2024 RELEASE

Nester's Microbiology

A HUMAN PERSPECTIVE

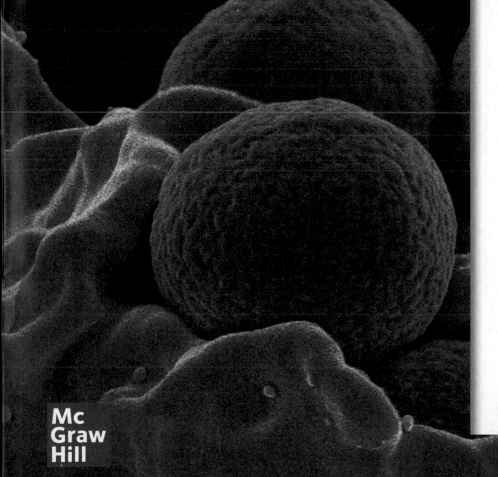

Denise Anderson | Sarah Salm | Mira Beins | Ann Auman | Jennifer Walker

Denise G. Anderson
UNIVERSITY OF WASHINGTON

Sarah N. Salm
BOROUGH OF MANHATTAN
COMMUNITY COLLEGE

Mira Beins
UNIVERSITY OF WASHINGTON

Ann Auman
PACIFIC LUTHERAN UNIVERSITY

Jennifer R. Walker
UNIVERSITY OF GEORGIA

Eugene W. Nester
UNIVERSITY OF WASHINGTON

Mc
Graw
Hill

NESTER'S MICROBIOLOGY

Published by McGraw Hill LLC, 1325 Avenue of the Americas, New York, NY 10019.

Some ancillaries, including electronic and print components, may not be available to customers outside the United States.

This book is printed on acid-free paper.

1 2 3 4 5 6 7 8 9 LWI 29 28 27 26 25 24

ISBN 978-1-266-86755-2
MHID 1-266-86755-4

Cover Image: *National Institute of Allergy and Infectious Diseases (NIAID)*

mheducation.com/highered

Brief Contents

About the Authors

The Nester Team:

Different Perspectives, One Vision, One Voice

The authors of this edition may be a set of individuals with different insights and unique experiences, but their cooperative relationship defines the word "team." What drives them is a single shared goal: to create the most learning-friendly and up-to-date microbiology textbook available. Each chapter was edited with students in mind, using simpler words where appropriate while maintaining the scientific rigor so important for today's healthcare professionals.

Courtesy of Richard Moore

Denise Anderson

Denise Anderson is a Senior Lecturer Emeritus in the Department of Microbiology at the University of Washington, where for over 30 years she taught a variety of courses, including general microbiology, medical bacteriology laboratory, recombinant DNA techniques, and medical mycology/parasitology laboratory. Equipped with a diverse educational background, including undergraduate work in nutrition and graduate work in food science and in microbiology, she first discovered a passion for teaching when she taught microbiology laboratory courses as part of her graduate training. Her enthusiastic teaching style, fueled by regular doses of Seattle's famous coffee, received high reviews from her students.

Denise now relaxes in the Yorkshire Dales of England, where she lives with her husband, Richard Moore. When not editing textbook chapters, she can usually be found walking scenic footpaths, chatting with friends, fighting weeds in her garden, or enjoying a fermented beverage at the local pub.

Courtesy of Sandy Coetzee

Sarah Salm

Sarah Salm, the digital author, is a Professor at the Borough of Manhattan Community College (BMCC) of the City University of New York, where she primarily teaches microbiology. She earned her undergraduate and doctoral degrees at the University of the Witwatersrand in Johannesburg, South Africa. She moved to New York, where she did postdoctoral work at the NYU Grossman School of Medicine. Her research background is diverse and includes plant virology, prostate cancer, and bacteria in contaminated water. Sarah frequently escapes the busy city to hike with friends and family.

Courtesy of Mira Beins

Mira Beins

Mira Beins is an Associate Teaching Professor in the Department of Microbiology at the University of Washington, where she teaches general microbiology, medical bacteriology, and medical mycology/parasitology. She completed her undergraduate studies in molecular biology and biotechnology at the University of the Philippines before moving to Wisconsin for graduate work in microbiology. Her graduate and postdoctoral research both focused on virology, which solidified her belief that viruses are amazing—although she now begrudgingly admits that bacteria, fungi, and eukaryotic parasites are pretty cool, too.

Mira lives in Seattle with her husband Mike and two kids, Maya and Noah. When she's not busy teaching or driving the kids to their many activities, she enjoys reading books, watching movies, hanging out with friends and family, and planning the next family trip (which Denise hopes will be to the Yorkshire Dales!).

Courtesy of John Froschauer/PLU

Ann Auman

Ann Auman is a Professor of Biology at Pacific Lutheran University (PLU) in Tacoma, WA. After earning her undergraduate degrees in microbiology and molecular and cell biology from the Pennsylvania State University, Ann completed a PhD in microbiology at the University of Washington. There, her thesis research included analyzing microbial communities using culture-independent methods. During her 20+ year teaching career, Ann has primarily taught microbiology and introductory biology courses. As a microbial ecologist, Ann's professional interests focus on understanding microbes' many contributions to global processes as well as the products they make that may be of biotechnological significance.

Ann lives in Kent, WA, with her partner Jeff and is now preparing for an empty nest as her two kids Rebecca and Josh transition into their next phases of life. Away from campus, she enjoys hanging out with her kids, going on walks with her partner and friends, reading and discussing books, and traveling (which Denise hopes will include a visit to the Yorkshire Dales!).

Jennifer Walker

Courtesy of Kate Walker

Jennifer Walker is a Senior Lecturer and the Undergraduate Coordinator for the Microbiology Department at the University of Georgia. Jennifer earned her B.S. in biology at UGA and then fell in love with microbiology after working as a lab technician for a year. She promptly returned to UGA to earn her PhD while studying peptide drug stability motifs and the Gram-negative biotin uptake system as a potential transporter for peptide antimicrobials. While working as a teaching assistant, Jennifer discovered her passion for education and thus combined her love of microbiology with instruction. Along with teaching introductory and upper-level undergraduate microbiology courses, Jennifer works with an amazing team of faculty and staff that mentor and challenge microbiology students to pursue their dreams.

Jennifer lives in Watkinsville with her husband, John, and her four teenage daughters. You can find Jennifer and her family in the fall rooting for the Dawgs at Saturday football games. She also enjoys sewing and working on her yard if she's not attending a pole vault meet, musical performance, volleyball scrimmage, or church event with her family.

Eugene Nester

Courtesy of Eugene Nester

Gene (Eugene) Nester was instrumental in establishing the text's reputation for excellence over the decades. Although no longer an active member of the author team, he wrote the original version of the present text with Evans Roberts and Nancy Pearsall more than 30 years ago. That text, *Microbiology: Molecules, Microbes and Man,* pioneered the organ system approach to the study of infectious disease and was developed specifically for allied health sciences.

Digital Tools for Your Success

Presentation Tools Allow Instructors to Customize Lecture

Enhanced Lecture Presentations contain lecture outlines, art, photos, tables, and animations embedded where appropriate. Fully customizable, but complete and ready to use, these presentations will enable you to spend less time preparing for lecture!

Animations—More than 100 animations bring key concepts to life; available for instructors and students.

Accessible PPTs—Our lecture presentations are formatted per the latest accessibility guidelines. Alternative text, written by our textbook author team, is included for all images and static tables.

Take Your Course Online—Easily—with One-Click Digital Lecture Capture

Tegrity in Connect® is a tool that makes class time available 24/7 by automatically capturing every lecture. With a simple one-click start-and-stop process, you capture all computer screens and corresponding audio in a format that is easy to search, frame by frame. Students can replay any part of any class with easy-to-use, browser-based viewing on a PC, Mac, or mobile device.

Educators know that the more students can see, hear, and experience class resources, the better they learn. In fact, studies prove it. Tegrity's unique search feature helps students efficiently find what they need, when they need it, across an entire semester of class recordings. Help turn your students' study time into learning moments immediately supported by your lecture. With Tegrity, you also increase intent listening and class participation by easing students' concerns about note-taking. Using Tegrity in Connect will make it more likely you will see students' faces, not the tops of their heads.

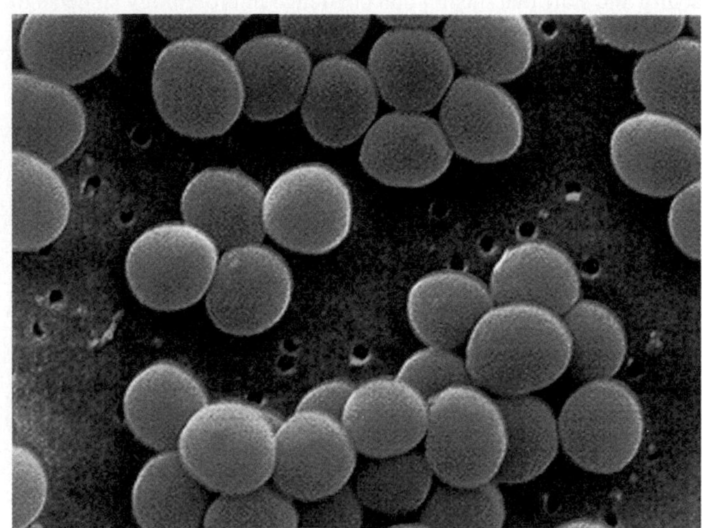

Janice Haney Carr/CDC

Remote Proctoring and Browser-Locking Capabilities

Remote proctoring and browser-locking capabilities, hosted by Proctorio within Connect, provide control of the assessment environment by enabling security options and verifying the identity of the student.

Seamlessly integrated within Connect, these services allow instructors to control the assessment experience by verifying identification, restricting browser activity, and monitoring student actions.

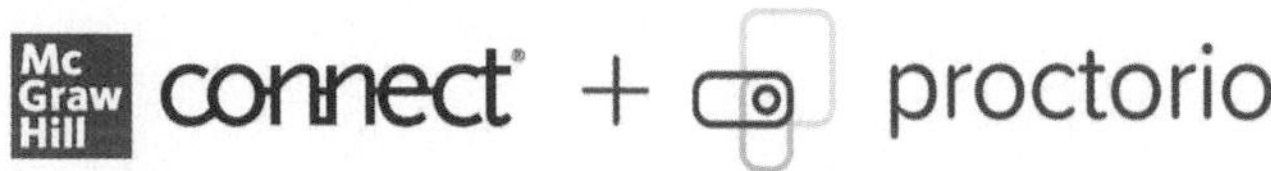

Instant and detailed reporting gives instructors an at-a-glance view of potential academic integrity concerns, thereby supporting evidence-based claims and avoiding personal bias.

Polling

Every learner has unique needs. Uncover where and when you're needed with the new Polling tool in McGraw Hill Connect! Polling allows you to discover where students are in real time. Engage students and help them create connections with your course content while gaining valuable insight during lectures. Leverage polling data to deliver personalized instruction when and where it is needed most.

Microbiology NewsFlash

NewsFlash brings the real world into your classroom! These activities in Connect tie current news stories to key concepts. After interacting with a contemporary news story, students are assessed on their ability to make the connections between real-life events and course content.

Relevancy Modules for Microbiology

With the help of our Relevancy Modules within McGraw Hill Connect, students can see how microbiology actually relates to their everyday lives. For select microbiology titles, students and instructors can access the Relevancy Modules eBook at no additional cost. Auto-graded assessment questions which correlate to the modules are also available within Connect. Each module consists of videos, an overview of basic scientific concepts, and then a closer look at the application of these concepts to the relevant topic. Some topics include microbes and cancer, fermentation, vaccines, biotechnology, global health, SARS-CoV-2, antibiotic resistance, and several others.

Virtual Labs

Virtual Labs and Lab Simulations

While the biological sciences are hands-on disciplines, instructors are often asked to deliver some of their lab components online: full online replacements, supplements to prepare for in-person labs, or make-up labs.

These simulations help each student learn the practical and conceptual skills needed, then check for understanding and provide feedback. With adaptive pre-lab and post-lab assessment available, instructors can customize each assignment.

From the instructor's perspective, these simulations may be used in the lecture environment to help students visualize complex scientific processes, such as DNA technology or Gram staining, while at the same time providing a valuable connection between the lecture and lab environments.

ReadAnywhere®

Read or study when it's convenient for you with McGraw Hill's free ReadAnywhere® app. Available for iOS or Android smartphones or tablets, ReadAnywhere gives users access to McGraw Hill tools including the eBook and SmartBook® 2.0 or Adaptive Learning Assignments in Connect. Take notes, highlight, and complete assignments offline—all of your work will sync when you open the app with Wi-Fi access. Log in with your McGraw Hill Connect username and password to start learning—anytime, anywhere!

OLC-Aligned Courses: Implementing High-Quality Instruction and Assessment through Preconfigured Courseware

In consultation with the Online Learning Consortium (OLC) and our certified Faculty Consultants, McGraw Hill has created preconfigured courseware using OLC's quality scorecard to align with best practices in online course delivery. This turnkey courseware contains a combination of formative assessments, summative assessments, homework, and application activities, and can easily be customized to meet each instructor's needs and desired course outcomes. For more information, visit https://www.mheducation.com/highered/olc.

Test Builder in Connect

Available within Connect, Test Builder is a cloud-based tool that enables instructors to format tests that can be printed, administered within a Learning Management System, or exported as Word documents. Test Builder offers a modern, streamlined interface for easy content configuration that matches course needs, without requiring a download.

Test Builder allows you to:

- access all test bank content from a particular title.
- easily pinpoint the most relevant content through robust filtering options.
- manipulate the order of questions or scramble questions and/or answers.
- pin questions to a specific location within a test.
- determine your preferred treatment of algorithmic questions.
- choose the layout and spacing.
- add instructions and configure default settings.

Test Builder provides a secure interface for better protection of content and allows for just-in-time updates to flow directly into assessments.

Create: Your Book, Your Way

McGraw Hill's Content Collections Powered by Create® is a self-service website that enables instructors to create custom course materials—print and eBooks—by drawing upon McGraw Hill's comprehensive, cross-disciplinary content. Choose what you want from our high-quality textbooks, articles, and cases. Combine it with your own content quickly and easily, and tap into other rights-secured, third-party content such as readings, cases, and articles. Content can be arranged in a way that makes the most sense for your course, and you can include the course name and information as well. Choose the best format for your course: color print, black-and-white print, or eBook. The eBook can be included in your Connect course and is available on the free ReadAnywhere® app for smartphone or tablet access as well. When you are finished customizing, you will receive a free digital copy to review in just minutes! Visit McGraw Hill Create®—www.mcgrawhillcreate.com—today and begin building!

Reflecting the Diverse World Around Us

McGraw Hill believes in unlocking the potential of every learner at every stage of life. To accomplish that, we are dedicated to creating products that reflect, and are accessible to, all the diverse, global customers we serve. Within McGraw Hill, we foster a culture of belonging, and we work with partners who share our commitment to equity, inclusion, and diversity in all forms. In McGraw Hill Higher Education, this includes, but is not limited to, the following:

- Refreshing and implementing inclusive content guidelines around topics including generalizations and stereotypes, gender, abilities/disabilities, race/ethnicity, sexual orientation, diversity of names, and age.
- Enhancing best practices in assessment creation to eliminate cultural, cognitive, and affective bias.
- Maintaining and continually updating a robust photo library of diverse images that reflect our student populations.
- Including more diverse voices in the development and review of our content.
- Strengthening art guidelines to improve accessibility by ensuring that meaningful text and images are distinguishable and perceivable by users with limited color vision and moderately low vision.

FOCUS ON UNDERSTANDING . . .

Student-Friendly Illustrations

Introduce the "big picture"

Focus figures provide an overview or highlight a key concept.

Keep the big picture in focus

A highlighted mini-version of the overview figure is often incorporated into the upper left corner of subsequent figures, helping students see how those figures fit into the big picture.

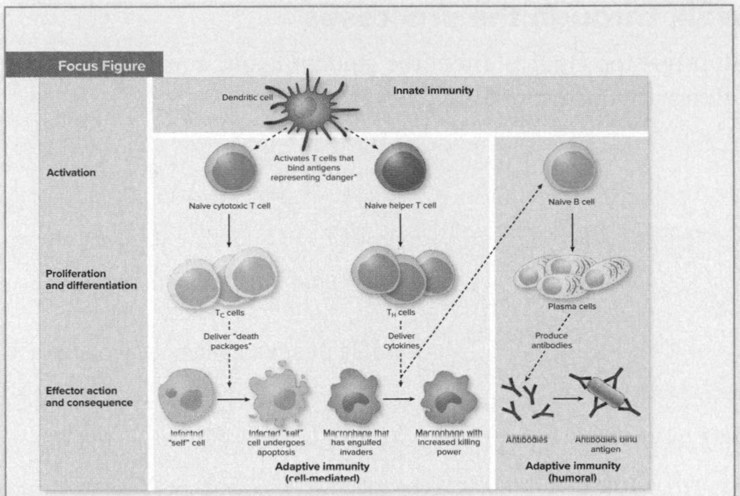

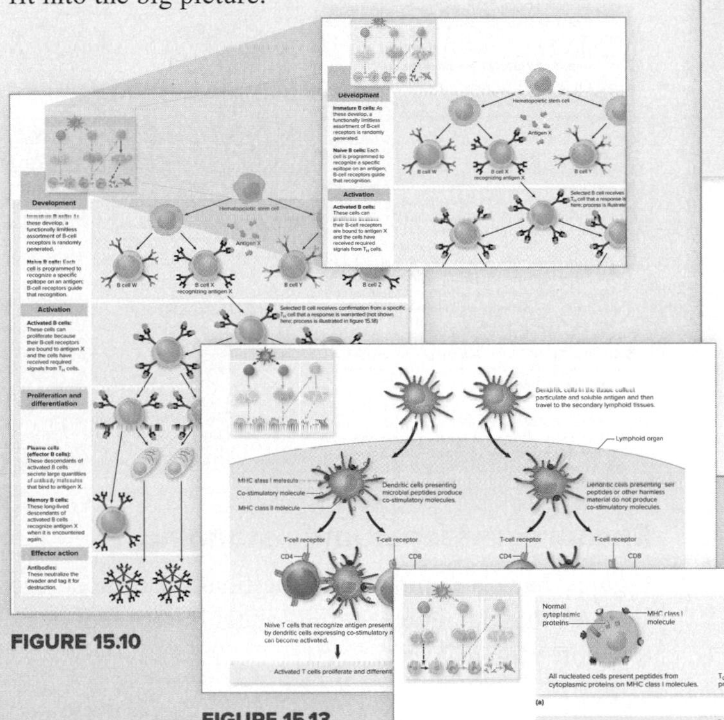

FIGURE 15.10

FIGURE 15.13

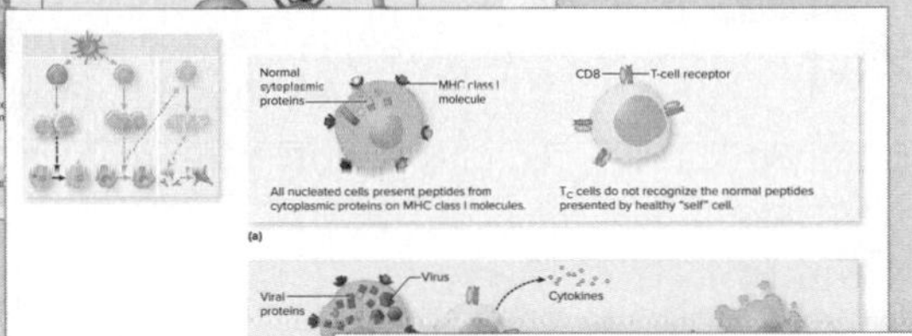

FIGURE 15.14

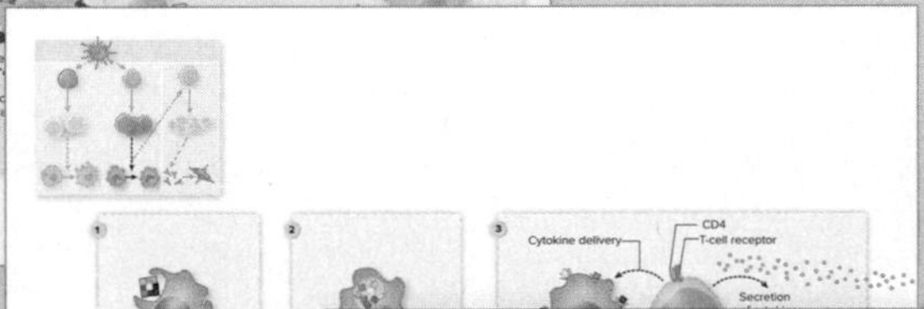

FIGURE 15.16

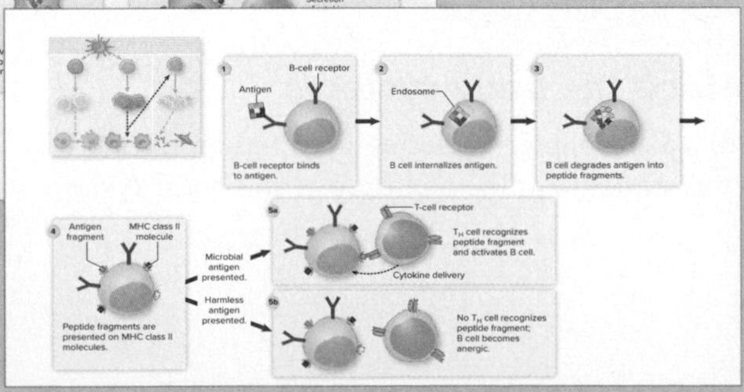

FIGURE 15.18

Focus Figure

FIGURE 15.1 Overview of the Adaptive Immune Response Cell-mediated immunity protects against antigens within host cells (intracellular antigens); humoral immunity protects against antigens in blood and tissue fluid (extracellular antigens). In this diagram, solid arrows represent the path of a cell or molecule; dashed arrows represent a cell's interactions and effector functions; antigen receptors and memory cells are not shown.

How does cell-mediated immunity eliminate intracellular antigens?

site is responsible for that recognition (**figure 15.2**). The antigen receptors on a single lymphocyte are identical and therefore recognize the same antigen, but because the body has hundreds of millions of different B cells and T cells, the immune system can recognize a nearly infinite assortment of antigens. General characteristics of the receptors are as follows:

- **T-cell receptors (TCRs).** These are on T cells. Conventional TCRs only bind an antigen "presented" by one of the body's own cells, an interaction guided by a surface molecule called a CD marker (CD stands for cluster of differentiation to reflect that scientists use the molecules to distinguish different groups of cells). Cytotoxic T cells have a CD marker called CD8, and the cells are sometimes referred to as CD8 T cells or CD8+ T cells; in contrast, helper T cells have a CD marker called CD4, and the cells are sometimes referred to as CD4 T cells or CD4+ T cells.
- **B-cell receptors (BCRs).** These are on B cells. BCRs are essentially membrane-anchored versions of the Y-shaped antibody molecules that the B cell is programmed to make. Unlike T-cell receptors, BCRs bind free antigens (in other words, antigens not presented by one of the body's own cells). The two arms of the BCR are identical to each other, resulting in two antigen-binding sites.

Cell-mediated and humoral immunity are both powerful and, if misdirected, can damage the body's own tissues. To provide the immune tolerance necessary to prevent inappropriate responses, two sequential processes are used:

- **Central tolerance.** This takes place as lymphocytes mature (T cells in the thymus and B cells in the bone

386

"Provides a logical unfolding conceptual framework that fosters better understanding."
—Jamal Bittar, University of Toledo

FOCUS ON UNDERSTANDING . . .

Walk through the processes

Step-by-step figures direct the student using numbered icons, often with corresponding icons in the text.

"The text and illustrations are 'tight' and give each other good support."

—Richard Shipee, Vincennes University

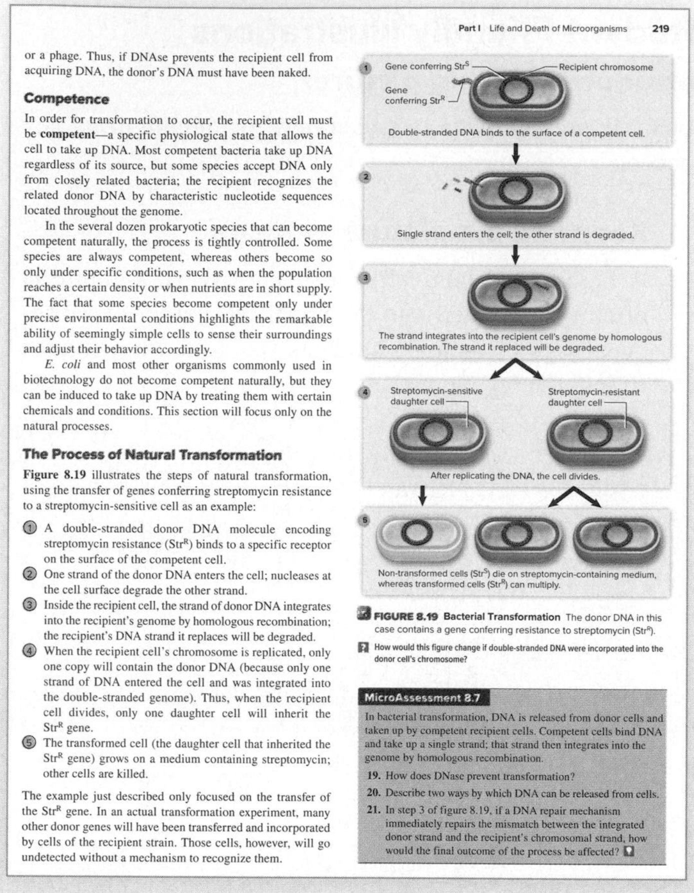

or a phage. Thus, if DNAse prevents the recipient cell from acquiring DNA, the donor's DNA must have been naked.

Competence

In order for transformation to occur, the recipient cell must be **competent**—a specific physiological state that allows the cell to take up DNA. Most competent bacteria take up DNA regardless of its source, but some species accept DNA only from closely related bacteria; the recipient recognizes the related donor DNA by characteristic nucleotide sequences located throughout the genome.

In the several dozen prokaryotic species that can become competent naturally, the process is tightly controlled. Some species are always competent, whereas others become so only under specific conditions, such as when the population reaches a certain density or when nutrients are in short supply. The fact that some species become competent only under precise environmental conditions highlights the remarkable ability of seemingly simple cells to sense their surroundings and adjust their behavior accordingly.

E. coli and most other organisms commonly used in biotechnology do not become competent naturally, but they can be induced to take up DNA by treating them with certain chemicals and conditions. This section will focus only on the natural processes.

The Process of Natural Transformation

Figure 8.19 illustrates the steps of natural transformation, using the transfer of genes conferring streptomycin resistance to a streptomycin-sensitive cell as an example:

1. A double-stranded donor DNA molecule encoding streptomycin resistance (Str^R) binds to a specific receptor on the surface of the competent cell.
2. One strand of the donor DNA enters the cell; nucleases at the cell surface degrade the other strand.
3. Inside the recipient cell, the strand of donor DNA integrates into the recipient's genome by homologous recombination; the recipient's DNA strand it replaces will be degraded.
4. When the recipient cell's chromosome is replicated, only one copy will contain the donor DNA (because only one strand of DNA entered the cell and was integrated into the double-stranded genome). Thus, when the recipient cell divides, only one daughter cell will inherit the Str^R gene.
5. The transformed cell (the daughter cell that inherited the Str^R gene) grows on a medium containing streptomycin; other cells are killed.

The example just described only focused on the transfer of the Str^R gene. In an actual transformation experiment, many other donor genes will have been transferred and incorporated by cells of the recipient strain. Those cells, however, will go undetected without a mechanism to recognize them.

FIGURE 8.19 Bacterial Transformation The donor DNA in this case contains a gene conferring resistance to streptomycin (Str^R).

How would this figure change if double-stranded DNA were incorporated into the donor cell's chromosome?

MicroAssessment 8.7

In bacterial transformation, DNA is released from donor cells and taken up by competent recipient cells. Competent cells bind DNA and take up a single strand; that strand then integrates into the genome by homologous recombination.

19. How does DNase prevent transformation?

20. Describe two ways by which DNA can be released from cells.

21. In step 3 of figure 8.19, if a DNA repair mechanism immediately repairs the mismatch between the integrated donor strand and the recipient's chromosomal strand, how would the final outcome of the process be affected?

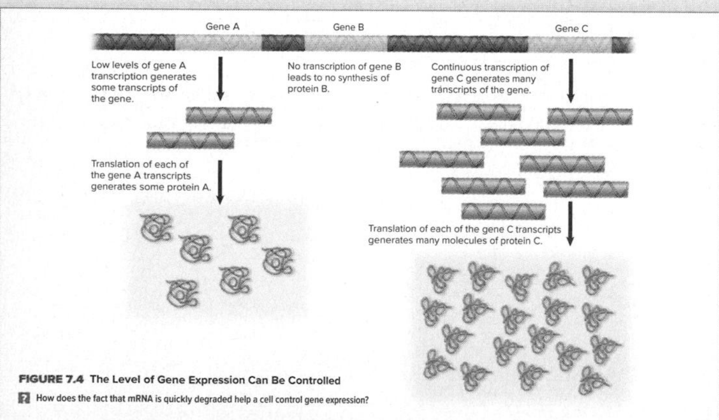

FIGURE 7.4 The Level of Gene Expression Can Be Controlled

How does the fact that mRNA is quickly degraded help a cell control gene expression?

Encourage deeper understanding

Figures have accompanying questions that encourage students to think more carefully about the concept illustrated in a figure.

Introduce the body systems

Each disease chapter includes a stunning figure that introduces the students to the anatomy of the body system.

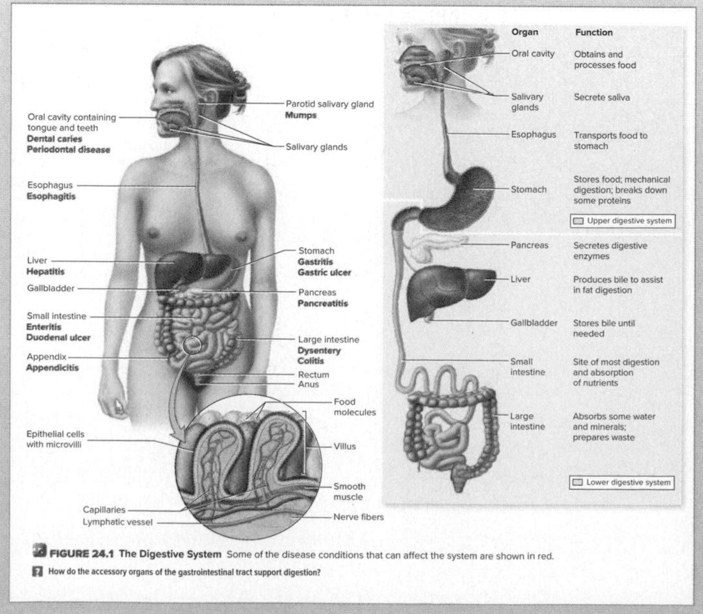

FIGURE 24.1 The Digestive System Some of the disease conditions that can affect the system are shown in red.

How do the accessory organs of the gastrointestinal tract support digestion?

Student-Friendly Chapter Features

Provide the tools for understanding

Key Terms for each chapter are defined on the opening page.

Share the history

A **Glimpse of History** opens each chapter, featuring engaging stories about the people who pioneered the field of microbiology.

Define the expectations

Learning outcomes are found at the beginning of each numbered section, allowing organization, evaluation, and assessment of instruction.

24 | Digestive System Infections

Intestinal microvilli (color-enhanced TEM). *Steve Gschmeissner/SPL/Science Source*

KEY TERMS

Cirrhosis Scarring of the liver that interferes with normal liver function.

Dental Caries Damage to tooth enamel resulting from acids produced as microbes in dental plaque ferment sugars; tooth decay.

Dental Plaque A biofilm on a tooth surface.

Dysbiosis An imbalance in the normal microbiota.

Dysentery A serious form of diarrhea characterized by blood, pus, and mucus in the feces.

Gastritis Inflammation of the lining of the stomach.

Gastroenteritis Inflammation of the lining of the stomach and intestines; the syndrome of nausea, vomiting, diarrhea, and abdominal pain.

Gingivitis Inflammation of the gums.

Hemolytic Uremic Syndrome (HUS) Serious condition characterized by red blood cell breakdown and kidney failure.

Hepatitis Inflammation of the liver.

Oral Rehydration Therapy (ORT) A treatment used to replace fluid and electrolytes lost due to diarrhea.

Periodontitis Inflammation of the periodontium (tissues supporting the teeth).

A Glimpse of History

Cholera is a very old disease, thought to have originated in the Far East thousands of years ago. With the increased shipping of goods and the mobility of people during the nineteenth century, cholera spread from Asia to Europe and then to North America. The disease caused major epidemics in the nineteenth century.

John Snow (1813–1858), a London physician, demonstrated that cholera was transmitted by contaminated water. He observed that almost all people who contracted the disease during an outbreak in 1854 got their water from a well on Broad Street, whereas neighbors who got their water elsewhere were unaffected. Even though the germ theory of disease had not yet been described, Snow was able to persuade local authorities to remove the pump handle from the suspected well so that people were forced to get their water elsewhere. Although the number of new cases decreased, Snow's explanation that cholera was a waterborne disease was not accepted by most doctors and government officials, partly because the outbreak had already begun subsiding before the handle was removed. By 1866, however, it was obvious that cholera occurred in areas where water had been contaminated with sewage, just as Snow had proposed. Public health agencies then played a major role in preventing epidemic cholera. In 1884 Robert Koch provided convincing evidence for the germ theory of disease after isolating *Vibrio cholerae*, the bacterium that causes cholera.

In late 2016, the largest cholera epidemic ever recorded started in Yemen. By April 2021, the country had reported over 2.5 million suspected cases and nearly 4,000 deaths. Conditions for the crisis were created by an ongoing war that devastated Yemen's health services and infrastructure, leaving more than 14 million people without access to safe drinking water. Although the WHO manages a stockpile of an oral cholera vaccine for use during epidemics, Yemen's unstable situation caused significant delays in initiating and then maintaining the vaccination programs. In addition, the sheer size of the epidemic and mass migration of people created further problems for disease control. Several humanitarian agencies responded to the crisis by opening up cholera treatment facilities, running awareness campaigns for disease prevention, delivering tanks of clean water to people, and training healthcare workers across the country to deal with the huge number of cholera cases—a multidisciplinary effort that demonstrates the importance of the One Health approach to disease control described in section 19.3. These efforts slowed the spread of disease, but the long-term solution requires a stable source of clean water.

24.1 ▪ Anatomy, Physiology, and Ecology of the Digestive System

Learning Outcomes

1. Describe the characteristics and functions of the digestive system components.
2. Describe the significance of the normal intestinal microbiota.

The main purpose of the digestive system is to convert the food we eat into a form that the body's cells can use as a

629

Assess understanding

A **MicroAssessment** at the end of each numbered section summarizes the concepts and includes review questions, usually featuring one that stimulates critical thinking (indicated by a light bulb icon).

MicroAssessment 3.2

Peptidoglycan is a molecule unique to bacteria that provides strength to the cell wall. The Gram-positive cell wall is composed of a relatively thick layer of peptidoglycan as well as teichoic acids. The Gram-negative cell wall has a thin layer of peptidoglycan and a lipopolysaccharide-containing outer membrane. Penicillin and lysozyme interfere with the structural integrity of peptidoglycan. *Mycoplasma* species lack a cell wall. Although archaea have a variety of cell wall types, most have S-layers.

4. What is the significance of lipid A?
5. How does the action of penicillin differ from that of lysozyme?
6. Explain why penicillin kills only actively multiplying cells, whereas lysozyme kills cells in any stage of growth.

Engage the reader

MicroBytes found throughout the chapter provide small "bytes" of information, capturing the reader's attention.

MicroByte

There are more bacteria in one person's mouth than there are people in the world!

FOCUS ON UNDERSTANDING . . .

Highlight the relevance

Focus on a Case boxes describe realistic clinical, veterinary, or environmental situations, along with questions and discussions designed to highlight the relevance of the information.

FOCUS ON A CASE 14.1

A 9-year-old boy with cystic fibrosis—a genetic disease that causes a number of problems, including the buildup of thick, sticky mucus in the lungs—complained of feeling tired, out of breath, and always coughing. When his mother took him to the doctor, she mentioned that his cough was productive, meaning that it contained sputum (pronounced *spew-tum*). She was particularly concerned that the sputum was a blue-green color. His doctor immediately suspected a *Pseudomonas aeruginosa* lung infection—a common complication of cystic fibrosis. A sputum sample was collected and sent to the clinical laboratory.

In the clinical laboratory, the sample was plated onto MacConkey agar and blood agar and incubated. Mucoid colonies surrounded by a bluish-green color grew on both types of agar media. The colonies on MacConkey had no pink coloration, so the medical tech-

The patient was treated with antibiotics, with only limited success. Like most cystic fibrosis patients, he developed a chronic lung infection that required repeated treatments.

1. What role did cystic fibrosis play in the disease process?
2. What is the significance of the mucoid phenotype of the colonies?
3. How would the siderophore (the iron-binding compound) benefit the bacterium?
4. Why would the boy's lung infection make his pre-existing respiratory problems even worse?

Discussion

1. Cystic fibrosis patients often have an accumulation of thick mucus in their

Pseudomonas aeruginosa cells to form biofilms. The biofilm protects the bacterial cells from various components of the immune system, including host defense peptides (antimicrobial peptides) and phagocytes. Bacteria growing within a biofilm are much more difficult for the immune system to destroy.

3. Siderophores help the bacterium obtain iron from the host. Recall that the body's iron-binding proteins (lactoferrin and transferrin) prevent microbes from using the host's iron supply and thereby limit their growth. Microorganisms that make siderophores essentially engage in a "tug-of-war" with the body over iron. That tug-of-war is especially important for *P. aeruginosa* because iron levels influence biofilm formation. When iron is limiting, *P. aeruginosa* cells are motile

Provide perspective

Focus Your Perspective boxes show how microorganisms and their products influence our lives in many different ways.

FOCUS YOUR PERSPECTIVE 9.1

The COVID-19 Response—The Power of Biotechnology

The COVID-19 response is an excellent illustration of the power of biotechnology. Because of several technologies described in this chapter, the pandemic's global outcome—although devastating—resulted in fewer deaths than feared or predicted.

SARS-CoV-2, the virus that causes COVID-19, has an RNA genome. When the virus was first discovered in China, researchers used the enzyme reverse transcriptase to make a cDNA copy of its genome. That cDNA was then cloned and sequenced, and the information was shared with scientists around the world, initiating a global effort to control the disease.

A major part of any disease control effort is diagnosing patients who have the disease.

CRISPR-Cas-based tests that give results in under an hour. While the first version of this test was authorized for use only in certified laboratories, researchers are also developing instrument-free versions for on-site use.

Data obtained via high-throughput sequencing were used to track the global spread of SARS-CoV-2. The tracking methods rely on detecting spontaneous mutations that inevitably occur as the virus replicates; these mutations serve as evolutionary markers. For example, viral genomes from a cluster of early cases in the Seattle area all shared the same unusual mutation, indicating that they all descended from the same source (see Focus Your Perspective 21.1). Using

addition, the knowledge gained can hopefully be used to prevent a similar pandemic in the future.

The fact that the genome sequence of SARS-CoV-2 was known early on facilitated research aimed at developing targeted antiviral therapies as well as vaccines, as described in Focus on the Future 20.1. By analyzing the viral genome, scientists determined the amino acid sequences of key proteins essential for viral replication. Relatively soon thereafter, the three-dimensional structures of three of those proteins were determined—one that the virus uses to attach to and then enter host cells (its spike protein), one it uses to replicate its genome (its

Introduce the concepts

Focus on a Disease boxes introduce a general category of disease (pneumonia, diarrheal disease, meningitis, sexually transmitted infections), giving students a framework for understanding specific diseases.

FOCUS ON PNEUMONIA

Pneumonia is a disease of the lower respiratory tract caused by bacterial, viral, or fungal infection of the lungs; the infection elicits an inflammatory response, resulting in the alveoli (air sacs) filling with fluids such as pus and blood. Pneumonia is the leading cause of death due to infectious disease in the United States.

Signs and Symptoms

The signs and symptoms of pneumonia generally include cough, chills, shortness of breath, fever, and chest pain. In severe cases, the patient may develop cyanosis (bluish skin color) due to poor blood oxygenation. Pneumonia ranges from mild to life-threatening, depending largely on the causative agent but also on any underlying health problems of the patient. It is often accompanied by a productive cough, meaning that a pus- and mucus-containing fluid called **sputum** comes up from the lungs.

Some pathogens cause what are referred to as atypical pneumonias, not because the diseases are uncommon, but because the symptoms or the treatments are slightly different than those of the more established types. "Walking pneumonia" is a term often used to describe some atypical pneumonias because of the milder symptoms observed.

To diagnose pneumonia, a physician uses a stethoscope to listen for a characteristic crackling or bubbling sound that occurs in the lungs as air passes by fluid in the alveoli. A chest X ray will likely be done to determine which parts of the lung are infected; areas of infection usually appear as white shadows. The patient may also be asked to give a sputum sample, which can be examined microscopically and inoculated onto appropriate laboratory media as part of the process to identify a bacterial or fungal cause of pneumonia.

Pathogenesis

Various bacteria, viruses, and fungi can all cause pneumonia, but typically only when the respiratory tract defenses are not func-

The damage from pneumonia is largely a result of the inflammatory response. As the capillaries become leaky during inflammation, fluids collect in the alveoli and interfere with O_2 and CO_2 exchange. In addition, phagocytes and other leukocytes are recruited to the site of infection, and mucus production increases. In severe cases, accumulating leukocytes and mucus create a thick substance that may clog the alveoli, a condition called consolidation. The inflammatory response seen in severe pneumonia often affects nerve endings in the pleura, causing pain. Fatal respiratory failure occurs when the lungs can no longer adequately oxygenate the blood or expel CO_2.

Epidemiology

Pneumonias are often categorized as either community-acquired, meaning they develop in members of the general public, or healthcare-associated, meaning they develop in hospitalized patients or other people within the healthcare system. Some types of community-acquired pneumonia (CAP) are contagious. Most, however, originate from the patient's own upper respiratory microbiota. These organisms may gain access to the lungs when a person inadvertently inhales his or her own throat secretions. As with CAPs, healthcare-associated pneumonias (HCAPs) often occur when the patient inhales his or her own upper respiratory microbiota. Patients at particular risk are those on mechanical ventilators used to help breathing, because the ventilator tube provides a portal for microbes to enter the lower airways. Pneumonias that develop this way are further classified as ventilator-associated pneumonias (VAPs).

Treatment and Prevention

Bacterial and fungal pneumonias are treated with antimicrobial medications, chosen according to the susceptibility of the causative agent. Unfortunately, bacteria that cause healthcare-associated

- **Summary** briefly reviews the key points.
- **Short Answer** questions review major chapter concepts.
- **Multiple Choice** questions allow self-testing; answers are provided in Appendix IV.
- **Application** questions provide an opportunity to use knowledge of microbiology to solve real-world problems.
- **Critical Thinking** questions encourage practice in analysis and problem solving that can be used by the student in any subject.

Review the information

End-of-chapter review encourages students to revisit the information.

Build the story

Logical chapter order helps students understand and connect the concepts.

The Immune Wars
Innate immunity (chapter 14)
Adaptive immunity (chapter 15)
→
The Pathogens Fight Back
Pathogenesis (part of chapter 16)
→
The Return of the Humans (Knowledge is Power)
Immunization and immunotherapy (chapter 17)
Epidemiology (chapter 19)
Antimicrobial medications (chapter 20)

FIGURE 17.1 The Host-Pathogen Trilogy

? How does immunization prevent disease?

Student-Friendly Descriptions

Include analogies

WHY? Analogies provide students a comfortable framework for making sense of difficult topics. Here's an example from chapter 14.

> **Innate Immunity:** *The innate immune system is easiest to understand by considering it as three general interacting components: first-line defenses, sensor systems, and innate effector actions. As a useful analogy, think of the defense systems of a high-security building or compound: The first-line defenses are the security walls surrounding the property; the sensor systems are the security cameras scattered throughout the property, monitoring the environment for signs of invasion; and the effector actions are the security teams sent to remove any invaders that have been detected, thereby eliminating the threat **(figure 14.1a).***

Steve Cole/E+/Getty Images

Image Source

Moodboard/Brand X Pictures/Getty Images

Emphasize the logic

WHY? Descriptions that emphasize the logic of processes make it easier for students to understand and retain the information. Here's an example from chapter 6.

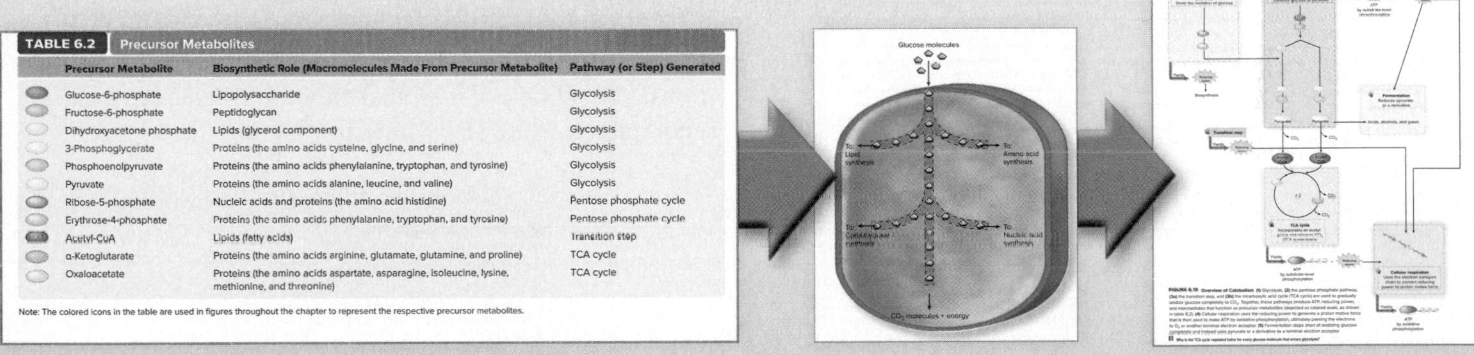

TABLE 6.2 Precursor Metabolites

Precursor Metabolite	Biosynthetic Role (Macromolecules Made From Precursor Metabolite)	Pathway (or Step) Generated
Glucose-6-phosphate	Lipopolysaccharide	Glycolysis
Fructose-6-phosphate	Peptidoglycan	Glycolysis
Dihydroxyacetone phosphate	Lipids (glycerol component)	Glycolysis
3-Phosphoglycerate	Proteins (the amino acids cysteine, glycine, and serine)	Glycolysis
Phosphoenolpyruvate	Proteins (the amino acids phenylalanine, tryptophan, and tyrosine)	Glycolysis
Pyruvate	Proteins (the amino acids alanine, leucine, and valine)	Glycolysis
Ribose-5-phosphate	Nucleic acids and proteins (the amino acid histidine)	Pentose phosphate cycle
Erythrose-4-phosphate	Proteins (the amino acids phenylalanine, tryptophan, and tyrosine)	Pentose phosphate cycle
Acetyl-CoA	Lipids (fatty acids)	Transition step
α-Ketoglutarate	Proteins (the amino acids arginine, glutamate, glutamine, and proline)	TCA cycle
Oxaloacetate	Proteins (the amino acids aspartate, asparagine, isoleucine, lysine, methionine, and threonine)	TCA cycle

Note: The colored icons in the table are used in figures throughout the chapter to represent the respective precursor metabolites.

Introduce the players: *Certain intermediates of catabolic pathways can be used in anabolic pathways, linking these two types of pathways. These intermediates—**precursor metabolites**—serve as carbon skeletons from which subunits of macromolecules can be made **(table 6.2).***

Reinforce the concept: *A cell's metabolic pathways make it easy for that cell to use a single type of substrate like glucose for multiple purposes. Think of the cells as extensive biological recycling centers that routinely process millions of glucose molecules **(figure 6.9).** Molecules that remain on the central deconstruction line are oxidized completely to CO_2, releasing the maximum amount of energy. Some breakdown intermediates, however, can exit that line to be used in biosynthesis.*

Put the pieces together: *Three key metabolic pathways—**the central metabolic pathways**—gradually oxidize glucose to CO_2, summarized as follows:*

$$C_6H_{12}O_6 + 6\,O_2 \rightarrow 6\,CO_2 + 6\,H_2O$$
(glucose) (oxygen) (carbon dioxide) (water)

The pathways are catabolic, but the precursor metabolites and reducing power they generate can also be diverted for use in biosynthesis.

FOCUS ON UNDERSTANDING . . .

Student-Friendly Disease Presentations

Help students think like experts

Within each body system chapter, diseases are separated by major taxonomic category (bacteria, viruses, fungi, protozoa). This organization reflects a major consideration with respect to treatment options, an important consideration for students going into healthcare-related fields.

22.4 ■ Fungal Diseases of the Skin

Learning Outcomes

9. Describe the characteristics of superficial cutaneous mycoses, including the role of dermatophytes in these infections.
10. Compare and contrast the roles of *Malassezia furfur* and *Candida albicans* in disease.

Diseases caused by fungi are called mycoses. Several fungi are responsible for mild to serious infections of the skin. The severity of most fungal infections is influenced by various host factors, including age and overall health.

Superficial Cutaneous Mycoses

A group of molds called **dermatophytes** can invade hair, nails, and the keratinized layer of the skin. The resulting mycoses have common names such as jock itch, athlete's foot, and ringworm, as well as Latin names that describe their location: tinea capitis (scalp), tinea barbae (beard), tinea axillaris (armpit), tinea corporis (body), tinea cruris (groin), and tinea pedis (feet), to list a few. Tinea simply means "worm," which probably reflects early misconceptions about the cause of these diseases.

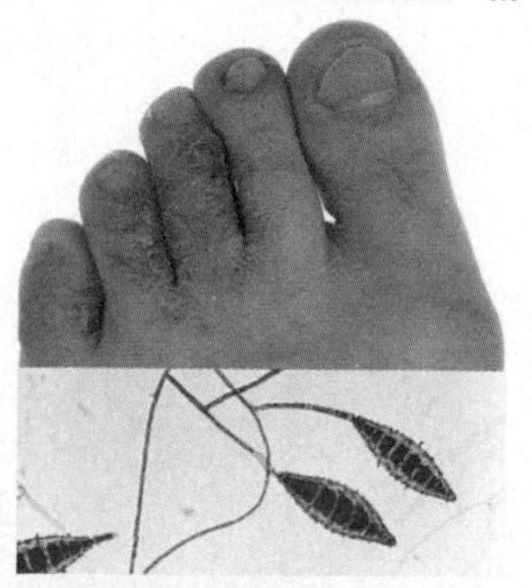

FIGURE 22.19 Dermatophytes (a) Tinea pedis, usually caused by species of *Trichophyton*. (b) Large boat-shaped conidia of *Microsporum gypseum*, a cause of scalp ringworm in children.
a: carroteater/Getty Images; b: Dr. Lucille K. Georg/CDC

? What is the common name for tinea pedis?

Signs and Symptoms

Most people colonized by dermatophytes have no signs or symptoms. Others complain of itching, a bad odor, or a rash. In ringworm, the rash at the site of infection appears as a scaly area surrounded by redness at the outer edge, producing irregular rings or a lacy pattern on the skin. On the scalp, patchy areas of hair loss can occur, leaving a fine stubble of short hair behind. Infected nails become thick and brittle and may separate from the nailbed. Sometimes, a rash consisting of fine papules and vesicles develops away from the infected area. This rash is referred to as a dermatophytid, or "id" reaction, a result of allergic reactions to products of the infecting fungus.

Causative Agents

Dermatophytes are a group of skin-invading molds that includes members of the genera *Epidermophyton, Microsporum,* and *Trichophyton* (**figure 22.19**). They can be grown on culture media especially designed for molds and are usually identified by their colony and microscopic morphologies. Biochemical tests and polymerase chain reaction (PCR) may be used as well. However, identification is not always necessary, because treatment for all dermatophyte infections is essentially the same.

Pathogenesis

The normal skin is generally resistant to fungal invasion, but dermatophytes can invade keratin-containing cells and structures in areas of the body where skin is moist (groin, armpits). The fungi produce an enzyme called keratinase that breaks down the protein, allowing them to use it as a nutrient source. Hair is invaded at the follicle, which is relatively moist.

Fungal products diffuse into the dermis and provoke an immune reaction, which probably explains why adults tend to be more resistant to infection than children. Children are more likely to have hypersensitivities, like eczema (a blistery skin rash that leaks fluid before forming crusts) and asthma.

Epidemiology

Patient age, the virulence of the infecting fungal strain, and moisture availability are important factors in determining the course of infection. Common causes of excessive moisture include obesity where folds of skin lie together, tight clothing, and plastic or rubber footwear. Potentially pathogenic molds may be present in soil and on pets such as cats and dogs; fungi acquired from these sources tend to cause more noticeable signs and symptoms in humans.

Treatment and Prevention

Several over-the-counter and prescription medications, such as clotrimazole (an azole) and terbinafine (an allylamine), can be used to treat superficial skin infections. Nail infections

Provide a consistent conceptual framework

Disease discussions are separated into consistent subsections, providing a conceptual framework and breaking the material into "bite-sized" pieces.

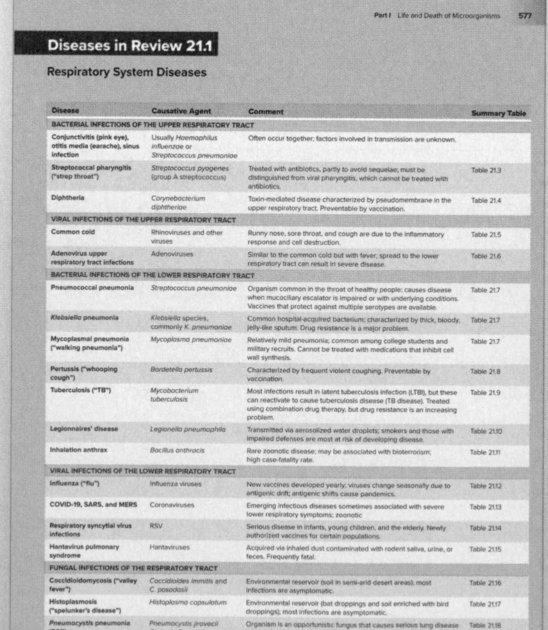

Diseases in Review 21.1

Respiratory System Diseases

Disease	Causative Agent	Comment	Summary Table
BACTERIAL INFECTIONS OF THE UPPER RESPIRATORY TRACT			
Conjunctivitis (pink eye), otitis media (earache), sinus infection	Usually *Haemophilus influenzae* or *Streptococcus pneumoniae*	Often occur together; factors involved in transmission are unknown.	
Streptococcal pharyngitis ("strep throat")	*Streptococcus pyogenes* (group A streptococcus)	Treated with antibiotics, partly to avoid sequelae; must be distinguished from viral pharyngitis, which cannot be treated with antibiotics.	Table 21.3
Diphtheria	*Corynebacterium diphtheriae*	Toxin-mediated disease characterized by pseudomembrane in the upper respiratory tract. Preventable by vaccination.	Table 21.4
VIRAL INFECTIONS OF THE UPPER RESPIRATORY TRACT			
Common cold	Rhinoviruses and other viruses	Runny nose, sore throat, and cough are due to the inflammatory response and cell destruction.	Table 21.5
Adenovirus upper respiratory tract infections	Adenoviruses	Similar to the common cold but with fever; spread to the lower respiratory tract can result in severe disease.	Table 21.6
BACTERIAL INFECTIONS OF THE LOWER RESPIRATORY TRACT			
Pneumococcal pneumonia	*Streptococcus pneumoniae*	Organism common in the throat of healthy people; causes disease when mucociliary escalator is impaired or with underlying conditions. Vaccines that protect against multiple serotypes are available.	Table 21.7
***Klebsiella* pneumonia**	*Klebsiella* species, commonly *K. pneumoniae*	Common hospital-acquired bacterium; characterized by thick, bloody, jelly-like sputum. Drug resistance is a major problem.	Table 21.7
Mycoplasmal pneumonia ("walking pneumonia")	*Mycoplasma pneumoniae*	Relatively mild pneumonia; common among college students and military recruits. Cannot be treated with medications that inhibit cell wall synthesis.	Table 21.7
Pertussis ("whooping cough")	*Bordetella pertussis*	Characterized by frequent violent coughing. Preventable by vaccination.	Table 21.8
Tuberculosis ("TB")	*Mycobacterium tuberculosis*	Most infections result in latent tuberculosis infection (LTBI), but these can reactivate to cause tuberculosis disease (TB disease). Treated using combination drug therapy, but drug resistance is an increasing problem.	Table 21.9
Legionnaires' disease	*Legionella pneumophila*	Transmitted via aerosolized water droplets; smokers and those with impaired defenses are most at risk of developing disease.	Table 21.10
Inhalation anthrax	*Bacillus anthracis*	Rare zoonotic disease; may be associated with bioterrorism; high case-fatality rate.	Table 21.11
VIRAL INFECTIONS OF THE LOWER RESPIRATORY TRACT			
Influenza ("flu")	Influenza viruses	New vaccines developed yearly; viruses change seasonally due to antigenic drift; antigenic shifts cause pandemics.	Table 21.12
COVID-19, SARS, and MERS	Coronaviruses	Emerging infectious diseases sometimes associated with severe lower respiratory symptoms; zoonotic	Table 21.13
Respiratory syncytial virus infections	RSV	Serious disease in infants, young children, and the elderly. Newly authorized vaccines for certain populations.	Table 21.14
Hantavirus pulmonary syndrome	Hantaviruses	Acquired via inhaled dust contaminated with rodent saliva, urine, or feces. Frequently fatal.	Table 21.15
FUNGAL INFECTIONS OF THE RESPIRATORY TRACT			
Coccidioidomycosis ("valley fever")	*Coccidioides immitis* and *C. posadasii*	Environmental reservoir (soil in semi-arid desert areas); most infections are asymptomatic.	Table 21.16
Histoplasmosis ("spelunker's disease")	*Histoplasma capsulatum*	Environmental reservoir (bat droppings and soil enriched with bird droppings); most infections are asymptomatic.	Table 21.17
***Pneumocystis* pneumonia (PCP)**	*Pneumocystis jirovecii* (formerly *P. carinii*)	Organism is an opportunistic fungus that causes serious lung disease in immunocompromised people, such as those with HIV/AIDS.	Table 21.18

Summarize each disease's characteristics

Summary tables serve as brief reminders of the important features of each disease. Major diseases are represented with an enhanced summary table that includes an outline of the disease process keyed to a human figure, showing the entry and exit of the pathogen.

Review the diseases as a group

Each disease chapter ends with a table that summarizes the key features of the diseases discussed in that chapter.

UPDATES—Maintaining the Cutting Edge

Evergreen

Content and technology are ever-changing, and keeping your course up to date with the latest information and assessments is important. That is why we want to deliver the most current and relevant content for your course, hassle-free.

Microbiology, A Human Perspective is moving to an Evergreen delivery model, which means it has content, tools, and technology that is updated and relevant, with updates delivered directly to your existing McGraw Hill Connect® course. Engage students and freshen up assignments with up-to-date coverage of select topics and assessments, all without having to switch editions or build a new course.

Global Changes in This Release

- Updated the disease statistics, vaccine recommendations, treatments, and terminology
- Increased the focus on worldwide effects of infectious diseases
- Aligned the descriptions of diseases that have similar symptoms or epidemiology so that comparisons can be made more easily
- Emphasized that cytokine release and the inflammatory response can lead to signs and symptoms of certain diseases
- Decreased the level of detail about non-specific disease treatments
- Deleted the Focus on the Future boxes, often moving the information into the narrative or using it to create new Focus Your Perspective boxes
- Modified the figures for increased accessibility
- Continued "wordsmithing" to improve the clarity and readability of the descriptions

Key Changes in Individual Chapters

Chapter 1 – Humans and the Microbial World

- Added mpox to the list of emerging infectious diseases and updated the COVID-19 death statistics; updated the information about wheat blast and African swine fever; added information about the recent spread of a novel avian influenza virus (section 1.2)
- Added *Thiomargarita magnifica* to Focus Your Perspective 1.1 (section 1.3)

Chapter 2 – The Molecules of Life

- Changed the name and narrowed the focus of the subsection now called *Triglycerides (Fats)* (was *Simple Lipids*); changed the heading and narrowed the focus of the subsection now called *Phospholipids* (was *Compound Lipids*); moved the information about *trans* fatty acids from a MicroByte into the main narrative; modified a MicroByte to emphasize the importance of artificial intelligence tools in predicting protein shapes based on DNA sequences (section 2.4)

Chapter 3 – Cells and Methods to Observe Them

- Added the terms *primary active transport* and *secondary active transport* (section 3.1)
- Added the term *haploid*; expanded on the functions of storage granules and encapsulin nanocompartments (section 3.4)
- Moved discussion of vesicles from section 3.5 to section 3.7
- Reorganized the discussion of eukaryotic membrane-bound organelles to separate those involved with the endomembrane system from those that are not, updating table 3.4 accordingly (section 3.7)
- Improved the description that distinguishes between resolving power and resolution; simplified the discussion of transmission electron microscopy sample preparation methods (section 3.8)

Chapter 4 – Dynamics of Microbial Growth

- Updated the description of alpha-hemolysis; added a Focus Your Perspective on Agar Art (section 4.6)

Chapter 5 – Control of Microbial Growth

- Added the terms *antimicrobial* and *biocide* to the list of key terms
- Introduced the term *cold pasteurization* (sections 5.1 and 5.3)
- Removed the description of depth filters; updated the discussion of ultraviolet radiation to include the use of UVC for ultraviolet germicidal irradiation; added the term *pascalization* to describe high pressure treatment (section 5.3)

Chapter 6 – Microbial Metabolism: Fueling Cell Growth

- Added the mention of Marjory Stephenson to the Glimpse of History
- Defined the term *intermediates*; reworded the definition of substrate-level phosphorylation (section 6.1)
- More explicitly compared the Embden-Meyerhof and Entner-Doudoroff glycolytic pathways (section 6.3)
- Added the mention of Jennifer Moyle's role in developing the chemiosmotic theory; reworked the detailed description of aerobic respiration in prokaryotes (section 6.4)
- Added a brief discussion of the different meanings of the term *fermentation* (section 6.5)
- Reduced the coverage of chemolithotrophs, referring the reader to chapter 11 for more information (section 6.7)
- Significantly revised the discussion of photosynthesis for better flow and increased understanding; added a brief description of phototrophy involving rhodopsin pigments (section 6.8)

Chapter 7 – The Blueprint of Life, from DNA to Protein

- Changed the Glimpse of History topic to tell the story of the Tsuneko Okasaki
- Clarified the role of DNA helicase (section 7.2)
- Changed the order of the paragraphs that introduce the components of translation so that tRNA is now described before ribosomes (section 7.3)
- Modified the description of quorum sensing (section 7.5)
- Revised the discussion of gene regulation so that the terms *constitutive*, *inducible*, and *repressible* are applied to genes rather enzymes (section 7.6)

Chapter 8 – Bacterial Genetics

- Replaced the term *base substitution* with *base-pair substitution*; redesigned figure 8.2 to clarify the process that leads to base-pair substitution (section 8.2)
- Replaced the critical thinking question in MicroAssessment 8.3 (section 8.3)
- Converted the descriptions of replica plating and Ames test into bullet lists; deleted the section on penicillin enrichment (section 8.5)
- Updated and added "titles" to the bullets and the accompanying figure that describe CRISPR (section 8.11)

Chapter 9 – Biotechnology

- Added clear descriptions of the terms *in vivo* and *in vitro* (section 9.1)
- Revised the discussions of gel electrophoresis and generating a recombinant DNA molecule (sections 9.1 and 9.2, respectively)
- Added the mention of Nobel Prize winners Jennifer Doudna and Emmanuelle Charpentier (section 9.3)
- Removed the description of RNA-seq (section 9.4)
- Changed the title and focus of the section now called *Considerations of Genetic Engineering* (was *Concerns Regarding Genetic Engineering*); added a discussion of gene therapy (section 9.7)

Chapter 10 – Identifying and Classifying Microorganisms

- Added a mention of point-of-care testing (section 10.1)
- Updated the description of prokaryotic nomenclature by indicating that the name of the phylum ends in the suffix *-ota*; updated table 10.1 and figure 10.1 to reflect the recent changes to bacterial phylum names (section 10.1)
- Added Nextstrain to the list of programs that track genomic changes of pathogens (section 10.4)
- Updated figure 10.14 to reflect the recent changes to bacterial phylum names (section 10.5)

Chapter 11 – The Diversity of Bacteria and Archaea

- Revised the discussion of anoxygenic phototrophs to remove much of the overlap with the improved section 6.8, which covers photosynthesis (section 11.2)
- Added the term *nuisance bloom* (section 11.3)
- Added the potential use of magnetosomes for targeted drug delivery systems; added a mention of *Thiomargarita magnifica* (section 11.7)
- Updated tables 11.1 through 11.3 to include the new bacterial phylum names

Chapter 12 – The Eukaryotic Members of the Microbial World

- Explained the use of the term *binary fission* to describe protozoan cell division, particularly how that division is different from the simple binary fission of prokaryotic cells; expanded the discussion of sexual life cycles of eukaryotes to include the terms *haploid-dominant*, *diploid-dominant*, and *alternation of generations* (chapter introduction)

- Added the term *harmful algal bloom* (section 12.3)
- Mentioned the association between the bites of certain types of ticks and the development of allergies to meats and animal products (section 12.5)

Chapter 13 – Viruses, Viroids, and Prions

- Updated the taxonomic classification of rubella virus to be under the family Matonaviridae; updated Focus Your Perspective 13.1 by including the term *nucleocytoplasmic large DNA viruses* (section 13.1)
- Updated and emphasized the discussion of phage therapy by creating Focus Your Perspective 13.2, which mentions the involvement of SEA-PHAGES (Science Education Alliance Phage Hunters Advancing Genomics and Evolutionary Science) (section 13.2)
- Streamlined the coverage by creating a new section (*Methods Used to Study Viruses*), which consolidates two previous sections (*Methods Used to Study Bacteriophages* and *Cultivating and Quantitating Animal Viruses*) (section 13.7)

Chapter 14 – The Innate Immune Response

- Updated the term *antimicrobial peptides (AMPs)* to *host defense peptides (HDPs)* (section 14.2)
- Added information about extracellular soluble pattern recognition molecules (section 14.5)
- Modified the description of phagocytosis to indicate that phagolysosome formation is part of phagosome maturation (section 14.7)
- Expanded the discussion of damaging effects of inflammation (section 14.8)

Chapter 15 – The Adaptive Immune Response

- Converted the descriptions of B-cell receptors and T-cell receptors to a bullet list (15.1)

Chapter 16 – Host-Microbe Interactions

- Converted what were headings under *Beneficial Roles of the Human Microbiome* to a bullet list; added a discussion of the gut microbiome–brain connection (section 16.2)
- Replaced the electron micrograph of a type III secretion system in figure 16.4 (section 16.5)
- Modified figure 16.12 to emphasize the involvement of cytokines in the response to superantigens (section 16.8)

Chapter 17 – Applications of Immune Responses

- Added a discussion of mRNA vaccines, viral vector vaccines, and DNA vaccines in a new subsection, *Nucleic Acid–Based Vaccines* changed the name and expanded the focus of the subsection now called *Vaccination Uses and Benefits* (was *Importance of Vaccines*); added information about ring vaccination; added a discussion about the novel oral polio vaccine type 2 (nOPV2) (section 17.2)

Chapter 18 – Immunological Disorders

- Revised the bullet list that describes the events leading to type I hypersensitivity to more closely align with figure 18.1; simplified table 18.2, which compares the major types of hypersensitivity reactions (section 18.1)
- Deleted Focus Your Perspective 18.1 (*The Fetus as an Allograft*)

Chapter 19 – Epidemiology

- Added information about surveillance case definitions, added the term *infection-fatality rate*, and rearranged the order of the other topics in the discussion (section 19.1)
- Updated the information about droplet transmission and airborne transmission to reflect the insights gained from the COVID-19 pandemic (section 19.2)
- Added information about the One Health approach to disease control (sections 19.3 and 19.6)
- Updated the information about nationally notifiable infectious diseases; added information about reportable infectious diseases: added information about the CDC's National Wastewater Surveillance System (NWSS); mentioned WHO's role of drawing attention to neglected tropical diseases (NTDs); added a Focus Your Perspective box on Bioterrorism Surveillance; moved the information on reduction and eradication of disease to section 19.6 (section 19.5)
- Changed the title and expanded the scope of the section now called *Trends in Infectious Diseases* (was *Emerging Infectious Diseases*); added dengue fever to the list of diseases likely to increase as a result of climate change (section 19.6)
- Moved the information about Standard Precautions into the main narrative (section 19.7)

Chapter 20 – Antimicrobial Medications

- Added mentions of the contributions of the "penicillin girls" and Dorothy Hodgkin Crowfoot to penicillin discovery and production; added the mention of the contributions of Albert Schatz and Elizabeth Bugie to the discovery of streptomycin; added the subsection *Rational Drug Design* (section 20.1)

- Reorganized the discussion of protein synthesis inhibitors, separating the drugs according to the ribosomal subunit they bind; updated the discussion of antimicrobials that act against *Mycobacterium tuberculosis* (section 20.3)
- Added a bullet list that describes the sources of antimicrobial resistance genes; updated the information about resistant strains of *M. tuberculosis* and *Neisseria gonorrhoeae* (section 20.5)
- Expanded the section on antiviral and antifungal drugs to include recent examples (sections 20.6 and 20.7, respectively)
- Added the mention of neglected tropical diseases; added an explanation of why antibacterial drugs can often be used to treat diseases caused by apicomplexan protozoa; updated the drug options listed in table 20.5 (section 20.8)

Chapter 21 – Respiratory System Infections

- Reorganized the anatomy/physiology descriptions to emphasize characteristics in common between the upper respiratory tract and the lower respiratory tract (section 21.1)
- Reorganized the discussion of tuberculosis treatment, using a bullet list to emphasize the two-phase treatment approach (section 21.4)
- Substantially revised and updated the influenza coverage; moved the coverage of COVID-19, SARS, and MERS forward to increase the emphasis; updated the coverage of COVID-19 to include emerging variants, new treatments, and vaccines; added information about the new respiratory syncytial virus (RSV) vaccines (section 21.5)

Chapter 22 – Skin Infections

- Added a discussion of acne as an inflammatory disease of the pilosebaceous unit (sebaceous gland and associated arrector pili), a term now introduced in figure 22.1 (section 22.2)
- Added a new Focus Your Perspective 22.1 on the mpox outbreak (section 22.3)

Chapter 23 – Wound Infections

- Expanded the discussion of gas gangrene to include two forms: traumatic and spontaneous (section 23.3); revised the discussion of bacterial infections of bite wounds (section 23.4)

Chapter 24 – Digestive System Infections

- Changed the headings and modified the coverage of the periodontal disease descriptions to reflect updates in terminology: *Periodontal Disease* is now *Dental Plaque-Induced Periodontal Diseases,* and *Acute Ulcerative Gingivitis* is now *Necrotizing Periodontal Diseases* (section 24.2)
- Revised the signs and symptoms discussion of oral herpes to emphasize the differences between the initial infection and recurrences (section 24.3)
- Rearranged some information in the disease discussions to emphasize similarities among the diseases; added information on XDR *Salmonella* Typhi as well as the new typhoid conjugate vaccine used in countries where typhoid is endemic (section 24.4)
- Added CDC's 2023 recommendation for hepatitis B screening; added the 2023 discovery that HCV uses flavin adenine dinucleotide (FAD) as 5′ cap on its RNA thus preventing its genome from being recognized by pattern recognition receptors in infected cells; moved the graph that compares the incidence of acute hepatitis A, B, and C to the end of the hepatitis coverage (section 24.6)

Chapter 25 – Cardiovascular and Lymphatic System Infections

- Changed the chapter title to emphasize the separate components of the cardiovascular and lymphatic systems; increased the emphasis on the terms *cytokine storm* and *sepsis* by introducing them in the chapter introduction
- Reorganized the anatomy/physiology descriptions to emphasize the separate components of the cardiovascular and lymphatic systems (section 25.1)
- Moved the description of bacterial sepsis forward to increase the emphasis, explained that sepsis can result from any systemic microbial infection (bacterial, viral, or fungal), and revised the bacterial sepsis pathogenesis discussion to include a bullet list; revised the Lyme disease epidemiology discussion, using a bullet format to describe and illustrate the *Ixodes scapularis* life cycle; converted the brucellosis "disease person" table to a standard table (section 25.2)
- Mentioned that viral infections can lead to sepsis; rearranged and aligned the coverage of the arboviral diseases (dengue and severe dengue, chikungunya, Zika virus disease, and yellow fever) to emphasize the similarities; added information about the new dengue vaccine used in countries where dengue is endemic (section 25.3)

- Simplified the description and the illustration of the malaria life cycle, and also added information about the new malaria vaccines; added a section on leishmaniasis (section 25.4)

Chapter 26 – Nervous System Infections

- Simplified the anatomy/physiology discussion (section 26.1)
- Updated the discussion of circulating vaccine-derived poliovirus strains (cVDPV) (section 26.3)
- Added *Mycobacterium lepromatosis* as another causative agent of leprosy; added descriptions of two additional forms of botulism: adult intestinal toxemia botulism and iatrogenic botulism; added mention that certain strains of *C. butyricum* and *C. baratii* produce botulinum toxin (section 26.7)

Chapter 27 – Genitourinary Tract Infections

- Added a description of how glycoprotein in the urine protects against urinary tract infections (section 27.2)
- Modified the coverage of leptospirosis to focus on mild versus severe disease symptoms (section 27.2)
- Distinguished between the terms *sexually transmitted infection (STI)* and *sexually transmitted disease (STD)* in the Focus on a Disease box; increased the emphasis on the importance of pelvic inflammatory disease (PID) (section 27.4)
- Moved the coverage of human papilloma STIs forward; substantially revised the HIV/AIDS coverage and, as part of that, included an updated replication cycle figure and mention of the new capsid inhibitor used for treatment (section 27.5).

Chapter 28 – Microbial Ecology

- Introduced the term *harmful algal bloom* (section 28.3)

Chapter 29 – Environmental Microbiology: Treatment of Water, Wastes, and Polluted Habitats

- Modified the definition of wastewater, and added the definition of sewage; added the terms *nuisance bloom* and *harmful algal bloom* (section 29.1)

Chapter 30 – Food Microbiology

- Modified the description of FoodNet (section 30.4)

McGraw Hill connect®

A complete course platform

Connect enables you to build deeper connections with your students through cohesive digital content and tools, creating engaging learning experiences. We are committed to providing you with the right resources and tools to support all your students along their personal learning journeys.

65%
Less Time Grading

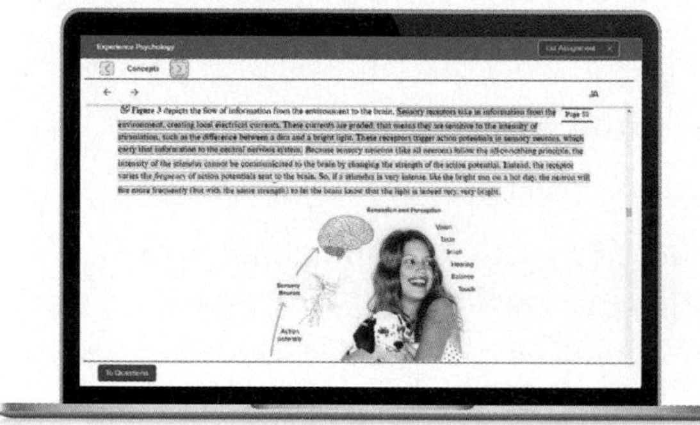

Laptop: Getty Images; Woman/dog: George Doyle/Getty Images

Every learner is unique

In Connect, instructors can assign an adaptive reading experience with SmartBook® 2.0. Rooted in advanced learning science principles, SmartBook 2.0 delivers each student a personalized experience, focusing students on their learning gaps, ensuring that the time they spend studying is time well spent. **mheducation.com/highered/connect/smartbook**

Study anytime, anywhere

Encourage your students to download the free ReadAnywhere® app so they can access their online eBook, SmartBook® 2.0, or Adaptive Learning Assignments when it's convenient, even when they're offline. And since the app automatically syncs with their Connect account, all of their work is available every time they open it. Find out more at **mheducation.com/readanywhere**

"I really liked this app— it made it easy to study when you don't have your textbook in front of you."

Jordan Cunningham, a student at *Eastern Washington University*

Effective tools for efficient studying

Connect is designed to help students be more productive with simple, flexible, intuitive tools that maximize study time and meet students' individual learning needs. Get learning that works for everyone with Connect.

Education for all

McGraw Hill works directly with Accessibility Services departments and faculty to meet the learning needs of all students. Please contact your Accessibility Services Office, and ask them to email **accessibility@mheducation.com**, or visit **mheducation.com/about/accessibility** for more information.

Affordable solutions, added value

Make technology work for you with LMS integration for single sign-on access, mobile access to the digital textbook, and reports to quickly show you how each of your students is doing. And with our Inclusive Access program, you can provide all these tools at the lowest available market price to your students. Ask your McGraw Hill representative for more information.

Solutions for your challenges

A product isn't a solution. Real solutions are affordable, reliable, and come with training and ongoing support when you need it and how you want it. Visit **supportateverystep.com** for videos and resources both you and your students can use throughout the term.

Updated and relevant content

Our new Evergreen delivery model provides the most current and relevant content for your course, hassle-free. Content, tools, and technology updates are delivered directly to your existing McGraw Hill Connect® course. Engage students and freshen up assignments with up-to-date coverage of select topics and assessments, all without having to switch editions or build a new course.

Acknowledgments

First and foremost, special thanks goes to Gene Nester, who led the team that wrote the first version of what became *Microbiology, A Human Perspective.* His efforts helped pioneer a new type of introductory microbiology textbook—with disease chapters organized by body systems—designed specifically for students entering healthcare-related fields. The editions since then have proudly built on that original vision.

Special thanks also go to our families, friends, and colleagues for picking up the many hairs we tore out while working on the textbook. Revising a textbook is an all-consuming task—from the initial development stage to proofing the pages during production—and numerous people have acted as advisors and cheerleaders throughout. This text would not exist without the contributions of our strong group of supporters.

We would also like to thank the reviewers and other instructors who guided us as we developed this release, as well as those whose input has helped the text evolve over the years. Deciding what to eliminate, what to add, and what to rearrange is always difficult, so we appreciate your suggestions.

Past students have been incredibly helpful as well. Every question helps us decide which parts of the textbook need more clarification, and every compliment lets us know when we are on the right track.

A list of acknowledgments is not complete without thanking our fearless leaders and friends at McGraw Hill. Our product portfolio manager Lauren Vondra stood by us, cheering us on and providing advice and support when needed. Product developer Monica Toledo kept us on task, making sure that everything was turned in on time. Content project manager Paula Patel showed tremendous patience as we kept adding the latest and greatest until the very end, and copy editor Kevin Campbell caught the typos we introduced while doing that. Executive marketing manager Tami Hodge helped ensure that word got out about this new release, allowing it to find its way into your hands. Additionally, we would like to thank digital content project manager Rachael Hillebrand for helping produce our digital resources that support the text and Lisa Burgess, who provided many wonderful photographs.

We hope that this text will be interesting and educational for students and helpful to instructors. Our goal is excellence, so with that in mind we would appreciate any comments and suggestions from our readers.

Denise Anderson
Mira Beins
Ann Auman
Jennifer Walker
Sarah Salm, digital author

Reviewers for the Tenth Edition

Andrea R. Beyer, *Virginia State University*
Bruce Bleakley, *South Dakota State University*
Anar A. Brahmbhatt, *San Diego Mesa College*
Linda D. Bruslind, *Oregon State University*
Carron Bryant, *East Mississippi Community College*
Matthew B. Crook, *Weber State University*
Jeremiah Davie, *D'Youville College*
Karim Dawkins, *Broward College*
Matthew Dodge, *Olympic College*
Robert A. Holmes, *University of Missouri–Kansas City*
Joshua Hughes, *Dakota State University*
Ilko B. Iliev, *Southern University at Shreveport*
Karen Kowalski, *Tidewater Community College*
Ruhul Kuddus, *Utah Valley University*
Lorie Lana, *Stevenson University*
Eddystone C. Nebel, *Delgado Community College*
Olabisi Ojo, *Albany State University*
Jennifer L. B. Roshek, *Stevenson University*
Dan Smith, *Seattle University*
Renato V. Tameta, *Schenectady County Community College*
Krystal Taylor, *Beaufort County Community College*
Roger Wainwright, *University of Central Arkansas*

Contents

5 Control of Microbial Growth 117

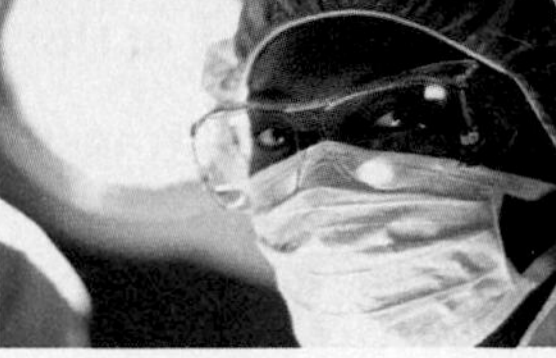
Arthur Tilley/Getty Images

6 Microbial Metabolism: Fueling Cell Growth 137

Comstock/PunchStock

7 The Blueprint of Life, from DNA to Protein 175

Molekuul/SPL/age fotostock

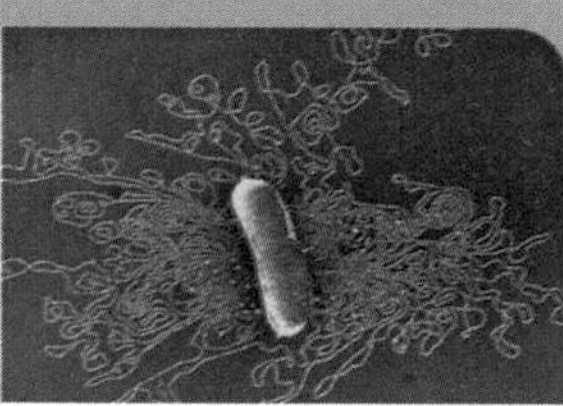
Dr. Gopal Murti/Science Source

atic12/123RF

Diane Keough/Moment/Getty Images

Heather Davies/Science Photo Library/Getty Images

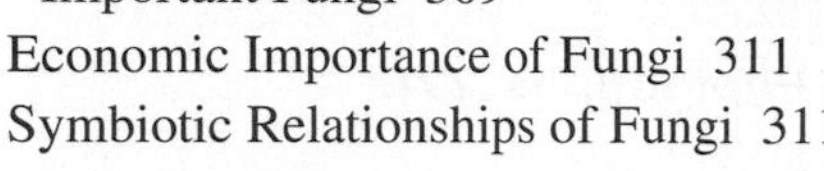

Steve Gschmeissner/Science Photo Library/Getty Images

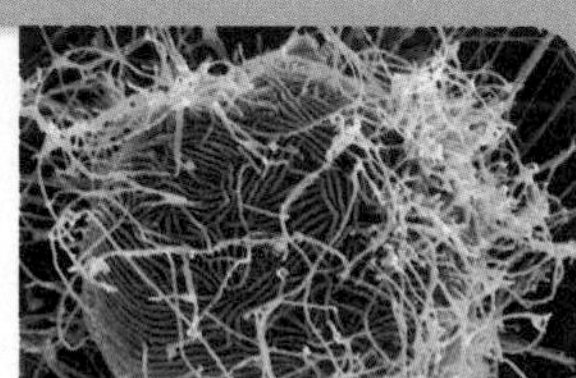

National Institute of Allergy and Infectious Diseases (NIAID)/CDC

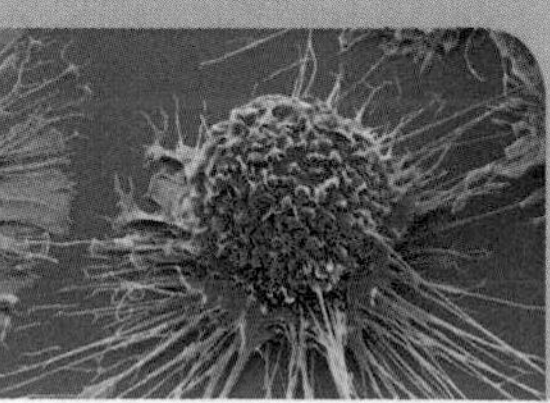

Steve Gschmeissner/Science Photo Library/Alamy Stock Photo

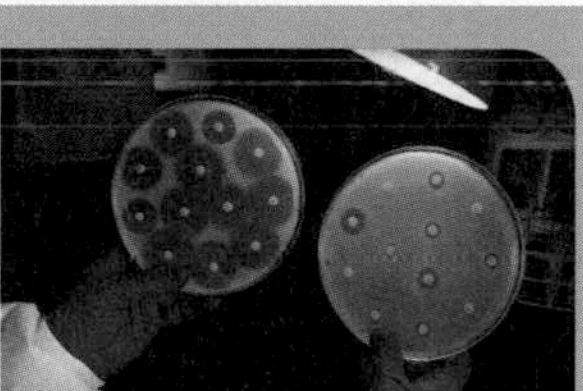
James Gathany/CDC

PART IV
Infectious Diseases

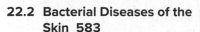

Steve Gschmeissner/SPL/Science Source

Steve Gschmeissner/Science Photo Library/Alamy Stock Photo

Robert Glusic/Getty Images

Images by Tang Ming Tung/Getty Images

1 Humans and the Microbial World

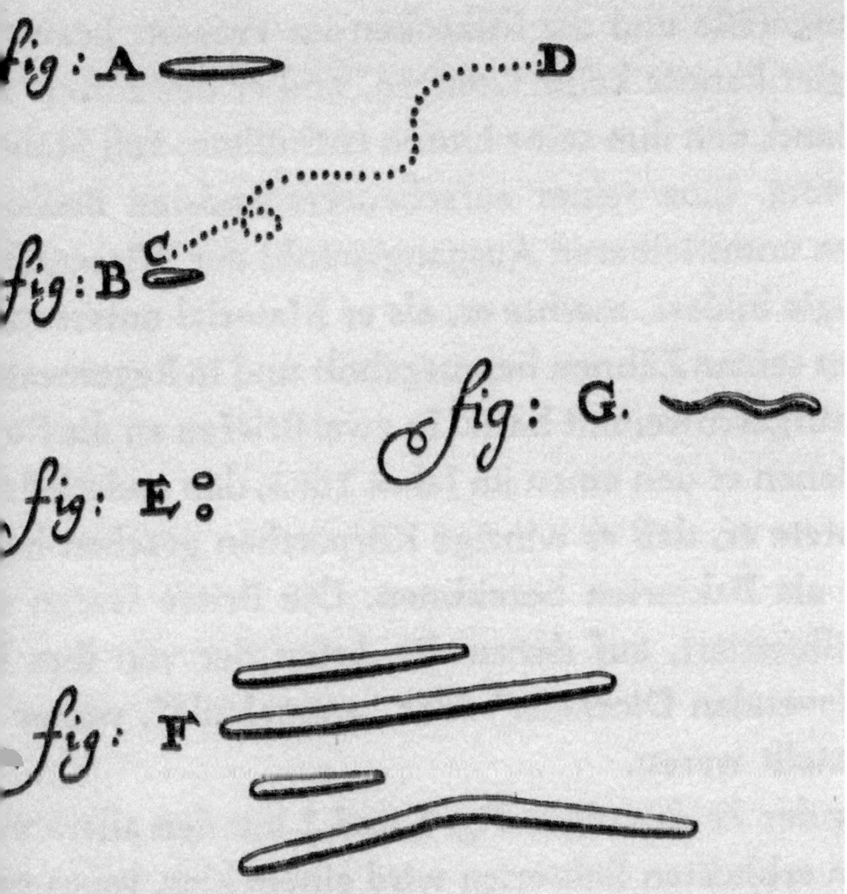

Drawings that van Leeuwenhoek made in 1683 of microorganisms he saw through his single-lens microscope. He also observed organism B moving from position C to D. *(Interfoto/Alamy Stock Photo)*

KEY TERMS

Domain The highest level in biological classification. There are three domains: Bacteria, Archaea, and Eukarya.

Eukaryote Organism composed of one or more eukaryotic cells; members of the domain Eukarya are eukaryotes.

Eukaryotic Cell Cell type characterized by a membrane-bound nucleus.

Prion An acellular infectious agent consisting only of protein.

Prokaryote Single-celled organism consisting of a prokaryotic cell; members of the domains Bacteria and Archaea are prokaryotes.

Prokaryotic Cell Cell type characterized by the lack of a membrane-bound nucleus.

Viroid An acellular infectious agent consisting only of RNA.

Virus An acellular infectious agent consisting of nucleic acid surrounded by a protein coat.

A Glimpse of History

Microbiology as a science was born in 1674 when Antonie van Leeuwenhoek, an inquisitive Dutch fabric merchant, looked at a drop of lake water through a glass lens he had carefully made. Although many people before him had used curved glass to magnify objects, Leeuwenhoek's skilled hands made a lens that uncovered a startling and amazing sight—the world of microbes. As van Leeuwenhoek wrote in a letter to the Royal Society of London, he saw

> Very many little animalcules, whereof some were roundish, while others a bit bigger consisted of an oval. On these last, I saw two little legs near the head, and two little fins at the hind most end of the body. Others were somewhat longer than an oval, and these were very slow a-moving, and few in number. These animalcules had diverse colours, some being whitish and transparent; others with green and very glittering little scales, others again were green in the middle, and before and behind white; others yet were ashed grey. And the motion of most of these animalcules in the water was so swift, and so various, upwards, downwards, and round about, that 'twas wonderful to see.

Before van Leeuwenhoek made these observations, Robert Hooke, an English microscopist, saw another kind of microorganism. In 1665, he described what he called a "microscopical mushroom." His drawing was so accurate that his specimen could later be identified as a common bread mold. Hooke also described how to make the kind of microscope that van Leeuwenhoek constructed almost 10 years later. Both men deserve equal credit for revealing the world of microbes—a world you are about to study.

Microbiology is the study of an amazing world made up of members too small to be seen without the aid of a microscope. Antonie van Leeuwenhoek described this world when he observed what he called "animalcules" through his simple microscope (**figure 1.1**). What he saw were **microorganisms** (organisms too small to see with the naked eye), including bacteria, protozoa, and some fungi and algae. The microbial world also includes viruses and other infectious agents that are not considered organisms because they are not composed of cells; they are acellular. When referring to general members of the microbial world, the term **microbe** is often used.

Microorganisms are the foundation for all life on Earth. They have existed on this planet for about 3.5 billion years, and over this time, plants, animals, and modern microorganisms have evolved from them. Even today, they continue to be a driving force in the evolution of all living things. Microorganisms may be small, but as you are about to learn, our life depends on their activities.

FIGURE 1.1 Model of van Leeuwenhoek's Microscope The original made in 1673 could magnify an object almost 300 times. Tetra Images/Alamy Stock Photo

? **What kinds of organisms did van Leeuwenhoek observe through his microscope?**

1.1 ■ The Dispute over Spontaneous Generation

Learning Outcomes

1. Describe the key experiments of scientists who disproved spontaneous generation.
2. Explain how the successful challenge to the idea of spontaneous generation led to the Golden Age of Microbiology.
3. Describe the scientific method, using Pasteur's swan-necked flask experiment as an example.

The discovery of microorganisms in various specimens raised an interesting question: "Where did these microscopic forms originate?" Some people believed that the tiny worms and other life-forms they saw arose from non-living material in a process known as **spontaneous generation.** This was challenged by an Italian biologist and physician, Francesco Redi. In 1668, he used a simple experiment to show that worms found on rotting meat originated from fly eggs, not from the decaying meat as supporters of spontaneous generation believed. In his experiment, Redi covered the meat with fine gauze that prevented flies from depositing their eggs; when he did this, no worms appeared. Despite Redi's work, it took more than 200 years and many experiments to amass conclusive evidence that microorganisms did not arise by spontaneous generation.

Early Experiments

In 1749, John Needham, a scientist and Catholic priest, showed that flasks containing various broths (made by soaking a nutrient source such as hay or chicken in water) gave rise to microorganisms even when the flasks were boiled and sealed with a cork. At that time, brief boiling was thought to kill all organisms, so this suggested that microorganisms did indeed arise spontaneously.

In 1776, the animal physiologist and priest Lazzaro Spallanzani obtained results that contradicted Needham's experiments; no bacteria appeared in Spallanzani's broths after boiling. His experiments differed from Needham's in two significant ways: Spallanzani boiled the broths for longer periods, and he sealed the flasks by melting their glass necks closed. Using these techniques, he repeatedly demonstrated that broths remained sterile (free of microorganisms). However, if the neck of the flask cracked, the broth rapidly became cloudy due to the growth of organisms. Spallanzani concluded that microorganisms had entered the broth with the air, and the corks used by Needham and other investigators did not keep them out.

Spallanzani's experiments did not stop the controversy. Some people argued that the heating process destroyed a "vital force" in the air that was necessary for spontaneous generation, and so the debate continued.

Experiments of Pasteur

One giant in science who helped disprove spontaneous generation was Louis Pasteur, the French chemist considered by many to be the founder of modern microbiology. In 1861, he did a series of clever experiments. First, he demonstrated that air contains microorganisms. He did this by filtering air through a cotton plug, trapping microorganisms. He then examined the trapped microorganisms with a microscope and found that many looked identical to those described by others who had been studying broths. When Pasteur dropped the cotton plug into a sterilized broth, the broth became cloudy from the growth of these microorganisms.

Most important, Pasteur demonstrated that sterile broths in specially constructed swan-necked flasks remained sterile even when left open to air (**figure 1.2**). Microorganisms from the air settled in the bends of the flask necks, never reaching the broth. Only when the flasks were tipped would microorganisms enter the broth and grow. Pasteur's simple and elegant experiments ended the arguments that unheated air or the broths themselves contained a "vital force" necessary for spontaneous generation. They led to the theory of **biogenesis,** the production of living things from other living things (*bio* means "life"; *genesis* means "to create").

Experiments of Tyndall

Although most scientists were convinced by Pasteur's experiments, some remained skeptical because they could not reproduce his results. An English physicist, John Tyndall, finally explained the conflicting data and, in turn, showed that Pasteur was correct. Tyndall found that some broths were sterilized by boiling for 5 minutes, whereas others, most notably broths made from hay, still contained living microorganisms even after boiling for 5 hours! Even when hay was merely present in the laboratory, broths that had previously been sterilized by boiling for

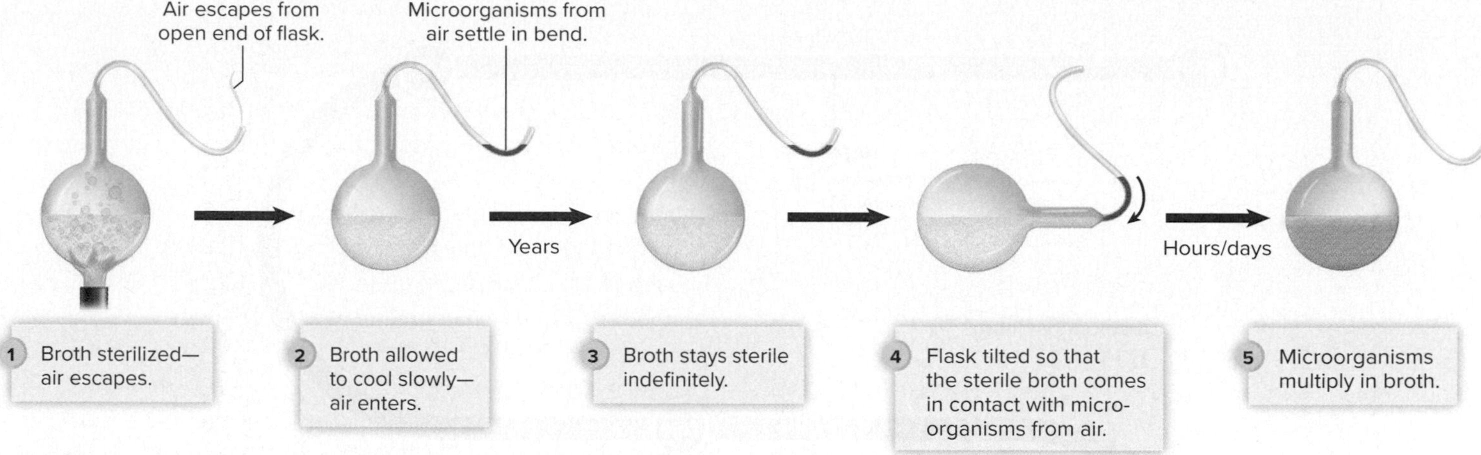
Air escapes from open end of flask.
Microorganisms from air settle in bend.
Years
Hours/days
1 Broth sterilized—air escapes.
2 Broth allowed to cool slowly—air enters.
3 Broth stays sterile indefinitely.
4 Flask tilted so that the sterile broth comes in contact with microorganisms from air.
5 Microorganisms multiply in broth.

FIGURE 1.2 **Pasteur's Experiment with the Swan-Necked Flask**

? **How did this experiment end arguments that a "vital force" in the air was necessary for spontaneous generation?**

5 minutes could not be sterilized by boiling for several hours. What was going on? Tyndall finally realized that the hay contained heat-resistant forms of microorganisms. When hay was brought into the laboratory, dust particles must have transferred these heat-resistant forms to the broths. Tyndall concluded that some microorganisms exist in two forms: a cell easily killed by boiling, and one that is heat resistant. In the same year (1876), a German botanist, Ferdinand Cohn, discovered **endospores,** the heat-resistant forms of some bacteria.

The extreme heat resistance of endospores explains the differences between Pasteur's results and those of other investigators. Organisms that produce endospores are commonly found in soil and were likely present in broths made from hay. Pasteur used only broths made with sugar or yeast extract, so his experiments probably did not have endospores. Scientists at the time did not appreciate the importance of the source of the broth, but in hindsight, the source was critical. This points out an important lesson for all scientists: When repeating an experiment, all conditions must be reproduced as closely as possible. What may seem like a trivial difference might be extremely important.

The Golden Age of Microbiology

The work of Pasteur and others in disproving spontaneous generation started an era called the Golden Age of Microbiology, during which time the field of microbiology blossomed. Many important advances were made during this period, including discoveries that led to the acceptance of the suggestion that microorganisms cause certain diseases, a principle now called the Germ Theory of Disease.

Figure 1.3 lists some of the important advances in microbiology made over the years in the context of other historical events. Rather than cover more history now, we will return to many of these milestones in brief stories called "A Glimpse of History" that open each chapter.

The Scientific Method

The dispute over spontaneous generation offers an excellent example of the process of science. This process, called the **scientific method,** separates science from intuition and beliefs. The scientific method involves a series of steps, including:

- **Making an observation and asking a question about that situation.** An example from this chapter was the observation that microorganisms were present in various examined specimens. This observation led to the question, "Where did the microorganisms originate?"
- **Developing an explanation and then devising an experiment that tests the explanation.** A testable explanation of an observation is called a **hypothesis,** and experiments are done to test the hypothesis. The dispute over spontaneous generation led to two opposing hypotheses: biogenesis and spontaneous generation. Various people designed different experiments to test the hypotheses.
- **Doing the experiment, collecting the data, and drawing a conclusion.** Experiments such as the one illustrated in figure 1.2 provided data about the growth of microorganisms in previously sterile broths. In doing a scientific experiment, a crucial component is a **control.** A control helps rule out alternative explanations of the results by showing that the only variable feature in the experiment was the characteristic being tested. Pasteur's swan-necked flask experiment was brilliantly designed because it provided the following control: After showing that the fluid in the swan-necked flasks remained sterile even when opened to air, he tipped the flasks so that bacteria could enter the fluid. By doing this, he showed that nothing in his original setup would have prevented bacteria from growing in the broth.
- **Communicating the methods, results, and conclusions.** Scientists share their work by publishing it in scientific

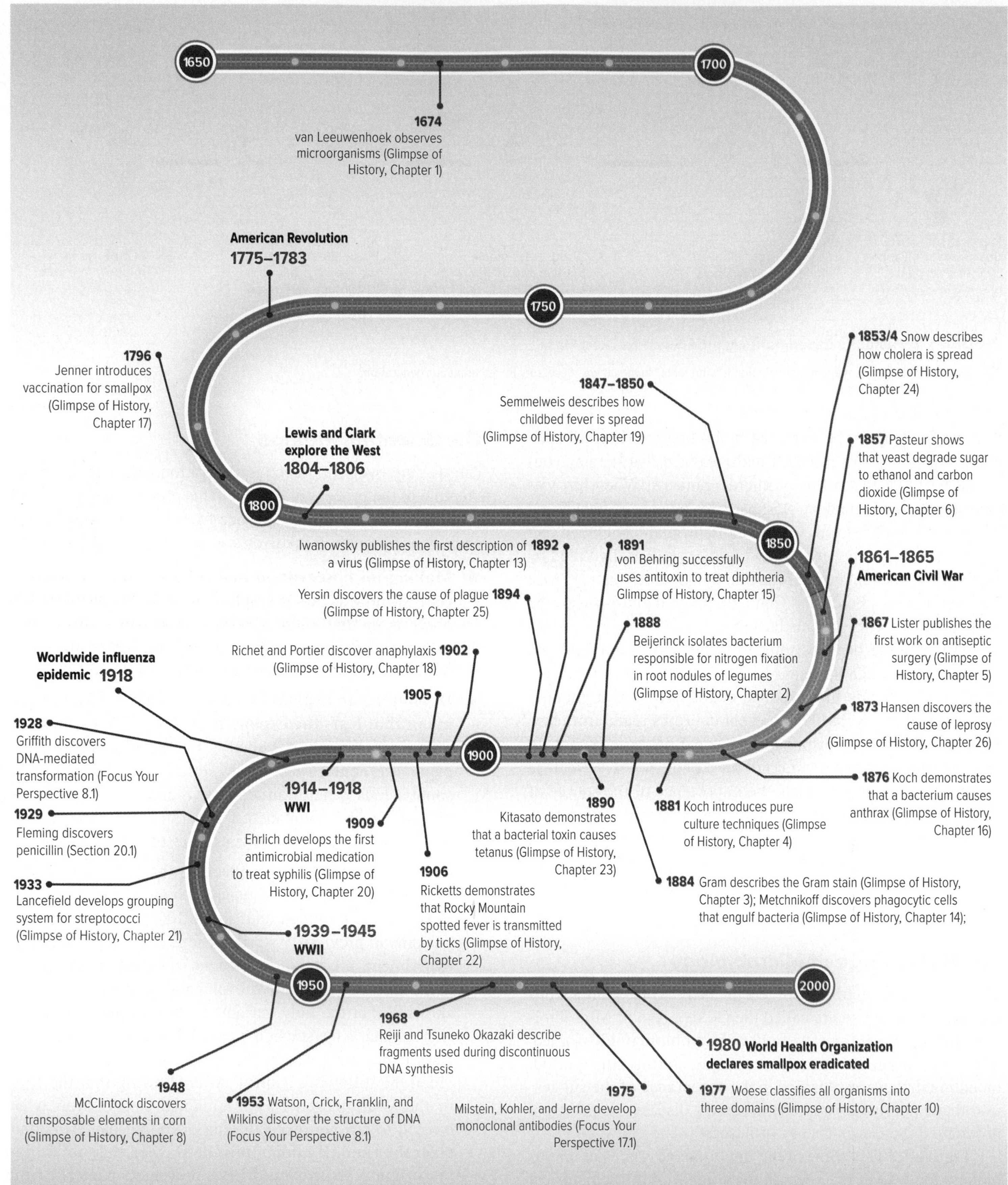

FIGURE 1.3 Historical Events in Microbiology Some major milestones in microbiology—and their timeline in relation to other historical events. The gold band indicates the Golden Age of Microbiology.

? **What is the Golden Age of Microbiology?**

journals. This step is particularly important because it allows other scientists to repeat the experiment to ensure the validity of the findings. Today, the respected scientific journals use a review process in which other experts in the field read communications before they are published. If deficiencies or flaws are noticed, the reviewers give suggestions for improving the experiments.

When an extensive amount of experimental evidence supports a hypothesis, that explanation may become a scientific **theory,** such as the Germ Theory of Disease. Note that the scientific meaning of the word *theory* is far different from the meaning of the word in common language, which is "a speculation or guess."

As you read the information in this textbook, continually challenge yourself by asking questions about what you have learned. If you find yourself asking a question such as, "How does that happen?" try to develop a hypothesis and then devise an experiment. As you do this, consider the controls you could use. Start learning to think like a scientist!

MicroAssessment 1.1

Experiments of Pasteur and Tyndall helped disprove spontaneous generation by showing that life arises from life. Many important discoveries were made during the Golden Age of Microbiology, including ones that led to the acceptance of the Germ Theory of Disease. The scientific method uses experimental evidence, including proper controls, to support or refute hypotheses.

1. Describe Pasteur's experiment that disproved the idea that a "vital force" in air was responsible for spontaneous generation.
2. How is the meaning of the word "theory" in science different from its meaning in everyday conversation?
3. Why is it important for scientists to repeat the experiments of others?

1.2 ■ Microbiology: A Human Perspective

Learning Outcomes

4. Explain the importance of microorganisms in the health of humans and the surrounding environment.
5. List three commercial benefits of microorganisms.
6. Describe why microorganisms are useful research tools.
7. Describe the role of microbes in disease, including examples of past triumphs and remaining challenges.

Microorganisms have a tremendous impact on all living things. We could not survive without them, and they also make our lives much more comfortable. At the same time, microbes can be harmful, and they have killed far more people than have ever been killed in war.

The Human Microbiome

Your body carries an enormous population of microorganisms—tens of trillions of bacterial cells alone. Many sources claim that the body carries 10 times as many microbial cells as human cells, but recent and probably more accurate estimates indicate that the ratio is likely closer to 1:1. Regardless, scientists have known for years that these microorganisms, collectively referred to as the **normal microbiota,** play an essential role in human health. For example, they prevent disease by competing with disease-causing microbes, help to degrade foods that the body otherwise could not digest, and promote the development of the immune system. In fact, studies indicate that early exposure to certain common microorganisms lessens the likelihood that an individual will develop allergies, asthma, and some other diseases. According to what is sometimes referred to as the "Old Friends" hypothesis, this early exposure helps the immune system learn to distinguish "friendly" microbes from those that can cause severe disease. In addition, animal studies suggest that the composition of the normal microbiota can affect brain chemistry and behavior, as well as the tendency to gain or lose weight.

The important role of the normal microbiota became even more obvious in recent years, thanks in part to the **Human Microbiome Project.** This coordinated set of studies used DNA sequencing technologies to characterize the microbial communities that inhabit the human body. The term **microbiome** has two overlapping meanings: (1) the total genetic content of a microbial community and (2) the microbial community itself. While the different meanings might seem confusing, they are actually quite similar because at this point the communities must be examined by studying their genetic material. The reason for this is that less than 1% of microorganisms can currently be grown in the laboratory, so for every microbe that had been studied in the laboratory, more than 99 others can only be characterized using DNA sequencing technologies.

The Human Microbiome Project changed the way scientists view the human body and also revealed how much more there is to discover about our microbial partners. To understand their significance, think of Earth's ecosystems (the environments and their interacting inhabitants). Over time, an interacting assortment of organisms has evolved to live in a given environment, resulting in a relatively stable community. Sudden changes can alter individual populations, often with negative consequences to the community as a whole. In turn, a disturbance in one ecosystem can affect the overall health of the planet. The human body, like a planet, is composed of various ecosystems—for example, the desert-like dry areas of the skin, and the nutrient-rich environment of the intestinal tract. An important part of these ecosystems is a population of interacting microbes. Disturbances in a microbial population can create an imbalance that may have negative consequences to that community, which, in turn, can harm a person's health. Observations such as these have led some scientists to suggest that the human body be considered a superorganism, meaning that our own cells interact with the body's normal microbiota to form a single cooperative unit.

The human microbiome's effect on health and disease is an exciting area of active research, but the results must be interpreted with caution. For example, researchers have found that the intestinal microbiome of people diagnosed with depression differs from those who report a good quality of life, but this correlation could be an effect of mood—perhaps even dietary changes associated with certain moods—rather than a cause. Likewise, bacterial species associated with gum disease have been found in the brains of people with Alzheimer's disease, but again, this could be effect rather than cause. Continuing studies aim to clarify the situation.

MicroByte

The Global Microbiome Conservancy is collecting fecal samples from people around the world in an effort to study and preserve the diversity of intestinal bacteria.

Microorganisms in the Environment

Microorganisms are the masters of recycling, and without them we would run out of certain nutrients. For instance, humans and other animals all require nitrogen, an essential part of nucleic acids and proteins. A plentiful source of nitrogen is N_2—the most common gas in the atmosphere—yet neither plants nor animals can use it. Instead, we depend on certain microbes that convert N_2 into a form of nitrogen that other organisms can use, a process called nitrogen fixation. Without nitrogen-fixing microbes, life as we know it would not exist.

Microorganisms are also important because they can degrade certain materials that other organisms cannot. Cellulose (an important component of plants) is an excellent example. Although humans and other animals cannot digest cellulose, certain microorganisms can, which is why leaves and fallen trees do not pile up in the environment. A group of cellulose-degrading microorganisms live in the specialized stomachs of ruminants (a group of plant-eating animals that includes cattle, sheep, and deer), helping those animals digest plant material; without the assistance of microbes, the ruminants would starve.

In recognition of the important role that microorganisms play in all aspects of life, additional programs promise to expand the scope of existing DNA-based studies. The National Microbiome Initiative (NMI) supports research on the microbiomes of humans as well as the surrounding environment, and an international effort called the Earth BioGenome Project aims to determine the DNA sequence of all known animal, plant, protozoan, and fungal species.

Commercial Benefits of Microorganisms

In addition to the crucial roles microorganisms play in our very existence, they also have made life more comfortable for humans over the centuries.

Food Production

Microorganisms have been used in food production since ancient times. In fact, Egyptians used yeast to make bread and beer. Virtually every population that raised milk-producing animals such as cows and goats also developed procedures to ferment milk, giving rise to foods such as yogurt, cheeses, and buttermilk. Today, the bacteria added to some fermented milk products are advertised as probiotics (live microorganisms that provide a health benefit), protecting against digestive disruptions.

Biodegradation

Microorganisms play essential roles in degrading various environmental pollutants. These include materials in wastewater, as well as toxic chemicals in contaminated soil and water. Bacteria also lessen the damage from oil spills. In some cases, microorganisms are added to pollutants to hasten their decay, a process called **bioremediation.**

Commercially Valuable Products from Microorganisms

Microorganisms synthesize a wide variety of commercially valuable products. Examples include antibiotics used to treat infectious diseases; ethanol used as a biofuel; hydrogen gas and certain oils potentially used as biofuels; amino acids used as dietary supplements; insect toxins used in insecticides; cellulose used in headphones; and polyhydroxybutyrate used in the manufacture of disposable diapers and plastics.

Biotechnology

Biotechnology—the use of microbiological and biochemical techniques to solve practical problems—relies on members of the microbial world. Information learned by studying microorganisms led to easier production of many medications, including the insulin used to treat diabetes. In the past, insulin was isolated from the pancreatic glands of cattle and pigs, but now certain microorganisms have been genetically engineered to make human insulin. The microbe-produced insulin is easier to obtain, and patients who use it have fewer allergic reactions than occurred with the animal-derived product. Biotechnology also allows scientists to genetically engineer plants to give them desirable qualities.

Microbes as Research Tools

Microorganisms are wonderful model organisms to study because they have the same fundamental metabolic and genetic properties as higher life-forms. All cells are composed of the same chemical elements, and they synthesize their cell structures by similar mechanisms. They all duplicate their DNA, and when they degrade foods to harvest energy, they do so via the same metabolic pathways. To paraphrase a Nobel Prize–winning microbiologist, Dr. Jacques Monod: What is true of elephants is also true of bacteria, and bacteria are much easier to study! In addition, bacteria can be used to obtain results very quickly because they grow rapidly and form billions of cells per milliliter on simple, inexpensive growth media. In fact, most major advances made in the last century toward understanding life have come through the study of microbes.

Microbes and Disease

Although most microbes are beneficial or not harmful, some are **pathogens,** meaning they can cause disease (a noticeable impairment in body function). The disease symptoms can result from damage caused by the pathogen's growth and products or by the body's immune system inadvertently damaging host tissues during the attempt to control the infection.

To appreciate the effect an infectious disease can have on a population, consider that more Americans died of influenza in 1918–1919 than were killed in World Wars I and II and the Korean, Vietnam, and Iraq wars combined. Estimates indicate that the COVID-19 pandemic resulted in the death of more than 15 million people worldwide by 2023, including over a million Americans.

Epidemics are not limited to human populations. The great famine in Ireland in the 1800s was due, in part, to a microbial disease of potatoes. A bacterial disease that kills olive trees was found in southern Italy in 2013 and has since spread to Spain, France, and Portugal, contributing to a recent worldwide drop in olive oil production. A fungal disease called "wheat blast" that devastated wheat crops in South America spread to Bangladesh in 2016, resulting in the loss of over 35,000 acres of crops that year. Two years later, it was found in Zambia, raising concerns about its potential spread to other parts of Africa. In 2001, a catastrophic outbreak of foot-and-mouth disease of livestock occurred in parts of England. To contain this viral disease, one of the most contagious known, almost 4 million pigs, sheep, and cattle were culled (selectively killed, in this case to limit spread of the disease). Since about 2007, African swine fever—a viral disease deadly to domestic and wild pigs—has gradually spread to over 70 countries, resulting in more than 800,000 pigs dying (from the disease or from culling). Most recently, a novel avian influenza virus that first appeared in 2021 is spreading around the world, killing wild birds and domestic poultry; in the United States alone, over 50 million poultry have died as a result (from the disease or from culling). Meanwhile, frog populations around the world have been decimated by chytridiomycosis, a fungal disease.

Past Triumphs

The Golden Age of Microbiology included an important period when scientists learned a great deal about pathogens. Between 1876 and 1918, most pathogenic bacteria were identified, and early work on viruses had begun. Once people realized that microbes could cause disease, they tried to prevent their spread. As illustrated in **figure 1.4,** the death rate due to infectious diseases has decreased dramatically since 1900,

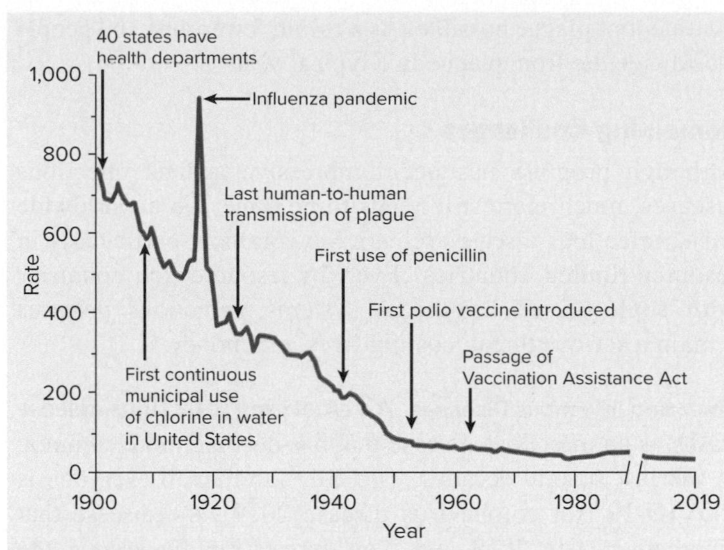

FIGURE 1.4 Trend in Death Rates Due to Infectious Diseases Crude death rate for infectious disease, United States, per 100,000 population per year.

? **Why would the creation of health departments lower the disease rate?**

due largely to preventing the spread of pathogens, developing vaccines to provide immunity, and using antibiotics to treat bacterial diseases when they do occur. To maintain this success, we must continue to develop new medications, vaccines, and disease-prevention strategies.

Perhaps the most significant triumph with respect to disease control was the eradication (elimination) of smallpox. This viral disease was one of the most devastating the world has ever known, killing about one-third of those infected. Survivors were sometimes blinded and often left with disfiguring scars. When Europeans carried the disease to the Americas, the effect on the populations of native inhabitants who had not been exposed before was catastrophic. A worldwide vaccination program eliminated the disease in nature, with no cases being reported since 1977. Laboratory stocks of the smallpox virus remain, however, raising the possibility that the virus could be used in bioterrorist attacks.

Polio, a disease that can cause paralysis and sometimes death, was once relatively common, but it has been nearly eliminated because of vaccination. In fact, the disease now occurs in only a few countries, and the goal is to eradicate it globally.

Plague is another major killer that has largely been brought under control. In the fourteenth century, one-third of the population of Europe, or approximately 25 million people, died of this bacterial disease in only 4 years (1347–1351). We now know that rodents can carry the bacterium, and their fleas can transmit the disease, so we take measures to control the rodent populations. We have also learned that the pneumonic form of the disease (meaning it is in the lungs) can spread from person to person through respiratory secretions, so special precautions are taken when a patient has pneumonic plague. In addition, the discovery of antibiotics in the twentieth century made treatment of plague possible. As a result, fewer than 100 people worldwide die from plague in a typical year.

Remaining Challenges

Although progress has been impressive against infectious diseases, much more still needs to be done. On a worldwide basis, infectious diseases remain too common, particularly in resource-limited countries. Even in resource-rich countries with sophisticated healthcare systems, infectious diseases remain a serious threat, costing lives and money.

Emerging Infectious Diseases An **emerging infectious disease (EID)** is an infectious disease that has become more common in the last several decades. The EID familiar to everyone is COVID-19 (for coronavirus disease 2019), the disease that emerged in late 2019 and then spread rapidly around the globe. COVID-19 is caused by a virus called SARS-CoV-2 (for severe acute respiratory syndrome coronavirus 2), which is commonly referred to as the COVID-19 virus. Some EIDs, such as malaria and tuberculosis, have been present for hundreds or thousands of years but have spread or become more common recently, often after disease-prevention methods controlled them previously. Others, like COVID-19, are new or newly recognized; examples include mpox (previously called monkeypox), Ebola virus disease (EVD), congenital Zika syndrome, *Candida auris* infection, hepatitis C, severe acute respiratory syndrome (SARS), Middle East respiratory syndrome (MERS), certain types of influenza, Lyme disease, acquired immunodeficiency syndrome (AIDS), mad cow disease (bovine spongiform encephalopathy), and hantavirus pulmonary syndrome (**figure 1.5**).

Some diseases arise as infectious agents evolve to infect new hosts, cause different types of damage, or become more difficult to treat because of antibiotic resistance. Genetic analysis indicates that the virus that causes COVID-19 arose from a strain that infects bats. HIV-1 (**h**uman **i**mmunodeficiency **v**irus type 1), the most common type of HIV to cause AIDS, arose from a virus that infects chimpanzees. A bacterium called *E. coli* O104:H4, which caused a severe foodborne diarrheal outbreak in Europe, appears to have gained the ability to make a specific toxin by acquiring genes from a related organism. Tuberculosis and malaria have increased in incidence in recent years, in part because the causative organisms developed resistance to many of the available medications.

As the rapid spread of COVID-19 around the globe certainly demonstrated, mobile populations can contribute to

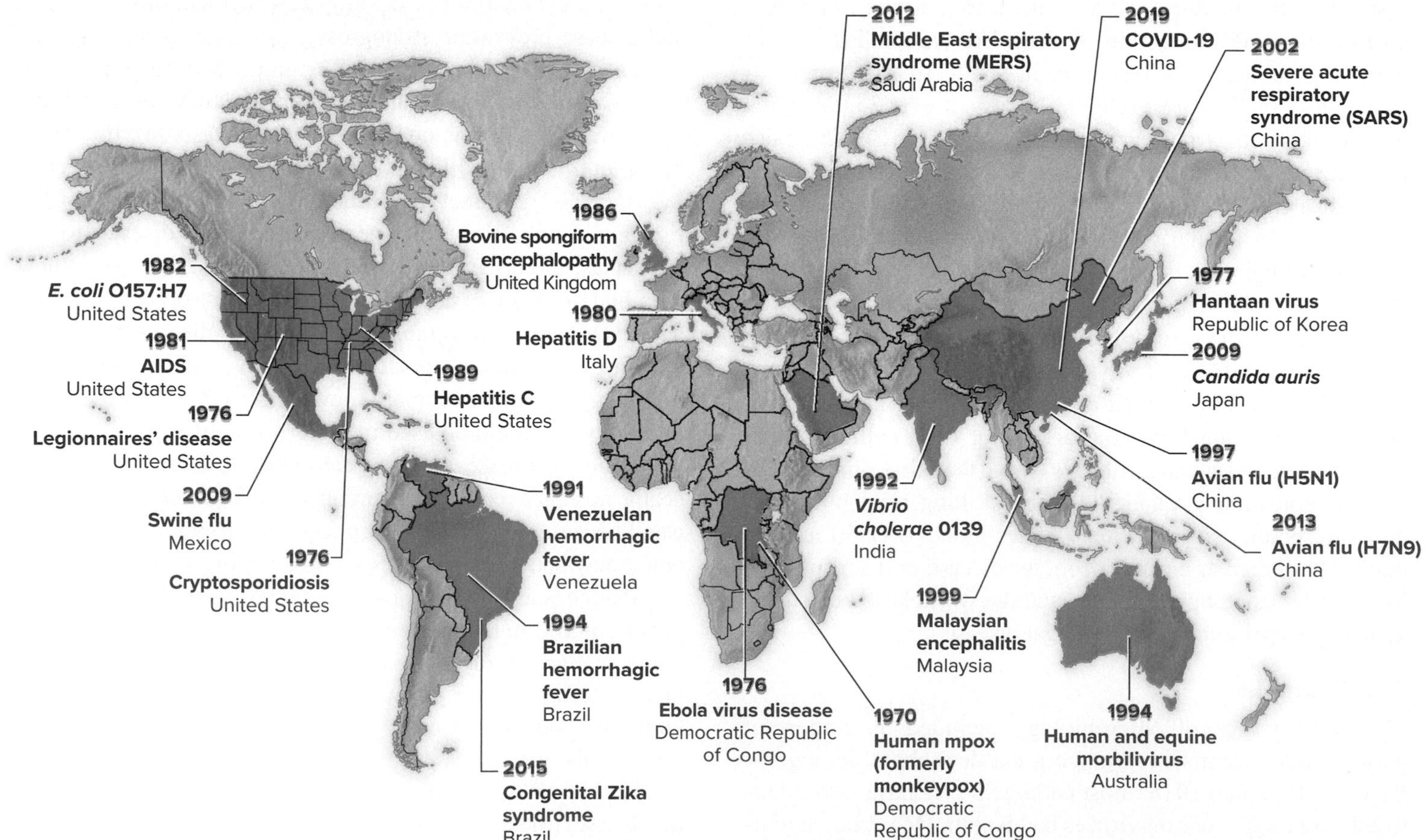

FIGURE 1.5 New and Newly Recognized Infectious Diseases or Disease Agents in Humans and Animals Since 1970 Countries where cases first appeared or were identified appear in a darker shade.

? Why might so many of the diseases first appear or be identified in the United States and other resource-rich countries?

FOCUS ON A CASE 1.1

A 24-year-old woman had suffered from recurrent severe episodes of an intestinal disorder called *Clostridioides* (formerly *Clostridium*) *difficile* infection (CDI) for the past 13 months. She routinely experienced profuse watery diarrhea, abdominal pain, and fever. In addition, she felt tired and hopeless because she did not seem to be getting well, despite long attempts at multiple different treatments.

As with most patients who develop CDI, the woman had been taking an oral antibiotic shortly before her symptoms began—in this case, to treat a tooth infection. The antibiotic had successfully killed the bacteria that caused her tooth infection, but it also killed some members of her normal intestinal microbiota. As a result, the bacterium *C. difficile*—often referred to simply as "*C. diff*"—thrived in her intestinal tract, growing to much higher numbers than it could before. The strain that caused her infection was able to make a toxin that damaged the lining of her intestinal tract.

When the patient first started experiencing CDI, her doctor told her to stop taking the antibiotic prescribed for her tooth infection, hoping that her CDI would resolve on its own. When that did not help, the doctor prescribed a different antibiotic that is often effective in treating CDI. The patient started feeling better, but the symptoms quickly returned when she stopped taking the medication. She also tried oral supplements containing *Lactobacillus* GG, a bacterium that may sometimes be effective in preventing antibiotic-associated diarrhea.

Because the patient's health was declining, doctors suggested a fecal microbiota transplant (FMT), an experimental procedure that involves inserting feces from a healthy person into the patient's intestinal tract in order to repopulate that environment with appropriate microbes. They chose to use her sister as a fecal donor, screening both the donor and the patient to ensure that neither was infected with certain pathogens. Approximately ¼ cup of fresh feces was mixed with 1 quart of water and delivered to the patient's intestinal tract via a colonoscope. Within days after the transplant, she began feeling better, and she soon recovered completely.

1. Why would certain oral antibiotics allow *C. difficile* to thrive in the intestinal tract?
2. Why would the doctors screen both the patient and the fecal donor for certain infectious agents?
3. Why would the doctors transplant feces rather than introducing isolated bacteria from feces to repopulate the colon?

Discussion

1. Antibiotics kill or inhibit not just pathogens, but also beneficial members of the normal microbiota, a group that protects against infection in at least two general ways: (1) using nutrients that would otherwise support the growth of *C. difficile* and other pathogens and (2) making compounds that are toxic or inhibitory to other organisms. The environment of the intestinal tract is quite complex, however, so other factors might also be playing a role.
2. Physicians screen the fecal donor to decrease the likelihood that disease-causing microbes could be transferred to the patient by the procedure. The doctors screen the patients to ensure that they are not already infected with the pathogens. For example, if this patient developed symptoms of a *Salmonella* infection after the procedure, how would the physicians know that she acquired the infection as a result of the procedure if they had not checked beforehand?
3. Feces contain many types of bacteria that cannot yet be grown in the laboratory. In addition, scientists do not yet know which types of fecal bacteria protect against CDI. In 2022, however, a new microbiota-based therapy called Rebyota—a rectally administered preparation of fecal microbiota—was approved by the U.S. Food and Drug Administration to prevent certain CDI recurrences.

disease emergence as people may inadvertently carry pathogens to different regions. Even diseases such as malaria, cholera, plague, and yellow fever that have largely been eliminated from developed countries can be carried to other places if travelers to regions where they still exist become infected and then move on before recovering. Meanwhile, as city suburbs expand into rural areas, human populations come into closer contact with animals as well as the mosquitoes and other arthropods that normally feed on those animals. Consequently, people are exposed to pathogens they might not have encountered previously.

The preventive measures used to control certain infectious diseases can become victims of their own success, a situation that can also lead to disease emergence. Decades of vaccination have nearly eliminated measles, mumps, and whooping cough in developed countries, so most people no longer have first-hand knowledge of the dangers of these diseases. Couple this with misinformation about vaccines, and some people develop irrational fears, falsely believing that vaccines are more harmful than the diseases they prevent. When this happens, parents often refuse to vaccinate their children appropriately, leading to situations where the diseases become more common again. Measles had been declared eliminated in the United States in 2000, but outbreaks in 2019 resulted in the highest number of cases in 25 years; the outbreaks generally start with unvaccinated travelers who bring the disease into the country, where it then spreads among others who are not vaccinated.

Chronic Diseases Some chronic illnesses once attributed to other causes may be due to microorganisms. Perhaps the best-known example is stomach ulcers, once thought to be due to stress. We now know that stomach ulcers are often caused by a bacterium (*Helicobacter pylori*) and are treatable with antibiotics. Chronic indigestion may be caused by the same bacterium. Another example is cervical cancer, which we now know is usually caused by human papillomavirus (HPV) infection; a vaccine against HPV helps prevent that cancer. Infectious microbes may play important roles in other chronic diseases as well.

Focus Figure

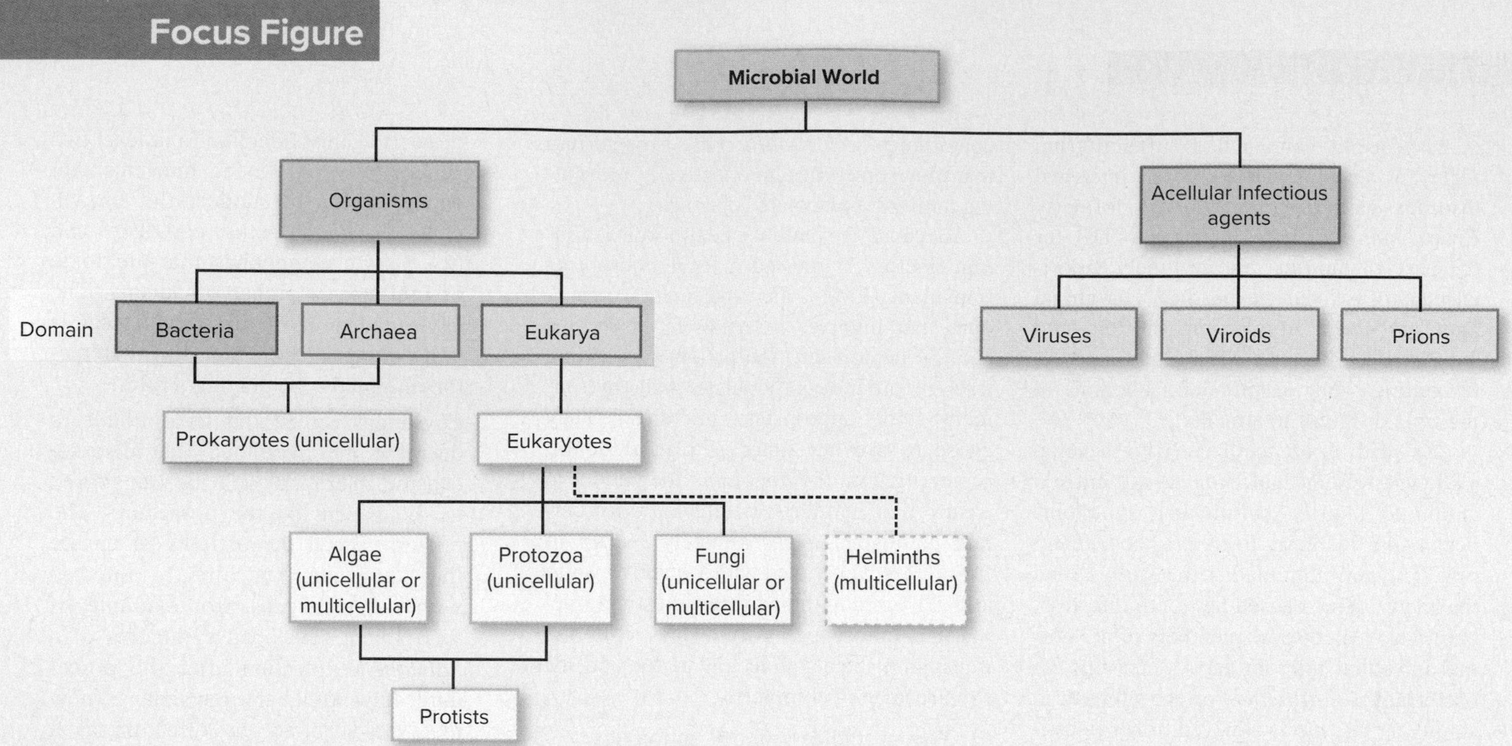

FIGURE 1.6 The Microbial World Although adult helminths (worms) can often be seen with the naked eye, some stages in the life cycle of helminths are microscopic.

Members of which two domains are prokaryotes?

MicroAssessment 1.2

Microbial activities are essential to human life as well as being commercially valuable. Microbes are important research tools. Although most microbes are beneficial or not harmful, some cause disease. Enormous progress has been made in preventing and curing infectious diseases, but some diseases are becoming more common.

4. Describe two microbial activities essential to life and three that make our lives more comfortable.
5. Describe three factors that cause certain infectious diseases to become more common.
6. During the COVID pandemic, results of some scientific studies were made available before they were peer-reviewed. What would be an advantage and a disadvantage of this approach?

1.3 ■ Members of the Microbial World

Learning Outcomes

8. Compare and contrast characteristics of members of the Bacteria, Archaea, and Eukarya.
9. Explain the features of an organism's scientific name.
10. Compare and contrast the algae, fungi, and protozoa.
11. Compare and contrast viruses, viroids, and prions.

Considering that small size is the only shared feature of all microbes, the group is tremendously diverse. If you look at the macroscopic world around you—the plants and animals—you should be impressed by the assortment of what you see. That range of types, however, is dwarfed by the huge variety of microbes! The extent of that diversity makes sense considering that microbes have inhabited this planet for billions of years and have evolved to thrive in every conceivable environment—from the hydrothermal vents at the bottom of the ocean to the icy tops of the highest mountains. Many people associate microbes with disease, but their contributions to our world go far beyond that. In fact, as section 1.2 describes, we could not survive without them.

Living organisms are all composed of cells with one of two basic structures—prokaryotic (*pro* means "prior to" and *karyote* means "nucleus") and eukaryotic (*eu* means "true"). **Prokaryotic cells** do not have a membrane-bound nucleus. Instead, the genetic material is located in a region called the nucleoid. In contrast, the genetic material in **eukaryotic cells** is contained within a membrane-bound nucleus, and the cells often have a variety of other membrane-bound organelles as well. Eukaryotic cells are also typically much larger and more complex than prokaryotic cells. Organisms that consist of one or more eukaryotic cells are called **eukaryotes,** whereas those composed of a prokaryotic cell are called **prokaryotes.** Prokaryotes fall into two very different groups—bacteria and archaea—as different from each other as they are from eukaryotes.

Because of the fundamental differences between bacteria, archaea, and eukaryotes, all living organisms are now classified into one of three **domains:** Bacteria, Archaea, and Eukarya (sometimes spelled Eucarya) (**figure 1.6**). Members of the

TABLE 1.1 Characteristics of Members of the Three Domains

Characteristic	Bacteria	Archaea	Eukarya
Cell type	Prokaryotic	Prokaryotic	Eukaryotic
Number of cells	Unicellular	Unicellular	Unicellular or multicellular
Membrane-bound organelles	No	No	Yes
Ribosomal RNA sequences unique to the group	Yes	Yes	Yes
Peptidoglycan in cell wall	Yes	No	No
Typical size range	0.3–2 μm	0.3–2 μm	5–50 μm

Bacteria and Archaea are referred to as bacteria and archaea, respectively, and are prokaryotes; members of the Eukarya are referred to as eukarya and are eukaryotes. **Table 1.1** compares some features of members of the three domains, but you will learn other important differences in later chapters.

The small size of microbes requires measurements not commonly used in everyday life (**figure 1.7**). Logarithms are extremely helpful in this regard, so you will find a brief discussion of them in Appendix I.

Scientific Names

When referring to microbes, we use their scientific names, which are written and pronounced in a Latin style; a pronunciation guide is in Appendix II, and an audio version is available on your Connect class site. The scientific names are assigned according to the binomial (two-part name) system of nomenclature developed by Carl Linnaeus in the 1700s. The first part of the name indicates the **genus,** with the first letter always capitalized; the second part indicates the specific epithet, or **species,** and is not capitalized. Both are usually italicized or underlined—for example, *Escherichia coli.* The part indicating the genus is commonly abbreviated, with the first letter capitalized—as in *E. coli.*

The origin of one or both parts of an organism's name often reflects a notable characteristic or honors a particular scientist (**table 1.2**). In the case of *Escherichia coli,* the name

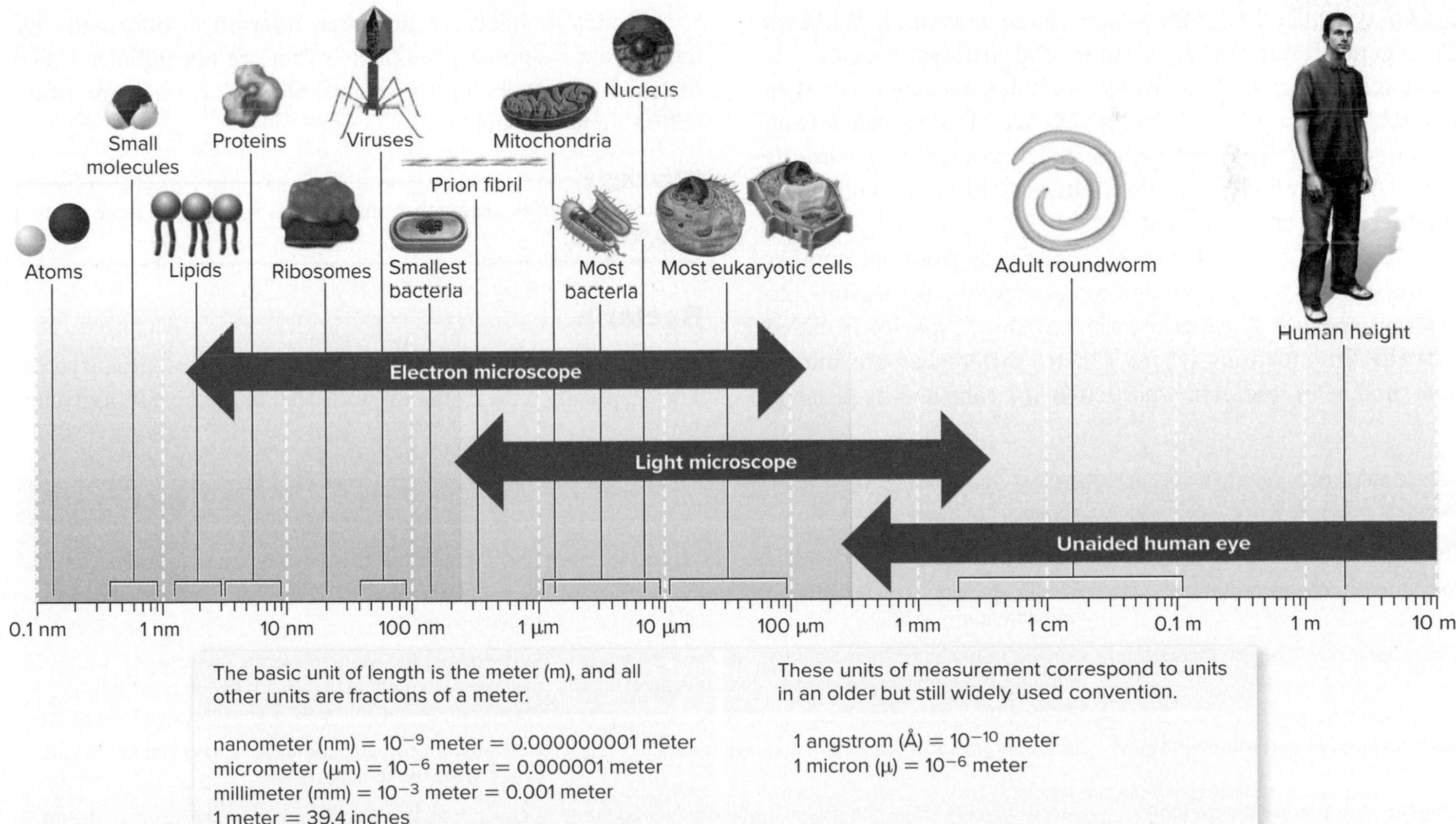

FIGURE 1.7 Sizes of Molecules, Non-Living Agents, and Organisms Note that the scale here is logarithmic (rather than linear), and each labeled increment increases by a factor of 10.

? **Why is a logarithmic scale useful when comparing sizes of members of the microbial world?**

FOCUS YOUR PERSPECTIVE 1.1

Every Rule Has an Exception

We might assume that because microorganisms have been so intensively studied, no major surprises are left to be discovered. This, however, is far from the truth. In the mid-1990s, an unusual organism was found in the intestinal tract of certain fish from both the Red Sea in the Middle East and the Great Barrier Reef in Australia. The organism's large size (600 μm long and 80 μm wide) makes it clearly visible without any magnification and suggested that it was a eukaryote. However, it does not have a membrane-bound nucleus. A chemical analysis of the cell confirmed that it is a prokaryote and a member of the domain Bacteria, and it was given the name *Epulopiscium fishelsoni* (*epulo* means "guest at a feast" and *piscium* means "fish").

In 1999, a spherical bacterium 70 times greater in volume than *E. fishelsoni* was isolated from the muck of the ocean floor off the coast of Namibia in Africa. Because it contains glistening globules of sulfur, it was named *Thiomargarita namibiensis,* meaning "sulfur pearl of Namibia." In 2022, an even larger species that has centimeter-long cells was described. This bacterium, now called *T. magnifica,* was discovered in a Caribbean mangrove swamp 12 years earlier, but it was thought to be fungus until researchers characterized it further.

In contrast to the examples of large bacteria, a unicellular alga found in the Mediterranean Sea is 1 μm in width. It is a eukaryote even though it is about the size of a typical bacterium.

How small can an organism be? A microbe discovered off the coast of Iceland is only about 400 nm (nanometers) wide and has one-tenth the amount of genetic information (DNA) compared to the common intestinal bacterium *Escherichia coli.* The tiny organism was found attached to a much larger microbe, a member of the Archaea growing in an ocean vent where the temperature was close to the boiling point of water. The larger organism is an *Ignicoccus* species (*igni* means "fire" and *coccus* means "sphere"). The tiny one, also a member of the Archaea, has been named *Nanoarchaeum equitans* (meaning "tiny archaea" and "rider"). *N. equitans* cannot be grown in the laboratory by itself, but *Ignicoccus* grows well without its *Nanoarchaeum* "rider."

These exceptions to long-standing rules point out the need to keep an open mind and not jump to conclusions! They also serve as excellent reminders that in a subject as complex as microbiology, there will almost always be exceptions!

---

of the genus honors Theodor Escherich, who discovered the bacterium; the species designation indicates the site where *E. coli* typically lives: the colon (large intestine). Within a given genus, there may be a number of different species. For example, the genus *Escherichia* includes species other than *E. coli*, such as *E. vulneris,* which was first isolated from human wounds (*vulneris* means "of a wound"). *E. vulneris* is genetically related to *E. coli,* but not closely enough to be considered the same species.

Members of the same species may vary from one another in minor ways, but not enough to separate the organisms into different species. A genetic variant within a species is called a **strain.** In situations where genetic differences are important, such as in research, a particular microbe and its progeny may be indicated with a strain designation—for example, *E. coli* B or *E. coli* K12.

Groups of microbes are often referred to informally by names that resemble genus names but are not italicized. For instance, members of the genus *Staphylococcus* are often called staphylococci.

MicroByte

Bacterial species outnumber mammalian species by more than 10,000-fold!

Bacteria

Bacteria (singular: bacterium) are single-celled prokaryotes. They typically have rigid cell walls that contain peptidoglycan,

TABLE 1.2 Origin of Various Scientific Names

Name	Genus Derivation	Species Derivation
Escherichia coli **(bacterium)**	Honors Theodor Escherich, the scientist who discovered the bacterium.	Derived from the word "colon," the body site inhabited by the bacterium.
Haemophilus influenzae **(bacterium)**	Derived from *haemo* (blood) and *phil* (loving), reflecting that the bacterium requires certain components of blood for growth.	Derived from the word "influenza," the disease mistakenly thought to be caused by the bacterium; we now know that influenza is caused by a virus.
Saccharomyces cereviseae **(fungus)**	Derived from *saccharo* (sugar) and *myces* (fungus).	Derived from *cerevisia* (beer), reflecting that the fungus (a yeast) is used to make beer.
Shigella dysenteriae **(bacterium)**	Honors Kiyoshi Shiga, the scientist who discovered the bacterium.	Derived from the word "dysentery," the disease caused by the bacterium.
Staphylococcus aureus **(bacterium)**	Derived from *staphylo* (bunch of grapes) and *kokkus* (berry), reflecting the grouping and shape of the cells.	Indicates the common color of visible masses of the cells (*aureus* means "golden").

a compound unique to bacteria. Many of the bacteria can move using flagella (singular: flagellum), filamentous appendages that extend from the cell.

Bacteria as a whole come in a variety of shapes, but cells of a given species are most often only one shape. Bacterial shapes include (**figure 1.8**):

- **Coccus (plural: cocci).** A spherical cell that may be flattened on one end or slightly oval.
- **Rod or bacillus (plural bacilli).** A cylindrical cell. One short enough to be confused with a coccus is called a coccobacillus. Note that the descriptive term "bacillus" should not be confused with *Bacillus,* the name of a genus. Although members of the genus *Bacillus* are rod-shaped, so are many other bacteria, including *Escherichia coli.*
- **Vibrio (plural: vibrios).** A short, curved rod.
- **Spirillum (plural: spirilla).** A curved rod long enough to form spirals.
- **Spirochete (plural: spirochetes).** A long, spiral-shaped cell with a flexible cell wall and a unique mechanism of motility.
- **Pleomorphic.** This is not an actual shape, but refers to bacteria that characteristically vary in their shape (pleo meaning "many" and morphic referring to shape).

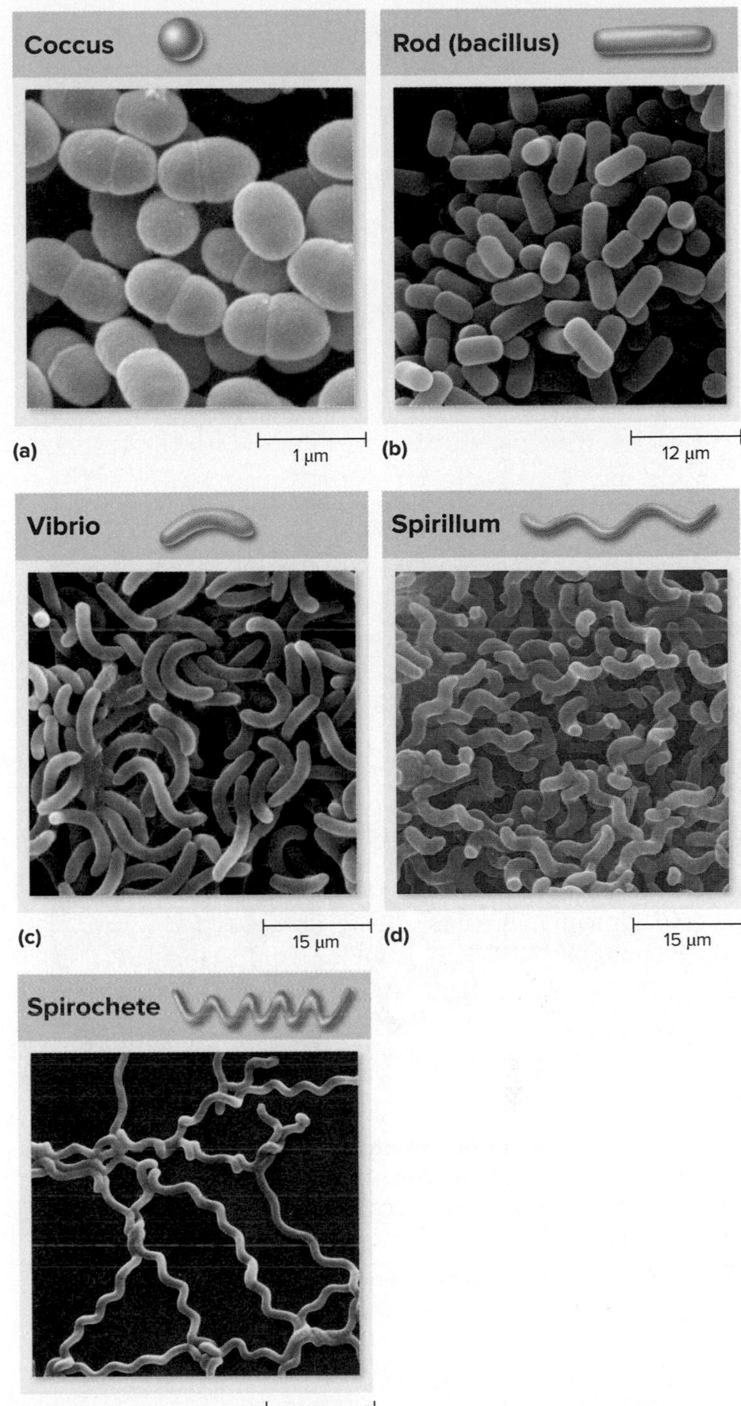

FIGURE 1.8 Shapes of Common Bacteria (a) Coccus; **(b)** rod; **(c)** vibrio; **(d)** spirillum; **(e)** spirochete. (SEM). a: SciMAT/Science Source; b–e: Dennis Kunkel Microscopy/SPL/Science Source

? **What are the two most common shapes of bacteria?**

Bacteria typically multiply by binary fission, a process in which one cell enlarges and then divides to form two cells, each equivalent to the original. The dividing cells often stick to each other, forming characteristic arrangements such as pairs, chains, or clusters, depending on the planes of division (**figure 1.9**). Cocci that typically remain as pairs are called diplococci; the characteristic diplococcus arrangement of *Neisseria gonorrhoeae* is an important clue in the identification of that bacterium. Division in one plane can also form long chains, a characteristic typical of some members of the genus *Streptococcus* (*strepto* means "twisted chain"). Cocci that divide in perpendicular planes form cubical packets; members of the genus *Sarcina* form such packets. Cocci that divide in several planes at random may form clusters; *Staphylococcus* species, which typically form grape-like clusters, are an example (*staphylo* means "bunch of grapes").

Some prokaryotes characteristically live as multicellular associations. Members of a group of bacteria called myxobacteria form swarms of cells that move as a pack, gliding over moist surfaces. The cells release enzymes, and, as a pack, they degrade organic material, including other bacterial cells. When water or nutrients become limiting, the cells come together to form a structure called a fruiting body, which is visible to the naked eye (see figure 11.14).

Many bacteria obtain energy from foods similar to those humans eat, but others can gain energy from seemingly unlikely sources such as hydrogen sulfide (a gas that smells like rotten eggs). Still others are photosynthetic, meaning they make cellular material using the radiant energy of sunlight.

Although most bacteria are beneficial, some cause diseases, and these will be a focus in the "disease chapters" of this textbook. Many of the early chapters will focus on bacteria in general, often with the aim of providing the necessary background for understanding infectious diseases.

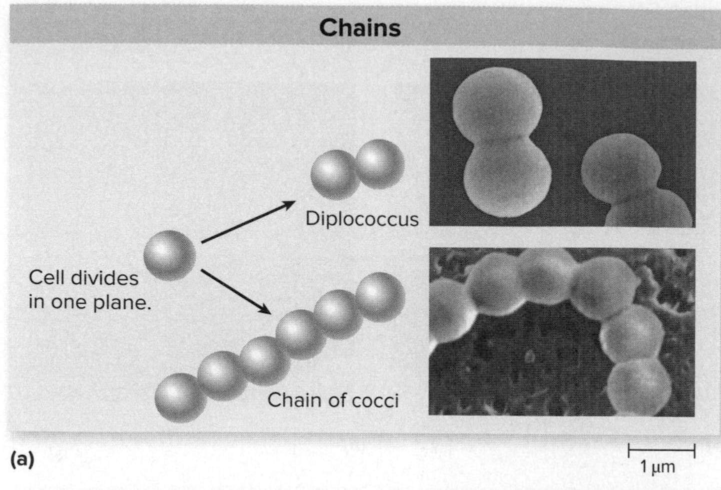

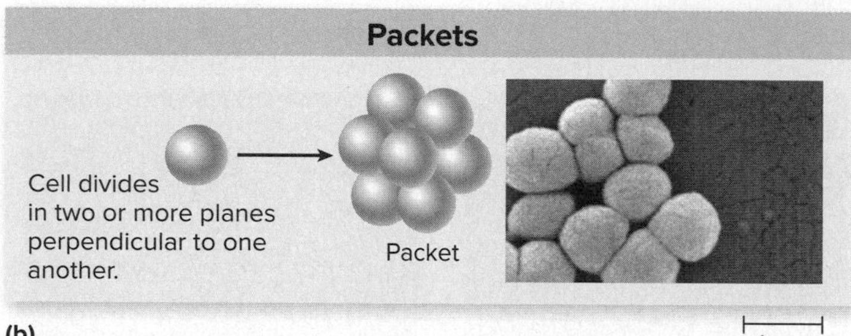

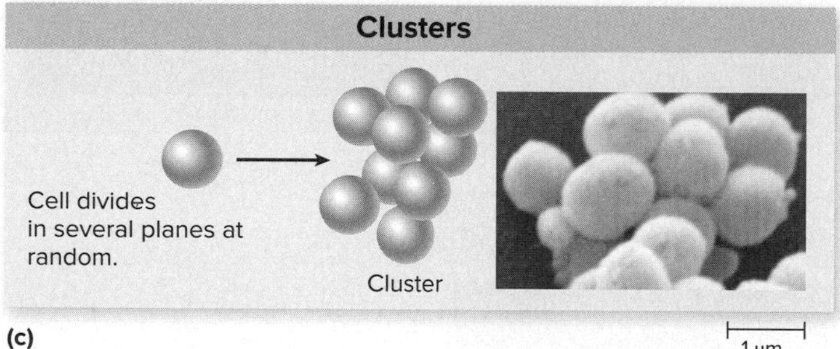

FIGURE 1.9 Common Cell Arrangements **(a)** Chains; **(b)** packets; **(c)** clusters. (SEM). a (top): Dennis Kunkel Microscopy/SPL/Science Source; a (bottom): BSIP SA/Alamy Stock Photo; b: CDC/Betsy Crane/Janice Haney Carr; c: Eye of Science/Science Source

? **How does bacterial cell division determine the characteristic cell arrangements?**

Archaea

Like bacteria, archaea (singular: archaeon) are single-celled prokaryotes. They have similar shapes, sizes, and appearances to bacteria. In addition, they also multiply by binary fission, move primarily by means of filamentous appendages, and have rigid cell walls. Like bacteria, different groups of archaea use different energy sources; some are photosynthetic, harvesting the energy of sunlight to make cellular material. Considering how much archaea look like bacteria, scientists initially believed they were closely related. We now know, however, that major differences exist between the two groups, and the groups are only distantly related to each other. In fact, you are more closely related to plants than archaea are to bacteria!

Archaea differ from bacteria in several of their structural and functional components. For example, the archaeal cell wall does not contain peptidoglycan, whereas the bacterial wall does. In addition, the composition and structural organization of the appendages the archaeal cells use for motility are different from bacterial flagella. Archaea also have characteristic nucleotide sequences in their ribosomal RNA (a molecule involved in protein synthesis) that differ significantly from those of bacteria. The discovery of the differences in ribosomal RNA sequences helped provide the basis for separating the two groups of prokaryotes into different domains.

An interesting feature of many archaea is their ability to grow in extreme environments in which most other organisms cannot survive. Some, for example, can grow in salt concentrations 10 times higher than that of seawater. These organisms grow in such habitats as the Great Salt Lake and the Dead Sea. Others grow best at extremely high temperatures. One archaeon can grow at a temperature of 122°C! (100°C is the temperature at which water boils at sea level.)

Although the archaea that grow in extreme environments are the most intensively studied, many others are common in moderate environments. They are widely distributed in soils, the oceans, and marshes, as well as in the intestinal tracts and oral cavities of animals.

Eukarya

Eukarya are eukaryotes; those studied by microbiologists include fungi, algae, protozoa, and helminths (worms). Algae and protozoa are also referred to as **protists.**

Fungi

Fungi (singular: fungus) are a diverse group of eukaryotes, ranging from single-celled yeasts that reproduce by budding to multicellular filamentous molds (**figure 1.10**). The microscopic filaments of molds, called hyphae (singular: hypha), can branch as well as twist and turn to form a visible mat called a mycelium. When you see moldy foods, you are looking at the mycelium, sometimes along with structures that give rise to microscopic reproductive forms called conidia (also referred to as spores). The conidia easily become airborne, allowing the fungus to spread. Some fungi make macroscopic reproductive structures called mushrooms.

Fungi harvest energy from organic materials. To do this, they secrete enzymes that degrade the organic material and then take in the nutrients that were released. Fungi are found in most places where organic materials, including dead plants and animals, are present.

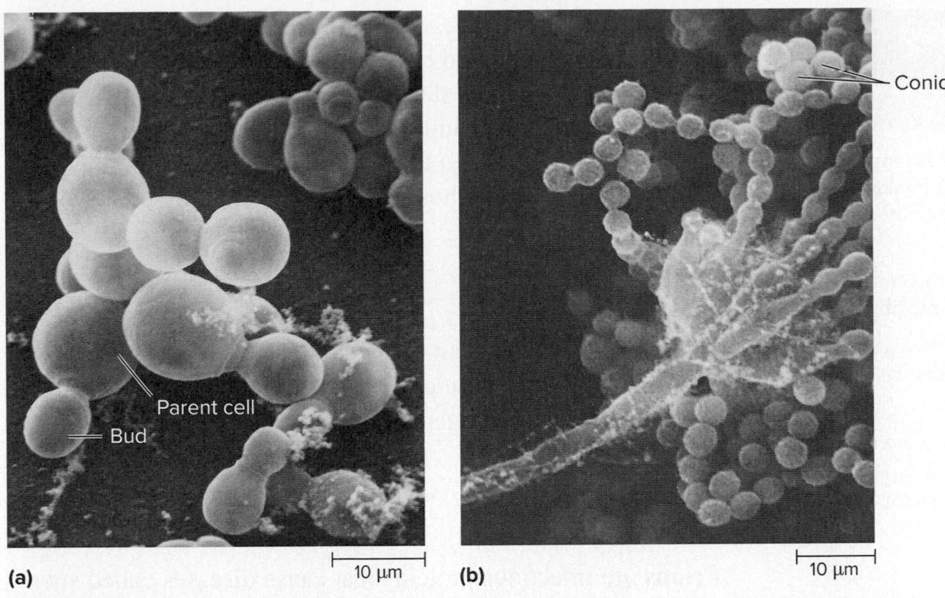

FIGURE 1.10 Fungi (a) Yeast, *Malassezia furfur.* **(b)** *Aspergillus,* a typical mold form whose reproductive structures rise above the mat of hyphae. Janice Haney Carr/CDC

? Molds and yeasts are made up of what type of cells?

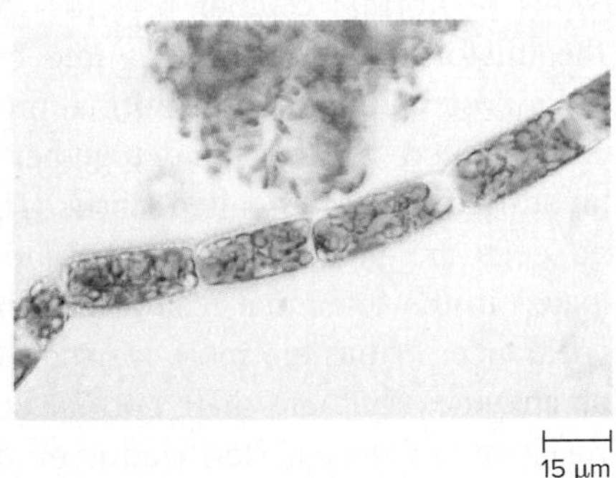

FIGURE 1.11 Alga *Ulothrix,* a filamentous green alga. Lisa Burgess/McGraw Hill

? What general features of algae distinguish them from other eukaryotic microorganisms?

Algae

Algae (singular: alga) are a diverse group of photosynthetic eukaryotes. Some are single-celled, whereas others are multicellular, such as seaweed (**figure 1.11**). Photosynthesis in algae occurs in chloroplasts, which have chlorophyll, a green pigment. Some algae also have other pigments as well, giving them characteristic colors. The pigments absorb the energy of light, which is used in photosynthesis.

Algae are usually found near the surface of either salt or fresh water or in moist terrestrial habitats. Their cell walls are rigid, but the chemical composition of the wall is quite distinct from that of bacteria and archaea. Many algae move by means of flagella, which are structurally far more complex and unrelated to flagella of prokaryotes.

Protozoa

Protozoa (singular: protozoan) are a diverse group of microscopic, single-celled eukaryotes that live in both aquatic and terrestrial environments. Although microscopic, they are very complex organisms and generally much larger than prokaryotes (**figure 1.12**). Unlike algae and fungi, protozoa do not have a rigid cell wall. Most protozoa are motile and ingest organic material as food sources.

Helminths

Parasitic **helminths** are worms that live at the expense of a host. They are an important cause of disease, particularly in developing countries. The adult worms are generally macroscopic, meaning they can be seen with the naked eye, and some are quite large, so technically they are not microorganisms. Microbiologists study them, however, because diagnosis of the diseases they cause often involves identifying their eggs and larval forms, which are microscopic. Helminths include roundworms, tapeworms, and flukes.

Table 1.3 summarizes the characteristics of eukaryotic organisms studied by microbiologists.

Acellular Infectious Agents

Viruses, viroids, and prions are acellular infectious agents, meaning that they are not composed of cells. They cannot reproduce independently and are considered non-living. By definition, an

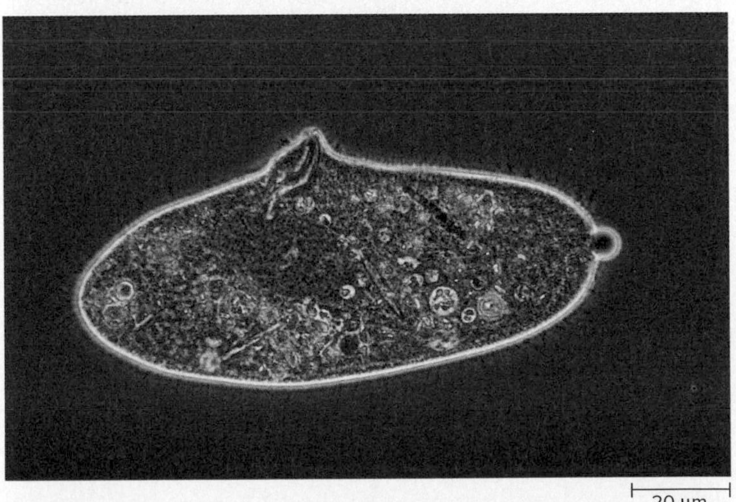

FIGURE 1.12 Protozoan A paramecium moves with the aid of hair-like appendages (called cilia) on the cell surface. Melba Photo Agency/Alamy Stock Photo

? How do protozoa differ from both fungi and algae?

TABLE 1.3 Eukaryotic Organisms Studied by Microbiologists

Organism	Characteristics
Fungi	Use organic material for energy. Sizes range from microscopic (yeasts) to macroscopic (molds); mushrooms are the reproductive structures of some fungi.
Algae	Use sunlight for energy. Sizes range from microscopic (single-celled algae) to macroscopic (multicellular algae).
Protozoa	Use organic material for energy. Single-celled microscopic organisms.
Helminths	Use organic material for energy. Adult worms are typically macroscopic and often quite large, but their eggs and larval forms are microscopic.

organism must be composed of one or more cells, so these acellular infectious agents are not microorganisms.

Viruses

Viruses consist of nucleic acid (either DNA or RNA) packaged within a protein coat (**figure 1.13**). To replicate, viruses infect living cells—referred to as **host cells**—and then hijack the machinery and nutrients of those cells to produce more phage particles. Outside the host cells, however, viruses are inactive. Thus, viruses are obligate intracellular agents, meaning that they cannot replicate outside a host cell.

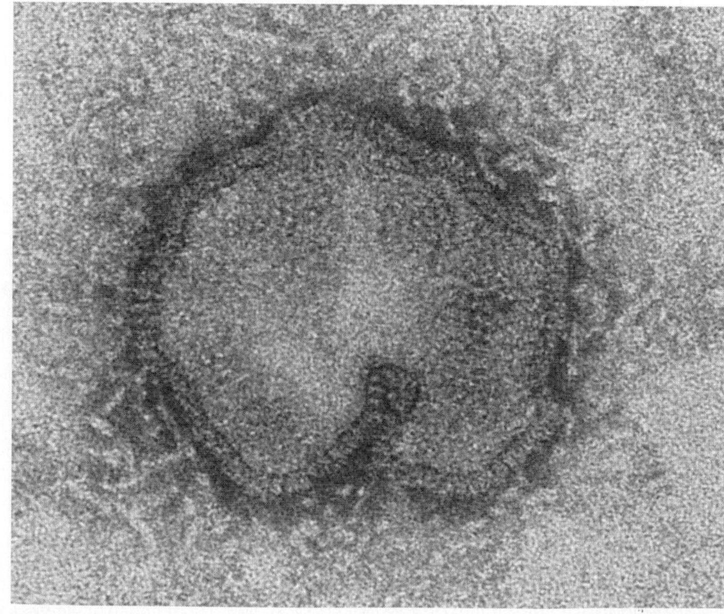

FIGURE 1.13 Virus Influenza virus, the cause of flu.
Cynthia S. Goldsmith and Thomas Rowe/CDC

? **Why can viruses be so much smaller than cells and still replicate?**

All forms of life, including bacteria, archaea, and eukarya, can be infected by viruses but of different types. The viruses frequently kill the cells in which they replicate, but some types have the additional possibility of remaining silent within a host cell; when this occurs, the viral genome is replicated along with the host cell genome and passed on to the cell's progeny.

Viroids

Viroids are simpler than viruses, consisting of only a single, short piece of ribonucleic acid (RNA). Like viruses, they are obligate intracellular agents. Viroids cause a number of plant diseases.

Prions

Prions are infectious proteins that cause diseases called spongiform encephalopathies, a name that reflects the sponge-like appearance of the affected brain tissue (*encephalo* means "brain" and *patho* means "disease"). Perhaps the most widely recognized example is bovine spongiform encephalopathy (BSE), commonly called mad cow disease. Prions are simply misfolded versions of normal cellular proteins found in the brain. When the misfolded version comes into contact with the normal cellular protein, it causes the normal protein to also misfold. These misfolded versions bind together within the cell to form thread-like structures called fibrils (**figure 1.14**). The fibril-filled cells are unable to function and eventually die, forming spaces in the brain that lead to the characteristic sponge-like appearance. Prions are more resistant to degradation by cellular enzymes than are their normal counterparts. They are also resistant to the usual sterilization procedures that destroy viruses and bacteria.

Characteristics of acellular infectious agents are listed in **table 1.4.**

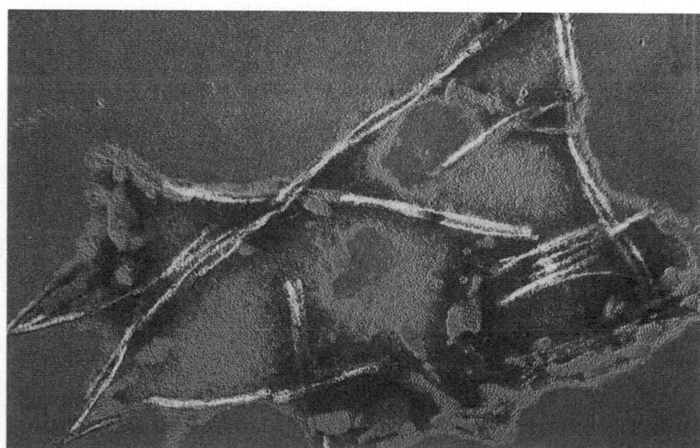

FIGURE 1.14 Prion Prion fibrils isolated from the brain of an infected cow. EM Unit, VLA/Science Source

? **How are prions different from the normal versions of the related proteins made by cells?**

MicroAssessment 1.3

Microbes are given a two-part name that indicates the genus and species. Three domains of life exist: Bacteria, Archaea, and Eukarya. Members of the Bacteria and Archaea are prokaryotes, but the two groups have significant differences. Members of the Eukarya are eukaryotes; within this group, microbiologists study algae, fungi, protozoa, and parasitic helminths. Viruses, viroids, and prions are acellular infectious agents.

7. List two features that distinguish prokaryotes from eukaryotes.
8. Describe the chemical composition of viruses, viroids, and prions.
9. Based on the genus name, what is the shape and arrangement of *Streptobacillus monoliformis* cells?

TABLE 1.4 Acellular Infectious Agents

Agent	Characteristic
Viruses	Consist of either DNA or RNA, surrounded by a protein coat. Obligate intracellular agents that use the machinery and nutrients of host cells to replicate.
Viroids	Consist only of RNA; no protein coat. Obligate intracellular agents that use the machinery and nutrients of host cells to replicate.
Prions	Consist only of protein; no DNA or RNA. Misfolded versions of normal cellular proteins that cause misfolding of the normal versions.

Summary

1.1 The Dispute over Spontaneous Generation

The belief in **spontaneous generation** was challenged by Francesco Redi in the seventeenth century.

Early Experiments

The experiments of John Needham supported the idea of spontaneous generation, while those of Lazzaro Spallanzani did not.

Experiments of Pasteur

The experiments of Louis Pasteur disproved spontaneous generation and supported what is now known as the theory of **biogenesis** (figure 1.2).

Experiments of Tyndall

John Tyndall showed that some microbial forms are not killed by boiling. He and Ferdinand Cohn discovered **endospores,** the heat-resistant forms of some bacteria.

The Golden Age of Microbiology

The field of microbiology blossomed after Pasteur and others disproved spontaneous generation, leading to the Golden Age of Microbiology. Discoveries during this time led to the acceptance of the Germ Theory of Disease.

The Scientific Method

The **scientific method** includes: (1) observing an occurrence and asking a question about that situation; (2) developing a **hypothesis** that explains the occurrence and devising an experiment that tests the hypothesis; (3) doing the experiment, collecting the data, and drawing conclusions; and (4) communicating the results, methods, and conclusions. A scientific **theory** is an explanation supported by a vast body of experimental evidence.

1.2 Microbiology: A Human Perspective

The Human Microbiome

The **normal microbiota** is essential to human health. The **microbiome** is an interacting community of microorganisms as well as their genetic information.

Microorganisms in the Environment

Without microorganisms, we would run out of certain nutrients; as one example, certain microorganisms convert the nitrogen gas in the air into a form that humans and other organisms can use. We also rely on them to degrade certain materials.

Commercial Benefits of Microorganisms

Microorganisms are used in the production of bread, wine, beer, and cheeses. They are also used to degrade toxic pollutants and to synthesize a variety of different useful products. **Biotechnology** depends on members of the microbial world.

Microbes as Research Tools

Microorganisms are wonderful model organisms to study because they have the same fundamental metabolic and genetic properties as higher life-forms, and they grow rapidly on simple, inexpensive growth media.

Microbes and Disease

Pathogens cause disease, but the death rate from infectious diseases has declined since 1900 as a result of disease-prevention efforts, including vaccination (figure 1.4). More needs to be done to prevent **emerging infectious diseases,** some of which are new or newly recognized (figure 1.5). Some chronic diseases are caused by microorganisms.

1.3 Members of the Microbial World

Considering that small size is the only shared feature of all microbes, the group is tremendously diverse (figure 1.6). All living organisms are classified into three **domains:** Bacteria, Archaea, and Eukarya (table 1.1). The small size of microbes requires measurements not commonly used in everyday life (figure 1.7).

Scientific Names

The first part of a scientific name indicates the **genus,** and the second part the **species;** these are written in italics or underlined (table 1.2). Members of the same species can vary, so strain designations are sometimes used.

Bacteria

Bacteria are single-celled **prokaryotes** that have peptidoglycan in their cell walls. Most bacteria are **cocci** or **rods;** other common shapes are **vibrios, spirilla,** and **spirochetes** (figure 1.8). Cells adhering to one another following division form characteristic arrangements such as chains, packets, and clusters (figure 1.9).

Archaea

Archaea are single-celled prokaryotes. Although they look like bacteria, there are significant differences. They do not contain peptidoglycan. Many archaea grow in extreme environments.

Eukarya

Eukarya are **eukaryotes** (table 1.3). **Fungi** include single-celled yeasts and multicellular molds and mushrooms; they use organic compounds as food (figure 1.10). **Algae** can be single-celled or multicellular, and they use sunlight as an energy source (figure 1.11). **Protozoa** are typically motile single-celled organisms that use organic compounds as food (figure 1.12). Parasitic **helminths** are worms that live at the expense of a host.

Acellular Infectious Agents

The non-living members of the microbial world are not composed of cells (table 1.4). **Viruses** consist of nucleic acid within a protein coat (figure 1.13). **Viroids** consist of a single, short RNA molecule. **Prions** consist only of protein; they are misfolded versions of normal cellular protein (figure 1.14).

Review Questions

Short Answer

1. How did Louis Pasteur help disprove spontaneous generation?
2. Describe the scientific method.
3. Explain why life could not exist without the activities of microorganisms.
4. How is the normal microbiota important to human health?
5. List four commercially important benefits of microorganisms.
6. What characteristics of microorganisms make them important research tools?
7. List three factors that contribute to the emergence of infectious diseases.
8. In the designation *Escherichia coli* B, which part of the name indicates the genus? Which part indicates the species? Which part indicates the strain?
9. Why are viruses not microorganisms?
10. Name three non-living groups in the microbial world, and describe their major properties.

Multiple Choice

1. The property of endospores that led to confusion in the experiments on spontaneous generation is their
   a) small size.
   b) ability to pass through cork stoppers.
   c) heat resistance.
   d) presence in all infusions.
   e) presence on cotton plugs.
2. The Golden Age of Microbiology was the time when
   a) microorganisms were first used to make bread.
   b) microorganisms were first used to make cheese.
   c) most pathogenic bacteria were identified.
   d) a vaccine against influenza was developed.
   e) antibiotics became available.
3. If all prokaryotes were eliminated from the planet,
   a) animals would thrive because there would be no disease.
   b) archaea would thrive because there would be no competition for nutrients.
   c) all animals would die.
   d) animals and archaea would thrive.
4. All of the following are emerging infectious diseases *except*
   a) smallpox.
   b) hepatitis C.
   c) Lyme disease.
   d) COVID-19.
   e) mad cow disease.
5. All of the following are biological domains *except*
   a) Bacteria.
   b) Archaea.
   c) Prokaryota.
   d) Eukarya.
6. Which name is written correctly?
   a) *staphylococcus aureus*
   b) *escherichia Coli*
   c) *Staphylococcus epidermidis*
   d) bacillus Anthracis
   e) Clostridium Botulinum
7. Members of which pairing are most similar in appearance to each other?
   a) Fungi and algae
   b) Algae and archaea
   c) Archaea and bacteria
   d) Bacteria and viruses
   e) Viruses and algae
8. If you wanted to increase your chances of obtaining a member of the Archaea (rather than a member of another domain), which would be the best site to obtain a sample?
   a) Intestine of an elephant
   b) Skin of an elephant
   c) A 95°C hot spring in Yellowstone National Park
   d) A 45°C hot spring in Hawaii
   e) A raw hamburger patty
9. Viruses
   a) contain both protein and nucleic acid.
   b) infect only eukaryotic cells.
   c) can grow in the absence of living cells.
   d) are generally the same size as prokaryotes.
   e) always kill the cells they infect.
10. Antonie van Leeuwenhoek could not have observed
   a) roundworms.
   b) *Escherichia coli.*
   c) yeasts.
   d) viruses.

Applications

1. The American Society for Microbiology is preparing a "Microbe-Free" banquet to emphasize the importance of microorganisms in food production. What foods could not be on the menu?
2. If you were asked to nominate one of the people mentioned in this chapter for the Nobel Prize, who would it be? Make a statement supporting your choice.

Critical Thinking

1. A microbiologist obtained two pure biological samples: one of a virus, and the other of a viroid. Unfortunately, the labels had been lost. The microbiologist felt she could distinguish the two by analyzing for the presence or absence of a single molecule. What molecule would she search for and why?
2. Why would archaea that grow in extreme environments be more intensively studied than those that do not?

2 The Molecules of Life

Space-filling models of water molecules. *(Lisa Burgess/McGraw Hill)*

KEY TERMS

Atom The basic unit of matter.

Buffer A chemical that stabilizes the pH of a solution.

Carbohydrate An organic compound composed of one or more simple sugars.

Covalent Bond A chemical bond formed when two atoms share electrons.

Hydrogen Bond The attraction between a hydrogen atom in a polar molecule and an electronegative atom in the same or another polar molecule.

Ion An atom or molecule that has gained or lost one or more electrons.

Ionic Bond A chemical bond resulting from the attraction between positively and negatively charged ions.

Lipid An organic molecule that is not soluble in water.

Molecule Two or more atoms held together by covalent bonds.

Nucleic Acid A macromolecule consisting of one or two nucleotide chains; DNA or RNA.

Organic Compound A compound that has a carbon atom covalently bonded to a hydrogen atom.

pH A measure of the hydrogen ion concentration or acidity of a solution on a scale of 0 to 14.

Protein A macromolecule consisting of one or more chains of amino acids.

A Glimpse of History

Farmers have understood for centuries that growing the same crop on the same land year after year reduces the crop yield. Allowing a field to remain unplanted for one or more seasons lets wild plants grow, and these appear to improve the soil. During the late 1880s, Martinus Beijerinck, a microbiologist from the Netherlands, helped to explain the science behind what the farmers already knew.

Beijerinck was described as a "keen observer" who was able "to fuse results of remarkable observations with a profound and extensive knowledge of biology and the underlying sciences." Fortunately for farmers, he disliked medical microbiology, preferring to study the agricultural applications of microbiology instead. Based on the work of other scientists, Beijerinck knew that certain plants improve the soil because microorganisms help them accumulate nitrogen from the air. All forms of life require nitrogen because it is an essential component of cellular material, but soils have only limited amounts. Almost 80% of Earth's atmosphere is nitrogen gas (N_2), so using that atmospheric nitrogen source might seem like an easy task, but it is far from it. N_2 is very stable, with strong chemical bonds holding the two nitrogen atoms together. Plants, animals, and most other organisms lack the ability to break those bonds, so it was unclear how microorganisms were helping plants use atmospheric nitrogen.

Beijerinck and other scientists studied nitrogen accumulation in a group of plants called legumes. Members of this group bear seeds in pods and include peas, beans, and clover, and their roots often have nodules (small growths that look like tumors) (see figure 11.17). It had been reported that the root nodules contained microorganisms, and that these allowed the plant to accumulate nitrogen. Beijerinck made a significant breakthrough by isolating a bacterium from inside a root nodule and then showing that it could convert atmospheric nitrogen into a form that can be incorporated into cellular material. This process is called nitrogen fixation, and the nitrogen is said to be "fixed." Then he showed that root nodules form when the nitrogen-fixing bacterium—a member of a group now called rhizobia—is incubated with seedlings of legumes. As those plants, or even parts of the plants, die and decay, the cellular material is released, enriching the soil with nitrogen-containing nutrients.

When farmers plant crops like soybeans or allow wild clover to grow in a field, rhizobia in the nodules fix nitrogen, which can then be used by the host plant. We now know that many types of rhizobia exist, each able to establish a relationship with only certain types of host plants. To honor Beijerinck for his achievements, a genus of nitrogen-fixing bacteria (*Beijerinckia*) is named for him.

Simply stated, chemistry is the study of matter, or the "stuff" of which the universe is composed. As you learn about microbiology, you will find that cells are masters at converting one set of chemicals to another and, in doing so, producing materials required to make new cells. This amazing ability to transform material is just one reason that microorganisms are crucial to life and that the principles described in this chapter are fundamental to information throughout the text.

2.1 ■ Elements and Atoms

Learning Outcomes

1. Describe the general structure of an atom and its isotopes.
2. Describe the importance of valence electrons.

Matter is categorized into **elements,** substances that have unique chemical properties and cannot be broken down by ordinary chemical means. An **atom** is the basic unit of all matter, and each element is composed of only one type of atom. Each element is indicated by a one- or two-letter symbol, with the first letter capitalized; for example, C represents carbon and Ca represents calcium. Not all chemical symbols are derived from English, as seen with the symbol Na, which stands for *natrium* in Latin, but which we know as sodium.

Of the 92 or so naturally occurring elements, biological systems are primarily composed of only six. As a way to remember these, think of the acronym CHONPS: carbon (C), hydrogen (H), oxygen (O), nitrogen (N), phosphorus (P), and sulfur (S). Other elements are also important in biological systems, including some referred to as trace elements, reflecting the fact that they are found in very small quantities.

Atomic Structure

Atoms consist of three major components:

- **Protons:** positively charged particles
- **Neutrons:** uncharged particles
- **Electrons:** negatively charged particles

Protons and neutrons together form the atomic nucleus (dense, central region of the atom), around which electrons move in a "cloud" (**figure 2.1**). Overall, the atom has no charge because the number of positive protons in the atomic nucleus is the same as the number of negative electrons in the cloud.

Each type of atom is distinguished by the number of protons in its nucleus, referred to as its **atomic number** (**table 2.1**). For example, a hydrogen atom has 1 proton, so its atomic number is 1; a carbon atom has 6 protons, so its atomic number is 6. Each atom also has a **mass number,** the sum of the number of protons and neutrons in the nucleus of that atom. A hydrogen atom with 1 proton and no neutrons has an atomic number of 1 as well as a mass number of 1. A carbon atom with 6 protons and 6 neutrons has an atomic number of 6 and a mass number of 12. Note that electrons are too light to contribute to the mass of an atom.

Atoms can be depicted in various ways to emphasize certain characteristics (**figure 2.2**). Some diagrams indicate the number of protons and neutrons in the nucleus as well as the number of electrons in the surrounding cloud. Frequently, however, the element symbol is used, with the atomic number in subscript to the left, and the mass number in superscript to the left. A depiction called a Lewis symbol uses the element symbol and indicates the number of valence electrons using dots (valence electrons are described next).

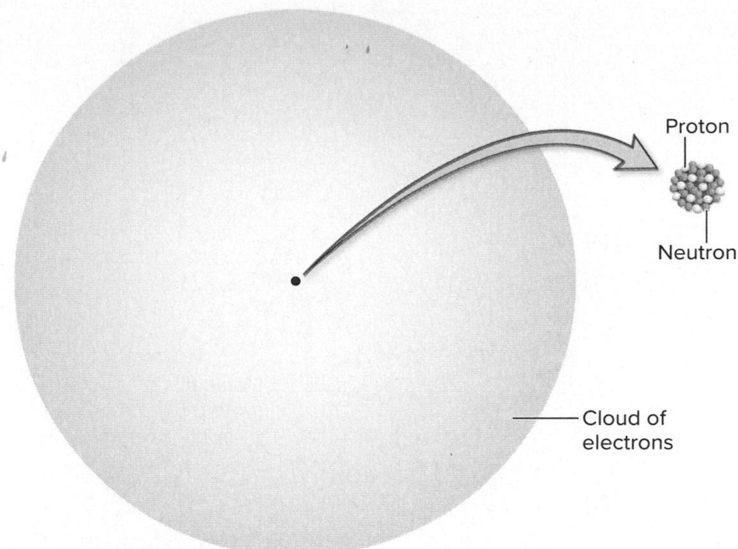

FIGURE 2.1 Atomic Structure A proton has a positive charge, a neutron has a neutral charge, and an electron has a negative charge. The electrons move around the nucleus in a cloud.

? **How does the number of electrons in an atom compare with the number of protons?**

TABLE 2.1 Characteristics of Atoms Common in Living Organisms

Atom	Symbol	Atomic Number	Mass Number
Hydrogen	H	1	1
Carbon	C	6	12
Nitrogen	N	7	14
Oxygen	O	8	16
Phosphorus	P	15	31
Sulfur	S	16	32

MicroByte

If the nucleus of an atom were a marble in the center of a football field, the electrons would occupy the entire space of the stadium.

The Role of Electrons

Electrons move in a cloud around an atom's nucleus, but at any point in time their precise location is impossible to determine (see figure 2.1). They are most likely to be in specific regions called shells, each of which is associated with a different energy level. The number of electrons a shell can hold is limited. Electrons are attracted to the protons in the nucleus, so shells closer to the nucleus are generally filled before electrons occupy other shells. The shell closest to

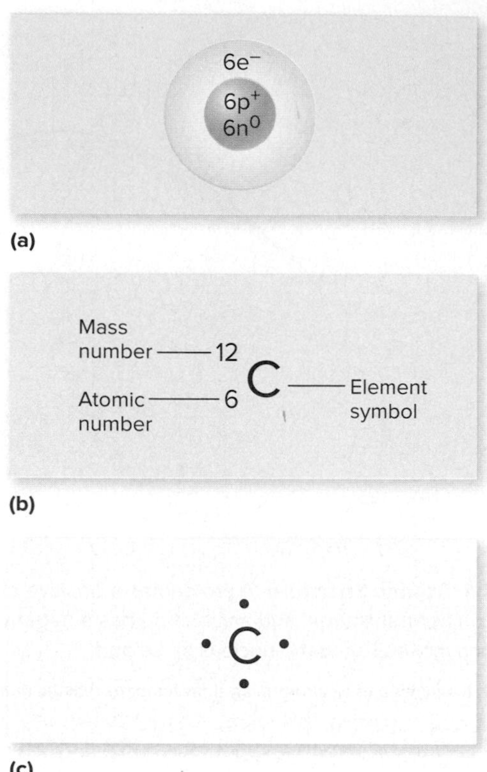

FIGURE 2.2 Depictions of a Carbon Atom (a) The number of protons (p^+) and neutrons (n^0) in the nucleus are indicated, along with the number of electrons (e^-) in the surrounding cloud. **(b)** The mass number and atomic number are indicated on the left of the element symbol. **(c)** The Lewis symbol uses dots to indicate the number of valence electrons.

? **Why are electrons not considered when determining the mass of an atom?**

the nucleus holds no more than 2 electrons; these electrons are most highly attracted to the nucleus and have the lowest energy level in that particular atom. Once that shell is filled, additional electrons occupy the next shell, which holds 8 electrons. Larger atoms have additional shells that hold even more electrons, but most atoms of biological significance follow the "octet rule," meaning they are most stable when their outer shell contains 8 electrons. An important exception is hydrogen, with a single shell; recall that the first shell has a limit of 2 electrons.

Electrons largely determine the chemical reactivity of an atom. An atom's **valence electrons,** meaning the electrons in the outermost shell (called the valence shell), are the most important in that regard. Atoms with the maximum number of electrons in the outer shell are very stable. Atoms with an unfilled outer shell tend to react with other atoms that have an unfilled outer shell by sharing or transferring electrons. Lewis symbols provide a simple visual representation of an atom's outer shell, with each dot representing one valence electron (see figure 2.2c); Lewis structures are similar, but they show the interaction of two or more atoms, with dots indicating shared and unshared valence electrons (see figure 2.6a).

Isotopes

All atoms of a given element have the same number of protons, but they can have different numbers of neutrons. That is, atoms of an element all have the same atomic number, but they can have different mass numbers. The various forms of atoms of an element are **isotopes** (*iso* means "same" and *tope* means "place"). For example, nearly 99% of naturally occurring carbon atoms have 6 neutrons, but some have 7 or 8. The fact that the atoms of an element can have different mass numbers is reflected in the term **atomic mass**—the average of the mass numbers of the atoms of the element, weighted according to the relative abundance of the naturally occurring isotopes. The atomic mass of carbon is 12.01.

Although the chemical behavior of different isotopes is similar to their more common counterparts, the structural differences can often be detected. Because of this, researchers and clinicians use isotopes to monitor the fate of specific atoms within a population of cells. Some isotopes are unstable and emit radiation that can be detected; others are tracked using special instruments that measure differences in mass. The energy emitted from radioactive isotopes (radioisotopes) is sometimes useful in medical diagnosis. For example, to evaluate proper functioning of the human thyroid gland (which produces the iodine-containing hormone thyroxine) doctors often administer radioactive iodine. The gland can then be scanned to determine if the amount and distribution of iodine is normal (**figure 2.3**).

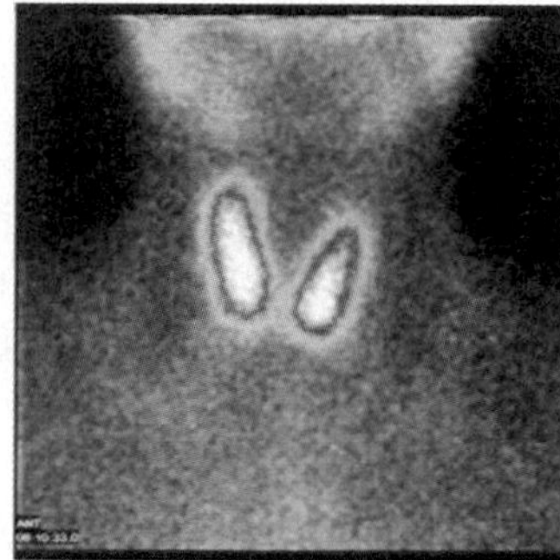

FIGURE 2.3 Radioisotopes A scan of the thyroid gland 24 hours after the patient received radioactive iodine. Stefania Arca/Shutterstock

? **Why is radioactive iodine concentrated in the thyroid gland?**

MicroAssessment 2.1

The six most common elements in biology are carbon, hydrogen, oxygen, nitrogen, phosphorus, and sulfur. The basic unit of all matter, the atom, is composed of protons, electrons, and neutrons. The reactivity of an atom is largely determined by the number of valence electrons. Isotopes are atoms of a given element that have a different number of neutrons.

1. What do the dots on a Lewis symbol represent?
2. What is the "octet rule" and its biologically important exception?
3. What do atoms of ^{14}C and ^{14}N have in common? How do they differ? **?**

2.2 ■ Chemical Bonds and Reactions

Learning Outcomes

3. Compare and contrast ionic bonds, covalent bonds, and hydrogen bonds.
4. Explain the role of an enzyme in a chemical reaction.

Atoms with an unfilled outer shell react with each other to lose, gain, or share their valence electrons. This allows them to achieve a more stable state and is the basis for chemical bond formation.

Ions and Ionic Bonds

An atom that gains or loses an electron is no longer neutral—it is an **ion** (**figure 2.4**). Atoms that gain an electron become negatively charged and are called **anions;** those that lose an electron become positively charged and are called **cations.** The type and amount of charge are indicated by a superscript to the right of the chemical symbol. For example, Na^+ indicates a sodium atom that has lost one electron and therefore carries a +1 (positive) charge; Mg^{2+} indicates a magnesium atom that has lost two electrons and therefore has a +2 charge. Note that a positive ion is formed by the loss of one or more valence electrons. The nucleus still contains the same number of protons, resulting in a positive charge. Na^+ cannot be formed by gaining a proton because gaining a proton would change the atomic number, and therefore the atom would no longer be sodium.

Ionic bonds form between cations and anions because of the attraction between positive and negative charges (see figure 2.4). The resulting product is called a salt. A common type of salt, sodium chloride (table salt), is composed of Na^+ (sodium cations) and Cl^- (chloride anions) and forms a solid crystal. The structure is highly ordered because the electrical attraction between positive and negative charges brings the ions together, but the like charges repel one another and are positioned as far apart as possible. Crystals continue to grow as new ions are added. Salts such as sodium chloride dissolve in water and are called **electrolytes,** meaning they conduct electricity.

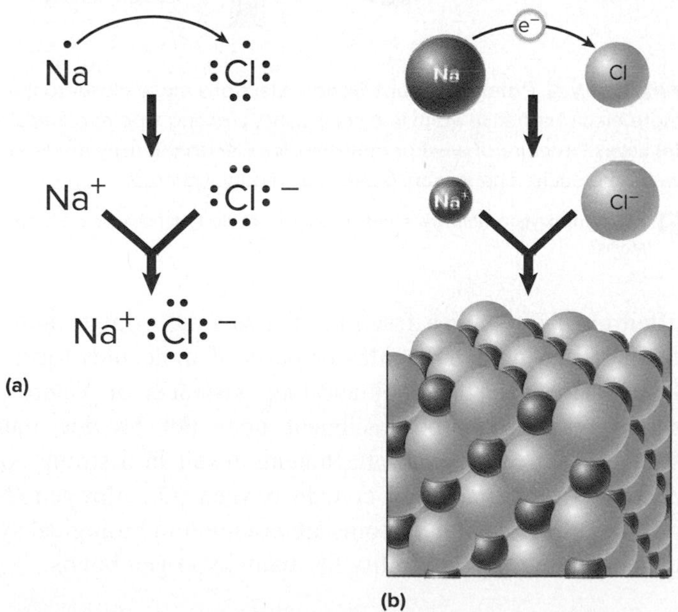

FIGURE 2.4 Ions and Ionic Bonds (a) Lewis symbols of sodium and chloride ions being formed, and an ionic bond between them. **(b)** Space-filling model of a salt crystal being formed by ionic bonding. Note that a cation is smaller than its neutral atom while an anion is larger.

? Which of the ions in this figure is an anion, and which is a cation?

MicroByte

Electrical charges from the heart are conducted by electrolytes and can be detected on the body surface as an electrocardiogram (ECG).

Covalent Bonds

Atoms do not always fill their valence shells (outer shells) by gaining or losing electrons. They may instead share pairs of valence electrons, forming **covalent bonds.** For example, a hydrogen atom (H) has one electron, so one additional electron is required to fill its valence shell; if each of two H atoms shares its single electron with the other, so that the atoms have two electrons between them, both atoms gain stability. The covalent bond between the two atoms is indicated by a dash, as H—H. Some atoms share more than one pair of electrons with each other, forming a double or triple covalent bond, indicated by a corresponding number of lines between the atoms. An oxygen atom needs two electrons to fill its valence shell; if each of two oxygen atoms shares two of its electrons with the other, then a double covalent bond is formed, represented as O=O. Double bonds are stronger than single bonds, meaning that more energy is required to break them, and triple bonds are stronger still.

Two or more atoms joined together by covalent bonds form a **molecule.** A molecule is represented by a molecular formula that indicates how many atoms of each type are present. Thus, a hydrogen molecule is represented by the formula H_2. If atoms that make up a molecule are different elements, the term **compound** may be used. (Salts such as NaCl are called ionic compounds.) Water is a compound that contains two hydrogen atoms and one oxygen atom; it is represented by the formula H_2O. The molecular mass of a molecule is based on the mass numbers of the component atoms. The molecular mass of most water molecules is 1 + 1 + 16, or 18.

Carbon (C) is a particularly important element in biological systems because its bonding properties provide the basis for many diverse structures. A carbon atom has four valence electrons, so it needs four more to fill its valence shell. Because of this, the atom forms four covalent bonds, typically with the main elements that make up cells: CHONPS (carbon, hydrogen, oxygen, nitrogen, phosphate, and sulfur). One C atom sharing electrons with four H atoms is methane (CH_4) (**figure 2.5**). Molecules that contain at least carbon and hydrogen are **organic compounds;** those that do not are inorganic compounds.

$4\ \dot{H}$ + $\cdot\dot{\underset{\cdot}{C}}\cdot$ → H:C:H (with H above and below) Methane

Each hydrogen atom needs one electron to fill its valence shell.

Carbon needs four electrons to fill its valence shell.

(a)

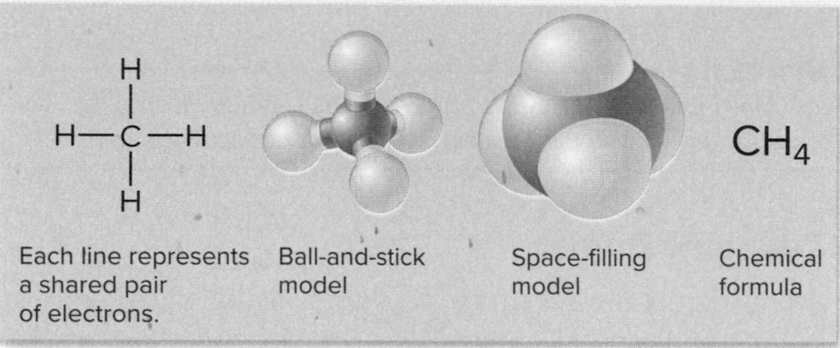

(b)

FIGURE 2.5 Covalent Bonds Covalent bonds are formed when atoms share electrons. **(a)** Methane is formed when a carbon atom fills its valence shell by sharing eight electrons—four belong to H atoms and four belong to the carbon atom. **(b)** Different ways of showing the methane molecule.

? **Is methane an organic molecule? Explain.**

TABLE 2.2	Electronegativity of Some Atoms Common in Biology
Element	**Electronegativity**
Hydrogen	2.1
Carbon	2.5
Nitrogen	3.0
Oxygen	3.5

TABLE 2.3	Non-Polar and Polar Covalent Bonds	
Type of Covalent Bond	**Atoms Involved and Charge Distribution**	
Non-polar	C—C C—H H—H	C and H atoms have similar electronegativities, so the charge on each atom is nearly equal.
Polar	O—H N—H O—C N—C	The O and N atoms are more electronegative than either C or H, so the O or N atoms have a slight negative charge, whereas the C or H atoms have a slight positive charge.

Electrons in a covalent bond may or may not be equally shared between the two atoms, depending on the electronegativity of the atoms involved; electronegativity is a measure of how strongly atoms attract electrons, with higher numbers indicating a stronger attraction (**table 2.2**). A **non-polar covalent bond** forms when electrons are shared equally, such as when identical atoms share electrons or when different atoms that have a similar electronegativity share electrons (**table 2.3**). In contrast, a **polar covalent bond** forms when electrons are shared unequally, which occurs when one atom is significantly more electronegative than the other.

The slight separation of charge resulting from a polar covalent bond is indicated by the Greek symbol delta (δ); the atom with a slight positive charge is δ^+ and the atom with the slight negative charge is δ^-. Consider the O—H bonds in water: The oxygen atom is more electronegative than the hydrogen atom, so the oxygen atom pulls the electrons toward it, giving it a slight negative charge (δ^-) and leaving each of the two hydrogen atoms with a slight positive charge (δ^+) (**figure 2.6**). Polar covalent bonds play a key role in biological systems because they often result in the formation of hydrogen bonds, discussed next.

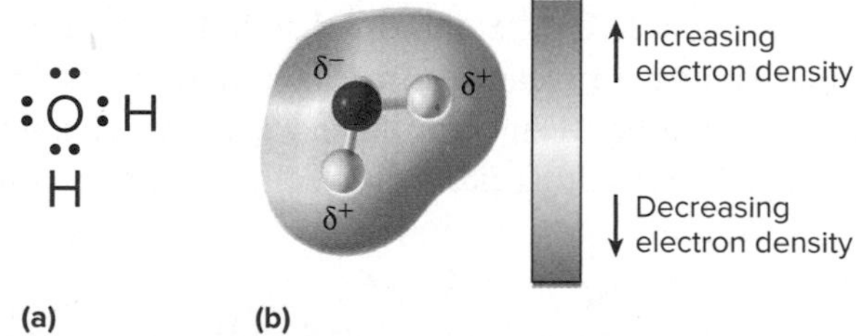

(a) (b)

FIGURE 2.6 Polar Covalent Bonds Electrons move closer to the more electronegative atom in a compound, creating a polar molecule. **(a)** Lewis structure of a water molecule. **(b)** Electron density model of a water molecule. The symbol δ indicates a partial charge.

? **Why is the oxygen atom in a water molecule more electron-rich than the hydrogen atoms?**

Hydrogen Bonds

Hydrogen bonds are weak bonds formed when a hydrogen atom in a polar molecule is attracted to an electronegative atom in the same or another polar molecule (**figure 2.7;** see also table 2.3). Although the individual bonds are weak and often exist for only a fraction of a second, a large number of them can hold molecules or parts of molecules together firmly. Consider the hook-and-loop fasteners of Velcro: A single hook-and-loop attachment does not provide much strength, but many such attachments result in a strong connection. Compounds that contain oxygen (O), nitrogen (N), or other electronegative atoms are common in biological systems, creating the possibility for many hydrogen bonds.

Molarity

A *mole* is a quantitative term used by chemists, much like a *dozen* is a quantitative term used by bakers. A baker may have a dozen cookies or a dozen bagels. A chemist may have a

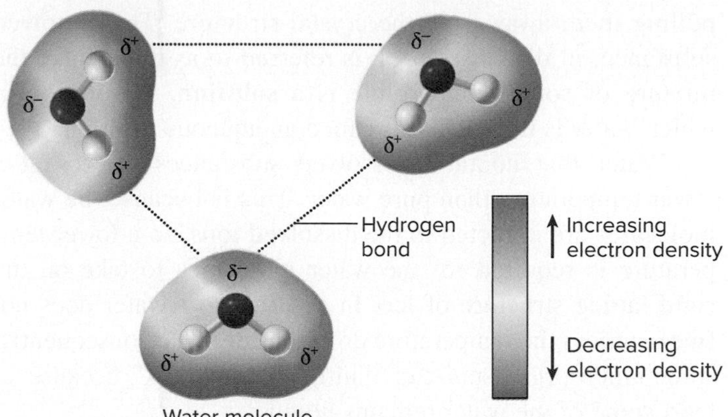

FIGURE 2.7 Hydrogen Bond Formation Hydrogen bonds form between water molecules because the electron-rich oxygen atom attracts electron-poor hydrogen atoms.

? **Explain why two identical atoms joined by a covalent bond cannot form a hydrogen bond.**

mole of glucose molecules or a mole of Na^+ ions. One **mole** is 6.022×10^{23} particles of a substance. That number is not important from a practical standpoint, but the concept is essential in chemistry—a mole of one substance has the same number of particles as a mole of any other. Scientists measure chemicals in much larger amounts than a single molecule or ion. Moles are measured in grams. One mole of sodium chloride (NaCl) weighs 58.44 grams. This is the sum of the atomic masses of the two elements (Na = 22.99 and Cl = 35.45) in grams.

The **molarity** (M) of a solution is defined as the number of moles of a compound dissolved in enough liquid to make 1 liter of solution. Therefore, a 1 M (read as "one molar") aqueous solution of NaCl has 58.44 grams of NaCl dissolved in enough water to bring the volume to 1 liter.

Chemical Reactions

Chemical reactions convert one or more substances (reactants) to others (products). During the course of the reaction, electrons are transferred, a process that often involves making and breaking bonds. Although various types of chemical reactions exist, the most common in biological systems include:

- **Synthesis reactions.** These combine two or more reactants to form a larger product.

$$\underset{\text{(reactants)}}{A + B} \longrightarrow \underset{\text{(product)}}{AB}$$

- **Decomposition reactions.** These split a reactant into one or more smaller products.

$$\underset{\text{(reactant)}}{AB} \longrightarrow \underset{\text{(products)}}{A + B}$$

- **Exchange reactions.** These couple synthesis and decomposition to trade one or more components of the reactants; as one bond is broken, another is formed. Two examples are shown below.

$$\underset{\text{(reactant)}}{AB + C} \longrightarrow \underset{\text{(products)}}{AC + B}$$

$$\underset{\text{(reactant)}}{AB + CD} \longrightarrow \underset{\text{(products)}}{AC + BD}$$

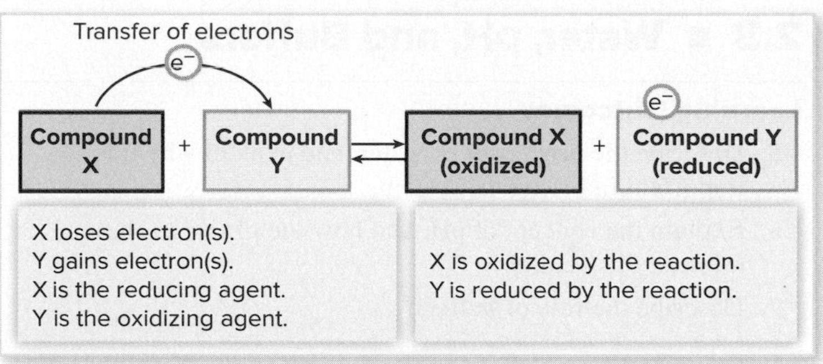

FIGURE 2.8. Oxidation-Reduction Reactions The substance that loses one or more electrons is oxidized by the reaction; the one that gains those electrons is reduced.

? **Hypochlorite (household bleach) is an oxidizing agent. Based on this information, is it more likely to donate electrons to other molecules, or remove them? Explain.**

Chemical reactions in which electrons are transferred from one reactant to the other are called **oxidation-reduction reactions** or **redox reactions** (**figure 2.8**). The reactant that loses the electron is **oxidized** by the reaction and the one that gained it is **reduced.** Reactants that typically lose electrons in a redox reaction are called **reducing agents,** whereas those that gain them are called **oxidizing agents.** As you will learn in later chapters, cells use redox reactions to convert energy into a usable form, and certain oxidizing agents have practical applications as disinfectants (chemicals that kill microbes).

Most biologically important chemical reactions would occur too slowly to be useful to a cell, so cells speed the rate of reactions by using biological catalysts called **enzymes.** The enzymes bind to one or more reactant molecules and position them in such a way that certain bonds are more likely to be broken or to form. The reactants in enzyme-catalyzed reactions are typically referred to as **substrates.**

MicroAssessment 2.2

Ionic bonds form between positively and negatively charged ions. Covalent bonds result from sharing electrons. Hydrogen bonds form between polar molecules or portions of molecules. Chemical reactions proceed from reactants to products. Enzymes speed up chemical reactions.

4. Compare the properties of covalent, hydrogen, and ionic bonds.
5. Which type of bond requires an enzyme to break it?
6. Why are hydrogen bonds associated with polar covalent bonds, but not with non-polar covalent bonds?

2.3 ■ Water, pH, and Buffers

Learning Outcomes

5. Describe the properties of water, and explain why it is so important in biological systems.
6. Explain the concept of pH, and how the pH of a solution relates to its acidity.
7. Describe the role of buffers.

Life on Earth has always been intimately associated with water, which makes up over half of the mass of a living organism and provides the medium in which countless chemical reactions fuel life processes.

Water

The unique properties of water are due to the features of a water molecule (H_2O). The oxygen atom is quite electronegative, so it pulls electrons away from the two hydrogen atoms, resulting in an asymmetrical polar molecule that readily forms hydrogen bonds (see figure 2.7). Neighboring water molecules form hydrogen bonds with one another, creating a network of interacting molecules. The extent and stability of this bonding depends on temperature. At room temperature, the molecules form a liquid as the hydrogen bonds continually break and re-form, allowing water molecules to slide past one another and move closer together. At freezing temperatures, however, the molecules slow down and form the maximum number of hydrogen bonds. This places each molecule a set distance away from its neighbors, producing a lattice-like structure called ice (**figure 2.9**). That structure makes ice less dense than liquid water, which is why ice floats.

Although hydrogen bonds between water molecules are constantly formed and broken at room temperature, the abundance of the bonds gives water its unique properties. For example, the hydrogen bonding between water molecules on the surface of liquid water creates the surface tension that prevents a drop of water from flattening and allows insects to "walk" on a pond. Hydrogen bonding of water molecules to other charged surfaces helps water travel through tiny vessels to the tops of trees. The world would look much different without the adhesive properties of water.

The polar nature of water molecules is why water is such an important **solvent,** meaning liquid in which substances dissolve. A crystal of table salt (NaCl) dissolves because the water molecule's slightly positively charged portion (hydrogen atoms) attracts the Cl^- ions, while the molecule's slightly negatively charged portion (oxygen atom) attracts the Na^+ ions (**figure 2.10**). As a result, water molecules surround the ions, pulling them away from the crystal structure. The dissolved substance, in this case NaCl, is referred to as the **solute;** the mixture of solvent and solute is a **solution.** A solution in which water is the solvent is called an aqueous solution.

Water that contains dissolved substances freezes at a lower temperature than pure water. This is because the water molecules are attracted to the dissolved ions, so a lower temperature is required for the water molecules to take on the rigid lattice structure of ice. In nature, most water does not freeze unless the temperature drops below 0°C. Consequently, some microorganisms can multiply below 0°C because at least some of the water remains liquid.

The terms hydrophilic and hydrophobic are used to describe a substance's interactions with water. Charged substances that are attracted to water are **hydrophilic,** meaning "water-loving." Non-polar substances that repel water are **hydrophobic,** meaning "water-fearing." Although hydrophobic molecules do not dissolve in water, they can be organized by the presence of water. For example, when oil drops are added to water, they simply float to the top and then often merge. This phenomenon occurs because oil molecules are pushed together as the water molecules form hydrogen bonds among themselves.

MicroByte

Astrobiologists who search for life elsewhere in the universe often concentrate on planets and moons where evidence of water exists.

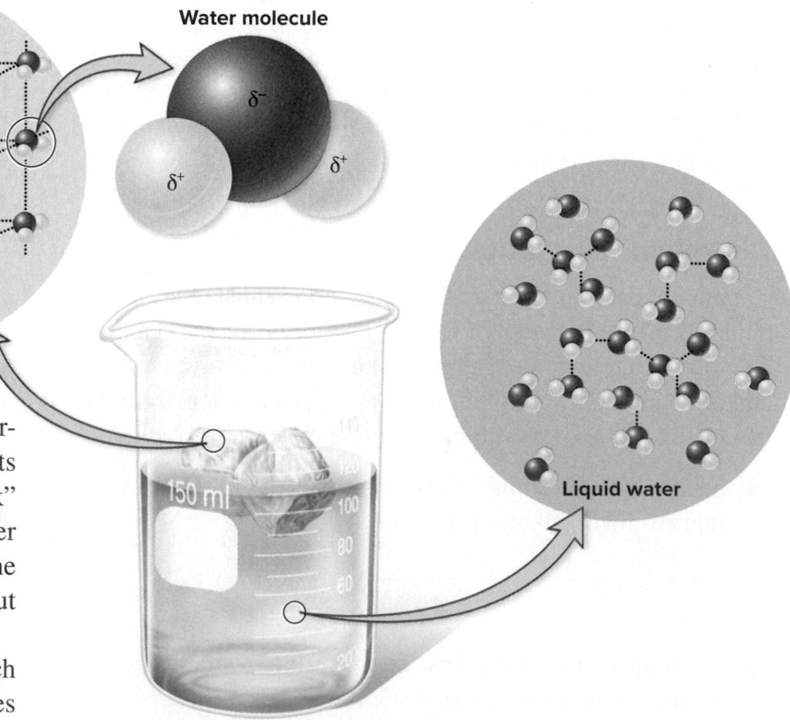

FIGURE 2.9 Water In liquid water, hydrogen bonds continuously break and re-form, allowing the molecules to move closer together. In ice, each H_2O molecule forms hydrogen bonds to other H_2O molecules, producing a rigid crystalline structure.

? **Why does ice float in water?**

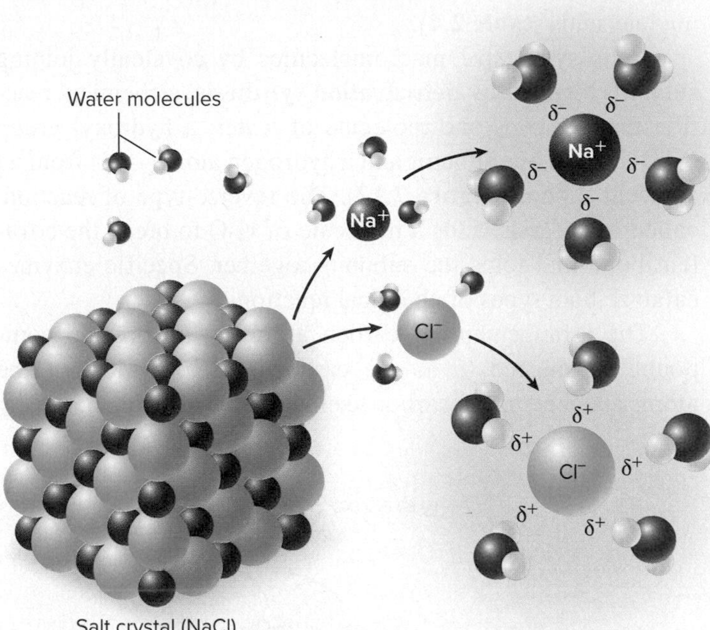

FIGURE 2.10 Salt (NaCl) Crystal Dissolving in Water In water, the Na^+ and Cl^- separate due to interaction with H_2O molecules. The Na^+ is attracted to the slightly negatively charged O^- portion of the water molecules, and the Cl^- is attracted to the slightly positively charged H^+ portion. In the absence of water, the salt is highly structured because of ionic bonds between Na^+ and Cl^- ions.

? If water were not polar, would it dissolve sodium chloride? Explain.

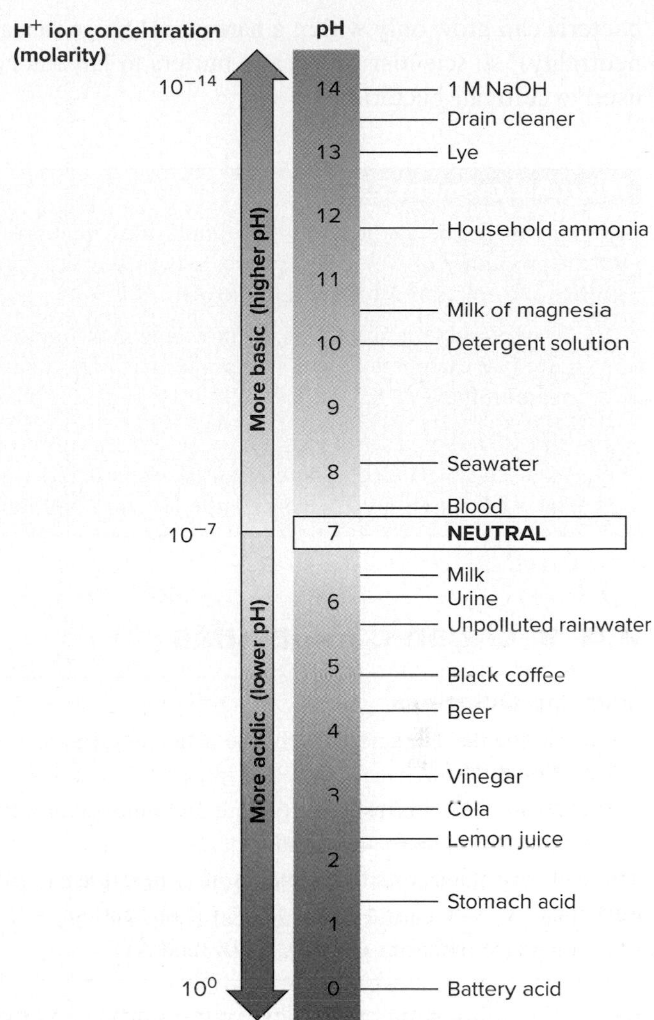

FIGURE 2.11 pH Scale The concentration of H^+ ions varies by a factor of 10 between each pH number since the scale is logarithmic.

? When the pH of a solution drops from 5 to 4, does the H^+ concentration increase, or decrease?

pH of Aqueous Solutions

pH is a measurement of the acidity of a solution. To understand the chemistry of what that means, it helps to know that water molecules dissociate (split apart) at a low rate because the covalent bond between the oxygen atom and a hydrogen atom in the molecule can break. When the bond breaks, the strongly electronegative oxygen atom takes the electron that had been shared by the weakly electronegative hydrogen atom. In a simplified view of the situation, this forms a proton (H^+; a hydrogen ion) and a hydroxide ion (OH^-). The reaction is reversible because the two ions can spontaneously rejoin to form a new water molecule (as indicated by the arrows in the following equation):

$$\underset{\text{(water)}}{H_2O} \rightleftarrows \underset{\text{(hydrogen ion)}}{H^+} + \underset{\text{(hydroxide ion)}}{OH^-}$$

In pure water, the concentration of H^+ and OH^- ions is equal, but when substances called acids or bases are added, the balance shifts. When an acid is added, the concentration of H^+ in the solution increases (acids dissociate to release H^+), but when a base is added, the concentration of H^+ decreases (it is easiest to think of bases as dissociating to release OH^-, which then combines with H^+ to make a water molecule, but the situation is not always that simple). The pH scale ranges from 0 to 14, with the values indicating the negative logarithm of the H^+ concentration (moles per liter); thus, the lower the pH value, the higher the H^+ concentration. Neutral solutions have a pH of 7 (equal amounts of H^+ than OH^-), acidic solutions have a pH less than 7, and basic solutions have a pH greater than 7 (**figure 2.11**). An important thing to remember is that because the scale is logarithmic, a change of one whole number represents a 10-fold difference. Thus, drinking a sugar solution that results in a drop of the pH of dental plaque (an accumulation of microbes on a tooth surface) from the typical value of about 7 to approximately 5 indicates a 100-fold increase in acidity (see figure 24.4).

Buffers

A **buffer** is a chemical that stabilizes the pH of a solution, thereby helping maintain a relatively constant pH. Buffers function by releasing H^+ ions to a solution when a base is added and combining with H^+ ions when an acid is added. Products of a cell's chemical reactions are often acidic or basic, yet most bacteria can grow only within a narrow pH range (usually near neutrality), so scientists often add buffers to laboratory media used to cultivate bacteria.

MicroAssessment 2.3

Weak hydrogen bonds between water molecules are responsible for the properties of water. The pH scale is used to indicate the acidity of a solution. Buffers stabilize pH.

7. What characteristics of a water molecule make it polar? Give two examples of why this property is important in microbiology.
8. How do buffers function?
9. What must be added to pure water to decrease the OH^- concentration? To decrease the H^+ concentration?

2.4 ■ Organic Molecules

Learning Outcomes

8. Describe the characteristics of the different types of carbohydrates.
9. Compare and contrast the structure and function of fatty acids, triglycerides (fats), and steroids.
10. Describe the factors that affect protein structure and function.
11. Compare and contrast the chemical compositions, structures, and major functions of DNA, RNA, and ATP.

Recall that all organic molecules contain carbon and hydrogen. Because each carbon atom forms four covalent bonds, an incredible assortment of organic molecules exists. In an organic compound with six carbon atoms, for example, those atoms may join to each other to form a linear chain, a branched chain, a ring, or any combination of these. In addition, the covalent bonds joining two carbon atoms within an organic molecule may be single, double, or even triple.

Organic molecules are often **macromolecules** (*macro* means "large"). Most macromolecules are **polymers** (*poly* means "many") formed by joining subunits or **monomers** (*mono* means "single"). Different classes of macromolecules are made up of different subunits. The four major classes of organic molecules are carbohydrates, lipids, proteins, and nucleic acids (**table 2.4**).

Cells synthesize macromolecules by covalently joining subunits together by **dehydration synthesis,** a chemical reaction that removes one molecule of water: a hydroxyl group (—OH) from one subunit and a hydrogen atom (—H) from an adjacent subunit (**figure 2.12**). The reverse type of reaction, called **hydrolysis,** adds a molecule of H_2O to break the covalent bond that joins the subunits together. Specific enzymes catalyze both types of chemical reactions.

The arrangement of carbon atoms in an organic compound is referred to as the carbon skeleton. Most of the atoms attached to the carbon skeleton are hydrogen, but many

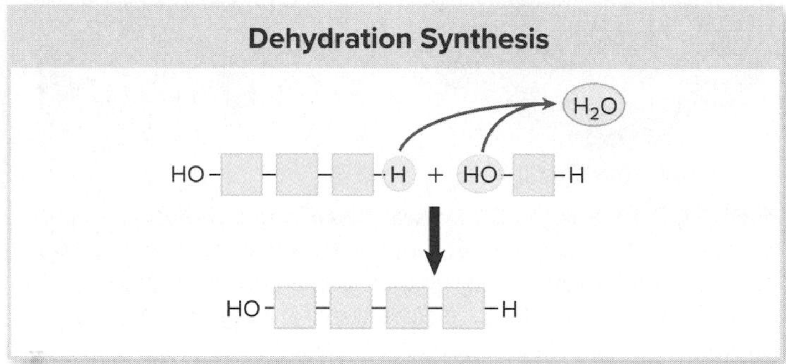

(a)

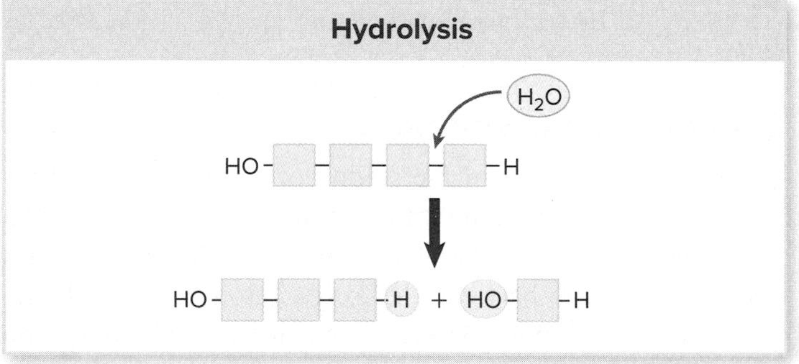

(b)

FIGURE 2.12 Synthesis and Breakdown of Macromolecules
(a) Dehydration synthesis is a chemical reaction in which water is removed as subunits are joined together (polymerized). **(b)** Hydrolysis is a chemical reaction in which a molecule of water is added to break the bond between subunits.

? What are the four major classes of macromolecules?

TABLE 2.4 Major Classes of Organic Molecules

Name	Subunits	Major Functions
Carbohydrates	Monosaccharides	Structural components of cell walls; energy sources
Lipids	Varies—subunits are not always similar	Some types are important components of cell membranes; energy storage
Proteins	Amino acids	Enzyme catalysts; structural portion of many cell components
Nucleic acids	Nucleotides	
DNA	Deoxyribonucleotides	Genetic information of a cell
RNA	Ribonucleotides	Various roles in protein synthesis; catalysis

TABLE 2.5 Biologically Important Functional Groups

Functional Group	Structure	Where Found
Aldehyde	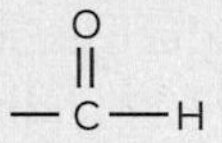 —C(=O)—H	Carbohydrates
Amino	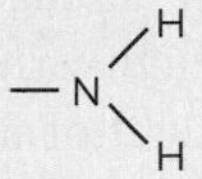 —N(—H)—H	Amino acids (the subunits of protein)
Carboxyl	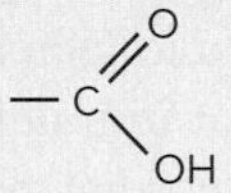 —C(=O)—OH	Organic acids, including amino acids and fatty acids
Hydroxyl	—OH	Carbohydrates; fatty acids; alcohol; some amino acids
Keto	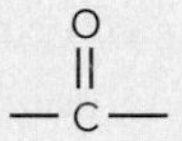 —C(=O)—	Carbohydrates; polypeptides
Methyl	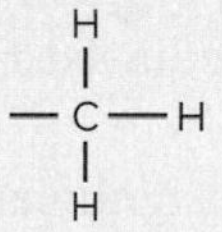 —C(H)(H)—H	Some amino acids; some nucleotides in DNA
Phosphate	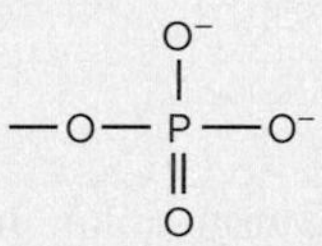 —O—P(=O)(O⁻)—O⁻	Nucleotides (the subunits of nucleic acids); ATP; signaling molecules
Sulfhydryl	—S—H	Part of the amino acid cysteine

organic molecules contain other atoms as well. Distinctive chemical arrangements called **functional groups** contribute to a molecule's properties (**table 2.5**). For example, a carboxyl group (—COOH) may release H^+ ions in solution, and so an organic molecule that contains it could act as an acid. Likewise, phosphate groups are usually ionized and carry a negative charge. As you read about the subunits that make up the macromolecules described in this section, try to identify the carbon skeletons and the various functional groups.

Carbohydrates

Carbohydrates are a diverse group of organic compounds composed of one or more simple sugars. Their biological roles include the following:

- **Energy source.** Organisms break down carbohydrates and harvest the energy they contain.
- **Energy storage.** Organisms can produce and store carbohydrates for later use.
- **Source of carbon for biosynthetic products.** Many microorganisms can make all of their cell components from a single carbohydrate: glucose.
- **Component of genetic material.** The subunits of DNA and RNA contain sugars.
- **Structural components of cells.** Cell walls of plants, fungi, and most bacteria contain carbohydrates.

Carbohydrates contain carbon, hydrogen, and oxygen atoms in an approximate ratio of 1:2:1. The general chemical formula of carbohydrate building blocks, CH_2O, reflects this ratio.

Monosaccharides

Monosaccharides, or simple sugars, are the basic units of a carbohydrate (*sacchar* means "sugar"). Most common monosaccharides have five or six carbon atoms, typically forming a ring; each carbon atom is numbered using a characteristic scheme, allowing scientists to describe the positions of various functional groups attached to the molecules.

Ribose and deoxyribose are 5-carbon monosaccharides (pentoses) found in the nucleic acids RNA and DNA, respectively (**figure 2.13**). These two sugars are identical, except that ribose has an oxygen atom that deoxyribose lacks (*de* means "away from"). Ribose has a hydroxyl group on the number 2 carbon (also called the 2 prime carbon, written 2′ carbon), whereas deoxyribose has only a hydrogen atom at that position.

Glucose, galactose, fructose, and mannose are all 6-carbon sugars (hexoses) with the molecular formula $C_6H_{12}O_6$. These are all **structural isomers,** meaning they contain the same atoms but in different chemical arrangements, much like the

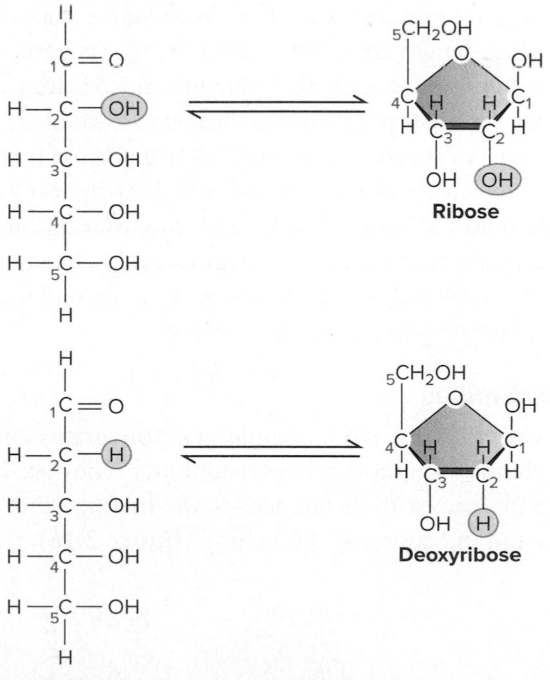

FIGURE 2.13 Ribose and Deoxyribose Note the difference at the number 2 carbon atoms. Although both linear and ring forms occur in the cell, the ring form predominates. In the diagram, the plane of the ring is at an angle to the plane of the paper with the thick line on the ring closest to the reader.

? What is the major chemical difference between ribose and deoxyribose?

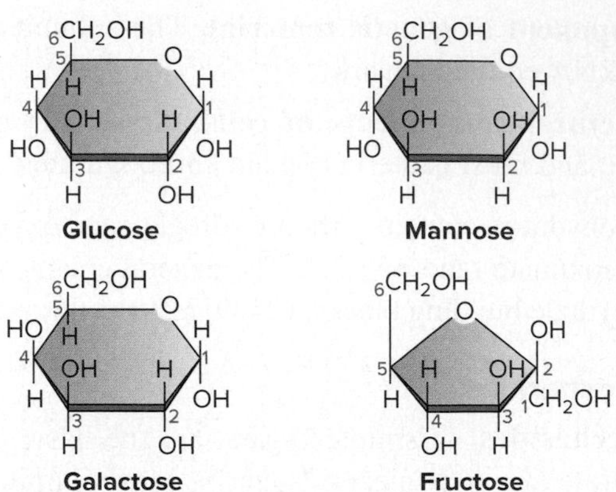

FIGURE 2.14 Common 6-Carbon Sugars These sugars are structural isomers with different properties.

What is a structural isomer?

words "list" and "slit" are composed of the same letters but in different arrangements (**figure 2.14**). Structural isomers result in distinct sugars with different properties. Glucose and mannose both have a sweet taste, for example, but mannose has a bitter aftertaste. Glucose is an important energy source for most cells. Mannose is medically significant because many pathogens have characteristic arrangements of mannose on their surface, allowing the body defenses to detect their presence, a topic discussed in chapter 14.

Disaccharides

A **disaccharide** is composed of two monosaccharides joined together by a covalent bond. The disaccharide sucrose (table sugar), which comes from sugar cane and sugar beets, is composed of the monosaccharides glucose and fructose. Lactose (milk sugar) consists of glucose and galactose. Maltose, a breakdown product of starch, is composed of two glucose molecules.

A disaccharide is formed when a dehydration synthesis reaction creates a bond between two monosaccharides, with the release of a water molecule (**figure 2.15**). The reaction is reversible, so hydrolysis, which adds a water molecule, produces the two original monosaccharides.

Polysaccharides

Polysaccharides are large complex carbohydrates composed of long chains of monosaccharide subunits. They are structurally very diverse, with differences in the linkages that join the subunits and the degree of branching (**figure 2.16**).

Cellulose, starch, glycogen, and dextran are all polymers of glucose. Although cellulose is the principal component of plant cell walls and the most abundant organic molecule on Earth, relatively few organisms can degrade it. Only certain bacteria and fungi produce the enzyme required to break the type of linkage between its glucose subunits, so these microbes play an important role in recycling organic material. In contrast, starch, the energy storage form produced by plants, can be degraded by a wide variety of organisms. Glycogen is an energy-storage product of animals and some bacteria. Dextran, a storage product of some microbes, is a component of certain products used to increase the volume of blood or deliver iron to the blood of iron-deficient patients.

Chitin and agar are other important polysaccharides in microbiology. Chitin, a polymer of the glucose derivative *N*-acetylglucosamine, is in the cell walls of fungi and is a major component in the exoskeletons of crustaceans (such as shrimp and crabs) and insects. Agar, a polymer of galactose, is found in the cell walls of certain types of algae; it is extensively used as a gelling agent in media used to grow microorganisms in the laboratory.

Table 2.6 summarizes the significance of common carbohydrates.

Lipids

Lipids are a diverse group of non-polar, hydrophobic molecules. Unlike other macromolecules, not all lipids are composed of similar subunits. Their single common feature is that they are insoluble or only slightly soluble in water.

Fatty Acids

Fatty acids are linear carbon skeletons with a carboxyl group (—COOH) at one end. Hundreds of different types exist, and they can be categorized based on the length of the carbon chain and the presence or absence of double bonds between carbon atoms (**figure 2.17**). Saturated fatty acids have no double bonds between carbon atoms; the term "saturated" indicates that they have the maximum number of hydrogen atoms. Unsaturated fatty acids contain one or more double bonds between carbon atoms.

Most naturally occurring unsaturated fatty acids are *cis*, meaning the hydrogen atoms attached to the double-bonded carbon atoms are on the same side of the bond. *Trans* fatty acids have hydrogen atoms on opposite sides of the double bond.

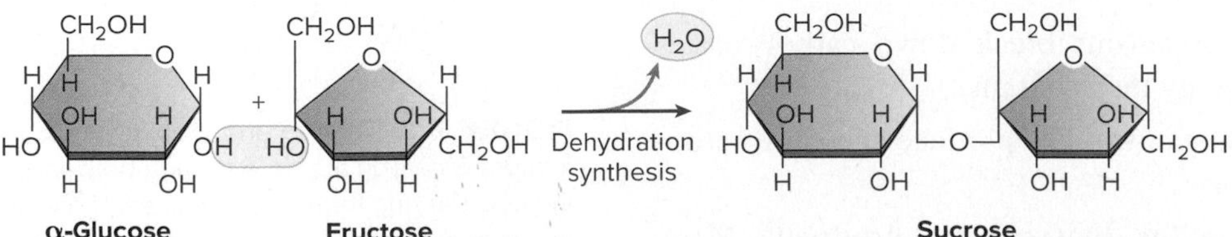

FIGURE 2.15 Formation of a Disaccharide Water is removed as glucose and fructose are joined to create a sucrose molecule.

What type of reaction would reverse the step shown in this diagram?

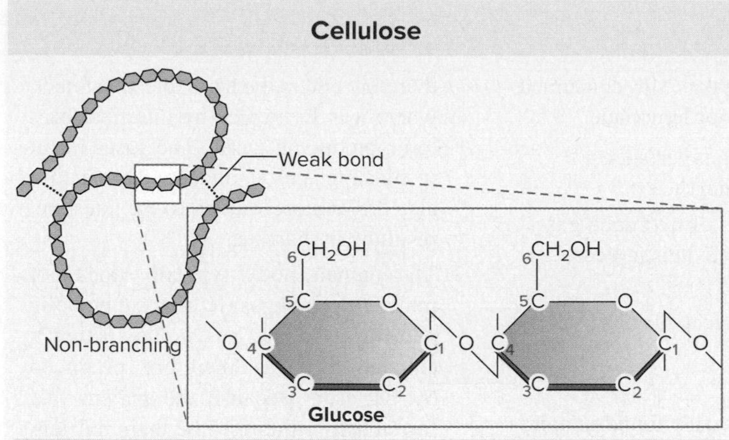

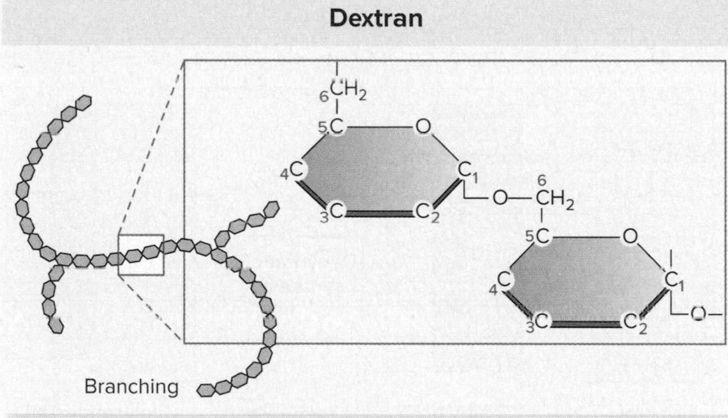

FIGURE 2.16 Three Important Polysaccharides The molecules shown have the same subunit (glucose), yet they are each distinct because of differences in the linkages that join the subunits and the degree of branching. Weak bonding forces are also involved.

? Considering that these three polysaccharides are all made from glucose subunits, how do they differ?

Cis *Trans*

Recent studies of the human microbiome highlight the importance of **short-chain fatty acids (SCFAs)** to human health. SCFAs contain five or fewer carbon atoms and are

TABLE 2.6 Common Monosaccharides, Disaccharides, and Polysaccharides

Name	Characteristics
Monosaccharides (5-carbon)	
Deoxyribose	Component of DNA
Ribose	Component of RNA
(6-carbon)	
Fructose	Fruit sugar; component of sucrose (table sugar)
Glucose	Common subunit of disaccharides and polysaccharides
Galactose	Component of lactose (milk sugar)
Mannose	Found in characteristic patterns on the surface of some microbes
Disaccharides	
Lactose	Glucose + galactose. Milk sugar
Maltose	Glucose + glucose. Breakdown product of starch
Sucrose	Glucose + fructose. Table sugar from sugar cane and beets
Polysaccharides	
Agar	Polymer of galactose. Gelling agent in bacteriological media; extracted from the cell walls of some algae
Cellulose	Polymer of glucose; relatively few organisms can break the type of linkage that joins the glucose subunits. Major structural polysaccharide in plant cell walls
Chitin	Polymer of *N*-acetyl-glucosamine. Major component in fungal cell walls and exoskeleton of insects and crustaceans
Dextran	Polymer of glucose. Storage product in some bacterial cells
Glycogen	Polymer of glucose. Major storage polysaccharide in animal and bacterial cells
Starch	Polymer of glucose. Major storage product in plants

Saturated fatty acid (palmitic acid)

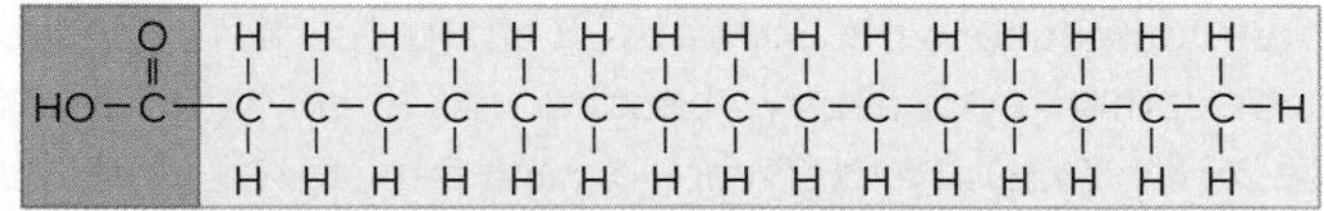

Unsaturated fatty acid (oleic acid)

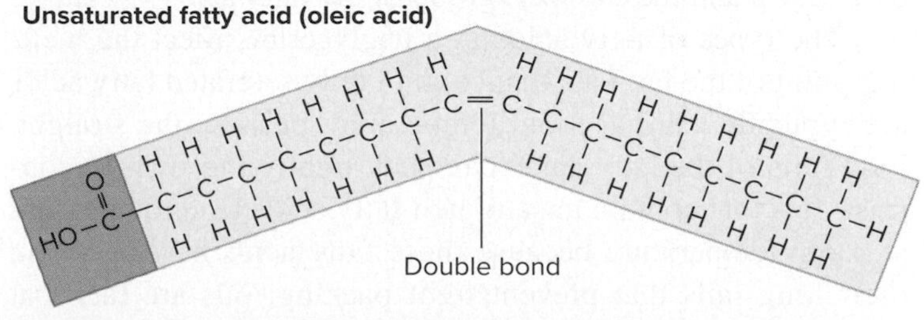

FIGURE 2.17 Fatty Acids Most fatty acids contain an even number of carbon atoms (commonly 16 or 18) and may be saturated or unsaturated.

? What distinguishes an unsaturated fatty acid from a saturated fatty acid?

FOCUS ON A CASE 2.1

The patient was a 15-year-old girl who had recently started dating. Most of the time she and her boyfriend went to a nearby diner, where they ate hamburgers and drank milkshakes. Once home, she routinely started feeling bloated and crampy, had to pass gas, and later suffered from diarrhea. She was embarrassed, and worried that she might be allergic to her new boyfriend. Her mother was afraid the girl had a nervous condition.

After taking a complete history, the doctor ordered a hydrogen breath test for the teen. The patient watched her diet for a few days, fasted the night before the test, and was given 25 grams (g) of lactose in water when arriving at the clinic. Several times during the next few hours, technicians measured the amount of hydrogen gas in the air she exhaled into a plastic bag. The teen entertained herself by reading magazines between tests, and as time passed, she began to feel bloated. When the doctor called the next day, she assured the mother that her daughter did not have a nervous condition. Instead, she suspected lactose intolerance and suggested that the girl replace the milkshakes she consumed on her dates with water or lemonade.

1. Why did the patient experience bloating, gas, and diarrhea?
2. How did high levels of hydrogen gas in the patient's exhaled air indicate lactose intolerance?
3. Was the patient allergic to the milkshakes?

Discussion

1. Lactose is a disaccharide found in dairy products such as milk, ice cream, and cheese. It is broken down into its constituent monosaccharides—glucose and galactose—by the enzyme lactase, which is produced in the small intestine. The monosaccharides are then absorbed into the bloodstream before reaching the large intestine. Babies, who derive much of their nutrition from milk, are generally born with the ability to produce plenty of lactase. By adulthood, however, many individuals produce much less of the enzyme and are unable to break down the amount of lactose in a small glass of milk. The disaccharide therefore enters the large intestine intact, where it is fermented by intestinal bacteria, producing gases. The gases result in bloating and flatulence. The lactose also draws more water into the intestine, resulting in diarrhea.
2. The human body typically does not make hydrogen gas (H_2) in its metabolic pathways, so the gas in the patient's exhaled air was therefore produced by bacteria. Because the patient had fasted before the test, the main nutrient source for the bacteria was the lactose ingested during the test. If the patient had had sufficient levels of lactase, her body would have broken the lactose into monosaccharides, and these would have been absorbed rather than passing through to her intestinal bacteria.
3. No, she was not suffering from an allergy because the immune system was not involved in her response. Her intolerance can be managed by simply avoiding milk products or supplementing her diet with an over-the-counter lactase supplement.

produced by bacteria in the large intestine as they degrade undigested plant material. Butyrate, a 4-carbon SCFA, is crucial for the normal growth and function of cells that line the large intestine (colonocytes); because of that role, butyrate helps prevent intestinal contents from leaking into the bloodstream and may also help prevent colon cancer.

Triglycerides (Fats)

Triglycerides (fats) consist of one molecule of glycerol—a 3-carbon compound with a hydroxyl group (—OH) attached to each carbon—covalently bonded to three fatty acids (**figure 2.18**). To form a triglyceride, each fatty acid is joined to glycerol via dehydration synthesis involving a hydroxyl group of glycerol and the carboxyl group of the fatty acid.

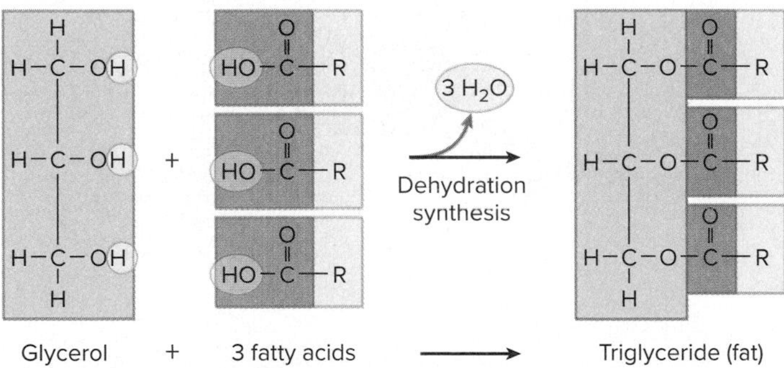

FIGURE 2.18 Formation of a Triglyceride Each fatty acid is joined to glycerol via dehydration synthesis involving a hydroxyl group of glycerol and the carboxyl group of the fatty acid; the R groups in the illustration are carbon-hydrogen chains, such as those shown in figure 2.17.

? **What would happen if triglycerides were placed in water?**

The types of fatty acids in a triglyceride affect the melting point of the fat. Fats that contain only saturated fatty acids are typically solid at room temperature because the straight, long tails of the fatty acids can pack tightly together. In contrast, fats that contain unsaturated fatty acids tend to be liquid at room temperature because these fatty acids have bends in their long tails that prevent tight packing; oils are fats that are liquid at room temperature. Manufacturing processes that hydrogenate oils convert many of the unsaturated fatty acids to saturated ones, thereby making the product solid at room temperature. Unfortunately, the process also sometimes converts *cis* fatty acids to *trans* fatty acids, and ingesting these has been linked to certain health problems. Diets rich in saturated fats and *trans* fats are associated with high blood cholesterol levels that may lead to clogged arteries.

Phospholipids

As the name implies, **phospholipids** contain a phosphate group as part of the structure. They are essential components of cell membranes, such as the cytoplasmic membrane that separates the interior of the cell from the outside environment. That membrane prevents cell contents from leaking out and keeps many molecules from entering cells.

Phospholipids of bacterial and eukaryotic cell membranes are similar to triglycerides, but replacing one of the fatty acids is a phosphate group linked to one of a variety of other polar groups (represented by "R"). The glycerol, phosphate group, and R group are polar, whereas the two fatty acids are non-polar. Thus, the molecule is amphiphilic (meaning "both loving") because it has polar and non-polar portions; the polar portion is referred to as a hydrophilic "head" whereas the non-polar portions are referred to as hydrophobic "tails" (**figure 2.19**). Because of their amphiphilic structure, the phospholipid molecules orient themselves in water to form a bilayer membrane: The hydrophobic tails face inward, toward their counterparts of the opposing layer, whereas the hydrophilic heads face outward, toward the aqueous environment. In other words, the bilayer resembles a sandwich, with the hydrophobic portions of the molecules buried between the hydrophilic portions. Water-soluble substances cannot pass through the hydrophobic region, so cells use transport mechanisms discussed in chapter 3 to move these across the membrane. Archaeal phospholipids have a structure similar to that of bacteria and eukaryotes, but their chemistry is unique.

Steroids

Steroids have a characteristic structure consisting of four connected rings (**figure 2.20**). They are quite different from fats or phospholipids but are classified as lipids because they too are poorly soluble in water. The hormones cortisol, estrogen, and testosterone are steroids. If a hydroxyl group (—OH) is attached to one of the rings, the steroid is a sterol. Cholesterol and ergosterol are sterols that provide rigidity to animal and fungal cell membranes, respectively.

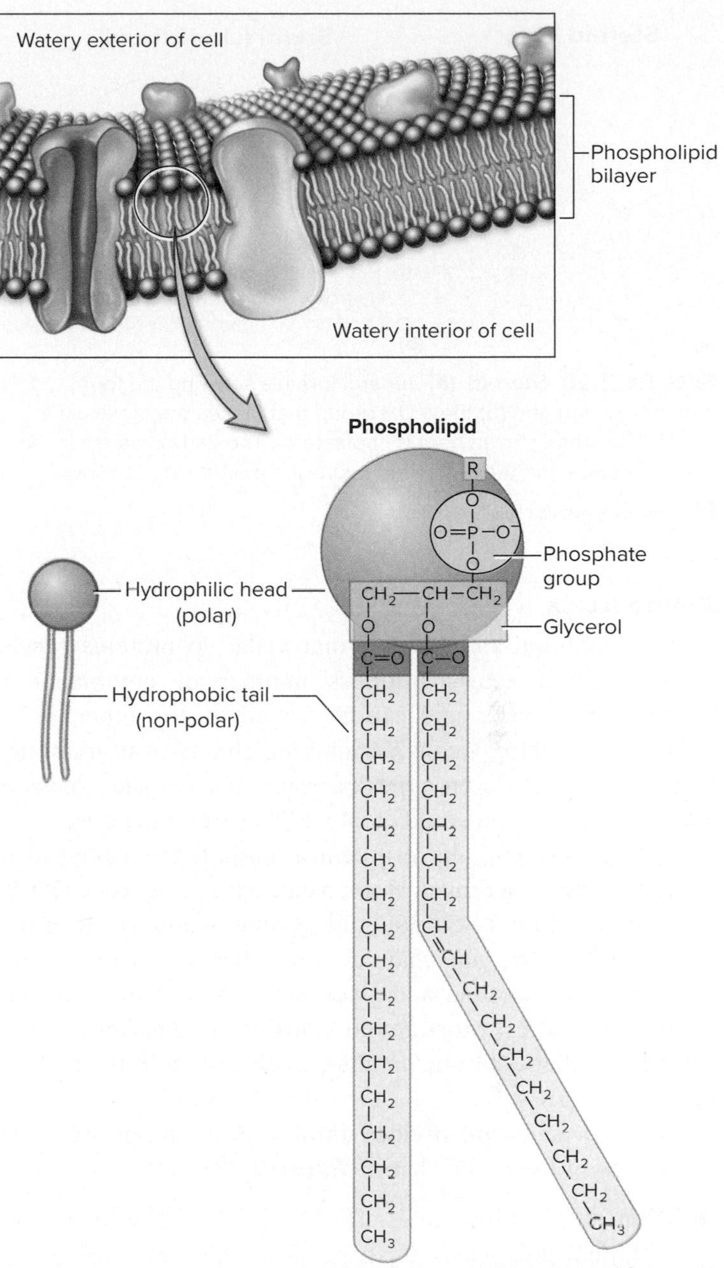

FIGURE 2.19 Phospholipids in Cell Membranes A phospholipid molecule has a "hydrophilic head" (glycerol, phosphate group, and R group) and two "hydrophobic tails" (fatty acids).

? What features of phospholipids make them ideal components of membranes?

Proteins

Even the simplest bacterial cell may contain several thousand different proteins. In spite of this great diversity, all proteins are composed of one or more chains of assorted subunits called **amino acids.** The chains vary in length among different proteins and are folded into complex three-dimensional shapes based on the sequence (order) of amino acids. Proper folding is crucial because even a slight change in shape can interfere with a protein's function, just as a slightly different key can no longer unlock a door. Protein functions include:

- **Enzyme catalysis.** Enzymes that speed up chemical reactions in a cell are typically proteins.
- **Transport.** Transport proteins move small substances, often across a cytoplasmic membrane into or out of cells.
- **Signal reception.** Receptor proteins on the cell surface recognize conditions in the external environment.
- **Regulation.** Some proteins serve as intercellular signals or bind to DNA and regulate gene expression.
- **Motility.** Proteins are essential components of flagella and cilia, structures that move cells.
- **Support.** Proteins make up the cytoskeleton, the structural framework of many cells.

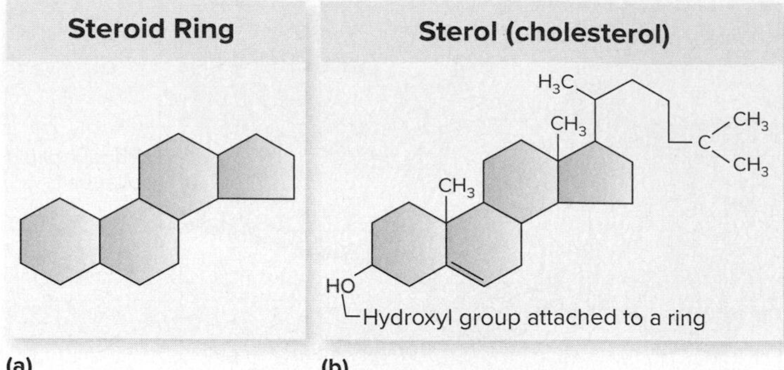

FIGURE 2.20 Steroid (a) General formula showing the four-membered ring and **(b)** the —OH group that makes the molecule a sterol. The sterol shown here is cholesterol. The carbon atoms in the ring structures and their attached hydrogen atoms are not shown.

? **Why are steroids classified as lipids?**

Amino Acids

The 20 common amino acids that make up proteins can be arranged in a nearly limitless number of combinations, because any amino acid can be joined to any other amino acid. As an analogy, consider making chains from unlimited numbers of 20 different types of beads. If you make chains of 100 beads each, you could create 20^{100} different chains!

All amino acids have a central carbon atom bonded to the following: a hydrogen atom; a carboxyl group (—COOH); an amino group (—NH_2); and a side chain or R group (**figure 2.21**). The side chain distinguishes the various amino acids from one another and gives each amino acid its characteristic properties. Most amino acids exist in two forms that are mirror images of one another, as described in Focus Your Perspective 2.1.

Amino acids are divided into several groups based on properties of their side chains (**figure 2.22**).

- **Non-polar amino acids.** The side chains of these have no polar covalent bonds. Cysteine, an amino acid in this group, is particularly important because its side chain has a sulfhydryl group (—SH), so a covalent bond can form between the sulfur atoms in different cysteine subunits; a bond between sulfur atoms is called a disulfide bond (S—S).

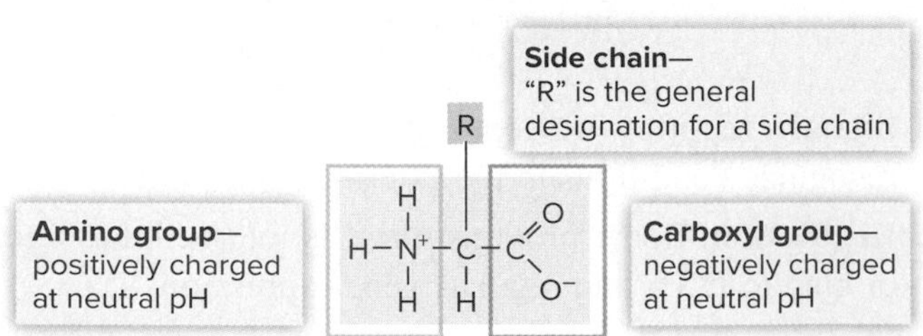

FIGURE 2.21 Generalized Amino Acid The different amino acids are distinguished from one another by their unique side chains (R groups).

? **Which portion of an amino acid is responsible for the unique properties of the molecule?**

- **Polar/hydrophilic (uncharged) amino acids.** The side chains of these contain a polar covalent bond.
- **Polar/hydrophilic (charged) amino acids.** The side chains of these include functional groups that can ionize, resulting in a positive or negative charge.

Amino Acid Chains

A short chain of amino acids is called a peptide and a longer one a **polypeptide,** but the distinction between the two is not precise. A **protein** is composed of one or more polypeptides folded to create a functional molecule. Note that the term *protein* implies a functional entity, but it is often used interchangeably with *polypeptide.*

The amino acids in a chain are joined together by covalent bonds called **peptide bonds.** These bonds form by a dehydration synthesis reaction between the carboxyl group (—COOH) of one amino acid and the amino group (—NH_2) of the neighboring amino acid (**figure 2.23**). The resulting molecule has an amino group at one end (called the N terminal or amino terminal end) and a carboxyl group at the other end (called the C terminal or carboxyl terminal end).

Certain proteins have other molecules covalently bonded to the side chains of some of their amino acids. If sugar molecules are bonded, the protein is a glycoprotein; if lipids are attached, the protein is a lipoprotein. Many of these proteins are found on the surfaces of cells.

Protein Structure

Proteins have up to four levels of structure (**figure 2.24**):

- **Primary structure.** This is the sequence of amino acids in the polypeptide, and it affects all other levels of protein structure. Thus, the primary structure directs the final shape of the protein and, as a result, is responsible for its properties.
- **Secondary structure.** This is a repeated coiling or folding in localized regions of a protein; a spiral or helical structure is an alpha (α)-helix, whereas parallel strands make up a beta (β)-pleated sheet. These characteristic patterns form due to weak hydrogen bonding between the carboxyl and amino groups of amino acids along the polypeptide chain.
- **Tertiary structure.** This is the overall three-dimensional shape of a folded polypeptide, and it results largely from the interactions between amino acid R groups (side chains). Non-polar amino acids have hydrophobic R groups, so water molecules tend push regions containing them to the interior of a protein. On the other hand, polar amino acids have hydrophilic R groups, and are typically located on the outside of a protein. Meanwhile, large R groups occupy more space than small ones, R groups of positively charged amino acids are attracted to those of negatively charged ones, and a disulfide bond can form between sulfur atoms

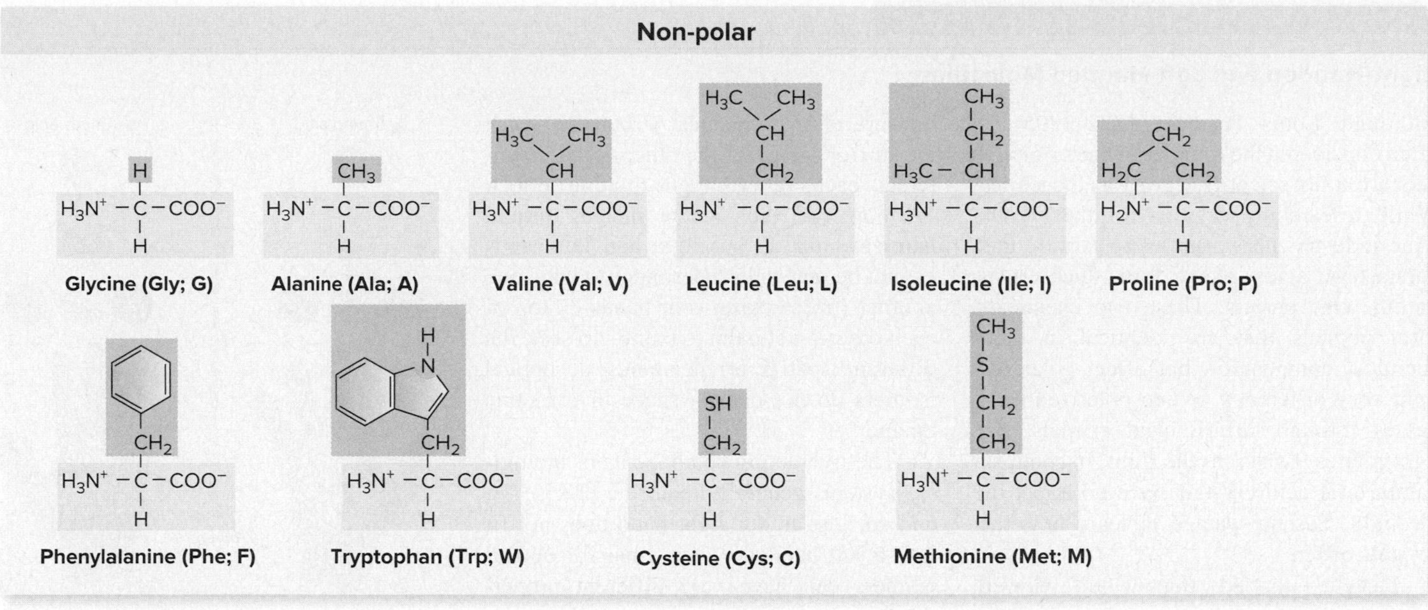

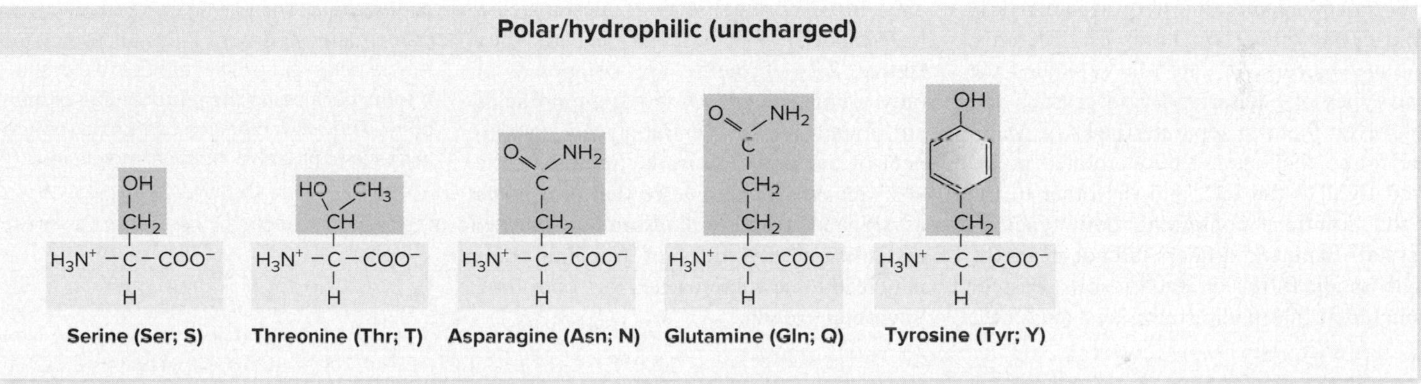

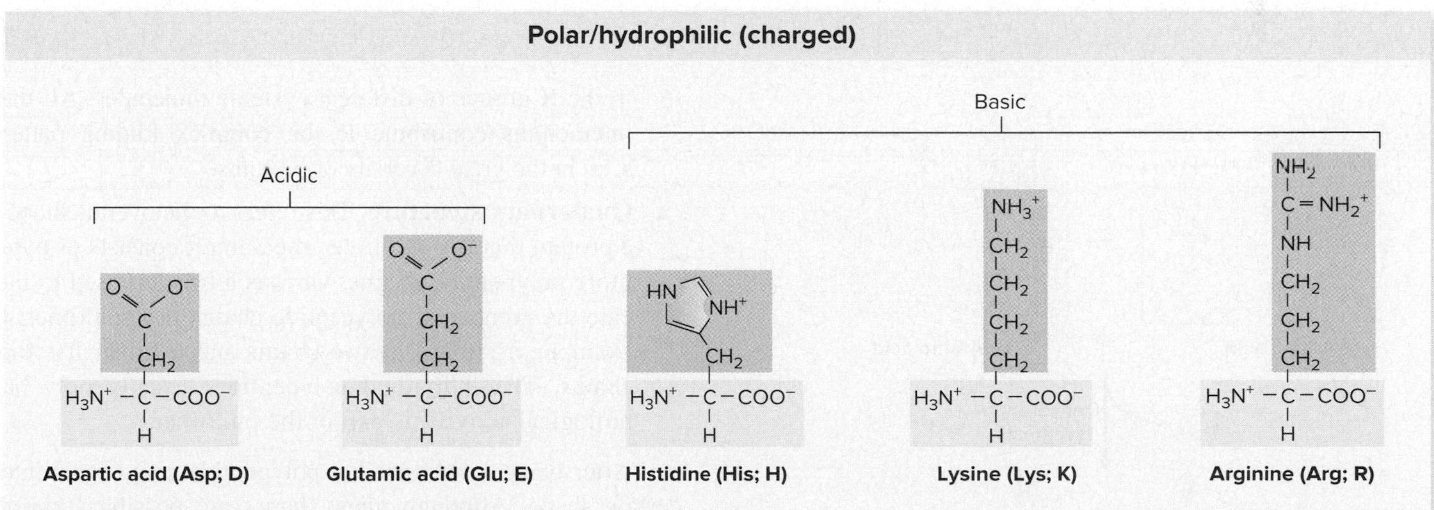

FIGURE 2.22 Common Amino Acids The groupings are based on the polarity and the overall charge of the amino acid. The basic and acidic amino acids have a net positive and a net negative charge, respectively. For simplicity, tyrosine is shown in only one group, but it has both non-polar and polar characteristics. The three-letter and single-letter code names for each amino acid are given.

? **What makes the side chain of cysteine unique?**

FOCUS YOUR PERSPECTIVE 2.1

Right-Handed and Left-Handed Molecules

Although Louis Pasteur (1822–1895) is often considered the founder of bacteriology, he started his scientific career as a chemist. While researching crystals for the French wine industry, he worked with tartaric and paratartaric acids, which form thick crusts within wine barrels. These two chemicals form crystals that are identical in their chemical composition but affect polarized light very differently. When polarized light passes through tartaric acid crystals, the light rotates (twists) to the right. In contrast, paratartaric acid crystals have no effect on the light. Pasteur wanted to learn how the crystals differ.

Pasteur noticed that when viewed using a microscope, all tartaric acid crystals looked identical, whereas paratartaric acid crystals had two kinds of structures. Using tweezers, he carefully separated the two types of paratartaric acid crystals and dissolved them in separate flasks of water. He found that one solution rotated polarized light to the left, and the other to the right; solutions containing both types of crystals in equal numbers did not rotate the light at all. Based on such results, Pasteur concluded that paratartaric acid must be a mixture of two chemical structures—each the mirror image of the other— that counteract each other's light-rotating effect. We now call such mirror images **optical isomers.** Just as our right and left hands cannot be precisely superimposed upon one another (try stacking your hands on top of each other—not palm to palm—to see), the different spatial arrangements of optical isomers do not occupy space in the same manner.

The interacting components of biological systems generally require a precise fit, and so, just as our right hand does not fit into a left-handed glove, a pair of optical isomers can have very different properties. Amino acids (except glycine) can exist in two optical isomers, designated L (left-handed) and D (right-handed) (box **figure 2.1**). Proteins are composed of only L-amino acids. A few D-amino acids are found in the peptidoglycan component of bacterial cell walls. Some bacteria also release D-amino acids that may serve as signals in cell wall assembly. Further insights into the roles of D-amino acids may lead to new approaches for controlling microbial growth.

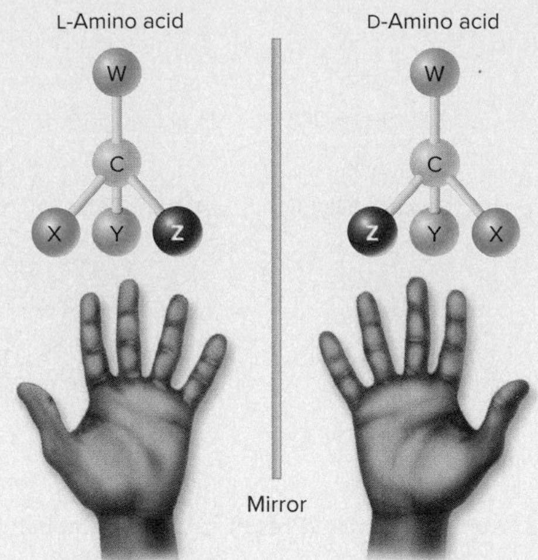

BOX FIGURE 2.1 Mirror Images of an Amino Acid The joining of a carbon atom to four different groups leads to asymmetry. Amino acids can exist in either the L or the D form, each being the mirror image of the other. The two molecules cannot be rotated in space to give two identical molecules.

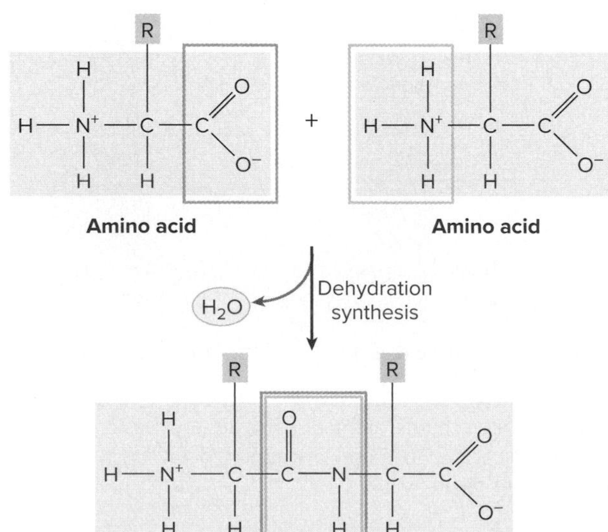

FIGURE 2.23 Peptide Bond Formation Dehydration synthesis forms the peptide bond, shown in red.

? **What two chemical groups are involved in the formation of a peptide bond?**

in the R groups of different cysteine molecules. All these interactions contribute to the complex folding patterns seen in the great diversity of proteins.

- **Quaternary structure.** This refers to the overall shape of a protein that is a multimer, meaning it consists of two or more polypeptide chains. Various terms are used to indicate the number of polypeptide chains in a multimer; for example, a dimer has two chains and a trimer has three chains. The individual polypeptides usually only have biological activity as part of the multimer.

After being synthesized, a polypeptide chain folds into a certain shape. Although many shapes are possible, typically only one functions properly. Most proteins fold spontaneously into their correct state, but special proteins called **molecular chaperones** are sometimes needed to help with the process.

MicroByte

Artificial intelligence programs have revolutionized scientists' ability to predict the three-dimensional shape of a given protein based on its DNA sequence (which determines the amino acid sequence).

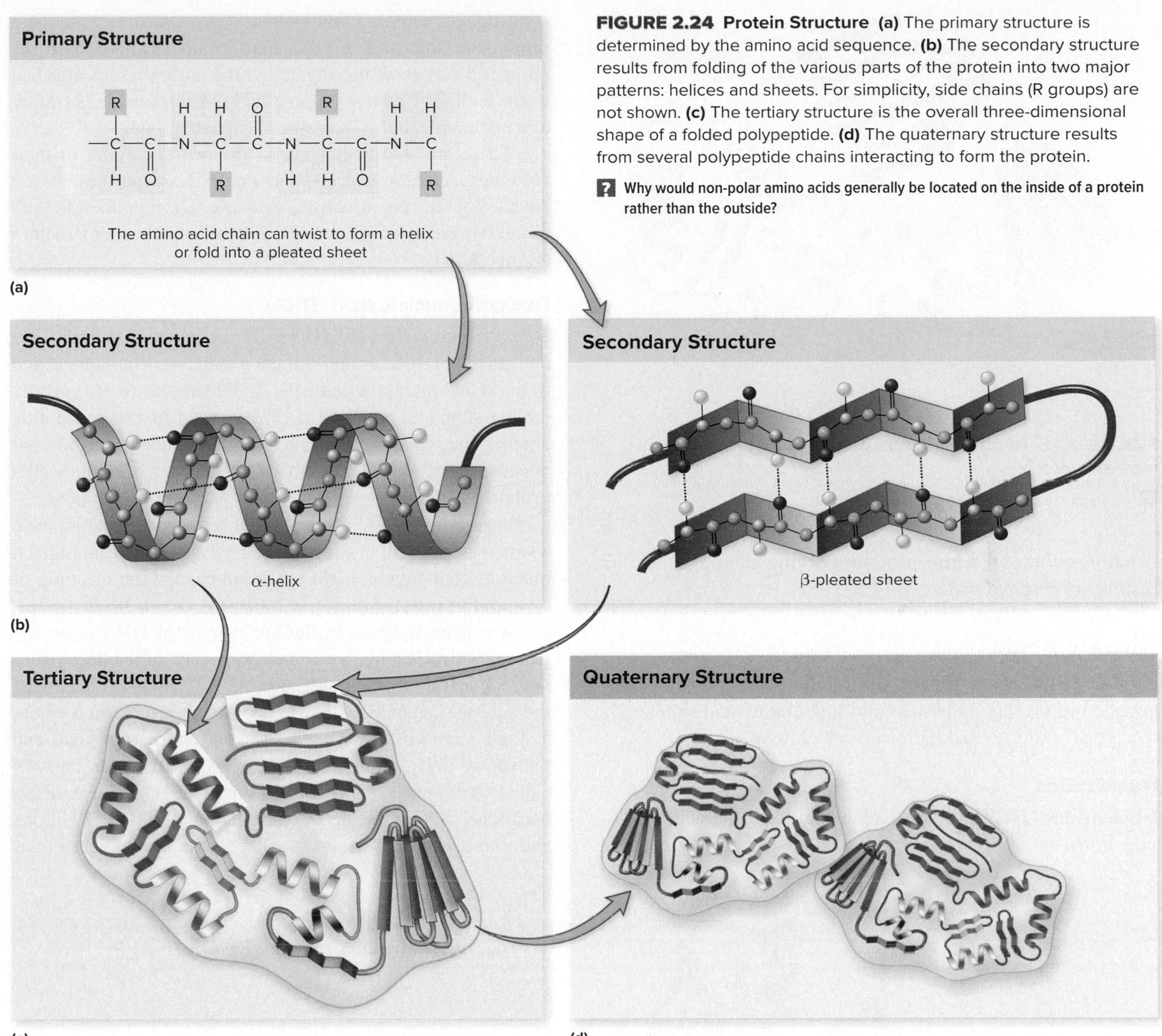

FIGURE 2.24 Protein Structure (a) The primary structure is determined by the amino acid sequence. **(b)** The secondary structure results from folding of the various parts of the protein into two major patterns: helices and sheets. For simplicity, side chains (R groups) are not shown. **(c)** The tertiary structure is the overall three-dimensional shape of a folded polypeptide. **(d)** The quaternary structure results from several polypeptide chains interacting to form the protein.

? Why would non-polar amino acids generally be located on the inside of a protein rather than the outside?

Protein Domains

As new technologies made it possible to study and compare the structures of a wide variety of different proteins, scientists noticed that proteins with similar functions often have one or more substructures in common. These substructures consist of sheets and helices that fold into a stable structure independently of other parts of the molecule. A substructure associated with a particular function is referred to as a **protein domain** (**figure 2.25**). For example, a certain domain may bind DNA; another may act as a catalyst. Once a known function can be attributed to a specific domain, the role of an unknown protein with that domain can be inferred. Large proteins sometimes have many domains and several different functions, whereas a small protein may have only one.

Protein Denaturation

High temperature, extreme pH, and certain solvents can affect the interactions between amino acids within a protein, causing it to **denature** or lose its characteristic shape (**figure 2.26**). As a result, it may become nonfunctional. This is why most organisms cannot grow in extreme environments—their enzymes denature and no longer catalyze chemical reactions. Denaturation may sometimes be reversible but not always. Boiling an egg, for example,

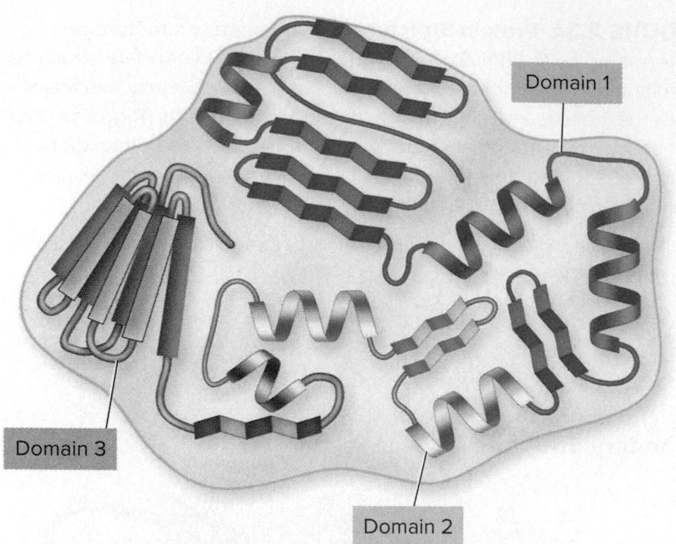

FIGURE 2.25 Domain Structure of a Protein Each domain has a different function.

? **Distinguish between a polypeptide and a domain.**

denatures the egg white protein; cooling the egg does not restore its original state.

Nucleic Acids

Nucleic acids carry genetic information. Cells decipher the encoded information and use it to link amino acid subunits in the proper order to make the various proteins.

Nucleotides

Nucleotides are the subunits of nucleic acids. Each nucleotide consists of a pentose sugar, a phosphate group, and a nitrogen-containing **nucleobase** (also called a base or a nitrogenous base). The phosphate group is attached to the number 5 carbon of the sugar, and the nucleobase is attached to the number 1 carbon (**figure 2.27**). A nucleoside is similar to a nucleotide, but it lacks the phosphate group.

There are five nucleobases, grouped according to their structure. Adenine (A) and guanine (G) are purines, which have a double-ring structure; cytosine (C), thymine (T), and uracil (U) are pyrimidines, which have a single-ring structure (**figure 2.28**).

Deoxyribonucleic Acid (DNA)

Deoxyribonucleic acid (DNA) is the genetic material of a cell; a cell's DNA contains all the genetic information needed to build and maintain that cell. A DNA molecule is typically double-stranded, meaning it is composed of two nucleotide chains. The sequence (order) of the nucleotides is significant because it encodes the amino acid sequences of all the cell's proteins, including enzymes and structural components; a change in the DNA sequence that results in changing even a single amino acid in a protein can have a significant impact, much as changing a single letter can change the meaning of a word. For lunch, would you rather eat rice or lice?

The pentose sugar in the nucleotides of DNA is deoxyribose (which is why these nucleotides are called deoxyribonucleotides); the nucleobases are A, T, C, and G. Because of the way the nucleotides are joined together to form a chain, a single strand of DNA has a series of alternating sugar and phosphate units, and these make up a structural framework called the sugar-phosphate backbone (**figure 2.29**). A linear backbone will always have a phosphate group (—PO_4) at one end and a hydroxyl group (—OH) at the other. These ends

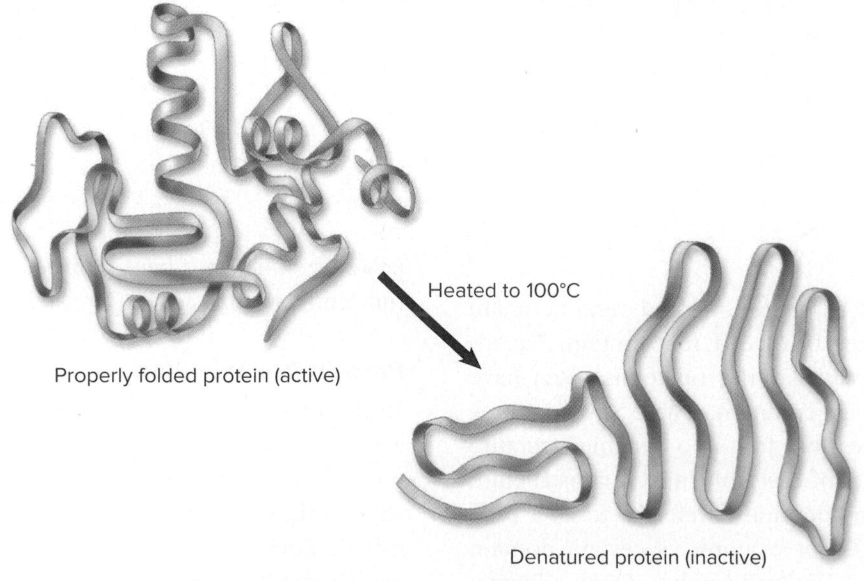

FIGURE 2.26 Denaturation of a Protein The denatured protein loses its function.

? **Describe two environmental conditions that can denature a protein.**

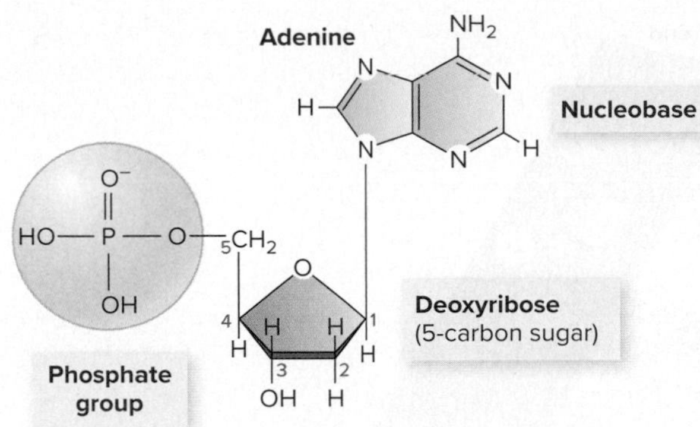

FIGURE 2.27 A Nucleotide A nucleotide consists of a pentose sugar, a phosphate group, and a nucleobase. In this illustration, the nucleobase is adenine, so the nucleotide is called deoxyadenosine-5′-phosphate.

? **What are the three components of a nucleotide?**

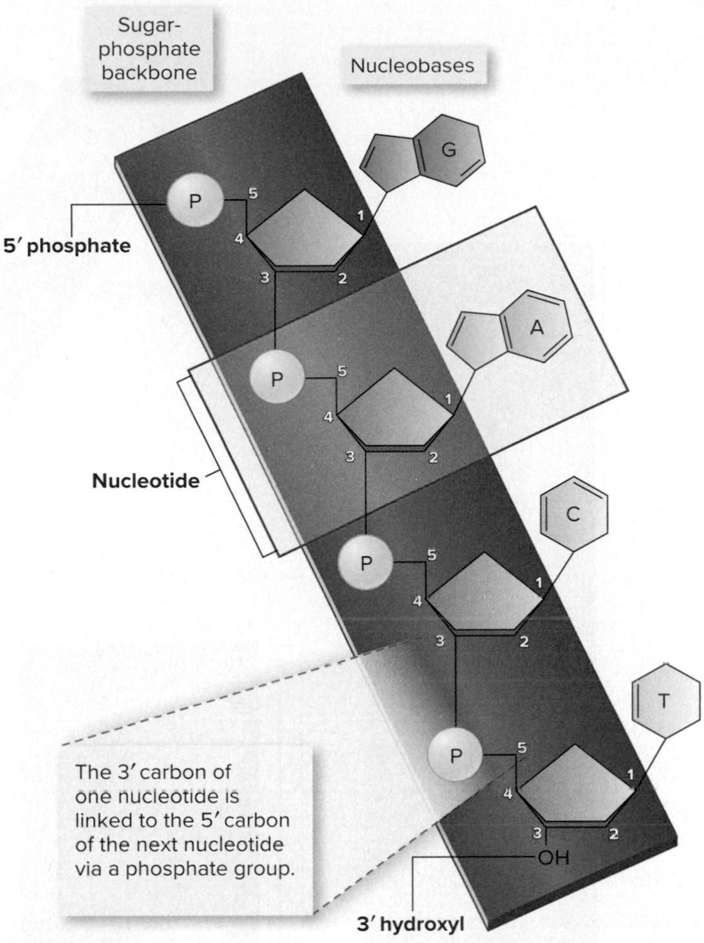

FIGURE 2.29 Single Strand of DNA Single chain of nucleotides in DNA showing the differences between the 5′ end and the 3′ end.

? **What functional group is at the 5' end of a DNA strand? What group is at the 3' end?**

are referred to as the **5′ end** (pronounced "5 prime end") and the **3′ end** ("3 prime end"), reflecting the positions at which the functional groups are attached to the carbon atoms of deoxyribose.

Double-stranded DNA is arranged somewhat like a spiral staircase with two railings, and is described as a double helix (**figure 2.30**). The "railings" represent the sugar-phosphate backbones of the two strands; the "stairs" are nucleobases attached to the railings and held together in pairs by hydrogen bonds. The two strands are **antiparallel,** meaning they are oriented in opposite directions; one strand is oriented in the 3′ to 5′ direction, whereas the opposite strand is in the 5′ to 3′ direction. The two strands are described as **complementary,** because wherever an A is in one strand, a T is in the other, and wherever a G is in one strand, a C is in the other. The opposing A-T bases are held together by two hydrogen bonds, whereas the G-C bases are held together by three hydrogen bonds. Although hydrogen bonds are quite weak, the double-stranded molecule is generally quite stable because of the many hydrogen bonds along its length. However, the two strands will quickly come apart if exposed to temperatures approaching 100°C. If the DNA pieces are short, the strands separate at lower temperatures.

The characteristic bonding of A to T and G to C is called **base-pairing** and is a fundamental characteristic of DNA. Because of the rules of base-pairing, one DNA strand can be used as a template to make the complementary strand.

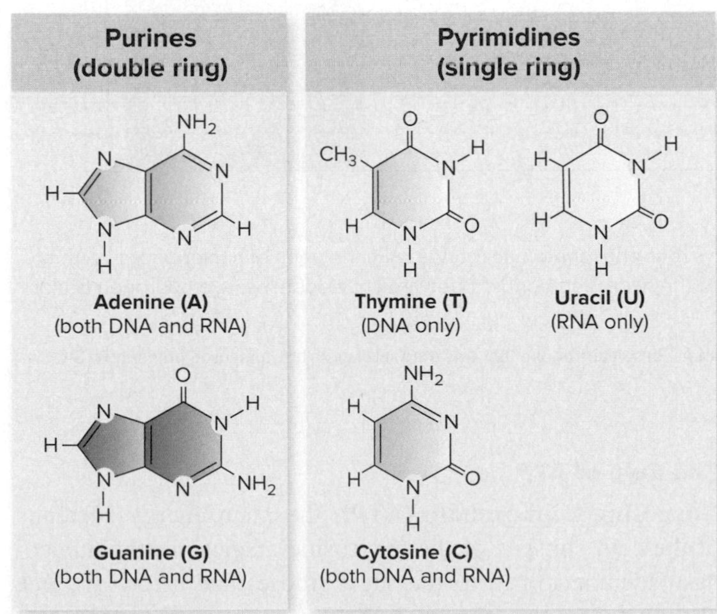

FIGURE 2.28 Purines and Pyrimidines Purines have a double-ring structure; pyrimidines have a single-ring structure.

? **Which of the nucleobases are found in DNA? In RNA?**

Ribonucleic Acid (RNA)

Several forms of **ribonucleic acid (RNA)** are involved in decoding the information in DNA to assemble a sequence of amino acids to build proteins. The basic structure of RNA is similar to that of DNA, but RNA contains the sugar ribose in place of deoxyribose. RNA also contains the nucleobase

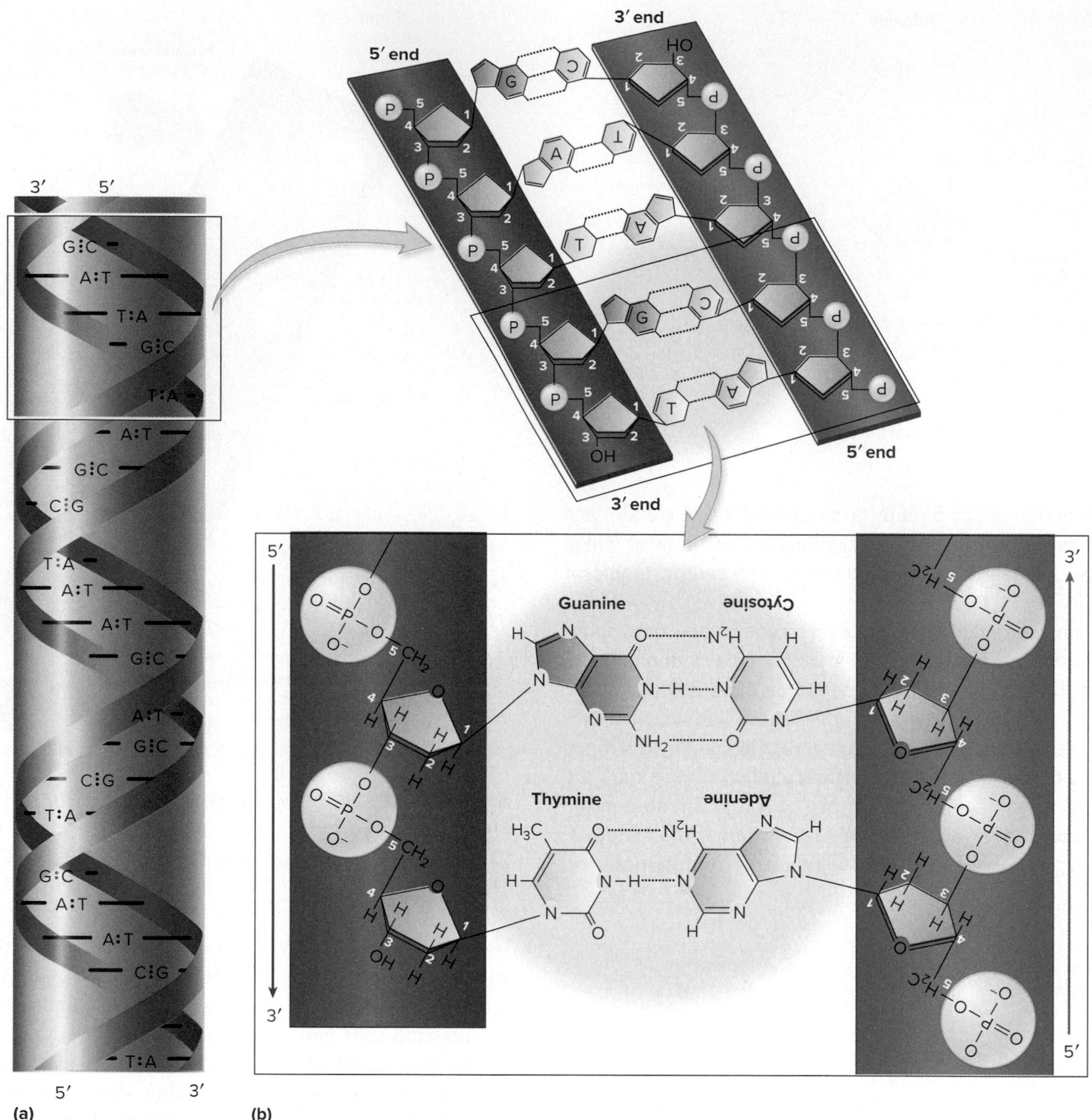

FIGURE 2.30 DNA Double-Stranded Helix **(a)** The "spiral staircase" of the sugar-phosphate backbones, with the pairs of nucleobases on the inside; the two strands are antiparallel. **(b)** The sugar-phosphate backbones and the hydrogen bonding between nucleobases; two hydrogen bonds form between adenine and thymine and three between guanine and cytosine.

? Which would require a higher temperature to denature—a DNA strand composed primarily of A-T base pairs or one that was the same length but composed primarily of G-C base pairs? Explain.

uracil (U) in place of thymine (T). Further, whereas DNA is a long, double-stranded helix, RNA is considerably shorter and generally exists as a single chain of nucleotides. Although RNA is typically single-stranded, some forms contain short double-stranded regions due to hydrogen bonding between complementary nucleobases in the single strand. The types of RNA important in protein synthesis are introduced in chapter 7.

The Role of ATP

Adenosine triphosphate (ATP), the main energy currency of the cell, consists of the nucleoside adenosine (the nucleobase adenine joined to the sugar ribose) and a row of three phosphate groups (**figure 2.31**). To understand the concept of "energy currency," consider the currency of the United States—the dollar. You can sell a valuable item for cash, thereby converting the value of the item into dollars. You can

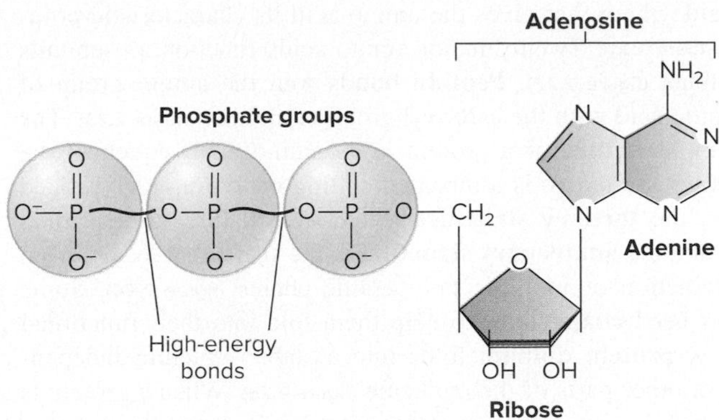

FIGURE 2.31 ATP Adenosine is a nucleoside composed of adenine and ribose. Adenosine triphosphate (ATP) serves as the energy currency of a cell. When the terminal phosphate bonds break, the energy released can be used to drive cellular reactions.

Why are the bonds between the phosphate groups of ATP "high energy"?

then spend the cash, converting the dollars into another item of value. Cells do something similar. They use energy available in sunlight or in molecules like sugars to make ATP, and then they "spend" the energy in ATP to drive chemical reactions in the cell. The energy in ATP is easily released because the three negatively charged phosphate groups repel one another, so the bonds joining them are inherently unstable. When these bonds break, the energy released can drive cellular processes. Because of the relatively high amount of energy released, the bonds are commonly referred to as **high-energy phosphate bonds,** indicated by the symbol ~. When the terminal phosphate bond of ATP breaks, inorganic phosphate and **adenosine diphosphate (ADP)** are formed. The processes cells use to capture the energy needed to make ATP are covered in chapter 6.

MicroAssessment 2.4

Carbohydrates serve as energy-storage molecules and structural material. Lipids are poorly soluble in water; phospholipids are important structural components of cell membranes. Proteins have complex structures based on the identity and sequence of amino acids. Many enzymes that catalyze chemical reactions are proteins. Nucleic acids, including DNA and RNA, are important in storing genetic information and synthesizing proteins. ATP is the main energy currency of a cell.

10. Name the subunits used to build a carbohydrate, a triglyceride, a protein, and a nucleic acid.

11. What level of protein structure is described as an α-helix?

12. If the DNA molecule were placed in boiling water, how would the molecule change?

Summary

2.1 ■ Elements and Atoms

Atomic Structure

Atoms are composed of **electrons, protons,** and **neutrons** (figures 2.1, 2.2). An **element** consists of a single type of atom.

The Role of Electrons

The reactivity of an atom is largely determined by the number of **valence electrons** in its outer shell.

Isotopes

Isotopes are atoms of a given element that have a different number of neutrons. Isotopes may emit energy that can be detected and tracked (figure 2.3).

2.2 ■ Chemical Bonds and Reactions

Ions and Ionic Bonds

Atoms that gain an electron become negatively charged **ions** called **anions;** those that lose an electron become positively charged and are called **cations. Ionic bonds** form between cations and anions because of the attraction between positive and negative charges (figure 2.4).

Covalent Bonds

Covalent bonds are formed by atoms sharing electrons (figure 2.5). Molecules that contain at least carbon and hydrogen are **organic compounds;** those that do not are inorganic compounds. When atoms have an equal attraction for electrons, a **non-polar covalent bond** is formed; when one atom is more electronegative, a **polar covalent bond** is formed (table 2.3; figure 2.6).

Hydrogen Bonds

Hydrogen bonds are weak bonds that result from the attraction of a hydrogen atom in a polar molecule to an electronegative atom in the same or another polar molecule (figure 2.7).

Molarity

A **mole** of one substance has the same number of particles as a mole of another substance.

Chemical Reactions

Chemical reactions convert one or more substances (reactants) to others (products). Chemical reactions in which electrons are transferred from one reactant to the other are called **oxidation-reduction reactions** or **redox reactions** (figure 2.8). **Enzymes** speed up chemical reactions.

2.3 ■ Water, pH, and Buffers

Water

Hydrogen bonding plays a very important role in the properties of water that make it essential for life (figures 2.9, 2.10). Water is an effective **solvent. Hydrophilic** substances are attracted to water; **hydrophobic** substances are repelled by water.

pH of Aqueous Solutions

pH is a measure of the acidity of a solution. A solution of pH 7 is neutral, whereas a solution with a pH less than 7 is acidic, and one with a pH higher than 7 is basic (figure 2.11).

Buffers

Buffers prevent a dramatic rise or fall of pH.

2.4 ■ Organic Molecules

Macromolecules are usually polymers of subunits. Different classes of macromolecules have different subunits (table 2.4). Macromolecules are made through **dehydration synthesis** and degraded by **hydrolysis** (figure 2.12). Organic molecules often contain **functional groups** that contribute to the molecule's properties (table 2.5).

Carbohydrates

Carbohydrates are a diverse group of organic compounds composed of one or more simple sugars. **Monosaccharides** include the 5-carbon sugars ribose and deoxyribose (figure 2.13) and the 6-carbon sugars glucose, galactose, fructose, and mannose (figure 2.14, table 2.6). **Disaccharides** include lactose, sucrose, and maltose (figure 2.15). **Polysaccharides** include cellulose, starch, glycogen, dextran, chitin, and agar (figure 2.16).

Lipids

Lipids are a diverse group of hydrophobic, non-polar molecules. Saturated fatty acids have no double bonds, whereas unsaturated fatty acids have one or more double bonds (figure 2.17). **Triglycerides** (fats) consist of fatty acids linked to glycerol and may be liquid or solid at room temperature (figure 2.18). **Phospholipids** are essential components of cytoplasmic membranes (figure 2.19). **Steroids** such as cholesterol have a characteristic structure consisting of four connected rings (figure 2.20).

Proteins

Proteins are made of amino acid subunits. **Amino acids** have a central carbon atom bonded to a carboxyl group, an amino group, and a side chain that gives the amino acid its characteristic properties (figure 2.21). Twenty major amino acids function as subunits of proteins (figure 2.22). **Peptide bonds** join the amino group of one amino acid with the carboxyl group of another (figure 2.23). The **primary structure** of a protein is its amino acid sequence; the **secondary structure** is a repeated coiling or folding of localized regions; the **tertiary structure** is the overall three-dimensional shape; and the **quaternary structure** is the shape that results from the interaction of multiple polypeptide chains (figure 2.24). Some proteins need **chaperones** to help them fold into their functional shape. A **protein domain** folds into a stable structure independently of other parts of the molecule (figure 2.25). When a protein is **denatured,** it may become non-functional (figure 2.26).

Nucleic Acids

Nucleic acids carry genetic information, which is decoded to produce proteins. **Nucleotides** are the subunits of nucleic acids and consist of a **nucleobase,** a pentose sugar, and a phosphate group (figures 2.27, 2.28, 2.29). The nucleotide sequence of **DNA (deoxyribonucleic acid)** codes for all of the proteins that a cell produces. DNA is a double-stranded helical molecule with a sugar-phosphate backbone (figure 2.30). The two strands of DNA are **antiparallel** and **complementary. RNA (ribonucleic acid)** is involved in the process that decodes the information contained in DNA. RNA is a single-stranded molecule, and its nucleotides contain uracil in place of thymine and ribose in place of deoxyribose (figure 2.28). **ATP (adenosine triphosphate)** carries energy in **high-energy phosphate bonds,** which, when broken, release the energy for use in the cell (figure 2.31).

Review Questions

Short Answer

1. Differentiate between an atom, a molecule, and a compound.
2. Why is water a good solvent?
3. Which solution is more acidic, one with a pH of 4 or one with a pH of 5? What is the concentration of H^+ ions in each?
4. Show how dehydration synthesis and hydrolysis reactions are related, using an example of each.
5. What is a structural isomer?
6. What is the significance of cellulose?
7. Name the major groups of lipids and give an example of each. What feature is common to all lipids?
8. List six functions of proteins.
9. What are the four levels of protein structure, and what is the distinguishing feature of each?
10. How do DNA and RNA differ from one another in composition and function?

Multiple Choice

1. Choose the list that goes from the lightest to the heaviest:
   a) proton, atom, molecule, electron.
   b) atom, proton, molecule, electron.
   c) electron, proton, atom, molecule.
   d) atom, electron, proton, molecule.
   e) proton, atom, electron, molecule.
2. An oxygen atom has an atomic number of 8. It typically forms
   a) no covalent bonds because it contains eight electrons.
   b) two covalent bonds because it contains six electrons in its valence shell.
   c) weak hydrogen bonds with two hydrogen atoms to form a water molecule.
   d) one covalent bond with a carbon atom to form carbon dioxide.
   e) ionic bonds with another oxygen atom.
3. Pure water has all of the following properties *except*
   a) polarity.
   b) ability to dissolve lipids.
   c) pH of 7.
   d) covalent joining of its atoms.
   e) ability to form hydrogen bonds.
4. When the pH of a solution changes from 3 to 8, the H^+ concentration
   a) decreases as the solution becomes more basic.
   b) decreases as the solution becomes more acidic.
   c) increases as the solution becomes more acidic.
   d) increases as the solution becomes more basic.
   e) does not change as the solution becomes more basic.
5. The function of a buffer is to
   a) bring the pH of a solution to neutral.
   b) speed up chemical reactions within a cell.
   c) interact with phospholipid molecules to form a cell membrane.

d) stabilize the pH of a solution.
e) provide energy for synthesis of ATP.

6. Dehydration synthesis is involved in the synthesis of all of the following *except*
   a) DNA.
   b) proteins.
   c) polysaccharides.
   d) lipids.
   e) monosaccharides.
7. The primary structure of a protein relates to its
   a) sequence of amino acids.
   b) length.
   c) shape.
   d) solubility.
   e) bonds between amino acids.
8. A shortage of nitrogen (N) would make it most difficult to construct a molecule of
   a) cellulose.
   b) cholesterol.
   c) an enzyme.
   d) a triglyceride.
   e) glucose.
9. Complementarity plays a major role in the structure of
   a) proteins.
   b) lipids.
   c) polysaccharides.
   d) DNA.
   e) RNA.
10. A bilayer is associated with
    a) proteins.
    b) DNA.
    c) RNA.
    d) polysaccharides.
    e) phospholipids.

Applications

1. A group of prokaryotes known as thermophiles thrive at high temperatures that would normally destroy other organisms, yet they cannot survive well at the lower temperatures normally found on Earth. Propose an explanation for this observation.
2. Microorganisms use hydrogen bonds to attach to surfaces, but many of the cells lose hold of the surface because of the weak nature of these bonds. Contrast the benefits and disadvantages of using covalent bonds as a means of attaching to surfaces.

Critical Thinking

1. What properties of the carbon atom make it ideal as the key atom for so many molecules in organisms?
2. A biologist determined the amounts of several amino acids in two separate samples of pure protein. The data are shown here:

Amino Acid	Leucine	Alanine	Histidine	Cysteine	Glycine
Protein A	7%	12%	4%	2%	5%
Protein B	7%	12%	4%	2%	5%

A student concluded that protein A and protein B were the same protein. Do you agree with this conclusion? Justify your answer.

3 Cells and Methods to Observe Them

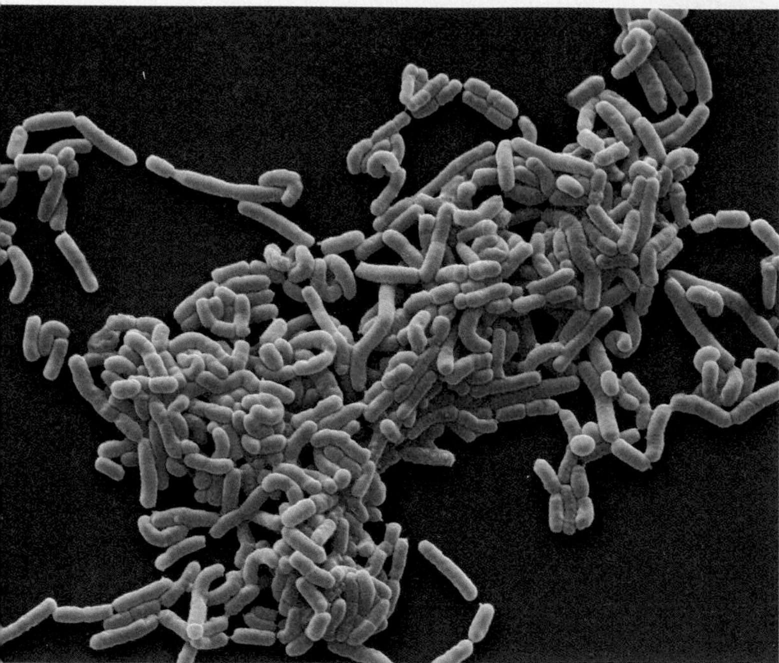
Bacterial cells (color-enhanced scanning electron micrograph). *(STEVE GSCHMEISSNER/Getty Images)*

KEY TERMS

Capsule A distinct, thick gelatinous material that surrounds some microorganisms.

Chemotaxis The movement of a cell toward or away from a certain chemical in the environment.

Chromosome A structure that carries an organism's hereditary information.

Cytoplasmic Membrane A phospholipid bilayer embedded with proteins that surrounds the cytoplasm and defines the boundary of the cell.

Endospore An extraordinarily resistant dormant cell produced by some types of bacteria.

Flagellum A type of structure used for cell movement.

Gram-Negative Bacteria Bacteria that have a cell wall characterized by a thin layer of peptidoglycan surrounded by an outer membrane; when Gram stained, these cells are pink.

Gram-Positive Bacteria Bacteria that have a cell wall characterized by a thick layer of peptidoglycan; when Gram stained, these cells are purple.

Lipopolysaccharide (LPS) The molecule that makes up the outer layer of the outer membrane of Gram-negative bacteria.

Peptidoglycan A macromolecule that provides strength to the cell wall; it is found only in bacteria.

Periplasm The gel-like material that fills the region between the cytoplasmic membrane and the outer membrane of Gram-negative bacteria and the cytoplasmic membrane and the peptidoglycan layer of at least some Gram-positive bacteria.

Pili Cell surface structures that allow cells to adhere to certain material; some types are involved in a mechanism of DNA transfer.

Plasmid An extrachromosomal DNA molecule that replicates independently of the chromosome.

Ribosome The structure involved in protein synthesis.

Transport Systems Mechanisms cells use to transport nutrients and other small substances across the cytoplasmic membrane.

A Glimpse of History

Hans Christian Joachim Gram (1853–1938) was a Danish physician working in a laboratory at the morgue of the City Hospital in Berlin, microscopically examining the lungs of patients who had died of pneumonia. He was working under the direction of Dr. Carl Friedländer, who was trying to identify the cause of pneumonia by studying patients who had died of it. Gram's task was to stain the infected lung tissue to make the bacteria easier to see under the microscope. Strangely, one of the methods he developed did not stain all bacteria equally; some types retained the first dye applied in this multistep procedure, whereas others did not. Gram's staining method revealed that two different kinds of bacteria were causing pneumonia, and these types retained the dye differently. We now recognize that this important staining method, called the Gram stain, efficiently identifies two large, distinct groups of bacteria: Gram-positive and Gram-negative. The staining outcome reflects a fundamental difference in the structure and chemistry of the cell walls of these two groups, which is why the Gram stain is a key test in the initial identification of a bacterium.

Microscopic study of cells has revealed two fundamental types: prokaryotic and eukaryotic. The cells of all bacteria and archaea are prokaryotic. In contrast, cells of all animals, plants, protozoa, fungi, and algae are eukaryotic. The similarities and differences between these two basic cell types are important from a scientific standpoint, and they also have significant consequences to human health. Bacterial cell components are particularly relevant because some are targets for antibacterial medications used to treat infectious diseases. By interfering with the function of components unique to bacteria, we can selectively kill or inhibit bacteria without harming the patient, as described in section 20.2. In addition, "alarm systems" in the body recognize compounds unique to bacteria or certain other microbial groups, alerting the body's immune system when invaders are present, as described in section 14.5.

Prokaryotic cells are generally much smaller than most eukaryotic cells—a trait that has certain advantages as well as disadvantages. Their small size gives the cells a high surface-area-to-volume ratio, making it easier for them to take in nutrients and excrete waste products. That small size, however, also makes the cells vulnerable to a variety of threats, including predators, parasites, and competitors. To cope, prokaryotes have evolved many unique features that increase their chances of survival.

Focus Figure

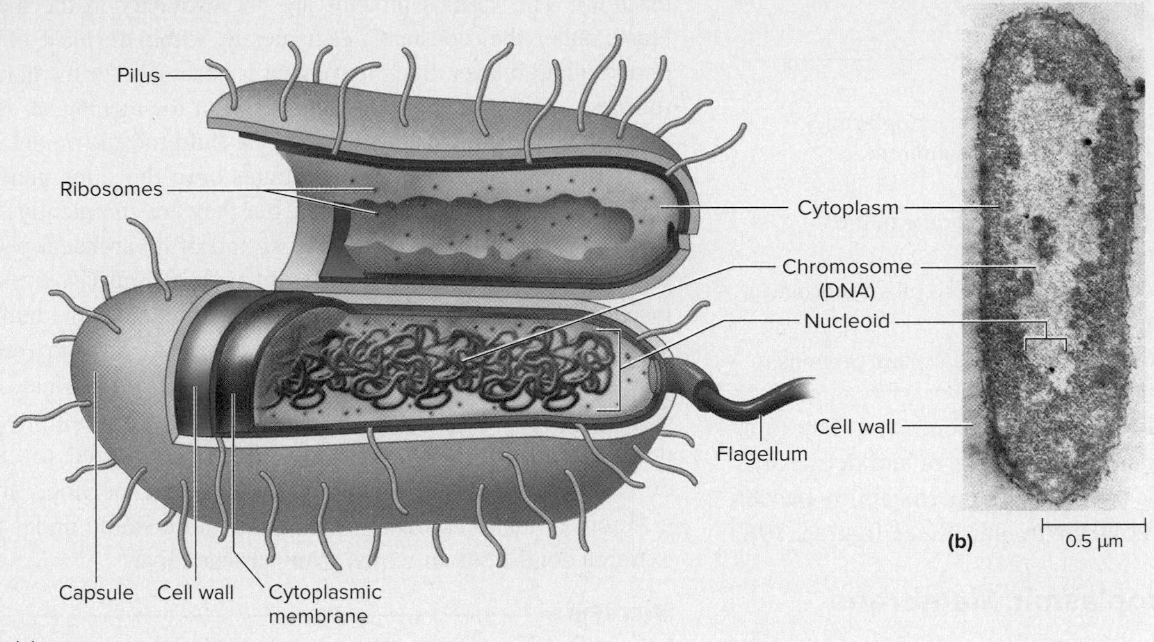

FIGURE 3.1 Typical Structures of a Prokaryotic Cell (a) Diagrammatic representation. Archaeal cells rarely have a capsule. **(b)** *Escherichia coli* cell (TEM). b: Science Source

Which layers make up the cell envelope?

Eukaryotic cells—which are defined by having a membrane-bound nucleus—are generally larger and much more complex than prokaryotic cells. In addition to the nucleus, they contain other membrane-bound compartments in which many of their cellular processes take place.

PROKARYOTIC CELL STRUCTURES AND THEIR FUNCTIONS

The surface layers of a prokaryotic cell, called the **cell envelope,** consist of the cytoplasmic membrane, the cell wall, and, if present, the capsule (**figure 3.1**). Although bacterial cells often have a capsule (a layer that helps protect the cell or allows it to attach to certain surfaces), archaeal cells rarely do. Enclosed by the envelope is the **cytoplasm,** a thick substance filled with nutrients, ribosomes, and enzymes; the fluid portion of the cytoplasm is called cytosol. The **nucleoid** is the gel-like region in the cytoplasm where the cell's chromosome is found. It is not enclosed by a membrane, which distinguishes the nucleoid from the nucleus that characterizes eukaryotic cells. The cell may also have appendages that provide useful traits, including motility and the ability to adhere to certain surfaces.

Although the general structure of bacterial and archaeal cells is the same, the components have several fundamental chemical differences. Because of this, and because bacteria and archaea are not closely related genetically, many scientists oppose lumping the two groups together under the general category "prokaryotic cells." Grouping them together is certainly convenient, however, particularly when trying to keep descriptions brief. With this in mind, we will use bacterial cells as a model when covering the general characteristics of prokaryotic cell structure. While doing this, we will also mention some of the most important ways in which archaeal cells differ. Bear in mind, though, that archaea have not been studied as extensively as bacteria, so scientists might still discover other significant differences.

We will start our discussion of prokaryotic cells by describing the cytoplasmic membrane because it defines the boundary of the cell; it serves as the cell's gatekeeper, determining what can enter or exit the cell. From there, we will work our way outward, next discussing the cell wall, an extremely important structure from a medical standpoint. Finally, we will describe the cell's internal components. Many of the cell parts are essential for growth and therefore are found in all prokaryotic cells, but others are optional. Although these optional parts are not required for growth in laboratory cultures, cells lacking them might not survive the competitive environment of the natural world. A table at the end of the discussion of prokaryotic cells summarizes the structures (see table 3.3).

MicroByte

Features found in only certain bacterial groups can be used to help identify a particular bacterium.

3.1 ■ The Cytoplasmic Membrane of Prokaryotic Cells

Learning Outcomes

1. Describe the structure and chemistry of the cytoplasmic membrane, focusing on how it relates to membrane permeability.
2. Describe the role of the cytoplasmic membrane in the development of a proton motive force.
3. Describe the systems prokaryotic cells use to move substances across the cytoplasmic membrane.
4. Explain why prokaryotic cells must secrete certain proteins.

The **cytoplasmic membrane** (or plasma membrane) is a thin, delicate structure that surrounds the cytoplasm and defines the boundary of the cell. It serves as the crucial permeability barrier between the cell and its external environment (see figure 2.19).

Structure of the Cytoplasmic Membrane

The prokaryotic cytoplasmic membrane is a typical biological membrane: a phospholipid bilayer embedded with proteins (**figure 3.2**). The phospholipid molecules are arranged in opposing layers so that their hydrophobic tails face inward, toward each other, and their hydrophilic heads face outward, interacting freely with aqueous solutions.

The proteins embedded in the cytoplasmic membrane have a variety of functions. Some act as selective gates, allowing nutrients or certain other substances to enter the cell and wastes to exit. Others serve as sensors of environmental conditions, giving the cell a mechanism to monitor and adjust to its surroundings. So, while the membrane functions as a permeability barrier, it also transmits information about the external environment to the inside of the cell. Membrane proteins can have other functions as well; for example, some are enzymes that catalyze essential chemical reactions. The various proteins are not stationary in the membrane; rather, they constantly drift laterally within the plane of the phospholipid bilayer. Such movement is necessary for membrane functions. The structure and composition of the membrane, with its resulting dynamic nature, is called the **fluid mosaic model.**

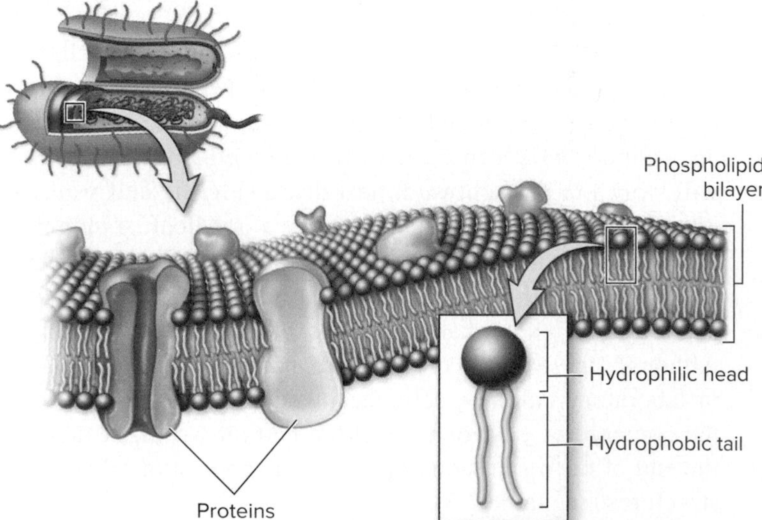

FIGURE 3.2 The Structure of the Cytoplasmic Membrane The membrane is a phospholipid bilayer embedded with proteins.

? Which part of the membrane is hydrophobic?

Archaeal cytoplasmic membranes have the same general structure as bacterial membranes, but they are chemically distinct. For example, the hydrophobic tail of an archaeal phospholipid is composed of a different type of unit (isoprenoid instead of fatty acid), and those units are linked to the hydrophilic head by a different type of chemical linkage (ether instead of ester). Additionally, some archaeal membranes are composed of monolayers, not bilayers; the phospholipids in these cases have two hydrophilic heads connected to each other by a membrane-spanning hydrophobic portion. Such differences make archaeal membranes more stable under the extreme conditions in which many archaea live.

MicroByte

E. coli's DNA encodes approximately 1,000 different membrane proteins, many of which are involved in transport.

Permeability of the Cytoplasmic Membrane

The cytoplasmic membrane is **selectively permeable,** meaning that only certain substances can cross it. Molecules that freely pass through the phospholipid bilayer include gases such as O_2, CO_2, and N_2, small hydrophobic compounds, and water (**figure 3.3**). Some cells facilitate water passage with **aquaporins,** pore-forming membrane proteins that specifically allow water molecules to pass through. Cells have transport systems to move certain other molecules across the membrane.

Simple Diffusion

Substances that freely pass through the phospholipid bilayer move in and out of the cell by **simple diffusion,** in which substances move from a region of high concentration to one of low concentration, until equilibrium is reached. The speed and direction of movement depend on the concentration gradient (the difference in concentrations of molecules on each side of the membrane); the greater the concentration difference, the higher the rate of diffusion. The rate slows as the gradient is reduced until the concentration is the same on both sides of the membrane. At equilibrium, substances still move back and forth across the membrane, but movement in one direction is balanced by movement in the other.

Osmosis

Osmosis is the diffusion of water across a selectively permeable membrane. It occurs when the concentrations of solute (dissolved substances like salts, sugars, and other molecules) on two sides of a membrane are unequal. Typical of diffusion, water moves down its concentration gradient from high water concentration (low solute concentration) to low water concentration (high solute concentration).

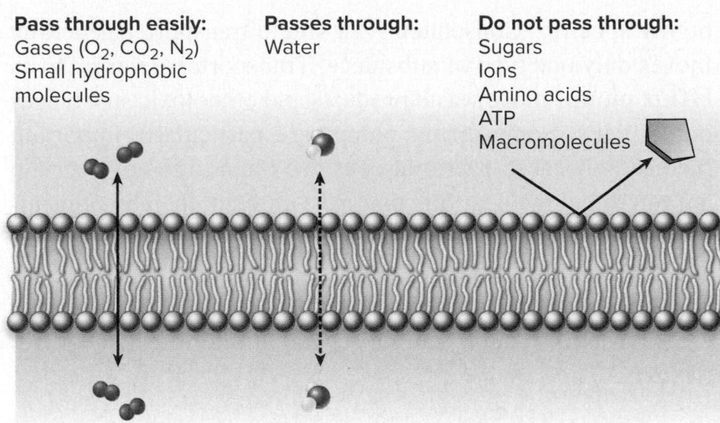

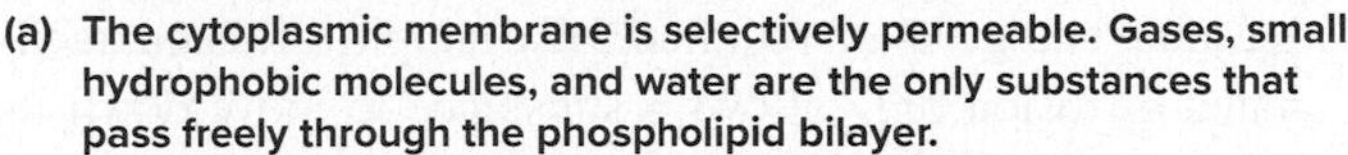

(a) The cytoplasmic membrane is selectively permeable. Gases, small hydrophobic molecules, and water are the only substances that pass freely through the phospholipid bilayer.

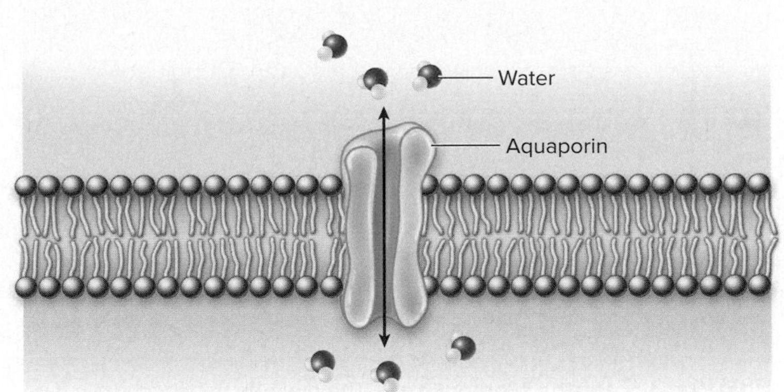

(b) Aquaporins allow water to pass through the cytoplasmic membrane more easily.

FIGURE 3.3 Permeability of the Phospholipid Bilayer **(a)** Most substances cannot pass through the phospholipid bilayer. **(b)** The role of aquaporins.

? **Why are ions unable to pass through the phospholipid bilayer?**

When describing osmosis, three terms are used to refer to the solutions on opposing sides of a membrane: hypotonic (*hypo* means "less"; *tonic* refers to solute), hypertonic (*hyper* means "more"), and isotonic (*iso* means "the same"). Water flows from the hypotonic solution to the hypertonic one. No net water movement occurs between isotonic solutions.

Osmosis has important biological consequences. The cytoplasm of a cell is a concentrated solution of inorganic salts, sugars, amino acids, and various other molecules. However, the environments in which bacteria and archaea normally grow are typically very dilute (hypotonic). Because water moves toward the high solute concentration, it flows from the surrounding medium into the cell (**figure 3.4a**). This inflow of water exerts tremendous osmotic pressure on the cytoplasmic membrane, much more than it typically can resist. However, the strong cell wall surrounding the membrane allows the cell to withstand increased osmotic pressure. The cytoplasmic membrane is forced against the wall but cannot balloon further. Damage to the cell wall weakens it, and cells may lyse (burst) as a result.

The Role of the Cytoplasmic Membrane in Energy Transformation

The cytoplasmic membrane of prokaryotic cells plays a crucial role in transforming energy—converting the energy of food or sunlight into ATP, the energy currency of a cell. This is an important difference between prokaryotic and eukaryotic cells; in eukaryotic cells, this process occurs in membrane-bound organelles, which are discussed in section 3.7.

As part of their energy-transforming processes, most prokaryotes have a series of protein complexes—collectively referred to as the **electron transport chain (ETC)**—embedded in their cytoplasmic membrane. Details of how the ETC works will be explained in chapter 6, but the net result is that protons are moved out of the cell. This creates an electrochemical gradient across the membrane—positively charged protons are concentrated immediately outside the membrane, whereas negatively charged hydroxide ions remain inside the cell (**figure 3.5**).

Water flows across a membrane toward the higher solute concentration.

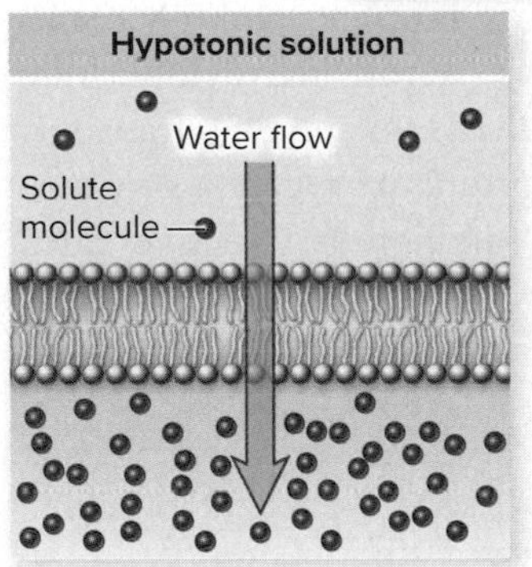

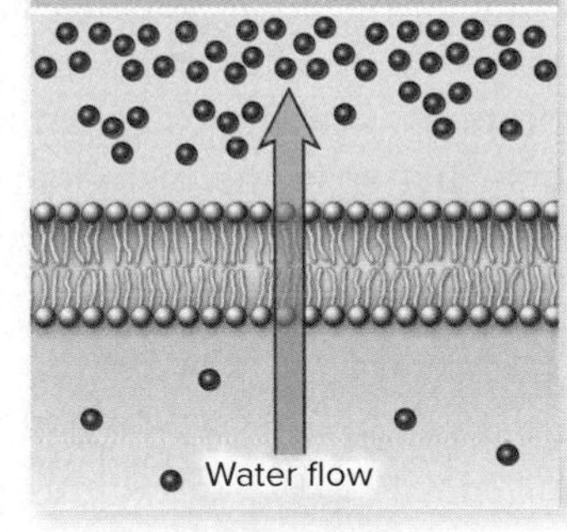

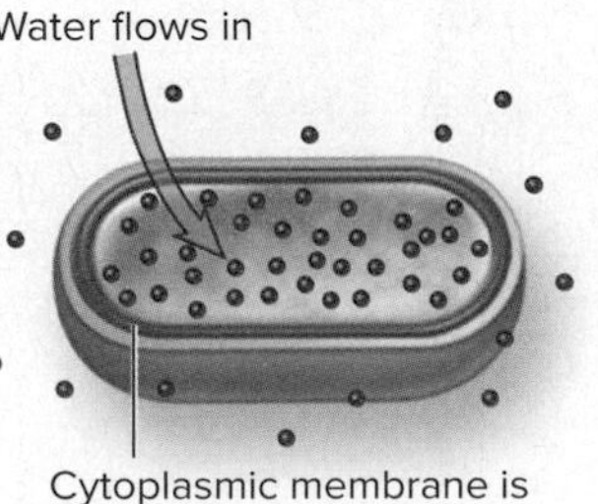

(a)

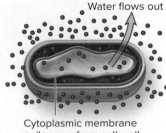

Water flows out

Cytoplasmic membrane pulls away from cell wall.

(b)

FIGURE 3.4 Osmosis **(a)** The effect of a hypotonic solution on a bacterial cell. **(b)** The effect of a hypertonic solution on a bacterial cell.

? **What might happen in part (a) if the cell wall were weakened?**

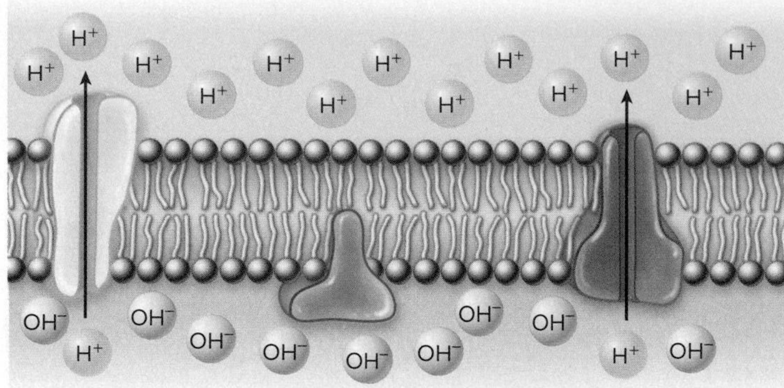
H+
OH−

FIGURE 3.5 Proton Motive Force The electron transport chain, a series of protein complexes within the membrane, moves protons out of the cell.

Why would the protons stay close to the membrane, rather than float away?

The charged ions attract each other, so they stay close to the membrane. This proton gradient is a form of energy called **proton motive force,** analogous to the energy stored in a battery.

The energy of the proton motive force is harvested by mechanisms that allow protons to move back into the cell. That energy is used to drive certain energetically unfavorable cellular processes (including ATP synthesis, some forms of motility, and one of the transport systems discussed next).

Transport of Small Substances Across the Cytoplasmic Membrane

Cells use **transport systems** to move nutrients and other substances (small molecules and ions) across the cytoplasmic membrane. These rely on specialized membrane transport proteins, sometimes called transporters, permeases, or carriers, that span the membrane and function as selective gates through which the substances pass (**figure 3.6**). The interaction between a transport protein and the substance it carries is highly specific. Consequently, a single transporter generally moves only one type of substance. Transporters referred to as **efflux pumps** move waste products and other toxic substances out of cells. Some efflux pumps are particularly important medically because bacterial cells use them to remove certain antimicrobial medications that have entered, thereby preventing the medications from reaching their intracellular targets and functioning (mechanisms of antimicrobial resistance are explained in more detail in section 20.5).

Transport systems can be separated into three categories (**figure 3.7**):

- **Facilitated diffusion.** This is a form of passive transport, meaning energy is not required. Like simple diffusion, facilitated diffusion only allows a substance to move down its concentration gradient (think of a twig floating downstream in a river). Unlike simple diffusion, however, facilitated diffusion requires membrane transport proteins through which substances move. The substances are transported from one side of the membrane to the other until their concentration is the same on both sides. Because prokaryotes typically grow in environments where the concentration of nutrients is lower outside the cell than inside, prokaryotic cells generally do not use facilitated diffusion to take in nutrients.
- **Active transport.** This moves a substance against a concentration gradient, a process that requires energy (think of swimming upstream). Prokaryotes routinely use active transport because nearly every organic molecule taken in by a cell moves against a concentration gradient. Primary active transport uses ATP as the energy source, while secondary active transport uses the energy stored within a concentration gradient like the proton motive force. Most transporters that use ATP as an energy source rely on specific binding proteins located immediately outside the cytoplasmic membrane to deliver molecules to the transport complex. Transporters that use proton motive force as an

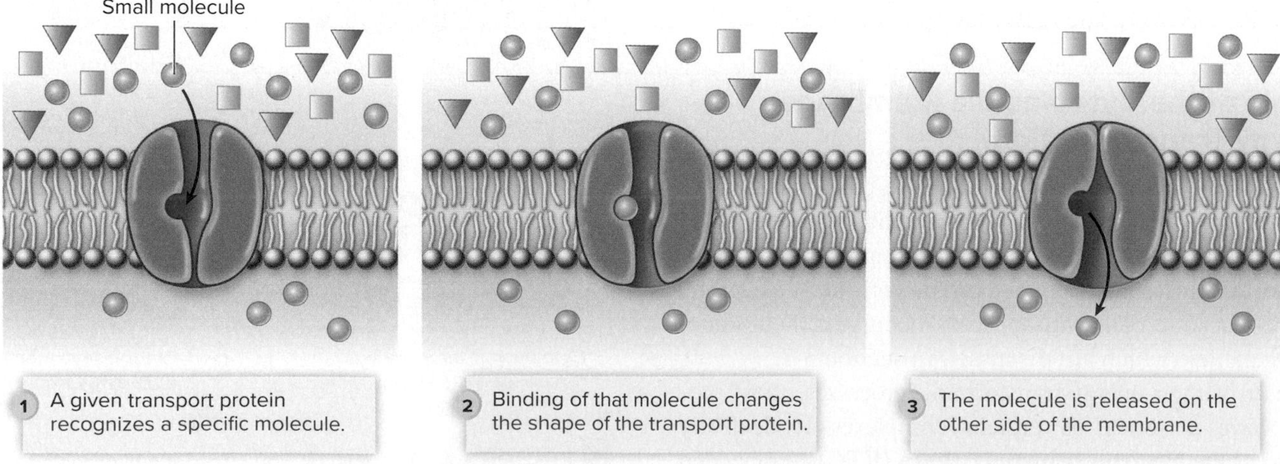
Small molecule
1 A given transport protein recognizes a specific molecule.
2 Binding of that molecule changes the shape of the transport protein.
3 The molecule is released on the other side of the membrane.

FIGURE 3.6 Transport Protein Only one type of substance is transported across the membrane through a given transport protein.

What are some examples of small substances that enter cells through transport proteins?

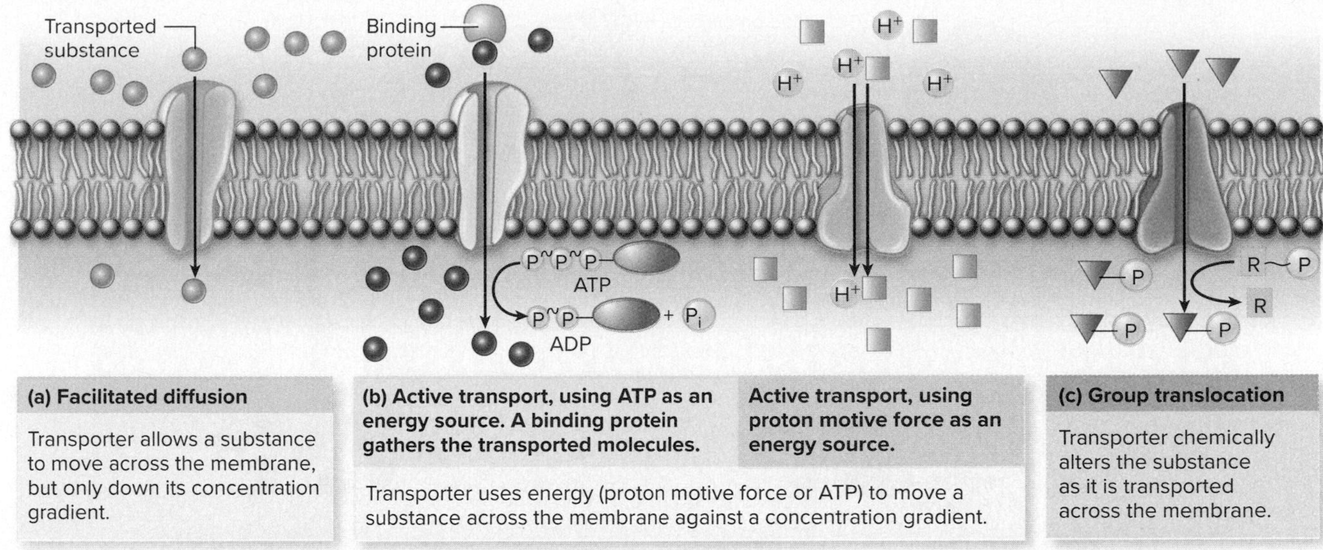

Transported substance
Binding protein
H+
P~P~P
ATP
P~P
ADP
+ Pi
P
R~P
R
(a) Facilitated diffusion
Transporter allows a substance to move across the membrane, but only down its concentration gradient.
(b) Active transport, using ATP as an energy source. A binding protein gathers the transported molecules.
Active transport, using proton motive force as an energy source.
Transporter uses energy (proton motive force or ATP) to move a substance across the membrane against a concentration gradient.
(c) Group translocation
Transporter chemically alters the substance as it is transported across the membrane.

FIGURE 3.7 Types of Transport Systems **(a)** Facilitated diffusion. **(b)** Active transport. **(c)** Group translocation.

In group translocation, how would the amount of unphosphorylated R group shown in the diagram indicate the quantity of transported substance being brought into the cell?

energy source allow a proton into the cell (down its concentration gradient) and simultaneously bring along or remove another substance (against its concentration gradient).

- **Group translocation.** This process chemically alters a substance during its passage through the cytoplasmic membrane. In the most well-understood examples, a phosphate group is added to a sugar such as glucose—a modification called phosphorylation. Although group translocation expends energy and can therefore be considered active transport, scientists typically consider it a separate category because the energy would be used anyway in the initial steps of sugar breakdown, a process described in chapter 6.

Characteristics of the transport mechanisms used by prokaryotic cells are summarized in **table 3.1.**

Protein Secretion

A cell moves certain proteins it has made to the outside of the cell by a process called **secretion.** As an example of a need for this, cells commonly make exoenzymes (extracellular enzymes) to break down polysaccharides or other large macromolecules into smaller subunits that can then be transported into the cell. Each of these exoenzymes, as well as any other protein destined for secretion, is synthesized within the cell as a form called a preprotein (precursor protein), which has a characteristic sequence of amino acids at one end of the molecule (**figure 3.8**). This sequence—a **signal sequence**—functions as a tag that directs the secretion machinery embedded in the cytoplasmic membrane to move the preprotein across the membrane. During the transport process, the signal sequence is removed and the protein ultimately folds into its functional shape outside the cytoplasmic membrane. Exoenzymes are not the only secreted proteins; others include the proteins that make up external structures such as flagella (appendages used for motility). Cells have a variety of different secretion systems, and some can even move molecules into other cells.

TABLE 3.1 Transport Mechanisms Used by Prokaryotic Cells

Transport Mechanism	Characteristics
Facilitated Diffusion	A substance moves down its concentration gradient through a membrane transport protein. No energy is used. Not commonly used by prokaryotes.
Active Transport	Energy in the form of ATP or a concentration gradient such as proton motive force is used to move a substance against a concentration gradient.
Group Translocation	The transported substance is chemically altered as it passes into the cell.

MicroAssessment 3.1

The cytoplasmic membrane is a phospholipid bilayer embedded with proteins. Substances that pass through the bilayer move in and out by simple diffusion; osmosis is the diffusion of water across the membrane. The electron transport chain generates a proton motive force, which is used to drive several cellular processes, including ATP synthesis, certain transport systems, and some types of motility. Transport systems move substances across the membrane using facilitated diffusion, active transport, or group translocation. Proteins destined for secretion have a characteristic signal sequence.

1. Explain why the cytoplasmic membrane is described as being selectively permeable.
2. Prokaryotes rarely use facilitated diffusion. Why is this so?
3. Membrane proteins move laterally in a phospholipid bilayer but generally not from one phospholipid layer to the other. Why would this be so?

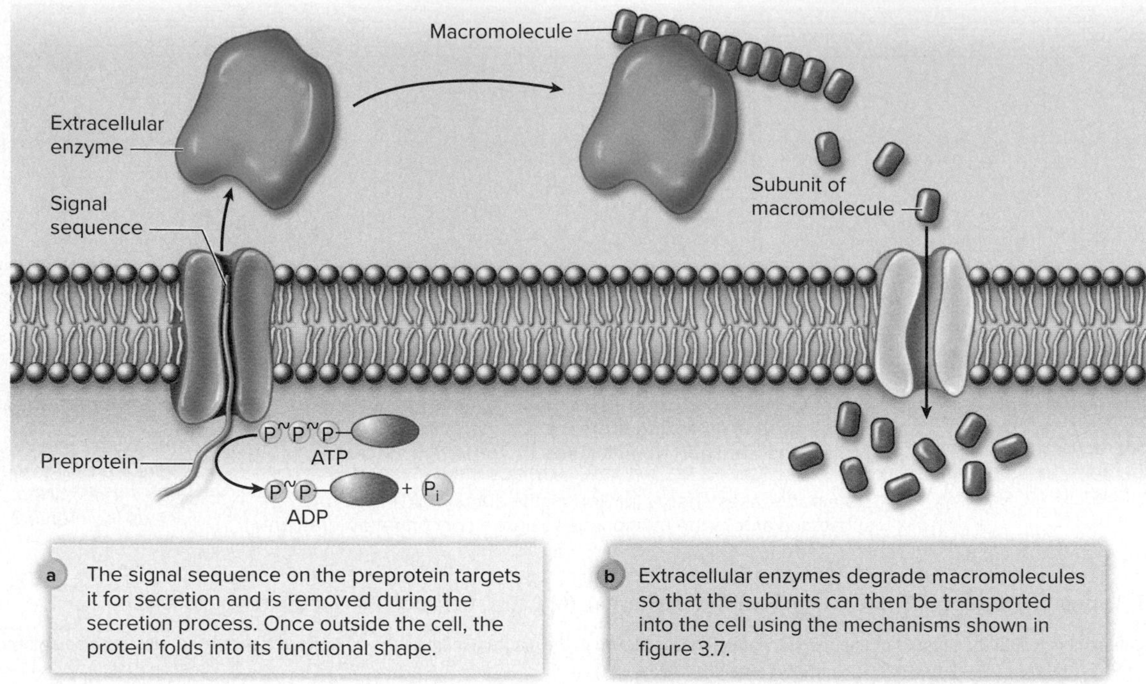
Macromolecule
Extracellular enzyme
Signal sequence
Subunit of macromolecule
Preprotein
ATP
ADP
+ P_i
a The signal sequence on the preprotein targets it for secretion and is removed during the secretion process. Once outside the cell, the protein folds into its functional shape.
b Extracellular enzymes degrade macromolecules so that the subunits can then be transported into the cell using the mechanisms shown in figure 3.7.

FIGURE 3.8 Protein Secretion (a) Generalized view of the protein secretion process. **(b)** One function of a secreted protein.

? **Why would a cell secrete enzymes rather than bring intact macromolecules into the cell?**

3.2 ■ The Cell Wall of Prokaryotic Cells

Learning Outcomes

5. Describe the chemistry and structure of peptidoglycan.
6. Compare and contrast the structure and chemistry of the Gram-positive and Gram-negative cell walls.
7. Explain the significance of lipid A and the O antigen of LPS.
8. Explain how the cell wall affects susceptibility to penicillin and lysozyme.
9. Explain how the cell wall affects Gram staining characteristics.
10. Describe the cell walls of archaea.

The prokaryotic cell wall is a strong, somewhat rigid structure that prevents the cell from bursting. Differences in its composition distinguish two main groups of bacteria: **Gram-positive** and **Gram-negative.**

Peptidoglycan

Gram-positive and Gram-negative cell walls both have a layer of **peptidoglycan,** a strong, mesh-like material found only in bacteria. Its basic structure is an alternating series of two major subunits related to glucose: *N*-acetylmuramic acid (NAM) and *N*-acetylglucosamine (NAG) (**figure 3.9**). These subunits are covalently joined to one another to form a linear polymer called a glycan chain (*glyco* means "sugar"), which serves as the backbone of the peptidoglycan molecule. Attached to each NAM molecule is a tetrapeptide chain (a string of four amino acids). These chains play an important role in the strength of peptidoglycan because they can connect to one another, thereby cross-linking adjacent glycan chains. Because of those connections, peptidoglycan is a single, very large three-dimensional molecule much like a flexible, multilayered chain-linked fence. Only a few types of amino acids, some of which are D-optical isomers (a form found in relatively few substances), make up these tetrapeptide chains. In Gram-negative bacteria, the connection between tetrapeptide chains is direct, whereas in Gram-positive bacteria, the connection is usually via a peptide interbridge (a series of amino acids). In either case, the connection requires an amino acid that binds to three other amino acids—rather than the normal two—so this amino acid requires two amino groups rather than only one. In Gram-negative bacteria, this diamino acid is commonly diaminopimelic acid, while in Gram-positive bacteria, it is commonly lysine. Diaminopimelic acid, which is related to lysine, has not been found in any other place in nature.

The Gram-Positive Cell Wall

A relatively thick layer of peptidoglycan characterizes the Gram-positive cell wall (**figure 3.10**). As many as 30 layers of interconnected glycan chains make up the polymer. Regardless of the layer's thickness, many small substances—including sugars and amino acids—can pass through. Within peptidoglycan are **teichoic acids** (from the Greek word *teichos,* meaning "wall") which may compose up to 60% of the cell wall by mass. They are

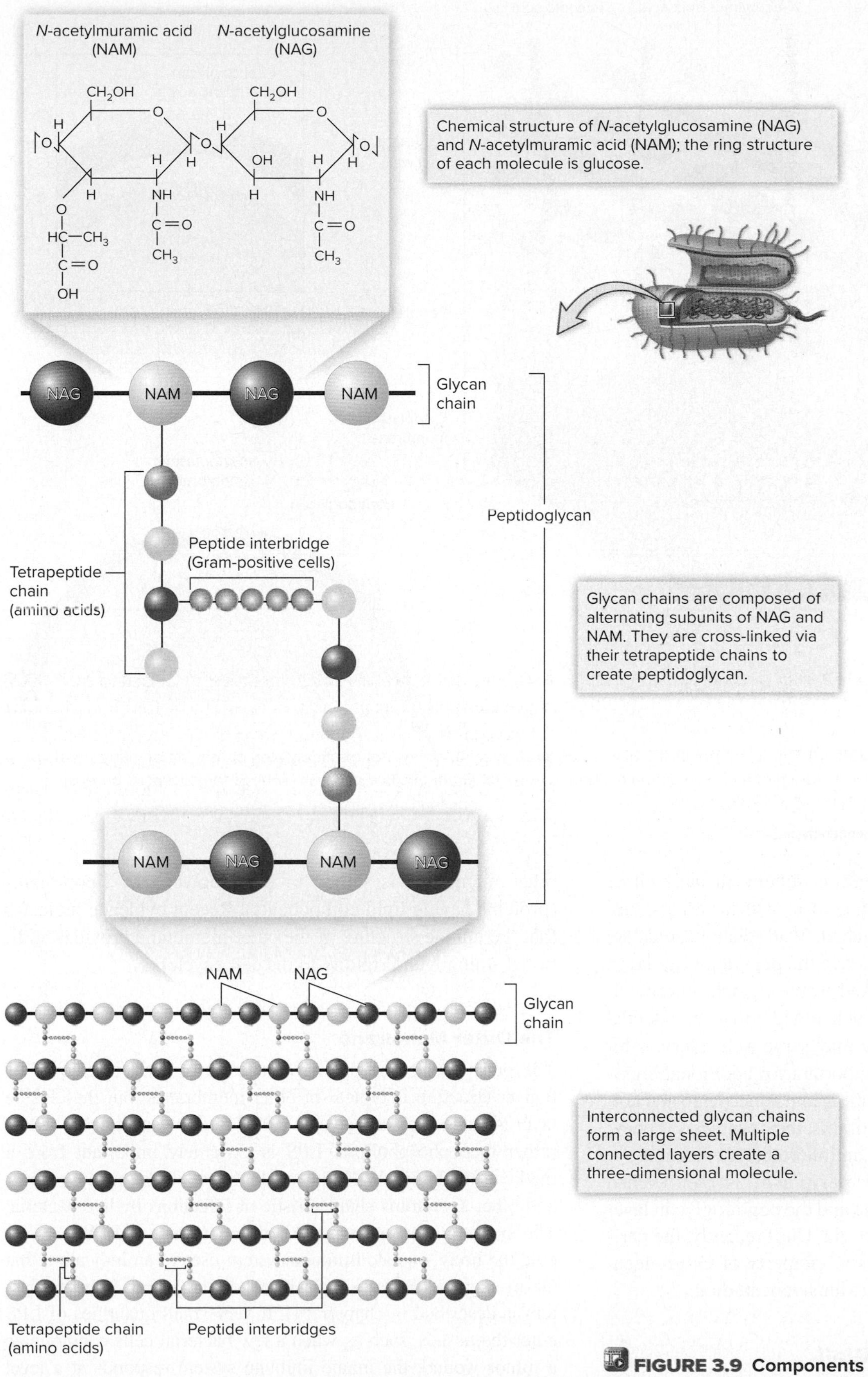
N-acetylmuramic acid (NAM)
N-acetylglucosamine (NAG)
Chemical structure of N-acetylglucosamine (NAG) and N-acetylmuramic acid (NAM); the ring structure of each molecule is glucose.
NAG
NAM
NAG
NAM
Glycan chain
Peptidoglycan
Peptide interbridge (Gram-positive cells)
Tetrapeptide chain (amino acids)
Glycan chains are composed of alternating subunits of NAG and NAM. They are cross-linked via their tetrapeptide chains to create peptidoglycan.
NAM
NAG
NAM
NAG
NAM
NAG
Glycan chain
Interconnected glycan chains form a large sheet. Multiple connected layers create a three-dimensional molecule.
Tetrapeptide chain (amino acids)
Peptide interbridges

FIGURE 3.9 Components and Structure of Peptidoglycan

Why might it be medically significant that peptidoglycan is found only in bacteria?

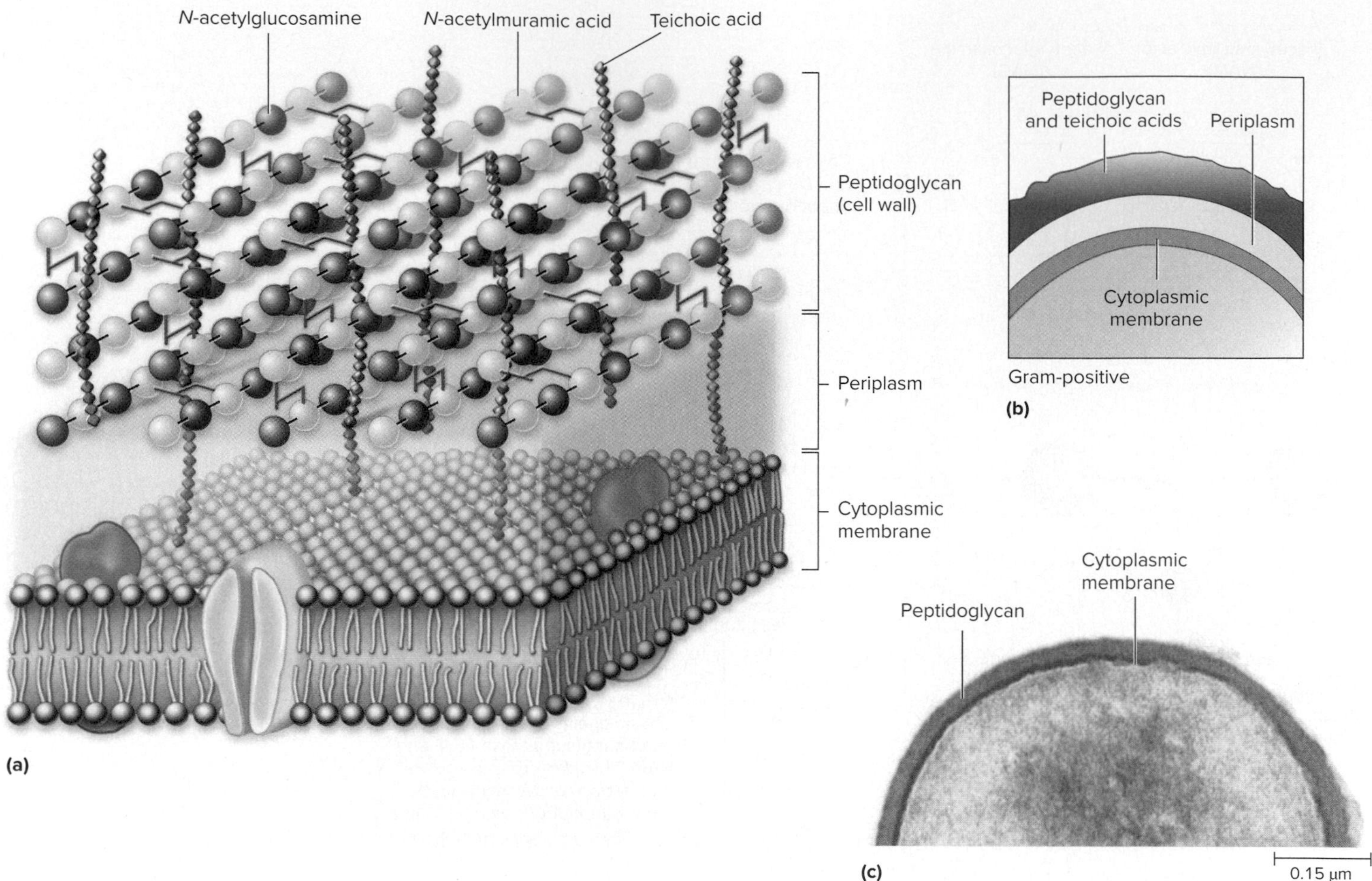

FIGURE 3.10 Gram-Positive Cell Wall (a) The Gram-positive cell wall has a relatively thick layer of peptidoglycan consisting of many sheets of interconnected glycan chains. **(b)** Simple diagram of a cross-section of the structure. **(c)** Gram-positive cell wall (TEM; *Staphylococcus aureus*).

c: Egbert Hoiczyk

? **What connects adjacent glycan chains in peptidoglycan?**

negatively charged chains of a common subunit (usually either ribitol-phosphate or glycerol-phosphate) to which various sugars and D-alanine are typically attached. Wall teichoic acids are covalently attached to and extend above the peptidoglycan layer, whereas lipoteichoic acids are linked to the cytoplasmic membrane. Teichoic acids bind cations such as Mg^{2+} that are essential for various cell functions, and they may serve as a reservoir for those cations. They may also be important for additional cross-linking, cell division, biofilm formation, and adhering to and colonizing host cell surfaces. Because of this, their synthesis is being investigated as a possible target for antimicrobial medications.

A gel-like substance called **periplasm** is sandwiched between the cytoplasmic membrane and the peptidoglycan layer of at least some Gram-positive bacteria. Until recently, the periplasm was thought to be an exclusive property of Gram-negative cells, where it has been more extensively studied.

The Gram-Negative Cell Wall

The Gram-negative cell wall contains only a thin layer of peptidoglycan (**figure 3.11**). Outside of that is the **outer membrane,** a unique lipid bilayer embedded with proteins. The outer membrane is joined to peptidoglycan by lipoproteins (proteins having lipid components). Recent evidence indicates that the unique structure of the outer membrane provides additional strength and rigidity to the cell envelope.

The Outer Membrane

The outer membrane of Gram-negative bacteria is unique. Its bilayer structure is typical of other membranes, but the outside layer is made up of a molecule called **lipopolysaccharide (LPS)** rather than phospholipid. LPS is extremely important from a medical standpoint. When injected into an animal, it causes fever and other symptoms characteristic of infections by live bacteria. The symptoms arise from the body's response to LPS, a molecule the body's innate immune system uses as an indication that Gram-negative bacteria have invaded (the innate immune system is described in chapter 14). If very small quantities of LPS enter the tissues, such as when a few bacterial cells contaminate a minor wound, the innate immune system responds at a level that can safely eliminate the invader. When significant amounts of the molecule spread throughout the body, however, such as when Gram-negative bacteria are growing in the bloodstream, the

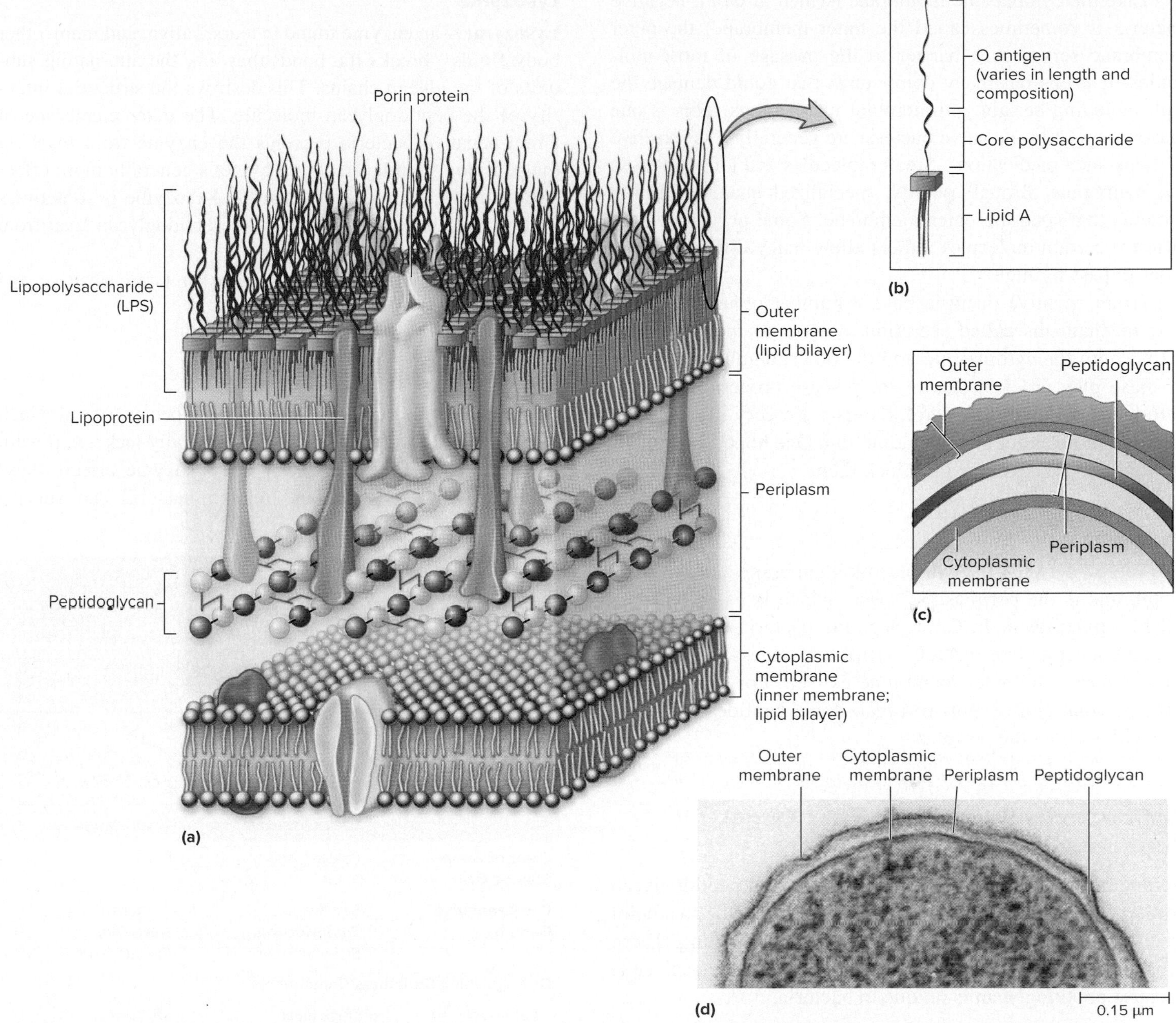

FIGURE 3.11 Gram-Negative Cell Wall (a) The Gram-negative cell wall has a thin layer of peptidoglycan made up of only one or two sheets of interconnected glycan chains. The outer membrane is similar to a typical phospholipid bilayer, but the outer layer is lipopolysaccharide. Porins span the membrane to allow specific molecules to pass through. Periplasm fills the region between the two membranes. **(b)** Structure of lipopolysaccharide. The sugars in the O antigen vary among bacterial species. **(c)** Simple diagram of a cross section of the structure. **(d)** Gram-negative cell wall (TEM; *Myxococcus xanthus*). d: Egbert Hoiczyk

? **How does the general structure of the Gram-negative cell wall differ from that of the Gram-positive cell wall?**

response can result in life-threatening *shock* (a systemic condition resulting from a sudden drop in blood pressure). Because of this effect, LPS is called **endotoxin** (*endo* meaning "inside," although LPS is actually a component of the envelope).

Two parts of the LPS molecule are particularly notable (see figure 3.11b):

- **Lipid A** anchors the LPS molecule in the lipid bilayer. This is the portion of the LPS molecule that the body recognizes as the sign of invading Gram-negative bacteria.
- **O antigen** is the part of LPS directed away from the membrane, at the end opposite lipid A. It is made up of a chain of sugar molecules, the number and type of which vary among different species. The differences can be used to identify certain species or strains.

MicroByte

The "O157" in *E. coli* O157:H7 refers to the characteristic O antigen.

Like the cytoplasmic membrane (which in Gram-negative bacteria is sometimes called the inner membrane) the outer membrane serves as a barrier to the passage of most molecules. It keeps out many compounds that could damage the cell, including certain antimicrobial medications. This is one reason why Gram-negative bacteria are generally less sensitive to many such medications. Small molecules and ions can cross the membrane through **porins,** specialized channel-forming proteins that span the outer membrane. Some porins are specific for certain molecules; others allow many different molecules to pass through.

Gram-negative bacteria have a number of unique transport mechanisms called secretion systems to move proteins across both the cytoplasmic and the outer membranes. Some of these play crucial roles in the disease process of certain pathogens, so medical microbiologists are very interested in learning more about how they function. One hope is that medications can be developed to block them.

Periplasm

The region between the cytoplasmic membrane and the outer membrane is the periplasmic space, which is filled with the gel-like **periplasm.** In Gram-negative bacteria, all exported proteins accumulate in the periplasm unless specifically moved across the outer membrane as well. For example, the enzymes that cells export to break down peptides and other molecules are in the periplasm.

Antibacterial Substances That Target Peptidoglycan

Compounds that interfere with the synthesis of peptidoglycan or alter its structural integrity weaken the molecule to a point where it can no longer prevent the cell from bursting. These substances have no effect on eukaryotic or archaeal cells because peptidoglycan is unique to bacteria.

Penicillin

Penicillin is the most thoroughly studied of a group of antibiotics that interfere with peptidoglycan synthesis. It functions by preventing the cross-linking of adjacent glycan chains (it inhibits enzymes that normally catalyze the cross-linking step).

Generally, but with notable exceptions, penicillin is far more effective against Gram-positive bacteria than against Gram-negative bacteria. This is because the outer membrane of Gram-negative cells prevents the medication from reaching the peptidoglycan layer. However, scientists have developed derivatives of penicillin (such as ampicillin) that pass through the outer membrane and can be effective in treating infections caused by some Gram-negative bacteria.

Lysozyme

Lysozyme—an enzyme found in tears, saliva, and many other body fluids—breaks the bonds that link the alternating subunits of the glycan chains. This destroys the structural integrity of the peptidoglycan molecule. The outer membrane of Gram-negative bacteria prevents the enzyme from reaching the peptidoglycan layer, so lysozyme is generally more effective against Gram-positive bacteria. Lysozyme is sometimes used in the laboratory to remove the peptidoglycan layer from bacteria for experimental purposes.

Table 3.2 compares the features of Gram-positive and Gram-negative bacteria.

Bacteria That Lack a Cell Wall

Certain bacteria, including *Mycoplasma* species (one of which causes a mild form of pneumonia), naturally lack a cell wall. As expected, neither penicillin nor lysozyme affects these organisms. *Mycoplasma* and related bacteria can survive

TABLE 3.2 Comparison of Features of Gram- Positive and Gram-Negative Bacteria

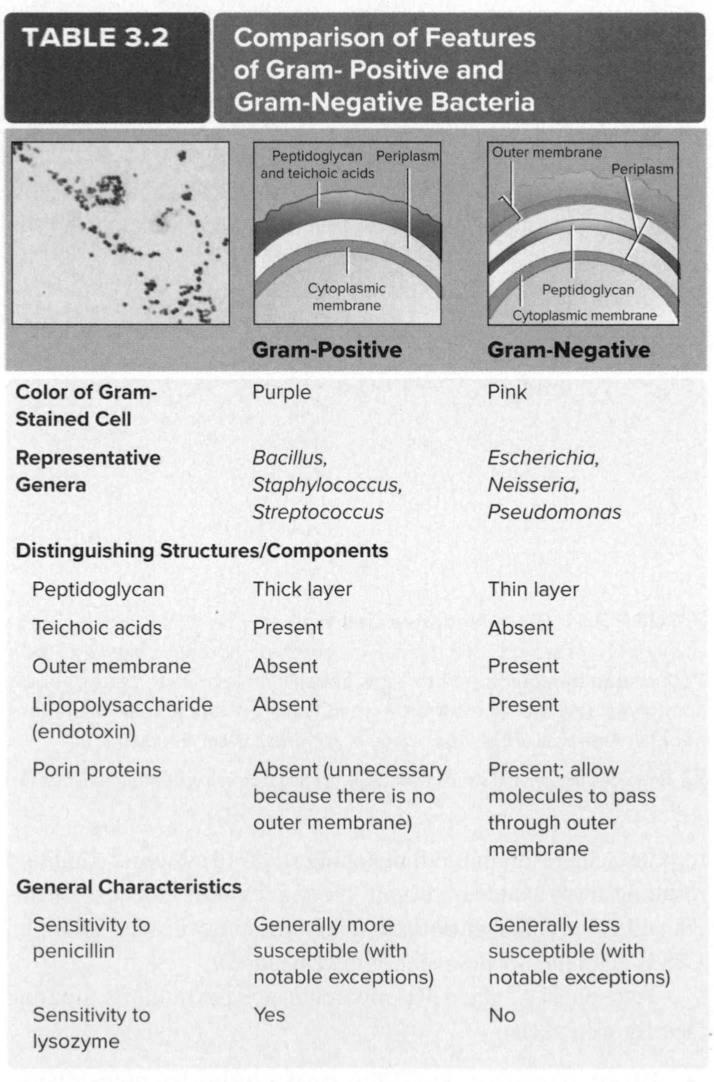

	Gram-Positive	Gram-Negative
Color of Gram-Stained Cell	Purple	Pink
Representative Genera	*Bacillus, Staphylococcus, Streptococcus*	*Escherichia, Neisseria, Pseudomonas*
Distinguishing Structures/Components		
Peptidoglycan	Thick layer	Thin layer
Teichoic acids	Present	Absent
Outer membrane	Absent	Present
Lipopolysaccharide (endotoxin)	Absent	Present
Porin proteins	Absent (unnecessary because there is no outer membrane)	Present; allow molecules to pass through outer membrane
General Characteristics		
Sensitivity to penicillin	Generally more susceptible (with notable exceptions)	Generally less susceptible (with notable exceptions)
Sensitivity to lysozyme	Yes	No

Lisa Burgess/McGraw Hill

FOCUS ON A CASE 3.1

Small outbreaks of *Staphylococcus aureus* skin infections were reported in people who had recently been tattooed. Investigation revealed 34 similar cases in three states over a 13-month period, involving 13 unlicensed tattooists. Most of the tattooists wore gloves while creating the tattoos, but they did not follow basic infection-control procedures, such as changing gloves between clients, handwashing, and disinfecting equipment. In several cases, the patient recalled seeing wounds on the tattooist's hands. In addition to the cases directly involving the tattooing procedure, an additional 10 people developed *S. aureus* infections as a result of contact with the tattoo recipients.

Treating the infections was complicated by the fact that the causative agent was resistant to several types of antibiotics. *S. aureus* is a Gram-positive coccus, and when penicillin was first used medically in 1941, most strains were quite sensitive to this life-saving antibiotic. Over time, however, some strains acquired the ability to produce penicillinase, an enzyme that degrades penicillin. Because the strains can destroy the medication, they are penicillin-resistant, meaning that the medication is no longer an effective treatment. To counteract the problem, scientists chemically modified penicillin to create methicillin and other derivatives not destroyed by penicillinase. Unfortunately, some bacteria have also gained the ability to resist the effects of these derivatives. In the case of *S. aureus*, the resistant bacteria are referred to as MRSA (pronounced mersa), which stands for **m**ethicillin-**r**esistant *Staphylococcus* ***a****ureus*. The strains in the outbreaks involving the tattoos were MRSA.

Fortunately, some treatment options were still available for the patients, so in most cases, the infections cleared with simple procedures (draining the wound and/or taking certain oral antimicrobial medications). Four patients, however, were hospitalized due to a MRSA bloodstream infection that required intravenous administration of vancomycin, an antimicrobial medication typically reserved as a last resort.

1. The tattooists wore gloves, so how could they have transmitted *S. aureus*?
2. Most *S. aureus* strains are now resistant to penicillin. Why would this be so?
3. Penicillin and its derivatives (including methicillin) all interfere with the cross-linking of adjacent glycan chains. How would this be lethal to bacterial cells?
4. MRSA strains have acquired the ability to produce a modified enzyme to make the cross-links between glycan chains. How would this allow them to be resistant to methicillin and other derivatives of penicillin?
5. Why would vancomycin be reserved only for use as a last resort?

Discussion

1. The tattooists could have contaminated their gloves in several ways. They might not have used proper procedures when putting on the gloves (specific procedures must be used to ensure that no part of the glove is touched with bare hands except the cuff). Even if the gloves had been put on properly, the tattooists might have touched contaminated surfaces with their gloved hands. Licensed tattooists are taught to use strict safety and sanitation procedures to avoid transmitting infections.
2. When penicillin is used, it kills only the penicillin-sensitive cells. The penicillin-resistant cells can still multiply, and in a penicillin-containing environment, they do so without competition. As penicillin is increasingly used, killing more of the sensitive cells, the resistant strains will predominate (see figure 20.13).
3. The cross-linking provides strength to the peptidoglycan molecule, just as the structure of a chain-link fence provides strength to that material. If the cross-links in peptidoglycan do not form properly, the cell wall will be too weak to prevent the cell from bursting.
4. Penicillin and its derivatives bind to the typical enzymes that form the cross-links in peptidoglycan, thereby interfering with their function. The antibiotics do not bind MRSA's modified enzyme.
5. When used to treat bloodstream or tissue infections, vancomycin must be given intravenously. Because of this, medications that can be given orally are usually a first choice. In addition, vancomycin sometimes has adverse side effects, so it should be avoided if safer alternatives are available.

Source: "Methicillin-Resistant *Staphylococcus aureus* Skin Infections Among Tattoo Recipients—Ohio, Kentucky, and Vermont, 2004–2005" Morbidity and Mortality Weekly Report June 23, 2006/55(24); 677–679. http://www.cdc.gov/mmwr/preview/mmwrhtml/mm5524a3.htm

without a cell wall because, unlike most other bacteria, their cytoplasmic membrane contains sterols, and these provide increased strength.

Cell Walls of Archaea

As a group, archaea have a variety of cell wall types. This is probably due to the fact that they inhabit a wide range of environments, including some that are extreme. Archaea have not been studied as extensively as bacteria, so less is known about their structures. Archaea do not have peptidoglycan in their cell walls, but some do have a similar molecule called pseudopeptidoglycan. Most have S-layers, which are sheets of flat protein or glycoprotein subunits. These subunits self-assemble, so they may have applications in nanotechnology—a branch of science that seeks to build functional items from molecules and atoms. Some bacterial cells also have S-layers, but the cell wall remains the primary structural component.

MicroAssessment 3.2

Peptidoglycan is a molecule unique to bacteria that provides strength to the cell wall. The Gram-positive cell wall is composed of a relatively thick layer of peptidoglycan as well as teichoic acids. The Gram-negative cell wall has a thin layer of peptidoglycan and a lipopolysaccharide-containing outer membrane. Penicillin and lysozyme interfere with the structural integrity of peptidoglycan. *Mycoplasma* species lack a cell wall. Although archaea have a variety of cell wall types, most have S-layers.

4. What is the significance of lipid A?
5. How does the action of penicillin differ from that of lysozyme?
6. Explain why penicillin kills only actively multiplying cells, whereas lysozyme kills cells in any stage of growth.

3.3 ■ Structures Outside the Cell Wall of Prokaryotic Cells

Learning Outcomes

11. Compare and contrast the structure and function of capsules and slime layers.
12. Describe the structure and arrangements of flagella, and explain how they are involved in chemotaxis.
13. Compare and contrast the structure and function of fimbriae and sex pili.

Many prokaryotes have structures outside the cell wall. These external structures are not essential to the life of microbes and are sometimes only produced under certain environmental conditions, but they provide a competitive advantage to cells that have them, including aiding in pathogenicity (ability to cause disease).

Capsules and Slime Layers

Bacteria often have a gel-like layer outside the cell wall that either protects the cell or allows it to attach to a surface (**figure 3.12**). A **capsule** is a distinct and gelatinous layer, whereas a **slime layer** is diffuse and irregular. Colonies that form either of these often appear moist and glistening.

Capsules and slime layers vary in their chemical composition, depending on the microbial species. Most are composed of polysaccharides and are commonly referred to as **glycocalyces** (singular: glycocalyx; *glyco* means "sugar" and *calyx* means "shell"). A few capsules consist of polypeptides made up of repeating subunits of only one or two amino acids. Interestingly, the amino acids are generally of the D-form, one of the few places this optical isomer is found in nature.

Some capsules and slime layers allow bacterial cells to adhere to specific surfaces, including teeth, epithelial cell layers, rocks, drainpipes, and other bacteria. Once attached, the cells can grow as a **biofilm,** a polymer-encased community of microbes. The polymers—collectively referred to as extracellular polymeric substances (EPS)—include the material that make up capsules and slime layers. A biofilm people encounter on a daily basis is dental plaque, which forms on teeth. When a person ingests sucrose (table sugar), *Streptococcus*

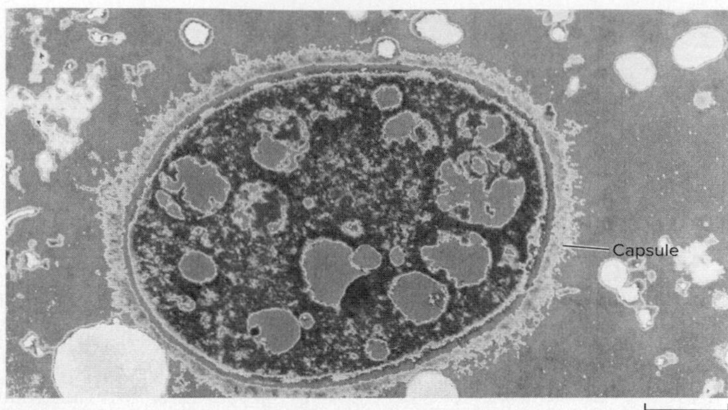

(a)

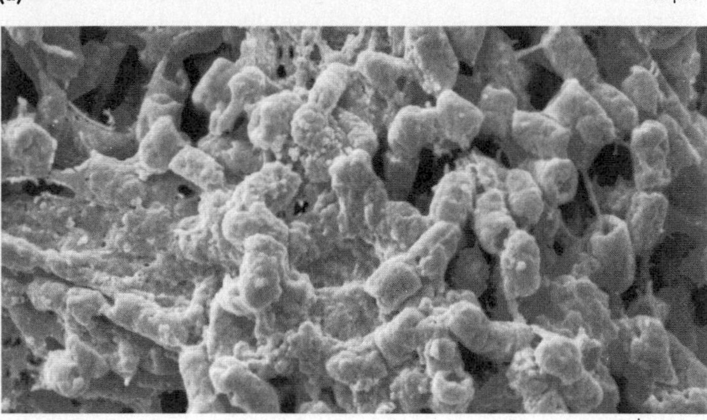

(b)

FIGURE 3.12 Capsules and Slime Layers (a) An encapsulated *Lactobacillus* species (TEM). **(b)** Masses of bacteria adhering in a layer of slime (SEM). a–b: Scimat/Science Source

? What are the functions of capsules and slime layers?

mutans (a common member of the oral microbiota) uses that to make a capsule. This allows *S. mutans* to attach to other bacterial cells on the teeth. In turn, additional bacteria adhere to the sticky layer, resulting in an even greater accumulation of plaque. Streptococci and other bacteria in the plaque produce acids that may then damage the tooth surface.

Some capsules allow bacteria to avoid the body's immune defenses that otherwise protect against infection. *Streptococcus pneumoniae,* an organism that causes bacterial pneumonia, can cause the disease only if it has a capsule. Unencapsulated cells are quickly engulfed and killed by phagocytes (a cell type of the body's immune system).

Flagella

Flagella (singular: flagellum) are long protein structures responsible for most types of bacterial motility in liquid (**figure 3.13**). They are anchored in the cytoplasmic membrane and cell wall and extend out from the cell surface. Flagella function by spinning like propellers, pushing the cell through liquid much as a ship is driven through water. That may sound like easy work, but water has the same relative viscosity to bacteria that molasses has to humans! The numbers and arrangement of flagella can be used to characterize flagellated bacteria. For example, some species have flagella distributed around the surface of a cell—an arrangement called

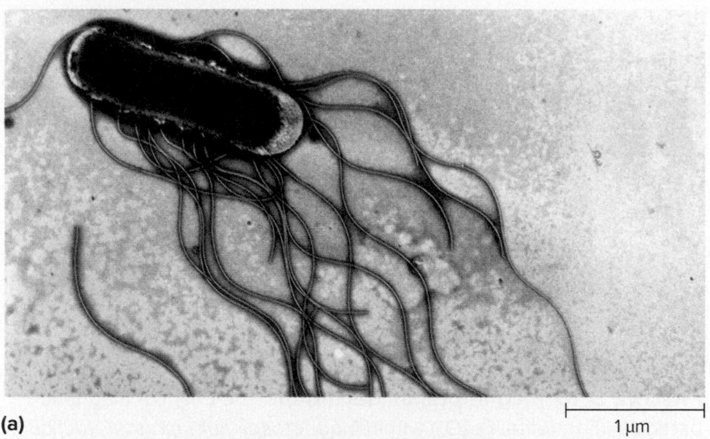

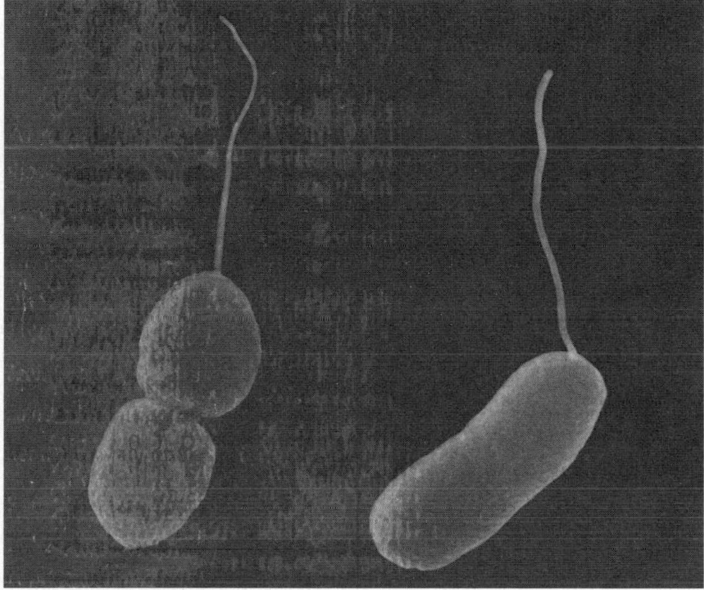

FIGURE 3.13 Flagella (a) Peritrichous flagella on *Salmonella enterica* (TEM). **(b)** Polar monotrichous flagellum on *Vibrio parahaemolyticus* (SEM). a: USDA/Science Source; b: Dennis Kunkel Microscopy/SPL/Science Source

How do flagella function?

peritrichous (*peri* means "around" and *trichous* means "hair-like"), whereas others have a single flagellum at one end—an arrangement called **polar monotrichous** (the structure is commonly referred to as a polar flagellum). Other arrangements also occur, including a tuft of flagella at one or both ends of the cell.

In some cases, flagella are important in disease. For example, *Helicobacter pylori,* the bacterium that sometimes causes gastric ulcers, has powerful multiple flagella at one end of the cell; these allow the bacterial cells to penetrate the thick mucous gel that coats the stomach epithelium.

MicroByte

Flagella can rotate over 100,000 rpm, propelling a bacterium 20 cell lengths per second. This is equivalent to a 6-foot-tall person running 82 mph!

Structure and Arrangement of Flagella

A bacterial flagellum has three basic parts: (**figure 3.14**).

- **Basal body.** This anchors the structure to the cell wall and cytoplasmic membrane; some of its components function together to sense the external environment, harvest the energy of the proton motive force, and cause the flagellum to rotate.
- **Hook.** This is a flexible curved segment that extends out from the basal body, connecting it to the filament.
- **Filament.** This extends out into the external environment and is made up of identical subunits of a protein called flagellin. These subunits form a chain that twists into a helical structure with a hollow core.

Archaea have structures called archaella (singular: archaellum) that are similar to flagella in function, but their composition and structural organization are distinctly different; they were previously called flagella, but the new name was given to emphasize their unique characteristics. Archaella are about half the diameter of flagella and do not have a hollow core, and the energy source used to power their rotation is ATP rather than the proton motive force.

Chemotaxis

Motile bacteria sense the presence of chemicals and respond by moving in a certain direction—a phenomenon called **chemotaxis.** If a chemical is a nutrient, it may serve as an

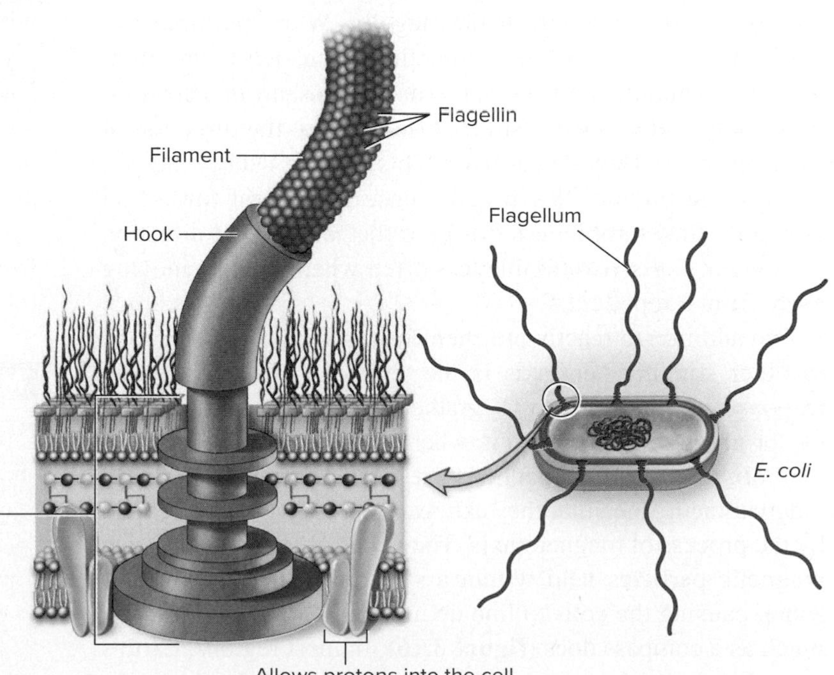

FIGURE 3.14 The Structure of a Flagellum in a Gram-Negative Bacterium The flagellum is composed of three basic parts—a filament, a hook, and a basal body.

What is the role of flagellin in the structure of a flagellum?

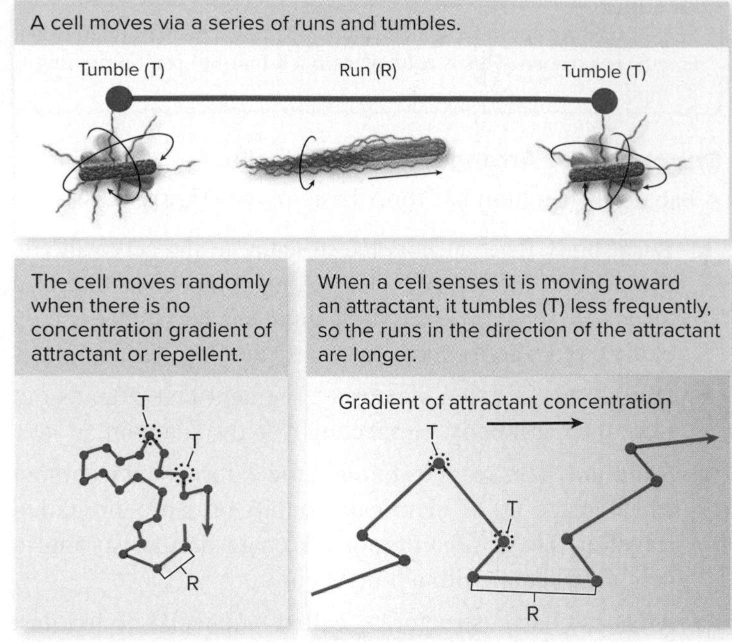

A cell moves via a series of runs and tumbles.
Tumble (T)
Run (R)
Tumble (T)
The cell moves randomly when there is no concentration gradient of attractant or repellent.
When a cell senses it is moving toward an attractant, it tumbles (T) less frequently, so the runs in the direction of the attractant are longer.
Gradient of attractant concentration
T
R

FIGURE 3.15 Chemotaxis

What mechanism causes a cell to tumble?

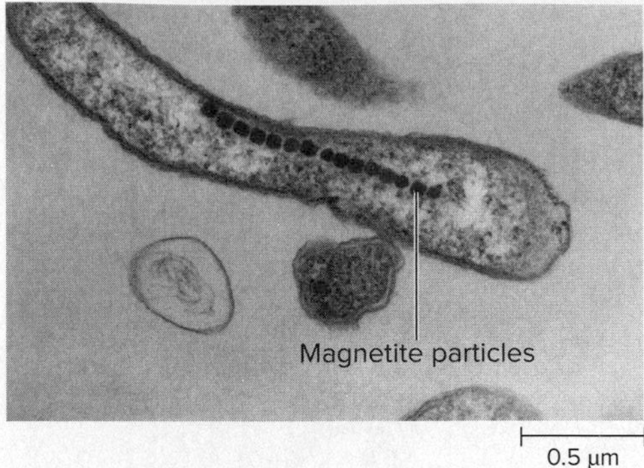

FIGURE 3.16 Magnetotactic Bacterium The chain of magnetic particles (magnetite: Fe_3O_4) within *Magnetospirillum magnetotacticum* helps align the cell along geomagnetic lines (TEM). Dennis Kunkel Microscopy/SPL/Science Source

How could magnetotaxis benefit a cell?

attractant, causing cells to move toward it. On the other hand, if the chemical is toxic, it may act as a repellent, causing cells to move away.

The movement of a bacterial cell toward an attractant is not in a straight line (**figure 3.15**). When an *E. coli* cell travels, it progresses in one direction for a short time, but then stops and tumbles for a fraction of a second. Consequently, the cell usually finds itself oriented in a completely different direction. It then moves in that direction for a short time and tumbles again. The seemingly odd pattern of travel is due to the coordinated rotation of the flagella. When multiple flagella on a single cell rotate counterclockwise, they form a tight propelling bundle, and the bacterium is pushed in a forward movement called a run. After a brief period, the direction of rotation of the flagella reverses. This change causes the cell to stop and tumble. When cells sense movement toward an attractant they tumble less often, so the runs in that direction are longer. Cells also tumble less often when they are moving away from a repellent.

In addition to reacting to chemicals, some bacteria respond to other stimuli. Aerotaxis is the term used to describe the response of organisms to O_2 gradients; organisms that require O_2 for growth move toward it, whereas bacteria that grow only in its absence are often repelled by it. Certain motile bacteria are magnetotactic, meaning they can react to Earth's magnetic field by the process of magnetotaxis. They contain a row of iron-rich magnetic particles held within a structure called a magnetosome, causing the cells to line up in a north-to-south direction much as a compass does (**figure 3.16**). In most regions, Earth's magnetic forces have a significant vertical component, so the magnetotactic cells move downward and into sediments where the O_2 concentration is low—the environment best suited for their growth. Some bacteria can move in response to variations in temperature (thermotaxis) or light (phototaxis).

Pili

Pili (singular: pilus) are considerably shorter and thinner than flagella, and their function is quite different. However, one part of their structure has a theme similar to the filament of a flagellum: a string of protein subunits arranged helically to form a long structure with a hollow core.

Bacteria as a group have several types of pili. Those sometimes referred to as common pili or **fimbriae** allow bacterial cells to attach to specific surfaces. For example, *E. coli* strains that cause watery diarrhea have pili that allow them to attach to epithelial cells that line the small intestine (**figure 3.17**); without the ability to attach, these bacterial cells would simply move along the intestinal tract with the other intestinal contents. Some types of pili help bacterial cells move on solid media. Twitching motility (characterized by jerking movements) and certain types of smooth, gliding motility involve pili. Another type of pilus, called a **sex pilus,** is used to join two bacterial cells as part of a process that transfers DNA from one to the other. This and other mechanisms of DNA transfer are described in chapter 8.

MicroAssessment 3.3

Capsules and slime layers allow organisms to adhere to surfaces and sometimes protect bacteria from certain immune defenses. Flagella allow for bacterial motility in liquid. Chemotaxis is the directed movement of cells toward an attractant or away from a repellent. Pili provide a mechanism for attachment to specific surfaces as well as for movement across surfaces.

7. How do capsules differ from slime layers?
8. *E. coli* cells have peritrichous flagella. What does this mean?
9. Explain why a sugary diet can lead to tooth decay.

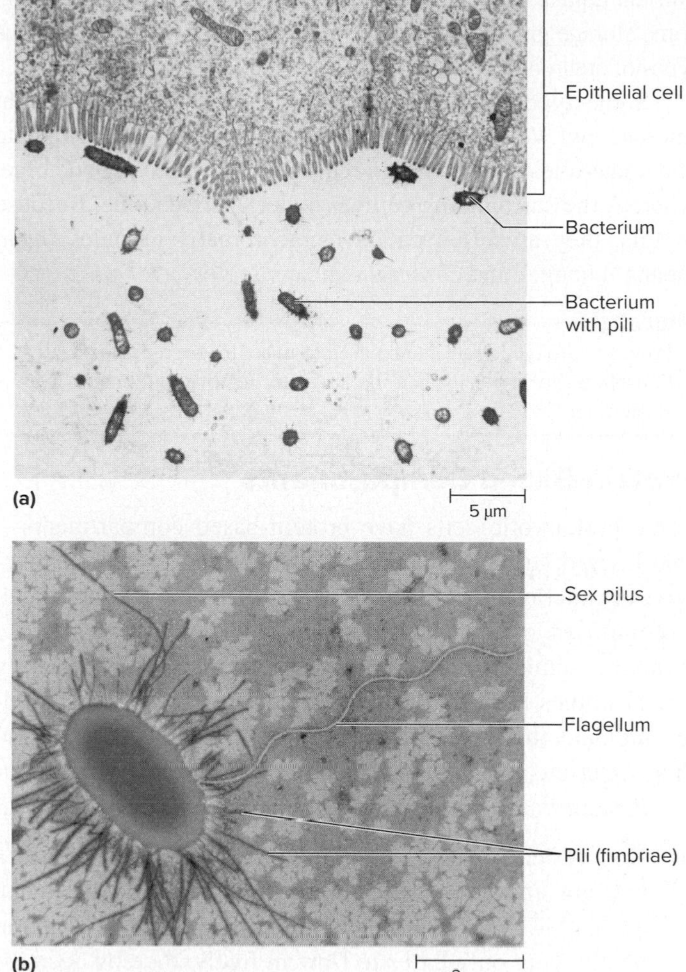

FIGURE 3.17 Pili (a) *Escherichia coli* attaching to epithelial cells in the small intestine of a pig (TEM). **(b)** Pili on an *E. coli* cell. The short pili (fimbriae) allow adherence, whereas the sex pilus is involved in DNA transfer (SEM). a: Harley W. Moon/U.S. Department of Agriculture; b: Dennis Kunkel/SPL/Science Source

? How does the function of pili compare to that of flagella?

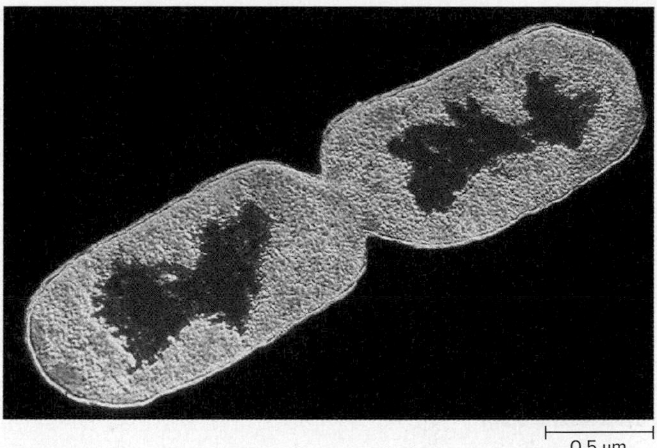

FIGURE 3.18 The Chromosome *Escherichia coli* undergoing cell division, with the DNA shown in red (TEM). CNRI/Science Source

? What is the gel-like region that contains the chromosome called?

3.4 ■ Internal Components of Prokaryotic Cells

Learning Outcomes

14. Describe the structure and function of the chromosome, plasmids, ribosomes, storage granules, protein-based compartments, and endospores.
15. Describe the significance and processes of sporulation and germination.

Prokaryotic cells have a variety of structures within the cell. Some, such as the chromosome and ribosomes, are essential for the life of all cells, whereas others give cells certain selective advantages.

Chromosome and Plasmids

The prokaryotic **chromosome** is typically a single, circular, double-stranded DNA molecule that contains all the genetic information required by a cell, as well as information that may be helpful but not required. Prokaryotic cells are considered to be haploid, a term that means they each carry only a single copy of their chromosome. Although the length of DNA in the prokaryotic chromosome greatly exceeds that of the cell (see chapter 8 opener), the chromosome folds and twists to form a tightly packed mass within the cytoplasm, creating a gel-like region called the **nucleoid** (**figure 3.18**). The compact shape is due partially to nucleoid-associated proteins that bind to DNA, creating a structure that bends and folds (eukaryotic cells have an analogous group of proteins, called histones). In addition, the DNA is twisted, or supercoiled. To understand what is meant by supercoiling, cut a rubber band and then twist one end several times before rejoining the cut ends. The twisting and coiling you see are analogous to supercoiling.

Plasmids have a structure similar to the chromosome, but they are smaller, replicate independently of the chromosome, and typically do not carry essential genetic information. For example, many plasmids code for the production of enzymes that destroy certain antibiotics, allowing the organism to resist the otherwise lethal effects of these medications. Although this genetic information is not useful to a bacterial cell growing in an antibiotic-free environment, it provides an advantage when antibiotics are present. A bacterium can sometimes transfer a copy of a plasmid to another bacterial cell, so this accessory genetic information can spread, which accounts in large part for the increasing frequency of antibiotic-resistant bacteria. A single cell can have more than one type of plasmid, and these can each be present in multiple copies. Additionally, plasmids typically code for multiple proteins.

Ribosomes

Ribosomes are involved in protein synthesis, where they facilitate the covalent joining of amino acids. Prokaryotic ribosomes are referred to as 70S, a term that reflects the relative size and density of the molecules. The "S" (for Svedberg) in the term is a unit used to indicate how fast particles settle when spun at very high speeds in an ultracentrifuge: The faster a particle moves toward the bottom, the higher the S value and the greater the size and density. Note that S units are not strictly additive.

The 70S ribosome is composed of a 30S and a 50S subunit (**figure 3.19**). Prokaryotic ribosomes differ from eukaryotic ribosomes, which are 80S. The fact that they are distinct is important medically because antibiotics that interfere with the function of the bacterial 70S ribosome have no effect on the 80S ribosome. Those antibiotics also have no effect on archaeal cells because archaeal ribosomes differ from their bacterial counterparts, even though they are both 70S.

Cytoskeleton

It was once thought that bacteria lacked a **cytoskeleton,** an interior protein framework that provides support and structure. However, several bacterial proteins that have similarities to those of the eukaryotic cytoskeletal proteins (actin and tubulin, described in section 3.5) have now been characterized, and these appear to be involved in cell division as well as in controlling cell shape.

Storage Granules

A **storage granule** is an accumulation of high-molecular-weight polymers synthesized from a nutrient that a cell has in relative excess. If nitrogen and/or phosphorus are lacking, for example, a cell cannot multiply even if glucose or another good carbon and energy source is plentiful. Rather than waste the carbon/energy source, cells may use it to produce glycogen (a glucose polymer). Later, when conditions are appropriate, cells degrade the glycogen and use the resulting glucose. Other bacterial species store carbon and energy as poly-β-hydroxybutyrate (PHB) or related compounds. Storing nutrients as polymers decreases the number of intracellular molecules, making the concentration gradient for nutrient uptake more favorable. Recent evidence also suggests that some storage granules may allow cells to better withstand various types of environmental stress.

Some types of granules can be easily detected by light microscopy. Volutin granules—a storage form of phosphate and sometimes energy—stain red with the dye methylene blue, whereas the surrounding cellular material stains blue. Because of this, they are often called metachromatic granules (*meta* means "change" and *chromatic* means "color").

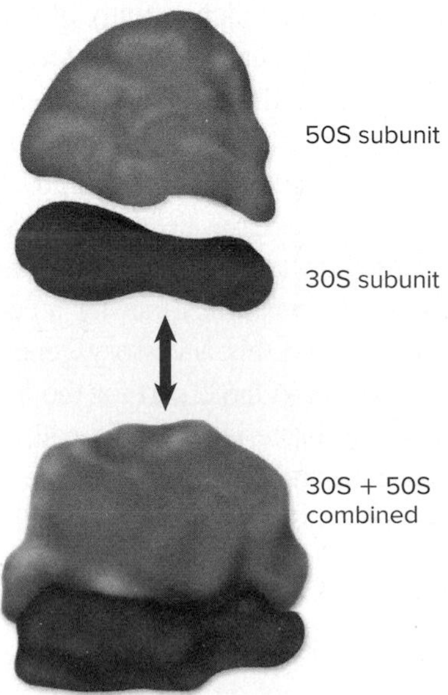

FIGURE 3.19 70S Ribosome The 70S ribosome is composed of 50S and 30S subunits.

? What is the function of ribosomes?

MicroByte

Poly-β-hydroxybutyrate can be used to make biodegradable plastics. Phosphate-storing cells can be used to remove pollutants from wastewater.

Protein-Based Compartments

Some prokaryotic cells have protein-based compartments—small, rigid structures that physically separate certain reactions or functions from the cell's cytosol. The structure's shell is composed of identical protein subunits that assemble to form the semipermeable compartment. Although relatively few examples have been characterized, recent genomic studies indicate that these structures are more widespread and their functions more diverse than previously recognized.

Protein-based compartments include the following:

- **Gas vesicles.** These provide aquatic prokaryotes with a mechanism of adjustable buoyancy, allowing cells to move up or down in the water column. The vesicles function by allowing only gases to flow in freely, thereby decreasing the density of the cell. By regulating the number of vesicles, a cell can float or sink to its ideal position with respect to conditions such as light or O_2 concentration.
- **Bacterial microcompartments (BMCs).** These contain enzymes required for certain metabolic reactions. By confining these in a compartment, the cell prevents unwanted side reactions and protects the cytosol from toxic metabolites. The most well-characterized of the BMCs are carboxysomes, within which carbon fixation occurs in cyanobacteria, a group of photosynthetic bacteria.
- **Encapsulin nanocompartments.** These are the most recently discovered type of compartment and appear to be widespread in bacteria and archaea. They are shaped like many bacterial viruses (bacteriophages), and their function is to hold certain proteins in isolation. For example, some nanocompartments contain proteins that bind free iron, thereby storing this essential mineral in a form that avoids cell damage (free iron can damage the cell). Scientists are exploring the possibility of using the nanocompartments as miniature delivery vehicles for materials, including medications and vaccines.

Endospores

An **endospore** is a unique type of dormant cell produced by certain bacterial species such as members of the genera *Bacillus* and *Clostridium* (**figure 3.20**). Endospores may remain

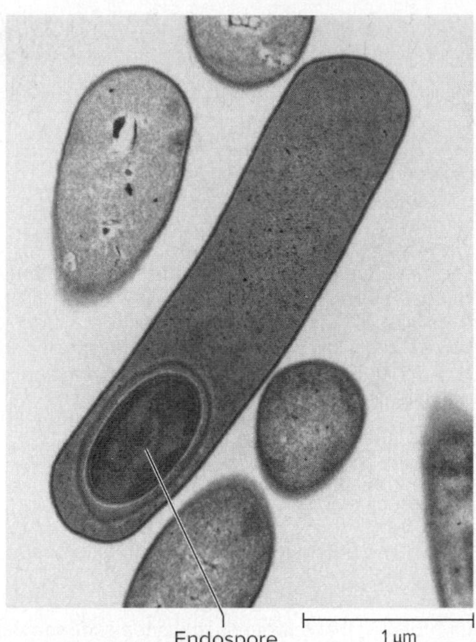

FIGURE 3.20 Endospore *Clostridioides difficile* forming an endospore (TEM). Dr. Kari Lounatmaa/Science Source

? **What is the function of an endospore?**

dormant for over 100 years, and they are extraordinarily resistant to damaging conditions including heat, desiccation, toxic chemicals, and ultraviolet (UV) light. Immersion in boiling water for hours may not kill them. When environmental conditions allow, an endospore can exit the dormant stage to become a typical multiplying cell, called a **vegetative cell,** which is more sensitive to damage from these stresses.

Because endospores can survive so long in a variety of conditions, they can be found virtually everywhere. They are common in soil, which can make its way into environments such as laboratories and hospitals and onto products such as medical devices, food, and media used to cultivate microbes. Keeping these environments and products microbe-free is very important, so special precautions must be taken to avoid or destroy endospores. A variety of diseases, including botulism, tetanus, gas gangrene, pseudomembranous colitis, and anthrax are all caused by endospore-formers.

Endospores are sometimes called spores. However, this latter term is also used to refer to structures produced by unrelated microbes such as fungi. Bacterial endospores are much more resistant to environmental conditions than are other types of spores.

MicroByte

A 500-year experiment started in 2014 is designed to study the survivability of *Bacillus subtilis* endospores.

Sporulation

Endospore formation, or **sporulation,** is a complex sequence of changes that begin when endospore-forming bacteria experience limiting amounts of carbon or nitrogen. The cells sense starvation conditions, which triggers them to begin the 8-hour sporulation process.

As sporulation begins, the cell stops growing, but it duplicates its DNA (**figure 3.21** ①). ② A septum forms, dividing the cell asymmetrically. ③ The larger compartment then engulfs the smaller compartment. These two compartments take on different roles in synthesizing the components that will make up the endospore. ④ As the developing endospore continues to mature, the smaller compartment develops into what is called the forespore, which will become the core of the endospore. Peptidoglycan-containing material is laid down between the two membranes that surround the forespore, forming the core wall and the cortex. Meanwhile, the larger

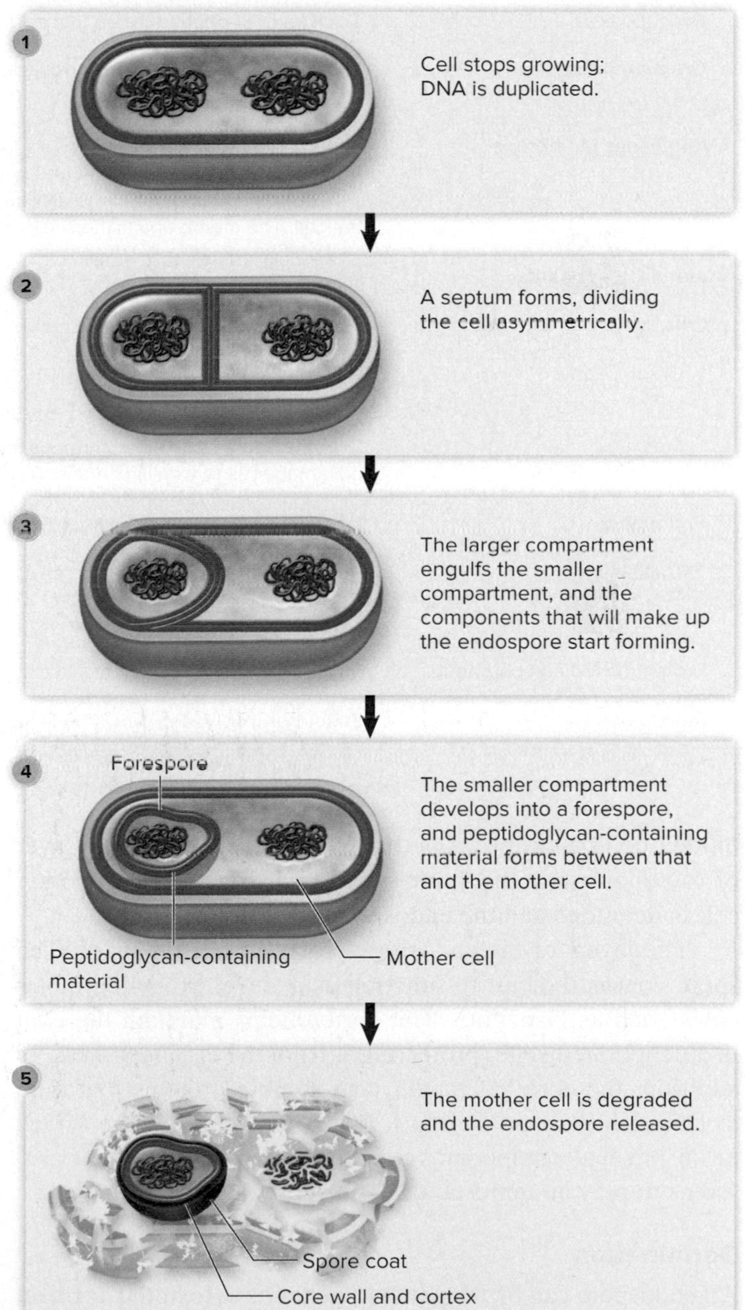

1
Cell stops growing; DNA is duplicated.
2
A septum forms, dividing the cell asymmetrically.
3
The larger compartment engulfs the smaller compartment, and the components that will make up the endospore start forming.
4
Forespore
The smaller compartment develops into a forespore, and peptidoglycan-containing material forms between that and the mother cell.
Peptidoglycan-containing material
Mother cell
5
The mother cell is degraded and the endospore released.
Spore coat
Core wall and cortex

FIGURE 3.21 The Process of Sporulation

? **Approximately how long does the sporulation process take?**

TABLE 3.3 Typical Bacterial Cell Structures

Structure	Characteristics
Structures Outside the Cell Wall	
Filamentous appendages	Long structures composed of protein subunits arranged helically to form a hollow chain.
Flagella	Appendages that provide the most common mechanism of motility.
Pili	Appendages that have different functions depending on the type. The common types, often called fimbriae, allow cells to adhere to surfaces. A few types are used for twitching or gliding motility. Sex pili are involved in DNA transfer.
Capsules and slime layers	Gel-like layers outside the cell wall, usually made of polysaccharide; allow bacteria to adhere to specific surfaces and are often involved in biofilm formation. Capsules are distinct and gelatinous and sometimes allow the bacterium to avoid the body's defenses and thus cause disease; slime layers are diffuse and irregular.
Cell Wall	Rigid layer that prevents the cells from lysing.
Gram-positive	Thick layer of peptidoglycan that contains teichoic acids and lipoteichoic acids.
Gram-negative	Thin layer of peptidoglycan surrounded by an outer membrane. The outer layer of the outer membrane is lipopolysaccharide.
Cytoplasmic Membrane	Biological membrane (phospholipid bilayer embedded with proteins) that surrounds the cytoplasm, separating it from the external environment and functioning as a permeability barrier. Some membrane proteins transport specific substances into and out of the cell, and others transmit information about the external environment to the inside of the cell. The membrane is the site of the electron transport chain.
Internal Components	
DNA	Macromolecule that carries the genetic information of the cell.
Chromosome	DNA molecule that carries the genetic information required by a cell. Typically a single, circular, double-stranded DNA molecule.
Plasmid	DNA molecule similar to a chromosome but generally smaller and carries only genetic information that may be advantageous to a cell in certain situations.
Endospore	Dormant cell type that is extraordinarily resistant to heat, desiccation, ultraviolet light, and toxic chemicals.
Cytoskeleton	Protein framework involved in cell division and control of cell shape.
Granules	High-molecular-weight polymers synthesized from a nutrient available in relative excess.
Protein-based compartments	Small, rigid structures that physically separate certain reactions or functions from the cell's cytosol.
Ribosomes	Structures involved in protein synthesis. Composed of two subunits, 30S and 50S, that join to form the 70S ribosome.

mother cell makes proteins that will form the outermost layer of the endospore, the spore coat. (5) Ultimately, the mother cell is degraded and the endospore is released.

The layers of the endospore protect it from damage. The spore coat is thought to function as a sieve, excluding molecules such as lysozyme. The cortex helps maintain the core in a dehydrated state, protecting it from the effects of heat. In addition, the core has small, acid-soluble proteins that bind to DNA, thereby protecting it from damage. The core is rich in an unusual compound (calcium dipicolinate), which also seems to play an important role in spore resistance.

Germination

An endospore can be triggered to **germinate** (exit the dormant stage and develop into a vegetative cell) by a brief exposure to heat or certain chemicals. Following such exposure, the endospore absorbs water and swells. The spore coat and cortex then crack open, and a vegetative cell emerges. As this happens, the core wall becomes the peptidoglycan layer of the vegetative cell. The process of germination is relatively fast in comparison to sporulation, taking 1–2 hours.

Endospore production and subsequent germination are not a means of cell reproduction. This is because one vegetative cell gives rise to one endospore, which then forms one vegetative cell. There is no increase in the number of cells; the process simply allows the organism to develop into a form that survives many types of adverse conditions commonly encountered in the environment.

Characteristics of typical bacterial cell structures are summarized in **table 3.3.**

MicroAssessment 3.4

The prokaryotic chromosome is usually a circular, double-stranded DNA molecule that contains all the genetic information required by a cell. Plasmids generally encode only information that is advantageous to a cell in certain conditions. Ribosomes facilitate the covalent joining of amino acids to form a protein. The cytoskeleton is an interior framework involved in cell division as well as controlling cell shape. Storage granules are polymers synthesized from nutrients a cell has in relative excess. Protein-based compartments provide a mechanism for the cell to physically separate certain reactions or functions from the cytosol. An endospore is a highly resistant dormant stage produced by certain bacterial species.

10. Explain how glycogen granules benefit a cell.
11. With respect to endospore-forming bacteria, contrast the outcomes of germination and sporulation.
12. What would be the consequence to a cell if its chromosome were damaged beyond repair, and how would the consequence differ if a plasmid were damaged instead?

EUKARYOTIC CELL STRUCTURES AND THEIR FUNCTIONS

Eukaryotic cells are generally much larger than prokaryotic cells, and their internal structures are far more complex (**figure 3.22**). One of their distinguishing characteristics is the presence of membrane-enclosed compartments, or **organelles.** The most significant of these is the nucleus, which contains the cell's genetic information (DNA). The organelles, which can occupy half the total cell volume, allow the cell to perform complex functions in physically separated regions. For example, when degradative enzymes are contained within an organelle, they can digest material without posing a threat to the integrity of the cell itself.

As a group, eukaryotic cells are highly variable. For example, protozoa, which are single-celled organisms, must each function exclusively as a self-contained unit that seeks and ingests food. A protozoan must be mobile and flexible to take in food particles, and it lacks a cell wall that would otherwise provide rigidity. Because animal cells must also be flexible to allow movement, they too lack cell walls. Fungi, on the other hand, are stationary, and a fungal cell benefits from the protection provided by a strong cell wall. Fungal cell walls are composed primarily of polysaccharides, including chitin (a polymer of *N*-acetylglucosamine), glucans (polymers of glucose), and mannans (polymers of mannose). Plants are also stationary, and plant cells each have a cell wall composed of cellulose (a polymer of glucose).

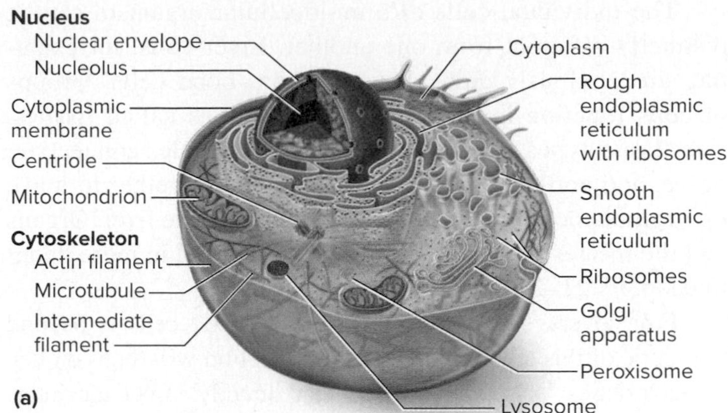

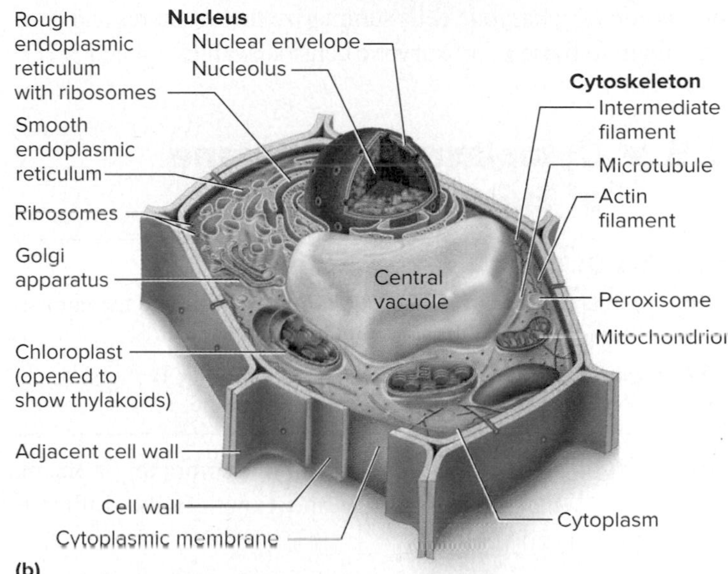

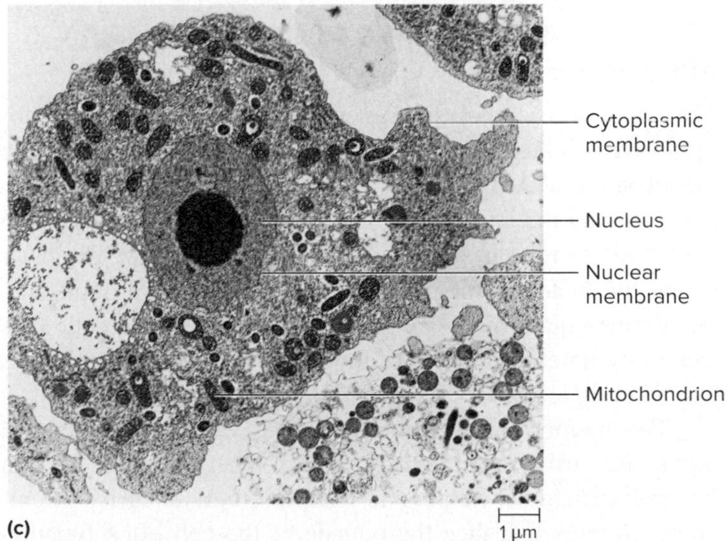

FIGURE 3.22 Eukaryotic Cells **(a)** Diagrammatic representation of an animal cell. **(b)** Diagrammatic representation of a plant cell. **(c)** Electron micrograph of an animal cell shows several membrane-bound structures, including a nucleus and mitochondria. A typical prokaryotic cell is about the size of a mitochondrion (TEM). c: Dr. Thomas Fritsche

Which organelle contains the cell's genetic information?

The individual cells of a multicellular organism can be distinctly different from one another. Liver cells, for example, are obviously quite different from bone cells. Groups of cells function in cooperative associations called **tissues.** The tissue types in your body include muscle, connective, nerve, and epithelial. Various tissues work together to make up **organs,** including the skin, the heart, and the liver. Organs and the tissues that compose them are covered in more detail in chapters 21–27.

Full coverage of all aspects of eukaryotic cells is beyond the scope of this textbook. Instead, this section will focus on key characteristics, particularly those that directly affect the interactions of microbes with human hosts. Tables at the end of the discussion of eukaryotic cells summarize the structures and compare them to those in prokaryotic cells (see tables 3.4 and 3.5).

3.5 ■ Cytoplasmic Membrane of Eukaryotic Cells

Learning Outcomes

16. Compare and contrast the eukaryotic cytoplasmic membrane with the prokaryotic counterpart.
17. Describe the mechanisms eukaryotic cells use to transfer molecules across the cytoplasmic membrane.

All eukaryotic cells have a cytoplasmic membrane, or plasma membrane, which is chemically and functionally similar to that of prokaryotic cells.

Structure and Function of the Cytoplasmic Membrane

The eukaryotic cytoplasmic membrane is a typical phospholipid bilayer embedded with proteins; the layer that faces the cytoplasm, however, has a very different lipid and protein composition than the layer facing the outside of the cell. The same is true for membranes that surround the organelles: The layer facing the inside of the organelle is similar to its cytoplasmic membrane counterpart that faces the outside of the cell. This lack of symmetry reflects the important role these membranes play in complex processes within the eukaryotic cell.

The membrane proteins perform a variety of functions. Some are involved in transport and others are attached to internal structures, helping to maintain cell integrity. Membrane proteins that face the outside of the cell often function as **receptors.** When a given receptor binds a specific molecule, referred to as its **ligand,** the receptor changes shape, which alters its interaction with intracellular components and may lead to a change in the cell's behavior. These receptor–ligand interactions are extremely important in multicellular organisms because they allow cells to communicate with each other—a process called signaling. For example, when certain cells of the human body encounter a compound unique to bacteria, they secrete specific proteins as a call for help in defending against the invader. Other cells of the body have surface receptors for these proteins, allowing them to recognize the signals and respond accordingly. Using this cell-to-cell communication, a multicellular organism can function as a cohesive unit.

The membranes of many eukaryotic cells contain sterols, which provide strength to the otherwise fluid structure. The sterol in animal cell membranes is cholesterol, whereas fungal membranes have ergosterol. This difference is important medically because some antifungal medications act by interfering with ergosterol synthesis or function.

Within the cytoplasmic membrane are distinct areas called lipid rafts (cholesterol-rich regions and associated proteins), which appear to be important in allowing the cell to detect and respond to signals in the external environment. From a microbiologist's perspective, they are also important because many viruses use these regions when they enter or exit a cell.

The cytoplasmic membrane of eukaryotes plays no role in ATP synthesis; instead, that task is performed within organelles called mitochondria. Even though the cytoplasmic membrane has no electron transport chain to generate a proton motive force, an electrochemical gradient across the membrane still exists. This is maintained by energy-requiring mechanisms that pump either sodium ions or protons to the outside of the cell.

Transfer of Molecules Across the Cytoplasmic Membrane

Various substances—including nutrients, signaling molecules, and waste products—must pass through the cytoplasmic membrane. Eukaryotic cells have several mechanisms to accomplish this.

Transport Proteins

The transport proteins of eukaryotic cells function as aquaporins, channels, or carriers. Although water diffuses across membranes directly, aquaporins increase the rate of water passage (as seen in some kidney cells specialized for water reabsorption). **Channels** form small pores in the membrane that allow small substances to diffuse through; the channel has a gate that can open or close in response to environmental conditions, allowing the cell to control passage. **Carriers** are analogous to the proteins in prokaryotic cells that function in facilitated diffusion and active transport. In contrast to the common situation for prokaryotes and other unicellular organisms, however, cells of multicellular organisms often take in nutrients by facilitated diffusion. This is because the organisms can often control the level of nutrients surrounding individual cells. In animals, blood typically has a high enough concentration of glucose for that sugar to enter cells by facilitated diffusion.

Endocytosis and Exocytosis

Many eukaryotic cells use **endocytosis** to take in material too large to fit through transport proteins. The types of endocytosis include the following (**figure 3.23**):

- **Pinocytosis.** This is used by cells to take in liquids along with any dissolved substances. The cell forms invaginations (inward folds) in its cytoplasmic membrane, which pinch off to form endocytic vesicles within the cell. These then fuse with membrane-bound compartments called **endosomes** to enter what is referred to as the endocytic pathway (a set of events that sort the material for either reuse or degradation). As a result, membrane receptors are sent back to the membrane for reuse, whereas foreign materials are sent to organelles called lysosomes (described later) for degradation.
- **Receptor-mediated endocytosis.** This is used by cells to take up extracellular ligands that bind to receptors on the cell's surface. When the receptors bind their ligands, the region is internalized to form a vesicle that contains the receptors along with their bound ligands. The fate of the vesicles is similar to that described for pinocytosis.
- **Phagocytosis.** This is used by phagocytes (a cell type of the body's immune system) and protozoa to ingest bacteria and other relatively large debris. The cells send out arm-like extensions called **pseudopods,** which surround and enclose extracellular material, including bacteria. This action brings the material into the cell in an enclosed compartment called a **phagosome,** the fate of which is similar to that of the vesicles just described.

Exocytosis is the reverse of endocytosis (**figure 3.24**). Secretory vesicles inside the cell fuse with the cytoplasmic membrane and release their contents to the outside of the cell.

MicroByte

Many viruses use receptor-mediated endocytosis to enter animal cells. Each mimics a specific receptor's ligand and, in doing so, "tricks" the cell into bringing it in.

Secretion

As in prokaryotic cells, proteins destined for secretion have a signal sequence: a characteristic amino acid sequence that functions as a tag. When synthesizing a protein with such a sequence, ribosomes attach to the membrane of an organelle called the endoplasmic reticulum (ER). Then, as the polypeptide is synthesized, it is threaded through the membrane and into the ER. From there, the protein can easily be transported by vesicles to a structure called the Golgi apparatus (described later) where the protein may be further modified. The Golgi apparatus then packages these proteins into secretory vesicles for transport to the outside of the cell. Proteins destined for various organelles also have specific amino acid tags.

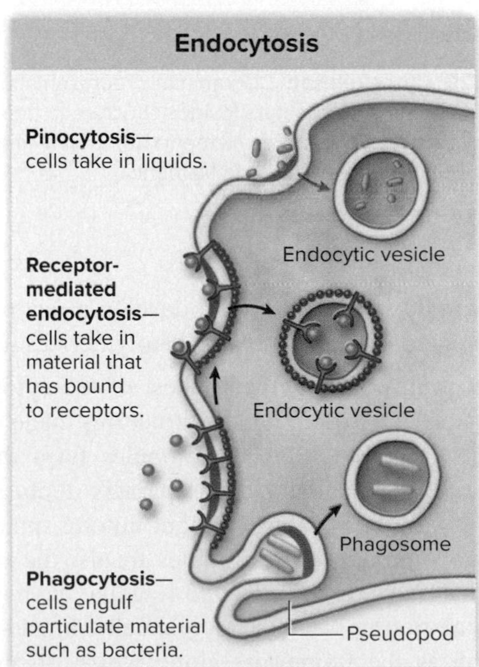

FIGURE 3.23 Endocytosis Cells use this process to take in substances too large to move through a transport protein.

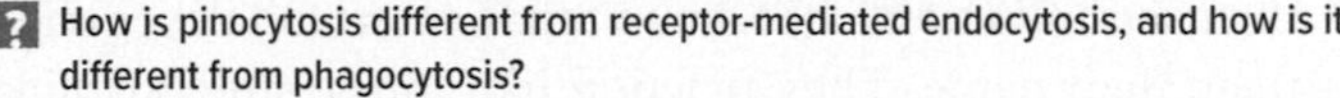

How is pinocytosis different from receptor-mediated endocytosis, and how is it different from phagocytosis?

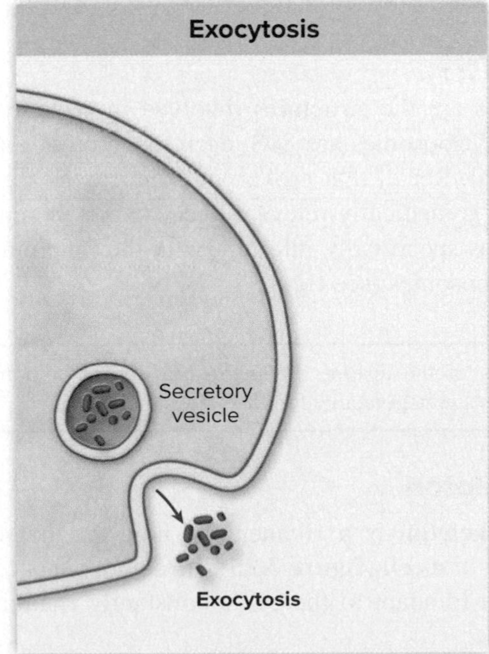

FIGURE 3.24 Exocytosis Cells use this process to remove contents of secretory vesicles from the cell.

Which layer of a secretory vesicle's bilayer membrane would be most similar to the layer of the cytoplasmic membrane that faces the outside of the cell?

MicroAssessment 3.5

The cytoplasmic membrane of eukaryotic cells is an asymmetrical phospholipid bilayer embedded with proteins. Specific receptors allow cell-to-cell signaling. Transport proteins function as either channels or carriers. Cells use pinocytosis to take in liquids along with any dissolved substances and use receptor-mediated endocytosis to bring in ligands bound to cell surface receptors. Protozoa and phagocytes internalize bacteria and debris by phagocytosis. Exocytosis is used to move material to the outside of the cell. Secreted proteins are first made by endoplasmic reticulum-associated ribosomes and then transported in vesicles to the Golgi apparatus. There, the proteins are modified and packaged into secretory vesicles that fuse with the cytoplasmic membrane, resulting in secretion.

13. What is the medical significance of ergosterol in fungal membranes?
14. How is the origin of an endocytic vesicle different from that of a phagosome?
15. How might a bacterium resist the killing effects of the environment within a phagolysosome?

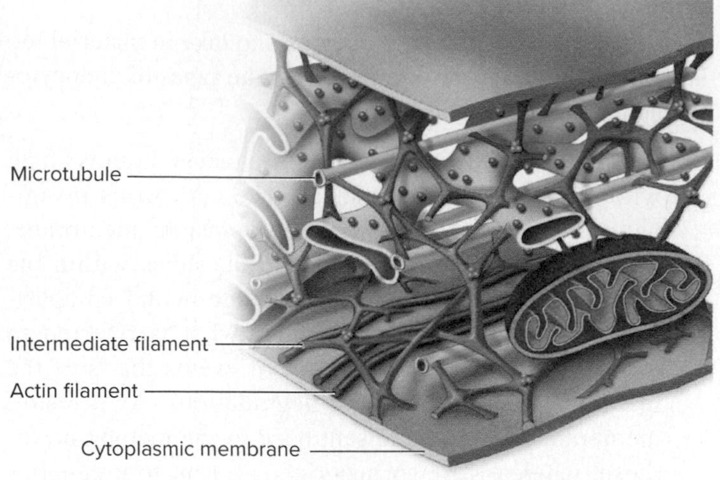

(a) Actin filament

(b) Microtubule

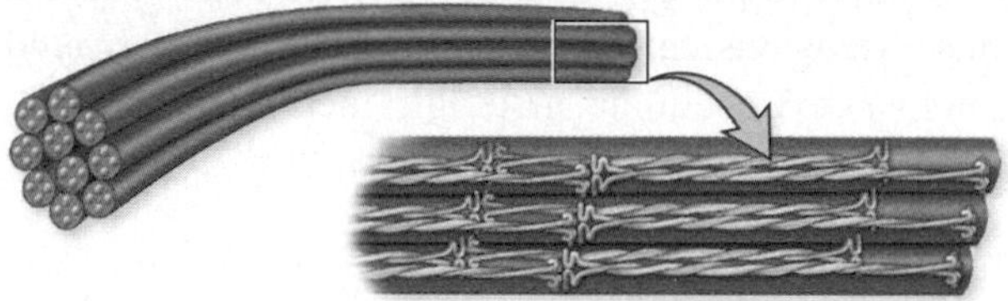

(c) Intermediate filament

FIGURE 3.25 Cytoskeleton Diagrammatic representation of the dynamic filamentous network that provides structure to the cell; the cytoskeleton is composed of three elements: **(a)** actin filaments, **(b)** microtubules, and **(c)** intermediate filaments.

What is the role of actin filaments?

3.6 ■ Protein Structures of Eukaryotic Cells

Learning Outcome

18. Describe the structure and function of eukaryotic ribosomes, the cytoskeleton, flagella, and cilia.

Eukaryotic cells contain a variety of important protein structures, including ribosomes, the cytoskeleton, flagella, and cilia.

Ribosomes

Ribosomes are the structures involved in protein synthesis. Eukaryotic ribosomes are 80S, each made up of a 60S and a 40S subunit. Recall that prokaryotic ribosomes are 70S. The difference is medically relevant because certain antibacterial medications specifically interfere with the function of bacterial 70S ribosomes (see figure 20.7).

MicroByte

Symptoms of the disease diphtheria are caused by a toxin that inactivates a protein required for 80S ribosome function.

Cytoskeleton

The **cytoskeleton** is a filamentous network that forms the framework of a cell (**figure 3.25**). Its components continually reconstruct to adapt to the cell's constantly changing needs, and include:

- **Actin filaments.** These allow the cell cytoplasm to move. The filaments are polymers of protein subunits called actin, which rapidly assemble and subsequently disassemble to cause motion. For example, the pseudopods that form during phagocytosis develop as actin polymerizes in one part of the cell and depolymerizes in another.
- **Microtubules.** These, the thickest of the cytoskeleton's components, are long, hollow structures made of protein subunits called tubulin. Microtubules form the mitotic spindles, the machinery that separates duplicated chromosomes as cells divide. Without mitotic spindles, cells could not reproduce. Microtubules are also the main structures that make up the cilia and flagella, the mechanisms of locomotion in some eukaryotic cells. In addition, they function as the framework along which organelles and vesicles move within a cell.
- **Intermediate filaments.** These filaments are thicker than actin filaments but thinner than microtubules, giving them their name. They function like ropes, strengthening

FOCUS YOUR PERSPECTIVE 3.1

Pathogens Hijacking Actin

Some pathogenic bacteria hijack the function of a eukaryotic cell's actin filaments for their own benefit. The pathogens make proteins that cause actin to polymerize or depolymerize on demand, allowing the bacteria to control movement of the eukaryotic cell's cytoplasm.

Salmonella species manipulate actin in intestinal epithelial cells (the cells that line the intestinal tract) to induce those cells to engulf them. The intestinal cells do not normally take in material using phagocytosis, but the bacteria trick them. The bacteria do this by injecting actin-rearranging proteins into an intestinal cell, which causes pseudopod-like membrane ruffles to form on the intestinal cell's surface (**box figure 3.1**). The ruffles extend outward and then enclose the bacteria, bringing them into the intestinal cell. Once inside an intestinal cell, the *Salmonella* can multiply, protected from components of the body's immune system.

Shigella species enter intestinal epithelial cells using a mechanism similar to that of *Salmonella* species. *Shigella* cells, however, have an additional trick. Once inside an intestinal epithelial cell, the bacteria direct their own transfer to adjacent cells. They do this by causing the host cell's actin to polymerize at one end of the bacterial cell. This forms an "actin tail" that propels the bacterium within the cell. The force of the propulsion is so great that the bacteria are often driven into the cytoplasm of neighboring cells. *Listeria monocytogenes* does the same thing (**box figure 3.2**).

Certain strains of diarrhea-causing *E. coli,* including *E. coli* O157:H7 and a group called enteropathogenic *E. coli,* attach to intestinal epithelial cells and then inject proteins that cause a thick structure—a "pedestal"—to form underneath the attached bacterial cell (**box figure 3.3**). The pedestal is the result of rearrangement of the actin just below the epithelial cell surface and appears to allow the multiplying bacteria to spread to adjacent cells. Unlike the *Salmonella, Shigella,* and *Listeria* previously discussed, however, the *E. coli* cells remain extracellular.

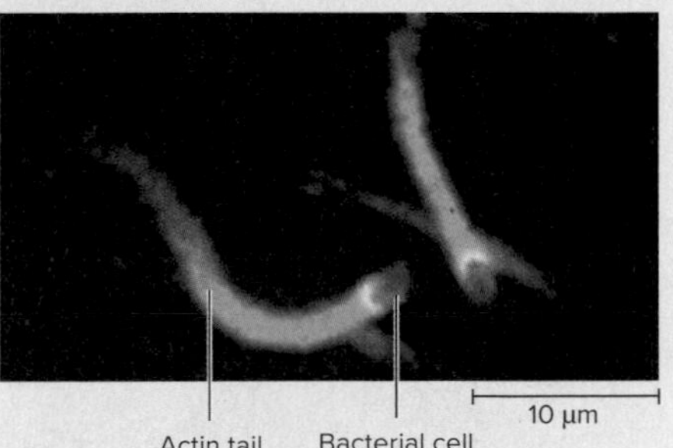

BOX FIGURE 3.2 Actin-Based Motility *Listeria* cells within host cells cause the host cell actin to polymerize at one end of the bacterial cell, propelling the bacterium with enough force to move it into the adjacent host cell. Pascale F. Cossart

BOX FIGURE 3.1 Ruffling *Salmonella* cells enter intestinal epithelial cells by inducing rearrangement of the cells' actin, causing ruffles to form that eventually enclose the bacterial cells. Mark A. Jepson

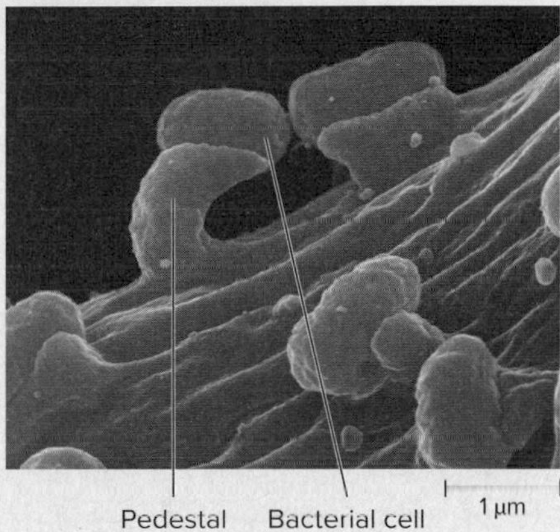

BOX FIGURE 3.3 Pedestal Formation Certain *E. coli* strains attach to intestinal epithelial cells and cause the host cell's actin to rearrange, forming a pedestal beneath the bacterial cell. SPL/Science Source

the cell mechanically, thereby allowing cells to resist physical stresses.

MicroByte

The antifungal medication griseofulvin interferes with the action of microtubules in some fungi.

Flagella and Cilia

Flagella and **cilia** are flexible structures that appear to project out of a cell yet are covered by extensions of the cytoplasmic membrane (**figure 3.26**). They are composed of long microtubules grouped in what is called a 9 + 2 arrangement: nine pairs of microtubules surrounding two individual ones. They originate from structures called basal bodies within the cell; the basal body has a slightly different arrangement of microtubules. Although eukaryotic flagella function in motility in liquid, they are very different from their prokaryotic counterparts both structurally and in how they move. Using ATP as a source of energy, they either propel the cell with a whip-like motion or thrash back and forth to pull the cell forward.

Cilia are shorter than flagella, often covering a cell and moving in synchrony (see figure 1.12). Their motion can move a cell forward in a liquid or propel surrounding material along a stationary cell. For example, epithelial cells that line the respiratory tract have cilia that beat together in a directed fashion. This moves the mucus film that covers those cells, directing it toward the mouth, where it can be swallowed. Because of this action, called the mucociliary escalator, most microbes that have been inhaled are removed before they can enter the lungs.

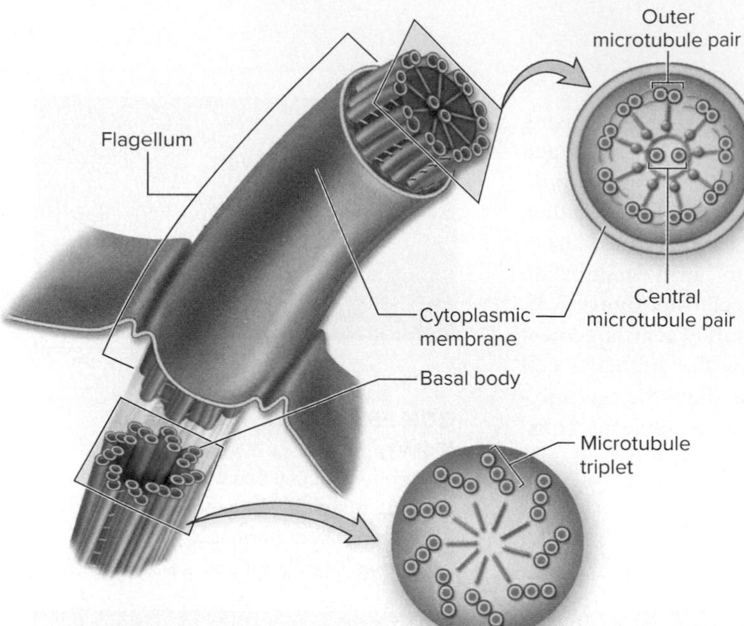

FIGURE 3.26 Flagellum A flexible structure involved in movement.

? How is the structure of a eukaryotic flagellum different from its prokaryotic counterpart?

MicroAssessment 3.6

The 80S eukaryotic ribosome is composed of a 60S and a 40S subunit. The cytoskeleton is a dynamic filamentous network that provides structure to the cell; it is composed of actin filaments, microtubules, and intermediate filaments. Flagella function in motility. Cilia either propel a cell or move material along a stationary cell.

16. Explain how actin filaments are related to phagocytosis.

17. How does the action of cilia on cells that line the respiratory tract prevent microbes from entering the lungs?

18. Why would 60S and 40S ribosomal subunits make an 80S ribosome rather than a 100S ribosome?

3.7 ■ Membrane-Bound Organelles of Eukaryotic Cells

Learning Outcomes

19. Describe the overall function of the endomembrane system and the specific functions of the organelles associated with it.

20. Describe the functions of mitochondria, chloroplasts, and peroxisomes.

Membrane-bound organelles are an important feature of eukaryotic cells that sets them apart from prokaryotic cells. The various membrane-bound organelles of eukaryotes can be divided into two groups: those involved with the endomembrane system and those that are not.

Nucleus and Organelles of the Endomembrane System

The nucleus and a series of organelles called the **endomembrane system** (the endoplasmic reticulum, the Golgi apparatus, lysosomes, and vesicles) work together to transport cellular materials (1) to other organelles within the cell or (2) to the cytoplasmic membrane for secretion from the cell. The materials are carried in membrane-bound vesicles that each form as a section of an organelle buds off, taking contents from the organelle's **lumen** (interior). When the vesicle encounters another organelle or the cytoplasmic membrane, the two membranes may fuse to become one unit, similar to two oil droplets merging. By doing so, the vesicle spills its contents into the lumen of the second organelle or to the outside of the cell (**figure 3.27**).

Nucleus

An important distinguishing feature of a eukaryotic cell is the **nucleus,** which contains the genetic information. The boundary of this structure is the nuclear envelope, composed of two phospholipid bilayer membranes: the inner and outer membranes (**figure 3.28**). Complex protein structures span the envelope, forming nuclear pores. These allow large molecules such as ribosomal subunits and proteins to be transported into and out of the nucleus. The nucleolus is a region of the nucleus where ribosomal RNAs are synthesized.

The nucleus contains multiple chromosomes, each one encoding different genetic information. Unlike the situation in most prokaryotic cells, the eukaryotic chromosomes are linear, and many eukaryotes are diploid, having two copies of each chromosome. The DNA within these chromosomes is wound around proteins called histones, a characteristic that adds structure and order to the long molecule.

Events that take place in the nucleus during cell division distinguish eukaryotes from prokaryotes. After DNA is replicated in eukaryotic cells, chromosomes go through a nuclear division process called mitosis, which ensures that the daughter cells receive the same number of chromosomes as the original parent. Because of mitosis, a cell that is diploid will generate two identical diploid daughter cells. In contrast, a process called meiosis involves two

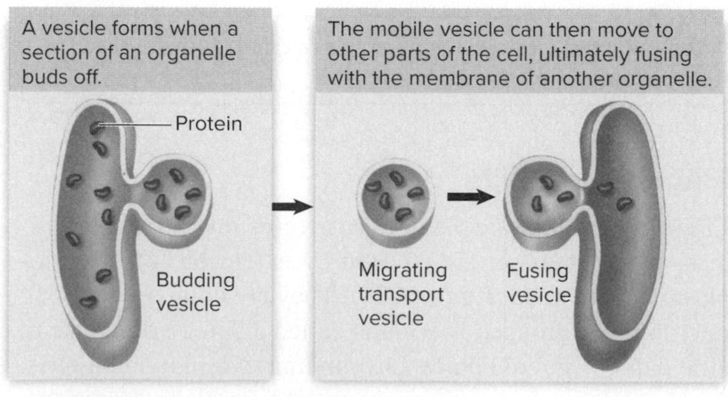

FIGURE 3.27 Vesicle Formation and Fusion

? What is the function of the endomembrane system?

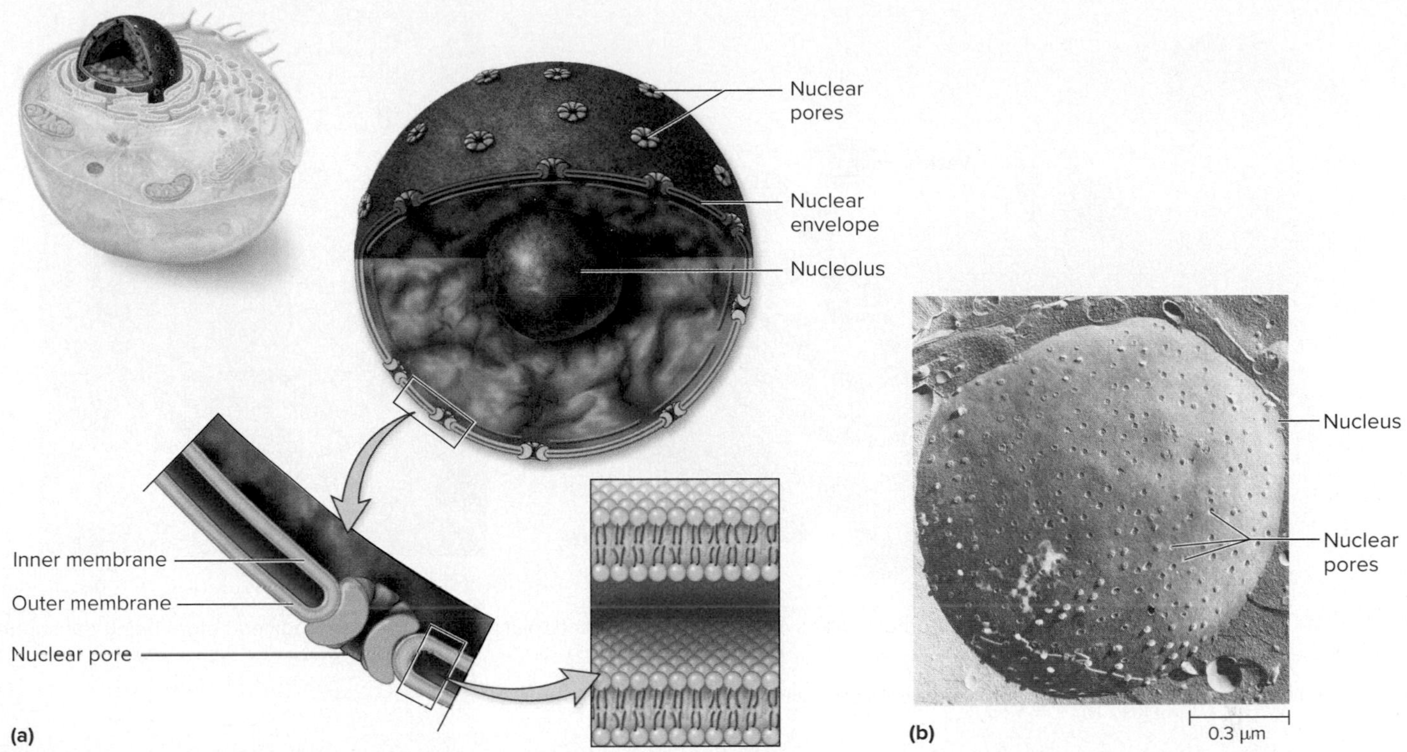

FIGURE 3.28 Nucleus Organelle that contains the cell's genetic information (DNA). **(a)** Diagrammatic representation. **(b)** Electron micrograph of a pig kidney cell by freeze-fracture technique. b: Biophoto Associates/Science Source

? What is the function of nuclear pores?

cell divisions to produce four distinct haploid daughter cells (meaning that each has only a single set of chromosomes). In sexual reproduction, two haploid cells fuse to become one diploid cell.

Some products made in the nucleus—including RNA molecules and partially assembled ribosomal subunits—are used elsewhere in the cell and must be transported through the pores of the nuclear envelope. Although these products are not transported in membrane vesicles, they are all needed for the synthesis of proteins that will be moved through the endomembrane system.

Endoplasmic Reticulum (ER)

The **endoplasmic reticulum (ER)** is a complex system of flattened sheets, sacs, and tubes (**figure 3.29**). The **rough endoplasmic reticulum** has a characteristic bumpy appearance due to the many ribosomes adhering to the surface. It

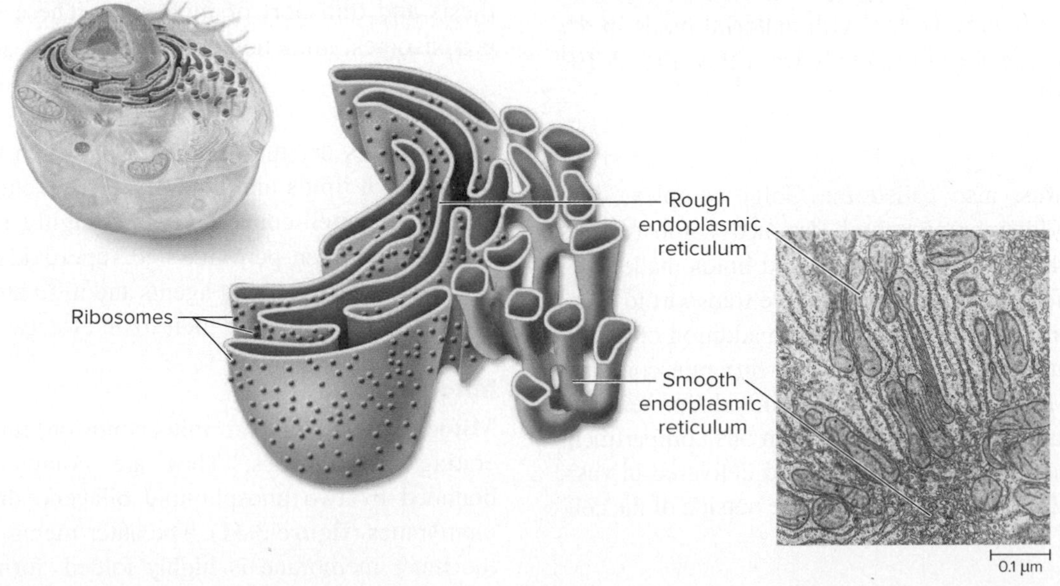

FIGURE 3.29 Endoplasmic Reticulum Site of synthesis of macromolecules destined for other organelles or the external environment. Don W. Fawcett/Science Source

? What causes the bumpy appearance of the rough endoplasmic reticulum?

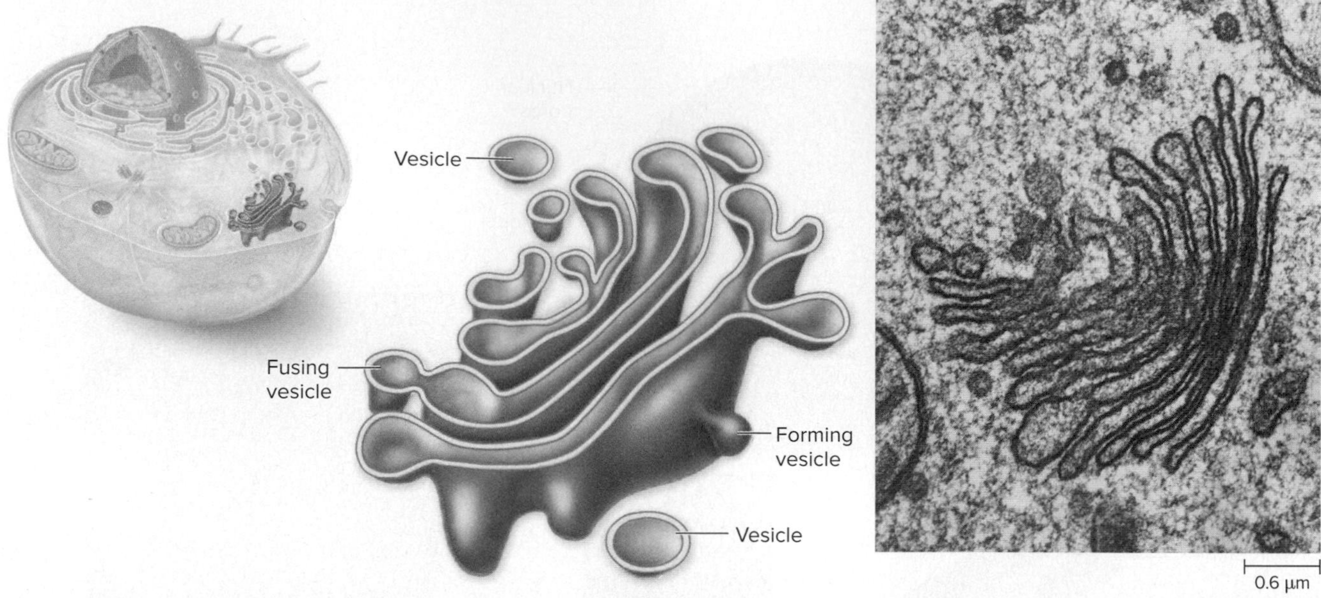

FIGURE 3.30 Golgi Apparatus Site where macromolecules synthesized from the endoplasmic reticulum are modified before being transported to other destinations. Biophoto Associates/Science Source

? How are the modified macromolecules transported from the Golgi apparatus to other sites?

is the site where proteins not destined for the cytoplasm are synthesized, including (1) proteins targeted for secretion, (2) proteins targeted for transfer to an organelle's lumen, and (3) membrane proteins such as receptors. When ribosomes begin making these proteins, they attach to the ER surface so that the polypeptides they are synthesizing can be threaded through gated pores in the membrane. This delivers the molecules to the ER's lumen, where they then fold to assume their three-dimensional shapes. Vesicles that bud off from the ER then transfer the newly synthesized proteins to the Golgi apparatus for further modification and sorting.

Some regions of the ER are smooth. This **smooth endoplasmic reticulum** functions in lipid synthesis and degradation and in calcium ion storage. As with material made in the rough ER, vesicles transfer compounds from the smooth ER to the Golgi apparatus.

Golgi Apparatus

The **Golgi apparatus,** also called the Golgi complex, consists of a series of membrane-bound flattened compartments (**figure 3.30**). It is the site where proteins and lipids made from the endoplasmic reticulum are modified before transport to other destinations. These modifications, such as the addition of carbohydrate and phosphate groups, take place in different compartments of the Golgi. Much like an assembly line, the protein and lipid molecules are transferred in vesicles from one compartment to another. The molecules are then sorted and delivered in vesicles to specific cellular compartments or to the outside of the cell.

Lysosomes

Lysosomes are organelles that contain a number of powerful degradative enzymes that could destroy the cell if not contained within the organelle. Endosomes and phagosomes fuse with lysosomes, allowing materials taken up by the cell to be degraded. In a similar manner, old organelles or vesicles formed around unneeded cellular components can fuse with lysosomes. This process, autophagy, digests and recycles the cell's own contents. The scientist who described the mechanism of autophagy (Yoshinori Ohsumi) was awarded a Nobel Prize in 2016.

Other Membrane-Bound Organelles

Several membrane-bound organelles found in eukaryotic cells are involved in other specific functions separate from the synthesis and transport of molecules. These organelles include peroxisomes, mitochondria, and chloroplasts.

Peroxisomes

Peroxisomes are the organelles in which O_2 is used to help break down lipids and detoxify certain chemicals. The organelle's enzymes convert O_2 into highly reactive molecules such as hydrogen peroxide and superoxide. The peroxisome contains these oxidizing agents and ultimately degrades them, protecting the cell from their toxic effects.

Mitochondria

Mitochondria (singular: mitochondrion) function as ATP-generating powerhouses. They are complex structures each bounded by two phospholipid bilayers: the outer and inner membranes (**figure 3.31**). The outer membrane is smooth, but the inner membrane is highly folded, forming invaginations called cristae. These folds increase the inner membrane's surface area, maximizing the ATP-generating capabilities of the

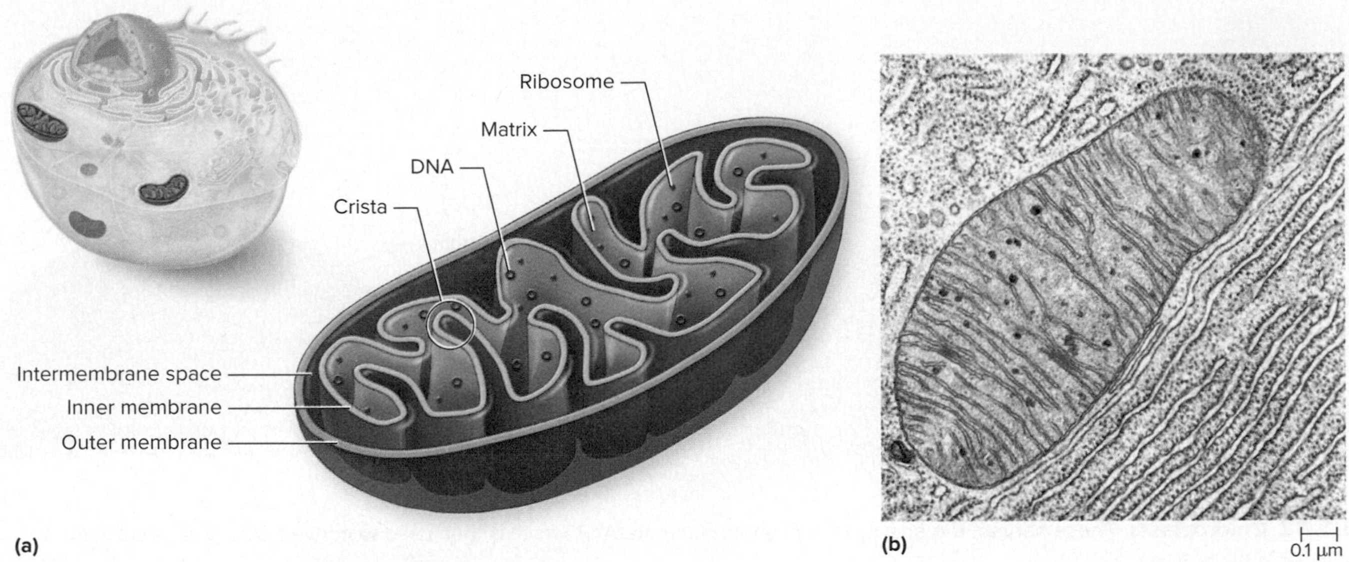

FIGURE 3.31 Mitochondria Organelles that harvest energy released during the degradation of organic compounds to synthesize ATP. **(a)** Diagrammatic representation. **(b)** Electron micrograph. b: Keith Porter/Science Source

? The matrix is which part of the mitochondrion?

organelle (these processes are discussed in chapter 6). The mitochondrial matrix—the gel-like material enclosed by the mitochondrial inner membrane—contains DNA, ribosomes, and other molecules necessary for protein synthesis.

The mitochondrial ribosomes are structurally similar to bacterial ribosomes. This observation, and the fact that mitochondria divide in a fashion similar to that of bacteria, were among the first pieces of evidence that led scientists like Lynn Marguilis to hypothesize that mitochondria evolved from bacterial cells. Significant additional evidence has now elevated that hypothesis to a scientific theory. The **endosymbiotic theory** states that the ancestors of mitochondria as well as chloroplasts (described next) were bacteria residing within other cells in a mutually beneficial partnership. The intracellular bacterium in such a partnership is called an endosymbiont. As time went on, each partner became indispensable to the other, and the endosymbiont eventually lost key features such as a cell wall and the ability to replicate independently.

Several early observations support the endosymbiotic theory. Mitochondria and chloroplasts, unlike other eukaryotic organelles, both carry some of the genetic information necessary for their function. This includes genes for certain components of their ribosomes. These ribosomes contrast with the typical 80S ribosomes that characterize eukaryotic cells and, in fact, are similar to the bacterial ribosomes. In addition, the double membrane that surrounds these organelles is similar to what is seen with present-day endosymbionts, which retain their cytoplasmic membranes and live within membrane-bound compartments in their eukaryotic host cells. Like these endosymbionts, as well as other bacteria, mitochondria and chloroplasts multiply by elongating and then dividing (binary fission).

Evidence in favor of the endosymbiotic theory continues to accumulate. Modern technology allows scientists to easily determine the nucleotide sequence of DNA molecules, making it possible to compare the organelle DNA with genomes of different bacteria. This led to the discovery that some mitochondrial DNA sequences bear a striking resemblance to DNA sequences of members of a group of obligate intracellular parasites, the rickettsias. These are probably relatives of modern-day mitochondria.

Chloroplasts

Chloroplasts, found exclusively in plants and algae, are the site of photosynthesis in eukaryotic cells. They harvest the energy of sunlight to generate ATP, which is then used to convert CO_2 to carbohydrates. Like mitochondria, chloroplasts are each bound by two membranes (**figure 3.32**). Within the chloroplast's stroma—the substance analogous to the mitochondrial matrix—are membrane-bound, disc-like structures called thylakoids. Chlorophyll and other pigments that capture light energy are embedded in the thylakoid membranes. As with mitochondria, substantial evidence indicates that chloroplasts evolved from endosymbiotic bacterial cells—in this case, a group of photosynthetic bacteria called cyanobacteria (see figure 12.14). Chloroplasts contain 70S ribosomes similar to those of bacteria and DNA. In addition, they elongate and divide, and they have photosynthetic mechanisms and DNA sequences similar to cyanobacteria.

Characteristics of eukaryotes are summarized in **table 3.4.** The functions of prokaryotic and eukaryotic cell components are compared in **table 3.5.**

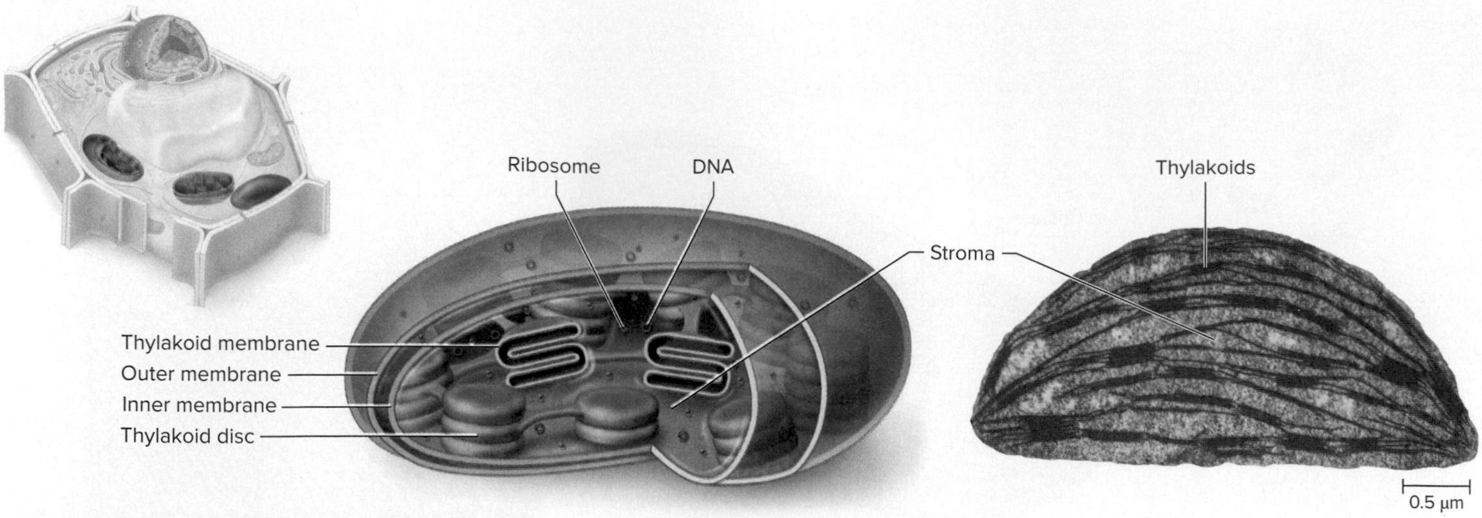

FIGURE 3.32 Chloroplasts These harvest the energy of sunlight to generate ATP, which is then used to convert CO_2 to an organic form.
Dr. Jeremy Burgess/Science Source

? **Scientific evidence indicates that chloroplasts evolved from which group of bacteria?**

TABLE 3.4 Typical Eukaryotic Cell Structures

Structure	Characteristics
Cytoplasmic Membrane	Asymmetrical phospholipid bilayer embedded with proteins; it serves as a permeability barrier, regulates transport, and is involved in cell-to-cell communication.
Protein Structures	
Cilia	Short appendages composed of microtubules covered by extensions of the cytoplasmic membrane that beat in synchrony to provide movement.
Cytoskeleton	Dynamic filamentous network that provides structure to the cell.
Flagella	Long appendages composed of microtubules covered by extensions of the cytoplasmic membrane that propel or push the cell with a whip-like or thrashing motion.
Ribosomes (80S)	Structures composed of two subunits, 60S and 40S, that function to covalently link amino acids.
Membrane-Bound Organelles Involved with the Endomembrane System	
Nucleus	Site where genetic information (DNA) is stored.
Endoplasmic reticulum	Site where proteins and lipids destined for other organelles, the cytoplasmic membrane, or the external environment are synthesized.
Rough	Structure with attached ribosomes that thread proteins they are synthesizing into the lumen of the organelle.
Smooth	Site where lipids are synthesized and degraded and of calcium ion storage.
Golgi apparatus	Site where proteins and lipids destined for other organelles are modified.
Lysosomes	Sites where macromolecules are digested.
Other Membrane-Bound Organelles	
Peroxisomes	Sites where oxidation of lipids and toxic chemicals occur.
Mitochondria	Sites where the energy released during the degradation of organic compounds is harvested to make ATP.
Chloroplasts	Sites where photosynthesis takes place: The energy of sunlight is harvested to generate ATP, which is then used to convert CO_2 to carbohydrates.

TABLE 3.5 Comparison of Typical Prokaryotic and Eukaryotic Cell Structures/Functions

	Prokaryotic	Eukaryotic
General Characteristics		
Size	Cells are generally 0.3–2 μm in diameter.	Cells are generally 5–50 μm in diameter.
Cell division	Chromosome replication is followed by binary fission.	Chromosome replication and mitosis is followed by division.
Chromosome number and location	Single, circular chromosome residing as an irregular mass in the nucleoid region, which is not membrane-bound.	Multiple, linear chromosomes with DNA wrapped around histones, all found in a membrane-bound nucleus.
Structures		
Cytoplasmic membrane	Bilayer that is relatively symmetrical with respect to the phospholipid content.	Bilayer that is highly asymmetrical, with the phospholipid composition of the outer layer differing significantly from that of the inner layer.
Cell wall	Layer composed of peptidoglycan (bacteria), with Gram-negative bacteria also having an outer membrane. Archaea have a variety of cell wall types.	Layer absent in animal cells; its composition in other cell types may include chitin, glucans, and mannans (fungi), and cellulose (plants).
Motility appendages used in liquid	Long protein structures (flagella in bacteria, archaella in archaea) attached to the cell envelope	Long (flagella) and short (cilia) structures each made up of a 9 + 2 arrangement of microtubules that are covered by an extension of the cytoplasmic membrane.
Membrane-bound organelles	Organelles absent.	Organelles present and include the nucleus, endoplasmic reticulum, Golgi apparatus, lysosomes, peroxisomes, mitochondria, and chloroplasts (only in plant cells).
Ribosomes	70S ribosomes, which are made up of 50S and 30S subunits.	80S ribosomes, which are made up of 60S and 40S subunits; mitochondria and chloroplasts have ribosomes similar to those of bacteria.
Functions		
Degradation of extracellular substances	Exoenzymes degrade macromolecules outside the cell. The resulting small molecules are transported into the cell.	Macromolecules can be brought into the cell by endocytosis. Lysosomes carry digestive enzymes.
Motility in liquid	Movement generally involves flagella (bacteria) or archaella (archaea), which rotate like propellers and are composed of protein subunits.	Movement involves cilia and flagella. Cilia move in synchrony; flagella propel a cell with a whip-like motion or thrash back and forth to pull a cell forward.
Protein secretion	Secretion systems transport proteins across the cytoplasmic membrane.	Secreted proteins are moved to the lumen of the rough endoplasmic reticulum as they are being synthesized. From there, they are transported to the Golgi apparatus for processing and packaging. Ultimately, secretory vesicles deliver them to the outside of the cell by the process of exocytosis.
Strength and rigidity	A cell wall and some cytoskeletal components provide these.	A cytoskeleton composed of microtubules, intermediate filaments, and microfilaments provide these; some cells also have a cell wall, and some have membrane sterols.
Transport	Active transport and group translocation are the main transport mechanisms.	Facilitated diffusion, active transport, and ion channels. Cells also take in material by endocytosis.

MicroAssessment 3.7

The endomembrane system is a series of membrane-bound organelles that work together to transport molecules destined for other organelles or for secretion. The nucleus, which contains the genetic information, is a distinguishing feature of a eukaryotic cell. The rough endoplasmic reticulum is the site where proteins not destined for the cytoplasm are synthesized. The smooth ER functions in lipid synthesis and degradation and calcium ion storage. The Golgi apparatus modifies and sorts molecules synthesized in the rough ER. Lysosomes are the organelles within which digestion takes place. Other membrane-bound organelles perform specific functions separate from the endomembrane system. In peroxisomes, O_2 is converted into highly reactive forms of oxygen that break down toxic substances. Mitochondria are ATP-generating powerhouses. Chloroplasts are the site of photosynthesis.

19. Describe the structure of the nucleus.
20. How does the function of the rough endoplasmic reticulum differ from that of the smooth endoplasmic reticulum?
21. Mitochondria have ribosomes similar to those of bacteria, yet antibiotics that target bacterial ribosomes are more toxic to bacteria than to humans. Why might this be so?

METHODS TO OBSERVE CELLS

3.8 ■ Microscopes

Learning Outcomes

21. Discuss the principles and importance of magnification, resolution, and contrast in microscopy.
22. Compare and contrast light microscopes, electron microscopes, and scanning probe microscopes.

One of the most important tools for studying microorganisms is the **light microscope,** which uses visible light and a series of lenses to magnify objects. These instruments are relatively easy to use and can magnify images approximately 1,000× (1,000-fold). They are routinely used in the laboratory to observe cell size, shape, and motility. The **electron microscope,** introduced in 1931, can magnify images in excess of 100,000×, revealing many fine details of cell structure. A major advancement came in the 1980s with the development of a scanning probe microscope, which allows scientists to produce images of individual atoms on a surface. Characteristics of the microscopes described here are summarized at the end of this section (see table 3.6).

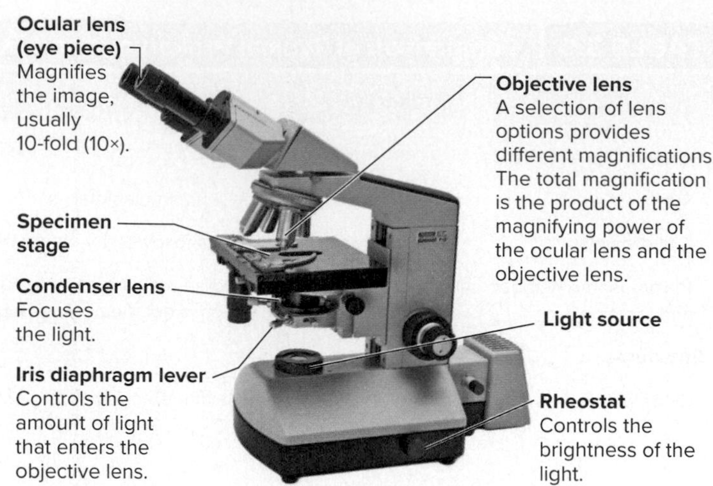
Ocular lens (eye piece)
Magnifies the image, usually 10-fold (10×).
Objective lens
A selection of lens options provides different magnifications. The total magnification is the product of the magnifying power of the ocular lens and the objective lens.
Specimen stage
Condenser lens
Focuses the light.
Light source
Iris diaphragm lever
Controls the amount of light that enters the objective lens.
Rheostat
Controls the brightness of the light.

FIGURE 3.33 A Modern Light Microscope The compound microscope uses a series of magnifying lenses. Scenics & Science/Alamy Stock Photo

? What are the two sets of magnifying lenses called, and how do these relate to total magnification?

Principles of Light Microscopy: Bright-Field Microscopes

In light microscopy, light passes through a specimen and then through a series of magnifying lenses before entering the observer's eye. The most common type of microscope—the **bright-field microscope**—uses light from below to evenly illuminate the field of view, resulting in a bright background against which a dark sample stands out.

Magnification

The modern light microscope, called a **compound microscope,** has multiple magnifying lenses. An **objective lens** is a lens system housed in a tube immediately above the object being viewed, whereas the **ocular lens,** or eyepiece, is a lens close to the eye (**figure 3.33**). Because the objective and ocular lenses are used in combination, the total magnification is equal to the product of their individual magnifications. For example, a structure is magnified 1,000× when viewed through a 10× ocular lens in series with a 100× objective lens. Most compound microscopes have a selection of objective lenses that are of different magnifications—typically 4×, 10×, 40×, and 100×—only one of which can be used at a time. The **condenser lens,** positioned between the light source and the specimen, does not magnify. It focuses the light on the specimen.

Resolution

The usefulness of a microscope depends primarily on its **resolving power,** which determines how much detail or resolution can be seen in the observed specimen. The resolving power of a microscope is a measure of its ability to distinguish two objects that are very close together. The **resolution** of an image refers to its quality as measured by the minimum distance between two points at which those points can still be observed as separate objects. A higher resolving power means that details of an image are clearer and the resulting image has higher resolution.

The resolving power of a microscope depends on the quality and type of lens, the wavelength of the light (shorter wavelengths give better resolution), the magnification, and how the specimen has been prepared. The maximum resolving power of the best light microscope is ordinarily 0.2 μm. This is sufficient to see the general morphology of a prokaryotic cell but too low to distinguish an object the size of most viruses. A recent and exciting advancement in light microscopy was the development of super-resolution microscopes, which use complex illumination mechanisms and merged data to construct an image with increased resolution, sometimes to 10 nm.

To obtain maximum resolution when using certain high-power objectives such as the 100× lens, immersion oil must be used to displace the air between the lens and the specimen. This prevents the refraction (bending of light rays) that occurs when light passes from glass to air (**figure 3.34**). Light rays bend when they pass from a medium of one refractive index (a measure of the relative speed of light as it passes through the medium) to another. If refraction occurs, some light rays will miss the relatively small openings of higher-power objective lenses, causing the image to look fuzzy. The immersion oil prevents refraction because it has nearly the same refractive index as glass, thus forming a contiguous column of material—glass of the slide, immersion oil, glass of the objective lens—all having the same refractive index.

Contrast

The amount of **contrast**—the difference in color intensity between an object and the background—affects how easily cells can be seen. As an example, colorless microorganisms do not stand out against a bright colorless background, and

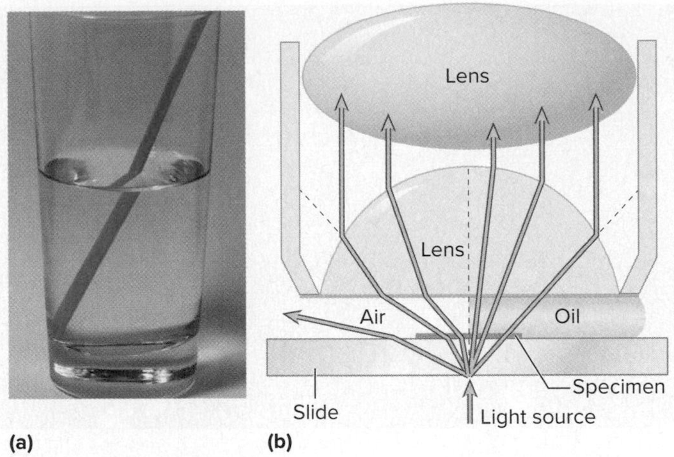

FIGURE 3.34 Refraction As light passes from one medium to another, the light rays will bend if the refractive indices of the two media are different. **(a)** The straw in water appears bent or broken because the refractive index of water is different from that of air. **(b)** Light rays bend as they pass from air to glass because of the different refractive indices of these media. Immersion oil and glass have the same refractive index, and therefore the light rays are not bent. Lisa Burgess/McGraw Hill

? A glass rod submerged in water is easier to see than one in immersion oil. Why?

that lack of contrast makes them harder to see (**figure 3.35**). One way to overcome this difficulty is to stain the cells with one of the various dyes that will be discussed in the next section. The staining process typically kills microbes, however, so it generally cannot be used to observe living cells.

Light Microscopes That Increase Contrast

Special light microscopes that increase the contrast between microorganisms and their surroundings overcome some of the difficulties of observing unstained cells. They allow scientists to easily see characteristics of living organisms such as motility. To view live organisms, the specimen is prepared as a wet mount—a drop of liquid on which a coverslip has been placed.

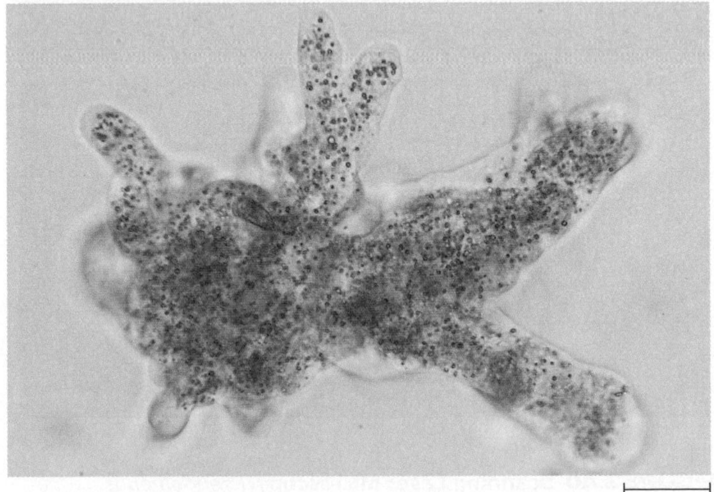

FIGURE 3.35 Bright-Field Microscopy Light illuminates the field evenly, producing a bright background. Note how difficult it is to see the transparent edges of the organism. *Amoeba proteus,* a protozoan. Lisa Burgess/McGraw Hill

? What can be done to increase the contrast before viewing microorganisms using bright-field microscopy?

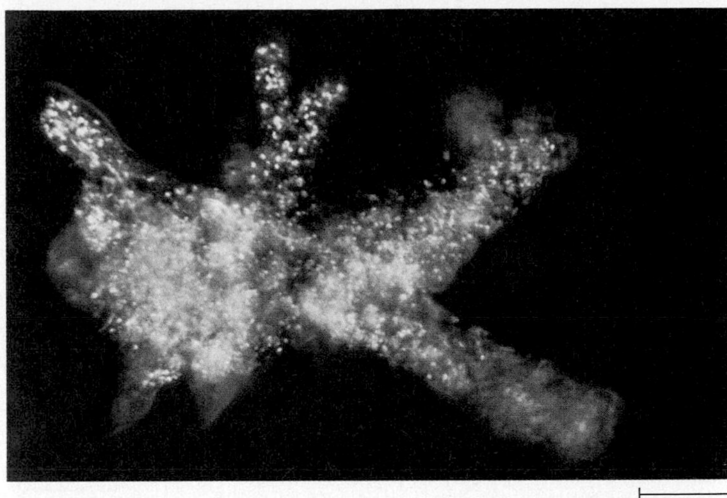

FIGURE 3.36 Dark-Field Microscopy Light enters at an angle, producing a dark background. *Amoeba proteus,* a protozoan. Lisa Burgess/McGraw Hill

? How does a dark-field microscope increase contrast?

Dark-Field Microscopes

Cells viewed through a **dark-field microscope** stand out as bright objects against a dark background (**figure 3.36**). The microscope works in the same way that a beam of light shining into a dark room makes dust visible. For dark-field microscopy, a special mechanism directs light toward the specimen at an angle, so that only light scattered by the specimen enters the objective lens. A simple attachment called a "dark-field stop" can be placed under the condenser lens of a bright-field microscope to temporarily convert it to a dark-field microscope.

MicroByte

Dark-field microscopy is often used to see *Treponema pallidum,* which causes syphilis; the cells can be difficult to see using bright-field microscopy because they are very thin and do not stain well with common dyes.

Phase-Contrast Microscopes

A **phase-contrast microscope** makes cells and other dense materials appear darker than normal (**figure 3.37**). It does this by amplifying the effects of the slight difference between the refractive index of dense material and that of the surrounding medium. As light passes through material, it is refracted a bit differently than when it passes through its surroundings. Special optical devices increase those differences, thereby enhancing the contrast.

Differential Interference Contrast (DIC) Microscopes

A **differential interference contrast (DIC) microscope** makes the image look three-dimensional (**figure 3.38**). This microscope, like the phase-contrast microscope, depends on differences in refractive indices as light passes through

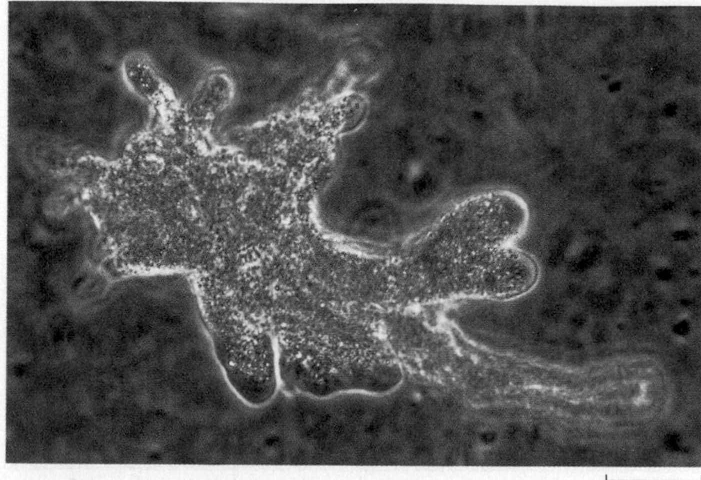

FIGURE 3.37 Phase-Contrast Microscopy Light is manipulated to darken dense material. *Amoeba proteus,* a protozoan. Lisa Burgess/McGraw Hilll

? How does a phase-contrast microscope increase contrast?

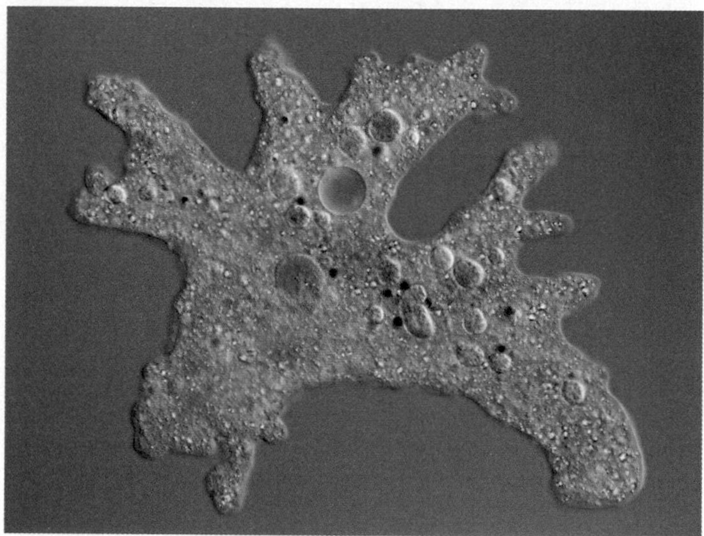

FIGURE 3.38 Differential Interference Contrast (DIC) Microscopy Light is manipulated to make the image appear three-dimensional. *Amoeba proteus,* a protozoan. Micro photo/iStockphoto/Getty Images

? How does a DIC microscope increase contrast?

different materials. It has a device for separating light into two beams that pass through the specimen and then recombine. The light waves are out of phase when they recombine, thus creating the three-dimensional appearance of the image.

Light Microscopes That Detect Fluorescence

Various light microscopes that detect fluorescence are useful in certain situations.

Fluorescence Microscopes

A **fluorescence microscope** is used to observe cells or other materials that are either naturally fluorescent or stained with

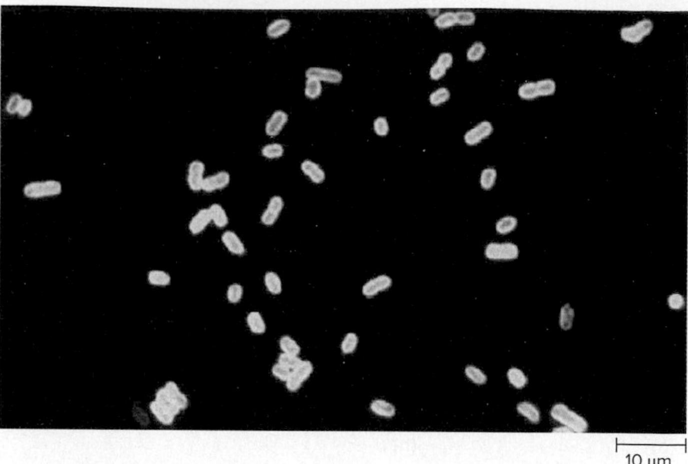

FIGURE 3.39 Fluorescence Microscopy Rod-shaped bacterial cells stained with a fluorescent molecule. Evans Roberts

? What is an epifluorescence microscope?

fluorescent dyes (**figure 3.39**). Fluorescent molecules absorb light at one wavelength (usually ultraviolet light) and then emit light of a longer wavelength. The microscope captures only the light emitted by the fluorescent molecules, allowing fluorescent cells to stand out as bright objects against a dark background. Most fluorescence microscopes used today are epifluorescence microscopes, meaning they project ultraviolet light onto the specimen rather than through it. Because the light does not need to travel through the specimen, cells attached to soil or other opaque particles can be observed.

Scanning Laser Microscopes (SLMs)

Scanning laser microscopes (SLMs) allow scientists to get detailed interior views of intact cells that have been stained with a fluorescent dye (**figure 3.40**). By using fluorescent

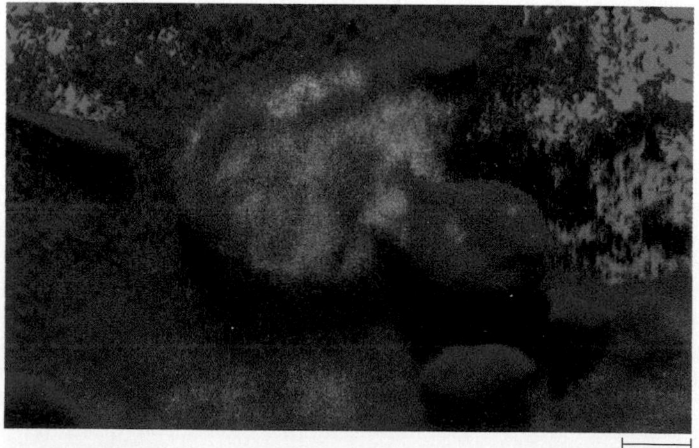

FIGURE 3.40 Scanning Laser Microscopy *Pseudomonas aeruginosa* biofilm (a polymer-encased community of microbial cells) stained with a fluorescent dye that differentiates live cells (green) and dead cells (red). The biofilm had been exposed to an antibacterial chemical, so this confocal image demonstrates that the chemical did not kill all cells in the biofilm. A. Harrer, B. Pitts, P. Stewart/MSU-CBE

? How is two-photon microscopy different from confocal microscopy?

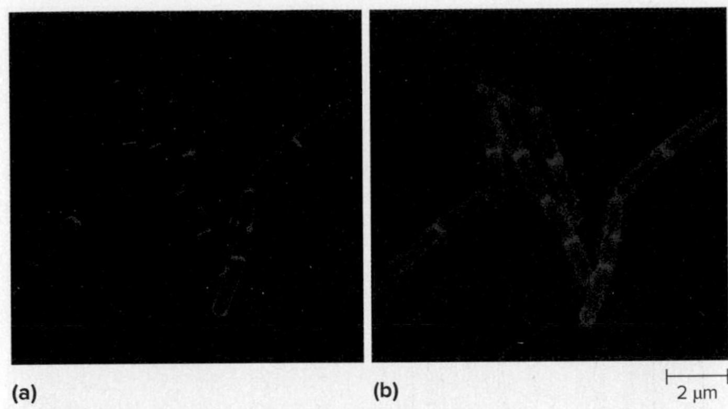

FIGURE 3.41 Super-Resolution Microscopy
Bacillus subtilis stained with a membrane dye (Nile red). **(a)** Super-resolution image. **(b)** Conventional fluorescence image. Dr. Henrik Strahl

? What factors normally limit the resolving power of a light microscope?

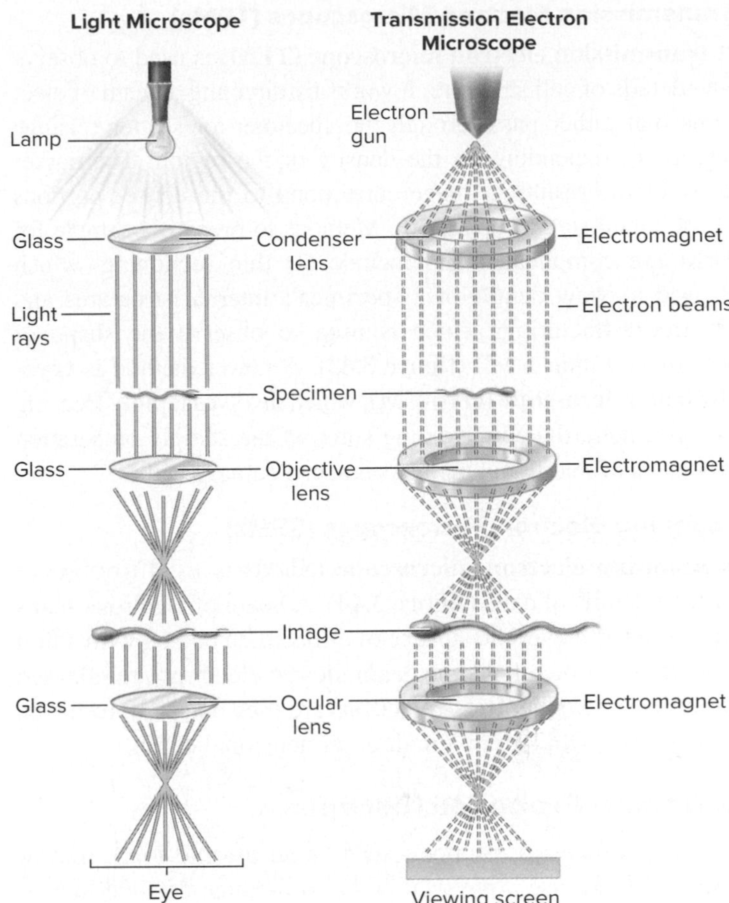

FIGURE 3.42 Principles of Light and Electron Microscopy For the sake of comparison, the light source for the light microscope has been inverted (the light is shown at the top and the ocular lens at the bottom).

? Some electron micrographs are "color enhanced." Why would this be done?

molecules that bind only to certain compounds, the precise intracellular location of those compounds can be determined. The lenses of a **confocal microscope** focus a laser beam to illuminate a given point on one vertical plane of a specimen. Mirrors then scan the laser beam across the specimen, illuminating successive regions and planes until the entire specimen has been scanned. Each plane corresponds to an image of one thin slice of the specimen. A computer then assembles the data and constructs a three-dimensional image, which is displayed on a screen. In effect, this microscope is a miniature computerized axial tomography (CAT) scan for cells. A **two-photon microscope** (also called multiphoton microscope) is similar to a confocal microscope, but lower-energy light is used. This light is less damaging to cells, so time-lapse images of live cells can be obtained. In addition, the light penetrates more deeply, making it possible to get interior views of relatively thick structures.

Super-Resolution Microscopes

As their name suggests, **super-resolution microscopes** achieve higher resolution than do other light microscopes (**figure 3.41**). In recognition of this revolutionary advancement, the developers (Eric Betzig, Stefan Hell, and William Moerner) were awarded a Nobel Prize in 2014. Multiple super-resolution techniques are now available. For example, one uses complicated mechanisms to separately illuminate fluorescent molecules that are otherwise too close together to be seen as distinct. The data obtained are then merged to create an image with increased resolution, sometimes reaching 10 nm.

Electron Microscopes

Electron microscopy is in some ways comparable to light microscopy, but it can clearly magnify an object 100,000×. Rather than using glass lenses, visible light, and the eye to observe the specimen, an electron microscope uses electromagnetic lenses, an electron beam, and a fluorescent screen to produce the magnified image (**figure 3.42**). In addition, rather than using a dye to stain cells, heavy metals such as lead or uranium are used instead. The resulting image can be photographed, creating a picture called an electron micrograph. The image is black-and-white, but it may be artificially colorized to make certain components stand out or to add visual interest. An electron beam has a wavelength about 1,000 times shorter than that of visible light, so the resolving power of electron microscopes is about 1,000× more than light microscopes—about 0.3 nanometers (nm) or 0.3×10^{-3} µm (see figure 1.7). Consequently, considerably more detail can be observed with electron microscopy.

Electron microscopes are complex instruments because the lenses and specimen must typically be in a vacuum to avoid air molecules that would otherwise interfere with the path of electrons. The need for a vacuum means that specimen preparation is complicated and makes it impossible to observe living cells. Techniques are being developed that do not require a vacuum, making it possible to view live cells, but images obtained are not as clear as with conventional methods.

Transmission Electron Microscopes (TEMs)

A **transmission electron microscope (TEM)** is used to observe fine details of cell structure. It works by directing a beam of electrons that either pass through the specimen or scatter (change direction), depending on the density of the region. The darker areas of the resulting image correspond to the denser portions of the metal-stained specimen. Methods to prepare a sample for TEM are complicated and include (1) thin sectioning, which is used to view details of a specimen's internal structures and (2) freeze-fracturing, which is used to observe the shape of structures within a cell (**figure 3.43**). A newer method is **cryo-electron microscopy (cryo-EM),** which involves rapidly freezing the specimen, thereby avoiding some of the sample preparation processes of traditional methods that can damage cells.

Scanning Electron Microscopes (SEMs)

A **scanning electron microscope (SEM)** is used to observe surface details of cells (**figure 3.44**). A beam of electrons scans back and forth over the surface of a specimen coated with a thin film of heavy metal. As the beam moves, electrons are released from the specimen. Those electrons are then detected, resulting in an image with a dramatic three-dimensional effect.

Scanning Probe Microscopes

Scanning probe microscopes, such as an **atomic force microscope (AFM),** use a physical probe to produce detailed images of surfaces (**figure 3.45**). An AFM has a resolving power much greater than that of an electron microscope and does not require special sample preparation. In fact, the instrument can inspect samples either in air or submerged in liquid. The mechanics of an AFM can be compared to that of a stylus mounted on the arm of an old-fashioned record player. A very sharp probe (stylus) moves across the sample's surface, "feeling" the bumps and valleys of the atoms. As the probe scans the sample, a laser measures its motion, and a computer produces a surface map of the sample.

Characteristics of the microscopes described in this section are summarized in **table 3.6.**

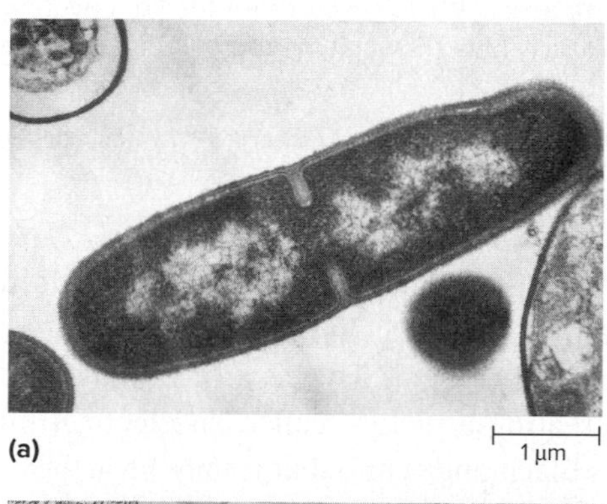

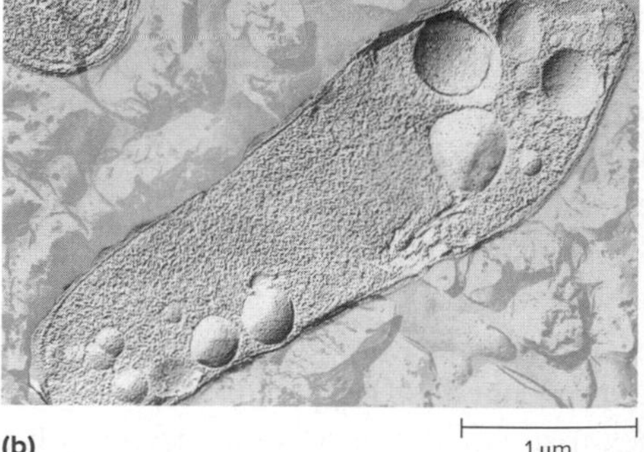

FIGURE 3.43 Transmission Electron Microscopy (TEM)
(a) *Bacillus licheniformis,* prepared by thin-sectioning;
(b) *Pseudomonas aeruginosa,* prepared by freeze-etching.
a: Lee D. Simon/Science Source; b: Dr. Tony Brain/Science Source

? How is thin-sectioning different from freeze-etching?

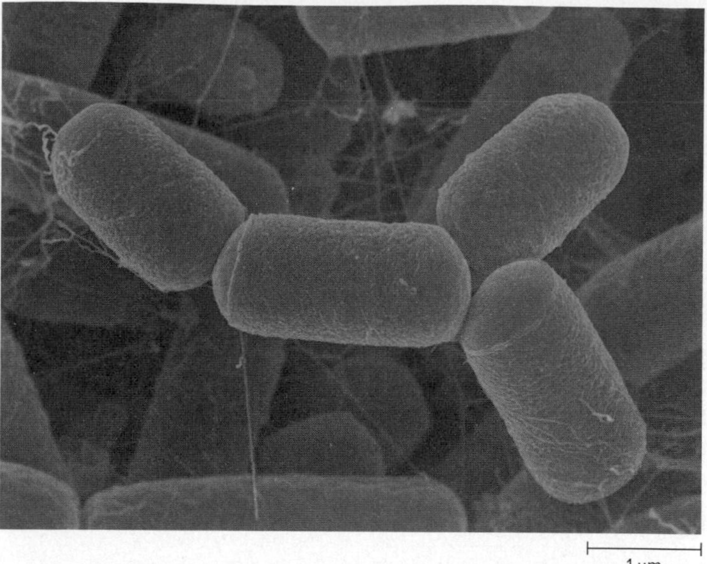

FIGURE 3.44 Scanning Electron Microscopy (SEM)
Methanobrevibacter smithii, an archaeon. Dennis Kunkel Microscopy/SPL/Science Source

? How is scanning electron microscopy different from transmission electron microscopy?

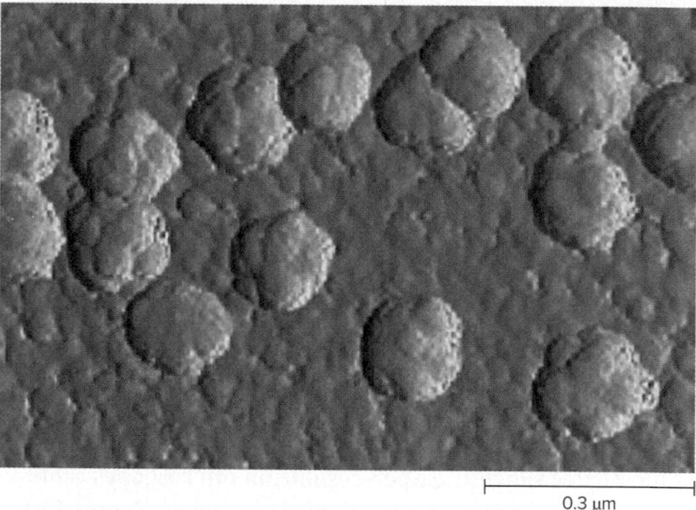

FIGURE 3.45 Scanning Probe Microscopy Atomic force micrograph. SARS virus exiting a host cell. Dr. Mary Ng Mah Lee, National University of Singapore, Singapore/CDC

? How does the resolving power of atomic force microscopy compare to that of electron microscopy?

TABLE 3.6 A Summary of Microscopic Instruments and Their Characteristics

Instrument	Mechanism	Comment
Light Microscopes	Visible light passes through a series of lenses to produce a magnified image.	They are relatively easy to use and much less expensive than electron and scanning probe microscopes.
Bright-field Lisa Burgess/McGraw Hill	The field of view is illuminated evenly, generating a bright background.	This is the most common type of microscope.
Dark-field Lisa Burgess/McGraw Hill	Light is directed toward the specimen at an angle.	Unstained organisms are easier to see because they stand out as bright objects against a dark background.
Phase-contrast Lisa Burgess/McGraw Hill	The differences in refractive indices are amplified, thus increasing the contrast.	Dense material appears darker than normal.
Differential interference contrast Gerd Guenther/Science Source	Two light beams pass through the specimen and then recombine.	The image of the specimen appears three-dimensional.
Fluorescence 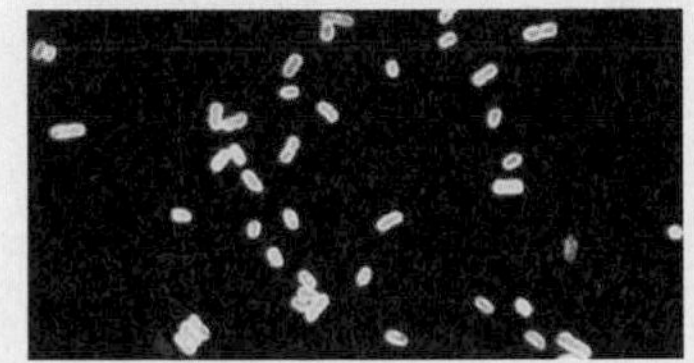 Evans Roberts	Ultraviolet light is projected, causing fluorescent molecules in the specimen to emit light of a longer wavelength.	Cells stained or tagged with a fluorescent dye are observed.

Instrument	Mechanism	Comment
Scanning laser A. Harrer, B. Pitts, P. Stewart/MSU-CBE	Mirrors scan a laser beam across successive regions and planes of a specimen. From that information, a computer constructs an image.	A three-dimensional image of a structure that has been stained with a fluorescent dye is produced, giving detailed sectional views of intact cells.
Super-resolution Dr. Henrik Strahl	Complex illumination mechanisms and merged data are used to construct the image.	A higher resolving power than that of a conventional fluorescence microscope is achieved.
Electron Microscopes	Electron beams are used in place of visible light to produce the magnified image.	Images magnified up to 100,000× are generated.
Transmission Lee D. Simon/Science Source	A beam of electrons is transmitted through a specimen.	Complicated specimen preparation is required.
Scanning Dennis Kunkel Microscopy/SPL/Science Source	A beam of electrons scans back and forth over the surface of a specimen.	Surface details are observed, producing a three-dimensional effect.
Scanning Probe Microscopes	A physical probe is used to produce detailed images of surfaces.	A map showing the bumps and valleys of the sample's surface is produced
Atomic Force Dr. Mary Ng Mah Lee, National University of Singapore, Singapore/CDC	A probe moves in response to even the slightest force between it and the sample.	No special sample preparation required; a three-dimensional effect is produced.

MicroAssessment 3.8

The usefulness of a microscope depends on its resolving power. Light microscopes can magnify objects 1,000×. The most common type of microscope is the bright-field microscope. Dark-field, phase-contrast, and DIC microscopes increase the contrast between a microorganism and its surroundings. Fluorescence microscopes are used to observe microbes stained with a fluorescent dye. Scanning laser microscopes are used to construct three-dimensional images of thick structures that have been stained with a fluorescent dye. Super-resolution microscopes have revolutionized fluorescence microscopy by improving the resolving power. Electron microscopes can magnify an object 100,000×; they include transmission electron microscopes (TEMs) and scanning electron microscopes (SEMs). Scanning probe microscopes, such as atomic force microscopes (AFMs), produce detailed images of surfaces.

22. Why must oil be used to obtain the best resolution with a 100× lens of a modern light microscope?

23. What are some drawbacks of electron microscopes?

24. If an object viewed using the phase-contrast microscope had the same refractive index as the background material, how would it appear?

3.9 ▪ Preparing Specimens for Light Microscopy

Learning Outcomes

23. Describe the principles of a wet mount, a simple stain, the Gram stain, and the acid-fast stain.

24. Describe the special stains used to observe capsules, endospores, and flagella.

25. Describe the benefits of staining with fluorescent dyes and tags.

One of the easiest ways to microscopically examine a specimen is to use a **wet mount.** This consists of a drop of a liquid specimen placed on a microscope slide and overlaid with a coverslip, sandwiching a thin film of liquid between the slide and the coverslip. Although a wet mount makes it possible to observe living, moving microorganisms, the cells can be difficult to see because they are usually colorless and often move around quickly. To avoid this problem, the cells can be immobilized on the slide and then stained with one or more dyes; stained cells are easily seen against a colorless background (imagine looking at a small red ruby versus a colorless diamond in a glass of water).

To prepare a specimen for staining, a drop of liquid containing the organism is placed on a microscope slide and allowed to dry (**figure 3.46**). The resulting specimen forms a film, or smear. The slide is then treated using heat or certain chemicals to fix (attach) the cells to the slide. Heat fixing can easily be done by passing the slide over a flame (smear side up); alternatively the slide can be placed on a slide warmer to both dry and fix the specimen. The cells are then stained, generally using one of the procedures described next; characteristics of the staining procedures are summarized in a table at the end of this section (see table 3.7).

Simple Staining

If only a single dye is used to stain a specimen, the procedure is called **simple staining.** A basic dye is typically used, meaning the dye carries a positive charge. Basic dyes stain cells because the positively charged dye particles are attracted to the many negatively charged cellular components, including peptidoglycan, teichoic acids, and lipopolysaccharides. Examples of basic dyes include methylene blue, crystal violet, safranin, and malachite green; staining protocols that use them are examples of **positive staining** because they actually stain cells. Although acidic dyes do not stain cells, they can be used for **negative staining,** a procedure that colors the background. The cells repel the negatively charged dye, allowing the colorless cell to stand out against the background. An advantage of negative staining is that it can be done as a wet mount, avoiding the heat-fixing step that can distort a cell's shape.

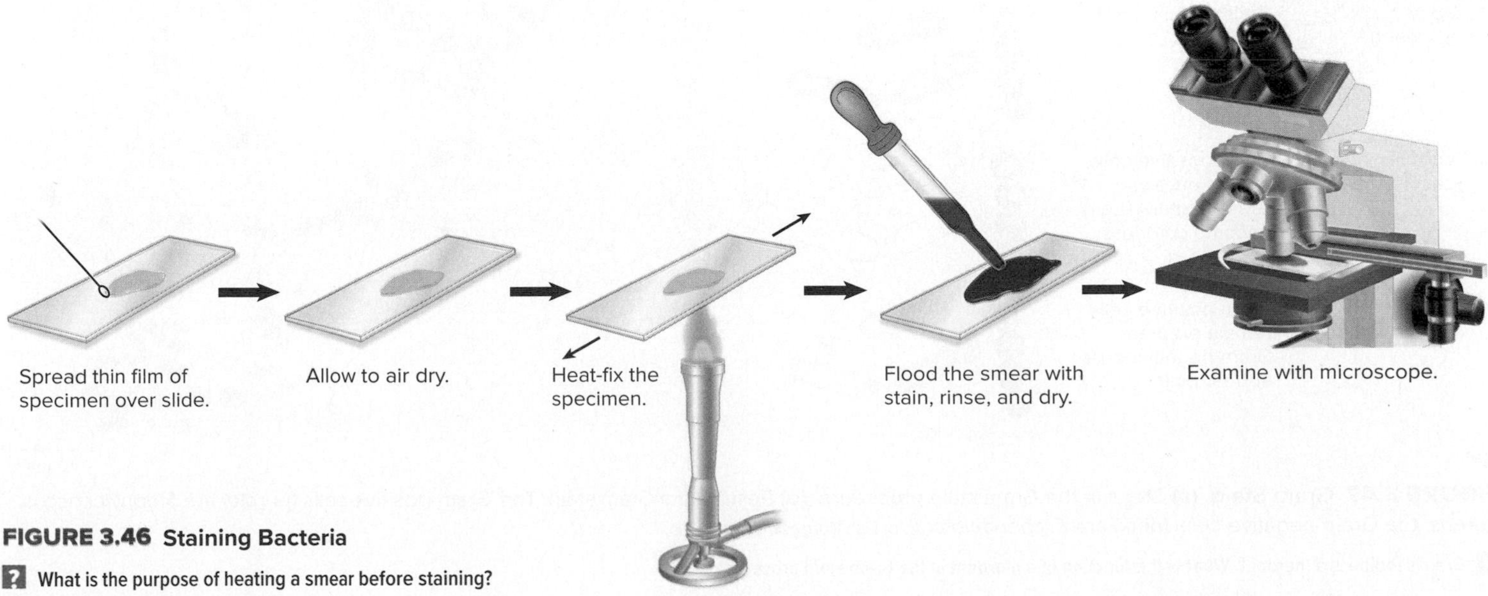

Spread thin film of specimen over slide.
Allow to air dry.
Heat-fix the specimen.
Flood the smear with stain, rinse, and dry.
Examine with microscope.

FIGURE 3.46 Staining Bacteria

What is the purpose of heating a smear before staining?

Differential Staining

Differential staining is used to distinguish different groups of bacteria. The two most frequently used differential staining techniques are the Gram stain and the acid-fast stain.

Gram Stain

The **Gram stain** is by far the most widely used procedure for staining bacteria. The basis for it was developed over a century ago by Dr. Hans Christian Joachim Gram (see **A Glimpse of History,** this chapter). He showed that bacteria can be separated into two major groups: **Gram-positive** and **Gram-negative.** We now know that the difference in the staining properties of these two groups reflects a fundamental difference in the structure of their cell walls, as discussed in section 3.2.

Gram staining involves four steps (**figure 3.47**).

1. The smear is flooded with the **primary stain,** crystal violet in this case. The primary stain is the first dye applied in differential staining and generally stains all cells.
2. The smear is rinsed to remove excess dye and then flooded with a solution called Gram's iodine. The iodine is a mordant, meaning it reacts with the dye to form a complex that is less likely to be washed from the cell.
3. The stained smear is rinsed, and then a **decolorizing agent**—usually 95% alcohol or an acetone alcohol mixture—is briefly added. This quickly removes the dye–iodine complex from the Gram-negative, but not the Gram-positive, bacteria.
4. The smear is rinsed and then a secondary stain, or **counterstain,** is applied to give a different color to the now colorless Gram-negative bacteria. For this purpose, the red dye safranin is used, which stains the cells pink. This dye enters Gram-positive cells as well, but because they are already purple, it makes little difference to their appearance.

The type of bacterial cell wall accounts for the Gram stain reaction, but most studies indicate that it is the inside of the cell, not the wall, that is stained by crystal violet (**figure 3.48**). Gram-positive cells retain the dye because their cell wall prevents the crystal violet–iodine complex from being washed out during decolorization. The decolorizing agent dehydrates the thick layer of peptidoglycan, and in this desiccated state the wall acts as a permeability barrier that prevents the dye from leaving the cell. In contrast, the solvent action of the decolorizing agent easily damages the outer membrane of Gram-negative bacteria. The relatively thin layer of peptidoglycan, which the decolorizing agent also damages, cannot hold back the dye complex. This model was recently challenged by a group of scientists who suggested that the crystal violet dye does not cross the cytoplasmic membrane to enter cells, but instead slowly enters peptidoglycan, where it becomes trapped. However, they used a substantially lower concentration of crystal violet than that used in the standard Gram staining protocol.

To obtain reliable results, the Gram stain must be done properly. One common mistake is to decolorize a smear for the incorrect amount of time. Even Gram-positive cells can lose the crystal violet–iodine complex during prolonged decolorization. When this happens, over-decolorized Gram-positive cells will appear pink after counterstaining. In contrast, under-decolorizing results in Gram-negative cells appearing purple. Another important consideration is the age of the culture. As bacterial cells age, they lose their ability to

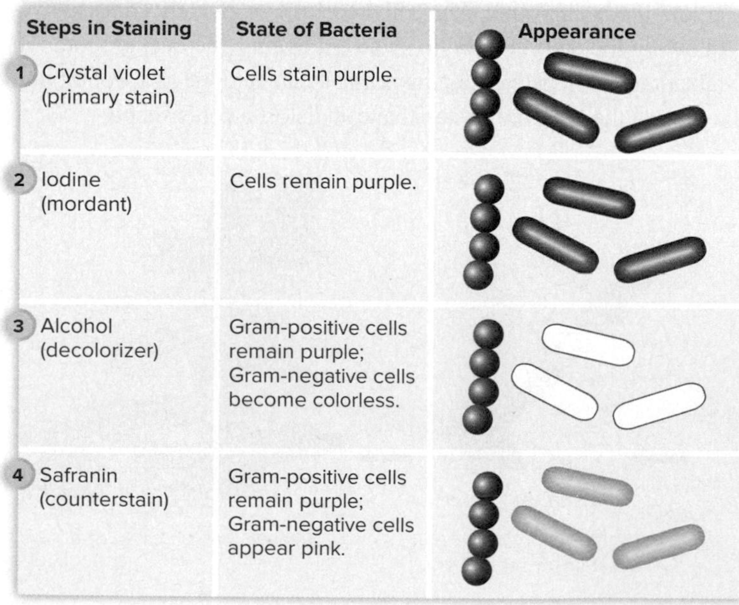

Steps in Staining
State of Bacteria
Appearance
1 Crystal violet (primary stain)
Cells stain purple.
2 Iodine (mordant)
Cells remain purple.
3 Alcohol (decolorizer)
Gram-positive cells remain purple; Gram-negative cells become colorless.
4 Safranin (counterstain)
Gram-positive cells remain purple; Gram-negative cells appear pink.
(a)

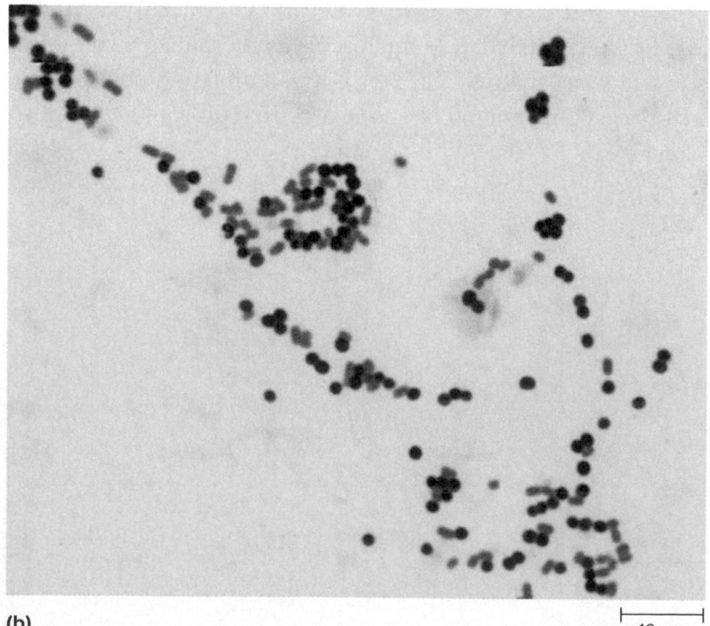

FIGURE 3.47 Gram Stain (a) Steps in the Gram stain procedure. **(b)** Results of a Gram stain. The Gram-positive cells (purple) are *Staphylococcus aureus;* the Gram-negative cells (pink) are *Escherichia coli.* b: Lisa Burgess/McGraw Hill

? **Gram's iodine is a mordant. What is the function of a mordant in the Gram stain procedure?**

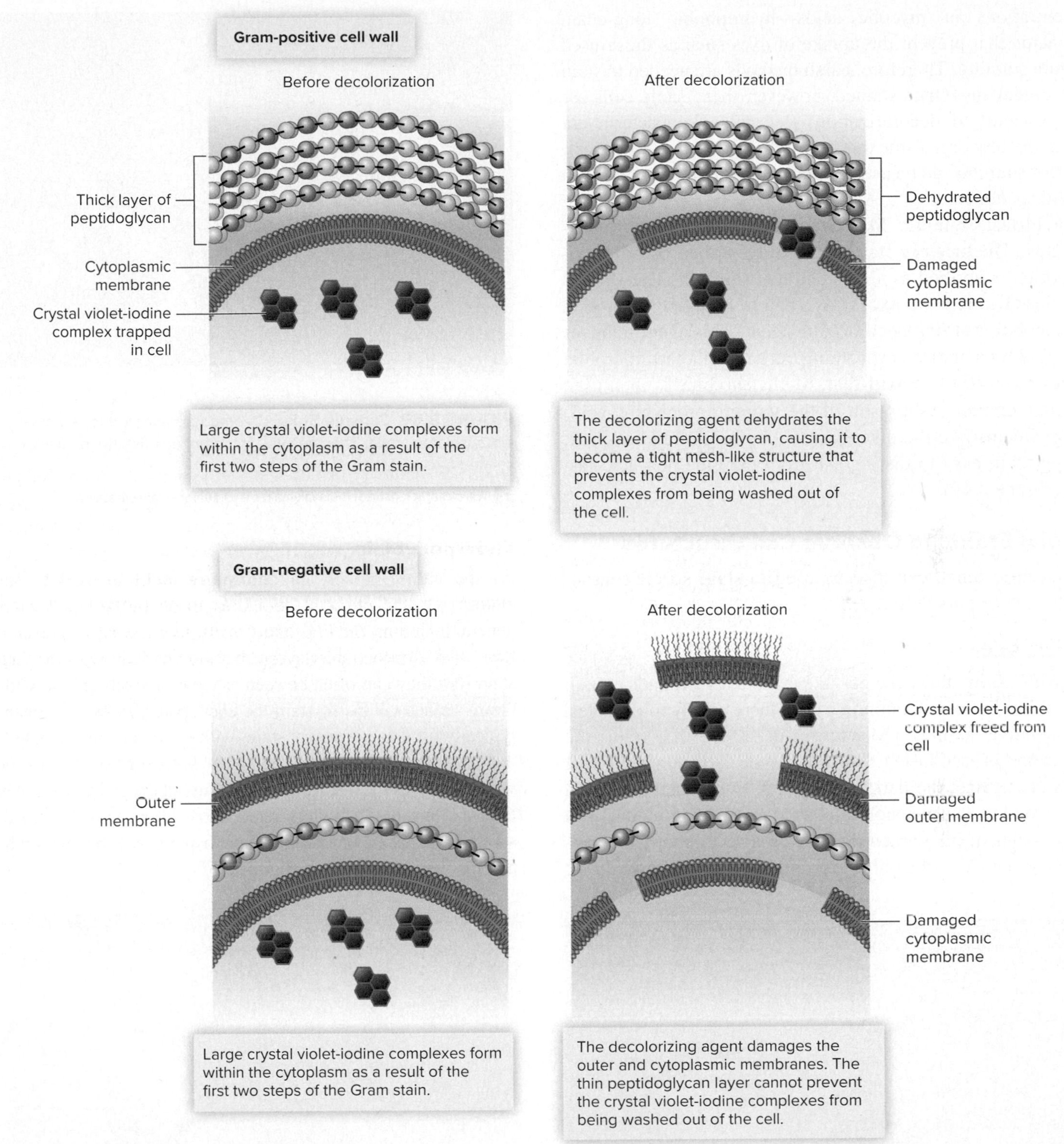

Gram-positive cell wall
Before decolorization
After decolorization
Thick layer of peptidoglycan
Cytoplasmic membrane
Crystal violet-iodine complex trapped in cell
Dehydrated peptidoglycan
Damaged cytoplasmic membrane
Large crystal violet-iodine complexes form within the cytoplasm as a result of the first two steps of the Gram stain.
The decolorizing agent dehydrates the thick layer of peptidoglycan, causing it to become a tight mesh-like structure that prevents the crystal violet-iodine complexes from being washed out of the cell.
Gram-negative cell wall
Before decolorization
After decolorization
Outer membrane
Crystal violet-iodine complex freed from cell
Damaged outer membrane
Damaged cytoplasmic membrane
Large crystal violet-iodine complexes form within the cytoplasm as a result of the first two steps of the Gram stain.
The decolorizing agent damages the outer and cytoplasmic membranes. The thin peptidoglycan layer cannot prevent the crystal violet-iodine complexes from being washed out of the cell.

FIGURE 3.48 Cell Wall Type and the Gram Stain

? The decolorizing agent damages the peptidoglycan layer. If cells of a Gram-positive bacterium were decolorized for a prolonged period, how might this affect the Gram stain results?

retain the crystal violet–iodine dye complex. As a result, cells from old cultures may appear pink. Thus, Gram staining fresh cultures (less than 24 hours old) gives more reliable results. Lastly, Gram staining works best using an even smear (not too thick nor too thin) so that the various reagents can access all cells equally and react with them similarly.

Acid-Fast Stain

The **acid-fast stain** is a procedure used to detect a small group of organisms that do not readily take up dyes. Among these are members of the genus *Mycobacterium,* including a species that causes tuberculosis and one that causes Hansen's disease (leprosy). The cell wall of these bacteria contains high concentrations of mycolic acids—hydrophobic long-chain fatty acids that prevent the uptake of dyes such as those used in Gram staining. Therefore, harsh methods are needed to stain these organisms. Once stained, however, these same cells are very resistant to decolorization. Because mycobacteria are among the few organisms that retain the dye in this procedure, acid-fast staining can be used to presumptively identify them in clinical specimens.

Acid-fast staining, like Gram staining, requires multiple steps. The primary stain is a red dye called carbol fuchsin, which is applied as a concentrated solution. The slide is then rinsed to remove excess dye before being flooded with acid-alcohol, a strong decolorizing agent. This agent removes the carbol fuchsin from everything but cells of a few microbial species referred to as **acid-fast.** Methylene blue is then used as a counterstain. As a result of the staining procedure, acid-fast organisms like the mycobacteria are bright reddish-pink, making them easy to distinguish from the blue non-acid-fast cells (**figure 3.49**).

Special Stains to Observe Cell Structures

Special procedures can also be used to stain specific structures inside or outside the cell.

Capsule Stain

A **capsule stain** allows the gel-like layer that surrounds certain microbes to be seen. Common dyes adhere poorly to capsules, so a negative staining procedure using a dye that stains the background is used instead; this makes the capsule visible as it stands out against the dark background. In one method, India ink is added to a suspension of cells to make a wet mount. The fine particles of ink darken the background, allowing the capsule to be seen as a clear area surrounding a cell (**figure 3.50**).

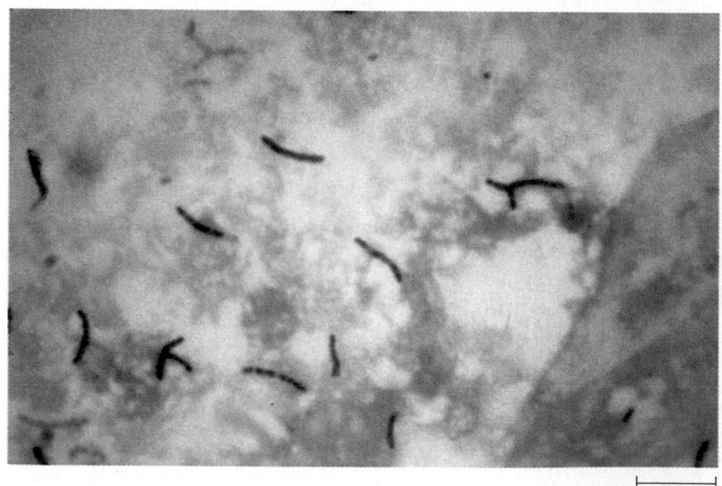

FIGURE 3.49 Acid-Fast Stain *Mycobacterium* cells retain the red primary stain, carbol fuchsin. Counterstaining with methylene blue colors the non-acid-fast cells blue. Dr. George P. Kubica/CDC

? What characteristic of *Mycobacterium* cells makes them acid-fast?

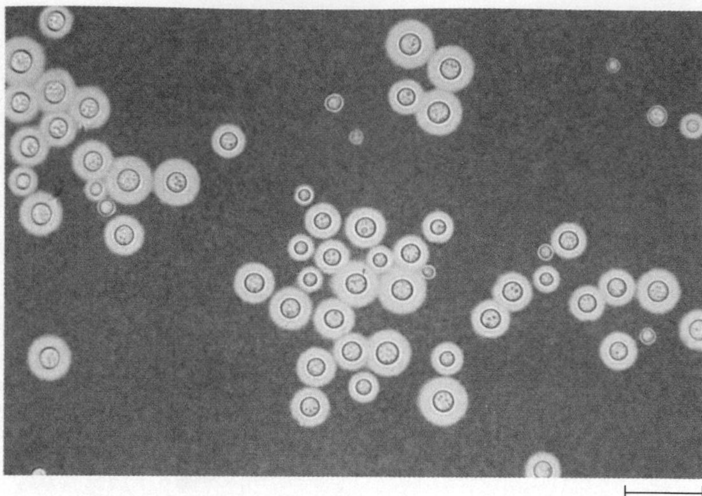

FIGURE 3.50 Capsule Stain *Cryptococcus neoformans,* an encapsulated yeast. The capsules stand out against the India ink–stained background. Dr. Leanor Haley/CDC

? What characteristic of India ink makes it useful for a capsule stain?

Endospore Stain

As the name implies, an **endospore stain** is used to stain endospores, the dormant structures made by certain bacterial genera, including *Bacillus* and *Clostridium.* Because endospores have thick impenetrable layers, they do not stain with the Gram stain (but they can often be seen as clear, smooth objects within Gram-stained cells). To stain the endospores, malachite green is typically used as a primary stain, with gentle heating to help the dye enter the endospores. The smear is then rinsed with water, which removes the dye from everything but the endospores. After that, the smear is counterstained, most often with the red dye safranin. Using this method, the endospores will be green while all other cells will be pink (**figure 3.51**).

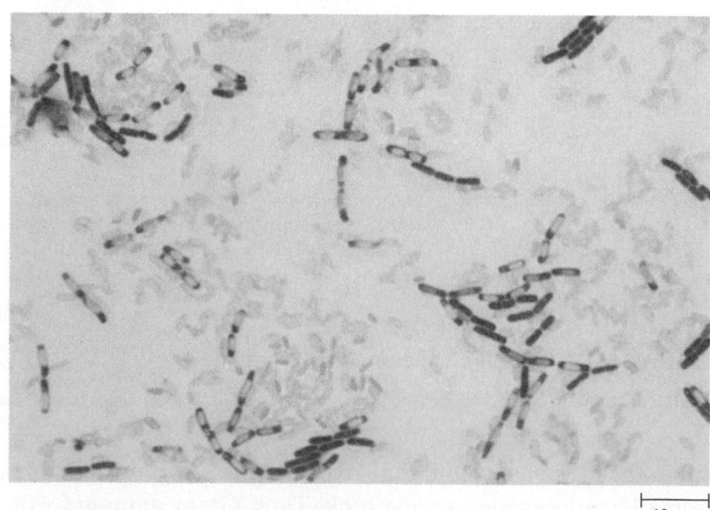

FIGURE 3.51 Endospore Stain Endospores retain the primary stain malachite green. Counterstaining with safranin colors other cells pink. Lisa Burgess/McGraw Hill

? What color would *Staphylococcus aureus* cells be following this endospore stain?

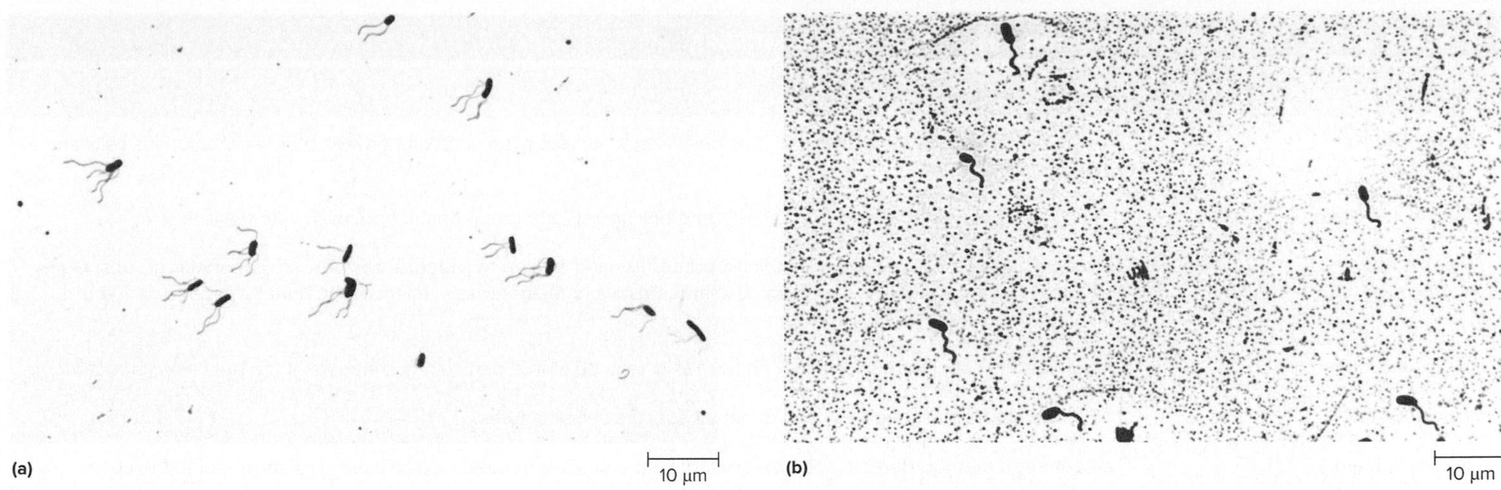

FIGURE 3.52 Flagella Stain The staining agent sticks to and coats the flagella. This increases their diameter so they can be seen with the light microscope. **(a)** *Bordetella bronchisepticum* (peritrichous flagella); **(b)** *Vibrio cholerae* (polar flagellum). Source: a-b: Dr. William A. Clark/CDC

? How can the flagella stain be helpful in identifying bacteria?

Flagella Stain

Flagella, the appendages that provide the most common mechanism of motility for bacterial cells, are ordinarily too thin to be seen with the light microscope. The **flagella stain** uses a substance that allows the staining agent to adhere to and coat the thin flagella, making them visible using light microscopy (**figure 3.52**). The presence and arrangement of flagella are distinguishing features that can be used to help identify an organism; recall that flagellar arrangements include peritrichous, polar monotrichous, and a tuft at one or both ends of a cell.

Fluorescent Dyes and Tags

Fluorescence can be used to observe total cells, a subset of cells, or cells with certain proteins on their surface, depending on the procedure (**figure 3.53**). One example is a fluorescent dye that binds to structures in all cells, so it can be used for determining the total number of microbial cells in a sample. Another fluorescent dye binds to all cells but is changed by cellular processes, so it can be used to distinguish between live and dead cells (see figure 3.40). Some fluorescent dyes bind to the mycolic acids in the cell walls of *Mycobacterium* species, making the dyes useful in a staining procedure similar to the acid-fast stain. A special technique called **immunofluorescence** is used to tag specific cell components with a fluorescent dye attached to an antibody (see figure 17.8). By tagging a protein unique to a given microbe, immunofluorescence can be used to detect and identify that organism. Antibodies, and how they are obtained, are described in chapters 15 and 17.

Characteristics of the stains discussed are summarized in **table 3.7.**

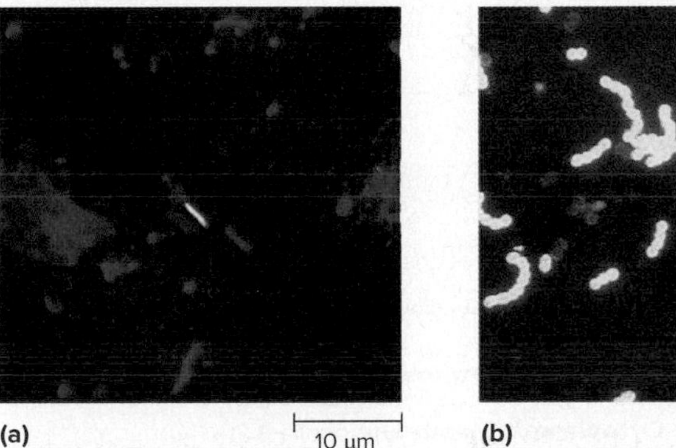

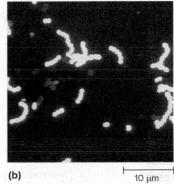

FIGURE 3.53 Fluorescent Dyes and Tags (a) To detect a *Mycobacterium* species in a specimen, a dye that binds mycolic acids and fluoresces yellow is used in a modification of the acid-fast technique. In this example, acridine orange was used to stain all other organisms. **(b)** Fluorescent antibodies tag specific molecules—in this case, the antibody binds to a molecule unique to *Streptococcus pyogenes.* a: CDC; b: Evans Roberts

? How can fluorescent dyes and tags be used to identify bacteria?

MicroAssessment 3.9

Dyes are used to stain cells so they can be seen against an unstained background. The Gram stain is the most commonly used differential stain. The acid-fast stain is used to detect *Mycobacterium* species. Specific dyes and techniques are used to observe cell structures such as capsules, endospores, and flagella. Fluorescent dyes and tags can be used to observe total cells, a subset of cells, or cells that have certain proteins on their surface.

25. What are the functions of a primary stain and a counterstain?
26. Describe one error in the staining procedure that would result in a Gram-positive bacterium appearing pink.
27. What color would a Gram-negative bacterium be in an acid-fast stain?

TABLE 3.7 A Summary of Stains and Their Characteristics

Stain	Characteristics
Simple Stains	A basic dye is used to stain cells and is an easy way to increase the contrast between otherwise colorless cells and a colorless background.
Differential Stains	A multistep procedure is used to stain cells and distinguish one group of microorganisms from another.
Gram stain	This is by far the most widely used staining procedure and is used to separate bacteria into two major groups: Gram-positive and Gram-negative. The staining characteristics of these groups reflect a fundamental difference in the chemical structure of their cell walls.
Acid-fast stain	This is used to detect organisms that do not easily take up stains, particularly members of the genus *Mycobacterium*.
Special Stains	These are staining procedures used to detect specific cell structures.
Capsule stain	The common method darkens the background, so the capsule stands out as a clear area surrounding the cell.
Endospore stain	This stains endospores, which do not readily take up stains. Members of the genera *Bacillus* and *Clostridium* are among the relatively few species that produce endospores.
Flagella stain	The staining agent adheres to and coats the otherwise thin flagella, making them visible with the light microscope.
Fluorescent Dyes and Tags	Some fluorescent dyes bind to compounds found in all cells; others bind to compounds specific to only certain types of cells. Antibodies to which a fluorescent molecule has been attached can be used to tag specific molecules.

Summary

PROKARYOTIC CELL STRUCTURES AND THEIR FUNCTIONS (figure 3.1; table 3.3)

3.1 ■ The Cytoplasmic Membrane of Prokaryotic Cells

Structure of the Cytoplasmic Membrane (figure 3.2)
The cytoplasmic membrane is a phospholipid bilayer embedded with a variety of different proteins.

Permeability of the Cytoplasmic Membrane (figure 3.3)
The cytoplasmic membrane is **selectively permeable.** A few substances pass through by **simple diffusion.** Some membrane proteins function as selective gates.

The Role of the Cytoplasmic Membrane in Energy Transformation (figure 3.5)
The **electron transport chain** generates an electrochemical gradient, a form of energy called **proton motive force.**

Transport of Small Substances Across the Cytoplasmic Membrane (figures 3.6, 3.7; table 3.1)
Facilitated diffusion moves substances through a membrane protein down a concentration gradient. **Active transport** moves substances against a concentration gradient and requires energy. **Group translocation** chemically modifies the molecules of a substance during its transport.

Protein Secretion (figure 3.8)
A characteristic signal sequence targets proteins for **secretion.**

3.2 ■ The Cell Wall of Prokaryotic Cells (table 3.2)

Peptidoglycan (figure 3.9)
Peptidoglycan provides rigidity to the bacterial cell wall. It is composed of glycan strands cross-linked via tetrapeptide chains.

The Gram-Positive Cell Wall (figure 3.10)
The Gram-positive cell wall contains a relatively thick layer of peptidoglycan. **Teichoic acids** can either project out of the peptidoglycan or anchor the peptidoglycan to the cytoplasmic membrane. A gel-like substance called **periplasm** may be sandwiched between the cytoplasmic membrane and the peptidoglycan layer.

The Gram-Negative Cell Wall (figure 3.11)
The Gram-negative cell wall has a relatively thin layer of peptidoglycan. An **outer membrane** contains **lipopolysaccharides.** LPS is called **endotoxin. Porins** permit small molecules to pass through the outer membrane. **Periplasm** fills the space between the inner and outer membranes.

Antibacterial Substances That Target Peptidoglycan
Penicillin interferes with peptidoglycan synthesis. **Lysozyme** destroys the structural integrity of peptidoglycan.

Bacteria That Lack a Cell Wall
Mycoplasma species are variable in shape and not affected by lysozyme or penicillin.

Cell Walls of Archaea
Archaea have a greater variety of cell wall types than do bacteria, with most having an S-layer; they all lack peptidoglycan.

3.3 ■ Structures Outside the Cell Wall of Prokaryotic Cells

Capsules and Slime Layers (figure 3.12)
Capsules and **slime layers** allow bacteria to adhere to surfaces. Some capsules allow pathogens to avoid the host immune system.

Flagella (figure 3.13)
Prokaryotic **flagella** are used for bacterial motility and function as propellers (figure 3.14); the functional counterparts in archaea

are called archaella and have a different structure and composition. **Chemotaxis** is movement toward an attractant or away from a repellent (figure 3.15). Cells use phototaxis, aerotaxis, magnetotaxis, and thermotaxis to move toward light, O_2, a magnetic field, and temperature, respectively.

Pili (figure 3.17)

Common **pili (fimbriae)** allow specific attachment of cells to surfaces. Other pili allow for surface motility. **Sex pili** are involved in a form of DNA transfer.

3.4 ■ Internal Components of Prokaryotic Cells

Chromosome and Plasmids

The **chromosome** forms a region called the **nucleoid** (figure 3.18); it contains the genetic information required by a cell. **Plasmids** usually encode only genetic information that may be helpful to a cell.

Ribosomes (figure 3.19)

Ribosomes are involved in protein synthesis. Prokaryotic ribosomes are 70S, and each is composed of a 50S and a 30S subunit.

Cytoskeleton

The **cytoskeleton** is involved in cell division and regulation of shape.

Storage Granules

Storage granules are synthesized from nutrients a cell has in excess.

Protein-Based Compartments

Protein-based compartments such as **gas vesicles, bacterial microcompartments,** and **encapsulin nanocompartments** physically separate certain reactions or functions from the cell's cytosol.

Endospores (figures 3.20, 3.21)

Endospores are resistant to heat, desiccation, toxic chemicals, and UV light; they can **germinate** to become **vegetative cells.**

EUKARYOTIC CELL STRUCTURES AND THEIR FUNCTIONS (figure 3.22; table 3.4)

3.5 ■ Cytoplasmic Membrane of Eukaryotic Cells

Structure and Function of the Cytoplasmic Membrane

The eukaryotic cytoplasmic membrane is an asymmetrical phospholipid bilayer embedded with proteins that are involved in transport, structural integrity, and signaling.

Transfer of Molecules Across the Cytoplasmic Membrane

Gated **channels** form pores in the membrane. **Carriers** function in facilitated diffusion and active transport. **Pinocytosis** is used by cells to take up liquids (figure 3.24). **Receptor-mediated endocytosis** is used by animal cells to take up material that binds to receptors. Protozoa and phagocytes take up material using **phagocytosis. Exocytosis** expels material (figure 3.25). Proteins destined for **secretion** are made by ribosomes attached to the endoplasmic reticulum.

3.6 ■ Protein Structures of Eukaryotic Cells

Ribosomes

Eukaryotic ribosomes are 80S, each composed of 60S and 40S subunits.

Cytoskeleton (figure 3.26)

The cytoskeleton is composed of **actin filaments, microtubules,** and **intermediate filaments.**

Flagella and Cilia (figure 3.27)

Flagella in eukaryotes propel a cell or pull the cell forward. **Cilia** move in synchrony to either propel a cell or move material along a stationary cell.

3.7 ■ Membrane-Bound Organelles of Eukaryotic Cells

Nucleus and Organelles of the Endomembrane System

The **nucleus** is bound by a nuclear envelope composed of two phospholipid bilayers and contains the cell's genetic information (DNA) (figure 3.28). Proteins not destined for the cytoplasm are synthesized by ribosomes that attach to the **rough endoplasmic reticulum**; the **smooth endoplasmic reticulum** is where lipids are synthesized and degraded, and calcium is stored (figure 3.29). The **Golgi apparatus** modifies, sorts, and packages proteins and lipids synthesized in the endoplasmic reticulum (figure 3.30). **Lysosomes** carry digestive enzymes.

Other Membrane-Bound Organelles

Peroxisomes are the organelles in which O_2 is converted into toxic compounds that oxidize and destroy certain substances. **Mitochondria** use the energy released during the degradation of organic compounds to generate ATP (figure 3.31). **Chloroplasts** capture the energy of sunlight and then use it to synthesize ATP, which is then used to convert CO_2 to an organic form (figure 3.32). The **endosymbiotic theory** states that mitochondria and chloroplasts each descended from endosymbiotic bacteria, which explains why these organelles share several characteristics with bacteria.

METHODS TO OBSERVE CELLS

3.8 ■ Microscopes (table 3.6)

Principles of Light Microscopy: Bright-Field Microscopes

The most commonly used type of microscope is the **bright-field microscope** (figure 3.33, figure 3.35). The usefulness of a microscope depends largely on its magnification and its **resolving power**. Modern bright-field microscopes are compound microscopes each having both an **ocular lens** and a changeable **objective lens**; the total magnification is the product of each lens's magnification. Staining specimens can be used to increase contrast.

Light Microscopes That Increase Contrast

Cells viewed through a **dark-field microscope** stand out against a dark background (figure 3.36). **Phase-contrast microscopes** increase differences in refraction (figure 3.37). **Differential interference contrast (DIC) microscopes** cause an image to appear three-dimensional (figure 3.38).

Light Microscopes That Detect Fluorescence

Fluorescence microscopes are used to observe cells stained with fluorescent dyes (figure 3.39). **Scanning laser microscopes (SLMs)** are used to obtain interior views of intact cells and three-dimensional images of thick structures (figure 3.40). **Super-resolution microscopes** use complex illumination mechanisms and merged data to achieve higher resolution than that of other light microscopes (figure 3.41).

Electron Microscopes

Transmission electron microscopes (TEMs) transmit a beam of electrons through a specimen stained with heavy metals (figure 3.43). **Scanning electron microscopes (SEMs)** scan a beam of electrons over the surface of a specimen stained with heavy metals (figure 3.44).

Scanning Probe Microscopes

Atomic force microscopes map surfaces on an atomic scale (figure 3.45).

3.9 ■ Preparing Specimens for Light Microscopy (table 3.7)

Simple Staining

Simple staining relies on a single dye. **Positive staining** procedures use basic dyes to stain cells while **negative staining** procedures use acidic dyes that stain the background.

Differential Staining

The **Gram stain** is widely used; **Gram-positive bacteria** stain purple and **Gram-negative bacteria** stain pink (figure 3.47, figure 3.48).

The **acid-fast stain** is used to stain *Mycobacterium* species; **acid-fast** organisms stain pink and all other organisms stain blue (figure 3.49).

Special Stains to Observe Cell Structures

The **capsule stain** is a negative staining procedure that allows a capsule to stand out as a clear zone around a cell (figure 3.50). The **endospore stain** uses heat to help the dye enter endospores, which are otherwise stain-resistant (figure 3.51). The **flagella stain** uses a substance that allows a stain to adhere to and coat the otherwise thin flagella (figure 3.52).

Fluorescent Dyes and Tags

Some fluorescent dyes bind compounds found in all cells, allowing for the cells to be counted; other fluorescent dyes bind compounds specific to certain cell types (figure 3.53). **Immunofluorescence** is used to tag proteins of interest with fluorescent compounds attached to antibodies.

Review Questions

Short Answer

1. Explain the medical significance of efflux pumps in bacteria.
2. Diagram the structure of peptidoglycan.
3. Give two reasons why the outer membrane of Gram-negative bacteria is medically significant.
4. Describe how a plasmid can help a cell.
5. How is receptor-mediated endocytosis different from phagocytosis?
6. What evidence led scientists to conclude that mitochondria evolved from bacterial cells?
7. Explain how the Golgi apparatus cooperatively functions with the endoplasmic reticulum.
8. Explain why resolving power is important in microscopy.
9. Describe the difference between a simple stain and a differential stain.
10. Describe what happens at each step in the Gram staining process.

Multiple Choice

1. Which bacterial cell component provides the best barrier for preventing most molecules from passing through it?
   a) Peptidoglycan
   b) Capsule
   c) Cytoplasmic membrane
   d) Slime layer
2. Endotoxin is associated with
   a) Gram-positive bacteria.
   b) Gram-negative bacteria.
   c) the cytoplasmic membrane.
   d) endospores.
3. The "O157" in the name *E. coli* O157:H7 refers to the type of O antigen. From this information you know that *E. coli*
   a) has a capsule.
   b) is a rod.
   c) is a coccus.
   d) is Gram-positive.
   e) is Gram-negative.
4. Eliminating which structure is *always* deadly to cells?
   a) Flagella
   b) Capsule
   c) Cell wall
   d) Cytoplasmic membrane
   e) Fimbriae
5. If you interfered with the ability of a *Bacillus* species to form endospores, what would be the result? The bacterium would no longer be
   a) able to multiply.
   b) antibiotic resistant.
   c) Gram-positive.
   d) Gram-negative.
   e) able to withstand boiling water.
6. If a virus mimics a ligand that normally participates in receptor-mediated endocytosis, the virus might
   a) prevent pinocytosis.
   b) be taken up by the cell.
   c) damage the cytoplasmic membrane.
   d) cause the cell to form pseudopods.
   e) damage the receptor.

7. The antibiotic erythromycin prevents protein synthesis in bacterial cells. Based on this information, which of the following might be targeted by the drug?
   1. 80S ribosomes
   2. 70S ribosomes
   3. 60S ribosomal subunit
   4. 50S ribosomal subunit
   5. 40S ribosomal subunit
      a) 1, 3
      b) 1, 4
      c) 2, 3
      d) 2, 4
      e) 2, 5
8. If a eukaryotic cell were treated with a chemical that destroys tubulin, all of the following would be directly affected *except*
   a) actin.
   b) cilia.
   c) eukaryotic flagella.
   d) microtubules.
   e) More than one of these
9. Which of the following is most likely to be used in a typical microbiology laboratory?
   a) Bright-field microscope
   b) Confocal scanning microscope
   c) Phase-contrast microscope
   d) Scanning electron microscope
   e) Transmission electron microscope
10. When a medical technologist wants to determine if a clinical specimen contains a *Mycobacterium* species, which should be used?
   a) Acid-fast stain
   b) Capsule stain
   c) Endospore stain
   d) Gram stain
   e) Simple stain

Applications

1. You are working in a laboratory producing new antibiotics for human and veterinary use. One compound with potential value inhibits the action of bacterial ribosomes. The compound, however, was shown to inhibit the growth of animal cells in culture. What is one possible explanation for its effect on animal cells?
2. A research laboratory is investigating environmental factors that inhibit the growth of archaea. The researchers wonder if penicillin would be effective in controlling their growth. Explain the probable results of an experiment in which penicillin is added to a culture of archaea.

Critical Thinking

1. This graph shows facilitated diffusion of a compound across a cytoplasmic membrane and into a cell. As the external concentration of the compound is increased, the rate of uptake increases until it reaches a point where it slows and then begins to plateau. This is not the case with passive diffusion, where the rate of uptake continually increases as the solute concentration increases. Why does the rate of uptake slow and then eventually plateau with facilitated diffusion?

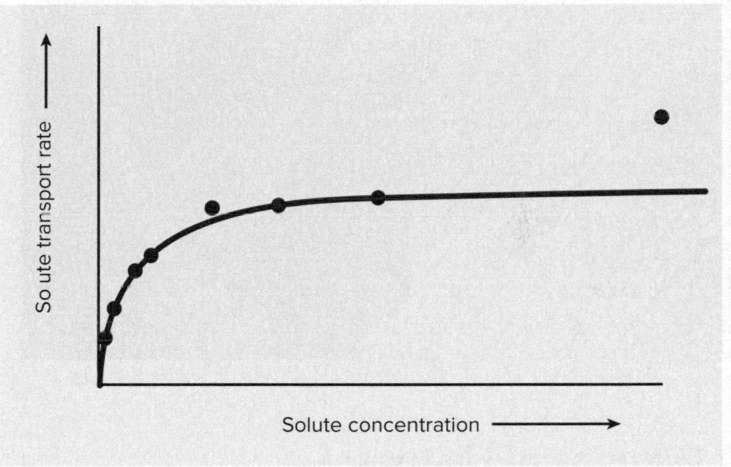

2. Most medically useful antibiotics interfere with either peptidoglycan synthesis or ribosome function. Why would the cytoplasmic membrane be a poor target for antibacterial medications?

www.mcgrawhillconnect.com

Enhance your study of this chapter with study tools and practice tests. Also ask your instructor about the resources available through Connect, including the media-rich eBook, interactive learning tools, and animations.

4 Dynamics of Microbial Growth

Escherichia coli growing on eosin methylene blue (EMB) agar. *Lisa Burgess/McGraw Hill*

KEY TERMS

Biofilm Polymer-encased community of microorganisms.

Chemically Defined Medium A culture medium of exact chemical composition; generally used only for specific experiments when nutrients must be precisely controlled.

Complex Medium A culture medium that contains protein digests, extracts, or other ingredients that vary in their chemical composition.

Differential Medium A culture medium with an ingredient that certain microorganisms change in a recognizable way; used to differentiate microbes based on their metabolic traits.

Exponential (Log) Phase Stage in the growth curve during which cells divide at a constant rate; generation time is measured during this period of active multiplication.

Facultative Anaerobe Organism that grows best if O_2 is available, but can also grow without it.

Generation Time The time it takes for a population to double in number.

Obligate Aerobe Organism that requires O_2.

Obligate Anaerobe Organism that cannot multiply, and often is killed, in the presence of O_2.

Pure Culture A population descended from a single cell.

Selective Medium A culture medium with an ingredient that inhibits the growth of microbes other than the one being sought.

A Glimpse of History

The greatest contributor to methods of cultivating bacteria was Robert Koch (1843–1910), a German physician who combined a medical practice with a productive research career for which he received a Nobel Prize. Koch was primarily interested in identifying disease-causing bacteria, but to do this, he needed simple methods to isolate and grow these particular species. Koch recognized that a single bacterial cell could multiply on a solid medium in a limited area to form a distinct visible mass of descendants, so he experimented with growing bacteria on the cut surfaces of potatoes. Some species would not grow, however, because the potatoes did not contain enough nutrients, so Koch experimented with methods to solidify any liquid nutrient medium. He used gelatin initially, but there were two major problems: Gelatin melts at the temperature preferred by many medically important microorganisms, and some bacteria can digest it. In 1882, Fannie Hesse, the wife of an associate of Koch, suggested using agar. This solidifying agent was used to harden jelly at the time and proved to be the perfect answer.

Today, we take pure culture techniques for granted because of their relative ease and simplicity. Their development, however, had a major impact on microbiology. By 1900, the agents causing most of the major bacterial diseases of humans had been isolated and characterized.

Microorganisms can be found growing even in the harshest climates and most severe conditions. Environments where no unprotected human could survive, such as the ocean depths, volcanic vents, and the polar regions, have thriving microbial species. Indeed many scientists believe that if life exists on other planets, it may resemble these organisms. Each species, however, has a limited set of environmental conditions in which it can grow; even then, it will grow only if specific nutrients are available. Some microorganisms can grow at temperatures above the boiling point of water but not at room temperature. Others can grow only within an animal host, and sometimes, only in specific areas of that host.

Because of the medical significance of some microbes, as well as the nutritional and industrial use of microbial by-products, scientists must be able to grow certain microorganisms in culture. This is why it is important to understand the basic principles involved in microbial growth, while recognizing that much information is yet to be discovered.

4.1 ■ Principles of Microbial Growth

Learning Outcome

1. Describe binary fission and how it relates to generation time and exponential growth.

Bacteria and archaea generally multiply by the process of **binary fission,** a process in which a cell increases its size, replicates its DNA, and then divides (**figure 4.1**). One cell

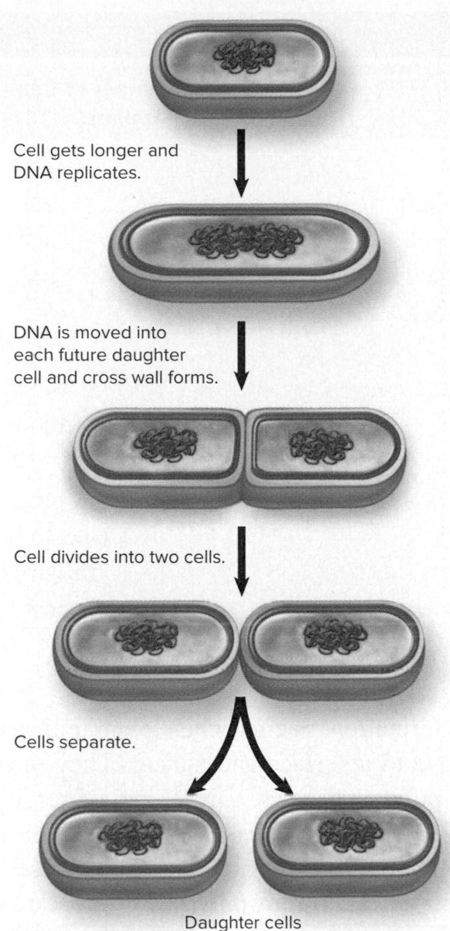

FIGURE 4.1 Binary Fission

How does the process of binary fission relate to the generation time?

divides into two, those two divide to become four, those four become eight, and so on. In other words, the increase in cell numbers is exponential. Because it is neither practical nor particularly meaningful to determine the relative size of the cells in a given population, microbial growth is defined as an increase in the number of cells in a population. The time it takes for a population to double in number is the **generation time.** This varies greatly from species to species and is influenced by the conditions in which the cells are grown. For example, the generation time of a given bacterium may be 30 minutes under ideal conditions, but when the conditions are not ideal, the organism will grow much more slowly, if at all.

The exponential multiplication of bacteria has important health consequences. For instance, a mere 10 cells of a foodborne pathogen in a potato salad, sitting for 4 hours in the warm sun at a picnic, may multiply to more than 40,000 cells. Although ingesting a few cells of the pathogen might not cause disease, consuming tens of thousands could be a different matter entirely. So keep exponential growth in mind, and your potato salad in the cooler, the next time you go to a picnic!

To calculate how many bacterial cells will be present in a product after a certain amount of time, two factors must be known: the initial number of cells in the original population and the number of times the cells will divide during the stated period. In the potato salad example, if the pathogen has a generation time of 20 minutes, then the number of cells in the population will double every 20 minutes. This means that in a single hour, the number of cells will double three times: The initial population of 10 cells will double to become 20 cells, which will then double to become 40 cells, which will then double to become 80 cells. As shown in **table 4.1,** over the course of 4 hours, the population will double 12 times, generating a population of 40,960 cells. The general equation illustrated by this example is $N_t = N_0 \times 2^n$, with N_t being the number of cells at a given time (in minutes), N_0 the initial number of cells and n the number of generations at that point.

MicroByte

Escherichia coli has a generation time of 20 minutes in ideal conditions; *Mycobacterium tuberculosis* needs at least 12 hours to double.

MicroAssessment 4.1

Most bacteria and archaea multiply by binary fission. The time required for a population to double in number is the generation time.

1. Explain why microbial growth refers to an increase in cell number rather than cell size.
2. If a bacterium has a generation time of 30 minutes, and you start with 100 cells at time 0, how many cells will you have in 30, 60, 90, and 120 minutes?

4.2 ▪ Microbial Growth in Nature

Learning Outcomes

2. Describe a biofilm, and explain why biofilms are important to humans.
3. Explain why microbes that grow naturally in mixed communities sometimes cannot be grown in pure culture.

Microorganisms have historically been studied by growing them in the laboratory. Scientists now recognize, however, that the changing and complex conditions of the natural environment, which differ greatly from the conditions in the laboratory, have significant effects on microbial growth and behavior. In fact, microbial cells can adjust to certain changes in their surroundings by sensing various chemicals and then

TABLE 4.1 Example of Exponential Growth

Time in Minutes *(t)*	Initial Population (N_0)	Number of Generations *(n)*	2^n	Number of Cells in the Population (N_t)
0	10	0		10
20	10	1	$2^1 (= 2)$	20
40	10	2	$2^2 (= 2 \times 2)$	40
60 (1 hour)	10	3	$2^3 (= 2 \times 2 \times 2)$	80
80	10	4	$2^4 (= 2 \times 2 \times 2 \times 2)$	160
100	10	5	$2^5 (= 2 \times 2 \times 2 \times 2 \times 2)$	320
120 (2 hours)	10	6	2^6	640
140	10	7	2^7	1,280
160	10	8	2^8	2,560
180 (3 hours)	10	9	2^9	5,120
200	10	10	2^{10}	10,240
220	10	11	2^{11}	20,480
240 (4 hours)	10	12	2^{12}	40,960

responding to that input by producing materials appropriate for the situation.

Biofilms

In nature, microorganisms can live suspended in an aqueous environment, but most attach to surfaces and live in polymer-encased communities called **biofilms (figure 4.2)**. Biofilms cause the slipperiness of rocks in a stream bed, the slimy "gunk" that coats kitchen drains, the scum that gradually accumulates in toilet bowls, and the dental plaque that forms on teeth.

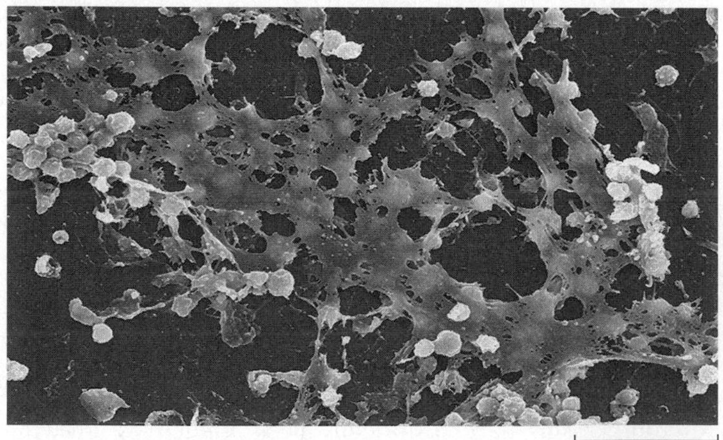

FIGURE 4.2 Biofilm on the Inside of an Indwelling Catheter
Rodney M. Donlan, Ph.D.; Janice Carr/CDC

? **Why would a biofilm within a catheter that drains urine from the bladder be a concern?**

Biofilm formation begins when planktonic (free-floating) cells move to a surface and adhere. They then multiply and release polysaccharides, DNA, and other hydrophilic polymers to which unrelated cells may attach and grow (**figure 4.3**). The mesh-like accumulation of these polymers, referred to as **extracellular polymeric substances (EPS),** gives a biofilm its characteristic slimy appearance and also forms a matrix that provides it with structure and stability. Biofilms are not random mixtures of microbes in a layer of EPS, but instead have characteristic formations with channels through which nutrients and wastes pass. Cells communicate with one another by synthesizing and responding to chemical signals—a process important in establishing a biofilm's architecture.

Biofilms are more than just an unsightly annoyance. Dental plaque leads to tooth decay and gum disease. Even troublesome, persistent ear infections and the complications of cystic fibrosis are due to biofilms. In fact, most bacterial infections involve biofilms. Treatment of these infections is difficult because microbes within the biofilm often resist the effects of antibiotics as well as the actions of the body's immune system. Biofilms are also important in industry, where their accumulations in pipes, drains, and cooling water towers can interfere with processes and damage equipment. Microbes within the biofilm are protected; thus they are much more resistant than their planktonic counterparts to disinfectants and dessication (drying).

Although biofilms can have negative effects, they also can be beneficial. Many bioremediation efforts, which use microbes to degrade harmful chemicals, are enhanced by biofilms. So, as some industries are exploring ways to destroy biofilms, others,

Planktonic bacteria move to the surface and adhere.
Bacteria multiply and produce extracellular polymeric substances (EPS).
Other bacteria may attach to the EPS and grow.
Cells communicate and create channels in the EPS that allow nutrients and waste products to pass.
Some cells detach and then move to other surfaces to create additional biofilms.

FIGURE 4.3 Development of a Biofilm

What are extracellular polymeric substances (EPS)?

such as wastewater treatment facilities, are looking for ways to encourage their development.

Interactions of Mixed Microbial Communities

Microorganisms in the environment regularly grow in close associations with many different species. Sometimes the interactions are cooperative, even helping the growth of certain species that otherwise could not survive. For example, organisms that cannot multiply in the presence of O_2 will grow in your mouth if neighboring microbial cells rapidly use that gas. Meanwhile, the metabolic wastes of one organism can serve as nutrients for another. Often, however, microbes in a community compete for nutrients, and some even resort to a type of biological warfare, synthesizing toxic compounds that inhibit competitors. Certain Gram-negative bacteria use a needle-like structure to inject toxic compounds directly into competing bacterial cells, a process called contact-dependent growth inhibition. Understandably, the conditions that characterize these associations of mixed microbial communities are extremely difficult to reproduce in the laboratory.

MicroAssessment 4.2

Biofilms have a characteristic architecture with channels through which nutrients and wastes can pass. In nature, prokaryotes often grow in associations with many different species.

3. Water bowls left out for pets sometimes develop a slimy layer if not washed regularly. What causes the slime?
4. Describe a situation in which the activities of one species benefit another.
5. Why is it more difficult to study microorganisms growing in their natural environment than to study them in the laboratory?

4.3 ■ Microbial Growth in Laboratory Conditions

Learning Outcomes

4. Describe how the streak-plate method is used to obtain a pure culture and how the resulting culture can be stored.
5. Describe the stages of a growth curve, and compare this closed system to colony growth and continuous culture.

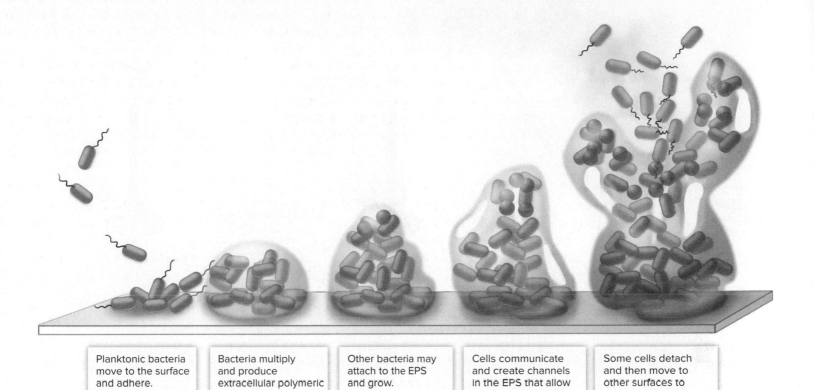

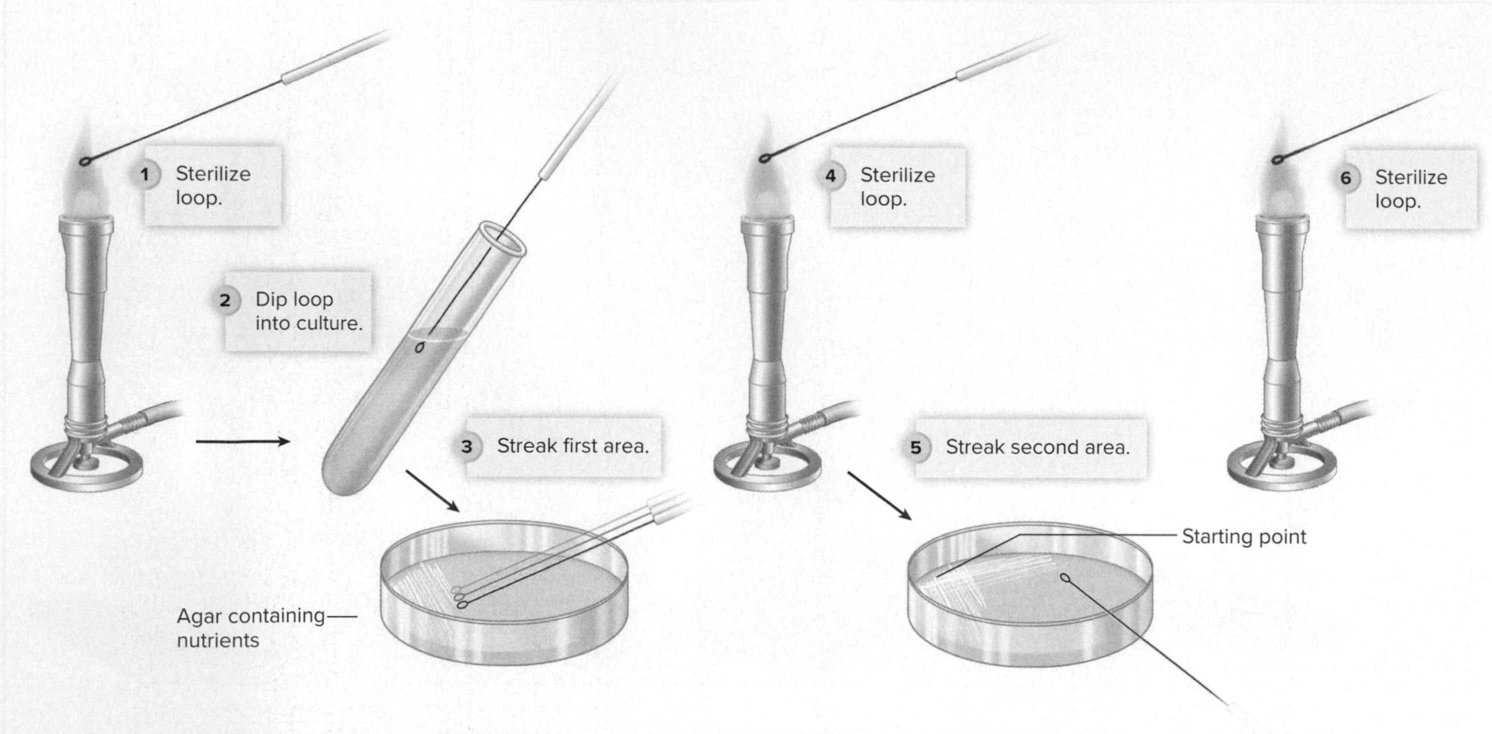

In the laboratory, bacteria and archaea are generally isolated and grown in pure culture. A **pure culture** is a population descended from a single cell and therefore contains only one species. Working with pure cultures makes it easier to identify and study the activities of a particular species. But although the results are much easier to interpret, organisms in pure cultures sometimes behave differently than they do in their natural environment, as just discussed. Another complicating issue is that with current methods, only an estimated 1% of all prokaryotes can successfully be grown in culture, which makes it more difficult to study the vast majority of microorganisms. Fortunately, most known medically significant bacteria can be grown in pure culture.

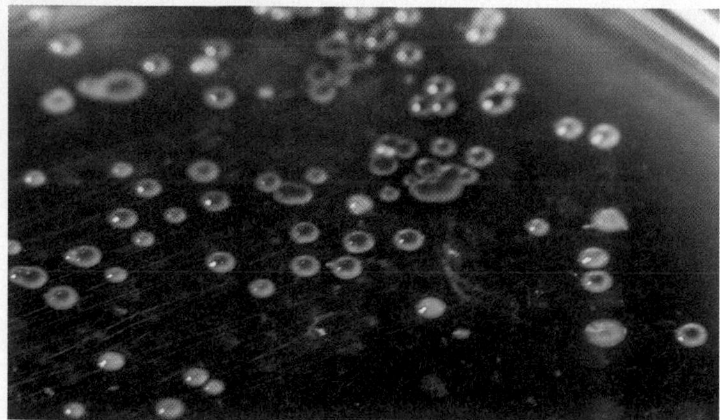

FIGURE 4.4 Colonies Growing on Agar Medium Lisa Burgess/McGraw Hill

What is the purpose of agar in the medium?

To work with a pure culture, all containers, media, and instruments must be sterile, or free of microbes, prior to use. These are then handled using **aseptic technique,** a set of procedures used to prevent the accidental introduction of unwanted microbes. The medium the cells are grown in, or on, is called a culture medium. It consists of nutrients dissolved in water, and it can be a liquid broth or a solid gel.

Obtaining a Pure Culture

The basic requirements for obtaining a pure culture are (1) a solid culture medium in a sterile container that prevents contaminating microbes from entering and (2) a method to separate individual microbial cells. A single microbial cell, supplied with the right nutrients and conditions, will multiply on the solid medium in a limited area to form a **colony,** a distinct mass of cells. About 1 million cells are required for a colony to be easily visible to the naked eye.

Agar, a polysaccharide extracted from certain types of seaweed, is used to solidify culture media (**figure 4.4**). Unlike gelatin and other gelling agents, agar is resistant to degradation by all but a very few microbes. It is not destroyed at high temperatures and can therefore be sterilized by heating—a process that also liquefies it. Melted agar will stay liquid until cooled to a temperature below 45°C, so nutrients that would be destroyed at high temperatures can be added at lower temperatures before the agar hardens. Once solidified, an agar medium will remain solid until heated above 95°C. Thus, unlike gelatin—which is liquid at 37°C—agar remains solid over the entire temperature range at which most microbes grow.

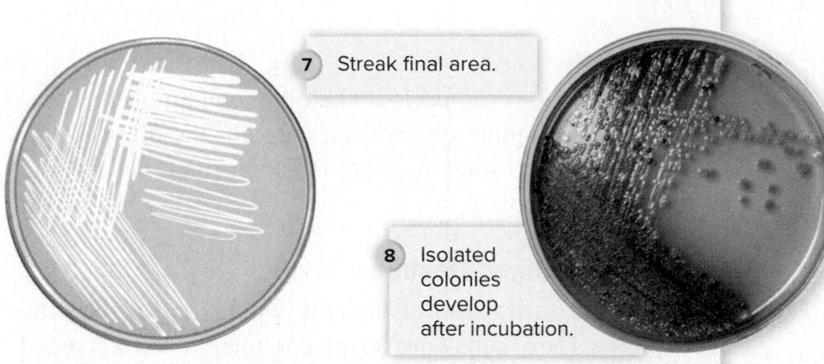

FIGURE 4.5 The Streak-Plate Method The successive streaks dilute the cells. By the third set of streaks, cells should be separated enough so that isolated colonies develop after incubation. Lisa Burgess/McGraw Hill

? **What is the purpose of obtaining isolated colonies?**

A solid culture medium is contained in a **Petri dish,** a two-part, covered container made of glass or plastic. Although not airtight, the dish excludes airborne contaminants. A culture medium in a Petri dish is commonly referred to as a plate of that medium—for example, a nutrient agar plate or, more simply, an **agar plate.**

MicroByte

Declining harvests of the seaweeds used to make agar have resulted in a worldwide agar shortage.

The Streak-Plate Method

The **streak-plate method** is the simplest and most commonly used technique for isolating microorganisms (**figure 4.5**). A sterile inoculating loop is dipped into a microbe-containing sample and then lightly drawn several times across the surface of an agar plate, creating a set of parallel streaks covering approximately one-third of the agar. The loop is then sterilized and a new series of parallel streaks is made across and at an angle to the previous ones, covering another surface section. This drags some of those cells streaked onto the first portion over to a fresh section, effectively inoculating it with a diluted sample. The loop is sterilized again, and another set of parallel streaks is made, dragging into a third area some of the organisms that had been moved into the second section. The goal is to reduce the number of cells being spread with each successive series of streaks. By the third set of streaks, cells should be separated enough so that distinct, well-isolated colonies will form.

Maintaining Stock Cultures

Once a pure culture has been obtained, it can be maintained as a stock culture, which is a culture stored for use as an inoculum in later procedures. Often, a stock culture is stored in the refrigerator as growth on the surface of an agar slant (a tube of agar that was held at an angle as it solidified). For long-term storage, stock cultures can be frozen at –70°C in a glycerol-containing solution that prevents ice crystals from forming and damaging cells. Alternatively, cells can be freeze-dried.

The Growth Curve

In the laboratory, many microorganisms are grown either on agar plates or in tubes or flasks of broth. These are considered **closed systems,** or batch cultures, because nutrients are not renewed, nor are wastes removed. As the cells grow in this type of system, the population increases in a distinct pattern of stages and then declines. The characteristic pattern observed in a broth culture is called a **growth curve,** and is characterized by five distinct stages: lag phase, exponential or log phase, stationary phase, death phase, and phase of prolonged decline (**figure 4.6**).

Lag Phase

When microorganisms are transferred into a different medium, the cells do not immediately begin dividing. Although there is no increase in cell number during this **lag phase,** cells begin synthesizing enzymes required for growth. The length of the lag phase depends on multiple factors. For example, if cells are transferred into a medium that contains fewer nutrients, the lag phase will be longer because the cells must first make components (such as amino acids) not supplied in the new medium.

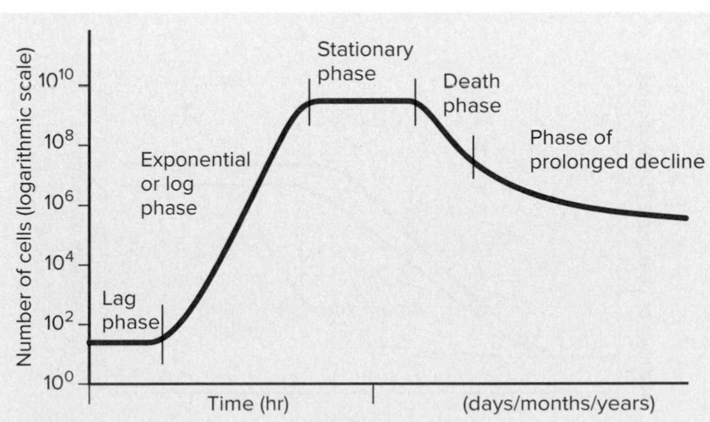

FIGURE 4.6 Growth Curve The growth curve is characterized by five distinct stages: lag phase, exponential or log phase, stationary phase, death phase, and phase of prolonged decline.

? **Generation time is measured during only one of the stages illustrated above. Considering the definition of generation time, during which stage would it be measured?**

Exponential Phase (Log Phase)

During the **exponential** or **log phase,** cells divide at a constant rate. Generation time is measured during this active growth phase. The phase is medically significant because bacteria are most sensitive to antimicrobial medications when cells are actively multiplying.

From a commercial standpoint, the exponential phase is important because some molecules made by growing cells are valuable. For instance, amino acids can be sold as nutritional supplements, and microbial waste products such as ethanol are used as biofuels. The small molecules made by cells as they multiply are called **primary metabolites.**

In the later stages of exponential growth, nutrients gradually become depleted and waste products accumulate. Cell activities shift as this occurs. Endospore formers such as *Bacillus* and *Clostridium* species may start the process of sporulation. Even cells that cannot form endospores often change their activities to prepare for starvation conditions. Microbial compounds that begin accumulating at this stage are made for purposes other than growth and are called **secondary metabolites** (**figure 4.7**). Commercially, the most valuable of these are antibiotics, which are produced by certain microorganisms.

Stationary Phase

Cells enter the **stationary phase** when the nutrient levels are too low to sustain growth. The total number of viable cells (cells capable of developing or multiplying) in the population remains relatively constant, but some cells are dying while others are multiplying. How can cells multiply when they have exhausted their supply of nutrients? Dead cells often burst, releasing nutrients that then fuel the growth of other cells.

During the stationary phase, the viable cells continue to synthesize secondary metabolites and maintain the altered properties they demonstrated in late log phase. The length of the stationary phase varies, depending on the species and environmental conditions. Some populations remain in the stationary phase for only a few hours, whereas others remain for days or longer.

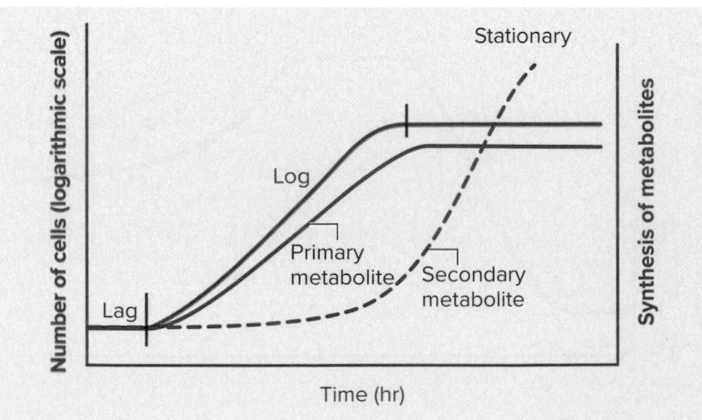

FIGURE 4.7 Primary and Secondary Metabolite Production
Primary metabolites are synthesized during the period of active multiplication. Secondary metabolites begin to be synthesized in late log phase.

? Which secondary metabolites are the most commercially valuable?

Death Phase

The **death phase** is the period when the total number of viable cells in the population decreases as cells die off at a constant rate. Like cell growth, death is exponential, but the rate is usually much slower.

Phase of Prolonged Decline

In many cases, a fraction of the cell population survives the death phase. These cells have adapted to tolerate the worsened conditions and can multiply for at least a short time using the nutrients released from the dead cells. As the conditions continue to deteriorate during this **phase of prolonged decline,** most of these survivors then die. However, the few progeny better equipped for survival can grow. This dynamic process generates successive waves of slightly modified populations, each more fit to survive than the previous ones. Thus, the statement "survival of the fittest" holds true even for closed cultures of microbes.

Colony Growth

The growth of bacteria on a solid medium is similar to that in a broth, but it has some important differences. After a lag phase, cells in a colony multiply exponentially and eventually compete with one another for available nutrients. Unlike the situation in a broth, however, the position of a single cell within a colony determines its environment. Cells multiplying on the edge of the colony face relatively little competition for O_2 and nutrients. In the center of the colony where the cell density is high, however, available O_2 and nutrients become depleted and harmful wastes such as acids accumulate. As a consequence, cells at the edge of the colony may be growing exponentially, while those in the center may be in the death phase. Cells in locations between these two extremes may be in the stationary phase.

Continuous Culture

Microbial cells can be kept in a state of continuous growth by using a **chemostat.** This device is an **open system,** meaning that nutrients can be added and waste products removed. The chemostat continually drips fresh medium into a broth culture contained in a chamber. With each drop that enters, an equivalent volume—containing cells, wastes, and spent medium—leaves through an outlet. By adjusting the nutrient content of the medium and the speed at which it enters the chamber, a constant growth rate and cell density can be maintained. This makes it possible to study a uniform population's response to different nutrient concentrations or environmental conditions. Chemostats are also used in industrial processes that harvest commercially valuable products made only during the log phase of microbial growth.

MicroAssessment 4.3

In the laboratory, the streak-plate method is used to obtain a pure culture. When grown in a closed system, a microbial population goes through distinct phases. Cells within a colony may be in any one of the phases, depending on their relative location. Chemostats are used to study uniform populations of growing cells.

6. To identify the causative agent of a given illness, a pure culture is often needed. How is a streak plate used to obtain a pure culture?
7. Explain the difference between the lag phase and the exponential phase.
8. Why would a compound that prevents bacteria from growing interfere with the function of the antibiotic penicillin?

4.4 ■ Environmental Factors That Influence Microbial Growth

Learning Outcomes

6. Describe the importance of a microorganism's requirements for temperature, O_2, pH, and water availability, and define the terms that indicate these requirements.
7. Explain the significance of reactive oxygen species, and describe the mechanisms cells use to protect against their effects.

As a group, microorganisms inhabit nearly every environment on Earth. Microbes we associate with disease and food spoilage live in habitats that humans consider quite comfortable. In contrast, microorganisms referred to as **extremophiles** (*phile* means "loving") live in harsh environments that would kill most other organisms; extremophiles are often archaea.

Recognizing the major environmental factors that affect growth (temperature, atmosphere, pH, and water availability) is essential for studying microbes and helps us understand their roles in the complex ecology of the planet. A table at the end of this section summarizes the environmental factors that influence microbial growth (see table 4.3).

Temperature Requirements

Each microbial species has a well-defined temperature range in which it grows. Within this range is the optimum growth temperature—the temperature at which the organism multiplies most rapidly. As a general rule, this optimum is close to the upper limit of the organism's temperature range.

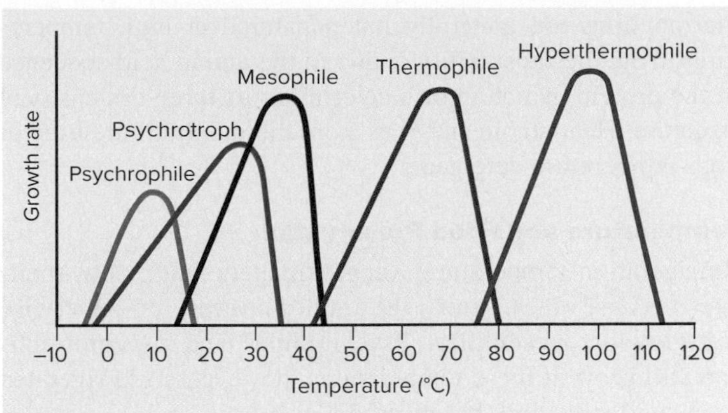

FIGURE 4.8 Temperature Requirements for Growth
Microorganisms are commonly divided into five groups based on their optimum growth temperatures. This graph shows a typical example of each group. The optimum temperature, the point at which the growth rate is highest, is near the upper limit of the range.

? **Most pathogens fall into which group on this chart?**

Microorganisms are commonly divided into five groups based on their optimum growth temperatures (**figure 4.8**). Note, however, that this merely represents a convenient organization scheme. In reality, no sharp dividing line exists between groups. Furthermore, not every organism in a group can grow in the entire temperature range typical for its group. The five groups are as follows:

- **Psychrophiles** (*psychro* means "cold") have an optimum between –5°C and 15°C. These organisms grow in the cold Arctic and Antarctic regions and in lakes fed by glaciers.
- **Psychrotrophs** (*troph* means "nourishment") have an optimum between 15°C and 30°C but grow well at lower temperatures. They are an important cause of spoilage in refrigerated foods.
- **Mesophiles** (*meso* means "middle") have an optimum between 25°C and about 45°C. *E. coli* and most other common bacteria are in this group. Pathogens, adapted to grow in the human body, typically have an optimum between 35°C and 40°C. Mesophiles that inhabit soil, a colder environment, generally have a lower optimum, closer to 30°C.
- **Thermophiles** (*thermo* means "heat") have an optimum between 45°C and 70°C. These organisms commonly live in hot springs and compost heaps.
- **Hyperthermophiles** (*hyper* means "excessive") have an optimum of 70°C or greater. These are usually archaea. An archeon isolated from the base of a hydrothermal vent deep in the ocean can grow at 122°C—the highest growth temperature yet recorded!

Why can some microbes withstand very high temperatures while most cannot? One reason is that proteins from thermophiles are generally not denatured at high temperatures. This thermostability is due to the amino acid sequence of the protein, which in turn determines its three-dimensional structure. Heat-stable enzymes from thermophiles are used in high-temperature detergents.

Temperature and Food Preservation

Refrigeration temperatures (approximately 4°C) slow spoilage because they limit the multiplication of otherwise fast-growing mesophiles. Psychrophiles and psychrotrophs can still grow at these temperatures, however, so refrigerated foods will still spoil, but more slowly.

Foods and other perishable products that withstand below-freezing temperatures can be frozen for long-term storage. It is important to recognize, however, that freezing is generally not an effective means of destroying microbes. Recall that freezing is routinely used to preserve stock cultures.

Temperature and Disease

The temperature of the heart, brain, and gastrointestinal tract is near 37°C, but that of the extremities is slightly lower. This is why some microbes cause disease more readily in certain parts of the body. Hansen's disease (leprosy) typically involves the coolest regions (ears, hands, feet, and fingers) because the causative bacterium *(Mycobacterium leprae)* grows best at these lower temperatures. The bacterium that causes syphilis *(Treponema pallidum)* also grows best at slightly cooler temperatures. Indeed, for more than 30 years the major treatment for syphilis was to deliberately induce high fever by introducing the organism that causes malaria.

Oxygen (O_2) Requirements

Like humans, some microorganisms have an absolute requirement for oxygen (O_2). They grow in **aerobic** environments, meaning that O_2 is present. Other microorganisms thrive in **anaerobic** environments, meaning that little or no O_2 is present. One way of determining an organism's O_2 requirement is to grow it in a shake tube. To do this, a tube of nutrient agar is boiled, which both melts the agar and drives off the O_2. The agar is then allowed to cool to just above its solidifying temperature. Next, the test organism is added and dispersed by gentle shaking or swirling. The agar medium is then allowed to harden and the tube is incubated. Because the solidified agar slows gas diffusion, the concentration of O_2 in the medium is relatively high at the top, whereas the bottom portion is anaerobic. The cells grow in the region that has a suitable O_2 level. Based on their O_2 requirements, microorganisms can be separated into these groups:

- **Obligate aerobes** have an absolute requirement for O_2. They use it in aerobic respiration, an energy-harvesting process. This and other ATP-generating pathways are discussed in chapter 6. An example of an obligate aerobe is *Micrococcus luteus,* which is common in the environment. Note that obligate aerobes are often simply called aerobes.
- **Facultative anaerobes** grow better if O_2 is present, but they can also grow without it. The term "facultative" means that the organism is flexible, in this case in its requirements for O_2. Facultative anaerobes use aerobic respiration if O_2 is available, but they resort to alternative types of metabolism if it is not. Growth is faster when O_2 is present because aerobic respiration produces the most ATP. *E. coli* is one of the most common facultative anaerobes in the large intestine.
- **Obligate anaerobes** cannot multiply if O_2 is present; in fact, they are often killed by even brief exposure to air. Obligate anaerobes harvest energy using processes other than aerobic respiration; the details of these are covered in chapter 6. Most inhabitants of the large intestine are obligate anaerobes, as is the bacterium that causes botulism, *Clostridium botulinum.* Note that obligate anaerobes are often simply called anaerobes.
- **Microaerophiles** require small amounts of O_2 (2% to 10%) for aerobic respiration; higher concentrations are inhibitory. An example is *Helicobacter pylori,* which can cause gastric and duodenal ulcers.
- **Aerotolerant anaerobes** are indifferent to O_2. They can grow in its presence, but they do not use it to harvest energy. They are also called **obligate fermenters** because fermentation is their only metabolic option. An example is *Streptococcus pyogenes,* which causes strep throat.

Reactive Oxygen Species (ROS)

When organisms use O_2 in aerobic respiration, harmful derivatives called **reactive oxygen species (ROS)** form as by-products. These molecules, which include superoxide (O_2^-) and hydrogen peroxide (H_2O_2), can damage cell components, so cells that grow aerobically must have mechanisms to protect against them; most obligate anaerobes lack these mechanisms (**table 4.2**). Virtually all organisms that grow in the presence of O_2 produce the enzyme **superoxide dismutase,** which inactivates superoxide by converting it to O_2 and hydrogen peroxide. Nearly all these organisms also produce the enzyme **catalase,** which converts hydrogen peroxide into O_2 and water. An important exception is the aerotolerant anaerobes; the fact that they do not produce catalase is useful in the laboratory. A simple test for the enzyme can be used to distinguish two groups of medically important Gram-positive cocci that grow aerobically: *Staphylococcus* species, which are catalase positive, and *Streptococcus* species, which are catalase negative.

TABLE 4.2 Oxygen (O_2) Requirements of Microorganisms

	Obligate Aerobe	Facultative Anaerobe	Obligate Anaerobe	Microaerophile	Aerotolerant Anaerobe
Pattern of Growth in a Shake Tube	Bacterial growth				
Growth Characteristics	Grows only when O_2 is available.	Grows best when O_2 is available, but also grows without it.	Cannot grow when O_2 is present.	Grows only if small amounts of O_2 are available.	Grows equally well with or without O_2.
Use of O_2 in Energy-Harvesting Processes	Requires O_2 for respiration.	Uses O_2 for respiration, if available.	Does not use O_2.	Requires O_2 for respiration.	Does not use O_2.
Typical Mechanisms to Protect Against Reactive Oxygen Species	Produces superoxide dismutase and catalase.	Produces superoxide dismutase and catalase.	Does not produce superoxide dismutase or catalase.	Produces some superoxide dismutase and catalase.	Produces superoxide dismutase but not catalase.

MicroByte

Over half of all the cytoplasm on Earth is probably in anaerobic microbes!

pH

Each microbial species can survive in a range of pH values, and within this range is its pH optimum. Despite the pH of the external environment, however, cells maintain a constant internal pH, typically near neutral. Many prokaryotes that grow in acidic environments quickly pump out protons (H^+) that enter the cell; recall that acidic environments have relatively high concentrations of protons. Prokaryotes that grow in alkaline conditions bring in protons.

Most microbes are **neutrophiles**—they live and multiply in the range of pH 5 (acidic) to pH 8 (basic) and have a pH optimum near neutral (pH 7). Food preservation methods such as pickling inhibit bacterial growth by increasing the acidity of the food.

Although most neutrophiles cannot withstand highly acidic conditions, one medically important bacterium has found a way. *Helicobacter pylori* grows in the stomach, sometimes causing ulcers. It decreases the acidity of its immediate surroundings by producing urease, an enzyme that splits urea into carbon dioxide and ammonia. The ammonia neutralizes any stomach acid surrounding the cell.

Acidophiles grow optimally at a pH below 5.5. *Picrophilus oshimae,* a member of the Archaea, has an optimum pH of less than 1! This organism, which was isolated from the dry, acidic soils of a gas-emitting volcanic fissure in Japan, has an unusual cytoplasmic membrane that is unstable at a pH above 4.0.

Alkaliphiles grow optimally at a pH above 8.5. They often live in alkaline lakes and soils.

Water Availability

All microorganisms require water for growth. Even if water is present, however, it may not be available in certain environments. Dissolved substances such as table salt (sodium chloride; NaCl) and sugars, for example, interact with water molecules, making them unavailable to the cell. In many environments, particularly in certain natural habitats such as salt marshes, microorganisms are faced with this situation. If the solute concentration is higher in the medium than in the cell, water diffuses out of the cell due to osmosis. This causes the cytoplasm to dehydrate and shrink from the cell wall, a phenomenon called **plasmolysis** (**figure 4.9**).

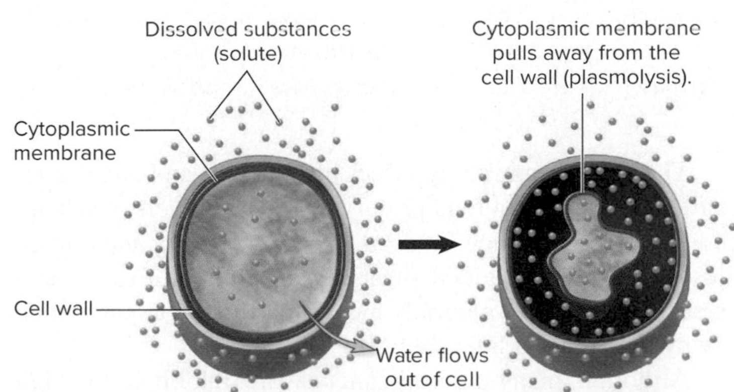

FIGURE 4.9 Effects of Solute Concentration on Prokaryotic Cells Water molecules pass freely through the cytoplasmic membrane. If the solute concentration is higher outside the cell, water moves out of the cell.

What is plasmolysis?

TABLE 4.3 Environmental Factors That Influence Microbial Growth

Environmental Factor/ Descriptive Terms	Characteristics
Temperature	Temperature tolerance partly reflects the thermostability of the organism's proteins.
Psychrophile	Optimum temperature between –5°C and 15°C.
Psychrotroph	Optimum temperature between 15°C and 30°C, but grows well at refrigeration temperatures.
Mesophile	Optimum temperature between 25°C and 45°C.
Thermophile	Optimum temperature between 45°C and 70°C.
Hyperthermophile	Optimum temperature of 70°C or greater.
Oxygen (O_2) Availability	Oxygen (O_2) requirement/tolerance reflects the organism's energy-harvesting mechanisms and its ability to inactivate reactive oxygen species.
Obligate aerobe	Requires O_2.
Facultative anaerobe	Grows best if O_2 is present, but can also grow without it.
Obligate anaerobe	Cannot grow in the presence of O_2.
Microaerophile	Requires small amounts of O_2, but higher concentrations are inhibitory.
Aerotolerant anaerobe (obligate fermenter)	Indifferent to O_2.
pH	Prokaryotes that live in pH extremes maintain a near-neutral internal pH by pumping protons out of or into the cell.
Neutrophile	Multiplies in the range of pH 5 to 8.
Acidophile	Grows optimally at a pH below 5.5.
Alkalophile	Grows optimally at a pH above 8.5.
Water Availability	Prokaryotes that can grow in high-solute solutions maintain the availability of water in the cell by increasing their internal solute concentration.
Halotolerant	Can grow in relatively high-salt solutions, up to approximately 10% NaCl.
Halophile	Requires high levels of sodium chloride.

The growth-inhibiting effect of high salt and sugar concentrations is used in food preservation. High levels of salt are added to preserve such foods as bacon, salt pork, and anchovies. Likewise, the added sugar in jams and jellies acts as a preservative. Honey naturally has a high sugar content, which is why it also has a long shelf life.

Although many microorganisms are inhibited by high salt concentrations, some withstand or even require them. Microbes that tolerate high salt concentrations, up to approximately 10% NaCl, are **halotolerant** (*halo* means "salt"). *Staphylococcus* species, bacteria that live on the dry, salty environment of the skin, are an example. **Halophiles** require high levels of NaCl. Many marine bacteria are mildly halophilic, requiring concentrations of approximately 3% NaCl. Certain archaea are extreme halophiles, requiring 9% NaCl or more. Extreme halophiles are found in environments such as the salt flats of Utah and the Dead Sea.

The environmental factors that influence microbial growth are summarized in **table 4.3.**

MicroAssessment 4.4

A microorganism can be categorized according to its optimum growth temperature as well as its O_2 requirements. Most microbes grow best at a near-neutral pH, although some prefer acidic conditions, and others grow best in alkaline conditions. Halophiles require high-salt conditions.

9. *Clostridium paradoxum* grows optimally at 55°C, pH 9.3; it will not grow in the presence of O_2. How should this bacterium be categorized with respect to its temperature, pH, and O_2 requirements?
10. What is the function of the enzyme catalase?
11. Why would hydrogen peroxide be an effective disinfectant?

4.5 ■ Nutritional Factors That Influence Microbial Growth

Learning Outcomes

8. List the required elements and give examples of common sources of required elements.
9. Explain the significance of a limiting nutrient.
10. Explain why fastidious microbes require growth factors.
11. Describe the energy and carbon sources used by photoautotrophs, chemolithoautotrophs, photoheterotrophs, and chemoorganoheterotrophs.

The growth of any microorganism depends not only on a suitable physical environment, but also on the availability of nutrients, which the cells use for biosynthesis. What sets bacteria and archaea apart from all other forms of life is their remarkable ability to use diverse sources of various chemical elements.

Required Elements

Chemical elements that make up cells are called **major elements;** these include carbon, oxygen, hydrogen, nitrogen, sulfur, phosphorus, potassium, magnesium, calcium, and iron. They are the essential components of proteins, carbohydrates, lipids, and nucleic acids.

Microorganisms can be categorized based on their source of carbon. **Heterotrophs** use organic carbon (*hetero* means "different" and *troph* means "nourishment"). Medically important bacteria are typically heterotrophs. **Autotrophs** (*auto* means "self") use inorganic carbon in the form of carbon dioxide (CO_2). They play a critical role in the cycling of carbon in the environment because they can convert inorganic carbon to an organic form, the process of **carbon fixation.** Without carbon fixation, the earth would quickly run out of organic carbon, which is essential to life.

Nitrogen is needed to make amino acids and nucleic acids. Some microorganisms use nitrogen gas (N_2) as a nitrogen source, converting it to ammonia and then incorporating that into cellular material. This process, **nitrogen fixation,** is unique to bacteria and archaea. Like carbon fixation, it is essential to life because once the nitrogen is incorporated into cellular material such as amino acids, other organisms can easily use it. Many microbes use ammonia as a nitrogen source. Some convert nitrate to ammonia, which is then incorporated into cellular material.

Sulfur is a component of some amino acids. Many microbes use inorganic sulfur sources such as sulfate, but others require organic sources such as sulfur-containing amino acids.

Phosphorus is a component of nucleic acids, membrane lipids, and ATP. As with sulfur, many organisms can use inorganic sources such as phosphate. Some, however, require organic sources, such as phosphorus-containing cell components.

Other elements, including potassium, magnesium, calcium, and iron, are required for some enzymes to function. A variety of inorganic and organic sources of these elements may be used by microbes.

Phosphorus and iron are important ecologically because they are often **limiting nutrients**—meaning they are available at the lowest concentration relative to need. To understand this concept, think about using a recipe to make chocolate chip cookies. Just as the quantity of chocolate chips available would determine the number of batches you could make (assuming the other ingredients were on hand), a limiting nutrient dictates the maximum level of microbial growth.

Trace elements are required in such small amounts that most natural environments, including water, have sufficient levels to support microbial growth. Trace elements include cobalt, zinc, copper, molybdenum, and manganese.

Representative functions of the major elements are summarized in **table 4.4.**

TABLE 4.4 Representative Functions of the Major Elements

Chemical	Function
Carbon, oxygen, and hydrogen	Component of amino acids, lipids, nucleic acids, and sugars
Nitrogen	Component of amino acids and nucleic acids
Sulfur	Component of some amino acids
Phosphorus	Component of nucleic acids, membrane lipids, and ATP
Potassium, magnesium, and calcium	Required for the functioning of certain enzymes; additional functions as well
Iron	Part of certain enzymes

Growth Factors

Some microbes cannot synthesize certain organic molecules such as amino acids, vitamins, and nucleotides. Therefore, these organisms can only grow if the molecules they cannot make are available in the surrounding environment; the molecules are called **growth factors.**

A microbe's growth factor requirements reflect its biosynthetic capabilities. Most *E. coli* strains, for example, can use glucose as the raw material to synthesize all of their cell components, so they do not need any growth factors. They multiply in a medium containing only glucose and several inorganic salts. In contrast, *Neisseria* species are less resourceful metabolically and require numerous growth factors, including vitamins and amino acids. Bacteria such as *Neisseria* species are **fastidious,** meaning they have complicated nutritional requirements.

Certain fastidious bacteria are used to measure the quantity of vitamins in food products. To do this, a well-characterized species that requires a specific vitamin is inoculated into a medium that lacks the vitamin but is supplemented with a measured amount of the food product. The extent of growth of the bacterium is related to the quantity of the vitamin in the product.

Energy Sources

Organisms harvest energy from either sunlight or chemical compounds, using processes discussed in chapter 6. **Phototrophs** obtain energy from sunlight (*photo* means "light"). They include plants, algae, and photosynthetic bacteria. **Chemotrophs** extract energy from chemical compounds (*chemo* means "chemical"). Mammalian cells, fungi, and many types of prokaryotes use organic chemicals such as sugars, amino acids, and fatty acids as an energy source. Some bacteria and archaea use inorganic chemicals such as hydrogen sulfide and hydrogen gas as an energy source, an ability that distinguishes them from eukaryotes.

Nutritional Diversity

Microbiologists often group microorganisms according to the energy and carbon sources they use (**table 4.5**):

- **Photoautotrophs** use the energy of sunlight to make organic compounds from CO_2. They are primary producers, meaning they support other forms of life by fixing carbon. Cyanobacteria are important primary producers that inhabit soil and aquatic environments. Many fix

FOCUS ON A CASE 4.1

The "patient" was Green Lake, a small lake in Seattle that sometimes had murky water. At its worst, a surface layer of greenish scum collected near the shore. In very hot weather, the area started smelling like rotting garbage—a characteristic that certainly discouraged people from enjoying the lake! The problems were due to the abundant growth of cyanobacteria, a group of photosynthetic bacteria.

Cyanobacterial blooms (accumulations of cyanobacteria), which occur most commonly in the summer months, are more than just an aesthetic annoyance. Some strains make toxins that can be deadly to animals that drink from the lake. Because of the potential danger, lakes may be closed to wading, swimming, and sailboarding when a cyanobacterial bloom occurs. In addition, dog owners are warned to keep their pets from drinking the lake water.

To control cyanobacterial growth in Green Lake, alum (aluminum sulfate) was added to the water. Aluminum combines with phosphorus and sinks to the bottom, removing phosphorus from the water column—an action that limits cyanobacterial growth. The treatment was successful, resulting in clear water throughout the year.

1. Why do cyanobacteria blooms occur in the summer months rather than year-round?
2. What caused the greenish scum to form?
3. Why does removing phosphorus from water limit cyanobacterial growth?
4. Why did the area start smelling like rotting garbage when the weather was hot?

Discussion

1. Cyanobacteria are photosynthetic, meaning they are photoautotrophs and therefore harvest their energy from sunlight. Summer months typically have longer days with more sunshine, providing photosynthetic organisms with plenty of their energy source. In addition, the warmer temperatures allow faster growth of the bacteria.
2. The greenish scum was an accumulation of cyanobacterial cells. The bacteria float to the surface because they have gas vesicles, which they use for buoyancy. Like plants, the bacteria are green because they contain chlorophyll. In fact, extensive scientific evidence indicates that chloroplasts—the photosynthetic organelles of plants—evolved from an ancestor of modern-day cyanobacteria.
3. Phosphorus is often the limiting nutrient for cyanobacteria. Carbon, which is frequently the limiting nutrient for heterotrophs, is not limiting for autotrophs because they obtain it from the CO_2 in the atmosphere. Nitrogen is often the limiting nutrient for autotrophs, but some cyanobacterial species can fix nitrogen, meaning that they can use atmospheric N_2. Considering that nitrogen-fixing cyanobacteria have readily available supplies of carbon and nitrogen, that leaves phosphorus as the most likely limiting nutrient. Without enough phosphorus, they cannot continue to synthesize their phosphorus-containing cell components, including DNA, RNA, membrane lipids, and ATP.
4. Some of the cyanobacterial cells died in the hot weather. Like any form of natural organic matter, the dead cells will decompose—a process that can generate unpleasant odors.

nitrogen as well, providing another indispensable role in the biosphere.

- **Photoheterotrophs** use the energy of sunlight and obtain their carbon from organic compounds. Some are facultative in their nutritional capabilities. For example, some members of a group called the purple non-sulfur bacteria grow anaerobically using light as an energy source and organic compounds as a carbon source (photoheterotrophs). They can also grow aerobically in the dark using organic sources of carbon and energy (chemoheterotrophs).
- **Chemolithoautotrophs** (*lith* means "stone"), often referred to simply as chemoautotrophs or chemolithotrophs, use inorganic compounds for energy and obtain their carbon from CO_2. These prokaryotes live in seemingly inhospitable places such as sulfur hot springs, which are rich in hydrogen sulfide, and other environments that have reduced inorganic compounds (see Focus Your Perspective 4.1). In regions of the ocean depths near hydrothermal vents, chemoautotrophs are the primary producers, supporting abundant life in these habitats that lack sunlight (see figure 28.12).

TABLE 4.5 Energy and Carbon Sources Used by Different Groups of Microorganisms

Type	Energy Source	Carbon Source
Photoautotroph	Sunlight	CO_2
Photoheterotroph	Sunlight	Organic compounds
Chemolithoautotroph	Inorganic chemicals (H_2, NH_3, NO_2^-, Fe^{2+}, H_2S)	CO_2
Chemoorganoheterotroph	Organic compounds (sugars, amino acids, etc.)	Organic compounds

FOCUS YOUR PERSPECTIVE 4.1

Can Microorganisms Live on Only Rocks and Water?

Microorganisms have been isolated from diverse environments that previously were thought to be incapable of sustaining life. For example, members of the Archaea have been isolated from environments 10 times more acidic than lemon juice. Other archaea have been isolated from oil wells a mile below the surface of the earth at temperatures of 70°C and pressures of 160 atmospheres (at sea level, the pressure is 1 atmosphere). The discovery of these microbes suggests that thermophiles may be widespread in the earth's crust.

Perhaps the most unusual environment from which prokaryotes have been isolated is volcanic rock 1 mile below the earth's surface near the Columbia River in Washington State. What do these organisms use for food? They apparently get their energy from the H_2 produced in a reaction between groundwater and the iron-rich minerals in the rock. The groundwater also contains dissolved CO_2, which the organisms use as a source of carbon. These prokaryotes apparently exist on nothing more than rocks and water!

- **Chemoorganoheterotrophs,** also referred to as chemoheterotrophs or chemoorganotrophs, use organic compounds for both energy and carbon. They are by far the most common group of microorganisms associated with humans and other animals. Individual species of chemoheterotrophs differ in the number of organic compounds they can use. Certain members of the genus *Pseudomonas* can obtain carbon and/or energy from more than 80 different organic compounds, including such unusual compounds as naphthalene (the ingredient associated with the smell of mothballs). At the other extreme, some organisms can degrade only a few compounds. *Bacillus fastidiosus* can use only urea and a few of its derivatives as sources of both carbon and energy.

MicroAssessment 4.5

Organisms require a source of major and trace elements. Heterotrophs use an organic carbon source, and autotrophs use CO_2. Phototrophs harvest energy from sunlight, and chemotrophs extract energy from chemicals.

12. To prevent excess phosphate from entering lakes and streams, certain laws govern the amount of phosphorus allowed in laundry and dishwasher detergents. What can happen if phosphorus levels in a lake increase?
13. How would your body's cells be categorized with respect to their carbon and energy sources?
14. Why would human-made materials (such as many plastics) be degraded only slowly or not at all?

4.6 ■ Cultivating Microorganisms in the Laboratory

Learning Outcomes

12. Compare and contrast complex, chemically defined, selective, and differential (conventional and chromogenic) media.
13. Explain how aerobic, microaerophilic, and anaerobic conditions can be provided.
14. Describe the purpose of an enrichment culture.

Cultivating microorganisms in the laboratory requires a suitable growth medium and an appropriate atmosphere.

General Categories of Culture Media

Considering the diversity of microorganisms, it should not be surprising that hundreds of different types of media are available. Even so, some medically important organisms and most environmental ones have not yet been grown in the laboratory.

Complex Media

A **complex medium** contains a variety of ingredients such as meat juices and digested proteins, forming what might be viewed as a tasty soup for microbes; it is easy to make and used for routine purposes. One common ingredient is peptone, a mixture of amino acids and short peptides produced by digesting any of a variety of different proteins. Extracts (the water-soluble components of a substance such as lean beef) are often included to provide vitamins and minerals. A common recipe for nutrient broth—a complex medium—uses peptone and beef extract in distilled water. If agar is added, then nutrient agar results.

Many medically important bacteria are fastidious, requiring a medium even richer than nutrient agar. Because of this, clinical laboratories often use blood agar. As the name implies, this contains red blood cells—a source of a variety of nutrients including hemin—as well as other ingredients. A medium used for even more fastidious bacteria is chocolate agar, named for its brownish appearance rather than its ingredients. Chocolate agar contains lysed red blood cells and additional supplements, thereby supplying fastidious bacteria the nutrients they need.

Even though a complex medium contains a specific amount of each ingredient, the exact chemical compositions of those ingredients can be highly variable. For example, each batch of peptone could have a different assortment of amino acids and other substances.

Chemically Defined Media

A **chemically defined medium** has a precise chemical composition because it is made of pure chemicals; it is generally used only for certain research experiments when the type and

TABLE 4.6 Ingredients in Two Types of Media That Support the Growth of *E. coli*

Nutrient Broth (Complex Medium)	Glucose-Salts Broth (Chemically Defined Medium)
Peptone	Glucose
Beef extract	Dipotassium phosphate
Water	Monopotassium phosphate
	Magnesium sulfate
	Ammonium sulfate
	Calcium chloride
	Iron sulfate
	Water

quantity of nutrients must be precisely controlled. A chemically defined medium called glucose-salts broth supports the growth of *E. coli* and contains only those chemicals listed in **table 4.6.** The cells grow more slowly in this medium than in nutrient broth because they must synthesize all of their components from glucose. A much longer recipe containing as many as 46 different ingredients must be used to make a chemically defined medium that supports the growth of the fastidious bacterium *Neisseria gonorrhoeae* (the cause of gonorrhea).

To keep the pH near neutral, buffers are often added to culture media. They are especially important in chemically defined media because some bacteria produce so much acid as a by-product of their metabolism that they inhibit their own growth. This is usually not a problem in complex media because the amino acids and other natural components provide at least some buffering action.

Special Types of Culture Media

To detect or isolate a particular species from a mixed population, special types of culture media are often used to make that species more prevalent or obvious. These special media can be either complex or chemically defined, depending on the needs of the microbiologist.

Selective Media

Selective media contain an ingredient that inhibits the growth of certain species in a mixed sample, while allowing the growth of the species of interest. Thayer-Martin agar, a medium used to isolate *Neisseria gonorrhoeae* from clinical specimens, is chocolate agar to which three or more antimicrobial drugs have been added. The antimicrobials inhibit fungi, Gram-positive bacteria, and Gram-negative rods, but not most *N. gonorrhoeae* strains, so the pathogen can grow with little competition. Modified versions of Thayer-Martin agar are also now used; the antimicrobials added have been modified to provide better coverage against current strains of *Neisseria gonorrhoeae*. MacConkey agar is used to isolate Gram-negative rods from various clinical specimens such as urine. This medium has a variety of nutrients that many bacteria can use, but it also includes two inhibitory components: crystal violet (a dye that inhibits Gram-positive bacteria) and bile salts (an ingredient that inhibits most non-intestinal bacteria).

Differential Media—Conventional

Differential media contain a substance that certain microbes change in a recognizable way. Blood agar is differential because bacteria that produce a hemolysin (a protein that causes red blood cells to burst) destroy red blood cells (RBCs) in the medium around the colonies, causing a zone of clearing called **beta hemolysis** (**figure 4.10**). This easily observable characteristic is important medically because some pathogens can be distinguished by their type of hemolysis. For example, colonies of *Streptococcus pyogenes* (the bacterium that causes strep throat) show beta hemolysis, which makes them stand out from other *Streptococcus* species that reside harmlessly in the throat. These other species may instead show **alpha hemolysis,** meaning their colonies are surrounded by a zone of greenish discoloration. The discoloration is caused by a change to hemoglobin in the RBCs (similar to what occurs with a bruise) rather than actual lysis of the RBCs; in fact, studies show that the RBC membrane remains intact, so alpha hemolysis is not true lysis. Other streptococci have no effect on red blood cells.

MacConkey agar is differential as well as selective (**figure 4.11**). In addition to its ingredients already mentioned, it contains lactose and a pH indicator. Bacteria that ferment the sugar produce acid, which turns the pH indicator pink-red. Therefore, colonies of lactose-fermenting bacteria are pink-red on MacConkey agar, whereas colonies of lactose-negative bacteria are colorless.

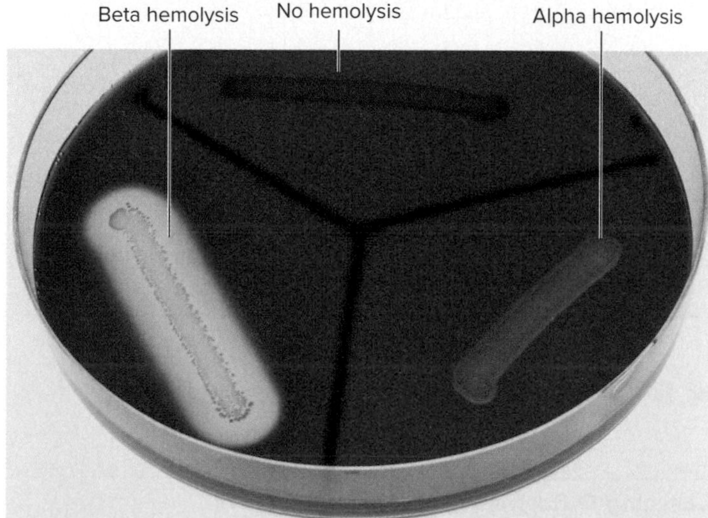

FIGURE 4.10 Blood Agar This complex medium is differential for hemolysis. A zone of colorless clearing around a colony growing on blood agar is called beta hemolysis; a zone of greenish discoloration is called alpha hemolysis. Lisa Burgess/McGraw Hill

? Which type of hemolysis characterizes *Streptococcus pyogenes*, the bacterium that causes strep throat?

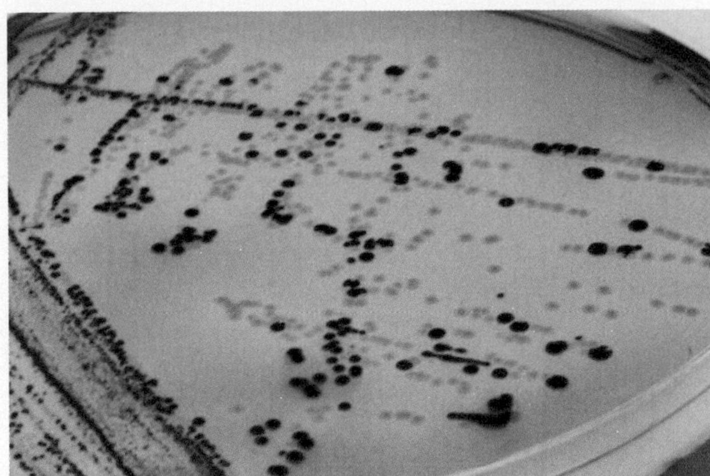

FIGURE 4.11 MacConkey Agar This complex medium is differential for lactose fermentation and selective for Gram-negative rods. Lisa Burgess/McGraw Hill

? **What specifically causes colonies of lactose fermenters to be pink-red on MacConkey agar?**

Figure 4.12 illustrates the function of the selective and differential media. Characteristics of various types of growth media are summarized in **table 4.7.**

TABLE 4.7 Characteristics of Representative Media Used to Cultivate Bacteria

Medium	Characteristic
Blood agar	Complex medium used routinely in clinical labs. Differential because colonies of hemolytic organisms are surrounded by a zone of red blood cell clearing or discoloration. Not selective.
Chocolate agar	Complex medium used to culture fastidious bacteria, particularly those found in clinical specimens. Not selective or differential.
Glucose-salts agar	Chemically defined medium. Used in laboratory experiments to study nutritional requirements of bacteria. Not selective or differential.
MacConkey agar	Complex medium used to isolate Gram-negative rods that typically reside in the intestine. Selective because bile salts and dyes inhibit Gram-positive organisms and Gram-negative cocci. Differential because the pH indicator turns pink-red when the sugar in the medium, lactose, is fermented.
Nutrient agar	Complex medium used for routine laboratory work. Supports the growth of a variety of nonfastidious bacteria. Not selective or differential.
Thayer-Martin agar	Complex medium used to isolate *Neisseria* species, which are fastidious. Selective because it contains antibiotics that inhibit most organisms except *Neisseria* species. Not differential.

Differential Media—Chromogenic

Chromogenic media provide a simple and colorful way to isolate precise microbial species or strains from a mixed microbial population. They make use of the same general principles as conventional differential agar media, but with an ingenious twist that uses multiple colors. A good example of the benefit involves the isolation of *Staphylococcus aureus*. Many conventional media used to isolate the bacterium rely on an agent that selects for halotolerant organisms such as *Staphylococcus,* as well as a pH-induced color change to distinguish *S. aureus* colonies from those of other staph species. Chromogenic agar can go one step further by using one color change to make *S. aureus* colonies look different from other staph species and another to distinguish colonies of methicillin-resistant

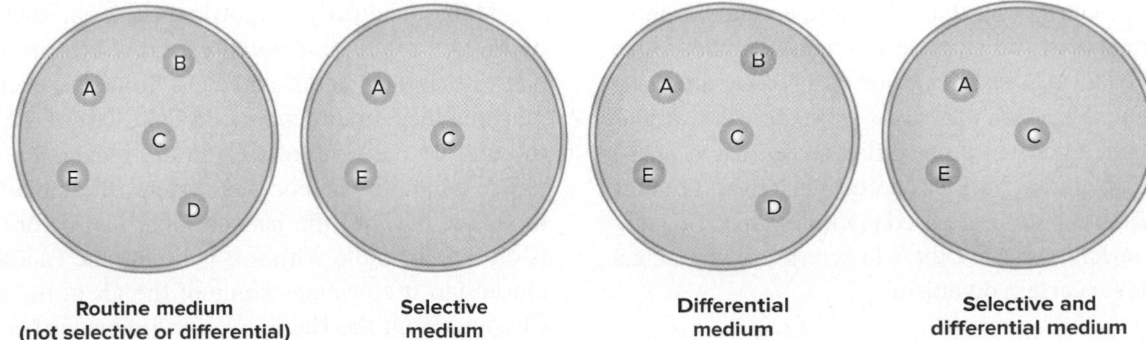

FIGURE 4.12 Selective and Differential Media A routine medium allows the growth of a variety of organisms; in this example, organisms A, B, C, D, and E grow, but all the colonies look similar. A selective medium contains an ingredient that inhibits the growth of certain organisms; in this example, only organisms A, C, and E can grow, but again all the colonies look similar. A differential medium contains an ingredient that certain microorganisms change in a recognizable way; in this example, the colonies of organisms B and E are pink, so they look different from the rest. A selective and differential medium inhibits certain organisms and causes colonies of some organisms to look different from others; in this example, only organisms A, C, and E can grow, and colonies of organism E are pink so they look different from colonies of organisms A and C.

? **How would you categorize nutrient agar to which penicillin had been added?**

FOCUS YOUR PERSPECTIVE 4.2

Agar Art: Creating Masterpieces with Microbes!

Every year since 2015, the American Society of Microbiology (ASM) has sponsored an Agar Art contest. Microorganisms such as bacteria and yeast are used as "paint" while different types of culture media become the artist's canvas (**box figure 4.1**). This process takes some planning and patience—the "paint" starts off as an invisible layer of microbes which will then grow, typically over the next few days. Both professional and budding microbiologists from all over the world get a chance to showcase their creative side by submitting a photo of their microbial masterpieces. Many of these creations feature organisms that are pigmented (meaning they make colored molecules) while others take advantage of color changes that occur when grown on differential media (either conventional or chromogenic). Several contest categories exist, depending on whether the art work was developed in a formal lab setting like in a university (professional) or in a more informal venue like a community lab (creator)—there is even a category for children! The contest is open for submissions in September, and winners for the different categories are announced in mid-November.

The contest is a great way to share your knowledge of microbes and showcase your talent as well—make sure to check it out!https://asm.org/Events/ASM-Agar-Art-Contest/Home

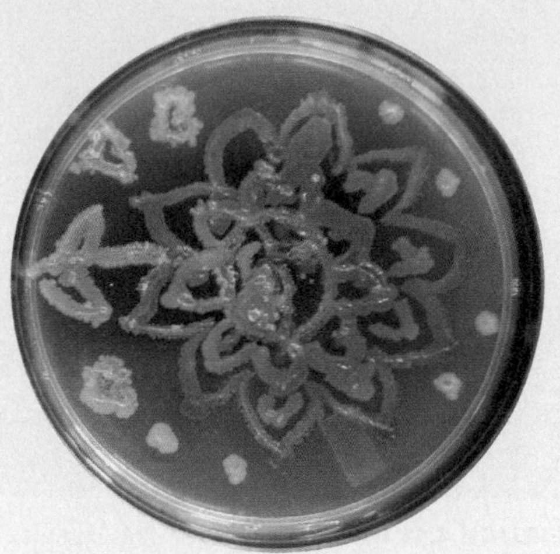

BOX FIGURE 4.1 "Flamingo Blooming," Agar Art by an Undergraduate Student
Jennifer Walker

S. aureus (MRSA) strains from those of their methicillin-susceptible counterparts (MSSA). Thus, the chromogenic agar helps identify the pathogen and simultaneously provides the patient care team with essential information to optimize treatment.

The feature that sets chromogenic agars apart from conventional differential media is the inclusion of one or more chromogens (*chrom* means "color" and *gen* means "generate"), which are colorless compounds that can be converted into colored ones. Scientists designed the chromogens so that only the microorganism of interest can catalyze the conversion. They did this by first identifying an enzyme that characterizes the microbial species or strains of interest, and then modifying the substrate of that enzyme to turn it into a chromogen. That modification involves attaching a chromophore (a molecule that functions as a dye) to the substrate in such a way that the substrate-chromophore complex is colorless, but the enzyme's action will release the chromophore to result in a distinct color change. By attaching a unique chromophore to each of several substrates, the various microbial colonies in a sample can be differentiated based on the enzyme expression patterns. Various chromogenic agars are now commercially available—some are used to identify specific pathogens in clinical specimens and others to screen environmental samples and foods for certain organisms.

Providing Appropriate Atmospheric Conditions

To grow microorganisms in culture, appropriate atmospheric conditions must be provided. Although many organisms can grow when incubated in air, others can only grow in the absence of O_2 or require a specific concentration of gases like O_2 or CO_2. Laboratories commonly use specific containers and techniques (described below) to achieve these atmospheric requirements. More recently, automated systems have also been developed where the concentrations of O_2, CO_2, or other gases in the container can be customized to fit the requirements of the organisms being grown. These systems provide both flexibility and reproducibility when culturing organisms in different atmospheric conditions.

Aerobic Conditions

Most obligate aerobes and facultative anaerobes do not need special atmospheric conditions to grow on an agar medium; they can be incubated in air (approximately 20% O_2). In a broth, however, the organisms grow best when the containers are shaken to provide maximum aeration.

Many medically important bacteria, including species of *Neisseria* and *Haemophilus,* grow best in aerobic atmospheres that have additional CO_2. Some are even capnophiles, meaning they require increased CO_2. One of the simplest ways to build up the level of CO_2 in the incubation atmosphere is to use a candle jar. For this system, the inoculated plates or tubes are put into the jar and then, just before the container is closed, a candle within it is lit. As the candle burns in the closed jar, it consumes some of the O_2 in the air, generating CO_2 and H_2O; the flame soon extinguishes because of insufficient O_2, but enough remains to support the growth of obligate aerobes and to prevent the growth of obligate anaerobes.

Microaerophilic Conditions

Microaerophilic organisms typically require O_2 concentrations lower than those achieved in a candle jar. These microbes are often incubated in a gas-tight container with a special

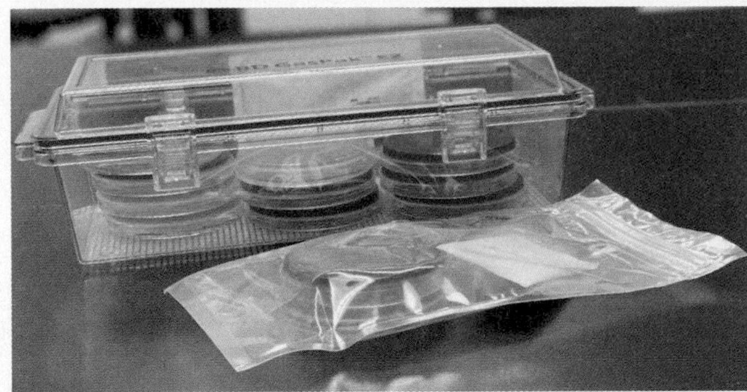

FIGURE 4.13 Anaerobic Containers A disposable packet contains chemicals that react with O_2, thereby producing an anaerobic environment. Mira Beins/McGraw Hill

Can an obligate anaerobe also grow in a candle jar?

disposable packet; the packet holds chemicals that react with O_2, reducing its concentration to approximately 5–15%.

Anaerobic Conditions

Obligate anaerobes are challenging to grow because of their sensitivity to O_2-containing environments. Those that can tolerate brief exposures to O_2 are often incubated in an **anaerobic container** (**figure 4.13**). This is the same type of container used to incubate microaerophiles, but the chemical composition of the disposable packet produces an anaerobic environment. An alternative method is to grow them in a tube of a semisolid culture medium that contains a reducing agent such as sodium thioglycolate, which reacts with O_2 to make water. In many cases, an O_2-indicating dye is included as well. The medium can be heated immediately before use to remove some dissolved O_2.

A more stringent method for working with anaerobes is to use an **anaerobic chamber,** an enclosed compartment maintained as an anaerobic environment (**figure 4.14**). A special port that can be filled with an inert gas such as N_2 is used to add or remove items. Airtight gloves allow researchers to handle items within the chamber.

Enrichment Cultures

An **enrichment culture** is used to isolate an organism present as only a very small fraction of a mixed population. It does this by providing conditions that preferentially enhance the growth of that particular species in a broth (**figure 4.15**). For instance, if the target microbe can grow using atmospheric nitrogen, then that element is left out of the medium. If it can use an unusual carbon source such as phenol, then that compound is added as the only carbon source. A selective agent such as bile salts may also be added. The culture is then incubated in conditions that preferentially promote the growth of the desired organism. As the microbes multiply, the relative concentration of the target organism can increase dramatically. A pure culture can then be obtained by streaking the enrichment onto an appropriate agar medium and selecting a single colony.

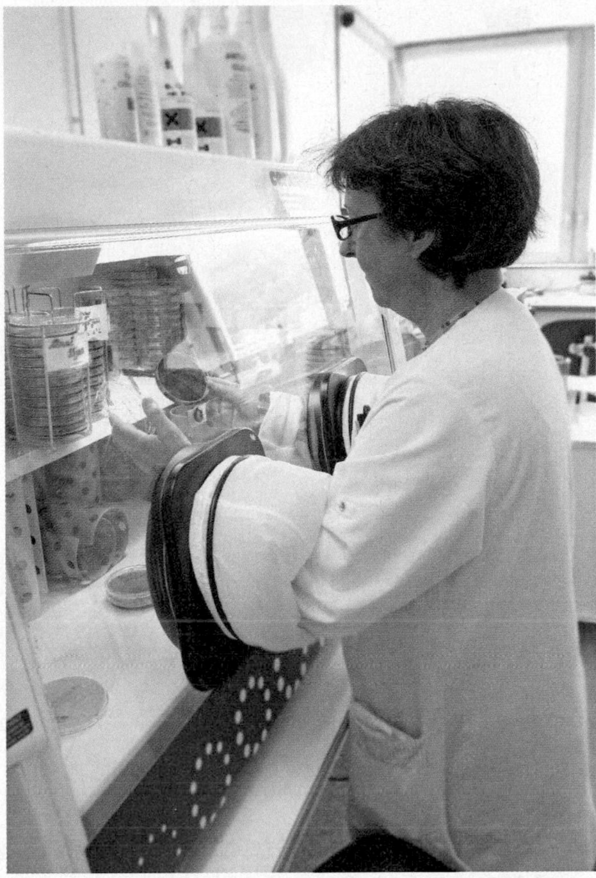

FIGURE 4.14 Anaerobic Chamber The enclosed compartment can be maintained as an anaerobic environment. A special port that can be filled with inert gas is used to add or remove items. The airtight gloves allow the researcher to handle items within the chamber. BSIP SA/Alamy Stock Photo

Why would an anaerobic chamber be preferable to an anaerobic container?

MicroAssessment 4.6

Culture media can be complex or chemically defined. Some media contain additional ingredients that make them selective or differential. Appropriate atmospheric conditions must be provided to grow microbes in culture. An enrichment culture increases the relative concentration of an organism growing in a broth.

15. Distinguish between complex and chemically defined media.

16. Describe two methods to create anaerobic conditions.

17. Would bacteria that cannot use lactose be able to grow on MacConkey agar?

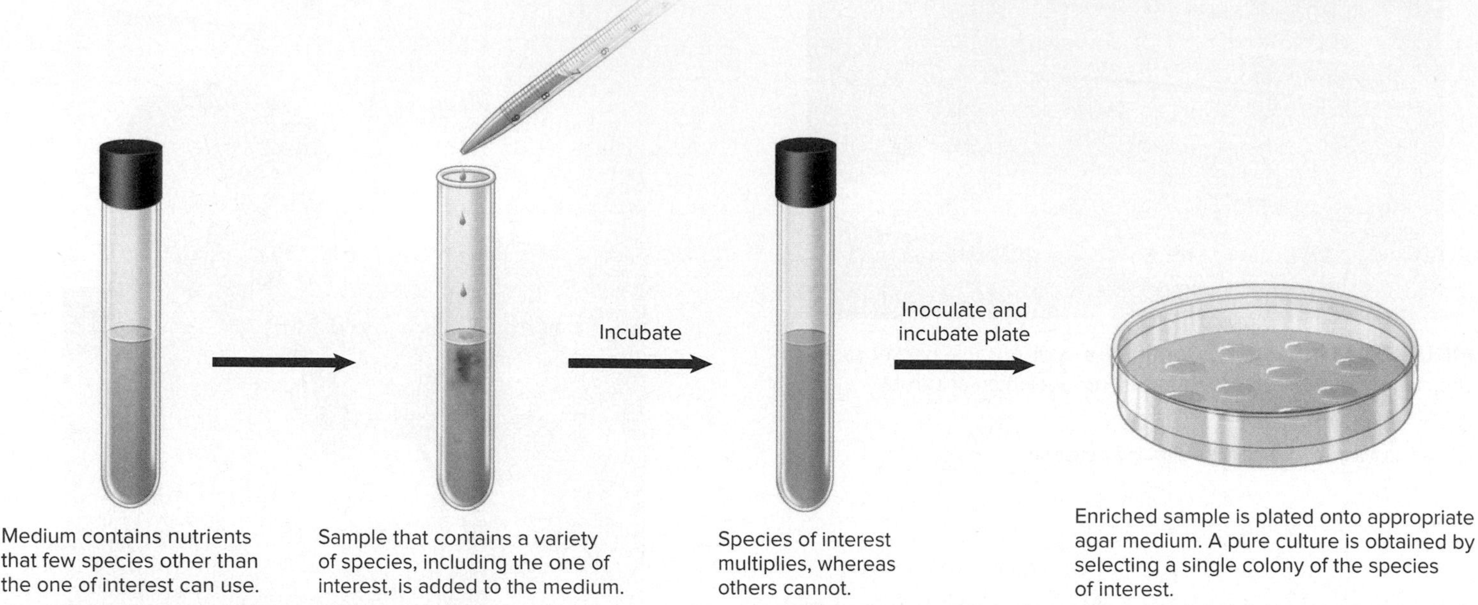

Incubate
Inoculate and incubate plate
Medium contains nutrients that few species other than the one of interest can use.
Sample that contains a variety of species, including the one of interest, is added to the medium.
Species of interest multiplies, whereas others cannot.
Enriched sample is plated onto appropriate agar medium. A pure culture is obtained by selecting a single colony of the species of interest.

FIGURE 4.15 Enrichment Culture Medium and incubation conditions favor growth of the desired species over other microorganisms in the same sample.

How would you enrich for an organism that can use phenol as a carbon source?

4.7 ▪ Methods to Detect and Measure Microbial Growth

Learning Outcome

15. Compare and contrast methods used for direct cell counts, viable cell counts, measuring turbidity, and detecting cell products.

A variety of methods can be used to monitor microbial growth; the choice depends on several factors including the nature of the sample and the expected number of microorganisms. A table at the end of this section summarizes the characteristics of methods used (see table 4.8).

Direct Cell Counts

Direct cell counts are particularly useful for determining the total numbers of microorganisms in a sample, including those that cannot be grown in culture. Unfortunately, the methods generally do not distinguish between living and dead cells in the specimen.

Direct Microscopic Count

One of the most rapid methods of determining the cell concentration in a suspension is the direct microscopic count (**figure 4.16**). A liquid specimen is added to a hemocytometer, a special glass slide designed specifically for counting cells. The slide has a thin chamber that holds a known volume of liquid atop a microscopic grid. The contents of the chamber

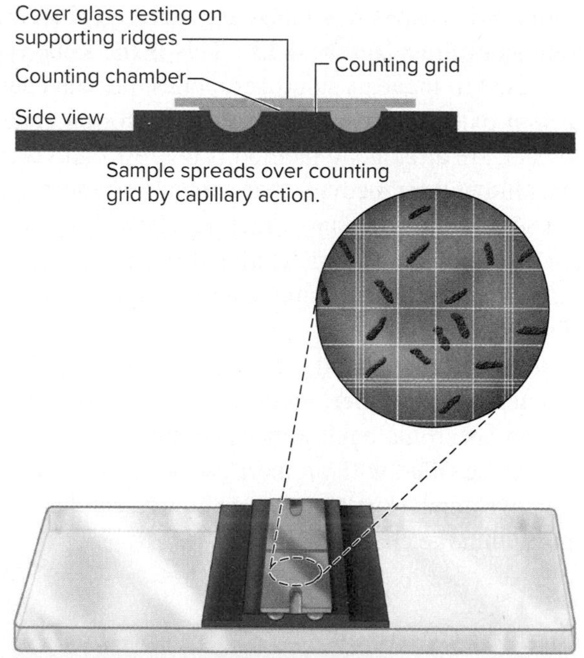

Cover glass resting on supporting ridges
Counting chamber
Counting grid
Side view
Sample spreads over counting grid by capillary action.
Using a microscope, the cells in several large squares like the one shown are counted and the results averaged. To determine the number of cells per ml, that number must be multiplied by 1/volume (in mL) held in the square. For example, if the square holds 1/1,250,000 mL, then the number of cells must be multiplied by 1.25 × 10⁶ mL.

FIGURE 4.16 Direct Microscopic Count The counting chamber holds a known volume of liquid. The number of microbial cells in that volume can be counted precisely.

What is one disadvantage of doing a direct microscopic count to determine the number of cells in a dilute suspension?

can be viewed under the light microscope, so the number of cells in a given volume can be counted precisely. A concentration of about 1 million (10^6) bacteria per milliliter (mL) is recommended to obtain an accurate count using this method, so a sample might need to be diluted or concentrated beforehand.

Cell-Counting Instruments

Cell-counting instruments are electronic devices that count cells in a suspension as they flow single file (one cell at a time) through a narrow channel past a detector. A **Coulter counter** counts the brief changes in resistance that occur when bacteria or other nonconducting particles in an electrically conducting fluid pass. A **flow cytometer** counts cells based on the light scattered as they pass one or multiple lasers (**figure 4.17**). Its advantage is that it can count total cells, cells of a particular size and granularity (internal complexity), or a specific subset that has been stained with a fluorescent dye or tag. A modification of flow cytometry is used to sort cells: As droplets of a suspension pass the detector, a device applies an electrical charge to any that contain a cell with a certain characteristic, and a mechanism then sorts the droplets based on their charge (see figure 17.12).

Viable Cell Counts

Viable cell counts determine the number of cells capable of multiplying. This type of count is particularly valuable when working with samples such as food and water that contain too few microbes for a direct microscopic count. In addition, by using appropriate selective and differential media, these methods can be used to count the cells of a particular microbial species.

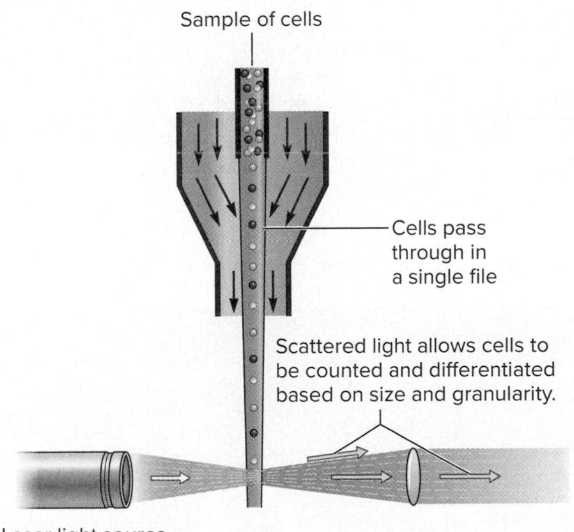
Sample of cells
Cells pass through in a single file
Scattered light allows cells to be counted and differentiated based on size and granularity.
Laser light source

FIGURE 4.17 A Flow Cytometer A cell suspension is focused into a single file by a specific type of fluid (arrows in blue region), and the individual cells are counted and differentiated as they scatter light from a laser.

? **How can a flow cytometer be used to count cells expressing a specific protein on its surface?**

Plate Counts

Plate counts measure the number of viable cells in a sample by taking advantage of the fact that an isolated microbial cell on an agar medium will give rise to one colony. A simple count of the colonies determines how many cells were in the initial sample. Plate counts generally are done only if a sample contains more than 100 organisms/mL. Otherwise, few if any cells will be transferred to the plates. In these situations, alternative methods give more reliable results.

When counting colonies, the ideal number on a plate is between 30 and 300; numbers outside that range are more likely to be inaccurate. Samples usually contain many more cells than that, so they generally must be diluted by a stepwise process called serial dilution (**figure 4.18**). This is done using a sterile liquid called the diluent, often physiological saline (0.85% NaCl in water). Dilutions are normally done in 10-fold increments, making the resulting math relatively simple.

Two techniques can be used to add samples to an agar plate (**figure 4.19**). In the **spread-plate method,** 0.1–0.2 mL of the diluted sample is transferred onto a plate of a solidified agar medium. It is then spread over the surface of the agar with a sterilized bent glass rod that resembles a miniature hockey stick. In the **pour-plate method,** 0.1–1.0 mL of the diluted sample is transferred to a sterile Petri dish; a melted agar medium cooled to 50°C is then added (at this temperature, agar is still liquid). The dish is then gently swirled to mix the microbial cells with the liquid agar. When the agar hardens, the individual cells become fixed in place and form colonies when incubated; colonies on the surface are larger than those embedded in the medium. Regardless of how the samples were added to the agar plate, the number of colonies formed after incubation is used to determine the concentration of the sample's **colony-forming units (CFUs)**—a measure that accounts for the fact that microbial cells often attach to one another and then grow to form a single colony. When calculating CFUs, three things must be considered: the number of colonies, the amount the sample was diluted before being plated, and the volume plated.

In a clinical lab, a simplified version of a plate count is done on urine specimens to help diagnose urinary tract infections (UTIs). The number of CFUs in these samples is important because small numbers are likely due to bacterial contamination of the sample during the collection process rather than to an actual infection. To determine the approximate number of CFUs per mL of urine, a calibrated loop is used to inoculate an agar plate with a tiny amount of specimen (as little as 0.001 mL). That same loop is then used to spread the sample, without flaming, to obtain isolated colonies. The labs have a set of criteria—(based on the urine collection method and the patient population)—for deciding how many CFUs in the specimen indicate a UTI.

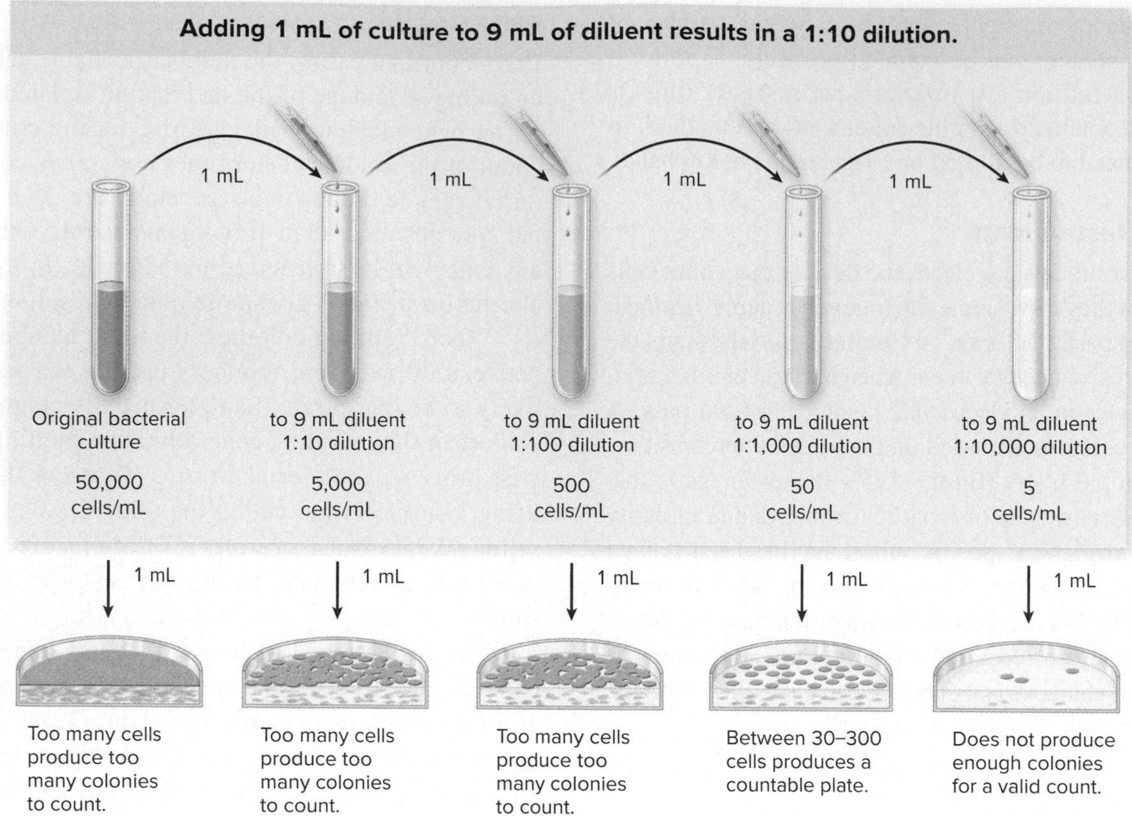
Adding 1 mL of culture to 9 mL of diluent results in a 1:10 dilution.
1 mL
1 mL
1 mL
1 mL
Original bacterial culture
50,000 cells/mL
to 9 mL diluent 1:10 dilution
5,000 cells/mL
to 9 mL diluent 1:100 dilution
500 cells/mL
to 9 mL diluent 1:1,000 dilution
50 cells/mL
to 9 mL diluent 1:10,000 dilution
5 cells/mL
1 mL
Too many cells produce too many colonies to count.
Too many cells produce too many colonies to count.
Too many cells produce too many colonies to count.
Between 30–300 cells produces a countable plate.
Does not produce enough colonies for a valid count.

FIGURE 4.18 Making Serial Dilutions To decrease the concentration of cells in a sample, 10-fold dilutions are often used.

? If the 1 mL sample had been added to 99 mL of diluent to make the first dilution (instead of 9 mL), how many cells would be in that particular dilution?

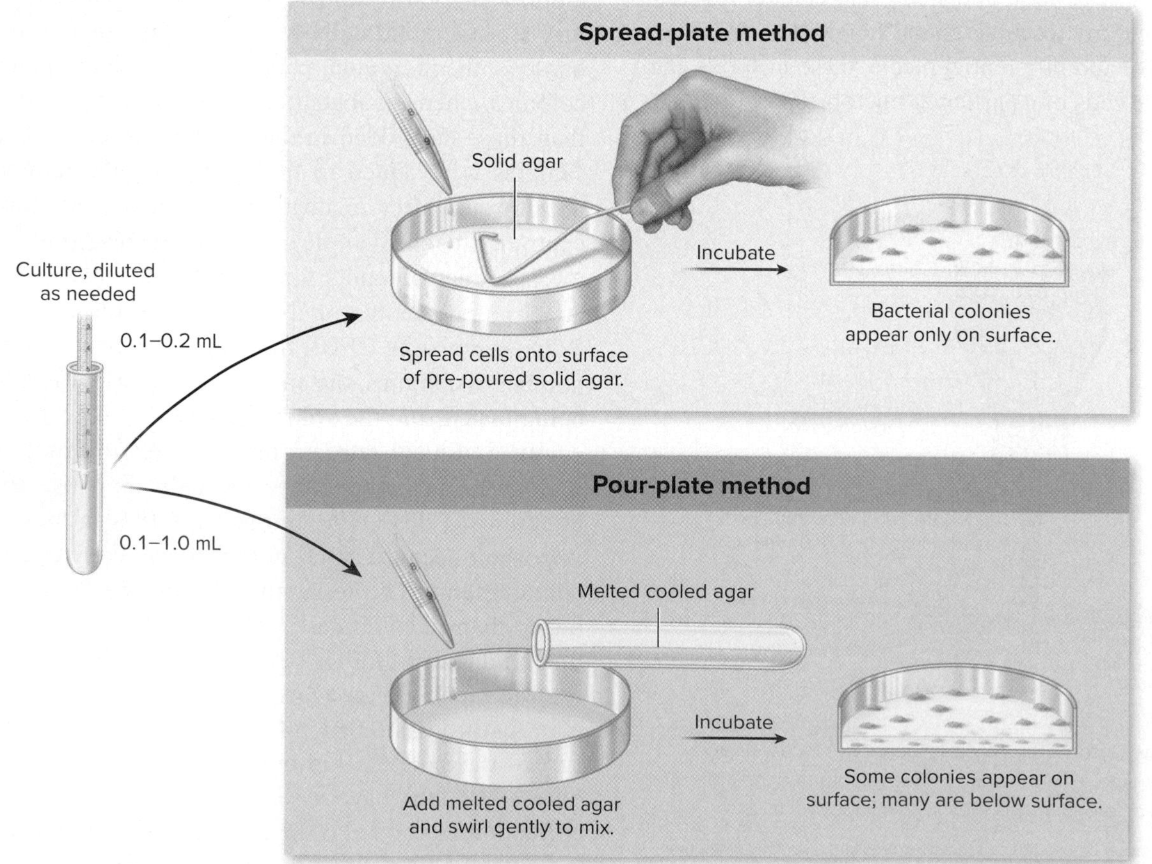
Spread-plate method
Solid agar
Incubate
Bacterial colonies appear only on surface.
Culture, diluted as needed
0.1–0.2 mL
Spread cells onto surface of pre-poured solid agar.
Pour-plate method
0.1–1.0 mL
Melted cooled agar
Incubate
Some colonies appear on surface; many are below surface.
Add melted cooled agar and swirl gently to mix.

FIGURE 4.19 Spread Plates and Pour Plates These techniques can be used for plate counts.

? Why are the results of plate counts expressed as the number of colony-forming units, rather than the number of cells?

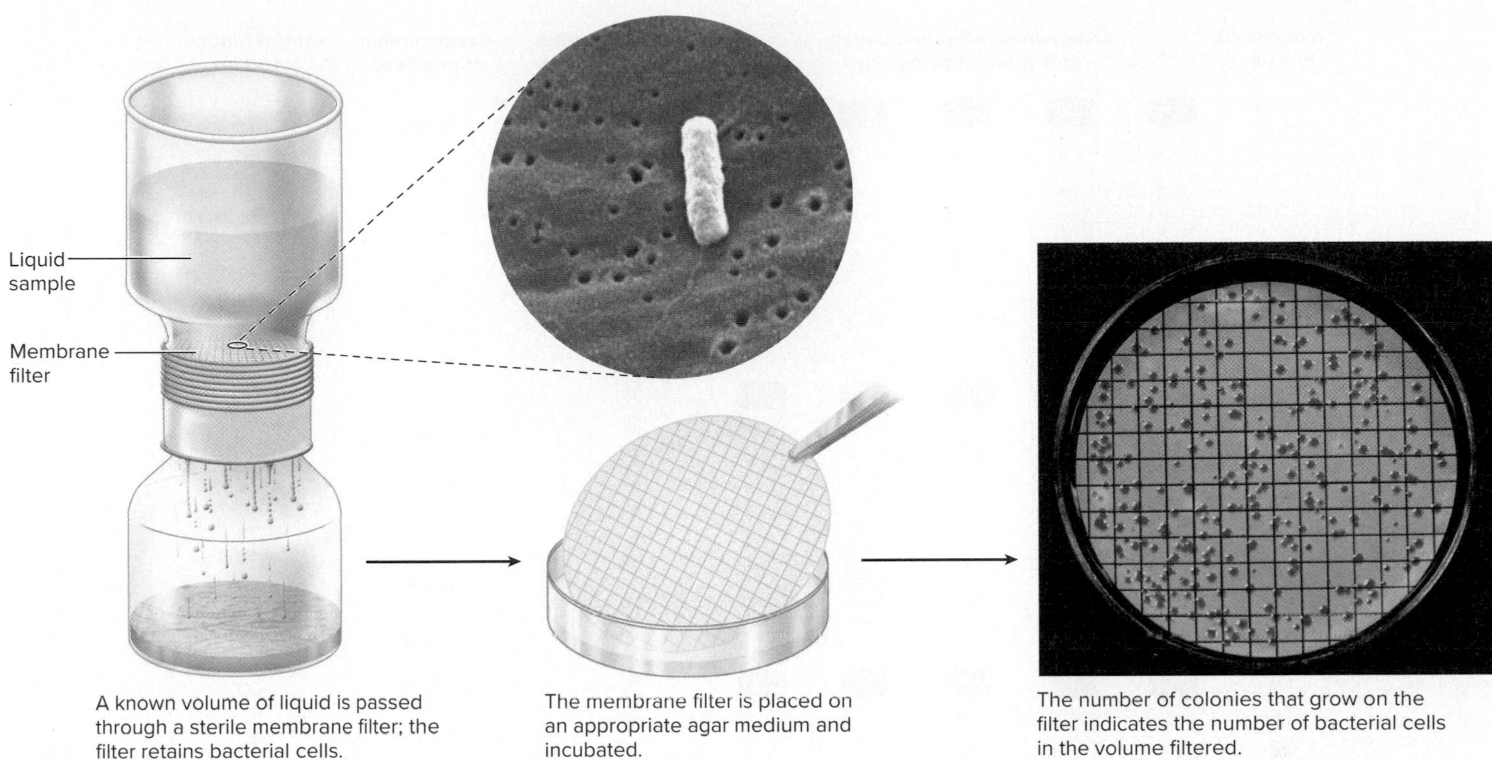

A known volume of liquid is passed through a sterile membrane filter; the filter retains bacterial cells.

The membrane filter is placed on an appropriate agar medium and incubated.

The number of colonies that grow on the filter indicates the number of bacterial cells in the volume filtered.

FIGURE 4.20 Membrane Filtration This technique catches the microbial cells on a membrane filter. top: Margaret Williams, PhD Claressa Lucas, PhD Tatiana Travis, BS/CDC; bottom: Lisa Burgess/McGraw Hill

? **In what situation would membrane filtration be used rather than a plate count?**

Membrane Filtration

Membrane filtration is used to count microorganisms in liquid samples that contain relatively few cells, as might occur in environments such as lakes or other natural bodies of water. A known volume of the liquid is passed through a sterile membrane filter that has a pore size small enough to prevent microorganisms from passing through (**figure 4.20**). By doing this, any microorganisms in the liquid are caught on the paper-thin filter. The filter is then placed on an appropriate agar medium and incubated. The number of colonies that form on the filter indicates the number of cells in the volume filtered.

Most Probable Number (MPN)

The **most probable number (MPN) method** provides an estimate of the concentration of cells in a specimen. The procedure uses a series of dilutions to determine the point at which additional dilutions receive no cells.

To determine the MPN, three sets of three or five tubes containing a growth medium are prepared (**figure 4.21**). Each set receives a measured amount of a sample such as water, soil, or food. The amount added is determined, in part, by the expected microbial concentration in that sample. What is important is that the second set receives 10-fold less than the first, and the third set 100-fold less. In other words, each set is inoculated with an amount 10-fold less than the previous set. After incubation, the presence or absence of turbidity (cloudiness) or other indication of growth is noted; the results are then compared against an MPN table, which gives a statistical estimate of the cell concentration.

Measuring Turbidity

Instead of directly counting the number of cells, the cloudiness or **turbidity** of a microbial suspension can be determined; it is proportional to the concentration of cells and is measured with a spectrophotometer (**figure 4.22**). This instrument shines light through a specimen and measures the percentage that reaches a light detector; the more cells present in the sample, the less light reaches the light detector. The amount of light that reaches the detector—called the transmittance—is inversely proportional to the amount absorbed, which is referred to as the optical density or absorbance. It is important to note that both viable and non-viable cells will be measured when determining optical density.

To use turbidity to estimate cell numbers, a one-time test must be done to determine the correlation between optical density and cell concentration for the specific organism and conditions under study. Once this correlation has been done—generally using a direct microscopic count or plate count to determine cell concentration—the turbidity measurement becomes a rapid and relatively accurate assay. One limitation of using turbidity to measure microbial growth is that the medium must contain a relatively high concentration of cells to be cloudy; a solution containing 1 million bacteria (10^6) per mL is still perfectly clear, and if it contains 10 times

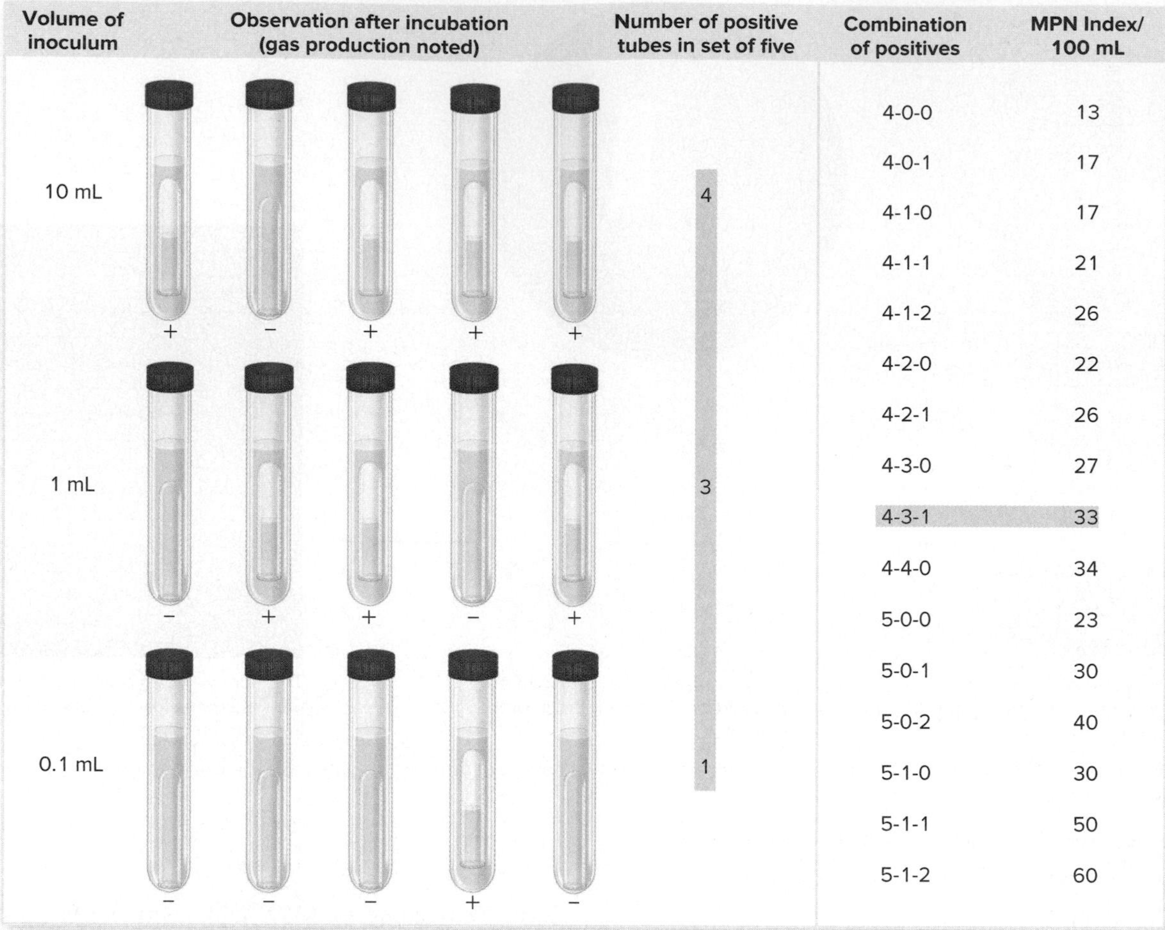

Volume of inoculum
Observation after incubation (gas production noted)
Number of positive tubes in set of five
Combination of positives
MPN Index/ 100 mL
10 mL
1 mL
0.1 mL
+
−
+
+
+
−
+
+
−
+
−
−
−
+
−
4
3
1
4-0-0 13
4-0-1 17
4-1-0 17
4-1-1 21
4-1-2 26
4-2-0 22
4-2-1 26
4-3-0 27
4-3-1 33
4-4-0 34
5-0-0 23
5-0-1 30
5-0-2 40
5-1-0 30
5-1-1 50
5-1-2 60

FIGURE 4.21 The Most Probable Number (MPN) Method In this example, three sets of five tubes containing the same growth medium were prepared. Each set received the indicated amount of inoculum. After incubation, the presence or absence of gas in each tube was noted. The results were then compared to an MPN table to get a statistical estimate of the concentration in the original sample of gas-producing bacteria that could grow in the medium.

If you were using a sample similar to the one illustrated, and all the tubes that received 10 mL had gas, but none that received 1.0 mL or 0.1 mL did, what would be the MPN index/100 mL?

that amount, it is barely turbid. It is important to remember that although a turbid culture indicates that microbes are present, a clear solution does not guarantee their absence.

In a clinical lab, a set of reference tubes called McFarland turbidity standards are used to prepare standardized inocula for certain tests such as antimicrobial susceptibility testing. By growing a bacterial culture to the same turbidity of a given McFarland turbidity standard, a known approximate cell concentration is achieved.

Detecting Cell Products

Microorganisms produce a variety of acids and, sometimes, gases as a result of their metabolism. Acids can be detected by including a pH indicator in the culture medium. Several pH indicators are available, and they differ in the pH value at which their color changes. To detect gas production, inverted tubes (Durham tubes) can be used to trap gas bubbles in broth cultures. Clinical labs use more sensitive methods to detect the slight amounts of CO_2 produced by bacteria growing in patient blood samples. One method uses a fluorescent sensor to detect the slight decrease in pH that accompanies the production of CO_2.

Characteristics of methods used to detect and measure microbial growth are summarized in **table 4.8.**

MicroAssessment 4.7

Direct microscopic counts and cell-counting instruments generally do not distinguish between living and dead cells. Plate counts determine the number of cells capable of multiplying; membrane filtration can be used to count microbial cells in dilute liquids. The most probable number (MPN) estimates cell concentration. Turbidity of a culture can be correlated to cell concentration. Microbial growth can be detected by the presence of cell products such as acids and gases.

18. When doing plate counts, why is it often necessary to dilute the sample?

19. Why is an MPN an estimate rather than an accurate number?

20. Water that accumulates in dirty bowls and dishes left in the sink over an extended period of time often becomes cloudy. Besides suspended food particles, what causes the turbidity?

(a) The cloudiness, or turbidity, of the liquid in the tube on the left is proportional to the concentration of cells.

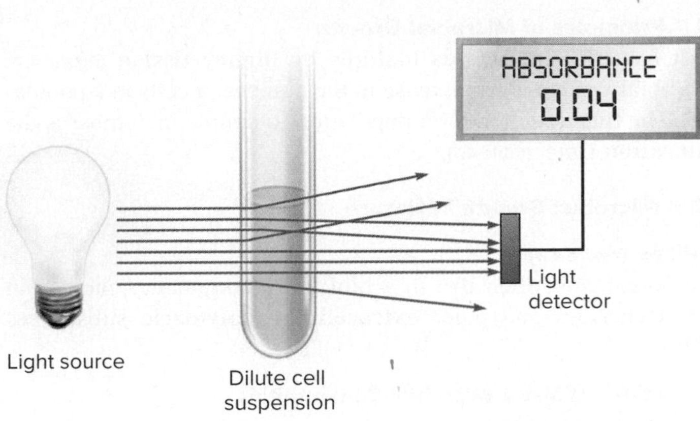

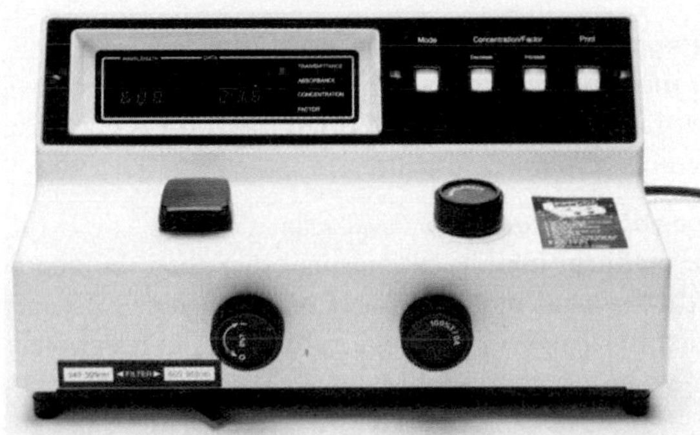

(b) A spectrophotometer is used to measure turbidity.

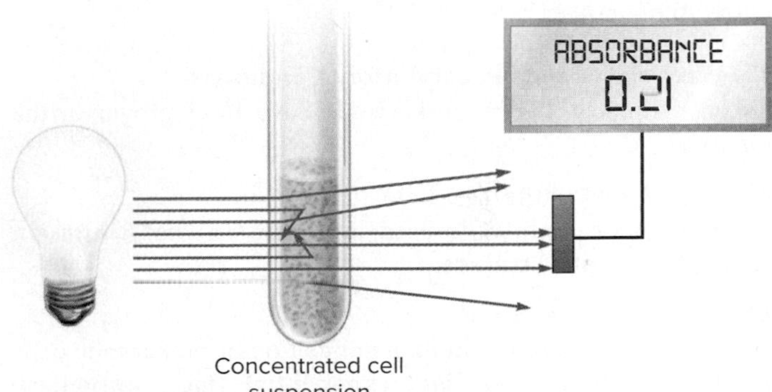

(c) The percentage of light that reaches the detector of the spectrophotometer is inversely proportional to the optical density (absorbance).

FIGURE 4.22 Measuring Turbidity with a Spectrophotometer a: Lisa Burgess/McGraw Hill; b: Martin Shields/Science Source

? Approximately how many bacterial cells must be in a suspension for it to be cloudy?

TABLE 4.8 Methods Used to Detect and Measure Microbial Growth

Method	Characteristics and Limitations
Direct Cell Counts	Used to determine total number of cells; counts include living and dead cells.
Direct microscopic count	Rapid, but sample might require dilution or concentration beforehand.
Cell-counting instruments	Coulter counters and flow cytometers count total cells in dilute solutions. Flow cytometers can also be used to differentiate and sort cells with certain characteristics.
Viable Cell Counts	Used to determine the number of cells capable of growing in a given set of conditions. Requires an incubation period of approximately 24 hours or longer. Selective and differential media can be used to determine the number of specific microbial species.
Plate count	Time-consuming but technically simple method that does not require sophisticated equipment. Generally used only if the sample has at least 10^2 cells/mL.
Membrane filtration	Used to count microorganisms in liquids that contain relatively few cells; the cells are caught on a paper-thin filter.
Most probable number	Estimates the likely cell concentration; it does not provide a precise count.
Measuring Turbidity	Cloudiness of cells in suspension is proportional to cell number. Very rapid method; used routinely. A one-time correlation with plate counts is required for turbidity to be used in determining the number of viable cells.
Detecting Cell Products	Acid and gas production can be used to detect growth.

Summary

4.1 ■ Principles of Microbial Growth

Most bacteria and archaea multiply by **binary fission** (figure 4.1). Microbial growth is an increase in the number of cells in a population. The time required for a population to double in number is the **generation time** (table 4.1).

4.2 ■ Microbial Growth in Nature

Biofilms (figure 4.3)
Microorganisms often live in a **biofilm,** a community encased in polysaccharides and other **extracellular polymeric substances (EPS).**

Interactions of Mixed Microbial Communities
Microorganisms often grow in close associations containing multiple different species. The activities of one organism often affect the growth of another.

4.3 ■ Microbial Growth in Laboratory Conditions

Only an estimated 1% of prokaryotes have been grown in the laboratory.

Obtaining a Pure Culture (figure 4.5)
The **streak-plate method** is used to isolate microorganisms in order to obtain a **pure culture.**

The Growth Curve (figure 4.6)
When grown in a **closed system,** a population of prokaryotic cells goes through five phases: **lag, exponential (log), stationary, death,** and **prolonged decline.**

Colony Growth
The position of a single cell within a colony influences its environment.

Continuous Culture
Microbes can be maintained in a state of continuous growth by using a **chemostat.**

4.4 ■ Environmental Factors That Influence Microbial Growth (table 4.3)

Temperature Requirements (figure 4.8)
Organisms can be grouped as **psychrophiles, psychrotrophs, mesophiles, thermophiles,** or **hyperthermophiles** based on their optimum growth temperatures.

Oxygen (O_2) Requirements (table 4.2)
Organisms can be grouped as **obligate aerobes, facultative anaerobes, obligate anaerobes, microaerophiles,** or **aerotolerant anaerobes** based on their oxygen (O_2) requirements.

pH
Organisms can be grouped as **neutrophiles, acidophiles,** or **alkaliphiles** based on their optimum pH.

Water Availability
Halophiles are adapted to live in high-salt environments.

4.5 ■ Nutritional Factors That Influence Microbial Growth

Required Elements (table 4.4)
The **major elements** make up cell components and include carbon, nitrogen, sulfur, and phosphorus. **Trace elements** are required in very small amounts.

Growth Factors
Microorganisms that cannot synthesize cell components such as amino acids and vitamins require these as **growth factors.**

Energy Sources
Organisms harvest energy either from sunlight or from chemical compounds.

Nutritional Diversity (table 4.5)
Photoautotrophs use the energy of sunlight along with the carbon in the atmosphere to make organic compounds. **Chemolithoautotrophs** use inorganic compounds for energy and derive their carbon from CO_2. **Photoheterotrophs** use the energy of sunlight and obtain their carbon from organic compounds. **Chemoorganoheterotrophs** use organic compounds for energy and as a carbon source.

4.6 ■ Cultivating Microorganisms in the Laboratory

General Categories of Culture Media (table 4.7)
A **complex medium** contains a variety of ingredients such as peptones and extracts. A **chemically defined medium** is composed of precise mixtures of pure chemicals.

Special Types of Culture Media (figure 4.12)
A **selective medium** inhibits organisms other than the one being sought. A **differential medium** contains a substance that certain microorganisms change in a recognizable way; conventional and chromogenic options are available.

Providing Appropriate Atmospheric Conditions
A candle jar provides increased CO_2, which enhances the growth of many medically important bacteria. Microaerophilic microbes are incubated in a gastight container along with a packet that generates low O_2 conditions. Anaerobes may be incubated in an **anaerobic container** or an **anaerobic chamber** (figures 4.13, 4.14).

Enrichment Cultures (figure 4.15)
An **enrichment culture** provides conditions in a broth that enhance the growth of one particular organism in a mixed population.

4.7 ■ Methods to Detect and Measure Microbial Growth (table 4.8)

Direct Cell Counts
Direct microscopic counts involve counting cells viewed through a microscope (figure 4.16). Both a **Coulter counter** and a **flow cytometer** can count cells as they pass through a narrow channel (figure 4.17).

Viable Cell Counts
Plate counts measure the number of viable cells by taking advantage of the fact that an isolated cell growing on an agar medium will form a single colony (figures 4.18, 4.19). **Membrane filtration** catches microbial cells in a liquid sample on a paper-thin filter; the filter is then incubated on an agar medium and the colonies counted (figure 4.20). The **most probable number (MPN) method** provides a statistical estimation of the cell concentration (figure 4.21).

Measuring Turbidity
Turbidity of a culture can be correlated with the concentration of cells; a spectrophotometer is used to measure turbidity (figure 4.22).

Detecting Cell Products
Products including acids and gases can indicate microbial growth.

Review Questions

Short Answer

1. Describe a detrimental effect and a beneficial effect of biofilms.
2. Define a *pure culture*.
3. Explain what occurs during each of the five phases of microbial growth.
4. Explain how the environment of a colony differs from that of cells growing in a liquid broth.
5. List the five categories of optimum temperature, and describe a corresponding environment in which a representative might thrive.
6. Why would botulism be a concern with canned foods?
7. Explain why O_2-containing atmospheres kill some microbes.
8. Explain why photoautotrophs are primary producers.
9. Distinguish between a selective medium and a differential medium.
10. If the number of microorganisms in lake water were determined using both a direct microscopic count and a plate count, which method would most likely give a higher number? Why?

Multiple Choice

1. If there are 10^3 cells per mL at the middle of log phase, and the generation time of the cells is 30 minutes, how many cells will there be 2 hours later?
   a) 2×10^3
   b) 4×10^3
   c) 8×10^3
   d) 1.6×10^4
   e) 1×10^7
2. Compared with their growth in the laboratory, bacteria in nature generally grow
   a) more slowly.
   b) faster.
   c) at the same rate.
3. Cells are most sensitive to penicillin during which phase of the growth curve?
   a) Lag
   b) Exponential
   c) Stationary
   d) Death
   e) More than one of these
4. Which best describes the intestinal tract?
   a) An open system with a chemically defined medium
   b) An open system with a complex medium
   c) A closed system with a chemically defined medium
   d) A closed system with a complex medium
5. *E. coli,* a facultative anaerobe, is grown for 24 hours on the same type of solid medium, but under two different conditions: one aerobic, the other anaerobic. The size of the colonies will be
   a) the same under both conditions.
   b) larger when grown under aerobic conditions.
   c) larger when grown under anaerobic conditions.
6. The generation time of a bacterium was measured at two different temperatures. Which results would be expected of a thermophile?
   a) 20 minutes at 10°C; 220 minutes at 37°C
   b) 220 minutes at 10°C; 20 minutes at 37°C
   c) No growth at 10°C; 20 minutes at 37°C
   d) 20 minutes at 45°C; 220 minutes at 65°C
   e) 220 minutes at 37°C; 20 minutes at 65°C
7. Which of the following is *false*?
   a) *E. coli* grows faster in nutrient broth than in glucose-salts medium.
   b) Organisms require nitrogen to make amino acids.
   c) Some eukaryotes can fix N_2.
   d) An organism that grows on ham is halotolerant.
   e) Blood agar is used to detect hemolysis.
8. If the pH indicator were left out of MacConkey agar, the medium would be
   a) complex.
   b) differential.
   c) defined.
   d) defined and differential.
   e) complex and differential.
9. A soil sample is placed in liquid and the number of bacteria in the sample determined in two ways: (1) colony count and (2) direct microscopic count. How would the results compare?
   a) Methods 1 and 2 would give approximately the same results.
   b) Many more bacteria would be estimated by method 1.
   c) Many more bacteria would be estimated by method 2.
   d) Depending on the soil sample, sometimes method 1 would be higher and sometimes method 2 would be higher.
10. If the concentration of *E. coli* in a broth is between 10^4 and 10^6 cells per mL, the best way to determine the precise number of living cells in the sample would be to
    a) use a counting chamber.
    b) plate out an appropriate dilution of the sample on nutrient agar.
    c) determine cell number by using a spectrophotometer.
    d) Any of these three methods would be satisfactory.
    e) None of these three methods would be satisfactory.

Applications

1. You are a microbiologist working for a pharmaceutical company and discover a new metabolite that can serve as a medication. You now must oversee its production. What are some factors you must consider if you need to grow extremely large (5,000 L) cultures of bacteria?
2. High-performance boat manufacturers know that microbes can collect on a boat, ruining its hydrodynamic properties. A boat-manufacturing facility recently hired you to help with this problem because of your microbiology background. What strategies other than routine cleaning would you pursue to come up with a long-term remedy for the problem?

Critical Thinking

1. This figure shows a growth curve plotted on a non-logarithmic, or linear, scale. Compare this with figure 4.6. In both figures, the number of cells increases dramatically during the log or exponential phase. In this phase, the cell number increases more and more rapidly (this effect is more apparent in the accompanying figure). Why should the increase be speeding up?

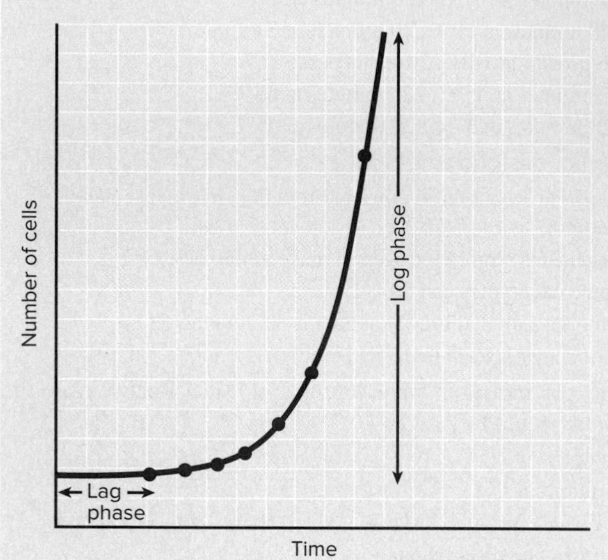

2. In question 1, how would the curve appear if the availability of nutrients were increased?

5 | Control of Microbial Growth

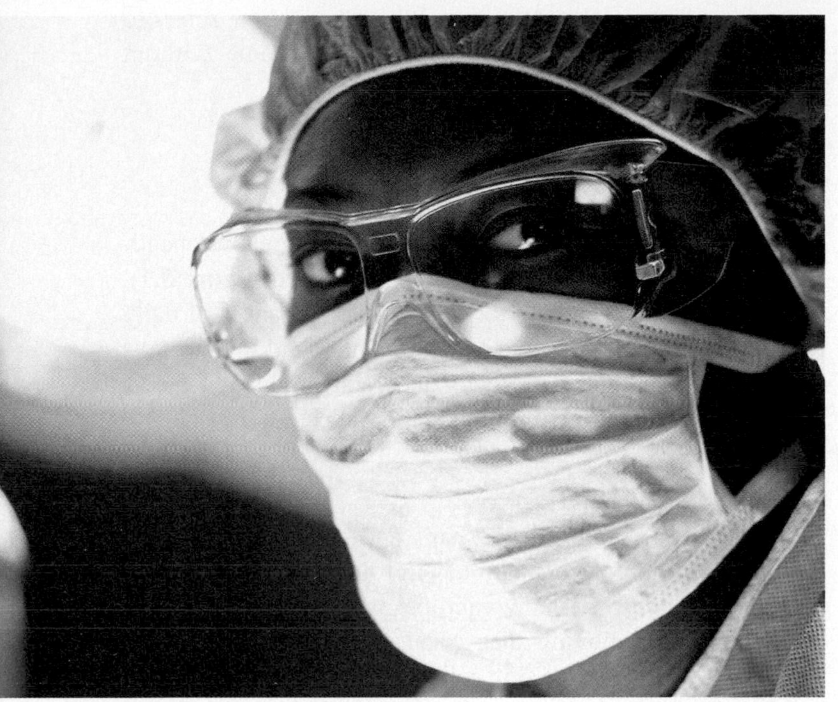

Medical settings warrant a high level of microbial control. *Arthur Tilley/Getty Images*

KEY TERMS

Antimicrobial A chemical or other agent that kills microbes or inhibits their growth.

Antiseptic An antimicrobial chemical nontoxic enough to be used on skin.

Aseptic Technique Set of procedures used to help prevent the accidental introduction of unwanted microbes.

Bactericide A chemical or other agent that kills bacteria.

Bacteriostatic Describes a chemical or other agent that stops the growth of bacteria without killing them.

Biocide A chemical or other agent that kills or inactivates living cells.

Disinfectant A chemical that destroys many, but not all, microbes on an inanimate surface or object.

Pasteurization A brief heat treatment that reduces the number of spoilage microorganisms and destroys pathogens in a product; a related term, cold pasteurization, refers to physical treatments that have the same outcome without using heat.

Preservation A treatment that inhibits microbial growth to delay spoilage.

Sterile Describes a product that is free of all viable microbes, including endospores and viruses.

Sterilization The destruction or removal of all microbes on or from a product.

A Glimpse of History

The British Medical Journal stated that the British physician Joseph Lister (1827–1912) "saved more lives by the introduction of his system than all the wars of the 19th century together had sacrificed." Lister revolutionized surgery by introducing methods that prevent wounds from becoming infected.

Impressed with Pasteur's work on fermentation (said to be caused by "minute organisms suspended in the air"), Lister wondered if "minute organisms" might also be responsible for the pus that forms in infected wounds. He then experimented by applying carbolic acid—a toxic compound which today is more commonly called phenol—directly onto damaged tissues and found it prevented infections. Carbolic acid wound dressings became standard in his practice, and he took pride in the fact that his patients no longer developed gas gangrene, a serious type of bacterial wound infection. His work also provided impressive evidence for the Germ Theory of Disease, even though microorganisms specific for various diseases were not identified for another decade.

Later, Lister improved his methods even further by using surgical procedures that excluded bacteria from wounds. These procedures included sterilizing instruments before use and maintaining a clean environment in the operating room. Preventing infection was preferable to killing the bacteria after they entered wounds because the toxic effects of the disinfectant could be avoided.

The oral antiseptic Listerine was named for Joseph Lister in 1879 when it was introduced as a surgical antiseptic.

Up until the late nineteenth century, patients having even minor surgeries were at great risk of developing fatal infections due to unsanitary medical practices and hospital conditions. Physicians would not accept that their hands could pass diseases from one patient to the next (see A Glimpse of History, chapter 19), nor did they understand that airborne microscopic organisms could infect open wounds. Fortunately, today's modern hospitals use strict procedures to avoid microbial contamination, allowing most surgeries to be performed with relative safety.

Microbial growth affects more than our health. Manufacturers of a wide variety of goods—ranging from food to lumber—recognize that microorganisms can reduce product quality by causing undesirable changes in the appearance, taste, odor, or function of products.

This chapter covers methods to control microbial growth on inanimate objects and some body surfaces. The methods often use **antimicrobials,** meaning chemicals or other agents that kill microbes or inhibit their growth. The antimicrobials described in this chapter are **biocides** (*bio* means "life," and *cida* means "to kill"), damaging all forms of life. In contrast, antibiotics and other antimicrobial medications are specifically toxic to microbes, which is why they are so valuable in treating infectious diseases. They will be discussed in chapter 20.

5.1 ■ Approaches to Control

Learning Outcomes

1. Explain the principles of sterilization, disinfection, pasteurization, decontamination, sanitization, and preservation.
2. Compare and contrast the methods used to control microbial growth in daily life, healthcare settings, microbiology laboratories, food and food production facilities, water treatment facilities, and other industries.

The various processes used to control microbes are either physical or chemical, or a combination of both. Physical methods include heat treatment, irradiation, filtration, high pressure, and mechanical removal (washing). Chemical methods use any of a variety of biocidal chemicals that damage cells by denaturing proteins, altering DNA, or disrupting cell membranes. The method chosen depends on the circumstances and level of control required.

Principles of Control

Sterilization is the removal or destruction of all microorganisms and viruses on or in a product. The destruction of microorganisms means they are no longer viable (cannot be "revived" to multiply even when transferred from the sterilized product to an ideal growth medium). A **sterile** item is free of all viable microbes, including endospores. Note, however, that the term *sterile* does not consider prions. These infectious protein particles are not destroyed by standard sterilization procedures.

Disinfection is the elimination of most or all pathogens on or in a material. In practice, the term generally implies the use of biocides. Unlike with sterilization, some viable microbes may remain after disinfection. **Disinfectants** are chemicals used for disinfecting inanimate objects. As they are effective against harmful microbes, they are often called **germicides. Bactericides** kill bacteria, **fungicides** kill fungi, and **virucides** inactivate viruses. **Antiseptics** are antimicrobial chemicals non-toxic enough to be used topically (on the skin or other body surfaces). These are routinely used to decrease microbial numbers on skin before invasive procedures such as surgery.

Decontamination reduces the number of pathogens on or in a product to a safe level, one below that required to initiate infection. The treatment can be as simple as thorough washing, or it may involve the use of heat or disinfectants. **Sanitization** generally implies a process that substantially reduces the microbial population to meet accepted health standards to minimize the spread of disease. Most people expect a sanitized object to look clean. This term does not indicate any specific level of control.

Preservation is the process of delaying spoilage of perishable products. One way to do this is to choose storage conditions that slow microbial growth. Alternatively, chemical preservatives can be added to a product; these are **bacteriostatic,** meaning they inhibit the growth of bacteria but do not kill them. **Pasteurization** is a brief heat treatment that reduces the number of spoilage organisms and destroys pathogens without changing the characteristics of the treated product; a related term, cold pasteurization, refers to physical treatments that have the same outcome without using heat.

Situational Considerations

Methods used to control microbial growth vary greatly depending on the situation and level of control required (**figure 5.1**). Control measures adequate for routine circumstances of daily life are usually not sufficient for situations found in hospitals, microbiology laboratories, foods and food production facilities, water treatment facilities, and other industries.

Daily Life

Washing and scrubbing with soaps and detergents is generally sufficient to control microbes in routine situations. Soap itself does not destroy many microbes, but it helps wash them off while also removing dirt, organic material, and—when used on the body—some skin cells. Regular handwashing and bathing do not harm the beneficial normal skin microbiota, which reside more deeply on underlying layers of skin cells and in hair follicles. Other methods used to control microorganisms in daily life include cooking foods, cleaning surfaces, and refrigeration.

MicroByte

Thorough handwashing with soap and water is the most important step in stopping the spread of many infectious diseases.

Hospitals and Other Healthcare Facilities

Controlling microbes in healthcare settings is particularly important because of the danger of **healthcare-associated infections (HAIs).** Patients in healthcare facilities, particularly hospitals, are often more susceptible to infectious agents because of their underlying conditions. In addition, pathogens are more likely to be found in such settings because of the high concentrations of patients with infectious disease. These patients can shed pathogens in their feces, urine, respiratory droplets, or other body secretions. Hospitals must be especially careful to control microorganisms in operating rooms (**figure 5.2**). Instruments used in invasive surgical procedures must be sterile to avoid introducing even normally harmless microbes into deep body tissue where they could easily cause infection. Appropriate care of surgical wounds after the procedure is also important for preventing infections.

(a) (b) (c) (d) (e)

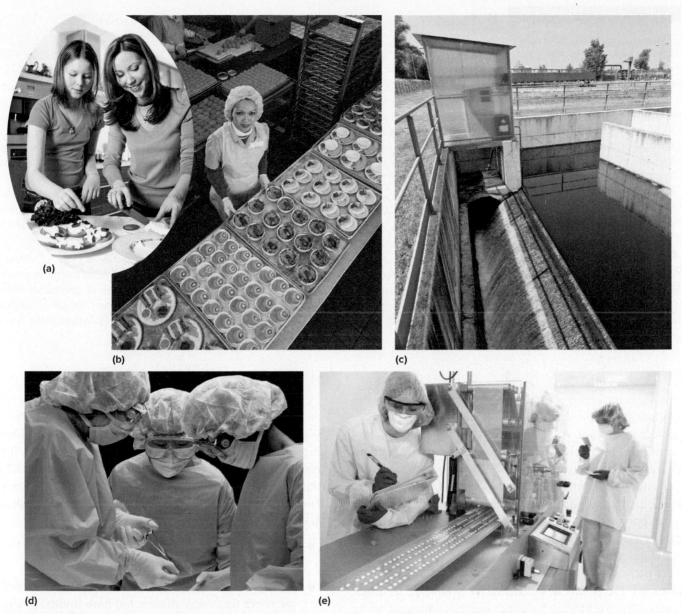

FIGURE 5.1 Situations That Warrant Different Levels of Microbial Control **(a)** Daily home life; **(b)** foods and food production facilities; **(c)** water treatment facilities; **(d)** hospitals; **(e)** other industries. a: Anna Bryukhanova/E+/Getty Images; b: Don Tremain/Getty Images; c: Marka/Alamy Stock Photo; d: Ryan McVay/Getty Images; e: Javier Larrea/age fotostock

? **Why would a hospital require a more stringent level of microbial control than daily home life?**

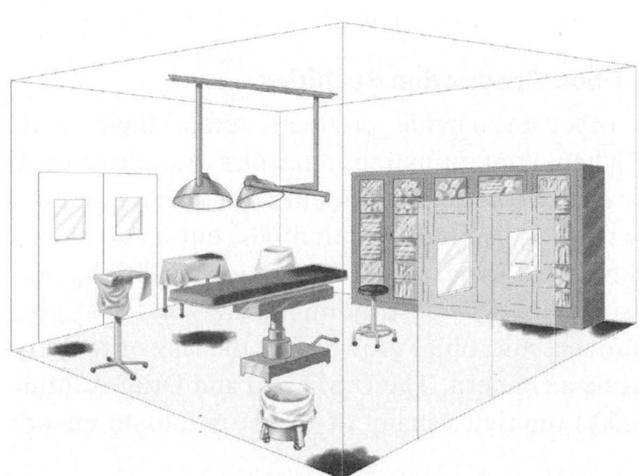

FIGURE 5.2 Contaminated Operating Room The reddish areas indicate places that were contaminated with *Pseudomonas aeruginosa* after treatment of a burn patient. Patient care rooms must be thoroughly cleaned after each use to prevent the spread of infection to others.

? **Will the operating room be sterilized between patients? Explain.**

FOCUS ON A CASE 5.1

The 88-year-old patient was asleep in the emergency department with several family members sitting nearby. She had broken her wrist in a fall but waited two days before calling for help. Doctors thought the wrist would heal with no complications, but the patient had become dehydrated, so they wanted to admit her to the hospital. The patient's granddaughter resisted, however, fearing that the woman would be exposed to dangerous infections. A doctor and nurse entered the room, passed by the germicide dispenser near the door, and approached the patient. The granddaughter stood up and insisted that the two stop and wash their hands.

1. Why was the granddaughter concerned about exposure to infections in the hospital?
2. Would use of the germicide be as effective at preventing the spread of infection as handwashing?
3. The doctor was wearing a necktie. Should the granddaughter be concerned about that?

Discussion

1. Healthcare facilities house a high concentration of patients with infectious disease, including some diseases caused by microbes that have become resistant after repeated exposure to certain antimicrobial medications. Although hospitals use rigorous cleaning techniques to control microbes, vulnerable patients may still have increased risk of infection. Up to 5% of patients who enter a hospital are infected during their stay.
2. Handwashing with soap and water is the best way to avoid the spread of disease. Proper handwashing physically removes microorganisms from the skin and sends them down the drain. Experts recommend singing *Happy Birthday* twice while washing so that the process lasts at least 20 seconds. When handwashing is not practical, the use of an alcohol-based germicide can effectively kill most microbes on the hands' surfaces.
3. Possibly. If the necktie is allowed to drape over a patient during examination, it could carry microorganisms from one patient to another, which is why many male-identifying doctors prefer bow ties.

In addition to protecting patients, healthcare facilities must protect personnel. As the COVID-19 pandemic unfortunately reminded us, healthcare workers are at risk of contracting infectious disease from patients. To minimize the risks, practices called Standard Precautions are used in patient care as a means to prevent infection of both patients and personnel. A set of supplementary measures called Transmission-Based Precautions is used in addition if a patient is, or might be, infected with a highly transmissible or epidemiologically important pathogen.

Prions are a relatively new concern for healthcare facilities. Fortunately, disease caused by these agents is thought to be exceedingly rare—less than 1 case per 1 million persons per year. Healthcare facilities, however, must take special precautions when handling tissue that may be contaminated with prions because these infectious particles are very difficult to destroy.

Microbiology Laboratories

Microbiology laboratories routinely work with microbial cultures and consequently must use rigorous methods to control microorganisms. To ensure that a culture remains pure, all media and instruments that contact it must first be sterilized to avoid contaminating the culture with environmental microbes. After use, all microbial cultures and contaminated items must be treated before disposal to avoid spread of microbes to workers and the environment. The set of procedures used to prevent microorganisms from contaminating an environment is called **aseptic technique.** Although all microbiology laboratory personnel must use these methods, those who work with known pathogens must be even more careful.

The U.S. Centers for Disease Control and Prevention (CDC) has established precautionary guidelines known as biosafety levels (BSLs) for laboratories working with microorganisms. The first level specifies standard aseptic procedures; each higher one includes all of the safety precautions from the previous level as well as additional precautions. The BSL levels include:

- **BSL-1.** Procedures for work with microbes not known to cause disease in healthy people.
- **BSL-2.** Procedures for work with moderate-risk microbes that cause disease but have limited potential for transmission.
- **BSL-3.** Procedures for work with pathogens that cause serious or potentially fatal disease through inhalation.
- **BSL-4.** Procedures for work with easily transmitted deadly pathogens.

Foods and Food Production Facilities

Foods and other perishable products retain their quality longer when contaminating microbes are destroyed, removed, or inhibited. Heat treatment is the most common and reliable method used to kill microbes, but it can alter a product's flavor and appearance. Irradiation and high pressure can also be used to destroy microbes. Chemical preservatives prevent microbial growth, but the risk of toxicity must always be a concern. The U.S. Food and Drug Administration (FDA) regulates many of these options to ensure safety.

Food-processing facilities need to keep surfaces relatively free of microorganisms to avoid contamination. If machinery used to grind meat is not cleaned properly, for example, it can create an environment in which microbes multiply, eventually contaminating large quantities of product.

Water Treatment Facilities

Water treatment facilities need to ensure that drinking water is free of pathogenic microbes. Chlorine has traditionally been used to disinfect water, saving hundreds of thousands of lives by preventing the spread of waterborne illnesses such as cholera. Chlorine and other disinfectants, however, can react with naturally occurring chemicals in the water (such as organic matter) to form **disinfection by-products (DBPs)**, some of which have been linked to long-term health risks (kidney, liver, and central nervous system problems; the development of cancer). In addition, certain intestinal pathogens like *Cryptosporidium hominis* produce structures called oocysts, each having a thick wall that allows them to survive traditional disinfection procedures. To address these problems, facilities are now required to minimize the level of both DBPs and *C. hominis* oocysts in treated water.

Other Industries

Many diverse industries have specialized concerns regarding microbial growth. Manufacturers of pharmaceuticals, cosmetics, deodorants, or any other product that will be ingested, injected, or applied to the skin must avoid microbial contamination that could affect the product's quality or safety.

MicroAssessment 5.1

The methods used to control microbial growth depend on the situation and the level of control required.

1. How is sterilization different from disinfection?
2. Why is it important for a microbiology laboratory technician to use aseptic technique?
3. How would a company that makes lipsticks determine if a batch had been contaminated by microorganisms during the production process?

5.2 ■ Selecting an Antimicrobial Procedure

Learning Outcome

3. Explain how the type and number of microbes, environmental conditions, risk for infection, and composition of the item influence the selection of an antimicrobial procedure.

Selecting an effective antimicrobial procedure is complicated by the fact that every option has disadvantages that limit its use. An ideal, multipurpose, non-toxic method simply does not exist. The ultimate choice depends on many factors, including the type and number of microbes to be controlled, environmental conditions, risk for infection, and the composition of the item to be treated.

Types of Microbes

One of the most critical considerations in selecting an antimicrobial procedure is the composition of the microbial population to be controlled. Products contaminated with microbes highly resistant to killing require a more rigorous treatment. Highly resistant microbes include the following:

- **Bacterial endospores.** The endospores of *Bacillus, Clostridium,* and related genera are the most resistant forms of life typically encountered. Only extreme heat or chemical treatment ensures their complete destruction.
- **Protozoan cysts and oocysts.** Cysts and oocysts are stages in the life cycle of certain intestinal protozoan pathogens such as *Giardia lamblia* and *Cryptosporidium hominis.* These disinfectant-resistant forms appear in the feces of infected humans and other animals, and they can cause diarrheal disease if ingested. Unlike endospores, protozoan cysts and oocysts are easily destroyed by boiling.
- ***Mycobacterium* species.** The hydrophobic cell walls of mycobacteria make them resistant to many chemical treatments. Because of this, stronger, more toxic chemicals must be used to disinfect environments that may contain *Mycobacterium* species, including *M. tuberculosis,* the cause of tuberculosis.
- ***Pseudomonas* species.** These common environmental organisms are not only resistant to some solutions of disinfectants, but in some cases actually grow in them. *Pseudomonas* species can cause serious healthcare-associated infections.
- **Non-enveloped viruses.** Viruses such as poliovirus that lack a lipid envelope are relatively resistant to disinfectants. Conversely, enveloped viruses, such as HIV and the virus that causes COVID-19, tend to be very sensitive to these chemicals.

Number of Microbes

The time it takes for heat or chemicals to destroy a microbial population depends in part on the number of cells or viral particles present; the larger the population, the more time it takes because only a fraction of cells or viral particles are destroyed during a given time interval. Thus, removing most of the microbes by washing or scrubbing can minimize the time necessary to sterilize or disinfect a product. Scrubbing also helps to remove biofilms, which is important because

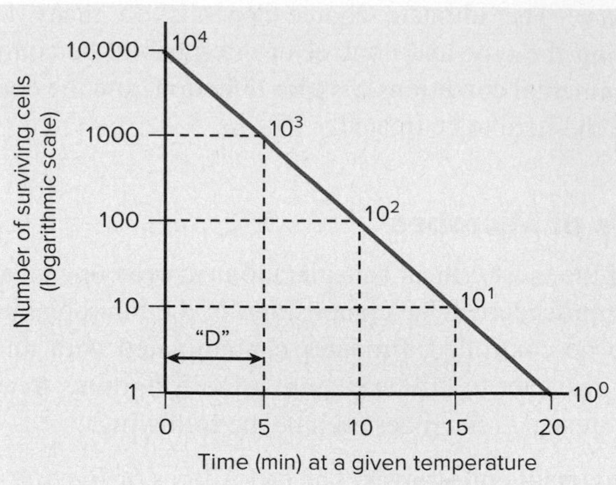

FIGURE 5.3 D Value The decimal reduction time (D value) is the time it takes for an antimicrobial treatment to kill 90% of a microbial population under specified conditions. In this example, the D value is 5 minutes. A population of 10,000 (10^4) organisms would be reduced to only one (10^0) organism after 4 D values, or 20 minutes. A larger population requires more time for elimination under the same conditions.

According to the data in the figure, how many organisms die in the interval from 10–15 minutes after the application of heat?

organisms growing within a biofilm are more resistant than their free-floating counterparts to chemical disinfectants.

The **decimal reduction time,** or **D value,** is the time in minutes it takes for an antimicrobial treatment to kill 90% of a microbial population under specific conditions. In the example illustrated in **figure 5.3,** 90% of a population of 10,000 (10^4) cells is killed in the first 5 minutes at a given temperature, so that 1,000 (10^3) cells remain. During the next 5 minutes, 90% of those cells are killed, leaving 100 (10^2) cells. After another 5 minutes, only 10 (10^1) cells remain. After another 5 minutes, only one (10^0) cell remains. Thus, in this example, it would take 20 minutes (4 D values) to reduce a population of 10^4 cells to only one survivor. Although the concept can be applied to chemical disinfection, it is primarily used by the commercial canning industry to describe adequate processing times at a specified temperature.

Environmental Conditions

Dirt, grease, and body fluids such as blood can interfere with heat penetration and the action of chemical disinfectants. For this reason alone, it is important to thoroughly clean items before disinfection or sterilization.

Temperature and pH also influence the effectiveness of disinfection techniques on microbial death rates. A solution of sodium hypochlorite (household bleach), for example, can kill a suspension of *M. tuberculosis* at a temperature of 55°C in half the time it would take if the suspension were held at 50°C. The hypochlorite solution is even more effective at a lower pH.

Risk for Infection

To guide the selection of germicidal procedures, medical instruments are categorized according to their risk for transmitting infectious agents. Those items that pose greater threats require more rigorous germicidal procedures before use. The categories are as follows:

- **Critical instruments.** These come into direct contact with body tissues; they include needles and scalpels. Critical instruments must be sterile.
- **Semicritical instruments.** These come into contact with mucous membranes but do not penetrate body tissue; they include gastrointestinal endoscopes and endotracheal tubes. Semicritical instruments must be free of all microorganisms and viruses, with the exception that small numbers of endospores may remain. Low numbers of common endospores pose little risk for infection because intact mucous membranes are generally effective barriers against their entry into deeper tissue.
- **Non-critical instruments and surfaces.** These come into contact only with unbroken skin, so they pose little risk for infection. Stethoscopes, blood pressure cuffs, and countertops are examples of non-critical items and surfaces.

Composition of the Item

Some sterilization and disinfection procedures are inappropriate for certain types of material. Heat treatment can damage many types of plastics and other materials. Although irradiation provides an alternative to heat, the process damages some types of plastics. Moist heat (such as boiling water) and liquid chemical disinfectants cannot be used to treat moisture-sensitive materials.

MicroAssessment 5.2

The types and numbers of microbes initially present, environmental conditions, the potential risks associated with use of the item, and the composition of the item must all be considered when determining which sterilization or disinfection procedure to use.

4. Describe three groups of microbes that are resistant to certain chemical treatments.
5. Why must critical instruments be sterile, whereas small numbers of endospores are permitted on semicritical instruments?
6. Is it safe to say that if all bacterial endospores have been killed, then all other medically important microorganisms also have been killed?

5.3 ■ Physical Methods Used to Destroy or Remove Microorganisms and Viruses

Learning Outcomes

4. Compare and contrast the methods and outcomes of boiling, pasteurization, sterilization using pressurized steam, the commercial canning process, and dry heat.
5. Describe how membrane filters and HEPA filters are used to remove microbes.
6. Compare and contrast the use of gamma radiation, ultraviolet radiation, microwaves, and high pressure for destroying microorganisms.

Heat treatment is one of the most useful methods of microbial control because it is reliable, safe, relatively fast, and inexpensive. Other physical methods including filtration, irradiation, and high-pressure treatment can be used to destroy or remove microbes from materials that cannot withstand heat treatment. A table at the end of this section summarizes the physical methods used to destroy or remove microorganisms and viruses (see table 5.1).

Moist Heat

Moist heat destroys microorganisms and viruses primarily by denaturing their proteins. Examples of moist heat treatment include boiling, pasteurization, and pressurized steam.

Boiling

Boiling (100°C at sea level) easily destroys most microorganisms and viruses. Because of this, drinking water that might be contaminated during floods or other emergency situations should be boiled for at least 5 minutes. Boiling is not a method of sterilization, however, because endospores can survive the process.

Pasteurization

Louis Pasteur developed the brief heat treatment we now call pasteurization as a way to prevent spoilage of wine without changing its flavor. Today, pasteurization is still used to destroy heat-sensitive spoilage organisms in foods and beverages, increasing the product's shelf life without significantly altering its quality. More importantly, pasteurization is widely used to destroy pathogens in milk and juices, protecting consumers from diseases such as brucellosis, salmonellosis, and typhoid fever.

Today, most pasteurization is by the **high-temperature–short-time (HTST) method.** With this treatment, milk is heated to 72°C and held for 15 seconds, but the parameters must be adjusted for different products. For example, ice cream is rich in fats, so it is pasteurized at 82°C for about 20 seconds.

Boxed juices and the single-serving containers of creamer served in restaurants require no refrigeration because they are treated using the **ultra-high-temperature (UHT) method** of processing ("ultra-pasteurization"). This destroys microorganisms that can grow under normal storage conditions. The UHT process for milk requires rapidly heating the milk to 140°C, holding it at that temperature for a few seconds, and then quickly cooling it. The product is then aseptically packaged in sterile containers.

MicroByte

Clothes can be pasteurized by regulating the temperature in a washing machine.

Sterilization Using Pressurized Steam

Heat- and moisture-tolerant items such as surgical instruments, most microbiological media, and reusable glassware are sterilized using an **autoclave.** This device is similar in principle to a pressure cooker, using steam under pressure to achieve high temperatures. Steam is pumped into the autoclave chamber as the air is removed. As steam continues to be pumped in, the pressure within the chamber increases as a result (**figure 5.4**). The higher pressure, in turn, increases the temperature at which steam forms. Steam at atmospheric pressure never exceeds 100°C, but steam at 15 psi (pounds per square inch) above that is 121°C, a temperature that kills even endospores. The pressure itself plays no direct role in the killing. The typical sterilization process is 121°C/15 psi

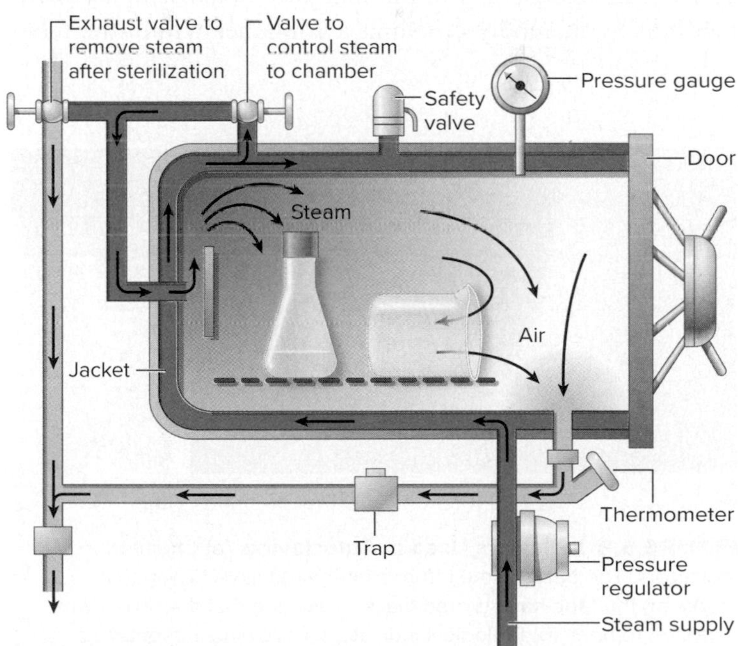

FIGURE 5.4 Autoclave In the illustrated device, steam first travels in an enclosed layer, or jacket, surrounding the chamber. It then enters the chamber, displacing the air downward and out through a port in the bottom of the chamber.

? **Why is autoclaving more effective than boiling?**

above atmospheric pressure for 15 minutes; longer treatment is necessary for large volumes because it takes more time for heat to completely penetrate the substance. When rapid processing is important, such as when sterile instruments must be consistently available in an operating room, flash sterilization with a higher temperature for a shorter time can be used. To destroy prions, the CDC recommends following World Health Organization (WHO) guidelines, which describe three rigorous treatments involving immersion in 1 M (molar) sodium hydroxide (NaOH) while or before autoclaving.

Autoclaving is consistently effective in sterilizing most objects, if done correctly. The temperature and the pressure gauges of the autoclave should be monitored to ensure proper operating conditions. Steam must enter items to displace air, so long, thin containers should be placed on their sides, and containers should never be closed tightly.

When using an autoclave, two types of indicators are important for quality control. Tape with a heat-sensitive indicator that turns black at a high temperature should be attached to items before autoclaving. This provides a visual signal that the items have been properly processed (**figure 5.5a**). A changed indicator, however, does not mean that the object is sterile, only that it has been heated. Biological indicators are used to ensure that an autoclave is working properly (**figure 5.5b**). A tube containing the heat-resistant endospores of *Geobacillus stearothermophilus* can be placed near the center of an autoclave load. After that load is autoclaved, the endospores in the tube are mixed with a growth medium by manually crushing a container within the tube. If the medium changes color after incubation of the tube, microbial growth occurred, indicating an unsuccessful autoclaving process.

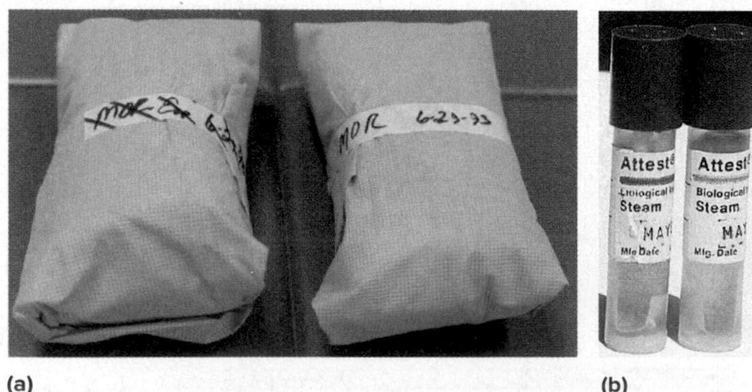

(a) (b)

FIGURE 5.5 Indicators Used in Autoclaving (a) Chemical indicators. The pack on the left has been autoclaved. Diagonal marks on the tape have turned black, indicating that the object was exposed to heat. **(b)** Biological indicators. Following incubation, a change of color to yellow indicates growth of the endospore-forming organism. (a&b): Evans Roberts

? **Why would a biological indicator be better than other indicators to determine if the sterilization procedure was effective?**

The Commercial Canning Process

Commercial canning uses an industrial-size autoclave (called a retort). For low-acid foods (having a pH above 4.6), the process ensures the destruction of all *Clostridium botulinum* endospores when done properly. This is critical because a canned food is anaerobic, and *C. botulinum* is an obligate anaerobe commonly found in the environment; this bacterium produces one of the most potent toxins known—botulinum toxin. If *C. botulinum* endospores survive the canning process in a low-acid food, they can germinate there, and the resulting vegetative cells grow and produce the toxin.

In destroying endospores of *C. botulinum,* the canning process also kills all other organisms that grow under normal storage conditions. Canned foods are **commercially sterile,** meaning that the endospores of some thermophiles may survive. These are usually not a concern, however, because they grow only at temperatures well above those of normal storage.

Several factors dictate the time and temperature of the canning process. First, as discussed earlier, the higher the temperature, the shorter the time needed to kill all organisms. Second, the higher the concentration of organisms, the longer the heat treatment required to kill them all. The commercial canning process for low-acid foods is designed to reduce a population of 10^{12} *C. botulinum* endospores to only one spore. In other words, it is a 12 D process. Because it is virtually impossible for a food to have this many endospores, the process provides a wide safety margin.

Dry Heat

Incineration (destruction by burning) oxidizes cell components to ashes. In microbiology laboratories, the wire loops continually reused to transfer bacterial cultures are sterilized by flaming—heating them in a flame until they are red hot. Alternatively, they can be heated to the same point in a benchtop incinerator designed for this purpose. Incineration is also used to destroy medical wastes and contaminated animal carcasses.

Temperatures achieved in hot air ovens kill microbes by destroying cell components and denaturing proteins. The process requires a higher temperature and a longer time than that for moist heat because dry heat takes longer to penetrate materials and is less efficient at killing microbes. Glass Petri dishes and glass pipettes are sterilized in ovens with non-circulating air at temperatures of 160°C to 170°C for 2 to 3 hours. Ovens with a fan that circulates the hot air can sterilize in a shorter time because they transfer heat more efficiently. Oils and dry materials such as powders are also sterilized in hot ovens.

Filtration

Recall that membrane filtration, used to determine the number of bacteria in a liquid medium, retains microbes on the surface of a filter while allowing the fluid to pass through. That same principle can be used to physically remove microbes from liquids or air.

Filtration of Fluids

Filtration is used extensively to remove organisms from heat-sensitive fluids such as sugar solutions, beer, wine, and some medications. Specially designed filtration units are also used by backpackers and campers to remove *Giardia* cysts and bacteria from natural waters (rivers, streams, and lakes).

Paper-thin **membrane filters** or **microfilters** have microscopic pores that allow liquid to flow through while trapping microbes and other particles too large to pass through (**figure 5.6**). A vacuum is commonly used to help pull the liquid through the filter; alternatively, pressure may be applied to push the liquid. The filters are made of nitrocellulose or other polymers and come in a variety of different pore sizes, including those smaller than any known viruses. Pore sizes smaller than necessary should be avoided, however, because they slow the flow of liquid through the filter. Filters with a pore size of 0.2 micrometers (μm) are commonly used to remove bacteria from a solution.

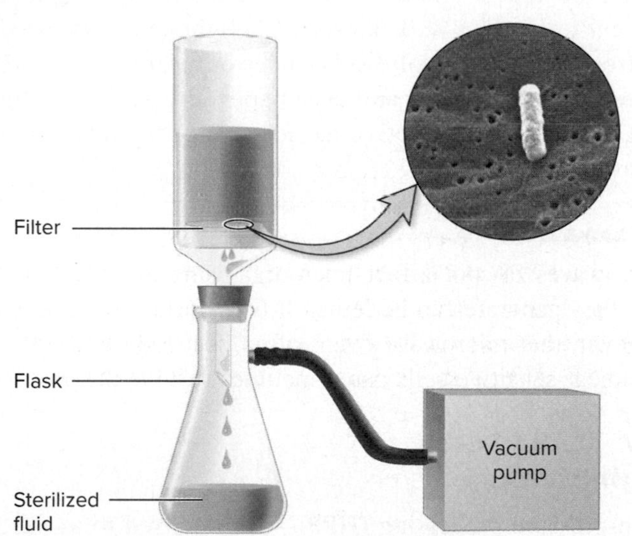

FIGURE 5.6 Sterilization of Fluids Using a Membrane Filter The liquid to be sterilized is pulled through the filter by means of a vacuum produced by a pump. The inset in the figure is a scanning electron micrograph, showing a bacterial cell retained by a membrane filter. Margaret Williams, Ph.D.; Claressa Lucas, Ph.D.; Tatiana Travis, BS/CDC

? **Why would a pore size smaller than necessary be a poor choice?**

Filtration of Air

High-efficiency particulate air (HEPA) filters remove nearly all airborne particles 0.3 μm or larger. These filters are used for keeping microorganisms out of specialized hospital rooms designed for patients extremely susceptible to infection. The filters are also used in biological safety cabinets in which laboratory personnel work with airborne pathogens such as *Mycobacterium tuberculosis.* A continuous stream of incoming and outgoing air flows through HEPA filters to hold microbes within the cabinet. The cabinets are also used to protect samples from contamination.

MicroByte

HEPA filters are used in passenger aircraft (to remove microbes from the recirculated cabin air) and in vacuum cleaners.

Irradiation

Radio waves, microwaves, visible and ultraviolet (UV) light rays, X rays, and gamma rays are all examples of **electromagnetic radiation,** a form of energy that travels in waves. The amount of energy is related to the wavelength, which is the distance from crest to crest (or trough to trough) of a wave. The wavelength is inversely proportional to the frequency (the number of waves per second). Radiation with short waves, and therefore high frequency, has more energy than radiation that has long waves with low frequency. The full range of wavelengths is called the electromagnetic spectrum (**figure 5.7**).

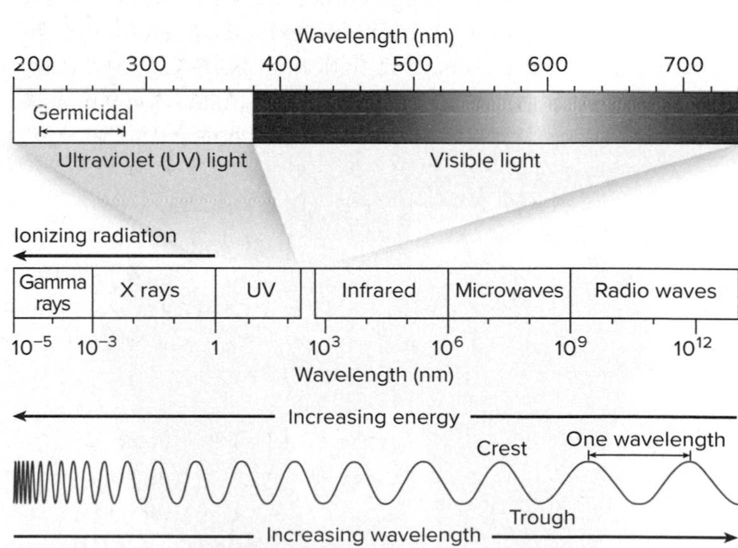

FIGURE 5.7 The Electromagnetic Spectrum Visible wavelengths include the colors of the rainbow.

? **Which types of radiation are ionizing?**

Ionizing Radiation

Ionizing radiation harms cells directly by destroying DNA and damaging cytoplasmic membranes. It also causes indirect damage, reacting with O_2 to produce reactive oxygen species (ROS) such as hydrogen peroxide. Bacterial endospores are among the most radiation-resistant microbial forms, whereas Gram-negative bacteria such as *Salmonella* and *Pseudomonas* species are among the most susceptible. Two important forms of ionizing radiation are gamma rays (emitted from decaying radioisotopes such as cobalt-60) and X rays, both of which penetrate surfaces and can be used to sterilize packaged materials. Ionizing radiation is used to sterilize medical items (equipment, disposable surgical supplies, gloves, and drugs such as penicillin) and laboratory supplies (Petri plates, plastic inoculating loops) as well as food products.

Many countries allow use of irradiation to control microorganisms on or in certain types of foods. The U.S. Food and Drug Administration (FDA) has approved the irradiation of spices, herbs, meats, fruits, vegetables, and grains. Irradiated foods sold in U.S. stores can be identified by a green radura symbol (**figure 5.8**). The number of microbes destroyed by irradiating a food product depends on the dose applied. Treatments designed to sterilize food can cause unwanted flavor changes, however, so lower doses are used to accomplish cold pasteurization (meaning the process eliminates pathogens and decreases the numbers of spoilage organisms without using heat).

Many consumers refuse to accept irradiated products, even though federal and global officials have endorsed the technique. Some people mistakenly believe that irradiated products are radioactive. Others think that irradiation-induced toxins or carcinogens are present in the foods, even though the FDA monitors the effects of radiation on both food and packaging to ensure that consumption of irradiated food is safe.

FIGURE 5.8 Radura Symbol The FDA requires display of this international symbol on foods that have been irradiated. The color and details vary among countries. H. S. Photos/Science Source

? **Why is irradiation not commonly used to sterilize foods?**

Another argument raised against irradiation is that it will allow people to ignore other important food-safety practices. Irradiation, however, is intended to complement, not replace, proper food-handling procedures by producers, processors, and consumers.

Ultraviolet Radiation

Ultraviolet (UV) light is non-ionizing radiation having wavelengths of approximately 220–300 nm. Wavelengths between 220–280 nm, which is within a range referred to as UVC, are most effective for killing microbes. These germicidal wavelengths are effective because they damage DNA (as described in section 8.3) and denature proteins. Actively multiplying organisms are the most easily killed, whereas bacterial endospores are the most UV-resistant. Using UV light as a germicide—referred to as UVGI (ultraviolet germicidal irradiation)—requires safety measures because UV exposure can damage the skin and eyes and promote skin cancer development. When used to disinfect rooms, for example, the germicidal UV lamps are on only when the room is vacant.

UVGI is used extensively to destroy microbes in drinking water and for air and surface disinfection in various settings, including hospitals and rooms where laboratory media is prepared. UVC emitting bulbs are now available for home use as well—for disinfecting surfaces, air, toothbrushes, and other items. UV light penetrates surfaces poorly, however, which limits its use; even a thin film of grease on the UV bulb or the material being treated can make it less effective. Likewise, UV light cannot be used to destroy microbes in solid substances or turbid liquids. Most types of glass, plastic, and even paper screen out UV light, so UVGI is most effective at close range against exposed microorganisms.

Microwaves

Microwaves do not affect microorganisms directly, but the heat they generate can be lethal. It is important to remember, however, that microwave ovens often heat food unevenly, so even heat-sensitive cells can sometimes survive the process.

High Pressure

High-pressure processing (HPP)—also referred to as pascalization or cold pasteurization—kills microbes by using pressures that may approach 120,000 psi. High temperatures are not required, so treated food products such as fruit juices and guacamole retain higher nutrient levels as well as the colors and flavors associated with fresh foods. The high pressure is thought to destroy microbes by denaturing proteins and altering cell permeability.

Table 5.1 summarizes the physical methods used to destroy or remove microorganisms and viruses.

TABLE 5.1 Physical Methods Used to Destroy or Remove Microorganisms and Viruses

Method	Characteristic	Use
Moist Heat	Denatures proteins. Is relatively fast, reliable, safe, and inexpensive.	
Boiling	Destroys most microorganisms and viruses when done for at least 5 minutes, with the notable exception of endospores.	Used to treat drinking water in emergency situations.
Pasteurization	Significantly decreases the numbers of heat-sensitive microorganisms, including spoilage microbes and pathogens; does not destroy endospores.	Used to pasteurize milk and juices.
Pressurized steam (autoclaving)	Destroys endospores during a typical treatment of 15 psi above atmospheric pressure/121°C for at least 15 minutes.	Used to sterilize microbiological media, laboratory glassware, surgical instruments, and other items that steam can penetrate. Used during the canning process to render foods commercially sterile.
Dry Heat	Destroys cell components and denatures proteins.	
Incineration	Burns cell components to ashes.	Used for sterilizing wire inoculating loops (flaming) and destroying medical wastes.
Hot air (ovens)	Requires longer times and higher temperatures than moist heat because it is less efficient.	Used in ovens to sterilize laboratory glassware, oils, and dry materials.
Filtration	Retains microbes while letting the suspending fluid or air pass through small holes.	
Filtration of fluids	Physically removes microbes larger than the pore size used; 0.2 μm pore-sized filters are commonly used to remove bacteria.	Used to remove microbes from heat-sensitive liquids such as beer, wine, and some medications.
Filtration of air (HEPA filters)	Removes microbes that have a diameter of 0.3 μm or greater.	Used in biological safety cabinets, specialized hospital rooms, and airplanes. Also used in some vacuum cleaners and home air purification units.
Irradiation	Damages cells in different ways, depending on the wavelength of radiation used.	
Ionizing radiation	Destroys DNA and possibly damages cytoplasmic membranes. Produces reactive molecules that damage other cell components. Allows items to be sterilized after packaging.	Used to sterilize heat-sensitive materials, including medical equipment, disposable surgical supplies, and drugs such as penicillin. Also used to destroy microbes in spices, herbs, meats, fruits, vegetables, and grains.
Ultraviolet radiation	Damages DNA. Penetrates poorly.	Used to destroy microbes in the air and drinking water, and to disinfect surfaces.
High Pressure	May denature proteins and alter the permeability of cells with treatments up to 120,000 psi.	Used to extend the shelf life of certain food products such as juices and guacamole.

MicroAssessment 5.3

Moist heat such as boiling water destroys most microbes. Pasteurization significantly reduces the numbers of heat-sensitive organisms. Autoclaves use pressurized steam to achieve high temperatures that kill microbes, including endospores. Dry heat takes longer than moist heat to kill microbes. Filters can be used to remove microorganisms and viruses from liquids and air. Ionizing radiation can be used to sterilize products and to decrease the number of microorganisms in foods. Ultraviolet light can be used to disinfect surfaces and air. Microwaves do not kill microbes directly but destroy them by the heat they generate. Extreme pressure can kill microorganisms.

7. Why is it important for the commercial canning process to destroy the endospores of *Clostridium botulinum*?
8. Describe two methods of cold pasteurization. How do they each kill microbes?
9. Why does a cell die when its proteins are denatured?

5.4 ■ Chemical Methods Used to Destroy Microorganisms and Viruses

Learning Outcomes

7. Describe the difference between sterilants, high-level disinfectants, intermediate-level disinfectants, and low-level disinfectants.
8. Describe the important factors to consider when selecting an appropriate germicidal chemical.
9. Compare and contrast the characteristics and uses of alcohols, aldehydes, biguanides, ethylene oxide gas, halogens, metals, ozone, peroxygens, phenolic compounds, and quaternary ammonium compounds as germicidal chemicals.

Germicidal chemicals are used to disinfect and, in some cases, sterilize. Most are biocidal and react irreversibly with proteins, DNA, cytoplasmic membranes, and/or viral envelopes, but their exact mechanisms of action are often poorly understood. They are generally less reliable than heat but useful for treating large surfaces and heat-sensitive items. Some are non-toxic enough to be used as antiseptics.

In the United States, the FDA is responsible for ensuring that chemicals used to disinfect or sterilize medical devices work as claimed and that antiseptics are safe and effective. As a result of recent FDA rulings, various active ingredients previously allowed in certain non-prescription antiseptic products must now be reevaluated before manufacturers can continue using them. The products affected include consumer antiseptic washes (antimicrobial soaps) and antiseptic rubs (hand sanitizers) as well as topical antiseptic products used by healthcare professionals (including hand washes, hand rubs, and products used to disinfect patients' skin in preparation for injections or surgery). In most cases, however, use of the ingredients had already been or were in the process of being phased out. The rulings do not cover antiseptics used in first-aid products or those used by the food industry. Most other chemical disinfectants are considered pesticides and, as such, are regulated by the U.S. Environmental Protection Agency (EPA).

Selecting the Appropriate Germicidal Chemical

Selecting the appropriate germicide is a complex decision. The chemical needs to kill the target microbes without damaging the material being treated. Ideally, the germicide is not dangerous to handle or inhale, but often the most effective options are quite toxic. Also, people generally object to bad odors, so the smell can be a factor. Points to be considered when choosing a germicide include:

- **Toxicity.** Germicides are at least somewhat toxic to humans and the environment. Therefore, the benefit of disinfecting or sterilizing an item or surface must be weighed against the risks associated with using the chemical. The risk of being exposed to pathogens in a hospital often calls for using the most effective germicides, even considering the potential risks of their use. Conversely, typical household and office situations do not justify using many of those same chemicals. To encourage the use of less toxic options, the EPA allows manufacturers to place a specially designed "Safer Choice" label on products considered the least hazardous (**figure 5.9**).
- **Activity in the presence of organic matter.** Hypochlorite (bleach) and many other germicides react with organic matter, losing their effectiveness in doing so. Other chemicals such as phenolics tolerate the presence of some organic matter.
- **Compatibility with the material being treated.** Items such as electrical equipment often cannot withstand liquid chemical germicides, and so gaseous alternatives must be used. Likewise, corrosive germicides such as hypochlorite often damage some metals and rubber.

FIGURE 5.9 Safer Choice Label The EPA allows manufacturers to use this label when all of the ingredients in a product have been tested to assure that they pose the least concern to health and to the environment. H. S. Photos/Alamy Stock Photo

? What characteristics would make a germicidal chemical safer for the environment?

- **Residue.** Many chemical germicides leave a toxic or corrosive residue. If such a germicide is used to treat an item, the residue must be removed by washing.
- **Cost and availability.** Some germicides are less expensive and more readily available than others. For example, hypochlorite can easily be purchased in the form of household bleach. In contrast, ethylene oxide gas is not only expensive, but its use requires a special chamber, which affects its cost and practicality.
- **Storage and stability.** Some germicides are sold as concentrated stock solutions, decreasing the required storage space. The stocks are simply diluted according to the manufacturer's instructions before use. Some have a limited shelf life once prepared.
- **Environmental risk.** Germicides that retain their antimicrobial activity after use can interfere with wastewater treatment systems. The activity of those germicides must be neutralized before disposal.

Categories of Germicidal Potency

Many different germicidal chemicals are marketed for medical and industrial use under a variety of trade names. Often they contain more than one antimicrobial chemical as well as buffers and/or other ingredients that can affect their antimicrobial activity. Germicides are grouped according to their potency:

- **Sterilants.** These destroy all microbes when used properly, including endospores and viruses. They are also called sporocides. Destruction of endospores usually requires a 6- to 10-hour treatment. Sterilants are used to treat heat-sensitive critical instruments (instruments that come into contact with body tissues).
- **High-level disinfectants.** These destroy all viruses and vegetative microorganisms, but they do not reliably kill endospores. Most are simply sterilants used for time periods too short to ensure endospore destruction. They are used to treat semicritical instruments such as gastrointestinal endoscopes.
- **Intermediate-level disinfectants.** These destroy all vegetative bacteria (including mycobacteria), as well as fungi, and most viruses. They do not kill endospores even with prolonged exposure. This group of germicides is used to disinfect non-critical instruments such as stethoscopes.
- **Low-level disinfectants.** These destroy fungi, vegetative bacteria except mycobacteria, and enveloped viruses. They do not kill endospores, nor do they always destroy non-enveloped viruses. Intermediate-level and low-level disinfectants are also called general-purpose disinfectants. In hospitals, they are used for disinfecting furniture, floors, and walls.

To perform properly, germicides must be used strictly according to the manufacturer's instructions as indicated on their labels, especially the instructions for dilution, temperature, and the contact time or dwell time (the amount of time the germicide must be in contact with an object being treated). It is extremely important to thoroughly clean objects before treatment, both to decrease the number of microbes present and to remove organic material that might interfere with the chemical's activity.

Classes of Germicidal Chemicals

Germicides are represented in a number of chemical families. Each type has characteristics that make it more or less appropriate for specific uses; these are described next and summarized at the end of the section (see table 5.2).

Alcohols

Aqueous solutions of 60–80% ethyl or isopropyl alcohol quickly destroy vegetative bacteria, enveloped viruses, and fungi. They do not, however, reliably destroy bacterial endospores and non-enveloped viruses. Alcohol denatures essential proteins such as enzymes and damages lipid membranes. Proteins are more soluble and therefore denature more easily in alcohol mixed with water, which is why the aqueous solutions are more effective than pure alcohol.

Alcohol solutions are commonly used as antiseptics to clean skin before procedures such as injections that break intact skin. In addition, the CDC recommends that alcohol-based hand sanitizers be used routinely by healthcare personnel as a way to protect patients. The hand sanitizers also provide an alternative for the general public when handwashing with soap and water is not practical.

Alcohol solutions are also used as disinfectants for treating instruments and surfaces. They are relatively non-toxic and inexpensive and do not leave a residue. Unfortunately, they evaporate quickly, which limits their contact time and, consequently, their germicidal effectiveness. In addition, they can damage rubber, some plastics, and other material.

Antimicrobial chemicals such as iodine are sometimes dissolved in alcohol. These alcohol-based solutions, called tinctures, can be more effective than the corresponding aqueous solutions.

Aldehydes

The aldehydes glutaraldehyde, *ortho*-phthalaldehyde (OPA), and formaldehyde destroy microorganisms and viruses by forming chemical bonds that cross-link and inactivate proteins and nucleic acids. A 2% solution of alkaline glutaraldehyde can be used as a liquid chemical sterilant for treating heat-sensitive medical items. Immersion in this solution for 10–12 hours destroys all microorganisms (including endospores) and viruses. Vegetative cells can be killed with soaking times as short as 10 minutes. Glutaraldehyde is toxic, however, so treated items must be thoroughly rinsed with sterile water before use.

Ortho-phthalaldehyde is a relatively new type of disinfectant that provides an alternative to glutaraldehyde. It requires shorter processing times and is less irritating to eyes and nasal passages, but it stains proteins gray, including those of the skin.

Formaldehyde is an extremely effective germicide that is used either as a gas or as formalin, an aqueous 37% solution. Formalin is used to kill bacteria and to inactivate viruses and toxins for use as vaccines, and it can also be used to preserve biological specimens. Formaldehyde's use is limited, however, because it releases irritating vapors and is a probable carcinogen.

Biguanides

Chlorhexidine, the most effective of a group of chemicals called biguanides, reacts with cell membranes to disrupt their integrity. It stays on skin and mucous membranes, is of relatively low toxicity, and destroys a wide range of microbes, including vegetative bacteria, fungi, and some enveloped viruses. Chlorhexidine is used in a variety of antimicrobial products, including disinfectant cleaning solutions, antiseptics available for use by healthcare professionals, and prescription mouthwashes. Chlorhexidine-impregnated catheters and implanted surgical mesh are used in medical procedures. Even tiny chips that slowly release chlorhexidine have been developed; these are inserted into periodontal pockets to treat gum disease. Adverse side effects of chlorhexidine are rare, but severe allergic reactions have been reported.

Ethylene Oxide

Ethylene oxide is an explosive but extremely effective gaseous sterilizing agent that destroys all microorganisms (including endospores) and viruses by chemically modifying proteins and nucleic acids. As a gas, it penetrates well into fabrics, equipment, and implantable devices such as pacemakers and artificial hips. It is particularly useful for sterilizing heat- or moisture-sensitive items such as electrical equipment, pillows, and mattresses. Many disposable laboratory items, including plastic Petri dishes and pipettes, are also sterilized with ethylene oxide.

A special chamber that resembles an autoclave is used to sterilize items with ethylene oxide. This allows careful control of factors such as temperature, relative humidity, and ethylene oxide concentration, all of which influence the effectiveness of the gas. Because ethylene oxide is explosive, it is generally mixed with a non-flammable gas such as carbon dioxide. Under these carefully controlled conditions, objects can be sterilized in 3–12 hours. The toxic and potentially carcinogenic ethylene oxide must then be eliminated from the treated material using heated forced air for 8–12 hours. This is important because the chemical irritates tissues and has a persistent antimicrobial effect, which, in the case of Petri dishes and other items used for growing microbes, is unacceptable.

Halogens

Halogens are a group of elements that include the common disinfectants chlorine and iodine, which are highly reactive oxidizing agents that damage proteins and other essential cell components.

Chlorine Chlorine can destroy all types of microorganisms (including endospores) and viruses if adequate concentrations and appropriate contact times are used. Chlorine-releasing compounds such as sodium hypochlorite can be used to disinfect waste liquids, swimming pool water, instruments, and surfaces. At much lower concentrations, they can be used to disinfect drinking water. Chlorine, unlike iodine, is too irritating to skin and mucous membranes to be used as an antiseptic.

Chlorine solutions are inexpensive, readily available disinfectants (**figure 5.10**). An effective solution can easily be made by diluting liquid household bleach (5.25% sodium hypochlorite) 1:100 in water, resulting in a solution of 500 ppm (parts per million) chlorine. This concentration is several hundred times the amount required to kill most pathogens, but usually necessary for fast, reliable effects. In situations when excessive organic material is present, a 1:10 dilution of bleach may be required. This is because chlorine readily reacts with organic compounds and other impurities in water, disrupting its germicidal activity. The use of such high concentrations should be avoided when possible, however, because chlorine is both corrosive and toxic. Diluted bleach deteriorates over time, so fresh solutions should be prepared regularly.

FIGURE 5.10 Chlorine in the Form of Household Bleach Jill Braaten/McGraw Hill

? **Why should bleach concentrations higher than necessary be avoided?**

Chlorine is used at very low levels to disinfect drinking water. It is important to remember, however, that these levels are not effective against *Cryptosporidium hominis* oocysts and *Giardia lamblia* cysts, both of which may be found in clear mountain streams and other natural waters. Also, the effective amount of chlorine depends on the amount of organic material in the water. The presence of organic compounds is a problem not only because of their interference with the germicide's activity, but also because chlorine can react with some organic compounds to form trihalomethanes (disinfection by-products that are potential carcinogens).

Chlorine dioxide (ClO_2) is a strong oxidizing agent increasingly being used as a disinfectant and sterilant. Its advantage is that it does not react with organic compounds to form trihalomethanes or other toxic chlorinated products. Compressed chlorine dioxide gas, however, is explosive, and liquid solutions decompose readily, so it must be generated on-site. It is used to treat drinking water, wastewater, and swimming pools.

MicroByte

In emergencies, drinking water can be disinfected by adding two drops of plain bleach per liter of water, stirring, and then allowing it to sit for 30 minutes before ingesting.

Iodine Iodine, unlike chlorine, does not reliably kill endospores. It may be used as a tincture—an alcohol-based solution—or more commonly as an iodophor, in which the iodine is linked to a carrier molecule that slowly releases free (unbound) iodine. Iodophors are not as irritating to the skin as tincture of iodine, nor are they as likely to stain. Iodophors used as disinfectants contain more free iodine (30–50 ppm) than do those used as antiseptics (1–2 ppm). Stock solutions must be strictly diluted according to the manufacturer's instructions because dilution affects the amount of free iodine available.

Surprisingly, some *Pseudomonas* species can survive in concentrated stock solutions of iodophors. The reasons are unclear, but it could be due to inadequate levels of free iodine in concentrated solutions; iodine may be released from the carrier only with dilution. *Pseudomonas* species also can form biofilms, within which cells are more resistant to chemicals. Healthcare-associated infections can result if a *Pseudomonas*-contaminated iodophor is unknowingly used to disinfect instruments.

Metal Compounds

Metal compounds kill microorganisms by combining with sulfhydryl groups (—SH) of enzymes and other proteins, thereby interfering with their function. High concentrations of most metals are too toxic to human tissue to be used medically. However, heavy metals like copper, nickel, and zinc are sometimes incorporated into surfaces of items to limit microbial growth. For example, copper can be used to line incubators to reduce contamination, and some metals are used to coat certain fixtures in healthcare facilities to minimize microbial spread.

Silver is one of the few metals still used as a disinfectant. Creams containing silver sulfadiazine (a combination of silver and a sulfa drug) are applied topically to prevent infection of second- and third-degree burns. Commercially available bandages with silver-containing pads can be used on minor scalds, cuts, and scrapes. For many years, doctors were required by law to add drops of another silver compound, 1% silver nitrate, into the eyes of newborns. This was to prevent ophthalmia neonatorum, an eye infection caused by *Neisseria gonorrhoeae,* acquired from infected mothers during the birth process; drops of antibiotics have now largely replaced the use of silver nitrate because they are less irritating to the eye. Silver is also sometimes incorporated into fabrics—such as those used to make certain types of cleaning cloths or athletic apparel—to inhibit the growth of odor-causing bacteria.

Compounds of mercury, tin, arsenic, copper, and other metals were once widely used as preservatives in industrial products and to prevent microbial growth in recirculating cooling water. Their extensive use resulted in serious pollution of natural waters, which has prompted strict controls.

Ozone

Ozone (O_3), an unstable gaseous form of oxygen, is a powerful oxidizing agent. It decomposes quickly, however, so it must be generated on-site, usually by passing air or O_2 between two electrodes. Ozone is used as an alternative to chlorine for disinfecting drinking water and wastewater.

Peroxygens

Peroxygens are powerful oxidizing agents that include hydrogen peroxide and peracetic acid. They can be used as sterilants under controlled conditions, are readily biodegradable, and, in normal concentrations of use, appear to be less toxic than the traditional alternatives (ethylene oxide and glutaraldehyde).

Hydrogen Peroxide The effectiveness of hydrogen peroxide (H_2O_2) as a germicide depends in part on whether it is used on living tissue or an inanimate object. This is because all cells that use aerobic metabolism (including tissue cells of the human body) produce catalase, the enzyme that inactivates hydrogen peroxide by breaking it down to water and O_2. Thus, when a solution of 3% hydrogen peroxide is applied to a wound, our cellular enzymes quickly break it down. When the same solution is used on an inanimate surface, however, it overwhelms the relatively low concentration of catalase produced by contaminating microbial cells.

Hydrogen peroxide is particularly useful as a disinfectant because it leaves no residue and does not damage stainless steel, rubber, plastic, or glass. Hot solutions are commonly used in the food industry to produce commercially sterile containers for aseptically packaged juices and milk. Vapor-phase hydrogen peroxide is more effective than liquid solutions and can be used as a sterilant.

Peracetic Acid Peracetic acid is an even more potent germicide than hydrogen peroxide. A 0.2% solution of peracetic acid, or a combination of peracetic acid and hydrogen peroxide, can be used to sterilize items in less than 1 hour. It is effective in the presence of organic compounds, leaves no residue, and can be used on a wide range of materials. It has a sharp, strong odor, however, and like other oxidizing agents, it irritates the skin and eyes.

Phenolic Compounds (Phenolics)

Phenol (carbolic acid) is important historically because it was one of the earliest disinfectants (see A Glimpse of History), but it has an unpleasant odor and irritates the skin. Phenolics are derivatives of phenol that have greater germicidal activity; because they are so effective, more dilute and therefore less irritating solutions can be used. Phenolics are the active ingredients in Lysol, a household disinfectant.

Phenolics destroy cytoplasmic membranes of microorganisms and denature proteins. They kill most vegetative bacteria and, in high concentrations (from 5–19%), many can kill *Mycobacterium* species. They do not, however, reliably inactivate all groups of viruses. The major advantages of phenolic compounds include their wide range of activity, reasonable cost, and ability to remain effective in the presence of detergents and organic contaminants. They also leave an active antimicrobial residue, which in some cases is desirable.

The phenolics hexachlorophene and triclosan were common in antiseptic products in the past, but safety concerns now limit their use. Hexachlorophene is effective against *Staphylococcus aureus,* the leading cause of wound infections, but its use has been associated with symptoms of neurotoxicity, so skin cleansers containing it are available only with a prescription. Triclosan had been widely used in a variety of personal care products, but the recent FDA rulings requiring manufacturers of antiseptic products to show they are safe and effective now limit its use.

Quaternary Ammonium Compounds (Quats)

Quaternary ammonium compounds (quats) are cationic (positively charged) detergents non-toxic enough to be used to disinfect food preparation surfaces. They are economical, effective, and widely used to disinfect clean inanimate objects and to preserve non-food substances. Some personal care products (such as shampoos and facial cleansers) contain the quat benzalkonium chloride.

Like all detergents, quats have both hydrophilic and hydrophobic properties. Because of this, they reduce the surface tension of liquids and help wash away dirt and organic material along with any attached microbes. Unlike most common household soaps and detergents—which are anionic (negatively charged) and therefore repelled by the negatively charged microbial cell surface—quats are attracted to the cell surface. They react with membranes, destroying many vegetative bacteria and enveloped viruses. Quats also enhance the effectiveness of some other disinfectants. Cationic soaps and organic materials such as cotton gauze, however, can neutralize their activity. In addition, *Pseudomonas,* a troublesome cause of healthcare-associated infections, resists the effects of quats and can even grow in solutions preserved with them.

Table 5.2 summarizes the characteristics and uses of antimicrobial chemicals used to treat non-food substances.

TABLE 5.2 Chemicals Used in Sterilization, Disinfection, and Preservation of Non-Food Substances

Chemical (Examples)	Characteristics	Use
Alcohols (ethanol and isopropanol)	Readily available and inexpensive. Limited in contact time due to rapid evaporation.	Aqueous solutions of alcohol are used as antiseptics to clean skin in preparation for procedures that break intact skin, and as disinfectants for treating instruments.
Aldehydes (glutaraldehyde, *ortho*-phthalaldehyde, and formaldehyde)	Capable of destroying all microbes. Irritating to the respiratory tract, skin, and eyes.	Glutaraldehyde and *ortho*-phthalaldehyde are used to sterilize medical instruments. Formalin is used in vaccine production and to preserve biological specimens.
Biguanides (chlorhexidine)	Relatively safe due to low toxicity. Capable of destroying a wide range of microbes. Able to adhere to and persist on skin and mucous membranes.	Chlorhexidine is used in a variety of products, including disinfectant cleaning solutions, antiseptics used by healthcare professionals, and prescription mouthwashes. It is also incorporated into catheters and surgical mesh.
Ethylene Oxide Gas	Capable of easily penetrating hard-to-reach places and fabrics. Can be used on moisture-sensitive material. Toxic, explosive, and possibly carcinogenic.	Ethylene oxide is commonly used to sterilize medical devices.
Halogens (chlorine and iodine)	Chlorine: inexpensive and readily available, but organic compounds and other impurities neutralize the activity; may react with organic compounds, forming toxic chlorinated products. Iodine: more expensive than chlorine; does not reliably kill endospores.	Chlorine solutions are widely used to disinfect inanimate objects, surfaces, drinking water, and wastewater. Tincture of iodine and iodophors can be used as disinfectants or antiseptics.
Metals (silver and others)	Too toxic to be used medically.	Silver sulfadiazine is used in topical dressings to prevent infection of burns; silver is also incorporated into some fabrics. Some metals are also used to line or coat instruments or fixtures in laboratory and/or healthcare settings.
Ozone	Unstable, readily breaking down to an ineffective form.	Ozone is used to disinfect drinking water and wastewater.
Peroxygens (hydrogen peroxide and peracetic acid)	Readily biodegradable and less toxic than traditional alternatives. Hydrogen peroxide: limited in effectiveness as an antiseptic because the enzyme catalase breaks it down. Peracetic acid: more potent as a germicide than hydrogen peroxide.	Hydrogen peroxide is used to sterilize containers for aseptically packaged juices and milk. Peracetic acid is widely used to disinfect and sterilize medical devices.
Phenolic Compounds (triclosan and hexachlorophene)	Widely ranging in activity. Relatively inexpensive. Effective in the presence of detergents and organic contaminants, leaving an active antimicrobial residue.	Phenolics are components of some household disinfectants. Hexachlorophene and triclosan were once common in antiseptic products, but safety concerns now limit their use.
Quaternary Ammonium Compounds (benzalkonium chloride and others)	Non-toxic enough to be used on food preparation surfaces. Inactive in presence of anionic soaps and detergents.	Quats are widely used to disinfect inanimate objects and to preserve non-food substances.

FOCUS YOUR PERSPECTIVE 5.1

Too Much of a Good Thing?

In our complex world, addressing one challenge may inadvertently create another. Scientists have long been pursuing less toxic alternatives to many traditional germicidal chemicals. For example, chlorhexidine is now generally used in place of hexachlorophene, and glutaraldehyde has largely replaced the more toxic formaldehyde. Meanwhile, ozone and hydrogen peroxide, which are both readily biodegradable, may eventually replace glutaraldehyde. While these less toxic alternatives are better for human health and the environment, their widespread acceptance and use may be contributing to an additional problem: the overuse and misuse of germicidal chemicals. Many products, including soaps, toothbrushes, and even clothing and toys are marketed with the claim of containing antimicrobial ingredients. Bacterial resistance to some of the chemicals included in these products already has been reported.

The issues surrounding the excessive use of antimicrobial chemicals are complicated. On one hand, there is no question that some microbes cause disease. Even those that are not harmful to human health can be troublesome because their metabolic end products (such as acids and/or gases) can ruin the quality of perishable products. Based on that information, it seems wise to destroy microbes or inhibit their growth whenever possible. The role of microbes in our life, however, is not that simple. Our bodies may harbor at least as many microbial cells as human cells, and this normal microbiota plays an important role in maintaining our health. Excessive use of antiseptics or other antimicrobials may actually predispose a person to infection by damaging the normal microbiota.

An even greater concern is that overuse of germicides will select for germicide-resistant microorganisms, a situation similar to our current problems with antibiotic resistance. By using antimicrobial chemicals indiscriminately, we may eventually make these useful tools obsolete. The excessive use of germicides may even be contributing to the problem of antibiotic resistance. The efflux pumps that bacteria use to remove certain germicides from the cell may also eject antibiotics; this was one of the factors that supported reconsideration of the widespread use of triclosan.

Other concerns about the overuse of antimicrobial chemicals involve the misguided belief that "non-toxic" or "biodegradable" chemicals cause no harm and the common notion that "if a little is good, more is even better." Concentrated solutions of hydrogen peroxide, though biodegradable, can cause serious damage, even death, when used improperly. Other chemicals, such as chlorhexidine, can elicit severe allergic reactions in some people. As less toxic germicidal chemicals are developed, people must be educated on the appropriate use of these alternatives.

MicroAssessment 5.4

Germicidal chemicals are used to disinfect and, in some cases, sterilize items, but they are less reliable than heat. They are especially useful for destroying microorganisms and viruses on heat-sensitive items and large surfaces.

10. Describe the factors to consider when selecting a germicide.

11. Explain why it is essential to dilute iodophors properly.

12. Why would a heavy metal be a more serious pollutant than most organic compounds?

5.5 ■ Preservation of Perishable Products

Learning Outcome

10. Compare and contrast chemical preservatives, low-temperature storage, and reducing the available water as methods to preserve perishable products.

Preventing or slowing the growth of microorganisms extends the shelf life of products such as food, medicines, deodorants, cosmetics, and contact lens solutions. Common methods of slowing microbial growth include adding chemical preservatives, storing products at low temperatures, and reducing available water.

Chemical Preservatives

Some of the germicidal chemicals described in section 5.4 can be used to preserve non-food items. For example, mouthwashes often contain a quaternary ammonium compound, and leather belts may be treated with one or more phenol derivatives. Food preservatives, however, must be non-toxic for safe ingestion.

Benzoic, sorbic, and propionic acids are weak organic acids sometimes added to foods such as breads, cheeses, and juices to prevent microbial growth. In acidic conditions, these chemicals affect cell membrane functions. Although this action inhibits the growth of many microbes, the reduced pH itself prevents the growth of most bacteria. Therefore, the primary benefit of adding the preservatives is that they inhibit molds, which otherwise grow at acidic pH.

Nitrate and its reduced form, nitrite, serve a dual purpose in processed meats. From a microbiological viewpoint, their most important function is to inhibit the germination of endospores and subsequent growth of *Clostridium botulinum*. If low levels of nitrate or nitrite are not added to cured meats such as ham, bacon, bologna, and smoked fish, *C. botulinum* may grow and produce deadly botulinum toxin. At higher concentrations than required for preservation, nitrate and nitrite react with myoglobin in the meat to form a stable pigment, giving fresh meat its desirable pink color. The preservatives pose a potential hazard, however, because they can be converted to nitrosamines—some of which are carcinogens—by the metabolic activities of intestinal bacteria or during cooking.

MicroByte

Swiss cheese naturally contains propionic acid, and cranberries have benzoic acid.

Low-Temperature Storage

Refrigeration inhibits the growth of many pathogens and spoilage microorganisms by slowing or stopping critical enzyme reactions. Psychrotrophic and some psychrophilic organisms, however, can grow at refrigeration temperatures.

Freezing preserves foods and other products by stopping microbial growth. The ice crystals that form kill some of the microbial cells, but those that survive can grow and spoil foods once thawed.

Reducing the Available Water

Salting, sugaring, and drying are ancient food preservation techniques that decrease the availability of water in food below the limits required for the growth of most microbes. The high-solute environment also causes plasmolysis, which damages microbial cells (see figure 4.9).

Adding Sugar or Salt

Sugar and salt draw water out of cells, dehydrating them. High concentrations of these solutes are added to many foods as preservatives. For example, fruit is made into jams and jellies by adding sugar, and fish and meats are cured by soaking them in salty water, or brine. It is important to note, however, that the food-poisoning bacterium *Staphylococcus aureus* can grow in relatively high-salt conditions.

Drying Food

Historically, food was dried in the sun, but now food dehydrators are commonly used. Dried foods often have added salt, sugar, or chemical preservatives. For example, meat jerkies usually have added salt and sometimes sugar.

Lyophilization (freeze-drying) is widely used for preserving foods such as coffee, milk, meats, and vegetables. In the process of freeze-drying, the food is first frozen and then dried in a vacuum. When water is added to the lyophilized material, it reconstitutes. The quality of the reconstituted product is often much better than that of products dried using ordinary methods. The light weight and stability without refrigeration of freeze-dried foods make them popular with hikers.

Although drying stops microbial growth, it does not reliably kill bacteria and fungi in or on foods. For example, cases of salmonellosis have been traced to dried eggs. Eggshells and even egg yolks may be contaminated with *Salmonella* species from the gastrointestinal tract of the hen, so some states have laws requiring dried eggs to be pasteurized before they are sold.

MicroAssessment 5.5

Preservation techniques slow or halt the growth of microorganisms to delay spoilage and prevent foodborne illness.

13. What is the most important function of nitrate in cured meat?

14. What is the primary benefit of adding benzoic acid to foods?

15. Preservation by freezing is sometimes compared to drying. Why would this be so?

Summary

5.1 Approaches to Control

The methods used to destroy microbes or remove them from a product can be physical—such as heat treatment, irradiation, and filtration—or chemical.

Principles of Control

A **sterile** item is free of viable microbes, including endospores and viruses. **Disinfection** is the elimination of most or all pathogens on or in a material. **Antiseptics** are antimicrobial chemicals non-toxic enough to be used on body tissue.

Situational Considerations (figure 5.1)

Situations encountered in daily life, hospitals, microbiology laboratories, food production facilities, water treatment facilities, and other industries warrant different degrees of microbial control. Healthcare facilities must be especially careful to control microorganisms to prevent **healthcare-associated infections** (figure 5.2).

5.2 Selecting an Antimicrobial Procedure

Types of Microbes

One of the most critical considerations in selecting a method of destroying microorganisms and viruses is the type of microbial population thought to be present on or in the product.

Number of Microbes

The amount of time it takes for heat or chemicals to kill a population of microorganisms is dictated in part by the number of cells initially present. Microbial death generally occurs at a constant rate. The **D value,** or **decimal reduction time,** is the time it takes to kill 90% of a microbial population under specific conditions (figure 5.3).

Environmental Conditions

Factors such as pH and the presence of organic materials influence microbial death rates.

Risk for Infection

Medical instruments are categorized as **critical, semicritical,** and **non-critical** according to their risk of transmitting infectious agents.

Composition of the Item

Some sterilization and disinfection procedures are inappropriate for certain types of material.

5.3 Physical Methods Used to Destroy or Remove Microorganisms and Viruses (table 5.1)

Moist Heat

Moist heat destroys microorganisms by irreversibly coagulating their proteins. **Pasteurization** uses a brief heat treatment to destroy spoilage and disease-causing organisms. Pressure cookers and **autoclaves** use pressurized steam to achieve temperatures that can kill endospores (figure 5.4, figure 5.5). Commercial canning uses an industrial-sized autoclave called a retort; for low-acid foods, the canning process is designed to destroy all *Clostridium botulinum* endospores.

Dry Heat

Incineration burns cell components to ashes. Temperatures achieved in hot air ovens destroy cell components and irreversibly denature proteins.

Filtration (figure 5.6)

Membrane filters retain microorganisms while letting the suspending fluid pass through. **High-efficiency particulate air (HEPA) filters** remove nearly all microorganisms from air.

Irradiation (figure 5.7, figure 5.8)

Ionizing radiation destroys cells by damaging their structures and producing reactive oxygen species. Ultraviolet light damages the structure and function of nucleic acids. Microwaves kill microorganisms by generating heat.

High Pressure

High pressure is thought to destroy microorganisms by denaturing proteins and altering the permeability of the cell.

5.4 ■ Chemical Methods Used to Destroy Microorganisms and Viruses

Selecting the Appropriate Germicidal Chemical

Factors that must be included in the selection of an appropriate germicidal chemical include toxicity, activity in the presence of organic matter, compatibility with the material being treated, residue, cost and availability, storage and stability, and environmental risk.

Categories of Germicidal Potency

Germicides are grouped according to their potency as **sterilants, high-level disinfectants, intermediate-level disinfectants,** or **low-level disinfectants.**

Classes of Germicidal Chemicals (table 5.2, figure 5.9)

Solutions of 60–80% ethyl or isopropyl alcohol in water rapidly destroy vegetative bacteria, enveloped viruses, and fungi by coagulating enzymes and other essential proteins, and by damaging lipid membranes. Glutaraldehyde, *ortho*-phthalaldehyde, and formaldehyde destroy microorganisms and viruses by inactivating proteins and nucleic acids. Chlorhexidine is a biguanide extensively used in antiseptic products. Ethylene oxide is a gaseous sterilizing agent that destroys microbes by reacting with proteins. Sodium hypochlorite (liquid bleach) is one of the least expensive and most readily available forms of chlorine. Chlorine dioxide is used as a sterilant and disinfectant. Iodophors are iodine-releasing compounds used as antiseptics. Metals interfere with protein function. Silver-containing compounds are used to prevent wound infections. Ozone is used as an alternative to chlorine in the disinfection of drinking water and wastewater. Peroxide and peracetic acid are both strong oxidizing agents that can be used alone or in combination as sterilants. Phenolics destroy cytoplasmic membranes and denature proteins. Quaternary ammonium compounds are cationic detergents; they are non-toxic enough to be used to disinfect food preparation surfaces.

5.5 ■ Preservation of Perishable Products

Chemical Preservatives

Benzoic, sorbic, and propionic acids are sometimes added to foods to prevent microbial growth. Nitrate and nitrite are added to some foods to inhibit the germination of endospores and subsequent growth of *Clostridium botulinum.*

Low-Temperature Storage

Low temperatures above freezing inhibit microbial growth. Freezing essentially stops all microbial growth.

Reducing the Available Water

Sugar and salt draw water out of cells, preventing the growth of microorganisms. Lyophilization is used for preserving food. The food is first frozen and then dried in a vacuum.

Review Questions

Short Answer

1. How is preservation different from pasteurization?
2. What is the most chemically resistant non-spore-forming bacterial pathogen?
3. Explain why it takes longer to kill a population of 10^9 cells than it does to kill a population of 10^3 cells.
4. What is the primary reason that wine is pasteurized?
5. What is the primary reason that milk is pasteurized?
6. When canning, why are low-acid foods processed at higher temperatures than high-acid foods?
7. How are heat-sensitive liquids sterilized?
8. How does microwaving a food product kill bacteria?
9. How is an iodophor different from a tincture of iodine?
10. Name two products commonly sterilized using ethylene oxide gas.

Multiple Choice

1. Unlike a disinfectant, an antiseptic
   a) sanitizes objects rather than sterilizes them.
   b) destroys all microorganisms.
   c) is non-toxic enough to be used on human skin.
   d) requires heat to be effective.
   e) can be used in food products.
2. The D value is defined as the time it takes to kill
   a) all microbial cells in a population.
   b) all pathogens in a population.
   c) 99.9% of a microbial population.
   d) 90% of a microbial population.
   e) 10% of a microbial population.
3. Which of the following is the most resistant to destruction by chemicals and heat?
   a) Bacterial endospores
   b) Fungal spores
   c) *Mycobacterium tuberculosis*
   d) *E. coli*
   e) HIV
4. Ultraviolet light kills bacteria by
   a) generating heat.
   b) damaging DNA.

c) inhibiting protein synthesis.
d) damaging cell walls.
e) damaging cytoplasmic membranes.

5. Which concentration of alcohol is the most effective germicide?
a) 100%
b) 75%
c) 50%
d) 25%
e) 5%
6. All of the following could be used to sterilize an item *except*
a) boiling.
b) incineration.
c) irradiation.
d) sporocides.
e) filtration.
7. All of the following are routinely used to preserve foods *except*
a) high concentrations of sugar.
b) high concentrations of salt.
c) benzoic acid.
d) freezing.
e) ethylene oxide.
8. Aseptically boxed juices and cream containers are processed using which of the following heating methods?
a) Commercial canning
b) High-temperature–short-time (HTST) method
c) Autoclaving
d) Ultra-high-temperature (UHT) method
e) Boiling
9. Commercial canning processes are designed to ensure destruction of which of the following?
a) All vegetative bacteria
b) All viruses
c) Endospores of *Clostridium botulinum*
d) *E. coli*
e) *Mycobacterium tuberculosis*
10. Which of the following is *false*?
a) A high-level disinfectant cannot be used as a sterilant.
b) Critical items must be sterilized before use.
c) Low numbers of endospores may remain on semicritical items.
d) Standard sterilization procedures do not destroy prions.
e) Quaternary ammonium compounds can be used to disinfect food preparation surfaces.

Applications

1. An agriculture extension agent is preparing pamphlets on preventing the spread of disease. In the pamphlet, the appropriate situations for using disinfectants around the house must be explained. What situations should the agent discuss?
2. As a microbiologist representing a food corporation, you have been asked to serve on a panel to debate the need for chemical preservatives in foods. Your role is to prepare a statement that compares the benefits of chemical preservatives to the risks. What points must you bring up that indicate the benefits of chemical preservatives?

Critical Thinking

1. This graph shows the time it takes to kill populations of the same microorganism under different conditions. What conditions would explain the differences in lines a, b, and c?

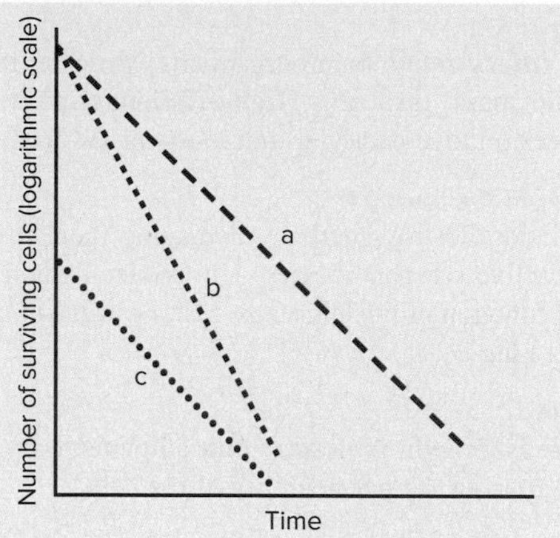

2. This diagram shows the filter paper method used to evaluate the inhibitory effect of chemical agents, heavy metals, and antibiotics on bacterial growth. A culture of a test bacterium is spread uniformly over the surface of an agar plate. Small filter paper discs containing the material to be tested are then placed on the surface of the medium. A disc that has been soaked in sterile distilled water is sometimes added as a control. After incubation, a lawn (film of growth) will cover the plate, but a clear zone will surround those discs that contain an inhibitory compound. The size of the zone reflects several factors, one of which is the effectiveness of the inhibitory agent. What are two other factors that might affect the size of the zone of inhibition? What is the purpose of the control disc? If a clear area were apparent around the control disc, how would you interpret the observation?

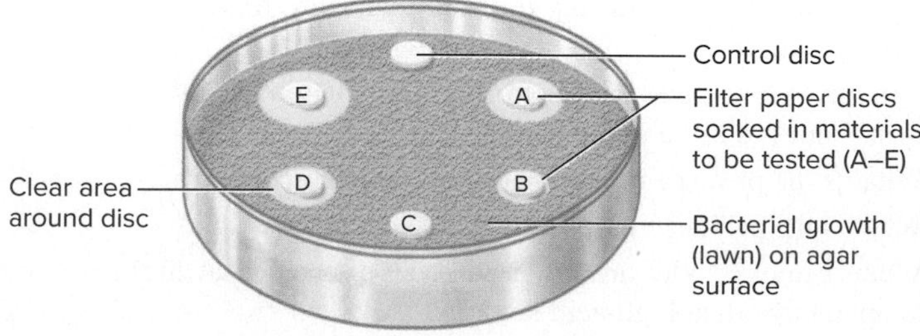

6 Microbial Metabolism: Fueling Cell Growth

Wine—a beverage produced using microbial metabolism. *Comstock/PunchStock*

KEY TERMS

Adenosine Triphosphate (ATP) The energy currency of cells. The bonds between its phosphate groups can be broken to release energy that powers endergonic (energy-consuming) reactions.

Anabolism Metabolic processes that synthesize the subunits of macromolecules and assemble them, using the energy of ATP; biosynthesis.

Catabolism Metabolic processes that break down compounds such as glucose in order to release their energy.

Cellular Respiration A series of metabolic processes that collectively transfer electrons stripped from a chemical energy source to an electron transport chain, generating a proton motive force that is used to synthesize ATP.

Electron Transport Chain (ETC) A group of membrane-embedded electron carriers that pass electrons from one to another, which results in protons being moved across the membrane to create a proton motive force.

Enzyme A molecule, usually a protein, that functions as a catalyst, speeding up a biological reaction.

Fermentation A metabolic process that stops short of oxidizing glucose or other organic compounds completely, using an organic compound as a terminal electron acceptor.

Oxidative Phosphorylation The synthesis of ATP using the energy of a proton motive force created by harvesting chemical energy; the enzyme ATP synthase catalyzes the reaction.

Photophosphorylation The synthesis of ATP using the energy of a proton motive force created by harvesting the energy of light.

Precursor Metabolites Metabolic intermediates that link catabolic and anabolic pathways because they can either be broken down to release more of their energy or be used to make the subunits of macromolecules.

Proton Motive Force A form of energy generated as an electron transport chain moves protons across a membrane to create a chemiosmotic gradient.

Substrate-Level Phosphorylation The synthesis of ATP using the energy of an exergonic (energy-releasing) chemical reaction to transfer a phosphate group from a donor molecule to ADP.

Terminal Electron Acceptor A chemical that is ultimately reduced as a consequence of fermentation or respiration.

A Glimpse of History

In the 1850s, Louis Pasteur tried to determine how alcohol forms from grape juice. Biologists had already noticed that when the juice is held in large vats, alcohol and CO_2 are produced and the numbers of yeast cells increase. They concluded that the multiplying cells were converting sugar in the grape juice to alcohol and CO_2. Pasteur agreed, but he could not convince two very powerful and influential German chemists, Justus von Liebig and Friedrich Wöhler, who refused to believe that microorganisms caused the breakdown of sugar. Both scientists mocked the hypothesis, publishing pictures of yeast cells looking like miniature animals taking in grape juice through one end and releasing CO_2 and alcohol through the other.

Pasteur studied the relationship between yeast and alcohol production using a strategy commonly employed by scientists today—that is, simplifying the experimental system so that relationships can be more easily identified. First, he prepared a solution of sugar, ammonia, mineral salts, and trace elements. He then added a few yeast cells. As the yeast multiplied, the sugar level decreased and the alcohol level increased, indicating that the sugar was being converted to alcohol as the cells grew. This strongly suggested that living cells caused the chemical transformation. Liebig, however, still would not believe that the process was occurring inside microorganisms. To convince him, Pasteur tried to extract something from inside the yeast cells that would convert the sugar. He failed, like many others before him.

In 1897, Eduard Buchner, a German chemist, showed that crushed yeast cells could convert sugar to ethanol and CO_2. We now know that the cells' enzymes carried out this transformation. For these pioneering studies, Buchner was awarded a Nobel Prize in 1907. In 1928, Marjory Stephenson, a prominent British biochemist, was the first to isolate a bacterial enzyme.

Microbial cells, like all cells, need to accomplish two fundamental tasks to grow. They must continually synthesize new parts—such as cell walls, membranes, ribosomes, and nucleic acids—that allow the cells to enlarge and eventually divide. In addition, cells need to harvest energy and convert it to a form that can power the various energy-consuming reactions, including those used to make new parts. The sum of all chemical reactions in a cell is called **metabolism.**

Microbial metabolism is important to humans for a number of reasons. As scientists look for new supplies of energy, some are investigating biofuels—fuels made from renewable biological sources such as plants and organic wastes. Microbes can produce biofuels, breaking down solid materials such as corn stalks, sugar cane, and wood to make ethanol (see Focus Your Perspective 6.1). Microbial metabolism is also important in food and beverage production; cheese-makers add *Lactococcus* and *Lactobacillus* species to milk because the metabolic wastes of these bacteria contribute to the flavor and texture of various cheeses. Bakers, brewers, vintners, and distillers use the yeast *Saccharomyces cereviseae* to make bread, beer, wine, and distilled spirits, respectively. In addition, products characteristic of specific microbes can be used as identifying markers in the laboratory. The metabolic processes of organisms such as *E. coli* also serve as an important model for studying similar processes in eukaryotic cells, including those of humans. Understanding metabolism is also medically relevant because reactions unique to microbes are potential targets for antimicrobial medications.

6.1 ■ Overview of Microbial Metabolism

Learning Outcomes

1. Compare and contrast catabolism and anabolism.
2. Describe the energy sources used by photosynthetic organisms and chemoorganoheterotrophs.
3. Describe the components of metabolic pathways (enzymes, ATP, chemical energy sources and terminal electron acceptors, and electron carriers) and the role of precursor metabolites.
4. Describe the roles of the three central metabolic pathways.
5. Distinguish between cellular respiration and fermentation.

Microbial metabolism (the sum of all chemical reactions in a cell) is easier to understand if you know the general principles and how they fit into the "big picture." With that in mind, this section will introduce you to the key concepts of metabolism and present an overview of the topic; later sections will then build on that information.

Metabolism can be separated into two components—catabolism and anabolism—each of which is required for cellular reproduction (**figure 6.1**). **Catabolism** is the set of chemical reactions that degrade compounds, releasing their energy. Cells capture that energy and use it to make ATP, the energy currency of the cell (see figure 2.31). **Anabolism, or biosynthesis,** is the set of chemical reactions that cells use to synthesize and assemble the subunits of macromolecules, using ATP to drive these energetically unfavorable processes. As described in chapter 2, the subunits of macromolecules include amino acids, nucleotides, monosaccharides, and fatty acids.

Although catabolism and anabolism are often discussed separately, they are intimately linked. As mentioned, ATP made during catabolism is used in anabolism. In addition, some of the compounds produced during catabolism are precursor metabolites, which are chemicals that link catabolic

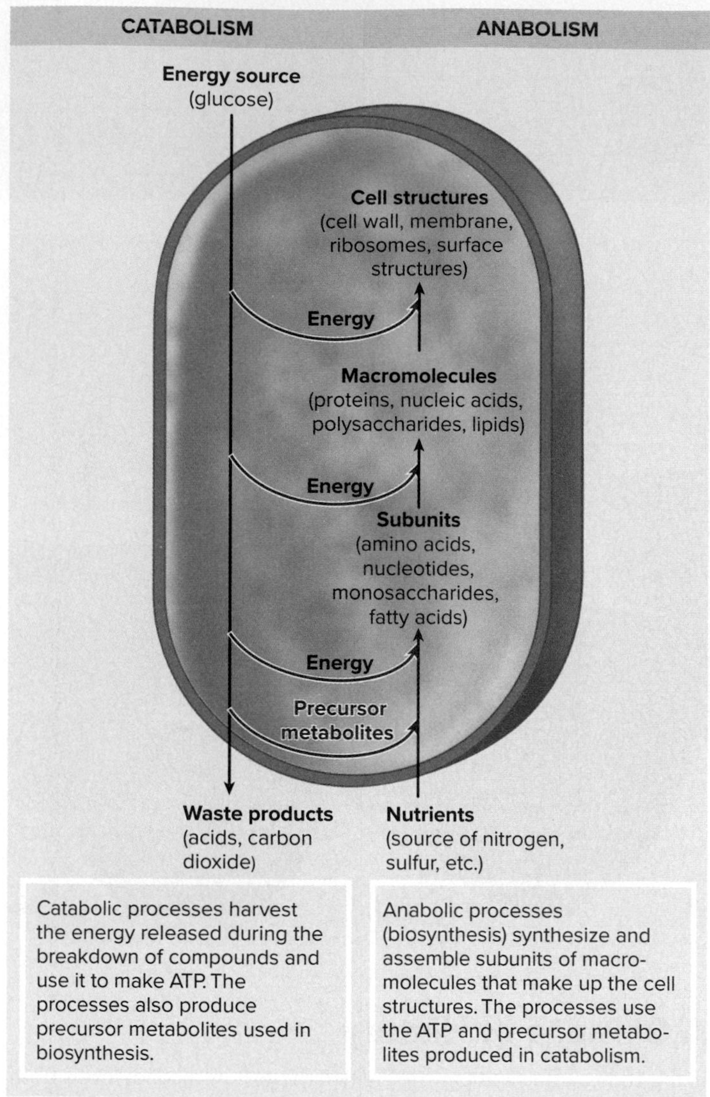
CATABOLISM
ANABOLISM
Energy source
(glucose)
Cell structures
(cell wall, membrane, ribosomes, surface structures)
Energy
Macromolecules
(proteins, nucleic acids, polysaccharides, lipids)
Energy
Subunits
(amino acids, nucleotides, monosaccharides, fatty acids)
Energy
Precursor metabolites
Waste products
(acids, carbon dioxide)
Nutrients
(source of nitrogen, sulfur, etc.)
Catabolic processes harvest the energy released during the breakdown of compounds and use it to make ATP. The processes also produce precursor metabolites used in biosynthesis.
Anabolic processes (biosynthesis) synthesize and assemble subunits of macromolecules that make up the cell structures. The processes use the ATP and precursor metabolites produced in catabolism.

FIGURE 6.1 The Relationship Between Catabolism and Anabolism Precursor metabolites are compounds produced during catabolism that can either be further degraded or be used in anabolism to make the subunits of macromolecules.

? **Which subunits make up proteins? Which make up nucleic acids?**

and anabolic processes; they can be further broken down to capture their energy (via catabolism), or they can be used to make certain subunits of macromolecules (via anabolism). As an analogy, think of what happens in a recycling center that breaks down electronic devices. Once certain components are freed from the devices, those components can be further broken down—generating raw materials such as gold that can be sold for cash—or they can be used to build refurbished devices.

Energy

Energy is the capacity to do work. It can exist as potential energy (stored energy) or as kinetic energy (energy of motion) (**figure 6.2**). Potential energy can be stored in various forms,

FIGURE 6.2 Forms of Energy Potential energy is stored energy, such as water held behind a dam. Kinetic energy is the energy of motion, such as movement of water through a specific point in the dam. Radoslaw Lecyk/Shutterstock

What is energy?

including chemical bonds, a rock on a hill, or water behind a dam.

Energy in the universe cannot be created or destroyed; however, it can be changed from one form to another. In other words, although new energy cannot be made, potential energy can be converted to kinetic energy and vice versa, and one form of potential energy can be converted to another. Hydroelectric dams release the potential energy of water stored behind a dam, creating the kinetic energy of moving water. This can be captured to generate an electrical current, which can then be used to charge a battery.

Metabolism involves processes that transform energy, converting it from one form into another. **Photosynthetic** organisms harvest the energy of sunlight, using it to power the synthesis of organic compounds from CO_2. By doing so, they convert the kinetic energy of photons (particles that travel at the speed of light) to the potential energy of chemical bonds. **Chemoorganotrophs** obtain energy by degrading organic compounds; they then use some of that energy to make other organic compounds. In other words, they take the potential energy of certain chemical bonds and use it to create other ones. Because chemoorganotrophs depend on a constant source of organic compounds, they generally rely on the metabolic activities of photosynthetic organisms (**figure 6.3**). In some environments, however, chemolithoautotrophs play the most significant role in synthesizing organic compounds (see figure 28.12).

The amount of energy released by breaking down a compound can be explained by the concept of free energy, the energy available to do work. From a biological perspective, this is the energy released when a chemical bond is broken. In a chemical reaction, some bonds are broken and others are formed. If the starting compounds have more free energy than

Sunlight
Photosynthetic organisms harvest the energy of sunlight and use it to power the synthesis of organic compounds from CO_2. This converts light energy to chemical energy.
Chemical energy (organic compounds)
Chemoorganotrophs degrade organic compounds, harvesting chemical energy.

FIGURE 6.3 Most Chemoorganotrophs Depend on Photosynthetic Organisms
middle: Robert Glusie/Photodisc/Getty Images; bottom: Digital Zoo/Digital Vision/PunchStock

Why do chemoorganotrophs require the activities of photosynthetic organisms?

the products, energy is released; the reaction is **exergonic.** In contrast, if the products have more free energy than the starting compounds, the reaction requires an input of energy and is **endergonic.**

The change in free energy for a given reaction is the same regardless of the number of steps involved. For example, converting glucose to CO_2 and water in a single step releases the same amount of energy as degrading it in a series of steps. Cells take advantage of this by degrading compounds step by step, carefully controlling the process to harvest the energy released in an exergonic reaction and having it power an endergonic one; this linking of an exergonic reaction to an endergonic one is called energy coupling.

Components of Metabolic Pathways

A **metabolic pathway** is a series of chemical reactions that converts one or more starting substrates to end product(s); recall that a substrate is a reactant in an enzyme-catalyzed reaction (see section 2.2). When the product of one reaction serves as the substrate for the next, it is called a metabolic intermediate. The pathways can be linear, branched, or cyclical (**figure 6.4**). Like the flow of rivers controlled by dams, metabolic pathways can be controlled at key points, allowing cells to regulate certain processes. This ensures that specific molecules are produced in precise quantities when needed. If a metabolic step is blocked, products "downstream" of that blockage will not be made.

To understand what metabolic pathways accomplish, it is important to be familiar with their essential components. These include enzymes, ATP, the chemical energy source and terminal electron acceptor, and electron carriers.

The Role of Enzymes

An **enzyme** is a molecule (usually a protein) that functions as a biological catalyst, speeding up the conversion of substrate(s) to product(s). A specific enzyme facilitates each step of a metabolic pathway (**figure 6.5a**). Without enzymes, energy-yielding reactions would still occur, but at rates so slow they would be insignificant. An enzyme catalyzes a chemical reaction by lowering the **activation energy**—the energy it takes to start a reaction (figure 6.5b); the activation energy serves to stress and break chemical bonds in the substrate(s) so that new bond(s) can more easily form to make product(s). Even an exergonic reaction has an activation energy, and by lowering this barrier, an enzyme speeds the reaction. Enzymes will be described in more detail later in the chapter.

The Role of ATP

Adenosine triphosphate (ATP) is the main energy currency of cells, serving as the ready and immediate donor of free energy. The molecule is composed of ribose, adenine, and three phosphate groups arranged in a row (see figure 2.31).

Cells produce "energy currency" by using energy to add a phosphate group to **adenosine diphosphate (ADP),** forming ATP. That energy currency can then be "spent" by removing the phosphate group, thereby releasing energy and converting the molecule back to ADP (**figure 6.6**). Cells produce ATP during catabolism and then use it to power anabolic reactions.

Chemoorganotrophs use two different processes to make ATP:

- **Substrate-level phosphorylation.** ATP is synthesized using the energy of an exergonic chemical reaction to transfer a phosphate group from a donor compound to ADP.

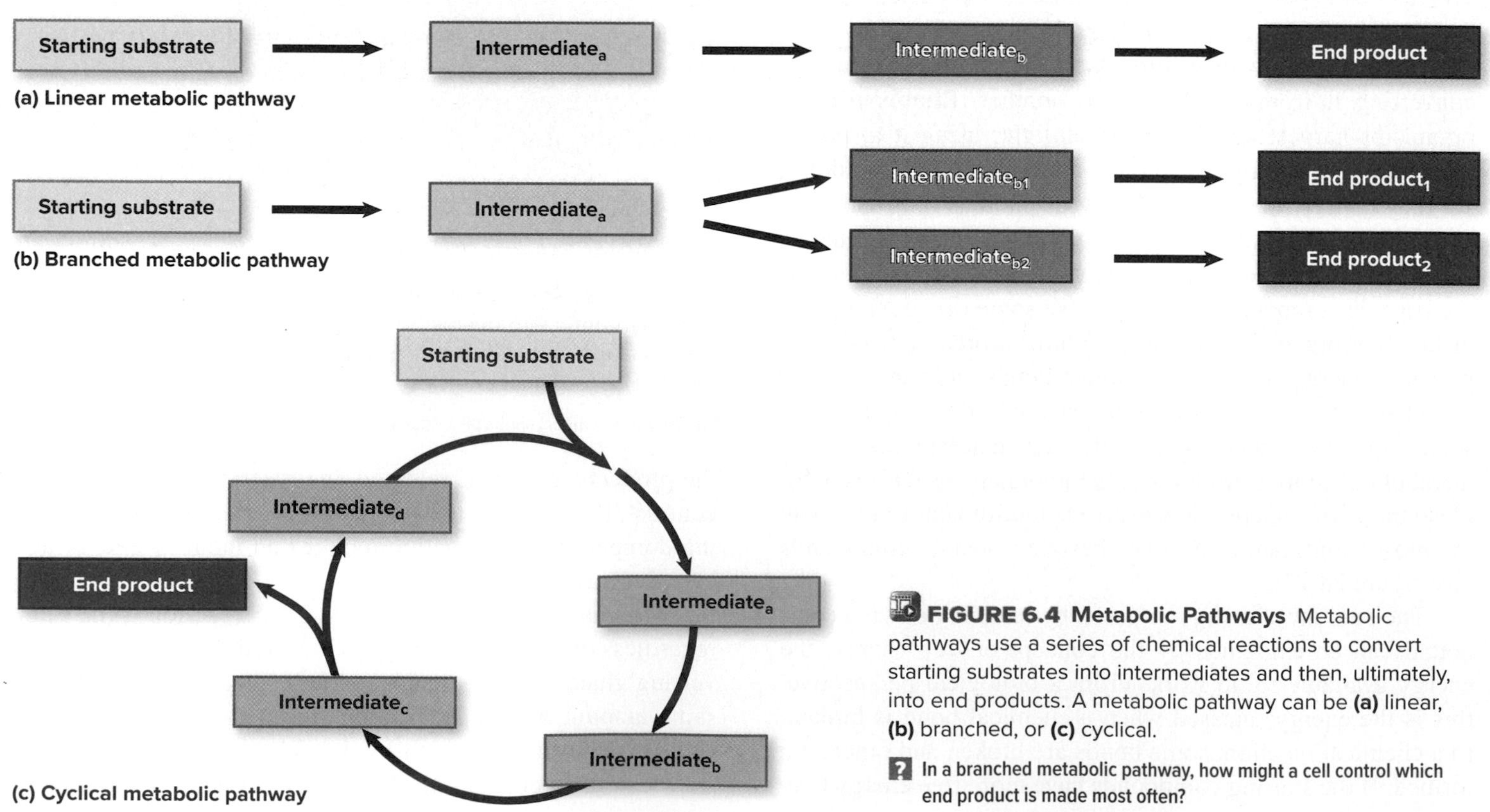

FIGURE 6.4 Metabolic Pathways Metabolic pathways use a series of chemical reactions to convert starting substrates into intermediates and then, ultimately, into end products. A metabolic pathway can be **(a)** linear, **(b)** branched, or **(c)** cyclical.

? **In a branched metabolic pathway, how might a cell control which end product is made most often?**

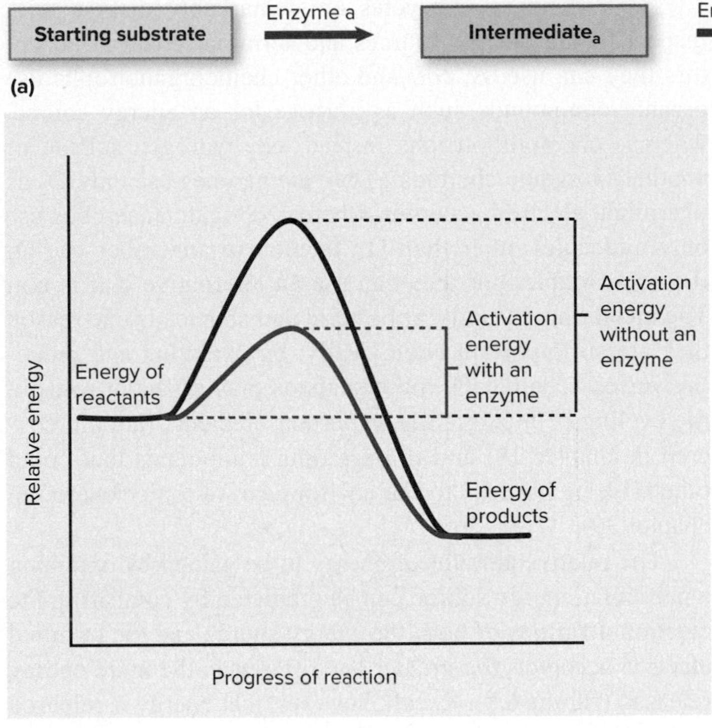

FIGURE 6.5 The Role of Enzymes Enzymes function as biological catalysts. **(a)** A specific enzyme facilitates each step of a metabolic pathway. **(b)** An enzyme catalyzes a chemical reaction by lowering the activation energy of the reaction.

? What is activation energy?

- **Oxidative phosphorylation.** ATP is synthesized using the energy of a proton motive force to add an inorganic phosphate group (Pi) to ADP; the reaction is catalyzed by an enzyme called ATP synthase. Recall that **proton motive force** is the form of energy that results from an electrochemical gradient across a membrane being established by the electron transport chain (see figure 3.5).

Photosynthetic organisms can generate ATP by **photophosphorylation,** which is similar to oxidative phosphorylation except that the energy of light (a type of radiant energy) is used to create the proton motive force.

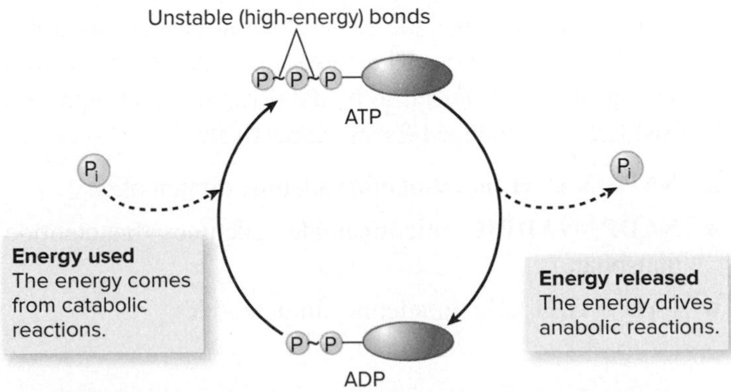

FIGURE 6.6 ATP Cells produce ATP by using energy to add a phosphate group to ADP. The symbol P_i indicates inorganic phosphate, but organic forms of phosphate can also be used. Energy is released in the reverse reaction, which breaks the phosphate bond to convert ATP to ADP + P_i.

? What structural characteristic of ATP makes its phosphate bonds "high energy"?

The Role of the Chemical Energy Source and the Terminal Electron Acceptor

To understand how cells obtain energy, it is helpful to recall that certain atoms are more electronegative than others, meaning they have a greater affinity (attraction) for electrons (see table 2.2). Likewise, certain molecules have a greater affinity for electrons than do other molecules, a characteristic that relates to the electronegativities of the atoms that make up the molecules. When electrons move from a molecule that has a relatively low electron affinity (tends to give up, not accept, electrons) to one that has a higher electron affinity (tends to accept, not give up, electrons), energy is released. This is how cells obtain the energy used to make ATP. They remove electrons from glucose or another low electron affinity chemical, and donate them to a molecule such as O_2 that has a higher electron affinity. The chemical that serves as the electron donor is the **energy source,** and the one that ultimately accepts those electrons is the **terminal electron acceptor.**

Cells remove electrons from the energy source through a series of **oxidation-reduction reactions,** or **redox reactions.** The substance that loses electrons is **oxidized** by the reaction; the one that gains those electrons is **reduced** (**figure 6.7**). When electrons are removed from a biological molecule, protons (H^+) often follow. The result is the removal of an electron-proton pair, or hydrogen atom. Thus, dehydrogenation (the removal of a hydrogen atom) is an oxidation. Correspondingly, hydrogenation (the addition of a hydrogen atom) is a reduction. Note that oxidation and reduction are linked in biological systems—it is not possible to have one without the other because electrons cannot exist on their own in cells.

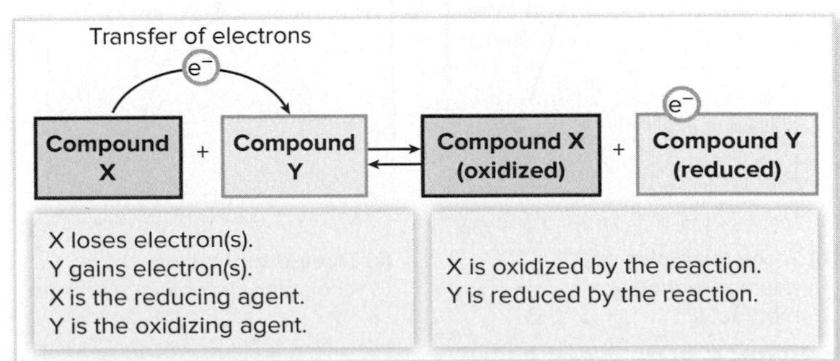

FIGURE 6.7 Oxidation-Reduction Reactions The substance that loses one or more electrons is oxidized by the reaction; the one that gains those electrons is reduced.

? Why is hydrogenation a reduction?

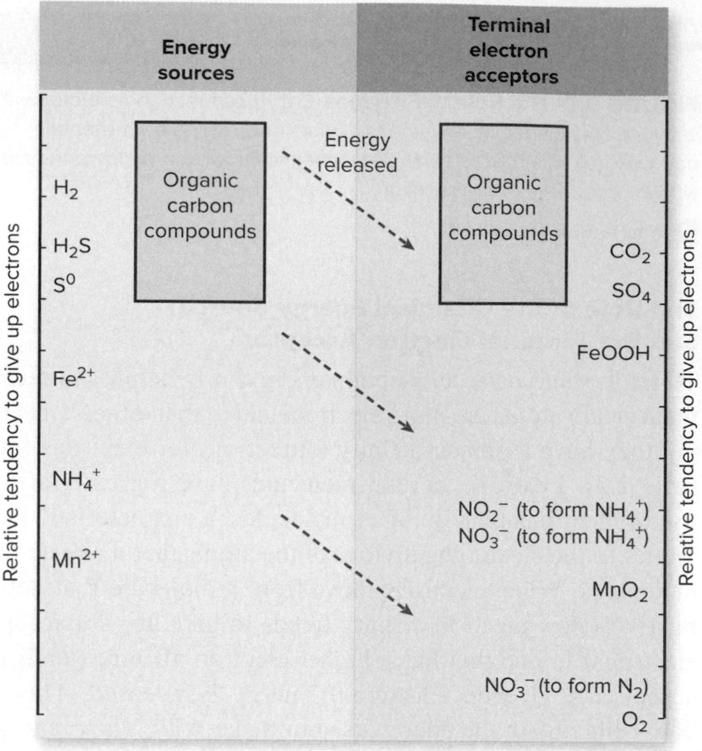

(a) Energy is released when electrons are moved from an energy source with a low affinity for electrons to a terminal electron acceptor with a higher affinity.

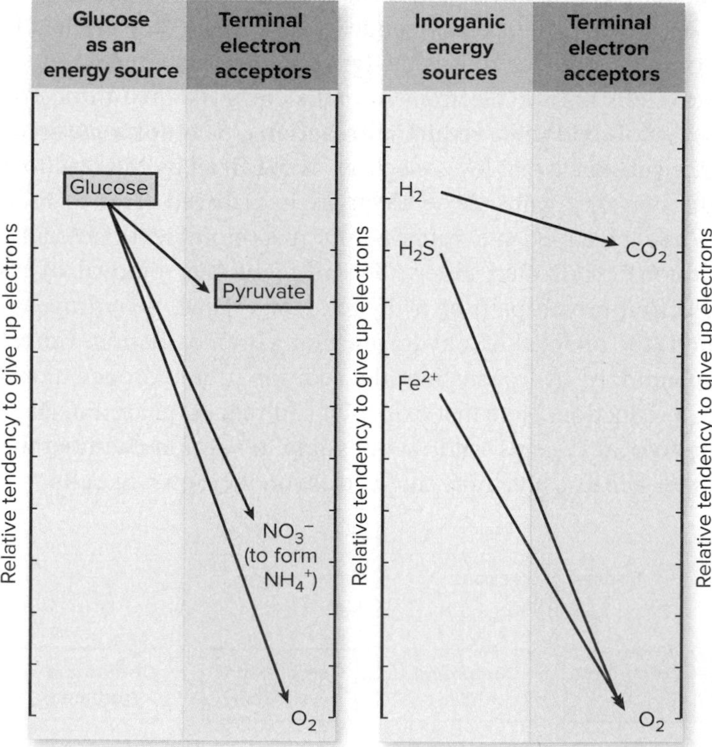

(b) Three examples of chemoorganotrophic metabolism

(c) Three examples of chemolithotrophic metabolism

FIGURE 6.8 Chemical Energy Sources and Terminal Electron Acceptors Catabolic pathways can proceed as long as the energy source has a lower electron affinity than the terminal electron acceptor. As a group, bacteria and archaea use a wide variety of energy sources and terminal electron acceptors, which is why the organisms are so important ecologically.

? *E. coli* **can use any of the examples shown in part (b) of figure 6.8. Considering this, why would it preferentially use O_2 rather than NO_3^- as a terminal electron acceptor? Why would it preferentially use NO_3^- rather than pyruvate as a terminal electron acceptor?**

As a group, prokaryotes are remarkably diverse with respect to the energy sources and terminal electron acceptors they can use. *E. coli* and other chemoorganotrophs use organic compounds such as glucose for an energy source, whereas chemolithotrophs instead use hydrogen sulfide or another inorganic chemical. Obligate aerobes use only O_2 as a terminal electron acceptor, whereas obligate anaerobes use only molecules other than O_2; facultative anaerobes use O_2 if it is available, but they can use an alternative if it is not. The metabolic diversity of bacteria and archaea is one reason they are so important ecologically; by oxidizing and reducing various chemicals, the organisms play essential roles in the cycling of biologically important elements (a topic covered in chapter 28) and in degrading compounds that could otherwise be harmful to the environment (a topic covered in chapter 29).

The relative amount of energy to be gained by oxidizing a particular energy source can be predicted by comparing the electron affinities of both the energy source and the terminal electron acceptor; the greater the difference, the more energy released (**figure 6.8**). Recall, however, that energy is released only when electrons are transferred from a compound with a relatively low electron affinity (high tendency to give up electrons) to one with a higher electron affinity (low tendency to give up electrons).

The Role of Electron Carriers

When cells remove electrons from an energy source, they do not remove them all at once, nor do they transfer them directly to the terminal electron acceptor. Instead, the electrons are initially transferred to **electron carriers** (**table 6.1**). Electron carriers can also be considered hydrogen carriers because, along with electrons, they carry protons. Note, however, that the location of protons in biological reactions is often ignored because in aqueous solutions protons—unlike electrons—do not require carriers but instead associate with water molecules. Cells have only a few different types of electron carriers—each with a distinct role—but the same electron carrier can assist many different enzymes. The three main types of electron carriers are usually referred to by abbreviations that represent their oxidized and reduced forms, respectively:

- **NAD^+/NADH** (nicotinamide adenine dinucleotide)
- **$NADP^+$/NADPH** (nicotinamide adenine dinucleotide phosphate)
- **FAD/$FADH_2$** (flavin adenine dinucleotide)

TABLE 6.1 Electron Carriers

Carrier	Oxidized Form (Accepts Electrons)		Reduced Form (Donates Electrons)	Typical Fate of Electrons Carried
			Electron / Electron carrier	
Nicotinamide adenine dinucleotide (carries 2 electrons and 1 proton)	$NAD^+ + 2\ e^- + 2\ H^+$	⇋	$NADH + H^+$	Used to generate a proton motive force that can drive ATP synthesis
Flavin adenine dinucleotide (carries 2 electrons and 2 protons; i.e., 2 hydrogen atoms)	$FAD + 2\ e^- + 2\ H^+$	⇋	$FADH_2$	Used to generate a proton motive force that can drive ATP synthesis
Nicotinamide adenine dinucleotide phosphate (carries 2 electrons and 1 proton)	$NADP^+ + 2\ e^- + 2\ H^+$	⇋	$NADPH + H^+$	Used to power anabolic reactions (biosynthesis)

Reduced electron carriers represent **reducing power** because they can easily transfer their electrons to another chemical that has a higher electron affinity. By doing so, they not only reduce the recipient molecule, they also raise its energy level. NADH and $FADH_2$ transfer their electrons to the electron transport chain, which then uses the energy to generate a proton motive force across the membrane. In turn, this can drive the synthesis of ATP in the process of oxidative phosphorylation. Ultimately the electrons are transferred to the terminal electron acceptor. The electrons carried by NADPH have an entirely different fate; they are used to reduce compounds during anabolic reactions. Note, however, that many microbial cells can use NADH to reduce $NADP^+$ (generating NADPH) and vice versa.

Precursor Metabolites

Certain intermediates of catabolic pathways can be used in anabolic pathways, linking these two types of pathways. These intermediates—**precursor metabolites**—serve as carbon skeletons from which subunits of macromolecules can be made (**table 6.2**). For example, the precursor metabolite pyruvate can be converted to any one of three amino acids: alanine, leucine, or valine.

Recall from chapter 4 that *E. coli* can grow in glucose-salts medium, which contains only glucose and a few inorganic salts. This means that the glucose is serving two purposes in the cell: (1) the energy source, and (2) the starting point from which all cell components are made—including proteins, lipids, carbohydrates, and nucleic acids. When *E. coli* cells degrade glucose molecules, the pathways they use release energy and also form a dozen or so precursor metabolites. Other organisms use the same catabolic pathways but sometimes lack the ability to convert a certain precursor metabolite into a compound needed for biosynthesis. Any essential compounds that a cell cannot synthesize must be provided from an external source.

TABLE 6.2 Precursor Metabolites

Precursor Metabolite	Biosynthetic Role (Macromolecules Made From Precursor Metabolite)	Pathway (or Step) Generated
Glucose-6-phosphate	Lipopolysaccharide	Glycolysis
Fructose-6-phosphate	Peptidoglycan	Glycolysis
Dihydroxyacetone phosphate	Lipids (glycerol component)	Glycolysis
3-Phosphoglycerate	Proteins (the amino acids cysteine, glycine, and serine)	Glycolysis
Phosphoenolpyruvate	Proteins (the amino acids phenylalanine, tryptophan, and tyrosine)	Glycolysis
Pyruvate	Proteins (the amino acids alanine, leucine, and valine)	Glycolysis
Ribose-5-phosphate	Nucleic acids and proteins (the amino acid histidine)	Pentose phosphate cycle
Erythrose-4-phosphate	Proteins (the amino acids phenylalanine, tryptophan, and tyrosine)	Pentose phosphate cycle
Acetyl-CoA	Lipids (fatty acids)	Transition step
α-Ketoglutarate	Proteins (the amino acids arginine, glutamate, glutamine, and proline)	TCA cycle
Oxaloacetate	Proteins (the amino acids aspartate, asparagine, isoleucine, lysine, methionine, and threonine)	TCA cycle

Note: The colored icons in the table are used in figures throughout the chapter to represent the respective precursor metabolites.

A cell's metabolic pathways make it easy for that cell to use a single type of substrate like glucose for multiple purposes. Think of the cells as extensive biological recycling centers that routinely process millions of glucose molecules (**figure 6.9**). Molecules that remain on the central deconstruction line are oxidized completely to CO_2, releasing the maximum amount of energy. Some breakdown intermediates, however, can exit that line to be used in biosynthesis. The exit points are located at the steps immediately after a precursor metabolite is made. So once a precursor metabolite is made in catabolism, it can be further oxidized to release energy, or it can be used in biosynthesis.

Catabolism

Catabolism of glucose, the preferred energy source of many cells, includes two key sets of processes:

- Oxidizing glucose molecules to generate ATP, reducing power (NADH, $FADH_2$, and NADPH), and precursor metabolites; this is accomplished in a series of reactions called the central metabolic pathways.

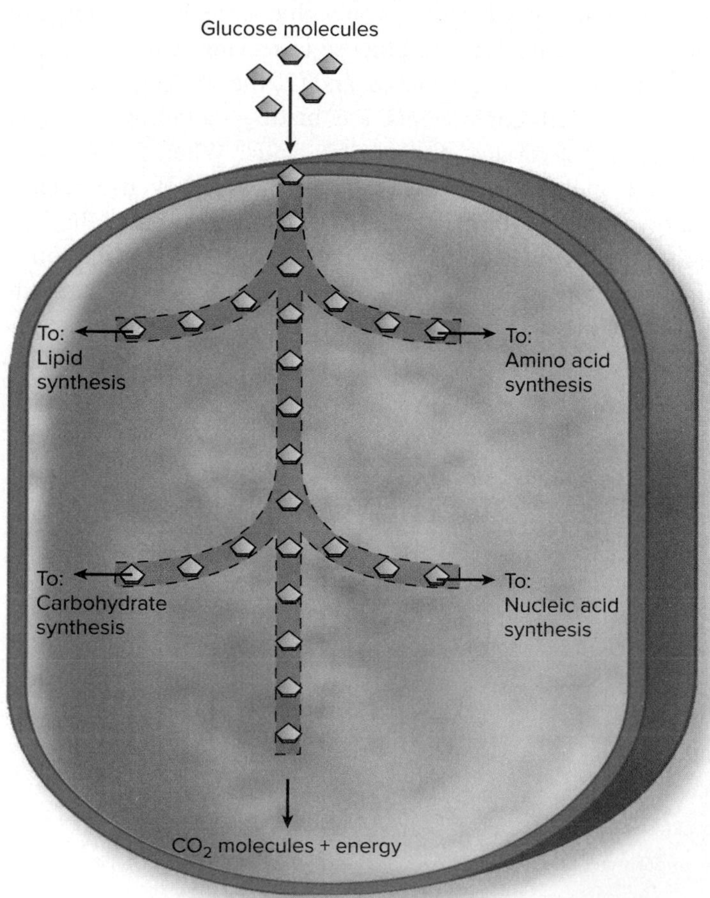

FIGURE 6.9 Cells Use Glucose for Multiple Purposes The millions of glucose molecules that continually enter a cell can have different fates. Some may be oxidized completely to release the maximum amount of energy, and others will be used in biosynthesis.

? What is the role of precursor metabolites in this process?

- Transferring the electrons carried by NADH and $FADH_2$ to the terminal electron acceptor, which occurs as part of either cellular respiration or fermentation (recall that the electrons carried by NADPH are used in biosynthesis).

Central Metabolic Pathways

Three key metabolic pathways—the **central metabolic pathways**—together completely oxidize glucose to CO_2, an outcome summarized as follows:

$$\underset{\text{(glucose)}}{C_6H_{12}O_6} + \underset{\text{(oxygen)}}{6\,O_2} \rightarrow \underset{\text{(carbon dioxide)}}{6\,CO_2} + \underset{\text{(water)}}{6\,H_2O}$$

The pathways are catabolic, but the precursor metabolites and reducing power they generate can also be diverted for use in biosynthesis. To reflect the dual role of these pathways, they are sometimes called amphibolic pathways (*amphi* meaning "both kinds"). The central metabolic pathways include the following (**figure 6.10**):

- **Glycolysis.** (1) This splits glucose and gradually oxidizes it to form two molecules of pyruvate. It provides the cell with a small amount of energy in the form of ATP, some reducing power, and six precursor metabolites.
- **Pentose phosphate pathway.** (2) This also breaks down glucose, but its primary role is to produce compounds used in biosynthesis, including two precursor metabolites as well as reducing power in the form of NADPH. A product of the pathway feeds into glycolysis.
- **Tricarboxylic acid (TCA) cycle.** This is also called the citric acid cycle or the Krebs cycle. (3a) Just before this cycle, a single reaction called the transition step (or the preparatory reaction) converts the pyruvate from glycolysis into acetyl-CoA. One molecule of CO_2 is released as a result. (3b) The TCA cycle then accepts the 2-carbon acetyl group, ultimately oxidizing it to release two molecules of CO_2. The transition step and the TCA cycle together generate the most reducing power of all the central metabolic pathways; they also produce three precursor metabolites and ATP.

As glucose is oxidized, a relatively small amount of ATP is made by substrate-level phosphorylation during glycolysis and the TCA cycle. The reducing power accumulated during the oxidation steps, however, can be used in cellular respiration to generate much more ATP by oxidative phosphorylation.

Cellular Respiration

(4) **Cellular respiration,** also simply called **respiration,** involves transferring the electrons from NADH and $FADH_2$ (which are originally derived from glucose) to the electron transport chain (ETC), which ultimately donates them to a terminal electron acceptor. The ETC uses the electrons to generate the proton motive force—the form of energy cells use to make ATP by oxidative phosphorylation. In **aerobic respiration,**

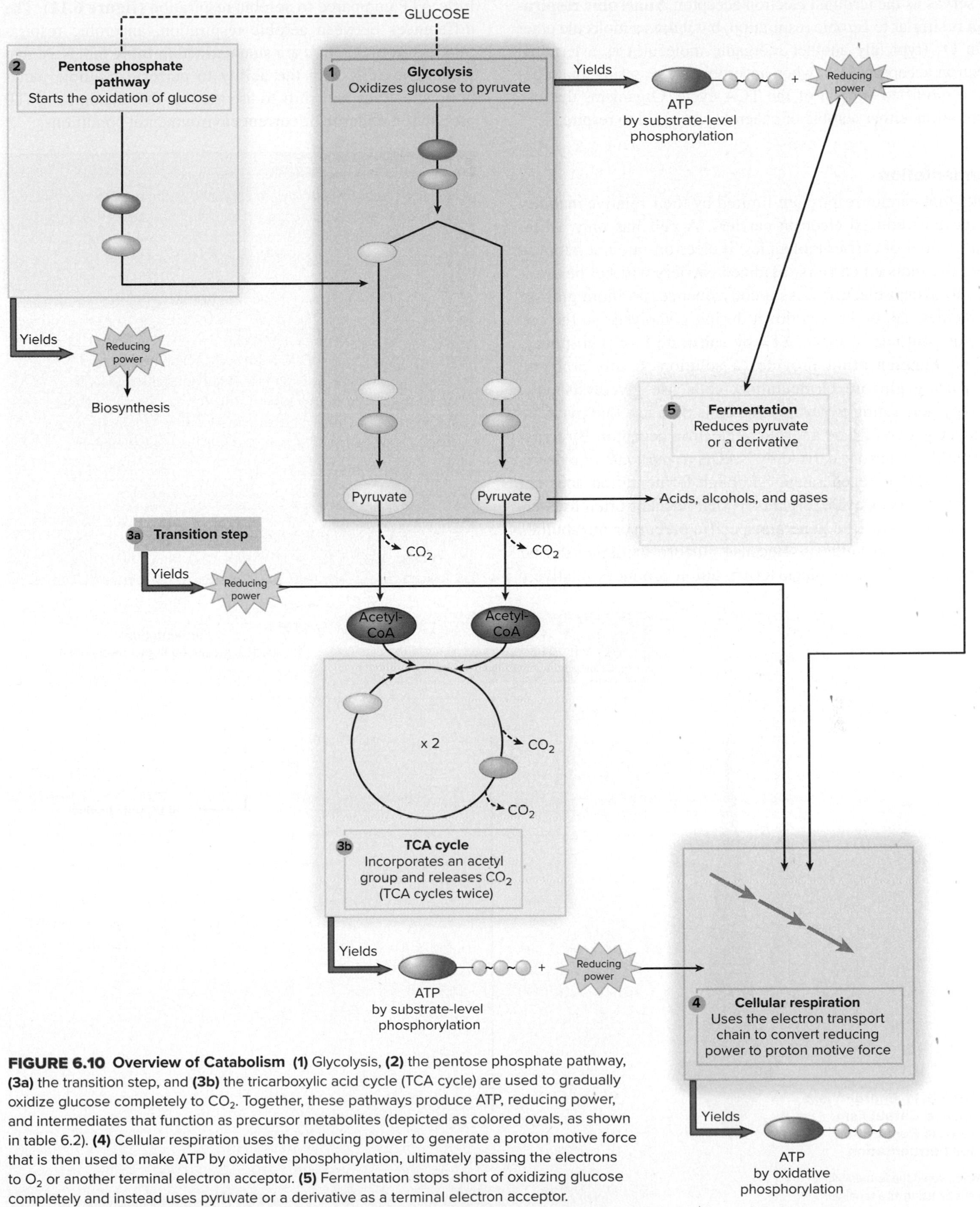

FIGURE 6.10 Overview of Catabolism **(1)** Glycolysis, **(2)** the pentose phosphate pathway, **(3a)** the transition step, and **(3b)** the tricarboxylic acid cycle (TCA cycle) are used to gradually oxidize glucose completely to CO_2. Together, these pathways produce ATP, reducing power, and intermediates that function as precursor metabolites (depicted as colored ovals, as shown in table 6.2). **(4)** Cellular respiration uses the reducing power to generate a proton motive force that is then used to make ATP by oxidative phosphorylation, ultimately passing the electrons to O_2 or another terminal electron acceptor. **(5)** Fermentation stops short of oxidizing glucose completely and instead uses pyruvate or a derivative as a terminal electron acceptor.

? Why is the TCA cycle repeated twice for every glucose molecule that enters glycolysis?

O_2 serves as the terminal electron acceptor. **Anaerobic respiration** is similar to aerobic respiration, but it uses a molecule other than O_2 (typically another inorganic molecule) as a terminal electron acceptor. Also, when anaerobically respiring, microbes use a modified version of the TCA cycle. Organisms that use respiration, either aerobic or anaerobic, are said to respire.

Fermentation

Cells that cannot respire are limited by their relative inability to recycle reduced electron carriers. A cell has only a limited number of carrier molecules; if electrons are not removed from the reduced carriers, oxidized carriers will not be available to accept electrons. As a consequence, no more glucose molecules can be broken down during glycolysis, so the cell cannot continue to make ATP by substrate-level phosphorylation. **Fermentation** provides a solution to this problem. (5) During glucose fermentation, cells use glycolysis only, thereby generating pyruvate. The cells then use that pyruvate or a derivative of it as a terminal electron acceptor. By transferring the electrons carried by NADH to pyruvate or a derivative, NAD^+ is regenerated. Although fermentation does not involve the TCA cycle, organisms that ferment often use certain key steps of it to generate specific precursor metabolites required for biosynthesis. Because glucose oxidation during fermentation is incomplete, fermentation produces relatively little ATP compared to aerobic respiration (**figure 6.11**). The differences between aerobic respiration, anaerobic respiration, and fermentation are summarized in **table 6.3.** Note that facultative cells with the ability to perform multiple metabolic strategies will shift to use the one that maximizes ATP production under their current environmental conditions.

MicroAssessment 6.1

Cells use catabolic pathways to gradually oxidize an energy source and capture the released energy. Each step of a metabolic pathway is catalyzed by a specific enzyme. Substrate-level phosphorylation uses the energy of exergonic chemical reactions to transfer a phosphate group from a donor molecule to ADP, synthesizing ATP; oxidative phosphorylation uses the energy of the proton motive force to add an inorganic phosphate group to ADP to do the same. Reducing power in the form of NADH and $FADH_2$ is used to generate the proton motive force; the reducing power of NADPH is used in biosynthesis. Precursor metabolites are metabolic intermediates that can be either further broken down to capture energy or used in biosynthesis. The central metabolic pathways generate ATP, reducing power, and precursor metabolites.

1. How does the fate of electrons carried by NADPH differ from the fate of electrons carried by NADH?
2. Why are the central metabolic pathways called amphibolic?
3. Why does fermentation supply less energy than cellular respiration?

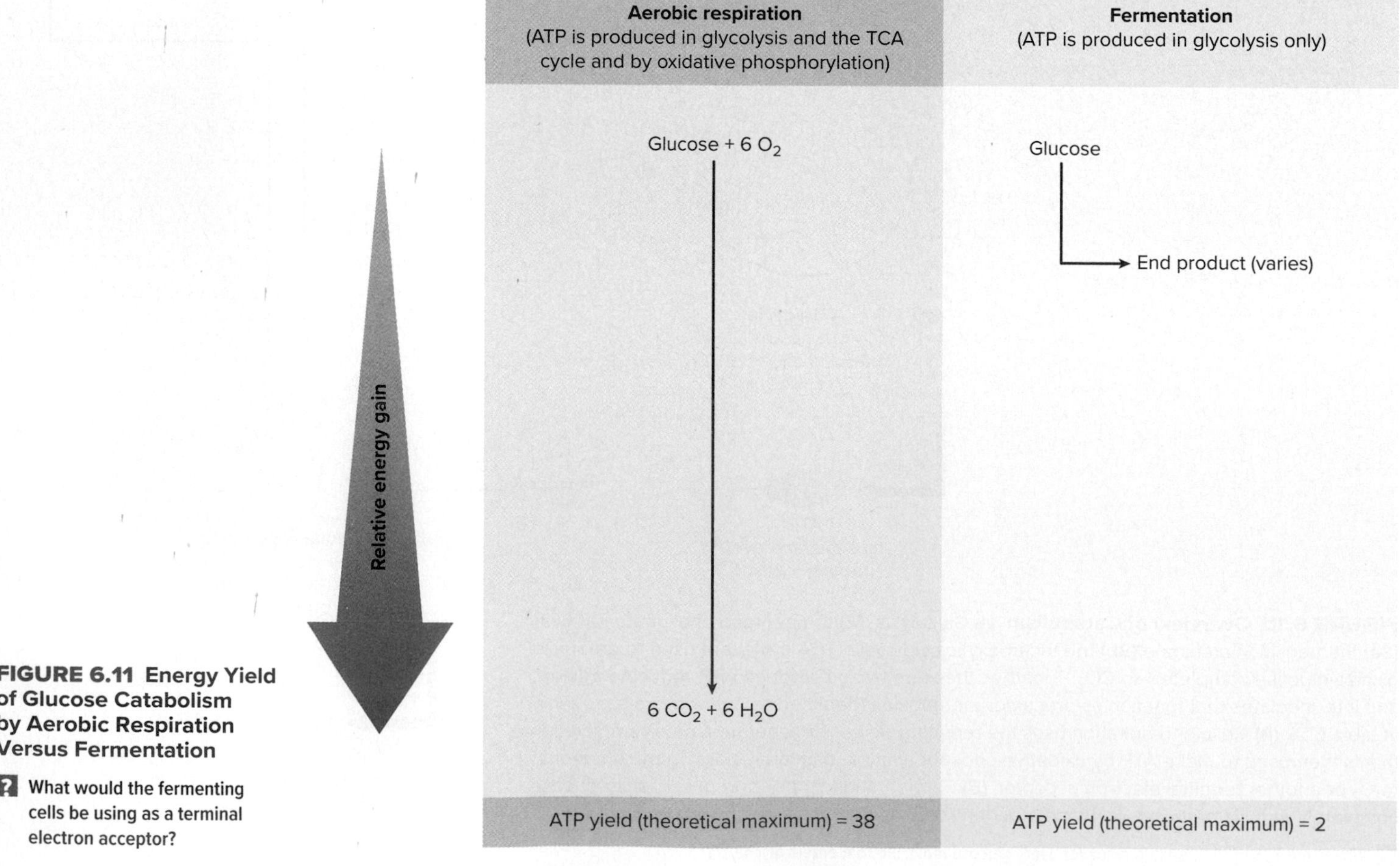

FIGURE 6.11 Energy Yield of Glucose Catabolism by Aerobic Respiration Versus Fermentation

? What would the fermenting cells be using as a terminal electron acceptor?

TABLE 6.3 ATP-Generating Processes of Prokaryotic Chemoorganoheterotrophs

Metabolic Process	Uses an Electron Transport Chain	Terminal Electron Acceptor	ATP Generated by Substrate-Level Phosphorylation (Theoretical Maximum)	ATP Generated by Oxidative Phosphorylation (Theoretical Maximum)	Total ATP Generated (Theoretical Maximum)
Aerobic respiration	Yes	O_2	2 in glycolysis (net) 2 in the TCA cycle 4 total	34	38
Anaerobic respiration	Yes	Molecule other than O_2 such as nitrate (NO_3^-), nitrite (NO_2^-), sulfate (SO_4^{2-})	Number varies; however, the ATP yield of anaerobic respiration is less than that of aerobic respiration but more than that of fermentation.		
Fermentation	No	Organic molecule (pyruvate or a derivative)	2 in glycolysis (net) 2 total	0	2

6.2 ■ Enzymes

Learning Outcomes

6. Describe the active site of an enzyme, and explain how it relates to the enzyme-substrate complex.
7. Compare and contrast cofactors and coenzymes.
8. List two environmental factors that influence enzyme activity.
9. Describe allosteric regulation.
10. Compare and contrast competitive and non-competitive enzyme inhibition.

Recall that enzymes are biological catalysts, meaning they increase the rates at which substrates are converted into products (see figure 6.5). They do this with extraordinary specificity and speed, usually acting on only one or a few structurally similar substrates. They are neither consumed nor permanently changed during a reaction, so a single enzyme molecule can function repeatedly. More than a thousand different enzymes exist in a cell, and most are proteins.

The name of an enzyme usually reflects its function and ends with the suffix *-ase*. For example, isocitrate dehydrogenase removes a hydrogen atom (a proton-electron pair) from isocitrate. Some groups of enzymes are referred to by their general function—for example, proteases degrade proteins.

MicroByte

In only one second, the fastest enzymes can convert more than 10^4 substrate molecules into products.

Mechanisms and Consequences of Enzyme Action

Each enzyme has an **active site,** typically a relatively small crevice for binding one or more substrates by weak forces (**figure 6.12**). The binding of a substrate to the active site causes the shape of the flexible enzyme to change slightly. This mutual interaction, or induced fit, results in a temporary intermediate called an enzyme-substrate complex. The substrate is held within this complex in a specific orientation so that existing bonds are destabilized and new ones can easily form, lowering the activation energy of the reaction. Then, after releasing their products, enzymes are left unchanged and are free to combine with new substrate molecules. Note that while some enzymes catalyze reactions involved in the breakdown of a substrate into smaller products, other enzymes join substrates to create larger products. Theoretically, all enzyme-catalyzed reactions are reversible, but the free energy change of certain reactions makes them effectively irreversible.

The interaction of an enzyme with its substrate is very specific. The substrate fits into the active site like a hand into a glove. Not only must it fit spatially, but appropriate chemical interactions such as hydrogen and ionic bonding between the active site and the substrate need to occur to induce the fit. This requirement for a precise fit and interaction explains why, with few exceptions, a unique enzyme is required to catalyze each reaction in a cell. Very few molecules of any particular enzyme are needed, however, as each can be used repeatedly.

Cofactors

Some enzymes require the assistance of an attached non-protein component called a **cofactor** (**figure 6.13**). Magnesium, zinc, copper, and other trace elements often function as inorganic cofactors. **Coenzymes** are a subset of cofactors; they are loosely attached non-protein organic compounds that help some enzymes transfer certain molecules or electrons from one compound to another. The various coenzymes function in different ways. Some remain bound to the enzyme during the transfer process, whereas others separate from the enzyme, carrying the substance being transferred along with them. The same coenzyme can assist many different enzymes. Because of this, far fewer different coenzymes are required than enzymes. Like enzymes, coenzymes can be reused and, consequently,

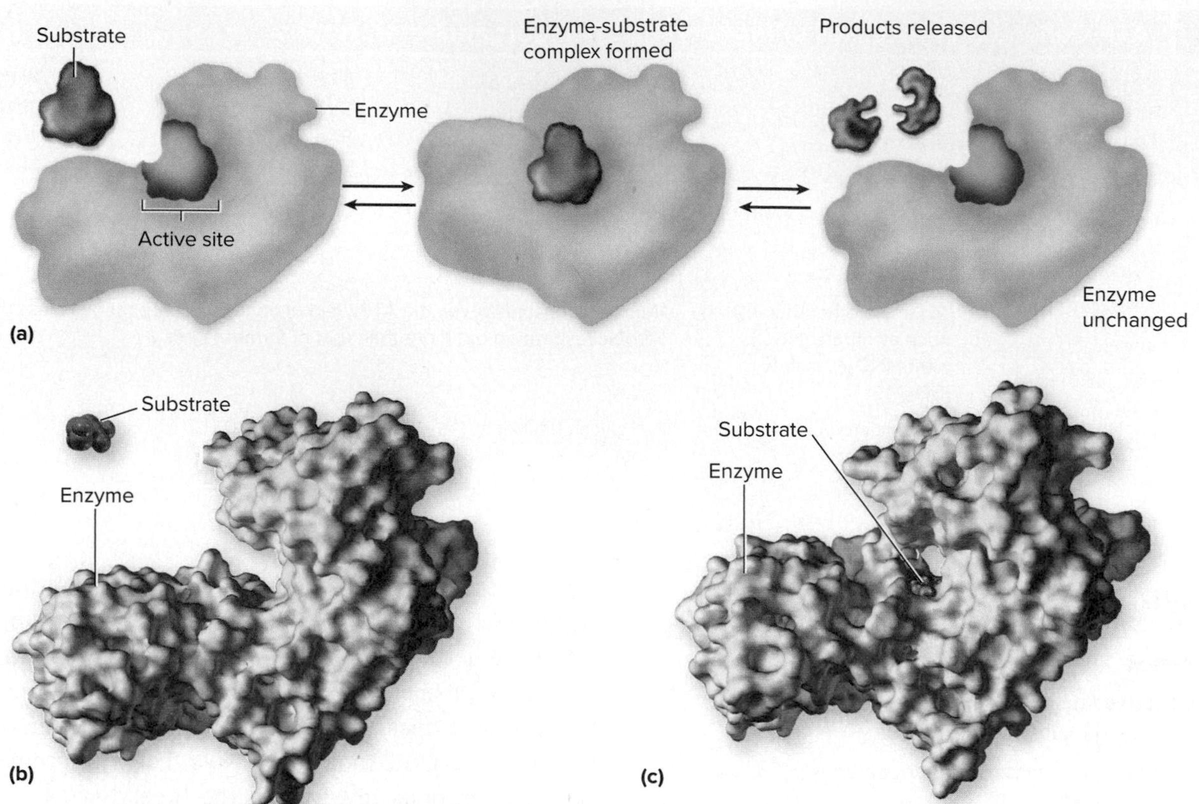

FIGURE 6.12 Mechanism of Enzyme Action (a) The substrate binds to the active site, forming an enzyme-substrate complex. The products are then released, leaving the enzyme unchanged and free to combine with new substrate molecules. (b) A model showing an enzyme and its substrate. (c) The binding of the substrate to the active site causes the shape of the flexible enzyme to change slightly.
b–c: Kenneth Eward/BioGrafx/Science Source

What does the enzyme called succinate dehydrogenase likely do?

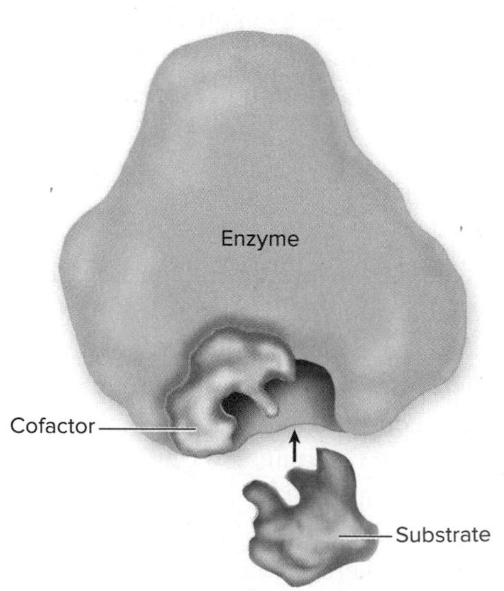

FIGURE 6.13 Some Enzymes Require the Assistance of a Cofactor Cofactors are non-protein components, either coenzymes or trace elements.

What is the function of the coenzyme FAD?

are needed only in very small quantities. The electron carriers FAD, NAD^+, and $NADP^+$ are coenzymes that play an important role in enzyme-catalyzed oxidation-reduction reactions.

Most coenzymes are derived from certain vitamins, which are organic compounds necessary for life but which the human body cannot synthesize. Without a given vitamin, any enzyme that requires the corresponding coenzyme for its activity cannot function (**table 6.4**). Thus, a single vitamin deficiency has serious health consequences. Most vitamins must be ingested as part of the diet, but certain ones made by the intestinal microbiota can be absorbed.

Environmental Factors That Influence Enzyme Activity

Several environmental factors influence how well enzymes function (**figure 6.14**). Each enzyme has a narrow range of conditions—including temperature, pH, and salt concentration—within which it functions best. A 10°C rise in temperature approximately doubles the speed of enzymatic reactions, until optimal activity is reached; this explains why bacteria tend to grow more rapidly at higher temperatures. If the temperature is

TABLE 6.4 Some Vitamins and the Uses of Their Derived Coenzymes

Vitamin	Coenzyme	Function of Coenzyme
Folate/folic acid	Tetrahydrofolate	Transfers 1-carbon compounds in nucleotide synthesis
Niacin	Nicotinamide adenine dinucleotide (NAD^+)	Carries reducing power
Pantothenic acid	Coenzyme A	Carries the acetyl group that enters the TCA cycle
Pyridoxine (vitamin B_6)	Pyridoxal phosphate	Transfers amino groups in amino acid synthesis
Riboflavin (vitamin B_2)	Flavin adenine dinucleotide (FAD)	Carries reducing power
Thiamine (vitamin B_1)	Thiamin pyrophosphate	Helps remove CO_2 from pyruvate in the transition step

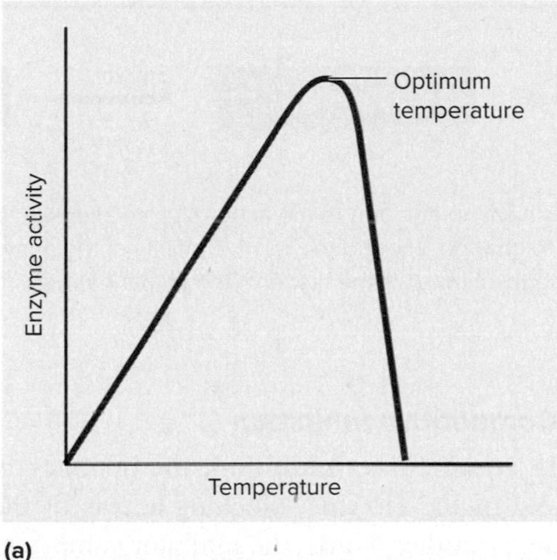

(a)

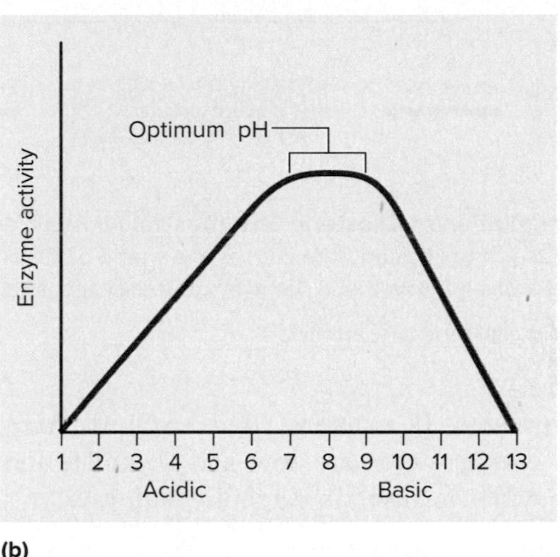

(b)

FIGURE 6.14 Environmental Factors That Influence Enzyme Activity (a) A rise in temperature increases the speed of enzymatic activity until the optimum temperature is reached. If the temperature gets too high, the enzyme denatures and no longer functions. **(b)** Most enzymes function best at pH values slightly above 7.

? **Why would an enzyme no longer function once it denatures?**

too high, however, proteins will denature (see figure 2.26) and no longer function. Most enzymes function best at low salt concentrations and at pH values slightly above 7. As a result, most microbes grow fastest under these same conditions. Some prokaryotes, however, particularly certain archaea, thrive in environments where conditions are extreme. They may require high salt concentrations, very acidic conditions, and/or temperatures near boiling. Enzymes from these organisms can be very important commercially because they function in harsh conditions that are often typical of industrial settings.

Allosteric Regulation

Cells can adjust the flow of metabolic pathways by regulating the activity of certain key enzymes. Enzymes that can be controlled are **allosteric** (*allo* means "other"; *stereos* means "shape"), meaning they each have one or more allosteric sites separate from the active site (**figure 6.15**). When a regulatory molecule binds to an allosteric site, the shape of the enzyme changes, altering the relative affinity (chemical attraction) of the enzyme for its substrate. An allosteric inhibitor decreases the affinity, while an allosteric activator increases it. Either way, the binding of the regulatory molecule is reversible, so the effect is only temporary.

Allosteric enzymes generally catalyze the first step of a pathway. If the pathway is biosynthetic, the end product generally acts as the allosteric inhibitor—a mechanism called feedback inhibition (see figure 6.15c). This allows the cell to shut down a pathway when the product begins accumulating. For example, the amino acid isoleucine is an allosteric inhibitor of the first enzyme of the pathway that converts threonine to isoleucine. When the level of isoleucine is relatively high, the pathway is shut down. The binding of the isoleucine is reversible, however, so the enzyme becomes active again if isoleucine levels decrease. Cells can also control the amount of enzyme they synthesize, a topic discussed in chapter 7.

Compounds that reflect a cell's relative energy supply often regulate allosteric enzymes of catabolic pathways. This allows cells to adjust the flow of metabolites through these pathways in response to changing energy needs. High levels of ATP inhibit certain enzymes and, as a consequence, slow

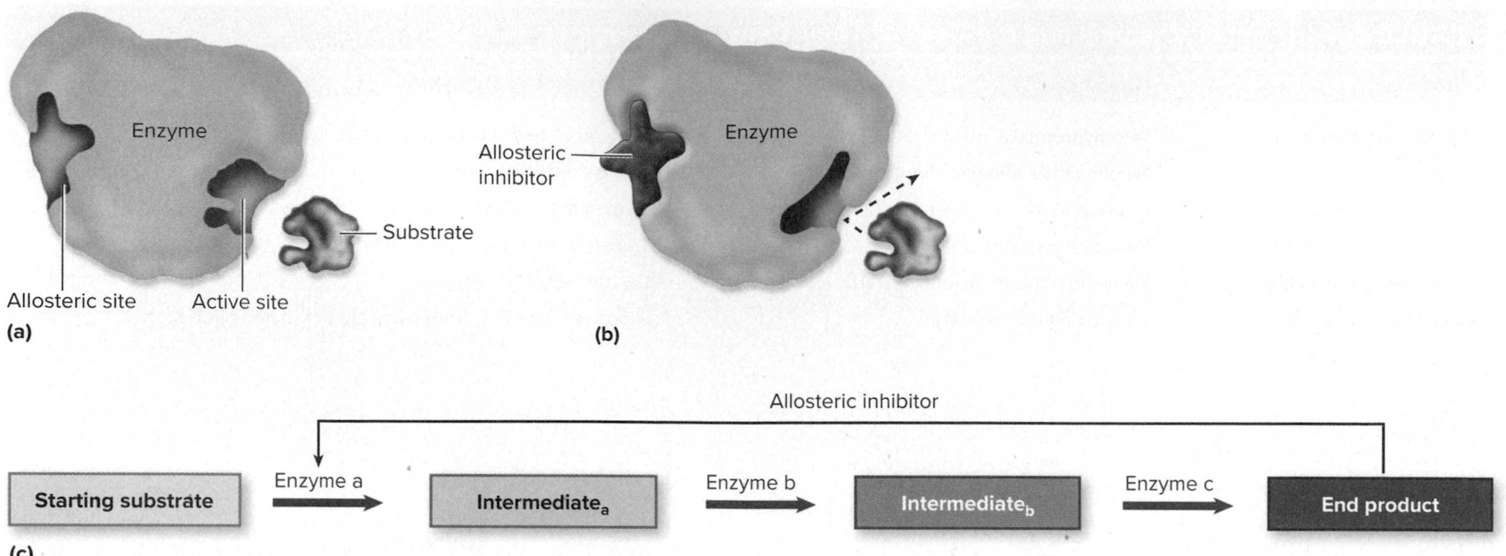

FIGURE 6.15 Regulation of Allosteric Enzymes **(a)** Allosteric enzymes have, in addition to the active site, an allosteric site. **(b)** The binding of a regulatory molecule to the allosteric site causes the shape of the enzyme to change, altering the relative affinity of the enzyme for its substrate. **(c)** The end product of a given biosynthetic pathway generally acts as an allosteric inhibitor of the first enzyme of that pathway.

Why would a cell need to regulate enzyme activity?

down catabolic processes. In contrast, high levels of ADP warn that a cell's energy stores are low and stimulate the activity of some enzymes involved in catabolic pathways.

Enzyme Inhibition

Enzymes can be inhibited by a variety of compounds other than the regulatory molecules just described. These inhibitory compounds can prevent microbial growth, so some are medically and commercially valuable. The site on the enzyme to which an inhibitor binds determines whether it functions as a competitive or non-competitive inhibitor.

Competitive Inhibition

In **competitive inhibition,** the inhibitor binds to the active site of the enzyme, blocking access of the substrate to that site; in other words, the inhibitor competes with the substrate for the active site. Generally the inhibitor has a chemical structure similar to that of the normal substrate.

A good example of competitive inhibition is the action of the group of antimicrobial medications called sulfa drugs (**figure 6.16**). These inhibit an enzyme in the pathway that many bacteria use to synthesize the vitamin folate; *E. coli* and others lack transporters to bring the vitamin into cells, so

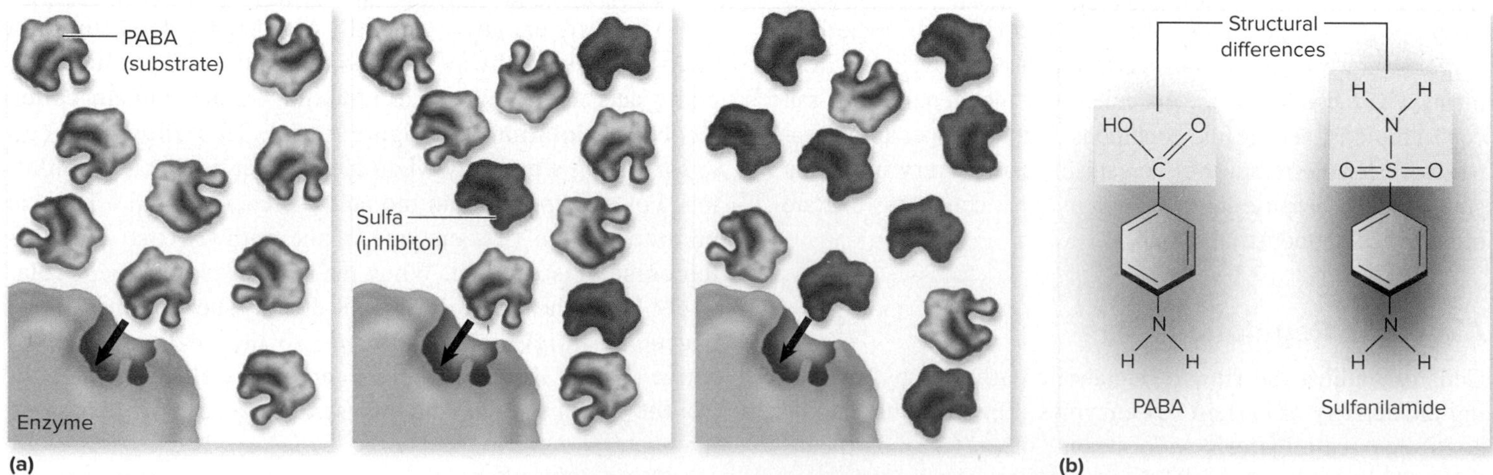

FIGURE 6.16 Competitive Inhibition by Sulfa Drugs **(a)** The sulfa drug (an inhibitor) competes with PABA (the normal substrate) for binding to the enzyme's active site. The greater the number of sulfa drug molecules relative to PABA molecules, the more likely it is that the active site of the enzyme will be occupied by a sulfa drug molecule. **(b)** A competitive inhibitor (such as sulfanilamide, a sulfa drug) generally has a chemical structure similar to an enzyme's normal substrate (such as PABA).

Will the illustrated enzyme function if sulfa is removed?

TABLE 6.5 Characteristics of Enzyme Inhibitors

Type	Characteristics
Competitive inhibition	Inhibitor binds to the active site of the enzyme, blocking access of the substrate to that site. Competitive inhibitors such as sulfa drugs are used as antibacterial medications.
Non-competitive inhibition (by regulatory molecules)	Inhibitor binds to a site other than the active site and changes the shape of the enzyme, so that the substrate can no longer bind the active site. This is a reversible action that cells use to control the activity of allosteric enzymes.
Non-competitive inhibition (by enzyme poisons)	Inhibitor binds to a site other than the active site and permanently changes the shape of the enzyme, making the enzyme non-functional. Enzyme poisons such as mercury are used in certain antimicrobial compounds.

they require the enzyme's activity. The drug does not affect human metabolism because humans cannot synthesize folate, so they must consume it in foods or in nutritional supplements (as folic acid). Sulfa drugs have a structure similar to *para*-aminobenzoic acid (PABA), an intermediate in the bacterial pathway for folate synthesis. Because of this, the drugs fit into the active site of the bacterial enzyme that normally uses PABA as a substrate. By doing so, they prevent the enzyme from binding PABA. The greater the number of drug molecules relative to PABA molecules, the more likely it is that a drug molecule will occupy the enzyme's active site. Once the sulfa drug is removed, however, the enzyme functions normally with PABA as the substrate.

Non-Competitive Inhibition

Non-competitive inhibition occurs when the inhibitor binds to a site other than the active site; the binding changes the shape of the enzyme so that the substrate can no longer bind the active site (see figure 6.15). The allosteric inhibitors that cells produce to regulate enzyme activity are non-competitive inhibitors that have a reversible action. In contrast, the effect of other non-competitive inhibitors is permanent. For example, mercury oxidizes the S–H groups of the amino acid cysteine in proteins. This converts cysteine to cystine, which cannot form the important covalent disulfide bond (S–S). As a result, the enzyme shape changes, making it non-functional; the inhibitor "poisons" the enzyme.

Table 6.5 summarizes the characteristics of enzyme inhibitors.

MicroAssessment 6.2

Enzymes catalyze chemical reactions without being consumed or permanently changed. The activity of some enzymes requires a cofactor. Environmental factors influence enzyme activity and, by doing so, determine how rapidly microorganisms multiply. The activity of allosteric enzymes can be regulated. Certain compounds inhibit enzymes, competitively or non-competitively.

4. Describe the function of a coenzyme.
5. Explain why sulfa drugs prevent bacterial growth without harming the human host.
6. Why is it important for a cell that allosteric inhibition be reversible?

6.3 ■ The Central Metabolic Pathways

Learning Outcome

11. Compare and contrast each of the central metabolic pathways with respect to the yield of ATP, reducing power, and number of different precursor metabolites.

The three central metabolic pathways—glycolysis, the pentose phosphate pathway, and the tricarboxylic acid cycle (TCA cycle)—degrade organic molecules in a step-wise fashion, generating:

- **ATP** by substrate-level phosphorylation
- **Reducing power** in the forms of NADH, $FADH_2$, and NADPH
- **Precursor metabolites** (see table 6.2)

This section describes how a molecule of glucose is broken down in the central metabolic pathways. Bear in mind, however, that many millions of molecules of glucose enter a cell, and different molecules can have different fates (see figure 6.9). For example, a cell might oxidize one glucose molecule completely to CO_2, thereby producing the maximum amount of ATP. Alternatively, another glucose molecule might enter glycolysis or the pentose phosphate pathway, only to be siphoned off as a precursor metabolite for use in biosynthesis. The step and rate at which the various intermediates are removed for biosynthesis will dramatically affect the overall energy gain of catabolism. This is generally overlooked in descriptions of the ATP-generating functions of these pathways for the sake of simplicity. However, it is important to recall that because these pathways serve more than one function, the calculated energy yields are only theoretical maximums.

The intermediates and end products of metabolic pathways are sometimes organic acids, which are weak acids. Depending on the pH of the environment, these occur as either the undissociated or the dissociated (ionized) form. Biologists often use the names of the two forms interchangeably—for example, pyruvic acid and pyruvate (an ion).

Glycolysis

Glycolysis is a series of steps that begins the oxidation of glucose, converting the 6-carbon compound to two molecules of the 3-carbon pyruvate. The most common glycolytic pathway is called the Embden-Meyerhof (EM) pathway, named after its discoverers. The steps of this pathway are easier to follow in the context of a diagram, so **figure 6.17** illustrates the process and includes a description of each step. As you will see by studying that diagram, for every glucose molecule that enters glycolysis, the steps produce:

- **ATP:** 2 molecules, net gain; 2 are spent early in the pathway (① and ③), but 4 are made by substrate-level phosphorylation after the 6-carbon molecule is split in half (⑦ and ⑩). Note that in bacteria, the first phosphate is added as glucose gets transported into the cell via group translocation; that phosphate is equivalent to one from ATP even though it comes from a different compound, so for simplicity its actual source is generally ignored.
- **Reducing power:** 2 molecules of NADH ⑥
- **Precursor metabolites:** Six different precursor metabolites (glucose-6-phosphate ①; fructose-6-phosphate ②; dihydroxyacetone phosphate ④; 3-phosphoglycerate ⑦; phosphoenolpyruvate ⑨; and pyruvate ⑩; see table 6.2)

Glycolysis is sometimes considered to be synonymous with the EM pathway, but some microbes use a more ancient form of glycolysis called the Entner-Doudoroff (ED) pathway. Although the ED pathway generates half as much ATP, it is particularly useful for microbes that inhabit the large intestine because it degrades gluconate, an acidic glucose derivative commonly found there; in fact, some organisms have the ED pathway in addition to (rather than instead of) the EM pathway. The ED pathway is also useful for bacterial identification because it involves unique enzymes and generates distinct intermediates that can be detected in the laboratory.

Pentose Phosphate Pathway

The pentose phosphate pathway also uses glucose as the starting substrate, but the oxidized product is glyceraldehyde-3-phosphate (G3P), a compound that then enters a step in glycolysis for further breakdown. Unlike glycolysis, however, this pathway does not make ATP by substrate-level phosphorylation, nor does it make reducing power that can be used to make ATP by oxidative phosphorylation. Instead, the pathway is important because of its contribution to biosynthesis: It generates reducing power in the form of NADPH and also produces two important precursor metabolites (as intermediates), releasing CO_2 as metabolic waste. The pathway is complex, with several alternative steps, so the yield varies depending on which alternatives are taken. In general, however, it produces:

- **Reducing power:** A variable amount of reducing power in the form of NADPH. Recall that NADPH is used in biosynthesis.
- **Precursor metabolites:** Two different precursor metabolites (ribose-5-phosphate and erythrose-4-phosphate; see table 6.2).

Transition Step and Tricarboxylic Acid (TCA) Cycle

The transition step and the tricarboxylic acid cycle together complete the oxidation of glucose. In prokaryotic cells, all the central metabolic pathways occur in the cytoplasm. In eukaryotic cells, however, the enzymes of glycolysis and the pentose phosphate pathway are located in the cytoplasm, whereas those of the transition step and the TCA cycle are within the mitochondrial matrix. Because of this, eukaryotic cells must transport pyruvate molecules into mitochondria for the transition step to occur.

Transition Step

The transition step oxidizes the end product of glycolysis (pyruvate), converting it to the compound used to begin the TCA cycle (acetyl-CoA). It therefore links glycolysis with the TCA cycle and occurs twice per glucose molecule (recall that glycolysis splits a glucose molecule, producing 2 pyruvate molecules per glucose). CO_2 is also produced as metabolic waste during the step. For each glucose that enters glycolysis, the transition step generates:

- **Reducing power:** 2 molecules of NADH
- **Precursor metabolites:** One of the precursor metabolites (acetyl-CoA; see table 6.2)

TCA Cycle

The tricarboxylic acid (TCA) cycle oxidizes the acetyl group of the end product of the transition step (acetyl-CoA), "turning" once to produce 2 molecules of CO_2 (released as waste); thus, the cycle "turns" twice for each glucose that enters glycolysis to complete that sugar's oxidation. As with glycolysis, the steps are easier to follow in the context of a diagram, so **figure 6.18** illustrates the process and includes a description of each step. Two "turns" of the TCA cycle produce:

- **ATP:** 2 molecules ⑤ made by substrate-level phosphorylation. In eukaryotic cells, this step produces a molecule related to ATP (GTP) that can be used to make ATP. However, *E. coli* produces ATP at this step.
- **Reducing power:** 6 molecules of NADH (③, ④, ⑧) and 2 molecules of $FADH_2$ (⑥).
- **Precursor metabolites:** Two different precursor metabolites (α-ketoglutarate ③ and oxaloacetate ⑧; see table 6.2).

The central metabolic pathways, along with the transition step that links glycolysis and the TCA cycle, are compared in **table 6.6.** The entire pathways with chemical formulas and enzyme names are illustrated in Appendix III.

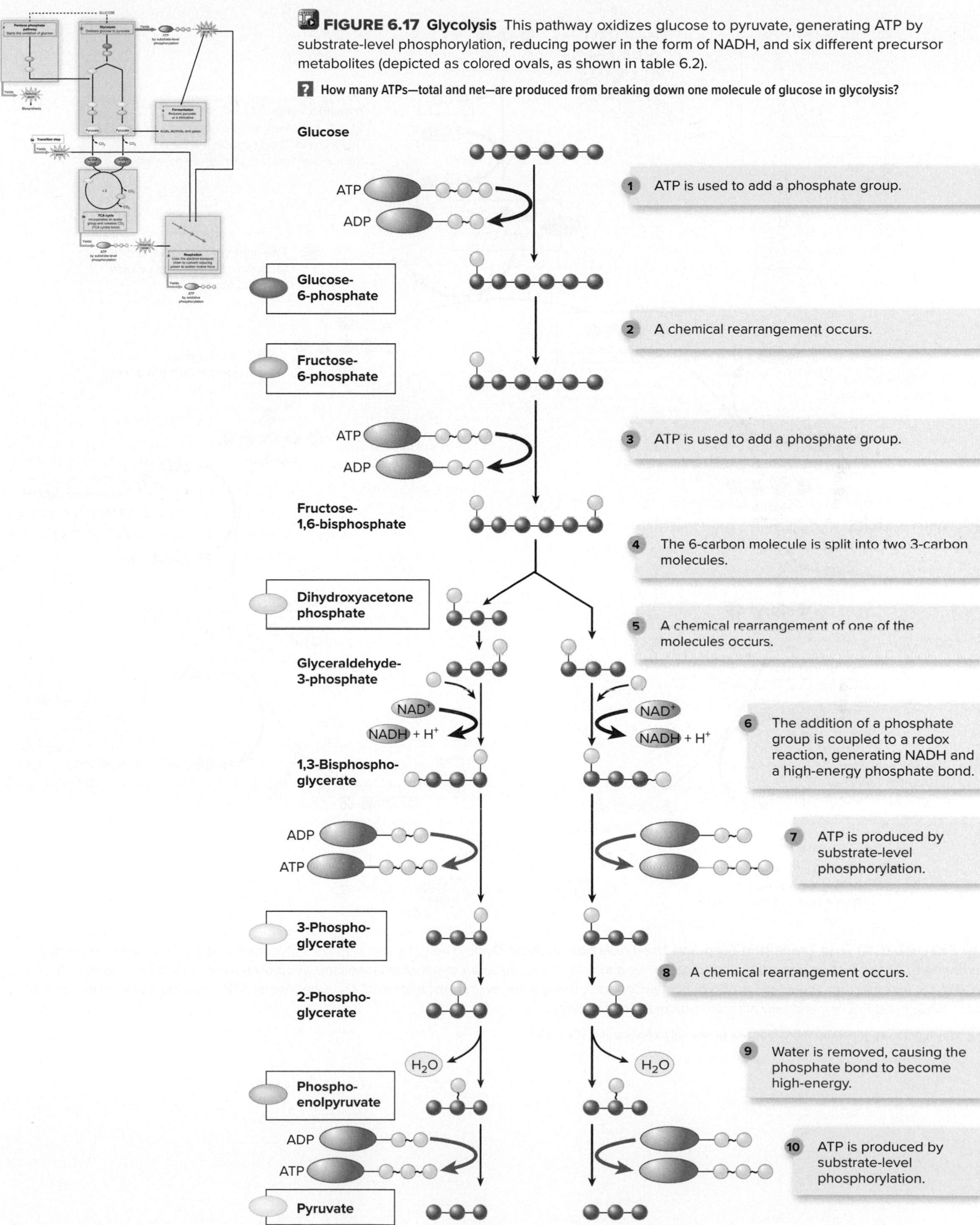

FIGURE 6.17 Glycolysis This pathway oxidizes glucose to pyruvate, generating ATP by substrate-level phosphorylation, reducing power in the form of NADH, and six different precursor metabolites (depicted as colored ovals, as shown in table 6.2).

How many ATPs—total and net—are produced from breaking down one molecule of glucose in glycolysis?

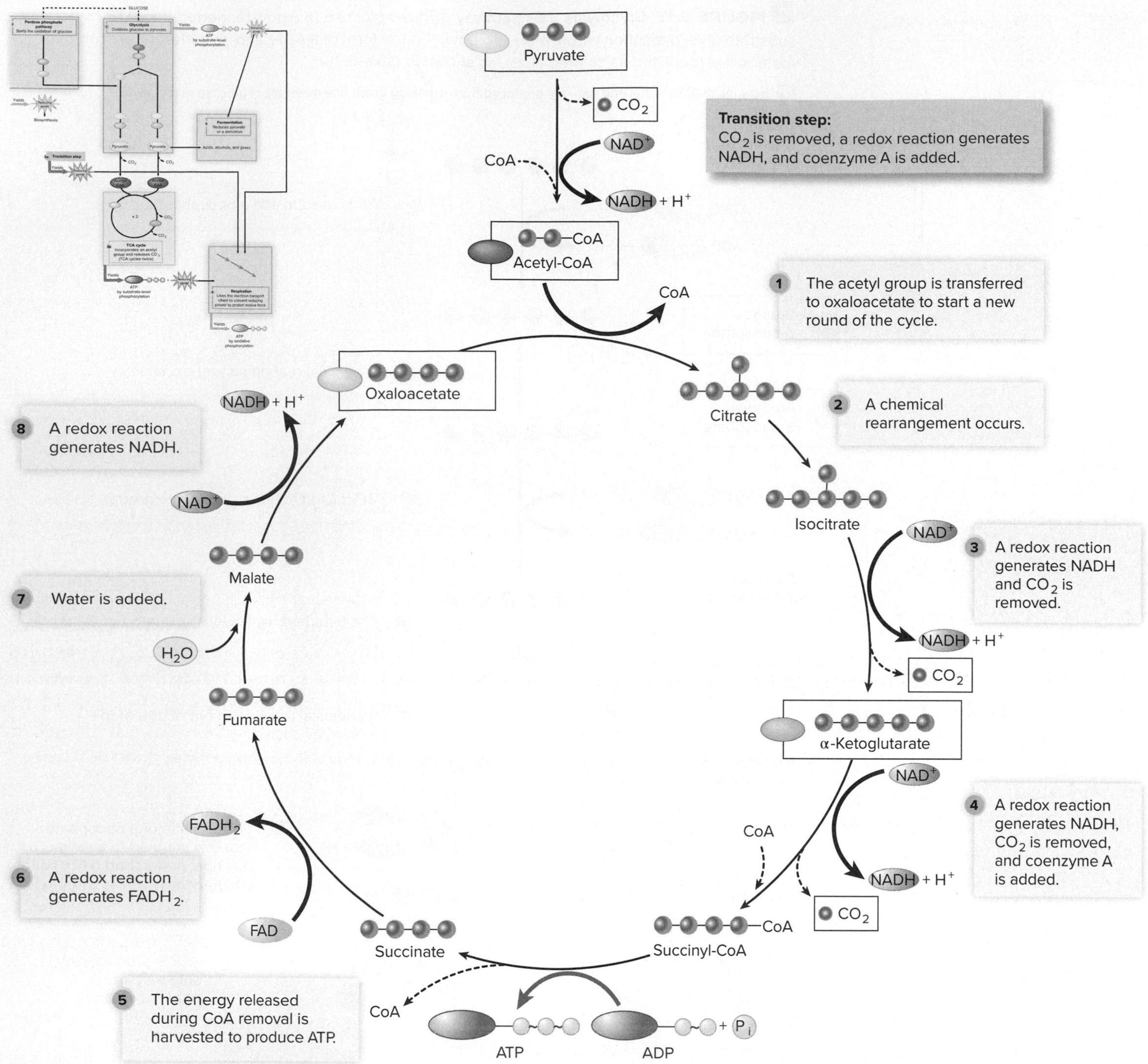
Pyruvate
CO_2
Transition step:
CO_2 is removed, a redox reaction generates NADH, and coenzyme A is added.
CoA
NAD^+
NADH + H^+
CoA
Acetyl-CoA
CoA
1 The acetyl group is transferred to oxaloacetate to start a new round of the cycle.
Oxaloacetate
Citrate
2 A chemical rearrangement occurs.
NADH + H^+
8 A redox reaction generates NADH.
NAD^+
Isocitrate
NAD^+
3 A redox reaction generates NADH and CO_2 is removed.
Malate
7 Water is added.
NADH + H^+
H_2O
CO_2
Fumarate
α-Ketoglutarate
NAD^+
4 A redox reaction generates NADH, CO_2 is removed, and coenzyme A is added.
$FADH_2$
CoA
6 A redox reaction generates $FADH_2$.
NADH + H^+
CO_2
FAD
Succinate
Succinyl-CoA
5 The energy released during CoA removal is harvested to produce ATP.
CoA
ATP
ADP
+ P_i

FIGURE 6.18 The Transition Step and the Tricarboxylic Acid Cycle The transition step links glycolysis and the TCA cycle, converting pyruvate to acetyl-CoA; it generates reducing power and one of the precursor metabolites (depicted as colored ovals, as shown in table 6.2). The TCA cycle incorporates the acetyl group of acetyl-CoA and, using a series of steps, releases CO_2; it generates ATP, reducing power in the form of both NADH and $FADH_2$, and two different precursor metabolites.

? Which pathway generates more reducing power—glycolysis or the TCA cycle?

TABLE 6.6 Comparison of the Central Metabolic Pathways

Pathway	Outcome Per Glucose Molecule
Glycolysis	Oxidizes glucose, converting it to 2 molecules of pyruvate and generating: • 2 ATP (net) by substrate-level phosphorylation • 2 NADH • six different precursor metabolites (one molecule of each)
Pentose phosphate cycle	Oxidizes glucose, converting it to an intermediate of glycolysis and generating: • NADPH (amount varies) • two different precursor metabolites (numbers of molecules of each vary)
Transition step	Oxidizes pyruvate, converting it to acetyl-CoA and in the process releasing CO_2 (a waste product). The step is repeated twice for each glucose molecule, generating: • 2 NADH • one of the precursor metabolites (two molecules)
TCA cycle	Oxidizes the acetyl group of acetyl-CoA, converting it to 2 molecules of CO_2 (a waste product). The cycle is repeated twice for each glucose molecule, generating: • 2 ATP by substrate-level phosphorylation (may involve conversion of GTP) • 6 NADH • 2 $FADH_2$ • two different precursor metabolites (two molecules of each)

MicroAssessment 6.3

Glycolysis oxidizes glucose to pyruvate, yielding some ATP by substrate-level phosphorylation, NADH, and six different precursor metabolites. The pentose phosphate pathway also oxidizes glucose, but more importantly, it produces two different precursor metabolites and NADPH for biosynthesis. The transition step and the TCA cycle, each repeated twice per starting glucose molecule, complete the oxidation of glucose, yielding some ATP by substrate-level phosphorylation, a great deal of reducing power, and three different precursor metabolites.

7. Considering that 4 ATP molecules are produced in glycolysis, why is the net ATP yield of the pathway only 2?
8. Which central metabolic pathway produces the greatest number of different precursor metabolites?
9. Which compound contains more free energy—glucose or an intermediate of the TCA cycle? On what did you base your conclusion?

6.4 ■ Cellular Respiration

Learning Outcomes

12. Describe the components of the electron transport chain and how they generate a proton motive force.
13. Compare and contrast the electron transport chains of eukaryotes and prokaryotes.
14. Describe how a proton motive force is used to synthesize ATP and how the ATP yield of aerobic respiration is calculated.

Cellular respiration uses the reducing power generated in glycolysis, the transition step, and the TCA cycle to synthesize ATP. The mechanism, **oxidative phosphorylation,** involves two separate processes:

- **Generating a proton motive force.** The electron transport chain uses the reducing power of NADH and $FADH_2$ to generate a proton motive force.
- **Synthesizing ATP.** The enzyme **ATP synthase** uses the energy of the proton motive force to drive the synthesis of ATP.

The association between the electron transport chain and ATP synthesis was proposed by the British scientist Peter Mitchell in 1961. His hypothesis, now called the **chemiosmotic theory,** was widely dismissed initially. After years of self-funded research in collaboration with Jennifer Moyle, they were finally able to convince others of its validity; he received a Nobel Prize in 1978 in recognition of this work.

The Electron Transport Chain (ETC)—Generating a Proton Motive Force

The **electron transport chain** (ETC) is a series of membrane-embedded electron carriers; it accepts electrons from NADH and $FADH_2$ and then passes those electrons from one carrier to another with each successive carrier having a stronger attraction (higher affinity) for electrons. The transfer of electrons can be likened to a ball falling down a set of stairs; energy is released as the electrons are passed (**figure 6.19**). That energy release allows the ETC to pump protons across the membrane, generating an electrochemical gradient called a proton motive force (see figure 3.5). In prokaryotic cells, the ETC is in the cytoplasmic membrane; in eukaryotic cells, it is in the inner membrane of each mitochondrion.

Components of an Electron Transport Chain

Most carriers in the ETC are grouped into several large protein complexes that function as proton pumps. Others move electrons from one complex to the next. Three general types of electron carriers in the ETC are notable: quinones, cytochromes, and flavoproteins. **Quinones** are lipid-soluble organic molecules that move freely in the membrane, transferring

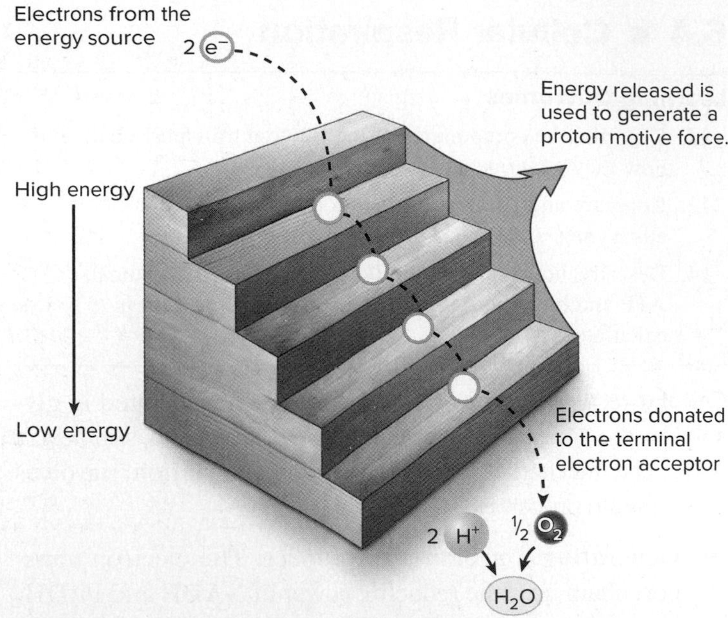

FIGURE 6.19 Electron Transport As electrons are passed along the electron transport chain, the energy released is used to establish a proton gradient.

? O_2 is serving as the terminal electron acceptor in this diagram; is it being oxidized or reduced?

electrons between certain protein complexes. Several types of quinones exist, one of the most common being ubiquinone (meaning "ubiquitous quinone"). Menaquinone, a quinone used in the ETC of some prokaryotes, serves as a source of vitamin K for humans and other mammals. This vitamin is required for proper blood coagulation, and mammals obtain much of their requirement by absorbing menaquinone produced by bacteria growing in the intestinal tract. **Cytochromes** are proteins that contain heme, a molecule that holds an iron atom in its center (heme is also found in red blood cells). Several different cytochromes exist, each designated with a letter, for example, cytochrome *c*. The presence of some cytochromes can be used to distinguish certain groups of bacteria. For instance, the oxidase test—which is used in the identification process for *Neisseria, Pseudomonas,* and *Campylobacter* species—detects the activity of cytochrome *c* oxidase (see table 10.3). **Flavoproteins** are proteins to which an organic group called a flavin is attached. FAD and other flavins are synthesized from the vitamin riboflavin (see table 6.4).

General Mechanisms of Proton Pumps

An important characteristic of the electron carriers in the ETC is that some accept only hydrogen atoms (proton-electron pairs), whereas others accept only electrons. The spatial arrangement of these two types of carriers causes protons to be moved from one side of the membrane to the other. This occurs because a hydrogen carrier receiving electrons from an electron carrier must pick up protons, which come from inside the cell (or matrix of the mitochondrion) due to the hydrogen carrier's relative location in the membrane. Conversely, when a hydrogen carrier passes electrons to a carrier that accepts electrons, but not protons, the protons are released to the outside of the cell (or intermembrane space of the mitochondrion). The net effect of these processes is that the ETC pumps protons from one side of the membrane to the other, generating the concentration gradient across the membrane. Note that the gradient could not be established if energy were not released during electron transfer.

The Electron Transport Chain of Mitochondria

The ETC of mitochondria has four different protein complexes, three of which function as proton pumps. In addition, two electron carriers (ubiquinone and cytochrome *c*) shuttle electrons between the complexes (**figure 6.20**). The protein complexes are:

- **Complex I** (also called NADH dehydrogenase complex). This accepts electrons from NADH, ultimately transferring them to ubiquinone (also called coenzyme Q); in the process, four protons are moved across the membrane.
- **Complex II** (also called succinate dehydrogenase complex). This accepts electrons from the TCA cycle when $FADH_2$ is formed during the oxidation of succinate (see figure 6.18, step 6). Electrons are then transferred from complex II to ubiquinone. Note that the electrons carried by $FADH_2$ enter the electron transport chain "downstream" (or "downstairs") of those carried by NADH. Because of this, a pair of electrons carried by NADH results in more protons being pumped across the membrane than does a pair carried by $FADH_2$.
- **Complex III** (also called cytochrome bc_1 complex). This accepts electrons from ubiquinone, which has carried them from either complex I or complex II. Complex III pumps four protons across the membrane before transferring the electrons to cytochrome *c*.
- **Complex IV** (also called cytochrome *c* oxidase complex). This accepts electrons from cytochrome *c* and pumps two protons across the membrane. Complex IV is a terminal oxidoreductase, meaning it transfers the electrons to the terminal electron acceptor, which in this case is O_2.

The Electron Transport Chains of Prokaryotes

Considering the versatility and diversity of prokaryotes, it should not be surprising that the types and arrangements of their electron transport components vary tremendously. In fact, a single species can have several alternative carriers, allowing cells to cope with ever-changing growth conditions.

The ETC of *E. coli* provides an excellent example of the versatility of some prokaryotes. This bacterium uses aerobic respiration when O_2 is available, but in the absence of O_2 it can switch to anaerobic respiration if a suitable electron acceptor such as nitrate (NO_3^-) is present. The *E. coli* ETC serves as a model for both aerobic and anaerobic respiration in bacteria.

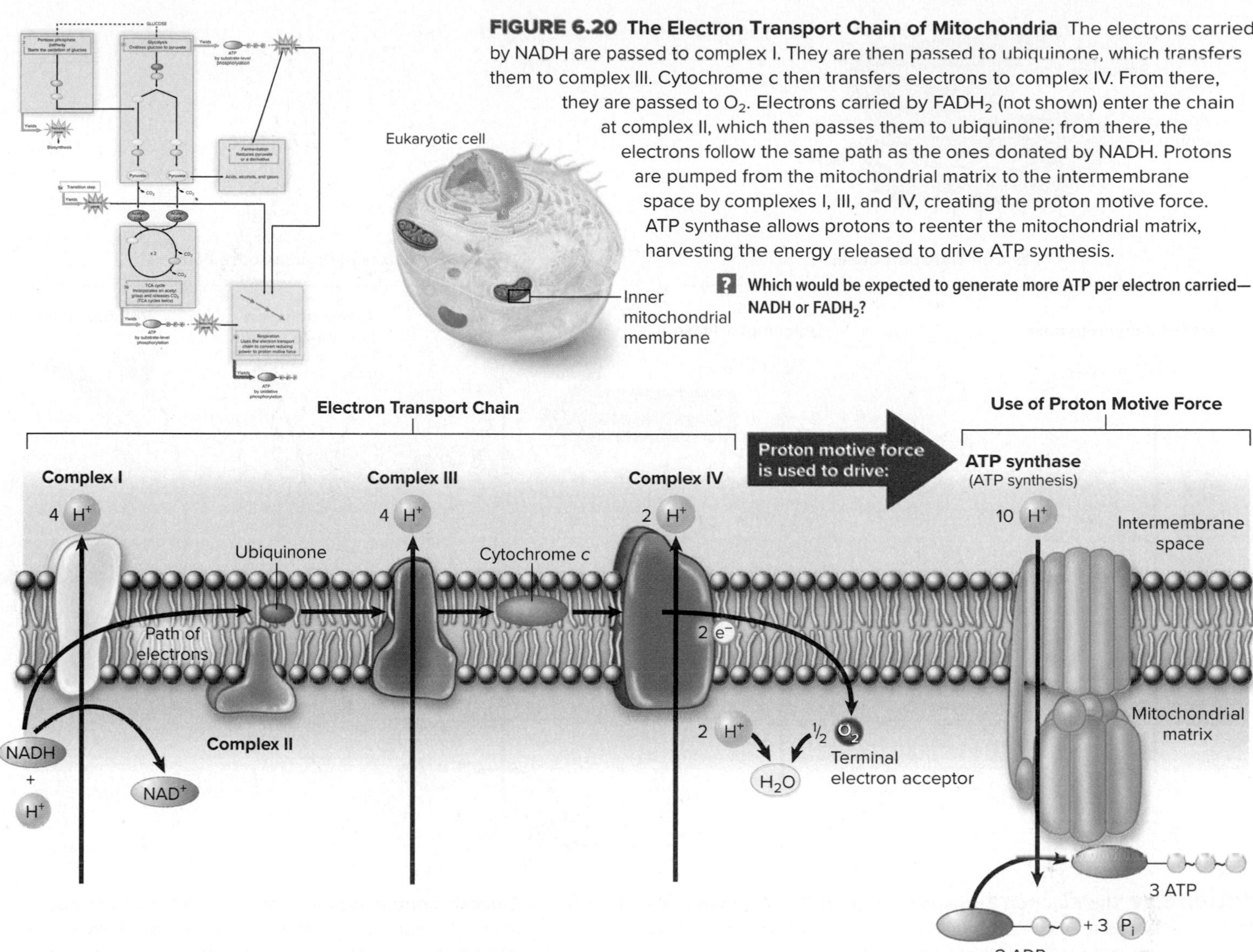

FIGURE 6.20 The Electron Transport Chain of Mitochondria The electrons carried by NADH are passed to complex I. They are then passed to ubiquinone, which transfers them to complex III. Cytochrome c then transfers electrons to complex IV. From there, they are passed to O_2. Electrons carried by $FADH_2$ (not shown) enter the chain at complex II, which then passes them to ubiquinone; from there, the electrons follow the same path as the ones donated by NADH. Protons are pumped from the mitochondrial matrix to the intermembrane space by complexes I, III, and IV, creating the proton motive force. ATP synthase allows protons to reenter the mitochondrial matrix, harvesting the energy released to drive ATP synthesis.

? **Which would be expected to generate more ATP per electron carried—NADH or $FADH_2$?**

Aerobic Respiration When growing aerobically using glucose, *E. coli* has two different NADH dehydrogenases, one of which is functionally equivalent to mitochondrial complex I, a proton pump (**figure 6.21**). The bacterium also has a functional equivalent to mitochondrial complex II (succinate dehydrogenase). *E. coli* can also make alternative complexes that allow it to use additional energy sources like hydrogen gas (H_2). Because the bacterium lacks equivalents to complex III and cytochrome *c,* quinones move the electrons directly to one of two ubiquinol oxidase variations, which each then pass the electrons to O_2. One variation works best at high O_2 concentrations and pumps out four protons, generating more proton motive force; the other has higher affinity for O_2 and is more effective at low O_2 concentrations, but only pumps out two protons.

Anaerobic Respiration Anaerobic respiration harvests less energy than aerobic respiration because of the lower electron affinities of the terminal electron acceptors used (see figure 6.8), and some of the ETC components are different from those of aerobic respiration. When *E. coli* cells anaerobically respire, which they do when O_2 is unavailable but nitrate is, they synthesize a terminal oxidoreductase that uses nitrate as a terminal electron acceptor. This produces nitrite (NO_2^-), which the cells then convert to ammonia to avoid the toxic effects of nitrite. Other bacteria can reduce nitrate even further, forming compounds such as nitrous oxide (N_2O) and nitrogen gas (N_2).

A group of obligate anaerobes called the sulfate-reducers use sulfate (SO_4^{2-}) as a terminal electron acceptor, producing hydrogen sulfide as an end product. The diversity and ecology of sulfate-reducing bacteria is discussed in chapter 11.

ATP Synthase—Using the Proton Motive Force to Synthesize ATP

Just as energy is required to establish a concentration gradient, energy is released as the gradient is removed or reduced. The enzyme complex ATP synthase uses the energy of the proton motive force to synthesize ATP. It does this by allowing protons to flow across the membrane down their concentration gradient back into the prokaryotic cell (or mitochondrial matrix) in

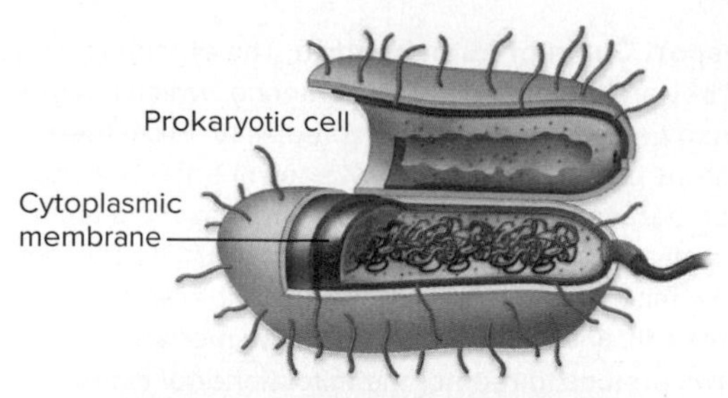

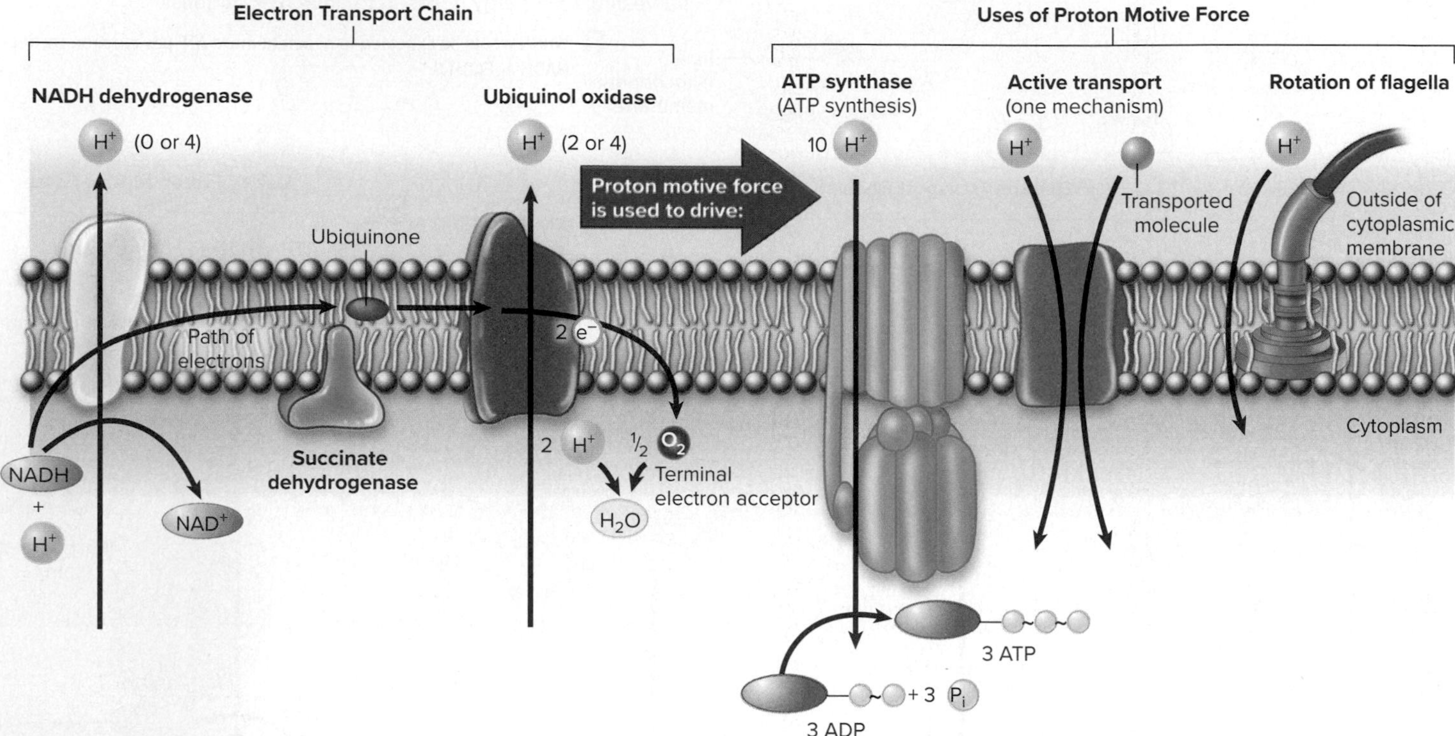

FIGURE 6.21 The Electron Transport Chain of *E. coli* Growing Aerobically in a Glucose-Containing Medium The electrons carried by NADH are passed to one of two different NADH dehydrogenases. They are then passed to ubiquinone, which transfers them to one of two ubiquinol oxidases. From there they are passed to O_2. The electrons carried by $FADH_2$ (not shown) enter the chain at succinate dehydrogenase, which then transfers them to ubiquinone; from there, the electrons follow the same path as the ones donated by NADH. Protons are pumped out by one of the two NADH dehydrogenases and both ubiquinol oxidases, creating the proton motive force. ATP synthase allows protons to reenter the cell, using the energy released to drive ATP synthesis. The proton motive force is also used to drive one form of active transport and to power the rotation of flagella. *E. coli* has other components of the electron transport chain that function under different growth conditions.

? Succinate dehydrogenase is equivalent to which component of the mitochondrial electron transport chain?

a controlled manner, using the energy released to add a phosphate group to ADP. One molecule of ATP forms from the flow of approximately three protons across the membrane.

Theoretical ATP Yield of Oxidative Phosphorylation

By calculating the ATP yield of oxidative phosphorylation, the relative energy gains of respiration and fermentation can be compared. It is not a straightforward comparison, however, because oxidative phosphorylation has so many variables. This is particularly true for prokaryotic cells because they use the proton motive force to drive processes other than ATP synthesis. In addition, as previously described, prokaryotic cells, as a group, use different carriers in their electron transport chain and pump out a variable number of protons per pair of electrons passed.

The basis for calculating the ATP yield of oxidative phosphorylation relies on experimental studies using rat mitochondria. These studies indicate that approximately 2.5 ATP are made per pair of electrons transferred to the electron transport chain by NADH; about 1.5 ATP are made per pair transferred by $FADH_2$. For simplicity we will use whole numbers (3 ATP/NADH and 2 ATP/$FADH_2$) in calculations. Using these numbers, the maximum theoretical energy yield for oxidative phosphorylation in a prokaryotic cell (assuming the electron transport chain is similar to that of mitochondria) is as follows:

From glycolysis:

- 2 NADH → 6 ATP (assuming 3 for each NADH)

From the transition step:

- 2 NADH → 6 ATP (assuming 3 for each NADH)

From the TCA cycle:

- 6 NADH → 18 ATP (assuming 3 for each NADH)
- 2 $FADH_2$ → 4 ATP (assuming 2 for each $FADH_2$)

Total maximum ATP yield from oxidative phosphorylation = 34

The ATP gain as a result of oxidative phosphorylation will be slightly less in eukaryotic cells than in prokaryotic cells because of the fate of the reducing power (NADH) generated during glycolysis. Recall that in eukaryotic cells, glycolysis takes place in the cytoplasm, whereas the electron transport chain is located in the mitochondria. Consequently, the electrons carried by cytoplasmic NADH must be moved across the mitochondrial membrane before they can enter the electron transport chain. This uses approximately 1 ATP per NADH generated during glycolysis.

ATP Yield of Aerobic Respiration in Prokaryotes

Now that the ATP-yielding components of the central metabolic pathways have been considered, we can calculate the theoretical maximum ATP yield of aerobic respiration in a prokaryotic cell (**figure 6.22**):

Substrate-level phosphorylation:

- 2 ATP (from glycolysis; net gain)
- <u>2 ATP</u> (from the TCA cycle)
- 4 total

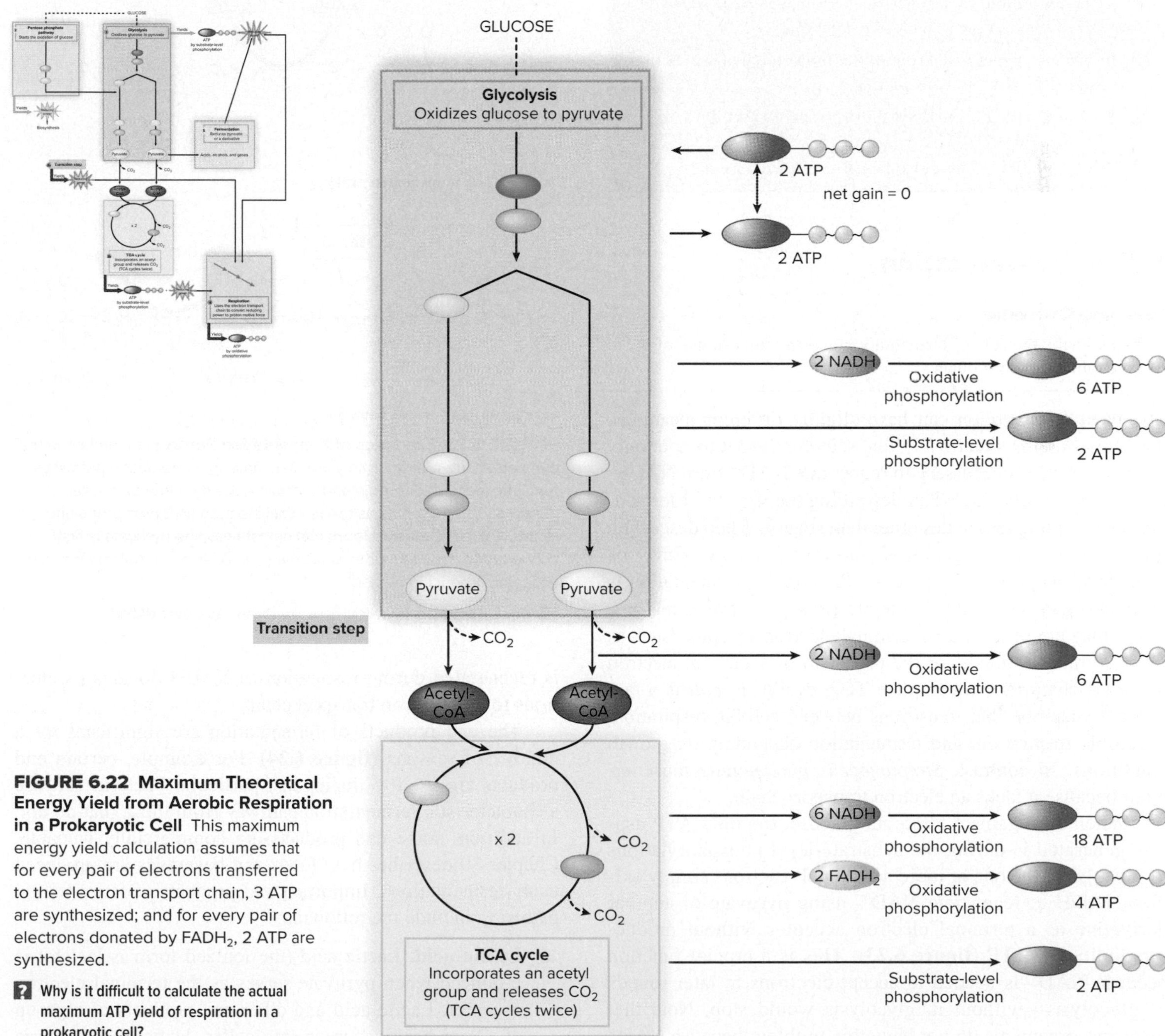

FIGURE 6.22 Maximum Theoretical Energy Yield from Aerobic Respiration in a Prokaryotic Cell This maximum energy yield calculation assumes that for every pair of electrons transferred to the electron transport chain, 3 ATP are synthesized; and for every pair of electrons donated by $FADH_2$, 2 ATP are synthesized.

? **Why is it difficult to calculate the actual maximum ATP yield of respiration in a prokaryotic cell?**

Oxidative phosphorylation:

- 6 ATP (from the reducing power gained in glycolysis)
- 6 ATP (from the reducing power gained in the transition step)
- 22 ATP (from the reducing power gained in the TCA cycle)
- 34 total

Total ATP gain (theoretical maximum) = 38

MicroAssessment 6.4

Cellular respiration uses the reducing power (NADH and $FADH_2$) generated collectively in glycolysis, the transition step, and the TCA cycle to synthesize ATP. The electron transport chain converts reducing power into a proton motive force. ATP synthase then harvests that energy to synthesize ATP by oxidative phosphorylation. In aerobic respiration, O_2 serves as the terminal electron acceptor; anaerobic respiration uses a molecule other than O_2.

10. In bacteria, what is the role of the molecule that serves as a source of vitamin K for humans?
11. Why is the overall ATP yield in aerobic respiration only a theoretical number?
12. Why could an oxidase also be called a reductase?

6.5 ■ Fermentation

Learning Outcome

15. Describe the role of fermentation and the importance of the common end products.

The term *fermentation* can have slightly different meanings, depending on the situation. Often, scientists use it to refer only to those reactions necessary to regenerate NAD^+ from NADH; this is particularly true when describing the variety of fermentations involving molecules other than sugars. When describing fermentations involving glucose, however, glycolysis may be considered as part of the process. Regardless, fermentation is used by organisms unable to respire for any of several reasons. There may not be a suitable terminal electron acceptor (such as O_2 or nitrate) available, or the organism may lack an electron transport chain (and a complete TCA cycle). *E. coli* is a facultative anaerobe that transitions between aerobic respiration, anaerobic respiration, and fermentation depending on growth conditions. In contrast, *Streptococcus pneumoniae* must ferment because it lacks an electron transport chain.

When an organism ferments glucose, the only ATP usually generated is during the substrate-level phosphorylations of glycolysis; the one or more additional reactions simply oxidize NADH to regenerate NAD^+, using pyruvate or another derivative as a terminal electron acceptor without producing additional ATP (**figure 6.23**). This is a crucial function because NAD^+ is needed to accept electrons in later rounds of glycolysis—without it, glycolysis would stop. Note that respiring organisms do not have this problem because NAD^+

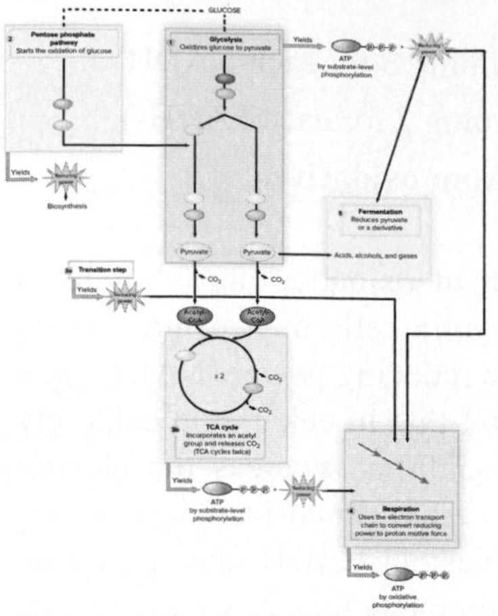

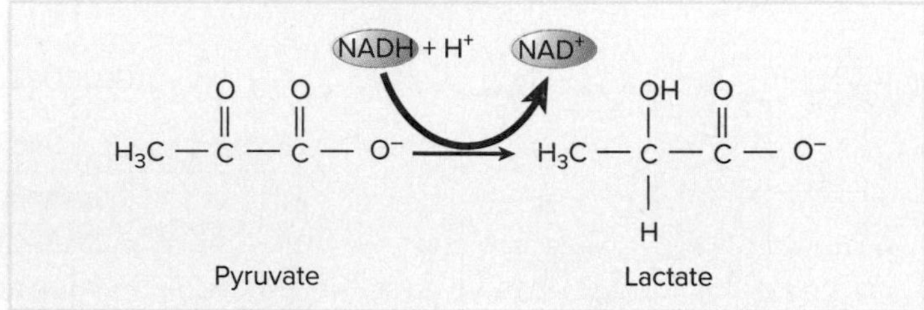

(a) Lactic acid fermentation pathway

(b) Ethanol fermentation pathway

FIGURE 6.23 Examples of Fermentation Pathways The fate of the pyruvate made during glycolysis differs among fermentation pathways. **(a)** In the lactic acid fermentation pathway, the pyruvate generated during glycolysis serves as the terminal electron acceptor, producing lactate. **(b)** In the ethanol fermentation pathway, the pyruvate is first converted to acetaldehyde, which then serves as the terminal electron acceptor, producing ethanol.

? Why is it important for cells to have a mechanism to oxidize NADH?

is regenerated during respiration as NADH donates its electrons to the electron transport chain.

The end products of fermentation are significant for a number of reasons (**figure 6.24**). For example, certain end products are used to distinguish particular bacteria that use a characteristic fermentation pathway from those that do not. In addition, some end products are commercially valuable. Chapter 30 describes how foods and beverages are produced using fermentations. Important end products of fermentation pathways include the following:

- **Lactic acid.** Lactic acid (the ionized form is lactate) is produced when pyruvate serves as the terminal electron acceptor. Lactic acid and other end products of a group of Gram-positive organisms called lactic acid bacteria contribute to the flavor and texture of cheese, yogurt, pickles, cured sausages, and other foods. Yet lactic acid production may also spoil food or cause tooth decay when made by bacteria living on the teeth.

- **Ethanol.** Ethanol is produced in a pathway that first removes CO_2 from pyruvate, generating acetaldehyde, which then serves as the terminal electron acceptor. The end products of these reactions—which are used in making wine, beer, spirits, and bread—are ethanol and CO_2. Ethanol is also an important biofuel. *Saccharomyces* (yeast) and *Zymomonas* (bacteria) use this pathway.
- **Butyric acid.** Butyric acid (the ionized form is butyrate) and a variety of other end products are produced in a complex multistep pathway used by *Clostridium* species, which are obligate anaerobes. Under certain conditions, some organisms use a variation of this pathway to produce the organic solvents butanol and acetone.
- **Propionic acid.** Propionic acid (the ionized form is propionate) is made in a multistep pathway that first adds CO_2 to pyruvate, producing a compound that then serves as a terminal electron acceptor. After this is reduced by NADH, it is further modified to form propionate. *Propionibacterium* species use this pathway, and their growth is encouraged for Swiss cheese production. The CO_2 they make forms the holes, and propionic acid gives the cheese its unique flavor.
- **Mixed acids.** These are produced in a multistep branching pathway, generating various different fermentation products including lactic acid, succinic acid (the ionized form is succinate), ethanol, acetic acid (the ionized form is acetate), and gases. This pathway differentiates certain members of the family Enterobacteriaceae. The methyl-red test detects the resulting low pH, distinguishing members that use this pathway, such as *E. coli,* from those that do not, such as *Klebsiella* and *Enterobacter* (see table 10.3).
- **2,3-Butanediol.** 2,3-Butanediol is produced in a multistep pathway that uses two molecules of pyruvate to generate acetoin and two molecules of CO_2. Acetoin is then used as the terminal electron acceptor. This is another pathway that differentiates members of the family Enterobacteriaceae—the Voges-Proskauer test detects acetoin, distinguishing members that use this pathway (such as *Klebsiella* and *Enterobacter*) from those that do not (such as *E. coli;* see table 10.3).

MicroAssessment 6.5

Fermentation of glucose stops short of the TCA cycle, using pyruvate or a derivative of it as a terminal electron acceptor. Many end products of fermentation are commercially valuable.

13. Why would a cell ferment rather than respire?
14. How do the methyl-red and Voges-Proskauer tests differentiate between certain members of the Enterobacteriaceae?
15. Fermentation is used as a means of preserving foods. Why would it slow spoilage?

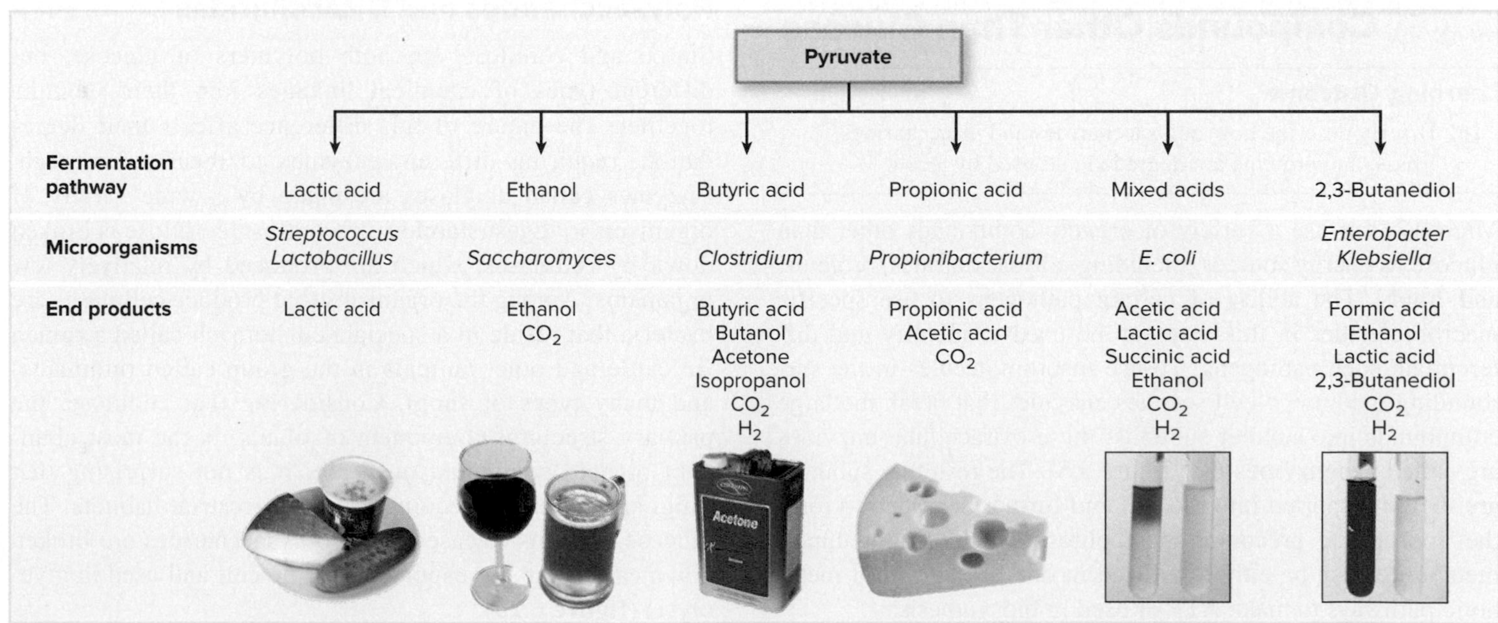

FIGURE 6.24 End Products of Fermentation Pathways

Lactic acid, ethanol, and butyric acid photos: Brian Moeskau; Propionic acid: Photographer's Choice/Getty Images; Mixed acids and 2,3-Butanediol photos: Auburn University Photographic Services/McGraw Hill

How can end products of fermentation help identify a bacterium?

FOCUS ON A CASE 6.1

A 9-year-old girl experiencing diarrhea, abdominal pain, fever, and nausea was taken to the emergency department at a local hospital. When her parents were questioned, they listed the foods she had eaten in recent days. The doctor also asked them about animals she had touched, and the parents mentioned that she had recently been given a pet turtle. A stool sample from the child was submitted to the clinical laboratory.

In the laboratory, the stool sample was inoculated onto various types of bacteriological media, including MacConkey agar, MacConkey-sorbitol agar, and Hektoen enteric agar. These are selective and differential media used to isolate common bacterial intestinal pathogens, allowing laboratory technicians to detect *Salmonella enterica, E. coli* O157:H7, and *Shigella* species. In addition, the specimen was inoculated into tetrathionate broth, a selective enrichment for *S. enterica.* After 24 hours of incubation, the agar plates were examined. Small colonies were present, but none had the colony morphology of any common intestinal pathogens. Meanwhile, the enrichment broth was turbid, indicating growth, so it was inoculated onto an agar plate designed to isolate *S. enterica.* After 24 hours of incubation, colonies of suspected *S. enterica* were seen. Further tests confirmed the identification of *S. enterica.*

1. Why was it important that the child had recently been given a pet turtle?
2. Considering that the child had symptoms typical of *Salmonella* gastroenteritis, why was it necessary for the clinical lab to examine the stool specimen?
3. Tetrathionate is an oxidized form of sulfur. Why might this ingredient enrich for *S. enterica* in a broth culture?
4. Why would *S. enterica* benefit from using tetrathionate?

Discussion

1. Pet turtles are a common source of *Salmonella enterica,* although the bacterium is most often acquired from contaminated foods. It is often found on poultry, so undercooked chicken, turkey, and eggs are common sources.
2. Many intestinal pathogens cause diarrhea, abdominal pain, fever, and nausea. In order to determine the precise cause of the illness, laboratory culture is generally done. Identifying the pathogen aids the physician in prescribing the proper treatment and also helps health officials notice disease trends—a cluster of similar cases could indicate an outbreak.
3. Tetrathionate broth has been used for many years to isolate *S. enterica,* but the mechanism of how it enriches for the bacterium has only recently been determined. Unlike *E. coli* and most of its other competitors, *Salmonella* species can use tetrathionate as a terminal electron acceptor for anaerobic respiration, which gives them an advantage in a tetrathionate broth culture. In a typical broth culture, the dissolved O_2 is quickly used up by aerobically respiring bacteria. At that point, most bacteria must switch to fermentation in order to continue growing (standard bacteriological media do not contain nitrate or any other terminal electron acceptor commonly used for anaerobic respiration). When tetrathionate is present, however, *Salmonella* species can anaerobically respire, which produces more ATP per glucose molecule than does fermentation. The anaerobically respiring *Salmonella* can therefore grow faster than the fermenting competitors.
4. Researchers have recently discovered that tetrathionate is produced in the intestinal tract as a result of inflammation, one of the body's defenses against infection. When a *Salmonella* species infects the intestinal tract, inflammation results, which then generates a terminal electron acceptor that *Salmonella* can use for anaerobic respiration. Just as in the tetrathionate broth used for enrichment, this gives the bacterium a selective advantage.

6.6 ■ Catabolism of Organic Compounds Other Than Glucose

Learning Outcome

16. Briefly describe how polysaccharides and disaccharides, lipids, and proteins are degraded and used by a cell.

Microbes can use a variety of organic compounds other than glucose as energy sources, including polysaccharides, proteins, and lipids. The ability of certain pathogens to use specific macromolecules in this way can be used to identify and differentiate such pathogens. To use macromolecules in the surrounding medium, a cell secretes enzymes that break the large compounds into smaller subunits; these extracellular enzymes are called **exoenzymes** (see figure 3.8). The resulting subunits are then transported into the cell and further degraded to form the appropriate precursor metabolites. Recall that precursor metabolites can be either oxidized in one of the central metabolic pathways to make ATP or used in biosynthesis.

Polysaccharides and Disaccharides

Starch and cellulose are both polymers of glucose, but different types of chemical linkages join their subunits together. The nature of this difference affects their degradation, requiring different enzymes to break down each. Enzymes called amylases are made by a wide variety of organisms to digest starches. In contrast, cellulose is broken down by cellulases, which are produced by relatively few organisms. Among the organisms that produce cellulases are bacteria that reside in a specialized stomach called a rumen (in cattle and other animals in the group called ruminants) and many types of fungi. Considering that cellulose, the primary structural component of plants, is the most abundant organic compound on Earth, it is not surprising that fungi are important decomposers in terrestrial habitats. The glucose subunits released when polysaccharides are broken down can then be transported into the cell and used in glycolysis (**figure 6.25**).

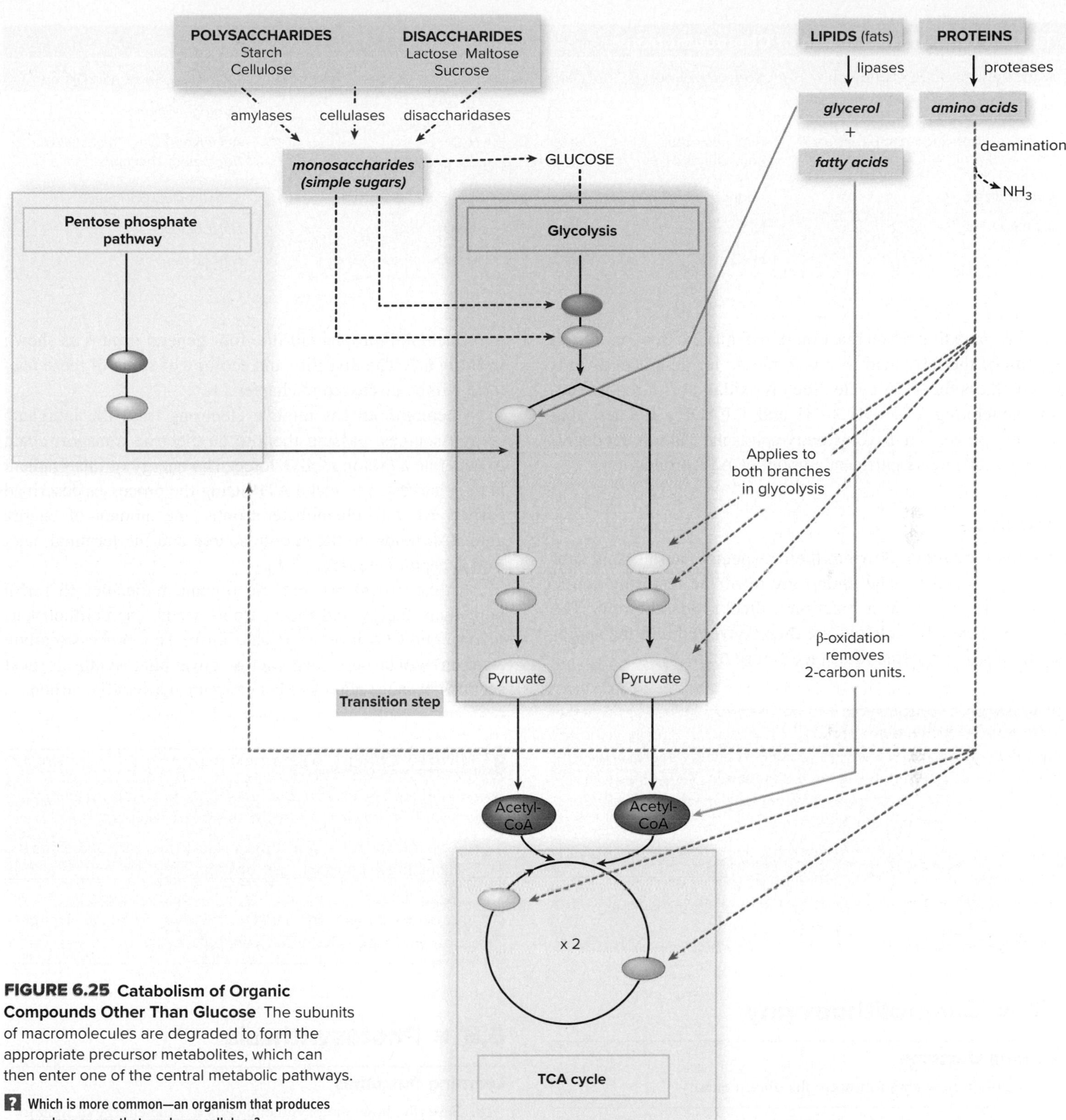

FIGURE 6.25 Catabolism of Organic Compounds Other Than Glucose The subunits of macromolecules are degraded to form the appropriate precursor metabolites, which can then enter one of the central metabolic pathways.

? **Which is more common—an organism that produces amylase or one that produces cellulase?**

Disaccharides including lactose, maltose, and sucrose can be transported into the cell, where specific disaccharidases then split them into their monosaccharide components. For example, the enzyme β-galactosidase breaks down lactose in the cell, forming glucose and galactose. Glucose can enter glycolysis directly, but other monosaccharides must first be converted to one of the precursor metabolites.

Lipids

Fats, the most common simple lipids, are broken down by lipases into their fatty acid and glycerol components. Glycerol is then converted to the precursor metabolite dihydroxyacetone phosphate, which enters glycolysis (see figure 6.25). The fatty acids are degraded using a series of reactions collectively

TABLE 6.7 Metabolism of Chemolithotrophs

Common Name of Organism	Source of Energy	Oxidation Reaction(s) (Energy Yielding)	Common Genera in Group
Hydrogen oxidizers	H_2	$H_2 + \frac{1}{2}O_2 \rightarrow H_2O$	*Hydrogenomonas*
Sulfur oxidizers (non-photosynthetic)	Reduced sulfur compounds (H_2S)	$H_2S + \frac{1}{2}O_2 \rightarrow H_2O + S$ $S + 1\frac{1}{2}O_2 + H_2O \rightarrow H_2SO_4$	*Acidithiobacillus, Thiobacillus, Beggiatoa, Thiothrix*
Iron oxidizers	Reduced iron (Fe^{2+})	$2\,Fe^{2+} + \frac{1}{2}O_2 + H_2O \rightarrow 2\,Fe^{3+} + 2\,OH^-$	*Sphaerotilus, Gallionella*
Nitrifiers	NH_3	$NH_3 + 1\frac{1}{2}O_2 \rightarrow HNO_2 + H_2O$	*Nitrosomonas*
	HNO_2	$HNO_2 + \frac{1}{2}O_2 \rightarrow HNO_3$	*Nitrobacter*

called β-oxidation. Each reaction transfers a 2-carbon unit from the end of the fatty acid to coenzyme A, forming acetyl-CoA, which enters the TCA cycle. Each β-oxidation is a redox reaction, generating 1 $NADH + H^+$ and 1 $FADH_2$. Because that reducing power can drive oxidative phosphorylation, fat degradation results in a significant amount of ATP production.

Proteins

Proteins are broken down to their respective amino acid subunits by proteases. The amino group of the resulting amino acids is removed by a reaction called a deamination. The remaining carbon skeletons are then converted into the appropriate precursor metabolites (see figure 6.25).

MicroAssessment 6.6

In order for macromolecules to be used as energy sources, they must be first broken down to smaller subunits. The subunits are then converted to the appropriate precursor metabolites, which can enter a central metabolic pathway.

16. Why do cells secrete enzymes that degrade macromolecules?

17. Explain the process used to degrade fatty acids.

18. How would cellulose-degrading bacteria in the rumen of a cow benefit the animal?

6.7 ▪ Chemolithotrophy

Learning Outcome

17. Explain how chemolithotrophs obtain energy.

Chemolithotrophic prokaryotes use reduced inorganic chemicals such as hydrogen sulfide (H_2S) and ammonia (NH_3) as energy sources (electron donors). They generally thrive in very specific environments where reduced inorganic compounds are found. Note that these compounds are the very ones produced as a result of anaerobic respiration, when inorganic molecules such as sulfate and nitrate serve as terminal electron acceptors. This is one important example of how nutrients are cycled; the waste products of some organisms serve as energy sources for others.

Chemolithotrophs fall into four general groups as shown in **table 6.7.** The diversity and ecology of some of these bacteria will be discussed in chapter 11.

Chemolithotrophs remove electrons from the inorganic energy sources, passing them to an electron transport chain to generate a proton motive force. The energy of this gradient is then harvested to make ATP, using the processes described earlier. As with chemoheterotrophs, the amount of energy gained depends on the energy source and the terminal electron acceptor (see figure 6.8).

Unlike organisms that use organic molecules to fulfill both their energy and their carbon needs, chemolithotrophs incorporate CO_2 into an organic form. This process (carbon fixation) will be described after we cover photosynthesis (next section) because photosynthetic organisms also fix carbon.

MicroAssessment 6.7

Chemolithotrophs use reduced inorganic compounds as energy sources. They use carbon dioxide as a carbon source.

19. Describe the roles of hydrogen sulfide and carbon dioxide in chemolithoautotrophic metabolism.

20. Which energy source, Fe^{2+} or H_2S, would result in the greatest energy yield when O_2 is used as a terminal electron acceptor? (*Hint:* Refer to figure 6.8)

6.8 ▪ Photosynthesis

Learning Outcomes

18. Describe how photosynthetic cells use the energy of light to power ATP synthesis.

19. Compare and contrast anoxygenic and oxygenic photosynthesis, including characteristics of their photosystems.

20. Describe the role of rhodopsin pigments in phototrophy.

Phototrophs, including plants, algae, and several groups of bacteria, harvest the energy of light and convert it into chemical energy through the process of **photosynthesis.** Photosynthesis includes two distinct stages: the light-dependent reactions and the light-independent reactions. During the **light-dependent reactions**, photosynthetic pigments capture light energy and use it to make ATP and reducing power (NADPH or NADH, depending on the organism). These products may be then used in the light-independent reactions to make organic compounds from CO_2; this latter process is used by both photosynthetic organisms and chemolithotrophs and is discussed in more detail in the next section.

Capturing the Energy of Light

Photosynthesis relies on various pigments that each absorb light at specific wavelengths and reflect light at another; they appear the color of light they reflect (see figure 5.7). For instance, plants are green because the cells' chloroplasts contain chlorophyll, a pigment that absorbs blue and red light, but not green. Colors reflected by different photosynthetic pigments include yellow, orange, red, and blue.

Photosynthetic pigments are organized into **photosystems** embedded within photosynthetic membranes—cytoplasmic membranes in photosynthetic bacteria or thylakoid membranes in the chloroplasts of eukaryotic cells. Each photosystem has two general pigment-containing components (**figure 6.26**):

- **Light-harvesting complex.** This functions as a light-collecting funnel whereby the pigments absorb the energy of light and pass the energy (but not electrons) among the pigment molecules within the photosystem. The primary pigment type within the complex is either bacteriochlorophyll or chlorophyll, depending on the organism. Other pigments, referred to as accessory pigments, capture additional wavelengths not absorbed by the primary pigments.

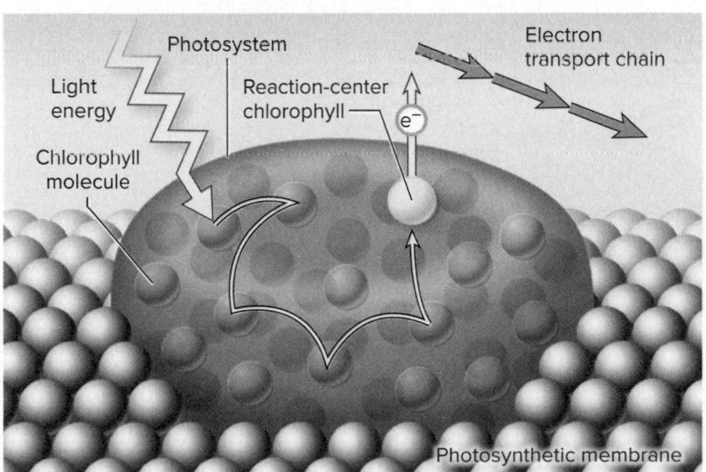

FIGURE 6.26 Photosystem Chlorophyll and other light-harvesting pigments capture the energy of light, which is then transferred among the pigment molecules (the transfers are indicated by the yellow lines) before it is transferred to a pigment in the reaction center. This pigment then emits an electron that is then passed to an electron transport chain.

? **What is the role of an electron transport chain in photosynthesis?**

- **Reaction center.** This receives energy collected by the light-harvesting complex. Significantly, it has pigment molecules arranged such that they emit high-energy electrons when they accumulate enough energy. Those electrons can then be transferred to an electron transport chain (ETC) embedded in the photosynthetic membrane.

Once high-energy electrons are passed from a reaction center in a photosystem to an ETC, they move through the ETC to generate a proton motive force, which ATP synthase then uses to make ATP. This mechanism of ATP synthesis, called **photophosphorylation** to reflect its dependence on light energy, is analogous to the oxidative phosphorylation process described for cellular respiration.

Electrons emitted by the reaction center must be replenished. Electrons that have been passed along the ETC in its entirety may return to the same reaction center that emitted them; this process is called cyclic photophosphorylation because the electrons follow a cyclical path leading to ATP production. Sometimes, however, photosynthetic organisms must use the high-energy electrons for other purposes, such as to generate reducing power to support biosynthetic pathways. In this case, the electrons follow a non-cyclical path, so they must be replenished from another source; the source of replenishing those electrons is what distinguishes anoxygenic photosynthesis from oxygenic photosynthesis.

Anoxygenic Photosynthesis

The modern photosynthesizing bacteria most closely related to the earliest examples use **anoxygenic photosynthesis,** meaning that their photosynthetic mechanisms do not generate O_2. The primary pigments in their light-harvesting complexes, as well as in their associated reaction centers, are bacteriochlorophylls. These pigments are structurally similar to chlorophylls, but absorb different wavelengths of light, thus allowing the bacteria to grow where oxygenic photosynthetic organisms cannot.

The two general types of anoxygenic photosynthetic bacteria each have a distinct version of their single type of photosystem:

- **Green bacteria.** These bacteria each have a photosystem that raises the energy level of the electrons from the reaction center to a high enough level that they can either be (1) passed along an ETC and then be used to generate proton motive force to make ATP (cyclical electron path) or (2) used to make reducing power (non-cyclical electron path).
- **Purple bacteria.** The single photosystem of these bacteria cannot raise the energy emitted from the reaction center to as high of a level as that of the green bacteria. Thus, although the electrons emitted from the reaction

center can be used to power photophosphorylation (cyclical electron path), their energy level is not high enough level to make reducing power; instead, the bacteria use a process called reversed electron transport—running the ETC in reverse or "uphill" to reduce NAD^+.

Oxygenic Photosynthesis

The cells of cyanobacteria, algae, and plants use **oxygenic photosynthesis** which, as the name implies, produces O_2. These cells each have two different types of photosystems—photosystem I and photosystem II—named in the order of their discovery. Photosystem I is similar to the photosystem of green bacteria, whereas photosystem II is similar to the photosystem of purple bacteria; like the photosystem of green bacteria, photosystem I raises the emitted electrons to a higher energy level than photosystem II does. Instead of bacteriochlorophyll, the photosystems of oxygenic photosynthesizers use chlorophyll *a* (the various chlorophylls are indicated by different letters).

Because of the combined actions of the dual photosystems (I and II), cyanobacterial, algal, and plant cells can make reducing power while also making ATP. In doing so, the electrons follow what is described as a Z pathway, starting with photosystem II. The reaction center of that photosystem loses high-energy electrons that are passed to an ETC to generate a proton motive force, which is harvested to make ATP. The electrons that have been passed along the ETC, however, are not returned to photosystem II; instead, they are donated to photosystem I to replace its high-energy electrons, which are used to make reducing power. To replace the ones lost by the photosystem II reaction center, cells use H_2O as an electron donor. To do this, an enzyme splits H_2O, releasing O_2 as metabolic waste (**figure 6.27**). This combined use of the two photosystems is referred to as non-cyclic photophosphorylation because the electrons follow a non-cyclic route to simultaneously drive the synthesis of both reducing power and ATP.

When cells need to make ATP but not reducing power, only photosystem I is used; the electrons emitted from this photosystem's reaction center are ultimately returned to it after being passed along the ETC. Because the electrons follow a cyclical path, this process is referred to as cyclic photophosphorylation.

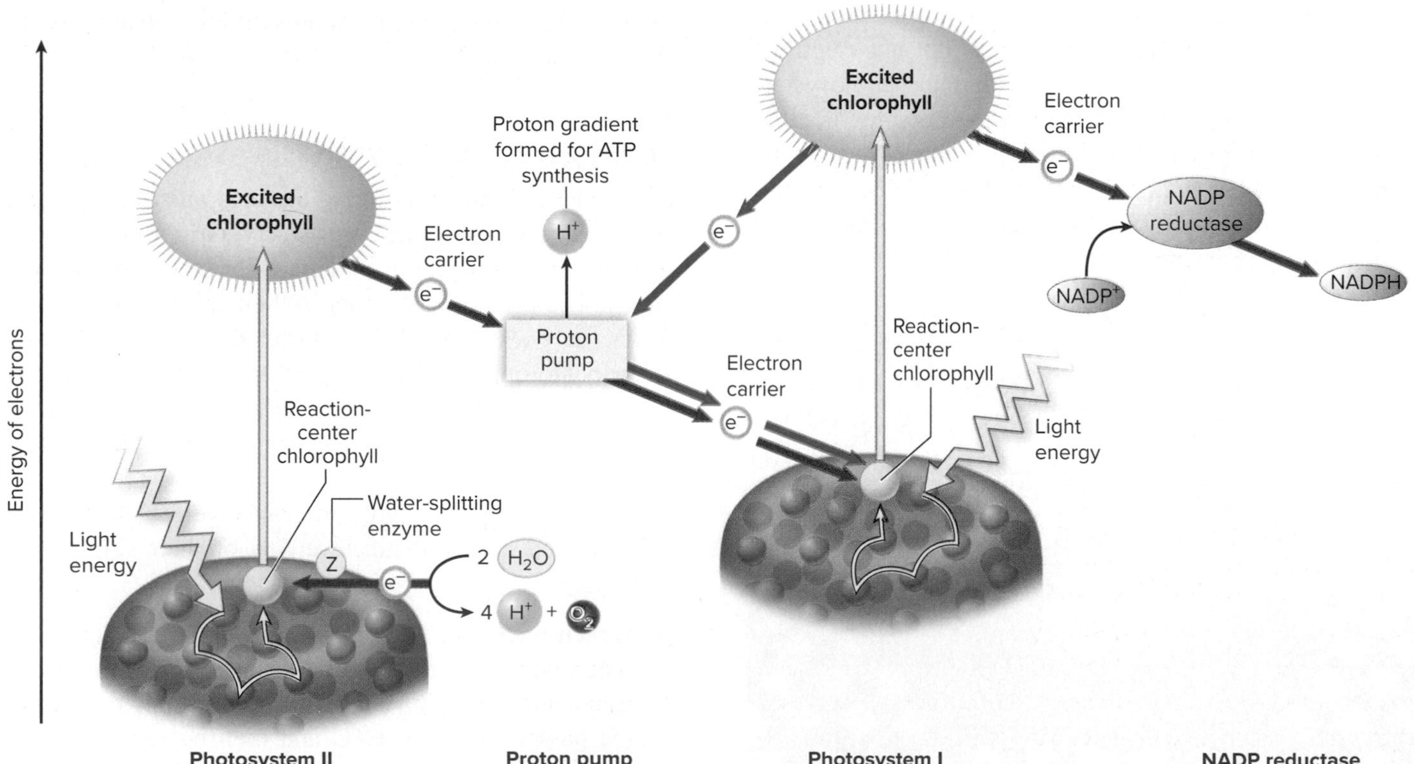

FIGURE 6.27 The Two Photosystems of Cyanobacteria and Chloroplasts Light energy captured by the pigments excites a reaction-center pigment, causing it to emit a high-energy electron, which is then passed to an electron transport chain. In cyclic photophosphorylation, electrons emitted by photosystem I are returned to that photosystem; the path of the electrons is shown in green arrows. In non-cyclic photophosphorylation, the electrons used to replenish photosystem I are donated by photosystem II's reaction center via the ETC; the path of these electrons is shown in orange arrows. Photosystem II replenishes its own electrons by stripping them from water, producing O_2.

When do the cells need to use non-cyclic photophosphorylation?

TABLE 6.8 Comparison of the Photosynthetic Mechanisms Used by Different Organisms

	Anoxygenic Photosynthesis		Oxygenic Photosynthesis	
	Purple Bacteria	**Green Bacteria**	**Plants, Algae**	**Cyanobacteria**
Location of the Photosystem	Within the cytoplasmic membrane	Primarily within the cytoplasmic membrane	In membranes within chloroplasts	In membranes within the cell
Type of Photosystem	Similar to photosystem II	Similar to photosystem I	Photosystem I and photosystem II	
Primary Pigments	Bacteriochlorophylls	Bacteriochlorophylls	Chlorophyll *a*	Chlorophyll *a*
Mechanism for Generating Reducing Power	Reversed electron transport	Non-cyclic use of the photosystem	Non-cyclic use of both photosystems: electrons from photosystem I are used to make reducing power and are replaced by ones from photosystem II	
Source of Electrons While Producing Reducing Power	Varies among the organisms in the group; may include H_2S, H_2, or organic compounds		H_2O	H_2O
Method of CO_2 Fixation	Calvin cycle	Reversed TCA cycle	Calvin cycle	Calvin cycle

Yield of Photosynthesis

Photosynthesis—including both the light-dependent and light-independent reactions—can be summarized as shown:

$$6\ CO_2 + 12\ H_2X \xrightarrow{\text{light energy}} C_6H_{12}O_6 + 12\ X + 6\ H_2O$$

For oxygenic photosynthesis, the "X" in the equation is an atom of oxygen, so "12 H_2X" is 12 molecules of H_2O (water) and "12 X" is 6 molecules of O_2. In contrast, anoxygenic photosynthesis uses an electron donor such as H_2S, H_2, or even reduced organic compounds instead of water, so the X is a molecule other than O_2.

Characteristics of the various photosynthetic mechanisms are summarized in **table 6.8,** and the types of photosynthetic bacteria are described in more detail in chapter 11.

Phototrophy Involving Rhodopsin Pigments

An additional prokaryotic mechanism for capturing light energy and converting it to chemical energy involves the use of prokaryotic forms of rhodopsin, a type of light-sensitive protein long known for its role in light detection in the eyes of humans and other animals. Bacteriorhodopsin was first discovered in extreme halophilic archaea (described in section 11.9), but sequences for comparable pigments, called proteorhodopsins, have been found in a wide range of taxonomically diverse bacteria in marine ecosystems globally.

Bacteriorhodopsin and proteorhodopsin are transmembrane proteins that each are bound to a pigment molecule called retinal. When light activates the retinal, the protein is able to pump protons across the membrane, resulting in the generation of a proton motive force. This proton motive force can then be used to make ATP by oxidative phosphorylation as previously described.

MicroAssessment 6.8

Photosynthesis harvests the energy of light and uses it to make organic compounds from CO_2. Photosystems capture the energy to drive cyclic photophosphorylation when only ATP is needed, but non-cyclic electron flow occurs when both ATP and reducing power are needed. Anoxygenic photosynthesis uses a reduced compound other than water as an electron source during non-cyclic electron flow, and O_2 is not generated. Oxygenic photosynthesis uses water as an electron source during non-cyclic photophosphorylation, and O_2 is generated. Prokaryotes that have bacteriorhodopsin or proteorhodopsin can also use light to drive ATP synthesis.

21. How is photophosporylation similar to oxidative phosphorylation?

22. What is the advantage to a cell of having two photosystems that work together?

23. Energy is required to reverse the flow of the electron transport chain in purple bacteria. Why is this so?

6.9 ■ Carbon Fixation

Learning Outcome

20. Describe the three stages of the Calvin cycle.

Chemolithoautotrophs and photoautotrophs incorporate carbon dioxide (CO_2) into organic compounds, the process of **carbon fixation.** In photosynthetic organisms, the steps used to do this are called the **light-independent reactions.** The process consumes a great deal of ATP and reducing power, which should not be surprising considering that the reverse process (oxidizing those same compounds to CO_2) releases a great deal of energy. The Calvin cycle is the most common pathway used to fix carbon, but some prokaryotes use other mechanisms. For example, the green bacteria and some archaea use a pathway that reverses the steps of the TCA cycle. Note that although heterotrophs use and therefore benefit from the organic compounds that autotrophs produce, the autotrophs make those compounds for their own use (including both biosynthesis and catabolism), not for the purpose of feeding other organisms.

Calvin Cycle

The **Calvin cycle,** or Calvin-Benson cycle, named in honor of the scientists who described much of it, is a complex cycle. The easiest way to understand the outcome is to consider six "turns" of the cycle; together, these turns use the following to produce one molecule of the 6-carbon sugar fructose-6-phosphate:

- 18 ATP
- 12 NADPH
- 6 CO_2

There are three stages of the Calvin cycle (**figure 6.28**):

① **Incorporating CO_2 into an organic compound.** Carbon dioxide enters the cycle when an enzyme commonly called rubisco (ribulose bisphosphate carboxylase) joins it to ribulose-1,5-bisphosphate (RuBP). The resulting compound spontaneously hydrolyzes to produce 2 molecules of 3-phosphoglycerate (3PG). Considering six turns of the cycle, the incorporation of 6 CO_2 molecules through their joining to 6 molecules of the 5-carbon RuBP results in the formation of 12 molecules of 3PG.

② **Reducing the resulting molecule.** An input of energy (12 ATP) and then reducing power (12 NADPH) converts the 12 molecules of 3PG to 12 molecules of glyceraldehyde-3-phosphate (G3P). This molecule is also a precursor metabolite formed as an intermediate in glycolysis and can have any of a variety of different fates: It can be used in biosynthesis, oxidized to make other precursor compounds, or combined with another G3P molecule to be converted to a 6-carbon sugar. An important aspect of the Calvin cycle, however, is that most of the G3P must be used to regenerate RuBP for the cycle to continue. Consequently, a maximum of one molecule of a 6-carbon sugar can be produced in six turns of the cycle.

③ **Regenerating the starting compound.** A series of complex reactions requiring the energy of 6 ATP converts the remaining 10 molecules of G3P to 6 RuBP molecules.

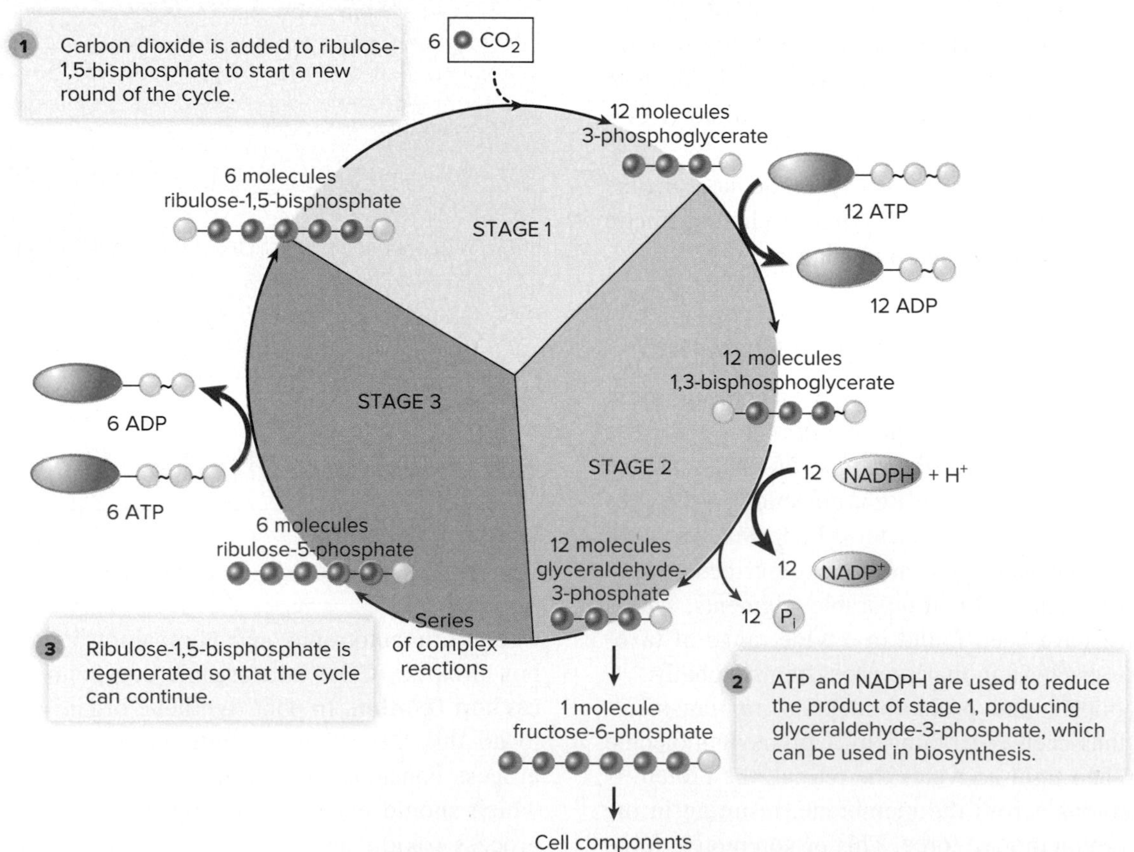

FIGURE 6.28 The Calvin Cycle The Calvin cycle has three essential stages: **(1)** Incorporating CO_2 into an organic compound, **(2)** reducing the resulting molecule, and **(3)** regenerating the starting compound.

? How much ATP and NADPH must be spent to synthesize one molecule of fructose?

MicroByte

Although rubisco is unique to autotrophs, it is probably the most abundant enzyme on Earth!

MicroAssessment 6.9

The process of carbon fixation consumes a great deal of ATP and reducing power. The Calvin cycle is the most common pathway used to incorporate inorganic carbon into an organic form.

24. What is the role of rubisco?

25. What would happen if ribulose-1,5-bisphosphate (RuBP) were depleted in a cell?

6.10 ■ Anabolic Pathways—Synthesizing Subunits from Precursor Molecules

Learning Outcome

21. Describe the synthesis of lipids, amino acids, and nucleotides.

Prokaryotes, as a group, are highly diverse with respect to their energy sources but similar in their biosynthetic processes. Starting with products of the central metabolic pathways (precursor metabolites, ATP, and reducing power in the form of NADPH), they can synthesize the necessary subunits of lipids (glycerol and fatty acids), proteins (amino acids), nucleic acids (nucleotides), and carbohydrates (sugars) (**figure 6.29**).

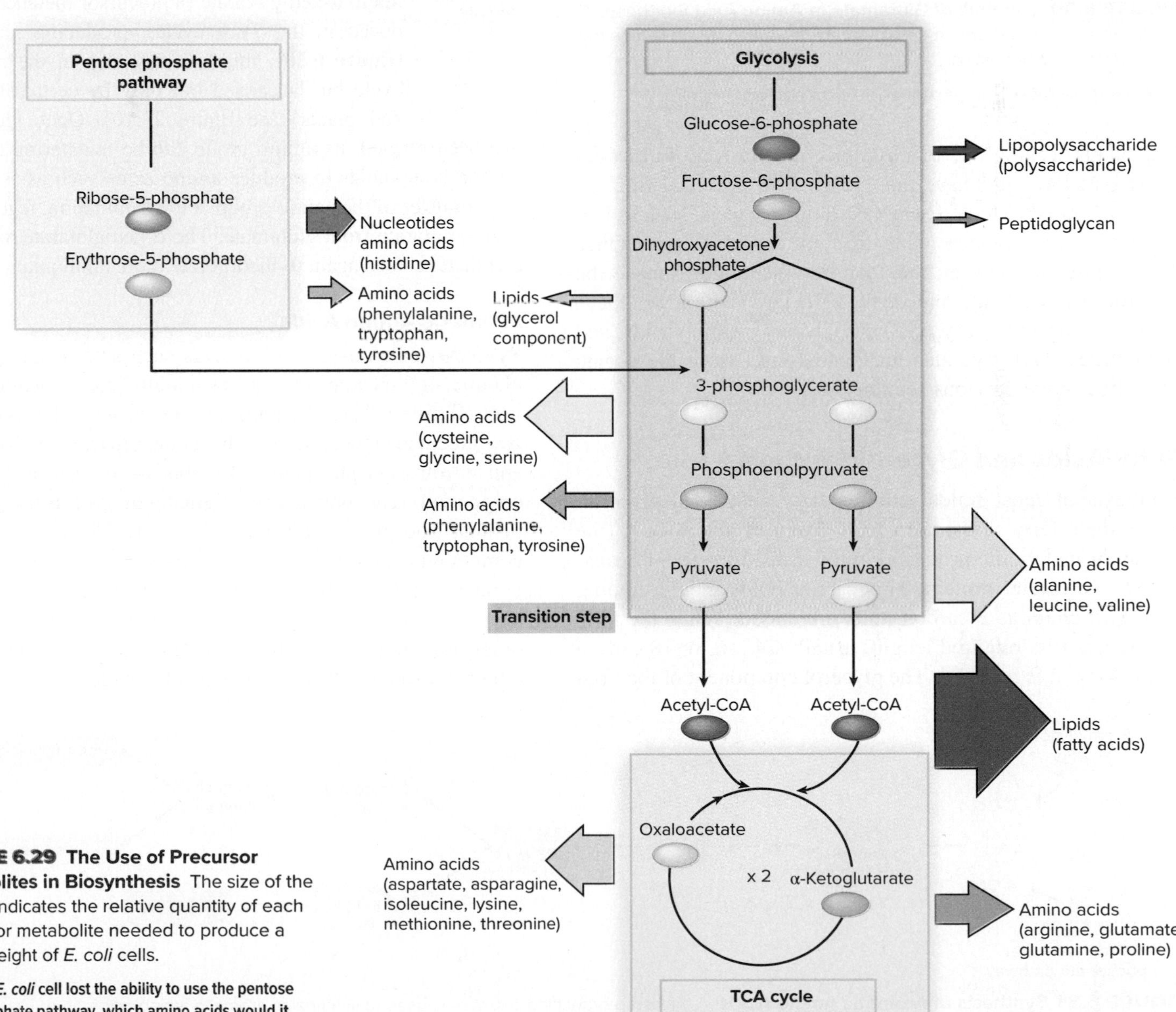

FIGURE 6.29 The Use of Precursor Metabolites in Biosynthesis The size of the arrows indicates the relative quantity of each precursor metabolite needed to produce a given weight of *E. coli* cells.

If an *E. coli* cell lost the ability to use the pentose phosphate pathway, which amino acids would it require as growth factors?

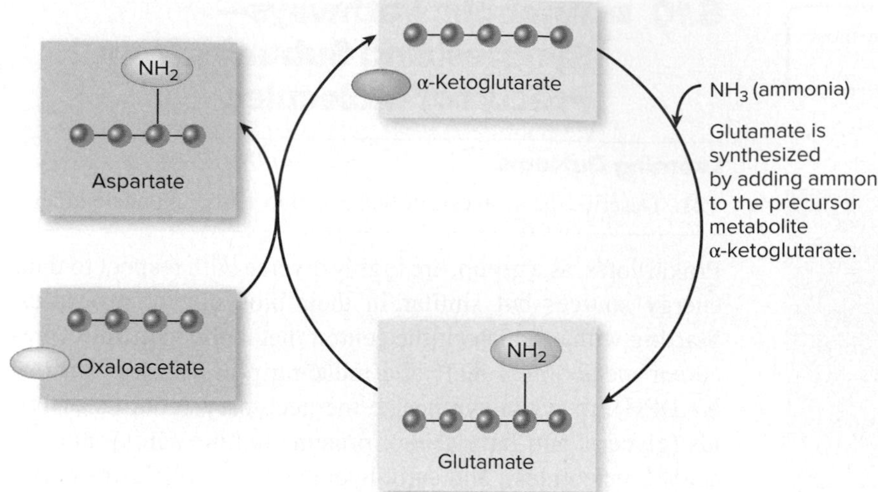

FIGURE 6.30 The Role of Glutamate in Amino Acid Synthesis Once glutamate has been synthesized from α-ketoglutarate, its amino group can be transferred to produce other amino acids.

? Alpha-ketoglutarate is produced in which central metabolic pathway?

is synthesized from dihydroxyacetone phosphate, a precursor metabolite generated in glycolysis.

Amino Acid Synthesis

Proteins are composed of various combinations of usually 20 different amino acids. These amino acids can be grouped into structurally related families that share common pathways of biosynthesis.

Glutamate

Glutamate synthesis is especially important because it is the primary mechanism for converting inorganic nitrogen, such as ammonium (NH_4^+), into an organic form—a process called nitrogen assimilation. This single-step reaction adds ammonia to α-ketoglutarate (a precursor metabolite produced in the TCA cycle), producing glutamate (**figure 6.30**); this is a key step in the nitrogen cycle but is carried out only by certain bacteria and plants (see figure 28.10). Once glutamate has been formed, its amino group can be transferred to other carbon compounds to produce amino acids such as aspartate. This transfer of the amino group, a transamination, regenerates α-ketoglutarate from glutamate. The α-ketoglutarate molecule can then be used again to incorporate more ammonia.

Organisms lacking one or more enzymes in a given biosynthetic pathway must have the end product provided from an external source (which is why fastidious bacteria, such as lactic acid bacteria, require many different growth factors). Once the subunits are present, they can be assembled to make the macromolecules that compose the cell. This section will focus on the synthesis of fatty acids and glycerol, amino acids, and nucleotides; carbohydrates are synthesized from the sugars described in the previous sections.

Fatty Acids and Glycerol Synthesis

Synthesis of most lipids requires fatty acids and glycerol. To produce fatty acids, the acetyl group of acetyl-CoA (the precursor metabolite produced in the transition step) is transferred to a carrier protein. This carrier holds the developing fatty acid chain as 2-carbon units are added. When the fatty acid reaches its required length, usually 14, 16, or 18 carbon atoms long, it is released. The glycerol component of the lipid

Aromatic Amino Acids

Synthesis of aromatic amino acids such as tyrosine, phenylalanine, and tryptophan requires a multistep, branching pathway (**figure 6.31**). The pathway begins with the joining of two precursor metabolites—phosphoenolpyruvate (3-carbon) and erythrose-4-phosphate (4-carbon)—to form a 7-carbon compound. The precursors originate in glycolysis and the pentose phosphate pathway, respectively. Then the 7-carbon compound is modified through a series of steps until a branch point is reached, where two options are possible. If synthesis proceeds in one direction, tryptophan is produced. In the other direction, another branch point is reached; from there, either tyrosine or phenylalanine can be made.

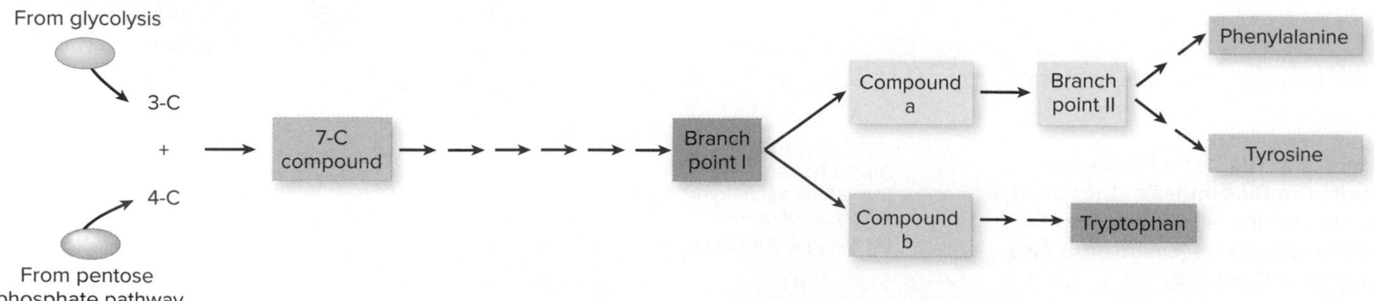

FIGURE 6.31 Synthesis of Aromatic Amino Acids A multistep branching pathway is used to synthesize aromatic amino acids.

? The synthesis of aromatic amino acids requires precursor metabolites made in which central metabolic pathways?

FOCUS YOUR PERSPECTIVE 6.1

Fueling the Future

Considering that microbes are masters at getting energy from seemingly unlikely sources, it should not be surprising that they are being used to help us meet our future energy needs. Bacteria, yeast, and algae play starring roles in most of the current research efforts toward developing cost-effective methods to produce **biofuels,** which are fuels made from renewable biological resources. One goal in producing and using these is to be carbon neutral, meaning that the amount of carbon released to the atmosphere when fuel is burned is the same as what goes into producing the fuel.

First-generation biofuels are made using standard technologies (such as fermentations) and readily available material (including corn and sugar cane). One example is using yeast to ferment the sugars in corn to produce bioethanol. Although this provides an alternative to fossil fuels, it uses an edible resource, contributing to food shortages and increased food prices. In addition, the farming practices used to grow corn on a large scale can harm the environment, and the process for making bioethanol uses fossil fuels, so the product is not carbon neutral.

In response to the problems with first-generation biofuels, newer technologies are being developed to produce what are collectively referred to as advanced biofuels. These include the following:

- **Bioethanol.** Rather than using edible foods, bioethanol produced with advanced technologies is made from cornstalks or other plant waste materials. One problem is that the main component of plant wastes is cellulose, and relatively few microbes degrade this carbohydrate. Microbes that degrade cellulose generally are more difficult to grow on a large-scale basis than yeast, so technical hurdles must be overcome for this to be a practical option.
- **Biodiesel.** Composed of fatty acid methyl esters, biodiesel has an advantage over bioethanol in that it has a higher energy content and is more compatible with current fuel storage and distribution systems. Algae and cyanobacteria—both of which are photosynthetic—capture radiant energy to make lipids that can be used to produce biodiesel. This provides a particularly attractive option because it bypasses the need for the two-stage process used to create bioethanol (plants convert sunlight to chemical energy, and then microbes ferment the plant material to produce ethanol). A problem with using photosynthetic microbes to produce biofuels, however, is that areas with enough sunlight to support their abundant growth generally lack sufficient water for the final steps of fuel production. Another option is to use a genetically engineered easy-to-grow bacterium such as *E. coli.* An *E. coli* strain has been engineered not only to produce biodiesel but also break down cellulose, so it can use cellulose-containing waste materials as an energy source. The efficiency is still low, however, so improvements are still necessary. Also, the engineered strain still relies on a second organism—in this case, plants—to convert radiant energy into chemical energy.
- **Biohydrogen.** Some cyanobacteria and anoxygenic phototrophs make hydrogen gas as a by-product of nitrogen fixation and grow using little more than sunlight and water. An advantage of biohydrogen is that it can be easily captured from a culture. As with the other advanced biofuels, however, there are still technical obstacles that must be overcome before the gas can be produced and stored on a large-scale basis.

Will microbes end our reliance on fossil fuels? Only time will tell, but the possibilities are exciting!

This pathway serves as an excellent illustration of many important features of the regulation of amino acid synthesis. When an amino acid is provided to a cell, it would be a waste of carbon, energy, and reducing power for that cell to continue synthesizing it. To regulate the pathway's activity, the cell controls the enzymes that catalyze the branch points and the first step of the pathway. Tryptophan is a feedback inhibitor of the enzyme that directs the branch to its synthesis; this sends the pathway to the steps leading to the synthesis of the other amino acids (tyrosine and phenylalanine). Both tyrosine and phenylalanine also each inhibit the first enzyme of the branch leading to their synthesis. The three amino acids that control the branch points also control the first step of the pathway: the formation of the 7-carbon compound. In *E. coli,* three different enzymes can catalyze this step; each has the same active site, but they have different allosteric sites. Each aromatic amino acid acts as a feedback inhibitor for one of the enzymes. If all three amino acids are present in the environment, then very little of the 7-carbon compound will be synthesized. If only one or two of those amino acids are present, then proportionally more of the compound will be synthesized. In addition to controlling enzyme activity, cells have mechanisms to regulate enzyme synthesis, as described in section 7.6.

Nucleotide Synthesis

Nucleotide subunits of DNA and RNA are composed of three parts: a 5-carbon sugar, a phosphate group, and a nucleobase, either a purine or a pyrimidine. They are synthesized as ribonucleotides, but these can then be converted to deoxyribonucleotides by replacing the hydroxyl group on the 2′ carbon of the sugar with a hydrogen atom.

The purine (double-ring) and pyrimidine (single-ring) nucleotides are synthesized in distinctly different manners. The starting compound of purine synthesis is ribose-5-phosphate, a precursor metabolite generated in the pentose phosphate pathway. Then, in a highly ordered sequence, atoms from the other sources are added to form the purine ring. This can then be converted to a purine nucleotide. To synthesize pyrimidine nucleotides, the pyrimidine ring is made first and then attached to ribose-5-phosphate. After one pyrimidine nucleotide is formed, the nucleobase component can be converted into any of the other pyrimidines.

MicroAssessment 6.10

Biosynthetic processes of different organisms are remarkably similar, using precursor metabolites, NADPH, and ATP to form subunits. Synthesis of the amino acid glutamate provides a mechanism for bacteria to incorporate nitrogen in the form of ammonia into organic material. Synthesis of aromatic amino acids involves branching pathways. The purine nucleotides are synthesized in a very different manner from the pyrimidine nucleotides.

26. Explain why glutamate synthesis is particularly important for a cell.

27. What three general products of the central metabolic pathways does a cell require to carry out biosynthesis?

28. With a branched biochemical pathway, why would it be important for a cell to shut down the first step as well as branching steps?

Summary

6.1 ■ Overview of Microbial Metabolism

Catabolism is the set of processes that capture and store **energy** by breaking down complex molecules. **Anabolism** includes processes that use energy to make and assemble the building blocks of a cell (figure 6.1).

Energy

Photosynthetic organisms harvest the energy of sunlight, using it to power the synthesis of organic compounds. **Chemoorganotrophs** harvest energy contained in organic compounds (figure 6.3). **Exergonic** reactions release energy; **endergonic** reactions use energy.

Components of Metabolic Pathways

A specific **enzyme** facilitates each step of a **metabolic pathway** (figure 6.5). **ATP** is the energy currency of the cell. A **redox** reaction is a pair of reactions: an oxidation and a reduction. In metabolism, the **energy source** is **oxidized** to release its energy (figure 6.8), while an **electron carrier** is **reduced** (figure 6.7). **NAD^+/NADH, $NADP^+$/NADPH,** and **FAD/$FADH_2$** are electron carriers (table 6.1).

Precursor Metabolites

Precursor metabolites may be used to make the subunits of macromolecules, or they may instead be oxidized to generate energy (table 6.2).

Catabolism (figure 6.10)

The **central metabolic pathways** are **glycolysis,** the **pentose phosphate pathway,** and the **tricarboxylic acid cycle (TCA cycle).** The **transition step** links glycolysis and the TCA cycle. **Cellular respiration** uses the reducing power accumulated in the central metabolic pathways to produce proton motive force which can be used to make ATP by oxidative phosphorylation. **Aerobic respiration** uses O_2 as a terminal electron acceptor; **anaerobic respiration** uses a molecule other than O_2 as a terminal electron acceptor (table 6.3). **Fermentation** of glucose uses pyruvate or another derivative as a terminal electron acceptor; this recycles the reduced electron carrier NADH.

6.2 ■ Enzymes

Enzymes function as biological catalysts; they are neither consumed nor permanently changed during a reaction.

Mechanisms and Consequences of Enzyme Action (figure 6.12)

One or more substrates bind to the **active site,** forming an enzyme-substrate complex that lowers the activation energy of the reaction.

Cofactors (figure 6.13; table 6.4)

Enzymes sometimes act with the assistance of **cofactors** such as **coenzymes** and certain trace elements.

Environmental Factors That Influence Enzyme Activity (figure 6.14)

The environmental factors most important in influencing enzyme activities are temperature, pH, and salt concentration.

Allosteric Regulation (figure 6.15)

Cells can fine-tune the activity of an allosterically regulated enzyme by using a molecule that binds to the **allosteric** site of the enzyme.

Enzyme Inhibition

Non-competitive inhibition occurs when the inhibitor and the substrate act at different sites on the enzyme. **Competitive inhibition** occurs when the inhibitor competes with the normal substrate for active site binding (figure 6.16).

6.3 ■ The Central Metabolic Pathways (table 6.6)

Glycolysis (figure 6.17)

Glycolysis converts one molecule of glucose into two molecules of pyruvate; the Embden-Meyerhof pathway produces 2 ATP (net) by substrate-level phosphorylation and 2 NADH. Six different precursor metabolites are made.

Pentose Phosphate Pathway

The **pentose phosphate pathway** forms NADPH and two different precursor metabolites.

Transition Step and Tricarboxylic Acid (TCA) Cycle (figure 6.18)

The **transition step** converts pyruvate to acetyl-CoA, a precursor metabolite. Repeated twice per glucose, this step produces 2 NADH. The **TCA cycle** completes the oxidation of glucose; two "turns" produce 6 NADH and 2 FADH2, in addition to 2 ATP by substrate-level phosphorylation. Two intermediates are precursor metabolites.

6.4 ■ Cellular Respiration

The Electron Transport Chain (ETC)—Generating a Proton Motive Force

The **electron transport chain (ETC)** is a series of membrane-embedded electron carriers that move protons across the membrane to generate a **proton motive force.** In the mitochondrial ETC, three different complexes (complexes I, III, and IV) function as proton pumps (figure 6.20). Prokaryotes vary with respect to the types and arrangements of their electron transport components (figure 6.21). Some prokaryotes can use molecules other than O_2 as terminal electron acceptors, a process known as anaerobic respiration; this harvests less energy than aerobic respiration.

ATP Synthase—Using the Proton Motive Force to Synthesize ATP

ATP synthase permits proton flow back across the membrane down their concentration gradient, harvesting the energy released to fuel the synthesis of ATP.

ATP Yield of Aerobic Respiration in Prokaryotes (figure 6.22)

The theoretical maximum ATP yield of aerobic respiration is 38 ATP.

6.5 ■ Fermentation

During the fermentation of sugars, ATP is usually only produced in the glycolytic pathway; steps after glycolysis provide a mechanism for recycling NADH back to NAD^+ (figure 6.23). Some end products of fermentation are commercially valuable (figure 6.24). Certain end products help in bacterial identification.

6.6 ■ Catabolism of Organic Compounds Other Than Glucose

Specific exoenzymes break down macromolecules extracellularly into their respective subunits that can then enter the cell (figure 6.25).

Polysaccharides and Disaccharides

Amylases digest starch, releasing glucose subunits, and are produced by many organisms. Cellulases degrade cellulose. Sugar subunits released when polysaccharides are broken down can then enter glycolysis.

Lipids

Fats are broken down by lipases, releasing glycerol and fatty acids. Glycerol is converted to the precursor metabolite dihydroxyacetone phosphate; fatty acids are degraded by β-oxidation, generating reducing power and the precursor metabolite acetyl-CoA.

Proteins

Proteins are broken down to their amino acid subunits by proteases. Deamination removes the amino group; the remaining carbon skeleton is then converted into the appropriate precursor molecule.

6.7 ■ Chemolithotrophy

Prokaryotes, as a group, are unique in their ability to use reduced inorganic compounds such as hydrogen sulfide (H_2S) and ammonia (NH_3) as energy sources. Chemolithotrophs are autotrophs.

6.8 ■ Photosynthesis

Capturing the Energy of Light

During the **light-dependent reactions,** phototrophs use pigments arranged in **photosystems** to capture light energy and convert it to chemical energy in the form of ATP. The high-energy electrons emitted by reaction centers are passed to an electron transport chain, which uses them to generate a proton motive force (figure 6.26). That is then harvested to make ATP by the process of **photophosphorylation.** During cyclic photophosphorylation, the high-energy electrons emitted by a reaction center return to it. However, when reducing power is needed in addition to ATP, an external electron source replaces electrons lost from the reaction center. Depending on the electron source, the process can be oxygenic or anoxygenic. The light-independent reactions then use the ATP and reducing power from the light-dependent reactions to synthesize organic compounds from CO_2.

Anoxygenic Photosynthesis

Purple and green bacteria each use only a single photosystem. They use non-cyclic electron flow when producing reducing power, and they obtain electrons from a reduced compound other than water, so they do not generate O_2.

Oxygenic Photosynthesis

Cyanobacteria, plants, and algae each have two types of photosystems that work together to make both ATP and reducing power simultaneously during non-cyclic photophosphorylation (figure 6.27). When high-energy electrons emitted from photosystem I are used to make reducing power, they are replaced by those lost from photosystem II (via the electron transport chain). Those emitted by photosystem II are then replaced by splitting water, generating O_2.

Yield of Photosynthesis

Photosynthesis uses light energy to convert 6 CO_2 + 12 H_2X to glucose ($C_6H_{12}O_6$) + 12 X + 6 H_2O. In oxygenic photosynthesis, X is O_2, while in anoxygenic photosynthesis, it is not.

Phototrophy Involving Rhodopsin Pigments

Some prokaryotes use membrane-embedded forms of rhodopsin that function as light-driven proton pumps; the energy of the resulting proton motive force is used to make ATP.

6.9 ■ Carbon Fixation

Calvin Cycle (figure 6.28)

The most common pathway used to incorporate CO_2 into an organic form is the **Calvin cycle.**

6.10 ■ Anabolic Pathways—Synthesizing Subunits from Precursor Molecules (figure 6.29)

Fatty Acids and Glycerol Synthesis

The fatty acid components of lipids are synthesized by adding 2-carbon units sequentially. The glycerol component is synthesized from dihydroxyacetone phosphate.

Amino Acid Synthesis

Synthesis of glutamate from α-ketoglutarate and ammonia provides a mechanism for cells to incorporate inorganic nitrogen into organic molecules (figure 6.30). Synthesis of aromatic amino acids requires a multistep branching pathway. Allosteric enzymes regulate key steps of the pathway (figure 6.31).

Nucleotide Synthesis

Purines and pyrimidine nucleotides are made in distinctly different manners.

Review Questions

Short Answer

1. Explain the difference between catabolism and anabolism.
2. How does ATP serve as a carrier of free energy?
3. How do enzymes catalyze chemical reactions?
4. Explain how precursor molecules are involved in catabolic as well as anabolic pathways.
5. How do cells regulate enzyme activity?
6. Why do the electrons carried by $FADH_2$ result in less ATP production than those carried by NADH?
7. Name three food products produced with the aid of fermenting microorganisms.
8. In photosynthesis, what is encompassed by the term "light reactions"?
9. Unlike the cyanobacteria, the anoxygenic photosynthetic bacteria do not produce O_2. Why not?
10. What is the role of transamination in amino acid biosynthesis?

Multiple Choice

1. Which of these factors do/does not affect enzyme activity?
   a) Temperature
   b) Inhibitors
   c) Coenzymes
   d) Humidity
   e) pH
2. Which of the following statements is *false*? Enzymes
   a) bind to substrates.
   b) lower the energy of activation.
   c) convert coenzymes to products.
   d) speed up biochemical reactions.
   e) can be named after the kinds of reaction they catalyze.
3. Based on the name, NADH dehydrogenase is
   a) a vitamin that oxidizes NADH.
   b) a vitamin that reduces NADH.
   c) a vitamin that produces NADH.
   d) an enzyme that oxidizes NADH.
   e) an enzyme that reduces NADH.
4. What is the end product of glycolysis?
   a) Glucose
   b) Citrate
   c) Oxaloacetate
   d) α-Ketoglutarate
   e) Pyruvate
5. The central metabolic pathway(s) is/are
   a) glycolysis and the TCA cycle only.
   b) glycolysis, the TCA cycle, and the pentose phosphate pathway.
   c) glycolysis only.
   d) glycolysis and the pentose phosphate pathway only.
   e) the TCA cycle only.
6. Which of these pathways gives a cell the potential to produce the most ATP?
   a) TCA cycle
   b) Pentose phosphate pathway
   c) Lactic acid fermentation
   d) Glycolysis
7. In fermentation, the terminal electron acceptor is
   a) oxygen (O_2).
   b) hydrogen (H_2).
   c) carbon dioxide (CO_2).
   d) an organic compound.
8. In the process of oxidative phosphorylation, the energy of a proton motive force is used to generate
   a) NADH.
   b) ADP.
   c) ethanol.
   d) ATP.
   e) glucose.
9. If a bacterium loses the ability to produce $FADH_2$, which of the following cannot continue?
   a) Biosynthesis
   b) Glycolysis
   c) Fermentation
   d) The TCA cycle
10. Degradation of fats as an energy source involves all of the following *except*
    a) β-oxidation.
    b) acetyl-CoA.
    c) glycerol.
    d) lipase.
    e) transamination.

Applications

1. A worker in a cheese-making facility argues that whey, a nutrient-rich by-product of cheese, should be dumped in a nearby pond where it could serve as fish food. Explain why this proposed action could actually kill the fish by depleting the O_2 in the pond.
2. Scientists working with DNA in the laboratory often store it in solutions that contain EDTA, a chelating agent that binds magnesium (Mg^{2+}). This is done to prevent enzymes called DNases from degrading the DNA. Explain why EDTA would interfere with enzyme activity.

Critical Thinking

1. A student argued that aerobic and anaerobic respiration should produce the same amount of ATP. He reasoned that they both use basically the same process; only the terminal electron acceptor is different. What is the primary error in this student's argument?
2. Chemolithotrophs near hydrothermal vents support a variety of other life-forms there. Explain how their role is analogous to that of photosynthetic organisms in terrestrial environments.

7 The Blueprint of Life, from DNA to Protein

Model of DNA double helix. *Molekuul/SPL/age fotostock*

KEY TERMS

Codon A series of three nucleotides that codes for a specific amino acid.

DNA Polymerase Enzyme that synthesizes DNA, using an existing strand as a template to make a new complementary strand.

DNA Replication Duplication of a DNA molecule.

Gene The functional unit of the genome; it encodes a product, most often a protein.

Genome Complete set of genetic information in a cell or a virus.

Messenger RNA (mRNA) Type of RNA molecule that contains the genetic information decoded during translation.

Promoter Nucleotide sequence of DNA to which RNA polymerase binds to start transcription.

Ribosomal RNA (rRNA) Type of RNA molecule present in ribosomes.

Ribosome Structure that facilitates the joining of amino acids during translation; composed of ribosomal RNA (rRNA) and protein.

RNA Polymerase Enzyme that synthesizes RNA using one strand of DNA as a template.

Transcription The process by which the information encoded in DNA is copied into RNA.

Transfer RNA (tRNA) Type of RNA molecule involved in interpreting the genetic code during translation; each tRNA molecule carries a specific amino acid.

Translation The process by which the information carried by mRNA is used to synthesize the encoded protein.

A Glimpse of History

Born in 1933, Tsuneko Okazaki is part of the first generation of women in Japan allowed to earn a university education. Upon completing her degree at Nagoya University in 1956, Tsuneko entered graduate school there, wanting to learn more about the process by which cells copy their DNA. That same year she met and married a fellow scientist, Reiji, and they soon became a husband-wife research team.

The conditions in post–World War II Japan made research difficult; many of the buildings had been destroyed, scientific journals were hard to obtain, and funding was scarce. In 1960, the Okazakis traveled to study in the United States, where they were able to expand their research, using methods and machinery that were state-of-the art at the time. One of the labs they worked in was that of Arthur Kornberg, who discovered DNA polymerase, the enzyme that synthesizes DNA from existing DNA.

After three years in the United States, the Okazakis returned to Nagoya University, where Reiji accepted a faculty position and Tsuneko completed her PhD. There, they worked to understand a particularly perplexing question related to how double-stranded DNA is replicated. DNA polymerase can only synthesize DNA in one direction (5′ to 3′), yet the two strands of DNA are antiparallel, meaning they are oriented in opposite directions (see figure 2.30). So how can both strands be synthesized simultaneously? In 1968, the Okazaki lab group published a paper that answered the question. When DNA polymerase synthesizes the problematic strand, it makes short DNA pieces, all in the 5′ to 3′ direction, and then joins these together to make one long strand. The term "Okazaki fragments" is now used to refer to the short pieces.

Sadly, Reiji passed away in 1975, meaning that Tsuneko lost not only her close collaborator but also her husband and the father to her young children. She considered giving up research, but at the urging of colleagues from all over the world, including Kornberg, she found the courage and help to continue. She and her lab later identified the small pieces of RNA that serve to start the process of DNA synthesis. Tsuneko Okazaki's dedicated efforts helped reveal the DNA replication mechanisms described in this chapter.

Consider for a moment the incredible diversity of life-forms in our world—from the remarkable variety of microorganisms, to the plants and animals consisting of many different specialized cells. Each of these cells' characteristics, from the shape to the function, is dictated by information within their deoxyribonucleic acid (DNA). DNA is the "blueprint," providing instructions for building an organism's components.

DNA itself is a simple structure—a long linear or circular molecule composed of only four different **nucleotides,** each containing a particular **nucleobase** (also called a nitrogenous base or simply a base): adenine (A), thymine (T), cytosine (C), or guanine (G). A set of three nucleotides encodes a specific amino acid; in turn, a string of amino acids makes up a protein,

Focus Figure

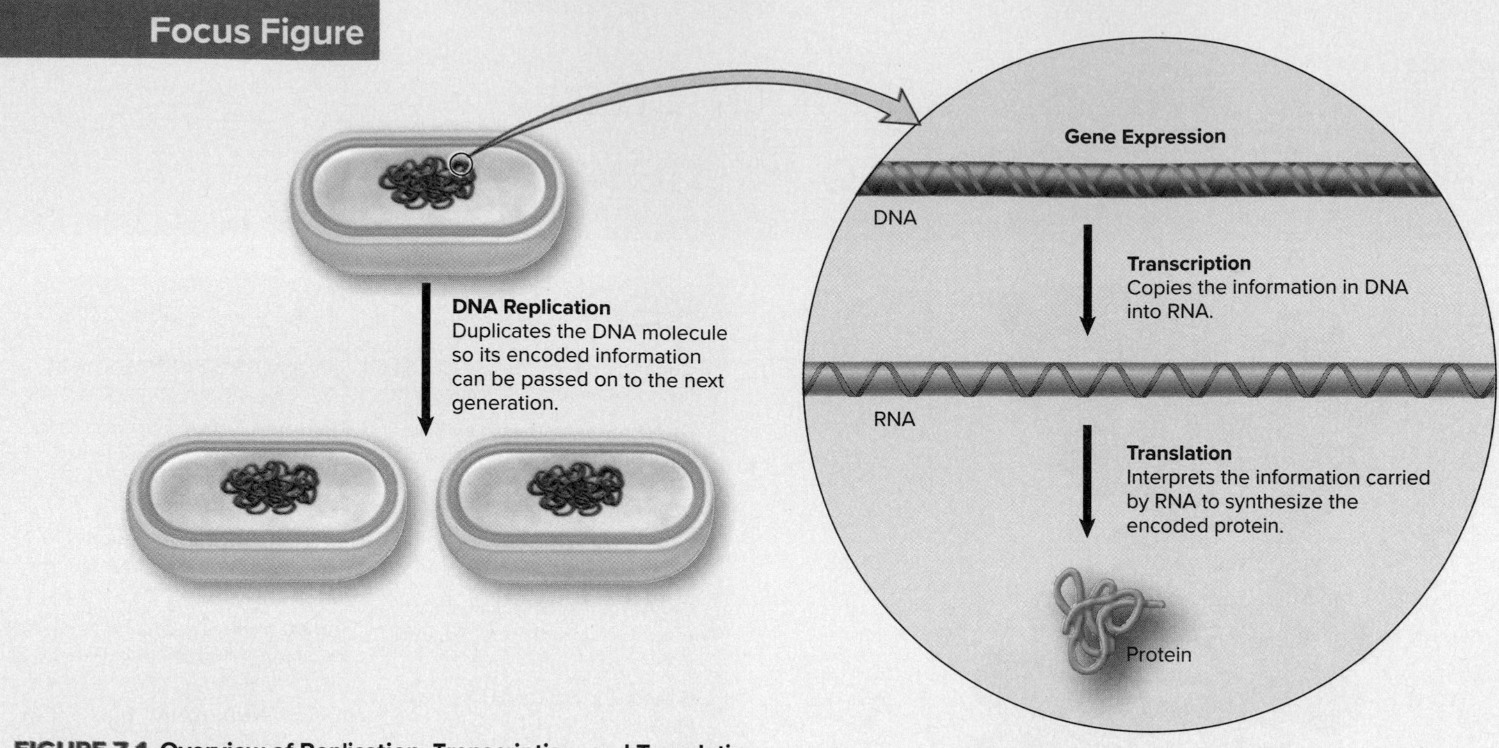

FIGURE 7.1 Overview of Replication, Transcription, and Translation

Which would result in more long-term consequences to a cell or its progeny: A mistake made during DNA synthesis or during RNA synthesis?

the structure and function of which is dictated by the order of the amino acid subunits. Some proteins serve as structural components of a cell. Others, such as enzymes, direct cellular activities, including biosynthesis and energy conversion. Together, all these proteins control the cell's structure and activities, thereby determining the overall characteristics of that cell.

Although at first it might seem unlikely that the vast array of life-forms could be encoded by a molecule consisting of only four different nucleotides, think about how much information can be transmitted by binary code, the language of all computers. Using only a simple series of ones and zeros, binary can code for each letter of the alphabet. String enough of these together in the right sequence and the letters become words. With longer and longer strings, the words can become complete sentences, chapters, books, or even whole libraries.

This chapter will focus on the processes bacteria use to replicate their DNA and convert the encoded information into proteins. The mechanisms used by eukaryotic cells have many similarities but are considerably more complicated, and they will be discussed only briefly. Some processes in archaea are similar to those of bacteria, but many others resemble those of eukaryotic cells.

7.1 ■ Overview

Learning Outcomes

1. Compare and contrast the characteristics of DNA and RNA.
2. Explain why gene regulation is important to the cell.

The complete set of genetic information of a cell—both the chromosome and any plasmids—is referred to as its **genome.** The genome of all cells is composed of DNA, but some viruses have an RNA genome. The functional unit of the genome is a **gene.** A gene encodes a product (called the gene product), most commonly a protein. The study and analysis of the nucleotide sequence of DNA is called **genomics.**

To multiply, all cells must (1) make a copy of their DNA and (2) use the information in their DNA to make functional products (**figure 7.1**). The double-stranded DNA must be duplicated before cell division so that its encoded information can be passed to the next generation. This is the process of **DNA replication.** Meanwhile, the information encoded in the DNA must be used to synthesize the necessary gene products. This process, **gene expression,** involves transcription and translation. **Transcription** is the process by which the information encoded in DNA is copied into a slightly different molecule: single-stranded RNA. In **translation,** the information carried by that RNA is interpreted and used to synthesize the encoded protein. The flow of information from DNA → RNA → protein is often referred to as the central dogma of molecular biology.

Characteristics of DNA

DNA is usually a double-stranded, helical structure. As described in chapter 2, each strand is a chain of deoxyribonucleotide subunits, more commonly called nucleotides (see figure 2.27). Each nucleotide consists of a 5-carbon sugar (deoxyribose), a phosphate group, and one of four different nucleobases (A, T, G, or C) (see figure 2.28). Recall that the carbon atoms on sugar molecules are numbered, thereby allowing scientists to describe the positions of various attached functional groups (figure 2.13). In a nucleotide, the phosphate group is attached to the 5′ (5 prime) carbon and is therefore referred to as the 5′-PO_4 (5 prime phosphate). The nucleotides of a DNA molecule are joined together by a covalent bond between the 5′-PO_4 of one nucleotide and the 3′-OH (3 prime hydroxyl) of the next. Joining nucleotides together this way creates a string of alternating sugar and phosphate units, called the sugar-phosphate backbone (see figure 2.29). Because of the chemical structure of nucleotides and how they are joined to each other, a single linear strand of DNA will always have a 5′-PO_4 at one end and a 3′-OH at the other. These ends are often referred to as the **5′ end** (5 prime end) and the **3′ end** (3 prime end).

The two strands of DNA are **complementary** and are held together by hydrogen bonds between the nucleobases (**figure 7.2**). Wherever an adenine (A) is in one strand, a thymine (T) is in the other; these opposing A-T bases are held together by two hydrogen bonds. Similarly, wherever a guanine (G) is in one strand, a cytosine (C) is in the other. These G-C bases are held together by three hydrogen bonds, a slightly stronger attraction than that of an A-T pair. The characteristic bonding of A to T and G to C is called **base-pairing** and is a fundamental characteristic of DNA. Because of the rules of base-pairing, one strand can always be used as a template for the synthesis of the opposing strand. Although the two strands are complementary, they are also **antiparallel.** That is, they are oriented in opposite directions. One strand is oriented in the 5′ to 3′ direction, and its complement is oriented in the 3′ to 5′ direction.

The duplex structure of double-stranded DNA is generally quite stable because of the numerous hydrogen bonds holding the strands together. Short DNA fragments have correspondingly fewer hydrogen bonds, so they are easily separated into single strands. Separating the two strands is called melting, or denaturing.

Characteristics of RNA

RNA, like DNA, is a chain of nucleotides. It is similar to DNA in many ways, with a few important exceptions. One difference is that the sugar in the nucleotides of RNA is ribose, not deoxyribose; ribose has an oxygen atom that deoxyribose lacks

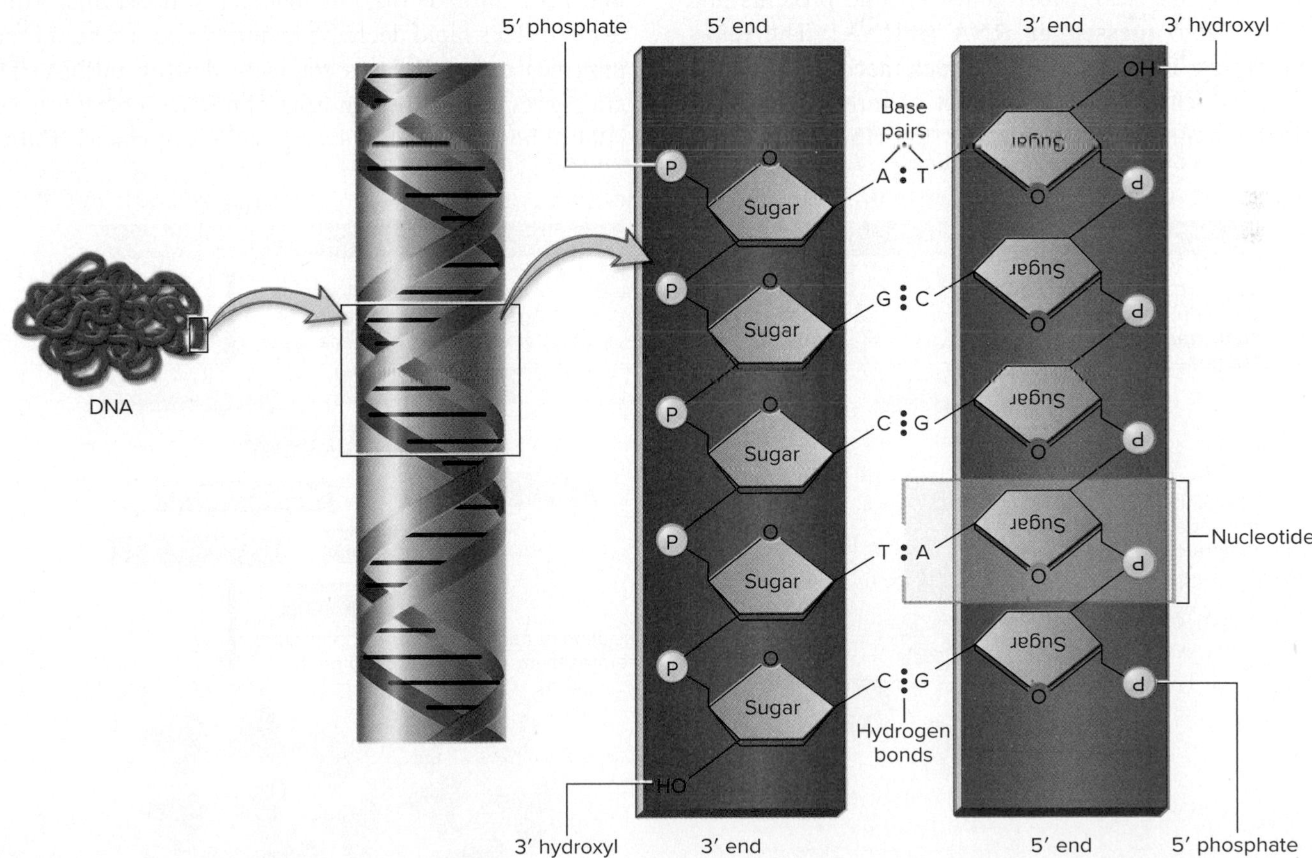

FIGURE 7.2 The Structure of DNA The two strands in the double helix are complementary. Three hydrogen bonds form between a G-C base pair and two between an A-T base pair. The strands are antiparallel; one is oriented in the 5′ to 3′ direction, and its complement is oriented in the 3′ to 5′ direction.

If a 100 base-pair double-stranded DNA fragment has 40 cytosines, how many adenines does it contain?

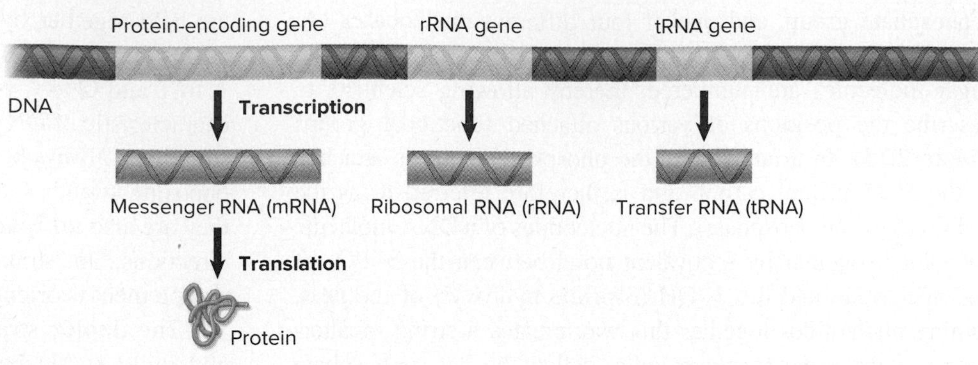

FIGURE 7.3 Three Functional Types of RNA Molecules The different functional types of RNA required for gene expression—messenger RNA (mRNA), ribosomal RNA (rRNA), and transfer RNA (tRNA)—are transcribed from different genes. The mRNA is translated, and the tRNA and rRNA fold into characteristic three-dimensional structures that each play a role in protein synthesis.

? **Ribosomal RNA is a component of ribosomes. What are ribosomes?**

(see figure 2.13). Another distinction is that RNA contains the nucleobase uracil in place of the thymine found in DNA (see figure 2.28). Also, RNA is usually a single-stranded linear molecule much shorter than DNA.

RNA is synthesized using a region of one of the two DNA strands as a template. In making the RNA molecule, or **transcript,** the base-pairing rules apply except that uracil, rather than thymine, pairs with adenine. The interaction of DNA and RNA is only temporary, however, and the transcript quickly separates from the template.

Three different functional types of RNA are required for gene expression, and these are transcribed from different sets of genes (**figure 7.3**). Most genes encode proteins and are transcribed into **messenger RNA (mRNA).** The information encrypted in mRNA is decoded according to the genetic code, which correlates each set of three nucleotides to a particular amino acid. The genes for **ribosomal RNA (rRNA)** and **transfer RNA (tRNA)** are never translated into proteins; instead, the RNA molecules themselves are the final products, and each type plays a different but critical role in protein synthesis.

Regulating Gene Expression

Although a cell's DNA can encode thousands of different proteins, not all of them are needed at the same time or in equal quantities. Because of this, cells require mechanisms to regulate the expression of certain genes. When transcription of a gene is turned "on," transcripts are available for translation, but if it is turned "off," the number of transcripts will rapidly decline. This rapid decrease in number occurs because cellular enzymes called RNases routinely destroy mRNA. Thus, by simply regulating the synthesis of mRNA molecules, a cell can control how much of a given protein is produced (**figure 7.4**).

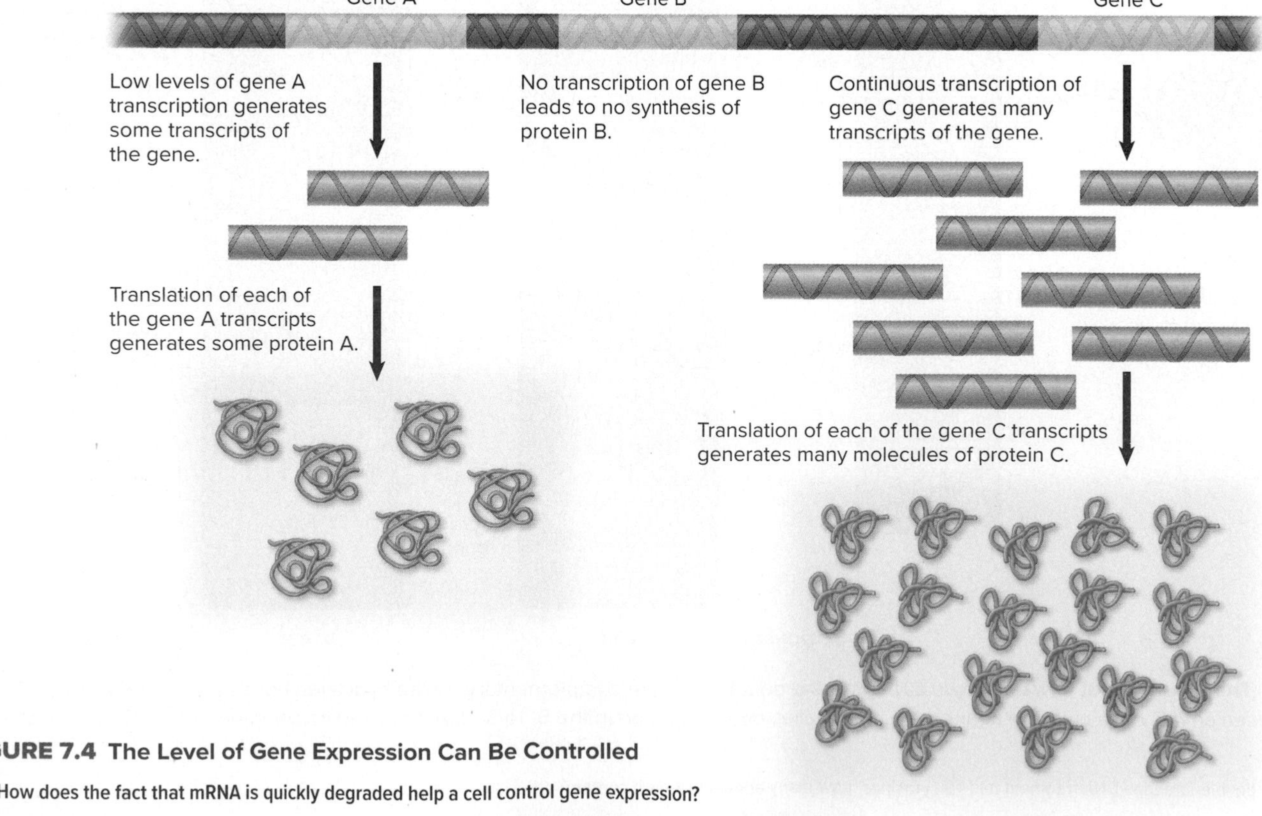

FIGURE 7.4 The Level of Gene Expression Can Be Controlled

? **How does the fact that mRNA is quickly degraded help a cell control gene expression?**

MicroAssessment 7.1

Replication is the process of duplicating a DNA molecule. Transcription is the process of copying the information encoded in DNA into RNA. Translation is the process in which the information carried by mRNA is used to synthesize the encoded protein.

1. How does the 5′ end of a DNA strand differ from the 3′ end?
2. What are the base-pairing rules?
3. If the nucleotide sequence of one strand of DNA is 5′ ACGTTGCA 3′, what is the sequence of the complementary strand?

7.2 ■ DNA Replication

Learning Outcome

3. Describe the bacterial DNA replication process.

DNA is replicated so that each of the two cells generated during binary fission can receive one complete copy of the genome. The replication process is generally bidirectional, meaning it proceeds in both directions from a specific starting point—a certain DNA sequence called the **origin of replication** (**figure 7.5**). Bidirectional replication allows a chromosome to be duplicated in half the time it would take if the process were unidirectional. The progression of bidirectional replication around a circular DNA molecule creates two advancing forks where DNA synthesis occurs. These regions, called **replication forks,** ultimately meet at a terminating site when the process is complete. Each of the two DNA molecules created through replication contains one original strand paired with a newly synthesized strand. Because one strand of the original molecule is conserved in each molecule, replication is said to be **semiconservative.**

The process that starts DNA replication is referred to as initiation. This begins when specific proteins recognize and bind to the origin of replication; bacterial chromosomes and plasmids typically contain only one origin of replication, and a DNA molecule that lacks the sequence will not be replicated. One type of protein that binds to the bacterial origin of replication is DNA helicase, an enzyme that unwinds the double helix and separates the DNA strands at that site, exposing single-stranded regions of DNA that can act as templates. Enzymes called primases then synthesize short stretches of RNA complementary to the exposed templates. These small fragments, called **primers,** are important in the steps of replication described next. Eukaryotic cells use initiation processes similar to those of bacteria, but with significant differences. For example, eukaryotic chromosomes typically have multiple origins of replication, and the proteins involved in the replication process are different. Likewise, the DNA replication machinery of archaea is also distinct.

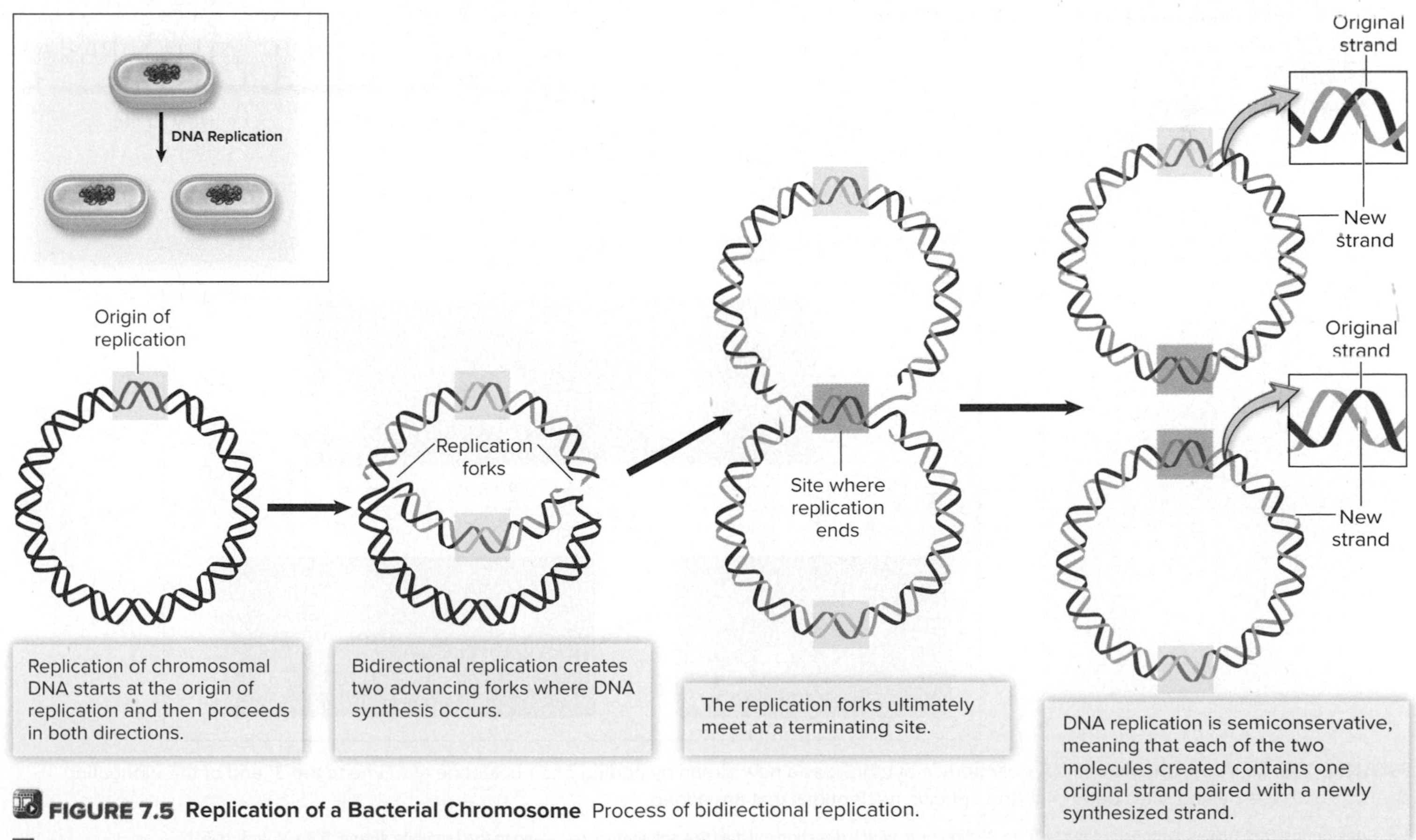
DNA Replication
Origin of replication
Replication forks
Site where replication ends
Original strand
New strand
Original strand
New strand
Replication of chromosomal DNA starts at the origin of replication and then proceeds in both directions.
Bidirectional replication creates two advancing forks where DNA synthesis occurs.
The replication forks ultimately meet at a terminating site.
DNA replication is semiconservative, meaning that each of the two molecules created contains one original strand paired with a newly synthesized strand.

FIGURE 7.5 Replication of a Bacterial Chromosome Process of bidirectional replication.

? Explain why the term "fork" is used to describe the separated DNA strands during replication.

Once replication is initiated, enzymes called **DNA polymerases** synthesize DNA in the 5′ to 3′ direction, using one parent strand as a template to make the complement (**figure 7.6**). To do this, a DNA polymerase adds nucleotides onto the 3′ end of the new strand, powering the reaction with the energy released when a high-energy phosphate bond of the incoming nucleotide is hydrolyzed. DNA polymerases add nucleotides only onto an existing nucleotide strand, so they cannot initiate synthesis. This explains why RNA primers are required at the origin of replication: They provide the DNA polymerase with a molecule to which it can add nucleotides.

The overall process of DNA replication is extraordinarily complicated, with most of the crucial enzymes and other proteins functioning together in a DNA-synthesizing protein complex called a **replisome** that simultaneously synthesizes both strands. The result is a twisted three-dimensional arrangement of the replication fork, much more complex than the flat layout depicted in the simplified diagrams used to explain the steps. DNA replication can be summarized as follows (**figure 7.7**):

1. For replication to progress, helicase molecules must progressively unwind the double helix and "unzip" the DNA strands at each replication fork to reveal additional template sequences. Simultaneous to this activity, the enzyme DNA gyrase breaks, uncoils, and reseals the parent DNA strands ahead of the fork to relieve the tension caused by the unwinding of the duplex DNA.
2. Synthesis of one new strand proceeds continuously as fresh template is exposed because the DNA polymerase simply adds nucleotides to the 3′ end. This strand is called the **leading strand.**
3. Synthesis of the other strand, the **lagging strand,** is more complicated. This is because DNA polymerases cannot add nucleotides to the 5′ end of a nucleotide chain, so synthesis must be reinitiated regularly as additional template is exposed. Each time synthesis is reinitiated, another RNA primer must be made first. The result is a series of small DNA fragments, each of which has a short stretch of RNA at its 5′ end. These fragments are called **Okazaki fragments** (see Glimpse of History, this chapter).
4. As the DNA polymerase adds nucleotides to the 3′ end of one Okazaki fragment, it eventually reaches the 5′ end of another. A different type of DNA polymerase then removes the RNA primer nucleotides and simultaneously replaces them with deoxynucleotides.

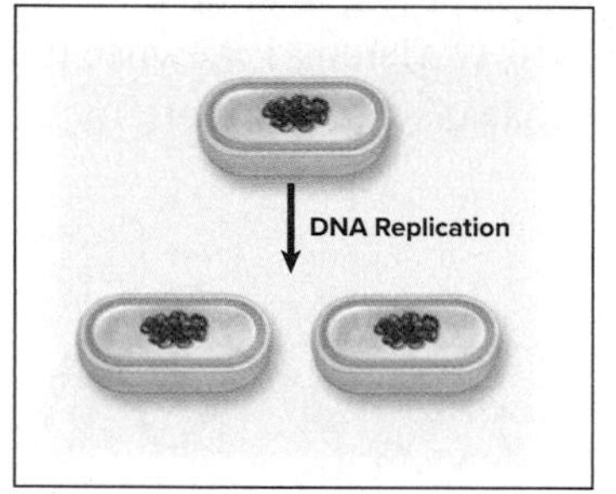

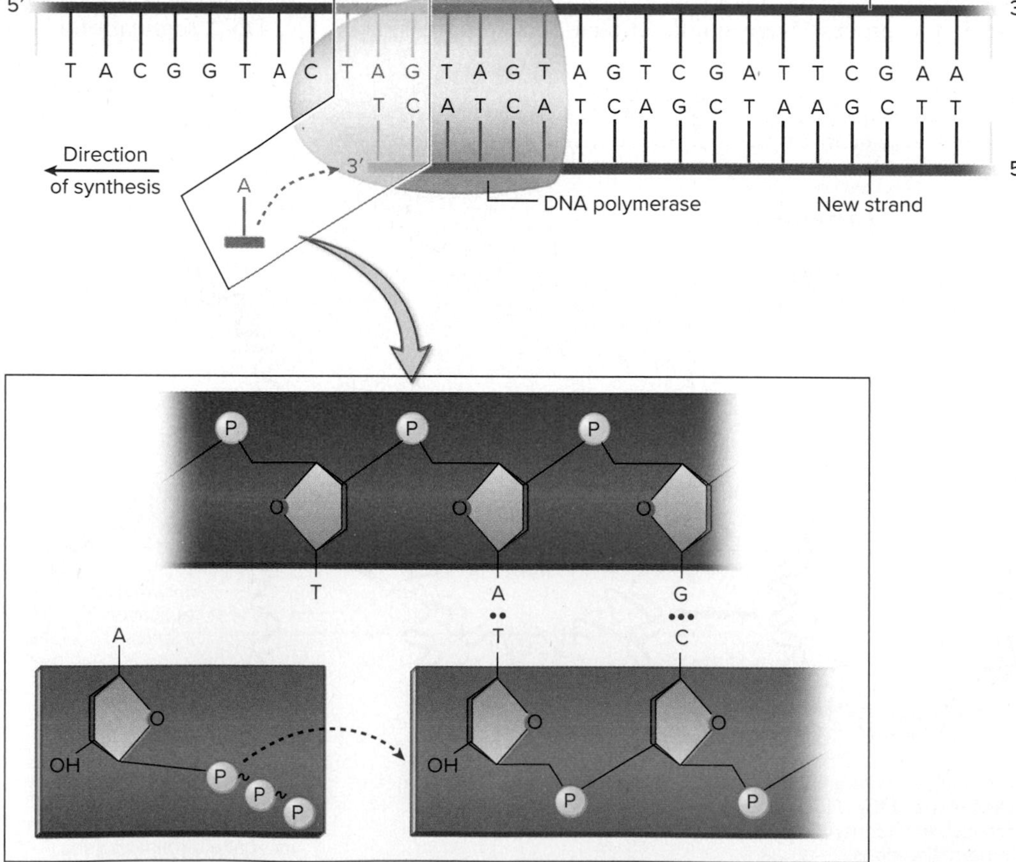

FIGURE 7.6 DNA Synthesis A DNA polymerase synthesizes a new strand by adding one nucleotide at a time to the 3′ end of the elongating strand. The base-pairing rules determine the specific nucleotides that are added.

? Considering that DNA is synthesized in the 5′ to 3′ direction, which direction must a DNA polymerase travel along the template strand: 5′ to 3′ or 3′ to 5′?

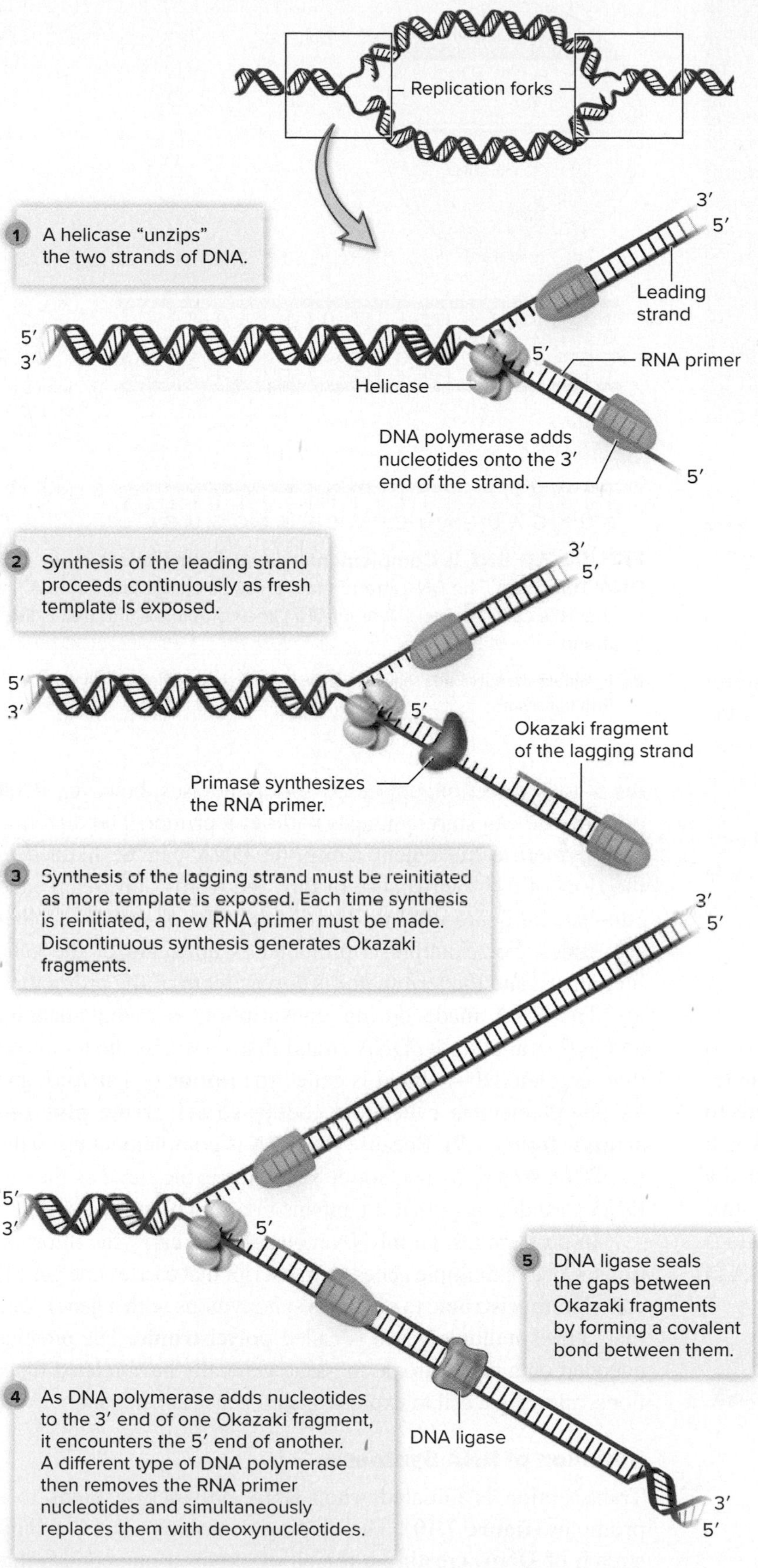
Replication forks
1 A helicase "unzips" the two strands of DNA.
3′
5′
Leading strand
5′
3′
5′
RNA primer
Helicase
DNA polymerase adds nucleotides onto the 3′ end of the strand.
5′
2 Synthesis of the leading strand proceeds continuously as fresh template is exposed.
3′
5′
5′
3′
5′
Okazaki fragment of the lagging strand
Primase synthesizes the RNA primer.
3 Synthesis of the lagging strand must be reinitiated as more template is exposed. Each time synthesis is reinitiated, a new RNA primer must be made. Discontinuous synthesis generates Okazaki fragments.
3′
5′
5′
3′
5′
5 DNA ligase seals the gaps between Okazaki fragments by forming a covalent bond between them.
4 As DNA polymerase adds nucleotides to the 3′ end of one Okazaki fragment, it encounters the 5′ end of another. A different type of DNA polymerase then removes the RNA primer nucleotides and simultaneously replaces them with deoxynucleotides.
DNA ligase
3′
5′

FIGURE 7.7 The Replication Fork This diagram is simplified to highlight the key differences between synthesis of the leading and lagging strands. Note that DNA gyrase is not shown.

? **Synthesis of which strand requires the repeated action of DNA ligase?**

⑤ The enzyme DNA ligase then seals the gaps between fragments by forming a covalent bond between the adjacent nucleotides. Replication continues around the circular chromosome.

When a circular bacterial chromosome is replicated, the two replication forks eventually meet at a site somewhat opposite the origin of replication. Two complete DNA molecules have been produced at this point, and these can be passed on to the two daughter cells. It takes approximately 40 minutes for the *E. coli* chromosome to be replicated. How, then, can the organism have a generation time of only 20 minutes? This happens because under favorable growing conditions, a cell initiates a new round of replication before the preceding one is complete. In this way, each of the two daughter cells will get one complete chromosome that has already started another round of replication.

Table 7.1 summarizes the key components of the replication process in bacteria.

TABLE 7.1 Components of DNA Replication in Bacteria

Component	Comment
DNA gyrase	Enzyme that temporarily breaks the strands of DNA, relieving the tension caused by unwinding the two strands of the DNA helix.
DNA helicase	Enzyme that unwinds the DNA helix and separates DNA strands.
DNA ligase	Enzyme that joins two DNA fragments together by forming a covalent bond between the sugar and phosphate residues of adjacent nucleotides.
DNA polymerases	Enzymes that synthesize DNA; they use one strand of DNA as a template to make the complementary strand. Nucleotides can be added only to the 3′ end of an existing fragment—therefore, synthesis always occurs in the 5′ to 3′ direction.
Okazaki fragment	Nucleic acid fragment produced during discontinuous synthesis of the lagging strand of DNA.
Origin of replication	Distinct region of a DNA molecule at which replication is initiated.
Primase	Enzyme that synthesizes small fragments of RNA to serve as primers for DNA synthesis.
Primer	Fragment of nucleic acid to which DNA polymerase can add nucleotides (the enzyme can add nucleotides only to an existing fragment).
Replisome	The complex of enzymes and other proteins that synthesize DNA.

MicroAssessment 7.2

DNA replication begins at the origin of replication and then proceeds bidirectionally, creating two replication forks. DNA polymerases synthesize DNA in the 5′ to 3′ direction, using one strand as a template to generate the complementary strand. The leading strand of DNA is synthesized continuously; the lagging strand is synthesized as fragments that DNA ligase then joins together.

4. Why is a primer required for DNA synthesis?
5. How does synthesis of the lagging strand differ from that of the leading strand?
6. Eukaryotic chromosomes have multiple origins of replication. Why would this be the case?

7.3 ■ Gene Expression in Bacteria

Learning Outcomes

4. Describe the process of transcription.
5. Describe the process of translation.

Recall that gene expression involves two separate but interrelated processes: transcription and translation. Transcription is the process of synthesizing RNA from a DNA template. During translation, information encoded by an mRNA transcript is used to synthesize a protein. "Polypeptide" would be a more accurate term, but the word "protein" is often used in this context for simplicity. The distinction between these two words is subtle: A polypeptide is simply a chain of amino acids, whereas a protein is a functional molecule made up of one or more polypeptides.

Transcription

In **transcription,** the enzyme **RNA polymerase** synthesizes single-stranded RNA using DNA as a template. Specific nucleotide sequences in the DNA direct the polymerase where to start and where to stop (**figure 7.8**). A DNA sequence to which RNA polymerase can bind and initiate transcription is called a **promoter;** a sequence that stops the process is a **terminator.** Like DNA polymerases, RNA polymerase can add nucleotides only to the 3′ end of a chain, and therefore it synthesizes RNA in the 5′ to 3′ direction; unlike DNA polymerases, however, RNA polymerase can start synthesis without a primer. The direction of polymerase movement along the DNA can be likened to the flow of a river. Because of this, the words "upstream" and "downstream" are used to describe relative positions of other sequences. For example, a promoter is upstream of the gene it controls, and the terminator is downstream of the promotor.

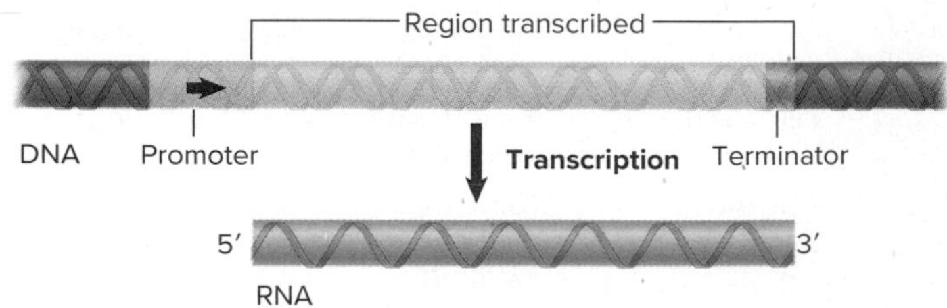

FIGURE 7.8 Nucleotide Sequences in DNA Direct Transcription The promoter is a DNA sequence to which RNA polymerase can bind to initiate transcription. The terminator is a sequence at which transcription stops.

In which direction is RNA synthesized: 5′ to 3′ or 3′ to 5′?

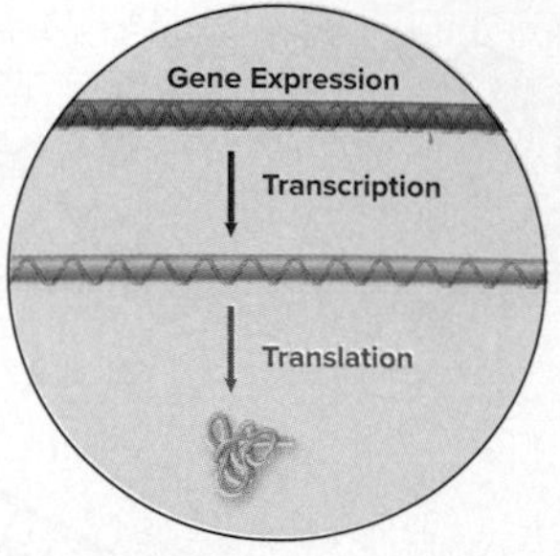

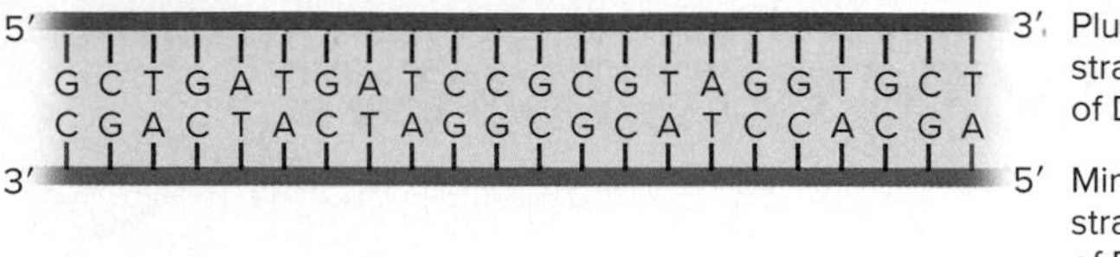

FIGURE 7.9 RNA Is Complementary and Antiparallel to the DNA Template The DNA strand that serves as a template for RNA synthesis is called the (–) strand of DNA; the complement to that is the (+) strand.

How does the nucleotide sequence of the (+) DNA strand differ from that of the RNA transcript?

The RNA made during transcription is complementary and antiparallel to the DNA strand that served as the template; that template DNA strand is called the **minus (–) strand,** and its complement is called the coding strand or the **plus (+) strand** (**figure 7.9**). Because the RNA is complementary to the (–) DNA strand, its nucleotide sequence is the same as the (+) DNA strand, except that it contains uracil in place of thymine.

In prokaryotes, an mRNA molecule can carry the information for one or multiple genes. A transcript that carries one gene is called **monocistronic** (a cistron is synonymous with a gene). One that carries multiple genes is called **polycistronic.** The proteins encoded on a polycistronic message generally have related functions, allowing a cell to express related genes as one unit.

Initiation of RNA Synthesis

Transcription is initiated when RNA polymerase binds to a promoter (**figure 7.10**). The binding denatures (melts) a short stretch of DNA, creating a region of exposed nucleotides that serves as a template for RNA synthesis. The part of RNA polymerase that recognizes the promoter is a loosely attached subunit called **sigma (σ) factor,** which usually separates from the polymerase after initiation is complete. A cell can produce

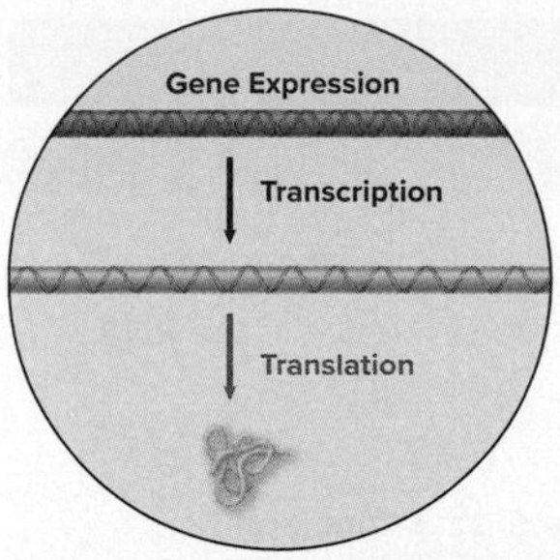

FIGURE 7.10 The Process of RNA Synthesis Bacterial RNA polymerases have a detachable subunit called sigma factor that allows them to recognize a promoter (as illustrated); the RNA polymerases of eukaryotic cells and archaea use transcription factors to recognize promoters.

? **Which component of the bacterial RNA polymerase recognizes the promoter?**

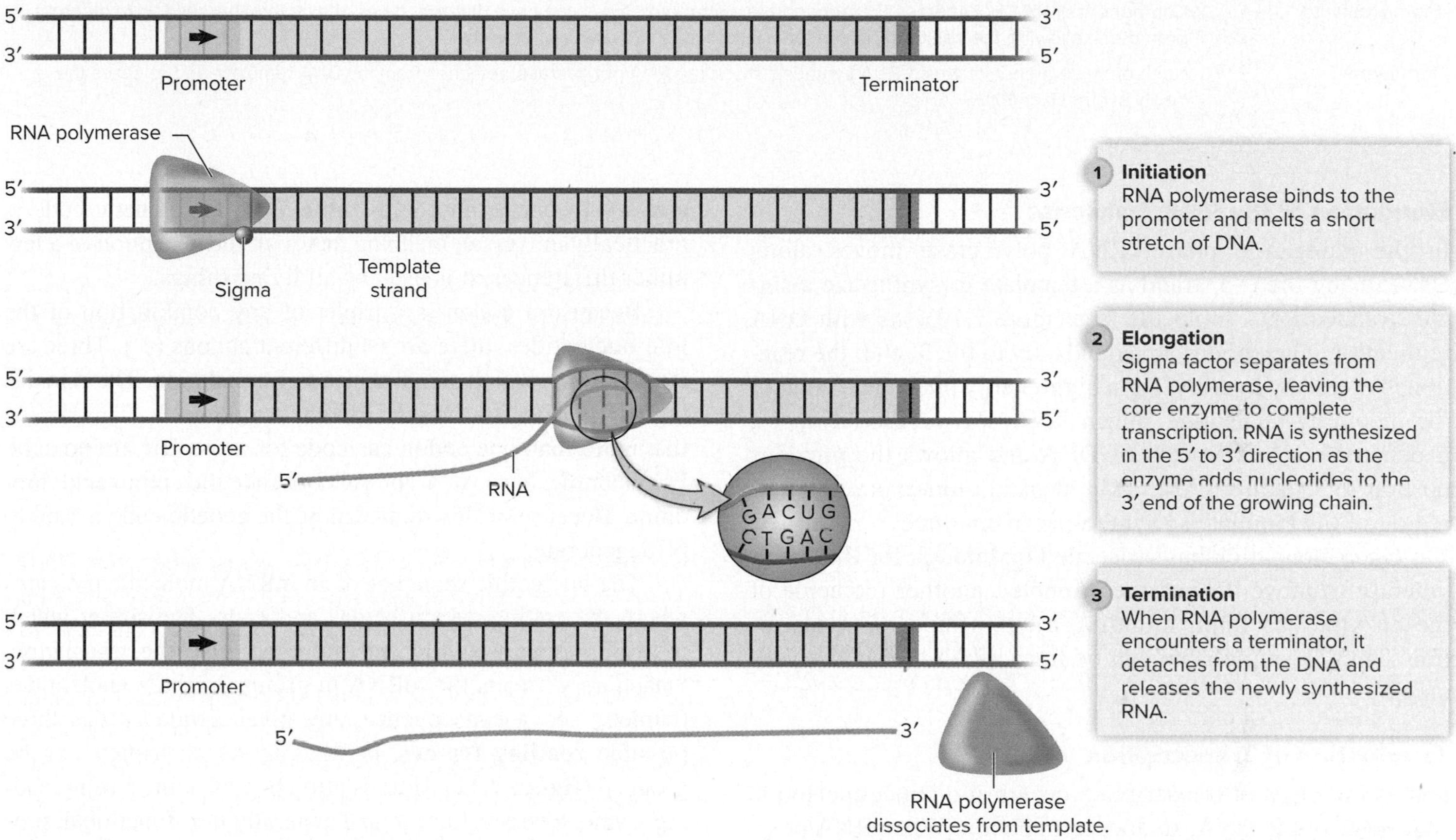

1 Initiation
RNA polymerase binds to the promoter and melts a short stretch of DNA.

2 Elongation
Sigma factor separates from RNA polymerase, leaving the core enzyme to complete transcription. RNA is synthesized in the 5′ to 3′ direction as the enzyme adds nucleotides to the 3′ end of the growing chain.

3 Termination
When RNA polymerase encounters a terminator, it detaches from the DNA and releases the newly synthesized RNA.

RNA polymerase dissociates from template.

various types of σ factors, each recognizing different promoters. By controlling which σ factors are made, cells can transcribe specialized sets of genes as needed. The RNA polymerases of eukaryotic cells and archaea use proteins called transcription factors to recognize promoters.

In identifying a region at which transcription begins, a promoter positions the RNA polymerase in one of the two possible orientations on the DNA molecule. In turn, this dictates the direction of transcription and thereby determines which strand will be used as a template (**figure 7.11**).

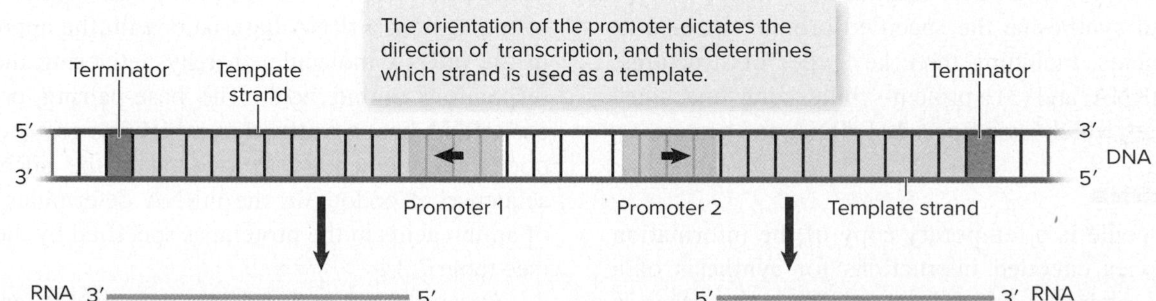

FIGURE 7.11 The Promoter Positions RNA Polymerase The direction of the promoter dictates which DNA strand the RNA polymerase will use as the template. In this diagram, the light blue RNA was transcribed from the red DNA strand (and is therefore analogous in sequence to the blue DNA strand), whereas the pink RNA was transcribed from the blue DNA strand (and is therefore analogous in sequence to the red DNA strand).

? **The light blue RNA strand is complementary to which DNA strand in this figure? To which DNA strand is the pink RNA strand complementary?**

TABLE 7.2 Components of Transcription in Bacteria

Component	Comment
(–) strand of DNA	Strand of DNA that serves as the template for RNA synthesis; the resulting RNA molecule is complementary to this strand.
(+) strand of DNA	Strand of DNA complementary to the one that serves as the template for RNA synthesis; the nucleotide sequence of the RNA molecule is the same as this strand, except it has uracil rather than thymine.
Promoter	Nucleotide sequence to which RNA polymerase binds to initiate transcription.
RNA polymerase	Enzyme that synthesizes RNA using one strand of DNA as a template; synthesis always occurs in the 5′ to 3′ direction.
Sigma (σ) factor	Component of RNA polymerase that recognizes a promoter. A cell can have different types of σ factors that recognize different promoters, allowing the cell to transcribe specialized sets of genes as needed.
Terminator	Nucleotide sequence at which RNA synthesis stops; the RNA polymerase detaches from the DNA template and releases the newly synthesized RNA.

Elongation of the RNA Transcript

In the elongation phase, RNA polymerase moves along DNA, using the (–) strand as a template to synthesize a single-stranded RNA molecule (see figure 7.10). As with DNA replication, nucleotides are added only to the 3′ end; the reaction is fueled by hydrolyzing a high-energy phosphate bond of the incoming nucleotide. When RNA polymerase advances, it denatures a new stretch of DNA and allows the previous portion to renature (close). The denaturation exposes a new region of the template so synthesis can continue.

Once elongation has proceeded far enough for RNA polymerase to move beyond the promoter, another molecule of the enzyme can bind, initiating a new round of transcription. Thus, a single gene can be transcribed repeatedly very quickly.

Termination of Transcription

Just as initiation of transcription occurs at distinct nucleotide sequences of the DNA, so does termination. When RNA polymerase encounters a sequence called a terminator, it detaches from the DNA and releases the newly synthesized RNA.

Table 7.2 summarizes the components of translation.

Translation

Translation is the process of decoding the information carried in the mRNA to synthesize the specified protein. It requires various components, including three key types of structures: (1) mRNA, (2) tRNA, and (3) a protein-synthesizing "machine" called a ribosome, which is composed of rRNA and protein.

The Role of mRNA

An mRNA molecule is a temporary copy of the information in DNA; it carries encoded instructions for synthesis of a specific protein, or in the case of a polycistronic message, a specific group of proteins. Recall that mRNA is composed of nucleotides, whereas proteins are composed of amino acids. Cells decode the information in mRNA using the **genetic code,** which correlates a series of three nucleotides—a **codon**—to one amino acid (**table 7.3**). The genetic code is practically universal, meaning that with the exception of a few minor differences, it is used by all living things.

Because a codon is a triplet of any combination of the four nucleotides, there are 64 different codons (4^3). Three are **stop codons,** which signal the end of translation. The remaining 61 translate to the 20 different amino acids. This means that more than one codon can code for a specific amino acid. For example, both ACA and ACG encode the amino acid threonine. Because of this redundancy, the genetic code is said to be degenerate.

The nucleotide sequence of an mRNA molecule indicates where the coding region begins and ends. The site at which it begins is particularly important because the translation, "machinery" reads the mRNA in groups of three nucleotides (triplets). As a consequence, any given sequence has three possible **reading frames,** or ways in which triplets can be grouped (**figure 7.12**). If translation begins in the wrong reading frame, a very different, and generally non-functional, protein is synthesized.

The Role of Transfer RNAs

A tRNA is an RNA molecule that carries an amino acid to be used in translation. Each tRNA is folded into a three-dimensional shape held together by hydrogen bonds; a specific amino acid is attached at one end (**figure 7.13**). As part of translation, a tRNA base-pairs with the appropriate codon in the mRNA molecule, thereby delivering the correct corresponding amino acid. The base-pairing occurs because each tRNA has an **anticodon,** which is a group of three nucleotides complementary to a codon in the mRNA. Thus, the sequence of codons in the mRNA determines the sequence of amino acids in the protein, as specified by the genetic code (see table 7.3).

Once a tRNA molecule has delivered its amino acid during translation, it can be recycled. An enzyme in the cytoplasm recognizes the tRNA and then attaches the appropriate amino acid; the process is called charging, and a tRNA not carrying an amino acid is referred to as *uncharged.*

TABLE 7.3 The Genetic Code

First Letter	Second Letter: U		C		A		G		Third Letter
U	UUU	Phe Phenylalanine	UCU	Ser Serine	UAU	Tyr Tyrosine	UGU	Cys Cysteine	U
	UUC		UCC		UAC		UGC		C
	UUA	Leu Leucine	UCA		UAA	"Stop"	UGA	"Stop"	A
	UUG		UCG		UAG	"Stop"	UGG	Trp Trytophan	G
C	CUU	Leu Leucine	CCU	Pro Proline	CAU	His Histidine	CGU	Arg Arginine	U
	CUC		CCC		CAC		CGC		C
	CUA		CCA		CAA	Gln Glutamine	CGA		A
	CUG		CCG		CAG		CGG		G
A	AUU	Ile Isoleucine	ACU	Thr Threonine	AAU	Asn Asparagine	AGU	Ser Serine	U
	AUC		ACC		AAC		AGC		C
	AUA		ACA		AAA	Lys Lysine	AGA	Arg Arginine	A
	AUG	Met Methionine; "Start"	ACG		AAG		AGG		G
G	GUU	Val Valine	GCU	Ala Alanine	GAU	Asp Aspartate	GGU	Gly Glycine	U
	GUC		GCC		GAC		GGC		C
	GUA		GCA		GAA	Glu Glutamate	GGA		A
	GUG		GCG		GAG		GGG		G

Note: The genetic code correlates a codon (a series of three nucleotides) to an amino acid, as shown in the table. In some cases, the codon encodes a "stop" signal instead of an amino acid. Many amino acids are specified by more than one codon. For example, threonine is specified by four codons, which differ only in the third nucleotide (ACU, ACC, ACA, and ACG).

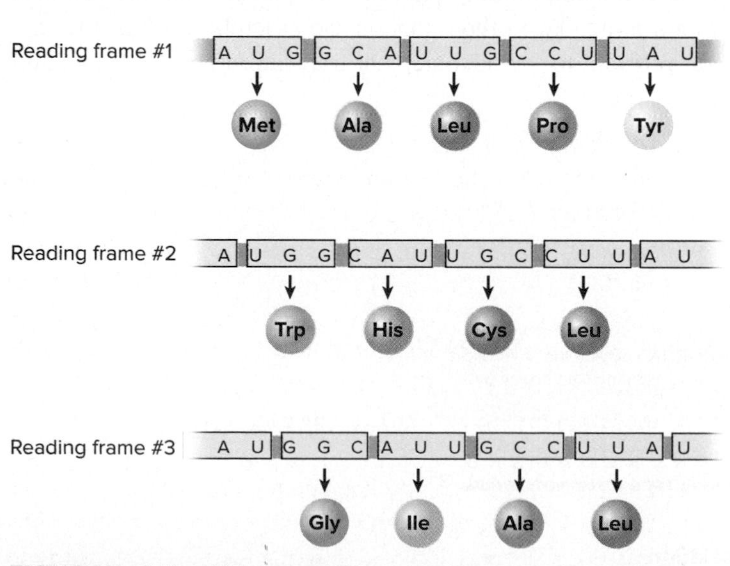

FIGURE 7.12 Reading Frames A nucleotide sequence has three potential reading frames, but only one is typically used for translation.

? Why is it important that the correct reading frame is used?

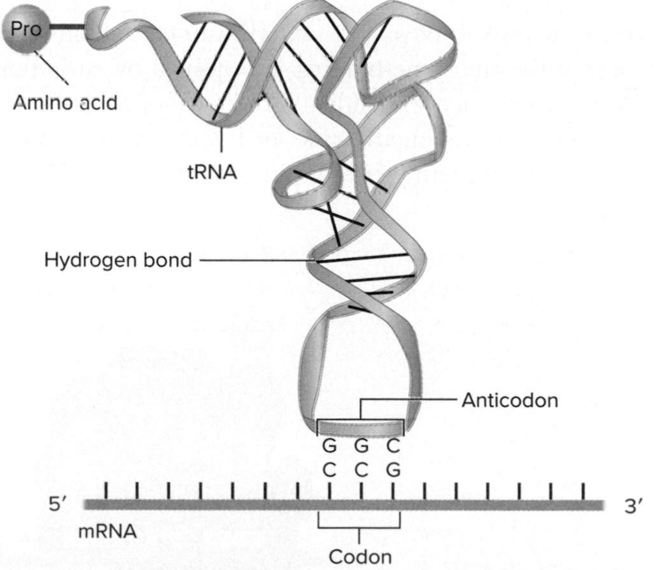

FIGURE 7.13 The Structure of Transfer RNA (tRNA) Three-dimensional illustration of tRNA. The amino acid that the tRNA carries is dictated by its anticodon. The tRNA that recognizes the mRNA codon CCG carries the amino acid proline—as specified by the genetic code.

? A tRNA that has the anticodon GAG carries which amino acid?

The Role of Ribosomes

A **ribosome** serves as a decoding site for mRNA, functioning as a protein-synthesizing machine where tRNAs interact with codons. As part of the process, the ribosome aligns two amino acids (carried by specific tRNAs) and then catalyzes the formation of a peptide bond between them.

Ribosomes also locate key sequences in the mRNA molecule as the points at which synthesis of a protein should start and stop (**figure 7.14**). In prokaryotes, the complex process of ribosome assembly begins at a sequence in mRNA called the **ribosome-binding site,** and translation begins just downstream from that at a **start codon** (usually the first AUG after the ribosome-binding site). The ribosome then moves along the mRNA in the 5′ to 3′ direction, "presenting" each codon in sequential order for decoding by an appropriate tRNA while maintaining the correct reading frame. Translation ends at a stop codon.

Prokaryotic ribosomes are each composed of a 30S subunit and a 50S subunit, both made up of protein and ribosomal RNA (rRNA) (see figures 3.19 and 10.10); the "S" stands for Svedberg unit—a measure of size. Svedberg units are not additive, which is why the 70S ribosome can have 30S and 50S subunits.

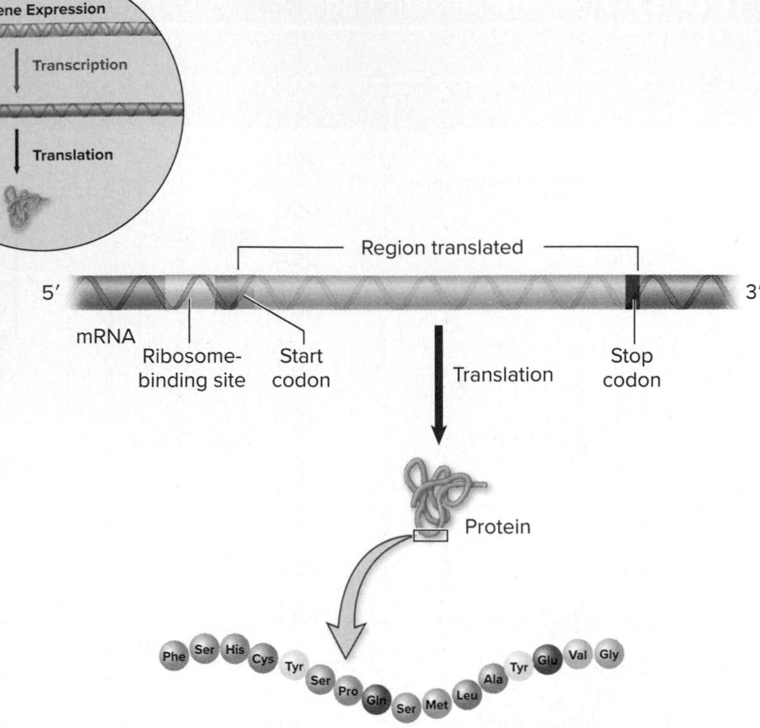

FIGURE 7.14 Nucleotide Sequences in mRNA Directs Translation The complete ribosome forms at the ribosome-binding site and starts translating at the start codon. Translation ends at a stop codon.

? In which direction does the ribosome move along RNA?

MicroByte

Several types of antibiotics, including tetracycline and azithromycin, interfere with the function of the bacterial 70S ribosome.

The Process of Translation

As with transcription, translation can be separated into three phases: initiation, elongation, and termination. The process as a whole requires significant energy expenditure.

Initiation In prokaryotes, the two ribosomal subunits come together at the ribosome-binding site, joined by an initiating tRNA that carries a chemically altered form of the amino acid methionine (*N*-formylmethionine, or f-Met). As the ribosome assembles, the initiating tRNA binds to the first AUG after the ribosome-binding site and occupies a ribosomal region called the P-site (**figure 7.15**). Thus, the initiating tRNA helps position the ribosome relative to the first AUG, which is important because the position determines the reading frame used for translating the remainder of that polypeptide. Note that AUG functions as a start codon only when preceded by a ribosome-binding site; at other sites, it simply encodes methionine. Two other sites on the assembled ribosome that are relevant to the next steps of translation are the A-site and the E-site.

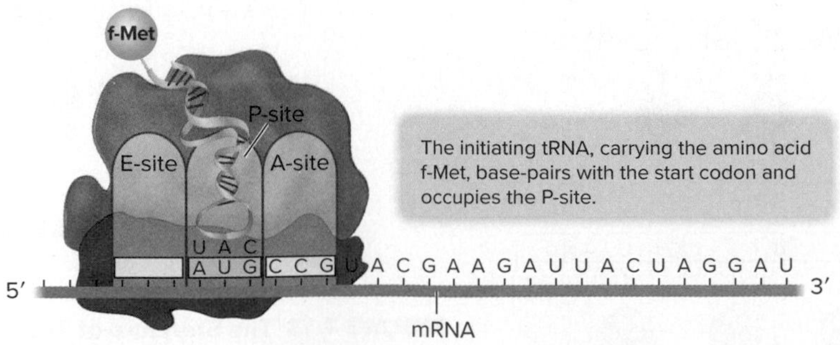

FIGURE 7.15 The Initiation Phase of Translation

? What nucleotide is encoded after the start codon in the illustrated mRNA? (Refer to table 7.3)

Elongation The elongation phase creates the full-length polypeptide and can be summarized as follows (**figure 7.16**):

1. At the start of elongation, the initiating tRNA (carrying the f-Met) still occupies the P-site. A tRNA that recognizes the next codon on the mRNA then enters the unoccupied A-site.
2. The ribosome catalyzes the joining of the amino acids carried by the tRNAs and, in doing so, transfers the amino acid from the initiating tRNA in the P-site to the one carried by the tRNA in the A-site.
3. The ribosome advances a distance of one codon, a process called translocation, moving along the mRNA in the 5′ to 3′ direction. As this happens, the uncharged initiating tRNA is released through the E-site to be recycled. The remaining tRNA, which now carries both amino acids, occupies the P-site. A tRNA that base-pairs with the codon in the A-site then quickly attaches there. Next, the amino acid chain carried by the tRNA now in the P-site will be transferred to the amino acid carried by the tRNA that just entered the A-site.
4. The elongation process continues to repeat as the ribosome moves along the mRNA in the 5′ to 3′ direction one codon at a time. After each translocation, the uncharged tRNA is released through the E-site, a tRNA enters the A-site, and the polypeptide chain carried by the tRNA in the P-site is transferred to the amino acid carried by the tRNA that just entered the A-site. These combined actions elongate the polypeptide chain. Once the ribosome has progressed far enough to clear the initiating sequences, another ribosome can bind. Thus, at any one time, multiple ribosomes can be translating a single mRNA molecule, maximizing the use of that template. The assembly of multiple ribosomes attached to a single mRNA molecule is called a polyribosome, or a polysome.

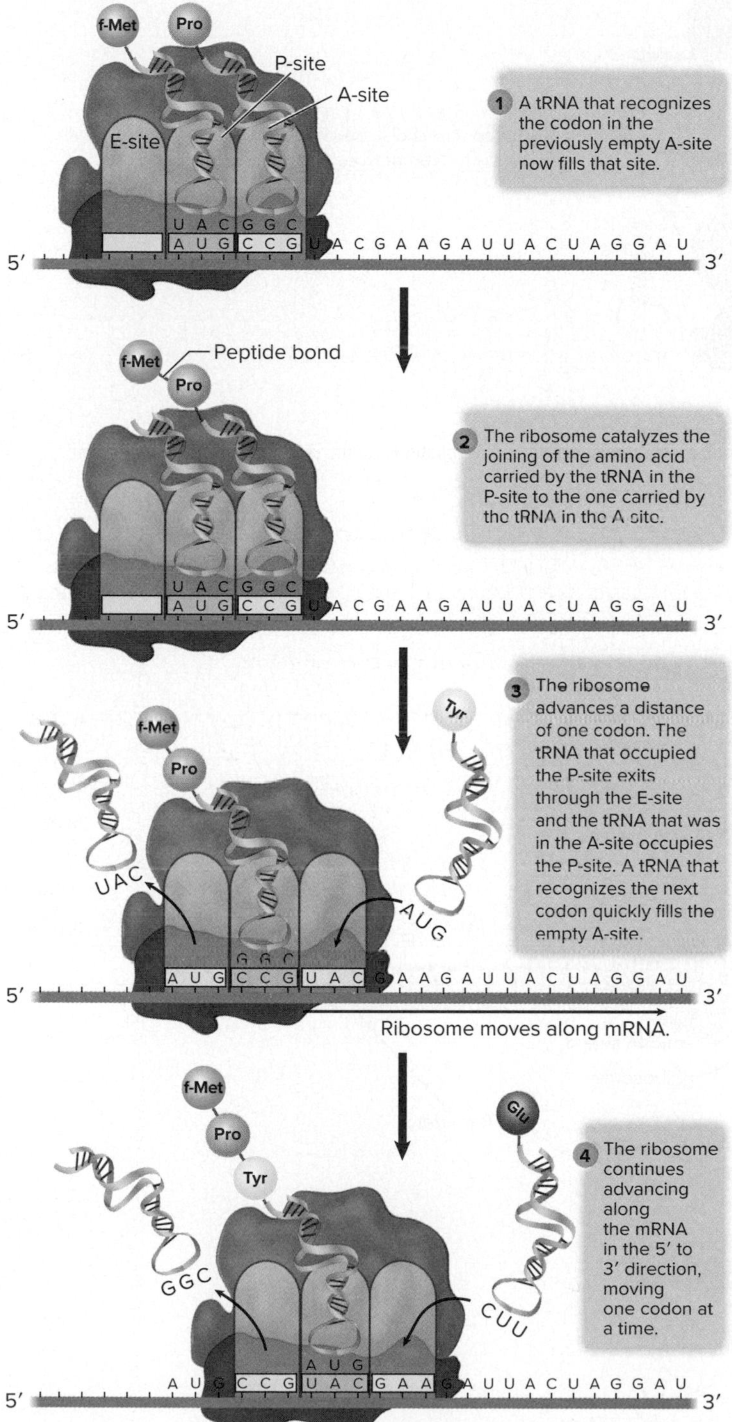

f-Met
Pro
P-site
A-site
E-site
1 A tRNA that recognizes the codon in the previously empty A-site now fills that site.
U A C G G C
A U G C C G U A C G A A G A U U A C U A G G A U
5′
3′
Peptide bond
2 The ribosome catalyzes the joining of the amino acid carried by the tRNA in the P-site to the one carried by the tRNA in the A-site.
Tyr
3 The ribosome advances a distance of one codon. The tRNA that occupied the P-site exits through the E-site and the tRNA that was in the A-site occupies the P-site. A tRNA that recognizes the next codon quickly fills the empty A-site.
UAC
AUG
Ribosome moves along mRNA.
Glu
4 The ribosome continues advancing along the mRNA in the 5′ to 3′ direction, moving one codon at a time.
GGC
CUU

FIGURE 7.16 The Elongation Phase of Translation

? What is the function of the anticodon on the tRNA?

Termination The process of translation terminates when the ribosome reaches a stop codon, a codon not recognized by a tRNA (**figure 7.17**). At this point, assisting enzymes free the polypeptide by breaking the covalent bond that joins it to the tRNA. The ribosome falls off the mRNA, dissociating into its two component subunits (30S and 50S). The subunits can then be reused to initiate translation at other sites.

Simultaneous Transcription and Translation

In prokaryotes, gene expression is particularly efficient because translation begins while the mRNA molecule is still being synthesized (**figure 7.18**). Also, as mentioned previously, multiple ribosomes can be translating the same mRNA molecule. (The latter trait is shared with eukaryotic cells.)

The components of translation are summarized in **table 7.4.**

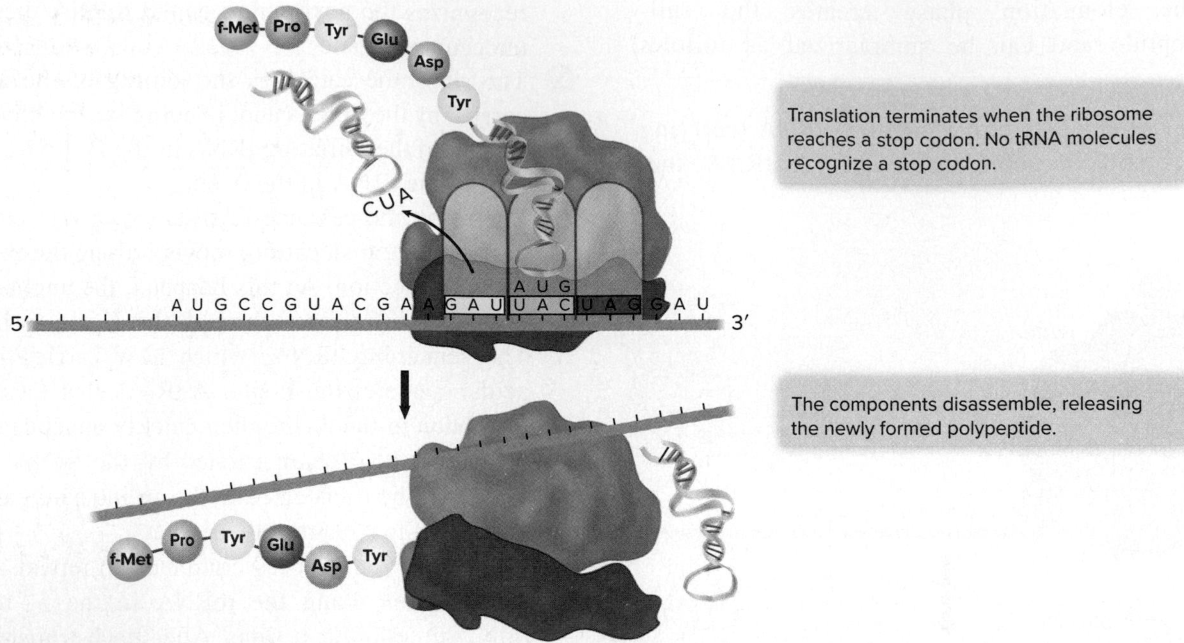

FIGURE 7.17 Termination of Translation

If you wanted to modify this illustration but still retain the functional components, what nucleotide sequences could you substitute for the stop codon shown here? (Refer to table 7.3)

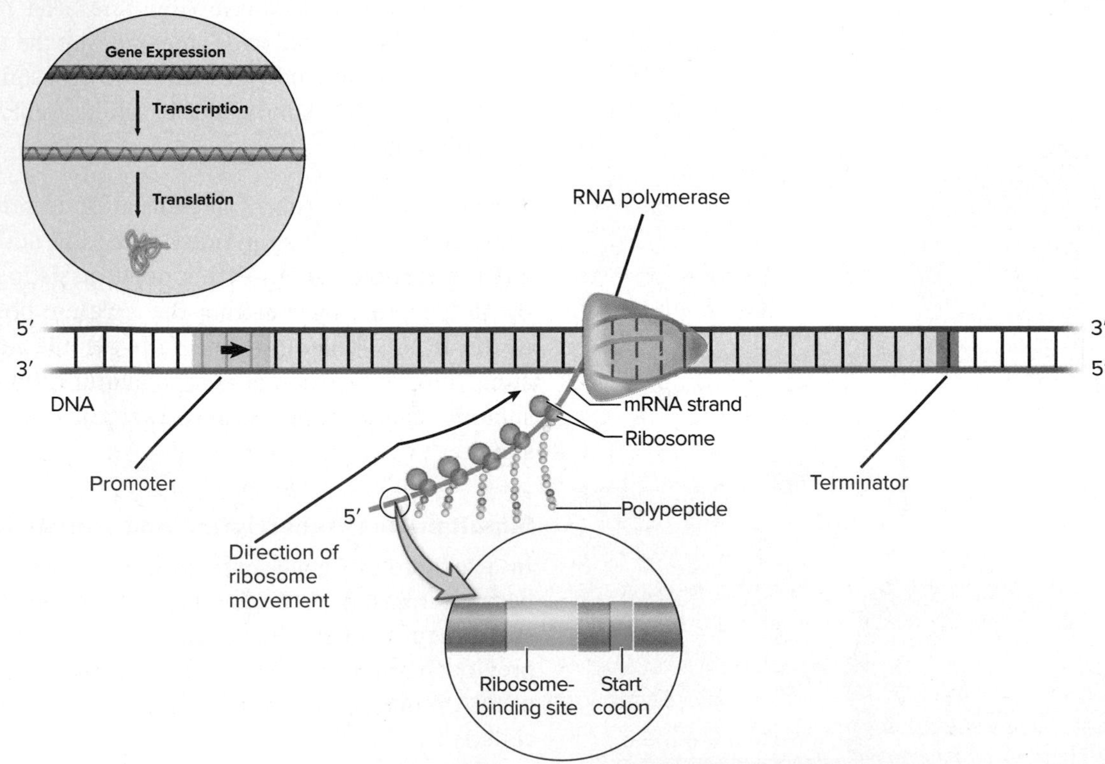

FIGURE 7.18 In Prokaryotes, Translation Begins as the mRNA Molecule Is Still Being Synthesized Ribosomes begin translating the mRNA molecule before transcription is complete. More than one ribosome can be translating the same mRNA molecule.

Why is the position of the first AUG after the ribosome-binding site important?

TABLE 7.4 Components of Translation in Bacteria

Component	Comment
Anticodon	Sequence of three nucleotides in a tRNA molecule that is complementary to a particular codon in mRNA. The anticodon allows the tRNA to recognize and bind to the appropriate codon.
mRNA	Type of RNA molecule that contains the genetic information decoded during translation.
Reading frame	Stretch of nucleotides grouped into sequential triplets that code for amino acids; an mRNA molecule has three potential reading frames, but only one is typically used in translation.
Ribosome	Structure that facilitates the joining of amino acids during the process of translation; composed of protein and ribosomal RNA. The prokaryotic ribosome (70S) consists of a 30S and a 50S subunit.
Ribosome-binding site	Sequence of nucleotides in mRNA to which a ribosome binds; the first time the codon for methionine (AUG) appears after that site, translation generally begins.
rRNA	Type of RNA molecule present in ribosomes.
Start codon	Codon at which translation is initiated; typically the first AUG after a ribosome-binding site.
Stop codon	Codon that terminates translation, signaling the end of the protein; there are three stop codons.
tRNA	Type of RNA molecule involved in interpreting the genetic code during translation; each tRNA molecule carries a specific amino acid dictated by its anticodon.

MicroAssessment 7.3

Gene expression involves transcription and translation. In transcription, RNA polymerase synthesizes RNA in the 5′ to 3′ direction, using one strand of DNA as a template. In translation, ribosomes synthesize proteins, using the nucleotide sequence of an mRNA molecule to determine the amino acid sequence of the encoded protein. The correct amino acids for given codons are carried by tRNAs.

7. How does a promoter dictate which DNA strand is used as the template?
8. What is the role of tRNA in translation, and how is tRNA different from rRNA?
9. Could two mRNA molecules have different nucleotide sequences and yet code for the same protein? Explain your answer.

7.4 ■ Differences Between Eukaryotic and Prokaryotic Gene Expression

Learning Outcome

6. Describe four differences between prokaryotic and eukaryotic gene expression.

Eukaryotes differ significantly from prokaryotes in several aspects of transcription and translation. Eukaryotic mRNA, for example, is synthesized in a precursor form, called **pre-mRNA.** The pre-mRNA must be processed (altered) both during and after transcription to form mature mRNA (**figure 7.19**). Shortly after transcription begins, a cap (a methylated guanine derivative) is added to the 5′ end of a pre-mRNA, a process called **capping.** This cap binds specific proteins that

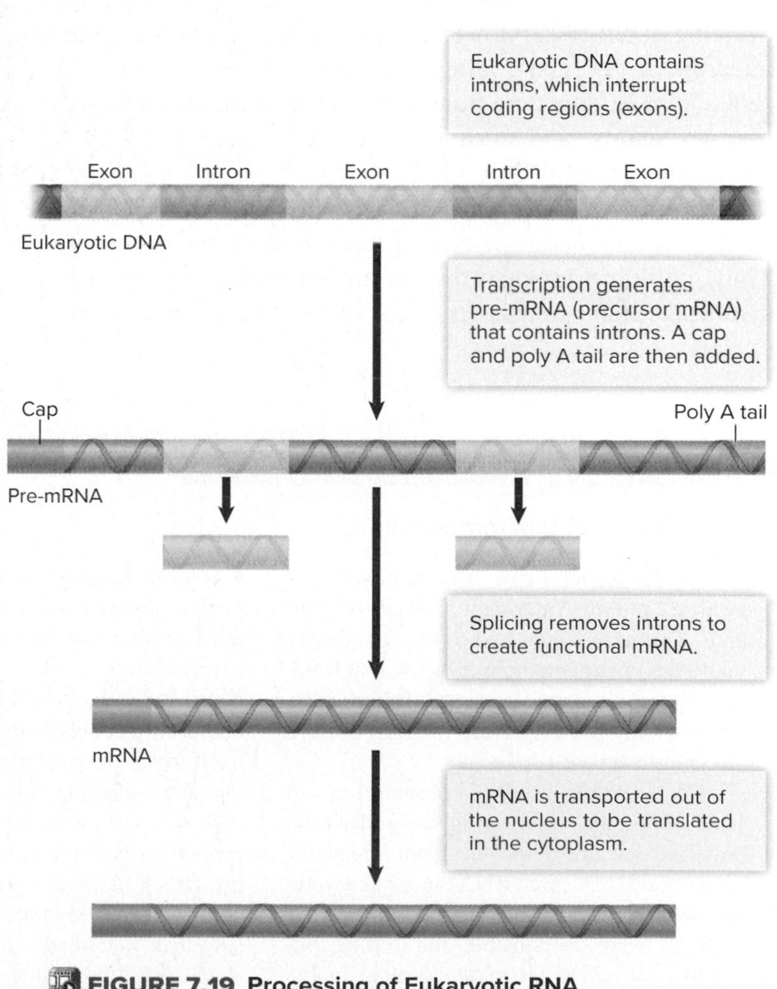

Eukaryotic DNA contains introns, which interrupt coding regions (exons).
Exon
Intron
Exon
Intron
Exon
Eukaryotic DNA
Transcription generates pre-mRNA (precursor mRNA) that contains introns. A cap and poly A tail are then added.
Cap
Poly A tail
Pre-mRNA
Splicing removes introns to create functional mRNA.
mRNA
mRNA is transported out of the nucleus to be translated in the cytoplasm.

FIGURE 7.19 Processing of Eukaryotic RNA

Introns are intervening sequences in genes; what are exons?

TABLE 7.5 Major Differences Between Prokaryotic and Eukaryotic Gene Expression

Prokaryotes	Eukaryotes
mRNA does not have a cap or a poly A tail.	Processing of the transcript (pre-mRNA) results in mRNA with a cap at the 5′ end and a poly A tail at the 3′ end.
Transcript (mRNA) does not contain introns.	Transcript (pre-mRNA) contains introns, which are removed by splicing.
Translation of mRNA begins as it is being transcribed.	mRNA is transported out of the nucleus so that it can be translated in the cytoplasm.
mRNA is often polycistronic; translation usually begins at the first AUG codon that follows a ribosome-binding site.	mRNA is monocistronic; translation begins at the first AUG.
Ribosomes are 70S.	Ribosomes are 80S.

stabilize the transcript and enhance translation. The 3′ end of the molecule is also modified, even before transcription has been terminated. This process, **polyadenylation,** cleaves the transcript at a specific sequence and then adds about 200 adenine derivatives to the new 3′ end. This creates a poly A tail, which is thought to stabilize the transcript as well as enhance translation. Another important modification is **splicing,** which removes specific segments of the transcript. Splicing is necessary because eukaryotic genes are often interrupted by non-coding sequences. These intervening sequences, **introns,** are transcribed along with the expressed regions, **exons,** and must be removed from pre-mRNA to create functional mRNA.

The mRNA in eukaryotic cells must be transported out of the nucleus before it can be translated in the cytoplasm. Thus, unlike in prokaryotes, an mRNA molecule cannot be synthesized and translated at the same time or even in the same cellular location. The mRNA of eukaryotes is generally monocistronic, and translation of the message typically begins at the first AUG in the molecule.

The ribosomes of eukaryotes differ from those of prokaryotes. Whereas the prokaryotic ribosome is 70S, made up of 30S and 50S subunits, the eukaryotic ribosome is 80S, made up of 40S and 60S subunits. The differences in ribosome structure are medically important because certain antibiotics bind to and inactivate bacterial 70S ribosomes, but not 80S ribosomes. This explains why those antibiotics kill or inhibit bacteria, usually without causing significant harm to mammalian cells.

The major differences between prokaryotic and eukaryotic gene expression are summarized in **table 7.5.**

MicroByte

Some viruses "steal" the 5′ cap from their host cell's mRNA, using it to make functional viral mRNA.

FOCUS YOUR PERSPECTIVE 7.1

RNA: The First Macromolecule?

The 1989 Nobel Prize in Chemistry was awarded to two Americans, Sidncy Altman and Thomas Cech, who independently made the unexpected observation that RNA molecules can act as enzymes. Before their studies, only proteins were thought to have enzymatic activity.

Cech made a key observation in 1982 when he was trying to understand how introns are removed from precursor ribosomal RNA (pre-rRNA) in a eukaryotic protozoan. Because he was convinced that proteins were responsible for cutting out introns, he added all of the protein in the cells' nuclei to a suspension of pre-rRNA. As expected, the introns were cut out. For a control, he used pre-rRNA to which no nuclear proteins were added, fully expecting that nothing would happen. Much to his surprise, the introns were removed in the control as well. Based on these results, Cech could only conclude that the RNA acted on itself to cut out the introns.

The studies of Altman and his colleagues showed that RNA has catalytic properties beyond cutting out introns from pre-rRNA. Altman's group found that RNA could convert a precursor transfer RNA (pre-tRNA) molecule to its final functional state. Additional studies have shown that enzymatic reactions in which catalytic RNAs, or ribozymes, play a role are very widespread. Ribozymes have been found in the mitochondria of eukaryotic cells and shown to catalyze other reactions that resemble the polymerization of RNA.

These observations have profound implications for a long-standing question in evolutionary biology: Which came first, proteins or nucleic acids? The answer seems to be nucleic acids, specifically RNA, which acted both as a carrier of genetic information and as an enzyme. Billions of years ago, before the present universe in which DNA, RNA, and protein are found, the only macromolecule was probably RNA. Once tRNAs became available, they carried amino acids to nucleotide sequences on a strand of RNA.

MicroAssessment 7.4

Eukaryotic pre-mRNA must be processed, which involves capping, polyadenylation, and splicing. In eukaryotic cells, the mRNA must be transported out of the nucleus before it can be translated in the cytoplasm. Eukaryotic mRNA is monocistronic.

10. What is an intron?

11. Would a deletion of two base pairs have a greater consequence if it occurred in an intron rather than in an exon?

7.5 ■ Sensing and Responding to Environmental Fluctuations

Learning Outcomes

7. Describe how quorum sensing and two-component regulatory systems allow cells to adapt to fluctuating environmental conditions.
8. Compare and contrast antigenic variation and phase variation.

Microorganisms are constantly faced with rapidly changing environmental conditions, and they must quickly adapt to the fluctuations if they are to survive. Consider the situation of bacterial communities in the intestinal tract of mammals. In this habitat, microbes must cope with alternating periods of feast and famine. For a limited time after a mammal eats, the bacterial cells thrive, bathed in the mixture of amino acids, vitamins, and other nutrients. Some bacteria actively take up these compounds they would otherwise need to synthesize. Simultaneously, the cells shut down their biosynthetic pathways, channeling the conserved energy into the rapid production of cell components such as DNA and protein. Famine, however, follows the feast. Between meals—which can be many days in the case of some mammals—the rich source of nutrients is depleted. Now the bacterial cells' biosynthetic pathways must be activated, using energy and slowing cell growth. Cells dividing several times an hour in a nutrient-rich environment might divide only once every 24 hours in a starved mammalian gut. When the animal defecates, some of the bacterial cells are eliminated in the feces. Outside the mammalian host, they must cope with yet a completely different set of conditions to survive.

Signal Transduction

Signal transduction is the process that transmits information from outside a cell to the inside. It allows cells to monitor and react to environmental conditions. Examples of mechanisms that bacterial cells use to sense and respond to environmental stimuli include quorum sensing and two-component systems.

Quorum Sensing

Some organisms can "sense" their own population density, a phenomenon called **quorum sensing,** which allows them to coordinate the activation of genes required for a particular activity. As an example, the cooperative activities leading to biofilm formation are controlled by quorum sensing. Some pathogens use the mechanism to coordinate expression of genes involved with the infection process.

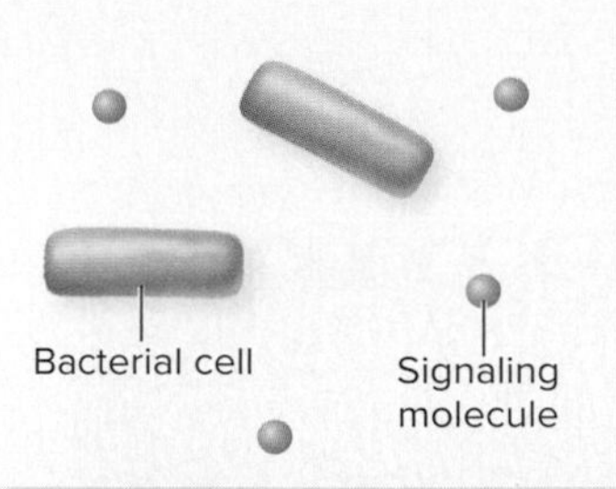

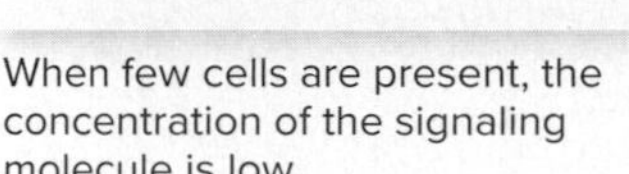

When few cells are present, the concentration of the signaling molecule is low.

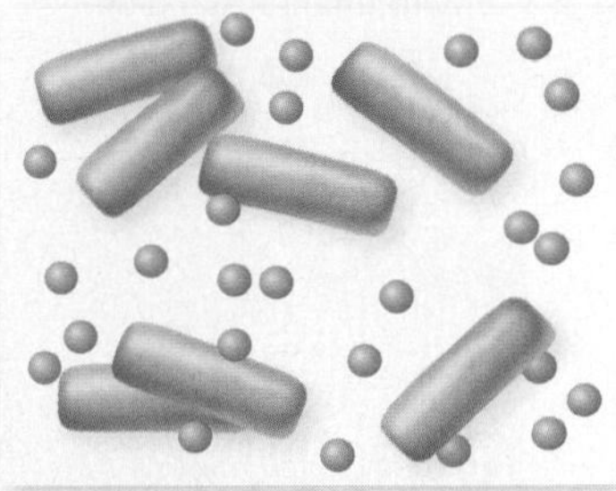

When many cells are present, the signaling molecule reaches a concentration high enough to induce the expression of certain genes.

FIGURE 7.20 Quorum Sensing

Why would it be beneficial for cells to wait until a critical population density is present before expressing certain genes?

Quorum sensing involves a process that allows bacteria to "talk" to each other by synthesizing one or more varieties of extracellular signaling molecules. When few cells are present, the concentration of a given signaling molecule is very low. As the cells multiply in a confined area, however, the concentration of that molecule increases proportionally. Only when a signaling molecule reaches a certain level does it change the expression of specific genes (**figure 7.20**). Some types of bacteria can detect and even interfere with the signaling molecules produced by other species. This allows them to "eavesdrop" on and even obstruct "conversations" of other bacteria.

Two-Component Regulatory Systems

A **two-component regulatory system** (**figure 7.21**) consists of two different proteins: a membrane-spanning sensor and a response regulator. When specific environmental variations occur, the sensor chemically modifies a region on its internal portion, usually by phosphorylating (adding a phosphate group to) a specific amino acid. The phosphate group is then transferred to a response regulator. When phosphorylated, the response regulator turns certain genes on or off. *E. coli* controls the expression of genes for its alternative types of metabolism by using a two-component regulatory system. When nitrate is present in anaerobic conditions, the cells activate genes required to use that compound as the terminal electron acceptor. Some pathogens use the system to activate appropriate genes for helping them avoid host defenses once they have invaded a host cell. They recognize their location by sensing magnesium concentrations, which are generally lower in certain host cells than in the extracellular environment.

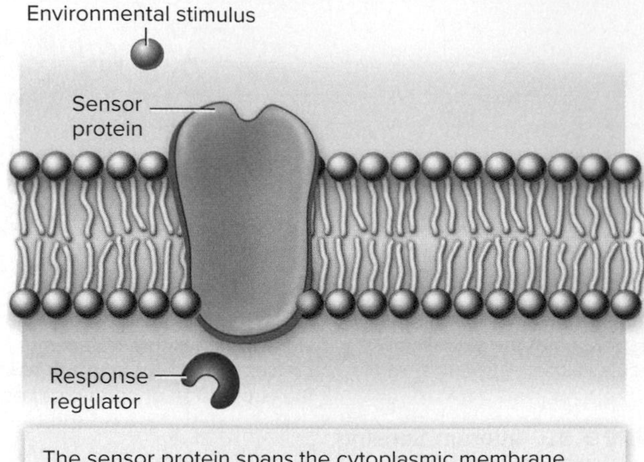

The sensor protein spans the cytoplasmic membrane. The response regulator is a protein inside the cell.

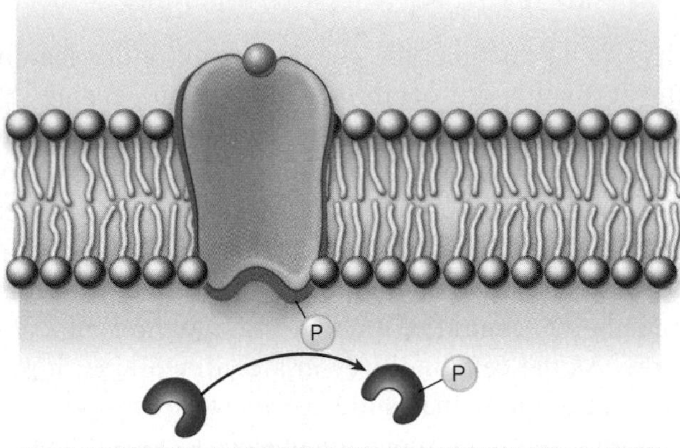

In response to a specific change in the environment, the sensor phosphorylates a region on its internal portion. The phosphate group is transferred to the response regulator, which can then turn genes on or off, depending on the system.

FIGURE 7.21 Two-Component Regulatory System

? *E. coli* cells use a two-component system to sense nitrate in the environment. What would happen if they lost the ability to sense that compound?

Natural Selection

Natural selection, the survival and growth of cells best adapted to live in a particular environment, can also play a role in gene expression. The expression of some genes changes randomly in cells, increasing the survival of at least a part of a population.

The role of natural selection is readily apparent in bacteria that undergo **antigenic variation,** an alteration in the characteristics of certain surface proteins. Pathogens that do this can stay one step ahead of the body's defenses by altering the very molecules our immune systems must learn to recognize. One of the most well-characterized examples is *Neisseria gonorrhoeae.* This bacterium has many different genes for pilin, the protein subunit that makes up pili, yet most of these genes are silent (not expressed). The only one expressed is in a particular chromosomal location called an expression locus. *N. gonorrhoeae* cells have a mechanism to randomly shuffle the pilin genes, moving different ones in and out of the expression locus. In a population of 10^4 cells, at least one is expressing a different type of pilin. During an infection, the body's immune system will begin responding to the dominant pilin type, but the bacterial cells that have already "switched" to produce a different pilin type will survive and then multiply. Eventually, the immune system learns to recognize those, but by that time, another subpopulation will have switched its pilin type.

Another mechanism of randomly altering gene expression is **phase variation,** the routine switching on and off of certain genes. In *E. coli,* certain types of pili required for attachment to mucosal epithelial cells undergo phase variation. Some of the bacterial cells adhering to an epithelial surface will spontaneously turn off the genes required for pili synthesis, causing those cells to detach from that surface. The process is reversible, so the detached cells will later turn the genes on again, allowing the cells to colonize epithelial cells elsewhere. By altering the expression of genes such as these, at least a part of the population is ready for change.

FOCUS ON A CASE 7.1

In October 2010, about 9 months after a serious earthquake devastated Haiti, an outbreak of severe diarrhea and vomiting was reported. Patients were producing several liters of liquid stool in a day, resulting in signs of dehydration, including sunken eyes, muscle cramps, and in a few cases, convulsions and death. The illness was quickly identified as cholera—a disease caused by the bacterium *Vibrio cholerae.* Finding that organism in Haiti was a surprise because cholera had not been reported there for at least a century. The outbreak was devastating—hundreds of thousands of people became ill, and thousands of patients died.

As with most cholera outbreaks, the disease spread in Haiti through contaminated drinking water. The bacteria survive in aqueous environments by producing extracellular polymeric substances (EPS) that allow them to form protective biofilms (see figure 4.3). Interestingly, although large numbers of live *V. cholerae* cells may be seen in biofilm samples from contaminated waters, relatively few of those cells grow in laboratory culture, suggesting that some enter a dormant stage.

After *V. cholerae*–contaminated water or food is ingested, the bacteria begin expressing different genes. They stop producing the EPS required for biofilm formation and instead make pili that they use to attach to the epithelial cells that line the small intestine. As *V. cholerae* cells multiply in the small intestine, they make cholera toxin (CT), a toxin that causes much more fluid to leave epithelial cells than can normally be reabsorbed. The

continued

toxin's effect results in the severe diarrhea that characterizes the disease.

1. Like many pathogens, *V. cholerae* uses quorum sensing as one of the regulatory mechanisms to control gene expression. Why would a bacterium use quorum sensing to control the expression of toxins or other molecules that allow the organism to cause disease?
2. Unlike pathogens that respond to high population density by turning on genes that allow them to cause disease, *V. cholerae* cells in a dense population turn off the genes associated with attachment and toxin production. As those genes are turned off, the cells turn on the genes required for biofilm production. How might this benefit the organism?
3. How could a better understanding of the signaling molecules involved in quorum sensing lead to new methods to control pathogens?
4. Researchers are investigating the use of quorum sensing signaling molecules of other bacteria to improve the isolation rate of *V. cholerae* from environmental samples. Why might these molecules induce *V. cholerae* cells to exit a dormant stage?

Discussion

1. Many bacterial pathogens do not produce their key virulence factors (the specific substances that allow the organism to cause disease) until the bacterial population densities are high enough to overwhelm the host's defense system. The bacterial cells produce signaling molecules, and once these reach a certain level, indicating that the population density is high, the cells produce and release the various virulence factors simultaneously. When large numbers of bacteria quickly release a toxin, for example, the host does not have time to mount an effective immune response, thereby allowing the pathogen to survive and multiply, using the affected host tissues as a source of nutrients.
2. *V. cholerae* uses quorum sensing in the opposite way from many other pathogens. In low population densities, the cells express genes for making attachment pili as well as cholera toxin (CT). This allows the bacteria to attach to the epithelial cells that line the small intestine, where they grow and produce CT with little competition (most members of the normal intestinal microbiota reside in the large intestine rather than the small intestine). Once the bacteria reach a high population density, they turn off the genes for toxin and pilus production. At the same time, the bacterial cells turn on genes for proteins that help them detach from the epithelial surface as well as genes for producing extracellular polymeric substances (EPS). The detachment increases the chance that the bacterial cells will be passed along the intestinal tract, eventually being shed in the feces, which can then contaminate water. Production of EPS allows the organisms to form biofilms, thereby increasing their chance of survival in the external environment.
3. By knowing more about the mechanisms bacteria use to signal one another, scientists might be able to use the signaling molecules to "fool" specific pathogens. For example, by forcing organisms to turn off essential virulence genes, scientists can control the pathogenicity and survival of those organisms.
4. *V. cholerae* might "eavesdrop" on the quorum-sensing communication among other bacteria as a means to assess environmental conditions. It appears that when neighboring cell populations are high, *V. cholerae* cells respond by exiting their dormant state. This is an area of study currently being explored, and there is much more to be learned!

MicroAssessment 7.5

Quorum sensing and two-component regulatory systems allow a microbial cell to respond to changing environmental conditions. The expression of some genes changes randomly, increasing the chances of survival of at least a subset of a cell population under varying environmental conditions.

12. Explain how certain bacteria "sense" the density of cells.

13. Describe antigenic variation.

14. In quorum sensing, why might a bacterium synthesize more than one type of signaling molecule?

7.6 ■ Bacterial Gene Regulation

Learning Outcomes

9. Explain how gene expression is regulated at the transcriptional level.

10. Using the *lac* operon as a model, explain the role of inducers, repressors, and carbon catabolite repression.

In bacterial cells, many genes are **constitutively** (routinely) expressed, but others are regulated in response to environmental conditions. Note that scientists describe these regulated genes as capable of being turned on or off, but in reality, there are no absolutes. In a population of cells, a gene that is off may still be expressed at very low levels.

A regulatory mechanism sometimes controls the transcription of only a limited number of genes, but in other cases, it controls many genes simultaneously. A set of regulated genes transcribed as a single mRNA molecule, along with the sequences that control its expression, is called an **operon.** One of the most well-characterized examples is the *lac* operon (see figure 7.25), which encodes proteins required for transporting and hydrolyzing the disaccharide lactose. Separate operons controlled by a single regulatory mechanism constitute a **regulon.** The simultaneous regulation of numerous genes is called **global control.**

When describing regulated genes (or operons), scientists group them according to their type of expression (**figure 7.22**):

- **Inducible.** An inducible gene is not routinely transcribed; instead, a regulatory molecule called an **inducer** can turn

Inducible genes
These are not routinely expressed, but mechanisms can turn on transcription for as long as needed; for example, when the substrate for an encoded enzyme is present.

Repressible genes
These are routinely expressed, but mechanisms can turn off transcription for as long as necessary; for example, when the product of an encoded enzyme is in sufficient quantity.

FIGURE 7.22 Principles of Regulation

? Why would a cell's genes that encode biosynthetic enzymes be repressible rather than constitutive or inducible?

the gene on when needed. The proteins encoded by these genes are often involved in the transport and breakdown of specific energy sources that are not always available. An example of such a protein is β-galactosidase, the enzyme that hydrolyzes lactose into its component monosaccharides—glucose and galactose. The gene that encodes this enzyme is part of the *lac* operon, which can only be turned on if lactose is present. This makes sense because a cell would waste precious resources if it expressed the operon's genes when lactose was not available.

- **Repressible.** Repressible genes are routinely transcribed but can be turned off when the gene product is not required. The proteins encoded by these genes are generally enzymes involved in biosynthetic (anabolic) pathways, such as pathways that synthesize amino acids. Cells need enough of a given amino acid to multiply, so it must be produced by the cell unless it is available in the environment. When the amino acid is available, however, synthesis of the enzymes used in its production would waste energy.

Mechanisms to Control Transcription

The methods a cell uses to prevent or facilitate transcription must be readily reversible, allowing cells to control the relative number of transcripts made. Two of the most common regulatory mechanisms are alternative sigma factors and DNA-binding proteins.

Alternative Sigma Factors

As described earlier, sigma factor is a detachable subunit of RNA polymerase that functions in recognizing specific promoters. Standard sigma factors recognize promoters for genes that need to be expressed during routine growth conditions, but a cell can also produce **alternative sigma factors.** These recognize different sets of promoters, thereby controlling the expression of specific groups of genes. In the endospore former *Bacillus subtilis,* the sporulation process is controlled by a number of different alternative sigma factors. One of these controls the steps at the beginning of sporulation, whereas others then guide the stages of development in the mother cell and spore. A cell can also express anti-sigma factors, which inhibit the function of specific sigma factors.

DNA-Binding Proteins

Transcription is often controlled by proteins that bind to specific DNA sequences. When a regulatory protein attaches to DNA, it can act as either a repressor (blocks transcription) or an activator (facilitates transcription).

Repressors A **repressor** is a regulatory protein that can block transcription (negative regulation). It does this by binding to an **operator,** a specific DNA sequence usually located immediately downstream of a promoter. When a repressor is bound to an operator, RNA polymerase cannot progress past that DNA sequence. Thus, the gene is "off." Repressors are allosteric proteins, however, meaning that specific molecules can attach to them and change their shape; the shape is important because it affects the repressor's ability to bind to the operator. As shown in **figure 7.23,** repressors can play a role in two types of transcriptional regulatory systems:

- **Induction.** The repressor is synthesized as a form that binds to the operator, blocking transcription. When an inducer attaches to the repressor, the shape of the repressor changes so that it can no longer attach to the operator. With the repressor unable to bind to DNA, RNA polymerase may transcribe the gene, and thus the gene would be "on."
- **Repression.** The repressor is synthesized as a form that cannot bind to the operator. However, when a molecule termed a **corepressor** attaches to the repressor, the corepressor-repressor complex can then bind to the operator, blocking transcription, and thus the gene would be "off."

Activators An **activator** is a regulatory protein that facilitates transcription (positive regulation). Genes controlled by an activator have an ineffective promoter preceded by an **activator-binding site.** The binding of the activator to that site enhances the ability of RNA polymerase to initiate transcription from the promoter. Like repressors, activators are allosteric proteins. When an inducer binds to an activator, the shape of the activator changes so that it can then bind to the activator-binding site (**figure 7.24**). Thus, the term "inducer" applies to any molecule that turns on transcription by (1) stimulating the function of an activator or (2) interfering with the function of a repressor.

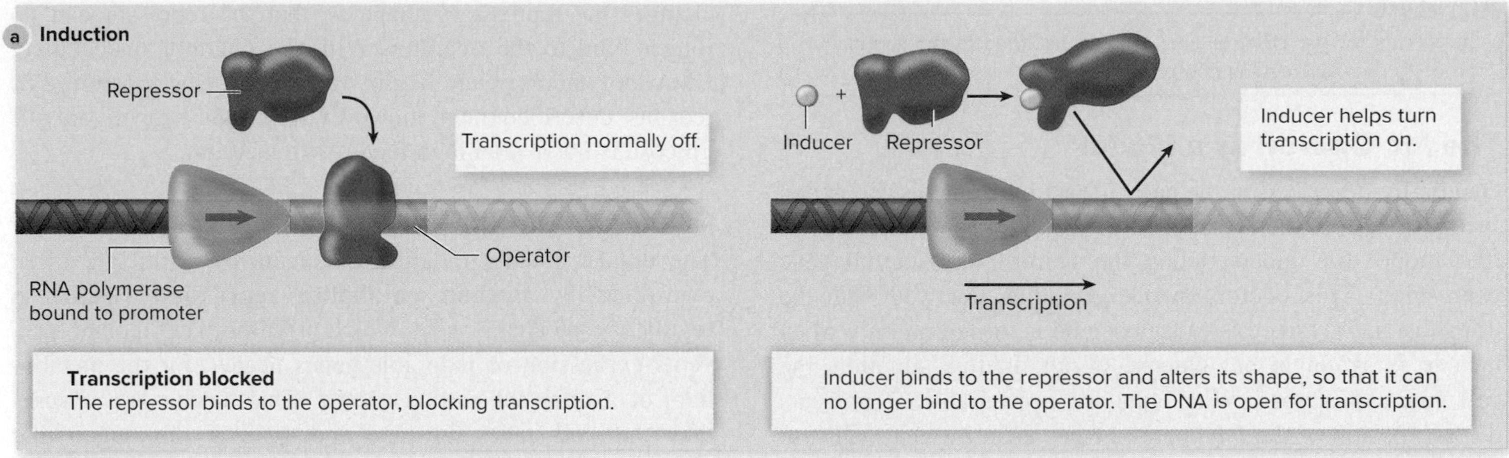
a Induction
Repressor
Transcription normally off.
Operator
RNA polymerase
bound to promoter
Transcription blocked
The repressor binds to the operator, blocking transcription.
Inducer
Repressor
Inducer helps turn
transcription on.
Transcription
Inducer binds to the repressor and alters its shape, so that it can
no longer bind to the operator. The DNA is open for transcription.

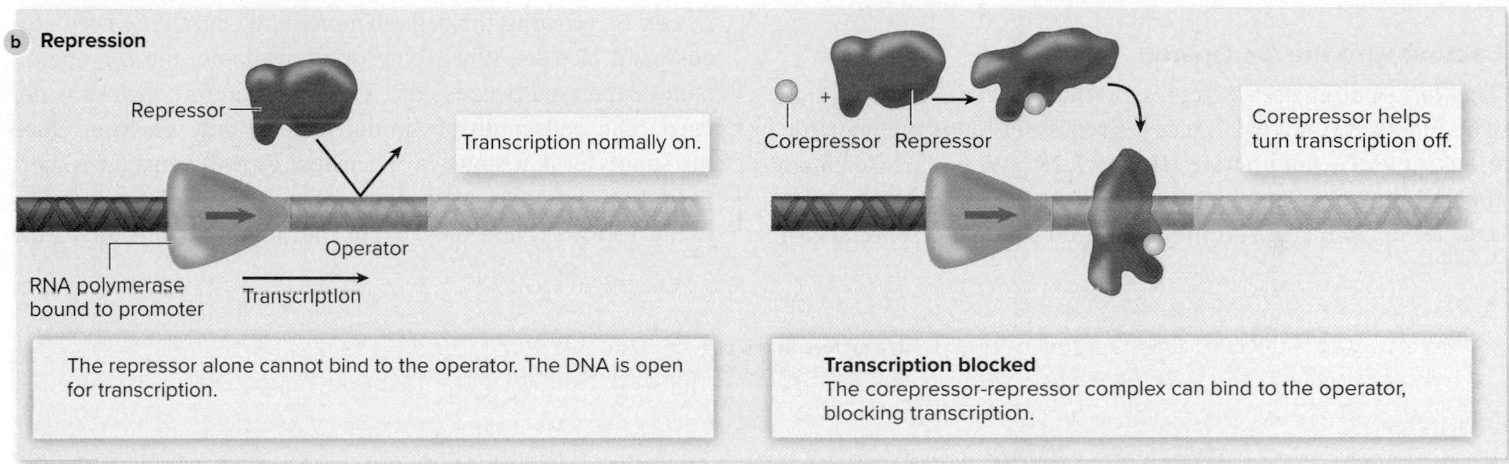
b Repression
Repressor
Transcription normally on.
Operator
RNA polymerase
bound to promoter
Transcription
The repressor alone cannot bind to the operator. The DNA is open
for transcription.
Corepressor
Repressor
Corepressor helps
turn transcription off.
Transcription blocked
The corepressor-repressor complex can bind to the operator,
blocking transcription.

FIGURE 7.23 The Role of Repressors in Transcriptional Regulation (a) Induction. **(b)** Repression.

How is a corepressor different from an inducer? How is it similar to an inducer?

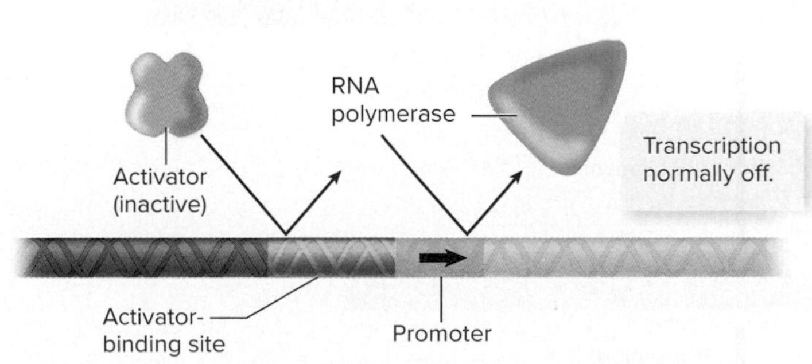
Activator
(inactive)
RNA
polymerase
Transcription
normally off.
Activator-
binding site
Promoter
RNA polymerase cannot bind to the promoter unless the activator
is bound to the activator-binding site, but the activator is in an
inactive form.

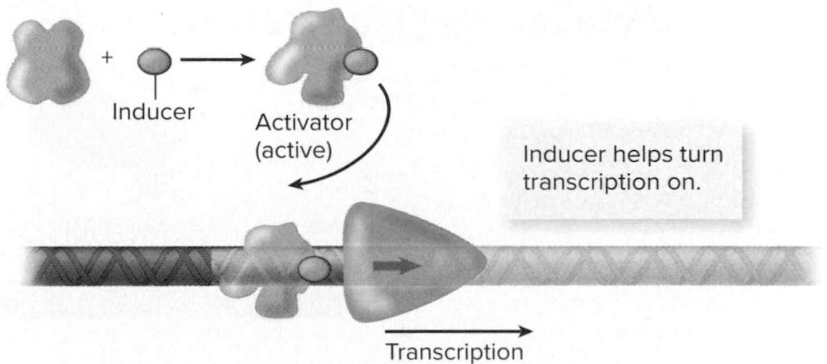
Inducer
Activator
(active)
Inducer helps turn
transcription on.
Transcription
Inducer binds to the activator and changes its shape, allowing the
activator to bind to the site. RNA polymerase can then bind to the
promoter and initiate transcription.

FIGURE 7.24 Transcriptional Regulation by Activators

How do activators facilitate transcription?

MicroByte

In *B. subtilis,* the DNA-binding protein that triggers the sporulation process controls a regulon of over 120 genes.

The *lac* Operon as a Model

Originally described in the early 1960s by François Jacob and Jacques Monod, the ***lac* operon** of *E. coli* serves as an important model for understanding the control of bacterial gene expression. This operon encodes proteins involved with the transport and hydrolysis of lactose and is turned on only when lactose is available but glucose is not. In this situation, the cell can use lactose. When glucose is available, it prevents the expression of the *lac* operon genes, ensuring that cells use the most efficiently metabolized carbon source (glucose) first.

Lactose and the *lac* Operon

The *lac* operon uses a repressor that prevents transcription when lactose is not available; the repressor binds the operator, blocking RNA polymerase (**figure 7.25**). When lactose enters the cell, however, some of it is converted to allolactose, which acts as an inducer. As allolactose binds to the repressor, it changes the repressor's shape so that the repressor can no longer bind to the operator. With the operator unoccupied, RNA polymerase can begin transcribing the operon. For reasons described next, however, this can happen only if glucose is not available in the growth medium.

Glucose and the *lac* Operon

The uptake and degradation of sugars by a cell are often controlled by **carbon catabolite repression (CCR),** a regulatory mechanism in which a carbon compound prevents expression of inducible genes needed for the metabolism of a different carbon source. This mechanism ensures that the cell takes up and metabolizes the sugar that allows it the fastest growth. CCR can be demonstrated in *E. coli* by growing the cells in a medium containing both glucose and lactose. When glucose is available, the *lac* operon is not expressed because of CCR, and therefore lactose is not used. The cells multiply initially using only glucose. Once the supply of that sugar is exhausted, growth stops for a short period as the cells synthesize enzymes needed for metabolizing lactose. Then they begin multiplying again, this time

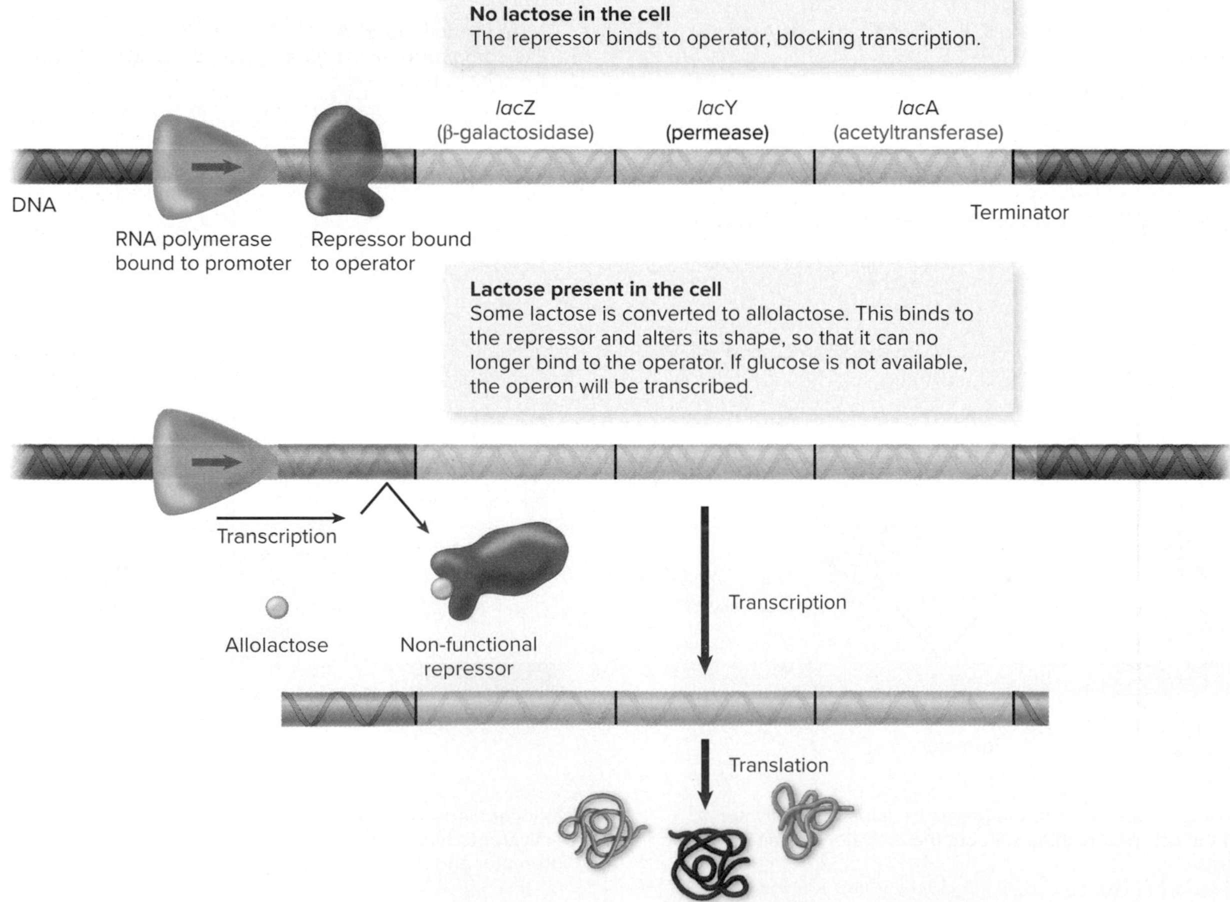
No lactose in the cell
The repressor binds to operator, blocking transcription.
lacZ (β-galactosidase)
lacY (permease)
lacA (acetyltransferase)
DNA
Terminator
RNA polymerase bound to promoter
Repressor bound to operator
Lactose present in the cell
Some lactose is converted to allolactose. This binds to the repressor and alters its shape, so that it can no longer bind to the operator. If glucose is not available, the operon will be transcribed.
Transcription
Allolactose
Non-functional repressor
Transcription
Translation

FIGURE 7.25 Lactose and the *lac* Operon β-Galactosidase is the enzyme that hydrolyzes lactose into its monosaccharide components. Permease is the transporter that brings lactose into the cell. The function of acetyltransferase in lactose metabolism is not fully understood.

? When would the *lac* operon in *E. coli* be repressed, and why?

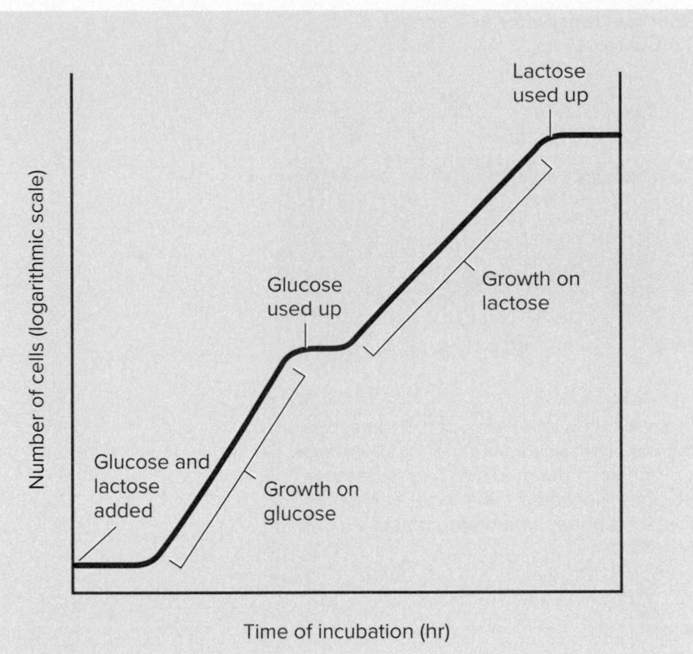

FIGURE 7.26 Diauxic Growth Curve of *E. coli* Growing in a Medium Containing Glucose and Lactose Cells preferentially use glucose. Only when the supply of glucose is used up do cells start metabolizing lactose. Note that the growth on lactose is slower than on glucose.

? **Why does bacterial growth stop temporarily when the glucose supply in the medium is used up?**

using lactose to fuel their growth. This characteristic two-phase growth pattern is called diauxic growth (**figure 7.26**).

Carbon catabolite repression is a global control system that allows glucose to regulate the expression of the *lac* operon as well as other sets of genes. Glucose does not act directly in the regulation, however. Instead, the cell's glucose transport system serves as a sensor of glucose availability. When the transport system is moving glucose molecules into the cell, catabolite repression prevents the *lac* operon from being expressed. When the transport system is idle, indicating that glucose is not available, then the *lac* operon can be turned on. How does the cell use the transporter to sense glucose levels? The transport process involves a relay system that donates a phosphate group to each incoming glucose molecule (group translocation; see figure 3.7c). When glucose levels are high, the unphosphorylated form of that transporter component predominates (**figure 7.27a**). Conversely, when glucose levels are low, the transporter component retains its phosphate group, indicating to the cell that other carbon sources need to be used (see figure 7.27b).

One mechanism of carbon catabolite repression involves an activator called CAP (catabolite activator protein), which is required for transcription of the *lac* operon. To be functional, the activator must be bound by an inducer—in this case, an ATP derivative called cAMP (cyclic AMP). The inducer is made only when extracellular glucose levels are low because the enzyme required for its synthesis is activated by the idle form of the glucose transporter component (see figure 7.27c).

Although a great deal of attention has been paid to the role of the activator in carbon catabolite repression, another mechanism of regulation called **inducer exclusion** might be more significant in *E. coli*. In this mechanism, when glucose is being moved into the cell, a glucose transport component binds to the lactose transporter (permease), temporarily locking it in a non-functional position. The locked permease cannot move lactose into the cell, so the *lac* operon will not be induced. Once the glucose supply diminishes, the glucose transporter becomes idle, so lactose can then be brought into the cell (see figure 7.27d). **Table 7.6** summarizes the effect of lactose and glucose levels on control of the *lac* operon.

TABLE 7.6 Glucose and the *lac* Operon

Control of the *lac* Operon			
Glucose Level	**Lactose Level**	***lac* Operon**	**Mechanism(s)**
High	Low/Absent	OFF	1. Because glucose is available, the activator (CAP) is not active, so the *lac* operon is not transcribed. 2. Because glucose is available, inducer exclusion prevents lactose from entering the cell. 3. Because lactose is absent, the repressor is active, thereby blocking transcription of the *lac* operon.
High	High	OFF	1. Because glucose is available, the activator (CAP) is not active, so the *lac* operon is not transcribed. 2. Because glucose is available, inducer exclusion prevents lactose from entering the cell. 3. Although lactose is present in the environment, inducer exclusion prevents it from entering the cell; because of this, the repressor is active, thereby blocking transcription of the *lac* operon.
Low/Absent	Low/Absent	OFF	1. Because lactose is absent, the repressor is active, thereby blocking transcription of the *lac* operon.
Low/Absent	High	ON	1. Because glucose is not available, the activator (CAP) is active, so it facilitates transcription of the *lac* operon. 2. Because glucose is not available, inducer exclusion does not occur and lactose enters the cell. 3. Because lactose is available in the cell, the repressor is not active, so transcription of the *lac* operon can proceed.

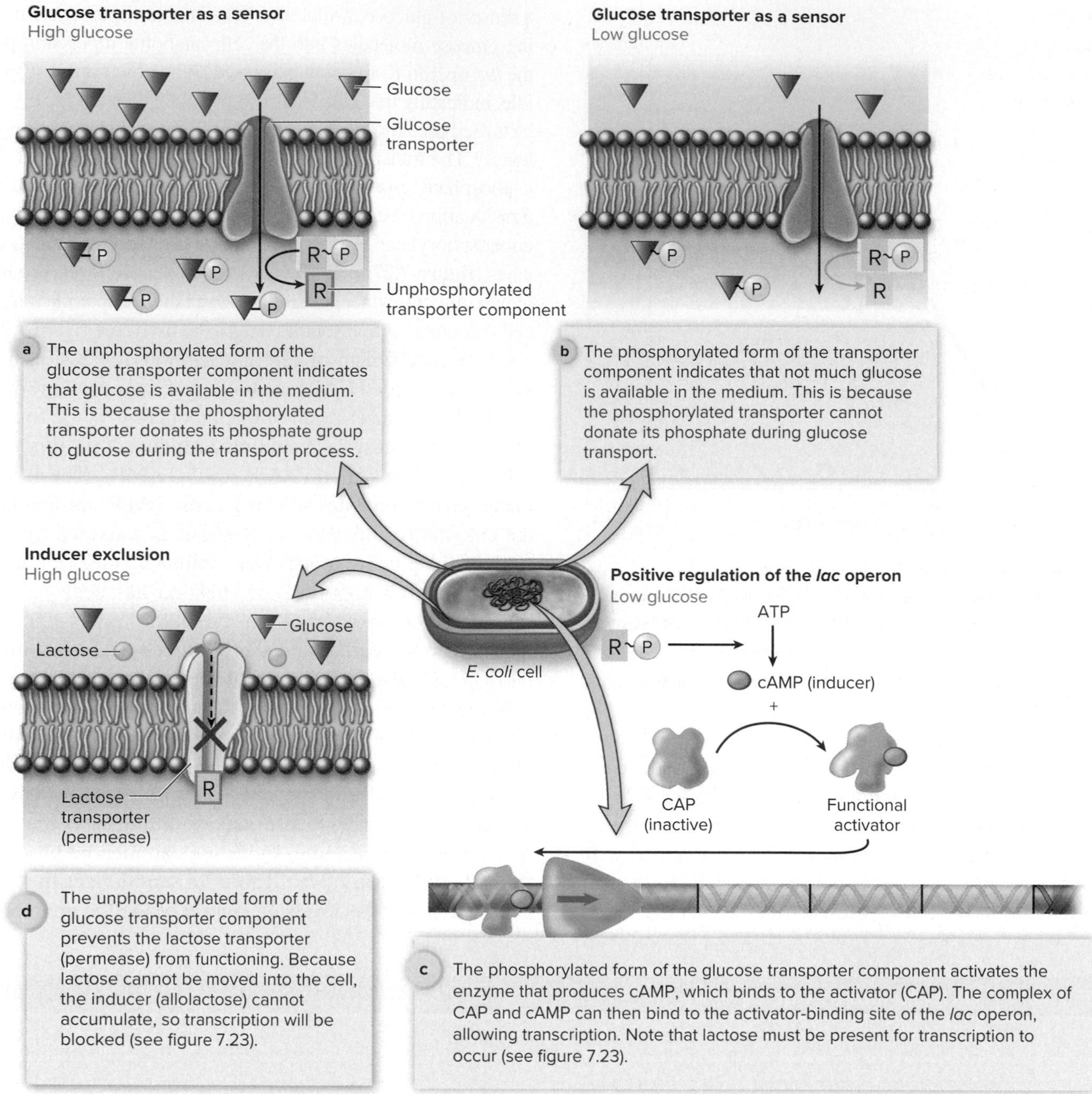

Glucose transporter as a sensor
High glucose
Glucose
Glucose transporter
R~P
R
P
Unphosphorylated transporter component
a The unphosphorylated form of the glucose transporter component indicates that glucose is available in the medium. This is because the phosphorylated transporter donates its phosphate group to glucose during the transport process.
Glucose transporter as a sensor
Low glucose
R~P
R
P
b The phosphorylated form of the transporter component indicates that not much glucose is available in the medium. This is because the phosphorylated transporter cannot donate its phosphate during glucose transport.
Inducer exclusion
High glucose
Lactose
Glucose
Lactose transporter (permease)
R
E. coli cell
Positive regulation of the lac operon
Low glucose
ATP
R~P
cAMP (inducer)
+
CAP (inactive)
Functional activator
d The unphosphorylated form of the glucose transporter component prevents the lactose transporter (permease) from functioning. Because lactose cannot be moved into the cell, the inducer (allolactose) cannot accumulate, so transcription will be blocked (see figure 7.23).
c The phosphorylated form of the glucose transporter component activates the enzyme that produces cAMP, which binds to the activator (CAP). The complex of CAP and cAMP can then bind to the activator-binding site of the lac operon, allowing transcription. Note that lactose must be present for transcription to occur (see figure 7.23).

FIGURE 7.27 Glucose and the *lac* Operon

? Why would it be advantageous for a cell to use glucose before lactose?

MicroAssessment 7.6

Many genes are constitutive, but some are inducible or repressible. A repressor blocks transcription when it binds to an operator. An activator enhances transcription when it binds to an activator-binding site. Inducers bring about gene expression by binding either to repressors (disabling them) or to activators (allowing them to attach to the activator-binding site). The *lac* operon, a model for regulation, is controlled by a repressor, an activator, and carbon catabolite repression.

15. Explain the difference between a constitutive gene and an inducible gene.
16. Explain how glucose prevents expression of the *lac* operon.
17. Why would it be advantageous for a cell to control the activity of an enzyme as well as its synthesis?

7.7 ■ Eukaryotic Gene Regulation

Learning Outcome

11. Describe how RNA interference silences genes.

Considering the complexity of eukaryotic cells and the diversity of cell types found in multicellular organisms, it is not surprising that eukaryotic gene regulation is much more complicated than that of prokaryotic organisms. Eukaryotic cells use a variety of control methods, including modifying the structure of the chromosome, regulating the initiation of transcription, and altering pre-mRNA processing and modification. We will focus only on a process called **RNA interference (RNAi),** a Nobel Prize–winning discovery that revolutionized views on gene regulation. Cells routinely use RNAi to destroy specific RNA transcripts, and scientists can manipulate the process to silence select genes.

In RNAi, a cell produces short, single-stranded RNA pieces to locate specific RNA transcripts targeted for destruction. To function in RNAi, a short RNA strand joins a multi-protein unit called an RNA-induced silencing complex (RISC). Within a RISC, the short RNA strand serves as the probe that allows the complex to locate a specific nucleotide sequence on mRNA molecules. The short RNA strand does this by binding to complementary sequences on an mRNA molecule, tagging that transcript for destruction by enzymes in the RISC (**figure 7.28**). The components of the RISC are not destroyed in the process, so the complex is catalytic, providing a rapid and effective means of silencing genes that have already been transcribed. Two different types of RNA molecules are used in RNAi: microRNA (miRNA) and short interfering RNA (siRNA). Although these differ in how they are produced, they are each about two dozen nucleotides in length and functionally equivalent.

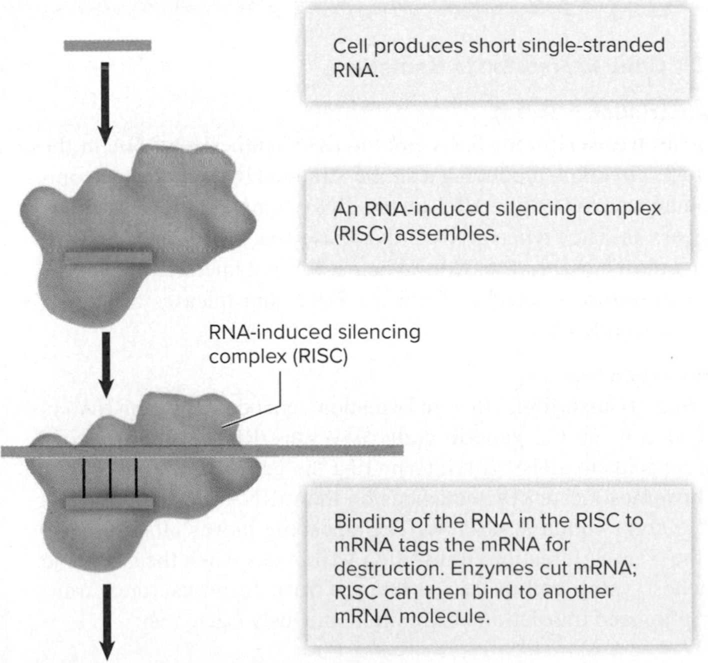

Cell produces short single-stranded RNA.
An RNA-induced silencing complex (RISC) assembles.
RNA-induced silencing complex (RISC)
Binding of the RNA in the RISC to mRNA tags the mRNA for destruction. Enzymes cut mRNA; RISC can then bind to another mRNA molecule.

FIGURE 7.28 RNA Interference (RNAi)

How might RNAi be used to treat an infectious disease?

MicroByte

The first medication based on RNAi was approved in 2018 to silence the gene that results in a condition called hereditary transthyretin amyloid.

MicroAssessment 7.7

Eukaryotic RNA interference uses short strands of RNA to identify specific RNA transcripts targeted for destruction.

18. What is the role of miRNA and siRNA in the regulation of gene expression?

7.8 ■ Genomics

Learning Outcomes

12. Explain how protein-encoding regions are found when analyzing a DNA sequence.
13. Describe metagenomics and the information it can provide.

Genomics is the study of genes—how they are arranged in a genome, what they encode, and how genetic information changes. In 1995, the nucleotide sequence of the chromosome of the bacterium *Haemophilus influenzae* was published, marking the first complete genomic sequence ever determined. Since then, sequencing microbial genomes has become relatively common, leading to many exciting advances, including a better understanding of the complex relationships between microbes and humans. In fact, many of the recent findings described in this textbook have been discovered through genomics.

Modern sequencing methods are rapidly generating data, but analyzing that information is far more difficult than it might seem. Imagine trying to determine the amino acid sequence of a protein encoded by a stretch of DNA without knowing anything about (1) the position of the gene's promoter or (2) the reading frame used for translation. Because either strand of the DNA molecule could potentially be the template strand, two entirely different RNA sequences must be considered as protein-encoding candidates. In turn, each of those two candidates has three potential reading frames, for a total of six possible reading frames. Yet only one of these actually codes for the protein. Understandably, computers are an invaluable aid and are used extensively to decipher the meaning of raw sequence data. As a result, a new field has emerged—**bioinformatics**—which uses computer technology to store, retrieve, and analyze biological data.

Analyzing a Prokaryotic DNA Sequence

When analyzing a DNA sequence, the (+) strand is used to represent the sequence of the corresponding RNA transcript. As an example, an ATG in the (+) strand of DNA indicates a possible start codon. Computers help locate protein-encoding regions in DNA by searching for **open reading frames (ORFs),** stretches of nucleotide sequences generally longer than 300 bp that begin with a start codon and end with a stop codon. An ORF potentially encodes a protein. Other characteristics such as an upstream sequence that can serve as a ribosome-binding site also suggest that an ORF encodes a protein.

The nucleotide sequence of an ORF can be compared with other known sequences by searching computerized databases of published sequences. Similar searches can be done using the amino acid sequence of the encoded protein. Not surprisingly, as genomes of more organisms are being sequenced, information contained in these databases is growing at a remarkable rate. If the encoded protein shows certain amino acid similarities to other characterized proteins, a presumed function can sometimes be assigned. For example, proteins that bind DNA have similar amino acid sequences in certain regions. Likewise, regulatory regions in DNA such as promoters can sometimes be identified based on similarities to known sequences.

Metagenomics

Metagenomics is the analysis of the total genomes (mostly microbial) in an environment. With metagenomics, researchers can study all the microorganisms and viruses in a community, not just the relatively few that can be isolated in the laboratory (see figure 28.5). Imagine the insights being gained from this relatively new approach: studying a person's microbiome over time to monitor changes during health and disease; comparing the microbiomes of different body sites; and even comparing microbiomes of different people around the world! Metagenomics is also being used to study microbial life in the open oceans and in soils. Analyzing these sequences gives an entirely new perspective on the extent of biodiversity and will probably lead to the discovery of new antibiotics and other medically useful compounds.

MicroByte

Metagenomic analysis of the human microbiome indicates that the human body carries 100 times more microbial DNA than human DNA.

MicroAssessment 7.8

Sequencing methods are rapid, but analyzing the data and extracting the pertinent information is difficult.

19. What is an open reading frame?
20. Describe two things that you can learn by searching a computerized database for sequences that have similarities to a newly sequenced gene.
21. There are characteristic differences in the nucleotide sequences of the leading and lagging strands. Why might this be so?

Summary

7.1 ■ Overview (figure 7.1)

Characteristics of DNA (figure 7.2)
DNA is a double helix, containing two **complementary** and **antiparallel** strands of nucleotides; a linear strand has a **5′ end** and a **3′ end.**

Characteristics of RNA
RNA, a single-stranded molecule, is transcribed from one of the two strands of DNA. Three different functional types of RNA molecules are required for gene expression: **messenger RNA (mRNA), ribosomal RNA (rRNA),** and **transfer RNA (tRNA)** (figure 7.3).

Regulating Gene Expression
Protein synthesis is generally controlled by regulating the synthesis of mRNA (figure 7.4).

7.2 ■ DNA Replication

DNA replication begins at the **origin of replication** and is **semiconservative.** The bidirectional progression of replication around a circular DNA molecule creates two **replication forks** (figure 7.5). A **DNA polymerase** synthesizes DNA in the 5′ to 3′ direction, using one strand as a **template** to generate the complementary strand (figures 7.6, 7.7).

7.3 ■ Gene Expression in Bacteria

Transcription (table 7.2)
During **transcription, RNA polymerase** synthesizes RNA in the 5′ to 3′ direction, producing a single-stranded RNA molecule complementary and antiparallel to the DNA template (figure 7.9). The process initiates when RNA polymerase recognizes and binds to a **promoter** (figures 7.10, 7.11). When RNA polymerase encounters a **terminator,** it detaches from the DNA and releases the newly synthesized RNA.

Translation (table 7.4)
During **translation,** the information encoded by mRNA is decoded using the **genetic code** (table 7.3). tRNAs carry specific amino acids to assist with interpreting the genetic code (figure 7.13). **Ribosomes** locate key sequences on the mRNA molecule including a **start codon** (figure 7.14). The ribosome moves along mRNA in the 5′ to 3′ direction; translation terminates when the ribosome reaches a **stop codon** (figures 7.15, 7.16, 7.17). In prokaryotes, transcription and translation occur simultaneously (figure 7.18).

7.4 ■ Differences Between Eukaryotic and Prokaryotic Gene Expression (table 7.5)

Eukaryotic mRNA is synthesized as **pre-mRNA** that must be processed to generate a functional molecule. Processing includes

capping, polyadenylation, and **splicing;** splicing removes **introns** (figure 7.19). In eukaryotic cells, the mRNA must be transported out of the nucleus before it can be translated in the cytoplasm.

7.5 ■ Sensing and Responding to Environmental Fluctuations

Signal Transduction

Signal transduction allows cells to monitor and react to environmental conditions. Bacteria use **quorum sensing** to activate genes that are useful only when expressed by a critical population density (figure 7.20). **Two-component regulatory systems** use a sensor that recognizes changes outside the cell and then transmits that information to a response regulator inside the cell (figure 7.21).

Natural Selection

Antigenic variation is a random alteration in the characteristics of certain surface proteins. **Phase variation** is the random switching on and off of certain genes.

7.6 ■ Bacterial Gene Regulation

Constitutive genes are constantly expressed. **Inducible genes** can be turned on when an **inducer** is present; **repressible genes** can be turned off by certain conditions (figure 7.22).

Mechanisms to Control Transcription

Repressors block transcription (figure 7.23). **Activators** enhance transcription (figure 7.24).

The *lac* Operon as a Model

A repressor prevents expression of the ***lac* operon** when lactose is not available (figure 7.25). **Carbon catabolite repression (CCR)** prevents transcription of the operon when glucose is available (figures 7.26, 7.27, table 7.6).

7.7 ■ Eukaryotic Gene Regulation

Regulation in eukaryotic cells is much more complicated than that in prokaryotic cells. In **RNA interference (RNAi),** a cell synthesizes short, single-stranded RNA pieces to locate specific RNA transcripts destined for destruction (figure 7.28).

7.8 ■ Genomics

Analyzing a Prokaryotic DNA Sequence

When analyzing a DNA sequence, the (+) strand is used to represent the sequence of the corresponding RNA transcript; computers are used to search for **open reading frames (ORFs).**

Metagenomics

Metagenomics allows researchers to study all organisms and viruses in a community, not just the relatively few that grow in culture.

Review Questions

Short Answer

1. Explain what *semiconservative* means with respect to DNA replication.
2. What is an origin of replication?
3. Why are primers required in DNA replication but not in transcription?
4. What is polycistronic mRNA?
5. Explain why knowing the orientation of a promoter is critical when determining the amino acid sequence of an encoded protein.
6. What is the function of a sigma factor?
7. What is the fate of a protein that has a signal sequence?
8. Explain how some bacteria sense the density of cells in their own population.
9. Compare and contrast regulation by a repressor to regulation by an activator.
10. Explain why locating protein-encoding regions in a genomic sequence can be difficult.

Multiple Choice

1. All of the following are involved in transcription *except*
   a) polymerase.
   b) a primer.
   c) a promoter.
   d) a sigma factor.
   e) uracil.
2. All of the following are involved in DNA replication *except*
   a) a polysome.
   b) gyrase.
   c) polymerase.
   d) primase.
   e) a primer.
3. All of the following are directly involved in translation *except*
   a) a promoter.
   b) ribosomes.
   c) a start codon.
   d) a stop codon.
   e) tRNAs.
4. Using the portion of DNA shown here as a template, what will be the sequence of the RNA transcript?

   5′ GCGTTAACGTAGGC 3′
   promoter → 3′ CGCAATTGCATCCG 5′

   a) 5′ GCGUUAACGUAGGC 3′
   b) 5′ CGGAUGCAAUUGCG 3′
   c) 5′ CGCAAUUGCAUCCG 3′
   d) 5′ GCCUACGUUAACGC 3′
5. A ribosome binds to the following mRNA at the site indicated by the dark box. At which codon will translation most likely begin?

   5′ ■ GCCGGAAUGCUGCUGGC

   a) GCC
   b) GGC
   c) AUG
   d) AAU
6. Which of the following statements about gene expression is *false*?
   a) More than one RNA polymerase can be transcribing a specific gene at a given time.
   b) More than one ribosome can be translating a specific transcript at a given time.

c) Translation begins at a site called a promoter.
d) Transcription stops at a site called a terminator.
e) Some amino acids are coded for by more than one codon.

7. The operon that encodes the enzymes used to synthesize the amino acid tryptophan is most likely
   a) constitutive.
   b) inducible.
   c) repressible.
   d) a and b
8. Under which of the following conditions will transcription of the *lac* operon occur?
   a) Lactose present/glucose present
   b) Lactose present/glucose absent
   c) Lactose absent/glucose present
   d) Lactose absent/glucose absent
   e) a and b
9. All of the following are characteristics of eukaryotic gene expression *except*
   a) a 5′ cap is on the mRNA.
   b) a poly A tail is on the 3′ end of mRNA.
   c) introns must be removed to create the mRNA that is translated.
   d) the mRNA is often polycistronic.
   e) translation begins at the first AUG.
10. Which of the following statements is *false*?
    a) A derivative of lactose serves as an inducer of the *lac* operon.
    b) Signal transduction provides a mechanism for a cell to sense the conditions of its external environment.
    c) Quorum sensing allows bacterial cells to sense the density of like cells.
    d) An example of a two-component regulatory system is the lactose operon, which is controlled by a repressor and an activator.
    e) An ORF is a stretch of DNA that may encode a protein.

Applications

1. A graduate student is trying to identify the gene coding for an enzyme found in a bacterial species that degrades trinitrotoluene (TNT). The student is frustrated to find that the organism does not produce the enzyme when grown in nutrient broth, making it difficult to collect the mRNA needed to help identify the gene. What could the student do to potentially increase the amount of the desired enzyme?
2. To isolate eukaryotic mRNA from cell lysates, a researcher has purchased magnetic beads that have short polymers of thymine nucleotides (T) attached. How could the beads be used to separate the mRNA from other cell components, including pre-mRNA?

Critical Thinking

1. Transcription of "Operon A" is regulated by Protein A. When the sugar arabinose is present, it attaches to Protein A, forming a complex that binds near the operon's promoter. This allows RNA polymerase to bind the promoter and start transcription. When arabinose is not present, Protein A will bind to a different region of DNA and, in doing so, block RNA polymerase. How would you classify Protein A in regard to its regulatory functions?
2. You have a solution containing an equal mixture of bacterial and eukaryotic mRNA. To separate the two kinds of mRNA, a tube containing cellulose beads that have long chains of thymine nucleotides can be used. When the solution is poured through the column, the solution that comes out the bottom will have a greater proportion of bacterial mRNA. Why would that be the case?

8 | Bacterial Genetics

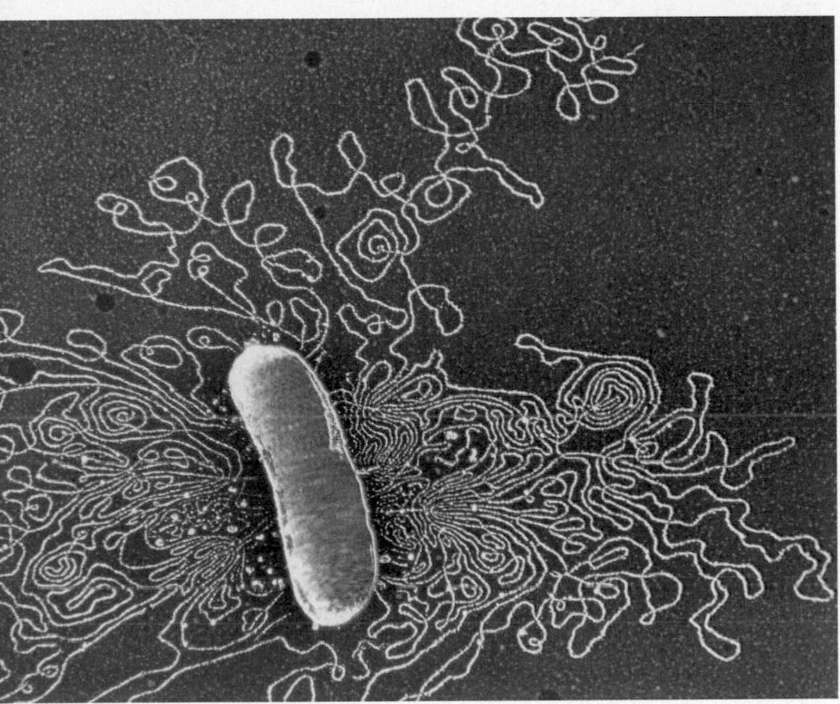

Chromosome being released from a gently lysed *E. coli* cell. *Dr. Gopal Murti/Science Source*

KEY TERMS

Auxotroph A microorganism that requires an organic growth factor.

Bacterial Transformation Mechanism of horizontal gene transfer in which naked DNA is taken up by the recipient cell.

Conjugation Mechanism of horizontal gene transfer in which the donor cell physically contacts the recipient cell.

Genotype The sequence of nucleotides in an organism's DNA.

Horizontal Gene Transfer Transfer of DNA from one organism to another by a process other than reproduction.

Mutation A change in the nucleotide sequence of a cell's DNA that is passed on to daughter cells.

Phenotype The observed characteristics of a cell.

Plasmid A DNA molecule that replicates independently of the chromosome.

Prototroph A microorganism that does not require any organic growth factors.

Transduction Mechanism of horizontal gene transfer in which bacterial DNA is transferred inside a phage coat.

Transposable Element Segment of DNA that can move from one site to another in a cell's genome.

Vertical Gene Transfer Transfer of DNA from parent to offspring.

A Glimpse of History

Barbara McClintock (1902–1992) was a remarkable scientist who made several very important discoveries in genetics well before the era of large research teams and sophisticated molecular techniques. With curiosity and determination, she eagerly worked 12-hour days, 6 days a week doing what she loved in a small laboratory at Cold Spring Harbor on Long Island, New York.

McClintock studied color variation in corn kernels. She noticed that the kernel colors were not inherited in a predictable manner. In fact, the colors varied from one ear of corn to another. Based on extensive data, McClintock concluded that segments of DNA, now called transposable elements (TEs), were moving into and out of genes involved with kernel color. These moving DNA segments destroyed the function of the genes, thereby changing kernel color.

At the time that McClintock published her results in 1950, most scientists believed that chromosomal DNA was very stable and changed only through recombination. Consequently, geneticists were skeptical of her conclusions. It was not until the late 1970s that her earlier ideas began to be accepted; by that time, TEs had been discovered in many organisms, including bacteria. In 1983, at age 81, McClintock was awarded a Nobel Prize for her discovery of TEs, popularly called "jumping genes."

In the ever-changing conditions that characterize most environments, organisms need to adapt in order to survive and multiply. If they fail to do this, competing organisms more fit to thrive in the new setting will soon predominate. This is the process of **natural selection.** Bacteria have two general means by which they routinely adjust to new circumstances: regulating gene expression (discussed in chapter 7) and genetic change, the focus of this chapter.

Genetic change is the raw material for evolution and is fundamental to all life on Earth. To study it, scientists often use *E. coli* as a model system because the cells grow rapidly in small volumes of simple, inexpensive media, accumulating in very large numbers. This makes it easier to examine rare events that give rise to different **strains**, which are genetic variants within a species.

8.1 ■ Genetic Change in Bacteria

Learning Outcome

1. Define the terms *mutation, horizontal gene transfer, genotype, phenotype, auxotroph,* and *prototroph.*

Genetic change in bacteria occurs by two mechanisms: mutation and horizontal gene transfer (**figure 8.1**). **Mutation** is a change in the nucleotide sequence of a cell's DNA that is then passed on to the cell's progeny ("daughter cells") through **vertical gene transfer.** The modified organism and progeny are referred to as mutants. **Horizontal gene transfer (HGT)** is the transfer of DNA from one organism to another by a process other than reproduction. Like mutations, the changes are then passed on to the progeny by vertical transfer.

A change in an organism's DNA alters its **genotype:** the sequence of nucleotides in the DNA. In bacterial cells, such a change can have a significant effect because bacteria are haploid, meaning that they contain only a single set of genes. There is no "backup copy" of a gene in a haploid organism. Because of this, a change in genotype often alters the organism's observable characteristics, or **phenotype.** Note, however, that the phenotype involves more than just the organism's genetic makeup; it can also be influenced by environmental conditions. For example, colonies of *Serratia marcescens* are red when incubated at 22°C but white when incubated at 37°C. In this case, the phenotype, but not the genotype, has changed. However, if the genes responsible for pigment production are removed, the organism's phenotype and genotype both change.

Scientists often study genetic changes that alter a microorganism's nutrient requirements. For instance, if a gene required for biosynthesis of the amino acid tryptophan is deleted, then the organism can multiply only if tryptophan is supplied in the growth medium. The same occurs if the gene is disrupted so that the protein it encodes no longer functions properly. A mutant that requires an amino acid or other growth factor is an **auxotroph** (*auxo* means "increase," and *troph* means "nourishment"). This is in contrast to a **prototroph,** which does not require growth factors (*proto* means "earliest form of"). Wild-type *E. coli* is a prototroph ("wild-type" refers to the typical phenotype of strains isolated from nature). By convention, a strain's characteristics are designated by three-letter abbreviations, with the first letter capitalized; for example, a strain that cannot make tryptophan is designated Trp^{-}. For simplicity, only if a growth factor is required is it indicated. A strain's resistance to antibiotics is also indicated; for example, streptomycin resistance is noted as Str^{R}.

MicroAssessment 8.1

The properties of bacteria can change through either mutation or horizontal gene transfer.

1. How is mutation different from horizontal gene transfer?
2. Contrast the meaning of the terms *auxotroph* and *prototroph*.
3. Which has a more lasting effect on a cell—a change in the genotype or a change in the phenotype?

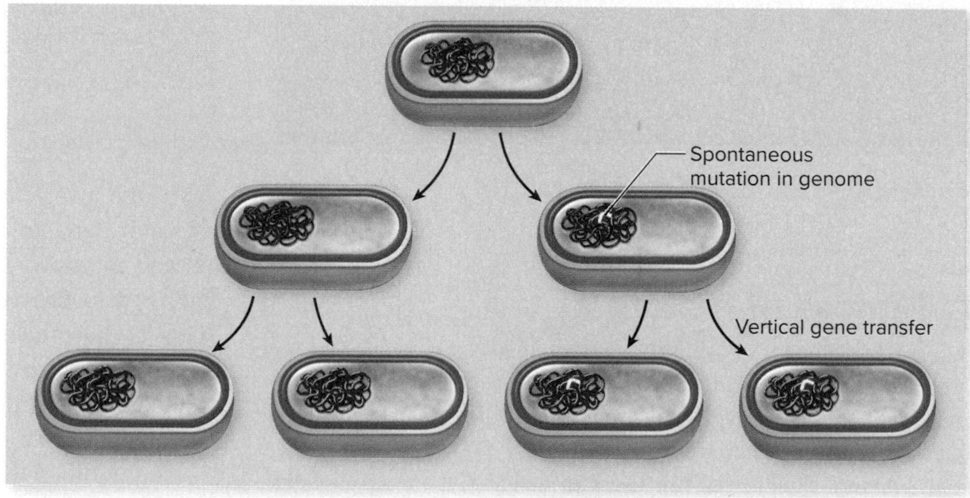

(a) Mutation

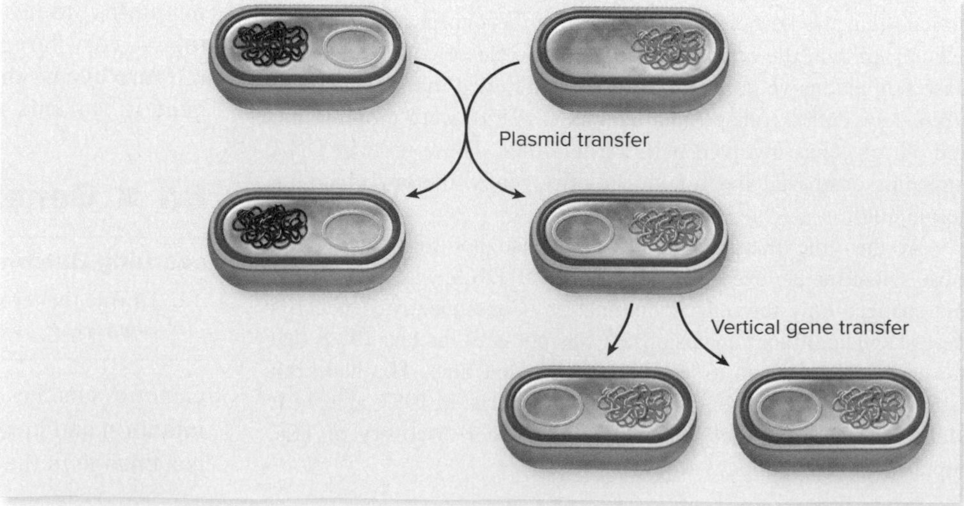

(b) Horizontal gene transfer

FIGURE 8.1 Mechanisms of Genetic Change in Bacteria (a) Mutation. **(b)** Horizontal transfer of plasmid-encoded genes; other DNA can be transferred horizontally as well.

Which would have a greater effect on the nucleotide sequence of an organism's DNA—mutation or horizontal gene transfer?

MUTATION AS A MECHANISM OF GENETIC CHANGE

8.2 ▪ Spontaneous Mutations

Learning Outcomes

2. Describe three outcomes of base-pair substitutions.
3. Describe the consequences of removing or adding nucleotide pairs.
4. Explain how transposable elements cause mutations.

Spontaneous mutations are random genetic changes that result from normal cell processes and are passed on to a cell's progeny. Because spontaneous mutations occur routinely yet infrequently, every large population contains mutants, so cells in a colony are not necessarily identical. The environment does not cause the mutations but selects those cells that can grow under its conditions. For example, a bacterial cell with a spontaneous mutation that results in resistance to an antimicrobial medication will become dominant in an environment where the medication is present because the sensitive cells are killed or inhibited, allowing the resistant cells to grow without competition (see figure 20.13).

A given gene mutates spontaneously at an infrequent but characteristic rate, defined as the probability that a mutation will occur in the gene. The mutation rate of different genes usually varies between 10^{-4} and 10^{-12} per cell division. In other words, the chance that a gene will undergo a mutation when a cell replicates its DNA prior to cell division is between one in 10,000 (10^{-4}) and one in a trillion (10^{-12}). A mutation can also change back to its original, non-mutated state; this change is called a **reversion,** or back mutation, and like the original mutation, occurs spontaneously at low frequencies.

A single spontaneous mutation is a rare event, so two such mutations developing simultaneously in a cell is even more unlikely. Doctors take advantage of this to prevent pathogens from becoming resistant to certain antimicrobial medications. In tuberculosis treatment, for example, two or more antimicrobial medications are given at the same time. The chance that a single cell will develop resistance through spontaneous mutation to both medications is the product of the mutation rates of the two genes (calculated by taking the sum of the exponents). For example, if the mutation rate for "antibiotic X" resistance is 10^{-6} per cell division and the mutation rate for "antibiotic Y" resistance is 10^{-8}, then the probability that both mutations will develop within a cell is $10^{-6} \times 10^{-8}$, or 10^{-14}.

Base-Pair Substitution

Base-pair substitution is the replacement of one base pair with another and can occur as a result of DNA polymerase incorporating an incorrect nucleotide during DNA synthesis. Recall that the only difference between the four nucleotides in DNA is their nucleobase (base), so although the error involves a nucleotide substitution, the relevant change is the base, which is why discussions emphasize that. As shown in **figure 8.2,** an error by DNA polymerase simply creates a mismatch between bases; only when another round of replication occurs—using the altered strand as template for synthesis—will the resulting DNA copy have a different base pair than the parent DNA. Because only one base pair is changed, it is termed a **point mutation.** As described in section 8.4, cells have mechanisms to correct mismatches between nucleobases, but those are not perfect, which is why mutations occur.

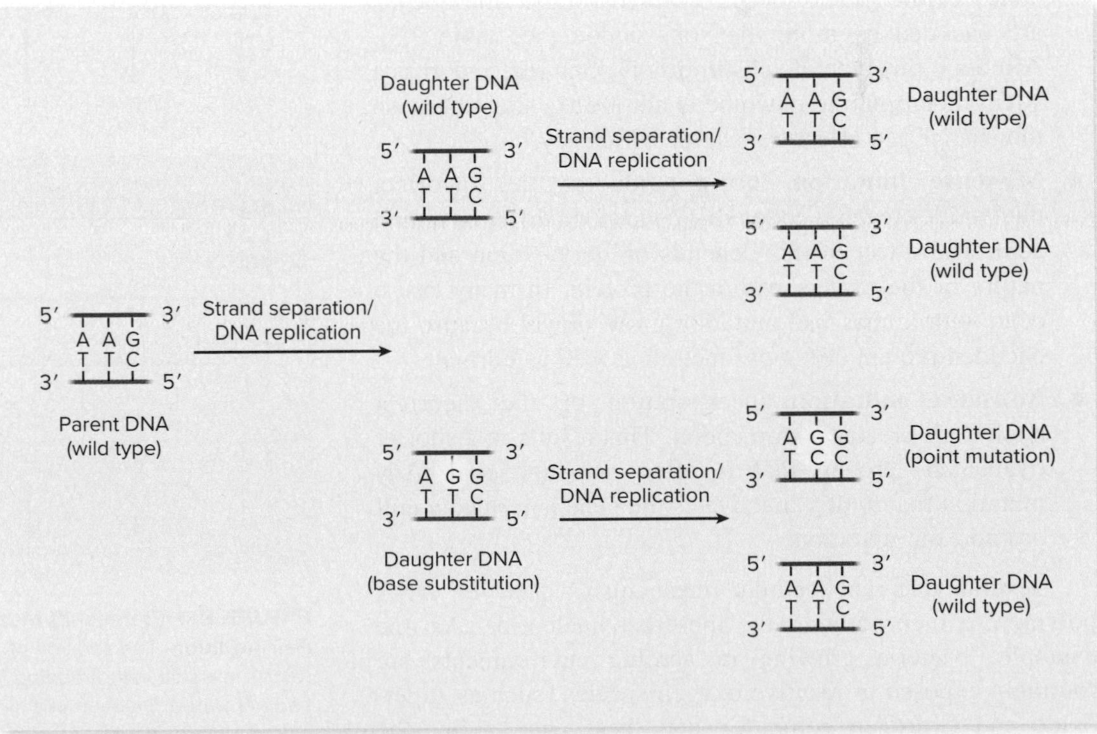

FIGURE 8.2 Base-Pair Substitution A replication error results in mismatched bases between the two DNA strands. Subsequent DNA replication using the altered strand as a template results in a point mutation.

? **What enzyme incorporated the incorrect nucleotide?**

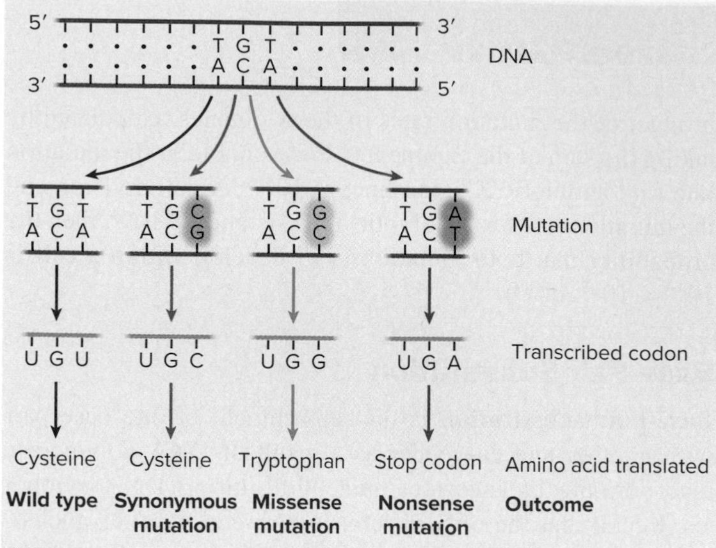

FIGURE 8.3 Potential Outcomes of a Base-Pair Substitution Outcomes include synonymous, missense, and nonsense mutations.

? **Why would some missense mutations have no effect on the encoded protein but other missense mutations in the same gene result in a partially functional or even non-functional protein?**

A base-pair substitution in a protein-encoding gene can lead to three possible mutation outcomes (**figure 8.3**):

- **Synonymous mutation.** Incorporation of the incorrect nucleotide creates a codon that encodes the same amino acid as the original. This can occur because of the redundancy of the genetic code; recall that most amino acids are encoded by more than one codon (see table 7.3). Although this type of substitution is often referred to as a silent mutation, the outcome is not always silent because the change can affect translation efficiency.
- **Missense mutation.** Incorporation of the incorrect nucleotide creates a codon that codes for a different amino acid. The effect of this depends on the position and the nature of the change within the protein. In many cases, cells with a missense mutation grow slowly because the encoded protein does not function as well as normal.
- **Nonsense mutation.** Incorporation of the incorrect nucleotide creates a stop codon. This results in a shorter (truncated) protein that is often non-functional. Any mutation that totally inactivates the gene is termed a null or knockout mutation.

Several different cellular mechanisms lead to DNA polymerase incorporating the incorrect nucleotide. As one example, bacteria growing in aerobic environments are routinely exposed to reactive oxygen species (such as superoxide and hydrogen peroxide), and these can oxidize the nucleobase guanine. DNA polymerases often mispair oxidized guanine with adenine rather than with cytosine.

Addition or Deletion of Nucleotide Pairs

Errors during DNA synthesis can also lead to the addition or deletion of nucleotide pairs; the consequence depends on how many pairs are involved. If three nucleotide pairs are added (or deleted), this adds (or deletes) one codon. When the gene is expressed, one additional (or one less) amino acid will be in the resulting protein; the seriousness of that change depends on its location in the encoded protein. In contrast, adding or subtracting one or two nucleotide pairs causes a **frameshift mutation,** which changes the reading frame of the corresponding mRNA molecule so that an entirely different set of codons is translated (**figure 8.4**). Frequently, one of the resulting downstream codons will be a stop codon; as a consequence, a frameshift mutation likely results in a shortened, non-functional protein—a knockout mutation.

Transposable Elements

Transposable elements (TEs), or jumping genes, are pieces of DNA that can move from one location to another in a cell's

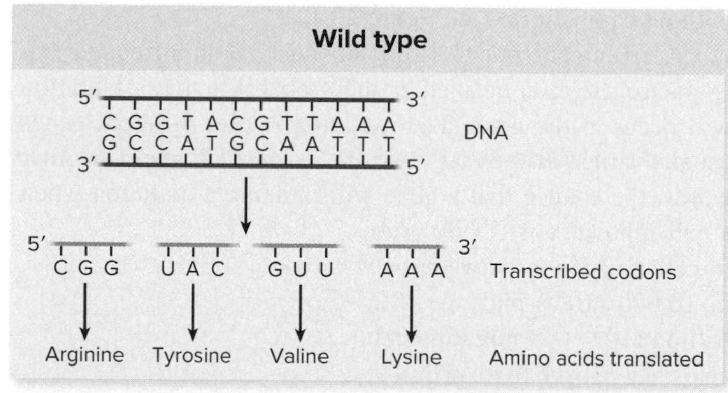

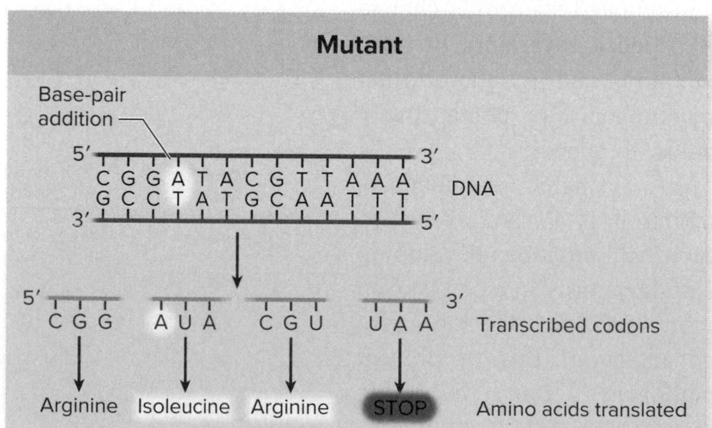

FIGURE 8.4 Frameshift Mutation as a Result of a Nucleotide-Pair Addition The addition of a nucleotide pair (base pair) to the DNA results in a shift in the reading frame when the sequence is transcribed and translated. Deletion of a single nucleotide pair in the DNA would have a similar effect.

? **This figure shows a single nucleotide-pair addition. What would happen if three nucleotide pairs were added?**

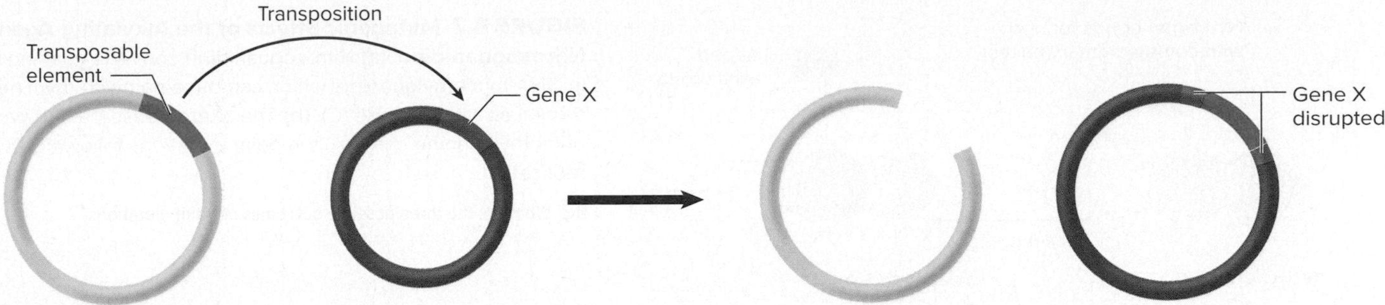

FIGURE 8.5 Transposition A transposable element has the ability to "jump" (transpose) from one piece of DNA into another. Host enzymes may repair the gap left by the departing transposon.

What effect does the transposable element have on gene X's function in this figure?

genome, a process called transposition (**figure 8.5**). They integrate into their new location through a process that does not require a similar nucleotide sequence in the region of recombination (the process is called non-homologous recombination); they simply insert into a stretch of DNA and do not replace the existing sequences. The gene into which a TE jumps is disrupted by the event, so the gene becomes nonfunctional, an outcome called **insertional inactivation.** Most TEs contain transcriptional terminators, so the expression of downstream genes in the same operon will stop as well. The structure and biology of TEs will be described in section 8.10.

The classic studies of transposition were carried out by Dr. Barbara McClintock (see A Glimpse of History, this chapter). She observed corn kernel color variations that resulted from TEs moving into and out of genes controlling pigment synthesis (**figure 8.6**).

FIGURE 8.6 Transposition Detected by Changes in Corn Kernel Color The variegation is caused by the insertion of transposable elements into genes involved in pigment synthesis. Bob Ross/Shutterstock

MicroAssessment 8.2

Spontaneous mutations happen during normal cell processes and can change the properties of the cell. Base-pair substitutions that occur during DNA synthesis can lead to synonymous, missense, and nonsense mutations. Removing or adding nucleotides can cause a frameshift mutation. Transposable elements can move from one location to another in a cell's genome.

4. Which is generally more serious to a cell: a missense mutation or a nonsense mutation?
5. What is the likely consequence of a frameshift mutation?
6. Is it as effective to take two antibiotics sequentially as it is to take them simultaneously, as long as the total length of time that they are both taken is the same? Explain.

8.3 ■ Induced Mutations

Learning Outcomes

5. Describe the three general groups of chemical mutagens.
6. Explain how X rays and UV light damage DNA.

Induced mutations are genetic changes that occur due to an influence outside of a cell, such as exposure to a chemical or radiation. An agent that induces the change is a **mutagen.** Geneticists—who depend on mutants to study cellular processes—often use mutagens to increase the mutation rate in bacteria, making mutants easier to find. The process is referred to as mutagenesis.

Chemical Mutagens

Some chemical mutagens cause base-pair substitutions, and others cause frameshift mutations.

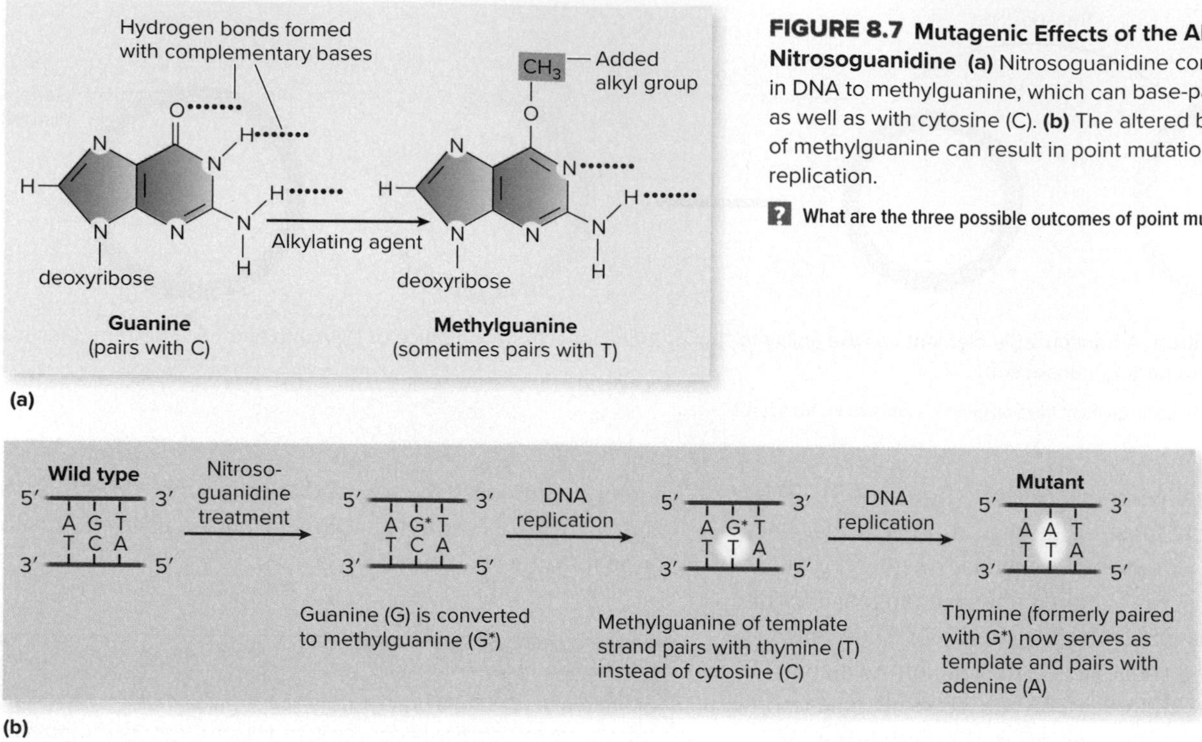

FIGURE 8.7 Mutagenic Effects of the Alkylating Agent Nitrosoguanidine **(a)** Nitrosoguanidine converts guanine bases in DNA to methylguanine, which can base-pair with thymine (T) as well as with cytosine (C). **(b)** The altered base-pairing property of methylguanine can result in point mutations following DNA replication.

? **What are the three possible outcomes of point mutations?**

Chemicals That Modify Nucleobases

A number of different chemicals modify the nucleobases in DNA, changing their base-pairing properties. By doing so, the chemicals increase the chance that an incorrect nucleotide will be incorporated during DNA replication. Chemicals called alkylating agents add a methyl group (CH_3) or other alkyl group onto nucleobases; when the alkylating agent nitrosoguanidine adds a methyl group to guanine, the modified nucleobase sometimes base-pairs with thymine (**figure 8.7**).

MicroByte

Many chemical mutagens are used in cancer therapy to kill rapidly dividing cancer cells. Unfortunately, they also damage DNA in normal cells.

Base Analogs

Base analogs structurally resemble nucleobases but have different hydrogen-bonding properties. The analogs can be mistakenly used in place of the nucleobases when the cells make nucleotides, and DNA polymerases then incorporate these into DNA. When the complementary strand is synthesized, the wrong nucleotide may be incorporated opposite the base analog. For example, 5-bromouracil resembles thymine, but it often base-pairs with guanine (G) instead of with adenine (A); similarly, 2-amino purine resembles adenine but often pairs with cytosine (C) instead of with thymine (T) (**figure 8.8**).

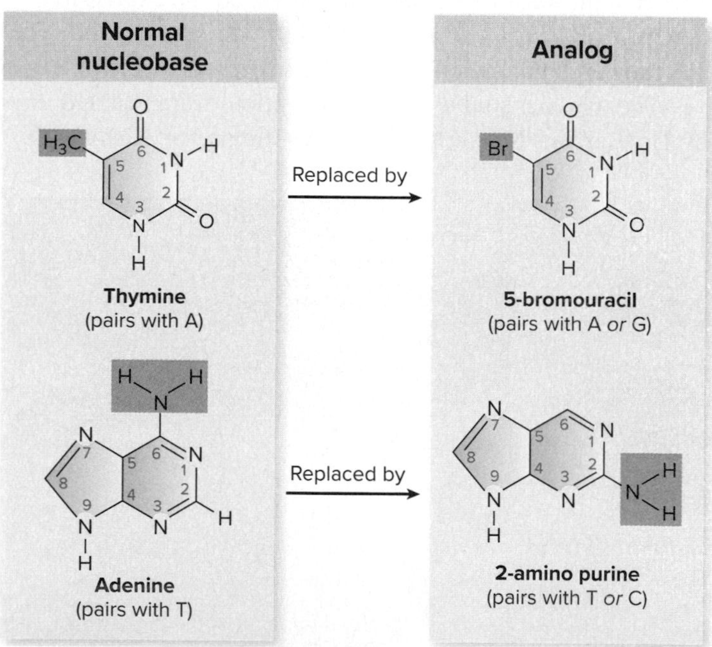

FIGURE 8.8 Two Base Analogs Used in Mutagenesis Some normal nucleobases can be replaced by analogs. The altered base-pairing properties of the analogs can lead to point mutations by the same mechanism shown in figure 8.7b.

? **What base-pair substitution would 5-bromouracil generate?**

Intercalating Agents

Intercalating agents increase the frequency of frameshift mutations because they can insert (intercalate) between adjacent bases of DNA. This pushes the nucleotides apart, producing a space between bases and thereby increasing the chance that insertions or deletions will be made during DNA replication. As in spontaneous frameshift mutations, this often results in the premature generation of a stop codon, giving rise to a shortened protein. Chemicals used to stain DNA in the laboratory are intercalating agents.

Transposons

Transposons, a type of transposable element, can be introduced intentionally into a cell in order to generate mutations; the gene into which a transposon jumps is usually inactivated (see figure 8.5). The structure and biology of transposons is described in section 8.10.

Radiation

Two kinds of radiation are commonly used as mutagens: ultraviolet (UV) light and X rays.

Ultraviolet Light

Ultraviolet light exposure causes covalent bonds to form between adjacent thymine nucleobases on a DNA strand, producing **thymine dimers** (**figure 8.9**). These dimers distort the DNA molecule because they cannot fit properly into the double helix; replication and transcription stop at the distortion, and as a result, the cells will die if the damage is not repaired. How, then, can UV light be mutagenic? Its major mutagenic action is indirect, resulting from the cell's attempt to repair the damage by SOS repair, described in the next section.

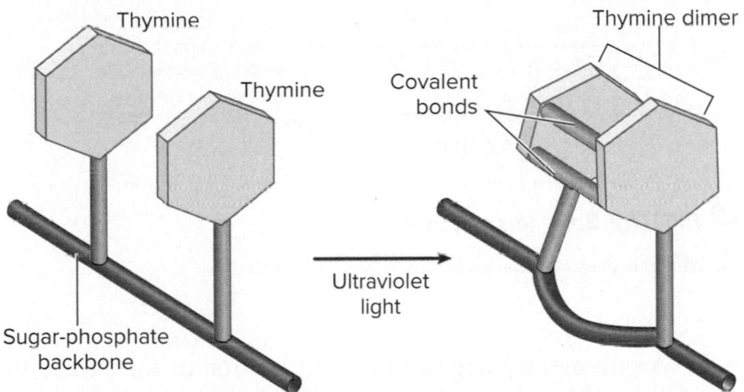

FIGURE 8.9 Thymine Dimer Formation UV light causes covalent bonds to form between adjacent thymine nucleobases on the same strand of DNA, distorting the shape of the DNA.

? What effect does a thymine dimer have on DNA synthesis?

X Rays

X rays cause single- and double-strand breaks in DNA, and they damage nucleobases. Double-strand breaks often result in deletions that are lethal to the cell.

Common mutagens and their characteristics are summarized in **table 8.1.**

MicroAssessment 8.3

Mutagens increase the frequency of mutations. Some chemical mutagens cause base-pair substitutions, but intercalating agents cause frameshift mutations. Mutations from UV light are due to SOS repair; X rays cause breaks in DNA strands and alter nucleobases.

7. How does an intercalating agent cause mutations?
8. What mutagen causes thymine dimers, and why does it kill cells?
9. Why would some bacterial species be more likely than others to develop UV-induced mutations?

8.4 ■ Repair of Damaged DNA

Learning Outcomes

7. Explain how cells repair errors in nucleotide incorporation by DNA polymerases, and the role of methylation in the process.
8. Explain how cells repair damage to nucleobases.
9. Describe three mechanisms cells use to repair thymine dimers.

Spontaneous and mutagen-induced damage to DNA, if not repaired, can quickly lead to cell death and, in animals, cancer. In humans, two genes associated with breast cancer code for enzymes that repair damaged DNA. Mutations that inactivate either gene result in a high probability (about 45–80%) of developing breast cancer.

Mutations are usually rare because cells have multiple mechanisms to repair damaged DNA before errors are passed on to progeny. The mechanisms are not perfect, however, and certain types of DNA damage such as insertional inactivation caused by transposons cannot be repaired.

TABLE 8.1 Common Mutagens

Agent	Action	Result
Chemical Mutagens		
Chemicals that modify nucleobases	Chemical modifications change base-pairing properties of nucleobases	Base-pair substitution
Base analogs	Base-pairing properties differ from those of nucleobases normally found in DNA	Base-pair substitution
Intercalating agents	Insert between base pairs, pushing them apart	Addition or deletion of nucleotides
Transposons	Randomly insert into DNA	Insertional inactivation
Radiation		
Ultraviolet (UV) light	Causes thymine dimers to form	Errors during repair process
X rays	Cause single- and double-strand breaks in DNA	Deletions

Repair of Errors in Nucleotide Incorporation

DNA polymerases sometimes incorporate the wrong nucleotide as they synthesize DNA (see figure 8.2). The resulting mispairing of nucleobases results in a slight distortion in the DNA helix, which can be recognized by enzymes within the cell that then repair the mistake. By repairing the error before the DNA is replicated, the cell prevents the mutation. Two mechanisms for this are proofreading by DNA polymerases and mismatch repair.

Proofreading by DNA Polymerases

DNA polymerases not only synthesize DNA, but most types also check the accuracy of their actions—a process called **proofreading.** Each of these enzymes can back up and excise (remove) a nucleotide not correctly hydrogen-bonded to the opposing nucleobase in the template strand and then insert the correct nucleotide. Although this proofreading function is very effective, it is not perfect, and some mutations remain.

Mismatch Repair

Mismatch repair fixes errors missed by the proofreading of DNA polymerases. A specific protein binds to the site of the mismatched nucleobase, directing an enzyme to cut the sugar-phosphate backbone of the new DNA strand. Another enzyme then degrades a short region of that DNA strand, thereby removing the misincorporated nucleotide. How does the cell know which strand is the new one? This is an important question because if the enzyme were to cut the template strand and not the new one, then the misincorporated nucleotide would remain. The key to the answer lies in methylation of the DNA nucleobases. Soon after a DNA strand is synthesized, an enzyme adds methyl groups (CH_3) to certain nucleobases. This takes time, however, so the new strand is still unmethylated immediately after it is synthesized. Because the template strand is methylated, whereas the new strand is not, the repair enzyme can distinguish between the two (**figure 8.10**). After the nucleotides are removed from the new strand, the combined actions of a DNA polymerase and DNA ligase then fill in that section and seal the gap.

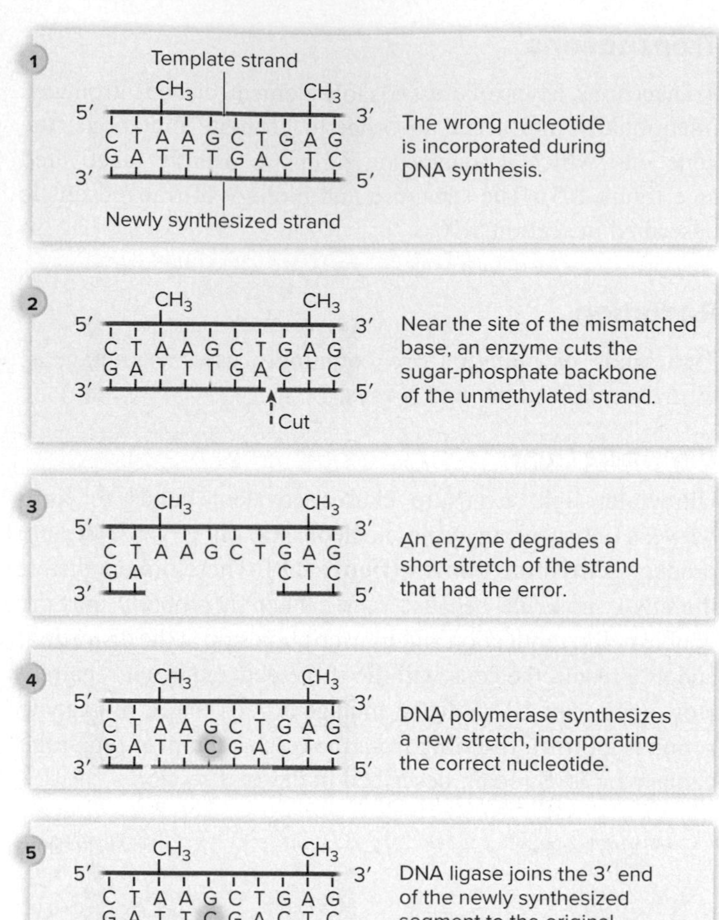

FIGURE 8.10 Mismatch Repair

What role does methylation play in mismatch repair?

Repair of Damage to Nucleobases

Oxidation or other damage to nucleobases in DNA can result in base-pair substitutions if not repaired before the DNA is replicated. A mechanism called **base excision repair** uses a type of enzyme called DNA glycosylase to remove the damaged nucleobase from the sugar-phosphate backbone (**figure 8.11**). Another enzyme then recognizes the space left by the missing nucleobase and cuts the DNA at this site. A DNA polymerase degrades a short section of this strand to remove the damage. This same enzyme synthesizes another strand with the proper nucleotides, and DNA ligase seals the gap in the single-stranded DNA.

Repair of Thymine Dimers

Organisms have mechanisms to prevent the DNA-damaging effects of UV light, a component of sunlight. Mechanisms used by bacteria include (**figure 8.12**):

- **Photoreactivation.** This mechanism, also called light repair, relies on an enzyme that uses the energy of visible light to break the covalent bonds of the thymine dimer. By breaking the bond, the DNA is restored to its original state.
- **Nucleotide excision repair.** This mechanism, also called dark repair, does not require energy from visible light. A specific enzyme recognizes the major distortions in DNA

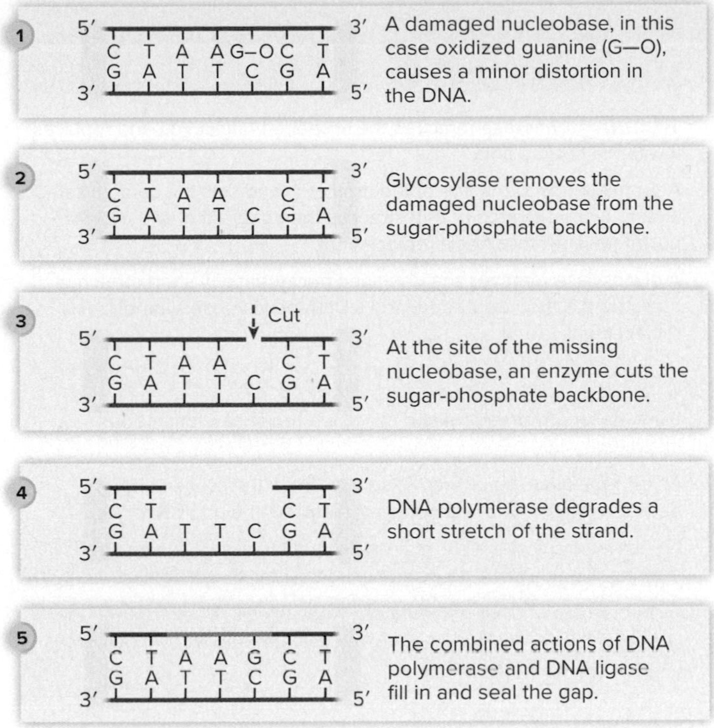

FIGURE 8.11 Base Excision Repair

What would be the effect on cells if they did not have the glycosylase enzyme?

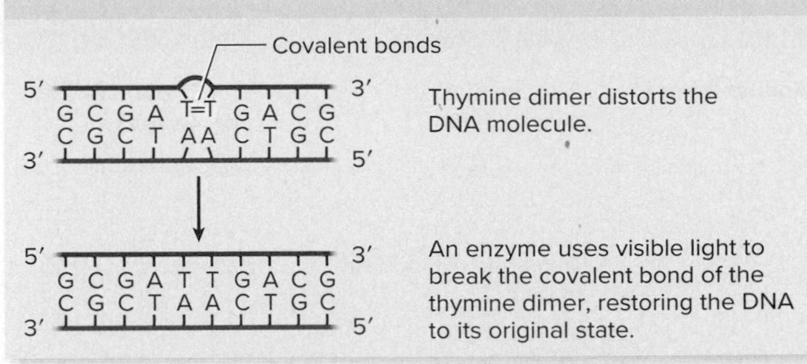

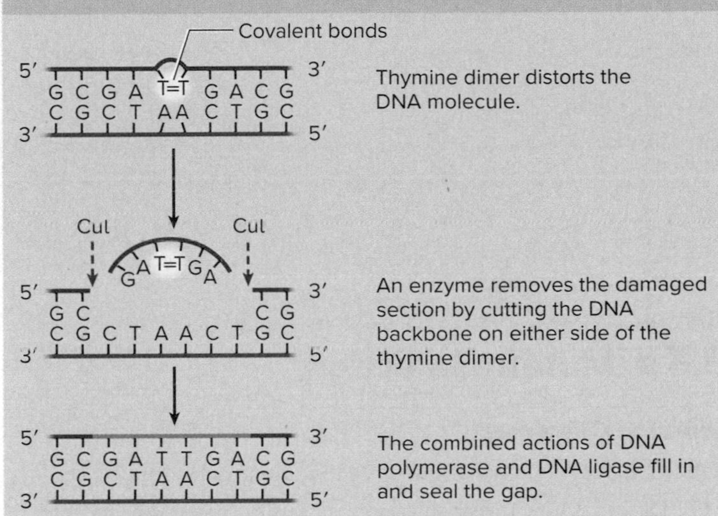

FIGURE 8.12 Repair of Thymine Dimers In photoreactivation, an enzyme uses the energy of light to break the covalent bonds. In nucleotide excision repair, the section containing the dimer is replaced.

How is the mechanism of nucleotide excision repair similar to that of mismatch repair?

that result from thymine dimer formation and removes the DNA strand with the damaged region. A DNA polymerase and DNA ligase then fill in and seal the gap left by the removal of the segment.

MicroByte

In humans, certain defects in the nucleotide excision repair system lead to a 1,000-fold increase in the rate of UV-induced skin cancers.

SOS Repair

SOS repair is a last-effort attempt that bacteria use to repair extensively damaged DNA, such as when UV damage is so great that the thymine-dimer repair mechanisms are overwhelmed. The enzymes that carry out SOS repair are induced when DNA polymerase stalls at a site of significant DNA damage; if not for SOS repair, the cell would be unable to replicate or transcribe its DNA and would therefore die. One component of the SOS system is a special type of DNA polymerase that synthesizes DNA even in extensively damaged regions. Unlike the standard DNA polymerases, however, the SOS DNA polymerase has no proofreading ability, so it often inadvertently introduces errors.

Mechanisms used to repair DNA are summarized in **table 8.2.**

MicroAssessment 8.4

Most DNA polymerases have proofreading ability. Mismatch repair fixes errors missed by the proofreading mechanism; methylation distinguishes the template strand. Base excision repair uses DNA glycosylases to remove damaged nucleobases. Thymine dimers can be repaired through photoreactivation and nucleotide excision repair; severe damage induces the SOS repair system.

10. Distinguish between photoreactivation and nucleotide excision repair.

11. How does UV light cause mutations?

12. To maximize the number of mutations following UV irradiation, should the irradiated cells be incubated in the light or in the dark, or does it make any difference? Explain your answer.

TABLE 8.2 Repair of Damaged DNA

	Type of Defect	Repair Mechanism	Characteristics of Repair
Spontaneous	Error in nucleotide incorporation	Proofreading by DNA polymerases	The DNA polymerase removes the mispaired nucleotide and replaces it with the correct one.
		Mismatch repair	A protein binds to the site of the mismatch and cuts the unmethylated strand. A short stretch of that strand is then degraded, and a DNA polymerase synthesizes a replacement.
	Damaged nucleobase in DNA	Base excision repair	A glycosylase removes the damaged nucleobase. A short stretch of that strand is then degraded, and a DNA polymerase synthesizes a replacement.
Mutagen-Induced			
Chemical	Error in nucleotide incorporation	Proofreading and mismatch repair	Same as for proofreading and mismatch repair of spontaneous mutations.
UV light	Thymine dimer	Photoreactivation (light repair)	An enzyme uses the energy of light to break the covalent bond between the thymine nucleobases, restoring the original molecule.
		Nucleotide excision repair (dark repair)	An enzyme removes a short stretch of the strand containing the thymine dimer; a DNA polymerase then synthesizes a replacement.
		SOS repair	A special type of DNA polymerase synthesizes DNA even when the template is extensively damaged; the cell may survive, but numerous mutations are introduced.

8.5 ■ Mutant Selection

Learning Outcomes

10. Compare and contrast direct and indirect selection.
11. Describe how direct selection is used to screen for possible carcinogens.

Mutations are rare events, even when mutagens are used. This presents a challenge to a scientist who wants to isolate a desired mutant. In a culture containing several billion cells, perhaps only one cell has the mutation of interest, making it very difficult to find. Two methods—direct and indirect selection—are used to isolate such mutants.

Direct Selection

Mutants that can grow under conditions in which the parent cells cannot are usually easy to isolate by **direct selection.** In this method, cells are inoculated onto an agar medium that supports the growth of the mutant but does not allow the parent to grow. For example, antibiotic-resistant mutants can be easily selected by inoculating cells onto a medium containing the antibiotic. Only the resistant cells will form colonies (**figure 8.13**).

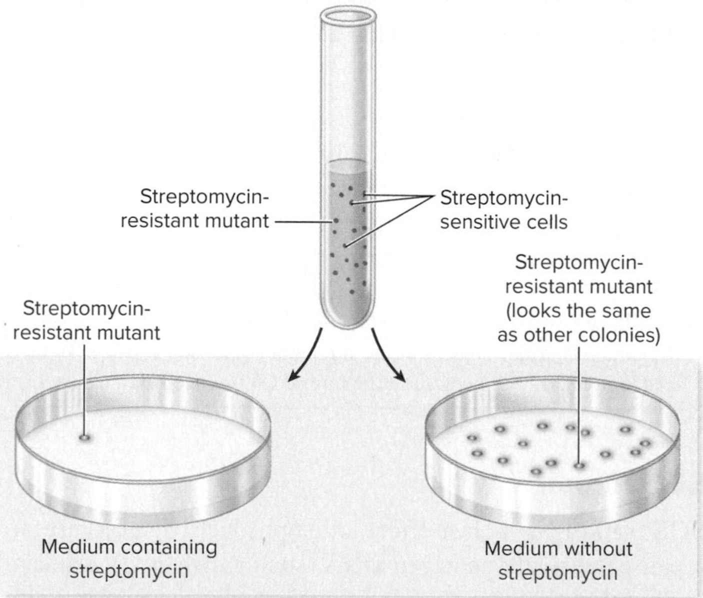

FIGURE 8.13 Direct Selection of Mutants Only cells carrying a mutation that confers resistance to streptomycin can grow on the selective medium used here.

Direct selection cannot be used to isolate auxotrophic mutants in a culture of prototrophic cells. Why not?

Indirect Selection

Indirect selection is used to isolate an auxotrophic mutant from a prototrophic parent strain. This is more difficult than direct selection because no medium allows growth of auxotrophs but not prototrophs. For example, Trp$^-$ mutants (auxotrophs) can grow only on a complex medium such as nutrient agar that supplies their required tryptophan, but Trp$^+$ parent cells (prototrophs) also grow on that medium. To deal with this problem, a clever indirect selection method called **replica plating** can be used. It relies on the use of a master plate—a plate that has colonies of both auxotrophic mutants and prototrophs—made by spreading the bacterial culture onto a nutrient agar plate and then incubating it to obtain isolated colonies. The steps from there include (**figure 8.14**):

① The master plate is pressed onto sterile velvet (a fabric with tiny threads that stand on end like small bristles), which picks up some cells of every colony on the plate.

② That velvet is then used to create two replica (copy) plates—one nutrient agar and one glucose-salts agar (a minimal medium lacking added nutrients; see table 4.6). This is done by pressing the plates in succession and in the same orientation onto the velvet, thereby inoculating both with cells from the master plate; a mark on the plates is used to maintain a consistent orientation. When the plates are incubated, prototrophs grow on both types of media, but auxotrophs grow only on the nutrient agar.

③ Auxotrophic mutants can be recognized because they form colonies on the nutrient agar plate but not on the glucose-salts agar replica. The growth factor required by a mutant can then be determined by adding various nutrients individually to a glucose-salts medium and finding out which one promotes growth.

Screening for Possible Carcinogens

Many cancers appear to be caused by chemicals that are **carcinogens** (meaning "cancer-generating"), and most carcinogens

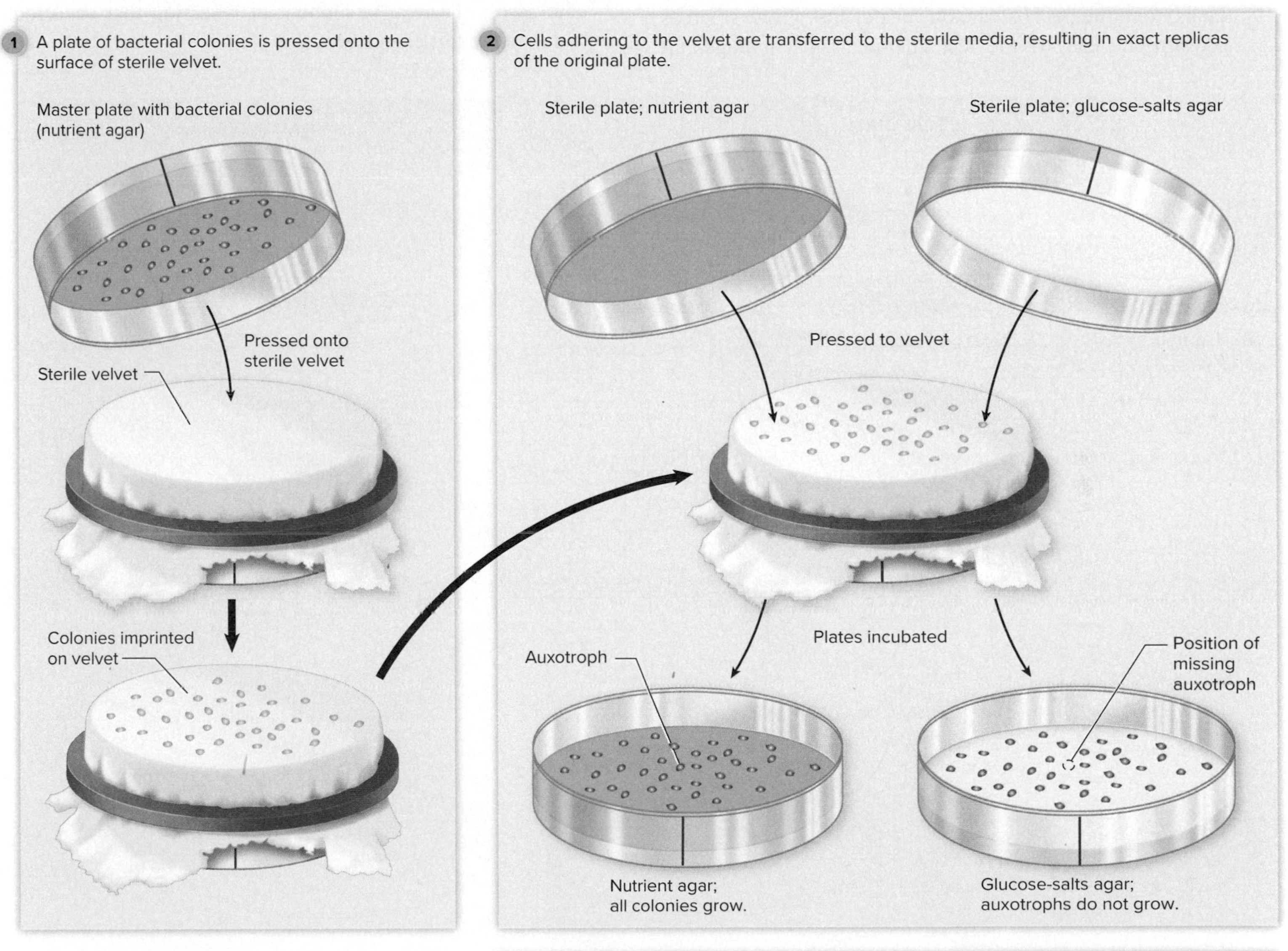
1 A plate of bacterial colonies is pressed onto the surface of sterile velvet.
Master plate with bacterial colonies (nutrient agar)
Pressed onto sterile velvet
Sterile velvet
Colonies imprinted on velvet
2 Cells adhering to the velvet are transferred to the sterile media, resulting in exact replicas of the original plate.
Sterile plate; nutrient agar
Sterile plate; glucose-salts agar
Pressed to velvet
Plates incubated
Auxotroph
Position of missing auxotroph
Nutrient agar; all colonies grow.
Glucose-salts agar; auxotrophs do not grow.
3 Auxotrophic mutants form colonies on the nutrient agar but not on the glucose-salts agar.

FIGURE 8.14 Indirect Selection of Mutants by Replica Plating The procedure shown was used by Joshua and Esther Lederberg, who developed the method in the 1950s, and it continues to be used today in many laboratories. Mutants are identified by comparing the growth of colonies on the two plates.

? Why go to the trouble of creating a master plate (why not simply plate the initial culture on both nutrient agar and glucose-salts agar)?

are mutagens. Thousands of common chemicals are potentially dangerous and must be tested for carcinogenic activity. Testing in animals takes several years and can be a costly way to test a single compound. To simplify the process, several quick, inexpensive methods have been developed to screen for possible carcinogens by examining the mutagenic effect of a given chemical on a microbiological system. The **Ames test,** devised by Bruce Ames and his colleagues over 50 years ago, illustrates the concept; it uses a histidine-requiring auxotroph (His^-) of *Salmonella enterica* serovar Typhimurium to inoculate two different plates (**figure 8.15**):

- **Test plate.** To do the test, the His^- bacteria are inoculated onto a glucose-salts agar plate, and liquid containing the chemical to be screened is then added. This agar medium lacks histidine, so His^- auxotrophs cannot grow, whereas prototrophs can. If the test chemical is a mutagen, it will induce mutations; some of the mutations will revert the His^- cells to the prototrophic phenotype, and these revertants will grow to form colonies on the agar. Thus, the number of colonies growing on the plate reflects the relative number of mutations that have occurred in the His^- culture.
- **Control plate.** As with any scientific procedure, a control must be done as part of the test. The need for this is obvious when you consider that spontaneous mutations occur routinely (at a low rate), so at least a few revertants will likely be present even if the chemical being tested is not a mutagen. For the control, the same procedure just described is used, but without the test chemical added. After incubation, the control and test plates are compared to assess the relative amount of reversion that occurred due to the test chemical.

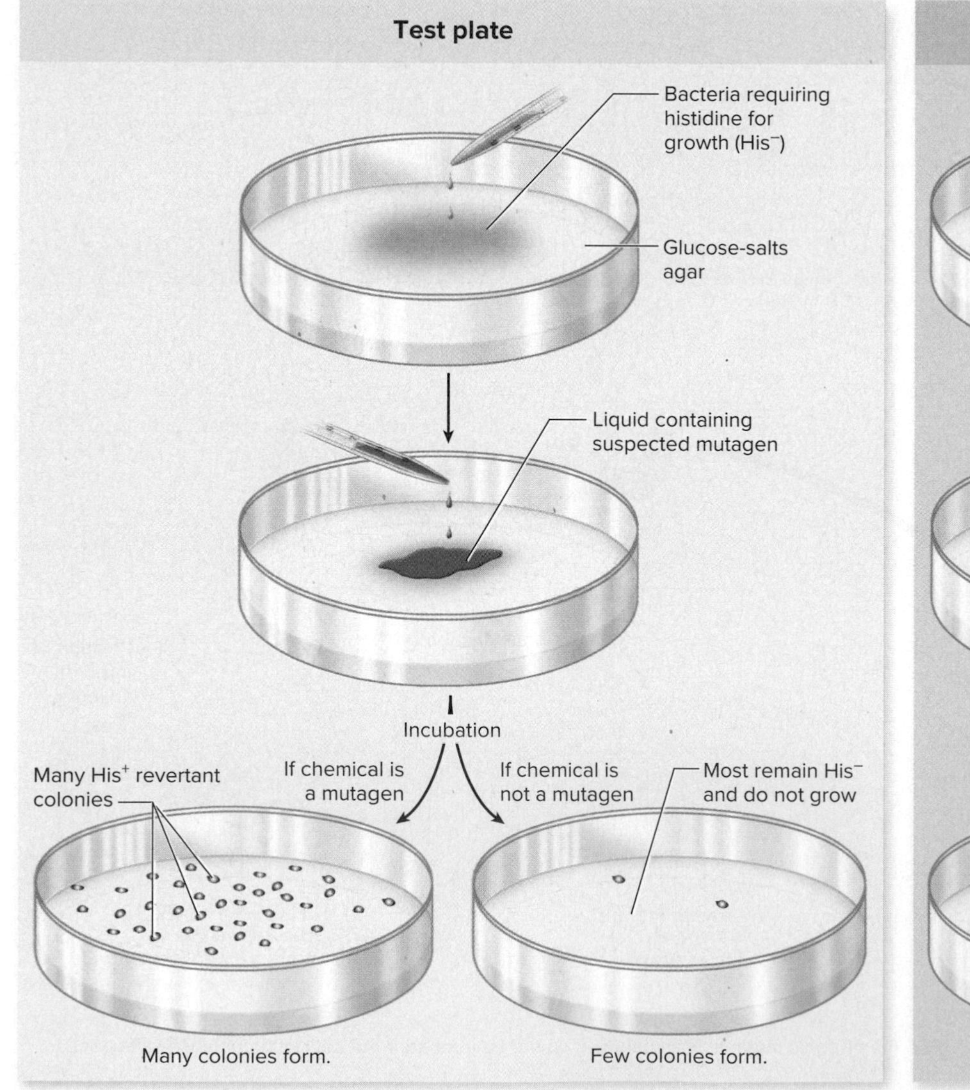

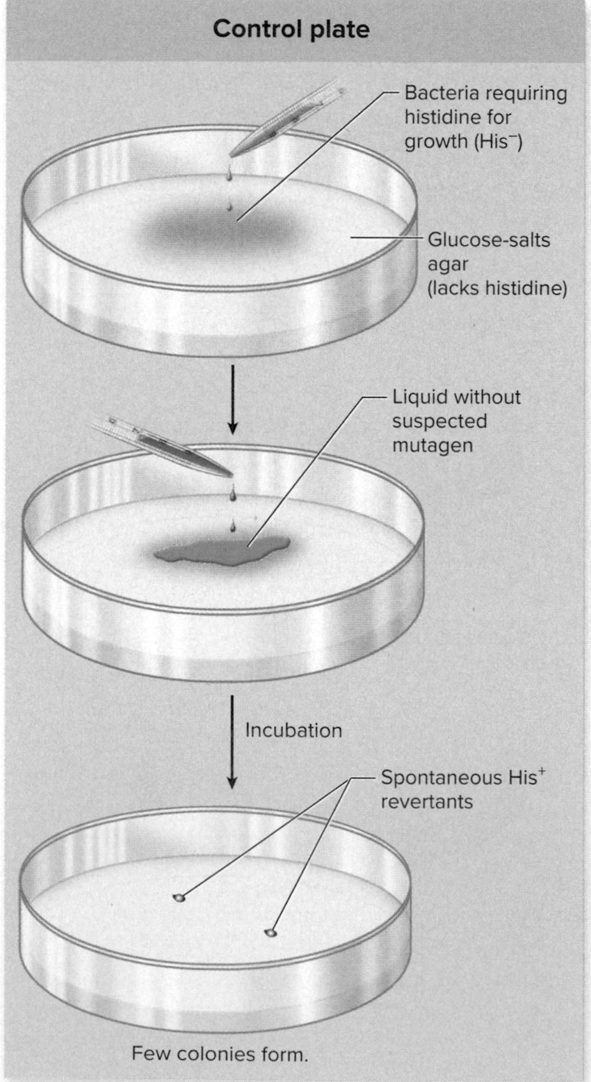

FIGURE 8.15 Ames Test to Screen for Mutagens If the chemical tested is a mutagen, it will increase the rate of His^+ reversion mutations; a control plate determines the spontaneous rate of reversion to His^+.

Why do some cells grow on the control plate in the absence of a mutagen?

Animal liver extract may also be added as part of the Ames test. This is because enzymes produced by the liver modify certain chemicals, and in some cases, this inadvertently transforms the chemicals into a mutagenic form. That is, the chemical would not normally be mutagenic, but once inside the body, the liver enzymes inadvertently convert it into a form that causes mutations.

If the test chemical causes mutation in the His^- cells, additional testing must be done in animals to determine if it is a carcinogen. This step is necessary because although most carcinogens are mutagens, not all mutagens are carcinogens. Thus, the Ames test is simply a valuable screening tool.

MicroAssessment 8.5

Mutants can be selected using either direct or indirect techniques. Replica plating is used for indirect selection. The Ames test is used to screen chemicals to determine which ones are mutagens and therefore possible carcinogens.

13. Distinguish between the kinds of mutants that can be isolated by direct selection and those for which indirect selection is needed.
14. In the Ames test, what information does the control provide?
15. Why might a chemical that causes mutations in bacteria not be cancer-causing in animals?

HORIZONTAL GENE TRANSFER AS A MECHANISM OF GENETIC CHANGE

8.6 ■ Overview of Horizontal Gene Transfer

Learning Outcomes

12. Distinguish between the terms *bacterial transformation*, *conjugation*, and *transduction*.
13. Explain why DNA that enters a bacterium must either be a replicon or become part of a replicon in order to be passed on to progeny.

Microorganisms commonly acquire genes from other cells through horizontal gene transfer (HGT). Not only does HGT allow organisms to change and adapt, scientists can use the transfer processes in the laboratory to intentionally move DNA from one organism (the donor) to another (the recipient). To determine if the recipient has indeed acquired new characteristics, it must be genetically different from the donor. Recipient cells will be **recombinants,** meaning they have properties of each of the original strains.

Figure 8.16 illustrates how HGT can be demonstrated. Two bacterial strains are used, neither of which can grow on a glucose-salts medium because of multiple growth factor requirements. Strain A is His^-, Trp^- (meaning it cannot make histidine or tryptophan and therefore requires those amino acids in the medium); strain B is Leu^-, Thr^- (meaning it cannot make leucine or threonine and therefore requires those amino acids in the medium). Neither population is likely to give rise to a spontaneous mutant that grows on the glucose-salts agar because two simultaneous mutations in the same cell would be required. When the strains are mixed and then spread on a glucose-salts agar plate, colonies should form only if cells of one strain acquired genes from cells of the other strain.

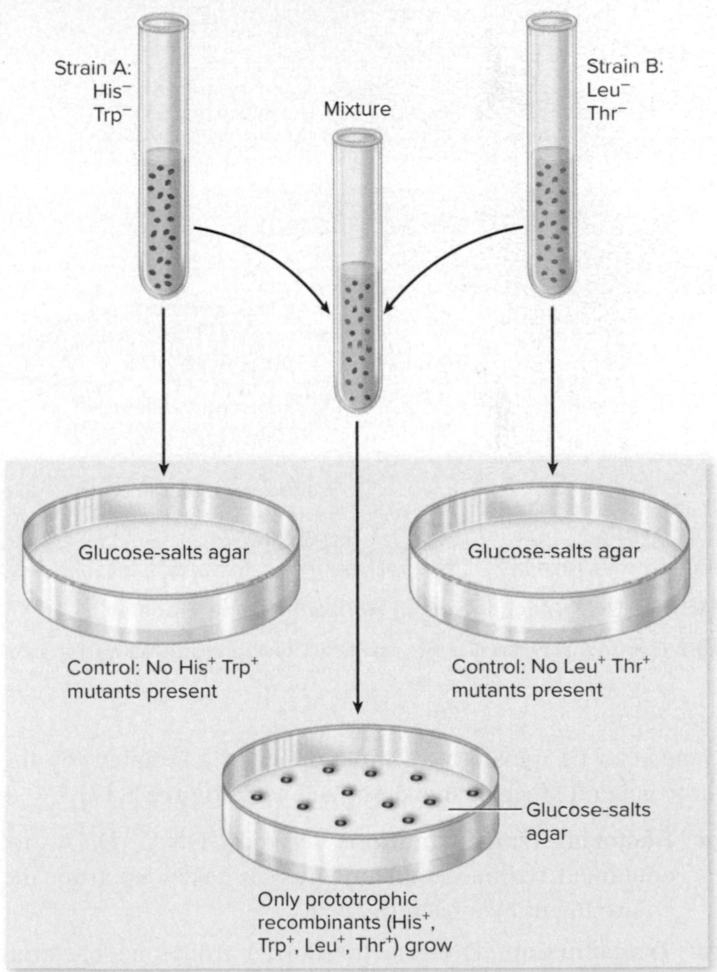

FIGURE 8.16 Experimental Demonstration of Horizontal Gene Transfer (HGT) in Bacteria Recombinant colonies have genetic traits from both strains present in the mixture. Control plates demonstrate that these colonies are not a result of spontaneous mutation.

When demonstrating HGT, why is it important to use strains that each require at least two different amino acids?

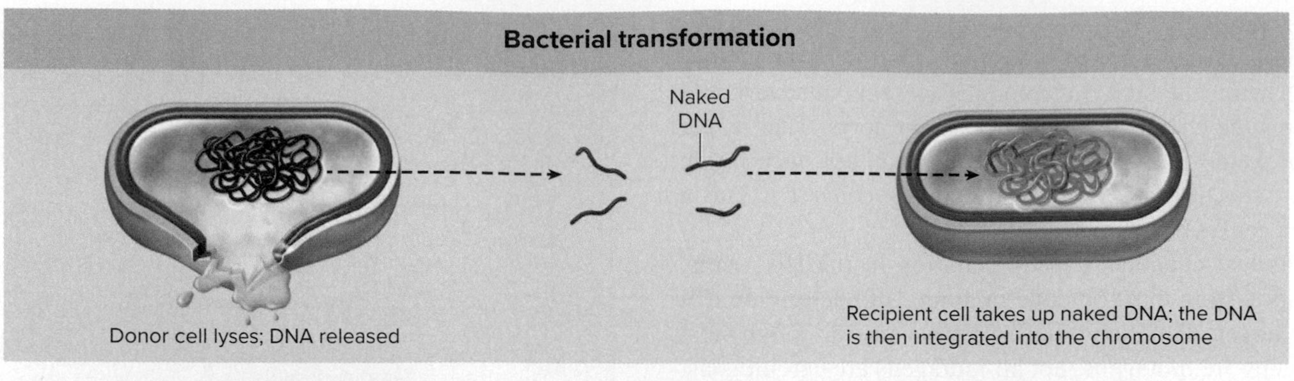

Donor cell lyses; DNA released

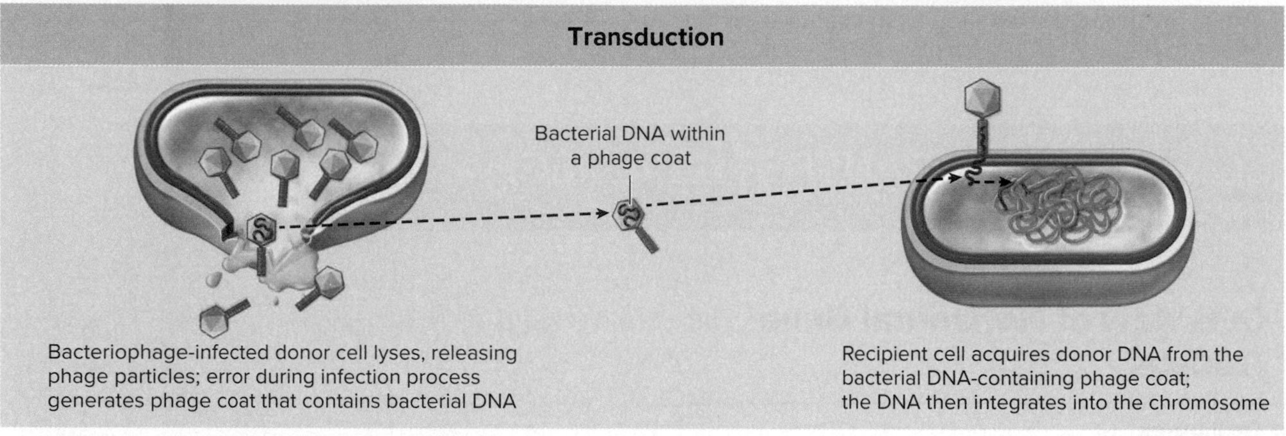

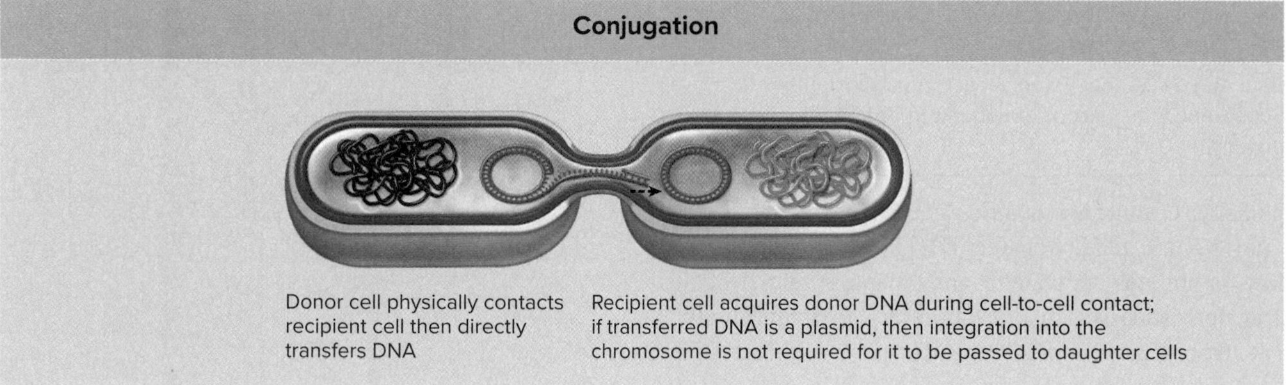

Bacterial DNA within a phage coat
Bacteriophage-infected donor cell lyses, releasing phage particles; error during infection process generates phage coat that contains bacterial DNA

Conjugation
Donor cell physically contacts recipient cell then directly transfers DNA
Recipient cell acquires donor DNA during cell-to-cell contact; if transferred DNA is a plasmid, then integration into the chromosome is not required for it to be passed to daughter cells

FIGURE 8.17 Mechanisms of Horizontal Gene Transfer

? DNAse (an enzyme that degrades naked DNA) would prevent which mechanism of DNA transfer?

Genes can be transferred from a donor to a recipient by the three general mechanisms described here (**figure 8.17**):

- **Bacterial transformation.** "Naked" DNA (DNA not contained within a cell or virus) is taken up from the environment by a bacterial cell.
- **Transduction.** DNA is transferred from one bacterial cell to another by a bacteriophage (a virus that infects bacteria; also called a phage).
- **Conjugation.** DNA is transferred during cell-to-cell contact.

Following gene transfer, a recipient cell must replicate the DNA to pass it on to daughter cells. This can happen if the transferred DNA is a **replicon,** meaning it has an origin of replication. Plasmids and chromosomes are replicons, but fragments of chromosomal DNA are not. If a chromosomal fragment is transferred, then it must become integrated into a replicon to be maintained in a population (**figure 8.18**). This involves a process called **homologous recombination,** which can happen only if the donor DNA is similar in nucleotide sequence to a region in the recipient cell's genome. In homologous recombination, a series of enzymes position the donor DNA next to a similar sequence in the recipient's genome and then exchange the segments, replacing the recipient DNA—which will be degraded—with donor DNA.

FOCUS YOUR PERSPECTIVE 8.1

The Biological Function of DNA: A Discovery Ahead of Its Time

In the 1920s, Frederick Griffith, an English bacteriologist, was studying *Streptococcus pneumoniae* ("pneumococcus"), a bacterial species that commonly causes pneumonia. These bacteria are pathogenic only if they are encapsulated (make a capsule). Unencapsulated pneumococci are not pathogenic. Griffith decided to do a four-part experiment, using mice (**box figure 8.1**): (1) He injected encapsulated bacteria into mice, and the mice died; (2) he injected unencapsulated bacteria into mice, and the mice remained healthy; (3) he injected heat-killed encapsulated bacteria into mice, and those mice also lived; and (4) he mixed heat-killed encapsulated (pathogenic) bacteria with live, unencapsulated (non-pathogenic) bacteria and injected them into mice. Surprisingly, the mice died. Griffith isolated live, encapsulated pneumococci from these mice.

Two years after Griffith reported these findings, another investigator, Martin H. Dawson, lysed heat-killed encapsulated pneumococci and passed the suspension of ruptured cells through a very fine filter, through which only the cytoplasmic contents of the bacteria could pass. When he mixed the filtrate (the material that had passed through the filter) with living bacteria unable to make a capsule, some bacteria began making a capsule. Moreover, the progeny of these bacteria could also make a capsule. Something in the filtrate was "transforming" the harmless unencapsulated bacteria into ones that could make a capsule.

What was this transforming principle? In 1944, after years of painstaking chemical analysis of lysates capable of transforming pneumococci, three investigators, Oswald T. Avery, Colin MacLeod, and Maclyn McCarty, purified the active compound and then wrote one of the most important papers ever published in biology. In it, they reported that the transforming molecule was DNA. The significance of their discovery was not appreciated at the time, and scientists were slow to recognize its significance. None of the three investigators received a Nobel Prize, although many scientists believe that they deserved it. Their studies pointed out that DNA is a key molecule in the scheme of life and led to James Watson and Francis Crick's determination of its structure, which they published in 1953. The understanding of the structure and function of DNA revolutionized the study of biology and started the era of molecular biology.

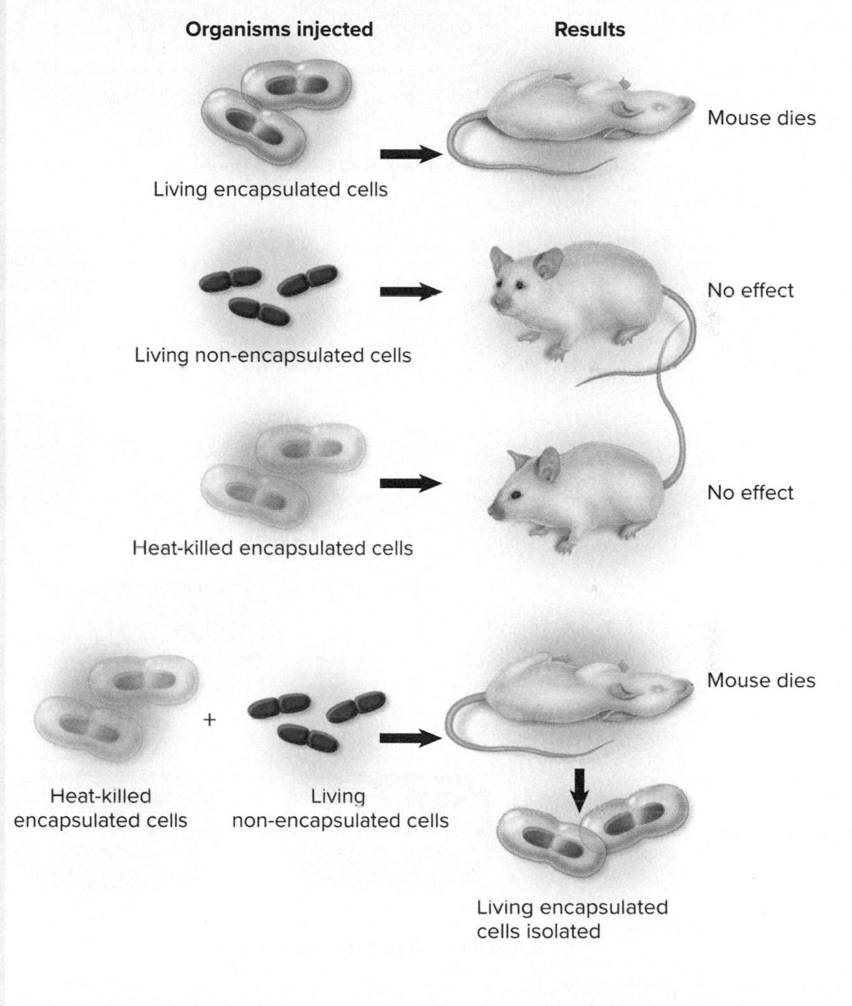

BOX FIGURE 8.1 Griffith's Demonstration of Genetic Transformation

MicroAssessment 8.6

Genes can be transferred from a donor cell to a recipient cell by three different mechanisms: bacterial transformation, transduction, and conjugation. To replicate in a recipient cell, the transferred DNA must either be a replicon or integrate into that cell's genome.

16. Which mechanism of horizontal gene transfer requires that the donor and recipient cells physically contact each other?
17. What is a replicon?
18. The donor cell usually survives which DNA transfer process?

FIGURE 8.18 DNA Must Be Part of a Replicon to Be Maintained in a Population **(a)** DNA without an origin of replication will not be passed on to any additional cells during growth of the population. **(b)** If the DNA becomes integrated within a replicon of the cell, it will be inherited by all daughter cells.

If the DNA fragment encoded penicillin resistance, how could you experimentally distinguish between (a) and (b)?

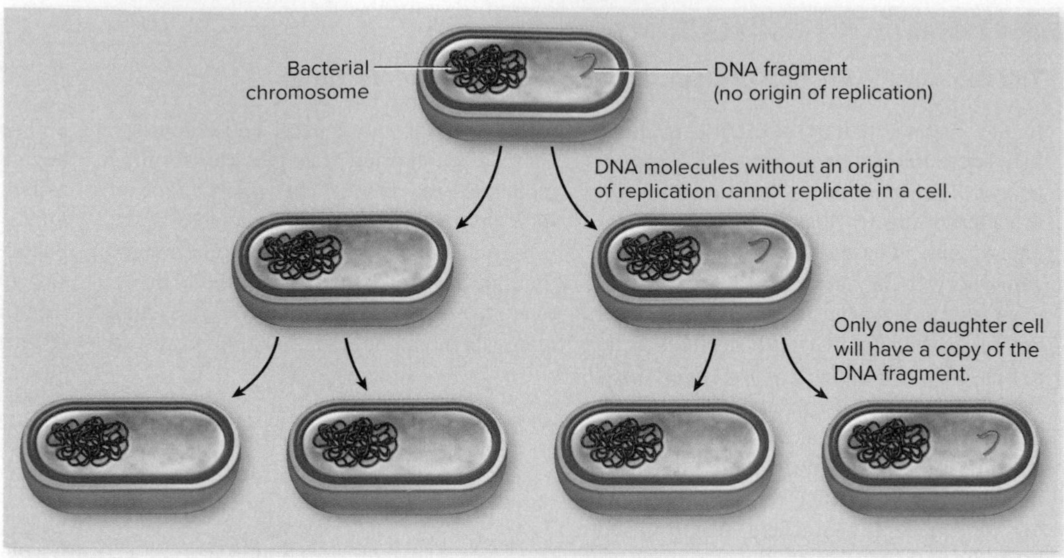

(a) Non-integrated DNA fragment

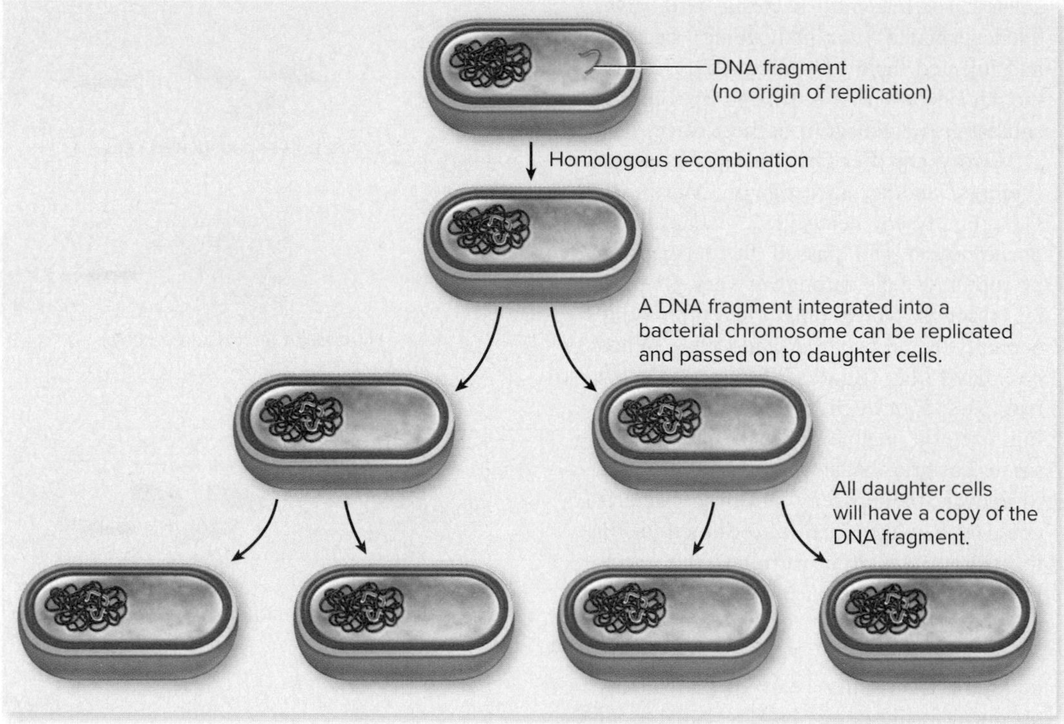

DNA fragment (no origin of replication)
Homologous recombination
A DNA fragment integrated into a bacterial chromosome can be replicated and passed on to daughter cells.
All daughter cells will have a copy of the DNA fragment.

(b) Integrated DNA fragment

8.7 ■ Bacterial Transformation

Learning Outcome

14. Describe the process of bacterial transformation, including the role of competent cells.

Bacterial transformation, also referred to as DNA-mediated transformation or simply transformation, involves the uptake of "naked" DNA by recipient cells. Naked DNA is free DNA in the cells' surroundings; it is not contained within a cell or a virus. Naked DNA often originates from cells that have died. As a cell bursts, its long and tightly packed chromosome breaks up into hundreds of small pieces. Another source of naked DNA is certain bacterial species that secrete small pieces of DNA, presumably as a means of promoting transformation.

The fact that naked DNA is used for bacterial transformation can be demonstrated by adding DNase (an enzyme that degrades DNA) to the mixture of donor and recipient cells. DNase will destroy only DNA that is free in the medium; it cannot access/degrade DNA within a cell or a phage. Thus, if DNAse prevents the recipient cell from acquiring DNA, the donor's DNA must have been naked.

Competence

In order for transformation to occur, the recipient cell must be **competent**—a specific physiological state that allows the cell to take up DNA. Most competent bacteria take up DNA regardless of its source, but some species accept DNA only from closely related bacteria; the recipient recognizes the related donor DNA by characteristic nucleotide sequences located throughout the genome.

In the several dozen prokaryotic species that can become competent naturally, the process is tightly controlled. Some species are always competent, whereas others become so only under specific conditions, such as when the population reaches a certain density or when nutrients are in short supply. The fact that some species become competent only under precise environmental conditions highlights the remarkable ability of seemingly simple cells to sense their surroundings and adjust their behavior accordingly.

E. coli and most other organisms commonly used in biotechnology do not become competent naturally, but they can be induced to take up DNA by treating them with certain chemicals and conditions. This section will focus only on the natural processes.

The Process of Natural Transformation

Figure 8.19 illustrates the steps of natural transformation, using the transfer of genes conferring streptomycin resistance to a streptomycin-sensitive cell as an example:

1. A double-stranded donor DNA molecule encoding streptomycin resistance (Str^R) binds to a specific receptor on the surface of the competent cell.
2. One strand of the donor DNA enters the cell; nucleases at the cell surface degrade the other strand.
3. Inside the recipient cell, the strand of donor DNA integrates into the recipient's genome by homologous recombination; the recipient's DNA strand it replaces will be degraded.
4. When the recipient cell's chromosome is replicated, only one copy will contain the donor DNA (because only one strand of DNA entered the cell and was integrated into the double-stranded genome). Thus, when the recipient cell divides, only one daughter cell will inherit the Str^R gene.
5. The transformed cell (the daughter cell that inherited the Str^R gene) grows on a medium containing streptomycin; other cells are killed.

The example just described only focused on the transfer of the Str^R gene. In an actual transformation experiment, many other donor genes will have been transferred and incorporated by cells of the recipient strain. Those cells, however, will go undetected without a mechanism to recognize them.

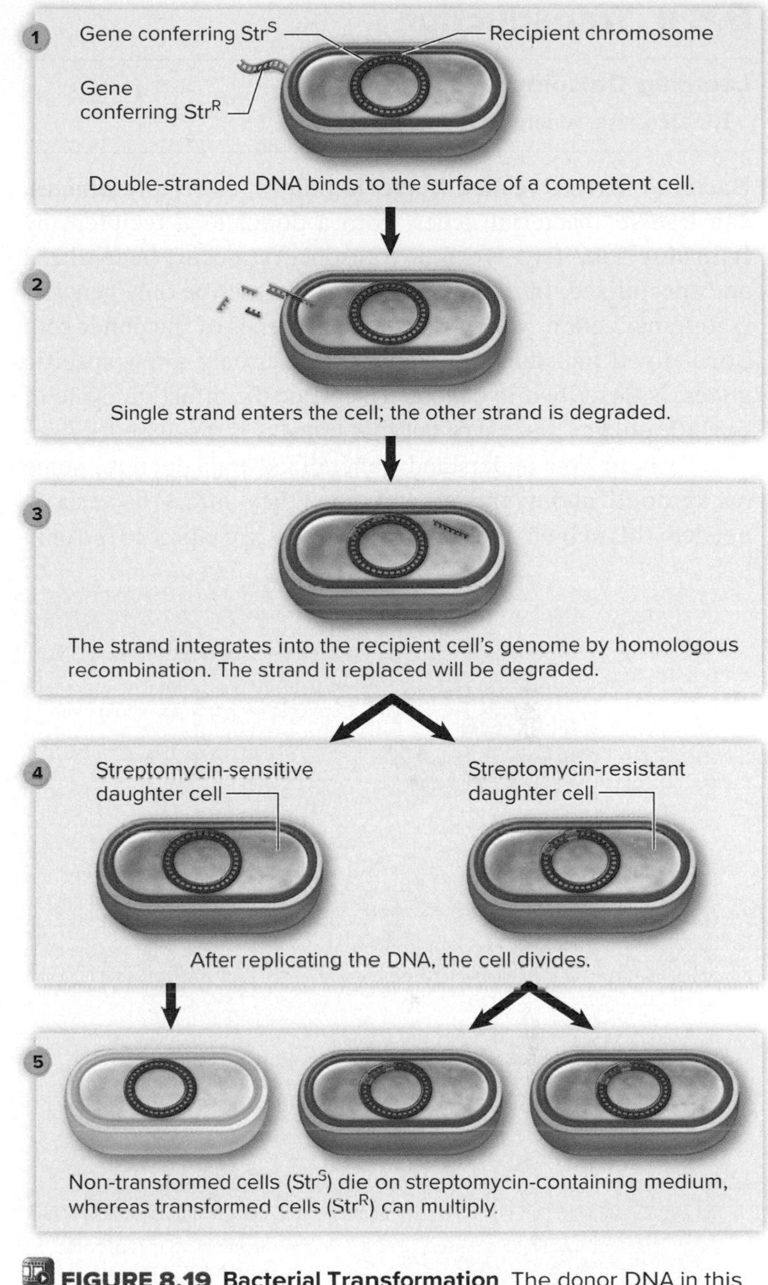
1
Gene conferring Str^S
Recipient chromosome
Gene conferring Str^R
Double-stranded DNA binds to the surface of a competent cell.
2
Single strand enters the cell; the other strand is degraded.
3
The strand integrates into the recipient cell's genome by homologous recombination. The strand it replaced will be degraded.
4
Streptomycin-sensitive daughter cell
Streptomycin-resistant daughter cell
After replicating the DNA, the cell divides.
5
Non-transformed cells (Str^S) die on streptomycin-containing medium, whereas transformed cells (Str^R) can multiply.

FIGURE 8.19 Bacterial Transformation The donor DNA in this case contains a gene conferring resistance to streptomycin (Str^R).

How would this figure change if double-stranded DNA were incorporated into the donor cell's chromosome?

MicroAssessment 8.7

In bacterial transformation, DNA is released from donor cells and taken up by competent recipient cells. Competent cells bind DNA and take up a single strand; that strand then integrates into the genome by homologous recombination.

19. How does DNase prevent transformation?

20. Describe two ways by which DNA can be released from cells.

21. In step 3 of figure 8.19, if a DNA repair mechanism immediately repairs the mismatch between the integrated donor strand and the recipient's chromosomal strand, how would the final outcome of the process be affected?

8.8 ■ Transduction

Learning Outcome

15. Describe generalized transduction.

Bacterial viruses, called **bacteriophages,** or simply **phages,** can transfer bacterial genes from a donor to a recipient by **transduction.** Two types of transduction occur: generalized and specialized. In this chapter, we will describe only generalized transduction, which transfers any genes of the donor cell. Specialized transduction, which transfers only a few specific genes, is described in chapter 13—after the infection cycle of bacteriophages that carry out this process is discussed.

To help you understand generalized transduction, some background about phages and how they infect bacteria is needed; this subject is covered more fully in chapter 13, so only essential details are described here. Phages consist of genetic material, either DNA or RNA (never both), surrounded by a protein coat. A phage infects a bacterium by attaching to the cell and then injecting its nucleic acid into that cell. Enzymes encoded by the phage genome then cut the bacterial DNA into small pieces. Next, the bacterial cell's enzymes replicate the phage nucleic acid and synthesize proteins that make up the phage coat. The phage nucleic acid then enters the phage coat, and the various components assemble to produce complete phage particles. These new phage particles are released from the bacterial cell, usually as a result of host cell lysis. The phage particles then attach to other bacterial cells and begin new cycles of infection.

Generalized transduction results from a rare error that sometimes occurs during the construction of phage particles (**figure 8.20**). A fragment of bacterial DNA (produced

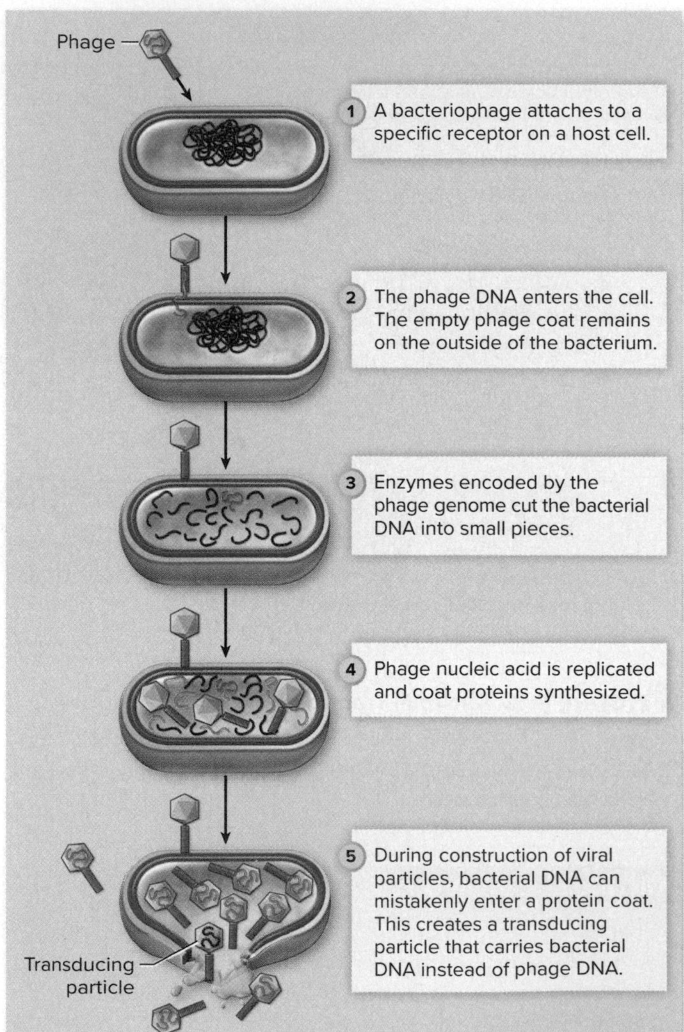

(a) Formation of a transducing particle

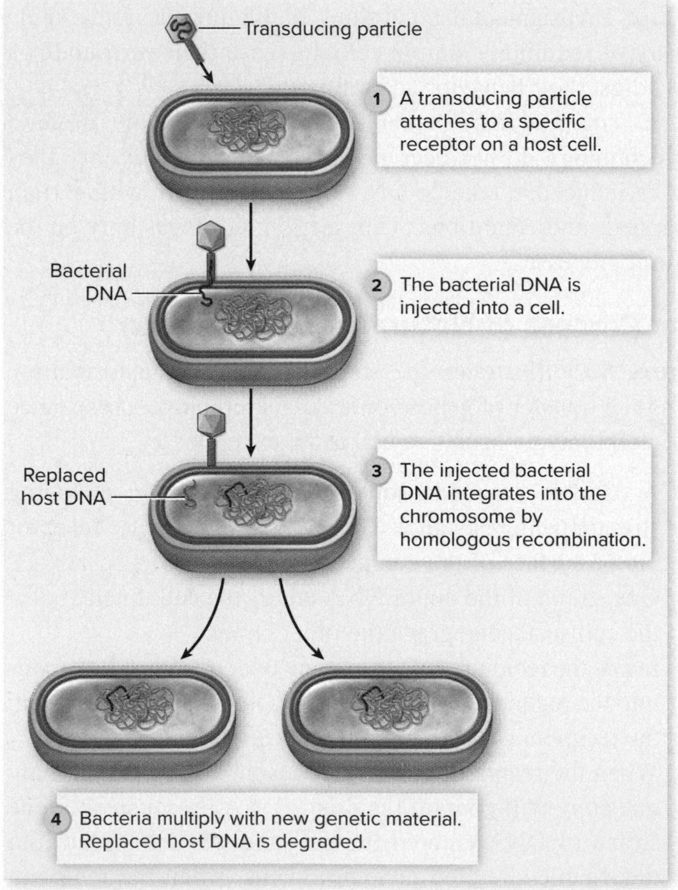

(b) The process of transduction

FIGURE 8.20 Generalized Transduction (a) An error during construction of phage particles produces a transducing particle, which contains bacterial DNA instead of phage DNA. **(b)** The bacterial DNA carried by the transducing particle is injected into a new host, resulting in generalized transduction. Essentially any bacterial gene can be transferred this way.

? After a phage injects its DNA into a bacterial cell, the cell begins making phage proteins. Why are phage proteins not made after a generalized transducing particle injects the DNA it carries?

when the phage-encoded enzyme cuts the bacterial genome) mistakenly enters the phage protein coat. This error creates what is called a **transducing particle,** which carries no phage DNA and therefore is not a phage. Like phage particles, a transducing particle will attach to another bacterial cell and inject the DNA it contains. The transducing particle, however, injects only bacterial DNA because that is all it contains. The bacterial DNA may then integrate into the recipient's chromosome by homologous recombination.

MicroAssessment 8.8

Transduction is the transfer of bacterial DNA from one cell to another by means of a bacteriophage. It results from an error that occurs during the infection cycle of the bacteriophage.

22. What error in the phage replication cycle leads to generalized transduction?
23. What is a transducing particle?
24. Two bacterial genes are transduced simultaneously. What does this suggest about the location of the two genes relative to each other?

8.9 ■ Conjugation

Learning Outcome

16. Compare and contrast conjugation involving the following donor cells: F⁺, Hfr, and F′.

Conjugation is a complex process that requires contact between donor and recipient bacterial cells. Gram-positive and Gram-negative bacteria can both transfer DNA this way, but the process is quite different in the two groups. For simplicity, we will describe conjugation only in Gram-negative bacteria.

Plasmid Transfer

Plasmids are most frequently transferred to other cells by conjugation. These DNA molecules are replicons, so they can be replicated in a cell without integrating into the recipient's chromosome.

Conjugative plasmids direct their own transfer from donor to recipient cells. The most thoroughly studied example is the **F plasmid** (F stands for fertility) of *E. coli.* Although this plasmid does not encode any notable characteristics except those required for transfer, other conjugative plasmids encode resistance to certain antibiotics, which explains how such resistance can easily spread among a population of cells. *E. coli* cells that contain the F plasmid are designated **F^+,** whereas those that do not are **F^-.** The F plasmid encodes several proteins required for conjugation, including the **F pilus,** also referred to as the sex pilus (**figure 8.21**).

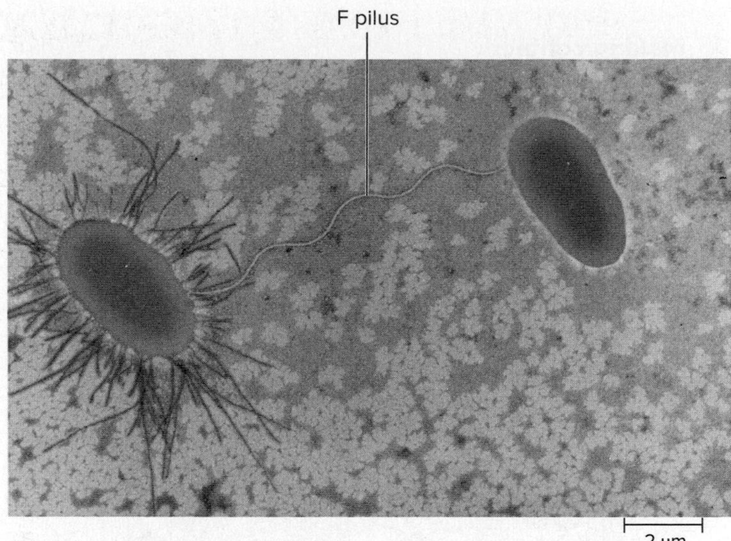

FIGURE 8.21 F Pilus Joining a Donor and Recipient Cell
Dennis Kunkel/SPL/Science Source

What are the hair-like appendages on the cell on the left?

Plasmid transfer involves a series of steps (**figure 8.22**):

1. **Making contact.** The F pilus of the donor cell binds to a specific receptor on the outer membrane of the recipient.
2. **Initiating transfer.** After contact, the F pilus retracts, pulling the two cells together. Meanwhile, a plasmid-encoded enzyme cuts one strand of the F plasmid at a specific nucleotide sequence, the origin of transfer.
3. **Transferring DNA.** A single strand of the F plasmid enters the F^- cell. Once inside the recipient cell, that strand serves as a template for synthesis of the complementary strand, generating an F plasmid. Likewise, the strand that remains in the donor serves as a template for DNA synthesis, regenerating the F plasmid. The transfer takes only a few minutes.
4. **Transfer complete.** Both the donor and the recipient cells are now F^+, so they can act as donors of the F plasmid.

Chromosome Transfer

Chromosomal DNA transfer is less common than plasmid transfer and involves **Hfr cells** (meaning **h**igh **f**requency of **r**ecombination cells). These are strains in which the F plasmid has integrated into the chromosome by homologous recombination, which happens on rare occasions. As shown in **figure 8.23,** integration of the F plasmid is reversible; the same process that generates an Hfr cell also allows the integrated plasmid to excise from the chromosome. In some cases, a mistake occurs during the excision process, creating what is called an F′ cell; the significance of this will be described when we discuss F′ donors.

When an Hfr cell transfers chromosomal DNA, the process involves the same general steps as for the F plasmid:

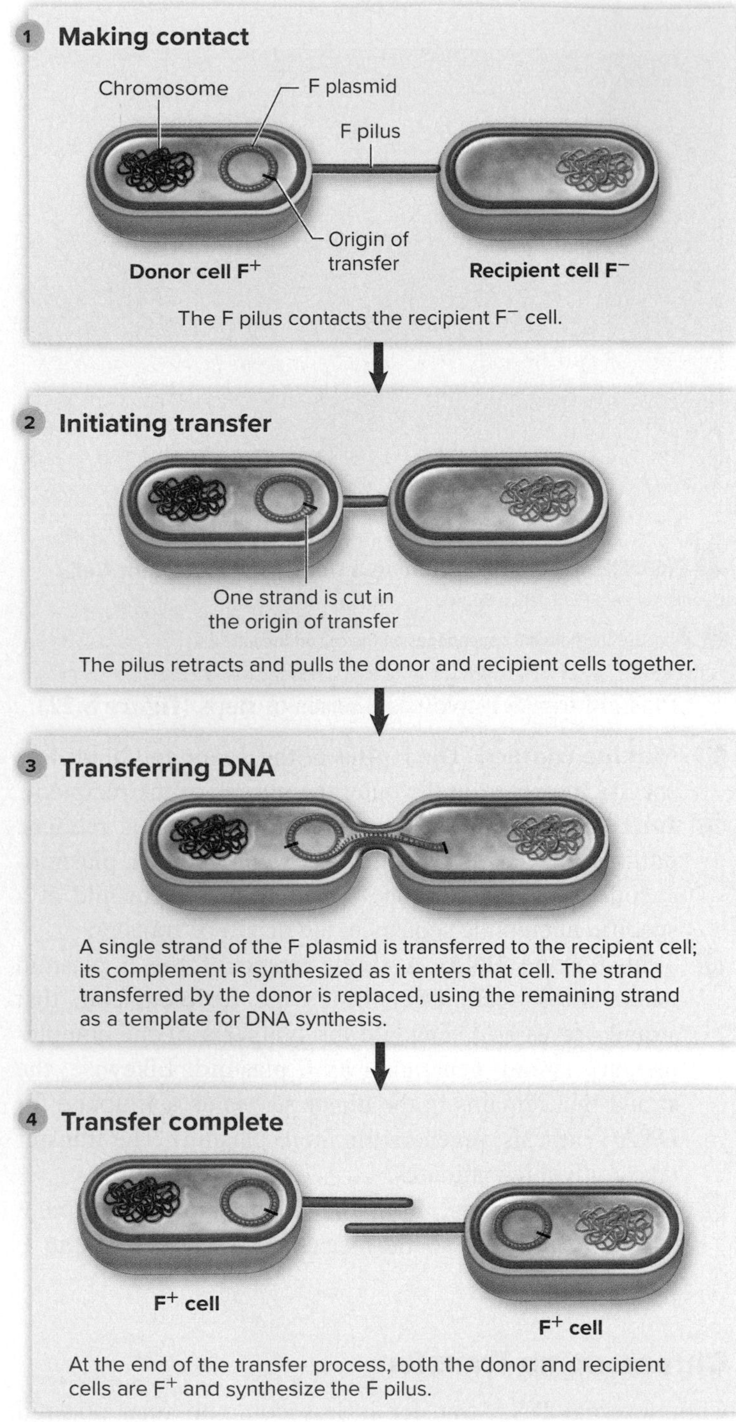
1 Making contact
Chromosome
F plasmid
F pilus
Origin of transfer
Donor cell F$^+$
Recipient cell F$^-$
The F pilus contacts the recipient F$^-$ cell.
2 Initiating transfer
One strand is cut in the origin of transfer
The pilus retracts and pulls the donor and recipient cells together.
3 Transferring DNA
A single strand of the F plasmid is transferred to the recipient cell; its complement is synthesized as it enters that cell. The strand transferred by the donor is replaced, using the remaining strand as a template for DNA synthesis.
4 Transfer complete
F$^+$ cell
F$^+$ cell
At the end of the transfer process, both the donor and recipient cells are F$^+$ and synthesize the F pilus.

FIGURE 8.22 Conjugation—F Plasmid Transfer

How does the recipient cell change as a result of conjugation?

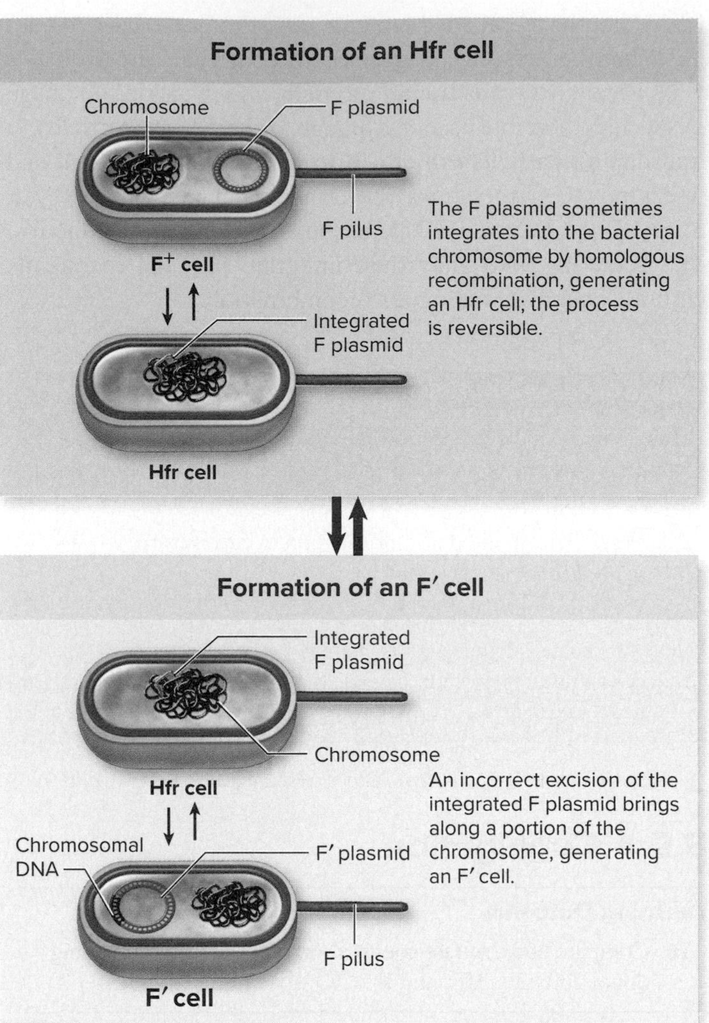
Formation of an Hfr cell
Chromosome
F plasmid
F pilus
F$^+$ cell
Integrated F plasmid
Hfr cell
The F plasmid sometimes integrates into the bacterial chromosome by homologous recombination, generating an Hfr cell; the process is reversible.
Formation of an F′ cell
Integrated F plasmid
Chromosome
Hfr cell
Chromosomal DNA
F′ plasmid
F pilus
F′ cell
An incorrect excision of the integrated F plasmid brings along a portion of the chromosome, generating an F′ cell.

FIGURE 8.23 Formation of Hfr and F′ Cells An Hfr cell is created when the plasmid integrates into the chromosome as a result of homologous recombination; note that the process is reversible. An F′ cell is created when certain recombination events result in an incorrect excision that removes a piece of the chromosome along with the F plasmid; this process is also reversible.

When the F plasmid integrates into a chromosome, it can only do so at specific locations. Why would this be the case?

Hfr cells produce an F pilus and the F plasmid DNA directs its movement to the recipient cell. Because the F plasmid DNA is integrated into the chromosome, however, chromosomal DNA accompanies it, beginning with the genes on one side of the origin of transfer (**figure 8.24**). The entire chromosome is generally not transferred because it would take approximately 100 minutes for this to occur, an unlikely event considering that the connection between the two cells usually breaks sooner than that. Because the entire integrated F plasmid is not transferred, the recipient remains F$^-$. The chromosomal DNA that enters the recipient is not a replicon, so it will be maintained only if it integrates into that cell's chromosome through homologous recombination. Unincorporated DNA will be degraded.

F′ Donors

Hfr strains can revert to F$^+$ because the process of F plasmid integration is reversible (see figure 8.23). In some instances, however, an error occurs during excision, and a piece of the bacterial chromosome is removed along with the F plasmid DNA. This action brings a chromosomal fragment into the F plasmid, producing a plasmid called **F′ (F prime).** Like the

FOCUS YOUR PERSPECTIVE 8.2

Bacteria Can Conjugate with Plants: A Natural Case of Genetic Engineering

The bacterium *Agrobacterium tumefaciens* uses a process analogous to conjugation to transfer certain genes into plant cells, including those of tobacco, carrots, and cedar trees. By doing so, the bacterium causes a common plant disease called crown gall, which is characterized by large tumors ("galls") on the crown of the plant (the portion above the soil line). Scientists studying crown gall tissues in the lab found that the tissue cells multiply in the absence of added plant hormones normally required for growth. In addition, crown gall tissues synthesize large amounts of amino acid derivatives called opines, which are not produced by normal tissues. Although the bacterium is required to start the altered growth, it is not needed to maintain the changes in the plant cells.

How does this bacterium permanently alter plant cells? *A. tumefaciens* strains that cause crown gall tumors contain a large plasmid called the Ti (tumor-inducing) plasmid. A specific piece of that plasmid called T-DNA (transferred DNA) moves from the bacterial cell to the plant cell, where it becomes incorporated into the plant cell chromosome **(box figure 8.2)**. The T-DNA carries genes for the synthesis of plant hormones as well as for opines. Those genes can be expressed in plant cells but not in *A. tumefaciens,* because their promoters resemble those of plant cells rather than those of bacteria. Once the plant cells begin expressing the transferred genes, the cells will be producing their own source of plant hormones—explaining why they can grow in the absence of added hormones—as well as opines.

How does *A. tumefaciens* benefit by altering plant cells? This bacterium can use opines as a source of carbon, nitrogen, and energy, whereas most other bacteria in the soil, as well as plants, cannot. In other words, *A. tumefaciens* genetically engineers the plant cells to multiply and produce food that only *A. tumefaciens* cells can use.

The *Agrobacterium*–crown gall system is important for several reasons. For one thing, it shows that DNA can be transferred from prokaryotes to eukaryotes. Many people believed that such transfer would be impossible in nature and could occur only in the laboratory. This system has also spawned an industry of plant biotechnology. The tumor formation genes in the Ti plasmid can be replaced with beneficial genes, and those genes will then be transferred and incorporated into the plant genome. Examples of genes that have been transferred into plants include those conferring resistance to viral pathogens, insects, and different herbicides. The *Agrobacterium*–crown gall system is an excellent example of how basic science research can lead to major industrial applications.

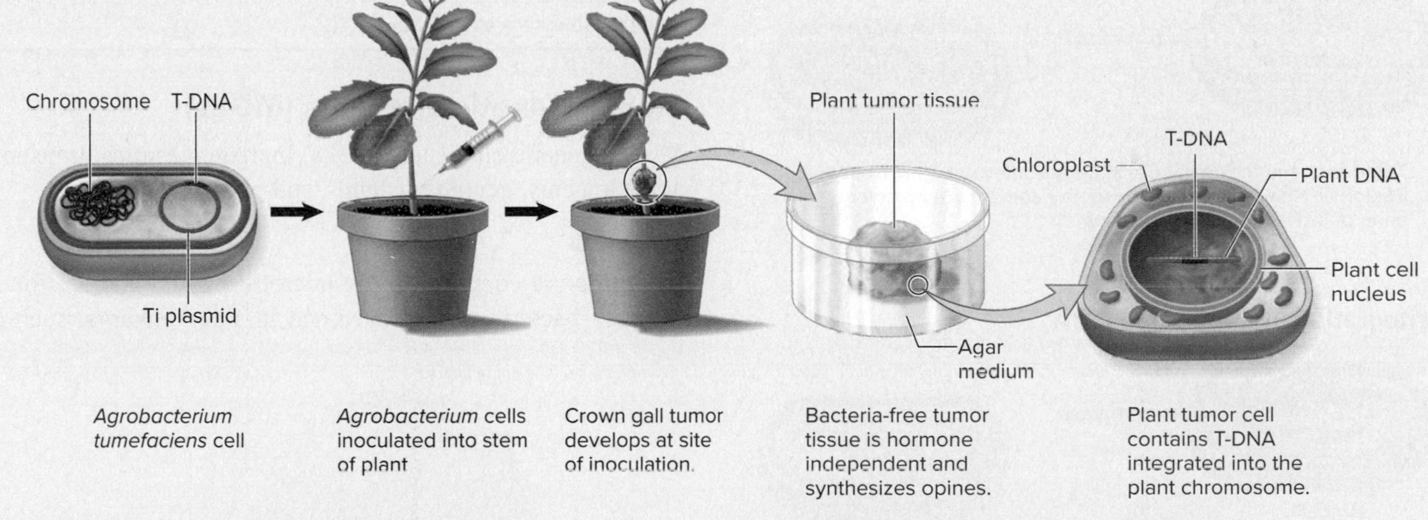

BOX FIGURE 8.2 ***Agrobacterium tumefaciens*** **Causes Crown Gall by Transferring Bacterial DNA to Plant Cells**

F plasmid, F′ is a replicon that is rapidly and efficiently transferred to F^- cells. In the case of F′, however, the chromosomal fragment is transferred as well.

MicroAssessment 8.9

Conjugation requires contact between donor and recipient cells. A donor cell that synthesizes an F pilus transfers the DNA to one that does not. Both plasmid and chromosomal DNA can potentially be transferred this way. Following transfer, plasmids replicate, but chromosomal DNA must be integrated into a replicon to replicate.

25. Why are conjugative plasmids that encode antibiotic resistance medically important?

26. Describe the outcomes of the three types of conjugation ($F^+ \times F^-$, Hfr $\times F^-$, and $F' \times F^-$).

27. Would you expect transfer of chromosomal DNA by conjugation to be more efficient if cells were plated together on solid medium (agar) or if they were mixed together in a liquid in a shaking flask? Explain.

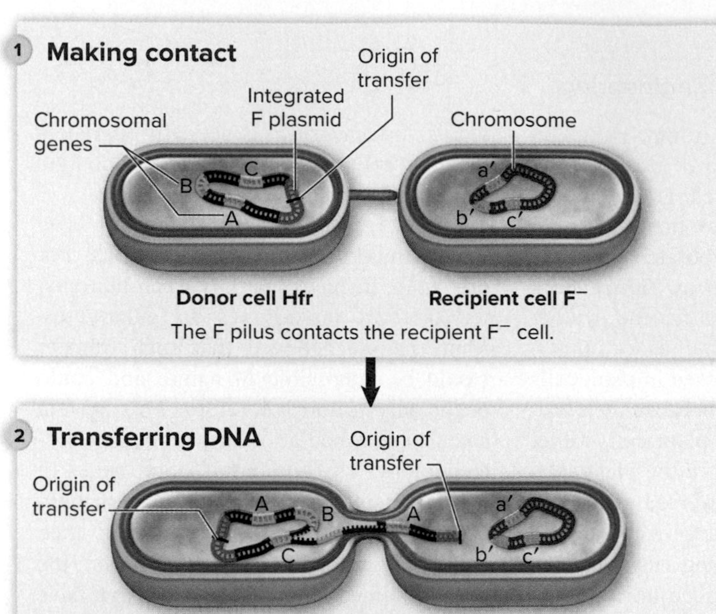

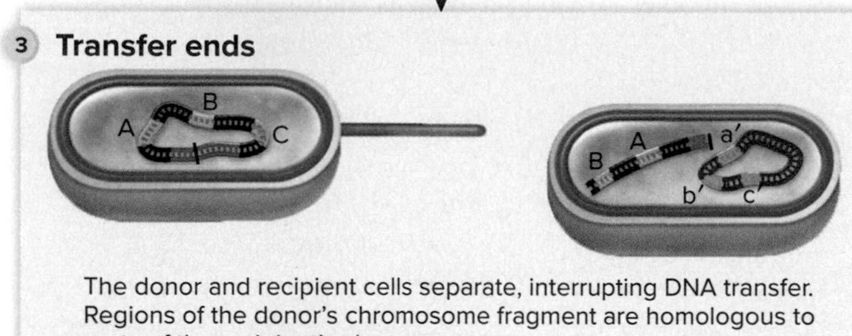

3 Transfer ends
The donor and recipient cells separate, interrupting DNA transfer. Regions of the donor's chromosome fragment are homologous to parts of the recipient's chromosome.

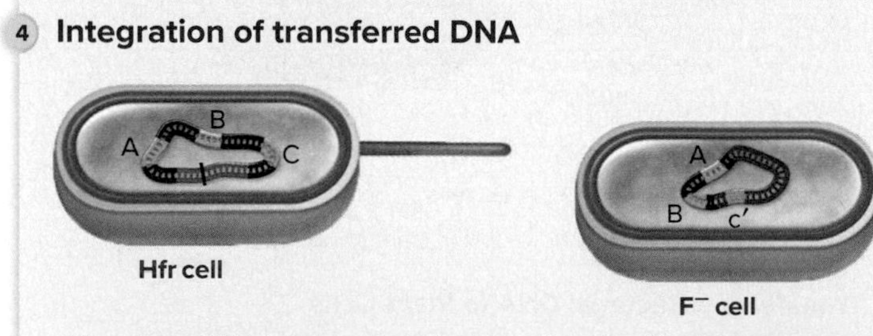

FIGURE 8.24 Conjugation—Chromosomal DNA Transfer The letters A, B, and C indicate genes in the Hfr cell; the letters a′, b′, and c′ indicate homologous but slightly different genes in the F^- cell.

? **Why is the entire donor chromosome seldom transferred?**

8.10 ▪ Genome Variability

Learning Outcomes

17. Describe the concepts of a pan-genome and a core genome.
18. Give four examples of mobile genetic elements (MGEs) and describe their significance.

Advances in genomics have uncovered surprising variability in the set of genes encoded by different strains of even a single species. In fact, less than half of an *E. coli* isolate's genes are shared by all strains of that species! This variation among strains has led to the concept of a **pan-genome**—the sum total of genes encoded by the various strains of a given species (**figure 8.25**). The pan-genome consists of three sets of genes: the **core genome** (common to all strains of the species), the accessory genome (present in more than one but not all strains of the species), and unique genes (found in only one strain of the species). Significant contributors to the accessory genome are segments of DNA called **mobile genetic elements (MGEs)**, which can move from one DNA molecule to another. The total set of MGEs in a genome is called a mobilome.

MicroByte

Almost 1,400 genes of the pathogen *E. coli* O157:H7 are not found in the laboratory strain *E. coli* K-12.

Mobile Genetic Elements (MGEs)

Mobile genetic elements (MGEs) include plasmids, transposable elements, genomic islands, and phage DNA.

Plasmids

Plasmids are common in the microbial world and are found in many bacteria and archaea and in some eukarya, such as

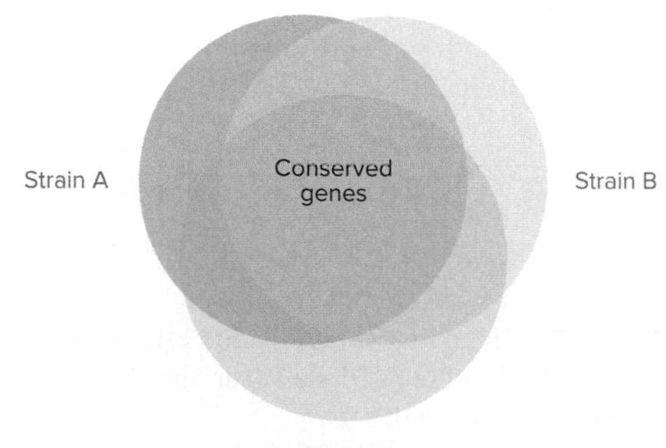

FIGURE 8.25 The Pan-Genome The pan-genome is the sum total of genes present in all strains of a given species. The core genome includes only those genes shared by all strains of that species (conserved genes).

? **What types of genes likely make up the core genome?**

TABLE 8.3 Some Plasmid-Encoded Traits

Trait	Organisms in Which Trait Is Found
Antibiotic resistance	Many
Antibiotic synthesis	*Streptomyces* species
Gas vacuole production	*Halobacterium* species
Increased virulence	*Yersinia* and *Shigella* species
Insect toxin synthesis	*Bacillus thuringiensis*
Nitrogen fixation	*Rhizobium* species
Oil degradation	*Pseudomonas* species
Pilus synthesis	*E. coli, Pseudomonas* species
Toxin production	*Bacillus anthracis*
Tumor formation in plants	*Agrobacterium* species (see Focus Your Perspective 8.2)

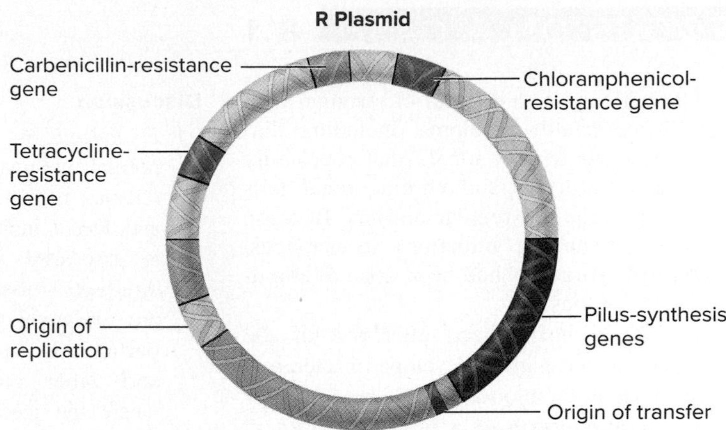

FIGURE 8.26 An R Plasmid The illustrated R plasmid carries multiple genes for antibiotic resistance as well as genes required for conjugation.

? How is an R plasmid similar to an F plasmid?

certain yeasts. Like chromosomes, most plasmids are circular double-stranded DNA molecules. They have an origin of replication and therefore can be replicated in a cell and passed to progeny cells; that is, they are replicons. Plasmids generally do not encode information essential to the life of a cell, and therefore cells can survive their loss. They are important, however, because they provide cells with the ability to survive in a particular environment (**table 8.3**).

Plasmids vary with respect to their properties. Some are small and carry only a few genes; others are large and carry many genes. Low-copy-number plasmids occur in only one or a few copies per cell, whereas high-copy-number plasmids are present in many copies, perhaps 500. Most plasmids have a narrow host range, meaning they can replicate in perhaps only one species. Broad host range plasmids replicate in many different species, sometimes including both Gram-negative and Gram-positive bacteria.

Many bacterial plasmids are readily transferred by conjugation. Conjugative plasmids are self transmissible, meaning they carry all of the genetic information needed for transfer, including an origin of transfer. In contrast, mobilizable plasmids encode an origin of transfer but lack other genetic information required for transfer. However, when a conjugative plasmid is in the same cell as a mobilizable plasmid, either plasmid can be transferred. Some plasmids can be transferred to various unrelated species. Genes carried on one type of bacterial plasmid can even be transferred to plant cells by a process analogous to conjugation (see Focus Your Perspective 8.2).

Resistance or **R plasmids** are particularly important medically because they encode resistance to antimicrobial chemicals, including antibiotics (**figure 8.26**). Many of these plasmids are conjugative, carrying not only genes encoding resistance (R genes), but also genes for pilus synthesis and other properties required for conjugation. Because many R plasmids have a broad host range and carry multiple resistance genes, they can allow a wide variety of organisms to become resistant to many different antimicrobials. Members of the normal microbiota that carry R plasmids can potentially transfer them to pathogens.

Transposable Elements (TEs)

As described in section 8.2, transposable elements (TEs) are pieces of DNA that can move from one location to another within a cell's genome. Because they can move into other replicons, including conjugative and mobilizable plasmids, they can potentially be transferred to other cells.

Several types of TEs exist, varying in their structural complexity. The simplest is an insertion sequence (IS), which typically encodes only transposase, the enzyme responsible for transposition; on each side of that gene are inverted repeats (sequences that are identical when read in the 5′ to 3′ direction) (**figure 8.27**). A type of TE called a composite transposon consists of one or more genes flanked by ISs. Composite transposons can move to another DNA site; their movement is easy to track when the genes they carry encode antibiotic resistance. Those transposons are particularly important medically because they can potentially transfer resistance genes to other bacteria.

Genomic Islands

Genomic islands are large DNA segments in a cell's genome that originated in another species. This conclusion is based on the fact that their nucleotide composition is quite different from the rest of the cell's genome. In general, each bacterial species has a characteristic proportion of G-C (guanine-cytosine) base pairs, so a large segment of DNA that has a very different G-C ratio suggests that the segment originated from a foreign source and was transferred to the cell through horizontal gene transfer.

The characteristics encoded by genomic islands include the use of specific energy sources, acid tolerance, the development

FOCUS ON A CASE 8.1

The patient was a 40-year-old woman with multiple health problems, including diabetes, chronic foot ulcers that repeatedly became infected, and chronic renal failure that required regular dialysis. Because of the recurring infections in her toes, multiple surgeries had been done to amputate those toes.

While hospitalized after one of the surgeries, the patient developed bacteremia (bacteria in the blood) as well as an abscess as a complication of a procedure used in dialysis. The causative agent in both infected sites was found to be methicillin-resistant *Staphylococcus aureus* (MRSA). Relatively few options existed for treating MRSA infections at the time, so in cases such as this, doctors often used the antibiotic vancomycin as a medication of "last resort." Past medical history indicated that the patient's foot ulcers had been treated with multiple courses of antibiotics, including vancomycin, over a span of a year before the surgery.

The patient's bacteremia was successfully treated with vancomycin and another antimicrobial medication, but she then developed an infection of the dialysis catheter exit-site. Vancomycin-resistant *S. aureus* (VRSA) and vancomycin-resistant *Enterococcus faecalis* (VRE) were isolated from the catheter tip, and then from two of her foot ulcers.

1. Based on the name *E. faecalis*, where would you think the bacterium is usually found?
2. How might the MRSA strain have become vancomycin-resistant?

Discussion

1. *E. faecalis* is part of the normal intestinal microbiota (*entero* means "intestine," and *faecal* indicates that it is associated with feces). Antibiotic-resistant strains are relatively common, particularly in hospitals and other environments where the medications are frequently used.
2. The only obvious difference between the patient's two *S. aureus* strains (MRSA and VRSA) was that the latter had a gene encoding resistance to vancomycin inserted into a plasmid present in both strains. Further studies showed that the vancomycin resistance gene of VRSA was part of a transposon that was identical to one in the VRE isolated from the same patient. In both VRSA and VRE, the transposon was integrated into a plasmid, but the plasmids in the two organisms were different. It appears that VRE transferred its transposon-containing plasmid to the vancomycin-sensitive *S. aureus* by conjugation (**box figure 8.3**). This entering plasmid was then destroyed by enzymes in *S. aureus,* but before that happened the transposon jumped to the plasmid already in the *S. aureus* cell.

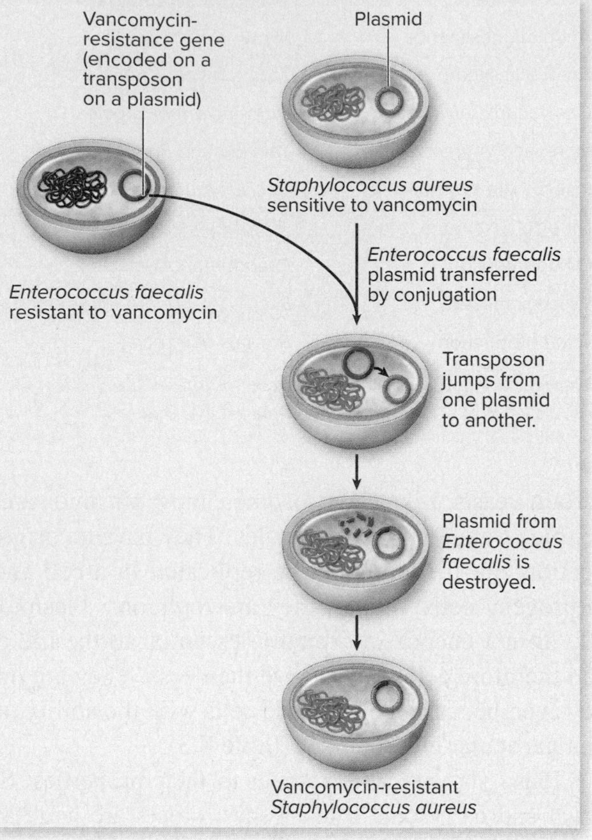

BOX FIGURE 8.3 Transfer of Vancomycin Resistance. The transfer involved both a plasmid and a transposon.

? What role did the transposon play in the transfer of vancomycin resistance?

Source: MMWR: 51(26):565–567 ["*Staphylococcus aureus* Resistant to Vancomycin—United States, 2002"] *Morbidity and Mortality Weekly Report* 51(26), July 5, 2002.

of symbiosis, and the ability to cause disease. Genomic islands that encode the latter are called **pathogenicity islands.**

Phage DNA

As described in chapter 13, certain types of phages can insert their DNA into the host cell chromosome. When this occurs, the phage DNA becomes part of the host cell's genome, which will be replicated and passed on to progeny cells. When inserted into the host cell's DNA, that phage DNA is called a **prophage.**

Mobile genetic elements and their characteristics are summarized in **table 8.4.**

TABLE 8.4 Mobile Genetic Elements and Their Characteristics

Mobile Genetic Element	Characteristics
Plasmid	Replicon that is independent of the chromosome and generally encodes only non-essential genetic information
Transposable element	Segment of DNA that directs its own movement to another location in chromosomal or plasmid DNA
Genomic Island	Large DNA segment in a cell's genome that originated in another species
Phage DNA	Phage genome that sometimes carries additional genes

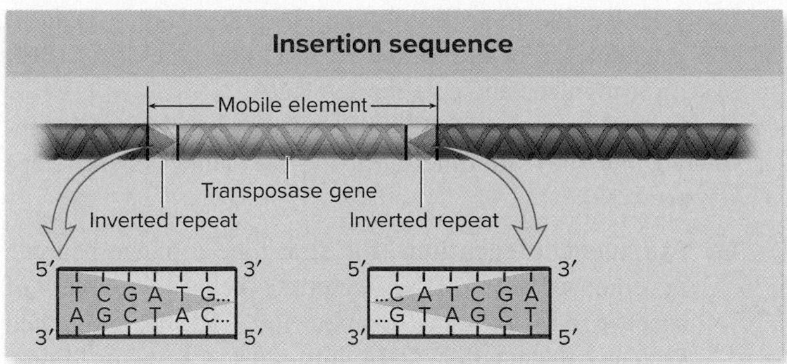

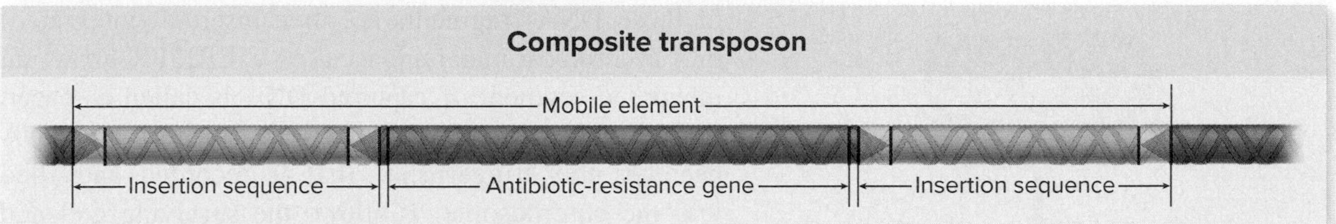

FIGURE 8.27 Transposable Elements The borders of insertion sequences are defined by inverted repeats 15–20 nucleotides in length. The first six nucleotides are shown here in expanded view to demonstrate their inverted orientation. Composite transposons, which move as single units, consist of two IS elements and the DNA between them.

Why are some transposons medically important?

MicroAssessment 8.10

The pan-genome is the sum total of genes encoded by the various strains of a given species; the core genome is the set of genes shared by all strains of a species. Mobile genetic elements (MGEs) are significant contributors to genome variation. MGEs include plasmids, transposable elements, genomic islands, and phage DNA.

28. What functions must a plasmid encode to be self-transmissible?
29. What characteristic of a genomic island suggests that it originated in another species?
30. Considering that *Staphylococcus epidermidis* does not typically cause disease in a healthy person, why would it be significant if it carried an R plasmid?

8.11 ▪ Bacterial Defenses Against Invading DNA

Learning Outcome

19. Explain how restriction modification systems and CRISPR systems recognize and destroy invading DNA.

Bacteria have various systems that recognize and destroy invading DNA (meaning foreign DNA that enters the cell). Although these mechanisms likely evolved as defenses against phages, any invading DNA may be destroyed. As described in chapter 9, the systems are also important in biotechnology because they allow scientists to cut DNA at precise nucleotide sequences. Once a cut has been made, the nucleotide sequence at that site can be manipulated.

Restriction-Modification Systems

Restriction-modification systems were discovered by scientists studying why certain bacterial strains are relatively resistant to phage infection. The scientists' research showed that those strains degrade foreign DNA through the combined action of two types of enzymes: a restriction enzyme and a modification enzyme. The restriction enzyme recognizes a specific short nucleotide sequence within the cell and then cuts the DNA molecule at that sequence. The modification enzyme protects that cell's own DNA from the action of the restriction enzyme by adding methyl groups to certain nucleobases in sequences recognized by the restriction enzyme. Because restriction enzymes cannot degrade methylated DNA, a restriction enzyme will destroy incoming unmethylated DNA but not the host DNA. Occasionally, the modification enzyme will methylate the incoming foreign DNA before the restriction enzyme has acted so that the invading DNA will not be degraded (**figure 8.28**). Different bacteria have different versions of restriction enzymes and modification enzymes, so hundreds of varieties exist, each recognizing different sequences in DNA.

CRISPR Systems

CRISPR systems (named for the characteristic **c**lustered **r**egularly **i**nterspersed **s**hort **p**alindromic **r**epeats in the bacterial genome) were discovered when scientists recognized that certain bacterial genomes include very small pieces of phage DNA. The bacteria had survived certain phage infections and retained small segments of the invaders' DNA, using them as "phage fingerprints" that function as a form of memory. The "fingerprints" are used by the bacterial cell

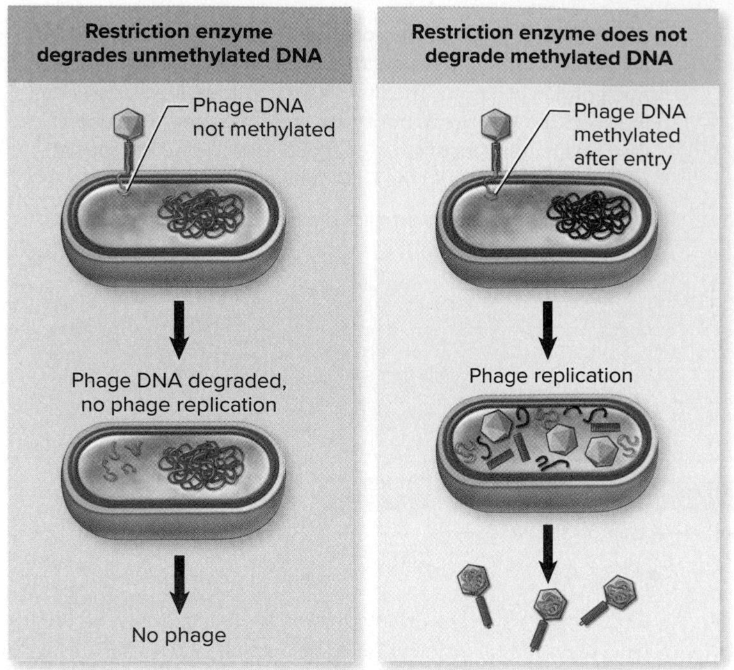

FIGURE 8.28 Restriction-Modification System

? How is the DNA modified by this system?

to recognize and then destroy the specific invading DNA if it is encountered in the future. At least one type of CRISPR system recognizes and cuts foreign RNA.

Several versions of CRISPR systems have been characterized, but they all function using the same general steps (**figure 8.29**):

1. **Fragment integration.** The first time a phage genome or other invading DNA enters a cell, a complex of specific Cas proteins (named for **C**RISPR-**as**sociated **s**equences) cuts that DNA into short fragments. Some of those DNA fragments are then inserted (integrated) into a chromosomal region called a CRISPR array; an integrated segment of captured DNA is called a spacer. In the cases of phage infections, the spacer DNA becomes the "fingerprint" that is recorded and filed into the chromosome. It allows the surviving cell and its descendants to recognize specific phage DNA if it is encountered again.
2. **crRNA expression.** The cell transcribes the CRISPR array—including the spacer sequences—and then processes that transcript to generate small RNAs called

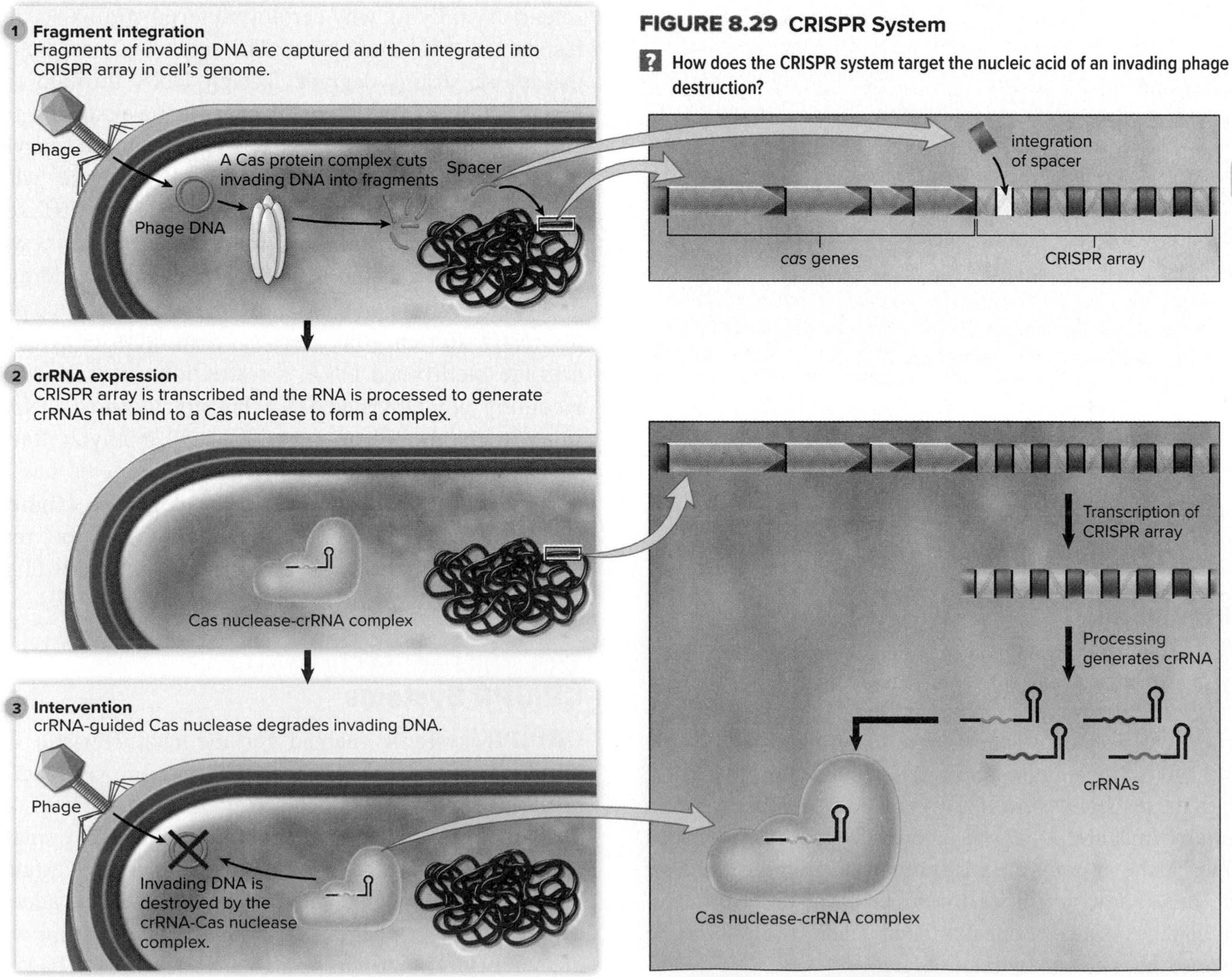

1 Fragment integration
Fragments of invading DNA are captured and then integrated into CRISPR array in cell's genome.

Phage
A Cas protein complex cuts invading DNA into fragments
Spacer
Phage DNA
integration of spacer
cas genes
CRISPR array

2 crRNA expression
CRISPR array is transcribed and the RNA is processed to generate crRNAs that bind to a Cas nuclease to form a complex.

Cas nuclease-crRNA complex
Transcription of CRISPR array
Processing generates crRNA
crRNAs
Cas nuclease-crRNA complex

3 Intervention
crRNA-guided Cas nuclease degrades invading DNA.

Phage
Invading DNA is destroyed by the crRNA-Cas nuclease complex.

FIGURE 8.29 CRISPR System

? How does the CRISPR system target the nucleic acid of an invading phage for destruction?

crRNAs, one for each spacer. Each crRNA binds to a Cas nuclease (a DNA-cutting enzyme) to form a complex. Within the complex, crRNA functions as a guide; if crRNA base-pairs with invading DNA, the Cas nuclease will cut that DNA.

③ **Intervention.** If the bacterial cell or one of its descendants re-encounters the same phage genome or other invading DNA, a Cas nuclease-crRNA complex will recognize and destroy that DNA.

MicroAssessment 8.11

Bacteria use restriction-modification systems and CRISPR systems to defend against invading DNA.

31. How do modification enzymes protect host cell DNA from restriction enzymes?
32. How does a bacterial cell acquire a historical record of phage infections?
33. How is the mechanism of a CRISPR system similar to the mechanism of RNA interference?

Summary

8.1 ■ Genetic Change in Bacteria

The **genotype** of bacteria can change through either **mutation** or **horizontal gene transfer** (figure 8.1). Bacteria are haploid, so any changes in DNA can easily alter the **phenotype.** A mutant that requires a growth factor is an **auxotroph;** in contrast, a **prototroph** does not require growth factors.

MUTATION AS A MECHANISM OF GENETIC CHANGE

8.2 ■ Spontaneous Mutations

Spontaneous mutations occur as a result of normal cell processes. They are stable but occasionally revert back to the non-mutant form. The chance of spontaneous mutations occurring in more than one gene in a cell is the product of the genes' individual mutation rates.

Base-Pair Substitution

A **base-pair substitution** occurs as a result of an error during DNA synthesis (figure 8.2). If the substitution occurs in a protein-encoding region, it results in one of three possible outcomes: a **synonymous mutation,** a **missense mutation,** or a **nonsense mutation** (figure 8.3).

Addition or Deletion of Nucleotide Pairs

Adding or deleting one or two nucleotide pairs in a protein-encoding region causes a **frameshift mutation** (figure 8.4). This often results in a shortened non-functional protein.

Transposable Elements

Transposable elements can move from one location to another in a cell's genome. The gene into which a transposon jumps is insertionally inactivated by the event (figure 8.5).

8.3 ■ Induced Mutations

Induced mutations are caused by **mutagens** (table 8.1).

Chemical Mutagens

Some chemicals modify nucleobases, altering their hydrogen-bonding properties (figure 8.7). **Base analogs** can be mistakenly incorporated in place of the usual nucleobases, and they have different hydrogen-bonding properties (figure 8.8). **Intercalating agents** insert between adjacent bases in the double helix and push nucleotides apart, resulting in frameshift mutations.

Transposons

Transposons, a type of transposable element, can be introduced intentionally into a cell in order to inactivate genes.

Radiation

Exposure to ultraviolet light results in **thymine dimer** formation (figure 8.9). The repair mechanism can cause mutations. X rays cause single- and double-strand breaks.

8.4 ■ Repair of Damaged DNA (table 8.2)

Repair of Errors in Nucleotide Incorporation

DNA polymerases have a **proofreading** function. **Mismatch repair** removes a portion of the strand that has a misincorporated nucleotide. A new DNA strand is then synthesized (figure 8.10).

Repair of Damage to Nucleobases

Base excision repair uses specific DNA glycosylases to remove damaged nucleobases in DNA (figure 8.11).

Repair of Thymine Dimers

In **photoreactivation,** an enzyme uses the energy of light to break the bonds of the thymine dimer (figure 8.12). In **nucleotide excision repair,** the damaged single-stranded segment is removed and replaced.

SOS Repair

SOS repair is a last-effort repair mechanism that uses a special DNA polymerase to bypass the damaged DNA but has no proofreading ability. Consequently, the newly synthesized DNA has many mutations.

8.5 ■ Mutant Selection

Direct Selection

Direct selection is used to obtain mutants that grow under conditions in which the parent cell cannot (figure 8.13).

Indirect Selection

Indirect selection uses **replica plating** to isolate an auxotroph from a prototrophic parent strain (figure 8.14).

Screening for Possible Carcinogens

The **Ames test** is used to determine if a chemical is a mutagen and therefore a possible **carcinogen** (figure 8.15).

HORIZONTAL GENE TRANSFER AS A MECHANISM OF GENETIC CHANGE

8.6 ■ Overview of Horizontal Gene Transfer (figure 8.17)

For newly acquired DNA to replicate in a cell, it must either be a replicon or integrate into the cell's genome (figure 8.18).

8.7 ■ Bacterial Transformation

Bacterial transformation transfers "naked" DNA.

Competence

A cell must be **competent** to take up DNA, and only certain species naturally become competent.

The Process of Natural Transformation

Short strands of double-stranded DNA bind to competent cells, but only one strand enters (figure 8.19).

8.8 ■ Transduction

Transduction is the transfer of bacterial DNA by a **bacteriophage.** There are two types: **generalized transduction** and **specialized transduction** (figure 8.20).

8.9 ■ Conjugation

Conjugation requires cell-to-cell contact.

Plasmid Transfer

F^+ cells synthesize an **F pilus,** encoded on an **F plasmid;** the F plasmid is transferred from an F^+ cell to an **F^- cell** (figure 8.22).

Chromosome Transfer

Hfr cells have the F plasmid integrated into the chromosome (figure 8.23). When the F plasmid is transferred, chromosomal DNA moves into a recipient cell along with it (figure 8.24).

F′ Donors

An **F′** donor carries a modified F plasmid that contains a piece of chromosomal DNA.

8.10 ■ Genome Variability

The sum total of genes encoded by the various strains of a given species is the **pan-genome.** Genes common to all strains of the species make up the **core genome** (figure 8.25).

Mobile Genetic Elements (MGEs) (table 8.4)

Mobile genetic elements (MGEs), including **plasmids, transposable elements, genomic islands,** and **phage** DNA, are significant contributors to genome variability.

8.11 ■ Bacterial Defenses Against Invading DNA

Restriction-Modification Systems

Restriction-modification systems use restriction enzymes to quickly distinguish and destroy invading DNA; modification enzymes protect the host's DNA by adding methyl groups to certain nucleobases (figure 8.28).

CRISPR Systems

CRISPR systems allow bacterial cells to recognize and destroy invading DNA that has been encountered previously (figure 8.29).

Review Questions

Short Answer

1. How is an auxotroph different from a prototroph?
2. Why does deleting one nucleotide pair generally have a more severe consequence than deleting three?
3. What type of mutation in an operon is most likely to affect the synthesis of more than one protein?
4. What is meant by "proofreading" with respect to DNA polymerases?
5. Why would a cell use SOS repair, considering that it introduces mutations?
6. Why is replica plating used to isolate an auxotrophic mutant from a prototrophic parent?
7. What is transduction?
8. How is an F^+ strain different from an Hfr strain?
9. Name four mobile genetic elements.
10. Why are R plasmids important?

Multiple Choice

1. UV light exposure forms
   a) covalent bonds between the two strands of DNA.
   b) hydrogen bonds between the two strands of DNA.
   c) covalent bonds between adjacent thymine nucleobases on the same strand of DNA.
   d) covalent bonds between adjacent guanine and cytosine nucleobases on the same strand of DNA.
   e) hydrogen bonds between adjacent guanine and cytosine nucleobases on the same strand of DNA.
2. If cells were exposed to UV light, the highest frequency of mutations would be obtained if the cells were immediately
   a) placed in the dark.
   b) exposed to visible light.
   c) shaken vigorously.
   d) incubated at a temperature below their optimum for growth.
3. If a nonsense mutation occurred in the first gene of a four-gene operon, how many functional proteins would be synthesized?
   a) 1
   b) 4
   c) 0
   d) 3
4. Which of the following repair mechanisms is the most error-prone?
   a) SOS repair
   b) mismatch repair
   c) proofreading by DNA polymerases
   d) light repair
   e) nucleotide excision repair

5. You are trying to isolate a mutant of wild-type *E. coli* that requires histidine for growth. This can best be done using
   a) direct selection.
   b) replica plating.
   c) the Ames test.
   d) reversion.
6. In the Ames test, the purpose of adding liver extract is to
   a) increase the mutation rate of bacteria.
   b) provide nutrients for the auxotrophs.
   c) provide nutrients for the prototrophs.
   d) provide mammalian enzymes that sometimes alter chemicals.
   e) provide the necessary DNA repair mechanisms.
7. Adding DNase to a mixture of donor and recipient cells will prevent gene transfer via
   a) bacterial transformation.
   b) conjugation.
   c) generalized transduction.
   d) specialized transduction.
8. The donor cell typically survives which DNA-transfer process?
   a) bacterial transformation
   b) conjugation
   c) specialized transduction
   d) generalized transduction
9. A plasmid that can replicate in *E. coli* and *Pseudomonas* is most likely a/an
   a) broad host range plasmid.
   b) self-transmissible plasmid.
   c) high-copy-number plasmid.
   d) essential plasmid.
   e) low-copy-number plasmid.
10. The frequency of transfer of an F′ molecule by conjugation is closest to the frequency of transfer of
    a) chromosomal genes by conjugation.
    b) an F plasmid by conjugation.
    c) an F plasmid by transformation.
    d) an F plasmid by transduction.
    e) an R plasmid by DNA transformation.

Applications

1. Some bacteria may have higher mutation rates than others following exposure to UV light. Discuss a reason why this might be the case. What experiments could you do to determine whether this is a likely possibility?
2. A pharmaceutical researcher is disturbed to discover that the major ingredient of a new drug formulation causes frameshift mutations in bacteria. What other information would the researcher want before looking for a substitute chemical?

Critical Thinking

1. You have the choice of different kinds of mutants for use in the Ames test to determine the frequency of reversion by suspected carcinogens. You can choose a deletion, a point mutation, or a frameshift mutation. Would it make any difference which one you chose? Explain.
2. You isolate an *E. coli* strain (strain X) that is resistant to four antibiotics (penicillin, streptomycin, chloramphenicol, and tetracycline). You then mix strain X with *E. coli* strain Y, which is sensitive to the same four antibiotics. After mixing the two strains, you find that cells of strain Y are now resistant to streptomycin, penicillin, and chloramphenicol (but still sensitive to tetracycline). Explain what microbial process might be occurring and why this could result in strain Y showing resistance to only three out of the four antibiotics.

9 | Biotechnology

Scientist working at a bench in a DNA laboratory. *atic12/123RF*

KEY TERMS

CRISPR-Cas Technologies Protein-based systems used to locate specific nucleotide sequences in vivo; some versions cut DNA as part of gene editing.

Colony Blotting A technique that uses a nucleic acid probe to determine which colonies on an agar plate contain a given nucleotide sequence.

DNA Microarray A nucleic acid probe-based in vitro technique used to study gene expression patterns.

DNA Probe A single-stranded piece of DNA, tagged with an identifiable marker, that is used to detect a complementary nucleotide sequence.

DNA Sequencing The process of determining the nucleotide sequence of a given DNA molecule.

Fluorescence In Situ Hybridization (FISH) A technique that uses a fluorescently labeled nucleic acid probe to detect a given nucleotide sequence within intact cells on a microscope slide.

Gel Electrophoresis An in vitro procedure used to separate large macromolecules, such as nucleic acids or proteins, by size.

Genetic Engineering The process of deliberately altering an organism's genetic information using in vitro techniques.

Molecular Cloning The in vitro process of inserting a DNA fragment into a vector and then transferring the resulting recombinant molecule into a cell where it will be replicated and passed on to the cell's progeny.

Polymerase Chain Reaction (PCR) An in vitro technique used to rapidly amplify (repeatedly duplicate) a specific region of a DNA molecule, increasing the number of copies exponentially.

Recombinant DNA Molecule A DNA molecule created by joining DNA fragments from two different sources.

Restriction Enzyme A type of enzyme used in vitro to cut a specific nucleotide sequence in DNA.

Vector A DNA molecule, often a plasmid, used in molecular biology that functions as a carrier of cloned DNA.

A Glimpse of History

In 1976, Argentine newspapers reported a violent shootout between soldiers and the occupants of a suburban Buenos Aires home, leaving the five "extremists" dead: a young couple and their three children, ages 6 years, 5 years, and 6 months. Over the next 7 years, similar scenarios recurred as the military junta that ruled Argentina killed thousands of citizens. This "Dirty War" finally ended in 1983 with the election of a new democratic government. The new leaders opened previously sealed records that confirmed what many had already suspected: More than 200 children had survived the bloodshed but had been kidnapped and placed with families that supported the junta.

American geneticist Dr. Mary-Claire King was enlisted to help in the effort to return the children to the surviving members of their biological families. Dr. King and others recognized that DNA technology could be used for this important humanitarian cause. Blood and tissue samples from one individual can be distinguished from those of another by comparing certain DNA sequences. These same principles can also be used to show that a particular child is the progeny of a given set of parents. Because a person has two copies of each chromosome—one inherited from each parent—half of a child's DNA will represent maternal sequences and the other half will represent paternal sequences. The case of the Argentine children was complicated, however, because most of the parents were dead or missing. Often, the only surviving relatives were aunts and grandmothers, and it is difficult to use chromosomal DNA to show genetic relatedness between a child and such relatives. Dr. King decided to investigate using mitochondrial DNA (mtDNA) instead. This organelle DNA, unlike chromosomal DNA, is inherited only from the mother. A child will have the same nucleotide sequence of mtDNA as any siblings, the mother and her siblings, and the maternal grandmother.

By comparing the nucleotide sequences of mtDNA in different individuals, Dr. King located key positions that varied extensively among unrelated people, but were similar in maternal relatives. Dr. King's technique, developed to help reunite families victimized by war, has found many uses. Today her lab still uses molecular biology techniques for humanitarian efforts, such as identifying the remains of victims of massacres around the world.

Biotechnology uses microbiological and biochemical procedures to solve practical problems and make useful products. In the traditional sense, biotechnology includes searching for naturally occurring mutants that have desirable characteristics, but today it is generally associated with procedures that involve manipulating DNA.

Modern biotechnology blossomed in the 1970s with the development of laboratory-based methods to deliberately alter a bacterium's genetic information, a process called **genetic engineering.** Scientists can now genetically alter not only bacteria, but also yeast and certain types of mammalian cells grown in culture, thereby providing important laboratory models for studying cell functions. Microorganisms and cultured animal cells can also be genetically engineered to produce commercially valuable proteins. Doing so creates protein sources that are reliably available and less expensive than traditional sources, particularly for proteins otherwise harvested from animal tissues. Plants and animals can also be genetically engineered, making what are commonly referred to as genetically modified organisms (GMOs). Technically, however, a GMO is any organism that has been genetically engineered, including bacteria and yeasts.

The insights gained through biotechnology gave rise to new fields to study the functions of certain components of biological systems. These fields include:

- **Genomics.** This is the study and analysis of genomes, as described in section 7.8. Advances in DNA sequencing technologies have played a tremendous role in the growth of genomics, which, in turn, spurred interest in other "omics" fields.
- **Transcriptomics.** This is the study of RNA molecules and their functions; the term commonly refers to the study of mRNA but also includes the analysis of non-coding RNAs. Transcriptomics is gaining more importance as scientists study gene expression to determine how cancer cells escape the mechanisms that normally limit multiplication of the body's cells. It is also tremendously important in the study of the host-pathogen interactions that result in disease.
- **Proteomics.** This is the study of proteins, their functions, and their expression profiles (meaning the conditions that promote the expression of various proteins by a cell or group of cells). Like transcriptomics, proteomics is playing an increasingly important role in cancer studies and infectious disease research.

The information and innovations generated through biotechnology affect society in many ways—from agricultural practices and medical diagnoses to evidence used in courtrooms. The applications seem limited only by the imagination and are far too numerous to list here, but some examples include:

- **Cancer treatment.** The revolutionary cancer treatments described in section 17.3 involve DNA manipulation.
- **Diagnostic tests.** Many tests used in disease diagnosis rely on the detection of nucleotide sequences unique to the relevant pathogen, as described in section 10.3. Methods under development aim to screen for dozens of pathogens simultaneously.
- **DNA typing.** The forensic DNA profiling method ("DNA fingerprinting") described in Focus on a Case 9.1 relies on the polymerase chain reaction (PCR), the topic of section 9.5.
- **Food production.** One of the most widely used proteins produced by genetically engineered microorganisms is chymosin (rennin), an enzyme used in cheese production; it was traditionally obtained from the stomachs of calves. A relatively new meat substitute, found in products such as the Impossible Burger, gets its more authentic flavor from soy leghemoglobin (a hemoglobin-like protein made by legume roots) produced by genetically engineered microorganisms.
- **Gene therapy.** Recent trials have shown tremendous promise in correcting certain life-threatening genetic defects, including myotubular myopathy (MTM), a disease that causes extreme muscle weakness.
- **Precision medicine.** This medical approach often uses a patient's genetic makeup as a basis for treatment.
- **Research.** The majority of updates in this textbook involve discoveries made through research that relied on genetic engineering or other biotechnologies.
- **Therapeutic proteins.** Human insulin, used in treating diabetes, was one of the first pharmaceutical proteins to be produced through genetic engineering. The original commercial product was extracted from pancreatic glands of cattle and pigs, and it sometimes caused allergic reactions in people receiving it.
- **Transgenic crops.** Various crops have been genetically engineered to resist certain insect pests by making a biological insecticide (Bt toxin) naturally produced by the bacterium *Bacillus thuringiensis.* Plants have also been engineered to resist the effects of the biodegradable herbicide glyphosate. In addition, plants have been engineered to improve their nutritional quality; "golden rice" carries the gene for the biosynthesis of β-carotene, a precursor of vitamin A. The development of transgenic plants started with basic research studying *Agrobacterium tumefaciens*, described in Focus Your Perspective 8.2.
- **Vaccines.** A vaccine protects against a disease by exposing a person's immune system to a killed or weakened form of the causative agent, or to parts of that agent. A genetically engineered microorganism can be designed to produce a protein from a pathogen for use in a vaccine; this type of vaccine is currently used to prevent hepatitis B, cervical cancer, and influenza.

In addition to the applications just mentioned, several biotechnologies play crucial roles in the global response to COVID-19. Specific examples will be described in Focus Your Perspective 9.1 after the relevant technologies are explained in more detail. **Table 9.1** summarizes the general applications of the DNA-based technologies covered in this chapter.

MicroByte

A bacterial enzyme that has been engineered to efficiently break down a common type of plastic may revolutionize the recycling of plastic bottles.

TABLE 9.1 Applications of DNA-Based Technologies

Technology	Application
Molecular Cloning	Used to genetically engineer microbes to: • make medically and commercially valuable proteins. • produce specific DNA sequences. • study gene expression.
CRISPR-Cas Technologies	Used to: • edit genomes. • study gene expression. • detect pathogens for rapid disease diagnosis.
DNA Sequencing	Used to: • identify microorganisms. • track the spread of disease. • trace the source of disease outbreaks. • determine the genetic relatedness of different microbial isolates. • characterize microbial community composition. • identify cancer types to allow for more precise treatment. • diagnose certain genetic disorders.
Polymerase Chain Reaction (PCR)	Used to: • detect pathogens for rapid disease diagnosis. • type DNA ("fingerprinting") for forensic evidence. • study gene expression.
Probe Technologies	
Colony blots	Used to detect colonies with a specific DNA sequence.
Fluorescence in situ hybridization (FISH)	Used to identify cells directly in a specimen.
DNA microarrays	Used to study gene expression.

9.1 ■ Fundamental Tools Used in Biotechnology

Learning Outcome

1. Describe the role of restriction enzymes, reverse transcriptase, and gel electrophoresis in biotechnology.

Biotechnology allows researchers to manipulate natural biological processes for beneficial purposes. Today, molecular biologists have a set of basic procedures in their "tool kit." Procedures or experiments that involve whole living organisms are called in vivo (Latin for "within the living"), whereas those that occur in artificial settings—such as test tubes, cell culture flasks, or Petri plates—are called in vitro (Latin for "in glass").

As we describe these basic procedures, remember that diagrams typically focus on only one or a few DNA molecules to illustrate what is happening at a molecular level. In reality, scientists are often working with millions of molecules simultaneously.

Restriction Enzymes

Restriction enzymes (also called restriction endonucleases) are powerful tools in biotechnology because they allow scientists to easily cut DNA in vitro in a predictable and controlled manner. Recall that these enzymes are produced by bacteria to destroy invading DNA, particularly phage genomes (see figure 8.28).

A restriction enzyme cuts DNA within or near a specific nucleotide sequence called a recognition sequence, generating a set of fragments called restriction fragments (**figure 9.1a**). Researchers use the enzymes not only to cut DNA, but also to

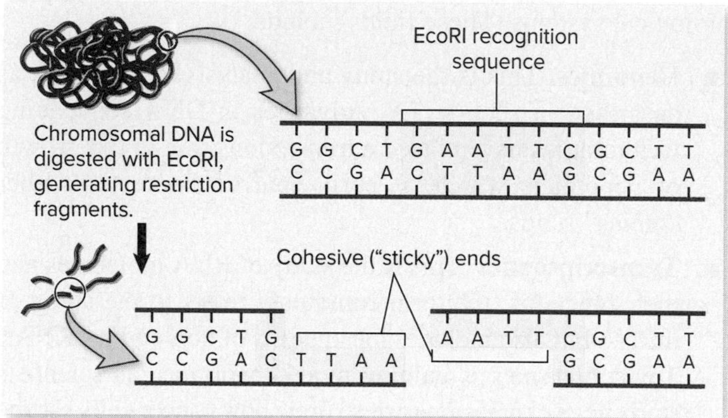

(a)

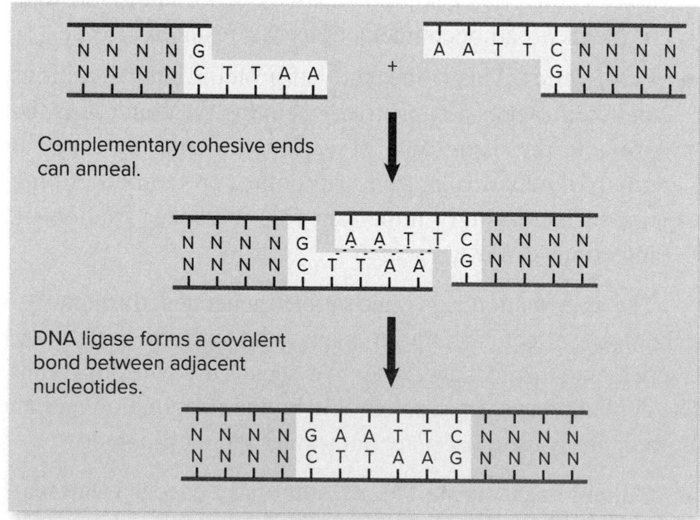

(b)

FIGURE 9.1 Action of Restriction Enzymes (a) In vitro digestion of DNA with a restriction enzyme generates restriction fragments. **(b)** Fragments that have complementary cohesive ends can anneal, regardless of their original source (N = nucleotide, meaning that any of the four nucleobases could be in that position as long as base-pairing rules are followed).

? How do restriction enzymes make it easier for scientists to create recombinant DNA molecules?

create recombinant DNA molecules (molecules made by joining DNA from two different sources). Most of the common restriction enzymes used to digest (cut) DNA make staggered cuts in the recognition sequence, meaning the cuts are not directly across from each other. The resulting fragments have short, single-stranded overhangs—referred to as sticky ends or cohesive ends—that can "stick" to one another. They can do this because their complementary nucleotide sequences are able to **anneal,** meaning they can form base pairs with one another (figure 9.1b). Any two complementary cohesive ends can anneal, even those from two different organisms. The enzyme DNA ligase is used to form a covalent bond between adjacent nucleotides, joining (ligating) the two molecules. Thus, if restriction enzymes are viewed as scissors that cut DNA into fragments, then DNA ligase is the glue that pastes the fragments together.

Each restriction enzyme recognizes a specific 4- to 6-base-pair nucleotide sequence (**table 9.2**). The sequences are typically palindromes, meaning they are the same on both strands when read in the 5′ to 3′ direction. The name of a particular restriction enzyme represents the bacterium from which it was first isolated. The first letter is the first letter of the genus name, and the next two are the first letters of the species name. Other numbers or letters indicate the strain and order of discovery; for example, EcoRI is from *E. coli* strain RY13. Although most restriction enzymes generate sticky ends, some make blunt ends. Due to the lack of overhangs, blunt ends are more difficult to ligate.

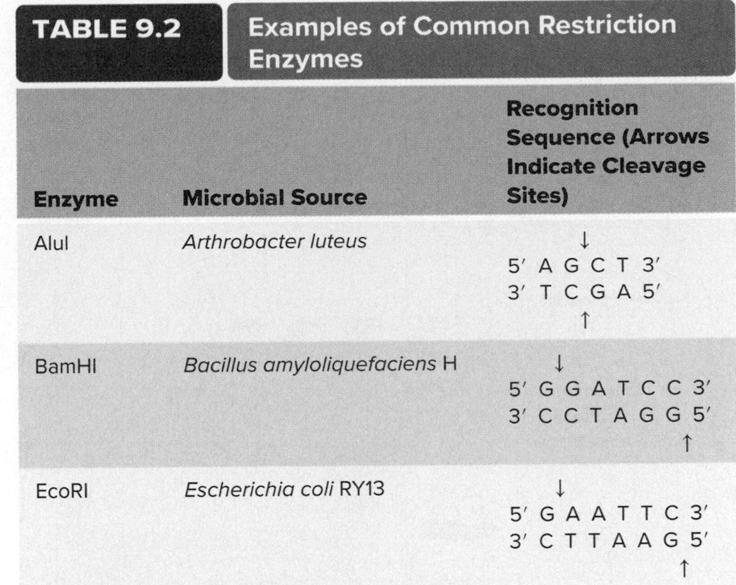

TABLE 9.2 Examples of Common Restriction Enzymes

Enzyme	Microbial Source	Recognition Sequence (Arrows Indicate Cleavage Sites)
AluI	*Arthrobacter luteus*	↓ 5′ A G C T 3′ 3′ T C G A 5′ ↑
BamHI	*Bacillus amyloliquefaciens* H	↓ 5′ G G A T C C 3′ 3′ C C T A G G 5′ ↑
EcoRI	*Escherichia coli* RY13	↓ 5′ G A A T T C 3′ 3′ C T T A A G 5′ ↑

Reverse Transcriptase

Reverse transcriptase is an enzyme that uses an RNA template to make DNA through a process called reverse transcription. The enzyme is naturally produced by retroviruses and other reverse-transcribing viruses.

Using RNA as a template, reverse transcriptase generates a single-stranded product referred to as **cDNA,** meaning complementary DNA; if double-stranded cDNA is needed, a DNA polymerase synthesizes the other strand. One of the earliest and still relevant in vitro uses of reverse transcriptase is to create DNA molecules that lack the introns common in eukaryotic genes. Recall that when eukaryotic genes are expressed, introns are removed during splicing of the pre-mRNA (see figure 7.19). The cDNA made from mRNA encodes the same protein as the original DNA but lacks introns (**figure 9.2**).

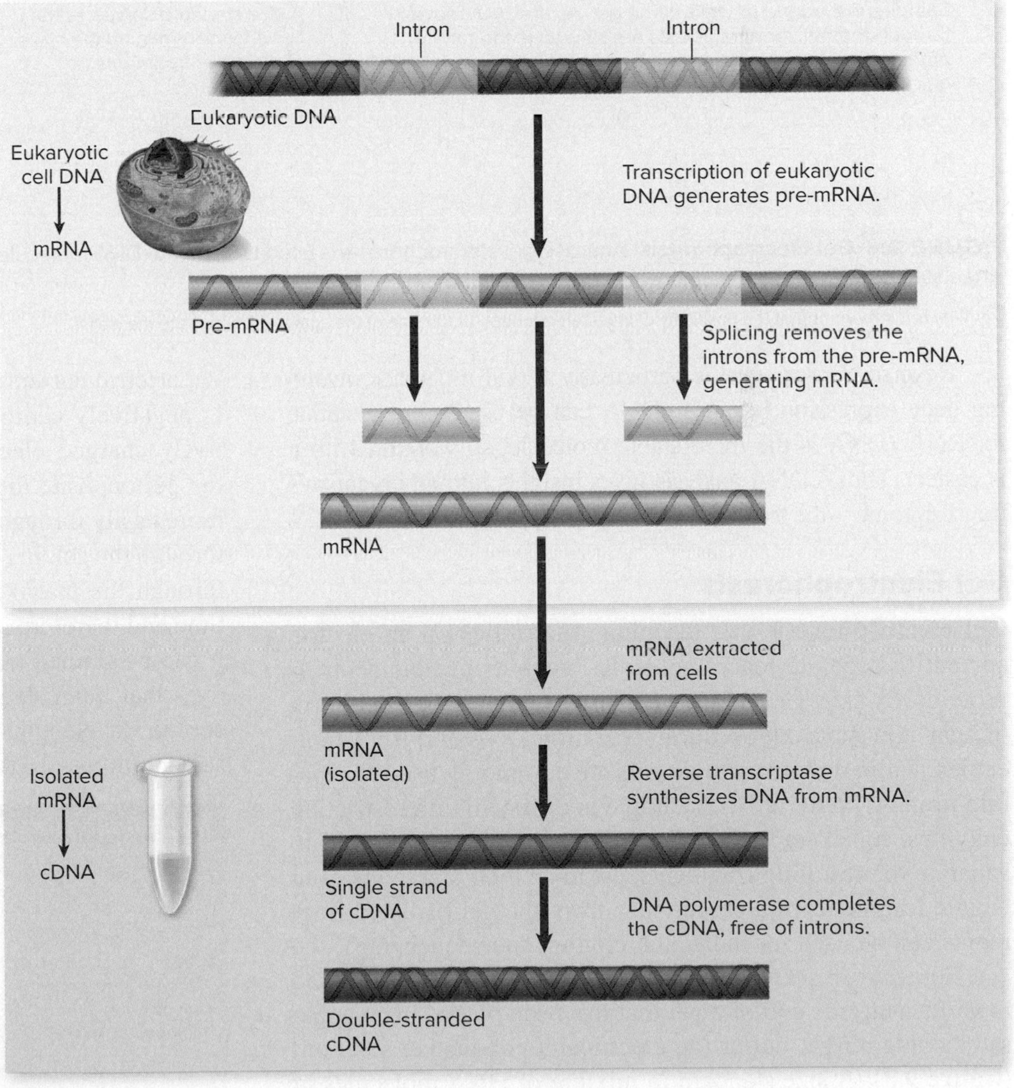

FIGURE 9.2 Making cDNA from Eukaryotic mRNA The cDNA encodes the same protein as the original DNA but lacks introns.

? **Why would the enzyme that uses RNA as a template to make DNA be called reverse transcriptase?**

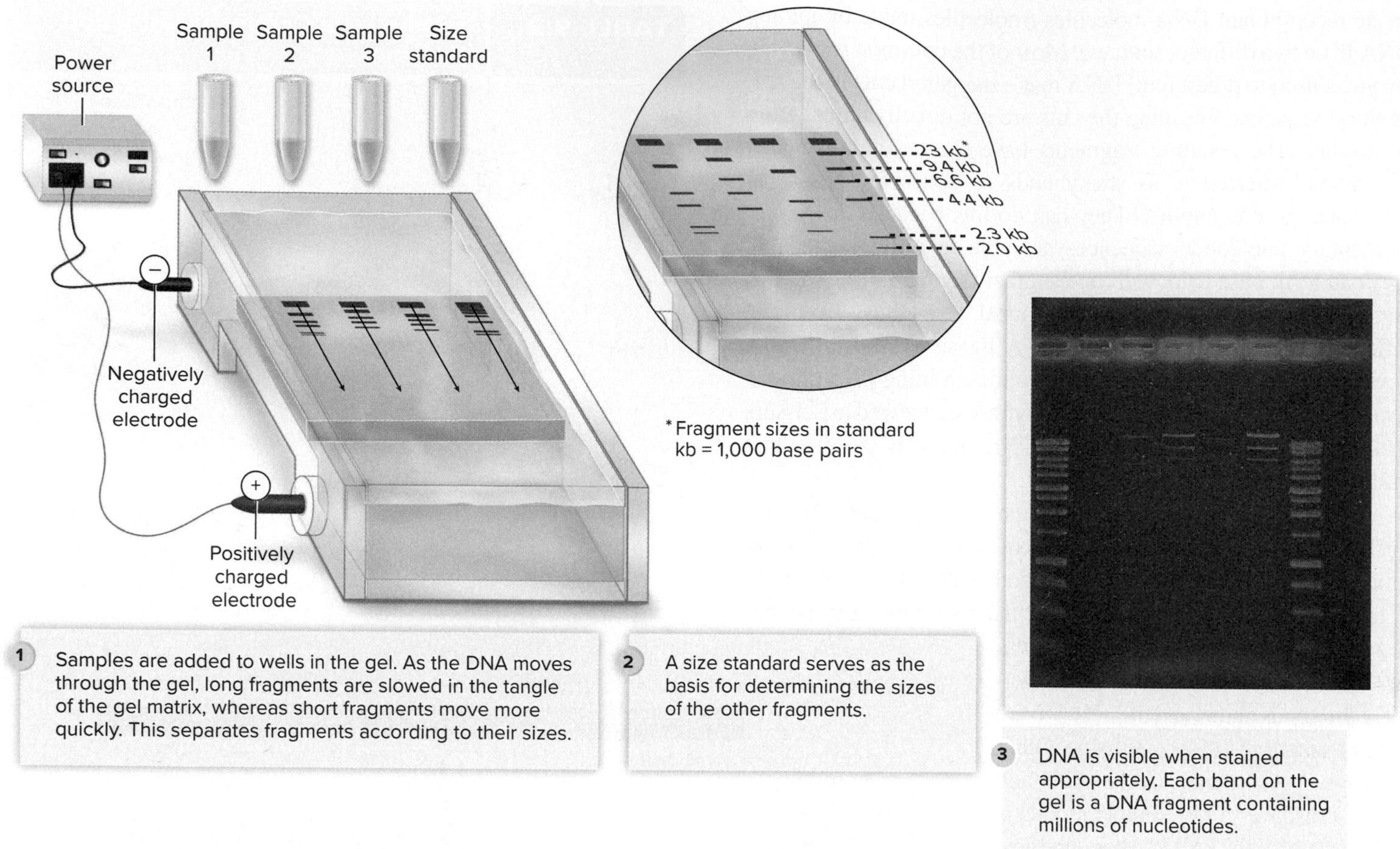

Power source
Sample 1
Sample 2
Sample 3
Size standard
Negatively charged electrode
Positively charged electrode
23 kb*
9.4 kb
6.6 kb
4.4 kb
2.3 kb
2.0 kb
*Fragment sizes in standard kb = 1,000 base pairs
1 Samples are added to wells in the gel. As the DNA moves through the gel, long fragments are slowed in the tangle of the gel matrix, whereas short fragments move more quickly. This separates fragments according to their sizes.
2 A size standard serves as the basis for determining the sizes of the other fragments.
3 DNA is visible when stained appropriately. Each band on the gel is a DNA fragment containing millions of nucleotides.

FIGURE 9.3 Gel Electrophoresis Agarose gel electrophoresis is used to separate DNA molecules according to size. DNA in the gel is visible when stained with a dye that fluoresces under UV light. Lisa Burgess/McGraw Hill

? Why is it important that the positively charged electrode is on the side of the agarose gel opposite the wells?

Reverse transcriptase is particularly useful in studies involving gene expression because cDNA can be used as a substitute for RNA (DNA is the more stable molecule, so working with it is easier). Thus, cDNA analysis gives insights into an organism's transcriptome—the total collection of RNA molecules in a cell.

Gel Electrophoresis

Gel electrophoresis is a technique that relies on an electric current to separate macromolecules, such as nucleic acids or proteins, by size. Gels made of agarose (a highly purified form of agar) are typically used for separating DNA and RNA molecules, while polyacrylamide gels are commonly used to separate proteins. After a DNA sample is digested with restriction enzymes, for example, agarose gel electrophoresis is used to separate the resulting fragments, estimate their sizes, and even isolate fragments of a certain size from the gel (isolated fragments can be used for molecular cloning and sequencing).

For agarose gel electrophoresis of DNA, samples are added to wells near one end of a gel that has been placed in an apparatus containing a buffer (an electrically conductive solution) (**figure 9.3**). A size standard (a mixture of DNA molecules of known sizes) is added to one or more unfilled wells; this will serve as a basis for comparison, allowing the researcher to estimate the sizes of the various DNA molecules in the samples. An electric current is then run through the gel. Because DNA is negatively charged, the molecules move toward the positively charged electrode, which is positioned at the end of the gel opposite the wells. Since short DNA molecules move more easily through the gel matrix, they migrate more quickly through the gel. In contrast, larger molecules move more slowly through the matrix, so they do not migrate as far in a given time and therefore remain closer to the wells. To see the DNA, it can be stained using one of several commercially available dyes that intercalate into the DNA helix and fluoresce under certain wavelengths of light. Each visible band in the gel represents millions of DNA molecules of a specific size.

MicroAssessment 9.1

Restriction enzymes recognize specific nucleotide sequences and then cut the DNA at or near those sequences, generating restriction fragments. Reverse transcriptase uses RNA to make cDNA. Gel electrophoresis is used to separate macromolecules, such as DNA fragments, according to their size.

1. What is the importance of sticky ends in genetic engineering?
2. How does gel electrophoresis separate differently sized DNA fragments?
3. What should a restriction enzyme isolated from *Staphylococcus aureus* strain 3A be called? ?

9.2 ▪ Molecular Cloning

Learning Outcomes

2. Outline the simplified steps of molecular cloning.
3. Describe the applications of molecular cloning.
4. Describe the process used to create a DNA library.

Molecular cloning (also called DNA cloning, or simply cloning) is the procedure used to introduce foreign or altered DNA into a recipient cell in a form that will be replicated and passed on to the cell's progeny. Recall, however, that DNA introduced into a cell will not be replicated by that cell unless it is a replicon (see figure 8.18). The discovery of restriction enzymes made it possible to insert a DNA segment into a plasmid or other small replicon, creating a **recombinant DNA molecule** that can be replicated in a cell.

The plasmid or other replicon into which the DNA fragment of interest is inserted functions as a carrier of that DNA, and is referred to as a cloning vector, or more simply a **vector.** A variety of vectors are now commercially available, each genetically modified to provide characteristics that make molecular cloning easier.

The Cloning Process—A Simplified View

A simplified view of the cloning process involves a series of steps including (**figure 9.4**):

① Isolate DNA containing the fragment of interest from the original organism.
② Use a restriction enzyme to cut that DNA into smaller fragments.
③ Use that same restriction enzyme to cut the vector to form a linear molecule.
④ Join (ligate) the DNA fragment of interest with the vector, creating a recombinant DNA molecule. The DNA fragment is said to have been inserted into the vector, and the fragment itself is referred to as an insert.
⑤ Introduce the recombinant molecule into a recipient cell (a host).

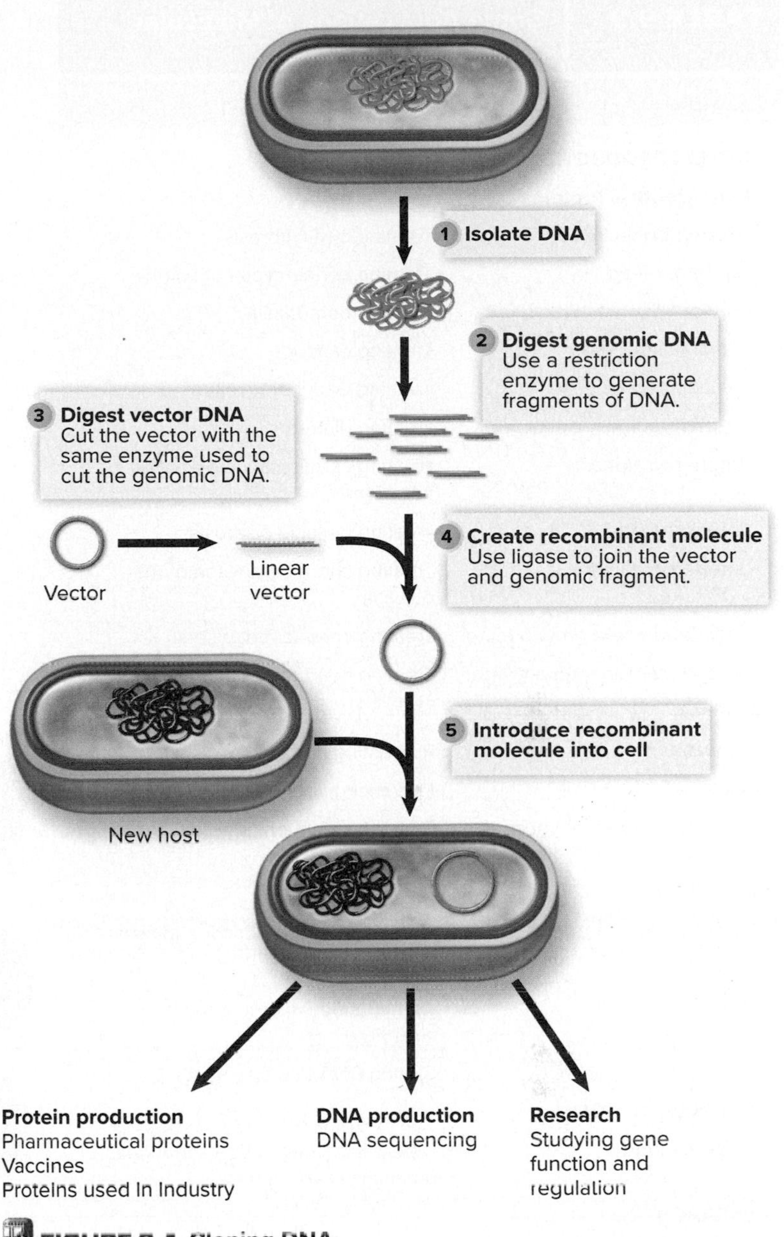

1 Isolate DNA
2 Digest genomic DNA
Use a restriction enzyme to generate fragments of DNA.
3 Digest vector DNA
Cut the vector with the same enzyme used to cut the genomic DNA.
Vector
Linear vector
4 Create recombinant molecule
Use ligase to join the vector and genomic fragment.
5 Introduce recombinant molecule into cell
New host
Protein production
Pharmaceutical proteins
Vaccines
Proteins used in industry
DNA production
DNA sequencing
Research
Studying gene function and regulation

FIGURE 9.4 Cloning DNA

Why is the same restriction enzyme used to cut both the vector and the insert in cloning?

Applications of Molecular Cloning

Molecular cloning was the groundbreaking process that allowed bacteria, yeast, and other cells that grow in culture to be genetically engineered. Once genes for commercially valuable proteins are introduced into cells, those cells can be grown to high densities; the cells express the newly inserted genes and the protein products can then be harvested. Likewise, genetically engineered cells can be used to produce DNA for study, and they are also valuable research tools (**table 9.3**).

MicroByte

The insulin yield from a 2,000-liter culture of *E. coli* genetically engineered to produce human insulin is the same as that from 1,600 pounds of pancreatic glands!

Creating a DNA Library—A Detailed View of the Cloning Process

An approach often used to clone a specific gene into a bacterium such as *E. coli* is to make a **DNA library.** This is a collection of cells, each containing a cloned DNA fragment from an organism; together, the cells carry the organism's entire genome (**figure 9.5**). To make a gene library, restriction enzymes are used to cut the DNA of the organism being studied, and then the entire set of restriction fragments is cloned into a population of *E. coli* cells. Although each cell in the resulting population contains only one fragment of the genome, the entire genome is represented in the population as a whole. Once a DNA library has been prepared, colony blots (described later) can be used to determine which cells contain the gene of interest.

TABLE 9.3 Some Applications of Molecular Cloning

Example	Use
PROTEIN PRODUCTION	
Pharmaceutical Proteins	
Deoxyribonuclease	Treating cystic fibrosis
Erythropoietin	Treating certain types of anemia
Factor VIII	Treating hemophilia
Gamma interferon	Treating cancer
Glucocerebrosidase	Treating Gaucher disease
Insulin	Treating diabetes
Interferon alpha	Treating certain cancers and viral infections
Interferon beta	Treating multiple sclerosis
Interferon gamma	Treating chronic granulomatous disease
Platelet-derived growth factor	Treating chronic foot ulcers
Tissue plasminogen activator	Treating heart attacks and strokes
Vaccines	
Hepatitis B	Preventing hepatitis
HPV	Preventing cervical cancer
Influenza (egg-free vaccine version)	Preventing influenza
Other Proteins	
Bovine somatotropin	Increasing milk production in cows
Chymosin	Cheese-making
Leghemoglobin	Contributing flavor to a meat substitute
Restriction enzymes	Cutting DNA into fragments
DNA PRODUCTION	
DNA for study	Determining nucleotide sequences; obtaining DNA probes
RESEARCH	
Studying gene function and regulation	Determining outcomes of turning genes on or off

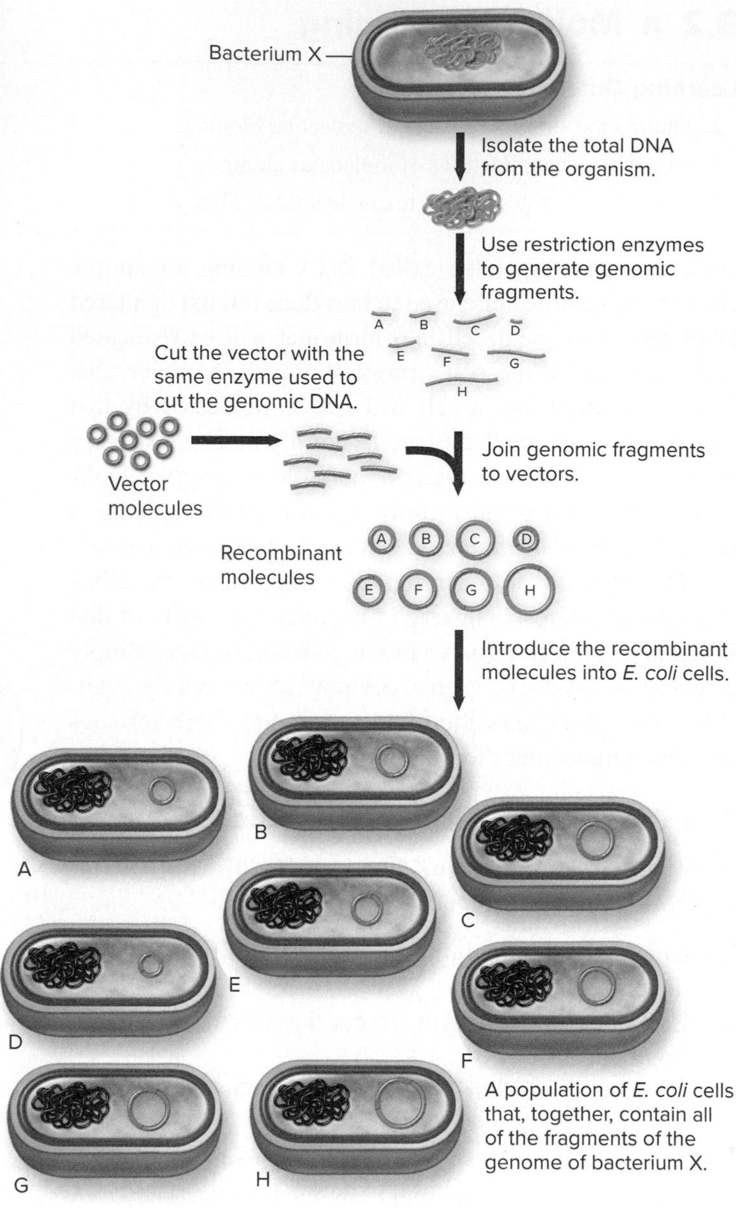

FIGURE 9.5 A DNA Library Each cell contains one fragment of a given genome.

What enzyme is used to join the genomic fragments to vector molecules?

Obtaining DNA for Cloning

The first step of cloning a bacterial gene is to obtain genomic DNA containing the gene of interest from the original bacterium. This is done by adding a detergent to cells in a broth culture to lyse them. The released genomic DNA is then separated from the rest of the cellular debris by precipitation using salt and alcohol.

When cloning a eukaryotic gene into bacteria for protein production, a researcher must first obtain a copy of the DNA of interest that lacks introns. To do this, the researcher isolates mature mRNA from the appropriate cells and then uses the combination of reverse transcriptase and a DNA polymerase to make double-stranded cDNA (see figure 9.2).

Generating a Recombinant DNA Molecule

A restriction enzyme is used to cut (1) the genome containing the DNA of interest and (2) the vector molecules. The two types of DNA are then combined, along with DNA ligase, allowing recombinant molecules to form.

In addition to having an origin of replication, each vector in use today typically has additional features to make the cloning process easier (**figure 9.6**):

- **Selectable marker.** This is usually a gene that encodes resistance to an antibiotic such as ampicillin which allows the researcher to eliminate any cells that have not taken up molecules containing vector sequences. Eliminating

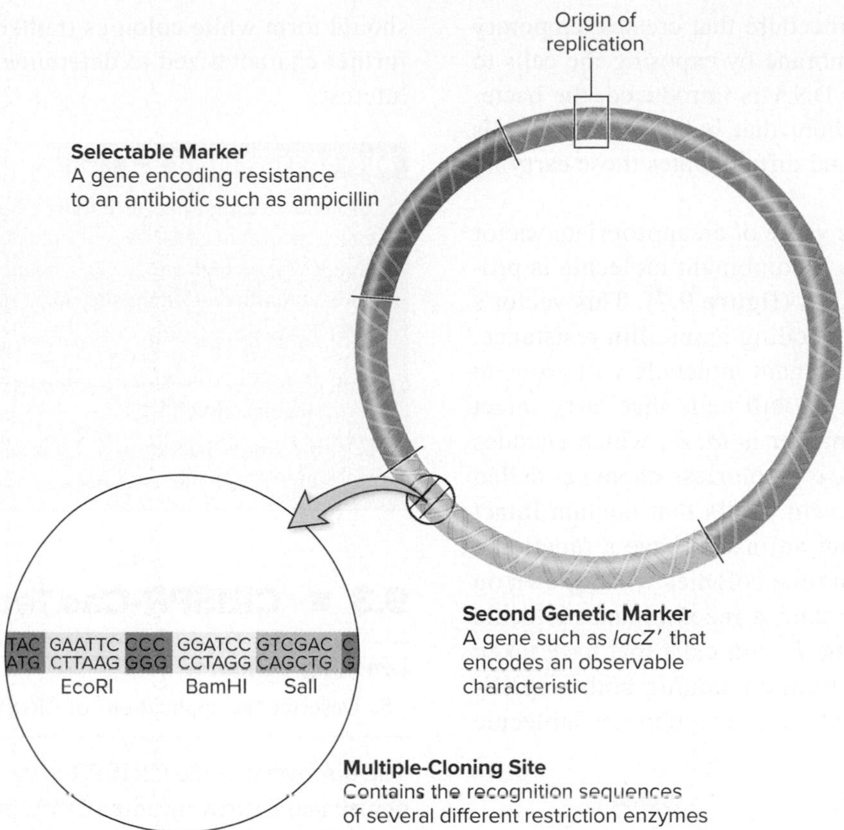

FIGURE 9.6 Typical Properties of an Ideal Vector In addition to having an origin of replication, most vectors have a selectable marker and a multiple-cloning site into which the DNA of interest can be inserted. The multiple-cloning site is usually situated within a second genetic marker and allows the researcher to identify cells containing a recombinant vector.

When a DNA fragment has been successfully joined with a vector, will both the selectable marker and the second genetic marker still be functional? Explain.

those cells is important because even under ideal conditions, relatively few cells in a population take up DNA. In the cloning process, only cells that have taken up a vector (with or without an insert) can grow on a selective medium containing the antibiotic.

- **Multiple-cloning site.** This is a short DNA region where the DNA of interest will be inserted to form the recombinant molecule; it has been engineered to include the recognition sequences of several different restriction enzymes that cut nowhere else in the vector. Its value is versatility—it can be used in different experiments for cloning various genes of interest by digesting the genomic DNA with different restriction enzymes. Significantly, the multiple-cloning site is located within the second genetic marker, described next.
- **Second genetic marker.** This gene encodes an observable characteristic that allows researchers to easily distinguish colonies of cells that contain recombinant molecules from those that do not. Doing so is important because when vector molecules are mixed with the DNA fragments to be cloned, most of the vector molecules will simply religate on themselves, recircularizing to regenerate the original (intact) version; cells containing those intact vector molecules can grow on the selective medium mentioned above but are an unwanted product. To distinguish cells containing intact vector from those containing recombinant molecules, the vector is designed such that its multiple cloning site is positioned within the second genetic marker but does not interfere with that marker's expression in an intact vector. Because of this positioning, the marker is functional in cells that contain intact vector but non-functional in cells that have a recombinant plasmid (because insertion of a fragment into the multiple cloning site inactivates the marker). Thus, the second genetic marker provides an easy way to identify the rare colonies of cells each containing a recombinant molecule. A common second genetic marker is a gene called *lacZ′*, which will be described shortly.

Obtaining Cells That Contain Recombinant DNA Molecules

A common method of introducing DNA into a host is bacterial transformation. *E. coli* cells do not naturally take up DNA, and therefore they must be treated with certain chemicals to make them competent. An alternative technique is to introduce the

DNA by electroporation, a procedure that creates temporary pores in the cytoplasmic membrane by exposing the cells to an electric current. Once the DNA is introduced, the bacteria are then grown on a medium that both selects for cells containing vector sequences and differentiates those carrying recombinant molecules.

A good illustration of the value of an appropriate vector in obtaining cells that have a recombinant molecule is provided by a vector called pUC18 (**figure 9.7**). This vector's selectable marker is a gene encoding ampicillin resistance. Thus, cells that carry a recombinant molecule will grow on media containing ampicillin, as will cells that carry intact vector. The second genetic marker is *lacZ′*, which encodes a polypeptide that helps cleave a colorless chemical called X-gal to produce a blue pigment. Cells that contain intact vector (circular vector without an insert) have a functional *lacZ′* gene and therefore form blue colonies when grown on X-gal, whereas those that contain a recombinant molecule form white colonies. When the *E. coli* cells that have taken up DNA are grown on a medium containing both ampicillin and X-gal, those that contain a recombinant molecule should form white colonies (rather than blue); they are then further characterized to determine if they carry the gene of interest.

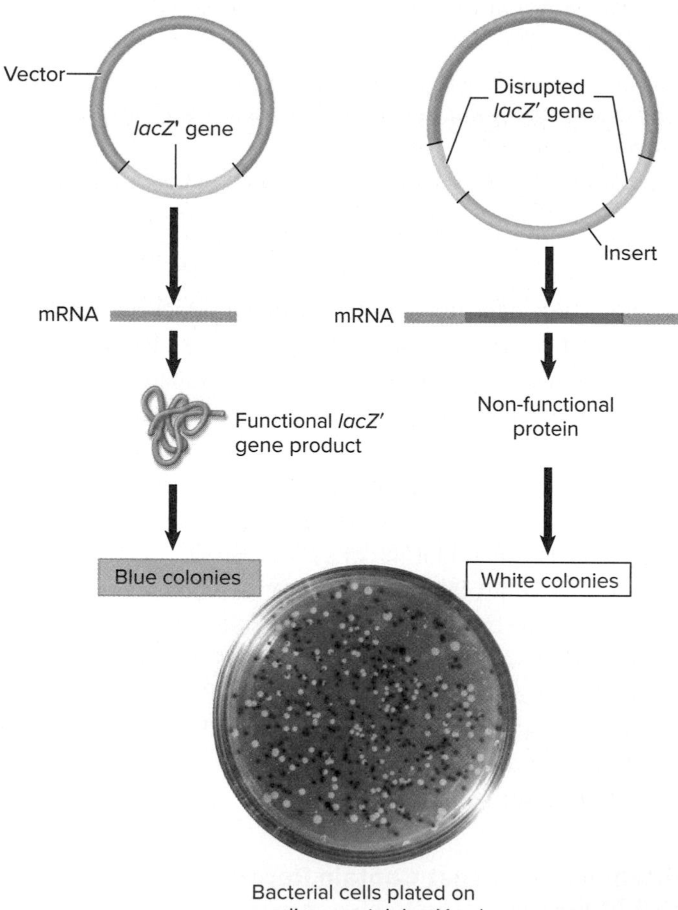

FIGURE 9.7 The Function of the *lacZ′* Gene in a Vector The *lacZ′* gene is used to differentiate cells that contain recombinant molecules from those that contain vector alone. Edvotek

? **What color colonies will cells that contain a recombinant molecule form?**

MicroAssessment 9.2

Molecular cloning involves joining a fragment of DNA to a vector to create a recombinant DNA molecule that is then moved into a cell, where it will be replicated. Most cloning vectors have a selectable marker, a multiple-cloning site, and a second genetic marker.

4. What is the function of a vector in molecular cloning?
5. Explain the role of ampicillin and the *lacZ′* gene in molecular cloning.
6. What would happen if a bacterial cell took up a fragment of foreign DNA that was not inserted into a vector?

9.3 ■ CRISPR-Cas Technologies

Learning Outcome

5. Describe the applications of CRISPR-Cas technologies.

The discovery of the CRISPR systems that bacteria use to recognize and destroy invading DNA, as described in section 8.11, gave scientists an entirely new range of biotechnology tools. What makes the systems so useful is that the Cas nucleases (nucleic acid-cutting enzymes) function within a living cell.

Scientists who developed what are called **CRISPR-Cas technologies** initially worked with a Cas nuclease called Cas9, but now various types of Cas nucleases have been engineered to accomplish different tasks. Like the original Cas nucleases, the modified forms use a piece of single-stranded RNA as a guide to recognize a particular DNA or RNA sequence within a cell; that guide RNA is referred to as gRNA. The modified Cas and gRNA can be introduced into a cell as a complete complex, or the genes for them can be moved in.

Applications of CRISPR-Cas Technologies

Various modified forms of Cas nucleases are used for **gene editing,** a process that introduces targeted changes into the existing nucleotide sequence of a cell's genome (**figure 9.8**). Directed by the gRNA, the nuclease makes a double-stranded cut in the DNA; scientists can control where the cut occurs simply by synthesizing a specific RNA molecule to act as the gRNA. Once the cut is made, the nucleotide sequence at that site can be changed, using processes that manipulate the cell's normal DNA repair mechanisms. Gene editing is now widely used to precisely modify a laboratory organism's DNA as a way to study gene function. Clinical trials are currently underway to assess its use in gene therapy, a way of correcting certain genetic disorders, including thalassemia, sickle cell disease, and a form of blindness. Because of the significant impact of

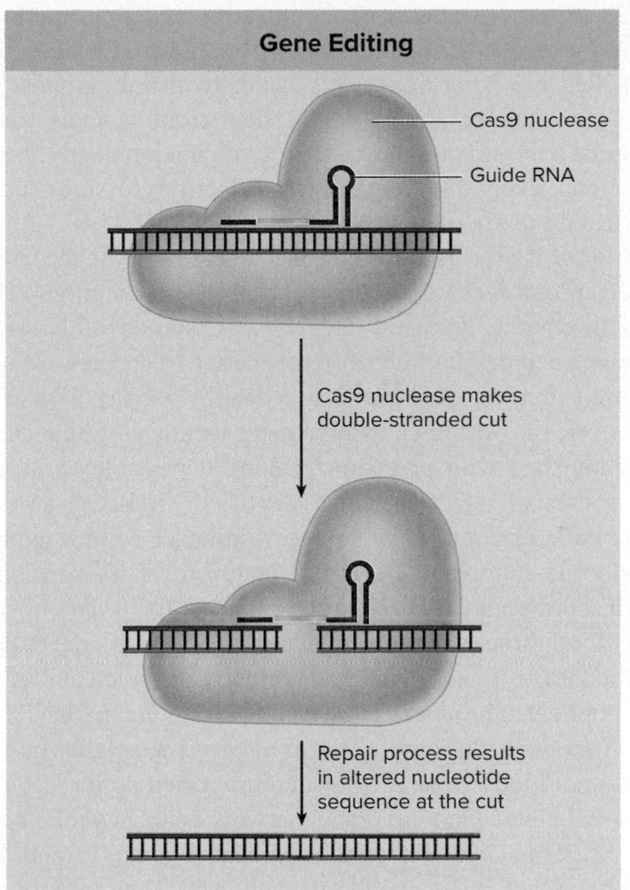

FIGURE 9.8 Using CRISPR for Gene Editing Cas9 nuclease, along with guide RNA, can be used to change a specific DNA sequence within a cell.

? What is the function of guide RNA?

CRISPR-Cas gene editing technologies, Jennifer Doudna and Emmanuelle Charpentier were awarded the 2020 Nobel Prize in Chemistry for developing the technologies.

Modified Cas9 nucleases called dCas9 (for "dead Cas") systems are used to study gene function. These modified forms are inactivated, meaning they do not cut DNA. Like the other Cas systems, however, a dCas uses a gRNA to locate and bind to a specific DNA sequence within a cell. One version of dCas binds DNA and physically blocks RNA polymerase from transcribing a gene, thereby turning the gene off; this allows researchers to study gene function **(figure 9.9a)**. Other versions have specific molecules attached to them, allowing those molecules to be delivered to precise locations in a cell's genome (figure 9.9b). Examples include: a dCas that carries an activator to turn a gene on, allowing researchers to study gene function and regulation; a dCas that adds methyl groups to DNA, allowing researchers to study the effects of DNA modification; and a dCas engineered to carry a fluorescent marker, allowing researchers to observe the location of certain DNA sequences within a cell.

One exciting application of Cas systems is rapid diagnosis of certain types of infectious disease or cancer. Various diagnostic kits are being developed, but they all rely on at least two components: (1) a Cas nuclease coupled with gRNA to recognize a nucleotide sequence for a specific disease, and (2) a mechanism to visibly detect that recognition. In 2020, the U.S. Food and Drug Administration (FDA) gave emergency-use authorization for a rapid COVID-19 diagnostic test that relies on CRISPR-Cas technologies; results can be obtained in about an hour (see Focus Your Perspective 9.1).

MicroAssessment 9.3

CRISPR-Cas technologies are used for gene editing, studies of gene regulation, and as a component of rapid diagnostic tests.

7. What is the natural function of the first CRISPR systems discovered in bacteria?
8. How can dCas be used to prevent a gene from being expressed?
9. Considering that restriction enzymes and Cas nucleases can both be used to cut DNA, why is Cas such an important new DNA-cutting tool for researchers?

Studies of Gene Function - Blocking Transcription

dCas9

dCas9 blocks transcription, turning the gene off

(a)

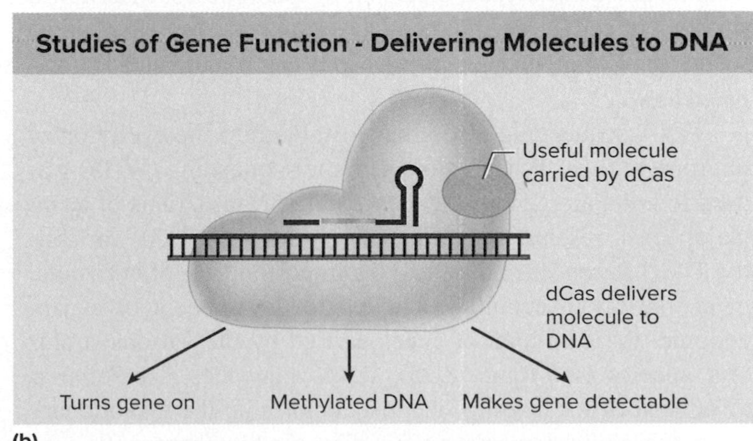

(b)

FIGURE 9.9 Using CRISPR to Study Gene Function **(a)** dCas can turn off a gene's expression by blocking transcription. **(b)** Versions of dCas that carry certain molecules can be used to deliver those molecules to specific chromosomal locations.

? How is a dCas different from a Cas?

9.4 ■ DNA Sequencing

Learning Outcomes

6. Describe the applications of DNA sequencing.
7. Describe the principles of high-throughput DNA sequencing.

The technologies for **DNA sequencing,** the process of determining the order of the nucleotides in a DNA molecule, have advanced rapidly over the last several decades. Today, thanks to highly automated and efficient methods, large-scale projects that would have been unheard of in the past are now a reality. The revolution started with the 1990 launch of the Human Genome Project, the effort that sequenced the human genome. In turn, that success led to the Human Microbiome Project, started in 2007 to use DNA sequencing to characterize the microbial communities that inhabit the human body. The most ambitious project yet, the Earth BioGenome Project, was launched in 2018 to sequence all known eukaryotic species.

In addition to providing valuable information for researchers, the large-scale sequencing projects encouraged significant advances in sequencing technologies. To put that into context: Bacterial genomes can now be sequenced in a matter of hours, whereas the first one, published in 1995, took about a year! The rapid explosion of data that followed gave rise to a new field—bioinformatics—which uses computer-based methods to organize and analyze sequence information and other biological data.

Applications of DNA Sequencing

With the advances in sequencing technologies, entire genomes can be sequenced with ease. Once the nucleotide sequence of a genome is known, the amino acid sequences of the encoded proteins can be determined. In addition, by comparing the genome sequences of various organisms, the roles and relevance of encoded proteins might be revealed. This, in turn, can give insights into an individual organism's capabilities, such as how it survives in certain environments or causes disease. Non-coding sequences and mobile genetic elements can be studied and compared as well.

DNA sequencing also gives insights into the evolution of microorganisms. When whole genome sequencing (WGS) was used to compare genome sequences of various strains of a single species, researchers sometimes found significant variability. The differences highlighted the important role of horizontal gene transfer in evolution and led to the concept of a pan-genome, the sum total of genes carried by the various strains of a species (see figure 8.25). DNA sequences also serve as a framework for helping scientists determine the evolutionary relatedness of organisms, a topic discussed in chapter 10.

The advances in sequencing technologies have made it feasible for WGS to be used for tracking pathogens around the globe, thus providing insights about their spread. For example, WGS of various strains of the globally emerging multi-drug resistant fungus *Candida auris*, an organism first described in 2009, showed at least five genetically distinct groups. This indicates that the strains do not stem from a single source; instead, they emerged in at least five geographically distinct locations, likely due to worldwide selective pressures such as the extensive use of antimicrobial medications.

WGS is also used to trace the source of pathogens implicated in outbreaks of healthcare-associated infections (HAI) and foodborne diseases. Tracking the source of HAIs has always been difficult because the causative species are often common in the environment or on healthy people. With WGS, however, the various phenotypically identical strains can be distinguished with precision, making it possible to identify the source of infection. Once identified, steps can be taken to remove or contain the source. Similar principles apply to foodborne outbreaks. A U.S. Centers for Disease Control (CDC) program called PulseNet catalogs WGS data of common foodborne pathogens, making it easier to connect cases of foodborne illness. After an outbreak is detected and traced, the implicated product can be recalled from the market. Using WGS to track the outbreaks can also give insights into the likely antibiotic resistance of the implicated strain. An FDA program called Genome Trakr Network enables public health organizations and universities to share genomic and geographical data about foodborne pathogens; data submitted to PulseNet are shared with Genome Trakr.

Advances in DNA sequencing have also given rise to precision medicine, an approach that tailors disease prevention and treatment to the individual, often using the patient's genetic makeup as a basis. As an example, breast cancer patients who have a type of cancer that results from over-expression of a protein called HER2 (human epidermal growth factor receptor 2) may now be treated with a medication that specifically blocks the protein rather than one that simply kills all rapidly dividing cells. In order to develop new targeted treatments and learn more about individual factors that correlate to disease risk, the White House established a research program in 2015 now known as *All of Us*. So far, nearly a million volunteers from across the United States, about 50% of which identify as belonging to historically underrepresented racial or ethnic groups, have provided DNA samples for analysis, along with personal information including medical history and intimate lifestyle details such as data from activity-tracking devices. The aim of the program is to build an extensive health database that allows researchers to study connections between individual factors (including lifestyle, biology, and environment) and health.

Because of efforts such as *All of Us*, your medical records might one day include a complete genetic profile, allowing for more targeted disease prevention and treatment. Concerns exist, however, regarding the appropriateness and confidentiality of information gained by analyzing a person's DNA. Will it be in a person's best interests to be told of a genetic predisposition for a life-terminating disease for which there is no intervention or treatment? And although the Genetic Information Nondiscrimination Act (GINA), passed in 2008, protects Americans against discrimination in healthcare coverage and employment based on their genome, could such information eventually be used to deny someone their rights?

High-Throughput Sequencing Methods

The term **high-throughput sequencing** refers to several highly automated methods that generate huge amounts of DNA sequence data relatively quickly. Also referred to as next-generation ("next-gen") sequencing, these methods have revolutionized biology research because their speed and efficiency lowers costs, thereby making large-scale sequencing projects more feasible.

Most high-throughput sequencing methods analyze millions of small but overlapping DNA fragments to determine their sequences. Computers are then used to align and merge those data to create one long sequence—a process called sequence assembly. **Figure 9.10** uses an analogy to show the concept of sequence assembly. The assembly is much easier if a reference genome has been completed; for example, once the DNA sequence of one strain of *E. coli* was determined, it became easier to assemble the sequences of other strains because the likely order of the fragments was known.

A relatively new high-throughput method is nanopore sequencing, also called third-generation sequencing, which measures brief changes in an electrical current to determine the nucleotide sequence of long fragments of DNA as they pass through microscopic protein pores in a membrane. Nanopore sequencing has been used to sequence microbial DNA aboard the International Space Station and has also been a valuable tool in sequencing and identifying strains of SARS-CoV-2 (following reverse transcription of their genomes to cDNA) during the COVID-19 pandemic.

Errors are common in high-throughput sequencing, so each genome must be analyzed multiple times to increase the reliability of the results. For example, if the identity of a particular nucleotide is determined 10 times, and one time it is shown to be an A, but the other nine times it is a T, then it most likely is a T.

Data	Alignment	Result
reading th keep read ing the textbook	reading th keep read ing the textbook	keep reading the textbook

FIGURE 9.10 Analogy for DNA Sequence Assembly Here, letters and spaces are being aligned and merged to form a phrase—an analogy for the process used in DNA sequence assembly.

? Why is it useful to have a reference sequence when assembling DNA sequencing data?

MicroAssessment 9.4

Efficient DNA sequencing methods have fueled the rapidly growing field of genomics. High-throughput methods generate huge amounts of data very quickly.

10. Describe three applications of DNA sequencing.

11. What is meant by the term *high-throughput sequencing*?

12. After several cholera outbreaks in Africa, whole genome sequencing was used to track the global spread of disease. How might this information be used to prevent future outbreaks in Africa?

9.5 ■ Polymerase Chain Reaction (PCR)

Learning Outcomes

8. Describe the applications of PCR.

9. Explain how PCR can be used to exponentially amplify a select region of DNA.

The **polymerase chain reaction (PCR)** allows for the creation of more than a billion copies of a given region of DNA—referred to as the target DNA or target sequence—in a matter of hours. Those copies, the **PCR product,** are generated in sufficient quantities to be detectable when DNA molecules in the sample are separated by gel electrophoresis, stained with a fluorescent dye, and visualized using UV light. Kary Mullis received a Nobel Prize for the invention.

Applications of PCR

One important application of PCR is disease diagnosis. For example, if a person is suspected of having the sexually transmitted disease gonorrhea, which is caused by *Neisseria gonorrhoeae*, PCR can be used to detect that bacterium in a specimen from the infected site. This is done by treating the specimen to release the DNA in cells and then using PCR to amplify (increase the amount of) a selected sequence unique to *N. gonorrhoeae*; successful amplification indicates that the pathogen is present (**figure 9.11**). The use of PCR to detect unique DNA sequences as a means to identify microorganisms is discussed in section 10.3.

Since PCR was introduced, numerous variations on the standard method have been developed, each with specific applications. Among the most widely used are:

- **RT-PCR (reverse-transcription PCR).** This is used to detect specific RNA sequences in a sample. It relies on standard PCR, but first the enzyme reverse transcriptase is used to synthesize cDNA from the RNA template. The cDNA then serves as the template for amplification.
- **qPCR (quantitative PCR).** Also called real-time PCR, this uses the standard method to amplify the target sequence, and it also labels the PCR product with a

FOCUS YOUR PERSPECTIVE 9.1

The COVID-19 Response—The Power of Biotechnology

The COVID-19 response is an excellent illustration of the power of biotechnology. Because of several technologies described in this chapter, the pandemic's global outcome—although devastating—resulted in fewer deaths than feared or predicted.

SARS-CoV-2, the virus that causes COVID-19, has an RNA genome. When the virus was first discovered in China, researchers used the enzyme reverse transcriptase to make a cDNA copy of its genome. That cDNA was then cloned and sequenced, and the information was shared with scientists around the world, initiating a global effort to control the disease.

A major part of any disease control effort is diagnosing patients who have the disease. Diagnosis not only guides treatment for an individual patient, but it also reveals the extent of the disease's spread in a population. Because researchers had access to the genome sequence of SARS-CoV-2 early on, they quickly developed real-time RT-PCR diagnostic tests, amplifying certain parts of the SARS-CoV-2 genome for detection in patient specimens. Test results were often available within a day. Countries with the most aggressive testing measures, as well as follow-up procedures to track contacts, were most successful in limiting the disease's spread. As the disease continues to cost lives, researchers are developing more rapid diagnostic tests, including CRISPR-Cas-based tests that give results in under an hour. While the first version of this test was authorized for use only in certified laboratories, researchers are also developing instrument-free versions for on-site use.

Data obtained via high-throughput sequencing were used to track the global spread of SARS-CoV-2. The tracking methods rely on detecting spontaneous mutations that inevitably occur as the virus replicates; these mutations serve as evolutionary markers. For example, viral genomes from a cluster of early cases in the Seattle area all shared the same unusual mutation, indicating that they all descended from the same source (see Focus Your Perspective 21.1). Using sequences of SARS-CoV-2 genomes from around the world, an open-source online project called Nextstrain created interactive visuals that resemble a family tree to illustrate viral evolution. Researchers involved with the project also determined that only some introductions of the virus into the United States were directly from China; others were from countries where the virus had been circulating after being introduced from China. While understanding the global spread did not stop the pandemic, it provided advance warning about the spread of COVID-19 so that preventive measures could be implemented. In addition, the knowledge gained can hopefully be used to prevent a similar pandemic in the future.

The fact that the genome sequence of SARS-CoV-2 was known early on facilitated research aimed at developing targeted antiviral therapies as well as vaccines, as described in Focus on the Future 20.1. By analyzing the viral genome, scientists determined the amino acid sequences of key proteins essential for viral replication. Relatively soon thereafter, the three-dimensional structures of three of those proteins were determined—one that the virus uses to attach to and then enter host cells (its spike protein), one it uses to replicate its genome (its polymerase), and one it uses to process and generate its proteins (its protease). Knowing those protein structures helped scientists to focus their efforts on developing compounds to specifically target these proteins. Several effective vaccine options that elicit an immune response against the spike protein were produced in record time, and the widely used medication for treatment, Paxlovid, includes a protease inhibitor. Although COVID-19 continues to be problematic, biotechnology has and continues to play an indispensable role in the prevention and treatment of this disease and has saved countless lives.

fluorescent marker so that its accumulation can be monitored during the course of the reaction. The amount of fluorescence increases as the PCR product accumulates, thereby allowing the researcher to (1) track the amplification in "real-time," rather than using gel electrophoresis at the end; and to (2) determine the relative amount of target DNA initially in the sample, which is reflected by the amount of fluorescence generated.

- **Real-time RT-PCR.** This combines the methods of RT-PCR and qPCR in order to detect and determine the relative amount of a given RNA sequence in a sample without the need for gel electrophoresis. Researchers often use it to study gene expression. It also plays an important role in COVID-19 diagnosis (see Focus Your Perspective 9.1).

The PCR Method

A standard polymerase chain reaction starts with a double-stranded DNA molecule that serves as a template from which more than a billion copies of the target sequence can be rapidly produced. The process involves DNA synthesis reactions and the following key ingredients:

- **Double-stranded DNA.** This contains the target sequence and serves as a template for DNA synthesis.
- ***Taq* polymerase.** This heat-stable DNA polymerase is from the thermophile *Thermus aquaticus*.
- **Primers.** Oligonucleotides—short, synthetically made single-stranded DNA molecules—are used as PCR primers. They are designed or chosen by the researcher and determine which portion of the template DNA is amplified.
- **Deoxynucleotides.** The four deoxynucleotides used in DNA synthesis—dATP, dGTP, dCTP, and dTTP—are required to make copies of the target.

The Three-Step Amplification Cycle

PCR uses a repeating cycle consisting of three steps (**figure 9.12**):

1. The sample is heated to near-boiling (about 95°C) in order to denature the DNA.

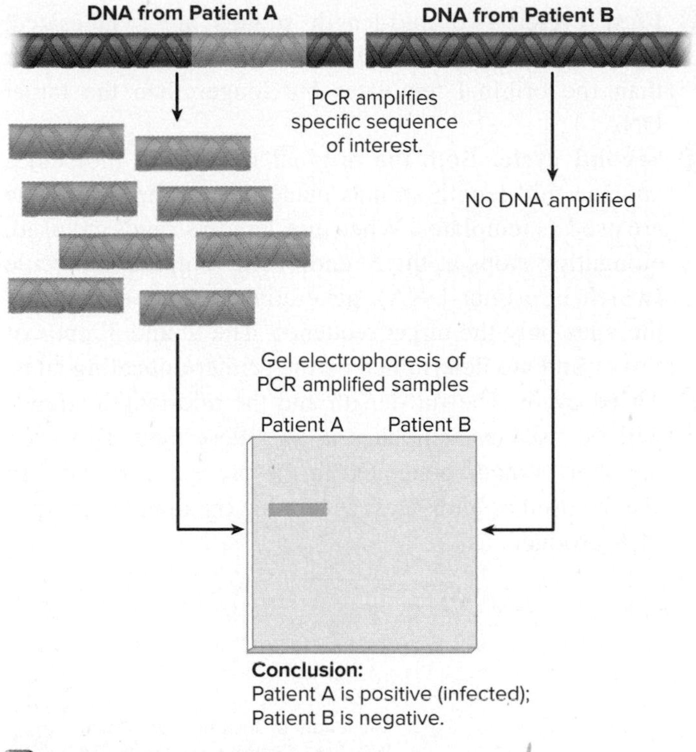
DNA from Patient A
DNA from Patient B
PCR amplifies specific sequence of interest.
No DNA amplified
Gel electrophoresis of PCR amplified samples
Patient A
Patient B
Conclusion:
Patient A is positive (infected);
Patient B is negative.

FIGURE 9.11 PCR Amplifies a Selected Sequence

How can PCR be used to diagnose an infectious disease?

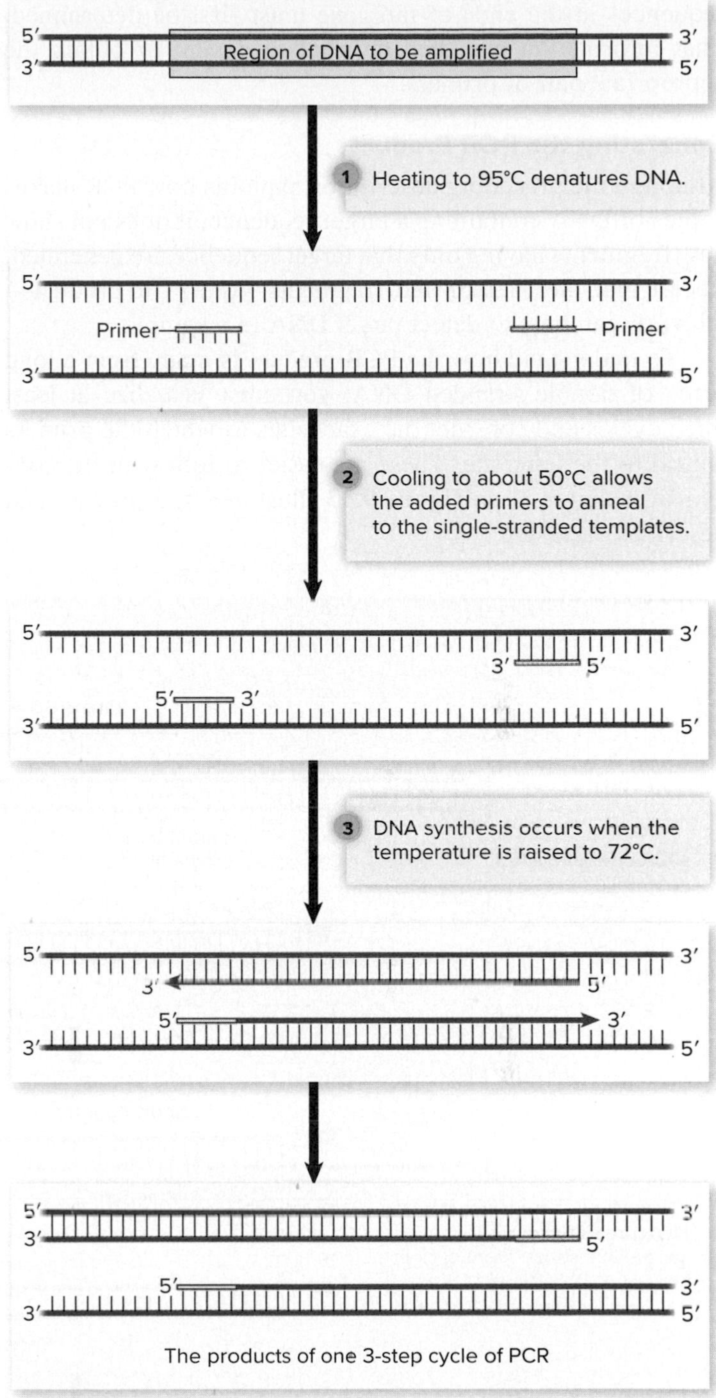
Region of DNA to be amplified
1 Heating to 95°C denatures DNA.
Primer
Primer
2 Cooling to about 50°C allows the added primers to anneal to the single-stranded templates.
3 DNA synthesis occurs when the temperature is raised to 72°C.
The products of one 3-step cycle of PCR

FIGURE 9.12 Steps of a Single Cycle of PCR

Considering that PCR can be used to make over a billion copies of target DNA in only a matter of hours, approximately how many molecules of primers must be included in the reaction?

2. The temperature is lowered to about 50°C; the exact temperature required depends on the lengths and nucleotide compositions of the primers. Within seconds, the primers anneal to their complementary sequences on the denatured target DNA.
3. The temperature is raised to the optimal temperature of *Taq* DNA polymerase (about 70°C), allowing DNA synthesis to occur. After one three-step cycle, a region containing the target DNA in the template is duplicated.

In subsequent amplification cycles, the original DNA strands serve as templates again, and so do the newly synthesized strands. Because the number of template molecules doubles with each cycle, PCR amplifies the DNA exponentially. After a single cycle of the three-step reaction, there will be two double-stranded DNA molecules for every original; after the next cycle, there will be four; after the next there will be eight; and so on.

A critical factor in PCR is *Taq* polymerase, the heat-stable DNA polymerase of *Thermus aquaticus*. This polymerase, unlike the DNA polymerase of *E. coli*, is not destroyed at the high temperature used to denature the DNA in the first step of each amplification cycle. If a heat-stable polymerase were not used, fresh polymerase would need to be added after that step in each cycle of the reaction. The discovery and characterization of *T. aquaticus* through basic research was key to developing this widely used and commercially valuable method.

Another critical factor in PCR is the selection of primer pairs used in the reaction because they dictate which part of the DNA is amplified. Each primer must be complementary to one end of the target DNA so that DNA synthesis will extend across that stretch of DNA (see figure 9.12). To amplify a DNA sequence that encodes a specific protein, the nucleotide sequences at the ends of the gene must first be determined. That information can then be used to design or obtain the appropriate pair of primers.

Generating the PCR Product

Although the preceding description explains how PCR makes copies of DNA containing a target sequence, it does not show how fragments having only that target sequence are generated. Recall that these amplified fragments—the PCR product—allow researchers to detect target DNA in a sample.

To understand how the PCR product is made from a long piece of double-stranded DNA, you must visualize at least three cycles and consider the exact sites to which the primers anneal to the template. These are easier to follow in the context of a diagram, so **figure 9.13** illustrates the process and describes each step.

① **First cycle.** Two mid-length strands are synthesized. Their 5′ ends are primer DNA, and the strands are shorter than the original templates but longer than the target DNA.

② **Second cycle.** Both the original full-length molecules and the mid-length strands made during the first cycle are used as templates. When mid-length strands are used, elongation stops at the 5′ end of the template molecule (which is primer DNA), generating a short strand that includes only the target sequence. The 5′ and 3′ ends of this strand are determined by the primer-annealing sites.

③ **Third cycle.** The full-length and the mid-length strands will be used as templates as just described. However, the short strands generated in the preceding round will also be used as templates, generating the double-stranded PCR product.

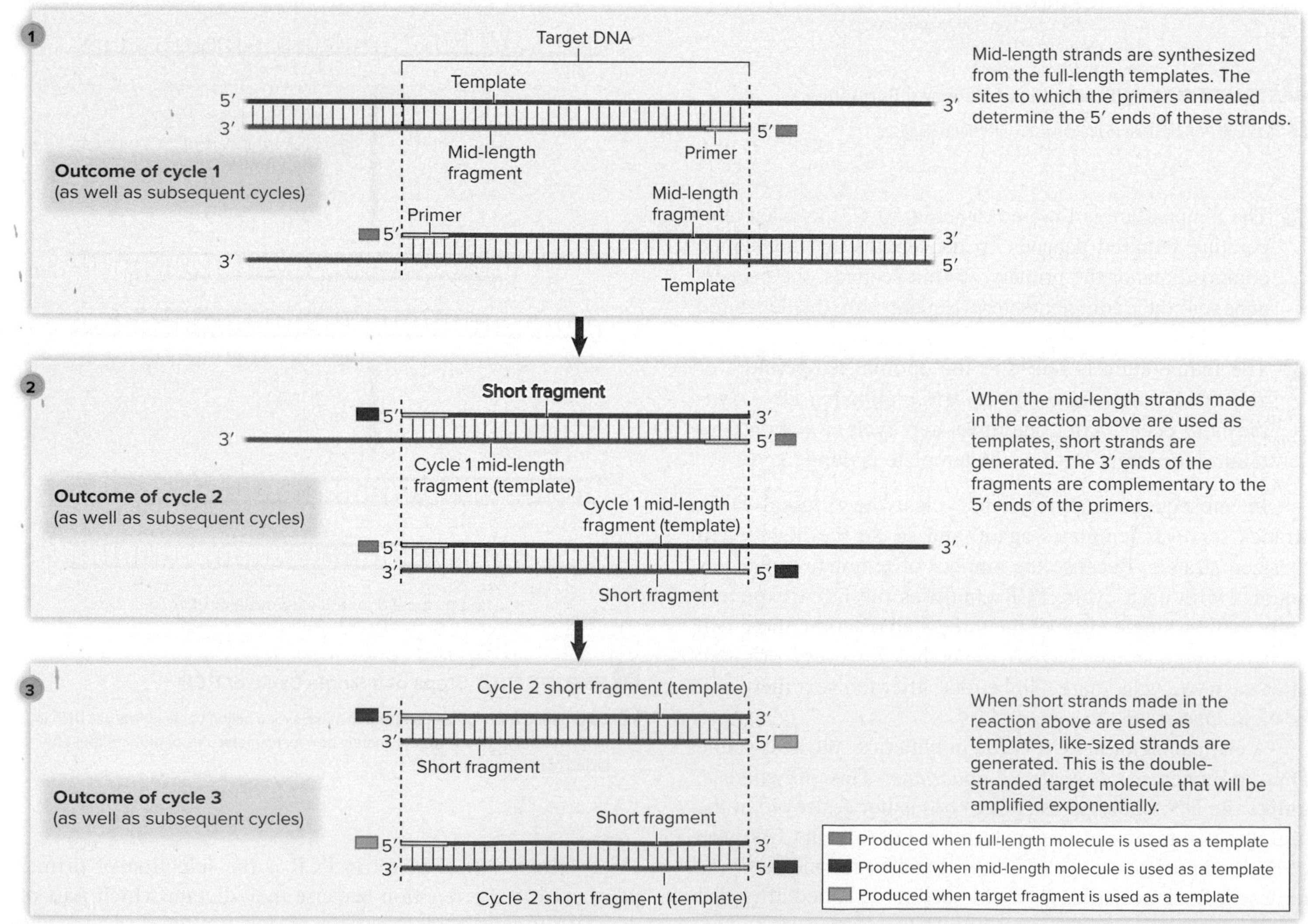

FIGURE 9.13 Generating the First Copies of the PCR Product. The positions to which the primers anneal to the template dictate the size and sequence of the fragment amplified exponentially.

? If two mid-length strands are present at the end of cycle 1, how many will be present at the end of cycle 3?

FOCUS ON A CASE 9.1

After serving more than 10 years on a rape charge, a wrongfully convicted man was released from prison when a new PCR-based technique for DNA typing (also called DNA profiling or DNA fingerprinting) indicated that he was not guilty. It showed that DNA from the semen taken from the victim did not match that of the accused man. Indeed, a DNA database showed a match with another man currently in prison for an unrelated rape charge.

The PCR-based method of DNA typing amplifies certain chromosomal regions that contain short tandem repeats (STRs). STRs are short nucleotide sequences (each typically two to six base pairs long) that repeat consecutively a variable number of times, usually within an intron or other untranslated region. For example, the sequence AATG repeats consecutively 5 to 14 times on chromosome 2 in an intron within the thyroid peroxidase gene. In one person, there may be 9 of these STRs in one copy of that chromosome and 7 in the other, whereas another person may have 11 and 5. This variation (called a polymorphism) makes the repeats a useful genetic marker for distinguishing individuals. The FBI now catalogs PCR-based DNA profiles from unsolved crimes and convicted violent offenders, making it easier to track or link the crimes of serial offenders. The database, called CODIS (**Co**mbined **D**NA **I**ndex **S**ystem), now uses the amplification pattern of 20 different STR loci (chromosomal locations). It is nearly impossible for two people to share the same number of repeats at all of these loci, unless they are identical twins.

1. Why would a PCR-based method for DNA typing be valuable in forensics?
2. How is PCR-based typing used to detect STR polymorphisms?
3. How can the method analyze 20 different loci simultaneously?

Discussion

1. Results can be obtained in less than 5 hours from a sample as small as a drop of body fluid the size of a pinhead.
2. Primers that bind regions on each side of the STRs are used to determine the number of repeats (**box figure 9.1**). An amplified fragment that contains 9 repeats, for example, will be longer than one that contains only 7. The PCR-amplified fragments can be quickly separated using a rapid type of gel electrophoresis called capillary electrophoresis. A size standard is included so that the sizes of the PCR-amplified fragments can be determined.
3. Commercially available kits contain fluorescently labeled primers that allow simultaneous amplification and subsequent recognition of each of the 20 loci. A laser detects the color of each amplified fragment as it moves out of a capillary gel, and computer analysis generates a pattern of peaks that reflect the STR profile of the DNA sample.

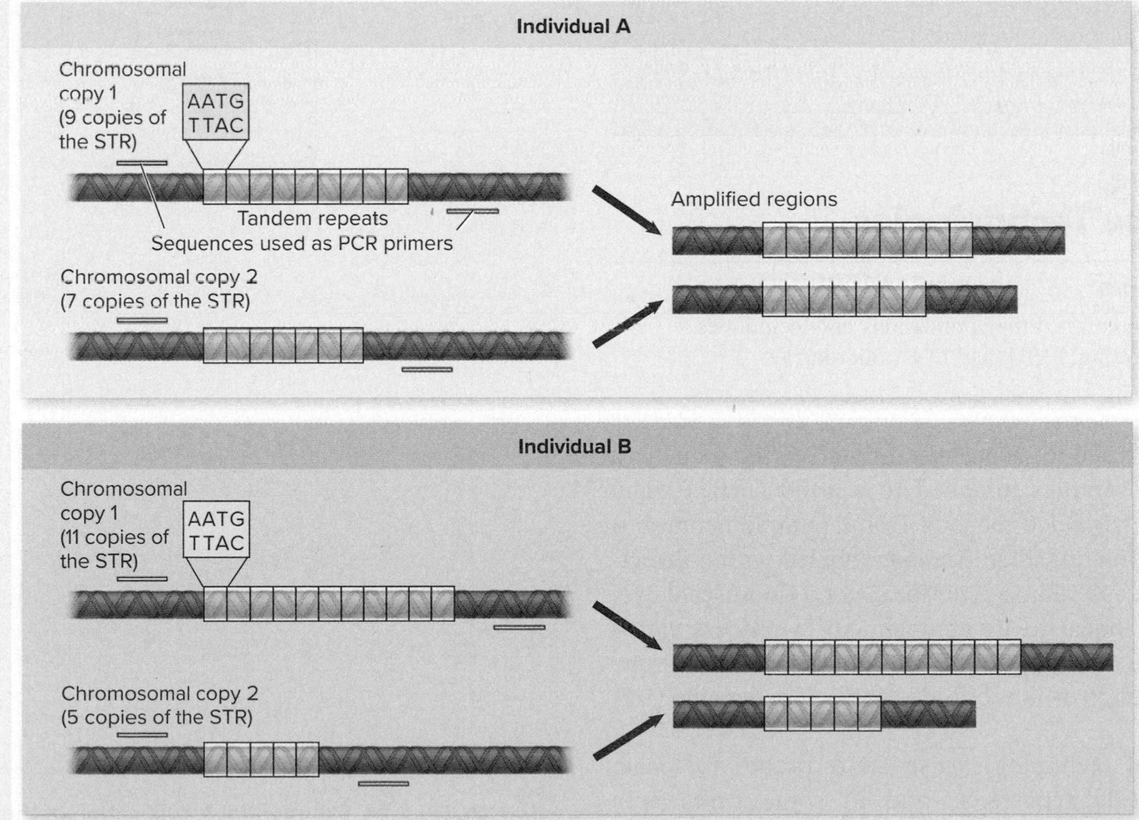

BOX FIGURE 9.1 Using PCR as Part of DNA Typing PCR is used to amplify certain chromosomal regions containing short tandem repeats (STRs). The number of copies of a given STR varies among people, resulting in corresponding differences in the length of the amplified fragments. Typically, at least 20 different STR locations are analyzed.

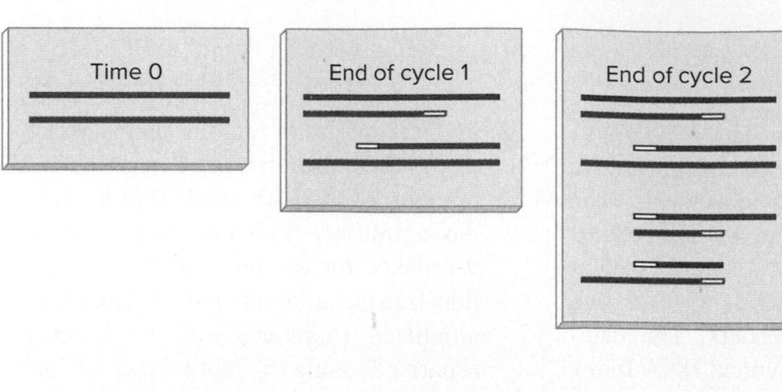

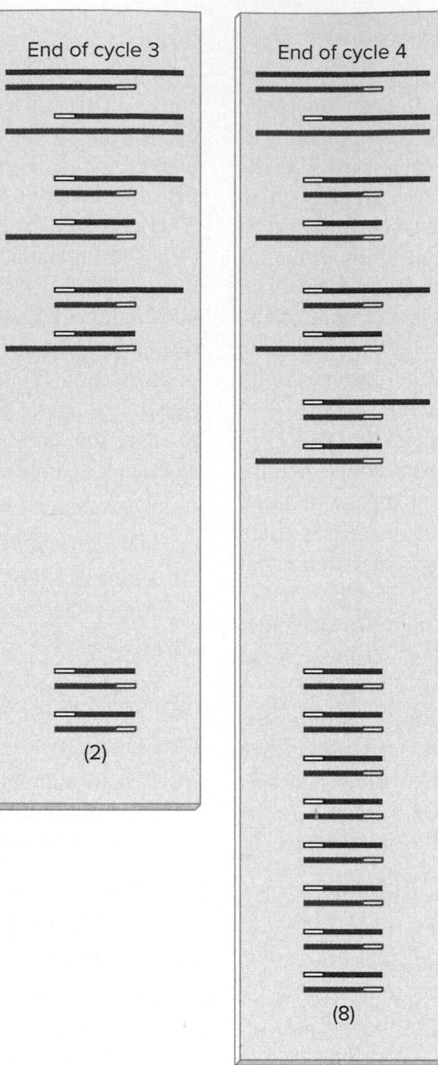

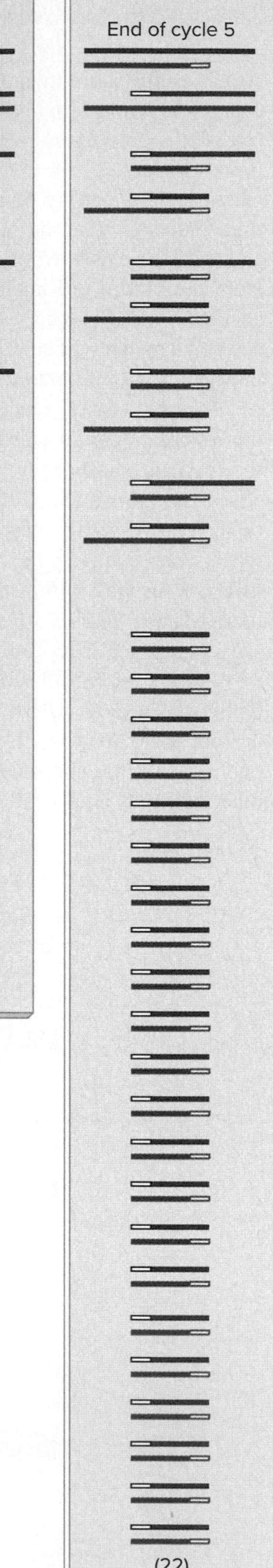

FIGURE 9.14 Exponential Amplification of the PCR Product
During PCR, mid-length fragments are amplified linearly (arithmetically), whereas the PCR product is amplified exponentially. After 30 cycles of PCR, more than a billion molecules of the PCR product will have been synthesized.

How many cycles are required to generate the first two copies of the PCR product?

Continuing to follow the events in further rounds of PCR reveals that the PCR product, which originated in the third cycle, is exponentially amplified (**figure 9.14**). Over one billion copies of PCR product can be obtained after 30 PCR cycles from a single template molecule.

MicroAssessment 9.5

The polymerase chain reaction (PCR) is used to rapidly increase the amount of a specific segment of DNA in a sample. Variations of the conventional methods can be used to study gene expression.

13. What is the role of the enzyme obtained from *Thermus aquaticus* in PCR?
14. Explain why it is important to use a polymerase from a thermophile in the PCR reaction.
15. Sequencing reactions can be done using PCR. In this case, would two primers be necessary? Explain.

9.6 ■ Probe Technologies

Learning Outcome

10. Compare and contrast the applications and techniques of colony blotting, FISH, and DNA microarray technologies.

DNA probes are used to locate specific nucleotide sequences in nucleic acid samples attached to a solid surface. The probe is a single-stranded piece of DNA, complementary to the sequence of interest, that has been labeled with a detectable marker such as a radioactive isotope or a fluorescent dye. The probe will anneal to its complement, a process called **hybridization.** By hybridizing to its complement, the probe "finds" the sequence of interest and makes it detectable (**figure 9.15**).

A variety of technologies use DNA probes to locate specific nucleotide sequences, and in some cases help identify a particular organism or group of organisms (see figure 10.9). Technologies that use probes include colony blotting, fluorescence in situ hybridization (FISH), and DNA microarrays.

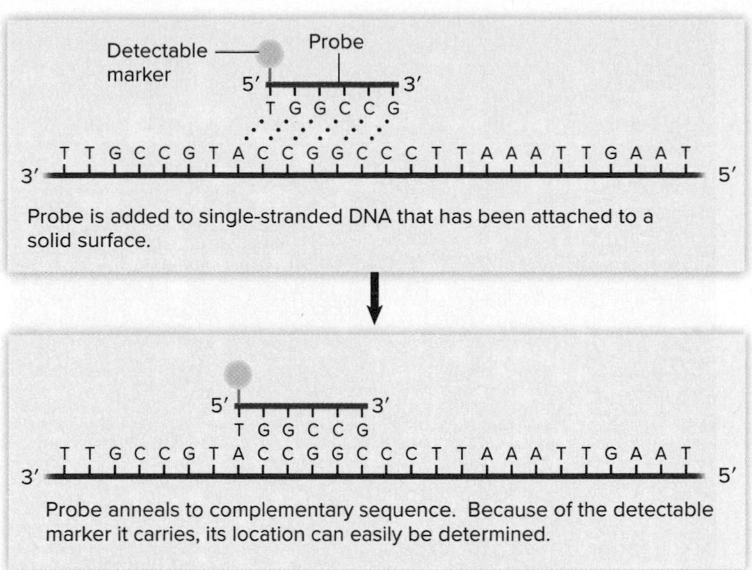

FIGURE 9.15 DNA Probes These single-stranded pieces of DNA, labeled with a detectable marker, are used to detect complementary nucleotide sequences in DNA or RNA samples that have been attached to a solid surface.

? **Why is it important that the probe be labeled?**

Colony Blotting

Colony blotting uses probes to detect specific DNA sequences in colonies grown on agar plates (**figure 9.16** ①). This method is commonly used to determine which clones in a DNA library or other collection contain a sequence being studied. ② The term "blot" in the name reflects the fact that the colonies are transferred in place ("blotted") onto a nylon membrane, retaining the spatial orientation of the colonies on the original plate. The membrane serves as a durable support for the cells of the colonies and their DNA. ③ After the transfer, the membrane is soaked in an alkaline solution to simultaneously lyse the cells and denature their DNA, generating single-stranded DNA molecules. ④ A solution containing the probe is then added to the membrane and incubated under conditions that allow the probe to hybridize to complementary sequences on the filter. Any probe that has not bound is then washed off. ⑤ The appropriate method is then used to detect the labeled probe. The positions to which the probe hybridized indicate colonies that have the DNA of interest.

Fluorescence in Situ Hybridization (FISH)

Fluorescence in situ hybridization (FISH) has two characteristics reflected in the name: (1) a fluorescently labeled probe is used to detect specific nucleotide sequences, and (2) the sequences are within intact cells, in this case attached to a microscope slide (in situ is Latin for "in the original place"). Cells containing the hybridized probe can then be observed using a fluorescence microscope, as described in section 3.8. To study microorganisms, a probe that hybridizes to sequences on ribosomal RNA (rRNA) is generally used. This is because multiplying cells can have thousands of copies of rRNA, increasing the technique's sensitivity. Other characteristics of rRNA that make it useful for identifying microorganisms are described in chapter 10.

FISH is revolutionizing microbial ecology research and holds great promise in clinical laboratories. It provides a way to rapidly identify microorganisms directly in a specimen, bypassing the need to grow them in culture. FISH can be used to detect either a group of related organisms or a specific species, depending on the nucleotide sequence of the probe. By using separate probes—each one specific for a different group of microorganisms and labeled with a different color of fluorescent marker—FISH can be used to determine the relative proportion of different groups of microorganisms in a specimen (**figure 9.17**). It can also be used in clinical studies to identify certain pathogens, such as *Mycobacterium tuberculosis* (the bacterium that causes tuberculosis) in a sputum specimen.

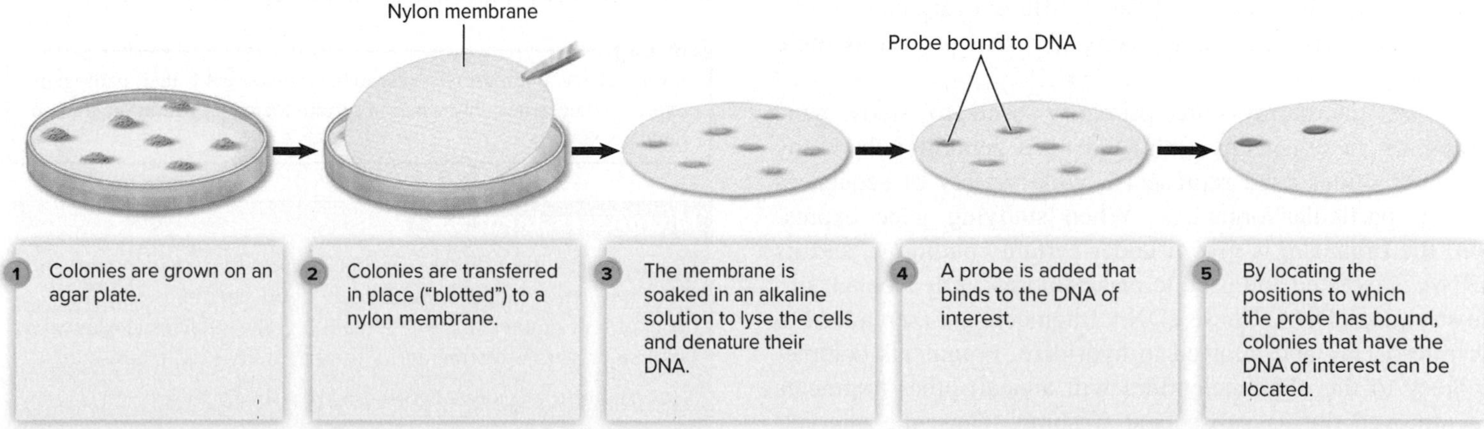

Nylon membrane
Probe bound to DNA
1 Colonies are grown on an agar plate.
2 Colonies are transferred in place ("blotted") to a nylon membrane.
3 The membrane is soaked in an alkaline solution to lyse the cells and denature their DNA.
4 A probe is added that binds to the DNA of interest.
5 By locating the positions to which the probe has bound, colonies that have the DNA of interest can be located.

FIGURE 9.16 Colony Blotting This technique is used to determine which colonies on an agar plate contain a given DNA sequence.

? **What is the purpose of soaking the membrane in an alkaline solution?**

FIGURE 9.17 Fluorescence In Situ Hybridization (FISH) Different probes, each labeled with a uniquely colored fluorescent marker, were used to distinguish various types of bacteria in an earthworm gut.
Seana Davidson

Why does FISH typically require a probe that hybridizes to rRNA?

To analyze a sample using FISH, the sample must first be treated with chemicals to: preserve the shape of the cells, inactivate enzymes that might otherwise degrade the nucleic acid, and make the cells more permeable so that the labeled probe molecules can easily enter. The treated specimen is then put on a glass slide, bathed with a solution containing the labeled probe, and incubated under conditions that allow hybridization to occur. Unbound probe is then washed off. Finally, the specimen is viewed using a fluorescence microscope.

DNA Microarrays

A **DNA microarray** (also called a gene chip) is a glass slide or other small solid support carrying an arrangement of tens or hundreds of thousands of DNA spots, each of which contains many copies of a specific oligonucleotide (a short single-stranded DNA sequence); each oligonucleotide functions like a probe. Each spot has a different oligonucleotide, so the microarray carries as many different probes as there are spots.

DNA microarrays are primarily used to study gene expression in organisms with sequenced genomes; the many oligonucleotides used represent a wide variety of sequences from a particular organism. When studying gene expression, the organism is grown under certain conditions, and its mRNA is isolated and used to make fluorescently labeled, single-stranded cDNA. These cDNA fragments are then added to the microarray and allowed to hybridize. Fragments complementary to the oligonucleotides will anneal; other fragments are removed during subsequent washing. Thus, the oligonucleotides to which the labeled cDNAs hybridize correspond to genes that the organism was expressing. The locations of those molecules on the array are identified and analyzed using

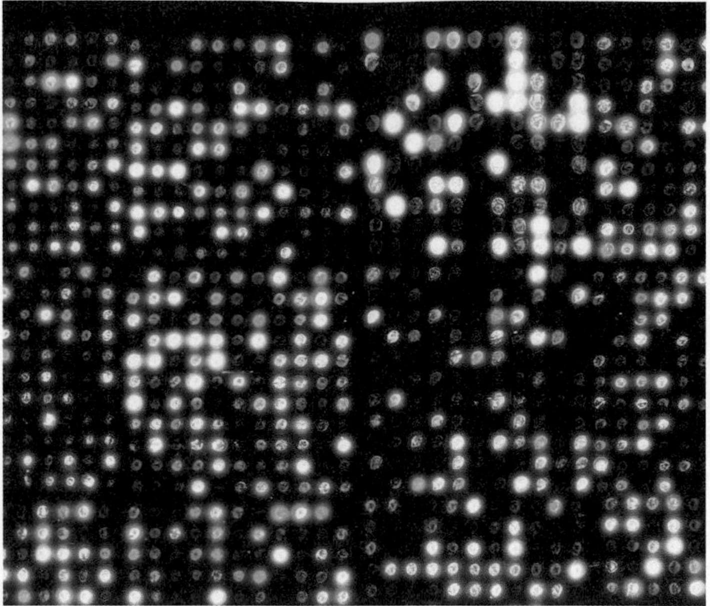

FIGURE 9.18 A DNA Microarray This is an example of an array used to study gene expression. Two different cDNA samples (one labeled with a red fluorescent marker and the other with a green fluorescent marker) were simultaneously hybridized to fragments in the microarray. Red dots indicate the positions to which one sample hybridized, and green dots indicate the positions to which the other hybridized. Yellow dots indicate that both samples hybridized.
Deco/Alamy Stock Photo

How can DNA microarrays be used to determine which genes of a bacterial pathogen are expressed both inside and outside a host cell?

a computerized scanner. Variations in gene expression can be revealed by doing the experiment using cDNA from cultures grown under different conditions and labeling each cDNA preparation with a different fluorescent marker (**figure 9.18**).

Versions of microarrays referred to as biochips are being developed to identify a variety of different bacterial pathogens in clinical specimens. The goal is to allow technicians to screen for dozens of disease-causing microorganisms simultaneously and to predict antibiotic susceptibility.

MicroByte

Using DNA microarrays, researchers discovered that pathogens express some genes only when in certain locations within the human body.

MicroAssessment 9.6

Colony blotting is used to identify colonies that contain a given DNA sequence. Fluorescence in situ hybridization is used to observe individual cells that contain a given nucleotide sequence. DNA microarrays are primarily used to study gene expression.

16. What role does colony blotting play in cloning?

17. Why is a probe that binds to rRNA used in FISH?

18. In FISH, what would happen if unbound probe molecules were not washed off?

9.7 ■ Considerations of Genetic Engineering

Learning Outcome

11. Discuss advantages and disadvantages of GMOs and gene therapy.

Any new technology should be tested to ensure it is safe and effective. When recombinant DNA methods first made gene cloning possible over 40 years ago, many concerns were raised about their use and possible abuse. Even the scientists who developed the technologies were uneasy about potential dangers. In response, guidelines for conducting research involving recombinant DNA were created when the National Institutes of Health (NIH) formed the Recombinant DNA Advisory Committee, now called the Novel and Exceptional Technology and Research Advisory Committee (NExTRAC). Today we are enjoying the benefits of many of those technologies. It should be noted, however, that although the technologies can be used to make life-saving products, there is no guarantee they will not be used for malicious purposes. It is a disturbing possibility that new infectious agents could be created for the purpose of bioterrorism.

Genetically Modified Organisms (GMOs)

Biotechnology has allowed for the creation of genetically modified organisms (GMOs), including transgenic crops that carry genes originally derived from other organisms. The development of GMOs has raised concerns—some logical and others not. Some people have expressed fear over the fact that GM foods "contain DNA." Considering that DNA is consumed routinely in every plant and animal we eat, this is an irrational concern. Others worry that unanticipated allergens could be introduced into food products, posing a health threat. To address this issue, the FDA has implemented strict guidelines, such as requiring producers to show that GM products intended for human consumption do not cause unexpected allergic reactions. Despite precautions, however, GM corn not approved for human consumption has been detected in products such as tortilla chips, and this causes continued unease about the effectiveness of regulatory control. However, to date, there has been no scientific evidence that humans or animals have experienced any negative health effects from consuming GM foods.

Another concern about GM crops is their possible unintended effects on the environment. Some laboratory studies have shown that pollen from plants genetically modified to produce Bt toxin can inadvertently kill monarch butterflies; other studies, however, have refuted the evidence. In addition, there are indications that herbicide-resistance genes can be transferred to weeds, decreasing the usefulness of the herbicide. Despite these issues, the benefits of GM foods may outweigh any negative consequences, real or perceived. GM foods can be designed to have increased nutritional value, such as the "golden rice" mentioned earlier. Introducing genes that encode pharmaceutical products, such as vaccines, into crops is also being investigated as an effective and inexpensive method for the mass manufacture of these products. GM crops containing genes for pesticides and/or herbicide resistance lead to reduced pesticide and herbicide use, respectively, reducing the exposure of farm workers and the overall environment to these chemicals, while also saving money. Overall, growing GM crops may reduce costs for consumers and increase agricultural productivity, which may be especially important as global climate change expands drought-prone regions and pest ranges. However, as with any new technology, the impact of GMOs will need to be carefully scrutinized to avoid negative consequences.

Gene Therapy

Gene therapy is the use of genetic engineering technologies to manipulate DNA in humans in an effort to treat various genetic diseases. Early strategies involved introducing functional copies of genes associated with various genetic diseases, such as certain primary immunodeficiencies, as described in section 18.3. Unfortunately, some patients in early gene therapy trials had unanticipated immune responses that led to organ failure, and a few patients developed cancer. More recently, scientists have developed newer, safer strategies for gene therapy. For example, a new cancer treatment involves collecting a patient's blood and genetically engineering the white blood cells to recognize and attack tumor cells, as described in section 17.3. With the development of new methods like CRISPR-Cas technologies, the possibility of not just introducing functional copies of genes but editing human DNA to fix the non-functional ones is within reach. This provides a great deal of promise for curing a wide range of human genetic diseases, including sickle-cell anemia, cystic fibrosis, muscular dystrophy, and many others.

While gene therapy has great potential, the risks need to be weighed carefully and explained to patient candidates. In the United States, the oversight of gene therapy is coordinated through collaboration between the FDA and NIH's NExTRAC. Additionally, because many gene therapy treatments are still largely experimental, they are often not covered by standard health insurance; this makes them very expensive for most people and only accessible to those who can afford them.

Overall, the use of gene therapy needs to be carefully considered. When it comes to using these technologies to manipulate a human genome, which interventions are ethical and which are not? Who gets access to these therapies? And who gets to decide? As technologies advance, it is important that ongoing robust discussions about these complex issues continue.

MicroAssessment 9.7

Genetic engineering holds great promise for society, but there are also many concerns about its use; they include potential adverse impacts of genetically modified organisms on human health and the environment, as well as ethical issues involving the manipulation of human DNA.

19. Describe two benefits and two concerns regarding GMOs.
20. Describe both the benefits and concerns regarding gene therapy.

Summary

9.1 Fundamental Tools Used in Biotechnology

Restriction Enzymes (figure 9.1)

Restriction enzymes cut DNA into fragments. Cohesive (sticky) ends will **anneal** to one another, making it possible to join DNA from two different organisms.

Reverse Transcriptase (figure 9.2)

Reverse transcriptase uses an RNA template to make DNA; the product, referred to as **cDNA,** is more stable than RNA.

Gel Electrophoresis (figure 9.3)

Agarose gel electrophoresis separates DNA fragments by size.

9.2 Molecular Cloning (table 9.3)

The Cloning Process—A Simplified View (figure 9.4).

Molecular cloning inserts a DNA fragment of interest into a **vector** to create a recombinant molecule that will be replicated in a recipient and passed on to that cell's progeny.

Applications of Molecular Cloning

Bacteria and other cells that grow in culture can be engineered for uses including protein production, DNA production, and research.

Creating a DNA Library—A Detailed View of the Cloning Process

A **DNA library** is made by cloning the entire set of restriction fragments obtained by digesting DNA from an organism of interest (figure 9.5). To obtain eukaryotic DNA without introns, reverse transcriptase is used to make a cDNA from an mRNA template (figure 9.2). Common cloning vectors have a selectable marker, a multiple cloning site, and a second genetic marker (figure 9.6). After the recombinant molecules are introduced into the new host using transformation or electroporation, the cells are grown on a medium that both selects for cells containing vector sequences and differentiates those that carry recombinant molecules (figure 9.7).

9.3 CRISPR-Cas Technologies

Applications of CRISPR-Cas Technologies

CRISPR-Cas technologies, which rely on genetically modified Cas nucleases of CRISPR systems, can be used to modify genomes, study gene function, and detect pathogens for rapid disease diagnosis (figures 9.8, 9.9).

9.4 DNA Sequencing

The technologies for **DNA sequencing** have advanced rapidly over the last several decades, making large-scale sequencing projects possible.

Applications of DNA Sequencing

A DNA sequence can be used to decipher the amino acid sequences of the encoded proteins. DNA sequences also give insights into the evolutionary relatedness of organisms. Whole genome sequencing (WGS) can be used to track disease spread and trace outbreaks. Personalized medicine also allows for the treatment of disease based upon an individual's own DNA sequence information.

High-Throughput Sequencing Methods

High-throughput sequencing methods allow for faster, less-expensive sequencing (figure 9.10).

9.5 Polymerase Chain Reaction (PCR)

PCR is used to exponentially amplify target DNA (figure 9.11). More than a billion copies of the target can be generated in a matter of hours. Researchers can then detect that **PCR product.**

Applications of PCR

One important application of PCR is disease diagnosis. **RT-PCR (reverse transcription PCR)** is used to detect a given RNA sequence in a sample. **qPCR (quantitative PCR)** (also called real-time PCR) allows researchers to track the PCR product accumulation during the course of a PCR reaction. **Real-time RT-PCR** combines both methods.

The PCR Method

PCR amplifies target DNA by repeatedly (1) denaturing double-stranded DNA, (2) allowing specific primers to anneal to their complementary sequences, and (3) synthesizing DNA (figures 9.12, 9.13, 9.14). The primers dictate which part of the DNA is amplified.

9.6 Probe Technologies

DNA probes are used to locate specific nucleotide sequences (figure 9.15).

Colony Blotting

Colony blotting uses a probe to identify colonies that contain a DNA sequence of interest (figure 9.16).

Fluorescence In Situ Hybridization (FISH)

Fluorescence in situ hybridization (FISH) uses a fluorescently labeled probe to detect specific nucleotide sequences within intact cells attached to a microscope slide (figure 9.17).

DNA Microarrays

DNA microarrays contain tens or hundreds of thousands of oligonucleotides that each function in a manner analogous to a probe (figure 9.18).

9.7 Considerations of Genetic Engineering

Genetic engineering can be used to make life-saving products. Significant oversight and guidelines for conducting research involving recombinant DNA minimize the possibility that it will be used for malicious purposes.

Genetically Modified Organisms (GMOs)

Genetically modified organisms hold much promise, but concerns exist about the inadvertent introduction of allergens into a food product and adverse effects on the environment. To date, there is no scientific evidence showing that humans or animals have experienced any negative health effects from consuming GM foods.

Gene Therapy

Gene therapy also has great potential for the treatment and cure of a wide variety of human genetic diseases, but its use must be considered carefully.

Review Questions

Short Answer

1. Describe three uses of biotechnology.
2. Why are restriction enzymes useful in biotechnology?
3. What is cDNA? Why is it so useful when studying gene expression?
4. What is a DNA library?
5. In a CRISPR-Cas system, how is a Cas used for gene editing different from one used to study gene function?
6. Why is whole genome sequencing (WGS) so useful for tracking infectious disease outbreaks?

7. How many different temperatures are used in each cycle of the polymerase chain reaction?
8. How does PCR eventually generate a discrete-sized fragment from a much longer piece of DNA?
9. What is the function of a DNA probe?
10. How does a DNA microarray function as a set of probes?

Multiple Choice

1. Basic research studying *Agrobacterium tumefaciens* led to methods for genetically engineering which of the following cell types?
   a) Animals
   b) Bacteria
   c) Plants
   d) Yeast
   e) All of these
2. In molecular cloning, what is the function of a vector?
   a) Destroys cells that do not contain cloned DNA
   b) Allows cells to take up foreign DNA
   c) Carries cloned DNA, allowing it to replicate in cells
   d) Encodes herbicide resistance
   e) Encodes Bt toxin
3. Which of the following can be used to generate a DNA library?
   a) PCR
   b) Sequencing
   c) Colony blotting
   d) Microarrays
   e) Molecular cloning
4. An ideal vector has all of the following *except*
   a) an origin of replication.
   b) a gene encoding a restriction enzyme.
   c) a gene encoding resistance to an antibiotic.
   d) a multiple-cloning site.
   e) the *lacZ'* gene.
5. Which of the following describes the function of the *lacZ'* gene in a cloning vector?
   a) Means of selecting for cells that contain vector sequences
   b) Means of distinguishing cells that have taken up recombinant molecules
   c) Site required for the vector to replicate
   d) Mechanism by which cells take up the DNA
   e) Gene for a critical nutrient required by transformed cells
6. Which is used for cloning eukaryotic genes but not prokaryotic genes?
   a) Restriction enzymes
   b) DNA ligase
   c) Reverse transcriptase
   d) Vector
   e) Selectable marker
7. The function of a labeled dCas molecule along with gRNA could best be compared to which of the following molecules?
   a) Restriction enzyme
   b) DNA probe
   c) Reverse transcriptase
   d) Ligase
   e) Vector
8. All of the following statements about high-throughput DNA sequencing are true *except*
   a) A bacterial genome can be sequenced in a matter of hours.
   b) The systems rarely make errors.
   c) Sequence assembly is easier if a reference genome has been completed.
   d) The methods are highly efficient and relatively inexpensive.
   e) c and d
9. The polymerase chain reaction uses *Taq* polymerase rather than a DNA polymerase from *E. coli* because *Taq* polymerase
   a) introduces fewer errors during DNA synthesis.
   b) is heat-stable.
   c) can initiate DNA synthesis at a wider variety of sequences.
   d) can denature a double-stranded DNA template.
   e) is easier to obtain.
10. The polymerase chain reaction generates a fragment of a distinct size even when an intact chromosome is used as a template. What determines the boundaries of the amplified fragment?
    a) The concentration of one particular deoxynucleotide in the reaction
    b) The duration of the elongation step in each cycle
    c) The position of a termination sequence, which causes the *Taq* polymerase to fall off the template
    d) The sites to which the primers anneal
    e) The temperature of the elongation step in each cycle

Applications

1. Two students in a microbiology class are arguing about the origins of biotechnology. One student argued that biotechnology started with the advent of genetic engineering. The other student disagreed, saying that biotechnology was as old as ancient civilization. What was the rationale for the argument by the second student?
2. A student wants to clone gene X. On both sides of the gene are the recognition sequences for AluI and BamHI (see table 9.2). Which enzyme would be easier to use for the cloning experiment and why?

Critical Thinking

1. Discuss some potential issues regarding gene editing to correct genetic defects in humans.
2. An effective DNA probe can sometimes be developed based on the amino acid sequence of the protein encoded by the gene. A student argued that this is too time-consuming since the complete amino acid sequence must be determined in order to create the probe. Does the student have a valid argument? Why or why not?

10 | Identifying and Classifying Microorganisms

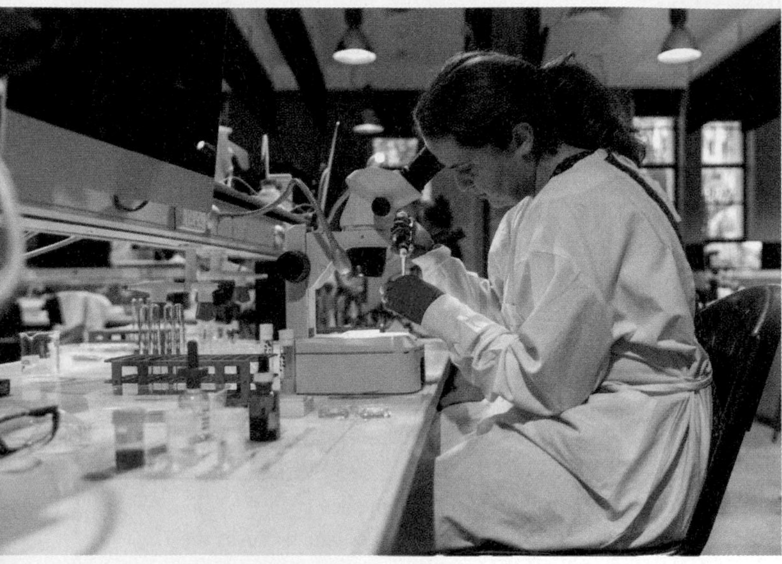

A scientist using biochemical tests to identify bacterial isolates. *Diane Keough/Moment/Getty Images*

KEY TERMS

Classification The process of arranging organisms into similar or related groups (taxa), primarily to make it easier to identify them for study.

Domain A collection of similar kingdoms; there are three domains: Bacteria, Archaea, and Eukarya.

Genus A collection of related species.

Identification The process of characterizing an isolate in order to determine the group (taxon) to which it belongs.

Nomenclature The system of assigning names to organisms.

Phylogeny Evolutionary relatedness of organisms.

Species A group of closely related strains; the basic unit of taxonomy.

Strain A pure culture isolate; subgroup within a species.

Taxonomy The science of characterizing organisms in order to arrange them into hierarchical groups (taxa); involves three interrelated areas: identification, classification, and nomenclature.

A Glimpse of History

In the early 1870s, the German botanist Ferdinand Cohn wrote several papers on bacterial classification, in which he grouped microorganisms according to shape—spherical, short rods, elongated rods, and spirals. That scheme was not adequate, however, because there were too many different kinds of bacteria with similar shapes.

The second major attempt at bacterial classification was initiated by the Danish scientist Sigurd Orla-Jensen. In 1908, he proposed that bacteria be classified according to their physiological properties. The idea was promising, but organisms classified as similar based on certain physiological traits would appear unrelated when comparing other traits.

A quarter of a century later, two Dutch microbiologists, Albert Kluyver and C. B. van Niel, proposed classification systems based on evolutionary relationships. They faced a serious problem, however, because they could not distinguish "resemblance" from "relatedness." The fact that two prokaryotes look or act alike does not mean they are genetically related.

In 1970, Roger Stanier, a microbiologist at the University of California, Berkeley, pointed out that relationships could be determined by comparing either physical traits, such as proteins and cell walls, or nucleotide sequences. At that time, most microbiologists, including Stanier, assumed that all prokaryotes are basically similar. However, chemical analysis of various prokaryotic structures—including the cell wall, cytoplasmic membrane, and ribosomal RNA—showed that this was not the case.

In the late 1970s, Carl Woese and his colleagues at the University of Illinois determined the nucleotide sequence of ribosomal RNA from a wide variety of organisms. Based on the data, they recognized that prokaryotes could be divided into two major groups that differ from each other as much as they differ from eukaryotic cells. This led to a revolutionary system of classification that separates prokaryotes into two domains: Archaea and Bacteria. All eukaryotic organisms, including humans, are in a third domain: Eukarya.

Information that is logically organized is easier to use. Newspapers, for instance, do not scatter various subjects throughout the paper; instead, the information is grouped into general topics, such as local news, sports, and entertainment. A large library would be extremely difficult to use if the locations of the many books were not organized by subject matter. Likewise, scientists have sorted living organisms into different groups, to better show the relationships among the species.

Take a moment to think about how you would group bacteria if you were to create a classification system. Would you group them according to shape? Or would it make more sense to consider their motility? Perhaps you would group them according to their medical significance. But then, how would you classify two apparently identical organisms that differed in their pathogenicity? This chapter describes some of the methods currently used by microbiologists to classify microorganisms.

10.1 ■ Principles of Taxonomy

Learning Outcome

1. Describe how microorganisms are identified, classified, and assigned names.

Taxonomy is the science of characterizing and naming organisms in order to arrange them into hierarchical groups (taxa). Organisms with similar properties are grouped together and separated from ones that are different. Taxonomy can be viewed as three separate but interrelated areas:

- **Identification.** The process of characterizing an isolate (a population of cells descended from a single cell) to determine the group (taxon) to which it belongs.
- **Classification.** The process of arranging organisms into similar or related groups, primarily to make it easier to identify them for study.
- **Nomenclature.** The system of assigning names to organisms.

Strategies Used to Identify Microorganisms

In practical terms, identifying the genus and species of a microorganism may be more important than understanding its genetic relationship to other microbes. A food manufacturer is most interested in detecting microbial contaminants that can spoil a food product. In clinical situations, identifying pathogens quickly and accurately is crucial so that patients can be treated appropriately.

To identify microorganisms, many different procedures may be used, including microscopic examination, culture characteristics, biochemical tests, and nucleic acid analysis. Immunological testing is also used, but those methods are covered in section 17.4, after the chapters that describe the immune responses. In a clinical laboratory, the patient's symptoms play an important role in the identification process. Pneumonia in an otherwise healthy adult is typically caused by *Streptococcus pneumoniae*, a bacterium easily differentiated from others using a few specific tests. In contrast, diagnosing the cause of a wound infection is often more difficult because several different microorganisms could be involved. If the specimen is from a body site that contains normal microbiota, then the strategy is to detect microbes known to cause the symptoms in question, rather than to conclusively identify each and every organism in the sample. For instance, a fecal specimen from a patient with nausea and diarrhea would generally be tested only for the presence of organisms that cause those symptoms. In most cases, tests used to identify pathogens must be done in laboratories, but some can be done at or near the site of patient care, a feature referred to as **point-of-care testing (POCT).**

Strategies Used to Classify Microorganisms

Understanding the **phylogeny** (evolutionary history) of microorganisms is important in creating a classification scheme that reflects their evolutionary relatedness. Such a scheme is more useful than one that simply groups organisms by random characteristics because it is less prone to the bias of human perceptions. It also makes it easier to classify newly recognized species and allows scientists to make predictions, such as the likelihood that genes will be acquired from a given organism.

Unfortunately, determining the phylogeny of microorganisms is more difficult than doing so for plants and animals. Not only do microorganisms have few differences in size and shape, but they do not undergo sexual reproduction. In higher organisms such as plants and animals, the basic taxonomic unit, a **species,** is generally considered to be a group of morphologically similar organisms capable of interbreeding to produce fertile offspring. Obviously, it is not possible to apply these same criteria to bacteria, archaea, and other microorganisms that reproduce by binary fission rather than sexual reproduction.

Historically, taxonomists have relied on phenotypic characteristics (such as cell wall type and ability to degrade certain compounds) to classify prokaryotes. The development and application of molecular techniques such as nucleotide sequencing, however, now make it possible for scientists to more accurately classify microorganisms based on their evolutionary history.

Taxonomic Hierarchies

Taxonomic classification categories are arranged in a hierarchical order, with the species being the basic unit. The species designation gives a formal taxonomic status to a group of related isolates, or **strains,** which, in turn, allows for their identification. Without classification, scientists and others would not be able to communicate efficiently about organisms. The names of groups within a given hierarchical level usually end with the same suffix, making it easier to recognize the level of a particular name. Taxonomic classification categories include:

- **Species.** A group of closely related strains or individuals. Note that members of a species are not all identical. With respect to microbiology, the difficulty for the taxonomist is to decide how different two isolates must be in order to be classified as separate species rather than strains of the same species. In certain cases, a species may be divided into two or more subspecies; this is particularly true for the commercially valuable microorganisms used to make fermented foods.
- **Genus.** A collection of similar species.
- **Family.** A collection of similar genera. Family names end in -aceae.
- **Order.** A collection of similar families. Order names end in -ales.
- **Class.** A collection of similar orders. Class names usually end in -ia.
- **Phylum** (sometimes called a division). A collection of similar classes. Phylum names usually end in -ota.

TABLE 10.1 Taxonomic Ranks of the Bacterium *Escherichia coli*

Formal Rank	Example
Domain	Bacteria
Phylum	Pseudomonadota (Proteobacteria)
Class	Gammaproteobacteria
Order	Enterobacteriales
Family	Enterobacteriaceae
Genus	*Escherichia*
Species	*coli*

- **Kingdom.** A collection of similar phyla or divisions.
- **Domain.** A collection of similar kingdoms. The domain reflects the characteristics of the cells that make up the organism (see A Glimpse of History).

Note, however, that microbiologists often group microorganisms into informal categories based on one or more distinctive characteristics rather than using the higher taxonomic ranks such as order, class, and phylum. Examples of informal groupings of bacteria include the lactic acid bacteria, the anoxygenic phototrophs, the endospore formers, and the sulfate reducers. Organisms within these groupings share similar phenotypic and physiological characteristics, but may not be genetically related.

Table 10.1 shows how a particular bacterial species is classified (the phylum name was recently changed so the former name is indicated in parentheses here to ease the transition). The rank of kingdom is not shown because that level of classification is not currently used for the bacteria and archaea.

Classification Systems

Classification systems change over the years as new information is discovered. There is no "official" classification system, and as new ones are introduced, others become outdated. The most widely used classification scheme today is the three-domain system, which designates all organisms as belonging to one of the three domains: Bacteria, Archaea, and Eukarya. The system is based on comparisons of nucleotide sequences in ribosomal RNA (rRNA) from a wide variety of organisms. The rRNA data are consistent with other differences between bacteria and archaea, including the chemical compositions of their cytoplasmic membranes and cell walls (**table 10.2**). Before the three domains were recognized, the most widely accepted system separated organisms into five kingdoms: Plantae, Animalia, Fungi, Protista (mostly single-celled eukaryotes), and Prokaryotae. Although this system recognizes the obvious morphological differences between plants and animals, it does not reflect the genetic insights of the ribosomal data, which indicate that plants and animals are more closely related to each other than archaea are to bacteria.

Figure 10.1a shows a tree of life (a diagram that illustrates the relatedness of organisms) based on rRNA sequence data that support the three-domain system. **Figure 10.1b** shows a newer and expanded version of the tree that was developed by comparing the amino acid sequences of ribosomal proteins. Because this latest version includes many bacteria that have not yet been grown in pure culture, it highlights the extraordinary diversity of the microbial world, particularly within the Bacteria. As metagenomics allows additional novel organisms to be characterized, the tree is expected to change even more.

Today, microbiologists generally rely on the phylogenetic classifications listed in the five-volume reference text *Bergey's Manual of Systematic Bacteriology;* an updated online version entitled *Bergey's Manual of Systematics of Archaea and Bacteria* is also available. In addition to containing descriptions of all known bacterial and archaeal species and their phylogenetic groupings, *Bergey's* includes information on the ecology, methods of enrichment, culture, and isolation of the organisms as well as methods for their maintenance and preservation.

Nomenclature

Bacteria and archaea are given names according to a set of internationally recognized rules: the *International Code of*

TABLE 10.2 A Comparison of Properties Typical of Members of the Three Domains

Cell Feature	Archaea	Bacteria	Eukarya
Cytoplasmic Membrane Lipids	Hydrocarbons (not fatty acids) linked to glycerol by ether linkage	Fatty acids linked to glycerol by ester linkage	Fatty acids linked to glycerol by ester linkage
Membrane-Bound Nucleus	No	No	Yes
Peptidoglycan Cell Wall	No	Yes	No
Presence of Introns	Sometimes	No	Yes
Ribosomes	70S	70S	80S

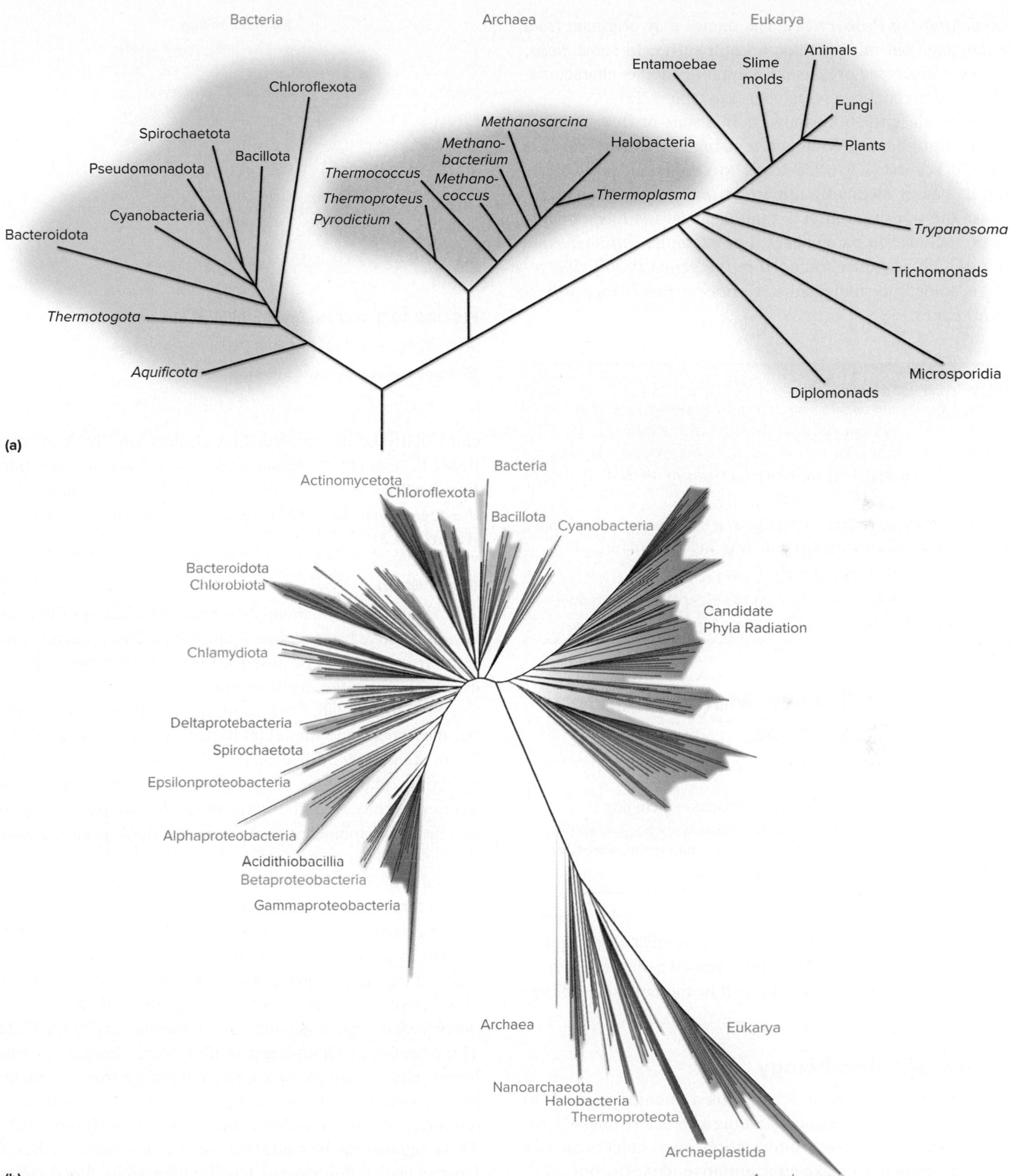

FIGURE 10.1 Tree of Life **(a)** Based on ribosomal RNA data. **(b)** Based on ribosomal protein sequence data, including information from many organisms that have not been grown in pure culture. The colors of the branches are arbitrary. For simplicity, only some of the names of the various organisms have been included here. The group of bacteria referred to as Candidate Phyla Radiation (CPR) have small genomes and appear to lack crucial genes required to multiply independently of a host cell. (b) Modified from: Hug et al., "A New View of the Tree of Life." *Nature Microbiology* 1, March 2016.

? Why would members of the Candidate Phyla Radiation (CPR) be difficult to grow in pure culture?

Nomenclature of Prokaryotes. The names may originate from any language but must include a Latin suffix. In some cases, the name reflects the organism's habitat or another characteristic, but it often honors a researcher (see table 1.2).

Just as classification is always in a state of flux, so is the assignment of names. Although revising names may increase scientific accuracy, it often leads to confusion, particularly when names of medically important organisms are changed. To ease the transition after a name change, the former name may be included in parentheses. For example, *Cutibacterium acnes,* which was once included in the genus *Propionibacterium,* is sometimes indicated as *Cutibacterium (Propionibacterium) acnes.*

MicroAssessment 10.1

Taxonomy consists of three interrelated areas: identification, classification, and nomenclature. In clinical laboratories, identifying the genus and species of an isolate is more important than understanding its evolutionary relationship to other organisms.

1. Why might it be easier to determine the cause of pneumonia than to determine the cause of a wound infection?
2. What is *Bergey's Manual*?
3. Some biologists were reluctant to accept the three-domain system. Why might this be the case?

10.2 ■ Identification Methods Based on Phenotype

Learning Outcome

2. Describe how phenotypic characteristics—including microscopic morphology, culture characteristics, metabolic capabilities, serological characteristics, and protein profile—can be used to identify microorganisms.

Phenotypic characteristics can be used to help identify microorganisms, without the need for sophisticated equipment. These identification methods, as well as those based on genotype, are summarized at the end of the next section (see table 10.4).

Microscopic Morphology

An important initial step in identifying a microorganism is to determine its size, shape, and staining characteristics. Microscopic examination gives information very quickly and is sometimes enough to make a presumptive identification.

Size and Shape

The size and shape of a microorganism can easily be determined microscopically. Based only on that information, the microbial group (such as prokaryote, fungus, or protozoan) can usually be determined. In a clinical lab, this can sometimes be enough to diagnose certain eukaryotic infections. For example, a wet mount of stool (feces) is examined for the eggs of parasites when certain roundworms are suspected (**figure 10.2**).

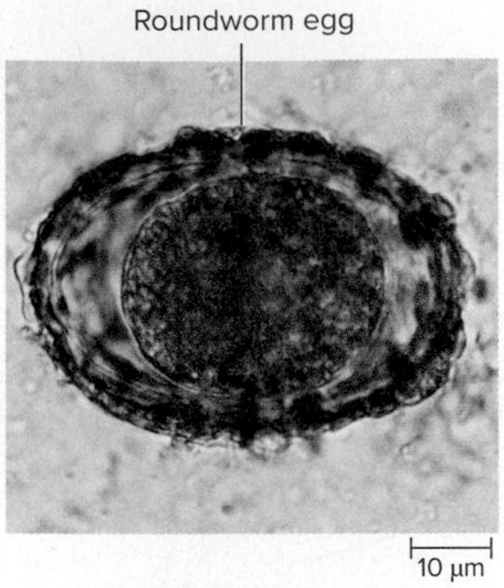

Roundworm egg

10 µm

FIGURE 10.2 Wet Mount of a Clinical Specimen Roundworm *(Ascaris)* egg in a stool specimen. BSIP/UIG/Getty Images

? Roundworms belong to which domain?

Gram Stain

The Gram stain distinguishes between Gram-positive and Gram-negative bacteria (see figure 3.47). This relatively rapid test narrows the list of possible identities of a bacterium, an essential step in the identification process.

In a clinical lab, Gram stain results are extremely useful, but most medically important bacteria cannot be identified by Gram stain alone. For example, *Streptococcus pyogenes,* the bacterium that causes strep throat, cannot be distinguished microscopically from the streptococci that are part of the normal throat microbiota. A Gram stain of a stool specimen cannot distinguish *Salmonella enterica* from *E. coli.* These organisms generally must be isolated in pure culture and tested by other means for accurate identification.

In certain cases, the Gram stain result gives enough information to start appropriate antimicrobial treatment. A Gram stain of sputum showing numerous white blood cells and Gram-positive diplococci is highly suggestive of *Streptococcus pneumoniae,* a bacterium that causes pneumonia (**figure 10.3a**). The presence of Gram-negative diplococci clustered in white blood cells in a sample of a urethral discharge from a male may be considered diagnostic for gonorrhea, the sexually transmitted infection caused by *Neisseria gonorrhoeae* (**figure 10.3b**). This diagnosis can be made because *N. gonorrhoeae* is the only Gram-negative diplococcus found within white blood cells in the male urethra.

Special Stains

Certain microorganisms have unique characteristics that can be detected with special staining procedures. If a

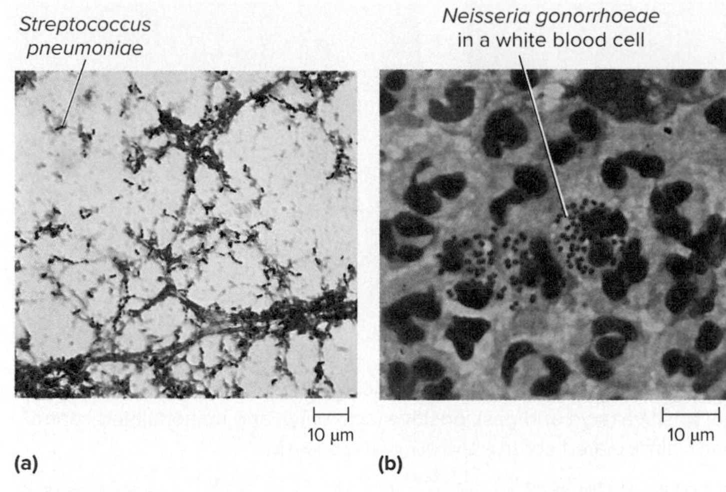

FIGURE 10.3 Gram Stains of Clinical Specimens (a) Sputum showing Gram-positive *Streptococcus pneumoniae* and **(b)** male urethral discharge showing Gram-negative *Neisseria gonorrhoeae* inside white blood cells. a: Dr. Mike Miller/CDC; b: Bill Schwartz/CDC

? **What disease does the patient in (a) have? What disease does the patient in (b) have?**

patient has symptoms of tuberculosis, then an acid-fast stain will be done on a sputum sample to help identify *Mycobacterium tuberculosis*; members of the genus *Mycobacterium* are some of the few acid-fast microorganisms (see figure 3.49).

Culture Characteristics

Culture characteristics can give initial clues to the identity of the organism. Colonies of the bacterium *Serratia marcescens* are often red when incubated at 22°C due to the production of a pigment. Cultures of the bacterium *Pseudomonas aeruginosa* have a distinct fruity odor; in addition, they often have a greenish tinge due to a soluble greenish pigment produced by the organism (see figure 11.11).

In a clinical lab, where rapid but accurate diagnosis is essential, specimens are inoculated onto differential media as a preliminary step in the identification process. A specimen taken by swabbing the throat of a patient complaining of a sore throat is inoculated onto blood agar. This makes it possible to detect the characteristic β-hemolytic colonies of *Streptococcus pyogenes,* the bacterium that causes strep throat (see figure 4.10). Urine collected from a patient suspected of having a urinary tract infection (UTI) is plated onto MacConkey agar; *E. coli,* the most common cause of UTIs, can grow on that medium, where it ferments lactose to form pink colonies (see figure 4.11).

Metabolic Capabilities

Various biochemical tests can be used to determine the metabolic capabilities of a microorganism—such as the types of sugars it ferments or the end products it makes. Their use is somewhat analogous to identifying a person based on characteristics such as hair color, food preferences, and hobbies: The traits are simply easy-to-detect features.

FIGURE 10.4 Catalase Test Catalase positive (left) and negative (right). Denise Anderson

? **What causes the bubbles to form in the test?**

One of the easiest and fastest biochemical tests is an assay for the enzyme catalase (**figure 10.4**); recall that many organisms produce catalase to protect against hydrogen peroxide (see table 4.2). To detect catalase, a small portion of a colony is transferred to a microscope slide or the inside of a Petri dish, and then a drop of hydrogen peroxide (H_2O_2) is added. If catalase is present, it immediately breaks down the hydrogen peroxide to form O_2 and water; the O_2 can be observed as bubbles in the reagent. Most bacteria that grow in the presence of O_2 are catalase-positive. Important exceptions are *Streptococcus* species and others in the group called lactic acid bacteria. Thus, if β-hemolytic colonies grow from a throat culture, but testing reveals they are all catalase-positive, then *Streptococcus pyogenes* has been ruled out.

Most biochemical tests rely on a chemical indicator that changes color when a compound is degraded, usually during an incubation period of at least 18 hours. For example, the test for the ability of an organism to ferment a given sugar involves adding the organism to a broth growth medium containing the sugar and a pH indicator and incubating overnight; if the organism ferments the sugar, acid is produced as the organism grows, which lowers the pH, resulting in a color change; an inverted tube traps any gas produced (**figure 10.5a**). A medium designed to detect urease (an enzyme that degrades urea to produce carbon dioxide and ammonia) contains urea and a pH indicator (**figure 10.5b**). This and other biochemical methods used to identify bacteria are summarized in **table 10.3.**

The basic strategy for identifying bacteria based on biochemical tests relies on a **dichotomous key,** a series of alternative choices that lead to an identification (**figure 10.6**). Because each biochemical test often requires an incubation period, however, it would be too time-consuming to proceed one step at a time. In addition, relying on a single result at each step could lead to misidentification of a strain that lost the ability to produce a single key enzyme. Therefore, several different biochemical tests are inoculated at the same time in order to identify the organism faster and more conclusively.

In certain cases, biochemical testing can be done without culturing the organism. *Helicobacter pylori,* the cause of most stomach ulcers, can be detected using the urea breath test (UBT), which assays for the presence of urease. The patient drinks a solution containing urea labeled with an isotope of carbon (see Focus on a Case 24.1). If *H. pylori* is present, its urease breaks down the urea, releasing labeled CO_2 that can be detected in the breath. This test is less invasive and, consequently, much cheaper and faster than the stomach biopsy that would be needed to culture the organism.

(a) (b)

FIGURE 10.5 Biochemical Tests that use a pH Indicator Results, left to right: **(a)** Sugar fermentation negative, positive (acid and gas), positive (acid only), and uninoculated control. **(b)** Urease positive and uninoculated control. Lisa Burgess/McGraw Hill

? **What causes the color to change in the tests?**

Several less labor-intensive commercial variations of traditional biochemical tests are available. As an example,

TABLE 10.3 Characteristics of Some Important Biochemical Tests

Biochemical Test	Principle	Positive Reaction
Catalase	Rapidly detects the activity of catalase, an enzyme that breaks down hydrogen peroxide to form O_2 and water.	Bubbles form.
Citrate	Determines whether or not citrate can be used as a sole carbon source.	Growth is evident, usually accompanied by the color change of a pH indicator.
Gelatinase	Detects enzymatic breakdown of gelatin.	The solid gelatin is converted to liquid.
Hydrogen Sulfide Production	Detects H_2S released as sulfur-containing amino acids are degraded.	A black precipitate forms due to the reaction of H_2S with iron salts in the medium.
Indole	Detects the enzymatic removal of the amino group from tryptophan.	The product, indole, reacts with an added chemical reagent, turning the reagent a deep red color.
Lysine Decarboxylase	Detects the enzymatic removal of the carboxyl group from lysine.	The medium becomes more alkaline, causing a pH indicator to change color.
Methyl Red	Detects mixed acids, the characteristic end products of a particular fermentation pathway.	The medium becomes acidic (pH below 4.5); a red color develops upon the addition of a pH indicator.
Oxidase	Rapidly detects the activity of cytochrome c oxidase, a component of the electron transport chain of specific organisms.	A dark color develops after a specific reagent is added.
Phenylalanine Deaminase	Detects the enzymatic removal of the amino group from phenylalanine.	The product of the reaction, phenylpyruvic acid, reacts with ferric chloride to give the medium a green color.
Sugar Fermentation	Detects the acidity resulting from fermentation of a specific sugar incorporated into the medium; also detects gas production.	The medium becomes acidic, causing a pH indicator to change color. An inverted tube traps any gas produced.
Urease	Detects the enzymatic degradation of urea to carbon dioxide and ammonia.	The medium becomes alkaline, causing a pH indicator to change color.
Voges-Proskauer	Detects acetoin, an intermediate of the fermentation pathway that leads to 2,3-butanediol production.	A red color develops after chemicals that detect acetoin are added.

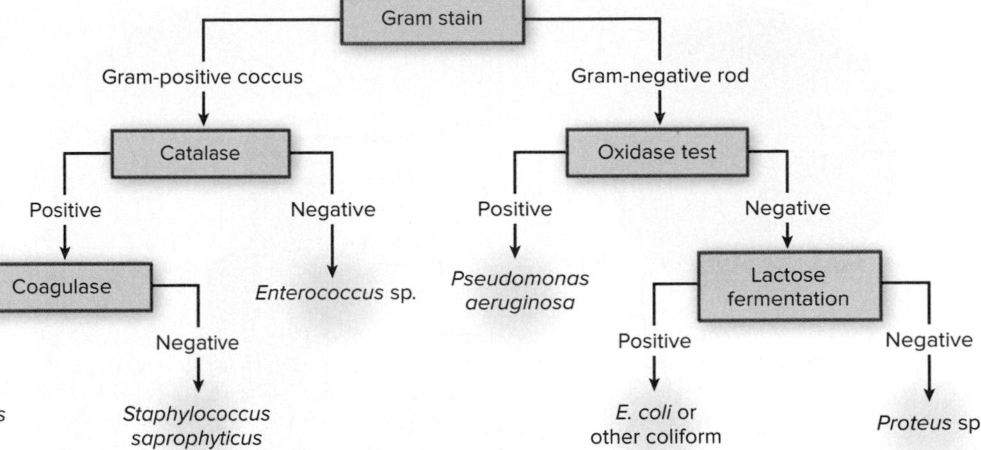

FIGURE 10.6 Dichotomous Key This shows an example of steps that can be used to distinguish some of the common causes of urinary tract infections. Additional tests may be done to confirm the identity of the pathogen.

? **When identifying organisms, why would a lab technician inoculate most biochemical tests at the same time, rather than waiting for one result before starting the next test?**

API test kits have a strip holding a series of tiny cups that contain dried media. A liquid suspension of the test bacterium is added to each compartment, inoculating as well as rehydrating the media. The media are similar to those used in traditional tests, giving rise to comparable color changes (**figure 10.7**). After incubation of the inoculated strip, the results are determined by inspection. The pattern of results is converted to a numerical score, which is then entered into a computer to identify the organism. A system by Biolog uses a microtiter plate, a small tray containing nearly 100 wells, to assay simultaneously an organism's ability to use a wide variety of carbon sources. Modifications of these plates allow researchers to characterize the metabolic capabilities of microbial communities, such as those in soil, water, or wastewater.

Highly automated systems also are available. The VITEK 2 system uses a miniature card that contains multiple wells with different types of dried media. After a relatively short incubation period (at least several hours, but sometimes longer), a computer reads the growth pattern in the wells.

Serological Characteristics

Serological testing uses antibodies to detect specific proteins and polysaccharides, a topic covered in chapter 17. A microbial cell's proteins and polysaccharides (particularly those that make up surface structures including the cell wall, capsule, flagella, and pili) are sometimes characteristic enough to be used as identifying markers. For example, certain *Streptococcus* species contain a unique carbohydrate as part of their cell wall, and antibodies can be used to detect this molecule. Some of the serological methods, such as those used to confirm the identity of *S. pyogenes,* are quite simple and rapid.

Protein Profile

A relatively new technology that determines a microorganism's protein profile is revolutionizing microbial identification because of its speed: microorganisms in a colony can often be identified in less than 15 minutes. The technology, called **MALDI-TOF MS** (**m**atrix-**a**ssisted **l**aser **d**esorption

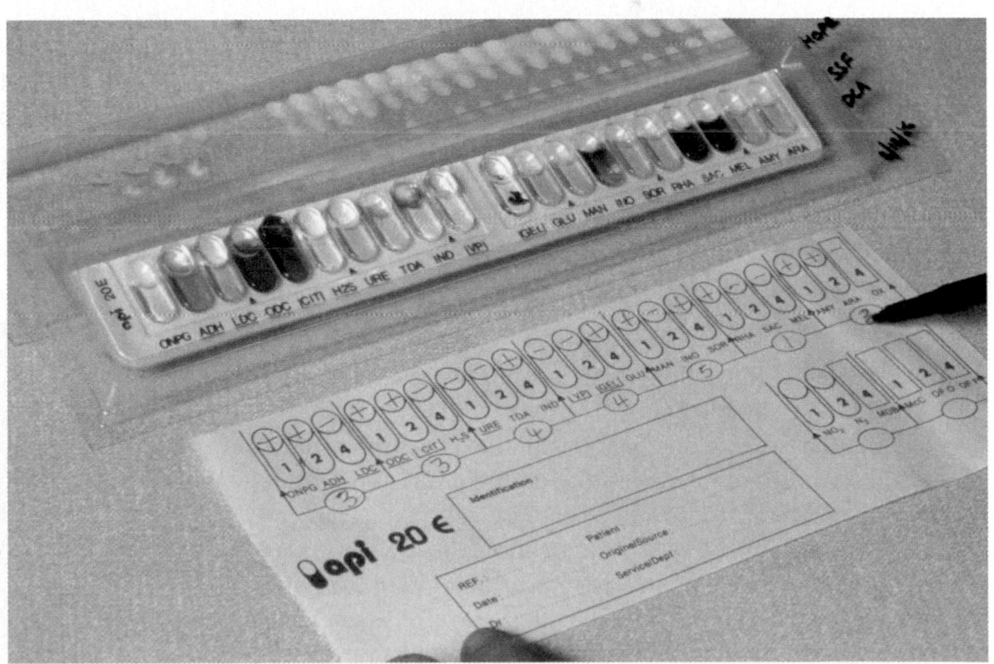

FIGURE 10.7 API Test Strip Each small cup contains a dried medium similar in formulation to the traditional tests. The pattern of results is converted to a numerical score, which is then entered into a computer to identify the organism. John Watney/Science Source

? **What advantage do commercial variations of biochemical tests have over traditional methods?**

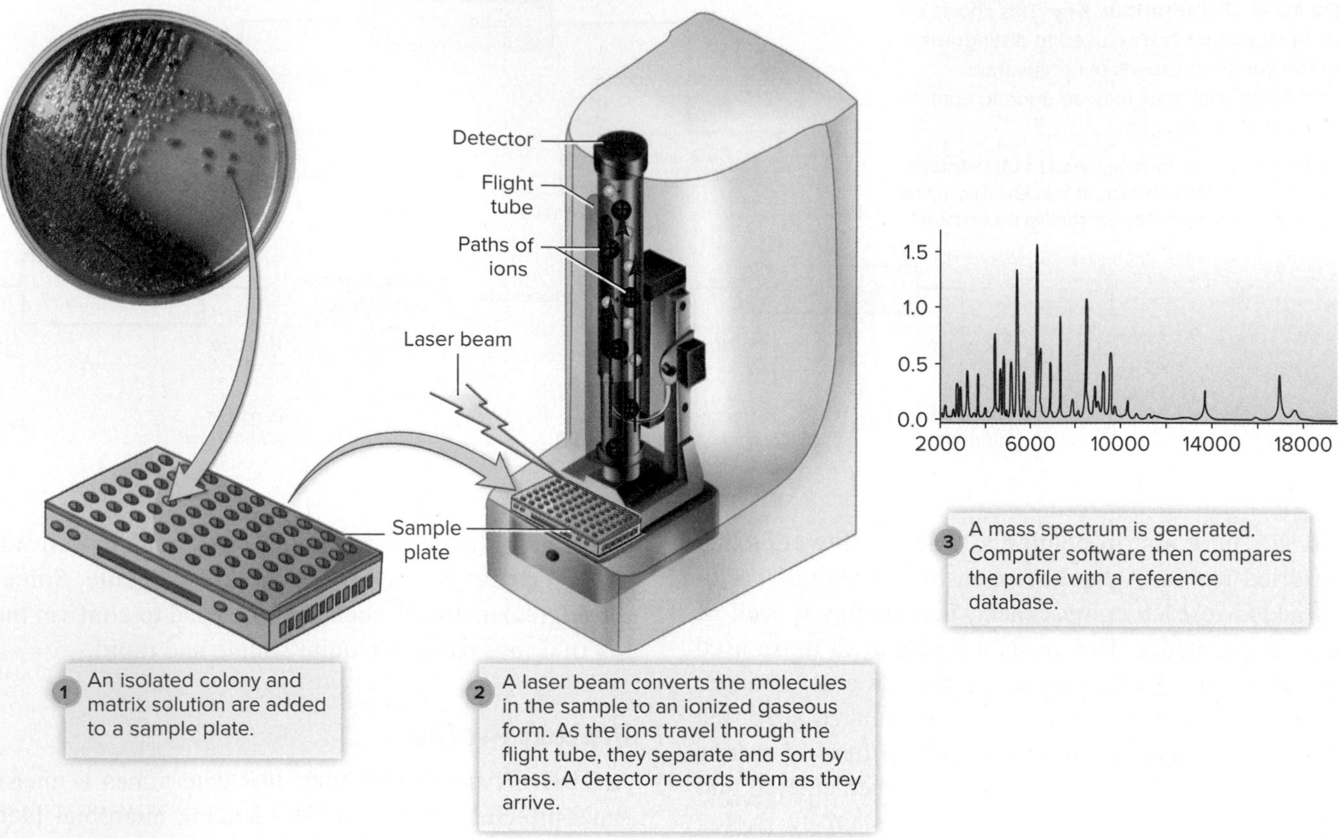

Detector
Flight tube
Paths of ions
Laser beam
Sample plate
1.5
1.0
0.5
0.0
2000
6000
10000
14000
18000
1 An isolated colony and matrix solution are added to a sample plate.
2 A laser beam converts the molecules in the sample to an ionized gaseous form. As the ions travel through the flight tube, they separate and sort by mass. A detector records them as they arrive.
3 A mass spectrum is generated. Computer software then compares the profile with a reference database.

FIGURE 10.8 MALDI-TOF The procedure separates and sorts a microorganism's proteins by mass to generate a profile that provides a fast way to identify a colony. Lisa Burgess/McGraw Hill

? In a clinical lab, what advantage does MALDI-TOF have over traditional biochemical testing?

ionization **t**ime **o**f **f**light **m**ass **s**pectrometry), is particularly important in clinical labs where rapid identification of bacterial isolates is crucial for patient care.

MALDI-TOF, a type of mass spectrometry, is used to determine the chemical composition of a sample by measuring the masses of the various components. The process is not only faster than other rapid identification methods, but it also requires less technical skill, as illustrated in the following steps (**figure 10.8**):

① An isolated colony, along with a solution called a matrix, is added to a sample plate.

② The plate is set into an apparatus that exposes the sample to a laser beam, a treatment that both converts molecules to a gaseous state (desorption) and ionizes them. The procedure is referred to as "soft ionization" because it leaves the large molecules relatively intact, rather than breaking them into small pieces. The ions then travel through what is called a flight tube, and because small ions travel faster than larger ones, this separates and sorts them by mass ("time of flight"). A detector records the ions as they arrive.

③ The detector generates a pattern of peaks called a mass spectrum, essentially a "fingerprint" or profile of the proteins and other macromolecules in the cell, particularly ribosomal proteins. Computer software then compares that profile with a reference database (results from known microorganisms) as a means to identify the microbe.

MicroAssessment 10.2

The size, shape, and staining characteristics of a microorganism give important clues to its identity. Culture characteristics provide additional information, but conclusive identification often relies on multiple biochemical tests. The proteins and polysaccharides that make up a bacterium are sometimes unique enough to be considered identifying markers. MALDI-TOF mass spectrometry characterizes an organism's proteins, and the profile generated can be used as an identifying trait.

4. How does MacConkey agar help identify the cause of a urinary tract infection?
5. Describe two methods to test for the enzyme urease.
6. A sample must contain many microbial cells for any to be seen by microscopic examination. Why?

10.3 ■ Identification Methods Based on Genotype

Learning Outcome

3. Describe how NAATs, nucleic acid probes, and sequencing 16S rRNA genes can be used to identify microorganisms.

Many of the technologies discussed in chapter 9 can be used to identify a microorganism based on its genotype. Some of these methods even make it possible to identify organisms that cannot yet be grown in culture.

Detecting Specific Nucleotide Sequences

Nucleic acid amplification tests (NAATs) and DNA probes can both be used to detect nucleotide sequences unique to a given species or related group. A significant limitation, however, is that each amplification or probe detects only a single possibility. If the organism in question could be one of five different species or related groups, then five distinguishable amplifications or probes would be needed.

Nucleic Acid Amplification Tests (NAATs)

PCR and other methods collectively referred to as **nucleic acid amplification tests (NAATs)** can be used to increase the number of copies of specific nucleotide sequences that serve as identifying markers. This allows researchers to detect the sequences in samples such as body fluids, soil, food, and water. The methods can be used to detect microorganisms even if present in extremely small numbers, including those that cannot yet be grown in culture. For using PCR to detect a DNA sequence of interest, a sample is treated to release and denature the DNA, and then specific primers and other ingredients are added (see figure 9.12). After 30 amplification cycles, the amount of target DNA will have been increased about a billion-fold (see figure 9.14). The method used to detect the amplified fragment depends on the procedure used.

DNA Probes

A **DNA probe** is a single-stranded piece of DNA that has been tagged with a detectable marker such as a radioisotope or a fluorescent dye. The probe is complementary to a nucleotide sequence of interest, so it binds to that sequence, bringing the detectable marker with it (**figure 10.9**). By doing so, it can locate a nucleotide sequence that characterizes a particular species or group. Oftentimes the methods rely on a preliminary step that increases the amount of DNA in the sample. This can be done by inoculating the specimen onto an agar medium so that each microbial cell multiplies, forming a colony. Alternatively, a preliminary in vitro DNA amplification step such as PCR can be done.

Fluorescence in situ hybridization (FISH), a technique used to detect a given nucleotide sequence within intact cells on a microscope slide, often relies on a probe that binds 16S ribosomal RNA (rRNA). An amplification step is not needed because numerous copies of rRNA are naturally present in multiplying cells. Various different probes that bind rRNA are available, each specific for a given **signature sequence**—a nucleotide sequence in rRNA that characterizes either a certain species or a group of related organisms (see figure 9.17).

Sequencing Ribosomal RNA Genes

The nucleotide sequence of ribosomal RNA molecules **(rRNAs),** or the DNA that encodes them **(rDNAs),** can be used to identify microbes (**figure 10.10**). rRNAs are useful in microbial classification and identification because they are present in all organisms and their sequences are relatively stable; the ribosome would not function with too many

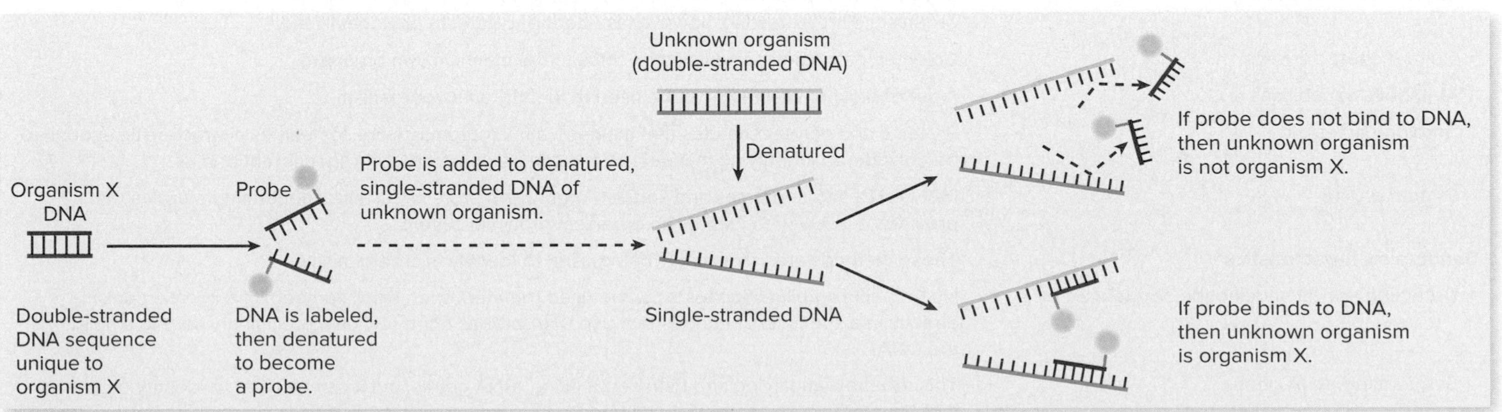

FIGURE 10.9 Nucleic Acid Probes Detect Specific DNA Sequences The probe, a single-stranded piece of nucleic acid labeled with a detectable marker, is used to locate a unique nucleotide sequence that identifies a particular microbial species.

? **What type of label is used for fluorescence in situ hybridization (FISH) probes?**

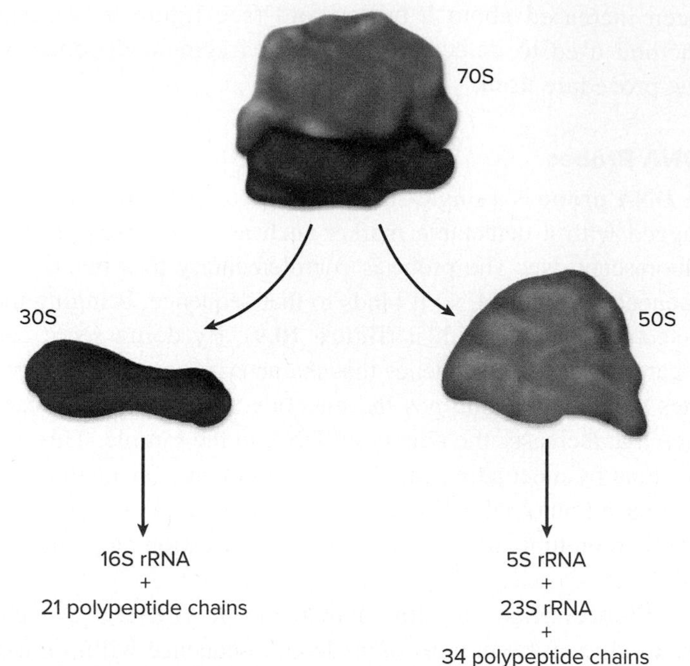

FIGURE 10.10 Ribosomal Components The 70S ribosome of prokaryotes is composed of three types of rRNA (5S, 16S, and 23S) and an assortment of different ribosomal proteins.

? Which type of ribosomal RNA is most often used in taxonomy?

mutations. Not all regions of the ribosome are equally stable, however: Some have relatively few changes, but others are more variable. Regions that are constant make it possible to design universal primers, meaning primers that will bind to the rRNA of most organisms. Regions that are variable are useful for distinguishing different species.

Of the different prokaryotic rRNAs (5S, 16S, and 23S), the 16S molecule is the most useful in taxonomy because of its moderate size (approximately 1,500 nucleotides). 16S RNA and its eukaryotic counterpart, 18S RNA, are called small subunit (SS or SSU) rRNAs because they are part of the small subunit of the ribosomes. Once the nucleotide sequence of an unknown organism's SSU rRNA has been determined, a researcher can compare it with sequences of known organisms by searching extensive computerized databases.

Using rDNA to Identify Uncultivated Organisms

The vast majority of microbes cannot yet be grown in culture. However, the DNA from these organisms can be amplified, cloned, and sequenced, making it possible to detect and identify them. The bacterium that causes Whipple disease, a rare intestinal illness, was identified this way. The organism was given the name *Tropheryma whipplei,* and a specific probe was then developed to detect it in intestinal tissue, well before it could be grown in culture.

Whole Genome Sequencing

As DNA sequencing technologies have advanced, whole genome sequencing (WGS) has become a feasible method for identifying a given isolate, particularly when other methods have failed to provide a reliable result. An additional advantage of WGS is that the data can be used to predict the antibiotic resistance of the isolate. WGS is also used to study disease outbreaks; the importance of this is discussed in the next section.

The identification methods based on phenotype and genotype are summarized in **table 10.4.**

TABLE 10.4 Methods Used to Identify Microorganisms

Method	Comments
Phenotypic Characteristics	Most of these methods do not require sophisticated equipment and can easily be done anywhere in the world.
Microscopic morphology	Size, shape, and staining characteristics can give suggestive information as to the identity of the organism. Further testing, however, is needed to confirm the identification.
Culture characteristics	Colony morphology can give initial clues to the identity of an organism.
Metabolic capabilities	A set of biochemical tests can be used to identify a microorganism.
Serological testing	Proteins and polysaccharides that make up a microorganism are sometimes characteristic enough to be considered identifying markers. These can be detected using specific antibodies.
Protein profile	MALDI-TOF MS separates and sorts an organism's proteins by mass, generating a profile that provides a fast way to identify an organism grown in culture.
Genotypic Characteristics	These methods are increasingly being used to identify microorganisms.
Detecting specific nucleotide sequences	Nucleic acid amplification tests can be used to detect even small numbers of a microorganism directly in a specimen. Methods that use DNA probes often rely on a preliminary step to amplify the DNA.
Sequencing rRNA genes	This requires amplifying and then sequencing rRNA genes, but it can be used to identify organisms that have not yet been grown in culture.
Whole genome sequencing	This is useful for identifying an isolate when other methods have failed; also used to study disease outbreaks.

MicroAssessment 10.3

A microorganism can be identified by using a NAAT or a probe to detect a nucleotide sequence unique to that organism. Ribosomal RNA genes can be sequenced to identify organisms, including those that cannot be grown in culture. Whole genome sequencing can also be used for identification.

7. When using a probe to detect an organism, why is it necessary to have some idea as to the organism's identity?
8. Why is 16S RNA also referred to as SSU RNA?
9. Why are molecular methods especially useful when bacteria are difficult to grow?

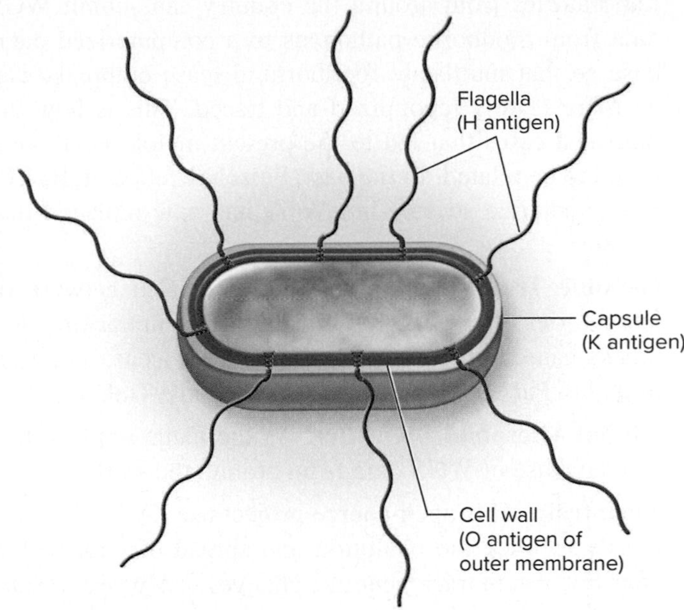

FIGURE 10.11 Serotypes The cell structures used to distinguish different strains of members of the family Enterobacteriaceae are shown.

What structures are reflected in the "O157:H7" of *E. coli* O157:H7?

10.4 ■ Characterizing Strain Differences

Learning Outcome

4. Describe five distinct methods to distinguish different strains.

The ability to distinguish different strains of a given species is useful in some situations, particularly when tracking the source of certain infections. In 2019, for example, 137 reported cases of diarrheal disease across 10 states in the United States involved a specific strain of *Salmonella enterica.* The outbreak was associated with eating pre-cut melons from a particular supplier, leading to a recall of the products. Linking the 137 cases among the many thousands of diarrheal cases that occur nationwide each year would be impossible without methods to distinguish different strains. Likewise, the ability to distinguish among strains of a common microbe is important for outbreak management in a healthcare setting.

Characterizing strain differences also plays an important role in forensic investigations of bioterrorism and other biocrimes, and in diagnosing certain diseases. The methods used to characterize different strains are summarized at the end of this section (see table 10.5).

Biochemical Typing

Biochemical tests are used to identify various species of bacteria, but they can also be used to distinguish strains. A group of strains that have a characteristic biochemical pattern is called a **biovar,** or a **biotype.** A biochemical variant of *Vibrio cholerae* called El Tor caused a worldwide epidemic of cholera beginning in 1961. Because this biovar can be readily distinguished from other strains, its spread can be traced.

Serological Typing

Proteins and carbohydrates that vary among strains can be used as distinguishing markers. Different strains of *E. coli* and related bacteria can be distinguished by the antigenic type of their flagella, capsules, and lipopolysaccharide molecules (**figure 10.11**). The "O157:H7" designation of *E. coli* O157:H7 refers to the antigenic type of its lipopolysaccharide (the O antigen) and flagella (the H antigen). A group of strains that have cell surface antigens different from other strains is called a **serovar,** or a **serotype.**

Whole Genome Sequencing

Advances in DNA sequencing methods have made it much easier to detect subtle differences among phenotypically identical strains. If fact, whole genome sequencing (WGS) has now largely replaced a method that compares patterns of fragment sizes called restriction fragment polymorphisms (RFLPS), which are produced when the same restriction enzyme is used to digest an isolate's DNA.

Various surveillance networks and other programs now use WGS data to help track infectious disease outbreaks both nationally and around the world. These include:

- **PulseNet.** This is a Centers for Disease Control (CDC) surveillance network that makes it easier for public health and food regulatory agencies to track foodborne outbreaks.

Laboratories from around the country can submit WGS data from foodborne pathogens to a computerized database so that multistate foodborne disease outbreaks can be more readily recognized and traced. This is how the diarrheal cases that led to the pre-cut melon recall were found to be related. In the past, PulseNet relied on RFLPs to characterize strains, but WGS has now replaced that system.

- **Genome Trakr.** This is an FDA-associated network of laboratories that collect WGS data for use in tracking outbreaks caused by various foodborne pathogens; data submitted to PulseNet are shared with Genome Trakr.
- **Global Microbial Identifier.** As the name implies, this is a database of WGS data from around the world.
- **Nextstrain.** This open-source project uses WGS data primarily to track the evolution and spread of viral pathogens but also to track genomic changes in *Mycobacterium tuberculosis*.

Phage Typing

Strains of a given species may differ in their susceptibility to bacteriophages. **Bacteriophages,** or phages, are viruses that infect bacteria, often lysing them; they will be described in more detail in chapter 13. The susceptibility of a bacterium to a particular type of phage can be easily determined. First, a culture of the test organism is inoculated into melted, cooled nutrient agar and poured onto the surface of an agar plate, creating a uniform layer of cells. Drops of different types of bacteriophage are then placed on the surface of the agar. During incubation, the bacterial cells multiply, forming a visible layer of turbidity. If the bacterial strain is susceptible to a specific type of phage that lyses its host, a clear area will form at the spot where the drop was added. The patterns of clearing indicate the susceptibility of the test organism to different phages (**figure 10.12**). Bacteriophage typing has now largely been replaced by other methods, but it is still a useful tool, particularly for laboratories that lack equipment to do molecular typing.

Antibiogram Typing

Antimicrobial susceptibility patterns, or **antibiograms,** can be used to distinguish different strains. As with phage typing, this method has largely been replaced by other methods. To determine the susceptibility pattern, a culture is uniformly inoculated onto the surface of nutrient agar. Paper discs containing different antibiotics or other antimicrobial medications are then placed on the agar. After incubation, clear areas (zones of inhibition) will be visible around discs of antimicrobials that inhibit or kill the organism (**figure 10.13**).

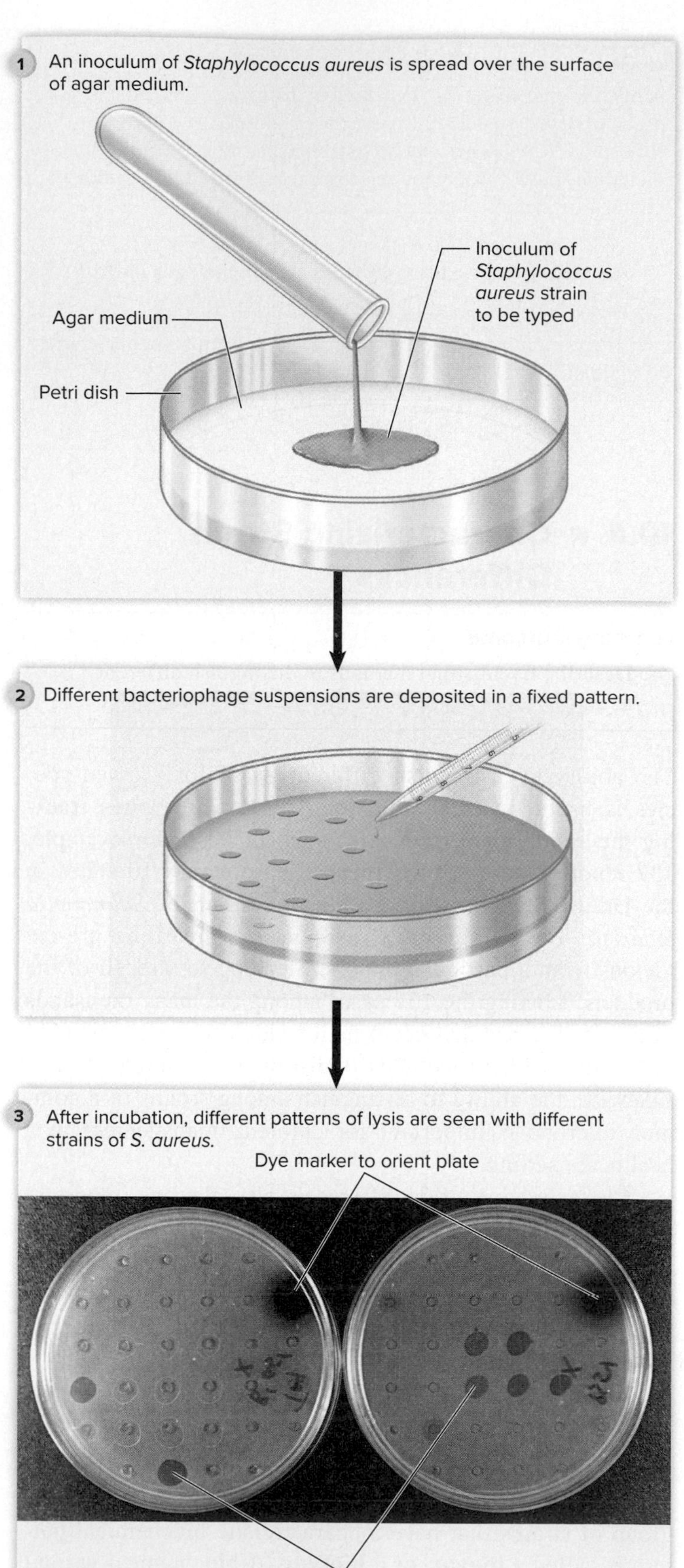
1 An inoculum of *Staphylococcus aureus* is spread over the surface of agar medium.
Inoculum of *Staphylococcus aureus* strain to be typed
Agar medium
Petri dish
2 Different bacteriophage suspensions are deposited in a fixed pattern.
3 After incubation, different patterns of lysis are seen with different strains of *S. aureus*.
Dye marker to orient plate
Lysis

FIGURE 10.12 Phage Typing

Evans Roberts

? Why would a lab technician do phage typing rather than molecular typing?

FOCUS ON A CASE 10.1

Wisconsin state health authorities alerted the CDC about an *E. coli* O157:H7 outbreak. Soon thereafter, officials from Oregon and New Mexico reported a similar outbreak. Within days, the CDC determined that the strains from Wisconsin matched those of the other two states, and that the patients recalled eating pre-packaged fresh spinach. The CDC then issued a press release advising people not to eat bagged fresh spinach, and a company that produces several brands of bagged spinach announced a voluntary recall of all fresh spinach products. In the end, the implicated strain was found to have caused 205 cases of illness in 26 states, resulting in three deaths.

1. How could the clinical laboratory separate and distinguish *E. coli* O157:H7 from the common *E. coli* strains that normally inhabit the human intestine?
2. How could the CDC determine that the strains from the three states originated from the same source?

Discussion

1. To identify *E. coli* O157:H7 in a stool specimen, the sample is inoculated onto a special agar medium designed to distinguish it from strains that typically inhabit the large intestine. One such medium is sorbitol-MacConkey, a modified version of MacConkey agar that contains the carbohydrate sorbitol in place of lactose. On this medium, most *E. coli* O157:H7 isolates are colorless because they do not ferment sorbitol. In contrast, common strains of *E. coli* ferment sorbitol, giving rise to pink colonies. Serological testing is then used to determine if the colorless *E. coli* colonies are serotype O157; strains that test positive are then generally tested to confirm they are serotype H7.
2. Today, whole genome sequencing (WGS) would be done. In the past, however, the restriction fragment polymorphism (RFLP) pattern of the isolates would have been determined.

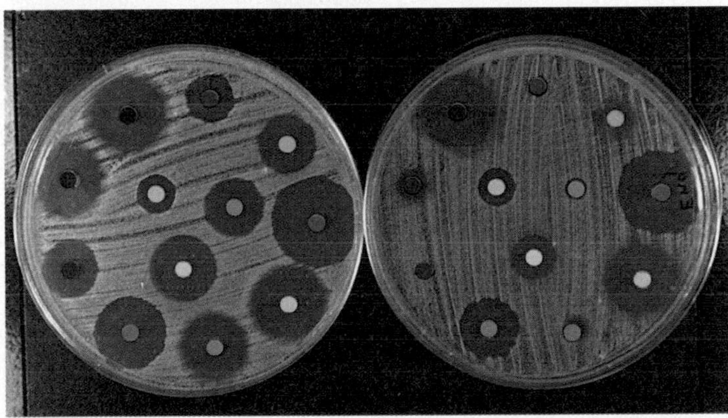

FIGURE 10.13 Antimicrobial Susceptibility Patterns In this example, 12 different antimicrobial medications incorporated in paper discs have been placed on two plates containing different cultures of *Staphylococcus aureus*. Clear areas are where bacterial growth has been inhibited. Evans Roberts

? How can you tell that these *S. aureus* isolates are different strains?

The methods used to characterize different strains are summarized in **table 10.5.**

MicroAssessment 10.4

Strains of a given species may differ in phenotypic characteristics such as biochemical capabilities and protein and polysaccharide components. Whole genome sequencing is increasingly being used to detect differences between strains. Phage typing and antibiogram typing can be used when other methods are not an option.

10. Explain the difference between a biotype and a serotype.

11. What is the function of PulseNet?

TABLE 10.5 Summary of Methods Used to Characterize Different Strains

Method	Comment
Biochemical Typing	Biochemical tests are most commonly used to identify bacteria, but in some cases they can be used to distinguish different strains. A group of strains that have a characteristic biochemical pattern is called a biovar or a biotype.
Serological Typing	Proteins and carbohydrates that vary among strains can be used to differentiate strains. A group of strains that have a characteristic serological type is called a serovar or a serotype.
Whole Genome Sequencing (WGS)	Various surveillance networks and other programs now use WGS data to help track outbreaks both nationally and around the world.
Phage Typing	Strains of a given species sometimes differ in their susceptibility to various types of bacteriophages.
Antibiogram Typing	Antibiotic susceptibility patterns can be used to characterize strains.

10.5 ■ Classifying Microorganisms

Learning Outcome

5. Describe how (1) sequence analysis of ribosomal components, (2) DNA-DNA hybridization, (3) sequence analysis of genomes, and (4) GC content are used to classify microorganisms.

Because the phylogeny (evolutionary history) of prokaryotes was difficult to determine before DNA sequencing techniques and other molecular methods were developed, classification schemes historically grouped these organisms by shape, staining characteristics, metabolic capabilities, and other phenotypic traits. Although this was convenient, phenotypic differences can be due to only a few gene products, and a single mutation resulting in a non-functional enzyme can change an organism's capabilities. In addition, phenotypically similar organisms may be only distantly related; conversely, those that appear dissimilar may be closely related.

DNA sequencing technologies revolutionized classification by providing insight into the phylogeny of different organisms. Random mutations cause an organism's nucleotide sequences to change over time, so the more time that has passed since two organisms diverged, the greater the differences in their sequences. By comparing DNA sequences of different organisms, scientists can construct a **phylogenetic tree,** a diagram that depicts the evolutionary history of organisms (**figure 10.14**).

Sequence Analysis of Ribosomal Components

The sequences of ribosomal components (ribosomal RNA and ribosomal proteins) are considered reliable indicators of phylogeny because ribosomes are present in all organisms and perform crucial and functionally constant tasks. Only a limited number of changes in the components can occur without affecting an organism's survival. Because of this, some of the

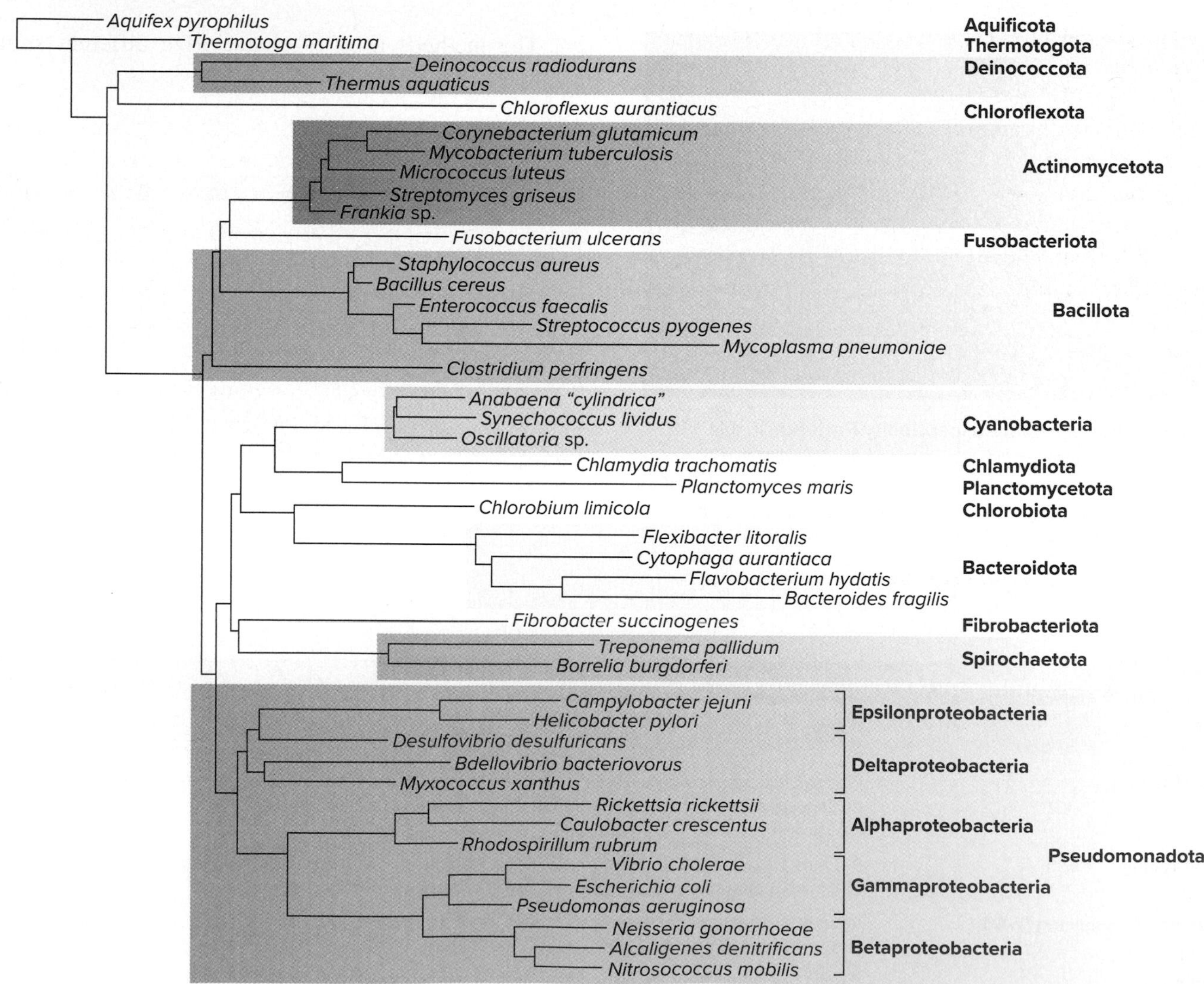

FIGURE 10.14 A Phylogenetic Tree of Bacteria Each branch represents the evolutionary distance between two species.

? Based on this tree, which is most closely related to *Bacillus cereus: Staphylococcus aureus* or *Clostridium perfringens*?

sequences will still be similar even in groups of organisms that diverged long ago. In addition, the ribosomal genes are not commonly horizontally transferred—an event that would complicate an analysis of evolutionary history.

Ribosomal RNA (rRNA)

Nucleotide sequence analysis of **rRNAs** (ribosomal RNAs) revolutionized microbial classification by giving rise to the three-domain system. With respect to classification, small ribosomal subunit rRNAs (SSU rRNAs) are the most useful because of their moderate size (see figure 10.10). Recall that in bacteria and archaea, the SSU rRNA is 16S; in eukaryotes it is 18S. Today, **rDNAs** (the genes that encode rRNAs) are usually studied because of the ease of DNA sequencing methods.

Even microorganisms that cannot yet be grown in culture can be tentatively classified by SSU rDNA sequence analysis. The SSU rDNA in a DNA sample extracted from an environmental sample such as soil or water can be amplified (using PCR), cloned, and sequenced. The sequences can then be compared with databases containing SSU rDNA sequences of known organisms.

Although SSU rDNA sequence analysis has been very helpful in determining evolutionary relationships among distantly related organisms, it is often unreliable when assessing closely related species. This is because closely related but genetically distinct microorganisms can have identical SSU rDNA sequences. In these cases, DNA-DNA hybridization (discussed shortly) is a better tool to assess relatedness.

Ribosomal Proteins

With the widespread availability of whole and partial genome sequences, scientists can now use the deduced amino acid sequences of various ribosomal proteins as a basis for determining relatedness. An advantage of this approach is that it accounts for novel organisms not detected by standard techniques used to amplify SSU rDNA. In addition, each organism has an assortment of different ribosomal proteins (see figure 10.10), so more gene products can be compared. The greatly expanded tree of life shown early in the chapter is based on comparisons of the amino acid sequences of 16 different ribosomal proteins from thousands of different species (see figure 10.1b). As more microbial genome sequences are determined, the tree is likely to become even more complex.

DNA-DNA Hybridization (DDH)

The extent of nucleotide sequence similarity between two organisms can be determined by measuring how completely single strands of their DNA hybridize to each other, meaning their strands come together and form base pairs. The extent of DNA-DNA hybridization (DDH) reflects the degree of sequence similarity. Two bacterial strains that show at least 70% similarity are generally considered to be members of the same species. Surprisingly, DDH has shown that members of the genera *Shigella* and *Escherichia,* which are quite different based on biochemical tests, should actually be grouped in the same species. Note that human and chimpanzee DNA have approximately 99% similarity by DDH. Therefore, by the criteria used to classify bacteria, humans and chimpanzees would be members of the same species!

Sequence Analysis of Genomes

Although DNA-DNA hybridization is a routine method for determining genetic relatedness, the accumulating data from whole genome sequencing (WGS) may eventually prove to be a more valuable tool. To assess the relatedness of two isolates, shared genes of genome sequences are compared, and a measure called the average nucleotide identity (ANI) is then calculated.

GC Content

The **GC content** (also called the G+C content), which is the percentage of G-C base pairs in an organism's DNA, can be used to roughly compare relatedness. If the GC content of two organisms differs by more than a small percentage, then the organisms cannot be closely related. A similarity of base compositions, however, does not necessarily mean that the organisms are related, because the nucleotide sequences and genome sizes could differ greatly.

The GC content is often measured by determining the temperature at which the double-stranded DNA denatures, or melts (**figure 10.15**). DNA that has a high GC content

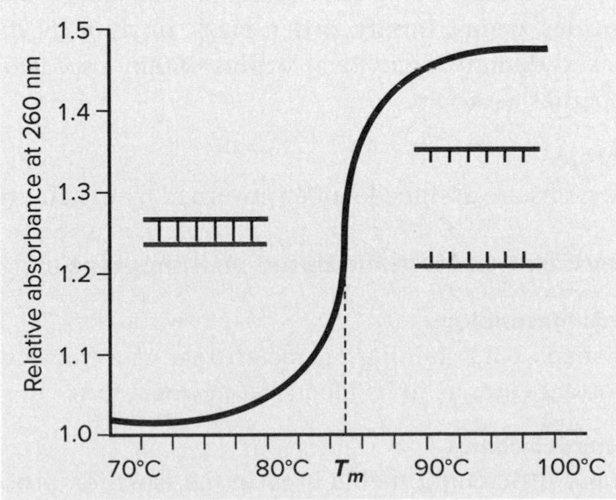

FIGURE 10.15 A DNA Melting Curve The absorbance (relative absorbance at 260 nm) rapidly increases as double-stranded DNA denatures (melts).

What characteristic of a DNA sequence does the T_m reflect?

TABLE 10.6 Methods Used to Determine the Relatedness of Different Microorganisms

Method	Comment
Sequence analysis of ribosomal components	Certain regions of the SSU rDNA (16S in bacteria and archaea; 18S in eukaryotes) can be used to determine distant relatedness of diverse organisms; other regions can be used to determine more recent divergence. Amino acid sequences of ribosomal proteins can also be compared.
DNA-DNA hybridization	The extent of nucleotide sequence similarity between two isolates can be determined by measuring how completely single strands of their DNA hybridize to one another.
Sequence analysis of genomes	Nucleotide sequences of shared genes can be compared, and the average nucleotide identity (ANI) then calculated.
GC content	GC content comparisons offer a crude comparison of genomes, but organisms with identical GC contents can be entirely unrelated.

melts at a higher temperature because G-C base pairs are held together by three hydrogen bonds, whereas A-T pairs are held together by only two hydrogen bonds. The melting temperature can be determined by measuring the absorbance of UV light by a DNA solution as it is heated; the absorbance rapidly increases as the DNA denatures.

Table 10.6 summarizes some of the methods used to determine the relatedness of microorganisms for the purpose of classification.

MicroAssessment 10.5

Analyzing and comparing the sequences of ribosomal RNA components has revolutionized the classification of microorganisms. Other characteristics used to determine relatedness include DNA-DNA hybridization, sequence analysis of genomes, and GC content.

12. Explain why ribosomal components are useful for determining phylogeny.
13. Explain the role of DNA-DNA hybridization in classification.
14. Why would it be easier to sequence rDNA than to sequence rRNA?

Summary

10.1 ■ Principles of Taxonomy

Taxonomy consists of three interrelated areas: **identification, classification,** and **nomenclature.**

Strategies Used to Identify Microorganisms

To characterize and identify microorganisms, a wide assortment of technologies may be used, including microscopic examination, cultural characteristics, biochemical tests, and nucleic acid analysis.

Strategies Used to Classify Microorganisms

Taxonomic classification categories are arranged in a hierarchical order, with the species being the basic unit. Taxonomic categories include **species, genus, family, order, class, phylum** (or division), **kingdom,** and **domain.** Individual **strains** within a species vary in minor properties (table 10.1).

Nomenclature

Microorganisms are assigned names governed by official rules.

10.2 ■ Identification Methods Based on Phenotype (table 10.4)

Microscopic Morphology

The size, shape, and staining characteristics of a microorganism yield important clues as to its identity (figures 10.2, 10.3).

Culture Characteristics

Selective and differential media used in the isolation process can provide information that helps to identify an organism.

Metabolic Capabilities

Most biochemical tests rely on a pH indicator or chemical reaction that shows a color change when a compound is degraded. The basic strategy for identification using biochemical tests relies on the use of a **dichotomous key** (figures 10.4, 10.5, 10.6; table 10.3).

Serological Characteristics

Proteins and polysaccharides that make up a prokaryote's surface are sometimes characteristic enough to be identifying markers.

Protein Profile

MALDI-TOF MS (matrix-assisted laser desorption ionization time of flight mass spectrometry) generates a profile of a colony's proteins and macromolecules, which can be used to identify the organism (figure 10.8).

10.3 ■ Identification Methods Based on Genotype (table 10.4)

Detecting Specific Nucleotide Sequences

Nucleic acid amplification tests (NAATs) such as PCR increase the number of copies of a specific nucleotide sequence, thereby determining if a given microorganism is present. A probe complementary to a sequence unique to a given microbe can also be used to detect that organism (figure 10.9).

Sequencing Ribosomal RNA Genes

The nucleotide sequence of ribosomal RNA molecules **(rRNAs),** or the DNA that encodes them **(rDNAs),** can be used to identify microbes. Microorganisms that cannot yet be cultivated can be identified by amplifying, cloning, and then sequencing specific regions of rDNA.

Whole Genome Sequencing (WGS)

Whole genome sequencing (WGS) is particularly helpful for identifying isolates when other methods have failed.

10.4 ■ Characterizing Strain Differences (table 10.5)

Biochemical Typing
A group of strains that has a characteristic biochemical variation is called a **biovar,** or a **biotype.**

Serological Typing
A group of strains that differs serologically from other strains is called a **serovar,** or a **serotype** (figure 10.11).

Whole Genome Sequencing
Whole genome sequencing (WGS) can be done to distinguish between isolates.

Phage Typing
The susceptibility to various types of bacteriophages can be used to demonstrate strain differences (figure 10.12).

Antibiogram Typing
Antimicrobial susceptibility patterns can be used to distinguish strains (figure 10.13).

10.5 ■ Classifying Microorganisms (table 10.6)

DNA sequences can be used to construct a **phylogenetic tree** (figure 10.14).

Sequence Analysis of Ribosomal Components
Analyzing and comparing the nucleotide sequences of **rRNA** and, more recently, **rDNA** has revolutionized the classification of organisms. The amino acid sequences of ribosomal proteins can also be compared.

DNA-DNA Hybridization (DDH)
The extent of nucleotide similarity between two organisms can be determined by measuring how completely single strands of their DNA will anneal to each other.

Sequence Analysis of Genomes
The average nucleotide identity (ANI) of shared genes can be used to measure relatedness.

GC Content
The **GC content** can be measured by determining the temperature at which double-stranded DNA melts (figure 10.15).

Review Questions

Short Answer

1. Name and describe the three areas of taxonomy.
2. Explain the basis for the three-domain systems of classification.
3. Describe how a dichotomous key is used when identifying bacteria.
4. Why is MALDI-TOF MS particularly important in clinical microbiology?
5. Describe the difference between using PCR and using a probe to detect a specific nucleotide sequence.
6. Explain how signature sequences are used in bacterial identification.
7. Explain why whole genome sequencing is so useful in tracking a foodborne outbreak.
8. What is a phylogenetic tree?
9. List two advantages of using sequence analysis of ribosomal components in classification.
10. Describe how the GC content of DNA can be measured.

Multiple Choice

1. Which of the following is the newest taxonomic unit?
   a) Strain
   b) Family
   c) Order
   d) Species
   e) Domain
2. If an acid-fast bacterium is detected in a clinical sample, then the organism could be
   a) *Cryptococcus neoformans.*
   b) *Mycobacterium tuberculosis.*
   c) *Neisseria gonorrhoeae.*
   d) *Streptococcus pneumoniae.*
   e) *Streptococcus pyogenes.*
3. The "breath test" for *Helicobacter pylori* infection detects the presence of which of the following?
   a) Antigens
   b) Catalase
   c) Hemolysis
   d) Lactose fermentation
   e) Urease
4. The "O157:H7" of *E. coli* O157:H7 refers to the
   a) biotype.
   b) serotype.
   c) phage type.
   d) antibiogram type.
5. When hydrogen peroxide is placed on a colony of an unknown bacterium, bubbles form. Based on this information, you can conclude that the bacterium
   a) is *Staphylococcus epidermidis.*
   b) is a lactic acid bacterium.
   c) is beta-hemolytic.
   d) is catalase-positive.
   e) can cause strep throat.
6. Which of the following is best to use for determining the evolutionary relatedness of organisms?
   a) Ability to form endospores
   b) 16S ribosomal RNA sequence
   c) Sugar degradation
   d) Motility
7. If the GC content of each of two organisms is 70%, the
   a) organisms are definitely related.
   b) organisms are definitely not related.
   c) AT content is 30%.
   d) organisms likely have extensive DNA homology.
   e) organisms likely have many characteristics in common.

8. If two microbial isolates have similar but different 16S rRNA sequences, they are probably
   a) both motile.
   b) both pathogens.
   c) both cocci.
   d) members of the same domain.
   e) the same strain.
9. The sequences of which ribosomal genes are most commonly used for establishing phylogenetic relatedness?
   a) 5S
   b) 16S
   c) 23S
   d) All of these are commonly used.
10. Which of the following is false?
    a) *Tropheryma whipplei* could be identified before it had been grown in culture.
    b) The GC content of an organism can be measured by determining the temperature at which its DNA melts.
    c) Sequence differences between organisms can be used to assess their relatedness.
    d) Based on DNA homology studies, members of the genus *Shigella* should be in the same species as *Escherichia coli.*
    e) The polymerase chain reaction (PCR) is used to determine the serotype of an organism.

Applications

1. Microbiologists debate the use of biochemical similarities and cell features as a way of determining the taxonomic relationships among prokaryotes. Explain why some microbiologists believe these similarities and differences are a powerful taxonomic indicator, whereas others think they are not very useful for that purpose.
2. A researcher interested in investigating the genetic relationship of mitochondria to bacteria must decide on the best method to study this. What advice would you give the researcher?

Critical Thinking

1. In figure 10.15, how would the curve appear if the GC content of the DNA sample were increased? How would the curve appear if the AT content were increased?
2. When DNA probes are used to detect specific sequence similarities in bacterial DNA, the probe is heated and the two strands of DNA are separated. Why must the probe DNA be heated?

11 The Diversity of Bacteria and Archaea

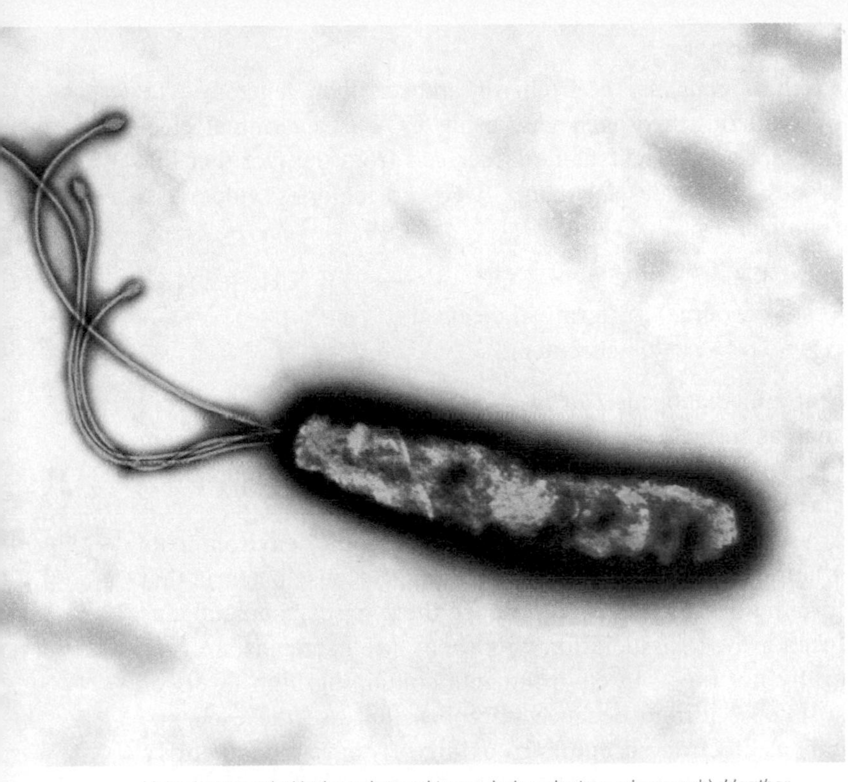

Helicobacter pylori (color-enhanced transmission electron micrograph). *Heather Davies/Science Photo Library/Getty Images*

KEY TERMS

Anoxygenic Phototrophs Photosynthetic organisms that do not produce O_2.

Chemolithotrophs Organisms that harvest energy by oxidizing inorganic chemicals.

Chemoorganotrophs Organisms that harvest energy by oxidizing organic chemicals.

Cyanobacteria Gram-negative oxygenic phototrophs; genetically related to chloroplasts.

Lactic Acid Bacteria Gram-positive bacteria that produce lactic acid as a major end product of their fermentative metabolism.

Methanogens Archaea that obtain energy by oxidizing hydrogen gas, using CO_2 as a terminal electron acceptor, thereby generating methane.

Myxobacteria Gram-negative bacteria that form complex multicellular structures called fruiting bodies.

Nitrifiers Gram-negative bacteria that obtain energy by oxidizing inorganic nitrogen compounds such as ammonium or nitrite.

Oxygenic Phototrophs Photosynthetic organisms that produce O_2.

Spirochetes Long, helical bacteria that have flexible cell walls and endoflagella.

Sulfur-Oxidizing Bacteria Gram-negative bacteria that obtain energy by oxidizing elemental sulfur and reduced sulfur compounds, generating sulfuric acid.

A Glimpse of History

Cornelius B. van Niel (1897–1985) earned his PhD. from the Technological University in Delft, Holland, the home of an approach to microbiology now commonly referred to as "the Delft School." The outstanding program there was chaired in succession by two well-known microbiologists: Martinus Beijerinck and Albert Kluyver.

As Kluyver's student, van Niel was influenced by his mentor's belief that biochemical processes were fundamentally the same in all cells and that microorganisms could be important research tools, serving as a model to study biochemical processes. Thirty years later, Kluyver and van Niel presented lectures that would be published in the book *The Microbe's Contribution to Biology*.

Shortly after completing his dissertation in 1928, van Niel accepted a position at the Hopkins Marine Station in California. There he continued the work he had begun under Kluyver's direction on the photosynthetic activities of the brightly colored purple bacteria. He demonstrated that these microbes require light for growth, yet, unlike plants and algae, they do not produce O_2. He also showed that the purple bacteria oxidize hydrogen sulfide as they fix CO_2. Furthermore, van Niel noted that the photosynthetic reactions in all photosynthetic organisms are remarkably similar, except the purple bacteria use hydrogen sulfide in place of water and produce oxidized sulfur compounds instead of O_2. This finding raised the possibility that O_2 generated by plants and algae did not come from carbon dioxide, as was believed at the time, but rather from water.

In addition to his scientific contributions, van Niel was an outstanding teacher. During the summers at Hopkins Marine Station, he taught a bacteriology course, inspiring many microbiologists with his enthusiasm for the diversity of microorganisms and their importance in nature. His sharp memory and knowledge of the literature, along with his appreciation for the remarkable abilities of microbes, allowed him to convey the wonders of the microbial world to his students.

Scientists are only beginning to understand the vast diversity of microbial life. Although well over a million species of prokaryotes are thought to exist, only a fraction of them have been described and classified. Traditional culture and isolation techniques have not supported the growth and subsequent study of the vast majority. This situation is changing as new molecular techniques make it easier to discover and characterize previously unrecognized species. The sheer volume of information uncovered by modern technologies, however, can be daunting for scientists and students alike.

This chapter covers a variety of prokaryotes, focusing primarily on their extraordinary diversity rather than the phylogenetic relationships discussed in chapter 10. Note, however, that no single chapter could describe all known prokaryotes and, consequently, only a relatively small selection is presented.

METABOLIC DIVERSITY

As a group, prokaryotes use an impressive range of mechanisms to harvest energy in order to produce ATP. This section highlights that metabolic diversity by describing select prokaryotes; a table at the end of the discussion summarizes the characteristics of the archaea and bacteria used as examples (see table 11.1).

11.1 ■ Anaerobic Chemotrophs

Learning Outcome

1. Compare and contrast the characteristics and habitats of methanogens, sulfur- and sulfate-reducing bacteria, *Clostridium* species, lactic acid bacteria, and *Propionibacterium* species.

For approximately the first 1.5 billion years that prokaryotes inhabited Earth, the atmosphere was **anoxic,** meaning it lacked O_2. In that anaerobic environment, some early **Chemotrophs** (organisms that harvest energy by oxidizing chemicals) probably used pathways of anaerobic respiration, using terminal electron acceptors such as carbon dioxide or elemental sulfur, which were plentiful in the environment. Others may have used fermentation, passing the electrons to an organic molecule such as pyruvate.

Today, anaerobic habitats are still common. Mud and tightly packed soil limit the diffusion of gases, and any O_2 that penetrates is quickly converted to water by aerobically respiring organisms. This creates anaerobic conditions just below the surface. Aquatic environments may also become anaerobic if they contain nutrients that promote the rapid growth of O_2-consuming microbes. This is evident in polluted lakes, where fish may die because of a lack of dissolved O_2. The human body also provides many anaerobic environments, such as in the intestinal tract. Even the skin and the oral cavity, which are routinely exposed to O_2, have anaerobic microenvironments—created as aerobes in a microbial community consume the available O_2.

MicroByte

Approximately 99% of the prokaryotes that inhabit the intestinal tract are obligate anaerobes.

Anaerobic Chemolithotrophs

Chemolithotrophs obtain energy by oxidizing inorganic chemicals such as hydrogen gas (H_2). Those growing anaerobically obviously cannot use O_2 as a terminal electron acceptor and instead must use an alternative such as carbon dioxide or sulfur. Relatively few anaerobic chemolithotrophs have been discovered, and most are archaea. Some bacterial examples that inhabit aquatic environments will be discussed later.

Methanogens

Methanogens are a group of archaea that generate ATP by oxidizing hydrogen gas, using CO_2 as a terminal electron acceptor. Their name is derived from the fact that this process generates methane (CH_4), a colorless, odorless, flammable gas:

$$\underset{\text{(energy source)}}{4\,H_2} + \underset{\text{(terminal electron acceptor)}}{CO_2} \longrightarrow CH_4 + 2\,H_2O$$

Many methanogens can also use alternative energy sources such as formate, methanol, or acetate. Representative genera of methanogens include *Methanobrevibacter* (see figure 3.44) and *Methanosarcina* (**figure 11.1**).

Methanogens are found in anaerobic environments where H_2 and CO_2 are both available. Because bacteria that ferment organic material produce these gases, methanogens often grow in association with them. Methanogens are generally not found in environments containing high levels of sulfate or nitrate because microbes that use these chemicals as electron acceptors to oxidize H_2 have a competitive advantage; the use of CO_2 as an electron acceptor releases comparatively little energy (see figure 6.8). Environments where methanogens are commonly found include swamps, marine sediments, rice paddies, and the digestive tracts of humans and other animals. The methane produced can be

FIGURE 11.1 Methanogen *Methanosarcina mazei* (SEM). Eye of Science/Science Source

What roles do hydrogen gas and carbon dioxide play in the metabolism of methanogens?

seen as bubbles rising in swamp waters and is part of the 10 cubic feet of gas discharged from a cow's digestive system each day. As a by-product of wastewater treatment plants, methane gas can be collected and used for heating and generating electricity.

Studying methanogens is challenging because they are very sensitive to O_2. Special techniques, including anaerobic chambers, are used to culture them.

Anaerobic Chemoorganotrophs—Anaerobic Respiration

Chemoorganotrophs oxidize organic compounds such as glucose to obtain energy. Those that grow anaerobically often use sulfur or sulfate as a terminal electron acceptor.

Sulfur- and Sulfate-Reducing Bacteria

When sulfur compounds are used as terminal electron acceptors, they become reduced to form hydrogen sulfide, the compound responsible for the rotten-egg smell of many anaerobic environments. An example of this reaction is:

organic compounds (energy source)	+	S (terminal electron acceptor)	⟶	$CO_2 + H_2S$

In addition to its unpleasant odor, H_2S is a problem to industry because it reacts with metals, thus corroding pipes and other structures. Ecologically, however, prokaryotes that reduce sulfur compounds are an essential component of the sulfur cycle.

Sulfate- and sulfur-reducing bacteria generally live in mud that has organic material and oxidized sulfur compounds. The H_2S they produce causes mud and water to turn black when it reacts with iron molecules. At least a dozen genera are recognized in this group; *Desulfovibrio* species, which are Gram-negative curved rods, are the most extensively studied.

Some archaea also use sulfur compounds as terminal electron acceptors, but the characterized examples generally do not inhabit the same environments as their bacterial counterparts. Although most of the sulfur-reducing bacteria are either mesophiles or thermophiles, the characterized sulfur-reducing archaea are hyperthermophiles, inhabiting extreme environments such as hydrothermal vents. They are discussed in section 11.9.

Anaerobic Chemoorganotrophs—Fermentation

Many types of anaerobic bacteria obtain energy by fermentation, producing ATP only by substrate-level phosphorylation. There are many variations of fermentation, using different organic energy sources and producing characteristic end products, but one example is:

glucose (energy source)	⟶	pyruvate (terminal electron acceptor)	⟶	lactic acid

The Genera *Clostridium* and *Clostridioides*

Members of the genera *Clostridium* and *Clostridioides,* collectively referred to as clostridia, are Gram-positive rods that can form endospores (see figure 3.20). They are common soil inhabitants, and the vegetative cells live in the anaerobic microenvironments created when aerobic organisms consume available O_2. Clostridial endospores can tolerate O_2 and survive for long periods by withstanding levels of heat, drying, chemicals, and irradiation that would kill vegetative bacteria. When conditions become favorable, these endospores germinate, and the resulting vegetative bacteria multiply. Examples of diseases caused by pathogenic *Clostridium* species include tetanus (*C. tetani*), gas gangrene (*C. perfringens*), and botulism (*C. botulinum*). *Clostridioides difficile,* which causes an antibiotic-associated diarrheal disease referred to as *C. difficile* infection (CDI), was included in the genus *Clostridium* up until recently. Some clostridia are normal inhabitants of the intestinal tract of humans and other animals.

As a group, clostridia ferment a wide variety of compounds, including sugars and cellulose. Some of the end products are commercially valuable—for example, *Clostridium acetobutylicum* produces acetone and butanol. Some species can ferment amino acids by an unusual process that oxidizes one amino acid, using another as a terminal electron acceptor. This generates a variety of foul-smelling end products associated with rotting flesh.

Lactic Acid Bacteria

Gram-positive bacteria that produce lactic acid as a major end product of their fermentative metabolism make up a group called the **lactic acid bacteria.** This includes members of the genera *Streptococcus, Enterococcus, Lactococcus, Lactobacillus,* and *Leuconostoc.* Most can grow in aerobic environments, but they typically only carry out fermentation. They can be easily distinguished from other bacteria that grow in the presence of O_2 because they lack the enzyme catalase (see figure 10.4).

Streptococcus species are cocci that typically grow in chains of varying lengths (**figure 11.2**). They inhabit the oral cavity, generally as part of the normal microbiota. *S. thermophilus* is important to the food industry because it is used to make yogurt. Some streptococci, however, are pathogens. One of the most important is *S. pyogenes* (group A strep), which causes pharyngitis (strep throat) and other diseases. Unlike the streptococci that typically inhabit the throat, *S. pyogenes* is β-hemolytic, an important characteristic used to distinguish it from most members of the normal microbiota (see figure 4.10).

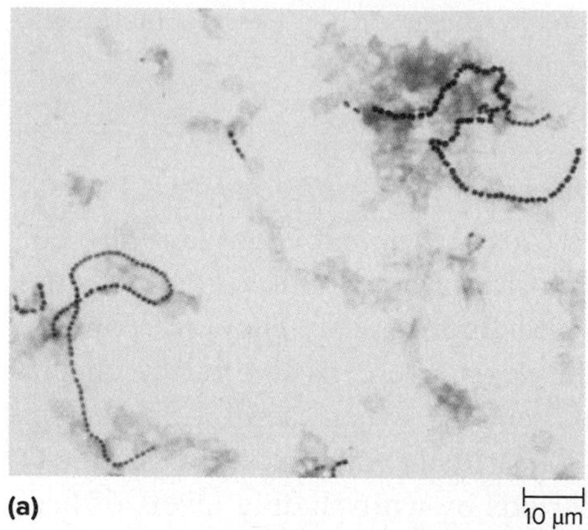

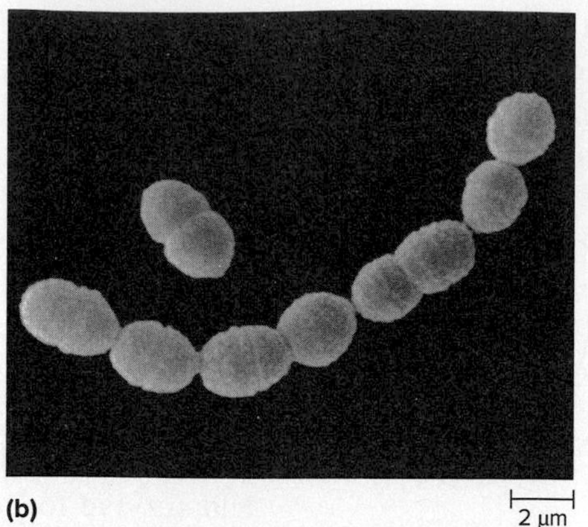

FIGURE 11.2 ***Streptococcus* Species** **(a)** Gram stain. **(b)** Scanning electron micrograph. a: Lisa Burgess/McGraw Hill; b: SCIMAT/Science Source

? **What is the major metabolic end product of *Streptococcus* species?**

Species of *Lactococcus* and *Enterococcus* were at one time included in the genus *Streptococcus.* The genus *Lactococcus* now includes species used to make cheeses. *Enterococcus* species, commonly referred to as enterococci, typically inhabit the intestinal tract of humans and other animals.

Members of the genus *Lactobacillus* are rod-shaped bacteria. They are common members of the microbiota in the mouth and the healthy vagina during childbearing years. In the vagina, they break down glycogen that has been deposited in the vaginal lining in response to estrogen (the female sex hormone). The resulting low pH helps prevent vaginal infections. Lactobacilli are also often present in milk and other dairy products as well as decomposing plant material. Like some other lactic acid bacteria, they are important in the production of fermented foods (**figure 11.3**).

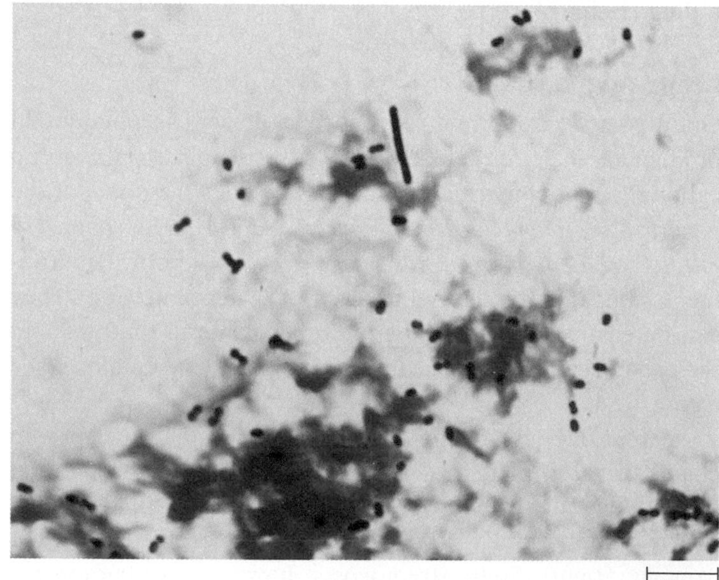

FIGURE 11.3 Lactic Acid Bacteria in Yogurt This Gram stain shows the Gram-positive rod *Lactobacillus delbrueckii* subsp. *bulgaricus* and the Gram-positive coccus *Streptococcus thermophilus*. Lisa Burgess/McGraw Hill

? ***Lactobacillus* species are common members of the normal microbiota of which human body sites?**

The Genus *Propionibacterium*

Propionibacterium species are Gram-positive pleomorphic (irregular-shaped) rods that produce propionic acid as their main fermentation end product. They can also ferment lactic acid, and so can extract energy from a waste product of other bacteria.

Propionibacterium species are valuable to the dairy industry because their fermentation end products are important in Swiss cheese production. The propionic acid gives the typical flavor of the cheese, and CO_2, also a product of the fermentation, creates the characteristic holes. Skin microbiota that were included in the genus *Propionibacterium* have been reclassified into a new genus, *Cutibacterium*.

MicroAssessment 11.1

Methanogens are archaea that oxidize hydrogen gas, using CO_2 as a terminal electron acceptor, to generate methane. The sulfur- and sulfate-reducing bacteria oxidize organic compounds, with sulfur or sulfate serving as a terminal electron acceptor, to generate hydrogen sulfide. Clostridia, the lactic acid bacteria, and *Propionibacterium* species oxidize organic compounds, with an organic compound serving as a terminal electron acceptor.

1. What metabolic process creates the rotten-egg smell characteristic of many anaerobic environments?
2. Describe two beneficial contributions of the lactic acid bacteria.
3. Relatively little is known about many obligate anaerobes. Why would this be so?

11.2 ■ Anoxygenic Phototrophs

Learning Outcome

2. Compare and contrast the characteristics of the purple bacteria and the green bacteria.

The earliest photosynthesizing organisms were anoxygenic phototrophs. As described in section 6.8, these bacteria use **anoxygenic photosynthesis,** meaning that they do not generate O_2 because they use hydrogen sulfide or organic compounds (rather than water) when making reducing power. For example:

$$\underset{\text{(carbon source)}}{6\ CO_2} + \underset{\text{(electron source)}}{12\ H_2S} \longrightarrow C_6H_{12}O_6 + 12\ S + 6\ H_2O$$

The modern-day anoxygenic phototrophs are a phylogenetically diverse group of bacteria that live in environments that have adequate light but little or no O_2. Typical habitats include bogs, lakes, and the upper layer of muds.

Purple Bacteria

The purple bacteria are Gram-negative organisms that appear red, orange, or purple due to their photosynthetic pigments. Unlike other anoxygenic phototrophs, their photosystems are contained within an internal membrane system created through folding of the cytoplasmic membrane. The purple bacteria can be separated into two groups: purple sulfur bacteria and purple non-sulfur bacteria.

Purple Sulfur Bacteria

Purple sulfur bacteria can sometimes be seen growing as colored masses in sulfur-rich aquatic habitats (**figure 11.4**). The cells are relatively large, sometimes larger than 5 μm in diameter, and some are motile by flagella. They may also have gas vesicles, allowing them to move up or down to their preferred level in the water column. Most species store sulfur in intracellular granules, but some form them outside the cell. They preferentially use hydrogen sulfide while making reducing power, although some can use other inorganic molecules (such as H_2) or organic compounds (such as pyruvate). Many are strict anaerobes, but some can grow aerobically in the dark, oxidizing reduced inorganic or organic compounds as a source of energy. Representative genera include *Chromatium, Thiospirillum,* and *Thiodictyon.*

FIGURE 11.4 Purple Sulfur Bacteria Photograph of purple sulfur bacteria growing in a bog. Johnny Madsen/Alamy Stock Photo

? **What causes the purple color of the bacterial masses?**

Purple Non-Sulfur Bacteria

The purple non-sulfur bacteria are found in a wide variety of aquatic habitats, including moist soils, bogs, and rice paddies. An important characteristic that distinguishes them from the purple sulfur bacteria is their preferential use of a variety of organic molecules rather than hydrogen sulfide as a source of electrons when making reducing power. They lack gas vesicles, and if they store sulfur, the granules form outside the cell. Many are remarkably versatile metabolically, growing as anaerobic phototrophs using organic or inorganic molecules as a source of electrons or growing aerobically in the dark as chemotrophs. Representative genera include *Rhodobacter* and *Rhodopseudomonas*.

Green Bacteria

The green bacteria are Gram-negative organisms that are typically green or brownish in color. Here we will focus on the green sulfur bacteria, which are similar to the purple sulfur bacteria with respect to habitats, source of electrons while making reducing power, and formation of extracellular sulfur granules. A distinguishing characteristic is that some of their photosynthetic pigments are located in structures called chlorosomes. The cells lack flagella, but many have gas vesicles. All are strict anaerobes, and none can use a chemotrophic metabolism. Representative genera include *Chlorobium* and *Pelodictyon*.

Filamentous Anoxygenic Phototrophic Bacteria

Filamentous anoxygenic phototrophic (FAP) bacteria are Gram-negative organisms that form multicellular arrangements and exhibit gliding motility. The most thoroughly studied of this group are members of the genus *Chloroflexus,* particularly the thermophilic strains that grow in hot springs. Many of the FAP bacteria have chlorosomes, which initially led scientists to believe they were related to the green sulfur bacteria, but their 16S rDNA sequences indicate otherwise. As a group, FAP bacteria are diverse metabolically. Some preferentially use organic compounds to generate reducing power and can also grow in the dark aerobically using chemotrophic metabolism.

Other Anoxygenic Phototrophs

Although the purple, green, and FAP bacteria have been studied most extensively, other types of anoxygenic phototrophs exist. Among these are members of the genus *Heliobacterium,* Gram-positive endospore-forming rods related to members of the genus *Clostridium.*

MicroAssessment 11.2

Anoxygenic phototrophs harvest the energy of sunlight, but they do not generate O_2. The purple sulfur bacteria and the green sulfur bacteria use hydrogen sulfide as a source of electrons to generate reducing power; the purple non-sulfur bacteria and many of the filamentous anoxygenic phototrophs preferentially use organic compounds.

4. Describe a structural characteristic that distinguishes the purple sulfur bacteria from the green sulfur bacteria.
5. What is the function of gas vesicles?
6. Given the metabolic capabilities of the different groups of anoxygenic phototrophs, which would you expect to find in the widest range of habitats and why?

11.3 ■ Oxygenic Phototrophs

Learning Outcome

3. Describe the characteristics of cyanobacteria, including how nitrogen-fixing species protect their nitrogenase enzyme from O_2.

Nearly 3 billion years ago, the Earth's atmosphere began changing as O_2 was gradually introduced to the previously anoxic environment. This was probably due to the evolution of the cyanobacteria, thought to be the earliest oxygenic phototrophs. As described in section 6.8, these organisms use **oxygenic photosynthesis,** meaning they generate O_2 because they use water as a source of electrons when making reducing power:

$$\underset{\text{(carbon source)}}{6\,CO_2} + \underset{\text{(electron source)}}{6\,H_2O} \longrightarrow C_6H_{12}O_6 + 6\,O_2$$

Cyanobacteria

Cyanobacteria are a diverse group of photosynthetic Gram-negative bacteria that inhabit a wide range of environments, including freshwater and marine habitats, soils, and the surfaces of rocks. They were initially thought to be algae and were called blue-green algae until electron microscopy revealed their prokaryotic structure. Many are able to convert nitrogen gas (N_2) to ammonia, which can then be incorporated into cell material. This process, called **nitrogen fixation,** is an exclusive ability of prokaryotes.

General Characteristics of Cyanobacteria

Cyanobacteria are morphologically diverse. Some are unicellular, with typical prokaryotic shapes such as cocci, rods, and spirals. Others form filamentous multicellular associations called trichomes that may or may not be enclosed within a sheath (a tube that holds and surrounds a chain of cells) (**figure 11.5**). Motile trichomes glide as a unit.

FIGURE 11.5 Trichome of *Oscillatoria* Note the arrangement of the individual cells in the trichome (DIC photomicrograph). M.I. Walker/Science Source

What is a trichome?

Cyanobacteria that live in aquatic environments often have gas vesicles, allowing them to move vertically within the water column. When large numbers of cyanobacteria accumulate in stagnant lakes or other freshwater habitats, it is called a bloom (**figure 11.6**). In the bright, hot conditions of summer, the buoyant cells lyse and decay, creating a foul-smelling scum referred to as a nuisance bloom. The ecological effects of these blooms on aquatic habitats are discussed in chapter 28.

The photosystems of the cyanobacteria are like those contained within the chloroplasts of algae and plants. This is not surprising in light of the genetic evidence indicating that chloroplasts evolved from a species of cyanobacteria that once resided as an endosymbiont within eukaryotic cells. In addition to light-harvesting chlorophyll pigments, cyanobacteria

FIGURE 11.6 Cyanobacterial Bloom Excessive growth of cyanobacteria in an aquatic environment leads to buoyant masses of cells rising to the surface. Wayne Carmichael (Wright State University), Mark Schneegurt (Wichita State University), and Cyanosite (www-cyanosite.bio.purdue.edu)

What allows cyanobacterial cells to rise to the surface?

have phycobiliproteins. These pigments absorb energy from wavelengths of light not well absorbed by chlorophyll and contribute to the blue-green, or sometimes reddish, color of the bacteria.

Nitrogen-Fixing Cyanobacteria

Nitrogen-fixing cyanobacteria are very important ecologically. They can incorporate both N_2 and CO_2 into organic material, so they generate a form of these nutrients that can then be used by other organisms. Thus, their activities can ultimately support the growth of a wide range of organisms in environments that would otherwise lack usable nitrogen and carbon. Also, like all cyanobacteria, they help limit atmospheric CO_2 buildup by using the gas as a carbon source.

Nitrogenase, the enzyme complex that catalyzes nitrogen fixation, is irreversibly inactivated by O_2; therefore, nitrogen-fixing cyanobacteria must protect the enzyme from the O_2 they generate. *Synechococcus* species fix nitrogen only in the dark. Consequently, nitrogen fixation and photosynthesis occur at different times of the day. *Anabaena* species, which are filamentous, isolate nitrogenase by restricting the process of nitrogen fixation to specialized thick-walled cells called heterocysts that appear to either not generate O_2 or quickly remove it; the heterocysts form at relatively regular intervals within the filaments, reflecting the ability of cells within a trichome to communicate (**figure 11.7**). One species of *Anabaena, A. azollae,* forms an intimate relationship with the water fern *Azolla*; the bacterium grows and fixes nitrogen within the protected environment of a special sac in the fern, providing *Azolla* with a source of available nitrogen.

Other Notable Characteristics of Cyanobacteria

Cyanobacteria have various other notable characteristics—some beneficial, others damaging. Filamentous cyanobacteria are responsible for maintaining the structure and productivity of soils in cold desert areas such as the Colorado Plateau. Their sheaths persist in soil, creating a sticky fibrous network that prevents erosion. In addition, these bacteria provide an important source of nitrogen and organic carbon in otherwise nutrient-poor soils. On the negative side, some cyanobacteria produce geosmin, a chemical that has a distinctive "earthy" odor and makes drinking water taste odd. Some aquatic species, such as *Microcystis aeruginosa,* produce toxins that can be deadly to humans and other animals.

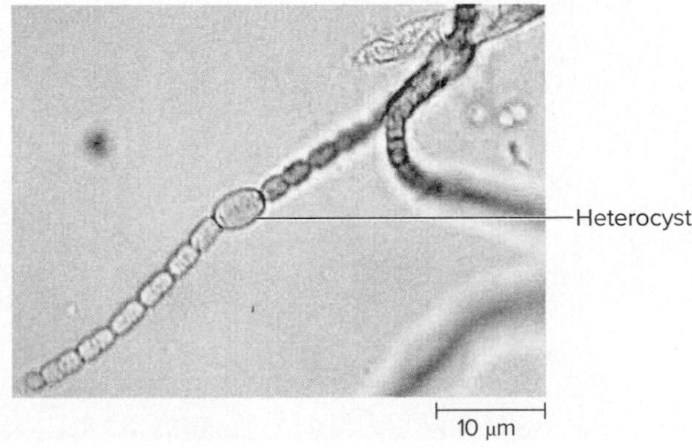

FIGURE 11.7 Heterocyst of an *Anabaena* Species Nitrogen fixation occurs within these specialized cells. Roger Burks (University of California at Riverside), Mark Schneegurt (Wichita State University), and Cyanosite (www-cyanosite.bio.purdue.edu)

? Why is it important for heterocysts to minimize O_2 production?

MicroAssessment 11.3

The photosystems of cyanobacteria generate O_2 and are similar to those of algae and plants. Many cyanobacteria can fix nitrogen.

7. What is the function of a heterocyst?
8. How do cyanobacteria prevent erosion in cold desert regions?
9. How could heavily fertilized lawns foster the development of cyanobacterial blooms?

11.4 ■ Aerobic Chemolithotrophs

Learning Outcome

4. Compare and contrast the characteristics of sulfur-oxidizing bacteria, nitrifiers, and hydrogen-oxidizing bacteria.

Aerobic chemolithotrophs obtain energy by oxidizing reduced inorganic chemicals, using O_2 as a terminal electron acceptor.

Sulfur-Oxidizing Bacteria

The **sulfur-oxidizing bacteria** are Gram-negative rods or spirals that sometimes grow as filaments. They obtain energy by oxidizing elemental sulfur and reduced sulfur compounds, including hydrogen sulfide and thiosulfate. O_2 serves as a terminal electron acceptor, generating sulfuric acid. An example of this reaction is:

$$\underset{\text{(energy source)}}{S} + \underset{\text{(terminal electron acceptor)}}{1\tfrac{1}{2}O_2} + H_2O \longrightarrow H_2SO_4$$

These bacteria are important in the sulfur cycle.

Filamentous Sulfur Oxidizers

Beggiatoa and *Thiothrix* species are filamentous sulfur oxidizers that live in sulfur springs, in sewage-polluted waters, and on the surfaces of marine and freshwater sediments. They store sulfur, depositing it as intracellular granules, but differ in certain characteristics of their multicellular filaments (**figure 11.8**). The filaments of *Beggiatoa* species move by gliding motility, a mechanism that does not require flagella.

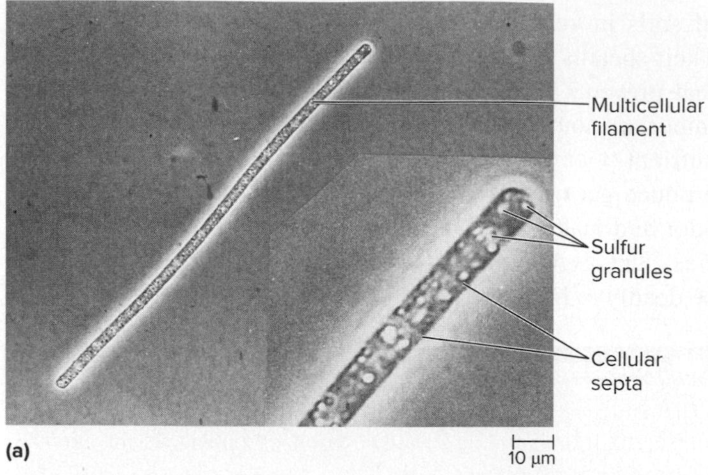

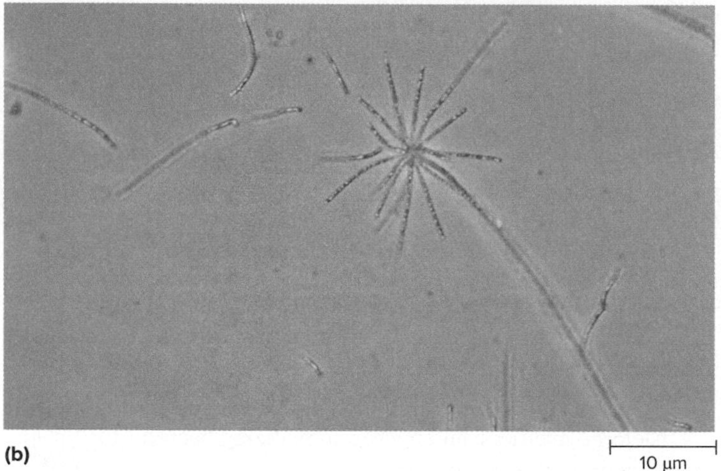

FIGURE 11.8 Filamentous Sulfur-Oxidizing Bacteria Phase-contrast photomicrographs. **(a)** Multicellular filament of a *Beggiatoa* species; the septa separate the cells. **(b)** Multicellular filaments of a *Thiothrix* species, forming a rosette arrangement. James T. Staley

? What is the role of sulfur in the metabolism of sulfur-oxidizing bacteria?

In contrast, the filaments of *Thiothrix* species are immobile; they fasten at one end to rocks or other solid surfaces, and often they attach to other cells, forming characteristic rosette arrangements of filaments. Progeny cells detach from the ends of these immobile filaments and use gliding motility to move to new locations, where they form additional filaments.

Overgrowth of *Beggiatoa* and *Thiothrix* in wastewater at treatment facilities causes a problem called bulking. Because the masses of filamentous organisms do not settle easily, bulking interferes with the process that separates the solid and liquid portions of the waste.

Unicellular Sulfur Oxidizers

Acidithiobacillus species are found in both terrestrial and aquatic habitats, where their ability to oxidize metal sulfides can be used to recover metals (see Focus Your Perspective 6.1). The bacteria oxidize insoluble metal sulfides such as gold sulfide, producing sulfuric acid. This lowers the pH, which converts the metal to a soluble form. *Acidithiobacillus* species can also be used to remove sulfur from coals and oils—an important step in preventing the acid rain that can result from burning those fuels. To do this, the bacteria are allowed to oxidize the sulfur to sulfate, a form that can then be extracted.

Acidithiobacillus species can also cause severe environmental problems. The strip mining of coal exposes metal sulfides, which the bacteria can then oxidize to produce sulfuric acid; some species produce enough acid to lower the pH to 1.0. The resulting runoff can acidify nearby streams, killing trees, fish, and other wildlife (**figure 11.9**). The runoff may also contain toxic metals made soluble by the bacteria.

Nitrifiers

Nitrifiers are a diverse group of Gram-negative bacteria that obtain energy by oxidizing inorganic nitrogen compounds such as ammonium or nitrite. These bacteria are a concern to farmers who fertilize their crops with compounds containing ammonium, a form of nitrogen retained by soils because its positive charge causes it to adhere to negatively charged soil particles. The potency and longevity of the fertilizer are affected by nitrifying bacteria converting the ammonium to nitrate. Although plants use the nitrate more easily, it is rapidly washed out of soils.

Nitrifying bacteria are also an important consideration when disposing of wastes that have a high ammonium concentration. As nitrifying bacteria oxidize nitrogen compounds, they use O_2, so waters polluted with nitrogen-containing wastes can quickly become hypoxic (low in dissolved O_2).

FIGURE 11.9 Acid Drainage from a Mine Sulfur-oxidizing bacteria oxidize exposed metal sulfides, generating sulfuric acid. The yellow-red color is due to insoluble iron oxides. Ronald Karpilo/Alamy Stock Photo

? How can the metabolic activities that result in this acid drainage be used in a commercially valuable manner?

The nitrifiers include two metabolically distinct groups that typically grow in close association. Together, they can oxidize ammonium to form nitrate. The ammonia oxidizers, which include the genera *Nitrosomonas* and *Nitrosococcus,* convert ammonium to nitrite in the following reaction:

$$\underset{\text{(energy source)}}{NH_4^+} + \underset{\text{(terminal electron acceptor)}}{1\tfrac{1}{2}O_2} \longrightarrow NO_2^- + H_2O + 2\,H^+$$

The nitrite oxidizers, which include the genera *Nitrobacter* and *Nitrococcus,* then convert nitrite to nitrate as follows:

$$\underset{\text{(energy source)}}{NO_2^-} + \underset{\text{(terminal electron acceptor)}}{\tfrac{1}{2}O_2} \longrightarrow NO_3^-$$

The latter group is particularly important in preventing the buildup of nitrite in soils, which is toxic and can contaminate groundwater. The oxidation of ammonium to nitrate (nitrification) is an important part of the nitrogen cycle.

Hydrogen-Oxidizing Bacteria

Members of the Gram-negative genera *Aquifex* and *Hydrogenobacter* are among the few hydrogen-oxidizing bacteria that are obligate chemolithotrophs. An example of the reaction in their metabolism is:

$$\underset{\text{(energy source)}}{H_2} + \underset{\text{(terminal electron acceptor)}}{\tfrac{1}{2}O_2} \longrightarrow H_2O$$

These related organisms are thermophilic and typically live in hot springs. Some *Aquifex* species have a maximum growth temperature of 95°C, the highest of any bacteria. Based on 16S rRNA studies, hydrogen-oxidizing bacteria were one of the earliest bacterial forms to exist on Earth. The fact that they require O_2 seems contradictory to their evolutionary position, but the low amount they require might have been available early on in certain niches due to photochemical processes that split water.

MicroAssessment 11.4

Sulfur oxidizers use sulfur compounds as energy sources, generating sulfuric acid. The nitrifiers oxidize nitrogen compounds such as ammonium or nitrite. Hydrogen-oxidizing bacteria oxidize H_2.

10. Describe two beneficial roles of sulfur oxidizers.

11. Why would farmers be concerned about nitrifying bacteria?

12. Why would sulfur-oxidizing bacteria store sulfur?

11.5 ■ Aerobic Chemoorganotrophs

Learning Outcomes

5. Describe the representative genera of obligate aerobes and facultative anaerobes.
6. Describe the family Enterobacteriaceae, and explain what distinguishes coliforms from other members of this family.

Aerobic chemoorganotrophs oxidize organic compounds to obtain energy, using O_2 as a terminal electron acceptor:

$$\underset{\text{(energy source)}}{\text{organic compounds}} + \underset{\text{(terminal electron acceptor)}}{O_2} \longrightarrow CO_2 + H_2O$$

They include a wide assortment of bacteria, ranging from some that inhabit very specific environments to others that are ubiquitous. This section will profile only representative genera found in a variety of different environments. Later sections will describe examples that thrive in specific habitats.

Obligate Aerobes

Obligate aerobes obtain energy using respiration exclusively; none of them can ferment.

The Genus *Micrococcus*

Members of the genus *Micrococcus* are Gram-positive cocci. The number of species assigned to the genus has decreased substantially due to reclassification, but as a group the remaining members have been isolated from a range of habitats, including a wastewater treatment plant and a research station in Antarctica. The species most commonly encountered in everyday life is *M. luteus,* which can be found in soil and on dust particles, various foods, and mammalian skin. Because that species is so common and the cells can easily become airborne, it is often encountered as a contaminant on bacteriological media, where it typically forms distinct yellow colonies ("luteus" means *yellow*) (**figure 11.10**). The bacterium tolerates dry conditions and can grow in salty environments such as 7.5% NaCl.

The Genus *Mycobacterium*

Mycobacterium species are widespread in nature and include harmless saprotrophs, which use nutrients from dead and decaying organic matter, as well as pathogens. They stain poorly because of mycolic acids (hydrophobic long-chain fatty acids) in their unusual lipid-rich cell wall, but special procedures can be used to increase the penetration of certain dyes. Once stained, the cells resist destaining, even with acidic decolorizing solutions. Because of this, *Mycobacterium*

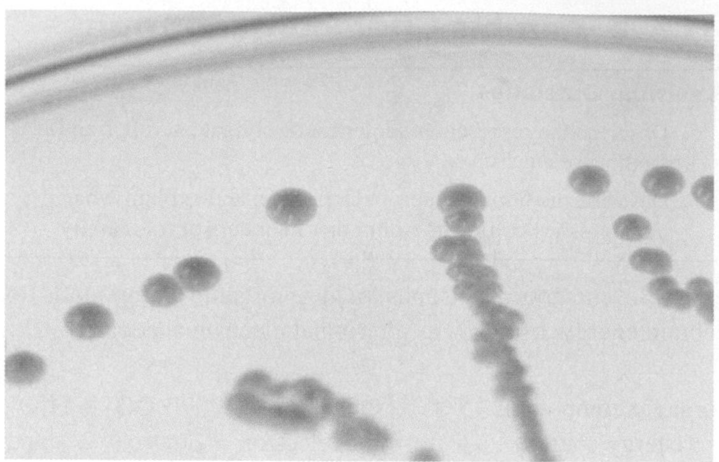

FIGURE 11.10 ***Micrococcus luteus* Colonies** Lisa Burgess/McGraw Hill

? Why is *Micrococcus luteus* a common contaminant on bacteriological media?

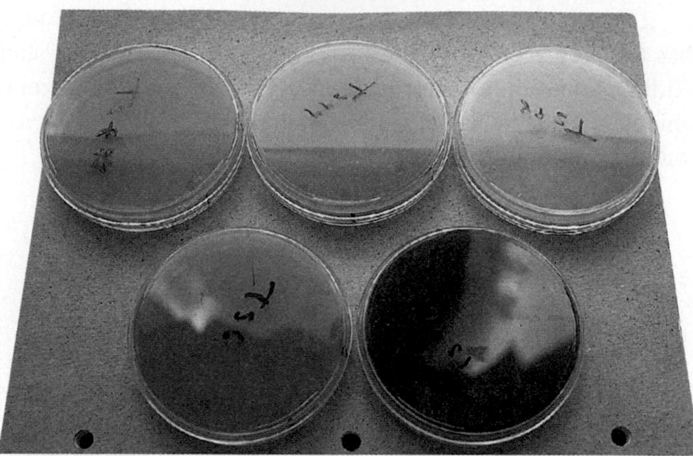

FIGURE 11.11 Pigments of *Pseudomonas* Species Cultures of different strains of *Pseudomonas aeruginosa*. Note the different colors of the water-soluble pigments. Evans Roberts

? The oxidase test detects the activity of a component used in which metabolic process?

species are called **acid-fast,** and the acid-fast staining procedure is an important step in identifying them (see figure 3.49). *Nocardia* species, a related group of bacteria common in soil, are also acid-fast.

Mycobacterium species are generally pleomorphic rods; they often occur in chains that sometimes branch, or bunch together to form cord-like groups. Several species are notable for their effect on human health, including *M. tuberculosis,* which causes tuberculosis, and *M. leprae,* which causes Hansen's disease (leprosy). *Mycobacterium* species are more resistant to disinfectants than most other vegetative bacteria. In addition, they are resistant to many of the most common antimicrobial medications.

The Genus *Pseudomonas*

Pseudomonas species are Gram-negative rods that typically have polar monotrichous flagella and often produce pigments (**figure 11.11**). Although most are strict aerobes, some can grow anaerobically if nitrate is available as a terminal electron acceptor. They are oxidase positive and do not ferment, characteristics that help distinguish them from members of the family Enterobacteriaceae, including *E. coli.*

As a group, *Pseudomonas* species have extremely diverse biochemical capabilities. Some can metabolize more than 80 different substrates, including unusual sugars, amino acids, and compounds containing aromatic rings. Because of this, *Pseudomonas* species play an important role in the degradation of many synthetic and natural compounds that resist breakdown by most other microorganisms. The ability to carry out some of these degradations is encoded on plasmids.

Pseudomonas species are widespread, typically inhabiting soil and water. Although most are harmless, some cause disease in plants and animals. Medically, the most significant species is *P. aeruginosa.* It is a common opportunistic pathogen, meaning that it primarily infects people who have underlying medical conditions. Unfortunately, it can grow in nutrient-poor environments, such as water used in mechanical ventilators that help patients breathe, and it is resistant to many disinfectants and antimicrobial medications. Because of this, hospitals must be very careful to prevent it from infecting patients.

The Genera *Thermus* and *Deinococcus*

Thermus and *Deinococcus* are related genera that have scientifically and commercially important characteristics. *Thermus* species are thermophilic, as their name implies, and this trait is valuable because of their heat-stable enzymes. The bacteria have an unusual cell wall and stain Gram-negative.

Deinococcus species' unusual cell wall has multiple layers, and they stain Gram-positive. They are unique in their extraordinary resistance to the damaging effects of gamma radiation. *D. radiodurans* can survive a radiation dose several thousand times higher than that which would be lethal to a human being. The dose literally shatters the organism's genome into many fragments, yet enzymes in the cells can repair the extensive damage. Scientists anticipate that through genetic engineering, *Deinococcus* species may eventually help clean up soil and water that have been contaminated by radioactive wastes.

MicroByte

The DNA polymerase of *T. aquaticus* (*Taq* polymerase) plays a key role in the polymerase chain reaction (PCR).

Facultative Anaerobes

Facultative anaerobes preferentially use aerobic respiration if O_2 is available. As an alternative, however, they can ferment.

The Genus *Corynebacterium*

Members of the genus *Corynebacterium* are widespread in nature. They are Gram-positive pleomorphic rods, often club-shaped (*koryne* is Greek for "club") and arranged to form V shapes or palisades (side-to-side stacks) (**figure 11.12**). Bacteria that exhibit this characteristic morphology are referred to as **coryneforms** or **diphtheroids.** *Corynebacterium* species are generally facultative anaerobes, but some are strict aerobes. Many *Corynebacterium* species live harmlessly in the throat, but toxin-producing strains of *C. diphtheriae* can cause the disease diphtheria.

The Family Enterobacteriaceae

Members of the family Enterobacteriaceae, often referred to as **enterics** or enterobacteria, are Gram-negative rods. Their name reflects the fact that most are found in the intestinal tract of humans and other animals (in Greek, *enteron* means "intestine"), although some thrive in rich soil. Enterics that are part of the normal intestinal microbiota include *Enterobacter, Klebsiella,* and *Proteus* species as well as most strains of *E. coli.* Enterics that cause diarrheal disease include *Shigella* species, *Salmonella enterica,* and some strains of *E. coli.* Life-threatening systemic diseases include typhoid fever, caused by *Salmonella enterica* serotype Typhi, and both the bubonic and pneumonic forms of plague, caused by *Yersinia pestis.*

Members of the Enterobacteriaceae are facultative anaerobes that ferment glucose and, if motile, generally have peritrichous flagella. The family includes over 40 recognized genera that can be distinguished using biochemical tests. Within a given species, many different strains have been described. These are often distinguished using serological tests that detect differences in cell walls, flagella, and capsules (see figure 10.11).

Enteric bacteria that characteristically ferment lactose are included in a group called **coliforms.** This is an informal grouping of certain common intestinal inhabitants such as *E. coli* that are easy to detect in food and water; for years, regulatory agencies have considered them to be an indicator of fecal pollution. Their presence indicates a possible health risk because fecal-borne pathogens might also be present.

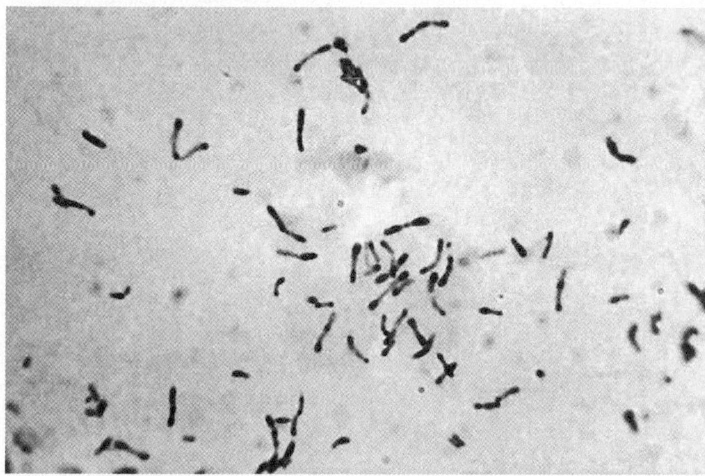

FIGURE 11.12 ***Corynebacterium*** The Gram-positive pleomorphic rods are often arranged to form V shapes or palisades. Dr. P.B. Smith/CDC

? **How does the name *Corynebacterium* reflect the shape of the bacterial cells?**

The Genus *Vibrio*

Members of the genus *Vibrio* are typically found in marine environments because most species require at least low levels of Na^+ for growth. They are Gram-negative straight or slightly curved rods and are facultative anaerobes. Pathogens include *V. cholerae,* which causes the severe gastrointestinal disease cholera, *V. parahaemolyticus,* which causes a milder gastrointestinal disease, and *V. vulnificus,* which can cause a systemic illness, particularly in patients who have an underlying illness such as liver disease. Some *Vibrio* species are bioluminescent, and these are described in section 11.7.

Table 11.1 summarizes characteristics of the archaea and bacteria used as examples in the discussion of metabolic diversity (many phylum names were recently changed, so in each case the previous name is included in parentheses).

MicroAssessment 11.5

Micrococcus, Mycobacterium, Pseudomonas, Thermus, and *Deinococcus* species are obligate aerobes that harvest energy by degrading organic compounds, using O_2 as a terminal electron acceptor. Most *Corynebacterium* species and all members of the family Enterobacteriaceae and the genus *Vibrio* are facultative anaerobes.

13. What unique characteristic makes members of the genus *Deinococcus* noteworthy?
14. What is the significance of finding coliforms in drinking water?
15. Why would it be an advantage for a *Pseudomonas* species to encode enzymes for degrading certain compounds on a plasmid rather than on the chromosome?

TABLE 11.1 Metabolic Diversity of Prokaryotes

Group/Genera	Characteristics	Phylum (Previous Name)
Anaerobic Chemolithotrophs		
Methanogens—*Methanospirillum, Methanosarcina*	Archaea that oxidize hydrogen gas, using CO_2 as a terminal electron acceptor to generate methane.	Euryarchaeota
Anaerobic Chemoorganotrophs—Anaerobic Respiration		
Sulfur- and sulfate-reducing bacteria—*Desulfovibrio*	Use sulfate as a terminal electron acceptor, generating hydrogen sulfide. Found in anaerobic muds rich in organic material. Gram-negative.	Pseudomonadota (Proteobacteria)
Anaerobic Chemoorganotrophs—Fermentation		
Clostridium and *Clostridioides*	Endospore-forming obligate anaerobes. Inhabitants of soil. Gram-positive.	Bacillota (Firmicutes)
Lactic acid bacteria—*Streptococcus, Enterococcus, Lactococcus, Lactobacillus, Leuconostoc*	Produce lactic acid as the major end product of their fermentative metabolism. Aerotolerant anaerobes. Several genera are used by the food industry. Gram-positive.	Bacillota (Firmicutes)
Propionibacterium	Obligate anaerobes that produce propionic acid as their main fermentation end product. Used in the production of Swiss cheese. Gram-positive.	Actinomycetota (Actinobacteria)
Anoxygenic Phototrophs		
Purple sulfur bacteria—*Chromatium, Thiospirillum, Thiodictyon*	Form colored masses in sulfur-rich aquatic habitats and use sulfur compounds as a source of electrons when making reducing power. Gram-negative.	Pseudomonadota (Proteobacteria)
Purple non-sulfur bacteria—*Rhodobacter, Rhodopseudomonas*	Grow in aquatic habitats, preferentially using organic compounds as a source of electrons for reducing power. Many are metabolically versatile. Gram-negative.	Pseudomonadota (Proteobacteria)
Green bacteria—*Chlorobium, Pelodictyon*	Some grow in habitats similar to those preferred by the purple sulfur bacteria. Gram-negative.	Chlorobiota (Chlorobi)
Filamentous anoxygenic phototrophic bacteria—*Chloroflexus*	Characterized by their filamentous growth. Gram-negative.	Chloroflexota (Chloroflexi)
Others—*Heliobacterium*	Have not been studied extensively.	Bacillota (Firmicutes)
Oxygenic Phototrophs—Cyanobacteria		
Anabaena, Synechococcus	Important primary producers. Some fix N_2. Gram-negative.	Cyanobacteria
Aerobic Chemolithotrophs		
Filamentous sulfur oxidizers—*Beggiatoa, Thiothrix*	Oxidize sulfur compounds as energy sources. Found in sulfur springs and sewage-polluted waters. Gram-negative.	Pseudomonadota (Proteobacteria)
Unicellular sulfur oxidizers—*Thiobacillus, Acidithiobacillus*	Oxidize sulfur compounds as energy sources. Some species produce enough acid to lower the pH to 1.0. Gram-negative.	Pseudomonadota (Proteobacteria)
Nitrifiers—*Nitrosomonas, Nitrosococcus, Nitrobacter, Nitrococcus*	Oxidize ammonia or nitrite as energy sources. This converts certain fertilizers to a form easily leached from soils, and it depletes O_2 in waters polluted with ammonia-containing wastes. Genera that oxidize nitrite prevent the toxic buildup of nitrite. Gram-negative.	Pseudomonadota (Proteobacteria)
Hydrogen-oxidizing bacteria—*Aquifex, Hydrogenobacter*	Thermophilic bacteria that oxidize hydrogen gas as an energy source. One of the earliest bacterial forms to exist on earth.	Aquificota (Aquificae)
Aerobic Chemoorganotrophs—Obligate Aerobes		
Micrococcus	Widely distributed; common laboratory contaminants. Gram-positive.	Actinomycetota (Actinobacteria)
Mycobacterium	Fatty-acid rich hydrophobic cell wall resists staining; acid-fast.	Actinomycetota (Actinobacteria)
Pseudomonas	Common environmental bacteria that, as a group, can degrade a wide variety of compounds. Gram-negative.	Pseudomonadota (Proteobacteria)
Thermus	*Thermus aquaticus* is the source of *Taq* polymerase (used in PCR). Stains Gram-negative.	Deinococcota (Deinococcus-Thermus)
Deinococcus	Resistant to the damaging effects of gamma radiation. Stains Gram-positive.	Deinococcota (Deinococcus-Thermus)
Aerobic Chemoorganotrophs—Facultative Anaerobes		
Corynebacterium	Widespread in nature. Gram-positive.	Actinomycetota (Actinobacteria)
The Enterobacteriaceae—*Escherichia, Enterobacter, Klebsiella, Proteus, Salmonella, Shigella, Yersinia*	Most reside in the intestinal tract. Those that ferment lactose are coliforms; their presence in water serves as an indicator of fecal pollution. Gram-negative.	Pseudomonadota (Proteobacteria)
Vibrio	Typically found in marine environments because most species require at least low levels of Na^+ for growth. Gram-negative.	Pseudomonadota (Proteobacteria)

ECOPHYSIOLOGICAL DIVERSITY

As a group, prokaryotes show remarkable diversity in their physiological adaptations to a wide range of habitats. Bacteria and archaea have evolved to live in virtually all environments —from the hydrothermal vents of deep oceans to the frozen expanses of Antarctica. This section will highlight the physiological mechanisms bacteria and archaea use to live in terrestrial and aquatic environments; the study of these adaptations is called **ecophysiology.** The section will also describe some examples of bacteria that use animals as habitats.

11.6 ■ Thriving in Terrestrial Environments

Learning Outcomes

7. Describe the bacterial groups that form resting stages.
8. Compare and contrast *Agrobacterium* species and rhizobia.

Microorganisms that live in soil must tolerate a variety of conditions. Daily and seasonally, soil can routinely alternate between wet and dry as well as warm and cold. Nutrient availability can also cycle from plentiful to scarce. To thrive in this ever-changing environment, microbes have evolved mechanisms to cope with unfavorable conditions and to use plants as sources of nutrients. A summary table at the end of section 11.7 includes the bacterial examples used here (see table 11.2).

Bacteria That Form a Resting Stage

Several genera that live in soil can form a resting stage that allows them to survive the dry periods that occur in many soils. Of the various types of dormant cells, endospores are by far the most resistant to environmental extremes.

Endospore Formers

Bacillus species and clostridia are the most common Gram-positive rod-shaped bacteria that form endospores; the position of the spore in the cell can help in identification (**figure 11.13**).

Clostridia, which are obligate anaerobes, were discussed earlier. *Bacillus* species include both obligate aerobes and facultative anaerobes, and some are medically important. *B. anthracis* causes the disease anthrax, which can be acquired from contacting its endospores in soil or in animal hides or wool. Unfortunately, the spores have also been used as an agent of domestic bioterrorism.

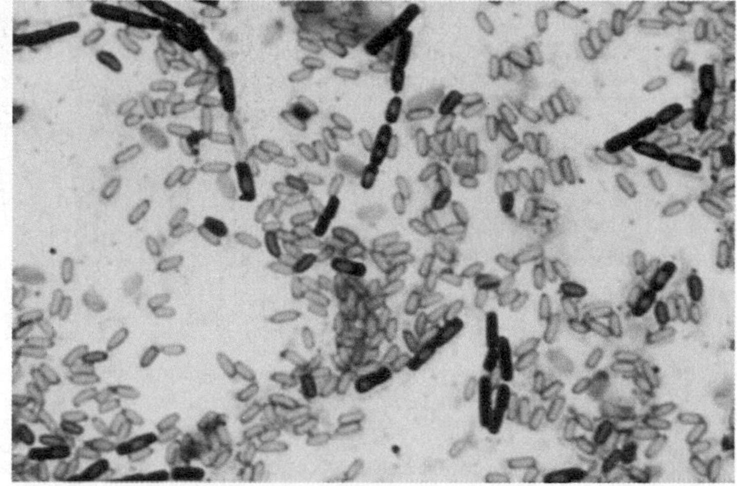

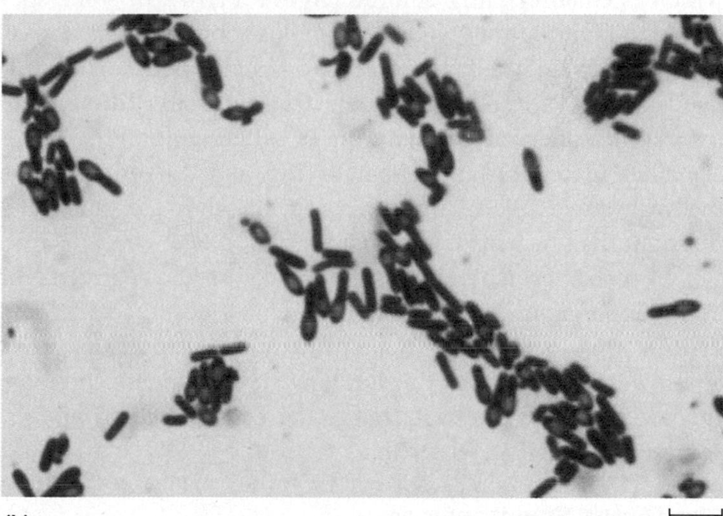

FIGURE 11.13 Endospore Formers (a) Endospores forming in the mid-portion of cells *(Bacillus megaterium).* **(b)** Endospores forming at the ends of cells *(Clostridium tetani).* Lisa Burgess/McGraw Hill

? Members of which genus are endospore-forming obligate anaerobes?

The Genus *Azotobacter*

Azotobacter species are Gram-negative pleomorphic, rod-shaped bacteria that live in soil. They can form a type of resting cell called a cyst that can withstand drying and ultraviolet radiation but not high heat.

Azotobacter species are also notable for their ability to fix nitrogen in aerobic conditions; recall that the enzyme nitrogenase is inactivated by O_2. Apparently, the extremely high respiratory rate of *Azotobacter* species consumes O_2 so rapidly that a low-O_2 environment is maintained inside the cell. In addition, a protein in the cell binds nitrogenase in a way that protects the enzyme from O_2 damage.

Myxobacteria

The **myxobacteria** are a group of aerobic Gram-negative rods that have a unique developmental cycle as well as a resting stage. When conditions are favorable, cells secrete a slime layer that other cells then follow, creating a swarm of cells. But then, when nutrient levels are low, the behavior of the

FIGURE 11.14 Myxobacterial Fruiting Body
Mary F. Lampe

How do fruiting bodies help myxobacteria survive unfavorable conditions?

group changes. The cells begin to come together and then pile up to form a multicellular structure called a **fruiting body,** which is often brightly colored (**figure 11.14**). In some species, the fruiting body is quite complex, consisting of a mass of cells elevated and supported by a stalk made of a hardened slime. The cells within the fruiting body differentiate to become spherical, dormant forms called microcysts. These are much more resistant to heat, drying, and radiation than are the vegetative cells of myxobacteria, but they are much less resistant than bacterial endospores.

Myxobacteria are important in nature as degraders of complex organic substances; they can digest bacteria and certain algae and fungi. Scientifically, these bacteria serve as an important model for studying developmental biology. Included in the myxobacteria are the genera *Chondromyces, Myxococcus,* and *Stigmatella.*

The Genus *Streptomyces*

Members of the genus *Streptomyces* are aerobic Gram-positive bacteria that resemble fungi in their pattern of growth. Like the fungi, they form a mycelium—a mass of interwoven branching filaments called hyphae, some of which grow up into the air (aerial hyphae). Growth of the mycelium gives rise to the characteristic dry, chalky appearance of *Streptomyces* colonies (**figure 11.15a**). At the tips of the aerial hyphae, chains of dormant cells called conidia develop (**figure 11.15b**). Conidia are resistant to drying and are easily spread in air currents. Note that even though this pattern of growth resembles fungi, which are eukaryotes, *Streptomyces* species have much smaller cells and are prokaryotes.

Streptomyces species produce a variety of extracellular enzymes that allow them to degrade a wide range of organic compounds. They also produce an array of medically useful antibiotics, including streptomycin, tetracycline, and erythromycin, and these give the organisms a competitive advantage in the natural microbial communities in soil. In addition, *Streptomyces* species are responsible for the characteristic "earthy" odor of soil because, like the cyanobacteria, they produce geosmin. One *Streptomyces* species, *S. somaliensis,* can cause an infection of subcutaneous tissue called an actinomycetoma.

Bacteria That Associate with Plants

Members of two related genera use very different means to obtain nutrients from plants. *Agrobacterium* species are plant pathogens that cause tumor-like growths, whereas *Rhizobium* species form a mutually beneficial relationship with certain types of plants.

The Genus *Agrobacterium*

Agrobacterium species are Gram-negative rod-shaped bacteria that have an unusual mechanism of gaining a competitive advantage in soil. They cause plant tumors, the outcome of their ability to genetically alter plants for their own benefit (**figure 11.16**). They do this by attaching to wounded plant tissue, and then transferring a part of a plasmid to a plant cell; in *A. tumefaciens* the plasmid is called the **Ti plasmid**

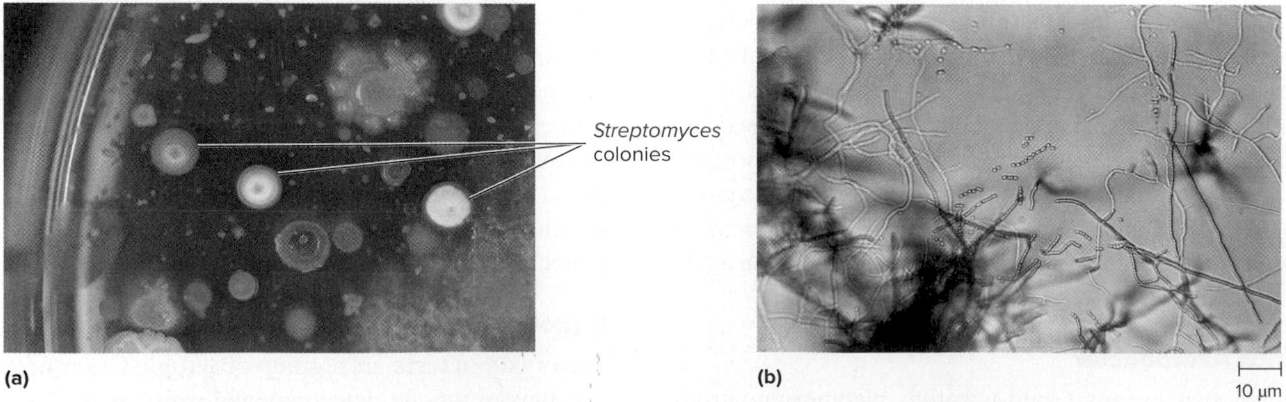

FIGURE 11.15 ***Streptomyces*** **(a)** Colonies, which typically have a chalky appearance and may be white or colored. **(b)** A photomicrograph showing the spherical conidia at the ends of the filamentous hyphae. a: Lisa Burgess/McGraw Hill; b: Dr. David Berd/CDC

How does conidia production benefit *Streptomyces* species?

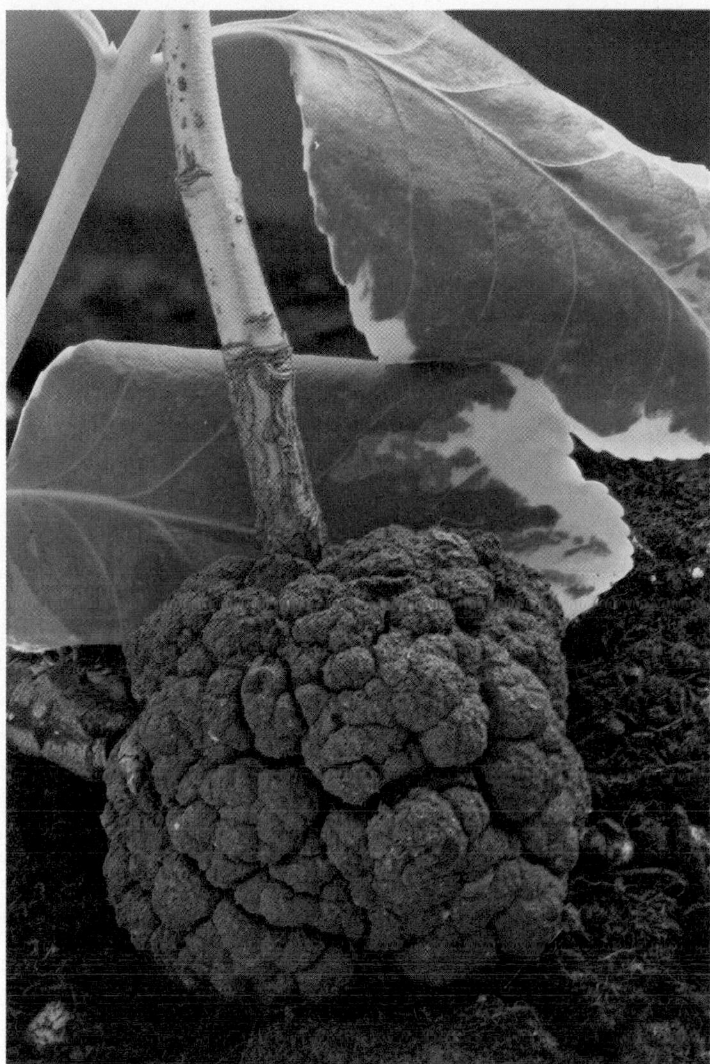

FIGURE 11.16 Plant Tumor Caused by *Agrobacterium tumefaciens* Custom Life Science Images/Alamy Stock Photo

? **What is the role of the Ti plasmid in plant tumor formation?**

(for "tumor-inducing"). The transferred DNA encodes the ability to synthesize plant growth hormones, causing uncontrolled growth of the plant tissue and resulting in a tumor. The transferred DNA also encodes enzymes that direct the synthesis of opines—a group of unusual amino acid derivatives that *A. tumefaciens* can use as a nutrient source (see Focus Your Perspective 8.2).

MicroByte

Scientists have modified the Ti plasmid, turning it into a commercially valuable tool used to genetically engineer plant cells.

Rhizobia

Rhizobia are a group of nitrogen-fixing Gram-negative rod-shaped bacteria that form intimate relationships with legumes (plants that bear seeds in pods), living within cells of root nodules (growths on plant roots) (**figure 11.17**). The plant synthesizes leghemoglobin, an O_2-binding protein that maintains low levels of O_2 in the nodules, and the bacteria fix nitrogen within the resulting microaerobic environment (see figure 28.14). Because the intracellular rhizobia benefit the plant (by providing a source of nitrogen), they are examples of endosymbionts.

MicroAssessment 11.6

Bacillus species and clostridia make endospores, the most resistant type of dormant cell known. *Azotobacter* species, myxobacteria, and *Streptomyces* species all produce dormant cells that tolerate some unfavorable conditions but are less resistant than endospores. *Agrobacterium* species and rhizobia obtain nutrients from plants, but the former are plant pathogens and the latter benefit the plant.

16. Why are myxobacteria important in nature?

17. How does *Agrobacterium* benefit from inducing a plant tumor?

18. If you wanted to determine the number of endospores in a sample of soil, what could you do before plating it?

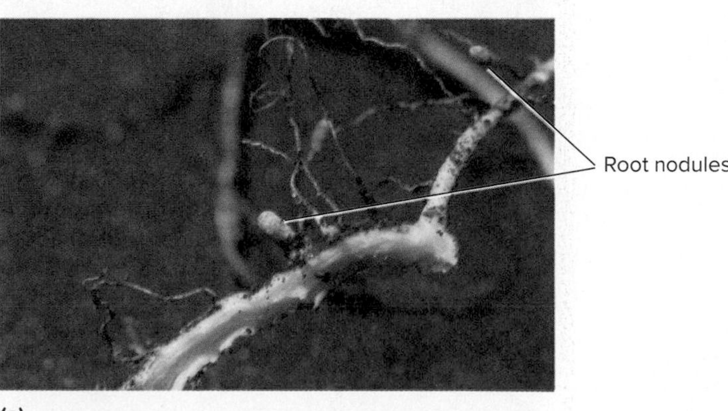

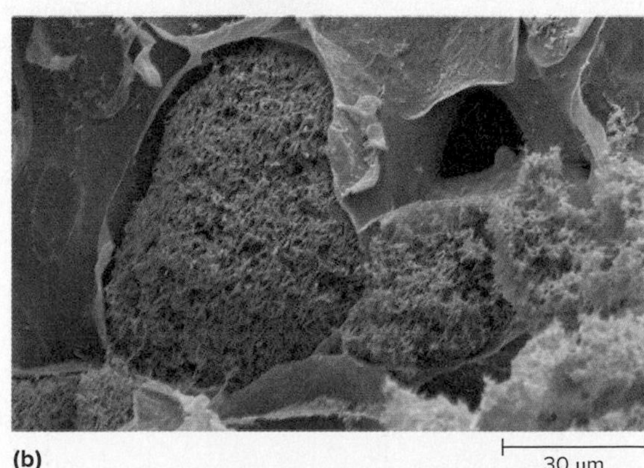

FIGURE 11.17 Symbiotic Relationship Between Rhizobia and Certain Plants (a) Root nodules. **(b)** Rhizobial cells within cells of a nodule (color-enhanced SEM). a: Lisa Burgess/McGraw Hill; b: Andrew Syred/Science Source

? **How do rhizobia benefit plants?**

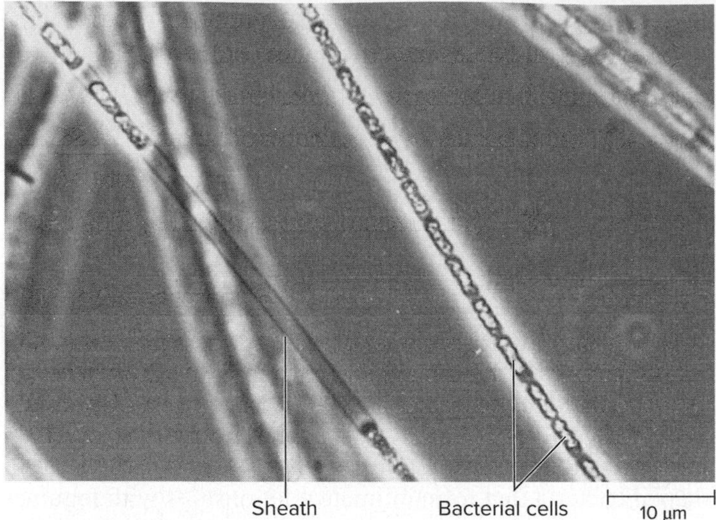

FIGURE 11.18 Sheathed Bacteria Phase-contrast photomicrograph of a *Sphaerotilus* species. James T. Staley

? In the case of sheathed bacteria, what is the role of the sheath?

11.7 ■ Thriving in Aquatic Environments

Learning Outcome

9. Describe examples of how bacteria that thrive in aquatic environments obtain and store nutrients.

Most aquatic environments lack a steady supply of nutrients. The bacteria that live in these habitats have evolved various mechanisms for obtaining and storing nutrients more efficiently.

Sheathed Bacteria

Sheathed bacteria form chains of cells encased within a tube, or sheath (**figure 11.18**). The sheath plays a protective role, helping the bacteria attach to solid objects in favorable habitats while sheltering them from attack by predators. Masses of filamentous sheaths can often be seen streaming from rocks in flowing water polluted by nutrient-rich wastes. They often interfere with sewage treatment and other industrial processes by clogging pipes. Sheathed bacteria include species of *Sphaerotilus* and *Leptothrix,* which are Gram-negative rods.

Sheathed bacteria spread by forming motile cells called swarmer cells that exit through the unattached end of the sheath. These then move to a new solid surface, where they attach. If enough nutrients are present, they can multiply and form a new sheath, which gets longer as the chain of cells grows.

Prosthecate Bacteria

The prosthecate bacteria are a diverse group of Gram-negative bacteria that have projections called prosthecae, which are extensions of the cytoplasm and cell wall. These extensions provide increased surface area to facilitate the absorption of nutrients. Some prosthecae allow the organisms to attach to solid surfaces.

The Genus *Caulobacter*

Because of their remarkable life cycle, *Caulobacter* species serve as a model for research on cellular differentiation. Entirely different events occur in an orderly fashion at opposite ends of the cell.

Caulobacter cells have a single polar prostheca, commonly called a stalk (**figure 11.19**). At the tip of the stalk is an adhesive holdfast, a structure that provides a mechanism for attachment. To multiply, the cell elongates and divides by binary fission, producing a motile swarmer cell at the end opposite to the stalk. This swarmer cell has a flagellum, located at the pole opposite the site of division. The swarmer cell detaches and

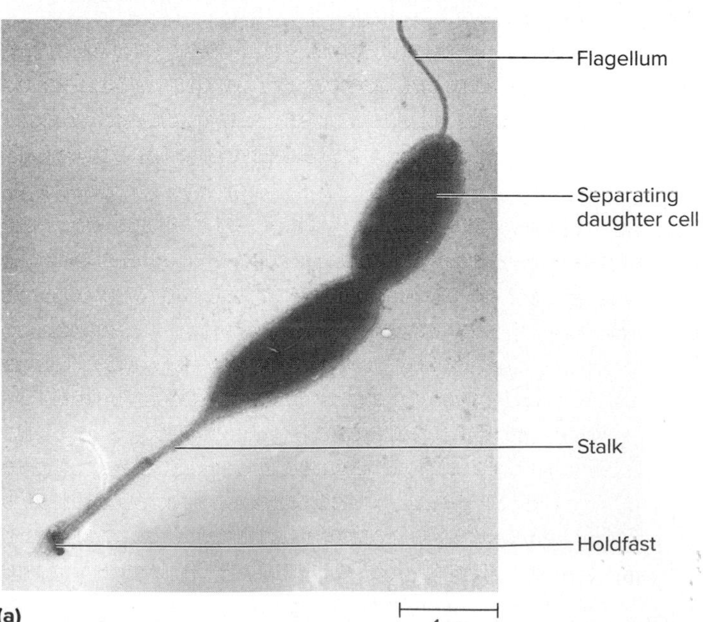

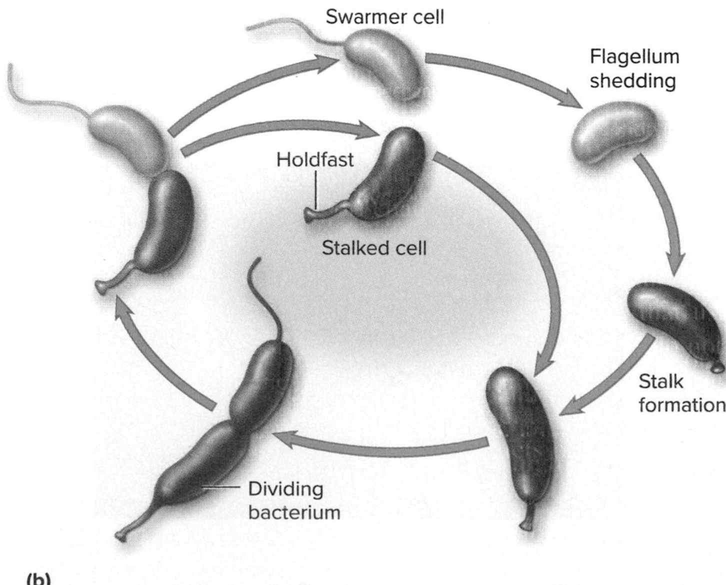

FIGURE 11.19 ***Caulobacter*** **(a)** Transmission electron micrograph. **(b)** Life cycle. a: James T. Staley

? What characteristic of *Caulobacter* species makes them important research models?

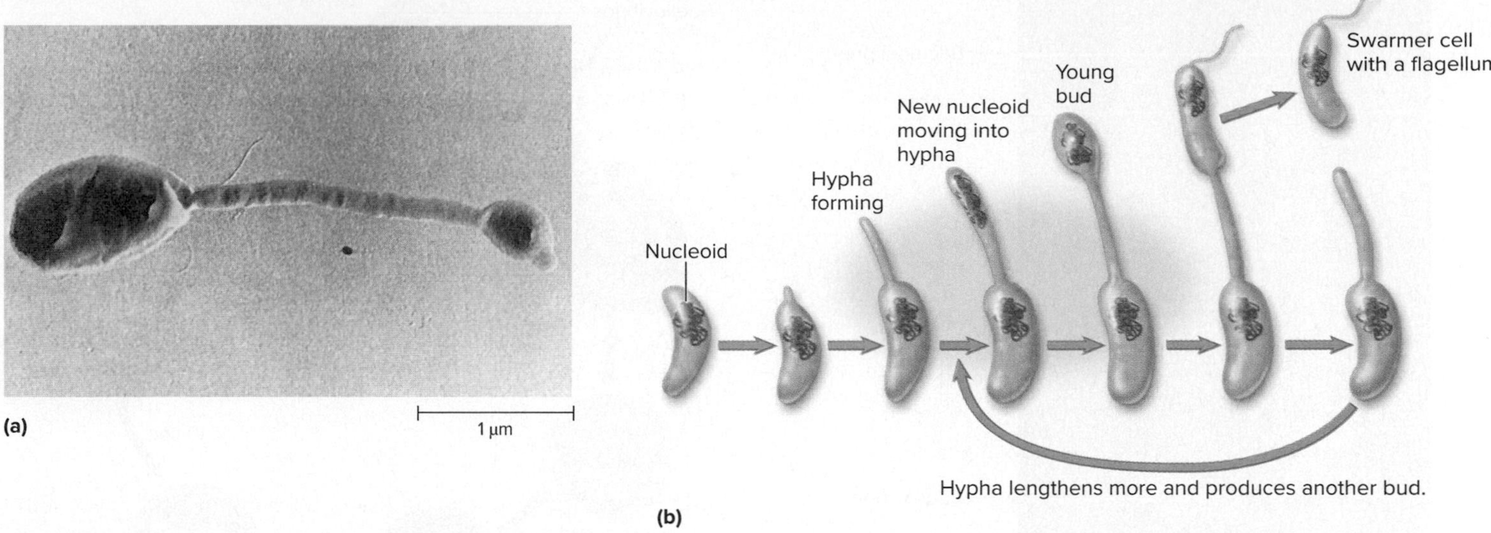

FIGURE 11.20 ***Hyphomicrobium*** **(a)** Electron micrograph. Note the bud forming at the tip of the polar prostheca (hypha). **(b)** Life cycle.
a: James T. Staley

What will happen to the swarmer cell?

moves to a new location, where it adheres via a holdfast near the base of its flagellum. It then loses its flagellum, replacing it with a stalk. Only then can the daughter cell replicate its DNA and repeat the process. In favorable conditions, a single cell divides and produces daughter cells many times.

The Genus *Hyphomicrobium*

Hyphomicrobium species are in many ways similar to *Caulobacter* species, except they have a distinct method of reproduction. The single polar prostheca (hypha) of the parent cell enlarges at the tip to form a bud (**figure 11.20**). This continues enlarging and also develops a flagellum, eventually giving rise to a motile daughter cell. The daughter cell (swarmer cell) then detaches and moves to a new location, eventually losing its flagellum and forming a polar prostheca at the opposite end to repeat the cycle. As with *Caulobacter* species, a single cell can repeatedly produce daughter cells.

Bacteria That Derive Nutrients from Other Organisms

Some bacteria obtain nutrients directly from other organisms. Examples include *Bdellovibrio* species, bioluminescent bacteria, *Epulopiscium* species, and *Legionella* species.

The Genus *Bdellovibrio*

Bdellovibrio species (*bdello,* from the Greek word for "leech") are highly motile Gram-negative curved rods that prey on *E. coli* and other Gram-negative bacteria (**figure 11.21**). When a *Bdellovibrio* cell attacks, the strike is so forceful that it propels the prey a short distance. The parasite then attaches to the prey and rotates with a spinning motion. At the same time, it makes digestive enzymes that break down lipids and peptidoglycan, eventually forming a hole in the prey's cell wall. This allows the parasitic bacterium to penetrate the peptidoglycan, lodging in the periplasm. There, over a period of several hours, the *Bdellovibrio* cell degrades and utilizes the prey's cellular contents. It derives energy by aerobically oxidizing amino acids and acetate. The parasite increases in length, ultimately dividing to form several motile daughter cells. When the host cell lyses, the *Bdellovibrio* progeny are released to find new hosts, repeating the cycle.

Bioluminescent Bacteria

Certain bacteria are **bioluminescent,** meaning they emit light (**figure 11.22**). This ability plays an important role in the mutually beneficial relationship between some of these bacteria and specific types of fish and squid. Certain types of squid have a specialized organ within their ink sac that is colonized by the bioluminescent bacterium *Aliivibrio (Vibrio) fischeri.* The light produced by the bacterial cells in the organ is thought to serve as a type of camouflage, masking the squid's contrast against the light from above and any shadow it might otherwise cast. Meanwhile, the squid provides nutrients to the bacteria in the organ. Another example is the flashlight fish, which carries bioluminescent bacteria in a light organ below its eye. By opening and closing a lid that covers the light organ, the fish can control the amount of light released, a tactic believed to confuse predators and prey.

Luminescence is catalyzed by the enzyme luciferase, which is expressed only when the density of the bacterial population is high. Studies of this cell-density-dependent system led to the discovery of quorum sensing, now recognized as an important mechanism by which a variety of different bacteria regulate the expression of certain genes.

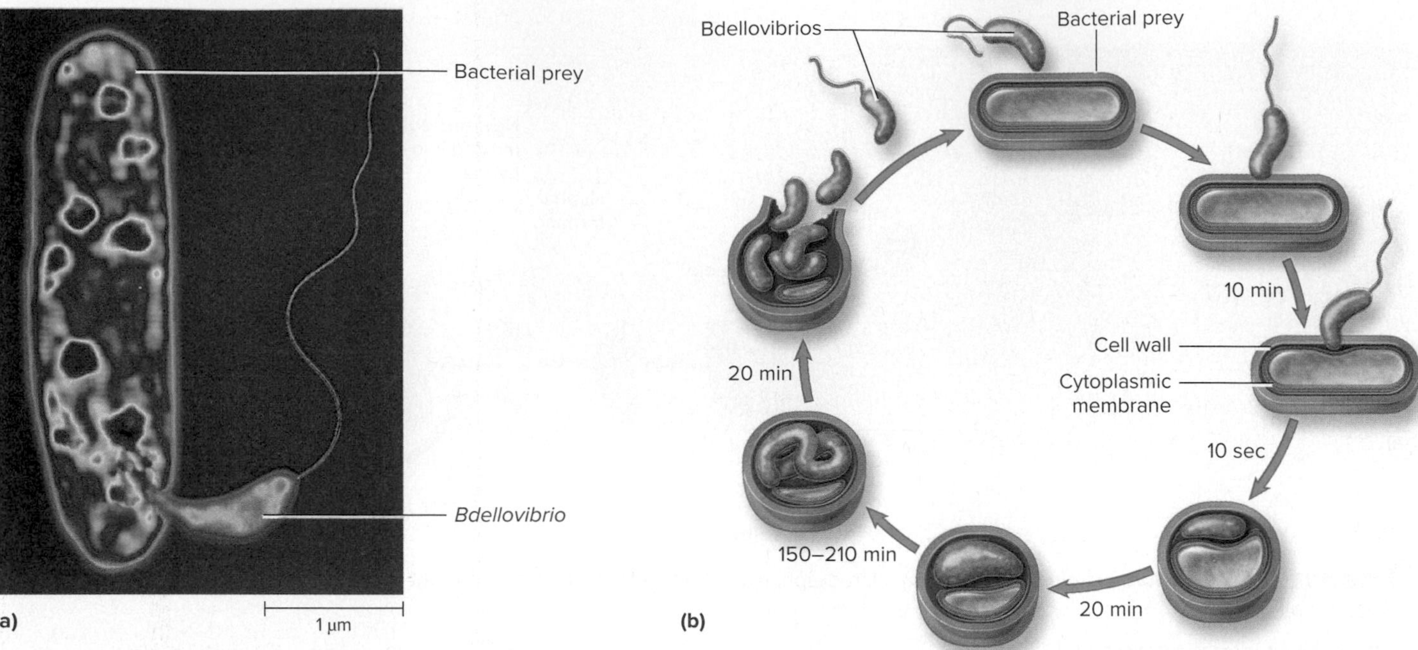

FIGURE 11.21 ***Bdellovibrio*** **(a)** Color-enhanced transmission micrograph of a *Bdellovibrio* cell attacking its prey. **(b)** Life cycle of *Bdellovibrio*. Note that the diagram exaggerates the size of the space in which *Bdellovibrio* multiplies. a: Alfred Pasieka/Photolibrary/Getty Images

How does a *Bdellovibrio* cell penetrate the prey?

In addition to *Aliivibrio fischeri,* other examples of bioluminescent bacteria include *Vibrio harveyi* and *Photobacterium phosphoreum.* All of these are Gram-negative, straight or curved rods and are facultative anaerobes. Bioluminescent bacteria typically inhabit marine environments.

The Genus *Legionella*

Legionella species are Gram-negative obligate aerobes found in a variety of aquatic environments, where they often live within protozoa; they have even been isolated from water in air conditioners and produce misters. The bacteria use amino acids, but not carbohydrates, as a source of carbon and energy. *Legionella pneumophila* can cause Legionnaires' disease, a type of pneumonia, when inhaled in aerosolized droplets.

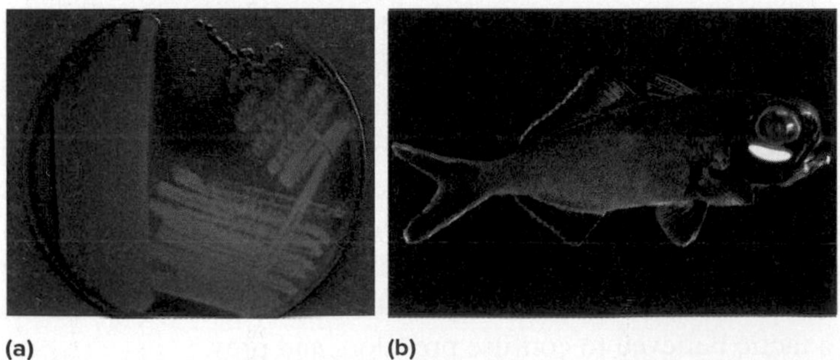

FIGURE 11.22 **Bioluminescent Bacteria (a)** Plate culture. **(b)** Flashlight fish; under the eye is a light organ colonized with bioluminescent bacteria. a: Amy Cheng Vollmer; b: Dantè Fenolio/Science Source

What benefit do bioluminescent bacteria in the ink sac of squid provide their host?

The Genus *Epulopiscium*

Epulopiscium species are Gram-positive cigar-shaped bacteria that live in the intestinal tract of surgeonfish. They are much larger than most prokaryotes (600 µm × 80 µm), and each cell has thousands of copies of the chromosome scattered throughout the cell. This means that the cell has thousands of copies of each gene, allowing the necessary proteins to be synthesized even in the far reaches of the large cell.

Epulopiscium species have an unusual life cycle. Rather than undergoing typical binary fission, they get very large, finally lysing to release up to seven daughter cells. They have not yet been grown in culture.

Bacteria That Move by Unusual Mechanisms

Some bacteria have unique mechanisms of motility that allow them to easily move to desirable locations. These organisms include the spirochetes and the magnetotactic bacteria.

Spirochetes

The **spirochetes** (Greek *spira* for "coil" and *chaete* for "hair") are a group of Gram-negative bacteria with a spiral shape and a unique motility mechanism that allows them to move through thick, viscous substances such as mud. Distinguishing characteristics include their spiral shape, flexible cell wall, and motility by means of endoflagella (also called axial filaments). Unlike typical flagella, endoflagella are contained within the periplasm. Either a single

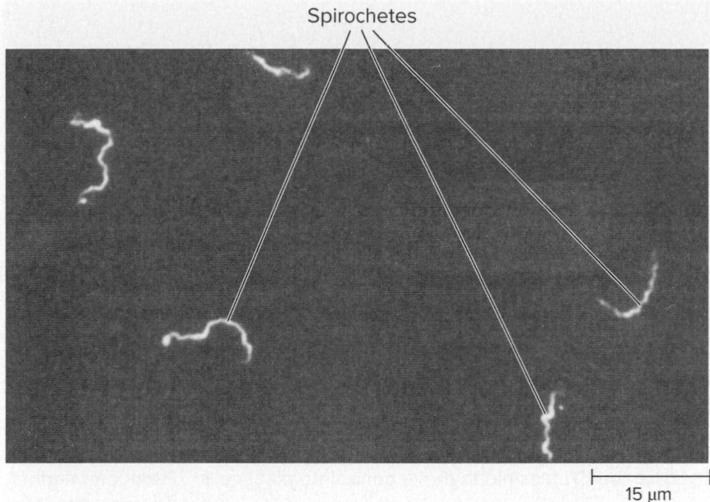

FIGURE 11.23 Spirochetes Dark-field photomicrograph of spirochetes. CDC/ C.W. Hubbard

? Why are many spirochetes difficult to see with bright-field microscopy?

endoflagellum or a tuft originates at each end of the cell, and the structures extend toward each other, overlapping in the mid-region of the cell. Rotation of the endoflagella within the limited area between the cytoplasmic and outer membranes causes the cell to move like a corkscrew, sometimes bending and twisting. Many spirochetes are difficult or impossible to grow in culture. They are often very slender and can be seen only by using special methods such as dark-field microscopy (**figure 11.23**).

Spirochetes include free-living species that inhabit aquatic environments, as well as ones that reside on or in animals. *Spirochaeta* species are anaerobes or facultative anaerobes that live in muds and anaerobic waters. *Leptospira* species are aerobic; some are free-living in aquatic environments, whereas others grow within animals. *L. interrogans* causes the disease leptospirosis, which can be transmitted in the urine of infected animals. Spirochetes adapted to live in body fluids of humans and other animals are discussed in the next section.

Magnetotactic Bacteria

Magnetotactic bacteria such as the Gram-negative, spiral-shaped *Magnetospirillum magnetotacticum* contain a string of magnetic crystals, held within a structure called a magnetosome, that align cells with the Earth's magnetic field (see figure 3.16). This allows the cells to move up or down in the water or sediments and is thought to allow the cells to locate the microaerophilic habitats they require. Scientists are exploring ways to use magnetosomes to help transport certain medications to specific sites within the human body; for example, by attaching magnetosomes to cancer-destroying medications, a magnetic field may then be used to guide the medications to specific tumors.

Bacteria That Form Storage Granules

A number of aquatic bacteria form granules that store nutrients. Recall that anoxygenic phototrophs often store sulfur granules, which can later be used as a source of electrons for reducing power. Some bacteria store phosphate, and others store compounds that can be used to generate ATP.

The Genus *Spirillum*

Spirillum species are large, Gram-negative, spiral-shaped, microaerophilic bacteria, often over 20 µm in length. In wet mounts, they may be seen moving to a narrow zone near the edge of the coverslip, where O_2 is available in the optimum amount. *Spirillum volutans* forms volutin granules, which are storage forms of phosphate; these are sometimes called metachromatic granules to reflect their characteristic staining with the dye methylene blue.

The Genera *Thioploca* and *Thiomargarita*

Thioploca and *Thiomargarita* species live in marine environments and store both sulfur (their energy source) and nitrate (their terminal electron acceptor). This allows the bacteria to compensate for the fact that their energy source and terminal electron acceptor are not usually found in the same location: Reduced sulfur compounds are often plentiful in anaerobic marine sediments, whereas nitrate is available in the waters above. By storing both sulfur and nitrate, the cells always have a supply of each.

Thioploca cells "commute" between the sulfur-rich sediments and the nitrate-rich waters, gathering and storing their energy source and terminal electron acceptor as needed. They do this by forming long sheaths within which the cells move between the two environments.

The cells of the huge bacterium *Thiomargarita namibiensis* ("sulfur pearl of Namibia") are a pearly white color due to globules of sulfur in their cytoplasm (see Focus Your Perspective 1.1). Each cell contains a large nitrate storage vacuole that takes up about 98% of the cell volume. These organisms, which can reach a diameter of 0.75 mm, are not motile and instead rely on storms or other disturbances to bring them into contact with nitrate-rich waters. An even larger member of the genus, with cells up to a centimeter long, was described in 2022. This bacterium, *T. magnificum,* is so large that it was originally thought to be a fungus.

Table 11.2 summarizes characteristics of the bacteria used as examples of diversity in terrestrial and aquatic environments (many phylum names were recently changed, so in each case the previous name is included in parentheses).

MicroByte

Thiomargarita namibiensis can store a 3-month supply of both its energy source (sulfur) and its terminal electron acceptor (nitrate).

TABLE 11.2 Ecophysiological Diversity

Group/Genera	Characteristics	Phylum (Previous Name)
Thriving in Terrestrial Environments		
Endospore formers—*Bacillus*, clostridia	*Bacillus* species include both obligate aerobes and facultative anaerobes; clostridia are obligate anaerobes. Gram-positive.	Bacillota (Firmicutes)
Azotobacter	Form cysts. Notable for their ability to fix nitrogen in aerobic conditions. Gram-negative.	Pseudomonadota (Proteobacteria)
Myxobacteria—*Chondromyces, Myxococcus, Stigmatella*	Cells come together to form a fruiting body, within which the cells differentiate to become dormant microcysts. Gram-negative.	Pseudomonadota (Proteobacteria)
Streptomyces	Resemble fungi in their pattern of growth; produce antibiotics. Gram-positive.	Actinomycetota (Actinobacteria)
Agrobacterium	Cause plant tumors. Scientists use their Ti plasmid to move genes into plant cells. Gram-negative.	Pseudomonadota (Proteobacteria)
Rhizobia—*Rhizobium, Sinorhizobium, Bradyrhizobium, Mesorhizobium, Azorhizobium*	Fix nitrogen; form a symbiotic relationship with legumes. Gram-negative.	Pseudomonadota (Proteobacteria)
Thriving in Aquatic Environments		
Sheathed bacteria—*Sphaerotilus, Leptothrix*	Form chains of cells enclosed within a protective sheath. Swarmer cells move to new locations. Gram-negative.	Pseudomonadota (Proteobacteria)
Prosthecate bacteria—*Caulobacter, Hyphomicrobium*	Appendages increase their surface area. Gram-negative.	Pseudomonadota (Proteobacteria)
Bdellovibrio	Predators of other bacteria. Gram-negative.	Pseudomonadota (Proteobacteria)
Bioluminescent bacteria—*Aliivibrio fischeri, Vibrio harveyi*, and *Photobacterium phosphoreum*	Some form symbiotic relationships with specific types of squid and fish. Gram-negative.	Pseudomonadota (Proteobacteria)
Legionella	Often reside within protozoa. Gram-negative.	Pseudomonadota (Proteobacteria)
Epulopiscium	Very large cigar-shaped bacteria that multiply by releasing several daughter cells; each cell has thousands of copies of the genome. Gram-positive.	Bacillota (Firmicutes)
Free-living spirochetes—*Spirochaeta, Leptospira* (some species)	Long, spiral-shaped bacteria that move by means of endoflagella. Gram-negative.	Spirochaetota (Spirochaetes)
Magnetospirillum	Magnetic crystals allow them to move in water and sediments. Gram-negative.	Pseudomonadota (Proteobacteria)
Spirillum	Spiral-shaped, microaerophilic bacteria. Gram-negative.	Pseudomonadota (Proteobacteria)
Thioploca, Thiomargarita	Use novel mechanisms to compensate for the fact that their energy source (reduced sulfur compounds) and terminal electron acceptor (nitrate) do not coexist.	Pseudomonadota (Proteobacteria)

MicroAssessment 11.7

Sheathed bacteria attach to solid objects in favorable locations. Prosthecate bacteria produce extensions that maximize the absorptive surface area. *Bdellovibrio* species, bioluminescent bacteria, and *Legionella* species use nutrients from other organisms. Spirochetes and magnetotactic bacteria move by unusual mechanisms. Some organisms form storage granules.

19. What role does quorum sensing play in the behavior of bioluminescent bacteria?
20. What is the habitat of *Legionella* species?
21. The genomes of free-living spirochetes are larger than those of ones that live within an animal host. Why would this be so?

11.8 ■ Animals as Habitats

Learning Outcome

10. Compare and contrast the examples of bacteria that use animals as habitats.

The bodies of animals, including humans, provide a wide variety of ecological habitats for microbes—from dry, O_2-rich surfaces to moist, anaerobic folds and depressions.

Bacteria That Inhabit the Skin

The skin is typically dry and salty, providing an environment inhospitable to many microorganisms. Members of the genus *Staphylococcus*, however, thrive under these conditions.

FOCUS ON A CASE 11.1

The "patient" was an enormous region of Lake Erie. An inches-thick green scum covered the surface during the summer months, and fish and other aquatic animals in the area were dying. Microscopic analysis revealed that the green scum was a bloom of cyanobacteria (a group of photosynthetic bacteria). The aquatic animals were dying because the waters had insufficient dissolved O_2, causing the animals to suffocate. In turn, this created a large "dead zone"—a region that lacks animal life.

The source of the problem was high levels of phosphorus in the lake. When excess phosphorus is available, cyanobacteria may grow to high numbers, an event referred to as a bloom. Potential sources of this type of pollution range from industries to individuals. For example, manure and chemical fertilizers used by farms and homeowners to promote plant growth often contain phosphate, a source of phosphorus for both plants and other organisms, including cyanobacteria. Fertilizers applied to soils can be washed into the lake by rain. Another source is wastewater treatment plants; detergents as well as biological materials can contain phosphate, and most wastewater treatment processes do not remove this substance. Liquid from septic systems also contains phosphate.

The obvious solution to the problem of excess cyanobacterial growth is to identify and control the most significant levels of phosphorus pollution. Although that might seem simple, it requires that communities and individuals recognize and accept responsibility for a widespread problem, and then work together to develop a solution.

1. Would "dead zones" lack all life, or only animal life?
2. Why would sunlight in combination with phosphorus-containing pollution promote the growth of cyanobacteria?
3. Why would cyanobacterial growth lead to decreased levels of O_2 in the water?
4. How could microorganisms be used to remove phosphate in wastewater?
5. What health risks could cyanobacterial blooms pose for residents of cities that draw their drinking water from the lake?

Discussion

1. Dead zones still have living organisms, just not those that depend on O_2 for respiration. For example, fermenting and anaerobically respiring microorganisms can survive in environments that have low or no O_2.
2. Cyanobacteria are photosynthetic, so they harvest energy from the radiant energy of the sun and use CO_2 as a carbon source. Many cyanobacterial species can fix nitrogen, converting N_2 in the atmosphere into forms that cells can incorporate into organic material. Considering that nitrogen-fixing cyanobacteria have virtually unlimited supplies of carbon and nitrogen, as well as plenty of their energy source in the summer, phosphorus is often the limiting nutrient.
3. Some of the cyanobacteria eventually burst, releasing organic materials that can be used as an energy source by various chemoheterotrophic microorganisms in the water. When these organisms aerobically respire, they use the dissolved O_2 in water as a terminal electron acceptor, reducing it to H_2O. This lowers the concentration of dissolved O_2, eventually to the point that not enough is available to support the requirements of fish and other aquatic animals.
4. *Spirillum volutans* and some other bacteria make volutin granules (storage forms of phosphate). These bacteria could be added to wastewater, where they would take up and accumulate the phosphate. The bacterial cells could then be removed, thereby removing the phosphate. Chemical methods for removing phosphate are also available, and these are increasingly being used as an additional step in wastewater treatment.
5. Some types of cyanobacteria make toxins. For a few days in August 2014, over 400,000 residents of Toledo, Ohio, were told not to use their tap water because it contained unsafe toxin levels due to a cyanobacterial bloom in Lake Erie.

The Genus *Staphylococcus*

Staphylococcus species are facultatively anaerobic Gram-positive cocci. Most, such as *S. epidermidis,* reside harmlessly as part of the normal microbiota of the skin. Like other bacteria that aerobically respire, *Staphylococcus* species are catalase-positive. This distinguishes them from *Streptococcus, Enterococcus,* and *Lactococcus* species, which are also Gram-positive cocci but lack the enzyme catalase. Several species of *Staphylococcus* are notable for their medical significance. *Staphylococcus aureus* causes a variety of diseases, including skin and wound infections, as well as food poisoning. *Staphylococcus saprophyticus* causes urinary tract infections.

Bacteria That Inhabit Mucous Membranes

Mucous membranes of the respiratory, genitourinary, and intestinal tracts provide a habitat for numerous kinds of bacteria, many of which have already been discussed. For example, *Streptococcus* and *Corynebacterium* species reside in the respiratory tract, *Lactobacillus* species inhabit the vagina, and *Clostridium* species and members of the family Enterobacteriaceae thrive in the intestinal tract. Some of the other genera are discussed next.

The Genus *Bacteroides*

Bacteroides species are small, strictly anaerobic, Gram-negative rods and coccobacilli that inhabit the mouth, intestinal tract, and genital tract of humans and other animals. They play an important role in digestion and make up about a third of the bacterial population in human feces. In addition, they are often responsible for abscesses and bloodstream infections that follow appendicitis and abdominal surgery. Many are killed by brief exposure to O_2, so they are difficult to study.

The Genus *Bifidobacterium*

Bifidobacterium species are Gram-positive, irregular, rod-shaped anaerobes that reside primarily in the intestinal tract of humans and other animals. They are the most common members of the intestinal microbiota of breast-fed infants and are thought to provide a protective function by excluding disease-causing bacteria. Formula-fed infants are also colonized with members of this genus, but generally the concentrations are lower.

The Genus *Bordetella*

Bordetella species are small, Gram-negative coccobacilli that only grow aerobically. Special media must be used for their cultivation because they are nutritionally fastidious. The most significant species medically is *Bordetella pertussis,* which causes whooping cough, a highly contagious infectious disease of humans. *Bordetella bronchiseptica* causes a respiratory infection in dogs, an illness commonly called "kennel cough."

The Genera *Campylobacter* and *Helicobacter*

Members of the genera *Campylobacter* and *Helicobacter* are microaerophilic curved Gram-negative rods. *Campylobacter jejuni* causes diarrheal disease in humans; it typically lives in the intestinal tract of domestic animals, particularly poultry. *Helicobacter pylori* inhabits the stomach, where it can cause peptic ulcers (lesions in the stomach and duodenum); it has also been linked to stomach cancer. An important factor in its ability to survive in the stomach is its production of the enzyme urease, which converts urea in gastric juices to ammonia, an alkaline compound that neutralizes stomach acid in the cell's immediate surroundings.

The Genus *Haemophilus*

Haemophilus species are Gram-negative coccobacilli that, as their name reflects, are "blood loving." They require hematin and/or NAD, which are found in blood. Many species are common microbiota of the respiratory tract. *H. influenzae* causes ear infections, respiratory infections, and meningitis, primarily in children. *Haemophilus ducreyi* causes the sexually transmitted infection (STI) chancroid.

The Genus *Mycoplasma*

Members of the genus *Mycoplasma* lack a cell wall, but most have sterols in their membrane to provide strength and rigidity. They are among the smallest forms of life, and their genomes are thought to be about the minimum size for encoding a free-living organism's essential functions. Because the bacteria lack a cell wall, they are unaffected by penicillin and other antibiotics that interfere peptidoglycan synthesis.

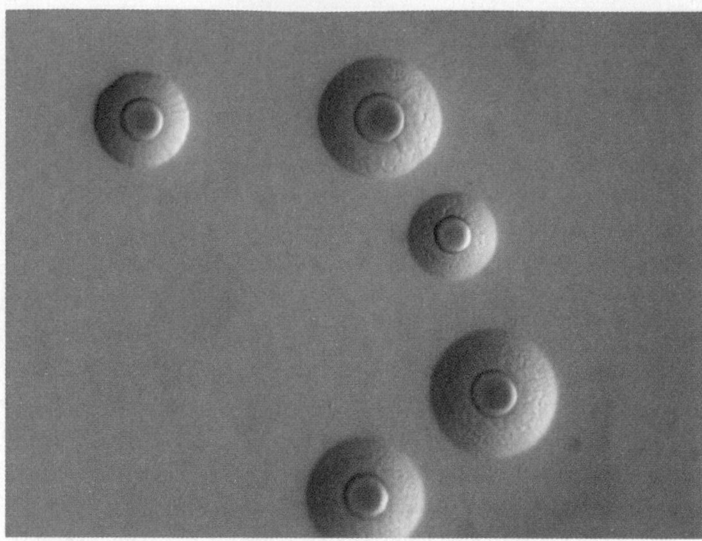

FIGURE 11.24 ***Mycoplasma*** **Colonies** The characteristically small colonies have a dense center, resulting in a "fried egg" appearance. Christine Citti

? Why are *Mycoplasma* species not affected by penicillin?

Mycoplasmas are a particular concern in laboratories studying viruses and eukaryotic cell physiology because the small bacteria can pass through the filters used to sterilize tissue culture media; they easily contaminate the media and thereby compromise experimental results. Mycoplasmas can grow to very high numbers in media without causing cloudiness, so contamination might not be noticed. When grown on agar media, *Mycoplasma* cells in the center of a colony grow into the medium, producing a dense central area that gives the characteristically small colony a "fried egg" appearance (**figure 11.24**).

Medically, one of the most significant mycoplasmas is *M. pneumoniae,* which causes a form of pneumonia often referred to as "walking pneumonia" because it tends to be mild. *M. genitalium* causes an emerging STI.

MicroByte

The genome of *Mycoplasma genitalium* is only 5.8×10^5 base pairs, approximately one-eighth the size of the *E. coli* genome.

The Genus *Neisseria*

Neisseria species are Gram-negative bacteria, typically kidney-bean-shaped cocci in pairs. They are common microbiota of animals including humans, growing on mucous membranes. *Neisseria* species are typically aerobes, but some can grow anaerobically if a suitable terminal electron acceptor such as nitrite is present. Those noted for their medical significance include *N. gonorrhoeae,* which causes the STI gonorrhea, and *N. meningitidis,* which causes meningitis; both are nutritionally fastidious.

The Genera *Treponema* and *Borrelia*

Members of the genera *Treponema* and *Borrelia* are spirochetes that typically inhabit body fluids and mucous membranes of humans and other animals. Recall that spirochetes are characterized by their corkscrew shape and endoflagella. Although they have a Gram-negative cell wall, they are often too thin to be seen using standard microscopy and staining methods.

Treponema species are obligate anaerobes or microaerophiles that often inhabit the mouth and genital tract. The species that causes the STI syphilis, *T. pallidum,* is difficult to study because it does not grow using standard culture techniques. Genome sequence studies indicate that the organism is a microaerophile with a metabolism highly dependent on its host; it lacks critical enzymes of the TCA cycle and a variety of other metabolic pathways.

Three *Borrelia* species are pathogens, transmitted by arthropods such as ticks and lice. *B. recurrentis* and *B. hermsii* both cause relapsing fever; *B. burgdorferi* causes Lyme disease. An unusual feature of *Borrelia* species is their genome—a linear chromosome and many linear and circular plasmids.

Obligate Intracellular Parasites

Obligate intracellular parasites cannot reproduce outside a host cell. By living within host cells, the parasites are supplied with a source of compounds they would otherwise need to synthesize for themselves. As a result, most intracellular parasites have lost the ability to make substances needed for extracellular growth. Examples described next are all tiny Gram-negative rods or coccobacilli.

The Genera *Rickettsia, Orientia,* and *Ehrlichia*

Species of *Rickettsia, Orientia,* and *Ehrlichia* are responsible for several serious human diseases spread by blood-sucking arthropods such as ticks and lice. *Rickettsia rickettsii* causes Rocky Mountain spotted fever, *R. prowazekii* causes epidemic typhus, *O. tsutsugamushi* causes scrub typhus, and *E. chaffeensis* causes human ehrlichiosis.

The Genus *Coxiella*

The only characterized species of *Coxiella, C. burnetii,* is an obligate intracellular bacterium that produces a spore-like form called a small-cell variant (SCV) that survives in the outside environment (**figure 11.25**). SCVs, however, lack the extreme resistance to heat and disinfectants characteristic of endospores. *Coxiella burnetii* causes Q fever of humans, a disease most often acquired by inhaling bacteria shed from infected animals. This is particularly a problem when animals give birth because high numbers of the bacteria can be found in the placenta of infected animals.

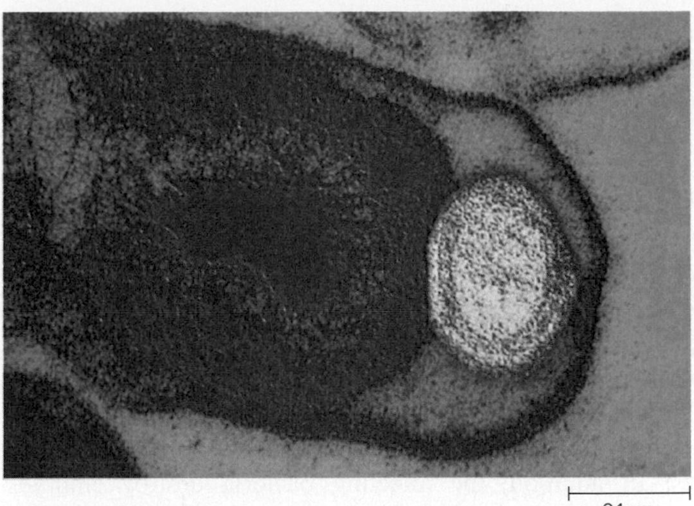

FIGURE 11.25 ***Coxiella*** Color-enhanced TEM of *C. burnetii.* The oval orange-colored object is the spore-like form. Alfred Pasieka/Science Photo Library/Science Source

? What disease does *C. burnetii* cause?

The Genera *Chlamydia* and *Chlamydophila*

Chlamydia and *Chlamydophila* species are transmitted directly from person to person and have a unique growth cycle involving two forms: a reticulate body (RB) and a smaller dense-appearing elementary body (EB) (**figure 11.26**). Within the host cell, the non-infectious RBs reproduce by binary fission, and these then differentiate into infectious

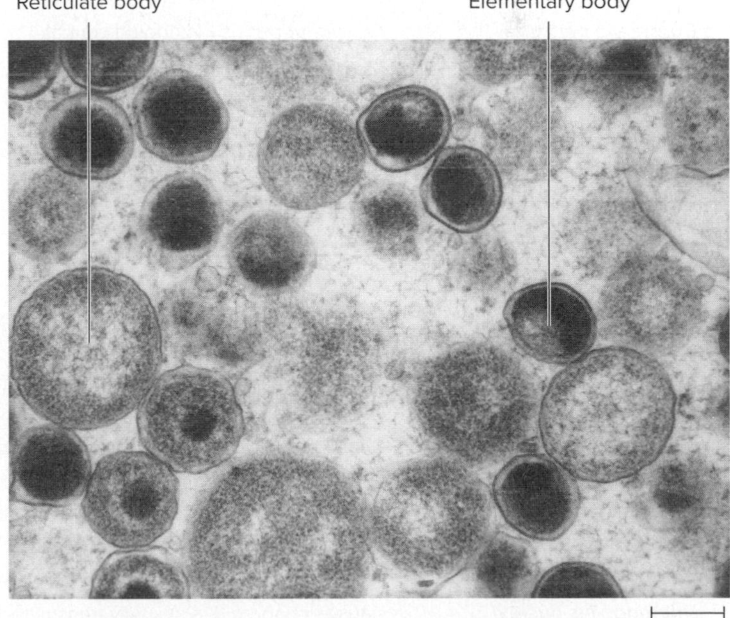

FIGURE 11.26 ***Chlamydia*** **Growing in Tissue Cell Culture** Reticulate bodies and elementary bodies are shown (TEM). David M. Phillips/Science Source

? Which form of *Chlamydia*—the reticulate body or elementary body—is infectious?

EBs that are released when the host cell ruptures. *Chlamydia trachomatis* causes eye infections and an STI that mimics gonorrhea; *Chlamydophila pneumoniae* causes atypical pneumonia; and *Chlamydophila psittaci* causes psittacosis, a form of pneumonia.

The Genus *Wolbachia*

Members of the genus *Wolbachia* infect arthropods (including insects, spiders, and mites) and parasitic worms. They are primarily transmitted maternally, via the eggs of infected females to their offspring. In arthropods, the bacteria use unique strategies to increase the overall population of infected females, including killing male embryos, allowing infected females to reproduce asexually, and causing infected males to gain female traits. In addition, the infecting bacteria destroy embryos resulting from the mating of an infected male with either an uninfected female or a female infected with a different strain.

Wolbachia species do not infect humans or other mammals, but *W. pipientis* is medically important because it resides within filarial worms that cause the diseases river blindness and elephantiasis; the chronic and debilitating inflammation associated with these diseases appears to result from the immune response directed against the bacterial cells in the invading worms. Scientists are now exploring treatments that not only kill the filarial worms but also lessen disease symptoms by eliminating the bacteria within the worms.

Wolbachia-infected mosquitoes have recently gained attention as a biological mechanism to prevent the spread of the viruses that cause dengue, chikungunya, Zika virus disease, and yellow fever; these viruses replicate poorly within the infected mosquitoes and thus are less likely to be transmitted to humans. Although *W. pipientis* naturally infects a wide range of insects, it must be introduced artificially into *Aedes aegypti* (a mosquito species capable of transmitting the viruses); once the mosquitoes are infected, they can breed with wild mosquitoes, and *W. pipientis* is then passed to the resulting offspring. As part of a study, *Wolbachia*-infected *A. aegypti* mosquitoes were released in an Australian city, resulting in a dramatic decrease in dengue cases. Similar decreases were observed when *Wolbachia*-infected mosquitoes were released in certain areas of Brazil, Indonesia, and other countries.

Table 11.3 lists the medically important bacteria covered in this and other sections of the chapter (many phylum names were recently changed, so in each case the previous name is included in parentheses).

TABLE 11.3 Medically Important Bacteria

Organism	Medical Significance	Phylum (Previous Name)
Gram-Negative Rods		
Bacteroides species	Obligate anaerobes that commonly inhabit the mouth, intestinal tract, and genital tract. Cause abscesses and bloodstream infections.	Bacteroidota (Bacteroidetes)
Bordetella pertussis	Causes pertussis (whooping cough).	Pseudomonadota (Proteobacteria)
Enterobacteriaceae	Most live in the intestinal tract.	Pseudomonadota (Proteobacteria)
Enterobacter species	Normal microbiota of the intestinal tract.	
Escherichia coli	Normal microbiota of the intestinal tract. Some strains cause urinary tract infections; some strains cause specific types of intestinal disease; some cause meningitis in newborns.	
Klebsiella pneumoniae	Normal microbiota of the intestinal tract. Causes pneumonia.	
Proteus species	Normal microbiota of the intestinal tract. Cause urinary tract infections.	
Salmonella enterica serotype Enteritidis	Causes gastroenteritis. Grows in the intestinal tract of infected animals; acquired by consuming contaminated food.	
Salmonella enterica serotype Typhi	Causes typhoid fever. Grows in the intestinal tract of infected humans; transmitted in feces.	
Shigella species	Cause dysentery. Grow in the intestinal tract of infected humans; transmitted in feces.	
Yersinia pestis	Causes bubonic plague, which is transmitted by fleas, and pneumonic plague, which is transmitted in respiratory droplets of infected individuals.	
Haemophilus influenzae	Causes ear infections, respiratory infections, and meningitis in children.	Pseudomonadota (Proteobacteria)
Haemophilus ducreyi	Causes chancroid, a sexually transmitted infection (STI).	Pseudomonadota (Proteobacteria)
Legionella pneumophila	Causes Legionnaires' disease, a respiratory disease. Grows within protozoa; acquired by inhaling contaminated water droplets.	Pseudomonadota (Proteobacteria)
Pseudomonas aeruginosa	Causes burn, urinary tract, and bloodstream infections. Common in the environment. Grows in nutrient-poor aqueous solutions. Resistant to many disinfectants and antimicrobial medications.	Pseudomonadota (Proteobacteria)

TABLE 11.3 Medically Important Bacteria (*continued*)

Organism	Medical Significance	Phylum (Previous Name)
Gram-Negative Rods—Obligate Intracellular Parasites		
Chlamydophila (Chlamydia) pneumoniae	Causes atypical pneumonia, or "walking pneumonia." Acquired from an infected person.	Chlamydiota (Chlamydiae)
Chlamydophila (Chlamydia) psittaci	Causes psittacosis, a form of pneumonia. Transmitted by birds.	Chlamydiota (Chlamydiae)
Chlamydia trachomatis	Causes an STI. Also causes conjunctivitis in newborns and trachoma (a serious eye infection).	Chlamydiota (Chlamydiae)
Coxiella burnetii	Causes Q fever. Acquired by inhaling organisms shed by infected animals.	Pseudomonadota (Proteobacteria)
Ehrlichia chaffeensis	Causes human ehrlichiosis. Transmitted by ticks.	Pseudomonadota (Proteobacteria)
Orientia tsutsugamushi	Causes scrub typhus. Transmitted by mites.	Pseudomonadota (Proteobacteria)
Rickettsia prowazekii	Causes epidemic typhus. Transmitted by lice.	Pseudomonadota (Proteobacteria)
Rickettsia rickettsii	Causes Rocky Mountain spotted fever. Transmitted by ticks.	Pseudomonadota (Proteobacteria)
Wolbachia pipientis	Resides within the filarial worms that cause river blindness and elephantiasis.	Pseudomonadota (Proteobacteria)
Gram-Negative Curved Rods		
Campylobacter jejuni	Causes gastroenteritis. Grows in the intestinal tract of infected animals; acquired by consuming contaminated food.	Pseudomonadota (Proteobacteria)
Helicobacter pylori	Causes peptic ulcers. Neutralizes stomach acid by producing urease.	Pseudomonadota (Proteobacteria)
Vibrio cholerae	Causes cholera, a severe diarrheal disease; acquired by drinking contaminated water.	Pseudomonadota (Proteobacteria)
Vibrio parahaemolyticus	Causes gastroenteritis. Acquired by consuming contaminated seafood.	Pseudomonadota (Proteobacteria)
Vibrio vulnificus	Causes a systemic disease, particularly in people who have liver failure or other underlying complications.	Pseudomonadota (Proteobacteria)
Gram-Negative Cocci		
Neisseria meningitidis	Causes meningitis.	Pseudomonadota (Proteobacteria)
Neisseria gonorrhoeae	Causes gonorrhea, an STI.	Pseudomonadota (Proteobacteria)
Gram-Positive Rods		
Bacillus anthracis	Causes anthrax. Acquired by inhaling endospores in soil, animal hides, and wool. Also acquired by touching or ingesting the endospores. Bioterrorism agent.	Bacillota (Firmicutes)
Bifidobacterium species	Predominant member of the intestinal tract in breast-fed infants. Thought to play a protective role in the intestinal tract by excluding pathogens.	Actinomycetota (Actinobacteria)
Clostridium botulinum	Causes botulism. Foodborne disease results from ingesting toxin-contaminated foods, typically canned foods that have been improperly processed.	Bacillota (Firmicutes)
Clostridium perfringens	Causes gas gangrene. Acquired when soil-borne endospores contaminate a wound.	Bacillota (Firmicutes)
Clostridium tetani	Causes tetanus. Acquired when soil-borne endospores are inoculated into deep tissue.	Bacillota (Firmicutes)
Clostridioides (Clostridium) difficile	Causes *C. difficile* infection (CDI), which is associated with antibiotic use and can result in severe diarrhea.	Bacillota (Firmicutes)
Corynebacterium diphtheriae	Toxin-producing strains cause diphtheria.	Actinomycetota (Actinobacteria)
Gram-Positive Cocci		
Enterococcus species	Normal microbiota of the intestinal tract. Cause urinary tract infections.	Bacillota (Firmicutes)
Micrococcus species	Found on skin as well as in a variety of other environments; often contaminate bacteriological media.	Actinomycetota (Actinobacteria)
Staphylococcus aureus	Leading cause of wound infections. Causes boils, carbuncles, food poisoning, and staphylococcal toxic shock syndrome.	Bacillota (Firmicutes)

continued

TABLE 11.3 Medically Important Bacteria (*continued*)

Organism	Medical Significance	Phylum (Previous Name)
Staphylococcus epidermidis	Normal microbiota of the skin.	Bacillota (Firmicutes)
Staphylococcus saprophyticus	Causes urinary tract infections.	Bacillota (Firmicutes)
Streptococcus pneumoniae	Causes pneumonia and meningitis.	Bacillota (Firmicutes)
Streptococcus pyogenes	Causes pharyngitis (strep throat), rheumatic fever, wound infections, glomerulonephritis, and streptococcal toxic shock.	Bacillota (Firmicutes)
Acid-Fast Rods		
Mycobacterium tuberculosis	Causes tuberculosis.	Actinomycetota (Actinobacteria)
Mycobacterium leprae	Causes Hansen's disease (leprosy); peripheral nerve invasion is characteristic.	Actinomycetota (Actinobacteria)
Spirochetes	Characterized by spiral shape, flexible cell wall, and endoflagella.	
Treponema pallidum	Causes syphilis, an STI. The organism has never been grown in culture.	Spirochaetota (Spirochaetes)
Borrelia burgdorferi	Causes Lyme disease, a tick-borne disease.	Spirochaetota (Spirochaetes)
Borrelia recurrentis and *B. hermsii*	Causes relapsing fever. Transmitted by arthropods.	Spirochaetota (Spirochaetes)
Leptospira interrogans	Causes leptospirosis, a waterborne disease. Excreted in urine of infected animals.	Spirochaetota (Spirochaetes)
Cell Wall-less		
Mycoplasma pneumoniae and *M. genitalium*	*M. pneumoniae* causes atypical pneumonia ("walking pneumonia"); M. *genitalium* causes an STI. Not susceptible to penicillin because they lack a cell wall.	Mycoplasmatota (Tenericutes)

MicroAssessment 11.8

Staphylococcus species thrive in the dry, salty conditions of the skin. *Bacteroides* and *Bifidobacterium* species live in the gastrointestinal tract; *Campylobacter* and *Helicobacter* species can cause disease when they reside there. *Neisseria* species, mycoplasmas, and spirochetes inhabit other mucous membranes. Obligate intracellular parasites—including *Rickettsia, Orientia, Ehrlichia, Coxiella, Chlamydia, Chlamydophila,* and *Wolbachia* species—are unable to reproduce outside a host cell.

22. How do *Helicobacter pylori* cells withstand stomach acidity?
23. What characteristic of *Mycoplasma* species separates them from other bacteria?
24. Why would breast feeding affect the composition of a baby's intestinal microbiota?

11.9 ■ Archaea That Thrive in Extreme Conditions

Learning Outcome

11. Compare and contrast the characteristics and habitats of the extreme halophiles and the extreme thermophiles.

Characterized archaea typically thrive in extreme environments, including conditions of high heat, acidity, alkalinity, and salinity. An exception is the methanogens, discussed earlier in the chapter; they inhabit anaerobic niches shared with bacteria. In addition to these archaea, many others have been detected in a variety of non-extreme environments using molecular techniques.

Extreme Halophiles

Extreme halophiles are found in high numbers in salty environments such as salt lakes, soda lakes, and brines used for curing fish. Most require a minimum of about 9% NaCl and grow well in saturated salt solutions (32% NaCl). Because they produce pigments, their growth can be seen as red patches on salted fish and pink blooms in concentrated saltwater ponds (**figure 11.27**).

Extreme halophiles are aerobic or facultatively anaerobic chemoheterotrophs, but some also use the light-sensitive pigment bacteriorhodopsin to absorb energy from light. As described in section 6.8, the cells then use that energy to pump protons from the cell, thus creating a proton gradient across the cytoplasmic membrane that can be used to synthesize ATP or drive the rotation of flagella. The organisms come in a variety of shapes, including rods, cocci, discs, and triangles. They include genera such as *Halobacterium, Halorubrum, Natronobacterium*, and *Natronococcus;* members of these latter two genera are extremely alkaliphilic as well as halophilic.

FIGURE 11.27 Salt Evaporation Ponds Chris Sattlberger/Blend Images

What causes the pink color in the salt evaporation ponds?

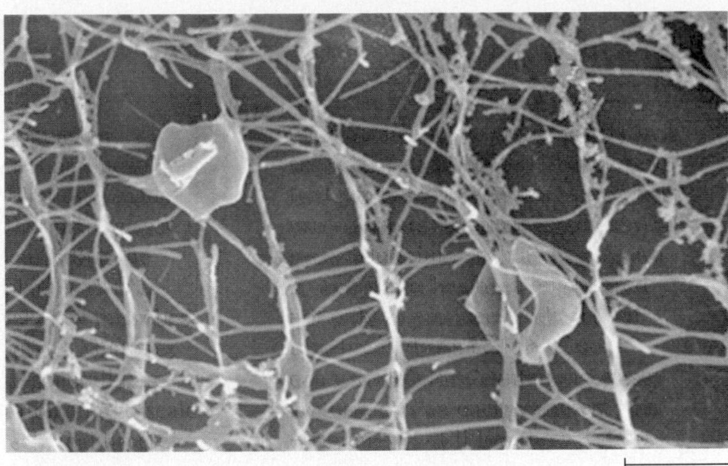

FIGURE 11.28 *Pyrodictium* The disc-shaped cells are connected by hollow tubes (SEM). Dr. Karl O. Stetter

How would *Pyrodictium* species be grouped with respect to their temperature preference?

Extreme Thermophiles

The extreme thermophiles (hyperthermophiles) are found near volcanic vents and fissures that release sulfurous gases and other hot vapors. Because these regions are thought to closely mimic early Earth's environment, scientists are interested in studying the prokaryotes that live there. Others are found in hydrothermal vents in the deep sea and hot springs.

Methane-Generating Hyperthermophiles

In contrast to the mesophilic methanogens discussed earlier, some methanogens are extreme thermophiles. *Methanothermus* species, which can grow in temperatures as high as 97°C, grow optimally at approximately 84°C. Researchers have grown *Methanopyrus kandleri* strain 116 at 122°C, currently the highest recorded growth temperature.

Sulfur-Reducing Hyperthermophiles

The sulfur-reducing hyperthermophiles are obligate anaerobes that use sulfur as a terminal electron acceptor, generating H_2S. They harvest energy by oxidizing organic compounds and/or H_2. These archaea can be isolated from hot sulfur-containing environments such as sulfur hot springs and hydrothermal vents. They include some of the most thermophilic organisms known, a few even growing above 100°C. One example, *Pyrolobus fumarii,* was isolated from a "black smoker" 3,650 m (about 12,000 feet) deep in the Atlantic Ocean. It grows between 90°C and 113°C. Another hydrothermal vent isolate, *Pyrodictium occultum,* has an optimum temperature of about 105°C and cannot grow below 82°C. Its disc-shaped cells are connected by hollow tubes, forming a web-like network (**figure 11.28**). A hydrothermal vent isolate called "strain 121" (also referred to as *Geogemma barossii*) grows at 121°C, which was previously the highest recorded growth temperature.

Nanoarchaea

The discovery of an archaeum so unique that it represents an entirely new phylum, *Nanoarcheota* ("tiny archaea"), was made possible by the earlier discovery of a new genus of sulfur-reducing hyperthermophiles, *Ignicoccus* ("the fire sphere"). *Nanoarchaeum equitans* ("rider") grows as 400 nm spheres attached to the surface of—presumably parasitizing—an *Ignicoccus* species.

Sulfur Oxidizers

Sulfolobus species are obligate aerobes found at the surface of acidic sulfur-containing hot springs such as many of those found in Yellowstone National Park (**figure 11.29**). They oxidize sulfur compounds, using O_2 as a terminal electron acceptor to generate sulfuric acid, and are thermoacidophilic, only growing above 50°C and at a pH between 1 and 6.

FIGURE 11.29 Typical Habitat of *Sulfolobus* Sulfur hot spring in Yellowstone National Park. blanscape/Getty Images

Why must *Solfolobus* species be able to tolerate acidic conditions?

TABLE 11.4 Archaea

Group/Genera	Characteristics	Phylum
Methanogens—*Methanospirillum, Methanosarcina*	Generate methane when they oxidize hydrogen gas as an energy source, using CO_2 as a terminal electron acceptor.	Euryarchaeota
Extreme halophiles—*Halobacterium, Halorubrum, Natronobacterium, Natronococcus*	Found in salt lakes, soda lakes, and brines. Most grow well in saturated salt solutions.	Euryarchaeota
Extreme thermophiles—*Methanothermus, Methanopyrus, Pyrodictium, Pyrolobus, Sulfolobus, Thermophilus, Picrophilus, Nanoarchaeum*	Found near hydrothermal vents and in hot springs; some grow at temperatures above 100°C. Includes examples of methane-generating, sulfur-reducing, and sulfur-oxidizing archaea, as well as extreme acidophiles.	Thermoproteota (was Crenarchaeota), Euryarchaeota, and Nanoarchaeota

Thermophilic Extreme Acidophiles

Members of two genera, *Thermoplasma* and *Picrophilus,* are notable for growing in extremely acidic, hot environments. *Thermoplasma* species grow optimally at pH 2; in fact, *T. acidophilum* lyses at neutral pH. It was originally isolated from a coal waste pile. *Picrophilus* species tolerate conditions even more acidic, growing optimally at a pH below 1. Two species isolated in Japan inhabit acidic areas in regions that spew sulfurous gases.

Characteristics of the archaea discussed in this chapter are summarized in **table 11.4**.

MicroAssessment 11.9

Many characterized archaea inhabit extreme environments. These include conditions of high salinity, heat, acidity, and alkalinity.

25. What is the habitat of *Nanoarchaeum equitans*?

26. At which relative depth in a sulfur hot spring would a sulfur oxidizer likely be found?

27. What characteristic of the methanogens makes it logical to discuss them with the bacteria rather than with the archaea described in this section?

FOCUS YOUR PERSPECTIVE 11.1

Astrobiology: Searching for Life Beyond Earth

If life as we know it exists elsewhere in the universe, it will likely be microbial. The task, then, is to figure out how to find and detect such extraterrestrial microorganisms. Considering how relatively little we know about the microbial life on our own planet, coupled with the extreme difficulty of obtaining or testing extraterrestrial samples, this is a daunting challenge with many unanswered questions. What is the most likely source of life beyond Earth? What is the best way to preserve specimens for study on Earth? What culture conditions should be used to grow such organisms?

Astrobiology, the study of life in the universe, is a relatively new field that brings together scientists from a wide range of disciplines, including microbiology, geology, astronomy, biology, and chemistry. The goal is to determine the origin, evolution, distribution, and destiny of life in the universe. Astrobiologists are also developing lightweight, dependable, and meaningful testing devices to be used in future space missions.

To prepare for researching life on other planets and their moons, microbiologists have turned to some of the most extreme environments here on Earth. These include glaciers and ice shelves, hot springs, deserts, volcanoes, deep ocean hydrothermal vents, and subterranean features such as caves. Because select microorganisms can survive in these environs, which are similar to conditions expected beyond Earth, they are good testing grounds for the technology to be used on future missions.

Summary

METABOLIC DIVERSITY (table 11.1)

11.1 ■ Anaerobic Chemotrophs

Anaerobic Chemolithotrophs

The **methanogens** are a group of archaea that harvest energy by oxidizing H_2, using CO_2 as a terminal electron acceptor (figure 11.1).

Anaerobic Chemoorganotrophs—Anaerobic Respiration

Desulfovibrio species reduce sulfur compounds to form hydrogen sulfide.

Anaerobic Chemoorganotrophs—Fermentation

Clostridia form endospores. The **lactic acid bacteria** produce lactic acid as their primary fermentation end product (figures 11.2, 11.3).

Propionibacterium species produce propionic acid as their primary fermentation end product.

11.2 ■ Anoxygenic Phototrophs

Purple Bacteria

The purple bacteria appear red, orange, or purple; the components of their photosynthetic apparatus are all within the cytoplasmic membrane.

Green Bacteria

The green bacteria are typically green or brownish. Their accessory pigments are located in chlorosomes.

Filamentous Anoxygenic Phototrophic Bacteria

The filamentous anoxygenic phototrophic bacteria form multicellular arrangements and exhibit gliding motility. Many have chlorosomes.

Other Anoxygenic Phototrophs

Other anoxygenic phototrophs have been discovered, including some that form endospores.

11.3 ■ Oxygenic Phototrophs

Cyanobacteria (figures 11.5, 11.6)

Genetic evidence indicates that chloroplasts evolved from a species of **cyanobacteria.** Nitrogen-fixing cyanobacteria are important ecologically because they provide an available source of both carbon and nitrogen (figure 11.7). Filamentous cyanobacteria maintain the structure and productivity of some soils. Some cyanobacteria produce toxins that can be deadly to animals that ingest contaminated water.

11.4 ■ Aerobic Chemolithotrophs

Sulfur-Oxidizing Bacteria

The filamentous sulfur oxidizers *Beggiatoa* and *Thiothrix* live in sulfur springs, in sewage-polluted waters, and on the surface of marine and freshwater sediments (figure 11.8). *Acidithiobacillus* species are found in both terrestrial and aquatic habitats (figure 11.9).

Nitrifiers

Ammonia oxidizers convert ammonium to nitrite; they include *Nitrosomonas* and *Nitrosococcus.* Nitrite oxidizers convert nitrite to nitrate; they include *Nitrobacter* and *Nitrococcus.*

Hydrogen-Oxidizing Bacteria

Aquifex and *Hydrogenobacter* species are thought to be among the earliest bacterial forms to exist on Earth.

11.5 ■ Aerobic Chemoorganotrophs

Obligate Aerobes

Micrococcus species are found in soil and on dust particles, various foods, and skin (figure 11.10). *Mycobacterium* species are widespread in nature; they are **acid-fast.** *Pseudomonas* species are widespread in nature and have extremely diverse metabolic capabilities (figure 11.11). *Thermus aquaticus* is the source of *Taq* polymerase, a component of PCR. *Deinococcus radiodurans* can survive radiation exposure several thousand times the level that would be lethal to a human being.

Facultative Anaerobes

Corynebacterium species are widespread in nature (figure 11.12). Members of the family Enterobacteriaceae typically inhabit the intestinal tract of animals, although some reside in rich soil. **Coliforms** are used as indicators of fecal pollution. *Vibrio* species typically live in marine environments.

ECOPHYSIOLOGICAL DIVERSITY (table 11.2)

11.6 ■ Thriving in Terrestrial Environments

Bacteria That Form a Resting Stage

Bacillus and clostridia form endospores, the most resistant dormant form known (figure 11.13). *Azotobacter* species form cysts. Cells of **myxobacteria** come together to form a fruiting body, within which the cells become dormant microcysts (figure 11.14). *Streptomyces* species form chains of conidia at the end of hyphae. Many species naturally produce antibiotics (figure 11.15).

Bacteria That Associate with Plants

Agrobacterium species cause plant tumors (figure 11.16). They transfer a portion of the **Ti plasmid** to plant cells, genetically engineering the plant cells to produce opines and plant growth hormones. **Rhizobia** reside as endosymbionts in nodules on the roots of legumes, fixing nitrogen (figure 11.17).

11.7 ■ Thriving in Aquatic Environments

Sheathed Bacteria

Sheathed bacteria attach to solid objects in favorable habitats; the sheath shelters them from attack by predators (figure 11.18).

Prosthecate Bacteria

Caulobacter species have a single polar prostheca called a stalk; at the tip of the stalk is a holdfast. The cells divide by binary fission (figure 11.19). *Hyphomicrobium* species divide by forming a bud at the tip of their single polar prostheca (figure 11.20).

Bacteria That Derive Nutrients from Other Organisms

Bdellovibrio species prey on other bacteria (figure 11.21). Certain species of bioluminescent bacteria form symbiotic relationships with specific types of squid and fish (figure 11.22). *Epulopiscium* species live within the intestinal tract of surgeonfish. *Legionella* species often reside within protozoa; they can cause respiratory disease when inhaled.

Bacteria That Move by Unusual Mechanisms

Spirochetes move by means of endoflagella (figure 11.23). Magnetotactic bacteria contain a string of magnetic crystals that allow them to move up or down in water or sediments to the microaerophilic niches they require.

Bacteria That Form Storage Granules

Spirillum volutans form volutin granules, a storage form of phosphate. *Thioploca* species "commute" to nitrate-rich waters. *Thiomargarita namibiensis,* the largest bacterium known, stores sulfur and has a nitrate-containing vacuole.

11.8 ■ Animals as Habitats (table 11.3)

Bacteria That Inhabit the Skin

Staphylococcus species are facultative anaerobes.

Bacteria That Inhabit Mucous Membranes

Bacteroides species inhabit the mouth, intestinal tract, and genital tract of humans and other animals. *Bifidobacterium* species reside in the intestinal tract of animals, including humans, particularly breast-fed infants. *Campylobacter* and *Helicobacter* species are microaerophilic. *Haemophilus* species require compounds found in blood. *Neisseria* species are nutritionally fastidious aerobes that grow in the oral cavity and genital tract. *Mycoplasma* species lack a cell wall; they often have sterols in their membrane that provide strength and rigidity (figure 11.24). *Treponema* and *Borrelia* species are spirochetes that typically inhabit mucous membranes and body fluids of humans and other animals.

Obligate Intracellular Parasites

Species of *Rickettsia, Orientia,* and *Ehrlichia* are spread when a flea or tick transfers bacteria during a blood meal. *Coxiella burnetii* survives well outside the host due to spore-like structures (figure 11.25). *Chlamydia* and *Chlamydophila* species are transmitted directly from person to person (figure 11.26). *Wolbachia* species alter the reproductive biology of infected arthropods; although *W. pipientis* does not infect mammals, it lives within the filarial worms that cause river blindness and elephantiasis.

11.9 ■ Archaea That Thrive in Extreme Conditions

Extreme Halophiles

Extreme halophiles are found in salt lakes, soda lakes, and brines used for curing fish (figure 11.27).

Extreme Thermophiles

Methanothermus and *Methanopyrus* species are hyperthermophiles that make methane. Sulfur-reducing hyperthermophiles are obligate anaerobes that use sulfur as a terminal electron acceptor. Nanoarchaea grow as spheres attached to an *Ignicoccus* species. Sulfur-oxidizing hyperthermophiles use O_2 as a terminal electron acceptor, generating sulfuric acid. Thermophilic extreme acidophiles have an optimum pH of 2 or below.

Review Questions

Short Answer

1. What kind of bacteria might compose the purplish subsurface scum of polluted ponds?
2. What kind of bacterium might be responsible for plugging the pipes in a sewage treatment facility?
3. Give three examples of energy sources used by chemolithotrophs.
4. Name two genera of endospore-forming bacteria. How do they differ?
5. How is the life cycle of *Epulopiscium* species unusual?
6. What unique motility structure characterizes the spirochetes?
7. In what way does the metabolism of *Streptococcus* species differ from that of *Staphylococcus* species?
8. How have species of *Streptomyces* contributed to the treatment of infectious diseases?
9. What characteristics of *Azotobacter* species protect their nitrogenase enzyme from inactivation by O_2?
10. Compare and contrast the relationships of *Agrobacterium* and *Rhizobium* species with plants.

Multiple Choice

1. A catalase-negative colony growing on a plate that was incubated aerobically could be which of these genera?
   a) *Bacillus*
   b) *Escherichia*
   c) *Micrococcus*
   d) *Staphylococcus*
   e) *Streptococcus*
2. Members of all of the following genera are spirochetes *except*
   a) *Borrelia.*
   b) *Caulobacter.*
   c) *Leptospira.*
   d) *Spirochaeta.*
   e) *Treponema.*
3. If you examined the acidic runoff from a coal mine, which of the following would you most likely find growing there?
   a) *Clostridium*
   b) *Escherichia*
   c) Lactic acid bacteria
   d) *Thermus*
   e) *Acidithiobacillus*
4. The dormant forms of which set of genera or groups are the most resistant to environmental extremes?
   a) *Azotobacter* and *Bacillus*
   b) *Bacillus* and *Clostridium*
   c) *Clostridium* and myxobacteria
   d) Myxobacteria and *Streptomyces*
   e) *Azotobacter* and *Streptomyces*
5. If you read that coliforms had been found in a lake, the report could have been referring to which of the following genera?
   a) *Bacteroides*
   b) *Bifidobacterium*
   c) *Clostridium*
   d) *Escherichia*
   e) *Streptococcus*
6. Members of which of the following genera prey on other bacteria?
   a) *Bdellovibrio*
   b) *Caulobacter*
   c) *Hyphomicrobium*
   d) *Photobacterium*
   e) *Sphaerotilus*
7. Members of all of the following genera are obligate intracellular parasites *except*
   a) *Chlamydia.*
   b) *Coxiella.*
   c) *Ehrlichia.*
   d) *Mycoplasma.*
   e) *Rickettsia.*
8. Members of all of the following genera fix nitrogen *except*
   a) *Anabaena.*
   b) *Deinococcus.*
   c) *Azotobacter.*
   d) *Rhizobium.*

9. Members of which of the following archaeal genera would most likely be found coexisting with bacteria?
   a) *Nanoarchaeum*
   b) *Halobacterium*
   c) *Methanococcus*
   d) *Picrophilus*
   e) *Sulfolobus*
10. *Thermoplasma* and *Picrophilus* grow best in which of the following extreme conditions?
    a) Low pH
    b) High salt
    c) High temperature
    d) Both a and c
    e) Both b and c

Applications

1. A student argues that it makes no sense to be concerned about coliforms in drinking water because they are harmless members of our normal microbiota. Explain why regulatory agencies are concerned about coliforms.
2. Friends who cherish the lush green lawn at their lakefront property complain of the foul-smelling green scum on the lake each summer. Explain how the lawn might be contributing to the problem.

Critical Thinking

1. Soil often goes through periods of extreme dryness and extreme wetness. What characteristics of *Clostridium* species make them well suited for these conditions?
2. Some organisms use sulfur as an electron donor (a source of energy), whereas others use sulfur as an electron acceptor. How can this be if there must be a difference between the electron affinity of electron donors and acceptors for an organism to obtain energy?

www.mcgrawhillconnect.com

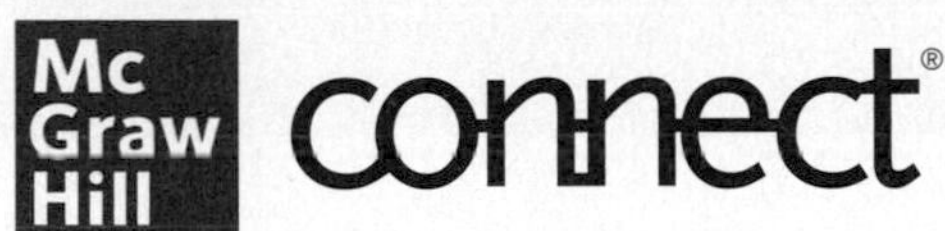

Enhance your study of this chapter with study tools and practice tests. Also ask your instructor about the resources available through Connect, including the media-rich eBook, interactive learning tools, and animations.

12 | The Eukaryotic Members of the Microbial World

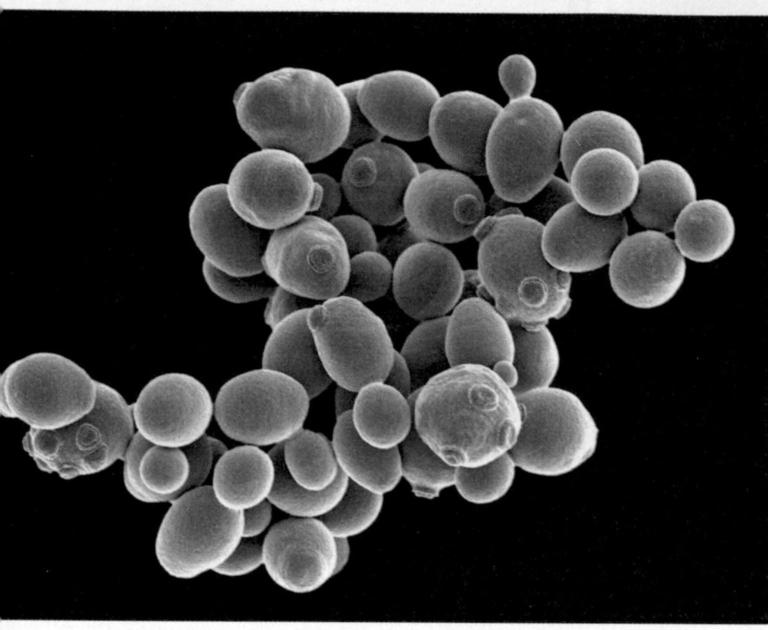

Budding yeast cells (color-enhanced scanning electron micrograph)
Steve Gschmeissner/Science Photo Library/Getty Images

KEY TERMS

Algae Eukaryotic photosynthetic organisms with relatively simple reproductive structures and no organized vascular system.

Arthropod Animal with an external skeleton and jointed appendages such as an insect or an arachnid; may act as a vector in disease transmission.

Definitive Host Organism in which a parasite undergoes sexual reproduction or matures to its adult form.

Fungus Heterotrophic eukaryotic organism with a chitin-containing cell wall.

Helminth A worm; parasitic helminths often have complex life cycles.

Intermediate Host Organism in which a parasite undergoes asexual reproduction or is found only in its immature form.

Mycosis Disease caused by fungal infection.

Protists Eukaryotes that are usually single-celled and are not fungi, plants, or animals.

Protozoa Protists that are not algae.

Saprotroph Organism that takes in nutrients from dead and decaying matter.

Yeasts Unicellular fungi that reproduce by budding.

A Glimpse of History

Sometimes, a single event can change the course of history. A water mold, *Phytophthora infestans,* which causes a disease called late blight in potatoes, contributed to just such an event—the Irish Potato Famine. Beginning in 1845, the disease ruined much of the potato crop in Ireland, and the Irish faced starvation. By time the famine ended 7 years later, an estimated 1 million people had died, while another 1 million moved to other countries; the population of Ireland fell by about 25%.

Two hundred years before the famine, Spaniards had brought potatoes to Europe from South America. The potato became the main source of nutrition for many Irish people because it was easy to grow and provided a food source that could be stored for months. Irish dependence on this single food left them open to a major disaster.

Phytophthora infestans affects every part of the potato plant, including the leaves, stems, and tubers (potatoes); the infected potatoes turn to a black, mushy mess. Because the tuber is underground, an infection is often not noticed until it has destroyed the entire plant. The pathogen easily contaminates other fields as it can be spread by the wind and also by people handling plants.

Plant breeders and scientists still battle late blight, and the disease results in global losses of billions of dollars per year. It is particularly devastating for developing countries where many people still rely on potatoes as their main source of nutrition.

All eukaryotic organisms are in the domain Eukarya, a diverse group ranging from microscopic members to plants and animals. Microscopic eukaryotes were traditionally classified based on anatomical characteristics, but that approach is problematic because physical features do not necessarily reflect evolutionary relationships. DNA sequencing now makes it possible to examine these organisms at the molecular level, but developing an accurate classification scheme is still challenging, particularly considering the rate at which sequencing data is being generated. Imagine trying to develop a logical organization in a bookstore when new books are being delivered daily, and not all fit neatly into the existing categories. At what point do you separate a rapidly expanding accumulation of cookbooks into subcategories? And should a book entitled *Cooking for Engineers* be placed in the cookbook section, or the engineering section? The task of classification is not easy!

Partly because of the rapid changes in classification schemes, some microscopic eukaryotes are discussed in informal groups that do not reflect their evolutionary relatedness. For example, the **fungi** are heterotrophic organisms with chitin in their cell wall, and **protozoa** are heterotrophic unicellular organisms that are not fungi. The term **algae** is used to collectively refer to simple autotrophic (photosynthetic) eukaryotes. Protozoa and algae are often grouped together as **protists,** a catch-all category of eukaryotes that are usually single-celled and are not fungi, plants, or animals. This chapter also covers certain multicellular worms and arthropods that are medically significant and often have forms that are indeed microscopic.

Eukaryotic cells are characterized by the presence of a membrane-bound nucleus (*eukaryote* means "true nucleus"), which contains their chromosomes. Because the DNA is located in the nucleus whereas the ribosomes are in the cytoplasm, the processes of transcription and translation cannot occur simultaneously as they do in prokaryotic cells. Eukaryotic cells also have lysosomes, Golgi, and other membrane-bound organelles that allow compartmentalization of cell processes. Energy transformation in eukaryotic cells primarily occurs in mitochondria and chloroplasts where electron transport chains (ETCs) operate; in prokaryotic cells, ETCs are typically in the cytoplasmic membrane. Many eukaryotic cells do not have a cell wall, and when they do, the wall lacks peptidoglycan, the molecule that characterizes the bacterial cell wall. Finally, most eukaryotic cells have a well-developed cytoskeleton that is important in movement and maintenance of structure (see figure 3.25).

Eukaryotic genomes each have multiple chromosomes, and the cells usually contain two sets of these (recall that bacteria typically have a single chromosome). Cells that carry two sets of the chromosomes are diploid, indicated as $2n$. Some eukaryotic cells, however, are haploid, meaning they carry a single set of their chromosomes, indicated as n.

The process of cell division is fundamentally different between prokaryotes and eukaryotes. Recall that a prokaryotic cell reproduces asexually using a relatively simple process of binary fission (see figure 4.1). In contrast, a eukaryotic cell must replicate all of the multiple chromosomes within its nucleus before dividing. To do this, a eukaryotic cell uses a complex nuclear division process called **mitosis**, which produces two identical nuclei, each containing the same number and type of chromosomes as the original parent nucleus. In most cases, the cell then divides to form two identical daughter cells, each with a single nucleus. In some microorganisms, however, that division is not equal; for example some fungal cells undergo a type of asymmetric division, giving rise to a small daughter cell that buds from the mother cell (see chapter opening photograph). Additionally, in some protozoa, mitosis occurs repeatedly before division, after which a process called multiple fission creates many daughter cells essentially simultaneously. To distinguish symmetric division from budding and multiple fission, scientists who study eukaryotic microorganisms sometimes use the term *binary fission* to describe symmetric division; note, however, that binary fission in eukaryotic cells is preceded by mitotic division and is thus very different from the simple binary fission process used by prokaryotic cells. Regardless of the characteristics of eukaryotic cell division, as a result of mitosis, a diploid cell produces diploid daughter cells, whereas a haploid cell produces haploid daughter cells.

Eukaryotic cells involved in sexual reproduction use a nuclear division called **meiosis,** which allows a diploid cell to undergo two divisions in a way that produces four genetically distinct haploid daughter cells (**figure 12.1**). The production of haploid cells is necessary because sexual reproduction involves fusion of two distinct haploid cells—one from each of two parents—that carries a chromosome set from each. The sexual life cycles of eukaryotes can be categorized into three general groups:

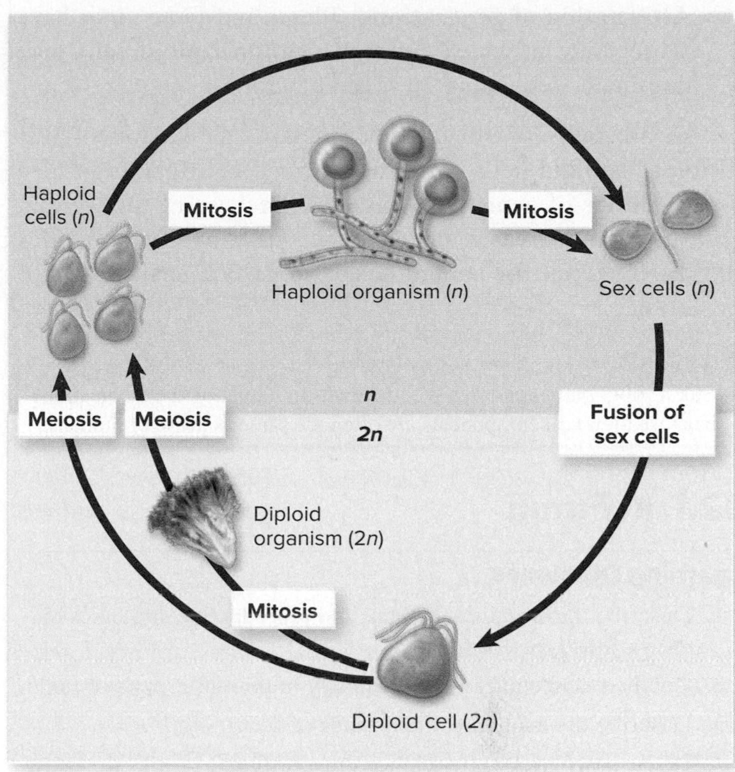

FIGURE 12.1 Cell Division Meiosis gives rise to haploid cells that can be used as sex cells (gametes) in sexual reproduction or, in some species, grow into a haploid organism by mitosis. Fusion of gametes or other sex cells forms a diploid cell that can grow into a diploid organism by mitosis or can undergo meiosis to form haploid cells.

? **In humans, is a gamete diploid? Explain.**

- **Diploid-dominant.** Many eukaryotic organisms, including humans and other animals as well as some protozoa, have this type of life cycle; the only haploid cells are sex cells called **gametes.** In animals, male gametes are called sperm and female gametes are called ova (or eggs). During sexual reproduction, male and female gametes fuse during fertilization to form a diploid cell called a **zygote.** This cell can then divide asexually, creating more diploid cells that, in the case of multicellular organisms, compose the organism's body.
- **Haploid-dominant.** Some eukaryotic microorganisms, including most fungi and some protozoa, have this type of life cycle; the only diploid cell that forms is the zygote arising from the fusion of two haploid cells. That zygote then undergoes meiosis to produce haploid cells, which each then divide asexually to produce more haploid cells. In these organisms, the haploid cells that fuse during fertilization are formed by the mitosis of haploid cells, not by meiosis.

- **Alternation of generations.** Plants and some algae have a life cycle in which either diploid or haploid cells may dominate, depending on several factors.

In all cases of sexual reproduction, the fusion of two different haploid cells to form a diploid cell results in new combinations of genetic information. Because of this, sexual reproduction increases the genome variation that provides the raw material for natural selection, which is the basis of evolution.

MicroByte

Eukaryotic pathogens can be difficult to target with medication because their cell components are often the same as those of humans.

12.1 ■ Fungi

Learning Outcomes

1. Describe the structure, habitat, and reproductive strategies of the various types of fungi.
2. Compare and contrast the medically important groups of fungi.
3. Describe the economic importance of fungi.
4. Discuss two symbiotic relationships of fungi.

The study of fungi (singular: fungus) is known as **mycology.** Forms of fungi include yeasts, molds, and mushrooms. These terms are unrelated to fungal classification and instead refer to morphological forms (**figure 12.2**):

- **Yeasts** are single-celled fungi.
- **Molds** are multicellular filamentous fungi.
- **Mushrooms** are simply the reproductive structures of certain fungi, similar to a peach on a peach tree. Some mushrooms are edible, but others are poisonous.

Fungi are heterotrophs and are among the main decomposers of organic materials, including lignin and cellulose (the main components of wood). In their key role as decomposers, fungi are **saprotrophs,** meaning they use nutrients from dead or decaying organic matter. They do this by secreting enzymes into the environment to break large molecules into smaller ones that they can then absorb. The overall decomposition process releases carbon dioxide and nitrogen compounds, which are then taken up by various organisms and again converted into organic compounds. Without this recycling, the earth would quickly be overrun with organic waste.

Some fungi can absorb nutrients from living tissue, acting as parasites (meaning they live at the expense of their host). The fungus *Batrachochytrium dendrobatidis* infects the skin of frogs and is believed to be responsible for the catastrophic decline in frog populations over the past two decades. Relatively few fungi infect humans—although athlete's foot and vaginal yeast infections are fairly common—but fungal diseases of plants can cause devastating destruction. More often, fungi form mutually beneficial partnerships with other organisms; for example, lichens are an association of fungi and algae that can grow on surfaces where neither partner can survive alone.

Characteristics of Fungi

Fungi are characterized by a cell wall that contains chitin—the same polysaccharide found in the exoskeleton of insects. Their wall is somewhat stronger than the cellulose-based cell wall of plants and is chemically distinct from the peptidoglycan cell wall of bacteria. In addition to chitin, the fungal cell wall typically contains glucan, a polymer of glucose. Glucan synthesis is the target of certain antifungal medications.

Fungal membranes generally contain ergosterol, distinguishing them from animal cell membranes, which contain cholesterol. Ergosterol is also the target for many antifungal medications.

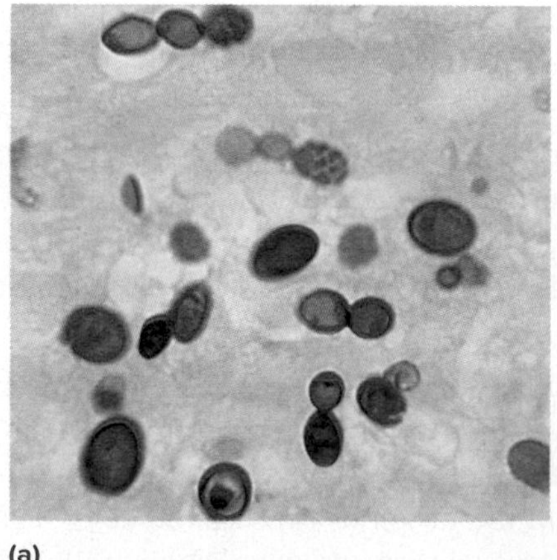
(a)

(b)

(c)

FIGURE 12.2 Morphology of Fungi (a) Photomicrograph of stained tissue containing *Histoplasma capsulatum.* **(b)** Mold on an orange. **(c)** *Amanita muscaria,* a highly poisonous mushroom. a: Dr. Libero Ajello/CDC; b: Fotosr52/Shutterstock; c: Jorgen Bausager/Folio Images/Getty Images

? Are fungi protists? Explain.

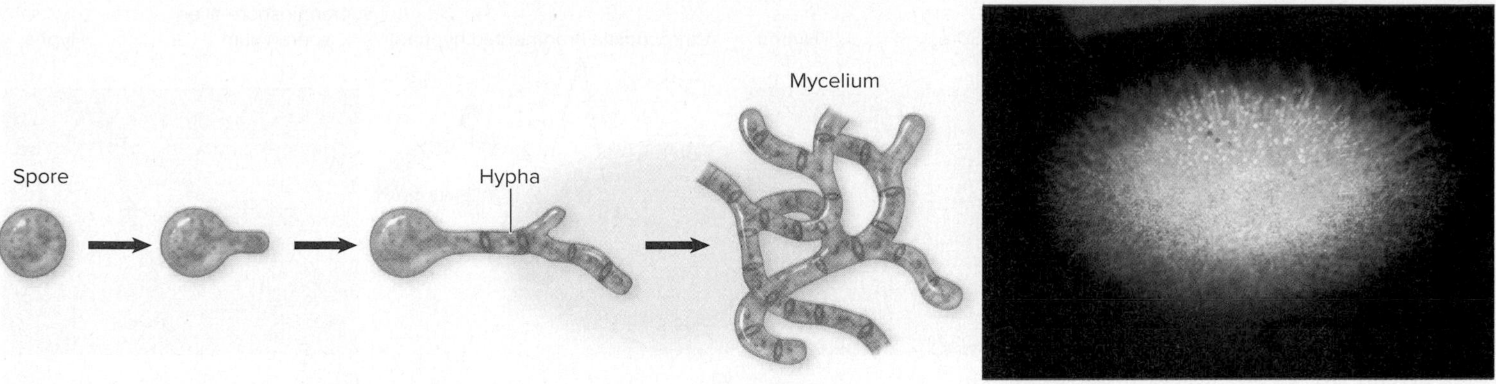

FIGURE 12.3 Hyphae and Mycelium Fungal spores (reproductive structures) germinate and then elongate to form hyphae that intertwine to form a mycelium. The white mass in the photograph is a mycelium. Don Rubbelke/McGraw Hill

? **What chemical in the fungal membrane is a target of many antifungal medications?**

Fungal Structure

Most fungi grow only as molds, although some grow only as yeasts. Fungi referred to as **dimorphic fungi** can transition between two morphologies—usually molds and yeasts—generally depending on the environmental conditions. Molds are characterized by intertwined thread-like multicellular filaments called **hyphae** (singular: hypha) (**figure 12.3**). A visible mass of hyphae is a **mycelium** (plural: mycelia). Most fungal species have septate hyphae, meaning that cross walls (septa) separate the cells; pores in the septa allow cytoplasm to flow from cell to cell. In some species, septa are rare or absent because the cells undergo mitosis without accompanying cell division, resulting in a filament with many nuclei. Because fungi are generally not motile, they cannot move toward a food; instead, the tips of what are referred to as vegetative hyphae respond to a nutrient source by growing in that direction. The surface growth you see on moldy food is composed of aerial hyphae, which give rise to spores (a generic term for the reproductive cells). The powdery or fuzzy and sometimes colorful appearance of molds is due to the aerial hyphae and spores (figure 12.2b).

MicroByte

The largest organism on Earth, called the humongous fungus, has a mycelium that spans over 2,200 acres in Oregon.

Fungal Habitats

Fungi are mainly terrestrial organisms but are found in nearly every habitat on Earth, including the thermal pools at Yellowstone National Park, volcanic craters, and bodies of water with high salt content, such as the Great Salt Lake and the Dead Sea. Some species can grow in concentrations of salts, sugars, or acids that inhibit the growth of most bacteria, so they are often responsible for spoiling pickles, jams, and other foods. They also grow better than most bacteria on acidic fruits and vegetables and on relatively dry foods like bread. Although fungi typically grow best between 20°C and 35°C, they can easily survive at refrigeration temperatures and below. They are most successful in moist environments, which is why molds become problems in damp rooms.

As a group, fungi can degrade leather, cork, hair, wax, ink, jet fuel, carpet, drywall, and even certain synthetic plastics. Some fungi are widespread because they can use a variety of carbon and energy sources; others occur only on a particular strain of one genus of plants. Most fungi are obligate aerobes, but some yeasts are facultative anaerobes, producing ethanol by fermentation. An unusual group of obligately anaerobic fungi live in the specialized stomach that characterizes cows and other mammalian herbivores called ruminants, helping those animals digest grasses.

Fungal Reproduction

The reproductive forms of fungi are very important in identification and classification. In medical microbiology, the forms that develop asexually are the most useful, because they can be seen and identified in pure cultures grown in the laboratory. The sexual forms play an important role in fungal classification.

The asexual reproductive cells of most molds are called **conidia,** but in one group they are referred to as **sporangiospores.** (However, the term *spore* is often used generically to refer to reproductive cells formed either sexually or asexually.) The conidia of some fungal species are held on conidia-bearing structures called conidiophores (*phore* means "to bear"). Other fungal species produce what are referred to as arthroconidia, which result from fragmentation of hyphae. In contrast, sporangiospores form within a sporangium (plural: sporangia). The conidia (including arthroconidia) and sporangiospores are small and easily carried by wind or water. They are somewhat resistant to drying and other unfavorable conditions, and they can persist for years until conditions improve; upon germination, they begin to grow hyphae in the direction of a food source. Laboratory identification of a mold often involves microscopically examining a culture

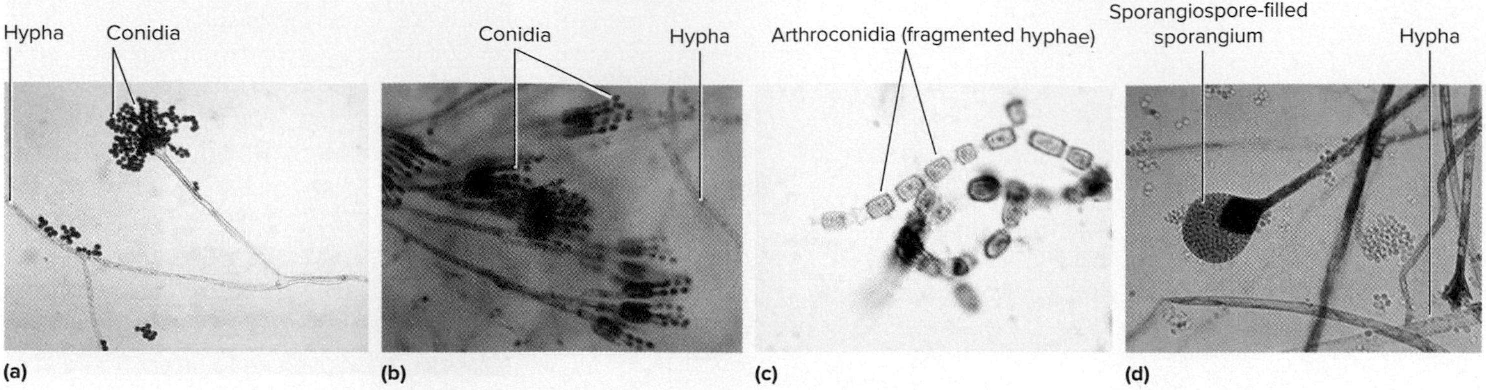

FIGURE 12.4 Asexual Reproductive Structures of Molds **(a)** *Aspergillus fumigatus;* **(b)** *Penicillum* species; **(c)** *Coccidioides immitis;* **(d)** *Mucor* species. a: CDC; b: Michael Abbey/Science Source; c: Dr. Hardin/CDC; d: Dr. Lucille K. Georg/CDC

? Why are fungal asexual reproductive structures particularly important in a clinical laboratory?

to observe the characteristic asexual reproductive structures (**figure 12.4**).

Yeast cells reproduce asexually by budding (see figure 12.2a). In this process, the nucleus divides by mitosis and one nucleus migrates into a smaller daughter cell, or bud (**figure 12.5**). The daughter cell then pinches off from a larger parent cell.

Sexual reproduction of fungi results when hyphae from two different mating types grow toward one another and fuse, creating a cell that contains both haploid nuclei. Because the two mating types have no obvious differences, mycologists refer to them as "+" and "–." In the case of certain fungi, the haploid nuclei then divide with each cell division, producing hyphae composed of cells each containing two genetically distinct haploid nuclei; those nuclei can eventually fuse and then undergo meiosis, forming haploid spores. Note that fungal spores are quite different from bacterial endospores, which are much more resistant to environmental conditions and are not a means of reproduction. Some of the sexual forms are commercially valuable; mushrooms, for example, are structures some fungi produce to hold their spores (see figure 12.2c).

The sexual stage of some fungi has never been seen, which made it difficult to classify these organisms before DNA sequencing methods were developed. Further complicating matters, many fungi have been inadvertently "discovered" twice—once based on the sexual reproductive forms and once based on the asexual forms—so they were known by two different names!

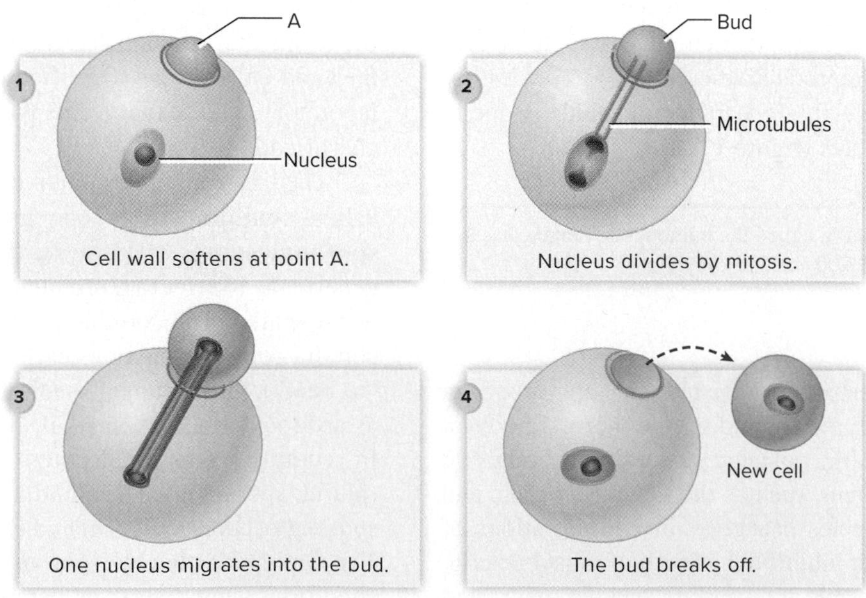

FIGURE 12.5 Budding in Yeast The nucleus divides by mitosis, and one nucleus moves into the bud. A cell wall is laid down beneath the bud. When the bud breaks off, it forms a new cell.

? What information in this figure indicates that the process generates a bud that has the same number of chromosomes as the original cell?

Classification of Fungi

Classification of fungi is in a state of flux and will continue to be so as the genomes of many more species are sequenced. About 100,000 fungal species have been described, and the genomes of over 3,000 have been sequenced, but studies indicate that there are perhaps 1.5 million species in nature. We will focus on only the three major fungal phyla that have medically important examples.

Ascomycota

Sexual reproduction of ascomycetes, or sac fungi, gives rise to ascospores, which are held within a sac called an ascus. In many cases, the sexual form of an ascomycete has never been seen, but DNA evidence allows classification.

In laboratory cultures, some ascomycetes grow as yeasts, some grow as molds with septate hyphae, and some are dimorphic, depending on the environment. Yeasts are typically identified by their metabolic traits, but some have a characteristic appearance that guides identification; molds are usually identified based on the appearance of their asexual reproductive structures (see figure 12.4).

Considering that about 75% of all known fungi are ascomycetes, it should not be surprising that the group includes most of the medically important fungi (described shortly). Other ascomycetes include the source of the first antibiotic discovered (penicillin, produced by a *Penicillium* species); fungi that make reproductive structures highly prized for their flavor (morels and truffles); and plant pathogens such as those that cause Dutch elm disease.

Basidiomycota

Sexual reproduction of basidiomycetes, or club fungi, gives rise to basidiospores, which are held on a club-like structure called a basidium. In the case of basidiomycetes that form mushrooms, the basidiospores are produced on the gills found on the underside of the mushroom cap (**figure 12.6**).

Many types of mushrooms are collected and grown for food, although some are deadly. Basidiomycetes also include the plant pathogens referred to as rusts and smuts—terms that reflect the appearance of their growth on the infected plant. Rusts and smuts cause significant losses in wheat, rye, and corn crops.

Mucoromycota

Sexual reproduction of mucoromycetes, commonly referred to as zygomycetes to reflect the previous classification, gives rise to zygospores, which are held within a zygosporangium that forms between hyphae of different strains. Characteristics that aid in laboratory identification of the mucoromycetes are their relatively wide hyphae with few or no septa and the production of sporangiospore-containing sporangia (see figure 12.4d). The common black bread mold *Rhizopus stolonifer* is a mucoromycete.

Characteristics of the major groups of fungi are summarized in **table 12.1.**

Groups of Medically Important Fungi

Relatively few fungal species infect humans. Life-threatening fungal infections are rare in otherwise healthy people, but in immunocompromised patients they are much more common and devastating. Fungal infection is called **mycosis.** Disease-causing fungi are described in more detail in the disease chapters of the textbook, but this section serves as a short introduction.

Medical mycologists separate pathogenic fungi into various groups based on the types of diseases they cause (**table 12.2**):

- **Superficial fungi.** These infect skin, hair, or nails. Fungi referred to as **dermatophytes** infect keratinized tissues, resulting in conditions such as athlete's foot and jock itch (see figure 22.19). Another superficial fungus is *Malassezia furfur,* which causes patches of skin discoloration called tinea versicolor (see figure 22.20).

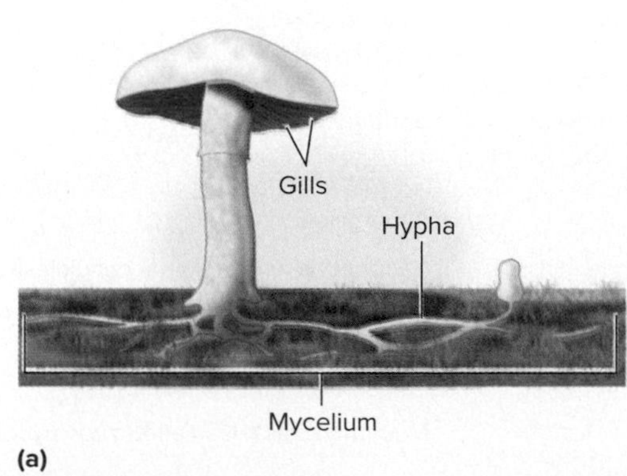

(a)

(b)

FIGURE 12.6 Mushrooms (a) Mushrooms are fruiting bodies produced by some basidiomycetes. **(b)** The basidospore-bearing gills are on the underside of the mushroom cap. b: Simon McGill/Getty Images

? With respect to fungal classification, what is the significance of mushroom gills?

TABLE 12.1 Characteristics of Major Phyla of Fungi

Phylum	Characteristics	Asexual Reproduction	Sexual Reproduction
Ascomycota	Includes about 75% of characterized fungi; some grow as yeasts, some grow as molds that have septate hyphae, and some are dimorphic	Yeasts bud; molds produce conidia on conidiophores or fragment to produce arthroconidia	Ascospore-containing ascus (sac)
Basidiomycota	Includes fungi that form mushrooms	Commonly absent; yeasts bud	Basidiospores borne on club-shaped structures
Mucoromycota (formerly Zygomycota)	Includes fungi that grow as molds characterized by wide hyphae with few or no septa	Sporangiospores within a sporangium	Zygospores within a zygosporangium

- **Subcutaneous fungi.** These infect deeper layers of the skin. An example is *Sporothrix schenkii,* which causes sporotrichosis ("Rose gardener's disease").
- **Opportunistic fungi.** These cause disease only under certain conditions. Some, such as *Candida albicans* and *Pneumocystis jirovecii,* are members of the normal microbiota. Others, including *Aspergillus* species, *Cryptococcus neoformans,* and *Rhizopus* species, are commonly found in the environment.
- **Systemic fungi.** These can cause systemic disease even in otherwise healthy people. Most of the systemic fungi are dimorphic; they typically grow in the soil as the mold form, and their conidia easily become airborne. When inhaled into the warm, moist environment of the lungs, the conidia develop into another form (most examples develop into a yeast, but one becomes what is called a spherule). *Coccidioides immitis* and *Histoplasma capsulatum* are examples of dimorphic fungi that cause systemic diseases.

In addition to the fungal groups that cause various infections, some fungi produce toxins. **Aflatoxins,** produced by *Aspergillus* species, are considered carcinogenic, so the U.S. Food and Drug Administration (FDA) monitors their levels in foods such as grains and peanuts. The rye mold *Claviceps purpurea,* also known as ergot, produces a toxin with hallucinogenic properties. Strange behavior in people eating contaminated rye may have led to the accusations of witchcraft in Salem, Massachusetts, in the late 1600s. The chemical has been purified to yield ergotamine, a drug that decreases blood flow and is used to relieve migraine headaches. *Amanita* species produce powerful toxins that can cause fatal liver damage (see figure 12.2c).

Some fungi are associated with hypersensitivity reactions, meaning an allergic reaction to the fungal components. People with allergies often monitor published pollen and mold counts and avoid unnecessary outdoor activity when levels are high.

TABLE 12.2 Examples of Medically Important Fungi

Group	Typical Growth	Disease Example(s)	Phylum
Superficial Fungi			
Dermatophytes (*Epidermophyton, Microsporidium,* and *Trichophyton*)	Mold	Athlete's foot, jock itch	Ascomycota
Malassezia furfur	Yeast in culture; yeast and hyphae in tissue	Tinea versicolor	Basidiomycota
Subcutaneous Fungi			
Sporothrix schenkii	Mold in the environment; yeast in tissue	Sporotrichosis	Ascomycota
Opportunists			
Aspergillus species	Mold	Aspergillosis	Ascomycota
Candida albicans	Yeast in culture; yeast and hyphae in tissue	Candidiasis (vulvovaginal candidiasis; thrush; bloodstream infections)	Ascomycota
Cryptococcus neoformans	Yeast (encapsulated)	Cryptococcocal meningitis	Basidomycota
Pneumocystis jirovecii	Does not grow in culture; cysts in tissue	Pneumocystis pneumonia	Ascomycota
Rhizopus species	Mold	Mucormycosis (also called zygomycosis)	Mucoromycota
Systemics			
Coccidioides immitis	Mold in the environment; spherules in tissue	Coccidiomycosis	Ascomycota
Histoplasma capsulatum	Mold in the environment; yeast in tissue	Histoplasmosis	Ascomycota

Economic Importance of Fungi

Many fungi are important commercially. Some, for example, synthesize life-saving antimicrobial medications like penicillin. In addition, yeasts have been genetically engineered to produce a variety of medically useful molecules, such as human insulin and a vaccine against hepatitis B. Fungi are also useful tools for studying complex eukaryotic events—such as cancer and aging—within a simple cell. The first eukaryotic genome sequenced was that of the yeast *Saccharomyces cerevisiae,* a model organism used in a variety of genetic and biochemical studies. *S. cerevisiae,* also known as brewer's yeast or baker's yeast, has long been used in the production of wine, beer, and bread. Other fungal species are used to make the large variety of cheeses produced throughout the world. Ironically, fungi are also among the greatest spoilers of food products; millions of tons of food are thrown away each year because fungal growth has made it inedible.

Fungi cause a number of plant diseases that result in billions of dollars worth of losses due to crop damage and expenditures on preventive measures. Dutch elm disease, caused by the fungus *Ophiostoma ulmi* (*Ulmus* is a genus of elm), dramatically changed the landscape of many U.S. cities when the elm trees lining streets and surrounding public buildings were killed. The cost of removing these diseased and dying trees was significant. In 2019, a devastating fungal disease of banana plants that has been slowly spreading around the world was detected in Columbia, one of several major banana-growing countries in Latin America. That disease—caused by a strain of *Fusarium oxysporum* referred to as TR4 (for tropical race 4)—can wipe out entire banana plantations. As a result, Columbia called a state of emergency in an effort to stop the disease's spread. Grain crops are particularly vulnerable to fungal infection. A fungal disease called "wheat blast" that devastated wheat crops in South America spread to Bangladesh in 2016, resulting in the loss of over 35,000 acres of crops that year; it has since been found in Zambia, raising concerns about its potential spread to other parts of Africa.

Symbiotic Relationships of Fungi

A **lichen** is a symbiotic life-form that results from the intimate association of a fungus with a photosynthetic microorganism such as an alga or a cyanobacterium. The fungal partner in the lichen absorbs water and minerals for the symbiotic pair, in addition to providing protection; the photosynthetic member supplies the fungus with organic nutrients (**figure 12.7**). Recent molecular and genetic studies of lichens have revealed that the associations may include yeasts and non-photosynthetic bacteria in addition to the already recognized fungal and photosynthetic partners. Lichens are found in terrestrial habitats all over the world and are often the first life to appear on bare rock, where they can begin the process of soil formation. They can grow in extreme ecosystems where neither partner could survive alone, such as in sub-Arctic tundra where they are the primary diet of reindeer. In spite of their hardiness in certain environments, lichens are easily killed by polluted air because the organisms absorb—but cannot excrete—common contaminants such as toxic metals, sulfur dioxide, and ozone; thus, they are often a good indicator of air quality and are not common in industrial cities.

Some fungi grow in a mutually beneficial association with plant roots, forming **mycorrhizas** (**figure 12.8**). The high surface area of fungal hyphae increases the plant's ability to absorb water and minerals. The fungus also supplies

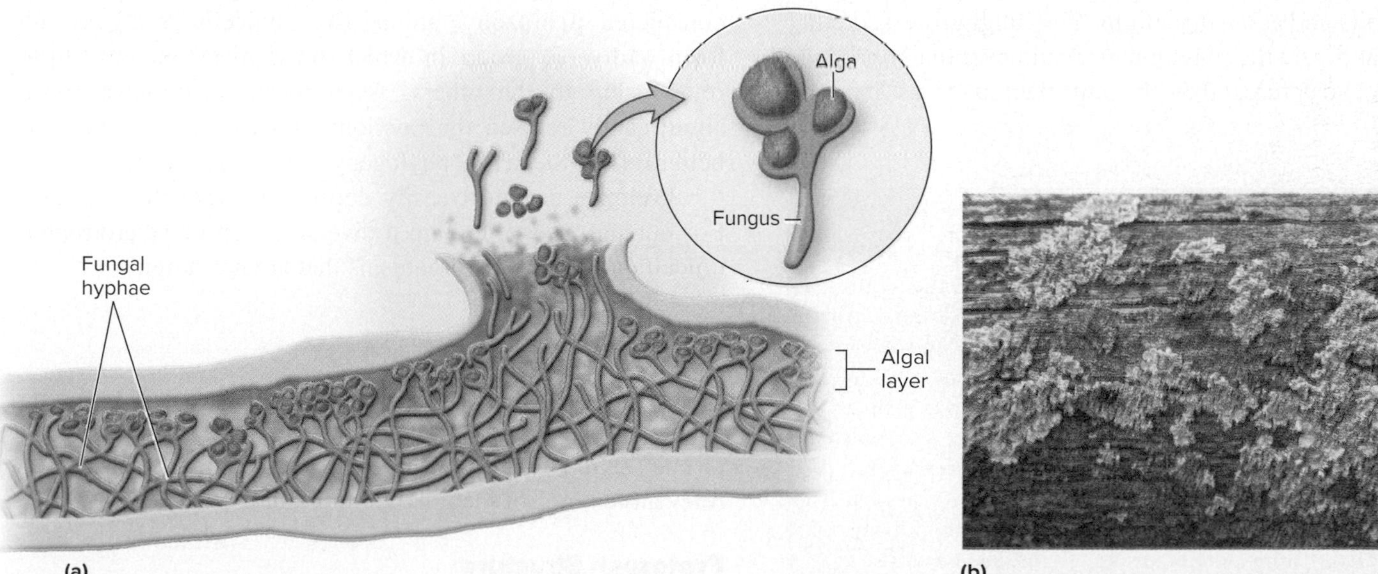

FIGURE 12.7 Lichens (a) A lichen includes photosynthetic cells intertwined within the hyphae of a fungal partner. **(b)** Lichens on a fallen tree.
Núria Talavera/Getty Images

? **What does the fungus contribute to the lichen?**

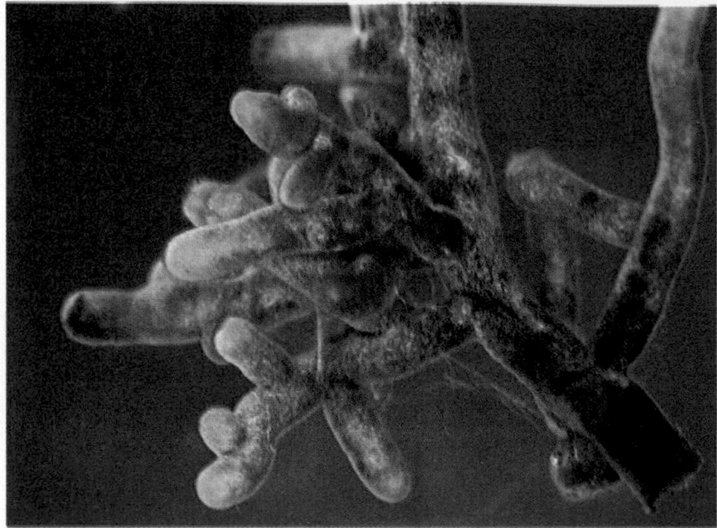

FIGURE 12.8 Mycorrhizas Fungi (seen here in white) form intimate relationships with the roots of most green plants. They supply the plant with nitrogen and phosphorus and increase the plant's ability to absorb water. Ellen Larsson

Are mycorrhizas considered parasites? Explain.

the plant with nitrogen and phosphorus from the breakdown of organic material in the soil. The plant, in return, supplies the fungus with organic compounds. An estimated 80% of vascular plants have some type of mycorrhizal association, and plants with mycorrhizas grow better than those without. Most orchids cannot germinate without mycorrhizas that help provide nutrients for the young plant.

Certain insects also depend on symbiotic relationships with fungi. Leaf-cutter ants farm their own fungal gardens (**figure 12.9**). The ants cannot eat tropical vegetation because the leaves are often poisonous. Instead, the ants chop the plant leaves into bits and add a mycelium. The fungi grow, secreting enzymes that digest the plant material and eventually produce reproductive structures that the ant then uses as its food source.

FIGURE 12.9 Leaf-Cutter Ants These ants carry food for fungi, whose reproductive structures then serve as a food source for the ants. imageBROKER/Alamy Stock Photo

Do leaf-cutter ants eat leaves? Explain.

MicroAssessment 12.1

As saprotrophs, fungi are important recyclers of carbon and other elements. Fungi typically have chitin and glucan in their cell walls, and their membrane contains ergosterol. Most fungi grow only as molds, but some grow only as yeasts; some species are dimorphic. Fungi are classified based on characteristics of their sexual reproductive forms, but fungal identification is primarily based on the asexual reproductive forms. The phyla Ascomycota, Basidiomycota, and Mucoromycota include some medically significant members. Many fungi are commercially important, but some cause devastating plant diseases. Some fungi form symbiotic relationships with other organisms, such as those of lichens or mycorrhizas.

1. What growth feature characterizes most fungi that cause systemic disease in otherwise healthy people?
2. Explain how the symbiotic association between fungi and green plants benefits each partner.
3. Fungi increase their conidia formation when food supplies are diminishing. Why would this be the case?

12.2 ■ Protozoa

Learning Outcomes

5. Describe the structure and habitats of protozoa.
6. Compare and contrast the groups of medically important protozoa.
7. Explain why slime molds, dinoflagellates, and euglenids highlight the difficulties of classification.

DNA sequencing studies show that the organisms classically considered **protozoa** ("animal-like" unicellular organisms) form a diverse group in which most members bear little relationship to the others. As a result, taxonomists have significantly revised the previous classification schemes to better reflect the actual phylogeny of the organisms.

Protozoa are most easily defined by what they are not. For our purposes, they are a diverse group of heterotrophic unicellular eukaryotic organisms that are not fungi.

Characteristics of Protozoa

The diversity of protozoa is easily seen in their wide-ranging characteristics, which are far too varied to adequately cover in one section. Thus, our discussion will mainly focus on the relevant features of medically important protozoa.

Protozoan Structure

Most protozoa lack a cell wall, although certain groups common in marine environments produce either calcium carbonate shells or silica exoskeletons. Many protozoa have specialized structures or mechanisms for movement, such as cilia, flagella, or pseudopods ("false feet"), and these are often used for identification, particularly in a clinical lab. As described in chapter 3, eukaryotic flagella and cilia are distinctly different from prokaryotic flagella (compare figures 3.14 and 3.26).

Several types of protozoa lack mitochondria, apparently as a result of gene loss; this characteristic is most common in protozoa adapted to live in anaerobic environments, such as the colon. Some of these protozoa have alternative organelles that play a role in energy transformation.

Certain types of protozoa can exist both as a **trophozoite** (growing, feeding form) and as a **cyst** (infectious survival form). In the case of intestinal examples, the trophozoites feed and multiply within the intestinal tract and then develop into cysts as they slowly exit in feces (figure 24.21). Unlike trophozoites, cysts can survive in the relatively harsh environmental conditions outside of the host. When ingested by a host, cysts survive passage through the stomach and then excyst to release trophozoites. Thus, the infectious form (the one transmitted from one host to the next) is the cyst, not the trophozoite.

A group of intestinal protozoa called coccidia produce an environmentally resistant form called an **oocyst.** Like the cysts just described, this is an infectious form that can be transmitted to the next host. Oocysts, however, develop as part of the sexual stage that characterizes the coccidia and related organisms. In some cases, the coccidian oocysts require time outside the host to mature into the infectious form.

Protozoan Habitats

The majority of protozoa are free-living aquatic organisms that are essential decomposers in many ecosystems. They make up part of the **zooplankton** (*zoo* means "animal" and *planktos* means "drifting"), the small animals and heterotrophic microorganisms that drift in oceans and other large bodies of water. On land, protozoa are abundant in soil as well as in or on plants and animals. Other protozoan habitats include the guts of termites, cockroaches, and ruminants such as cattle. Some protozoan species are parasitic, living at the expense of a host.

Protozoa are an important part of the food chain. They help maintain ecological balance by devouring large numbers of bacteria and algae and, in turn, serve as food for larger species. A single paramecium (a type of protozoan) can ingest as many as 5 million bacteria in one day.

Groups of Medically Significant Protozoa

The majority of protozoa do not cause disease, but those that do have significant impacts on global health. They are described in more detail in the disease chapters of the textbook, but this section serves as a short introduction. A table at the end of the discussion summarizes the characteristics of the protozoa described here (see table 12.3).

Protozoa have traditionally been grouped primarily based on their mode of locomotion, but DNA sequencing data show that separate events in evolutionary history can give rise to similar structures or characteristics; thus, organisms that share a common method of movement (such as pseudopods, flagella, or cilia) are not necessarily closely related. Classifications have been updated to reflect the sequencing data, but although the new groupings are helpful with respect to understanding the evolution of the organisms, they pose difficulties for medical microbiologists accustomed to the traditional ones. This is particularly true when phenotypic characteristics provide the basis for laboratory identification of a given pathogen. With that in mind, this section groups medically significant protozoa according to characteristics emphasized in a clinical lab. A figure at the end of the next section illustrates a current phylogenetic grouping of the protists (protozoa and algae) described in the chapter (see figure 12.18).

Amoebae

Amoebae (singular: amoeba) move by extending and retracting portions of their cytoplasm; the combined actions send out pseudopods that not only move the cell forward, but also allow it to engulf particles. The shape of the cell changes as it moves, a characteristic termed amoeboid. Amoebae reproduce by the asexual process of mitosis followed by binary fission, which we will refer to as *mitotic binary fission.*

Intestinal amoebae are typically able to form cysts, allowing them to survive in the environment between hosts (**figure 12.10**). If these are ingested, the trophozoites will be released in the intestinal tract, where they then multiply. The cells lack mitochondria, so they rely only on fermentation for energy transformation, but some have structures called mitosomes; these appear to be relics of mitochondria, but

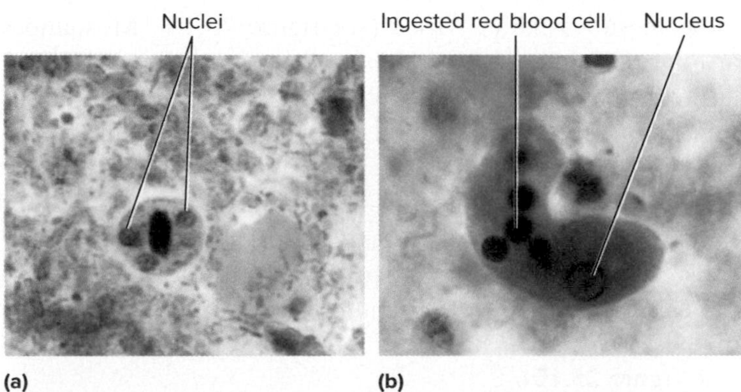

FIGURE 12.10 ***Entamoeba histolytica*** **(a)** Cyst, which typically has four nuclei; the large mass in the middle is a chromotoid body—an accumulation of RNA. **(b)** Trophozoite; ingested red blood cells are an identifying characteristic. a: Dr. Mae Melvin/CDC; b: Dr. N.J. Wheeler, Jr./CDC

? **Which form shown above can survive outside of the intestinal tract, and why is that important?**

their function is not understood. Most intestinal amoebae are commensals, meaning that they live in the intestinal tract without harming the host. *Entamoeba histolytica* is an exception; it is a pathogen that causes disease ranging from mild diarrhea to severe dysentery (diarrhea characterized by blood and mucus in the stool).

Naegleria fowleri is normally free-living in warm waters, but if it gains access to the sinuses, it can then invade and destroy brain tissue. It has three forms: an amoeboid trophozoite; a temporary flagellated cell that can swim to new locations; and a cyst that can survive environmental stress.

Apicomplexans

Apicomplexans are sometimes referred to as sporozoa, a term that reflects an earlier classification scheme. The cells have a characteristic structure called an apical complex at one end, which helps the organism penetrate the cytoplasmic membrane of host cells. Most also have an essential structure called an apicoplast, which appear to have evolved from an endosymbiotic bacterium; it is significant medically because it makes apicomplexans susceptible to certain antibacterial medications. Medically important apicomplexans are obligate intracellular parasites, meaning they require a host cell in which to replicate.

Apicomplexans have complicated life cycles that involve sexual and asexual reproduction, often in different hosts. If different hosts are required, the one in which sexual reproduction occurs is called the **definitive host;** any other host is referred to as an **intermediate host.** The sexual and asexual reproductive processes often involve multiple fission, resulting in many progeny being produced simultaneously.

Apicomplexans include the *Plasmodium* species that cause malaria, making them one of the most significant causes of infectious disease in the world. The World Health Organization estimates that in 2021, 247 million people suffered from malaria, resulting in 619,000 deaths, most of them in sub-Saharan Africa (see figure 25.16). Mosquitoes are the definitive host for *Plasmodium* species; humans are an intermediate host, with the parasites undergoing multiple rounds of asexual replication as they repeatedly infect and lyse red blood cells (RBCs). Within an infected RBC, the malarial parasite enlarges, undergoes multiple rounds of mitosis, and then divides by multiple fission to produce many cells called **merozoites** (**figure 12.11**). These are released as the RBC lyses, and they go on to infect other RBCs. Details of the *Plasmodium* life cycle will be covered in chapter 25 (see figure 25.15).

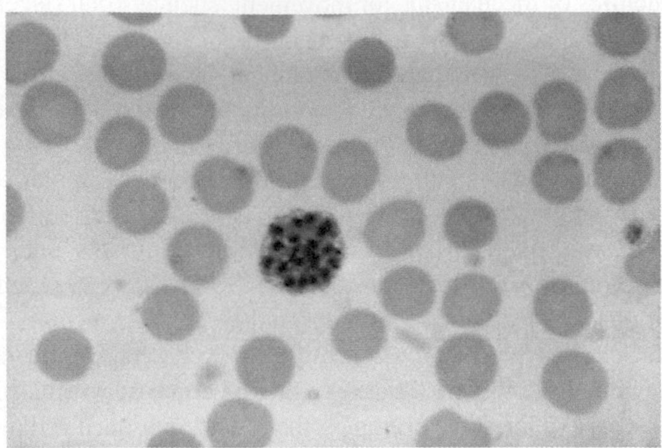

FIGURE 12.11 ***Plasmodium vivax*** **Within an Infected Red Blood Cell** This infected red blood cell contains 18 merozoites, a form that results from asexual reproduction. Dr. Mae Melvin/CDC

? What human disease do Plasmodium species cause?

Apicomplexans include the coccidia, the group of intestinal parasites mentioned earlier that produce environmentally resistant oocysts. The entire life cycle of two examples of coccidia—*Cryptosporidium hominis* and *Cyclospora cayetanensis*—can be completed in one host. Transmission, however, requires that oocytes produced in the intestinal tract be passed in feces. Once a mature oocyst is ingested by the next host, a form emerges that can penetrate and multiply in the intestinal cells of that host, often causing diarrhea. Many animals, including humans, can serve as hosts for *C. hominis*, but humans are the only host for *C. cayetanensis*.

Cats are the definitive host for another coccidian parasite, *Toxoplasma gondii;* oocytes are shed in the feces of an infected cat. Various animals, including humans, can serve as intermediate hosts that become infected by ingesting mature oocysts or the tissue cysts (described shortly). Once in that host, the parasite can enter and multiply in nearly any tissue, but it often infects brain cells. In humans this can lead to severe outcomes, including: (1) encephalitis in people who are immunocompromised, and (2) fetal infections that lead to stillbirth. The parasite multiplies rapidly in tissues until an immune response develops, at which point it forms a walled structure called a tissue cyst. A tissue cyst contains numerous *T. gondii* cells that can survive for the life of the host and reactivate if the immune response lessens. Cats become infected when they eat tissues of an infected intermediate host such as a mouse, thus completing the life cycle (see figure 26.14). Because cats do not usually eat people, humans are a dead-end host, meaning that the parasite cannot complete its life cycle.

Hemoflagellates

Hemoflagellates are a group of flagellated protozoa that multiply asexually by mitotic binary fission and live in the bloodstream for part of their life cycle. They all require a blood-feeding insect as part of that cycle.

Hemoflagellates are part of a related group called kinetoplastids. Members of this group are characterized by a single large mitochondrion that has an unusual feature called a kinetoplast—a distinctive mass of thousands of interlocking molecules of circular DNA. Because the structure is unique to the kinetoplastids, it may be a useful target for medications to treat diseases caused by this group.

Several hemoflagellates cause devastating diseases. *Trypanosoma brucei* is transmitted by tsetse flies and causes African trypanosomiasis (sleeping sickness) (see figure 26.13). *T. cruzi* is transmitted by "kissing" bugs and causes American trypanosomiasis (Chagas disease). *Leishmania* species, which are transmitted by sandflies, cause leishmaniasis.

Lumen-Dwelling Flagellates

Lumen-dwelling flagellates colonize the hollow inside portion (the lumen) of either the intestinal tract or the genital tract, using flagella for movement. The medically important examples reproduce asexually by mitotic binary fission.

Giardia lamblia, a flagellate that lives in the intestinal tract, is among the leading causes of diarrhea worldwide. Like the intestinal amoeba, *G. lamblia* forms infectious cysts that survive in the environment (see figure 24.21). Because the cysts can be found in streams and lakes, people who drink directly from those water sources can become infected. After a cyst is ingested, characteristic leaf-shaped flagellated trophozoites emerge and multiply (see figure 24.20). The cells lack mitochondria but have mitosomes (structures that appear to be remnants of mitochondria).

Trichomonas vaginalis is a sexually transmitted flagellate that colonizes the genital tract. It lacks mitochondria but forms a unique structure called a hydrogenosome, which produces some ATP while generating molecular hydrogen (H_2).

Characteristics of some medically important protozoa are summarized in **table 12.3.**

Other Protozoan Groups

Several protozoan groups are not only interesting, but they also highlight the difficulties of classification.

Slime Molds

Slime molds were once considered types of fungi. Although they may look and act like fungi, at a cellular and molecular level, they are more closely related to *Entamoeba* species.

Social Amoebae (Cellular Slime Molds) Social amoebae are also known as cellular slime molds. These amoebae live as single cells that ingest bacteria growing on decaying vegetation, but when nutrients are scarce, they aggregate into a moving mass of cells referred to as a slug. Some of the cells then form a fruiting body, while others differentiate into spores (**figure 12.12**). Scientists use one species, *Dictyostelium discoideum,* as a model for studying cell aggregation and multicellular development.

TABLE 12.3 Examples of Medically Important Protozoa

Group/Genus of Pathogen	Characteristics	Disease
Amoebae	Cells move using pseudopods	
Entamoeba histolytica	Intestinal parasite; forms trophozoites and cysts	Amebiasis
Naegleria fowleri	Normally free-living but can invade through the sinuses to access and destroy brain tissue	Primary amebic meningoencephalitis
Apicomplexans	Cells have a characteristic apical complex that helps the parasite attach to or penetrate host cells; complex life cycle, involving sexual and asexual stages	
Cryptosporidium hominis	Many animal hosts, including humans	Cryptosporidiosis
Cyclospora	Humans are the only host	Cyclosporiasis
Toxoplasma gondii	Cats are the definitive host; many animals can be an intermediate host; humans are a dead-end intermediate host	Toxoplasmosis
Plasmodium species	Mosquitoes are the definitive host; humans are an intermediate host	Malaria
Hemoflagellates	Cells use flagella to move and are cell characterized by a kinetoplast	
Leishmania species	Transmitted by sandflies	Leishmaniasis
Trypanosoma brucei	Transmitted by tsetse flies	African trypanosomiasis (sleeping sickness)
Trypanosoma cruzi	Transmitted by kissing bugs	American trypanosomiasis (Chagas disease)
Lumen-Dwelling Flagellates	Cells use flagella to move	
Giardia lamblia	Intestinal parasite; forms trophozoites and cysts	Giardiasis
Trichomonas vaginalis	Sexually transmitted	Trichomoniasis

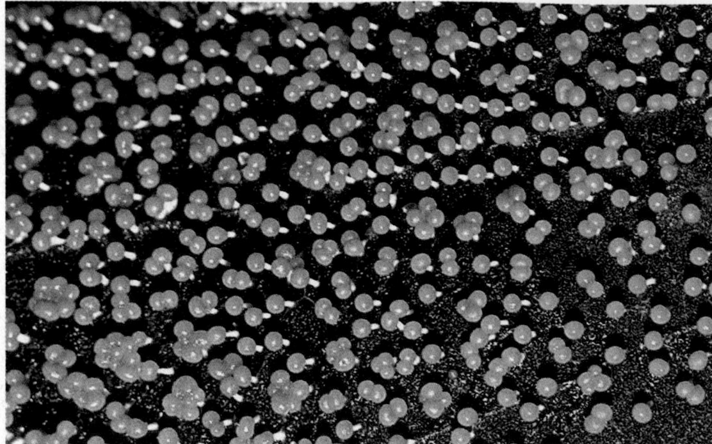

FIGURE 12.12 Fruiting Bodies of a Social Amoeba Richard Gross/McGraw Hill

? **Why are scientists particularly interested in social amoebae?**

Plasmodial Slime Molds Plasmodial slime molds are large, multinucleated "super-amoebae" that may easily reach 0.5 m in diameter. They are widespread and readily visible in their natural environment due to their large size and often bright color (**figure 12.13**). Following the germination of haploid spores, the cells fuse to form a diploid cell in which the nucleus divides repeatedly, forming a multinucleated stage called a **plasmodium.** The plasmodium oozes over the surface of decaying wood and leaves, ingesting organic debris and microorganisms. When food or water is in short supply, the plasmodium is stimulated to form spore-bearing fruiting bodies, and the process begins again.

Dinoflagellates

Most dinoflagellates are single-celled marine organisms; some are heterotrophs, but others are photosynthetic. Photosynthetic microorganisms are traditionally considered algae, but DNA studies indicate that dinoflagellates are more closely related to the apicomplexa than to algae. The photosynthetic dinoflagellates apparently arose when an ancestor engulfed an algal cell.

Regardless of their taxonomic placement, dinoflagellates are medically important because some species produce neurotoxins. Under warm and nutrient-rich conditions in coastal waters, the toxin-producing photosynthetic dinoflagellate *Karenia brevis* may grow to such high concentrations (a "bloom" or "harmful algal bloom") that the waters may take on a reddish coloration, referred to as a **red tide.** During such blooms, the toxin (brevetoxin) is produced in high

FOCUS ON A CASE 12.1

The patient was Dudley, a Golden Retriever mix brought to the veterinarian's office by his owner, Sam. "The Dude" was usually playful, but for several days he had simply sprawled in front of the sliding glass door and watched squirrels scamper across the patio. Sam was concerned when Dudley did not want to eat, and alarmed by Dudley's recent explosive diarrhea.

The vet asked if Dudley had done anything unusual in the past few weeks. Sam explained that they had gone camping a few weeks earlier. Further questioning revealed that while hiking, Dudley had lapped water directly from a clear stream. Sam groaned as the vet nodded his head and asked the technician to gather supplies to take a stool sample from Dudley. The vet suspected that "The Dude" might be suffering from giardiasis caused by ingesting cysts of *Giardia lamblia,* a protozoan sometimes found in clear streams.

1. What happens to ingested cysts of *G. lamblia*?
2. How can a stool sample confirm the vet's suspicions that Dudley was infected with *G. lamblia*?
3. Once the diagnosis of giardiasis was confirmed, the vet wrote a prescription for metronidazole, a medication used to treat infections caused by anaerobic organisms. Why would this medication be a common choice for treating this disease?

Discussion

1. The cysts, which are surrounded by a thick wall that protects them from the acidity of stomach secretions, pass through the stomach. Once in the small intestine, trophozoites are released, and these attach to the intestinal wall and multiply, often causing diarrhea and cramping. When trophozoites detach and slowly pass through the intestine, they develop into the cyst form. The cysts contained in feces are highly infectious and, if ingested (such as in contaminated water or food), begin the cycle in a new host.
2. If Dudley has diarrhea due to rapid movement of waste materials through the digestive tract, microscopic examination of a fecal smear may reveal the trophozoite form of the parasite. If waste materials are moving slowly enough to form solid stools, then the trophozoites will have time to develop into cysts before exiting the body. A fecal flotation test can reveal the presence of cysts. Feces are mixed with a test solution that is denser than the cysts; the cysts then float to the top of the mixture where they can be skimmed off and examined under the microscope. Shedding of the parasite is often intermittent, so microscopic examination of a stool sample may not reveal its presence. A test called ELISA can also be used to detect the parasite in the sample (see figure 17.10).
3. *G. lamblia* lacks mitochondria and therefore does not use typical aerobic cellular respiration. Unlike most other eukaryotes, it produces an enzyme similar to one made by anaerobic bacteria that results in the transfer an electron to metronidazole; this activates the medication, which then binds to and damages DNA, thus killing the parasite.

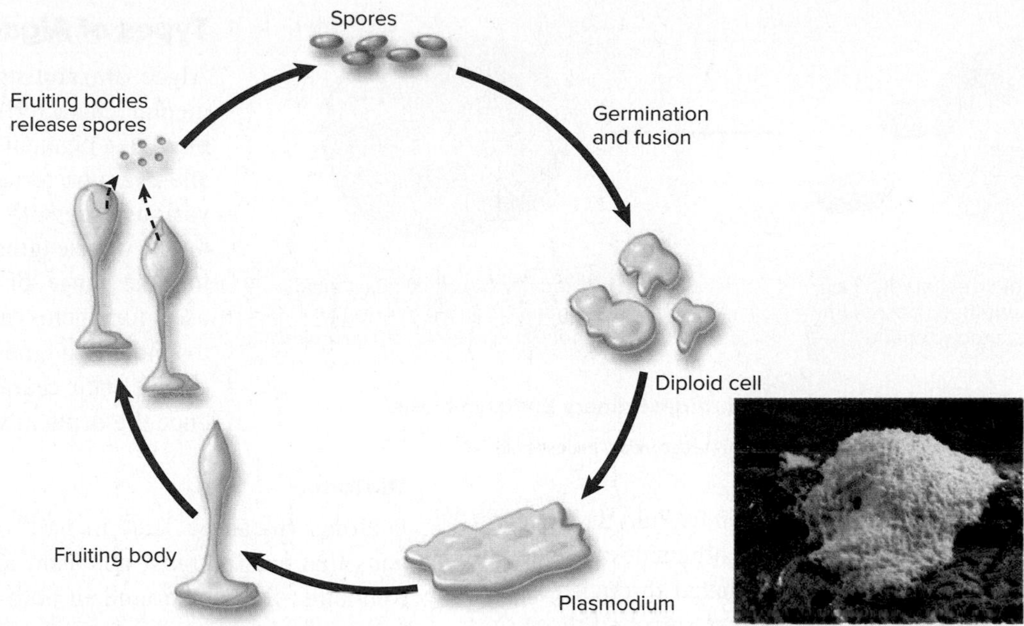

FIGURE 12.13 Plasmodial Slime Mold Haploid spores germinate and fuse to form a diploid cell in which the nucleus divides repeatedly, resulting in a multinucleated plasmodium. Laurie Knight/Getty Images

? **What are the activities of this type of slime mold when it is in the plasmodial form?**

enough quantities to kill fish and other marine life. The toxin is also concentrated in shellfish, posing a risk of neurotoxic shellfish poisoning (NSP) to people who eat them. The contaminated shellfish appear normal, and cooking does not destroy the toxin. Symptoms include gastrointestinal distress, numbness of parts of the face, and dizziness. A similar situation can lead to paralytic shellfish poisoning (PSP), caused by a toxin (saxitoxin) produced by several genera of photosynthetic dinoflagellates.

Euglenids

Euglenids are another group in which some members are heterotrophic and others are photosynthetic. As with the photosynthetic dinoflagellates, the photosynthetic euglenids appear to have arisen when an ancestor engulfed an algal cell. DNA studies indicate that euglenids are more closely related to trypanosomes and other kinetoplastids than to algae.

MicroAssessment 12.2

Protozoa are a diverse group of single-celled organisms. They occupy a variety of habitats and are a very important part of food chains. Groups of medically important protozoa include amoebae, apicomplexa, hemoflagellates, and lumen-dwelling flagellates. Some protozoa, including slime molds, dinoflagellates, and euglenids, highlight the difficulties of classification.

4. How is a cyst different from an oocyst?
5. What is the function of the apical complex of apicomplexans?
6. *Trichomonas vaginalis* does not produce a cyst form. What does this suggest about its transmission?

12.3 ■ Algae

Learning Outcomes

8. Compare primary and secondary endosymbiosis.
9. Describe the structure and habitat of algae.
10. List the general characteristics of diatoms, brown algae, green algae, and red algae.
11. Explain why water molds are often included in discussions of algae.

The term **algae** refers to simple photosynthetic eukaryotes. Unlike plants, algae lack an organized vascular system and have relatively simple reproductive structures. Sexual reproduction is common in algae, and many alternate between a haploid generation and a diploid generation. Some algae reproduce asexually.

Characteristics of Algae

Photosynthesis in algae occurs within chloroplasts; recall that these evolved from ancestors of modern-day cyanobacteria. In certain algae, the chloroplasts originated through primary endosymbiosis: a non-photosynthetic eukaryotic cell engulfed a cyanobacterium (**figure 12.14**). In other algal types, the chloroplasts originated through secondary endosymbiosis, meaning that a non-photosynthetic eukaryotic cell engulfed a photosynthetic eukaryotic cell, which itself had arisen through primary endosymbiosis; recall that the photosynthetic dinoflagellates and euglenas described in section 12.2 appear to have acquired their chloroplasts this way. The number of endosymbiotic

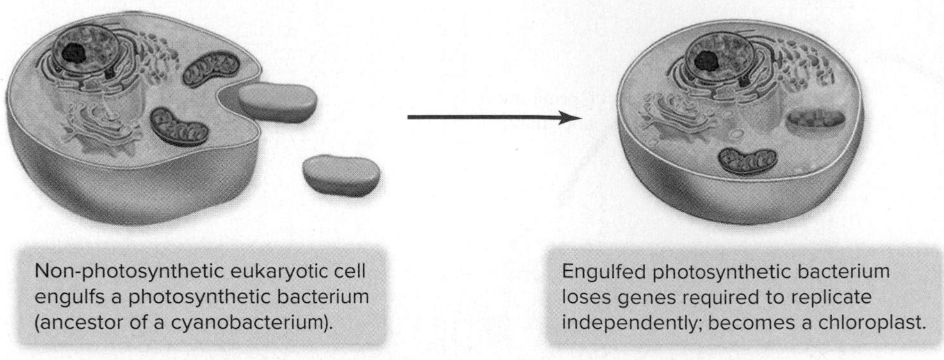

Non-photosynthetic eukaryotic cell engulfs a photosynthetic bacterium (ancestor of a cyanobacterium).
Engulfed photosynthetic bacterium loses genes required to replicate independently; becomes a chloroplast.

FIGURE 12.14 Chloroplasts Originated Through Primary Endosymbiosis

? How would this figure be changed if it were to illustrate secondary endosymbiosis?

events involved in chloroplast acquisition explain the different numbers of membranes surrounding the organelles: Two membranes surround chloroplasts that originated through primary endosymbiosis, whereas three membranes surround those that originated through secondary endosymbiosis. Some chloroplasts appear to have arisen through tertiary endosymbiosis.

Algal Structure

Algae can be either unicellular or multicellular and usually have rigid cell walls mostly composed of cellulose. The unicellular algae are well adapted to an aquatic environment—their single cells provide relatively large absorptive surfaces, allowing efficient use of dilute nutrients. Some single cells are propelled by flagella, whereas others simply float. Certain microscopic algae grow in long multicellular chains or filaments; others form colonies that can be visible to the naked eye. Macroscopic multicellular algae are often called seaweed, and they typically have structures called holdfasts that allow them to adhere to solid surfaces.

Algal Habitats

Algae are found in both fresh and salt water—at depths that allow penetration of light—as well as in moist soil. Because the oceans cover more than 70% of the earth's surface, aquatic algae are major producers of O_2 as well as important users of CO_2.

Unicellular algae make up a significant part of **phytoplankton** (*phyton* means "plant" and *planktos* means "drifting"), the photosynthetic microorganisms that drift in large bodies of water. Phytoplankton form the base of aquatic food chains; zooplankton graze on phytoplankton, and then both become food for other organisms, including filter-feeding whales, some of the world's largest mammals.

Algal growth in aquatic environments is typically limited by low nitrogen and phosphorus levels. When those nutrients are added, such as when fertilizers wash into a body of water, algal growth can be excessive, resulting in what is commonly referred to as a *bloom*; the term is also used to refer to excessive growth of other photosynthetic aquatic organisms, including dinoflagellates and cyanobacteria (see figure 11.6).

Types of Algae

Algae are protists that share some fundamental characteristics. All contain chlorophyll *a,* a pigment also found in green plants and cyanobacteria. In addition, they contain various accessory pigments that absorb different wavelengths of light, thereby extending the range of light waves that can be used for photosynthesis. The types of photosynthetic pigments give the different algal groups their characteristic colors and influence the depth at which the algae can grow.

Diatoms

Diatoms (meaning "cut in half") are single-celled organisms that are the most abundant algal type in aquatic environments; they are found in both saltwater and freshwater habitats, as well as terrestrial environments. Diatoms are distinguished by intricate, rigid, silica-containing external structures called frustules. The arrangement of a frustule is a bit like a Petri dish, enclosing the cell with an overlapping top and bottom layer, although many diatoms are elongated rather than round (**figure 12.15**). Diatoms are sometimes called "jewels of the sea" due to the optical properties of this structure. They are usually a golden brown color, due to the relatively abundant quantities of the photosynthetic pigment fucoxanthin (a carotenoid) that masks other pigments.

When diatoms die in a body of water, they sink to the bottom, where their remains accumulate because the cell wall does not readily decompose. Deposits of fossilized remains are mined for use as a substance known as **diatomaceous earth,** used for filtering systems, abrasives in polishes, and

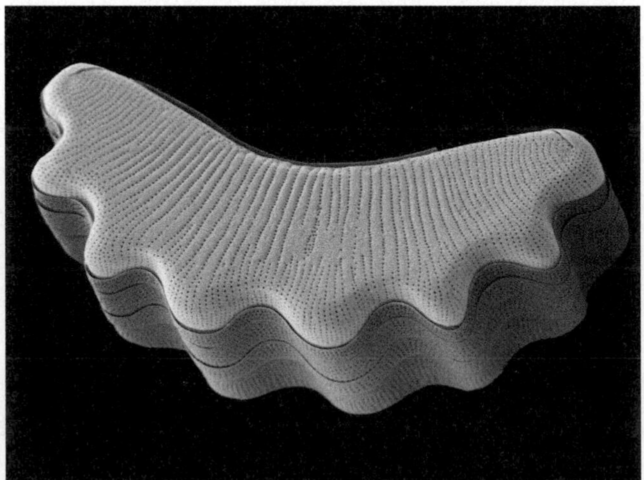

FIGURE 12.15 Diatom Color-enhanced SEM Steve Gschmeissner/Science Photo Library/Alamy Stock Photo

? The name "diatom" reflects what characteristic of the organisms? (Think about the arrangement of the frustule.)

many other purposes. Diatoms that sank to the ocean floor millions of years ago have become a major source of crude oil and natural gas.

MicroByte

Forensic scientists sometimes analyze diatom populations in the lungs of someone who has drowned, to determine where the incident occurred.

Brown Algae

Brown algae are a large group of multicellular algae that typically grow in cold salt water. Kelps are the largest, growing to about 75 meters long in tidal areas, forming underwater "forests" that provide food and habitat for other marine organisms **(figure 12.16).** Like diatoms, brown algae have relatively abundant quantities of the photosynthetic pigment fucoxanthin, which is responsible for their brownish color. Some brown algae are commercially valuable because they are the source of alginate, a thickening agent used in ice cream, toothpaste, and various other products.

Blooms of *Sargassum* species, brown algae that grow in warmer waters, have resulted in environmental and economic damage. Free-floating varieties may accumulate in masses large enough to be tracked by satellites. While small amounts of the seaweed provide important habitats for marine life, the huge mats can entangle marine animals, smother corals, and interfere with boat traffic. When the seaweed washes up on beaches, its decay results in a foul-smelling mess.

Green Algae

Green algae grow in a wide variety of saltwater and freshwater environments, as well as on rocks, trees, and other terrestrial surfaces. Most are single-celled, but some are multicellular. *Volvox* is a green alga that forms colonies (**figure 12.17**).

FIGURE 12.16 Kelp Forest The kelp grow as high as 75 meters in tidal areas. Douglas Klug/Flickr/Moment/Getty Images

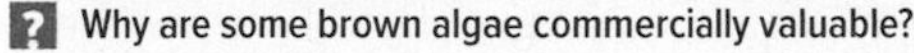
? Why are some brown algae commercially valuable?

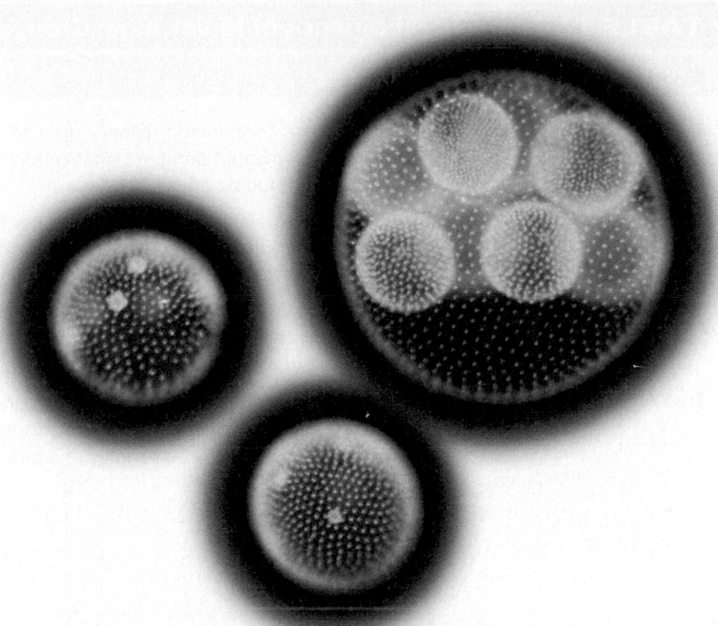

FIGURE 12.17 Volvox A colony of cells forms a hollow sphere. Daughter colonies develop within the spheres. Stephen Durr

? What pigments give the green color to ***Volvox?***

Like plants, green algae have chlorophyll *a* and *b;* in fact, green algae are closely related to plants. Not all algae in this group are green; the accessory pigments of some give rise to different colors.

Some green algae have symbiotic relationships with other organisms. In lichens, they often serve as the photosynthetic member (see figure 12.7). They also live within certain protists, providing the host with energy from photosynthesis; the chloroplasts in euglenids appear to have arisen this way.

Red Algae

Red algae grow in warm salt water, particularly in tropical coastal areas. Some are single-celled, but others are multicellular. Their characteristic reddish color is due to certain accessory pigments (phycoerythrin and phycocyanin) that absorb wavelengths of light that penetrate to great depths, allowing the algae to grow in waters as deep as 70 meters—much deeper than other algae.

Some red algae are commercially useful. The gelling agents carrageenan and agar are polysaccharides extracted from certain types. Nori, the dark covering on sushi rolls, is a red alga. A group of red algae called corallines secrete calcium carbonate, a hard substance that has medical and dental uses associated with bone grafting.

Characteristics of major algal groups are summarized in **table 12.4.**

TABLE 12.4 Characteristics of Major Algal Groups

Example	Characteristics	Comments
Diatoms	Single cells within silica-containing frustules; grow in a variety of environments including saltwater, freshwater, and terrestrial; golden brown due to the photosynthetic pigment fucoxanthin	Diatomaceous earth (the remains of diatoms) is used in filtering systems and as an abrasive
Brown Algae	Multicellular; include large kelp that form underwater forests; typically grow in cold salt water, although *Sargassum* species grow in warmer salt water; brown color due to the photosynthetic pigment fucoxanthin	Source of the thickening agent alginate; huge blooms of *Sargassum* species are damaging
Green Algae	Most are single-celled, but some are multicellular; grow in a variety of environments, including saltwater, freshwater, and terrestrial; green color due to chlorophyll a and b, but other colors occur due to various pigments	Closely related to plants; some are involved in symbiotic relationships
Red Algae	Some are single-celled and others are multicellular; grow in warm salt water, often relatively deep; reddish color due to the photosynthetic pigments phycoerythrin and phycocyanin, which absorb wavelengths of light that penetrate to great depths	Source of the gelling agents agar and carrageenan

Exceptions to the Rule

The water molds, or oomycetes, are not photosynthetic and they lack chloroplasts. They were once considered fungi because of certain shared characteristics, but DNA analysis indicates that they are more closely related to brown algae and diatoms. The shared characteristics of fungi and water molds are good examples of convergent evolution—a process that occurs when organisms develop similar characteristics independently as they adapt to similar environments.

Water molds form masses of white threads on decaying material. Like fungi, they secrete digestive enzymes onto a substrate and absorb small molecules for nutrients. The cytoplasm in their filaments is continuous with many nuclei. However, water molds have cellulose in their cell walls rather than chitin.

Oomycetes cause some serious diseases of food crops, such as late blight of potatoes and downy mildew of grapes. Late blight was a factor in the potato famine in Ireland in the 1840s that sent waves of immigrants to other countries (see A Glimpse of History). Oomycetes are also important fish pathogens.

Figure 12.18 illustrates the major phylogenetic groupings of the protists described in this and the previous section.

MicroAssessment 12.3

Algae are aquatic eukaryotic organisms that have chlorophyll *a* and carry out photosynthesis; they form the base of aquatic food chains and produce much of our atmospheric O_2. Oomycetes are genetically related to certain types of algae but are not photosynthetic.

7. How is primary endosymbiosis different from secondary endosymbiosis?
8. What commercially valuable products are derived from algae?
9. Why do algae have a greater variety of photosynthetic pigments than land plants?

12.4 ▪ Multicellular Parasites: Helminths

Learning Outcomes

12. Explain how disease-causing helminths can be transmitted to humans.
13. Compare and contrast the structures of a roundworm, a tapeworm, and a fluke; describe one disease caused by each of these.

Alveolata	Stramenophila	Archaeplastida	Discoba	Metamonada	Amoebozoa
Apicomplexans Dinoflagellates	Brown algae Diatoms Water molds	Green algae Red algae	Euglenids Haemoflagellates *Naegleria*	*Giardia* *Trichomonas*	*Entamoeba* Plasmodial slime molds Social amoebae

FIGURE 12.18 Major Phylogenetic Groups of Some Protozoa and Algae These groups are sometimes referred to as supergroups or superclades.

? What accounts for some dinoflagellates being photosynthetic?

Helminths are worms, a type of animal. Some are parasites of humans or other animals, causing tissue damage or robbing the body of nutrients. Although not microorganisms, parasitic helminths are often identified using microscopic and immunological techniques familiar to microbiologists. The microscopic eggs (ova) of each helminth species are typically distinctive and can be identified by size, shape, and other features. The helminths are divided into two general groups: **roundworms (nematodes)** and **flatworms (platyhelminths).** The flatworms are further divided into **tapeworms (cestodes)** and **flukes (trematodes).** Multicellular parasites have been largely controlled in industrialized nations, but they still cause suffering and the death of many millions each year in developing parts of the world.

Life Cycles and Transmission of Helminths

Some helminths have complex life cycles involving one or more **intermediate hosts** within which a sexually immature stage of the parasite develops. Sexual reproduction takes place in the **definitive host.**

Helminths enter the body in a number of ways. The larvae (immature forms) of hookworms, a type of nematode, live in the soil and can burrow through human skin. The adult worms live in the human digestive tract, where the female worm lays eggs that are eliminated with feces. The eggs hatch in the soil, giving rise to more larvae. When sanitation is poor and people are barefooted, the parasite is easily transmitted. An estimated 576 to 740 million people are infected with hookworms, mostly in tropical and subtropical regions.

Some helminths are inadvertently eaten with food, as when the larva of the nematode *Trichinella spiralis* is ingested in the meat of a carnivore (a meat-eating animal). Eating undercooked pork is one of the most common causes of trichinellosis (also called trichinosis). More often, helminth eggs are ingested on contaminated foods. Children with pinworms *(Enterobius vermicularis),* for example, may pick up eggs by touching their anus and then transmit those eggs to a surface; a food handler who does not use proper handwashing may then inadvertently transfer the eggs from that surface to the food.

Some helminths are transmitted through insect bites. *Wuchereria bancrofti,* a type of filarial (thin or thread-like) nematode, is transmitted by mosquitoes. The adult worms live in the lymphatic vessels, eliciting an inflammatory response that interferes with lymphatic drainage; the resulting fluid buildup can cause massive swelling in various parts of the body, a condition called elephantiasis (**figure 12.19**). *Onchocerca volvulus,* the filarial worm that causes river blindness, is transmitted by a type of biting fly (see Focus Your Perspective 12.1).

Roundworms (Nematodes)

A **roundworm (nematode)** has a cylindrical, tapered body with a digestive tract that extends from the mouth to the anus. The nematode *Caenorhabditis elegans* is a model eukaryotic organism that has been the subject of numerous studies in genetics and development because it matures quickly, its genome has been sequenced, and all of its 959 body cells can be identified. Many nematodes are free-living in soil and water. Others are parasites and cause serious disease.

FOCUS YOUR PERSPECTIVE 12.1

What Causes River Blindness?

Female black flies, sometimes called buffalo gnats, swarm around their hosts and bite repeatedly to obtain blood needed to produce eggs. They require rapidly flowing water for larval development and so are most often found near rivers. These flies are associated with a disease called river blindness (onchocerciasis) that affects at least 18 million people, 99% of whom live in Africa; the World Health Organization estimates that about 270,000 are blind as a result of their infection and up to another 1 million people have impaired vision. River blindness is the second leading cause of infectious blindness (the leading cause is trachoma caused by the bacterium *Chlamydia trachomatis*). But are the black flies the cause of this disease?

When biting, black flies may transmit larvae of a filarial nematode, *Onchocerca volvulus,* to a human host, eventually leading to river blindness. In a human host, the larvae reside in nodules and mature to adulthood in about a year, at which time adult females produce millions of microfilariae. These can migrate through the skin, where they can be picked up by another black fly as it bites, thus continuing the life cycle. When infections are heavy, the microfilariae can also be found in the blood and in the eye. So are the *Onchocerca* larvae the real cause of the disease?

As microfilariae move throughout the body of their host, they carry a bacterial population of *Wolbachia pipientis* that is necessary for fertility and viability of the worms. When microfilariae die, they release the *Wolbachia*. The bacteria cause an inflammatory response that can damage sensitive host tissues. When this happens in the eye, the result is vision impairment and blindness. Ultimately, it is the bacteria carried by *Onchocerca* larvae that lead to the symptoms of river blindness.

Efforts to control river blindness have focused on eliminating the black fly vector and providing a medication called ivermectin that targets the worm. Ivermectin reduces the number of microfilariae for a few months but does not destroy the adults, so repeated medication is needed. An alternative approach is to use ivermectin in combination with doxycycline, an antibacterial medication that targets *Wolbachia*. The doxycyline not only reduces the effects of the disease, but it sterilizes the worms so that they can no longer produce microfilariae, thus disrupting the life cycle.

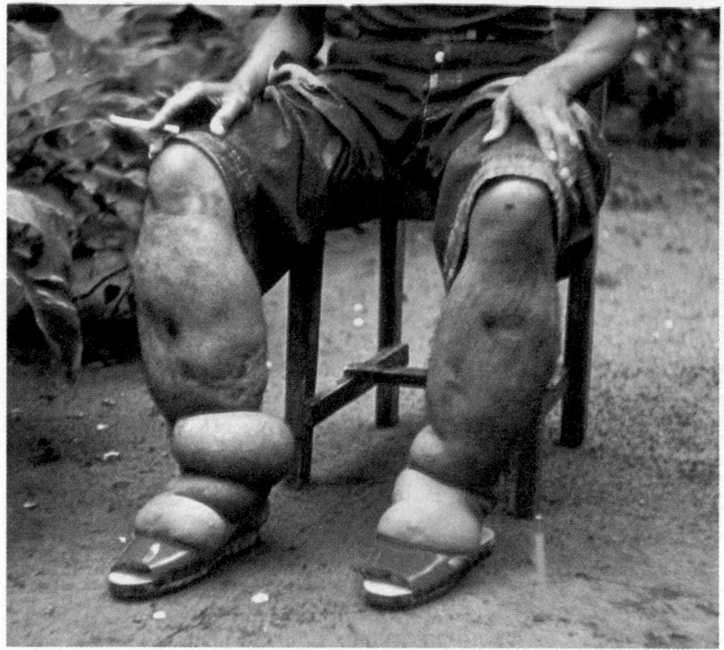

FIGURE 12.19 Elephantiasis Buildup of fluid has occurred as a result of adult worms living in the lymphatic vessels leading from the limbs. CDC

? What type of worm causes elephantiasis?

Ascariasis, caused by *Ascaris lumbricoides,* is the most common human disease caused by roundworms. Females are larger than males and may exceed 30 cm long while producing more than 200,000 eggs per day that are eliminated in the feces. Ingested eggs hatch in the digestive tract, releasing immature worms that burrow into the bloodstream (**figure 12.20**). After they reach the lungs, they can be coughed up and swallowed. When the immature worms again reach the intestine, they grow and begin producing eggs that will be released with the feces. Although the worms do not feed on human tissue, they rob the body of nutrients by feeding on material that passes through the digestive tract. In addition, they may cause choking and pulmonary symptoms when they enter the respiratory tract.

MicroByte

Over 1 billion people worldwide are believed to carry the roundworm *Ascaris lumbricoides.*

Tapeworms (Cestodes)

Tapeworms (cestodes) have flat, ribbon-shaped bodies. They have no digestive system and do not feed directly on host tissues. Rather, adult tapeworms attach within the definitive host's intestines, where their bodies absorb predigested nutrients. Some types reach over 15 meters in length while living for over a decade in the host intestine. Depending on the species of the worm, the **scolex** (head end) may have

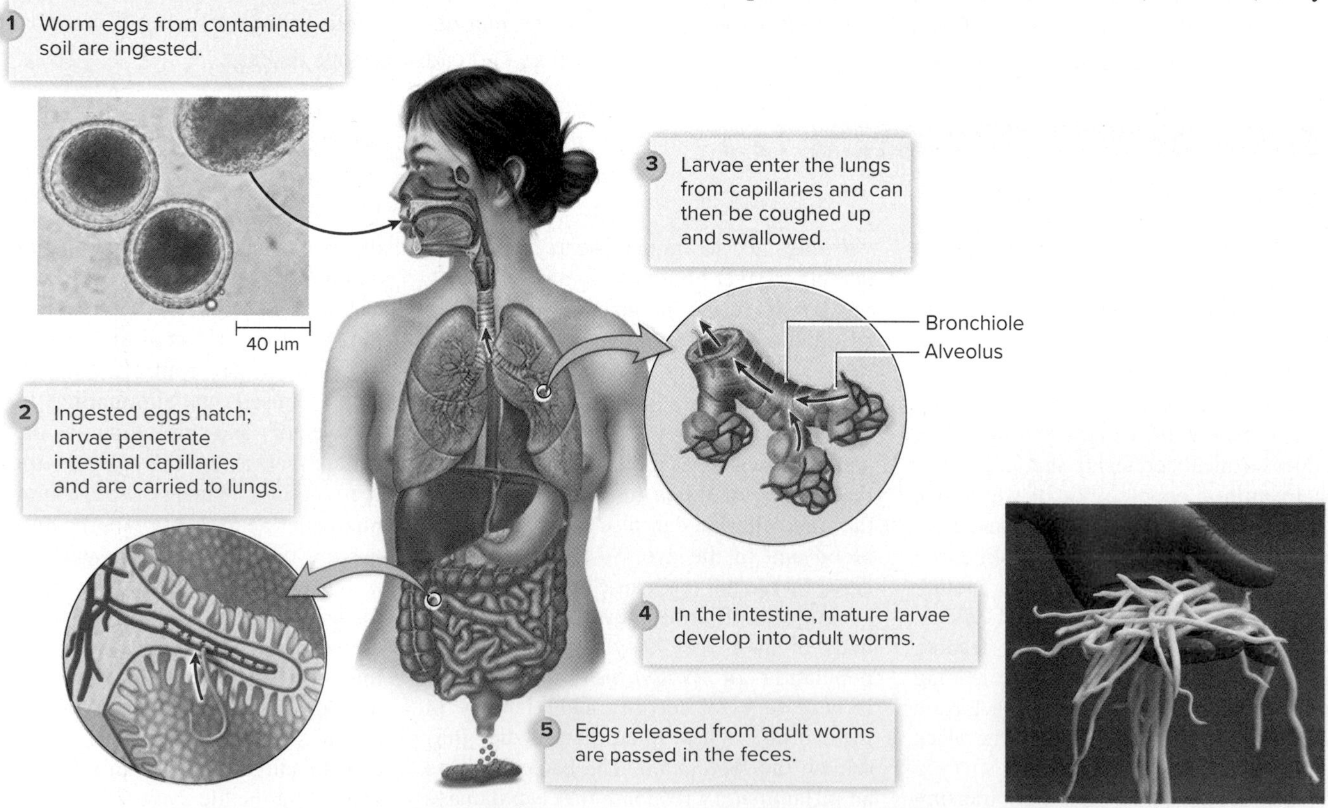

1 Worm eggs from contaminated soil are ingested.
40 µm
2 Ingested eggs hatch; larvae penetrate intestinal capillaries and are carried to lungs.
3 Larvae enter the lungs from capillaries and can then be coughed up and swallowed.
Bronchiole
Alveolus
4 In the intestine, mature larvae develop into adult worms.
5 Eggs released from adult worms are passed in the feces.

FIGURE 12.20 Life cycle of *Ascaris lumbricoides* Larvae hatching in the intestine migrate through the lungs and back to the intestine before maturing to adulthood. (both Photos): Lisa Burgess/McGraw Hill

? How do *Ascaris lumbricoides* larvae get from the lungs to the intestines?

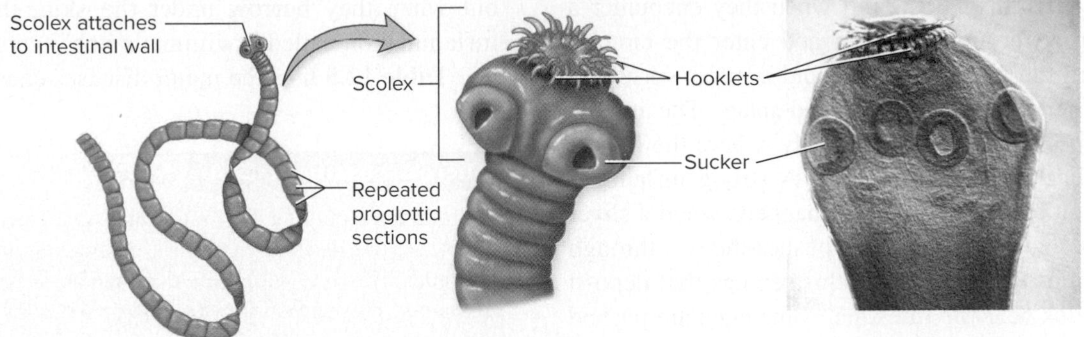

FIGURE 12.21 Tapeworm The scolex holds the tapeworm to the intestinal surface. Proglottids contain reproductive structures and are shed in the feces as the tapeworm elongates by adding new segments. Eric Grave/SPL/Getty Images

? **Why can a tapeworm live without a digestive system?**

suckers and hooks for attachment (**figure 12.21**). Connected to the scolex is the neck and then a number of segments called **proglottids** that contain both male and female reproductive structures. The proglottids farthest from the scolex contain fertilized eggs. As the worm grows, these segments break off and are eliminated in the feces along with the eggs. When a suitable intermediate host ingests the eggs, the eggs hatch, releasing larvae; these then penetrate the intestinal wall and migrate into tissues, where they develop into larval cysts called cysticerci (singular: cysticercus), which are infectious when consumed by a definitive host.

The most common tapeworms of humans have intermediate hosts of cattle, pigs, and fish. Humans (the definitive host) become infected when they eat raw or undercooked meat containing the larval forms. Unfortunately, humans can also serve as the accidental intermediate host of the pork tapeworm (*Taenia solium*). If someone inadvertently ingests *T. solium* eggs, the eggs can hatch in the intestinal tract, releasing larvae. These migrate to the tissues and form cysticerci, resulting in the disease called cysticercosis. If cysticerci form in the brain, serious neurological symptoms may result.

Flukes (Trematodes)

Flukes (trematodes) are flat, leaf-shaped worms with two suckers that the adult worms use to both attach to and move along a surface of their definitive host. The worms have a mouth but no anus, so nutrients enter and wastes exit from the same opening. The parasites typically have a complex life cycle that involves at least two hosts: a snail in which the organism asexually reproduces, and a mammal or other vertebrate in which the organism sexually reproduces. There are two general categories of flukes: tissue flukes and blood flukes. The tissue flukes are hermaphroditic (have both sex organs in the same worm), whereas the blood flukes have separate sexes.

Although many flukes are medically important, *Schistosoma* species (blood flukes) are particularly significant because of the devastating disease they cause—schistosomiasis (also called bilharzia). Spread of that disease requires a suitable snail host, along with sewage-contaminated fresh water. Schistosome-infected snails release thousands of cercariae, the parasite's larval form (**figure 12.22**). These swim in search

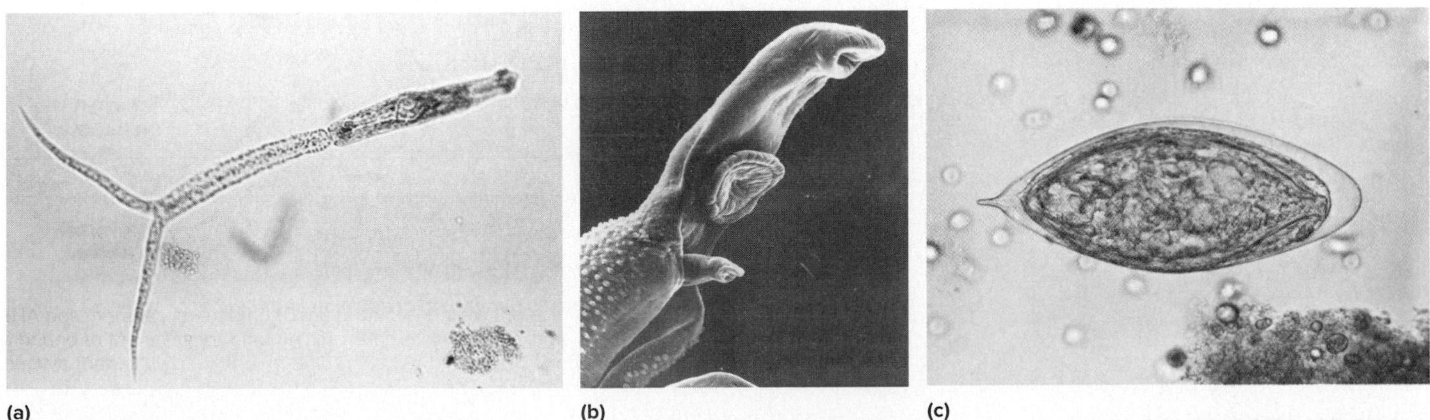

FIGURE 12.22 Schistosome **(a)** Cercaria. **(b)** Male schistosome embracing the much smaller female worm; note the suckers on the male (SEM); **(c)** characteristic egg of *Schistosoma hematobium*. a: Melissa Rethlefsen and Marie Jones, Prof. William A. Riley/Minnesota Department of Health, R.N. Barr Library/CDC; b: Bruce Wetzel and Harry Schaefer/National Cancer Institute (NCI); c: CDC

? **Considering the two general requirements for the spread of schistosomiasis, what could be done to prevent disease?**

of an appropriate definitive host, and when they encounter a human, they burrow through the skin and enter the circulatory system. There the worms mature, and a male worm and a female worm then find each other and mate. The mating pair moves to a specific region of the body, where the female deposits her fertilized ova in tiny veins. A strong inflammatory response develops, similar to what happens when a sliver lodges in the skin, and this sometimes pushes the ova through the walls of the vein. In the case of schistosomes that deposit their ova in the veins near the intestine, some eggs are pushed into the intestinal tract, where they are then eliminated with feces. For the schistosome that deposits its ova in the veins near the bladder, some eggs are pushed into the bladder, where they are eliminated with urine. In either case, if viable ova reach fresh water, they hatch to release ciliated larvae that infect specific snail species, completing the life cycle.

The most severe outcomes of schistosomiasis are due to eggs not released into the intestinal tract or the bladder. As the inflammatory response to the ova continues, normal tissues are destroyed and replaced with scar tissue (see Focus Your Perspective 14.1). A particularly damaging situation occurs if the blood circulatory system moves ova deposited near the intestinal tract to the liver. There, the inflammatory response gradually destroys liver tissue, leading to life-threatening liver damage.

Swimmers are sometimes infected with the larvae of schistosomes that typically complete their life cycles in fish or water birds. These parasites cannot mature in humans, but when they burrow under the skin, they cause a local inflammation called "swimmer's itch" (cercarial dermatitis).

Table 12.5 lists the major diseases caused by helminths.

MicroAssessment 12.4

Helminths, including the roundworms, tapeworms, and flukes, cause serious diseases in humans. Many helminths have a complex life cycle with more than one host. They may enter a human host by ingestion with food or water, by an insect bite, or by burrowing through skin.

10. What are the major differences among nematodes, cestodes, and trematodes?
11. Differentiate between a definitive host and an intermediate host.
12. As an intermediate host for *Taenia solium*, why would humans be considered a dead-end host?

12.5 ■ Arthropods

Learning Outcomes

14. Describe two ways in which arthropods are related to disease in humans.
15. Give an example of a disease transmitted by each of the following: mosquitoes, flies, fleas, lice, and ticks.

TABLE 12.5 Nematodes, Cestodes, and Trematodes

Infectious Agents	Disease and Its Characteristics
Nematodes (roundworms)	
Ascaris lumbricoides (large roundworm)	Ascariasis—abdominal pain, vomiting, intestinal blockage
Enterobius vermicularis (pinworm)	Enterobiasis—anal itching, restlessness, irritability, poor sleep
Necator americanus and *Ancylostoma duodenale* (hookworms)	Hookworm disease—anemia, weakness, fatigue; the associated nutrient loss can result in physical and intellectual disability in children
Onchocerca volvulus	Onchocerciasis (river blindness)—severe itching, thickening of skin, vision loss
Strongyloides stercoralis	Strongyloidiasis—rash at site of penetration, cough, abdominal pain, weight loss
Trichinella spiralis	Trichinellosis (trichinosis); fever—swelling of upper eyelids, muscle soreness
Trichuris trichiura (whipworm)	Trichuriasis—abdominal pain, bloody stools, weight loss
Wuchereria bancrofti and *Brugia malayi* (filarial worms)	Lymphatic filariasis—swelling of lymphatic structures, genitals, and extremities, fever
Cestodes (tapeworms)	
Diphyllobothrium latum (fish tapeworm)	Diphyllobothriasis (fish tapeworm disease)—few or no symptoms, sometimes anemia
Taenia saginata (beef tapeworm)	Taeniasis (beef tapeworm disease)—few or no symptoms
Taenia solium (pork tapeworm)	Taeniasis (pork tapeworm disease)—few or no symptoms
	Cysticercosis—variable symptoms depending on location and number of cysticerci in the body
Trematodes (flukes)	
Schistosoma species (blood flukes)	Schistosomiasis—liver damage, malnutrition, weakness, and accumulation of fluid in the abdominal cavity
Schistosomes that infect birds and other animals	Swimmer's itch—inflammation of the skin, itching

Arthropods are a group of invertebrate animals that includes insects (such as flies, mosquitoes, lice, and fleas) and arachnids (such as ticks and mites). Some are **vectors** that can transmit pathogens to humans. An arthropod may act as a mechanical vector that simply transfers a microbe from one surface to another, or it may be a biological vector within which the microbe develops or multiplies. For example, *Plasmodium* species that cause malaria multiply within *Anopheles* mosquitoes during their life cycle, and species of trypanosomes that cause African sleeping sickness multiply within the tsetse fly (*Glossina* species).

When an arthropod vector feeds on an infected host, it may pick up a pathogen that can be transferred to a human in a later bite. Some arthropods bite only one type of host. Certain mosquitoes that carry *Plasmodium* species bite only humans; typically, only female mosquitoes take a blood meal. In contrast, fleas that carry *Yersinia pestis,* the bacterium responsible for plague, bite both humans and small mammals such as rats.

The incidence of vector-borne diseases can be decreased by controlling the vector or the infected hosts. The risk of mosquito-borne diseases can be minimized by eliminating standing water or using insect repellent to reduce the incidence of vector transmission. Plague can be controlled by eliminating rodent populations that may infect their fleas with *Y. pestis.* Examples of some important arthropods, the agents they transmit, and the resulting diseases are shown in **table 12.6.**

Some arthropods cause disease even when they do not act as vectors. For example, the pubic louse *(Phthirus pubis),* commonly transmitted during sexual intercourse, is not a vector of infectious agents but can cause an unpleasant itch associated with "crabs." Similarly, dust mites do not transmit infectious disease, but inhalation of the mites and their waste products can sometimes trigger asthma. The larvae of some mites are called "chiggers" and may cause intense itching where they attach and feed on fluids within skin cells. Scabies, a disease caused by another mite, is easily transmitted by personal contact. Allergic reactions to female mites that have burrowed into the outer layers of skin are largely responsible for the itchy rash of scabies. Bites of certain ticks, particularly lone star ticks, are associated with the development of alpha gal syndrome—a condition that results in potentially life-threatening allergies to certain meats and animal products.

MicroAssessment 12.5

Arthropods such as flies, mosquitoes, fleas, lice, and ticks act as vectors for the spread of disease. Infestations of mites and lice may cause itching.

13. How can arthropods spread disease in humans?

14. What causes the itching of scabies?

15. Why are diseases transmitted by insect vectors more common in the summer than in the winter?

TABLE 12.6 Some Arthropods That Transmit Infectious Agents

Arthropod	Infectious Agent	Disease and Characteristic Features
Insects		
Black fly (*Simulium* species)	*Onchocerca volvulus*	Onchocerciasis (river blindness)—rash, itching, visual impairment
Flea (*Xenopsylla cheopis*)	*Yersinia pestis*	Plague—fever, headache, confusion, enlarged lymph nodes, skin hemorrhage
Louse (*Pediculus humanus*)	*Rickettsia prowazekii*	Typhus—fever, hemorrhagic rash, confusion
Mosquito (*Aedes* species)	Chikungunya virus	Chikungunya—fever, headache, rash, severe joint pain that can become chronic
	Dengue virus	Dengue and severe dengue—fever; headache; joint pain; bleeding and shock can occur.
	Yellow fever virus	Yellow fever—fever, vomiting, jaundice, hemorrhagic rash
	Zika virus	Zika virus disease—mild fever and joint pain; can cause congenital Zika syndrome
Mosquito (*Anopheles* species)	*Plasmodium* species	Malaria—bouts of recurring fever and chills
Mosquito (*Culex* species)	West Nile virus and others	Viral encephalitis—fever, nausea, convulsions, coma
Sand fly (*Phlebotomus* species)	*Leishmania* species	Leishmaniasis—ulcerated skin lesions; mucosal lesions in the nose, mouth, and throat; fever; enlarged spleen and liver
Tsetse fly (*Glossina* species)	*Trypanosoma brucei*	African trypanosomiasis (sleeping sickness)—sleepiness, headache, coma
Arachnids		
Tick (*Dermacentor* species)	*Rickettsia rickettsii*	Rocky Mountain spotted fever—fever, hemorrhagic rash, confusion
Tick (*Ixodes* species)	*Borrelia burgdorferi* and some other species	Lyme disease—fever, rash, joint pain, nervous system impairment

Summary

Algae, fungi, and **protozoa** are informal groupings that do not reflect evolutionary relatedness. Eukaryotes are characterized by a membrane-bound nucleus. Reproduction may be asexual using **mitosis** or sexual using **meiosis,** which forms **gametes** (figure 12.1).

12.1 ■ Fungi

Yeast, mold, and **mushroom** are common terms that indicate morphological forms of fungi (figure 12.2).

Characteristics of Fungi (table 12.1)

Fungi cell walls contain chitin and typically have glucans. Fungal membranes typically contain ergosterol. **Dimorphic fungi** can grow either as yeasts or as molds. Molds are characterized by a mass of **hyphae** that form a **mycelium** (figure 12.3). The asexual reproductive cells of most molds are called **conidia** or, in one fungal group, **sporangiospores**; the characteristics of these forms are used to identify fungi in the laboratory (figure 12.4). Yeast cells reproduce asexually by budding (figure 12.5). Fungal classification relies on sexual forms.

Classification of Fungi (table 12.1)

Major phyla of fungi include Ascomycota, Basidiomycota, and Mucoromycota.

Groups of Medically Important Fungi (table 12.2)

Medical mycologists separate fungi that infect humans into the following groups: superficial fungi, subcutaneous fungi, opportunistic fungi, and systemic fungi. Some fungi produce toxins, and others are associated with hypersensitivity reactions.

Economic Importance of Fungi

Commercially important fungi include fungi that synthesize antimicrobial medications, and the yeast *Saccharomyces cerevisiae.* Fungi also cause several significant plant diseases.

Symbiotic Relationships of Fungi

Lichens result from the association of a fungus with a photosynthetic microorganism (figure 12.7). A **mycorrhiza** is an intimate association of a fungus and the roots of a plant (figure 12.8). Leaf-cutter ants grow gardens of fungus for food (figure 12.9).

12.2 ■ Protozoa

Characteristics of Protozoa

Most protozoa lack a cell wall. Certain protozoa can exist as **trophozoites** or as **cysts**. Some protozoa form **oocysts** that develop as part of the sexual stage. Most protozoa are free-living, but some are parasitic.

Groups of Medically Significant Protozoa (table 12.3)

Medical microbiologists typically rely on traditional methods of grouping protozoa that emphasize methods of locomotion. These include **amoebae** *(Entamoeba* and *Naegleria),* **apicomplexans** *(Plasmodium, Toxoplasma, Cryptosporidium, Cyclospora),* **hemoflagellates** *(Trypanosoma, Leishmania),* and lumen-dwelling flagellates *(Giardia* and *Trichomonas)* (figures 12.10, 12.11).

Other Protozoan Groups

Several protozoan groups, including slime molds (figures 12.12, 12.13), dinoflagellates, and euglenids, are examples of why classification can be difficult.

12.3 ■ Algae

Characteristics of Algae

Photosynthesis in algae occurs within chloroplasts that evolved from ancestors of modern-day cyanobacteria (figure 12.14). Algae usually have rigid cell walls mostly composed of cellulose. Algae are found in fresh and salt water as well as in soil. Unicellular algae make up a significant part of the **phytoplankton.**

Types of Algae (table 12.4)

Types of algae include diatoms (figure 12.15), brown algae (figure 12.16), green algae (figure 12.17), and red algae. These differ in their photosynthetic pigments, but all contain chlorophyll *a*.

Exceptions to the Rule

Water molds (oomycetes), which are not photosynthetic, were once considered fungi because of certain shared characteristics, but DNA analysis indicates that they are more closely related to brown algae and diatoms. They can cause serious diseases of food crops.

12.4 ■ Multicellular Parasites: Helminths (table 12.5)

Helminths are worms, a type of animal. Some are parasitic.

Life Cycles and Transmission of Helminths

Some helminths have complex life cycles with asexual stages occurring in one or more **intermediate hosts** and the sexual or adult stage occurring in the **definitive host.** Helminths may enter a human host by ingestion, by an insect bite, or by penetrating the skin.

Roundworms (Nematodes)

Many **roundworms (nematodes)** are free-living, but some are parasites that cause serious diseases; ascariasis is an example (figure 12.20).

Tapeworms (Cestodes)

Tapeworms (cestodes) have segmented bodies (figure 12.21). Most tapeworm infections occur in persons who eat uncooked or undercooked meats.

Flukes (Trematodes)

Flukes (trematodes) often have complicated life cycles that require more than one host. The larvae of *Schistosoma* species can penetrate the skin of persons wading in infected waters and cause serious disease (figure 12.22).

12.5 ■ Arthropods

Arthropods can act as **vectors** for disease (table 12.6). Some arthropods can infest the body and cause itching and a rash.

Review Questions

Short Answer

1. What are the major differences between a prokaryotic cell and a eukaryotic cell?
2. What are the differences among a yeast, a mold, and a mushroom?
3. What is a mycosis? Give an example.
4. In what ways are fungi economically important?
5. Why are few lichens found in industrial areas?
6. Compare and contrast the organisms that cause malaria and African trypanosomiasis (sleeping sickness) and their transmission.
7. What characteristics do all algae have in common?
8. Name a disease for which humans are an intermediate host and another for which humans are a definitive host. Give an example of a disease in which humans are a dead-end host.
9. Describe the life cycle of *Schistosoma* species.
10. Explain the difference between a mechanical vector and a biological vector.

Multiple Choice

1. Cell walls of this group contain chitin.
   a) Algae
   b) Protozoa
   c) Fungi
   d) Helminths
2. Certain members of this group are the source of diatomaceous earth.
   a) Algae
   b) Protozoa
   c) Fungi
   d) Helminths
3. Which of the following statements about fungi is *false*?
   a) Dimorphic fungi can transition between morphologies—usually molds and yeasts.
   b) Some fungi produce life-threatening toxins.
   c) Fungal spores may cause allergic responses in humans.
   d) Systemic mycoses are common in otherwise healthy adults.
4. Which of the following statements regarding protozoa is *false*?
   a) Intestinal amoebae are typically able to form cysts.
   b) A defining characteristic of protozoa is the presence of mitochondria.
   c) Some protozoa are parasitic.
   d) Malaria is caused by a protozoan.
5. Which of the following is mismatched?
   a) *Plasmodium*—malaria
   b) Trypanosomes—dysentery
   c) Dinoflagellates—red tide
   d) Nematode—trichinellosis
6. Which of the following is mismatched?
   a) Trematode—fluke
   b) Mosquito—malaria transmission
   c) Baker's yeast—algae
   d) Apicomplexan—protozoa
7. The current phylogeny of eukaryotes is based upon
   a) rRNA sequence comparisons.
   b) possession of photosynthetic pigments.
   c) mode of reproduction.
   d) mode of locomotion.
8. All algae
   a) have chlorophyll *a*.
   b) have cell walls that contain agar.
   c) make holdfasts.
   d) are multicellular.
9. Which of the following statements regarding protists is *false*?
   a) They include both autotrophic and heterotrophic organisms.
   b) They include both microscopic and macroscopic organisms.
   c) They are an intermediate host in schistosomiasis.
   d) They include algae and protozoa.
10. Which of the following statements regarding beef tapeworms is *false*?
    a) They absorb nutrients from the host through their body wall.
    b) They complete their life cycle in a single host.
    c) They are hermaphroditic.
    d) They cannot be transmitted from human to human.

Applications

1. A student in a clinical microbiology class wanted to see examples of *Entamoeba histolytica* cysts, and was frustrated to find that nearly all of the examples of the parasite in a stained slide of a stool specimen were trophozoites. What might account for the lack of cysts?
2. A paper recycling company refuses to collect products that are contaminated with food or have been sitting wet for several days. A college student who runs a recycling program on campus wants to know why. What reason did the company scientist probably give?

Critical Thinking

1. If you discovered a new type of nucleated cell in a lake near your home, how would you determine whether the cell was from a fungus, an alga, or a protozoan?
2. Fungi are known for growing in a wide range of environmental extremes in temperature, pH, and osmotic pressure. What does this tolerance for extremes indicate about fungal enzymes?

13 | Viruses, Viroids, and Prions

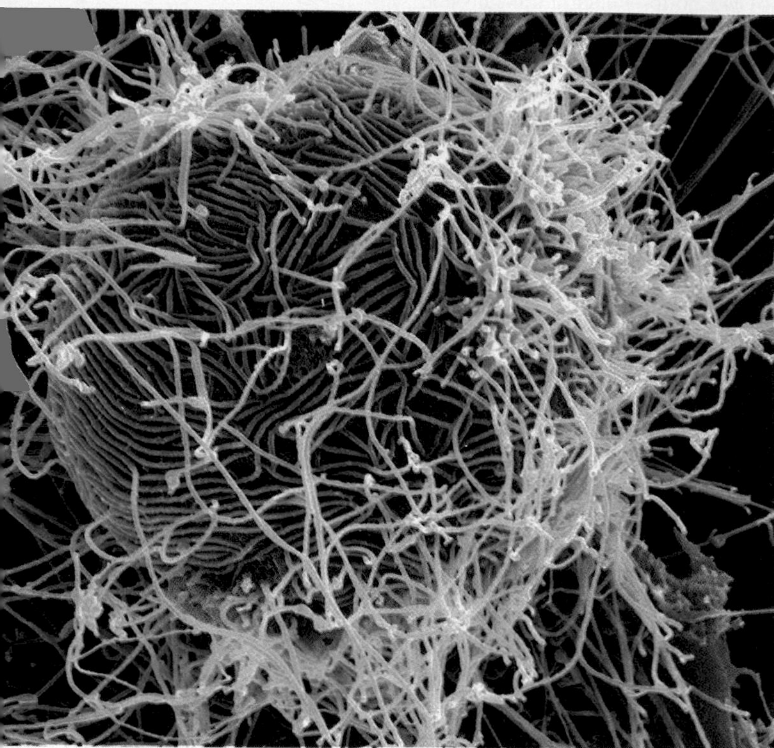

Ebola virus particles budding from a monkey kidney epithelial cell (color-enhanced scanning electron micrograph). *Source: National Institute of Allergy and Infectious Diseases (NIAID)/CDC*

KEY TERMS

Bacteriophage A virus that infects bacteria; often called a *phage*.

Latent Infection With respect to viral infections, an infection in which the viral genome persists silently in the host cell, so new viral particles are not being produced.

Lysogen A bacterium that carries phage DNA (a prophage) integrated into its genome.

Lysogenic Conversion A change in one or more of the properties of a bacterium upon acquiring a prophage.

Lytic Infection A productive infection that causes the host cell to burst.

Prion An infectious protein that causes a neurodegenerative disease.

Productive Infection Viral infection in which more viral particles are produced.

Viroid An infectious agent of plants that consists only of RNA.

Virion A complete virus in its inert non-replicating form outside a host cell; also referred to as a viral particle.

A Glimpse of History

During the late nineteenth century, many bacteria, fungi, and protozoa were identified as infectious agents. Most of these organisms could be easily seen using a light microscope and could be grown in the laboratory. In the 1890s, however, Dimitri Iwanowsky and Martinus Beijerinck discovered that a mosaic disease of tobacco plants was caused by an unusual agent. The agent was too small to be seen with the light microscope. Furthermore, it passed through filters that retained most known bacteria, and it could be grown only in media that contained living cells. Beijerinck called the agent a "filterable virus." About 10 years later, Frederick W. Twort and Félix d'Herelle discovered a filterable virus that destroyed bacteria. With time, the adjective filterable was dropped, and now we use only the word *virus* to describe these agents.

In 1935, Wendell Stanley demonstrated that viruses have features more characteristic of complex chemicals than of cells. He was able to crystallize the virus now called tobacco mosaic virus (TMV), which causes a serious disease in tobacco plants, resulting in a mosaic pattern and ring lesions on the tobacco leaves (see figure 13.22a). His finding indicated that the physical and chemical properties of viruses differ from those of cells, which cannot be crystallized. Surprisingly, the crystallized TMV could still cause disease.

In the simplest of terms, viruses can be viewed as genetic information—either DNA or RNA—contained within a protective protein coat. They are inert particles incapable of metabolism, replication, or motility on their own. When a viral genome enters a host cell, however, it can hijack that cell's replication machinery, inducing the cell to produce more viral particles. In other words, viruses straddle the definition of life: Outside a host cell they are inert particles, but inside a host cell they direct activities that can have a profound effect on that cell. So although viruses are infectious agents, they are not organisms.

Viruses can be broadly grouped into two general types based on the category of cells they infect. Some infect eukaryotic cells, and others infect prokaryotic cells. Viruses that infect bacteria are also referred to as **bacteriophages,** or simply **phages** (*phage* means "to eat").

The fact that viruses are obligate intracellular parasites makes them very difficult to study. Unlike most bacteria and eukaryotic cells, which can be grown in pure culture, viruses require live organisms as hosts. In addition, the vast majority of viruses are too small to be seen with a light microscope and can only be visualized with an electron microscope.

The first section of this chapter describes the general characteristics of viruses. The primary focus of the remaining sections is on bacteriophages and animal viruses. Animal viruses are obviously important because of the diseases they cause, but why learn about bacteriophages? We study them partly because they are an easy-to-cultivate model for understanding animal viruses and their relationships with cells they infect. Much of what you learn about them will apply to similar relationships between the medically important viruses and their hosts. Bacteriophages are also important because they serve as a vehicle for horizontal gene transfer in bacteria, which is described in chapter 8. In addition, because bacteriophages kill bacteria, they reduce bacterial populations, which is significant ecologically and also has medical applications. In recent years, the Food and Drug Administration (FDA) has approved the use of species-specific preparations of bacteriophages for controlling bacteria during food production and storage. These preparations have been used to prevent the growth of food-contaminating pathogens such as *Listeria monocytogenes, Salmonella enterica,* and *E. coli* O157:H7 on various food products. Phage therapy, the use of bacteriophages to treat bacterial infections, has been used in rare cases when treatment with antibiotics has failed; it is discussed in Focus Your Perspective 13.2.

MicroByte

An estimated 10^{31} bacteriophages are on Earth, making them the most numerous of all biological entities.

13.1 ■ General Characteristics of Viruses

Learning Outcomes

1. Describe the general features of viral structure.
2. Describe how viruses are classified and named.

Most viruses are extremely small (**figure 13.1**). In fact, they are approximately 100 to 1,000 times smaller than the host cells they infect. The smallest viruses are about 10 nm in diameter and contain very little nucleic acid, with perhaps as few as 10 genes. The largest known virus is about 1,500 nm long, the size of many bacterial cells. Indeed, when giant viruses were first discovered they were often mistaken for bacteria (see Focus Your Perspective 13.1).

Viral Structure

At a minimum, a **virion** (viral particle) consists of nucleic acid surrounded by a protein coat. Virions contain only a single type of nucleic acid—either RNA or DNA—but never both. This provides a useful method for classifying viruses, which are often referred to as either RNA or DNA viruses. The genome may be linear or circular, and either single-stranded or double-stranded. The protein coat, called the **capsid,** protects the viral nucleic acid from enzymes and toxic chemicals in the environment; it is composed of precisely arranged identical protein subunits called capsomeres. The capsid together with the nucleic acid it encloses is called the **nucleocapsid.**

Viruses have specific protein components that allow them to attach to specific receptor sites on host cells. Phages, for example, have tail fibers for attachment, and many animal viruses have protein structures called **spikes** on their surface.

Some viruses have an outer lipid bilayer called an envelope. **Non-enveloped viruses** (also called naked viruses) do not have that layer (**figure 13.2**). **Enveloped viruses** obtain their outer layer from the host cell; in addition, they have a viral matrix protein that links the envelope to the nucleocapsid. In general, enveloped viruses are more susceptible to soaps, detergents, and disinfectants because these chemicals damage the envelope, thereby removing the spikes and making the viruses non-infectious. The virus that causes COVID-19 is enveloped, which is why frequent hand-washing with soap and water helps limit the spread of the disease. Phages are typically non-enveloped.

MicroByte

A phage's nucleic acid is so tightly packed inside the capsid that the internal pressure is 10 times higher than the pressure in a champagne bottle!

A virus is generally one of three different shapes: icosahedral, helical, or complex (**figure 13.3**). Icosahedral viruses appear spherical when viewed with an electron microscope, but their surface is actually 20 flat triangles arranged in a manner somewhat similar to that of a soccer ball. Helical viruses appear cylindrical when viewed with an electron microscope; their capsomeres are arranged in a helix, somewhat similar to a spiral staircase. Some helical viruses are short and rigid, whereas others are long and filamentous. Complex viruses have more complicated structures. Phages are the most common examples of this, many having an icosahedral nucleocapsid, referred to as the head, with a long helical protein portion called the tail; tail fibers and other components are attached to the tail.

Viral Taxonomy

Although viruses are not living organisms, they are biological entities that are classified to provide easy identification and study (**table 13.1**). The International Committee on Taxonomy of Viruses (ICTV) keeps an online database and publishes a report describing the classification, nomenclature, and key features of recognized viruses; the report published in 2021 describes 10,434 viral species belonging to 2,606 genera. Up until recently, the classification scheme focused on grouping closely related viruses, primarily based on features such genome structure (type of nucleic acid and strandedness) and hosts infected (bacteria, archaea, animals, or plants); a variety of other traits were also considered, including virus shape. The ICTV has now expanded the classification framework to allow consideration of more distant evolutionary relationships. The new system includes taxonomic hierarchies

FIGURE 13.1 Relative Sizes of Viruses

Why are viruses so much smaller than the cells they infect?

(a) Non-enveloped virus

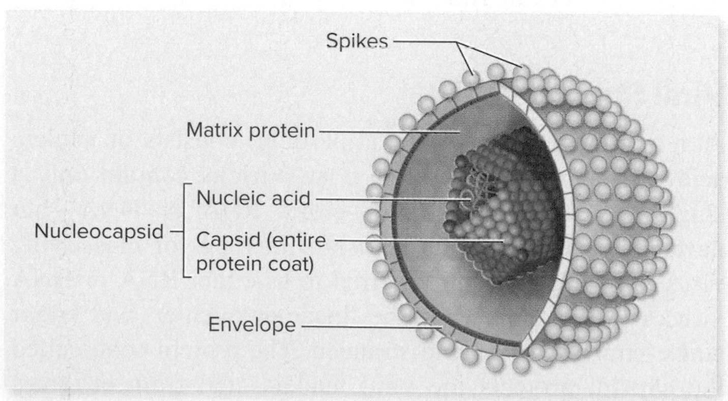

(b) Enveloped virus

FIGURE 13.2 Two Different Types of Viral Structure

Why would a disinfectant that destroys the envelope (thereby removing the spikes from the particle) make the remaining particle non-infectious?

similar to those used for other biological disciplines, with eight principal ranks—from species through realm (realm is the most inclusive rank, analogous to domain).

Our discussion of classification will focus on the viruses that infect animals. The names of virus families are derived from a variety of sources, but they all end in the suffix *-viridae*. These names follow no consistent pattern. In some cases, the name indicates the appearance of the viruses in the family—for example, *Coronaviridae* comes from *corona,* which means "crown." In other cases, the virus family was named for the geographic area from which a member was first isolated. *Bunyaviridae* is derived from Bunyamwera, a town in Uganda. However, as of 2015, viruses are no longer named after the location of their

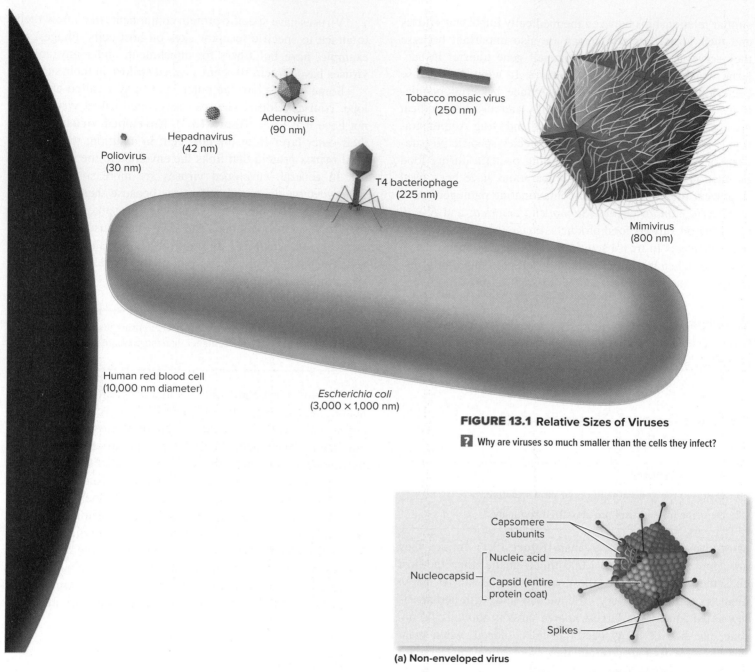

Icosahedral

Nucleic acid

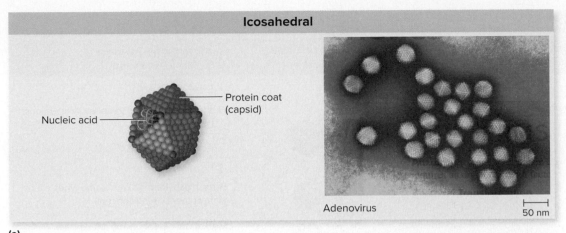

Protein coat (capsid)

Adenovirus

50 nm

(a)

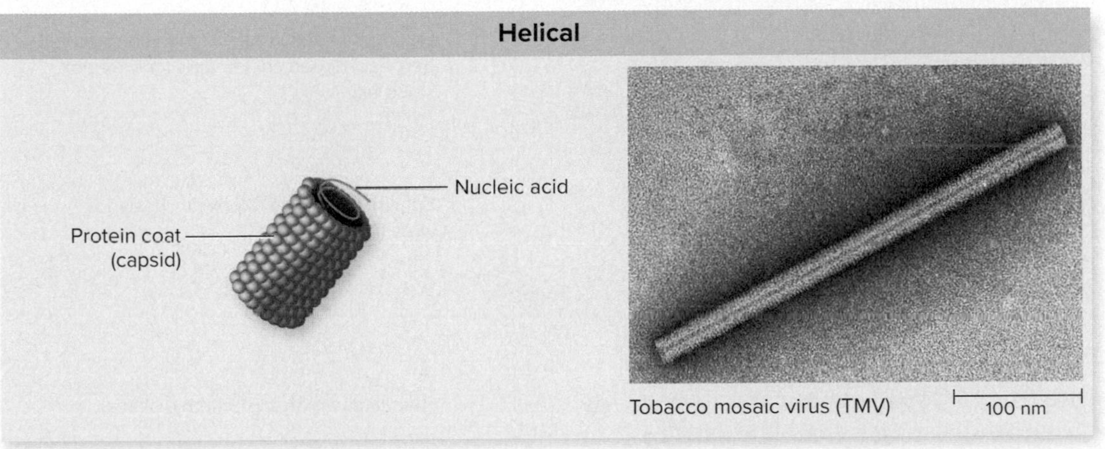

(b)

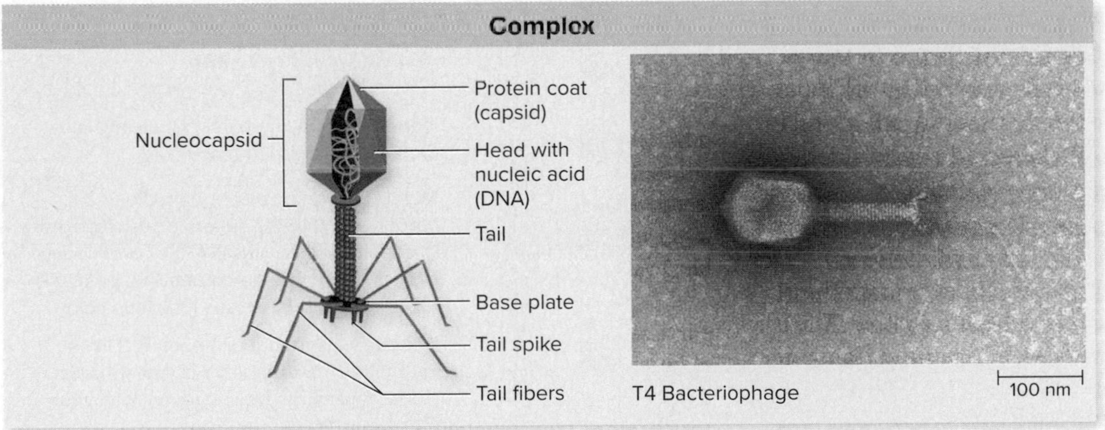

(c)

FIGURE 13.3 Common Shapes of Viruses

a: Eye of Science/Science Source; b: Science Source; c: Biophoto Associates/Science Source

? **What determines the shape of the virus?**

first isolation. Each family contains numerous genera, the names of which end in *-virus,* making it a single word—such as *Enterovirus.* The species name is often the name of the disease the virus causes—for example, *poliovirus* causes polio. All levels of formal taxonomic names of viruses are written in italics, unlike the nomenclature rules for other biological disciplines.

In contrast to bacterial nomenclature, in which an organism is usually referred to by its genus and species name, viruses are commonly referred to informally only by their species name or by another name, neither of which is capitalized or italicized—for example, rhabdovirus. In addition, informal terms are often used to refer to groups of animal viruses that are not taxonomically related but share critical characteristics, such as the primary route of transmission (**table 13.2**). Viruses transmitted via the fecal-oral route are referred to as enteric viruses (enteric means relating to the intestines), and viruses transmitted through the respiratory route are called respiratory viruses. Zoonotic viruses cause zoonoses, which are diseases transmitted from a non-human

TABLE 13.1 Classification of Some Human Viruses

Nucleic Acid	Outer Covering	Family	Drawing of Virion (Not to Scale)	Representative Member (and Disease Caused)
DNA Viruses				
Double-stranded DNA	Non-enveloped	*Adenoviridae*		Human adenoviruses (respiratory infections)
		Papillomaviridae		Human papillomaviruses (some types cause warts; others cause cancers)
		Polyomaviridae		Merkel cell polyomavirus (Merkel skin cancer)
	Enveloped	*Herpesviridae*		Herpes simplex viruses (cold sores, genital herpes); herpes zoster virus (chickenpox, shingles)
		Poxviridae		Smallpox virus (smallpox); monkeypox virus (mpox)
Single-stranded DNA	Non-enveloped	*Parvoviridae*		Human parvovirus B19 (fifth disease)
RNA Viruses				
Double-stranded	Non-enveloped	*Reoviridae*		Human rotaviruses (gastroenteritis)
Single-stranded (plus strand)	Non-enveloped	*Caliciviridae*		Norovirus (gastroenteritis)
		Picornaviridae		Hepatitis A virus (hepatitis A); polioviruses (polio); rhinovirus (common cold)
	Enveloped	*Coronaviridae*		Middle East respiratory syndrome coronavirus (MERS); severe acute respiratory syndrome coronavirus (SARS); severe acute respiratory syndrome coronavirus 2 (COVID-19); human coronaviruses (common cold)
		Flaviviridae		Dengue virus (dengue); hepatitis C virus (hepatitis C); West Nile virus (encephalitis); yellow fever virus (yellow fever); Zika virus (Zika virus disease)
		Matonaviridae		Rubella virus (rubella)
		Togaviridae		Chikungunya virus (Chikungunya)
Single-stranded (minus strand)	Enveloped	*Arenaviridae*		Lassa virus (Lassa fever)
		Bunyaviridae		Hantavirus (hantavirus pulmonary syndrome)
		Filoviridae		Ebola virus (Ebola virus disease); Marburg virus (Marburg virus disease)

TABLE 13.1 Classification of Some Human Viruses (*Continued*)

Nucleic Acid	Outer Covering	Family	Drawing of Virion (Not to Scale)	Representative Member (and Disease Caused)
		Orthomyxoviridae		Influenza virus (influenza)
		Paramyxoviridae		Measles virus (measles); mumps virus (mumps)
		Pneumoviridae		Respiratory syncytial virus (RSV bronchiolitis, croup)
		Rhabdoviridae		Rabies virus (rabies)
Reverse Transcribing Viruses				
DNA	Enveloped	*Hepadnaviridae*		Hepatitis B virus (hepatitis B)
RNA	Enveloped	*Retroviridae*		Human immunodeficiency virus (AIDS)

TABLE 13.2 Grouping of Human Viruses Based on Route of Transmission

Virus Group	Mechanism of Transmission	Common Viruses Transmitted (and Diseases Caused)
Enteric	Fecal-oral route	Enteroviruses (polio); noroviruses and rotaviruses (diarrhea)
Respiratory	Respiratory or salivary route	Influenza viruses (influenza); measles virus (measles); rhinoviruses (colds); coronaviruses (colds, COVID-19)
Sexually Transmitted	Sexual contact	Herpes simplex virus type 1 and type 2 (genital herpes); HIV (HIV/AIDS)
Zoonotic	Vector (such as arthropods)	Dengue viruses (dengue and severe dengue), West Nile virus (encephalitis); Zika virus (Zika virus disease)
	Animal to human directly	Rabies virus (rabies)

animal to a human. One group of viruses, the **arboviruses** (meaning **ar**thropod **bo**rne), is so named because they are spread by arthropods such as mosquitoes, ticks, and sandflies. The arthropods are vectors—they do not cause disease themselves, but they can transmit pathogens from one host to another; this occurs when a virus-infected arthropod bites an animal and transmits the virus to that animal. In many cases, arboviruses can infect widely different species. If a virus-infected arthropod bites birds, reptiles, and mammals, for example, it can transfer the virus among those different groups. Arboviruses cause important diseases such as West Nile encephalitis, La Crosse encephalitis, yellow fever, and dengue fever.

MicroAssessment 13.1

Viruses consist of nucleic acid (RNA or DNA) surrounded by a protein coat and sometimes an outer lipid layer (envelope). Viruses are typically icosahedral, helical, or complex in shape. Almost all bacteriophages are non-enveloped, whereas animal viruses are either non-enveloped or enveloped. Viruses are classified based primarily on the characteristics of their genome, such as type of nucleic acid and strandedness. Viruses are often grouped by their route of transmission. They replicate only inside living cells.

1. Why are some viruses more easily destroyed by soaps, detergents, and disinfectants than others?
2. List three ways in which viruses can be transmitted from one organism to another.
3. Most enteric viruses are non-enveloped. Why would this be so?

FOCUS YOUR PERSPECTIVE 13.1

Microbe Mimicker

Like bacteria, viruses vary tremendously in size and complexity. This is becoming increasingly apparent as viruses from a range of hosts are studied in greater detail. A group of viruses called the giant viruses is changing the views of scientists on the origin of viruses and possibly the history of life. One of the largest giant viruses characterized came from *Acanthamoeba polyphaga,* a free-living amoeba. The virus was first thought to be a Gram-positive bacterium because of its size, staining characteristics, and hairy appearance, with many fibrils sticking out from its capsid (see figure 13.1). Its name, *Acanthamoeba polyphaga mimivirus* (APMV) reflects this original misconception (**mim**icking **mi**crobe). APMV has a diameter of approximately 800 nm; this is twice the size of a small bacterium, and these virus particles can be observed using a light microscope. Its genome size is enormous for a virus—1.2 million base pairs, which is large enough to encode over 1,000 proteins. Many of the encoded proteins have never been found in viruses before; they include enzymes of nucleic acid synthesis, DNA repair, translation, and polysaccharide biosynthesis. However, the virion does not contain ribosomes, and it cannot replicate outside a host cell. Mimivirus belongs to a taxonomic group called nucleocytoplasmic large DNA viruses (NCLDVs), members of which infect distinctly different organisms, including vertebrates and algae. These viruses typically induce the formation of "viral factories," areas in the host cytoplasm where organelles are excluded and new virions are assembled.

After the discovery of mimivirus, a mimivirus-infected amoeba was found infected with an additional virus. Its discoverers named this new virus Sputnik and called it a virophage. Virophages are tiny and have a very small genome (about 18,000 base pairs); they can replicate in a host cell only when another virus (called a helper virus) has infected the same host cell. The scientists who discovered Sputnik named its helper virus mamavirus because it was larger than the original mimivirus (**box figure 13.1**). Coinfection of a host amoeba with a Sputnik and mamavirus results in production of fewer mamavirus particles, and those particles display unusual morphologies. Thus, Sputnik behaves as a true parasite.

Mimivirus and mamavirus are not the only unusual giant amoeba-infecting viruses. Several other giant viruses have been described. Now two new giant viruses called tupanviruses have been discovered. Tupanviruses have a genome of approximately 1.4 to 1.5 million base pairs, and they contain nearly all the genes necessary for protein production—in fact, if they had ribosomes, they could synthesize proteins without the need for a host cell. The discovery and characterization of giant viruses expands our knowledge of viruses. Their genetic complexity also challenges our definition of what is living and non-living—and perhaps will provide an answer to the question: "Where did viruses originate?"

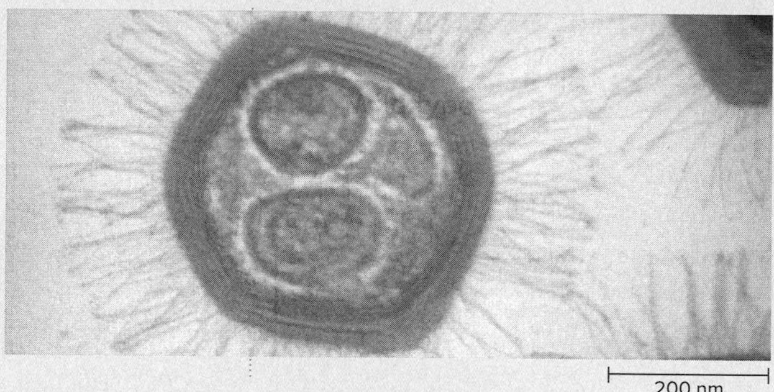

BOX FIGURE 13.1 Mamavirus (Large Mimivirus) Dr. Raoult/Science Source

13.2 ■ Bacteriophages

Learning Outcomes

3. Compare and contrast lytic, temperate, and filamentous phage infections.
4. Describe two consequences of lysogeny.

An enormous variety of phages exists, each with a characteristic shape, size, genome structure, and replication strategy. The general strategies for phage replication result in two possible outcomes: (1) a **productive infection,** in which new viral particles are produced, and (2) a **latent state,** in which the viral genome remains silent within the cell but is replicated along with the host cell genome (**figure 13.4**). In a productive infection, some types of viruses kill their host, but others do not.

Lytic Phage Infections: T4 Phage as a Model

By definition, **lytic phages** (also called **virulent phages**) exit the host at the end of the infection by lysing the cell, an action that kills the cell. An example of a lytic phage is T4, a double-stranded DNA phage that infects *E. coli*; the steps of its replication cycle are similar to those of other lytic phages (**figure 13.5**):

① **Attachment.** Phages collide with their host cells by chance. On contact, the phage attaches to a receptor on the host cell surface or an appendage such as a pilus. In the case of T4, the phage tail fibers bind to receptors on the outer membrane of Gram-negative bacteria. The receptors used by phages for attachment normally perform important functions for the cell—the phages merely take advantage of the molecules for their own use. Cells that lack the receptor used by a particular phage are resistant to infection by that phage.

Focus Figure

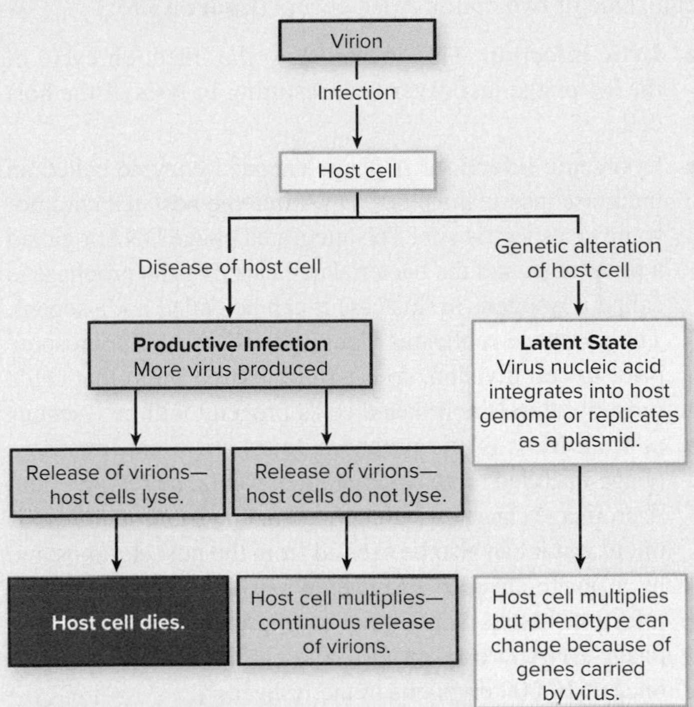

FIGURE 13.4 Major Types of Relationships Between Viruses and the Cells They Infect

Which of the illustrated interactions is least harmful to the host? Explain.

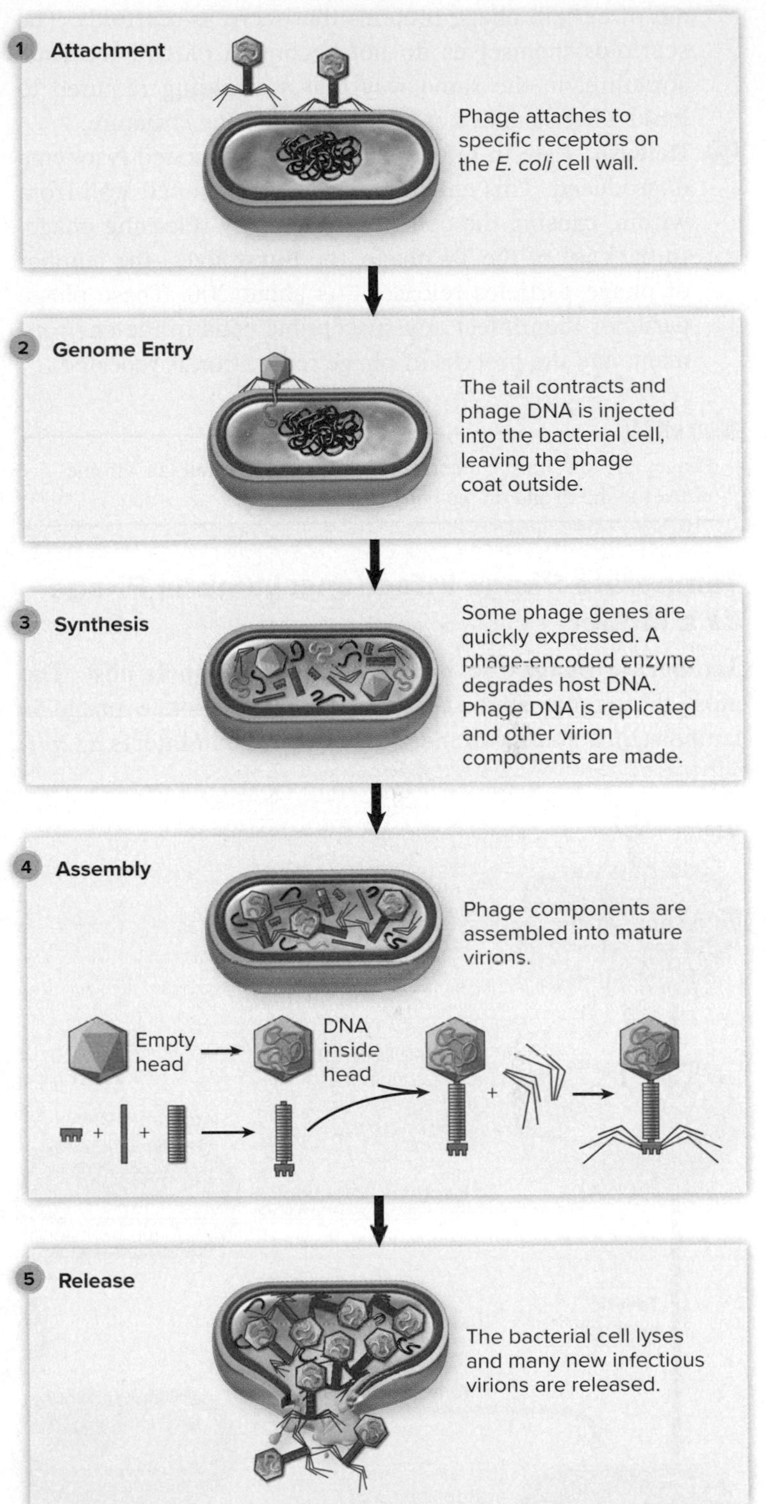

FIGURE 13.5 Steps in the Replication of the Lytic Phage T4 in *E. coli*

If one phage particle infects one bacterial cell, how many particles are inside the cell 5 minutes after infection? Explain.

2. **Genome entry.** Following attachment, the phage injects its genome into the cell. T4 does this by degrading a small part of the bacterial cell wall, using an enzyme located in the tip of its tail. This enzyme, called T4 lysozyme, is functionally similar to the lysozyme found in tears and other body secretions, and it degrades peptidoglycan. The tail then contracts so that the phage particle appears to squat on the cell surface. This action injects the phage DNA through the host's cell wall and cytoplasmic membrane, and into the cell. The phage capsid remains on the outside of the host cell.
3. **Synthesis of phage proteins and genome.** Within minutes after the phage DNA is injected into a host cell, that cell's machinery transcribes and translates some of the phage genes. The genes are expressed in a specific sequence to control the course of infection. First, the proteins needed to start phage replication are made; one of these is a nuclease that degrades the host cell's DNA. The phage genome is then synthesized. Toward the end of the replication cycle, phage structural proteins are made, including those that make up the capsid, the tail, and the tail fibers.
4. **Assembly (maturation).** This complex, multistep process assembles the various phage components made in the previous step to form new phage particles. After the phage head is formed, DNA is packed into it, and then the tail is attached; the tail fibers then attach to the tail structure. Some of these components self-assemble, meaning that the proteins join together spontaneously to form a specific structure. Other components assemble with the

aid of certain phage proteins that serve as scaffolds. The scaffolds themselves do not become a part of the final structure, in the same way that scaffolding required to build a house does not become part of the structure.

5 **Release.** Late in infection, the phage-encoded lysozyme is produced. This enzyme digests the host cell wall from within, causing the cell to lyse, thereby releasing phage. In the case of the T4 phage, the **burst size**—the number of phage particles released—is about 200. These phage particles then infect any susceptible cells in the environment, and the process of phage replication is repeated.

MicroByte

Phages are the most numerous members of the human virome—viruses found in and on the body.

Temperate Phage Infections: Lambda Phage as a Model

Temperate phages do not necessarily lyse their host. The most thoroughly studied example of a temperate phage is lambda (λ), a double-stranded DNA phage that infects *E. coli.* It has a linear chromosome, but once that genome is injected into a cell, the ends join together to form a circular DNA molecule. One of two options then occurs (**figure 13.6**):

- **Lytic infection.** This is similar to the infection cycle of the T4 phage just described, resulting in lysis of the host cell.
- **Lysogenic infection.** A phage-encoded enzyme called an integrase inserts the phage DNA into the host cell chromosome at a specific site. The integrated phage DNA is called a **prophage,** and the bacterial cell carrying the prophage is called a **lysogen** (in this case it can be called a λ lysogen). The prophage replicates along with the host chromosome prior to cell division, so it will be passed on to that cell's progeny; thus, the infected cell's progeny will be lysogens as well. Most of the prophage genes are silent because a phage-encoded repressor protein prevents their expression. A prophage can remain integrated in this latent state indefinitely, but it can also be excised from the host chromosome by a phage-encoded enzyme. When this happens, a lytic infection begins. Under ordinary growth conditions, the phage DNA is excised from the chromosome only about once in 10,000 divisions of the lysogen.

Which type of infection occurs initially—lytic or lysogenic—is mostly random, but the metabolic state of the host cell has an effect. For example, if a bacterial cell is growing slowly because of nutrient limitation, then a lysogenic infection is more likely to occur.

Whether the prophage persists or the lytic cycle begins depends on a complex series of events involving the phage-encoded repressor and other regulatory proteins. The repressor prevents expression of the gene required for excision and is therefore essential for maintaining the lysogenic (latent) state. The repressor does this by binding to the operator which controls the expression of phage genes that direct a lytic infection. Excision can be induced by treating a lysogenic culture with a DNA-damaging agent such as ultraviolet light. As a result of the damage, the SOS repair system activates an enzyme that inadvertently destroys the repressor. Once the repressor is gone, the prophage will be excised from the chromosome—allowing

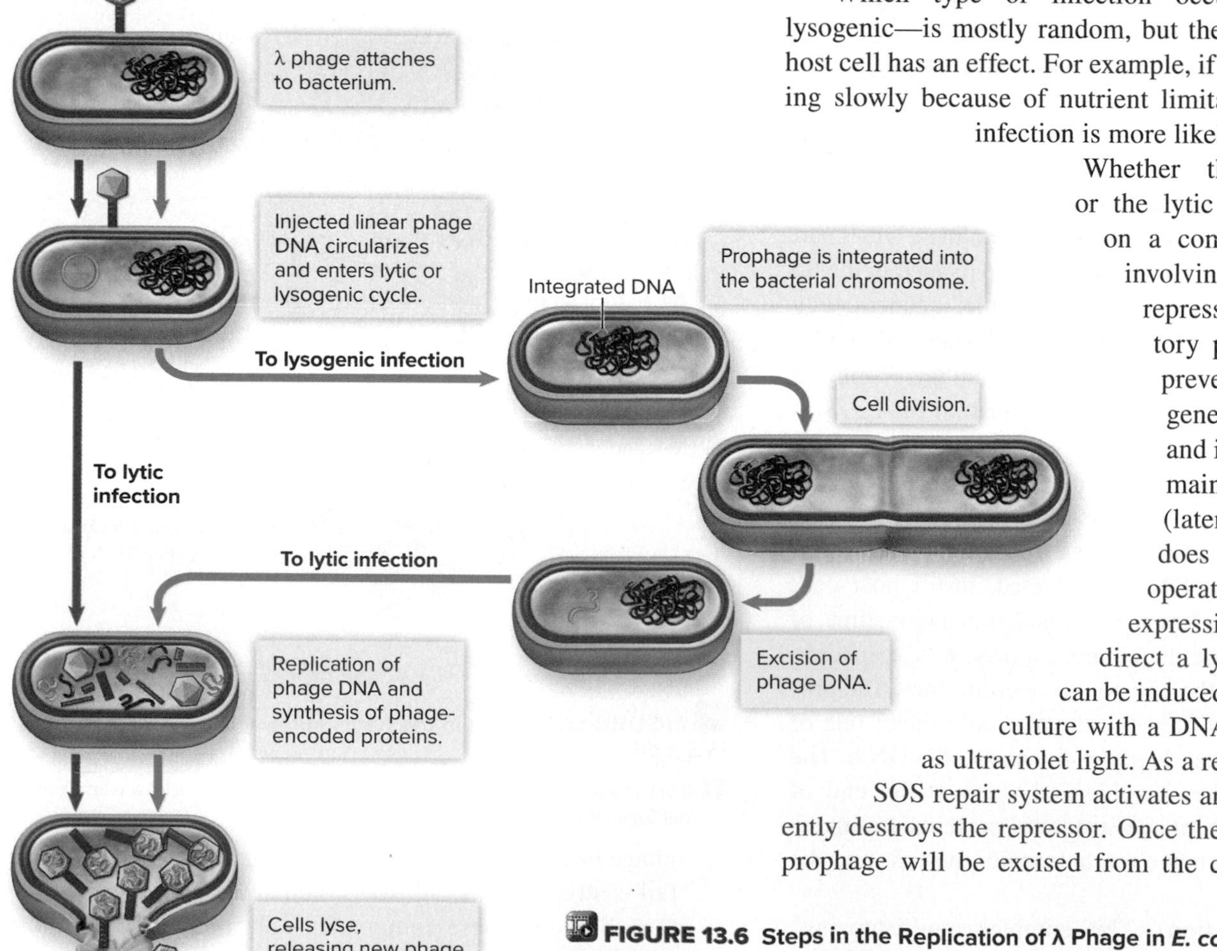
λ phage attaches to bacterium.
Injected linear phage DNA circularizes and enters lytic or lysogenic cycle.
To lysogenic infection
Integrated DNA
Prophage is integrated into the bacterial chromosome.
Cell division.
To lytic infection
To lytic infection
Excision of phage DNA.
Replication of phage DNA and synthesis of phage-encoded proteins.
Cells lyse, releasing new phage.

FIGURE 13.6 Steps in the Replication of λ Phage in *E. coli*

When is it advantageous for a temperate phage to excise its DNA from the host chromosome?

the phage to enter the lytic cycle—and a productive infection results. This process, **phage induction,** lets the phage escape from a damaged host.

Consequences of Lysogeny

Although a lysogen is morphologically identical to an uninfected cell, lysogeny has consequences. These include:

- **Immunity to superinfection.** This refers to the blocking of an additional infection of a lysogen by the same type of phage. It occurs because the repressor that maintains the prophage in the integrated state will also bind to the operator on incoming phage DNA. Thus, if another λ phage injects its DNA into a λ lysogen, that DNA will not be expressed. Note that this type of immunity is different from that of the human immune system, which is discussed in chapters 14 and 15.
- **Lysogenic conversion.** This is a change in the phenotype of a lysogen as a consequence of the specific prophage it carries. For example, only strains of *Corynebacterium diphtheriae* that are lysogenic for a certain phage can synthesize the toxin that causes diphtheria. If a toxin is encoded exclusively by phage genes, then only bacterial strains that carry the prophage will synthesize the toxin. Some additional examples of lysogenic conversion are given in **table 13.3.**

Filamentous Phage Infections: M13 Phage as a Model

Filamentous phages are single-stranded DNA phages that look like long fibers (filaments). They cause productive infections, but the process does not kill the host cells. Infected cells, however, grow more slowly than uninfected cells.

M13 is a filamentous phage that initiates infection by attaching to a protein on the F pilus of *E. coli*. Once the phage genome enters a bacterial cell, the host cell DNA polymerase synthesizes the complementary strand to create a double-stranded DNA molecule referred to as the replicative form (RF). One strand of the RF is then used as a template to make mRNA as well as multiple copies of the phage genome (**figure 13.7**).

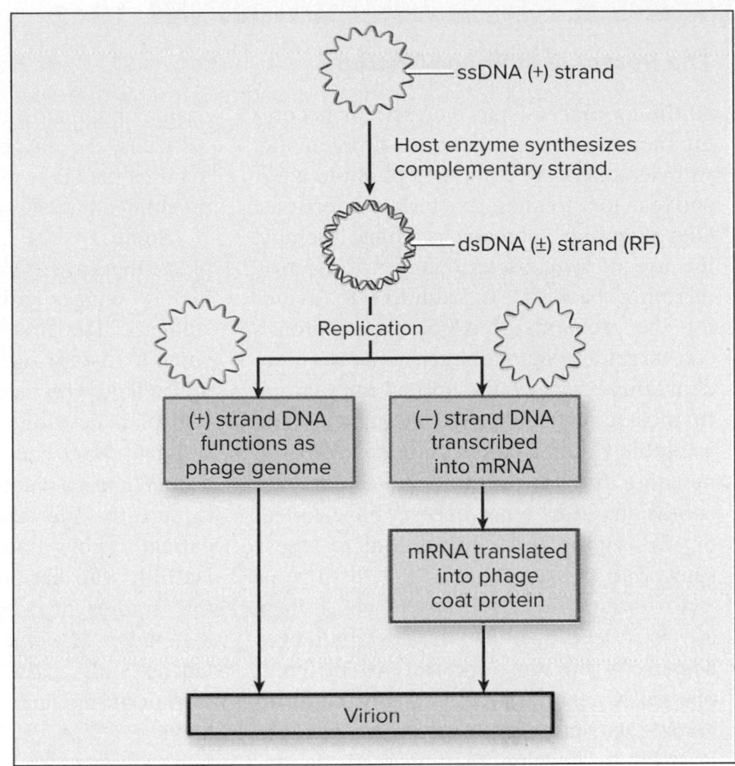

FIGURE 13.7 Macromolecule Synthesis in Filamentous Phage Replication

Which host enzyme synthesizes the complement to the ssDNA to make the RF?

Once phage mRNA has been made, the genes are expressed to produce the various viral proteins. Phage particles are assembled as they exit the cell, a process called extrusion. For this to occur, phage capsomeres insert into the host cell's cytoplasmic membrane as certain other phage proteins assemble to create pores that span the cell envelope (**figure 13.8**). Then, as phage DNA is extruded through the pores, the capsomeres coat that single-stranded genome to form a nucleocapsid.

TABLE 13.3 Some Properties Encoded by Prophages (Lysogenic Conversion)

Microorganism	Medical Importance	Property Encoded by Prophage
Clostridium botulinum	Causes botulism	Botulinum toxin (interferes with nerve impulse transmission and thus muscle contraction)
Corynebacterium diphtheriae	Causes diphtheria	Diphtheria toxin (interferes with protein synthesis, causing cell death)
Escherichia coli O157:H7	Causes hemolytic uremic syndrome	Shiga toxin (interferes with protein synthesis, causing cell death particularly in blood vessel lining)
Streptococcus pyogenes	Causes streptococcal toxic shock syndrome	Streptococcal pyrogenic exotoxins (SPEs) (result in a cytokine storm due to inappropriate activation of certain cells of the immune system)
Vibrio cholerae	Causes cholera	Cholera toxin (results in the outpouring of fluid from intestinal cells)

FOCUS YOUR PERSPECTIVE 13.2

The Potential of Phage Therapy

Antimicrobial resistance (AMR) is becoming increasingly prevalent and problematic, so researchers are exploring possible alternatives for treating bacterial infections. One potential approach is phage therapy, the use of lytic bacteriophages to destroy infecting bacteria. In addition to avoiding the problems of AMR, this approach can target a given pathogen, thus avoiding damaging effects to the normal microbiota. In most countries, phage therapy is only available (1) on a compassionate use basis, meaning as a last resort for life-threatening conditions after other options have failed, or (2) as part of a clinical trial to assess safety and efficacy.

Although the specificity of phage therapy is a benefit, it also creates challenges because a mixture (cocktail) of different phages is generally required. For compassionate use cases, a personalized approach is used to develop that cocktail: Scientists screen a variety of different phages against the patient's infecting bacterial strain to determine which options lyse the strain, and they then use that information to decide on the most effective mix. For clinical trials, the phage cocktail must be able to target the many different bacterial strains infecting a patient population.

Some of the recent successes with phage therapy have relied on cocktails that include phages isolated by undergraduate students. The first of these was in 2018, when a 15-year-old cystic fibrosis patient in England who recently had a double lung transplant developed a systemic multidrug-resistant *Mycobacterium abscessus* infection. When conventional treatments failed to cure the life-threatening infection, the patient's physicians turned to Graham Hatfull, who heads a research laboratory that studies mycobacteriophages (phages that target *Mycobacterium* species). That lab not only maintains an extensive collection of mycobacteriophages, but it also helps run SEA-PHAGES (Science Education Alliance Phage Hunters Advancing Genomics and Evolutionary Science), a program that aims to increase undergraduate engagement by helping students discover and characterize bacteriophages. Hatfull's group was sent a culture of the patient's infecting *M. abscessus* strain, which was then used to screen for phages that targeted it. Based on the findings, the group then optimized a phage cocktail to be used for treatment. One of the phages selected, called Muddy, had been discovered and named by the undergraduate student in South Africa who isolated it from a decaying eggplant! After six weeks of intravenous and topical treatment with the phage cocktail, the patient's infection had mostly cleared, and she was eventually able to return to school. Since then, Muddy has played a role in other successful cases of compassionate use phage therapy.

Many scientists are very optimistic about using phage therapy to combat "superbugs," and considering the estimated 10^{31} phages on Earth, the potential supply of these agents is enormous. Programs like SEA-PHAGES allow a path toward discovery of potentially life-saving tools while also educating and training young scientists about microbiology research.

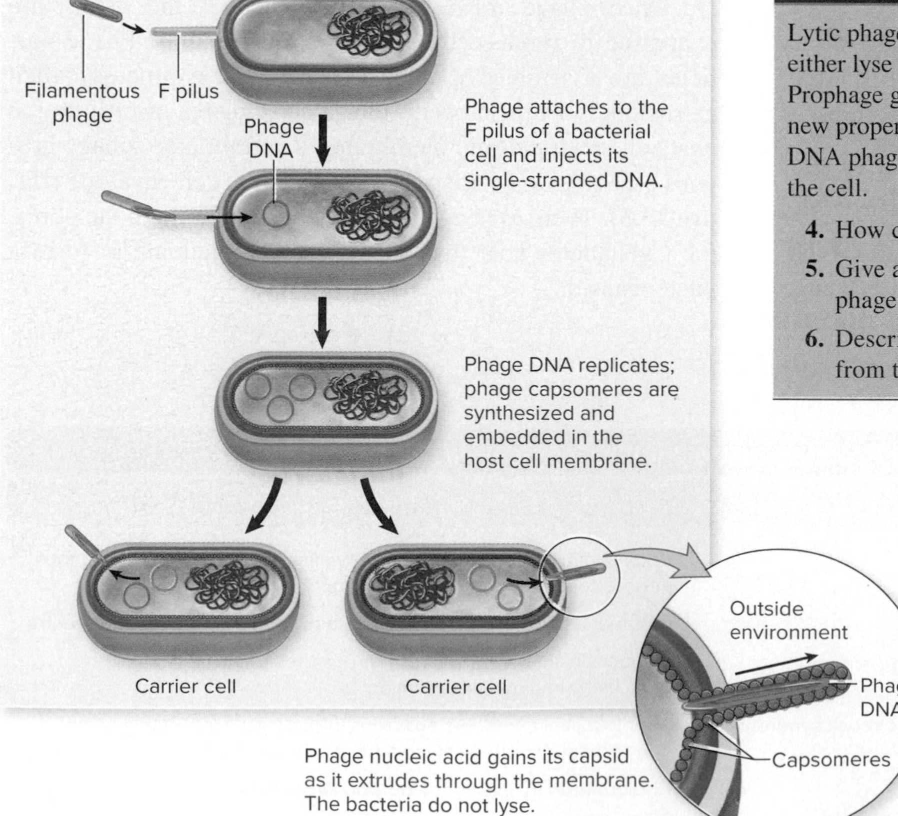
Filamentous phage
F pilus
Phage DNA
Phage attaches to the F pilus of a bacterial cell and injects its single-stranded DNA.
Phage DNA replicates; phage capsomeres are synthesized and embedded in the host cell membrane.
Carrier cell
Carrier cell
Outside environment
Phage DNA
Capsomeres
Phage nucleic acid gains its capsid as it extrudes through the membrane. The bacteria do not lyse.

FIGURE 13.8 Replication of a Filamentous Phage

? **Why would infected cells grow more slowly than uninfected cells?**

MicroAssessment 13.2

Lytic phages lyse their host cells, whereas temperate phages either lyse their host or exist within the host as a prophage. Prophage genes sometimes code for products that confer new properties on the host cell. Filamentous single-stranded DNA phages are extruded from the host cell without killing the cell.

4. How can a productive phage infection not kill a host cell?
5. Give an example of a virulent phage and of a temperate phage.
6. Describe how the replication cycle of T4 phage is different from that of lambda (λ).

13.3 ■ The Roles of Bacteriophages in Horizontal Gene Transfer

Learning Outcome

5. Compare and contrast generalized and specialized transduction.

Errors during the phage replication process play an important role in horizontal gene transfer in bacteria. As briefly discussed in chapter 8, phages can transfer DNA from one bacterial cell (the donor) to another (the recipient) in the process called transduction. The two types of transduction—generalized and specialized—are due to two different types of errors.

Generalized Transduction

Generalized transduction results from a packaging error during phage assembly. Lytic phages degrade the bacterial chromosome into many fragments during infection, and any of these short bacterial DNA fragments can be mistakenly packaged into the phage head in place of phage DNA (see figure 8.20). Such an error creates what is called a **transducing particle,** which carries bacterial DNA but no phage DNA and therefore is not a phage.

Like phage particles, a transducing particle will attach to another bacterial cell and inject the DNA it contains. The transducing particle, however, injects only bacterial DNA because that is all it contains. The bacterial DNA may then integrate into the recipient's chromosome by homologous recombination. Any gene of the donor cell can be transferred this way, which is why this mechanism is called generalized transduction.

Specialized Transduction

Specialized transduction results from an excision mistake made by a temperate phage during its transition from a lysogenic cycle to a lytic cycle. In contrast to generalized transduction, only bacterial genes adjacent to the integrated phage DNA can be transferred. **Figure 13.9** illustrates the steps that lead to specialized transduction:

1. As with the start of any lysogenic cycle, a temperate phage first injects its DNA into a bacterial host.
2. That phage DNA then integrates into the host chromosome to become a prophage.
3. As the prophage is excised from the host cell chromosome to begin the lytic cycle, a rare mistake occurs: Rather than the prophage being excised precisely from the bacterial chromosome, a short piece of bacterial DNA on one side of the phage DNA is taken, and a piece of phage DNA is left behind.
4. The excised DNA—containing both bacterial and phage DNA—replicates and then becomes incorporated into phage heads during assembly. This results in the production and release of defective phage particles that do not carry the complete set of phage genes.

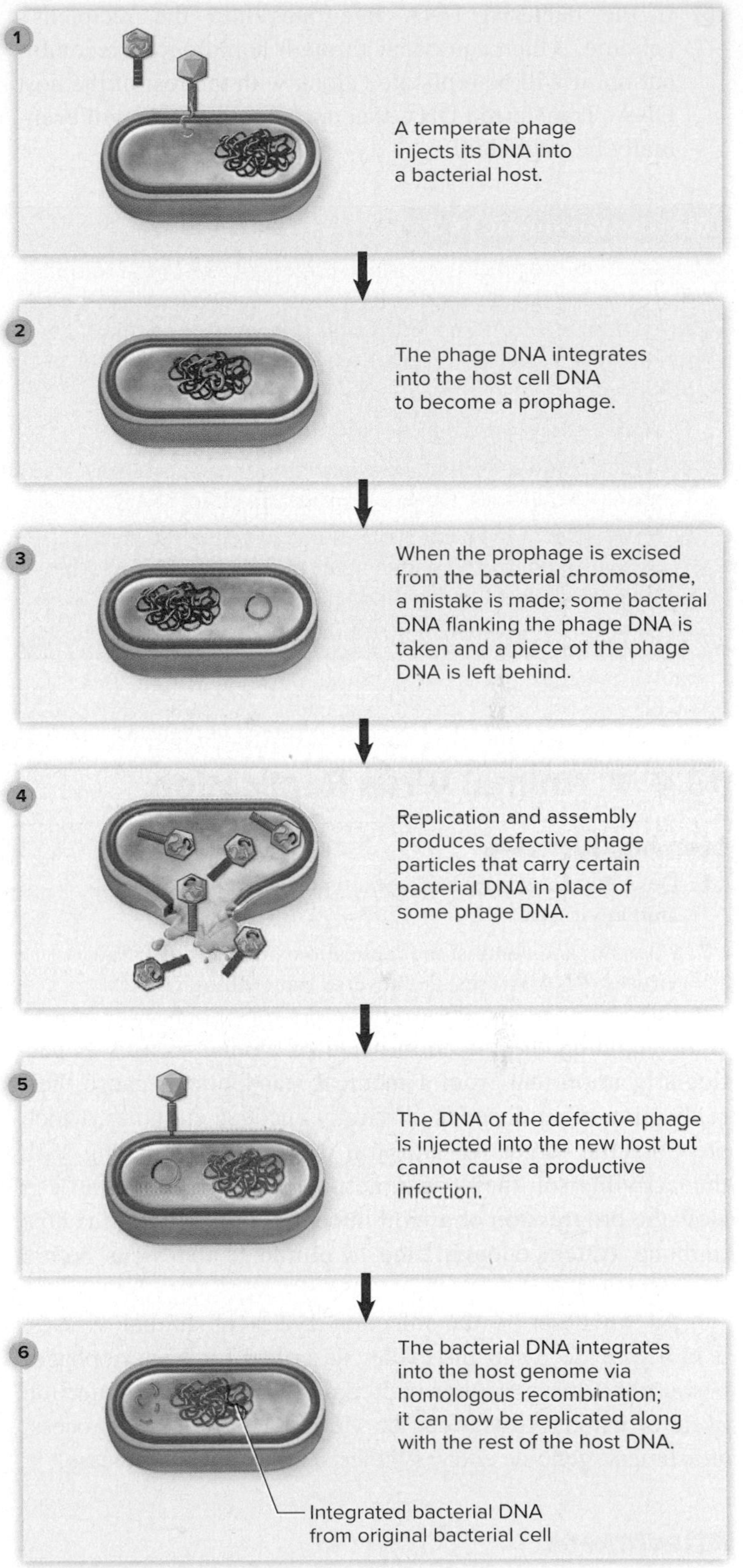

1
A temperate phage injects its DNA into a bacterial host.
2
The phage DNA integrates into the host cell DNA to become a prophage.
3
When the prophage is excised from the bacterial chromosome, a mistake is made; some bacterial DNA flanking the phage DNA is taken and a piece of the phage DNA is left behind.
4
Replication and assembly produces defective phage particles that carry certain bacterial DNA in place of some phage DNA.
5
The DNA of the defective phage is injected into the new host but cannot cause a productive infection.
6
The bacterial DNA integrates into the host genome via homologous recombination; it can now be replicated along with the rest of the host DNA.
Integrated bacterial DNA from original bacterial cell

FIGURE 13.9 Specialized Transduction

? What mistake in the temperate phage replication cycle leads to specialized transduction?

5. The defective phage injects the DNA it contains into another bacterial cell; thus, both phage DNA and bacterial DNA enter that cell.

6. If the bacterial DNA integrates into the recipient's genome, which can occur through homologous recombination, it will be replicated along with the rest of the host DNA. Transferred DNA that does not integrate will eventually be degraded.

MicroAssessment 13.3

Transduction is a process by which bacterial DNA is transferred from one bacterium to another by a phage. Generalized transduction results from a DNA packaging error, whereas specialized transduction results from an error during excision of a prophage.

7. With respect to transduction, what is a defective phage?
8. Which is more likely to be a specialized transducing phage—a lytic or a temperate phage? Explain.
9. Most types of temperate phages integrate into the host chromosome, whereas some replicate as plasmids. Which kind of relationship do you think would be more likely to maintain the phage in the host cell? Explain.

13.4 ▪ Animal Virus Replication

Learning Outcomes

6. Describe the general steps of a typical infection cycle of animal viruses.
7. Compare and contrast the replication strategies of DNA viruses, RNA viruses, and reverse-transcribing viruses.

Understanding the infection cycle of animal viruses is particularly important from a medical standpoint because their replication often depends on virally encoded enzymes, which are potential targets of antiviral drugs. By interfering with the activities of these enzymes, antiviral medications can slow the progression of a viral infection, often giving the host immune system enough time to eliminate the virus before illness occurs.

As we describe the infection cycles of animal viruses, it is helpful to recall the cycles described for bacteriophages because they have features in common. A generalized infection cycle of animal viruses can be viewed as a five-step process: attachment, genome entry, synthesis, assembly, and release.

Attachment

The process of attachment is basically the same in all virus–cell interactions. Animal viruses have attachment proteins (spikes) on their surfaces (see figure 13.2). The receptors these viral proteins bind are usually glycoproteins on the host cell cytoplasmic membrane, and often more than one receptor is required for effective attachment. HIV, for example, must bind to two different molecules before it can enter the cell. As with the receptors used by phages, the normal function of these receptors is completely unrelated to their role in virus attachment.

Because a virion must bind to specific receptors, a particular virus may be able to infect only one or a limited number of cell and tissue types; this is called tissue tropism. Likewise, most viruses can infect only a single species; the range of cell types, tissues, and species that a virus can infect—referred to as its host range—explains the resistance that some animals have to certain diseases. For example, dogs do not contract measles from humans, because dog cells do not have receptors for measles virus; likewise, humans do not contract distemper from cats. Some viruses, such as influenza virus, however, can infect unrelated animals such as pigs, birds, and humans.

MicroByte

Each cell can have tens to hundreds of thousands of receptor copies on its surface, providing many opportunities for a virion to encounter a receptor.

Entry and Uncoating

The mechanism an animal virus uses to enter its host cell depends in part on whether the virion is enveloped or non-enveloped. In all cases, the entire virion is taken into the cell. This differs from phages, where only nucleic acid enters the bacterial cell and the capsid (protein coat) stays outside.

Enveloped viruses enter the host cell by one of two mechanisms (**figure 13.10**):

- **Fusion with the host membrane.** The lipid envelope of the virion fuses with the cytoplasmic membrane of the host cell after the virion has attached to the host cell receptors. This process is similar to the way drops of oil can fuse together when they come into contact with each other. As a result of fusion, the nucleocapsid is released directly into the host cell's cytoplasm.
- **Endocytosis.** This involves receptor-mediated endocytosis, a normal mechanism by which a cell brings certain extracellular material into its cytoplasm (see figure 3.23). The viral particle binds to the receptors that normally facilitate endocytosis, causing the cell to take it up. After a virion is taken into the cell, the nucleocapsid is released into the cytoplasm; in many cases this occurs via fusion of the viral envelope with the endosomal membrane.

Non-enveloped viruses, which have no lipid envelope, cannot fuse with host membranes to enter cells. Therefore, the virions can only enter via endocytosis.

In most viruses, the nucleic acid separates from its capsid before the start of replication. This process, called **uncoating,** may occur either simultaneously with entry of the virion

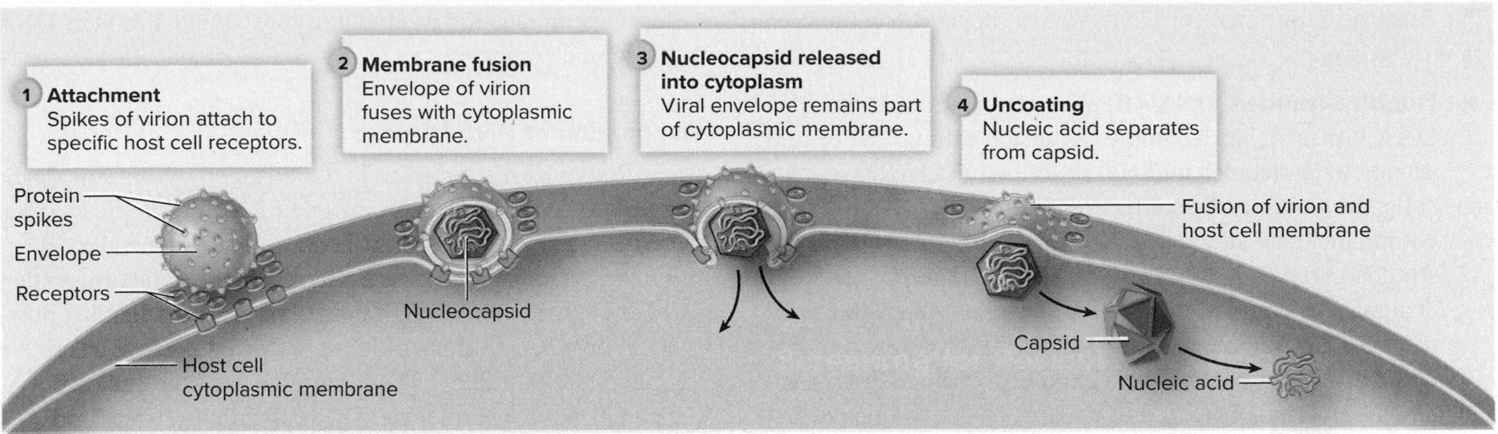
1 Attachment
Spikes of virion attach to specific host cell receptors.
2 Membrane fusion
Envelope of virion fuses with cytoplasmic membrane.
3 Nucleocapsid released into cytoplasm
Viral envelope remains part of cytoplasmic membrane.
4 Uncoating
Nucleic acid separates from capsid.
Protein spikes
Envelope
Receptors
Nucleocapsid
Host cell cytoplasmic membrane
Fusion of virion and host cell membrane
Capsid
Nucleic acid

(a) Entry by membrane fusion

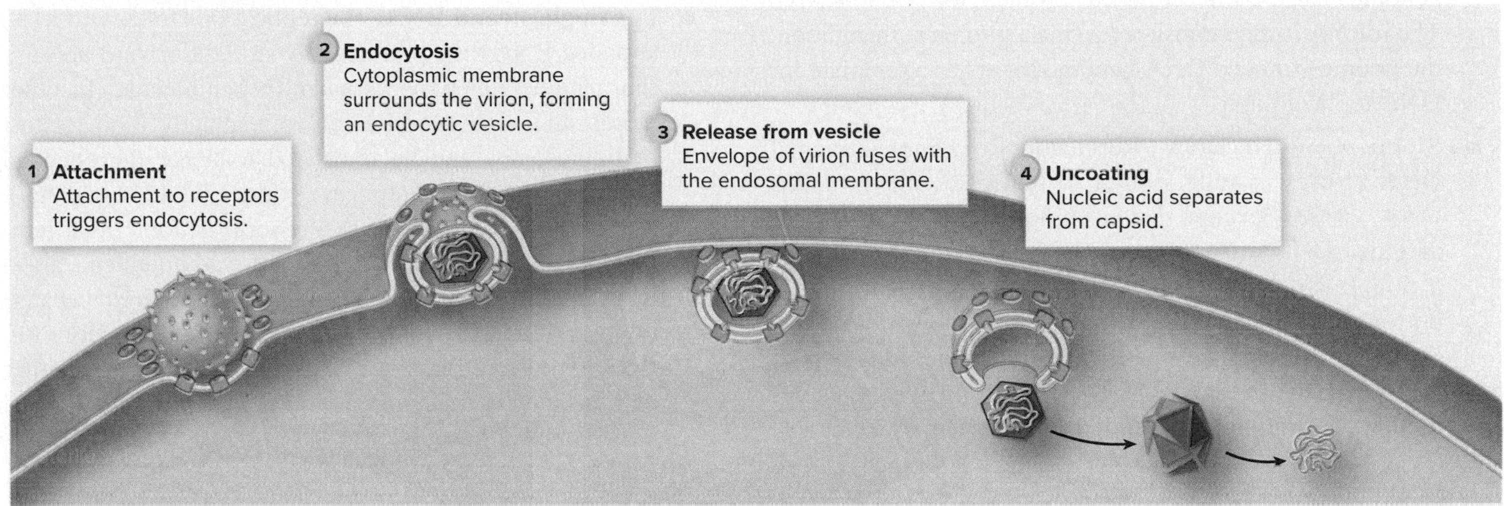
1 Attachment
Attachment to receptors triggers endocytosis.
2 Endocytosis
Cytoplasmic membrane surrounds the virion, forming an endocytic vesicle.
3 Release from vesicle
Envelope of virion fuses with the endosomal membrane.
4 Uncoating
Nucleic acid separates from capsid.

(b) Entry by endocytosis

FIGURE 13.10 Entry of Enveloped Animal Viruses into Host Cells **(a)** Entry following membrane fusion. **(b)** Entry by endocytosis.

? Could phages enter bacteria by the process of membrane fusion? Explain.

into the cell or after the nucleocapsid is near the intracellular site of viral replication. Most DNA viruses multiply in the nucleus; their genome, perhaps still protected by capsid components, enters through nuclear pores using viral proteins that have nuclear localization signals.

Replication

Production of viral particles in an infected cell requires two distinct but interrelated events:

- Expression of viral genes to produce structural and catalytic proteins, such as capsid proteins and any enzymes required for replication.
- Synthesis of multiple copies of the viral genome.

The viral proteins are sometimes synthesized as a polyprotein that is subsequently cleaved into individual proteins by viral proteases, which are potential targets for antiviral medications. For example, several medications used to treat HIV infection inhibit the HIV protease. Likewise, some medications for treating hepatitis C inhibit the HCV protease.

The replication schemes of viruses can be divided into three general categories: those used by (1) DNA viruses, (2) RNA viruses, and (3) reverse-transcribing viruses. As you will see in the following discussion, the type of genome has a significant influence on the viral replication scheme, a primary reason that virus classification takes genome structure into account (see table 13.1).

Replication of DNA Viruses

Most DNA viruses replicate in the nucleus of the host cell and use host cell components for DNA synthesis and gene expression. These viruses often encode their own DNA polymerase, however, which allows them to replicate even if the host cell is not actively duplicating its own chromosome.

The replication scheme for DNA viruses depends on the type of viral genome:

- **Double-stranded DNA.** Replication of double-stranded DNA viruses is fairly simple because it follows the central dogma of molecular biology, described in chapter 7. Like cells, the genome of double-stranded DNA viruses has two complementary strands. Recall that complementary DNA strands can be referred to as (+) strands and (–) strands, designations that indicate which strand is used as the template for RNA synthesis (see figure 7.9). As shown in **figure 13.11** (top panel), the genome of a double-stranded DNA virus can be referred to as (+/–) to indicate that it has both a (+) strand and a (–) strand. These (+/–) genomes can be transcribed to produce mRNA (+RNA). The mRNA is then translated to make proteins. In addition, the double-stranded DNA genome serves as a template for DNA replication.
- **Single-stranded DNA.** Replication of single-stranded DNA viruses is quite similar to that of double-stranded DNA viruses, except that a complement to the single-stranded DNA molecule must be synthesized to generate a double-stranded (+/–) DNA molecule (figure 13.11, bottom panel). Once that has occurred, the genes can be expressed to produce the encoded proteins. Meanwhile, the newly synthesized DNA strand—which is complementary to the single-stranded DNA genome—also acts as a template for producing more single-stranded DNA genome copies.

Replication of RNA Viruses

Most RNA viruses replicate in the host cell cytoplasm. Their replication requires a virally encoded RNA polymerase, often called a **replicase.** The replicase is an RNA-dependent RNA polymerase, meaning that it uses an RNA template to synthesize a new strand of RNA. Compare this with the RNA polymerase involved in transcription—that enzyme synthesizes a strand of RNA from a DNA template and is thus a DNA-dependent RNA polymerase. The replication scheme for RNA viruses depends on the type of viral genome:

- **(+) single strand RNA.** The replication strategy of a (+) stranded RNA virus is relatively straightforward because the genome functions as an mRNA molecule. In other words, the viral genome can immediately bind to host cell ribosomes and be translated to make the encoded proteins, including a replicase (**figure 13.12**, top panel). The replicase will then use the viral genome as a template to make multiple copies of the complement: a (–) RNA molecule. These molecules are required because they serve as a template for replicase to make more (+) RNA. The resulting (+) RNA molecules can then either be translated to make more viral proteins or be packaged as genomes of new virions.

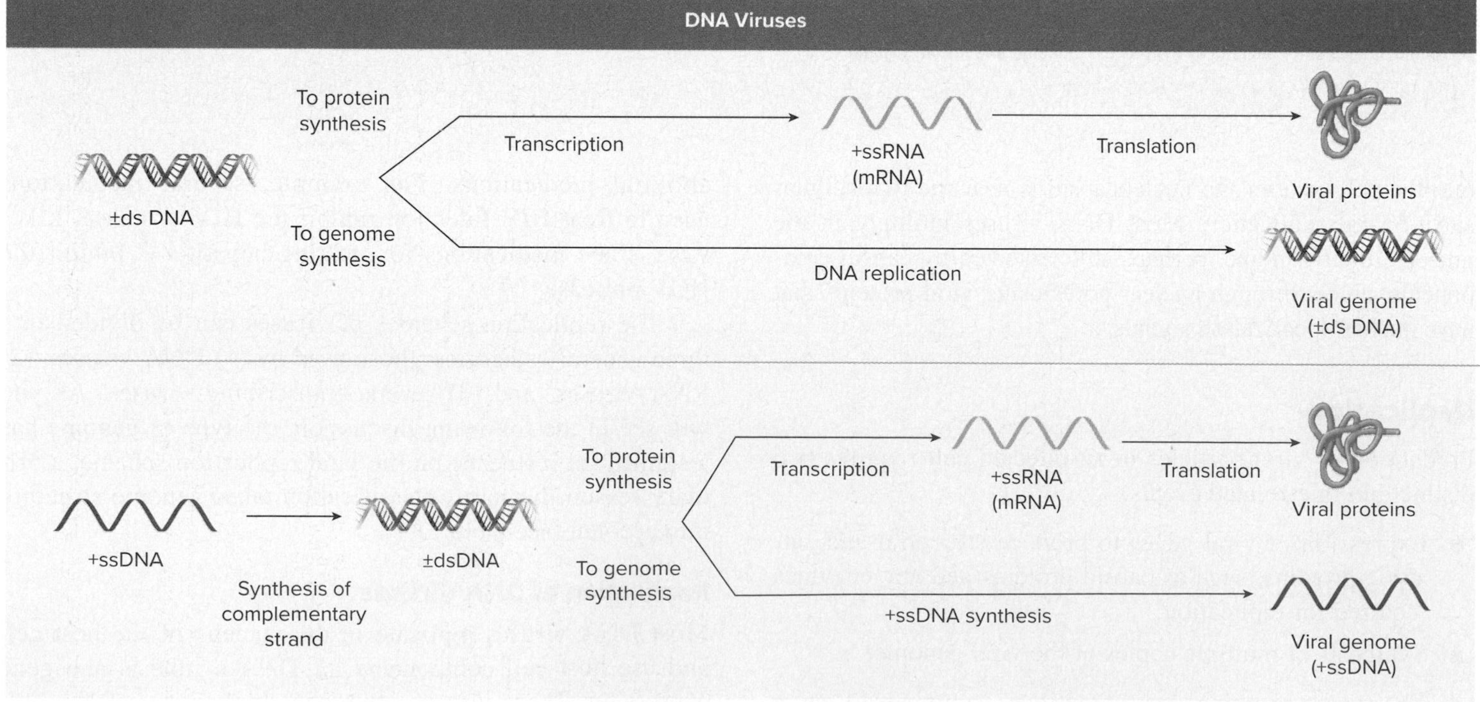

FIGURE 13.11 Replication of DNA Viruses

? **Why is it advantageous for DNA viruses to carry their own DNA polymerase?**

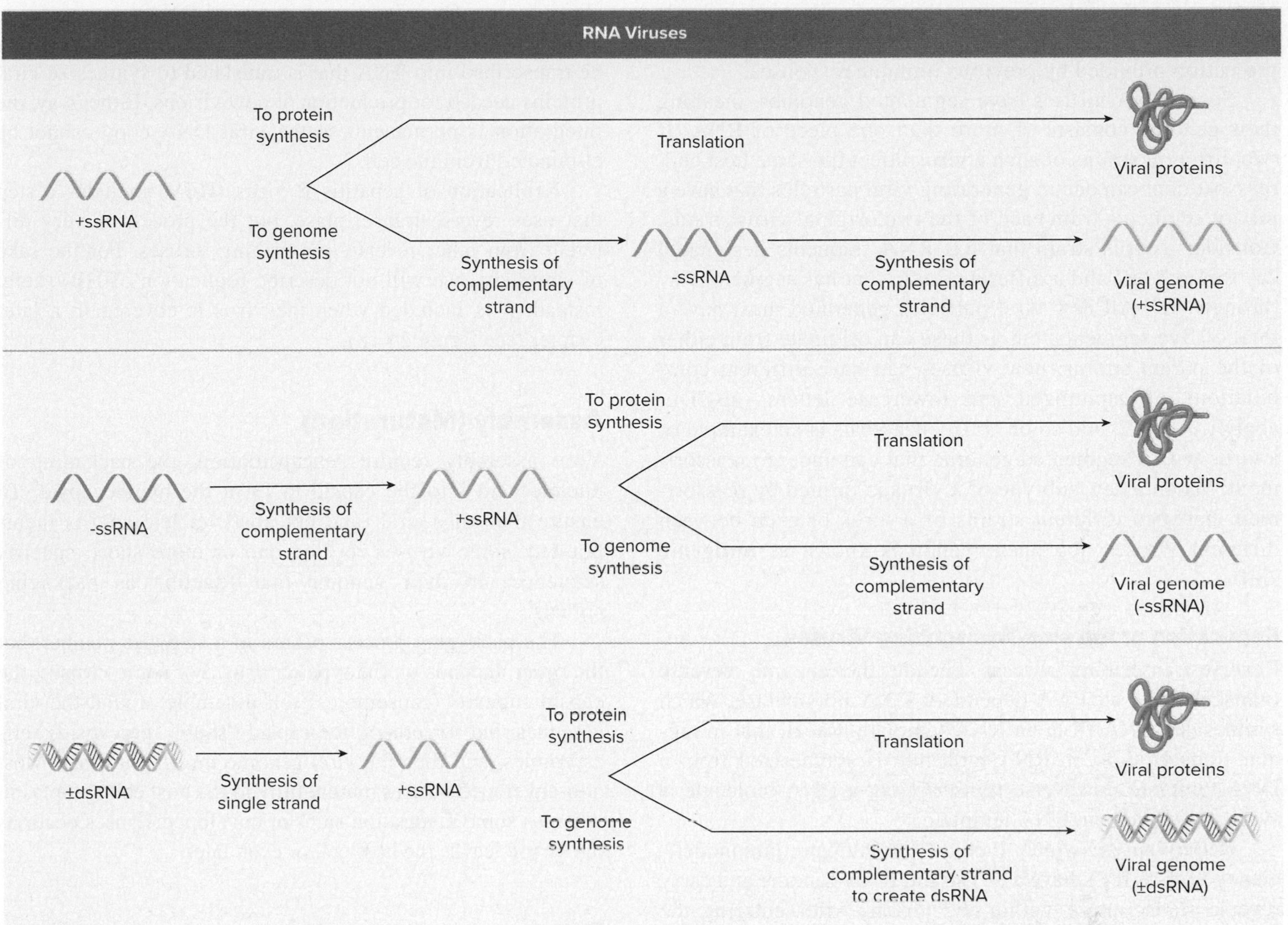

FIGURE 13.12 Replication of RNA Viruses A virally encoded replicase is required to synthesize RNA from an RNA template.

Why do RNA viruses need to encode a polymerase in order to replicate their genome, while DNA viruses do not?

- **(−) single strand RNA.** The replication strategy of a single-stranded (−) RNA virus is more complicated because the viral genome cannot be translated directly. Instead, it must first be used as a template to create a (+) RNA strand (see figure 13.12, middle panel). This requires a replicase, however, so that enzyme must be carried into the host cell as part of the viral particle. Once (+) RNA has been produced, it can either be translated to make viral proteins, including replicase, or be used as a template for synthesizing new (−) RNA strands. As the new viral particles are assembled, a molecule of the replicase is packaged along with the genome into the capsid.
- **Double-stranded RNA.** These viruses must also carry their own replicase because the host cell machinery cannot translate double-stranded RNA. The replicase immediately uses the (−) RNA strand of the viral genome as a template to make (+) strand RNA (see figure 13.12, bottom panel). This molecule is then translated to make more replicase, and the infection cycle can continue. Double-stranded RNA viruses are relatively uncommon.

Most replicases lack proofreading ability and therefore make more mistakes than DNA polymerases, generating mutations during replication. These mutations can lead to antigenic variation and allow some RNA viruses to adapt to selective pressures. Influenza viruses, for example, exhibit a type of antigenic variation called **antigenic drift.** This occurs as mutations accumulate in genes encoding key viral surface proteins recognized by the immune system. Because of such changes, people whose immune response protected them against influenza virus one year may not be fully protected against the influenza virus strain that circulates the next year. A similar situation occurs with SARS-CoV-2.

Mutations in the gene encoding the viral spike protein leads to emergence of variants able to at least partially escape the protection provided by previous immune responses.

Some RNA viruses have segmented genomes, meaning their genome consists of more than one piece of RNA. If two different strains of such a virus infect the same host cell, reassortment can occur, generating viral particles that have a mix of segments from each of the two original virus strains. Consider a virus strain that has RNA segments designated "a" through "e" and a different strain that has segments "A" through "E." All new viral particles generated must have a total of five segments, but as these can originate from either of the parent strains, new viruses can have different combinations of capitalized and lowercase letters—aBCDE, abcDE, AbcDE, and so on. Influenza virus is an example of a virus with a segmented genome that can undergo reassortment. When a new subtype of a virus is formed by reassortment between different strains of a virus or even between different viruses, the phenomenon is known as **antigenic shift.**

Replication of Reverse-Transcribing Viruses

Reverse-transcribing viruses encode the enzyme **reverse transcriptase,** an RNA-dependent DNA polymerase, which synthesizes DNA from an RNA template. Recall that in regular transcription, an RNA molecule is synthesized from a DNA template. In reverse transcription, a DNA molecule is synthesized from an RNA template.

Retroviruses, which include the human immunodeficiency virus (HIV), have a (+) strand RNA genome and carry reverse transcriptase within the virion. After entering the host cell, the reverse transcriptase uses the RNA genome as a template to make one strand of DNA (**figure 13.13**). The complement to that DNA strand is then synthesized to make double-stranded DNA, which integrates into the host cell chromosome. Once integrated, the viral DNA may remain in a latent state (similar to what occurs with λ phage), or it may be transcribed into RNA that is translated to synthesize viral proteins needed for production of new virions. Either way, the integration is permanent, so the viral DNA copy cannot be eliminated from the cell.

Replication of hepatitis B virus (HBV) includes a step that uses reverse transcriptase, but the process is very different from other reverse-transcribing viruses. For the sake of simplicity, we will not describe replication of HBV here; instead, it is included when the virus is covered in a later chapter (see figure 24.18).

Assembly (Maturation)

Viral assembly requires encapsidation, the packaging of nucleic acid into the capsid to form the nucleocapsid. To ensure that only viral (and not host) nucleic acid is incorporated, some viruses rely on one or more short, specific sequences in their genome that function as packaging signals.

The packaging process occurs in a stepwise manner, but the order depends on the type of virus. For some viruses, the capsid subunits (capsomeres) self-assemble around the viral genomes, but for others, the capsid "shell" (procapsid) self-assembles, and then the viral genome inserts into that. Most non-enveloped viruses mature fully in the host cell cytoplasm, whereas some maturation steps of enveloped viruses occur as the virion leaves the host cell or even later.

Release

Just as the entry mechanism into a host cell depends on whether the virion is enveloped or non-enveloped, so does its release. Most enveloped viruses are released by **budding,** a

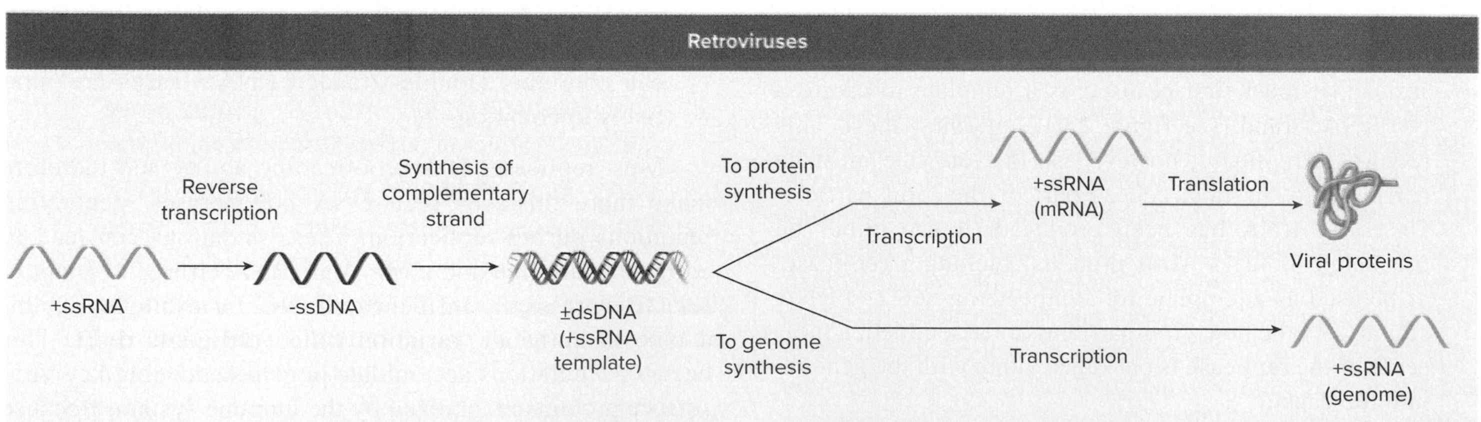

FIGURE 13.13 Replication of Reverse Transcribing Viruses The virally encoded enzyme reverse transcriptase is required to synthesize DNA from an RNA template.

? What is the role of reverse transcriptase in retroviruses such as HIV?

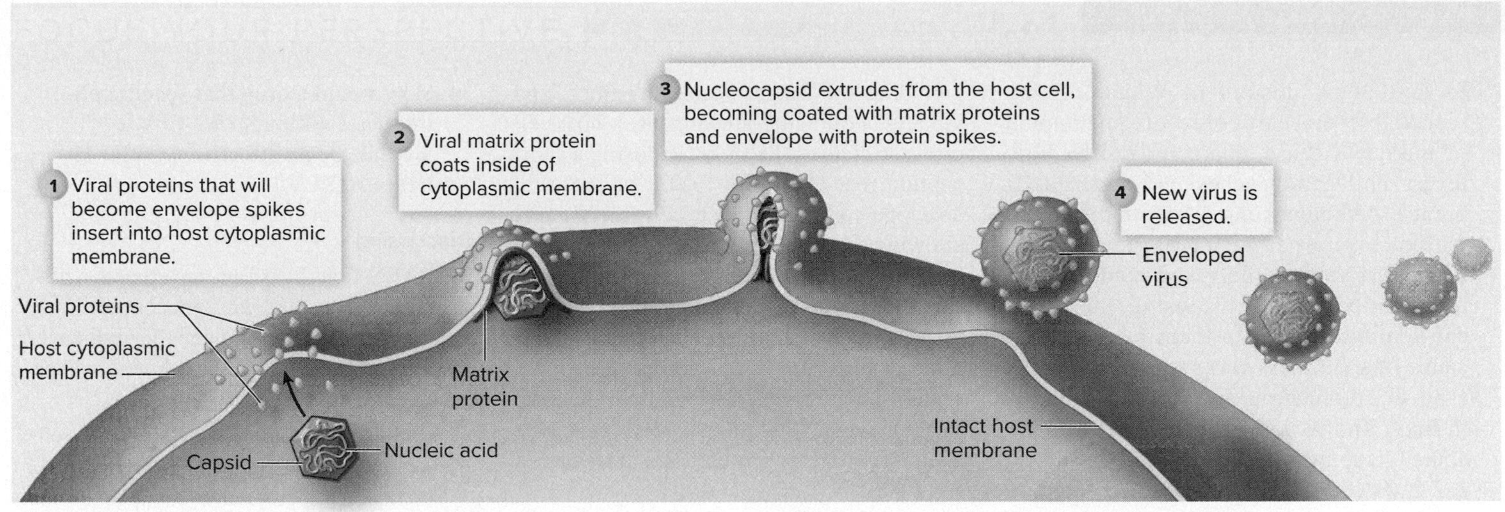
1 Viral proteins that will become envelope spikes insert into host cytoplasmic membrane.
2 Viral matrix protein coats inside of cytoplasmic membrane.
3 Nucleocapsid extrudes from the host cell, becoming coated with matrix proteins and envelope with protein spikes.
4 New virus is released.
Enveloped virus
Viral proteins
Host cytoplasmic membrane
Matrix protein
Capsid
Nucleic acid
Intact host membrane

(a)

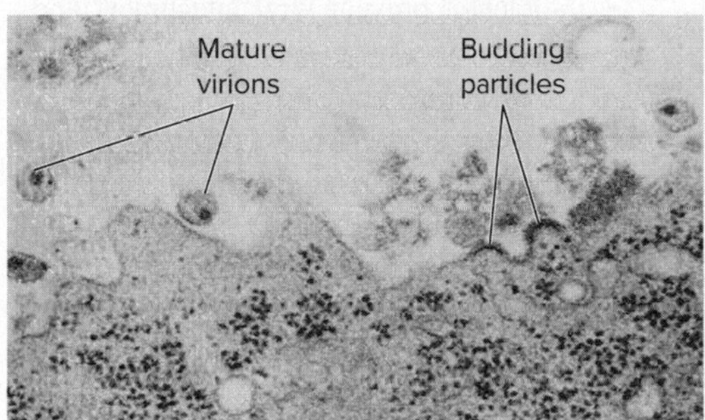
Mature virions
Budding particles

(b)

FIGURE 13.14 Mechanisms for Releasing Enveloped Virions (a) Process of budding. **(b)** Virions budding from the surface of a human cell (TEM). b: CDC

What component of the virion is gained in the process of budding?

process whereby the virus acquires its envelope (**figure 13.14**). Before budding occurs, virally encoded protein spikes insert into specific regions of the host cell's membrane. Matrix protein accumulates on the inside surface of those same regions. Assembled or partially assembled nucleocapsids are then extruded from the cell at these regions, becoming covered with a layer of matrix protein and lipid envelope in the process. Not all enveloped viruses have cytoplasmic membrane-derived envelopes, however. Some obtain their envelope from the membrane of an organelle such as the Golgi apparatus or the rough endoplasmic reticulum. They do this by budding into the organelle. From there, they are transported in vesicles to the outside of the cell. Budding may not destroy the cell because the membranes can be repaired after the viral particles exit.

Non-enveloped viruses are released when the host cell dies. How this occurs, however, is sometimes quite different from how bacteriophages kill their hosts. Many viruses trigger a normal programmed cell death called apoptosis. The virions released from the dead cells may then invade any healthy cells in the area.

To be maintained in nature, infectious virions must leave one animal host and be transmitted to another. Viral particles may be released in feces, urine, genital secretions, blood, or mucus and saliva. From there, the virus enters the next host to begin another round of infection.

MicroAssessment 13.4

Animal viruses and bacteriophages share similar features in their infection cycle. Because many animal viruses are enveloped and phages are not, differences exist in the way they enter and exit a host cell. The genomes of animal viruses are diverse, influencing viral replication strategies.

10. Why are virally encoded enzymes medically important?

11. How do enveloped viruses exit a cell?

12. Why can viruses not replicate independently of living cells?

13.5 ■ Categories of Animal Virus Infections

Learning Outcome

8. Compare and contrast acute infections and the two types of persistent infections caused by animal viruses.

Animal virus infections can be divided into two major categories: acute and persistent (**figure 13.15**). **Acute infections** are characterized by sudden onset and a relatively short duration. In contrast, **persistent infections** can continue for years, or

FOCUS ON A CASE 13.1

In late 2019, doctors in Wuhan, China reported several cases of severe pneumonia of unknown cause, characterized by high fevers, chills, body aches, and shortness of breath. Although some patients recovered within several days, an alarming number deteriorated rapidly and died, despite interventions such as using mechanical ventilators to help them breathe. The cause of the illness was soon identified as a novel coronavirus that likely originated in bats. The virus was later named severe acute respiratory syndrome coronavirus 2 (SARS-CoV-2) because its genome sequence showed its close evolutionary relationship to SARS-CoV, the coronavirus that caused the SARS outbreak in 2003. The disease was named coronavirus disease 2019, or COVID-19. Within weeks, COVID-19 spread globally, and scientists worldwide raced to learn more about the causative agent.

SARS-CoV-2 is an enveloped, single-stranded positive-sense RNA virus; characteristic spikes on the surface create a crown-like appearance (*corona* means "crown"). During infection, these spikes attach to a host membrane protein called angiotensin-converting enzyme 2 (ACE2) that is found on cells in various organs, including the lungs, heart, intestines, and kidneys. Following attachment, a host cell protease cleaves the spike, exposing a short peptide that facilitates fusion of the viral envelope with the host cell membrane, allowing the virus to enter the cell. Once in the host cell, the virus begins replicating. The SARS-CoV-2 proteins are synthesized as polyproteins that are then cut by viral proteases to produce various functional proteins. Using information about the viral structure and infection cycle, scientists began developing prevention and treatment strategies.

1. To prevent transmission of COVID-19, people are advised to wash their hands frequently and thoroughly with soap and water. Why is it not surprising that soap is an effective means to destroy SARS-CoV-2?
2. Why might a drug that specifically binds to the SARS-CoV-2 spike be an effective COVID-19 treatment?
3. Why would a drug that specifically interferes with the SARS-CoV-2 RNA-dependent RNA polymerase be an effective COVID-19 treatment?
4. Why would a drug that specifically interferes with a SARS-CoV-2 protease be an effective COVID-19 treatment?

Discussion

1. SARS-CoV-2 is an enveloped virus, meaning it has an outer lipid layer. Soaps can easily damage the viral envelope along with the attachment spikes, making the virus non-infectious.
2. Entry of SARS-CoV-2 into host cells requires attachment of the spike to ACE2 on host cell membranes. Compounds that bind to the spike in a way that blocks this interaction prevent viral attachment and subsequent entry into host cells.
3. SARS-CoV-2 is an RNA virus and, like all RNA viruses, it encodes an RNA-dependent RNA polymerase to make copies of its genome. Interfering with the activity of that enzyme therefore prevents replication of the genome.
4. The viral proteases are required to cleave the viral polyproteins into their functional components; thus, a drug that inhibits a SARS-CoV-2 protease prevents the production of more viral particles.

even for the life span of the host. Persistent infections can be chronic or latent.

Although we often categorize viral infections as either acute or persistent, they do not always fall neatly into a single category. For instance, when a person is first infected with HIV, the virus replicates to high levels, often causing an acute illness characterized by fever, fatigue, swollen lymph nodes, and headache. The immune system soon eliminates most virions, and the obvious symptoms subside. However, the DNA copy of the viral genome integrates into the host cell chromosome, resulting in a persistent infection that can lead to the development of acquired immunodeficiency syndrome (AIDS) after many years. Thus, HIV infection has features of both acute and persistent infections.

Acute Infections

Acute infections result in a burst of virions being released from infected host cells. Although the virus-infected cells often die, the host may survive. This is because the immune system of the animal host may gradually eliminate the virus over a period of days to months. Examples of diseases due to acute viral infections include influenza, mumps, and polio.

Persistent Infections

Persistent infections remain for years, or even the lifetime of the host—sometimes without any symptoms. In laboratory studies, two general types of persistent infections can be identified: chronic and latent. These categories may overlap in an actual infection, however, with different cells experiencing different types of infection. Chronic and latent infections are described next, and examples are listed in **table 13.4.**

Chronic infections are characterized by the continuous low-level production of viral particles. In some cases, the infected cell survives and slowly releases viral particles. In other cases, the infected cell lyses, but only a small proportion of cells is infected at any given time. From a practical standpoint, the important aspect of a chronic infection

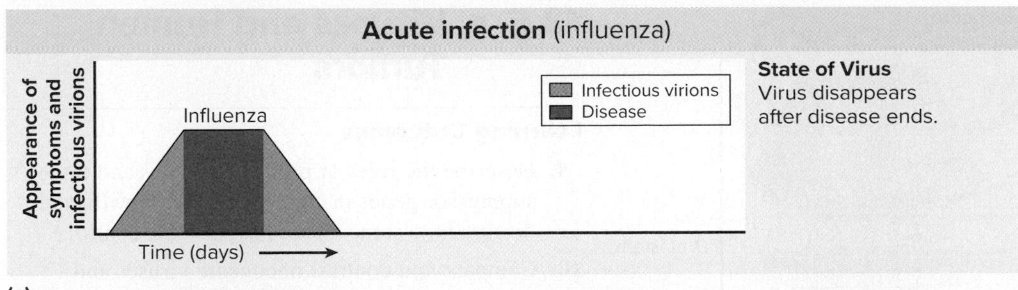

(a)

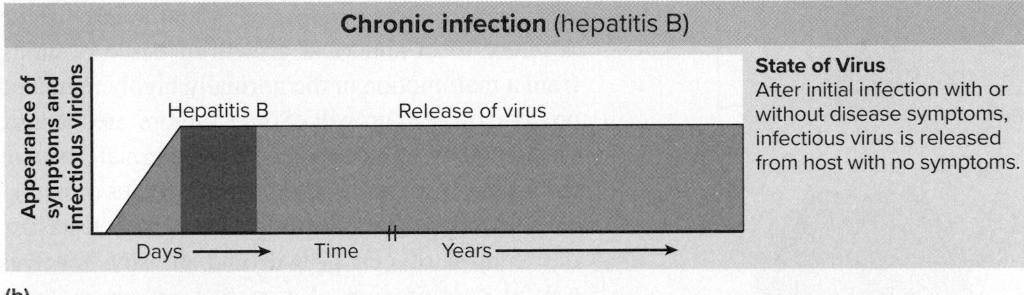

(b)

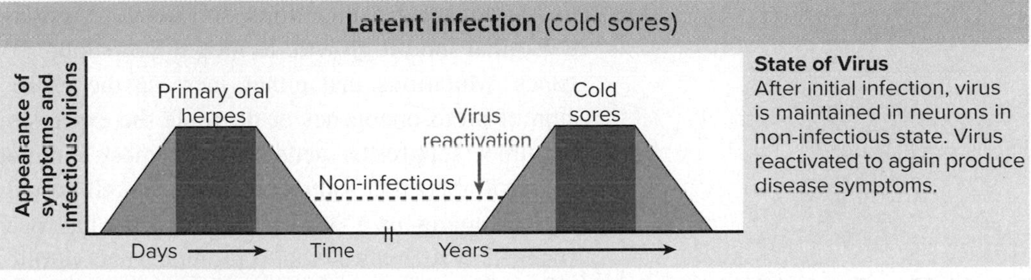

(c)

FIGURE 13.15 Types of Infection by Animal Viruses (a) Acute infection. **(b)** Persistent infection—chronic. **(c)** Persistent infection—latent.

? **Into which category would you classify the common cold, and why?**

is the continuous production of infectious viral particles, often in the absence of disease symptoms. Consequently, a person who is asymptomatic (meaning without symptoms) can still transmit the virus to others. For example, some people infected with hepatitis B virus develop a chronic infection; they become carriers of the virus, able to pass it to other people through blood and body fluids.

In **latent infections,** the viral genome remains silent within a host cell, yet it can reactivate to cause a productive infection. In some cases, the viral genome is integrated into the host cell chromosome and is called a **provirus.** Other viruses do not integrate their genome; rather, they replicate independently of the host genome, much like a plasmid. The fact that a latent virus cannot be eliminated from the body means that the disease can recur even after an extended period without symptoms. Consider the disease shingles. A person can develop shingles only if that individual previously had chickenpox—a disease caused by varicella zoster virus (VZV). Following the acute infection that caused chickenpox, the VZV genome goes dormant in nerve ganglia (groups of nerve cell bodies). Under certain circumstances, the virus can reactivate—giving rise to new VZV particles that move down peripheral nerves and spread locally, causing the painful skin lesions characteristic of shingles. Latent infection also explains the recurrence of cold sores, which are usually caused by herpes simplex virus type 1 (HSV-1). HSV-1

TABLE 13.4 Examples of Persistent Infections

Virus	Type	Cells Involved	Disease
Hepatitis B virus	Chronic	Hepatocytes (liver cells)	Hepatitis, cirrhosis, liver cancer
Hepatitis C virus	Chronic	Hepatocytes (liver cells)	Hepatitis, cirrhosis, liver cancer
Herpesviruses			
Cytomegalovirus (CMV)	Latent	Salivary glands, kidney epithelium, leukocytes	CMV pneumonia, eye infections, congenital CMV infection
Epstein-Barr virus (EBV)	Latent	B cells, which are involved in antibody production	Mononucleosis, Burkitt's lymphoma
Herpes simplex virus type 1 (HSV-1)	Latent	Neurons of sensory ganglia	Primary oral herpes (viral reactivation results in cold sores)
Herpes simplex virus type 2 (HSV-2)	Latent	Neurons of sensory ganglia	Genital herpes (viral reactivation results in recurrent outbreaks)
Varicella zoster virus (VZV)	Latent	Satellite cells of sensory ganglia	Chickenpox (viral reactivation results in shingles)
Human immunodeficiency virus (HIV)	Mixed	Chronic: Activated helper T cells, macrophages Latent: Memory helper T cells	AIDS

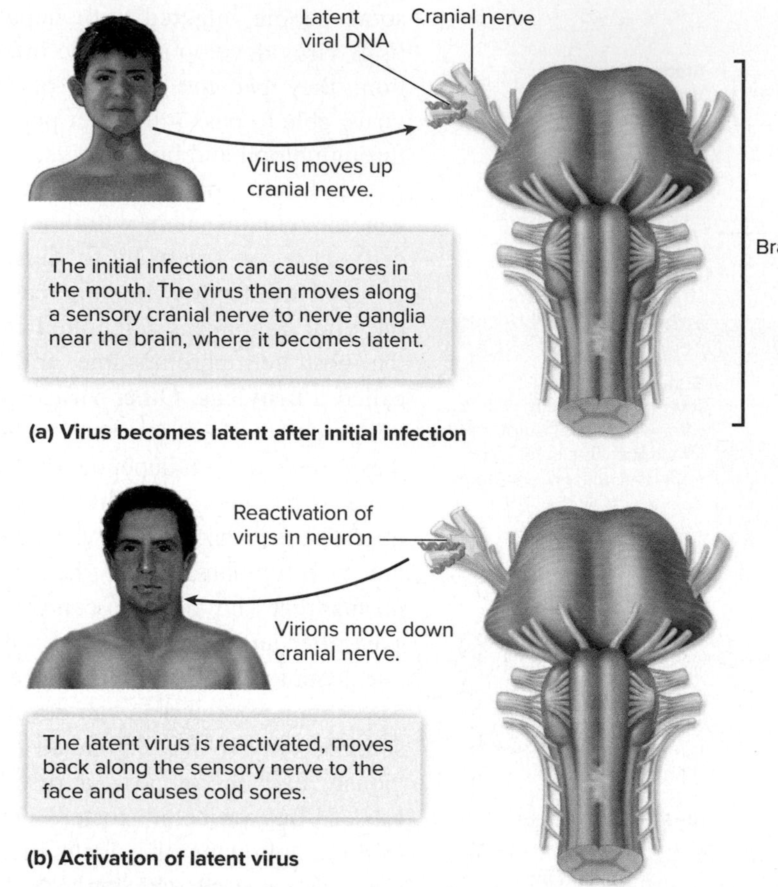
Latent viral DNA
Cranial nerve
Virus moves up cranial nerve.
The initial infection can cause sores in the mouth. The virus then moves along a sensory cranial nerve to nerve ganglia near the brain, where it becomes latent.
Brainstem
(a) Virus becomes latent after initial infection
Reactivation of virus in neuron
Virions move down cranial nerve.
The latent virus is reactivated, moves back along the sensory nerve to the face and causes cold sores.
(b) Activation of latent virus

FIGURE 13.16 **Infection Cycle of Herpes Simplex Virus Type 1, HSV-1**

? What is a provirus?

causes an initial oral infection that is often asymptomatic but sometimes leads to painful sores in the mouth. From there, the virus spreads to sensory nerve cells, where it remains latent. After that, the latent virus can repeatedly reactivate to cause recurrent episodes (outbreaks) of cold sores (**figure 13.16**). What reactivates the virus is not clear, but certain physiological or immunological changes in the host are often involved.

MicroAssessment 13.5

Many viruses cause acute infections in which viruses multiply and spread rapidly in the host. Some viruses cause persistent infections; these long-term infections can be chronic or latent.

13. Distinguish between acute and persistent viral infections at the cellular level.
14. How are latent viral infections different from chronic infections?
15. Could the same type of virus cause both an acute and a persistent infection? Explain.

13.6 ▪ Viruses and Human Tumors

Learning Outcomes

9. Describe the roles of proto-oncogenes and tumor suppressor genes in controlling cell growth, and discuss how some viruses bypass this control.
10. Compare and contrast oncogenic viruses and oncolytic viruses.

A tumor is an abnormal growth of tissue resulting from a malfunction in the normally highly regulated process of cell growth. Some tumors are benign, meaning they do not invade nearby normal tissue or metastasize (spread). Others are cancerous or malignant, meaning they have the potential to metastasize.

Control of cell growth and division involves careful coordination of two sets of genes: those that stimulate growth, called proto-oncogenes, and those that inhibit growth, termed tumor suppressor genes. Mutations that either increase the expression of proto-oncogenes or decrease the expression of tumor suppressor genes are the most common cause of abnormal and/or uncontrolled cell growth. An **oncogene** is a proto-oncogene that has been changed in such a way that it promotes uncontrolled growth, leading to tumor formation. Numerous events, including spontaneous and induced mutations, can lead to conversion of a host's normal proto-oncogene into an oncogene. A single change in the DNA sequence of regulatory genes is probably not enough to cause a tumor; rather, multiple changes at different sites are required.

Most tumors are caused by mutations in host genes that regulate cell growth, but some are caused by viruses. In contrast to the cancer-causing viruses, other viruses offer potential for treating cancers.

Cancer-Causing Viruses

Viruses that can cause cancer in humans are called **oncoviruses** or oncogenic viruses. The cancer may arise as a direct result of viral infection, such as when a viral gene functions as an oncogene, or when the viral genome inserts into the host cell chromosome in such a way that a proto-oncogene is converted to an oncogene. Alternatively, the cancer may arise indirectly as a result of DNA damage that occurs during long-term inflammation associated with chronic virus infections.

Most virus-induced tumors are associated with certain DNA viruses, although some RNA viruses are also oncogenic. Seven human oncogenic viruses are currently recognized (**table 13.5**). Cancers caused by these viruses do not develop immediately, but instead become apparent between 15 and

TABLE 13.5 Viruses Associated with Cancers in Humans

Virus	Nucleic Acid Type	Kind of Tumor
Epstein-Barr virus	DNA	Burkitt's lymphoma; nasopharyngeal carcinoma; B-cell lymphoma
Hepatitis B virus	DNA	Hepatocellular carcinoma (liver cancer)
Human herpesvirus type 8	DNA	Kaposi's sarcoma
Human papillomaviruses	DNA	Different kinds of tumors, caused by different HPV types
Merkel cell polyomavirus	DNA	Merkel cell carcinoma (skin cancer)
Hepatitis C virus	RNA	Hepatocellular carcinoma (liver cancer)
Human T-lymphotropic virus type 1	RNA (retrovirus)	Adult T-cell leukemia

40 years after infection. Only a small percentage of oncovirus-infected people develop the associated cancer, however, indicating that other important factors must be involved. Some virus-associated cancers are preventable—at least 13 serotypes of human papillomaviruses (HPVs) increase the risk of cervical and some other cancers, but a vaccine is available that protects against nine of those serotypes.

The various effects that animal viruses can have on their host cells are shown in **figure 13.17.**

Cancer-Fighting Viruses

In contrast to oncogenic viruses, **oncolytic viruses** specifically target and kill cancer cells, a characteristic that makes them potentially useful in cancer treatment. Some oncolytic viruses destroy the cancer cells directly by multiplying within them and inducing lysis; others act indirectly by stimulating the host's cancer-fighting immune cells. Although naturally occurring oncolytic viruses were described decades ago, scientists are now genetically engineering other well-characterized viruses to develop oncolytic versions for therapy.

The first oncolytic viral therapy approved by the U.S. Food and Drug Administration (FDA) is a genetically engineered herpes simplex virus type 1 used to treat advanced inoperable melanoma (an aggressive form of skin cancer). The virus has been genetically engineered to (1) preferentially infect and replicate in the tumor cells and (2) encode production of GM-CSF (granulocyte-macrophage colony-stimulating factor), a protein that stimulates an anti-tumor immune response. Thus, when the genetically modified virus is injected directly into melanoma lesions, it destroys tumor cells and also triggers a robust immune response.

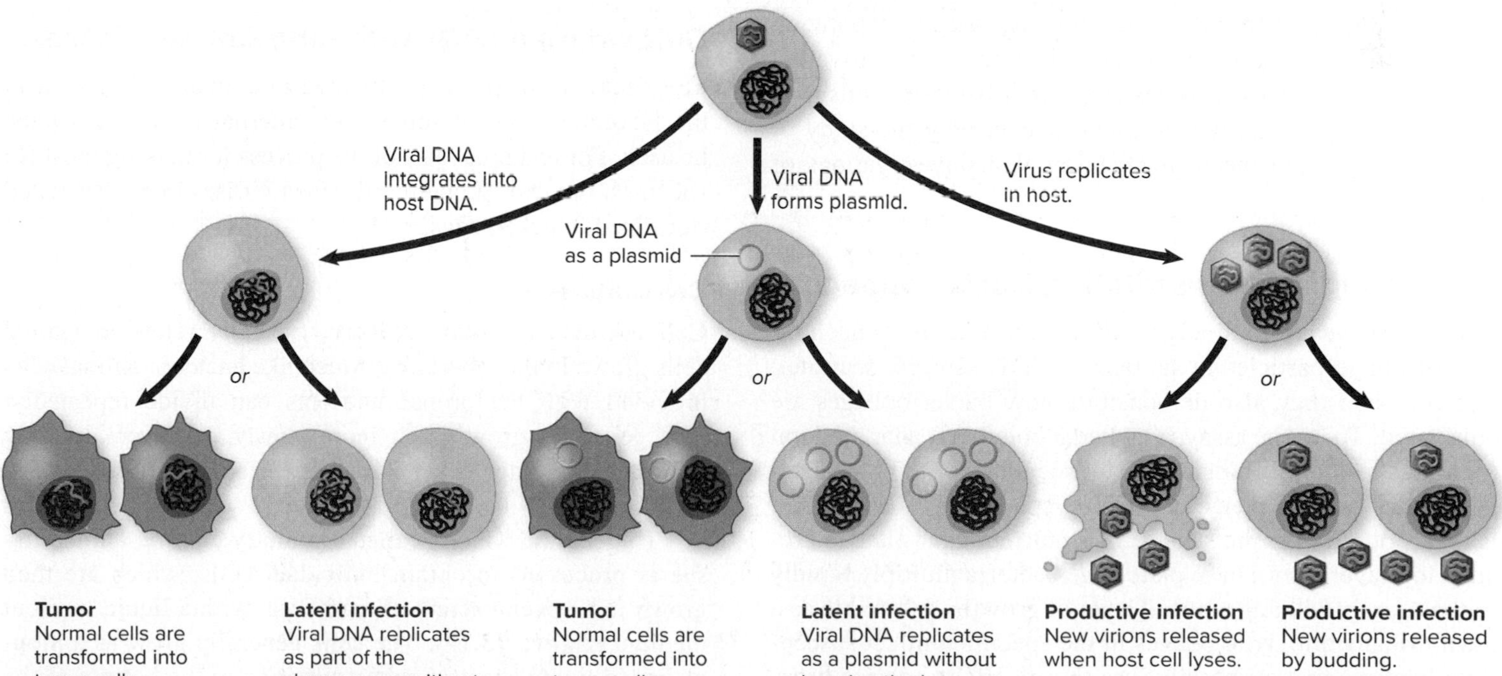
Viral DNA integrates into host DNA.
Viral DNA forms plasmid.
Virus replicates in host.
Viral DNA as a plasmid
or
or
or
Tumor
Normal cells are transformed into tumor cells.
Latent infection
Viral DNA replicates as part of the chromosome without harming the host.
Tumor
Normal cells are transformed into tumor cells.
Latent infection
Viral DNA replicates as a plasmid without harming the host.
Productive infection
New virions released when host cell lyses.
Productive infection
New virions released by budding.

FIGURE 13.17 Various Effects of Animal Viruses on the Cells They Infect

? **Which of the illustrated relationships is the most destructive to the infected host cell? Explain.**

Other engineered oncolytic viruses show promise in clinical trials—a modified poliovirus that infects cancerous neuronal cells is being investigated as a possible treatment for glioblastoma, a type of brain cancer. The virus has been given "breakthrough therapy" designation by the FDA, which means research and clinical trials using the virus can be advanced more quickly.

MicroAssessment 13.6

Viruses that lead to the development of cancer are called oncogenic viruses or oncoviruses. Most tumors are not caused by viruses, however; they are caused by mutations in certain host genes that regulate cell growth. Oncolytic viruses target and destroy cancer cells.

16. What is the function of a proto-oncogene?
17. Explain why a vaccine can prevent cervical cancer.
18. Why would it be advantageous to a virus to interfere with the function of proto-oncogenes or tumor suppressor genes?

13.7 ■ Methods Used to Study Viruses

Learning Outcomes

11. Describe how a plaque assay is used to detect and estimate the numbers of phage particles.
12. Describe the methods used to cultivate and quantitate animal viruses.

Viruses can replicate only inside live, actively metabolizing cells, so studying them in the laboratory requires cultivating appropriate host cells. It is easier to grow bacterial cells than to grow animal cells, which is one reason why the study of bacteriophages advanced much faster than investigations of animal viruses.

Cultivating and Quantitating Bacteriophages

Plaque assays are routinely used to determine the concentration of phage particles in samples such as sewage, seawater, and soil, and they also demonstrate how bacteriophages are cultivated. To do the assay, a melted, cooled soft agar medium (contains less solidifying agent) is inoculated with both a bacterial host and the phage-containing specimen. That mixture is then poured over the surface of a nutrient agar plate to create a top layer over a base plate. The bacteria multiply rapidly in the agar, producing a turbid layer of growth, referred to as a lawn. Meanwhile, lytic phages in the specimen infect susceptible bacteria and eventually lyse those cells to release progeny phage, which then diffuse through the soft agar to infect and lyse neighboring bacterial cells. As a result, circular zones of clearing called **plaques** develop in the lawn; these result from lysis of the bacterial host cells, with each plaque

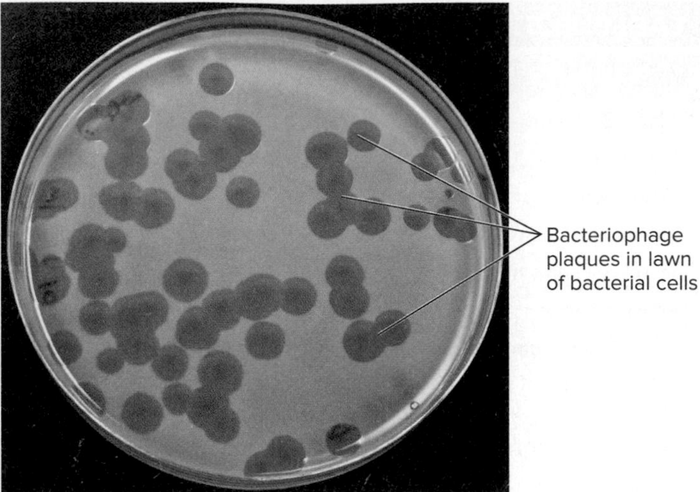

FIGURE 13.18 Phage Plaques Lisa Burgess/McGraw Hill

Plaques are limited in size; that is, they do not keep getting larger. Why would this be so?

representing a plaque-forming unit (PFU) initiated by a single phage particle infecting a cell (**figure 13.18**). In the case of temperate phages, the plaques are cloudy in appearance rather than clear because only some of the bacterial hosts are lysed.

The assay is done using several dilutions of the phage suspension so that one will yield a statistically valid number of plaques to be accurately counted; the number of plaques is used to determine the titer, which is the concentration of infectious viral particles in the original suspension.

Cultivating and Quantitating Animal Viruses

Most animal viruses are cultivated in animal cells grown in the laboratory, but in some cases alternative methods must be used. For example, part of the process for making most flu vaccines, involves growing influenza viruses in embryonated (fertilized) chicken eggs.

Cell Culture

Cell culture, or **tissue culture,** typically refers to animal cells grown in the laboratory. Much like bacteria, animal cells provided with the proper nutrients can divide repeatedly; they generally grow much more slowly, however, dividing no faster than once every 24 hours (*E. coli* can divide every 20 minutes).

To generate what is called a primary culture, animal tissue is processed to obtain individual cells, which are then grown in a special screw-capped flask with a liquid nutrient medium (**figure 13.19**). The cells generally grow as a monolayer (a single sheet of cells), adhering to the bottom of the flask; an exception is white blood cells, which do not adhere but grow as single cells in suspension. Normal cells can divide only a limited number of times, however, so new primary cultures must be regularly generated. As an alternative

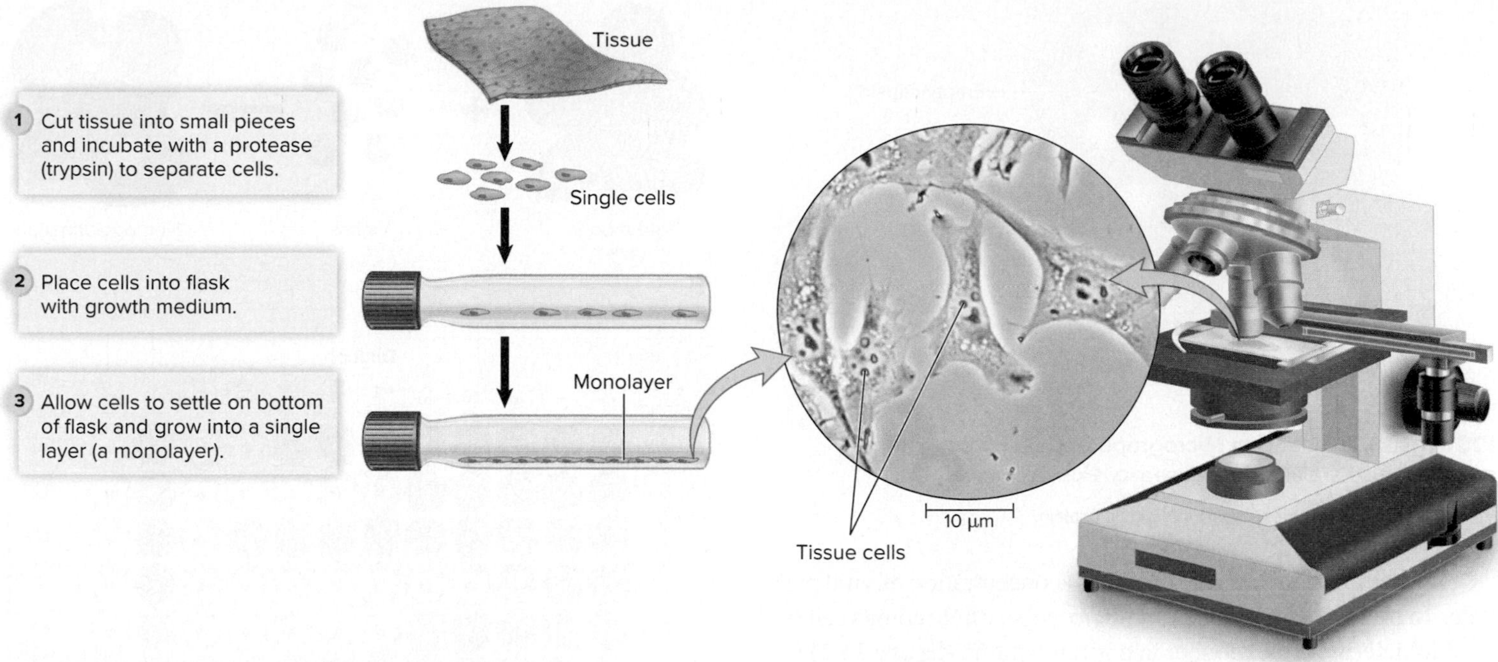
Tissue
1 Cut tissue into small pieces and incubate with a protease (trypsin) to separate cells.
Single cells
2 Place cells into flask with growth medium.
Monolayer
3 Allow cells to settle on bottom of flask and grow into a single layer (a monolayer).
10 µm
Tissue cells

FIGURE 13.19 Preparation of a Primary Cell Culture Dr. Cecil Fox/National Cancer Institute (NCI)

? **Why must primary cell cultures be remade every so often?**

to normal cells, tumor cells—which can multiply indefinitely in vitro—are often used to create what is referred to as an established cell line.

Many viruses can be detected by their effect on cells in culture. A virus propagated in cell culture often causes distinct morphological changes in infected cells, called **cytopathic effects** (**figure 13.20**). The host cells, for example, may change shape, detach from the surface, or lyse. Several viruses, such as HIV and measles, can cause infected cells to fuse into a giant multinuclear cell (called a syncytium), a mechanism of viral spread. Some viruses, including rabies, cause an infected cell to form a distinct region called an **inclusion body,** the site of viral replication; the position of an inclusion body in a cell depends on the type of virus. Because cytopathic effects are often characteristic of a particular virus, they are useful to scientists studying and identifying viruses.

MicroByte

The commonly used HeLa cell line was derived from cervical cancer cells taken from a patient named **He**nrietta **La**cks, who died from the cancer at age 31.

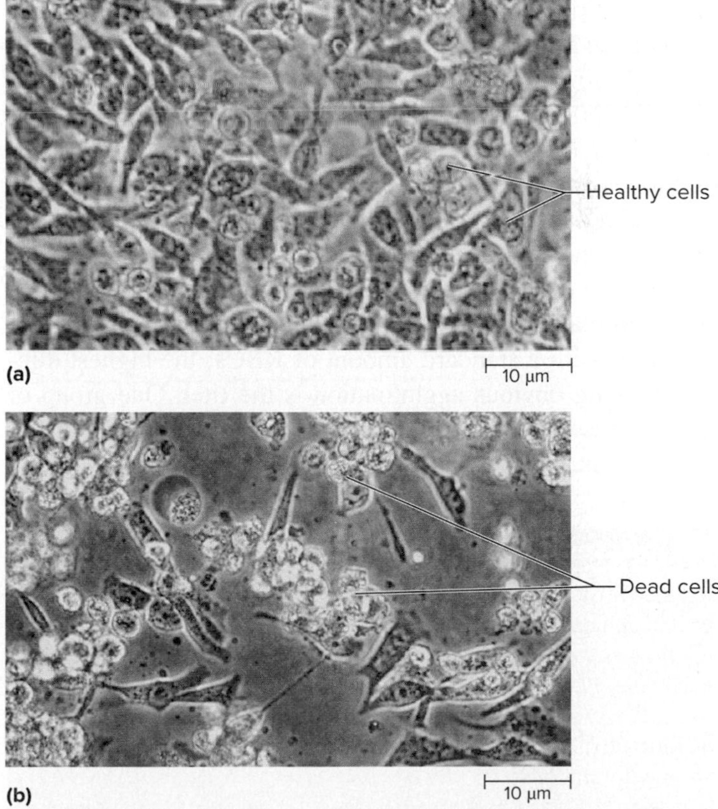
Healthy cells
(a)
10 µm
Dead cells
(b)
10 µm

FIGURE 13.20 Cytopathic Effects in Virally Infected Cells in Culture **(a)** Healthy HeLa cells. **(b)** HeLa cells infected with a bunyavirus. Modified from "The Genome Sequence of Lone Star Virus, a Highly Divergent Bunyavirus Found in the *Amblyomma americanum Tick*", PLOS ONE, April 2013, vol.8 , Issue 4, e62083, with permission from University of California, San Francisco and the Centers for Disease Control and Prevention.

Quantitating Animal Viruses

A plaque assay, similar in principle to that described to quantitate bacteriophages, is one of the most precise methods for determining the animal virus concentration in a sample. This is done using tissue culture, generating countable clear zones (plaques) surrounded by uninfected cells.

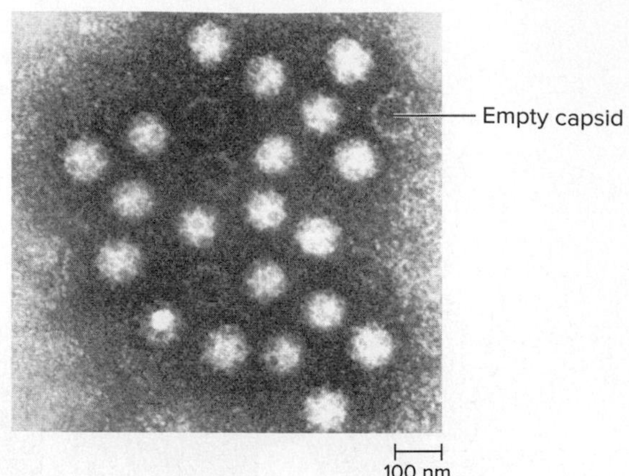

FIGURE 13.21 Electron Micrograph of Calicivirus Note that empty capsids are also easy to see. Dr. Erskine Palmer/CDC

Is a virion with an empty capsid infectious? Explain.

If a sample contains a high enough concentration of viral particles to be seen with an electron microscope, direct counts can be used to determine the number in a given volume (**figure 13.21**).

A viral titer can be estimated using a **quantal assay.** In this method, several dilutions of the virus preparation are administered to a number of animals, cells, or chick embryos, depending on the host specificity of the virus. The titer of the virus is the dilution at which 50% of the inoculated hosts are infected or killed. This is reported as either the $\mathbf{ID_{50}}$ (infective dose) or the $\mathbf{LD_{50}}$ (lethal dose).

High concentrations of certain viruses cause red blood cells (RBCs) to clump, a phenomenon called hemagglutination, so for those viruses a **hemagglutination assay** can be used to determine the relative concentration of viral particles. Hemagglutination occurs when individual viral particles attach to surface molecules of multiple RBCs simultaneously, connecting the cells to form an aggregate (**figure 13.22**). To do a hemagglutination assay, serial dilutions of the viral suspension are mixed with a standard amount of RBCs; the highest dilution showing obvious agglutination is the titer. One group of animal viruses that can agglutinate red blood cells is the orthomyxoviruses, of which influenza virus is a member.

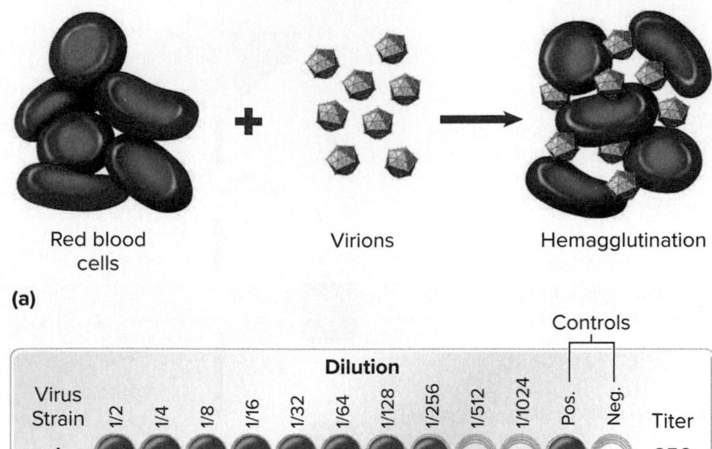

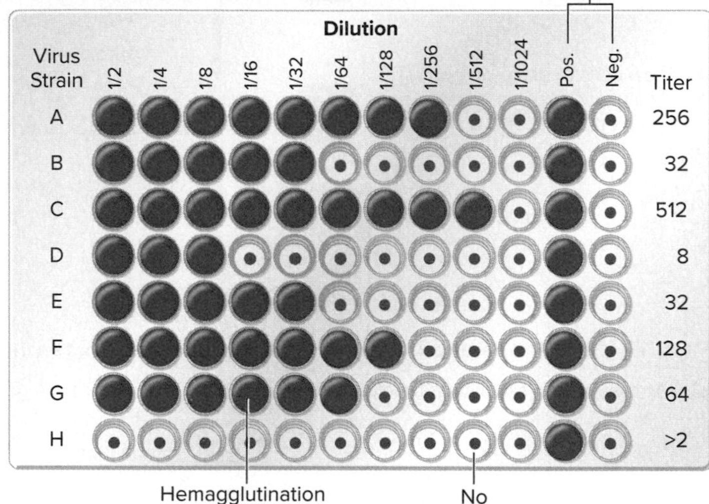

FIGURE 13.22 Hemagglutination Assay (a) Diagram showing virions combining with red blood cells, resulting in hemagglutination. **(b)** Viral titer determination using the hemagglutination assay. Red blood cells normally form a pellet when they sink to the bottom of the wells, whereas agglutinated ones do not.

What is the titer of virus strain F?

MicroAssessment 13.7

Various hosts are required to cultivate different viruses; bacteriophages require bacterial host cells while animal viruses are typically cultivated in cell culture using the appropriate host cells. Plaque assays are used to quantitate bacteriophages and animal viruses that lyse their host cells. Other methods to quantitate animal viruses include direct counts, quantal assays, and hemagglutination.

19. What are plaque-forming units?

20. Why is it advantageous to use tumor cells rather than normal cells in tissue culture?

21. Is it important to have fewer phages than bacterial host cells when doing a quantitative plaque assay? Explain.

13.8 ■ Plant Viruses

Learning Outcomes

13. Compare and contrast the mechanisms by which plant and animal viruses enter host cells.

14. Discuss the ways that plant viruses can be transmitted to their hosts.

Viral diseases of plants are economically important, particularly when they affect crop plants such as corn, wheat, rice, and soybeans. Over half of the crop yields can be lost when a serious plant virus spreads.

Viral infection of plants can be recognized through various visible signs, including yellowing of foliage with irregular lines appearing on leaves and fruit (**figure 13.23**). Individual cells or specialized organs of the plant may die, and tumors may appear. Usually, infected plants are stunted in their growth, although in a few cases growth is stimulated, leading to deformed structures. Plants generally do not recover from viral infections because, unlike animals, plants

(a)

(b)

FIGURE 13.23 Signs of Viral Diseases of Plants (a) Typical mosaic pattern and ring lesions on a tobacco leaf resulting from infection by tobacco mosaic virus. **(b)** Wheat plant showing yellowing caused by wheat streak mosaic virus. a: Nigel Cattlin/Alamy Stock Photo b: Nigel Cattlin/Science Source

? How do most plant viruses enter plants?

are not capable of developing specific immunity to rid themselves of invading viruses. In severely infected plants, virions may accumulate in enormous numbers. As much as 10% of the dry weight of a tobacco mosaic virus-infected plant may consist of virus. The virus, which causes a serious disease of tobacco, is unusually difficult to eliminate from a contaminated area because the virions remain infective for at least 50 years. Other plant viruses are also extraordinarily stable in the environment.

Certain plants have been purposely maintained in a virus-infected state. The most well-known example involves tulips, in which a virus transmitted through the bulbs can cause a desirable color variegation of the flowers (**figure 13.24**). The infecting virus was transmitted through bulbs for a long time before the cause of the variegation was even suspected.

FIGURE 13.24 Tulip with Symptoms Resulting from a Viral Infection Anna Yu/Photodisc/Getty Images

? Why are tulips maintained in a virus-infected state?

In contrast to phages and animal viruses, when plant viruses infect a cell they do not attach to specific receptors. Instead, they enter through wound sites in the cell wall, which is otherwise very tough and rigid. From there, they can then spread from cell to cell through channels that interconnect cells (the plasmodesmata) and then to the rest of the plant via its vascular tissue (the phloem).

Plant viruses can be transmitted multiple ways. Insects are important transmitters, so insect control is a critical tool for preventing viral spread. Humans and other animals who contact diseased plants can also spread the viruses; for example, growers who have been handling infected plants can inadvertently transmit viruses to healthy seedlings; even healthy leaves that rub up against virally infected ones can transmit viruses. Some plant viruses can also be transmitted through contaminated seeds, tubers, soil, or even airborne pollen.

MicroByte

Tobacco in cigarettes, chewing tobacco, and cigars may carry the tobacco mosaic virus. Smokers who work with plants should wash their hands thoroughly before handling them.

MicroAssessment 13.8

Plant viruses cause many plant diseases and are of major economic importance. They invade through wound sites in the cell wall and are spread through soil, insects, and growers.

22. How does a plant virus penetrate the tough outer coat of the plant cell?

23. How are plant viruses transmitted?

24. Why is it especially important for plant viruses that they remain stable outside the plant?

13.9 ■ Other Infectious Agents: Viroids and Prions

Learning Outcome

15. Compare and contrast the characteristics of viroids and prions.

Although viruses are composed of only nucleic acid surrounded by a protective protein coat, viroids and prions have an even simpler structure.

Viroids

A **viroid** consists solely of a small single-stranded circular non-coding RNA molecule, about one-tenth the size of the smallest viral RNA genome known. All known viroids infect only plants, where they cause serious diseases such as potato spindle tuber, chrysanthemum stunt, citrus exocortis, and cadang-cadang. Like plant viruses, viroids enter plants through wound sites and then spread from cell to cell via the plasmodesmata and to the rest of the plant through the phloem. Although many questions about these unusual members of the microbial world still remain, scientists have learned a great deal about viroid structure and its interactions with host factors. These studies not only shed light on how viroids cause disease, but could also impact our understanding of the origin and roles of other non-coding RNAs such as those used in RNA interference.

Prions

Prions are composed only of protein, which is reflected in the name (derived from proteinaceous infectious agent); they have no nucleic acid. These agents have been linked to a number of slow, degenerative diseases that are always fatal. Human prion diseases include Creutzfeldt-Jakob disease, fatal familial insomnia, and kuru; animal diseases include scrapie (sheep and goats), mad cow disease or bovine spongiform encephalopathy (cattle), and chronic wasting disease (deer and elk) (**table 13.6**). In all of these diseases, prion proteins accumulate in neural tissue. For unknown reasons, neurons die and brain function deteriorates as the tissues develop characteristic holes (**figure 13.25**). The resulting sponge-like appearance of the brain tissues gave rise to the general term **transmissible spongiform encephalopathies,** which refers to all prion diseases.

Considering that prions lack nucleic acid, the mechanism by which they accumulate in tissue and cause disease has long been an intriguing question. An answer began to emerge with the discovery of a normal neuronal protein identical in amino acid sequence to a prion protein. The infectious protein is simply a misfolded version of the normal protein, but that alteration has significant consequences. For one thing, it makes the infectious protein resistant to degradation by the host cell proteases, which normally destroy older proteins as new ones are synthesized. The misfolding also causes the molecules to become insoluble, resulting in their aggregation in cells. The infectious prion proteins are now referred to as **PrP^{SC}** (for **pr**ion **p**rotein, **sc**rapie) and the normal counterparts as **PrP^{C}** (for **pr**ion **p**rotein, **c**ellular). A significant factor in PrP^{SC} accumulation is that it somehow interacts with and changes the folding of PrP^{C} molecules, thus converting them to PrP^{SC} (**figure 13.26**).

In most cases, a prion disease is transmitted only to members of the same species. Exceptions exist, however, such as when the prion strain that caused mad cow disease in England also led to fatal disease in more than 170 people, presumably after they ate contaminated beef; the symptoms in humans were very similar to those of Creutzfeldt-Jakob disease, so the illness was called variant Creutzfeldt-Jakob disease. So far, no human deaths have been attributed to eating deer or elk that have chronic wasting disease or sheep that have scrapie.

Chronic wasting disease (CWD) provides a good example of how a prion disease can spread. Deer at a research facility

TABLE 13.6 Prion Diseases

Disease	Host
Scrapie	Sheep and goats
Bovine spongiform encephalopathy (mad cow disease)	Cattle
Chronic wasting disease	Deer and elk
Transmissible mink encephalopathy	Ranched mink
Exotic ungulate encephalopathy	Antelope in South Africa
Feline spongiform encephalopathy	Cats
Kuru	Humans (transmitted by cannibalism)
Variant Creutzfeldt-Jakob disease	Humans (transmitted by consumption of prion-contaminated beef)
Creutzfeldt-Jakob disease	Humans (inherited)
Gerstmann-Sträussler-Scheinker syndrome	Humans (inherited)
Fatal familial insomnia	Humans (inherited)

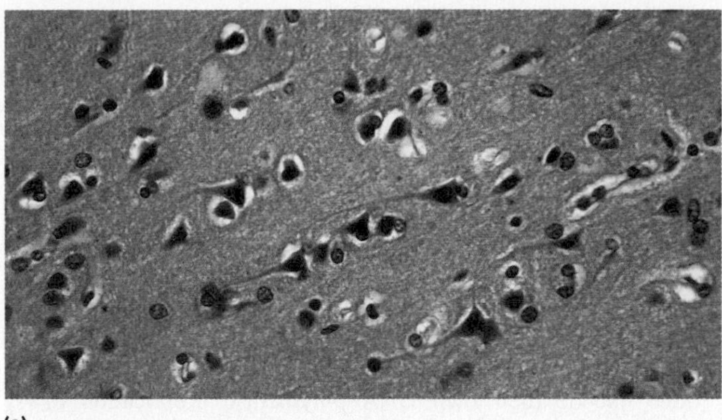

(a)

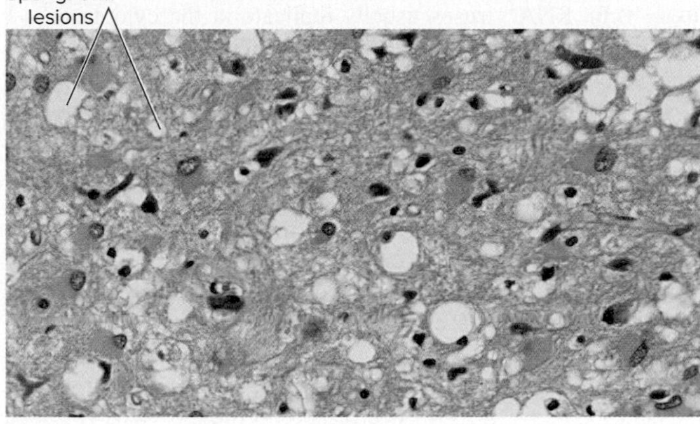
Spongiform
lesions

(b)

FIGURE 13.25 Appearance of a Brain with Spongiform Encephalopathy (a) Normal brain section. **(b)** Brain section of patient with spongiform encephalopathy. a: M. Abbey/Science Source; b: Sherif Zaki; MD; PhD; Wun-Ju Shieh; MD; PhD; MPH/CDC

From these photos, why has this disease been given the name it has?

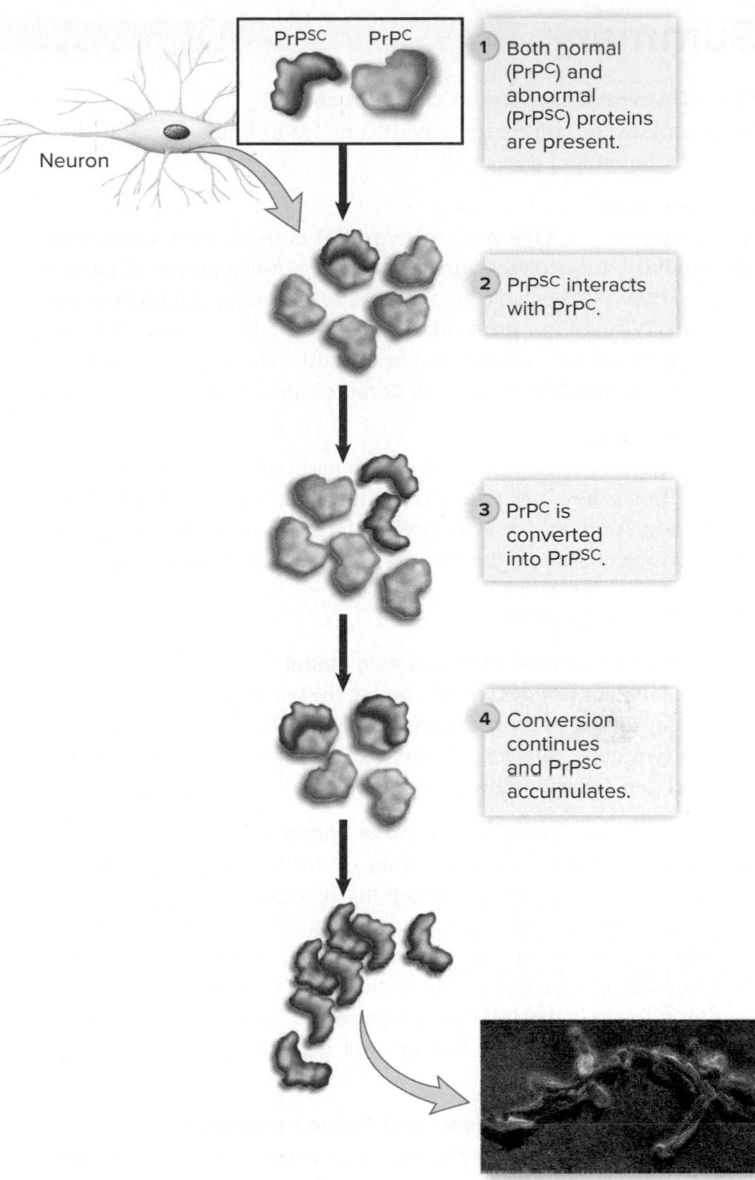
PrPSC
PrPC
Neuron
1 Both normal (PrPC) and abnormal (PrPSC) proteins are present.
2 PrPSC interacts with PrPC.
3 PrPC is converted into PrPSC.
4 Conversion continues and PrPSC accumulates.

FIGURE 13.26 Proposed Mechanism by Which Prions Propagate b: Eye of Science/Science Source

Why would PrP^{SC} accumulate when PrP^{C} does not?

in Colorado were the first animals to show signs of CWD in 1967. Within 40 years, the disease had spread from Colorado to wild deer populations in at least two dozen states in the United States and to parts of Canada, through natural deer migration and as a result of transporting captive farmed animals. The disease is transmitted through direct animal-to-animal (nose-to-nose) contact and indirect exposure to prions in the environment. Feed, water sources, and soil may be contaminated with prions shed in saliva, urine and feces, or decomposing carcasses.

Prions are unusually resistant to heat and chemical treatments, making them difficult, if not impossible, to eliminate from the environment. Additional information about the role of prions in human disease is covered in chapter 26.

MicroAssessment 13.9

Two infectious agents that are structurally simpler than viruses are viroids and prions. Viroids contain only single-stranded RNA and no protein; prions contain protein and no nucleic acid.

25. Distinguish between a viroid and a prion in terms of structure and hosts.

26. Discuss how prion proteins accumulate in nervous tissue.

27. Must all prion diseases result from eating infected food? Explain.

Summary

13.1 ■ General Characteristics of Viruses

Most viruses are approximately 100 to 1,000 times smaller than the cells they infect (figure 13.1).

Viral Structure

At a minimum, a **virion** (viral particle) consists of nucleic acid surrounded by a protein **capsid.** Capsids are composed of capsomeres. Some viruses have an envelope surrounding the **nucleocapsid;** other viruses are **non-enveloped** or naked (figure 13.2). Viruses contain either RNA or DNA, but never both. The shape of a virus is generally icosahedral, helical, or complex (figure 13.3).

Viral Taxonomy

Viruses are classified primarily on the basis of their genome structure and hosts they infect (table 13.1). The names of virus families end in *-viridae*. Viruses are also given informal names and are sometimes grouped based on their routes of transmission (table 13.2).

13.2 ■ Bacteriophages

Lytic Phage Infections: T4 Phage as a Model

Lytic or **virulent phages** exit the host at the end of the cycle by lysing the host, resulting in a **productive infection.** The infection proceeds through five steps: attachment, genome entry, synthesis of phage proteins and genome, assembly (maturation), and release (figure 13.5).

Temperate Phage Infections: Lambda Phage as a Model

Temperate phages have the option of either directing a productive infection or initiating a **lysogenic infection** (figure 13.6); in the case of a lysogenic infection, the infected cell is called a **lysogen.** A **repressor** maintains the **prophage** in an integrated state, but the prophage can be excised to initiate a lytic infection. Lysogens are immune to superinfection. **Lysogenic conversion** occurs if a prophage carries genes that change the phenotype of the host cell (table 13.3).

Filamentous Phage Infections: M13 Phage as a Model

Filamentous phages cause productive infections, but the viral particles are continually extruded from the host in the assembly process, and the cells are not killed (figure 13.8).

13.3 ■ The Roles of Bacteriophages in Horizontal Gene Transfer

Generalized Transduction

Generalized transduction results from a packaging error during phage assembly. Generalized transducing particles can transfer any gene of a donor cell to a recipient cell.

Specialized Transduction

Specialized transduction results from an excision mistake made by a temperate phage during its transition from a lysogenic cycle to a lytic cycle (figure 13.9). Only genes located near the site at which the temperate phage integrates are transduced.

13.4 ■ Animal Virus Replication

The infection cycle of animal viruses typically involves a five-step process.

Attachment

Attachment proteins or spikes on the viral particle attach to specific receptors on the cell surface.

Entry and Uncoating

In the case of animal viruses, the entire virion enters the cell. Enveloped viruses either fuse with the host membrane or are taken in by receptor-mediated endocytosis (figure 13.10). Non-enveloped virions enter by receptor-mediated endocytosis. **Uncoating** releases the nucleic acid from the protein coat.

Replication

DNA viruses generally replicate in the nucleus and use the host cell machinery for DNA synthesis as well as gene expression, although they often encode their own DNA polymerase. Replication of single-stranded DNA viruses is similar to that of double-stranded DNA viruses, but the complementary strand must be synthesized first (figure 13.11). RNA viruses usually replicate in the cytoplasm. Replication requires a virally encoded **replicase** to synthesize the complementary RNA strand. This enzyme lacks proofreading ability and thus is relatively error-prone compared to DNA polymerases (figure 13.12). Reverse-transcribing viruses encode **reverse transcriptase,** which synthesizes DNA from an RNA template. As with replicases, these enzymes are error-prone (figure 13.13).

Assembly (Maturation)

Capsids are formed, and then the genome and any necessary proteins are packaged within it.

Release

Enveloped virions most often exit by **budding** (figure 13.14). Non-enveloped virions are released when the host cell dies.

13.5 ■ Categories of Animal Virus Infections (figure 13.15)

Acute infections are characterized by sudden onset and a relatively short duration. **Persistent infections** can continue for years, or even the lifetime of the host.

Acute Infections

Acute infections result in a burst of virions being released from infected host cells; although the cells often die, the host can survive because of the immune response.

Persistent Infections

Chronic infections are characterized by the continuous production of low levels of viral particles; in **latent infections** the viral genome remains silent in the host cell, but it can reactivate to cause a productive infection.

13.6 ■ Viruses and Human Tumors

Cancer-Causing Viruses

Oncoviruses are viruses that can cause cancer; in some cases they carry **oncogenes** that interfere with the ability of the cell to control growth. Most virus-induced tumors are caused by certain DNA viruses (table 13.5). A vaccine against human papillomaviruses (HPVs) prevents many cervical cancers.

Cancer-Fighting Viruses

Oncolytic viruses target and destroy cancer cells.

13.7 ■ Methods Used to Study Viruses

Viruses need to be cultivated and studied in an appropriate host cell. Bacterial cells are easier to cultivate than animal cells, so most studies on viruses have been performed on bacteriophages.

Cultivating and Quantitating Bacteriophages

Plaque assays are used to quantitate the phage particles in samples (figure 13.18). Each plaque represents a plaque-forming unit (PFU) initiated by a single phage particle infecting a cell.

Cultivating and Quantitating Animal Viruses

Cell culture or **tissue culture** is commonly used to cultivate most viruses (figure 13.19). Virus propagated in cell culture often cause a **cytopathic effect** (figure 13.20). Plaque assays are used to determine the concentration of animal viruses in a sample. In some cases, virions can be counted with an electron microscope (figure 13.21). **Quantal assays** determine the $\mathbf{ID_{50}}$ or $\mathbf{LD_{50}}$. For certain viruses, a **hemagglutination assay** can be used to determine the relative concentration of viral particles (figure 13.22).

13.8 ■ Plant Viruses

Viral infection of plants can be recognized by outward signs such as yellowing foliage, stunted growth, and tumor formation; plants usually do not recover from the infection (figure 13.23). Virions do not bind to receptor sites on plant cells but enter through wound sites. Insects are probably the most important transmitter of plant viruses.

13.9 ■ Other Infectious Agents: Viroids and Prions

Viroids

Viroids are plant pathogens that consist of small, circular, single-stranded non-coding RNA molecules.

Prions

Prions are composed solely of protein and cause a number of **transmissible spongiform encephalopathies** (table 13.6). Prions accumulate by converting PrP^{C} to PrP^{SC}, proteins that are less susceptible to proteases and form aggregates (figure 13.26).

Review Questions

Short Answer

1. Why are non-enveloped viruses generally more resistant to disinfectants than are enveloped viruses?
2. How is the replication cycle of lambda phage different from that of T4?
3. What is lysogenic conversion?
4. How is specialized transduction different from generalized transduction?
5. What is the difference between acute and persistent infections?
6. Why must (−) strand but not (+) strand RNA viruses bring their own replicase into a cell?
7. Why are RNA viruses and retroviruses more error-prone in their replication than DNA viruses?
8. What is the role of proviruses in persistent infections?
9. How do oncogenes differ from proto-oncogenes?
10. Describe how prions propagate.

Multiple Choice

1. Capsids are composed of
   a) DNA.
   b) RNA.
   c) protein.
   d) lipids.
   e) polysaccharides.
2. The tail fibers on phages are associated with
   a) attachment.
   b) penetration.
   c) transcription of phage DNA.
   d) assembly of a virus.
   e) lysis of the host.
3. Classification of viruses is based on all of the following *except*
   a) type of nucleic acid.
   b) shape of virus.
   c) size of virus.
   d) host infected.
   e) strandedness of nucleic acid.
4. Temperate phages can do all of the following *except*
   a) lyse their host cells.
   b) change the properties of their hosts.
   c) integrate their DNA into the host DNA.
   d) bud from their host cells.
   e) become prophages.
5. All phages must be able to
   1. inject their nucleic acid into a host cell.
   2. kill the host cell.
   3. multiply in the absence of living bacteria.
   4. lyse the host cell.
   5. have their nucleic acid replicate in the host cell.
      a) 1, 2
      b) 2, 3
      c) 3, 4
      d) 4, 5
      e) 1, 5
6. Filamentous phages
   a) infect animal and bacterial cells.
   b) cause their host cells to grow more quickly.
   c) are extruded from the host cell.
   d) undergo assembly in the cytoplasm.
   e) degrade the host cells' DNA.
7. Influenza vaccines must be changed yearly because the amino acid sequence of the viral proteins changes gradually over time. Based on this information, which is the most logical conclusion? The influenza virus
   a) is enveloped.
   b) is non-enveloped.
   c) has a DNA genome.
   d) has an RNA genome.
   e) causes a persistent infection.

8. Acute infections of animals
   1. are a result of productive infection.
   2. generally lead to long-lasting immunity.
   3. result from the integration of viral nucleic acid into the host.
   4. are usually followed by chronic infections.
   5. often lead to tumor formation.
   a) 1, 2
   b) 2, 3
   c) 3, 4
   d) 4, 5
   e) 1, 5
9. Determining viral titers of both phage and animal viruses frequently involves
   a) plaque formation.
   b) quantal assays.
   c) hemagglutination.
   d) determining the ID_{50}.
   e) counting of virions by microscopy.
10. Prions
   a) contain only nucleic acid without a protein coat.
   b) replicate like HIV.
   c) integrate their nucleic acid into the host genome.
   d) cause diseases of humans.
   e) cause diseases of plants.

Applications

1. A public health physician isolated large numbers of phages from rivers used as a source of drinking water in western Africa. The physician is very concerned that humans might become ill from drinking this water, although she knows that phages specifically attack bacteria. Why is she concerned?
2. SARS-CoV-2 and related viruses are unusual among RNA viruses because their replicase enzymes have proofreading ability. How will this affect the development of vaccines and therapies that target COVID-19?

Critical Thinking

1. Viruses that infect bacterial cells do not infect human cells, and viruses that infect human cells do not infect bacterial cells. Explain why this should be the case.
2. Why is it virtually impossible to eradicate (eliminate) a disease caused by a zoonotic virus?

14 | The Innate Immune Response

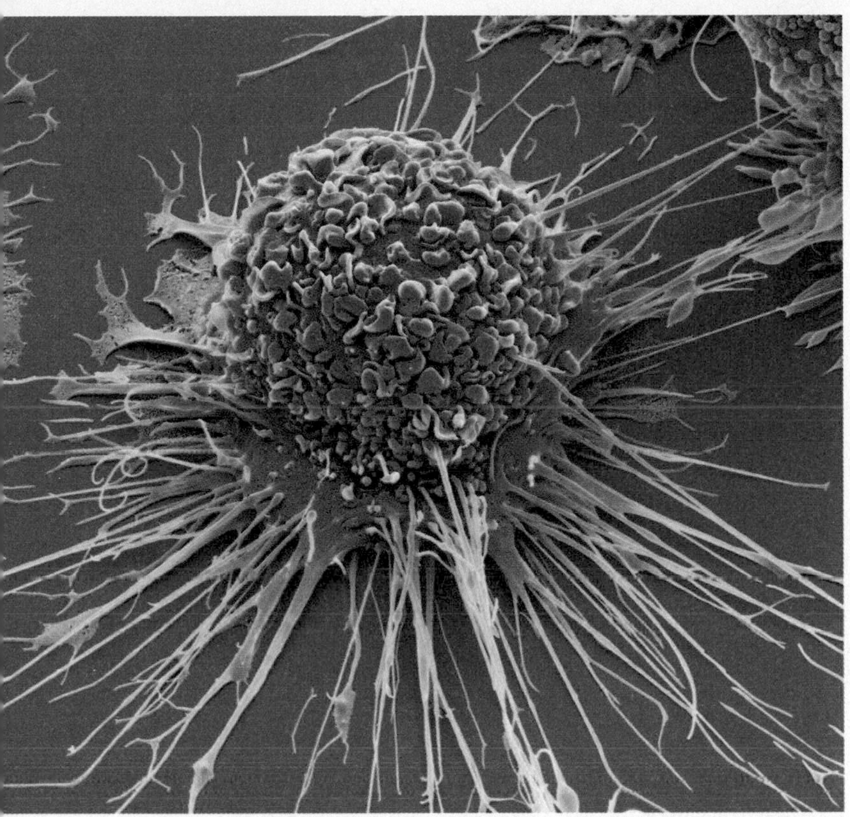

Activated macrophage engulfing material (color-enhanced scanning electron micrograph). *Steve Gschmeissner/Science Photo Library/Alamy Stock Photo*

KEY TERMS

Apoptosis Type of programmed death of "self" cells that does not cause inflammation.

Complement System Series of proteins in blood and tissue fluids that can be activated to help destroy and remove invading microbes.

Cytokines Proteins that function as chemical messengers, allowing the cells involved in host defenses to communicate.

Inflammatory Response Coordinated innate response with the purpose of containing a site of damage, localizing the response, eliminating the invader, and restoring tissue function.

Innate Immunity Protection provided by routine immune responses present at birth; involves anatomical barriers, sensor systems that recognize patterns associated with microbes or tissue damage, phagocytic cells, the inflammatory response, and fever.

Macrophage Type of phagocytic cell that wanders or resides in tissues; it has multiple roles, including scavenging debris and producing pro-inflammatory cytokines.

Membrane Attack Complexes (MACs) Group of complement system components assembled to form pores in membranes of invading cells.

Neutrophil Major type of phagocytic cell in blood; neutrophils quickly move to infected tissues, where they use multiple mechanisms to destroy invading microbes.

Opsonization Coating of an object with molecules for which phagocytes have receptors, thereby making it easier for a phagocyte to engulf the object.

Pattern Recognition Receptors (PRRs) Proteins that recognize specific compounds unique to microbes or tissue damage, thus allowing the body's cells to "see" signs of microbial invasion and respond accordingly.

Phagocyte Cell type that specializes in engulfing and digesting microbes and cell debris (a process called phagocytosis).

A Glimpse of History

Once microorganisms were shown to cause disease, scientists worked to explain how the body defends itself against their invasion. Ilya Metchnikoff, a Russian-born scientist, hypothesized that specialized cells in the body destroy invading organisms. His ideas came about while he was studying the larval form of starfish. As he looked at the larvae under the microscope, he could see amoeba-like cells within the bodies. He described his observations:

> . . . I was observing the activity of the motile cells of a transparent larva, when a new thought suddenly dawned on me. It occurred to me that similar cells must function to protect the organism against harmful intruders. . . . I thought that if my guess was correct a splinter introduced into the larva of a starfish should soon be surrounded by motile cells much as can be observed in a man with a splinter in his finger. No sooner said than done. In the small garden of our home . . . I took several rose thorns that I immediately introduced under the skin of some beautiful starfish larvae which were as transparent as water. Very nervous, I did not sleep during the night, as I was waiting for the results of my experiment. The next morning, very early, I found with joy that it had been successful.

Metchnikoff reasoned that certain cells in animals can ingest and destroy foreign material. He called these cells phagocytes ("cells that eat") and proposed that they were primarily responsible for the body's ability to destroy invading microbes. He then studied the process by watching phagocytes ingest and destroy invading yeast cells in transparent water fleas. In 1884, Metchnikoff published a paper supporting his belief that phagocytic cells were primarily responsible for destroying disease-causing organisms. He spent the rest of his life studying this process and other biological phenomena. Metchnikoff was awarded a Nobel Prize in 1908 for his studies of immunity.

From a microorganism's standpoint, the tissues and fluids of the human body are much like a culture flask filled with a warm nutrient-rich solution, but guarded by armies of cells. Those armies generally keep the interior of the body—including blood, muscles, and bones—sterile. If this were not the case, microbes would simply degrade our tissues, just as they readily decompose dead animals. When the body's defense systems work to eliminate an invader, however, they must minimize unintentional damage to the body's own tissues. Likewise, they must avoid waging war with the normal microbiota that lives on the body's surfaces, because the resulting constant conflict would damage the tissues. The system must maintain a delicate balancing act, destroying pathogens while maintaining relatively stable conditions within the human body—a state called homeostasis (*homeo* means "similar" and *stasis* means "standing still"). Failure to do that can be deadly.

Like other multicellular organisms, our bodies have several mechanisms of defense. **Innate immunity** is the routine protection present at birth and involves anatomical barriers as well as certain cell types and chemicals. In addition to innate immunity, vertebrates have evolved a more specialized defense system, providing protection called **adaptive immunity.** This develops throughout life as a result of exposure to microbes or certain other types of foreign material, and it substantially increases the body's ability to defend itself. A substance capable of eliciting an immune response is called an **antigen.** Each time the body reacts to an antigen, the adaptive defense system first "learns" and then "remembers" the most effective response to that specific antigen; it then reacts accordingly if the antigen is encountered again. An important action of the adaptive immune response is the production of proteins called **antibodies.** These bind specifically to antigens, thereby targeting them for destruction or removal by other immune defenses. The adaptive immune response can also destroy the body's own cells—referred to as **host cells** or **self cells**—that are infected with a virus or other invader.

To simplify the description of the immune system, we will consider it as a series of individual parts; this chapter will focus almost exclusively on the components of innate immunity. Remember, however, that although the various parts are discussed separately, their actions are connected and coordinated. In fact, as described in chapter 15, certain components of the innate defenses educate the adaptive defenses, helping them to recognize that a particular antigen represents a microbial invader.

14.1 ■ Overview of the Innate Immune Defenses

Learning Outcome

1. Outline the fundamental components of the innate defenses.

The innate immune system is easiest to understand by considering it as three general interacting components: first-line defenses, sensor systems, and innate effector actions. As a useful analogy, think of the defense systems of a high-security building or compound: The first-line defenses are the security walls surrounding the property; the sensor systems are the security cameras scattered throughout the property, monitoring the environment for signs of invasion; and the effector actions are the security teams sent to remove any invaders that have been detected, thereby eliminating the threat (**figure 14.1a**).

The **first-line defenses** prevent microbes and other foreign material from entering the body's tissues. These defenses include the physical barriers provided by the skin and the mucous membranes, along with the antimicrobial substances that bathe them (**figure 14.1b**). Members of the normal microbiota residing on the surfaces also provide protection.

Sensor systems allow the immune system to recognize when the first-line defenses have been breached (see figure 14.1b). Any microbe that passes through the first-line defenses and into tissue is perceived by the immune system as an invader. Certain host cells serve as **sentinel cells** (lookouts or guards), positioned at strategic sites in the body to detect invading microbes in blood or tissue fluids. The sentinel cells recognize microbes by detecting their unique components, such as peptidoglycan, using a special group of receptors called **pattern recognition receptors (PRRs).** Some PRRs are located on the surfaces of sentinel cells, allowing the cells to detect surrounding invaders; others are within the sentinel cells' endosomes or phagosomes, allowing the cells to determine what they have engulfed. In addition to the PRRs of sentinel cells, many cell types have a distinct PRR set in their cytoplasm, allowing those cells to detect the presence of an intracellular microbial invader. Some PRRs are soluble extracellular molecules that bind to invaders, thereby targeting them for elimination by other components of the immune system. A very different type of sensor circulates in blood and tissue fluids; this sensor, called the **complement system,** was named because it can "complement" (act in combination with) the adaptive immune defenses. The complement system is a series of proteins that circulate in an inactive form but become activated in response to certain stimuli, setting off a chain of events that results in the removal and destruction of invading microbes.

Innate effector actions help eliminate invaders (see figure 14.1b). When one or more of the sensor systems just described detects an invading microbe, various effector mechanisms may be called into action. Because the sensors typically recognize patterns associated with certain groups of microbes, the effector actions can be tailored to defend against those groups. For example, when a host cell recognizes that it is infected by a virus, the cell produces an **interferon (IFN)**, a type of protein that warns nearby cells about the virus. Those neighboring cells react by preparing to shut down their biosynthetic activities if they too become infected. By doing so, the cells can deprive the virus of a mechanism to replicate. In response to sensor signals that indicate a bacterial infection or tissue damage, **phagocytes** are recruited to the site of invasion or damage; these cells specialize in engulfing and digesting microbes and cell debris, a process called **phagocytosis** (see figure 3.23). Some complement system proteins and soluble PRRs also play a role in the

Focus Figure

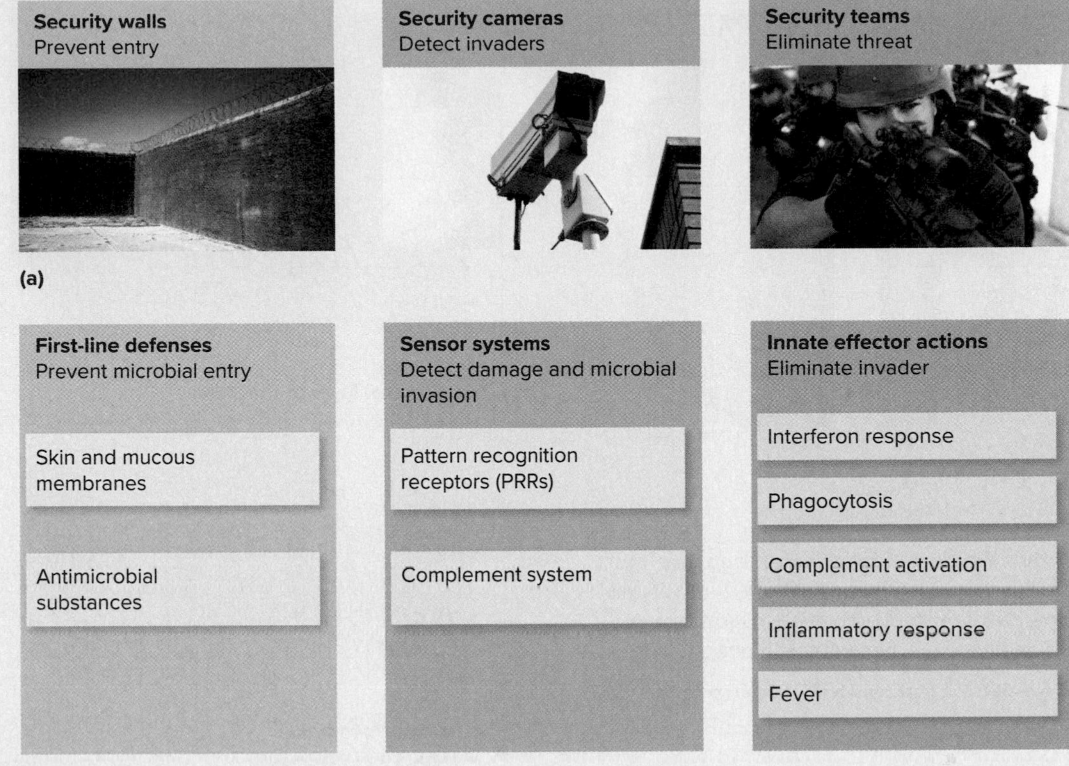

FIGURE 14.1 Defense Systems (a) Systems that protect a high-security compound. **(b)** Components of innate immunity that protect the body against infection. a (left): Steve Cole/E+/Getty Images; a (middle): Image Source; a (right): Moodboard/Brand X Pictures/Getty Images

What is the role of the sensor systems in innate immunity?

process by binding to foreign material—an action that makes it easier for phagocytes to engulf the material. A coordinated set of events called the **inflammatory response** occurs as a general response to microbial invasion or tissue damage. As part of this, local blood vessels become leaky, allowing fluids containing complement system proteins and other protective substances to enter tissues; phagocytic cells move from the bloodstream and into tissues as well. Another innate response is **fever**, a higher-than-normal body temperature. Fever interferes with the growth of some pathogens and can enhance the effectiveness of other responses.

MicroAssessment 14.1

First-line defenses prevent microbes from entering tissues. Sensor systems recognize invading microbes. Innate effector actions eliminate the invader.

1. List two sensor systems of the innate defenses.
2. Describe three innate effector actions.
3. In addition to peptidoglycan, which molecules unique to bacteria might pattern recognition receptors recognize?

14.2 ▪ First-Line Defenses

Learning Outcome

2. Describe the first-line defenses, including the physical barriers, antimicrobial substances, and normal microbiota.

The body's borders are the first line of defense against invading microbes (**figure 14.2**). Some of these borders are thought of as being "inside" the body, but they directly contact the external environment. For example, the digestive tract, which begins at the mouth and ends at the anus, is simply a hollow tube that runs through the body, allowing intestinal cells to absorb nutrients from food that passes through (see figure 24.1); the respiratory tract is an open system that allows O_2 and CO_2 to be exchanged (see figure 21.1).

In this section, we will describe the general physical and chemical aspects of the anatomical barriers, as well as the protective contributions of the normal microbiota. These are described in more detail in the chapters dealing with each body system.

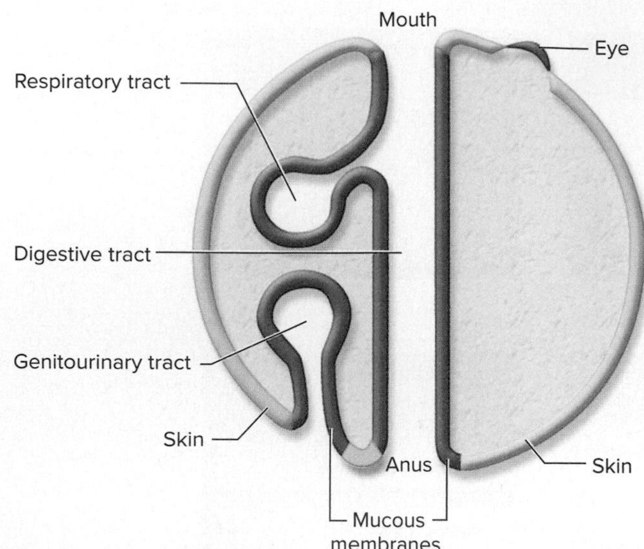

FIGURE 14.2 The Body's Borders These borders separate the interior of the body from the surrounding environment; they are the initial obstacles microorganisms must overcome to invade tissues. The skin is shown in tan, and mucous membranes in pink.

? Why are the contents of the digestive tract considered to be in contact with the external environment?

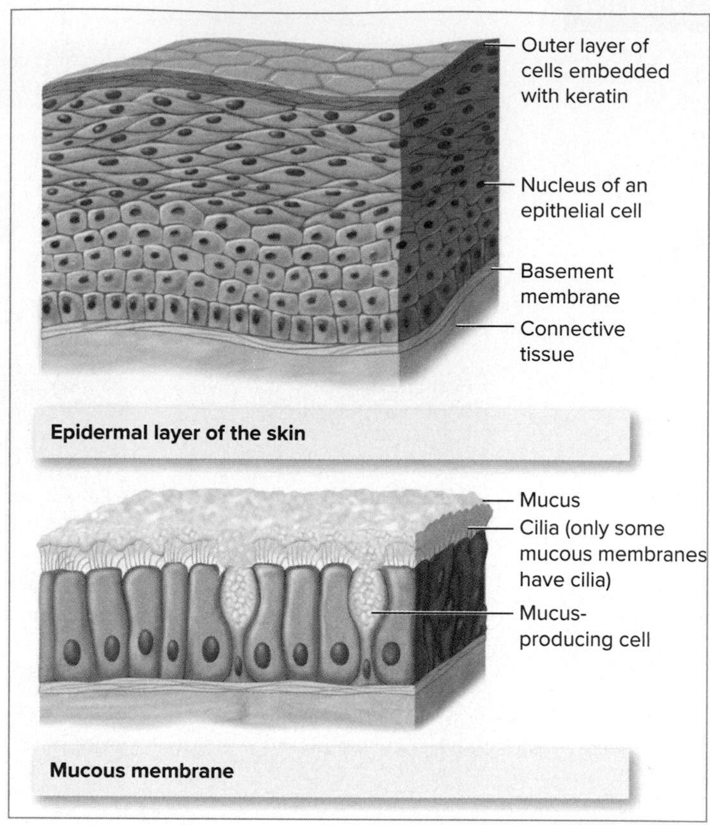

FIGURE 14.3 Epithelial Barriers Cells of these barriers are tightly packed together and rest on a layer of thin fibrous material, the basement membrane

? What is the purpose of the cilia on the respiratory epithelium?

Physical Barriers

All exposed surfaces of the body are lined with tightly packed epithelial cells that rest on a thin layer of fibrous material, the basement membrane. The two general types of barriers include (**figure 14.3**):

- **Skin.** This obvious visible barrier is the most difficult for microbes to penetrate. It is composed of two main layers: the epidermis and the dermis (see figure 22.1). The epidermis, the surface layer, is composed of many sheets of epithelial cells. The outermost sheets are made up of flattened dead cells filled with a water-repelling protein called keratin, resulting in a dry surface environment; those dead cells continually flake off, taking with them any microbes that might be adhering. The dermis contains tightly woven fibrous connective tissue, making it extremely tough and durable (the dermis of cattle is used to make leather).
- **Mucous membranes.** These moist membranes line the digestive tract, respiratory tract, and genitourinary tract and are constantly bathed with mucus or other secretions that help wash microbes from the surface. Most mucous membranes have mechanisms that move microbes toward areas where they can be eliminated. Peristalsis—wave-like contractions of the intestinal tract—propels food and liquid toward the anus and also helps remove microbes. The respiratory tract is lined with ciliated cells; the hair-like cilia constantly beat in a synchronized manner, moving materials away from the lungs to the throat, where they can then be swallowed. This movement out of the respiratory tract is referred to as the **mucociliary escalator.** As with skin cells, mucosal epithelial cells are constantly shed, taking attached microbes with them.

MicroByte

A person sheds approximately 1 billion intestinal cells per hour.

Antimicrobial Substances

Skin and mucous membranes are protected by a variety of substances that inhibit or kill microorganisms. For example, the salty residue that accumulates on skin as perspiration (sweat) evaporates inhibits all but salt-tolerant microbes. Other antimicrobial substances include (**figure 14.4**):

- **Lysozyme.** This peptidoglycan-degrading enzyme is in tears, saliva, and mucus. It is also found in phagocytic cells, blood, and the fluid that bathes tissues.
- **Peroxidases.** These are part of systems that produce antimicrobial compounds, using hydrogen peroxide (H_2O_2) in the process. Microorganisms that make the enzyme catalase are less susceptible to the lethal effects of peroxidase systems because they can potentially convert hydrogen peroxide to water and O_2 before peroxidases have a

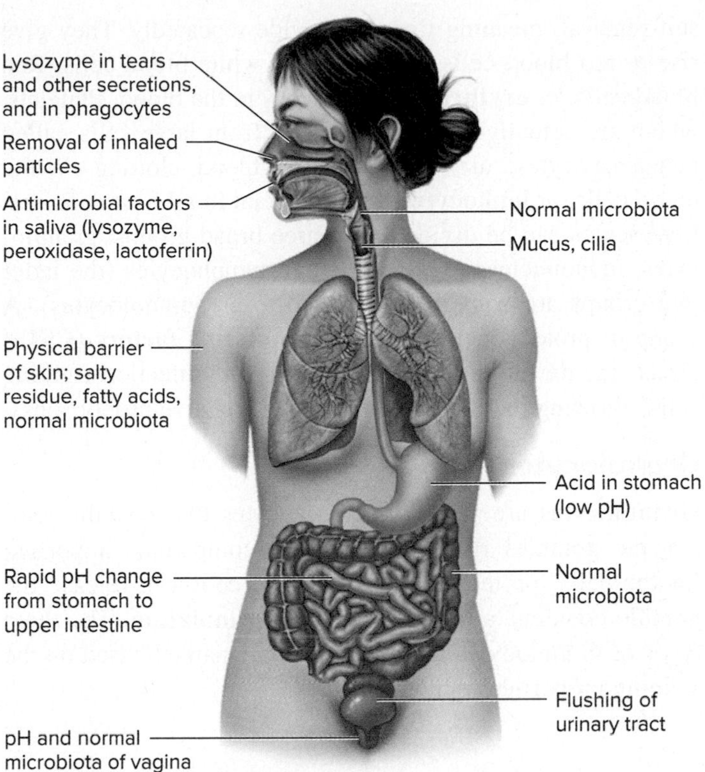

FIGURE 14.4 Antimicrobial Substances and the Normal Microbiota These play important roles in protecting the body's borders.

? **How is lysozyme antibacterial?**

chance to use it. Peroxidase systems are found in saliva, milk, body tissues, and phagocytes.

- **Lactoferrin.** This is an iron-binding protein in saliva, mucus, milk, and some types of phagocytes. A similar compound, **transferrin,** is in blood and tissue fluids. By binding to iron, these proteins make it unavailable to microorganisms. Recall that iron is one of the major elements required by organisms, so withholding it prevents microbial growth. Some microorganisms can capture iron from the host, however, counteracting this defense.
- **Host defense peptides (HDPs).** Also referred to as antimicrobial peptides, these short chains of amino acids (usually 15–20 amino acids long) have antimicrobial and other protective activities and are produced by a wide range of organisms. A group of positively charged HDPs called **defensins** insert into microbial membranes, forming pores that damage cells. Certain epithelial cells produce and release defensins, preventing the invasion of the skin and mucous membranes; production increases when microbial invasion is detected, thereby helping the body to eliminate the infection. Defensins are also produced by phagocytes, which use them to destroy microorganisms they have ingested. In addition to directly killing invading microorganisms, defensins and other HDPs also promote and guide various immune responses. Vitamin D plays a role in regulating the expression of some HDPs, which might explain why people who have a vitamin D deficiency are more susceptible to certain diseases.

Normal Microbiota

The **normal microbiota** of humans is the population of microorganisms that routinely inhabit the skin and mucous membranes of healthy people (see figure 16.1); the community as a whole is referred to as the human microbiome, and some of its beneficial roles are discussed further in section 16.2. Although these organisms are not technically part of the immune system, they provide considerable protection. As an example, they prevent pathogens from adhering to host cells by covering binding sites that might otherwise be used for attachment. The population also consumes available nutrients that could otherwise support the growth of less desirable organisms. Some members of the normal microbiota produce compounds toxic to other bacteria. In the hair follicles of the skin, for instance, *Cutibacterium* species degrade lipids, releasing fatty acids that inhibit the growth of many pathogens. In the gastrointestinal tract, some strains of *E. coli* synthesize colicins, a group of proteins toxic to certain bacteria. *Lactobacillus* species growing in the vagina produce lactic acid as a fermentation end product, resulting in an acidic pH that inhibits the growth of some pathogens.

The normal microbiota is also essential to the development of the immune system. As certain microbes are encountered, the system learns to distinguish harmless ones from pathogens. An inability to tolerate harmless microbes can result in chronic inflammatory conditions such as Crohn's disease.

Disruption of the normal microbiota can result in **dysbiosis**—an imbalance in the population—which can predispose a person to various infections and other conditions. Antibiotic use, for example, can cure infectious diseases but can also predispose a patient to developing antibiotic-associated diarrhea and pseudomembranous colitis, caused by the growth of toxin-producing strains of *Clostridioides* (*Clostridium*) *difficile* in the intestine. In women, antibiotic use and hormonal changes are associated with vulvovaginitis, caused by excessive growth of *Candida albicans* in the vagina.

MicroAssessment 14.2

Physical barriers that prevent microbes from entering the body include skin and mucous membranes. Antimicrobial substances, including lysozyme, peroxidases, lactoferrin, and defensins, are on body surfaces. The normal microbiota excludes pathogens and promotes immune system development.

4. How does peristalsis protect against intestinal infection?
5. What is the role of lactoferrin and transferrin?
6. How would damage to the ciliated cells of the respiratory tract predispose a person to infection?

14.3 ■ The Cells of the Immune System

Learning Outcome

3. Describe the characteristics and roles of granulocytes, mononuclear phagocytes, dendritic cells, and lymphocytes.

The cells of the immune system can move from one part of the body to another, traveling through the body's circulatory systems like vehicles on an extensive interstate highway system. Most types are always found in normal blood, but their numbers usually increase during an infection, recruited from reserves of immature cells in the bone marrow. They can leave the blood circulatory system and move to various tissues, particularly during an infection.

The formation and development of blood cells is called **hematopoiesis** (Greek for "blood" and "to make"). All blood cells, including those important in the body's defenses, originate from the same cell type, the **hematopoietic stem cell,** found in the bone marrow (**figure 14.5**). As with other types of stem cells, hematopoietic cells are capable of long-term self-renewal, meaning they can divide repeatedly. They give rise to red blood cells, platelets, and white blood cells. Red blood cells, or **erythrocytes,** carry O_2 in the blood. Platelets, which are actually fragments arising from large cells called megakaryocytes, are important for blood clotting. White blood cells, or **leukocytes,** are important in all host defenses. Leukocytes can be divided into three broad groups: granulocytes, mononuclear phagocytes, and lymphocytes (the latter two groups are sometimes referred to as agranulocytes). A group of proteins called **colony-stimulating factors (CSFs)** directs the development of these important infection-fighting cells, allowing the body to control their relative proportions.

Granulocytes

Granulocytes are a group of leukocytes that contain cytoplasmic granules filled with various compounds important for the cells' protective functions; the cells can release the granule contents, a process called **degranulation.** The three types of granulocytes found in blood are named based on the staining properties of their granules:

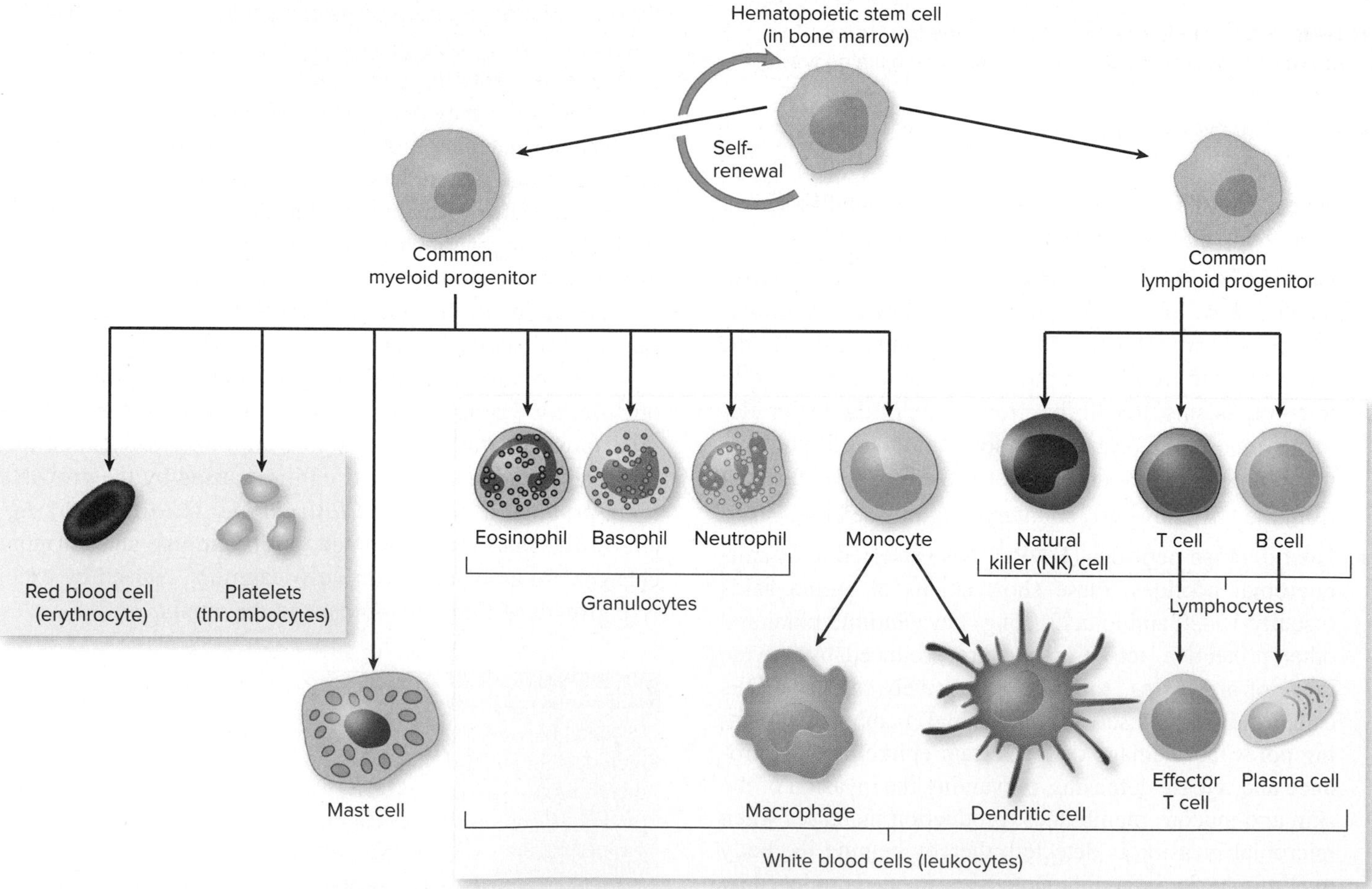

FIGURE 14.5 Blood Cells and Their Derivatives All descend from hematopoietic stem cells found in the bone marrow. Not all steps are shown; multiple steps occur between the hematopoietic stem cell and the final cells produced.

? **Certain colony-stimulating factors (CSFs) may be administered to cancer patients to treat chemotherapy-induced neutropenia—a low concentration of neutrophils (a type of leukocyte) in the blood. Why would a CSF be effective in doing this?**

- **Neutrophils.** These are phagocytes that have multiple mechanisms for destroying microbial invaders. They are essentially microbe-destroying machines—their granules, which stain poorly, contain many destructive enzymes and antimicrobial substances. When a neutrophil engulfs microbes, the granule contents then help destroy the microbes as part of the usual phagocytic process. Neutrophils can also degranulate to kill nearby microbes, or they can behave as mobile grenades, bursting in an area of infection to release not only the contents of their granules, but also cellular DNA. This burst creates what is called a neutrophil extracellular trap (NET): The DNA strands ensnare microbes so that they can be more easily destroyed by the granule contents. Neutrophils are the most numerous and important granulocytes of the innate responses, normally accounting for over half of the circulating leukocytes; their numbers typically increase during bacterial infections. Few are present in tissues, except during inflammation. Neutrophils are also called polymorphonuclear leukocytes, polys, or PMNs, names that reflect the appearance of the multiple lobes of their single nucleus. We will describe their role in phagocytosis in more detail in section 14.7.
- **Eosinophils.** These are important in ridding the body of parasitic worms. They are also involved in allergic reactions, causing some of the symptoms associated with allergies, but reducing others. Their granules, which stain red with the acidic dye eosin, contain various antimicrobial substances that can be released by degranulation.
- **Basophils.** These are involved in allergic reactions and inflammation. Their granules, which stain dark purplish-blue with the basic dye methylene blue, contain histamine and other chemicals involved with inflammation. When released, the chemicals increase capillary permeability. **Mast cells** are similar in appearance and function to basophils but are found in tissues rather than blood. They are an important type of sentinel cell that detects tissue damage and degranulates in response—an action that induces an inflammatory response.

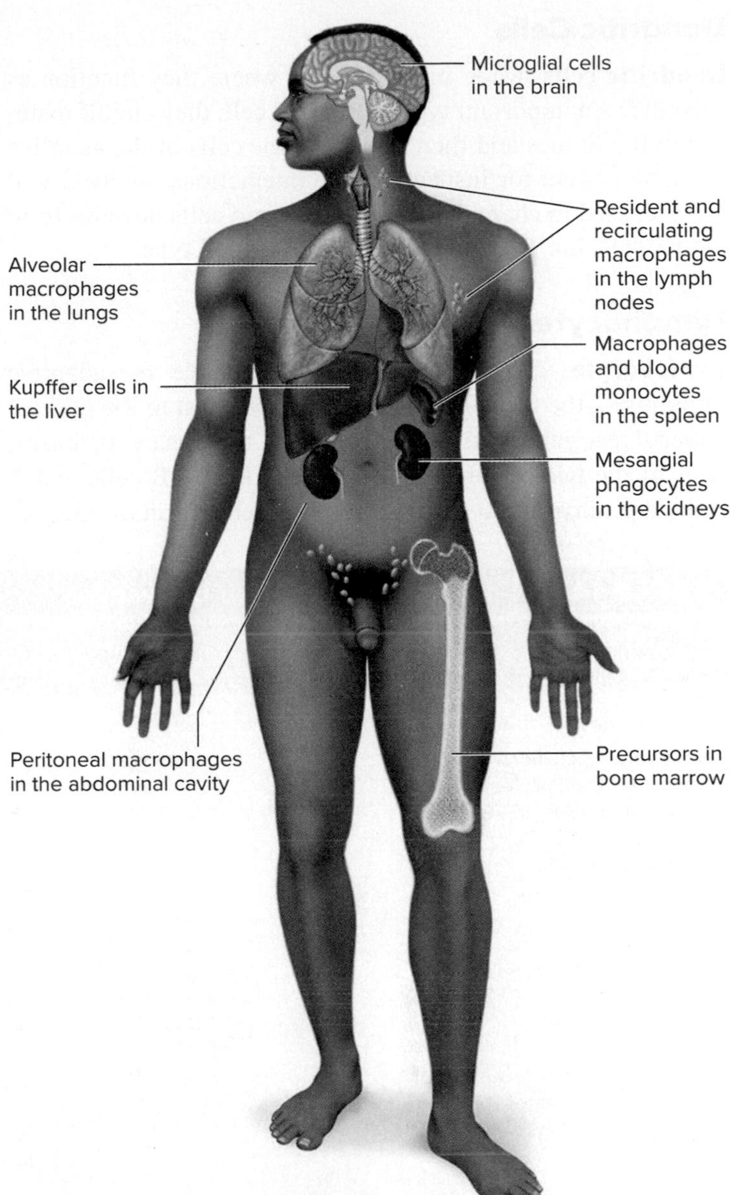

FIGURE 14.6 Mononuclear Phagocytes Resident macrophages sometimes have special names depending on their location—for example, Kupffer cells (in the liver) and alveolar macrophages (in the lung).

? **Macrophages develop from which type of blood cell?**

Mononuclear Phagocytes

Mononuclear phagocytes include (**figure 14.6**):

- **Monocytes.** These circulate in the bloodstream. They can move into tissues, where they develop into the macrophages and dendritic cells described next.
- **Macrophages.** The name means "large eater," reflecting their role as an important type of phagocyte. Some are referred to as wandering macrophages because they move around in tissues. Others, collectively referred to as resident macrophages, live permanently in tissues; they are particularly abundant in the liver, spleen, lymph nodes, lungs, and the peritoneal (abdominal) cavity. Resident macrophages are sometimes given different names based on their location—for example, Kupffer cells (in the liver) and alveolar macrophages (in the lung). Scientists initially thought that all macrophages originated from monocytes in the bloodstream, but recent evidence indicates that many resident macrophages self-renew by dividing. In addition to the role of macrophages in phagocytosis, they are an important type of sentinel cell, detecting tissue invaders and then alerting other components of the host defenses. Macrophages, along with neutrophils, will be discussed in more detail in section 14.7.
- **Dendritic cells.** These engulf material, but their role is quite different from that of other phagocytic cells, so they are described separately (next).

Dendritic Cells

Dendritic cells reside in the tissues, where they function as "scouts." An important type of sentinel cell, they engulf material in the tissues and then bring it to the cells of the adaptive immune system for inspection; the interactions involved will be discussed in chapter 15. Most dendritic cells develop from monocytes, but some develop from other cell types.

Lymphocytes

Lymphocytes are the leukocytes responsible for adaptive immunity—the focus of chapter 15. In contrast to the generic pattern recognition that characterizes the innate defenses, cells of the two major groups of lymphocytes, **B cells** and **T cells,** are remarkably specific in their recognition of antigen. As we will describe in chapter 15, these cell types generally reside in lymph nodes and other lymphatic tissues (e.g., lymph nodes, spleen, appendix, tonsils).

A group of lymphocytes called **innate lymphoid cells (ILCs)** differ from B and T cells in that they lack specificity in their mechanism of antigen recognition. Several subsets of ILCs have been identified quite recently; they are common near mucous membranes and appear to have multiple roles that help promote a balanced inflammatory response. One type of ILC, called a **natural killer (NK) cell,** has been recognized for quite some time. As its name implies, it kills certain types of cells—a role described in chapter 15.

Characteristics of leukocytes and their derivatives are summarized in **table 14.1.**

TABLE 14.1 Leukocytes and Their Derivatives

Cell Type (Percentage of Blood Leukocytes)	Major Function and Other Characteristics
Granulocytes	
Neutrophils (polymorphonuclear neutrophilic leukocytes or PMNs, often called polys; 55–65%)	Phagocytosis; they can also degranulate or release neutrophil extracellular traps (NETs). Most abundant leukocyte in blood.
Eosinophils (2–4%)	Degranulate to release chemicals that destroy eukaryotic parasites. Found mainly in tissues below the mucous membranes.
Basophils (0–1%), mast cells	Degranulate to release histamine and other inflammation-inducing chemicals. Basophils are found in blood, whereas mast cells are present in most tissues and are an important sentinel cell.
Mononuclear Phagocytes	
Monocytes (3–8%)	Phagocytosis. Found in blood; they differentiate into either macrophages or dendritic cells when they migrate into tissues.
Macrophages	Phagocytosis; they are an important type of sentinel cell in tissues. Some are wandering macrophages and others are resident macrophages; resident macrophages are sometimes known by different names based on the tissue in which they are found.
Dendritic cells	Function as "scouts"; they are an important type of sentinel cell. Engulf material in the tissues and then bring it to cells of the adaptive immune system for inspection.
Lymphocytes (25–35%)	
B and T cells	Responsible for adaptive immunity. Found in lymphoid organs (e.g., lymph nodes, spleen, appendix, tonsils, thymus, bone marrow); also in blood.
Innate lymphoid cells (NK cells are an example)	Various subsets have different roles and different locations.

MicroAssessment 14.3

Granulocytes include neutrophils, eosinophils, basophils, and mast cells. Mononuclear phagocytes include monocytes and macrophages. Dendritic cells function as scouts for the adaptive immune system. Lymphocytes are responsible for adaptive immunity.

7. Which type of granulocyte is the most abundant?
8. How are most dendritic cells related to macrophages?
9. Why can bone marrow transplants be used to replace defective lymphocytes?

14.4 ■ Cell Communication

Learning Outcome

4. Describe the characteristics and roles of surface receptors, cytokines, and adhesion molecules in innate immunity.

Immune cells must communicate with each other in order to mount a coordinated response to microbial invasion. They do this through surface receptors, cytokines, and adhesion molecules.

Surface Receptors

Surface receptors can be viewed as the "eyes" and "ears" of a cell. They are proteins that generally span the cytoplasmic membrane, connecting the outside of the cell with the inside, allowing the cell to sense and respond to external signals. Each receptor is specific with respect to the compound or compounds it will bind; a molecule that can bind to a given receptor is called a **ligand** for that receptor. When a ligand binds to its surface receptor, the internal part of the receptor is modified, triggering a specific response by the cell. Cells can alter the types and numbers of surface molecules they make, allowing them to respond to signals relevant to their immediate situation.

Cytokines

Cytokines can be viewed as the "voices" of a cell. They are secreted ligands that act at extremely low concentrations, having local, regional, or systemic effects. A cytokine produced by one cell diffuses to another and binds to the appropriate cytokine receptor of that cell, triggering a specific response by the cell. Cytokines include:

- **Chemokines.** These are important in chemotaxis of immune cells. Certain types of cells have receptors for chemokines, allowing those cells to sense and move to the relative location where they are needed, such as an area of inflammation.
- **Colony-stimulating factors (CSFs).** As mentioned in section 14.3, these are important in the differentiation of leukocytes (see figure 14.5). When more leukocytes are needed during an immune response, a variety of different colony-stimulating factors direct immature cells into the appropriate maturation pathways.
- **Interferons (IFNs).** These were discovered because of their antiviral effects, but they have several roles in the host defenses. They are important in a number of regulatory mechanisms, stimulating the responses of some cells and inhibiting others. In section 14.5, we will focus on the effector action of IFNs during viral infection.
- **Interleukins (ILs).** These are produced by a variety of cells and have diverse, often overlapping, functions. As a group, they are important in both innate and adaptive immunity.
- **Tumor necrosis factor (TNF).** This multi-functional cytokine was discovered because of its involvement in killing tumor cells, a characteristic reflected by the name, but its main role appears to be initiation of the inflammatory response; it can also cause certain host cells to self-destruct, a process called programmed cell death that is discussed in section 14.8.

Groups of cytokines often act together or in sequence to generate a response. A group referred to as **pro-inflammatory cytokines** (TNF, IL-1, IL-6, and others) contribute to inflammation. Other functions of cytokine groups include promoting antibody responses (IL-4 and others) and stimulating certain types of T cells (IL-2, an IFN, and others). The sources and effects of some cytokines are listed in **table 14.2.**

Although cytokines play an essential role in helping the body eliminate invaders, they can also lead to damaging responses. The situation can be likened to a crowded chaotic emergency situation where everyone is shouting different instructions; the reactions will likely be counterproductive and possibly harmful. Likewise, an excessive release of cytokines, called a **cytokine storm,** can result in tissue-damaging responses. This is particularly likely to occur during a systemic infection (meaning throughout the body). As scientists learn more about the roles of cytokines in disease outcomes, various medications are being developed to modify the responses, an approach called immunotherapy that is described in section 17.3.

MicroByte

HIV takes advantage of two chemokine receptors, CCR5 and CXCR4, using them as attachment sites for infection.

Adhesion Molecules

Adhesion molecules on the surface of cells allow those cells to "grab" other cells. When phagocytic cells in the blood are needed in tissues, the endothelial cells that line the blood vessels synthesize adhesion molecules that bind to passing phagocytic cells. This slows the rapidly moving phagocytes, allowing them to then leave the bloodstream. Cells also use adhesion molecules to attach to other cells; by doing this, a cell can deliver cytokines or other molecules directly to another cell.

TABLE 14.2 Some Important Cytokines

Cytokine	Source	Effect
Chemokines	Various cells	Chemotaxis
Colony-Stimulating Factors (CSFs)	Various cells	Direct the differentiation of leukocytes
Interferons	Various cells	Regulate immune responses; antiviral
Interleukins (ILs)		
IL-1	Macrophages, epithelial cells	Promotes inflammation; induces fever; macrophage activation; T-cell activation
IL-2	T cells	T-cell proliferation
IL-4	Mast cells, T cells,	Promotes antibody responses
IL-6	T cells, macrophages	Contributes to inflammation; induces fever; T- and B-cell proliferation
Tumor Necrosis Factor (TNF)	Macrophages, T cells, NK cells	Promotes inflammation; induces programmed cell death; regulates certain immune functions

MicroAssessment 14.4

Surface receptors allow a cell to detect molecules that are outside of that cell. Cytokines provide cells with a mechanism of communication. Adhesion molecules allow one cell to adhere to another.

10. What is a ligand?

11. How do cytokines function?

12. Why might some cytokines have overlapping functions?

14.5 ▪ Pattern Recognition Receptors (PRRs)

Learning Outcomes

5. Describe the significance of pattern recognition receptors (PRRs) in the immune response.
6. Compare and contrast the various PRRs that monitor a cell's surroundings, the material ingested by the cell, and the cell's cytoplasm.
7. Explain how the interferon response prevents viral replication.

Pattern recognition receptors (PRRs) are proteins that recognize specific compounds unique to microbes or tissue damage, allowing the body's cells to "see" signs of microbial invasion and respond accordingly. Furthermore, the signals from different PRRs can complement each other when an invader is detected, thereby provoking a stronger response. Many PRRs detect components of certain groups of microbes—for example, cell wall–associated compounds (peptidoglycan, teichoic acids, lipopolysaccharide, and lipoproteins), flagellin subunits, and microbial nucleic acid. These compounds are called **pathogen-associated molecular patterns (PAMPs),** reflecting the fact that they are characteristics of pathogens. They are not exclusive to pathogens, however, which is why many microbiologists prefer the term **microbe-associated molecular patterns (MAMPs).** Some PRRs recognize **damage-associated molecular patterns (DAMPs),** meaning molecules that indicate host cell damage.

The discovery of PRRs changed the way that immunologists view the immune system. In the past, the innate immune response was considered "non-specific," but we now know that the PRRs help the body's cells recognize the general category of an infectious agent, thereby playing an important role in shaping the overall response to that agent. If a macrophage's PRRs detect bacterial products, then that cell produces pro-inflammatory cytokines, leading to an inflammatory response. Likewise, if a dendritic cell's PRRs detect bacterial products, then the cell relays that information to lymphocytes, allowing those cells to mount an appropriate response. If a virally infected cell's PRRs detect viral nucleic acid, then that cell produces an interferon; the interferon alerts immune cells to the presence of a virus, and also promotes an antiviral response in nearby tissue cells.

The outcome of certain diseases is influenced by PRR-generated signals from sentinel cells and infected cells. In most cases the signals induce a protective response, but sometimes the response can be excessive and therefore damaging. For example, people who have mutations in the genes encoding certain PRRs are more likely to develop inflammatory diseases or autoimmune diseases. Scientists are now working to learn more about the mechanisms and outcomes of pattern recognition, in hopes that a better understanding will lead to new treatment options for a wide range of disease states—from illnesses caused by pathogens, to those resulting from immune system dysfunction.

This section will focus on PRRs found in three distinct locations on or in cells: (1) on the cell surface, (2) in endosomes and phagosomes, and (3) free in the cytoplasm (**figure 14.7**). Because of these locations, they provide cells with important information about the invading microbe's relative location (inside versus outside the host cell). In addition to the PRRs located on or in cells, some pattern recognition

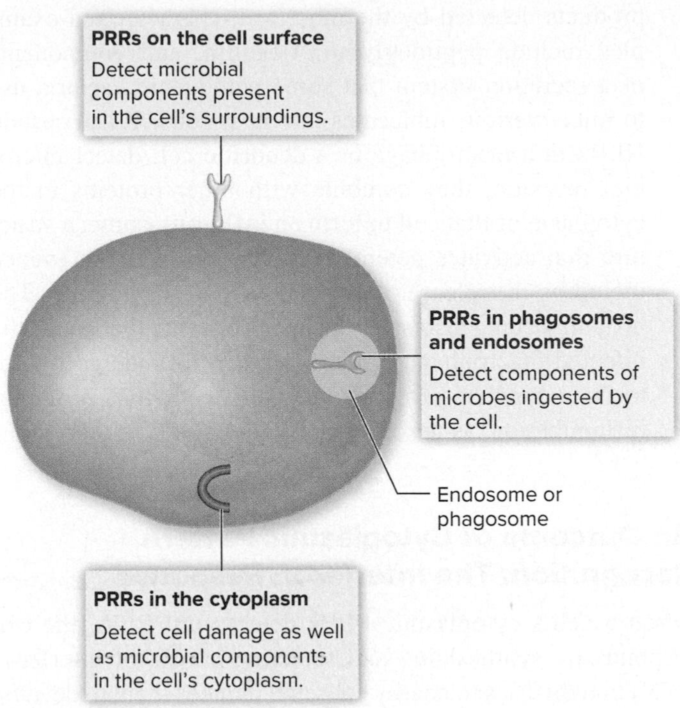
PRRs on the cell surface
Detect microbial components present in the cell's surroundings.
PRRs in phagosomes and endosomes
Detect components of microbes ingested by the cell.
Endosome or phagosome
PRRs in the cytoplasm
Detect cell damage as well as microbial components in the cell's cytoplasm.

FIGURE 14.7 Cellular Locations and Corresponding Roles of Pattern Recognition Receptors (PRRs)

? Why would a macrophage produce pro-inflammatory cytokines when its PRRs detect bacterial components?

receptors are extracellular soluble molecules. These soluble molecules bind to certain PAMPs and DAMPs on cells, thus targeting the cells for elimination by phagocytes or other components of the immune response.

Pattern Recognition Receptors (PRRs) That Monitor a Cell's Surroundings

Sentinel cells such as phagocytes and cells that line blood vessels and other sterile body sites have PRRs on their surfaces (anchored in their cytoplasmic membranes), allowing the cells to detect invaders in the surrounding environment. The most well-characterized of these PRRs are the **toll-like receptors (TLRs).** A number of different TLRs have been described (at least 10 in humans); those on the cell surface are considered here, but others (discussed shortly) are in endosomes and phagosomes. Each TLR recognizes a distinct compound or group of compounds associated with microbes (**figure 14.8**). TLRs on the cell surface generally detect components of the outermost layers of microbial cells, including lipopolysaccharide (LPS), lipoproteins, and flagellin. In addition, a group of cell-surface PRRs called C-type lectin receptors (CLRs) bind to certain carbohydrate molecules often found on microorganisms. Dendritic cells have TLRs as well as CLRs, so they can gather a great deal of information about invaders they encounter. They then pass that information on to lymphocytes, thereby helping to shape the adaptive immune response.

MicroByte

"Toll-like" reflects the reaction of a researcher who exclaimed "toll"—German slang for "awesome"—upon discovering the gene for a similar receptor in a fruit fly.

Pattern Recognition Receptors (PRRs) That Monitor Material Ingested by a Cell

Phagocytic cells have PRRs in their phagosomes and endosomes, allowing the cells to inspect ingested materials; these TLRs are anchored in the phagosomal and endosomal membranes, oriented so that they face the contents in the interior (lumen) of the compartments. The TLRs typically recognize characteristics of nucleic acids that indicate a microbial origin. Although it might seem surprising that a cell can recognize microbial nucleic acid, several features distinguish it from normal host cell DNA or RNA. For one thing, certain nucleotide sequences are much more common in bacterial or viral nucleic acids than in normal host cell nucleic acids. In addition, the genome of RNA viruses is often double-stranded (dsRNA) during the viral replication cycle, whereas host cell RNA is single-stranded (ssRNA). Even DNA viruses may generate long dsRNA because the genes sometimes overlap

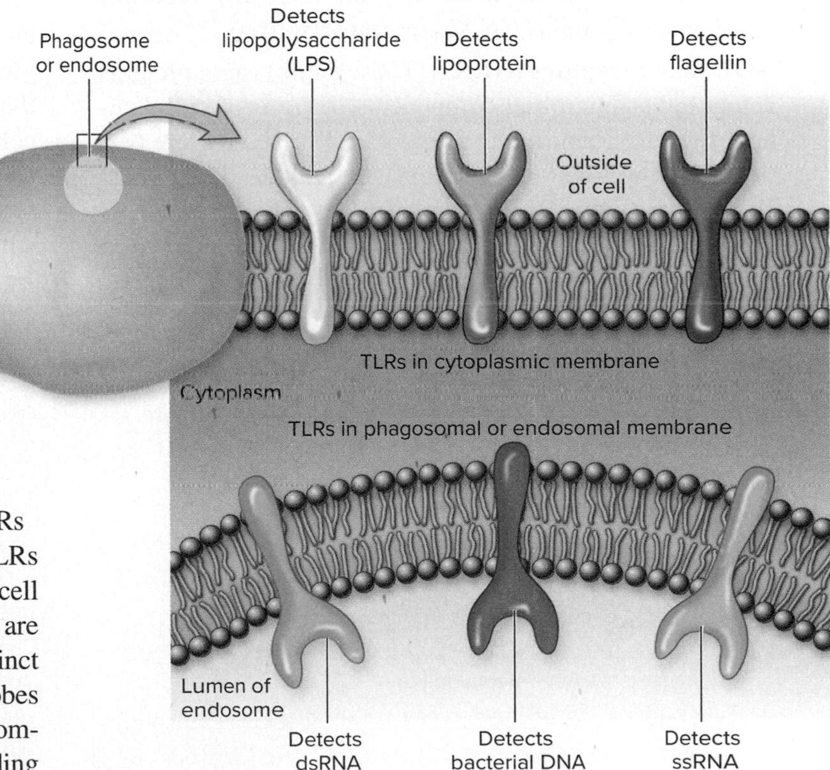
Phagosome or endosome
Detects lipopolysaccharide (LPS)
Detects lipoprotein
Detects flagellin
Outside of cell
TLRs in cytoplasmic membrane
Cytoplasm
TLRs in phagosomal or endosomal membrane
Lumen of endosome
Detects dsRNA
Detects bacterial DNA
Detects ssRNA

FIGURE 14.8 Toll-Like Receptors (TLRs) These pattern recognition receptors are anchored in membranes of sentinel cells, allowing these cells to "see" microbial compounds that originated outside the cell. Not all of the compounds recognized by TLRs are shown in this figure.

? From the standpoint of defending the human body, why would it be beneficial for the cells that line the blood vessels to have TLRs on their surface?

and are oriented in different directions, such that both strands of DNA may be used as templates for transcription, leading to production of complementary mRNA molecules; this is not a characteristic of host cell genes. Certain TLRs recognize these distinct differences between host and microbial nucleic acids.

Pattern Recognition Receptors (PRRs) That Monitor a Cell's Cytoplasm

Most host cells have PRRs in their cytoplasm, allowing the cells to monitor their own cytoplasmic contents for signs of invasion. Examples of cytoplasmic PRRs include (**figure 14.9**):

- **RIG-like receptors (RLRs).** These cytoplasmic proteins detect viral RNA and are found in most cell types. Because RLRs are so widespread, they represent a very important early-warning system for viral infections—nearly any virally infected cell can alert neighboring cells that a virus is present. RLRs can distinguish viral RNA from normal cellular RNA because of at least two of its distinct characteristics: (1) viral RNA is often double-stranded (dsRNA) and (2) viral RNA often lacks a cap (recall that a process called capping normally modifies the 5′ end of cellular RNA after transcription).
- **NOD-like receptors (NLRs).** These cytoplasmic proteins detect either microbial components or signs of cell damage; they are found in a variety of cell types but are particularly important in macrophages and dendritic cells. Over 20 different NLRs have been described, but details about their roles are still being uncovered; the microbial products detected by the most well-characterized examples include peptidoglycan, flagellin, and components of a secretion system that some pathogenic bacteria use to inject various molecules into host cells. When certain NLRs in a macrophage or a dendritic cell detect microbial invasion, they combine with other proteins in the cytoplasm of that cell to form an **inflammasome,** a structure that activates potent inflammation-inducing events, including the release of pro-inflammatory cytokines. The inflammasome also starts a chain of events that cause the infected macrophage or dendritic cell to undergo pyroptosis, a type of programmed cell death that promotes inflammation, as described in section 14.8).

An Outcome of Cytoplasmic Pattern Recognition: The Interferon Response

When a cell's cytoplasmic PRRs detect viral RNA, the cell responds by synthesizing and secreting a type of interferon (IFN) that diffuses to nearby cells and induces them to develop an antiviral state (**figure 14.10**). The IFN molecules attach to specific receptors on cells, causing the cells to express what can be viewed as inactive "suicide enzymes." For convenience, we will refer to these enzymes collectively as inactive antiviral proteins (iAVPs). These iAVPs can be activated by viral dsRNA to become antiviral proteins (AVPs). The AVPs then degrade mRNA and stop protein synthesis, inducing the cell to self-destruct—a type of programmed cell death called apoptosis, described in section 14.8. A key feature of interferon response is that the iAVPs are activated by long dsRNA, which is typically found only in virally infected cells. Thus,

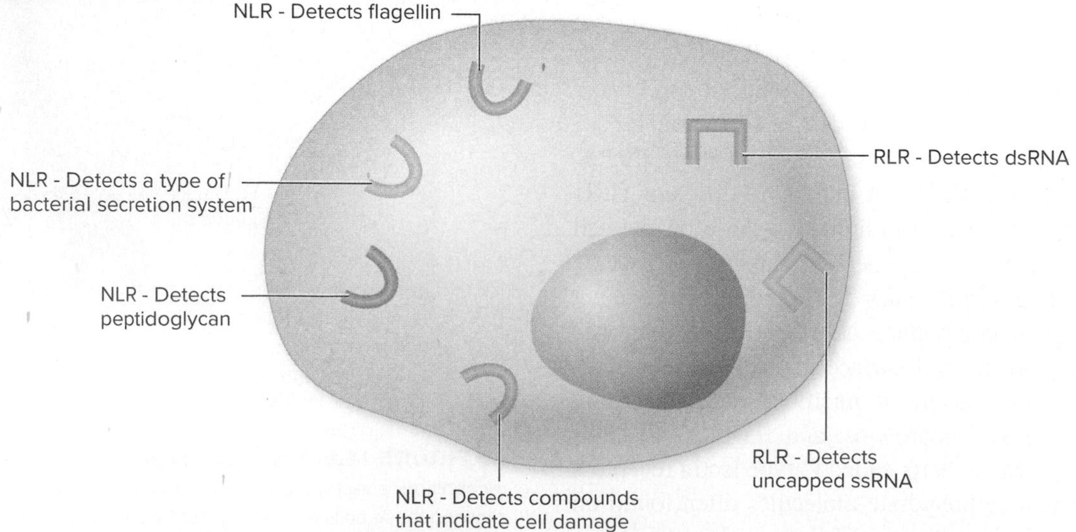

FIGURE 14.9 NOD-Like Receptors (NLRs) and RIG-Like Receptors (RLRs) These pattern recognition receptors are found within cells and detect either microbial components or signs of cell damage. Not all NLRs and RLRs are shown in the illustration.

? **With respect to the source of microbial compounds detected, how do NLRs and RLRs differ from TLRs?**

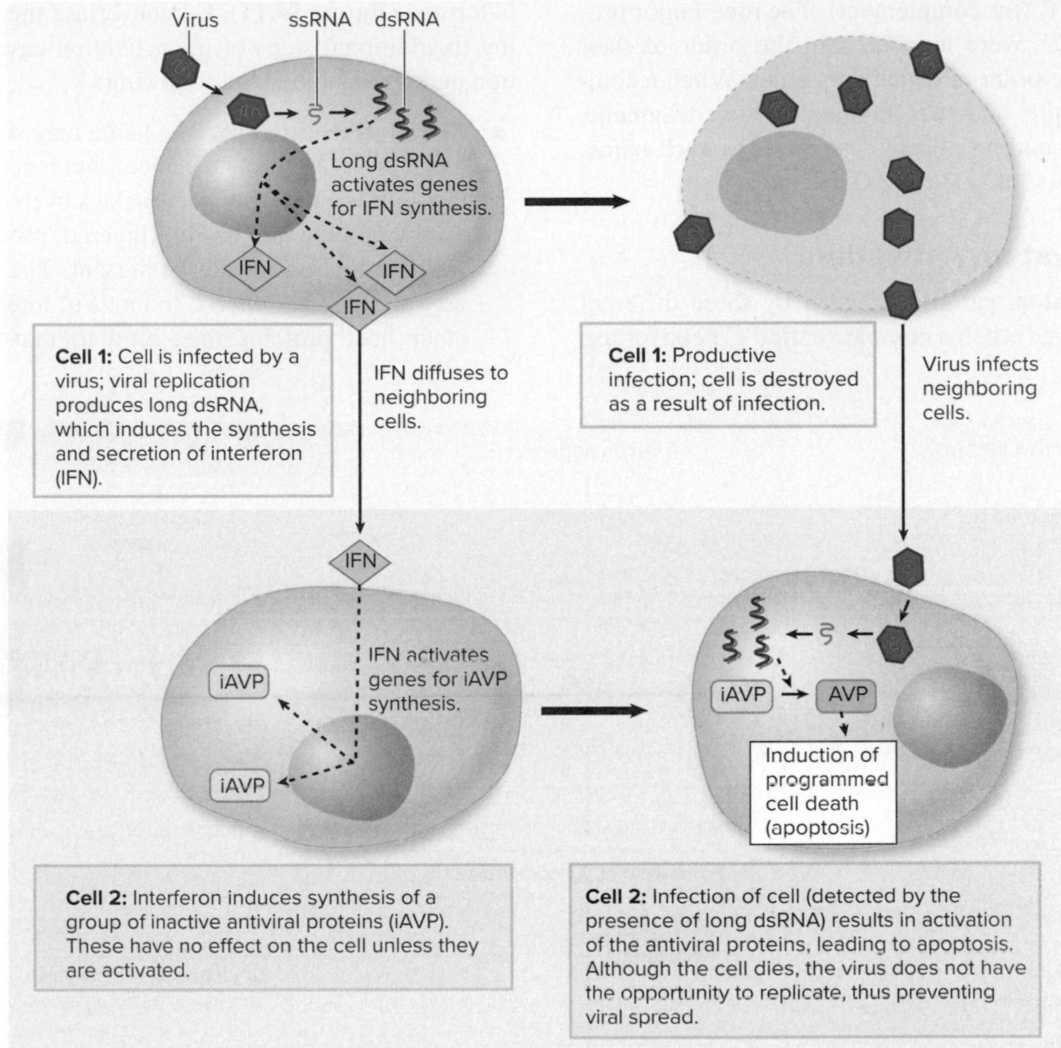

FIGURE 14.10 Antiviral Effects of Interferon

? **Why would it be beneficial to a host for a virally infected cell to undergo apoptosis?**

when interferon binds cells, only the infected cells will be sacrificed; their uninfected counterparts remain functional but are prepared to undergo apoptosis should they become infected.

MicroAssessment 14.5

Sentinel cells use pattern recognition receptors (PRRs) to detect microbial components in the surroundings and in ingested material. Many cells use PRRs to determine if they are infected. Viral RNA triggers an interferon response.

13. Give three examples of PAMPs (also called MAMPs).
14. If a cell produces antiviral proteins (AVPs), what happens to that cell when those proteins encounter long dsRNA?
15. Why would the discovery of TLRs alter the view that innate immunity is non-specific?

14.6 ■ The Complement System

Learning Outcome

8. Describe the three pathways that lead to complement system activation and the three outcomes of activation.

The **complement system,** often simply referred to as complement, is a series of proteins that routinely circulate in the blood and the fluid that bathes the tissues. The proteins are produced in an inactive form, but certain signals that indicate the presence of microbial invaders start a reaction cascade that rapidly activates the system. The activated complement proteins have specialized functions that help remove and destroy the invader.

The name of the complement system is derived from the observation that the system "complements" the function of antibodies, a component of adaptive immunity. Each of the major complement system proteins has been given a number along with the letter C (for complement). The nine major proteins, C1 through C9, were numbered in the order of their discovery and not the order in which they react. When a complement protein is split into two fragments, those fragments are distinguished by adding a lowercase letter to each name. For example, C3 splits into C3a and C3b.

Complement System Activation

The complement system can be activated by three different pathways that converge when a complex called C3 convertase is formed **(figure 14.11)**. C3 convertase then splits C3, leading to additional steps of the activation cascade. The activation pathways include the following:

- **Alternative pathway.** The name may seem to imply that the pathway is "second choice," but it actually reflects the fact that the pathway was not discovered first. The pathway is quickly and easily triggered, providing vital early warning that an invader is present. The alternative pathway is triggered when C3b binds to foreign cell surfaces; other host proteins then bind to that C3b, eventually

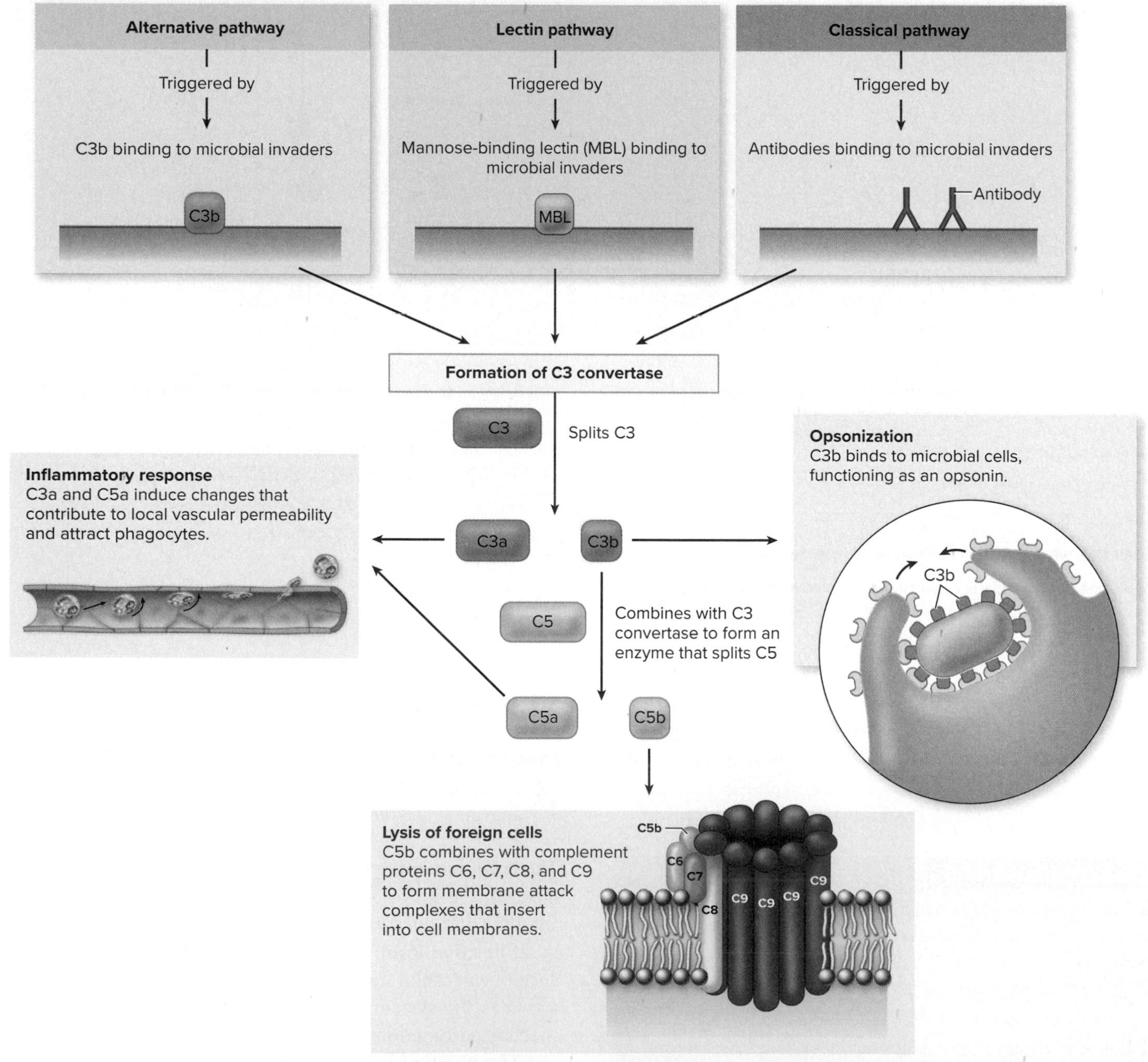

FIGURE 14.11 Complement System The three pathways of complement system activation converge with the formation of C3 convertase, leading to the same three outcomes (the inflammatory response, lysis of foreign cells, opsonization). Not all of the steps in these pathways are shown.

? How can C3b be both a product of complement activation and an activator of the complement system?

forming the C3 convertase. What might seem confusing is the fact that C3b is a product of complement activation, yet it also triggers the alternative pathway. How can it be both a product and a trigger? This can occur because C3 is somewhat unstable, and at a low rate spontaneously splits even when the complement system has not been activated (this is referred to as *tickover*). The C3a and C3b formed this way are rapidly inactivated by regulatory proteins, but some C3b is always present to trigger the alternative pathway when needed.

- **Lectin pathway.** Activation of the complement system via this pathway involves a type of soluble pattern recognition molecule called mannose-binding lectin (MBL; also called mannan-binding lectin), which binds to certain arrangements of mannose, a monosaccharide commonly found on the surface of bacteria and fungi. Once an MBL attaches to a surface, it interacts with complement system components to form a C3 convertase.
- **Classical pathway.** Complement system activation by this pathway requires antibodies. When multiple antibody molecules bind to an antigen (forming an antigen-antibody complex, also called an immune complex), they interact with the same complement system components involved in the lectin pathway to form a C3 convertase.

Effector Functions of the Complement System

Activation of the complement system eventually leads to three major protective outcomes (see figure 14.11):

- **Opsonization.** The C3b concentration increases substantially when the complement system is activated, and these molecules bind to microbial cells or other foreign particles. This has two effects: (1) continued complement activation via the alternative pathway, and (2) **opsonization.** Material that has been opsonized (meaning "prepared for eating") is easier for phagocytes to bind to and engulf because the phagocytic cells have receptors that attach specifically to molecules referred to as **opsonins** (in this case, C3b).
- **Inflammatory response.** The complement component C5a is a potent chemoattractant, drawing phagocytes to the area where the complement system has been activated. In addition, C3a and C5a induce changes in the endothelial cells that line the blood vessels, contributing to the vessel leakiness associated with inflammation. They also cause mast cells to degranulate, releasing various pro-inflammatory cytokines and other chemicals that contribute to inflammation.
- **Lysis of foreign cells.** Complexes of complement system proteins (C5b, C6, C7, C8, and multiple C9 molecules) spontaneously assemble in cell membranes, forming doughnut-shaped structures called **membrane attack complexes (MACs)** (**figure 14.12**). This creates pores in the membrane, causing the cells to lyse. MACs have little effect on Gram-positive bacteria because the cells' thick peptidoglycan layer prevents the components from reaching their cytoplasmic membranes. In contrast, MACs damage both the outer and the cytoplasmic membranes of Gram-negative bacteria.

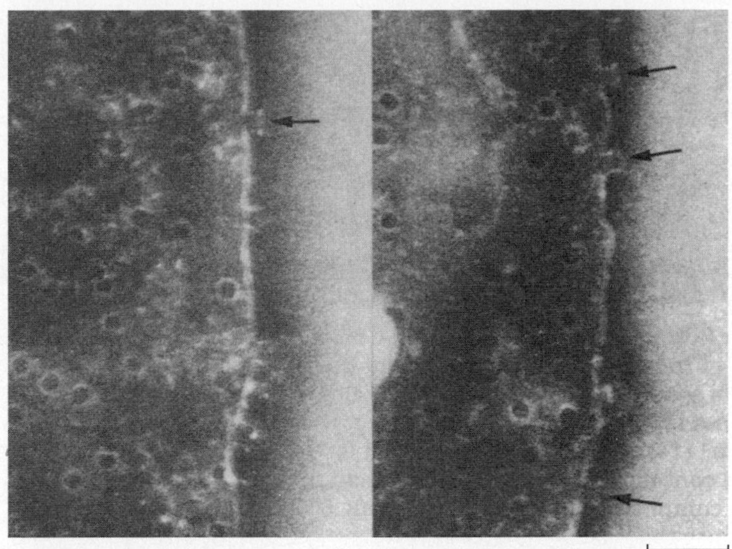

FIGURE 14.12 Membrane Attack Complexes (MACs) Electron micrographs of MACs in membranes of two different cells (arrows point to side views; dark dots are head-on views). Sucharit Bhakdi

? How do MACs cause cells to lyse?

MicroByte

Think of opsonization as coating a microbe with one of the two opposing strips of Velcro. In this analogy, phagocytes are coated with the other strip, so they can easily attach to microbes.

Regulation of the Complement System

Various control mechanisms protect host cells from the complement system. One example involves molecules in host cell membranes that bind C3b-inactivating regulatory proteins (**figure 14.13**). Not only does this prevent host cell surfaces from triggering the alternative pathway of complement regulation, it also prevents host cells from being opsonized. Most microbial cell surfaces do not bind the regulatory proteins, but as discussed in chapter 16, some pathogens have mechanisms to hijack the host's protective mechanisms.

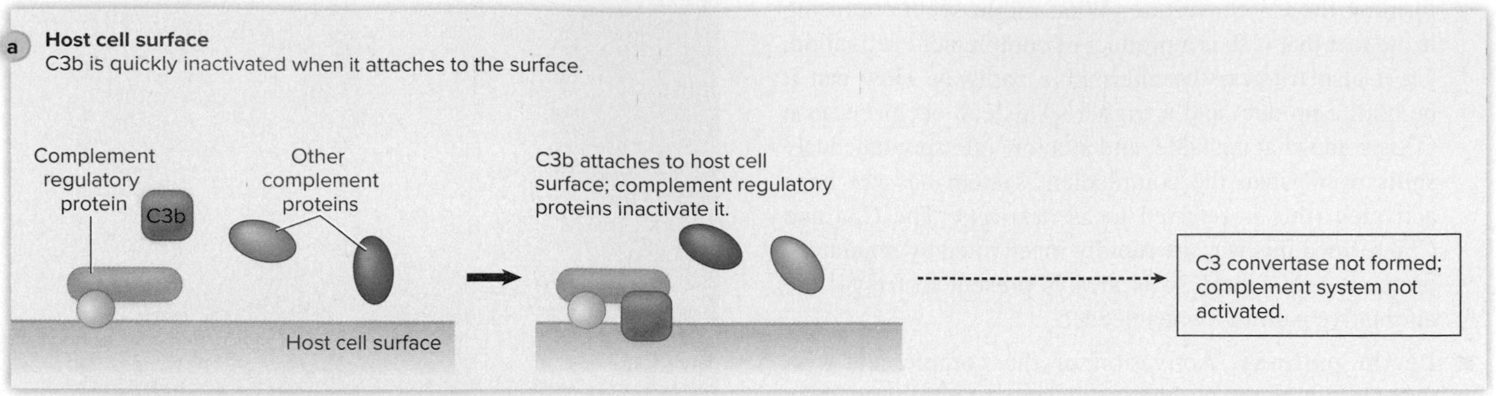
a
Host cell surface
C3b is quickly inactivated when it attaches to the surface.
Complement regulatory protein
Other complement proteins
C3b
Host cell surface
C3b attaches to host cell surface; complement regulatory proteins inactivate it.
C3 convertase not formed; complement system not activated.

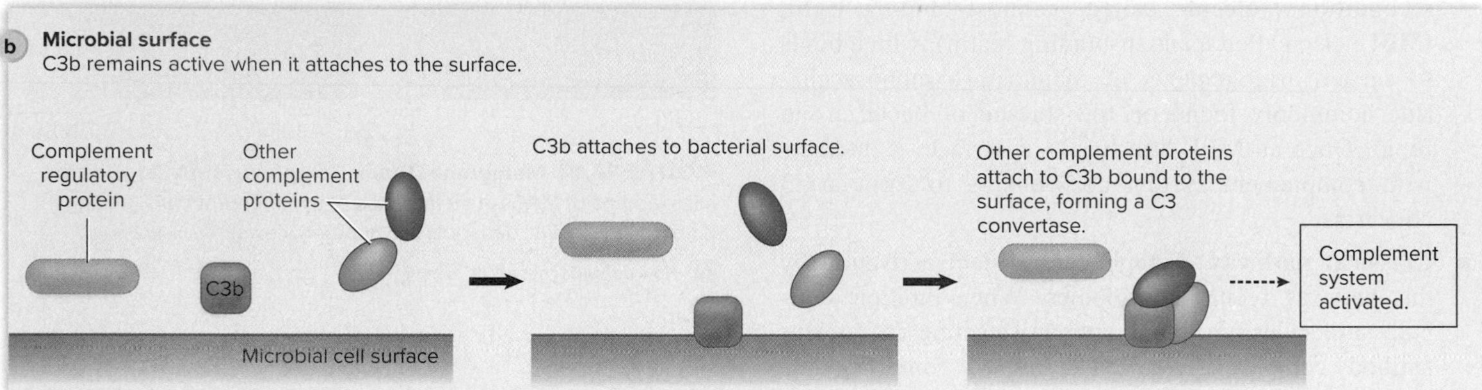
b
Microbial surface
C3b remains active when it attaches to the surface.
Complement regulatory protein
Other complement proteins
C3b
Microbial cell surface
C3b attaches to bacterial surface.
Other complement proteins attach to C3b bound to the surface, forming a C3 convertase.
Complement system activated.

FIGURE 14.13 Complement Regulatory Proteins Inactivate C3b (a) Molecules in host cell membranes bind regulatory proteins that quickly inactivate C3b. **(b)** Most microbial cell surfaces do not bind the regulatory proteins, so they trigger the alternative pathway of complement activation.

? **Some pathogens attract complement regulatory proteins to their surfaces. How would this help the pathogens avoid destruction?**

MicroAssessment 14.6

The complement system can be activated by three different pathways that each lead to opsonization, an inflammatory response, and lysis of foreign cells.

16. Describe the outcome of opsonization.
17. The body's own cells do not trigger the alternative pathway of complement system activation. Explain why this is so.
18. Some pathogens produce C5a peptidase, an enzyme that destroys C5a. How would this benefit the pathogen?

14.7 ■ Phagocytosis

Learning Outcomes

9. Outline the steps of phagocytosis.
10. Compare and contrast the roles of macrophages and neutrophils.

Phagocytes routinely engulf and digest material, including invading microbes. In routine situations, such as when microbes enter through a minor skin wound, resident macrophages in the tissues destroy the relatively few invaders that enter. If the microbes are not rapidly cleared, however, macrophages produce cytokines to recruit additional phagocytes—particularly neutrophils—for extra help.

The Steps of Phagocytosis

The steps of **phagocytosis** are particularly important medically because most pathogens have evolved the ability to evade one or more of them, a topic explored in chapter 16. The steps include the following (**figure 14.14**):

1. **Chemotaxis.** Phagocytes are recruited to the site of infection or tissue damage by chemicals that act as chemoattractants. These include products of microorganisms, materials released by injured host cells, chemokines, and the complement system component C5a.
2. **Recognition and attachment.** Various receptors on phagocytes bind invading microbes either directly or indirectly. Direct binding occurs when a phagocyte's receptors bind mannose molecules on the microbe's surface. Indirect binding happens when a phagocyte binds **opsonins** that have attached to the invader. Recall that opsonins bind to microbes and other foreign particles, making it easier for phagocytes to attach to and then engulf them; opsonins include the complement component C3b, some

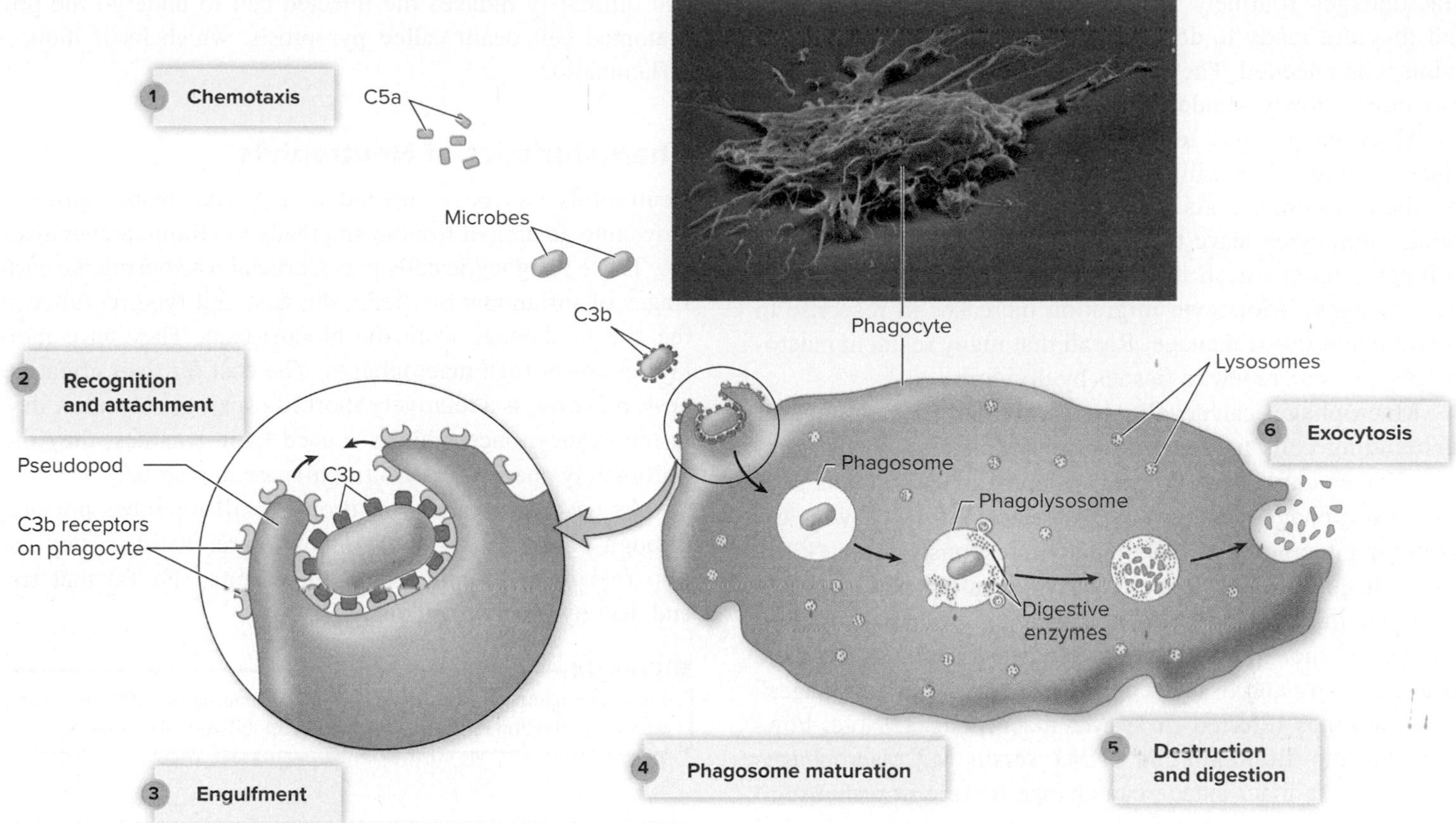

FIGURE 14.14 Phagocytosis This diagram shows a microbe that has been opsonized by the complement protein C3b; certain classes of antibodies can also function as opsonins. Eye of Science/Science Source

? **What would happen if a bacterium prevented phagosome maturation?**

soluble pattern recognition molecules, and certain classes of antibody molecules.

③ **Engulfment.** Once a phagocyte has attached to a particle, it sends out pseudopods that surround and engulf the material. This action brings the material into the cell, enclosed in a **phagosome** (see figure 3.23). If a phagocyte encounters an invader too large to engulf, it may release its toxic contents as a means of destroying it.

④ **Phagosome maturation.** Initially, a phagosome has no antimicrobial capabilities, but it matures to develop these; for example, the pH within the phagosome becomes progressively more acidic. Maturation is a complex, highly regulated process, with events tailored to the type of material ingested; if a phagocyte's toll-like receptors (TLRs) indicate that a phagosome contains microbial components, for example, then that phagosome will have a different fate than if the contents are only normal host debris. As part of the process, the phagosome fuses with enzyme-filled lysosomes (and granules, in the case of neutrophils), and in doing so becomes a **phagolysosome.**

⑤ **Destruction and digestion.** A number of factors within the phagolysosome work together to destroy an engulfed invader, and these differ somewhat between macrophages and neutrophils. In general, however, the many enzymes contributed by the lysosomes (or granules) degrade various bacterial cell components, including peptidoglycan. Host defense peptides (antimicrobial peptides) damage the invader's membranes, and lactoferrin limits microbial growth by binding iron. Meanwhile, special pumps move protons into the phagolysosome, further lowering the pH, and an enzyme converts O_2 to toxic reactive oxygen species (ROS). Another enzyme makes nitric oxide, which reacts with ROS to produce additional toxic compounds. The net outcome of these actions is a very toxic environment for any ingested microbes.

⑥ **Exocytosis.** The phagolysosome releases undigested debris to the outside of the cell by fusing with the phagocyte's cytoplasmic membrane (see figure 3.24). As you will learn in chapter 15, a special process in macrophages puts some of the ingested material on the cell's surface as a way of displaying bits of invaders to certain cells of the adaptive immune system.

Characteristics of Macrophages

Macrophages are the everyday protectors of tissues, playing an essential role in every major tissue in the body. As an analogy, they would be the "beat cops" that protect city streets.

Macrophages routinely phagocytize dead cells and debris, and they are ready to destroy invaders and call in reinforcements when needed. They are always present in tissues, where they either slowly wander or remain stationary.

Macrophages live for weeks to months, or even longer. They maintain their killing power by continually regenerating their lysosomes. As wandering macrophages die, circulating monocytes leave the blood and migrate to the tissues to replace them (recall that monocytes can differentiate into macrophages). Monocyte migration increases in response to invasion and tissue damage. Recall that many resident macrophages can self-renew in tissues by dividing.

Macrophages can develop into **activated macrophages** if surrounding cells produce certain cytokines or other chemicals. For example, certain pro-inflammatory cytokines activate macrophages, giving them greater killing power and promoting a continued inflammatory response; these "killer macrophages" are sometimes referred to as M1 macrophages. Other chemicals cause macrophages to produce cytokines that lessen the inflammatory response, as well as promote wound healing and tissue repair; these "healer macrophages" are sometimes referred to as M2 macrophages. Note, however, that the differentiation of M1 versus M2 macrophages is not fixed; a macrophage can change its role in response to environmental cues, so presenting the types as two polarized alternatives is not entirely accurate. Nevertheless, the relative activities of macrophages seem to drive certain disease states. In type 2 diabetes, for example, an overabundance of inflammation-promoting (M1) macrophages in the adipose tissue decreases the sensitivity of various tissues to insulin. In contrast, anti-inflammatory macrophages (M2) promote an environment where tissue cells respond to insulin.

If activated macrophages fail to destroy microbes, the phagocytes can fuse together to form **giant cells.** Macrophages, giant cells, and T cells form concentrated groups called **granulomas** that wall off and retain organisms or other material that cannot be destroyed; again, this is an example of the cooperation between defense systems. Granulomas, which are part of the disease process in tuberculosis and several other illnesses, prevent the microbes from escaping to infect other cells (see figure 21.20). Unfortunately, they also harm the host because they interfere with normal tissue function.

In addition to their phagocytic activities, macrophages are very important sentinel cells. They are well equipped with pattern recognition receptors (PRRs), allowing them to "see" microbes in the surrounding environment, in material they have ingested, and in their cytoplasm. If a macrophage detects a microbe that has invaded the body, it produces cytokines that alert and stimulate various other cells of the immune system. If a macrophage's cytoplasmic PRRs detect microbial components, indicating that the phagocytic cell itself is infected, an inflammasome forms; recall that this cytoplasmic protein structure triggers a strong inflammatory response and ultimately induces the infected cell to undergo the programmed cell death called pyroptosis, which itself induces inflammation.

Characteristics of Neutrophils

Neutrophils can be compared to a SWAT team—quick to move into an area of trouble and ready to eliminate the invaders. These phagocytic cells play a crucial role during the early stages of inflammation, being the first cell type recruited to the site of damage from the bloodstream. They have more killing power than macrophages. The cost for their effectiveness, however, is a relatively short life span of only a few days in the tissues; once they have used their granules, they die. Fortunately, many more neutrophils are in reserve.

As described earlier, neutrophils kill microbes not only through phagocytosis, but also by degranulating. They can also release neutrophil extracellular traps (NETs) that trap and destroy microorganisms.

MicroByte

For every neutrophil in the circulatory system, about 100 more are waiting in the bone marrow, ready to be mobilized when needed.

MicroAssessment 14.7

The process of phagocytosis includes chemotaxis, recognition and attachment, engulfment, phagosome maturation, destruction and digestion, and exocytosis. Macrophages are long-lived and are always present in tissues; they can be activated to have more power. Neutrophils are highly active, short-lived cells that must be recruited to the site of damage.

19. How does a phagolysosome differ from a phagosome?
20. Tuberculosis is characterized by granulomas called tubercles. What is a granuloma?
21. What could a microorganism do to avoid engulfment by a phagocyte?

14.8 ■ The Inflammatory Response

Learning Outcomes

11. Describe the vascular and cellular changes associated with inflammation.
12. Compare and contrast apoptosis, pyroptosis, and necroptosis.

When microbes are introduced into normally sterile body sites, or when tissues are damaged, **inflammation** occurs. The purpose is to contain a site of damage, localize the reaction, eliminate the invader, and restore tissue function. Everyone has experienced inflammation; in fact, the Roman physician Celsus described what are referred to as its four cardinal signs about 2,000 years ago: swelling, redness, heat, and pain. A fifth sign, loss of function, is sometimes present.

The inflammatory response is a complex and highly regulated process that involves various cell types that recognize microbial invasion or tissue damage. Recall that many cells—particularly sentinel cells such as mast cells and macrophages—have various pattern recognition receptors (PRRs) that allow them to detect pathogen-associated molecular patterns (PAMPs) and damage-associated molecular patterns (DAMPs). When a mast cell detects tissue damage, for example, it degranulates to release histamine and various other **inflammatory mediators** (a collective term for signaling molecules involved in the inflammatory response); some of these induce changes in local blood vessels, such as increased permeability. Likewise, when a macrophage detects microbial products, it produces various pro-inflammatory cytokines (another group of inflammatory mediators); some of these induce the liver to synthesize acute-phase proteins, a group of proteins that facilitate phagocytosis and complement activation. Meanwhile, microbial surfaces trigger complement activation, also leading to an inflammatory response. If blood vessels are injured, two enzymatic cascades are activated. One is the coagulation cascade, which results in blood clotting, and the other produces several molecules that increase blood vessel permeability.

MicroByte

Aspirin and other non-steroidal anti-inflammatory drugs (NSAIDs) interfere with the production of prostaglandins, a group of inflammatory mediators.

The Inflammatory Process

The inflammatory process involves a series of events that result in dilation of small blood vessels, leakage of fluids from those vessels, and the migration of leukocytes out of the bloodstream and into the tissues (**figure 14.15**). Two general types of changes occur during the process:

- **Vascular changes.** The diameter of local blood vessels increases due to the action of histamine and other inflammatory mediators. This results in greater blood flow to the area, causing the heat and redness associated with inflammation. It also slows the blood flow in the capillaries. At the same time, changes in the endothelial cells that line the capillaries create small gaps in the normally tight junctions between the cells, increasing the vascular permeability and thus allowing more fluid to leak from the blood vessels and into the tissue. This protein-rich fluid, referred to as exudate, contains transferrin, complement system proteins, antibodies, and other substances that help counteract invading microbes. The accumulation of exudate in the tissues causes the swelling and pain associated with inflammation. Pain also results from the direct effect of certain inflammatory mediators on sensory nerve endings.
- **Cellular changes.** Some of the pro-inflammatory cytokines cause endothelial cells of the local blood vessels to produce adhesion molecules that loosely "grab" phagocytes in the bloodstream. The phagocytes normally flow rapidly through the vessels but slowly tumble to a halt as the adhesion molecules attach to them. The phagocytic cells themselves begin producing a different type of adhesion molecule that strengthens the attachment. Then, in response to various chemoattractants, the phagocytes leave the blood vessels and move into the surrounding tissues. They do this by squeezing between the cells of the dilated vessel, a process called diapedesis. Neutrophils are the first to arrive at the site of infection, and they actively phagocytize foreign material. Monocytes (which mature into macrophages at the site of infection) and lymphocytes arrive later. Clotting factors in the exudate initiate clotting reactions in the surrounding area, walling off the site of infection; this helps prevent bleeding and stops the spread of invading microbes. As the inflammatory process continues, large quantities of dead neutrophils accumulate. Those dead cells, along with tissue debris, make up pus. A localized collection of pus within a tissue is called an abscess (see figure 22.2).

The extent of inflammation varies depending on the nature of the injury, but the response is typically localized, begins immediately upon injury, and increases rapidly. A short-term inflammatory response is called **acute inflammation** and is characterized by an abundance of neutrophils. When the infection is brought under control, resolution of inflammation occurs. Neutrophils stop entering the area, and macrophages clean up the damage by ingesting dead cells and debris. As the area heals, new capillaries grow, destroyed tissues are replaced, and scar tissue forms.

If the body's defenses cannot limit the infection, **chronic inflammation** results. This is a long-term inflammatory process that can last for years. In chronic inflammation, macrophages and giant cells accumulate, and granulomas form.

Damaging Effects of Inflammation

The inflammatory process can be compared to a sprinkler system that prevents fire from spreading in a building. Although the process usually limits damage and restores tissue function, the actions themselves—including increased leakiness of blood vessels and the release of enzymes and toxic products contained within phagocytic cells—can cause significant damage. If inflammation is limited, such as in a response to a cut finger, the damage is usually minimal and can be repaired by the normal healing processes. In a delicate system, however, such as the meninges (membranes that surround the brain and spinal cord), the consequences can be severe, even life-threatening. As you learn more about infectious diseases, you will notice that many of the severe effects

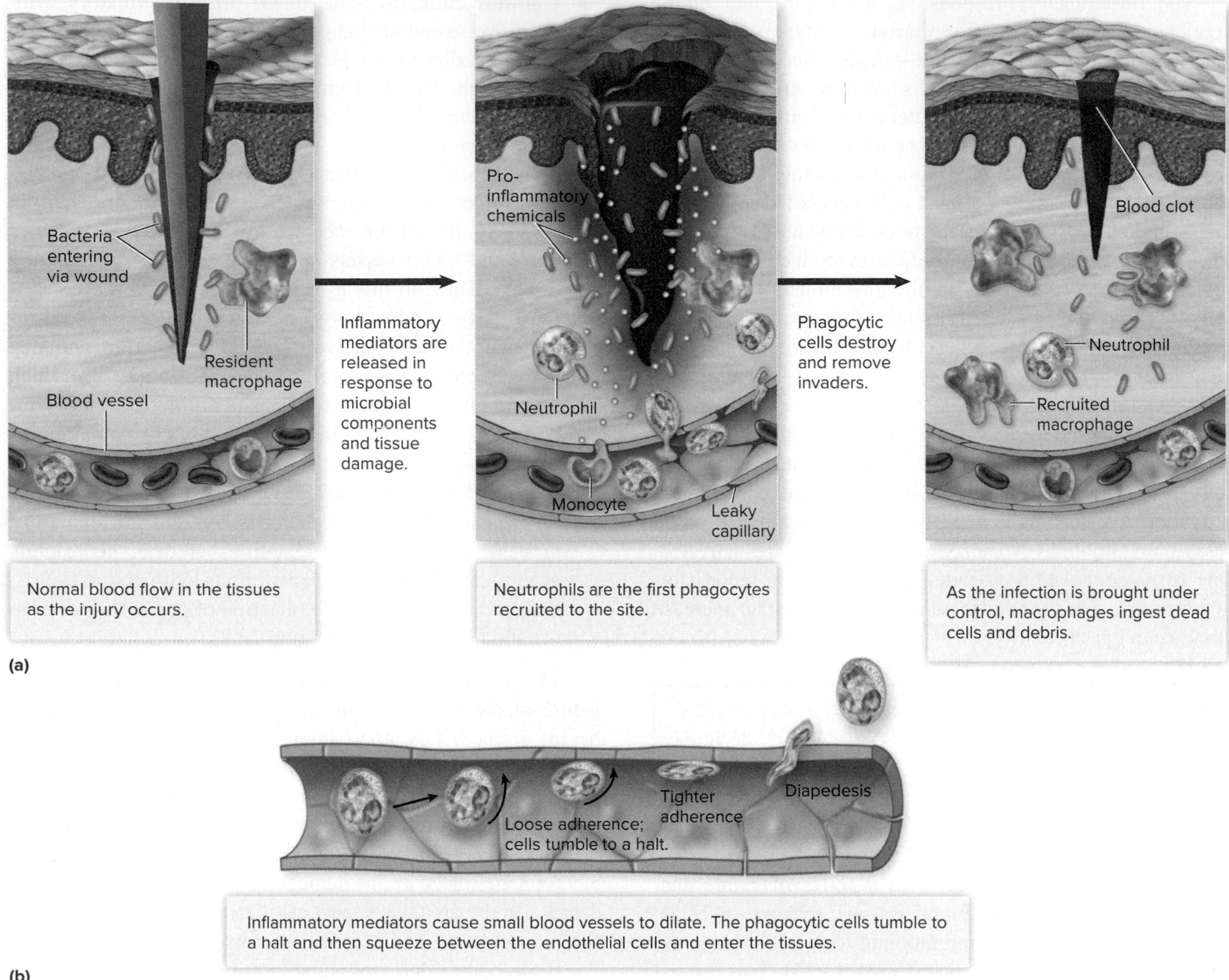
Bacteria entering via wound
Resident macrophage
Blood vessel
Normal blood flow in the tissues as the injury occurs.
Inflammatory mediators are released in response to microbial components and tissue damage.
Pro-inflammatory chemicals
Neutrophil
Monocyte
Leaky capillary
Neutrophils are the first phagocytes recruited to the site.
Phagocytic cells destroy and remove invaders.
Blood clot
Neutrophil
Recruited macrophage
As the infection is brought under control, macrophages ingest dead cells and debris.
(a)
Loose adherence; cells tumble to a halt.
Tighter adherence
Diapedesis
Inflammatory mediators cause small blood vessels to dilate. The phagocytic cells tumble to a halt and then squeeze between the endothelial cells and enter the tissues.
(b)

FIGURE 14.15 The Inflammatory Response (a) The process of inflammation. **(b)** Phagocytes leave the blood vessels and move to the site of infection.

? Which type of phagocyte is the first to be recruited to a site of inflammation?

of infection result from the inflammatory response. The suffix *-itis* in many disease names means inflammation—for example, meningitis (inflammation of the meninges), encephalitis (inflammation of the brain), colitis (inflammation of the colon), and gingivitis (inflammation of the gums around the base of the teeth). As you encounter those terms, consider the potential consequences of inflammation in those body sites.

Inflammation triggered throughout the body, such as that which can occur during a bloodstream infection, can be devastating. The response may result in widespread capillary dilation and leakiness, leading to inadequate blood flow to the organs. It may also trigger an excessive release of pro-inflammatory cytokines—a cytokine storm—resulting in further damage that amplifies the problem. The net result of a systemic infection is sometimes **sepsis,** a life-threatening condition characterized by organ damage due to a dysfunctional immune response to an infection (see figure 25.3). If the blood pressure becomes too low support adequate blood flow to organs, a condition called **shock** develops; shock resulting from sepsis is referred to as **septic shock.**

Chronic inflammatory conditions, particularly autoimmune diseases, lead to unwarranted damaging responses. For instance, the inflammation associated with the autoimmune disease rheumatoid arthritis can eventually lead to disabling joint damage (see figure 18.7). Chronic inflammation is also associated with diabetes, some cancers, certain types of obesity, and various other conditions. The triggers for the inflammation in these cases are not well understood, nor is it known if inflammation is a cause, effect, or simply a correlation, so scientists are currently trying to understand the complex factors involved.

FOCUS YOUR PERSPECTIVE 14.1

For *Schistosoma*, the Inflammatory Response Delivers

Schistosoma species, the parasitic flatworms that cause the disease schistosomiasis (also called bilharzia), use the immune response to assist them in completing one portion of their complex life cycle.

A person can become infected with schistosomes by wading or swimming in water that contains a larval form of the parasite called cercariae, which are released from infected snails (see figure 12.22a). Cercariae penetrate skin by burrowing through with the aid of digestive enzymes. They then move into the bloodstream, where they mature into adult worms that can live for over 25 years. Adult worms mask themselves from the immune system by coating themselves with various blood proteins, an ability that provides them with a primitive stealth "cloaking device."

Schistosoma species have separate sexes, and the male and female worms manage to locate each other in the bloodstream. The male's body has a deep longitudinal groove in which he clasps his female partner in life-long embrace (schistosoma means "split-body," referring to the long slit) (see figure 12.22b). To reproduce, the worms migrate to the tiny veins of either the intestines or the bladder (depending on the schistosome species), where the female lays hundreds of ova per day (see figure 12.22c).

In contrast to the adult schistosomes—which effectively hide from the immune system—the eggs provoke a strong inflammatory response. This pushes the eggs to the closest body surface, in a manner similar to what is experienced as a sliver in the skin works its way out. In the case of schistosome species that deposit ova in veins near the intestine, the eggs are pushed out into the intestinal tract, where they are eliminated in feces. Ova of the species that deposits its eggs near the bladder are pushed into the bladder, where they are eliminated in urine. If untreated sewage that contains schistosome eggs reaches fresh water, the ova can hatch, releasing a ciliated larval stage that infects and multiplies asexually in a specific freshwater snail host. The infected snail then releases large numbers of cercariae, which can infect a human host to complete the parasite's life cycle.

The symptoms of schistomiasis are due to the many ova that are not expelled. If these ova are swept to the liver by the bloodstream, the resulting inflammatory process and granuloma formation gradually destroy liver cells. The cells are replaced with scar tissue, causing the liver to malfunction. In turn, this results in a fluid buildup in the abdominal cavity, as well as malnutrition. Chronic schistosomiasis can also damage the lungs and bladder, and occasionally, the central nervous system.

Despite their complex life cycle, *Schistosoma* species are highly successful parasites. Not only do they avoid certain immune responses that would otherwise lead to their destruction, but they also exploit the inflammatory response for their own spread.

Programmed Cell Death and the Inflammatory Response

In addition to traumatic cell death (necrosis) that results from tissue damage, host cells can self-destruct, an action referred to as programmed cell death. This capability allows the body to eliminate any cells no longer needed, and it also serves as a mechanism for sacrificing "self" cells that might otherwise spread an infection. One type of programmed cell death avoids an inflammatory response, whereas other types promote one.

Apoptosis (*apo* means "off"; *ptosis* means "falling") is a programmed cell death that does not trigger inflammation. During apoptosis, the dying cells undergo certain changes: The cell shape changes, enzymes cut the DNA, and pieces of the cell bud off, effectively shrinking the cell. Some changes appear to serve as a signal to macrophages that the remains of the cell are to be engulfed without the events associated with inflammation.

In contrast to apoptosis, **pyroptosis** (*pyro* means "fire") and **necroptosis** (*necro* means "tissue death") are types of programmed cell death that induce an inflammatory response. The triggers and the signaling pathways of these two types of cell death differ, but the outcome is the same: The infected cell is sacrificed so that it cannot play host to an invader, and its death releases damage-associated molecular patterns (DAMPs) and cytokines. As a result, various components of the immune system are recruited to the region.

MicroAssessment 14.8

The inflammatory response is initiated when microbes invade or tissues are damaged. The outcome is dilation of small blood vessels, leakage of fluids from those vessels, and migration of leukocytes out of the bloodstream and into the tissue. Inflammation helps contain an infection, but the response itself can be damaging. Apoptosis destroys "self" cells without initiating inflammation; pyroptosis and necroptosis both trigger an inflammatory response.

22. How do the vascular changes associated with inflammation assist the cellular changes?
23. How is apoptosis different from pyroptosis?
24. Infection of the fallopian tubes can lead to infertility. Why would this be so?

14.9 ■ Fever

Learning Outcome

13. Describe the induction and outcomes of fever.

Fever is an important host defense mechanism and a strong indication of infectious disease, particularly a bacterial one. Human body temperature is normally kept around 37°C by a

FOCUS ON A CASE 14.1

A 9-year-old boy with cystic fibrosis—a genetic disease that causes a number of problems, including the buildup of thick, sticky mucus in the lungs—complained of feeling tired, out of breath, and always coughing. When his mother took him to the doctor, she mentioned that his cough was productive, meaning that it contained sputum (pronounced *spew-tum*). She was particularly concerned that the sputum was a blue-green color. His doctor immediately suspected a *Pseudomonas aeruginosa* lung infection—a common complication of cystic fibrosis. A sputum sample was collected and sent to the clinical laboratory.

In the clinical laboratory, the sample was plated onto MacConkey agar and blood agar and incubated. Mucoid colonies surrounded by a bluish-green color grew on both types of agar media. The colonies on MacConkey had no pink coloration, so the medical technologist concluded that the cells did not ferment lactose. She noted the blue-green color on the agar plates and in the sputum, knowing that *P. aeruginosa* makes several pigmented compounds that give rise to colors ranging from yellow to blue. One of the pigments functions as a siderophore, which is a molecule that binds iron. Another is important for biofilm formation. Further testing showed that the bacterium was an oxidase-positive, Gram-negative rod, consistent with the physician's initial suspicions.

The patient was treated with antibiotics, with only limited success. Like most cystic fibrosis patients, he developed a chronic lung infection that required repeated treatments.

1. What role did cystic fibrosis play in the disease process?
2. What is the significance of the mucoid phenotype of the colonies?
3. How would the siderophore (the iron-binding compound) benefit the bacterium?
4. Why would the boy's lung infection make his pre-existing respiratory problems even worse?

Discussion

1. Cystic fibrosis patients often have an accumulation of thick mucus in their lungs, which interferes with the mucociliary escalator and other first-line defenses. With a compromised (weakened) mucociliary escalator, inhaled microbes are not easily removed. In addition, the accumulated mucus serves as a nutrient source for bacteria.
2. The mucoid colonies suggest that the bacterium produces an extracellular material that forms a capsule or a slime layer. This material, also referred to as extracellular polymeric substances (EPS), allows *Pseudomonas aeruginosa* cells to form biofilms. The biofilm protects the bacterial cells from various components of the immune system, including host defense peptides (antimicrobial peptides) and phagocytes. Bacteria growing within a biofilm are much more difficult for the immune system to destroy.
3. Siderophores help the bacterium obtain iron from the host. Recall that the body's iron-binding proteins (lactoferrin and transferrin) prevent microbes from using the host's iron supply and thereby limit their growth. Microorganisms that make siderophores essentially engage in a "tug-of-war" with the body over iron. That tug-of-war is especially important for *P. aeruginosa* because iron levels influence biofilm formation. When iron is limiting, *P. aeruginosa* cells are motile and do not initiate biofilm formation.
4. In response to a bacterial infection in the lungs, an inflammatory response develops. The capillaries in the lungs become leaky, allowing fluids from the blood to enter the local tissues; those fluids cover the respiratory surfaces, thus interfering with gas exchange. In addition, inflammation recruits neutrophils to the area, some of which release the destructive enzymes in their granules.

temperature-regulation center in the brain. During an infection, the regulating center "sets" the body's thermostat at a higher level. An oral temperature above 37.8°C is regarded as fever.

A higher temperature setting results when macrophages release pro-inflammatory cytokines in response to microbial products. The cytokines act as messages carried in the bloodstream to the brain, where the temperature-regulating center raises the body temperature in response. The rise in temperature prevents microbes with lower optimum temperatures from growing, giving the immune system time to eliminate them before they cause too much harm. A moderate fever has also been shown to enhance several protective processes, including the inflammatory response, phagocytic activity, lymphocyte multiplication, release of neutrophil-attracting substances, and interferon and antibody production. In addition, fever stimulates the release of leukocytes from the bone marrow into the bloodstream.

Fever-inducing cytokines and other substances are **pyrogens** (*pyro* means "fire" and *gen* means "to generate"). Pyrogenic cytokines are endogenous pyrogens, meaning the body makes them, whereas microbial products such as lipopolysaccharide (LPS) are exogenous pyrogens, meaning they are introduced from external sources.

MicroAssessment 14.9

Fever results when macrophages release pro-inflammatory cytokines; this occurs when macrophages detect microbial products.

25. What is a pyrogen?

26. How does fever inhibit the growth of pathogens?

27. Syphilis was once treated by infecting the patient with the parasite that causes malaria, a disease characterized by repeated cycles of fever, shaking, and chills. Why would this treatment control syphilis?

Summary

14.1 ■ Overview of the Innate Immune Defenses

First-line defenses prevent microbial entry into the body's tissues, sensor systems detect invasion, and effector mechanisms destroy and remove invaders (figure 14.1).

14.2 ■ First-Line Defenses (figure 14.2)

Physical Barriers

The skin is composed of two main layers—the epidermis and the dermis. Mucous membranes are constantly bathed with mucus and other secretions that help wash microbes from the surfaces (figure 14.3).

Antimicrobial Substances

Lysozyme, peroxidases, lactoferrin, and **host defense peptides** inhibit or kill microorganisms (figure 14.4).

Normal Microbiota

Members of the **normal microbiota** competitively exclude pathogens and stimulate the host defenses.

14.3 ■ The Cells of the Immune System (figure 14.5, table 14.1)

Granulocytes

Granulocytes include **neutrophils, eosinophils, basophils,** and **mast cells.**

Mononuclear Phagocytes

Monocytes circulate in blood; **macrophages** are in tissues (figure 14.6).

Dendritic Cells

Dendritic cells develop from monocytes; some have other origins.

Lymphocytes

Lymphocytes, which include **B cells, T cells,** and **innate lymphoid cells (ILCs),** are involved in adaptive immunity.

14.4 ■ Cell Communication

Surface Receptors

Surface receptors bind **ligands,** allowing the cell to detect certain substances.

Cytokines (table 14.2)

Cytokines include **chemokines, colony-stimulating factors (CSFs), interferons (IFNs), interleukins (ILs),** and **tumor necrosis factor (TNF).**

Adhesion Molecules

Adhesion molecules allow cells to adhere to other cells.

14.5 ■ Pattern Recognition Receptors (PRRs) (figure 14.7)

Pattern recognition receptors (PRRs) are sensors that allow the body's cells to "see" signs of microbial invasion. Many PRRs detect **pathogen-associated molecular patterns (PAMPS),** also called **microbe-associated molecular patterns (MAMPs),** and some detect **damage-associated molecular patterns (DAMPs).** The outcome of certain diseases is influenced by PRR-generated signals from sentinel cells and infected cells.

Pattern Recognition Receptors (PRRs) That Monitor a Cell's Surroundings

Some **toll-like receptors (TLRs)** are anchored in the cytoplasmic membranes of sentinel cells such as phagocytes and cells that line blood vessels (figure 14.8). These TLRs detect certain microbial surface components. Membrane-anchored C-type lectin receptors (CLRs) detect certain carbohydrate molecules found on the surface of some microbial cells.

Pattern Recognition Receptors (PRRs) That Monitor Material Ingested by a Cell

Some toll-like receptors (TLRs) are anchored in endosomal and phagosomal membranes. These TLRs typically recognize characteristics of microbial nucleic acid.

Pattern Recognition Receptors (PRRs) That Monitor a Cell's Cytoplasm

RIG-like receptors (RLRs) detect viral RNA in a cell's cytoplasm. **NOD-like receptors (NLRs)** detect microbial components or signs of damage in a cell's cytoplasm (figure 14.9). Certain NLRs in macrophages and dendritic cells allow formation of an **inflammasome.**

An Outcome of Cytoplasmic Pattern Recognition: The Interferon Response

Virally infected cells respond to the infection by making interferons, causing nearby cells to prepare to undergo apoptosis if they become infected with a virus (figure 14.10).

14.6 ■ The Complement System

Complement System Activation

The **complement system** is activated when certain of its components detect microbial cells or antibodies bound to antigens (figure 14.11).

Effector Functions of the Complement System

The major protective outcomes of complement system activation include **opsonization,** an inflammatory response, and lysis of foreign cells (figure 14.12).

Regulation of the Complement System

Complement regulatory proteins prevent host cell surfaces from activating the complement system via the alternative pathway (figure 14.13).

14.7 ■ Phagocytosis

The Steps of Phagocytosis (figure 14.14)

The steps of **phagocytosis** include chemotaxis, recognition and attachment, engulfment, **phagosome** maturation and **phagolysosome** formation, destruction and digestion, and exocytosis.

Characteristics of Macrophages

Macrophages are always present in tissues to some extent but can call in reinforcements when needed. A macrophage can become an **activated macrophage.** Macrophages, **giant cells,** and T cells form **granulomas** that wall off and retain material that cannot be destroyed. Macrophages are important sentinel cells.

Characteristics of Neutrophils

Neutrophils are the first cell type recruited from the bloodstream to the site of damage.

14.8 ■ The Inflammatory Response

Swelling, redness, heat, and pain are the signs of **inflammation,** the body's attempt to contain a site of damage, localize the response, eliminate the invader, and restore tissue function. Inflammation is initiated when tissue damage occurs, or when microbes are detected by pattern recognition receptors (PRRs) or the complement system.

The Inflammatory Process

The inflammatory process involves vascular and cellular changes that result in dilation of small blood vessels, leakage of fluids from those vessels, and movement of leukocytes from the bloodstream into the tissues (figure 14.15). **Acute inflammation** is characterized by an abundance of neutrophils; **chronic inflammation** is characterized by macrophage and giant cell accumulation and granuloma formation.

Damaging Effects of Inflammation

The inflammatory response can be damaging to the host, and in some cases this is life-threatening.

Programmed Cell Death and the Inflammatory Response

Apoptosis is a mechanism of eliminating "self" cells without triggering an inflammatory response; **pyroptosis** and **necroptosis** both trigger an inflammatory response.

14.9 ■ Fever

Fever results when macrophages release certain pro-inflammatory cytokines. It inhibits the growth of many pathogens and increases the rate of various body defenses.

Review Questions

Short Answer

1. Describe how the skin protects against infection.
2. What characteristics of saliva help protect against microbes?
3. Why is iron availability important in body defenses?
4. Name two categories of cytokines and give their effects.
5. What is the function of pattern recognition receptors?
6. Contrast the pathways of complement activation.
7. How do complement proteins cause foreign cell lysis?
8. How do phagocytes enter tissues during an inflammatory response?
9. How is acute inflammation different from chronic inflammation?
10. Describe the function of apoptosis.

Multiple Choice

1. Lysozyme does which of the following?
   a) Disrupts cell membranes
   b) Hydrolyzes peptidoglycan
   c) Waterproofs skin
   d) Propels gastrointestinal contents
   e) Propels the cilia of the respiratory tract
2. The hematopoietic stem cells in the bone marrow can develop into which of the following cell types?
   1. Red blood cell
   2. T cell
   3. B cell
   4. Monocyte
   5. Macrophage

   a) 2, 3
   b) 2, 4
   c) 2, 3, 4, 5
   d) 1, 4, 5
   e) 1, 2, 3, 4, 5
3. All of the following refer to the same type of cell *except*
   a) macrophage.
   b) neutrophil.
   c) poly.
   d) PMN.
4. Considering the role of pattern recognition receptors (PRRs), which of the following are they least likely to detect?
   a) Peptidoglycan
   b) Glycolysis enzymes
   c) Lipopolysaccharide
   d) Flagellin
   e) Certain nucleotide sequences
5. The direct/immediate action of interferon on a cell is to
   a) interfere with the replication of the virus.
   b) prevent the virus from entering the cell.
   c) stimulate synthesis of inactive antiviral proteins.
   d) stimulate the immune response.
   e) stop the cell from dividing.
6. A pathogen that can avoid the complement component C3b would directly protect itself from
   a) opsonization.
   b) triggering inflammation.
   c) lysis.
   d) inducing interferon.
   e) antibodies.
7. Which of the following statements about phagocytosis is *false*?
   a) Phagocytes move toward an area of infection by chemotaxis.
   b) Digestion of invaders occurs within a phagolysosome.
   c) Phagocytes have receptors that recognize C3b bound to bacteria.
   d) Phagocytes have receptors that recognize antibodies bound to bacteria.
   e) Macrophages die after phagocytizing bacteria, but neutrophils regenerate their lysosomes and survive.
8. If you are analyzing the cell types in a granuloma, which of the following are you least likely to find?
   a) Neutrophils
   b) Macrophages
   c) Giant cells
   d) T cells
9. All of the following trigger an inflammatory response *except*
   a) engagement of PRRs.
   b) complement system activation.
   c) interferon induction of antiviral protein synthesis.
   d) tissue damage.
10. Which of the following statements about inflammation is *false*?
    a) Vasodilation results in leakage of blood components.
    b) The process can damage host tissue.
    c) Neutrophils are the first to migrate to a site of inflammation.
    d) Apoptosis induces inflammation.
    e) The signs of inflammation are redness, swelling, heat, and pain.

Applications

1. Patients who have recently had a bone marrow transplant are extremely susceptible to infection. Why would this be so?
2. A cattle farmer inspects a sore on the leg of one of his cows and notices that the area just around the sore is warm to the touch. A veterinarian examines the wound and explains that the warmth may be due to inflammation. The farmer asks about the difference between the localized warmth and fever. What explanation would the vet give?

Critical Thinking

1. Why would it benefit the body to have an adaptive immune system in addition to an innate immune system?
2. Some bacterial cells avoid the killing effects of activated complement proteins. How might they do this?

15 | The Adaptive Immune Response

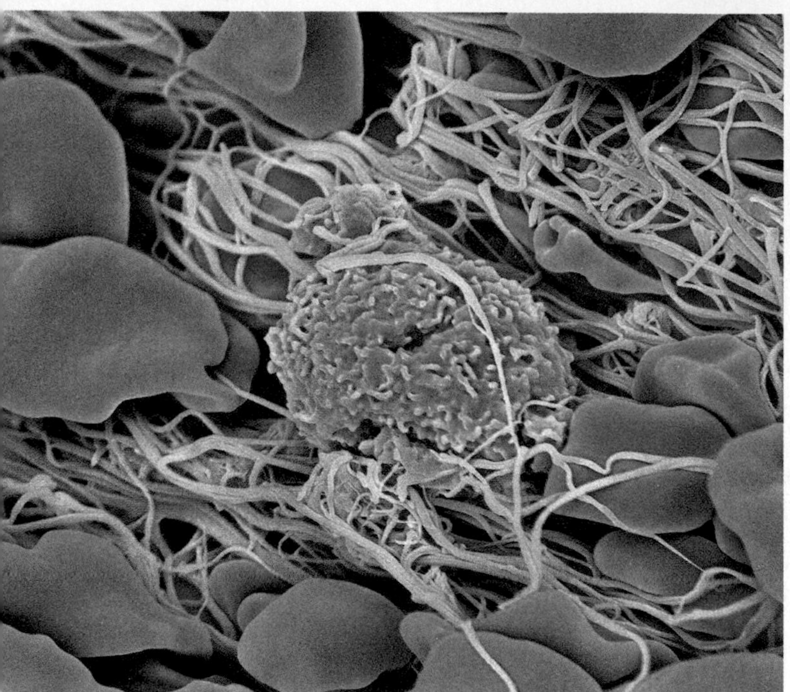

Blood clot, with erythrocytes, fibrin filaments, and a lymphocyte (color-enhanced scanning electron micrograph). *Science Photo Library/Getty Images*

A Glimpse of History

Near the end of the nineteenth century, diphtheria was a terrifying disease that killed many infants and small children. The first symptom was a sore throat, often followed by the development of a gray membrane that could come loose and block the airway. Frederich Loeffler, working in Robert Koch's laboratory in Berlin, found club-shaped bacteria growing in the throats of people with the disease but not elsewhere in their bodies. He hypothesized that the organisms were making a poison that spread through the bloodstream. In Paris, at the Pasteur Institute, Emile Roux and Alexandre Yersin followed up by growing the bacteria and extracting the poison, or toxin, from culture fluids. When the toxin was injected into guinea pigs, it generally killed them. Back in Berlin, Emil von Behring injected the toxin into guinea pigs that had recovered from lab-induced diphtheria. These animals did not become ill, suggesting that something in their blood protected them; von Behring called it *antitoxin*. To test this idea, he mixed toxin with serum (the liquid portion of clotted blood) from a guinea pig that had recovered from diphtheria, and injected it into one that had not had the disease. The animal remained well. In further experiments, he cured animals with diphtheria by giving them antitoxin.

The effectiveness of antitoxin was put to the test in late 1891, when a diphtheria epidemic broke out in Berlin. On Christmas night, antitoxin was first given to an infected child, who then recovered from the dreaded disease. The substances with antitoxin properties were then given the name *antibodies*, and materials that induced antibody production were called *antigens*.

KEY TERMS

Adaptive Immunity Protection provided by immune responses that improve due to exposure to antigens; involves B cells and T cells.

Antibody Y-shaped protein that binds antigen.

Antigen Molecule that reacts specifically with either an antibody or an antigen receptor on a lymphocyte.

Antigen-Presenting Cells (APCs) Cells such as dendritic cells, B cells, and macrophages that can present exogenous antigens to T cells.

B Cell Type of lymphocyte programmed to make antibody molecules.

Cell-Mediated Immunity (CMI) Immunity involving a T-cell response.

Clonal Selection Process in which a lymphocyte's antigen receptor binds to an antigen, allowing the lymphocyte to multiply.

Cytotoxic T Cell Type of lymphocyte programmed to destroy infected or cancerous "self" cells.

Dendritic Cell Cell type responsible for activating naive T cells.

Effector Lymphocyte Differentiated descendant of an activated lymphocyte; its actions help eliminate antigen.

Helper T Cell Type of lymphocyte programmed to activate B cells and macrophages and assist other parts of the adaptive immune response.

Humoral Immunity Immunity involving B cells and an antibody response.

Lymphocytes A group of white blood cells (leukocytes) involved in adaptive immunity; B cells and T cells are examples.

Major Histocompatibility Complex (MHC) Molecules Host cell surface proteins that present antigens to T cells.

Memory Lymphocytes Long-lived descendants of activated lymphocytes that can quickly respond if a specific antigen is encountered again.

Plasma Cell Effector form of a B cell; it functions as an antibody-secreting factory.

T_C Cell Effector form of a cytotoxic T cell; it induces apoptosis in infected or cancerous "self" cells.

T_H Cell Effector form of a helper T cell; it activates B cells and macrophages and releases cytokines that stimulate other cells of the immune system.

Emil von Behring received the first Nobel Prize in Medicine in 1901 for his work on antibody therapy. It took many more decades of investigation to reveal the biochemical nature of antibodies. In 1972, Rodney Porter and Gerald Edelman were awarded a Nobel Prize for their part in determining the structure of antibodies.

Whereas the innate immune response described in chapter 14 is always ready to react to patterns that signify microbial invasion or host cell damage, the adaptive immune response improves with each exposure to an **antigen** (a microbial invader or another foreign material). When an antigen is first encountered, certain **lymphocytes**—the main participants in adaptive immunity—recognize it and then proliferate (multiply quickly). This dramatically increases the number of the most effective lymphocytes, allowing efficient removal of the invader.

The protection provided by this response is called **adaptive immunity.** An important characteristic of the adaptive immune response is molecular specificity, meaning that the recognition of the antigen is precise. So if a person had the viral disease measles, for example, the response that eliminated the measles virus does not protect against a different disease such as mumps.

On first exposure to a given microbe or other antigen, adaptive immunity takes a week or more to build. During this delay, the host depends on innate immunity for protection, which may not be sufficient to prevent disease. In some cases, a person will not survive long enough for the adaptive response to reach a helpful level. When the response is successful, however, components are retained so that a faster and more effective reaction occurs upon re-exposure; in essence, the system "learns" how to effectively protect against the pathogen. This phenomenon, called **immunological memory,** is why someone who survived a disease such as measles, mumps, or diphtheria generally never develops that same disease again. Today, vaccination prevents many diseases by exposing a person's immune system to relatively harmless forms of a pathogen or its products. The vaccine triggers an adaptive immune response, so if the vaccine recipient is later exposed to the actual pathogen, the memory response eliminates the agent before disease develops. Certain diseases can be acquired multiple times, but that is generally due to the pathogen's ability to avoid recognition by the host defenses, a topic discussed in chapter 16.

Another important aspect of adaptive immunity is **immune tolerance,** which is the ability to ignore any given molecule. Most significantly, the immune system can distinguish between normal host cells and invading microbes, an ability referred to as self versus non-self recognition. A more accurate description might be "healthy self" versus "dangerous," the latter including invading microbes as well as cancerous or other "corrupt" self cells. Regardless, the ability to develop immune tolerance is crucial because without it the immune system would routinely turn against the body's own cells, attacking them just as it does invading microbes. It would also regularly attack harmless substances such as pollen. Immune tolerance is not a fail-safe system, however, which explains the occurrence of autoimmune diseases and hypersensitivity reactions such as allergies and gluten intolerance.

The adaptive immune system is extraordinarily complex, involving a network of cells, cytokines, and other compounds. The need for such complexity is logical given the system's task, but it makes explaining and understanding the system difficult! Although any description should seemingly start at the beginning and then continue in a linear progression, that approach can be confusing to someone new to the subject. As an analogy, imagine explaining how to build a car to someone who has never even seen a car, much less driven one; the person would have no idea why certain parts were being put together. With that in mind, we have chosen to start this chapter with a general overview of the adaptive immune response, focusing on the essential characteristics of the process and how the response eliminates invading microbes. After that fundamental groundwork has been laid, we will then progress to the more complex aspects of the system.

15.1 ■ Overview of the Adaptive Immune Response

Learning Outcomes

1. Compare and contrast the general characteristics of cell-mediated and humoral immunity.
2. Explain the importance of central tolerance and peripheral tolerance.
3. Define the terms *primary immune response* and *secondary immune response*.
4. Describe six protective outcomes of antibody-antigen binding.
5. Compare and contrast the roles of lymphatic vessels, the primary lymphoid organs, and the secondary lymphoid organs.
6. Diagram a simplified "big picture" summary of the immune response, including the roles of dendritic cells, helper T cells, cytotoxic T cells, and B cells.

To help simplify something as complex as adaptive immunity, scientists often regard it as two interacting mechanisms for eliminating foreign material in the body (**figure 15.1**):

- **Cell-mediated immunity (CMI),** or cellular immunity, deals with invaders that are within a "self" cell, meaning within one of the body's own cells. These invaders include viruses and bacteria replicating within a self cell. CMI relies on **T lymphocytes,** or **T cells,** a name that reflects the fact that they mature in the thymus. Two types of T cells are involved with eliminating antigen: **Cytotoxic T cells** are responsible for inducing apoptosis in self cells infected with viruses or are otherwise "corrupt," and **helper T cells** are responsible for directing and assisting the various immune responses.
- **Humoral immunity** (*humor* means "fluids") eliminates microbial invaders and toxins that are outside of a self cell; in other words, invaders in the blood or in tissue fluids. It involves **B lymphocytes,** or **B cells,** a cell type that in mammals develops in the bone marrow. B cells are programmed to produce Y-shaped proteins called **antibodies.** These bind to specific antigens and, in doing so, mark them as invaders to be eliminated.

Each T cell and each B cell has thousands of copies of a receptor on its surface that allows it to recognize a specific antigen; a region of the receptor called an antigen-binding

Focus Figure

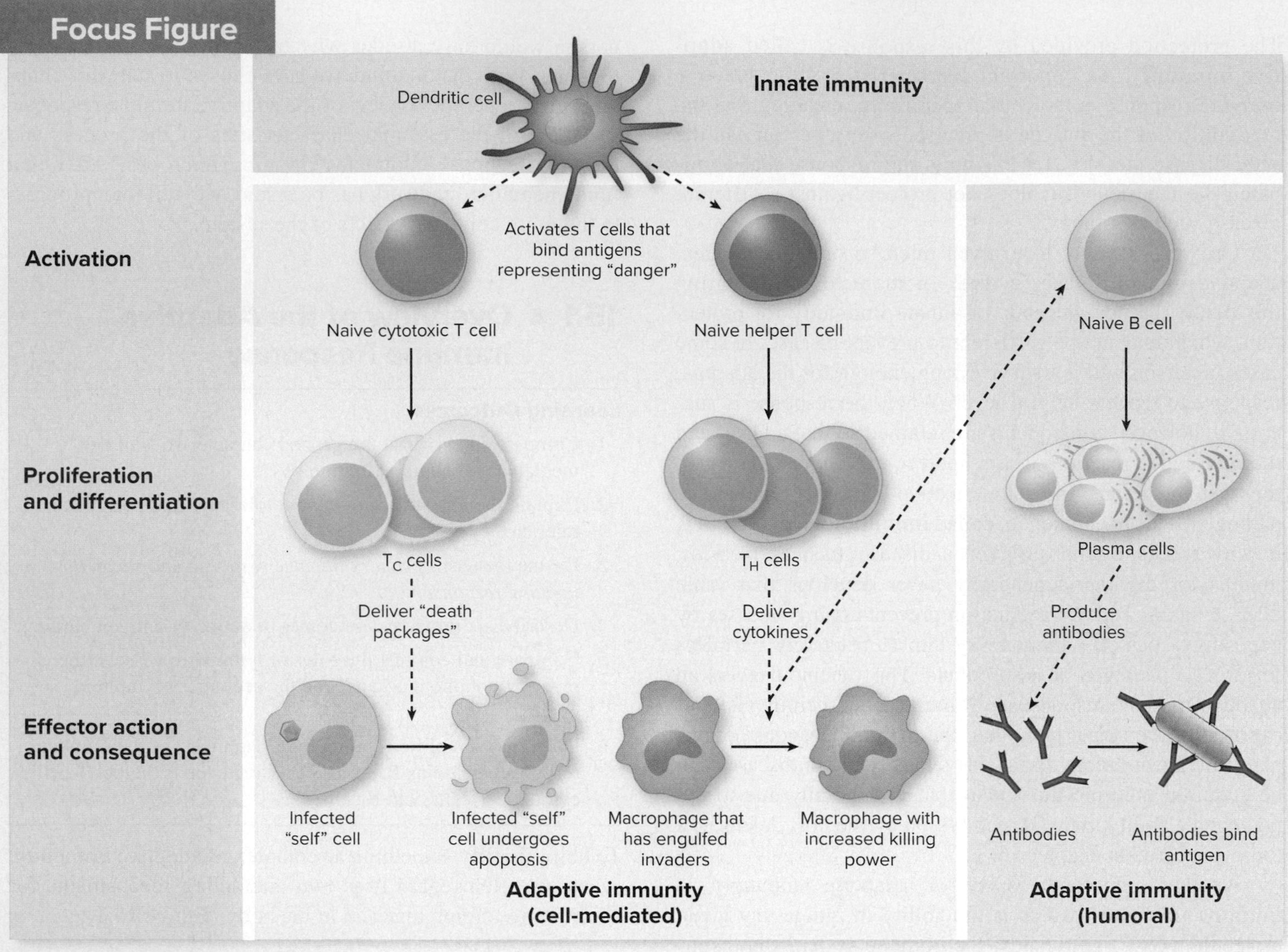

FIGURE 15.1 Overview of the Adaptive Immune Response Cell-mediated immunity protects against antigens within host cells (intracellular antigens); humoral immunity protects against antigens in blood and tissue fluid (extracellular antigens). In this diagram, solid arrows represent the path of a cell or molecule; dashed arrows represent a cell's interactions and effector functions; antigen receptors and memory cells are not shown.

? How does cell-mediated immunity eliminate intracellular antigens?

site is responsible for that recognition (**figure 15.2**). The antigen receptors on a single lymphocyte are identical and therefore recognize the same antigen, but because the body has hundreds of millions of different B cells and T cells, the immune system can recognize a nearly infinite assortment of antigens. General characteristics of the receptors are as follows:

- **T-cell receptors (TCRs).** These are on T cells. Conventional TCRs only bind an antigen "presented" by one of the body's own cells, an interaction guided by a surface molecule called a CD marker (CD stands for **c**luster of **d**ifferentiation to reflect that scientists use the molecules to distinguish different groups of cells). Cytotoxic T cells have a CD marker called CD8, and the cells are sometimes referred to as CD8 T cells or CD8+ T cells; in contrast, helper T cells have a CD marker called CD4, and the cells are sometimes referred to as CD4 T cells or CD4+ T cells.
- **B-cell receptors (BCRs).** These are on B cells. BCRs are essentially membrane-anchored versions of the Y-shaped antibody molecules that the B cell is programmed to make. Unlike T-cell receptors, BCRs bind free antigens (in other words, antigens not presented by one of the body's own cells). The two arms of the BCR are identical to each other, resulting in two antigen-binding sites.

Cell-mediated and humoral immunity are both powerful and, if misdirected, can damage the body's own tissues. To provide the immune tolerance necessary to prevent inappropriate responses, two sequential processes are used:

- **Central tolerance.** This takes place as lymphocytes mature (T cells in the thymus and B cells in the bone

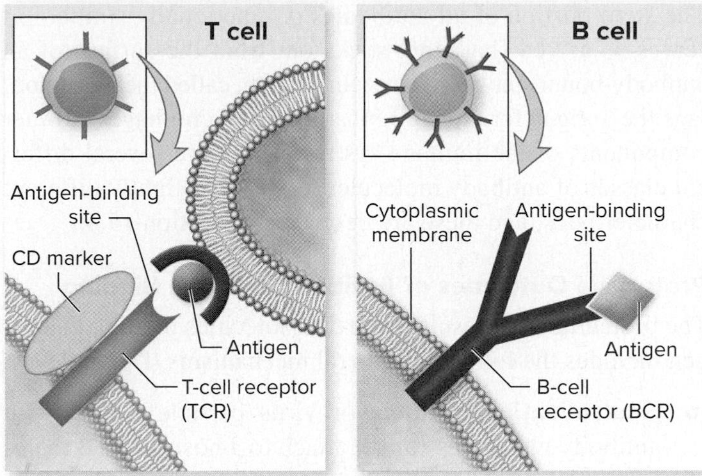

FIGURE 15.2 Antigen Receptors on Lymphocytes T cells and B cells have surface receptors that allow them to recognize specific antigens. T-cell receptors only bind antigens presented by another cell; B-cell receptors bind free antigens.

? How is a B-cell receptor similar to an antibody?

marrow); it eliminates immature T and B cells found to recognize certain "self" molecules.

- **Peripheral tolerance.** This occurs after the lymphocytes mature; it prevents any T and B cells not eliminated during central tolerance from reacting against self or other harmless molecules. As an example, a **naive lymphocyte** (meaning a mature lymphocyte that has not encountered antigen previously) cannot react to an antigen until specific signals from another cell confirm that the antigen is a microbe or other potentially harmful material. This form of peripheral tolerance can be likened to the two actions needed to move a parked car forward: Before the car will move, the brake must be released and the gas pedal depressed. In the case of a naive T cell or B cell, the cell's antigen receptors must bind the antigen (which "releases the brake"), and the confirmatory cell must provide the signal to indicate that a response is warranted (which "depresses the gas pedal"). Some mechanisms of peripheral tolerance interfere with the body's ability to destroy cancer cells, and medications that block these are now being used to treat certain types of cancer (see section 17.3).

Once a naive lymphocyte receives the signals that a response is needed, it becomes activated, meaning it can proliferate (see figure 15.1). Some descendants of activated lymphocytes differentiate to become **effector lymphocytes**—short-lived cells that can perform certain actions to help eliminate invaders. Other descendants become **memory lymphocytes,** long-lived cells that can be activated more quickly if the antigen is encountered again. The first response to a particular antigen is called the **primary immune response;** any additional encounters with the same antigen result in a faster and more effective reaction called the **secondary immune response.** Memory lymphocytes are responsible for the effectiveness of the secondary immune response.

Cell-Mediated Immunity

As just described, an immune response cannot begin until a lymphocyte becomes activated. **Dendritic cells,** the scouts of innate immunity, help activate the appropriate naive T cells (see figure 15.1). To do this, dendritic cells in the tissues first collect various antigens, including material that might have originated from invading microbes. They then travel to regions where naive T cells gather, and they interact with those cells in a way that passes on information about the antigen. Dendritic cells do this by simultaneously:

- presenting various pieces of the antigen, and
- producing surface proteins called co-stimulatory molecules if the antigen being presented is microbial or otherwise represents "danger"; in essence, these molecules function as "emergency lights."

If a T cell's TCR binds any part of an antigen presented by a dendritic cell that is also expressing co-stimulatory molecules, a lengthy interaction between the two cells may result in T-cell activation. This requirement, an example of peripheral tolerance, helps ensure that a response is warranted. The response that occurs depends on the type of T cell.

Once a cytotoxic T cell is activated, it proliferates, and then some of the clones (cell copies) differentiate to become an effector form called a **T_C cell** (also referred to as a CTL, for **c**ytotoxic **T** **l**ymphocyte). These T_C cells then search for certain infected self cells—specifically, any cell infected with the invader (antigen) the T_C recognizes. If one is encountered, the T_C cell delivers a "death package" to that infected self cell, inducing it to undergo apoptosis (see figure 15.1). The intracellular invader is thus deprived of a host in which to multiply. Meanwhile, other clones become memory cytotoxic T cells that can respond quickly upon re-exposure to the same antigen.

Likewise, once a helper T cell is activated, it proliferates. In this case, however, some of the clones differentiate to become an effector form called a **T_H cell,** which delivers cytokines to certain macrophages and B cells, thus activating them (see figure 15.1). T_H cells also produce cytokines that direct and support other cells, including various T cells. Other clones become memory helper T cells that respond quickly if the same antigen is encountered again.

T cells in a third subset, **regulatory T cells** (formerly T suppressor cells), are similar to the other T cells in that they have TCRs, but their role and development are entirely different. Instead of aiding an immune response, they help prevent one. This is another mechanism of peripheral tolerance, stopping the immune system from overreacting and responding to harmless substances. Regulatory T cells (T_{reg}), often referred to as "T regs," are involved in some of the disorders described in chapter 18. The topics covered in this chapter primarily involve helper and cytotoxic T cells.

Humoral Immunity

The activation of B cells is quite different from that of T cells. In most cases, T_H cells are required for activation of naive B cells (see figure 15.1). When a naive B cell encounters an antigen that its B-cell receptor (BCR) binds, the B cell takes up that antigen, degrades it, and then presents pieces of it to T_H cells. If a T_H cell's antigen receptor (TCR) binds to any of the pieces, the T_H cell delivers cytokines to the B cell, activating it. Again, this is an example of a mechanism of peripheral tolerance, because the T_H cell's existence helps confirm that the antigen represents "danger"; recall that a T_H cell develops only if a dendritic cell has activated its naive predecessor.

Once a B cell becomes activated and proliferates, many of the clones differentiate to become **plasma cells,** which are effector B cells (see figure 15.1). Plasma cells make the Y-shaped proteins called **antibodies,** which are secreted versions of the BCR; thus, the antibodies have the same binding specificity as the BCR of the responding B cell. When antibodies bind to the surfaces of antigens, including microbial cells, toxins, and viruses, they protect the body against the effects of those antigens. For example, antibodies that bind diphtheria toxin essentially coat the toxin so that it can no longer bind to host cells, thus protecting the patient from the effects of the toxin (see A Glimpse of History, this chapter); they also target it for elimination. Some of the B cell clones instead become memory B cells that respond quickly upon re-exposure to the same antigen.

The structure of an antibody molecule accounts for its ability to protect against an invader. An antibody has two functional regions: the two identical arms and the single stem of the Y-shaped molecule (**figure 15.3**). The arms, called the **Fab region** (or Fab regions), are the portions that attach to an antigen. That binding is very specific, so the immune system must be prepared to produce a wide variety of different antibody molecules, each with a slightly different set of "arms."

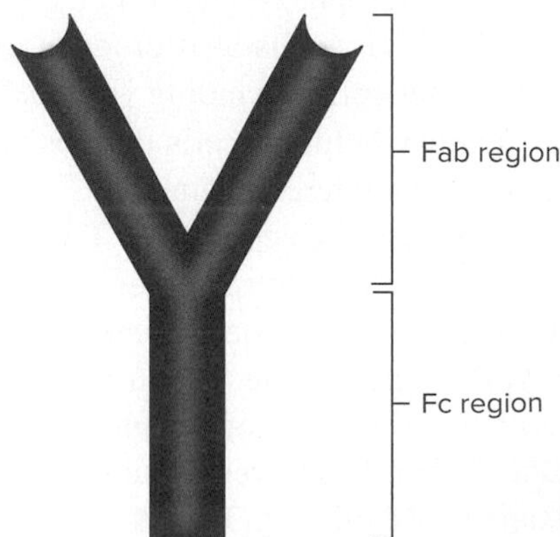

FIGURE 15.3 Simplified Y-Shaped Antibody Structure The arms of the Y make up the Fab region, and the stem is the Fc region.

How does the function of the Fc region differ from that of the Fab region?

The stem portion of all antibodies is functionally similar—it serves as a "red flag" that sticks out from the surface of an antibody-bound antigen. That stem portion, called the **Fc region,** tags the antigen for rapid elimination by macrophages or other components of the immune system. There are several different classes of antibody molecules, each with slightly different characteristics, and these are described in section 15.4.

Protective Outcomes of Antibody-Antigen Binding

The protection that results from the molecules binding to antigen includes the following general mechanisms (**figure 15.4**):

- **Neutralization.** A toxin or virus particle coated with antibody molecules cannot attach to a host cell and therefore cannot damage that cell.
- **Opsonization.** Phagocytic cells have receptors for the Fc portion of certain classes of antibody molecules, making it easier for the phagocyte to engulf antibody-coated antigens. Recall from chapter 14 that the complement protein C3b opsonizes antigens; antibody molecules have a similar effect.
- **Complement system activation.** When multiple molecules of certain antibody classes are bound to a cell surface or other antigen, a specific complement system protein attaches to side-by-side Fc regions. This triggers the classical pathway of complement system activation, leading to production of the opsonin C3b, initiation of an inflammatory response, and formation of membrane attack complexes.
- **Immobilization and prevention of adherence.** Binding of antibodies to flagella interferes with a microbe's ability to move, and binding to pili prevents a bacterium from attaching to surfaces. These capabilities are often necessary for a pathogen to infect a host, so antibodies that bind to flagella or pili prevent infection.
- **Cross-linking.** The two arms of an antibody can bind separate but identical antigen molecules, linking them. The overall effect is that large antigen-antibody complexes form, creating big "mouthfuls" of antigens for phagocytic cells to engulf.
- **Antibody-dependent cellular cytotoxicity (ADCC).** When multiple molecules of certain classes of antibodies bind to a virally infected cell or a tumor cell, that cell becomes a target for destruction by natural killer (NK) cells. The NK cell attaches to the Fc regions and then kills that cell.

The Nature of Antigens

The term *antigen* was first used to refer to compounds that induce antibody production; it is derived from the descriptive expression ***antibody generator***. Today, the term is used more

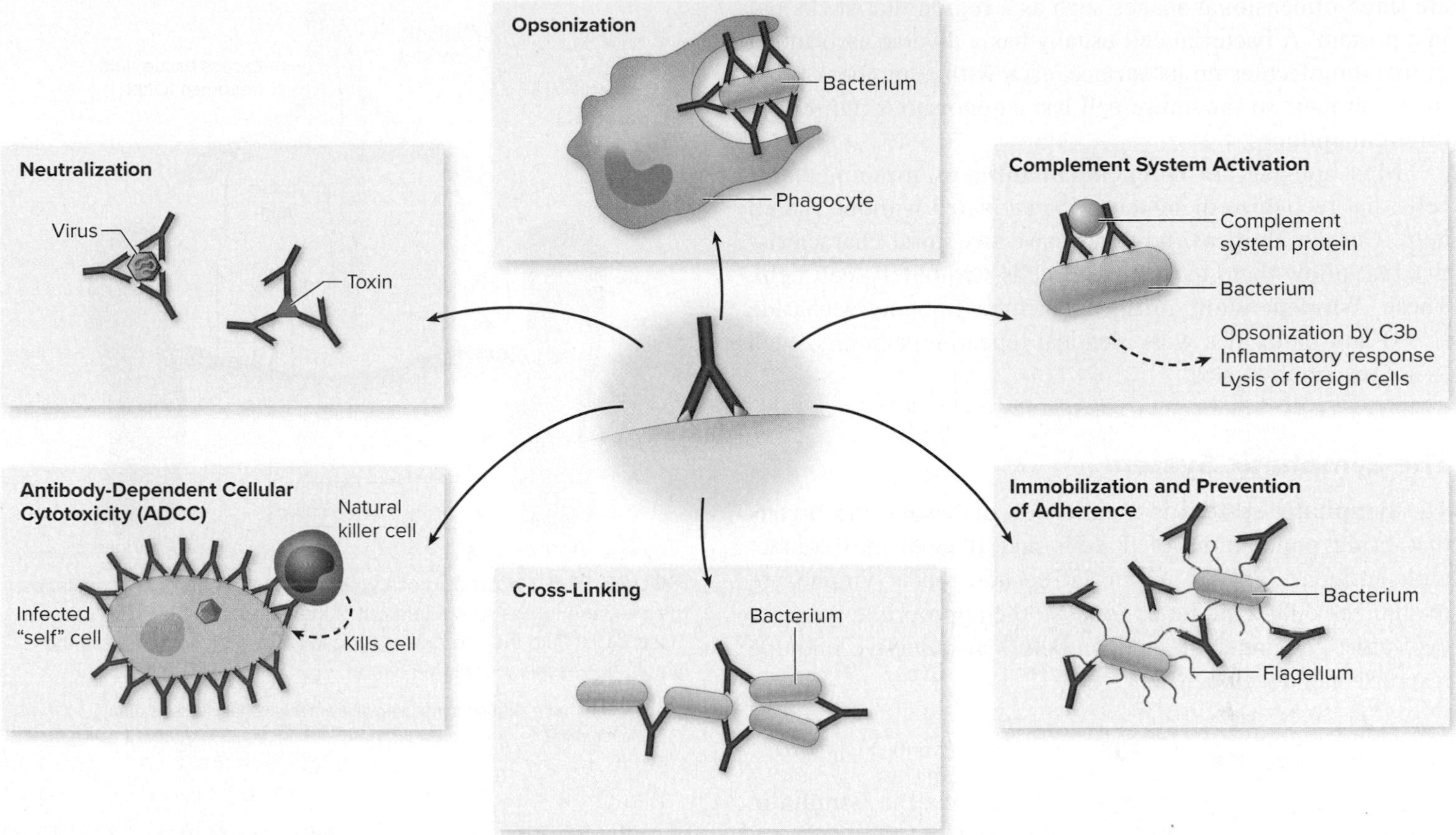

FIGURE 15.4 Protective Outcomes of Antibody-Antigen Binding

? **Which pathway of complement activation is depicted in this figure?**

broadly to describe any molecule that reacts specifically with a T-cell receptor (TCR), a B-cell receptor (BCR), or an antibody; it does not necessarily imply that the molecule induces an immune response.

Antigens include an enormous variety of materials, from invading microbes and their parts and products to plant pollens. The various antigens differ in their effectiveness in stimulating an immune response. Proteins generally induce a strong response, whereas lipids and nucleic acids often do not. The term **immunogenic** is used to describe the relative ability of an antigen to elicit an immune response. Small molecules are usually not immunogenic, meaning that they do not elicit a response; exceptions are small molecules called haptens, which become immunogenic once attached to a carrier protein.

Although antigens are generally large molecules, the adaptive immune system recognizes distinct regions of the molecule known as **epitopes,** or antigenic determinants (**figure 15.5**). Thus, when an antibody, a BCR, or a TCR binds to an antigen, it is actually binding to an epitope of the antigen (even so, discussions regarding the interactions usually use the term *antigen* rather than *epitope of an antigen*). Some epitopes are stretches of 10 or so amino acids, whereas others

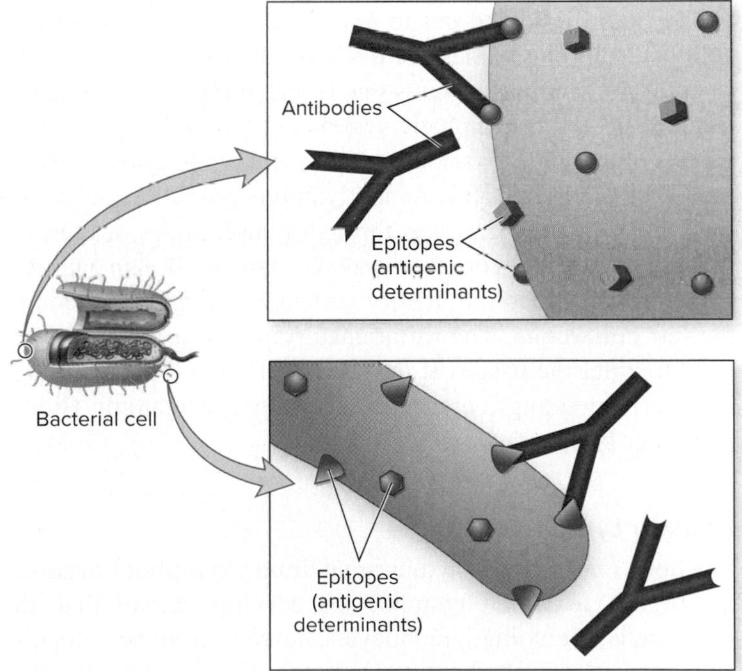

FIGURE 15.5 Antibodies Binding to Epitopes When an antibody binds to an antigen, it is actually binding to an epitope of the antigen (even so, discussions regarding the interactions usually use the term *antigen* rather than *epitope of an antigen*).

? **Explain why it is essentially impossible for a pathogen to have only a single epitope.**

are three-dimensional shapes such as a region that sticks out in a protein. A bacterial cell usually has a diverse assortment of macromolecules on its surface, each with a number of distinct epitopes, so the entire cell has an enormous number of different epitopes.

Most antigens are **T-dependent antigens,** meaning that B cells that recognize them cannot be activated without T_H cell help. Certain antigens, however, have structural characteristics that allow them to activate B cells without T_H cell help. These **T-independent antigens** include lipopolysaccharide (LPS) and molecules with identical repeating subunits, such as some carbohydrates.

The Lymphatic System

The **lymphatic system** is a collection of tissues and organs that bring populations of T cells and B cells into contact with antigens. This is important because each lymphocyte recognizes only a single epitope, so the appropriate lymphocyte must encounter the antigen before an effective immune response can develop.

Lymphatic Vessels

Flow within the lymphatic system occurs via the **lymphatic vessels,** which carry **lymph**—a colorless fluid derived from tissue fluid that forms as a result of the body's cardiovascular system (see figure 25.1). As blood enters capillaries, some of the liquid is forced out to join the tissue fluid (the extracellular fluid that bathes the tissues). Most of the liquid will re-enter the capillaries, but some is left in the tissues. Excess tissue fluid enters lymphatic vessels to become lymph, which carries along with it various antigens and other material from the tissues (**figure 15.6**). As the lymph is pushed through the lymphatic vessels, it passes through lymph nodes, a type of secondary lymphoid organ (discussed shortly). It then empties back into the blood circulatory system at a large vein behind the left collarbone. The inflammatory response causes more fluid to enter the tissues at the site of inflammation, resulting in a corresponding increase in the antigen-containing fluids that enter lymphatic vessels.

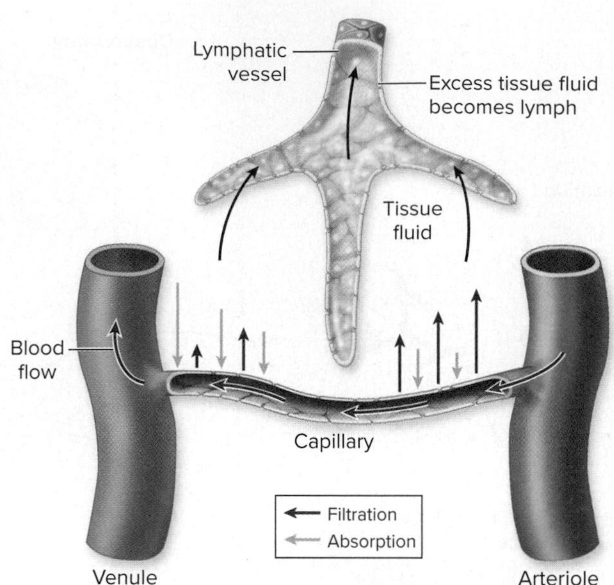

FIGURE 15.6 Formation of Lymph As blood enters the capillaries, the pressure forces some fluid out of the vessels and into the tissues; most of that fluid then re-enters the capillaries, but some enters the lymphatic vessels, becoming lymph.

? **Why does more antigen-containing fluid enter lymphatic vessels during inflammation?**

Primary Lymphoid Organs

The bone marrow and thymus are **primary lymphoid organs,** the organs in which lymphocytes develop. Recall that all blood cells, including lymphocytes, descend from hematopoietic stem cells in the bone marrow (see figure 14.5). B cells mature in the bone marrow, whereas T cells mature in the thymus.

As the lymphocytes develop, they acquire their ability to recognize distinct epitopes. This complex process involves rearrangement of DNA in the B-cell receptor (BCR) and T-cell receptor (TCR) genes, a process described later in the chapter. Some of the rearrangements inevitably give rise to receptors that recognize **autoantigens** (self molecules), but central tolerance eliminates developing lymphocytes found to be self-reactive. This occurs by exposing the lymphocytes to various autoantigens; if a developing lymphocyte binds an autoantigen in such a way that would elicit an unnecessary immune response, that cell is then modified or eliminated (a process referred to as negative selection). Not all self-reactive lymphocytes can be removed this way, though, which is why peripheral tolerance is also important.

Once mature, lymphocytes leave the primary lymphoid organs and move via the lymphatics to the secondary lymphoid organs.

Secondary Lymphoid Organs

Secondary lymphoid organs, including the lymph nodes, spleen, and tonsils, are the sites where antigens that have entered the body are brought into contact with dense populations of lymphocytes (**figure 15.7**). Lymph nodes, for example, capture materials from the lymphatics, and the spleen collects materials from blood. The organs are like busy, highly organized lymphoid coffee shops where many cellular meetings take place. The anatomy of these organs provides a structured center for various cells of the immune system to interact and transfer cytokines. No other places in the body do this, so these organs are the only sites where adaptive immune responses can be initiated.

Some secondary lymphoid organs are less organized in structure than the lymph nodes and spleen, but their purpose is the same—to capture antigens and bring them into contact with lymphocytes. Among the most important are the **Peyer's patches,** tissues in the intestinal walls where samples of intestinal contents are inspected. Specialized intestinal epithelial cells called M cells transfer material from the intestinal lumen (the hollow inside portion of the intestinal tract) to the Peyer's patches (**figure 15.8**). Dendritic cells in and near the Peyer's patches can also reach through the epithelial layer and grab material in the intestinal lumen to present it to naive T cells. Peyer's patches are part of a network of lymphoid tissue called **mucosa-associated lymphoid tissue (MALT),** which is distributed in various body sites including the respiratory, gastrointestinal, and urogenital tracts; MALT plays a crucial role in **mucosal immunity,** the immune response that prevents microbes from invading the body via the mucous membranes.

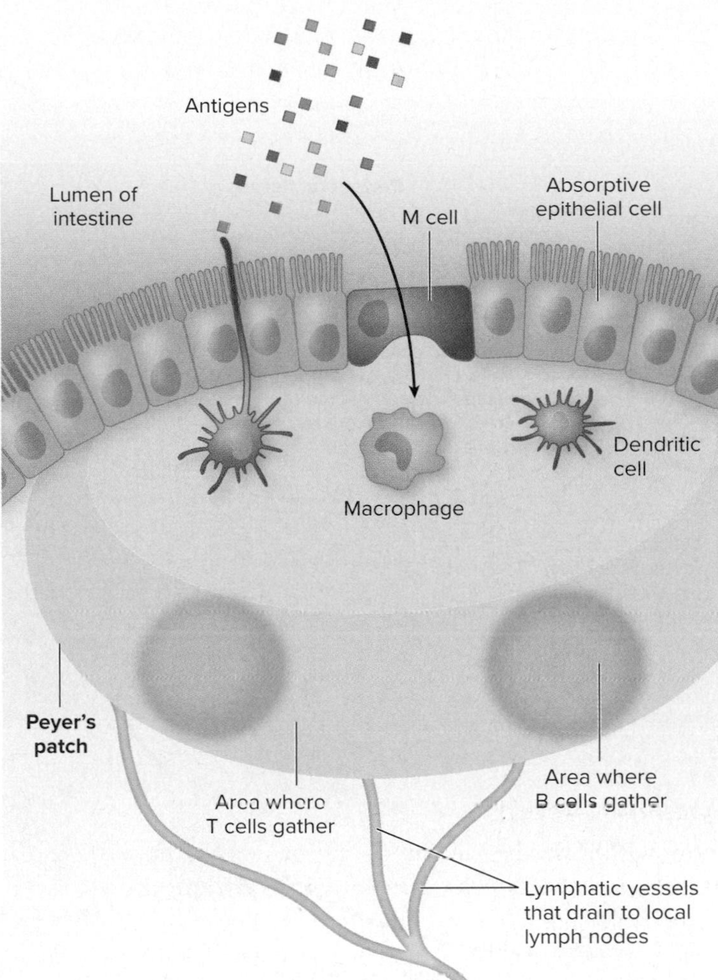

FIGURE 15.8 Peyer's Patch M cells transfer samples of intestinal contents to lymphocytes that reside in Peyer's patches. Dendritic cells can reach through the epithelial layer and grab material in the intestine to present it to naive T cells.

? What is mucosal immunity?

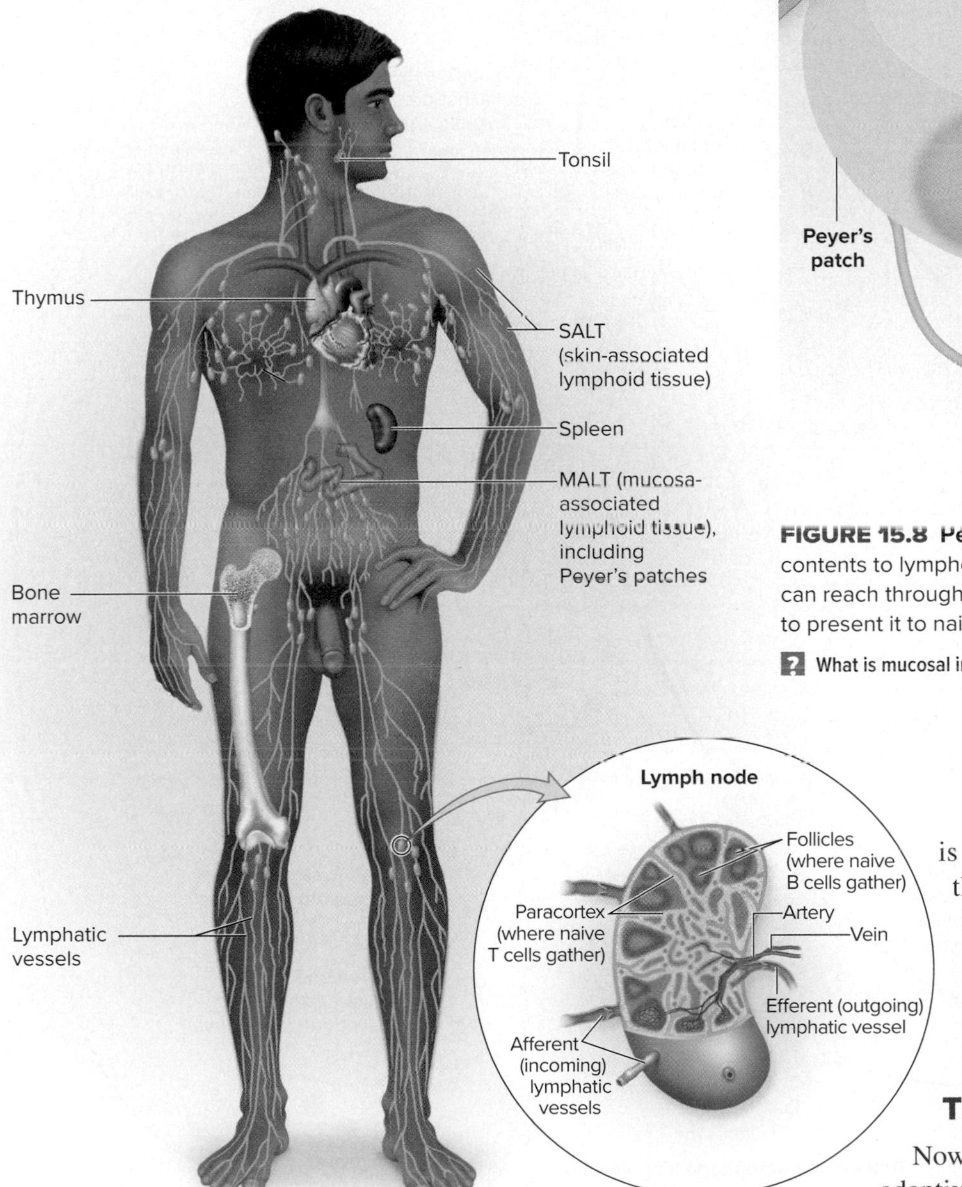

FIGURE 15.7 Anatomy of the Lymphatic System Lymph flows through a system of lymphatic vessels, passing through lymph nodes and lymphoid tissues.

? What is the function of the lymph nodes?

The Big Picture Summary

Now that we have completed an overview of the adaptive immune response, take a moment to review the general concepts, and use **figure 15.9** to follow the events that take place in the lymphoid organs and the peripheral tissues (meaning tissues other than the lymphoid organs). Once you are confident with that "big picture," then continue on to learn more advanced, yet significant, details.

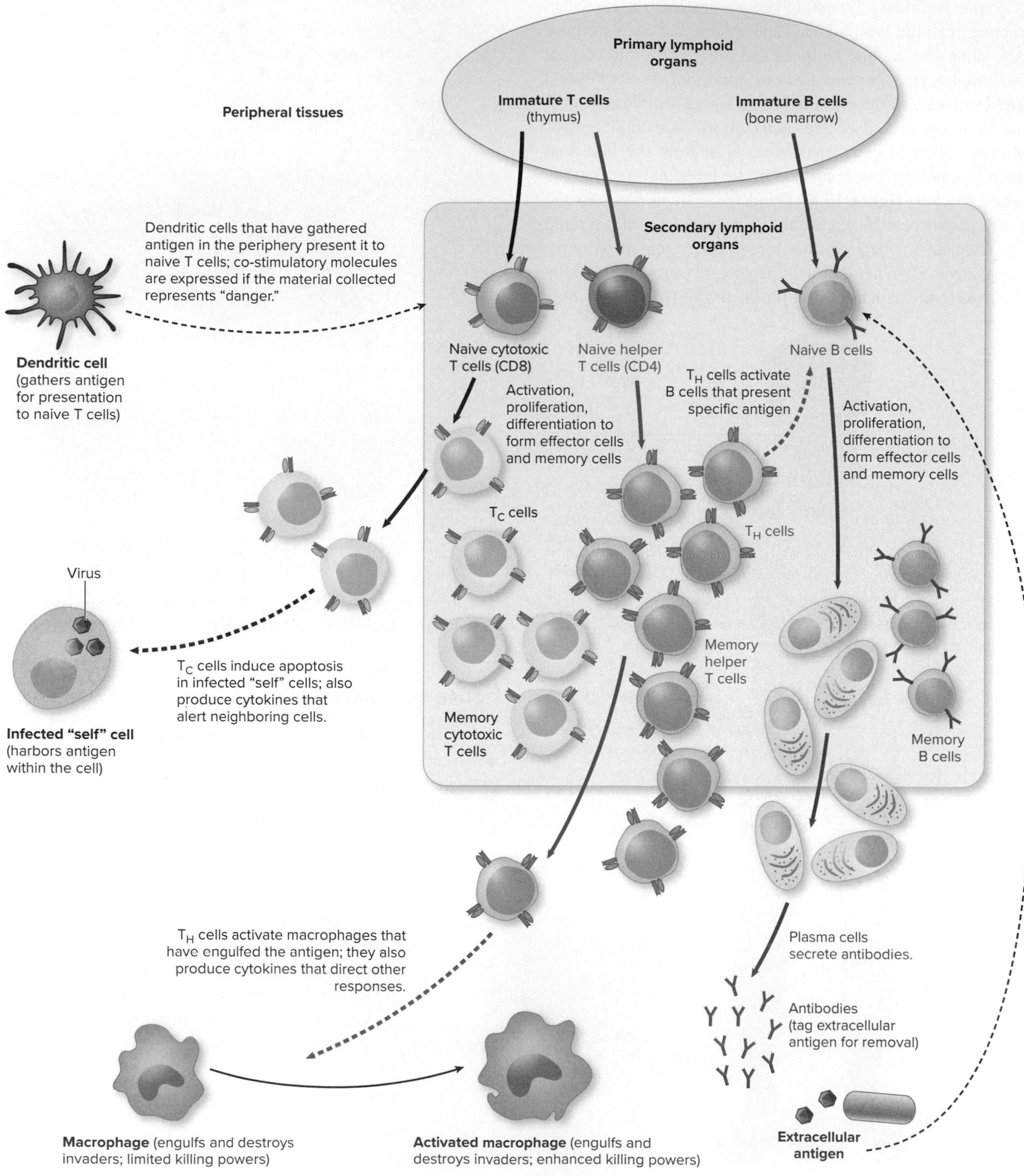

Primary lymphoid organs
Immature T cells (thymus)
Immature B cells (bone marrow)
Peripheral tissues
Secondary lymphoid organs
Dendritic cells that have gathered antigen in the periphery present it to naive T cells; co-stimulatory molecules are expressed if the material collected represents "danger."
Dendritic cell (gathers antigen for presentation to naive T cells)
Naive cytotoxic T cells (CD8)
Naive helper T cells (CD4)
Naive B cells
T_H cells activate B cells that present specific antigen
Activation, proliferation, differentiation to form effector cells and memory cells
Activation, proliferation, differentiation to form effector cells and memory cells
T_C cells
T_H cells
Virus
T_C cells induce apoptosis in infected "self" cells; also produce cytokines that alert neighboring cells.
Infected "self" cell (harbors antigen within the cell)
Memory helper T cells
Memory cytotoxic T cells
Memory B cells
T_H cells activate macrophages that have engulfed the antigen; they also produce cytokines that direct other responses.
Plasma cells secrete antibodies.
Antibodies (tag extracellular antigen for removal)
Extracellular antigen
Macrophage (engulfs and destroys invaders; limited killing powers)
Activated macrophage (engulfs and destroys invaders; enhanced killing powers)

FIGURE 15.9 Summary of the Adaptive Immune Response

? How would the adaptive immune response be affected if memory cells could not be produced?

MicroAssessment 15.1

Cell-mediated immunity involves T cells; humoral immunity involves B cells. T cells have T-cell receptors (TCRs) to recognize antigens; B cells have B-cell receptors (BCRs) to do the same. The processes of central tolerance and peripheral tolerance help prevent inappropriate immune responses. Before a naive T cell or B cell can respond to an antigen it recognizes, a confirming signal from another cell is usually required. Cytotoxic T cells are programmed to destroy infected "self" cells as a means of eliminating intracellular antigens. Helper T cells are programmed to direct and assist the immune response. B cells are programmed to make antibodies that help eliminate extracellular antigens. Antibody-antigen binding results in neutralization, immobilization and prevention of adherence, cross-linking, opsonization, complement activation, and antibody-dependent cytotoxicity. When an antibody, a BCR, or a TCR binds to an antigen, it is actually binding to an epitope of the antigen. Lymphatic vessels carry antigen-containing fluid collected from tissues to the lymph nodes. Primary lymphoid organs are where T and B cells develop. Secondary lymphoid organs are where antigens that have entered the body are brought into contact with dense populations of lymphocytes.

1. How is the cell-mediated response different from the humoral response?
2. Why is it important that T cells and B cells become activated before they can begin multiplying in response to an antigen?
3. How would you expect a T_C cell to respond if it encountered a T_H cell that was infected with a virus?

15.2 ■ Clonal Selection and Expansion of Lymphocytes

Learning Outcome

7. Outline the process of clonal selection and expansion.

The **clonal selection theory** explains how the adaptive immune system can respond to a seemingly unlimited range of antigens by using a pre-existing population of lymphocytes, each one already programmed to recognize a specific epitope on an antigen. When the immune system encounters a given antigen, only the lymphocytes recognizing it can multiply. Thus, the antigen determines which lymphocytes multiply.

Although clonal selection applies to both T and B cells, it is perhaps easier to understand using B cells as a model. During lymphocyte development, a given B cell is programmed to make only a single specificity of antibody (**figure 15.10**). Thus, each naive B cell in the secondary lymphoid organs is waiting for the "antigen of its dreams"—an antigen that has the epitope to which that particular B cell is programmed to respond. When an antigen is introduced, only the B cells that have a B-cell receptor (BCR) that binds the antigen have the possibility of becoming activated; other B cells remain inactive. In most cases, the B cells must present the antigen to T_H cells to receive the signals to become activated, but once this occurs the activated B cells multiply, generating a population of B clones—copies of the specific B cells capable of making the appropriate specificity of antibody. The antibody has the same binding specificity as the BCR.

The activities of individual lymphocytes change as they encounter antigens. As a means of clarifying discussions of lymphocyte characteristics, descriptive terms are sometimes used:

- **Immature lymphocytes.** The antigen-specific receptors on these are not yet fully developed.
- **Naive lymphocytes.** These have antigen receptors but have not yet encountered the antigen to which they are programmed to respond.
- **Activated lymphocytes.** These are able to proliferate; they recognize the presence of a specific antigen because their antigen receptor has attached to it, and they have received any necessary signals confirming that the antigen represents "danger."
- **Effector lymphocytes.** These are descendants of activated lymphocytes, armed with the ability to produce specific cytokines or other protective substances. Plasma cells are effector B cells, T_C cells are effector cytotoxic T cells, and T_H cells are effector helper T cells.
- **Memory lymphocytes.** These are long-lived descendants of activated lymphocytes; they can quickly become activated when an antigen is encountered again. Memory lymphocytes are responsible for the speed and effectiveness of the secondary immune response.

In the first (primary) exposure to an antigen, development of a protective response usually takes about 10 to 14 days. During this delay, the person might experience signs and symptoms of an infection, which could be life-threatening. The immune system, however, is actively responding, generating effector and memory cells. By the time the antigen is cleared, there will be many more memory cells than the relatively few naive cells that recognized the antigen originally. Those memory cells can quickly develop into effector cells, resulting in a stronger and more rapid response to future encounters with the antigen.

MicroAssessment 15.2

In response to an antigen, only those lymphocytes that recognize the antigen multiply. Depending on their functional characteristics, lymphocytes may be referred to as immature, naive, activated, effector, or memory cells.

4. Describe the clonal selection theory.
5. How does a naive lymphocyte differ from an activated one?
6. What would happen if the body lost the ability to make memory lymphocytes?

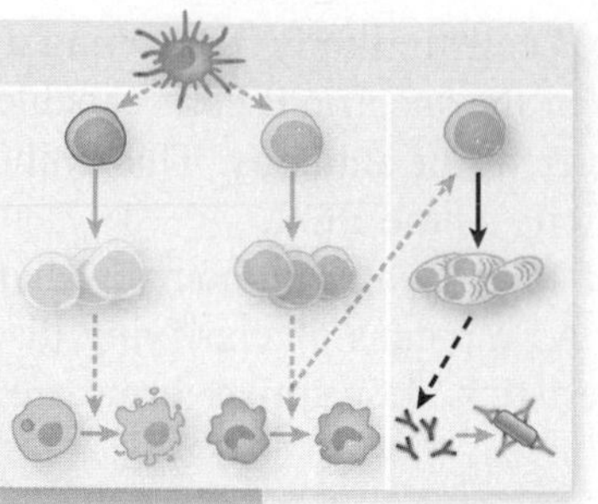

FIGURE 15.10 Clonal Selection and Expansion During the Antibody Response The antibodies shown are a class composed of five Y-shaped subunits.

? What is meant by clonal selection?

Development

Immature B cells: As these develop, a functionally limitless assortment of B-cell receptors is randomly generated.

Naive B cells: Each cell is programmed to recognize a specific epitope on an antigen; B-cell receptors guide that recognition.

Activation

Activated B cells: These cells can proliferate because their B-cell receptors are bound to antigen X and the cells have received required signals from T_H cells.

Proliferation and differentiation

Plasma cells (effector B cells): These descendants of activated B cells secrete large quantities of antibody molecules that bind to antigen X.

Memory B cells: These long-lived descendants of activated B cells recognize antigen X when it is encountered again.

Effector action

Antibodies: These neutralize the invader and tag it for destruction.

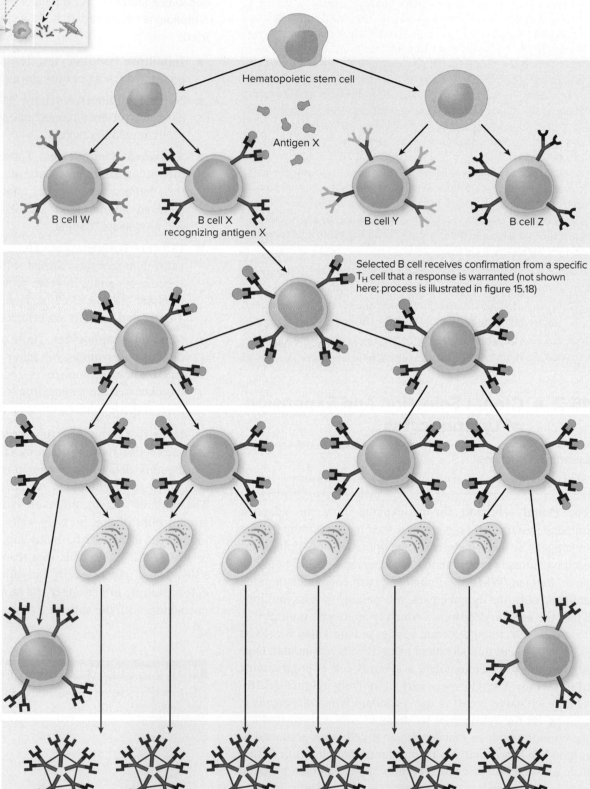

Hematopoietic stem cell

Antigen X

B cell W

B cell X recognizing antigen X

B cell Y

B cell Z

Selected B cell receives confirmation from a specific T_H cell that a response is warranted (not shown here; process is illustrated in figure 15.18)

15.3 ■ The T-Cell Response: Cell-Mediated Immunity

Learning Outcomes

8. Describe the importance of T-cell receptors, MHC molecules, and CD markers.
9. Describe the role of dendritic cells in T-cell activation.
10. Compare and contrast T_C and T_H cells with respect to antigen recognition and the response to antigen.

General Characteristics of T Cells

As mentioned in the overview, each T cell has thousands of identical copies of a **T-cell receptor (TCR)** on its surface that allow the cell to recognize and bind to a specific epitope on an antigen. Conventional TCRs, however, do not interact with free antigen. Instead, the antigen must be "presented" by another host cell, an action called **antigen presentation.** The host cell does this by (1) partly degrading (processing) the antigen's proteins into short peptide fragments and then (2) placing individual peptide fragments into the groove of surface proteins called **major histocompatibility complex (MHC) molecules.**

When a T cell recognizes and binds a presented antigen, the TCR is actually binding both the antigen (peptide fragment) and the MHC molecule simultaneously; thus, the TCR is binding antigen:MHC (**figure 15.11a**). Conventional T cells have a TCR composed of two different polypeptide chains, called alpha and beta, connected by a disulfide bond (**figure 15.11b**). Certain regions of the polypeptide chains are folded, so that each chain has two characteristic protein domains referred to as immunoglobulin domains. When TCRs that bind different antigens are compared, tremendous variation is seen in the amino acid sequences of the domains farthest from the cell surface; this part of a TCR is referred to as the variable region, and it accounts for the antigen:MHC-binding specificity. The domains closest to the cell surface have a consistent amino acid composition and make up the constant region. It is worth noting that a group of T cells called gamma delta T cells have a receptor composed of polypeptide chains called gamma and delta, and they behave quite differently from other T cells. Gamma delta T cells are relatively uncommon except near the intestinal tract, but they appear to play an important role in certain immune responses. They are not well understood, however, so we will focus our attention exclusively on conventional T cells.

Two types of MHC molecules are used to present antigen: MHC class I and MHC class II (**figure 15.12**). Cytotoxic T cells only recognize antigens presented on **MHC class I molecules,** and helper T cells only recognize antigens presented on **MHC class II molecules** (the significance of this difference will become clear when we describe the effector functions of the different cell types). The peptide-binding groove of both types of MHC molecules is shaped somewhat like an elongated bun and holds the peptide fragment lengthwise, like a bun holds a hot dog. The variable region of the TCR recognizes the "whole sandwich"—the peptide: MHC complex.

Recall that cytotoxic and helper T cells are identical microscopically, so scientists distinguish them based on the cells' **CD markers** (**c**luster of **d**ifferentiation markers). Cytotoxic T cells typically have the CD8 marker, whereas helper T cells typically have the CD4 marker; those CD markers

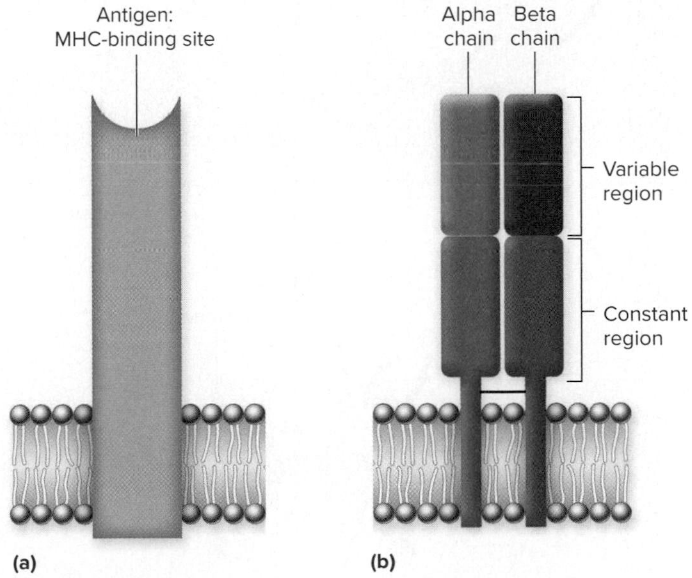

FIGURE 15.11 Structure of a T-Cell Receptor (a) Simplified structure, showing the antigen:MHC-binding site. **(b)** The molecule is made of two different chains, linked by a disulfide bond, and has a variable region and a constant region.

? **Why is it logical that the amino acid sequences of the TCR domains farthest from the cell surface are variable?**

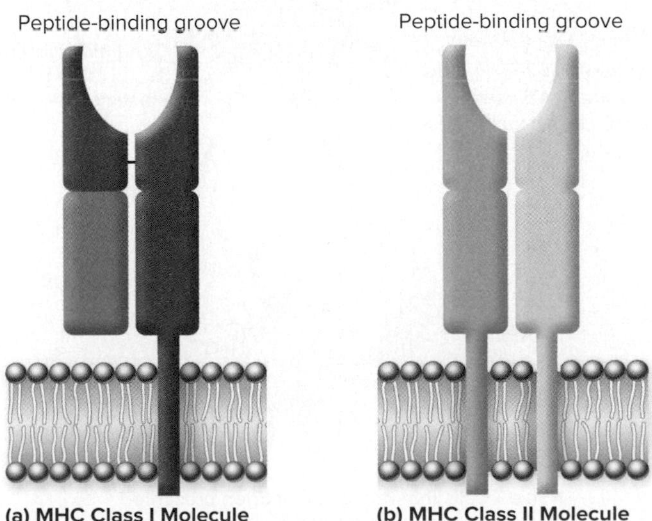

FIGURE 15.12 MHC Molecules (a) MHC class I molecule. **(b)** MHC class II molecule.

? **Which T cell type recognizes peptide fragments presented on MHC class I molecules? Which cell type recognizes peptide fragments presented on MHC class II molecules?**

help the respective T cell types interact with the appropriate MHC class. Note that CD4 is also a receptor for HIV, which explains why the virus infects helper T cells.

Activation of T Cells

Dendritic cells activate T cells. First, however, immature dendritic cells residing in the skin and other peripheral tissues must gather various materials from those areas; the cells use both phagocytosis and pinocytosis to take up particulate and soluble material that could contain foreign proteins. Dendritic cells located just below the mucosal barriers can even send tentacle-like extensions between the epithelial cells of the barriers. Using this action, they gather material from the respiratory tract and the lumen of the intestine. The dendritic cells have TLRs (toll-like receptors) and other pattern recognition receptors that allow them to recognize microbes or other signs of "danger." If such material is detected or if the cells are at the end of their life span, they take up even more material and then enter lymphatic vessels, which transport them to secondary lymphoid organs where naive T cells gather. En route, the dendritic cells mature into a form that can present antigen to naive T cells; any dendritic cells that detected "danger" also produce **co-stimulatory molecules** to communicate the significance of the material to the T cells. Unlike most cells, dendritic cells can present peptide fragments from material they have taken up on both types of MHC molecules—class I and class II; this ability, called cross-presentation, allows them to present those antigens to cytotoxic T cells as well as helper T cells.

In the secondary lymphoid organs, the dendritic cells present antigen to naive T cells (**figure 15.13**). Naive T cells that recognize antigen presented by dendritic cells displaying co-stimulatory molecules can become activated. In contrast, naive T cells that recognize antigen presented by a dendritic cell not displaying co-stimulatory molecules often become anergic (unresponsive) and eventually undergo apoptosis; this eliminates lymphocytes that might recognize autoantigens or other harmless material and is an important mechanism of peripheral tolerance. Instead of becoming anergic, some of the responding $CD4^+$ T cells will become regulatory T cells

FIGURE 15.13 T-Cell Activation

? What causes a dendritic cell to produce co-stimulatory molecules?

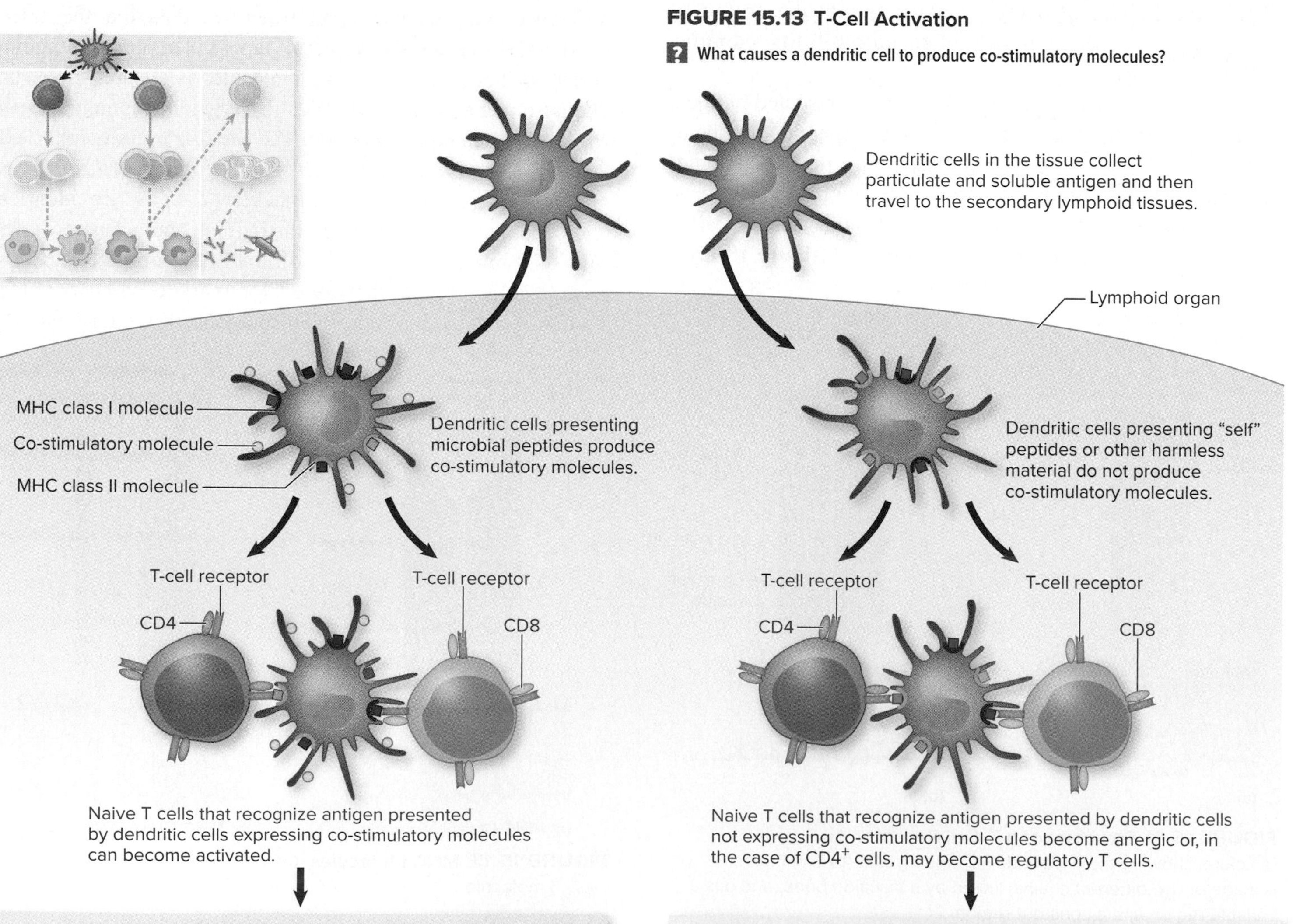
Dendritic cells in the tissue collect particulate and soluble antigen and then travel to the secondary lymphoid tissues.
Lymphoid organ
MHC class I molecule
Co-stimulatory molecule
MHC class II molecule
Dendritic cells presenting microbial peptides produce co-stimulatory molecules.
Dendritic cells presenting "self" peptides or other harmless material do not produce co-stimulatory molecules.
T-cell receptor
T-cell receptor
CD4
CD8
T-cell receptor
T-cell receptor
CD4
CD8
Naive T cells that recognize antigen presented by dendritic cells expressing co-stimulatory molecules can become activated.
Naive T cells that recognize antigen presented by dendritic cells not expressing co-stimulatory molecules become anergic or, in the case of CD4+ cells, may become regulatory T cells.
Activated T cells proliferate and differentiate.
Anergic T cells cannot respond and eventually undergo apoptosis. Regulatory T cells prevent certain immune responses.

(T_{reg} cells). The role of T_{reg} cells is to prevent certain immune responses, so this outcome provides another mechanism of peripheral tolerance.

Once activated, a T cell proliferates as described previously, eventually giving rise to effector cells. Some of these leave the secondary lymphoid organ to circulate in the bloodstream, and from there they can enter tissues at sites of infection or move to other secondary lymphoid organs; their role is to directly interact with other cells—target cells—to cause distinct changes in those cells. Other activated T cells give rise to long-lived memory cells, and these can respond quickly if the antigen is encountered again—an important characteristic of the secondary immune response.

Effector Functions of T_C (CD8) Cells

T_C cells induce apoptosis in infected or "corrupt" self cells. How do T_C cells distinguish such cells from their normal counterparts? The answer lies in the significance of antigen presentation on MHC class I molecules. All nucleated cells routinely degrade a portion of the proteins produced inside the cell and then load the resulting peptide fragments into the groove of MHC class I molecules to be delivered to the cell surface. Because the proteins originate from inside the cell, they are called endogenous antigens (*endo* means "inside" and *gen* means "generate"). What this does is allow circulating T_C cells to inspect samples of proteins made within the cell.

If a self cell is producing only normal proteins, then none of the T_C cells should recognize any of the peptide fragments being presented (**figure 15.14a**); recall that central and peripheral tolerance generally eliminate any lymphocytes that might mount a response against autoantigens. If, however, the self cell is infected with a replicating virus or microorganism, then some of the peptides presented on MHC class I molecules will be from the invader, making that self cell a target for the effector actions of a T_C cell (**figure 15.14b**). When a T_C cell binds to a presented peptide, that cell releases several proteases along with perforin, a molecule that forms pores in the target cell membrane. The proteases then enter the target cell through the pores and cause reactions that induce the cell to undergo apoptosis. In addition, a specific molecule on the T_C cell can engage a "death receptor" on the target cell, also initiating apoptosis. Killing the target cell by inducing apoptosis rather than lysis minimizes the number of intracellular microbes that might spill into the surrounding area and infect other cells. Most microbes remain in the cell remnants until they are ingested by macrophages. The T_C cell survives and can go on to kill other targets.

T_C cells can potentially induce apoptosis in cancerous self cells because these cells often make abnormal proteins; however, peripheral tolerance mechanisms referred to as immune checkpoints often stop that from happening. The checkpoints, and cancer therapies that block their action, are described in section 17.3.

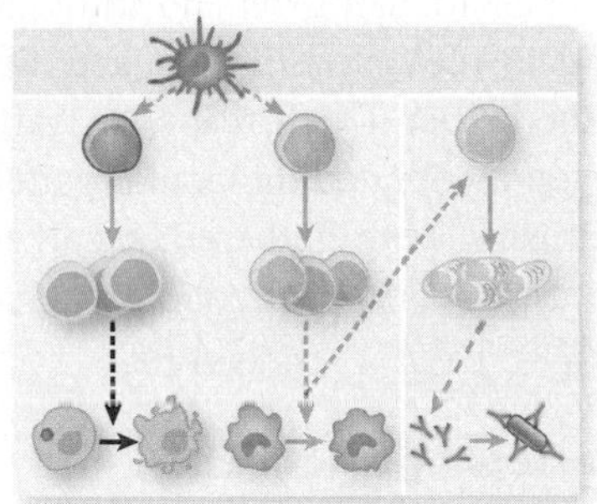

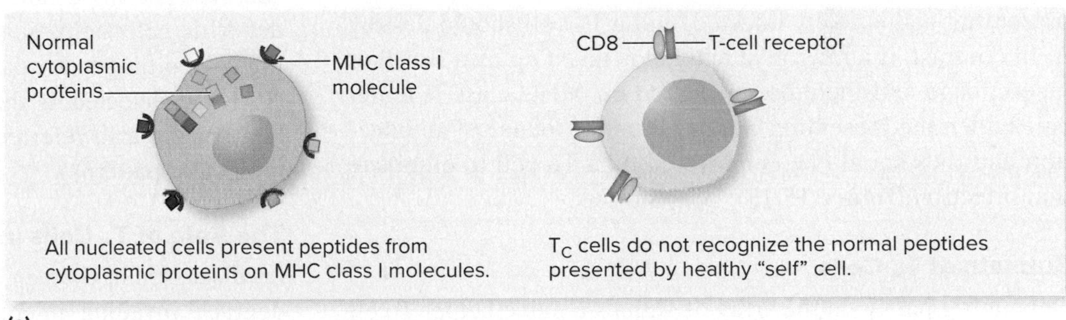

Normal cytoplasmic proteins
MHC class I molecule
All nucleated cells present peptides from cytoplasmic proteins on MHC class I molecules.
CD8
T-cell receptor
T_C cells do not recognize the normal peptides presented by healthy "self" cell.
(a)

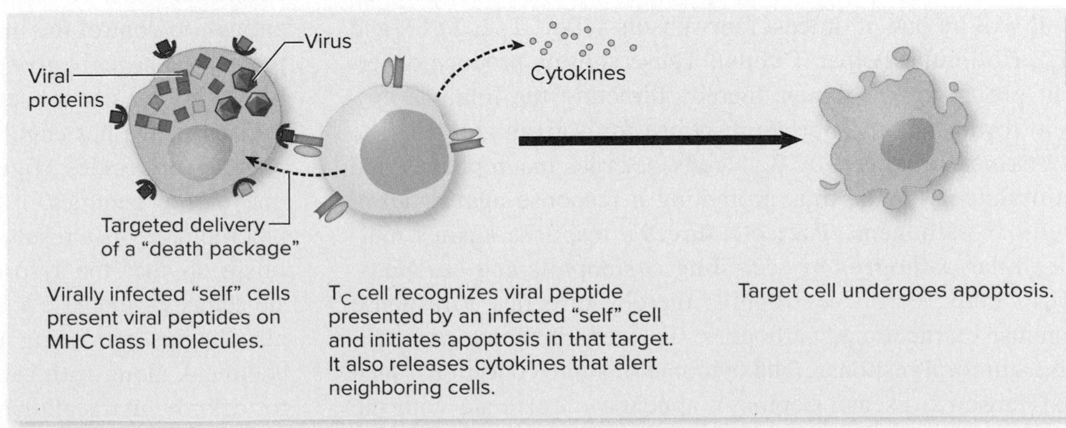

Viral proteins
Virus
Cytokines
Targeted delivery of a "death package"
Virally infected "self" cells present viral peptides on MHC class I molecules.
T_C cell recognizes viral peptide presented by an infected "self" cell and initiates apoptosis in that target. It also releases cytokines that alert neighboring cells.
Target cell undergoes apoptosis.
(b)

FIGURE 15.14 Functions of T_C Cells **(a)** T_C cells ignore healthy "self" cells. **(b)** T_C cells induce apoptosis in virally infected "self" cells.

Could a T_C cell induce apoptosis in a "self" cell that lost the ability to produce MHC class I molecules?

In addition to inducing apoptosis in the target cells, the T_C cell also produces various cytokines that strengthen the "security system." One cytokine increases antigen processing and presentation in nearby cells, making it easier for T_C cells to find other infected cells. Another cytokine activates local macrophages whose toll-like receptors (TLRs) have been triggered. Note that a more efficient mechanism of macrophage activation involves T_H cells and will be discussed shortly.

Effector Functions of T_H (CD4) Cells

T_H cells orchestrate the immune response by activating B cells and macrophages and by producing cytokines that direct the activities of other cells involved in the immune response, including B cells, macrophages, and T cells.

T_H cells recognize antigens presented on MHC class II molecules. Only certain cell types (dendritic cells, B cells, and macrophages), collectively referred to as **antigen-presenting cells (APCs),** make MHC class II molecules. With the exception of cross-presentation by dendritic cells during T-cell activation, a crucial difference between MHC class I and MHC class II molecules is the origin of the antigens presented on them. Whereas MHC class I molecules are used to present antigens made by the self cell (endogenous antigens), MHC class II molecules are used to present antigens taken up by the self cell from its surrounding environment (the antigens are referred to as exogenous antigens; *exo* means "outside"). Take a minute to think about the significance of these differences. If a microbial antigen is made within a cell and therefore presented on MHC class I molecules, then the presenting cell is likely infected, and a T_C cell should destroy it. In contrast, if a microbial antigen is taken up by a B cell or macrophage and therefore presented on MHC class II molecules, then the presenting cell has found evidence of an infection and thus should be empowered by a T_H cell to eliminate that infection (**figure 15.15**).

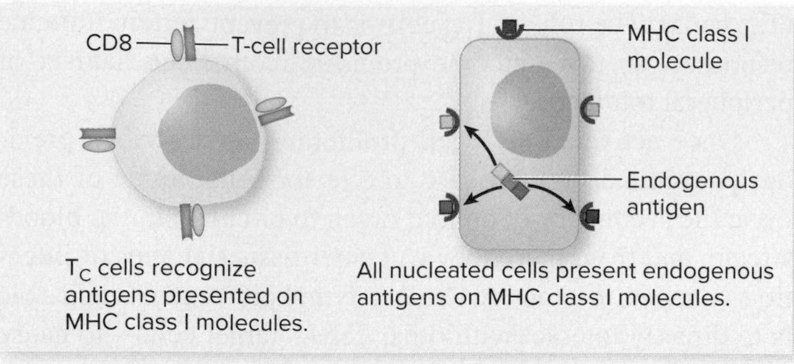
CD8
T-cell receptor
MHC class I molecule
Endogenous antigen
T_C cells recognize antigens presented on MHC class I molecules.
All nucleated cells present endogenous antigens on MHC class I molecules.

(a)

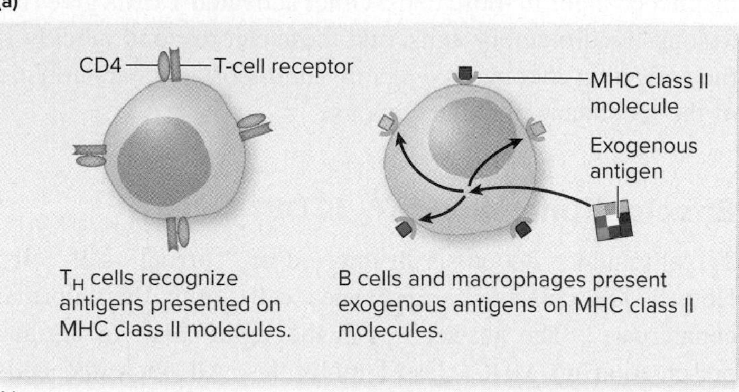
CD4
T-cell receptor
MHC class II molecule
Exogenous antigen
T_H cells recognize antigens presented on MHC class II molecules.
B cells and macrophages present exogenous antigens on MHC class II molecules.

(b)

FIGURE 15.15 Antigen Recognition by Effector T Cells (a) Cytotoxic T (T_C) cell. **(b)** Helper T (T_H) cell.

? What is the fate of a cell that presents antigen recognized by a T_C cell? What is the fate of a cell that presents antigen recognized by a T_H cell?

Subsets of T_H Cells

Depending partly on the array of signals a naive helper T cell receives from an activating dendritic cell, the resulting T_H cell will be one of at least four subsets—T_H1, T_H2, T_H17, and T_{FH} (follicular helper T cells). These subsets produce different groups of cytokines, thereby directing the immune system toward an appropriate response for a given antigen. The cytokines produced by T_H1 cells activate macrophages and stimulate T_C cells, thus promoting a response against intracellular pathogens. T_H2 cells direct a response against multicellular pathogens by recruiting eosinophils and basophils. T_H17 cells recruit neutrophils, thereby directing a response against extracellular pathogens. T_{FH} cells promote an effective humoral response. The outcome of some conditions, such as Hansen's disease (leprosy), appears to correlate with the type of helper T-cell response.

For simplicity, we will consider the general functions of T_H cells as a group, rather than focusing on the individual subsets. The important thing to recognize is that activating dendritic cells helps shape the helper T cell response, tailoring it according to the pathogen-associated molecular patterns (PAMPs) detected as the antigens were gathered; recall that PAMPs are also referred to as MAMPs (microbe-associated molecular patterns).

The Role of T_H Cells in Macrophage Activation

As discussed in chapter 14, macrophages routinely engulf and degrade invading microbes, clearing most organisms even before an adaptive response is mounted. If this alone is not enough to control the invader, however, T_H cells can activate the macrophages to give them more potent destructive capabilities. This can occur because peptide fragments from material a macrophage has engulfed are routinely presented on MHC class II molecules (**figure 15.16**). If a T_H cell recognizes one of the peptides, it delivers cytokines that activate the macrophage. As a result, the macrophage increases its metabolism so that the lysosomes—which contain antimicrobial substances—increase in number. The activated macrophage also begins producing nitric oxide, a potent antimicrobial chemical, along with various compounds that can be released to destroy extracellular microorganisms. In addition, the T_H cell releases cytokines that stimulate nearby T_C cells.

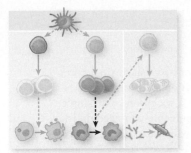

FIGURE 15.16 The Role of T_H Cells in Macrophage Activation

? How does macrophage activation help prevent disease?

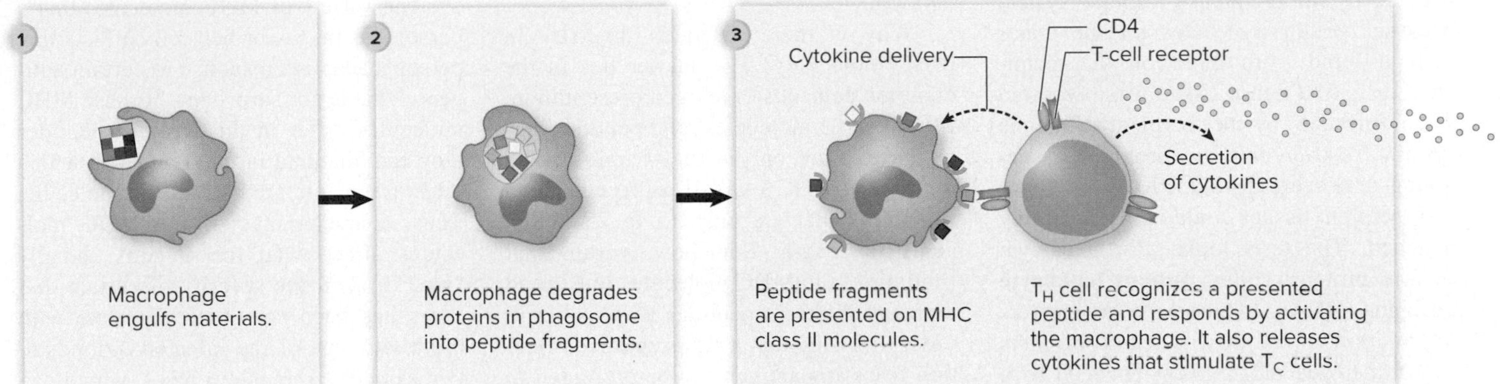
1
2
3
Cytokine delivery
CD4
T-cell receptor
Secretion of cytokines
Macrophage engulfs materials.
Macrophage degrades proteins in phagosome into peptide fragments.
Peptide fragments are presented on MHC class II molecules.
T_H cell recognizes a presented peptide and responds by activating the macrophage. It also releases cytokines that stimulate T_C cells.

If the immune response is still not sufficient to control the infection, activated macrophages fuse together, forming **giant cells.** These, along with other macrophages and T cells, can form granulomas that wall off the offending agent, preventing infectious microbes from escaping to infect other cells. Activated macrophages are important in the response against the bacterium that causes tuberculosis *(Mycobacterium tuberculosis)* and other microorganisms that can survive within regular macrophages.

The Role of T_H Cells in B-Cell Activation

In the case of most antigens, T_H cells are required for the activation of B cells. The process is similar to that described for macrophage activation, and details are described in the next section.

The characteristics of cytotoxic and helper T cells are summarized in **table 15.1.**

MicroAssessment 15.3

Dendritic cells expressing co-stimulatory molecules activate T cells that recognize the presented antigen. T_C (CD8) cells recognize antigen presented on MHC class I molecules; they induce apoptosis in target cells and produce cytokines that increase the level of surveillance. T_H (CD4) cells recognize antigen presented on MHC class II molecules (found on B cells and macrophages); they activate the target cells and secrete various cytokines that orchestrate the immune response.

7. If an effector CD8 cell recognizes antigen presented on an MHC class I molecule, how should it respond?
8. If an effector CD4 cell recognizes antigen presented on an MHC class II molecule, how should it respond?
9. Why would a person who has AIDS be more susceptible to the bacterium that causes tuberculosis?

TABLE 15.1 Characteristics of T Cells

T Cell Type/CD Marker	Antigen Recognition	Effector Form	Potential Target Cells	Effector Function	Source of Peptide Recognized by Effector Cell
Cytotoxic/CD8	Peptide fragments presented on MHC class I molecules	T_C cell	All nucleated cells	Induces target cell to undergo apoptosis	Endogenous antigen (produced within the target cell)
Helper/CD4	Peptide fragments presented on MHC class II molecules	T_H cell	B cells, macrophages	Activates target cell	Exogenous antigen (produced outside of the target cell)

FOCUS YOUR PERSPECTIVE 15.1

What Flavors Are Your Major Histocompatibility Complex (MHC) Molecules?

The major histocompatibility complex (MHC) molecules were discovered through research into tissue and organ transplantation, long before scientists knew about their essential role in adaptive immunity. Scientists noticed that transplants were generally rejected by the recipient's immune system because certain molecules on the donor cells differed from those on the recipient's cells; the donor tissue was perceived as an "invader" by the recipient's immune system. To overcome this problem, tissue-typing tests were developed so that donor and recipient tissues could be more closely matched. The tests looked for leukocyte surface proteins called **human leukocyte antigens (HLAs),** which serve as markers for tissue compatibility. Later, researchers determined that HLAs were encoded by a cluster of genes, now called the major histocompatibility complex. Unfortunately, the terminology can be confusing because the proteins that transplant biologists refer to as HLAs are called MHC molecules by immunologists.

It is highly unlikely that two random individuals will have identical MHC molecules. This is because the genes encoding them are polygenic, meaning they are encoded by more than one locus (position on the chromosome), and each locus is highly polymorphic (multiple forms). As an analogy, if MHC molecules were candy, each cell would be covered with pieces of chocolate, taffy, and lollipop. The type of chocolate could be dark, white, milk, or a number of various flavors; the taffy could be peppermint, raspberry, or cinnamon, and so on. As you might imagine, the number of different possible combinations is enormous.

There are three loci (specific physical locations) of MHC class I genes, designated HLA-A, HLA-B, and HLA-C (**box figure 15.1**); there are at least 890 alleles (forms) for HLA-A, 1,400 for HLA-B, and 620 for HLA-C. In addition, the loci are all co-dominantly expressed. In other words, the set of the three MHC class I genes you inherited from your mother and the set you inherited from your father are both expressed. Putting this all together, your cells most likely express six different varieties of MHC class I molecules—two of the over 890 known HLA-A possibilities, two of the over 1,400 HLA-B possibilities, and two of the over 620 HLA-C possibilities. Thus, the chance that anyone you know will also have that exact same combination of MHC class I molecules is extremely low, unless you have an identical twin.

Why is there so much diversity in MHC molecules? The answer lies in the complex demands of antigen presentation. MHC class I molecules bind peptide fragments that are only 8 to 10 amino acids in length; MHC class II molecules bind fragments that are only 13 to 25 amino acids in length. Somehow, within that limitation, the MHC molecules must bind as many different peptides as possible in order to ensure that a representative selection from any antigen can be presented to T cells. The ability to bind a wide variety of peptides is particularly important considering how readily microbes evolve in response to selective pressure. For example, if a single alteration in a viral protein prevented all MHC molecules from presenting peptides from that antigen, then a virus with that mutation could overwhelm the defenses. The ability to bind different peptides is not enough, however, because, ideally, a given peptide should be presented in several slightly different orientations so that distinct parts of the three-dimensional structures can be inspected by T-cell receptors. No single variety of MHC molecule can accomplish all of these aims, which explains the need for their diversity.

The variety of MHC molecules that a person has on his or her cells affects that person's adaptive response to certain antigens. This is not surprising because MHC molecules differ in the array of peptides they can bind and in the manner in which those fragments are held in the molecule. Thus, characteristics of the MHC molecules affect what the T cells actually "see." In fact, the severity of certain diseases has been shown to correlate with the MHC type of the infected individual. For example, rheumatic fever, which can occur as a consequence of *Streptococcus pyogenes* infection, develops more often in people with certain MHC types. The most serious outcomes of schistosomiasis have also been shown to correlate with certain MHC types. Epidemics of life-threatening diseases such as plague and smallpox appear to have dramatically altered the relative proportion of MHC types in certain populations, killing those whose MHC types ineffectively present peptides from the causative agent.

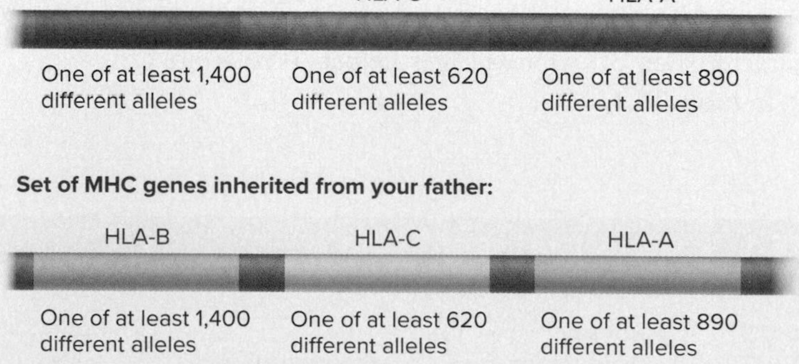

BOX FIGURE 15.1 MHC Polymorphisms The order of the MHC class I genes on the chromosome is B, C, and A.

15.4 ■ The B-Cell Response: Humoral Immunity

Learning Outcomes

11. Describe the role of T_H cells in B-cell activation.
12. Compare and contrast the five classes of immunoglobulins.
13. Compare and contrast the primary and the secondary immune responses to T-dependent antigens.
14. Explain why T-independent antigens are important medically.

This section will mainly focus on the B-cell response to T-dependent antigens, because that is most common. The response to T-independent antigens will be described at the end of the discussion.

General Characteristics of B cells

As mentioned in the overview, each B cell has thousands of copies of a **B-cell receptor (BCR)** on its surface that allow the cell to recognize and bind to a specific epitope on an antigen. The receptor is essentially a membrane-bound version of the Y-shaped antibody that the B cell is programmed to make (**figure 15.17**). It is composed of four polypeptide chains—two duplicate copies of a heavy chain and two duplicate copies of a light chain (the names reflect the relative molecular weights)—linked together by disulfide bonds. This creates the characteristic Y-shaped structure with two identical arms and a stem; at the junction of the arms and the stem is a flexible stretch. As with the T-cell receptor (TCR), certain regions of the polypeptide chains are folded to form characteristic domains. In fact, from a structural standpoint, one arm of the BCR can be compared to a TCR: The part farthest from the cell surface is a variable region that binds to the antigen, and the part closest to the cell surface is the constant region. The BCR's stem portion is also part of the constant region. A crucial difference between a BCR and a TCR, however, is that the BCR recognizes free antigen (meaning antigen not presented on an MHC molecule).

B-Cell Activation

Naive B cells gather in the secondary lymphoid organs to encounter antigens. When a naive B cell's receptors bind an antigen, the B cell internalizes that antigen and then degrades it into various peptide fragments to be presented on MHC class II molecules (**figure 15.18**). T_H cells, which also gather in the secondary lymphoid organs, scan the peptide fragments presented by naive B cells. If a T_H cell's T-cell receptor (TCR) binds any of the presented fragments, that T cell then delivers cytokines to the B cell to activate it. Note that the T_H cell does not need to recognize the same epitope as the B cell; in fact, a responding B cell probably recognized an epitope on a pathogen's surface, whereas a T_H cell could very well recognize a peptide fragment from within the pathogen. If no T_H cells recognize the peptides presented by a B cell, that B cell may become anergic, an important mechanism of peripheral tolerance. Once a B cell is activated, it proliferates as described previously, eventually giving rise to antibody-secreting plasma cells as well as memory cells (see figure 15.10).

Characteristics of Antibodies

All antibodies, also called immunoglobulins, have the same basic Y-shaped structure, called an antibody monomer (**figure 15.19**). Like the B-cell receptor (BCR), an antibody monomer is made of heavy chains and light chains folded into domains and linked by disulfide bonds. Consistent with the structure of the BCR, a variable region at the end of each arm gives the antibody its antigen-binding specificity; the remaining part of the antibody is the constant region. As described in the overview,

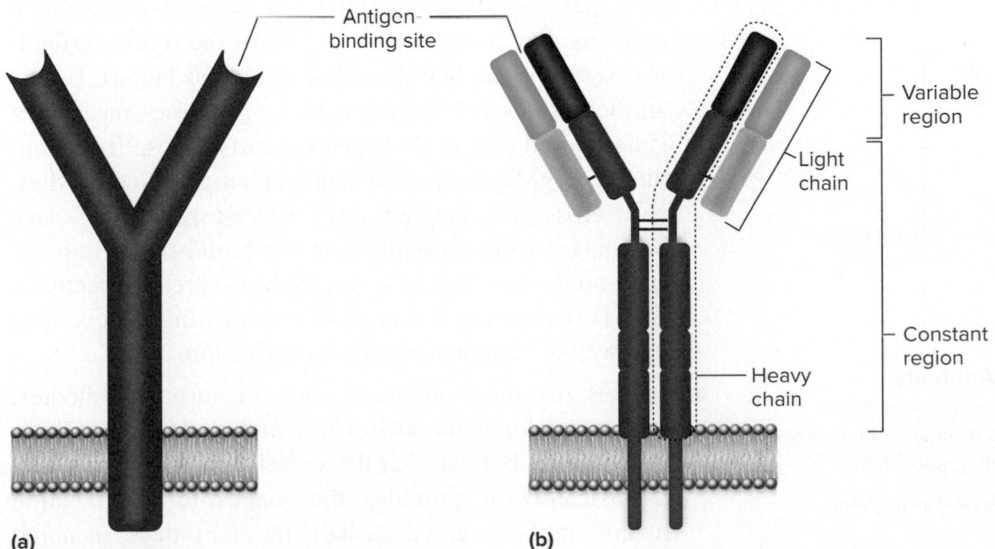

FIGURE 15.17 Structure of a B-Cell Receptor (BCR) (a) Simplified structure, showing the antigen-binding site. **(b)** The B-cell receptor is made up of two identical heavy chains and two identical light chains. Disulfide bonds link the chains. The amino acid sequence of the variable region accounts for the antigen-binding specificity.

? **How is the B-cell receptor similar to a T-cell receptor? How is it different?**

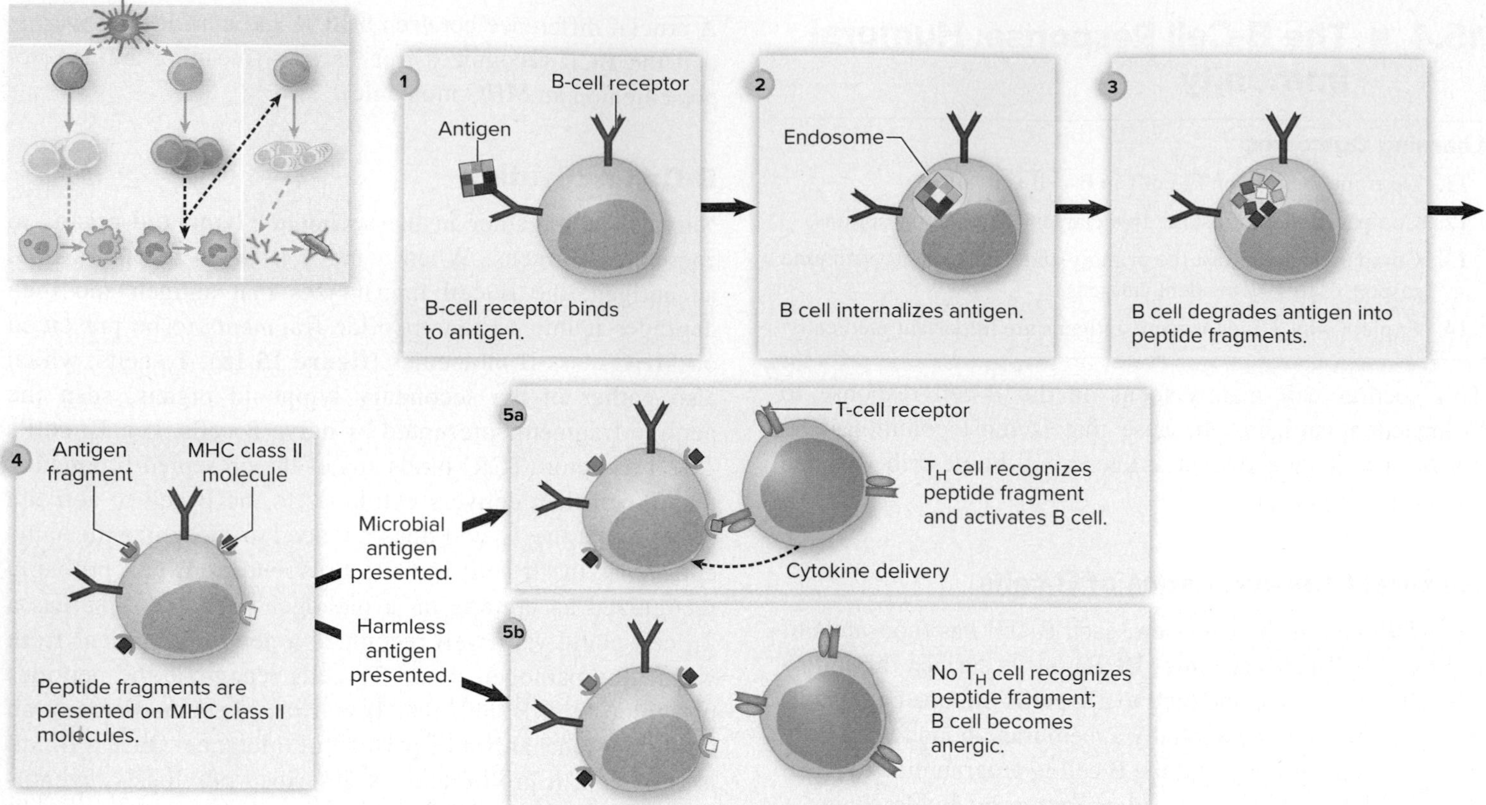

FIGURE 15.18 B-Cell Activation The B cell processes the antigen and presents it to T_H cells. If a T_H cell recognizes the antigen, it activates the B cell, allowing it to undergo clonal expansion.

? Why is it beneficial to the body for a B cell to become anergic if no T_H recognizes an antigen fragment that the B cell presents?

the antibody monomer has two general parts: The two identical arms are called the **Fab region** (or Fab regions) and the stem is the **Fc region.** These names were assigned following early studies that showed that enzymatic digestion of antibodies yielded two types of fragments—fragments that were antigen-binding (Fab) and fragments that could be crystallized (Fc). The junction of the Fab and Fc regions is a flexible stretch called the hinge region.

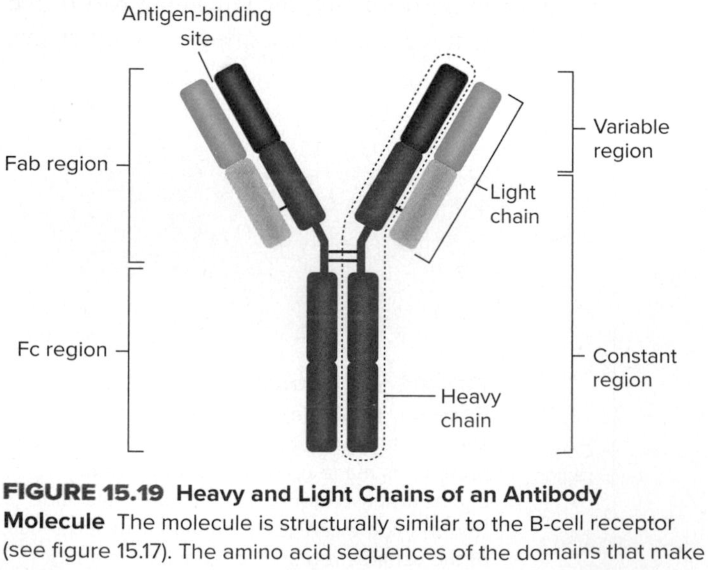

FIGURE 15.19 Heavy and Light Chains of an Antibody Molecule The molecule is structurally similar to the B-cell receptor (see figure 15.17). The amino acid sequences of the domains that make up the variable region account for the antigen-binding specificity.

? Why would cutting the molecule with some enzymes yield two Fab fragments whereas cutting with others yields only one?

Five general types of constant regions correspond to the major classes of immunoglobulin (Ig) molecules. Each class (also called an isotype) has distinct functions and properties:

- **IgM** is the first antibody class produced during the primary immune response (as shown in figure 15.10) and the principal one made in response to some T-independent antigens. It accounts for 5% to 13% of the serum antibodies (serum is the liquid portion of clotted blood). Unlike antibodies of other classes, IgM is a pentamer, meaning it is composed of five Y-shaped subunits. Those five subunits give IgM a total of 10 identical antigen-binding sites, so it cross-links antigens very effectively. Its large size prevents it from crossing from the bloodstream into tissues, so its main role is to control bloodstream infections. IgM is the most efficient class in triggering the classical pathway of complement system activation.
- **IgG** is the most abundant class of serum antibodies, accounting for about 80% to 85% of the total. It circulates in the blood but can exit the vessels to enter the tissues. Significantly, it provides the longest-term protection of any antibody class; its half-life is 21 days, meaning

that the number of IgG molecules in a given set will be reduced by about 50% after 21 days. IgG is generally the first and most abundant serum antibody produced during the secondary immune response, with protective mechanisms including neutralization, opsonization, complement activation, immobilization and prevention of adherence, cross-linking, and ADCC (see figure 15.4). Another important feature is that IgG can cross the placenta, so it protects a developing fetus against infections. Women who are not immune to a particular pathogen lack IgG against that microbe, so they are warned to take extra precautions during pregnancy to avoid pathogens that can infect and damage a fetus. Maternal IgG protects not only the fetus but also the newborn. The maternal antibodies present at birth gradually degrade over a period of about 6 months, but during this time the infant begins producing protective antibodies (**figure 15.20**). IgG is also in colostrum, the first form of breast milk produced after giving birth; the newborn's intestinal tract absorbs these antibodies.

- **IgA** is the most abundant antibody class produced. It is not common in serum, however, because the vast majority is a form called **secretory IgA (sIgA),** a dimer transported across mucosal membranes. During the transport, a polypeptide called the secretory component is added to each dimer; this not only attaches the sIgA to the layer of mucus that coats the mucosal surface but also protects the molecule from destruction by enzymes there. Secretory IgA provides **mucosal immunity** because it accumulates on mucous membranes that line the gastrointestinal, genitourinary, and respiratory tracts and is also in secretions such as saliva and tears. In addition, sIgA is in breast milk, thereby protecting breast-fed infants against intestinal pathogens. Protection by sIgA is primarily due to the direct effect of its binding, an action that neutralizes toxins and viruses and interferes with the attachment of microbes to host cells. IgA is mainly produced by plasma cells that descended from B cells in the mucosa-associated lymphoid tissues (MALT); those B cells would have encountered antigens entering through mucous membranes, so an sIgA response against them is appropriate. Relatively little IgA remains in the serum; only about 10% to 13% of antibodies in the serum are IgA, and these are the monomeric form.
- **IgD** accounts for less than 1% of all serum immunoglobulins. It is involved with the development and maturation of the antibody response, but its functions in serum have not been clearly defined.
- **IgE** is barely detectable in serum because rather than being free in circulation, most is tightly bound via the Fc region to basophils and mast cells, allowing these cells to detect and respond to antigens. When adjacent IgE molecules carried by a basophil or a mast cell bind to an antigen, the cell degranulates to release histamine and other inflammatory mediators; this IgE-associated action helps eliminate parasites, particularly parasitic worms (helminths). Unfortunately for allergy sufferers, basophils and mast cells also degranulate when the IgE they carry binds to normally harmless materials such as foods, dusts, and pollens, leading to immediate reactions such as coughing, sneezing, and tissue swelling. In some cases these allergic, or hypersensitivity, reactions can be life-threatening.

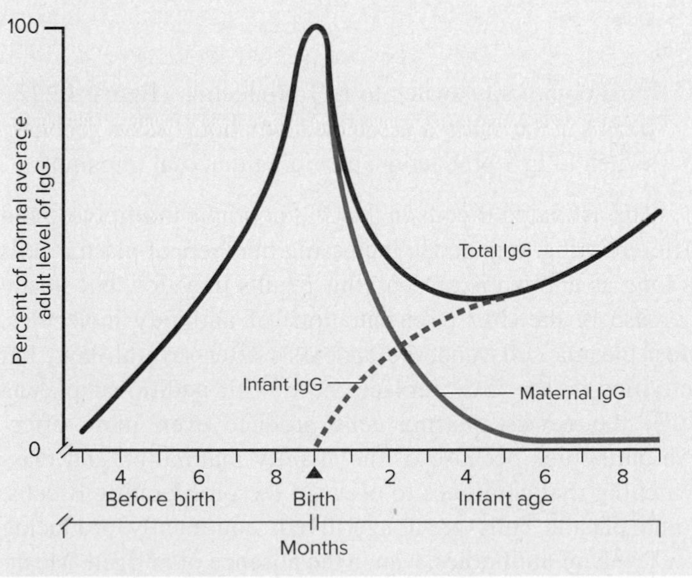

FIGURE 15.20 Immunoglobulin G Levels in the Fetus and Infant As the fetus develops, maternal IgG is transported across the placenta to provide protection. After birth, the infant begins producing antibodies.

? **Why does the level of maternal antibodies in an infant decrease over time?**

The characteristics of the main classes of antibody molecules are summarized in **table 15.2.**

Evolution of the Humoral Response to T-Dependent Antigens

During a humoral response to a given T-dependent antigen, changes occur in the responding B cell population that result in even more effective antibodies.

Primary Immune Response

When the relatively few naive B cells that recognize a particular antigen are activated, they multiply to generate a population of clones. Some of the clones differentiate to form antibody-secreting plasma cells, ultimately resulting in the first detectable level of IgM production; others, however, move within a secondary lymphoid organ to form a region called a **germinal center (GC).** The GC can be viewed as a B-cell "training camp"—a concentrated region of activity

TABLE 15.2 Characteristics of the Main Classes of Antibodies

Class and Molecular Weight (Daltons)	Structure	Percentage of Total Serum Immunoglobulin (Half-Life in Serum)	Properties and Functions
IgM 970,000		5–13% (10 days)	■ First antibody class made during the primary immune response. ■ Principal class made in response to T-independent antigens. ■ Provides direct protection by neutralizing viruses and toxins, immobilizing motile organisms, preventing microbes from adhering to cell surfaces, and cross-linking antigens. ■ Binding of IgM to antigen leads to activation of the complement system (classical pathway).
IgG 146,000		80–85% (21 days)	■ Most abundant class in the blood and tissue fluids. ■ Provides longest-term protection because of its long half-life. ■ Provides direct protection by neutralizing viruses and toxins, immobilizing motile organisms, preventing microbes from adhering to cell surfaces, and cross-linking antigens. ■ Binding of IgG to antigen facilitates phagocytosis of that antigen, leads to activation of the complement system (classical pathway), and allows antibody-dependent cellular cytotoxicity (ADCC). ■ Transported across the placenta, providing protection to a developing fetus; long half-life extends the protection through the first several months after birth.
IgA monomer 160,000; secretory IgA 390,000		10–13% (6 days)	■ Most abundant class produced, but the majority of it is secreted into mucus, tears, and saliva, providing mucosal immunity. ■ Protects mucous membranes by neutralizing viruses and toxins, immobilizing motile organisms, and preventing attachment of microbes to cell surfaces. ■ Component of breast milk; protects the intestinal tract of breast-fed infants.
IgD 184,000		<1% (3 days)	■ Involved in the development and maturation of the antibody response. Its functions in blood are not well understood.
IgE 188,000		<0.01% (2 days)	■ Binds via the Fc region to mast cells and basophils. This bound IgE allows those cells to detect parasites and other antigens and respond by releasing their granule contents. ■ Involved in many allergic reactions.

where various cell types, including T_H cells, assist and direct the activated B cells to optimize the response. As a result, the B cells undergo changes including:

- **Affinity maturation.** This is a form of natural selection among proliferating B cells (**figure 15.21**). As activated B cells multiply, spontaneous mutations commonly occur in certain regions of the B-cell receptor (BCR) genes. Some of these result in slight changes in the antigen-binding site of the BCRs. B cells that are most effective at binding antigen are most likely to proliferate and, in turn, the plasma cells that descend from these cells secrete antibodies that bind the antigen more effectively (recall that antibodies are essentially secreted forms of the BCRs).
- **Class switching.** All B cells are initially programmed to differentiate into plasma cells that secrete IgM. As the activated B cells multiply, however, some are induced to switch that genetic program through loss of DNA, causing them to differentiate into plasma cells that secrete other antibody classes. B cells in the lymph nodes most commonly switch to IgG production (**figure 15.22**). B cells in the mucosa-associated lymphoid tissues generally switch to IgA production, providing mucosal immunity.

The activated B cells in the GCs continue multiplying and differentiating to generate increasing numbers of plasma cells as long as antigen is present; this results in a slow but steady increase in the titer (concentration) of antibody molecules. Most plasma cells undergo apoptosis after several days, but activities in the GCs replace these with additional plasma cells; the newest plasma cells produce even more effective antibodies because of the affinity maturation and class switching that continues to occur in the proliferating B cells. Some plasma cells are longer-lived, continually producing low levels of antibodies even in the absence of antigen. Meanwhile, some of the B-cell clones become long-lived memory B cells, which can quickly become activated if the antigen is encountered again. In addition, a specialized cell type in the germinal center gathers and retains some of the antigens, helping to maintain memory of the response; that cell type

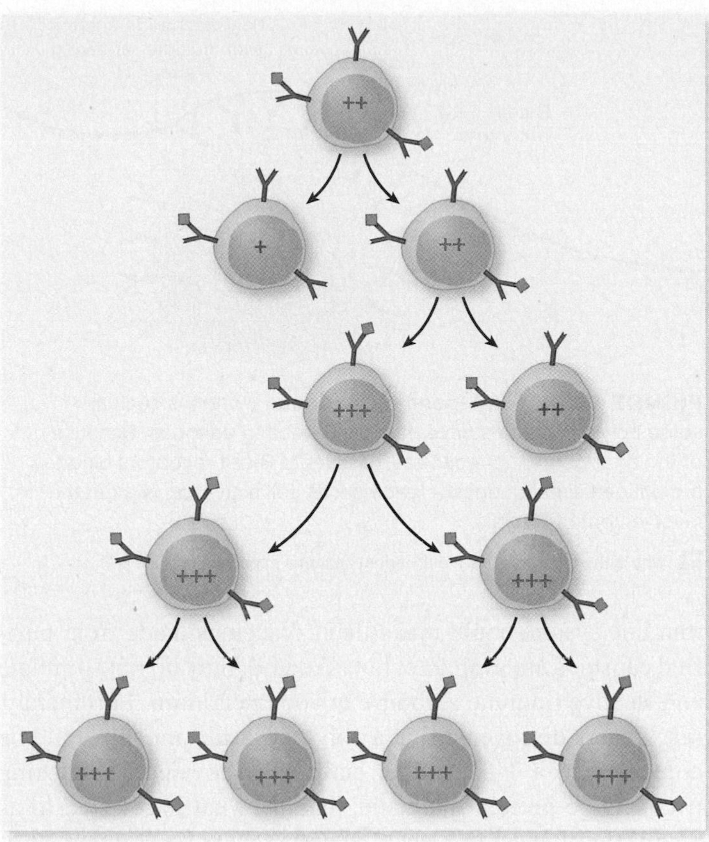

FIGURE 15.21 Affinity Maturation B cells that bind antigen most effectively are the most likely to proliferate. The plus signs indicate the relative quality of binding of the B-cell receptors (and therefore the antibodies) to the antigen; those in green indicate the most "fit" to continue proliferating.

? **What accounts for the change in a B cell's ability to bind antigen?**

also produces cytokines that promote affinity maturation and class switching to IgG.

Once the antigen is cleared, the antibody response begins to decrease. As fewer molecules of antigen remain to stimulate the lymphocytes, the activated lymphocytes undergo apoptosis. The long-lived plasma cells continue producing antibodies, however, resulting in a steady but low supply.

Secondary Immune Response

Because of memory lymphocytes, the secondary immune response is much faster and more effective than the primary immune response. In fact, repeat invaders are generally eliminated before they cause noticeable harm. This is why a person who has recovered from a particular disease typically has long-lasting immunity to the causative agent; vaccination takes advantage of this naturally occurring phenomenon. Although the secondary immune response involves both cell-mediated immunity and humoral immunity, the outcome is easier to demonstrate and monitor with humoral immunity because antibody levels are relatively easy to measure.

Memory B cells can scavenge even low concentrations of antigen because their receptors have been fine-tuned through affinity maturation during the primary immune response (see figure 15.21). When the memory cells bind to antigen,

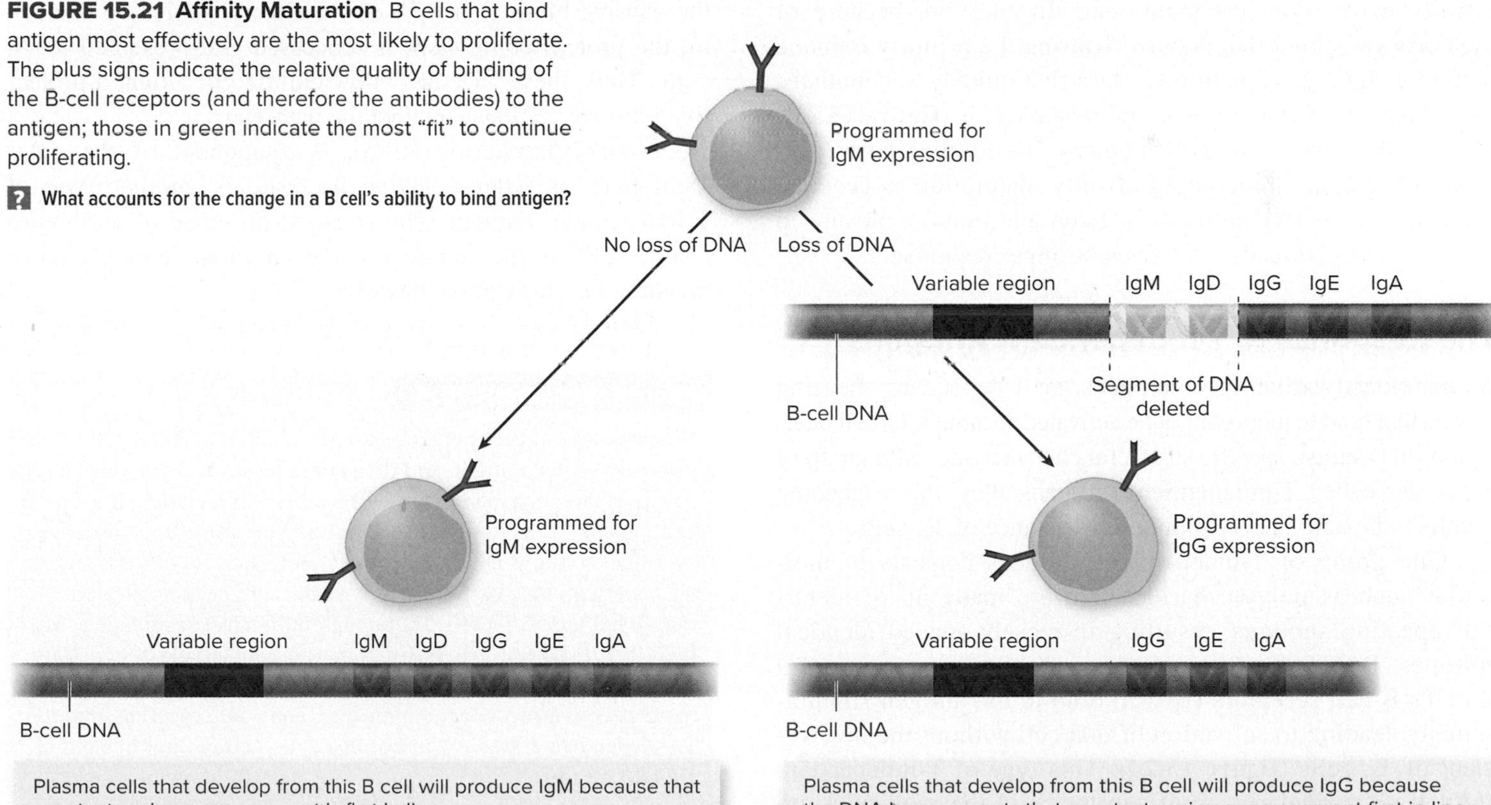

FIGURE 15.22 Class Switching Naive B cells are programmed to produce IgM antibodies. Activated B cells undergo class switching. Class switching does not alter the variable region and therefore has no effect on the antibody specificity. The plasma cells descended from B cells in lymph nodes most commonly produce IgG after class switching.

? **How would this illustration change if it showed class switching by B cells residing in Peyer's patches?**

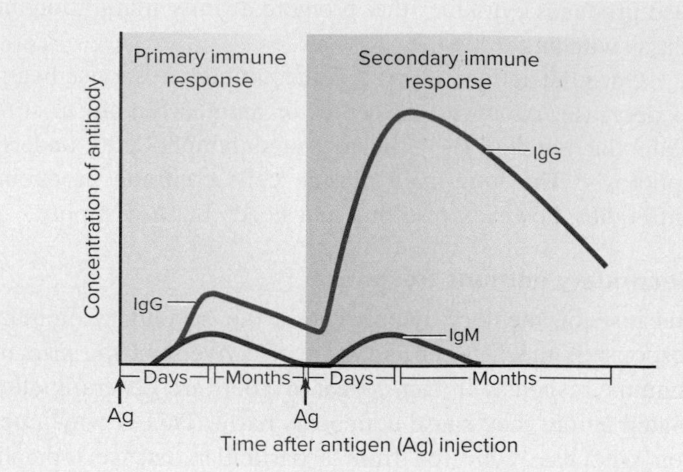

FIGURE 15.23 The Primary and Secondary Immune Responses to Antigen The first exposure to antigen elicits relatively low amounts of IgM, followed by IgG. The second exposure, which characterizes the memory of the adaptive immune system, elicits rapid production of relatively large quantities of IgG.

How would this graph be different if it illustrated antibody levels on a mucosal surface?

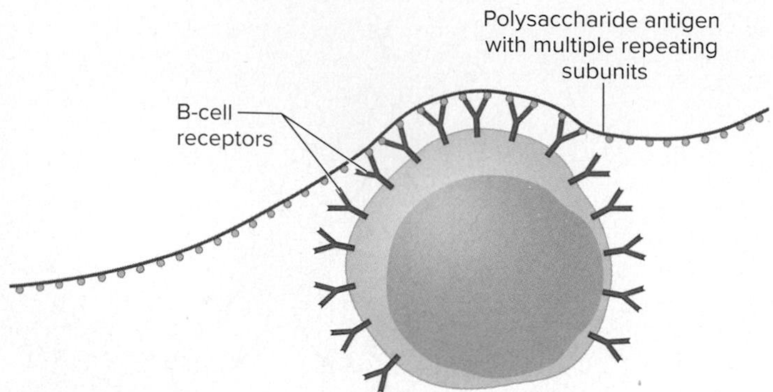

FIGURE 15.24 T-Independent Antigens Antigens such as some polysaccharides have multiple repeating epitopes. Because of the arrangement of epitopes, clusters of B-cell receptors bind to the antigen simultaneously, leading to B-cell activation without the involvement of T_H cells.

Why is the response to T-independent antigens important medically?

some of them quickly differentiate to form plasma cells, resulting in the rapid production of antibodies; these antibodies are also highly effective in binding antigen because of the affinity maturation just mentioned. In addition, because of the class switching that occurred during the primary immune response, IgG is the antibody class that quickly accumulates to highest levels in the blood and tissue fluids (**figure 15.23**). Meanwhile, other activated memory B cells begin proliferating, once again undergoing affinity maturation to generate even more effective antibodies. Each additional exposure to the same antigen leads to an even stronger response.

The Response to T-Independent Antigens

As mentioned earlier, most antigens are T-dependent, meaning B cells that bind to them cannot be activated without a T_H cell interaction. In contrast, specific structural characteristics of a group of molecules called **T-independent antigens** allow the responding B cells to be activated without the assistance of T_H cells.

One group of T-independent antigens consists of molecules such as polysaccharides that are made up of identical repeating subunits, resulting in closely spaced identical epitopes. Because of the epitope arrangement, clusters of a B cell's B-cell receptors (BCRs) bind to the antigen simultaneously, leading to activation of that cell without the involvement of T_H cells (**figure 15.24**). This type of T-independent antigen is particularly significant because the molecules are not very immunogenic in young children. As a result, children less than 2 years of age are more susceptible to pathogens such as *Streptococcus pneumoniae* and *Haemophilus influenzae*, which coat themselves in polysaccharide capsules. Antibodies that bind the capsules would be protective if the child's immune system could make them. Vaccines made from purified capsules are available, but, likewise, they do not stimulate an effective immune response in young children. Fortunately, researchers discovered that a polysaccharide antigen could be converted to a T-dependent antigen by covalently attaching it to a large protein molecule, making what is referred to as a **conjugate vaccine.** Once the polysaccharide component of the vaccine binds to a BCR and the entire molecule is taken in, the protein component is processed and presented to T_H cells. Thus, these vaccines allow children to mount a protective immune response against the pathogens.

Lipopolysaccharide (LPS), a component of the outer membrane of Gram-negative bacteria, is another type of T-independent antigen. The constant presence of antibodies against LPS is thought to provide an early defense against invading Gram-negative bacteria.

MicroAssessment 15.4

When a naive B cell's receptors bind an antigen, the B cell internalizes that antigen and then degrades it into various peptide fragments to be presented on MHC class II molecules. If a T_H cell recognizes any of the presented peptide fragments, it delivers cytokines to the B cell to activate it. Activated B cells proliferate, ultimately giving rise to antibody-secreting plasma cells and long-lived memory cells. Immunoglobulin classes include IgM, IgG, IgA, IgD, and IgE. Affinity maturation and class switching occur in the primary immune response; these allow a swift and more effective secondary immune response. B cells that bind to T-independent antigens can become activated without the aid of T_H cells.

10. Describe the significance of class switching.
11. How does the ability to bind antigen increase as B cells multiply?
12. Why should B cells residing in the mucosa-associated lymphoid tissues produce IgA?

FOCUS ON A CASE 15.1

A 34-year-old man who had recently returned from a safari in Africa experienced what he assumed was severe jet lag, feeling dizziness, a headache, and body aches. His condition gradually worsened over several days as he developed a fever and became nauseated. At his family's urging, he visited his primary care physician, who questioned him about the trip. The patient reported that he had visited a travel clinic beforehand, where he was vaccinated against several diseases, including polio, typhoid fever, meningococcal meningitis, yellow fever, and hepatitis A. He was also given information about measures to prevent malaria and was prescribed an anti-malarial medication as a prophylaxis (preventive measure) but had stopped taking it after only a few days. He did, however, follow the clinic's advice to use an insect repellent while he was there because mosquitoes transmit the disease, but he ignored the suggestion of using a product that contained DEET (*N, N*-diethyl-*meta*-toluamide). Instead, he chose a natural plant-based product that was not very effective.

Based on the patient's symptoms and travel history, the physician suspected malaria. This disease is caused by several different species of *Plasmodium,* a type of eukaryotic parasite that infects red blood cells (RBCs). The physician took a blood sample and immediately sent it to the clinical lab for testing. When the technician examined the stained specimen under the microscopic, she saw many *Plasmodium*-infected RBCs; she also noticed banana-shaped *Plasmodium* gametocytes—a morphology that identifies the parasite as *P. falciparum.* She quickly reported the results to the physician, recognizing that malaria caused by this species can be a medical emergency.

The patient was admitted to the hospital, where he was immediately started on an intravenous antimalarial medication. His condition continued to deteriorate, however, and he began experiencing episodes of chills and fever, and was often drenched with sweat. Within a matter of days, he developed neurological symptoms, including confusion, anxiety, and seizures. He soon slipped into a coma and died.

Autopsy results indicated that the patient had developed cerebral malaria, a deadly complication that sometimes occurs with malaria caused by *P. falciparum.* This species makes a protein that inserts into the cytoplasmic membrane of infected RBCs, causing those cells to stick to capillary walls, blocking blood flow and depriving local tissues of O_2; cerebral malaria is the result of capillaries being blocked in the brain

1. *Plasmodium* species are characterized by a complex life cycle that includes multiple antigenically distinct stages. How would the parasite benefit by having several different antigenic forms?
2. What is the advantage to the parasite of multiplying within RBCs?
3. What is the advantage to the parasite of making a protein that causes RBCs to stick to the capillary walls?
4. Why might the patient have died even though he was being treated?

Discussion

1. The multiple antigenic forms make it more difficult for the immune response to eliminate the parasite. By the time an effective adaptive response has developed against one stage in the parasite's life cycle, the organism has progressed to the next form. Over a period of weeks, however, an effective immune response against the form that infects RBCs limits disease progression.
2. By hiding within one of the body's own cells, the parasite avoids humoral immunity. Multiplying within RBCs provides an additional advantage because the cells do not make MHC molecules, so the infected cells are not a target of cytotoxic T cells.
3. By sticking to the capillaries, the RBCs do not circulate in the bloodstream. Circulating blood passes through the spleen, a secondary lymphoid organ that removes old and damaged RBCs. The parasite has multiple genes for the protein that attaches to the walls of the capillaries, so it can produce dozens of antigenically distinct varieties; as the immune response begins to make antibodies that recognize one version of the protein, some of the parasites will have switched to make a different antigenic type.
4. The damage had probably already started before treatment was begun. Once the RBCs clog the capillaries, the obstruction deprives brain cells of O_2. An inflammatory response develops in the damaged tissues, which makes the problem even worse. The infecting parasite might also have been resistant to the medications being delivered; drug resistance in *Plasmodium* species is an increasing problem.

15.5 ■ Lymphocyte Development

Learning Outcomes

15. Describe the roles of gene rearrangement, imprecise joining, and combinatorial associations in the generation of diversity.
16. Describe positive and negative selection of lymphocytes.

As descendants of hematopoietic stem cells develop into B cells and T cells, they acquire their ability to recognize distinct epitopes. B cells undergo the developmental stages in the bone marrow; T cells begin their development in the bone marrow, but then move to the thymus where further maturation processes occur.

Generation of Diversity

Studies using B cells uncovered the genetic mechanisms that give rise to a population of lymphocytes that together produce a seemingly limitless assortment of antibodies and antigen-specific receptors. The processes in B cells are very similar to those in T cells, so we will use B cells as a general model to describe this generation of diversity.

Each B cell responds to only one epitope, yet the population of B cells within the body appears able to respond to more than 100 million different epitopes. Based on this number and the information presented in chapter 7, it might seem logical to assume that the human genome has over 100 million different antibody genes, each encoding specificity for a single epitope. This is impossible, however, because the human genome has only 3 billion nucleotides and contains about 25,000 genes.

The question of how such tremendous diversity in antibodies could be generated puzzled immunologists until Dr. Susumu Tonegawa solved the mystery. For this work, he was awarded a Nobel Prize in 1987.

Gene Rearrangement

A primary mechanism for producing a wide variety of different antibodies using a limited-size region of DNA relies on a strategy similar to that of a practical and well-dressed traveler living out of a small suitcase. A traveler can create a wide variety of unique outfits from a limited number of components by mixing and matching different shirts, pants, and shoes. Likewise, a B cell forms a nearly unique variable region of an antibody heavy chain by "mixing and matching" gene segments—using one each from sets of DNA segments called V (variable), D (diversity), and J (joining). Recall that a membrane-bound form of an antibody also serves as the B-cell receptor.

With respect to the DNA that encodes the variable region of an antibody heavy chain, a human hematopoietic stem cell has about 40 different V segments, 25 different D segments, and 6 different J segments (**figure 15.25**). As a stem cell

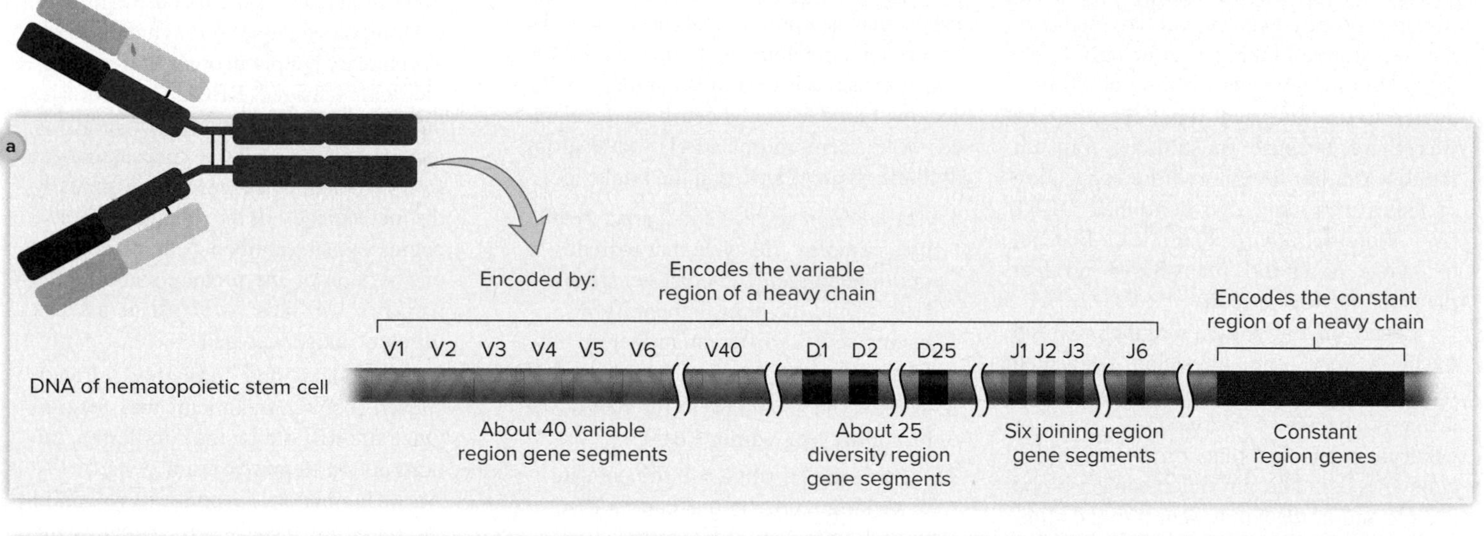

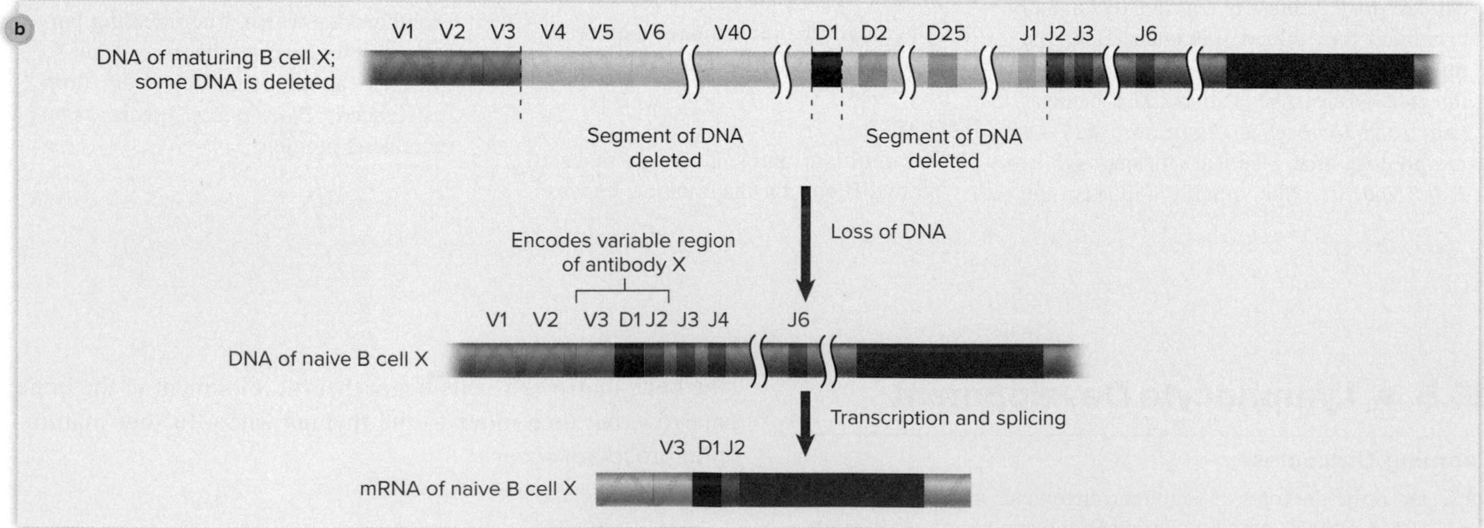

FIGURE 15.25 Antibody Diversity (a) The variable region of an antibody heavy chain is encoded by one each of three gene segments: V (variable), D (diversity), and J (joining). **(b)** As a hematopoietic stem cell differentiates to become a B cell, some of the DNA in the antibody-encoding region is lost, resulting in the joining of distinct V, D, and J segments. Together, the joined segments encode the variable region of the antibody heavy chain that the given B cell is programmed to make. This diagram shows only the gene segments for the heavy chain and is not drawn to scale.

? Considering that a similar process occurs in the light chain genes, how does the random joining of a light chain and heavy chain contribute to antibody diversity?

differentiates to become a B cell, two large regions of that DNA are permanently removed, thereby joining distinct V, D, and J segments. Those joined segments encode the antibody heavy chain that the mature B cell is programmed to make. Thus, one B cell could express the combination V3, D1, and J2 to produce its heavy chain, whereas another B cell might use V19, D25, and J6; each V, D, and J combination would result in a unique specificity with respect to the epitope recognized. In addition to the rearrangements that occur in the heavy chain gene segments, a similar process happens in the light chain genes as well.

Imprecise Joining

As the DNA segments are joined together during the gene rearrangement just described, nucleotides are often deleted or added between them. This imprecise joining changes the reading frame of the encoded polypeptide so that two B cells that have the same V, D, and J segments for their heavy chain could potentially give rise to antibodies with very different specificities. Likewise, the segments that encode the light chain often join imprecisely.

Combinatorial Associations

The variable regions of the heavy and light chains independently acquire diversity through the mechanisms just described; the randomly generated heavy chain and light chain are then combined to create an antibody molecule. This substantially increases the potential for antibody diversity because of the many light chain/heavy chain combination possibilities; recall that the antigen-binding site of an antibody molecule involves the variable domains of both chains (see figure 15.19).

Negative Selection of Self-Reactive B Cells

Once a maturing B cell has developed its antigen receptor (B-cell receptor), it undergoes a process called negative selection, which eliminates any B cell that binds "self." Most developing B cells fail negative selection and, as a consequence, are induced to undergo apoptosis. If these cells are not eliminated, then the immune system may attack "self" substances by mistake, resulting in autoimmune disease.

Positive and Negative Selection of Self-Reactive T Cells

The fate of developing T cells rests on two phases of trials: positive and negative selection. Positive selection permits only those T cells that recognize MHC to develop further. Recall that the T-cell receptor, unlike the B-cell receptor, recognizes a peptide:MHC complex. The T-cell receptor, therefore, must show at least some recognition of an MHC molecule regardless of the peptide fragment being carried. T cells that show insufficient recognition fail positive selection and, as a consequence, are eliminated. Each T cell that passes positive selection is also subjected to negative selection. T cells that recognize "self" peptides presented on MHC molecules are eliminated. Positive and negative selection processes are so strict that over 95% of developing T cells undergo apoptosis in the thymus.

MicroAssessment 15.5

Diversity of antigen specificity in lymphocytes is generated through rearrangement of gene segments, imprecise joining of those segments, and combinatorial associations of heavy chains and light chains. Negative selection eliminates B cells and T cells that recognize normal "self" molecules. Positive selection permits only those T cells that recognize the MHC molecules to develop further.

13. What three gene segments encode the variable region of the heavy chain of an antibody molecule?
14. Why is negative selection important?
15. How is imprecise joining similar to a frameshift mutation?

15.6 ■ Natural Killer (NK) Cells

Learning Outcome

17. Describe two distinct protective roles of NK cells.

Natural killer (NK) cells are innate lymphoid cells (ILCs), a group of lymphocytes lacking the antigen-specific receptors that characterize B cells and T cells. The activities of NK cells assist the adaptive immune responses.

NK cells induce apoptosis in antibody-bound self cells. This process, **antibody-dependent cellular cytotoxicity (ADCC)**, allows NK cells to destroy self cells that have viral or other foreign proteins inserted into their membrane (see figure 15.4). NK cells can do this because they have Fc receptors for IgG molecules on their surface; recall that Fc receptors bind the "red flag" portion of antibody molecules. The NK cell attaches to the antibodies on the self cell and then delivers perforin- and protease-containing granules directly to that cell, initiating apoptosis.

NK cells also recognize and destroy stressed self cells that do not have MHC class I molecules on their surface (**figure 15.26**). This is important because some viruses have evolved mechanisms to avoid the action of cytotoxic T cells by interfering with the process of antigen presentation; cells infected with such a virus will essentially lack MHC class I molecules and thus cannot be a target of cytotoxic T cells.

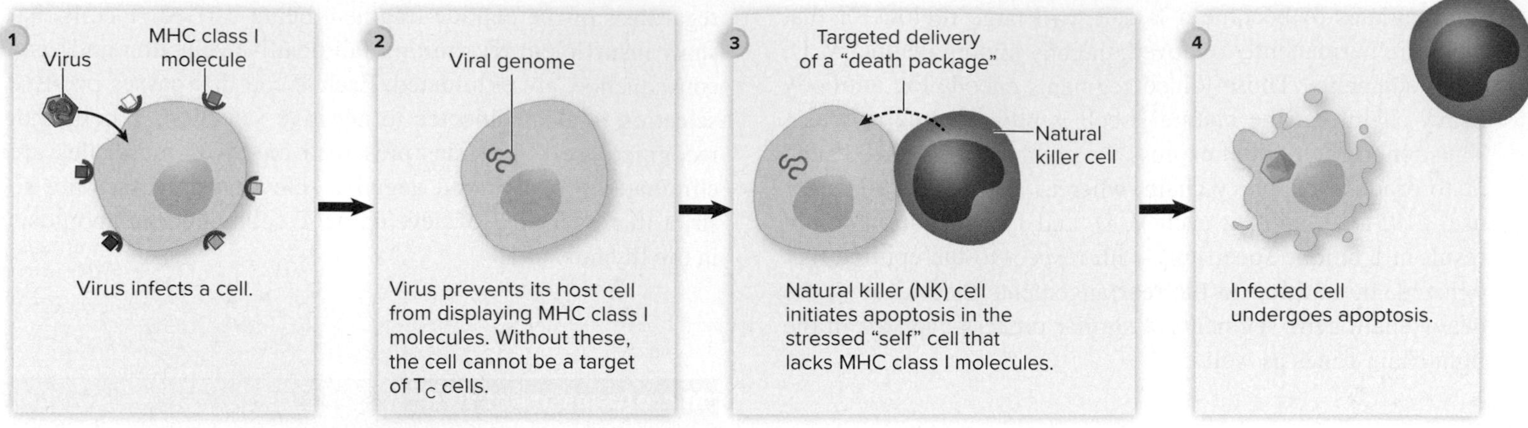

FIGURE 15.26 Natural Killer (NK) Cells Destroy Stressed "Self" Cells That Lack MHC Class I Molecules

? Why would interfering with a host cell's production of MHC class I molecules be an advantage to a virus?

The NK cells recognize the absence of MHC class I molecules on those cells, along with the presence of certain molecules that indicate the cells are under stress, and induce the infected cells to undergo apoptosis.

Recent evidence indicates that NK cells are more than killing machines. For example, they produce cytokines that help regulate and direct certain immune responses. Unfortunately, studying these actions is difficult because there are different subsets of NK cells, and the activities of the various subsets are influenced by cues in their local environment.

MicroAssessment 15.6

Natural killer (NK) cells can kill antibody-bound cells by antibody-dependent cellular cytotoxicity (ADCC). NK cells also kill stressed cells that lack MHC class I molecules on their surface.

16. What mechanism do NK cells use to kill self cells?

17. What can cause a self cell to not display MHC class I molecules?

18. Why might a virus that interferes with the production of the host's MHC class I molecules encode a non-functional version?

Summary

15.1 ■ Overview of the Adaptive Immune Response (figure 15.1)

Cell-mediated immunity (CMI) deals with intracellular antigens; it involves **T cells (T lymphocytes). Humoral immunity** works to eliminate extracellular antigens; it involves **B cells (B lymphocytes).** The antigen-binding site of a T cell's **T-cell receptor (TCR)** and B cell's **B-cell receptor (BCR)** allows the respective cells to recognize antigens (figure 15.2). As part of peripheral tolerance, naive lymphocytes generally cannot respond until they receive signals from another cell type to become activated. An activated lymphocyte can proliferate, ultimately giving rise to effector lymphocytes and memory lymphocytes. As a result of the **primary immune response** to an antigen, the **secondary immune response** is more effective. Memory lymphocytes are responsible for the effectiveness of the secondary immune response.

Cell-Mediated Immunity

In response to intracellular antigens, **cytotoxic T cells** proliferate, and then many of the clones differentiate into **T_C cells** that induce apoptosis in "self" cells harboring the intruder. Memory cytotoxic T cells are also formed. **Helper T cells** proliferate, and then many of the clones differentiate to form **T_H cells** that help orchestrate the various responses of humoral and cell-mediated immunity. Memory helper T cells are also formed.

Humoral Immunity

In response to extracellular antigens, B cells proliferate, and then many of the clones differentiate into **plasma cells** that function as antibody-producing factories. **Antibodies** are Y-shaped proteins with an antigen-binding site at the end of each arm (figure 15.3). The tail of the Y is the **Fc region.** Memory B cells are also formed. Antibodies binding to antigens result in neutralization, **opsonization,** complement activation, immobilization and prevention of adherence, cross-linking, and **antibody-dependent cellular cytotoxicity (ADCC)** (figure 15.4).

The Nature of Antigens

Antigens are molecules that react specifically with an antibody or lymphocyte. The immune response is directed to **epitopes** (antigenic determinants) on the antigen (figure 15.5).

The Lymphatic System (figure 15.7)

Lymph flows in the **lymphatic vessels** to the lymph nodes (figure 15.6). **Primary lymphoid organs** are the sites where B cells and T cells mature. **Secondary lymphoid organs** are the sites at which lymphocytes gather to contact antigens (figure 15.8).

The Big Picture Summary (figure 15.9)

15.2 ■ Clonal Selection and Expansion of Lymphocytes

When an antigen enters a secondary lymphoid organ, only the lymphocytes that specifically recognize that antigen via their antigen receptor will respond (figure 15.10). Lymphocytes may be immature, naive, activated, effector, or memory cells.

15.3 ■ The T-Cell Response: Cell-Mediated Immunity

General Characteristics of T Cells (table 15.1, figure 15.15)

Cytotoxic (CD8) T cells recognize antigen presented on **major histocompatibility complex (MHC) class I molecules.** Helper (CD4) T cells recognize antigen presented on **MHC class II molecules** (figure 15.12).

Activation of T Cells (figure 15.13)

Dendritic cells sample material in tissues and then travel to secondary lymphoid organs to present antigens to naive T cells. The dendritic cells that detect molecules associated with danger produce **co-stimulatory molecules** and are able to activate both subsets of T cells.

Effector Functions of T_C (CD8) Cells

T_C cells induce apoptosis in cells that present antigens they recognize on MHC class I molecules; they also produce cytokines that increase the level of surveillance (figure 15.14). All nucleated cells present endogenous antigens on MHC class I molecules.

Effector Functions of T_H (CD4) Cells

T_H cells activate cells that present antigens they recognize on MHC class II molecules; various cytokines are released, depending on the subset of the responding T_H cell (figures 15.16, 15.18). Macrophages and B cells present exogenous antigens on MHC class II molecules. Subsets of T_H cells direct the immune system to an appropriate response for a given antigen.

15.4 ■ The B-Cell Response: Humoral Immunity

Most antigens are **T-dependent antigens,** meaning the B cells that recognize them require help from T_H cells.

General Characteristics of B cells

The B-cell receptor recognizes free antigen—that is, antigen not presented on an MHC molecule (figure 15.17).

B-Cell Activation

B cells present peptide fragments from T-dependent antigens to T_H cells for inspection. If a T_H cell recognizes a peptide, it delivers cytokines to the B cell, initiating the process of clonal expansion, which ultimately gives rise to antibody-producing plasma cells (figure 15.18).

Characteristics of Antibodies

The antibody monomer is composed of two identical **heavy chains** and two identical **light chains.** The **variable region** contains the antigen-binding site; the **constant region** encompasses the entire Fc region as well as parts of the Fab regions. Antibody classes include IgM, IgG, IgA, IgD, and IgE (table 15.2).

Evolution of the Humoral Response to T-Dependent Antigens

During the primary immune response, the expanding B-cell population undergoes **affinity maturation** (figure 15.21). Under the direction of T_H cells, **class switching** and memory cell formation also occur (figure 15.22). Memory cells are responsible for the swift and effective secondary immune response, eliminating invaders before they cause noticeable harm (figure 15.23).

The Response to T-Independent Antigens

T-independent antigens include polysaccharides that have multiple identical evenly spaced epitopes, and LPS (figure 15.24).

15.5 ■ Lymphocyte Development

Generation of Diversity

Mechanisms that generate the diversity of antigen specificity in lymphocytes include rearrangement of gene segments, imprecise joining of those segments, and combinatorial associations of heavy and light chains (figure 15.25).

Negative Selection of Self-Reactive B Cells

Negative selection occurs as B cells develop in the bone marrow; if the B-cell receptor of a developing B cell binds material in the bone marrow, that B cell undergoes apoptosis.

Positive and Negative Selection of Self-Reactive T Cells

Positive selection permits only those T cells that show moderate recognition of the MHC molecules to develop further. Negative selection also occurs.

15.6 ■ Natural Killer (NK) Cells

NK cells mediate **antibody-dependent cellular cytotoxicity (ADCC).** NK cells also induce apoptosis in host cells not bearing MHC class I molecules on their surface (figure 15.26).

Review Questions

Short Answer

1. Diagram the basic structure of an antibody molecule and label (a) the Fc region and (b) the areas that combine with antigen.
2. Describe the protective outcomes of antibody-antigen binding.
3. What is a secondary lymphoid organ?
4. Describe clonal selection and expansion in the immune response.
5. Describe the role of dendritic cells in T-cell activation.
6. What are antigen-presenting cells (APCs)?
7. Which antibody class is the first produced during a primary immune response?
8. Which antibody class neutralizes viruses in the intestinal tract?
9. How do T-independent antigens differ from T-dependent antigens?
10. How does the role of natural killer cells differ from that of cytotoxic T cells?

Multiple Choice

1. Which of the following specifically refers to an effector lymphocyte?
   a) B cell
   b) CD8 T cell
   c) CD4 T cell
   d) Plasma cell

2. Which markers are found on all nucleated cells?
   a) MHC class I molecules
   b) MHC class II molecules
   c) CD4
   d) CD8
3. Which of the following are examples of antigen-presenting cells (APCs)?
   a) Macrophage and neutrophil
   b) Macrophage and B cell
   c) Neutrophil and T cell
   d) B cell and plasma cell
   e) Macrophage, neutrophil, and plasma cell
4. Which is an appropriate response if an antigen is presented on MHC class II molecules?
   a) An effector CD8 cell kills the presenting cell.
   b) An effector CD4 cell kills the presenting cell.
   c) An effector CD8 cell activates the presenting cell.
   d) An effector CD4 cell activates the presenting cell.
5. The variable regions of antibodies are located in the
   a) Fc region and light chain only.
   b) Fc region, light chain, and heavy chain.
   c) Fab region and light chain only.
   d) Fab region and heavy chain only.
   e) Fab region, light chain, and heavy chain.
6. Which of the following statements about antibodies is *false*?
   a) If you removed the Fc portion, antibodies would no longer be capable of opsonization.
   b) If you removed the Fc portion, antibodies would no longer be capable of activating the complement system.
   c) If you removed an Fab portion, an antibody would no longer be capable of cross-linking antigen.
   d) If IgG were a pentamer, it would bind antigens more efficiently.
   e) If IgE had a longer half-life, it would protect newborn infants.
7. Which class of antibody can cross the placenta?
   a) IgA
   b) IgD
   c) IgE
   d) IgG
   e) IgM
8. A person who has been vaccinated against a disease should have primarily which of these types of serum antibodies against that agent 2 years later?
   a) IgA
   b) IgD
   c) IgE
   d) IgG
   e) IgM
9. Which of the following statements about B cells/antibody production is *false*?
   a) A B cell initially has the potential to make more than one class of antibody.
   b) In response to antigen, all B cells located close to the antigen begin dividing.
   c) Each B cell is programmed to make a single specificity of antibody.
   d) The B-cell receptor allows B cells to detect antigen.
   e) The cell type that makes and secretes antibody is called a plasma cell.
10. Which term describes the loss of specific heavy chain genes?
   a) Affinity maturation
   b) Apoptosis
   c) Clonal selection
   d) Class switching

Applications

1. Many dairy operations keep cow's milk for sale and use formula and feed to raise any calves. One farmer noticed that calves raised on the formula and feed needed to be treated for diarrhea more frequently than calves left with their mothers to nurse. Tests run on the diets showed no differences in the calories or nutritional content. The farmer called a veterinarian and asked him to explain the observations. What was the vet's response?
2. What kinds of diseases would be expected to occur as a result of a lack of T or B lymphocytes?

Critical Thinking

1. The primary immune response to an antigen may take a week or more to develop fully, whereas the secondary immune response is rapid and relies on the activation of high numbers of memory cells. Would it be better if high numbers of reactive cells were available regardless of prior exposure, so that the body could respond rapidly to *any* antigen exposure? Why or why not?
2. Two patients had high levels of IgG antibodies that recognize a given SARS-CoV-2 antigen, but only one of the patients also had high levels of IgM antibodies recognizing the same antigen. What is the most likely explanation for these results?

16 | Host-Microbe Interactions

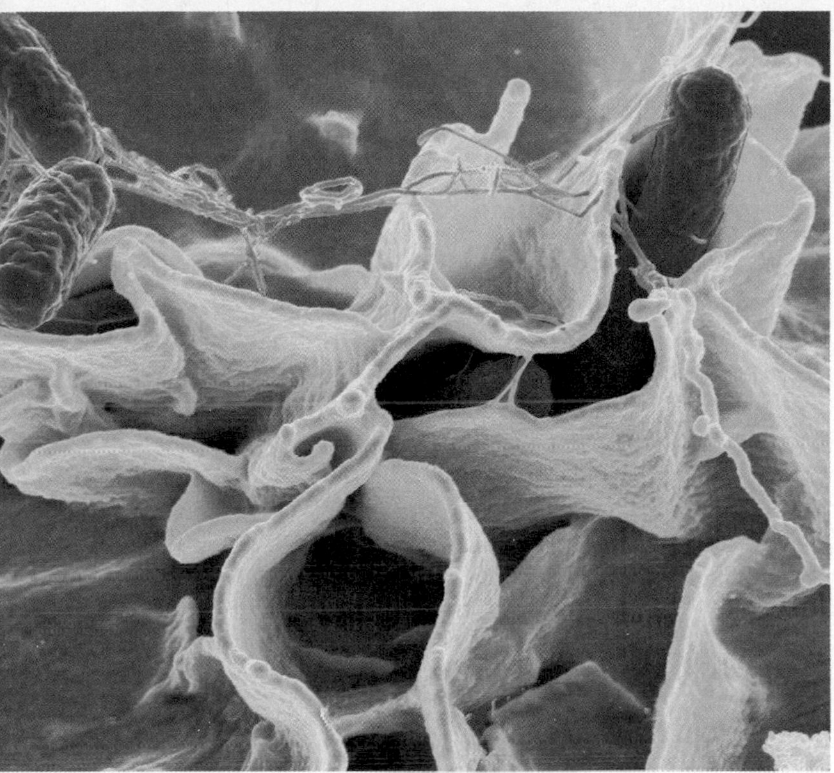

Salmonella enterica serotype Typhimurium invading cultured human cells (color-enhanced SEM). *Rocky Mountain Laboratories/NIH/NIAID*

KEY TERMS

Acute Infection An infection characterized by symptoms that develop fairly quickly and last a relatively short time.

Chronic Infection An infection that generally develops slowly and lasts for months or years.

Colonization Establishment and growth of a microorganism on a surface.

Endotoxin The lipopolysaccharide (LPS) component of the outer membrane of Gram-negative bacteria; lipid A is responsible for the toxic properties of LPS.

Exotoxin A toxic protein produced by a microorganism; often simply referred to as a *toxin.*

Immunocompromised Having a weakness or defect in the innate or adaptive defenses.

Infection Colonization by a pathogen on or within the body.

Infectious Disease An infection that prevents the body from functioning normally.

Latent Infection Infection in which the infectious agent is present but not causing symptoms.

Microbiome The total genetic information of a community of microorganisms in a given environment; also the community itself, in which case the term is often used synonymously with normal microbiota.

Normal Microbiota The group of microorganisms that routinely colonize the body of a healthy individual.

Opportunistic Pathogen A microbe that causes disease only when introduced into an unusual location or into an immunocompromised host.

Primary Pathogen A microbe able to cause disease in an otherwise healthy individual.

Virulence Factors Traits of a microbe that promote pathogenicity

A Glimpse of History

Throughout history, people have tried to understand the spread of diseases. In ancient times, people thought that illnesses were divine punishments for their sins. Gradually it became obvious, however, that at least some diseases could be transmitted by contact with people who had the disease. By the Middle Ages, many people accepted this and fled cities to escape diseases.

With van Leeuwenhoek's discovery of microorganisms in the late seventeenth century, people began to suspect that microorganisms might cause disease, but the techniques of the times could not prove this. It was not until 1876 that Robert Koch offered convincing evidence of what is now known as the germ theory of disease. He showed that *Bacillus anthracis* causes anthrax, an often fatal disease of humans, sheep, and other animals. With his microscope, Koch observed *B. anthracis* cells in the blood and spleen of dead sheep. He then inoculated mice with the infected sheep blood and recovered *B. anthracis* from the blood of those mice. In addition, he grew the recovered bacteria in pure culture and showed that they caused anthrax when injected into healthy mice. From these experiments and later work with *Mycobacterium tuberculosis,* Koch formalized what is now known as Koch's postulates—a group of criteria for establishing the cause of an infectious disease.

Every day we come into contact with an enormous number and variety of microorganisms. Some enter our respiratory system as we breathe; others are ingested with each bite of food or sip of drink; and still more adhere to our skin whenever we touch an object or surface. It is important to recognize, however, that the vast majority of these microbes cause no harm. Some may colonize the body surfaces, taking up residence with the variety of other harmless microbes that live there; others are shed with dead epithelial cells. Most of those swallowed are either killed in the stomach or eliminated in feces.

Relatively few microbes cause noticeable damage to the human body; those that can are called **pathogens.** They have distinct characteristics that allow them to avoid at least some of the body's defenses. Research into how these microbes evade our innate and adaptive defenses is leading to many fascinating discoveries; the knowledge gained will hopefully make it possible to develop therapies targeted at specific pathogens.

This chapter will explore some of the ways in which microbes colonize the human host, living either as members of the normal microbiota or causing disease. It will also describe how pathogens evade or overcome the immune responses and damage the host.

MICROBES, HEALTH, AND DISEASE

Many people think of microorganisms as "germs" that should routinely be killed or avoided. Most microbes are harmless, however, and many are beneficial. It is a delicate balancing act, though, because even typically harmless microbes can cause disease if the opportunity arises. Weaknesses or defects in the innate or adaptive defenses can leave people vulnerable to invasion; such individuals are said to be **immunocompromised.** Factors that can cause a person to become immunocompromised include malnutrition, cancer, AIDS, surgery, wounds, genetic defects, alcohol or drug abuse, and immunosuppressive therapy that accompanies procedures such as organ transplants. Pregnant females, while not considered immunocompromised, have an altered immune system that makes them more susceptible to some infections and more likely to develop a severe form of certain diseases, including COVID-19.

16.1 ■ The Anatomical Barriers as Ecosystems

Learning Outcome

1. Compare and contrast mutualism, commensalism, and parasitism.

The skin and mucous membranes are barriers against invading microorganisms, but they are also part of a complex ecosystem—an interacting biological community along with the environment that shapes it. The intimate relationships between the microorganisms and the human body are an example of **symbiosis,** meaning "living together."

Microorganisms can have a variety of symbiotic relationships with one another and with the human host. These relationships may take on different characteristics depending on the closeness of the association and the relative advantages to each partner. Symbiotic associations can take several forms, and these may change, depending on the state of the host and the traits of the microbes:

- **Mutualism.** This is an association in which both partners benefit. Members of the human normal microbiota contribute to our health in many ways; as just one example, some bacteria in the large intestine synthesize vitamin K and certain B vitamins, which the body can then absorb. The microbes benefit as well, as they are supplied with warmth and a variety of different energy sources.
- **Commensalism.** This is an association in which one partner benefits but the other remains unharmed. Many microbes living on the skin are neither harmful nor helpful to the human host, but they obtain nutrients from the host.
- **Parasitism.** This is an association in which one organism, the parasite, benefits at the expense of the other. All pathogens are parasites, but medical microbiologists often reserve the word *parasite* for eukaryotic pathogens such as certain protozoa and helminths (worms).

MicroAssessment 16.1

The symbiotic relationship between a host and a microbe can be described as mutualism, commensalism, or parasitism, depending on the relative benefit to each partner.

1. How is mutualism different from commensalism?

16.2 ■ The Human Microbiome

Learning Outcomes

2. Describe how the composition of a person's microbiome can change over time.
3. Describe (1) five beneficial roles of the normal human microbiota and (2) the gut–microbiome brain connection.

The **normal microbiota** of humans is the group of microorganisms routinely found growing on the body surfaces (skin and mucous membranes) and in the gut (intestinal tract) of healthy individuals (**figure 16.1**). The community itself is called the **microbiome** (the term is also used to refer to the total

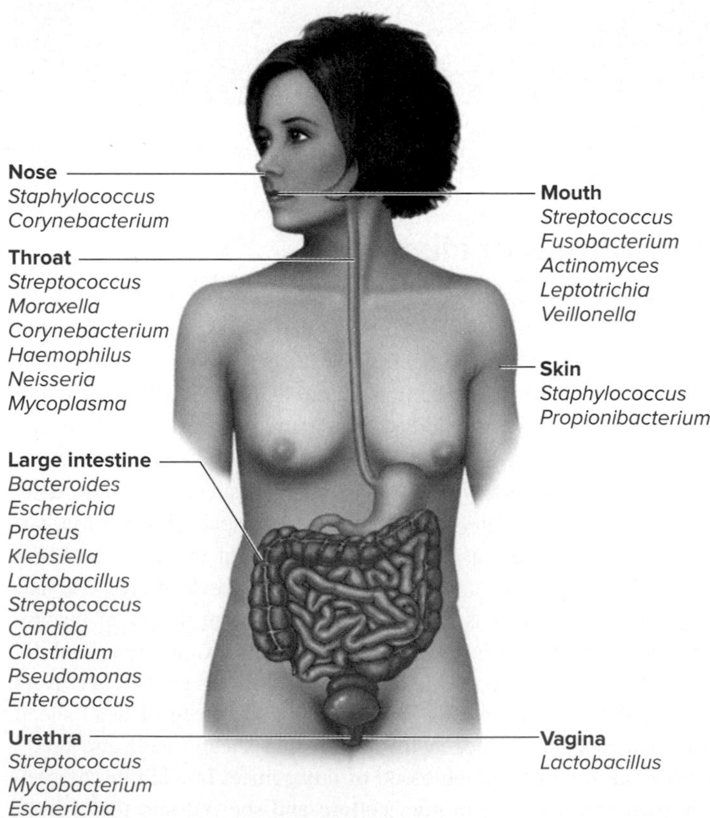

FIGURE 16.1 Normal Microbiota This shows only some of the common genera; many others may also be present.

What might happen if members of the intestinal normal microbiota are killed or their growth is suppressed?

genetic information of the community). Microbes that typically inhabit body sites for extended periods are resident microbiota, whereas temporary occupants are transient microbiota.

Considering how important the microbiome is to human health, relatively little is known about its members. As described in chapter 1, however, that is quickly changing as large-scale research projects focus on this diverse population. The studies typically rely on **metagenomics,** the sequence analysis of the DNA extracted directly from a given environment, which allows scientists to investigate all microbes in a sample, including those that have not yet been grown in culture.

MicroByte

There are more bacteria in one person's mouth than there are people in the world!

Composition of the Microbiome

Babies begin acquiring their microbiota at birth, when they first encounter a wide variety of microorganisms. During passage through the mother's birth canal, the baby is exposed to lactobacilli and an assortment of other microbes that then take up residence in its digestive tract and on the skin. A baby delivered by cesarean section is not exposed to the genital fluids, so microbes from the mother's skin and the environment establish themselves as residents on the newborn instead. Preterm infants, whether delivered vaginally or by cesarean section, have a different microbiome from that of full-term infants.

Breastfeeding also influences the composition of a baby's microbiome. Recent studies show that breast milk contains a remarkable variety of microorganisms as well as certain carbohydrates that appear to specifically nourish a healthy microbiome in an infant's intestine. Babies who are breastfed have higher concentrations of *Bifidobacterium* species and lactic acid bacteria in their intestinal tract.

The composition of a baby's microbiome changes over time, as microbes encountered in various foods, on pets, on other humans, and in the environment establish themselves. Starting at 1 year of age and continuing over a period of about 2 years, the makeup of a child's intestinal microbiome slowly begins to resemble that of an adult.

An average adult carries over 100 trillion microorganisms, and considerable diversity is seen in the composition of the microbiome, both among different individuals and in one person over time. Comparisons of microbial populations, however, reveal certain consistencies, such as the dominant presence in the intestinal tract of members of the phyla Bacillota (includes *Clostridium* and *Bacillus* species) and Bacteroidota (includes *Bacteroides* species); the variety of species within these phyla, however, can vary significantly.

The makeup of the microbiome represents a balance of many forces that can alter the microbial population's quantity and composition. Changes occur in response to physiological variations within the host (such as hormonal changes) and as a direct result of the activities of the human host (such as consuming food). Although intriguing results have been observed, additional studies are needed as confirmation. As an example of the importance of further studies, early research indicated that proportions of certain microbial phyla differ in obese people compared to lean individuals, but follow-up studies have not consistently supported those observations.

Researchers are currently attempting to determine if certain microbiome compositions result in a given disease state. For example, there seems to be a link between intestinal **dysbiosis** (an imbalance in the microbiome) and a number of diseases, including diabetes and inflammatory bowel disease. Although some relationships have been observed, it is too early to tell if these simply represent correlation rather than showing cause and effect.

Beneficial Roles of the Human Microbiome

Scientists have long recognized the importance of the normal human microbiome in human health, and studies continue to give new insights into its significance. Examples of the beneficial roles include:

- **Protecting against infection.** As discussed in chapter 14, the normal microbiota excludes pathogens by (1) covering binding sites used for attachment, (2) consuming available nutrients, and (3) producing compounds toxic to other bacteria. When members of the normal microbiota are killed or their growth suppressed, as can happen during antibiotic treatment, pathogens may colonize and cause disease. For instance, vulvovaginal candidiasis can result from antibiotics that inhibit *Lactobacillus* species that normally suppress growth of the yeast *Candida albicans* in the vagina. Oral antibiotics can also inhibit members of the normal intestinal microbiota, allowing the overgrowth of *Clostridioides difficile* strains that cause antibiotic-associated diarrhea and colitis.
- **Stimulating adaptive immunity.** When small numbers of bacteria that normally live on the skin enter tissues through scrapes and cuts, the body develops an immune response against them; surface proteins of these organisms are often similar to those of pathogens, so antibodies against skin bacteria will bind to pathogens as well. The importance of the microbiome in stimulating the adaptive immune system can be shown in mice reared in a microbe-free environment; these "germ-free" animals have underdeveloped mucosa-associated lymphoid tissue (MALT). Certain members of the intestinal microbiota also appear to protect against infection and cancer by increasing the patrolling of certain T cell subsets in the intestine.
- **Promoting immune tolerance.** The microbiome appears to play an important role in the development of immune tolerance. In a complex series of events, our defenses

FOCUS YOUR PERSPECTIVE 16.1

The Potential of Probiotics

Considering the important protective roles of the normal microbiota in human health, it is not surprising that ingesting live beneficial microbes—referred to as **probiotics**—has been suggested as therapy for diarrheal and other diseases. Researchers are studying the effects of probiotics on various diseases associated with digestion such as antibiotic-associated diarrhea (AAD), inflammatory bowel disease (IBD), and inflammatory bowel syndrome (IBS), sometimes with conflicting results. For example, several studies indicate that ingestion of *Lactobacillus rhamnosus* GG can shorten the duration of AAD, but studies using a mixture containing *L. bulgaricus* and *L. acidophilus* generally showed no helpful effect. Meanwhile, some studies examining the effects of probiotic supplements on mental health, using human as well as non-human animal subjects, indicate a correlation with reduced anxiety symptoms; however, the specific bacterial strains, dosage, and delivery (pills versus fermented foods) have yet to be determined. Although current research certainly highlights the potential of probiotics for treating various medical conditions, larger and well-controlled scientific studies are needed to provide conclusive evidence.

As researchers investigate the potential medical uses of probiotics, it is important to recognize that most of the commercially available products are considered "dietary supplements" and not subject to certain U.S. Food and Drug Administration (FDA) regulations; only if a probiotic is marketed as a medical treatment for a disease must it be proven safe and effective. Thus, a product claimed to contain "live active cultures" may have no health benefit whatsoever. For one thing, positive effects appear to be strain-specific, and those strains beneficial for one condition might not help against another. In addition, the strains in a supplement might not be able to survive passage through the stomach, and even if they do, they might not be able to colonize the intestinal tract. The number of viable microbial cells in a supplement portion can also vary, particularly if the organisms die over time during storage. To help ensure at least some consistency in probiotic supplements, the FDA is now considering requiring manufacturers to label the live microbial quantity, expressed as colony-forming units (CFUs). Note, however, that products with high CFUs are not necessarily superior.

learn to lessen the immune response to the many microbes that routinely inhabit the gut, as well as to foods that pass through. Recent studies into the actions of regulatory T cells indicate that early and consistent exposure to certain microbes in the gut stimulates these T cells, thereby preventing the immune system from overreacting to harmless microbes and substances. This idea is the basis of the **hygiene hypothesis,** which proposes that insufficient exposure to microbes can lead to allergies and autoimmune diseases. It is a fine balance, however, because contact with certain pathogens can be deadly.

- **Aiding digestion.** The human genome encodes relatively few enzymes that degrade complex carbohydrates, so the body relies on microorganisms to break down those compounds. In the anaerobic environment of the intestinal tract, for example, microbes ferment dietary fibers (plant material that the body cannot digest), making short-chain fatty acids (SCFAs) that the body can then absorb. Of the various microbes that inhabit the intestinal tract, *Bacteroides* species appear to make the greatest variety of carbohydrate-degrading enzymes.
- **Producing substances important for human health.** As mentioned earlier, bacteria in the intestinal tract produce vitamin K and certain B vitamins that can be used by the host. In addition, bacterial fermentation of dietary fiber produces butyrate, a type of SCFA and an important energy source for the epithelial cells that line the large intestine. By nourishing those epithelial cells, butyrate helps maintain the mucosal barrier, thereby preventing leakage of damaging microbial products into the bloodstream.

The Gut Microbiome–Brain Connection

Increasing evidence points to the gut microbiome not only playing a part in digestive health but also in brain function. The variety of intestinal microbes releasing compounds into the gut may have an effect on neurological processes. Scientists believe the vagus nerve which runs from the base of the brain to body organs, including the large intestines, may be sensing bacterial metabolites such as SCFAs. Research using animal models indicates that some types of behavior may be correlated with changes in gut microbiota. Some studies are investigating the types of intestinal microbes associated with depression symptoms, while other studies are exploring the gut microbiome's influence on host immune and neurological functions associated with Parkinson's disease (a brain disorder resulting in uncontrollable movements). This exciting field of study continues to reveal the importance of the gut microbiome in human health.

MicroAssessment 16.2

The microbiome protects against potentially harmful organisms, stimulates the immune system, promotes oral tolerance, aids digestion, and produces substances important for human health. Increasing evidence points to the gut microbiome having some effect on brain health.

2. What factor favors the overgrowth of *Clostridioides difficile* in the intestine?
3. What role does dietary fiber play in maintaining the lining of the large intestine?
4. Some research suggests that babies delivered by cesarean section and not breastfed are more prone to developing allergies. What could explain this effect?

16.3 ▪ Principles of Infectious Disease

Learning Outcomes

4. Define the terms *primary pathogen, opportunistic pathogen,* and *virulence.*
5. Describe the characteristics of infectious diseases, including the course of disease, duration of symptoms, and distribution of the pathogen.

The term **colonization** refers to the establishment and growth of a microbe in a particular environment. If the microbe has a parasitic relationship with the host, then the term **infection** can be used. That is, a member of the normal microbiota is said to have colonized the host, but a pathogen is described as having either colonized or infected the host. Infection does not always lead to illness. It can be **subclinical,** meaning that symptoms either do not appear or are mild enough to go unnoticed.

An infection that results in disease (a condition that prevents the body from functioning normally) is called an **infectious disease.** Diseases are characterized by symptoms and signs; **symptoms** are the subjective effects experienced by the patient, such as pain and nausea, whereas **signs** are the objective evidence, such as rash, pus formation, and swelling.

Effects of one disease may leave a person predisposed to developing another. For example, a respiratory illness that damages the mucociliary escalator makes a person more likely to develop pneumonia; in other words, the **primary infection** (in this case the respiratory illness) may allow for a **secondary infection** (in this case pneumonia).

Pathogenicity

A pathogen is a disease-causing microbe, and **pathogenicity** is the ability to cause disease. A **primary pathogen** can cause disease even in an otherwise healthy person; diseases such as malaria, measles, influenza, strep throat, plague, tetanus, and tuberculosis are caused by primary pathogens. In contrast, an **opportunistic pathogen,** or opportunist, causes disease only when the body's defenses are weakened (such as through injury or the effects of another disease) or when relatively high numbers are introduced to an unusual site. Opportunists can be members of the normal microbiota or they can be from the environment. *Pseudomonas* species, for example, are common environmental bacteria that routinely come into contact with healthy individuals without harmful effects, yet they can cause sometimes-fatal opportunistic infections. In 2023, an outbreak of *Pseudomonas aeruginosa* eye infections, some resulting in eye loss or even death, was linked to contaminated eye drops. The bacterium has long been a concern in healthcare facilities because it causes pneumonia in individuals who have the genetic disease cystic fibrosis and readily infects wounds, particularly thermal (burn) wounds (see figure 23.6). Ironically, as healthcare systems improve, extending the life span of patients through surgery and immunosuppressive medications, diseases caused by opportunists are becoming more common; also, many organisms previously considered non-pathogenic have been known to cause disease in severely immunocompromised patients.

The term **virulence** refers to the degree of pathogenicity of an organism. An organism described as highly virulent is more likely to cause disease—particularly severe disease—than might otherwise be expected. *Streptococcus pyogenes* causes strep throat, but certain strains are particularly virulent, causing diseases such as necrotizing fasciitis ("flesh-eating disease"). **Virulence factors** are the traits of a microorganism that specifically allow it to cause disease. The genes encoding these traits can sometimes be transferred horizontally.

Characteristics of Infectious Disease

Infectious diseases that spread from one host to another are called **communicable** or **contagious diseases.** Some communicable diseases, such as colds and measles, are easily transmitted. The ease with which a communicable disease spreads partly reflects the **infectious dose**—the number of microbes necessary to establish an infection. The intestinal disease shigellosis is quite contagious in humans because only 10 to 100 cells of a *Shigella* species need to be ingested to establish an infection; in contrast, salmonellosis does not spread as easily because as many as 10^6 cells of *Salmonella enterica* serotype Enteritidis must be consumed to cause illness. The difference in these infectious doses reflects, in part, the pathogen's ability to survive the acidic conditions encountered as the cells pass through the stomach. Generally, the infectious dose is expressed as the ID_{50}, an experimentally derived figure that indicates the number of microbial cells administered to test subjects that resulted in disease in 50% of that population.

Progression of Infectious Disease

The progression of an infectious disease includes several stages:

- **Incubation period.** This is the time between the introduction of a microbe to a susceptible host and the onset of signs and symptoms. It varies considerably, from only a few days (for the common cold), to several weeks (for hepatitis A), to many months (for rabies), and even years (for Hansen's disease). The length of the incubation period depends on a variety of factors, including the growth rate of the pathogen, the host's condition, and the number of infectious cells or virions encountered.
- **Illness.** During this period, a person experiences the signs and symptoms of the disease. In some cases, onset of illness is preceded by a **prodromal phase**—a period of early, vague symptoms such as malaise (a general feeling of illness) and headache.
- **Convalescence.** This is the stage of recuperation and recovery from the disease.

Even though there may be no indication of infection during the incubation and convalescent periods, many infectious agents can still be spread during these stages. Some individuals, called **carriers,** harbor an infectious agent for months or years and continue to spread the pathogen, even though they appear healthy.

Following recovery from infection, the person's immune system usually has produced enough protective antibodies and memory lymphocytes to prevent reinfection with the same microbe. This is true after vaccination as well.

Duration of Symptoms

Infections and the associated diseases are often described according to the timing and duration of the symptoms (**figure 16.2**):

- **Acute infections** are characterized by symptoms that develop quickly but last a relatively short time; an example is strep throat.
- **Chronic infections** develop slowly and last for months or years; an example is hepatitis C.
- **Latent infections** are never completely eliminated; the microbe continues to exist in host tissues, held in check by the immune system without causing any symptoms. If immunity decreases sometime later, the latent infection may reactivate and become symptomatic. For example, the initial infection by the varicella-zoster virus results in the characteristic symptoms of chickenpox. That illness is stopped by an effective immune response, but the virus is not completely eliminated; instead, the viral genome remains silent in sensory nerves. Later in life, infectious viral particles may be produced again, causing the disease shingles (herpes zoster). Other diseases in which the causative agent becomes latent include tuberculosis, cold sores, and genital herpes.

Distribution of the Pathogen

Infections are often described according to the distribution of the causative agent in the body. In a **localized infection,** the microbe is limited to a small area; an example is a hair follicle infection caused by *Staphylococcus aureus*. In a **systemic infection,** the infectious agent disseminates (spreads) throughout the body; an example is Lyme disease. Systemic infections often include a characteristic set of signs and symptoms—such as fever, fatigue, and headache—that result from the systemic immune response to the infecting agent.

The *suffix -emia* means "in the blood." Thus, **bacteremia** indicates that bacteria are circulating in the bloodstream. Note that this term does not necessarily imply a disease state. A person can have temporary bacteremia after forceful tooth brushing; on the other hand, a person with a bloodstream infection has potentially life-threatening bacteremia. **Toxemia** indicates that toxins are circulating in the bloodstream; for instance, the organism that causes tetanus produces a localized infection, yet its toxins circulate in the bloodstream. The term **viremia** indicates that viral particles are circulating in the bloodstream.

Incubation period → Illness → Convalescence

Acute. Illness is short term because the pathogen is eliminated by the host defenses; person is usually immune to reinfection.

Incubation period → Illness (long lasting)

Chronic. Illness persists over a long time period.

Incubation period → Illness → Convalescence → Latency → Recurrence

Latent. Illness may recur if immunity weakens.

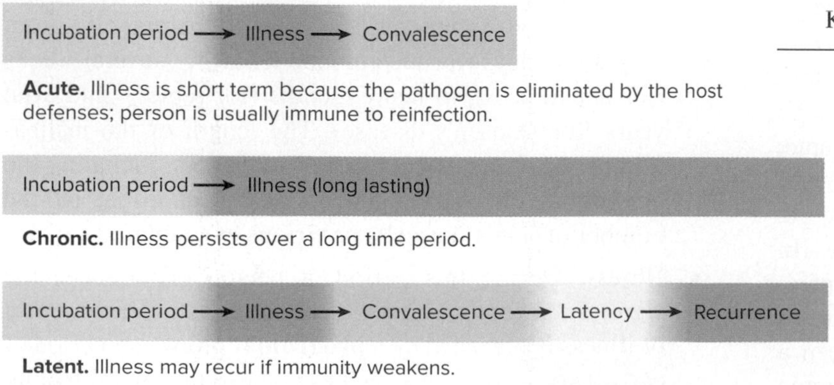

FIGURE 16.2 Duration of Symptoms Infections can be acute, chronic, or latent.

? Some diseases include a prodromal phase. Where would this phase fit in the figure?

MicroAssessment 16.3

A primary pathogen can cause disease in an otherwise healthy individual; an opportunist causes disease when host immune defenses are weakened or when introduced into an unusual site. The course of infectious disease includes an incubation period, illness, and a period of convalescence. Infections can be acute, chronic, or latent; they can be localized or systemic.

5. Why are diseases caused by opportunists becoming more frequent?
6. Give an example of a microbe that causes a latent infection.
7. What factors might contribute to a long incubation period?

16.4 ■ Determining the Cause of an Infectious Disease

Learning Outcome

6. List Koch's postulates, and compare them to the molecular Koch's postulates.

Criteria are needed to guide scientists as they try to determine the cause of an infectious disease, and they can also be helpful when studying the disease process.

Koch's Postulates

The steps that Robert Koch used to show that *Bacillus anthracis* causes anthrax (see **A Glimpse of History**) are now known as **Koch's postulates.** Although they were never meant to be applied rigidly, they still provide scientists with a logical framework for establishing that a given microbe causes a certain infectious disease (**figure 16.3**):

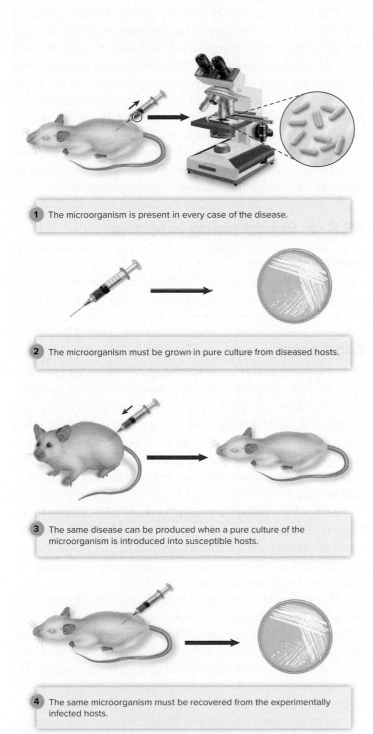

1. The microorganism is present in every case of the disease.
2. The microorganism must be grown in pure culture from diseased hosts.
3. The same disease can be produced when a pure culture of the microorganism is introduced into susceptible hosts.
4. The microorganism must be recovered from the experimentally infected hosts.

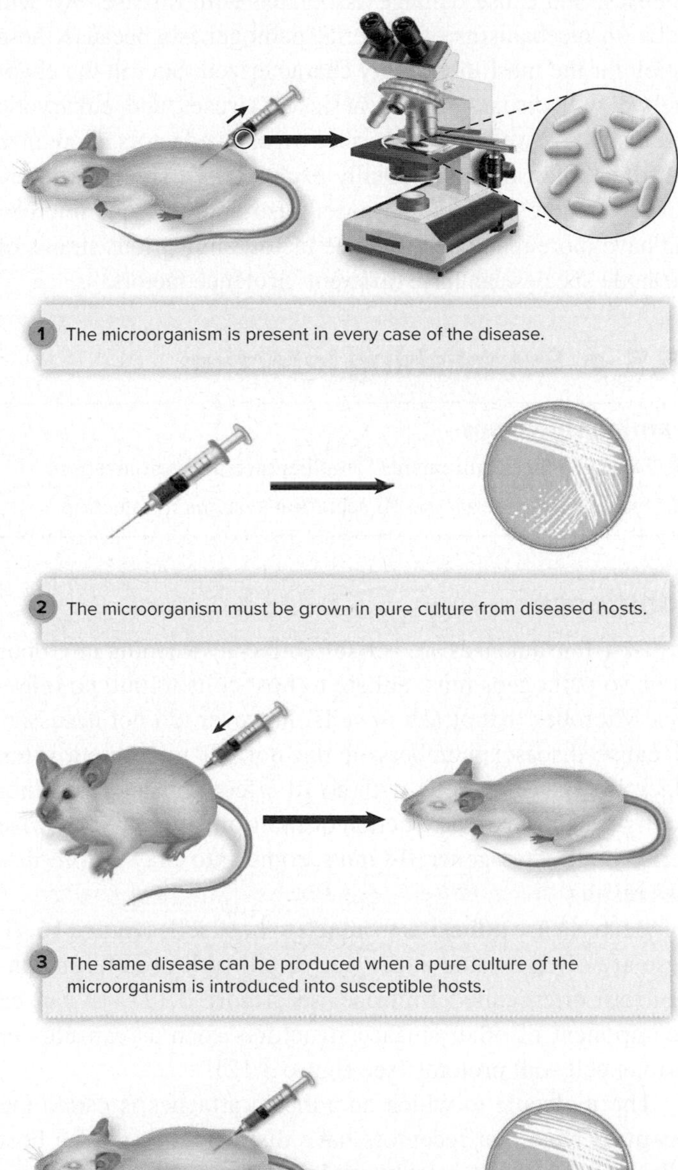

1 The microorganism is present in every case of the disease.
2 The microorganism must be grown in pure culture from diseased hosts.
3 The same disease can be produced when a pure culture of the microorganism is introduced into susceptible hosts.
4 The same microorganism must be recovered from the experimentally infected hosts.

FIGURE 16.3 Koch's Postulates These criteria provide a foundation for establishing that a given microbe causes a specific disease.

? **To fulfill Koch's postulates, why must an organism suspected of causing the disease be able to grow in laboratory media?**

Although the postulates serve an important function, they have limitations. The second postulate cannot be fulfilled for organisms that do not grow in laboratory media, which was the case for *Treponema pallidum* (causes syphilis) until recently. The third postulate does not necessarily hold true for diseases such as polio in which many infected people do not have signs or symptoms of the disease. Also, some diseases, including periodontal diseases, are polymicrobial, meaning that multiple species act together to cause the illness. In the case of certain diseases, suitable experimental animal hosts are not available, and it would be unethical to test the postulates on humans because of safety concerns.

Molecular Koch's Postulates

Molecular Koch's postulates are similar in principle to Koch's postulates, but they rely on molecular techniques to study a microbe's virulence factors. They are particularly relevant in the study of pathogens such as *E. coli* and *Streptococcus pyogenes,* which can cause several different diseases depending on the virulence factors of a given strain. Molecular Koch's postulates are as follows:

1. The virulence factor gene or its product should be found in pathogenic strains of the microorganism.
2. Mutating the virulence gene to disrupt its function should reduce the virulence of the pathogen.
3. Reversion of the mutated virulence gene or replacement with a wild-type version should restore virulence to the strain.

As with the traditional Koch's postulates, it is not always possible to apply all of these criteria, but they provide an approach to studying how infectious agents cause disease. Perhaps most importantly, both sets of postulates remind us that scientific evidence should be the foundation for determining the cause of disease.

MicroAssessment 16.4

Koch's postulates can be used to establish that a given microbe causes a specific infectious disease. Molecular Koch's postulates are used to identify the virulence factors responsible for disease.

8. How were Koch's postulates used to prove the cause of anthrax?

9. Why are Koch's postulates not suitable for identifying the cause of diseases that involve polymicrobial infections?

MECHANISMS OF PATHOGENESIS

From a microbe's perspective, the interior of the human body is a rich source of nutrients guarded by innate and adaptive defenses. The ability to get past these defenses and cause damage is what distinguishes pathogens from other microbes. Understanding how they do this helps illustrate why only certain microbes can cause disease in a healthy host. Pathogenic mechanisms generally follow one of several patterns:

- **Produce toxin (that is then ingested).** The microbe does not grow on or in the host, so this is not an infection but rather a foodborne intoxication, a form of food poisoning. The only relevant virulence determinant in this situation is toxin production. Relatively few bacteria cause foodborne intoxication; these include *Clostridium botulinum* (causes botulism) and toxin-producing strains of *Staphylococcus aureus* (cause staphylococcal food poisoning).
- **Colonize mucous membranes and produce toxin.** The microbe adheres to a mucous membrane (such as the lining of the intestinal tract or the upper respiratory tract), multiplies to high numbers, and produces a toxin that interferes with host cell function. Examples of bacteria that do this include *Vibrio cholerae* (causes cholera) and *Corynebacterium diphtheriae* (causes diphtheria).
- **Multiply within host tissues.** The microbe penetrates the first-line defenses and then multiplies within the tissues. Organisms that do this generally have mechanisms to avoid destruction by macrophages; some also have mechanisms to avoid antibodies. Examples of bacteria that multiply within host tissues include *Neisseria meningitidis* (causes meningococcal disease) and *Yersinia pestis* (causes plague).
- **Multiply within host tissues and produce toxin.** Microbes that do this are similar to those in the previous category, but in addition to multiplying in host tissues, they also make toxins. Examples include *Shigella dysenteriae* (causes shigellosis) and *Clostridium tetani* (causes tetanus).

A successful pathogen needs to overcome the host defenses only long enough to multiply and then exit the host. In fact, a pathogen that completely overwhelms the host defenses is actually at a disadvantage because it will likely kill the host. If the host dies, the pathogen loses an exclusive source of nutrients and perhaps the opportunity to be transmitted.

Pathogens and their hosts generally evolve over time to a state of **balanced pathogenicity.** The pathogen becomes less virulent while the host becomes less susceptible. This was demonstrated when the myxoma virus was intentionally introduced into Australia in the early 1950s to kill the rapidly increasing rabbit population. As expected, the rabbit population dropped dramatically after the virus was introduced. Eventually, however, the number of rabbits began rising. Viruses isolated from these rabbits were shown to be less virulent than the original strain, and the rabbits were more resistant to the original virus strain.

The next sections will describe how pathogens adhere to and colonize host tissue, avoid innate defenses, avoid adaptive defenses, and cause damage associated with disease. We will focus on mechanisms of bacterial pathogenesis because these are by far the most thoroughly characterized; later in the chapter, we will discuss pathogenesis of viruses and eukaryotic organisms. As we describe various virulence factors, recognize that their roles are not mutually exclusive—a single structure can serve more than one purpose. Also note that one microbe can have more than one virulence factor, and various strains of the same species can have different virulence factors.

16.5 ▪ Establishing Infection

Learning Outcomes

7. Describe the requirements for adherence and colonization.
8. Explain the role of type III secretion systems in infection.

Adherence

The first-line defenses are very effective at sweeping microbes away, so pathogens must adhere to host cells to initiate infection. Microbes that attach to cells, however, do not necessarily cause disease; members of the normal microbiota often adhere to epithelial cells with no ill effect whatsoever. Other factors such as toxin production or multiplication in otherwise sterile body sites generally must come into play before disease results.

Bacteria use **adhesins** to attach to host cells (**figure 16.4**); these are often located at the tips of pili (pili used for attachment are often called fimbriae; see figure 3.17) but can be a component of other surface structures such as capsules or various cell wall proteins (see figure 3.12).

The molecule to which an adhesin attaches is called the **receptor.** Note that receptors have distinct roles for the host cells; the microbes merely exploit the molecules for their own use. The normal role of the receptor used by *Neisseria gonorrhoeae* is to help protect host cells from damage by the complement system. Receptors are typically glycoproteins or glycolipids, and the adhesin binds to the sugar portion.

Adhesin-receptor binding is highly specific, dictating the types of cells to which the bacterium can attach. The adhesin of common *E. coli* strains allows them to attach to cells that line the large intestine, where the strains multiply as part of the normal microbiota. Pathogenic *E. coli* strains have additional adhesins, broadening the range of tissues to which they can attach; strains that cause urinary tract infections have pili that attach to cells lining the bladder, and strains that cause watery diarrhea have pili that adhere to cells lining the small intestine.

Colonization

A microorganism must multiply in order to colonize the host. In many cases, pathogens grow on host surfaces as a biofilm (a polymer-encased microbial community).

To colonize a mucosal surface, the pathogen must deal with the host's defenses that protect those surfaces. Recall, for example, that the body uses lactoferrin and transferrin to bind iron, thereby limiting the growth of microbes. Some pathogens respond by producing their own iron-binding molecules, called **siderophores;** others can use the iron bound to the host proteins.

Secretory IgA also protects mucosal surfaces. Many pathogens, however, have evolved mechanisms to avoid those antibodies; mechanisms include rapid turnover of pili (to shed any bound antibody), antigenic variation, and **IgA proteases** (enzymes that cleave IgA antibodies).

If the body site has normal microbiota, a new arrival must compete for space and nutrients. It must also tolerate any toxic products such as fatty acids produced by competitors.

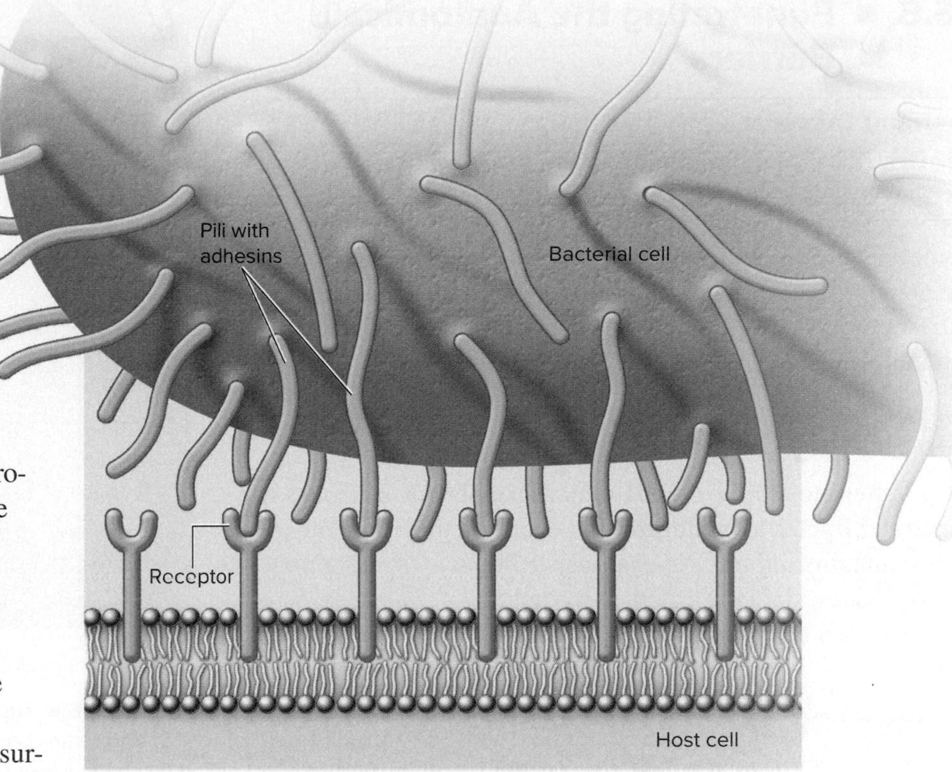

FIGURE 16.4 Adhesins In this illustration, adhesins at the tips of pili (fimbriae) attach to receptors on the host cell surface.

What would happen if a pathogen lost the ability to produce adhesins?

Delivering Effector Proteins to Host Cells

Some Gram-negative pathogens deliver molecules directly into host cells using one of several types of secretion systems. A **type III secretion system,** or injectisome, is a syringe-like structure that injects proteins into eukaryotic cells (**figure 16.5**). The injected proteins, referred to as effector proteins, induce changes such as altering the cell's cytoskeleton structure. Some effector proteins direct the host cell to engulf the bacterial cell, a process discussed in the next section.

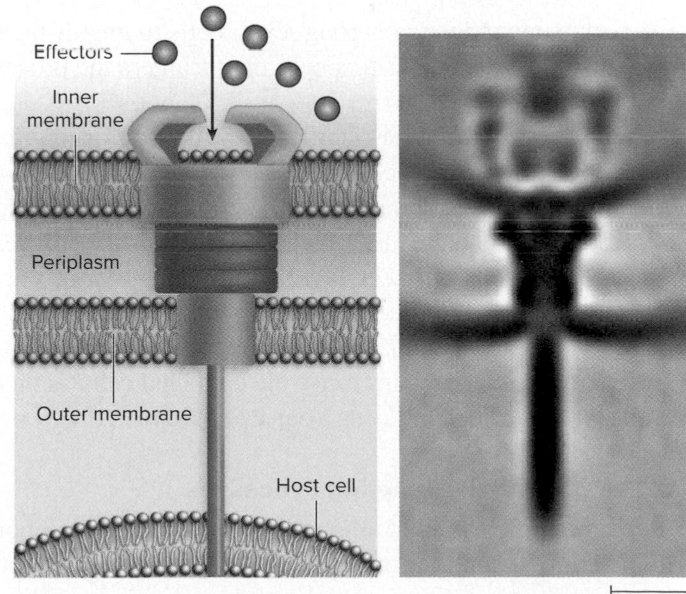

FIGURE 16.5 Type III Secretion Systems Gram-negative bacteria use type III secretion systems to deliver certain molecules directly to host cells, inducing changes in those cells. Peptidoglycan is not shown in the drawing. Electron micrograph shows a type III secretion system in *Salmonella enterica*. Donghyun Park, Maria Lara-Tejero, M Neal Waxham, Wenwei Li, Bo Hu, Jorge E Galán, Jun Liu (2018) Visualization of the type III secretion mediated Salmonella–host cell interface using cryo-electron tomography

What bacterial structure do type III secretion systems resemble?

MicroAssessment 16.5

Pathogens use adhesins, often on pili, to bind to a body surface. To colonize a surface, the pathogen must often compete with the normal microbiota, prevent binding of secretory IgA, and obtain iron. Some bacteria deliver effector proteins to epithelial cells, inducing a specific change in those cells.

10. What are siderophores?
11. What is a type III secretion system?
12. From a microbe's perspective, why is it a good strategy to adhere to a receptor that plays a critical function for a host cell?

16.6 ■ Penetrating the Anatomical Barriers

Learning Outcome

9. Describe the mechanisms pathogens use to penetrate the skin and mucous membranes.

Some bacterial pathogens cause disease while remaining on the mucosal surfaces, but many others penetrate the anatomical barriers. By crossing the epithelial barrier, invading microbes can multiply in the nutrient-rich tissues without competition.

Penetrating the Skin

Skin is the most difficult anatomical barrier for microbes to penetrate. Bacterial pathogens that enter via this route rely on skin-damaging injury. As examples, *Staphylococcus aureus* enters tissues via a cut or other wound and *Yersinia pestis* is injected via a bite from an infected flea.

Penetrating Mucous Membranes

Mucous membranes are the entry points for most pathogens, but the penetration processes are complex and difficult to study. It appears, however, that at least two mechanisms can be used to cross the membranes: directed uptake by cells and exploiting antigen-sampling processes.

Directed Uptake by Cells

Some pathogens induce non-phagocytic cells to engulf them. The pathogen first attaches to a cell, then triggers the process of endocytosis.

Gram-negative bacteria often inject effector proteins that induce engulfment by host cells. *Salmonella enterica*, for example, uses a type III secretion system to deliver specific proteins into intestinal epithelial cells, causing pseudopod-like "ruffles" to appear on the cells' surfaces (**figure 16.6**); the ruffles eventually enclose the bacterial cells, bringing them into the intestinal cells. This phenomenon, called **membrane ruffling,** occurs because the bacterial proteins cause the epithelial cells' actin to rearrange, as described in Focus Your Perspective 3.1.

Exploiting Antigen-Sampling Processes

As described in chapter 15, the mucosa-associated lymphoid tissue (MALT) monitors material on the surface of mucous membranes for signs of microbial invaders, using specialized antigen-sampling cells to transfer the material to the underlying phagocytes and lymphocytes. Some pathogens exploit this process to cross the membranes.

Shigella cells use M cells to cross the intestinal barrier. Recall that M cells transport material from the lumen of the intestine to the Peyer's patches (see figure 15.8). Most microbes delivered this way are destroyed by macrophages in the Peyer's patches, but certain pathogens have mechanisms to avoid this fate. When *Shigella* cells are transferred to the macrophages, for instance, the bacteria not only survive but even multiply. Eventually, this triggers lysis of the host cell, which frees the bacterial cells (**figure 16.7**). The freed bacterial cells then bind to the base of the mucosal epithelial cells and cause these non-phagocytic cells to engulf them, using a mechanism similar to that just described for *Salmonella.*

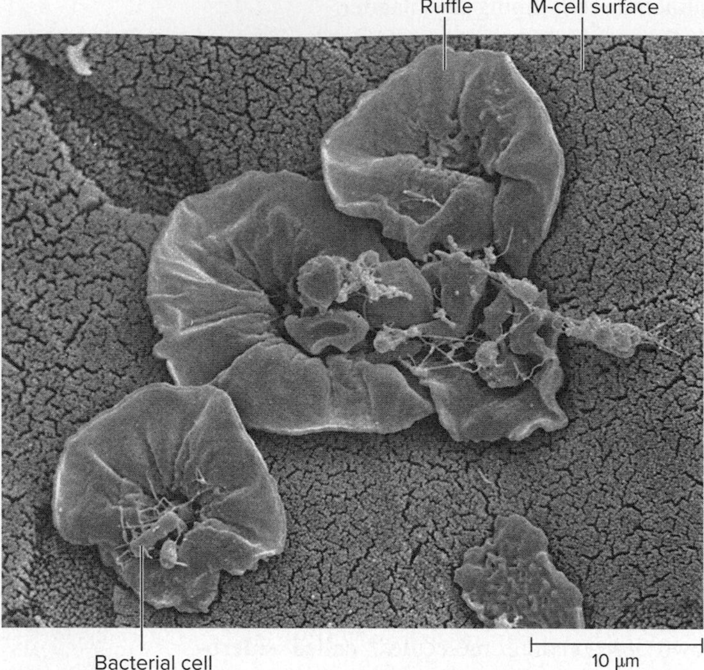

FIGURE 16.6 Ruffling *Salmonella enterica* serotype Typhimurium inducing ruffles on an M cell (a specialized intestinal epithelial cell), leading to uptake of the bacterial cells. Mark A. Jepson

? How do *Salmonella* cells induce ruffling?

MicroAssessment 16.6

Skin is the most difficult barrier for microbes to penetrate. Some pathogens cross the mucous membrane barrier by inducing mucosal epithelial cells to engulf them or by taking advantage of antigen-sampling processes.

13. How does *Salmonella enterica* enter intestinal epithelial cells?
14. How do *Shigella* species enter intestinal epithelial cells?
15. Why would it be difficult to study the invasion of mucous membranes?

16.7 ■ Avoiding the Host Defenses

Learning Outcome

10. Describe mechanisms that bacteria use to avoid destruction by phagocytes, killing by complement system proteins, and recognition by antibodies.

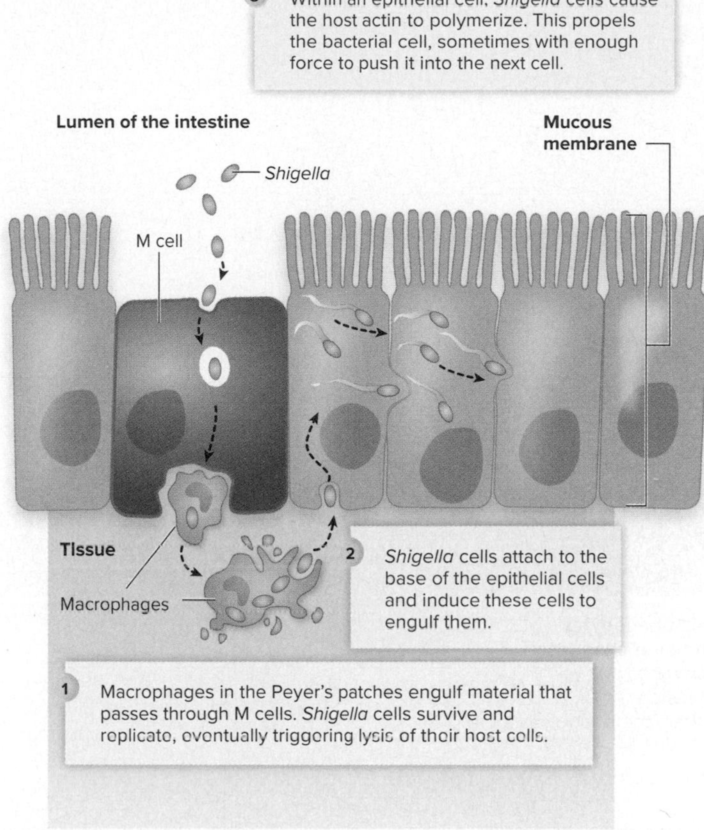

FIGURE 16.7 Antigen-Sampling Processes Provide a Mechanism for Invasion *Shigella* species use M cells to move across the intestinal epithelial barrier. Once the bacterial cells are on the other side, macrophages ingest them, but the bacteria are able to escape and then infect other cells.

? **What is the normal function of M cells?**

Once within body tissues, microorganisms encounter the innate and adaptive immune defenses. Pathogens as a group have evolved a wide variety of mechanisms to avoid the otherwise lethal effects of these defenses.

Hiding Within a Host Cell

Some pathogens enter host cells, where they hide from complement proteins, phagocytes, and antibodies. Once a *Shigella* cell is within an intestinal epithelial cell, for example, it directs its own transfer to adjacent cells (see figure 16.7). It does this by causing the host cell actin to polymerize at one end of the bacterial cell, thereby forming an "actin tail" that propels the bacterium within the cell. The force of the propulsion is so great that the bacterial cells are often pushed into neighboring host cells. *Listeria monocytogenes* (causes meningitis) does the same thing (**figure 16.8**).

Avoiding Destruction by Phagocytes

Phagocytes destroy microbes using a process that involves multiple steps: chemotaxis, recognition and attachment, engulfment, and fusion of the phagosome with lysosomes (see figure 14.14). Pathogens have evolved several mechanisms to avoid destruction (**figure 16.9**).

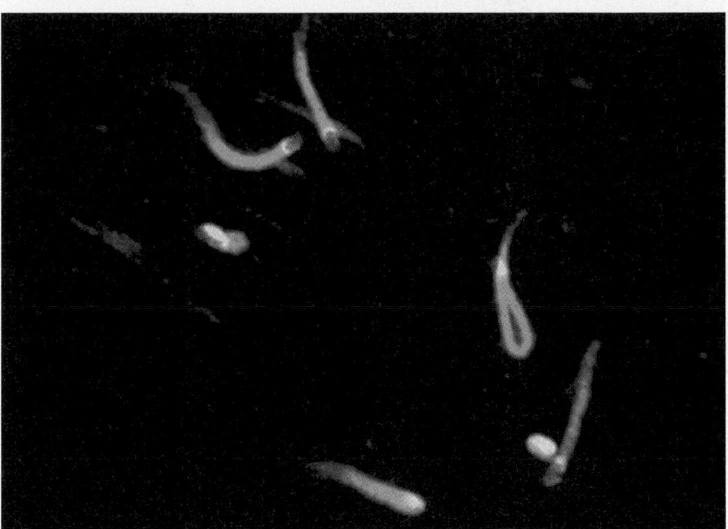

FIGURE 16.8 Actin Tail of Intracellular *Listeria monocytogenes* Rapid polymerization of host cell actin (green) at one end of the bacterial cell (orange) propels the bacterium within the host cell. Pascale F. Cossart

? **How does an actin tail benefit a bacterial cell?**

① Preventing Encounters with Phagocytes

Some pathogens prevent phagocytosis by avoiding macrophages and neutrophils altogether. The mechanisms include:

- **C5a peptidase.** This enzyme degrades the complement system component C5a, a chemoattractant that recruits phagocytic cells. *Streptococcus pyogenes* (causes strep throat) makes C5a peptidase.
- **Membrane-damaging toxins.** These kill phagocytes and other cells, often by forming pores in their membranes. The leaky cells swell and then lyse. *S. pyogenes* makes a membrane-damaging toxin called streptolysin O.

② Avoiding Recognition and Attachment

Some pathogens avoid being recognized by phagocytes. Recall that phagocytes recognize and attach to foreign material more efficiently if opsonins such as C3b or antibodies coat it. Mechanisms that bacteria use to avoid opsonization include:

- **Capsules.** Some capsules bind the host's complement regulatory proteins that inactivate C3b—in other words, the pathogen hijacks the mechanism that the body uses to protect its own cells from the effects of complement (see figure 14.13). Not only is the inactivated C3b no longer an effective opsonin, but it also cannot activate the complement system by the alternative pathway (see figure 14.11). *Streptococcus pneumoniae* (causes pneumonia) is an example of a pathogen that uses a capsule to avoid phagocytosis.

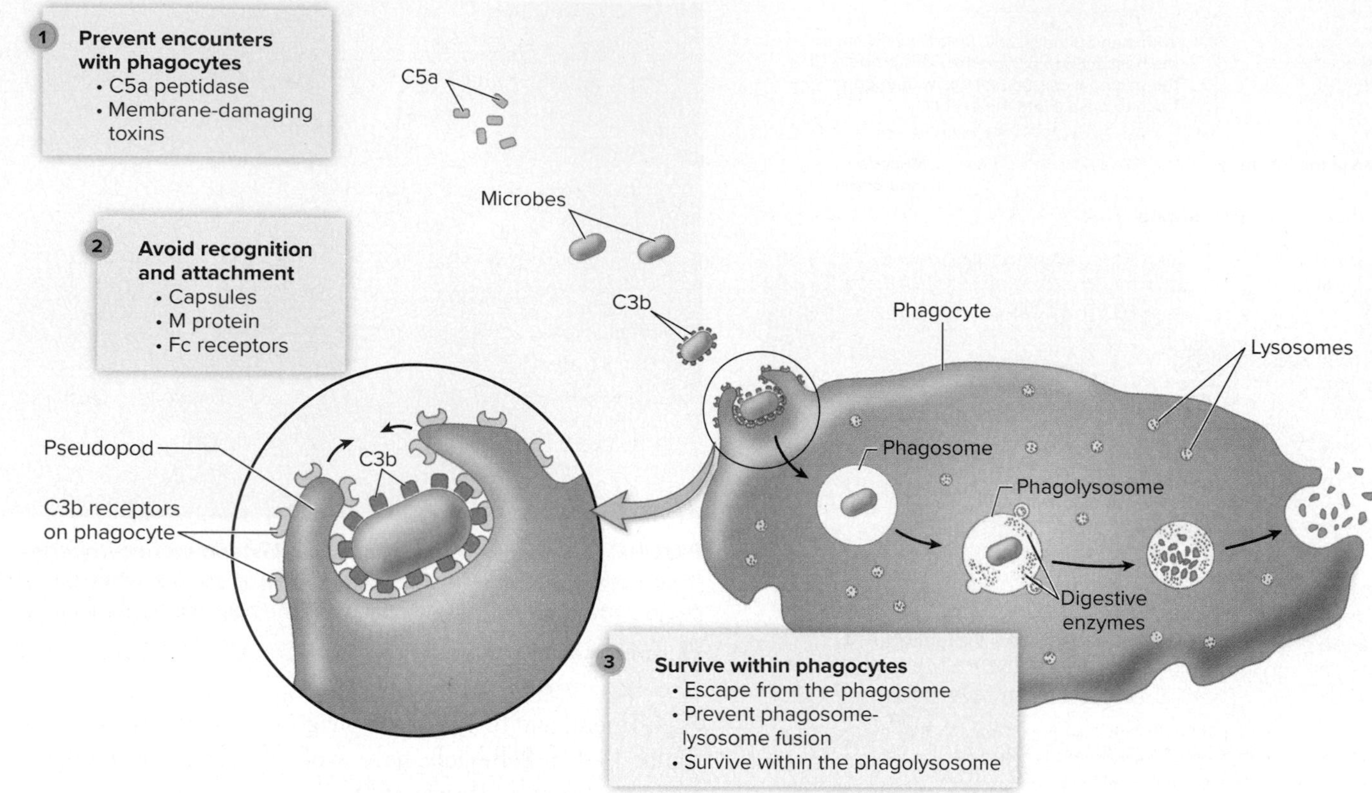

FIGURE 16.9 Avoiding Destruction by Phagocytes

? In addition to avoiding destruction, what other advantages would a bacterium have by surviving within a phagocyte?

- **M protein.** This component of the cell wall of *Streptococcus pyogenes* functions in a manner similar to that just described for capsules: It binds a complement regulatory protein that inactivates C3b, thereby preventing C3b from being an effective opsonin and from activating the complement system.
- **Fc receptors.** These cell surface proteins bind the Fc region of antibodies, interfering with their function as opsonins (**figure 16.10**). Recall that antibodies have two parts: the Fab region, which binds specifically to antigens, and the Fc region, which functions as a "red flag" (see figure 15.3). In a normal situation, when antibody molecules are bound to a microbial cell, the Fc region sticks out, serving as a marker to the immune system that the cell is an invader. Bacterial cells that have Fc receptors on their surface bind the Fc portion of random antibodies, orienting the molecules so that the Fab region sticks out; a phagocytic cell has no mechanism for recognizing the Fab regions, so this masks the bacterial cell from phagocytes. *Staphylococcus aureus* cells have protein A, a type of Fc receptor.

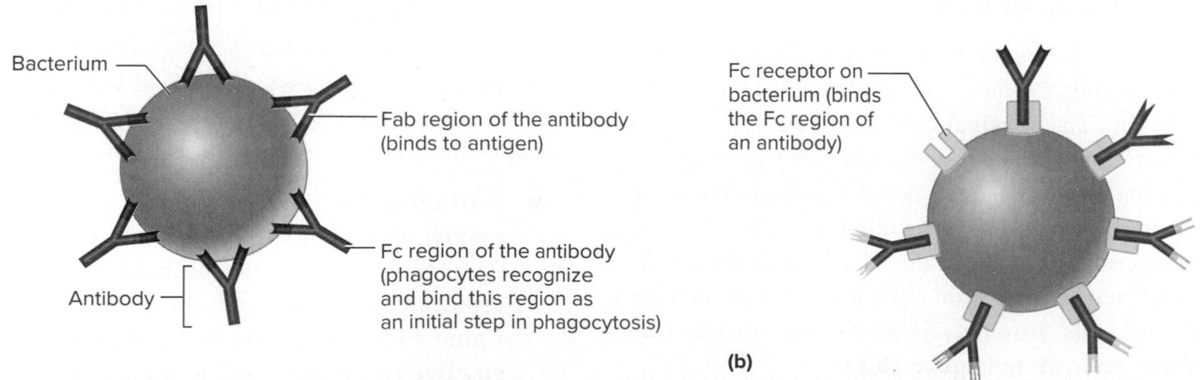

FIGURE 16.10 Fc Receptors Prevent Opsonization by Antibodies **(a)** The normal orientation of antibody molecules on the surface of a bacterium; note that the Fc region of the antibody sticks out from the bacterial cell, making it available for a phagocyte to recognize and bind. **(b)** The effect of Fc receptors on a bacterial cell's surface; the receptors bind the Fc portion of antibodies, regardless of the specificity of the antibodies.

? Are the antibodies bound by Fc receptors on a bacterial surface specific for that particular bacterial cell? Explain.

③ Surviving Within Phagocytes

Some bacteria make no attempt to avoid engulfment by phagocytes, instead using it as an opportunity to hide from antibodies and be transported to other locations in the body. Mechanisms used to survive within phagocytes include:

- **Escape from the phagosome.** Some pathogens escape from the phagosome before it fuses with lysosomes. The bacteria then multiply within the cytoplasm of the phagocyte, protected from other host defenses. *Listeria monocytogenes* produces a molecule that forms pores in the phagosomal membrane, allowing the bacterial cells to escape. *Shigella* species lyse the phagosome.
- **Preventing phagosome-lysosome fusion.** Bacteria that prevent phagosome-lysosome fusion avoid the otherwise inevitable exposure to the destructive components of lysosomes. *Salmonella enterica* cells can sense that they have been ingested by a macrophage and respond by producing a protein that blocks the fusion process.
- **Surviving within the phagolysosome.** *Coxiella burnetii* (causes Q fever) is one of the few microbes that can survive the destructive environment within the phagolysosome. Once the organism has been ingested by a macrophage, it delays fusion of the phagosome with the lysosome, allowing additional time for the microbe to equip itself for growth within the phagolysosome.

Avoiding Killing by Complement System Proteins

As described in chapter 14, one of the three primary outcomes of activation of the complement system is lysis of foreign cells by membrane attack complexes (MACs) (see figure 14.11). Gram-negative bacteria are susceptible to MACs because their outer and cytoplasmic membranes serve as targets, whereas the thick layer of peptidoglycan of Gram-positive bacteria prevents the MAC components from reaching the cytoplasmic membrane. Gram-negative bacteria referred to as **serum-resistant** have mechanisms to prevent complement-mediated killing; for example, strains of *Neisseria gonorrhoea* that cause disseminated gonococcal infection hijack the mechanism that host cells use to prevent their own surfaces from activating the complement system (see figure 14.13). By binding to the host's complement regulatory proteins, they avoid complement activation by the alternative pathway, thereby postponing MAC formation.

Avoiding Recognition by Antibodies

Pathogens that survive the innate defenses soon encounter an additional obstacle, the adaptive defenses. For bacteria, the most important of these are usually antibodies. Mechanisms for avoiding them include:

- **IgA protease.** This enzyme—produced by *Neisseria gonorrhoeae* and a variety of other pathogens—cleaves IgA, the class of antibody found in mucus and other secretions. The enzyme also cleaves certain proteins that help bacteria survive within host cells.
- **Antigenic variation.** Some pathogens routinely alter the structure of their surface antigens, allowing them to stay ahead of antibody production by altering the very molecules that antibodies would otherwise recognize. *N. gonorrhoeae* can vary the antigenic structure of its pili; antibodies produced by the infected host in response to one variation of the pili cannot bind effectively to another.
- **Mimicking host molecules.** Pathogens sometimes produce and cover themselves with molecules similar to those normally found in the host. This molecular mimicry takes advantage of the fact that the immune system typically does not mount an attack against "self" molecules. Certain strains of *Streptococcus pyogenes* have a capsule composed of hyaluronic acid, a polysaccharide found in human tissues.

MicroAssessment 16.7

Mechanisms bacteria use to avoid destruction by phagocytes include preventing encounters with phagocytes, avoiding recognition and attachment, and surviving within phagocytes. Serum-resistant bacteria avoid the killing effects of complement system proteins. Mechanisms for avoiding antibodies include IgA protease, antigenic variation, and mimicking host molecules.

16. Describe how Fc receptors help a microbe avoid phagocytosis.
17. Describe three mechanisms pathogens may use to survive within phagocytic cells.
18. Encapsulated organisms can be phagocytized once antibodies against the capsule have been produced. Why would this be so?

16.8 ■ Damage to the Host

Learning Outcomes

11. Describe the difference between exotoxins and endotoxins.
12. Compare and contrast neurotoxins, enterotoxins, and cytotoxins, giving two examples of each.
13. Explain how inflammation and antibodies can cause damage.

Damage due to infection can be the result of a pathogen's direct effects, such as toxins produced, or indirect effects, such as the immune response. In many cases, the damage helps the organism exit the host, allowing it to spread to others. *Bordetella pertussis* (causes whooping cough) causes severe bursts of coughing, propelling respiratory pathogens into the air. *Vibrio cholerae* (causes cholera) induces watery diarrhea—up to

20 liters of microbe-containing fluid in one day! In areas of the world without adequate sewage treatment, this can lead to contaminated water supplies and widespread outbreaks.

Exotoxins

A number of Gram-positive and Gram-negative pathogens make **exotoxins**—proteins that have very specific and harmful effects; the toxins are often a major cause of damage to an infected host. Exotoxins are either secreted by the bacterium or leaked into the surrounding fluid following lysis of the bacterial cell. In most cases, the pathogen must colonize a body surface or tissue to produce enough toxin to cause damage. With foodborne intoxication, however, the bacterial cells multiply in a food product, where they produce toxin that is then consumed. In the case of botulism, ingestion of even tiny amounts of botulinum toxin is sufficient to cause paralysis. Like most other exotoxins, botulinum toxin can be destroyed by heating.

Exotoxins can act locally, or they may be carried in the bloodstream throughout the body, causing systemic effects. *Corynebacterium diphtheriae* (causes diphtheria) grows and releases its exotoxin in the throat. There, the toxin destroys local cells, leading to the accumulation of dead host cells, pus, and blood. This forms a "pseudomembrane" in the throat that sometimes dislodges and blocks the airway. In addition, the toxin can be absorbed and carried to the heart, nervous system, and other organs, causing systemic damage.

Because exotoxins are proteins, the immune system can generally produce neutralizing antibodies against them (see figure 15.4). Unfortunately, many exotoxins are so powerful that fatal damage occurs before an adequate immune response is mounted. Vaccination, however, prevents otherwise common and often fatal toxin-mediated diseases such as tetanus and diphtheria (see table 17.3). The vaccines against tetanus and diphtheria are **toxoids,** which are inactivated toxins; the toxoid induces production of specific antibodies that immediately bind the toxin if encountered later. If a person develops symptoms of a toxin-mediated disease, he or she can often be treated with **antitoxin,** a suspension of neutralizing antibodies.

Most exotoxins can be grouped into functional categories according to the tissues they affect (see table 16.1).

- **Neurotoxins.** These damage the nervous system, causing symptoms such as paralysis.
- **Enterotoxins.** These cause symptoms associated with intestinal disturbance, such as diarrhea and vomiting.
- **Cytotoxins.** These damage a variety of different cell types, either by interfering with essential cellular mechanisms or by lysing cells.

Exotoxins usually fall into three general categories that reflect their structure and general mechanism of action: A-B toxins, membrane-damaging toxins, and superantigens.

A-B Toxins

A-B toxins consist of two parts: The A subunit is the toxic (active) portion and the B subunit binds to a specific surface molecule on cells (**figure 16.11**). In other words, the A subunit, usually an enzyme, is responsible for the effects of the toxin, whereas the B subunit dictates the type of cell to which the toxin is delivered. The structure of A-B toxins offers novel approaches for the development of vaccines and therapies. For example, the B subunit of cholera toxin is used as a component of an orally administered cholera vaccine available in some countries; the vaccine elicits production of antibodies that bind the subunit, and these protect the vaccine recipient because they prevent cholera toxin from attaching to intestinal cells. Researchers are now experimenting with joining medications to B subunits, allowing those beneficial compounds to be delivered specifically to the cell type targeted by the B subunit.

Membrane-Damaging Toxins

Membrane-damaging toxins are cytotoxins that disrupt eukaryotic cytoplasmic membranes, causing the cell to lyse. Many lyse red blood cells, causing hemolysis that can be observed when the organisms are grown on blood agar; these toxins are often referred to as **hemolysins.**

General types of membrane-damaging toxins include:

- **Pore-forming toxins.** These insert into the phospholipid bilayer of membranes, forming channels that allow uncontrolled passage of fluids. An example of a pore-forming toxin is streptolysin O, the compound responsible for the characteristic β-hemolysis of *Streptococcus pyogenes* grown anaerobically on blood agar (see figure 4.10). Recall that streptolysin O also helps *S. pyogenes* avoid phagocytosis.

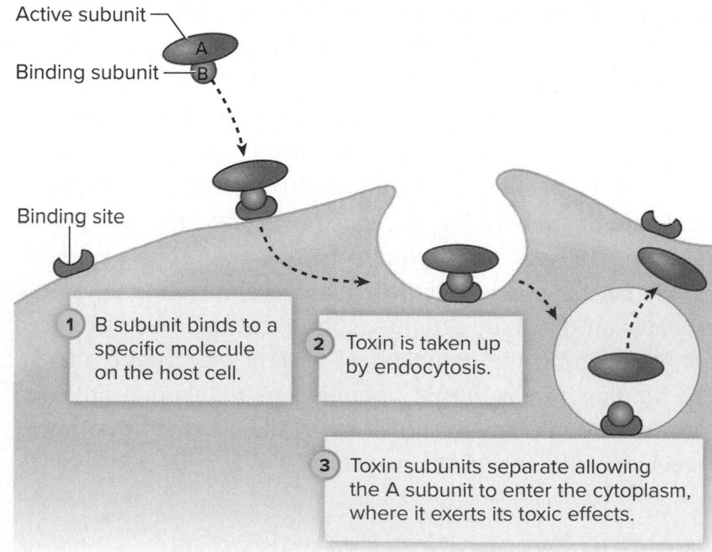

Active subunit
A
Binding subunit
B
Binding site
1 B subunit binds to a specific molecule on the host cell.
2 Toxin is taken up by endocytosis.
3 Toxin subunits separate allowing the A subunit to enter the cytoplasm, where it exerts its toxic effects.

FIGURE 16.11 The Action of an A-B Toxin The structure and mechanisms of uptake of different A-B toxins may vary slightly.

? **Would an antibody response against the B subunit protect against the effects of the toxin? Why or why not?**

- **Phospholipases.** These hydrolyze phospholipids in the cytoplasmic membrane. The α-toxin of *Clostridium perfringens* (causes gas gangrene) is an example.

Superantigens

Superantigens are exotoxins that stimulate an abnormally high number of T_H cells (effector helper T cells), causing a massive release of cytokines (a "cytokine storm"). This leads to fever, nausea, vomiting, and diarrhea. The effects of a cytokine storm can be life-threatening, leading to organ failure and circulatory collapse. Examples of superantigens include toxic shock syndrome toxin (TSST) as well as several other toxins produced by *Staphylococcus aureus* and *Streptococcus pyogenes.*

Superantigens function by overriding the normal specificity of helper T cells' antigen recognition. They do this by binding simultaneously to the outer portion of the major histocompatibility (MHC) class II molecule on antigen-presenting cells and the T-cell receptor (**figure 16.12**). The T-cell machinery interprets the binding to mean that the T-cell receptor recognizes the antigen presented on the MHC molecule, when it probably does not. An antigen usually stimulates about one in 10,000 helper T cells, whereas superantigens stimulate as many as one in five. Following that stimulation, the T cells often undergo apoptosis, thereby suppressing the immune response. Superantigens are also suspected of contributing to autoimmune diseases; by overriding the normal control mechanisms of adaptive immunity, superantigens may promote the proliferation of T cells that respond to healthy "self."

The exotoxins produced by *S. aureus* strains that cause foodborne intoxication are superantigens. Although they result in nausea and vomiting and are therefore referred to as enterotoxins, their structure and action are very different from the enterotoxins of *V. cholerae* and pathogenic *E. coli* strains. The mechanism by which they induce vomiting is poorly understood. Unlike most other exotoxins, the enterotoxins produced by *S. aureus* are heat-stable; thorough cooking of foods contaminated with these toxins will not prevent illness.

MicroByte

Botox, used for medical and cosmetic treatments, is a dilute suspension of botulinum toxin (see Focus Your Perspective 26.1).

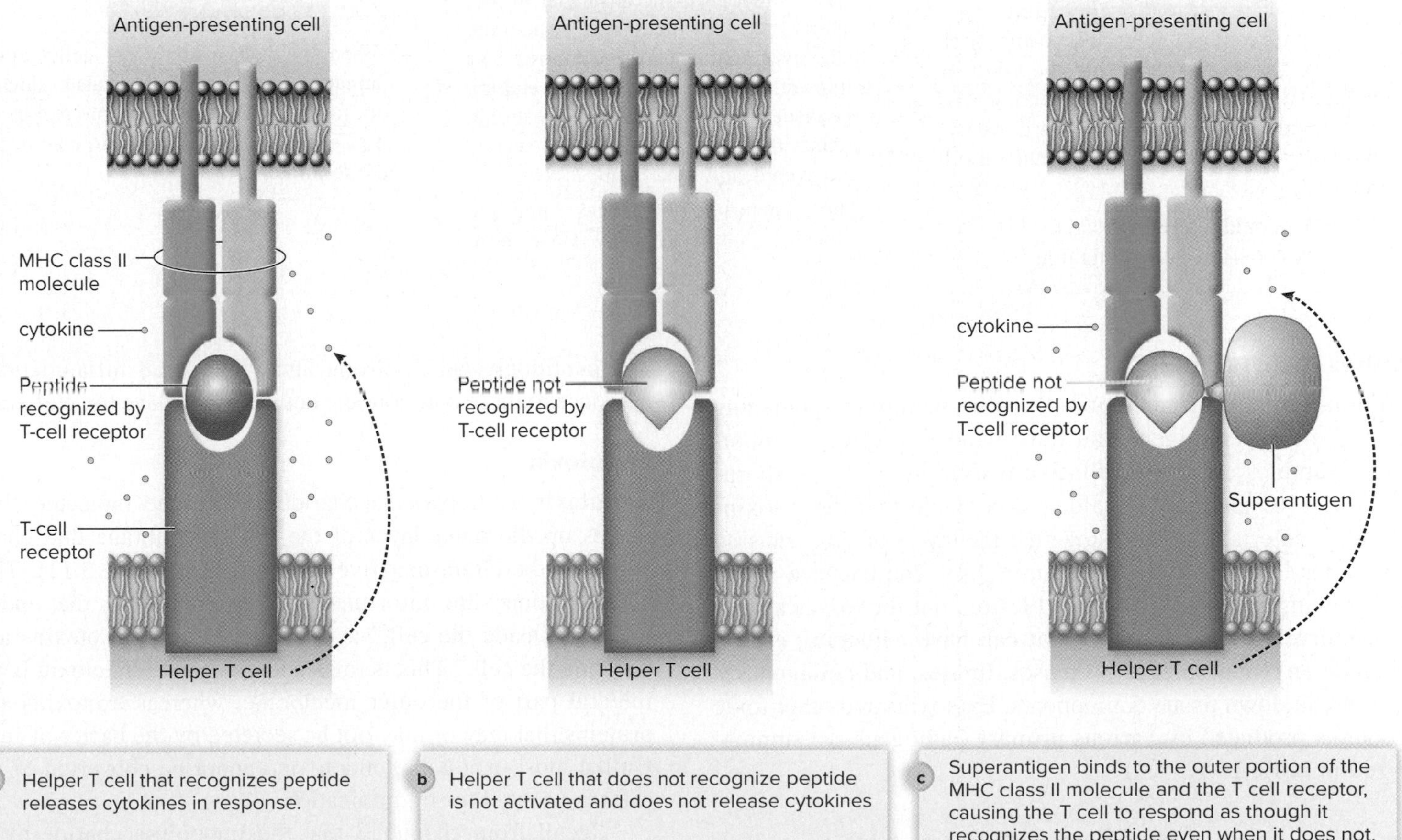

FIGURE 16.12 Superantigens (a) Normal situation: antigen recognition. **(b)** Normal situation: lack of antigen recognition. **(c)** Effect of a superantigen.

? **What specifically causes the toxic effects of superantigens?**

FOCUS ON A CASE 16.1

The patient was a 14-year-old boy who had been coughing frequently for about 2 weeks. He initially developed what his parents assumed was the common cold, with symptoms of a runny nose, malaise, low fever, and mild cough. His cough gradually worsened and was particularly severe at night. Some of his coughing episodes were so intense that they caused him to vomit.

The physician considered the possibility that the boy had pertussis (whooping cough), a disease caused by the bacterium *Bordetella pertussis*. The boy had been vaccinated against the disease as a child, but the physician knew that recent outbreaks suggest that the vaccine used since the 1990s might not provide long-lasting immunity. The physician swabbed the boy's nasopharynx (where the nasal passage connects with the throat) and sent the sample to the clinical lab for culture and PCR testing. In talking to the boy's parents about the next steps, she told them to watch him for signs such as a spike in fever, which could indicate a secondary infection.

Culture and PCR results came back positive for *B. pertussis,* and the boy was treated with antibacterial medication. His cough persisted for several more weeks, but he gradually improved and made a full recovery.

1. Why would it be beneficial to *B. pertussis* to cause coughing?
2. The bacterium colonizes the ciliated cells of the upper respiratory tract. How might it avoid being swept away by the mucociliary escalator?
3. The bacteria do not invade the epithelial cells, yet they damage those cells. What might cause the damage?
4. Why would the physician be concerned about a secondary infection?

Discussion

1. Coughing propels the bacterial cells into the air, where another person can inhale them. In this way, the bacterium can move to the next susceptible host. Because whooping cough is easily transmitted, it spreads quickly among unvaccinated populations, so contacts of people who have the disease should be notified. Signs and symptoms of pertussis are often mild in adults, mimicking the common cold, so people with mild infections can unknowingly serve as a source of *B. pertussis* to others. Infants cannot receive their first whooping cough vaccination until age 2 months, so they are at particular risk for developing the disease. This is a concern because in infants the disease can be quite severe, and even deadly (see figure 21.16).
2. The bacterium produces multiple adhesins that specifically allow it to colonize ciliated cells. The attachment structures include fimbriae and a protein called filamentous hemagglutinin (FHA), named for its ability to agglutinate red blood cells.
3. *B. pertussis* produces several toxins that damage cells. One of them, tracheal cytotoxin (TCT), is a fragment of peptidoglycan that *B. pertussis* specifically releases during growth and is toxic to ciliated epithelial cells. Another is pertussis toxin (PT), an A-B toxin that modifies a regulatory protein in respiratory cells, causing the accumulation of respiratory secretions and mucus. The combination of increased fluid production and decreased ciliary action results in the characteristic cough of pertussis, because only the cough reflex remains for clearing respiratory secretions. The classical "pertussis cough" includes a rapid series of coughs—a reflex for clearing fluid from the airways—followed by a whooping sound as the patient gasps for air.
4. When ciliated respiratory cells are damaged, the mucociliary escalator does not function normally, therefore making it easier for other pathogens to colonize the respiratory tract.

Other Toxic Proteins

Various proteins that are not A-B toxins, membrane-damaging toxins, or superantigens can have damaging effects. Important examples are the **exfoliative toxins** produced by strains of *S. aureus* that cause scalded skin syndrome; these toxins destroy material that binds together the layers of skin, causing the outer layer to peel (see figure 22.4). The bacteria might be growing in a small, localized lesion, but the toxins spread systemically. Other proteins that can have damaging effects include enzymes such as proteases, lipases, and collagenases that break down tissue components. Exotoxins and other toxic proteins produced by various primary pathogens are summarized in **table 16.1.**

Endotoxin and Other Bacterial Cell Wall Components

The host defenses are primed to respond to various bacterial cell wall components, including lipopolysaccharide and peptidoglycan. A strong and widespread inflammatory response to these compounds, however, can have toxic effects.

Endotoxin

Endotoxin is lipopolysaccharide (LPS), the molecule that makes up the outer layer of the outer membrane that characterizes the Gram-negative cell wall (see figure 3.11). The name is somewhat unfortunate because it implies that endotoxin is "inside the cell," and, conversely, that exotoxins are "outside the cell." This is misleading because endotoxin is an integral part of the outer membrane, whereas exotoxins are proteins that may or may not be secreted by the bacterial cell. Unlike most exotoxins, endotoxin cannot be converted to an effective toxoid for immunization.

Recall from chapter 3 that the lipopolysaccharide molecule contains lipid A, which the body recognizes as an indication that Gram-negative bacteria have invaded. When lipid A is present in a localized region, the response helps clear an infection, but when the infection is systemic, as in

TABLE 16.1 Exotoxins and Other Toxic Proteins Produced by Various Primary Pathogens

Example	Name of Disease; Name of Toxin	Characteristics of the Disease	Mechanism
A-B TOXINS—Composed of two subunits, A and B. The A subunit is the toxic, or active, part; the B subunit binds to the target cell.			
Neurotoxins			
Clostridium botulinum	Botulism; botulinum toxin	Flaccid paralysis	Blocks transmission of nerve signals to muscles by preventing the release of neurotransmitters.
Clostridium tetani	Tetanus; tetanus toxin	Spastic paralysis	Blocks the action of inhibitory neurons by preventing the release of neurotransmitters.
Enterotoxins			
Enterotoxigenic *E. coli* (ETEC)	Traveler's diarrhea; heat-labile enterotoxin (cholera-like toxin)	Severe watery diarrhea	Modifies a regulatory protein in intestinal cells, causing an outpouring of water and electrolytes from the cells.
Vibrio cholerae	Cholera; cholera toxin	Severe watery diarrhea	Modifies a regulatory protein in intestinal cells, causing an outpouring of water and electrolytes from the cells.
Cytotoxins			
Bacillus anthracis	Anthrax; anthrax toxin (includes edema factor and lethal factor)	Inhaled form—overwhelming systemic illness; cutaneous form—skin lesions	Edema factor modifies a regulatory protein in cells, causing accumulation of fluids in the tissues. Lethal factor inactivates proteins involved in cell signaling.
Bordetella pertussis	Pertussis (whooping cough); pertussis toxin	Sudden bouts of violent coughing	Modifies a regulatory protein in respiratory cells, causing accumulation of respiratory secretions and mucus. Other factors also contribute to the symptoms.
Corynebacterium diphtheriae	Diphtheria; diphtheria toxin	Pseudomembrane in the throat; heart, nervous system, kidney damage	Inhibits protein synthesis by inactivating an elongation factor of eukaryotic cells. Kills local cells (in the throat) and is carried in the bloodstream to various organs.
Shiga toxin–producing *E. coli* (STEC)	Bloody diarrhea, hemolytic uremic syndrome; Shiga toxin	Diarrhea that may be bloody; kidney damage	Inactivates eukaryotic ribosomes, stopping protein synthesis.
Shigella dysenteriae	Dysentery, hemolytic uremic syndrome; Shiga toxin	Diarrhea that contains blood, pus, and mucus; kidney damage	Inactivates eukaryotic ribosomes, stopping protein synthesis.
MEMBRANE-DAMAGING TOXINS (cytotoxins)—Disrupt plasma membranes, causing leakiness that results in cell lysis.			
Clostridium perfringens	Gas gangrene; α-toxin	Extensive tissue damage	Removes the polar head group on the phospholipids in the membrane, damaging the membrane structure.
Staphylococcus aureus	Wound and other infections; α-toxin, Panton-Valentine leukocidin (PVL)	Accumulation of pus	Inserts into membranes, forming pores that allow fluids to enter the cells.
Streptococcus pyogenes	Pharyngitis and other infections; streptolysin O	Accumulation of pus	Inserts into membranes, forming pores that allow fluids to enter the cells.
SUPERANTIGENS—Override the specificity of the T-cell response.			
Staphylococcus aureus (certain strains)	Foodborne intoxication; staphylococcal enterotoxins	Nausea and vomiting	Not well understood with respect to how the ingested toxins lead to the characteristic symptoms of foodborne intoxication.
Staphylococcus aureus (certain strains)	Staphylococcal toxic shock; toxic shock syndrome toxin-1	Fever, vomiting, diarrhea, muscle aches, rash, low blood pressure	Systemic toxic effects due to the resulting massive release of cytokines.
Streptococcus pyogenes (certain strains)	Streptococcal toxic shock; streptococcal pyrogenic exotoxins	Fever, vomiting, diarrhea, muscle aches, rash, low blood pressure	Systemic toxic effects due to the resulting massive release of cytokines.
OTHER TOXIC PROTEINS			
Staphylococcus aureus	Scalded-skin syndrome; exfoliative toxins	Separation of the outer layer of skin	Destroys material that holds the layers of skin together.
Various organisms	Various diseases; proteases, lipases, and other hydrolases	Tissue damage	Degrades proteins, lipids, and other compounds that make up tissues.

a bloodstream infection, the result can be devastating. Imagine inflammation occurring throughout the body, with permeable blood vessels leaking fluids and recruited neutrophils damaging tissues. This is the response that can lead to the life-threatening tissue damage and organ dysfunction that characterizes bacterial **sepsis** (see figure 25.3). Such an overwhelming systemic response can result in a dramatic drop in blood pressure, leading to **shock,** a life-threatening condition resulting from insufficient blood flow. Widespread blood clotting may also occur, leading to disseminated intravascular coagulation (DIC).

Lipid A is part of the Gram-negative outer membrane and causes a response only when released from the cell. This occurs primarily when a bacterium lyses, which can happen as a result of phagocytosis, activation of the complement system (due to membrane attack complexes), or treatment with certain types of antibiotics. Once released from a cell, the LPS molecules can activate the innate and adaptive defenses by a variety of mechanisms. Monocytes, macrophages, and other cells have toll-like receptors (TLRs) that detect LPS, inducing the cells to produce pro-inflammatory cytokines. LPS also functions as a T-independent antigen; at high concentrations, it activates a variety of different B cells, regardless of the specificity of their B-cell receptor.

Endotoxin is heat-stable; it is not destroyed by autoclaving. Thus, although autoclaved material is sterile, it may still contain endotoxin. Disastrous outcomes, including death, have resulted from administering intravenous (IV) fluids contaminated with endotoxin. To verify that such fluids are not contaminated, a very sensitive test known as the Limulus amoebocyte lysate (LAL) assay is done. This uses proteins extracted from blood of the horseshoe crab *(Limulus polyphemus)* that form a gel-like clot when exposed to even minute amounts of endotoxin. Horseshoe crabs are one of this planet's more unique and ancient life-forms. The essential role they play in this test has led to an increased awareness of the importance of their habitat. A non-lethal system of capture, blood sampling, and release has been developed.

Table 16.2 summarizes some of the differences between exotoxins and endotoxin.

Other Bacterial Cell Wall Components

Peptidoglycan and other bacterial cell wall components can cause symptoms similar to those that characterize the response to endotoxin. The systemic response can lead to sepsis, shock, and DIC.

Damaging Effects of the Immune Response

Although the immune response eliminates invading microbes, it can inadvertently damage host tissues as well; the reactions to endotoxin and other cell wall components are examples of damaging effects of the immune response but are typically considered a toxic effect of the bacterium because the reactions can be immediate and overwhelming. Some of the damaging responses discussed next take longer to appear.

Damage Associated with Inflammation

The inflammatory response itself can be damaging whether in a local region or widespread throughout the body. Phagocytic cells recruited to an area release enzymes and toxic products, and these may damage surrounding tissues. The life-threatening aspects of bacterial meningitis, for example, are due to the inflammatory response itself. Complications of certain sexually transmitted infections are also due to the damage associated with inflammation. If *Chlamydia trachomatis* or *Neisseria gonorrhoeae* infections involve the fallopian tubes, the inflammatory response can lead to scarring that obstructs the tubes, either preventing fertilization or predisposing a woman to an ectopic pregnancy (meaning the fertilized egg

TABLE 16.2 Comparison of Exotoxins and Endotoxin

Property	Exotoxins	Endotoxin
Bacterial Source	Gram-positive and Gram-negative species	Gram-negative species only
Location in the Bacterium	Synthesized in the cytoplasm; may or may not be secreted	Component of the outer membrane that characterizes the Gram-negative cell wall
Chemical Nature	Protein	Lipopolysaccharide (the lipid A component)
Ability to Form a Toxoid	Generally, yes	No
Heat Stability	Generally inactivated by heat	Heat-stable
Mechanism	A distinct toxic mechanism for each	Innate immune response; systemic response can lead to sepsis, shock, and disseminated intravascular coagulation.
Toxicity	Generally very potent; some are among the most potent toxins known	Small amounts in a localized area lead to an appropriate immune response that helps clear an infection, but systemic distribution can be deadly.

implants outside the uterus). When inflammation occurs in the lungs, fluids leaking from the local capillaries may collect in the alveoli (air sacs) and interfere with O_2 and CO_2 exchange, such as in severe COVID-19.

Bloodstream infections can cause extensive harm due to the systemic inflammatory response triggered by LPS (endotoxin), peptidoglycan, or other bacterial components. The widespread capillary leakiness can lead to a blood volume too low to support adequate flow to organs, and a cytokine storm (an excessive and potentially damaging cytokine response) may develop, amplifying the problem. A systemic infection sometimes leads to sepsis (see figure 25.3).

Damage Associated with Adaptive Immunity

The adaptive immune response can also lead to damaging effects. Mechanisms include:

- **Immune complexes.** When antibodies bind to antigens, the complexes can settle in the kidneys and joints, where they activate the complement system, causing destructive inflammation. An example is post-streptococcal glomerulonephritis, a complication that can follow skin and throat infections caused by *Streptococcus pyogenes;* the immune complexes that form as a result of the infection trigger a response that damages kidney structures called glomeruli.
- **Cross-reactive antibodies.** Certain antibodies produced in response to an infection bind to the body's own tissues, promoting an autoimmune response. Acute rheumatic fever—a complication that can follow strep throat—appears to be the result of antibodies against *S. pyogenes* binding to normal tissue proteins. This occurs most often in people with certain MHC types (see Focus Your Perspective 15.1).

MicroAssessment 16.8

Damage can be due to exotoxins, including A-B toxins, superantigens, membrane-damaging toxins, and other toxic proteins. Endotoxin and other cell wall components in the bloodstream can trigger a widespread inflammatory response. Inflammation can cause local tissue damage that leads to scarring; a systemic inflammatory response can result in sepsis. Immune complexes can trigger damaging inflammation in the kidneys and joints; cross-reactive antibodies can lead to an autoimmune response.

19. Cholera toxin is an A-B toxin. What does that tell you about its structure?
20. How is an enterotoxin different from endotoxin?
21. Home-canned foods should be boiled before consumption to prevent botulism. Considering that boiling does not destroy endospores, why would it prevent the disease?

16.9 ■ Mechanisms of Viral Pathogenesis

Learning Outcomes

14. Describe how viruses bind to host cells and how they spread to other cells.
15. Describe how viruses avoid immune responses and ultimately damage cells.

To infect a host, a virus must enter an appropriate cell, use the cell's machinery for replication, keep the host from recognizing and destroying the infected cell, and then move to new cells or hosts. Damage to the host can occur as a direct result of infection or from the immune response.

Binding to Host Cells and Invasion

As discussed in chapter 13, viruses attach to target cells via specific receptors on those cells. Only cells that have the receptor can be infected, so this influences the host range and tissue specificity of a particular virus. HIV's receptor is CD4, a molecule found on helper T cells and macrophages. The coronaviruses that cause COVID-19 and SARS bind to ACE2 (**a**ngiotensin-**c**onverting **e**nzyme 2), an enzyme found on a variety of cell types, including lung cells. Depending on the type of virus, either receptor-mediated endocytosis or fusion with the cytoplasmic membrane is used to enter a host cell (see figure 13.10).

Virions released from a cell can either infect neighboring cells or spread to other tissues via the bloodstream or lymphatic system. Polioviruses initially infect cells in the throat and intestinal tract; upon release from these cells, some viral particles enter the bloodstream and may then spread to infect motor nerve cells of the brain and spinal cord, causing paralysis.

Avoiding Immune Responses

To replicate in a host, viruses must avoid at least some of the defenses that detect and eliminate the invaders. These include interferons, cell-mediated responses, and antibodies.

Avoiding the Antiviral Effects of Interferons

Interferons normally limit viral spread in a host by causing cells to produce enzymes that, when activated, induce infected cells to undergo apoptosis (see figure 14.10). As a group, viruses use various mechanisms to avoid this, and many viruses use multiple mechanisms. Some viruses coat their RNA with a virally encoded protein, hiding it from the host cell's pattern recognition receptors that detect viral nucleic acid. Viruses may also shut down host gene expression, thereby preventing the synthesis of proteins involved in the interferon response. Some viruses interfere with activation of the enzymes used by infected cells to inhibit viral activities. Papillomaviruses, which cause various types of warts, interfere with the normal function of p53, a protein that regulates apoptosis in cells; tumors sometimes develop as a result of this inhibition.

Avoiding the Cell-Mediated Immune Response

To avoid being detected by the cell-mediated immune response, many types of viruses interfere with antigen presentation by MHC class I molecules. Herpesviruses, for example, block the movement of these MHC molecules to the surface of the infected cell, so that T_C cells (effector cytotoxic T cells) cannot inspect the proteins being made. The immune system, however, is prepared for such a strategy. Natural killer (NK) cells recognize stressed cells that lack MHC class I molecules and destroy them (see figure 15.26). Perhaps not surprisingly, some viruses have methods to trick NK cells. Cytomegalovirus (CMV), which causes disease in immunocompromised people, encodes the production of "fake" MHC class I molecules. These decoy versions are displayed on the surface of the host cell, tricking the immune system into believing that all is well in the cell (**figure 16.13**).

Avoiding Antibodies

Antibodies generally control the spread of viruses by neutralizing extracellular virions (see figure 15.4). Some viruses move directly from one cell to its immediate neighbors, thus avoiding antibodies. Other viruses remain intracellular by forcing cellular neighbors to fuse, forming a large multinucleated cell called a **syncytium;** HIV induces syncytia formation.

The surface antigens of some viruses change rapidly, outpacing the body's capacity to produce effective neutralizing antibodies. This rapid evolution typically occurs because many replicases of RNA viruses and the reverse transcriptases of retroviruses have no proofreading ability, so mutations occur fairly often; thus, as the viruses replicate, they give rise to pools of genetically altered virions. An immune response against essential proteins that cannot tolerate extensive mutations may eventually clear the viruses. In the case of HIV, however, an infected person produces many anti-HIV antibodies, but those antibodies are not effective in eliminating the virus.

Some viral infections are actually aided by antibodies—a process called antibody-dependent enhancement of infection (ADE). This is thought to occur when antibodies bind to viral particles but fail to neutralize them; instead, the bound antibodies facilitate uptake of the particles by phagocytes, within which the virus replicates. Severe dengue is a disease that involves ADE.

Damage to the Host

As viruses replicate, they often damage host cells. In addition, the immune response to the infection can also be damaging. Many viruses cause what are referred to as **flu-like symptoms**—malaise, fever, and body aches—which are a result of the body's response to the infection. Some viruses, including the one that causes COVID-19, can elicit the sometimes-deadly overproduction of cytokines called a **cytokine storm.**

MicroAssessment 16.9

Viruses attach to host cells via receptors on those cells. Viruses, as a group, have several mechanisms for avoiding the effects of interferon, avoiding the cell-mediated immune response, and avoiding antibodies.

22. From a virus's perspective, why would it be beneficial to prevent apoptosis?

23. How do cytomegaloviruses avoid the cell-mediated immune response?

24. Why would various unrelated viral infections often include a similar set of symptoms (fever, headache, fatigue, and runny nose)?

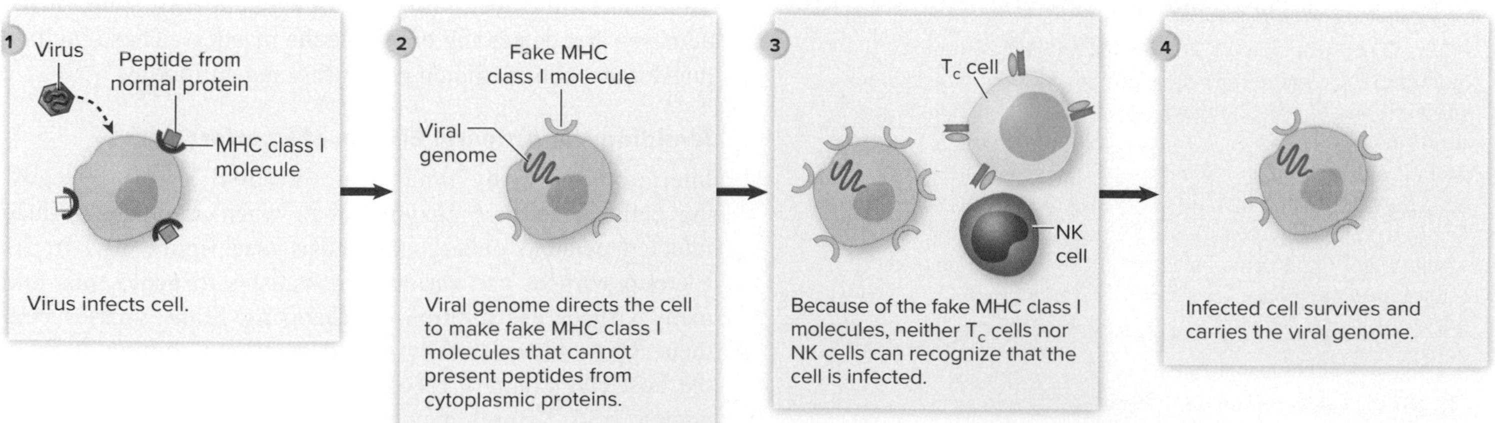

FIGURE 16.13 The Role of MHC Class I Decoys in Viral Pathogenesis

What is the role of MHC class I molecules in adaptive immunity?

16.10 ■ Mechanisms of Eukaryotic Pathogenesis

Learning Outcomes

16. Describe the mechanisms of pathogenesis of dermatophytes, *Candida albicans,* and dimorphic fungi.
17. Compare and contrast the mechanisms of pathogenesis of protozoa and helminths.

Pathogenesis of eukaryotic cells, including fungi and protozoa, involves the same basic scheme as that of bacterial pathogens: colonization, evasion of host defenses, and damage to the host. The mechanisms, however, are often not as well understood.

Fungi

Members of a group of fungi referred to as dermatophytes cause superficial infections of hair, skin, and nails, but do not invade deeper tissues. These fungi have keratinases, enzymes that break down the keratin in superficial tissues, allowing the fungi to use this protein as a nutrient source. Dermatophytes cause diseases such as ringworm and athlete's foot.

Fungi in the normal microbiota, especially the yeast *Candida albicans,* can cause disease in immunocompromised hosts. Factors that can lead to excessive growth of *C. albicans* include AIDS, uncontrolled diabetes, severe burns, and inhibition of normal microbiota due to hormonal influences or antibiotic treatment. *C. albicans* generally infects the mucous membranes, causing thrush (an infection of the throat and mouth) or vulvovaginal candidiasis.

Cryptococcus species make a very large capsule. This capsule, like those produced by certain bacteria, interferes with phagocytosis.

Most serious fungal infections are caused by dimorphic fungi, which can transition between two forms depending on the environmental conditions. These fungi grow as molds in the soil, producing small conidia (reproductive structures) that easily become airborne; when these are inhaled, they lodge deep within the lungs and then develop into other forms, usually yeasts. The immune system generally controls the infection, so most cases are asymptomatic. If the person is immunocompromised, however, the infections can become life-threatening.

Some fungi produce toxins, collectively referred to as mycotoxins. *Aspergillus flavus,* a fungus that grows on certain grains and nuts, produces aflatoxin; when ingested, this toxin can damage the liver and increase the risk of liver cancer. Fungal conidia and other products can cause hypersensitivities in some people.

Protozoa and Helminths

Most pathogenic protozoa and helminths either live within the intestinal tract or enter the body's tissues via the bite of an arthropod. *Schistosoma* species, however, can enter the skin directly (see Focus Your Perspective 14.1).

Like bacteria and viruses, eukaryotic parasites attach to host cells via specific receptors. *Plasmodium vivax,* one of the two most common causes of malaria, typically attaches to the Duffy blood group antigen on red blood cells. Most people of West African ancestry lack this antigen and are therefore resistant to infection by the common strains of *P. vivax. Giardia lamblia* cells use a disc that functions as a suction cup to attach to intestinal cells; an adhesin associated with the disc helps the initial attachment.

Protozoa and helminths use a variety of mechanisms to avoid antibodies. Some hide within cells, thus avoiding exposure to antibodies as well as certain other defenses. Malarial parasites, for example, infect red blood cells (RBCs), a cell type that does not present antigens to T_C cells. *Plasmodium falciparum* makes proteins that insert into the infected RBCs, causing the RBCs to stick to the lining of the capillaries rather than circulating through the spleen where they might otherwise be destroyed. *Leishmania* species survive and multiply within macrophages when phagocytized. Cells of the parasite that causes African trypanosomiasis (*Trypanosoma brucei*) routinely change their surface antigens to prevent antibodies from recognizing them. *Schistosoma* species coat themselves with host proteins, thereby disguising themselves. Many parasitic worms appear to suppress immune responses by secreting certain molecules—sometimes carried within extracellular vesicles (membrane-bound particles)—to moderate the inflammatory response.

The extent and type of damage caused by parasites vary tremendously. In some cases, the parasites compete for nutrients in the intestinal tract, contributing to malnutrition of the host. Helminths that accumulate in high enough numbers or grow quite long can block the intestines or other organs. Parasites that produce tissue-digesting enzymes cause direct damage, and those that elicit a strong immune response cause indirect damage; the latter accounts for the high fevers that characterize malaria and the granulomas that form around *Schistosoma* eggs.

MicroAssessment 16.10

Pathogenic mechanisms of fungi, protozoa, and helminths involve the same basic scheme as described for bacteria: colonization, evasion of host defenses, and damage to the host.

25. What is the importance of keratinase?
26. How does the parasite that causes African trypanosomiasis avoid the effects of antibodies?
27. Why would relatively few people of West African ancestry have the Duffy blood group antigen?

Summary

MICROBES, HEALTH, AND DISEASE

16.1 ■ The Anatomical Barriers as Ecosystems

In **mutualism,** both partners benefit; in **commensalism,** one partner benefits while the other is unaffected; and in **parasitism,** the parasite benefits at the expense of the host.

16.2 ■ The Human Microbiome (figure 16.1)

Composition of the Microbiome

Babies begin acquiring their microbiome during delivery and feeding. The microbial community of adults shows considerable diversity. Bacillota and Bacteroidota are commonly represented phyla in the intestinal tract of adults.

Beneficial Roles of the Human Microbiome

The human microbiome protects against infection by excluding pathogens, priming the adaptive immune system to react against pathogens, helping the immune system to develop immune tolerance to harmless substances, aiding digestion, and producing certain vitamins and other substances important for health.

The Gut Microbiome-Brain Connection

Increasing evidence points to the gut microbiome playing a part not only in digestive health but also in brain function.

16.3 ■ Principles of Infectious Disease

Infectious diseases have characteristic **signs** and **symptoms.** A **secondary infection** can occur as the result of a **primary infection.**

Pathogenicity

Pathogenicity is the ability to cause disease. A **primary pathogen** causes disease in even otherwise healthy individuals; an **opportunistic pathogen** causes disease only when introduced into an unusual site or when the body's defenses are weakened. **Virulence** refers to the degree of pathogenicity of an organism.

Characteristics of Infectious Disease

Communicable or **contagious diseases** spread from one host to another; ease of spread partly reflects the **infectious dose.** Stages of infectious disease include the **incubation period, illness,** and **convalescence;** during the illness a person experiences signs and symptoms of the disease (figure 16.2). **Infections** can be described as **acute**, **chronic,** or **latent,** depending on the timing and duration of symptoms. Infections can be **localized** or **systemic.**

16.4 ■ Determining the Cause of an Infectious Disease

Koch's Postulates

Koch's postulates provide a foundation for determining the cause of an infectious disease (figure 16.3).

Molecular Koch's Postulates

Molecular Koch's postulates provide a foundation for identifying virulence factors that contribute to disease.

MECHANISMS OF PATHOGENESIS

16.5 ■ Establishing Infection

Adherence

Bacteria use **adhesins** to bind to host cells (figure 16.4).

Colonization

Rapid turnover of pili, antigenic variation, and **IgA proteases** allow bacteria to avoid the effects of secretory IgA. **Siderophores** enable microbes to scavenge iron.

Delivering Effector Proteins to Host Cells

Type III secretion systems of Gram-negative bacteria allow them to deliver proteins directly to host cells (figure 16.5).

16.6 ■ Penetrating the Anatomical Barriers

Penetrating the Skin

Some pathogens enter tissues via a bite or other skin-damaging injury.

Penetrating Mucous Membranes

Some pathogens induce mucosal epithelial cells to engulf bacterial cells; others exploit antigen-sampling processes (figures 16.6, 16.7).

16.7 ■ Avoiding the Host Defenses

Hiding Within a Host Cell

Some bacteria can evade innate defenses, as well as certain adaptive defenses, by entering host cells.

Avoiding Destruction by Phagocytes (figure 16.9)

Mechanisms to prevent encounters with phagocytes include C5a peptidase and **membrane-damaging toxins.** Mechanisms to avoid recognition and attachment by phagocytes include **capsules,** M protein, and **Fc receptors** (figure 16.10). Mechanisms to survive within the phagocyte include escape from the phagosome, preventing phagosome-lysosome fusion, and surviving within the phagolysosome.

Avoiding Killing by Complement System Proteins

Certain bacteria that are **serum resistant** postpone the formation of membrane attack complexes by interfering with activation of the complement system via the alternative pathway.

Avoiding Recognition by Antibodies

Mechanisms to avoid antibodies include **IgA protease, antigenic variation,** and mimicking "self."

16.8 ■ Damage to the Host

Exotoxins (table 16.1)

Exotoxins are proteins that have very specific damaging effects; they may act locally or cause dramatic systemic effects. Many can be grouped into categories such as **neurotoxins, enterotoxins,** or **cytotoxins.** The toxic activity of **A-B toxins** is mediated by the A subunit; the binding to specific cells is mediated by the B subunit (figure 16.11). **Membrane-damaging toxins** disrupt cell membranes either by forming pores or by removing the polar head group on phospholipids in the membrane. **Superantigens** override the specificity of the T-cell response, causing systemic effects due to the massive release of cytokines (figure 16.12).

Endotoxin and Other Bacterial Cell Wall Components

The symptoms associated with **endotoxin** are due to a vigorous host response. Lipid A of lipopolysaccharide is responsible for its toxic properties. Peptidoglycan and certain other components induce various cells to produce pro-inflammatory cytokines.

Damaging Effects of the Immune Response

The release of enzymes and toxic products from phagocytic cells can damage tissues. **Immune complexes** can cause kidney and joint damage; cross-reactive antibodies can result in an autoimmune response.

16.9 ■ Mechanisms of Viral Pathogenesis

Binding to Host Cells and Invasion

Viruses attach to specific receptors on the target cell.

Avoiding Immune Responses

Many viruses can avoid the effect of interferons; some avoid the cell-mediated immune response (figure 16.13). To avoid antibodies, some viruses transfer directly from cell to cell; some change their surface antigens quickly, outpacing the production of antibodies.

Damage to the Host

Viruses often damage host cells directly. The immune response to many viruses causes **flu-like symptoms;** some viruses elicit a potentially deadly **cytokine storm.**

16.10 ■ Mechanisms of Eukaryotic Pathogenesis

Fungi

Saprophytes are generally opportunists; dermatophytes cause superficial infections of skin, hair, and nails. The most serious fungal infections are caused by dimorphic fungi.

Protozoa and Helminths

Eukaryotic parasites attach to host cells via specific receptors. They use a variety of mechanisms to avoid antibodies; the extent and type of damage they cause vary tremendously.

Review Questions

Short Answer

1. Describe three types of symbiotic relationships.
2. Describe two situations that can lead to changes in the composition of a person's microbiome.
3. How are acute, chronic, and latent infections different from one another?
4. Why are Koch's postulates not sufficient to establish the cause of all infectious diseases?
5. Describe the four general mechanisms by which microorganisms cause disease.
6. Describe two mechanisms that bacteria use to invade via mucous membranes.
7. Explain how a capsule can allow an organism to be serum resistant and to avoid phagocytosis.
8. Give an example of a neurotoxin, an enterotoxin, and a cytotoxin.
9. Describe two mechanisms a virus might use to prevent the induction of apoptosis in an infected cell.
10. How do *Schistosoma* species avoid antibodies?

Multiple Choice

1. Opportunistic pathogens are *least* likely to affect which of the following groups?
   a) AIDS patients
   b) Cancer patients
   c) College students
   d) Drug addicts
   e) Transplant recipients
2. Capsules and M protein are thought to interfere with which of the following?
   a) Opsonization by complement proteins
   b) Opsonization by antibodies
   c) Recognition by T cells
   d) Recognition by B cells
   e) Phagosome-lysosome fusion
3. The C5a peptidase enzyme of *Streptococcus pyogenes* breaks down C5a, resulting in
   a) lysis of *S. pyogenes* cells.
   b) lack of opsonization of *S. pyogenes* cells.
   c) killing of phagocytes.
   d) decreased accumulation of phagocytes.
   e) inhibition of membrane attack complexes.
4. All of the following are known microbial mechanisms of avoiding the effects of antibodies *except*
   a) antigenic variation.
   b) mimicking "self."
   c) synthesis of an Fc receptor.
   d) synthesis of IgG protease.
   e) remaining intracellular.
5. Which of the following statements about diphtheria toxin is *false*?
   a) It is an example of an endotoxin.
   b) It is produced by a species of *Corynebacterium.*
   c) It inhibits protein synthesis.
   d) It can cause local damage to the throat.
   e) It can cause systemic damage (that is, to organs such as the heart).
6. Which of the following statements about botulism is *true*?
   a) It is caused by *Bacillus botulinum,* an obligate aerobe.
   b) The toxin is heat-resistant, withstanding temperatures of 100°C.
   c) The organism that causes botulism can cause disease without encountering the immune response.
   d) Vaccinations are routinely given to prevent botulism.
7. Superantigens
   a) are exceptionally large antigen molecules.
   b) cause a very large antibody response.
   c) elicit a response from a large number of T cells.
   d) attach non-specifically to B-cell receptors.
   e) assist in a protective immune response.

8. An endotoxin is
   a) an A-B toxin.
   b) a component of Gram-positive bacteria.
   c) a substance that can be converted to a toxoid.
   d) heat-stable.
   e) a substance that causes T cells to release cytokines.
9. The tissue damage caused by *Neisseria gonorrhoeae* is primarily due to
   a) cross-reactive antibodies.
   b) exotoxins.
   c) hydrolytic enzymes.
   d) the inflammatory response.
   e) all of these.
10. Which of the following statements about viruses is *false*?
   a) They may colonize the skin.
   b) They may enter host cells by endocytosis.
   c) They may enter host cells by fusion of the viral envelope with the cell membrane.
   d) They may prevent apoptosis of infected host cells.
   e) They may suppress the expression of MHC class I molecules on host cells.

Applications

1. A group of smokers suffering from *Staphylococcus aureus* infections is suing the cigarette companies, claiming that smoking aggravated the disease. The group is citing studies indicating that compounds in cigarette smoke harm phagocytes. A statement prepared by their lawyers states that *S. aureus* would not have caused such a severe disease if the phagocytes were functioning properly. During the proceedings, a microbiologist was called in as a professional witness for the court. What were her conclusions about the validity of the claim?
2. A microbiologist wrote a grant proposal to identify molecules that jam type III secretion systems. Her principal rationale for the research was that interfering with secretion would disrupt a pathogen's ability to cause disease without harming the normal microbiota. Is this a reasonable proposal? Why or why not?

Critical Thinking

1. A student argued that no distinction should be made between commensalism and parasitism. Even in commensalism, the microorganisms are gaining some benefit (such as nutrients) from the host, and this represents a loss to the host. In this sense, the host is damaged. Does the student have a valid argument? Why or why not?
2. A microbiologist argued that there is no such thing as "normal" microbiota of the human body because the microbiome is dynamic and constantly changing, depending on diet and external environment. What would be an argument against this microbiologist's view?

www.mcgrawhillconnect.com

Enhance your study of this chapter with study tools and practice tests. Also ask your instructor about the resources available through Connect, including the media-rich eBook, interactive learning tools, and animations.

17 | Applications of Immune Responses

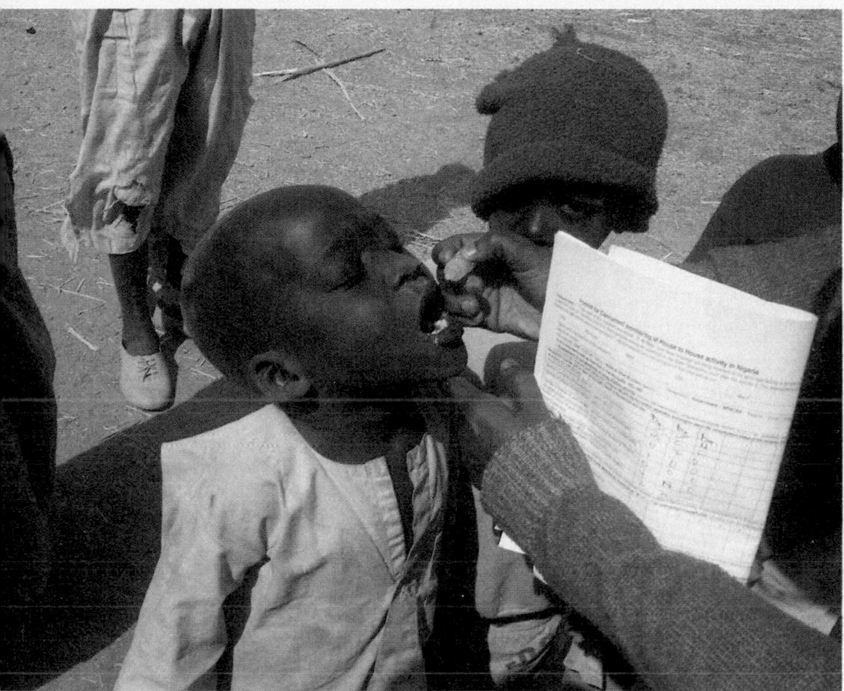

Child receiving oral polio vaccine. *Binta Bako Sule, Nigeria/CDC*

KEY TERMS

Active Immunity Immunity that results from an immune response upon exposure to an antigen.

Antiserum A preparation of serum that contains protective antibodies.

Attenuated Vaccine Vaccine composed of a weakened form of the pathogen that is generally unable to cause disease.

Enzyme-Linked Immunosorbent Assay (ELISA) Technique that uses enzyme-labeled antibodies to detect given antigens or antibodies.

Fluorescent Antibody (FA) Test Technique that uses fluorescently labeled antibodies to detect specific antigens on cells attached to a microscope slide.

Herd Immunity Protection of an entire population based upon a critical concentration of immune hosts that prevents the spread of an infectious agent.

Immunoassay A test that takes advantage of the specificity of antigen-antibody interactions by using known antibodies or antigens to detect or quantify given antigens or antibodies.

Immunotherapy A medical intervention that modifies specific immune responses as a means to treat certain diseases.

Inactivated Vaccine Vaccine composed of killed bacterial cells, inactivated viral particles, fractions of the pathogen, or inactivated toxin.

Passive Immunity Immunity that results when antibodies are transferred to an individual.

Vaccine A preparation of a pathogen or its products used to induce active immunity.

Western Blotting Procedure that uses labeled antibodies to detect specific antigens in a mixture of proteins separated according to their molecular weight.

A Glimpse of History

Long before people knew that microbes cause disease, they recognized that individuals who recovered from a disease such as smallpox rarely got it a second time. Old Chinese writings dating from the Sung dynasty (960–1280) describe a procedure known as variolation, in which small amounts of the powdered scabs from smallpox lesions were inhaled or scratched into the skin. The resulting disease was usually mild, and the person was then immune to smallpox. Occasionally, however, severe disease developed, often resulting in death. In addition, the person became contagious, so the disease could spread.

Although variolation was practiced in China and the Mideast a thousand years ago, it was not widely used in Europe until after 1719. At that time, Lady Mary Wortley Montagu, wife of the British ambassador to Turkey, had their children immunized against smallpox in this way. Variolation then became popular in Europe. Because of the dangers, however, and the fact that the procedure was expensive, many people remained unprotected.

As an apprentice physician, Edward Jenner noted that milkmaids who had recovered from cowpox (a disease of cows that caused few or no symptoms in humans) rarely got smallpox. Then, in 1796, long before viruses had been discovered, he conducted a classic experiment in which he deliberately transferred material from a cowpox lesion on the hand of a milkmaid, Sarah Nelmes, to a scratch he made on the arm of a young boy named James Phipps. Six weeks later, Jenner inoculated Phipps with pus from a smallpox victim, but the boy did not develop smallpox. Phipps had been made immune to smallpox when he was inoculated with pus from the cowpox lesion.

Using the less dangerous cowpox material in place of the scabs from smallpox cases, Jenner and others worked to spread the practice of variolation. Later, Pasteur used the word vaccination (from the Latin *vacca* for "cow") to describe any type of protective inoculation. By the twentieth century, most of the industrialized world was generally free of smallpox as the result of routine vaccination.

In 1967, the World Health Organization (WHO) started a program of intensive smallpox vaccination. Because there were no animal hosts and no non-immune humans to whom it could be spread, the disease died out. The last naturally contracted case occurred in Somalia, Africa, in 1977, and 2 years later WHO declared the world free of smallpox. Nevertheless, a few laboratories around the world still have the virus. In this age of bioterrorism concerns, some see smallpox as a major threat should the deadly virus ever be released into the largely unprotected populations of the world. Because of this, large supplies of the vaccine have been stockpiled in the United States and other countries to be used in case of an emergency.

Chapters 14 and 15 discussed the innate and adaptive defense systems, describing the protective functions of antibodies and lymphocytes. This chapter will consider how **immunization,** the process of inducing immunity, can be

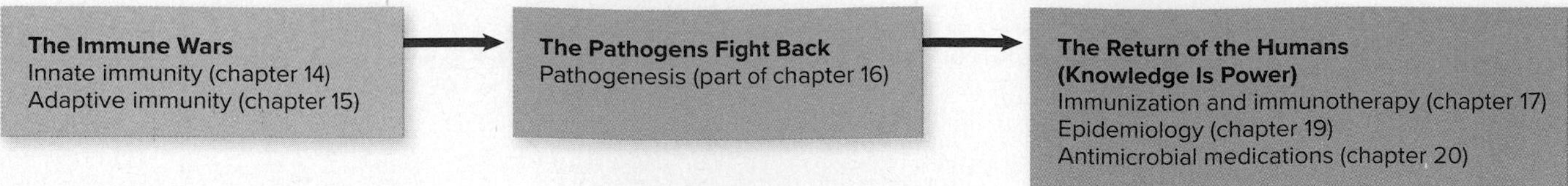

FIGURE 17.1 The Host-Pathogen Trilogy

How does immunization prevent disease?

used to prevent certain diseases. In fact, immunization has probably had the greatest impact on human health of any medical procedure, and it is just one example of how knowledge is power with respect to fighting disease (**figure 17.1**). We will also discuss the applications of **immunotherapies,** methods designed to either enhance or suppress specific immune responses as a means to treat certain diseases. Finally, we will explore some useful applications of immunological reactions in diagnostic tests.

IMMUNIZATION AND IMMUNOTHERAPY

17.1 ■ Principles of Immunization

Learning Outcome

1. Compare and contrast naturally acquired active immunity, artificially acquired active immunity, naturally acquired passive immunity, and artificially acquired passive immunity.

Naturally acquired immunity is the adaptive immunity an individual gains through normal events, such as infection by a pathogen. Immunization mimics those same events, preventing disease by inducing **artificially acquired immunity.** The protection provided by immunization can be either active or passive.

Active Immunity

Active immunity is the result of an immune response in a person who has been exposed to antigen. As a result of the exposure, specific B and T cells are activated and they then multiply, giving the person lasting protection due to immunological memory. Active immunity can develop either naturally from an actual infection or artificially from vaccination.

Passive Immunity

Passive immunity results when protective antibodies are transferred to someone, which occurs naturally during pregnancy and through breast feeding (**figure 17.2**). During pregnancy, the mother's IgG antibodies cross the placenta and protect the fetus; these antibodies remain active in the newborn during the first few months of life, when his or her own immune responses are still developing (see figure 15.20). This is why a number of infectious diseases are generally not observed until a baby is 3 to 6 months old, after the maternal antibodies have been degraded. When an infant is breastfed, the secretory IgA in the milk protects the digestive tract of the child. Artificial passive immunity results from an injection of antibodies produced by other people, animals, or even modified B cells growing in culture. It is used to prevent disease immediately before or after likely exposure to a pathogen or its toxins. Note that passive immunity provides no memory; once the transferred antibodies are degraded, the protection is lost.

A preparation of serum (the fluid portion of blood that remains after blood clots) containing protective antibodies is referred to as **antiserum.** An antiserum that protects against a toxin is called an **antitoxin.** Two types of antisera (or antitoxins) can be used to provide passive immunity immediately after a person is exposed to certain diseases. **Immune globulin,** the IgG fraction of blood plasma pooled from many donors, has a variety of antibodies due to typical infections and vaccinations experienced by the donors; it is used to protect unvaccinated people who have been recently exposed to certain diseases, including measles and hepatitis A, and to protect immunosuppressed people who have low antibody levels. **Hyperimmune globulin** is prepared from the sera of donors with high amounts of antibodies to certain disease agents and is used to protect against those specific agents. Examples of hyperimmune globulin include tetanus immune globulin (TIG), rabies immune globulin (RIG), and hepatitis B immune globulin (HBIG).

MicroAssessment 17.1

Active immunity occurs naturally in response to infections, and artificially in response to vaccination. Passive immunity occurs naturally from maternal antibodies transferred during pregnancy and breast feeding, and artificially through administration of immune globulin or hyperimmune globulin.

1. How is naturally acquired active immunity different from artificially acquired active immunity?
2. What is antitoxin?
3. If an unvaccinated person suffers a deep puncture wound that puts him or her at risk for tetanus, why would TIG be advised rather than simply vaccinating the individual?

Focus Figure

Active Immunity	Passive Immunity

Natural Active Immunity

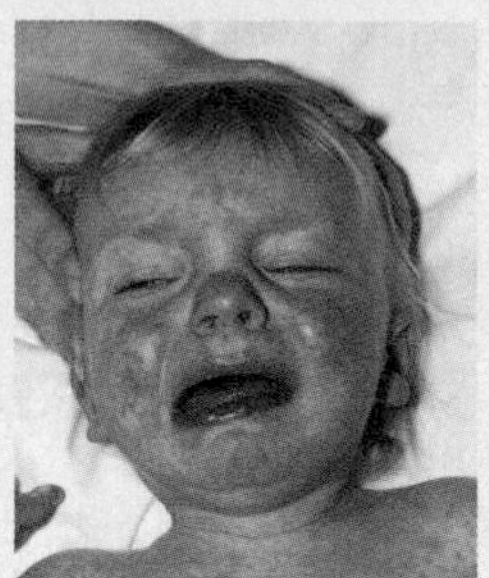

Immunity that results from an immune response in an individual after exposure to an infectious agent.

Natural Passive Immunity

Immunity that results when antibodies from a woman are transferred to her developing fetus during pregnancy or to an infant during breastfeeding.

Artificial Active Immunity

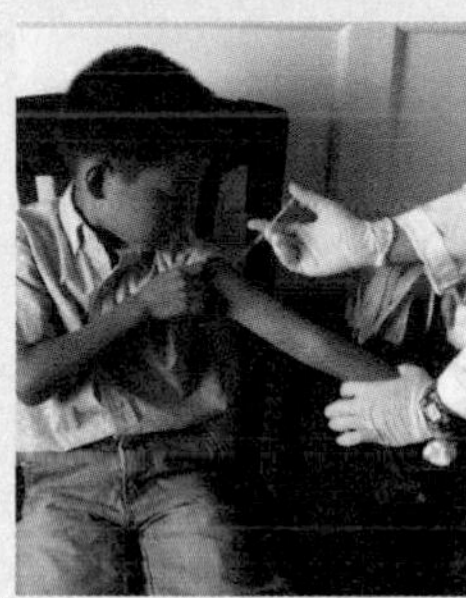

Immunity that results from an immune response in an individual after vaccination.

Artificial Passive Immunity

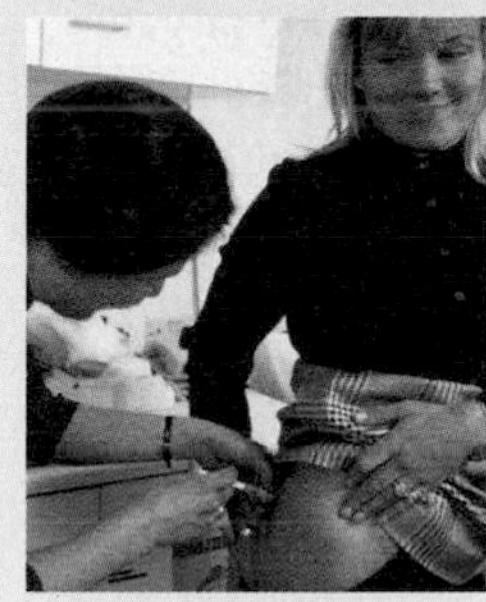

Immunity that results when antibodies contained in the serum of other people or animals are injected into an individual.

FIGURE 17.2 Acquired Immunity Acquired immunity can be natural or artificial, active or passive. (top left): SPL/Science Source; (top right): Jose Luis Pelaez Inc/Blend Images LLC; (bottom left): Jill Braaten/McGraw Hill; (bottom right): SPL/Science Source

? **Why does active immunity last longer than passive immunity?**

17.2 ■ Vaccines and Immunization Procedures

Learning Outcomes

2. Describe the role of vaccines in providing herd immunity.
3. Compare and contrast the characteristics of attenuated, inactivated, and nucleic acid–based vaccines.
4. Describe the benefits and uses of routine and non-routine vaccinations.
5. Compare and contrast the inactivated polio vaccine (IPV) and the oral polio vaccine (OPV).

A **vaccine** is a biological preparation used to prevent disease by inducing active immunity against a particular pathogen. Effective vaccines should be safe, with few side effects, while giving lasting protection against the illness. They should induce protective antibodies or immune cells, or both, as appropriate for controlling the infection. Ideally, vaccines should be inexpensive, stable with a long shelf life, and easy to administer.

If a vaccine prevents infection as well as disease, then widespread vaccination may prevent that disease from spreading in a population. This phenomenon, called **herd immunity,** occurs when the pathogen has too few hosts in which to multiply and then be transmitted to another susceptible host. Herd immunity is important because it protects the few individuals who were either unable to get vaccinated or were vaccinated but did not mount a strong enough immune response to prevent disease. Vaccination-induced herd immunity is responsible for dramatic declines in childhood diseases, both in the United States and in other countries. Unfortunately, some of these diseases reappear and spread as a direct consequence of parents' failure to have their children vaccinated.

Because of the time it takes for active immunity to develop, vaccines are typically given to an individual well before possible exposure to a pathogen. In the few situations where the incubation period of a particular pathogen is quite long, such as in the case of rabies where the incubation period is 2 to 3 months, the vaccine can even be protective if given soon after exposure. The purpose of **post-exposure prophylaxis (PEP)** vaccination is to prevent an already infected individual from becoming sick or to at least lessen the disease severity. When someone has been bitten by or exposed to an animal that has rabies, for example, several doses of the rabies vaccine are administered to the exposed person, usually along with human rabies immunoglobulin (HRIG). Post-exposure prophylaxis is also used if someone is exposed to hepatitis B.

Vaccines have traditionally fallen into two general categories—attenuated and inactivated—based on whether the immunizing agent can replicate and thereby amplify the amount of antigen (attenuated agents can replicate, whereas inactivated agents cannot); each type has characteristic advantages and disadvantages. An exciting new development is the availability of a third type—nucleic acid–based vaccines—which plays a crucial role in preventing severe COVID-19.

Attenuated Vaccines

An **attenuated vaccine** is a weakened form of the pathogen that generally cannot cause disease. The attenuated strain replicates in the vaccine recipient, causing an infection with undetectable or mild symptoms that typically results in long-lasting immunity. Because infection with the attenuated strain mimics that of a wild-type strain, it induces the appropriate type of immune response. For instance, attenuated vaccines given orally induce mucosal immunity (a secretory IgA response), protecting against pathogens that infect via the gastrointestinal tract. Some attenuated vaccines can stimulate cytotoxic T cells, inducing cell-mediated immunity.

Attenuated strains are often produced by growing a microbe under conditions that cause mutations to accumulate, thereby making the microbe less pathogenic. Viruses that infect humans can sometimes be attenuated by growing them in cells of a different animal species; the mutations that allow the viruses to multiply in the other animal cells often cause them to grow poorly in human cells. Genetic manipulation is now also being used to produce strains of pathogens with low virulence; specific genes are mutated and used to replace wild-type genes. The inserted mutant genes are engineered so that the microbes cannot revert back to the wild type version.

Attenuated vaccines have several advantages compared to their inactivated counterparts. For one thing, often only one or two doses of an attenuated agent are enough to induce relatively long-lasting immunity. This might be because the microbe multiplies in the body, causing the immune system to be exposed to the antigen for a longer period and in greater amounts than with inactivated agents. In addition, the vaccine strain has the added potential of being spread from an individual being immunized to other non-immune people, thereby immunizing the contacts of the vaccine recipient.

The disadvantages of attenuated agents are that they sometimes cause disease in immunosuppressed people and that they can occasionally mutate to become pathogenic again. Attenuated vaccines are generally not advised for pregnant women because of the possibility that the vaccine strain may cross the placenta and damage the developing fetus. Another disadvantage of attenuated vaccines, especially in developing countries where they are desperately needed, is that refrigeration is usually required to keep them active.

Attenuated vaccines currently in widespread use include those against measles, mumps, rubella, chickenpox, and rotavirus gastroenteritis. The adenovirus vaccine given to military recruits to protect them from severe respiratory disease contains non-attenuated replicating viruses, but the dose is taken orally, a route that does not result in disease.

Inactivated Vaccines

An **inactivated vaccine** is unable to replicate but retains the immunogenicity of the pathogen or toxin. The advantage of inactivated vaccines is that they cannot cause infections or revert to pathogenic forms. Because they do not replicate, however, there is no amplification of the dose in vivo, so the magnitude of the immune response is limited. To compensate, regular booster doses are usually needed to maintain sufficient immunity to be protective. Some inactivated vaccines include the whole infectious agent, and others include only fractions of the agent. Examples currently in use include:

- **Inactivated whole agent vaccines.** These contain killed microorganisms or inactivated viruses. The vaccines are made by treating the pathogen with formalin or another chemical that does not significantly change the surface epitopes. The treatment leaves the agent immunogenic even though it cannot reproduce. Vaccines in this category include those against influenza, rabies, and hepatitis A.
- **Toxoid vaccines.** These are inactivated toxins used to protect against diseases caused by bacterial toxins. They are prepared by treating the toxins to destroy the toxic part of the molecules while retaining the antigenic epitopes. Diphtheria and tetanus vaccines are toxoids.
- **Subunit vaccines.** These consist of key protein antigens or antigenic fragments of a pathogen. Obviously, they can be developed only after research has revealed which of the microbe's components are most important in triggering a protective immune response. Their advantage is that parts that may cause undesirable side effects are not included. The vaccine currently used to prevent whooping cough (pertussis) is a subunit vaccine, referred to as the acellular pertussis (aP) vaccine. It does not cause the side effects

that sometimes occurred with the killed whole-cell vaccine used previously. Recombinant subunit vaccines are produced using genetically engineered microorganisms. The hepatitis B virus vaccine, for example, is made using yeast cells engineered to produce a viral surface protein.

- **VLP (virus-like particle) vaccines.** These contain empty capsids. Laboratory microorganisms are genetically engineered to produce the major capsid proteins of a virus, which then self-assemble. The human papillomavirus (HPV) vaccine is a VLP vaccine.
- **Polysaccharide vaccines.** These contain the polysaccharides that make up the capsules of certain organisms. They are not effective in young children because polysaccharides are T-independent antigens; recall that these antigens generally elicit a poor response in this age group. One of the pneumococcal vaccines given to adults is a polysaccharide vaccine.
- **Conjugate vaccines.** These are polysaccharides linked to proteins, a modification that converts the polysaccharides into T-dependent antigens. The first conjugate vaccine developed was against *Haemophilus influenzae* type b (Hib), and it has nearly eliminated Hib meningitis in children. The conjugate vaccine developed against certain *Streptococcus pneumoniae* serotypes promises to do the same for a variety of infections caused by those serotypes.

Many inactivated vaccines contain an **adjuvant,** a substance that enhances the immune response to antigens (*adjuvare* means "to help"). These are necessary additives because purified antigens such as toxoids and subunit vaccines are often poorly immunogenic by themselves because they lack the "danger" signals—the PAMPs (pathogen-associated molecular patterns) that characterize invading microbes. Recall that the PAMPs, also referred to as MAMPs (microbe-associated molecular patterns), cause dendritic cells to produce costimulatory molecules, allowing them to activate helper T cells, which, in turn, activate B cells (see figure 15.13). Adjuvants are thought to function by providing the danger signals to dendritic cells. Some adjuvants appear to adsorb the antigen, releasing it at a slow but constant rate to the tissues and surrounding blood vessels. Unfortunately, many effective adjuvants trigger an intense inflammatory response, making them unsuitable for use in vaccines for humans. Alum (aluminum hydroxide and aluminum phosphate) is the most common adjuvant used, but others, including one that uses a derivative of lipid A, have also been developed.

Nucleic Acid–Based Vaccines

A **nucleic acid–based vaccine** consists of segments of nucleic acid—either DNA or mRNA—that encode key antigens of an infectious agent. When the nucleic acid is delivered into the vaccine recipient's cells, the microbial genes are temporarily expressed, resulting in production of microbial proteins (antigens) that induce an immune response. As with many inactivated vaccines, some nucleic acid–based vaccines require an adjuvant. Although the first examples of this type of vaccine have only recently been approved for human use, scientists have been working to develop and refine the technology for over 30 years. Given the success of the current versions, many more will likely be developed in the years to come. General types of nucleic acid–based vaccines include:

- **mRNA vaccines.** These are segments of antigen-encoding mRNA molecules within fatty bubbles called lipid nanoparticles (LNPs). The LNPs protect the mRNA molecules from enzymatic degradation and serve as a mechanism for delivery of the mRNA into host cells. When the vaccine is injected into tissues, the LNPs are taken up by host cells through endocytosis and then fuse with endosomal membranes, thus releasing the mRNA into the cytoplasm where it can be translated. The manufacturing process for mRNA vaccines is relatively fast and adaptable, so the mRNA sequences can easily be changed if new variants of the target pathogen emerge. The vaccines need to be stored at very low temperatures (−20 to −70°C) to maintain stability, however, making it a challenge to deliver the vaccines to many areas of the world. Two COVID-19 vaccines are examples (Pfizer–BioNTech and Moderna).
- **Viral vector vaccines.** These are modified viruses, usually a replication-defective adenovirus, to which key antigen-encoding viral genes have been added. The viral vector simply functions as a delivery system: Once the vector enters the cell, the antigen gene(s) it carries can be expressed. Advantages of these vaccines are that they are inexpensive to manufacture and can be kept at room temperature, making them easy to store and distribute globally. Some individuals have existing immunity to the vector, however, which can interfere with the immune response and potentially result in rare adverse reactions. Two COVID-19 vaccines are examples of this type of vector vaccine (Jannsen/Johnson & Johnson and Oxford/AstraZeneca). The new vaccine to prevent Ebola virus disease uses a different vector, attenuated vesicular stomatitis virus (VSV), which delivers the antigen-encoding genes in a replicating form.
- **DNA vaccines.** These are antigen-encoding recombinant plasmid DNA molecules. Although none are currently approved in the United States for use in humans, a few are approved for veterinary use, including a vaccine to prevent West Nile virus disease in horses. In addition, one was recently approved in India to prevent COVID-19; it uses a needle-free device to deposit the vaccine into the skin, where the plasmids are taken up by local

TABLE 17.1 Characteristics of Attenuated, Inactivated, and Nucleic Acid-Based Vaccines

Characteristic	Attenuated Vaccines	Inactivated Vaccines	Nucleic Acid–Based Vaccines
Types	Attenuated viruses, attenuated bacteria	Inactivated whole agents, toxoids, subunits, VLPs (virus-like particles), polysaccharides, conjugates	Antigen-encoding DNA or RNA segments
Route of administration	Injection, oral, or nasal	Injection	Injection
Need for adjuvant	No	Often	Variable
Antibody response (memory)	IgG; secretory IgA if administered orally or nasally	IgG	IgG
Cell-mediated immune response	Good	Poor	Good
Relative duration of protection	Longer	Shorter	Shorter
Number of doses	Usually one or two	Multiple	Multiple
Risk of mutation to virulence	Very low	Absent	Absent
Risk to immunocompromised recipient	Can be significant	Absent	Absent
Stability in warm temperatures	Poor	Good	Variable

immune cells and expressed. As with mRNA vaccines, the sequence of the DNA can be easily changed.

Table 17.1 compares the characteristics of attenuated, inactivated, and nucleic acid–based vaccines.

Vaccination Uses and Benefits

Vaccination has significantly improved the health of individuals as well as communities. Before vaccines were available to prevent common diseases, thousands of people, mainly children, died or were permanently disabled from diseases that are now preventable (**table 17.2**). The introduction of a routine vaccination schedule for children from the time they are born has dramatically reduced the number of childhood deaths.

TABLE 17.2 The Effectiveness of Universal Immunization in the United States

Disease	Cases per Year Before Immunization	Decrease After Immunization
Diphtheria	175,885 (1920–1922)	Nearly 100%
Haemophilus influenzae type b invasive disease	20,000 (estimated)	Over 99%
Measles	503,282 (1958–1962)	Over 99%
Mumps	152,209 (1968)	Over 99%
Pertussis (whooping cough)	147,271 (1922–1925)	90%
Polio	16,316 (1951–1954)	100%
Rubella (German measles)	50,230 (1966–1969)	Nearly 100%
Smallpox	48,164 (1900–1904)	100%
Tetanus	1,314 (1922–1926)	98%

Unfortunately, many people still become ill or even die from vaccine-preventable diseases. One reason is that some parents refuse to have their children vaccinated, fearing that vaccination might be harmful. In situations such as this, vaccines have become victims of their own success. They have been so effective at preventing diseases that many people no longer realize how serious these diseases can be. Reports of adverse effects of vaccination have led some people to falsely believe that the risk of vaccination is greater than the risk of diseases.

There will always be at least some risk associated with almost any medical procedure, but there is no question that the benefits of routine vaccinations greatly outweigh the very slight risks. Data show that a child with measles has a 1:2,000 chance of developing serious brain inflammation, compared with a 1:1,000,000 chance from the measles vaccine. Between 1989 and 1991, measles immunization rates dropped 10% and an outbreak of 55,000 cases occurred, with 120 deaths (see figure 22.13). Once the vaccination rates increased again, measles outbreaks became rare. The suggestion that certain vaccines are associated with autism again threatened the acceptance of immunization, resulting in another series of disease outbreaks. It is important to note, however, that numerous scientific studies show no evidence of a link between the two.

Routine Vaccines

Some vaccines are considered routine, meaning they are part of recommended vaccination schedules. **Table 17.3** lists bacterial diseases prevented by routine vaccines; **table 17.4** lists viral diseases prevented by routine vaccines. The routine vaccines administered in the United States are generally covered by the National Vaccine Injury Compensation Program, a no-fault mechanism for resolving vaccine injury

TABLE 17.3 Bacterial Diseases Prevented by Routine Vaccines

Disease	Vaccine Type (Vaccine Acronym)	Persons Who Should Receive the Vaccine
Diphtheria	Toxoid (the "D" in DTaP, and the "d" in Tdap and Td; the lowercase letter indicates a lower dose)	DTaP: infants, toddlers, and preschoolers Tdap: adolescents, pregnant women, and anyone who did not receive it as an adolescent Td or Tdap: adults, every 10 years as a booster
Haemophilus influenzae type b disease	Conjugate (Hib)	Infants and toddlers; older children and adults with certain medical conditions that put them at increased risk
Meningococcal disease	Conjugate against serogroups ACWY (MenACWY) Recombinant subunit against serogroup B (MenB)	MenACWY: adolescents age 11 or 12, with booster at age 16; anyone at increased risk due to certain medical conditions or potential exposures MenB: people 10 years or older who are at increased risk due to certain medical conditions or potential exposure
Pertussis (whooping cough)	Subunit (the "aP" in DTaP and the "ap" in Tdap; the lowercase letter indicates a lower dose)	DTaP: infants, toddlers, and preschoolers Tdap: adolescents, pregnant women, and anyone who did not receive it as an adolescent; an alternative to Td as a booster every 10 years
Pneumococcal disease	Conjugate (PCV13; PCV15; PCV 20) Polysaccharide (PPSV23)	PCV13 or PCV15: infants and children under 5; children 5 to 18 years with certain health conditions PCV15 or PCV 20: adults age 65 and over; adults with certain health conditions PPSV23: adults age 65 and over; anyone 2 years or over with certain health conditions
Tetanus	Toxoid (the "T" in DTaP, Tdap, and Td)	DTaP: infants, toddlers, and preschoolers Tdap: adolescents, pregnant women, and anyone who did not receive it as an adolescent Td or Tdap: adults, every 10 years as a booster

TABLE 17.4 Viral Diseases Prevented by Routine Vaccines

Disease	Vaccine Type (Vaccine Acronym)	Persons Who Should Receive the Vaccine
COVID-19	mRNA; a recombinant subunit may be used as an alternative in certain situations	Children 6 months and older; adults
Hepatitis A	Inactivated virus (HepA)	Toddlers; older children and adolescents who have not received the vaccine; adults who want the vaccine or are at increased risk due to potential exposure
Hepatitis B	Recombinant subunit (HepB)	Newborns, infants, and toddlers; older children and adolescents who have not received the vaccine; adults who are at increased risk due to certain medical conditions or potential exposure
Human papillomavirus (HPV)–related diseases	VLPs of 9 serotypes (9vHPV)	Adolescents age 11 or 12 (may be given ages 9 through 26)
Influenza	Inactivated virus (IIV is produced in chicken eggs; ccIV is from cell culture) Recombinant subunit (RIV) Attenuated virus (LAIV), taken as a nasal mist	People 6 months of age or older (vaccine type used depends on patient's age, allergies, or other characteristics), yearly because the antigens of the virus change frequently
Measles	Attenuated virus (an "M" in MMR and MMRV)	MMR: Toddlers and preschoolers; infants who will be traveling outside the U.S.; older children, adolescents, and adults who are not already immune MMRV: An alternative to MMR for ages 1 through 12
Mumps	Attenuated virus (an "M" in MMR and MMRV)	MMR: Toddlers and preschoolers; infants who will be traveling outside the U.S.; older children, adolescents, and adults who are not already immune MMRV: An alternative to MMR for ages 1 through 12
Polio	Inactivated virus (IPV); an attenuated virus (OPV) is used in some parts of the world	IPV: Infants, toddlers, and preschoolers; older children who have not received the vaccine; adults at increased risk due to potential exposure
Rotavirus gastroenteritis	Attenuated virus (RV)	Infants
Rubella (German measles)	Attenuated virus (the "R" in MMR and MMRV)	MMR: Toddlers and preschoolers; infants who will be traveling outside the U.S.; older children, adolescents, and adults who are not already immune MMRV: An alternative to MMR for ages 1 through 12
Shingles	Recombinant virus (RZV) Attenuated virus (HZV)	RZV: Adults age 50 and over (even if previously vaccinated with HZV) HZV: An alternative to RZV, adults age 60 and over
Varicella-zoster (chickenpox)	Attenuated virus (VAR; also the "V" in the MMRV vaccine)	VAR or MMRV: Toddlers and preschoolers; older children, adolescents, and adults who are not already immune

claims that was established to stabilize the vaccine market; a tax on every vaccine dose purchased funds the program. Some routine vaccines are administered in combination, thereby minimizing the number of injections required. For example, MMRV is a vaccine used to immunize against measles, mumps, rubella, and varicella. DTaP-IPV-Hib is used to immunize against diphtheria, tetanus, pertussis, polio, and *Haemophilus influenzae* type B disease; a similar combination vaccine adds hepatitis B (HepB) to that mix.

The U.S. Centers for Disease Control and Prevention (CDC) regularly publishes recommended immunization schedules for children, adolescents, and adults. These are updated regularly as vaccines are developed and modified. Because of the complexity of the schedules and how often they are updated, it is important to know how to access the most current versions, which are available at the following CDC website: http://www.cdc.gov/vaccines/schedules/.

Non-Routine Vaccines

Certain vaccines have been approved by the CDC but only for use in particular situations. For example, some are approved only for people who live in or travel to areas of the world where certain diseases are endemic. Others are approved only for people whose occupation puts them at increased risk of exposure to a dangerous or deadly disease. **Table 17.5** lists diseases that can be prevented by non-routine vaccines.

If an outbreak of a vaccine-preventable communicable disease not controlled by routine vaccination occurs, a strategy called **ring vaccination** is sometimes used to limit the spread. As an example, rather than using precious resources for a mass Ebola vaccination campaign during a localized Ebola virus disease outbreak, only the people most likely to be infected are vaccinated: the contacts of identified cases as well as the contacts of those contacts. Thus, a vaccination "ring" is created around the cases. Ring vaccination was instrumental in the final stages of the campaign that successfully eradicated smallpox; the last few smallpox outbreaks were unable to spread because of ring vaccination.

An Example of Vaccination Strategy—The Campaign to Eliminate Polio (Poliomyelitis)

Vaccines against polio (poliomyelitis) provide an excellent illustration of the complexity of vaccination strategies. The virus that causes this disease enters the body orally, infects the cells that line the throat and intestinal tract, and then invades the bloodstream. From there, it can invade nerve cells and cause the disease polio (see figure 26.8). Any of the three serotypes of poliovirus (designated types 1, 2, and 3) can cause the disease. The Salk vaccine, developed in the mid-1950s, consists of inactivated virus particles of all three serotypes. This vaccine, now called the **inactivated polio vaccine (IPV),** dramatically lowered the rate of the disease but had the disadvantage of requiring

TABLE 17.5 Diseases Prevented by Non-Routine Vaccines

Disease	Vaccine Type	Persons Who Should Receive the Vaccine
Adenovirus respiratory infection	Two serotypes of active virus (taken orally)	Military recruits entering basic training and certain other military personnel at higher risk for infection
Anthrax	Subunit	Adults in occupations that put them at risk of exposure, such as some military personnel; unvaccinated people who have recently been exposed
Cholera	Attenuated bacterium (taken orally)	Adults ages 18 through 64 traveling to endemic areas
Severe dengue	Attenuated virus	Children ages 9 through 16 with laboratory confirmed previous dengue virus infection and are living in an endemic area
Ebola virus disease	Attenuated viral vector	Healthcare personnel working at Ebola treatment centers or responding to outbreaks; laboratory workers at risk of exposure
Japanese encephalitis	Inactivated virus	People 2 months old or over who are traveling to or living in countries where the disease occurs; laboratory workers at risk of exposure
Rabies	Inactivated virus	People at high risk for exposure, such as veterinarians and other animal handlers; people who have recently been exposed
Smallpox	Vaccinia virus	Used only in response to biological warfare threat; laboratory workers at risk of exposure
Tuberculosis	Attenuated bacterium (BCG strain)	Used only in special circumstances in the United States; widely used in other countries
Typhoid fever	Two forms—attenuated bacterium (taken orally) and polysaccharide	People traveling to parts of the world where typhoid fever is common; people in close contact with a typhoid carrier; laboratory personnel who work with *Salmonella* Typhi
Yellow fever	Attenuated virus	People ages 9 months through 59 years traveling to or living in parts of the world where the disease is known to exist; laboratory workers at risk of exposure

a series of injections for maximum protection. In 1961, the Sabin vaccine became available. This vaccine, now called the **oral polio vaccine (OPV)**, consists of one or more attenuated strains that replicate in cells that line the throat and intestinal tract. This vaccine has the advantage of cheaper oral administration.

OPV and IPV both cause the immune system to produce antibodies that protect against polio by preventing the virus from invading the central nervous system. OPV, however, has a distinct advantage over IPV in that it induces better mucosal immunity (secretory IgA response). Thus, the antibodies in the throat and intestinal tract can neutralize wild poliovirus before it infects cells of those linings, thereby preventing the virus from replicating in the cells and then being transmitted in feces. Because of this, OPV provides better herd immunity. A disadvantage of OPV is that the attenuated viruses can mutate, giving rise to virulent versions referred to as vaccine-derived poliovirus (VDPV), which can cause the same disease as wild poliovirus and lead to outbreaks in communities with low vaccination rates. An obvious way to avoid vaccine-related polio is to abandon OPV in favor of IPV; however, the situation is not as simple as it might seem. If only IPV is used, wild poliovirus can still replicate in the vaccine-recipient's throat and intestinal lining and then be transmitted to others in feces, thereby spreading in a population. Therefore, polio eradication (complete elimination) depends on OPV because it prevents the spread of wild poliovirus.

A campaign to eliminate polio using OPV was so successful that by 1980 the United States was free of wild poliovirus (see figure 26.9). By 1991, the virus had been eliminated from the Western Hemisphere. Because of the continued risk of vaccine-associated polio, a vaccine strategy that attempted to capture the best of both vaccines was adopted in the United States. Children first received doses of IPV, protecting them from the disease. Following these doses, OPV was given, providing mucosal protection while also boosting immunity. In 2000, the routine use of OPV was discontinued altogether in the United States. Today, most countries use only IPV.

The original goal of global eradication of polio by 2000 was not achieved, but substantial progress has been made, and efforts continue. For these eradication programs, OPV must be used in regions where wild poliovirus might still be present because that vaccine prevents transmission of the virus. Using this strategy, two of the three serotypes of wild poliovirus have now been declared globally eradicated—type 2 in 2015 and type 3 in 2019. Unfortunately, type 2 VDPV continues to circulate in some parts of the world, so a new vaccine is authorized for emergency use by the World Health Organization (WHO) to help prevent its spread; the vaccine, novel oral polio vaccine type 2 (nOPV2), consists of a more genetically stable modification of the attenuated strain such that it is less likely to mutate and become virulent. Chapter 26 has more information about polio and the progress toward its elimination.

MicroAssessment 17.2

An attenuated vaccine is a weakened form of the pathogen. An inactivated vaccine is unable to replicate but retains the immunogenicity of the pathogen or toxin; examples include killed microorganisms, inactivated viruses, and fractions of the agents, including toxoids. Routine childhood immunizations have prevented millions of cases of disease and many deaths during the past decades.

4. What is the difference between an attenuated and an inactivated vaccine?
5. Childhood diseases such as measles and mumps are rare now, so why is it important for children to be immunized against them?
6. In 2004, Israel switched from using OPV to using IPV. Then, after wild-type poliovirus was found to be circulating in the population in 2013, a supplementary OPV program was begun for children who had received only IPV. What would be the rationale for an OPV supplementary program?

17.3 ■ Immunotherapies

Learning Outcomes

6. With respect to cancer immunotherapies, describe the following: (1) immunomodulators, (2) monoclonal antibodies, (3) checkpoint inhibitors, (4) CAR T cells, (5) oncolytic viruses, and (6) therapeutic vaccines.
7. Give two examples of immunotherapies used for immunological disorders.

Immunotherapies are medical interventions that modify specific immune responses as a means to treat diseases. The term is primarily used in the context of cancer treatments, but the approaches are used to treat other diseases as well. Recent advances in immunotherapies have been dramatic, with dozens of treatment options currently approved by the U.S. Food and Drug Administration (FDA). Many more are still in the development pipeline—a process that includes various clinical trials to demonstrate that the benefit of a treatment outweighs the risk. Oftentimes, the therapies have significant side effects, in which case their use is limited to only disabling or life-threatening diseases.

Immunotherapies for Cancer

Cancers are "self" cells that multiply and spread without control, leading to disease. Cancerous cells often abnormally express certain proteins referred to as tumor antigens, so it seems they should be eliminated by the immune system; however, they often have ways to evade detection or destruction. For example, some cancer cells interfere with an effective immune response by exploiting the peripheral tolerance mechanisms the immune system uses to avoid responses against "self" cells. They also down-regulate the production of MHC (major histocompatibility complex) molecules so that less antigen is presented. Cancer immunotherapies attempt to overcome those evasion mechanisms.

Immunomodulators

Immunomodulators are therapies that either boost or suppress the immune response. Those approved by the FDA to treat certain cancers enhance the anti-cancer response and include:

- **Cytokines.** As described in chapter 14, these direct certain immune cell activities. Interleukin 2 (IL-2) stimulates the proliferation of T cells, and interferon alpha increases tumor cell surveillance by T cells.
- **Adjuvants.** These boost the immune system. The only one currently approved for cancer treatment targets a toll-like receptor pathway; it is used to treat patients with a certain type of skin cancer.
- **Checkpoint inhibitors.** These have revolutionized the treatment of certain types of cancer and will be discussed in a separate subsection.

Monoclonal Antibodies

A **monoclonal antibody (mAb)** is a preparation of antibody molecules produced by clones of a single B cell, so that all molecules in the preparation recognize only a single epitope. They are a fundamental part of some immunoassays, as we will describe in the next section, but more recently they have become an important part of immunotherapies. Monoclonal antibodies are obtained through a complicated process that involves taking B cells from an immunized animal, and then fusing those short-lived B cells with other cells that will divide repeatedly in culture (see Focus Your Perspective 17.1). Medications composed of monoclonal antibodies have names ending in "-mab." The mAbs used therapeutically are generally "humanized," meaning that recombinant DNA techniques have been used to replace most of the animal-derived antibody molecule with human equivalents. In humans, these antibodies have a longer half-life than standard mAbs because the immune system is less likely to recognize them as foreign and destroy them. Medications composed of humanized monoclonal antibodies have names ending in "-zumab."

Examples and mechanisms of mAbs approved by the FDA for use in certain cancer immunotherapies include:

- **Naked antibodies.** These are simply mAbs without attached components (in contrast to the others discussed next). In certain cases, by binding to cancer cells, the mAbs mark the cells for destruction via antibody-dependent cellular cytotoxicity (ADCC). An example is rituximab, which is used to treat some types of B-cell cancers. It binds to a B-cell surface protein called CD20, thereby marking B cells for destruction. Some naked antibodies interfere with chemicals called growth factors that are required for cancer cell proliferation. An example is trastuzumab, which is used to treat certain breast cancers. These cancers express abnormally high amounts of a protein called human epidermal growth factor receptor 2 (HER2), allowing them to better receive the growth-stimulating signal. By binding to HER2, the mAb interferes with the cancer cells' growth.
- **Conjugated antibodies.** These have been constructed to deliver a toxin or other molecule to a cancerous cell. Some are derivatives of the naked antibodies just described; for example, rituximab has been conjugated to a cell-damaging radioactive molecule so that the B cell is destroyed when the conjugated antibody binds to it. Trastuzumab has been conjugating to a cytotoxin; once the trastuzumab component has bound to an HER2 expressing cell, the cytotoxin enters the cell by receptor-mediated endocytosis, where it interferes with the function of tubulin, the protein that makes up microtubules.
- **Bi-specific antibodies.** These have been constructed to have two different antigen-binding sites. An example called blinatumomab is used to treat some types of acute lymphocytic leukemias: It has one antigen-binding site that binds to a protein found on T cells and another that binds to a protein found on some cancerous B cells involved in the leukemia. Thus, it connects the two cell types, which increases the chance that the T cell will destroy the cancerous B cell; this bi-specific antibody is also referred to as a bi-specific T-cell engager (BiTE).
- **Checkpoint inhibitors.** These are naked antibodies that are also immunomodulators, and they are described next.

Checkpoint Inhibitors

Checkpoint inhibitors interfere with peripheral tolerance mechanisms called immune checkpoints that normally help prevent cytotoxic T cells from killing "self" cells. The checkpoints are cell surface proteins that can block inappropriate responses that could otherwise lead to autoimmunity, inflammatory damage, and allergies; in doing so, however, they also interfere with the immune system's ability to destroy cancer cells. In recognition of the importance of checkpoint inhibitors in cancer therapy, two scientists who pioneered the field, James Allison and Tasuku Honjo, were awarded a Nobel Prize in 2018. Although the checkpoint inhibitors allow the immune system to attack cancer cells, they also increase the likelihood of attack on normal host cells.

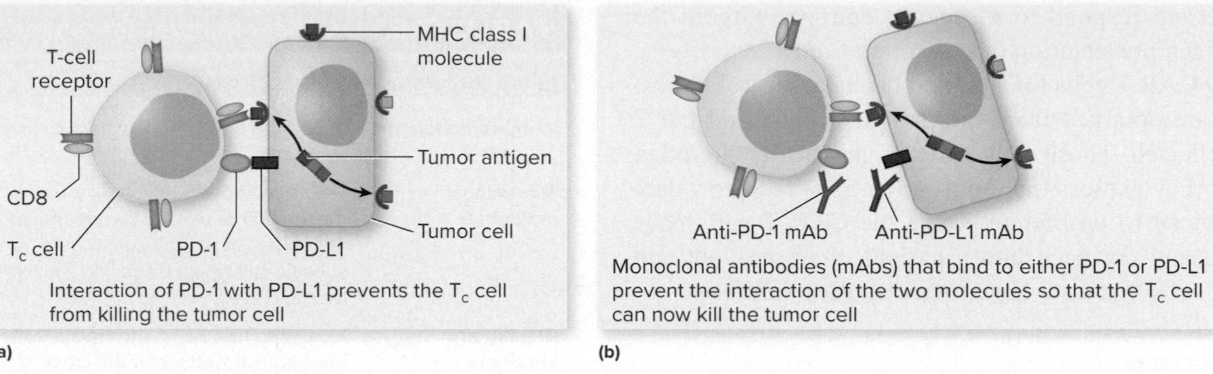

FIGURE 17.3 Checkpoint Inhibitor (a) Normal interaction of PD-1 with PD-L1 serves as a checkpoint that promotes peripheral tolerance. **(b)** mAbs that bind either PD-1 or PD-L1 are checkpoint inhibitors.

? **Describe two other anti-tumor mechanisms of mAbs used in cancer therapy.**

The FDA-approved checkpoint inhibitors are monoclonal antibodies that fall into two groups:

- **Inhibitors of the PD-1/PD-L1 pathway.** These block a peripheral tolerance pathway that involves a surface protein on effector cytotoxic T cells (T_C cells). When that protein, called programmed death receptor 1 (PD-1), interacts with a protein called PD-L1 on the surface of another cell, the T_C cell's ability to kill the other cell is suppressed; thus the PD-1/PD-L1 interaction acts as a T_C cell "off switch" (**figure 17.3**). Notably, cancer cells often increase their production of PD-L1, which protects them from immune attack. Several checkpoint inhibitors are monoclonal antibodies that bind to PD-1 on the T-cell surface, thereby preventing the interaction between PD-1 and PD-L1; other options bind to PD-L1 on the interacting cells.
- **Inhibitors of the CTLA-4 pathway.** These block a peripheral tolerance pathway that involves a T-cell surface receptor called cytotoxic T-lymphocyte-associated protein 4 (CTLA-4). This receptor acts as a competitor of a different receptor on a T cell that normally binds co-stimulatory molecules on antigen-presenting cells (see figure 15.13). When CTLA-4 binds the co-stimulatory molecules, it effectively hides them from the receptor that would otherwise bind to them as part of T-cell activation. Thus, by blocking CTLA-4, the checkpoint inhibitor makes T-cell activation more likely to occur.

Chimeric Antigen Receptor (CAR) T Cells

Chimeric antigen receptor (CAR) T cells are a patient's own T cells that have been genetically engineered to express a modified T-cell receptor. The intracellular portion of the chimeric antigen receptor is the signaling domain of a T-cell receptor, often with other signaling components added, and the extracellular part is derived from one "arm" of a monoclonal antibody (**figure 17.4**). Thus, a CAR T cell is engineered

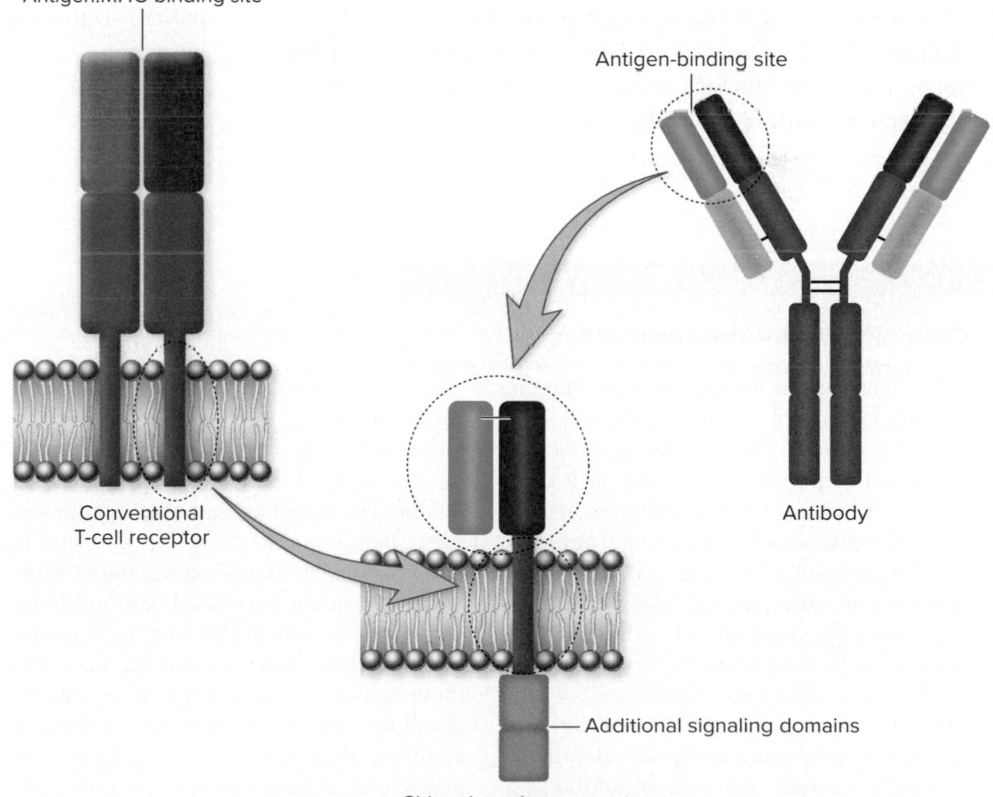

FIGURE 17.4 Components of a Chimeric Antigen Receptor on a CAR T Cell The intracellular portion of the receptor is the signaling domain of a T-cell receptor, often with other signaling components added; the extracellular part is derived from one "arm" of a monoclonal antibody.

? **In cancer therapy, what is the advantage of a CAR T cell over a conventional T cell?**

to recognize and respond to a specific epitope, without the need for antigen presentation or co-stimulation.

Creating CAR T cells for a patient can take several weeks. Leukocytes are obtained from the patient's blood and the T cells then collected. Those cells are then genetically altered in vitro so that they express the modified antigen receptor; they are then induced to proliferate. Once the CAR T cells have been created and put back into a patient, they seek out and destroy host cells that have the epitope on their surface.

Oncolytic Viruses

As discussed in chapter 13, a genetically modified virus has been approved by the FDA for use in cancer therapy. Not only does it destroy cancerous cells directly, but it also encodes the production of GM-CSF (granulocyte-macrophage colony-stimulating factor), a protein that stimulates an anti-tumor immune response. Its presence also increases the antigenicity of the tumors, increasing the likelihood that the immune system will attack the tumor cells.

Therapeutic Vaccines

The only FDA-approved therapeutic cancer vaccine is used to treat prostate cancer, and it relies on an in vitro step to direct the patient's immune cells to attack the cancer cells. To do this, leukocytes are collected from the patient's blood and then exposed to a genetically engineered protein consisting of two critical parts: One is GM-CSF (granulocyte-macrophage colony-stimulating factor), which promotes the development of antigen-presenting dendritic cells; the other is an antigen that characterizes prostate cancer. Thus, some of the leukocytes develop into dendritic cells that will present the prostate cancer antigen to T cells. These are then infused back into the patient so that they can activate any T cells that recognize the antigen.

Examples of cancer immunotherapies are summarized in **table 17.6.**

TABLE 17.6 Cancer Immunotherapies

Immunotherapy	Characteristics
Immunomodulators	Molecules that enhance the anti-cancer immune response.
Monoclonal antibodies	Naked antibodies that mark a target cell for destruction or block a certain receptor; conjugated antibodies deliver a toxin or other molecule to specific cell types; bispecific antibodies connect two cell types.
Checkpoint inhibitors	Molecules that overcome mechanisms cells normally use to prevent attack by effector cytotoxic T cells.
Chimeric antigen receptor (CAR) T cells	A patient's own T cells that have been genetically engineered to express an antigen receptor that allows the cells to recognize and respond to a given antigen without the need for antigen presentation or co-stimulation.
Oncolytic virus	A genetically modified virus that attacks cancer cells and encodes a protein that stimulates an antitumor response.
Therapeutic vaccine	A patient's own immune cells that have been exposed to a genetically modified protein to enhance the anti-cancer response.

FOCUS YOUR PERSPECTIVE 17.1

Obtaining Monoclonal Antibodies

When an animal is injected with an antigen, its immune system responds to the antigen's different epitopes. So even though there is a single antigen, a variety of different B cells respond (each to a single epitope), resulting in the production of polyclonal antibodies. Unfortunately, this makes it difficult to standardize experimental results because the antibody composition is different each time the antiserum is made.

In 1975, Georges Köhler and Cesár Milstein overcame the problem of variable antiserum preparations by developing a technique to make monoclonal antibodies. These are antibodies produced by a single B clone, so all molecules in a preparation will have the same constant and variable regions and, thus, the same functional characteristics and epitope specificity. With such consistency, tests can be standardized more easily and with greater reliability.

To make monoclonal antibodies, a laboratory animal is immunized with the agent being studied, and that animal's B lymphocytes are then isolated (**box figure 17.1**). These are then fused with myeloma cells, which are malignant (cancerous) plasma cells. Unlike normal plasma cells, these myeloma cells can divide repeatedly in culture and do not make antibodies. In addition, they have lost the capacity to produce a critical enzyme, so they cannot grow in a medium that contains the drug aminopterin. When the B cells and the myeloma cells are mixed and grown in a medium that contains aminopterin, only the fusion products, called hybridomas, can proliferate. A hybridoma cell retains critical traits from the fused cells: The B cell supplies the genes for the specific antibody production, and the myeloma cell supplies the cellular machinery for producing the antibodies and multiplying indefinitely.

In the laboratory, monoclonal antibodies are the basis of a number of diagnostic tests. For example, monoclonal antibodies against a hormone can detect pregnancy only 10 days after conception. Specific monoclonal antibodies are used for rapid diagnosis of a wide variety of infectious diseases and are now being used in immunotherapies as well. Köhler and Milstein were awarded a Nobel Prize in 1984 for their work.

continued

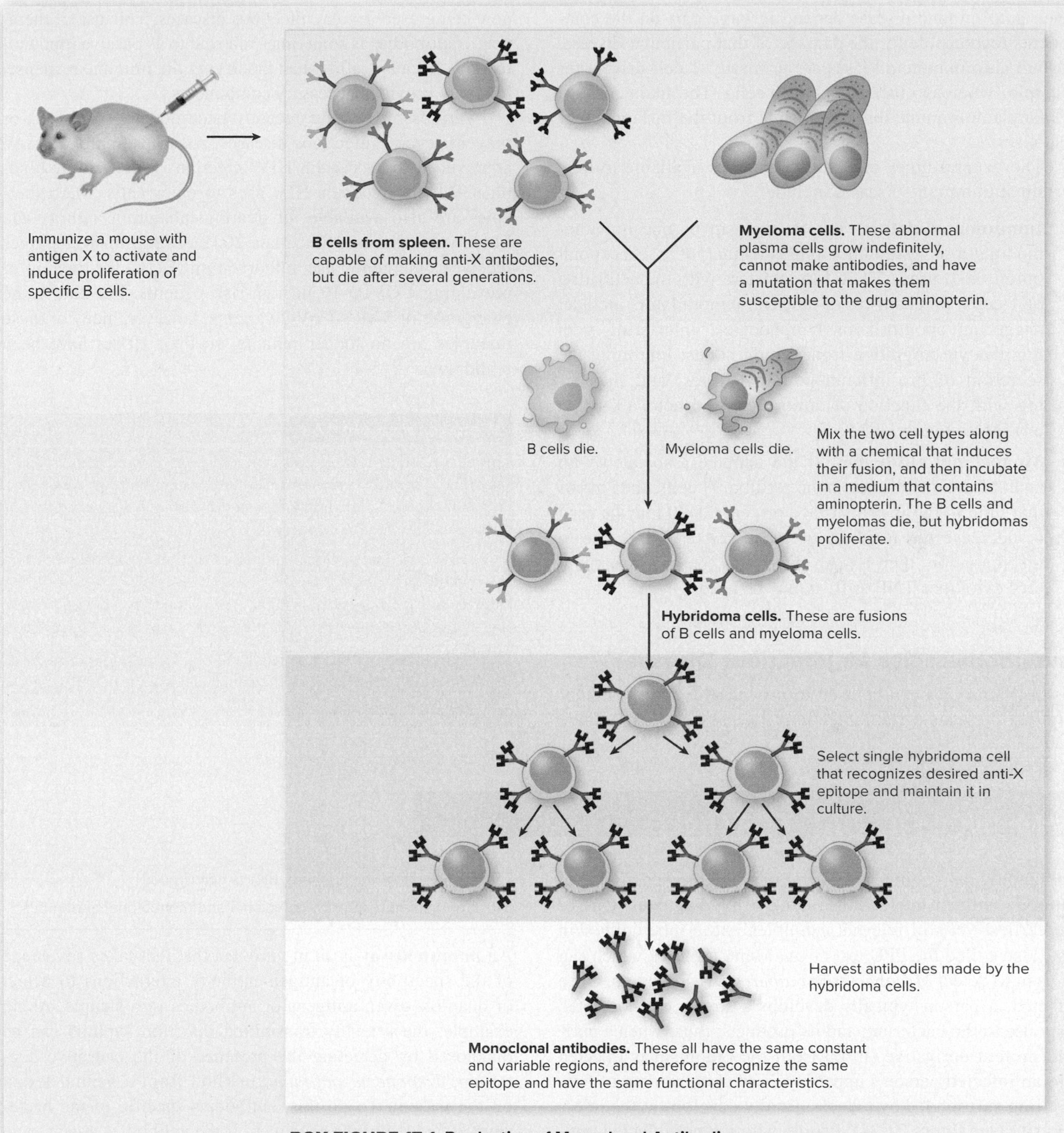

Immunize a mouse with antigen X to activate and induce proliferation of specific B cells.
B cells from spleen. These are capable of making anti-X antibodies, but die after several generations.
Myeloma cells. These abnormal plasma cells grow indefinitely, cannot make antibodies, and have a mutation that makes them susceptible to the drug aminopterin.
B cells die.
Myeloma cells die.
Mix the two cell types along with a chemical that induces their fusion, and then incubate in a medium that contains aminopterin. The B cells and myelomas die, but hybridomas proliferate.
Hybridoma cells. These are fusions of B cells and myeloma cells.
Select single hybridoma cell that recognizes desired anti-X epitope and maintain it in culture.
Harvest antibodies made by the hybridoma cells.
Monoclonal antibodies. These all have the same constant and variable regions, and therefore recognize the same epitope and have the same functional characteristics.

BOX FIGURE 17.1 Production of Monoclonal Antibodies

Immunotherapies for Immunological Disorders

Immunotherapies are currently used to treat some immunological disorders. This section will focus only on the treatment of autoimmune diseases; those used to treat hypersensitivities will be described in chapter 18, after the mechanisms of the hypersensitivities are explained.

Autoimmune diseases result from immune reactions that attack "self" molecules or cells. The immunotherapies for these diseases suppress certain components of the immune response (in contrast to the cancer immunotherapies just described, which enhance the immune response). A variety of immunotherapies are available, but the ones suitable for a given autoimmune disease depend in large part on the components responsible for the damage of that particular disease. Certain autoimmune diseases are primarily T cell driven, for example, whereas others involve B cells. The main damage of many autoimmune diseases results from the inflammatory response.

The general types of immunotherapies available to treat certain autoimmune diseases include:

- **Immunomodulators.** The mechanism of one immunomodulator is essentially opposite that of a checkpoint inhibitor: It uses CTLA-4 to interfere with the activation of T cells. Mechanisms of other immunodulators include: triggering apoptosis of lymphocytes; interfering with lymphocyte migration from lymph nodes; inhibiting the secretion of pro-inflammatory cytokines; and interfering with the function of tumor necrosis factor (TNF), a pro-inflammatory cytokine.
- **Monoclonal antibodies.** If the damage associated with an autoimmune disease is the result of B cells, then mAbs that bind to the B-cell surface protein CD20 may be used to decrease the numbers of those cells. If it is due to inflammation, then a mAb that binds the pro-inflammatory cytokines TNF or IL-6 may be used.

Immunotherapies for Infectious Diseases

Just as antibodies can be used to provide the passive immunity that prevents certain infectious diseases, some mAbs are now being used to treat infectious diseases. This use of therapeutic antibodies is sometimes referred to as passive immunotherapy, because rather than modifying the immune response, it simply provides necessary components.

Various mAbs have recently been developed to treat or prevent several infectious diseases. As an example, an mAb approved to treat certain HIV infections functions by binding CD4, the receptor HIV uses to enter cells. MAb therapies are also available for treating inhalation anthrax and Ebola virus disease. Since late 2021, several mAb therapies received emergency use authorization (EUA) for treating or preventing COVID-19 in high-risk patients, but due to the emergence of SARS-CoV-2 variants, however, many of these therapies are no longer reliable, so their EUAs have been withdrawn.

MicroAssessment 17.3

Immunotherapies for cancers generally boost components of the immune system, whereas the therapies for autoimmune diseases typically suppress the immune system. Antibodies are used to treat certain infectious diseases.

7. How can a monoclonal antibody function as a checkpoint inhibitor?
8. How can a monoclonal antibody be used to decrease inflammation?
9. Why would a cancer patient be at risk for developing an autoimmune disease when using a checkpoint inhibitor as therapy?

IMMUNOLOGICAL TESTING

Immunological testing takes advantage of the specificity of antibody-antigen interactions, using it for diagnosis. One of the earliest yet still relevant examples is the tuberculin skin test (also called the PPD test or the Mantoux test), which can be used to detect *Mycobacterium tuberculosis* infection. Once infected, a person typically develops a strong cell-mediated response to the bacterium and its products; thus, when a purified protein derivative (PPD) from the organism is injected into an infected person's upper layers of skin, an area of firm swelling surrounded by redness usually develops at the injection site (see figure 21.21). People who are not infected typically show little, if any, response (unless they have received the BCG vaccine against tuberculosis; see table 17.5).

17.4 ■ Principles of Immunoassays

Learning Outcomes

8. Describe how known antibodies can be used to identify an unknown antigen, and vice versa.
9. Explain how the antibody titer is determined.
10. Compare and contrast polyclonal and monoclonal antibodies.

An **immunoassay** is an in vitro test that that takes advantage of the specificity of antigen-antibody interactions to detect or quantify given antigens or antibodies in a sample. As an example, the sexually transmitted infection syphilis can be diagnosed by detecting the presence of the causative bacterium, *Treponema pallidum,* in fluid from a genital lesion on the patient. To do this, antibodies specific to the bacterium are added to the specimen; if the antibodies attach to an organism in the specimen, then the organism is *T. pallidum* (**figure 17.5**). In the reverse situation, the patient can also be tested for antibodies that bind specifically to *T. pallidum.* If antibodies are present, then the patient's immune system must have responded to the microbe at some point, indicating either previous or current infection.

To determine if a person has certain specific antibodies in their blood, either the serum or the plasma is tested. **Serum**

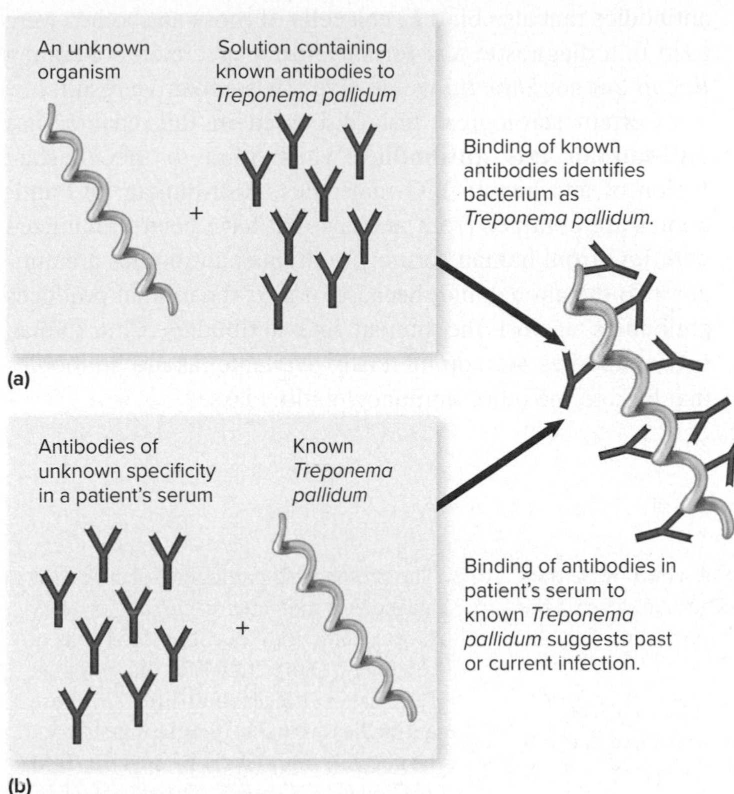
(a)
An unknown organism
Solution containing known antibodies to *Treponema pallidum*
+
Binding of known antibodies identifies bacterium as *Treponema pallidum*.
(b)
Antibodies of unknown specificity in a patient's serum
Known *Treponema pallidum*
+
Binding of antibodies in patient's serum to known *Treponema pallidum* suggests past or current infection.

FIGURE 17.5 Principles of Immunoassays These assays can be used to **(a)** detect and therefore identify unknown bacteria (or other antigens); **(b)** detect specific antibodies.

? Other than a current infection, what else might account for a person having antibodies to a specific infectious agent?

is the fluid portion that remains after blood clots; **plasma** is the fluid portion of blood treated with an anticoagulant to prevent clotting. Because serum is so often used as a source of antibodies, the study of in vitro antibody-antigen interactions is referred to as **serology,** and immunoassays are also referred to as serological tests. In general, serological testing implies examining a patient's blood for specific antibodies.

A person who has not been exposed to a given pathogen typically lacks specific antibodies against that microbe and is referred to as seronegative. Once infected, that person will begin producing detectable levels of specific antibodies about a week or two later, becoming seropositive. This change from seronegative to seropositive is referred to as **seroconversion.** As the infection progresses, increasing amounts of specific antibodies are produced. A rise in the concentration (titer) of specific antibodies is characteristic of an active infection. In contrast, low but steady levels of specific antibodies indicate a previous infection or vaccination.

Quantifying Antigen-Antibody Reactions

The concentration of antibody molecules in a specimen such as serum is usually determined by making serial dilutions, similar to what is done when counting bacterial cells (see figure 4.18). A series of 2-fold or 10-fold dilutions is used to dilute the specimen, and then the antigen is added to each dilution. The **titer** (concentration) is expressed as the reciprocal of the last dilution that gives a detectable antigen-antibody reaction. Thus, if a positive reaction is observed in the dilution 1:256 but not in 1:512, then the antibody titer is 256.

Immunoassays can be done in test tubes, but this requires many tubes and large amounts of reagents, so the assays are usually done using microtiter plates, which are plastic plates that have many tiny wells (**figure 17.6**). The volumes used in each well are a mere fraction of those needed for even a small test tube, so tests can be done on very small samples. Special equipment can be used to mix the reagents and read the results.

Obtaining Known Antibodies

Depending on the situation, the antibodies used in immunoassays can be either monoclonal or polyclonal. As described in the previous section, a **monoclonal antibody** recognizes only a single epitope because it was produced by clones of a single B cell (see Focus Your Perspective 17.1). In contrast, **polyclonal antibodies** recognize multiple epitopes on the same antigen because they are produced by clones of the collection of B cells that responded to the antigen.

The first step in producing polyclonal antibodies is similar to the initial step of making monoclonal antibodies: An animal such as a rabbit or goat is immunized with the agent. In the case of polyclonal antibodies, however, the animal serum is later collected to harvest the antibodies. Multiple naive B cells of the animal will have responded to the immunization, giving rise to a mixture of antibodies that together bind a variety of the antigen's epitopes. Immunizing the animal with a whole agent

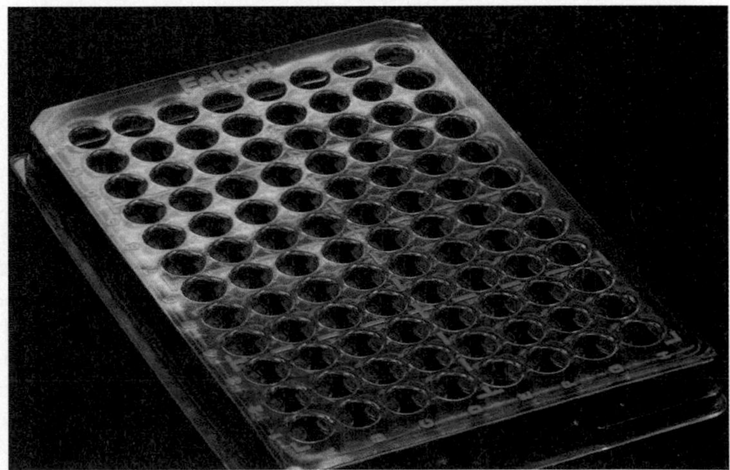

FIGURE 17.6 Microtiter Plate Immunoassays can be done in the wells of these small plates. Lisa Burgess/McGraw Hill

? What is an advantage of doing immunoassays in a microtiter plate?

results in a wider range of antibody specificities than immunizing with only part of the agent because of the greater number of epitopes on a whole agent. As with monoclonal antibodies, a variety of polyclonal antibody preparations are commercially available from laboratory supply companies.

Although polyclonal antibodies are less expensive to produce than monoclonal antibodies and sometimes give better results, one problem is that some may bind to closely related organisms, resulting in false-positive reactions. *Shigella* species have outer membrane proteins that are highly homologous to ones from *E. coli,* for example, so an animal immunized with whole *Shigella* cells would produce some antibodies that also bind *E. coli* cells. If those antibodies were used in a diagnostic test for *Shigella,* a specimen containing *E. coli* but not *Shigella* would give a false-positive result.

Certain serological tests discussed in this chapter use **anti-human IgG antibodies,** which bind to the constant region of any human IgG molecules. Anti-human IgG antibodies are obtained from animals that have been immunized with IgG from human serum. The human antibodies are antigenic to the animal and, because of this, the animal produces antibodies against the human IgG antibodies. Anti-human IgG antibodies are commercially available, as are antibodies that bind to the other immunoglobulin classes.

FOCUS ON A CASE 17.1

Patient A, an unvaccinated university student, had recently traveled to Europe. He became worried when he developed a fever and swelling on half his face, so he visited the campus clinic. He was diagnosed with a bacterial infection and prescribed antibiotics. A week later, Patient A returned to the clinic, this time reporting pain in his testicles. At that point, he was suspected of having mumps, a viral disease that often affects the parotid glands (the largest salivary glands) and sometimes other sites, including the testicles. There is no effective treatment for mumps, but the patient was referred for serological testing. He did not follow through.

A month later, the roommate of Patient A visited the clinic, also complaining of pain, fever, and swelling on half his face. Although he had been vaccinated, mumps was suspected, and serological testing was done. Those results were inconclusive: The test for mumps-specific IgM antibodies was negative, but the test for mumps-specific IgG antibodies was positive. A short time later, three other students visited the clinic, experiencing the same symptoms, and again, mumps was suspected. Their diagnosis was confirmed using polymerase chain reaction (PCR).

Health officials were concerned because the mumps cases appeared to be linked, so an investigation was done to determine the extent of the outbreak. A total of 29 cases were found, approximately half of which were confirmed by PCR; one was confirmed by finding mumps-specific IgM antibodies. Seven of those who were ill either were unvaccinated or had not received the recommended number of MMR (measles, mumps, and rubella) doses.

All students at the university were then encouraged to get an additional dose of the MMR vaccine, and vaccine clinics were set up around the campus.

1. Why was it significant that Patient A had not been vaccinated and had traveled to Europe?
2. What is the significance of finding mumps-specific IgM antibodies?
3. Why might the roommate's serology test results have been negative for mumps-specific IgM and positive for mumps-specific IgG?
4. What might explain the fact that some students became ill with mumps even though they had been adequately vaccinated?

Discussion

1. Without vaccination against mumps, Patient A was susceptible to the disease. He was unlikely to contract the disease in the United States, because of herd immunity—widespread vaccination has made most people immune to the disease, and the lack of susceptible hosts prevents the virus from spreading. Unfortunately, unfounded concerns over vaccination have decreased the vaccination rate in some countries, leading to outbreaks of vaccine-preventable diseases, including mumps. Patient A had traveled to a country that was experiencing an outbreak, and he likely contracted the disease there, bringing the infecting virus back with him when he returned to the United States.
2. The presence of pathogen-specific IgM antibodies indicates a current or very recent infection, because IgM has a relatively short half-life. In contrast, IgG has a longer half-life, so those antibodies remain circulating longer. In general, high levels of specific IgM antibodies indicate a current infection or recent vaccination, whereas high levels of specific IgG antibodies indicate a current infection, past infection, or vaccination.
3. The roommate had been vaccinated, so he likely had at least some circulating mumps-specific IgG antibodies as a result. In addition, vaccination would have induced the production of mumps-specific memory B cells, so more IgG would have been rapidly produced upon exposure to the virus. Some IgM was also likely made upon infection, but levels may not have been high enough to be detectable when he was tested. Although the patient had circulating IgG antibodies against the mumps virus, the quantities were not sufficient to protect him from the disease.
4. Mumps cases in vaccinated people could be due to waning immunity. Although mumps vaccination is part of the childhood vaccination schedule, the resulting immunity lessens over time. In addition, some individuals never received the recommended number of doses. Finally, some vaccine recipients do not mount a strong enough response to completely protect them from the disease.

Source: "Mumps Outbreak on a University Campus—California, 2011" *Morbidity and Mortality Weekly Report (MMWR)* December 7, 2012 /61(48); 986–989. Centers for Disease Control and Prevention http://www.cdc.gov/mmwr/preview/mmwrhtml/mm6148a2.htm?s_cid=mm6148a2_e

MicroAssessment 17.4

Immunoassays are used to detect or quantify given antigens or antibodies in a sample. The sample may be serially diluted to determine the antibody titer. Antibodies used in immunoassays can be either polyclonal or monoclonal.

10. What is the significance of a rise in titer of specific antibodies in serum samples taken at different times?

11. How are polyclonal antibodies different from monoclonal antibodies?

12. Would antibodies produced by a patient in response to infection be monoclonal or polyclonal?

17.5 ■ Common Types of Immunoassays

Learning Outcomes

11. Explain how labeled antibodies are used in direct and indirect tests.

12. Compare and contrast fluorescent antibody tests, ELISAs, and Western blots.

13. Describe how the fluorescence-activated cell sorter is used in immunoassays.

14. Compare and contrast precipitation reactions and agglutination reactions.

Antibodies are small molecules, so they cannot be seen using conventional microscopy. Because of this limitation, many clever assays have been developed that allow scientists to monitor antigen-antibody interactions. The assays fall into two general categories:

1. **Immunoassays that use labeled antibodies.** When antibodies are labeled with a detectable marker such as an enzyme, a fluorescent dye, or a radioactive isotope, their location can be tracked. Scientists can then determine if the antibodies are attached to an antigen. An advantage of these tests is that they are relatively sensitive, meaning they are more likely than other methods to detect if the antigen in question is present in a sample. Various labeled antibodies are extensively used in research and clinical laboratories, so many classes and specificities are commercially available.
2. **Immunoassays that involve visible antigen-antibody aggregates.** Many tests in this category were extensively used in the past, but they have now largely been replaced by newer, more sensitive tests that use labeled antibodies. The older tests are relatively inexpensive and technically simple, however, so some are still extremely useful in certain situations.

Immunoassays That Use Labeled Antibodies

Diagnostic tests that use labeled antibodies are routine in clinical laboratories, so understanding the basic principles is important. We will start by describing those principles, and then we will focus on specific tests and their applications. A table at the end of the discussion summarizes the characteristics of the various tests (see table 17.7).

Basic Principles

Labeled antibodies can be used to detect a given bacterium or other antigen as a way of identifying that antigen. Alternatively, they can be used to detect certain antibodies as a way of confirming that a patient's immune system has mounted a response against a given antigen. These types of tests are classified as either direct or indirect:

- **Direct immunoassays.** These are typically used to identify an unknown antigen (for example, a certain pathogen) in a clinical specimen. In the example illustrated in **figure 17.7a,** the unknown antigen is suspected to be "antigen X." The specimen containing the unknown antigen is first attached to a solid surface, and then labeled antibodies that will bind only antigen X are added. A washing step removes all unbound antibodies. The bound antibodies are then located by testing for the label. If the label is detected, then the specimen contained antigen X; conversely, if the label is not detected, then the specimen did not contain antigen X.
- **Indirect immunoassays.** These are used to detect antibodies of a given specificity in a patient's serum. The tests are called indirect because they require a labeled secondary antibody to detect the unlabeled first (primary) antibody. In the example illustrated in **figure 17.7b,** the patient's serum is being tested for IgG antibodies against "antigen X." To detect these, known antigen X must be attached to a solid surface. The patient's serum (containing antibodies) is then added; any anti-X antibodies in the serum will bind the antigen X. A washing step then removes any unbound antibodies. The next steps are aimed at detecting any bound IgG antibodies (primary antibodies). To do this, labeled anti-human IgG antibodies (secondary antibodies) are used; these bind to any human IgG molecules, regardless of their source or specificity. A washing step then removes any unbound molecules. In the final step, a test is done to detect the label. If it can be detected, then the labeled secondary antibodies are present. That, in turn, indicates that the primary antibodies are present.

Fluorescent Antibody (FA) Test

The **fluorescent antibody (FA) test** uses fluorescence microscopy to locate fluorescently labeled antibodies bound to antigens fixed to a microscope slide. Several different fluorescent dyes, including fluorescein (fluoresces green) and rhodamine (fluoresces red), can be used to label the antibodies. By using various fluorescent dyes, different antigens in the same preparation can be located. The fact that the results are examined by microscopy is an advantage over other methods because the antibody-bound structures can be observed;

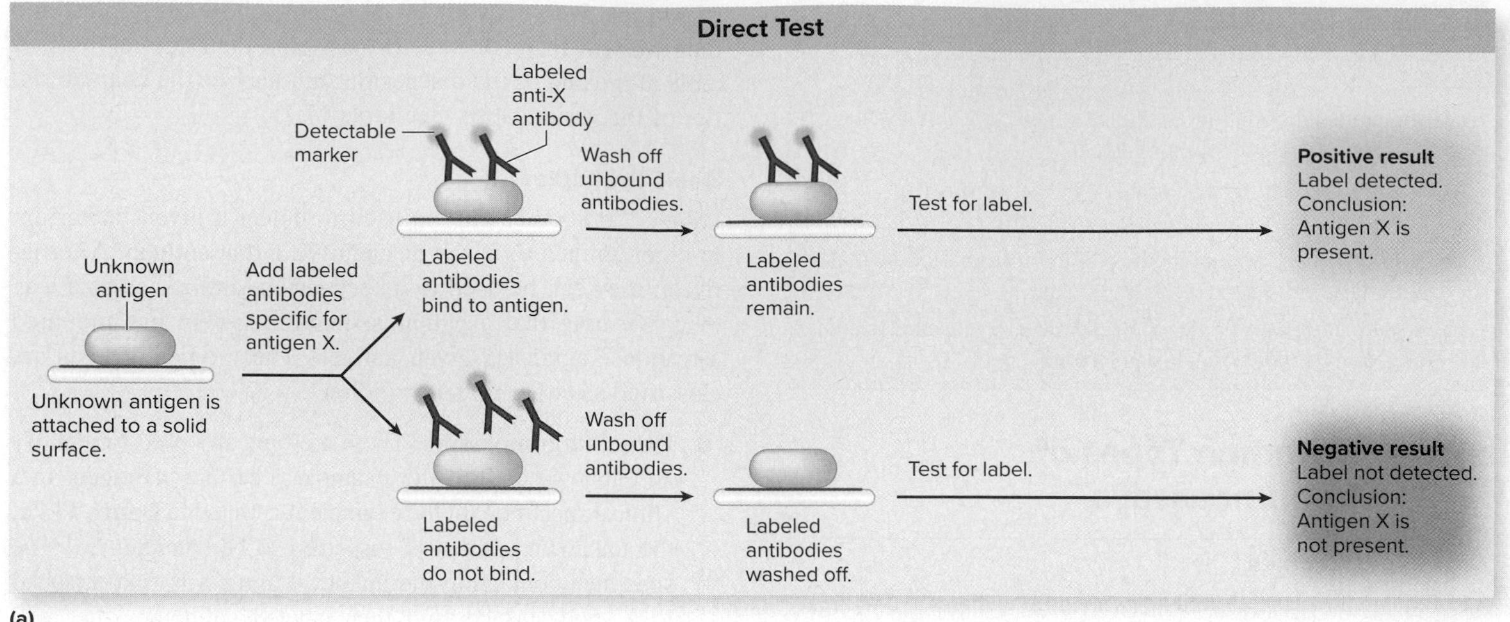
Direct Test
Labeled anti-X antibody
Detectable marker
Wash off unbound antibodies.
Test for label.
Positive result
Label detected.
Conclusion: Antigen X is present.
Unknown antigen
Add labeled antibodies specific for antigen X.
Labeled antibodies bind to antigen.
Labeled antibodies remain.
Unknown antigen is attached to a solid surface.
Wash off unbound antibodies.
Test for label.
Negative result
Label not detected.
Conclusion: Antigen X is not present.
Labeled antibodies do not bind.
Labeled antibodies washed off.
(a)

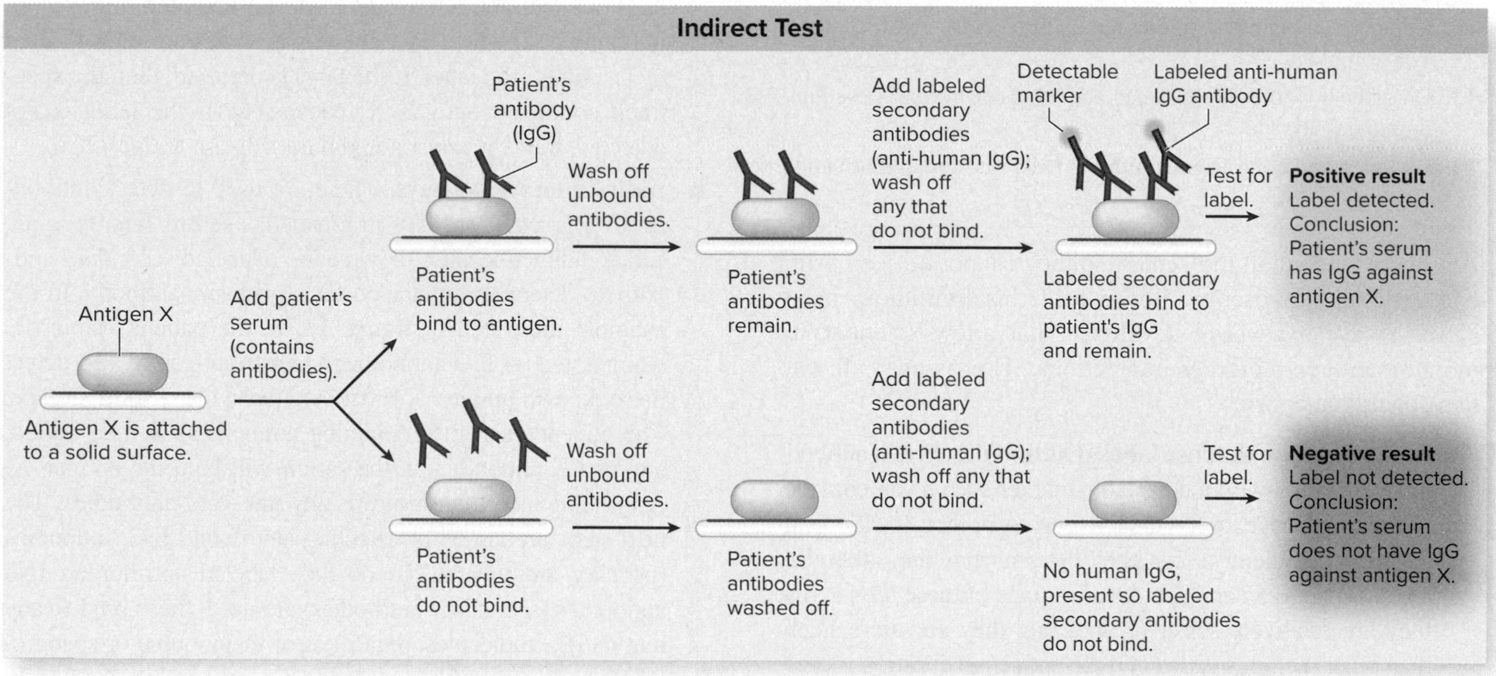
Indirect Test
Patient's antibody (IgG)
Wash off unbound antibodies.
Add labeled secondary antibodies (anti-human IgG); wash off any that do not bind.
Detectable marker
Labeled anti-human IgG antibody
Test for label.
Positive result
Label detected.
Conclusion: Patient's serum has IgG against antigen X.
Patient's antibodies bind to antigen.
Patient's antibodies remain.
Labeled secondary antibodies bind to patient's IgG and remain.
Antigen X
Add patient's serum (contains antibodies).
Antigen X is attached to a solid surface.
Wash off unbound antibodies.
Add labeled secondary antibodies (anti-human IgG); wash off any that do not bind.
Test for label.
Negative result
Label not detected.
Conclusion: Patient's serum does not have IgG against antigen X.
Patient's antibodies do not bind.
Patient's antibodies washed off.
No human IgG, present so labeled secondary antibodies do not bind.
(b)

FIGURE 17.7 Basic Principles of Tests That Use Labeled Antibodies to Detect Antigen-Antibody Interactions (a) Positive and negative results of a direct test. **(b)** Positive and negative results of an indirect test.

? **In (b), why would it be more efficient to use labeled anti-human IgG rather than label the patient's antibodies?**

this makes it possible to visualize individual cells and even sub-cellular structures. A disadvantage is that the procedure is relatively time-consuming because each specimen must be individually examined using a fluorescence microscope.

The FA test can be direct or indirect:

- **Direct FA test.** This is used to detect certain antigens in a sample (**figure 17.8a**). If the antigen is a microorganism, the researcher can see its size and shape. Viruses are too small to be seen as individual particles, but infected cells will fluoresce. The direct FA test is used to detect rabies virus-infected cells in brain tissue.
- **Indirect FA test.** This is used to detect certain antibodies in a sample, usually as a way of confirming diagnosis of a disease such as syphilis or toxoplasmosis (**figure 17.8b**).

Enzyme-Linked Immunosorbent Assay (ELISA)

As the name implies, **enzyme-linked immunosorbent assays (ELISAs)** use enzyme-labeled antibodies—in other

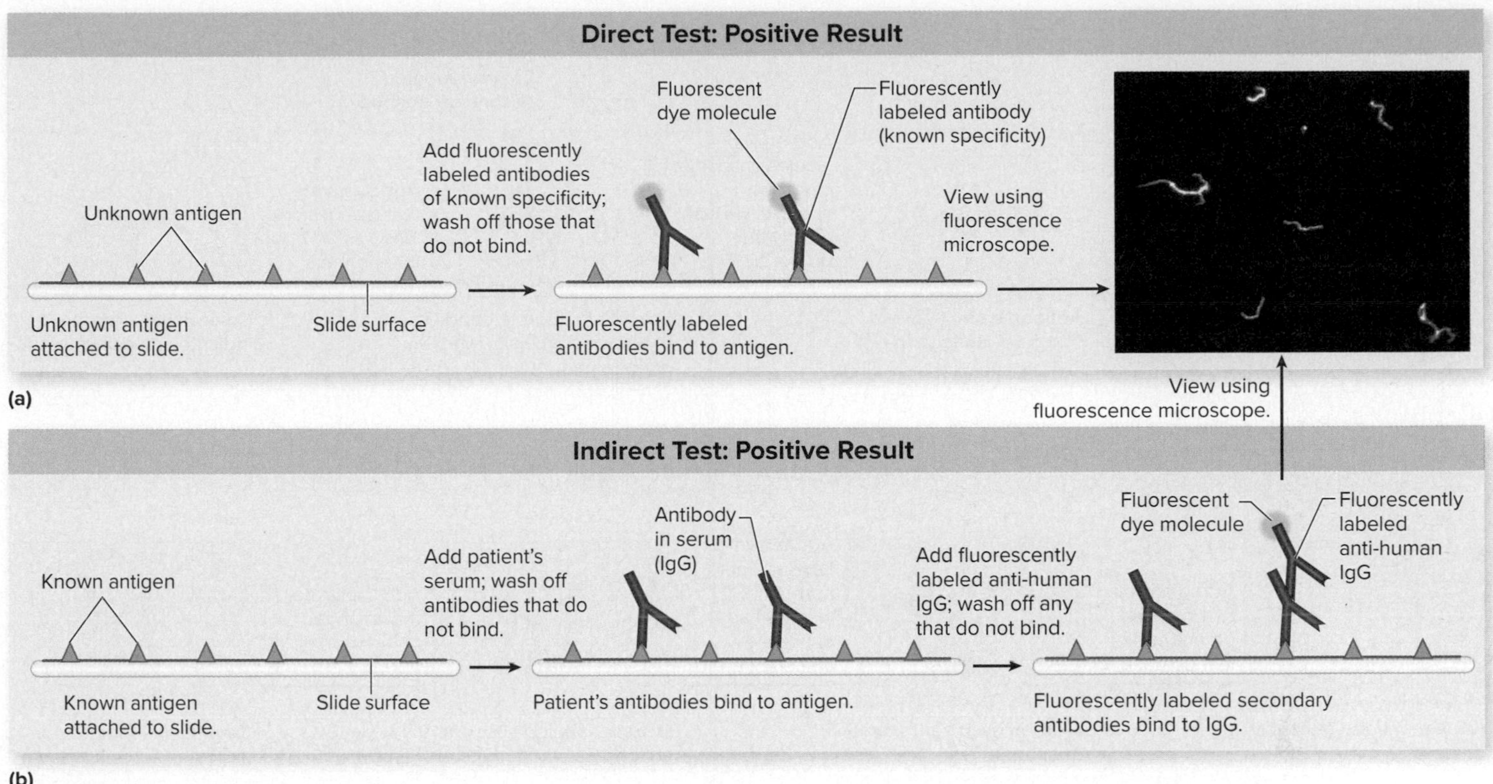
Direct Test: Positive Result
Fluorescent dye molecule
Fluorescently labeled antibody (known specificity)
Add fluorescently labeled antibodies of known specificity; wash off those that do not bind.
Unknown antigen
View using fluorescence microscope.
Unknown antigen attached to slide.
Slide surface
Fluorescently labeled antibodies bind to antigen.
(a)
View using fluorescence microscope.
Indirect Test: Positive Result
Fluorescent dye molecule
Fluorescently labeled anti-human IgG
Antibody in serum (IgG)
Add patient's serum; wash off antibodies that do not bind.
Add fluorescently labeled anti-human IgG; wash off any that do not bind.
Known antigen
Known antigen attached to slide.
Slide surface
Patient's antibodies bind to antigen.
Fluorescently labeled secondary antibodies bind to IgG.
(b)

FIGURE 17.8 Fluorescent Antibody (FA) Test **(a)** Positive direct FA test. **(b)** Positive indirect FA test. Russell/CDC

? **In this figure, why is anti-human IgG used in the indirect test but not in the direct test?**

words, antibodies with an enzyme attached; the assays are also referred to as enzyme immunoassays (EIAs). An enzyme commonly used to label antibodies is peroxidase from the horseradish plant. To detect the enzyme, a colorimetric assay is used to measure the enzymatic conversion of a colorless substrate into a colored product. The assays have advantages over other common immunoassays in that they are easy to do and often require minimal technical skills. In fact, they provide the basis for several simple strip tests, including the rapid group A strep tests done in doctor's clinics and the at-home COVID-19 tests and pregnancy tests (**figure 17.9**). More sensitive detection methods are also available in which fluorescent or luminescent (light-producing) substrates are used instead.

Clinical labs often do ELISAs as part of the diagnoses of a variety of different diseases and conditions, including COVID-19, Lyme disease, West Nile virus disease, giardiasis, measles, hepatitis B and C, and HIV infection. In addition, the tests are used to screen donated blood for antibodies that suggest the presence of certain pathogens, including HIV, hepatitis B virus, and hepatitis C virus. ELISAs are often done in microtiter plates, allowing multiple samples to be tested all at once, and can be direct or indirect:

- **Direct ELISA.** This is often done using what is called a "sandwich method" (**figure 17.10a**). With this method, specific antigens in the sample are "captured" by antibodies that have been attached to the inside surface of the well; the captured antigens are then detected, and therefore identified, using labeled antibodies. To diagnose giardiasis (a diarrheal disease caused by *Giardia lamblia*), commercially available microtiter plates that have wells coated with anti-*G. lamblia* antibodies may be used. When a stool sample is added to a well, the antibodies capture the *G. lamblia* antigens; enzyme-labeled antibodies are then used to detect those antigens.

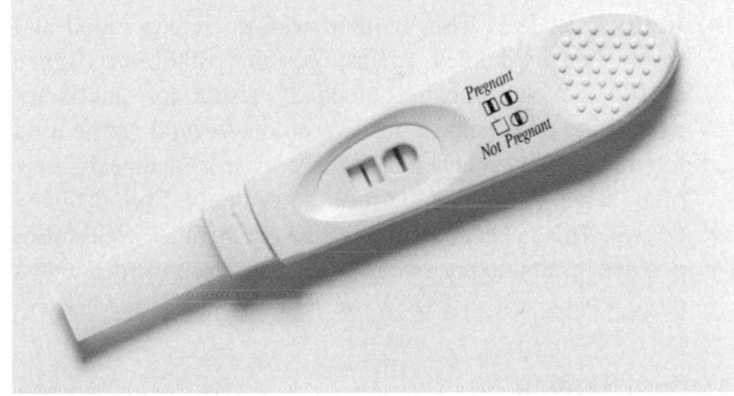
Pregnant
Not Pregnant

FIGURE 17.9 ELISA Test for Pregnancy The test detects human chorionic gonadotropin (HCG), an antigen present only in pregnant females. A urine sample is applied on the left. Two pink lines indicate reaction of HCG with antibodies, a positive test. A single line indicates absence of HCG in the urine, a negative test. Raimund Koch/The Image Bank/Getty Images

? **What is the purpose of the line that forms even if the test is negative?**

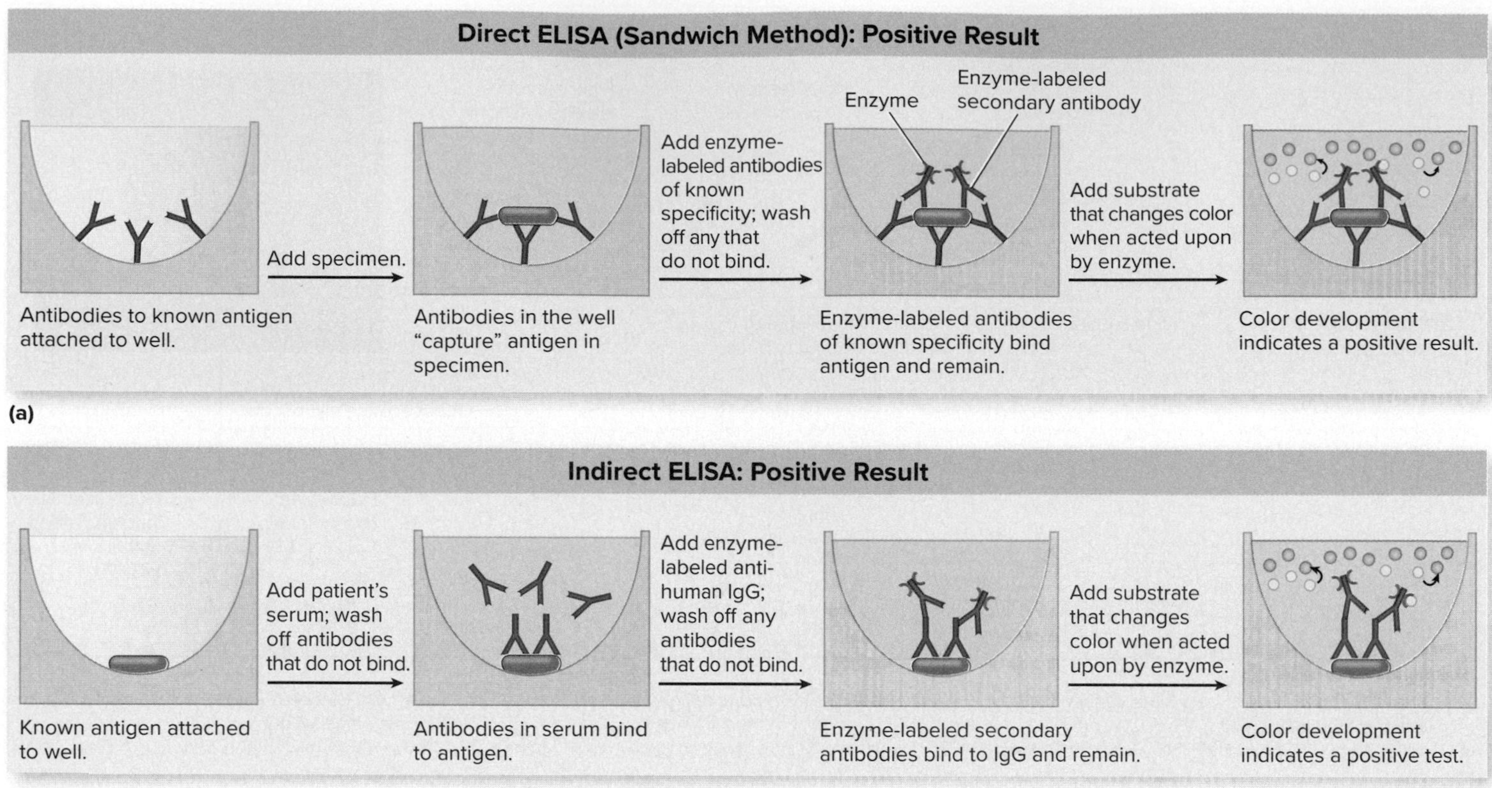
Direct ELISA (Sandwich Method): Positive Result
Add specimen.
Add enzyme-labeled antibodies of known specificity; wash off any that do not bind.
Enzyme
Enzyme-labeled secondary antibody
Add substrate that changes color when acted upon by enzyme.
Antibodies to known antigen attached to well.
Antibodies in the well "capture" antigen in specimen.
Enzyme-labeled antibodies of known specificity bind antigen and remain.
Color development indicates a positive result.
(a)
Indirect ELISA: Positive Result
Add patient's serum; wash off antibodies that do not bind.
Add enzyme-labeled anti-human IgG; wash off any antibodies that do not bind.
Add substrate that changes color when acted upon by enzyme.
Known antigen attached to well.
Antibodies in serum bind to antigen.
Enzyme-labeled secondary antibodies bind to IgG and remain.
Color development indicates a positive test.
(b)

FIGURE 17.10 Enzyme-Linked Immunosorbent Assay (ELISA) **(a)** Positive direct ELISA. **(b)** Positive indirect ELISA.

In the first panel of part (a), what is the purpose of the antibodies attached to the well?

- **Indirect ELISA.** This is often used to screen blood and serum for antibodies against certain pathogens (**figure 17.10b**). When donated blood is tested for antibodies against the pathogens, a positive product would not be used for transfusion. When ELISAs are used in diagnosis, they often serve as a screening tool rather than a final diagnostic test. This is because ELISAs can sometimes yield false-positive results, so positive results are often confirmed with a more reliable test such as Western blotting (described next).

Western Blotting

In the **Western blotting** technique, the various proteins that make up a sample are separated by size before reacting them with antibodies. This makes it possible to determine exactly which proteins the antibodies are recognizing, an essential part of accurate diagnostic testing. A drawback of the procedure is that relatively few samples can be tested at a time.

The basic steps of the Western blotting procedure are illustrated in **figure 17.11a.** First, a special type of gel electrophoresis is used to separate the proteins. To do this, samples are loaded onto a polyacrylamide gel matrix, and an electrical current is then run through the gel. Smaller proteins move faster through the gel matrix than the larger ones, so the proteins separate on the basis of size. The separated proteins in the gel are then transferred ("blotted") to a membrane; this immobilizes them in the same positions they were in the gel, creating what is referred to as a blot. The steps after that are very similar to those of an ELISA. To determine if a patient's serum has antibodies specific for any of the proteins in the sample, some of that serum is added to the blot, after which unbound antibodies are washed off. Enzyme-labeled anti-human IgG antibodies are then added, and these secondary antibodies bind to any serum antibodies attached to the proteins; unbound secondary antibodies are then washed off. Finally, the label is detected, generating a bar code–like pattern that reflects which proteins on the blot were recognized by the patient's antibodies (**figure 17.11b**). The detection method depends on the type of molecule carried by the secondary antibodies.

Fluorescence-Activated Cell Sorter (FACS)

A **fluorescence-activated cell sorter (FACS)** can be used for a variety of purposes, including diagnosing and monitoring the progression of diseases involving white blood cells. It is used to characterize different types and stages of lymphomas and leukemias as well as to diagnose immunodeficiency diseases. In addition, it can be used to track the progression of HIV infection by determining the serum levels of CD4 T cells.

A FACS is a specialized version of a flow cytometer that sorts and counts cells labeled with fluorescent antibodies (**figure 17.12**). Perhaps more important, it also measures the amount of fluorescence, along with cell size and granularity (internal complexity), allowing researchers to compare the concentrations and characteristics of the labeled cells or other

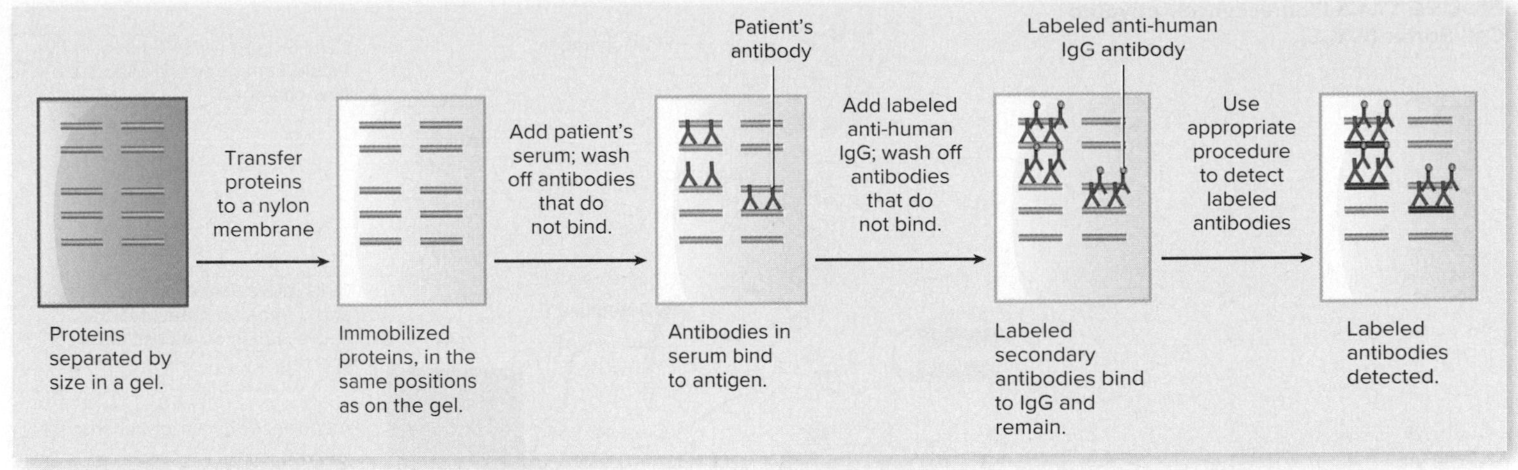

(a)

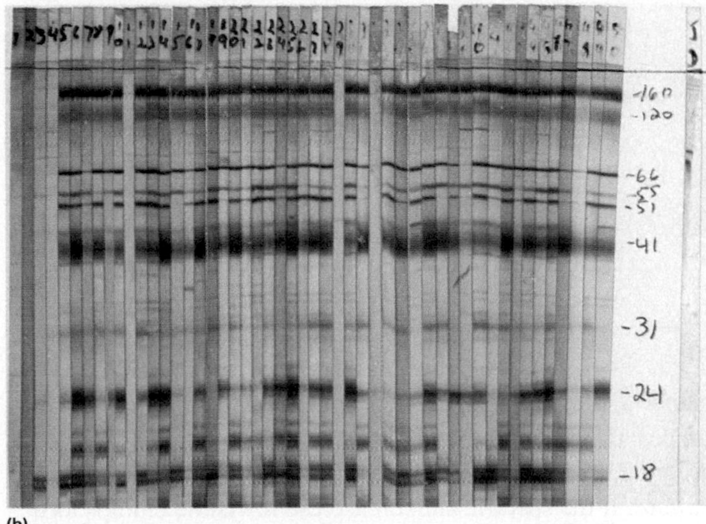

(b)

FIGURE 17.11 Western Blotting (a) Procedure. **(b)** Results of HIV testing. Each vertical strip is from a different patient. The dark bands represent proteins recognized by the patients' antibodies, and the numbers to the right indicate protein sizes. b: Hank Morgan/Science Source

? **Why are the results of a Western blot generally more reliable than those of an ELISA?**

particles. This is why it is so useful for monitoring differences in various types of white blood cells.

Characteristics of immunoassays that use labeled antibodies are summarized in **table 17.7.**

TABLE 17.7 Characteristics of Immunoassays That Use Labeled Antibodies

Method	Characteristics and Uses
Fluorescent antibody (FA) tests	Fluorescence microscopy is used to locate fluorescently labeled antibodies bound to antigens. Individual cells and even sub-cellular structures can be visualized, but each specimen needs to be examined individually. Used in clinical labs to diagnose certain diseases, including rabies and syphilis.
Enzyme-linked immunosorbent assay (ELISA)	A colorimetric assay is used to locate enzyme-labeled antibodies bound to antigens. When done in microtiter plates, large numbers of specimens can be screened all at once, but false positives can be a problem. Commercially available options are easy to perform and widely used. Used in at-home pregnancy tests and COVID-19 tests; in physicians' offices to diagnose strep throat; In clinical labs as part of the diagnosis of a wide range of diseases and conditions, including COVID-19, Lyme disease, West Nile virus disease, hepatitis B and C, and HIV infection. Also used to screen donated blood for antibodies that suggest the presence of HIV, hepatitis B virus, or hepatitis C virus.
Western blot	Various methods are used to detect labeled antibodies bound to proteins that have been separated by size. The results are generally more reliable than with the ELISA because they provide more information, but fewer samples can be tested at a time. Used in clinical labs to confirm certain positive ELISA tests, such as for Lyme disease.
Fluorescence-activated cell sorter	The device is used to separate and determine the relative concentrations and certain characteristics of fluorescent antibody-labeled cells. Widely used to diagnose and monitor the progression of diseases involving white blood cells. Used in clinical labs to track the progression of HIV infection, to characterize different types and stages of lymphomas and leukemias, and to diagnose immunodeficiency diseases.

Immunoassays That Involve Visible Antigen-Antibody Aggregates

As described in chapter 15, antibodies can cross-link antigens, thereby creating large "mouthfuls" for phagocytic cells (see figure 15.4). These clumped antigen-antibody complexes can be observed in agglutination and precipitation reactions.

Agglutination Reactions

Agglutination reactions take advantage of the visible clumping that occurs when antibodies cross-link relatively large particles such as cells. The reactions are used in red blood cell (RBC) typing and to identify certain types of microorganisms.

In a direct agglutination test (DAT), an antibody suspension is mixed with the large antigen molecules (**figure 17.13a**). If the antibodies bind to the antigens, visible clumping will occur—a positive test. The agglutination of RBCs by antibody binding or other means is referred to as hemagglutination and

FIGURE 17.12 Fluorescence-Activated Cell Sorter (FACS)

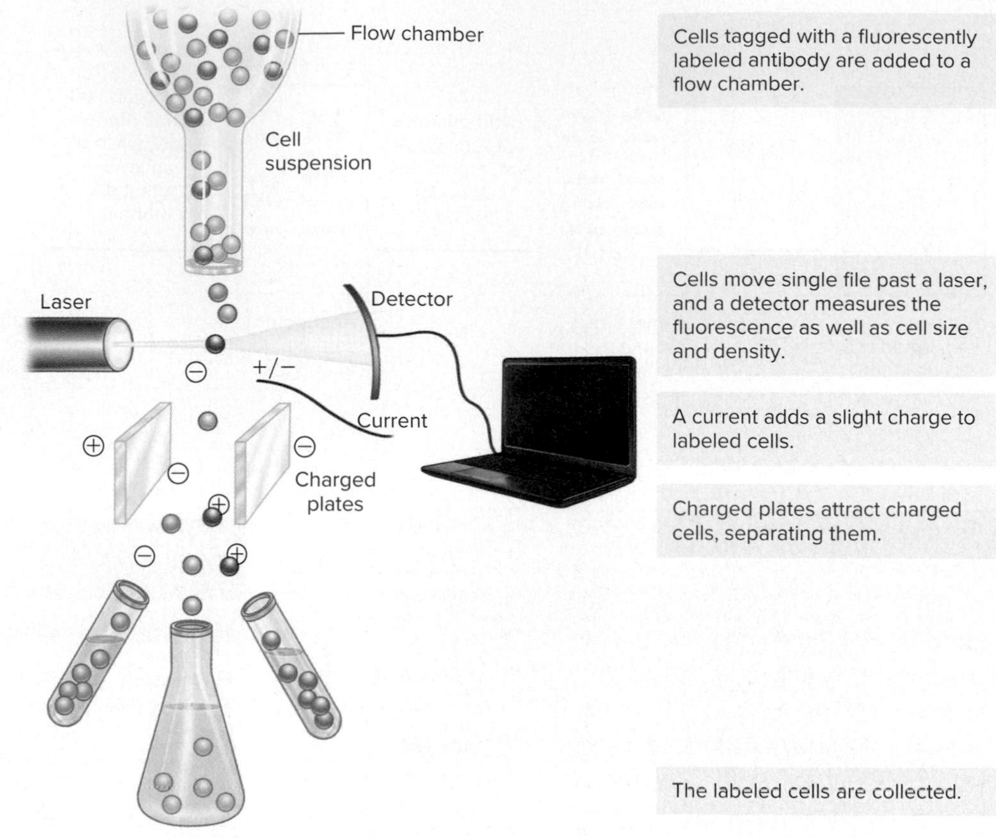

Flow chamber
Cell suspension
Laser
Detector
+/−
Current
Charged plates
Waste
Cells tagged with a fluorescently labeled antibody are added to a flow chamber.
Cells move single file past a laser, and a detector measures the fluorescence as well as cell size and density.
A current adds a slight charge to labeled cells.
Charged plates attract charged cells, separating them.
The labeled cells are collected.

is used in blood typing (**figure 17.13b**). Blood typing involves detecting certain antigens on RBCs and is important for avoiding hemolytic transfusion reactions (see table 18.1).

A passive agglutination test is used instead of a DAT if the antigens are relatively small. In the passive test, either the antibodies or the antigens are attached to particles such as latex beads to make the aggregates larger and therefore easier to see (**figure 17.14**). Latex beads to which specific antibodies have been attached are available commercially to identify various bacteria, fungi, viruses, and parasites, and to detect certain hormones, drugs, and other substances in body fluids. The beads are mixed with a drop of a body fluid or suspended microbial culture. If the specific antigen is present, visible clumps form.

Precipitation Reactions

When antibodies bind to soluble antigens—that is, molecules that are dissolved rather than suspended as particles—large lattice-like complexes may form that precipitate out of solution

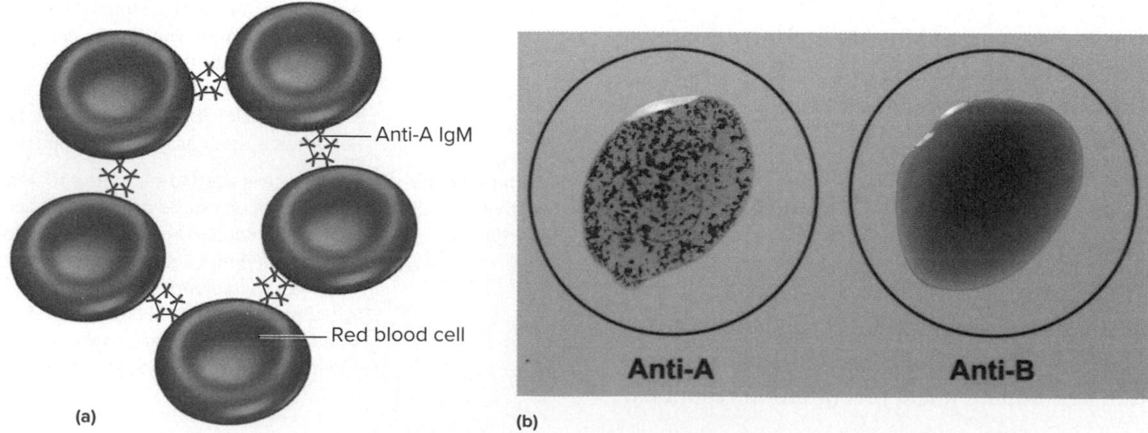

FIGURE 17.13 Direct Agglutination In this example, red blood cells (RBCs) are being tested for ABO blood groups by mixing them with separate antibody suspensions specific for either A or B antigens. **(a)** If antibodies bind an antigen on the RBCs, the resulting cross-linking will cause agglutination. **(b)** The RBCs agglutinated when mixed with anti-A antibodies, but did not agglutinate when mixed with anti-B antibodies, indicating that the blood group is type A. b. Lisa Burgess/McGraw Hill

? If the RBCs agglutinated when mixed with both anti-A and anti-B antibodies, what would the blood group be?

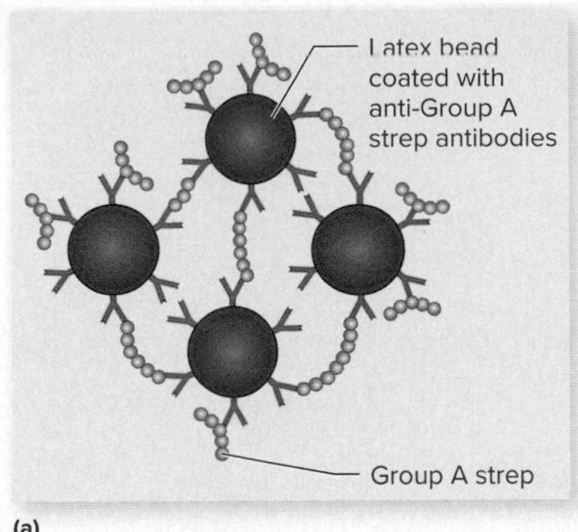

(a)

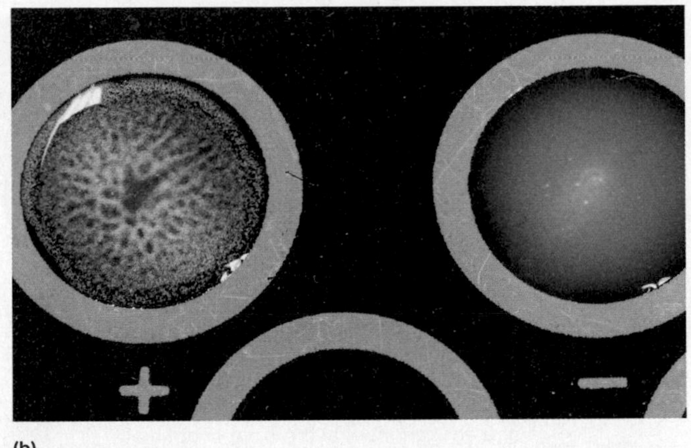

(b)

FIGURE 17.14 Passive Agglutination In this example, antibody-coated latex beads are used to identify *Streptococcus pyogenes.* **(a)** The latex beads are coated with antibodies that bind specifically to cell wall antigens of the bacteria. **(b)** Visible clumping (shown on the left) confirms that the organism is *S. pyogenes.* Negative test results are shown on the right. b: Lisa Burgess/McGraw Hill

Why use antibodies attached to latex beads instead of simply mixing them directly with the bacterial suspension?

(**figure 17.15**). This is the basis of **precipitation reactions.** The complexes can take several hours to form and develop only at certain relative concentrations of antibody and antigen molecules. The easiest way to get the proper concentrations is to place the antigen and antibody suspensions near each other in a gel and let the molecules diffuse toward each other—a process called immunodiffusion. A precipitate will form in a distinct region called the zone of optimal proportions.

Precipitation reactions are not commonly used in diagnosis today because they have largely been replaced by methods that rely on labeled antibodies. However, the principle of the reactions is nicely demonstrated by the Ouchterlony technique, which can be done in a Petri dish (**figure 17.16**). Antigen and antibody solutions are placed into separate wells cut in the gel contained in the dish. The two solutions will gradually diffuse outward, meeting between the wells. If the antibody molecules recognize the antigen, a line of precipitation will form at the zone of optimal proportions. The Ouchterlony test can be used to detect autoantibodies associated with certain connective tissue disorders.

MicroAssessment 17.5

Antibodies labeled with a detectable marker can be used to identify an antigen (direct test) or to detect a patient's antibodies to a known antigen (indirect test). Examples of methods that use labeled antibodies include ELISA, Western blotting, fluorescence antibody tests, and fluorescence-activated cell sorters. Agglutination and precipitation reactions both depend on the formation of visible antigen-antibody complexes.

13. Why are ELISAs more commonly used than Western blots in diagnosing diseases?
14. How is a direct agglutination test different from a passive agglutination test?
15. Why is a false positive more significant in HIV testing of patients than in screening donated blood for transfusions?

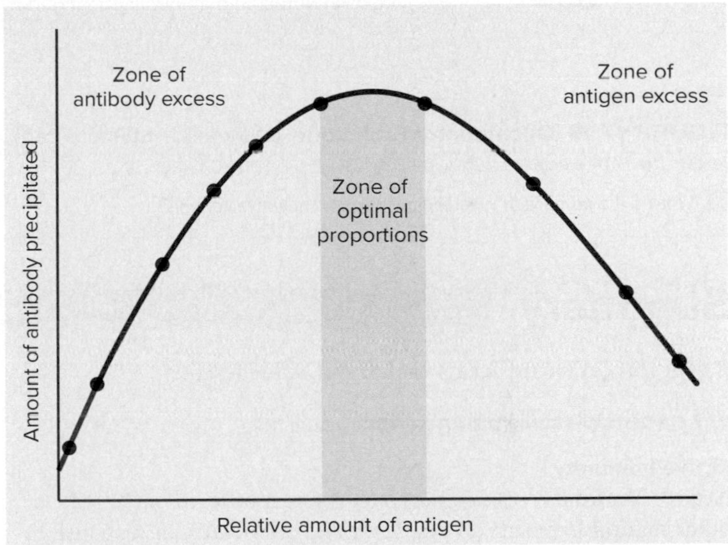

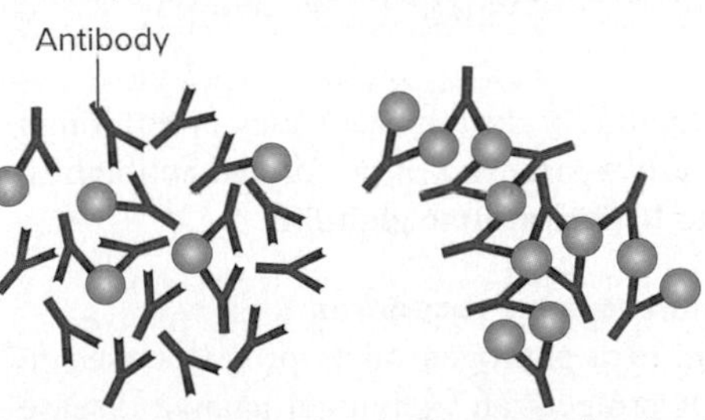

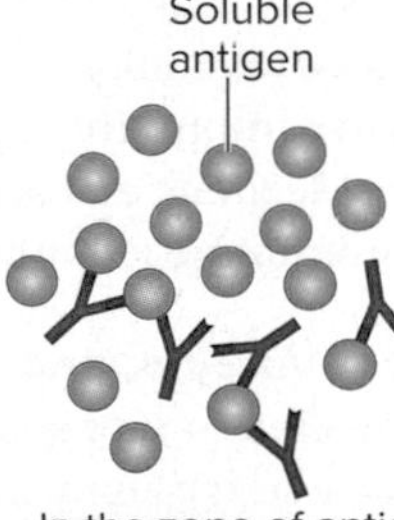

Soluble antigen
In the zone of antigen excess, little or no cross-linking occurs; no visible precipitate forms.

FIGURE 17.15 Antigen-Antibody Precipitation Reactions The maximum amount of precipitate forms in the zone of optimal proportion.

It takes fewer molecules of IgM than of IgG to cause precipitation. Why would this be so?

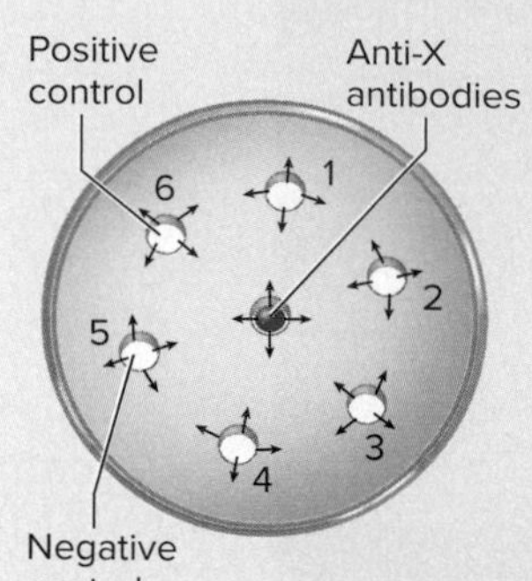

Antibody and antigen solutions are placed into separate wells cut into the gel. The antigens and antibodies diffuse toward each other. In this example, antibodies that bind antigen X (called anti-X antibodies) were added to the well in the center of the gel. Samples that contain unknown antigens are added to wells 1 through 4. A negative control (contains no antigen X) was added to well 5, and a positive control (contains known antigen X) was added to well 6.

When antibody molecules that recognize the antigen meet at the zone of optimal proportions, antigen-antibody complexes precipitate out of solution, forming a visible line. In this example, a line has formed between the center well and well 6 (the positive control). A line has also formed between the center well and the sample in well 4, indicating that the sample contains antigen X. The other samples do not contain detectable amounts of antigen X.

(a)

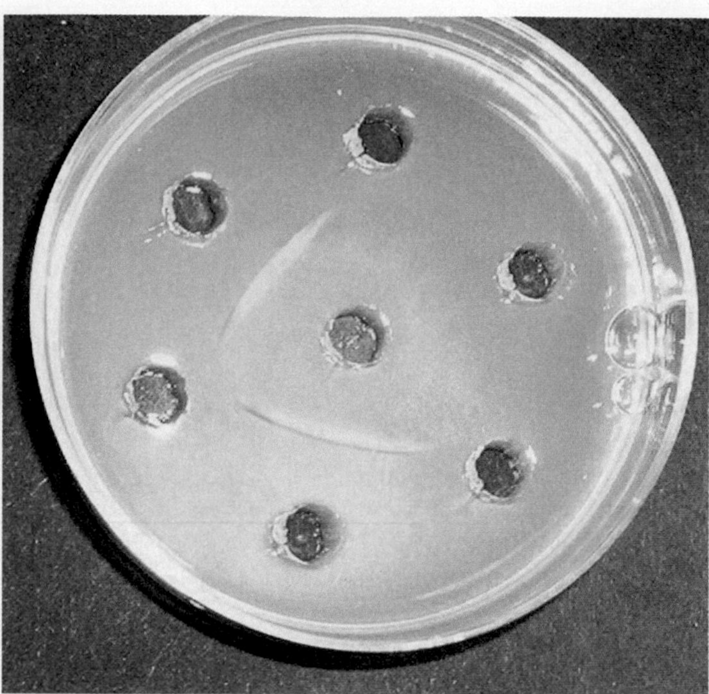

(b)

FIGURE 17.16 Ouchterlony Technique (a) Method. **(b)** Photograph of results. Schadler, D. L. 2003. Antigen-antibody testing: A visual simulation or virtual reality. *The Plant Health Instructor*. DOI: 10.1094/PHI-K-2003-0224-01

? **What is the purpose of including positive and negative controls?**

Summary

IMMUNIZATION AND IMMUNOTHERAPY

17.1 ■ Principles of Immunization (figure 17.2)

Active Immunity

Active immunity occurs naturally in response to infections or other natural exposure to antigens, and artificially in response to vaccination.

Passive Immunity

Passive immunity occurs naturally during pregnancy and through breast feeding, and artificially by transfer of preformed antibodies, as in **immune globulin** and **hyperimmune globulin.**

17.2 ■ Vaccines and Immunization Procedures (table 17.1)

A **vaccine** is a preparation of a pathogen or its products used to induce active immunity. It protects an individual against disease and may also provide **herd immunity.**

Attenuated Vaccines

An **attenuated vaccine** is a weakened form of the pathogen that can replicate but is generally unable to cause disease.

Inactivated Vaccines

Inactivated vaccines are unable to replicate but retain the immunogenicity of the infectious agent or toxin. They include inactivated whole agents, toxoids, subunits, VLPs, polysaccharides, and conjugates. **Adjuvants** increase the intensity of the immune response to the antigen in a vaccine.

Nucleic Acid–Based Vaccines

Nucleic acid–based vaccines consist of nucleic acid segments encoding key antigens of an infectious agent. Microbial proteins (antigens) that induce an immune response are produced when the nucleic acid is delivered into the vaccine recipient's cells. They include mRNA vaccines, viral vector vaccines, and DNA vaccines.

Vaccination Uses and Benefits (tables 17.3, 17.4, 17.5)

Routine childhood immunizations have prevented millions of cases of disease and many deaths (table 17.2). Non-routine vaccines are recommended only in situations where there is an increased risk of acquiring the infection.

An Example of Vaccination Strategy—The Campaign to Eliminate Polio (Poliomyelitis)

Vaccines against polio provide an excellent illustration of the complexity of vaccine strategies.

17.3 ■ Immunotherapies

Immunotherapies are methods designed to either enhance or suppress specific immune responses as a means to treat certain diseases. The term is primarily used in the context of cancer treatments, but the approaches are used to treat other diseases as well.

Immunotherapies for Cancer (table 17.6)

Immunotherapies for cancer include **immunomodulators, monoclonal antibodies, checkpoint inhibitors, chimeric antigen receptor (CAR) T cells**, an engineered **oncolytic virus,** and a therapeutic vaccine (figures 17.3, 17.4).

Immunotherapies for Immunological Disorders

The immunotherapies for immunological disorders include immunomodulators and monoclonal antibodies.

Immunotherapies for Infectious Diseases

Just as antibodies can be used to provide the passive immunity that prevents certain diseases, some are now used to treat infectious diseases.

IMMUNOLOGICAL TESTING

17.4 ■ Principles of Immunoassays (figure 17.5)

A seronegative individual becomes seropositive after initial infection with an agent; this **seroconversion** usually takes about a week or two. To determine if a patient has antibodies in the blood against a specific infectious agent, the patient's **serum** or **plasma** is tested.

Quantifying Antigen-Antibody Reactions

The concentration of antibody molecules in a specimen is usually determined by making serial dilutions; the last dilution that gives a detectable antigen-antibody reaction reflects the **titer.**

Obtaining Known Antibodies

Polyclonal antibodies recognize multiple epitopes, whereas **monoclonal antibodies** recognize only a single epitope. **Anti-human IgG antibodies** are used to detect IgG molecules in a patient specimen. Recombinant DNA techniques have been used to produce humanized monoclonal antibodies.

17.5 ■ Common Types of Immunoassays

Immunoassays That Use Labeled Antibodies (table 17.7)

Direct immunoassays are typically used to identify unknown antigens; **indirect immunoassays** are typically used to detect antibodies of a given specificity in a patient's serum (figure 17.7). The **fluorescent antibody (FA) test** relies on fluorescence microscopy to locate fluorescently labeled antibodies bound to antigens fixed to a microscope slide (figure 17.8). The **enzyme-linked immunosorbent assay (ELISA)** uses antibodies labeled with a detectable enzyme (figure 17.9, 17.10). In the **Western blot** technique, the various proteins that make up a sample are separated by size before reacting them with antibodies (figure 17.11). The **fluorescence-activated cell sorter (FACS)** can be used to count and separate antigens labeled with fluorescent antibodies, as well as to determine concentrations and characteristics of the labeled particles (figure 17.12).

Immunoassays That Involve Visible Antigen-Antibody Aggregates

Antibodies bound to particulate antigens cause obvious aggregates to form. Examples of **agglutination reactions** include direct agglutination tests and passive agglutination tests (figures 17.13, 17.14). Antibodies bound to soluble antigens may form complexes that precipitate out of solution (figure 17.15); an example of a test that involves a **precipitation reaction** is the Ouchterlony technique (figure 17.16).

Review Questions

Short Answer

1. How is immune globulin different from hyperimmune globulin?
2. Describe two advantages of an attenuated vaccine over an inactivated one.
3. Describe two advantages of an inactivated vaccine over an attenuated one.
4. What is herd immunity?
5. Describe how both active and passive immunization can be used to combat tetanus.
6. With respect to immunotherapies, what advantage do humanized monoclonal antibodies have over non-engineered versions?
7. What are CAR T cells?
8. What is the purpose of anti-human IgG antibodies in immunological testing?
9. What is an advantage of the fluorescent antibody test over an ELISA?
10. Is blood typing an example of a precipitation reaction or an agglutination reaction? Explain.

Multiple Choice

1. Which is an example of immunization that elicits active immunity?
   a) Giving antibodies against diphtheria
   b) Immune globulin injections to prevent hepatitis
   c) Inactivated polio vaccine
   d) Rabies immune globulin
   e) Tetanus immune globulin
2. Breastfeeding provides which of the following to an infant?
   a) Artificial active immunity
   b) Artificial passive immunity
   c) Natural active immunity
   d) Natural passive immunity
3. Vaccines ideally should be all of the following *except*
   a) effective in protecting against the disease.
   b) inexpensive.
   c) stable.
   d) living.
   e) easily administered.
4. Severely immunosuppressed people should *not* receive the measles vaccine. Based on this information, the vaccine is likely
   a) an inactivated whole agent.
   b) a toxoid.
   c) a subunit vaccine.
   d) a genetically engineered vaccine against hepatitis B.
   e) an attenuated vaccine.
5. All of the following diseases have attenuated vaccines *except*
   a) chickenpox.
   b) mumps.
   c) rubella.
   d) pertussis.
   e) measles.

6. *Haemophilus influenzae* is an encapsulated bacterium that causes infant meningitis. Considering this, which vaccine would best protect infants against disease caused by this organism?
   a) Conjugate vaccine
   b) Polysaccharide vaccine
   c) Attenuated vaccine
   d) Subunit vaccine
   e) Toxoid
7. In quantifying antibodies in a patient's serum,
   a) total protein in the serum is measured.
   b) the antibody is usually measured in grams per mL.
   c) the serum is serially diluted.
   d) both antigen and antibody are diluted.
   e) the titer refers to the amount of antigen added.
8. Which of the following statements about immunological testing is *false*?
   a) Polyclonal antibody preparations recognize multiple epitopes.
   b) Monoclonal antibodies recognize a single epitope.
   c) Serum and plasma can both be tested for antibodies.
   d) The direct ELISA uses anti-human IgG antibodies.
   e) A rise in specific antibody titer indicates an active infection.
9. All of the following are matching pairs *except*
   a) ELISA—radioactive label.
   b) fluorescence-activated cell sorter—flow cytometry.
   c) fluorescent antibody test—microscopy.
   d) Western blot—gel electrophoresis.
10. Which of the following would be most useful for screening thousands of specimens for antibodies that indicate a certain disease?
    a) Western blot
    b) Fluorescent antibody
    c) ELISA
    d) All of the above
    e) None of the above

Applications

1. A new parent asks you which vaccines the CDC recommends for a 2-month-old infant. Based on the CDC's latest recommended immunization schedules found online, what is your answer?
2. In tests to determine if a patient has measles, the laboratory looks for IgM. Why would finding IgM be more significant than finding IgG?

Critical Thinking

1. For many years, the main method of HIV testing involved screening patients using the ELISA test and then confirming any positive results using the Western blot test. Why would the tests not be performed the other way around, with the Western blot first?
2. *Staphylococcus aureus* makes a protein called protein A, which binds to the Fc region of antibody molecules from a wide variety of species. How could protein A be exploited in immunoassays?

18 Immunological Disorders

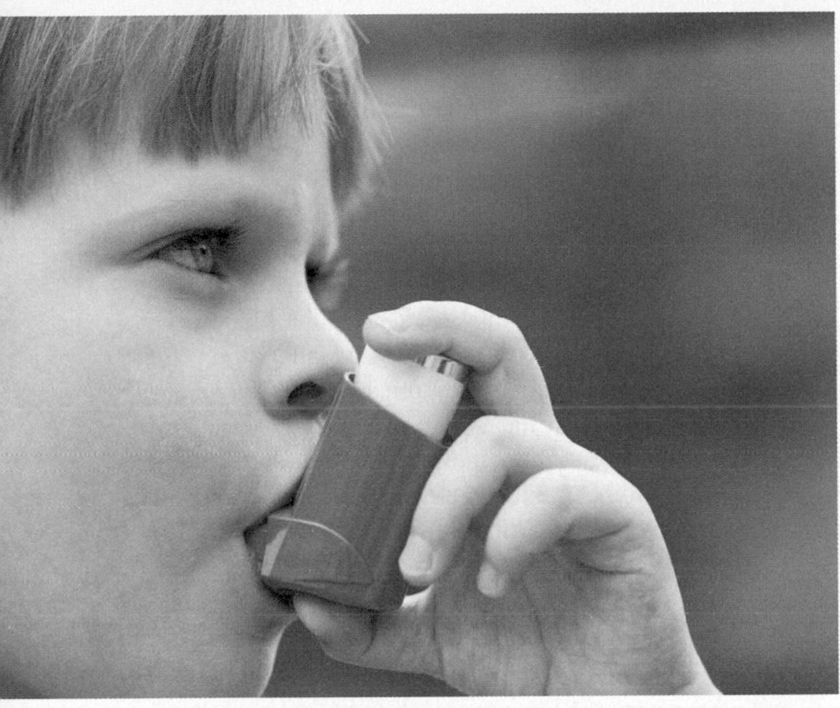
Asthmatic individual using an inhaler. *bubutu/Shutterstock*

KEY TERMS

Allergy A damaging immune response to a typically harmless environmental antigen (allergen).

Autoimmune Disease Damaging reaction of the immune system against "self" antigens.

Delayed-Type Hypersensitivity Exaggerated immune response of antigen-specific T cells.

Desensitization Allergy treatment that induces IgG production by gradual exposure to small amounts of allergen; IgG competes with IgE.

Hypersensitivity Exaggerated immune response that damages tissue.

Immune Complex Combination of antibody and soluble antigen capable of triggering the classical pathway of complement system activation.

Immunodeficiency A condition in which the immune system does not provide an adequate response, leaving a person vulnerable to infection.

Systemic Anaphylaxis Potentially life-threatening immediate hypersensitivity reaction caused by IgE attached to circulating basophils.

A Glimpse of History

Pasteur is widely quoted as saying, "Chance favors the prepared mind." This was the case when Charles Richet (1850–1935) discovered hypersensitivities, sometimes called allergies. Richet was a French physiologist who performed early experiments on toxins and the immune responses to them. He and his colleague, Paul Portier, were cruising on Prince Albert of Monaco's yacht when they hypothesized in 1901 that the Portuguese man-of-war jellyfish must produce a toxin that causes the swollen painful reactions to its stings. Prince Albert encouraged them to study the reactions, so they made an extract of the jellyfish tentacles and showed that it could indeed be toxic.

Upon returning to France, Richet and Portier continued their work by studying the effects of the toxin from a common sea anemone. When they tested it on dogs, some survived the first exposure to the potent toxin, but when given a small second dose at least 3 weeks later, the dogs died within minutes. Richet and Portier had worked with toxins earlier, so they had prepared minds. They recognized that the dogs' reactions were probably caused by an immune response. Unlike protective immune responses they had seen before, however, this one was destructive. They called the dangerous response anaphylaxis, indicating it was the opposite of phylaxis, the Greek term for protection. Richet received a Nobel Prize in 1913 for his work on anaphylaxis.

As you learned in chapters 14 and 15, the host defenses involve complex interactions between many types of cells and chemical messengers. When those interactions are coordinated and appropriate, the innate and adaptive immune responses provide excellent protection against microbial invaders. In some cases, however, the immune system fails to work appropriately, resulting in disease. This chapter covers three general categories of immune system disorders: exaggerated immune responses that damage normal host tissue (hypersensitivities), misdirected immune responses that attack normal host tissues (autoimmune diseases), and lack of adequate immune responses that leave an individual vulnerable to infection (immunodeficiencies).

18.1 ■ Hypersensitivities

Learning Outcomes

1. Compare and contrast the immunological reactions involved in types I, II, III, and IV hypersensitivities.
2. Describe two examples of each type of hypersensitivity.

The immune system does an excellent job of protecting the body from infection, but these same protective mechanisms can be harmful if uncontrolled. Just as a building's sprinkler system that protects against fire can cause significant water damage if set off by accident, inappropriate immune responses can also become destructive. An exaggerated immune response that injures tissue is called **hypersensitivity,** and it can be categorized into one of four groups according to the mechanisms and timing of the response.

Type I Hypersensitivities: Immediate IgE-Mediated

A **type I hypersensitivity** involves IgE and is commonly referred to as an allergic reaction, or **allergy.** This is a rapid, exaggerated immune response to an **allergen**—an antigen that is usually a harmless environmental substance. Although any type of hypersensitivity can technically be called an allergy, we will follow the common practice of using the term to refer only to type I hypersensitivities.

Allergic reactions begin within minutes of exposure to the allergen. A variety of common substances can trigger the reactions, including the following:

- Inhaled substances such as pollen, pet dander, or mold
- Injected substances such as insect venom or certain medications such as penicillin
- Ingested substances such as peanuts or seafood

Allergic reactions occur only in people who have been sensitized due to a previous exposure to the specific allergen. The process can be summarized as follows (**figure 18.1**):

1. **B-cell activation.** When the first contact with the allergen occurs, naive B cells that bind the antigen are activated by effector helper T cells (T_H cells).
2. **Proliferation, class switching, and differentiation.** The activated B cells begin multiplying, and some then differentiate into IgM-producing plasma cells. As other B cells continue multiplying, however, they undergo class switching (see figure 15.22). Type I hypersensitivities result from class switching that gives rise to IgE-producing plasma cells; the resulting IgE antibodies are key players in the allergic reaction.
3. **Sensitization.** Fc receptors on either mast cells or basophils bind to the Fc region of the IgE molecules, positioning the antibodies so that their antigen-binding sites are available to interact with the allergen if it is encountered again.
4. **Cross-linking of cell-bound IgE.** On subsequent exposures to the allergen, the IgE antibodies captured by mast cells and basophils bind to the antigen. As a result, adjacent captured IgE antibodies are cross-linked, which causes the cells to degranulate. Recall that mast cells and basophils contain granules holding powerful inflammatory mediators such as histamine, leukotrienes, and prostaglandins.
5. **Degranulation and release of mediators.** As a result of degranulation, the inflammatory mediators are released.
6. **Development of allergy symptoms.** The inflammatory mediators cause various allergy symptoms, including fluid accumulation (due to leaky, dilated blood vessels), itching and pain (due to effects on nerve endings), smooth muscle contraction, and increased mucus production. Note that the degranulation response that leads to allergy symptoms plays an important role in the defense against parasitic worms that are too large to be engulfed by phagocytes.

Localized Allergic Reactions

Localized allergic reactions often occur in the skin or in the respiratory tract when IgE-sensitized mast cells in the area degranulate, resulting in the release of histamine and other inflammatory mediators. Outcomes include hives, hay fever, and asthma, depending on the site of the reaction:

- **Hives (urticaria).** This is an allergic skin condition characterized by the formation of a wheal and flare, a reaction that can also be seen in positive skin tests for allergens; the wheal is a raised (swollen) itchy area surrounded by redness—the flare (**figure 18.2**). Hives can occur in response to a number of different allergens, including food, medications, and insect stings. In the case of hives associated with a food or oral medication, allergens absorbed from the intestinal tract enter the bloodstream and are then transported to the skin; the allergic reaction occurs there, resulting in the release of inflammatory mediators that cause local swelling as fluid leaks from dilated capillaries. The condition is often self-limiting, but antihistamine medications may be used to relieve symptoms by blocking binding sites for histamine.
- **Hay fever (allergic rhinitis).** This results from an immediate allergic reaction in the upper respiratory tract when pollen of ragweed or other plants is inhaled. When the pollen grains contact the mucous membranes of the nose and upper parts of the airway, they trigger the allergic reaction, resulting in the release of inflammatory mediators that dilate capillaries under the mucous membranes. This allows fluid leakage, leading to the familiar signs of hay fever: teary eyes, a runny nose, and sneezing (see figure 18.1). Allergic rhinitis can be treated with antihistamines or glucocorticoid nasal sprays, which lessen the inflammatory response.
- **Asthma.** This usually involves an immediate allergic reaction in the lower respiratory tract, resulting in the release of inflammatory mediators that interfere with breathing in two ways: (1) immediate spasms of smooth muscle tissue lining the bronchial tubes and (2) increased mucus production. At the onset of an asthma attack, people may use an inhaler to self-deliver a bronchodilator medication that relaxes the constricted muscles and helps open the airways. To prevent asthma attacks, inhalers are used that contain glucocorticoids to reduce inflammation and possibly a long-acting bronchodilator as well (see the chapter-opening photo).

Systemic Anaphylaxis

Systemic anaphylaxis is a rare but serious form of IgE-mediated allergy that can occur when an antigen enters the bloodstream and spreads throughout the body. When the antigen binds to IgE on mast cells and circulating basophils, the cells release their inflammatory mediators systemically,

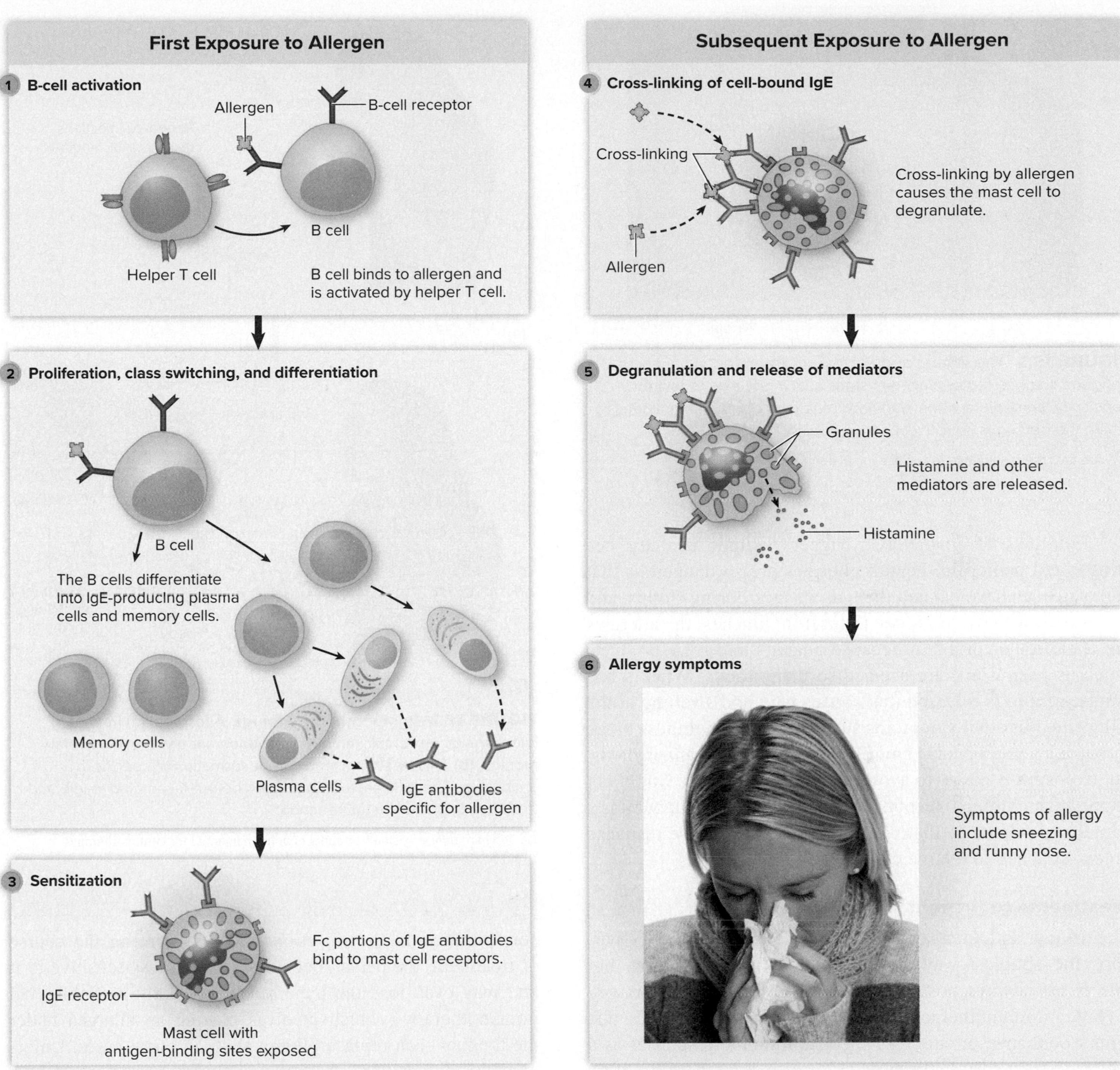

FIGURE 18.1 Type I Hypersensitivity: Immediate IgE-Mediated First exposure to an allergen can lead to sensitization; subsequent exposures can result in allergy symptoms. 6. Image Source

? Why are antihistamines sometimes used to reduce allergy symptoms?

resulting in extensive blood vessel dilation and loss of blood volume. This causes a severe drop in blood pressure that may lead to heart failure and insufficient blood flow to the brain and other vital organs, a condition termed *shock*; when this occurs as a result of an allergic reaction, it is called **anaphylactic shock.** When tissues lining the airways swell and the bronchial tubes constrict, the person may suffocate. Systemic anaphylaxis is usually controlled by immediate injection of epinephrine, which temporarily stops fluid leakage from the bloodstream and dilates the bronchial tubes. People who know they are prone to anaphylaxis often carry an auto-injector containing epinephrine that temporarily reverses the anaphylactic reaction. Otherwise, anaphylactic reactions can be fatal within minutes (see A Glimpse of History, this chapter).

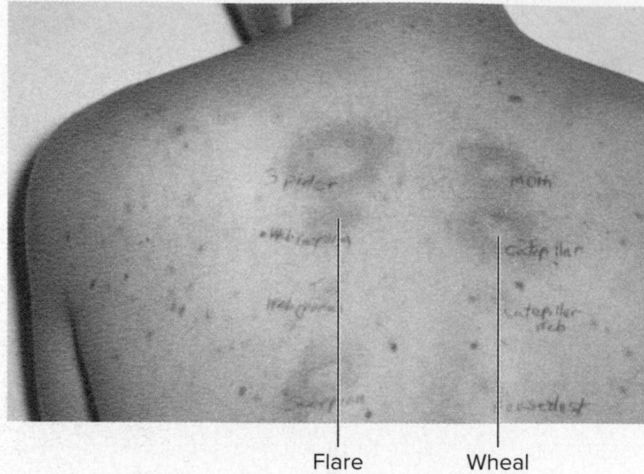

FIGURE 18.2 The Wheal and Flare Skin Reaction A variety of antigens are injected or placed in small cuts in the skin to test for sensitivity. Immediate wheal and flare reactions occur with antigens to which the person is sensitive. Dr. Frank Perlman, M.A. Parson/CDC

? Are the results shown here positive for all antigens tested? Explain.

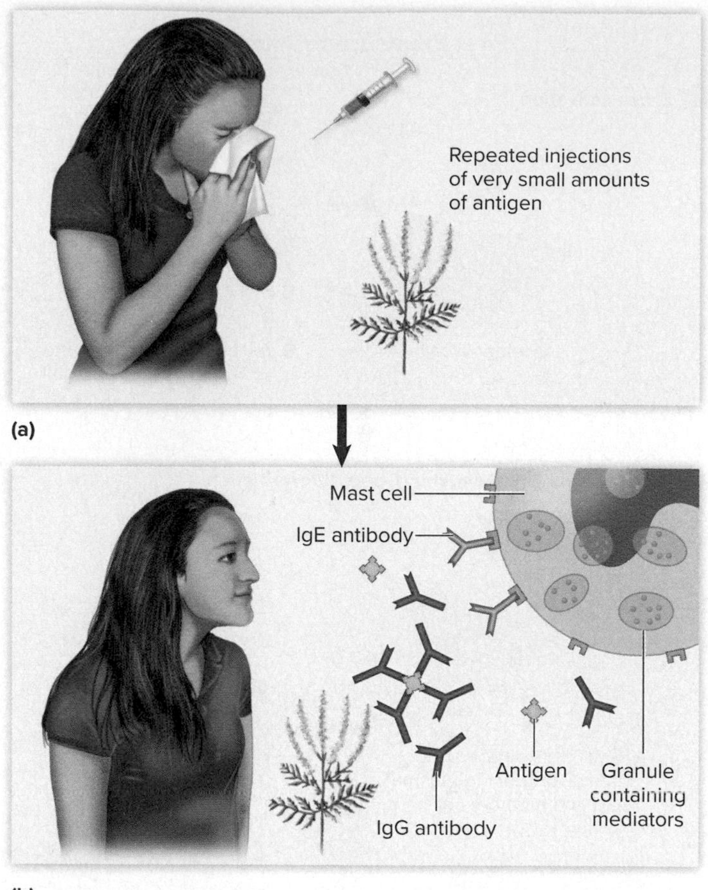

FIGURE 18.3 Immunotherapy for IgE Allergies (a) Repeated injections of very small amounts of antigen are given over several months. **(b)** This treatment leads to the formation of specific IgG antibodies. The IgG reacts with antigen before it can bind to IgE, and therefore it blocks the IgE reaction.

? Why are only very small amounts of antigen injected for immunotherapy?

Examples of anaphylaxis triggers include peanuts, bee stings, and penicillin. Peanut allergies are so dangerous that some airlines have banned them as a snack during flights, and many schools have forbidden them from lunches; the allergies are a particular problem because peanuts and their products (such as peanut oil) are used in so many foods. When a bee stings, venom is released that causes pain and swelling at the sting site, but other symptoms such as hives or tightness in the throat may also occur. Penicillin can also cause anaphylaxis, but exposure is easier to avoid. Penicillin normally would not stimulate an immune response because it is so small, but it is a hapten—a substance that does not elicit an immune response unless attached to a large carrier protein.

Treatments to Prevent Allergic Reactions

Identifying and then avoiding the relevant allergen is perhaps the simplest way to prevent an allergic reaction, but this is not always possible or convenient. Another strategy involves immunotherapy; as described in section 17.3, this aims to enhance or suppress specific immune responses as a means to treat certain diseases.

A common immunotherapy for allergies is **desensitization,** a procedure that causes the immune system to produce IgG against the allergen. The IgG antibodies probably protect the patient by binding to the offending antigen, thus coating it and facilitating its removal before it can attach to bound IgE on mast cells or basophils. Regulatory T cells may also play a role through the release of cytokines that suppress the IgE response. Desensitization therapy often involves "allergy shots"—injecting the person with the allergen in extremely dilute, but gradually increasing, concentrations over a period of several months (**figure 18.3**). The antigen must be diluted enough to avoid an anaphylactic reaction. As the antigen concentration in the injections increases during the course of treatment, the person becomes less and less sensitive to it and may even lose the hypersensitivity entirely. Sublingual immunotherapy—which involves placing an allergen under the tongue—can replace allergy shots in some cases. Unfortunately, desensitization does not always work, and it may require continued therapeutic exposure over several years to maintain effectiveness. Also, desensitization does not work well for food allergies.

Monoclonal antibodies (mAbs), described in section 17.3, may be used to treat severe asthma that does not respond well to other therapies; their use is restricted, however, because they are expensive and can have significant side effects. Examples include omalizumab, which binds specifically to the Fc portion of IgE molecules (thereby blocking the site that would otherwise attach to mast cells and basophils) and mepolizumab, which targets a specific interleukin associated with inflammatory eosinophil activity.

Type II Hypersensitivities: Cytotoxic

A **type II hypersensitivity reaction** results when antibodies bind to molecules on the surface of a normal host cell and trigger its destruction by the complement system or by antibody-dependent cellular cytotoxicity (ADCC). Because the responses destroy cells, type II reactions are called cytotoxic hypersensitivities. Hemolytic transfusion reactions, hemolytic disease of the newborn, and some autoimmune diseases involve type II reactions.

Hemolytic Transfusion Reaction

A **hemolytic transfusion reaction** occurs when a person receives a transfusion of erythrocytes (red blood cells; RBCs) with antigens different from his or her own, and it results in destruction of the cells. Human ABO blood types, for example, reflect the presence of A or B antigens on RBC membranes (**table 18.1**). People with A antigens on their RBC surfaces are blood type A; people with B antigens are blood type B; people with both antigens are blood type AB; and people with neither are blood type O. The different blood types can be determined using hemagglutination reactions to identify the antigens present (see figure 17.13). What is significant with respect to hemolytic transfusion reactions is that a person's plasma (the liquid portion of blood) contains antibodies to the A or B antigens not on that individual's own RBCs. This means that people with blood type A have anti-B antibodies; people with blood type B have anti-A antibodies; people with blood type AB have neither; and people with blood type O have both anti-A and anti-B antibodies. The antibodies are called natural antibodies because they occur without obvious pre-exposure; they generally appear within 6 months of birth, probably due to multiple exposures to bacteria, dust, and food that have components similar to blood group antigens. Because the A and B antigens are polysaccharides (T-independent antigens) the antibodies against them are mostly IgM, which cannot cross the placenta.

If a person receives a transfusion of RBCs that have ABO blood group antigens different from that person's own, natural antibodies will bind to the transfused cells, marking them for destruction. The complement system may be activated in response, resulting in the formation of membrane attack complexes that rapidly lyse the RBCs; the released hemoglobin and damaged membranes can block blood vessels and initiate clotting reactions. In addition, activated complement components trigger an inflammatory response that attracts cytokine-releasing phagocytes. The overall result includes kidney damage, shock, and disseminated intravascular coagulation (DIC), a condition in which clots form in small blood vessels, leading to failure of vital organs.

An RBC surface protein called RhD can also be involved in hemolytic transfusion reactions similar to those described for the ABO group. A person who has RhD is referred to as Rh-positive; someone without it is Rh-negative. In contrast to the ABO system, however, Rh-negative individuals do not have natural antibodies against the RhD antigen; the antibodies develop only when an Rh-negative individual is exposed to Rh-positive cells.

Because of the potentially life-threatening consequences of giving the wrong blood to a patient, extensive testing of the patient blood is done before a transfusion to determine not only the type, but also the presence of antibodies that could cause hemolysis of transfused blood. The donor blood is also typed. Just before the transfusion, the donor and recipient blood are cross-matched. This is done by mixing the cells from the donated blood with the serum of the recipient. Any clumping indicates an incompatibility between the two blood samples in any of the many antigens on the RBC surface; if clumping occurs, further testing may then be done, but in most cases the blood would not be used in the transfusion.

Hemolytic Disease of the Newborn

Like hemolytic transfusion reactions, **hemolytic disease of the newborn (HDN)** involves the destruction of RBCs, but it stems from an entirely different set of circumstances and usually involves RhD. If an Rh-negative female carries an Rh-positive fetus, she is likely to develop antibodies to the RhD antigen if fetal blood enters her circulation, which often occurs to at least some extent immediately before or during childbirth. Fetal blood can also enter maternal circulation during pregnancy or after a miscarriage or an induced abortion. The maternal antibodies will usually not affect the woman's

TABLE 18.1 Antigens and Antibodies in Human ABO Blood Groups

			Incidence of Blood Type in United States			
Blood Type	**Antigen on RBC**	**Antibody in Plasma**	**Among Asians**	**Among Blacks**	**Among Caucasians**	**Among Hispanics**
A	A	Anti-B	28%	26%	40%	31%
B	B	Anti-A	25%	19%	11%	10%
AB	A and B	Neither anti-A nor anti-B	7%	4%	4%	2%
O	Neither	Both anti-A and anti-B	40%	51%	45%	57%

first baby because IgM, which is produced upon first exposure to an antigen (including RhD), cannot cross the placenta. If the woman carries a second Rh-positive fetus, however, her anti-RhD IgG antibodies can cross the placenta and damage the RBCs of the developing fetus. The affected fetus responds by producing immature red blood cells called erythroblasts, so the disease is also called erythroblastosis fetalis. The fetus is monitored to determine the degree of anemia, and in utero transfusions of Rh-negative blood are given if needed. HDN generally becomes most severe soon after birth. Immediate treatment is sometimes required to correct anemia and prevent permanent brain damage.

HDN is rare today because of a medication called RhoGAM, which contains anti-RhD antibodies; Rh-negative women carrying an Rh-positive fetus are injected with RhoGAM during pregnancy and again shortly after delivery (**figure 18.4**). The anti-RhD antibodies bind to any Rh-positive RBCs that may have entered the mother's circulation from the

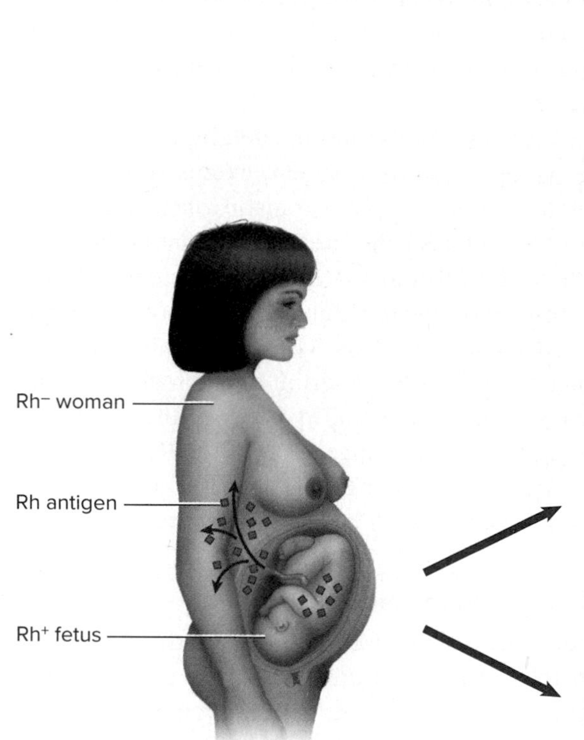

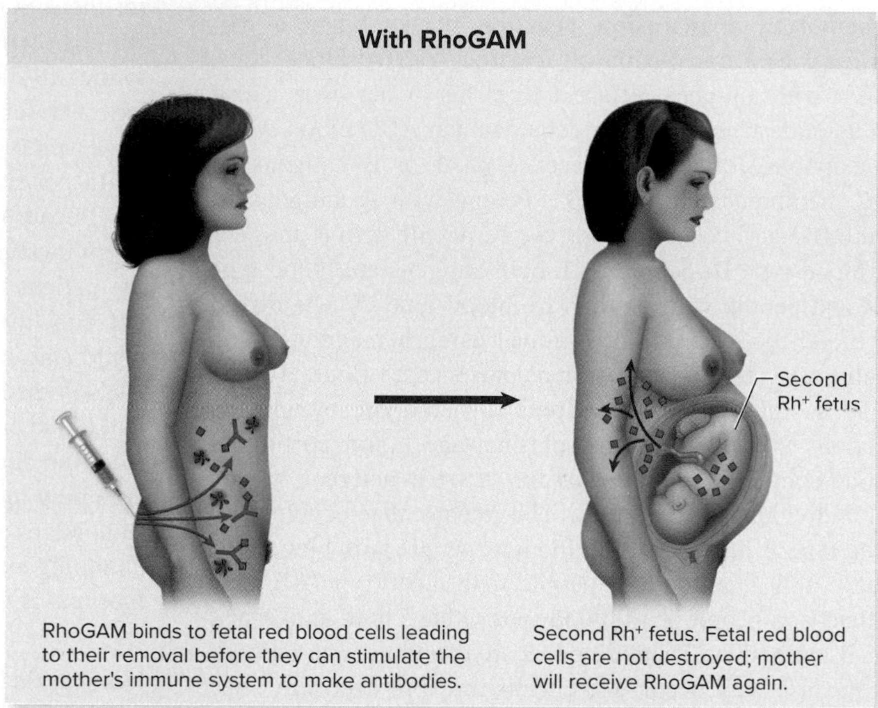

With RhoGAM
Second Rh⁺ fetus
RhoGAM binds to fetal red blood cells leading to their removal before they can stimulate the mother's immune system to make antibodies.
Second Rh⁺ fetus. Fetal red blood cells are not destroyed; mother will receive RhoGAM again.

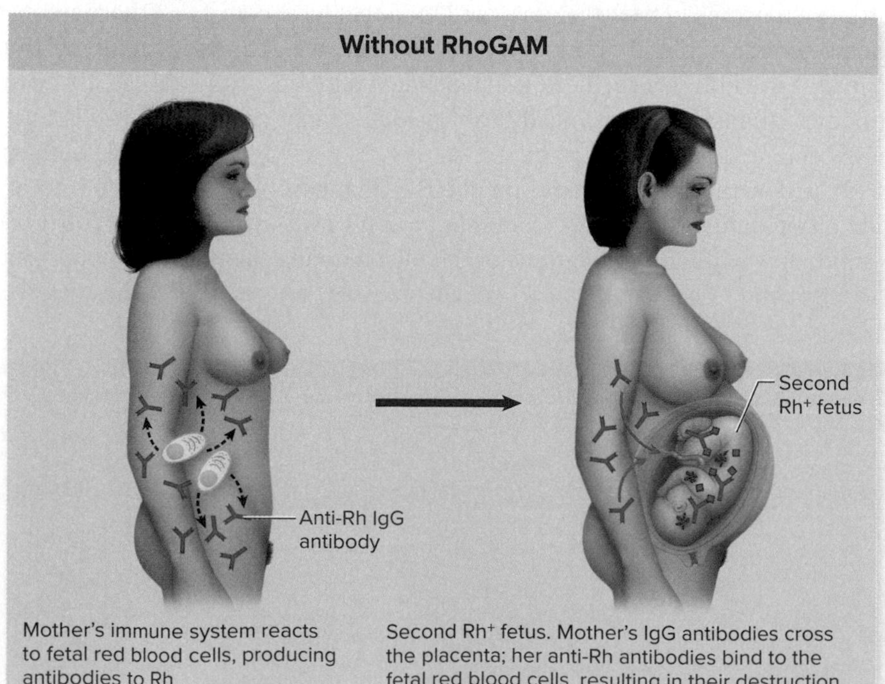

Without RhoGAM
Second Rh⁺ fetus
Anti-Rh IgG antibody
Mother's immune system reacts to fetal red blood cells, producing antibodies to Rh.
Second Rh⁺ fetus. Mother's IgG antibodies cross the placenta; her anti-Rh antibodies bind to the fetal red blood cells, resulting in their destruction.

FIGURE 18.4 Hemolytic Disease of the Newborn (HDN) RhoGAM contains antibodies that bind to Rh antigens on fetal RBCs, preventing them from stimulating a primary immune response in the mother.

? Would a mother need RhoGAM again with a third Rh⁺ fetus? Explain.

fetus, thereby preventing those RBCs from stimulating a primary immune response in the mother. If the mother does not receive RhoGam in a timely manner and has already formed memory cells, injecting anti-RhD antibody is not effective.

Type III Hypersensitivities: Immune Complex–Mediated

Type III hypersensitivities are due to small **immune complexes** (also referred to as antibody-antigen complexes) that form when the two arms of an antibody bind separate but identical soluble antigen molecules, linking them. The size of immune complexes depends on the relative proportion of antibody and antigen molecules, as described in the discussion of in vitro precipitation reactions (see figure 17.15). If the complexes are large, phagocytes quickly engulf them, removing them. However, if the complexes are small—which occurs when there is more antigen than antibody—they are less likely to be engulfed by phagocytes, so they remain in tissues where they were formed or circulate in the bloodstream. Circulating immune complexes can become trapped in the walls of small blood vessels in various tissues including the skin, joints, or in organs such as kidneys. Once stuck, they can activate the complement system, causing inflammation that damages the surrounding tissue (**figure 18.5**). When the complexes are deposited in the kidneys, they can cause glomerulonephritis (inflammation in the kidney glomeruli, the tufts of tiny blood vessels where blood is filtered for urine formation; see figure 21.7). The complexes are

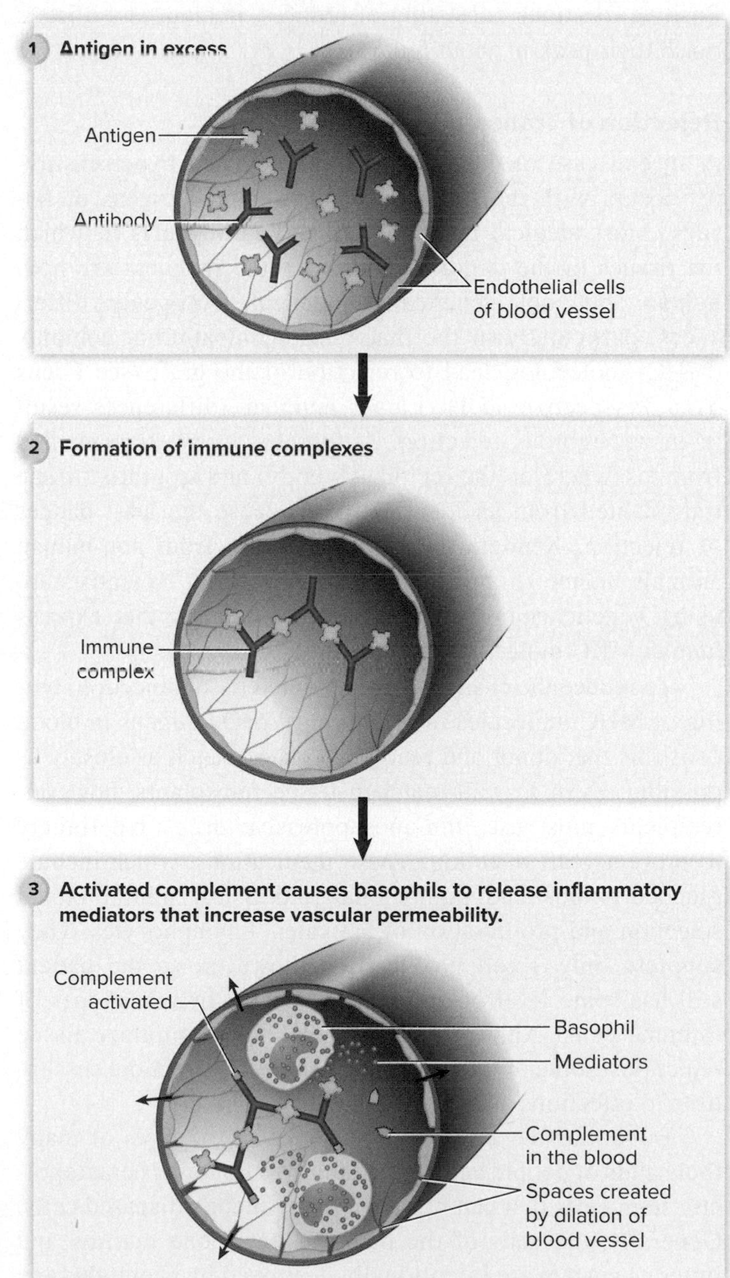

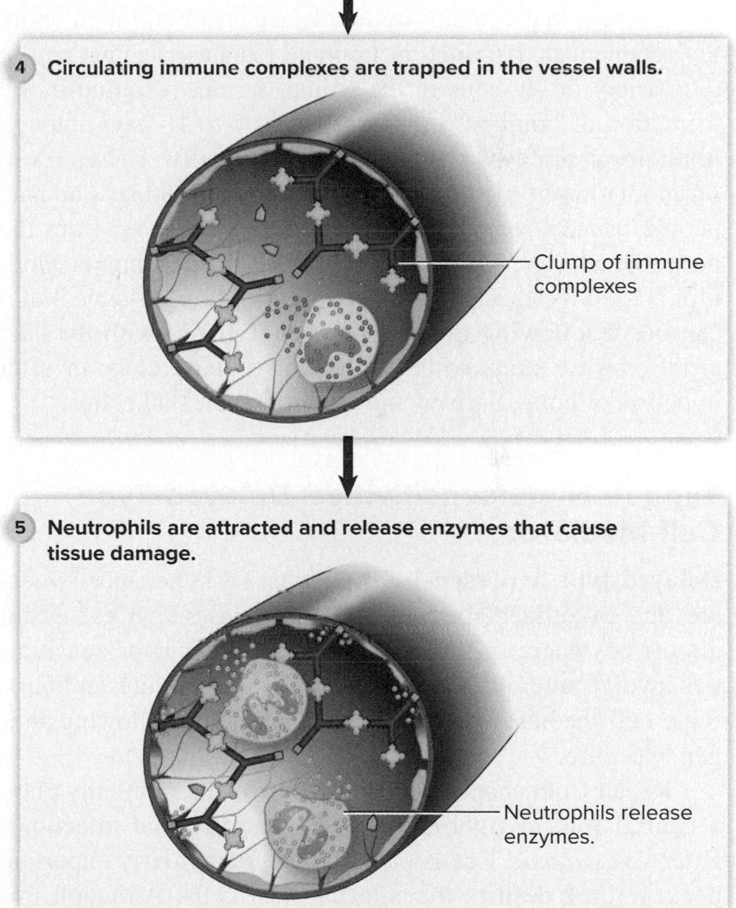

FIGURE 18.5 Type III Hypersensitivity: Immune Complex–Mediated

? Why do immune complexes form only when an antigen is soluble?

also responsible for the rashes, joint pains, and other symptoms seen in certain diseases, such as infective endocarditis and the autoimmune diseases systemic lupus erythematosus and rheumatoid arthritis.

The **Arthus reaction** is a localized immune complex reaction. If antigen is injected into a previously immunized person who already has high levels of circulating specific antibody—for example, if the tetanus-diphtheria booster vaccine is given to a person too frequently—immune complexes can form at the site of injection where antigen levels are high. The immune complexes form in skin tissues, where they activate complement, resulting in a local inflammatory response characterized by severe pain, swelling, and redness that peaks in 6 to 12 hours.

Serum sickness is a systemic immune complex disease. It can occur following injection of antibodies from a non-human animal into a person to provide immediate (but short-term) passive immunity. For example, if botulism is treated using horse serum that contains antibodies against botulinum toxin, the recipient may mount an immune response against any of the variety of antigens in the animal serum, resulting in the formation of immune complexes. After 7 to 10 days, enough immune complexes circulate to potentially cause fever, inflammation of blood vessels, arthritis, and kidney damage; people usually recover as the antigens are cleared from the body. Serum sickness is rare now because hyperimmune globulin is used instead of animal serum whenever possible, but it can occur following treatment of heart attacks with the bacterial enzyme streptokinase to dissolve blood clots, or after injection of horse-derived antivenom to treat snakebites.

Type IV Hypersensitivities: Delayed-Type Cell-Mediated

Delayed-type hypersensitivities (type IV hypersensitivities) are due to antigen-specific T-cell responses and can occur almost anywhere in the body. Whereas other hypersensitivities involve antibodies and can occur within minutes or hours, these cell-mediated reactions peak 2 to 3 days following antigen exposure.

Recall from chapter 15 that cell-mediated immunity plays a central role in fighting intracellular microbial infections. Effector cytotoxic T cells (T_C cells) are particularly important because they destroy the infected host cells. Although this prevents the infection from spreading, it also damages tissues. With chronic infections, delayed-type hypersensitivity causes extensive host cell destruction and progressive loss of tissue function. These reactions can also be seen in some autoimmune diseases and in rejection of transplanted tissue.

Contact Hypersensitivities

Contact hypersensitivity, also called **allergic contact dermatitis,** is due to the effector T-cell response to small molecules that penetrate intact skin and act as sensitizing agents. Substances involved include the nickel of metal jewelry, the chromium salts in certain leather products, components of some cosmetics, latex in products like gloves or condoms, and oils on the surface of poison ivy and poison oak leaves (**figure 18.6**). All these materials shed small chemicals that act as haptens and combine chemically with proteins in the skin. Dendritic cells take up these hapten:protein complexes, process them, and then present the resulting hapten:peptides to T cells in nearby lymph nodes. The resulting activated T cells form memory cells, sensitizing the individual to the substance. Once the person has been sensitized, a second exposure causes activation of memory T cells and a damaging cell-mediated response, resulting in irritating rashes and sometimes blisters. The substance causing a contact hypersensitivity is commonly identified by patch tests, in which suspect substances are applied to the skin under a bandage. Positive reactions consisting of redness, itching, and blisters reach their peak in about 3 days.

Rejection of Transplanted Tissues

A special case of delayed-type cell-mediated hypersensitivity occurs with the rejection of transplanted organs or tissues. Most medical transplants involve **allografts** in which the tissues of the donor and those of the recipient are both human, but not genetically identical. Antigenic differences, particularly in the major histocompatibility complex (MHC) molecules, lead to rejection of the graft (see Focus Your Perspective 15.1). Larger antigenic differences result in more vigorous rejection. Autografts (tissue transplanted from elsewhere on the recipient's body) and isografts (tissue transplanted from an identical twin) cause the least danger of rejection. Xenografts involving tissues from non-human animals induce strong immune responses, but scientists are using genetic engineering to develop animals that express human MHC molecules.

To reduce the chance of transplanted tissue rejection, typing of MHC molecules in tissues and ABO antigens in blood confirms that donor and recipient tissues match as closely as possible. Even in well-matched tissue transplants, however, recipients must use immunosuppressive drugs indefinitely to prevent graft rejection. These medications, which include glucocorticoids and monoclonal antibodies, inhibit clonal selection and proliferation of activated T lymphocytes. They suppress only T-cell proliferation, however, so the patient still has some level of immunity mediated by other types of immune cells. Although these medications minimize tissue rejections, some of them also make the recipient more susceptible to infections and cancer.

Bone marrow transplants have saved the lives of many thousands of people because the marrow contains hematopoietic stem cells that can replace the recipient's diseased cells. Generally, the cells of the patient's own bone marrow and immune system are intentionally destroyed by chemotherapy

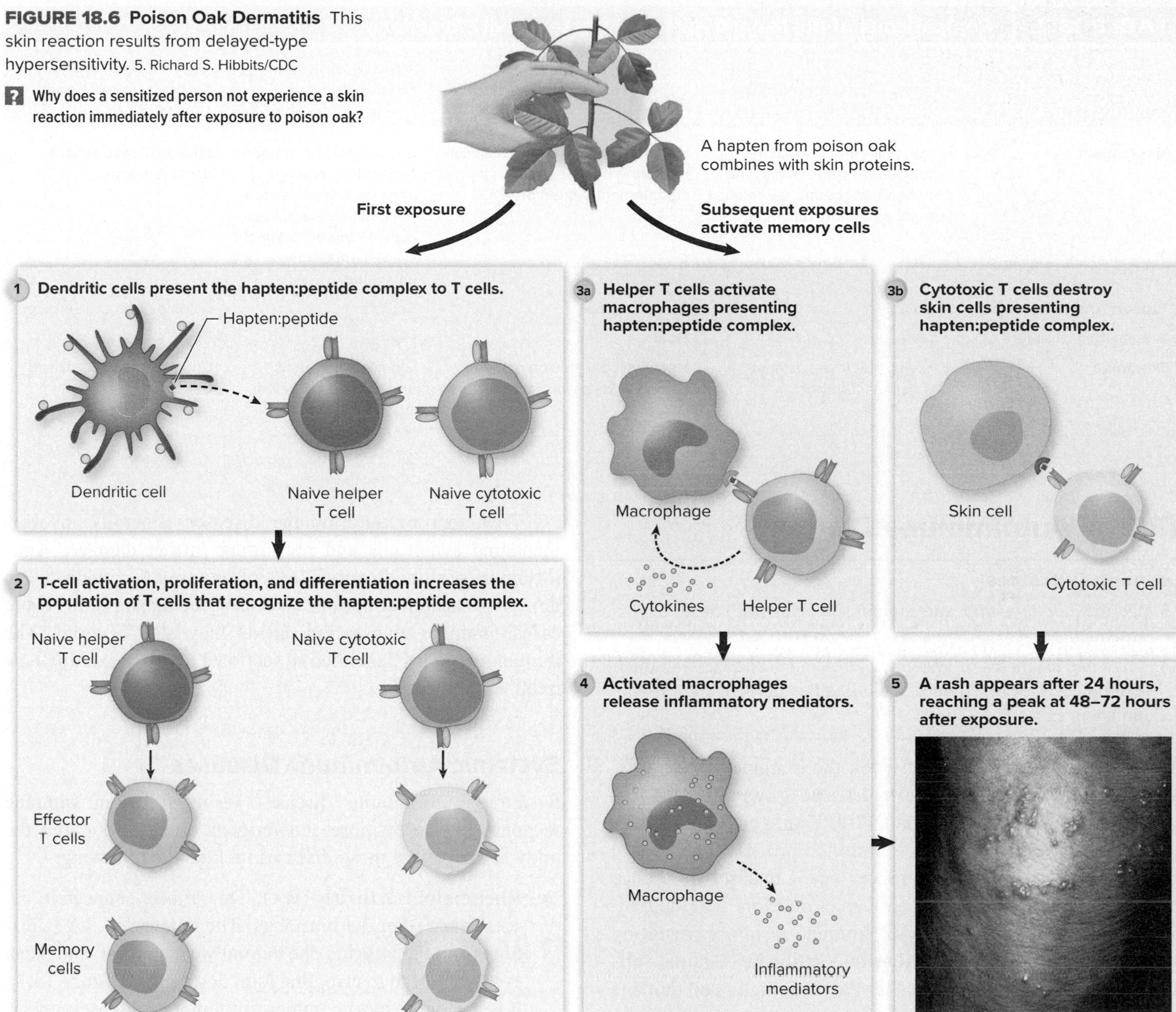
A hapten from poison oak combines with skin proteins.
First exposure
Subsequent exposures activate memory cells
1 Dendritic cells present the hapten:peptide complex to T cells.
Hapten:peptide
Dendritic cell
Naive helper T cell
Naive cytotoxic T cell
2 T-cell activation, proliferation, and differentiation increases the population of T cells that recognize the hapten:peptide complex.
Naive helper T cell
Naive cytotoxic T cell
Effector T cells
Memory cells
3a Helper T cells activate macrophages presenting hapten:peptide complex.
Macrophage
Cytokines
Helper T cell
3b Cytotoxic T cells destroy skin cells presenting hapten:peptide complex.
Skin cell
Cytotoxic T cell
4 Activated macrophages release inflammatory mediators.
Macrophage
Inflammatory mediators
5 A rash appears after 24 hours, reaching a peak at 48–72 hours after exposure.

FIGURE 18.6 Poison Oak Dermatitis This skin reaction results from delayed-type hypersensitivity. 5. Richard S. Hibbits/CDC

Why does a sensitized person not experience a skin reaction immediately after exposure to poison oak?

and possibly radiation before the transplant. Donor bone marrow cells are then infused through the subclavian vein under the left collarbone (clavicle). The hematopoietic stem cells circulate until they reach the recipient's bone marrow, where they leave the bloodstream and begin multiplying. Close monitoring is required for months after a transplant because the recipient is at risk of infection and bleeding while the grafted marrow restores the supply of red and white blood cells and platelets. Recovery may be complicated by a graft-versus-host reaction in which the graft-derived immune system recognizes the patient's own tissues as foreign and attacks them.

Table 18.2 compares the major types of hypersensitivity reactions.

MicroAssessment 18.1

Hypersensitivities are exaggerated immune reactions that cause tissue damage. Type I hypersensitivities, commonly called allergies, occur immediately after exposure to antigen. Type II hypersensitivities result in cell destruction. Type III hypersensitivities arise when immune complexes in the tissues activate the complement system, triggering an inflammatory response. Type IV delayed-type hypersensitivities are due to the actions of effector T cells.

1. Describe the process of desensitization that reduces the IgE response to an allergen.
2. Describe the events that result in a rash after exposure to poison ivy.
3. Why is the recipient of a kidney transplant unlikely to suffer from a graft-versus-host reaction?

TABLE 18.2 Comparisons of the Major Types of Hypersensitivity Reactions

Characteristic	Type I Hypersensitivity: Immediate IgE-Mediated	Type II Hypersensitivity: Cytotoxic	Type III Hypersensitivity: Immune Complex–Mediated	Type IV Hypersensitivity: Delayed-Type Cell-Mediated
Mechanism	IgE molecules carried by basophils or mast cells bind antigen, triggering the release of inflammatory mediators	IgG or IgM binds normal host cells, triggering their destruction by the complement system or ADCC	IgG or IgM bind antigen molecules, forming immune complexes that deposit in tissues; these activate the complement system, triggering inflammation	Effector T cells actions damage tissues
Type of Antigen	Soluble	On cell surface	Soluble	Soluble or on cell surface
Onset After Exposure	Minutes	Hours	Hours	Days
Examples	Hives, hay fever, asthma, anaphylactic shock	Hemolytic transfusion reaction, hemolytic disease of the newborn	Glomerulonephritis, arthus reaction, serum sickness	Contact dermatitis, tissue transplant rejection

18.2 ■ Autoimmune Disease

Learning Outcomes

3. Define *autoimmunity,* and explain some possible causes and treatments.
4. Give two examples of systemic autoimmune diseases and three examples of localized ones, including the mechanism of tissue injury in each.

Autoimmune disease results when the immune system mistakenly attacks the body's normal tissues. Lymphocytes that could react against autoantigens ("self" antigens) are usually eliminated or silenced by the combined processes of central tolerance and peripheral tolerance, but if that self-tolerance is faulty, an autoimmune disease may develop. The autoantigens may stimulate a humoral immune response, resulting in the production of **autoantibodies** (antibodies against self-antigens) or they may activate T cells, resulting in cell damage and inflammation in the body's tissues. Sometimes autoantigens do both. Autoimmune diseases can be organ-specific or systemic, depending on whether the autoantigens are localized to an organ or found throughout the body.

The cause of an autoimmune disease likely involves multiple factors, but these are not always clear. Genetic factors may be involved because the diseases are often associated with certain types of major histocompatibility complex (MHC) molecules (see Focus Your Perspective 15.1). Environmental factors, including certain infections, may also play a role. For example, a pathogen may cover itself with molecules that resemble those on host tissues in order to evade the immune system. An immune response against that pathogen may inappropriately attack healthy tissues as well, causing inflammation and damage even in the absence of the pathogen. In other cases, the action or control of T_{reg} cells may be deficient.

Treatment of autoimmune diseases generally involves managing symptoms and preventing further damage. Anti-inflammatory medications are often used, in addition to non-pharmaceutical options such as reducing stress. In severe cases, immune-suppressing agents may also be used. The immunotherapies described in section 17.3 offer exciting new treatment options.

Systemic Autoimmune Diseases

Systemic autoimmune diseases result from an immune response to one or more autoantigens found throughout the body. Examples of these diseases include the following:

- **Rheumatoid arthritis (RA).** This affects connective tissues, most often within joints. The damage of RA is distinct from the arthritis due to joint wear and tear, and it can eventually lead to crippling joint destruction (**figure 18.7**). It is one of the most common autoimmune diseases, occurring most often in women ages 30 to 50, although it can begin at any age. As is the case with several other autoimmune diseases, RA is characterized by symptomatic episodes often referred to as flare-ups or simply flares (a term that has no relation to the wheal and flare seen in hives). Multiple mechanisms of damage are involved, but T cells appear to be an important factor; they accumulate in the joints and then release inflammatory cytokines when stimulated by specific antigens. In addition, autoantibodies to soluble substances in the joint fluids form immune complexes characteristic of type III hypersensitivities, causing further damage. Several of the medications used to prevent further RA damage target the cytokines that would otherwise trigger an inflammatory response.
- **Systemic lupus erythematosus (SLE).** This can affect many different body systems, resulting in a wide range

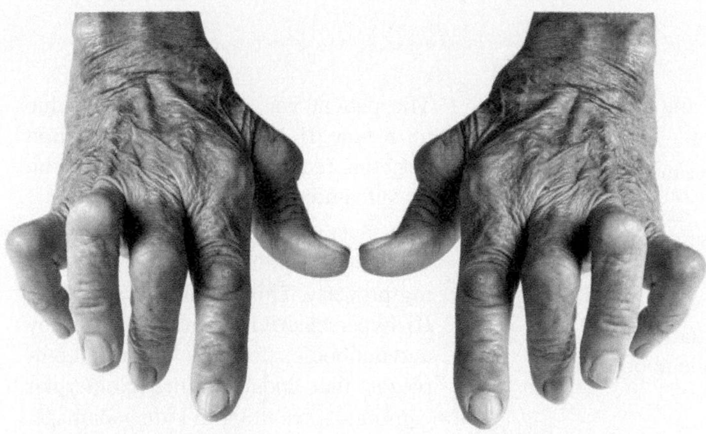

FIGURE 18.7 Rheumatoid Arthritis Autoimmune reactions cause chronic inflammation and destruction of the joints. chaowalit407/iStock/Getty Images

? **How do autoantibodies contribute to damage to the joints in these hands?**

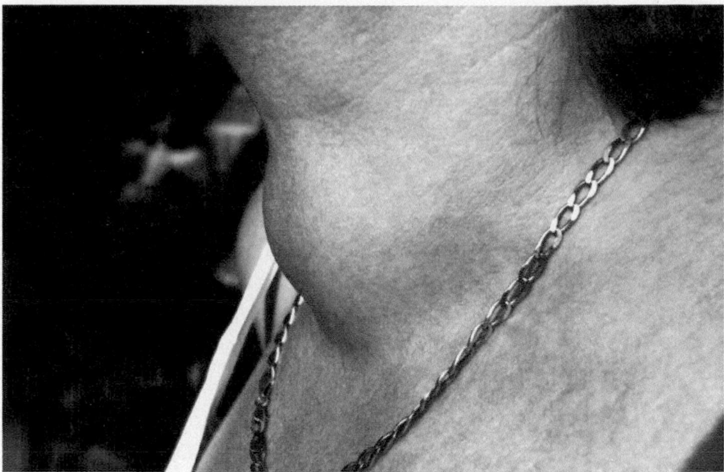

FIGURE 18.8 Goiter A goiter is an enlarged thyroid gland that may result from overstimulation of the gland. Chris Pancewicz/Alamy Stock Photo

? **What receptors in the thyroid gland are stimulated in Graves' disease?**

of possible signs and symptoms; it mostly affects women of childbearing age. As with rheumatoid arthritis, multiple mechanisms are likely involved, but a significant factor is the formation of immune complexes in different tissues. The disease is also characterized by a variety of destructive autoantibodies that target DNA and other molecules found in cell nuclei. The signs and symptoms of SLE vary considerably but typically include fatigue, fever, and rashes, along with joint pain and swelling. Flares are often characteristic. In severe cases, kidneys or other organs may be damaged. Treatment of SLE varies based on disease manifestations and severity but is aimed at preventing flares, in addition to managing symptoms and reducing organ damage.

Organ-Specific Autoimmune Diseases

Organ-specific autoimmune diseases result from autoantigens found only within a specific organ. Examples of the diseases include the following:

- **Graves' disease.** This causes an over-activation of the thyroid gland (hyperthyroidism) and results from an autoantibody that binds to the gland's receptor for thyroid-stimulating hormone (TSH). The binding mimics that of TSH, resulting in the continuous production of thyroid hormones. Metabolic rate increases in affected individuals, resulting in weight loss, fatigue, irritability, heat intolerance, and rapid heartbeat. Bulging eyes and a goiter (enlarged thyroid gland) may develop (**figure 18.8**). Effective treatment options include: (1) medications that block thyroid hormone synthesis and release, (2) destruction of thyroid tissue using radioactive iodine, and (3) surgical removal of the thyroid gland. The latter two are usually followed by lifelong treatment with replacement thyroid hormone.
- **Myasthenia gravis (MG).** The name means "grave muscle weakness" and reflects the fact that this disease involves a disruption in nerve impulse transmission to muscles. Normally, a muscle cell contracts when it receives an electrical signal (the nerve impulse) from a nearby nerve cell. The two cells do not touch; instead the impulse passes from the nerve cell to the muscle cell at the neuromuscular junction (where the two cells meet) by means of a chemical called acetylcholine. That chemical binds to receptors on the muscle cell membrane, initiating muscle cell contraction. In MG, autoantibodies bind to the acetylcholine receptors, blocking the nerve impulses. The binding may also activate the complement system, which then damages membranes at the junction. Medications that inhibit the enzyme cholinesterase (an enzyme that degrades acetylcholine) lessen the effects of the diseases by allowing acetylcholine to accumulate at the neuromuscular junction. Other treatments are used for severe or rapidly worsening cases, particularly when muscles that control swallowing or breathing are affected.
- **Multiple sclerosis (MS).** This affects the central nervous system, specifically the protective layer (myelin sheath) that surrounds certain nerves. The inflammatory damage associated with the disease compromises nerve impulse conduction, leading to fatigue and weakness, and in some people, eventual complete paralysis. As with several other autoimmune diseases, MS is more common in women. Most types of MS are characterized by flares that occur months or years apart, but some are slowly progressing without flares. Disease management involves treating or relieving symptoms during flares, and slowing disease progression.

FOCUS ON A CASE 18.1

A 28-year-old female reported to her physician that she had no energy and was tired all the time. She had a slight fever and could not shake a cough that was disturbing her sleep at night. Her joints ached, and her hands and feet were often swollen. The patient has two older sisters. One suffers from Graves' disease, as did their mother, and the other has rheumatoid arthritis.

The patient's white blood cell (WBC) count was normal, but a differential WBC count revealed 18% lymphocytes (the normal range is 25–33%). Her red blood cell count was 3.5 million/mm^3, which is below the normal range of 4–6 million/mm^3. The levels of complement proteins in her blood were below normal, but she had significant protein levels in her urine. A positive antinuclear antibody (ANA) test indicated the presence of autoimmune disease.

After reviewing the patient's history as well as her physical exam and test results, the physician diagnosed her as having systemic lupus erythematosus (SLE). The patient was given a prescription for prednisone, an immunosuppressant; a flu shot and a vaccination against pneumococcal pneumonia was ordered.

1. What evidence led the physician to diagnose SLE?
2. How might SLE have caused the fatigue in this patient?
3. What is a likely cause of the abnormal protein levels in the urine and blood of this patient?
4. Why is it important that this patient get a flu shot and a pneumococcal vaccination?

Discussion

1. Systemic lupus erythematosus (SLE) is a widespread disorder of connective tissue that may affect many different organs. It is difficult to diagnose because it mimics a number of other conditions. The ANA test is not specific for lupus, but a positive test is seen in 97% of those with the disorder. When combined with physical symptoms such as joint pain, and a family history of autoimmune disorders, such as rheumatoid arthritis and Graves' disease, the results are more reliable.
1. The patient was anemic, possibly due to a type II hypersensitivity reaction targeting red blood cells for destruction by autoantibodies.
2. Elevated protein levels in the urine indicate that the kidneys are not functioning properly. This could indicate a type III hypersensitivity reaction caused by autoantibodies forming immune complexes that lodge in the glomerular capillaries, resulting in kidney damage. Low levels of complement protein in the blood are consistent with systemic inflammation caused by SLE.
3. The differential WBC count indicates that the immune system could also be targeting lymphocytes for destruction, which would make the patient more vulnerable to infections. Moreover, prednisone further suppresses immunity. The patient's cough indicated that she may already have some fluid in her lungs. The vaccinations will minimize the patient's risk for common respiratory infections.

- **Type 1 diabetes mellitus.** Once called insulin-dependent or juvenile diabetes, this is caused by the destruction of pancreatic β cells by cytotoxic T cells. Pancreatic β cells produce insulin, a hormone that allows cells to take up glucose from the bloodstream. Lack of insulin increases blood glucose levels, drawing water from the cells and leading to increased thirst and urination. Because the cells do not take in glucose that would be used as their energy source, symptoms also include extreme hunger, fatigue, and weight loss. People with diabetes are vulnerable to high blood pressure, stroke, blindness, kidney disease, and conditions that lead to the need for limb amputation. Insulin injections are used to treat the disease.

Characteristics of these autoimmune diseases are summarized in **table 18.3.**

TABLE 18.3 Characteristics of Some Autoimmune Diseases

Disease	Specificity	Major Mechanism of Tissue Damage
Systemic		
Rheumatoid arthritis	Systemic, especially joints	T-cell-directed inflammatory destruction of joint tissues; immune complexes also involved
Systemic lupus erythematosus	Systemic	Immune complexes deposited in multiple tissues; autoantibodies target DNA and other nuclear components
Organ-Specific		
Graves' disease	Thyroid	Autoantibodies bind thyroid-stimulating hormone receptors, overstimulating the thyroid
Multiple sclerosis	Brain and spinal cord	Inflammatory response in the central nervous system damages the myelin sheath of nerve cells, interfering with nerve impulse conduction
Myasthenia gravis	Muscle	Autoantibodies bind to acetylcholine receptors on muscle cell membranes, preventing muscle contraction
Type 1 diabetes mellitus	Pancreas	T-cell destruction of pancreatic β cells, preventing insulin production

MicroAssessment 18.2

Autoimmune disease can result when the immune system attacks autoantigens. Some autoimmune diseases are systemic, but others are organ-specific. Treatment often involves anti-inflammatory medications, but immune-suppressing medications may be used for severe cases.

4. What might make an autoimmune reaction general rather than tissue-specific?
5. Describe the three treatments used for Graves' disease.
6. How could viruses or bacteria potentially trigger autoimmune disease?

18.3 ■ Immunodeficiency Disorders

Learning Outcomes

5. Compare and contrast primary and secondary immunodeficiencies.
6. Explain how immunodeficiency can lead to multiple and unusual infections.

In contrast to hypersensitivities and autoimmune diseases, which arise from an exaggerated or inappropriate immune response, **immunodeficiencies** develop when the body cannot initiate or sustain an immune response. People with immunodeficiencies tend to have repeated infections; the types of infections often depend on which part of the immune system is affected. For example, a B-cell deficiency will result in less antibody production and more infections by bacteria that travel in the bloodstream.

Primary Immunodeficiencies

Primary immunodeficiencies, which are present from birth, are generally rare. They can occur at any point in the many steps that lead to an effective immune response and may affect B cells, T cells, natural killer (NK) cells, phagocytes, or complement components. Scientists know many of the gene defects that cause the deficiencies, and work is underway to correct them. Examples of primary immunodeficiencies include the following:

- **Antibody deficiencies.** One of the most common primary immunodeficiencies is selective IgA deficiency, in which very little or no IgA is produced. People with this disorder may appear healthy, but they often develop repeated infections of the respiratory, gastrointestinal, and genitourinary tracts, where secretory IgA normally protects against colonization by pathogens. Another antibody deficiency is agammaglobulinemia, a disease in which few or no antibodies are produced.
- **Lymphocyte deficiencies.** Two examples of lymphocyte deficiencies are severe combined immunodeficiency (SCID) and DiGeorge syndrome. SCID is a group of rare disorders that result from mutations in genes involved in lymphocyte differentiation and proliferation. Infants with SCID have few functioning T and B cells, and sometimes few NK cells as well; they often die of infectious disease before their first birthday unless successfully treated. Until recently, the only cure was a stem cell transplant, a procedure that uses hematopoietic (blood-forming) stem cells from a healthy donor to replace the patient's own stem cells. The donor stem cells populate the patient's bone marrow and give rise to functioning lymphocytes (see figure 14.5). Recently, however, scientists successfully used gene therapy to restore immunity in children with certain types of SCID. The patient's own hematopoietic stem cells were modified in vitro, using a lentivirus (a type of retrovirus) to insert a normal copy of the defective gene, and the modified cells were then introduced into the patient. DiGeorge syndrome results when the thymus fails to develop, so cell-mediated immunity is deficient. Affected individuals have other developmental defects as well, such as heart and blood vessel abnormalities, and the inability to maintain proper levels of calcium and phosphorus. Those with severe thymic deficiency may be treated with a transplant of thymus tissue. As expected from a lack of T cells, affected people are susceptible to a variety otherwise uncommon infections.
- **Defects in phagocytic cells.** Chronic granulomatous disease (CGD) is caused by a defect that results in the failure of phagocytes to produce reactive oxygen species (ROS) within a phagolysosome. Surprisingly, these patients can mount an immune response to most types of infections, but they cannot respond effectively to a few common pathogens such as *Staphylococcus aureus* and molds of the genus *Aspergillus*. Patients also produce an abundance of non-functional phagocytes, forming granulomas. Treatments for CGD include using antibiotics to reduce the chance of infection and administering interferons to boost the immune response.
- **Defects in complement system components.** People with deficiencies in the early components of the complement system (such as C1 and C2) may develop immune complex diseases because these components normally help clear circulating complexes. People who lack late components (C5, C6, C7, C8) have recurrent *Neisseria* infections; typically *Neisseria* cells are destroyed by membrane attack complexes. People who lack certain important regulatory proteins of the complement system experience uncontrolled complement activation, resulting in fluid accumulation and potentially fatal tissue swelling, a condition called hereditary angioedema.

MicroByte

SCID is sometimes called "bubble boy disease" after David Vetter, who survived with this condition for 12 years in an inflated sterile plastic bubble.

Secondary Immunodeficiencies

Secondary, or acquired, immunodeficiency diseases develop after birth. For example, cancers involving the lymphatic system often decrease effective antibody-mediated immunity. Multiple myeloma is a cancer arising from a single plasma cell that proliferates out of control and produces large quantities of a single specificity of antibody—an overproduction that occurs at the expense of others needed to fight infection.

Some viral infections deplete certain cells of the immune system, resulting in secondary immunodeficiency. AIDS (acquired immunodeficiency syndrome), caused by human immunodeficiency virus (HIV), is one of the most serious of these; AIDS occurs as result of HIV-associated destruction of helper T cells, leaving a person highly susceptible to various opportunistic infections (see table 27.14). An important cause of a temporary immunodeficiency is the measles virus, which replicates in various immune cells, killing many of them and leaving the body more vulnerable to other infections. Syphilis, leprosy, and malaria affect the T-cell population and also macrophage function, causing temporary defects in cell-mediated immunity.

Examples of immunodeficiency diseases are listed in **table 18.4.**

TABLE 18.4 Immunodeficiency Diseases

Disease	Immune Component Involved
Primary Immunodeficiencies	
Agammaglobulinemia	B cells (deficiency)
Chronic granulomatous disease (CGD)	Phagocytes (defect)
DiGeorge syndrome	T cells (deficiency)
Hereditary angioedema	Complement regulator (deficiency)
Selective IgA deficiency	B cells making IgA (deficiency)
Severe combined immunodeficiency (SCID)	Bone marrow stem cells (defect)
Secondary Immunodeficiencies	
Acquired immunodeficiency syndrome (AIDS)	Helper T cells (destroyed during HIV infection)
Multiple myeloma	Plasma cell (multiplies out of control)

MicroAssessment 18.3

Primary immunodeficiencies are present at birth and result from defects in components of the immune response. Secondary, or acquired, immunodeficiencies result from infection or malignancies.

7. Compare and contrast severe combined immunodeficiency disease (SCID) and chronic granulomatous disease (CGD).
8. How can a person with multiple myeloma be immunodeficient?
9. Why are people with B-cell deficiencies more prone to bacterial infections, whereas people with T-cell deficiencies are more prone to viral infections?

Summary

18.1 Hypersensitivities (table 18.2)

An exaggerated immune response that injures tissue is called **hypersensitivity.**

Type I Hypersensitivities: Immediate IgE-Mediated

Type I hypersensitivities occur when specific antigens react with IgE bound to mast cells or basophils, causing the cell to degranulate and release powerful mediators of the allergic reaction (figure 18.1). Localized allergic reactions include **hives** (urticaria), **hay fever** (allergic rhinitis), and **asthma** (figure 18.2). **Systemic anaphylaxis** is a rare but serious reaction that can lead to shock and death. **Desensitization** is a form of immunotherapy (figure 18.3). Monoclonal antibodies are used to treat certain cases of severe asthma.

Type II Hypersensitivities: Cytotoxic

Type II hypersensitivities, or cytotoxic reactions, occur when antibodies bound to the surface of a host cell trigger its destruction by complement proteins or by antibody-dependent cellular cytotoxicity (ADCC). The ABO blood group antigens are associated with **hemolytic transfusion reactions** (table 18.1). The rhesus blood group antigens (Rh) are usually involved with **hemolytic disease of the newborn** (figure 18.4). Injected anti-Rh antibody (RhoGAM) helps prevent Rh sensitization of Rh-negative mothers.

Type III Hypersensitivities: Immune Complex–Mediated

Type III hypersensitivities are mediated by small **immune complexes** that activate complement, attracting neutrophils, and contributing to inflammation (figure 18.5). This can result in inflammatory conditions such as glomerulonephritis and serum sickness.

Type IV Hypersensitivities: Delayed-Type Cell-Mediated

Delayed-type hypersensitivities are due to antigen-specific T-cell responses and can occur almost anywhere in the body. Contact hypersensitivity, or **allergic contact dermatitis,** is due to the response of effector T cells to small molecules that penetrate intact skin and act as sensitizing agents (figure 18.6). Transplantation rejection of **allografts** is caused largely by type IV hypersensitivities.

18.2 ■ Autoimmune Disease (table 18.3)

Responses against autoantigens can lead to **autoimmune diseases**. Some autoimmune diseases are caused by antibodies that target body components, and others by cell-mediated reactions or a combination of antibodies and immune cells.

Systemic Autoimmune Diseases

Systemic autoimmune diseases result from an immune response to one or more autoantigens found throughout the body. Examples of these diseases include rheumatoid arthritis and systemic lupus erythematosus (figure 18.7).

Organ-Specific Autoimmune Diseases

Organ-specific autoimmune diseases result from autoantigens found only within a specific organ. Examples of the diseases include Graves' disease (figure 18.8), myasthenia gravis, multiple sclerosis, and type I diabetes mellitus.

18.3 ■ Immunodeficiency Disorders (table 18.4)

Immunodeficiencies occur when the immune system does not provide an adequate response, leaving a person vulnerable to infection.

Primary Immunodeficiencies

Primary immunodeficiencies are genetic or developmental defects. B-cell immunodeficiencies result in diseases involving a lack of antibody production, such as selective IgA deficiency and agammaglobulinemias. Lymphocyte deficiencies result in diseases such as severe combined immunodeficiency (SCID) and DiGeorge syndrome. Defective phagocytes are found in chronic granulomatous disease. Defects in complement system components may result in immune complex diseases, recurrent *Neisseria* infections, or uncontrolled complement activation.

Secondary Immunodeficiencies

Acquired immunodeficiencies can result from cancers such as multiple myeloma or infectious diseases (such as AIDS).

Review Questions

Short Answer

1. How does treatment for hay fever differ from treatment for asthma?
2. List two physical responses of systemic anaphylaxis that can lead to rapid death.
3. What are the major differences between an IgE-mediated skin reaction, such as hives, and a delayed-type hypersensitivity reaction, such as rejection of transplanted tissue?
4. Compare and contrast the Arthus reaction and serum sickness.
5. What leads to hemolytic disease of the newborn?
6. Compare and contrast the autoimmune processes causing myasthenia gravis and Graves' disease.
7. Give an example of an organ-specific autoimmune disease and one that is systemic, involving a variety of tissues and organs.
8. Explain why a patient who received a successful lung transplant might subsequently die from pneumonia.
9. Why does an IgA deficiency predispose a person to diarrheal diseases?
10. What is the difference between a primary immunodeficiency and a secondary immunodeficiency? Give an example of each.

Multiple Choice

1. An IgE-mediated allergic reaction
   a) reaches a peak within minutes after exposure to antigen.
   b) occurs only in response to polysaccharide antigens.
   c) requires complement activation.
   d) requires considerable macrophage participation.
   e) is characterized by bleeding.
2. Which of the following statements is true of the ABO blood group system in humans?
   a) A and B antigens are present on type O red blood cells.
   b) People with blood group A have A antigens on their red blood cells and natural anti-A antibodies in their plasma.
   c) Natural anti-A and anti-B antibodies are of the class IgG.
   d) People with blood group O do not have natural antibodies against A and B antigens.
   e) In blood transfusions, incompatibilities cause complement-mediated lysis of red blood cells.
3. Repeated injections of very small amounts of antigen are an effective therapy for combating allergies because
   a) IgG production is stimulated; the IgG outcompetes IgE for the antigen.
   b) IgG binds to mast cells before IgE can do so.
   c) repeated exposure will cause production of IgE to decrease.
   d) antigen binds to mast cells and prevents their degranulation.
   e) repeated exposure to antigen prevents the production of histamine.
4. Which of the following could be the result of a type III hypersensitivity reaction?
   a) DiGeorge syndrome
   b) Glomerulonephritis
   c) Type I diabetes mellitus
   d) Hemolytic disease of the newborn
   e) Anaphylactic shock
5. Delayed-type hypersensitivity reactions in the skin
   a) are characterized by a wheal and flare reaction.
   b) peak at 4 to 6 hours after exposure to antigen.
   c) require complement activation.
   d) are due to an effector T cell response to an allergen.
   e) depend on activities of the Fc portion of antibodies.
6. Organ transplants, such as transplanted kidneys,
   a) are only available in a few clinical trials at present.
   b) can be successful only if there are exact matches between donor and recipient.
   c) are only successful if irradiated before the transplant.
   d) are only successful if B cells are suppressed in the recipient.
   e) may be rejected by a complex process in which cellular mechanisms predominate.

7. All of the following are true of autoimmune disease *except*
   a) some seem to have a genetic component.
   b) induction of tolerance may alleviate symptoms.
   c) damage to organs occurs due to long-term exaggerated production of IgE.
   d) disease may result from reaction to viral antigens that are similar to autoantigens.
   e) some are organ-specific and some are widespread in the body.
8. All of the following approaches are used to treat autoimmune diseases *except*
   a) immunosuppressant drugs.
   b) surgery.
   c) antibiotics.
   d) anti-inflammatory medications.
   e) replacement therapy, as with insulin in diabetes.
9. If the thymus is removed from a 2-year-old child, you might expect which of the following to occur?
   a) Increased success of an organ transplant and increased incidence of bacterial infection
   b) Increased success of an organ transplant and decreased incidence of bacterial infection
   c) Decreased success of an organ transplant and increased incidence of viral infection
   d) Decreased incidence of cancer and decreased incidence of viral infection
   e) Decreased success of an organ transplant and increased incidence of cancer
10. In which one of the following combinations of hypersensitivity reactions are B cells involved?
   a) Type I, but not types II, III, and IV hypersensitivity
   b) Type IV, but not types I, II, and III hypersensitivity
   c) Types I, II, and III, but not type IV hypersensitivity
   d) Types II, III, and IV, but not type I hypersensitivity
   e) B cells are involved in all types of hypersensitivity

Applications

1. Jack and Jill were badly burned in an accident at the well and both were taken to the burn unit of the local hospital. Although serious, their burns covered only a small area of skin, so grafts were prepared for both patients from the skin of Jack's thigh. Jack's graft was successful, and his burn healed completely. Jill, however, rejected the grafted skin. Explain the immune responses of both patients to these grafts. What treatments could have helped Jill to avoid rejection of her graft?
2. Horse serum containing specific antibody to snake venom has been a successful approach to treating snakebites in humans. How do you think this antivenom could be generated? What are some advantages of using horses instead of humans to produce the antibody? Why might it be unsafe to administer the antivenom to a person more than once?

Critical Thinking

1. Hypersensitivity reactions, by definition, lead to tissue damage. Can they also be beneficial? Explain.
2. Why might malnutrition result in immunodeficiencies?

19 Epidemiology

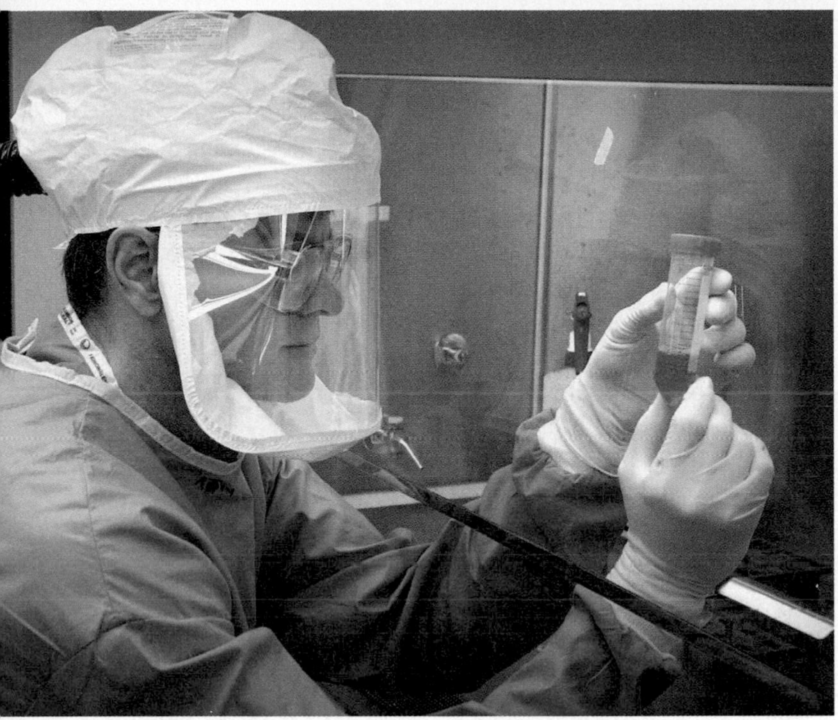
CDC microbiologist studies influenza virus. *James Gathany/CDC*

KEY TERMS

Attack Rate The proportion of susceptible persons developing illness in a population exposed to an infectious agent.

Case-Fatality Rate The proportion of persons diagnosed with a specific disease who die from that disease.

Communicable Disease An infectious disease that can be transmitted from one host to another.

Endemic A disease or other occurrence that is constantly present in a population.

Epidemic A disease or other occurrence that has a much higher incidence than usual.

Herd Immunity Protection of an entire population based upon a critical concentration of immune hosts that prevents the spread of an infectious agent.

Incidence The number of new cases of a disease or condition in a population at risk during a specified time period.

Morbidity Illness.

Mortality Death; often expressed as the rate of death in a defined population.

Outbreak A group of cases occurring during a brief time interval and affecting a specific population.

Pandemic An epidemic that has spread around the world.

Prevalence The total number of cases of a disease or condition in a given population at a point in time or over a specified time period.

Reservoir of Infection The natural habitat of a pathogen; sum of the potential sources of an infectious agent.

A Glimpse of History

Puerperal fever, a bacterial infection of the uterus following childbirth, rose to epidemic proportions when more women chose to deliver their babies in hospitals. By the mid-1800s in the hospitals of Vienna, the medical hub of the world at that time, about one of every eight women died of puerperal fever following childbirth.

In 1841, Ignaz Semmelweis, a Hungarian, traveled to Vienna to study medicine. After medical school, he worked for Professor Johann Klein, who, with a staff of medical students, ran one section of a maternity hospital. Midwives and midwifery students served a second section. Semmelweis noticed that the incidence of puerperal fever in Dr. Klein's section was as high as 18%, four times that in the section served by the midwives. During this period, a friend of Semmelweis was cut with a scalpel while supervising an autopsy and died of symptoms very similar to those of puerperal fever. Semmelweiss reasoned that the "poison" that killed his friend probably also contaminated the hands of the medical students who did autopsies. Perhaps these students were transferring the "poison" from the dead bodies to the women in childbirth. Midwives, after all, did not perform autopsies. This was before Pasteur and Koch established the germ theory of disease, so Semmelweis had no way of knowing that the "poison" being transferred was probably *Streptococcus pyogenes*, a common cause of many infections, including puerperal fever.

To test his hypothesis that physicians and students were transferring "poison" to patients, Semmelweis had them wash their hands with a strong disinfectant before attending patients. The incidence of puerperal fever dropped to one-third its previous level. Instead of appreciating these findings, Semmelweis's colleagues refused to admit responsibility for the deaths of so many patients. His work was so fiercely attacked that he was forced to leave Vienna and return to his native Hungary. Although his disinfection techniques achieved a remarkable reduction in the number of deaths from puerperal fever there as well, Semmelweis became increasingly outspoken and bitter. He was finally confined to a mental institution, where he died one month later of a generalized infection similar to the kind that had killed his friend and the many women who had contracted puerperal fever following childbirth.

As the global spread of COVID-19 began to dominate news stories in early 2020, anyone following the unfolding story learned a great deal about **epidemiology,** the study of the distribution and causes of disease in populations. Terms such as *pandemic, case-fatality rate, droplet spread, aerosol, herd immunity,* and *personal protective equipment (PPE)* were mentioned regularly in the news and soon made their way into normal conversations. Although epidemiology took on a special urgency as a result of the pandemic, many of our daily habits, from handwashing to waste disposal, are based on its principles.

19.1 ■ Basic Concepts of Epidemiology

Learning Outcome

1. Define the terms commonly used to describe the epidemiology of disease.

Epidemiologists are the "disease detectives" who collect and compile data about sources of disease and associated risk factors. Using that information, they design strategies to prevent or predict the spread of disease. Epidemiologists approach a disease outbreak much as criminal detectives describe the scene of a crime, using expertise in diverse disciplines including ecology, microbiology, sociology, statistics, and psychology.

Each disease of public health importance has a specific **case definition**—a set of diagnostic criteria that allows consistent reporting and counting. Based on which criteria are met, a case is classified as *suspect*, *probable*, or *confirmed.* With regard to COVID-19, for example, someone who was verified to be infected using a diagnostic laboratory test such as PCR (polymerase chain reaction) done by a certified laboratory is counted as a confirmed case. In contrast, someone who had characteristic COVID-19 symptoms and close contact with a confirmed case but no diagnostic laboratory evidence of infection is considered a probable case. Someone who tested positive using an at-home COVID-19 test but had no symptoms or laboratory confirmed evidence of infection is counted as a suspect case; such a case would often be followed up to determine probable or confirmed status. Note that case definitions may change over time to reflect new methodologies or knowledge. For example, the 2020 case definition for COVID-19 included only nucleic acid tests such as PCR as confirmatory laboratory evidence, whereas the 2021 case definition added genomic sequencing as a laboratory option. A searchable database of current surveillance case definitions established by the Centers for Disease Control and Prevention (CDC) is available at the following website: https://ndc.services.cdc.gov/.

Infectious diseases that can be transmitted from one host to another, such as measles, colds, influenza, and COVID-19, are **contagious,** or **communicable, diseases.** In contrast, **noncommunicable diseases** do not spread from one host to another; microorganisms that cause these diseases most often arise from an individual's normal microbiota or from the environment. Legionellosis, for example, which is caused by the bacterium *Legionella pneumophila,* can be contracted from warm natural waters or from a building's water systems. Unlike most other respiratory infections, it typically does not spread from person to person.

Some diseases are routinely present in a given region, whereas others are not. For instance, the common cold is **endemic** (meaning routinely found) in the United States, whereas malaria is not. An unusually large number of cases in a population constitutes an **epidemic.** Epidemics may be caused by diseases not normally present in a population (such as when COVID-19 emerged), or by fluctuations in the incidence of endemic diseases such as influenza and pneumonia (**figure 19.1**). The term **outbreak** describes a group of cases occurring during a brief time interval in a specific population; an outbreak may indicate the onset of an epidemic. An epidemic that spreads around the world, as COVID-19 has, is called a **pandemic.**

Epidemiologists generally are more concerned with the rate of disease than with the number of cases. Consider the difference between having 100 people develop genital herpes in a city of 1,000,000 versus the same number developing that disease in a town of 5,000; the second scenario would be of greater concern to an epidemiologist because of the higher rate. The **attack rate** describes the number of susceptible people who become ill in a population after exposure to an infectious agent. If 100 people at a party eat chicken contaminated with *Salmonella enterica,* and 10 people develop symptoms of salmonellosis, then the attack rate is 10%. The attack rate reflects many factors, including the infectious dose of the organism and the general health of the host population. Communicable diseases may have a significant secondary attack rate—the proportion of susceptible people in contact with an infected individual who then go on to develop the disease. A related term is the **basic reproductive number (R_0),** or R nought, which is the average number of secondary cases that develop from a single case in a susceptible population. The R_0 of a disease can change as interventions are implemented, so epidemiologists

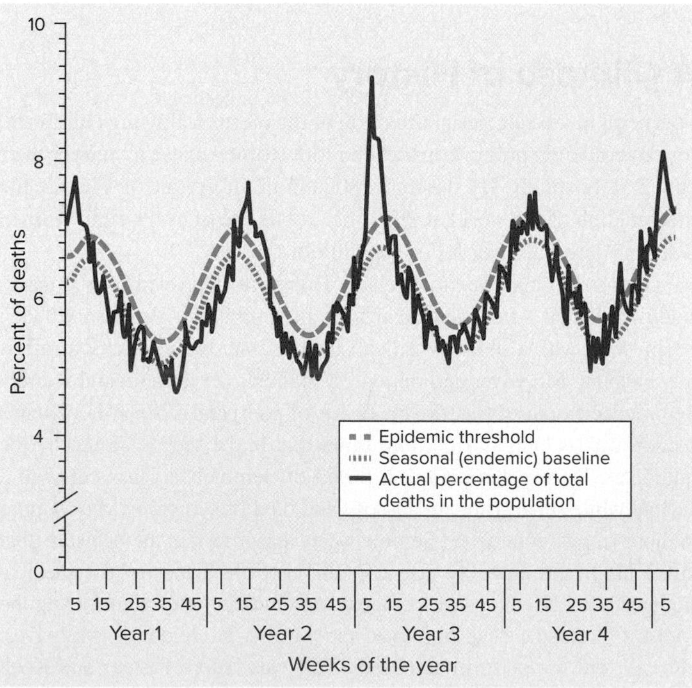

FIGURE 19.1 Endemic Disease Can Be Epidemic Example of yearly fluctuation of pneumonia and influenza deaths (expressed as a percentage of all deaths).

? **Based on this graph, during what season(s) of the year is a pneumonia epidemic most likely?**

focus on that number as they try to control a disease. If the R_0 for a given disease is greater than 1, then the number of new cases will increase over time; if it is less than 1, then they will decrease over time.

The terms *incidence* and *prevalence* are used to describe disease frequency: **Incidence** is the number of new cases in a population in a specific time period (often one year), whereas **prevalence** is the number of existing cases in a population at a point in time or over a specific time period. Because incidence considers only new cases, it provides a measure of the risk that a person will contract the disease. Prevalence includes all cases, so it reflects the overall impact of a disease on society, which is particularly important when considering chronic diseases. Incidence and prevalence are both usually expressed as the number per 100,000 people.

Various terms are helpful in describing a disease's effect on the population. **Morbidity** refers to illness. **Mortality** refers to death, and is often expressed as a rate—the number of people in a defined population who die during a given period. **Case-fatality rate** is the proportion of persons diagnosed with a specific disease who die from that disease; the case-fatality rate can change over time—particularly during an epidemic—as additional cases are identified, treatments improve, and case definitions change. A related term is **infection-fatality rate,** which is an estimate of the proportion of persons infected with a specific pathogen who die as a result; the number of persons infected (including identified cases as well as undiagnosed and asymptomatic individuals) is often estimated using serological data.

MicroByte

Measles is one of the most contagious diseases, with a secondary attack rate of over 90%.

MicroAssessment 19.1

Each disease of public health importance has a specific case definition. Communicable diseases can be transmitted from one host to another. An unusually large number of cases within a population constitutes an epidemic. Rates of disease within a population are a concern of epidemiologists, who study disease patterns.

1. Name two typical sources of non-communicable infectious diseases.
2. Explain the difference between the incidence of a disease and the prevalence of the disease.
3. Why might a disease be endemic in one region but not in another?

19.2 ▪ Chain of Infection

Learning Outcome

2. Describe the steps involved in the chain of infection.

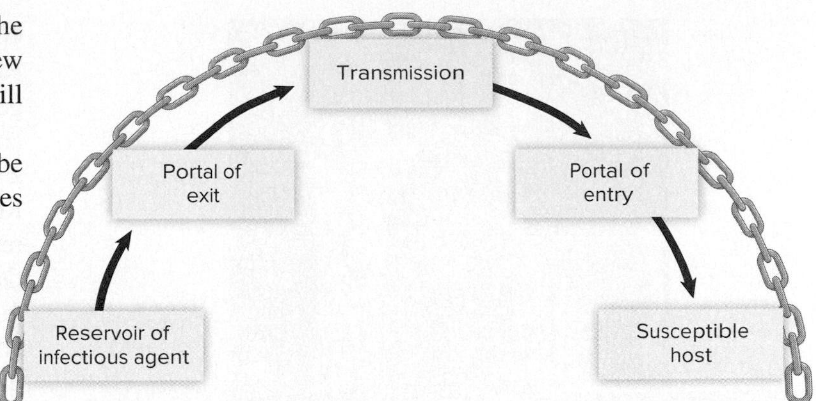

FIGURE 19.2 Chain of Infection If any link in this chain is broken, disease transmission is slowed or stopped.

? How could disease transmission be stopped at the portal of exit?

The spread of an infectious disease follows a series of steps often termed the chain of infection. First, there must be a source, or reservoir, of an infectious agent. If the reservoir is an infected host, the agent must leave that host through a portal of exit, be transmitted to a new host, and then enter the new host through a portal of entry (**figure 19.2**). Knowing the chain of infection for a given disease enables researchers and public health workers to determine where links in the chain can be broken, thereby stopping or slowing the spread of the disease.

Reservoirs of Infection

The natural habitat of a pathogen, the **reservoir of infection,** may be on or in an animal (including humans) or in an environment such as soil or water (**figure 19.3**). The reservoir of infection affects the extent and distribution of a disease. Once the reservoir is identified, susceptible people can be prevented from coming into contact with the disease source.

Human Reservoirs

Infected humans are a significant reservoir of most communicable diseases. Humans are often the only reservoir, but some pathogens can also colonize non-human animals or grow in the environment. When infected humans are the only reservoir, the disease can be easier to control because setting up prevention and control programs in humans is simpler than in wild animals. The eradication of smallpox is an excellent example: The combined effects of widespread vaccination (which resulted in fewer susceptible people) and the isolation of those who became infected eliminated circulating smallpox virus because it no longer had a reservoir in which to multiply.

Symptomatic Infections People with symptomatic illnesses are an obvious source of infectious agents. Ideally they understand the importance of taking precautions to avoid transmitting their illness to others. Resting at home while ill both helps the body to recover and protects others from exposure to the

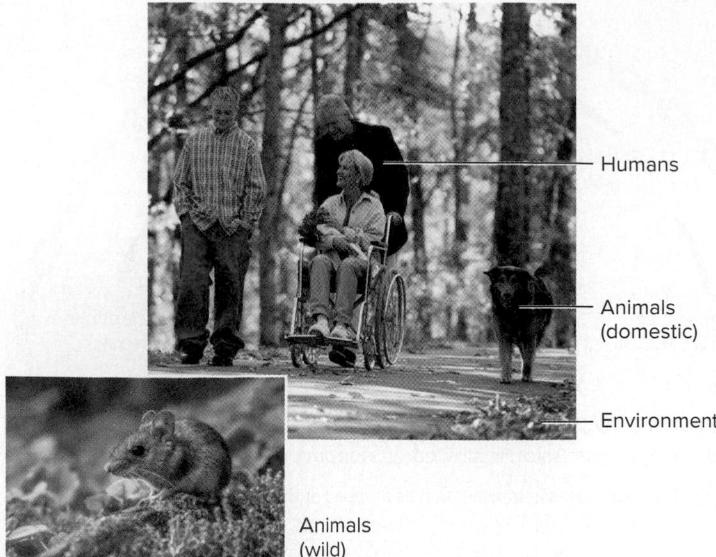

FIGURE 19.3 Reservoirs of Infection (top): Thinkstock/Getty Images; (bottom): Rudmer Zwerver/Shutterstock

? How might you be a reservoir of infection?

disease-causing agent. Even conscientious people, however, can unintentionally be a source of infection. For example, people who are in the incubation period of COVID-19 shed virus before symptoms appear.

Asymptomatic Carriers A person can harbor a pathogen with no ill effects, acting as a **carrier** of the disease agent. Depending on the situation, these people may shed the microbe intermittently or constantly for months, years, or even a lifetime.

Some carriers have an asymptomatic infection; their immune system is actively responding to the invading microbe, but they have no obvious clinical symptoms. Because these people often have no reason to consider themselves a reservoir, they move freely about, spreading the pathogen. A significant complicating factor in the control of COVID-19 is that many infected people are asymptomatic, which is why even healthy people were encouraged or ordered to stay home during certain phases of the pandemic. Sexually transmitted infections (STIs) such as chlamydia are also often asymptomatic; studies indicate that perhaps 80% of people infected with *Chlamydia trachomatis,* the bacterium that causes chlamydia, have no symptoms and therefore may unknowingly transmit the organism to their sexual partners.

Some potentially pathogenic microbes can colonize the skin or mucosal surfaces, establishing themselves as part of a person's microbiota. In fact, many people carry *Staphylococcus aureus* as a part of their nasal or skin microbiota. Carriers of *S. aureus* may never have any illness as a result of the colonization, but they remain a potential source of disease. Unfortunately, ridding a colonized carrier of the infectious organism is often difficult, even with the use of antimicrobial medications.

Non-Human Animal Reservoirs

Non-human animal reservoirs are the source of many pathogens. These can be very difficult to control, particularly in wild animal populations. In the United States, raccoons, skunks, and bats are reservoirs of the rabies virus (see figure 26.11). Rodents such as rats, rock squirrels, and prairie dogs are the reservoir of *Yersinia pestis*, the bacterium that causes plague. Occasional transmission of plague to humans is still reported in the western United States, but epidemics of that disease no longer occur, in part because of rodent control. Rodents, particularly the deer mouse, are also the reservoir for hantavirus. Poultry are a reservoir of gastrointestinal pathogens such as species of *Campylobacter* and *Salmonella.*

Diseases such as plague and rabies that can be transmitted to humans but exist primarily in other animals are called zoonotic diseases, or **zoonoses.** Zoonotic diseases are often more severe in humans than in the typical animal host because the infection in humans is accidental; there has been no evolution toward the balanced pathogenicity that normally exists between host and parasite.

Environmental Reservoirs

Some pathogens have environmental reservoirs. *Clostridium botulinum,* which causes botulism, and *Clostridium tetani,* which causes tetanus, are both widespread in soils. Pathogens that have environmental reservoirs are difficult or impossible to eliminate.

Portals of Exit

A pathogen must leave its reservoir to be transmitted to a susceptible host. If the reservoir is a human or other animal, the body orifice or surface from which a microbe is shed is called the **portal of exit** (**figure 19.4**). Intestinal pathogens such as *Vibrio cholerae* are shed in the feces; the massive volumes of watery diarrhea triggered by that bacterium can contaminate drinking water and food. Pathogens such as *Mycobacterium tuberculosis* and various respiratory viruses, including the virus that causes COVID-19, exit the body in droplets of saliva and mucus when people talk, laugh, sing, sneeze, or cough. Organisms that inhabit the skin are constantly shed on skin cells; even as you read this text you are shedding skin cells, some of which may have *Staphylococcus aureus* on their surface. Genital pathogens such as *Chlamydia trachomatis* can be carried in urethral and vaginal secretions.

Disease Transmission

An infectious agent must somehow be transmitted from its reservoir to the next host. **Vertical transmission** is the transfer of an infectious agent from a pregnant female to her fetus, or from a mother to her infant during childbirth or breast feeding. Prenatal care includes interventions to prevent vertical transmission of pathogens. **Horizontal transmission** refers to transfer of an infectious agent by any other method, including

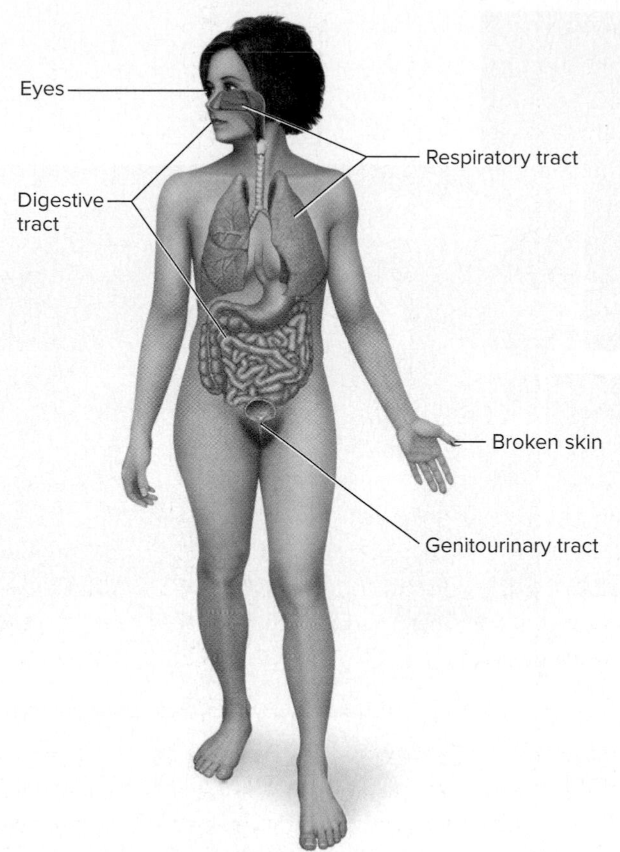

FIGURE 19.4 Portals of Exit or Entry A portal of exit is the route by which an infectious agent leaves the host. Some pathogens cause disease only when they enter a host through a specific portal of entry.

? **What two portals of exit are most likely used by organisms that infect the digestive tract?**

person to person or environment to person (**figure 19.5**). Transmission of the pathogen can be direct, indirect, or sometimes both.

Direct Transmission

Direct transmission typically involves immediate transfer of the infectious agent from a reservoir to an appropriate host; this could occur via direct physical contact or projection of respiratory droplets onto mucous membranes.

Direct Contact Touching is **direct contact.** It can be as simple as a handshake or as intimate as sexual intercourse. Organisms with a low infectious dose—meaning that very few microbial cells or viral particles are needed to initiate an infection—are often transmitted by direct contact. An example is *Shigella* species, which are easily transmitted by direct contact, particularly in day-care settings where young children play together. Routine handwashing with soap and water, which physically removes microbes and destroys enveloped viruses, is considered to be the single most important measure for preventing the spread of infectious diseases, particularly those transmitted by direct contact. When handwashing is not possible, the use of alcohol gels may provide protection, but they are not effective against all pathogens.

Pathogens that cannot survive for extended periods in the environment must generally be transmitted through direct contact. *Neisseria gonorrhoeae,* for example, which causes gonorrhea, dies quickly when exposed to a relatively cold or dry environment. Because of this, intimate sexual contact or other activity involving direct mucous-membrane-to-mucous-membrane contact is generally required to transmit the bacterium.

Droplet Transmission Droplet transmission results from inhaling relatively large microbe-containing respiratory droplets that are released as people talk, laugh, sing, sneeze, or cough. In most cases, the inhaled droplets only contact the upper respiratory tract because the mucus that lines the nose and throat usually traps them before they enter the lungs.

Droplet transmission typically occurs only if the infected person is very close because the droplets generally fall to the ground within about 1 meter (approximately 3 feet) from release, but it is particularly a concern in densely populated buildings such as schools and military barracks. Methods to prevent the transmission include encouraging people to cover their mouths (preferably not with bare hands) when coughing or sneezing and spacing desks or beds more than 1.5 m (about 5 feet) apart. Droplet transmission plays a role in the spread of COVID-19, which is why control mechanisms include social distancing (physical distancing) and face mask use; a properly worn face mask traps droplets released by the wearer and also prevents the wearer from inhaling droplets released by others.

Indirect Transmission

Indirect transmission of a pathogen from a reservoir to a host occurs via airborne particles, vehicles (such as inanimate objects, food, or water), or vectors (mosquitoes, flies, fleas, lice, or ticks).

Airborne Transmission Airborne transmission results from inhaling microbe-containing airborne particles that can remain suspended in air because of their small size. The suspended particles, often referred to as aerosols, are typically smaller than about 5 μm. They can originate from any number of sources, but those referred to as droplet nuclei are the remains of evaporated respiratory droplets and one of the concerns with respect to COVID-19 transmission, particularly indoors. Soil disturbed by wind, skin cells, and household dust can also become airborne.

Understandably, airborne transmission is very difficult to control. To prevent the buildup of airborne pathogens, modern public buildings have ventilation systems that constantly change the air. The air pressure in hospital microbiology laboratories can be lowered so that air flows in from the corridors, preventing microorganisms and viruses from being swept out of the lab to other parts of the building. Air in specialized hospital rooms, jetliners, and some laboratories is circulated through **high-efficiency particulate air (HEPA)** filters to

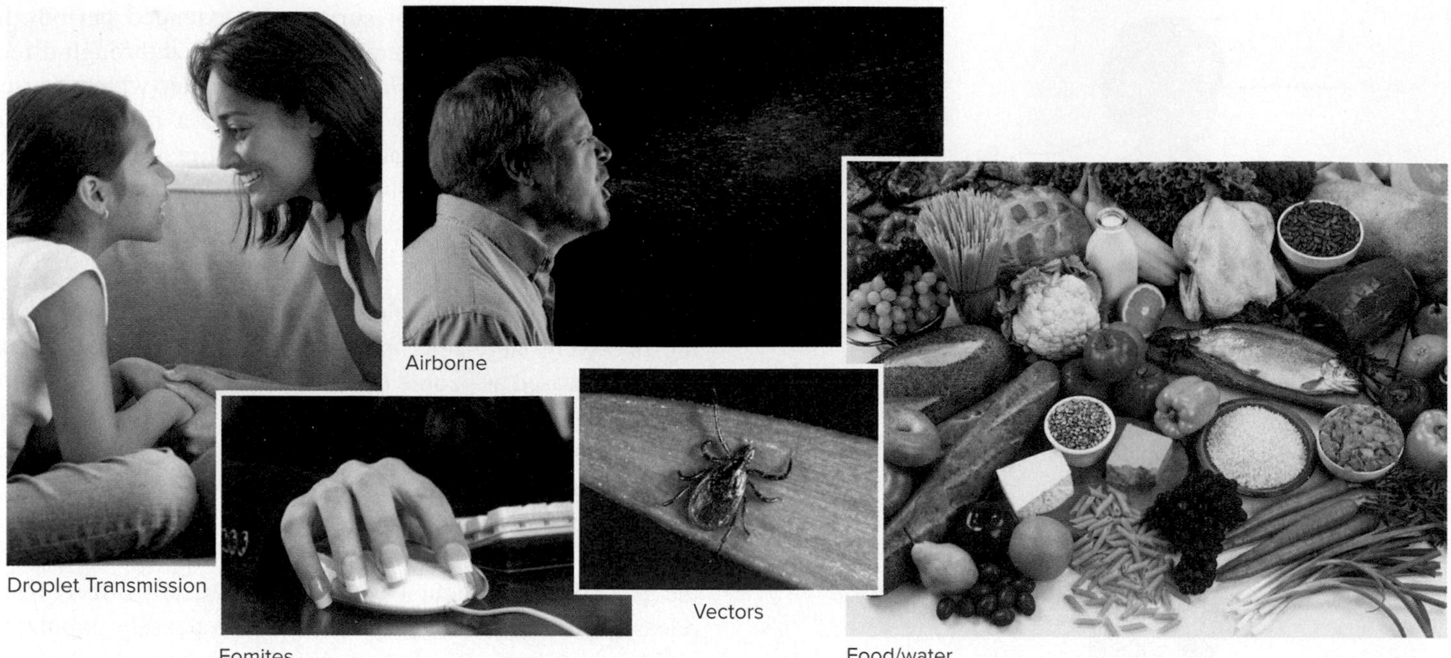

FIGURE 19.5 Horizontal Transmission of Pathogens (talking): Fuse/Getty Images; (computer mouse): Ingram Publishing/Alamy Stock Photo; (sneeze): James Gathany/CDC; (tick): Jim Gathany/CDC; (foods): Mitch Hrdlicka/Photodisc/Getty Images

? If vertical transmission were added to this figure, photographs showing what subjects might be added?

remove airborne organisms that may be present. In addition, face mask use helps prevent airborne transmission for the reasons described for droplet transmission; they are particularly important for trapping the respiratory droplets released by the wearer, because the droplets could otherwise evaporate to become droplet nuclei.

Vehicle-Borne Transmission Pathogens can be transmitted via vehicles such as **fomites** (inanimate objects), foods, and water.

Fomites, including clothing, keyboards, cell phones, doorknobs, and drinking glasses, can become contaminated when someone touches them or a respiratory droplet lands on them. When a carrier of *Staphylococcus aureus* touches a skin lesion or a colonized nostril, for example, the bacterial cells now on that person's fingers can be transferred to a fomite. Another person then handling a contaminated object can acquire the microbe. Handwashing helps prevent transmission via fomites.

Foods can become contaminated in a number of ways. Animal products such as meat and eggs may carry pathogens that originated from the animal's intestinal tract. This is the case with poultry contaminated with species of *Salmonella* or *Campylobacter,* and hamburger contaminated with *E. coli* O157:H7. Pathogens can also be unintentionally added during food preparation. *S. aureus* carriers who do not wash their hands before preparing food can easily contaminate the food. **Cross-contamination** results when pathogens from one food are transferred to another. A cutting board used first to chop raw chicken and then to cut salad ingredients can transfer *Salmonella* cells from the chicken onto the salad (**figure 19.6**). Because many foods are a rich nutrient source, microorganisms can multiply to high numbers if the contaminated food is not refrigerated. Sensible food-handling methods, including sanitary preparation as well as thorough cooking and proper storage, can prevent foodborne diseases.

Waterborne disease outbreaks can involve large numbers of people because municipal water systems distribute water to widespread areas. The 1993 waterborne outbreak of cryptosporidiosis from *Cryptosporidium parvum,* an intestinal parasite, in Milwaukee, Wisconsin, was estimated to have affected

FIGURE 19.6 Cross-Contamination Vegetables exposed to juices from raw chicken may be a source of foodborne infection. mauritius images GmbH/Alamy Stock Photo

? What is the source of *Salmonella* species in raw chicken?

more than 400,000 people. Preventing waterborne diseases requires disinfecting or filtering drinking water and properly treating sewage.

Vector-Borne Transmission A **vector** is any living organism that can transmit a disease-causing microbe, but most commonly the term is used to refer to arthropods such as mosquitoes, flies, fleas, lice, and ticks (see table 12.6). There are two general categories of arthropod vectors (**figure 19.7**):

- **Mechanical vector.** This simply carries a pathogen from one place to another. When a fly lands on feces, for example, an intestinal pathogen might stick to its legs; if that fly then lands on food, it can transfer the pathogen to that food.
- **Biological vector.** This transmits a pathogen that multiplies within it. Biological vectors typically feed on blood: They acquire the pathogen while feeding on an infected host, and—after the pathogen multiplies—they transmit it while feeding on another host. Plague, malaria, and Lyme disease are examples of diseases spread by biological vectors.

Preventing vector-borne disease relies largely on arthropod control. Malaria, once endemic in the continental United States, was successfully eliminated from the nation through a combination of mosquito control and prompt treatment of infected patients. Unfortunately, worldwide eradication efforts that initially showed great promise ultimately failed, in part due to the decreased vigilance that accompanied the dramatic but short-lived decline of the disease.

Portals of Entry

The last step in the chain of infection is colonization of a new host. In some cases, the pathogen grows on the outer surface of the new host, but often it must gain access to the host through a **portal of entry** (see figure 19.4). Respiratory pathogens released into the air during a cough generally cause disease only when someone inhales them; for these pathogens, the nose is the typical portal of entry. As described for droplet transmission, a properly worn face mask can help prevent inhalation of pathogen-containing respiratory particles. Intestinal pathogens such as *Vibrio cholerae* and *Shigella* species must usually be ingested; the mouth is the portal of entry. For *Shigella* species, the infectious dose is about 10 to 100 cells, a number easily transferred by casual contact; if cells of a *Shigella* species were transferred by touch, however, they would cause disease only if they were then transported to the mouth. When fecal organisms are inadvertently ingested, the transfer is called **fecal-oral transmission.**

Some organisms can cause disease if they enter one body site but are harmless if they enter another. For instance, *Enterococcus faecalis*, a member of the normal intestinal microbiota, can cause a bladder infection if it enters the urinary tract, but it is harmless when ingested.

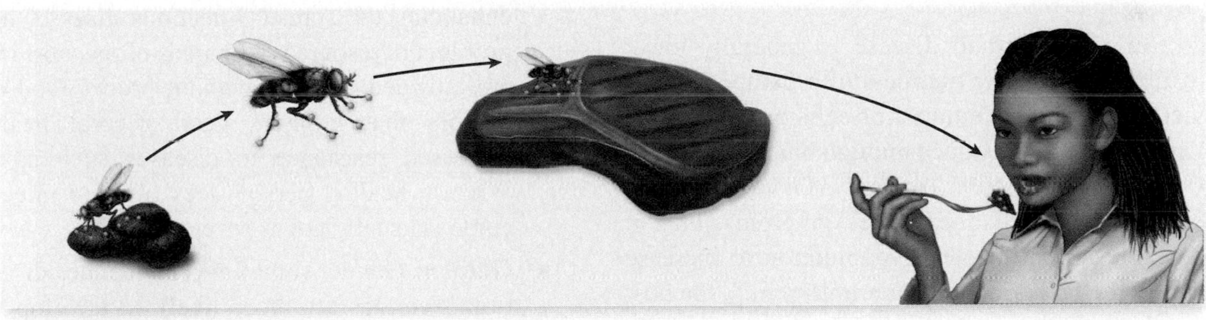

(a) Mechanical vector

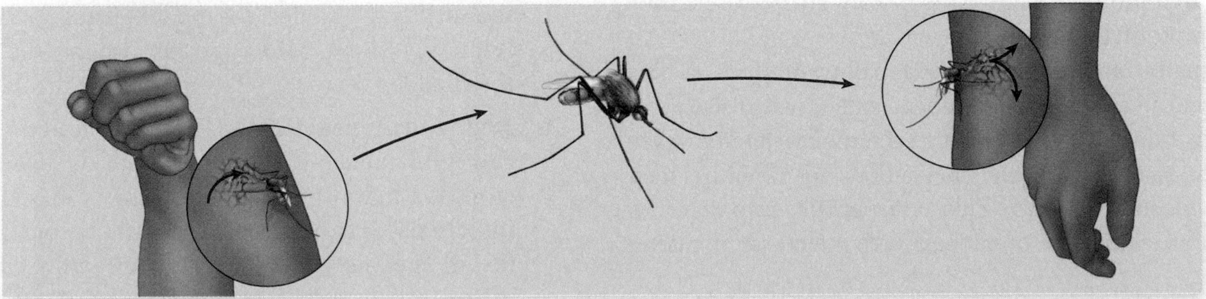

(b) Biological vector

FIGURE 19.7 Vector-Borne Transmission (a) A mechanical vector moves microbes from one place to another. **(b)** A biological vector participates in the life cycle of the pathogen and provides a place for it to multiply.

? **Why are biological vectors typically arthropods that take a blood meal?**

MicroAssessment 19.2

Spread of a disease can be prevented by breaking the chain of infection. The reservoir of a disease agent can be infected people, other animals, or the environment. To spread, infectious microbes must exit one reservoir, be transmitted to a susceptible host, and then enter that host. Handwashing and vector control can prevent many diseases; airborne transmission of pathogens is difficult to control.

4. How can an arthropod act as both a mechanical vector and a biological vector?
5. How does handwashing reduce the spread of disease?
6. Considering that circulating blood is not normally released from the body, describe how blood-borne microbes might exit.

19.3 ■ Factors That Influence the Epidemiology of Disease

Learning Outcome

3. Explain how the dose of the infecting agent, the incubation period, and characteristics of the host population can influence the epidemiology of a disease.

The host-microbe interactions described in chapter 16 influence the epidemiology of disease, as do various other factors, including the dose of the infecting agent, the incubation period, the characteristics of the host population, and the environment.

The Dose

The probability of infection and disease is generally lower when one is exposed to smaller numbers of a pathogen. This is because a certain minimum number of cells or particles of the pathogen are required to produce enough damage to cause disease. If 30 bacterial cells enter the body, for example, but 1,000,000 cells are required to produce symptoms, then it will take some time for the bacterial population to increase to that number. As the bacterial cells are multiplying, the host defenses are working to eliminate them and may successfully do so before symptoms appear. This is why small doses often result in asymptomatic infections.

An unusually large exposure to a pathogen, such as that which can occur in a laboratory accident, may cause disease in a person who is ordinarily immune to that microbe. Therefore, even immunized persons should take precautions to minimize their exposure to infectious agents. This is especially important for medical workers taking care of patients with infectious diseases.

The Incubation Period

The extent of the spread of an infectious agent is influenced by its incubation period. Diseases with long incubation periods can spread far and wide before the first symptomatic cases appear. An excellent example was a typhoid fever outbreak that originated at a ski resort in Switzerland in 1963. As many as 10,000 people were exposed to drinking water containing small numbers of the bacterium *Salmonella enterica* serotype Typhi. The long incubation period of typhoid fever, 10 to 14 days, allowed the organism to be spread by the skiers, who flew home to various parts of the world before they became ill. As a result, more than 430 cases of the disease appeared in at least six countries.

The Host Population

Some populations more than others are likely to be affected by a given pathogen. Population characteristics that influence the occurrence of disease include:

- **Immunity to the pathogen.** Previous exposure or immunization of a population to a disease agent decreases incidence of the disease. If a very high percentage of the population is immune to the pathogen, **herd immunity** protects non-immune individuals because not enough susceptible hosts are available to allow the pathogen to spread. Unfortunately, some infectious agents can undergo antigenic variation or have high mutation rates and thereby overcome herd immunity. Novel agents can become pandemic because the global population lacks immunity, a situation that occurred with the virus that causes COVID-19.
- **General health.** Malnutrition, overcrowding, and fatigue increase people's susceptibility to infectious diseases, enhancing the spread. Infectious diseases are more of a problem in resource-poor areas of the world where people are crowded together without proper food or sanitation. Factors that promote good general health result in increased resistance to diseases such as tuberculosis. When a healthy person does become infected, the outcome is usually not as severe.
- **Gender.** Gender sometimes influences disease distribution. Females are more likely to develop urinary tract infections because their urethra (the tube that connects the urinary bladder to the external environment) is relatively short. Microbes can ascend the urethra into the bladder.
- **Behavioral practices.** Certain practices can significantly influence the rate and type of disease transmission. An infant who is breast-fed is less likely to have infectious diarrhea because of the protective antibodies in the mother's milk; groups who eat traditional dishes made from raw freshwater fish are more likely to acquire the tapeworm *Diphyllobothrium latum,* a parasite normally killed by cooking; and people who have multiple sexual partners are more likely to acquire sexually transmitted infections.

- **Genetic background.** Natural immunity can vary with genetic background, but determining the relative importance of genetic, behavioral, and environmental factors can be difficult. In a few instances, however, the genetic basis for resistance to infectious disease is known. For example, many people of black African ancestry are not susceptible to malaria caused by common strains of *Plasmodium vivax* because they lack a specific red blood cell receptor used by the infecting organisms. Some populations of Northern European ancestry are less susceptible to HIV infection because they lack a certain receptor on their white blood cells.
- **Age.** The very young and the elderly are generally more susceptible to infectious agents. In the case of young children, the immune system is not fully developed and, consequently, they are vulnerable to certain diseases. Young children are particularly susceptible to meningitis caused by *Haemophilus influenzae;* fortunately, use of the Hib vaccine dramatically decreased the incidence of meningitis in this age group. The elderly are more prone to disease because immunity wanes over time. Influenza outbreaks in nursing homes often have high case-fatality rates; likewise, COVID-19 outbreaks in nursing homes have been devastating. Older adults are also less likely to update their immunizations, making them more susceptible to diseases such as tetanus that require regular booster doses (**figure 19.8**).

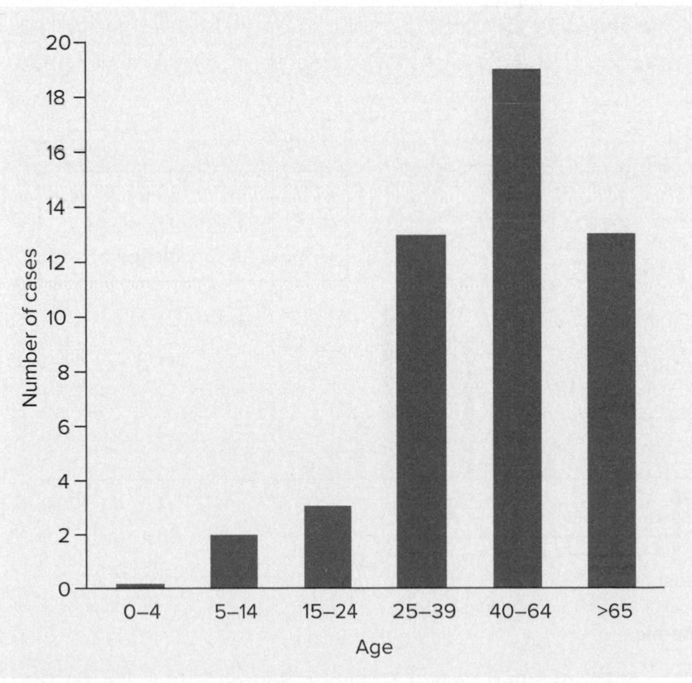

FIGURE 19.8 Typical Number of New Cases of Tetanus by Age Group

Why might people in the 40–64 age group develop tetanus more often than younger people?

The Environment

Factors in the physical environment, including temperature, water and nutrient supply, and the availability of light and O_2, determine which microbes can exist and reproduce in a given location. In addition to physical factors, the environment includes other organisms, including humans. Overcrowding in human populations is a major risk factor for infectious disease because transmission is often easier under crowded conditions, and poor sanitation creates opportunities for microbial spread and growth.

The spread of the mosquito *Aedes albopictus*, a vector for dengue virus, chikungunya, and other infectious agents, illustrates the interacting environmental factors that influence the epidemiology of disease. That mosquito is native to Southeast Asia, but human activities—including international transport of used tires that collect rainwater in which the mosquitoes lay eggs—have allowed its spread to parts of Europe, the Americas, Africa, and the Pacific; in turn, the diseases it transmits have spread accordingly. Scientists are concerned that climate change will increase the mosquito's range even further by expanding the territory in which it can survive and reproduce.

Recognition that disease spread involves such a wide variety of factors led to what is called the One Health approach. This relies on experts in a variety of fields—including public health, veterinary medicine, and environmental engineering—working together to develop solutions to the complex health problems facing the world today.

MicroAssessment 19.3

The epidemiology of disease is affected by the dose of the infecting agent, the incubation period, characteristics of the host population, and the environment.

7. Explain how a low dose of an infectious agent can result in an asymptomatic infection.
8. Why are diseases with long incubation periods more likely to result in an epidemic?
9. Why do influenza outbreaks in nursing homes typically have higher case-fatality rates than influenza outbreaks in a college dormitory?

19.4 ■ Epidemiological Studies

Learning Outcomes

4. Compare and contrast descriptive studies, analytical studies, and experimental studies.
5. Describe how a common-source epidemic can be distinguished from a propagated epidemic.

British physician John Snow illustrated the power of a well-designed epidemiological study over 150 years ago, long before the relationship between microbes and disease was accepted (see A Glimpse of History, chapter 24). During a cholera epidemic in London in 1854, he carefully compared the conditions of households that were affected by cholera to those that were not, eventually determining that the primary difference was their water supply. Preventing access to the contaminated water broke the chain of infection.

The urgency posed by COVID-19 illustrates the power of new technologies in studying widespread epidemics. Various smartphone apps that allow people to self-report their symptoms were developed to assist investigators studying not just the characteristics of the disease but also the spread of the causative agent. In addition, rapid genomic sequencing of viral isolates from patients around the world allows scientists to detect slight genomic differences that are then used to trace the virus's evolution and spread. Genomes of patients are being studied as well, in hopes of detecting genetic differences that influence disease outcome.

Descriptive Studies

When a disease outbreak occurs, epidemiologists conduct a **descriptive study** by gathering data about the time, the place, and the individuals affected. That information is used to compile a list of possible risk factors involved in the spread of disease.

The Person

Determining the profile of those who become ill is critical to defining the population at risk. Variables such as age, gender, ethnicity, occupation, personal habits, previous illnesses, socioeconomic class, and marital status may all yield clues about risk factors for developing the disease. In the Swiss ski resort epidemic of typhoid fever mentioned earlier, cases occurred only in tourists because the local people rarely drank water, preferring wine instead.

The Place

The location of disease occurrence helps pinpoint the exact source and may also give clues about potential reservoirs, vectors, or geographic boundaries that might affect the spread. For example, malaria can be transmitted only in regions that have an appropriate mosquito vector.

The Time

The timing of the symptom onset of cases during an outbreak yields important clues about the nature of the disease. To study this, epidemiologists make a graph that illustrates the time symptoms began for each case. The graph, called an **epi curve (epidemic curve),** helps reveal how the disease is being spread. The general categories of spread include (**figure 19.9**):

- **Propagated epidemic.** This occurs when a disease is contagious, with one person transmitting it to several others, who then transmit it to several more, and so on. In this situation, the number of ill people rises gradually, with higher peaks in the number of cases over time. The first case in such an outbreak is called the **index case.** The average time between symptom onset in the index case and symptom onset of the next cases reflects the average incubation period of the disease. As the disease continues to spread, the gaps between new cases disappear and

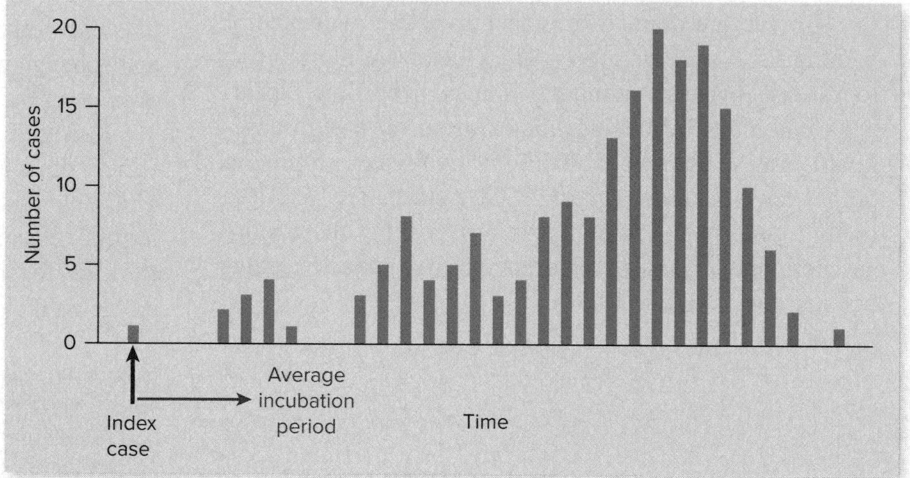

(a) Propagated Epidemic

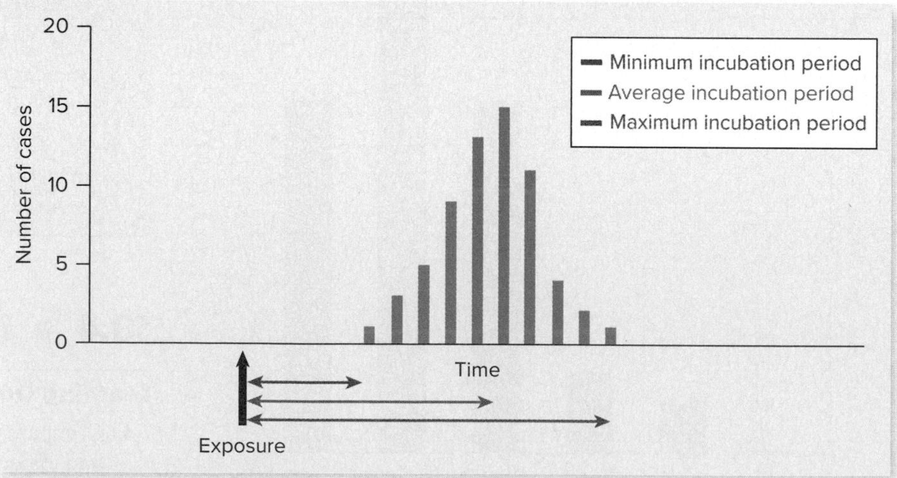

(b) Common-Source Epidemic

FIGURE 19.9 Comparison of Propagated Versus Common-Source Epidemics (a) The time between the index case and subsequent cases reflects the average incubation period of the disease. As the disease is propagated, this interval is blurred. **(b)** Cases in a common-source epidemic appear only within the incubation period.

? **Why does the number of cases increase over several incubation periods in a propagated epidemic?**

the peaks get higher. The phrase "flatten the curve"—an expression commonly used to describe the goal of lockdowns and social distancing protocols to slow COVID-19 spread—refers to the epi curve. By lowering the peak number of infections during a significant epidemic, the strain on healthcare systems is lessened.

- **Common-source epidemic.** This occurs when all the cases result from exposure to a single source of the infectious agent, such as eating contaminated food at a picnic. The resulting epi curve will show a rapid rise in the number of people who become ill. Not all common-source epidemics involve a single exposure to the infectious agent. In some instances, contact is continuous or intermittent, such as with a faulty drinking water disinfection system. Also, some epidemics start with a single source but become propagated when the first infected people transmit the disease to their contacts.

The season in which the epidemic occurs can also be significant. Respiratory diseases, including influenza, COVID-19, and the common cold, are more easily transmitted in crowded indoor conditions during the winter. Conversely, vector-borne and foodborne diseases are often transmitted in warm weather when people are more likely to be exposed to mosquitoes and ticks or to eat picnic food that has not been stored properly (**figure 19.10**).

Analytical Studies

Analytical studies are designed to determine which of the potential risk factors identified by the descriptive studies are actually relevant in the spread of the disease. They include three general types (**figure 19.11**):

- **Case-control.** These are retrospective studies, meaning they look back in an attempt to determine the chain of events leading to an outbreak. The investigator examines the past activities of the cases (people who developed the disease) and compares them with those of controls (similar people who remained healthy). If an activity is common among the cases, but not the controls, that activity may have been a factor in disease development. An important consideration in the study design is the selection of controls, which must be matched to cases with respect to several variables, including age, gender, and socioeconomic status. The aim is to ensure that the controls all had the potential to become cases.
- **Cross-sectional.** These studies do not try to establish the cause of a disease, but instead seek to identify risk factors for the disease. The investigator collects data about a population at a given point in time to get a snapshot of the population's characteristics—such as behaviors, exposure, and disease.
- **Cohort.** These are prospective studies, meaning they look ahead to see if previously identified risk factors actually predict a tendency to develop the disease. The investigator selects and then monitors cohort groups (sets of people whose exposure to the risk factor is known). The incidence of disease in those who were exposed to the risk factor and those who were not is compared over time, as a means to determine if the suspected cause does indeed correlate with the expected effect. Compared to retrospective studies, cohort studies are less likely to be influenced by poor memory, but they are generally more time-consuming and expensive, particularly when examining a disease with a long incubation period.

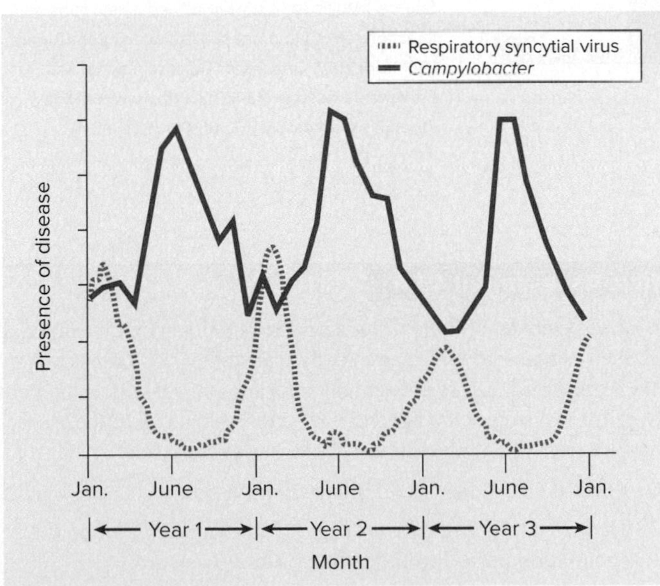

FIGURE 19.10 Seasonal Occurrence of Respiratory and Gastrointestinal Infections

Why does the level of foodborne disease increase in the summer?

Experimental Studies

An **experimental study** is sometimes used to judge the cause-and-effect relationship between the risk factors and

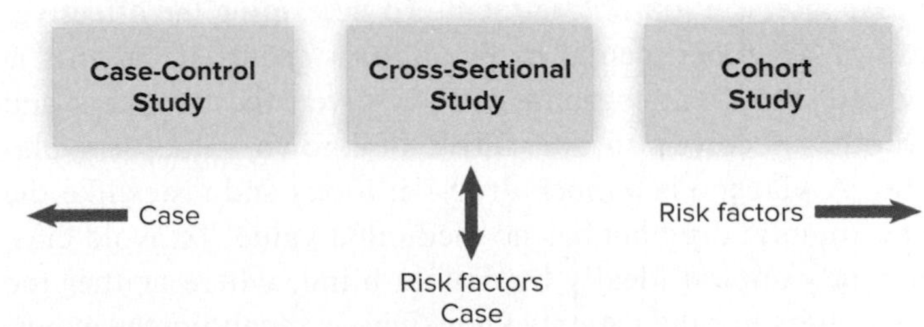

FIGURE 19.11 Analytical Studies Case-control studies are retrospective, cross-sectional studies gather information at one point in time, and cohort studies are prospective.

In a case-control study, how do the cases differ from the controls?

FOCUS ON A CASE 19.1

A 32-year-old female went to the emergency department complaining of fatigue, nausea, vomiting, and abdominal pain. The nurse recorded a fever of 100°F and noticed a yellowish skin color. Suspecting that it could be a type of hepatitis, a contagious viral disease that affects the liver, the physician ordered several tests, including one for liver enzyme activity and one that detects antibodies against hepatitis A virus (HAV). Results of the tests confirmed a diagnosis of hepatitis A, a disease spread through fecal-oral transmission. As required by law, the physician reported the case to the state health officials.

The source of hepatitis A can be difficult to trace because of the long incubation period—usually 3 to 5 weeks. After interviewing the patient, health officials determined that she was one of 162 people who were diagnosed with hepatitis A during a 4-month period. The source of infection was found to be a mixture of frozen fruits marketed as an organic antioxidant blend. The epi curve is shown in **box figure 19.1.**

1. Why is reporting the case to public health officials required by law?
2. How would public health officials determine the source of the infection?
3. The epi curve for this outbreak looks different from the common-source and propagated epi curves shown in figure 19.9. How can this be explained?

Discussion

1. By collecting the data, disease outbreaks can be detected early. If a commercial product is the source of infection, as it was in this situation, a recall can be issued to help control the outbreak. In addition, insights gained by analyzing the data can be used to develop programs to prevent future outbreaks.
2. Health officials gather and examine descriptive information about when and where cases occurred and who became ill. Hepatitis A is usually transmitted via contaminated food or water, so in this instance, officials asked people to list the foods they had eaten. The outbreak was traced to pomegranate seeds in the antioxidant product.
3. The implicated product was consumed over several weeks before the outbreak was noticed. Because of the long and variable incubation period of hepatitis A, cases occurred over a relatively long time compared with a typical common-source outbreak with an abrupt start. Also, hepatitis A is contagious and can be spread to secondary contacts, which complicates the situation.

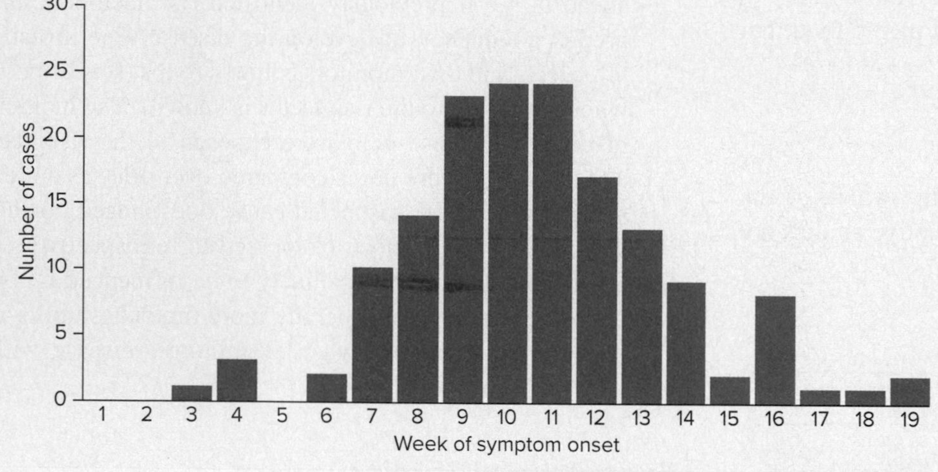

BOX FIGURE 19.1 Epi Curve The pattern of disease occurrence does not always clearly indicate whether an epidemic is propagated or common-source.
Source: "Multistate outbreak of hepatitis A virus infections linked to pomegranate seeds from Turkey (Final Update)," Centers for Disease Control and Prevention, October 23, 2013. http://www.cdc.gov/hepatitis/outbreaks/2013/a1b-03-31/epi.html

disease but is most often done to assess the value of a particular intervention or treatment. To determine the effectiveness of an experimental medication, a group of patients is divided into two subgroups; one is given the treatment and the other receives an alternative of known value or a placebo. A placebo is a mock drug—it looks and tastes like the experimental drug but has no medicinal value. To avoid bias, the study should ideally be double-blind, where neither the researchers nor the patients know who is receiving the experimental treatment until the study is over. Ethical issues sometimes make it necessary to use animals rather than human patients in experimental studies.

MicroAssessment 19.4

Descriptive epidemiological studies attempt to identify the potential risk factors that lead to disease. Analytical studies try to determine which suspected factors are actually relevant to disease development. Experimental studies are generally used to evaluate the effectiveness of a treatment or an intervention in preventing disease.

10. On what three factors does a descriptive study generally focus?

11. How is the timing of a propagated epidemic related to the incubation period of the pathogen?

12. Why must a scientific study to assess the benefit of a drug include an alternative medication or a placebo?

19.5 ■ Infectious Disease Surveillance

Learning Outcomes

6. Compare and contrast the roles of the Centers for Disease Control and Prevention, state public health departments, and the World Health Organization.

Infectious disease surveillance—which involves monitoring disease occurrence and then analyzing the data—is one of the most important aspects of disease prevention. By being aware of disease trends, the burden of a disease on a population and the effectiveness of preventive interventions can be evaluated. In addition, surveillance data can provide early warning of an outbreak.

National Disease Surveillance Network

Infectious disease control in the United States depends on a network of agencies across the country to monitor disease occurrence. It is partly because of this network that infectious diseases do not claim more lives.

Centers for Disease Control and Prevention (CDC)

The **Centers for Disease Control and Prevention (CDC)** in Atlanta, Georgia, is part of the U.S. Department of Health and Human Services. It provides support for public health departments in the United States and abroad, collects data on diseases that impact public health, and uses various means to distribute health information. As part of the CDC, the National Notifiable Diseases Surveillance System (NNDSS) collects and organizes data on *nationally notifiable diseases*—diseases for which state and local public health departments voluntarily report cases to the CDC (**table 19.1**); these diseases are typically of relatively high incidence or pose a potential danger to public health. The current data collected by the NNDSS, along with historical numbers to reflect any trends, are published online (https://www.cdc.gov/nndss/data-statistics/index.html).

A recent addition to the CDC's surveillance network is the National Wastewater Surveillance System (NWSS). This was developed to detect the presence of SARS-CoV-2 in wastewater as a means to track the virus's spread, but it also shows promise for detecting certain other pathogens as

TABLE 19.1 National Notifiable Infectious Diseases, 2023

- Anthrax
- Arboviral neuroinvasive and non-neuroinvasive diseases
- Babesiosis
- Botulism
- Brucellosis
- Campylobacteriosis
- *Candida auris,* clinical or screening
- Carbapenemase-producing organisms, clinical or screening
- Chancroid
- *Chlamydia trachomatis* infection
- Cholera
- Coccidioidomycosis
- Congenital syphilis, including stillbirth
- Coronavirus disease 2019 (COVID-19)
- Cryptosporidiosis
- Cyclosporiasis
- Dengue virus infections
- Diphtheria
- Ehrlichiosis and anaplasmosis
- Giardiasis
- Gonorrhea
- *Haemophilus influenzae,* invasive disease
- Hansen's disease
- Hantavirus infection
- Hantavirus pulmonary syndrome (HPS)
- Hemolytic uremic syndrome, post-diarrheal
- Hepatitis, viral (acute and chronic)
- HIV infection
- Influenza-associated pediatric mortality
- Invasive pneumococcal disease
- Legionellosis
- Leptospirosis
- Listeriosis
- Lyme disease
- Malaria
- Measles
- Melioidosis
- Meningococcal disease
- Mpox virus infection
- Mumps
- Novel influenza A infections
- Pertussis
- Plague
- Poliomyelitis, paralytic
- Poliovirus infection, nonparalytic
- Psittacosis
- Q fever
- Rabies, animal and human
- Rubella
- Rubella, congenital syndrome
- *Salmonella* Paratyphi infection
- *Salmonella* Typhi infection
- Salmonellosis
- Severe acute respiratory syndrome–associated coronavirus disease
- Shiga toxin–producing *Escherichia coli*
- Shigellosis
- Smallpox
- Spotted fever rickettsiosis
- Streptococcal toxic shock syndrome
- Syphilis
- Tetanus
- Toxic shock syndrome (other than streptococcal)
- Trichinellosis
- Tuberculosis
- Tularemia
- Vancomycin-intermediate *Staphylococcus aureus* (VISA) and vancomycin-resistant *Staphylococcus aureus* (VRSA) infection
- Varicella, including varicella deaths
- Vibriosis
- Viral hemorrhagic fever (VHF)
- Yellow fever
- Zika virus disease and Zika virus infection, congenital and non-congenital

FOCUS YOUR PERSPECTIVE 19.1

Bioterrorism Surveillance

Monitoring for indications of **bioterrorism**—the deliberate release of infectious agents or their toxins as a means to cause harm—is an important part of infectious disease surveillance. Prompt recognition of a bioterrorism attack, followed by rapid and appropriate isolation and treatment procedures, can help to minimize the consequences. The CDC, in cooperation with the Association for Professionals in Infection Control and Epidemiology (APIC), has prepared a bioterrorism readiness plan to be used as a template by healthcare facilities. Many of the recommendations are based on the Standard Precautions already used by hospitals to prevent the spread of infectious agents (see section 19.7).

The CDC separates bioterrorism agents into three categories based on the ease of spread and severity of disease. **Category A agents** pose the highest risk because they are easily spread or transmitted from person to person and result in high mortality. These agents include:

- ***Bacillus anthracis.*** Endospores of this bacterium are found in nature and are very stable. Inhalation anthrax, which results when an individual breathes in the airborne spores, can lead to a rapidly fatal systemic illness. Cutaneous anthrax, which occurs when the organism enters the skin, manifests as a blister that develops into a skin ulcer with a black center. Although the blister usually heals without treatment, the disease can potentially progress to a fatal bloodstream infection. Gastrointestinal anthrax results from consuming contaminated food, leading to bloody vomiting and severe diarrhea; it is not common but has a high case-fatality rate. Anthrax can be prevented by vaccination, but that option is not widely available. Preventive treatment with antimicrobial medications is possible for those who might have been exposed, but this requires prompt recognition of exposure. Fortunately, person-to-person transmission of the agent is not likely.
- ***Clostridium botulinum.*** Botulism occurs naturally if botulinum toxin produced by the organism is ingested. Any mucous membrane can absorb the toxin, so aerosolized toxin could be used as a weapon. The disease can be prevented by vaccination, but that option is not widely available; an antitoxin is also available in limited supplies. Botulism is not contagious.
- ***Yersinia pestis.*** Pneumonic plague, caused by inhalation of Yersinia pestis, is the most likely form of plague to result from a biological weapon. Although no effective vaccine is available, post-exposure preventive treatment with antimicrobial medications is possible. Special isolation precautions must be used for patients who have pneumonic plague because the disease is easily transmitted by respiratory droplets.
- **Smallpox virus.** Although a vaccine is available to prevent infection with this virus, routine immunization was stopped over 50 years ago because the natural disease has been eradicated. Special isolation precautions must be used for smallpox patients because the virus can be acquired through droplet, airborne, or contact transmission.
- ***Francisella tularensis.*** This bacterium, naturally found in animals such as rodents and rabbits, causes the disease tularemia; inhalation of even 50 cells can result in severe pneumonia. A vaccine is not available, but post-exposure preventive treatment with antimicrobial medications is possible. Fortunately, person-to-person transmission of the agent is not likely.
- **Filoviruses and arenaviruses.** These include the causative agents of Ebola virus disease and Marburg virus disease. Symptoms of the diseases vary, but severe cases may show signs of bleeding from many sites. Some of the viruses can be transmitted from person to person, so patient isolation in these cases is important. The only approved vaccine and treatments specifically target *Zaire ebolavirus* (commonly referred to as Ebola virus).

Category B agents pose moderate risk because they are relatively easy to spread and cause moderate morbidity. **Category C** agents are emerging pathogens that could be engineered for easy dissemination.

well. For example, some states are using wastewater testing to monitor the spread of the viruses that cause mpox and polio.

Each week, the CDC publishes the *Morbidity and Mortality Weekly Report (MMWR),* which includes articles relating to disease prevention efforts (http://www.cdc.gov/mmwr/). Potentially significant case reports—such as the 1981 report of a cluster of opportunistic infections in young homosexual men that marked the start of the AIDS epidemic—are also included in the *MMWR.* This publication is an invaluable aid to physicians, public health agencies, teachers, students, and others concerned about infectious disease or public health. In fact, many of the epidemiological charts and stories in this textbook are from the *MMWR.*

The CDC also conducts research relating to infectious diseases and can dispatch teams worldwide to assist with identifying and controlling epidemics. As the COVID-19 pandemic showed, an outbreak anywhere in the world is a potential threat to people living in the United States. The CDC also provides refresher courses for laboratory and infection control personnel.

MicroByte

Chlamydial infection is the most commonly reported sexually transmitted infection in the United States, with over 1.4 million cases reported to the CDC in 2022.

Public Health Departments

Each state has a state epidemiologist (or equivalent) who oversees a network of public health departments involved in various health-related activities, including infection surveillance and control. Individual states have the authority to mandate which diseases are *reportable diseases*—diseases for which physicians or other healthcare providers must report cases to the state. Once a case is reported, information is collected to help locate the source of the infection and then prevent the spread. The prompt response of health authorities in Washington State that helped stop a foodborne outbreak of *Escherichia coli* O157:H7 in 1993 was partly because Washington then was one of the few states with surveillance and reporting measures for that organism; the epidemic had started in other states but gone unrecognized. Many of the reportable diseases are also nationally notifiable, so the public health department then sends the relevant data, stripped of any personal identifying information, to the CDC.

Other Components of the Public Health Network

The public health network also includes public schools, which report absentee rates, and hospital laboratories, which report on the isolation of pathogens with epidemiological significance. In conjunction with these local activities, the news media alert the public to the presence of infectious diseases.

Worldwide Disease Surveillance

The **World Health Organization (WHO),** an agency of the United Nations, directs and coordinates health programs worldwide. One focus is to draw attention to what are referred to as neglected tropical diseases (NTDs)—a group of about 20 diseases that have devastating consequences primarily in resource-limited areas of the world. Like the CDC, the WHO collects data and provides health guidance, but the effort is maintained at a global level. The information provided by the WHO is distributed through a series of periodicals and books as well as through its website (www.who.int). For example, the *Weekly Epidemiological Record* reports timely information about epidemics of public health importance, particularly those of global concern.

MicroAssessment 19.5

In the United States, a network of agencies—including the Centers for Disease Control and Prevention (CDC) and state and local public health departments—monitor disease development. The World Health Organization (WHO) directs and coordinates worldwide health programs.

13. What is the *MMWR*?
14. Distinguish between a nationally notifiable disease and a reportable disease.
15. Explain why we have relatively accurate data on the number of cases of measles that occur in the United States but not on the number of cases of the common cold.

19.6 ▪ Trends in Infectious Diseases

Learning Outcomes

7. Describe the conditions that can allow eradication of an infectious disease.
8. Describe six factors that contribute to the emergence of infectious diseases.

Humans have been enormously successful in eliminating or reducing the occurrence of many diseases through improved sanitation, reservoir and vector control, vaccination, and antibiotic treatment (see figure 1.4). In fact, scientific advances made over the past several decades led to speculation that the war against infectious diseases, particularly those caused by bacteria, had been won. Microorganisms, however, have occupied this planet far longer than humans, evolving to reside in every habitat with the potential for life, including the human body. Thus, while certain diseases have been controlled, others—including new diseases and some that were previously contained—are emerging.

Reduction and Eradication of Infectious Diseases

In the United States and many other countries, certain diseases that were once common and claimed many lives are now relatively rare. Successful vaccination programs led to dramatic decreases in the number of deaths caused by *Haemophilus influenzae, Corynebacterium diphtheriae, Clostridium tetani, Bordetella pertussis,* and others. Meanwhile, controlling the source of diseases such as malaria, plague, and cholera effectively limited their spread.

One human disease—smallpox—has been globally eradicated, eliminating the natural occurrence of a disease that had a 25% case-fatality rate and disfigured many who survived (**figure 19.12**). Because humans were the only reservoir for smallpox virus, the disease could be eradicated through extensive vaccination (see A Glimpse of History, chapter 17); mass vaccination programs worldwide successfully led to dramatic reductions in the number of cases, and then ring vaccination was used for the final eradication steps. Dracunculiasis, caused by a helminth (worm) transmitted through contaminated drinking water, may be the next disease to be globally eradicated. An estimated 3.5 million cases occurred in 1986, but only 13 were reported in 2022. Meanwhile, rinderpest—a devastating disease of cattle and buffalo—is the only non-human animal disease that has been eradicated.

Even with the great successes in disease reduction, it is important to bear in mind that diseases previously controlled can rebound unless the pathogens are eliminated. Measles outbreaks associated with low vaccination rates are a sobering reminder of the need for continuous vigilance. Polio has been on the brink of eradication for several years, but outbreaks in countries where it had been eliminated are worrisome.

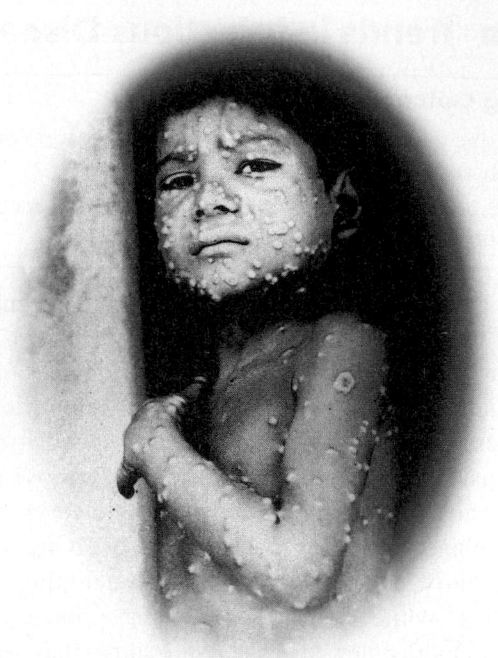

FIGURE 19.12 Child with Smallpox, a Disease That Has Since Been Eradicated World History Archive/Newscom

What characteristics make a disease a good candidate for eradication?

Emerging Infectious Diseases

Emerging infectious diseases are those that are novel or have increased in incidence in the last several decades. The one that everyone is now likely familiar with is COVID-19; other new or newly recognized diseases include Ebola virus disease, Zika virus disease, severe acute respiratory syndrome (SARS), Middle East respiratory syndrome (MERS), and *Candida auris* infection (see figure 1.5). Meanwhile, familiar diseases such as malaria, tuberculosis, and measles have increased in incidence in many areas of the world after years of decline. Some factors that contribute to the emergence of diseases include:

- **Microbial evolution.** Some diseases emerge due to the natural evolution of microbes. SARS-CoV-2, the virus that causes COVID-19, appears to have evolved from a virus that infects bats, as did the related viruses that cause SARS and MERS. Genomic sequencing studies of SARS-CoV-2 demonstrate that the virus continues to change, giving rise to new variants, some of which are more easily transmitted or escape previous immunity gained through vaccination or exposure. Strains of a relatively new serotype of *Vibrio cholerae*, designated O139, are closely related to those of an older serotype but gained the ability to produce a protective capsule; immunity against the older strains does not guarantee immunity to the newer ones. In the case of avian influenza, scientists are concerned that the virus will evolve to spread from person to person. Resistance to antimicrobial drugs is contributing to the reemergence of many diseases, including malaria.
- **Complacency.** As an infectious disease is controlled and therefore of less concern, complacency can develop, paving the way for resurgence of the disease. The early success of the plan to eliminate tuberculosis in the United States by the year 2000 caused news reports, education efforts, and research money to be diverted to more common diseases. At the same time, certain social welfare programs were cut, resulting in more people with poor health living in unstable conditions that put them at higher risk of developing active tuberculosis; in turn, the incidence of the disease increased. Fortunately, new public health measures brought the disease back under control in the United States.
- **Misinformation.** As the spread of misleading and false information about COVID-19 on social media demonstrated, misinformation can interfere with disease control. Measles, polio, and other vaccine-preventable diseases can resurge when misinformation leads to unfounded fears over vaccination.
- **Widespread use of antimicrobial medications.** Although antimicrobial medications can be life-saving, they can also create an environment that favors the growth of certain pathogens typically unable to compete with the normal microbiota. Compounding the problem is that people on such medications often have severe underlying conditions that makes them even more susceptible to infection. *Candida auris*, a yeast first isolated in 2009 in Japan, has recently caused severe infections in hospitalized patients around the globe. Treatment of those infections is difficult because some strains are resistant to the common antifungal medications.
- **Changes in human society.** Day-care centers, where diapered infants mingle, unmindful of sanitation and hygiene, are a relatively new component of society. For obvious reasons, these centers can be hotbeds of contagious diseases. This is particularly true for the intestinal diseases giardiasis and shigellosis, not only because the causative agents have a low infectious dose, but also because infants often explore through taste and touch and are thus likely to ingest fecal organisms (**figure 19.13**). Moreover, many young children have not yet acquired immunity to communicable diseases such as colds and diarrheal illnesses that are readily transmitted among this susceptible population.
- **Advances in technology.** Technology may make life easier, but it can inadvertently create new habitats for microorganisms. For example, the introduction of contact lenses gave microorganisms the opportunity to grow in a new location—the lenses and storage solutions of people who did not use proper disinfection techniques. In turn, this resulted in new types of eye infections.

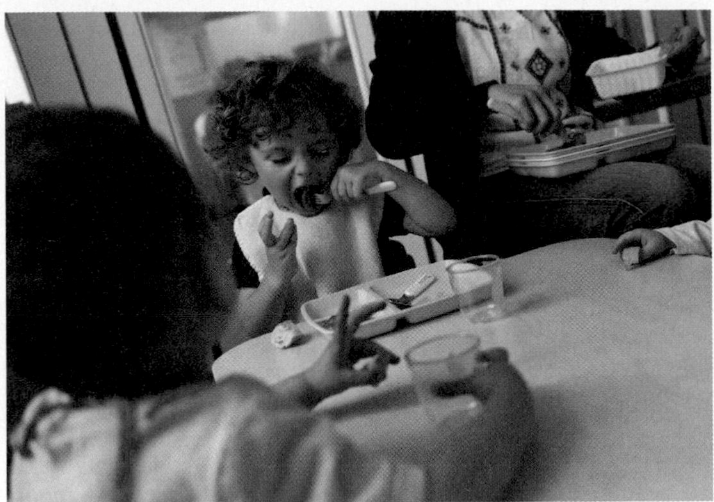

FIGURE 19.13 Children in a Day-Care Center Loic Venance/AFP/Getty Images

Why do shigellosis and giardiasis easily spread in day-care centers?

- **Population expansion.** As populations increase, people often move into areas where they are more likely to encounter reservoirs of disease. When new homes are built along the borders of forests, for example, humans are exposed to more deer that carry ticks, the host for *Borrelia burgdorferi,* the cause of Lyme disease.
- **Development.** Dams, which provide important sources of power necessary for economic development, have inadvertently extended the range of certain diseases. The life cycle of the parasite that causes the disease schistosomiasis involves an aquatic snail. Construction of dams such as the Aswan Dam on the Nile River has increased the habitat for the snail, thus extending the range of the parasite.
- **Mass production, widespread distribution, and importation of food.** Foodborne illness has always existed, but the ease with which we can now transport items worldwide from a single production plant can create new problems. Widespread distribution of foods contaminated with pathogens can result in similarly broad outbreaks of disease. In 2022, 17 countries reported cases of *Salmonella enterica* serotype Typhimurium linked to chocolates produced at a single factory.
- **War and civil unrest.** War and civil unrest can disrupt the infrastructure on which disease prevention relies. Refugee camps that crowd people into substandard living quarters lacking toilet facilities and safe drinking water are hotbeds of infectious diseases such as cholera and dysentery. Unfortunately, war also disrupts disease-eradication efforts.
- **Climate changes.** Warm temperatures favor the survival and spread of some arthropods, which can serve as vectors for diseases such as malaria, dengue, West Nile encephalitis, and Zika virus disease. Heavy rainfall and flooding often result in surges of cholera cases in developing countries.

The list of factors that contribute to disease emergence helps illustrate the importance of the One Health approach described in section 19.3. Because disease emergence involves complex interactions among humans, animals, and the environment, experts in a variety of fields must work together to devise and implement methods to control the occurrence.

MicroByte

About 75% of emerging infectious diseases affecting humans are of animal origin.

MicroAssessment 19.6

Humans have been successful in reducing or eliminating certain diseases, but people must remain vigilant to prevent their resurgence. A number of situations, including evolution of the pathogens, changes in human behavior, and climate change, can provide new opportunities for pathogens to thrive.

16. Explain why smallpox was successfully eradicated, but rabies probably never will be.
17. How would you expect a trend toward warmer climates to affect the spread of vector-borne disease?
18. What political and societal factors might lead to a decrease in childhood immunizations?

19.7 ■ Healthcare-Associated Infections

Learning Outcomes

9. Describe five reservoirs of infectious agents in healthcare settings and three mechanisms by which the agents can be transferred to patients.
10. Identify the most common types of healthcare-associated infections.
11. Describe the role of the Standard Precautions in preventing healthcare-associated infections.

Healthcare-associated infections (HAIs) are infections acquired while receiving treatment in a healthcare setting such as a hospital, urgent care facility, or long-term care residence; those acquired in the hospital may also be called **nosocomial infections** (derived from the Greek word for hospital). HAIs are a significant concern because they are a relatively common but largely preventable cause of death. Although tremendous progress has been made toward preventing the infections, more can still be done. In addition, vigilance is needed to maintain the progress made so far, as illustrated by HAI rates that increased when healthcare systems were strained during the height of the COVID-19 pandemic.

Reservoirs of Infectious Agents in Healthcare Settings

Many agents that cause HAIs seldom cause disease in healthy people, but a sick or immunocompromised person undergoing surgery or being treated with an invasive medical procedure is highly vulnerable to infection. The microbes involved—some of which are listed in **table 19.2**—can originate from a number of different sources, including:

- **Patient microbiota.** Nearly any invasive procedure can transmit an organism of a patient's normal microbiota to otherwise sterile body sites. When intravenous fluids are administered, *Staphylococcus epidermidis* or other skin microbiota can potentially gain access to the bloodstream. Similarly, some patients can inadvertently inhale their normal oral microbiota, resulting in healthcare-associated pneumonia. Severely immunocompromised patients, such as people who have undergone cancer chemotherapy or are on immunosuppressive drugs, are prone to activation of latent infections that their immune system previously controlled.
- **Healthcare environment.** Some Gram-negative rods—particularly the common opportunistic pathogen *Pseudomonas aeruginosa*—can thrive in moist healthcare environments such as sinks, ventilators, and toilets. Not only is *P. aeruginosa* resistant to many disinfectants and antimicrobial medications, it requires few nutrients, enabling it to multiply in environments containing little other than water. Many HAIs have been traced to soaps, disinfectants, and other aqueous solutions that were contaminated with the bacterium.
- **Other patients.** Patients who have an infectious disease, perhaps the reason for their hospitalization, may release pathogens into the environment via skin cells, respiratory droplets, feces, and other body secretions or excretions. Thorough cleaning and the use of disinfectants minimize the spread of these pathogens.
- **Visitors.** Patients can be exposed to infectious agents by visitors to their hospital rooms. To minimize this risk, hospitals often limit the number of visitors allowed and ask anyone who has symptoms of an infection to contact staff for advice before visiting. Also, visitors and anyone else should wash their hands or use alcohol gel before and after touching a patient. Healthcare facilities may also ban animals or live plants in some patient rooms.
- **Healthcare workers.** Outbreaks of HAIs are sometimes traced to infected personnel. Clearly, those who report to work with even a mild case of influenza can expose patients to an infectious agent that may have serious or fatal consequences to those in poor health. A more troublesome source of infection is a healthcare worker who is an asymptomatic carrier of a pathogen such as *Staphylococcus aureus*; these personnel may not recognize that they pose a risk to patients until they are implicated in an outbreak. Carriers who are members of a surgical team pose a particular threat because transfer of a pathogen directly into a surgical site can result in a systemic infection.

Transmission of Infectious Agents in Healthcare Settings

Transmission of infectious agents in healthcare settings can be direct or indirect and include:

- **Direct Transmission—Healthcare Personnel.** Healthcare personnel must be extremely vigilant to avoid

TABLE 19.2 Common Bacterial Causes of Healthcare-Associated Infections (HAIs)

Infectious Agent	Comments
Acinetobacter baumannii	Environmental bacterium found on skin of healthy people; causes a variety of HAIs; most strains are multi-drug resistant
Candida species (*C. albicans, C. auris,* and others)	Yeasts; some are part of the normal microbiota; common cause of bloodstream infections; some are resistant to antifungal medications
Clostridioides difficile	Endospore-former that is sometimes part of the intestinal microbiota; if antibiotic use suppresses competing bacteria, the organism grows to high numbers; toxin-producing strains cause diarrhea and colitis
Enterobacteriaceae (*E. coli, Klebsiella pneumoniae,* and others)	Part of the normal intestinal microbiota; cause a variety of HAIs; some strains are multidrug resistant, including to last-resort antibiotics
Enterococcus species	Part of the normal intestinal microbiota; cause a variety of HAIs; some strains are multi-drug resistant
Pseudomonas species	Environmental bacteria that grow in moist, nutrient-poor environments such as the humidifier of a mechanical ventilator; cause a variety of HAIs; some strains are multi-drug resistant
Staphylococcus aureus	Often part of the skin microbiota of healthy people; causes a variety of HAIs; some strains are multi-drug resistant

transmitting infectious disease agents, particularly from patient to patient. What Ignaz Semmelweis found to be true in the 1800s is equally true today—handwashing helps prevent the spread of disease (see A Glimpse of History). Healthcare personnel should routinely wash or disinfect their hands after touching one patient before going to the next. Patient rooms typically have alcohol gel dispensers for use upon entering, but soap and water is more effective in preventing the transmission of some agents, such as *Clostridioides difficile;* recall that this diarrhea-causing organism produces endospores that are not destroyed by alcohol but can be removed by washing. Healthcare professionals should perform a more thorough hand scrubbing (lasting 10 minutes with a suitable antiseptic) before participating in surgery or when working in a nursery or an intensive care or isolation unit. They should also wear gloves when contacting blood, mucous membranes, broken skin, or body fluids.

- **Indirect Transmission—Medical Devices.** Devices used for diagnostic and therapeutic procedures can inadvertently transmit infectious agents to patients; patients in intensive care units (ICUs) are particularly vulnerable because so many invasive mechanisms are needed to provide life-saving support (**figure 19.14**). Many hospitalized patients have a urinary catheter in place, which can potentially introduce microbes into the normally sterile urinary bladder, resulting in a catheter-associated urinary tract infection (CAUTI). Intravascular (IV) catheters used to deliver fluids and medications into the bloodstream also provide a route for microbial entry, potentially leading to catheter-related bloodstream infections (CRBSIs). A mechanical ventilator that assists a patient's breathing by pumping air directly into the trachea can also potentially deliver microbes that cause ventilator-associated pneumonia (VAP). Meanwhile, if inadequately sterilized instruments are used in an invasive procedure such as surgery or a biopsy, they can transmit infectious agents. Common types of healthcare-associated infections, many of which are transmitted by devices, are listed in **table 19.3.**

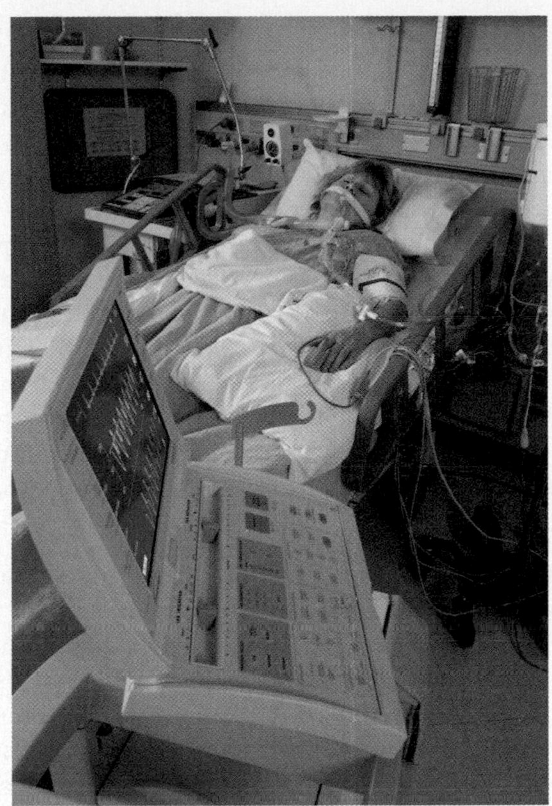

FIGURE 19.14 Patient in an Intensive Care Unit David Joel/Photographer's Choice/Getty Images

? To what sources of HAIs is this patient exposed?

- **Indirect Transmission—Airborne.** Most hospitals are designed to minimize the airborne spread of microbes. Airflow to operating rooms is usually supplied under slight pressure, thereby preventing contaminated air in the corridors from flowing into the room. Floors are washed with a damp mop or floor washer to avoid sweeping microbes into the air. High-efficiency particulate air (HEPA) filters, which remove most airborne particles, are used to exclude airborne microbes from rooms in which extremely susceptible patients reside.

TABLE 19.3 Common Healthcare-Associated Infections

Type	Comments
Urinary tract infections (UTIs)	Urinary catheter use can lead to catheter-associated urinary tract infection (CAUTI).
Bloodstream infections	Intravascular catheter use can lead to catheter-related bloodstream infection (CRBSI).
Pneumonia	Mechanical ventilator use can lead to ventilator-associated pneumonia (VAP).
Surgical site infections (SSIs)	Incisions made in skin or mucous membranes can lead to infection.
Gastrointestinal infections	Antibiotic use can lead to gastrointestinal infections caused by a variety of microbes, particularly *Clostridioides difficile*.

Preventing Healthcare-Associated Infections

One of the biggest challenges for a healthcare facility has always been to prevent the spread of disease within that confined setting. To develop solutions, nearly every hospital has an infection control committee composed of various personnel from within the facility, including nurses, physicians, dietitians, environmental services staff, epidemiologists, and medical laboratory scientists. Hospitals may employ an infection control practitioner (ICP) to perform active surveillance of the types and numbers of infections that occur. The infection control committee, along with the ICP, develops infection prevention and control policies based on the set of procedures referred to as **Standard Precautions,** which apply to all patient care and can be summarized as follows:

- **Hand hygiene.** Unnecessary touching of surfaces near a patient should be avoided. If hands are not visibly soiled, they should be decontaminated with an alcohol-based rub before direct contact with patients; after contact with blood, body fluids, secretions, excretions, or contaminated items; immediately after removing gloves; and between patient contacts. If hands are visibly dirty or contaminated or if contact with spores is likely to have occurred, hands should be washed with soap and water.
- **Personal protective equipment (PPE).** Gloves should be worn when touching blood, body fluids, secretions, excretions, mucous membranes, non-intact skin, and any contaminated items. A gown should be worn when contact of clothing or exposed skin with blood/body fluids, secretions, and excretions is anticipated. A mask, goggles, or face shield should be worn when splashes or sprays of blood, body fluids, or secretions are anticipated.
- **Respiratory hygiene/cough etiquette.** Symptomatic persons should cover their mouth/nose with tissue when sneezing/coughing, and then dispose of the tissue in a no-touch receptacle. Hand hygiene should be used if hands are contaminated with respiratory secretions. A surgical mask should be worn if tolerated; otherwise, spatial separation of more than 3 feet should be maintained, if possible.
- **Patient placement.** Certain high-risk patients, including those most likely to transmit a disease or at risk of a severe outcome if an HAI is acquired, should be prioritized for single-patient rooms.
- **Patient-care equipment and instruments/devices.** Soiled equipment should be handled in a manner that prevents transfer of microorganisms to others and to the environment; gloves should be worn if the item is visibly contaminated.
- **Textiles and laundry.** Textiles and laundry should be handled in a manner that prevents transfer of microorganisms to others and to the environment.
- **Safe injection practices.** Aseptic technique should be used to avoid contaminating sterile injection equipment. Specific precautions describe how medications and IV solutions are stored and administered.
- **Worker safety including proper handling of needles and other sharps.** Federal and state requirements for protection of healthcare personnel from exposure to bloodborne pathogens should be obeyed.

A set of supplementary measures called **Transmission-Based Precautions** is used in addition to the Standard Precautions if a patient is, or might be, infected with a highly transmissible or epidemiologically important pathogen. The Transmission-Based Precautions are separated into three sets—Airborne Precautions, Droplet Precautions, and Contact Precautions—that are used singly or in combination as appropriate. The set of precautions appropriate for a given patient are often posted outside the patient's door to inform staff and visitors of the policies.

To assist the work of hospitals, the CDC's National Healthcare Safety Network (NHSN) tracks and publishes HAI data, making it easier to identify problem areas as well as to assess progress in prevention. In addition, the Healthcare Infection Control Practices Advisory Committee (HICPAC) provides advice to hospitals and recommends guidelines for surveillance, prevention, and control of healthcare-associated infections.

MicroAssessment 19.7

Invasive treatments and resistant microorganisms contribute to the incidence of healthcare-associated infections (HAIs). These infections may originate from other patients, the healthcare environment, healthcare workers, or the patient's own normal microbiota. Diagnostic and therapeutic procedures can potentially transmit infectious agents. The most important steps in preventing HAIs are to first detect their occurrence and then establish policies to prevent their development.

19. Explain why an IV catheter poses a risk to a patient.
20. Describe two ways in which infectious agents can be transmitted to a patient.
21. Explain why the rate of HAIs is often relatively high in emergency room settings.

Summary

19.1 ■ Basic Concepts of Epidemiology

Epidemiologists study the frequency and distribution of disease in order to identify its cause, source, and route of transmission. Diseases of public health significance have a **case definition.** Some diseases are **endemic** to a given region. An **epidemic** can be caused by an endemic disease or one recently introduced to a region (figure 19.1). An epidemic may spread globally to become a **pandemic.** Epidemiologists focus on the rate and effects of disease, including the **attack rate**, the **case-fatality rate,** and the **infection-fatality rate**.

19.2 ■ Chain of Infection

The spread of a disease can be prevented by breaking the chain of infection (figure 19.2).

Reservoirs of Infection (figure 19.3)

Preventing susceptible people from coming in contact with a **reservoir of infection** can prevent infectious disease. People who have asymptomatic infections or are colonized with a pathogen are **carriers** of the infectious agent. **Zoonoses** can be transmitted to humans but exist primarily in other animals. Pathogens with environmental reservoirs are probably impossible to eliminate.

Portals of Exit (figure 19.4)

Pathogens may be shed in feces, in respiratory droplets, on skin cells, in genital secretions, and in urine.

Disease Transmission (figure 19.5)

Vertical transmission occurs between mother and fetus or infant. **Horizontal transmission** includes direct transmission and indirect transmission. **Direct transmission** includes direct contact and droplet spread. **Indirect transmission** may be airborne, vehicle-borne, or vector-borne. Airborne transmission is the most difficult to control. Vehicles of transmission include **fomites,** food, and water. A **mechanical vector** carries a microbe on its body; a pathogen can multiply within a **biological vector** (figure 19.7).

Portals of Entry

Some organisms can cause disease if they enter one body site but are harmless if they enter another.

19.3 ■ Factors That Influence the Epidemiology of Disease

The Dose

The probability of infection and disease is generally lower if an individual is exposed to small numbers of pathogens.

The Incubation Period

Diseases with long incubation periods can spread far and wide before the first cases appear.

The Host Population

Disease distribution can be influenced by various host population characteristics, including immunity to a pathogen, general health, gender, behavioral practices, genetic background, and age (figure 19.8).

The Environment

Environmental factors determine which organisms can exist, reproduce, and spread in a given location.

19.4 ■ Epidemiological Studies

Descriptive Studies

Descriptive studies are used to identify potential risk factors that correlate with disease development. Determining the time that symptoms began for each case helps distinguish a **common-source epidemic** from a **propagated epidemic** (figure 19.9). Some epidemics are seasonal (figure 19.10).

Analytical Studies (figure 19.11)

Analytical studies are designed to determine which risk factors are actually relevant to disease development. A **case-control study** compares the past activities of cases with those of controls to help determine the cause of the epidemic. **Cross-sectional studies** survey a range of people at a defined point in time. **Cohort studies** compare groups to determine if the identified risk factors predict a tendency to develop disease.

Experimental Studies

Experimental studies are generally used to evaluate the effectiveness of a treatment or intervention in preventing disease.

19.5 ■ Infectious Disease Surveillance

National Disease Surveillance Network

The **Centers for Disease Control and Prevention (CDC)** provides support for public health departments in the United States and abroad; as part of the CDC, the National Notifiable Diseases Surveillance System (NNDSS) collects and organizes data on nationally notifiable diseases (table 19.1). State public health departments are involved in infection surveillance and control (table 19.1).

Worldwide Disease Surveillance

The **World Health Organization (WHO)** directs and coordinates health programs worldwide.

19.6 ■ Trends in Infectious Diseases

Reduction and Eradication of Infectious Diseases

Smallpox is the only human disease that has been eradicated (figure 19.12). Rinderpest is the only non-human animal disease that has been eradicated.

Emerging Infectious Diseases

Emerging infectious diseases are new or newly recognized, or are reemerging after years of decline. Factors that contribute to disease emergence include: microbial evolution; complacency; spread of misinformation; widespread use of antimicrobial medications; changes in human society; advances in technology; population expansion; development; mass production, distribution and importation of food; war and civil unrest; and climate changes (figure 19.13).

19.7 ■ Healthcare-Associated Infections (table 19.2; table 19.3)

Healthcare-associated infections (HAIs) are acquired by individuals in a healthcare setting. **Nosocomial infections** are acquired in a hospital.

Reservoirs of Infectious Agents in Healthcare Settings

The organisms that cause HAIs can originate from a number of sources, including the patient's own microbiota, the healthcare environment, other patients, visitors, and healthcare workers.

Transmission of Infectious Agents in Healthcare Settings
Infectious agents can be inadvertently transmitted to patients by healthcare personnel, medical devices, and by airborne spread.

Preventing Healthcare-Associated Infections
Infection prevention and control policies rely on the set of procedures referred to as **Standard Precautions**, which apply to all patient care. **Transmission-Based Precautions** are used in addition to the Standard Precautions if a patient is, or might be, infected with a highly transmissible or epidemiologically important pathogen.

Review Questions

Short Answer

1. Compare the impact on a society of an endemic debilitating disease with high incidence to the impact on society of a similar disease with high prevalence.
2. What is the epidemiological significance of people who have asymptomatic infections?
3. Explain why zoonotic diseases are often severe in humans.
4. List the main portals of exit from the human body.
5. Name the most important control measure for preventing person-to-person transmission of a disease.
6. Describe the factors within a population that may make it more susceptible to infectious disease.
7. Draw representative epi curves (time versus number of people ill) depicting a propagated and a common-source epidemic.
8. Differentiate among a case-control study, a cross-sectional study, and a cohort study.
9. Describe the factors that contribute to the emergence or reemergence of disease.
10. What are the main reservoirs of healthcare-associated infections?

Multiple Choice

1. Which of the following is an example of a fomite?
   a) Table
   b) Flea
   c) *Staphylococcus aureus* carrier
   d) Water
   e) Air
2. Which of the following would be the easiest to eradicate?
   a) A pathogen that is common in wild animals but sometimes infects humans
   b) A disease that occurs exclusively in humans, always resulting in obvious symptoms
   c) A mild disease of humans that often results in no obvious symptoms
   d) A pathogen found in marine sediments
   e) A pathogen that readily infects both wild animals and humans
3. Which of the following methods of disease transmission is the most difficult to control?
   a) Airborne
   b) Foodborne
   c) Waterborne
   d) Vector-borne
   e) Direct person to person
4. Which of the following statements is *false*?
   a) A botulism epidemic that results from improperly canned green beans is an example of a common-source outbreak.
   b) Droplet nuclei fall quickly to the ground.
   c) Congenital syphilis is an example of a disease acquired through vertical transmission.
   d) Plague is endemic in the rock squirrel population in parts of the United States.
   e) The first case in an outbreak is called the index case.
5. Which of the following statements is *false*?
   a) A disease with a long incubation period might spread extensively before an epidemic is recognized.
   b) A person exposed to a low dose of a pathogen might not develop disease.
   c) The young and the aged are more likely to develop certain diseases.
   d) Malnourished populations are more likely to develop certain diseases.
   e) Herd immunity occurs when a population does not engage in a given behavior, such as eating raw fish, that would otherwise increase their risk of disease.
6. The purpose of an analytical study is to
   a) identify the person, place, and time of an outbreak.
   b) determine which potential risk factors result in high frequencies of disease.
   c) assess the effectiveness of preventive measures.
   d) determine the effectiveness of a placebo.
   e) None of the above
7. If you and your family all develop infectious diarrhea, the most likely portal of entry for the pathogen was the
   a) large intestine.
   b) mouth.
   c) skin.
   d) respiratory tract.
   e) nose.
8. All of the following are thought to contribute to the emergence of disease *except*
   a) advances in technology.
   b) breakdown of public health infrastructure.
   c) construction of dams.
   d) mass distribution and importation of food.
   e) widespread vaccination programs.
9. Which of the following common causes of healthcare-associated infections is an environmental organism that grows readily in nutrient-poor solutions?
   a) *Enterococcus*
   b) *Escherichia coli*
   c) *Pseudomonas aeruginosa*
   d) *Staphylococcus aureus*

10. An endemic disease
    a) is typically transmitted by asymptomatic carriers.
    b) is more prevalent in the winter months than in the summer months.
    c) requires transmission by a vector.
    d) has been eradicated.
    e) is always present at some level in a given population.

Applications

1. A news station reported about a potentially fatal epidemic disease occurring in a small African village. An epidemiologist from the CDC was interviewed to discuss the disease and was very distressed that it was not being contained. Why did the epidemiologist feel the disease was a concern for people in North America?
2. An international team was gathered to discuss how funding should be spent to eliminate human infectious disease. There is only enough funding to eliminate one disease. How would the scientists go about choosing the next disease to be eliminated from the planet?

Critical Thinking

1. A student disagreed with the presentation of the examples in figure 19.9. She claimed that the number of cases from a common-source outbreak could remain high over a much longer period of time in some cases and not decrease to zero. Is the student's claim reasonable? Why or why not?
2. As shown in the graphs that follow, the proportion of deaths due to cancer and heart disease in the United States in 2010 was much higher than it was in 1900. Explain the factors that most likely contribute to this difference.

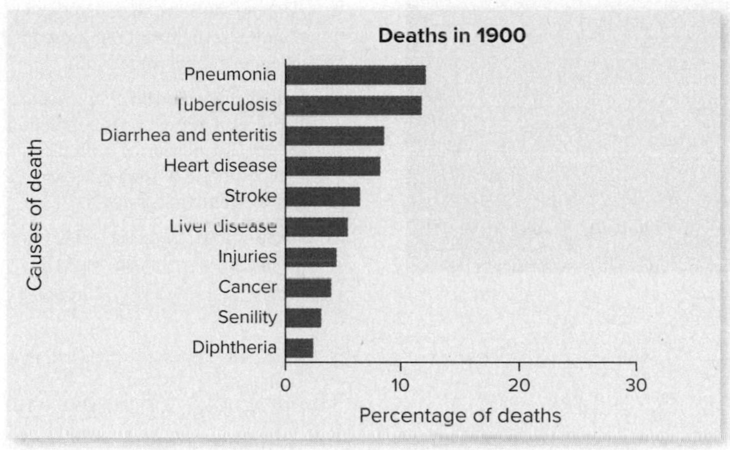

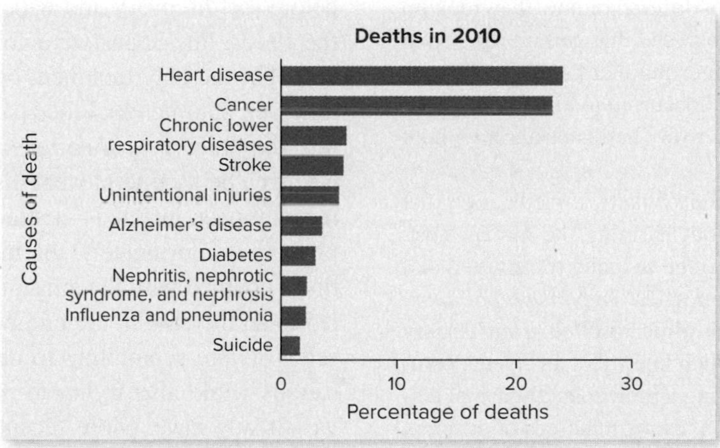

20 Antimicrobial Medications

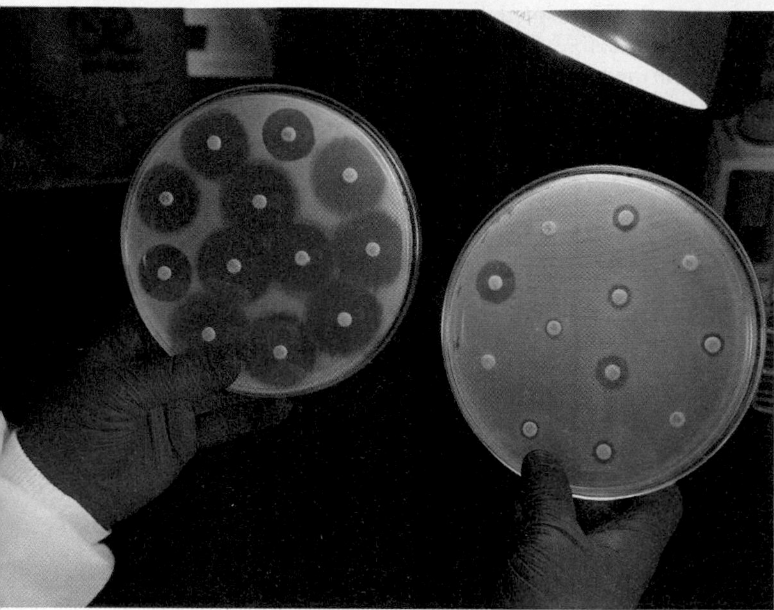

Antibiotic susceptibility testing. *(James Gathany/CDC)*

KEY TERMS

Acquired Resistance Resistance that develops due to genetic changes, including mutation and horizontal gene transfer.

Antibiotic A compound naturally produced by molds or bacteria that inhibits the growth of or kills other microorganisms.

Antimicrobial Medication An antibiotic or other chemical that kills microbes or inhibits their growth; also called an antimicrobial drug.

Antiviral Medication A chemical that is used to treat a viral infection and acts by interfering with the infection cycle of the virus; also called an antiviral drug.

Bactericidal Describes a chemical or other agent that kills bacteria.

Bacteriostatic Describes a chemical or other agent that stops the growth of bacteria without killing them.

Broad-Spectrum Antibiotic An antibiotic that is effective against a wide range of bacteria, generally including both Gram-positive and Gram-negative bacteria.

Chemotherapeutic Agent A chemical used to treat disease.

Intrinsic (Innate) Resistance Resistance due to inherent characteristics of an organism.

Narrow-Spectrum Antibiotic An antibiotic that is effective against a limited range of bacteria.

R Plasmid A plasmid that encodes resistance to one or more antimicrobial medications.

A Glimpse of History

Paul Ehrlich (1854–1915), a German physician and bacteriologist, noticed how various types of body cells differed in how they took up dyes and other substances. When he observed that certain dyes stain bacterial cells but not animal cells, suggesting that the two cell types are somehow fundamentally different, it occurred to him that it might be possible to find a chemical that selectively harms bacteria without affecting human cells.

Ehrlich began searching for a "magic bullet," a medication that would kill a microbial pathogen without harming the human host. He began looking for a chemical to cure the sexually transmitted disease syphilis, which is caused by the spirochete *Treponema pallidum*. Much of the mental illness during this time resulted from tertiary syphilis, a late stage of the disease. Ehrlich knew that an arsenic compound had shown some success in treating a protozoan disease of animals, and so he and his colleagues began testing hundreds of different arsenic compounds in search of a cure for syphilis. In 1909, the 606th compound tested, arsphenamine, was found to be highly effective in treating the disease in laboratory animals. Although the compound itself was potentially lethal for patients, it cured infections previously considered hopeless. The medication was given the name Salvarsan, a term derived from the words salvation and arsenic. Ehrlich's discovery proved that some chemicals could indeed selectively kill microbes.

Think back to the last time you were prescribed an **antimicrobial medication,** a chemical used to treat an infectious disease. Could you have recovered without the medication? The prognosis for people with common diseases such as bacterial pneumonia and severe staphylococcal infections was grim before the discovery and widespread availability of penicillin in the 1940s. Physicians were able to identify the cause of the disease, but the only treatment option was usually bed rest. Today, however, antimicrobial medications are routinely prescribed, and this simple cure is often taken for granted. In some cases, they can even be used to prevent disease in someone who might have recently been infected—a strategy known as post-exposure prophylaxis. Unfortunately, the misuse of these life-saving medications, coupled with the amazing ability of microbes to adapt, has led to an increase in the number of resistant strains. In response, scientists are scrambling to develop new varieties of the medications while also trying to maintain the effectiveness of ones we already have. More recently, the COVID-19 pandemic has spurred research toward expanding the relatively meager selection of medications available for treating viral diseases.

20.1 ■ History and Development of Antimicrobial Medications

Learning Outcomes

1. Describe the discovery of antimicrobial medications, including antibiotics.
2. Explain how new antimicrobial medications are developed.

To appreciate the important role of antimicrobial medications in modern life, it is necessary to understand the history and development of these life-saving remedies. The options used to treat bacterial diseases were discovered first and are by far the most numerous, so this section will focus mostly on those.

Discovery of Antimicrobial Medications

The development of Salvarsan by Paul Ehrlich was the first documented example of a chemical used successfully as an antimicrobial medication (see A Glimpse of History). The next breakthrough came almost 25 years later when the German chemist Gerhard Domagk discovered that the red dye Prontosil could be used to treat streptococcal infections in animals. Surprisingly, Prontosil had no effect on streptococci grown in test tubes. It was later discovered that enzymes in the animals' blood split the Prontosil molecule, releasing sulfanilamide, which acted against the infecting streptococci. Thus, the discovery of sulfanilamide, the first of a group of chemicals now called sulfa drugs, was based on both luck and scientific effort. If Prontosil had been screened only against bacteria in test tubes and not given to infected animals, its effectiveness might not have been discovered.

Salvarsan and Prontosil are **chemotherapeutic agents,** chemicals used to treat disease. Because they are used to treat microbial infections, they are more specifically called antimicrobial medications, antimicrobial drugs, or, more simply, antimicrobials.

Discovery of Antibiotics

In 1928, Alexander Fleming, a British scientist, was working with cultures of *Staphylococcus aureus* when he noticed that colonies growing near a contaminating mold appeared to be dissolving (**figure 20.1**). Recognizing that the mold might be secreting a substance that killed bacteria, he studied it more carefully. He identified the mold as a species of *Penicillium* and found that it was indeed producing a bacteria-killing substance; he called this penicillin. Even though Fleming could not purify penicillin, he showed that it was remarkably effective in killing many different bacterial species and did not cause adverse effects when injected into rabbits and mice. Fleming recognized the potential medical significance of his discovery but eventually stopped studying it after he became discouraged with his inability to purify the compound.

About 10 years after Fleming's discovery of penicillin, two other scientists working in England, Ernst Chain and Howard Florey, successfully purified the compound. In 1941, the chemical was tested for the first time on a police officer with a life-threatening *Staphylococcus aureus* infection. The patient improved so dramatically that within 24 hours his illness seemed under control. Unfortunately, the supply of purified penicillin ran out, and the man eventually died of the infection. Later, with greater supplies of penicillin, two deathly ill patients were successfully cured.

The need for effective antimicrobial medications to treat soldiers wounded in World War II caused British and American scientists to collaborate to develop the means for large-scale penicillin production and to determine its chemical structure. Under the direction of Florey, efforts to mass-produce penicillin at Oxford University were initiated in the late 1930s; a team of six women, nicknamed "the penicillin girls" were instrumental in this work. Penicillin production was greatly expanded during World War II across both nations. Meanwhile, in 1941, Oxford biochemist Dorothy Crowfoot Hodgkin began X-ray crystallography studies to determine penicillin's structure, a feat she completed in 1945. In recognition of this as well as other contributions, she received the Nobel Prize in Chemistry in 1964. Ultimately, several penicillins were found in the different *Penicillium* cultures and were designated alphabetically. Penicillin G (or benzyl penicillin) was found to be the most suitable for treating infections. This was the first of what we now call **antibiotics**—antimicrobial medications naturally produced by microorganisms.

Soon after the discovery of penicillin, Selman Waksman's research team screened soil bacteria (especially a type called actinomycetes) for antibiotic production. Waksman and colleagues Albert Schatz and Elizabeth Bugie ultimately discovered that the soil bacterium *Streptomyces griseus* produced a powerful antibiotic they called streptomycin. The realization that both bacteria and molds could produce antibiotics prompted researchers to begin screening hundreds of thousands of microbial strains for antibiotic production. Even today, pharmaceutical companies examine soil samples from around the world in hopes of finding microbes that produce previously undiscovered antibiotics.

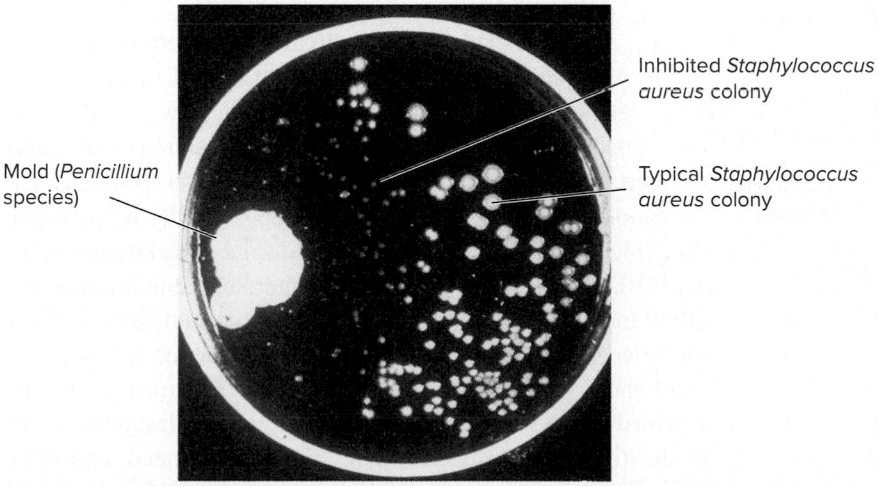

Mold (*Penicillium* species)

Inhibited *Staphylococcus aureus* colony

Typical *Staphylococcus aureus* colony

FIGURE 20.1 Mold Affecting the Growth of *Staphylococcus aureus* Alexander Fleming's photograph of his contaminated plate that led to the discovery of penicillin.
Biophoto Associates/Science Source

? **Why are only the colonies growing close to the mold inhibited?**

Development of New Antimicrobial Medications

Most antibiotics come from microorganisms that normally live in the soil, where many microbes produce chemicals including antibiotics to gain a competitive advantage in natural microbial communities. Species of *Streptomyces* and *Bacillus* (bacteria) and *Penicillium* and *Cephalosporium* (fungi) are known for their antibiotic-producing capabilities. To commercially produce an antibiotic, a carefully selected strain of the appropriate species is grown in a huge vat of broth medium. The antibiotic is then extracted from the medium and purified.

In the 1960s, scientists began altering the chemical structure of certain antibiotics to give them new properties (**figure 20.2**). For example, penicillin G, which is active mainly against Gram-positive bacteria, was altered to create ampicillin, which kills a variety of Gram-negative species as well. Another change to penicillin created methicillin, which is less susceptible to some bacterial enzymes that destroy penicillin. Today a variety of penicillin-like medications exist, referred to as the family of penicillins, or simply penicillins. Other unrelated antibiotics have also been altered to give them new characteristics. These chemically modified compounds are called semisynthetic. In some cases, the entire substance can be synthesized in the laboratory. By convention, these partially or totally synthetic chemicals are still called antibiotics because microorganisms can produce the core structure naturally.

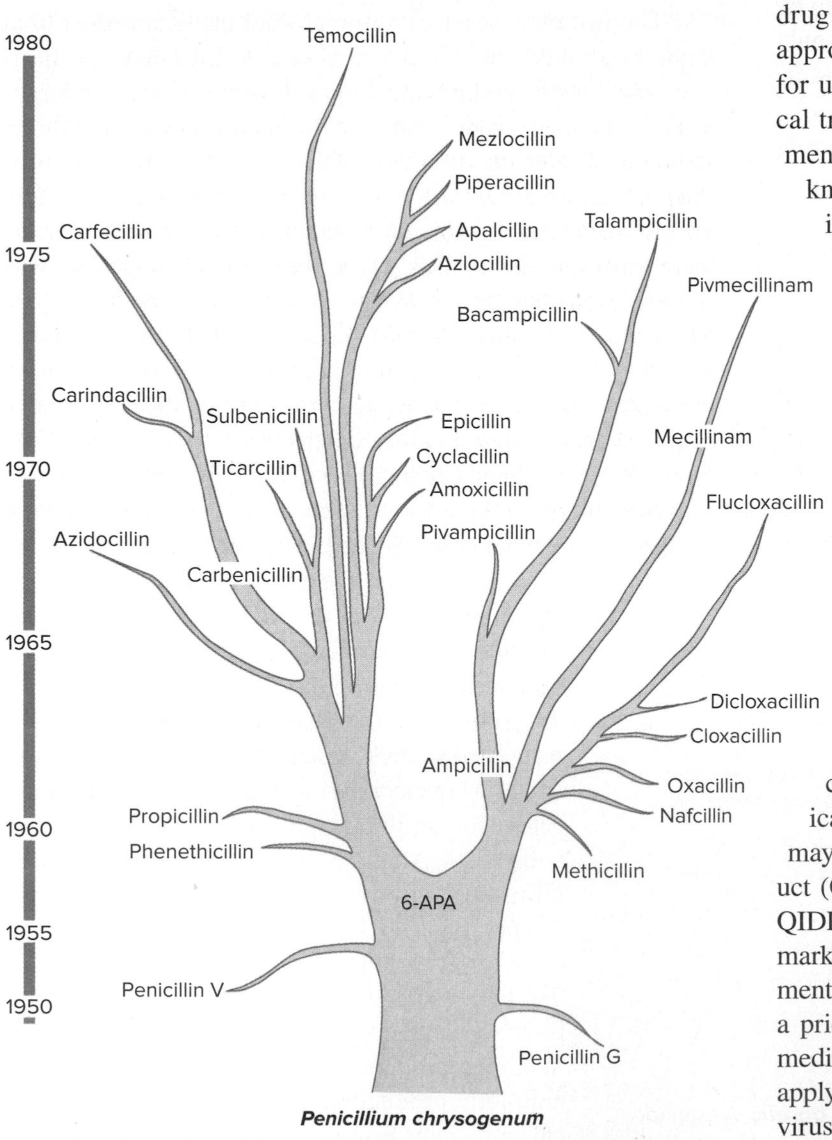

Penicillium chrysogenum

FIGURE 20.2 Family Tree of Penicillins All of the derivatives contain 6-aminopenicillanic acid (6-APA), the core portion of penicillin G.

? Why is it necessary to develop new generations of antimicrobial medications?

Microbes continually evolve, allowing pathogens to develop resistance to the available antimicrobial medications. As a result, scientists must constantly work to create new versions. This task has become increasingly difficult, however, because the obvious options have already been discovered. The problem is compounded by the fact that the process of developing new antimicrobial medications is financially risky—multiple stages of clinical trials are required before a drug can receive U.S. Food and Drug Administration (FDA) approval, a designation that means the drug can be marketed for use in treating a specific disease or condition. The clinical trials are used to show that the drug is an effective treatment and that the benefits of the drug's use outweigh the known risks for the intended population. Even if the effort is successful, pathogens will likely develop resistance, which lessens the return on investment. Also, the use of new antimicrobial medications is sometimes limited because they are reserved as last resort options for treating certain severe infections; this restricted use slows the evolution and spread of resistance to the potentially life-saving medications, but it also lessens the financial rewards for companies that develop them. Because of these challenges, everyone must cooperate to protect the medications we currently have and to promote research to develop new options.

In response to the increasing problem of antibiotic resistance, the U.S. government enacted a new law—Generating Antibiotics Incentives Now (GAIN)—to encourage companies to develop medications that target certain bacterial and fungal pathogens. Under GAIN, a medication under development that targets a qualifying pathogen may receive the designation Qualified Infectious Disease Product (QIDP). This adds potential commercial value because any QIDP medication that receives FDA approval can be exclusively marketed for an additional 5 years. To accelerate the development and approval of QIDP medications, a fast-track status and a priority review process are available. Other designations for medications in development have also been created, and these apply to a broad range of medications, including those that target viruses. A designation called breakthrough therapy is given to drugs shown to offer significant benefits over existing therapies for treating serious diseases; it provides additional FDA guidance to streamline the approval and development process. In addition, during significant threats such as the COVID-19 pandemic, the FDA may give unapproved medical treatments **Emergency Use Authorization (EUA)**. This allows the use of an unapproved medical product—including a medication, vaccine, or diagnostic test—for the treatment, prevention, or diagnosis of a serious disease or condition. This also allows for the use of a medical treatment previously approved for another purpose to be used for a different disease or condition when no other approved alternatives are available.

Rational Drug Design

Rational drug design, a targeted approach for developing medications, was pioneered by Gertrude Elion and George Hitchings, whose decades-long collaboration began in the 1940s. They reasoned that pathogens and tumors require high levels of nucleic acid synthesis to sustain their rapid proliferation, so they tried using compounds structurally similar to certain nucleotides as possible inhibitors. As a result, they discovered drugs effective in the treatment of leukemia, malaria, toxoplasmosis, bacterial infections, and many viruses. Elion later developed acyclovir, which has been used to treat herpesviruses including the Epstein-Barr virus, chickenpox, and shingles; her work also led to the discovery of azidothymidine (AZT), the first widely used drug therapy for HIV/AIDS. For their significant contributions in drug discovery, they were awarded the Nobel Prize in Physiology or Medicine in 1988.

MicroAssessment 20.1

Antimicrobials are chemotherapeutic agents that are effective in treating microbial infections. Antibiotics are antimicrobial chemicals naturally produced by microorganisms. Pathogens can become resistant to antimicrobials, so new options must continue to be developed. Rational drug design is an approach whereby drugs are screened for their ability to inhibit a function necessary for the proliferation of pathogens and tumors.

1. How is the microbe that makes penicillin different from the one that makes streptomycin?
2. Define and contrast the terms *chemotherapeutic agent, antimicrobial medication,* and *antibiotic.*
3. How might *Streptomyces griseus* cells protect themselves from the effects of streptomycin?

20.2 ■ Characteristics of Antimicrobial Medications

Learning Outcome

3. Describe the characteristics of antimicrobial medications including: (1) selective toxicity, (2) antimicrobial action, (3) spectrum of activity, (4) effects of combinations, (5) tissue distribution/metabolism/excretion, (6) adverse effects, and (7) resistance to antimicrobials.

Many different antimicrobial medications are available, each with characteristics that make it more or less suitable for a given clinical situation. Hundreds of thousands of tons of these chemicals, worth many billions of dollars, are now produced each year.

Selective Toxicity

Antimicrobial medications exhibit **selective toxicity,** which means they cause greater harm to microbes than to humans and other animal hosts. They do this by interfering with essential structures or biochemical processes that are common or accessible in microbes but not in host cells. With respect to antiviral medications, finding a target for selective toxicity is difficult because viruses rely on human cells for their replication.

Although the ideal antimicrobial medication is non-toxic to humans and other animals, most can be harmful at high concentrations. In other words, selective toxicity is relative. The toxicity of a given medication is expressed as the **therapeutic index,** which is the lowest dose toxic to the patient divided by the dose typically used for therapy. An antimicrobial that has a high therapeutic index is less toxic, often because the medication acts against an essential biochemical process that is unique to microbes. For example, penicillin G, which interferes with bacterial cell wall synthesis, has a very high therapeutic index. A related term is **therapeutic window,** which is the range between the dose used therapeutically and the toxic dose; a medication that has a high therapeutic index has a wide therapeutic window.

When using an antimicrobial with a low therapeutic index, the concentration in the patient's blood must be carefully monitored to ensure it does not reach a toxic level. Medications too toxic for systemic use are sometimes used topically (meaning applied to a body surface), such as in first-aid ointments used to prevent infections of minor skin injuries.

Antimicrobial Action

Some antimicrobial medications kill microbes, whereas others only inhibit their growth. Both actions are medically important. Antibacterial medications can be divided into two general groups, based on their actions:

- **Bacteriostatic.** These inhibit the growth of bacteria but do not kill them. A patient taking a bacteriostatic medication must rely on the body's defense systems to kill or eliminate the pathogen after its growth has been stopped. Sulfa drugs, for example, are often prescribed for treating urinary tract infections; they prevent bacteria in the bladder from growing so that urination can more effectively eliminate them.
- **Bactericidal.** These kill bacteria. These are particularly useful when the host defenses cannot reliably eliminate pathogens. It is important to note, however, that bactericidal medications may be only inhibitory when used at low concentrations or during certain stages of bacterial growth.

Spectrum of Activity

Antimicrobial medications vary with respect to the range of microbes they kill or inhibit. Antibiotics, for example, can be divided into two general groups:

- **Broad-spectrum antibiotics.** These affect a wide range of bacteria, generally including both Gram-positive and Gram-negative organisms, so they are important for treating acute life-threatening diseases when immediate antimicrobial treatment is essential and there is no time to culture and identify the pathogen. The disadvantage of broad-spectrum antibiotics is that by affecting a wide range of bacteria they also disrupt the microbiome.
- **Narrow-spectrum antibiotics.** These affect a limited range of bacteria, such as only Gram-positive bacteria, so they are less disruptive to the microbiome. Their use, however, depends on knowing the antimicrobial susceptibility of the pathogen. A patient may be started on a broad-spectrum antibiotic and then switched to a narrow-spectrum one once the relevant information has been determined.

Effects of Antimicrobial Combinations

Combinations of antimicrobials are sometimes used to treat infections, but these must be chosen carefully because some options counteract the effects of others. Bacteriostatic antimicrobials that prevent bacterial cell division, for example, interfere with the action of bactericidal medications that kill only actively dividing cells. Counteracting combinations are called antagonistic, while combinations in which the activity of one medication enhances the activity of the other are called synergistic. Combinations neither synergistic nor antagonistic are additive.

Tissue Distribution, Metabolism, and Excretion of the Medication

Antimicrobials differ not only in their action and activity, but also in how they are distributed in tissues, metabolized, and excreted by the body. Only some medications cross the blood-brain barrier (the layer of cells that restricts movement of substances from the bloodstream into the cerebrospinal fluid); this is an important factor in treating meningitis. Medications that are unstable at low pH are destroyed by stomach acid when swallowed, so these options are typically given by intravenous or intramuscular injection. Likewise, medications that are poorly absorbed from the intestinal tract must be administered by injection when used to treat systemic infections; however, they can be taken orally to treat localized intestinal infections.

Another important characteristic of an antimicrobial medication is its rate of elimination, expressed as its half-life—the time it takes for the chemical's concentration in the serum to decrease by 50%. This dictates the dosage frequency required to maintain an effective level in the body. Penicillin V, which has a very short half-life, needs to be taken four times a day, whereas azithromycin, which has a half-life of over 24 hours, is taken no more than once a day. Patients who have kidney or liver dysfunction often excrete or metabolize medications more slowly, and so the dosages must be adjusted to avoid toxic levels.

Adverse Effects

As with any medication, several concerns and dangers are associated with antimicrobials. It is important to remember, however, that the medications have saved countless lives when properly prescribed and used.

Allergic Reactions

Some people develop allergies to certain antimicrobials. An allergy to penicillin or a related medication usually results in a fever or rash but can abruptly cause life-threatening systemic anaphylaxis. For this reason, people who have an allergic reaction to a given antimicrobial must alert their healthcare providers so that an alternative can be prescribed. They should also wear a bracelet or necklace that conveys that information in case of emergency.

Toxic Effects

Several antimicrobials are toxic at high concentrations or occasionally cause adverse reactions. Aminoglycosides such as streptomycin can damage kidneys, impair the sense of balance, and even cause irreversible deafness. Patients taking these medications must be closely monitored because of the low therapeutic index. Some antimicrobials have such severe potential side effects or are so toxic that they are used only for life-threatening conditions when no other options are available. In rare cases, for example, chloramphenicol causes the potentially lethal condition aplastic anemia, in which the body is unable to make white and red blood cells. Polymyxin E (colistin) was once considered too toxic to be used systemically, but as bacteria have become resistant to other antibiotics, it is sometimes the only remaining choice. Certain antimicrobial medications should be avoided during pregnancy because of the potential adverse effects on the developing fetus.

Dysbiosis

Taking an antimicrobial can lead to **dysbiosis,** an imbalance in the microbiome. Among the many possible effects, one obvious outcome is that pathogens normally unable to compete may multiply to high numbers. For example, patients who take certain broad-spectrum antimicrobials orally sometimes develop diarrheal disease caused by *Clostridioides difficile,* a pathogen that generally cannot establish itself in the intestine due to competition from other bacteria. When the microbiome is disrupted, however, *C. difficile* can sometimes grow to high numbers and cause serious intestinal damage, resulting in symptoms that range from mild diarrhea to life-threatening colitis.

Focus Figure

Resistance to Antimicrobials

Because of their specific structural or other characteristics, certain organisms are inherently resistant to the effects of some antimicrobials, a trait called **intrinsic (innate) resistance.** *Mycoplasma* species lack a cell wall, so they are intrinsically resistant to penicillin and other antimicrobials that interfere with peptidoglycan synthesis. Many Gram-negative bacteria are resistant to certain medications because the outer membrane prevents the molecules from entering (see figure 3.11).

In contrast to intrinsic resistance, **acquired resistance** refers to the development of resistance in a previously sensitive organism. This occurs through spontaneous mutation or horizontal gene transfer (see figure 8.1). Acquired resistance is a significant and ongoing problem—pharmaceutical companies are creating a greater variety of antimicrobial medications, but microorganisms continue to evolve, developing mechanisms to avoid the drugs' effects. This very important topic will be discussed in section 20.5.

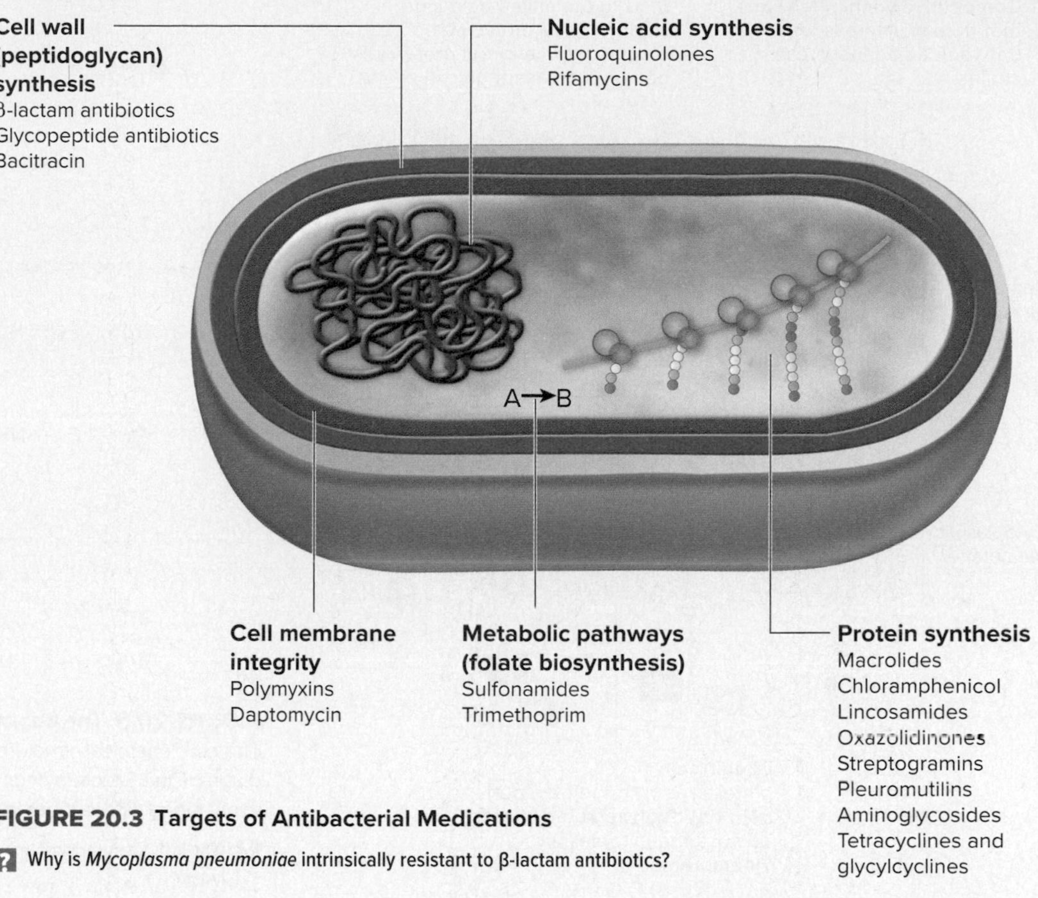

FIGURE 20.3 Targets of Antibacterial Medications

Why is *Mycoplasma pneumoniae* intrinsically resistant to β-lactam antibiotics?

MicroAssessment 20.2

When choosing an antimicrobial to prescribe, a variety of factors must be considered, including the therapeutic index, antimicrobial action, spectrum of activity, effects of combinations, tissue distribution, half-life, adverse effects, and resistance of the microbe.

4. Which would be safer to use: an antimicrobial that has a low therapeutic index or one that has a high therapeutic index? Why?
5. In what clinical situation is it most appropriate to use a broad-spectrum antimicrobial?
6. Why would antimicrobials that have toxic effects be used at all?

20.3 ■ Mechanisms of Action of Antibacterial Medications

Learning Outcomes

4. Describe the β-lactam antibiotics and other antimicrobials that inhibit cell wall synthesis.
5. Describe the antimicrobial medications that interfere with the following: (1) protein synthesis, (2) nucleic acid synthesis, (3) metabolic pathways, (4) cell membrane integrity.
6. Describe the antibacterial medications used to treat *Mycobacterium tuberculosis* infections.

This section describes the mechanisms of action of various classes of antibacterial medications, highlighting their bacterial targets (**figure 20.3**). A group of medications called β-lactam antibiotics will be covered in the greatest detail because their features illustrate some important general concepts. A table at the end of the section summarizes the characteristics of the various classes of medications.

Inhibit Cell Wall Synthesis

Bacterial cell walls are unique in that they contain peptidoglycan, which is composed of glycan chains that are cross-linked via peptide bridges between NAM molecules of adjacent chains (see figures 3.9, 3.10, and 3.11). Because peptidoglycan is so important for bacterial cell wall strength, antibacterial medications that interfere with its synthesis weaken the wall to the point where the cell may burst. Medications that do this include β-lactam antibiotics, glycopeptide antibiotics,

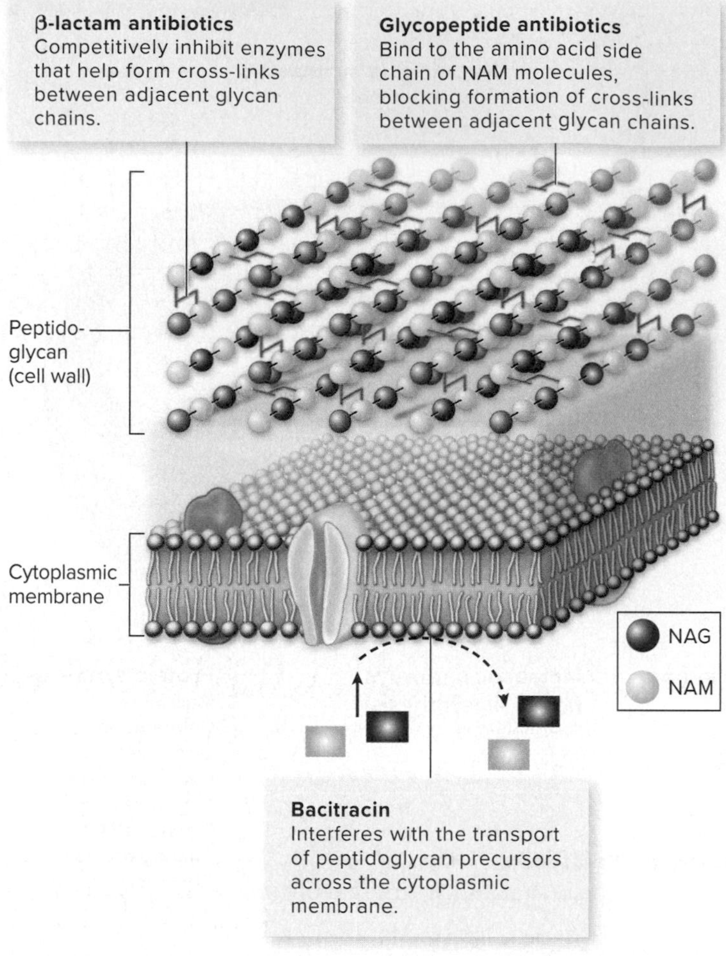
β-lactam antibiotics
Competitively inhibit enzymes that help form cross-links between adjacent glycan chains.
Glycopeptide antibiotics
Bind to the amino acid side chain of NAM molecules, blocking formation of cross-links between adjacent glycan chains.
Peptido-glycan (cell wall)
Cytoplasmic membrane
NAG
NAM
Bacitracin
Interferes with the transport of peptidoglycan precursors across the cytoplasmic membrane.

FIGURE 20.4 Antibacterial Medications That Interfere with Cell Wall Synthesis

? What is the function of peptidoglycan in a bacterial cell wall?

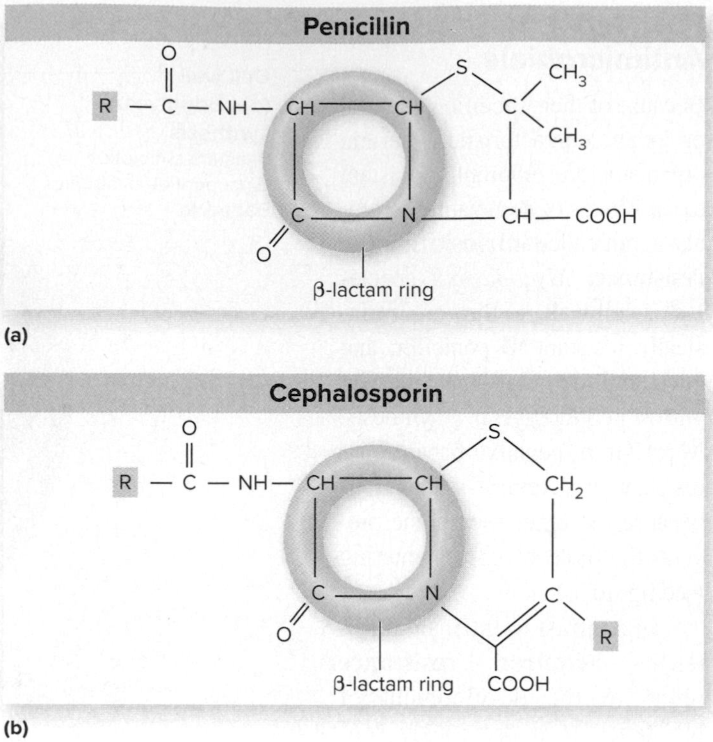

Penicillin

(a)

Cephalosporin

(b)

FIGURE 20.5 The β-Lactam Ring of Penicillins and Cephalosporins The core chemical structure of **(a)** a penicillin; **(b)** a cephalosporin. In the diagram, the β-lactam rings are indicated by an orange ring. The R groups vary among different penicillins and cephalosporins.

? Why is it not surprising that penicillins and cephalosporins have a high therapeutic index?

and bacitracin (**figure 20.4**). They are typically bactericidal only against growing bacteria because these cells continuously synthesize peptidoglycan.

β-Lactam Antibiotics

Penicillins, cephalosporins, carbapenems, and monobactams are **β-lactam antibiotics**—they share a chemical structure called a β-lactam ring (**figure 20.5**). Because β-lactam antibiotics each have a high therapeutic index, they are often the preferred option for treating bacterial infections. All β-lactam antibiotics interfere with peptidoglycan synthesis by competitively inhibiting a group of enzymes that catalyze the formation of the peptide bridges that cross-link adjacent glycan chains; these enzymes are commonly called **penicillin-binding proteins (PBPs),** a name that reflects how they were discovered rather than their normal function.

The different β-lactam antibiotics vary in their spectrums of activity. One reason for this is the cell wall structure of the bacteria. The peptidoglycan of Gram-positive bacteria is exposed to the outside environment, so the β-lactam antibiotics can directly contact the enzymes that synthesize the molecule. In contrast, the outer membrane of Gram-negative bacteria prevents some types of the medications from reaching their target. Another factor is the type of PBPs. The PBPs of Gram-positive bacteria differ somewhat from those of Gram-negative bacteria, and the PBPs of obligate anaerobes differ from those of aerobes. The various PBPs have different affinities (amounts of attraction) for the β-lactam antibiotics. Differences in affinity can even exist among related organisms.

Some bacteria resist the effects of certain β-lactam drugs by synthesizing a **β-lactamase,** an enzyme that breaks the β-lactam ring, thus destroying the activity of the antibiotic. Just as there are many β-lactam antibiotics, there are various β-lactamases, and these differ in the range of medications they destroy. Penicillinase is a β-lactamase that inactivates only members of the penicillin family. In contrast, extended-spectrum β-lactamases (ESBLs) inactivate a wide variety of β-lactam antibiotics, including penicillins, cephalosporins, and monobactams. Carbapenemases are β-lactamases that inactivate the greatest variety of β-lactam antibiotics; some inactivate all classes, including carbapenems, penicillins, cephalosporins, and monobactams. As a whole, Gram-negative bacteria produce a much wider variety of β-lactamases than Gram-positive organisms can.

Penicillins All **penicillins** share a common basic structure. This structure's side chain has been chemically modified to create penicillin derivatives, each with unique characteristics (**figure 20.6**). The derivatives can be loosely grouped into several categories:

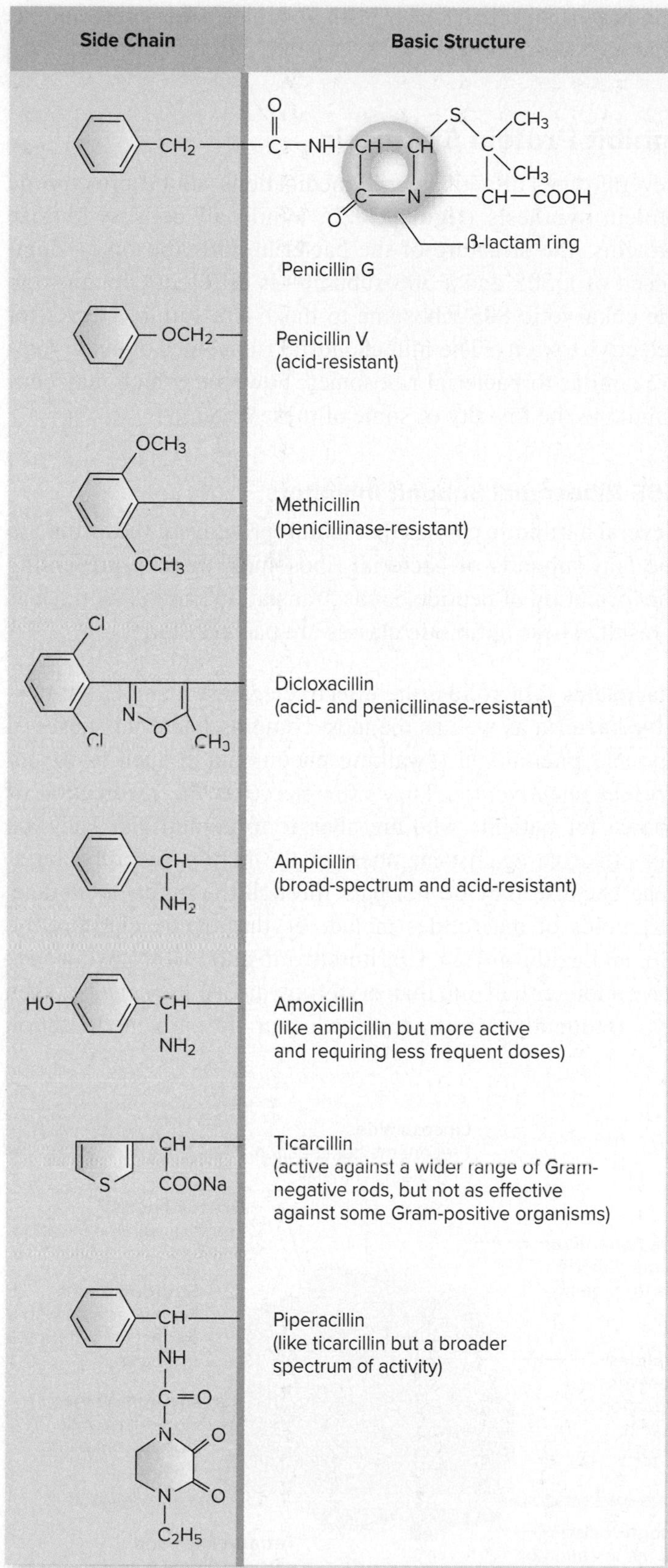

FIGURE 20.6 Chemical Structures and Properties of Representative Members of the Penicillin Family The entire structure of penicillin G and the side chains of other penicillins are shown.

? **Could penicillin be used to treat MRSA (methicillin-resistant *Staphylococcus aureus*) infections? Explain.**

- **Natural penicillins.** These are the original penicillins produced naturally by the mold *Penicillium chrysogenum.* They are narrow-spectrum antibiotics, effective against Gram-positive and a few Gram-negative bacteria. Penicillin V is more stable in acid and, therefore, better absorbed than penicillin G when taken orally. Bacteria that produce penicillinase are resistant to the natural penicillins.
- **Penicillinase-resistant penicillins.** Scientists developed these in response to the problem of penicillinase-producing *Staphylococcus aureus* strains. Penicillinase-resistant penicillins include methicillin and dicloxacillin. Unfortunately, some *S. aureus* strains acquired the ability to make an altered PBP (PBP2a) to which most β-lactam antibiotics do not bind as well. These strains are called MRSA (methicillin-resistant *S. aureus*).
- **Broad-spectrum penicillins.** These are active against not only penicillin-sensitive Gram-positive bacteria, but also many Gram-negative bacteria. Unfortunately, they can be inactivated by many β-lactamases. Broad-spectrum penicillins include ampicillin and amoxicillin.
- **Extended-spectrum penicillins.** These have greater activity against most of the Enterobacteriaceae as well as *Pseudomonas aeruginosa.* This is important because these Gram-negative bacteria are common causes of healthcare-associated infections and are often resistant to many other antimicrobial medications. Extended-spectrum penicillins, however, have less activity against Gram-positive bacteria. Like the broad-spectrum penicillins, many β-lactamases destroy them. Examples of extended-spectrum penicillins include ticarcillin and piperacillin.
- **Penicillins + β-lactamase inhibitor.** This is a combination of agents. The β-lactamase inhibitor interferes with the activity of some types of β-lactamases, thereby protecting the penicillin against enzymatic destruction. Augmentin, which is a combination of amoxicillin and clavulanic acid, is an example.

Cephalosporins The chemical structure of cephalosporins protects them from destruction by certain β-lactamases. By chemically modifying the basic structure, scientists created first-, second-, third-, and fourth-generation cephalosporins, each with a slightly different spectrum of activity. The later generations are generally less effective against Gram-positive bacteria but more effective against Gram-negative bacteria and less susceptible to destruction by some β-lactamases. Examples of cephalosporins include cephalexin and cefazolin (first-generation), cefaclor and cefprozil (second-generation), cefixime and ceftriaxone (third-generation), and cefepime (fourth-generation). Two newer cephalosporins, referred to as fifth-generation cephalosporins, are additionally active against MRSA; the only one currently approved in the United States is ceftaroline, although another (ceftobiprole) has QIDP status and will likely be approved soon. In addition, certain cephalosporins are now available in combination with a β-lactamase inhibitor.

Carbapenems Carbapenems are effective against a wide range of Gram-negative and Gram-positive bacteria. They are not inactivated by the extended-spectrum β-lactamases (ESBLs) produced by certain Gram-negative bacteria, so they are usually reserved as a last resort for treating severe diseases caused by ESBL-producing organisms. Bacteria that produce a carbapenemase are resistant to all carbapenems and often all other β-lactam antibiotics as well. These bacteria also usually have mechanisms to resist other antimicrobials, meaning that they are resistant to all, or nearly all, conventional antimicrobial medications; in some cases, the only treatment option is an antibiotic normally considered too toxic to use systemically. Carbapenems include imipenem, ertapenem, meropenem, and doripenem.

Monobactams Monobactams are antibiotics each containing a β-lactam ring but lacking another ring fused to it. The only monobactam used medically, aztreonam, is primarily effective against members of the family Enterobacteriaceae. Bacteria that produce an ESBL or certain carbapenemases are resistant to aztreonam.

Glycopeptide Antibiotics

Glycopeptide antibiotics interfere with peptidoglycan synthesis by binding to the amino side chain of NAM molecules, blocking cross-linking between adjacent glycan chains. These antibiotics are effective against only Gram-positive bacteria because the outer membrane of Gram-negative bacteria prevents them from reaching their target. They have a relatively low therapeutic index due to side effects unrelated to their activity against peptidoglycan, so they are usually reserved for treating serious infections caused by Gram-positive bacteria resistant to β-lactams.

In the United States, the most widely used glycopeptide is vancomycin. Acquired resistance is typically due to a change in the peptide side chain of the NAM molecule that prevents antibiotic binding. Vancomycin is poorly absorbed from the intestinal tract, so it must be administered intravenously except when used to treat intestinal infections. Newer glycopeptide antibiotics include telavancin, dalbavancin, and oritavancin; they not only share vancomycin's effect on peptidoglycan synthesis, but also insert into the bacterial membrane, causing the cell to become leaky.

Bacitracin

Bacitracin inhibits cell wall biosynthesis by interfering with the transport of peptidoglycan precursors across the cytoplasmic membrane. Its toxicity typically limits its use to topical applications (such as on the skin); however, it is a common ingredient in over-the-counter (non-prescription) first-aid skin ointments.

Inhibit Protein Synthesis

Several types of antibacterial medications inhibit prokaryotic protein synthesis (**figure 20.7**). While all cells synthesize proteins, the structure of the bacterial 70S ribosome—composed of a 50S and a 30S subunit—is different enough from the eukaryotic 80S ribosome to make it a suitable target for selective toxicity. The mitochondrial ribosomes of eukaryotes are similar to bacterial ribosomes, however, which may contribute to the toxicity of some of these drugs.

50S Ribosomal Subunit Inhibitors

Several antibiotic classes that inhibit protein synthesis bind to the 50S subunits of bacterial ribosomes, thereby preventing the formation of peptide bonds; translation cannot continue as a result. These antibiotic classes are bacteriostatic.

Macrolides Macrolides are effective against many Gram-positive bacteria as well as the most common bacterial causes of atypical pneumonia ("walking pneumonia"), such as *Mycoplasma pneumoniae*. They often serve as the medication of choice for patients who are allergic to penicillins. They are not effective against members of the family Enterobacteriaceae because they do not pass through the outer membrane. Examples of macrolides include erythromycin, clarithromycin, and azithromycin. Clarithromycin and azithromycin each have a longer half-life than erythromycin, so they can be taken less frequently. Resistance can occur through modification

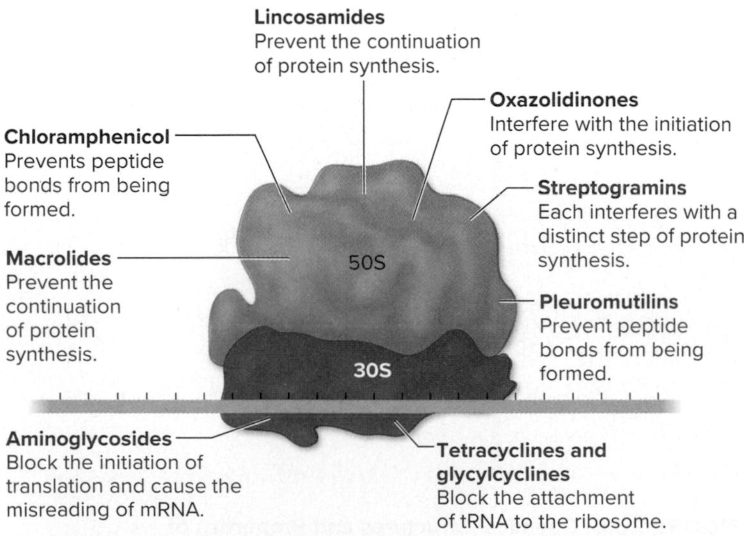
Lincosamides
Prevent the continuation of protein synthesis.
Oxazolidinones
Interfere with the initiation of protein synthesis.
Chloramphenicol
Prevents peptide bonds from being formed.
Streptogramins
Each interferes with a distinct step of protein synthesis.
Macrolides
Prevent the continuation of protein synthesis.
50S
Pleuromutilins
Prevent peptide bonds from being formed.
30S
Aminoglycosides
Block the initiation of translation and cause the misreading of mRNA.
Tetracyclines and glycylcyclines
Block the attachment of tRNA to the ribosome.

FIGURE 20.7 Antibacterial Medications That Inhibit Bacterial Protein Synthesis These medications bind to the bacterial ribosome.

? Some medications that inhibit bacterial protein synthesis have a low therapeutic index. Why would this be so?

of the ribosomal RNA target, production of an enzyme that chemically modifies the medication, or alterations that result in decreased uptake by the bacterial cell.

Chloramphenicol Chloramphenicol is active against a wide range of bacteria and was commonly used in the past to treat a variety of infections, but its use is now generally limited because even low doses may cause an extremely rare but lethal side effect—aplastic anemia (the inability of the body to form white and red blood cells). In the United States, its use is restricted to treating certain life-threatening diseases. In many areas of the world, however, it is also used topically as drops and ointments to treat eye and ear infections.

Lincosamides Lincosamides inhibit a variety of Gram-positive bacteria as well as some Gram-negative anaerobes. They are particularly useful for treating infections resulting from intestinal perforation because they inhibit *Bacteroides fragilis,* a member of the normal intestinal microbiota that is frequently resistant to other antimicrobials. Unfortunately, the risk of developing *Clostridioides difficile* infection (CDI) is greater for people taking lincosamides than some other antimicrobials because most *C. difficile* strains are resistant to the lincosamides. The most commonly used lincosamide is clindamycin. Resistance results from enzymatic modification of the 50S ribosome.

Other 50S Ribosomal Subunit Inhibitors Several other protein synthesis inhibitors that bind the 50S ribosomal subunit are available but are less commonly used than those just described; all are effective against a variety of Gram-positive bacteria. They include the following:

- **Oxazolidinones.** These are synthetic antimicrobials primarily used for treating infections caused by bacteria resistant to other options; linezolid and tedizolid are examples.
- **Streptogramins.** Two of these (quinupristin and dalfopristin) act synergistically by binding to two different sites on the 50S subunit; although individually, each drug is bacteriostatic, together they are bactericidal.
- **Pleuromutilins.** These have been used to treat animals for many years, but one (lefamulin) is now available to treat certain types of pneumonia in humans; another (retapamulin) is restricted to topical applications.

30S Ribosomal Subunit Inhibitors

Several other antibiotic classes function by binding specifically to the 30S ribosomal subunit of bacterial cells. Due to differences in binding locations, these inhibitors interfere with translation in different ways.

Aminoglycosides Aminoglycosides are bactericidal antibiotics that irreversibly bind to the 30S ribosomal subunit, thereby blocking the initiation of translation and also causing misreading of mRNA by ribosomes that pass the block. Unfortunately, the medications can cause severe side effects, including hearing loss and vertigo (dizziness) as a result of damage to sensory components of the inner ear. They can also cause kidney damage. Consequently, aminoglycosides are typically used only when less toxic alternatives are not available.

Aminoglycosides are generally not effective against anaerobes, enterococci, and streptococci because they enter bacterial cells by a process that requires respiratory metabolism. To extend their spectrum of activity, the aminoglycosides are sometimes used in a synergistic combination with a penicillin. The penicillin interferes with cell wall synthesis, allowing the aminoglycoside to enter cells that would otherwise be resistant.

Examples of aminoglycosides include streptomycin, gentamicin, and tobramycin. A form of tobramycin that can be inhaled makes treatment of *Pseudomonas aeruginosa* lung infections in cystic fibrosis patients safer and more effective. Another aminoglycoside, neomycin, is too toxic for systemic use; however, it is a common ingredient in over-the-counter first-aid skin ointments.

Tetracyclines and Glycylcyclines Tetracyclines reversibly bind to the 30S ribosomal subunit, blocking the attachment of tRNA and preventing translation from continuing. These bacteriostatic antibiotics are effective against certain Gram-positive and Gram-negative bacteria. Some tetracyclines, including doxycycline, have a longer half-life, allowing less-frequent doses. Resistance to the tetracyclines is due to a decrease in their accumulation by the bacterial cell—by either decreased uptake or increased excretion (efflux)—or to structural alteration of the target.

Glycylcyclines are functionally and structurally related to the tetracyclines, but they have a wider spectrum of activity. In addition, they are effective against many bacteria that have acquired resistance to the tetracyclines. Tigecycline is the only example currently approved. Glycylcyclines are relatively new, so acquired resistance is still rare, but resistance seems to be due to increased efflux.

MicroByte

Tetracyclines and glycylcyclines can cause discoloration in teeth when used by young children.

Inhibit Nucleic Acid Synthesis

Enzymes required for nucleic acid synthesis are the targets of some groups of antimicrobial medications.

Fluoroquinolones

Fluoroquinolones are synthetic compounds that inhibit one or more of a group of enzymes called topoisomerases, which maintain the supercoiling of DNA within the bacterial cell. One type of topoisomerase, DNA gyrase, breaks, uncoils, and rejoins strands to relieve the strain caused by the localized unwinding of DNA during replication and transcription. Consequently, inhibition of this enzyme prevents these essential cell processes.

The fluoroquinolones are bactericidal against a wide variety of bacteria, including both Gram-positive and Gram-negative organisms. They have been used extensively in the past, but their use is now limited due to severe side effects. Examples of fluoroquinolones include ciprofloxacin, levofloxacin, moxifloxacin, and ofloxacin. Acquired resistance is commonly due to an alteration in the DNA gyrase target.

Rifamycins

Rifamycins are antibiotics that block bacterial RNA polymerase from initiating transcription. Rifampin, the most widely used rifamycin, is bactericidal against many Gram-positive and some Gram-negative bacteria as well as members of the genus *Mycobacterium.* It is primarily used to treat tuberculosis and Hansen's disease (leprosy) and to prevent meningitis in people who have been exposed to *Neisseria meningitidis.* In some patients, a reddish-orange pigment appears in urine and tears. Bacteria quickly develop resistance to rifampin due to a mutation in the gene that encodes RNA polymerase. Certain poorly absorbed forms of rifamycin can be used orally to treat some types of traveler's diarrhea.

Fidaxomicin

Fidaxomicin is a relatively new bactericidal antibiotic that interferes with transcription by binding to RNA polymerase. It passes through the intestinal tract without being absorbed and is particularly useful for treating *C. difficile* infection.

Metronidazole

Metronidazole (Flagyl) is a synthetic compound that interferes with DNA synthesis and function, but only in anaerobic microorganisms. The selective toxicity is due to the fact that anaerobic metabolism is required to convert the medication to its active form. The active form then binds DNA, interfering with synthesis and causing damaging breaks. Metronidazole is used to treat bacterial vaginosis and *C. difficile* infection.

Interfere with Metabolic Pathways

Relatively few antibacterial medications interfere with metabolic pathways. Among the most useful are the folate synthesis inhibitors—sulfonamides and trimethoprim. These synthetic compounds each inhibit different steps in the pathway that leads initially to the synthesis of folate and ultimately to the synthesis of a coenzyme required for nucleotide biosynthesis. The combination of a sulfonamide and trimethoprim has a synergistic effect, so both are included in a medication called **co-trimoxazole.** Animal cells lack the enzymes for the early steps in the pathway, which is why folate is a dietary requirement for animals.

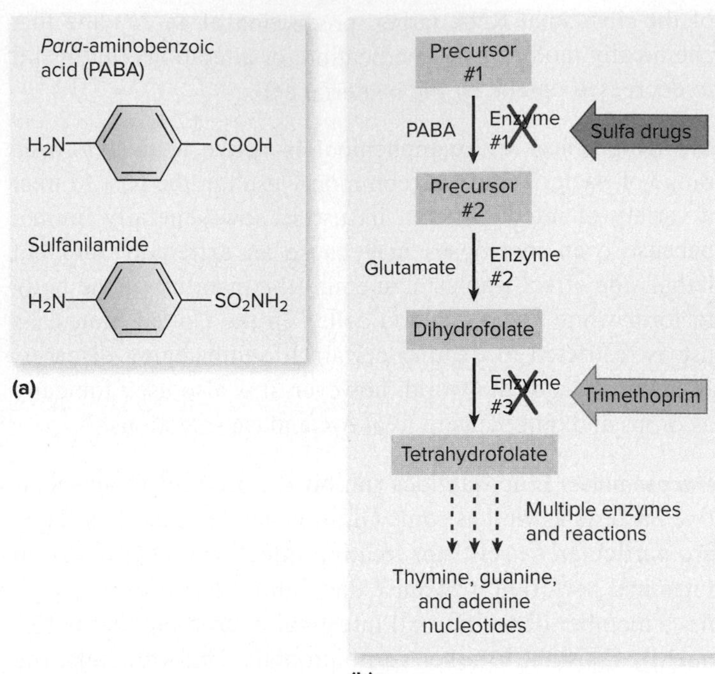

FIGURE 20.8 Inhibitors of the Folate Pathway (a) The chemical structure of PABA and a sulfa drug (sulfanilamide). (b) The sulfa drugs and trimethoprim interfere with different steps of the pathway.

? Sulfa drugs have a high therapeutic index. Why would this be so?

Sulfonamides

Sulfonamides and related compounds, collectively referred to as **sulfa drugs,** inhibit the growth of many Gram-positive and Gram-negative bacteria. They are structurally similar to *para*-aminobenzoic acid (PABA), a substrate in the folate biosynthesis pathway (**figure 20.8**). Because of this similarity, the enzyme that normally binds PABA binds sulfa drugs instead, an example of competitive inhibition (see figure 6.16). Human cells lack this enzyme, providing the basis for the selective toxicity of the medications. Resistance often results from acquiring a plasmid that encodes an enzyme the medications do not bind to as well.

Trimethoprim

Trimethoprim inhibits the bacterial enzyme that catalyzes a metabolic step following the one inhibited by sulfonamides. Fortunately, the medication has little effect on this enzyme's counterpart in human cells. The most common mechanism of resistance is a plasmid-encoded alternative enzyme that the medication does not bind to as well. Unfortunately, the genes encoding resistance to trimethoprim and to sulfonamide are often carried on the same plasmid.

Interfere with Cell Membrane Integrity

A few antibiotics damage bacterial membranes, causing cellular leakage leading to cell death. These include:

- **Daptomycin.** This inserts into bacterial cytoplasmic membranes and is used to treat certain infections caused by Gram-positive bacteria that are resistant to other medications. It is not effective against Gram-negative bacteria because it cannot pass through the outer membrane.
- **Polymyxins.** These bind to the membranes of Gram-negative cells. Unfortunately, these also bind to eukaryotic cells though to a lesser extent, which limits their use. Despite their toxicity, polymyxin B and polymyxin E, the latter of which is also called colistin, are each sometimes used to treat certain life-threatening bacterial infections that do not respond to other antimicrobials. Polymyxin B is also a common ingredient in first-aid skin ointments.
- **Glycopeptides.** As mentioned previously, the two newest glycopeptide antibiotics—albavancin and oritavancin—disrupt cell membranes in addition to interfering with peptidoglycan synthesis.

Act Against *Mycobacterium tuberculosis*

Relatively few antimicrobials are effective against *Mycobacterium tuberculosis.* This is due to several factors, including the organism's waxy cell wall (which prevents the entry of many drugs) and slow growth. A group of four medications serve as the **first-line drugs** for treating tuberculosis (TB), meaning they are preferred because they are the most effective as well as the least toxic. The treatment regimen used depends on the circumstances, but as an example, all four are given in combination for an initial 8-week treatment phase (aimed at rapidly reducing the number of infecting *M. tuberculosis* cells), and then two are continued for another 18 weeks (to ensure that all infecting cells are killed). This combination therapy decreases the chance that resistant mutants will develop; if some cells in the infecting population spontaneously develop resistance to one medication, another one will eliminate them. The **second-line drugs** are used for strains resistant to the first-line drugs; however, they either are less effective or have greater risk of toxicity.

The core first-line drugs for treating TB have different targets. Isoniazid (INH) is thought to inhibit the synthesis of mycolic acids, an important cell wall component. Ethambutol (EMB) inhibits enzymes required for synthesis of other mycobacterial cell wall components. The target of pyrazinamide (PZA) is not clear. Another first-line drug, rifampin (RIF), was discussed earlier in this section.

When *M. tuberculosis* strains become resistant to first-line drugs, second-line drugs must be used as substitutes; 10 or so second-line drugs are approved as options. A new combination for treating disease caused by the most resistant *M. tuberculosis* strains includes bedaquiline (specifically inhibits mycobacterial ATP synthase), pretomanid (inhibits a step in mycolic acid synthesis), and linezolid (a member of the oxazolidinones, discussed earlier).

Characteristics of the antimicrobial medications described in this section are summarized in **table 20.1.**

TABLE 20.1 Characteristics of Antibacterial Medications

Target/Class/Example(s)	Comments/Characteristics
Cell Wall Synthesis	
β-lactam antibiotics	Bactericidal against a variety of bacteria; interfere with peptidoglycan synthesis by inhibiting enzymes called penicillin-binding proteins (PBPs) that help form cross-links between adjacent glycan chains.
Penicillins Penicillin G, methicillin, dicloxacillin, ampicillin, amoxicillin, ticarcillin, piperacillin	Groups include natural penicillins, penicillinase-resistant penicillins, broad-spectrum penicillins, and extended-spectrum penicillins; these differ in spectrum of activity and susceptibility to β-lactamases.
Cephalosporins Cephalexin, cefazolin, cefaclor, cefprozil, cefixime, ceftriaxone, cefepime, ceftaroline	The later generations are usually more effective against Gram-negative bacteria and less susceptible to destruction by certain β-lactamases.
Carbapenems Imipenem, meropenem, ertapenem, and doripenem	These are broad-spectrum and are not destroyed by most β-lactamases, including extended-spectrum β-lactamases, but are susceptible to carbapenemases.
Monobactams Aztreonam	These are primarily active against members of the family Enterobacteriaceae and are not inactivated by most β-lactamases, but are susceptible to extended-spectrum β-lactamases and some carbapenemases.
Glycopeptide antibiotics Vancomycin, televancin, dalbavancin, and oritavancin	Bactericidal against Gram-positive bacteria; interfere with peptidoglycan synthesis by binding to the amino acid side chain of NAM molecules, blocking formation of cross-links between adjacent glycan chains. Dalbavancin and oritavancin also disrupt membrane integrity. Used to treat serious infections caused by Gram-positive bacteria that are resistant to most other options.
Bacitracin	Bactericidal against Gram-positive bacteria; interferes with the transport of peptidoglycan precursors across the cytoplasmic membrane. Common ingredient in non-prescription antibiotic ointments.

(continued)

TABLE 20.1 Characteristics of Antibacterial Medications (*Continued*)

Target/Class/Example(s)	Comments/Characteristics
Protein Synthesis	
50S ribosomal subunit inhibitors	Bacteriostatic; bind to the 50S ribosomal subunit, preventing the ribosome from continuing translation.
Macrolides Erythromycin, clarithromycin, azithromycin	Bacteriostatic against many Gram-positive bacteria as well as the most common causes of atypical pneumonia.
Chloramphenicol	Bacteriostatic and broad-spectrum; generally used only as a last resort for life-threatening infections.
Lincosamides Lincomycin, clindamycin	Bacteriostatic against a variety of Gram-positive and Gram-negative bacteria.
Oxazolidinones Linezolid, tedizolid	Bacteriostatic against a variety of Gram-positive bacteria. Useful for treating infections caused by bacteria that are resistant to other options.
Streptogramins Quinupristin, dalfopristin	Synergistically bactericidal when used in combination; the two drugs bind to two different sites on the 50S ribosomal subunit. Effective against a variety of Gram-positive bacteria, but generally reserved for strains that are resistant to other antimicrobials.
Pleuromutilins Lefamulin, retapamulin	Bacteriostatic against a variety of Gram-positive bacteria.
30S ribosomal subunit inhibitors	Bactericidal or bacteriostatic; bind to 30S ribosomal subunit, with effect on translation dependent on binding location.
Aminoglycosides Streptomycin, gentamicin, tobramycin, neomycin	Bactericidal against aerobic and facultative bacteria; binding to the 30S ribosomal subunit blocks the initiation of translation and causes the misreading of mRNA. Toxicity limits the use. Neomycin is commonly used in non-prescription topical antibiotic ointments.
Tetracyclines and glycylcyclines Tetracyline and doxycycline (tetracyclines), tigecycline (a glycylcycline)	Bacteriostatic against some Gram-positive and Gram-negative bacteria; bind to the 30S ribosomal subunit blocks the attachment of tRNA.
Nucleic Acid Synthesis	
Fluoroquinolones Ciprofloxacin, levofloxacin, moxifloxacin	Bactericidal against a wide variety of Gram-positive and Gram-negative bacteria; inhibit topoisomerases.
Rifamycins Rifampin	Bactericidal against Gram-positive and some Gram-negative bacteria. Bind RNA polymerase, blocking the initiation of RNA synthesis.
Fidaxomicin	Bactericidal; particularly useful for treating *Clostridioides difficile* infections. Interferes with RNA polymerase.
Metronidazole	Bactericidal against anaerobes. Activated by anaerobic metabolism and then binds DNA, interfering with synthesis and causing damaging breaks.
Folate Biosynthesis	
Sulfonamides	Bacteriostatic against a variety of Gram-positive and Gram-negative bacteria. Structurally similar to para-aminobenzoic acid (PABA) and therefore inhibit the enzyme for which PABA is a substrate.
Trimethoprim	Synergistically bacteriostatic when used with sulfonamide, a combination called co-trimoxazole. Inhibits the enzyme that catalyzes a step following the one inhibited by the sulfonamides.
Cell Membrane Integrity	
Daptomycin	Bactericidal against Gram-positive bacteria by damaging the cytoplasmic membrane.
Polymyxins Polymyxin B, polymyxin E (colistin)	Bactericidal against Gram-negative bacteria by damaging cell membranes. Toxicity limits their use.
Mycobacterium tuberculosis	
Ethambutol	Bacteriostatic by inhibiting the synthesis of a component of the mycobacterial cell wall.
Isoniazid	Bactericidal, possibly by inhibiting synthesis of mycolic acids.
Pyrazinamide	Not clear.
Bedaquiline	Bactericidal by inhibiting mycobacterial ATP synthase.
Pretomanid	Bactericidal by inhibiting synthesis of mycolic acids.

MicroAssessment 20.3

Antimicrobial medications may target steps in the pathways for synthesizing peptidoglycan, protein, nucleic acid, or folate; some instead interfere with membrane integrity. Medications used to treat tuberculosis often interfere with processes unique to *Mycobacterium tuberculosis.*

7. Why are β-lactam antibiotics bactericidal only to growing bacteria?
8. What is the target of the macrolides?
9. Considering that all β-lactam antibiotics have the same target, why do they vary in their spectrums of activity?

20.4 ■ Antimicrobial Susceptibility Testing

Learning Outcomes

7. Describe the Kirby-Bauer disc diffusion test.
8. Describe how the minimum inhibitory concentration (MIC) and the minimum bactericidal concentration (MBC) are determined.

When a pathogen's susceptibility to various antimicrobial medications is unpredictable, laboratory tests are used to guide the treatment choice. In our discussion, we will focus on methods used to determine the susceptibility of bacteria, but similar procedures are used for some fungi, particularly yeasts.

Conventional Disc Diffusion Method

The **Kirby-Bauer disc diffusion test** is a relatively simple method routinely used to determine the susceptibility of a given bacterial strain to a variety of antimicrobial medications. A standard inoculum of the strain is first uniformly spread on the surface of an agar plate. Then 12 or so discs, each containing a known amount of a different antimicrobial, are placed on the surface of the medium. During incubation, the various antimicrobials diffuse outward, forming a concentration gradient around each disc. Meanwhile the bacterial cells multiply, eventually forming a film of growth on the plate, except in regions around the discs where the bacteria were killed or their growth was inhibited. This clear area in which no visible growth occurs is called a **zone of inhibition.**

The size of the zone of inhibition around an antimicrobial disc reflects, in part, the degree of susceptibility of the organism to the medication (**figure 20.9**). The zone size is also influenced by characteristics of the chemical, including its molecular weight and stability, as well as the amount in the disc. Special charts have been prepared that correlate the diameter of the zone of inhibition around a particular antimicrobial disc to an organism's susceptibility. Based on the zone's size and information in the chart, a bacterial strain can be described as susceptible, intermediate, or resistant to

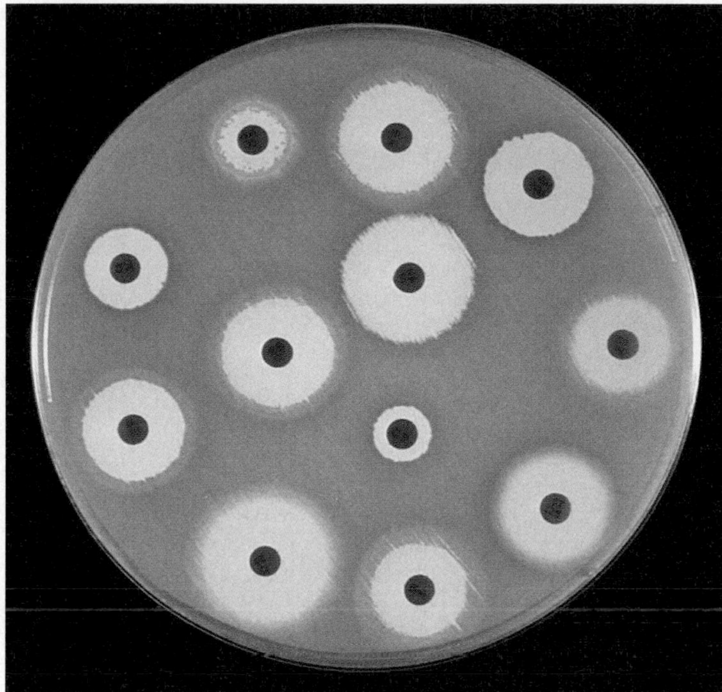

FIGURE 20.9 Kirby-Bauer Method for Determining Antimicrobial Susceptibility The size of the zone of inhibition surrounding the disc reflects, in part, the susceptibility of the bacterial strain to the medication. Gilda L. Jones/CDC

When using the Kirby-Bauer test to determine antimicrobial susceptibility, a chart must be consulted. Why not simply choose the option that gives the largest zone size?

a particular antimicrobial. The procedures discussed next are used to gain more precise information about an organism's susceptibility.

Minimum Inhibitory and Minimum Bactericidal Concentrations (MIC and MBC)

If the Kirby-Bauer test gives unusual results, or when a company is developing a new antimicrobial medication, then a broth dilution test to determine the **minimum inhibitory concentration (MIC)** may be necessary. The MIC is the lowest concentration of a specific antimicrobial medication needed to prevent the visible growth of a given microbial strain in vitro. It is determined by growing the test strain in broth cultures containing different concentrations of the antimicrobial (**figure 20.10**). To do this, serial dilutions are used to generate decreasing concentrations of the medication in tubes containing a suitable growth medium. Then, a standard inoculum of microbial cells is added to each tube. The tubes are incubated for at least 16 hours and then examined for turbidity (cloudiness), which indicates growth. The lowest concentration of the medication that prevents microbial growth is the medication's MIC for that particular microbial strain.

From a medical standpoint, a microbial strain is considered susceptible to a given medication only if the organism is inhibited by concentrations used clinically. For example, an organism with an MIC of 16 μg/mL would be considered resistant to an

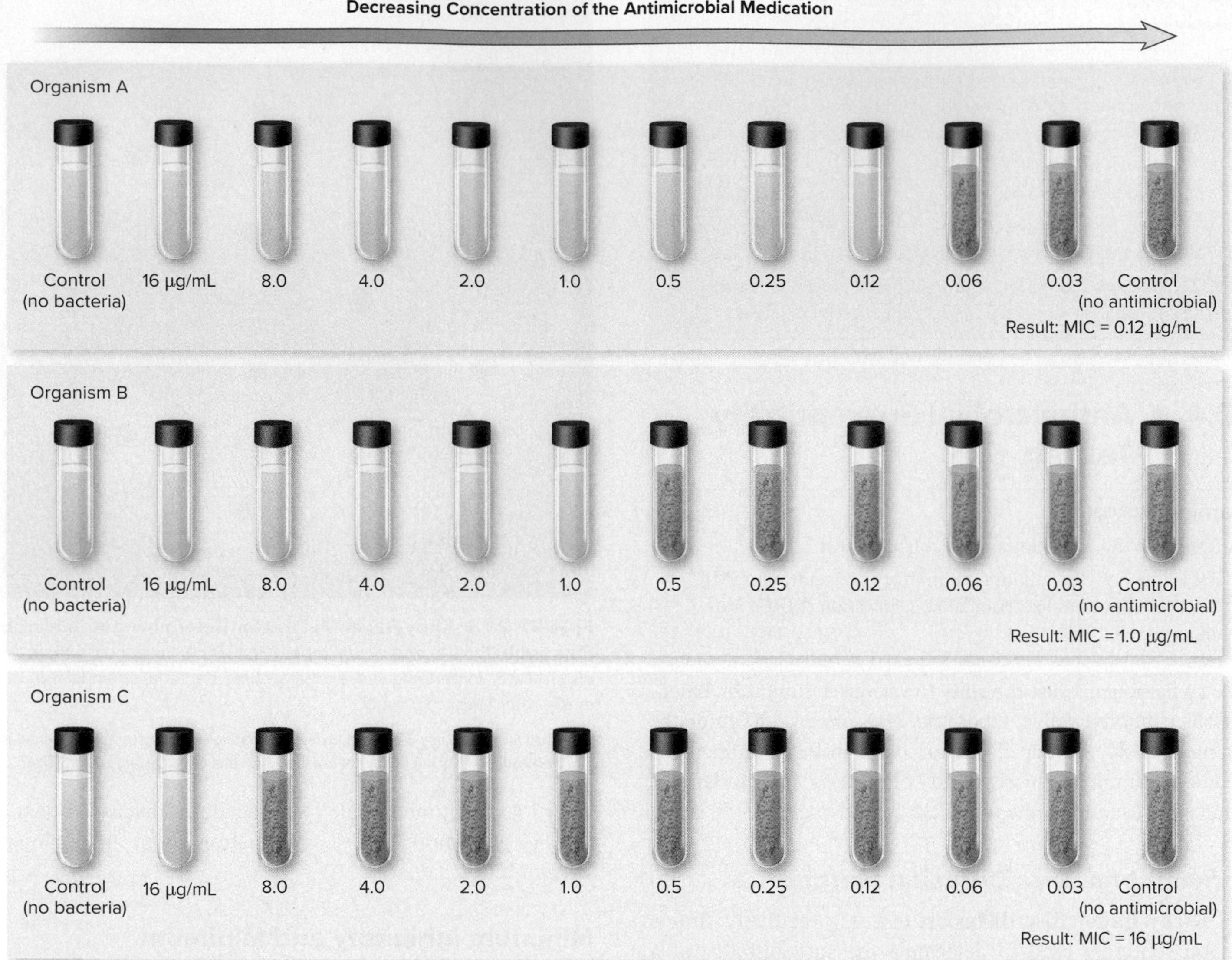

FIGURE 20.10 Determining the Minimum Inhibitory Concentration (MIC) of an Antibacterial Medication In the tubes in this diagram, the pale yellow color indicates clear broth (no visible bacterial growth) whereas the gold color with dots indicates turbid broth (visible bacterial growth). The lowest concentration of the antimicrobial medication that prevents growth of the test strain is the MIC for that particular strain.

? Considering that organism C does not grow in the tube that contains 16 μg/mL of the medication, why might the bacterium still be considered resistant to the medication?

antimicrobial if the highest level that could be achieved in vivo were less than that. Bacteria that have an MIC between susceptible (treatable) and resistant (untreatable) are called intermediate.

The **minimum bactericidal concentration (MBC)** is the lowest concentration of a specific antimicrobial medication that kills 99.9% of cells of a given bacterial strain in vitro. The MBC is determined by finding out how many live bacterial cells remain in tubes from the MIC test that showed no growth. A known volume from each of those tubes is transferred to a plate containing an antimicrobial-free agar medium; any colonies that form when the plate is incubated can be counted to determine how many cells in the sample survived that particular antimicrobial concentration. A similar test is sometimes done to determine the minimum fungicidal concentration (MFC) of a given fungal strain.

Determining the MIC and the MBC (or MFC) using these methods gives precise information regarding an organism's in vitro susceptibility. The techniques, however, are labor-intensive and therefore expensive. In addition, individual sets of tubes must be inoculated to determine susceptibility to each medication tested.

Commercial Modifications of Antimicrobial Susceptibility Testing

Commercial modifications of the conventional susceptibility testing methods offer certain advantages. They are less labor-intensive, and the results can be obtained in as little as 4 hours. One system uses a small card with tiny wells containing specific antimicrobial concentrations. The highly automated system inoculates and incubates the cards, determines

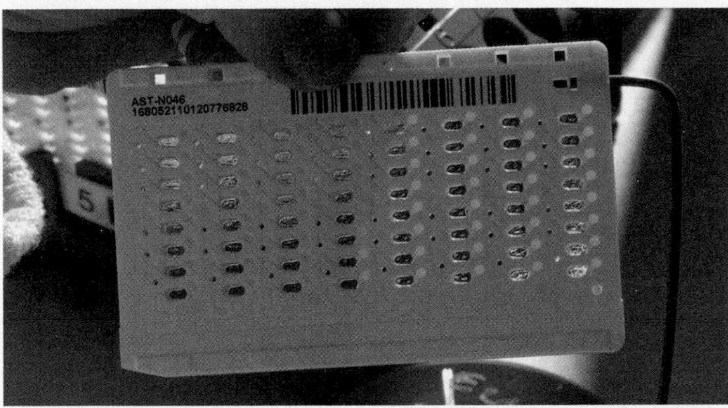

FIGURE 20.11 Automated Tests Used to Determine Antimicrobial Susceptibility The tiny wells in the card contain specific concentrations of an antimicrobial. An automated system inoculates and incubates the cards, determines the growth rate by reading turbidity, and uses mathematical formulas to interpret the results and derive the MICs. BSIP/UIG/Getty Images

What are two advantages of automated tests used to determine antimicrobial susceptibility?

the growth rate by reading the turbidity, and then uses mathematical formulas to interpret the results and determine the MICs, all within 5 to 15 hours (**figure 20.11**).

The E test, a modification of the disc diffusion test, uses a strip containing a gradient of concentrations of an antimicrobial medication. Multiple strips, each containing a different antimicrobial, are placed on the surface of an agar medium that has been uniformly inoculated with the test organism. During incubation, the test organism grows, and a zone of inhibition forms around the strip. Because of the gradient of antimicrobial concentrations, the zone of inhibition is shaped like a teardrop that intersects the strip (**figure 20.12**). The

FIGURE 20.12 The E Test The strip has a gradient of concentrations of a given antimicrobial medication. The MIC is determined by reading the number on the strip at the point where growth intersects the strip. Sirirat/Shutterstock

Why is the zone of inhibition larger at one end of the strip?

MIC is determined by reading the printed number at the point where microbial growth intersects the strip.

Newer rapid systems do not examine in vitro susceptibility and instead detect genes encoding antimicrobial resistance. Some of these systems use microarrays to detect the genes; others use PCR.

MicroAssessment 20.4

Disc diffusion tests can determine whether an organism is susceptible, intermediate, or resistant to a variety of different antimicrobials. The MIC and the MBC (or, in the case of fungi, MFC) are quantitative measures of a microbial strain's susceptibility to an antimicrobial medication. Commercial tests for determining antimicrobial susceptibility are less labor-intensive and often more rapid.

10. List two factors other than a strain's susceptibility that influence the size of the zone of inhibition around an antimicrobial disc.
11. Explain the difference between the MIC and the MBC.
12. Why would it be important for the Kirby-Bauer disc diffusion test to use a standard concentration of the bacterial strain being tested?

20.5 ■ Resistance to Antimicrobial Medications

Learning Outcomes

9. Describe four general mechanisms of antimicrobial resistance.
10. Describe how antimicrobial resistance can be acquired.
11. List five examples of emerging antimicrobial resistance.
12. Describe how the emergence and spread of antimicrobial resistance can be slowed.

After sulfa drugs and penicillin were introduced, people hoped that such medications would eliminate most bacterial diseases. We now realize, however, that resistance limits the usefulness of all known antimicrobials. The spread of antimicrobial resistance (AMR) is alarming because of the impact on the cost, complications, and outcomes of treatment. Dealing with the problem requires an understanding of the mechanisms and spread of resistance. This section focuses on bacterial resistance, but the principles largely also apply to fungi and other pathogenic microbes.

As antimicrobial medications are increasingly used and misused, resistant organisms have a selective advantage over their susceptible counterparts (**figure 20.13**). For example, when penicillin G was first introduced, less than 3% of *Staphylococcus aureus* strains were resistant to its effects. Heavy use of the antibiotic, measured in hundreds of tons per year, eliminated susceptible strains, so that 90% or more are now resistant.

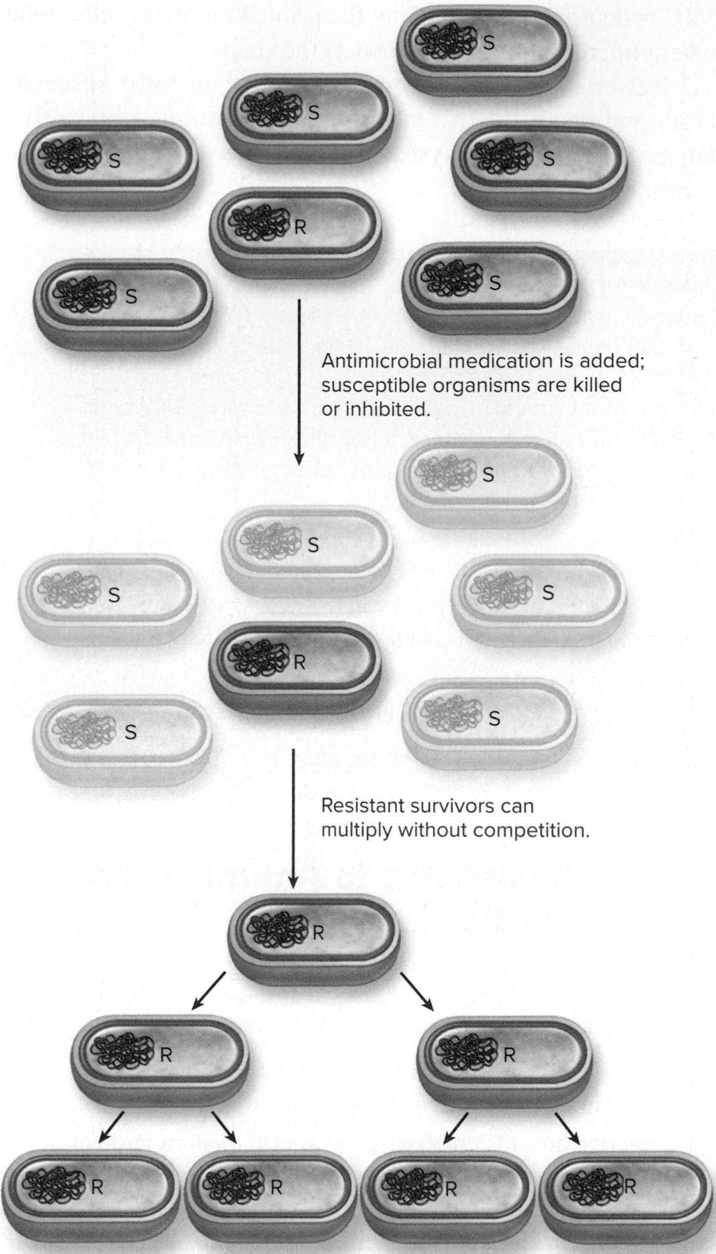

FIGURE 20.13 The Selective Advantage of Resistance to Antimicrobial Medications When antimicrobials are used, organisms that are resistant (R) to their effects have a selective advantage over their susceptible (S) counterparts.

? How does overuse of antibiotics contribute to increasing numbers of antibiotic-resistant bacteria?

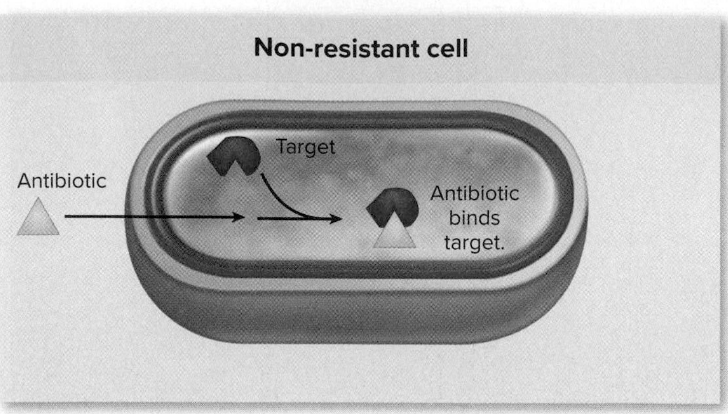

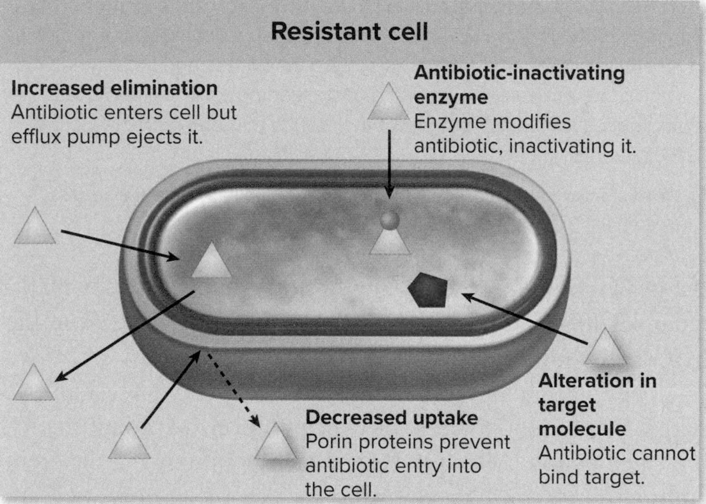

FIGURE 20.14 Common Mechanisms of Acquired Resistance to Antibiotics and Other Antimicrobial Medications

? As a group, bacteria use which three methods to avoid the effects of β-lactam antibiotics?

Mechanisms of Acquired Resistance

Common microbial mechanisms of acquired resistance to antimicrobial medications include the following (**figure 20.14**):

- **Medication-inactivating enzymes.** Some bacteria produce enzymes that chemically modify a specific medication, interfering with its function. One example is penicillinase, an enzyme that destroys penicillin. Another is the enzyme chloramphenicol acetyltransferase, which chemically alters the antibiotic chloramphenicol, making it ineffective. Some inactivating enzymes have an extended spectrum, meaning that they confer resistance to a wide variety of antibacterial medications. Extended-spectrum β-lactamases and carbapenemases inactivate many different β-lactam antibiotics.
- **Alteration in the target molecule.** Antimicrobial medications generally act by binding to specific target molecules in a microorganism, interfering with the target's function. Minor structural changes in the target can prevent the medication from binding, thereby protecting the organism from its effects. For example, modifications in the penicillin-binding proteins (PBPs) prevent β-lactam antibiotics from binding to them. Similarly, a change in ribosomal RNA (rRNA), the target for the macrolides, prevents those medications from interfering with ribosome function. Some bacteria have acquired the ability to produce an enzyme that adds a methyl group to an rRNA molecule of the 50S ribosome, thereby preventing macrolides, lincosamides, and streptogramins from binding to their target.

- **Decreased uptake of the medication.** Recall that porin proteins in the outer membrane of Gram-negative bacteria selectively permit small molecules to pass through that membrane and enter the cell's periplasm; from there the molecules may cross the cytoplasmic membrane to enter the cytoplasm. Changes in the porin proteins can therefore prevent certain antimicrobials from entering the cell's periplasm or cytoplasm. By stopping entry of an antimicrobial, an organism avoids its effects.
- **Increased elimination of the medication.** Microorganisms use **efflux pumps** to transport antimicrobials and other damaging compounds out of a cell. When a cell makes more of these pumps, the compound is ejected faster. In addition, structural changes in the pumps can influence the range of compounds that can be pumped out. In some cases, the organism becomes resistant to several different antimicrobials simultaneously.

Acquisition of Resistance

Antimicrobial resistance can be acquired by either spontaneous mutation, which alters existing genes, or gaining new genes (see figure 8.1).

Spontaneous Mutation

As cells replicate, spontaneous mutations happen at a relatively low rate. Those few mutations that occur, however, can have a significant effect on the resistance of a bacterial population to an antimicrobial medication. Acquired resistance to streptomycin (an aminoglycoside) is a good example. A single base-pair change in the gene encoding a ribosomal protein alters the target enough to make the cell streptomycin-resistant. When a streptomycin-susceptible strain is grown in streptomycin-free medium to a population of 10^9 cells, at least one cell in the population probably has that particular mutation. If streptomycin is then added to the medium, only that cell and its descendants will still be able to replicate, giving rise to a streptomycin-resistant population.

Medications for which a single point mutation confers resistance are sometimes used in combination with one or more other drugs, an approach called **combination therapy.** If any cell spontaneously develops resistance to one medication, another one will still kill it. Combination therapy is effective because the chance that a cell will simultaneously develop mutational resistance to multiple antimicrobial medications is extremely low.

When an antimicrobial medication has several different targets or can bind to multiple sites on a single target, resistance due to spontaneous mutation is less likely because multiple mutations would be required to prevent the antimicrobial from binding to its target. Consider a susceptible strain grown to a population of 10^9 cells: A few of the cells might have a mutation that allows them to be slightly less susceptible to the medication, but it is unlikely that a single cell could have accumulated the precise combination of mutations necessary to be resistant. Thus, appropriate levels of the medication for a long enough period of time would still be effective against all the cells in that particular population. If doses are skipped, however, the medication level may drop to sub-inhibitory levels. In this environment, the least susceptible cells may multiply, which increases the chance that they can accumulate the needed mutations to be resistant. Likewise, if the treatment is stopped too soon, the least susceptible cells may not have been eliminated, and when these cells multiply, additional mutations may make them even less susceptible. Progressive misuse of the medications can eventually lead to resistant strains.

Gene Transfer

Genes encoding resistance to antimicrobial medications can spread to different strains, species, and even genera. Original sources of those genes include:

- **Antibiotic producers.** Microbes that naturally produce an antibiotic can be a source of resistance to the corresponding antibiotic. Although some microbes are inherently resistant to the compounds they make (for example, fungi are naturally resistant to antibacterial chemicals), others have resistance-encoding genes. Certain *Streptomyces* species, for instance, have gene clusters that encode not only production of a given antibiotic but also resistance; in fact, the gene encoding an aminoglycoside-modifying enzyme likely came from the *Streptomyces* species that produced the drug.
- **Resistant mutants.** When a microbe develops resistance through spontaneous mutation, it can potentially transfer that resistance to other organisms.

Transfer of resistance genes is commonly through conjugative transfer of **R plasmids.** These plasmids often carry several different resistance genes, each one encoding resistance to a specific antimicrobial. Thus, when an organism acquires an R plasmid, it may become resistant to several different medications simultaneously.

Examples of Emerging Resistance

Some of the problems associated with the increasing AMR are highlighted by the following examples.

Enterococci

Enterococci are part of the normal intestinal microbiota and a common cause of healthcare-associated infections (HAIs). They are intrinsically less susceptible to many common antimicrobials. For example, their penicillin-binding proteins have low affinity for certain β-lactam antibiotics. In addition, many enterococci have R plasmids. Some strains, called **vancomycin-resistant enterococci (VRE),** are even resistant to vancomycin. Recall that this antibiotic is usually reserved as a last resort for treating serious infections caused by Gram-positive bacteria resistant to all β-lactam antibiotics. Because vancomycin resistance in VRE strains is encoded on a plasmid, the resistance is transferable to other organisms.

Enterobacteriaceae

Members of the family Enterobacteriaceae are intrinsically resistant to many antimicrobials because the outer membrane of these Gram-negative bacteria prevents the medications from entering cells. The situation became more complicated when some enterics developed the ability to produce a β-lactamase, allowing the strains to resist the effects of ampicillin and other penicillins. Some strains then developed the ability to produce extended-spectrum β-lactamases (ESBLs), making them resistant to most cephalosporins and monobactams, as well as to penicillins. More recently, strains referred to as **carbapenem-resistant Enterobacteriaceae (CRE)** have been discovered. Many of these produce an enzyme that inactivates carbapenems as well as all other β-lactam antibiotics. The strains have additional mechanisms to resist all or nearly all other routine medications, which is why colistin—a medication normally considered too toxic to be used systemically—is now sometimes administered. The recent discovery of plasmid-encoded resistance to colistin in some members of the family Enterobacteriaceae has caused understandable concern that a CRE strain will acquire resistance to that medication as well.

Mycobacterium tuberculosis

Mycobacterium tuberculosis can easily become resistant to the first-line anti-TB drugs (the preferred medications) through spontaneous mutation. Large numbers of bacterial cells are found in an active infection, so at least one of the cells has likely developed spontaneous resistance to a drug, which is why combination therapy is required. Tuberculosis (TB) treatment has always been a long and complicated process, requiring combinations of antimicrobial medications taken for a duration of at least 4 months but often much longer; the length of treatment is due in part to the very slow growth of *M. tuberculosis.* Many tuberculosis patients make the mistake of skipping doses or stopping treatment too soon. As a consequence, strains of *M. tuberculosis* develop resistance to the first-line drugs. This results in even longer, more expensive treatments that are also less effective.

Tuberculosis that is resistant to treatment by two of the first-line anti-TB medications—isoniazid and rifampin—is called **multi-drug-resistant tuberculosis (MDR-TB).** To prevent the emergence of resistant *M. tuberculosis* strains, some cities are using directly observed therapy (DOT); with DOT, healthcare workers watch patients swallow the prescribed pills to ensure treatment compliance. When strains develop resistance to any of the first-line drugs, options in a group of about 10 second-line drugs are substituted instead. **Extensively drug-resistant tuberculosis (XDR-TB)** is an even greater concern. This is defined as tuberculosis resistant to treatment with isoniazid and rifampin as well as three or more of the second-line anti-TB medications. In response to the problem of resistance, the FDA recently approved a new combination treatment specifically for XDR-TB: a new medication called pretomanid, to be taken orally with two second-line TB medications, bedaquiline and linezolid.

Neisseria gonorrhoeae

Gonorrhea, the sexually transmitted infection caused by *Neisseria gonorrhoeae,* was very easy to treat when penicillin was first introduced (about 80 years ago) because the bacterium was quite susceptible to the drug. Then, some strains developed resistance to penicillin through mutation, and others acquired a plasmid that encoded production of penicillinase. As these strains spread, penicillin could no longer reliably be used to treat the infection. Several other treatment options were available, including tetracyclines, macrolides, and fluoroquinolones, but as these antimicrobial medications were increasingly used, some *N. gonorrhoeae* strains gradually developed resistance to them as well. Today, only certain cephalosporins are reliably effective against most strains; gonorrhea is currently treated with a single intramuscular dose of ceftriaxone (a third-generation cephalosporin).

Staphylococcus aureus

Staphylococcus aureus, another common cause of HAIs, is becoming increasingly resistant to antimicrobials. Although nearly all strains were susceptible to penicillin when that antibiotic was first introduced, most are now resistant due to their acquisition of a gene that encodes penicillinase. Until recently, infections by these strains could be treated with methicillin or other penicillinase-resistant penicillins. New strains have emerged, however, that not only produce penicillinase, but also make a penicillin-binding protein called PBP2a, which has a low affinity for most β-lactam antibiotics. These strains, called **methicillin-resistant *S. aureus* (MRSA),** are resistant to methicillin as well as to nearly all other β-lactam antibiotics. Ceftaroline, a relatively new cephalosporin, is an important exception because it binds PBP2a.

MRSA strains are described as two categories: healthcare-associated (HA-MRSA) and community acquired (CA-MRSA). HA-MRSA strains are generally resistant to a wide range of antimicrobial medications, so severe infections are often treated with vancomycin. Some hospitals, however, have reported isolates that are no longer susceptible to normal levels of vancomycin. Infections caused by these **vancomycin-intermediate *S. aureus* (VISA)** and **vancomycin-resistant *S. aureus* (VRSA)** strains are relatively uncommon, and the optimal treatment is still uncertain. Fortunately, most CA-MRSA strains are currently susceptible to certain medications other than β-lactam antibiotics.

Streptococcus pneumoniae

Until recently, *Streptococcus pneumoniae,* the leading cause of pneumonia in adults, has remained very susceptible to penicillin. Some isolates, however, are now resistant to the antibiotic. This acquired resistance is due not to the production of a β-lactamase, but to changes in the chromosomal genes coding for the targets of penicillin—the penicillin-binding proteins. The modified targets have lower affinities for the medication. The nucleotide changes do not appear to have come from point mutations as one might expect; instead, they are due to the acquisition of chromosomal DNA from other species of *Streptococcus.* As you may recall from earlier reading, *S. pneumoniae* can acquire DNA through bacterial transformation. In addition to penicillin resistance, some *S. pneumoniae* strains have developed resistance to other antibiotics, including macrolides.

Preventing Resistance

Antimicrobial resistance has become such a concern that the CDC published a document, *Antibiotic Resistance Threats in the United States, 2019,* which lists the most serious threats—including antimicrobial-resistant bacteria and fungi—and categorizes them by level: urgent, serious, and concerning (**table 20.2**). These categories reflect multiple factors, including the public health significance of the resistant strains as well as the likelihood that the organisms will continue to spread. Unfortunately, as shared in the CDC's 2022 special

TABLE 20.2 Threats of Resistance to Antimicrobial Medications, 2019

Microorganism	Comments
CDC's Threat Level: Urgent	
Carbapenem-resistant *Acinetobacter*	*Acinetobacter* species are a common cause of healthcare-associated infections (HAIs). Carbapenems are often a last resort, so the infections are very difficult or impossible to treat effectively.
Candida auris (a fungus)	*C. auris*, first identified in 2009, has caused HAI outbreaks around the world. Some strains are resistant to all available antifungal medications.
Clostridioides (Clostridium) difficile	*C. difficile* causes antibiotic-associated intestinal infections that can sometimes be life-threatening; the organism is considered an urgent threat because *C. difficile* infection (CDI) is related to the use of antimicrobials.
Carbapenem-resistant Enterobacteriaceae (CRE)	Members of the family Enterobacteriaceae are a common cause of HAIs. Carbapenems are often a last resort, so CRE infections are very difficult or impossible to treat effectively.
Drug-resistant *Neisseria gonorrhoeae*	*N. gonorrhoeae* causes gonorrhea, a sexually transmitted infection. The current recommended treatment is a single intramuscular dose of ceftriaxone (a cephalosporin).
CDC's Threat Level: Serious	
Drug-resistant *Campylobacter*	*Campylobacter* species are a common cause of foodborne diarrhea; resistance to fluoroquinolones and macrolides is increasing.
Drug-resistant *Candida* (a fungus)	*C. albicans* and other species are a common cause of HAIs; relatively few antifungal medications are available, so acquired resistance to even one type is significant.
Extended-spectrum β-lactamase (ESBL) producing Enterobacteriaceae	Members of the family Enterobacteriaceae are a common cause of HAIs. ESBL-producing Enterobacteriaceae are resistant to all the β-lactam antibiotics, leaving carbapenem as the last treatment option.
Vancomycin-resistant *Enterococcus* (VRE)	*Enterococcus* species are a common cause of HAIs, including urinary tract and bloodstream infections. Enterococci are innately resistant to many antimicrobials, and VRE strains are resistant to nearly all antimicrobial medications.
Multi-drug-resistant *Pseudomonas aeruginosa*	*P. aeruginosa* is a common cause of HAIs; it is also a common cause of pneumonia in cystic fibrosis patients. *P. aeruginosa* is innately resistant to many antimicrobials, so multi-drug resistance makes a difficult situation even more challenging.
Drug-resistant non-typhoidal *Salmonella*	Non-typhoidal *Salmonella* strains are a common cause of foodborne diarrhea. Uncomplicated infections are not usually treated with antimicrobial medications, but infections that spread to the bloodstream can be life-threatening and should be treated.
Drug-resistant *Salmonella* Typhi	*Salmonella* Typhi causes typhoid fever, a life-threatening systemic disease. Most illnesses are travel-related, and strains are becoming increasingly resistant to certain treatment options.
Drug-resistant *Shigella*	*Shigella* species cause diarrheal diseases that spread easily in conditions of poor sanitation. Widespread antimicrobial resistance makes treatment more difficult.
Methicillin-resistant *Staphylococcus aureus* (MRSA)	*S. aureus* is a common cause of healthcare-associated (HA) and community-acquired (CA) skin and wound infections. MRSA is resistant to nearly all β-lactam antibiotics; HA-MRSA strains are generally resistant to many other antimicrobial medications as well, so severe infections are typically treated with vancomycin; CA-MRSA strains are usually susceptible to some other options.

(continued)

TABLE 20.2 Threats of Resistance to Antimicrobial Medications, 2019 (*Continued*)

Microorganism	Comments
Drug-resistant *Streptococcus pneumoniae*	*S. pneumoniae* causes pneumonia and meningitis, as well as bloodstream, ear, and sinus infections. Some strains are resistant to a variety of antibiotics, including penicillins and macrolides.
Drug-resistant tuberculosis	*M. tuberculosis* causes tuberculosis. Treatment has always been a long and complicated process because the organism is slow-growing and innately resistant to many antimicrobials. Treatment of infections caused by drug-resistant strains takes longer, is more expensive, and carries more health risks. Drug-resistant strains are an increasing problem worldwide, but are relatively rare in the United States.
CDC's Threat Level: Concerning	
Erythromycin-resistant Group A streptococcus	Group A streptococcus causes a variety of infections, including pharyngitis, impetigo, and tissue infections. A main concern regarding antibiotic resistance is invasive infections (infections of otherwise sterile body sites).
Clindamycin-resistant Group B streptococcus	Group B streptococcus is an important cause of neonatal meningitis; pregnant women who are colonized with the organism are treated to prevent infection of the newborn.

report *COVID-19: U.S. Impact on Antimicrobial Resistance*, the 2020 data for over half of the AMR threats was delayed or is unavailable because of the COVID-19 pandemic; further, the pandemic also led to significant increases in both infections and deaths from resistant healthcare-associated pathogens.

To reverse the alarming trend of increasing AMR, everyone must cooperate. On an individual level, physicians as well as the general public must take more responsibility for the appropriate uses of these life-saving medications. On a global scale, countries around the world need to make important policy decisions about what are, and what are not, appropriate uses of these medications.

The Responsibilities of Physicians and Other Healthcare Workers

Physicians and other healthcare workers need to increase their efforts to identify the cause of a given infection and, only if appropriate, prescribe suitable antimicrobials. Recent studies show that the rate at which physicians prescribe them varies widely across the United States, perhaps reflecting inappropriate use of these life-saving medications in certain regions (**figure 20.15**). In response to the problem of increasing AMR, hospitals and healthcare systems are implementing antimicrobial stewardship programs to help ensure appropriate treatments. Although these efforts may be more expensive in the short term, they will ultimately save both lives and money.

The Responsibilities of Patients

Patients need to carefully follow the instructions that accompany their prescriptions, even if those instructions seem inconvenient. When a patient skips a scheduled dose, the blood level of that medication may not remain high enough to inhibit the growth of the least-susceptible members of the microbial population. If these less-susceptible organisms then have a chance to grow, they will give rise to a population that is not as susceptible as the original. Likewise, failure to complete the prescribed course of treatment may not kill the least-susceptible organisms, allowing their subsequent multiplication. The misuse of antimicrobials by skipping doses or failing to complete the prescribed treatment increases the likelihood that resistant mutants will develop.

The Importance of an Educated Public

A greater effort must be made to educate people about the role and limitations of antimicrobial medications. First and foremost, people need to understand that antibiotics are not effective against viruses. Taking antibiotics will not cure the

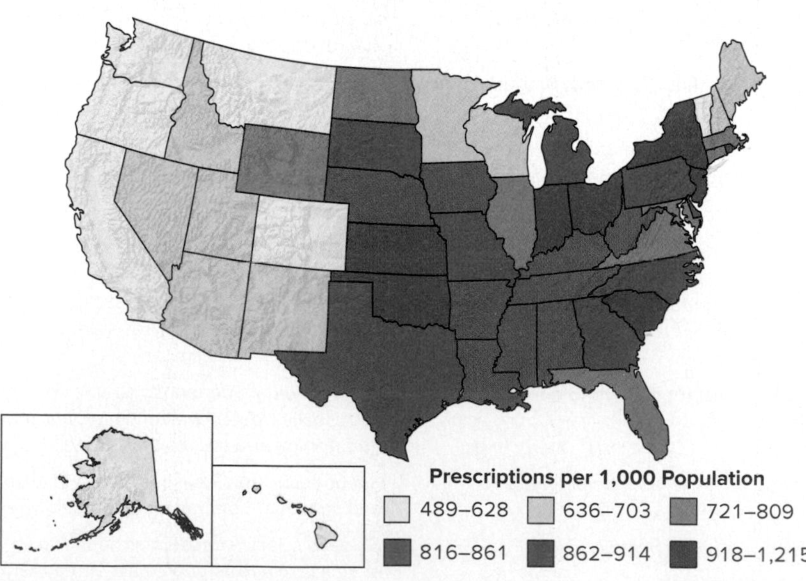

FIGURE 20.15 Outpatient Antimicrobial Prescriptions (per 1,000 Persons) in the United States, by State, 2017 Centers for Disease Control and Prevention. Outpatient antibiotic prescriptions — United States, 2017

? What suggestions might an antimicrobial stewardship program give to decrease high rates of antimicrobial prescriptions?

FOCUS ON A CASE 20.1

A 36-year-old woman came to the emergency department complaining of severe shortness of breath, cough, and chest pain. She said she had been feeling ill for 3 months and had lost 20 pounds as a result. When questioned about her travel history, she said she had immigrated to the United States from India when she was 10 and had since taken several trips back there to visit family members. A chest X ray was done, and lesions in one of her lungs were consistent with pulmonary tuberculosis, a lung infection caused by *Mycobacterium tuberculosis.*

The patient's cough was not productive (meaning it was dry), so physicians sampled material from her lower respiratory tract using bronchoalveolar lavage (BAL). This involves delivering sterile saline into the lungs via a long thin tube and then collecting the fluid. The BAL fluid was then sent to the clinical laboratory for examination.

In the clinical laboratory, a technician prepared a smear of the BAL fluid on a microscope slide, in preparation for doing an acid-fast stain. He also inoculated the fluid onto media that support the growth of *M. tuberculosis.* After doing the acid-fast stain, the technician carefully examined the stained specimen microscopically. He could not find any acid-fast bacilli (AFB), so he reported the result as negative. The technician then did a nucleic acid amplification test (NAAT) designed to detect *M. tuberculosis* DNA; that result was positive.

The patient was prescribed the standard four-drug treatment for tuberculosis disease: isoniazid (INH), rifampin (RIF), pyrazinamide (PZA), and ethambutol (EMB). After 3 weeks, she reported feeling slightly better, but was still coughing. A month later, the culture results were positive for *M. tuberculosis.* Antimicrobial susceptibility testing was done, and the infecting strain was found to be resistant to INH and RIF. Two other medications were then substituted, and the local public health office was contacted to provide directly observed therapy (DOT) for the remainder of her treatment.

1. Why was the patient's travel history important?
2. Why are *M. tuberculosis* infections more difficult to treat than most other bacterial infections?
3. Why was a combination of antimicrobial medications prescribed?
4. Why did it take so long for culture results to be available?
5. What is the significance of the fact that the strain was resistant to isoniazid and rifampin?
6. Why was directly observed therapy (DOT) used?

Discussion

1. Although tuberculosis is no longer endemic in the United States, it is still common in many parts of the world, including India. The patient could have contracted the infection as a child in India or during her trips to visit family members.
2. *M. tuberculosis* has a very waxy cell wall, which prevents many antibacterial medications from penetrating. In addition, it is extremely slow growing, and most antimicrobial medications are most effective against rapidly multiplying bacteria.
3. Combinations of antimicrobials are used to prevent the development of resistant mutants. Consider a situation where a single point mutation in a gene allows a bacterium to become resistant to a given antimicrobial, and the chance of that mutation occurring is one in a million (10^{-6}). If a susceptible bacterial cell is grown to a population of a billion cells, then about 1,000 cells will have spontaneously developed resistance to that medication. Meanwhile, if another medication is prescribed, but the rate of spontaneous mutations to resistance for that one is 10^{-7}, then 100 cells will have developed resistance to that option. However, the chance that a single cell will spontaneously develop resistance to both medications simultaneously is the product of the two rates, meaning 10^{-13}.
4. *M. tuberculosis* grows very slowly, with a generation time of over 16 hours, so it takes weeks for visible colonies to form. This is why detection methods that do not require culture are so important for early diagnosis.
5. Isoniazid and rifampin are first-line anti-tuberculosis drugs, the preferred options for treatment. When *M. tuberculosis* is resistant to the first-line drugs, then the patient must be treated with second-line drugs, and these are not as effective and are also more toxic and more expensive. Tuberculosis caused by strains resistant to isoniazid and rifampin is called MDR-TB (multidrug-resistant tuberculosis).
6. DOT ensures that patients comply with the prescribed treatment, thereby preventing additional resistance from developing. In addition, patients who comply with their treatment become non-infectious sooner, so there is less opportunity for the infectious agent to be transmitted. From a public health standpoint, this is extremely important because of the increased costs and less favorable outcomes associated with treating MDR-TB.

common cold, COVID-19, or any other viral illness. A few antiviral medications are available, but they are effective against only certain viruses. Unfortunately, far too many people mistakenly believe that antibiotics are effective against viruses, and often seek prescriptions to "cure" viral infections. This misuse only selects for antibiotic-resistant bacteria in the normal microbiota. Even though these organisms typically do not cause disease, they can serve as a reservoir for R plasmids, eventually transferring their resistance genes to an infecting pathogen.

Global Impacts of the Use of Antimicrobial Medications

The overuse of antibiotics and other antimicrobial medications is a worldwide concern. Countries may vary in their laws and customs, but antimicrobial resistance recognizes no political boundaries; an organism that develops resistance in one country can quickly be spread globally. In many parts of the world, particularly in resource-limited countries, antimicrobial medications are available on a non-prescription basis. Because of the consequences of inappropriate use, restricting or eliminating the over-the-counter availability of these medications should be considered.

Another worldwide concern is the use of antimicrobials in animal feeds. Low levels of these substances in feeds enhance the growth of animals, a seemingly attractive option. This use, like any other, however, selects for drug-resistant organisms, which has caused many scientists to question its ultimate wisdom. In fact, infections caused by drug-resistant *Salmonella* strains have been linked to animals whose feed was supplemented with those medications. In response to these concerns, the FDA developed and implemented a plan that has now phased out the growth-enhancing use of medically important antimicrobials in animals raised as food in the United States. In addition, there is growing pressure worldwide to ban the use of antimicrobials in animal feeds.

MicroAssessment 20.5

Mutations and transfer of genetic information allow microorganisms to become resistant to antimicrobial medications. Antimicrobial resistance affects the outcomes of medical treatment. Slowing the emergence and spread of resistant microbes involves the cooperation of healthcare personnel, patients, educators, and the general public.

13. Explain how using a combination of two antimicrobial medications helps prevent the development of spontaneously resistant mutants.
14. Explain the significance of a member of the normal microbiota that harbors an R plasmid.
15. A student argued that "spontaneous mutation" meant that an antimicrobial could cause mutations. Is the student correct? Why or why not?

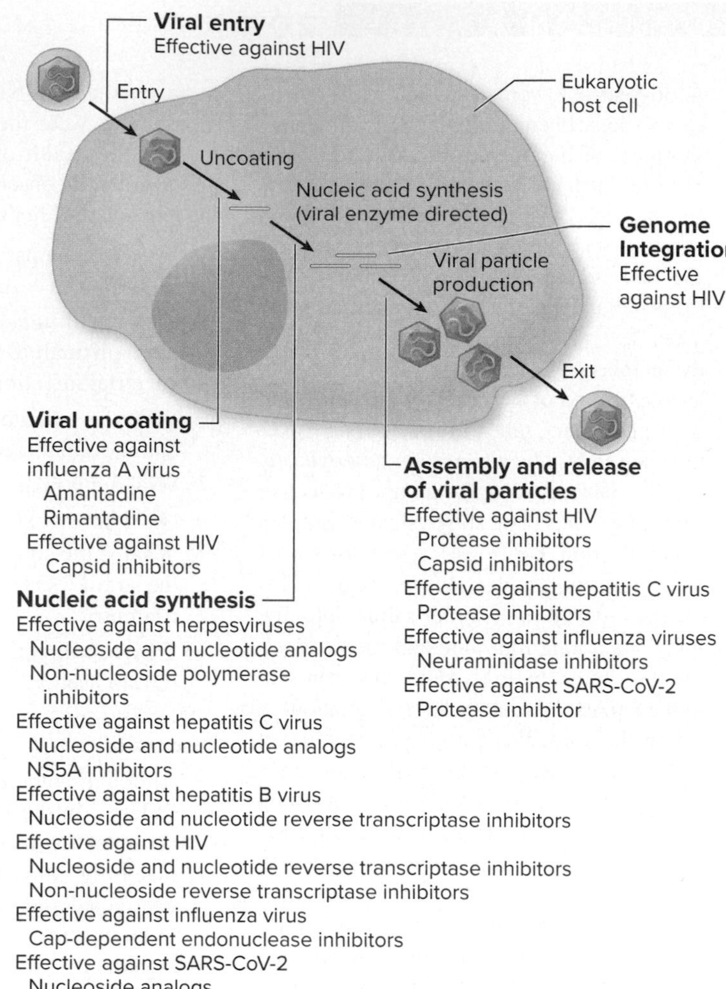

FIGURE 20.16 Targets of Antiviral Medications

Why are there relatively few options for antiviral medications?

20.6 ■ Mechanisms of Action of Antiviral Medications

Learning Outcome

13. Describe the antiviral medications that interfere with the following: (1) viral uncoating, (2) nucleic acid synthesis, (3) genome integration, (4) assembly and release of viral particles.

Viruses rely almost exclusively on the host cell's metabolic machinery for their replication, making it difficult to find a target for selective toxicity. They have no cell wall, ribosomes, or any other structure targeted by antibiotics. Because of this, viruses are completely unaffected by antibiotics. Many encode their own polymerases, however, and these, along with a few other unique viral proteins, are potential targets of **antiviral medications** (**figure 20.16**). These medications, also called antiviral drugs or antivirals, are generally effective against only a specific type of virus. The greatest variety of antivirals currently available are directed at HIV (human immunodeficiency virus). Fortunately, antivirals that target SARS-CoV-2, the virus that causes COVID-19, are now in use; the strategies for their development are described in Focus Your Perspective 20.1.

An important limitation of antivirals is that they are only effective against replicating viruses. Viruses such as herpesviruses and HIV can remain latent in cells, so the medications for treating them do not cure these infections. To treat a herpesvirus infection, a medication is generally taken for only a limited time period as a means to shorten the duration of a symptomatic episode; latent virus can still reactivate, causing symptoms to recur, in which case a course of the medication is taken again. With HIV, different combinations of appropriate medications are taken for the person's lifetime in order to slow disease progression; combination therapy must be used to prevent the virus from developing resistance.

For viruses such as HIV and hepatitis C virus (HCV) that evolve rapidly to develop resistance to single medications, combination therapy is typically used. Manufacturers have now developed a number of fixed-dose combinations that make it easier for patients to adhere to prescribed treatment protocols.

Prevent Viral Entry

Some medications effective against HIV are entry inhibitors, meaning they prevent the virus from entering host cells, either by preventing attachment or by preventing fusion of the viral envelope with the host cell membrane. They can be separated into four categories:

- **Attachment inhibitor.** A new medication, fostemsavir (FTR), binds to the HIV surface protein called Env that the virus uses to attach to the CD4 molecule on host cells.
- **Post-attachment inhibitor.** Ibalizumab is not a chemotherapeutic agent but a monoclonal antibody (mAb) used to treat certain multidrug-resistant HIV infections. It binds to CD4 molecules on host cells in a way that prevents any virus that binds the receptor from then binding to a co-receptor.
- **CCR5 antagonist.** Maraviroc (MCV) binds to the cell surface protein CCR5, a molecule that certain HIV strains use as a co-receptor to attach to cells. The medication cannot be used to treat infections caused by HIV strains that use the alternative co-receptor CXCR4.
- **Fusion inhibitor.** Enfuvirtide (ENF) binds to an HIV protein that otherwise promotes fusion of the viral envelope with the cell membrane, thereby blocking fusion.

Interfere with Viral Uncoating

Two groups of medications target the uncoating process that releases the nucleic acid of a viral particle from the protein coat. The medications amantadine and rimantadine interfere with uncoating of influenza A virus, but they are not currently used because of widespread resistance. An entirely new class of medication, capsid inhibitors, targets the HIV capsid.

Capsid Inhibitors

A recently approved drug for the treatment of HIV infections, lenacapavir, represents a new class of antiviral drugs—capsid inhibitors. Lenacapavir (LEN) works by binding to the protein that makes up the subunits (capsomers) that compose the viral capsid. This binding interferes with both capsid disassembly during viral uncoating and capsid assembly during viral maturation, thus targeting the virus during two separate stages of its replication cycle. Lenacapavir is particularly useful for treating patients with multi-drug resistant HIV infections.

Interfere with Nucleic Acid Synthesis

Many of the most effective antiviral medications target virally encoded enzymes used during replication of viral nucleic acid. These antivirals are generally limited to treating infections caused by herpesviruses, HBV (hepatitis B virus), HCV (hepatitis C virus), or HIV. Nucleic acid synthesis is also a target of certain medications used to treat COVID-19.

Nucleoside and Nucleotide Analogs

Several antiviral medications are nucleoside or nucleotide analogs—chemicals structurally similar to the building blocks of DNA and RNA that can interfere with nucleic acid synthesis. The two types of analogs are considered together because a nucleoside analog can be converted within a cell to a nucleotide analog—a process that involves phosphorylation by a virally encoded or a normal cellular enzyme. Many nucleotide analogs act as chain terminators during nucleic acid synthesis; when the analog is incorporated into a growing nucleotide chain, additional nucleotides cannot be added to that chain. The basis for the selective toxicity of most analogs is the fact that error-prone virally encoded polymerases are more likely than the host cell enzymes to incorporate them. Therefore, more damage is done to the rapidly replicating viral genome than to the host cell genome.

Acyclovir, famciclovir, and valacyclovir are nucleoside analogs used to speed the healing of the characteristic lesions associated with shingles, chickenpox, cold sores, and genital herpes—diseases all caused by herpesviruses. The medications cause little harm to uninfected cells because normal cellular enzymes are not likely to convert the chemicals into nucleotide analogs. Instead, a virally encoded enzyme found only in infected cells converts the medications into their active form.

Sofosbuvir is a relatively new medication that has revolutionized the treatment of HCV infections because it is so effective when used in combination with at least one other anti-HCV medication. This nucleotide analog is incorporated by the HCV's replicase (an RNA-dependent RNA polymerase), resulting in chain termination.

Recall that HIV and HBV are both reverse-transcribing viruses, meaning that a step in their replication cycle requires the virally encoded enzyme reverse transcriptase. Nucleoside and nucleotide analogs that interfere with the activity of reverse transcriptase are referred to as NRTIs (nucleoside/nucleotide reverse transcriptase inhibitors). To treat HIV infections, two NRTIs are often used in combination with at least one other antiretroviral medication; the NRTIs used for this include abacavir (ABC), emtricitabine (FTC), lamivudine (3TC), tenofovir disoproxil fumarate (TDF), and azidothymidine (AZT), also known as zidovudine (ZDV). The NRTIs used to treat chronic hepatitis B (caused by HBV) include tenofovir, adefovir, and lamivudine.

Certain nucleoside analogs are used only for severe infections because of their significant side effects. An example is ganciclovir, which is used to treat life- or sight-threatening cytomegalovirus (CMV) infections in immunocompromised patients.

As described in in Focus Your Perspective 20.1, two nucleoside analogs, remdesivir and molnupiravir, are now being used to treat certain COVID-19 patients.

Non-Nucleoside Polymerase Inhibitors

Non-nucleoside polymerase inhibitors are compounds that inhibit the activity of viral polymerases by binding to a site other than the nucleotide-binding site; like many antiviral medications, these are typically virus-specific. The non-nucleoside polymerase inhibitor dasabuvir is a component of a fixed-dose combination used to treat HCV infections. Foscarnet is used to treat severe infections caused by herpesviruses that are resistant to other medications.

Non-Nucleoside Reverse Transcriptase Inhibitors (NNRTIs)

Non-nucleoside reverse transcriptase inhibitors (NNRTIs) inhibit the activity of reverse transcriptase by binding to a site other than the nucleotide-binding site of the enzyme. NNRTIs used to treat HIV infections include doravirine (DOR), efavirenz (EFV), etravirine (ETR), nevirapine (NVP), and rilpivirine (RPV).

NS5A Inhibitors

NS5A inhibitors offer a relatively new option for treating HCV infections. As the name implies, they inhibit NS5A, an HCV-encoded protein required for replication of the viral genome. Examples include elbasvir, ombitasvir, and velpatasvir.

Cap-Dependent Endonuclease Inhibitors

Cap snatching is a process used by some viruses to "steal" the 5′ cap from host cell mRNA so that functional viral mRNA can be produced. The medication baloxavir marboxil (often simply called baloxavir) targets cap snatching in influenza viruses by interfering with the function of the viral endonuclease that catalyzes the process—polymerase acidic (PA) protein.

Prevent Genome Integration

Integrase strand transfer inhibitors (INSTIs), also simply called integrase inhibitors, interfere with the action of the HIV-encoded enzyme integrase, thereby preventing the virus from inserting the DNA copy of its genome into the host cell genome. Examples of INSTIs include dolutegravir (DTG), elvitegravir (EVG), raltegravir (RAL), and bictegravir (BIC).

Prevent Assembly and Release of Viral Particles

Virally encoded enzymes required for the assembly and release of viral particles are the targets of medications used to treat certain viral infections. As described earlier in the section, the new capsid inhibitor that targets HIV interferes with the uncoating process as well as viral maturation.

Protease Inhibitors

Protease inhibitors (PIs) are chemicals that inhibit virally encoded proteases. Recall that some viruses express certain groups of genes as a single unit to form a polyprotein. Those viruses then use virally encoded proteases to cleave the viral polyproteins in order to release the individual proteins. Viral proteases are virus-specific, so each type of protease inhibitor is virus-specific as well. Protease inhibitors used to treat HIV infections have the suffix -navir, and include atazanavir (ATV) and darunavir (DRV). Another HIV protease inhibitor, ritonavir (RTV), is used in combination with other protease inhibitors because it helps prevent their breakdown in the body. Protease inhibitors used to treat HCV infections have the suffix -previr and include grazoprevir, paritaprevir, and simeprevir. Additionally, as described in Focus your Perspective 20.1, a SARS-CoV-2 protease inhibitor is part of a medication recently developed to treat COVID-19.

Neuraminidase Inhibitors

Neuraminidase inhibitors inhibit neuraminidase, an influenza virus-encoded enzyme required for the release of viral particles from infected cells. Three neuraminidase inhibitors are currently available in the United States: oseltamivir (taken orally), zanamivir (inhaled), and peramivir (given by injection).

Characteristics of antiviral medications described in this section are summarized in **table 20.3.**

TABLE 20.3 Characteristics of Antiviral Medications

Target/Class/Example(s)	Comments/Characteristics
Viral Entry	
Attachment inhibitor	Binds to the HIV Env protein, which the virus uses to attach to the CD4 molecule on host cells.
Fostemsavir (FTR)	Used to treat HIV infections.
Post-attachment inhibitor	Binds to CD4 molecules on host cells in a way that prevents any viral particle that binds the receptor from then binding to a co-receptor.
Ibalizumab	Used to treat HIV infections.
CCR5 antagonist	Blocks the HIV co-receptor CCR5.
Maraviroc (MVC)	Used to treat HIV infections.
Fusion Inhibitors	Bind to an HIV protein that promotes fusion of the viral envelope with the cell membrane.
Enfuvirtide (ENF)	Used to treat HIV infections.

TABLE 20.3 Characteristics of Antiviral Medications (*Continued*)

Target/Class/Example(s)	Comments/Characteristics
Viral Uncoating	
Capsid inhibitors	Bind to protein subunits composing HIV's capsid, interfering with both uncoating and maturation.
Lenacapavir (LEN)	Used to treat multi-drug resistant HIV infections.
Uncoating inhibitors	Prevent release of viral nucleic acid from the protein coat.
Amantadine and rimantadine	Used in the past to reduce severity and duration of influenza A infections.
Nucleic Acid Synthesis	
Nucleoside and nucleotide analogs	Incorporated by error-prone virally encoded polymerases, resulting in defective viral nucleic acid.
Acyclovir, famciclovir, valacyclovir	Used to treat shingles, chickenpox, cold sores, and genital herpes—diseases all caused by herpesviruses.
Sofosbuvir	Used to treat HCV infections.
Nucleoside and nucleotide reverse transcriptase inhibitors (NRTIs): abacavir (ABC), emtricitabine (FTC), lamivudine (3TC), tenofovir disoproxil fumarate (TDF), and zidovudine (AZT, ZDV)	Used to treat HIV infections; some are used to treat chronic hepatitis B.
Ganciclovir	Used to treat life- or sight-threatening cytomegalovirus infections.
Remdesivir, molnupiravir	Used to treat COVID-19.
Non-nucleoside polymerase inhibitors	Inhibit the activity of viral polymerases by binding to a site other than the nucleotide-binding site.
Dasabuvir	Used in a fixed-dose combination to treat HCV infections.
Foscarnet	Used to treat severe infections caused by herpesviruses that are resistant to other medications.
Non-nucleoside reverse transcriptase inhibitors (NNRTIs)	Inhibit the activity of reverse transcriptase by binding to a site other than the nucleotide-binding site.
Doravirine (DOR), efavirenz (EFV), etravirine (ETR), nevirapine (NVP), and rilpivirine (RPV)	Used to treat HIV infections.
NS5A inhibitors	Inhibit NS5A, an HCV-encoded protein required for replication of the viral genome.
Elbasvir, ombitasvir, velpatasvir	Used to treat HCV infections.
Cap-dependent endonuclease inhibitors	Inhibit an enzyme used to make functional viral mRNA.
Baloxavir marboxil	Used to treat influenza.
Genome Integration	
Integrase strand transfer inhibitors	Interfere with the HIV-encoded enzyme integrase
Dolutegravir (DTG), elvitegravir (EVG), raltegravir (RAL), and bictegravir (BIC)	Used to treat HIV infections.
Assembly and Release of Viral Particles	
Protease inhibitors	Inhibit virally encoded proteases, which cleave viral polyproteins to release individual proteins.
Atazanavir (ATV), darunavir (DRV), and ritonavir (RTV)	Used to treat HIV infections.
Grazoprevir, paritaprevir, simeprevir	Used to treat hepatitis C.
Nirmatrelvir	Used in combination with ritonavir to treat COVID-19.
Neuraminidase inhibitors	Inhibit an influenza virus-encoded enzyme required for release of viral particles from infected cells.
Oseltamivir, zanamivir, peramivir	Used to treat influenza.
Capsid inhibitors	Described above because they also interfere with uncoating.

FOCUS YOUR PERSPECTIVE 20.1

The Race to Develop COVID-19 Treatments

Almost immediately after the emergence of COVID-19, scientists raced to find effective treatments. An early focus was on drug repurposing—the use of approved or investigational drugs for new therapeutic uses. Approved drugs are those that have undergone the testing required for the U.S. Food and Drug Administration (FDA) to authorize marketing of the drug; investigational drugs are experimental drugs that the FDA has authorized for testing in humans. An enormous advantage of a repurposed drug is that it has already gone through clinical trials to assess its safety in patients. Thus, a repurposed drug can often be brought into wide-scale use more quickly than a brand new medication.

The first example of a repurposed antiviral medication authorized by the FDA for treating COVID-19 was the nucleoside analog remdesivir, an investigational drug developed to treat various diseases caused by RNA viruses. Clinical studies using intravenously administered remdesivir to treat severe COVID-19 showed that it shortens the duration of symptoms. Because of this and the fact that no other antiviral treatment options were available early during the pandemic, the FDA first granted emergency authorization for its use in treating patients with severe COVID-19. Remdesivir is now approved for use in treating COVID-19 in adults and children at risk of progression to severe disease. Molnupiravir, another nucleoside analog, was originally shown to have activity against multiple RNA viruses and was in Phase I clinical trials for treating influenza when the pandemic started. Researchers then shifted their focus and showed that it was also active against SARS-CoV-2, the virus that causes COVID-19. The FDA has since granted molnupiravir emergency use authorization for the treatment of certain adult COVID-19 patients.

Armed with a better understanding of SARS-CoV-2, scientists are now using rational drug design to develop medications that specifically target key viral proteins. One drug designed using this approach, nirmatrelvir, is part of an oral medication that has received emergency use authorization for treating certain COVID-19 patients. Nirmatrelvir specifically inhibits a SARS-CoV2-encoded protease required for viral replication; it is given in combination with ritonavir, which helps prevent the breakdown of nirmatrelvir in the body.

The speed of the progress with which effective COVID-19 treatments have been developed is a testament to the effort and cooperation of countless scientists around the world. Despite the remarkable progress, however, the job is not finished. New variants continue to emerge, so long-term control of the disease might require a variety of drugs, each interfering with a different target.

MicroAssessment 20.6

Viral replication generally uses host cell machinery; because of this, there are few targets for selectively toxic antiviral medications. Available antivirals are virus-specific; targets include viral entry, viral uncoating, nucleic acid synthesis, genome integration, and the assembly and release of viral particles.

16. Explain why acyclovir has fewer side effects than do other nucleoside analogs.
17. How do protease inhibitors interfere with the production of infectious viral particles?
18. Why are nucleoside analogs active only against replicating viruses?

20.7 ■ Mechanisms of Action of Antifungal Medications

Learning Outcome

14. Describe the antifungal medications that interfere with the following: (1) fungal cytoplasmic membrane synthesis and function, (2) cell wall synthesis, (3) cell division, (4) nucleic acid synthesis, and (5) protein synthesis.

Eukaryotic pathogens such as fungi more closely resemble human cells than do bacteria, which is why there are so few targets for antifungal medications (**figure 20.17**). Because of the relative scarcity of treatment options, acquired resistance is a significant concern. Scientists are currently paying particularly close attention to infections caused by the fungus *Candida auris* because some strains of this emerging health-care associated pathogen are resistant to multiple antifungals (see table 20.2).

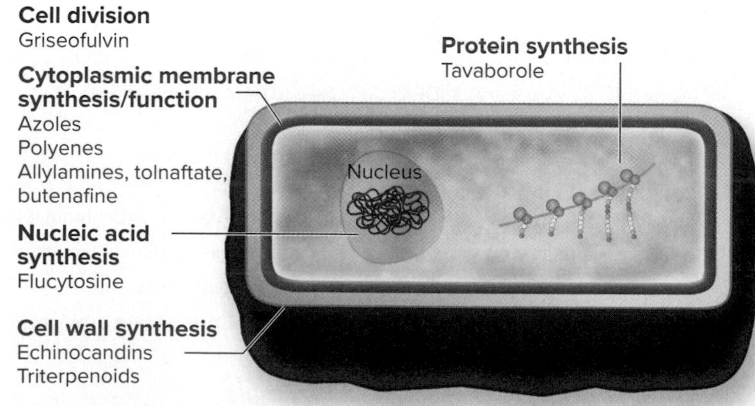

FIGURE 20.17 Targets of Antifungal Medications

What compound in the fungal cytoplasmic membrane is the target of most antifungal medications?

Interfere with Cytoplasmic Membrane Synthesis and Function

The target of most antifungal medications is ergosterol, a sterol found in the cytoplasmic membrane of fungal but not human cells.

Azoles

The azoles are a class of synthetic compounds that function by inhibiting ergosterol synthesis, resulting in defective fungal membranes that leak cytoplasmic contents. There are three families of antifungal azoles:

- **Imidazoles.** Two imidazoles—miconazole and clotrimazole—are commonly used in non-prescription creams, ointments, and suppositories to treat vulvovaginal candidiasis (vaginal yeast infections). They are also used on skin to treat dermatophyte infections.
- **Triazoles.** Triazoles have a variety of uses. Four examples—fluconazole, voriconazole, posaconazole, and isavuconazole—are extremely important for treating systemic fungal infections, and each has a slightly different spectrum of activity; fluconazole is additionally used to treat a wide variety of localized fungal infections, including vulvovaginal candidiasis and oropharyngeal candidiasis (thrush). Another triazole, eficonazole, is used for topically treating nail infections.
- **Tetrazoles.** Tetrazoles, a new family of azoles, are represented by oteseconazole. This drug has been approved for the oral treatment of complicated vulvovaginal candidiasis (chronic yeast infections) and has been shown to have higher specificity for ergosterol, resulting in fewer side effects.

Polyenes

The polyenes, a group of antibiotics produced by certain species of *Streptomyces,* bind to ergosterol. This binding disrupts the fungal membrane, killing the cell by allowing its cytoplasmic contents to leak out. Unfortunately, the polyenes are quite toxic to humans, which limits their use. The side effects of the polyene amphotericin B are so severe that the drug is usually reserved for treating life-threatening infections for which no other option is available. Newer lipid-based emulsions are less toxic but more expensive. Current formulations of another polyene, nystatin, are too toxic to be used systemically; however, because the medication is not absorbed, it can be used in skin creams, lozenges, and oral suspensions to treat *Candida albicans* infections.

Allylamines, Tolnaftate, Butenafine

The allylamines inhibit an enzyme in the pathway of ergosterol synthesis; tolnaftate and butenafine appear to do the same. They can be applied to the skin to treat dermatophyte infections, but terbinafine (an allyamine) can also be taken orally.

Interfere with Cell Wall Synthesis

Fungal cell walls contain β-1,3 glucan, a polysaccharide not produced by animal cells. Two families of antifungal medications—echinocandins and triterpenoids— interfere with β-1,3 glucan synthesis, causing the cells to burst.

- **Echinocandins.** Echinocandins are particularly useful for treating systemic *Candida* infections. The current examples—caspofungin, micafungin, and anidulafungin—are all administered intravenously, so they are primarily used in treating critically ill hospitalized patients. Some *C. auris* strains are resistant to these medications, however, which is one reason why the organism is a particular concern. Caspofungin can also be used for treating invasive aspergillosis that does not respond to other medications.
- **Triterpenoids.** The triterpenoids (also called the fungerps) are a new antifungal class represented by the oral medication ibrexafungerp. This drug has recently been approved for use in treating vulvovaginal candidiasis and is particularly useful when fluconazole is not an option. Ibrexafungerp is effective at treating infections caused by echinocandin-resistant strains because it binds to the enzyme responsible for β-1,3 glucan synthesis in a different location than the echinocandins.

Interfere with Cell Division

The target of one antifungal medication, griseofulvin, is cell division. Griseofulvin interferes with the action of tubulin, the protein that polymerizes to form microtubules. Because tubulin is a part of all eukaryotic cells, the selective toxicity of the medication may be due to its greater uptake by fungal cells. When taken orally for months, it is absorbed and concentrated in the dead keratinized layers of the skin. Fungi that then invade keratin-containing structures such as skin and nails take up the medication, which prevents them from multiplying. Griseofulvin is active only against fungi that invade keratinized cells and is used to treat skin and nail infections.

Interfere with Nucleic Acid Synthesis

Flucytosine (a synthetic derivative of the nucleobase cytosine) interferes with nucleic acid synthesis in yeast cells. It is taken up by those cells and then converted by yeast enzymes to a form that inhibits an enzyme required for nucleic acid synthesis. Unfortunately, resistant mutants are common, and therefore, flucytosine is used mostly in combination with amphotericin B or an alternative. Side effects can be significant, so it is used only for serious yeast infections.

TABLE 20.4 Characteristics of Antifungal Medications

Target/Class/Example(s)	Comments
Cytoplasmic Membrane	
Azoles	Interfere with ergosterol synthesis, leading to defective fungal membranes.
Imidazoles: miconazole, clotrimazole	Used in non-prescription creams, ointments, and suppositories to treat vaginal yeast infections; also used on skin to treat dermatophyte infections.
Triazoles: fluconazole, voriconazole, posaconazole, isavuconazole, and eficonazole	Most are extremely important for treating a variety of systemic fungal infections; fluconazole is additionally used to treat localized fungal infections, including vaginal yeast infections and thrush; eficonazole is only used topically.
Tetrazoles: oteseconazole	Used orally to treat some cases of vulvovaginal candidiasis.
Polyenes	Bind to ergosterol, causing the fungal membrane to leak. Very toxic.
Amphotericin B, nystatin	Amphotericin B is used as a last resort for treating life-threatening infections; nystatin is used topically.
Allylamines, tolnaftate, butenafine	Inhibit an enzyme in the pathway of ergosterol synthesis.
Naftifine, terbinafine (allylamines); tolnaftate; butenafine	Used topically to treat dermatophyte infections; terbinafine can be taken orally.
Cell Wall Synthesis	
Echinocandins	Interfere with β-1,3 glucan synthesis.
Caspofungin, micafungin, and anidulafungin	Used to treat systemic *Candida* infections; some are used to treat invasive aspergillosis that resists other medications.
Triterpenoids (fungerps)	Interfere with β-1,3 glucan synthesis.
Ibrexafungerp	Used orally to treat vulvovaginal candidiasis.
Cell Division	
Griseofulvin	Used to treat skin and nail infections. Active only against fungi that invade keratinized cells.
Nucleic Acid Synthesis	
Flucytosine	Used to treat systemic yeast infections; inhibits an enzyme required for nucleic acid synthesis.
Protein Synthesis	
Tavaborole	Used topically to treat nail infections.

Interfere with Protein Synthesis

The antifungal medication tavaborole inhibits protein synthesis by interfering with an enzyme that "charges" a tRNA molecule by attaching the correct amino acid. It is used topically to treat nail infections.

The characteristics of various antifungal medications are summarized in **table 20.4.**

MicroAssessment 20.7

Because fungi are eukaryotic cells, there are relatively few targets for selectively toxic antifungal medications. Most antifungal medications interfere with ergosterol function or synthesis. Other targets include cell wall synthesis, cell division, and nucleic acid synthesis.

19. Why is amphotericin B used only for treating life-threatening infections?

20. Why is flucytosine generally used only in combination with other drugs?

21. Why is it reasonable to assume that the therapeutic index of fluconazole is higher than that of nystatin?

20.8 ■ Mechanisms of Action of Antiprotozoan and Antihelminthic Medications

Learning Outcome

15. Describe the targets of most antiparasitic medications.

Most antiparasitic medications probably interfere with biosynthetic pathways of protozoan parasites or the neuromuscular function of worms. Unfortunately, compared with antibacterials, antifungals, and antivirals, relatively little research and development go into antiparasitic medications, mainly because most parasitic diseases are neglected tropical diseases (NTDs); recall that NTDs are concentrated in resource-limited areas of the world where people simply cannot afford to spend money on expensive medications. In many cases, the modes of action for available medications are not well understood, hindering efforts to develop more effective derivatives. Note that some antibacterial drugs (including doxycycline, tetracycline, and clindamycin) can also be used for treating diseases caused by apicomplexan protozoa—including

Plasmodium species (causes malaria) and *Toxoplasma gondii* (causes toxoplasmosis); these drugs seem to target structures called apicoplasts, which are of bacterial origin and thought to be a product of secondary endosymbiosis. Even so, the need for new medication options is becoming urgent, as the most prevalent *Plasmodium* species are developing resistance to common antimalarial drugs.

Some of the most important antiparasitic medications and their characteristics are summarized in **table 20.5.**

TABLE 20.5 Characteristics of Some Antiprotozoan and Antihelminthic Medications

Causative Agent/Medication	Comments
Intestinal Protozoa	
Iodoquinol	Used to eliminate the cysts of the protozoan that causes intestinal amebiasis; mechanism unknown.
Metronidazole, tinidazole	Used to treat amebiasis, giardiasis, and trichomoniasis; activated by the metabolism of anaerobic organisms, forming reactive compounds that alter DNA.
Nitazoxanide	Used to treat cryptosporidiosis and giardiasis; interferes with anaerobic energy metabolism.
Parmomycin	Used to eliminate the protozoan that causes intestinal amebiasis from the intestinal tract; mechanism of action against protozoa is unclear.
Plasmodium* (Malaria) and *Toxoplasma	
Artemisinin-based combination therapy (ACT)	Used to treat malaria; intended to prevent development of additional resistance. Combines a fast acting artemisinin derivative with one of various other types of anti-malarial medications. The artemisinin derivative is activated by the parasite's heme digestion products to form a molecule that damages parasite components; the companion medications differ in their actions. Lumefantrine, a medication commonly used with artemether (an artemisinin derivative), is thought to interfere with the parasite's ability to detoxify heme.
Artesunate	An artemisinin derivative that is administered intravenously to treat severe malaria.
Atovaquone-proguanil	A synergistic combination used to prevent and treat malaria, particularly in regions where chloroquine resistance occurs. Atovaquone interferes with mitochondrial electron transport, and proguanil disrupts folate synthesis.
Chloroquine, halofantrine, hydroxychloroquine, quinine	Used to treat malaria; thought to interfere with the parasite's ability to detoxify heme.
Mefloquine	Used to treat malaria, particularly that which is acquired in regions where chloroquine resistance occurs; interferes with the function of the parasite's 80S ribosome.
Primaquine, tafenoquine	Used to treat relapsing forms of malaria, with tafenoquine being longer-lasting; each is activated by the parasite to form a molecule thought to damage the parasite components; both kill the dormant liver stage, but tafenoquine may also be effective at killing some blood stage parasites.
Pyrimethamine- sulfonamide	A synergistic combination used to treat toxoplasmosis and malaria; interferes with folate metabolism.
Trypanosomes and *Leishmania*	
Amphotericin B	Used to treat some types of leishmaniasis.
Benznidazole	Used to treat acute Chagas' disease; enzymes in the parasite convert it into a reactive form that damages certain cell components.
Eflornithine	Used to treat the late stage of chronic human African trypanosomiasis; inhibits the enzyme ornithine decarboxylase.
Fexinidazole	New option for oral treatment of both stages of chronic human African trypanosomiasis; interferes with DNA synthesis.
Melarsoprol	Used to treat the late stage of chronic and acute human African trypanosomiasis, but treatment can be lethal; inactivates enzymes by binding sulfhydryl groups.
Miltefosine	Used to treat some types of leishmaniasis; mechanism of action likely involves interaction with lipids and interference with mitochondrial function.
Nifurtimox	Used to treat acute Chagas' disease and, in combination with other medications, the late stage of chronic human African trypanosomiasis; thought to form reactive compounds that damage certain cell components.
Parmomycin	Used to treat some types of leishmaniasis; mechanism of action against protozoa is unclear.
Pentamidine	Used to treat the early stage of chronic human African trypanosomiasis; mechanism is poorly understood.
Sodium stibogluconate	Used to treat some types of leishmaniasis; mechanism of action involves sulfhydryl groups of cell components.
Suramin	Used to treat the early stage of acute human African trypanosomiasis; mechanism is poorly understood.
Intestinal and Tissue Helminths	
Albendazole, mebendazole	Albendazole is used to treat a variety of helminth infections, including tissue and intestinal infections; mebendazole is poorly absorbed, so it is only used to treat intestinal helminths; both bind to tubulin of helminths, resulting in immobilization of the worm.

(continued)

TABLE 20.5 Characteristics of Some Antiprotozoan and Antihelminthic Medications (*Continued*)

Causative Agent/Medication	Comments
Diethylcarbamazine	Used to treat infections caused by certain filarial nematodes; results in damage to the outer surface of the microfilaria, making it easier for the immune system to eliminate them.
Ivermectin	Used to treat infections caused by *Strongyloides stercoralis* and certain tissue nematodes; interferes with nerve transmission in the worms, causing flaccid (limp) paralysis.
Moxidectin	Used to treat onchocerciasis (river blindness); thought to interfere with nerve transmission in the worms.
Praziquantel	Used to eliminate a variety of trematodes (flukes) and cestodes (tapeworms); results in sustained contractions of the worms.
Pyrantel pamoate	Used to eliminate certain intestinal nematodes (roundworms); interferes with neuromuscular activity of worms.
Triclabendazole	Used to eliminate liver flukes; interferes with microtubule formation in the worms.

Summary

20.1 ■ History and Development of Antimicrobial Medications

Discovery of Antimicrobial Medications

Salvarsan, developed by Paul Ehrlich, was the first documented example of an **antimicrobial medication.**

Discovery of Antibiotics

Alexander Fleming discovered that a species of the fungus *Penicillium* produces an antimicrobial chemical he called penicillin, the first example of an **antibiotic** (figure 20.1).

Development of New Antimicrobial Medications

Most modern antibiotics come from species of *Streptomyces* and *Bacillus* (bacteria) and *Penicillium* and *Cephalosporium* (fungi). Antimicrobial medications can be chemically modified to give them new properties (figure 20.2).

Rational Drug Design

Rational drug design is a targeted approach used to develop medications for treating a wide variety of conditions, including bacterial and viral infections.

20.2 ■ Characteristics of Antimicrobial Medications

Selective Toxicity

Medically useful antimicrobials have **selective toxicity;** the relative toxicity of a medication is expressed as the **therapeutic index.**

Antimicrobial Action

Bacteriostatic chemicals inhibit the growth of bacteria; chemicals that kill bacteria are **bactericidal.**

Spectrum of Activity

Broad-spectrum antibiotics affect a wide range of bacteria; those that affect a narrow range are called **narrow-spectrum.**

Effects of Antimicrobial Combinations

Combinations of antimicrobial medications can be synergistic, antagonistic, or additive.

Tissue Distribution, Metabolism, and Excretion of the Medication

Some antimicrobials cross the blood-brain barrier into the cerebrospinal fluid; these can be used to treat meningitis. Medications that are unstable in acid must be administered through injection. The half-life of a medication affects how frequently it must be taken.

Adverse Effects

Some people develop allergies to certain antimicrobials. Some antimicrobials can have potentially damaging side effects such as kidney damage. By disrupting the microbiome, antimicrobials can cause **dysbiosis.**

Resistance to Antimicrobials

Certain types of bacteria have **intrinsic (innate) resistance** to a particular antimicrobial medication. **Acquired resistance** is due to spontaneous mutation or the acquisition of new genetic information.

20.3 ■ Mechanisms of Action of Antibacterial Medications

(figure 20.3, table 20.1)

Inhibit Cell Wall Synthesis (figure 20.4)

The **β-lactam antibiotics,** which include penicillins, cephalosporins, carbapenems, and monobactams, inhibit enzymes that help form cross-links between adjacent glycan chains, ultimately leading to cell lysis. Other medications that interfere with cell wall synthesis include glycopeptide antibiotics and bacitracin.

Inhibit Protein Synthesis (figure 20.7)

Antibiotics that inhibit protein synthesis by binding to the bacterial 50S ribosome include macrolides, chloramphenicol, lincosamides, oxazolidinones, streptogramins, and pleuromutilins; those that bind the 30S ribosome include aminoglycosides, tetracyclines, and glycylcyclines.

Inhibit Nucleic Acid Synthesis

Antimicrobials that inhibit nucleic acid synthesis include fluoroquinolones, rifamycins, fidaxomicin, and metronidazole.

Interfere with Metabolic Pathways (figure 20.8)

Sulfa drugs and trimethoprim block different enzymes in a metabolic pathway required for nucleotide biosynthesis. The two medications are often used together as a synergistic combination called **co-trimoxazole.**

Interfere with Cell Membrane Integrity

Daptomycin and polymyxins damage bacterial membranes, as do two of the new glycopeptide antibiotics.

Act Against *Mycobacterium tuberculosis*

First-line drugs used to treat tuberculosis include isoniazid, ethambutol, and pyrazinamide. **Second-line drugs** are used for strains resistant to the first-line drugs. Bedaquiline, pretomanid, and linezolid (an oxazolidinone) are used in combination to treat the most resistant *M. tuberculosis* strains.

20.4 ■ Antimicrobial Susceptibility Testing

Conventional Disc Diffusion Method (figure 20.9)
The **Kirby-Bauer disc diffusion test** determines the susceptibility of a bacterial strain to a variety of antimicrobial medications.

Minimum Inhibitory and Minimum Bactericidal Concentrations (MIC and MBC) (figure 20.10)
The **minimum inhibitory concentration (MIC)** is the lowest concentration of a specific antimicrobial needed to prevent the growth of a bacterial strain in vitro. The **minimum bactericidal concentration (MBC)** is the lowest concentration of a specific antimicrobial that kills 99.9% of cells of a given bacterial strain in vitro.

Commercial Modifications of Antimicrobial Susceptibility Testing
Automated methods can determine antimicrobial susceptibility in as little as 4 hours (figure 20.11).

20.5 ■ Resistance to Antimicrobial Medications (figure 20.13)

Mechanisms of Acquired Resistance (figure 20.14)
Enzymes that chemically modify an antimicrobial destroy the activity of the medication. Structural changes in the target can prevent the medication from binding. Altered porin proteins prevent drugs from entering cells. **Efflux pumps** actively pump drugs out of cells.

Acquisition of Resistance
Resistance can be acquired through spontaneous mutation or horizontal gene transfer. The most common mechanism of transfer of antibiotic resistance genes is through the conjugative transfer of **R plasmids.**

Examples of Emerging Resistance (table 20.2)
Strains of many organisms, including enterococci, Enterobacteriaceae, *Mycobacterium tuberculosis, Neisseria gonorrhoeae, Staphylococcus aureus,* and *Streptococcus pneumoniae* are becoming increasingly resistant to antimicrobial medications.

Preventing Resistance
Physicians should prescribe antimicrobial medications only when appropriate. Patients need to carefully follow prescribed instructions when taking antimicrobials. The public must be educated about the appropriateness and limitations of antimicrobial therapy.

20.6 ■ Mechanisms of Action of Antiviral Medications (figure 20.16, table 20.3)

Prevent Viral Entry
Entry inhibitors are available to treat HIV infections; they include an attachment inhibitor, post-attachment inhibitor, a CCR5 antagonist, and a fusion inhibitor.

Interfere with Viral Uncoating
Amantadine and rimantadine block the uncoating of influenza A virus after it enters a cell. Capsid inhibitors interfere with HIV uncoating (they also interfere with the maturation of HIV particles).

Interfere with Nucleic Acid Synthesis
Many of the most effective antiviral medications target virally encoded enzymes used during replication of viral nucleic acid; these medications include nucleoside and nucleotide analogs, non-nucleoside polymerase inhibitors, non-nucleoside reverse transcriptase inhibitors, NS5A inhibitors, and cap-dependent endonuclease inhibitors. Virus-specific options are available to treat infections caused by herpesviruses, HCV, HIV, SARS-CoV-2, and influenza viruses.

Prevent Genome Integration
Integrase strand transfer inhibitors (INSTIs) interfere with the function of HIV integrase.

Prevent Assembly and Release of Viral Particles
Protease inhibitors bind to and inhibit virally encoded proteases; virus-specific options are used to treat HIV infection, HCV infection, and COVID-19. Neuraminidase inhibitors interfere with the release of influenza virus particles from a cell. The capsid inhibitors that target HIV interfere with the maturation of HIV particles (in addition to interfering with HIV uncoating).

20.7 ■ Mechanisms of Action of Antifungal Medications (figure 20.17, table 20.4)

Interfere with Cytoplasmic Membrane Synthesis and Function
Azoles interfere with ergosterol synthesis. Polyenes disrupt fungal cell membranes by binding to ergosterol. Allylamines, tolnaftate and butenafine all inhibit an enzyme in the pathway of ergosterol synthesis.

Interfere with Cell Wall Synthesis
Echinocandins and triterpenoids interfere with the synthesis of β-1,3 glucan.

Interfere with Cell Division
Upon oral administration, griseofulvin is concentrated in keratinized skin cells, where it inhibits fungal cell division.

Interfere with Nucleic Acid Synthesis
Flucytosine is taken up by yeast cells and converted by yeast enzymes to an active form.

Interfere with Protein Synthesis
Tavaborole inhibits protein synthesis by interfering with an enzyme that "charges" a tRNA molecule by attaching the correct amino acid.

20.8 ■ Mechanisms of Action of Antiprotozoan and Antihelminthic Medications (table 20.5)

Most antiparasitic medications are thought to interfere with biosynthetic pathways of protozoan parasites or the neuromuscular function of worms.

Review Questions

Short Answer

1. Describe the difference between the terms *antibiotic* and *antimicrobial.*
2. Define *therapeutic index* and explain its importance.
3. Explain how penicillin-binding proteins (PBPs) are involved in a bacterium's susceptibility to β-lactam antibiotics.
4. Name three classes of antimicrobial medications that target bacterial ribosomes.

5. Explain the roles of the first-line drugs versus the second-line drugs in the treatment of tuberculosis.
6. Compare and contrast the method for determining the minimum inhibitory concentration (MIC) with the Kirby-Bauer disc diffusion test.
7. Name three targets that can be altered sufficiently via spontaneous mutation to result in resistance to an antimicrobial medication.
8. What is MRSA? Why is it significant?
9. Why is it difficult to develop antiviral medications?
10. Explain the difference between the mechanism of action of an azole and that of a polyene.

Multiple Choice

1. Which of the following targets would you expect to be the most selective with respect to toxicity?
   a) Cytoplasmic membrane function
   b) DNA synthesis
   c) Glycolysis
   d) Peptidoglycan synthesis
   e) 70S ribosome
2. Penicillin has been modified to make derivatives that differ in all of the following *except*
   a) spectrum of activity.
   b) resistance to β-lactamases.
   c) potential for allergic reactions.
   d) a and c.
3. Which of the following is the target of β-lactam antibiotics?
   a) Peptidoglycan synthesis
   b) DNA synthesis
   c) RNA synthesis
   d) Protein synthesis
   e) Folate synthesis
4. Which of the following statements is *false*?
   a) A bacteriostatic chemical stops the growth of a microorganism.
   b) The lower the therapeutic index, the less toxic the medication.
   c) Broad-spectrum antibiotics are associated with the development of *Clostridioides difficile* infections.
   d) Azithromycin has a longer half-life than does penicillin V.
   e) Chloramphenicol can cause a life-threatening type of anemia.
5. All of the following interfere with the function of the bacterial ribosome *except*
   a) fluoroquinolones.
   b) lincosamides.
   c) macrolides.
   d) streptogramins.
   e) tetracyclines.
6. The target of the sulfonamides is
   a) cytoplasmic membrane proteins.
   b) folate synthesis.
   c) gyrase.
   d) peptidoglycan biosynthesis.
   e) RNA polymerase.
7. Routine antimicrobial therapy to treat tuberculosis involves taking
   a) one medication for 10 days.
   b) combinations of medications for 10 days.
   c) one medication for 4 months.
   d) combinations of medications for at least 4 months, but often longer.
   e) five medications for 2 years.
8. *Staphylococcus aureus* strains referred to as HA-MRSA are sensitive to
   a) methicillin.
   b) penicillin.
   c) most cephalosporins.
   d) vancomycin.
   e) none of the above
9. Acyclovir is a
   a) nucleoside analog.
   b) non-nucleoside polymerase inhibitor.
   c) protease inhibitor.
   d) none of the above
10. The antifungal medication griseofulvin is used to treat
    a) vaginal infections.
    b) systemic infections.
    c) nail infections.
    d) eye infections.

Applications

1. A physician was treating one young woman and one elderly patient for urinary tract infections caused by the same type of bacterium. Although the patients had similar body dimensions and weight, the physician gave a smaller dose of the antibacterial medication to the older patient. What was the physician's rationale for this decision?
2. Many scientists have criticized the use of low-dosage antibiotics and other antimicrobial agents to enhance the growth of cattle and chickens. Why would they be against this practice? Why would producers be reluctant to stop it?

Critical Thinking

1. Sulfonamides are not as effective in the presence of high-nutrient material such as pus or dead tissue. Why might this be the case?
2. Figure 20.12 shows the E-test procedure for determining an MIC value. How would the zone of inhibition appear if the medication concentrations in the strip were decreased slightly?

21 | Respiratory System Infections

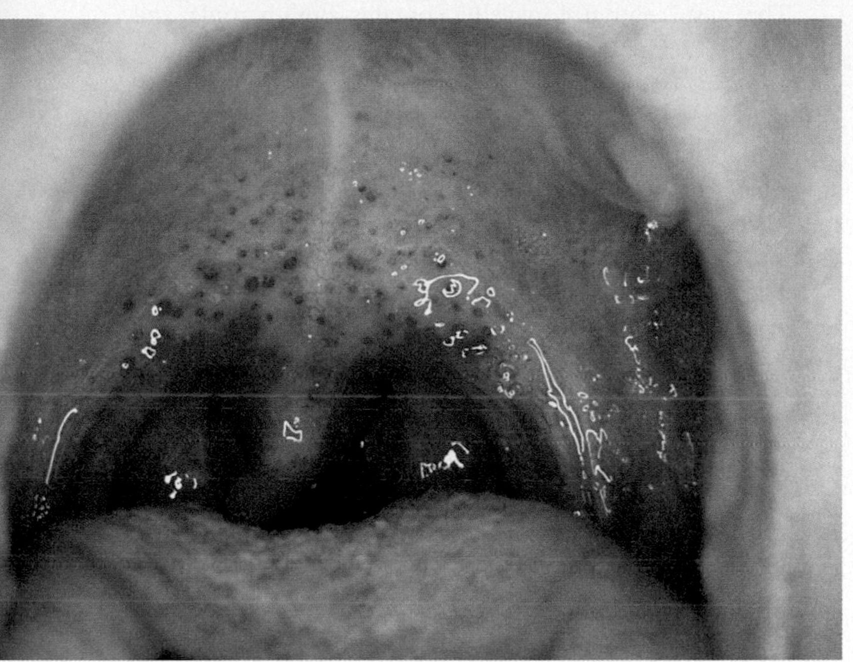

View of the throat in an individual with streptococcal pharyngitis. *Heinz F. Eichenwald, MD/CDC*

KEY TERMS

Antigenic Drift Minor changes that occur naturally in influenza virus antigens as a result of mutation.

Antigenic Shift Major changes in the influenza virus antigens that result from reassortment of viral RNA during infection of the same host cell by different viral strains.

Granuloma Collections of lymphocytes and macrophages found in a chronic inflammatory response, an attempt by the body to wall off and contain persistent organisms and antigens.

Mucociliary Escalator Moving layer of mucus propelled by cilia lining the respiratory tract that traps bacteria and other particles and carries them toward the throat.

Otitis Media Inflammation of the middle ear.

Pharyngitis Inflammation of the pharynx (throat).

Pneumonia Infection of the lungs that elicits an inflammatory response, resulting in the alveoli (air sacs) filling with fluids such as pus and blood.

Sputum Thick fluid containing mucus, pus, and other material coughed up from the lungs.

A Glimpse of History

Rebecca Lancefield (1895–1981) was a prominent microbiologist, famous for demonstrating that streptococci can be classified according to their cell wall carbohydrates. Her system, now known as "Lancefield grouping," is still used today.

Lancefield attended Wellesley College in 1912, and then went to Columbia University, where she obtained her master's degree. She discontinued her studies during World War I, but earned her PhD from Columbia in 1925 while maintaining a position at Rockefeller Institute of Medical Research (now Rockefeller University), where her work focused on the streptococci. At the time, streptococcal classification was largely based on the type of hemolysis displayed when organisms were grown on blood agar. β-hemolytic strep colonies produce a colorless clearing of red blood cells (β-hemolysis), whereas α-hemolytic strep colonies produce a zone of greenish discoloration of the red blood cells (α-hemolysis). A turning point in her research came when she recognized that streptococci have important antigens on their surface, some of which correlate with the disease-causing potential of a strain. She noticed that one particular antigen—a cell wall carbohydrate—could be used to separate the various streptococci into different groups. Many strains of β-hemolytic streptococci isolated from human infections have the same "A" cell wall carbohydrate; others have "B" carbohydrate; many from cattle, horses, and guinea pigs have "C" carbohydrate; some human normal microbiota streptococci have "D" carbohydrate; and so on. Lancefield grouping was a much better predictor of pathogenic potential than hemolysis on blood agar. Lancefield also discovered M protein on the surface of *Streptococcus pyogenes*, a group A streptococcus (GAS). M protein shows great variation among strains of *S. pyogenes* and plays a significant role in its virulence.

Rebecca Lancefield was the first female president of the American Association of Immunologists and in 1970 was elected to the prestigious National Academy of Sciences.

Respiratory infections include an enormous variety of illnesses ranging from the trivial to the fatal. They can be divided into infections of the upper respiratory tract (in the head and neck) and infections of the lower respiratory tract (in the chest). Common upper respiratory infections such as colds or sinus infections are uncomfortable, but they are not life-threatening. Lower respiratory infections such as pneumonia and tuberculosis, however, are often serious, and may be fatal.

21.1 ■ Anatomy, Physiology, and Ecology of the Respiratory System

Learning Outcomes

1. Outline the structures of the upper and lower respiratory tracts and their functions.
2. Describe the components of the mucociliary escalator and explain its importance.

The major function of the respiratory system is gas exchange. Each breath brings in O_2, which replenishes the supply in the blood, and then releases CO_2, the waste product of cellular metabolism. In addition, movement of air over the vocal cords makes sound (which allows us to speak) and sensors in the nose detect odors. Although the mouth can also be used for breathing, it is considered part of the digestive tract, so it is covered in chapter 24.

The respiratory system consists of the upper respiratory tract and the lower respiratory tract. We also include the eyes and the ears in our discussion. Even though these are not part of the respiratory tract, they are associated with ducts that open onto mucous membranes lining the nasal cavity and throat. Due to that connection, they can serve as portals of entry to the respiratory tract.

Mucous membranes line the respiratory tract. These are coated with mucus, a slimy glycoprotein material that traps airborne dust and other particles, including microorganisms. Mucus is produced by specialized cells called **goblet cells** that are scattered among the cells of the membrane. Their name reflects their shape, which is narrow at the base and wide at the surface, like a wine glass or goblet.

Ciliated epithelium makes up most mucous membranes of the respiratory system. Cells of this lining have cilia—tiny, hair-like projections—along their exposed free border. The cilia beat synchronously, continually propelling mucus away from the lungs. The mucus, along with any trapped microbes, is then swallowed and digested. This mechanism, the **mucociliary escalator,** is an important part of our innate defenses that help prevent inhaled microorganisms from reaching the lungs. Smoking, alcohol or narcotic abuse, and viral infections all impair ciliary movement and increase the chance of disease.

MicroByte

Cilia beat at a rate of about 1,000 times a minute, propelling a film of mucus along the mucociliary escalator.

The Upper Respiratory Tract

The upper respiratory tract includes the nose and nasal cavity, pharynx (throat), larynx (voice box), and epiglottis (**figure 21.1**). In addition to the ciliated mucous membranes that line these structures, tonsils provide another important defense for the respiratory tract. There are three pairs of these secondary lymphoid organs in different areas of the throat, positioned so that they come into contact with microbes entering the upper respiratory tract. They are important in the immune response but can also be the sites of infection, resulting in tonsillitis. Inflamed tonsils at the opening of the throat (palatine tonsils) are visible in the chapter opening photo. Tonsils in the upper part of the throat, behind the nasal cavity, are called adenoids.

The Nose and Nasal Cavity

Air enters the respiratory system at the nostrils and moves into the nasal cavity. When cold air enters the nose, local blood flow immediately increases due to nervous reflexes. This warms the air in the nose to near body temperature while passage over mucous membranes saturates it with water vapor.

The nasal entrance usually contains diphtheroids and staphylococci. Up to 30% of healthy people consistently carry *Staphylococcus aureus* in their nostrils; an even higher percentage of healthcare workers are likely to carry this important pathogen. Farther inside the nasal passages, the normal microbiota is similar to that of the throat. Infection of the nasal passages, usually by viruses, causes inflammation or rhinitis, resulting in the familiar "runny nose."

The skull contains air-filled cavities called sinuses that are lined with ciliated epithelium continuous with that of the nasal cavity. Several pairs of sinuses found in different areas of the skull are named based on their location. Infection of the sinuses is called rhinosinusitis, or sinusitis, and may cause a "sinus headache" or postnasal drip.

Pharynx (Throat), Larynx (Voice Box), and Epiglottis

From the nose, air moves to the throat (pharynx). Inflammation of the throat, **pharyngitis,** is commonly the result of viral infection. The larynx connects the throat to the lower respiratory tract and produces voice sounds when speaking or singing. Inflammation of the larynx, called laryngitis, results in hoarseness; the most common cause is viral infection, but it can also be due to other factors including voice overuse. The throat is shared by the respiratory system and the digestive system. During swallowing, a small flap called the epiglottis covers the glottis, the opening that allows air to pass between the upper and lower respiratory tracts. This prevents swallowed material from entering the lungs. Inflammation of the epiglottis, called epiglottitis, can be life-threatening because the swollen flap can block the glottis and close the airway.

Streptococci, including the viridans group streptococci (a group of mainly α-hemolytic streptococci; *viridis* means "green"), are common members of the normal microbiota of the throat. A variety of other bacteria are also found there, including *Moraxella* species, diphtheroids, and various obligate anaerobes. Some common opportunistic pathogens also colonize the throat, including *Streptococcus pneumoniae, Haemophilus influenzae,* and *Neisseria meningitidis.*

Eyes

The surface of the eyes and the lining of the eyelids are covered by mucous membranes called conjunctiva. Infection of the conjunctiva is called **conjunctivitis.** Surprisingly, even though the eyes are constantly exposed to large numbers of microorganisms, the conjunctiva of healthy people usually have very few bacteria. This is because the eye is bathed with tears rich in lysozyme and secretory IgA and is cleaned by

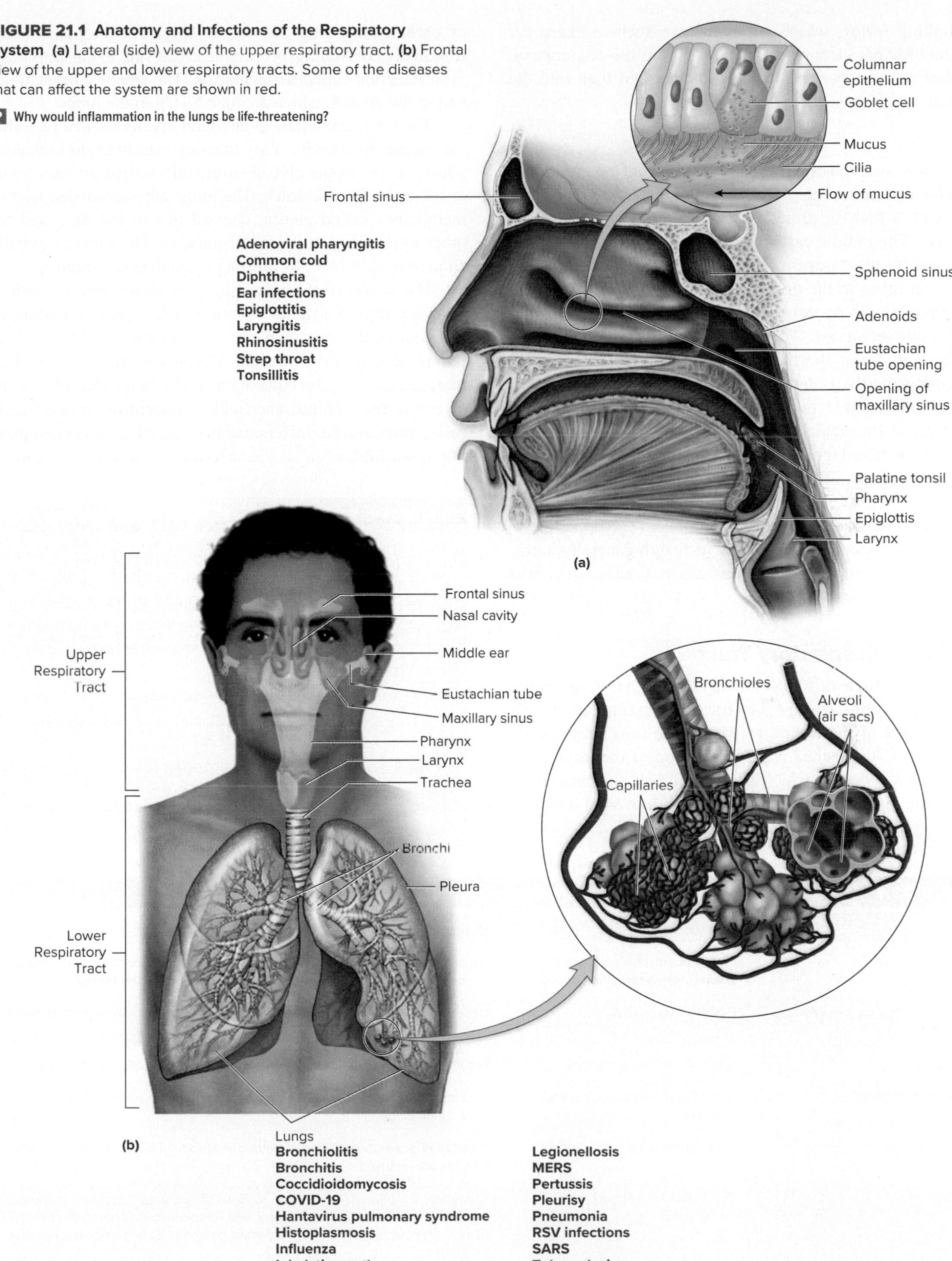

FIGURE 21.1 Anatomy and Infections of the Respiratory System **(a)** Lateral (side) view of the upper respiratory tract. **(b)** Frontal view of the upper and lower respiratory tracts. Some of the diseases that can affect the system are shown in red.

? **Why would inflammation in the lungs be life-threatening?**

the blinking reflex, which wipes the eye surface like a car windshield wiper cleans a windshield. From the conjunctiva, microbes may be swept into the tear duct and then into the nasal cavity.

Ears

The ear has three parts: external, middle, and inner ear. The external ear includes the outer portion of the ear and the auditory canal, which is protected from microbes by cerumen (ear wax). The middle ear, which is separated from the auditory canal by the eardrum, is sterile. It is connected by the Eustachian tubes to the upper portion of the pharynx. These tubes equalize the pressure in the middle ear and drain normal mucous secretions. They are also a route through which microbes can enter the middle ear, leading to an infection called **otitis media.** Enlargement of the adenoids can contribute to middle ear infections by interfering with normal drainage from the Eustachian tubes. The fluid-filled inner ear is also microbe-free. Occasionally, however, viruses or bacteria enter the inner ear, often following an upper respiratory tract infection.

Some of the bacterial genera that inhabit the upper respiratory tract are listed in **table 21.1.** Although generally harmless, they are opportunists that can cause disease when host defenses are impaired.

The Lower Respiratory Tract

The lower respiratory tract includes the trachea (windpipe), bronchi, and lungs (**figure 21.1b**). Some scientists include the lower part of the larynx as part of the lower respiratory tract as well. The trachea is a continuation of the larynx and branches into two bronchi. Inflammation of the bronchi is called bronchitis and is commonly the result of viral infection or chronic coughing due to smoking. The bronchi branch repeatedly, becoming bronchioles, the site of an important viral infection called **bronchiolitis.** The smallest bronchioles end in the alveoli, which are found deep in the lungs.

The lungs are a pair of air-filled organs located on either side of the chest cavity. They function mainly in gas exchange, which occurs in the alveoli—tiny, thin-walled air sacs found at the end of bronchioles. The lungs are surrounded by two membranes called pleura; one adheres to the lung and the other to the chest wall and diaphragm. The pleura normally slide over each other as the lung expands and contracts.

The action of the mucociliary escalator and the antimicrobial action of alveolar secretions help to keep microorganisms out of the lungs. In addition, macrophages in the lung tissues readily move into the alveoli and airways to engulf infectious agents. Inflammation of the lungs that causes the alveoli to fill with pus and fluid as a result of an infection is called **pneumonia.** Inflammation of the pleura is called pleurisy (pleuritis), which is characterized by severe chest pain.

MicroAssessment 21.1

The respiratory system provides a warm, moist environment for microorganisms. It is protected by tonsils and adenoids, and by the mucociliary escalator. The upper respiratory tract contains a highly diverse microbiota. Some members of the normal microbiota are opportunistic pathogens. The mucociliary escalator helps prevent inhaled microorganisms from reaching the lungs.

1. What is the normal function of the tonsils?
2. Describe the normal microbiota of the upper respiratory tract.
3. How would damage to the cilia affect the occurrence of pneumonia?

TABLE 21.1 Normal Microbiota of the Upper Respiratory System

Genus	Characteristics	Comments
Corynebacterium	Pleomorphic, often club-shaped Gram-positive rods; non-motile	Aerobic or facultatively anaerobic. Those that look similar to *C. diphtheriae*, the cause of diphtheria, are included in a group called diphtheroids.
Fusobacterium, Porphyromonas, Prevotella	Gram-negative rods	Obligate anaerobes. Most often found in the space between teeth and gums.
Haemophilus	Small, Gram-negative rods	Facultative anaerobes. Commonly include the potential pathogen *H. influenzae.*
Neisseria, Moraxella	Gram-negative diplococci and diplobacilli	Aerobic. Some microscopically resemble pathogenic *Neisseria* species such as *N. meningitidis.*
Staphylococcus	Gram-positive cocci in clusters	Facultative anaerobes. Potential pathogen *Staphylococcus aureus* commonly inhabits the nostrils.
Streptococcus	Gram-positive cocci in chains	Aerotolerant (obligate fermenters). Viridans group streptococci are common; many members are α-hemolytic (greenish discoloration). β-hemolytic (clear hemolysis) streptococci and the potential pathogen *S. pneumoniae* may also be present.

UPPER RESPIRATORY TRACT INFECTIONS

21.2 ■ Bacterial Infections of the Upper Respiratory System

Learning Outcomes

3. List the parts of the upper respiratory system commonly infected by *Streptococcus pneumoniae* and *Haemophilus influenzae*.
4. Compare the distinctive characteristics of strep throat and diphtheria.

Bacterial infections of the upper respiratory tract are fairly common and are caused by various species. Some of these organisms are members of the normal microbiota that access normally sterile sites following damage caused by a viral infection.

Pink Eye, Earache, and Sinus Infections

Bacterial **conjunctivitis** ("pink eye"; inflammation of the eye surface), **otitis media** (inflammation of the middle ear), and sinusitis (inflammation of the sinuses) often occur together and frequently have the same causative agent. Pink eye is easily spread and is therefore a concern when it occurs in a daycare facility or school. Otitis media is especially common in children and is responsible for millions of doctor visits per year in the United States. Sinusitis is common in both adults and children.

Signs and Symptoms

The signs and symptoms of acute bacterial conjunctivitis include increased tears, redness of the conjunctiva, swollen eyelids, sensitivity to bright light, and large amounts of pus (**figure 21.2**). Acute bacterial conjunctivitis differs from viral conjunctivitis, in which eyelid swelling and pus are usually minimal.

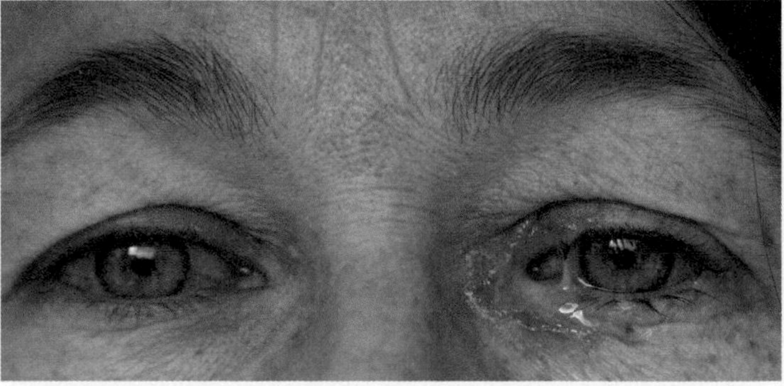

FIGURE 21.2 Bacterial Conjunctivitis Redness, swollen eyelids, and pus are characteristic of bacterial conjunctivitis. Anita van den Broek/Shutterstock

? What symptom shown here is usually minimal in viral conjunctivitis?

Otitis media causes a severe earache; the intense pain often causes poor feeding in young children. Ear tugging and difficulty hearing may occur as well. Fever may be mild or absent.

In sinusitis, facial pain and a pressure sensation develop in the region of the involved sinus. Headache, severe malaise, and a thick green nasal discharge that sometimes contains pus and blood may also occur.

Causative Agents

Pink eye, earache, and sinus infections are often caused by two common bacterial pathogens: (1) *Haemophilus influenzae,* a tiny Gram-negative rod; and (2) *Streptococcus pneumoniae,* a Gram-positive encapsulated diplococcus also known as pneumococcus. Strains that infect the conjunctiva have adhesins that allow them to attach firmly to the epithelium.

Conjunctivitis can also be caused by a variety of other organisms, including *Staphylococcus aureus*, certain *Moraxella* species, and enterobacteria. *Neisseria gonorrhoeae* can also cause a more acute form of conjunctivitis, while *Chlamydia trachomatis* can cause a more chronic, long-lasting disease.

Although not common, some eye infections are caused by environmental microbes that contaminate eye medications and contact lens solutions. When these infections occur, they are usually more serious because they affect the cornea (clear dome around the colored part of the eye) and can lead to permanent eye damage. People who wear contact lenses should make sure to strictly follow the instructions for using the cleaning solution; cloudy or outdated solutions and eye medications should be thrown away.

Otitis media and sinusitis can also be caused by *Mycoplasma pneumoniae, Streptococcus pyogenes, Moraxella catarrhalis,* and *Staphylococcus aureus.* About one-third of cases are caused by respiratory viruses; this explains why some infections do not respond to antibiotics, which have no effect on viruses.

Pathogenesis

Few details are known about the pathogenesis of acute bacterial conjunctivitis. Organisms are probably inoculated directly onto the conjunctiva from airborne respiratory droplets or transferred from contaminated hands. They resist destruction by lysozyme. Attachment may be aided by the degradation of mucin, a protective component of surface mucus. Following attachment, the bacteria release various tissue-damaging enzymes, sometimes combined with toxins, that further harm eye tissue.

Otitis media and sinusitis are usually preceded by infections of the nasal cavity and upper pharynx (see figure 21.1). Infection damages the ciliated cells, resulting in inflammation and swelling. When the Eustachian tube cannot drain secretions from the middle ear, fluid and pus collect behind the eardrum. This leads to a buildup of pressure, causing the ear to ache. The eardrum may perforate (burst), discharging blood or pus from the ear and giving immediate relief from pain. With treatment, holes in the eardrum usually heal quickly. In some people, the causative organisms create a biofilm, leading to chronic infections that may be difficult to treat. Fluid behind the eardrum may impair hearing ability, causing a delay in speech development in some young children. Both middle ear and sinus infections sometimes spread to the membranes covering the brain, causing meningitis.

Epidemiology

The epidemiological factors involved in the development and spread of the eye, ear, and sinus infections caused by *H. influenzae* and *S. pneumoniae* are largely unknown. Some people carry the bacteria without having any symptoms, so factors such as the virulence of the strains, crowding of potential hosts, and presence of respiratory viruses probably play important roles in outbreaks.

A preceding or simultaneous viral illness is common in otitis media and sinusitis; the virus probably damages the mucociliary mechanism that would normally protect against infection. Otitis media is rare in the first month of life but becomes very common in early childhood, particularly in children who use pacifiers after the age of 2. Conditions that cause inflammation of the nasal mucosa—including viral infections, nasal allergies, and exposure to air pollution or cigarette smoke—play a role in some cases. Children eventually develop immunity to *H. influenzae,* so the bacterium rarely causes otitis media in children over age 5. Sinusitis tends to affect adults and older children in whom the sinuses are more fully developed.

Treatment and Prevention

Bacterial conjunctivitis is effectively treated with eyedrops or ointments containing an antimicrobial medication such as erythromycin or ofloxacin, a fluoroquinolone. Antibiotics can be prescribed to treat otitis media, but in most cases physicians recommend waiting 2 to 3 days to see if symptoms improve on their own. Although pain- and fever-reducing medications can be used to help relieve symptoms, decongestants and antihistamines are generally ineffective and can even be harmful because they reduce the immune response. Bacterial sinusitis can be typically treated with amoxicillin; in areas where antibiotic-resistant strains of *H. influenzae* and *S. pneumoniae* are common, alternative medications are used instead. Unfortunately, if the causative agent forms a biofilm, the disease often recurs.

General preventive measures for conjunctivitis include handwashing and protecting the eyes from contamination by avoiding rubbing or touching them, particularly with shared towels. Bacterial conjunctivitis is highly contagious, so people suspected of having it should be kept home from school or day-care settings until treated. Otitis media during the "flu" season can be substantially decreased by vaccinating infants in day-care facilities against influenza. Surgical removal of enlarged adenoids improves drainage from the Eustachian tubes and can help prevent recurrences in certain patients. In those with chronically malfunctioning Eustachian tubes and hearing loss, tiny plastic ventilation tubes are often inserted through the eardrums so that pressure can equalize (**figure 21.3**). There are no proven preventive measures for sinusitis.

Streptococcal Pharyngitis ("Strep Throat")

Sore throat is one of the most common reasons that people in the United States seek medical care. Many of these visits are due to reasonable concerns about streptococcal pharyngitis ("strep throat"), because this can lead to complications that develop after the initial infection. These complications, referred to as post-streptococcal sequelae (*sequi* means "follow after") will be discussed shortly.

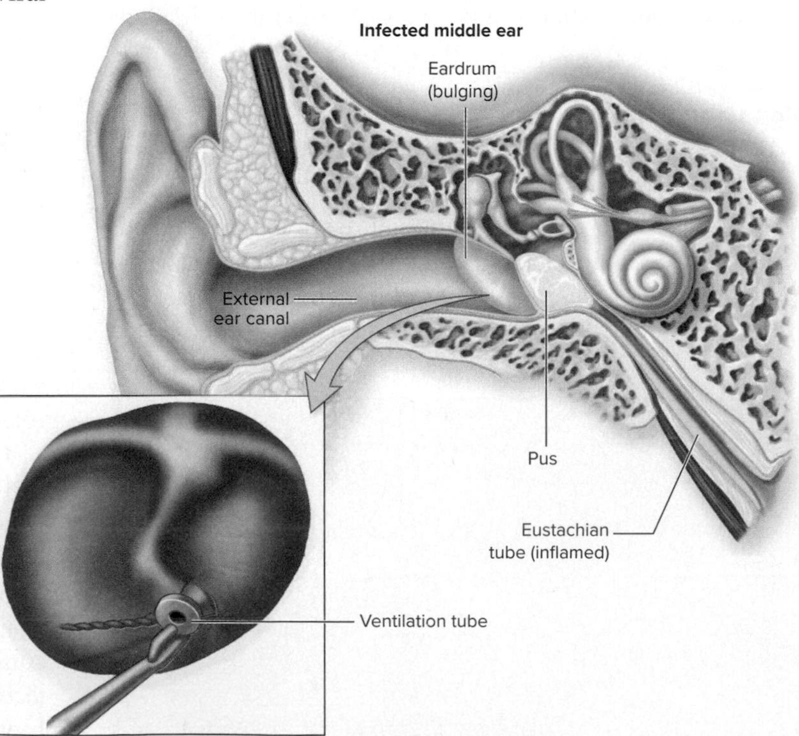

FIGURE 21.3 Otitis Media Inset shows a ventilation tube placed in the eardrum to equalize middle ear pressure in individuals with chronically malfunctioning Eustachian tubes.

? **Why is otitis media painful?**

Signs and Symptoms

Strep throat is characterized by a sore throat, difficulty swallowing, and fever, which develop after an incubation period of 2 to 5 days. The throat is red—with patches of pus and scattered tiny hemorrhages—and the lymph nodes in the neck are swollen and tender. Abdominal pain or headache may occur in older children and young adults. Patients do not usually have a cough, weepy eyes, or runny nose. Most patients with strep throat recover spontaneously after about a week; in fact, many infected people have only mild symptoms or no symptoms at all.

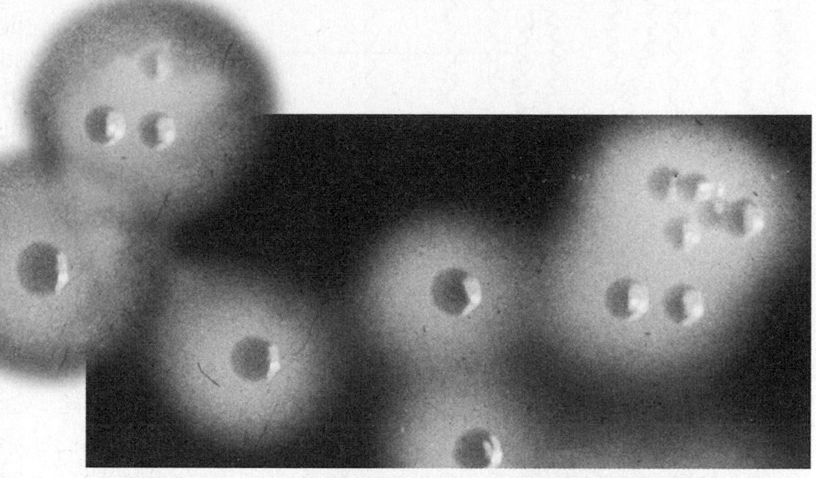

FIGURE 21.5 ***Streptococcus pyogenes* Colonies on Blood Agar** The colonies are surrounded by a wide clear zone of β-hemolysis. Evans Roberts

How can α-hemolysis be differentiated from β-hemolysis?

Causative Agent

Strep throat is caused by *Streptococcus pyogenes,* a Gram-positive coccus that grows in chains (**figure 21.4**). The organism can be differentiated from other streptococci that normally inhabit the throat by its colony morphology on blood agar—*S. pyogenes* colonies are surrounded by a characteristic clear zone of β-hemolysis (**figure 21.5**). In contrast, most species of *Streptococcus* that are typically part of the normal throat microbiota are either non-hemolytic or α-hemolytic, producing a zone of dark or greenish discoloration around colonies on blood agar. A few other streptococci and some other bacteria are also β-hemolytic, so further tests are needed to identify *S. pyogenes.*

Streptococcus pyogenes is commonly referred to as group A streptococcus (GAS), reflecting its Lancefield grouping (see A Glimpse of History in this chapter). This grouping system uses antibodies to distinguish the cell wall carbohydrates in streptococcal species. *S. pyogenes* is characterized by the "A" carbohydrate in its cell wall. Antibodies that bind to the group A carbohydrate are the basis for many of the rapid diagnostic tests done on throat specimens in a physician's office.

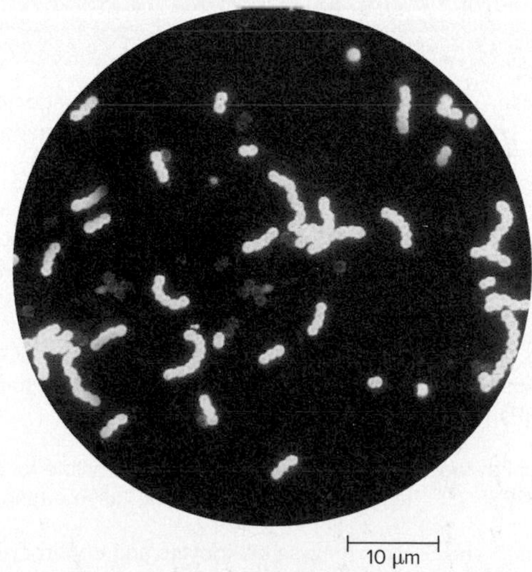

FIGURE 21.4 ***Streptococcus pyogenes*** Chain formation by *S. pyogenes*, revealed by fluorescence microscopy. Evans Roberts

What symptoms of strep throat accompany the typical sore throat?

Different *S. pyogenes* strains are distinguished by variations of a surface antigen called **M protein,** an important virulence factor. Strains with certain types of M protein are the cause of strep throat, whereas others with different M protein are more likely to cause skin infections.

Pathogenesis

Proteins in the cell wall of *S. pyogenes* allow the bacteria to attach to host cells (**figure 21.6**). M protein is an important adhesin and a key virulence factor—host antibodies that bind to it prevent infection. Unfortunately, more than 200 antigenic types of M protein occur among the many strains of *S. pyogenes,* and antibodies to one type do not prevent infection by a strain that has a different type. Another group of proteins called fibronectin-binding proteins (FBPs) allow *S. pyogenes* to attach to fibrin, a protein found on epithelial cells, including the ones in the throat. Once *S. pyogenes* colonizes host tissues, it produces various tissue-damaging enzymes such as hyaluronidase that break down connections between cells, allowing the organism to spread rapidly to other cells. This spread is assisted by streptokinase, which causes the breakdown of blood clots.

In addition to factors important for host attachment and colonization, *S. pyogenes* also has mechanisms for avoiding the host immune system. The cells have a hyaluronic acid capsule that prevents phagocytotosis; hyaluronic acid is a normal component of human tissue, so the capsule acts as a cloaking mechanism, making it difficult for the host immune system to detect the pathogen. M protein, in addition to aiding attachment to the host cell, also interferes with phagocytosis by preventing complement component C3b (an opsonin) from being deposited on the bacterial cell wall. The organism further inhibits phagocytosis by releasing C5a peptidase, an enzyme

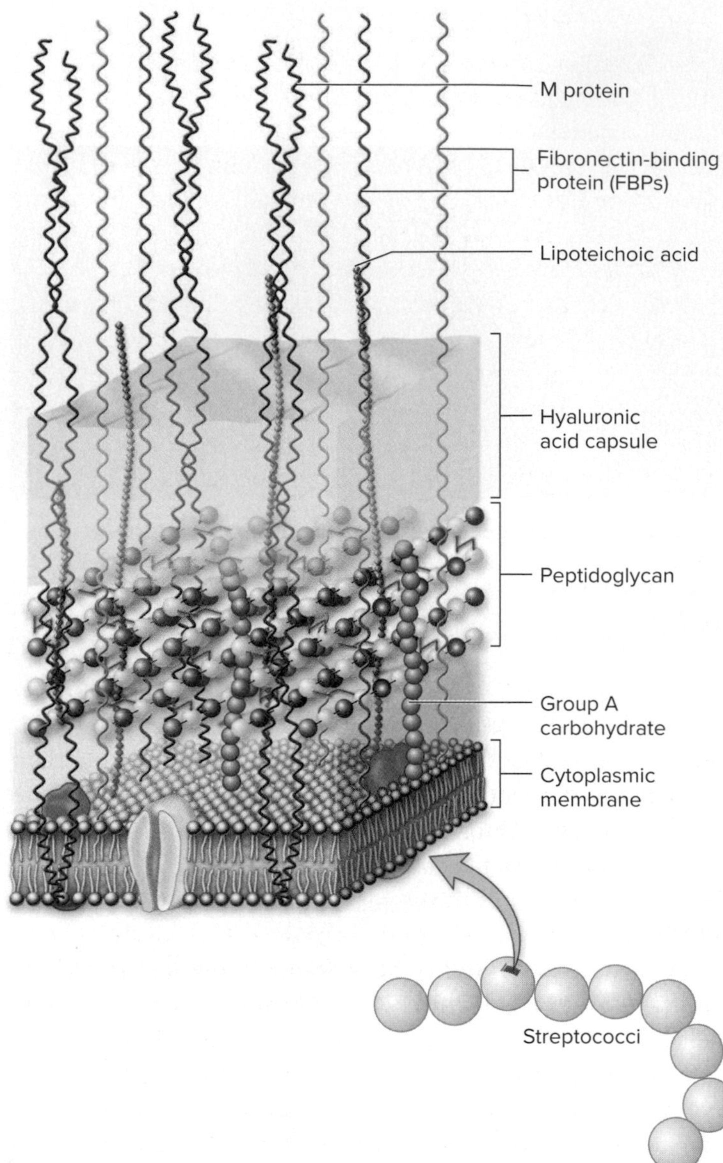

FIGURE 21.6 Virulence Factors Associated with the Cell Wall of *Streptococcus pyogenes*

? What is the role of M protein in pathogenesis?

that destroys complement component C5a (a chemoattractant for phagocytes). It also produces streptolysins O and S—enzymes that destroy blood cells by making holes in their cell membranes. Destruction of leukocytes (white blood cells) by the streptolysins inhibits the immune response; destruction of erythrocytes (red blood cells) causes the β-hemolysis exhibited by *S. pyogenes* (see figure 21.5).

A few strains of *S. pyogenes* produce **streptococcal pyrogenic exotoxins (SPEs),** a family of exotoxins that cause severe streptococcal diseases characterized by high fever (*pyro* means "fire"). Some SPEs are encoded by prophages (phage DNA inserted into a bacterial chromosome), and therefore their production is an example of lysogenic conversion. These toxins are superantigens, causing massive activation of T cells; the resulting uncontrolled release of cytokines (a cytokine storm) is probably responsible for the seriousness of these infections. A person with strep throat caused by an SPE-producing strain of *S. pyogenes* may develop scarlet fever, characterized by high fever and a red rash that feels rough (like sandpaper); the rash is found on the head, neck, chest, and thighs, but usually not around the mouth. The toxin also causes the tongue to look like a ripe strawberry—red and spotted (called a "strawberry tongue"). Although the toxin is an SPE, it is traditionally referred to as erythrogenic toxin (*erythro* means "red") because of the characteristic red rash and strawberry tongue. Some SPE-producing strains of *S. pyogenes* have additional virulence factors that allow them to cause severe invasive diseases such as streptococcal toxic shock syndrome and necrotizing fasciitis ("flesh-eating disease"). The virulence factors of *S. pyogenes* are summarized in **table 21.2.**

MicroByte

Streptococcal streptokinase is used as a clot-dissolving medication in treating some heart attacks and pulmonary embolisms.

Epidemiology

Streptococcus pyogenes naturally infects only humans. People with strep throat spread the organism easily in respiratory droplets generated by coughing and sneezing. Peak incidence of the disease occurs in winter or spring and is highest in children aged 5 to 15. Epidemics have also originated from food contaminated with *S. pyogenes.*

Some people become chronic carriers of *S. pyogenes,* usually harboring the organism in the throat. These asymptomatic carriers are unlikely to transmit the disease to others, perhaps because the colonizing organism becomes less

TABLE 21.2 Virulence Factors of *Streptococcus pyogenes*

Product	Effect
C5a peptidase	Inhibits recruitment of phagocytes by destroying complement component C5a
Hyaluronic acid capsule	Inhibits phagocytosis
M protein	Interferes with phagocytosis by causing inactivation of complement component C3b, an opsonin; involved in attachment to host cells
Fibronectin-binding proteins (FBPs)	Responsible for attachment to host cells via fibrin
Streptococcal pyrogenic exotoxins (SPEs)	Superantigens responsible for scarlet fever, toxic shock, "flesh-eating" fasciitis
Streptolysins O and S	Lyse leukocytes and erythrocytes
Tissue-damaging enzymes	Enhance spread of bacteria by breaking down proteins, hyaluronic acid, blood clots, and other components

virulent or is carried at a relatively low density. If the carrier develops viral pharyngitis, however, that person might be misdiagnosed as having strep throat and unnecessarily treated with antibacterial medications. *S. pyogenes* can occasionally colonize other body surfaces, and anal carriers have been implicated in healthcare-associated infections.

Treatment and Prevention

People with fever and sore throat should be seen by a physician so that a throat swab can be taken for a rapid strep test (RST) or rapid antigen detection test (RADT) and throat culture. Confirmed strep throat is usually treated with an oral penicillin, which eliminates the organism in most cases. Treatment given as late as 9 days after onset prevents certain post-streptococcal sequelae (covered next).

Ensuring adequate ventilation and avoiding crowds help to control the spread of streptococcal infections. No vaccine is available. **Table 21.3** summarizes important information about strep throat.

Post-Streptococcal Sequelae

Post-streptococcal sequelae—complications that can develop after strep throat or other streptococcal infections—are thought to be a result of immune responses to *Streptococcus pyogenes*. The specific complications that develop are related to the targets of the immune responses, including the heart, kidneys, joints, brain, or skin. Sequelae are uncommon in resource-rich countries because of quick treatment of *S. pyogenes* infections. Two of the most common are discussed here.

Acute Rheumatic Fever

Acute rheumatic fever (ARF) usually begins about 3 weeks after recovery from strep throat. Signs and symptoms include fever, joint pains, chest pain, rash, and nodules under the skin. Uncontrollable body movements (Syndenham chorea, also called "St. Vitus' dance") can occur and, when present, are major criteria for diagnosis. Carditis (inflammation of heart tissue), the most serious complication, develops in about 30 to 50% of patients. This can lead to rheumatic heart disease in which one or more of the heart valves are damaged, causing them to leak; this can result in heart failure later in life. The damaged valves are also prone to infection, usually by bacteria from the normal skin or mouth microbiota, potentially resulting in a heart valve infection called infective endocarditis.

ARF develops only in people who are genetically predisposed to it; some MHC class II alleles are involved in susceptibility (see Focus Your Perspective 15.1). It is thought to be an autoimmune response involving antibodies to *S. pyogenes* that cross-react with host tissue antigens similar to the bacterial antigens. For example, antibodies to streptococcal M protein may also recognize and bind to myosin in the heart. The heart tissue is then targeted for attack by the host's own immune system. As a result, effector helper T (T_H) cells release pro-inflammatory cytokines, causing inflammation that leads to permanent damage to the local tissues, particularly the heart valves.

The incidence of ARF has declined in the United States, probably as a result of prompt treatment of strep throat with antibiotics and decreased prevalence of the strains associated with the disease. The risk of developing ARF after severe

TABLE 21.3 Strep Throat (Streptococcal Pharyngitis) and Post-Streptococcal Sequelae

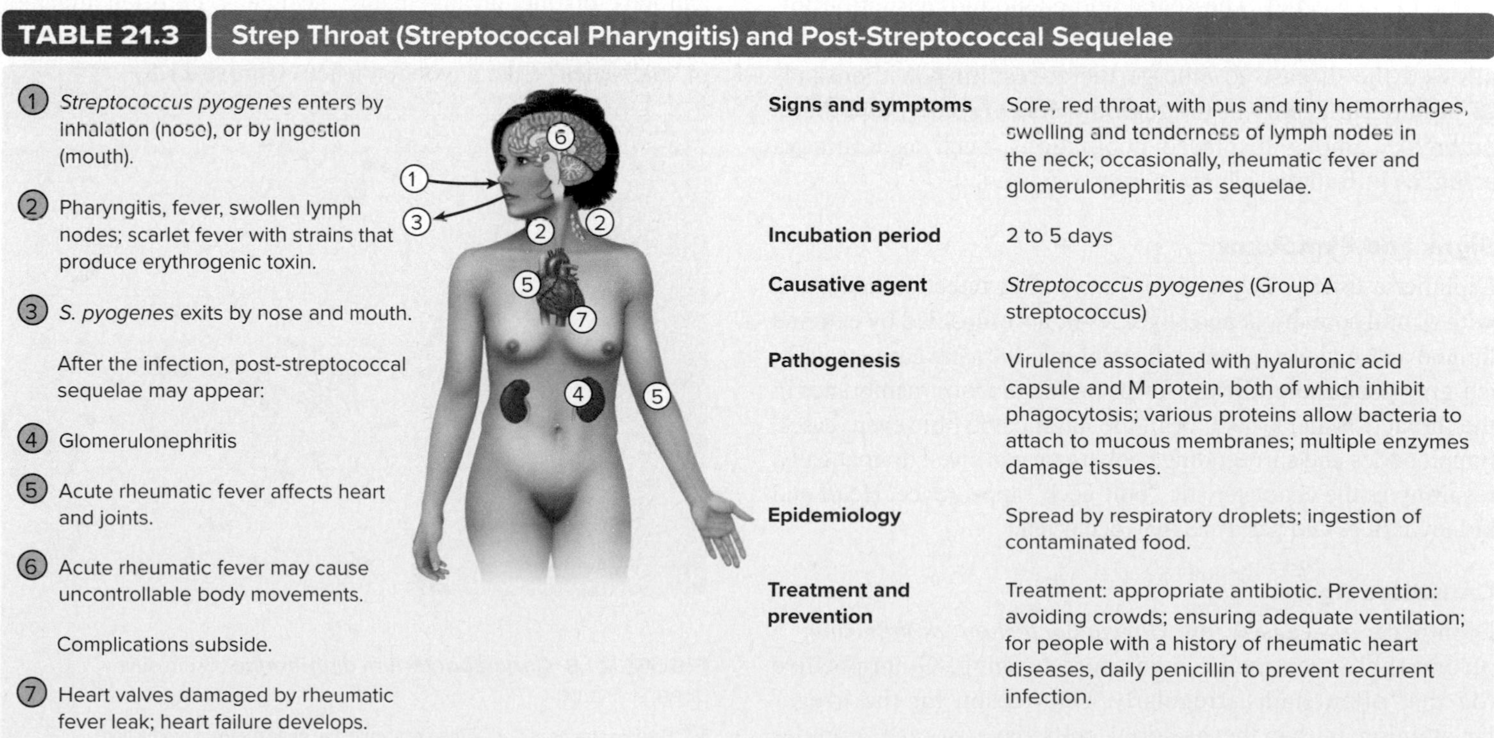

1 *Streptococcus pyogenes* enters by inhalation (nose), or by ingestion (mouth).
2 Pharyngitis, fever, swollen lymph nodes; scarlet fever with strains that produce erythrogenic toxin.
3 *S. pyogenes* exits by nose and mouth.
After the infection, post-streptococcal sequelae may appear:
4 Glomerulonephritis
5 Acute rheumatic fever affects heart and joints.
6 Acute rheumatic fever may cause uncontrollable body movements.
Complications subside.
7 Heart valves damaged by rheumatic fever leak; heart failure develops.
1
2
3
4
5
6
7

Signs and symptoms	Sore, red throat, with pus and tiny hemorrhages, swelling and tenderness of lymph nodes in the neck; occasionally, rheumatic fever and glomerulonephritis as sequelae.
Incubation period	2 to 5 days
Causative agent	*Streptococcus pyogenes* (Group A streptococcus)
Pathogenesis	Virulence associated with hyaluronic acid capsule and M protein, both of which inhibit phagocytosis; various proteins allow bacteria to attach to mucous membranes; multiple enzymes damage tissues.
Epidemiology	Spread by respiratory droplets; ingestion of contaminated food.
Treatment and prevention	Treatment: appropriate antibiotic. Prevention: avoiding crowds; ensuring adequate ventilation; for people with a history of rheumatic heart diseases, daily penicillin to prevent recurrent infection.

untreated strep throat is now 3% or less, and untreated mild cases carry an even lower risk. Nonetheless, around 500,000 new cases occur globally each year, mostly in resource-limited countries. Signs and symptoms usually decrease with rest and anti-inflammatory medicines. People with cardiac damage due to ARF typically take penicillin or an alternative daily for years to prevent a recurrence of strep infection.

Post-Streptococcal Glomerulonephritis

Post-streptococcal glomerulonephritis (PSGN) occasionally follows strep throat but is more often an aftermath of a strep skin infection. It typically begins 1 to 3 weeks after strep throat, and 3 to 6 weeks after a strep skin infection. Signs and symptoms include fever, fluid retention, high blood pressure, and blood and protein in the urine (making the urine look like tea or cola). There are no streptococci in the urine or the diseased kidney tissues, and the body's immune response has generally eliminated *S. pyogenes* from the throat before the symptoms of PSGN appear. Kidney damage is due to an inflammatory reaction caused by streptococcal antigens that accumulate in the kidney glomeruli (tufts of tiny blood vessels where blood is filtered for urine formation); antibodies bound to these antigens form immune complexes that activate the complement system (**figure 21.7**). Only a few strains of *S. pyogenes* cause PSGN; it is rare to have it twice.

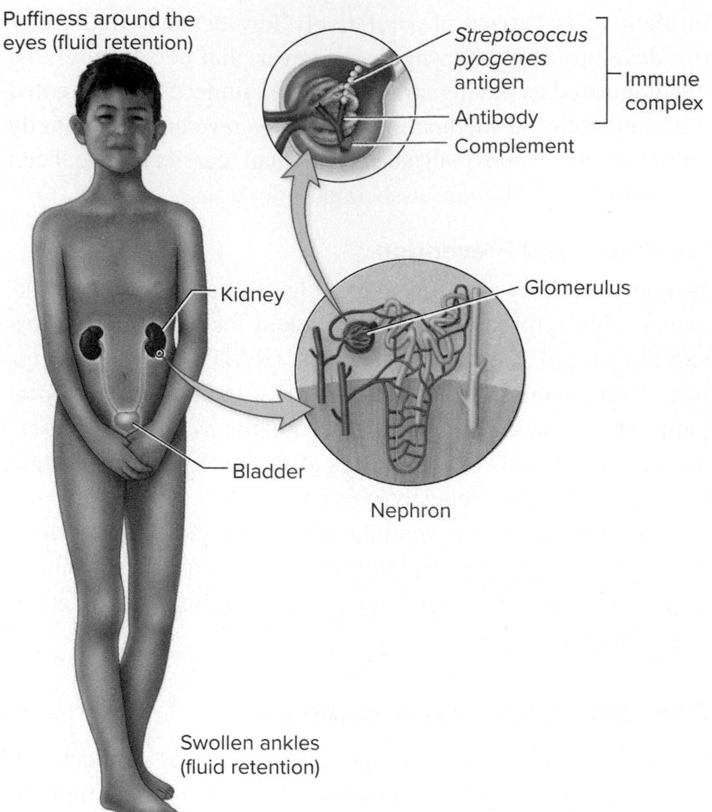

FIGURE 21.7 Pathogenesis of Post-Streptococcal Glomerulonephritis Immune complexes are deposited in the kidney glomeruli, causing an inflammatory response.

? Why might blood appear in the urine during glomerulonephritis?

Diphtheria

Diphtheria, a deadly toxin-mediated disease, is now rare in the United States because of childhood immunization. Events in other parts of the world, however, are a reminder of what can happen when vaccination and other public health measures are neglected. The social and economic disruption following the breakup of the Soviet Union in the early 1990s allowed the disease to emerge there, resulting in thousands of diphtheria deaths in the region. More recently, diphtheria outbreaks among displaced populations, such as Rohingya refugees in Bangladesh, have been reported.

Signs and Symptoms

Diphtheria usually begins 2 to 6 days after infection and starts with a mild sore throat and slight fever, accompanied by extreme fatigue and malaise (general discomfort). In many cases, a whitish-gray **pseudomembrane** forms on the mucous membranes in the throat, tonsils, larynx, or in the nasal cavity. In severe cases, lymph nodes and surrounding neck tissue can swell dramatically, resulting in the characteristic "bull-neck" appearance. Heart and kidney failure and paralysis may occur later.

Causative Agent

Diphtheria is caused by *Corynebacterium diphtheriae,* a pleomorphic, non-motile, non-spore-forming, Gram-positive rod that often stains irregularly. One reason for the irregular staining is that the bacterial cells have storage granules (metachromatic granules) at their poles that stain darkly. The cells are characteristically club-shaped (*koryne* means "club") and have distinct arrangements: Two cells are often attached at one end to form a "V" shape, or the cells align side by side in "palisades" (like a wooden fence) (**figure 21.8**).

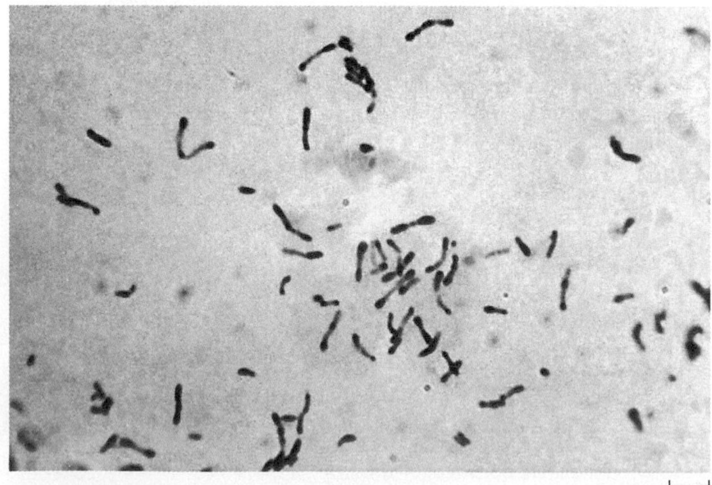

FIGURE 21.8 ***Corynebacterium diphtheriae*** Gram stain
Dr. P.B. Smith/CDC

? How do strains of *C. diphtheriae* acquire the gene for toxin production?

FOCUS ON A CASE 21.1

The patient was a college student who complained to her roommate about a sore throat. Although she had been previously diagnosed with strep throat just 2 years ago, the student was not too worried because she did not feel particularly ill, and she did not plan to go to the campus clinic. Her roommate, a microbiology student, argued that a visit to the clinic was important, noting that her instructor had said that strep throat should be treated with antibiotics. Eventually, the ill student agreed to visit the clinic.

At the clinic, the student was given a rapid strep test (RST)—the doctor swabbed the back of the student's throat and used a kit to immediately detect the presence of *Streptococcus pyogenes,* the causative agent of strep throat. The test was negative. The doctor told the student that he would send the throat swab for culture and would have a definitive answer for her soon. Meanwhile, he sent her home with a prescription for penicillin, but told her not to fill it until she heard back from him. A day later, the clinic doctor called the student and told her that she did indeed have strep throat and to begin taking the penicillin immediately.

1. Considering that the student had suffered from strep throat just 2 years previously, why would she develop it again?
2. Why would the RST be negative when the student did in fact have strep throat?
3. Why was the student given penicillin to treat her infection?
4. Why is it important to treat strep throat promptly?
5. Why are there no vaccines against *Streptococcus pyogenes*?

Discussion

1. Numerous antigenically distinct strains of *S. pyogenes* exist, and immunity to one strain does not protect against another. From an infection standpoint, the most important antigen of *S. pyogenes* is M protein, a major virulence factor that allows the bacterial cells to avoid phagocytosis. Antibodies to M protein prevent opsonization and protect against infection, but more than 200 different varieties of the protein are known to occur. M protein variation means that a person can get strep throat more than once—in this case, the student's current strep throat was likely due to a *S. pyogenes* strain that had an M protein type different from the strain that had infected her previously.
2. The rapid strep test (RST) is very specific but is not as sensitive as culture. Specificity indicates the reliability of a positive test; the high specificity of RST means that a positive result is likely a true positive—*S. pyogenes* is present in the specimen. Sensitivity indicates the reliability of a negative test; the low sensitivity of RST means that a negative result could be a false negative—*S. pyogenes* could be present in the specimen, but the test was simply unable to detect it. That is why laboratory culture, which has both high specificity and high sensitivity, is used as a backup for diagnosis. *S. pyogenes* is easily recognizable on blood agar by the zone of β-hemolysis surrounding the colonies on this medium. Other bacteria are also β-hemolytic, so catalase tests and group A–specific latex agglutination are done to confirm the identification.
3. The student was given penicillin because *S. pyogenes* is still sensitive to this antibiotic. Penicillin is generally the medication of choice for treating strep throat because it is inexpensive, safe, and effective. Had the student been allergic to penicillin, the doctor would have likely prescribed azithromycin, also effective in treating strep throat.
4. Although strep throat itself is not a very serious disease, it can lead to poststreptococcal sequelae or other complications. Treating with antibiotics may lessen the chance of these complications. The sequelae, including acute rheumatic fever (ARF) and post-streptococcal glomerulonephritis (PSGN), are due to the immune response against *S. pyogenes* and occur after the infection has resolved. ARF is an autoimmune reaction that results from molecular mimicry of *S. pyogenes*—the epitopes of some versions of M protein are similar to epitopes on self proteins. Thus, an immune response to the M protein may result in the production of antibodies that also attach to self cells, targeting them for destruction. Glomerulonephritis results from an inflammatory reaction in the glomeruli of the kidneys. In the case of PSGN, an accumulation of immune complexes consisting of antibodies bound to streptococcal antigens is thought to trigger the inflammation. Another concern regarding strep throat is that a few strains produce streptococcal pyrogenic exotoxins (SPEs), and these can cause certain severe diseases, including scarlet fever.
5. Developing a vaccine against *S. pyogenes* has not been possible because of the great variation in the M protein of these organisms. A vaccine against one M type might not protect against another. An additional complication is the risk of autoimmunity, such as that seen in ARF. An appropriate vaccine would have to protect against many different strains of *S. pyogenes* without causing an autoimmune response.

Most strains of *C. diphtheriae* release diphtheria toxin, a powerful cytotoxin that causes the serious symptoms of the disease. The toxin is encoded by a prophage (phage DNA inserted into a bacterial chromosome); its production is an example of lysogenic conversion.

Pathogenesis

Corynebacterium diphtheriae has little invasive ability, so it rarely enters the blood or tissues. The disease symptoms result from damage caused by the toxin released by the bacteria as they grow in the upper respiratory tract. The toxin kills eukaryotic cells, leading to the formation of a local pseudomembrane, made up of dead epithelial cells and clotted blood, along with fibrin and the leukocytes that accumulate during the resulting inflammatory response. If the pseudomembrane comes loose it can obstruct the airways, causing the patient to suffocate. In addition, the toxin may be absorbed into the bloodstream, allowing it to access and damage heart, nerve, and kidney tissues.

Diphtheria toxin is an A-B toxin (**figure 21.9**). The B subunit attaches to specific receptors on a host cell membrane, and the entire toxin molecule is taken into the cell by endocytosis. Cells lacking these receptors do not take up the toxin and are unaffected by it, explaining why the toxin only damages some tissues. Once the toxin is inside a cell, the A subunit separates from the B subunit. It then becomes a functional enzyme that inactivates elongation factor 2 (EF-2), which is required for movement of the eukaryotic ribosome on mRNA. This stops protein synthesis, and the cell dies. The toxin is extremely potent because the A subunit, an enzyme, is not used up in the reaction, so a single molecule can inactivate nearly all of the cell's EF-2.

Epidemiology

Humans are the primary reservoir for *Corynebacterium diphtheriae.* The organism is typically spread by inhalation of airborne droplets from infected people or through fomites. *C. diphtheriae* can also cause cutaneous diphtheria, resulting in skin ulcers that can become a source of infection to others.

Treatment and Prevention

Diphtheria is treated by injecting the patient with antiserum against diphtheria toxin once the disease is suspected; the toxin is so potent that delaying treatment to wait for culture results can be fatal. *C. diphtheriae* strains are sensitive to various antibiotics, including erythromycin and penicillin. Although antibiotic treatment can prevent further transmission, it has no effect on toxin that has already been absorbed. Even with treatment, about 10% of diphtheria patients die.

Because diphtheria results from toxin production rather than microbial invasion, immunization with a toxoid (inactivated toxin) can effectively prevent disease. The toxoid causes the body to produce antibodies that specifically neutralize the diphtheria toxin. The childhood vaccination DTaP consists of diphtheria and tetanus toxoids along with components of *Bordetella pertussis,* the bacterium that causes pertussis (whooping cough). Unfortunately, these immunizations are often neglected, and serious epidemics of diphtheria have occurred periodically. An active campaign in the United States during the 1980s led to a requirement that nearly all children entering school be immunized against diphtheria. As a result, the incidence of the disease has been reduced to only a handful in the past decades. Immunity decreases after childhood, however, so booster injections must be given every 10 years to maintain protection. **Table 21.4** summarizes the main features of diphtheria.

MicroAssessment 21.2

Conjunctivitis, otitis media, and sinusitis are common infections that often occur together and are caused by the same pathogens. *Streptococcus pyogenes* causes a sore throat commonly known as strep throat. Untreated *S. pyogenes* infections can sometimes cause serious diseases (sequelae) in other parts of the body long after the initial infections have resolved. *Corynebacterium diphtheriae* produces a toxin that kills host cells, resulting in symptoms characteristic of diphtheria.

4. Name two post-streptococcal sequelae.
5. How does diphtheria toxin kill cells?
6. How would adequate ventilation help to prevent the spread of streptococcal infections?

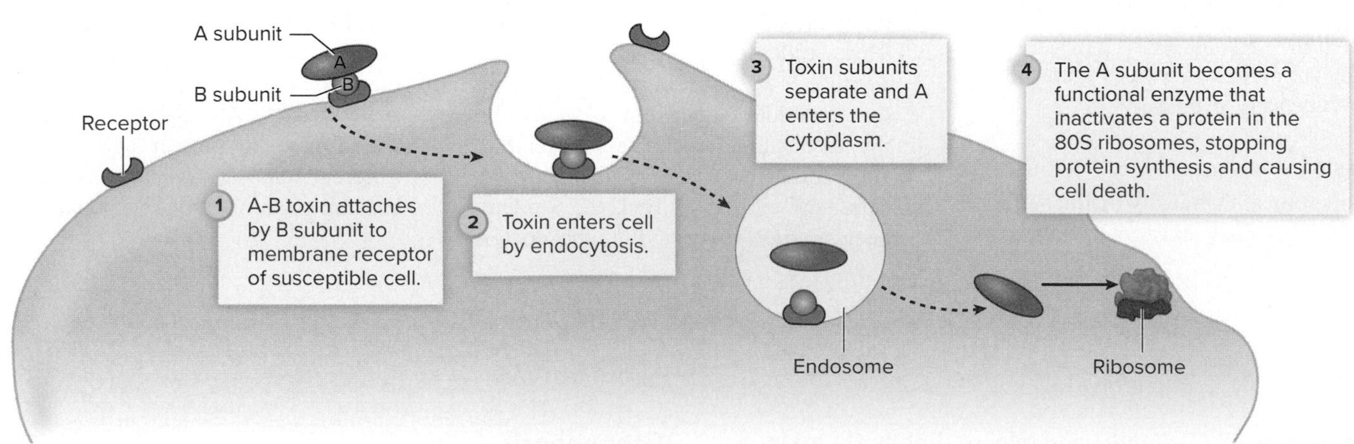

FIGURE 21.9 Mode of Action of Diphtheria Toxin

Why is diphtheria toxin so potent?

TABLE 21.4 Diphtheria

1. *Corynebacterium diphtheriae* enters by inhalation.
2. Infection established in throat.
3. Toxin released locally, pseudomembrane forms.
4. Circulating toxin may cause paralysis or damage heart muscle, kidneys, nerves.
5. Membrane may come loose and obstruct breathing.
6. Exit from body by respiratory secretions.

Signs and symptoms	Sore throat, fever, fatigue, and malaise; pseudomembrane forms on the throat, tonsils, larynx or in the nasal cavity; paralysis, heart and kidney failure
Incubation Period	2 to 6 days
Causative Agent	*Corynebacterium diphtheriae*, a toxin-producing, non-spore-forming Gram-positive rod
Pathogenesis	Infection in upper respiratory tract; diphtheria toxin is released locally and may be absorbed into the bloodstream; toxin kills cells by interfering with protein synthesis; affects cells that have receptors for the toxin—mainly heart, kidney, and nerve tissue.
Epidemiology	Inhalation of infectious droplets; indirect contact with fomites.
Treatment and Prevention	Treatment: antitoxin; appropriate antibiotic to prevent transmission. Prevention: toxoid vaccine; part of routine vaccination for infants and children; boosters for adults.

21.3 ■ Viral Infections of the Upper Respiratory System

Learning Outcomes

5. List the strategies helpful in avoiding common colds.
6. Give the distinctive characteristics of adenoviral respiratory tract infections.

The average person in the United States gets two to five viral upper respiratory infections each year. Although hundreds of different viruses cause these infections, they produce similar symptoms and generally resolve without treatment. However, they impair respiratory tract defenses and allow for more serious secondary bacterial infections.

The Common Cold

The common cold is the most frequent infectious disease in humans, accounting for more than half of upper respiratory tract infections every year. Colds are the leading cause of absences from school and result in people missing an estimated 150 million workdays per year in the United States.

MicroByte

The average adult gets two to three common colds a year, whereas children typically get more, as many as eight colds in this time.

Signs and Symptoms

Colds begin 1 to 2 days after infection and start with malaise, followed by a runny nose, sneezing, coughing, a mild sore throat, and hoarseness. The nasal secretions are initially profuse and watery but may thicken and become cloudy as the cold progresses. There is typically no fever (except in younger children) unless secondary bacterial infection occurs. Symptoms typically last about a week, but a mild cough may continue a bit longer.

Causative Agents

Viruses that cause the common cold are often simply called "cold viruses." Between 30% and 50% of colds are caused by the 100 or more types of human rhinoviruses (*rhino* means "nose," as in rhinoceros, or "horny nose") (**figure 21.10**). These are members of the family *Picornaviridae* (*pico* means "small" thus, "small RNA viruses"), a group of non-enveloped viruses with a single-stranded RNA genome. Other cold viruses include certain types of adenoviruses, enteroviruses, and coronaviruses.

Pathogenesis

Cold viruses typically bind to and infect nasal epithelial cells, causing them to release pro-inflammatory cytokines that recruit neutrophils to the area. Other chemical mediators are also released, with the overall result of sneezing, increased

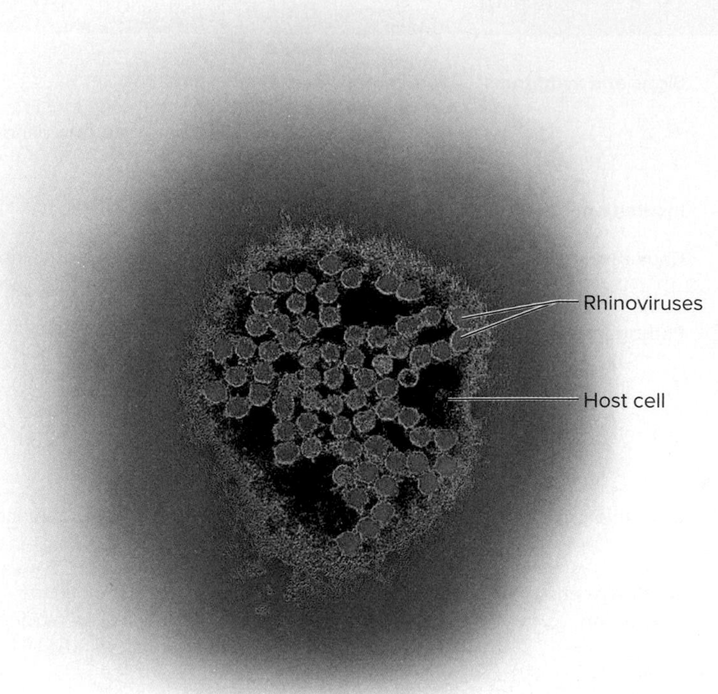

FIGURE 21.10 Rhinovirus-Infected Cell Color-enhanced TEM.
A.B. Dowsette/SPL/Science Source

? **There are no vaccines that prevent the common cold. Why do you think that is?**

nasal secretions, and tissue swelling that partially or completely obstructs the airways. The infection is stopped by immune responses, but it can spread into the ears, the sinuses, or even the lower respiratory tract before this occurs.

Epidemiology

Humans are the only source of cold viruses, which are spread when airborne virus-containing droplets are inhaled or when secretions from infected people are accidentally rubbed into the eyes or nose by contaminated hands. Young children transmit cold and other respiratory viruses very easily because they are often careless with their respiratory secretions. Infected people have high concentrations of virus in their nasal secretions and on their hands during the first 2 to 3 days of a cold and are most likely to transmit the virus then. Virus levels are often undetectable after 4 or 5 days, but low levels can be present for 2 weeks.

Treatment and Prevention

There are no proven treatments for the common cold. Viruses are not affected by antibiotics or other antibacterial medications. Analgesics (painkillers) and antipyretics (fever-reducers) such as aspirin and ibuprofen can help reduce symptoms, but children and teenagers should not be given aspirin because of the risk of developing **Reye's syndrome,** a rare, potentially fatal condition that can affect many organs but especially the brain and the liver.

One of the most important ways to prevent the spread of cold viruses is handwashing. Alcohol-based hand sanitizers have been shown to be effective as well. Other preventive measures include keeping hands away from the face and avoiding crowded places when respiratory diseases are prevalent. It is especially important to avoid people with colds during the first few days of their symptoms, when they are shedding high numbers of viral particles. Developing a vaccine has not been possible because such a large number of immunologically different viruses cause colds. The genomes of all known rhinovirus strains have been sequenced, however, revealing new possibilities for vaccines. A number of antivirals are in clinical trials as well. **Table 21.5** summarizes some important features of the common cold.

Adenovirus Respiratory Tract Infections

Human adenoviruses are widespread and can cause a variety of different types of infections, depending on the viral serotype. Several serotypes infect the upper respiratory tract, causing a sore throat accompanied by a fever. Human adenoviruses are also the most common viral cause of pink eye.

Signs and Symptoms

Adenovirus serotypes that infect the upper respiratory tract generally cause a runny nose and a sore throat, but unlike the common cold, fever is typically present. Signs and symptoms appear 5 to 10 days after infection. The throat is usually sore, with regions of gray-white pus on the pharynx and

TABLE 21.5	The Common Cold
Signs and Symptoms	Malaise, runny nose, sneezing, cough, sore throat, hoarseness
Incubation Period	1 to 2 days
Causative Agent	Mainly rhinoviruses—more than 100 types
Pathogenesis	Viruses attach to respiratory epithelium; some cells slough off; mucus secretion increases, and inflammatory reaction occurs.
Epidemiology	Transmitted through inhalation of infectious droplets or transfer of infectious mucus to nose or eye by contaminated fingers.
Treatment and Prevention	Treatment: no treatment except for control of symptoms. Prevention: handwashing, avoiding people with colds, and keeping hands away from face.

tonsils—characteristics that can be confused with those of strep throat. Lymph nodes in the neck can become swollen, and a mild cough is common. Patients sometimes develop a severe cough with chest pain suggestive of pneumonia, particularly if they are immunocompromised. Recovery usually takes about 1 to 3 weeks, although the virus has been shown to persist in certain cells and in lymphoid tissues such as the tonsils.

Causative Agent

Adenoviruses are non-enveloped, double-stranded DNA viruses discovered in cell lines derived from adenoid tissue (thus the name). More than 50 different serotypes infect humans, and many of these cause upper respiratory tract infections. Certain serotypes, however, are associated with other diseases such as conjunctivitis, gastrointestinal disease, or pneumonia, and a number of these have caused sporadic outbreaks over the years. Adenovirus serotypes that cause severe respiratory infections have been a particular concern for military recruits.

Pathogenesis

Adenoviruses attach to and enter epithelial cells. The association of different serotypes with certain diseases is partially explained by the specific cell types that they can infect, but host characteristics are also involved. Once inside the cell, the genome is transported to the host cell nucleus, where the virus multiplies. Adenoviruses have many mechanisms for avoiding host defenses, including delaying apoptosis, blocking interferon function, and interfering with antigen presentation by MHC class I molecules. Once replication is complete, virions are released after host cell lysis. In severe infections, extensive cell destruction and inflammation occur.

Epidemiology

Human adenoviruses are shed from the respiratory tract during acute illness and transmitted directly via droplet spread and indirectly through fomites. They can be shed in the feces for months after infection, so spread via the fecal-oral route is also possible. Because these viruses can persist in the environment and can potentially be transmitted by multiple routes, healthcare personnel must use appropriate precautions (contact precautions and/or droplet precautions) to prevent their transfer among patients. The viruses also commonly infect schoolchildren, occasionally causing outbreaks in winter and spring. Summer epidemics can occur when viruses are transmitted in inadequately chlorinated swimming pools. Respiratory disease can be a problem in military recruits and other people living together in crowded conditions. Asymptomatic adenovirus infections are common, which fosters epidemic spread.

TABLE 21.6 Adenovirus Respiratory Tract Infections

Signs and Symptoms	Fever, sore throat, cough, swollen lymph nodes of neck, pus on tonsils and throat
Incubation Period	5 to 10 days
Causative Agent	Adenoviruses—more than 50 serotypes—non-enveloped, double-stranded DNA genome
Pathogenesis	Virus multiplies in host cell nucleus; cell destruction and inflammation occur; different serotypes produce different symptoms.
Epidemiology	Inhalation of infectious droplets; possible spread from gastrointestinal tract.
Treatment and Prevention	Treatment: no treatment except for control of symptoms. Prevention: handwashing, avoiding. crowds; vaccine for military recruits

Treatment and Prevention

As with colds, there is no specific treatment for adenovirus disease, and most patients recover on their own. Secondary bacterial infections may occur, however, and these require treatment with an antibacterial medication. Immunocompromised patients who are more prone to severe, even fatal disease may be given an antiviral medication, although no drug is specifically approved for treating adenovirus infections.

Methods to prevent adenovirus infections are the same as described for the common cold, including handwashing and avoiding crowds. An orally administered vaccine against the two serotypes most likely to cause severe respiratory disease in military recruits is available. **Table 21.6** summarizes some important features of adenoviral infections.

MicroAssessment 21.3

The common cold may be caused by many different viruses, particularly rhinoviruses. Transmission requires close personal contact during which respiratory droplets from an infected person or respiratory secretions on contaminated hands are transferred to the nasal epithelium. Symptoms typically last for about a week before the disease is stopped by immune responses. Adenovirus infections may resemble colds, but fever is present.

7. Why are people most infectious during the first few days of a cold?
8. How is an adenovirus infection treated?
9. Contrary to your grandmother's warnings, going outside in cold weather without a coat is not likely to cause a cold. Why not?

LOWER RESPIRATORY TRACT INFECTIONS

FOCUS ON PNEUMONIA

Pneumonia is a disease of the lower respiratory tract caused by bacterial, viral, or fungal infection of the lungs; the infection elicits an inflammatory response, resulting in the alveoli (air sacs) filling with fluids such as pus and blood. Pneumonia is the leading cause of death due to infectious disease in the United States.

Signs and Symptoms

The signs and symptoms of pneumonia generally include cough, chills, shortness of breath, fever, and chest pain. In severe cases, the patient may develop cyanosis (bluish skin color) due to poor blood oxygenation. Pneumonia ranges from mild to life-threatening, depending largely on the causative agent but also on any underlying health problems of the patient. It is often accompanied by a productive cough, meaning that a pus- and mucus-containing fluid called **sputum** comes up from the lungs.

Some pathogens cause what are referred to as atypical pneumonias, not because the diseases are uncommon, but because the symptoms or the treatments are slightly different than those of the more established types. "Walking pneumonia" is a term often used to describe some atypical pneumonias because of the milder symptoms observed.

To diagnose pneumonia, a physician uses a stethoscope to listen for a characteristic crackling or bubbling sound that occurs in the lungs as air passes by fluid in the alveoli. A chest X ray will likely be done to determine which parts of the lung are infected; areas of infection usually appear as white shadows. The patient may also be asked to give a sputum sample, which can be examined microscopically and inoculated onto appropriate laboratory media as part of the process to identify a bacterial or fungal cause of pneumonia.

Pathogenesis

Various bacteria, viruses, and fungi can all cause pneumonia, but typically only when the respiratory tract defenses are not functioning optimally. Smoking, alcohol use, and narcotic use can all damage the mucociliary escalator, as can viral respiratory infections such as influenza. That is why patients who have illnesses that damage the respiratory tract or interfere with the patient's ability to cough are more susceptible to developing pneumonia. Bacterial pathogens that cause pneumonia often make a capsule. Because of the capsule, alveolar macrophages that normally destroy invading microbes cannot effectively eliminate the pathogen at first. Once opsonizing antibodies are produced during a B-cell response, however, phagocytes can remove the microbes.

The damage from pneumonia is largely a result of the inflammatory response. As the capillaries become leaky during inflammation, fluids collect in the alveoli and interfere with O_2 and CO_2 exchange. In addition, phagocytes and other leukocytes are recruited to the site of infection, and mucus production increases. In severe cases, accumulating leukocytes and mucus create a thick substance that may clog the alveoli, a condition called consolidation. The inflammatory response seen in severe pneumonia often affects nerve endings in the pleura, causing pain. Fatal respiratory failure occurs when the lungs can no longer adequately oxygenate the blood or expel CO_2.

Epidemiology

Pneumonias are often categorized as either community-acquired, meaning they develop in members of the general public, or healthcare-associated, meaning they develop in hospitalized patients or other people within the healthcare system. Some types of community-acquired pneumonia (CAP) are contagious. Most, however, originate from the patient's own upper respiratory microbiota. These organisms may gain access to the lungs when a person inadvertently inhales his or her own throat secretions. As with CAPs, healthcare-associated pneumonias (HCAPs) often occur when the patient inhales his or her own upper respiratory microbiota. Patients at particular risk are those on mechanical ventilators used to help breathing, because the ventilator tube provides a portal for microbes to enter the lower airways. Pneumonias that develop this way are further classified as ventilator-associated pneumonias (VAPs).

Treatment and Prevention

Bacterial and fungal pneumonias are treated with antimicrobial medications, chosen according to the susceptibility of the causative agent. Unfortunately, bacteria that cause healthcare-associated pneumonias are often multi-drug-resistant. With a few important exceptions, including pneumonia caused by COVID-19 or influenza, no effective treatments are currently available for viral pneumonias.

Vaccines are available to prevent pneumococcal disease but not other common types of pneumonia. Vaccines are also available for certain diseases that sometimes progress to pneumonia, including COVID-19 and influenza. Prevention by other means involves making healthy choices, such as not smoking, as well as avoiding other respiratory illnesses by staying out of crowded situations and by using good hygiene practices such as handwashing.

21.4 ■ Bacterial Infections of the Lower Respiratory System

Learning Outcomes

7. Compare the distinctive features of pneumococcal, *Klebsiella*, and mycoplasmal pneumonias.
8. Outline the pathogenesis and treatment of pertussis, tuberculosis, Legionnaires' disease, and inhalation anthrax.

Bacterial infections of the lower respiratory tract are less common than those of the upper respiratory tract, but are generally much more serious. An earache or a sore throat is unlikely to be life-threatening, but the same infectious agents can cause serious illnesses in the lungs.

Pneumonia is an infection of the lungs resulting in inflammation accompanied by filling of the air sacs with fluids such as pus and blood. It tops the list of fatal community-acquired infectious diseases in the United States and is also one of the most prevalent healthcare-associated infections. Most common types of pneumonia are bacterial, but some are fungal or viral; certain features are similar regardless of the causative agent so the general characteristics are highlighted in the box Focus on Pneumonia.

Pneumococcal Pneumonia

Pneumococci are an important cause of bacterial pneumonia, accounting for about 400,000 hospitalizations and about one-third of all community-acquired pneumonia cases in the United States each year.

Signs and Symptoms

Typical signs and symptoms of pneumococcal pneumonia start after an incubation of 1 to 3 days and begin abruptly with fever and shaking chills. Patients experience severe chest pain that is aggravated by each breath or cough, causing shallow, rapid breathing. Blood from the lungs makes the sputum pinkish or rust-colored (blood-tinged); poor blood oxygenation leads to cyanosis (bluish skin color). Most patients improve after 3 to 5 days of antimicrobial treatment, but full recovery can take weeks.

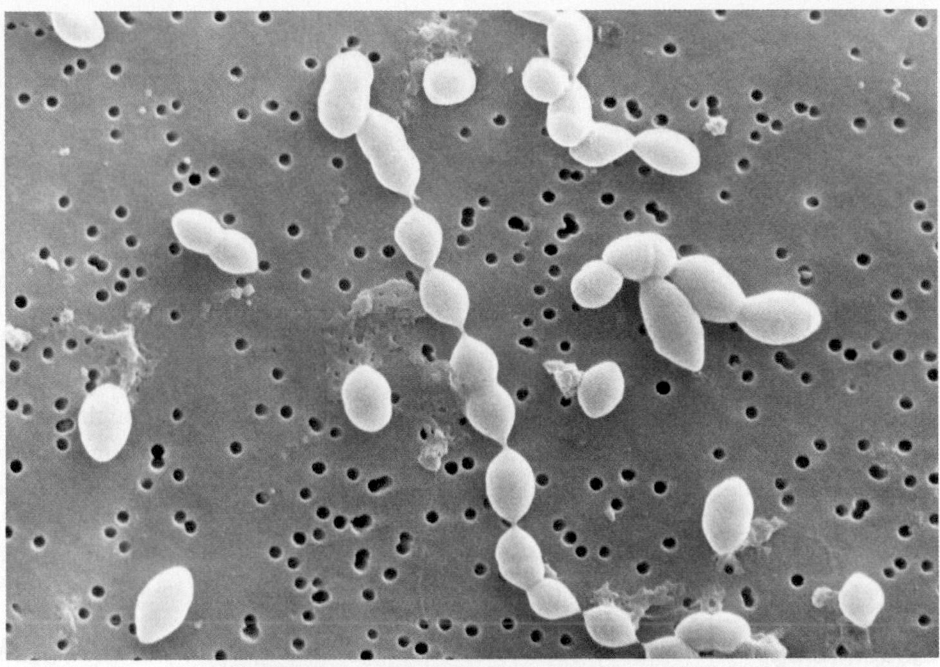

(a)

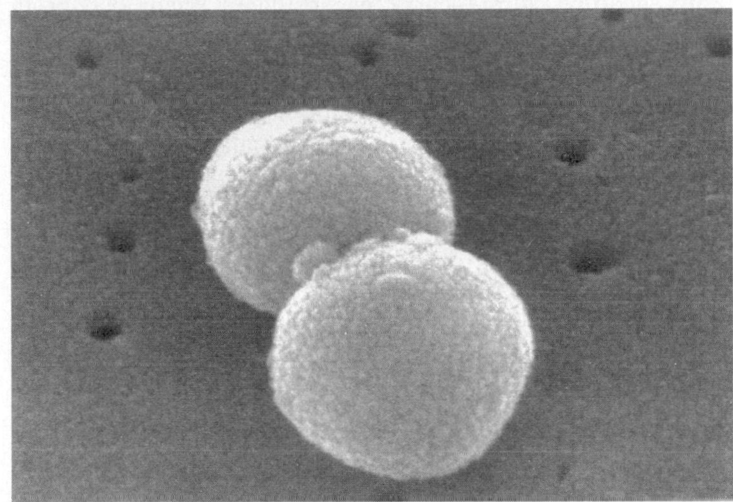

(b)

FIGURE 21.11 ***Streptococcus pneumoniae*** Scanning electron micrographs. **(a)** Emphasizing the lancet-shape of the cells. **(b)** Emphasizing the diplococcus arrangement (higher magnification).

a: BSIP/Science Source; b: Janice Haney Carr/CDC

? What color would *Streptococcus pneumoniae* cells be after a Gram stain?

Causative Agent

Pneumococcal pneumonia is caused by *Streptococcus pneumoniae,* a Gram-positive diplococcus known as pneumococcus. The cells, which are often elongated with a tapered end, are referred to as lancet-shaped (a lancet is a surgical instrument with a pointed end) and are typically arranged as diplococci (**figure 21.11**). The most striking characteristic of *S. pneumoniae* is its thick polysaccharide capsule, which is responsible for the organism's virulence. There are over 100 different serotypes of *S. pneumoniae,* each with different capsular antigens. Certain serotypes are more commonly associated with pneumonia and other serious complications, including sepsis and meningitis. Strains of the organism that lack a capsule do not typically cause serious disease.

Pathogenesis

The capsule of pneumococcus is a key virulence factor—unencapsulated strains do not cause pneumonia. Encapsulated bacteria are resistant to phagocytosis because their capsule interferes with the action of the complement system component C3b, an important opsonin. The bacteria also produce pneumolysin, a membrane-damaging toxin released when they die; this potent cytotoxin kills eukaryotic cells and also activates the inflammatory response. As with most severe pneumonias, the inflammatory response leads to an accumulation of fluid and phagocytic cells in the lung alveoli, causing difficulty breathing. This fluid can be seen as abnormal shadows on chest X rays of patients (**figure 21.12**).

Pneumococci may enter the bloodstream from the inflamed lungs, causing three often fatal complications: sepsis (due to a bloodstream infection); infective endocarditis (due to an infection of the heart valves); and meningitis (due to an infection of the membranes covering the brain and spinal cord). People who do not develop complications usually produce enough specific anti-capsular antibodies within about a week to allow phagocytosis and destruction of the pneumococci, resulting in complete recovery.

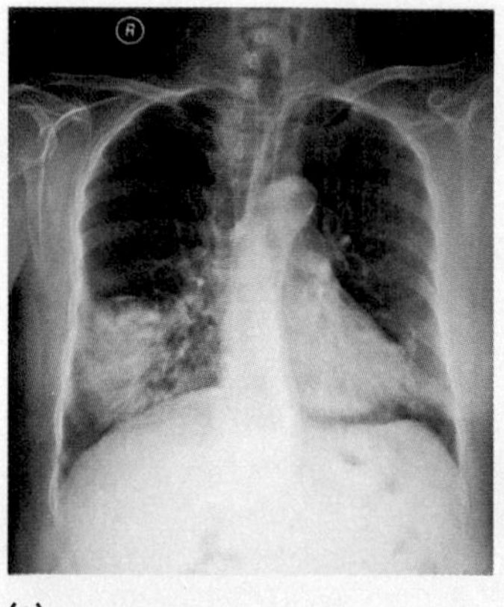

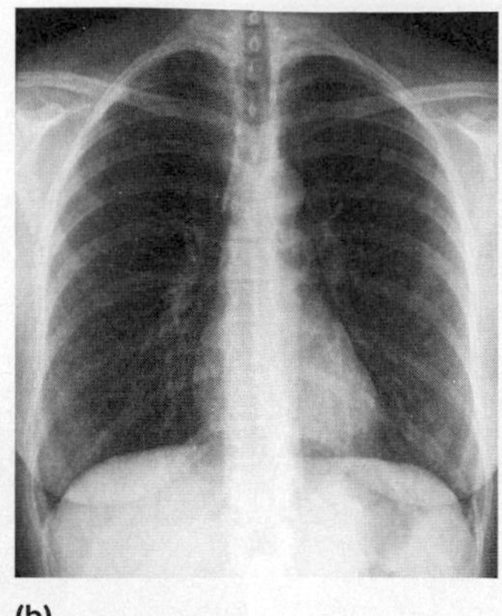

(a) (b)

FIGURE 21.12 Chest X Ray Appearance in Pneumococcal Pneumonia (a) Pneumonia. The right lung (left side of figure) appears white because the alveoli are filled with fluid. **(b)** Normal X ray. a: Hong xia/Shutterstock; b: skyhawk x/Shutterstock

? **Why does the sputum of a pneumonia patient become pinkish or rust-colored?**

Epidemiology

Many healthy people carry encapsulated pneumococci in their throat, but the bacteria seldom reach the lungs because the mucociliary escalator effectively removes them. When this defense mechanism is impaired, however, the risk of pneumococcal pneumonia rises dramatically. There is also an increased risk of the disease in people over 65, or in those with underlying heart or lung disease, diabetes, or cancer.

Treatment and Prevention

Most pneumococcal infections can be cured with penicillin or amoxicillin if given early in the illness, but antibiotic-resistant strains of pneumococci are becoming increasingly common.

Vaccines for preventing pneumococcal disease target the *S. pneumoniae* capsular polysaccharides and include two general types:

- **Pneumococcal conjugate vaccine (PCV).** This type contains polysaccharides attached to bacterial proteins. Recall that polysaccharides are T-independent antigens, but by attaching them to a carrier protein, they become T-dependent antigens; thus, the vaccines are effective even in young children. The three types of PCV currently available, PCV13, PCV15, and PCV20, use polysaccharides from 13, 15, and 20 pneumococcal serotypes, respectively. PCV13 or PCV15 is recommended for children under 5 years old and for high-risk children through age 18. PCV15 or PCV20 is recommended for adults 65 years or older and for some high-risk adults 19 years or older.
- **Pneumococcal polysaccharide vaccine (PPSV).** This type contains purified polysaccharides. The one version available, PPSV23, contains polysaccharides from 23 pneumococcal serotypes. It is recommended for high-risk children 2 through 18 years old and as a follow-up for anyone age 19 or older who had PCV15.

Klebsiella Pneumonia

Enterobacteria such as *Klebsiella* species and other Gram-negative rods can cause pneumonia, especially if host defenses are impaired. Enterobacteria are common hospital-acquired pathogens in the United States and cause most of the deaths from healthcare-associated infections. They are also significant causes of community-acquired pneumonia in parts of Asia and Africa.

Signs and Symptoms

The general signs and symptoms of *Klebsiella* pneumonia—cough, chills, shortness of breath, fever, chest pain, and cyanosis—are the same as those of pneumococcal pneumonia and also appear after an incubation of 1 to 3 days. The sputum of *Klebsiella* pneumonia patients, however, can be thick, bloody, and jelly like, which is different from the blood-tinged sputum seen with pneumococcal pneumonia.

Causative Agent

Klebsiella pneumoniae, a Gram-negative rod with a large capsule, produces big, noticeably mucoid colonies (**figure 21.13**). It is commonly part of the normal microbiota of the gastrointestinal tract and may be found in the mouth or the throat.

Pathogenesis

K. pneumoniae is contracted through secretions transmitted by person-to-person contact, or from medical equipment such as ventilators. Organisms first colonize the throat and gain access to the lung via inhaled air or mucus. Specific adhesins aid colonization. The capsule is an essential virulence factor, probably interfering with the action of complement system component C3b. In addition, the bacterium produces a specific siderophore with a very high affinity for iron, allowing it to "steal" the essential element from host cells. Recent studies indicate that iron depletion causes a type of cellular stress that induces

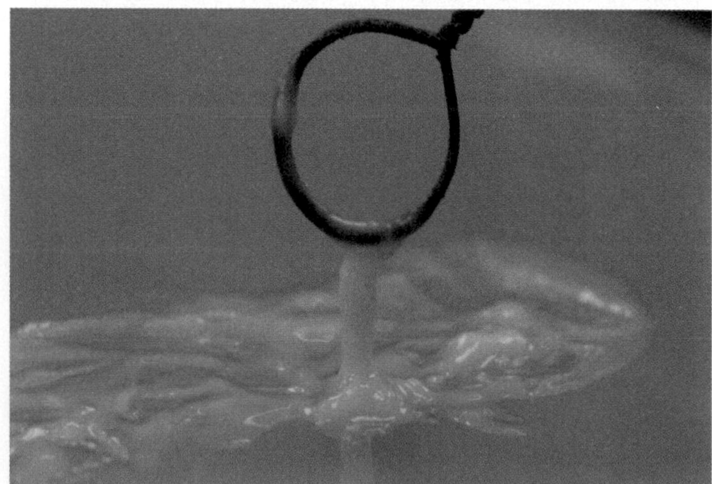

FIGURE 21.13 ***Klebsiella pneumoniae*** Growth on an agar plate. Notice the mucoid nature of the bacterial growth. Lisa Burgess/McGraw Hill

? **Why is *Klebsiella* pneumonia often fatal?**

inflammation and enhances the spread of the bacterium. *Klebsiella* pneumonia results in more tissue damage than pneumococcal pneumonia and in the rapid formation of lung abscesses. Therefore, even with effective antibacterial medication, the lung can be permanently damaged, and the patient may die. Infection often spreads to the bloodstream, leading to abscess formation in other tissues such as the liver and brain, and also causing septic shock.

Epidemiology

Klebsiella species are widespread in nature. In humans, they are commonly part of the normal intestinal microbiota. Typically, people who contract *Klebsiella* pneumonia are very old, are very young, or have a compromised immune system (such as alcoholics, or those in a hospital or other medical setting). The strains that circulate in hospitals and nursing homes are often resistant to antimicrobial medications and are increasingly becoming multidrug-resistant.

Treatment and Prevention

Klebsiella pneumonia is treated with antibiotics. Antimicrobial susceptibility testing must be carried out to determine which medications should be used. In seriously ill patients, immediate combination antibiotic therapy is given. Unfortunately, some of the medications used in these combinations, such as aminoglycosides, have severe side effects. Surgery may be required to drain abscesses.

Treatment of *Klebsiella* infections can be very challenging because strains are becoming increasingly antibiotic resistant. *Klebsiella* species commonly produce a plasmid-encoded β-lactamase, an enzyme that makes the bacteria resistant to certain β-lactam antibiotics (primarily the penicillins). Many strains also produce an extended-spectrum β-lactamase (ESBL), which makes them resistant to many of the cephalosporins as well. Some strains produce versatile β-lactamases called carbapenemases, which makes them resistant to carbapenems and other β-lactam drugs. These strains, referred to as CRKP (carbapenem-resistant *K. pneumoniae*), are included in the CRE (carbapenem-resistant Enterobacteriaceae), which the CDC lists as an urgent threat with respect to antimicrobial resistance (see table 20.2). Carbapenems are usually given as a last resort, meaning that very few effective treatment choices remain for CRKP infections. The case-fatality rate for CRKP pneumonia, even with treatment, is as high as 50%, and patients tend to die more quickly than other pneumonia patients.

There are no specific preventive measures such as vaccination for *Klebsiella* pneumonia. To prevent spread between patients, healthcare workers must follow infection control measures, such as washing hands and wearing gloves and gowns when in a room with a *Klebsiella* patient. Disinfecting the environment, using sterile respiratory equipment, and using antimicrobial medications only when necessary help control the organisms and their development of resistance in hospitals.

Mycoplasmal Pneumonia ("Walking Pneumonia")

Mycoplasmal pneumonia is one of the most common kinds of pneumonia in children and young adults. The risk is greatest in crowded conditions that might occur in a college dormitory or military barracks. The disease is generally mild (as reflected by its popular name "walking pneumonia") and seldom requires hospitalization. Because it does not show the severe signs and symptoms of most other pneumonias, it is considered an atypical pneumonia.

Signs and Symptoms

The incubation period for mycoplasmal pneumonia is typically 2 to 3 weeks, but symptoms may take longer to fully develop. The early symptoms are sore throat, chills, fever, headache, muscle pain, and fatigue. After several days, a dry cough begins which can worsen over weeks, a condition commonly called a "chest cold." In 1 out of 10 cases, additional symptoms consistent with a milder form of pneumonia develop, including chest pain and difficulty breathing. Mucoid sputum is produced in some cases.

Causative Agent

Mycoplasmal pneumonia is caused by *Mycoplasma pneumoniae,* a tiny, spindle-shaped, aerobic bacterium with no cell wall (**figure 21.14**). *M. pneumoniae* requires specialized media and grows very slowly, forming very tiny colonies in culture.

Pathogenesis

Only a few inhaled *M. pneumoniae* cells are required to start an infection. Cells attach by means of a specialized organelle

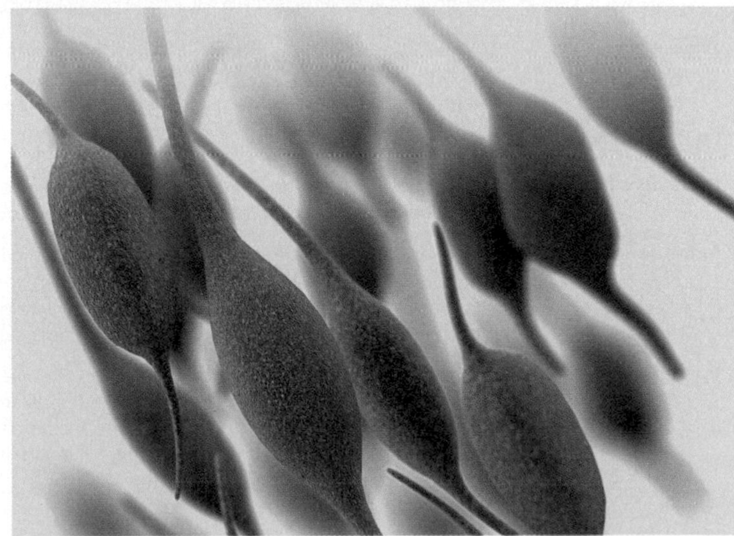

FIGURE 21.14 ***Mycoplasma pneumoniae*** Computer-generated illustration of spindle-shaped cells based on scanning electron microscopy. Sarah Bailey Cutchin/CDC

? **Why is penicillin ineffective in treating mycoplasmal pneumonia?**

to the base of the cilia on the respiratory epithelium. This close association allows the organisms to escape clearance by the mucociliary escalator. In addition, bacterial attachment damages the respiratory epithelium, causing the ciliated cells to slough off. The bacteria also produce a cytotoxin called CARDS (**c**ommunity **a**cquired **r**espiratory **d**istress **s**yndrome) toxin, which causes additional damage and induces the release of pro-inflammatory cytokines. The resulting inflammatory response leads to an accumulation of lymphocytes and phagocytes and a thickening of the walls of the bronchial tubes and alveoli.

Epidemiology

M. pneumoniae is spread by aerosolized droplets of respiratory secretions. The organisms are shed in these secretions for a long time period, ranging from about a week before symptoms begin to many weeks afterward, thereby increasing the likelihood of transmission. Infections tend to be more common in summer and early fall. Mycoplasmal pneumonia can account for up to one-fifth of bacterial pneumonias, with a peak incidence in young people. Immunity after recovery is not permanent, and repeat infections have occurred within 5 years.

Treatment and Prevention

M. pneumoniae lacks a cell wall, so antibiotics that act against bacterial cell wall synthesis are not effective. Macrolides shorten the illness if given early.

Preventive measures for mycoplasmal pneumonia are similar to those for other respiratory diseases and include handwashing as well as avoiding crowding in schools and military facilities. There are no vaccines to prevent mycoplasmal pneumonia. **Table 21.7** compares pneumococcal, *Klebsiella*, and mycoplasmal pneumonias.

Pertussis ("Whooping Cough")

Whooping cough is the common name for pertussis (*per* means "intensive"; *tussis* means "cough"). This vaccine-preventable disease is still endemic in many countries, including the United States. Worldwide, the disease causes up to half a million deaths yearly.

Signs and Symptoms

Pertussis has a typical incubation period of 7 to 10 days and has three stages:

- **Catarrhal stage** (meaning excessive mucus due to inflammation of the mucous membranes). This typically lasts 1 to 2 weeks, with signs and symptoms that resemble an upper respiratory tract infection—runny nose, sneezing, low fever, and mild cough.
- **Paroxysmal stage** (meaning repeated sudden attacks). This lasts for about 1 to 6 weeks, and each attack includes frequent bursts of violent, uncontrollable coughing. The cough is dry, but is severe enough to burst small blood vessels in the eyes. The coughing spasm is followed by a forceful attempt to inhale, causing the characteristic "whoop" of this disease. The patient may also become cyanotic (blue from lack of O_2) during the attack.

TABLE 21.7 Pneumococcal, *Klebsiella*, and Mycoplasmal Pneumonias Compared

	Pneumococcal Pneumonia	*Klebsiella* Pneumonia	Mycoplasmal Pneumonia
Signs and Symptoms	Cough, sudden chills and fever, shortness of breath, chest pain, cyanosis, rust-colored sputum (blood tinged)	Cough, repeated chills, fever, shortness of breath, chest pain, cyanosis, bloody jelly-like sputum	Gradual onset of dry cough, fever, fatigue, headache, and muscle aches
Incubation Period	1 to 3 days	1 to 3 days	2 to 3 weeks
Causative Agent	*Streptococcus pneumoniae* (pneumococcus); encapsulated strains	*Klebsiella pneumoniae*, encapsulated enterobacterium	*Mycoplasma pneumoniae*; lacks cell wall
Pathogenesis	Inhalation of infected droplets. Colonization of alveoli triggers an inflammatory response; fluid and inflammatory cells fill the alveoli.	Inhalation of infected droplets. Destruction of lung tissue and abscess formation common; infection spreads via blood to other body tissues.	Inhalation of infected droplets. Damage to respiratory epithelium; inflammatory response and destruction of cells via CARDS toxin.
Epidemiology	High carrier rates for *S. pneumoniae*. Risk of pneumonia increases with conditions such as alcoholism, narcotic use, and viral infections that impair the mucociliary escalator. Other risk factors are chronic heart or lung disease, diabetes, and cancer.	*Klebsiella* species and other Gram-negative rods are common causes of fatal healthcare-associated pneumonias. Often resistant to antibiotics.	Mild infections are common; infected people spread the disease.
Treatment and Prevention	Treatment: antibiotics. Prevention: PCV13, PCV15, and PCV20 (conjugate vaccines); PPSV23 (polysaccharide vaccine).	Treatment: a combination of antibiotics; resistance is a problem. Prevention: no vaccine available.	Treatment: antibiotics, excluding cell wall synthesis inhibitors. Prevention: no vaccine available; avoid crowding in schools and military facilities.

Exhaustion, vomiting, and seizures can occur during or after each attack. Patients may not look sick between coughing spasms.

- **Convalescent stage** (meaning recovery). During this stage the person is no longer contagious. The coughing attacks gradually become less frequent, and the person slowly recovers over a period of several weeks.

Causative Agent

Whooping cough is caused by *Bordetella pertussis,* a tiny, encapsulated, strictly aerobic, Gram-negative rod. The cells are sensitive to drying and sunlight, and they die quickly outside the host.

Pathogenesis

When *B. pertussis* is inhaled, it attaches specifically to ciliated cells of the respiratory epithelium, aided by several protein factors (filamentous hemagglutinin, pertactin, and others). The organism colonizes the upper throat, trachea, bronchi, and bronchioles, where it releases three toxins that play critical roles in the disease process:

- **Pertussis toxin (PT).** This is an A-B exotoxin (**figure 21.15**). The B subunits attach to receptors on the host cell surface, allowing the A subunit to move through the cytoplasmic membrane. As it does so, it becomes a functional enzyme that inactivates a regulatory protein that controls the production of cyclic adenosine monophosphate (cAMP), leading to increased production of this molecule. High levels of cAMP interfere with cell signaling pathways, resulting in a significant increase in mucus output, decreased killing ability of phagocytes, massive release of lymphocytes into the bloodstream, ineffectiveness of natural killer cells, and low blood sugar.

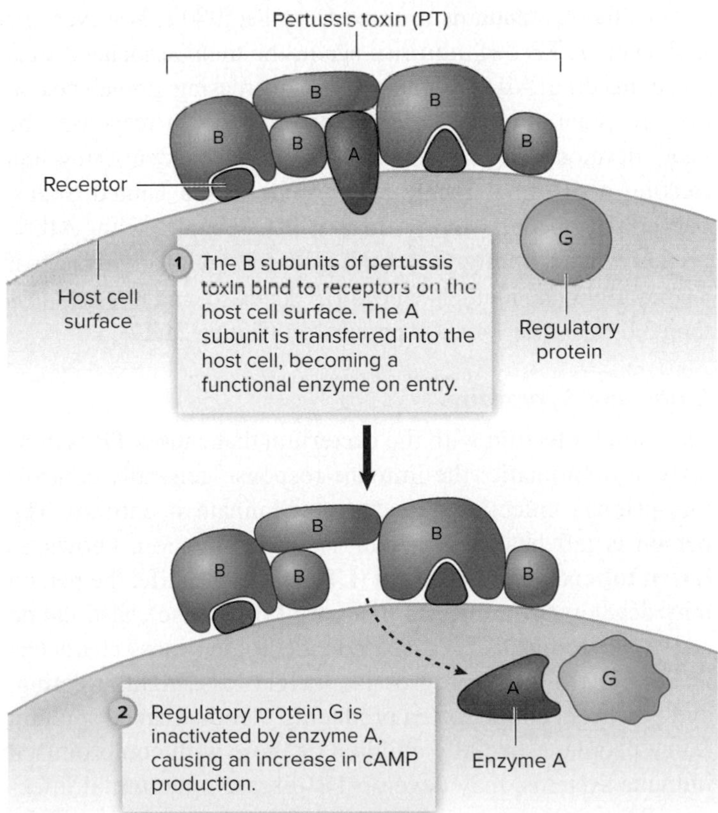

Pertussis toxin (PT)
B
B
B
B
A
B
Receptor
G
1 The B subunits of pertussis toxin bind to receptors on the host cell surface. The A subunit is transferred into the host cell, becoming a functional enzyme on entry.
Host cell surface
Regulatory protein
B
B
B
B
B
A
G
2 Regulatory protein G is inactivated by enzyme A, causing an increase in cAMP production.
Enzyme A

FIGURE 21.15 Mode of Action of Pertussis Toxin (PT)

? What is the result of increased cAMP production?

- **Adenylate cyclase toxin (ACT).** This is both an enzyme and a membrane-damaging toxin. It interferes with phagocytosis by causing lysis of accumulating leukocytes. Inside a host cell, it also catalyzes the reaction that converts ATP to cAMP.
- **Tracheal cytotoxin (TCT).** This is a fragment of peptidoglycan released during bacterial growth that causes host cells to release a fever-inducing cytokine (interleukin-1; IL-1). It is also toxic to ciliated epithelial cells, causing them to die and slough off, resulting in a rapid decline in ciliary action.

Increased mucus production along with decreased ciliary action results in the severe cough that characterizes pertussis because only the cough reflex remains for clearing secretions from the lungs. Some bronchioles are completely obstructed by mucus, resulting in small areas of collapsed lung. Coughing spasms or partial mucus blockage of bronchioles let air enter but not escape. Pneumonia due to *B. pertussis* or, more commonly, secondary bacterial infection may cause death.

Epidemiology

Pertussis is highly contagious; it spreads by direct contact with respiratory secretions or by inhaling airborne droplets of these secretions. Patients are most infectious during the catarrhal stage, when large numbers of organisms are present in mucus secretions. Pertussis is classically a disease of infants, and most fatal cases occur in infants under 1 year of age. Older children and adults often have milder symptoms that may be mistaken for a persistent cold, asthma, or bronchitis; these infected people may unknowingly serve as a source of infection for high-risk populations (such as infants who are not yet vaccinated), potentially leading to outbreaks of severe disease. Despite the availability of a successful vaccine, the reported incidence of pertussis has been increasing. Possible factors contributing to this trend include better surveillance and diagnostic techniques, suboptimal vaccination (especially in resource-limited countries), decreasing immunity in vaccinated people, and the development of *B. pertussis* strains for which the current vaccine is not as effective.

Treatment and Prevention

Macrolides such as erythromycin taken during the catarrhal stage usually eliminate *B. pertussis* from the respiratory secretions and limit its spread. Antibiotics are ineffective once the paroxysmal stage is reached. Pertussis in infants can be life-threatening; severe cases often require intensive supportive therapy (**figure 21.16**).

Pertussis is effectively prevented with a vaccine. The original vaccine, composed of whole *B. pertussis* cells, produced a dramatic decrease in pertussis cases but sometimes caused severe side effects. It has been replaced with a newer acellular (subunit) pertussis vaccine (aP) that includes only part of the bacterium instead of whole cells. This acellular version is given in combination with diphtheria and tetanus toxoids—a grouping referred to as the DTaP vaccine—at 2, 4, 6, and 15 to 18 months, and again at 4 years old. Evidence indicates that immunity induced by the acellular vaccine decreases over time. For this reason, adolescents should receive a booster with a lower dose of pertussis antigen, which is given in combination with a tetanus and diphtheria booster (a vaccine referred to as Tdap). That same booster is given to women in their third trimester of pregnancy as a means to provide passive immunity to the newborn. Tdap should also be administered to adults every 10 years to ensure continued protection against pertussis as well as tetanus and diphtheria. The main features of pertussis are shown in **table 21.8.**

TABLE 21.8	Pertussis ("Whooping Cough")
Signs and Symptoms	Characterized by three stages: catarrhal stage includes a runny nose, cough, and fever; paroxysmal stage consists of spasms of violent coughing, sometimes leading to vomiting and convulsions; convalescent stage is indicated by less frequent coughing as the person recovers.
Incubation Period	1 to 2 weeks
Causative Agent	*Bordetella pertussis*, a tiny Gram-negative rod
Pathogenesis	Colonization of ciliated respiratory tract surfaces; destructive toxins produced; increased mucus along with decreased ciliary action results in the severe coughing spasms; air can enter bronchioles but not escape.
Epidemiology	Direct contact with respiratory secretions or inhalation of infected droplets; older children and adults have mild symptoms.
Treatment and Prevention	Treatment: certain antibiotics given before coughing spasm stage begins. Prevention: acellular vaccine (DTaP) for immunization of infants and children; Tdap booster for adolescents, pregnant women, and for adults every 10 years.

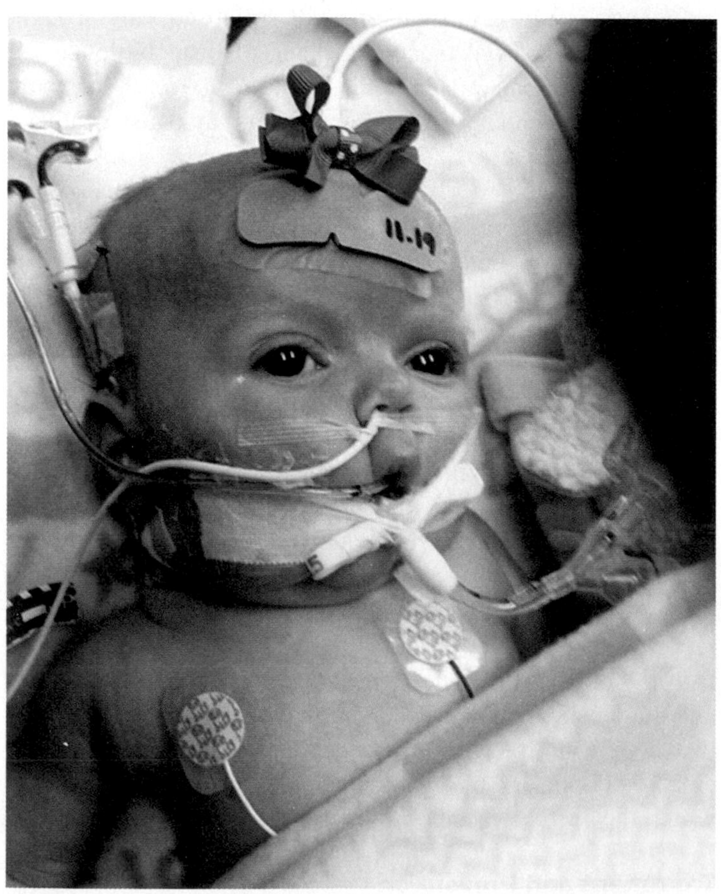

FIGURE 21.16 Infant with Severe Pertussis This baby's blood is being oxygenated outside the body while the lungs are impaired. CDC

? What is the effect of *B. pertussis* on ciliated respiratory epithelial cells?

Tuberculosis ("TB")

Tuberculosis (TB) was once a very common disease, but the number of cases gradually declined in resource-rich countries as living standards improved. In the 1980's, however, the incidence of TB began to rise again, the trend associated with the expanding AIDS epidemic and increasing prevalence of drug-resistant strains of the causative agent. In response, the CDC developed a plan to increase efforts in identifying and treating cases among high-risk groups in the United States, particularly people living in poverty, people with AIDS, prisoners, and immigrants from countries with high rates of TB; by 1993, the incidence began to decrease, and it is now less than 3.0 cases per 100,000 population (**figure 21.17**).

Signs and Symptoms

The initial infection with the bacterium that causes TB is typically asymptomatic: the immune response generally controls this primary infection but cannot eliminate it entirely. The person is left healthy but with a latent infection, known as **latent tuberculosis infection (LTBI).** Later in life, the person may develop **tuberculosis disease (TB disease),** also called active tuberculosis. TB disease is a chronic illness characterized by slight fever, progressive weight loss, night sweating, and persistent cough, often producing blood-streaked sputum. Some people, especially children or those with compromised immune systems, may develop TB disease upon initial infection. The causative agent mainly infects the lungs, but if the bacterial cells enter the bloodstream, they may spread to other parts of the body including kidney, bone, joints, and the

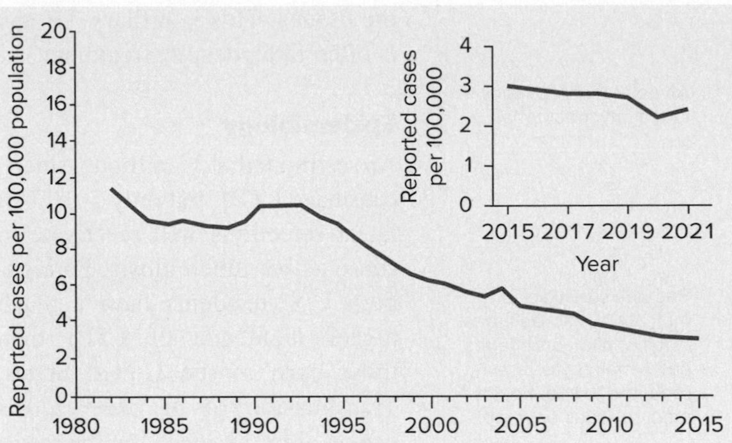

FIGURE 21.17 Incidence of TB Disease, United States, 1982–2021

? What caused the rise in TB disease incidence between 1988 and 1993?

central nervous system. This disseminated disease is called miliary TB because the lesions caused by the bacteria resemble millet seeds (grain) in the tissues.

Causative Agent

Tuberculosis is caused by *Mycobacterium tuberculosis*, commonly called the tubercle bacillus. The organism is a slender, acid-fast, rod-shaped bacterium (**figure 21.18**). It is a strict aerobe that grows very slowly, with a generation time of over 16 hours, making rapid TB diagnosis difficult. The organism has an unusual cell wall that contains a large number of complex glycolipids called mycolic acids. These make the cells unusually resistant to drying, disinfectants, and strong acids and alkali, although they are easily killed by pasteurization. The mycolic acids are also responsible for the bacterium's acid-fast staining. A group of related mycobacteria, including *M. bovis*, *M. africanum* and others, can also cause

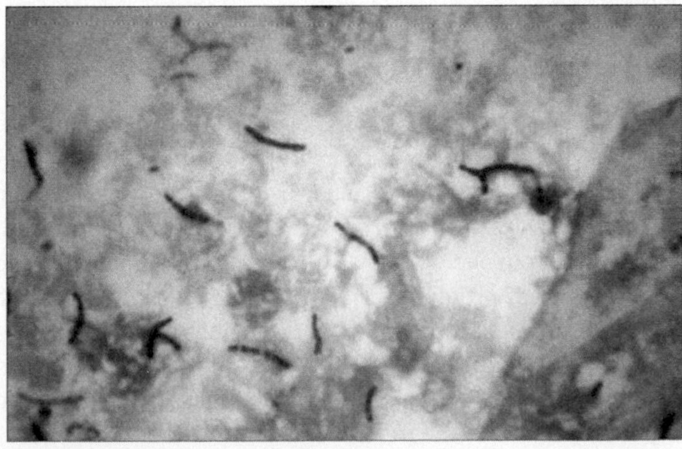

FIGURE 21.18 ***Mycobacterium tuberculosis*** The cells appear reddish after acid-fast staining. Dr. George P. Kubica/CDC

? Why is *M. tuberculosis* so resistant to drying?

tuberculosis in humans or other animals. These organisms, along with *M. tuberculosis*, are part of the *Mycobacterium tuberculosis* complex (MTBC).

Pathogenesis

When airborne *M. tuberculosis* cells are inhaled, they may enter the lungs, and a complex set of events then occurs, as described below (**figure 21.19**):

1. Alveolar macrophages engulf the bacteria but cannot destroy them. Instead, the ingested bacteria manipulate the phagocytic cell, creating an environment in which they can multiply.
2. The bacteria survive and multiply in the macrophages. They then recruit more macrophages to the site, thereby increasing the number of available host cells in which to multiply. Lymphocytes eventually collect around the infected macrophages, walling off the area from the surrounding tissue. This localized collection of immune cells (a granuloma) is the body's characteristic response to microorganisms and other foreign substances that resist destruction and removal; the granulomas of tuberculosis are called **tubercles.** Effector helper T cells release cytokines to activate the macrophages within the tubercle, but the mycobacteria will have already induced some of their host cells to accumulate large numbers of oil droplets, becoming foamy macrophages. The lipids in foamy macrophages are thought to help the intracellular bacteria survive in these cells.
3. A fibrous layer forms around the group of walled-off macrophages. The lesion then calcifies and can be seen on an X ray as a Ghon focus. If the adjacent lymph nodes are involved, the focus is called a Ghon complex. Some of the mycobacteria in the calcifying tubercles survive but cannot multiply because of conditions such as low pH and low available O_2. The bacteria remain in this state for many years, causing a latent TB infection (LTBI). People with LTBI are asymptomatic and non-infectious. In many cases, the infection resolves.
4. TB disease results if the immune response cannot contain the mycobacteria. This can occur during primary infection, but it can also happen in a person with LTBI—the infection reactivates (reactivation TB) if the person's immunity becomes impaired by stress, advanced age, or disease such as AIDS. Within the tubercle, macrophages containing mycobacteria die, releasing bacteria, enzymes, and cytokines. An area of necrosis (tissue death) is formed in the center of the tubercle; this has the texture of soft cheese and is referred to as caseous necrosis. It is thought that foamy macrophages play an important role in necrosis and that the caseum contains lipids from these cells.
5. The tubercle then ruptures, releasing the bacteria and dead material into the airways.

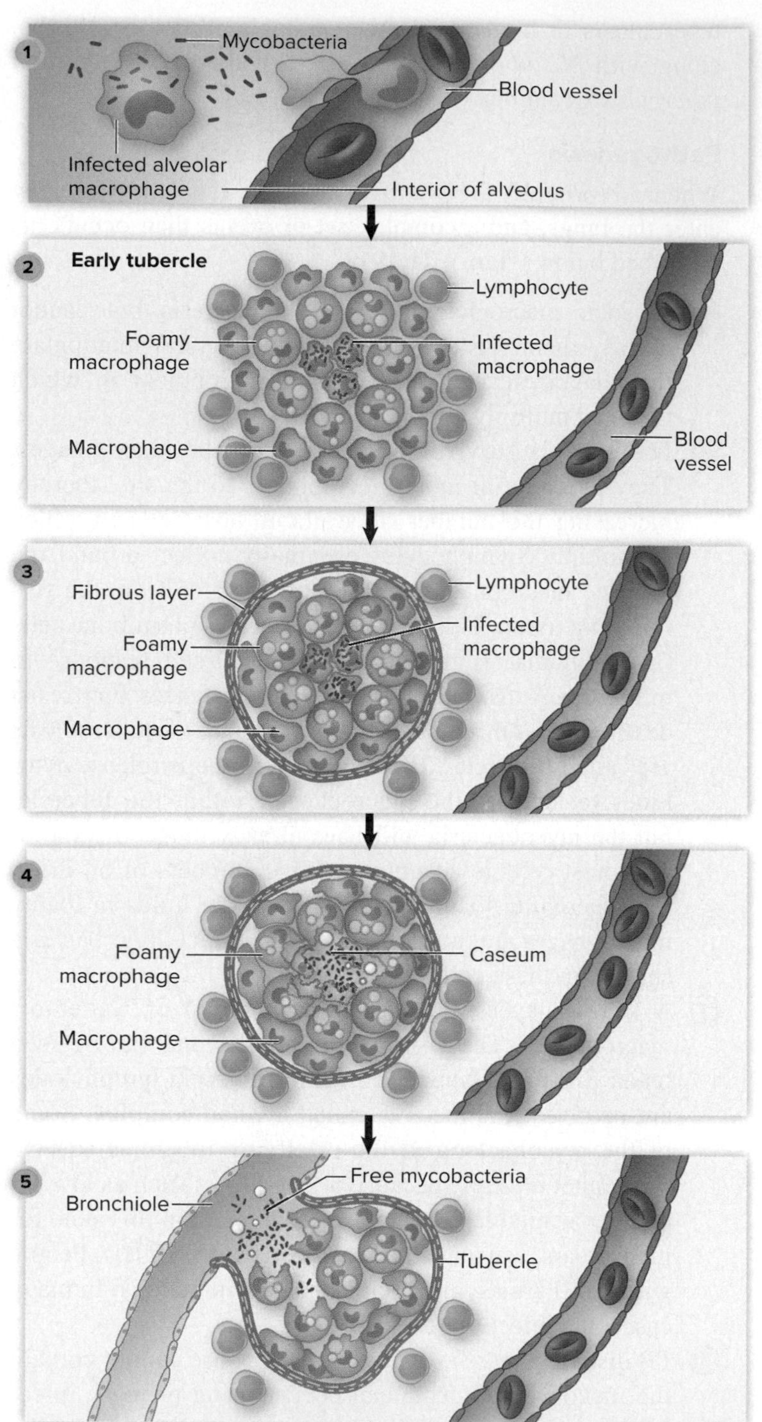

FIGURE 21.19 Pathogenesis of Tuberculosis

? What is a tubercle?

A rupturing tubercle causes a large lung defect called a tuberculous cavity that spreads the bacteria to other parts of the lung (**figure 21.20**). The cavities characteristically persist, slowly enlarging for months or years and shedding bacterial cells into the bronchi. The organisms can then be transmitted to other people by coughing and spitting. If the tubercle bacilli enter the bloodstream and are spread throughout the body, they cause lesions that look like small white granules in the tissues. This is miliary TB and is often fatal, despite treatment.

Epidemiology

An estimated 13 million Americans have LTBI, but only 5–10% of latent infections will reactivate to cause active tuberculosis. Foreign-born U.S. residents have a much higher incidence of LTBI than those born in the United States. Transmission of *M. tuberculosis* occurs almost entirely by the respiratory route; 10 or fewer inhaled cells are enough to cause infection if they reach the alveoli. Factors important in transmission include the frequency of coughing, the adequacy of ventilation (transmission is unlikely to occur outdoors), and the degree of crowding. Immunodeficiency increases reactivation of *M. tuberculosis* in those with LTBI, a significant problem in AIDS patients. TB disease in a person with HIV is an AIDS-defining condition.

The WHO began accelerating the fight against tuberculosis in 2015 by adopting a program called the *End TB Strategy*—a plan to end the global TB epidemic by 2035. Important milestones have been reached, but due to disruptions resulting from the COVID-19 pandemic, the number of deaths due to TB in 2021 increased for the first time since 2005.

The tuberculin skin test (TST), also known as the Mantoux (pronounced man-too) test, is an extremely important tool for studying the epidemiology of the disease and for identifying those who are infected with *M. tuberculosis.* The test is carried out by injecting into the skin a small amount of a sterile fluid called purified protein derivative (PPD), obtained from *M. tuberculosis* cultures. People infected with the bacterium develop redness and a firm swelling (induration) at the injection site that reaches a peak intensity after 48 to 72 hours (**figure 21.21**). This reaction is due to the accumulation of macrophages and T lymphocytes at the injection site, caused by a delayed-type hypersensitivity reaction to the injected antigens. The test is considered positive

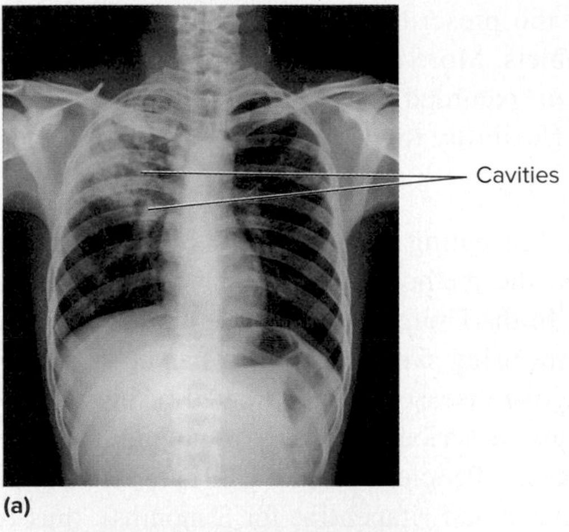

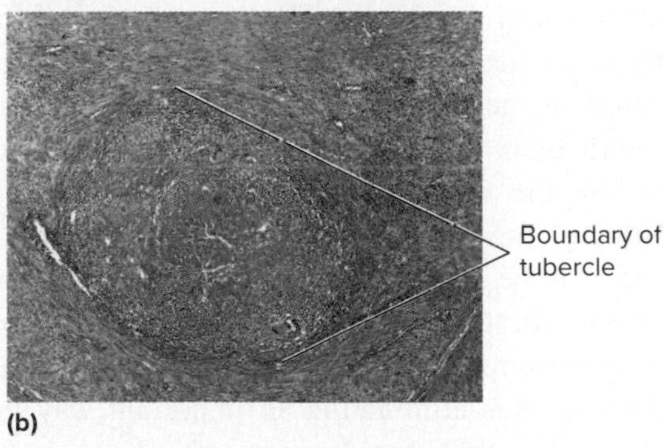

FIGURE 21.20 Lung Damage Caused by Tuberculosis (a) Chest X ray of a person with TB disease. **(b)** Stained lung tissue showing tubercle. In the center of the tubercle, most of the nuclei have disappeared because the cells are dead and the tissue has begun to liquefy. a: kaling2100/Shutterstock; b: Jose Luis Calvo/Shutterstock

? **What is a tubercle?**

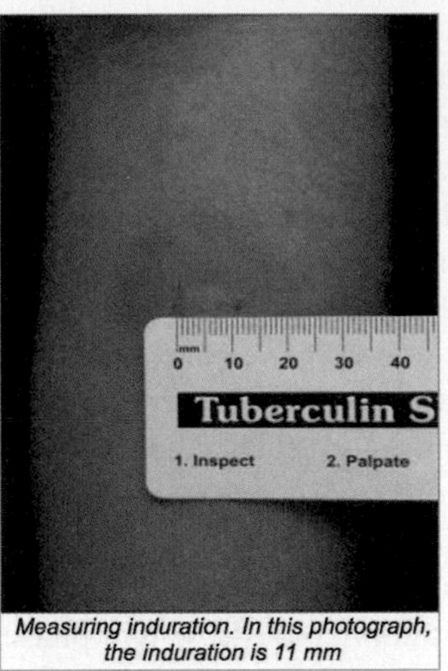

FIGURE 21.21 Tuberculin Skin Test This positive test is caused by a delayed-type hypersensitivity reaction to *Mycobacterium tuberculosis* antigens injected into the skin. CDC

? **Explain why a tuberculin skin test on an AIDS patient with tuberculosis may show a negative result.**

if the induration exceeds a certain diameter within 3 days. A positive reaction does not necessarily mean that the person has TB disease, only that the person may have been infected by *M. tuberculosis* at some time in the past; that infection could have resulted in either LTBI or TB disease.

Besides the tuberculin skin test, blood tests are available for diagnosing *M. tuberculosis* infection. Interferon-gamma release assays (IGRAs) detect interferon-gamma produced in response to *M. tuberculosis.* IGRAs generally give results similar to TST but more quickly and with greater specificity. This is the preferred type of test for people who have received a TB vaccine called the BCG vaccine (described shortly), or who cannot return for a TST follow-up.

A newer assay called the Xpert MTB/RIF is used to detect the DNA of *M. tuberculosis* and closely related species, as well as genes for resistance to the antibiotic rifampicin, a first-line medication used to treat TB disease. Nucleic acid amplification methods can be used on clinical specimens and are fast, sensitive, and accurate. Continued research is already producing additional rapid diagnostic assays.

MicroByte

One-fourth of the global population is believed to be infected with *Mycobacterium tuberculosis*.

Treatment and Prevention

Tuberculosis treatment can be complicated and requires close patient monitoring. Because of this, many efforts focus on preventing the disease.

Treatment Medications used to treat tuberculosis fall into two general groups:

- **First-line drugs.** These are the preferred options because they are most effective and least toxic. They have bactericidal effects against actively growing organisms, and some kill metabolically inactive intracellular ones as well. Standard first-line drugs include rifampin, isoniazid, pyrazinamide, and ethambutol. A somewhat newer option is rifapentine, a derivative of rifampin that has a longer half-life.
- **Second-line drugs.** This group of 10 or so drugs are typically less effective, more toxic, or more expensive than the first-line medications. They are important treatment options, however, particularly for infections caused by strains resistant to one or more of the first-line agents or in cases of drug intolerance.

Treatment of tuberculosis relies on combination therapy—an approach used to prevent the acquisition of microbial resistance through spontaneous mutation. As described in chapter 20, a bacterial cell might spontaneously develop resistance to a single medication but is unlikely to develop resistance to multiple drugs simultaneously. In the case of tuberculosis, combination therapy must be continued for months because *M. tuberculosis* grows very slowly and can evade destruction by body defenses. Unfortunately, if a patient fails to comply with the complex treatment regimen—for example, if the patient feels better and then becomes careless about continuing to take the prescribed medications—the bacterium may develop resistance.

To combat the emergence and spread of drug resistance, tuberculosis treatment strategies used today are tailored to individual patients. As such, the recommended treatment options vary according to patient characteristics (such as age, medical history, and stability of living situation) as well as the drug susceptibility of the infecting strain. In general, however, tuberculosis treatment consists of two phases, each involving different drug combinations and lasting at least 2 months:

- **Initial phase.** This is an intensive 8-week phase with the aim of rapidly reducing the number of infecting *M. tuberculosis* cells, thereby (1) improving the patient's health and overall prognosis, (2) making the patient less infectious, and (3) decreasing the likelihood of bacterial mutation to resistance. A combination of four medications—traditionally the standard first-line drugs—is used during this phase. A newer regimen option that uses a novel combination of four drugs has the advantage of allowing a shorter treatment period for the next phase (the continuation phase), but its use is currently limited to certain patient groups because of the potential for side effects.
- **Continuation phase.** As the name implies, this continues treatment following the initial phase; it lasts long enough to ensure that all *M. tuberculosis* cells are destroyed. The phase requires a combination of at least two drugs, usually taken for at least 18 weeks; an exception is when the newer initial phase regimen is used, in which case the continuation phase is only 9 weeks.

The problem of increasing drug resistance in *M. tuberculosis* became significant in the 1990s with the spread of strains resistant to treatment by two of the most effective first-line drugs, rifampin and isoniazid; tuberculosis caused by these strains is referred to as **multi-drug-resistant TB (MDR-TB)**. By the end of the decade, **extensively drug-resistant TB (XDR-TB)** emerged; this is defined as tuberculosis caused by strains resistant to the two most effective first-line drugs as well as many of the second-line options. Fortunately, a newer combination of second-line drugs (pretomanid, bedaquiline and linezolid) is effective in treating XDR-TB, and additional drugs are in clinical trials. Drug-resistant strains, however, threaten tuberculosis control efforts around the world.

A strategy called **DOT (directly observed therapy)** that ensures patient compliance has been very successful for TB control globally. With this strategy, healthcare workers supplying the prescribed medications watch patients swallow the tablets. More recently, electronic DOT (eDOT) uses real-time or recorded video to monitor the patients, thus providing flexibility to the process as well as making it more cost effective.

Prevention Preventing and controlling TB are global challenges, and the methods used depend partly on a country's resources. In the United States, prevention involves identifying TB cases using skin or blood tests and chest X rays. People with active disease are then treated (using the two-phase approach just described), thereby removing a source of the causative agent. People who have LTBI are also treated (often using isoniazid and rifapentine for 3 months), thus reducing the patient's risk of developing TB disease later in life.

Many countries lack the resources for the prevention method used in the United States, and instead rely on vaccination with BCG (Bacille Calmette-Guérin), an attenuated derivative of the cattle-infecting species, *Mycobacterium bovis*. Although the BCG vaccine prevents childhood TB disease, it seems to be ineffective in preventing LTBI, which can later reactivate. In the United States, use of the vaccine is discouraged because recipients usually develop a positive tuberculin skin test, thus eliminating an important way to detect TB. Additionally, BCG is also not safe to use in severely immunocompromised patients. Several new genetically engineered vaccines are being developed, many of which are currently being tested in TB-endemic regions. The main features of tuberculosis are shown in **table 21.9.**

Legionnaires' Disease (*Legionella* Pneumonia)

Legionnaires' disease was not discovered until 1976, when a number of people attending an American Legion convention in Philadelphia developed a mysterious pneumonia that was fatal in many cases. Months of scientific investigation eventually paid off when the cause was discovered to be a previously unknown bacterium commonly present in the natural environment.

Signs and Symptoms

Legionnaires' disease typically begins after an incubation of 2 to 10 days with headache, muscle aches, high fever, confusion, and shaking chills. A dry cough develops that later produces small amounts of sputum, sometimes containing blood. Shortness of breath is common. About 25% of the cases also have some digestive tract symptoms such as diarrhea, abdominal pain, and vomiting. Recovery is slow, and weakness and fatigue last for weeks.

Causative Agent

Legionnaires' disease is caused by *Legionella* species, most commonly *L. pneumophila*. These bacteria are also responsible for a variety of other diseases collectively called legionellosis—

TABLE 21.9 Tuberculosis ("TB")

1. Airborne *Mycobacterium tuberculosis* cells are inhaled and lodge in the lungs.
2. The bacteria are phagocytized by lung macrophages and multiply within them.
3. Lymphocytes eventually collect around the infected macrophages, walling off the area from the surrounding tissue and forming a tubercle.
4. Bacterial growth within the tubercle is controlled.
5. With uncontrolled or reactivated infection, infected macrophages die, releasing mycobacteria and enzymes that can cause caseous necrosis; cavities may form.
6. The bacterium may also spread in the bloodstream to infect and multiply within macrophages in various parts of the body such as the kidneys, brain, lungs, and lymph nodes, repeating the previous steps but in other parts of the body.
7. *M. tuberculosis* exits the body through the mouth with coughing.

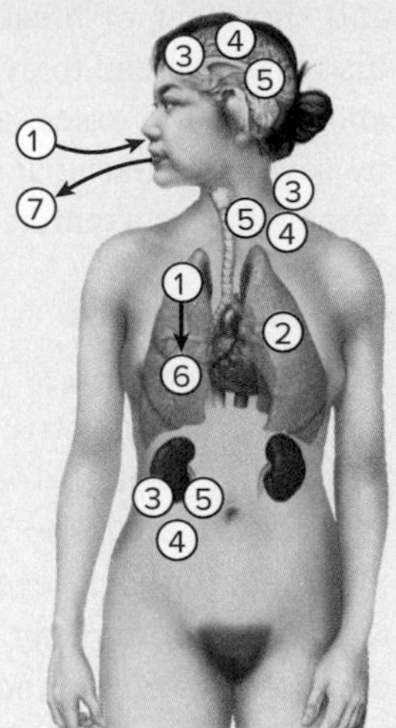

Signs and Symptoms	Chronic fever, weight loss, cough, sputum production
Incubation Period	2 to 10 weeks
Causative Agent	*Mycobacterium tuberculosis*; unusual cell wall with high lipid content
Pathogenesis	Bacterial cells survive and multiply within alveolar macrophages; organisms may be carried to other body tissues; granulomas form.
Epidemiology	Inhalation of airborne organisms; latent infections can reactivate.
Treatment and Prevention	Treatment: lengthy process requiring four and then at least two anti-TB medications; directly observed therapy (DOT). Prevention: detection tests including tuberculin (Mantoux) skin test and interferon-gamma release assays (IGRAs) allows early treatment to limit spread; treatment of latent infections.

including "Pontiac fever," a milder flu-like illness. *L. pneumophila,* an aerobic, fastidious Gram-negative rod, stains poorly with conventional dyes but can be detected using immunofluorescence (**figure 21.22**). It is a facultative intracellular parasite and survives well in various freshwater protozoa, including their cysts (survival forms). *Legionella* also persists in biofilms—if the biofilm is disturbed, huge numbers of the bacteria are released into the water.

Pathogenesis

Legionella pneumophila is acquired by inhaling aerosolized water contaminated with the organism. Infections in healthy people are often asymptomatic, and the disease is more common in smokers and in people with impaired host defenses.

Once the bacteria are in the lungs, they are taken up by alveolar macrophages. The bacteria survive within the macrophage by using a type IV secretion system to release various effector proteins that prevent phagosome-lysosome fusion. The effector proteins also recruit other cellular components that the bacteria use to alter the phagosome, remodeling it to create a *Legionella*-containing vacuole (LCV) where the bacteria can multiply intracellularly. The LCV expands as the bacteria divide, and eventually the host cell dies, releasing bacterial cells that can then infect nearby tissues or enter the bloodstream to involve other organs. In the lungs, the inflammatory response leads to an accumulation of phagocytic cells and fluids in the alveoli.

Epidemiology

Legionella pneumophila is widespread in warm natural waters containing other microorganisms such as protozoa, in which the bacteria live and multiply. The organism also survives in the water systems of buildings, particularly in hot water systems,

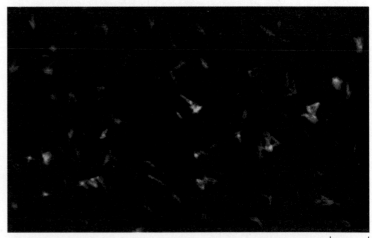

FIGURE 21.22 ***Legionella pneumophila* Detected Using Immunofluorescence** The bacterium does not stain well with common methods used to examine tissue or sputum.
CDC/ Dr. William Cherry

? **How do the organism's staining characteristics and fastidious nature help explain why it was unrecognized as a cause of disease for so long?**

where chlorine levels are generally low; even if higher chlorine levels are used, *L. pneumophila* cells inside protozoa might survive. Cases have originated from contaminated aerosols from large central air-conditioning systems, nebulizers, and sprays to freshen produce, as well as from showers and water faucets. In rare cases, the organism has been acquired via aspiration of contaminated drinking water. Direct person-to-person spread does not seem to occur.

Treatment and Prevention

Legionnaires' disease is typically treated with a macrolide or a fluoroquinolone. *L. pneumophila* produces a β-lactamase, which makes it resistant to many penicillins and some cephalosporins. In addition, these and other β-lactam antibiotics are not effective because they do not accumulate within the alveolar macrophages where *Legionella* bacteria multiply.

Most efforts to control Legionnaires' disease have focused on designing equipment to minimize the risk of infectious aerosols and on disinfecting procedures. The main features of the disease are given in **table 21.10.**

Inhalation Anthrax

Anthrax is primarily a disease of domestic livestock and other herbivores, but it also occurs in humans. It is particularly significant in the United States because of the 2001 domestic bioterrorism attack linked to mail containing purified *Bacillus anthracis* spores. Eleven people developed inhalation anthrax as a result of the attack, and five of those died; in addition, 11 people developed cutaneous anthrax. The causative agent, *B. anthracis,* is now listed as a Category A bioterrorism agent.

Signs and Symptoms

Inhalation anthrax typically begins with flu-like symptoms—discomfort, mild fever, and a nonproductive cough. These usually begin within a week after exposure, but may take up to 2 months to appear. Fever then increases within a few days, and shortness of breath and chest pain develop. Respiratory distress may lead to cyanosis as tissues are deprived of O_2. In about half the cases, anthrax meningitis develops, indicated by severe headache and evidence of blood when the cerebrospinal fluid is examined.

TABLE 21.10	Legionnaires' Disease (*Legionella* Pneumonia)
Signs and Symptoms	Muscle aches, headache, fever, cough, shortness of breath, chest and abdominal pain, diarrhea, vomiting
Incubation Period	2 to 10 days
Causative Agent	*Legionella* species, primarily *L. pneumophila*, a fastidious Gram-negative bacterium that stains poorly
Pathogenesis	Organism multiplies within phagocytes; released with death of the cell; necrosis of cells lining the alveoli; inflammation and formation of multiple small abscesses
Epidemiology	Originates mainly from warm water contaminated with protozoa, such as that found in air-conditioning systems
Treatment and Prevention	Treatment: antibiotics. Prevention: avoiding contaminated water aerosols; regular cleaning and disinfection of humidifying devices.

Causative Agent

Bacillus anthracis is an endospore-forming, non-motile, Gram-positive rod-shaped bacterium that forms non-hemolytic, rough-looking colonies on blood agar (**figure 21.23**). The endospores can remain in the environment indefinitely. Vegetative cells produce a capsule that protects them from phagocytosis, but unlike most capsules, it is composed of an amino acid polymer (poly-D-glutamic acid) rather than a polysaccharide.

Pathogenesis

Once inhaled deep into the lungs, *B. anthracis* endospores are taken up by alveolar macrophages and carried to lymph nodes in the chest, where they germinate. In some cases, the endospores may remain in the lungs for up to 60 days before germinating. In general, clumps of endospores do not cause disease when inhaled, because particles larger than 5 μm are trapped and removed by the mucociliary escalator or other defenses of the upper respiratory tract before reaching the lungs.

Vegetative *B. anthracis* cells produce anthrax toxin, a system of three interacting proteins: protective antigen (PA) and the enzymes edema factor (EF) and lethal factor (LF). PA serves as the delivery mechanism; it first binds to the host cell and then to either EF or LF. As a result, two types of toxic molecules form at the cell surface: edema toxin (PA + EF) and lethal toxin (PA + LF). These are taken up by endocy-

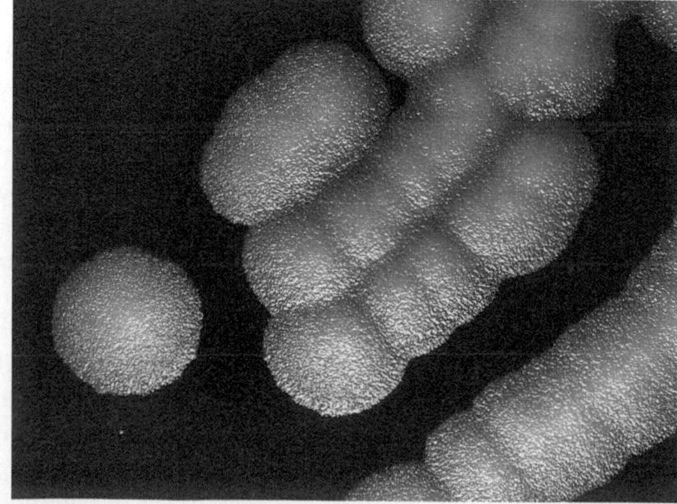

FIGURE 21.23 ***Bacillus anthracis*** Colonies grown on laboratory media have a rough surface. J. Todd Parker; PhD and Luis Lowe; MS; MPH/CDC

? **What color would the cells in these colonies be if Gram stained?**

tosis and then released from the endosome through the pore-forming action of PA. Once inside the cytosol, EF and LF disrupt critical cellular functions. Anthrax toxin appears to particularly target macrophages and dendritic cells, thus interfering with the host immune response and allowing the infecting vegetative bacteria to continue multiplying and to produce more toxin. The accumulating toxin eventually causes systemic damage.

In the later stages of infection, the toxin-producing encapsulated *B. anthracis* cells damage endothelial cells; this leads to vascular leakage and pulmonary edema. They spread into the bloodstream, overwhelming the body's defense mechanisms and damaging tissues. Without treatment, the inhalation anthrax case-fatality rate is near 95%; early treatment reduces it to about 45%.

Epidemiology

Anthrax is a zoonotic disease that can be transmitted to humans who work with animals or animal products, particularly in countries where *B. anthracis* is prevalent. People who work in slaughterhouses or leather tanneries, for example, are most likely to be affected. Although the disease occurs naturally in the United States, it is extremely rare, so bioterrorism is suspected whenever a case occurs. Fortunately, anthrax is not transmitted from person to person.

Treatment and Prevention

Inhalation anthrax should be treated immediately because the toxins released by vegetative bacteria can cause death within a few days. Bactericidal medications such as ciprofloxacin and a protein synthesis inhibitor like doxycycline are given along with a monoclonal antibody that binds to protective antigen (PA). The antibody-bound PA cannot attach to host cells, so LF and EF can no longer enter cells to cause further damage. Treatment may also include mechanical ventilation and drainage of fluid from the lungs.

Individuals who might have been exposed to *B. anthracis* endospores, like the postal workers in the 2001 attack, undergo a **post-exposure prophylaxis** regimen that includes antimicrobial treatment for 60 days in addition to the anthrax vaccine (discussed next). The duration of preventive antimicrobial treatment is important because inhaled endospores may remain in the lungs for some time before germinating. If administered before symptoms arise, antibiotic treatment generally prevents the development of the disease.

An important control mechanism is to reduce infection in livestock, particularly using an attenuated vaccine; note that the vaccine used for livestock is not appropriate for humans because of the risks. An acellular (subunit) anthrax vaccine is available for people at increased risk of contact with anthrax endospores, such as members of the U.S. military, people who handle certain animals or animal products, and some laboratory workers; this subunit vaccine is also used for post-exposure prophylaxis. The main features of inhalation anthrax are shown in **table 21.11.**

TABLE 21.11 Inhalation Anthrax

Signs and Symptoms	Fever, nonproductive cough; progressing to shortness of breath, chest pain; meningitis in about 50% of cases.
Incubation Period	3–60 days
Causative Agent	*Bacillus anthracis*, an endospore-forming Gram-positive rod.
Pathogenesis	Endospores engulfed by macrophages in lung germinate and produce toxins that weaken innate immunity, cause pulmonary edema, and damage respiratory cells.
Epidemiology	Rare zoonotic disease that may be transmitted by animals or animal products; no person to person transmission; possible agent of bioterrorism.
Treatment and Prevention	Treatment: antimicrobial medications; monoclonal antibody that binds to PA. Prevention: vaccinate livestock; acellular vaccine for people at increased risk of exposure; post-exposure prophylaxis.

MicroAssessment 21.4

Pneumococci are a common cause of community-acquired pneumonia. Pneumonia due to *Klebsiella* species is mainly hospital-acquired and is the leading cause of death from healthcare-associated infections. Symptoms of mycoplasmal pneumonia are usually relatively mild. Whooping cough (pertussis) is mainly a threat to infants but is commonly spread by adults; childhood immunization against the disease protects them, but immunity often does not persist to adulthood. Tuberculosis is a chronic disease; most infections become latent, posing the risk of reactivation throughout life. *Legionella pneumophila,* the bacterium that causes Legionnaires' disease, originates from water containing other microorganisms, where it can grow within protozoa. Inhalation anthrax is rare in the United States, but is significant because of bioterrorism concerns.

10. Bacteria that cause pneumonia often produce a capsule. How does this structure help the organisms?
11. Outline the pathogenesis of tuberculosis.
12. With respect to pneumococcal vaccines, why would PPSV23 be given to someone who has already had PCV-15?

21.5 ■ Viral Infections of the Lower Respiratory System

Learning Outcomes

9. Describe antigenic drift and antigenic shift, and discuss how they affect the epidemiology of influenza.
10. Compare the distinctive characteristics of COVID-19, SARS, MERS, respiratory syncytial virus infection, and hantavirus pulmonary syndrome.

DNA viruses such as the adenoviruses sometimes cause serious pneumonias, but RNA viruses are of greater overall importance because of the large number of people they infect and their potential for serious outcomes. The following section covers some diseases caused by RNA viruses.

Influenza ("Flu")

Influenza is a good example of the constantly changing interaction between people and infectious agents because antigenic changes in the influenza viruses lead to serious annual outbreaks of the disease. The four influenza virus types (A, B, C, and D) are classified based on their core proteins; only types A and B cause significant human disease. Influenza A virus causes the most serious disease and the most widespread epidemics. Influenza B virus outbreaks occur each year as well, but they are less extensive and the disease is not as severe. Influenza C virus only causes mild disease in humans, while type D has not been documented to cause human disease. This section will focus on disease caused by influenza A and influenza B viruses.

Signs and Symptoms

After a short incubation period averaging 2 days, influenza typically begins with headache and muscle aches, fever, sore throat, and fatigue, peaking in 6 to 12 hours. A dry cough develops and worsens over a few days. These symptoms usually last less than a week, but a lingering cough, fatigue, and generalized weakness can continue for days or weeks. Influenza viruses do not cause "stomach flu," and gastrointestinal symptoms are usually not associated with influenza infections in adults (but are in 10–20% of cases in children).

Causative Agent

Influenza viruses are enveloped single-stranded RNA viruses with segmented genomes; they belong to the family *Orthomyxoviridae*. The influenza A and influenza B virus genomes are both segmented into eight pieces, with each piece containing one or two genes (**figure 21.24**).

Embedded in the influenza A and influenza B viral envelopes are two kinds of glycoprotein spikes—hemagglutinin (HA) and neuraminidase (NA)—which both have roles in viral pathogenesis. The HA spikes recognize and attach to specific receptors on ciliated host epithelial cells, facilitating viral entry; they also provide a useful characteristic for virus identification and study because they cause red blood cells to stick together (hemagglutination; see figure 13.22). NA is an enzyme that plays a critical role in the release of newly formed virions from infected host cells. As new virions are made, they bud out of the host cell but remain attached to surface receptors in the cell membrane. NA destroys these receptors, allowing the virions to leave the infected host cell, and it aids in the spread of the virus to uninfected host cells.

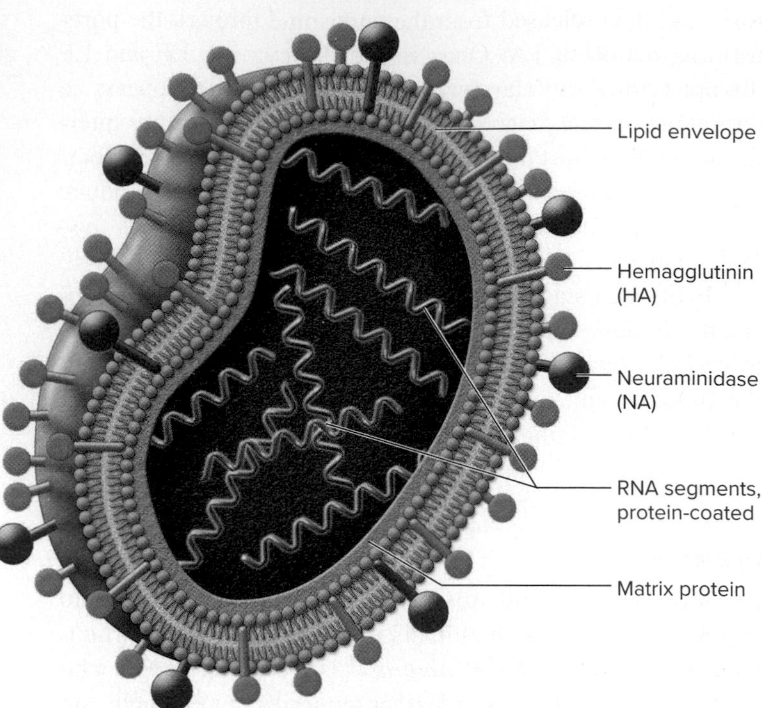

FIGURE 21.24 Structure of Influenza Virus

What are the functions of the HA and NA spikes?

Influenza A viruses are classified into subtypes based on the antigenically distinct HA and NA spikes on the viral surface, and the different subtypes are assigned numbers—H1, H2, N1, N2, and so on. For example, the "avian flu" epidemic of 1997 was caused by influenza A virus H5N1, whereas the "swine flu" epidemic of 2009 was caused by influenza A virus H1N1. Of the 18 HA and 11 NA subtypes, only H1, H2 and H3, and N1 and N2 typically spread from person to person.

The HA and NA spikes of influenza B viruses show less variation than those of influenza A viruses, so influenza B viruses are not classified into subtypes. They are, however, classified into two lineages (meaning a closely related group descended from a common ancestor): B/Yamagata and B/Victoria.

Pathogenesis

People acquire influenza by inhaling aerosolized respiratory secretions from someone who has the disease or from fomites. The virions attach via their HA spikes to specific receptors on ciliated respiratory epithelial cells and enter the cell by endocytosis. New viral parts are quickly synthesized, and regions of the host cell membrane become embedded with virally encoded HA and NA glycoproteins. Mature virions then bud from the host cell, acquiring host cell–derived membrane containing these HA and NA spikes as they do so. The virus spreads rapidly to nearby cells, including mucus-secreting cells and cells of the alveoli. The common flu signs and symptoms are caused by tissue damage as well as the effects of pro-inflammatory cytokines produced by virus-infected cells. Infected epithelial cells die and slough off, thus destroying the mucociliary escalator. The damage to this important first line defense makes the person susceptible to secondary respiratory infections. The immune response quickly controls the influenza virus infection in most cases, although complete recovery of the respiratory epithelium may take months.

Epidemiology

The case-fatality rate for influenza is typically very low, but many people fall ill during an epidemic, so the total number of deaths is high. Although influenza virus infection alone can kill otherwise healthy people, most deaths are due to secondary bacterial infections that lead to pneumonia. People are predisposed to these infections because of the virus-induced damage to the respiratory epithelium. In fact, the bacterium *Haemophilus influenzae* got its name because it was often found in the lungs of people who died after having influenza, leading to the incorrect conclusion that it caused the disease.

Influenza A outbreaks, and sometimes epidemics, occur every year. Pandemics occur periodically over the years, marked by rapid spread of the viruses around the globe and higher-than-normal morbidity. Several factors are involved in the spread of influenza A viruses, but major attention has focused on their antigenic changeability. Two types of variation occur—antigenic drift and antigenic shift:

- **Antigenic drift.** This is caused by minor mutations in the genes that code for the HA and NA antigens and is responsible for the yearly occurrence of influenza outbreaks, called **seasonal influenza.** The mutations happen during normal viral replication and often cause a change in only a single amino acid in the HA or NA spikes. They occur often, however, and are enough to make immunity developed to virus strains of previous years less effective. Strains that arise because of antigenic drift are named to indicate the year and location of isolation, along with the strain number and subtype. For example, A/Kansas/14/2017 (H3N2) is an influenza A subtype H3N2 virus isolated in Kansas in 2017 and designated strain 14, whereas A/Hong Kong/45/2019 (H3N2) was isolated in Hong Kong in 2019 and designated strain 45. The Hong Kong/2019 strain has minor mutations that distinguish its HA antigens from the Kansas/2017 strain, so antibodies produced by people who recovered from Kansas/2017 will not be fully protected against the Hong Kong/2019 strain. Thus, the newer Hong Kong strain might be able to cause an outbreak in populations previously exposed to the Kansas strain.
- **Antigenic shift.** This is an uncommon but more dramatic change that occurs as a result of influenza A viral genome reassortment and is the cause of **pandemic influenza.** Recall that the viral genome is segmented, meaning that viral proteins are encoded on eight different RNA segments rather than being encoded on one long molecule. When two different influenza A viruses infect a cell at the same time, the progeny released from that cell can have RNA segments from either of the viruses. From an infectious disease standpoint, this is particularly problematic when a genome segment from a virus strain that normally does not infect humans is acquired by a virus strain that does. For example, if a pig is simultaneously infected with two influenza A virus strains—one that infects only pigs and birds and another that infects pigs and humans—the viral particles that emerge will still have eight genome segments, but each segment may originate from either of the initial infecting strains. This can result in a strain that infects humans and has novel HA and/or NA antigens for which human populations have no immunity (**figure 21.25**). In 2009, an influenza A(H1N1) virus emerged that was found to have genes from avian, swine, and human flu viruses; to distinguish it from previous influenza A(H1N1) strains, it was assigned the name A(H1N1)pdm09. The pandemic that resulted from A(H1N1)pdm09 ended in 2010, but not before an estimated quarter of a million people died, mostly in Africa and Southeast Asia. Descendents of that strain replaced the previously predominant seasonal H1N1 strain and now circulate as seasonal influenza.

Ecological studies show that all known influenza A virus subtypes exist in ducks and other wild aquatic birds, generally causing asymptomatic or mild intestinal infections. The viruses circulate in these waterfowl but can be transmitted to various wild and domestic birds, sometimes causing the disease referred to as avian influenza ("bird flu"). Depending on the severity of disease they cause in domestic poultry, the viruses are categorized as either low pathogenic avian influenza (LPAI) or highly pathogenic avian influenza (HPAI): LPAI strains cause only asymptomatic infections or mild disease, but HPAI strains cause up to 100% mortality. Avian flu is a concern not only because of its effects on birds, but also because infected poultry can potentially transmit the disease to humans—particularly people who work closely with flocks—or to other animals. In addition, a constant concern is that an HPAI strain will evolve through antigenic drift or antigenic shift to be easily transmissible person-to-person. Outbreaks of HPAI avian influenza that resulted in poultry-to-human transmission include the following:

- **H5N1, Hong Kong, 1997.** This was the first avian flu outbreak in domestic poultry that spilled over into humans, resulting in several cases that were often fatal. Fortunately, the virus did not spread easily from person to person, and implementation of control measures limited the poultry-to-human transmission. The virus continued to cause outbreaks in wild birds and domestic poultry, while also evolving through mutation and genome reassortment; those outbreaks sometimes led to infections in people who had close contact with sick birds.
- **H7N9, China, 2013.** As with the other avian influenza outbreaks, the virus that caused this outbreak mostly affects birds, but people exposed to infected poultry

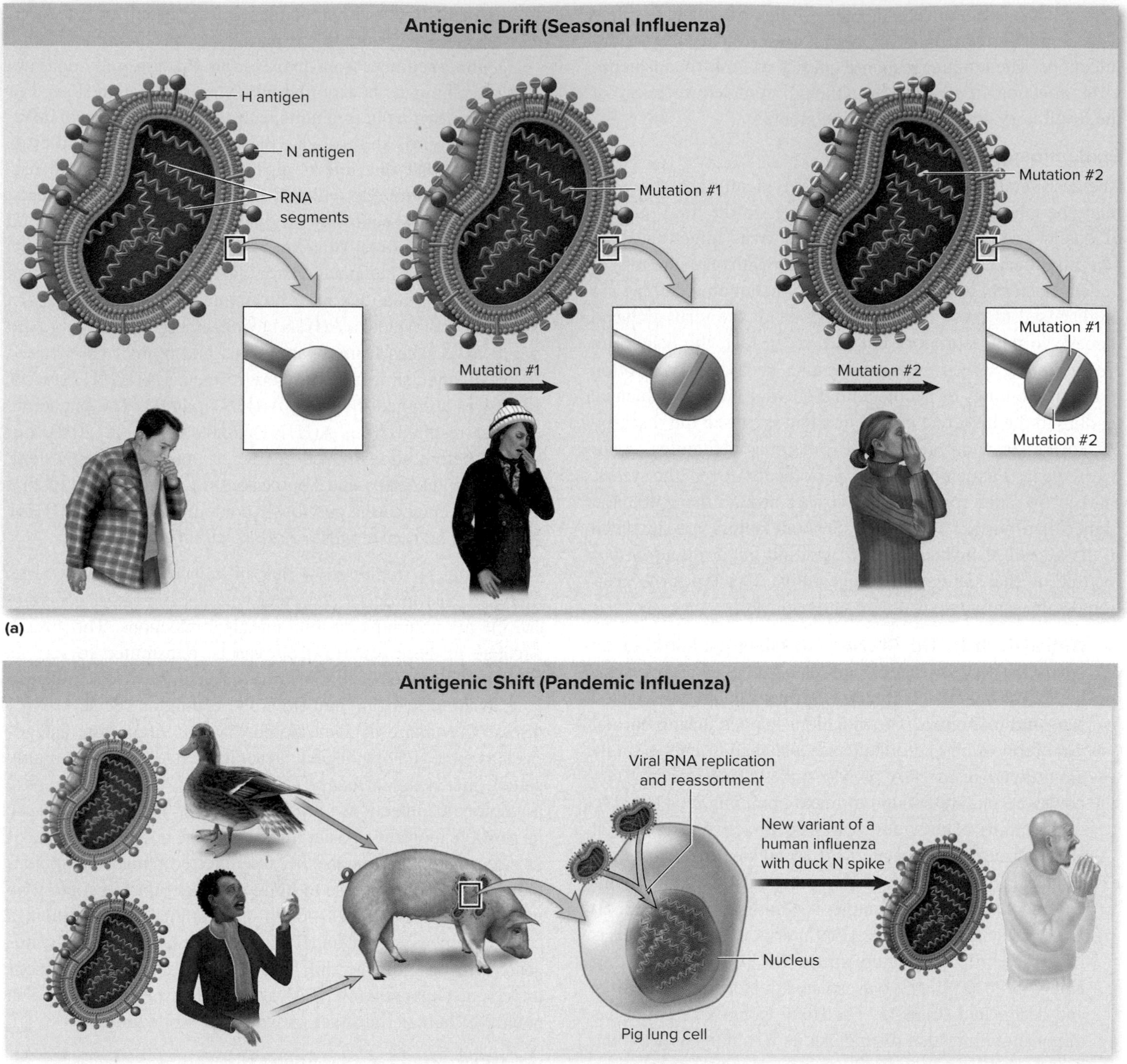

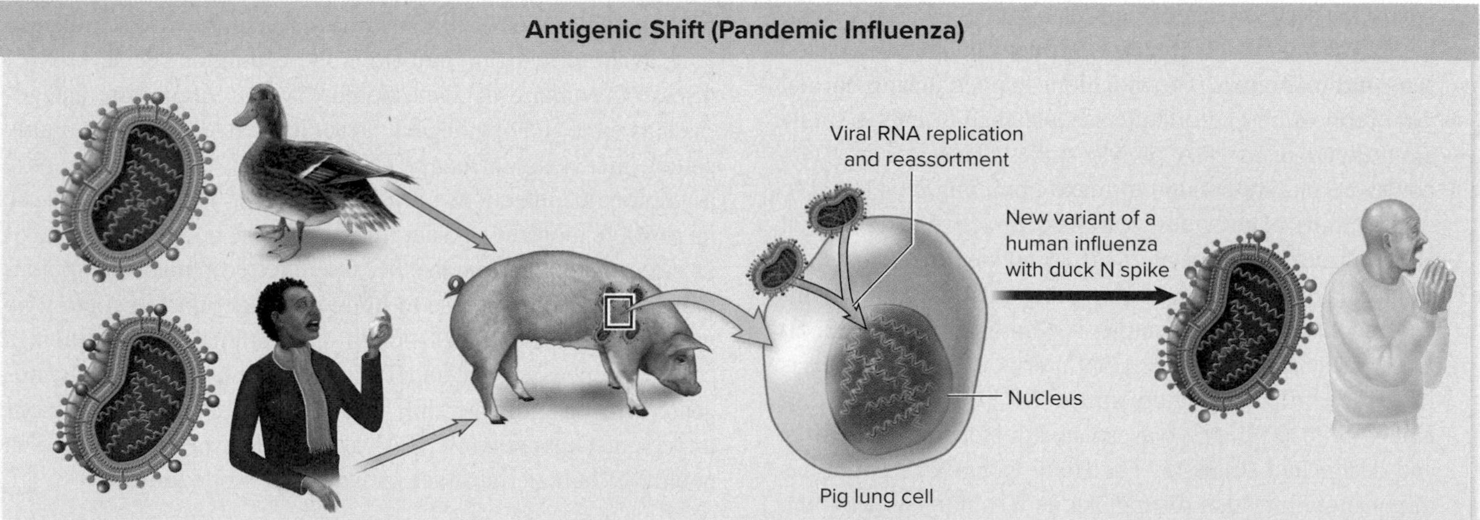

FIGURE 21.25 Influenza Virus: Antigenic Drift and Antigenic Shift With drift, repeated mutations cause a gradual change in the HA and/or NA spikes, so that antibodies against the original virus become progressively less effective. With shift, there is a sudden major change in the spikes because the virus acquires a new genome segment.

? Why can antigenic shifts cause pandemics?

sometimes develop disease that is often fatal. Sustained person-to-person transmission has not been observed, but annual outbreaks in poultry and sporadic human infections continued to occur for several years.

- **H5N1, global, current.** The virus causing this outbreak is a novel H5N1 variant that emerged in 2021 and quickly spread around the world. In the United States, outbreaks in poultry have been reported in at least 47 states, leading to the death of over 50 million domestic birds from disease or culling (selective slaughter, in this case to prevent further spread of the disease). In contrast to previous avian influenza virus variants, this one is also causing significant outbreaks in wild birds, particularly those that gather in large flocks. Although the virus has also caused outbreaks in certain mammals, it does not appear to easily infect people. The CDC surveillance system continues to monitor individuals—particularly those who work closely with poultry—for signs of infection.

MicroByte

An H1N1 influenza A virus caused the 1918 flu pandemic that killed 50–100 million people around the globe.

Treatment and Prevention

Like all viral infections, antibiotics are not effective for treating influenza. However, antiviral medications that specifically target influenza A and B viruses generally shorten the duration and severity of illness if given early. The medications include various neuraminidase inhibitors, which prevent newly synthesized virions from spreading within the body, and baloxavir, a cap-dependent endonuclease inhibitor that prevents the viruses from replicating. Two other medications (amantadine and rimantadine) target influenza A virus but are not currently recommended because of widespread resistance among circulating strains.

Although fever reducers and pain relievers are often given to help with signs and symptoms of influenza and most other viral infections, children and teenagers should not be given aspirin because of the risk of developing **Reye's syndrome.** This disease is a rare, potentially fatal condition that can affect many organs but especially the brain and the liver.

Vaccination is used to prevent influenza, but new vaccines must be developed annually because influenza A virus changes so quickly. The vaccines now used in the United States are quadrivalent, meaning they are designed to protect against four strains—two influenza A viruses (usually an H3N2 strain and an H1N1 strain) and two influenza B viruses (one Yamagata strain and one Victoria strain). Once the vaccine composition is chosen, the manufacturing process to make adequate amounts requires 6 to 9 months. A variety of different vaccine types are now available, including inactivated influenza vaccine (IIV), recombinant influenza vaccine (RIV), and live attenuated influenza vaccine (LAIV). The choice of which vaccine type to use depends on patient characteristics; for example, a high-dose vaccine is available for people over 65 and a recombinant vaccine (not made in eggs) for people allergic to eggs. In addition to vaccination, other prevention measures include handwashing, frequently disinfecting surfaces, and avoiding close contact with people who are ill. The main features of influenza are summarized in **table 21.12.**

COVID-19, SARS, and MERS

Most coronaviruses that cause disease in humans infect the upper respiratory tract, leading to mild symptoms, but three can infect the lower respiratory tract, leading to severe disease. The diseases—COVID-19 (**co**rona**vi**rus **d**isease-20**19**), SARS (**s**evere **a**cute **r**espiratory **s**yndrome), and MERS (**M**iddle **E**ast **r**espiratory **s**yndrome)—can be fatal, particularly in adults who have underlying health problems. The viruses that cause these severe coronavirus diseases are relatively new (likely evolved from strains that normally infect bats), so until they began spreading in human populations, people had no specific immunity against them.

COVID-19 was first reported in China in late 2019 and then spread rapidly around the globe (see Focus Your

TABLE 21.12 Influenza ("Flu")

1. Influenza virus is inhaled and carried to the lungs.
2. Viral hemagglutinin attaches to receptors on ciliated epithelial cells and the virus enters the cell by endocytosis.
3. Host cell synthesis is diverted to producing virions.
4. Newly formed virions bud from infected cells; they are released by viral neuraminidase and infect nearby cells.
5. Infected cells ultimately die and slough off; recovery of the mucociliary escalator may take weeks.
6. Secondary bacterial infection of the lungs, ears, and sinuses is common.
7. The virus exits with coughing.

Signs and Symptoms	Fever, muscle aches, lack of energy, headache, sore throat, nasal congestion, cough
Incubation Period	1 to 2 days
Causative Agent	Influenza A and influenza B viruses, enveloped single-stranded RNA viruses with a segmented genome
Pathogenesis	Infection of respiratory epithelium; cells destroyed and virus released to infect other cells. Secondary bacterial infection results from damaged mucociliary escalator.
Epidemiology	Antigenic drift is responsible for seasonal influenza; antigenic shift is responsible for pandemic influenza.
Treatment and Prevention	Treatment: antiviral medications somewhat effective for treatment when given early in the disease. Prevention: vaccines developed and given annually.

Perspective 21.1). The disease's emergence shared similarities to a brief SARS pandemic that began in China in 2003 and involved at least 27 countries, resulting in over 8,000 cases and almost 800 deaths; fortunately, international infection control efforts for SARS were so successful that no new cases have been reported since 2004. MERS was first identified in Saudi Arabia in 2012, and has now been reported in at least 27 countries, with over 2,600 cases and 900 deaths.

Signs and Symptoms

COVID-19, SARS, and MERS usually begin 2 to 14 days after infection, with typical flu-like signs and symptoms of fever, malaise, muscle aches, a non-productive cough, and shortness of breath. Abdominal symptoms such as nausea, vomiting, and diarrhea have also been reported. In some cases, particularly patients who are elderly or have other medical conditions, complications such as pneumonia and acute respiratory distress develop, often leading to death. Many people infected with SARS-CoV-2 are asymptomatic or have mild upper respiratory symptoms, including nasal congestion and sore throat. An unusual loss of the sensation of smell and/or taste may also occur. MERS has also been reported to cause only mild infections in a few patients. SARS is unusual because no upper respiratory symptoms appear during the early stage of illness.

Causative Agents

COVID-19 is caused by the virus SARS-CoV-2 (CoV indicates coronavirus); SARS is caused by SARS-CoV; and MERS is caused by MERS-CoV. These viruses, which we will refer to collectively as severe coronaviruses, belong to the *Coronaviridae* family. Members of this family are enveloped, single-stranded RNA viruses characterized by spikes on their surface that create a crown-like appearance (*corona* means "crown") (**figure 21.26**).

As SARS-CoV-2 continues to spread, mutations in the viral genome give rise to new viral variants, and these are assigned names based on a standard phylogenetic system (e.g. B.1.1.7); some are assigned a Greek letter that reflects their order of emergence, such as alpha, beta, gamma. Variants that show evidence of increased virulence, transmissibility, or evasion of previous immunity are classified as variants of concern (VOC).

Pathogenesis

Coronaviruses use their spike proteins to bind to cells in the respiratory tract, including bronchial and alveolar epithelial cells; SARS-CoV and SARS-CoV-2 spike proteins bind the same receptor (ACE2), but MERS-CoV uses a different one. Once attached to a receptor, the spike protein is cleaved by a host cell protease, allowing fusion of viral and host membranes, resulting in entry of the virus into the cell. The virus then induces changes in the host cell's endoplasmic reticulum membrane, resulting in the formation of double membrane vesicles (DMVs) in the cytoplasm, within which viral genome replication and transcription occur. Recall that these processes generate double-stranded RNA intermediates that would normally be detected by the cell's pattern recognition receptors, thereby triggering destruction of the cell; when confined within DMVs, however, these products remain hidden from those receptors. Pores in the DMVs allow the viral RNA to move to the cytoplasm for translation.

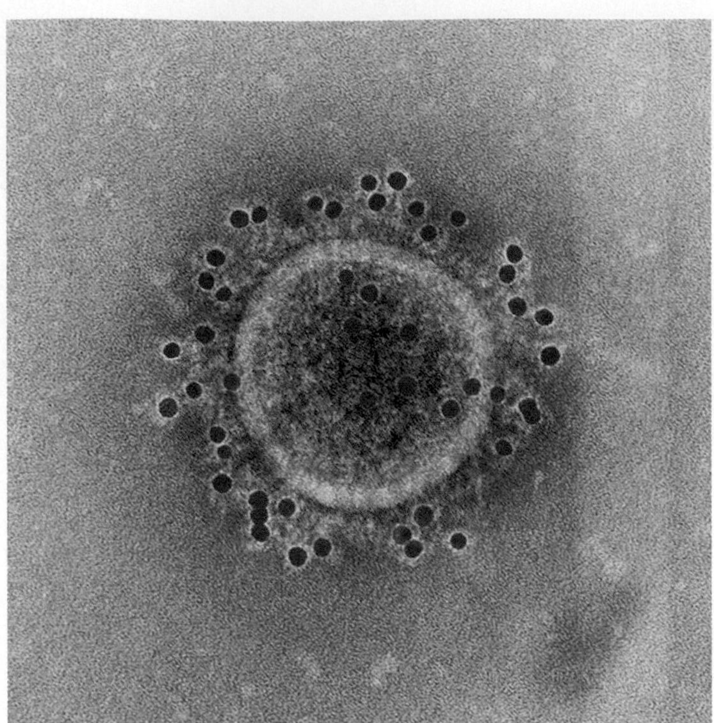

FIGURE 21.26 Coronavirus Color-enhanced TEM NIAID

? **Besides humans, what animals are believed to host the severe coronaviruses?**

Coronavirus infection directly damages the host cells and induces an inflammatory response that, coupled with an adaptive response, typically limits the infection. With severe coronavirus infection, however, the immune response can become dysregulated. One potential result is an excessive release of pro-inflammatory cytokines (a "cytokine storm"), which appears to be largely responsible for the most devastating effects of the disease.

Epidemiology

COVID-19, SARS, and MERS are zoonotic diseases that likely originated in bats, which are known to harbor several different coronaviruses (strains resembling the three severe viruses still circulate in bats). Researchers believe that in each instance the virus was initially transmitted from a bat to an intermediate animal host that has close contact with humans; the virus replicated to high numbers in the intermediate host, thereby making transmission to a human more likely. For SARS-CoV-2, the identity of the intermediate host has not been conclusively determined. For SARS-CoV however, genetic and epidemiological evidence indicates that a bat coronavirus was transmitted to a palm civet (a cat-like animal) or some other animal sold in live animal markets in China. For MERS, a bat coronavirus appears to have been transmitted to a dromedary camel (a camel with one hump) over 30 years ago; these camels now carry different variations of MERS-CoV, some of which can also infect humans.

Coronaviruses are transmitted via respiratory droplets during close person-to-person contact, but aerosolized droplets also likely play a role in COVID-19 transmission. At least some people with COVID-19 shed virus even when they have no apparent symptoms, although symptoms increase the shedding. This is in contrast to SARS and MERS, which are generally transmitted only by people with symptoms severe enough to require hospitalization. The difference in shedding explains why SARS and MERS primarily spread in hospital settings, whereas COVID-19 spread widely in the general population. As SARS-CoV-2 evolves, new variants arise, and certain variants—particularly those shed at higher levels before symptoms appear—spread most widely; fortunately, the most transmissible variants appear less likely to cause severe disease.

In elderly people who have underlying conditions, COVID-19, SARS, and MERS all have very high case fatality rates. It is important to recognize, however, that although severe COVID-19 is most likely to develop in older adults (over age 50) and people with certain medical conditions, it can sometimes occur in otherwise healthy young people. In the SARS epidemic, mainly hospitalized patients developed the disease, but healthcare workers who had close contact with those infected patients sometimes became ill. MERS is almost exclusively limited to hospitalized patients, suggesting that underlying health problems are an essential predisposing factor.

Treatment and Prevention

In response to the COVID-19 pandemic, a global effort focused on developing drugs that target SARS-CoV-2 (see Focus Your Perspective 20.1). A few drugs are now approved by the U.S. Food and Drug Administration (FDA) for treating COVID-19 patients at high risk of developing severe disease. An example is remdesivir, a nucleoside analog that causes the SARS-CoV-2 replicase (an RNA-dependent RNA polymerase) to prematurely terminate synthesis of RNA molecules. Another example is Paxlovid, which includes two components: (1) nirmatrelvir, an inhibitor of a SARS-CoV-2 viral protease, and (2) ritonavir, which helps prevent the breakdown of nirmatrelvir in the body. Several other medications currently have Emergency Use Authorization (EUA), meaning they can be used in certain situations when no approved option is available. An example is molnupiravir, a drug that causes the SARS CoV-2 replicase to introduce so many mutations that the viral genome copies are rendered useless. Although monoclonal antibodies (mAbs) that bind the spike protein and thus neutralize the virus were among the first COVID-19 treatments to receive EUA, that authorization was withdrawn in 2023 because the antibodies are unlikely to be effective against the circulating SARS-CoV-2 variants. Additional measures used for treating severe COVID-19 include medications that suppress the inflammatory response as well as supportive therapies, such as breathing support (supplementary oxygen or mechanical ventilation). No antivirals are currently approved to specifically target SARS-CoV or MERS-CoV.

Three vaccines that prevent or decrease the severity of COVID-19 are currently approved or under EUA in the United States for persons age 18 and older; some of these are approved for children and infants, Several other COVID-19 vaccines have been approved for use in other countries. The vaccines approved or authorized in the United States include the following types:

- **mRNA vaccine.** The Pfizer–BioNTech and Moderna vaccines are both mRNA vaccines that encode the SARS-CoV-2 spike protein; updated booster versions are bivalent, meaning that they encode two versions of the protein, including one of a widespread recent variant.
- **Subunit vaccine.** The Novavax vaccine is a subunit vaccine that contains part of the SARS-CoV-2 spike protein.

A viral vector vaccine (Janssen COVID-19 vaccine) was previously used under EUA but is no longer available.

No vaccines against MERS-CoV are currently available, but several are in clinical trials. Efforts to develop a vaccine for camels is also underway, which could help reduce the transmission potential of the virus from camels to humans.

In addition to vaccination, other prevention measures for the severe coronavirus diseases include handwashing, frequently disinfecting surfaces, and avoiding close contact with people who are ill. **Table 21.13** presents some important features of COVID-19, SARS, and MERS.

TABLE 21.13	COVID-19, SARS, and MERS
Signs and Symptoms	Fever, muscle aches, cough; in severe cases, eventual respiratory distress or pneumonia
Incubation Period	2 to 14 days
Causative Agent	Severe coronaviruses: SARS-CoV-2, SARS-CoV, and MERS-CoV
Pathogenesis	Alveolar damage, inflammation, fluid accumulation; cytokine storm may damage tissue.
Epidemiology	Zoonotic origin; spread by respiratory droplets; underlying conditions predispose people to severe disease.
Treatment and Prevention	Treatment: antivirals. Prevention: vaccines available to prevent COVID-19.

FOCUS YOUR PERSPECTIVE 21.1

A Global Lesson in Microbiology: The COVID-19 Pandemic

In December 2019, reports of pneumonia cases of an unknown cause began surfacing in Wuhan, the largest city in central China, with many of the initial cases traced to a market that sold live seafood and animals for consumption. The mysterious disease spread rapidly, and by late January, several thousand more cases were reported each day. As the outbreak progressed, it became clear that the disease was being transmitted person-to-person, with symptoms developing within 2 weeks of exposure.

Scientists in China quickly identified the causative agent as a coronavirus and determined the nucleotide sequence of the viral RNA genome. This allowed development of PCR-based diagnostic tests to detect what became commonly known as the novel coronavirus. The genome sequence also revealed the virus's close evolutionary relationship to the one that emerged in 2003 to cause the disease SARS (severe acute respiratory syndrome). Like the SARS virus, the new coronavirus appears to have originated from a bat coronavirus, possibly with the involvement of another animal host.

Cases of the novel coronavirus disease soon began appearing outside of China, carried by travelers. Because of the disease's rapid spread and severity, the World Health Organization (WHO) declared the outbreak a Public Health Emergency of International Concern in late January 2020. Meanwhile, the International Committee on Taxonomy of Viruses (ICTV) called a special meeting and chose the name SARS-CoV-2 to reflect the virus's similarity to the original SARS virus (SARS-CoV). At about the same time, the WHO named the disease COVID-19, for coronavirus disease-2019. In the weeks that followed, large outbreaks of the disease were reported in South Korea, Iran, and Italy.

The first reported COVID-19 cases in the United States occurred near Seattle, Washington. A man who had returned to the area from Wuhan in mid-January developed symptoms of pneumonia and was sent to the hospital with suspected COVID-19. The CDC confirmed that diagnosis within a week. A month later, a resident of a skilled nursing facility near Seattle who did not travel or have any contact with known COVID-19 patients was found to have the disease; additional testing later identified over 160 other cases linked to the facility. By mid-March, it was clear that the virus had been introduced into the United States multiple times by international travelers and had been circulating in all states without detection.

On March 11, with COVID-19 having spread to over 110 countries, the WHO declared the disease a pandemic. Six months later, over 30 million infections and 1 million deaths had been reported worldwide, staggering numbers yet likely underestimates as tests to confirm cases were often unavailable. As the number of cases increased around the world, governments ordered unprecedented shutdowns of schools, workplaces, and public events to "flatten the curve"—a reference to an epi curve (see figure 19.9). Social distancing measures such as home isolation for all but essential workers were established. In essence, the world went quiet as normal activities were shut down in an effort to control the pandemic.

Just as the rapid spread of COVID-19 was unprecedented, so was the speed of research efforts as scientists united to control the disease. Scientists around the globe sequenced the genome of SARS-CoV-2 isolates and then shared the data to track the virus's spread and mutation rate. Worldwide efforts to develop treatments and vaccines also began immediately. By mid-February 2020, scientists determined the three-dimensional structure of the SARS-CoV-2 spikes—the surface structures that not only give the virus its characteristic crown-like appearance but also are used by the virus to bind to and enter host cells. In late 2020, monoclonal antibody (MAb) treatments were given U.S. Food and Drug Administration (FDA) emergency use authorization (EUA) for treating COVID-19 patients most at risk of progressing to severe disease. These antibodies work by binding the spike proteins on the viral surface, effectively neutralizing the virus.

Efforts to develop a vaccine focused on eliciting the production of spike-blocking antibodies that neutralize the virus. By December 2020, three different SARS CoV-2 vaccines were given EUA. At no other time in history have vaccines been developed so quickly—the first doses were administered less than a year after the virus was discovered! Other vaccines soon became available for emergency use worldwide.

As vaccines and treatments became more readily available, many people hoped that the end of the pandemic was near. Soon, however, sequencing data revealed that SARS-CoV-2 variants—mutants of the original virus—began to emerge. Such evolution was not surprising given the sheer number of opportunities the virus has to replicate in the previously uninfected global population. Some newer variants spread more easily, and some have mutations in genes that encode the virus's spike protein, resulting in decreased effectiveness of both existing vaccines and of monoclonal antibodies available for treatment.

Although many of the SARS-CoV-2 variants can infect even vaccinated individuals, vaccination continues to prevent severe disease, thus decreasing hospitalization and saving lives. As more and more people are vaccinated and treatments for severe disease become more readily available, restrictions on social distancing and masking have relaxed. Still, however, scientists believe that we must remain vigilant. As the virus continues to evolve, so must our efforts to control infections.

Respiratory Syncytial Virus (RSV) Infections

Respiratory syncytial virus (RSV) infection is the leading cause of serious lower respiratory tract infections in infants and young children and one of the leading causes of death in infants worldwide. It is also responsible for serious disease in elderly people and for healthcare-associated epidemics.

Signs and Symptoms

Signs and symptoms of RSV infection begin after an incubation period of 4 to 6 days with runny nose that may be followed by fever, cough, wheezing, and difficulty breathing. The severity of symptoms depends on the patient's age, health status, and previous RSV exposure. Younger children infected with RSV for the first time (primary infection) often have more severe lower respiratory tract disease, while healthy older children and adults generally show symptoms of a bad cold. Patients can develop a bluish color around the lips, indicating that they are not getting enough O_2 mainly due to **bronchiolitis**, an inflammation of the small airways (bronchioles) and a common feature of RSV infection.

RSV is one of the causes of **croup,** a condition resulting from obstruction of the trachea and larynx that manifests as a loud, high-pitched cough and noisy inhalation due to airway obstruction.

Hospitalized infants seldom die from RSV infection, but it can be fatal for elderly patients with underlying diseases such as heart and lung disease, cancer, and immunodeficiency.

Causative Agent

RSV is a single-stranded, enveloped RNA virus of the family *Pneumoviridae.* It causes cells in culture to form **syncytia** (clumps of fused cells), thus the name of the virus. The two major subtypes of the virus, A and B, differ in their envelope glycoproteins; both subtypes can be circulating during an outbreak, but type A is associated with more severe disease.

Pathogenesis

RSV enters the body by inhalation and replicates in the nasopharynx. The infection spreads to the epithelial cells in the bronchioles, where the resulting inflammatory response causes the cells to die and slough off. Not only does this damage the mucociliary escalator, but the bronchioles become partially plugged by sloughed cells, mucus, and fluid that oozes from the bronchial walls. The initial obstruction causes wheezing when air rushes through the narrowed passageways, a manifestation that may be mistaken for an asthma attack. The obstruction often acts like a one-way valve, allowing air to enter the lungs but not leave. In many cases the inflammatory process extends into the alveoli, causing pneumonia. As with influenza, RSV increases the risk of secondary infection because of damage to the mucociliary escalator.

Epidemiology

RSV outbreaks are common from late fall to late spring, peaking in mid-winter. Healthy children and adults usually have mild illness and readily spread the virus to others. Most children have been infected by the time they are 2, but recovery produces only weak and short-lived immunity so re-infection is common and can recur throughout life. Many different RSV strains exist, and dominant strains often shift yearly.

Treatment and Prevention

There are no effective antiviral medications for treating RSV infections. Preventing healthcare-associated spread of the virus requires strict isolation techniques. To prevent RSV illness, two new vaccine options are approved for adults age 60 and older; one of those is recommended for pregnant women (to protect newborns). In addition, monoclonal antibody preparations are available to provide passive immunity to newborns and certain high-risk infants and children during the RSV season. The main features of RSV infections are summarized in **table 21.14.**

Hantavirus Pulmonary Syndrome

Hantavirus pulmonary syndrome (HPS) is a life-threatening, rodent-borne zoonotic viral infection. Although rare, it is considered worrisome because it can cause severe, acute disease in previously healthy individuals. It was discovered in the spring of 1993 in the Four Corners region of the southwestern United States (where Arizona, Colorado, New Mexico, and Utah come together) when an outbreak of unexplained respiratory illness occurred in a group of previously healthy people.

Signs and Symptoms

After an incubation period ranging from 7 days to 8 weeks, hantavirus pulmonary syndrome begins with fever, fatigue, and muscle aches (especially in the lower back and thighs), often followed by nausea, vomiting, and diarrhea. Unproductive cough and increasingly severe shortness of breath appear within a few days, often followed by shock and death.

Causative Agents

Hantavirus pulmonary syndrome is mainly caused by a hantavirus called the Sin Nombre virus (SNV)— sin nombre means "no name" in Spanish (**figure 21.27**). Hantaviruses are enveloped viruses of the family *Bunyaviridae*. Their genome consists of three segments of single-stranded RNA. In nature, these viruses primarily infect rodents, causing lifetime infections, without any apparent harm to the animals. Each type of hantavirus generally

TABLE 21.14	Respiratory Syncytial Virus (RSV) Infections
Signs and Symptoms	Runny nose, cough, fever, wheezing, difficulty breathing, bluish color
Incubation Period	1 to 4 days
Causative Agent	Respiratory syncytial virus (RSV), a pneumovirus that produces syncytia
Pathogenesis	Sloughing of respiratory epithelium and inflammatory response plug bronchioles, cause bronchiolitis; pneumonia from bronchiolar or alveolar inflammation; secondary infection.
Epidemiology	Yearly epidemics during the cool months; readily spread by otherwise healthy older children and adults with mild symptoms; no lasting immunity.
Treatment and Prevention	Treatment: no effective antiviral treatment. Prevention: vaccines for adults age 60 and older; vaccine for pregnant women (to protect newborns). For high-risk children including newborns, passive immunization with a monoclonal antibody.

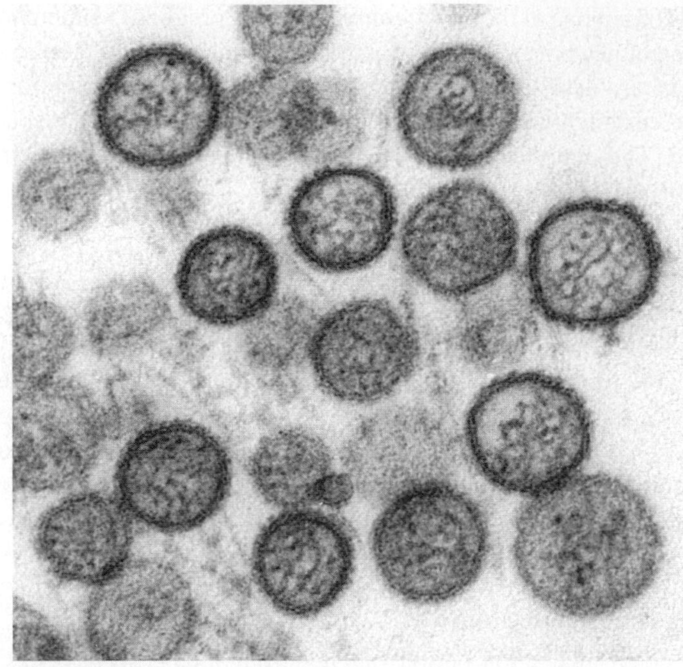

(a)

(b)

FIGURE 21.27 Hantavirus (a) Transmission electron micrograph of Sin Nombre virus. **(b)** Deer mouse (*Peromyscus maniculatus*), an important reservoir. a: Cynthia Goldsmith, Luanne Elliott/CDC; b: James Gathany/CDC

? Will deer mice infected with Sin Nombre virus exhibit the same symptoms as humans?

infects a specific rodent species. Sin Nombre virus infects deer mice, which transmitted the virus to humans during the initial outbreak in 1993 when the virus was discovered.

Pathogenesis

The hantavirus enters the body by inhalation of airborne dust contaminated with the urine, feces, or saliva of infected rodents. In the lungs, it is taken up by phagocytes and transported to the lymph nodes. It then enters the circulation and is carried throughout the body, infecting endothelial cells that line capillaries, particularly in the lung. Massive amounts of the viral antigen appear in lung capillaries, and an inflammatory response to that antigen causes those vessels to leak large amounts of fluid into the lungs. This suffocates patients and causes their blood pressure to fall, leading to shock and death in almost 40% of the cases. Fortunately, few mature infectious virions enter the air passages of the lung, so person-to-person transmission is rare.

Epidemiology

Hantavirus pulmonary syndrome is a zoonosis. Since the syndrome was described, cases have been identified from Canada to Argentina, in addition to the United States. The main risk factor for infection is cleaning poorly ventilated rodent-infested homes and buildings. Infections are also more prevalent in the fall when wild rodents seek shelter from the cold weather. Complex ecological factors that affect mouse population levels, such as numbers of predators (foxes, owls, etc.) and El Niño weather patterns (which affect food supply), can also affect transmission. The emergence of hantavirus pulmonary syndrome is a convincing example of how environmental change can result in infectious human disease.

Treatment and Prevention

There is no proven antiviral treatment for hantavirus pulmonary syndrome, a disease that is often fatal. Prevention is based on minimizing exposure to rodents as well as dusts contaminated by rodent urine, saliva, and feces. Rodent populations should be controlled by keeping foods in containers and making buildings as mouse-proof as possible. When cleaning an area where rodents are found, maximum ventilation should be ensured, and the area should be mopped with a disinfectant solution rather than being swept with brooms and vacuum cleaners that can stir up dust. The main features of hantavirus pulmonary syndrome are summarized in **table 21.15.**

TABLE 21.15	Hantavirus Pulmonary Syndrome
Signs and Symptoms	Fever, muscle aches, vomiting, diarrhea, cough, shortness of breath, shock
Incubation Period	7 days to 8 weeks
Causative Agent	Sin Nombre and related hantaviruses; enveloped viruses with segmented single-stranded RNA
Pathogenesis	Viral antigen localizes in capillary walls in the lungs; inflammation
Epidemiology	Zoonotic disease; epidemics associated with increases in mouse populations near housing; generally no person-to-person spread
Treatment and Prevention	Treatment: no antiviral treatment. Prevention: avoiding contact with rodents; sealing access to houses, food supplies; good ventilation; avoiding dust; using disinfectants in cleaning rodent-infested infested areas.

MicroAssessment 21.5

The rapid evolution of influenza viruses and the speed with which they can be spread around the world make influenza an extremely serious threat. COVID-19, SARS, and MERS are caused by novel coronaviruses. Respiratory syncytial virus is the leading cause of serious respiratory disease in infants and young children. Hantavirus pulmonary syndrome is often fatal.

13. Why are there so many deaths from influenza when it is generally a mild disease?
14. What is the source of the virus that causes hantavirus pulmonary syndrome?
15. Why might you expect an influenza epidemic to be more severe following an antigenic shift in the virus rather than after antigenic drift?

21.6 ■ Fungal Infections of the Lower Respiratory System

Learning Outcomes

11. Describe the pathogenesis of histoplasmosis.
12. Outline the epidemiology of coccidioidomycosis and *Pneumocystis* pneumonia.

Serious fungal infections are quite unusual in healthy, immunocompetent people, but coccidioidomycosis and histoplasmosis are important exceptions. The causative agents of these diseases are dimorphic (meaning two forms), growing in the soil as molds but differentiating during an infection into forms adapted to grow in the body. The diseases are limited to certain geographical regions so are often referred to as endemic mycoses; because the diseases can involve multiple body systems they are also called systemic mycoses. In contrast to these dimorphic fungi, the fungus that causes *Pneumocystis* pneumonia (PCP) is a widespread opportunist; it can cause serious disease in HIV patients and other immunocompromised individuals.

Coccidioidomycosis ("Valley Fever")

In the United States, most cases of coccidioidomycosis occur in Arizona and California. Farm workers, construction workers, and other people exposed to dust and soil that contains infectious spores are most likely to become infected, but fewer than half of people infected develop symptoms.

Signs and Symptoms

Signs and symptoms of coccidioidomycosis are similar to influenza or community-acquired pneumonia and can appear 1 to 3 weeks after exposure. Common manifestations include fever, fatigue, cough, and chest pain; night sweats, muscle and joint pain, rashes, and painful nodules sometimes occur. Although most patients recover spontaneously within a month, a small percentage develop chronic disease.

Causative Agent

Coccidioidomycosis is caused by *Coccidioides immitis* and *Coccidioides posadasii,* dimorphic fungi that grow in the soil as molds. Their hyphae give rise to numerous barrel-shaped structures called arthroconidia, which can become airborne and be inhaled (**figure 21.28a**). In infected tissues, the arthroconidia develop into thick-walled spherules that may contain several hundred small cells called endospores, not to be confused with bacterial endospores (**figure 21.28b**).

Pathogenesis

Arthroconidia enter the lungs with inhaled air and develop into spherules. The spherules eventually rupture, releasing their endospores, which then develop into new endospore-producing spherules, and the process is repeated. Every time this cycle occurs, an inflammatory response is provoked which can cause a local pulmonary lesion to form.

The organisms are usually eliminated by body defenses, but the large spherules can sometimes be resistant to phagocytosis, resulting in granulomas that resemble those seen

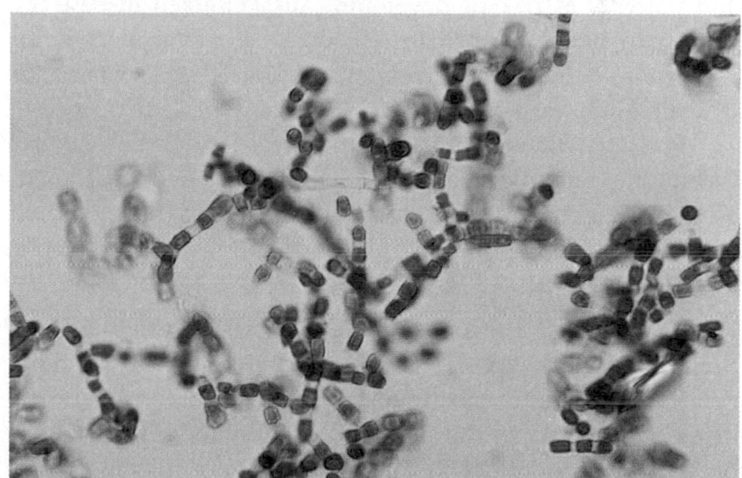

(a)

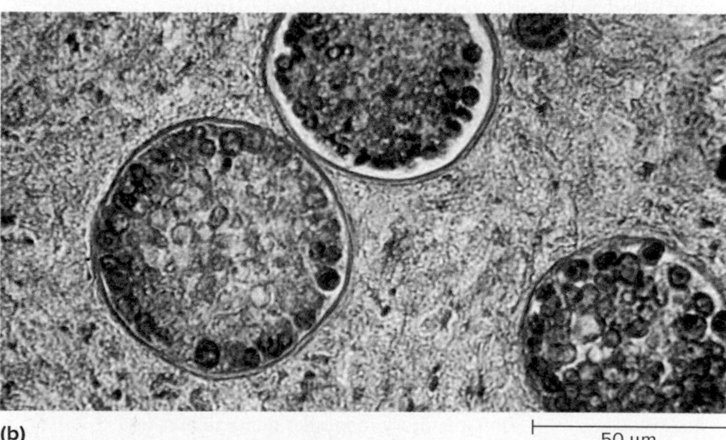

(b) 50 μm

FIGURE 21.28 ***Coccidioides immitis*** **(a)** Mold-phase hyphae giving rise to barrel-shaped arthroconidia. **(b)** Spherules in tissues. a: Dr. Lucille K. Georg/CDC; b: Lucille K. Georg/CDC

Which form of this fungus is found in host tissues?

in tuberculosis. Occasionally, recruited macrophages carry organisms to the lymph nodes; from there, the organisms can disseminate throughout the body. The disseminated form of the disease occurs more often in people with AIDS or other immunodeficiencies and is fatal without treatment.

Epidemiology

Coccidioides species grow only in semi-arid desert areas of the Western Hemisphere. In these areas, infections occur during the hot, dry, dusty seasons when airborne arthroconidia are easily dispersed from the soil. Rainfall encourages growth of the fungus, which then produces increased numbers of arthroconidia when dry conditions return. Infections often occur when dust is stirred up during earthquakes, construction, excavations, and dust storms. Travelers to these areas can inhale the arthroconidia and develop the disease upon their return home. Soil containing the fungus can unknowingly be transported to other areas, but the organism apparently cannot thrive in moist climates.

Treatment and Prevention

Most coccidioidal infections are not treated, but serious cases and patients at risk of developing disseminated disease are usually treated with an azole (fluconazole or itraconazole) for 3 months, or even longer. Amphotericin B is also used for treatment, but only in severe cases because of its toxicity.

Preventive measures in endemic areas include dust control (watering and planting vegetation) and avoiding dust. **Table 21.16** describes the features of coccidioidomycosis.

TABLE 21.16	Coccidioidomycosis ("Valley Fever")
Signs and Symptoms	Fever, fatigue, cough, chest pain; less frequently, night sweats, muscle and joint pain, rash and painful nodules on extremities
Incubation Period	1 to 3 weeks
Causative Agent	*Coccidioides immitis* and *C. posadasii*, dimorphic fungi
Pathogenesis	Inhaled arthroconidia develop into spherules that mature and rupture to release endospores, which can each develop into another spherule; inflammatory response damages tissue
Epidemiology	Occurs only in certain semi-arid regions of the Western Hemisphere
Treatment and Prevention	Treatment: antifungal medications. Prevention: dust control and avoiding dust.

Histoplasmosis ("Spelunker's Disease")

Histoplasmosis, like coccidioidomycosis, is often asymptomatic. When symptoms do occur, they are usually mild but occasionally mimic tuberculosis. Rare, serious forms of the disease suggest that the patient has an underlying immunodeficiency such as AIDS. The distribution of histoplasmosis is more widespread than that of coccidioidomycosis and is associated with different soil types and climate.

Signs and Symptoms

Signs and symptoms of histoplasmosis may develop a few days to weeks after exposure and are usually mild. Fever, cough, and chest pain are the most common, but shortness of breath, headaches, chills, fatigue, and body aches can also occur in some patients. These symptoms typically resolve after several weeks. The infection can sometimes persist in the lungs or even spread to other parts of the body in patients with weakened immune systems or in patients who were exposed to unusually high levels of the organism.

Causative Agent

Histoplasmosis is caused by the dimorphic fungus *Histoplasma capsulatum* (the name is misleading because the fungus does not have a capsule). In pus or tissue from people with active disease, *H. capsulatum* is a tiny oval yeast that grows within macrophages (**figure 21.29a**). The mold form of the organism characteristically produces two kinds of conidia: macroconidia, which often have numerous projecting knobs (**figure 21.29b**), and tiny microconidia.

Pathogenesis

Histoplasma capsulatum microconidia inhaled into the lungs are taken up by resident macrophages. The fungus then develops into the yeast form, which multiplies within the phagocytes. Granulomas develop, closely resembling those seen in tuberculosis, sometimes even showing caseous necrosis. Eventually, the lesions are replaced with scar tissue, and many calcify, becoming visible on X rays. In rare cases, particularly in those who are immunodeficient, the disease spreads throughout the body.

Epidemiology

The distribution of histoplasmosis is quite different from that of coccidioidomycosis. **Figure 21.30** shows histoplasmosis distribution in the United States, but the disease also occurs in tropical and temperate zones scattered around the world. Most cases in the United States have occurred in the Mississippi and Ohio River drainage areas and in South Atlantic states. Studies reveal that millions of people living in these areas have been infected. The organism grows well in soils enriched with bird droppings, so construction and excavation workers, farmers, and people who clean chicken coops are at risk in endemic areas. Cave explorers (spelunkers) are also at risk because the organism can be found in bat droppings (many caves have bats).

Treatment and Prevention

Treatment of histoplasmosis is similar to that of coccidioidomycosis. Itraconazole is used for treating mild to moderate disease. In severe cases, amphotericin B is given until the patient improves, at which point the less toxic itraconazole is used.

No proven preventive measures for histoplasmosis are known other than to avoid areas where soil is heavily enriched with bat and bird droppings, especially if the soil has been left

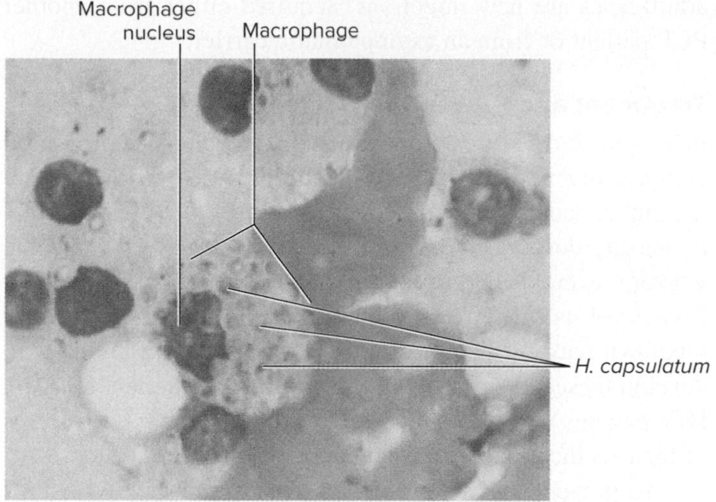

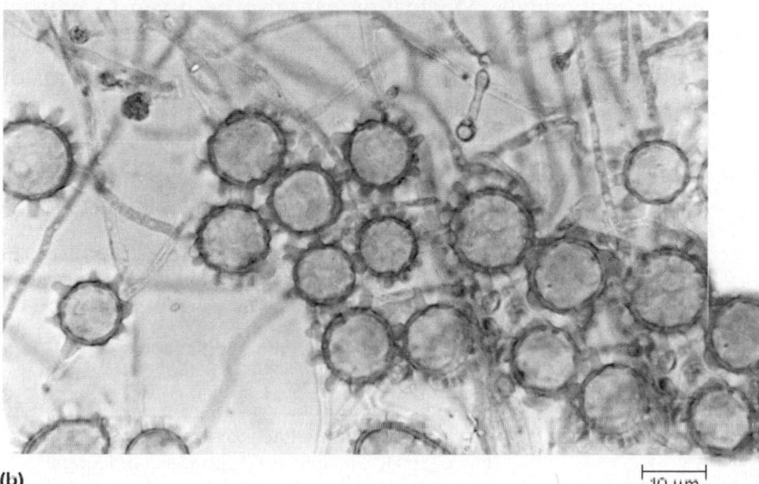

FIGURE 21.29 ***Histoplasma capsulatum*** **(a)** Yeast-phase organisms In the cytoplasm of a macrophage. **(b)** Mold phase, showing macroconidia. Evans Roberts

? Is this fungus encapsulated?

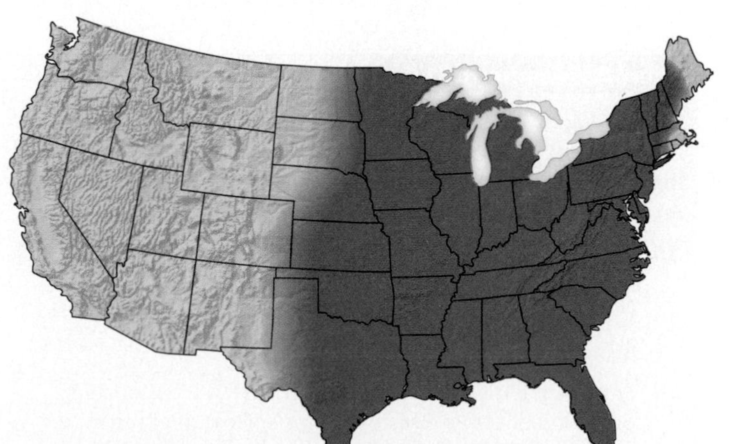

FIGURE 21.30 Geographic Distribution of *Histoplasma capsulatum* in the United States

? Why is histoplasmosis also called "spelunker's disease"?

undisturbed for a long period. Professional hazard removal companies should be used to clean areas that have large amounts of bird or bat droppings. Immunocompromised patients should also avoid activities associated with potentially contaminated areas such as exploring caves and cleaning chicken coops. The main features of histoplasmosis are described in **table 21.17.**

Pneumocystis Pneumonia (PCP)

Pneumocystis pneumonia (PCP), a severe, infectious lung disease, was recognized just after World War II in Europe when it killed malnourished, premature infants in hospitals. Subsequently, occasional cases were identified among immunodeficient patients. PCP incidence increased rapidly during the AIDS epidemic, and it is still a common opportunistic infection in AIDS patients who are not receiving preventive care, as well as in other severely immunocompromised patients.

Signs and Symptoms

Many immunocompetent people can be temporarily colonized by the fungus that causes PCP. It can live in the respiratory tract, causing few to no symptoms, and the immune system typically clears the organism after several months. In immunocompromised or immunosuppressed people, however, the organism causes disease. The signs and symptoms in HIV-infected individuals begin slowly after an incubation period of several weeks, with a mild fever, gradually increasing shortness of breath, rapid breathing, and a non-productive cough. As the disease progresses, a dusky coloration of the skin and mucous membranes appears and gradually worsens—this is caused by poor oxygenation of the blood that can become fatal. The disease

TABLE 21.17	Histoplasmosis ("Spelunker's Disease")
Signs and Symptoms	Mild respiratory symptoms, mainly fever, cough, and chest pain; also headaches, chills, fatigue, and body aches in some patients
Incubation Period	3 days to a few weeks
Causative Agent	*Histoplasma capsulatum*, a dimorphic fungus
Pathogenesis	Microconidia inhaled, change to yeast phase, multiply in macrophages; granulomas form; disease spreads throughout the body in people with AIDS or other immunodeficiencies.
Epidemiology	The fungus grows in soil contaminated with bird or bat droppings; most cases occur in the Ohio and Mississippi River valleys, and in the southeast United States. Found in many other countries. Spelunkers are at risk.
Treatment and Prevention	Treatment: antifungal medications. Prevention: avoiding soils contaminated with bird or bat droppings.

typically develops faster in patients who are immunosuppressed, such as transplant recipients or patients with certain blood cancers. In these patients, symptoms appear after only a few days and are usually accompanied by a high fever.

Causative Agent

Pneumocystis pneumonia is caused by *Pneumocystis jirovecii,* a tiny yeast-like fungus that was formerly classified as *P. carinii. P. jirovecii* differs from many fungi in the chemical composition of its cell wall and cytoplasmic membrane, so it is often resistant to medications typically used against fungal pathogens. The organism has a trophic (feeding) form that predominates during lung infection and several cyst forms at various stages of maturation; mature cysts are likely involved in transmission. Cysts have a characteristic appearance, which helps in identification (**figure 21.31**). The organism has not reliably been cultivated in vitro.

Pathogenesis

P. jirovecii cells are easily inhaled into lung tissue and attach to the alveolar walls. Attachment to the alveoli epithelium initiates an inflammatory response, and the alveoli fill with fluid, macrophages, and masses of *P. jirovecii* cells in various stages of development. Later, the alveolar walls become thickened and scarred, preventing the free passage of O_2. Activated alveolar macrophages are critical for clearance of *P. jirovecii,* and these are typically impaired in HIV-infected individuals.

Epidemiology

P. jirovecii is distributed worldwide, but PCP typically occurs only in people who are severely immunocompromised. The organism is transmitted via the respiratory route, and most people are exposed as children, developing an asymptomatic infection that is eventually cleared. PCP in adults was originally thought to be due to reactivation of a latent infection acquired during childhood, but scientists now believe that adult cases are new infections, acquired either from another PCP patient or from an asymptomatic carrier.

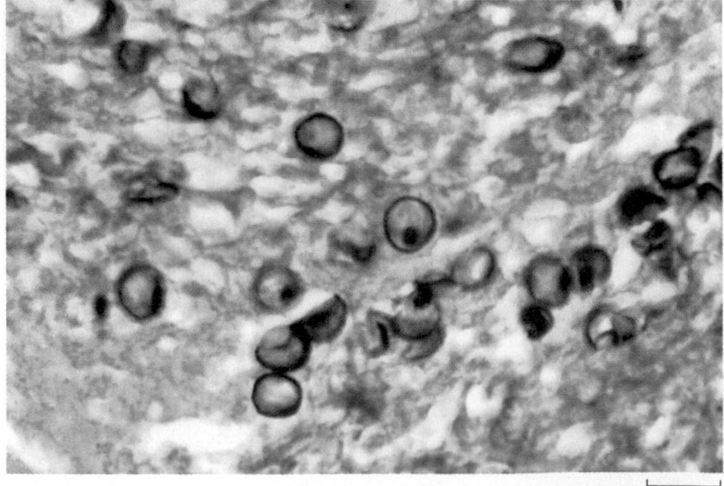

FIGURE 21.31 Methanamine Silver Stain of *Pneumocystis jirovecii* The dark circles and cup-like forms are *P. jirovecii* cysts.
Dr. Edwin P. Ewing, Jr./CDC

? **Why is *Pneumocystis* pneumonia known by the acronym PCP?**

Treatment and Prevention

PCP is most often treated with co-trimoxazole. The drug is so effective that patients with a mild allergy to the medication are recommended to undergo desensitization by using gradually increasing doses. Alternative medications are given to people who are severely allergic or who cannot tolerate co-trimoxazole because of its side effects (mainly rash, nausea, and fever). For unknown reasons, people with HIV disease are more likely to develop these side effects than others. After treatment for PCP, HIV patients and others must receive lower prophylactic doses of their medication to prevent recurrent infection.

To prevent the spread of PCP in hospitals, PCP patients should not be placed in a room with other immunocompromised individuals. Medication can also be given to prevent susceptible populations (HIV-infected patients, transplant recipients) from acquiring the infection. The main features of *Pneumocystis* pneumonia are presented in **table 21.18.**

TABLE 21.18	*Pneumocystis* Pneumonia (PCP)
Signs and Symptoms	Fever, shortness of breath, rapid breathing, nonproductive cough and dusky color of skin and mucous membranes
Incubation Period	Several weeks in HIV-infected patients
Causative Agent	*Pneumocystis jirovecii*
Pathogenesis	Inhaled fungal cells attach to alveolar walls, causing alveoli to fill with fluid, macrophages, and fungal cells. Alveolar walls thicken, impairing O_2 exchange.
Epidemiology	Most people become infected in early childhood, airborne transmission from other PCP patients, or asymptomatic carriers.
Treatment and Prevention	Treatment: antimicrobial medications. Prevention: prophylactic doses for certain immunosuppressed or immunodeficient patients.

MicroAssessment 21.6

Coccidioidomycosis and histoplasmosis are caused by fungi that live in the soil. The body responds to these infections in a way that mimics tuberculosis. Pneumocystis pneumonia (PCP) affects immunocompromised people.

16. Why should an immunodeficient person avoid traveling through hot, dry, dusty areas of the southwestern United States?

17. Why might cave exploration increase the risk of histoplasmosis?

18. Several students staying in a hotel next to a bulldozing operation developed histoplasmosis. How might the bulldozing explain the outbreak?

The key features of the diseases covered in this chapter are highlighted in the **Diseases in Review 21.1** table.

Diseases in Review 21.1

Respiratory System Diseases

Disease	Causative Agent	Comment	Summary Table
BACTERIAL INFECTIONS OF THE UPPER RESPIRATORY TRACT			
Conjunctivitis (pink eye), otitis media (earache), sinus infection	Usually *Haemophilus influenzae* or *Streptococcus pneumoniae*	Often occur together; factors involved in transmission are unknown.	
Streptococcal pharyngitis ("strep throat")	*Streptococcus pyogenes* (group A streptococcus)	Treated with antibiotics, partly to avoid sequelae; must be distinguished from viral pharyngitis, which cannot be treated with antibiotics.	Table 21.3
Diphtheria	*Corynebacterium diphtheriae*	Toxin-mediated disease characterized by pseudomembrane in the upper respiratory tract. Preventable by vaccination.	Table 21.4
VIRAL INFECTIONS OF THE UPPER RESPIRATORY TRACT			
Common cold	Rhinoviruses and other viruses	Runny nose, sore throat, and cough are due to the inflammatory response and cell destruction.	Table 21.5
Adenovirus upper respiratory tract infections	Adenoviruses	Similar to the common cold but with fever; spread to the lower respiratory tract can result in severe disease.	Table 21.6
BACTERIAL INFECTIONS OF THE LOWER RESPIRATORY TRACT			
Pneumococcal pneumonia	*Streptococcus pneumoniae*	Organism common in the throat of healthy people; causes disease when mucociliary escalator is impaired or with underlying conditions. Vaccines that protect against multiple serotypes are available.	Table 21.7
***Klebsiella* pneumonia**	*Klebsiella* species, commonly *K. pneumoniae*	Common hospital-acquired bacterium; characterized by thick, bloody, jelly-like sputum. Drug resistance is a major problem.	Table 21.7
Mycoplasmal pneumonia ("walking pneumonia")	*Mycoplasma pneumoniae*	Relatively mild pneumonia; common among college students and military recruits. Cannot be treated with medications that inhibit cell wall synthesis.	Table 21.7
Pertussis ("whooping cough")	*Bordetella pertussis*	Characterized by frequent violent coughing. Preventable by vaccination.	Table 21.8
Tuberculosis ("TB")	*Mycobacterium tuberculosis*	Most infections result in latent tuberculosis infection (LTBI), but these can reactivate to cause tuberculosis disease (TB disease). Treated using combination drug therapy, but drug resistance is an increasing problem.	Table 21.9
Legionnaires' disease	*Legionella pneumophila*	Transmitted via aerosolized water droplets; smokers and those with impaired defenses are most at risk of developing disease.	Table 21.10
Inhalation anthrax	*Bacillus anthracis*	Rare zoonotic disease; may be associated with bioterrorism; high case-fatality rate.	Table 21.11
VIRAL INFECTIONS OF THE LOWER RESPIRATORY TRACT			
Influenza ("flu")	Influenza viruses	New vaccines developed yearly; viruses change seasonally due to antigenic drift; antigenic shifts cause pandemics.	Table 21.12
COVID-19, SARS, and MERS	Coronaviruses	Emerging infectious diseases sometimes associated with severe lower respiratory symptoms; zoonotic	Table 21.13
Respiratory syncytial virus infections	RSV	Serious disease in infants, young children, and the elderly. Newly authorized vaccines for certain populations.	Table 21.14
Hantavirus pulmonary syndrome	Hantaviruses	Acquired via inhaled dust contaminated with rodent saliva, urine, or feces. Frequently fatal.	Table 21.15
FUNGAL INFECTIONS OF THE RESPIRATORY TRACT			
Coccidioidomycosis ("valley fever")	*Coccidioides immitis* and *C. posadasii*	Environmental reservoir (soil in semi-arid desert areas); most infections are asymptomatic.	Table 21.16
Histoplasmosis ("spelunker's disease")	*Histoplasma capsulatum*	Environmental reservoir (bat droppings and soil enriched with bird droppings); most infections are asymptomatic.	Table 21.17
***Pneumocystis* pneumonia (PCP)**	*Pneumocystis jirovecii* (formerly *P. carinii*)	Organism is an opportunistic fungus that causes serious lung disease in immunocompromised people, such as those with HIV/AIDS.	Table 21.18

Summary

21.1 ■ Anatomy, Physiology, and Ecology of the Respiratory System

Ciliated cells line much of the respiratory tract and remove microorganisms by constantly propelling mucus out of the respiratory system.

The Upper Respiratory Tract

The upper respiratory tract includes the nose and nasal cavity, pharynx (throat), and epiglottis (figure 21.1). A wide variety of microorganisms colonize parts of the tract (table 21.1).

The Lower Respiratory Tract

The lower respiratory tract includes the larynx, trachea, bronchi, and lungs. Pleural membranes surround the lungs. Innate immune defenses such as the mucociliary escalator normally keep microorganisms out of the lower respiratory tract.

UPPER RESPIRATORY TRACT INFECTIONS

21.2 ■ Bacterial Infections of the Upper Respiratory System

Pink Eye, Earache, and Sinus Infections

Conjunctivitis (pink eye) is usually caused by *Haemophilus influenzae* or *Streptococcus pneumoniae* (pneumococcus) (figure 21.2). **Otitis media** and sinusitis often develop when infection spreads from the upper portion of the pharynx (figure 21.3).

Streptococcal Pharyngitis ("Strep Throat") (tables 21.2, 21.3; figure 21.6)

Streptococcus pyogenes is a β-hemolytic Gram-positive coccus that causes strep throat (figures 21.4, 21.5). Strains that produce **streptococcal pyrogenic exotoxins (SPEs)** can cause serious diseases, including scarlet fever.

Post-Streptococcal Sequelae

Post-streptococcal sequelae, including rheumatic fever and glomerulonephritis, may follow strep throat and are due to the immune response (figure 21.7).

Diphtheria (table 21.4)

Diphtheria, caused by *Corynebacterium diphtheriae,* is a toxin-mediated disease that can be prevented by immunization (figures 21.8, 21.9).

21.3 ■ Viral Infections of the Upper Respiratory System

The Common Cold (table 21.5)

The common cold can be caused by many different viruses, rhinoviruses being the most common (figure 21.10).

Adenovirus Respiratory Tract Infections (table 21.6)

Adenoviruses cause illnesses that can resemble a common cold or strep throat, with symptoms varying from mild to severe.

LOWER RESPIRATORY TRACT INFECTIONS

Focus on Pneumonia

In **pneumonia**, infection causes the alveoli of the lungs to fill with fluids. Signs and symptoms generally include cough, chills, shortness of breath, fever, and chest pain. A cough often brings up **sputum** from the lungs. Pneumonia is often caused by opportunists that cause disease when the mucociliary escalator is not functioning optimally. Vaccines are available to prevent pneumococcal disease but not the other common types of pneumonia.

21.4 ■ Bacterial Infections of the Lower Respiratory System

Pneumococcal Pneumonia (table 21.7; figures 21.11, 21.12)

Streptococcus pneumoniae is a very common cause of community-acquired pneumonia.

***Klebsiella* Pneumonia** (table 21.7)

Klebsiella pneumonia causes many healthcare-associated pneumonias that result in permanent lung damage (figure 21.13). Serious complications such as lung abscesses and bloodstream infection are more common than with other bacterial pneumonias.

Mycoplasmal Pneumonia ("Walking Pneumonia") (table 21.7, figure 21.14)

Mycoplasmal pneumonia is often called walking pneumonia; serious complications are rare. Penicillins and cephalosporins are not useful in treatment because the causative agent, *Mycoplasma pneumoniae,* lacks a cell wall.

Pertussis ("Whooping Cough") (table 21.8; figures 21.15, 21.16)

Whooping cough is characterized by violent spasms of coughing. Immunization against the causative agent, *Bordetella pertussis,* prevents the disease.

Tuberculosis ("TB") (table 21.9; figures 21.17 21.18, 21.19, 21.20, 21.21)

Tuberculosis, caused by the acid-fast rod *Mycobacterium tuberculosis,* is a chronic disease; the initial infection is typically asymptomatic but becomes latent, presenting the risk of later reactivation. Other mycobacteria, including *M. bovis* and *M. africanum*, can also cause tuberculosis. Together with *M. tuberculosis*, they are part of the *Mycobacterium tuberculosis* complex (MTBC).

Legionnaires' Disease (*Legionella* Pneumonia) (table 21.10, figure 21.22)

Legionnaires' disease is most likely to occur in smokers or in those with an underlying lung disease. The cause, *Legionella pneumophila,* is common in the environment.

Inhalation Anthrax (table 21.11, figure 21.23)

Inhalation anthrax is an animal disease that rarely occurs in humans, but it has been associated with bioterrorism and is a Category A agent.

21.5 ■ Viral Infections of the Lower Respiratory System

Influenza ("Flu") (table 21.12; figures 21.24, 21.25)

Widespread epidemics are characteristic of influenza A viruses. **Antigenic drift** is responsible for **seasonal influenza**; **antigenic shift** can cause **pandemic influenza.** Deaths are usually (but not always) caused by secondary infection.

COVID-19, SARS, and MERS (table 21.13, figure 21.26)

COVID-19, SARS, and MERS are emerging infectious diseases sometimes associated with severe lower respiratory symptoms and caused by coronaviruses that likely originated in bats. International infection control efforts limited the spread of both SARS and MERS but not COVID-19.

Respiratory Syncytial Virus (RSV) Infections (table 21.14)

RSV is the leading cause of serious respiratory disease in infants and young children.

Hantavirus Pulmonary Syndrome (table 21.15, figure 21.27)
Hantavirus pulmonary syndrome is an often fatal disease contracted by inhalation of airborne dust contaminated with the urine of hantavirus-infected mice.

21.6 ■ Fungal Infections of the Lower Respiratory System

Coccidioidomycosis ("Valley Fever") (table 21.16, figure 21.28)
Coccidioidomycosis is caused by the dimorphic soil fungi *Coccidioides immitis* and *C. posadasii.* It occurs in semi-arid areas in the Western Hemisphere and is acquired by inhalation of airborne arthroconidia.

Histoplasmosis ("Spelunker's Disease") (table 21.17, figure 21.29)
Histoplasmosis is similar to coccidioidomycosis but occurs in tropical and temperate zones around the world. The causative fungus, *Histoplasma capsulatum,* is dimorphic and is found in bat droppings or soils contaminated by bird droppings (figure 21.30).

***Pneumocystis* Pneumonia (PCP)** (table 21.18, figure 21.31)
Pneumocystis pneumonia occurs as an opportunistic infection in immunocompromised people, including those with HIV/AIDS.

Review Questions

Short Answer

1. How can contamination of the eye lead to an upper respiratory infection?
2. After you recover from strep throat, can you get it again? Explain.
3. What is the source of the gene that encodes diphtheria toxin?
4. Why has no vaccine been developed for the common cold?
5. How do alcoholism and cigarette smoking predispose a person to pneumonia?
6. Describe a mechanism by which *Klebsiella pneumoniae* can become antibiotic-resistant.
7. Why does the incidence of whooping cough rise promptly when pertussis immunizations are stopped?
8. Why is combination therapy used to treat TB disease?
9. How can Legionnaires' disease be prevented?
10. Why is it important to get a new influenza vaccination every year?

Multiple Choice

1. The following are all complications of streptococcal pharyngitis *except*
   a) glomerulonephritis.
   b) scarlet fever.
   c) damaged heart valves.
   d) acute rheumatic fever.
   e) Reye's syndrome.
2. All of the following are true of diphtheria *except*
   a) a membrane that forms in the throat can cause suffocation.
   b) a toxin is produced that interferes with ribosome function.
   c) the causative organism typically invades the bloodstream.
   d) immunization with a toxoid prevents the disease.
   e) it may form chronic skin ulcers.
3. Adenoviral infections and the common cold are both
   a) caused by picornaviruses.
   b) often associated with fever.
   c) associated with severe sore throat.
   d) lower respiratory infections.
   e) avoided by handwashing.
4. All are true of mycoplasmal pneumonia *except*
   a) it is a mycosis.
   b) it usually does not require hospitalization.
   c) penicillin is ineffective for treatment.
   d) it is the leading cause of bacterial pneumonia in college students.
   e) the infectious dose of the causative organism is low.
5. All of the following are true of Legionnaires' disease *except*
   a) the causative organism can grow inside protozoa.
   b) it spreads readily from person to person.
   c) it is more likely to occur in long-term cigarette smokers than in non-smokers.
   d) it is often associated with diarrhea or other intestinal symptoms.
   e) it can be contracted from household water supplies.
6. Which of the following infectious agents is most likely to cause a pandemic?
   a) Influenza A virus
   b) *Streptococcus pyogenes*
   c) *Histoplasma capsulatum*
   d) Sin Nombre virus
   e) Respiratory syncytial virus
7. Respiratory syncytial virus
   a) is a leading cause of serious lower respiratory tract infections in infants.
   b) is an enveloped DNA virus of the adenovirus family.
   c) attaches to host cell membranes by means of neuraminidase.
   d) poses no threat to elderly people.
   e) mainly causes disease in the summer months.
8. In the United States, hantaviruses
   a) are limited to northwestern states.
   b) are carried only by deer mice.
   c) infect humans, resulting in life-threatening disease.
   d) were first identified in the early 1970s.
   e) are contracted mainly in bat caves.
9. All of the following are true of coccidioidomycosis *except*
   a) it is contracted by inhaling arthroconidia.
   b) it is caused by a dimorphic fungus.
   c) endospores are produced within a spherule.
   d) it is more common in Maryland than in California.
   e) it is often associated with fever, cough and chest pain.

10. The disease histoplasmosis
    a) is caused by an encapsulated bacterium.
    b) is contracted by inhaling arthroconidia.
    c) occurs mostly in hot, dry, and dusty areas of the American Southwest.
    d) is a threat to AIDS patients living in areas bordering the Mississippi River.
    e) is commonly fatal for pigeons and bats.

Applications

1. A physician is advising the family on the condition of a diphtheria patient. How would the physician explain why the disease affects some tissues and not others?
2. How should a physician respond to a mother who asks if her daughter can get pneumococcal pneumonia again?

Critical Thinking

1. If all transmission of *Mycobacterium tuberculosis* from one person to another were stopped, how long would it take for the world to be rid of the disease?
2. Medications that prevent and treat influenza by binding to neuraminidase on the viral surface act against all the kinds of influenza viruses that infect humans. What does this imply about the nature of the interaction between the medications and the neuraminidase molecules?

22 Skin Infections

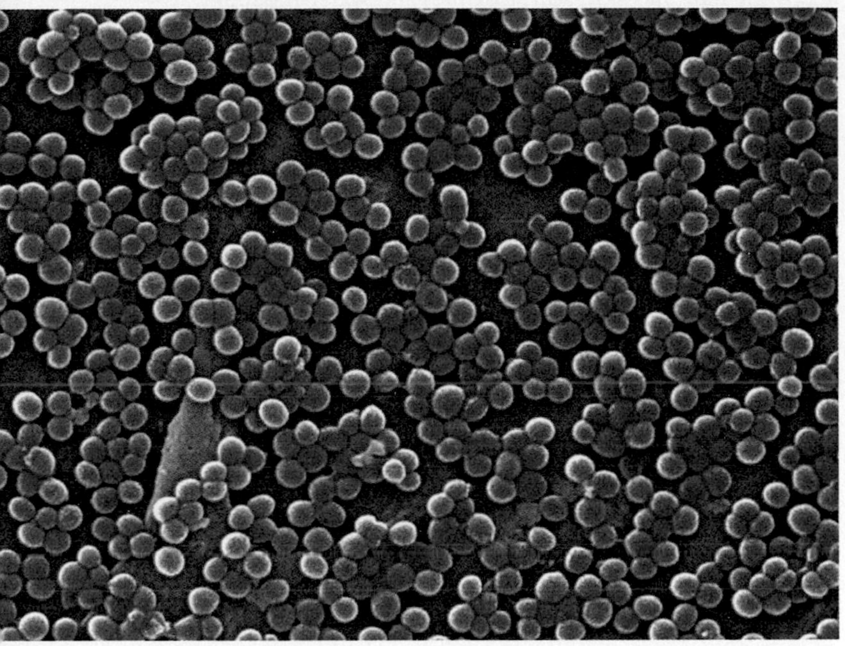

Methicillin-resistant *Staphylococcus aureus* (MRSA) (color-enhanced scanning electron micrograph). *Janice Carr/CDC*

KEY TERMS

Abscess A localized collection of pus within a tissue.

Carbuncle A large, red, swollen, and painful cluster of interconnected boils.

Exfoliative toxins Bacterial toxins that cause the outer epidermis to peel.

Folliculitis Inflammation of hair follicles.

Furuncle Also called a boil; a painful, localized, pus-filled lesion involving an infected hair follicle and surrounding tissue.

Pus Yellowish fluid composed of proteins, living and dead leukocytes, and tissue debris.

Pyoderma Any skin disease characterized by production of pus.

A Glimpse of History

Howard T. Ricketts (1871–1910) studied medicine in Chicago, and then specialized in pathology, the study of the nature of disease and its causes. In 1902, he was appointed to the faculty of the University of Chicago, where his research interests turned to Rocky Mountain spotted fever (RMSF), an often fatal disease characterized by a dramatic rash. RMSF was poorly understood at the time, but scientists suspected it was contracted from tick bites.

Ricketts found that by injecting blood from an ill patient into laboratory animals, he could transmit RMSF. Furthermore, he suspected that the tiny rod-shaped bacteria he saw in the blood of patients and the laboratory animals caused the disease. He then confirmed that certain species of ticks could transmit the disease. The infected ticks remained healthy, yet their eggs frequently contained large numbers of the same bacteria found in the patients' blood. Once fertilized, these eggs developed into infected ticks. Ricketts was never able to cultivate the bacteria for further studies, so he declined to give them a scientific name.

Ricketts later traveled to Mexico to study a very similar disease: louse-borne typhus. Unfortunately, he contracted that disease and died at the age of 39. A Czech scientist who was also studying typhus, Stanislaus Prowazek, met the same fate at almost the same age. The martyrdom of these two young scientists is memorialized in the name of the louse-borne typhus agent, *Rickettsia prowazekii*. Both the genus and the species names of the RMSF agent, *Rickettsia rickettsii*, recognize Howard Ricketts. We now know that these bacteria are obligate intracellular parasites. Because most culture media lack live cells, the bacteria did not grow in the culture media used by Ricketts and Prowazek.

Much of the human body's contact with the outside world occurs at the skin surface. Although this tough, flexible outer covering provides a remarkable barrier to invasion and infection, its exposed state also leaves it vulnerable to a variety of injuries. Cuts, punctures, burns, chemical injury, and insect or tick bites can break this barrier, giving pathogens a way to infect the skin and also access underlying tissues. Skin infections can also occur when microbes that enter the body via the respiratory system or another site are carried by the bloodstream to the skin.

22.1 ■ Anatomy, Physiology, and Ecology of the Skin

Learning Outcomes

1. Describe the functions of skin in health and disease.
2. Explain the role of normal skin microbiota in health and disease.

The skin is far more than a passive wrapping for the body; it functions as a barrier that prevents microbial entry, helps regulate body temperature, and limits the loss of tissue fluid. In addition, sensory receptors within it provide the central nervous system with information about the environment. The skin also plays an essential role in immune system function; if the skin barrier breaks, resident macrophages and dendritic cells produce cytokines that trigger an immune response.

The skin is composed of two layers: the **epidermis** and the **dermis** (**figure 22.1**). The subcutaneous layer, or hypodermis, lies beneath the dermis and supports the skin. Characteristics of these three layers are as follows:

- **Epidermis.** This is the surface layer, made up of multiple layers of flat epithelial cells. The outermost cells are dead and filled with keratin—a tough, water-resistant protein also found in hair and nails. These keratinized skin cells regularly flake off, taking attached microbes with them. Cells at the boundary between the epidermis and the dermis are actively dividing, giving rise to cells that migrate to the skin's surface. There, they become flattened and die as keratin fills their cytoplasm.
- **Dermis.** This layer, composed of connective tissue, is directly below the epidermis. It contains many tiny nerves, glands, blood vessels, and lymphatic vessels, and provides the epidermis with nutrients and removes wastes from it.
- **Subcutaneous tissue** (*cutaneous* relates to the skin). This supportive layer, directly below the dermis, is made up of fat and other cell types; it also contains blood vessels.

The outermost layers of skin are bathed in secretions produced by glands in the dermis. Sweat glands produce sweat, a mixture of mainly water and salt, which is then delivered to the skin surface; the water evaporates and the salty residue left behind inhibits the growth of many microbes. Sebaceous glands produce the oily secretion called **sebum.** These glands open into hair follicles, which along with the associated arrector pili (which causes the hair to stand when it contracts), form what is referred to as a pilosebaceous unit; the sebum flows into the follicles and spills out over the skin surface. Sebum keeps the hair and skin soft, flexible, and water-repellent.

In addition to being a physical barrier, the skin is a distinct ecological habitat. Compared with the moist, warm conditions in the respiratory tract, most areas of the skin are cool and dry. Members of the normal skin microbiota are uniquely suited to thrive in this environment, using the substances in sebum and sweat as nutrients to fuel their growth; in doing so, they inhibit other microbes. For example, resident bacteria degrade lipids in sebum, producing fatty acids that are toxic to many bacteria. In fact, the skin surface is an unfriendly habitat for most pathogens, being too dry, salty, acidic, and toxic for their survival. Those that tolerate the conditions are often shed with dead skin cells.

Different regions of the skin can be compared to unique neighborhoods, made up of distinct numbers and types of inhabitants. Depending on the body location and amount of moisture, the number of bacterial cells on the skin surface may range from only about 1,000 per square centimeter (on the back) to more than 10 million (in the groin and the armpit, where there is plenty of moisture).

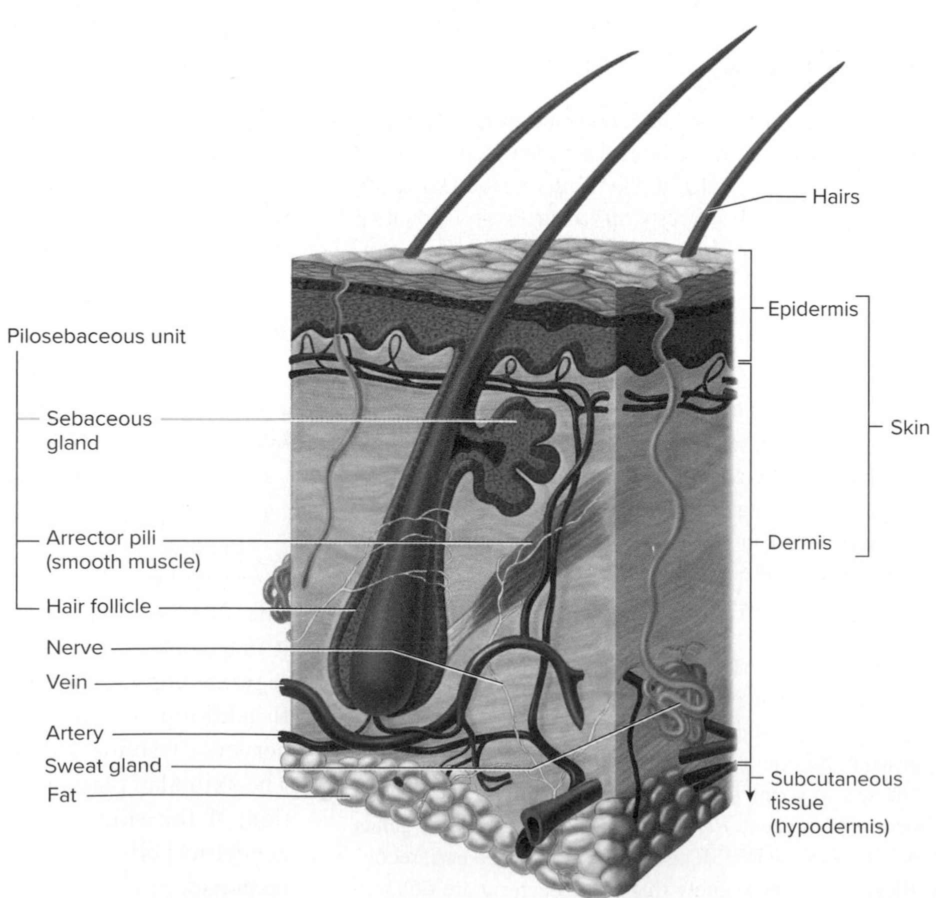

FIGURE 22.1 Microscopic Anatomy of the Skin

? What is the function of keratin in the epidermis?

TABLE 22.1 Principal Members of the Normal Skin Microbiota

Name	Characteristics
Diphtheroids	Variably shaped, non-motile, Gram-positive rods of the *Cutibacterium* and *Corynebacterium* genera
Staphylococci	Gram-positive cocci arranged in packets or clusters; facultatively anaerobic
Fungi	Small yeasts of the genus *Malassezia*; require oily substances for growth

Most of the microbial skin inhabitants can be categorized into three groups: diphtheroids, staphylococci, and fungi (**table 22.1**). Diphtheroids are most common in oily regions like the forehead, upper chest, and back. The name of this informal grouping refers to the fact that these Gram-positive bacteria physically resemble *Corynebacterium diphtheriae,* the pathogen that causes diphtheria. Among the most common diphtheroids on the skin are species of *Cutibacterium (*formerly *Propionibacterium)*—anaerobes that grow within hair follicles, where O_2 is limited. Staphylococci such as *Staphylococcus epidermidis* are salt-tolerant Gram-positive cocci that grow well on the dry environment of the skin surface. These bacteria use available nutrients there, preventing pathogen colonization; they also produce antimicrobial substances active against other Gram-positive bacteria, helping to maintain a balance among the microbial inhabitants of the skin ecosystem. *Malassezia* species are tiny lipid-dependent yeasts (fungi) that are also part of the normal human skin microbiota.

Although members of the normal microbiota play an essential protective role on the skin, they can also be a problem. Some are opportunistic pathogens, causing disease in people with impaired immune defenses. In addition, metabolic products of certain members of the normal microbiota are responsible for body odor: Sweat is odorless when first secreted, but the bacteria degrade some of its components, producing foul-smelling compounds. Most antiperspirants prevent body odor by decreasing the number and metabolic activity of bacteria at the site of application.

Infections of the skin often give rise to characteristic lesions and rashes, including:

- **bullae** (singular: bulla): large, fluid-filled blisters (examples: staphylococcal scalded skin syndrome and impetigo)
- **exanthem:** widespread rash that can cover the whole body (examples: chickenpox, measles)
- **macules:** flat, discolored areas that do not affect skin texture (examples: rubella and rubeola)
- **maculopapular rash:** rash with flat or slightly raised red bumps (examples: rubella and rubeola)
- **nodules:** relatively large, raised, firm bumps (example: acne)
- **papules:** small raised bumps that do not contain fluid (examples: acne and warts)
- **petechiae** (singular: petechia): small purplish spots caused by blood leakage from capillaries under the skin or mucous membranes (example: Rocky Mountain spotted fever)
- **pustules:** red, raised bumps containing **pus,** which is composed of living and dead neutrophils, bacteria, and tissue debris (example: acne)
- **vesicles:** small, elevated lesions that contain clear liquid (example: chickenpox)
- **vesicular rash:** rash containing multiple vesicles (examples: chickenpox and shingles)

MicroByte

The average person sheds about 40,000 skin cells daily, or approximately 1,600 during a typical 1-hour microbiology lecture.

MicroAssessment 22.1

The skin provides a physical barrier to infection and is a distinct ecological habitat; members of the normal skin microbiota help prevent colonization by pathogens. Resident microbes are responsible for body odor and may be opportunistic, causing disease when the host's defenses are impaired. Infections of the skin often give rise to characteristic lesions and rashes.

1. Describe four characteristics of skin that help it resist infection.
2. Name and describe three groups of microorganisms generally present on normal skin.
3. Which group would you expect to have larger numbers of bacteria living on their skin: people living in the tropics or people living in the desert? Explain.

22.2 ■ Bacterial Diseases of the Skin

Learning Outcomes

3. Compare and contrast acne and hair follicle infections.
4. Describe the characteristics of staphylococcal scalded skin syndrome and impetigo.
5. Compare and contrast the pathology and epidemiology of Rocky Mountain spotted fever with those of cutaneous anthrax.

Few bacterial species invade intact skin directly. However, hair follicles provide a route of invasion for pathogens.

Acne Vulgaris

Acne vulgaris, often simply referred to as acne, is an inflammatory disease of the pilosebaceous unit (see figure 22.1). The most common form of acne begins at puberty in association with a rise in sex hormones, although some adults have acne as well. Although it is a common skin condition, acne can have a profound effect on a person's self-esteem and can lead to depression and anxiety—particularly in those who have chronic versions of the disease.

Signs and Symptoms

Acne is characterized by enlarged sebaceous glands and increased sebum secretion, which clogs the hair follicle. This gives rise to lesions called **comedones,** of which there are various forms. The two general types of non-inflamed comedones are:

- **Whiteheads** (closed comedones). These appear as small, white or flesh-colored bumps; the hair follicle pore is completely blocked.
- **Blackheads** (open comedones). These are similar to whiteheads, but are larger and the follicle is open to the skin surface, so sebum and dead skin cells in the clogged follicle are exposed to air. This causes melanin (skin pigment) to oxidize and darken, resulting in the characteristic blackhead.

When comedones become inflamed, they take on different characteristics. Categories of inflamed comedones include:

- **Papules and pustules.** These are commonly referred to as pimples or zits. As described in the previous section, papules are small red bumps; when pus-filled, they are called pustules.
- **Nodules.** These are large, painful, inflamed lumps beneath the skin that are characteristic of more severe forms of acne; when pus-filled, they are sometimes referred to as cysts. Nodules can result in permanent skin discoloration and scarring.

Causative Agent

The species typically associated with acne is *Cutibacterium (Propionibacterium) acnes,* Gram-positive rods that grow anaerobically mainly in sebum-rich pilosebaceous units. Certain subtypes of *C. acnes* are part of the normal skin microbiota of most people throughout their lives.

Pathogenesis

Acne formation involves many factors. It begins when sebaceous glands enlarge and sebum production increases. The hair follicle epithelium thickens and sloughs off in clumps, gradually blocking the flow of sebum to the skin surface. Continued sebum production by the infected gland can force a plug of material to the surface, where it is visible as a comedone.

Various factors are involved with acne formation, but the cause-and-effect relationships are not clear. Dysbiosis (an imbalance in the normal microbiota) of the pilosebaceous unit seems to be involved—specifically increased numbers of certain types of *C. acnes*. Higher levels of keratinization in the follicle and sebum production by the sebaceous glands also occur. An inflammatory response results, attracting neutrophils to the area, sometimes leading to pus-filled comedones. The degree of inflammation and pus accumulation characterizes the various types of acne lesions. Follicle rupture can lead to scarring.

Epidemiology

Acne vulgaris is the most commonly diagnosed skin condition in the United States, affecting over 50 million people annually. It typically begins at puberty and affects a large percentage of adolescents and young adults, although some people have acne well into their 30s or 40s. The increased incidence of acne during puberty is likely due to excess sebum production caused by increased hormone levels, particularly testosterone. Infants may also have acne caused by maternal hormones that stimulate sebum production.

Treatment and Prevention

Treatment of acne depends on the type. Mild acne may be treated with various over-the-counter topical medications (meaning medications used on a body surface) that contain ingredients such as retinoids, benzoyl peroxide, and salicylic acid. These function together to unblock follicles, decrease inflammation, and inhibit bacterial growth. In more severe cases, oral medications are typically prescribed in addition to topical treatments. These include antibiotics such as doxycycline and, in cases where changes in hormone levels contribute to acne formation in females, certain oral contraceptives. Isotretinoin, an oral retinoid that reduces sebaceous gland function, is generally prescribed for only the most serious cases because of its potentially significant side effects. Squeezing acne lesions is ill-advised, because it may cause the follicles to rupture, leading to more acne scars. It can also introduce bacteria into the bloodstream, allowing their spread to multiple locations throughout the body. Acne can be prevented or reduced by keeping the skin and hair clean, minimizing the use of makeup and skincare products that contain oil, and maintaining a healthy diet. **Table 22.2** describes the main features of this disease.

Hair Follicle Infections

Hair follicle infections are generally mild and commonly clear up without treatment. In some cases, however, they may progress into severe or even life-threatening disease.

Signs and Symptoms

The three outcomes of hair follicle infections, described by increasing severity, include:

- **Folliculitis.** This is inflammation of the hair follicles, characterized by small red bumps in the skin. The hair can be pulled from its infected follicle, sometimes releasing a

TABLE 22.2	**Acne Vulgaris**
Signs and Symptoms	Enlarged and blocked sebaceous glands causing comedones; these can become inflamed and sometimes pus-filled.
Incubation Period	Variable
Causative Agent	*Cutibacterium acnes*
Pathogenesis	Increased sebum production and other factors lead to clogged hair follicles. Dysbiosis in the pilosebaceous unit contributes to inflammation and pus formation.
Epidemiology	Common during puberty because of increased hormone (especially testosterone) levels.
Treatment and Prevention	Treatment: over-the-counter medications; sometimes antibiotics, oral contraceptives and, in severe cases, sebum-reducing medications. Prevention: keeping the skin and hair clean, avoiding skin products that contain oil, maintaining a healthy diet.

small amount of pus. Folliculitis often resolves without further treatment.

- **Furuncle (boil).** This occurs when a hair follicle infection extends to adjacent tissues, causing localized redness, swelling, and pain. The pus-filled lump that characterizes a furuncle is a type of skin **abscess** (a collection of pus within a tissue); pus may need to be drained from the furuncle to aid healing.
- **Carbuncle.** This occurs when the infection expands to become a cluster of connected furuncles. It involves several hair follicles and results in a large area of redness, swelling, and pain, with multiple sites of draining pus; fever can be present, along with other signs of a serious infection. Carbuncles usually develop in areas of the body where the skin is thick, such as the back of the neck.

Causative Agent

Most furuncles and carbuncles, as well as many cases of folliculitis, are caused by *Staphylococcus aureus,* a Gram-positive coccus. The genus name of this bacterium derives from the Greek root *staphyle* or "a bunch of grapes," referring to the arrangement of the cells; the species name, *aureus,* or "golden," refers to the typical creamy color of the colonies. *S. aureus* colonizes the skin and mucous membranes of humans and other animals. Although the nostrils seem to be the preferred habitat of the organism (up to 30% of healthy adults carry the bacterium in their nose), moist areas of the skin are also frequent sites of colonization. Despite being part of the normal microbiota of many people, *S. aureus* is an extremely important pathogen and is mentioned often throughout this text as the cause of a number of medical conditions (**table 22.3**).

S. aureus can be distinguished from most other staphylococci because it is one of the few that produces **coagulase** and **clumping factor.** In the laboratory, when *S. aureus* is inoculated into plasma (the fluid portion of blood), its coagulase causes a large clot to form during several hours of incubation. In contrast, clumping factor causes rapid clotting when plasma is mixed into a concentrated suspension of the bacterial cells and is easier to test for in the lab; it is often called "slide coagulase." Both coagulase and clumping factor are important virulence factors for *S. aureus.*

Pathogenesis

Infection begins when *S. aureus* attaches to cells of a hair follicle, multiplies, and spreads to involve the sebaceous gland. This induces an inflammatory response with swelling and redness, followed by attraction and accumulation of neutrophils. If the infection continues, the follicle becomes a plug of inflammatory cells and dead tissue overlying a small abscess. The infection sometimes spreads deeper, reaching the subcutaneous tissue, where a large abscess forms, causing the painful localized swelling of a boil (**figure 22.2**). Without effective treatment, the abscess may expand to other hair follicles and form a carbuncle. If the bacteria enter the bloodstream, the infection can spread to other parts of the body, such as the heart, bones, or brain.

S. aureus strains can have many different virulence factors, although not all strains make the same ones. Nearly all strains have **protein A,** a cell wall component that interferes with phagocytosis by binding to the Fc portion of antibodies and preventing opsonization (see figure 16.10). Many also synthesize a polysaccharide capsule that inhibits phagocytosis (**figure 22.3**).

TABLE 22.3 Some Diseases Often Caused by *Staphylococcus aureus*

Disease	Comments
Food poisoning	Results from ingesting exotoxin in food
Hair follicle infections	Outcomes include folliculitis, furuncles, carbuncles
Impetigo	Superficial skin disease with pus production
Infective endocarditis	Infection of heart valves or inner lining of heart
Staphylococcal scalded skin syndrome	Toxin-mediated skin disease
Toxic shock syndrome	Superantigens causing low blood pressure and organ failure
Wound infections	Colonization of wounds; possible systemic complications

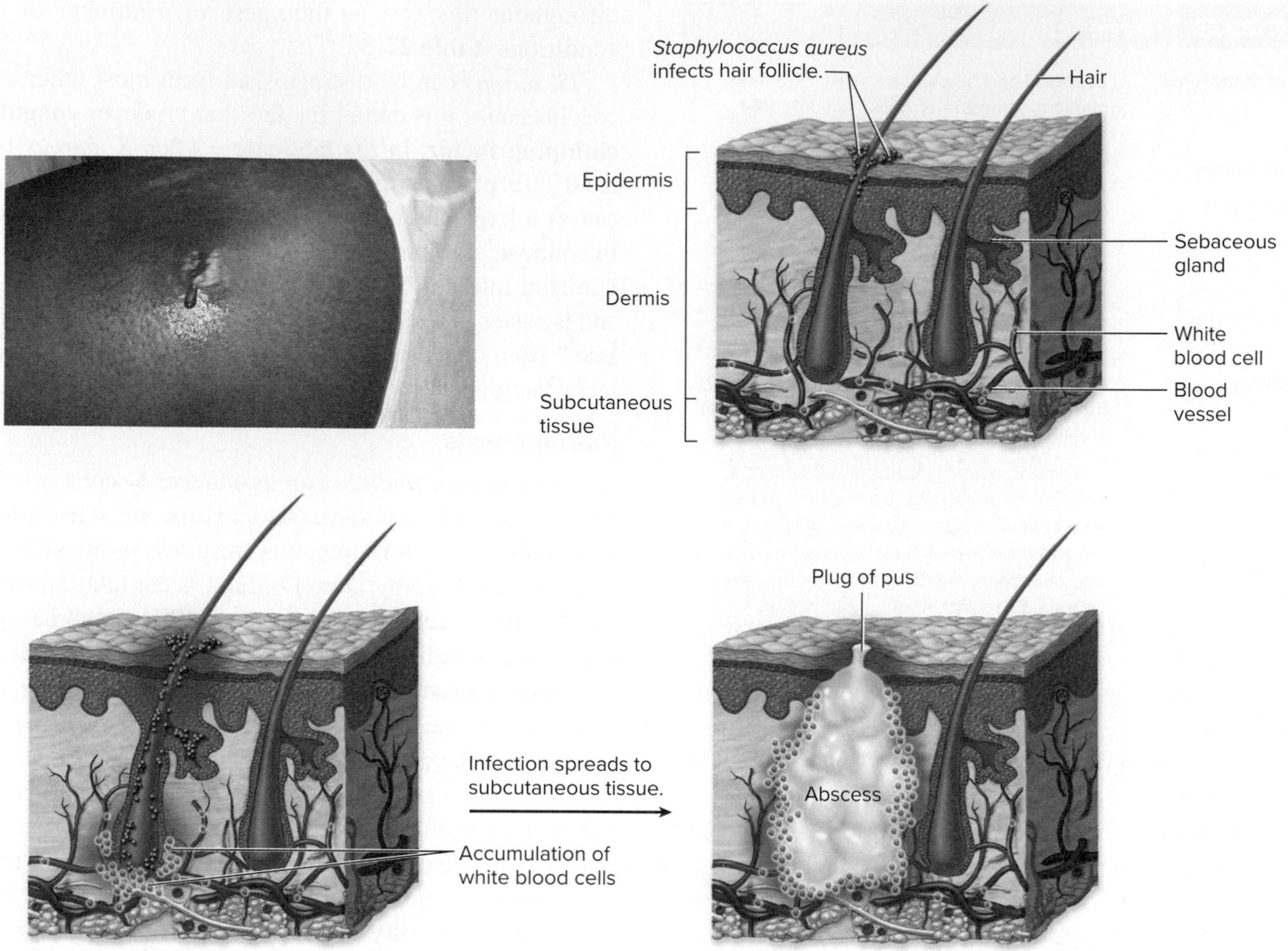

Staphylococcus aureus infects hair follicle.
Hair
Epidermis
Dermis
Subcutaneous tissue
Sebaceous gland
White blood cell
Blood vessel
Plug of pus
Infection spreads to subcutaneous tissue.
Abscess
Accumulation of white blood cells

FIGURE 22.2 Pathogenesis of a Boil (Furuncle) *Staphylococcus aureus* enters a hair follicle through its opening on the skin surface. The infection produces a plug of necrotic material, a small abscess in the dermis, and finally, a larger abscess in the subcutaneous tissue. Centers for Disease Control and Prevention

? What is the risk with an untreated abscess?

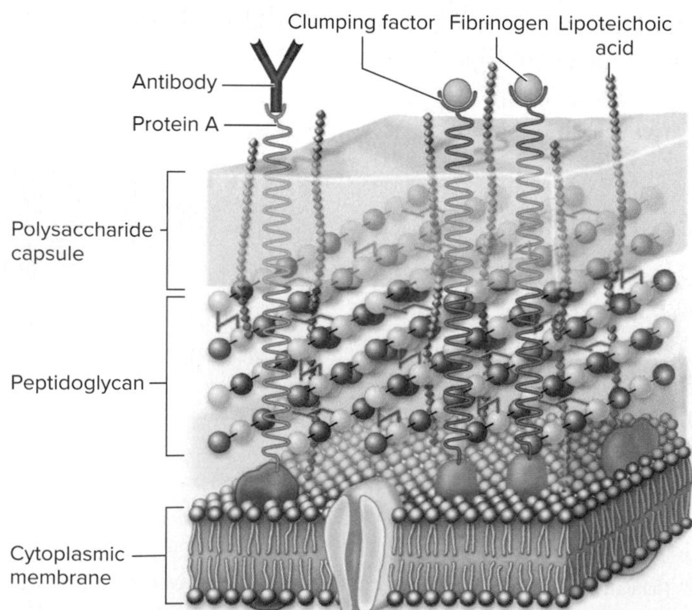

FIGURE 22.3 Virulence Factors Associated with the *Staphylococcus aureus* Cell Surface A polysaccharide capsule, protein A, and clumping factor contribute to the virulence of this organism.

? How does protein A help *Staphylococcus aureus* evade phagocytes?

The coagulase and clumping factor used to identify *S. aureus* are virulence factors that manipulate components of the host's blood clotting system. By doing so, they allow the organism to colonize soft tissues while avoiding the immune response. Coagulase triggers a series of reactions that convert a blood protein called fibrinogen into fibrin (a fibrous protein involved in blood clot formation). Most of the coagulase is released by the bacterial cells, resulting in blood clots that slow leukocyte movement to the site; some remains tightly bound to the cell surface, however, forming a fibrin coating that effectively disguises the cells so they are less likely to be detected by the immune system. Meanwhile, clumping factor on the bacterial cell surface attaches to fibrinogen and fibrin, thus helping the cells colonize the area. Additional cell-surface molecules called fibronectin-binding proteins adhere to fibrinogen as well as to fibronectin (another protein involved in clot formation). Plastic medical devices such as intravenous catheters and artificial heart valves can become coated with fibrinogen once in a patient, so those devices also become targets for colonization.

S. aureus secretes many tissue-damaging enzymes. Hyaluronidase degrades hyaluronic acid, a component of host tissue that helps hold the cells together. Proteases degrade various host proteins, including collagen (the white fibrous protein found in skin, tendons, and connective tissue). Lipases degrade lipids, forming fatty acids and glycerol that provide a nutrient source for *S. aureus* when it colonizes hair follicles.

Most strains of *S. aureus* also produce one or more toxins that damage or kill host cells. The membrane-damaging α-toxin is a pore-forming toxin that kills host cells, including red blood cells; it is responsible for the observed β-hemolysis of *S. aureus* colonies on blood agar. Some particularly virulent *S. aureus* strains also produce a toxin called Panton-Valentine leukocidin (PVL)—a pore-forming toxin that causes leukocyte destruction and tissue necrosis. The toxin is encoded by a specific prophage carried by the bacterium, an example of lysogenic conversion. Certain *S. aureus* strains produce one or more additional disease-specific toxins; for example, exfoliative toxins cause staphylococcal scalded skin syndrome, described shortly. **Table 22.4** summarizes properties of *S. aureus* implicated in its virulence.

Epidemiology

Hair follicle infections caused by *S. aureus* are relatively common—any carrier of the organism is at risk of infection. Carriers can also easily transfer the organism via their hands to fomites, food, or other people, which is one reason why handwashing is so important; this is particularly significant because staphylococci survive well in the environment, so they can be easily transferred from one host to another via fomites. People with boils and other staphylococcal infections shed large numbers of the organism and should not work with food, or near patients with surgical wounds or chronic illnesses.

Because *S. aureus* is so common, tracing the source of a staphylococcal outbreak requires precise identification of the strain involved. This can be done using techniques such as whole genome sequencing, phage typing, and antibiogram typing (see section 10.4).

Treatment and Prevention

Although folliculitis usually resolves on its own, boils and carbuncles may require minor surgery to drain the pus from the lesion. Afterward, patients are often given oral antibiotics.

Treating any type of *S. aureus* infection can be complicated because many strains are resistant to multiple antibacterial medications. Nearly all *S. aureus* strains were initially susceptible to penicillin. However, due to the rapid rise and spread of antimicrobial resistance, most isolates are now resistant, not just to penicillin, but to many of the other β-lactam drugs as well as other antimicrobials (see table 20.1). These strains are grouped as:

TABLE 22.4 Properties of *Staphylococcus aureus* Implicated in Its Virulence

Product	Effect
Capsule	Inhibits phagocytosis
Clumping factor	Attaches the bacterium to fibrin, fibrinogen, and plastic devices
Coagulase	Slows progress of leukocytes into infected area by producing clots in the surrounding capillaries
Enterotoxins	Superantigens that cause food poisoning if ingested, cause toxic shock if systemic
Exfoliative toxins (ET)	Destroy material that binds outer layers of the epidermis together, causing scalded skin syndrome
Fibronectin-binding protein	Attaches bacterium to endothelium, epithelium, acellular tissue substances such as clots, and indwelling devices
Tissue-damaging enzymes	Hyaluronidase breaks down hyaluronic acid component of tissue (allowing infection to spread); lipase breaks down fats; proteases degrade collagen and other tissue proteins.
Pore-forming toxins	Make holes in cell membranes of various host cells, including red blood cells (α-toxin) and leukocytes (PVL)
Protein A	Binds to Fc portion of antibody molecules, thereby interfering with opsonization that otherwise facilitates phagocytosis
Toxic shock syndrome toxin	Causes rash, diarrhea, and shock

- **Methicillin-resistant *S. aureus* (MRSA).** These produce PBP2a, a modified version of a penicillin-binding protein, the target of all β-lactam drugs. As a result, these strains are resistant to all β-lactam antibiotics except ceftaroline, a relatively new cephalosporin. When MRSA strains first appeared, most could be traced to hospitals and clinics. These strains are referred to as HA-MRSA (healthcare-associated MRSA). Many HA-MRSA strains are resistant to a number of other medications as well. More recently, completely different strains have become widespread among healthy carriers in the community; these strains are called CA-MRSA (community-acquired MRSA). In general, CA-MRSA strains are resistant only to β-lactam antibiotics (except ceftaroline) and macrolides.
- **Vancomycin-intermediate *S. aureus* (VISA) and vancomycin-resistant *S. aureus* (VRSA).** Most MRSA strains were reliably treated with vancomycin until the late 1990s when these strains were identified. Although newer medications such as daptomycin and linezolid are sometimes used, the optimal treatments for infections due to VISA or VRSA are uncertain.

Preventing staphylococcal skin disease is difficult because so many people carry the organism. The carrier state can potentially be eliminated—or at least the numbers of colonizing bacterial cells reduced—by applying an antibacterial cream to the nostrils and washing the skin using soaps or wipes containing an antibacterial agent such as chlorhexidine. Because *S. aureus* is most commonly transmitted by the hands, handwashing and regular use of hand sanitizers limit the spread of these bacteria. In an effort to control the spread of MRSA, many hospitals screen patients at admission. The patients who carry a MRSA strain are isolated and given appropriate antibacterial treatment to prevent the bacterium from spreading to other people. **Table 22.5** describes the main features of hair follicle infections caused by *S. aureus*.

TABLE 22.5 Hair Follicle Infections

Signs and Symptoms	Small red bumps that may progress to furuncles or carbuncles with swelling, redness, tenderness, pus, and sometimes fever.
Incubation Period	Variable
Causative Agent	*Staphylococcus aureus*
Pathogenesis	Bacteria enter a hair follicle, causing an inflammatory response. The follicle becomes plugged with white blood cells and dead tissue. Bacteria may invade subcutaneous tissue, causing an abscess.
Epidemiology	Many people are carriers. Person-to-person transmission; fomites
Treatment and Prevention	Treatment: sometimes minor surgery to drain the pus, followed by an oral antibiotic. Prevention: handwashing and hand sanitizers.

MicroByte

People colonized with *S. aureus* may have as many as 10^8 cells of the bacterium per nostril.

Staphylococcal Scalded Skin Syndrome

Staphylococcal scalded skin syndrome (SSSS) is a toxin-mediated disease that occurs mainly in infants.

Signs and Symptoms

As the name suggests, SSSS causes a patient's skin to look as though it has been burned with hot liquid or steam; the outer layer blisters and peels, resulting in significant tissue fluid loss and increased risk of secondary infections (**figure 22.4**). SSSS begins after a variable incubation period with a generalized redness of the skin, sometimes affecting the entire body. Other early signs and symptoms include malaise (a vague feeling of discomfort and uneasiness) as well as irritability and fever. The nose, mouth, and genitalia may be painful for one or more days before the typical features of the disease become apparent. Within 48 hours after the redness appears, the skin becomes wrinkled and then bullae (large blisters filled with clear fluid) develop. The skin is tender to the touch, peels easily, and feels like sandpaper.

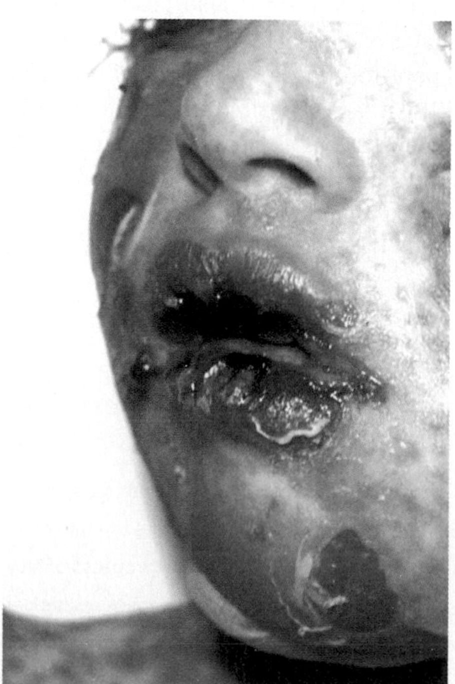

FIGURE 22.4
Staphylococcal Scalded Skin Syndrome (SSSS) Exfoliative toxins, which are produced by certain strains of *Staphylococcus aureus,* cause the outer layer of skin to blister and peel. BSIP SA/Alamy Stock Photo

? Why might this disease sometimes be fatal?

Causative Agent

Staphylococcal scalded skin syndrome is caused by *Staphylococcus aureus* strains that produce one or more **exfoliative toxins (ETs).** Only about 5% of *S. aureus* strains make an exfoliative toxin, and some of these produce more than one type.

Pathogenesis

Exfoliative toxins cause the epidermis to peel off just below the dead keratinized outer layer by digesting an important cell-to-cell adhesion molecule. Even though the bacteria generally colonize a relatively small area of the skin, that localized infection results in widespread skin damage because the toxin is carried throughout the body via the bloodstream. The skin damage predisposes the patient to significant fluid loss and secondary infections, which may prove fatal without treatment.

Epidemiology

Staphylococcal scalded skin syndrome can affect anyone, but it occurs most often in newborn infants; elderly and immunocompromised people are also at increased risk. Transmission is generally by direct person-to-person contact. The disease is usually sporadic, but small epidemics in nurseries sometimes occur.

Treatment and Prevention

Patients with SSSS are typically treated with intravenous (IV) antibiotic therapy and rehydration. Dead skin is removed (a procedure called debridement) to help prevent secondary infections. Although the disease can be fatal, quick treatment usually leads to full recovery. There are no preventive measures except to place SSSS patients in protective isolation, which limits the spread of the pathogen to others and helps prevent secondary infections in the isolated patient. **Table 22.6** describes the main features of this disease.

TABLE 22.6 Staphylococcal Scalded Skin Syndrome

Signs and Symptoms	Malaise, irritability, fever, sensitive red rash with sandpaper texture, large blisters, peeling of outer skin layers
Incubation Period	Variable, usually days
Causative Agent	Strains of *Staphylococcus aureus* that produce exfoliative toxins
Pathogenesis	Exfoliative toxins produced in a localized infection are carried throughout the body by the bloodstream, causing widespread blistering and peeling of the skin's outer layer; fluid loss and secondary infections contribute to mortality.
Epidemiology	Occurs mainly in infants; person-to-person transmission
Treatment and Prevention	Treatment: antibiotics; remove dead tissue. Prevention: isolate patient to limit spread of the pathogen to others and to prevent secondary infection.

Impetigo

Impetigo is the most common type of **pyoderma**—a superficial skin disease characterized by pus production (*pyo-* means pus)—and the terms are often used interchangeably. Impetigo is highly contagious and can sometimes occur after a cut, insect bite, or other skin injury that may not even be noticeable.

Signs and Symptoms

Signs and symptoms of impetigo usually develop up to 10 days after the causative agent colonizes either intact or broken skin, although not everyone who is colonized develops the disease. The resulting infection causes inflammation in patches of epidermis just beneath the dead, keratinized outer layer. There are two main types of impetigo:

- **Non-bullous impetigo.** This is the more common form. Small red papules form that develop into vesicles and then pustules; these break and ooze, giving rise to yellowish, crusty lesions of drying plasma (**figure 22.5**). The lesions are usually on the face or the extremities, and multiple lesions can sometimes develop.
- **Bullous impetigo.** This is a localized form of staphylococcal scalded skin syndrome, characterized by large, thin-walled blisters (bullae) that are initially filled with clear or yellowish fluid which gradually becomes dark. Yellow crusts do not form.

Causative Agent

Most cases of non-bullous impetigo are due to either *Staphylococcus aureus* or *Streptococcus pyogenes*, but sometimes both are involved. *S. aureus* was described earlier in this chapter, and *S. pyogenes* was described in chapter 21. Although these organisms are both Gram-positive cocci, they have significant differences, as summarized in **table 22.7**. Only exfoliative toxin-producing strains of *S. aureus* cause bullous impetigo.

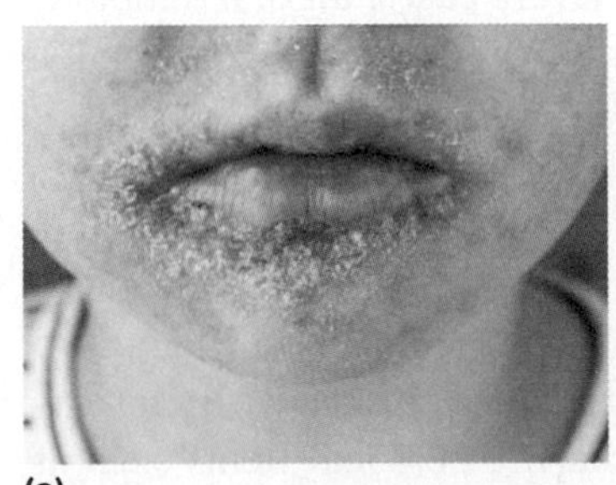
(a)

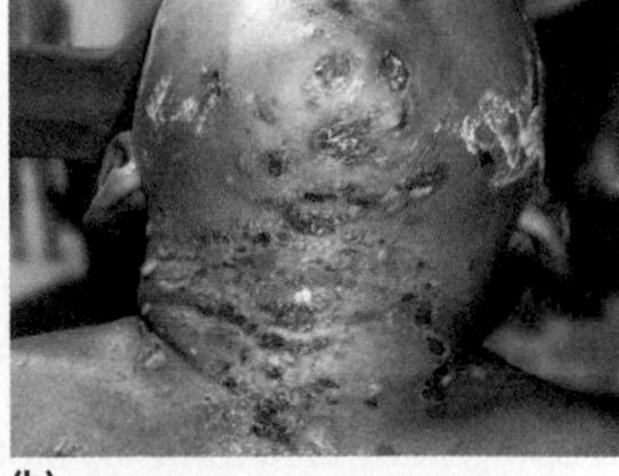
(b)

FIGURE 22.5 Patients with Impetigo (a) Non-bullous impetigo. **(b)** Bullous impetigo FotoHelin/Shutterstock; Zay Nyi Nyi/Shutterstock

? What causes the yellow crusts that develop in non-bullous impetigo?

TABLE 22.7 *Streptococcus pyogenes* Versus *Staphylococcus aureus*

	Streptococcus pyogenes	*Staphylococcus aureus*
Morphology	Chains of Gram-positive cocci	Clusters of Gram-positive cocci
Growth Characteristics	Beta-hemolytic colonies; catalase-negative, coagulase-negative, obligate fermenter	Golden, beta-hemolytic colonies, catalase-positive, coagulase-positive, facultative anaerobe
Virulence Factors	Bacterium produces hemolysins (streptolysins O and S), streptokinase, DNase, hyaluronidase, and others	Bacterium produces coagulase, pore-forming toxins, hyaluronidase, nuclease, protease, and others
Diseases	Strep throat, scarlet fever, non-bullous impetigo, wound infections, puerperal fever, streptococcal toxic shock syndrome, necrotizing fasciitis. Complications: glomerulonephritis and rheumatic fever	Hair follicle infections (folliculitis, furuncles, carbuncles), staphylococcal scalded skin syndrome, non-bullous and bullous impetigo, wound infections, abscesses, bone infections, staphylococcal toxic shock syndrome, and food poisoning

Pathogenesis

Minor injuries introduce *S. aureus* or *S. pyogenes* growing on superficial skin layers into deeper layers of epidermis, leading to infection. Scratching an infected area may spread the bacteria to other areas of the skin. As described previously and summarized in table 22.7, both organisms have numerous virulence factors that interfere with phagocytosis and cause tissue damage. Impetigo caused by *S. pyogenes* may lead to complications, particularly acute post-streptococcal glomerulonephritis (see figure 21.7).

Bullous impetigo is caused by strains of *S. aureus* that produce exfoliative toxins that weaken cell connections in the epidermis, resulting in the formation of large blisters.

Epidemiology

Impetigo is most common in children between 2 and 5 years old but can occur at any age; rates of the disease are highest in resource-limited settings in areas that are hot, humid, and crowded. Person-to-person contact spreads the disease, as does contact with fomites such as toys and towels.

Treatment and Prevention

Mild cases of impetigo are treated by gently cleaning the sores, removing any crusts, and applying a prescription topical antibiotic cream or ointment. In more severe cases, an oral antibiotic such as cephalexine (a cephalosporin) is given; the medication chosen should be effective against both *S. aureus* and *S. pyogenes* unless only a single causative agent is identified. In cases involving MRSA, non-β-lactam drugs such as doxycycline are used instead.

People with impetigo should avoid contact with others to avoid spreading the bacteria. Transmission can be limited by keeping skin clean and by not sharing personal items such as towels and washcloths. Antiseptics and wound cleansing also decrease the chance of infection. **Table 22.8** summarizes the main features of impetigo.

Rocky Mountain Spotted Fever

Rocky Mountain spotted fever (RMSF)—a disease first recognized in the Rocky Mountain area of the United States—is one of several diseases in a group called spotted fever rickettsioses. These arthropod-borne diseases are grouped together primarily for reporting purposes because they can be difficult to tell apart in their early stages and are caused by related pathogens. In the United States, RMSF is the most serious and common of the group.

Signs and Symptoms

RMSF begins after an incubation period of 3 to 12 days with a headache, fever, and muscle and joint pain. In most cases, a maculopapular rash characterized by flat pink spots appears

TABLE 22.8 Impetigo

Signs and Symptoms	Non-bullous impetigo: papules, vesicles, then pustules that break, releasing plasma which dries, forming yellowish crusts. Bullous impetigo: large, gradually darkening blisters (bullae), no crusts
Incubation Period	Up to 10 days
Causative Agents	*Staphylococcus aureus*, *Streptococcus pyogenes*
Pathogenesis	Organisms enter the skin, sometimes through minor breaks; some *S. pyogenes* strains that cause impetigo can also cause glomerulonephritis; exfoliative toxin-producing *S. aureus* causes bullous impetigo
Epidemiology	Spread by direct contact and fomites.
Treatment and Prevention	Treatment: appropriate antibiotic, topical or oral. Prevention: keeping skin clean; treating skin injuries.

on the wrists and ankles a few days after the onset of the fever and can spread up the arms and legs to the rest of the body (**figure 22.6**). The rash may also spread to the palms and soles of the feet, which is very characteristic of RMSF. Over time, small bumps and petechiae (which form when small blood vessels bleed under the skin) can appear. Damage to the heart, kidneys, brain, and other body tissues can cause a drop in blood pressure. Shock (a condition resulting from life-threatening low blood pressure) and death follow unless treatment is given promptly.

Causative Agent

RMSF is caused by *Rickettsia rickettsii,* a tiny Gram-negative coccobacillus (**figure 22.7**). The organism is an obligate intracellular parasite, and its dependence on host cells makes growing it in culture very difficult (see A Glimpse of History in this chapter).

Pathogenesis

RMSF is transmitted by the bite of a tick infected with *R. rickettsii.* The tick attaches and feeds, but the bite is usually painless and goes unnoticed. The bacteria are typically transmitted to the host after 4–10 hours of feeding. Once the bacteria are released into capillary blood, they induce the vessel endothelial cells to engulf them. Inside these cells, the bacteria leave the phagosome and multiply. Like several other pathogens, they cause host cell actin to polymerize, forming an "actin tail" that propels the bacterial cells into adjacent host cells (see Focus Your Perspective 3.1). Eventually, the host cell membrane becomes so damaged that the cell ruptures, releasing bacteria to travel through the bloodstream, infecting more cells.

Infection directly damages the endothelial cells that line blood vessels. The resulting inflammatory response causes an increase in vascular permeability as well as an accumulation of platelets (cells involved in blood clotting) in the affected areas. This process results in the characteristic rash and other symptoms seen in RMSF. In severe cases, the systemic inflammation results in damage to various organs including the brain, heart, and kidneys.

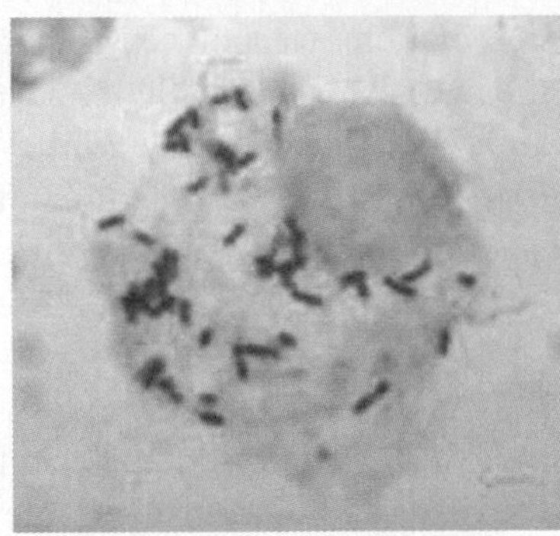

FIGURE 22.7 The Obligate Intracellular Bacterium *Rickettsia rickettsii* Centers for Disease Control and Prevention

? Can *Rickettsia rickettsii* be grown on blood agar? Explain.

Epidemiology

Despite its name, the highest incidence of RMSF in the United States has generally been in the south Atlantic and south-central states (**figure 22.8**). Five states (North Carolina, Oklahoma, Arkansas, Tennessee, and Missouri) account for 60% of cases. The disease is more common when ticks are most active: April to September.

RMSF is a zoonosis, maintained in nature in various species of ticks and mammals. Rickettsias generally do not cause disease in their natural hosts—the host and pathogen have co-evolved to reach a state of balanced pathogenicity. Because humans are accidental hosts, they can develop severe disease.

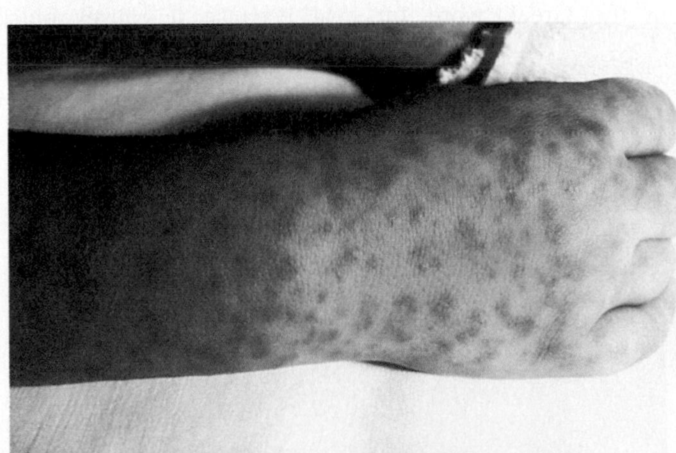

FIGURE 22.6 Rash Caused by Rocky Mountain Spotted Fever (RMSF) Early in the disease, most people have a rash of flat pink non-itchy spots. The rash typically begins on the wrists and ankles, then spreads to the rest of the body. Centers for Disease Control and Prevention

? How do petechiae form?

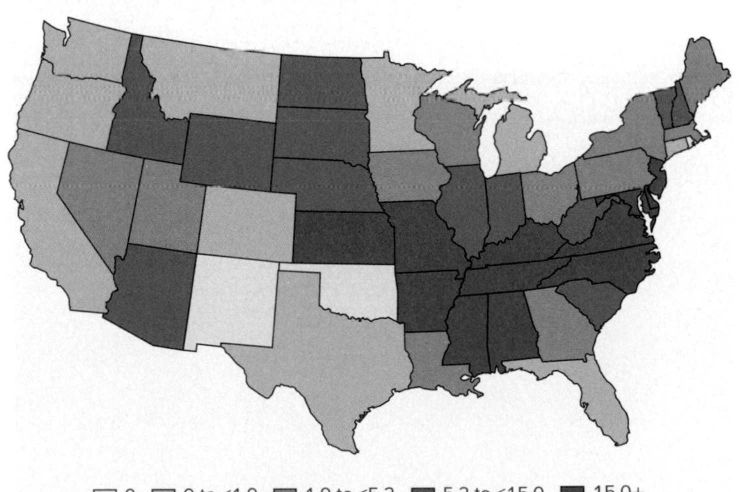

FIGURE 22.8 Annual Incidence of Spotted Fever Rickettsiosis Reported cases per million persons in the United States, 2019. "Tickborne Diseases of the United States: Rocky Mountain Spotted Fever," Centers for Disease Control and Prevention. www.cdc.gov/rmsf/stats/index.html

? Why do most cases of RMSF occur in the summer months?

Several species of ticks transmit RMSF to humans. The main vector in the western United States is the wood tick, *Dermacentor andersoni* (**figure 22.9**), while in the eastern United States it is the dog tick, *D. variabilis.* The brown dog tick, *Rhicephalus sanguineus*, serves as the primary vector in parts of the southwestern United States and Mexico.

Treatment and Prevention

The antibiotic doxycycline is very effective in treating RMSF if given early in the disease, before irreversible organ damage has occurred. Because of this, treatment is initiated once the disease is suspected based on signs and symptoms. Without treatment, the overall case-fatality rate is about 20%, but it can be much higher in elderly patients. With early diagnosis and treatment, the case-fatality rate is reduced to less than 5%.

There is no vaccine against RMSF. To prevent infection, people in regions where the tick vectors are present should (1) avoid grassy, bushy, and wooded areas; (2) use approved tick repellents such as those containing DEET (*N,N*-diethyl-*meta*-toluamide); (3) treat clothing and gear with 0.5% permethrin; and (4) carefully inspect themselves and their pets for ticks several times daily. When removing attached ticks, care should be taken not to crush them because that could contaminate the bite wound; instead, tweezers should be used to grasp the tick head and mouthparts, and then gently pull the tick away from the skin. An antiseptic should be applied to the bite site. The main features of RMSF are summarized in **table 22.9.**

TABLE 22.9	**Rocky Mountain Spotted Fever (RMSF)**
Signs and Symptoms	Headache, muscle and joint pain, and fever, followed by a rash that begins on the extremities
Incubation Period	3 to 12 days
Causative Agent	*Rickettsia rickettsii*, an obligate intracellular bacterium
Pathogenesis	Organisms multiply at site of tick bite, invade the bloodstream, and then infect endothelial cells; blood vessel involvement and systemic inflammatory response damage tissues.
Epidemiology	A zoonosis transmitted by bite of infected tick, usually *Dermacentor* species
Treatment and Prevention	Treatment: appropriate antibiotics. Prevention: avoid tick-infested areas; use tick repellent; remove attached ticks.

Cutaneous Anthrax

Anthrax is found in animal populations around the world but is rare in humans. It may, however, occur when endospores of *Bacillus anthracis* enter the body through the respiratory tract (inhalation anthrax), by ingestion (gastrointestinal anthrax), or through skin lesions (cutaneous anthrax). Cutaneous anthrax is the most common type in humans, but inhalation anthrax has a much higher case-fatality rate.

FIGURE 22.9 ***Dermacentor andersoni,* the Wood Tick** The wood tick is the main vector of RMSF in the western United States. James Gathany/CDC

? Humans are accidental hosts for *R. rickettsia*. What does this mean, and why is it a problem?

Signs and Symptoms

Within 1 to 7 days after skin exposure to *B. anthracis* endospores, a papule (small red bump) appears at the site of infection. This gradually develops vesicles or blisters and eventually forms a painless ulcer surrounded by swelling. A characteristic black eschar (dead tissue resembling a flat scab) covers the center of the area and may persist for several weeks (**figure 22.10**).

Causative Agent

B. anthracis is an endospore-forming, Gram-positive, non-motile, rod-shaped bacterium. Its species name is derived from the Greek word for coal because it causes black

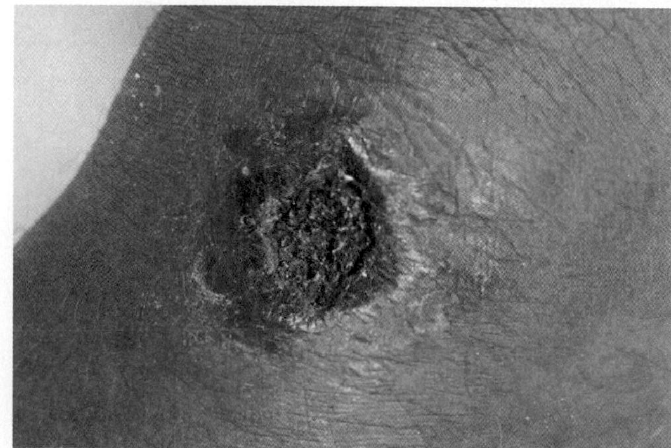

FIGURE 22.10 Anthrax Lesion This lesion on the forearm shows the distinctive tissue erosion and black eschar associated with *Bacillus anthracis* infection. James H. Steele/CDC

? What causes the characteristic eschar of cutaneous anthrax?

"coal-like" scabs to form on infected skin (see figure 22.10). The infectious form of the organism is an endospore that can survive in the environment for extended periods. In the host, the endospores germinate to become encapsulated vegetative cells capable of producing exotoxins.

Pathogenesis

Once introduced under the skin, *B. anthracis* endospores germinate, giving rise to vegetative cells that have an antiphagocytic capsule. The cells make three key proteins that assemble to become toxins that disrupt cell signaling pathways, inhibit immune cells, and eventually lead to tissue-damaging cell death. The interactions of the proteins are described in the discussion of inhalation anthrax in section 21.4. All types of anthrax have the potential to become systemic, but this is much less likely with cutaneous anthrax.

MicroByte

Intravenous drug abusers may develop injection anthrax, which resembles cutaneous anthrax but the infection can be deeper—sometimes involving muscle tissue—and may become systemic.

Epidemiology

Cutaneous anthrax accounts for over 90% of human anthrax cases. It is usually seen in people who work with animals or animal products such as farmers, veterinarians, leather tanners, or those who work with wool. Cases are usually mild and require no treatment. A small percentage of cases become systemic, most often leading to death. Routine person-to-person transmission of cutaneous anthrax does not occur.

Treatment and Prevention

Cutaneous anthrax can be treated with ciprofloxacin (or another fluoroquinolone) or doxycycline. The treatment has little effect on healing of the skin lesion, but it may prevent the organism from spreading systemically. Typical cases do not require hospitalization. As with inhalation anthrax, a vaccine is available but only given to people at greatest risk of exposure; these include people working with potentially infected animals or contaminated animal products, or certain laboratory personnel who work with the organism. The main features of cutaneous anthrax are summarized in **table 22.10.**

TABLE 22.10 Cutaneous Anthrax

Signs and Symptoms	Small red bump develops into painless black eschar
Incubation Period	Usually 1 to 7 days
Causative Agent	*Bacillus anthracis*, an endospore-forming, toxin-producing, encapsulated Gram-positive rod
Pathogenesis	Vegetative cells produce three proteins that assemble to become toxins that inhibit immune function and cause cell death, resulting in tissue damage.
Epidemiology	Occurs in people who work with animals or animal products; no person-to-person transmission
Treatment and Prevention	Treatment: antimicrobial medications. Prevention: vaccine for at-risk populations.

MicroAssessment 22.2

Acne is associated with *Cutibacterium acnes*, and while very common, varies in severity. Folliculitis, furuncles, and carbuncles are usually caused by *Staphylococcus aureus*. Some strains of *S. aureus* produce exfoliative toxin and cause staphylococcal scalded skin syndrome. Impetigo is a pyoderma often caused by *Streptococcus pyogenes* or *S. aureus*. Rocky Mountain spotted fever is caused by *Rickettsia rickettsii* and is transmitted by ticks. Cutaneous anthrax is caused by *Bacillus anthracis* and is rare in humans.

4. List four extracellular products of *S. aureus* that contribute to its virulence.
5. Describe the pattern of skin rash used to diagnose Rocky Mountain spotted fever.
6. Patients with staphylococcal scalded skin syndrome often do not have *S. aureus* growing in the affected areas. How can this be explained?

22.3 ■ Viral Diseases of the Skin

Learning Outcomes

6. Compare and contrast chickenpox and shingles, measles, and rubella.
7. Compare and contrast fifth disease, roseola, and hand-foot-and-mouth disease.
8. Describe warts and their treatment.

While some viruses that cause skin rashes enter through skin lesions, several others initially infect the respiratory tract and then are carried in the blood to the skin. Regardless of the entry route, the resulting rash is called a viral exanthem (from the Greek word exanthema which means "breaking out"). The appearance of the rash and other clinical findings are useful in diagnosis, but immunological testing can also be used when the disease is not typical.

Varicella (Chickenpox) and Herpes Zoster (Shingles)

Chickenpox is the popular name for varicella, a type of rash that was common in childhood before the varicella vaccine was introduced in 1995. The causative agent is a member of the *Herpesviridae* family. All herpesviruses produce latent infections that can reactivate long after recovery from the initial illness; in the case of chickenpox, that reactivation results in herpes zoster (**shingles**).

Signs and Symptoms

Most cases of childhood chickenpox are mild. Early signs and symptoms include fever, headache, and malaise. A characteristic pruritic (itchy) rash develops 10 to 21 days after infection, and lasts for 5 to 10 days. It begins with macules (small red spots) that progress to papules (little bumps). These become small vesicles (fluid-filled blisters), forming a vesicular rash, that then burst and crust over. All stages of the rash can be seen at the same time, and the lesions can develop anywhere on the body, but usually first appear on the head, chest, and back (**figure 22.11**). They also may occur on mucous membranes such as those lining the mouth and eyelids. Patients may have only a few lesions or as many as a few hundred. Scratching the lesions should be avoided because it can lead to serious, even fatal, secondary infection by *Streptococcus pyogenes* or *Staphylococcus aureus.*

The signs and symptoms of chickenpox tend to be more severe in older children and adults. Some adults can develop varicella pneumonia, characterized by rapid breathing, coughing, shortness of breath, and a dusky skin color from lack of O_2. The pneumonia subsides with the rash, but respiratory signs and symptoms often persist for weeks. Chickenpox is a threat to immunocompromised patients of any age—the virus damages the lungs, heart, liver, kidneys, and brain, resulting in death in about 20% of the cases.

Anyone who has been infected with the virus that causes chickenpox is at risk of later developing shingles, a disease that results from reactivation of the latent virus; it can occur at any age but becomes increasingly common later in life. Shingles begins with pain or tingling in the region supplied by a sensory nerve that houses the latent virus. After about 1 to 5 days, a chickenpox-like rash appears, but unlike with chickenpox, the rash is usually restricted to the area supplied by the branches of the sensory nerve involved (**figure 22.12**). These nerves most often serve the chest or abdomen, but shingles can appear on a limb or on the face. Facial shingles can pose a risk to the eyes. Unlike the chickenpox rash, the shingles rash is very painful. Although the lesions generally subside within a week, the pain may persist for months, or longer. In people with AIDS or other immunodeficiency, the rash may spread over the entire body.

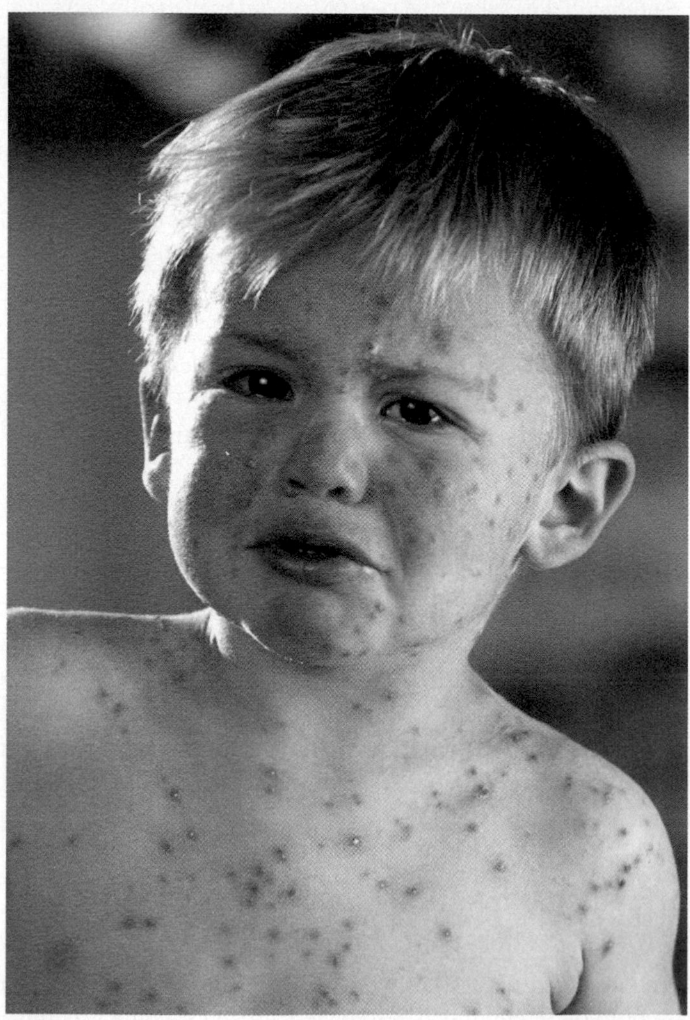

FIGURE 22.11 A Child with Chickenpox (Varicella) Characteristically, lesions in various stages of development—macules, papules, and vesicles—are present. SW Productions/Brand X Pictures/Getty Images

? **Why is scratching chickenpox lesions potentially risky?**

Causative Agent

Chickenpox and shingles are caused by varicella-zoster virus (VZV), a member of the *Herpesviridae* family. It is an enveloped, double-stranded DNA virus that becomes latent in the nerve cells of infected individuals.

Pathogenesis

VZV enters via the upper respiratory tract and replicates in local mucosal cells. Infected phagocytes carry the virus to the regional lymph nodes, where VZV infects T cells; these

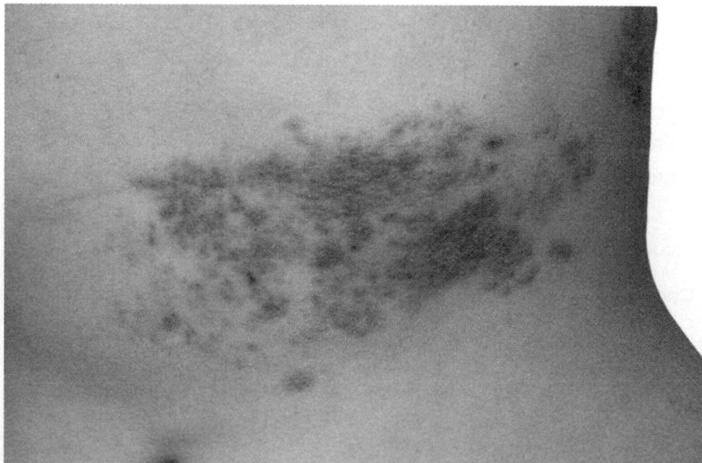

FIGURE 22.12 Shingles The painful rash typically is limited to the surface region of the body associated with a sensory nerve housing the latent virus. Franciscodiazpagador/iStock/Getty Images

? **Why is the shingles rash limited to a smaller area and not as widespread as the chickenpox rash?**

enter the bloodstream and allow the virus to replicate in other body sites. Eventually, the virus travels to the skin and once there, VZV spreads locally, resulting in development of the characteristic vesicular rash. Stained preparations of VZV-infected cells show inclusion bodies which function as virus-producing factories. Some of the infected cells fuse together, forming multinucleated giant cells which swell and eventually lyse.

During varicella infection, the virus can enter nerve endings and establish a latent infection in the ganglia (singular: ganglion), which are collections of nerve cell bodies (the part of a nerve cell that contains the cell nucleus) located near the spinal cord. Throughout the latent infection, the VZV genome is not replicated, and very few viral genes are expressed; the mechanism of suppression is not known but may involve regulation by immune cells.

Shingles is most likely to occur when cell-mediated immunity declines, allowing VZV to reactivate in an infected nerve cell. The virus replicates in the cell, and then moves via the nerve to the skin, where it can spread locally, causing skin lesions. Unlike with chickenpox, however, the replicating virus originates from a single infected nerve cell rather than from the bloodstream; thus, the rash of shingles is limited to only a specific region, as opposed to the widespread rash of chickenpox. An intense secondary (memory) immune response begins, leading to the disappearance of the lesions. Sometimes, however, the inflammatory response leads to scars and chronic pain called postherpetic neuralgia that lasts long after the rash has disappeared.

Epidemiology

Chickenpox is a nationally notifiable disease, requiring that confirmed cases are reported to the CDC; many cases are so mild, however, that they go unreported. In the early 1990s, before a vaccine was available, approximately 4 million cases of chickenpox occurred each year in the United States, resulting in 10,000 hospitalizations and approximately 100 to 150 deaths; within 10 years after the vaccine was approved, there were fewer than half a million cases of chickenpox and 8 deaths each year.

VZV is highly contagious and is mainly transmitted by direct contact with skin lesions or via inhalation of aerosolized respiratory secretions or skin lesion debris. As with many diseases transmitted through inhalation, most cases occur in the winter and spring. Humans are the only reservoir. People are infectious from 1 to 2 days before the rash appears until all the lesions have crusted (usually 4 days after the onset).

Chickenpox is a serious threat to babies. Without treatment, it has a case-fatality rate as high as 30% in newborn infants. A baby born to a woman who had chickenpox early in pregnancy may have congenital varicella syndrome, a rare disorder characterized by abnormalities such as cataracts and underdeveloped head and limbs.

The ability of VZV to form a latent infection allows it to persist within a population. A person who had chickenpox in the past can develop shingles, creating a source of the virus that can be transmitted to susceptible individuals. Those newly infected people will then develop chickenpox. Thus, epidemics of chickenpox can reappear from exposure to cases of shingles if enough susceptible people are present.

Treatment and Prevention

Antiviral medications, such as acyclovir and valacyclovir, are recommended for treating VZV infections in groups at high risk for complications. Certain over-the-counter medications may be given to ease fever, but aspirin should not be given to children or teenagers because it increases the risk of Reye's syndrome.

An attenuated chickenpox vaccine is recommended for all healthy children and adults without a history of chickenpox or evidence of immunity. By preventing chickenpox, the vaccine also reduces the chance of developing shingles. The first dose is usually given to children between 12 and 15 months old with a second dose between ages 4 and 6 years old. The doses are typically administered as a single-agent vaccine (VAR), but the combination vaccine MMRV that also immunizes against measles, mumps, and rubella (German measles) can be used instead. The chickenpox vaccine should not be given to certain people, including pregnant females and individuals with immunodeficiencies.

If a susceptible person has a significant exposure to VZV, post-exposure prophylaxis (PEP) can be used to prevent chickenpox or lessen its severity. Unless contraindicated, VAR should be administered within five days of exposure; for anyone not eligible to receive VAR, varicella-zoster immune globulin (VARIZIG; a hyperimmune globulin) can be administered to provide passive immunity.

To prevent shingles, two doses of a highly effective recombinant vaccine (RZV) is recommended for adults age 50 and over and for immunocompromised individuals 19 years old and older. The main features of chickenpox and shingles are summarized in **table 22.11.**

Rubeola (Measles)

Measles is a common name for rubeola. The dramatic reduction in measles cases as a result of immunizing children with an effective attenuated vaccine licensed in 1968 is a great success story: By the end of 2000, measles was no longer considered endemic in the United States (**figure 22.13**). Unfortunately, the number of significant outbreaks has increased in recent years, due in large part to some parents' failure to vaccinate their children. Multiple outbreaks in 2019, including a group of related ones in New York, led to 1,274 reported cases of measles in the United States—the highest in

TABLE 22.11 Varicella (Chickenpox) and Herpes Zoster (Shingles)

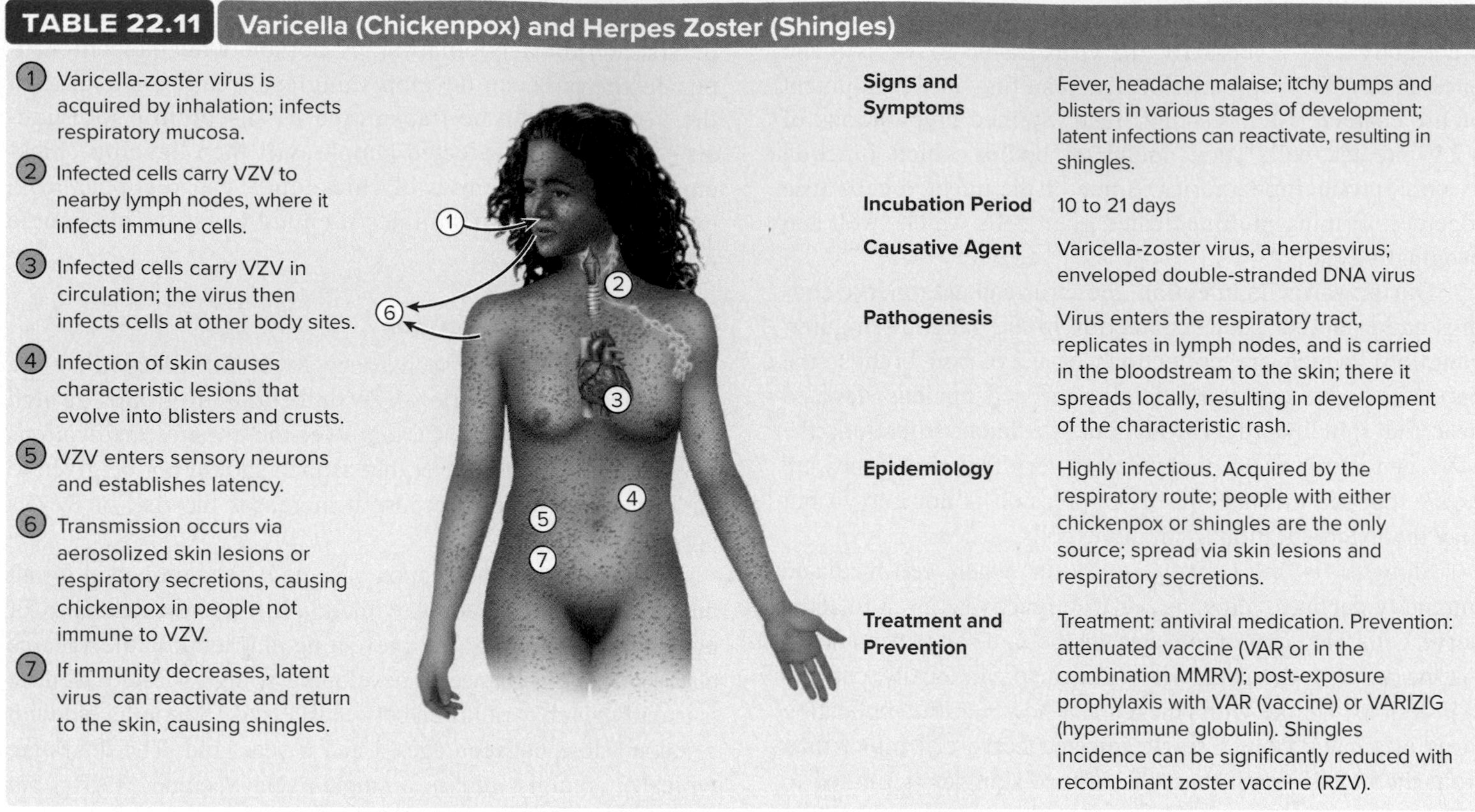

1 Varicella-zoster virus is acquired by inhalation; infects respiratory mucosa.
2 Infected cells carry VZV to nearby lymph nodes, where it infects immune cells.
3 Infected cells carry VZV in circulation; the virus then infects cells at other body sites.
4 Infection of skin causes characteristic lesions that evolve into blisters and crusts.
5 VZV enters sensory neurons and establishes latency.
6 Transmission occurs via aerosolized skin lesions or respiratory secretions, causing chickenpox in people not immune to VZV.
7 If immunity decreases, latent virus can reactivate and return to the skin, causing shingles.

Signs and Symptoms	Fever, headache malaise; itchy bumps and blisters in various stages of development; latent infections can reactivate, resulting in shingles.
Incubation Period	10 to 21 days
Causative Agent	Varicella-zoster virus, a herpesvirus; enveloped double-stranded DNA virus
Pathogenesis	Virus enters the respiratory tract, replicates in lymph nodes, and is carried in the bloodstream to the skin; there it spreads locally, resulting in development of the characteristic rash.
Epidemiology	Highly infectious. Acquired by the respiratory route; people with either chickenpox or shingles are the only source; spread via skin lesions and respiratory secretions.
Treatment and Prevention	Treatment: antiviral medication. Prevention: attenuated vaccine (VAR or in the combination MMRV); post-exposure prophylaxis with VAR (vaccine) or VARIZIG (hyperimmune globulin). Shingles incidence can be significantly reduced with recombinant zoster vaccine (RZV).

25 years. Almost 90% of the cases were in people who were unvaccinated or had an unknown vaccination status, and 10% required hospitalization. Measles incidence has been much lower more recently, likely as a result of lockdowns and social distancing associated with the COVID-19 pandemic.

Signs and Symptoms

Measles begins after an incubation period of 7 to 14 days with fever, runny nose, cough, and swollen, red, weepy eyes. **Koplik spots,** small white spots that look like grains of salt on red and rough oral mucosa, may appear in the mouth (**figure 22.14a**).

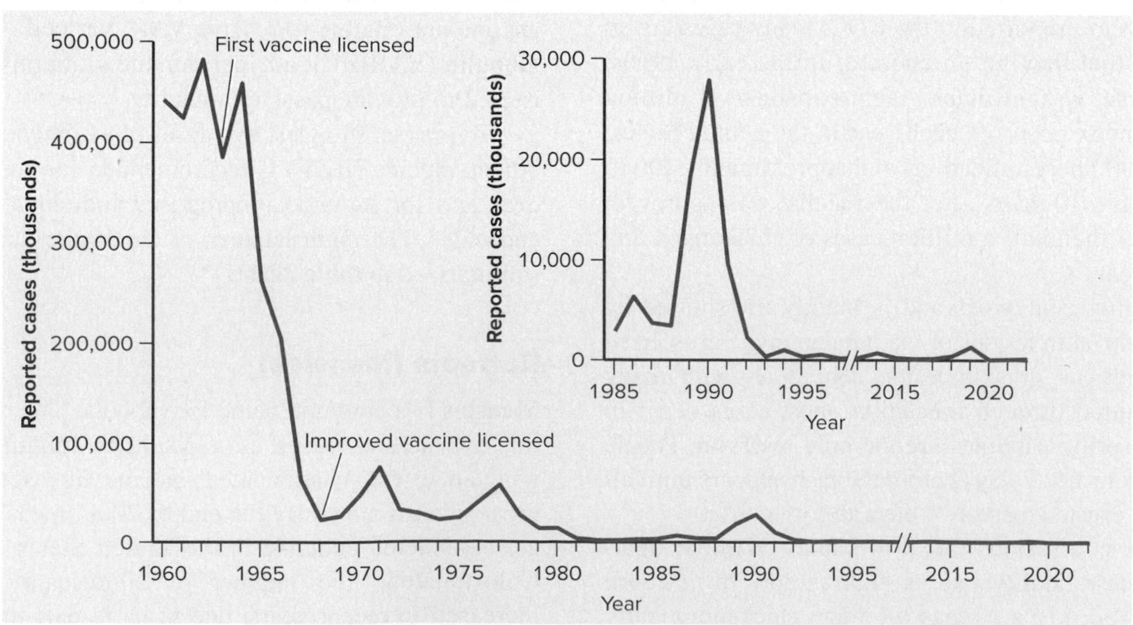

500,000
400,000
300,000
200,000
100,000
0
Reported cases (thousands)
First vaccine licensed
Improved vaccine licensed
1960 1965 1970 1975 1980 1985 1990 1995 2015 2020
Year
30,000
20,000
10,000
0
Reported cases (thousands)
1985 1990 1995 2015 2020
Year

FIGURE 22.13 Measles Cases in the United States, 1960-2022 Measles was declared eliminated in the United States in the year 2000, but outbreaks have occurred in recent years.

? **Why did the reported incidence of measles temporarily increase after each vaccine was licensed?**

FOCUS YOUR PERSPECTIVE 22.1

Mpox Outbreak

In May 2022, several European countries began reporting patients presenting with a characteristic pustular rash. These cases were soon identified as the viral disease monkeypox (now called mpox), a less severe and less contagious "cousin" of the deadly disease smallpox; recall that a successful vaccination campaign led to the elimination of smallpox, which was declared eradicated in 1980 by the World Health Organization (see A Glimpse of History, chapter 17). By August 2023, over 89,000 confirmed mpox cases and over 150 deaths had been reported in 114 countries, including the United States.

Mpox is a zoonotic disease—a disease of animals that can be transmitted to humans. The name was changed in 2022 because of the stigma associated with the original name, as well as to avoid a misconception about the natural source: The original name was given because the disease was first detected in laboratory monkeys, but the reservoir is thought to be rodents. Mpox is caused by the monkeypox virus, a large double-stranded DNA virus with over 200 genes. It is a member of the *Poxviridae*, a family that includes variola virus, which causes smallpox, and vaccinia virus, the vaccine used against smallpox.

The first recorded case of human mpox occurred in 1970, when an infant was admitted to the hospital in the Democratic Republic of Congo (DRC) with a smallpox-like disease. Since then, sporadic outbreaks originating from infected animals have occurred in the DRC as well as in tropical rainforest areas of Central and West Africa. Before the current outbreak, a few cases of mpox outside of endemic regions have occurred, with the largest outbreak in the United States in 2003. That outbreak, which had 47 confirmed and probable cases, resulted from exposure to infected prairie dogs purchased as pets. An animal dealer had housed those animals with various types of wild-caught rodents—later found to be infected—that had been imported from endemic areas to be sold as exotic pets.

In the last decade or so, the incidence of mpox in endemic areas has increased, likely for reasons that relate to smallpox vaccination: The smallpox vaccine offers cross-protection against mpox, but use of that vaccine was phased out in the 1970s as smallpox was eliminated. Thus, relatively few people now have smallpox vaccine-induced immunity to mpox, so populations are becoming increasingly susceptible to the disease.

Mpox is generally a self-limiting disease, although severe and occasionally fatal cases sometimes occur. After an incubation period of about 1 to 2 weeks, early non-specific symptoms such as a fever, malaise, and headaches may begin. Lymphadenopathy (swollen lymph nodes) is usually observed, which distinguishes mpox from smallpox. The characteristic pustular rash, most typically in the extremities, appears after the non-specific symptoms; lesions then dry and scab over within several weeks. Aside from contact with an infected animal, the disease can be transmitted person-to-person through close contact with lesions, body fluids, or respiratory droplets. In the outbreak that started in 2022, transmission has mainly been during sexual activity, through direct contact of skin or mucosal areas with lesions.

Fortunately, many of the prevention and treatment strategies developed for smallpox apply to mpox as well. The most commonly used vaccine for preventing mpox is an attenuated, non-replicating form of vaccinia virus that is approved in the United States for preventing smallpox and mpox in high-risk adults. Supplies are limited, however, so intradermal administration, which requires one-fifth the dose of the approved route (subcutaneous), was recently authorized. An alternative vaccine option can be used in some cases, but in the United States it is only approved for preventing smallpox in high-risk individuals. Although that vaccine, a modified version of the attenuated vaccine used to eradicate smallpox, is more readily available, it can have serious side effects. Because of the limited supplies, vaccination is available according to priority: (1) pre-exposure for those at highest risk, (2) post-exposure (within 4 days) for those who have been exposed, and (3) pre-exposure for those most likely to be exposed. Antivirals approved for treating smallpox are being used on an investigational basis for treating mpox; these include tecovirimat (blocks viral assembly) and brincidofovir (interferes with poxvirus DNA replication). Unlike SARS-CoV-2, monkeypox virus has a DNA genome and much more accurate proofreading and repair mechanisms, so viral variants are less likely to arise.

The spots are temporary and may go unnoticed as they can disappear within a day of developing. Within a few days, a fine red maculopapular rash appears on the face and spreads to the neck and body, and then to the extremities (**figure 22.14b**). Patients are usually contagious from 4 days before to 4 days after the rash appears. Unless complications occur, signs and symptoms generally disappear in about a week. Unfortunately, many cases are complicated by secondary bacterial infections; bacteria easily invade the body because the measles virus damages the respiratory mucous membranes and also temporarily suppresses cell-mediated immunity.

The most common complication of measles, particularly in children, is pneumonia; it can be caused by the measles virus itself or other infectious agents (bacterial or viral) as a result of the mucosal damage and immune suppression just mentioned. Encephalitis (inflammation of the brain) is another serious complication, which can sometimes result in permanent brain damage, with intellectual disability, deafness, and epilepsy.

Very rarely, a disease called subacute sclerosing panencephalitis (SSPE) can develop up to 10 years after a measles infection. It is marked by slow, progressive brain degeneration, generally resulting in death within 2 years. Defective measles virus particles can be detected in the brains of these patients along with high levels of measles antibodies in their blood. Immunization programs in the United States have decreased the incidence of SSPE, but recent measles outbreaks may lead to more SSPE in the future.

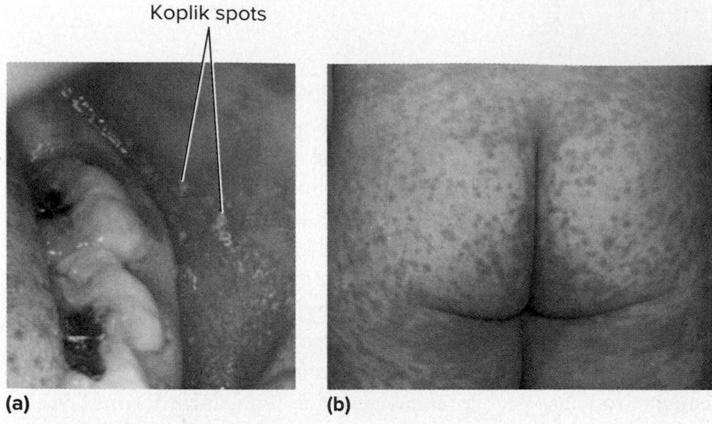

FIGURE 22.14 Characteristic Symptoms of Measles (Rubeola) (a) Koplik spots in the oral mucosa resemble grains of salt on a red base. **(b)** A child with the maculopapular rash covering the face, neck and upper body. a, b: Centers for Disease Control and Prevention

? Can someone have measles without developing Koplik spots?

Measles that occurs during pregnancy increases the risk of miscarriage, premature labor, and low-birth-weight babies. Birth defects, however, generally do not occur.

Causative Agent

Measles is caused by rubeola virus (commonly referred to as the measles virus), an enveloped, single-stranded RNA virus of the *Paramyxoviridae* family.

Pathogenesis

Rubeola virus is acquired by the respiratory route. It initially infects dendritic cells and macrophages in the respiratory tract. These cells migrate to regional lymph nodes where the virus infects lymphocytes. Infected immune cells move into the bloodstream, and infection spreads to other organs in the body, including the skin. The rash is caused by an immune response to infected skin cells.

Two biologically important proteins are found on the viral envelope: hemagglutinin and fusion protein. The virus uses hemagglutinin to attach to host cells and the fusion protein to fuse the viral envelope with the host cell's cytoplasmic membrane as part of viral entry into the cell. The fusion protein also causes adjacent infected host cells to join together, forming multinucleated giant cells.

Measles results in temporary immune suppression—the virus causes lymphocyte depletion, and in particular, reduces the number of memory T cells and B cells. This effect, which can last for 2 to 3 years, is called measles amnesia; it makes the person more susceptible to other diseases and raises the overall mortality associated with the disease; for example, latent tuberculosis infections may reactivate. The virus also damages the respiratory mucous membranes, thereby increasing the susceptibility of measles patients to secondary infections, especially of the lungs. Damage to the intestinal epithelium may explain why diarrhea—a contributing factor to high measles death rates in resource-limited countries—is a common complication of this disease.

Epidemiology

Humans are the only natural host of rubeola virus. Before vaccination became widespread, almost everyone got infected with this highly contagious virus, resulting in about 2.6 million deaths worldwide each year. Global vaccination efforts led to decreased measles incidence, but the disease remains one of the leading causes of death in children worldwide: In 2019, over 200,000 measles deaths were reported, most of them in children under 5 years old. As with many diseases, declining vaccination rates, particularly during the COVID-19 pandemic, increases the risk of disease resurgence.

Although endemic measles was eliminated in the United States in 2000, outbreaks can increase the incidence. The outbreaks in 2019 likely originated from non-immune individuals infected in other countries where measles is still endemic who then returned to or visited the United States. Once introduced into a population, measles can spread rapidly among non-immune individuals, such as those who are unvaccinated (including children too young to be vaccinated) and those who are inadequately vaccinated.

MicroByte

Measles is so contagious that a single cough from an infected person in a crowded place such as a subway car could infect up to 90% of the non-immune people in that area.

Treatment and Prevention

No antiviral treatment is approved for measles, but an attenuated vaccine can prevent the disease. The vaccine is given as a combination with mumps and rubella vaccines (MMR), and sometimes with varicella vaccine as well (MMRV), administered in two doses; the first dose is given at 12 to 15 months, while the second is given between 4 and 6 years old. People lacking presumptive evidence of immunity (documented proof of receiving the vaccine, laboratory-confirmed immunity or disease, or birth before 1957) should also be vaccinated. The vaccine can also be used as post-exposure prophylaxis (PEP) in non-immune persons within 72 hours of exposure to the virus. Immune globulin is also available for PEP, administered up to 6 days after exposure. Some features of measles are summarized in **table 22.12.**

Rubella (German Measles)

German measles is a common name for **rubella.** The term *German measles* arose because Germany is where the disease was first described as a separate disease from measles and scarlet fever. In contrast to other viral diseases with a rash (viral exanthem), rubella is typically mild and can be difficult to diagnose. It is concerning, however, because infection of pregnant women can have tragic consequences for the fetus.

TABLE 22.12 Rubeola (Measles)

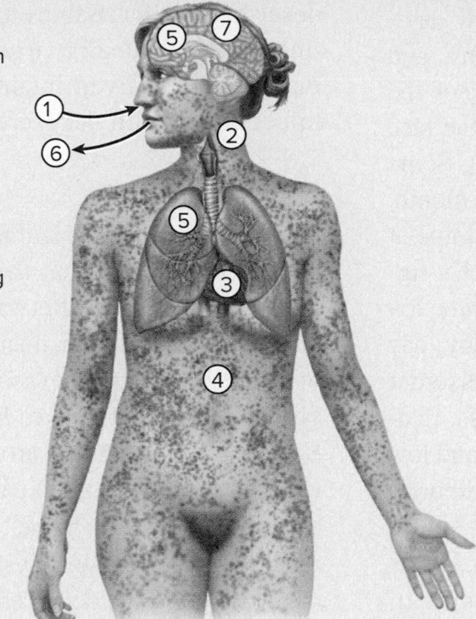

1. Rubeola virus is acquired by inhalation and infects immune cells in the respiratory tract.
2. Infected cells carry the virus to regional lymph nodes, where it infects lymphocytes.
3. Infected lymphocytes enter the bloodstream, carrying the virus to other organs in the body, including the skin.
4. Skin cells are infected with the virus, and the immune response to the infection causes the generalized rash.
5. Complications can include pneumonia and encephalitis.
6. Transmission is by respiratory secretions.
7. Rarely, subacute sclerosing panencephalitis can occur months or years later, likely resulting from defective viral particles persisting in the brain.

Signs and Symptoms	Koplik spots, fever, weepy eyes, cough, nasal discharge, rash
Incubation Period	7 to 14 days
Causative Agent	Rubeola virus, an enveloped single-stranded RNA virus
Pathogenesis	Virus multiplies in respiratory tract; spreads to lymph nodes, then to other parts of body; immune response to infection in skin cells leads to maculopapular rash.
Epidemiology	Acquired by respiratory route; highly contagious; humans are the only source
Treatment and Prevention	Treatment: no antiviral treatment. Prevention: attenuated vaccine (MMR or MMRV).

FOCUS ON A CASE 22.1

A 20-year-old man was immunized against measles as a requirement for starting college. He had received his first dose of the measles vaccine when he was approximately 1 year old. Past medical history revealed that he is positive for the human immunodeficiency virus (HIV). Laboratory tests showed that he had a very low $CD4^+$ lymphocyte count, indicating a severely damaged immune system.

About a month after his pre-college immunization, he developed *Pneumocystis* pneumonia, a lung infection characteristic of AIDS and was hospitalized. He had a good response to treatment given in the hospital, and was discharged. Ten months later, he was again hospitalized for symptoms of a severe lung infection. He had no rash. Multiple laboratory tests to determine the cause of his infection were negative. Finally, a lung biopsy was performed and revealed multinucleated giant cells. These results were highly suggestive of measles pneumonia, and measles virus was subsequently recovered from cell cultures of the biopsy material. Other studies showed it to be the vaccine strain of measles virus. The patient received intravenous immune globulin and improved. Subsequently, however, his condition deteriorated, and he died of presumed complications of AIDS.

1. Is measles immunization a good idea for people with immunodeficiency?
2. If immunocompromised patients such as the one in this case cannot be vaccinated with the measles vaccine, how can they be protected after an exposure?
3. Despite the severe infection, the patient did not have a rash. What is the most likely reason?

Discussion

1. Measles is often disastrous for people with AIDS or other immunodeficiencies. HIV patients should be immunized as soon as possible in their illness, before the immune system becomes so weakened it cannot respond effectively to the vaccine. Also, as this and other cases have shown, the vaccine strain itself (an attenuated virus) can cause disease when immunodeficiency is severe.
2. With the worldwide effort to eliminate measles, the risk of exposure to the wild-type measles virus is declining, but outbreaks in colleges and other institutions still occur. If a severely immunodeficient person is exposed to the wild-type virus, passive immunity can be provided by administering immune globulin.
3. Following acute infection, the measles virus floods the bloodstream and is carried to various tissues of the body, including the skin. The rash that characterizes measles is caused by an immune response against infected skin cells. Because the patient was immunocompromised, this likely affected the typical response in the skin during a measles infection, so the rash did not occur.

Signs and Symptoms

Characteristic signs and symptoms of German measles develop 2 to 3 weeks after infection and typically last only a few days. These include a low-grade fever, mild cold symptoms, and swollen lymph nodes behind the ears and on the back of the neck. After about a day, a faint pink rash appears over the face and then spreads to the rest of the body (**figure 22.15**). Some teenagers and adults can develop joint pain for up to a month. Up to 50% of infected individuals may show no symptoms at all. Although the disease is mild in most people, it poses a significant threat to a developing fetus, particularly if the infection happens during the first 12 weeks of pregnancy. The baby may be born with congenital rubella syndrome (CRS)—a disorder characterized by various abnormalities including deafness, cataracts and other eye defects, brain damage, heart defects, and low birth weight despite normal gestation. Babies may be stillborn.

Causative Agent

German measles is caused by the rubella virus, a member of the *Matonaviridae* family. The virus is an enveloped, single-stranded RNA virus.

Pathogenesis

The rubella virus enters the body via the respiratory tract and replicates in cells in the nasopharynx. It then spreads to local lymph nodes, causing a sustained viremia as it enters the bloodstream. Once in the bloodstream, the virus can access various body tissues, including the skin and the joints. Humoral and cell-mediated immunity develop against the virus, and the resulting immune response likely causes the rash and joint symptoms.

Rubella virus that crosses the placenta in a pregnant woman can infect virtually all types of fetal cells; some cells are killed, whereas others develop a persistent infection in which cell division is impaired and chromosomes are damaged, leading to the various developmental problems described earlier. Babies may be stillborn; those that live continue to shed rubella virus in throat secretions and urine for many months. Even infants who are apparently healthy can infect others with the excreted virus.

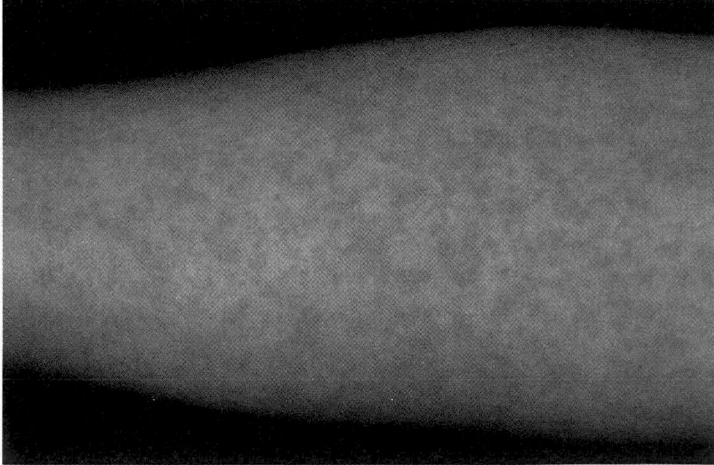

FIGURE 22.15 Adult with German Measles (Rubella) Symptoms are often very mild, but the effects on a fetus can be devastating.
Dr. P. Marazzi/Science Source

? **Is congenital rubella syndrome more likely to occur if a pregnant woman is infected early or late in the pregnancy?**

Epidemiology

Humans are the only natural host for rubella virus. Infected people can usually transmit the virus from 1 week before the rash appears until 1 week afterward. The disease is highly contagious, although less so than measles; estimates indicate that before widespread use of the vaccine began in the late 1960s rubella virus had infected 85% to 90% of the population in the United States. Major epidemics arose periodically, and a 1964 epidemic resulted in about 20,000 cases of congenital rubella syndrome, causing a "rubella bulge" of hearing-impaired children that affected the educational system for decades. Since 2004, rubella is no longer considered endemic in the United States, but it continues to circulate in other parts of the world. Therefore, as with measles, non-immune travelers to these regions can still become infected and spread the disease among unvaccinated populations upon their return. Cases of rubella and congenital rubella syndrome are both reported to the CDC.

Treatment and Prevention

No specific antiviral therapy for rubella is available. The disease can be prevented by an attenuated rubella virus vaccine given to babies at 12 to 15 months of age with a second dose at age 4 to 6 years (as part of the MMR or MMRV vaccine). The vaccine produces long-lasting immunity, and use of the vaccine has reduced the incidence of rubella in the United States to less than 10 cases per year (**figure 22.16**). Women planning to become pregnant are counseled to make sure they are immune to rubella. If they are not already immune, the vaccine can only be given to them before they become pregnant; the vaccine is an attenuated virus, so it should not be given to pregnant women for fear it might result in congenital defects. As an added precaution, women are advised not to become pregnant for at least 4 weeks after receiving the vaccine. Some features of German measles are summarized in **table 22.13.**

Other Viral Rashes of Childhood

Many different kinds of viruses cause childhood rashes; enteroviruses, for example, likely have around 50 types associated with skin lesions. In the early 1900s, the causes of common childhood rashes were largely unknown, and the practice was to number them 1 to 6 as follows: (1) rubeola; (2) scarlet fever; (3) rubella; (4) Duke's disease (also called fourth disease)—seldom reported and now thought to be misdiagnosis of scarlet fever or rubella; (5) erythema infectiosum (commonly called fifth disease); and (6) roseola (also called sixth disease).

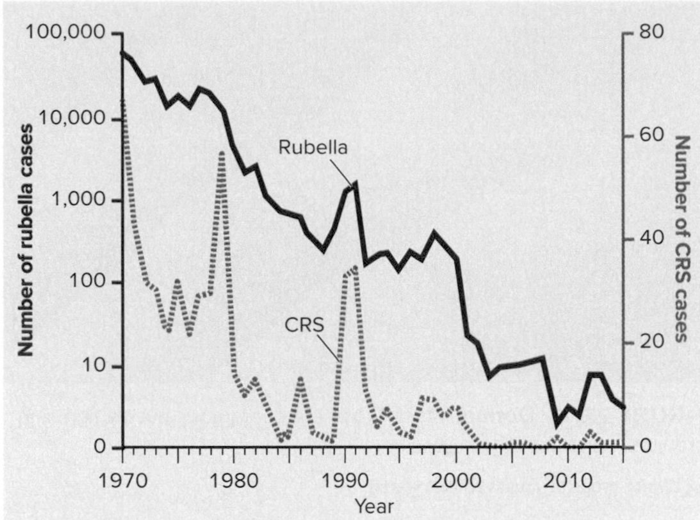

Number of rubella cases
100,000
10,000
1,000
100
10
0
Rubella
CRS
Number of CRS cases
80
60
40
20
0
1970
1980
1990
2000
2010
Year

FIGURE 22.16 Reported Cases of German Measles (Rubella) and Congenital Rubella Syndrome (CRS), United States, 1970–2015 The spikes that occurred in 1978 and 1991 resulted from failure to vaccinate in certain populations. After 2015, there continued to be less than 10 German measles cases per year (and less than 5 CRS cases per year).

? Why are there fewer cases of CRS than of German measles at every time tested?

Fifth Disease

Fifth disease (erythema infectiosum) occurs in both children and young adults. It begins with mild fever, runny nose, and headache. During this time the disease is spread by respiratory droplets. A diffuse redness appears on the cheeks, making the face appear as if it has been slapped ("slapped cheek" rash) (**figure 22.17**). The rash commonly spreads to other parts of the body, especially the extremities, and may come and go for 2 weeks or more before recovery. As the centers of the red blotches begin to clear, the rash appears "lacy." Adults with fifth disease often have joint pain. The disease is caused by parvovirus B19, a non-enveloped, single-stranded DNA virus that preferentially infects certain bone marrow cells, leading to a temporary reduction in red blood cell production; this is a major threat to people with sickle cell anemia or other anemias because these people are already deficient in red blood cells. Rarely, it can cause severe anemia in a developing fetus; up to 5% of women infected with the virus during pregnancy suffer a miscarriage.

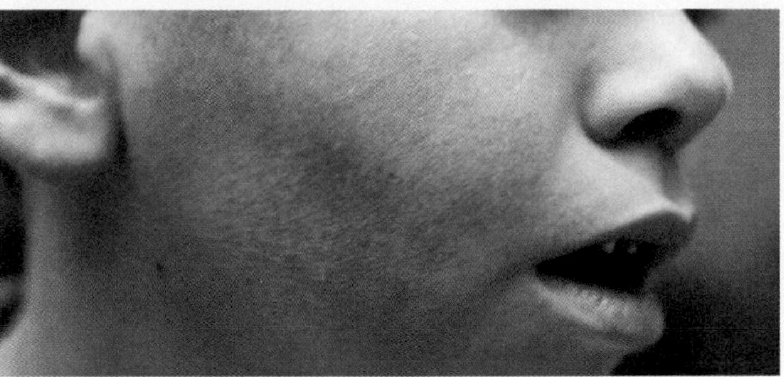

FIGURE 22.17 Fifth Disease "Slapped cheek" appearance of the rash on the face. Dr. Philip S. Brachman/CDC

? Why is fifth disease a major threat to individuals with sickle cell anemia?

Roseola

Roseola (also called roseola infantum) is a common but mild disease that typically affects children under 2 years old. Symptoms begin abruptly with a fever that may reach as high

TABLE 22.13 Rubella (German Measles)

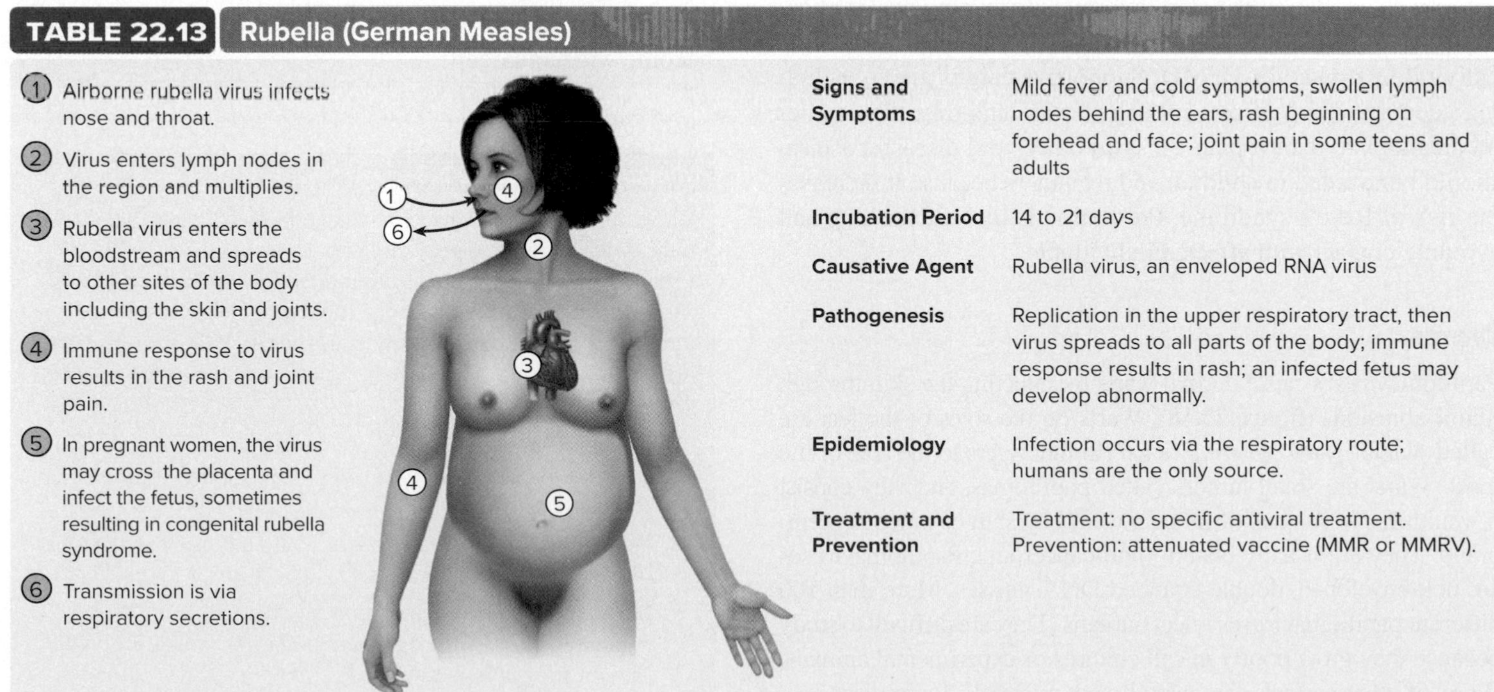

1 Airborne rubella virus infects nose and throat.
2 Virus enters lymph nodes in the region and multiplies.
3 Rubella virus enters the bloodstream and spreads to other sites of the body including the skin and joints.
4 Immune response to virus results in the rash and joint pain.
5 In pregnant women, the virus may cross the placenta and infect the fetus, sometimes resulting in congenital rubella syndrome.
6 Transmission is via respiratory secretions.
1
2
3
4
5
6

Signs and Symptoms	Mild fever and cold symptoms, swollen lymph nodes behind the ears, rash beginning on forehead and face; joint pain in some teens and adults
Incubation Period	14 to 21 days
Causative Agent	Rubella virus, an enveloped RNA virus
Pathogenesis	Replication in the upper respiratory tract, then virus spreads to all parts of the body; immune response results in rash; an infected fetus may develop abnormally.
Epidemiology	Infection occurs via the respiratory route; humans are the only source.
Treatment and Prevention	Treatment: no specific antiviral treatment. Prevention: attenuated vaccine (MMR or MMRV).

as 105°F; this can make parents anxious, especially because febrile seizures may occur in some children. The child may have a runny nose and eye redness, but generally does not appear ill. After several days, the fever goes away and a red rash appears mainly on the chest and abdomen but resolves in a few hours to 2 days. This disease is typically caused by human herpesviruses type 6B (HHV-6B) and less commonly, type 7 (HHV-7). There is no vaccine against the disease. No treatments are available except to reduce the risk of seizures by sponging with lukewarm water and using fever-reducing medication; as with other viral diseases, aspirin should not be given to children or teenagers because it increases the risk of Reye's syndrome.

Hand-Foot-and-Mouth Disease (HFMD)

Hand-foot-and-mouth disease (HFMD)—which is not the same as foot and mouth disease in animals—is a generally mild viral disease that typically affects infants and children under 5 years old, although it may also occur in older children and adults. Signs and symptoms typically start with a sore throat, fever, and general malaise. Sores can develop in the mouth, and these may blister and become very painful. In addition, a rash appears on the palms of the hands, soles of the feet, and sometimes also on the buttocks and genital area. In rare cases, a child with HFMD may develop viral meningitis or encephalitis. HFMD can be caused by different serotypes of enteroviruses, a family of non-enveloped, single-stranded RNA viruses. In the United States, most HFMD is caused by coxsackievirus A16; in parts of Asia, enterovirus A71 has caused outbreaks of severe disease. Infection occurs through the fecal-oral route, or by contact with skin lesions and oral secretions. The virus replicates in lymphoid cells, mainly in the gastrointestinal tract, and then in regional lymph nodes. It spreads to the bloodstream and then to different organs, including the skin and mucous membranes; replication at these sites leads to the characteristic lesions. There is no antiviral treatment for HFMD. Supportive care is given, including adequate fluids to prevent dehydration and over-the-counter medications to reduce pain. As with other viral diseases, aspirin should be avoided in children and teenagers because it increases the risk of Reye's syndrome. Prevention is by handwashing and avoiding contact with affected individuals.

Warts

Papillomaviruses cause dermal warts by infecting the skin through minor abrasions (**figure 22.18**). Warts on the soles of the feet are called plantar warts (*plantar* is a term that refers to the sole of the foot). Warts are small tumors called papillomas, and they consist of multiple protrusions of tissue covered by skin or mucous membrane. They are usually benign (non-cancerous). Papillomaviruses are non-enveloped, double-stranded DNA viruses. More than 100 different papillomaviruses infect humans. They are difficult to study because they grow poorly in cell cultures or experimental animals. Warts of other animals are generally not infectious for humans.

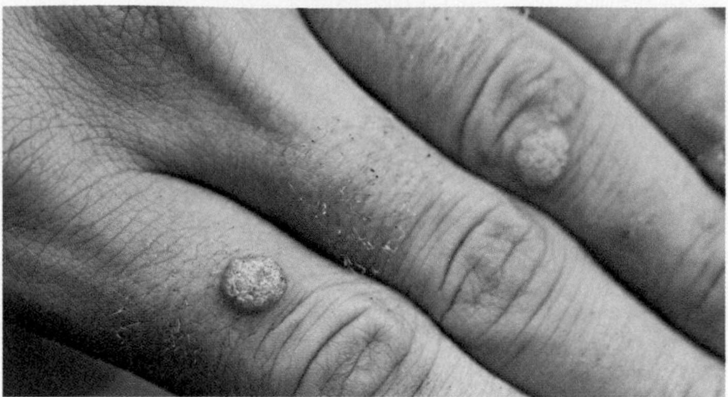

FIGURE 22.18 Dermal Wart Warts are contagious but benign skin growths. Marcel Jancovic/Shutterstock

? **Which group of viruses causes warts?**

Papillomaviruses can remain infectious on inanimate objects such as wrestling mats, towels, and shower floors, and infection can be acquired from such contaminated objects. The viruses infect the deeper cells of the epidermis; some of those infected cells grow abnormally, forming the wart. The incubation period ranges from 2 to 18 months. Infectious virus is present in the wart and can contaminate fingers or objects that pick or rub the lesions.

Dermal warts often disappear within 2 years without treatment. Like other tumors, however, warts can be treated effectively by killing or removing all of the abnormal cells. This can usually be done by several methods, including treatment with salicylic acid, freezing the wart with liquid nitrogen, or surgically removing the wart. Virus generally remains in the adjacent normal-appearing skin, however, and may cause additional warts. Plantar warts are more difficult to treat because the pressure of standing on them causes them to become wide and deep.

MicroAssessment 22.3

The virus that causes varicella (chickenpox) becomes latent and can reactivate to cause shingles. Measles (rubeola) makes a person more susceptible to other infections. Rubella (German measles) during early pregnancy can cause devastating fetal damage. Other common childhood rashes include fifth disease and roseola. Dermal warts are benign tumors that often resolve without treatment.

7. If a non-immune person is exposed to someone with shingles, which are they at risk of developing—shingles or chickenpox?
8. What important diagnostic sign is often present in the mouth of measles (rubeola) patients?
9. If a version of the combination MMR vaccine no longer contained the vaccine against rubella (R), would it now be safe for pregnant women?

22.4 ■ Fungal Diseases of the Skin

Learning Outcomes

9. Describe the characteristics of superficial cutaneous mycoses, including the role of dermatophytes in these infections.
10. Compare and contrast the roles of *Malassezia furfur* and *Candida albicans* in disease.

Diseases caused by fungi are called mycoses. Several fungi are responsible for mild to serious infections of the skin. The severity of most fungal infections is influenced by various host factors, including age and overall health.

Superficial Cutaneous Mycoses

A group of molds called **dermatophytes** can invade hair, nails, and the keratinized layer of the skin. The resulting mycoses have common names such as jock itch, athlete's foot, and ringworm, as well as Latin names that describe their location: tinea capitis (scalp), tinea barbae (beard), tinea axillaris (armpit), tinea corporis (body), tinea cruris (groin), and tinea pedis (feet), to list a few. Tinea simply means "worm," which probably reflects early misconceptions about the cause of these diseases.

Signs and Symptoms

Most people colonized by dermatophytes have no signs or symptoms. Others complain of itching, a bad odor, or a rash. In ringworm, the rash at the site of infection appears as a scaly area surrounded by redness at the outer edge, producing irregular rings or a lacy pattern on the skin. On the scalp, patchy areas of hair loss can occur, leaving a fine stubble of short hair behind. Infected nails become thick and brittle and may separate from the nailbed. Sometimes, a rash consisting of fine papules and vesicles develops away from the infected area. This rash is referred to as a dermatophytid, or "id" reaction, a result of allergic reactions to products of the infecting fungus.

Causative Agents

Dermatophytes are a group of skin-invading molds that includes members of the genera *Epidermophyton, Microsporum,* and *Trichophyton* (**figure 22.19**). They can be grown on culture media especially designed for molds and are usually identified by their colony and microscopic morphologies. Biochemical tests and polymerase chain reaction (PCR) may be used as well. However, identification is not always necessary, because treatment for all dermatophyte infections is essentially the same.

Pathogenesis

The normal skin is generally resistant to fungal invasion, but dermatophytes can invade keratin-containing cells and structures in areas of the body where skin is moist (groin, armpits). The fungi produce an enzyme called keratinase that breaks down the protein, allowing them to use it as a nutrient source. Hair is invaded at the follicle, which is relatively moist.

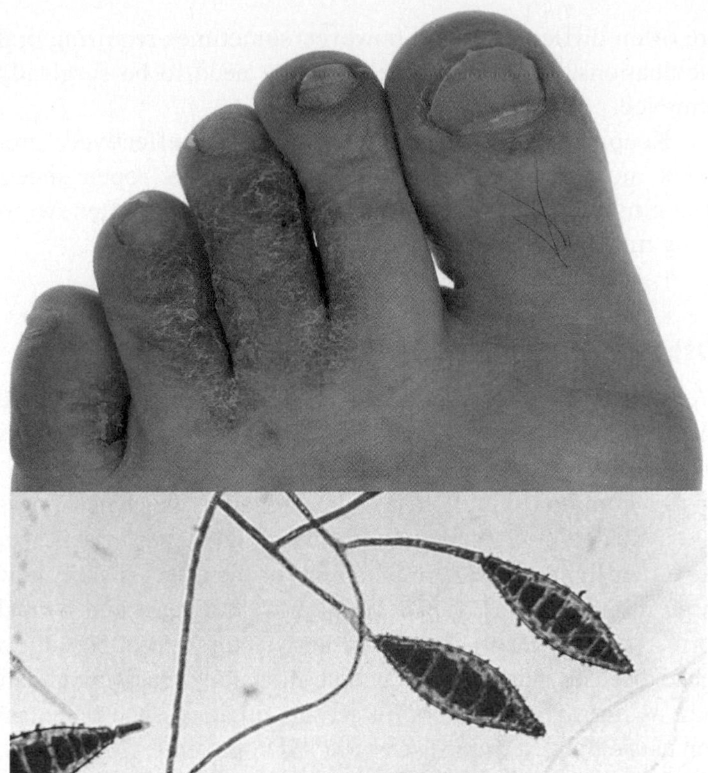

FIGURE 22.19 Dermatophytes (a) Tinea pedis, usually caused by species of *Trichophyton.* **(b)** Large boat-shaped conidia of *Microsporum gypseum,* a cause of scalp ringworm in children.
a: carroteater/Getty Images; b: Dr. Lucille K. Georg/CDC

? **What is the common name for tinea pedis?**

Fungal products diffuse into the dermis and provoke an immune reaction, which probably explains why adults tend to be more resistant to infection than children. Children are more likely to have hypersensitivities, like eczema (a blistery skin rash that leaks fluid before forming crusts) and asthma.

Epidemiology

Patient age, the virulence of the infecting fungal strain, and moisture availability are important factors in determining the course of infection. Common causes of excessive moisture include obesity where folds of skin lie together, tight clothing, and plastic or rubber footwear. Potentially pathogenic molds may be present in soil and on pets such as cats and dogs; fungi acquired from these sources tend to cause more noticeable signs and symptoms in humans.

Treatment and Prevention

Several over-the-counter and prescription medications, such as clotrimazole (an azole) and terbinafine (an allylamine), can be used to treat superficial skin infections. Nail infections are often difficult to cure, however, sometimes requiring oral medications for months; the nail may need to be surgically removed.

Keeping skin and nails clean and dry effectively prevents most dermatophyte infections. Powders, open shoes, changing of socks, and applying rubbing alcohol after swimming may help prevent toenail infections.

Other Fungal Diseases

Malassezia furfur is generally harmless and commonly found on the skin, but in some people it can cause skin conditions such as a scaly facial rash, dandruff, or tinea versicolor. The latter is a common skin disease characterized by patchy scaliness and a change in pigmentation, resulting in patches of skin that are either lighter or darker. Scrapings of the affected skin show large numbers of *M. furfur* cells—both the yeast and hyphal forms (**figure 22.20**). Unknown host factors are important in these diseases because most people carry the organism on their skin as normal microbiota. In certain situations, the organism can cause more serious disease; in AIDS patients, for example, it often causes a severe rash with pus-filled pimples, and in people receiving lipid-containing intravenous feedings it may even infect internal organs.

The yeast *Candida albicans* may live harmlessly among the normal skin microbiota, but it can also cause disease. It is a common cause of dermatitis in infants, a condition called diaper rash (**figure 22.21**). In some people, *C. albicans* invades the deep layers of the skin and subcutaneous tissues, from where it may enter and multiply in the bloodstream (candidemia), resulting in serious illness. In many people with candidal skin infections, the cause for the invasion cannot be determined.

The key features of the diseases covered in this chapter are highlighted in the Diseases in Review 22.1 table that follows.

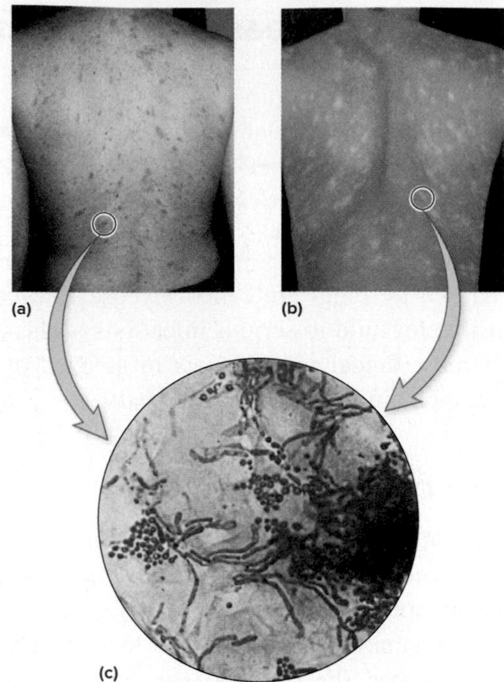

FIGURE 22.20 Tinea Versicolor Patches of **(a)** increased and **(b)** decreased pigmentation relative to surrounding skin. **(c)** Microscopic appearance of a stained skin scraping showing *Malassezia furfur* yeast and hyphae. a: Biophoto Associates/Science Source; b: Medical-on-Line/ Mediscan/ Alamy Stock Photo; c: BSIP/UIG/Getty Images

? What other conditions can be caused by this fungus?

MicroAssessment 22.4

The fungi that cause skin mycoses can commonly colonize skin without causing signs or symptoms. Molds called dermatophytes can invade the hair, nails, and skin. One of the best ways to prevent fungal skin infections is to keep the skin clean and dry.

10. What is a mycosis?

11. Name some of the superficial cutaneous mycoses caused by dermatophytes.

12. Why do you think it is so much more difficult to treat nail infections than to treat other superficial dermatophytoses?

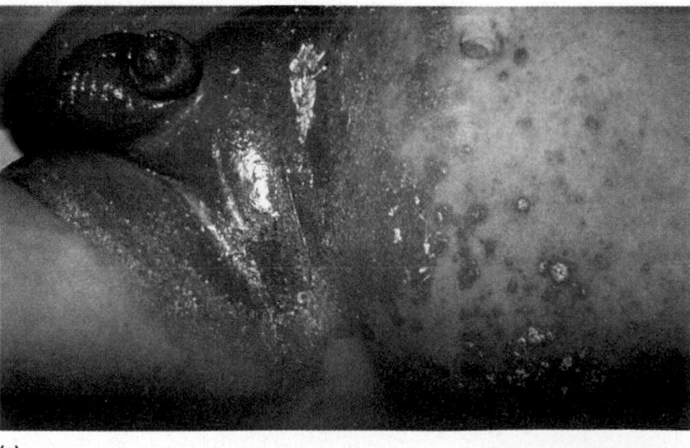

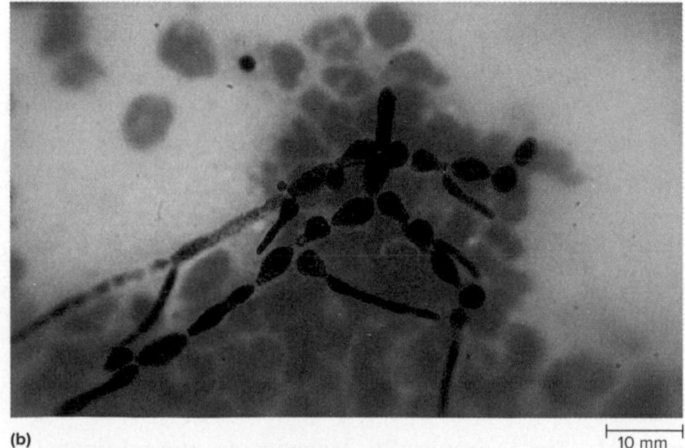

FIGURE 22.21 ***Candida albicans*** **(a)** Causing a diaper rash. **(b)** Gram stain of pus showing *C. albicans* yeast forms and filamentous forms called pseudohyphae. a: Centers for Disease Control and Prevention; b: Evans Roberts

? Would an antibiotic ointment containing penicillin be effective against this candidal infection? Explain your answer.

Diseases in Review 22.1

Common Bacterial, Viral, and Fungal Skin Diseases

Disease	Causative Agent	Comment	Summary Table
BACTERIAL SKIN DISEASES			
Acne vulgaris	*Cutibacterium* (formerly *Propionibacterium*) *acnes* commonly associated	Most common during puberty, probably due to excess sebum secretion in response to increased hormone levels.	Table 22.2
Hair follicle infections	*Staphylococcus aureus*	Causative agent commonly colonizes the nostrils and moist skin areas; skin infections include folliculitis, furuncles, and carbuncles. Organism is often resistant to multiple antibiotics.	Table 22.5
Staphylococcal scalded skin syndrome (SSSS)	Exfoliative toxin-producing strains of *S. aureus*	Characterized by peeling of the outer layer of skin; occurs in infants, as well as the elderly and immunocompromised.	Table 22.6
Impetigo	*S. aureus* or *Streptococcus pyogenes*	Non-bullous impetigo: papules, vesicles, then pustules that break, releasing plasma and forming yellowish crusts. Bullous impetigo: large, gradually darkening blisters (bullae), no crusts.	Table 22.8
Rocky Mountain spotted fever (RMSF)	*Rickettsia rickettsii*	Spread by ticks; characterized by rash; inflammatory response to damaged endothelium can lead to potentially fatal organ damage.	Table 22.9
Cutaneous anthrax	*Bacillus anthracis*	Zoonotic disease that is rare in humans; characterized by black eschar.	Table 22.10
VIRAL SKIN DISEASES			
Varicella (chickenpox) and herpes zoster (shingles)	Varicella-zoster virus (VZV)	Virus enters via the respiratory tract; infection characterized by itchy skin lesions; VZV becomes latent and can later reactivate to cause shingles. Preventable by vaccination.	Table 22.11
Rubeola (measles)	Rubeola virus	Virus enters via the respiratory tract; characterized by respiratory symptoms and a spreading rash; Koplik spots occur in mouth. Preventable by vaccination.	Table 22.12
Rubella (German measles)	Rubella virus	Virus enters via the respiratory tract, can cause mild symptoms (including rash) or no symptoms; can damage a developing fetus (congenital rubella syndrome). Preventable by vaccination.	Table 22.13
Other viral diseases (fifth disease, roseola, hand-foot-and-mouth disease)	Parvovirus B-19; human herpesvirus type 6B; coxsackievirus A16, enterovirus A71	Fifth disease is characterized by "slapped cheek" rash; roseola causes high fever, sometimes leading to convulsions; hand-foot-and-mouth disease is characterized by painful mouth lesions and a rash on palms and soles.	
Warts (dermal warts)	Papillomaviruses	Warts are benign skin tumors that can be removed by freezing, burning, surgery, or topical medication.	
FUNGAL SKIN DISEASES			
Superficial cutaneous mycoses	Usually *Epidermophyton*, *Microsporum*, or *Trichophyton* species	Fungi invade keratinized skin, causing conditions that are commonly known as athlete's foot, jock itch, and ringworm.	
Other fungal diseases	*Malassezia furfur*, *Candida albicans*	Both organisms are usually harmless on the skin, but *M. furfur* sometimes causes skin conditions such as scaly face rash, dandruff, or tinea versicolor. *C. albicans* sometimes invades deeper layers and subcutaneous tissues.	

Summary

22.1 ■ Anatomy, Physiology, and Ecology of the Skin

The skin prevents the entry of microbes, regulates body temperature, restricts the loss of fluid from body tissues, and plays an essential role in the function of the immune system. It is composed of the **epidermis** and the **dermis** (figure 22.1). Common members of the skin microbiota include diphtheroids, staphylococci, and fungi (table 22.1).

22.2 ■ Bacterial Diseases of the Skin

Acne Vulgaris (table 22.2)

Acne is characterized by enlarged sebaceous glands and increased secretion of **sebum,** which clogs the hair follicle. This gives rise to non-inflamed lesions called comedones—whiteheads and blackheads. *Cutibacterium acnes* may grow to high numbers in clogged follicles. Their metabolic products cause inflamed comedones, including pustules (pimples), nodules, and sometimes cysts.

Hair Follicle Infections (table 22.5)

Folliculitis, furuncles, and **carbuncles** are caused by *Staphylococcus aureus,* which is coagulase-positive and often resistant to penicillin and other antibiotics (figures 22.2 and 22.3; and tables 22.3 and 22.4). There are many different strains of *S. aureus*, and these vary in virulence. A carbuncle is more serious than folliculitis and furuncles because the infection is more likely to spread to the heart, bones, or brain.

Staphylococcal Scalded Skin Syndrome (table 22.6)

Staphylococcal scalded skin syndrome (SSSS) is caused by strains of *Staphylococcus aureus* that produce one or more **exfoliative toxins.** (figure 22.4).

Impetigo (table 22.8)

Impetigo is a pyoderma often caused by *Streptococcus pyogenes* or *S. aureus.* Two forms of the disease occur: non-bullous (most common) and bullous (a localized form of SSSS) (figure 22.5; table 22.7).

Rocky Mountain Spotted Fever (table 22.9; figure 22.6)

Rocky Mountain spotted fever is an often fatal disease, caused by the obligate intracellular bacterium *Rickettsia rickettsii* (figure 22.7); it is transmitted to humans by the bite of an infected tick (figure 22.8, figure 22.9).

Cutaneous Anthrax (table 22.10)

Cutaneous anthrax develops when *Bacillus anthracis* infects the skin, causing a small red bump that develops into an eschar resembling a flat black scab (figure 22.10). The disease is rare in humans; person-to-person transmission does not occur.

22.3 ■ Viral Diseases of the Skin

Varicella (Chickenpox) and Herpes Zoster (Shingles) (table 22.11)

Chickenpox, once a common disease of childhood, is caused by the varicella-zoster virus (figure 22.11). **Shingles** can occur months or years after chickenpox and is due to reactivation of the virus (figure 22.12). Shingles cases can be sources of chickenpox epidemics. Immunization with an attenuated virus protects against chickenpox; vaccines are also available to prevent shingles.

Rubeola (Measles) (table 22.12)

Measles (rubeola) can lead to serious secondary bacterial infections and fatal lung or brain damage (figures 22.14). Measles can be prevented by vaccinating with an attenuated virus (figures 22.13).

Rubella (German Measles) (figure 22.15, table 22.13)

German measles (rubella) is usually a mild disease, but if contracted by a woman early in pregnancy, the baby may be born with congenital rubella syndrome (CRS). Immunization with an attenuated virus protects against this disease (figures 22.16).

Other Viral Rashes of Childhood

Many childhood rashes are caused by viruses (figures 22.17). Examples include fifth disease (erythema infectiosum), roseola (sixth disease), and hand-foot-and-mouth disease (HFMD).

Warts

Warts are skin tumors caused by a number of papillomaviruses (figures 22.18). They are usually benign and may disappear without treatment. Warts on the soles of the feet are called plantar warts.

22.4 ■ Fungal Diseases of the Skin

Superficial Cutaneous Mycoses

Dermatophytes cause athlete's foot, ringworm, and invasions of the hair and nails (figures 22.19).

Other Fungal Diseases

Malassezia species can cause tinea versicolor and dandruff, as well as serious skin disease in AIDS patients (figures 22.20). *Candida albicans* may live harmlessly among the normal microbiota, but it can invade deeper layers of the skin and subcutaneous tissues (figures 22.21).

Review Questions

Short Answer

1. What is the difference between a furuncle and a carbuncle?
2. Why do only certain strains of *Staphylococcus aureus* cause scalded skin syndrome?
3. Compare the signs and symptoms of bullous and non-bullous impetigo.
4. How does the fact that Rocky Mountain spotted fever is a zoonosis relate to the relative severity of the disease symptoms?
5. Who is most at risk for contracting cutaneous anthrax?
6. Describe the progression of the rash of varicella.
7. What is the relationship between chickenpox (varicella) and shingles (herpes zoster)?
8. Why are many cases of measles complicated by secondary infections?
9. What is the significance of rubella viremia during pregnancy?
10. How does a person contract warts?

Multiple Choice

1. Which of the following conditions is important in the ecology of the skin?
   a. Temperature
   b. Salt concentration
   c. Lipids
   d. pH
   e. All of the above
2. *Staphylococcus aureus* can be responsible for all of the following conditions *except*
   a. impetigo.
   b. food poisoning.
   c. toxic shock syndrome.
   d. scalded skin syndrome.
   e. athlete's foot.
3. The main effect of staphylococcal protein A is to
   a. interfere with phagocytosis.
   b. enhance the attachment of the Fc portion of antibody to phagocytes.
   c. coagulate plasma.
   d. kill white blood cells.
   e. degrade collagen.
4. Which of the following contributes to the virulence of *Streptococcus pyogenes*?
   a. Protease
   b. Hyaluronidase
   c. DNase
   d. All of the above
   e. None of the above
5. Which of the following statements is true of streptococcal impetigo?
   a. It is caused by a Gram-negative rod.
   b. It cannot be transmitted from one person to another.
   c. Pathogenic streptococci all produce coagulase.
   d. All of the above.
   e. None of the above.
6. All of the following are true of Rocky Mountain spotted fever *except*
   a. the disease is most prevalent in the western United States.
   b. it is caused by an obligate intracellular bacterium.
   c. it is a zoonosis transmitted to humans by ticks.
   d. those with the disease characteristically develop a hemorrhagic rash.
   e. antibiotic therapy is usually curative if given early in the disease.
7. If a woman begins to develop shingles, which of the following can you conclude?
   a. She was recently exposed to someone who had chickenpox.
   b. She was recently exposed to someone who had shingles.
   c. She had chickenpox in the past.
   d. She had shingles in the past.
   e. The rash will likely spread over her entire trunk and extremities.
8. Which of the following statements is more likely to be true of measles (rubeola) than of German measles (rubella)?
   a. Koplik spots are present.
   b. It causes birth defects.
   c. It causes only a mild illness.
   d. Humans are the only natural host.
   e. Attenuated virus vaccine is available for prevention.
9. All of the following must be cultivated in cell cultures instead of cell-free media *except*
   a. *Rickettsia rickettsii.*
   b. rubella virus.
   c. varicella-zoster virus.
   d. *Candida albicans.*
   e. rubeola virus.
10. All of the following might contribute to the development of ringworm or other superficial cutaneous mycoses *except*
   a. obesity.
   b. playing with kittens.
   c. rubber boots.
   d. using skin powder.
   e. dermatophyte virulence.

Applications

1. A school administrator in a small Iowa community prohibited a child with chickenpox from attending school. He said this was the first case of chickenpox in the school in 6 years, and he did not want to have an outbreak. Several parents argued to the school board that an outbreak would benefit the school in the long term. Discuss the pros and cons of allowing this child to attend school.
2. A public health official was asked to speak about immunization during a civic group luncheon. One parent asked if rubella was still a problem. In answering, the official cautioned women planning to have another child to have their present children immunized against rubella. Why did the official suggest this?

Critical Thinking

1. In the past, teenagers with acne were told to stop eating chocolate and French fries and to vigorously scrub their skin. Explain how you think these recommendations would affect the teenagers' acne.
2. Why might it be more difficult to eliminate a disease like Rocky Mountain spotted fever from the earth than to eliminate rubeola or rubella?

23 | Wound Infections

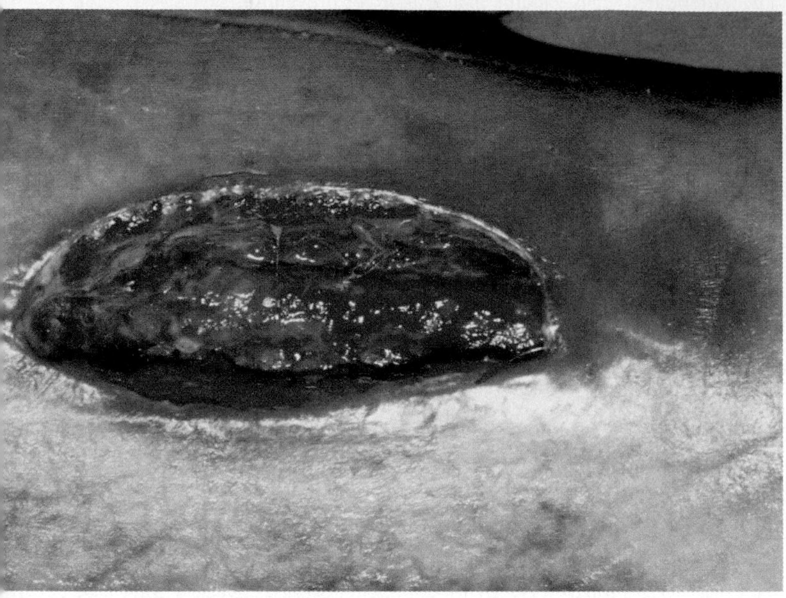

Infected leg wound. *(Garry Watson/Science Source)*

KEY TERMS

Abscess A localized collection of pus within a tissue.

Granulation Tissue New tissue formed during healing of an injury.

Necrotizing Fasciitis Tissue death resulting from inflammation of the fascia—bands of fibrous tissue that underlie skin and surround muscle and body organs.

Pus Yellowish fluid composed of proteins, living and dead leukocytes, and tissue debris.

Pyogenic Pus-producing.

Superantigens Molecules that bind to and stimulate helper T cells, resulting in overproduction of cytokines and sometimes fatal shock.

Synergistic Infection A co-infection in which the combined effect of two or more species is more damaging than the sum of each one acting alone.

A Glimpse of History

Endospores of *Clostridium tetani*, the bacterium that causes tetanus (lockjaw), are found in soil and dust virtually everywhere. Tetanus was common before its cause and pathogenesis were understood, often ending in an agonizingly painful death. Dr. Shibasaburo Kitasato (1853–1931), who was studying the disease in Robert Koch's laboratory in Germany, discovered that *C. tetani* is an obligate anaerobe. This crucial information helped him develop a method to grow the bacterium in pure culture, an essential step toward characterizing a pathogen and learning how it causes disease.

Kitasato showed that laboratory animals injected with *C. tetani* developed tetanus. He was puzzled, however, by a surprising finding: Although the animals died of generalized disease, the only *C. tetani* cells in the animal were at the injection site. By doing experiments in which he injected the tails of mice and then removed the inoculated tissue at hourly intervals, he showed that the animals developed tetanus only if the bacteria remained in the animals for more than an hour. He also showed that the bacterial cells stayed at the site of inoculation; at no time were they found in the rest of the body. Kitasato reasoned that something other than bacterial invasion was causing the disease.

While Kitasato was working with tetanus, another scientist, Emil von Behring, was busy investigating how *Corynebacterium diphtheriae* causes the disease diphtheria. Together, Kitasato and von Behring showed that toxins produced by the bacteria cause both diseases. The concept that a bacterial toxin could cause disease was an extremely important advancement in the understanding and control of infectious diseases.

Most people occasionally suffer wounds that cause breaks in the skin or the mucous membranes. Microorganisms almost always enter the wound, originating from either the damaging object or the environment. Infection may develop as a result, but this depends on several factors, including: (1) virulence of the microbes; (2) number of microbial cells in the wound; (3) status of the host's immune system; and (4) the type of wound. Wounds that contain dirt or other foreign material usually become infected and do not heal until the contaminating material is removed. This is because foreign substances create surfaces for biofilm development and provide places for microorganisms to multiply out of the reach of phagocytes and other immune defenses. They may also reduce available O_2, allowing the growth of anaerobic pathogens and interfering with phagocyte function. Clean wounds often heal without treatment despite microbial colonization, but sometimes even a minor wound can result in a severe, or possibly fatal, infection.

23.1 ■ Anatomy, Physiology, and Ecology of Wounds

Learning Outcomes

1. Name three tissue components exposed by wounds to which pathogens specifically attach.
2. Describe the beneficial and harmful aspects of abscess formation.

Wounds vary in their characteristics, severity, and associated risks. The general categories of wounds include:

- **Incisions:** Produced by a knife or other sharp object.
- **Punctures:** Result from penetration by a small, sharp object, such as a needle, nail, or animal tooth.

- **Lacerations:** Occur when the tissue is torn.
- **Contusions:** Produced by a blow that crushes tissue.
- **Abrasions:** Occur when the epidermis is scraped off.
- **Gunshot wounds:** Caused by bullets or other projectiles.
- **Burns:** Caused by heat (thermal burns), electricity, chemicals, radiation, or friction.

Wounds expose tissue usually protected by skin or mucous membranes, uncovering components such as collagen, fibronectin, fibrinogen, and fibrin that pathogens can attach to and then colonize. **Collagen** is a fibrous material, the main supportive protein of skin, tendons, and other body structures. **Fibronectin** is a fibrous glycoprotein found both in a circulating form and in tissue, where it binds cells and other tissue substances together. **Fibrinogen** is a blood protein; when a wound occurs, this protein is converted to **fibrin,** which forms clots in the damaged vessels. The clots plug the break and thereby help stop excessive bleeding, the first step in the wound repair process.

As wound healing begins, connective tissue cells (fibroblasts) multiply and new capillaries form in the area. This produces a light red, translucent (clear) fibrous material called **granulation tissue.** In clean wounds, granulation tissue fills the space created by the wound; this tissue gradually shrinks and is converted to collagen, a component of scar tissue that gradually retracts as epithelium is regenerated (**figure 23.1**). When a wound results in extensive tissue damage or severe infection, the dead, damaged, and infected tissue is removed—a medical procedure called **debridement**—to help wounds heal faster.

MicroByte

In maggot debridement therapy (MDT), live, sterile maggots (fly larvae) are placed on a wound to remove dead tissue and aid in wound healing.

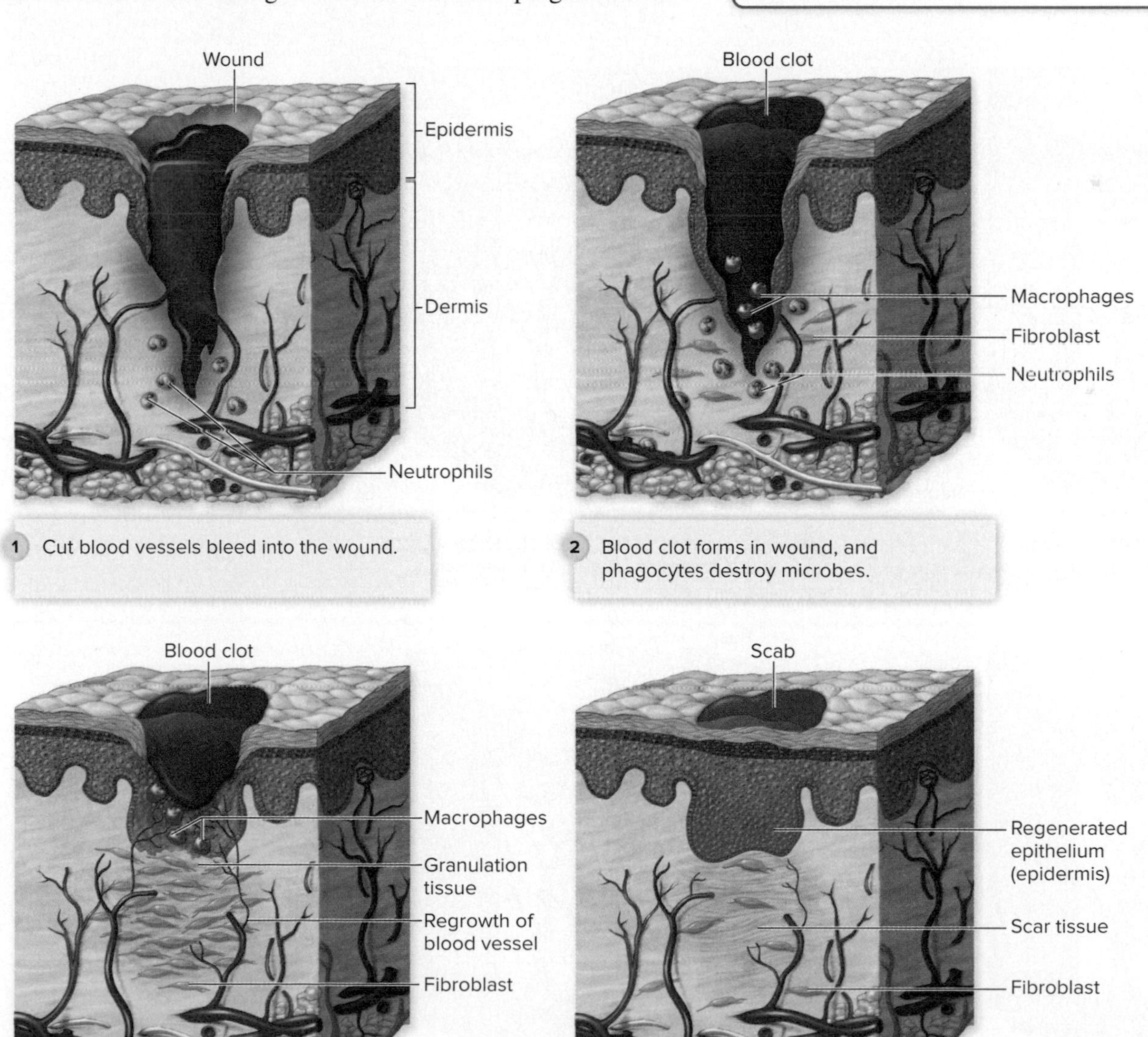

FIGURE 23.1 The Process of Wound Repair

What is the function of granulation tissue?

Wound Abscesses

An **abscess** is a localized collection of **pus**—a thick, yellowish fluid that contains leukocytes, tissue debris, and proteins—surrounded by inflamed tissue. Abscesses form as a result of the body's immune defenses and usually indicate an infection (**figure 23.2**). Although an abscess helps localize the infection and prevents its spread, it also indicates a potentially serious situation: If microbial cells escape the abscess, they can enter the blood or the lymph, leading to infection in other parts of the body.

The very nature of abscesses makes them difficult to treat. As pus forms and the abscess expands, blood vessels around the area are pushed aside and destroyed; additionally, adjacent blood vessels are often blocked by clots. This lack of blood circulation to the abscess means that microbes within it may not be exposed to antimicrobials used for treatment. Therefore, to be effectively treated, abscesses must either burst through toward a body surface or be drained surgically.

MicroAssessment 23.1

Wounds can be classified as incisions, punctures, lacerations, contusions, abrasions, gunshots, or burns. Wounds expose tissue components to which pathogens can attach. Granulation tissue fills a healing wound. An abscess helps isolate an infection and prevents its spread, but is difficult to treat.

1. Name and describe two substances in wounds to which pathogens attach.
2. Explain why an abscess might not respond to antibiotic treatment.
3. Why is it important that the granulation tissue shrinks after it is formed?

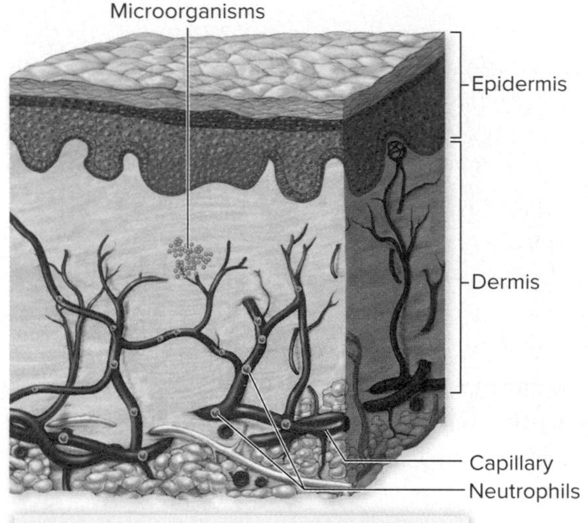

1 Microorganisms enter the tissue from a wound or from the bloodstream.

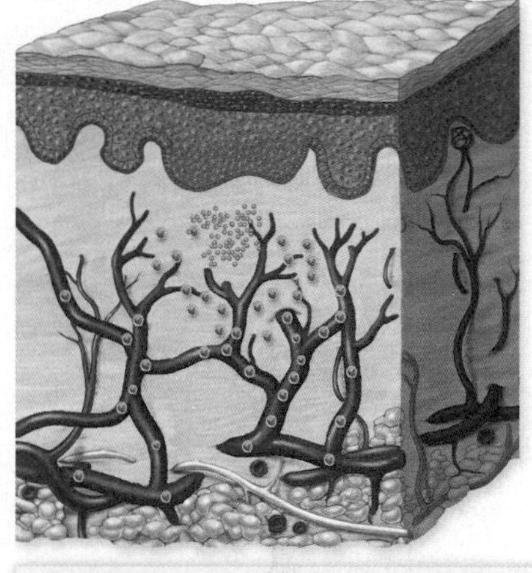

2 Blood vessels dilate, and leukocytes migrate to the area of the developing infection.

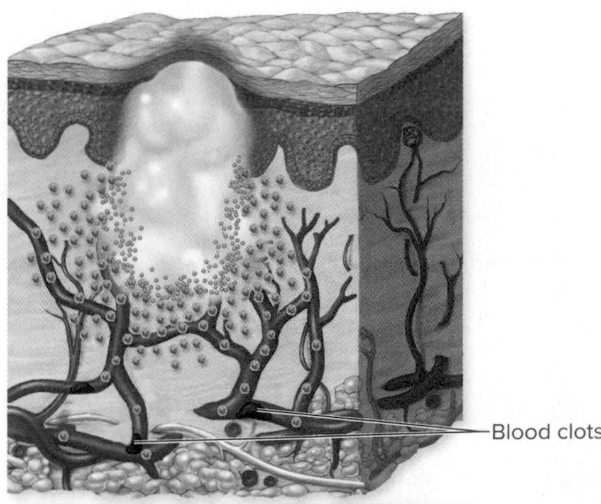

3 Pus forms and an abscess develops; clotting occurs in adjacent blood vessels.

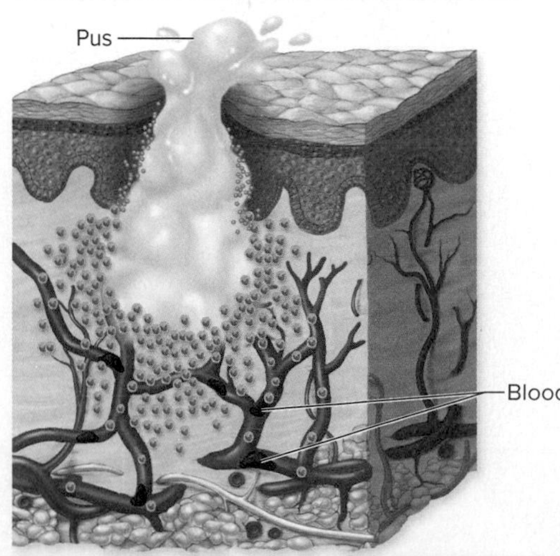

4 Buildup of pressure causes the abscess to expand in the direction of least resistance; if it reaches a body surface, it may rupture and discharge its contents.

FIGURE 23.2 Abscess Formation

? What is the composition of pus?

23.2 ■ Common Bacterial Infections of Wounds

Learning Outcomes

3. Describe the distinctive characteristics of wound infections caused by *Staphylococcus aureus, Streptococcus pyogenes,* and *Pseudomonas aeruginosa.*
4. Discuss how *Staphylococcus epidermidis* can form biofilms on medical devices such as artificial joints.

Infected wounds heal more slowly, can result in difficult-to-treat abscesses, and carry the risk of the bacteria or their toxins spreading to other body sites. If the bacteria spread to an implant such as an artificial hip or knee, the device may then have to be removed before the infection can be eliminated. Infected surgical wounds often split open as swelling causes the stitches to pull through tissues weakened by the infection.

Staphylococcal Wound Infections

Staphylococcus species, common inhabitants of the nostrils and the skin, are the leading causes of wound infections. Of the 40 or more recognized *Staphylococcus* species, only two cause most human wound infections: *Staphylococcus aureus* and *Staphylococcus epidermidis.* The more important of these is *S. aureus,* which was covered extensively in chapter 22.

Signs and Symptoms

Staphylococcus species are **pyogenic,** which means they cause pus production (*pyo* means "pus" and *genic* means "generating"). Staph infections are usually characterized by an inflammatory reaction, with swelling, redness, and pain (**figure 23.3**). Fever occurs if the infected area is large or if the infection has spread to the blood or the lymph.

Toxic shock syndrome can occur if the wound is infected with a toxin-producing *S. aureus* strain. Signs and symptoms of this include high fever, muscle aches, a life-threatening drop in blood pressure, and shock. Sometimes the infected person will also have a rash and diarrhea.

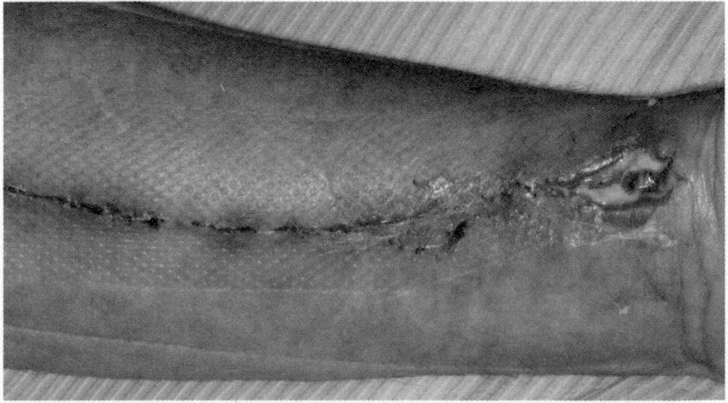

FIGURE 23.3 Surgical Wound Infection Dr. P. Marazzi/Science Source

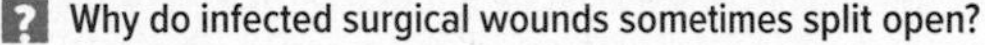

? Why do infected surgical wounds sometimes split open?

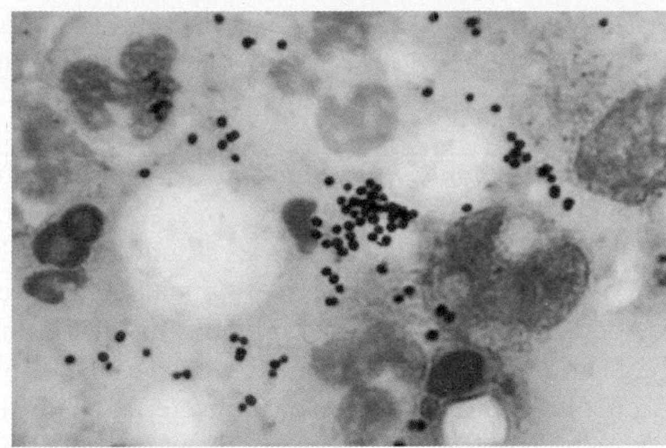

FIGURE 23.4 *Staphylococcus aureus* Gram stain of pus. Lisa Burgess-Scimeca

? What does the name *Staphylococcus* indicate about the typical growth arrangement of the bacterial cells?

Causative Agents

Staphylococcus aureus and *S. epidermidis* are Gram-positive cocci that grow in clusters (**figure 23.4**). They are facultative anaerobes and can grow on skin. They also survive well in the environment and are easily transferred from person to person.

The coagulase test is used to distinguish *S. aureus* from many other staphylococci. *S. aureus* is often referred to as "coag-positive staph" because it typically makes the protein coagulase, whereas the group of staphylococci that do not are collectively referred to as "coag-negative staph." Although *S. aureus* is the most common cause of serious wound infections, coag-negative staph, particularly *S. epidermidis,* can cause healthcare-associated infections, including those of surgical wounds.

Pathogenesis

Staphylococcus aureus *S. aureus* produces multiple virulence factors that can act together in the disease process (see table 22.4). These virulence factors are covered extensively in chapter 22, but it is important to recognize that they play a critical role in wound infections as well. Clumping factor and other proteins allow the bacteria to attach to clots and tissue components, an initial step in colonization. Lipases, proteases, and hyaluronidases together cause tissue damage. Capsules, coagulase, and protein A protect the cells from the complement system, phagocytes, and antibodies. Immunity to *S. aureus* infection is generally weak or non-existent, probably because the organism evades the immune defenses so effectively.

S. aureus infections sometimes cause systemic complications. Bacteria growing in a wound can enter the bloodstream and then spread to other parts of the body, leading to abscesses in the heart, bones, or other tissues. Some *S. aureus* strains produce **superantigens** that can enter the circulation and activate many helper T cells (see figure 16.12). The T cells then release large amounts of cytokines (a cytokine storm), which can lead to staphylococcal toxic shock syndrome—characterized by rapid onset of fever, a drop in blood pressure and in some cases life-threatening multi-organ failure.

Staphylococcus epidermidis *S. epidermidis* is not particularly virulent and cannot invade healthy tissues. However, the bacterium often causes minor abscesses around sutures. It can also form biofilms on medical devices—including indwelling catheters and artificial joints—by binding to the fibrinogen and fibronectin that quickly coat these materials. Biofilms are difficult to eliminate because antibacterial medications diffuse slowly into them, and even then, the medications might be ineffective because the bacteria within a biofilm are often metabolically inactive. The bacteria in biofilms can come loose and be carried by the bloodstream to the heart, potentially leading to infective endocarditis. In people with compromised immune systems, bloodstream spread can result in tissue abscesses in multiple sites or even sepsis.

MicroByte

More than 100,000 *S. aureus* cells need to be injected into skin to cause an abscess, but introducing only 100 into a suture site will do the same.

Epidemiology

Some people are *S. aureus* carriers, putting them at greater risk for surgical wound infections caused by this species. Other factors that increase a person's risk include advanced age, overall poor health, immunosuppression, an infection at another body site, and an extended stay in the hospital before surgery. Additional information about the epidemiology of *S. aureus* is discussed in chapter 22.

S. epidermidis is part of the normal microbiota on the skin and mucous membranes of most people. It is an opportunist that can cause disease in people who are immunocompromised.

Treatment and Prevention

Treating staphylococcal infections is often difficult because many strains are resistant to antibiotics. For example, **methicillin-resistant *S. aureus* (MRSA)** strains are resistant to all β-lactam antibiotics except ceftaroline, a newer cephalosporin. MRSA may be healthcare-associated (HA-MRSA) or community-acquired (CA-MRSA). Fortunately CA-MRSA can usually be treated successfully with sulfa drugs, tetracyclines, or clindamycin. HA-MRSA strains are sometimes resistant even to these and are often susceptible only to a few medications, including vancomycin, daptomycin, and ceftaroline. **Vancomycin-intermediate *S. aureus* (VISA)** and **vancomycin-resistant *S. aureus* (VRSA)** have emerged, making treatment of these infections more difficult. Although newer medications such as daptomycin and linezolid are sometimes used, the optimal treatments for infections due to VISA or VRSA are uncertain.

To reduce the chance of infection, wounds should be thoroughly cleaned and when necessary, quickly closed by sutures. Surgical wound infection rates can be greatly reduced if the patient is given an effective anti-staphylococcal medication immediately before and after surgery. **Table 23.1** summarizes the main features of staphylococcal wound infections.

Group A Streptococcal "Flesh-Eating Disease"

Streptococcus pyogenes—also called Group A Streptococcus, because it has the Lancefield group A cell wall carbohydrate on its surface—causes wound infections that are generally easy to treat. Occasionally, however, the infections can progress rapidly, even leading to death despite antimicrobial treatment. These more severe infections are called "invasive" because they spread from the wound into the bloodstream and then into tissues and organs, causing diseases such as meningitis, puerperal fever, streptococcal toxic shock syndrome, and fasciitis (inflammation of the fascia that surround muscles and body organs). This section will focus on **necrotizing fasciitis** ("flesh-eating disease"), a rare but serious complication of invasive *S. pyogenes* infection.

Signs and Symptoms

Signs and symptoms of necrotizing fasciitis are very serious. Severe pain develops at the site of the wound, which sometimes can be so minor that no break in the skin is even seen. Within a short time, swelling occurs, causing the overlying skin to become stretched and discolored (**figure 23.5**). The person develops fever and confusion and may also suffer from fatigue and vomiting. Necrotizing fasciitis can progress

TABLE 23.1 Staphylococcal Wound Infections

Signs and Symptoms	Redness, swelling, and pain at wound site; pus formation; sometimes fever; occasionally shock
Incubation Period	Variable
Causative Agent	*Staphylococcus aureus* and *S. epidermidis*
Pathogenesis	*S. aureus* produces many virulence factors that cause tissue damage and destruction; *S. epidermidis* is less virulent but can form biofilms.
Epidemiology	*S. aureus* carriers are at greater risk for wound infection; predisposing factors also include advanced age, overall poor health, immunosuppression, and a prolonged hospital stay. *S. epidermidis* is an opportunist and affects immunocompromised people.
Treatment and Prevention	Treatment: antibiotics; resistance a problem with *S. aureus*. Prevention: cleaning wounds, pre- and post-operative wound care.

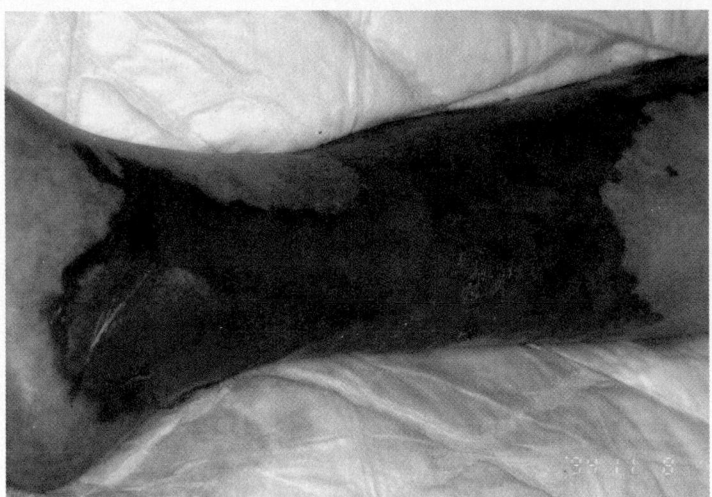

FIGURE 23.5 Individual with *Streptococcus pyogenes* "Flesh-Eating Disease" (Necrotizing Fasciitis) DermNet New Zealand

? Why does necrotizing fasciitis require immediate treatment?

very rapidly, sometimes within hours. Unless treatment is started immediately, the patient may quickly develop shock (life-threatening drop in blood pressure) and die.

Causative Agent

Streptococcus pyogenes is a Gram-positive coccus that typically forms long chains. It is an aerotolerant anaerobe, meaning that although it can grow in aerobic environments it does not use aerobic metabolism (it only ferments). As described in chapter 21, the organism causes streptococcal pharyngitis (strep throat) and is identified by the β-hemolytic colonies it forms on blood agar (see figure 21.5). Strains that cause invasive disease are more virulent because they produce various enzymes and toxins that cause severe tissue damage.

Pathogenesis

The pathogenesis of *S. pyogenes* is covered in detail in chapter 21. The bacterium produces adhesins that help it colonize the wound, including M protein and fibronectin binding proteins (see table 21.2). It also makes a variety of enzymes that help it spread by destroying the subcutaneous fatty tissue, fascia, and sometimes muscle tissue. These enzymes include streptokinase (also called fibrinolysin; breaks down blood clots), hyaluronidases (break down connections between cells), and streptolysins O and S (damage leukocytes and red blood cells).

In addition to aiding in attachment, M protein is shed by growing bacteria and causes neutrophils to release inflammatory molecules that increase vascular permeability. Blood vessels leak fluid, resulting in a drop in blood pressure and subsequent shock. M protein also helps the organism avoid phagocytosis by promoting inactivation of the complement component C3b.

Strains of *S. pyogenes* that cause necrotizing fasciitis produce a variety of streptococcal pyrogenic exotoxins (SPEs). SPEs A and C are superantigens that induce inflammation by nonspecifically activating helper T cells, causing them to release large amounts of cytokines (see figure 16.12). SPEs are encoded by prophages (phage DNA inserted into the bacterial chromosome), another example of lysogenic conversion. SPE B is a chromosome-encoded protease that causes tissue death and breakdown, leading to fluid accumulation in the area and intense swelling. The organisms multiply in the dead tissue, using the breakdown products as nutrients.

Epidemiology

Necrotizing fasciitis is a very rare and sporadic cause of death in the United States. The risk of developing any invasive *S. pyogenes* infection is increased with skin wounds (trauma, burns, surgery, and diseases such as chickenpox), chronic diseases (such as diabetes and alcoholism), and a weakened immune system (due to cancer, AIDS, or injection-drug abuse). Necrotizing fasciitis and other invasive infections seldom occur in healthy people with minor injuries.

Treatment and Prevention

The toxins produced by *S. pyogenes* spread with such speed that immediate surgery is often essential to reduce the pressure of the swollen tissue and to remove dead tissue (debridement). Amputation is sometimes necessary to quickly remove the source of toxins. A combination of broad-spectrum antibiotics is given intravenously, along with intensive supportive care.

No vaccines are currently available to prevent *S. pyogenes* infections, including necrotizing fasciitis. However, several vaccine candidates, including ones that target conserved regions of the M protein, have shown promise in animal studies. **Table 23.2** summarizes the main features of necrotizing fasciitis.

TABLE 23.2 Necrotizing Fasciitis

Signs and Symptoms	Severe pain and swelling at wound site; skin discoloration; fever, confusion, and shock
Incubation Period	Varies; several hours (in severe cases) to several days (in less severe cases)
Causative Agent	*Streptococcus pyogenes*, Lancefield group A β-hemolytic streptococci
Pathogenesis	Organism destroys muscle and organ fascia by releasing destructive enzymes and toxins; blood vessel leakiness and possible shock triggered by M protein.
Epidemiology	Very rare; most cases occur in people with skin wounds, chronic diseases, or a compromised immune system.
Treatment and Prevention	Treatment: aggressive treatment essential with debridement or amputation of affected tissue, and intravenous antibiotics. Prevention: no proven preventive measures.

Pseudomonas aeruginosa Infections

Pseudomonas aeruginosa, an opportunistic pathogen, is a significant cause of healthcare-associated infections. In hospitals, it is an important cause of lung infections and a common cause of wound infections, especially of thermal burns. Burns result in large exposed areas of dead tissue that are not protected by normal body defenses and are therefore highly susceptible to bacterial infection. Almost any opportunistic pathogen can infect burns, but *P. aeruginosa* is among the most common and most difficult to treat.

P. aeruginosa occasionally causes community-acquired infections, including: skin rashes and external ear canal infections from contaminated swimming pools and hot tubs; eye infections from contaminated contact lens solutions; bone infections from puncture wounds; heart valve infections due to injection-drug abuse; and soft tissue infections from procedures such as ear piercing. It also frequently forms biofilms in the lungs of people with the inherited disease cystic fibrosis, causing serious infection.

Signs and Symptoms

A very noticeable sign of *P. aeruginosa* infection of burns and other wounds is pus of a characteristic green color, caused by water-soluble pigments produced by the organism (**figure 23.6a;** see also figure 11.11). These pigments are virulence factors; they include yellow pyoverdine and blue pyocyanin, which together are green (**figure 23.6b**). If the organism enters the bloodstream, signs and symptoms include chills, fever, and shock.

Causative Agent

P. aeruginosa is a motile Gram-negative rod with a single polar flagellum (**figure 23.6c**). It is found in a wide variety of environments such as soil and water, where it grows quickly. The bacterium is classified as an aerobe, but it can respire anaerobically if nitrate is present—an important factor in the organism's ability to form biofilms.

Pathogenesis

P. aeruginosa has many virulence factors, expression of which is often regulated using quorum sensing. It uses its polar flagellum and pili for attachment and colonization; once established on tissue surfaces, quorum sensing signals can lead to biofilm formation. The organism also secretes an exotoxin (exotoxin A) that is taken up by host cells; the toxin blocks protein synthesis and kills the cells. The bacterium uses a type III secretion system (T3SS) to inject various effector proteins into cells; different strains produce different effector proteins, but those that cause severe disease generally produce exoenzyme S (ExoS). This protein interferes with the cell's regulatory proteins, eventually inducing apoptosis. *P. aeruginosa* also produces membrane-damaging toxins, including phospholipase C, which destroys lecithin, an

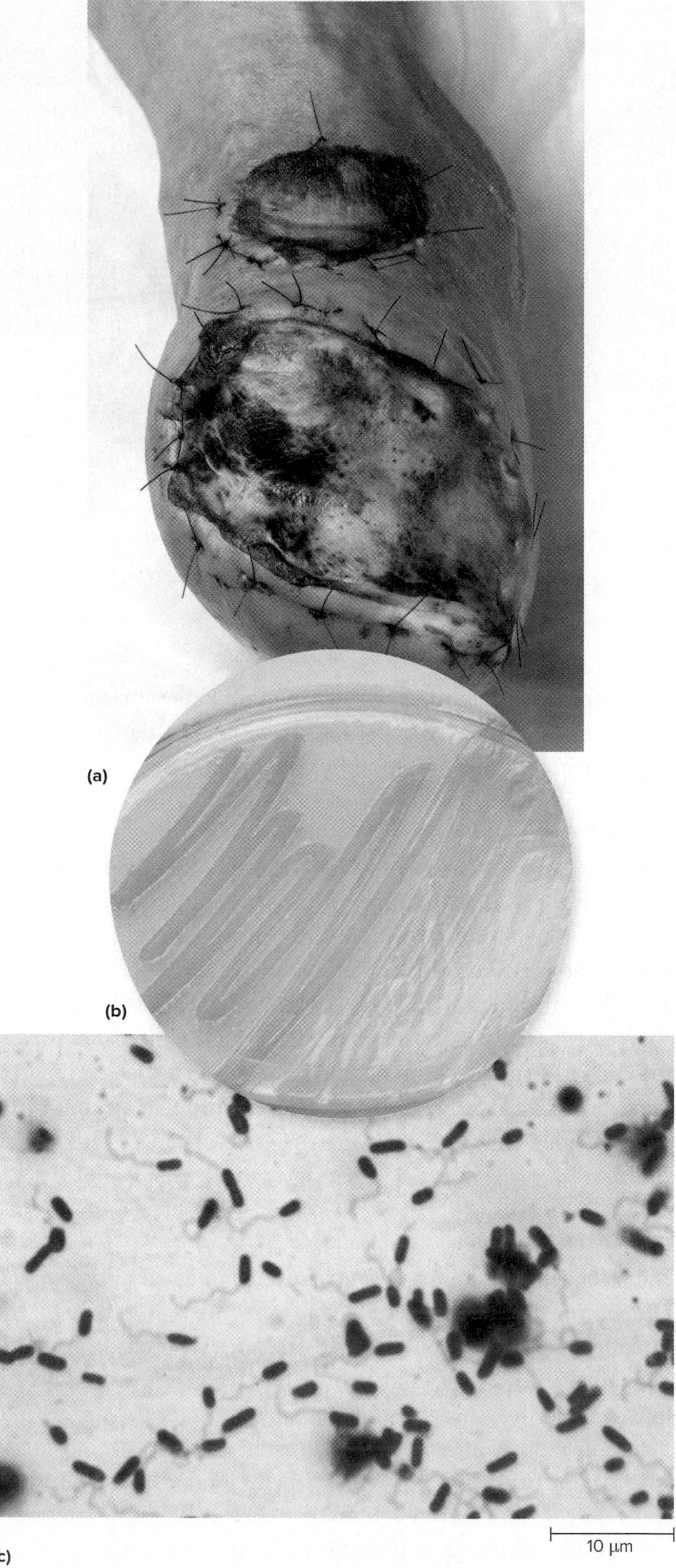

FIGURE 23.6 ***Pseudomonas aeruginosa*** **(a)** Infected skin graft. **(b)** Culture showing green discoloration of an agar medium. **(c)** Stained cells with their single polar flagellum. a: Matt Meadows/Science Source; b-c: Lisa Burgess/McGraw Hill

? What causes the green discoloration of the wound and the culture medium?

important lipid component of cell membranes. In addition, the organism makes proteases (such as elastases) that destroy tissue-binding components, allowing the organism to spread in tissue.

The pigments produced by *P. aeruginosa* contribute to its virulence. Pyocyanin plays a role in increasing oxidative stress in host cells as well as in biofilm formation, allowing the organism to avoid host defenses and antimicrobial medications. Pyoverdine is a siderophore that the bacterium uses to acquire iron, which is often a limiting nutrient in the host.

Epidemiology

P. aeruginosa is widespread in nature. It can grow in most moist places, including swimming pools and hot tubs as well as in soaps, ointments, eyedrops, contact lens solutions, cosmetics, disinfectants, and even distilled water. Because the organism can be on flowers, potted plants, and fruit baskets, hospital visitors are not allowed to take these items into burn wards or intensive care units. *P. aeruginosa* can also be found on many kinds of hospital equipment, on the soles of shoes, and in illegal injectable drugs, all of which have been sources of serious infections.

Treatment and Prevention

P. aeruginosa infections are difficult to treat because the organism is resistant to a wide range of antibiotics. The situation is made even worse by the fact that *P. aeruginosa* forms biofilms, and bacterial cells within biofilms are even more resistant. Only a few antibiotics are effective against this pathogen, including pipercillin, cefepime, and ciprofloxacin, and even these may not be effective against all strains. Antibiotic sensitivity tests are done to determine the most appropriate medication for treatment. For systemic infections, antibacterial medications must usually be given intravenously in high doses.

TABLE 23.3 *Pseudomonas aeruginosa* **Infections**

Signs and Symptoms	Green-colored pus; chills, fever, and shock in systemic infections
Incubation Period	Variable, depending on site of infection
Causative Agent	*Pseudomonas aeruginosa*, a Gram-negative motile rod that can respire both aerobically and anaerobically
Pathogenesis	Secretes exotoxin A; injects effector proteins that alter cell activities; produces toxins that disrupt host cytoplasmic membranes; pigments contribute to pathogenicity.
Epidemiology	Organism widespread in nature, carried on vegetation; difficult to avoid
Treatment and Prevention	Treatment: antibiotics, although resistance and biofilm formation are problematic. Prevention: eliminating bacterial source; proper wound care, including removing dead tissue.

TABLE 23.4 Leading Causes of Wound Infections

Causative Organism	Characteristics	Consequences
Staphylococcus aureus	Gram-positive cocci in clusters, coagulase-positive	Delayed healing; abscess formation; extension into tissues, artificial devices, or bloodstream; some strains can cause toxic shock syndrome
Streptococcus pyogenes	Gram-positive cocci in chains; Lancefield group A	Same as above, except some strains can cause "flesh-eating" necrotizing fasciitis
Pseudomonas aeruginosa	Gram-negative rods; pigments produced result in a green color	Delayed healing; abscess formation; extension into tissues, artificial devices, or bloodstream; shock

P. aeruginosa infections can be prevented by eliminating possible sources of the bacterium and by prompt wound care. In the case of burn wounds, antibacterial cream should be applied after dead tissue is removed. **Table 23.3** summarizes the main features of *P. aeruginosa* infections.

Table 23.4 summarizes the characteristics of the leading causes of wound infections.

MicroAssessment 23.2

Staphylococcus aureus is the most important cause of wound infections because it is commonly carried by humans, transfers easily from person to person, and has multiple virulence factors. *Staphylococcus epidermidis* forms biofilms on foreign materials, protecting the bacteria from body defenses and antibacterial medications. "Flesh-eating disease" is uncommon but can be life-threatening. *Pseudomonas aeruginosa,* a pigment-producing organism, is widespread in the environment and is a major cause of healthcare-associated infections.

4. Why are antibacterial medications alone not effective for treating necrotizing fasciitis?
5. Why do wound infections caused by *Pseudomonas aeruginosa* sometimes have green pus?
6. Why is it not surprising that staphylococci are the most common cause of wound infections?

23.3 ■ Diseases Due to Anaerobic Bacterial Wound Infections

Learning Outcomes

5. Describe the conditions that lead to the development of anaerobic wound infections.
6. Discuss why it is difficult to treat diseases resulting from wounds infected with toxin-producing bacteria.

An important feature of many wounds is that they may be hypoxic (lacking in O_2), thus allowing the growth of obligate anaerobes such as *Clostridium tetani.* Anaerobic conditions can result from numerous factors, including extensive tissue damage, contamination with dirt, and infections involving multiple types of organisms (polymicrobial infection) in which aerobic or facultative microorganisms use up the available O_2 during cellular respiration. This section describes two distinctive diseases that result from anaerobic bacterial wound infections: tetanus and gas gangrene.

Tetanus ("Lockjaw")

Tetanus is rare in resource-rich countries because of vaccination, but has a high case-fatality rate. Exposure to the causative organism cannot be avoided because it produces endospores that are widespread in dust and dirt, often contaminating clothing, skin, and wounds. Even a minor wound in a non-immunized person can result in tetanus if the wound provides conditions that allow the spores to germinate.

Signs and Symptoms

Tetanus is characterized by continuous, painful, and uncontrollable cramp-like muscle spasms; the spasms often begin with the jaw muscles, giving the disease the common name "lockjaw." Early signs and symptoms usually appear about a week after infection and include restlessness, irritability, difficulty swallowing, spasms of the jaw muscles, and sometimes seizures. As more muscles go into sustained contraction (called tetany), breathing becomes difficult, abnormal heart rhythms may occur, and in some cases bones can fracture. After a period of severe pain, the infected person often dies of pneumonia or from lung damage caused by regurgitation of stomach contents that are then aspirated (inhaled) into the lung. Surviving patients remain in the hospital for long periods of time, which increases their risk of healthcare-associated infections.

Newborns can also develop a form of tetanus called neonatal tetanus. About 4 to 14 days after birth, the baby becomes irritable and develops muscle rigidity (**figure 23.7**). As the muscles of the face and mouth are affected, the infant can no longer feed. Respiratory muscles are also affected, which can interfere with breathing.

Causative Agent

Tetanus is caused by *Clostridium tetani,* an anaerobic, Gram-positive rod that produces terminal endospores, meaning they form at the end of the cell (**figure 23.8**). The disease is typically diagnosed by the characteristic signs and symptoms that result from the bacterium's toxin, which plays a key role in pathogenesis (discussed next).

FIGURE 23.7 Infant with Neonatal Tetanus CDC

? What typically causes the death of an infant with tetanus?

Pathogenesis

C. tetani is not invasive, and colonization is generally localized to a wound. If the wound is anaerobic, contaminating endospores germinate, producing vegetative cells. As these grow they release tetanus toxin, an A-B toxin also called tetanospasmin, which binds to peripheral motor neurons (nerves that activate voluntary muscles, meaning the ones we can consciously control). From there, the toxin moves up the nerve and into the spinal cord and brainstem, where it enters other neurons. Normal muscle function involves two types of neurons: (1) motor neurons, which release neurotransmitters (chemical messengers) that cause muscles to contract, and (2) inhibitory interneurons, which release neurotransmitters that block the activity of motor neurons, thereby allowing the muscle to relax. Tetanus toxin prevents inhibitory neurons from releasing neurotransmitters, thereby resulting in sustained muscle contraction (a condition called spastic paralysis) and characteristic muscle spasms (**figure 23.9**).

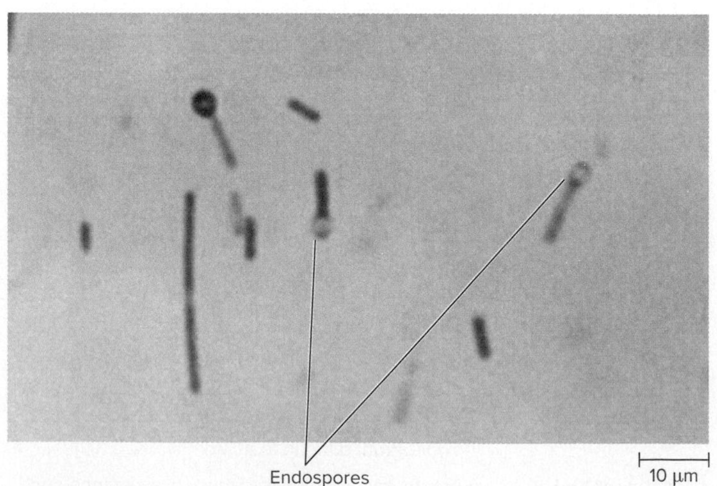

FIGURE 23.8 ***Clostridium tetani*** Terminal endospores are characteristic of this species. Dr. Holdeman/CDC

? How does this anaerobic species persist in an aerobic environment?

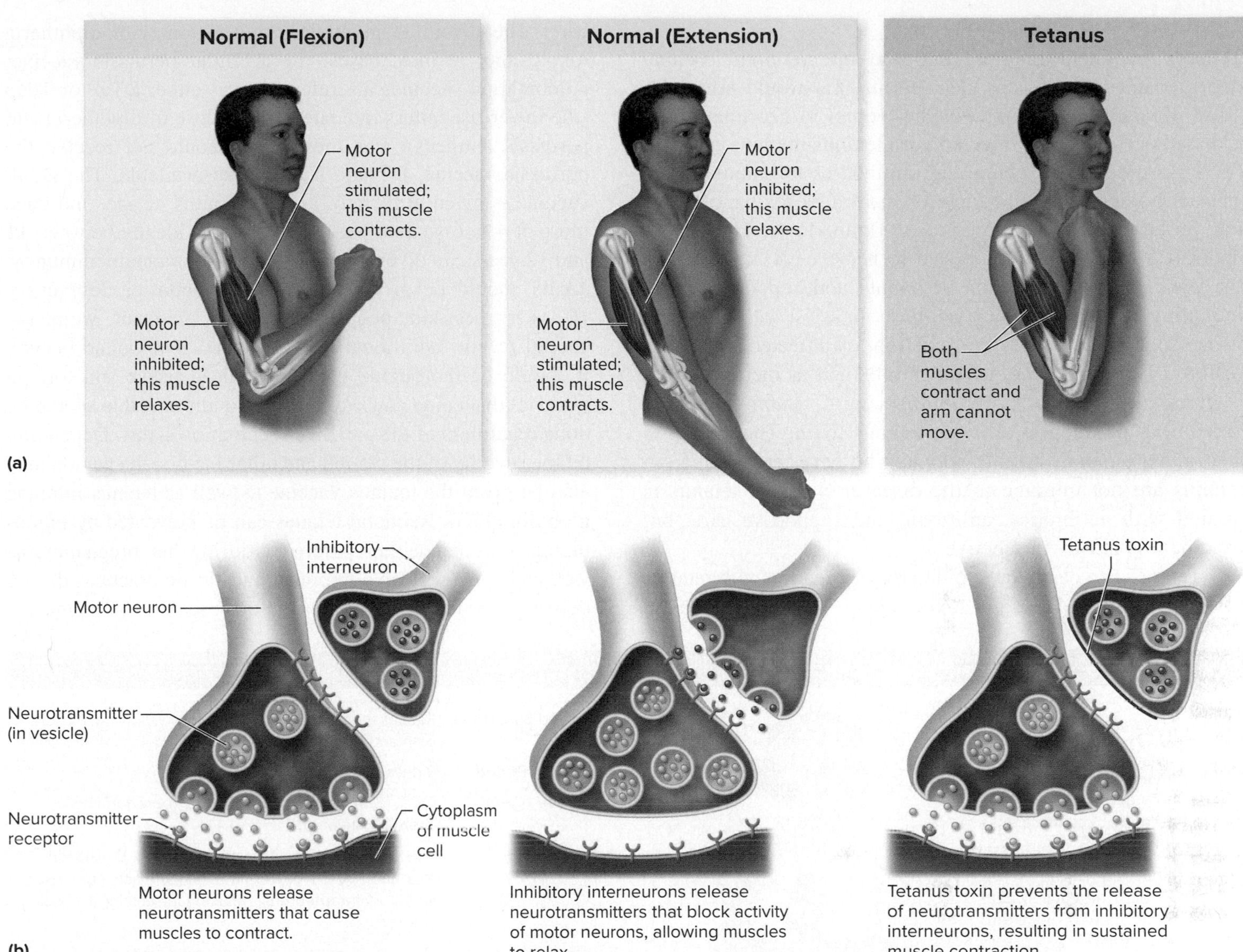

FIGURE 23.9 Tetanus and Inhibitory Neuron Function **(a)** Normal muscle function involves the release of stimulatory and inhibitory neurotransmitters that cause one muscle to contract while an opposing muscle relaxes. **(b)** In tetanus, tetanospasmin prevents the release of the inhibitory neurotransmitters, so both muscles contract simultaneously, causing spasms and cramps.

? **How does tetanus toxin cause sustained muscle contraction?**

Epidemiology

C. tetani spores are found not only in dirt and dust, but also in the gastrointestinal tract and feces of humans and other animals that have eaten foods containing its spores. Therefore, fecal contamination is a potential source of infection. Many cases of tetanus result from anaerobic puncture wounds, which occur from stepping on a sharp object (such as a nail), body piercing, tattooing, animal bites, splinters, injection-drug abuse, and insect stings. Cases can also occur following surgery. Even surface wounds such as scrapes and burns can result in tetanus if the presence of dirt or dead tissue gives rise to anaerobic conditions. Neonatal tetanus may occur if *C. tetani* spores are unintentionally introduced into the umbilical cord wound, either when the cord is cut with unsterilized instruments or when it is covered with a non-sterile dressing.

People who develop tetanus usually have inadequate immunity, meaning they never received a tetanus vaccine or did not receive a scheduled booster shot. In the case of neonatal tetanus, the woman who gave birth to the baby had inadequate immunity; recall that maternal antibodies remain active in the newborn, providing passive immunity for the first several months of life (see figure 15.20). Tetanus occurs most often in resource-limited countries because vaccination is not always available and access to adequate healthcare is often lacking.

MicroByte

In resource-limited countries, World Health Organization initiatives such as maternal immunization and improved hygiene during home births have reduced the cases of neonatal tetanus by over 95% in the last 30 years.

Treatment and Prevention

Tetanus treatment requires multiple interventions. Wound debridement removes any dead tissue that would otherwise allow *C. tetani* spores still in the wound to germinate. The affected person is also injected with tetanus immune globulin (TIG), a preparation containing antibodies to tetanus toxin. The antibodies bind to and therefore neutralize toxin molecules not yet attached to nerve cells (see figure 15.4). TIG cannot neutralize toxin already attached to nerve cells, however, so the patient is given muscle relaxants and supportive care, including being placed on a ventilator to assist with breathing if needed. Over time, the function of affected nerves is restored. An antibacterial medication such as metronidazole is given to kill any actively multiplying *C. tetani* cells, thus preventing production of more tetanus toxin. The person is also given tetanus vaccine; people who have recovered from tetanus are not immune to the disease. Neonatal tetanus is treated with antibiotics, antitoxin, and supportive care, but treatment is mostly ineffective.

Tetanus can easily be avoided by vaccination with tetanus toxoid (inactivated tetanus toxin) along with proper wound care. The toxoid is given in combination with diphtheria toxoid and, in most cases, the acellular pertussis vaccine; combination vaccines are referred to as either DTaP or Tdap (the lowercase letters indicate a lower dose of diphtheria and pertussis antigens). For those who should not receive the pertussis vaccine, DT and Td are also available. The DTaP vaccine is given at 2, 4, 6, and 18 months of age, and once more at 4 to 6 years of age. Adolescents, ideally between 11 and 12 years, are given a dose of Tdap. To maintain immunity, adults should be given a Tdap or Td booster dose every 10 years; certain people—including pregnant women—should receive additional booster doses. If a wound occurs, it should be thoroughly cleaned of dead tissue and foreign material that could cause anaerobic conditions; this should be done regardless of the patient's vaccination status. Depending on the severity of the wound and other factors, the person may also be given the tetanus vaccine as well as tetanus immune globulin (TIG). Neonatal tetanus can be prevented by immunizing a woman either before or during her pregnancy, as well as by using clean equipment and sterile practices during delivery. **Table 23.5** describes the main features of tetanus.

TABLE 23.5 Tetanus ("Lockjaw")

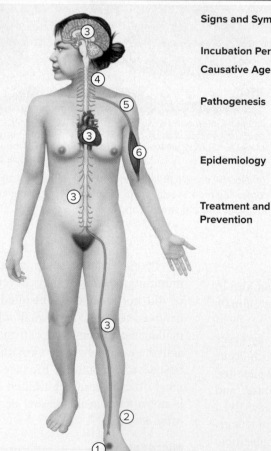

1. *Clostridium tetani* spores in dust or dirt enter a wound.
2. In anaerobic wounds, the spores germinate, and vegetative bacteria release tetanus toxin.
3. Tetanus toxin is carried to the central nervous system by motor nerves or via the bloodstream.
4. The toxin prevents any inhibitory interneurons it reaches from functioning.
5. Other neurons that cause muscle contraction act unopposed, causing muscle spasms.
6. The result is sustained, painful, cramp-like muscle spasms.

Signs and Symptoms	Restlessness, irritability, difficulty swallowing; muscle pain and spasm in jaw, abdomen, back, or entire body
Incubation Period	3 days to 3 weeks; average 8 days
Causative Agent	*Clostridium tetani*, an anaerobic, spore-forming, Gram-positive rod
Pathogenesis	Tetanus results from tetanus toxin, an A-B exotoxin produced by the bacterium. The toxin acts on inhibitory interneurons, resulting in sustained muscle contraction.
Epidemiology	Organisms common in soil; spores contaminate wounds, germinate under anaerobic conditions, such as in puncture wounds.
Treatment and Prevention	Treatment: wound debridement, tetanus immune globulin (TIG) and antibiotics. Prevention: toxoid vaccine; DT or DTaP for infants and children; Td or Tdap for adolescents and as boosters for adults.

Clostridial Myonecrosis ("Gas Gangrene")

Clostridial myonecrosis (commonly called "gas gangrene") is a serious, often life-threatening infection of the muscle caused by a few species of *Clostridium.* Two forms occur: traumatic and spontaneous. The traumatic form is mainly associated with neglected trauma wounds, and the much rarer spontaneous form is usually linked to a weakened immune system or to cancer of the colon or rectum.

Signs and Symptoms

Clostridial myonecrosis signs and symptoms appear dramatically within 6 hours to several days after infection. In the traumatic form, the infected wound quickly becomes extremely painful and the area begins to swell. The skin over the wound becomes discolored, starting out bronze and then progressing to red or purple. Large fluid-filled blisters (bullae) can also form on the skin surface. A thin, foul-smelling, bloody or brownish fluid may leak from the site; the fluid may look frothy because of gas bubbles released by the organism (thus the common name "gas gangrene"). The skin eventually turns black as the tissue dies (**figure 23.10**). If the bacteria enter the bloodstream, serious complications including shock and multi-organ failure can develop. Spontaneous gas gangrene occurs in the absence of any visible trauma or wound; symptoms include fever and an abrupt onset of severe muscle pain. Swelling and bullae similar to those seen in the traumatic form may eventually develop.

Causative Agent

The most common cause of traumatic gas gangrene is *Clostridium perfringens,* a Gram-positive, anaerobic, toxin-producing rod.

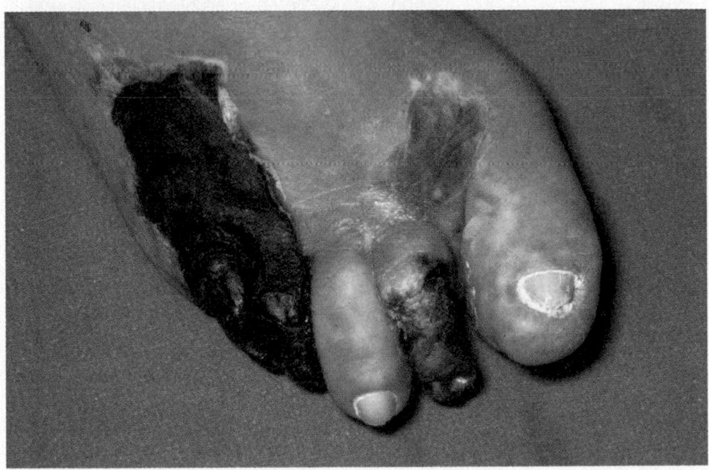

FIGURE 23.10 Foot of Person with Traumatic Gas Gangrene (Clostridial Myonecrosis) Central Manchester University Hospitals NHS Foundation Trust, UK/Science Source

? Why does the affected tissue turn black in gas gangrene?

Although *C. perfringens* is an endospore-forming organism, it usually does not produce spores in wounds or cultures. A related species, *C. septicum,* is the most common cause of the spontaneous form.

Pathogenesis

Two main factors lead to the development of traumatic clostridial myonecrosis: (1) the presence of dirt and dead tissue in the wound and (2) long delays before the wound is treated. *C. perfringens* cannot infect healthy tissue but grows easily in poorly oxygenated and dead tissues that provide a source of nutrients. The bacterium produces multiple toxins, two of which play key roles in disease development. Alpha-toxin (α-toxin) is an enzyme that hydrolyzes phospholipids in host cell membranes, leading to cell lysis; perfringolysin O (PFO) is a pore-forming toxin that works together (synergistically) with α-toxin. The toxins diffuse from the area of infection, killing leukocytes and tissue cells. Several enzymes produced by the pathogen, including collagenase and hyaluronidase, break down the host tissues. As the bacterial cells multiply, they ferment the tissue breakdown products, resulting in the release of large amounts of hydrogen and carbon dioxide. As these gases accumulate in the tissue, the local pressure increases, which helps spread the infection. Without quick surgical treatment, large amounts of the toxins diffuse into the bloodstream and destroy red and white blood cells, tissue capillaries, and various structures throughout the body; this systemic involvement leads to widespread cytokine release, resulting in shock and organ failure. In contrast to traumatic gas gangrene, the spontaneous form occurs when *C. septicum* enters the bloodstream through lesions formed in the gastrointestinal tract; it can then seed deep muscle tissues where it produces several toxins, including a toxin similar to PFO.

Epidemiology

C. perfringens is widespread in soil and is common in the intestinal tract of many animals, including humans, so it can easily infect trauma or surgical wounds. Diseases such as artherosclerosis or diabetes—which can lead to poor oxygenation of tissue by restricting blood flow—are predisposing factors. *C. perfringens* may be found in the vagina of healthy women, so gas gangrene of the uterus sometimes occurs after self-induced abortions, and it occasionally occurs after miscarriages and childbirth. *C. septicum* is also commonly found in soil, and it can be part of the intestinal microbiota of animals (including humans) as well. Lesions in the gastrointestinal tract allow the organism to enter the bloodstream, which can lead to the spontaneous form of the disease; predisposing factors include immune deficiency diseases (such as AIDS) and conditions that damage the intestinal epithelium (such as colon or rectal cancer and inflammatory bowel disease).

Treatment and Prevention

Treatment of gas gangrene includes prompt debridement of all dead and infected tissues and may require amputations. High doses of antibiotics—often a combination of penicillin and clindamycin—are given to help stop bacterial growth and toxin production, but the medications may not diffuse well into large areas of dead tissue; in addition, they do not inactivate α-toxin or perfringolysin.

There is no available vaccine for gas gangrene, but the traumatic form can be prevented by promptly and thoroughly cleaning wounds, including removing any foreign material as well as dead tissue. **Table 23.6** describes the main features of traumatic gas gangrene.

MicroAssessment 23.3

Anaerobic conditions in wounds are created by the presence of dead tissue and foreign material. Tetanus ("lockjaw") is caused by *Clostridium tetani,* an anaerobic, spore-forming bacterium with little invasive ability. The organism releases tetanus toxin that is carried to the nervous system, causing uncontrolled muscle spasms and possibly death. Traumatic gas gangrene (clostridial myonecrosis), is caused by another anaerobic spore-former, *C. perfringens.* The disease usually starts in neglected wounds containing dead tissue and foreign material and then spreads into normal muscle tissue. The spontaneous form of gas gangrene is caused by a related organism, *C. septicum.*

7. Why do many tetanus patients fail to show immediate improvement when given tetanus antitoxin?
8. What factors lead to the development of traumatic gas gangrene?
9. Do newborns need to be immunized against tetanus if their mother had been immunized against the disease? Explain.

TABLE 23.6 Traumatic Clostridial Myonecrosis ("Gas Gangrene")

1. *Clostridium perfringens* spores enter a wound that has two essential characteristics: dead tissue and anaerobic conditions.
2. The spores germinate, and the vegetative bacteria multiply in dead tissue, producing α-toxin and perfringolysin O.
3. The toxins diffuse into and destroy normal tissue. The infection spreads into the dead tissues as bacterial enzymes break down host cells.
4. Swelling occurs; gas produced by fermentation aids rapid progress of the infection.
5. The toxins diffuse into the bloodstream, destroying blood cells and other cells throughout the body. Systemic response to the damage leads to shock and organ failure.

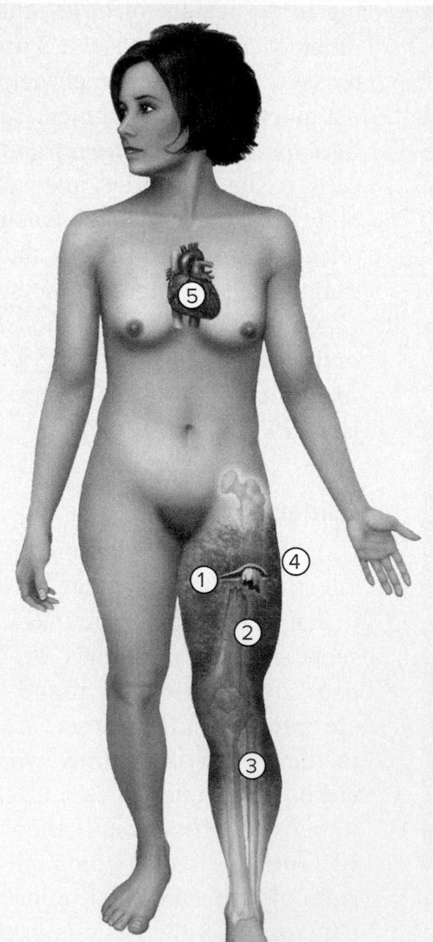

Signs and Symptoms	Severe pain; gas and fluid seep from wound; blackening of overlying skin; shock and multi-organ failure can develop.
Incubation Period	16 hours to several days
Causative Agent	Usually *Clostridium perfringens*
Pathogenesis	Organism grows in dead and poorly oxygenated tissue; releases toxins that kill leukocytes and tissue cells by degrading lecithin in their cell membranes; systemic response leads to shock and organ failure.
Epidemiology	Trauma or surgical wounds contaminated with spores; impaired circulation to tissue as in people with diabetes or artherosclerosis; self-induced abortions
Treatment and Prevention	Treatment: surgical removal of dead tissues; antibiotics to kill vegetative *Clostridium* cells. Prevention: prompt and thorough cleaning of wounds; no vaccine available.

FOCUS ON A CASE 23.1

A 63-year-old woman, healthy except for mild diabetes, had her gallbladder removed. The surgery went well, but within 72 hours the surgical incision became swollen and pale. Within hours, the swelling had spread, and the affected area developed a bluish discoloration. The surgeon suspected traumatic gas gangrene. Antibiotic treatment was started, and the patient was rushed back to the operating room, where the entire swollen discolored area, including the repaired operative incision, was surgically removed. After that, the wound healed normally, although the patient required a skin graft to close the large wound. High numbers of *Clostridium perfringens* were grown from the wound culture.

Six days later, a 58-year-old woman had surgery in the same operating room for a malignant tumor of the colon. The surgery was performed without difficulty, but 48 hours later she developed rapidly advancing swelling and bluish discoloration of her surgical wound. As with the first case, traumatic gas gangrene was suspected, and the patient was treated with antibiotics and surgical removal of the affected tissue. She also needed a skin graft. Her wound culture also showed *C. perfringens*.

Because none of the surgery department's patients had developed infections with *C. perfringens* before this, the hospital epidemiologist was asked to investigate. Among the findings of the investigation:

- Cultures of surfaces in the operating room grew large numbers of *C. perfringens*.
- Unknown to the medical staff, a worker had recently serviced a fan in the ventilation system of the operating room, and for a time air was allowed to flow into the operating room, rather than out of it.
- A road outside the hospital was being repaired using heavy machinery, creating clouds of dust.

As a result of these findings, the operating room and its ventilating system were cleaned and upgraded. No further cases of surgical wound gas gangrene developed.

1. Was the surgeon's diagnosis correct?
2. Many other patients had surgery in the same operating room. Why did only these two patients develop traumatic gas gangrene?
3. What could be done to help identify the source of the patients' infections?

Discussion

1. *C. perfringens* is commonly cultivated from wounds without any evidence of infection. However, in the cases presented here, not only were high numbers of the organism present, but the clinical picture also suggested traumatic gas gangrene. The surgeon's diagnosis was correct.
2. In both cases, an underlying condition increased the two patients' susceptibility to infection: cancer in one, and diabetes in the other. Moreover, both had a recognized source for the organism: Cultures of as many as 20% of diseased gallbladders are positive for *C. perfringens,* and the organism is commonly found in large numbers in the human intestine (a potential source in the case involving removal of the colon tumor).
3. The surgeon thought that the infecting organism might have come from the patients themselves because such strains are often much more virulent than strains that live and sporulate in the soil. On the other hand, the gross contamination of the operating room, as revealed by the cultures of its surfaces, could indicate a very large infecting dose at the operative site, possibly compensating for lesser virulence. In addition, no further cases occurred after the operating room was cleaned and the ventilation system serviced. Unfortunately, no cultures were done from the removed gallbladder or colon tumor, nor were the strains isolated from the wounds and the environment compared. Today, whole genome sequencing could be used to compare the isolates to help identify the source of the infections.

23.4 ■ Bacterial Infections of Bite Wounds

Learning Outcomes

7. Explain why human mouth microbiota can cause serious bite wound infections.
8. Describe two zoonotic wound infections.

Each year, about 2 to 5 million animal or human bite wounds occur in the United States. The most common complication after a bite is an infection at the wound site; this can be a superficial skin infection or one deeper in tissue. The severity can also depend on the types of pathogens introduced into the bite. Infections that result from animal bites are much more common than those from human bites and are considered **zoonoses**. The viral disease rabies, described in section 26.3, is the most concerning outcome of animal bites, although the risk of acquiring the disease is less in countries where vaccination of domesticated animals (such as dogs) is widespread.

Human Bite Infections

Wounds caused by human teeth (such as bites that break the skin, hitting the teeth of another person, or being injured with objects that have been in a person's mouth) are common and can result in very serious infections by normal mouth microbiota. Occasionally, diseases such as syphilis, hepatitis B, and hepatitis C are transmitted this way because the causative agents of these diseases can be found in the blood or saliva.

FOCUS YOUR PERSPECTIVE 23.1

Infection Caused by a Human "Bite"

A 26-year-old man injured a finger of his right hand during a bar fight when he punched someone in the mouth. The man did not seek medical help until more than 36 hours after the fight, at which point his entire hand was very red and tender, with swelling that was spreading to his arm. The surgeon cut open the infected tissues to allow pus to drain, removed the damaged tissue, and washed the wound with sterile fluid. Smears and cultures of the infected material showed aerobic and anaerobic bacteria characteristic of normal mouth microbiota, including species of *Bacteroides* and *Streptococcus*. The patient was given antibiotics, but the wound did not heal well and continued to drain pus. Several weeks later, X rays revealed that infection had spread to the bone at the base of the finger. To cure the infection, the finger had to be amputated.

Signs and Symptoms

The wound may appear minor at first, but within a few hours it becomes painful, swells massively, and often leaks foul-smelling pus (**figure 23.11**).

Causative Agents

The most frequent causes of infection in human bite wounds are members of the normal mouth microbiota, including streptococci, fusiforms, spirochetes, and *Bacteroides* species, often in association with *Staphylococcus aureus.*

Pathogenesis

Bite wound infections involve multiple species that together create a **synergistic infection,** meaning that the infecting organisms acting together results in greater damage than each pathogen causes individually. As the aerobes and facultative anaerobes grow, they consume available O_2, creating anaerobic microenvironments in which anaerobes can multiply. Meanwhile, the bacteria as a group produce a wide variety of damaging toxins and enzymes, including leukocidin, collagenase, hyaluronidase, and various proteases including enzymes that destroy antibodies and proteins of the complement system. Irreversible damage to tissues such as tendons can result in permanent loss of function.

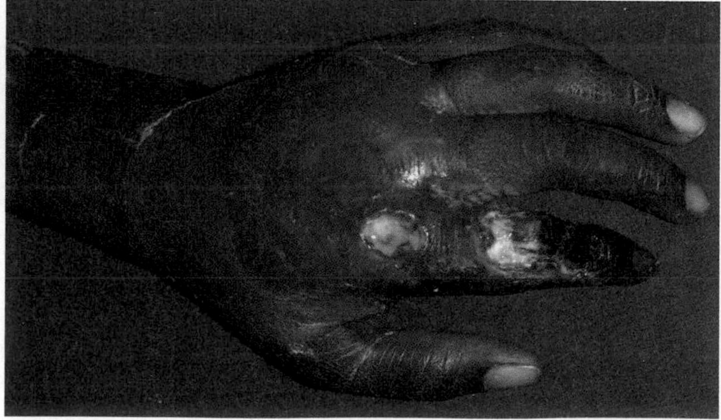

FIGURE 23.11 Infected Bite Injury Hand with infected bite injury to the index finger. Dr. M.A. Ansary/Science Source

? What is the likely source of the organisms infecting this wound?

Epidemiology

Most serious human bite infections occur because of fights or during forcible restraint, situations often encountered in law enforcement and in psychiatric hospitals. The risk is greatly increased when the person who bites has poor mouth care and extensive dental disease. Bites by young children are usually not serious, mostly because children seldom break the skin or cause crushing wounds when they bite.

Treatment and Prevention

Bite wound infections should be treated immediately to avoid serious complications, including infection of muscles and tendons. The infected area should be opened with a scalpel so that dirt, foreign matter (such as broken teeth), and dead tissue can be removed, and the wound should then be washed thoroughly with sterile fluid.

Preventing bite wound infections involves avoiding situations that lead to biting and hitting. If a bite wound occurs, it should be cleaned promptly with soapy water or an antiseptic solution to help prevent infection. An antimicrobial medication such as Augmentin (amoxicillin-clavulanic acid) is typically given prophylactically; alternatively, a combination of medications effective against both aerobes and anaerobes can also be given. The main features of human bite wound infections are presented in **table 23.7.**

TABLE 23.7 Human Bite Infections

Signs and Symptoms	Rapid onset, pain, massive swelling, drainage of foul-smelling pus
Incubation Period	Usually 6 to 24 hours
Causative Agent	Mixed mouth microbiota: anaerobic streptococci, fusiforms, spirochetes, anaerobic Gram-negative rods, often in association with *Staphylococcus aureus*
Pathogenesis	Various mouth bacteria act synergistically to destroy tissue.
Epidemiology	Fighting; forcible restraint; poor mouth care and extensive dental disease
Treatment and Prevention	Treatment: wound care and antibacterial medications. Prevention: clean bite wounds promptly.

Pasteurella multocida Infections

Infections that arise from animal bites involve normal microbiota of the animal's respiratory tract, particularly *Pasteurella multocida.*

Signs and Symptoms

Early signs of infection in wounds caused by animal bites develop within 24 hours and include redness, tenderness, and swelling, followed by pus discharge in some cases. Complications include abscess formation and bone infections. If the bacteria enter the bloodstream, endocarditis or meningitis may result.

Causative Agent

Pasteurella multocida is a Gram-negative, facultatively anaerobic encapsulated coccobacillus. The various strains are classified into different serogroups based on their capsule composition.

Pathogenesis

The details of *P. multocida* pathogenesis are not well understood. The cells have an antiphagocytic capsule that presumably acts by interfering with opsonization by complement proteins. Strains vary in their pathogenicity, however, so whole genome sequencing is used to identify characteristics shared by virulent strains. Some strains produce an A-B toxin called *Pasteurella multocida* toxin (PMT) that interferes with host cell signaling pathways.

Epidemiology

Animals such as cats, dogs, and monkeys carry *P. multocida* as normal oral and upper respiratory microbiota. Cats are more likely to carry *P. multocida* than dogs, so the infections are more frequently caused by cat bites. *P. multocida* also causes disease—and sometimes fatal outbreaks—in a number of other animal species.

Treatment and Prevention

Unlike many Gram-negative pathogens, *P. multocida* is susceptible to penicillin, so a medication that is a combination of a penicillin derivative (amoxicillin) and a β-lactamase inhibitor (clavulanate) is used. The clavulanate is helpful because many bite wounds often also contain strains of β-lactamase-producing *Staphylococcus aureus.* This and other antibacterial medications are effective if given early in the infection, so treatment is started even before the diagnosis has been confirmed by culturing the bacterium.

No vaccines for *P. multocida* are available for use in humans. Immediate cleaning of bite wounds and quick medical attention usually prevent the development of serious infection. **Table 23.8** gives the main features of *P. multocida* bite wound infections.

TABLE 23.8 *Pasteurella multocida* Bite Wound Infections

Signs and Symptoms	Spreading redness, tenderness, swelling, pus discharge
Incubation Period	24 hours or less
Causative Agent	*Pasteurella multocida*, a Gram-negative, facultatively anaerobic, encapsulated coccobacillus
Pathogenesis	*P. multocida* introduced by bite; capsule prevents phagocytosis; may produce an A-B toxin that kills host cells.
Epidemiology	Carried by many animals in their mouth or upper respiratory tract.
Treatment and Prevention	Treatment: antibiotics, effective if given promptly. Prevention: no vaccine available for humans; prompt wound care.

Bartonellosis ("Cat Scratch Disease")

Bartonellosis, also known as "cat scratch disease," is a bacterial disease that can be acquired after being bitten or scratched by a cat infected with *Bartonella henselae.* Cats become infected by the bite of fleas that carry the organism.

Signs and Symptoms

Cat scratch disease (CSD) typically begins within a week of a scratch or bite with the appearance of a papule or pustule at the site of injury. Painful enlargement of local lymph nodes (lymphadenopathy) develops in 1 to 3 weeks, and in some cases the lymph nodes become pus-filled and soft. Patients also experience fever, malaise, headache, and fatigue. In immunocompromised individuals, infections of the eye, liver, brain, bones, and heart valves can occur.

Causative Agent

CSD is caused by *Bartonella henselae,* a fastidious, slow growing, curved, Gram-negative rod.

Pathogenesis

The virulence factors of *B. henselae* and the process by which it causes disease are not well described. However, it uses an adhesin called BadA to attach to and enter host cells. It also uses a secretion system to deliver a variety of effector proteins into host cells. *B. henselae* cells are eventually carried to the lymph nodes, causing them to become swollen. The infection usually clears uneventfully, but in people with an impaired immune system the organism may spread via the bloodstream, causing serious complications such as bacillary angiomatosis (vascular lesions) or bacillary peliosis hepatis (liver vascular lesions).

Epidemiology

CSD occurs worldwide, often in children and young adults. The disease is transmitted to humans mainly through cat scratches and bites. Cat fleas carry the bacterium; transmission to humans occurs via a cat scratch contaminated with flea stool as well as through cat bites and when cats lick open wounds.

Treatment and Prevention

Severe *B. henselae* infections can usually be treated with antibiotics such as ampicillin, but some strains are resistant to this medication.

There are no proven preventive measures for CSD, other than to avoid handling stray cats and to make sure domestic cats use appropriate flea prevention products. Any cat scratch or bite should be promptly cleaned with soap and water and then treated with an antiseptic. If signs of infection develop, prompt medical evaluation is needed. The main features of cat scratch disease are presented in **table 23.9.**

Streptobacillus moniliformis Infections

Rat bites are fairly common in resource-limited large cities and among workers who handle laboratory rats. *Streptobacillus moniliformis,* a slow growing, pleomorphic, Gram-negative rod, is carried in the nose and throat of rats, mice, and other rodents. If the bacterium enters the human body—through a bite or scratch, or via food or water contaminated with rat feces—it can cause streptobacillary rat-bite fever, or streptobacillosis. This disease, which is rare in the United States, is characterized by fever, head and muscle aches, and vomiting. A petechial or maculopapular rash usually appears on the hands and feet after a few days, followed by joint pain. Serious and sometimes fatal complications such as abscesses in the abdominal cavity, meningitis, or heart valve infection can occur. Because of this, empiric treatment (therapy given without confirming the cause of infection) with either penicillin or ceftriaxone is usually started immediately as laboratory confirmation can take weeks. Prevention of this disease includes control of wild rat populations and care in handling laboratory rats and their feces.

TABLE 23.9 Bartonellosis ("Cat Scratch Disease")

Signs and Symptoms	Papule or pustule appears at the bite or scratch site, followed by local lymph node swelling; nodes may become pus-filled and soft.
Incubation Period	Usually within 1 week
Causative Agent	*Bartonella henselae*, a Gram-negative rod
Pathogenesis	Bacterial cells enter with cat bite or scratch and reach lymph nodes, where the disease is usually stopped; in immunocompromised people, the organism may spread, causing serious complications.
Epidemiology	Acquired from cats when they bite or scratch, or from their fleas.
Treatment and Prevention	Treatment: antibiotics. Prevention: promptly washing wounds, applying antiseptic; avoiding rough play with cats and keeping them free of fleas.

MicroAssessment 23.4

Human bite infections can be dangerous because certain generally harmless members of the mouth microbiota can invade and destroy tissue when growing synergistically. *Pasteurella multocida* can infect bite wounds caused by a number of different animals, especially cats. Cat bites and scratches can also transmit *Bartonella henselae,* the cause of cat scratch disease, which is typically characterized by local lymph node enlargement. *Streptobacillus moniliformis* is the cause of streptobacillary rat-bite fever, characterized by fever and a rash.

10. What Gram-negative organism commonly infects wounds caused by bites of any of a variety of different animals?

11. What are two potentially serious complications of cat scratch disease?

12. Why are members of the normal mouth microbiota a cause of serious infection in human bites?

23.5 ■ Fungal Wound Infections

Learning Outcome

9. Give the distinctive features of sporotrichosis.

Fungal infections of wounds are more serious than dermal mycoses. Typically, they are caused by soil fungi that enter through injuries that allow them to invade subcutaneous tissue.

Sporotrichosis ("Rose Gardener's Disease")

Sporotrichosis, also known as "rose gardener's disease," occurs around the world and is mostly associated with activities that lead to puncture wounds from plant material. Although many cases are sporadic, small outbreaks can occur among farmers, gardeners, and agricultural workers. In certain areas of the world, sporotrichosis can also be transmitted by infected animals, particularly cats; because of this, the disease is classified as an emerging zoonosis.

Signs and Symptoms

Signs and symptoms of sporotrichosis resemble those of a bacterial skin infection, often leading to misdiagnosis. The site of infection in most cases is the hand or arm, but other parts of the body can also be involved. Typically, a chronic ulcer forms at the wound site, followed by a slowly progressing series of ulcerating nodules that develop near the original lesion (usually along the path of a lymphatic vessel). Local lymph nodes enlarge, but patients generally do not feel ill. If a person has AIDS or another immunodeficiency, however, the fungus can disseminate, infecting joints and the central nervous system, and becoming life-threatening.

Causative Agent

Sporotrichosis is caused by species of the dimorphic fungus *Sporothrix,* which lives in soil and on vegetation (**figure 23.12**). Most cases worldwide are caused by *S. schenckii,* but zoonotic infections in some South American countries have been caused by *S. brasiliensis.*

Pathogenesis

S. schenckii conidia enter the body through an injury caused by plant material, while *S. brasiliensis* is transmitted mainly through contact with infected cats. After an incubation period that usually ranges from 1 to 3 weeks, the multiplying fungi cause a small nodule to form at the site of the injury. This slowly enlarges and ulcerates, producing a painless red lesion that bleeds easily. There is little to no pus unless secondary bacterial infections occur. After a week or longer, more lesions form—usually along a lymphatic vessel away from the extremities and toward the center of the body. In healthy people, the infection does not usually spread to other sites. *S. brasiliensis*, however, typically causes more severe skin lesions and is associated with an increased number of disseminated infections.

Epidemiology

Sporotrichosis is distributed worldwide, mostly in warmer regions but extending into temperate climates. It is an occupational disease of farmers, carpenters, gardeners, greenhouse workers, and others who work with plant materials. People in certain areas who handle infected animals are also at risk. Other risk factors for the disease include diabetes, immunosuppression, and alcoholism. Individuals with chronic lung disease can contract *S. schenckii* lung infections from inhaling contaminated dust from hay or cattle feed. Deaths from sporotrichosis are rare.

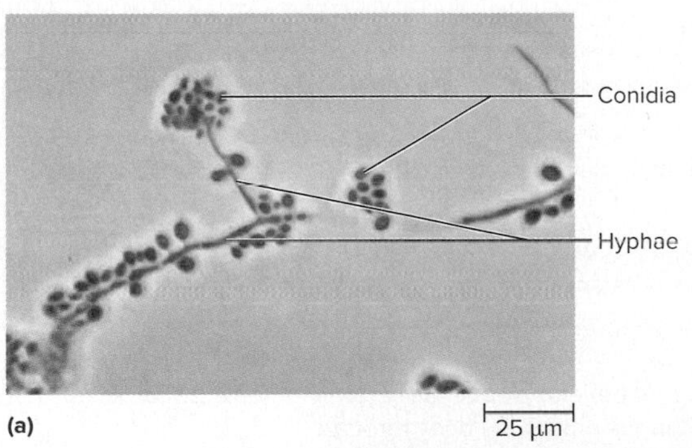

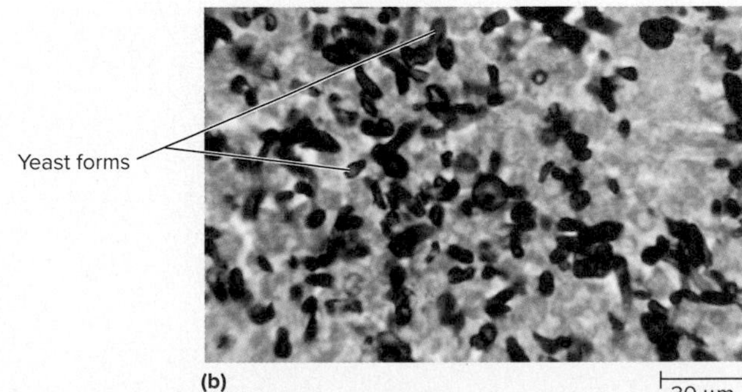

FIGURE 23.12 ***Sporothrix schenckii*** **(a)** Mold form. **(b)** Yeast form, as seen in infected tissue from a laboratory mouse. a: CDC; b: Libero Ajello, Ph.D./CDC

? *Sporothrix schenckii* is dimorphic; what does this mean?

TABLE 23.10 Sporotrichosis ("Rose Gardener's Disease")

Signs and Symptoms	Painless, ulcerating nodules appearing in sequence in a linear pattern
Incubation Period	Usually 1 to 3 weeks
Causative Agent	*Sporothrix schenckii*, a dimorphic fungus; also *S. brasiliensis* in certain regions
Pathogenesis	Fungal cells multiply at site of introduction, causing a small nodule that ulcerates. Fungi carried by lymph flow lead to additional lesions along the lymphatic vessel.
Epidemiology	Widespread distribution. People in certain occupations, such as those that require contact with sharp plant materials, are at particular risk.
Treatment and Prevention	Treatment: antifungal medications. Prevention: protective gloves and clothing when handling plant material, avoiding scratches by infected animals.

Treatment and Prevention

Itraconazole is typically used to treat sporotrichosis, with the addition of amphotericin B in certain severe cases. Oral potassium iodide (KI) was commonly used in the past, and it is still used in resource-limited countries, but it can cause uncomfortable side effects. Without treatment, the disease can become chronic.

Sporotrichosis can be prevented by wearing protective gloves and a long-sleeved shirt when working with soil and vegetation. People in areas shown to have zoonotic transmission should avoid scratches from infected animals, especially cats. **Table 23.10** presents some of the main features of sporotrichosis.

MicroAssessment 23.5

Rose gardener's disease (sporotrichosis), mainly caused by the dimorphic fungus *Sporothrix schenckii,* is widely distributed around the world. It generally affects those with occupations that expose them to splinters and sharp vegetation. Unlike most invasive fungal infections, sporotrichosis is usually easy to treat.

13. List some of the risk factors associated with contracting rose gardener's disease.

14. How is it possible for a person to contract *S. schenckii* lung infections?

15. Why is sporotrichosis often misdiagnosed?

The key features of the diseases covered in this chapter are highlighted in the **Diseases in Review 23.1** table that follows.

Diseases in Review 23.1

Wound Infections

Disease	Causative Agent	Comment	Summary Table
BACTERIAL INFECTIONS			
Staphylococcal infections	*Staphylococcus aureus* and *S. epidermidis*	*S. aureus* (described in chapter 22) is the most common cause of wound infections, but *S. epidermidis*, part of the normal skin microbiota, can form biofilms on medical devices.	Table 23.1
Group A streptococcal "flesh-eating disease"	SPE-producing strains of *Streptococcus pyogenes*	SPEs and enzymes of *S. pyogenes* (described in chapter 21) cause inappropriate T-cell activation, tissue breakdown, and toxic shock.	Table 23.2
***Pseudomonas aeruginosa* infections**	*Pseudomonas aeruginosa*	Widespread environmental organism that commonly infects burned tissues; produces pigments that result in a green color; resistant to a wide variety of antimicrobial medications.	Table 23.3
DISEASES DUE TO ANAEROBIC BACTERIAL WOUND INFECTIONS			
Tetanus ("lockjaw")	*Clostridium tetani*	Toxin-producing, endospore-forming anaerobe grows in wound-damaged tissues; tetanus toxin blocks the action of inhibitory interneurons, leading to continuous painful muscle contractions; typically fatal without treatment. Preventable by vaccination (DTaP and boosters).	Table 23.5
Clostridial myonecrosis ("gas gangrene")	Usually *Clostridium perfringens* (traumatic form); *C. septicum* (spontaneous form)	Toxin-producing, endospore-forming anaerobes; *C. perfringens* grows in necrotic and poorly oxygenated tissue; *C. septicum* enters through lesions in gastrointestinal tract; gases produced accumulate in tissues; typically fatal without treatment.	Table 23.6
BACTERIAL INFECTIONS OF BITE WOUNDS			
Human bite infections	Mixed aerobic and anaerobic species, often members of the normal mouth microbiota	Crushing nature of wound causes tissue damage and anaerobic conditions; synergistic infection adds to that tissue damage.	Table 23.7
***Pasteurella multocida* infections**	*Pasteurella multocida*	Characterized by abscess at the site of the wound; causative agent is a common member of the normal mouth microbiota of animals; results most often from cat bites.	Table 23.8
Bartonellosis ("cat scratch disease")	*Bartonella henselae*	Characterized by local lymph node enlargement; acquired via a cat bite or scratch.	Table 23.9
Streptobacillary rat-bite fever	*Streptobacillus moniliformis*	Acquired from rats, or food or water contaminated with rat feces; characterized by recurring fever and rash; uncommon and rarely serious.	
FUNGAL WOUND INFECTIONS			
***Sporotrichosis* ("rose gardener's disease")**	*Sporothrix schenckii; S. brasiliensis*	Characterized by painless ulcerating nodules that develop sequentially along a lymphatic vessel; environmental dimorphic fungus enters the body through an injury; zoonotic transmission via infected cats	Table 23.10

Summary

23.1 ■ Anatomy, Physiology, and Ecology of Wounds

Wounds expose tissue components to which pathogens specifically attach. Wounds heal by forming **granulation tissue,** which fills the space created by the wound; this gradually shrinks and is converted to collagen, a component of scar tissue that gradually retracts as epithelium is regenerated (figure 23.1).

Wound Abscesses

An **abscess**, composed of **pus,** localizes an infection within tissue, preventing its spread (figure 23.2).

23.2 ■ Common Bacterial Infections of Wounds (table 23.4)

Wound infections may lead to delayed healing, abscess formation, spread of bacteria or toxins throughout the body, and opening of surgical wounds.

Staphylococcal Wound Infections (figures 23.3, 23.4)

Staphylococcus aureus is the most common cause of wound infections and has many virulence factors (table 23.1). *Staphylococcus epidermidis* is less virulent but can form biofilms on catheters and other devices.

Group A Streptococcal "Flesh-Eating Disease" (figure 23.5)

Necrotizing fasciitis is caused by strains of *Streptococcus pyogenes* that produce a variety of toxins, including SPE B, a protease thought to be responsible for tissue destruction.

Pseudomonas aeruginosa Infections (figure 23.6)

Pseudomonas aeruginosa is widespread in the environment and a cause of both healthcare-associated and community-acquired infections. Infections are characterized by production of green pus.

23.3 ■ Diseases Due to Anaerobic Bacterial Wound Infections

Anaerobic conditions are likely to occur in wounds containing dead tissue or foreign material, and in wounds with limited exposure to the air.

Tetanus ("Lockjaw") (table 23.5)

Tetanus is characterized by sustained, painful muscle spasms (figure 23.7) caused by tetanus toxin, an exotoxin made by the endospore-forming anaerobe *Clostridium tetani* (figure 23.8). The toxin prevents release of neurotransmitters by inhibitory interneurons, causing uncontrolled muscle contraction (figure 23.9). The disease is prevented by vaccination with toxoid, followed by regular booster doses.

Clostridial Myonecrosis ("Gas Gangrene") (table 23.6)

Gas gangrene has two forms—traumatic and spontaneous—and is caused by certain endospore-forming anaerobes in the genus *Clostridium*. The traumatic form, usually caused by *C. perfringens*, results in dark, discolored skin and fluid-filled blisters that leak a thin, brown, bubbly fluid (figure 23.10); toxins cause tissue necrosis, and gas is produced as tissue nutrients are fermented. The spontaneous form is caused by *C. septicum,* which enters the bloodstream via gastrointestinal lesions and seeds deep muscle tissues; a toxin causes tissue damage. There is no vaccine for either form.

23.4 ■ Bacterial Infections of Bite Wounds

The kind of bite wound infection depends on the types of bacteria in the mouth of the biting animal and the nature of the wound—whether punctured, crushed, or torn.

Human Bite Infections (table 23.7)

Human bites can result in very serious infections, with pain, massive swelling, and foul-smelling pus (figure 23.11). Infections are usually caused by synergistic activity among normal mouth microbiota introduced into the wound, and are treated with antibiotics and prevented by prompt cleansing and use of an antiseptic.

Pasteurella multocida Infections (table 23.8)

Pasteurella multocida can infect bite wounds inflicted by a number of animal species that are asymptomatic carriers. Infection begins with spreading redness, tissue tenderness, and swelling followed by pus formation; abscesses are common.

Bartonellosis ("Cat Scratch Disease") (table 23.9)

Cat scratch disease is caused by *Bartonella henselae* and begins with a pimple at the site of a bite or scratch, followed by enlargement of local lymph nodes, which often become pus-filled.

Streptobacillus moniliformis Infections

Streptobacillus moniliformis can cause streptobacillary rat-bite fever, characterized by recurring fevers, head and muscle aches, and vomiting. A rash and joint pains often develop.

23.5 ■ Fungal Wound Infections

Fungal wound infections are usually more serious than dermal mycoses.

Sporotrichosis ("Rose Gardener's Disease") (table 23.10)

Sporotrichosis is characterized by painless, ulcerating nodules that develop along the path of a lymphatic vessel; it is caused by certain dimorphic fungi in the genus *Sporothrix* (figure 23.12). Most cases worldwide are caused by *S. schenckii,* a soil- and vegetation-borne organism that enters the body through an injury caused by plant material; *S. brasiliensis* is found in parts of South America, mainly transmitted by contact with infected cats.

Review Questions

Short Answer

1. What property of *Staphylococcus epidermidis* helps it to colonize plastic materials used in medical procedures?
2. How does the action of the superantigens made by some *S. aureus* strains result in toxic shock?
3. Name two underlying conditions that predispose a person to *Streptococcus pyogenes* "flesh-eating disease."
4. Give two sources of *Pseudomonas aeruginosa.*
5. Outline the pathogenesis of tetanus.

6. Explain why tetanus immunoglobulin (TIG) cannot reverse the muscle paralysis seen in tetanus.
7. What characteristics of bite wounds lead to anaerobic infections?
8. What is the causative agent of cat scratch disease? Why is it a threat to patients with an impaired immune system?
9. What is a synergistic infection? How might one be acquired?
10. Why is sporotrichosis sometimes called "rose gardener's disease"?

Multiple Choice

1. Which of the following about *Staphylococcus aureus* is *false*?
   a) It is often referred to as coag-positive staph ("coag" referring to coagulase).
   b) Its infectious dose is increased in the presence of foreign material.
   c) Some strains infecting wounds can cause toxic shock.
   d) Nasal carriers have an increased risk of surgical wound infection.
   e) It is pyogenic.
2. Which of these statements about *Streptococcus pyogenes* is *false*?
   a) It is a Gram-positive coccus occurring in chains.
   b) Some strains that infect wounds can cause toxic shock.
   c) Some strains that infect wounds can cause necrotizing fasciitis.
   d) It can cause puerperal fever.
   e) A vaccine is available for preventing *S. pyogenes* infections.
3. Choose the one *false* statement about *Pseudomonas aeruginosa.*
   a) It is widespread in nature.
   b) Some strains can grow in distilled water.
   c) It is a Gram-positive rod.
   d) It produces pigments that together are green.
   e) Under certain circumstances, it can grow anaerobically.
4. Which of these statements about tetanus is *true*?
   a) It can start from a bee sting.
   b) Immunization is carried out using tiny doses of killed *C. tetani.*
   c) Those who recover from the disease are immune for life.
   d) Tetanus immune globulin does not prevent the disease.
   e) It is easy to avoid exposure to spores of the causative organism.
5. Choose the one *true* statement about gas gangrene.
   a) It typically starts in neglected wounds with dead tissue.
   b) When treating the disease, it is best to rely on antibacterial medications and avoid disfiguring surgery.
   c) A toxoid is generally used to protect against the disease.
   d) Only one antitoxin is used for treating all cases of the disease.
   e) Avoiding spores of the causative agent is easy.
6. Which of the following statements about human bite wound infections is *false*?
   a) They are caused by synergistic bacterial infections.
   b) The causative agents are normal mouth microbiota.
   c) They are seldom serious and often resolve without treatment.
   d) They may sometimes be contaminated with debris.
   e) The causative agents include both aerobes and anaerobes.
7. Which of the following statements about *Pasteurella multocida* is *false*?
   a) Infections generally respond to a penicillin.
   b) It can cause oubreaks of fatal disease in animals.
   c) It is commonly found in the mouths of biting animals, including humans.
   d) A vaccine is used to prevent *P. multocida* disease in people.
   e) Cat bites are more likely than dog bites to result in *P. multocida* infections.
8. Which of these statements about cat scratch disease is *false*?
   a) It causes local lymph node enlargement.
   b) It is a serious threat to people with AIDS.
   c) Cat scratches are the only mode of transmission to humans.
   d) It can be treated with antibiotics.
   e) It can affect the brain or heart valves in a small percentage of cases.
9. The following statements about *Streptobacillus moniliformis* are all true *except*
   a) it can be transmitted by food contaminated with rodent feces.
   b) it may cause fatal complications involving the heart or brain.
   c) it can be transmitted by the bites of animals other than rats.
   d) human infection is characterized by irregular fevers, rash, and joint pain.
   e) it is a Gram-positive spore-forming rod.
10. Which statement concerning sporotrichosis is *false*?
   a) It is characterized by ulcerating lesions along the path of a lymphatic vessel.
   b) Person-to-person transmission is common.
   c) It can occur in groups of people.
   d) It can persist for years if not treated.
   e) The causative organism is a dimorphic fungus.

Applications

1. Clinicians become concerned when the laboratory reports that organisms capable of digesting collagen and fibronectin are present in a wound culture. What is the basis of their concern?
2. An army field nurse working at a mobile surgical hospital asks this question of all the ambulance drivers: "Was the soldier wounded while in a field with cows?" Why does the nurse ask this question?

Critical Thinking

1. In what way would the incidence of tetanus at various ages in a developing country differ from age incidence in developed countries?
2. Could colonization of a wound by a non-invasive bacterium cause disease? Explain your answer.

24 | Digestive System Infections

Intestinal microvilli (color-enhanced TEM). *Steve Gschmeissner/SPL/Science Source*

KEY TERMS

Cirrhosis Scarring of the liver that interferes with normal liver function.

Dental Caries Damage to tooth enamel resulting from acids produced as microbes in dental plaque ferment sugars; tooth decay.

Dental Plaque A biofilm on a tooth surface.

Dysbiosis An imbalance in the normal microbiota.

Dysentery A serious form of diarrhea characterized by blood, pus, and mucus in the feces.

Gastritis Inflammation of the lining of the stomach.

Gastroenteritis Inflammation of the lining of the stomach and intestines; the syndrome of nausea, vomiting, diarrhea, and abdominal pain.

Gingivitis Inflammation of the gums.

Hemolytic Uremic Syndrome (HUS) Serious condition characterized by red blood cell breakdown and kidney failure.

Hepatitis Inflammation of the liver.

Oral Rehydration Therapy (ORT) A treatment used to replace fluid and electrolytes lost due to diarrhea.

Periodontitis Inflammation of the periodontium (tissues supporting the teeth).

A Glimpse of History

Cholera is a very old disease, thought to have originated in the Far East thousands of years ago. With the increased shipping of goods and the mobility of people during the nineteenth century, cholera spread from Asia to Europe and then to North America. The disease caused major epidemics in the nineteenth century.

John Snow (1813–1858), a London physician, demonstrated that cholera was transmitted by contaminated water. He observed that almost all people who contracted the disease during an outbreak in 1854 got their water from a well on Broad Street, whereas neighbors who got their water elsewhere were unaffected. Even though the germ theory of disease had not yet been described, Snow was able to persuade local authorities to remove the pump handle from the suspected well so that people were forced to get their water elsewhere. Although the number of new cases decreased, Snow's explanation that cholera was a waterborne disease was not accepted by most doctors and government officials, partly because the outbreak had already begun subsiding before the handle was removed. By 1866, however, it was obvious that cholera occurred in areas where water had been contaminated with sewage, just as Snow had proposed. Public health agencies then played a major role in preventing epidemic cholera. In 1884 Robert Koch provided convincing evidence for the germ theory of disease after isolating *Vibrio cholerae*, the bacterium that causes cholera.

In late 2016, the largest cholera epidemic ever recorded started in Yemen. By April 2021, the country had reported over 2.5 million suspected cases and nearly 4,000 deaths. Conditions for the crisis were created by an ongoing war that devastated Yemen's health services and infrastructure, leaving more than 14 million people without access to safe drinking water. Although the WHO manages a stockpile of an oral cholera vaccine for use during epidemics, Yemen's unstable situation caused significant delays in initiating and then maintaining the vaccination programs. In addition, the sheer size of the epidemic and mass migration of people created further problems for disease control. Several humanitarian agencies responded to the crisis by opening up cholera treatment facilities, running awareness campaigns for disease prevention, delivering tanks of clean water to people, and training healthcare workers across the country to deal with the huge number of cholera cases—a multidisciplinary effort that demonstrates the importance of the One Health approach to disease control described in section 19.3. These efforts slowed the spread of disease, but the long-term solution requires a stable source of clean water.

24.1 ■ Anatomy, Physiology, and Ecology of the Digestive System

Learning Outcomes

1. Describe the characteristics and functions of the digestive system components.
2. Describe the significance of the normal intestinal microbiota.

The main purpose of the digestive system is to convert the food we eat into a form that the body's cells can use as a source of energy and raw materials for growth. It does this by physically breaking down food into small particles, chemically degrading those particles even further, and then absorbing the available nutrients. The waste material that remains is eliminated as feces.

The digestive system includes two general components: the digestive tract and the accessory organs (**figure 24.1**). The digestive tract is a hollow tube that starts at the mouth and ends at the anus. When referring only to the stomach and the intestines, the term *gastrointestinal tract* is often used. The accessory organs—which include the salivary glands, pancreas, liver, and gallbladder—support the digestive process by producing vital enzymes and other substances that help break down food.

Like the respiratory system and the skin, the digestive tract is one of the body's major boundaries with the environment. It is distinct, however, in that the digestive process provides a plentiful source of carbon and energy for microbial growth: Every mouthful of food you eat feeds not only yourself but also the microbial population in your digestive tract.

Most microbes that inhabit the digestive tract live in harmony with the host, but the balance in this complex ecosystem is delicate. In many parts of the tract, a mucous membrane only a single cell layer thick separates the microbial

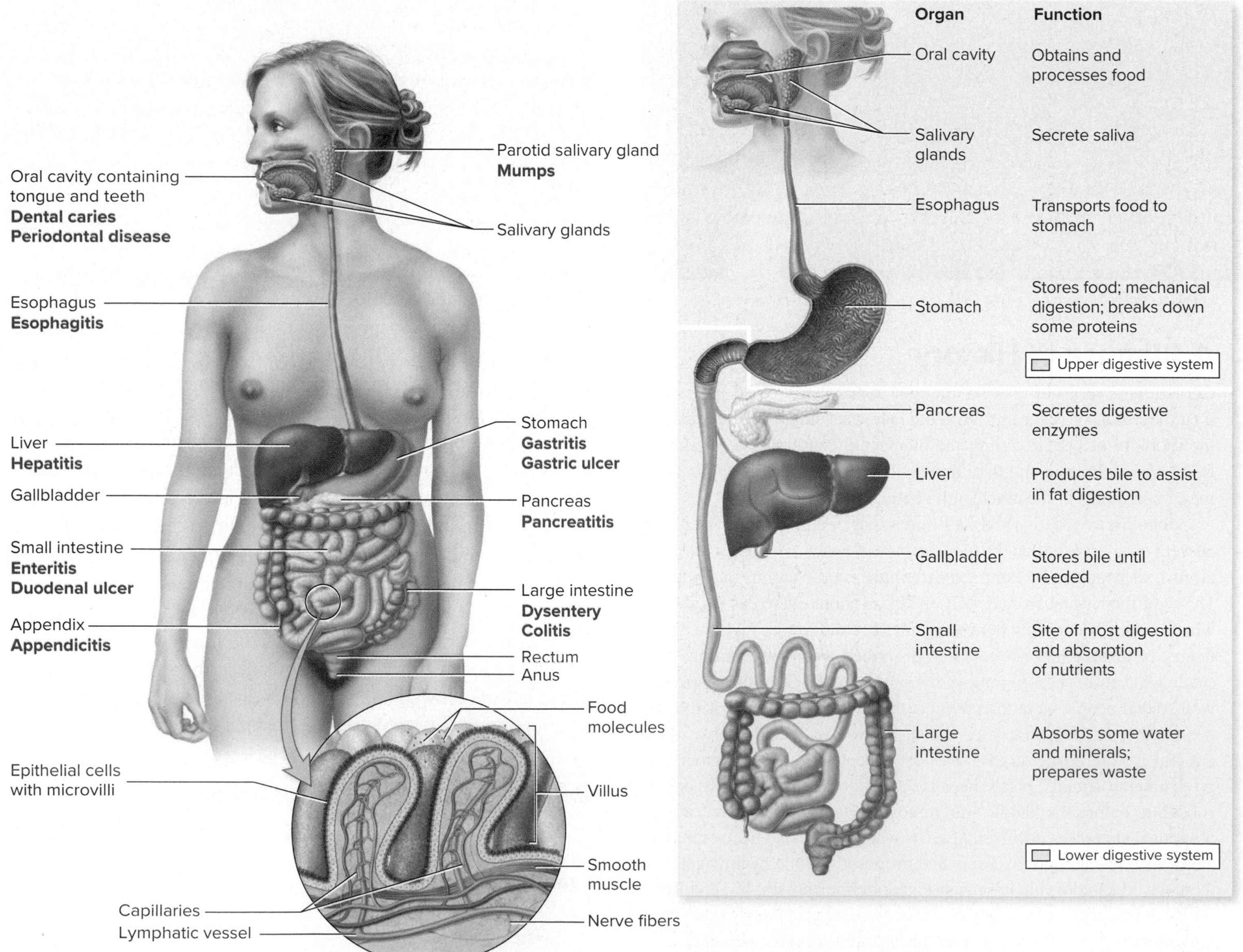

FIGURE 24.1 The Digestive System Some of the disease conditions that can affect the system are shown in red.

How do the accessory organs of the gastrointestinal tract support digestion?

population from underlying tissues, so damage to this layer allows resident microbes to penetrate deeper tissue. Ingested pathogens often have mechanisms to breach intact barriers.

The Upper Digestive System

The upper digestive system includes the mouth, salivary glands, esophagus, and stomach.

The Mouth and Salivary Glands

The mouth begins the physical and chemical digestion of food. The act of chewing allows the teeth to grind food into smaller pieces, while the saliva moistens the food and provides amylase, an enzyme that helps start the breakdown of starches.

Teeth are protected by a hard substance called enamel (**figure 24.2**). Proteinaceous material from the saliva adheres to the enamel, creating a thin film called a pellicle. Various types of bacteria can attach to receptors on the pellicle, colonizing the tooth surface to create a biofilm called **dental plaque** or simply plaque. Over time, calcium salts deposit in plaque, creating a hardened, crusty form of plaque referred to as dental calculus or tartar. Routine brushing and interdental cleaning such as flossing can remove plaque, but tartar removal requires professional cleaning.

When the tooth enamel deteriorates or is damaged, microorganisms can enter the substance of the tooth. **Dental caries** (tooth decay) can lead to holes called cavities in the hard tooth surface; dental fillings are used to repair these. If bacteria reach the root canal—the interior part of the tooth filled with a tissue called pulp—either the tooth must be removed or the pulp must be replaced with an inert substance (a procedure called a root canal treatment, or simply a root canal).

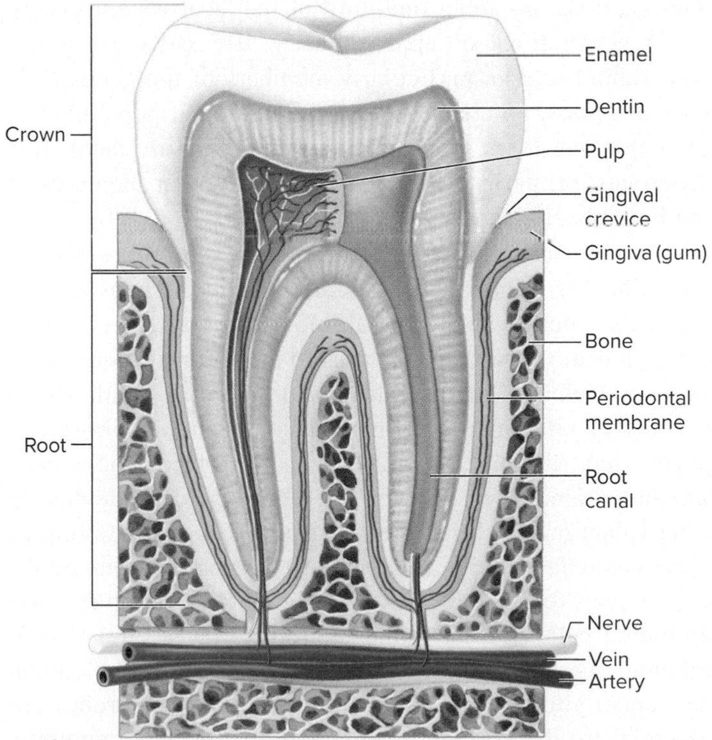

FIGURE 24.2 Structure of a Tooth and Its Surrounding Tissues

? What is the difference between dental plaque and tartar?

Another frequent site of microbial accumulation and infection is the gingival crevice, the space between the gum and a tooth. In response to accumulated plaque in that region, the gums become inflamed, a condition called **gingivitis.** The irritated gums may recede from the tooth root, allowing bacteria to reach that region as well, which can ultimately result in tooth loss.

Salivary glands produce saliva (about 1 liter per day) that flows via ducts into the mouth. In addition to aiding in digestion, the saliva helps protect the mouth because it contains secretory IgA (the antibodies that provide mucosal immunity) as well as antimicrobial substances including lysozyme, lactoferrin, and host defense peptides. People with poor saliva production, such as those undergoing radiation therapy for head and neck cancer, are at risk of severe tooth decay.

Even with the antimicrobial effects of saliva, over 700 microbial species are found in the mouth. They withstand the flushing effect of salivary flow as well as the scrubbing action of both food and the tongue by binding specifically to molecules on host tissues or to other attached microbes. Among the most numerous inhabitants are members of the genus *Streptococcus,* which are obligate fermenters. Obligate anaerobes are also surprisingly common in the mouth, thriving in complex communities where aerobic species consume available O_2. The foul-smelling end products of anaerobic metabolism are associated with bad breath, also known as halitosis.

The Esophagus

The esophagus is a muscular tube about 10 inches long that connects the mouth to the stomach. Food is pushed down to the stomach by peristalsis, the rhythmic muscular contractions of the tube. The esophagus has a relatively sparse microbial population, consisting mostly of bacteria from the mouth and the upper respiratory tract. One reason for the lack of microbes is the secretory IgA–containing mucus that bathes the lining of the tube; this traps microbes and then propels them through the esophagus, along with food and liquids. The esophagus rarely becomes infected except in people with AIDS or other immunodeficiencies.

The Stomach

The stomach is an expandable and muscular sac-like structure within which food particles are broken down and stored before they enter the small intestine. Specialized cells in the stomach produce highly acidic gastric juice that denatures proteins; the low pH also activates an acid-tolerant digestive enzyme that further breaks down proteins. A layer of gel-like mucus coats the lining of the stomach, protecting that mucous membrane from the harsh effects of the acid and enzymes. The conditions kill most microbial cells, however, so a normal empty stomach has few microorganisms.

The Lower Digestive System

The lower digestive system includes the small and large intestines, as well as the pancreas, liver, and gallbladder.

The Small Intestine

Digestive fluids from the pancreas and liver join the stomach contents as they enter the duodenum (the first part of the small intestine). The alkaline fluid from the pancreas neutralizes stomach acid and contains various digestive enzymes; the liver contributes **bile,** a fluid that contains bile salts, a type of emulsifying agent that helps physically break down fat globules into smaller droplets, thus increasing the surface area for better access by digestive enzymes. Many types of bacteria are killed by bile, but those adapted to live in the intestinal tract resist its bactericidal effects. This has practical applications in the laboratory because bile salts are used in selective media designed to isolate intestinal bacteria.

If the small intestine were a simple pipe, its 6-meter length would have an inside surface area smaller than that of a bath towel. Instead, that surface is approximately 30 square meters—about the size of three parking spaces! The enormous surface area results from many tiny, finger-like projections called **villi,** each lined with cells that have tiny cytoplasmic projections called **microvilli** (see figure 24.1 and chapter-opening image). These microvilli-covered epithelial cells produce digestive enzymes that are anchored in the cytoplasmic membranes; because a microvilli-covered surface is often called a brush border, the enzymes are referred to as brush border enzymes. At the base of the villi are pits called crypts—small glands that continuously secrete large amounts of enzyme-containing fluids and mucus into the lumen (the interior portion of an organ, in this case the small intestine). Certain cells in the crypts give rise to new intestinal epithelial cells. Intestinal epithelial cells are completely replaced every 9 days—one of the fastest cell turnover rates in the body.

A major role of the small intestine is nutrient and fluid absorption. Cells that line the villi take in amino acids and monosaccharides, bringing in sodium ions (Na^+) at the same time. Fatty acids, vitamins, and minerals such as iron are absorbed as well. An enormous volume of fluids is absorbed each day—approximately 9 liters from the combination of digestive juices, saliva, and the liquids taken in as food and drink. Understandably, disruptions that interfere with the balance between fluid secretion and absorption can result in diarrhea.

Few microbes reside in the upper small intestine because the flushing action of rapidly passing digestive juices limits their ability to colonize the surface. As the involuntary muscle action of peristalsis in the small intestine propels its contents toward the large intestine, the movement slows and the microbial population increases. The immune system monitors this population using the combined actions of dendritic cells that sample the intestinal contents and M cells that deliver small amounts of the contents to cells in the Peyer's patches (see figure 15.8).

The Large Intestine

A primary function of the large intestine, principally the colon (the main portion), is absorption of water as well as the vitamins produced by the resident microbiota. Because the small intestine absorbs so much, only 300 to 1,000 milliliters of fluid normally reach the large intestine per day, and most of this is also absorbed. The semisolid feces, composed of indigestible material and bacteria, remain in the portion of the large intestine called the rectum until defecation (the act of expelling feces) through the anus. Infection of the large intestine can interfere with absorption and can also stimulate the painful contractions known as "stomach cramps."

Microbes flourish in the large intestine, supported by the abundance of nutrients in undigested food material. In fact, bacteria make up about one-third of the fecal weight, reaching concentrations of approximately 10^{11} cells per gram! Anaerobic bacteria, particularly members of the genus *Bacteroides,* make up about 99% of the microbial population in the colon and feces. The remaining microbes are facultative anaerobes, primarily *Escherichia coli* and other members of the Enterobacteriaceae.

As described in section 16.2, intestinal microbes play an essential role in maintaining human health, perhaps including brain function; their crucial role became even more obvious in recent years, thanks in part to the Human Microbiome Project. Members of the intestinal microbiota synthesize a number of vitamins, including niacin, thiamine, riboflavin, pyridoxine, vitamin B_{12}, folic acid, pantothenic acid, biotin, and vitamin K (see table 6.4). They also degrade dietary fiber (plant material that the body cannot digest) to produce short-chain fatty acids, which serve as energy sources for colonocytes (epithelial cells of the colon); by assisting those epithelial cells, the intestinal microbes play a vital role in maintaining the mucosal barrier between the intestinal lumen and underlying tissues. In addition, intestinal microbes are essential for the normal development of mucosal immunity, and they also help prevent pathogens from colonizing the intestines. Antibiotic treatment, especially with broad-spectrum medications, disrupts the microbial population, creating an imbalance called **dysbiosis** that can lead to mild to severe **antibiotic-associated diarrhea.** In some cases, this is caused by toxin-producing strains of *Clostridioides difficile,* which readily colonize the colon of anyone whose normal intestinal microbiota has been reduced by antimicrobial chemotherapy. Suppression of intestinal microbiota with antibacterial medications also increases susceptibility to other pathogens such as *Salmonella enterica.*

Although intestinal microbes generally benefit host health, they can have detrimental effects as well. Many of them are opportunistic pathogens that cause disease if they gain access to another body site such as the urinary tract. Also, certain intestinal microbes appear to be associated with intestinal cancer risk; a number of species, for example, produce a toxin called colibactin that has been linked to the development and progression of colorectal cancer, and some microbial enzymes convert various food components into carcinogens. Meanwhile, if microbial fermentation of a food material produces large amounts of gas, abdominal discomfort and flatus (gas expelled via the anus) may result.

MicroByte

The Global Microbiome Conservancy currently stores around 11,000 bacterial strains collected from the feces of 40 people in 7 countries, in an effort to archive the diversity of human intestinal microbes.

The Pancreas

The pancreas—located behind the stomach—makes hormones (including insulin) and digestive enzymes. About 2 liters of alkaline pancreatic digestive juices empty into the upper portion of the small intestine each day, neutralizing the acidic material from the stomach.

The Liver and Gallbladder

The liver is a large organ located in the upper right part of the abdomen. It produces bile, which is then concentrated and stored in the sac-like gallbladder. The bile then flows through a system of tubes into the upper small intestine. Severe liver disease or obstruction of the bile ducts can cause **jaundice,** a yellow color of the skin and eyes caused by the buildup of a bile component (bilirubin) in the blood.

The liver also detoxifies substances in the bloodstream. As an example of the importance of this, the ammonia absorbed into the bloodstream as intestinal bacteria degrade amino acids could reach damaging levels if not detoxified by the liver. Many medications are also altered or removed by detoxification processes, but the standard prescribed doses account for this loss. If the liver is damaged, dose adjustments might be required to prevent an accidental overdose.

MicroAssessment 24.1

The digestive tract is a complex ecosystem and a major route for pathogens to enter the body. Bacteria colonizing the tooth surface can create a biofilm called dental plaque. Infections of the esophagus are so unusual that their occurrence suggests immunodeficiency. Acidic gastric fluid destroys most microbes before they reach the intestines. The small intestine carries out most of the digestion and absorption of nutrients; undigested material becomes feces in the large intestine. Disruption of the normal microbiota through antibiotic treatment can result in antibiotic-associated diarrhea.

1. What causes halitosis (bad breath)?
2. What makes the surface area of the small intestine so large, when it is only about 6 meters long?
3. Two patients recently had spinal surgery, but one is jaundiced. Which patient would likely need a lower dose of acetaminophen? Explain.

UPPER DIGESTIVE SYSTEM INFECTIONS

24.2 ■ Bacterial Diseases of the Upper Digestive System

Learning Outcomes

3. Compare and contrast dental caries, dental plaque-induced periodontal diseases, and necrotizing periodontal diseases.
4. Describe *Helicobacter pylori* gastritis and how it relates to peptic ulcers and stomach cancer.

Bacterial diseases of the upper digestive system often involve the teeth or gums, as well as the stomach. Local infections of the teeth and gums—which often go unnoticed for years—may have consequences for the rest of the body as well; some studies suggest that chronic gum infections contribute to arterial disease, rheumatoid arthritis, premature births, and dementia. In addition, members of the normal oral microbiota can enter the bloodstream during dental procedures and cause infective endocarditis.

Dental Caries (Tooth Decay)

Dental caries (tooth decay) is damage to tooth enamel resulting from the acids produced as microbes in the biofilm called dental plaque ferment sugars; it develops gradually and can eventually result in dental cavities (holes in the hard tooth surface). Although the term *caries* is often used synonymously with *cavities,* scientists studying tooth decay emphasize that the two terms are different: A cavity is a consequence of caries. Dental caries is one of the most common chronic diseases worldwide and the main reason for tooth loss.

Signs and Symptoms

Dental caries can become very advanced before signs and symptoms develop. The decay might result in sensitivity or changes to the tooth such as brown or black discoloration, roughness, or a visible hole; in some cases, a tooth can break during chewing. However, the severe, throbbing pain of a toothache is often the first symptom.

Causative Agents

Dental caries involves *Streptococcus mutans* and certain other Gram-positive bacteria that are **cariogenic** (meaning "caries generating"). These organisms thrive in the acidic conditions that result from their fermentative production of lactic acid (see figure 6.24); that is, they are acidogenic (produce acids) and aciduric (tolerate acidic conditions).

Pathogenesis

Dental caries results from acids produced by microbes in plaque on the tooth surface slowly eroding the tooth's enamel covering (**figure 24.3**). This is in contrast to the plaque-induced periodontal diseases, discussed next, in which the plaque accumulates in the gingival crevice.

The overall process of dental caries begins when oral streptococci adhere to the dental pellicle—a thin film of proteinaceous material that forms on tooth enamel. Other mouth microbiota then also attach, creating the dental plaque. If dietary sucrose is present, *S. mutans* exoenzymes first split the sucrose into its monosaccharide components, glucose and fructose, and then polymerize the glucose to make glucans; these polymers help create an even thicker biofilm by binding together a mixed population of microbes close to the tooth surface. Meanwhile, the bacteria ferment the fructose, producing lactic acid. Thus, the plaque acts as a tiny acid-soaked sponge closely applied to the tooth; the acid, rather than the plaque itself, causes tooth decay. Both *S. mutans* and a sucrose-rich diet are required for dental caries to occur on smooth surfaces of the teeth. In deep fissures (grooves) or pits, plaque can accumulate in the absence of *S. mutans,* so any acid–producing bacteria can cause caries if fermentable substances are present

FIGURE 24.3 Infections of the Teeth and Gums (a) Healthy tooth with resident microbiota, limited plaque formation, and mild inflammation (left); tooth with decay and plaque-induced periodontitis (right). **(b)** Dental X rays showing a dental cavity (top) and gum infection with bone loss, and a filled cavity (bottom). Dr. Larry Hoffman

? **How does brushing away plaque decrease the chance of developing caries?**

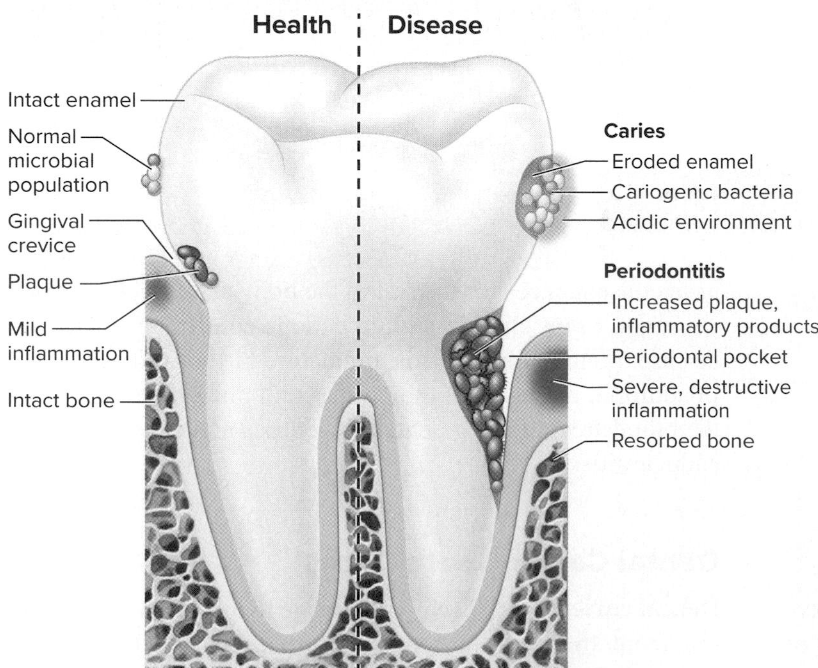

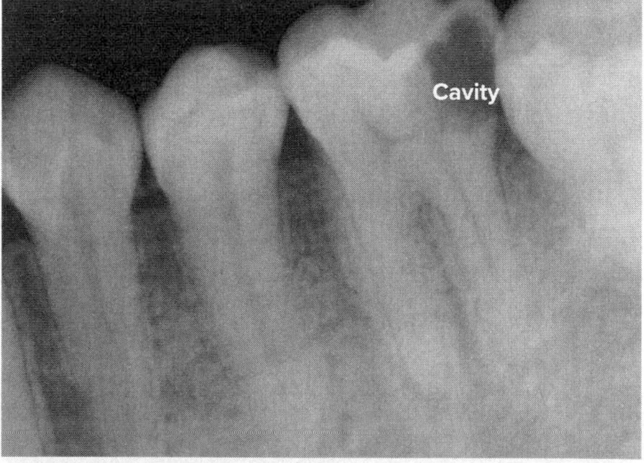

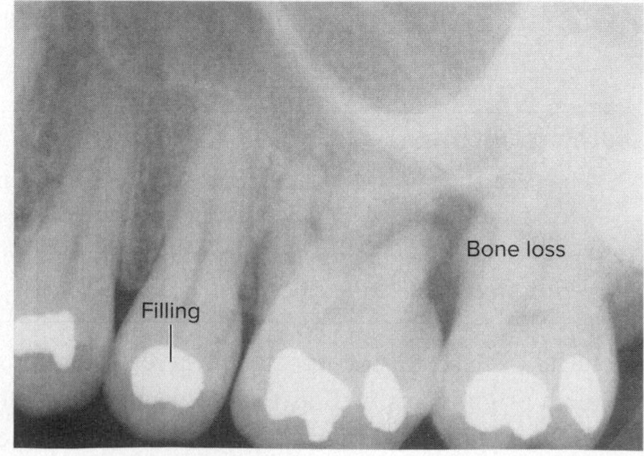

Dental caries is primarily associated with frequent sugar consumption. When sugar enters the mouth, the normally neutral pH of plaque rapidly drops as cariogenic bacteria ferment the sugar (**figure 24.4**). Tooth enamel then begins to dissolve at a pH of 5.5, through a process called demineralization. Once the sugar leaves the mouth, however, the buffering action of saliva will return the pH to neutral, and remineralization can restore the enamel. The problem with frequent sugar consumption is that the continuous acid production can overwhelm the buffering capacity of saliva; it also prevents remineralization and favors growth of the acid-tolerant cariogenic bacteria. Over time, continuous tooth erosion results in tiny holes developing in the enamel, eventually penetrating and then damaging the underlying softer material called dentin. When this happens, the enamel may collapse, resulting in a dental cavity. Dentin has tiny tubules that connect with the pulp-containing region, providing a passageway for microbes into the pulp; microbial invasion via those tubules leads to inflammation, and the resulting pressure on the local nerves causes pain and sensitivity.

Epidemiology

Dental caries occurs in people worldwide, but the progression varies significantly depending mainly on dietary sucrose intake and preventive dental care. Genetics also plays an important role—people differ in characteristics such as enamel development and degradation, saliva composition and flow rate, and morphological features of the teeth (pits and fissures). The incidence of dental cavities peaks during the teen years and decreases with age, probably because the pits and fissures where plaque accumulates wear down over time. Although older people have a lower incidence of fissure-related cavities, they are prone to receding gums that expose root surfaces, which become sites for dental caries.

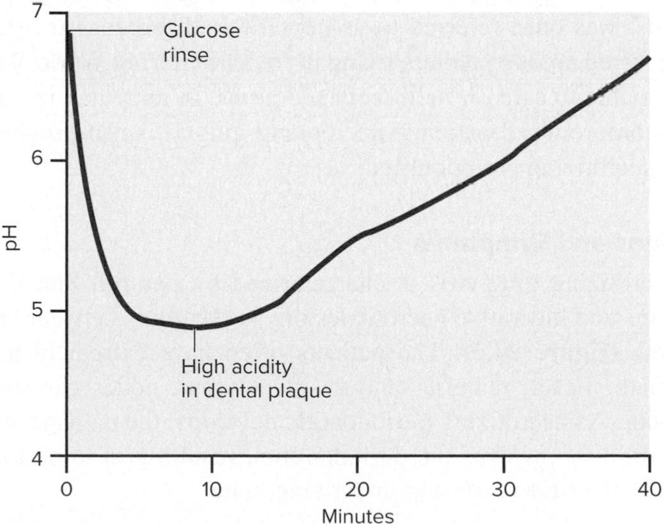

FIGURE 24.4 Acidity of Dental Plaque When the mouth is rinsed with a glucose solution, the pH of plaque can drop within minutes from its normal value of about 7 to below 5, a 100-fold increase in acidity. Tooth enamel begins to dissolve at about pH 5.5.

? Would you expect this dip in pH if *Streptococcus mutans* metabolized sugar via aerobic respiration rather than via fermentation? Explain.

Treatment and Prevention

Once dental cavities form, treatment may require restoration, which involves drilling out the cavity and then filling the defect with material such as composite resin (tooth-colored) or dental amalgam (silver-colored).

The most important method for controlling dental caries is restricting dietary sucrose and other refined carbohydrates. This reduces *S. mutans* colonization as well as acid production by cariogenic bacteria. However, the quantity of sugar in the diet is not the most important factor—the frequency of eating sugary foods and the length of time the food stays on the teeth are more critical. Chewing sugar-free gum reduces the incidence of dental cavities, probably because it increases saliva flow.

Trace amounts of fluoride prevent dental caries by making tooth enamel harder and more resistant to acid erosion. In the United States, nearly three quarters of the population is supplied with fluoridated public drinking water, resulting in significant reductions in dental cavities. In areas where fluoridated drinking water is not available, fluoride tablets or solutions can be used. To have optimum effect, children should begin receiving fluoride before their permanent teeth appear. Fluoride applied to tooth surfaces in the form of mouthwashes, gels, or toothpaste is generally less effective.

Daily toothbrushing and interdental cleaning such as flossing disrupt and remove plaque from most tooth surfaces. Toothbrush bristles, however, cannot remove plaque from the deep narrow pits and fissures normally present in children's teeth, which is where most childhood tooth decay starts. Application of a clear dental sealant to a tooth's chewing surface prevents cavities by filling the fissures and pits and forming a protective layer on the tooth (the sealant is usually used on the back teeth). Tartar (hardened calcified plaque) can only be removed from teeth by scraping it off, which is one reason why regular dental visits for professional cleaning are important.

Dental Plaque-Induced Periodontal Diseases

Dental plaque-induced periodontal diseases are the most common type of periodontal diseases (diseases of the tissues that support and anchor the teeth). **Gingivitis,** meaning inflammation of the gums, develops when dental plaque (including tartar) accumulates in the gingival crevice, especially in hard-to-clean areas between the teeth; it typically resolves once the plaque is removed. If not removed, however, the plaque gradually extends further into the crevice, eventually resulting in **periodontitis,** meaning inflammation of the periodontium (soft tissues and bone supporting the teeth). Unlike gingivitis, the damage associated with periodontitis is permanent, and ongoing treatment is required to prevent the condition from worsening. Advanced periodontitis is an important cause of tooth loss from middle age onward.

Signs and Symptoms

Gingivitis is marked by gums that are red, tender, and bleed when probed. As periodontitis develops, the red, bleeding gums begin to recede (pull away from the tooth), leaving periodontal pockets between the gums and teeth in which plaques accumulate (see figure 24.3). In turn, this leads to deeper periodontal pockets and bad breath. As the disease progresses, the teeth start looking longer as their bases become more exposed. Those exposed portions gradually become discolored and are more susceptible to dental caries. In the later periodontitis stages, the teeth become loose.

Causative Agents

Plaque-induced periodontal disease results from an inflammatory response to plaques composed primarily of Gram-negative anaerobes, which are thus quite different from those associated with dental caries. Moreover, a recent study based on 16S rDNA sequencing data showed that the plaques associated with gingivitis differ from those of more advanced stages of periodontitis. Determining the causative agents and their roles is complicated, however, because the infections are polymicrobial, meaning that multiple interacting microbial species are involved. Hundreds of different species have been identified in the plaques, most of which have not been cultivated; bacterial genera implicated include *Porphyromonas, Tannerella, Prevotella, Treponema, Selenomonas,* and *Aggregatibacter.*

Pathogenesis

The processes that lead to plaque-induced periodontal diseases appear to involve many factors, including host characteristics and the composition of the microbial population in the plaque. In general, when plaque accumulates in a gingival crevice, pattern recognition receptors of surrounding cells detect various bacterial products—including the lipopolysaccharide of the Gram-negative outer membrane (endotoxin)—leading to an inflammatory response. If the plaque levels remain small, the inflammation is limited and characteristic of gingivitis. When the microbial population is not controlled, however, periodontitis may slowly develop. As the disease progresses, tissue-degrading enzymes released by the plaque microbes cause the gingival crevice to widen and deepen. This allows the plaque to spread farther toward the root of the tooth, leading to the release of pro-inflammatory cytokines that induce an inflammatory response strong enough to damage tissues. With progressive damage, the attachment between the tooth root and the bone weakens, bone loss occurs, and the tooth becomes loose and may fall out.

Epidemiology

Plaque-induced periodontal disease can begin in childhood. After age 65, almost 70% of people have some degree of periodontitis. Smokers and those with immunodeficiency often have severe periodontitis that leads to tooth loss.

Treatment and Prevention

Plaque-induced periodontal disease can be treated in its early stages by deep cleaning any inflamed gingival crevices to remove plaque and tartar. In advanced cases, minor surgery is usually required to expose and clean the roots of the teeth. These physical treatment methods may be used along with antibiotic therapy in some patients.

Daily toothbrushing and interdental care such as flossing reduce the incidence of plaque-induced periodontal disease, especially when combined with regular dental visits for professional cleaning. These same oral hygiene procedures, often along with antiseptic mouthwashes, are also important for stopping the disease progression.

Necrotizing Periodontal Diseases

Necrotizing periodontal diseases (NPDs) involve tissue death often leading to ulcerations. Unlike the periodontal diseases just discussed, they have a rapid onset and involve processes other than routine plaque accumulation. Necrotizing gingivitis, until recently called acute necrotizing ulcerative gingivitis, was once referred to as trench mouth because it often occurred among soldiers living in trenches during World War I, unable to care for their teeth and gums. In untreated immunocompromised patients, necrotizing gingivitis can progress to necrotizing periodontitis.

Signs and Symptoms

Necrotizing gingivitis is characterized by painful, bleeding gums and ulcerative necrotic lesions in the gums between the teeth (**figure 24.5**). The patients often have extremely bad breath. Fever, malaise, and swollen lymph nodes can also occur. As necrotizing periodontitis develops, the damage and tissue loss involves the periodontium, resulting in tooth loss and sometimes exposing underlying bone.

Causative Agents

NPDs are associated with the heavy growth of various anaerobes at the gum line. Bacteria that appear to play a role in the polymicrobial infection include spirochetes (such as *Treponema* species), fusiforms (rods with tapered ends), and *Prevotella intermedia.*

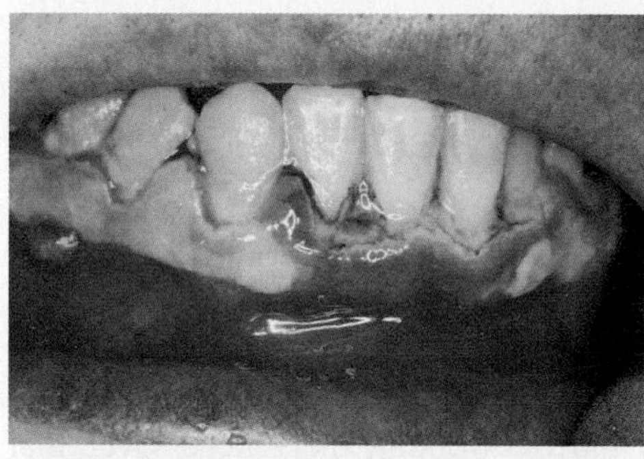

(a)

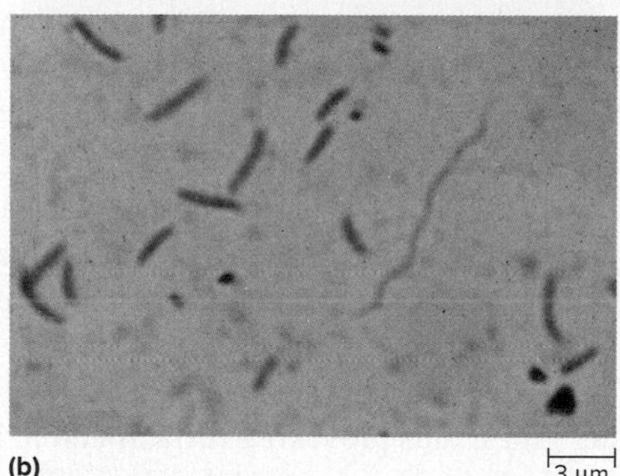

(b)

FIGURE 24.5 Necrotizing Gingivitis (a) Red, swollen gums with loss of tissue, especially between the teeth. **(b)** Gram stain of material from the gingival crevice showing a spirochete and rod-shaped bacteria. a: Biophoto Associates/Science Source; b: Evans Roberts

? **Why might necrotizing gingivitis patients have extremely bad breath?**

Pathogenesis

The spirochetes and other anaerobes associated with NPDs are presumed to act together to destroy tissues in the oral cavity, but the precise mechanisms are unknown. The spirochetes invade the tissue, causing necrosis and ulceration. Inflammation may lead to bone destruction around the teeth, with subsequent tooth loss.

Epidemiology

NPDs can occur at any age in association with poor oral hygiene, particularly when combined with poor nutrition, smoking, high-sugar diet, chronic stress, substance abuse, and immunodeficiency.

Treatment and Prevention

An antimicrobial medication such as metronidazole is typically used to treat NPDs. Because the bacteria involved are anaerobes, local hydrogen peroxide treatment in the form of mouthwash rapidly relieves the symptoms. In some cases, pain medication is also given. Necrotic tissue is removed by debridement. Prevention of NPDs begins with daily brushing, interdental cleaning such as flossing, and regular dental visits for professional cleaning.

The main features of teeth and gum infections are presented in **table 24.1.**

TABLE 24.1 Important Infections of the Teeth and Gums

	Dental Caries	Plaque-Induced Periodontal Diseases	Necrotizing Periodontal Diseases
Signs and Symptoms	None until advanced disease; then roughness, discoloration, broken tooth, throbbing pain	Gingivitis: tender, bleeding gums. Periodontitis: bad breath; red gums that bleed easily; loose teeth; exposed bases of teeth due to receding gums	Painful, bleeding gums, necrotic ulcerative lesions; extremely bad breath; bone and tooth loss
Incubation Period	Months before cavity is detectable	Gingivitis can begin within days, but periodontitis takes months or years to develop	Undetermined
Causative Agents	*Streptococcus mutans* and other cariogenic bacteria in dental plaque	Various Gram-negative anaerobic microbes in dental plaque	Probably a spirochete of the genus *Treponema* acting with other anaerobes, including *Prevotella intermedia*
Pathogenesis	Cariogenic bacteria in plaque ferment dietary sugars, producing acid that slowly dissolves the tooth enamel; sucrose is usually critical for plaque formation.	Bacterial products in plaque trigger inflammation of the gums (gingivitis) or the periodontium (periodontitis)	Spirochetes and certain other anaerobes act synergistically to destroy tissues; spirochetes invade tissue, causing necrosis and ulceration.
Epidemiology	Caries progression depends on dietary sucrose and preventive dental care; cavities common in the young.	Gingivitis can begin in childhood, but periodontitis primarily occurs in older people, particularly smokers and the immunodeficient.	Associated with poor oral hygiene, malnutrition, smoking, substance abuse, or immunodeficiency. All ages are susceptible.
Treatment and Prevention	Treatment: fill cavities. Prevention: restrict dietary sucrose; supplemental fluoride; oral hygiene aimed at removing plaque on teeth; dental sealant to protect tooth chewing surface; regular dental visits.	Treatment: for severe cases of periodontitis, minor surgery to clean tooth roots. Prevention: oral hygiene aimed at removing plaque near gums; regular dental visits.	Treatment: antibiotic therapy along with local treatment; tissue debridement. Prevention: general oral hygiene; regular dental visits.

Helicobacter pylori Gastritis and Peptic Ulcer Disease

Helicobacter pylori colonization of the stomach typically results in **gastritis** (inflammation of the stomach lining), sometimes leading to peptic ulcer disease (characterized by lesions in the lining of the stomach or duodenum) or even cancer. The association of *H. pylori* with gastritis was demonstrated in the 1980s when Barry Marshall, one of the Australian scientists who discovered the bacterium, intentionally drank a culture and soon developed gastritis. Marshall and his collaborator Robin Warren received a Nobel Prize in 2005 for their discovery of *H. pylori* and its association with gastritis and ulcers.

Signs and Symptoms

H. pylori infections are typically asymptomatic, but acute gastritis can result in abdominal pain, nausea, and sometimes vomiting. Without treatment, the infection becomes chronic but is often asymptomatic even if peptic ulcer disease develops. In some cases though, disease results in upper abdominal tenderness and burning pain, particularly when the stomach is empty, along with belching (burping) and loss of appetite; in severe cases, bleeding may occur, sometimes detected as dark red or black stool. In a small number of people, chronic gastritis leads to stomach cancer.

Causative Agent

H. pylori is a short, curved, Gram-negative, microaerophilic bacterium with multiple sheath-covered polar flagella.

Pathogenesis

H. pylori cells survive the acidic environment of the stomach by: (1) producing urease, which converts urea in gastric juices to ammonia, an alkaline compound that neutralizes stomach acid in the immediate microenvironment, and (2) using their flagella to burrow through the mucus layer that coats the stomach lining. Upon reaching the nearly neutral epithelial surface, the *H. pylori* cells then attach to the epithelial cells and multiply, protected by the mucus.

H. pylori is well equipped to manipulate the host's immune response and establish chronic infection. Its lipopolysaccharide (LPS) and flagella have modifications that prevent them from being recognized by certain pattern recognition receptors of the innate immune system, thus limiting the inflammatory response to infection. At the same time, the bacterium produces several proteins that damage host cells or alter their activities, thus provoking inflammation. For example, the toxin VacA (vacuolating cytotoxin) induces epithelial cell death and interferes with the function of various types of immune cells. Many strains additionally make the toxin CagA (cytotoxin-associated gene), which interferes with host cell signaling and promotes inflammation. The infection damages

FOCUS ON A CASE 24.1

A 35-year-old man consulted his physician because of upper abdominal pain, which he described as a steady burning or gnawing sensation, like severe hunger pain. Usually it came on 1.5 to 3 hours after eating and was generally relieved by eating or by taking antacid medicines.

On examination, the patient appeared well, without evidence of weight loss. The only notable finding was tenderness slightly to the right of the midline in the upper part of the abdomen. A test of his feces was positive for blood; the remaining laboratory tests were normal.

The patient was given a rapid, noninvasive test called the urea breath test (UBT), which is used to diagnose *Helicobacter pylori* infections. For this test, the patient drinks a solution containing urea labeled with an isotope of carbon; if *H. pylori* is present, labeled CO_2 is produced that can be detected in the breath. The patient's UBT was positive.

1. Why is labeled CO_2 produced if *H. pylori* is present?
2. How do flagella and the enzyme urease function as bacterial virulence factors in the development of peptic ulcers?
3. It took a long time for doctors to accept that peptic ulcers were caused by infection. Why?

Discussion

1. *H. pylori* uses the enzyme urease to break down urea into ammonia and CO_2. Because the urea is labeled with a carbon isotope, the CO_2 produced from it also has the label.
2. Flagella propel the bacterial cells through the mucus layer, moving them from the acidic environment of the lumen (approximately pH 2) to the epithelial surface, where the pH is nearly neutral. Non-motile strains do not cause disease because they are removed with mucus turnover. Urease breaks down urea in the stomach to produce ammonia, an alkaline compound that neutralizes the gastric fluid around the cell. This protects the organism until it can burrow below the mucus layer.
3. Claude Bernard, a scientist of Pasteur's time, put it this way: "It is that which we do know which is the greatest hindrance to our learning that which we do not know." In 1983, when Barry Marshall proclaimed before an international gathering of infectious disease experts that a bacterium caused stomach and duodenal ulcers, everyone "knew" it could not be true because they thought no organism could survive stomach acidity and enzymes. Indeed, almost everyone already "knew" the cause of ulcers to be psychological stress. There is much still to be learned about the cause of ulcers, however, and Bernard's statement remains relevant.

FIGURE 24.6 ***Helicobacter pylori* Pathogenesis** Heather Davies/Science Photo Library/Getty Images

? How does urease allow *H. pylori* to survive in the stomach?

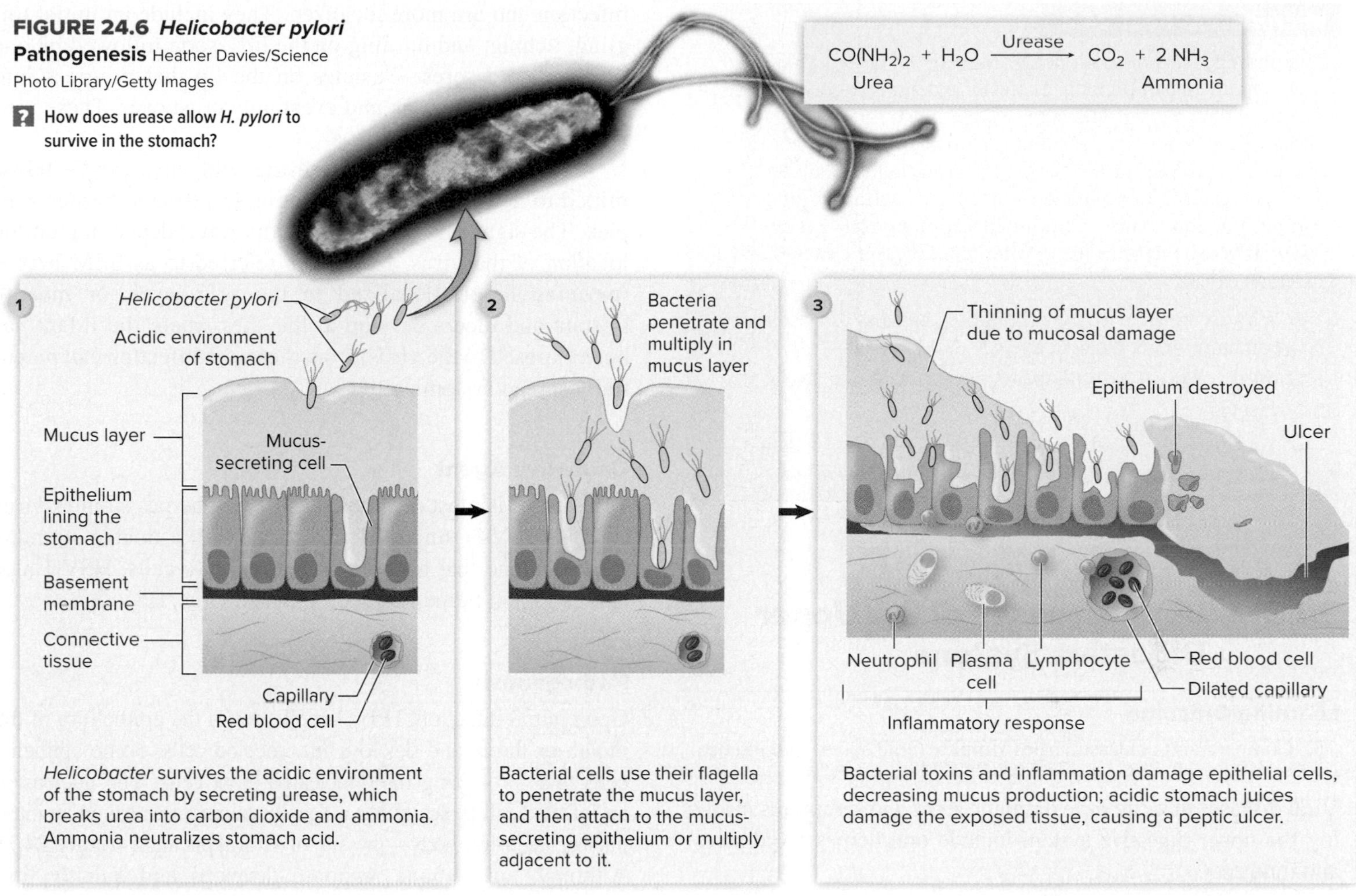

Urease
$CO(NH_2)_2 + H_2O \xrightarrow{\text{Urease}} CO_2 + 2\,NH_3$
Urea
Ammonia
1
Helicobacter pylori
Acidic environment of stomach
Mucus layer
Mucus-secreting cell
Epithelium lining the stomach
Basement membrane
Connective tissue
Capillary
Red blood cell
Helicobacter survives the acidic environment of the stomach by secreting urease, which breaks urea into carbon dioxide and ammonia. Ammonia neutralizes stomach acid.
2
Bacteria penetrate and multiply in mucus layer
Bacterial cells use their flagella to penetrate the mucus layer, and then attach to the mucus-secreting epithelium or multiply adjacent to it.
3
Thinning of mucus layer due to mucosal damage
Epithelium destroyed
Ulcer
Neutrophil
Plasma cell
Lymphocyte
Red blood cell
Dilated capillary
Inflammatory response
Bacterial toxins and inflammation damage epithelial cells, decreasing mucus production; acidic stomach juices damage the exposed tissue, causing a peptic ulcer.

gastric epithelial cells through the combined effects of chronic inflammation, *H. pylori* products, and stomach acid, resulting in decreased mucus production. In turn, the thinning of the protective mucus layer and host cell damage probably account for the development of peptic ulcers (**figure 24.6**).

Epidemiology

About half the world's population is thought to be infected with *H. pylori.* Infections tend to cluster in families, and rates are highest in low socioeconomic groups. Transmission of the bacterium probably occurs by the fecal-oral route, possibly via contaminated water or food; oral-oral transmission (contact with gastric secretions) might also occur. The infections persist for years, often for life, and about one in six infected people develop ulcers. Only a small percentage of chronically infected people develop stomach cancer, but most of those with stomach cancer are infected.

Treatment and Prevention

Several options are available for treating *H. pylori* infection, all of which use a combination of antibiotics along with a proton pump inhibitor (PPI, a medication that inhibits stomach acid production) and sometimes bismuth subsalicylate (Pepto Bismol). Any ulcers generally heal once the infection has cleared, but treatment failures sometimes occur. Because of this, post-treatment tests are done to confirm that the infection has been eliminated; if it has not, then one of a group of alternative combinations referred to as salvage therapy is used. No proven preventive measures exist. The main features of *H. pylori* gastritis are shown in **table 24.2.**

TABLE 24.2	*Helicobacter pylori* Gastritis and Peptic Ulcer Disease
Signs and Symptoms	Infections are often asymptomatic, but peptic ulcer disease can cause abdominal pain, tenderness, and bleeding. Stomach cancer occasionally develops.
Incubation Period	Usually undetermined
Causative Agent	*Helicobacter pylori,* a curved, Gram-negative, microaerophilic bacterium, with multiple sheath-covered polar flagella.
Pathogenesis	*H. pylori* cells survive stomach acidity by producing ammonia-generating urease and by burrowing into the stomach's mucus coating. Bacterial products and inflammation damage the mucosal layer, which can lead to peptic ulcers.
Epidemiology	Probably fecal-oral or oral-oral transmission. Incidence increases with age.
Treatment and Prevention	Treatment: a combination of antibiotics along with a proton pump inhibitor and sometimes bismuth subsalicylate. Prevention: no proven preventive.

MicroAssessment 24.2

The pathogenesis of tooth decay depends on both dietary sucrose and acid-producing bacteria in dental plaque on teeth. Plaque-induced gingivitis is reversible inflammation of the gums; plaque-induced periodontitis results in permanent tissue damage due to the inflammation. Necrotizing periodontal diseases result in non-plaque-induced destruction of gingival and periodontal tissues. Chronic infection by *Helicobacter pylori* is a key factor in the development of peptic ulcers and stomach cancer.

4. Why are new cavities less common, but loss of teeth more common, in people over age 65?
5. What is the relationship between *H. pylori* and stomach cancer?
6. Your parents probably told you that candy causes dental cavities. Is this claim true? Explain your answer.

24.3 ■ Viral Diseases of the Upper Digestive System

Learning Outcome

5. Compare and contrast herpes simplex (cold sores) and mumps.

Viral diseases that can have dramatic signs and symptoms involving the upper digestive system include oral herpes (cold sores) and mumps.

Oral Herpes (Cold Sores)

Oral herpes is known by several names that reflect the characteristic recurrent lesions that sometimes develop on the lips: cold sores, fever blisters, and herpes labialis (*labia* means "lip"). Although the disease is usually insignificant, it can sometimes have serious consequences.

Signs and Symptoms

Initial (primary) infection by the virus that causes oral herpes is often asymptomatic. Sometimes, however, gingivostomatitis (inflammation of the gums and of the mouth) develops within 2 to 12 days of infection; this often leads to small blisters or sores appearing around the mouth and on the gums, often accompanied by fever, headache, and swollen lymph nodes. The blisters break within a day or two, producing superficial ulcers that can be so painful they make it difficult to eat or drink. Although the lesions of gingivostomatitis typically heal without treatment within 2 weeks, the virus persists in a latent state (a latent infection) and can reactivate to cause a recurrence (an outbreak). Symptoms of recurrences are usually less severe than those of the primary infection and are more localized. They include an initial tingling, itching, and burning on the lips, soon followed by one or more cold sores—vesicles on the lip that progress into blisters that then break and eventually crust over. These usually heal within 7 to 10 days.

In rare cases, the virus causing cold sores can be transmitted to newborn infants, resulting in neonatal herpes simplex. The signs and symptoms of this vary, depending on the location of the virus. In what is referred to as SEM herpes (meaning herpes localized to the skin, eyes, or mouth), blisters and ulcers develop at the sites where the infant has been exposed to the virus; from there, the infection can potentially spread systemically.

Causative Agent

Oral herpes is most commonly caused by herpes simplex virus type 1 (HSV-1), an enveloped virus with a double-stranded DNA genome that becomes latent in nerve cells. HSV-1 also causes genital herpes, as does a related virus, HSV-2.

Pathogenesis

Upon initial infection, HSV-1 replicates in the epithelium of the mouth or throat and destroys the infected cells. Some epithelial cells fuse, producing multinucleated giant cells. The nucleus of an infected cell typically has a deeply staining area called an intranuclear inclusion body—the site of viral replication (**figure 24.7**). Although an immune response develops and quickly limits the infection, some virions enter sensory nerve cells in

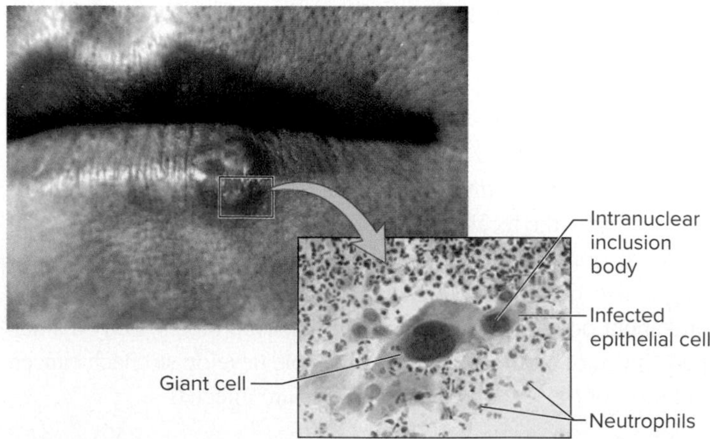

FIGURE 24.7 Oral Herpes (Cold Sores) The photomicrograph shows stained material from a herpes lesion, including a multinucleated giant cell and intranuclear inclusion bodies, the sites of viral replication. Centers for Disease Control and Prevention; (inset): Frederick C. Skvara, M.D.

How do the signs and symptoms of recurrent oral herpes differ from those associated with the primary infection?

the area, where HSV-1 persists in a latent non-infectious form. The latent virus can occasionally reactivate, however, giving rise to new viral particles that infect nearby skin or mucous membranes to produce cold sores (figure 13.16). Stresses that can trigger such recurrences include menstruation, sunburn, and any febrile (fever-associated) illness.

Epidemiology

HSV-1 is common worldwide, with approximately two-thirds of the global population infected. In the United States, approximately one in three people suffers from recurrent cold sores. Most infections are acquired during childhood, with transmission by close physical contact. The greatest risk of infection is from contact with lesions or saliva of a symptomatic person, but even asymptomatic people can be infectious, posing a risk to dentists and other healthcare workers.

HSV-1 can infect almost any body tissue. For example, nurses are at risk for herpetic whitlow (a painful finger infection). Wrestlers can develop infections at almost any skin site because saliva containing the virus can contaminate wrestling mats and get rubbed into abrasions. Blindness can result if the eyes are infected. Although uncommon, HSV-1 is the most frequently identified cause of sporadic viral encephalitis, a serious brain disease.

Treatment and Prevention

Medications such as acyclovir target the virus's DNA polymerase and are useful for treating severe cases and for preventing disabling recurrences. They do not affect the latent virus, however, and therefore cannot cure the infection. Sunlight exposure can trigger cold sore recurrences, so sunscreens are sometimes a helpful preventive. Some of the main features of oral herpes are shown in **table 24.3.**

TABLE 24.3 Oral Herpes (Cold Sores)

Signs and Symptoms	Initial infection: fever, sore throat, headache, blisters in the oral cavity; these break, leaving superficial, painful ulcers. Recurrences: itching, tingling, or pain on the lips, followed by blisters (cold sores); less severe.
Incubation Period	2 to 12 days
Causative Agent	Usually herpes simplex virus type 1 (HSV-1)
Pathogenesis	Virus replicates in the epithelium of the mouth and throat. An immune response limits the infection, but HSV remains latent in nerve cells. The virus can later reactivate, causing recurrent sores.
Epidemiology	Common worldwide; transmitted by close physical contact. Lesions and saliva of symptomatic people are most infectious, but saliva of asymptomatic people can also be infectious.
Treatment and Prevention	Treatment: no cure; anti-HSV medications can shorten duration of symptoms. Prevention: anti-HSV medications can prevent recurrences; sunscreen helps minimize recurrences.

Mumps

Mumps is an acute viral illness that often affects the parotid glands (the largest of the salivary glands). Formerly common in the United States, the disease is now relatively rare because of routine childhood immunization. Outbreaks do occur, however, mostly due to fading immunity in college students and other young adults.

Signs and Symptoms

Symptoms of mumps begin 15 to 21 days after infection, with fever, appetite loss, and headache. Painful swelling of one or both parotid glands may then occur (**figure 24.8**). Spasm of the underlying muscle makes it difficult to chew or talk, perhaps giving rise to the name of the disease (*mump* means "to mumble" or "whisper"). Symptoms usually last about a week.

Mumps can occasionally lead to complications with or without parotid swelling. About 25% of cases in post-pubertal males develop orchitis—a rapid, intensely painful and significant swelling of one or both testicles. Atrophy (shrinkage) of the involved testicles commonly occurs later as a result, and this occasionally decreases fertility. In post-pubertal females, ovarian involvement characterized by pelvic pain occurs in about 5% of cases, resulting in lower abdominal pain, vomiting, and fever. Neurological complications include meningitis (infection of the meninges, the coverings of the brain and spinal cord), encephalitis (infection of the brain), and possible deafness.

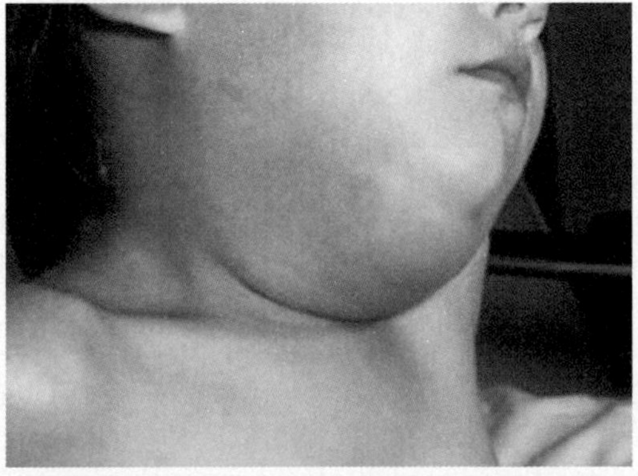

FIGURE 24.8 A Child with Mumps The swelling below the earlobe and under the chin is due to salivary gland infection. CDC

? **Why does mumps make it difficult to speak and chew?**

Causative Agent

Mumps virus is an enveloped single-stranded RNA virus of the family *Paramyxoviridae*. Only one serotype of the mumps virus is known.

Pathogenesis

The mumps virus enters the body when virus-containing droplets of saliva are inhaled. It replicates first in cells of the upper respiratory tract, then spreads throughout the body in the bloodstream. Symptoms begin only after certain tissues such as the parotid glands, meninges, ovaries, or testicles are infected, which is why the incubation period is relatively long.

In the parotid glands, the virus replicates in the epithelium of ducts that deliver saliva to the mouth. This destroys these cells, releasing enormous quantities of virus into the saliva and eliciting a strong inflammatory response, which causes the severe swelling and pain characteristic of the disease. A similar sequence of events occurs in the testicles, where the virus infects the tubules that convey the sperm. The swelling and pressure often impair the blood supply, which can lead to hemorrhage and death of testicular tissue.

Epidemiology

Humans are the only reservoir of the mumps virus. It is often spread by asymptomatically infected people who secrete the virus in their saliva and continue to contact other people. In patients who develop symptoms, the virus can be present in saliva from almost a week before symptom onset to 2 weeks afterward. Peak infectivity, however, is from 1 to 2 days before parotid swelling until the gland begins to return to normal size. People who recover from mumps appear to have lifelong immunity.

Treatment and Prevention

No antiviral treatment exists for mumps. An effective attenuated vaccine has been available in the United States for over 50 years; it is part of the measles, mumps, and rubella vaccine (MMR), as well as the measles, mumps, rubella, and varicella vaccine (MMRV) used for children. **Figure 24.9** shows how the incidence of mumps has generally declined, although it increased in the 1980s, associated with temporary funding cuts for vaccinations. Large outbreaks have still occurred since then, primarily in settings where people have close contact with one another, such as on college campuses. Following an outbreak of over 20,000 cases in 1986–1987, a two-dose vaccination schedule was introduced; an additional dose is sometimes recommended for those at risk of exposure during an outbreak. Mumps is a good candidate for eradication because it infects only humans, latent infections do not occur, and there is only one serotype. The main features of mumps are presented in **table 24.4.**

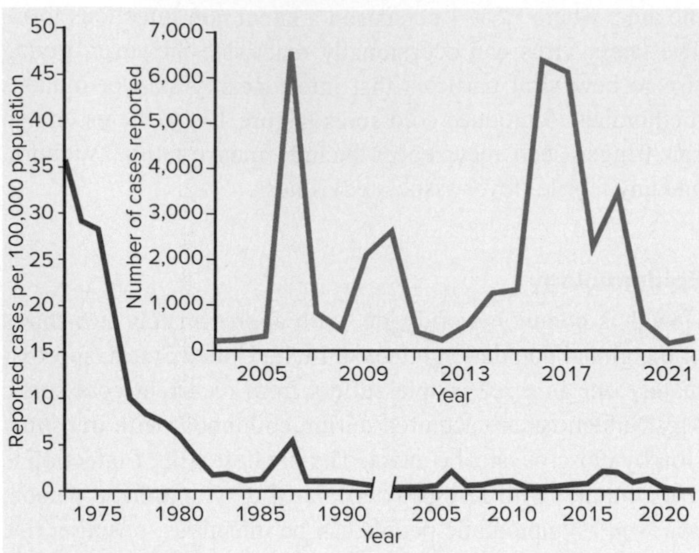

FIGURE 24.9 Reported Cases of Mumps, United States, 1973–2022 Mumps vaccine was licensed in 1967.

? **What events or factors might contribute to mumps outbreaks among college students?**

TABLE 24.4 Mumps

Signs and Symptoms	Fever, headache, loss of appetite, followed by painful swelling of parotid gland(s). Complications include painful enlargement of the testicles in men, pelvic pain in females, and neurological involvement.
Incubation Period	About 2 to 3 weeks
Causative Agent	Mumps virus, a single-stranded RNA virus
Pathogenesis	Viral replication destroys cells lining parotid gland ducts; inflammatory responses cause swelling and pain.
Epidemiology	Humans the only reservoir. Saliva of symptomatic and asymptomatically infected people is infectious.
Treatment and Prevention	Treatment: no antiviral therapy is available. Prevention: vaccination (MMR or MMRV).

MicroAssessment 24.3

Herpes simplex (cold sores) is characterized by acute disease followed by lifelong latency and the possibility of recurrences. Infectious virus is often present in saliva in the absence of symptoms. Mumps characteristically results in parotid gland enlargement but can also involve the testicles, ovaries, and brain.

7. In what type of cell does HSV-1 persist?
8. Why is mumps virus a good candidate for worldwide eradication?
9. Why would medication fail to cure HSV infections even though it prevents recurrent cold sores?

LOWER DIGESTIVE SYSTEM INFECTIONS

24.4 ■ Bacterial Diseases of the Lower Digestive System

Learning Outcomes

6. Describe the general characteristics of diarrheal diseases.
7. Compare and contrast cholera, shigellosis, the various types of *E. coli* gastroenteritis, *Salmonella* gastroenteritis, typhoid and paratyphoid fevers, campylobacteriosis, and *Clostridioides difficile* infection.

Diarrheal illness is a frequent result of lower intestinal tract infection, whether caused by bacteria, viruses, or protozoa. These infections have many features in common, particularly with respect to the epidemiology and prevention, so the important general characteristics are covered in the feature Focus on Diarrheal Diseases. The intestines also provide certain microbes a route of entry to other parts of the body.

This section focuses on bacterial intestinal infections; the mechanisms of cholera and shigellosis pathogenesis will be explained in the greatest detail because they illustrate principles shared by several other intestinal pathogens. In addition to the intestinal infections covered in this chapter, chapter 30 describes *Staphylococcus aureus* foodborne intoxication, which causes vomiting and diarrhea, and it provides more information about foodborne infections. Tapeworm and roundworm infections of the intestines are discussed in chapter 12.

FOCUS ON DIARRHEAL DISEASES

Hundreds of thousands of children around the world die of diarrheal illnesses each year. Most are infants, but all age groups are affected. In the United States and other resource-rich countries, fatalities are much less common and mostly occur among the elderly and those with AIDS or other immunodeficiencies, but millions of diarrhea cases still occur each year.

Some physicians refer to diarrheal disease as **gastroenteritis** (*gastro-*, "stomach," *entero-*, "intestine," *-itis*, "inflammation"), whereas others use the term "stomach flu" (but bear in mind that "stomach flu" has no relationship to "the flu," or influenza).

Signs and Symptoms

The signs and symptoms of gastroenteritis include diarrhea, loss of appetite, nausea, vomiting, and sometimes fever; the incubation period is typically a day or two. Diarrhea can be abundant and watery when infection involves the small intestine. Large intestine invasion typically results in frequent small amounts of loose stool that contains mucus and pus; if blood is present as well, the diarrhea is referred to as **dysentery.**

Pathogenesis

The pathogenesis of diarrheal disease varies according to the causative agent. In the case of bacteria, many of the responsible virulence genes are on plasmids, phages, or other mobile genetic elements, which can be transferred from one species or strain to another by horizontal gene transfer. Because of this, even strains within the same species can differ in the ways they cause disease. Common mechanisms involved include exotoxin production, epithelial cell invasion, and interfering with epithelial cell function. The exotoxins involved fall into two groups: **enterotoxins,** which typically cause water and electrolytes to flow from intestinal cells; and **cytotoxins**, which cause cell death. Some types of cytotoxins can be absorbed into the bloodstream, resulting in systemic effects.

The **infectious dose** of an intestinal pathogen (the number of cells or viral particles required to establish infection) is related to its ability to survive exposure to stomach acid. Acid-tolerant organisms are more likely to survive passage through the stomach, and the surviving cells then multiply in the intestines to high enough numbers to cause disease symptoms. In contrast, acid-sensitive organisms are often killed in the stomach, so only when millions of cells are ingested will enough survive to reach the intestines. People with low gastric acidity are more susceptible to intestinal infections, even by organisms that normally have a high infectious dose.

The location of the infection in the intestine influences the outcome. Infections of the small intestine generally lead to watery diarrhea due to disruptions in the fluid exchange that normally occurs there. The loss of water and electrolytes (salts) draws fluid from the bloodstream, resulting in dehydration. In severe cases, the reduced blood volume may not be sufficient to keep vital organs working properly, a potentially fatal situation. In contrast, damage due to infection of the large intestine often elicits a strong inflammatory response and leads to dysentery.

Epidemiology

Diarrheal diseases typically spread by **fecal-oral transmission,** meaning the pathogen is shed in the feces of one host and enters the mouth of the next. This commonly occurs via food or water contaminated with human or animal feces, but microbes with a low infectious dose—perhaps 100 cells—can additionally be transmitted through direct person-to-person contact. Pathogens that have a low infectious dose are a particular problem in day-care centers and other places where hygiene standards might be difficult to maintain. Sexual practices that lead to oral-anal contact can also transfer intestinal pathogens.

Treatment and Prevention

To counteract the loss of fluid and electrolytes due to any type of diarrhea, a method of fluid replacement called **oral rehydration therapy (ORT)** is often used. Although this involves giving liquids by mouth, plain water is not sufficient because the intestinal cells cannot absorb enough to keep pace with the rapid fluid loss.

continued

FOCUS ON DIARRHEAL DISEASES (continued)

The discovery that glucose increases the absorptive capacity of the cells led to the development of what is called oral rehydration salts (ORS) solution, a highly effective lifesaver for cases of severe diarrhea, regardless of the cause. ORS is a mixture of glucose and various salts (sodium chloride, potassium chloride, and trisodium citrate) commercially available as pre-measured packets to be dissolved in clean water. The World Health Organization (WHO) has also developed a list of recommended home fluids (RHF) that can prevent dehydration. If oral rehydration cannot be tolerated—for example, if the patient is vomiting—then intravenous hydration may be required.

Antimicrobial medications are not helpful in most intestinal infections, and they often prolong the illness because they suppress the normal microbiota. These drugs can be life-saving, however, if microorganisms invade beyond the intestine. For malnourished children in resource-limited countries, zinc supplements decrease the length and severity of diarrheal disease. No specific treatments are available for viral diarrhea.

Long-term control of diarrheal diseases requires breaking the transmission cycle by providing access to safe, clean water supplies, providing adequate sanitation (such as treating sewage), and promoting handwashing; the acronym WASH (water, sanitation, hygiene) is sometimes used by programs working to break the cycle. Only a few vaccines are available to prevent diarrheal diseases, and these are pathogen-specific.

People who have symptoms of an intestinal infection should be especially careful to take precautions aimed at preventing transmission of a possible causative agent. These precautions should be maintained until at least 24 hours after symptoms resolve; for a confirmed infection, a longer symptom-free period or laboratory testing might be needed to ensure that the infection has cleared. In the United States, surveillance using PulseNet—a DNA subtyping resource—helps track illness caused by specific intestinal pathogens; this helps public health agencies detect foodborne outbreaks so that intervention strategies can be implemented. Because food workers who have an infectious diarrheal disease can potentially spread the pathogens via the food they prepare, the U.S. Food and Drug Administration (FDA) publishes requirements about restrictions for infected employees of food establishments.

Cholera

Cholera causes diarrhea so severe that it can be fatal within hours. Seven cholera pandemics have occurred since the early 1800s; the last one began in 1961 and continues to this day, mostly in resource-limited regions of the world.

Signs and Symptoms

Cholera is a classic example of severe watery diarrheal disease. Signs and symptoms develop suddenly after an incubation period of about 12 to 48 hours. Huge volumes of watery diarrhea are produced—up to 20 liters a day—leading to severe dehydration that can result in shock (a drop in blood pressure that results in inadequate blood flow to the organs), multiple organ failure, and death. The diarrheal fluid is described as "rice water stool" because of its cloudy and light colored appearance that resembles water in which rice has been rinsed. Vomiting often occurs at the onset of the disease, and many people suffer muscle cramps caused by loss of fluid and electrolytes.

Causative Agent

Cholera is caused by *Vibrio cholerae,* a curved, Gram-negative rod with a single polar flagellum. Several different serotypes exist, grouped according to differences in their O antigen; the two that account for current cholera epidemics are O1 (which is pandemic) and O139 (which is largely limited to Asia). *V. cholerae* is halotolerant and can grow in alkaline conditions, characteristics used to design appropriate selective media.

Pathogenesis

V. cholerae is sensitive to acid and thus has a high infectious dose; only if large numbers of cells are ingested will enough survive passage through the stomach to colonize the intestine. Once in the small intestine, a bacterial cell will use its flagellum to swim through the mucus layer and then use its pili and other surface proteins to attach to the local epithelial cells. As the bacteria multiply, they cause no visible damage to the epithelial cells but produce cholera toxin, a diarrhea-inducing enterotoxin. That toxin is encoded by a prophage (phage DNA inserted into a bacterial chromosome; the latent form of a temperate phage), and therefore its production is an example of lysogenic conversion.

Cholera toxin is an A-B toxin composed of multiple copies of a B (binding) subunit and a single copy of an A (active) subunit. Its mechanism of action illustrates how certain exotoxins cause an increased flow of fluid out of cells (**figure 24.10**):

1. When cholera toxin contacts an intestinal epithelial cell, the B subunits attach to specific receptor molecules, causing the cell to take in the toxin. Once inside the cell, the A subunit is released.
2. The A subunit chemically alters a host cell G protein (a protein that acts as a molecular on/off switch to control intracellular activities). This change to the G protein essentially locks it in the "on" (active) position.
3. A normal function of the "on" form of the G protein is to activate adenylate cyclase, an enzyme that converts ATP to cyclic AMP (cAMP). As a result of the constant activity of the altered G protein, abnormally high levels of cAMP are produced.

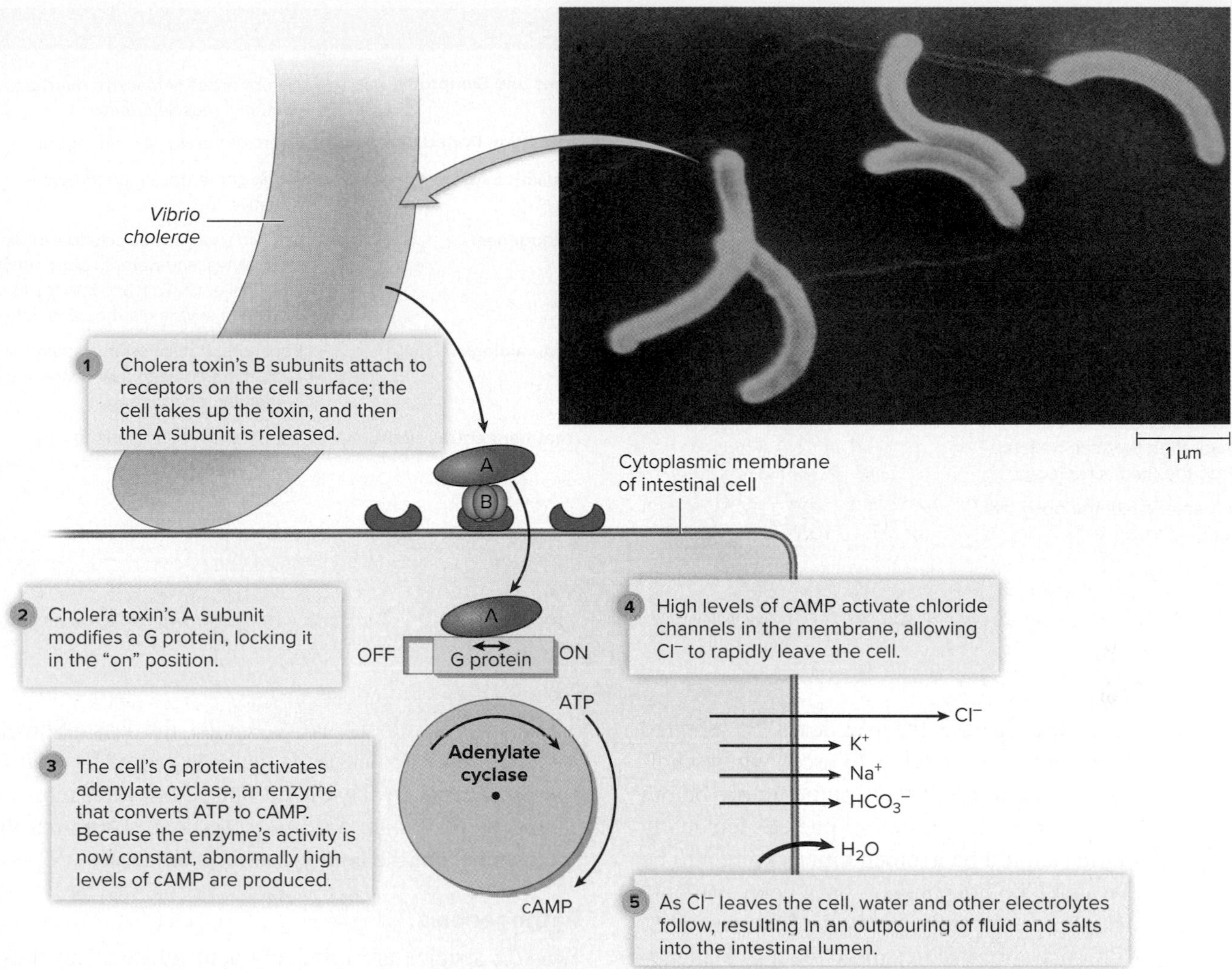

FIGURE 24.10 ***Vibrio cholerae*** **Pathogenesis** *Vibrio cholerae* attaches to the small intestine using pili and begins to produce cholera toxin, an A-B toxin. London School of Hygiene & Tropical Medicine/Science Source

? How would vaccination against the B subunit of cholera toxin help prevent disease?

④ High levels of cAMP activate chloride channels in the cell's cytoplasmic membrane, allowing chloride ions (Cl^-) to rapidly leave the cell.

⑤ As chloride ions leave the cell, other electrolytes and water follow due to osmosis, resulting in an outpouring of fluid and salts into the intestinal lumen. The volume of fluid released in the small intestine is beyond what the colon can absorb, leading to the severe diarrhea that results in potentially life-threatening dehydration. The normal turnover of intestinal cells eventually removes the toxin.

Epidemiology

Cholera spreads by fecal-oral transmission, most commonly through contaminated water, but also through foods such as contaminated crabs, oysters, and vegetables. Nearly all U.S. cases involve international travel. A person with cholera can discharge a million or more *V. cholerae* cells in each milliliter of feces, so untreated wastes can potentially spread the pathogen far and wide. Although cholera is relatively common in many areas of the world, effective water and wastewater treatment can stop its spread.

Treatment and Prevention

Treatment of cholera depends primarily on the rapid replacement of fluid and electrolytes before irreversible damage to vital organs can occur. If patients are vomiting or are too weak to drink oral rehydration solutions, then intravenous (IV) rehydration is started. The prompt administration of oral or IV rehydration therapy decreases the mortality of cholera from over 30% to less than 1%. In severe cases, antibiotics may be administered in addition to rehydration therapy.

Cholera prevention requires adequate sanitation and access to safe, clean water supplies. With that in mind, the Global Task Force on Cholera Control (a WHO-coordinated network) established a plan in 2017 that includes investing in safe water, sanitation, and hygiene in areas where cholera is most common, with the goal of decreasing cholera deaths by

TABLE 24.5 Cholera

① *Vibrio cholerae* enters the mouth with fecally contaminated food or water.

② The bacteria attach to epithelial cells of the small intestine.

③ Cholera toxin causes electrolyte-containing fluids to rapidly flow from cells.

④ The outpouring of fluids into the intestinal lumen causes watery diarrhea.

⑤ Fluid loss causes severe dehydration that can be fatal unless the fluid is replaced.

⑥ The bacteria exit the body with feces.

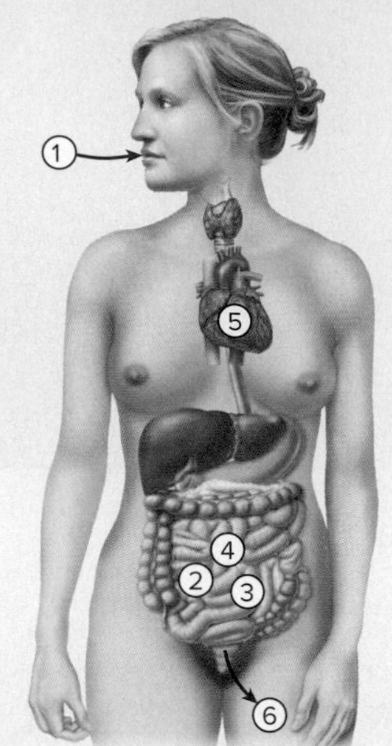

Signs and Symptoms	Abrupt onset of massive diarrhea, vomiting, muscle cramps
Incubation Period	Short, generally 12 to 48 hours
Causative Agent	*Vibrio cholerae*, a curved Gram-negative rod
Pathogenesis	Cholera toxin causes chloride, other electrolytes, and water to pour out of the intestinal epithelium and into the lumen, leading to severe diarrhea and dehydration
Epidemiology	Fecal-oral transmission, usually via contaminated water but sometimes contaminated foods
Treatment and Prevention	Treatment: rehydration therapy. Prevention: adequate sanitation and safe, clean water supplies; vaccination

90% and eliminating the disease from at least 20 affected countries by 2030. Meanwhile, travelers to areas where cholera is endemic are advised to cook food immediately before eating it. No fruit should be eaten unless peeled personally by the traveler, and ice should be avoided unless known to be made from boiled water. Cholera vaccines have been available in endemic countries since the early 1990s. In 2016, the U.S. Food and Drug Administration (FDA) approved a live attenuated oral vaccine against cholera caused by serogroup O1 for use by travelers between 18 and 64 years of age who are visiting endemic areas; its production was temporarily stopped, however, as a result of decreased international travel during the height of the COVID-19 pandemic. The main features of cholera are summarized in **table 24.5.**

Shigellosis

Shigellosis can be mild or life-threatening, depending largely on the infecting species. The disease is found all over the world, most commonly in crowded, resource-limited areas lacking adequate wastewater treatment.

Signs and Symptoms

Shigellosis is a classic example of a diarrheal disease that results in dysentery (bloody, mucoid diarrhea), but some *Shigella* species cause watery diarrhea. Signs and symptoms appear after an incubation period of 1 to 3 days and often include fever and vomiting, in addition to diarrhea.

Causative Agent

Shigellosis is caused by the four species of *Shigella: S. dysenteriae, S. flexneri, S. boydii,* and *S. sonnei.* Of these, *S. dysenteriae* is the most virulent, and *S. sonnei* the least. Shigellosis in resource-limited countries is typically caused by *S. dysenteriae* and *S. flexneri.* In the United States, however, most cases are caused by *S. sonnei. Shigella* species are Gram-negative rods and members of the family Enterobacteriaceae.

Pathogenesis

Shigella species are relatively acid tolerant and thus have a low infectious dose; even small numbers of ingested cells will likely survive passage through the stomach. Once in the large intestine, the cells invade epithelial cells, leading to a strong inflammatory response; their invasion process illustrates how some bacteria manipulate normal host cell functions as a means to gain access to a cell's cytosol (**figure 24.11**):

① To initiate invasion, *Shigella* cells take advantage of M cells, using them to cross the mucosal epithelium (see figure 16.7). Recall that the normal function of M cells is to sample microbe-containing material from the intestinal lumen and deliver it to macrophages within the Peyer's patches; the macrophages then destroy the material, and parts of it are presented to T cells (figure 15.8). When *Shigella* cells are delivered to a macrophage, however, they survive the destructive process by escaping from the phagosome, and they then multiply within the macrophage's cytosol.

② The *Shigella* cells inside a macrophage induce their host cell to undergo a type of programmed cell death (pyroptosis); this releases the bacteria, which then attach to specific receptors on the base of intestinal epithelial cells and inject proteins referred to as invasins. These cause rearrangement of the host cell actin, resulting in endocytosis of the bacterial cells. Once inside an epithelial cell, the *Shigella* cells escape from the endosome to access the host cell's cytosol, where they then multiply.

③ As the *Shigella* cells multiply within the epithelial cells, they make a protein that causes the host cell actin to polymerize (see Focus Your Perspective 3.1). As a result, an "actin tail" forms at one end of each bacterial cell, and this propels a bacterium within the host cell, sometimes with enough force to drive it into a neighboring cell.

④ Infected epithelial cells eventually die, resulting in patches of dead epithelium that slough off. This leaves intensely inflamed bare areas in the intestine that are covered with pus and blood, leading to dysentery.

In addition to invading cells, some strains of *Shigella dysenteriae* produce a potent cytotoxin known as Shiga toxin, a chromosomally encoded A-B toxin. The toxin enters the bloodstream, and then the B portion binds to endothelial cells that line the blood vessels. This allows the A subunit to then enter those cells, where it interacts with ribosomes, stopping protein synthesis and leading to cell death. Shiga toxin is responsible for **hemolytic uremic syndrome (HUS),** a life-threatening condition that can follow *Shigella dysenteriae* infection. In HUS, damage to the lumen of the tiny blood vessels of the kidneys results in (1) lysis of fragile red blood cells as they squeeze through and (2) clotting that blocks the kidneys' filtration system, which can lead to kidney failure. Symptoms of HUS sometimes include paralysis or other indications of nervous system injury. Shiga toxin is also produced by strains of *Escherichia coli* that cause HUS.

Epidemiology

Shigellosis is almost exclusively a disease of humans, spread by fecal-oral transmission via direct person-to-person contact or through contaminated food or water. As such, the disease spreads readily in overcrowded populations with poor sanitation—including refugee camps and day-care centers. People who engage in anal intercourse are also at risk of contracting the disease.

An estimated 450,000 shigellosis cases occur in the United States each year. Globally, an estimated 75 million cases occur annually, resulting in about 210,000 deaths.

Treatment and Prevention

Shigellosis generally resolves on its own, but antimicrobial medications are useful in severe cases because they shorten the duration of not only the symptoms but also the shedding of the bacterium in feces; the latter effect decreases transmission. Unfortunately, many strains have R plasmids that encode resistance to a variety of medications.

The spread of *Shigella* is controlled by sanitary measures that reduce fecal contamination of food and water supplies. Shigellosis can be tracked through PulseNet, making it easier for public health agencies to detect outbreaks. To prevent spread of the disease by food workers, FDA guidelines cover restrictions on food establishment employees who are exposed or infected. **Table 24.6** describes the main features of shigellosis.

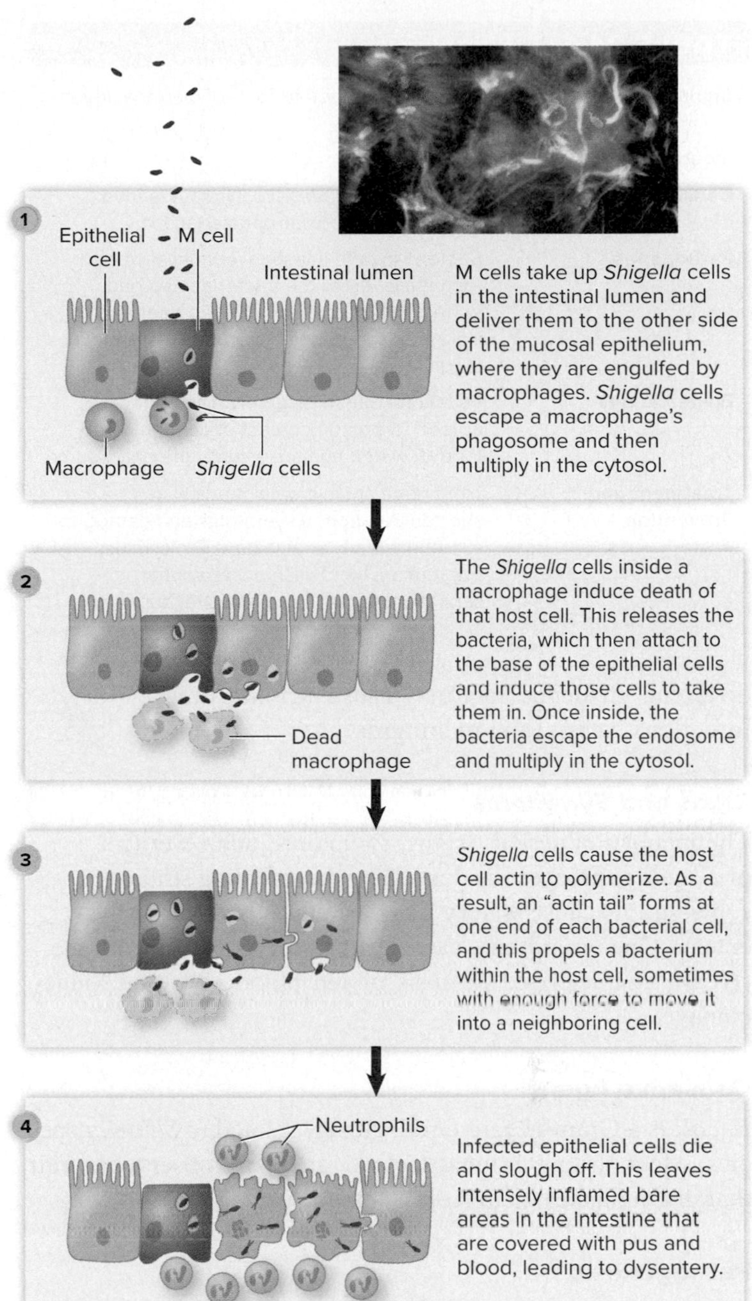

FIGURE 24.11 ***Shigella* Pathogenesis** The photomicrograph shows the actin tails (green) that form on intracellular *Shigella* cells (orange). Dr. Philippe J. Sansonetti

? **Why does shigellosis typically result in dysentery, whereas cholera is characterized by watery diarrhea?**

Escherichia coli Gastroenteritis

E. coli strains are intestinal residents of almost all humans and a number of other animals; most of the strains are harmless, but some produce specific virulence factors that allow them to cause intestinal disease. Other strains—with

TABLE 24.6 Shigellosis

Signs and Symptoms	Dysentery or another form of diarrhea, fever, vomiting
Incubation Period	1 to 3 days
Causative Agent	Four species of *Shigella,* Gram-negative, members of the Enterobacteriaceae
Pathogenesis	Bacteria invade intestinal epithelial cells; "actin tails" allow the bacteria to spread to neighboring cells, leading to death and sloughing of epithelium. Some strains make Shiga toxin.
Epidemiology	Fecal-oral transmission involving direct person-to-person contact or contaminated food or water; humans usually the only source
Treatment and Prevention	Treatment: antimicrobial medications shorten duration of symptoms and pathogen excretion; many strains have R plasmids. Prevention: handwashing, adequate sanitation, and safe, clean water supplies

different virulence factors—cause urinary tract infections, sepsis, and meningitis.

Signs and Symptoms

The incubation periods, signs, symptoms, and severity of *E. coli* gastroenteritis depend on the infecting strain—some strains cause watery diarrhea and others cause dysentery. One group can cause hemolytic uremic syndrome (HUS), characterized by lysis of red blood cells and kidney damage.

Causative Agent

E. coli is a Gram-negative rod, closely related to *Shigella* species. Most strains ferment lactose, an easily observable trait that distinguishes them from *Shigella* species.

Pathogenesis

The *E. coli* strains capable of producing gastroenteritis have a variety of pathogenic mechanisms (**figure 24.12**). Not all strains produce the same virulence factors, which accounts for the different symptoms associated with the various strains.

E. coli strains that cause intestinal disease can be grouped into six pathovars (pathogenic varieties), based on their array of virulence factors (**table 24.7**):

- **Shiga toxin–producing *E. coli* (STEC).** These strains, also referred to as EHEC or enterohemorrhagic *E. coli,* produce Shiga toxins, a family of functionally identical cytotoxins that includes the toxin of *Shigella dysenteriae.* In STEC strains, the genes for Shiga toxins are encoded by various related prophages, an example of lysogenic conversion. STEC strains typically colonize the large intestine, attaching to the epithelial cells there. After a bacterial cell attaches, it injects effector proteins into the epithelial cell; in turn, these proteins cause the epithelial cell's actin to rearrange, resulting in the formation of a pedestal-like structure directly under the bacterial cell (see box figure 3.3). This characteristic damage, referred to as **attaching and effacing (A/E) lesions**, leads to diarrhea. The feces becomes bloody (hemorrhagic diarrhea) due to the action of Shiga toxins on the local blood vessels; hemolytic uremic syndrome (HUS) develops in about 5% to 10% of infected people. Most of the strains identified in outbreaks belong to a single serotype, O157:H7, but important exceptions exist. The STEC strain responsible for the 2011 outbreak in Europe (serotype O104:H4) is unusual because it does not produce A/E lesions and has characteristics of the EAEC pathovar, which will be discussed shortly.
- **Enterotoxigenic *E. coli* (ETEC).** These strains make pili that allow them to attach to and colonize the small

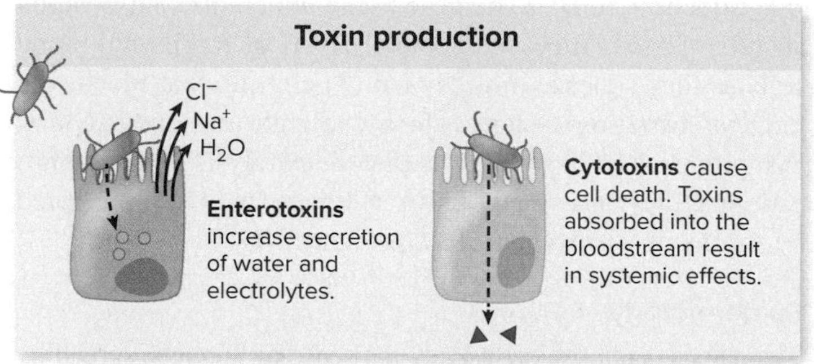

Toxin production
Cl^-
Na^+
H_2O
Enterotoxins increase secretion of water and electrolytes.
Cytotoxins cause cell death. Toxins absorbed into the bloodstream result in systemic effects.

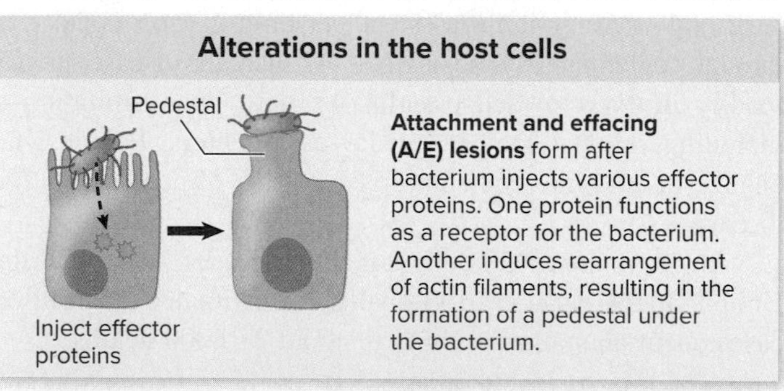

Alterations in the host cells
Pedestal
Inject effector proteins
Attachment and effacing (A/E) lesions form after bacterium injects various effector proteins. One protein functions as a receptor for the bacterium. Another induces rearrangement of actin filaments, resulting in the formation of a pedestal under the bacterium.

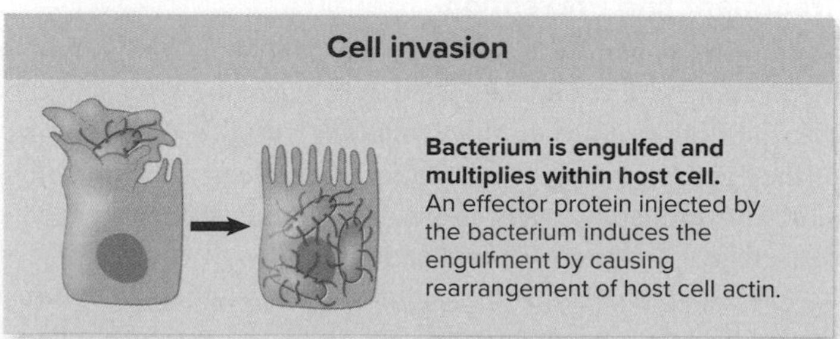

Cell invasion
Bacterium is engulfed and multiplies within host cell.
An effector protein injected by the bacterium induces the engulfment by causing rearrangement of host cell actin.

FIGURE 24.12 Common Pathogenic Mechanisms of *E. coli* Some other intestinal pathogens use these mechanisms.

? Why is adhesion generally a prerequisite for pathogenicity?

intestine. In addition, they secrete enterotoxins, one of which is nearly identical to cholera toxin. The genes for adhesin and toxin synthesis are on plasmids.

- **Enteroinvasive *E. coli* (EIEC).** These strains invade the intestinal epithelium, causing a disease similar to shigellosis.
- **Enteropathogenic *E. coli* (EPEC).** These strains produce pili that allow them to colonize the small intestine, where they inject effector proteins that cause A/E lesions.
- **Enteroaggregative *E. coli* (EAEC).** These strains produce pili that allow them to adhere to the intestinal epithelium. There, they grow in characteristic aggregations ("brick-like") in a thick mucus-associated biofilm. In addition, they produce enterotoxins and cytotoxins, damaging the intestinal cells and evoking an inflammatory response.
- **Diffusely adhering *E. coli* (DAEC).** These strains are similar to EAEC, but rather than forming aggregations, they grow as a diffuse layer.

Epidemiology

Just as the symptoms and pathogenesis of *E. coli* gastroenteritis depend on the infecting strain, so does the epidemiology.

STEC strains can be foodborne, and epidemics have involved contaminated ground beef, green leafy vegetables, and bean sprouts, as well as unpasteurized milk and apple juice. The initial source of infection is often untreated cow manure, reflecting the fact that cattle are an important reservoir. The infectious dose of STEC strains is typically very low, so in addition to foodborne transmission, the bacteria are easily spread by direct contact.

ETEC strains commonly cause diarrhea in infants in resource limited countries as well as in travelers visiting those regions. The infectious dose of these strains is relatively high, so the bacteria typically do not spread by direct contact. ETEC strains are host species–specific, and those that infect animals are responsible for significant mortality in young livestock.

EIEC strains primarily cause disease in young children in resource-limited countries. When compared with some other *E. coli* strains, they are not easily spread by direct contact, indicating that the infectious dose is relatively high. Humans appear to be the only source of infection.

EPEC strains are an important cause of chronic diarrhea in infants. The infection is uncommon in breast-fed infants, probably because of protective antibodies in breast milk. Outbreaks have occurred in hospital nurseries, so the infectious dose for infants is thought to be very low. A variety of animals have been shown to harbor EPEC, but their importance as a source of infection is not known.

EAEC strains cause diarrhea in children, travelers, and AIDS patients. A variety of animals harbor EAEC strains, but it is not clear that the animal strains cause human disease.

DAEC strains have caused diarrhea outbreaks in children, but relatively little is known about their epidemiology.

Treatment and Prevention

Treatment of *E. coli* gastroenteritis varies according to the symptoms and the infecting strain, but as with any disease, fluid lost from vomiting and diarrhea should be replaced. Antibacterial medications are typically not used because most cases are self-limiting. With respect to STEC infections, studies indicate that patients treated with antibiotics typically have an overall worse outcome.

Measures to prevent *E. coli* gastroenteritis include methods that reduce fecal contamination of food and water supplies, along with handwashing, cooking food thoroughly, and pasteurizing beverages. Infections caused by *E. coli* O157:H7

TABLE 24.7 Characteristics of Diarrhea-Causing *Escherichia coli*

Designation	Characteristic Features	Clinical Picture
Diffusely adhering *E. coli* (DAEC)	Grows as a diffuse layer in a thick mucus-associated biofilm on the intestinal epithelium; produces toxins	Diarrhea, particularly in children
Enteroaggregative *E. coli* (EAEC)	Grows in brick-like aggregations in a thick mucus-associated biofilm on the intestinal epithelium; produces toxins	Variable symptoms; nausea, watery diarrhea
Enteroinvasive *E. coli* (EIEC)	Invades the intestinal epithelium, causing a disease very similar to shigellosis	Fever, cramps, diarrhea containing blood and pus
Enteropathogenic *E. coli* (EPEC)	Colonizes the small intestine; induces changes in actin filaments, causing the microvilli to be replaced by pedestals under the bacterial cells (A/E lesions)	Fever, vomiting, watery diarrhea containing mucus
Enterotoxigenic *E. coli* (ETEC)	Colonizes the small intestine and produces toxins, one nearly identical to cholera toxin	Nausea, vomiting, abdominal cramps, massive watery diarrhea leading to dehydration
Shiga toxin–producing *E. coli* (STEC)	Same as EPEC, except it produces Shiga toxin and mainly colonizes the large intestine rather than the small intestine	Fever, abdominal cramps, bloody diarrhea without pus; some patients develop hemolytic uremic syndrome; most strains identified in outbreaks are serotype O157:H7

TABLE 24.8 *Escherichia coli* **Gastroenteritis**

① Pathogenic strain of *E. coli* enters directly from an infected person or via contaminated food or drink.

② Most strains colonize the small intestine and produce watery diarrhea.

③ Other strains invade the large intestine and cause dysentery.

④ Some strains produce Shiga toxin, which is absorbed by the bloodstream and causes hemolytic-uremic syndrome.

⑤ The bacteria exit the body in feces.

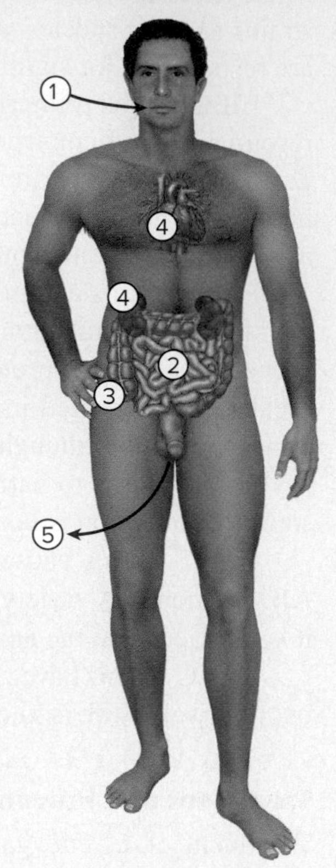

Signs and Symptoms	Watery diarrhea or dysentery
Incubation Period	2 hours to 6 days
Causative Agent	*Escherichia coli,* certain strains only; member of the Enterobacteriaceae
Pathogenesis	Various mechanisms; attachment to small intestinal cells allows colonization; some strains produce one or more enterotoxins; some strains invade the intestinal epithelium; others cause attachment and effacing (A/E) lesions, and may produce Shiga toxin
Epidemiology	Fecal-oral transmission: route depends on infecting strain; some strains common in travelers; some strains have an animal source
Treatment and Prevention	Treatment: rehydration therapy. Prevention: handwashing, adequate sanitation, and safe, clean water supplies; cooking foods thoroughly; pasteurizing beverages; bismuth compounds help prevent traveler's diarrhea

and other STEC strains can be tracked using PulseNet; in addition, FDA guidelines cover restrictions on food workers who are exposed or infected. Traveler's diarrhea can usually be prevented with bismuth preparations (such as Pepto-Bismol). Unfortunately, the widespread use of antibiotics to prevent diarrhea has promoted the development of resistant strains. Some features of *E. coli* gastroenteritis are summarized in **table 24.8.**

Salmonella Gastroenteritis

Salmonella gastroenteritis is caused by numerous serotypes of *Salmonella enterica* and can be acquired from many animal sources. Large outbreaks are usually due to commercially distributed foods contaminated by animal feces.

Signs and Symptoms

Signs and symptoms of *Salmonella* gastroenteritis include diarrhea (sometimes bloody), abdominal cramps, nausea, vomiting, and fever. The disease is often short-lived and mild, but that varies depending on the virulence of the infecting strain and the number of cells ingested. Similarly, the incubation period varies from 6 hours to 3 days.

Causative Agent

S. enterica is a Gram-negative rod and, like *Shigella* and *E. coli,* a member of the Enterobacteriaceae. The various *Salmonella* strains are subdivided into more than 2,400 serotypes based on differences in their cell wall (O), flagellar (H), and capsular (K) antigens (see figure 10.11). Each serotype was once considered a separate species and given a distinct name. However, DNA sequencing data indicate there are only two species: *S. enterica* and *S. bongori,* the latter of which is only rarely isolated from humans.

The serotype of a *Salmonella* strain is significant with respect to both the epidemiology and the disease; for instance, certain serotypes cause enteric fever (discussed next) instead of gastroenteritis. Because of the significance, the serotype is often included with the name—for example, *Salmonella enterica* serotype Dublin. For convenience, the name is often shortened, as in *Salmonella* Dublin. Note that the serotype is capitalized and not italicized, thus distinguishing the serotype from the species name. *Salmonella* Typhimurium and *Salmonella* Enteritidis are the most common serotypes in the United States.

Pathogenesis

Most *Salmonella* serotypes are sensitive to acid, so millions of cells must usually be ingested for enough to survive passage through the stomach to colonize the intestines. After reaching the distal small intestine (the region farthest from the stomach), some of the bacterial cells attach to specific receptors on the surface of the epithelial cells. That contact activates a type III secretion system in the bacterial cell, which then transfers bacterial effector proteins into the epithelial cell. Within minutes, these proteins cause the epithelial cell to take in the bacterial cells by endocytosis, and the bacteria begin multiplying within that cell (see figure 16.6). Some of the bacteria escape the endosome and multiply in the cytosol, but others are released from the base of the cell by exocytosis. Macrophages and neutrophils take up those released bacteria, but the macrophages are often destroyed as a result. The infection, however, usually remains localized. The host response increases epithelial cell fluid secretion, causing diarrhea.

Not all of the infecting *Salmonella* cells invade the epithelial cells. Those that do not can benefit from the inflammatory response caused by the invading bacteria. The oxidizing environment created by neutrophils recruited during inflammation generates tetrathionate—a compound that only *Salmonella* and a very small number of other intestinal organisms can use as a terminal electron acceptor for anaerobic respiration (see Focus on a Case 6.1). The ability to anaerobically respire gives *Salmonella* cells a competitive advantage in the densely populated intestinal tract where most organisms can only ferment; recall that the ATP yield of respiration is greater than that of fermentation (see table 6.3).

Epidemiology

Salmonella strains that cause gastroenteritis live in the intestinal tracts of a number of various types of domestic animals, and the disease spreads through fecal-oral transmission—usually via food or water contaminated with feces of those animals. The bacteria sometimes survive for months in soil and water, so they can be spread in untreated manure. Poultry often carry the bacteria, and eggs can be contaminated as well. A variety of other contaminated products—including alfalfa sprouts, protein supplements, raw eggs, and dry milk—have started outbreaks. Children are commonly infected through contact with seemingly healthy, but colonized, pets (particularly lizards, snakes, and turtles) that shed the bacteria in their feces. The disease has been on the rise, largely due to mass production and distribution of foods; an estimated 1.35 million cases occur in the United States each year, with about 400 deaths.

Treatment and Prevention

Most people with *Salmonella* gastroenteritis recover without antimicrobial treatment; in fact, antibiotics are not advised for treatment except in cases involving tissue or bloodstream invasion. *S. enterica* strains that cause gastroenteritis have shown increasing plasmid-mediated resistance to antimicrobial medications, and some are resistant to five or more medications. Resistance is likely due to the widespread use of subtherapeutic levels of antibiotics in animal feeds, which is why several countries have now banned this practice.

Control of *Salmonella* gastroenteritis depends on sanitary handling of animal carcasses, pasteurizing or irradiating animal products, and testing products for contamination. Surveillance and tracing contaminated sources using PulseNet are also important, and FDA guidelines cover restrictions on food workers who are exposed to or develop the disease. Adequate cooking kills *Salmonella* cells, but the center of cooked food does not always reach the temperature needed to kill foodborne bacteria (about 165° Fahrenheit; 74° Celsius). **Table 24.9** presents the characteristics of *Salmonella* gastroenteritis.

TABLE 24.9 *Salmonella* Gastroenteritis

Signs and Symptoms	Diarrhea, vomiting, abdominal pain, and fever
Incubation Period	Usually 6 to 72 hours
Causative Agent	*Salmonella enterica*, Gram-negative, member of the Enterobacteriaceae
Pathogenesis	Induced uptake by epithelial cells in the distal small intestine; bacteria multiply with those cells and then are discharged at the base of the infected cells; inflammatory response increases fluid secretion
Epidemiology	Fecal-oral transmission, usually via food contaminated by animal feces, especially poultry
Treatment and Prevention	Treatment: antimicrobial medication is generally not advised. Prevention: sanitary handling of animal carcasses; adequately cooking and properly handling food.

Enteric Fever (Typhoid and Paratyphoid)

Typhoid fever and paratyphoid fever are **enteric fevers**—systemic diseases that originate in the intestine. They are clinically indistinguishable and caused by related organisms. Although rare in the United States, millions of cases occur in resource-limited countries.

Signs and Symptoms

The signs and symptoms of enteric fever begin after an incubation period of 1 to 4 weeks, with larger doses usually leading to the shorter incubation periods. Patients develop a severe headache, constipation, abdominal pain, a fever that gradually increases over several days, and sometimes a characteristic rash with small red spots ("rose spots") on the chest and abdomen. In severe cases, intestinal rupture, internal bleeding, shock, and death may follow.

Causative Agent

Typhoid fever is caused by *Salmonella enterica* serotype Typhi, whereas paratyphoid fever is caused by *Salmonella enterica* serotype Paratyphi (these serotypes are often referred to as *S.* Typhi and *S.* Paratyphi); both are Gram-negative members of the Enterobacteriaceae. Because the infections are systemic, cases are confirmed by blood culture rather than stool culture.

Pathogenesis

The bacteria that cause enteric fever colonize the intestines, cross the mucous membrane (often via M cells), and then multiply within macrophages and dendritic cells in the Peyer's patches. They are then carried in the bloodstream to locations throughout the body. The systemic infection causes fever, abscesses, sepsis (life-threatening tissue damage and organ dysfunction resulting from an overwhelming host response to an infection), and shock, often with little or no diarrhea. The Peyer's patches are sometimes destroyed, leading to intestinal perforation, hemorrhage, and death. Surprisingly little is known about the bacterial mechanisms that lead to enteric fever; an A-B toxin (typhoid toxin) is produced, but studies using human volunteers suggest that it is not required for disease development.

Epidemiology

Humans are the only known reservoir for *S.* Typhi and *S.* Paratyphi, and the bacteria are spread from person to person through fecal-oral transmission—usually via contaminated food or water. Some patients surviving enteric fever remain colonized with the causative agents and are thus carriers of the disease. In these patients, the bacteria usually reside in the gallbladder, where they multiply without competition because most other bacteria are killed or inhibited by concentrated bile. Carriers can shed high numbers of the bacterial cells in feces for years. Mary Mallon ("Typhoid Mary"), an Irish cook living in New York State in the early 1900s, was a *S.* Typhi carrier; she was linked to at least 53 cases of typhoid fever over a 15-year period. In her day, about 350,000 cases occurred in the United States each year; the incidence is now less than 400 cases per year, and most of these are related to international travel.

Treatment and Prevention

Antimicrobial medications are used to treat enteric fever, but susceptibility testing must be done because multidrug-resistant strains are now found worldwide; strains of *S.* Typhi referred to as XDR (extensively drug-resistant) are resistant to all but one oral antimicrobial. The susceptibility patterns of *S.* Typhi and *S.* Paratyphi vary by geographic location, so until the results are known, a patient's travel history can be used to guide the treatment choice. Surgical removal of the gallbladder and months of antibiotic therapy are often necessary to rid a carrier of the infection.

As with other fecally transmitted diseases, prevention of enteric fever involves adequate sanitation and safe, clean water supplies. Although enteric fever is rare in the United States, FDA guidelines cover restrictions on food workers who are exposed to or infected with *S.* Typhi. Two typhoid fever vaccines, both about 50% to 75% effective, are available in the United States for people traveling to endemic countries. One is an attenuated live oral vaccine for people at least 6 years old and the other an injectable vaccine composed of *S.* Typhi capsular polysaccharide for people at least 2 years old. Those vaccines are also recommended by WHO, as is a new conjugate vaccine; this latest option, capsular polysaccharide attached to a tetanus toxoid carrier, can be used for children at least 6 months of age and adults up to age 45 in countries where typhoid is endemic. There is no vaccine against *S.* Paratyphi. **Table 24.10** summarizes some of the features of enteric fever.

TABLE 24.10	Enteric Fever (Typhoid and Paratyphoid)
Signs and Symptoms	Fever, severe headache, constipation, abdominal pain and rash; sometimes followed by intestinal rupture, internal bleeding, shock, and death
Incubation Period	1 to 4 weeks
Causative Agent	*Salmonella* Typhi (typhoid fever) and *Salmonella* Paratyphi (paratyphoid fever), Gram-negative, members of the Enterobacteriaceae
Pathogenesis	Bacteria cross the intestinal epithelium, multiply, and are then carried in the bloodstream throughout the body; destruction of Peyer's patches leads to intestine perforation and hemorrhage.
Epidemiology	Fecal-oral transmission, usually via contaminated food or water; humans are the only source of infection.
Treatment and Prevention	Treatment: antimicrobial medications. Prevention: vaccines for typhoid fever; no vaccine for paratyphoid fever; handwashing, adequate sanitation, and safe, clean water supplies.

Campylobacteriosis

Campylobacter jejuni was first isolated from a diarrheal stool in 1972, but development of a suitable method to selectively culture the organism then took 5 years. Once this was widely used, *C. jejuni* infection was found to be a common cause of diarrhea.

Signs and Symptoms

The incubation period of campylobacteriosis can vary from 1 to 11 days, but signs and symptoms usually appear in 2 to 5 days. These include fever, vomiting, diarrhea, and abdominal cramps. Dysentery occurs in about half the cases.

Causative Agent

C. jejuni is a curved, Gram-negative rod (**figure 24.13**). It can be cultivated from feces under microaerophilic conditions using a selective medium to suppress the growth of other intestinal organisms.

Pathogenesis

C. jejuni cells attach to and then invade the epithelial cells of the small and large intestines. The bacteria multiply within and beneath the epithelium and cause a localized inflammatory reaction.

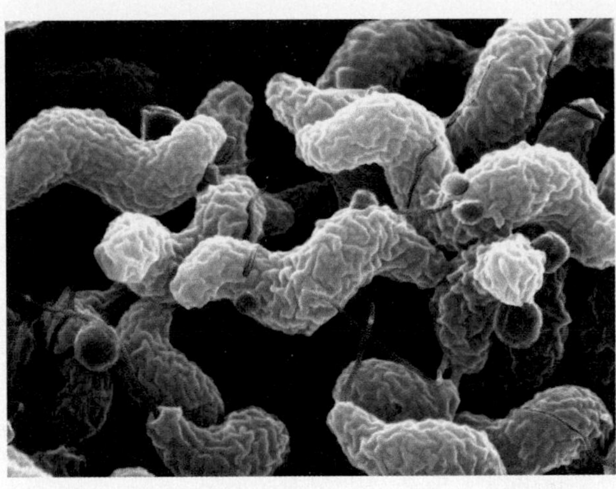

FIGURE 24.13 ***Campylobacter jejuni*** Color-enhanced scanning electron micrograph. De Wood, Chris Pooley/USDA

? **What general type of medium is used to culture this organism, and why?**

TABLE 24.11	**Campylobacteriosis**
Signs and Symptoms	Diarrhea, fever, abdominal cramps, vomiting, bloody stools
Incubation Period	Usually 2 to 5 days
Causative Agent	*Campylobacter jejuni*, a curved Gram-negative, microaerophilic rod
Pathogenesis	Low infectious dose; bacteria multiply within and beneath the epithelial cells, causing an inflammatory response. Complicated by Guillain-Barré syndrome on rare occasions.
Epidemiology	Fecal-oral transmission, usually via food contaminated by animal feces, especially poultry, but also contaminated water.
Treatment and Prevention	Treatment: antibiotics in severe cases. Prevention: adequately cooking and properly handling foods, particularly poultry; adequate sanitation and safe, clean water supplies.

In rare cases, individuals with campylobacteriosis go on to develop an autoimmune complication called Guillain-Barré syndrome in which the immune system damages certain nerves. The syndrome begins within about 10 days of the onset of diarrhea, with tingling in the feet followed by progressive paralysis in the legs, the arms, and the rest of the body. Most Guillain-Barré patients require hospitalization, but recover completely; about 5% of patients die despite treatment.

Epidemiology

C. jejuni lives in the intestines of a variety of domestic animals, including pets and migratory birds, and the disease spreads through fecal-oral transmission, usually via food or water contaminated with feces of those animals. Poultry is a common source of infection, and up to 90% of raw poultry products contain the bacterium. The infectious dose of *C. jejuni* is as low as 500 organisms, so one drop of juice from raw chicken meat can easily result in infection. Unpasteurized milk and non-chlorinated surface water have also started epidemics. Despite the relatively low infectious dose, direct person-to-person spread of *C. jejuni* is rare.

Campylobacteriosis is a leading bacterial diarrheal illness in the United States, with an estimated 1.5 million cases each year. Numerous foodborne and waterborne outbreaks have been reported, but most cases are sporadic.

Treatment and Prevention

C. jejuni gastroenteritis is typically self-limiting, but azithromycin or fluoroquinolones are used to treat severe cases.

Campylobacteriosis can be prevented by cooking and handling raw poultry properly to avoid cross-contamination. Chicken should be cooked until no longer pink (about 165° Fahrenheit; 74° Celsius). Pet owners should wash their hands after contact with animal feces, and outdoor play areas should be kept free of bird droppings. Additional measures include avoiding untreated water and unpasteurized milk. PulseNet helps track *Campylobacter* spread. The main features of campylobacteriosis are listed in **table 24.11.**

Clostridioides difficile Infection (CDI)

Clostridioides (*Clostridium*) *difficile* causes antibiotic-associated diarrheal disease that can sometimes be life-threatening. Because of the potential severity of *C. difficile* infection (CDI) and its relationship to antibiotic use, the CDC considers the bacterium to be an urgent health threat (see table 20.2).

Signs and Symptoms

C. difficile infection severity ranges widely. Some patients have only mild diarrhea, often accompanied by fever and abdominal pain and usually beginning less than a week after infection. In other cases, the disease progresses to severe colitis (inflammation of the lining of the colon), sometimes with patchy areas on the colon called pseudomembranes, which are composed of dead epithelium, inflammatory cells, and clotted blood; the formation of these is a characteristic of pseudomembranous colitis. In severe cases, CDI is life-threatening.

Causative Agent

C. difficile (commonly referred to as "*C. diff*"), is a Gram-positive, rod-shaped, endospore-forming obligate anaerobe. The endospores are highly resistant to common disinfectants and environmental conditions, making spread of the disease difficult to control. When the endospores are ingested, they can germinate, forming vegetative cells that can sometimes colonize the large intestine.

Many strains of *C. difficile* exist, and these can be separated into several large groups that differ in their pathogenicity. Strains that cause CDI produce one or more characteristic toxins, and tests that detect the toxins or toxin-encoding genes are often used to diagnose the disease. The greatest concern currently is a related group of hypervirulent strains that emerged just over 20 years ago and have been increasing in prevalence. Unlike most *C. difficile* strains, these are resistant to fluoroquinolones, so frequent use of the medications may have given them a selective advantage.

Pathogenesis

C. difficile can proliferate when antibiotics have disrupted the normal microbiota in the large intestine, creating dysbiosis. Pathogenic strains tested so far all produce produce toxin B (TcdB) and often toxin A (TcdA). Both of these toxins interfere with regulation of host cell actin polymerization and various other signaling pathways, resulting in the death of intestinal epithelial cells. The toxins also damage macrophages, causing them to release proinflammatory cytokines and undergo pyroptosis, which, in turn, induces a strong inflammatory response. In severe cases, the overall result is formation of pseudomembranes on the inner wall of the colon (**figure 24.14**). The hypervirulent *C. diff* strains produce two forms of toxin B and also make a third toxin, *C. difficile* transferase (CDT), sometimes referred to as binary toxin (a name that indicates it has two components); like toxins A and B, CDT interferes with actin polymerization.

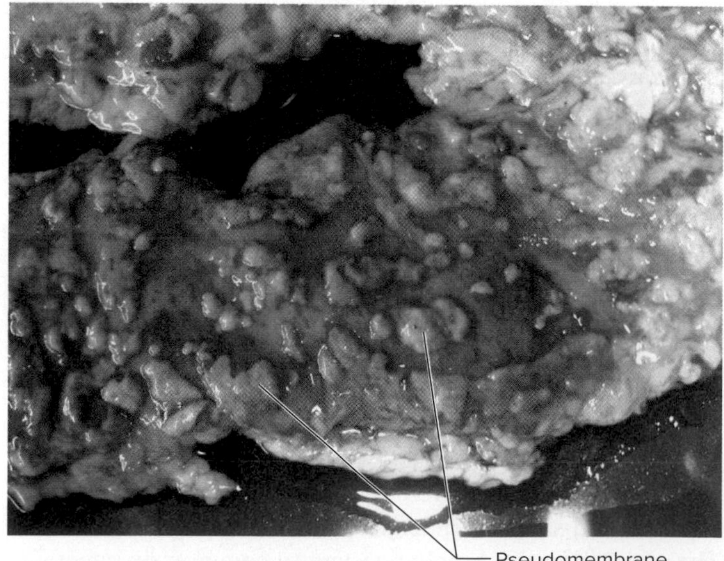

FIGURE 24.14 Pseudomembranes The pseudomembranes are on the inner wall of a section of colon.

? How does antibiotic therapy predispose a person to this condition?

Epidemiology

C. difficile infection primarily occurs in patients on antibiotic therapy, probably because the organism can only grow to high numbers when the normal microbiota has been disrupted. Some patients carry it as a minor component of their normal microbiota, but many acquire the bacterium while in the hospital or other healthcare setting—an example of a healthcare-associated infection. Infectious endospores are shed in the feces and can be transmitted through the environment or by contact with healthcare workers or other patients. An increasing number of CDI cases are community acquired, and many of these patients have an underlying chronic intestinal condition such as Crohn's disease.

Treatment and Prevention

If CDI occurs, any predisposing antibiotics the patient was taking are stopped when practical; if antibiotics are still necessary, options less likely to worsen CDI are substituted. To treat an initial CDI episode, oral fidamoxin or oral vancomycin is typically used. Although these medications kill vegetative bacteria, the endospores survive, so CDI often recurs. If treatment fails or if CDI recurs within 2 months, a different antibiotic course is chosen (either a different antibiotic or the same antibiotic but with a different dosage); a monoclonal antibody (bezlotoxumab) against the *C. difficile* toxin TcdB may be administered along with treatment to prevent additional recurrences. If CDI recurs yet again, a new microbiota-based therapy (Rebyota) may be used following antibiotic treatment to prevent recurrences. This biotherapeutic option, approved by the FDA in 2022, is a rectally administered preparation of fecal microbiota. By adding healthy, normal intestinal microbiota to the disrupted system, a balanced population that can compete with *C. difficile* may be created. The development of this biotherapy—the first of its kind—was spurred by the successes of an experimental treatment called fecal microbiota transplant (FMT), which uses feces from a healthy donor introduced into the patient's colon, usually via a colonoscope (see Focus on a Case 1.1). Success rates for FMT have been encouraging, but the procedure has risks, including the unintentional transmission of pathogens with the transplant material, which is why scientists have been exploring alternative options. Other biotherapies under development for treating CDI include orally administered capsules containing either lyophilized (freeze-dried) fecal microbiota or purified bacterial endospores isolated from feces.

Measures to prevent initial CDI include avoiding inappropriate antibiotic use and preventing *C. difficile* transmission. Methods to accomplish the latter are particularly important in healthcare settings and include handwashing with soap and water (hand sanitizers do not reliably kill endospores), wearing gloves, keeping surfaces disinfected, and isolating CDI patients. Characteristics of *C. difficile* infection are summarized in **table 24.12.**

TABLE 24.12 *C. difficile* Infection (CDI)

Signs and Symptoms	Variable; mild diarrhea to potentially fatal pseudomembranous colitis
Incubation Period	Usually less than a week
Causative Agent	*Clostridioides difficile,* a Gram-positive, rod-shaped, endospore-forming, toxin-producing anaerobe
Pathogenesis	Toxins disrupt host cell actin and cell signaling, causing lethal effects to the intestinal epithelium and inducing inflammation.
Epidemiology	Primarily occurs in patients on antibiotic therapy.
Treatment and Prevention	Treatment: for initial CDI, stop any predisposing antibiotics if possible; treat with an antibiotic that targets *C. difficile.* If CDI recurs, treat using an antibiotic course different from that used previously; a monoclonal antibody may be given as well. If CDI recurs again, a new microbiota-based biotherapy may be used following treatment to prevent an additional recurrence. Fecal microbiota transplants are used experimentally. Prevention: avoid inappropriate antibiotic use and prevent transmission by proper handwashing, wearing gloves, keeping surfaces disinfected, and isolating CDI patients.

MicroAssessment 24.4

Vibrio cholerae, Shigella, various *E. coli* strains, *Salmonella,* and *Campylobacter jejuni* account for most intestinal bacterial infections around the world. Pathogenic mechanisms of these bacteria include attachment, enterotoxin or cytotoxin production, cell invasion, and destruction of microvilli. Dehydration can be treated with oral rehydration therapy. Typhoid and paratyphoid fevers are systemic, life-threatening diseases. *Clostridioides difficile* causes antibiotic-associated diarrheal disease.

10. Explain how *Vibrio cholerae* causes cholera.
11. How does the epidemiology of *Salmonella* gastroenteritis differ from that of typhoid fever?
12. How would you devise a selective medium for cultivating *Vibrio cholerae*?

24.5 ■ Viral Diseases of the Lower Digestive System—Intestinal Tract

Learning Outcome

8. Compare and contrast the diseases caused by rotaviruses and noroviruses.

Viral infections of the lower digestive system are common in all age groups, resulting in millions of cases of gastroenteritis each year in the United States alone. As mentioned in the last section, infections of the lower digestive system have similar epidemiology and prevention, so the important general characteristics are covered in the box Focus on Diarrheal Diseases.

Rotavirus Gastroenteritis

Most cases of viral gastroenteritis in infants and children are caused by rotaviruses. Before the first attenuated vaccines became available in 2006, rotavirus infections in the United States resulted in over 55,000 hospital admissions each year. Worldwide, more than 100,000 children still die from rotavirus gastroenteritis each year.

Signs and Symptoms

After an incubation period of about 24 to 48 hours, rotaviral gastroenteritis begins abruptly with vomiting and slight fever, soon followed by profuse, watery diarrhea. Signs and symptoms often last less than a week, but fatal dehydration can occur if fluids are not replaced.

Causative Agent

Rotaviruses are non-enveloped viruses with a triple-layered capsid and a double-stranded, 11-segment RNA genome (**figure 24.15**). The viruses represent a major subgroup of the family *Reoviridae.*

Pathogenesis

Rotaviruses mainly infect the epithelial cells that line the upper part of the small intestine. Infection causes host cell death, leading to loss of microvilli along with the associated

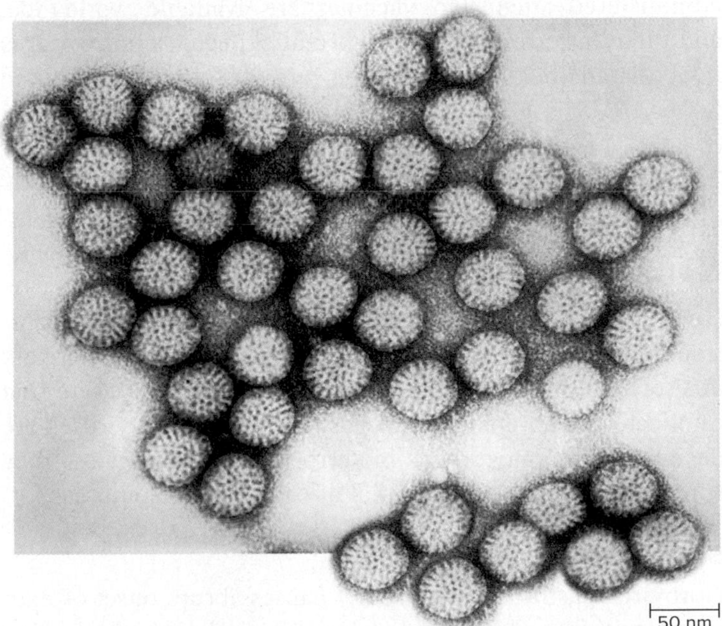

FIGURE 24.15 Rotavirus Color-enhanced electron micrograph shows a distinctive rim of radiating capsomere units that serve as spikes (*rota* means wheel). Dr. Erskine Palmer & Byron Skinner/CDC

What is unusual regarding the rotavirus capsid?

digestive enzymes (brush border enzymes). The damaged lining fails to absorb fluids, resulting in watery diarrhea. In addition, one of the viral proteins appears to function as an enterotoxin, causing fluid secretion in a manner somewhat similar to that of cholera toxin.

Epidemiology

Rotaviruses are spread by fecal-oral transmission via direct person-to-person contact or through contaminated food or water. The disease is highly contagious, reflecting the virus's relatively low infectious dose and stability in the environment. Childhood epidemics generally occur in winter in temperate climates, probably because children are often indoors in groups where the viruses spread easily. In regions where vaccination is not common, most children are infected before age 5. The infection results in some immunity, so the second and third infections cause much milder diarrhea than the first. Rotaviruses also cause about 25% of traveler's diarrhea cases.

Strains of rotavirus that infect wild and domestic animals typically do not infect humans, but reassortment of genetic segments has been shown to occur when a host contains two different viral strains.

Treatment and Prevention

Although no specific treatment of rotavirus infection is available, some infants and young children are hospitalized and given intravenous fluids to prevent dehydration.

Handwashing, disinfectant use, and other sanitary measures help limit the spread of rotaviruses. Various orally administered attenuated vaccines are available worldwide, and these have resulted in a substantial decrease in rotavirus infection and deaths. Two options are available in the United States: RV5 is given as a three-dose series (at 2, 4, and 6 months of age), whereas RV1 is given as a two-dose series (at 2 and 4 months of age).

Norovirus Gastroenteritis

Noroviruses (NoVs) are the most common cause of viral gastroenteritis worldwide. They were originally called "Norwalk viruses," from Norwalk, Ohio, the place where they were first implicated in an epidemic of gastroenteritis. The viruses are designated a Category B bioterrorism agent because they spread easily, with the potential of causing large outbreaks.

Signs and Symptoms

Norovirus gastroenteritis usually causes abrupt onset of nausea, vomiting, and watery diarrhea, after an incubation period of generally 12 to 48 hours. As with other gastrointestinal diseases, dehydration may occur. Vomiting, which is typically most severe in older children and adults, generally resolves within the first day or two. Other symptoms take several days to subside.

Causative Agent

NoVs are non-enveloped, single-stranded RNA viruses (**figure 24.16**). These viruses represent a group of gastroenteritis-producing viruses within the *Caliciviridae.* New strains emerge every few years, sometimes—but not always—causing a significant increase in the number of gastroenteritis cases.

Pathogenesis

NoVs infect the epithelium of the upper small intestine, causing cell death and intestinal damage similar to that of rotaviruses. Immunity to the causative virus is long-lasting, but because of strain diversity and viral evolution, individuals can have repeated infections. The virus has recently been cultivated in the laboratory, but the methods are still not suitable for wide-scale research; once they have been optimized, studying the mechanisms of pathogenesis and immunity should be easier, which will hopefully lead to new methods of treatment and prevention.

Epidemiology

NoV spreads primarily by fecal-oral transmission via direct person-to-person contact or through contaminated food or water; vomit also contains infectious viral particles, so transmission through aerosols and contaminated surfaces can also occur. The disease is highly contagious due to the virus's low infectious dose (less than 100 viral particles) and stability in the environment.

NoV epidemics are relatively common on cruise ships and in college residence halls, and easy spread of these viruses also makes them a concern for healthcare facilities. They also spread easily in families. More than 50% of foodborne disease outbreaks are due to norovirus, so in 2009 the

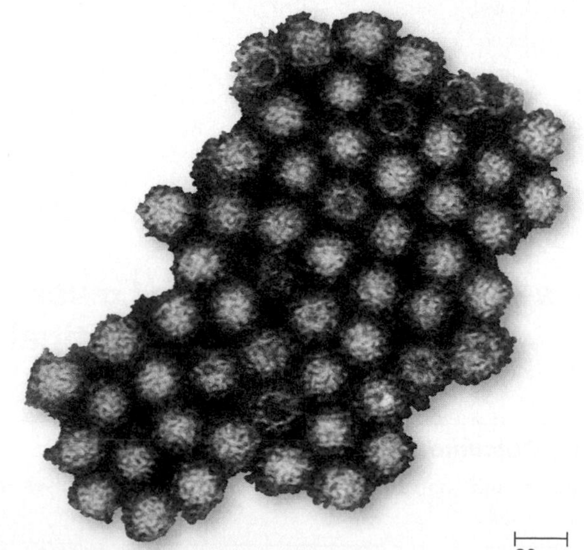

FIGURE 24.16 Norovirus Color-enhanced electron micrograph. Charles D. Humphrey/CDC

? **Which group of NoV-infected individuals is most likely to experience severe vomiting—young children or young adults?**

CDC started CaliciNet, a national surveillance network to trace NoV strains.

Treatment and Prevention

No proven anti-noroviral medications are available, and there is no vaccine. Because hand sanitizers do not reliably inactivate the viral particles, thorough handwashing using soap and water is a vital preventive measure, along with disinfectant use and other sanitary measures. To avoid transmitting norovirus, anyone having symptoms consistent with NoV infection should take appropriate precautions—such as not preparing food for others—until at least 2 days after symptoms subside. FDA guidelines cover restrictions on food workers exposed to or infected with NoV. **Table 24.13** compares noroviruses and rotaviruses.

TABLE 24.13 Rotavirus and Norovirus Gastroenteritis Compared

Causative Agent	Rotavirus	Norovirus
Virus Characteristics	Triple-walled capsid; double-stranded segmented RNA genome	Single-stranded RNA genome
Signs and Symptoms	Vomiting, abdominal cramps, diarrhea, lasting 3 to 8 days	Vomiting, abdominal cramps, diarrhea, lasting 1 to 3 days
Incubation Period	24 to 48 hours	12 to 48 hours
Epidemiology	Fecal-oral transmission involving direct person-to-person or contaminated food or water; highly contagious; milder disease in older children	Fecal-oral transmission involving direct person-to-person contact or through contaminated food or water; highly contagious; epidemics common
Prevention	Vaccine (RV1 or RV5); handwashing, disinfectant use, and other sanitary measures.	No vaccine; handwashing, disinfectant use, and other sanitary measures.

MicroAssessment 24.5

Rotaviruses are the leading cause of viral gastroenteritis in infants and children, and they also are a common cause of traveler's diarrhea. Noroviruses are the most common cause of viral gastroenteritis. They are highly infectious and easily spread.

13. Why are rotavirus infections common in young children but not in adults?
14. Why is handwashing an important means to control the spread of norovirus?
15. Why might it be more difficult to develop an effective vaccine against noroviruses than against rotaviruses?

24.6 ■ Viral Diseases of the Lower Digestive System—Liver

Learning Outcome

9. Compare and contrast hepatitis A, B, and C.

At least five different viruses can cause **hepatitis**—inflammation of the liver—but this discussion will focus on the three types that account for most cases in the United States: hepatitis A virus (HAV), hepatitis B virus (HBV), and hepatitis C virus (HCV). These viruses are unrelated to one another, but the liver damage they cause often results in similar signs and symptoms—the most noticeable being jaundice (yellowing of the skin and the whites of the eyes). Patients with any form of hepatitis, regardless of the cause, should avoid alcohol, acetaminophen, and other chemicals known to damage the liver.

Viral hepatitis can be classified as acute or chronic—terms that refer to the duration of infection without regard to the intensity of symptoms; acute viral hepatitis includes the first 6 months of infection, whereas chronic viral hepatitis refers to anything later. HAV infections resolve within 6 months and are therefore not chronic, whereas HBV and HCV infections can become chronic. For all the virus types and infection stages, at least some people are asymptomatic.

Hepatitis A

Hepatitis A is an acute disease spread by fecal-oral transmission. The reported incidence in the United States decreased significantly after an effective vaccine was introduced in 1995, but beginning in 2015 the incidence began increasing, peaking in 2019 (see figure 24.19). That increase was mostly associated with outbreaks of person-to-person transmission involving people living in unstable conditions associated with homelessness and drug abuse. Although the reported incidence then decreased markedly in 2020—perhaps due to decreased transmission and diagnosis during the height of the COVID-19 pandemic—it was still well above the 2015 level.

Signs and Symptoms

Signs and symptoms of hepatitis A begin after an incubation period of about 1 month (range of 15 to 50 days). They usually start abruptly with fever, fatigue, nausea, and vomiting; clay-colored feces and dark urine may become apparent shortly thereafter, with jaundice often developing a few days later. Patients typically recover within a few weeks, but recovery sometime takes up to several months. Older children and adults generally experience symptoms, but infections in most young children (less than 6 years old) and many older children (ages 6 to 14) are asymptomatic. About one in five infected adults develops disease severe enough to require hospitalization.

Causative Agent

Hepatitis A virus (HAV) is a non-enveloped, single-stranded RNA virus of the family *Picornaviridae*.

Pathogenesis

Following ingestion, HAV enters the bloodstream by an unknown mechanism and is then carried to the liver, the main site of viral replication and the only tissue known to be damaged by the infection. The virus is released into the bile and eliminated with the feces.

Epidemiology

HAV spreads by fecal-oral transmission involving direct person-to-person contact or through contaminated food or water. Groups at high risk for person-to-person transmission include people living in unstable conditions, children in day-care centers, nursing home residents, international travelers, and people having sexual contact with an infected person. Many foodborne outbreaks have been traced to restaurants where infected food handlers failed to wash their hands. Raw shellfish are also a frequent source of infection because they concentrate the virus from fecally polluted seawater. Because of the long incubation period, HAV can spread widely through a population before being detected (see Focus on a Case 19.1). Infected infants and children are often asymptomatic and can shed the virus in their feces for several months.

Treatment and Prevention

No antiviral treatment is available for hepatitis A, but supportive care helps relieve symptoms.

To prevent hepatitis A, an inactivated vaccine is given as a two-dose series. The vaccine is recommended for all children between 1 and 2 years of age, for children and adolescents ages 2 to 18 who have not received the vaccine, and for adults at risk of exposure or the development severe disease. Although unvaccinated adults are still at risk for the disease, the vaccine has resulted in a substantial decrease in the number of reported cases. If an unvaccinated person is exposed to HAV, then timely post-exposure prophylaxis (PEP) using the vaccine—sometimes along with immune globulin—prevents the disease. Recall that vaccination induces the long-term protection of active immunity, whereas immunoglobulin provides the immediate but short-term protection of passive immunity. Because the virus is transmitted through the fecal-oral route, handwashing is effective in preventing spread. FDA guidelines cover restrictions on food workers who are exposed to or develop hepatitis A.

Hepatitis B

Hepatitis B is transmitted through contact with body fluids and can be acute or chronic. Infections in infants and children are a particular concern because individuals in this age group are much more likely to develop a chronic infection even though the acute stage is most often asymptomatic. An estimated 860,000 people in the United States have a chronic infection, and about two-thirds of these people are unaware of their condition.

Signs and Symptoms

Signs and symptoms of acute hepatitis B are similar to those of other forms of acute hepatitis, ranging from mild to severe. The incubation period varies considerably—from about 2 to 5 months—depending on the viral dose received. The acute disease is rarely fatal, and the virus is usually cleared within weeks to months of initial symptoms; some people develop chronic HBV infection, however, and about one in five of these people eventually develop **cirrhosis** (scarring of the liver), liver failure, liver cancer, or other chronic liver disease.

Causative Agent

Hepatitis B virus (HBV) is an enveloped virus that has a mostly double-stranded DNA genome (a portion is single-stranded) and is a member of the family *Hepadnaviridae*. Unlike most enveloped viruses, HBV is remarkably resistant to environmental conditions—the virions can still be infectious after a week outside the body.

Three HBV components produced during viral replication are notable because they serve as useful markers of infection (**figure 24.17**):

- **Hepatitis B surface antigen (HBsAg).** This envelope protein begins appearing in the bloodstream days or weeks after infection, often long before signs of liver

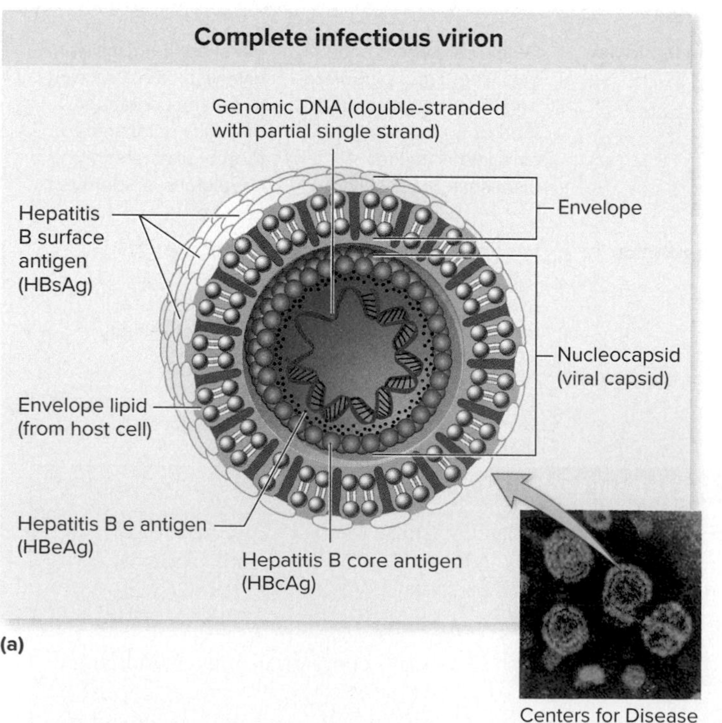

FIGURE 24.17 Hepatitis B Virus Diagrammatic representation of a viral particle showing important components Dr. Erskine Palmer/CDC

? **What does circulating HBeAg indicate about individuals with chronic HBV infection?**

damage are evident. Antibodies to HBsAg confer immunity to the virus.

- **Hepatitis B core antigen (HBcAg).** IgM antibodies against this antigen indicate active viral replication.
- **Hepatitis B e antigen (HBeAg).** High levels of this soluble component of the viral core correlate with increased risk of liver damage and increased risk of transmitting the disease.

Pathogenesis

Following its entry into the body, HBV is carried to the liver by the bloodstream. Surface antigen (HBsAg) allows the virus to attach to and enter host cells.

The next steps of HBV replication are complex, but they help explain why certain antiviral medications can reduce disease symptoms. After uncoating occurs, the viral genome is transported to the host cell nucleus and the single-stranded gap is then filled in by host cell machinery (**figure 24.18**). At this point, the genome is a double-stranded covalently closed circular DNA molecule (cccDNA). A host cell RNA polymerase uses the cccDNA as a template to produce mRNA molecules that are then translated in the cytoplasm to make the various viral proteins. The significant and unusual aspect of HBV is that an RNA molecule is produced to serve as a template for DNA synthesis; in other words, HBV is a reverse-transcribing virus. The RNA molecule—referred to as pregenomic RNA (pgRNA)—is packaged into the nucleocapsid along with an HBV-encoded **reverse transcriptase.** This enzyme uses the pgRNA as a template to make one strand of DNA; it then uses that DNA molecule as a template to synthesize the complementary strand. The process is not finished inside the virion, however, leaving the HBV genome only partly double-stranded.

The long-term liver damage from HBV infection is likely due to the cell-mediated immune response, as effector cytotoxic T cells attack infected liver cells. The role of the virus in liver cancer development is not well understood, but in most cases of the cancer, the viral DNA has integrated into the host cell genome.

Epidemiology

HBV can be transmitted in body fluids such as saliva, blood, blood products, and semen. Any activity that mixes these fluids from two persons is considered a risk factor, including sharing needles, toothbrushes, razors, or towels, and using unsterile tattooing or ear-piercing instruments. Unprotected sex is a particularly risky behavior, with nearly half of new hepatitis B cases in the United States acquired through sexual intercourse. The virus can also be transmitted vertically (mother to baby).

When HBV infection becomes chronic, the virus continues replicating, circulating in the blood for many years, even if the patient is asymptomatic. These people are extremely important in the spread of hepatitis B because they are often unaware of their infection. The failure to clear the virus is age related. More than 90% of infants who contract the virus from an infected mother at or shortly after birth (vertical

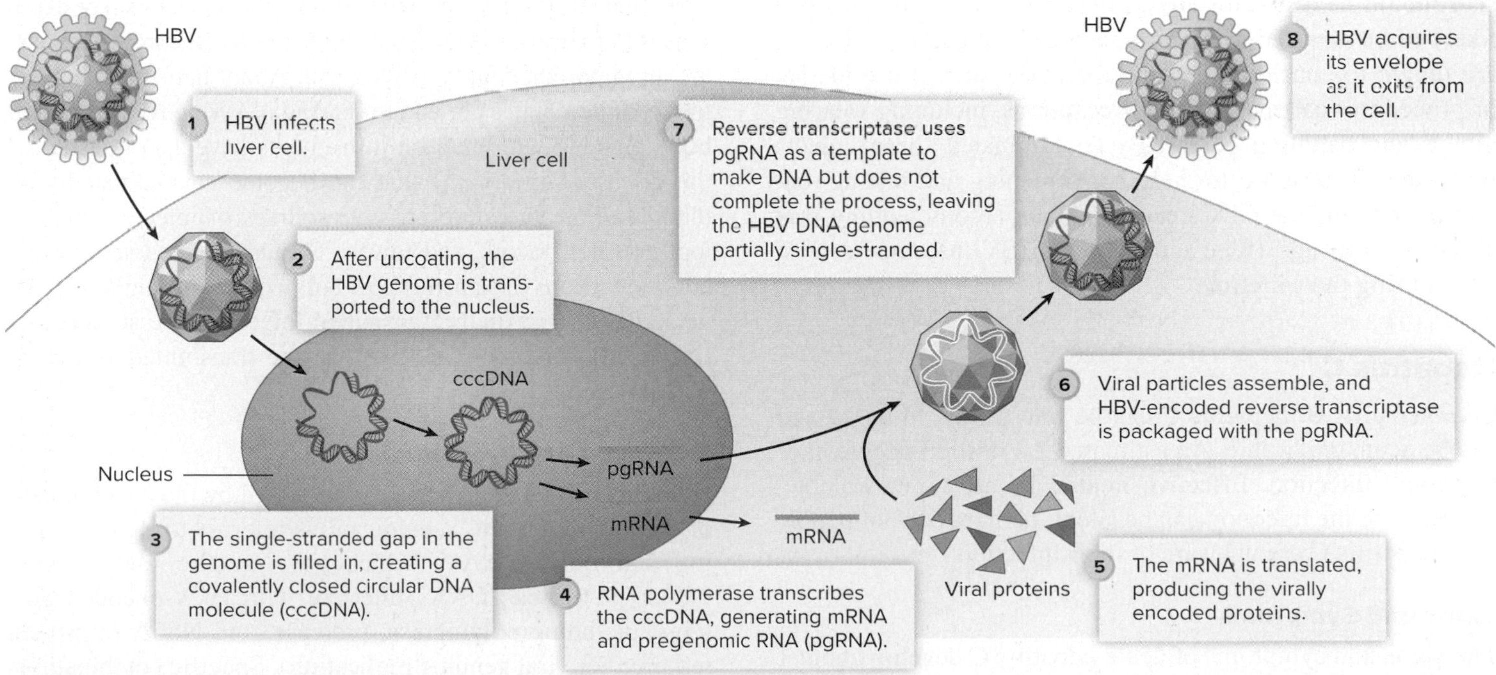

FIGURE 24.18 Replication of Hepatitis B Virus

? With respect to the replication strategy, what unusual feature does HBV have in common with HIV?

transmission) develop a chronic infection. In contrast, 25% to 50% of children who acquire the virus between the ages of 1 and 5 will develop a chronic infection, as will 6% to 10% of people infected as older children or adults.

Treatment and Prevention

No curative antiviral treatment currently exists for hepatitis B, so most acute cases are managed with only supportive care. Individuals with chronic infections show remarkable improvement when given certain reverse transcriptase inhibitors such as tenofovir; injections of genetically engineered interferon may also be used.

A vaccine to prevent hepatitis B was licensed in the early 1980s, but that was replaced with a more readily available recombinant subunit vaccine in 1986, leading to a significant decrease in the number of reported cases (see figure 24.19). Today, several versions of that vaccine are available, usually given as a three-dose series. For most babies, the first dose is recommended within 24 hours of birth. Pregnant women are screened for HBV infection and, if positive, then within 12 hours of birth the baby is given (1) the first vaccine dose and (2) hepatitis B immune globulin (HBIG, a hyperimmune globulin that provides passive immunity); infants born to infected mothers are also tested for HBV. For people not already vaccinated against hepatitis B, the vaccine is recommended for anyone through age 59 and is available to anyone 60 or older. Post-exposure prophylaxis (PEP) using the hepatitis B vaccine—sometimes along with HBIG—can also prevent disease. Additional measures to prevent hepatitis B include educating high-risk groups about protective measures, including proper use of condoms; anyone at increased risk should be tested for HBV infection periodically. Persons likely to be exposed to blood—such as healthcare workers—are taught to consider all blood infectious and to use the set of procedures called Standard Precautions, including wearing gloves and handling potentially contaminated sharp objects with care. In addition to the other screening and testing recommendations, the CDC recently began recommending that all adults over age 18 be screened for HBV infection at least once during their lifetime.

Hepatitis C

Like hepatitis B, hepatitis C can be transmitted in blood and can be acute or chronic. An estimated 2.4 million people have a chronic infection. Effective treatments are now available, but because the infection can be silent, perhaps 50% of people with hepatitis C are unaware of their infection.

Signs and Symptoms

The signs and symptoms of acute hepatitis C develop about 1 to 2 months after infection (with a range of 2 to 26 weeks) and are similar to those of acute hepatitis A and B but generally milder. Many HCV-infected individuals have no symptoms during the acute stage of the disease, however, but most infections become chronic; without treatment, 10% to 20% of patients with chronic HCV infection develop cirrhosis or liver cancer.

Causative Agent

Hepatitis C virus (HCV) is an enveloped, single-stranded RNA virus of the family *Flaviviridae*. Considerable genetic variability of the virus is seen, with types and subtypes differing in pathogenicity as well as in response to treatment.

Pathogenesis

HCV infection generally occurs from exposure to contaminated blood. After entering liver cells, the viral genome is translated as a single polyprotein that is then cleaved by an HCV protease to make multiple proteins. The infection triggers inflammatory and immune responses, and the disease process then starts and stops; at times liver function seems to return to normal and then, weeks or months later, signs of marked inflammation appear.

In 2023, researchers reported that HCV uses a previously undescribed mechanism to prevent its RNA genome from being recognized by pattern recognition receptors within infected cells: The virus uses the cellular metabolite flavin adenine dinucleotide (FAD) to create a 5′ cap on the RNA. This occurs because the viral replicase (an RNA-dependent RNA polymerase) routinely uses FAD as the starting nucleotide when initiating RNA synthesis.

Epidemiology

Although HCV is transmitted in blood, the mechanism of exposure is not always obvious; as with hepatitis B, chronic carriers are an important source of infection. About half the infections in the United States are due to sharing of syringes. Tattoos and body piercing with unclean instruments have also transmitted the disease. Other items that can become contaminated with blood and are therefore possible sources of infection include toothbrushes, razors, and towels. Sexual intercourse is a less common means of transmission, but people with multiple partners and other sexually transmitted infections are at increased risk for the disease. The virus can be transmitted vertically, with an overall risk of about 5%.

Treatment and Prevention

Hepatitis C treatment was revolutionized by the development of a group of anti-HCV medications referred to as direct-acting antivirals (DAAs), which have a cure rate of over 90%. The various DAAs interfere with HCV-encoded proteins, including polymerase, protease, and NS5A (a protein required for viral genome replication). Specific combinations of DAAs are taken orally for treatment, some of which are available in fixed-dose tablets. For example, one formulation includes sofosbuvir (a nucleotide analog) along with velpatasvir (an NS5A inhibitor); another includes elbasvir (an NS5A inhibitor) and grazoprevir (a protease inhibitor). Different treatment combinations are related to differences in HCV genotypes.

No vaccine is available for preventing hepatitis C. However, physicians often recommend vaccination against hepatitis A and B to help prevent a combination of infections that might severely damage the liver.

The trends in reported incidence of acute hepatitis A, B, and C are compared in **figure 24.19**; remember that many infections are unrecognized, however, so the estimated number of infections is significantly higher. The main features of hepatitis are given in **table 24.14.**

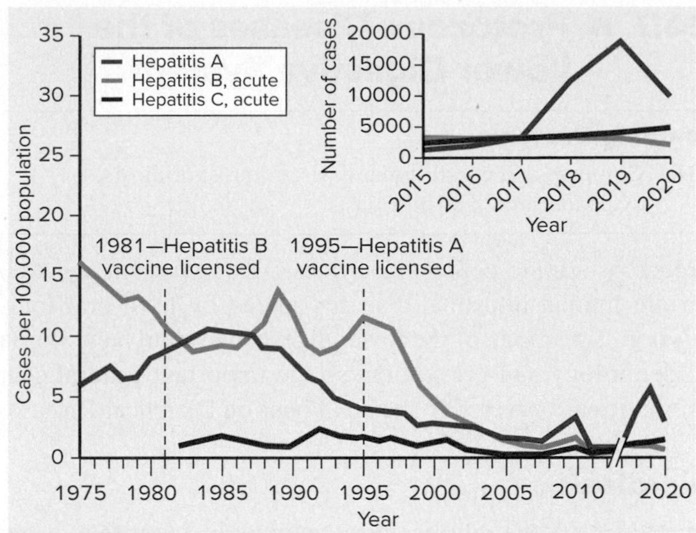

FIGURE 24.19 Reported Cases of Acute Viral Hepatitis in the United States, 1975–2020 The actual number of cases is significantly higher, a result of under-diagnosis and under-reporting. The hepatitis B vaccine licensed in 1981 was replaced in 1986 by a recombinant subunit version.

? **What is the most noticeable sign of hepatitis?**

MicroAssessment 24.6

Hepatitis viruses are a diverse, unrelated group that cause liver inflammation. Hepatitis A virus is transmitted by the fecal-oral route; infection does not become chronic. Hepatitis B virus is transmitted by exposure to body fluids; some infections, particularly those in infants and children, become chronic. Hepatitis C virus is transmitted by blood and occasionally by sexual intercourse; many infections become chronic. Chronic hepatitis can result in cirrhosis and liver cancer.

16. At what stage in the replication of hepatitis B virus does a reverse transcriptase inhibitor act?

17. What behaviors might favor transmission of hepatitis C?

18. Why would the first dose of the hepatitis B vaccine be given to newborns, rather than waiting until age 1 to 2 months so that it could be given with other common vaccines?

TABLE 24.14 Viral Hepatitis

	Hepatitis A	Hepatitis B	Hepatitis C	Hepatitis D	Hepatitis E
Signs and Symptoms	Fever, fatigue, nausea, vomiting, clay-colored feces, dark urine, and jaundice. No chronic stage	Acute stage: similar to hepatitis A but often more severe. Chronic stage: can lead to cirrhosis and cancer	Acute stage: usually few or no symptoms. Chronic stage: can lead to cirrhosis and cancer	Worsening hepatitis B symptoms (infection occurs only in patients infected with HBV)	Similar to hepatitis A, except severe in pregnant women. No chronic stage
Incubation Period	About 1 month	About 2 to 5 months	About 1 to 2 months	About 1 month	About 1 month
Causative Agent	Hepatitis A virus (HAV); non-enveloped, single-stranded RNA picornavirus	Hepatitis B virus (HBV); enveloped, double-stranded DNA hepadnavirus; reverse transcribing	Hepatitis C virus (HCV); enveloped, single-stranded RNA flavivirus	Hepatitis D virus (HDV); defective single-stranded RNA virus	Hepatitis E virus (HEV); non-enveloped, single-stranded RNA calicivirus
Epidemiology	Fecal-oral transmission	Transmitted in body fluids (blood, semen); chronic carriers are an important infection source; infection in infants and children most likely to become chronic, but adults also at risk.	Transmitted in blood, possibly semen; chronic carriers are an important infection source.	Transmitted in blood, semen.	Fecal-oral transmission
Treatment and prevention	Treatment: no specific treatment available. Prevention: vaccine; post-exposure prophylaxis; handwashing	Treatment: no curative treatment available, but chronic infections improve with an antiviral. Prevention: vaccine; post-exposure prophylaxis; avoid exposure to contaminated body fluids.	Treatment: highly effective treatments are newly available. Prevention: no vaccine; avoid exposure to contaminated blood.	Treatment: no specific treatment available. Prevention: protect against HepB infection.	Treatment: no specific treatment available. Prevention: no FDA-approved vaccine (licensed only in China).

24.7 ■ Protozoan Diseases of the Lower Digestive System

Learning Outcome

10. Compare and contrast giardiasis, cryptosporidiosis, cyclosporiasis, and amebiasis.

Protozoa—single-celled eukaryotes—are important causes of certain human intestinal diseases spread by fecal-oral transmission. Infections of the lower digestive system have similar epidemiology and prevention, so the important general characteristics are covered in the box Focus on Diarrheal Diseases.

Giardiasis

Giardiasis is one of the most commonly identified waterborne illnesses in the United States. It can be contracted from mountain streams, chlorinated but unfiltered city water, and person-to-person contact. The disease occurs worldwide and is responsible for many cases of traveler's diarrhea, but many people infected with the causative protozoan are asymptomatic.

Signs and Symptoms

The incubation period of giardiasis is generally 6 to 20 days, with symptoms ranging from mild (indigestion, "gas," and nausea) to severe (vomiting, explosive diarrhea, abdominal cramps, fatigue, and weight loss). Symptoms usually resolve in 1 to 3 weeks, but some cases become chronic.

Causative Agent

Giardia lamblia (also known as *G. duodenalis* and *G. intestinalis*) can exist in two forms: a growing, feeding **trophozoite** and a dormant **cyst.** The flagellated trophozoite is shaped like a pear cut lengthwise and has a distinctive appearance due to an adhesive disc on its undersurface and two side-by-side nuclei that resemble eyes (**figure 24.20**). Trophozoites multiply in the intestinal tract and may be present in fresh diarrhea, but they do not survive in the environment. The cysts, on the other hand, have a thick chitin cell wall that withstands harsh environmental conditions (see Focus on a Case 12.1). *G. lamblia* lacks mitochondria but contains structures called mitosomes that appear to be remnants of mitochondria, although their function is not clearly defined.

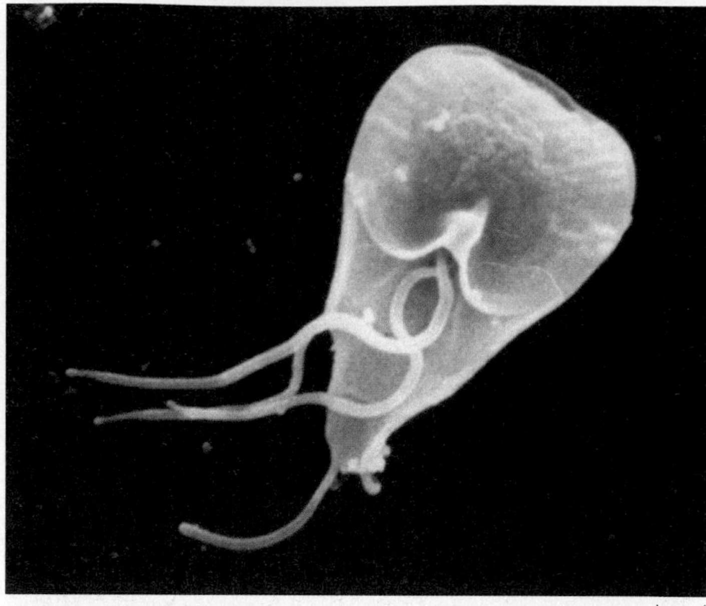

FIGURE 24.20 ***Giardia lamblia*** Scanning electron micrograph (SEM). Janice Haney Carr/CDC

? Does this image show a cyst or a trophozoite? Explain.

Pathogenesis

The cysts of *G. lamblia* are infectious because, unlike the trophozoites, they are resistant to acid so they can survive passage through the stomach. Once in the small intestine, trophozoites emerge from the cysts. Some of these attach to the epithelium (the brush border) by their adhesive discs; others use their flagella to move freely in the intestinal mucus. Some may even migrate up the bile duct to the gallbladder and cause cramping or jaundice.

The trophozoites attached to the brush border interfere with the intestine's ability to absorb nutrients and produce digestive enzymes. The result is fatty stool, intestinal gas (from bacterial fermentation of unabsorbed nutrients), and malnutrition. Some intestinal impairment may be the result of the immune response to the parasite.

When trophozoites detach from the epithelium, the intestinal contents carry them toward the large intestine. If the transit time is long enough, they develop into cysts. Thus, a person who has formed stools is more likely to pass cysts in the feces, whereas a person with diarrhea is apt to eliminate trophozoites (**figure 24.21**).

Epidemiology

Giardia lamblia is transmitted by the fecal-oral route and spreads easily because only about 10 cysts are needed to establish infection. Asymptomatically infected people can function as carriers, unknowingly excreting infectious cysts with their feces for months or longer. The cysts can remain viable in cold water for more than 2 months; water contaminated with human feces is a common source of infection, but feces from animals such as beavers, raccoons, dogs, and cats can also be implicated. Hikers who drink from streams, even in remote areas thought to contain safe water, are at risk of contracting giardiasis.

Although waterborne outbreaks are the most common, person-to-person contact can also transmit giardiasis. This is especially likely in day-care centers where workers' hands become contaminated while changing diapers. Sexual

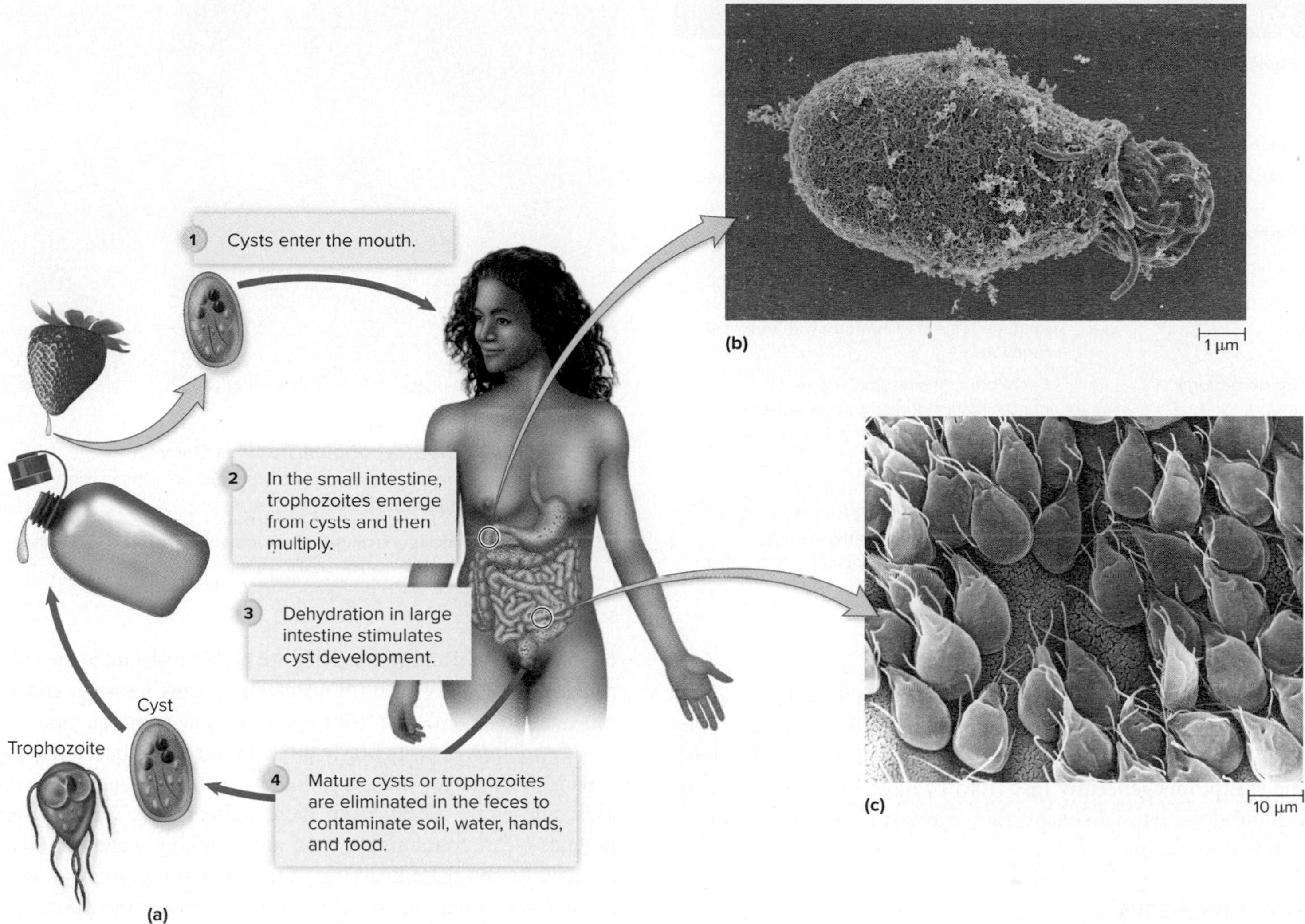

FIGURE 24.21 ***Giardia lamblia*** **(a)** Life cycle. **(b)** Trophozoite emerging from a cyst. **(c)** Trophozoites attached to the intestinal wall. (b, c) Dr. Stan Erlandsen/CDC

? **Why would an infected person with diarrhea be less likely to transmit giardiasis than one who has formed stools?**

practices that lead to oral-anal contact can also transmit the disease. Transmission by fecally contaminated food has also been reported. Good personal hygiene, especially handwashing, decreases the chance of spreading the infection.

MicroByte

The trophozoite form of a *G. lamblia* was first reported by Antonie van Leeuwenhoek when he examined his own diarrhea using a microscope he made.

Treatment and Prevention

Treatment of giardiasis is sometimes not necessary if symptoms are mild and transmission of the parasite to others is unlikely. When treatment is appropriate, however, several effective medications are available, including tinidazole, metronidazole (Flagyl), and nitazoxanide.

Measures to prevent giardiasis include avoiding fecal contamination of food and water supplies, along with handwashing, cooking food thoroughly, and pasteurizing beverages. The level of chlorine used to disinfect municipal water supplies does not destroy *Giardia* cysts, so the water must be treated using ultraviolet radiation, ozone, or certain filtration methods. For hikers, the most effective ways to eliminate the parasite in drinking water are to boil the water for 1 minute or use a portable filter with a pore size of 1 μm or smaller. Other methods, including adding commercial water-purifying tablets, tincture of iodine, or household bleach are much less reliable. **Table 24.15** describes the main features of giardiasis.

Cryptosporidiosis ("Crypto")

Cryptosporidiosis, commonly called "crypto," is one of the most common waterborne diseases in the United States. It was first identified as a threat to humans when the AIDS epidemic started in the early 1980s, but the disease is a health risk not only to those with an immunodeficiency but also to the general public. A 1993 waterborne outbreak in Milwaukee affected more than 400,000 people!

TABLE 24.15 Giardiasis

Signs and Symptoms	Mild: indigestion, intestinal gas, nausea. Severe: vomiting, diarrhea, abdominal cramps, weight loss.
Incubation Period	6 to 20 days
Causative Agent	*Giardia lamblia*, a flagellated pear-shaped protozoan with two nuclei
Pathogenesis	Ingested cysts survive stomach passage; trophozoites emerge from cysts in the small intestine, where some attach to the epithelium; adherent trophozoites and the immune response interfere with mucosal function.
Epidemiology	Fecal-oral transmission, usually via contaminated drinking water but also direct person-to-person contact; low infectious dose.
Treatment and Prevention	Treatment: several antimicrobial medication options. Prevention: avoiding fecal contamination; handwashing, adequate sanitation, and safe, clean water supplies

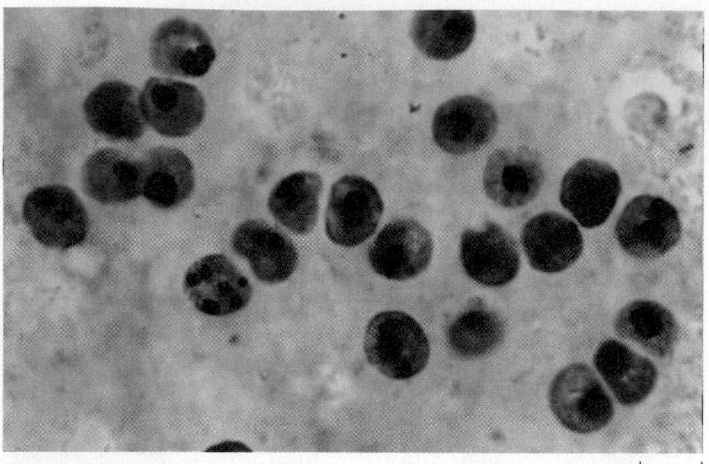

FIGURE 24.22 *Cryptosporidium hominis* Oocysts Feces contain oocysts. Person-to-person and waterborne transmission are common. Michael Abbey/Science Source

? **Based on the symptoms of cryptosporidiosis, cells of which part of the intestine do you think are affected—large or small? Explain.**

Signs and Symptoms

Signs and symptoms of cryptosporidiosis include fever, appetite loss, nausea, abdominal cramps, and profuse watery diarrhea, beginning after an incubation period of 4 to 12 days. The symptoms generally last 10 to 14 days, but in people with immunodeficiency diseases, they can last for months and may be life-threatening.

Causative Agent

Cryptosporidiosis is caused by *Cryptosporidium hominis* (formerly *parvum*), an apicomplexan. The oocyst stage is an environmentally resistant acid-fast sphere that contains four sporozoites (**figure 24.22**).

Pathogenesis

In the small intestine, sporozoites are released from ingested oocysts. The sporozoites then invade and grow in the epithelial cells of the small intestine, altering the epithelium and intestinal villi and causing inflammation. Water and electrolyte secretion increase, and nutrient absorption decreases. Cell-mediated immunity is important in controlling the infection.

Epidemiology

Oocysts of *C. hominis* passed in feces are immediately infectious. The infectious dose is as few as 10 oocysts, so direct person-to-person spread readily occurs with poor sanitation. The organism is responsible for many cases of traveler's diarrhea.

Infected people often have prolonged diarrhea and can continue to eliminate infectious oocysts for 2 weeks or more after the diarrhea ceases. These oocysts can survive for up to 6 months in food and water. They are highly resistant to the chlorine levels typically used for disinfecting drinking water and are too small to be removed from water by some filtration methods.

C. hominis is particularly difficult to control because it has a wide host range, infecting domestic animals such as dogs, pigs, and cattle. Feces from these animals—as well as from humans—can contaminate food and drinking water. Epidemics have arisen from drinking water, swimming pools, a water slide, a zoo fountain, day-care centers, and various foods and beverages including unpasteurized apple juice.

Treatment and Prevention

Nitazoxanide may be prescribed for people with healthy immune systems, although most otherwise healthy people will recover without treatment; the medication is ineffective in immunocompromised patients.

Measures for preventing cryptosporidiosis are similar to those described for giardiasis—avoiding fecal contamination of food and water supplies, along with handwashing, cooking food thoroughly, pasteurizing beverages, and treating drinking water. The main features of cryptosporidiosis are shown in **table 24.16.**

Cyclosporiasis

Cyclosporiasis first came to medical attention in the late 1980s, with widely scattered epidemics of severe diarrhea. More recently, it has made the news because of several outbreaks across the United States.

Signs and Symptoms

Cyclosporiasis begins after an incubation period of about 1 week, with fatigue, loss of appetite, slight fever, vomiting, and watery

TABLE 24.16 Cryptosporidiosis

Signs and Symptoms	Fever, loss of appetite, nausea, abdominal cramps, watery diarrhea
Incubation Period	4 to 12 days
Causative Agent	*Cryptosporidium hominis*, an apicomplexan
Pathogenesis	Ingested oocysts release sporozoites in the intestine. The sporozoites invade epithelial cells, causing an inflammatory response. Secretion of water and electrolytes increases; absorption of nutrients decreases.
Epidemiology	Oocysts of *C. hominis* are immediately infectious and have a low infectious dose. Fecal-oral transmission involvings direct person-to-person contact or contaminated food and water. The oocysts resist chlorination and pass through many municipal water filtration systems. People can excrete oocysts for weeks after diarrhea ends. Wide host range.
Treatment and Prevention	Treatment: antimicrobial medication. Prevention: avoiding fecal contamination; handwashing, adequate sanitation, and safe, clean water supplies; pasteurizing beverages.

diarrhea, followed by weight loss. The diarrhea usually subsides in 3 to 4 days, but relapses occur for up to 4 weeks.

Causative Agent

Cyclospora cayetanensis is an apicomplexan, and its oocysts are similar to those of *C. hominis* but larger. Unlike *C. hominis* oocysts, however, *C. cayetanensis* oocysts are not yet infectious when passed in the feces; in favorable conditions outside the body, they mature to become infectious. Mature oocysts contain sporozoites within structures called sporocysts.

Pathogenesis

Little is known about the pathogenesis of *Cyclospora cayetanensis* because there is no good animal model for the disease. No hosts other than humans are known.

Epidemiology

Because the oocysts of *C. cayetanensis* are not infectious when passed in feces, person-to-person spread does not occur. In temperate regions, most infections happen in spring and summer months and among travelers to tropical areas. This is likely because warm, moist conditions favor maturation of the oocysts. In the United States, outbreaks of cyclosporiasis have been associated with raspberries or other fresh produce imported from tropical regions; a 2023 outbreak appeared to be linked to imported broccoli.

Treatment and Prevention

Co-trimoxazole (trimethoprim plus sulfamethoxazole) effectively treats most cases of cyclosporiasis.

TABLE 24.17 Cyclosporiasis

Signs and Symptoms	Fatigue, loss of appetite, vomiting, watery diarrhea, and weight loss. Symptoms improve in 3 to 4 days, but relapses can occur for up to a month.
Incubation Period	Usually about 1 week
Causative Agent	*Cyclospora cayetanensis*, an apicomplexan
Pathogenesis	Little is known.
Epidemiology	Fecal-oral transmission, usually involving contaminated food. The oocysts of *C. cayetanensis* are not infectious when passed in the feces, so person-to-person spread does not occur. Travelers to tropical countries are at risk of infection. Fresh produce has been implicated in most North American outbreaks.
Treatment and Prevention	Treatment: antimicrobial medication. Prevention: avoiding fecal contamination; handwashing, adequate sanitation, and safe, clean water supplies; pasteurizing beverages.

Measures for preventing cyclosporiasis are similar to those described for giardiasis—avoiding fecal contamination of food and water supplies, along with handwashing, cooking food thoroughly, pasteurizing beverages, and treating drinking water. The main features of cyclosporiasis are presented in **table 24.17.**

Amebiasis

Amebiasis is caused by *Entamoeba histolytica,* a parasite believed to infect up to 50 million people worldwide, mostly in resource-limited countries. Although usually asymptomatic or mild, life-threatening disease can occur in areas where virulent strains spread easily in crowded, unsanitary living conditions. An estimated 55,000 or more people die from these strains every year.

Signs and Symptoms

Patients with *E. histolytica* infections are commonly asymptomatic. When symptoms do occur, the incubation period is usually 1 to 4 weeks. Mild disease can be self-limiting, but some patients suffer from chronic, mild diarrhea, lasting months or years. In the most severe cases, however, sometimes fatal dysentery occurs. The causative agent may also spread to the liver, causing liver abscesses that result in fever, nausea, vomiting, abdominal pain, and liver enlargement.

Causative Agent

E. histolytica is an amoeba that can sometimes lyse tissues (hence the species name). Like *G. lamblia*, its life cycle involves an active trophozoite stage and an infectious cyst; trophozoites may be present in fresh diarrhea, but they do not survive in the environment. Cysts—characterized by four nuclei—can survive weeks in the environment (**figure 24.23**).

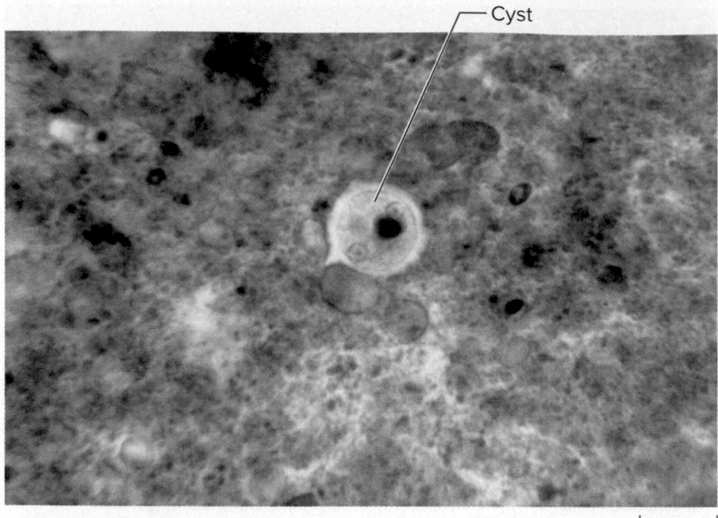

FIGURE 24.23 ***Entamoeba histolytica* Cyst** Not all of the nuclei can be seen in this image. CDC

? Why would the organism be transmitted in the cyst form and not as a trophozoite?

Pathogenesis

Ingested *E. histolytica* cysts survive passage through the stomach. Once in the small intestine, trophozoites emerge from the cysts; these then move to the large intestine, where they feed on mucus and intestinal bacteria to multiply. Virulent strains can attach to intestinal epithelial cells and kill them on contact, an activity referred to as contact-dependent cytotoxicity. Certain strains have also been shown to take "bites" out of live host cells, a phenomenon called trogocytosis (trogo means "gnaw"); the trophozoites then display on their surface certain proteins from the membranes of those host cells, an action that appears to prevent recognition by the complement system. The organisms may also invade the intestinal lining, causing ulcerations and bloody diarrhea, a condition called **amebic dysentery.** Sometimes the invading amoebae penetrate deeper, entering blood vessels that carry them to the liver or other organs. This extraintestinal spread results in tissue damage as the organisms multiply, leading to the development of amebic abscesses in those organs. Meanwhile, some of the *E. histolytica* trophozoites multiplying in the intestine may slowly develop into cysts. As with giardiasis, a person who has formed stools is more likely to eliminate cysts in the feces, whereas a person with diarrhea is apt to eliminate trophozoites.

Epidemiology

E. histolytica spreads by fecal-oral transmission, involving direct person-to-person contact or contaminated food or water. Because the cyst form is acid-tolerant and can easily survive passage through the stomach, it has a low infectious dose. The disease is more common in tropical areas where sanitation is poor. In the United States, cases occur mainly in poverty-stricken areas and among migrant farm workers and men who have sex with men. Humans are the only important reservoir.

TABLE 24.18 Amebiasis

Signs and Symptoms	Diarrhea, abdominal pain, blood in feces; occasionally liver abscesses
Incubation Period	Usually 1 to 4 weeks
Causative Agent	*Entamoeba histolytica*
Pathogenesis	Trophozoites that emerge from ingested cysts feed on mucus and bacteria in the large intestine; cytotoxic enzyme kills intestinal cells. The trophozoites may penetrate the intestinal wall and are sometimes carried to the liver and other organs, resulting in abscesses.
Epidemiology	Fecal-oral transmission involving direct person-to-person contact or contaminated food or water; associated with poverty, migrant workers, and men who have sex with men.
Treatment and Prevention	Treatment: antimicrobial medication. Prevention: handwashing, adequate sanitation, and safe, clean water supplies

Treatment and Prevention

The treatment choice for *E. histolytica* infections depends on the severity of the infection. For most infections, an orally administered medication that is poorly absorbed is used (an example is parmomycin); the action of this type of agent, called a luminal agent, is limited to the gut lumen. More severe infections, particularly those that are invasive, are typically treated with metronidazole, a medication that kills trophozoites; this is followed up by a luminal medication, often parmomycin, to ensure destruction of organisms in the intestine. Tinidazole is sometimes used as an alternative to metronidazole, again followed by paromomycin. Preventing amebiasis depends on sanitary measures and avoiding fecal contamination of foods and drinking water. **Table 24.18** gives the main features of amebiasis.

MicroAssessment 24.7

Giardiasis and cryptosporidiosis are common waterborne diseases that can also be transmitted person to person; cyclosporiasis is a similar disease but not transmissible person to person. Amebiasis is caused by an amoeba that ulcerates the large intestinal epithelium, resulting in dysentery.

19. What causes the excessive intestinal gas that characterizes giardiasis?

20. Why is cryptosporidiosis difficult to control?

21. How can the fact that *Cryptosporidium hominis* is acid-fast be used in diagnosis?

The key features of the diseases covered in this chapter are highlighted in the **Diseases in Review 24.1** table.

Diseases in Review 24.1

Digestive System Diseases

Disease	Causative Agent	Comment	Summary Table
BACTERIAL INFECTIONS OF THE UPPER DIGESTIVE SYSTEM			
Dental caries (tooth decay)	*Streptococcus mutans*	Cariogenic organisms adhering to teeth produce tooth-damaging acids; progression correlates with sugar consumption and lack of preventive dental care.	Table 24.1
Plaque-induced periodontal diseases	Certain groups of Gram-negative anaerobes	An inflammatory response to plaque and tartar accumulation in the gingival crevice results in gingivitis; an inflammatory response to the accumulation in periodontal pockets leads to periodontitis, which damages the structures that support the teeth, leading to tooth loss.	Table 24.1
Necrotizing periodontal diseases	*Treponema* species with other anaerobes	Characterized by painful, bleeding and ulcerated gums and extremely bad breath; associated with poor oral hygiene, particularly in combination with other stresses.	Table 24.1
***Helicobacter pylori* gastritis**	*Helicobacter pylori*	Can result in peptic ulcers; also associated with gastric cancers; *H. pylori* withstands stomach acid by producing ammonia from urea.	Table 24.2
VIRAL INFECTIONS OF THE UPPER DIGESTIVE SYSTEM			
Oral herpes simplex (cold sores)	Herpes simplex virus (usually type 1)	Initial infection typically occurs during childhood; latent virus reactivates, causing recurrent cold sores.	Table 24.3
Mumps	Mumps virus	Characterized by painful swelling of the parotid glands; virus enters via the respiratory tract and then spreads to glands; can cause infertility in males; preventable by vaccination.	Table 24.4
BACTERIAL INFECTIONS OF THE LOWER DIGESTIVE SYSTEM			
Cholera	*Vibrio cholerae*	Classic example of watery diarrhea; *V. cholerae* colonizes the small intestine and produces cholera toxin, causing secretion of water and electrolytes; rehydration is critical.	Table 24.5
Shigellosis	*Shigella species*	Classic example of bacterial dysentery; *Shigella* species invade cells of large intestine; low infectious dose; some species produce Shiga toxin.	Table 24.6
***Escherichia coli* gastroenteritis**	Certain strains of *E. coli*	Symptoms vary depending on the pathovar; STEC strains cause bloody diarrhea and HUS; symptoms of ETEC are similar to those of cholera; symptoms of EIEC are similar to those of shigellosis.	Table 24.7, Table 24.8
***Salmonella* gastroenteritis**	*Salmonella enterica* (not Typhi or Paratyphi)	Foodborne, various animals including poultry	Table 24.9
Enteric fever (typhoid and paratyphoid)	*Salmonella enterica* serotypes Typhi and Paratyphi	Life-threatening diseases caused by *Salmonella* serotypes Typhi and Paratyphi crossing the intestinal mucosa and then invading the bloodstream; vaccine for typhoid fever is available.	Table 24.10
Campylobacteriosis	*Campylobacter jejuni*	Foodborne, particularly poultry	Table 24.11
***Clostridioides difficile* infection (CDI)**	*Clostridioides difficile*	The infection typically occurs only in patients on antibiotic therapy; most common cause of diarrhea in healthcare settings.	Table 24.12
VIRAL INFECTIONS OF THE LOWER DIGESTIVE SYSTEM—INTESTINAL TRACT			
Rotavirus gastroenteritis	Rotaviruses	Most common cause of viral gastroenteritis in infants and children in the United States; preventable by vaccination.	Table 24.13
Norovirus gastroenteritis	Noroviruses	Most common cause of viral gastroenteritis in the United States; highly infectious; epidemics are common on cruise ships and in college dormitories.	Table 24.13
VIRAL INFECTIONS OF THE LOWER DIGESTIVE SYSTEM—LIVER			
Hepatitis A	Hepatitis A virus (HAV)	Spreads via fecal-oral route; usually mild symptoms; no chronic carriers; preventable by vaccination.	Table 24.14
Hepatitis B	Hepatitis B virus (HBV)	Spreads in blood and semen; reverse transcribing virus; chronic infections can result in cirrhosis and liver cancer; chronic carriers; preventable by vaccination.	Table 24.14
Hepatitis C	Hepatitis C virus (HCV)	Spreads in blood, possibly semen; chronic infections can result in cirrhosis and liver cancer; chronic carriers.	Table 24.14
PROTOZOAN INFECTIONS OF THE LOWER DIGESTIVE SYSTEM			
Giardiasis	*Giardia lamblia*	Common waterborne illness; low infectious dose; cysts are chlorine-resistant.	Table 24.15
Cryptosporidiosis	*Cryptosporidium hominis*	Wide host range; oocysts are chlorine-resistant; low infectious dose.	Table 24.16
Cyclosporiasis	*Cyclospora cayetanensis*	No direct person-to-person spread because oocysts must mature in the environment to become infectious; imported produce has been implicated in outbreaks.	Table 24.17
Amebiasis	*Entamoeba histolytica*	Mostly occurs in tropical regions where sanitation is poor; low infectious dose; amebic dysentery results from intestinal damage.	Table 24.18

Summary

24.1 ■ Anatomy, Physiology, and Ecology of the Digestive System

The digestive system encompasses the digestive tract and accessory organs (figure 24.1).

The Upper Digestive System

The upper digestive system includes the mouth and salivary glands, the esophagus, and the stomach. Bacteria growing on teeth attach to the pellicle that adheres to the enamel (figure 24.2), forming a biofilm called **dental plaque.** Peristalsis in the esophagus propels microbes to the stomach. The acidity and enzymes in the stomach destroy most bacterial cells.

The Lower Digestive System

The lower digestive system includes the small and large intestines, as well as the pancreas and the liver. As the stomach contents enter the small intestine, digestive fluids from the pancreas and the liver are mixed in. **Villi** and **microvilli** increase the surface area for absorption in the small intestine. The large intestine absorbs remaining nutrients and more water; feces are eliminated. Microbes make up about one-third of the weight of feces. The liver produces **bile,** which is stored in the gallbladder.

UPPER DIGESTIVE SYSTEM INFECTIONS

24.2 ■ Bacterial Diseases of the Upper Digestive System

Dental Caries (Tooth Decay) (table 24.1)

Dental caries is a chronic biofilm-mediated disease that damages tooth enamel, often resulting in dental cavities. Plaque that contains *Streptococcus mutans* or other cariogenic microbes can act as a tiny acid-soaked sponge applied closely to the tooth (figure 24.3, figure 24.4). Control of dental caries depends mainly on restricting dietary sucrose, supplying fluoride, flossing, and brushing teeth; dental sealants can also be applied.

Dental Plaque-Induced Periodontal Diseases (table 24.1)

Gingivitis is reversible inflammation of the gums due to plaque accumulation in the gingival crevice; **periodontitis** is inflammation of the tissues supporting the teeth and results in permanent damage (figure 24.3).

Necrotizing Periodontal Diseases (table 24.1)

Necrotizing periodontal diseases (NPD), including necrotizing gingivitis and necrotizing periodontitis, can occur at any age and are associated with poor oral hygiene (figure 24.5).

Helicobacter pylori Gastritis and Peptic Ulcer Disease (table 24.2)

Helicobacter pylori causes gastritis that can lead to peptic ulcers (figures 24.6). Treatment with antimicrobial medications can cure the infection and prevent peptic ulcer development.

24.3 ■ Viral Diseases of the Upper Digestive System

Oral Herpes (Cold Sores) (table 24.3; figure 24.7)

Oral herpes is caused by herpes simplex viruses, which can persist as a latent infection inside sensory nerves; the virus can reactivate to cause cold sores.

Mumps (table 24.4; figure 24.8)

Mumps virus infects the parotid glands and a variety of other body tissues; it generally causes more severe disease in individuals beyond the age of puberty. An attenuated vaccine can be used to prevent disease (figure 24.9).

LOWER DIGESTIVE SYSTEM INFECTIONS

Focus on Diarrheal Diseases

Infections of the small intestine typically cause watery diarrhea, whereas invasion of the large intestine causes **dysentery.** The causative agents are transmitted via the fecal-oral route, often through contaminated foods and water; those that have a low infectious dose can be transmitted by direct contact as well. Dehydration can be treated with **oral rehydration therapy (ORT)** using oral rehydration salts (ORS).

24.4 ■ Bacterial Diseases of the Lower Digestive System

Cholera (table 24.5)

Cholera is a severe form of watery diarrhea caused by a toxin of *Vibrio cholerae* that acts on the small intestinal epithelium (figure 24.10).

Shigellosis (table 24.6)

Shigella species invade the epithelium of the large intestine, causing dysentery (figure 24.11). *Shigella dysenteriae* produces Shiga toxin, which causes **hemolytic uremic syndrome (HUS).**

Escherichia coli Gastroenteritis (tables 24.7, 24.8; figure 24.12)

Escherichia coli strains that cause intestinal disease can be grouped into six pathovars: STEC (or EHEC), ETEC, EIEC, EPEC, EAEC, and DAEC. STEC strains cause hemolytic uremic syndrome.

Salmonella Gastroenteritis (table 24.9)

Most cases of *Salmonella* gastroenteritis originate from animals. The organisms are often foodborne, commonly in poultry and eggs.

Enteric Fever (Typhoid and Paratyphoid) (table 24.10)

Typhoid fever is caused by *Salmonella* Typhi, and paratyphoid fever is caused by *Salmonella* Paratyphi; both infect only humans. The diseases are characterized by high fever, headache, and abdominal pain. Untreated, they can be fatal. A vaccine is available to help prevent typhoid fever.

Campylobacteriosis (table 24.11)

Campylobacter jejuni is the most common bacterial cause of diarrhea in the United States; it usually originates from domestic animals (figure 24.13).

Clostridioides difficile Infection (CDI) (table 24.12)

Clostridioides difficile causes diarrhea and **pseudomembranous colitis,** primarily in patients on antibiotic therapy (figure 24.14).

24.5 ■ Viral Diseases of the Lower Digestive System—Intestinal Tract

Rotavirus Gastroenteritis (table 24.13)

Rotaviral gastroenteritis is the main diarrheal illness of infants and young children.

Norovirus Gastroenteritis (table 24.13)

Norovirus is the most common cause of viral gastroenteritis in the United States.

24.6 ■ Viral Diseases of the Lower Digestive System—Liver (figure 24.19)

By definition, acute hepatitis virus infection lasts 6 months or less, whereas chronic hepatitis virus infection lasts longer than 6 months.

Hepatitis A (table 24.14)
Hepatitis A is an acute disease spread by fecal-oral transmission. Infections in older children and adults generally lead to symptoms, but infections in most young children and many older children are asymptomatic.

Hepatitis B (table 24.14)
Hepatitis B can be acute or chronic and is spread in blood, blood products, and semen (figures 24.17, 24.18). Chronic infection can lead to **cirrhosis** and liver cancer.

Hepatitis C (table 24.14)
Hepatitis C is transmitted mainly in blood. The acute stage is often asymptomatic, and many infections become chronic, leading to cirrhosis and liver cancer.

24.7 ■ Protozoan Diseases of the Lower Digestive System

Giardiasis (table 24.15)
Transmission of *Giardia lamblia* is usually via drinking water contaminated by feces. It is a common cause of traveler's diarrhea (figures 24.20, 24.21). The **cysts** survive in chlorinated water but can be removed by filtration.

Cryptosporidiosis ("Crypto") (table 24.16)
Cryptosporidium hominis (formerly *parvum*) oocysts are infectious, resist chlorination, and are too small to be removed by some filters (figure 24.22). *C. hominis* is a cause of many waterborne and foodborne epidemics and traveler's diarrhea.

Cyclosporiasis (table 24.17)
Cyclospora cayetanensis is transmitted via fecally contaminated water or food; it causes traveler's diarrhea. Oocysts passed in feces are not immediately infectious, so there is no person-to-person spread; no hosts other than humans are known.

Amebiasis (table 24.18)
Entamoeba histolytica is an important cause of dysentery; infection can spread to the liver and other organs (figure 24.23). Cysts are infectious.

Review Questions

Short Answer

1. Describe two characteristics of *Streptococcus mutans* that contribute to its involvement in dental caries.
2. Describe the differences between gingivitis and periodontitis.
3. How does *Helicobacter pylori* cause stomach ulcers?
4. Besides the ducts of the salivary glands, what other body sites are sometimes infected by the mumps virus?
5. What characterizes the solutions used for oral rehydration therapy?
6. How do *Shigella* cells move from one host cell to another even though they are non-motile?
7. Name four different pathogenic groups of *Escherichia coli.*
8. What predisposes someone to a *Clostridioides difficile* infection?
9. Name two kinds of hepatitis preventable by vaccination.
10. Contrast the cause and epidemiology of giardiasis and amebiasis.

Multiple Choice

1. Which of the following about intestinal bacteria is *false*?
   a) They produce vitamins.
   b) They can produce carcinogens.
   c) They are mostly aerobes.
   d) They produce gas during the breakdown of certain foods.
   e) They include potential pathogens.
2. All of the following attributes of *Streptococcus mutans* are important in tooth decay *except*
   a) it produces endotoxin, which triggers an inflammatory response.
   b) it can grow at pH below 5.
   c) it produces lactic acid.
   d) it synthesizes glucans.
   e) it adheres to the dental pellicle.
3. *Helicobacter pylori* has all of the following characteristics *except*
   a) it is a helical bacterium with sheath-covered flagella.
   b) it is an obligate anaerobe.
   c) it produces urease.
   d) it causes infections that can last for years.
   e) it can cause stomach ulcers.
4. *Vibrio cholerae* pathogenesis involves all of the following *except*
   a) attachment to the small intestinal epithelium.
   b) production of cholera toxin.
   c) watery diarrhea.
   d) acid resistance.
5. Which of the following statements concerning *Salmonella enterica* serotype Typhi is *false*?
   a) It is commonly acquired from domestic animals.
   b) It can colonize the gallbladder.
   c) It is highly resistant to being killed by bile.
   d) It can destroy Peyer's patches.
   e) It causes typhoid fever.
6. Which statement about rotavirus gastroenteritis is *false*?
   a) A vaccine is available to prevent the disease.
   b) Fatal dehydration can occur if fluids are not replaced.
   c) Most cases of the disease occur in infants and children.
   d) The causative agent infects mainly the stomach.
   e) The disease is transmitted by the fecal-oral route.
7. Which of the following statements about noroviruses is *false*?
   a) They are the most common cause of viral gastroenteritis in the United States.
   b) They have a low infectious dose.
   c) They generally cause vomiting lasting 1 to 2 weeks.
   d) Immunity does not last long.
   e) They are a Category B bioterrorism agent.

8. Which of the following statements about hepatitis is *false*?
   a) Both RNA and DNA viruses can cause hepatitis.
   b) Some kinds of hepatitis can be prevented by vaccination.
   c) HCV infections are often associated with injected-drug abuse.
   d) Lifelong carriers of hepatitis A are common.
   e) Hepatitis A spreads by the fecal-oral route.
9. Which of the following statements about hepatitis B virus is *false*?
   a) Replication involves reverse transcriptase.
   b) Infected people may have large numbers of non-infectious viral particles circulating in their bloodstream.
   c) In the United States, infection rates have been steadily increasing over the last few years.
   d) Asymptomatic infections can last for years.
   e) Infection can result in cirrhosis.
10. Choose the most accurate statement about cryptosporidiosis.
   a) Waterborne transmission is unlikely.
   b) The host range of the causative agent is narrow.
   c) It is prevented by chlorination of drinking water.
   d) Person-to-person spread does not occur.
   e) The causative agent invades epithelial cells of the small intestine.

Applications

1. One concern with respect to chlorinating a water supply is that chlorine can react with substances in water or in the intestine to produce carcinogens. How do you assess the relative risks of chlorinating or not chlorinating drinking water?
2. A medical scientist is designing a research program to determine the effectiveness of hepatitis B vaccine in preventing liver cell cancer. Because liver cell cancer probably has multiple causes, how would you measure the success of an anti-cancer vaccination program?

Critical Thinking

1. Why does the lack of a brown color in feces indicate hepatitis?
2. Mutant strains of *Helicobacter pylori* that lack the ability to produce urease fail to cause infection when they are swallowed. Infection occurs, however, if a tube is used to introduce them directly into the layer of mucus that overlies the stomach epithelium. What does this imply about the role of urease in the bacterium's pathogenicity?

25 | Cardiovascular and Lymphatic System Infections

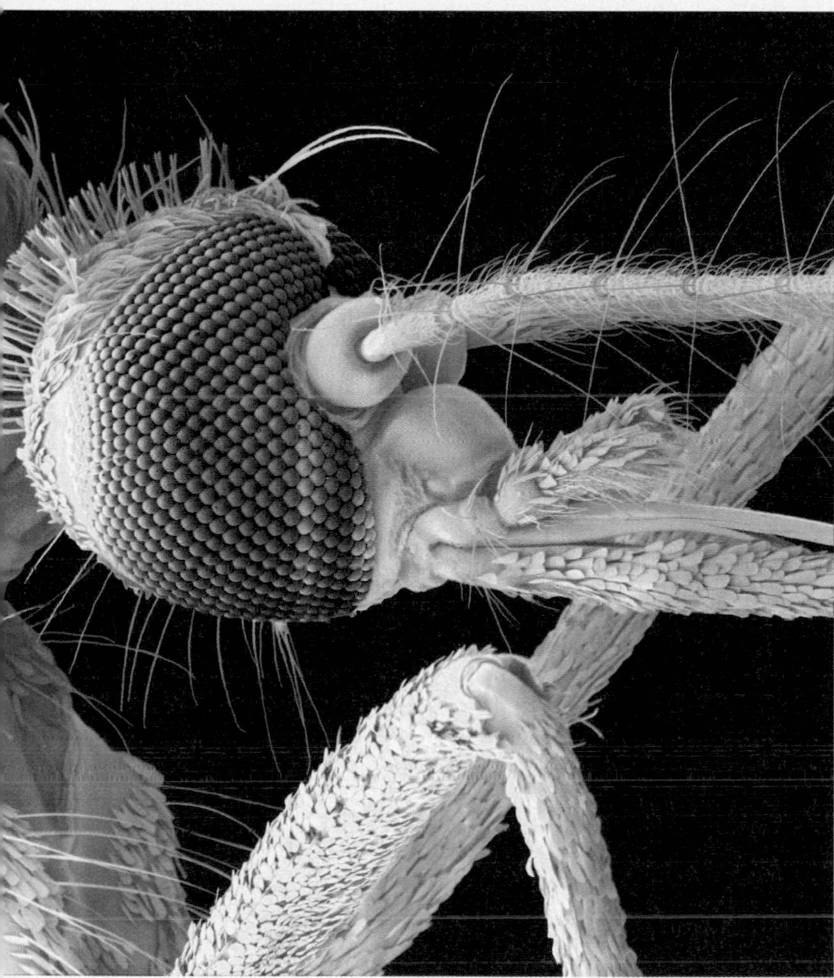

Color-enhanced SEM of mosquito head. *Steve Gschmeissner/Science Photo Library/ Alamy Stock Photo*

KEY TERMS

Bubo An enlarged, tender lymph node characteristic of plague and some sexually transmitted infections.

Cytokine Storm Excessive release of pro-inflammatory cytokines.

Disseminated Intravascular Coagulation (DIC) Systemic activation of the body's blood clotting system, potentially leading to clots in small blood vessels throughout the body; it can result in organ damage (due to blocked blood vessels) and hemorrhage (due to depletion of clotting proteins and platelets).

Infective Endocarditis Inflammation of the heart valves or lining of the heart chambers resulting from infection.

Lymphangitis Inflammation of lymphatic vessels.

Paroxysm A sudden recurrence or worsening of symptoms as seen in the cycle of chills, fever, and sweats of malaria.

Petechiae Purple spots on the skin and mucous membranes caused by blood leakage from small blood vessels.

Sepsis Life-threatening tissue damage and organ dysfunction resulting from an overwhelming host response to to an infection.

Shock A life-threatening condition resulting from inadequate blood flow to organs due to a significant drop in blood pressure.

A Glimpse of History

Alexandre Emile Jean Yersin (1863–1943) is one of the most interesting, though relatively unknown, contributors to the understanding of infectious diseases. He developed an interest in science as a young boy, when he found a microscope and dissecting instruments that had belonged to his late father. A local physician befriended Yersin and influenced him to study medicine. After becoming a physician, Yersin volunteered at the Pasteur Institute in Paris, where he was hired by Émile Roux, a coworker of Louis Pasteur. Yersin became well recognized for his work at the Institute but was bored with research and did not want to practice medicine because he felt it was wrong for physicians to make money from other people's sickness.

Yersin left the Pasteur Institute and was employed as a physician on a ship sailing from France to Vietnam. He became enchanted with Vietnam and its people, so made it his home for the rest of his life. There, Yersin studied diseases that affected the Vietnamese people, including plague. The cause and transmission of this serious disease were completely unknown at the time.

In 1894, Yersin went to study an outbreak of plague in Hong Kong. Unfortunately for him, Shibasaburo Kitasato, the famous colleague of Robert Koch, had arrived 3 days earlier with a large team of scientists. British authorities had given Kitasato and his colleagues access to patients and laboratory facilities. Yersin did not speak English, so it was difficult for him to communicate with the authorities. He set up a laboratory in a bamboo hut and was forced to do his research on samples from dead plague victims that he got from British soldiers, whose job it was to bury them. A week after arriving, Yersin reported his discovery of a bacillus that was always present in the swollen lymph nodes of plague victims. A pure culture of the bacterium caused plague-like disease when injected into rats, and the disease could be transmitted from one rat to another. Kitasato, working with blood from the hearts of plague victims, also announced that he had isolated the bacterium that caused plague. His initial reports on the organism were somewhat vague, probably because his samples were contaminated. Nonetheless, it is generally accepted that he did indeed work with the plague bacillus, and Kitasato is credited as its co-discoverer.

Yersin's plague bacillus was later named *Yersinia pestis*. It was used to make a vaccine, and later, antiserum prepared against the organism was used to cure a patient with plague—the first successful treatment of a plague victim.

The cardiovascular system (which circulates blood) and the lymphatic system (which carries lymph) are largely separate, but their routes involve essential interactions. Infections of these systems can be serious because the pathogens become **systemic,** meaning carried throughout the body. Once disseminated (spread), the microbes can infect one or more organs. In addition, when the innate defenses recognize a disseminated pathogen, they elicit an equally widespread response. That systemic response might eliminate the microbe, but in doing so it can also damage tissues and even result in life-threatening organ failure.

Separate but related terms are used to refer to damaging systemic immune responses: cytokine storm and sepsis. **Cytokine storm** is a general term used to describe an excessive release of pro-inflammatory cytokines. Recall that cytokines are powerful inflammatory mediators that cause an array of effects, including increased blood vessel permeability (due to vessel dilation) and adherence of phagocytes to vessel walls (see figure 14.15). Although those effects play an essential role in localized inflammatory responses, they can be disastrous when their actions are widespread (imagine leaky blood vessels throughout the body). **Sepsis** is life-threatening tissue damage and organ dysfunction resulting from an overwhelming host response to an infection; the mechanisms are complex, but the outcomes are clear.

25.1 ■ Anatomy, Physiology, and Ecology of the Cardiovascular and Lymphatic Systems

Learning Outcome

1. Describe the characteristics and functions of the cardiovascular and lymphatic system components.

The cardiovascular and lymphatic systems are typically considered sterile, but microbes can enter in various ways, including from other body sites or through bites of arthropods such as fleas and mosquitoes. Normally, the immune system quickly removes any microbes that enter, but disease may result if microbes overcome those defenses.

The Cardiovascular System

The cardiovascular system is composed of the heart and blood vessels, as well as the blood that flows within them.

Blood

Circulating blood provides cells with nutrients and O_2 and removes waste products; it also heats or cools the body. Blood is composed of erythrocytes (red blood cells; RBCs), leukocytes (white blood cells; WBCs), and cell fragments called platelets, all suspended in the nutrient-rich fluid called plasma. RBCs, the most abundant cell type, are filled with hemoglobin, an iron-containing protein that binds O_2 in the lungs and releases it in the tissues. Characteristics of the different kinds of WBCs are described in chapter 14 and summarized in table 14.1. Platelets are required for blood clotting.

When a microbe is in the bloodstream, the condition is given a name that specifies the type of microbe, as in **bacteremia, viremia, fungemia,** and **parasitemia.** These terms do not imply a disease state—in many cases, the circulating microbe causes no symptoms. For example, toothbrushing often leads to brief bacteremia because mouth bacteria enter small abrasions caused by the brush.

The Heart

The heart—a muscular organ enclosed in a fibrous sac called the pericardium—pumps the blood throughout the body. A wall of tissue called the septum separates the heart into right and left sides. Each side is divided into two chambers: (1) the atrium that receives blood, and (2) the ventricle that discharges it.

A crucial function of blood circulation is O_2 delivery to the tissues. Blood that has delivered O_2 is referred to as deoxygenated (O_2-poor), and it enters the right atrium of the heart. From there it flows to the right ventricle, which pumps it to the lungs where it becomes oxygenated (O_2-rich). The oxygenated blood then returns to the heart, entering the left atrium and passing into the left ventricle. That ventricle then pumps the blood to the arterial system (the series of blood vessels that carries oxygenated blood away from the heart and into the tissues). Valves at the entrance and exit of each ventricle prevent the blood flow from reversing. After O_2 is delivered to the tissues, the deoxygenated blood is carried back to the right atrium by a series of blood vessels called the venous system, thus completing the circuit (**figure 25.1**). Although infections of the pericardium, heart muscle, and heart valves are not common, they can be serious because of the effects on blood circulation.

Blood Vessels

Blood vessels include arteries (the arterial system), capillaries, and veins (the venous system). Arteries have thick muscular walls to withstand the high pressure of blood pumped by the ventricles. In contrast, microscopic capillaries have very thin walls, allowing gas and fluid exchange between the blood and the tissues. The walls of veins are thinner than those of arteries because the pressure within the veins is low. Blood continues its circular route through the venous system due to the combination of the (1) pumping action as the veins become compressed when surrounding skeletal muscles contract and (2) one-way valves in the veins that prevent flow from reversing.

The Lymphatic System

As described in chapter 15, the lymphatic system consists of lymphatic vessels and lymphoid organs (including lymph nodes, spleen, and tonsils), as well as the lymph that flows within them (see figure 15.7).

Lymph

Lymph, a clear leukocyte-containing fluid, plays an essential role in the immune response. Its formation is intimately associated with the cardiovascular system: Some of the excess tissue fluid—which originates from plasma that squeezes out of blood as it enters capillaries—enters lymphatic capillaries

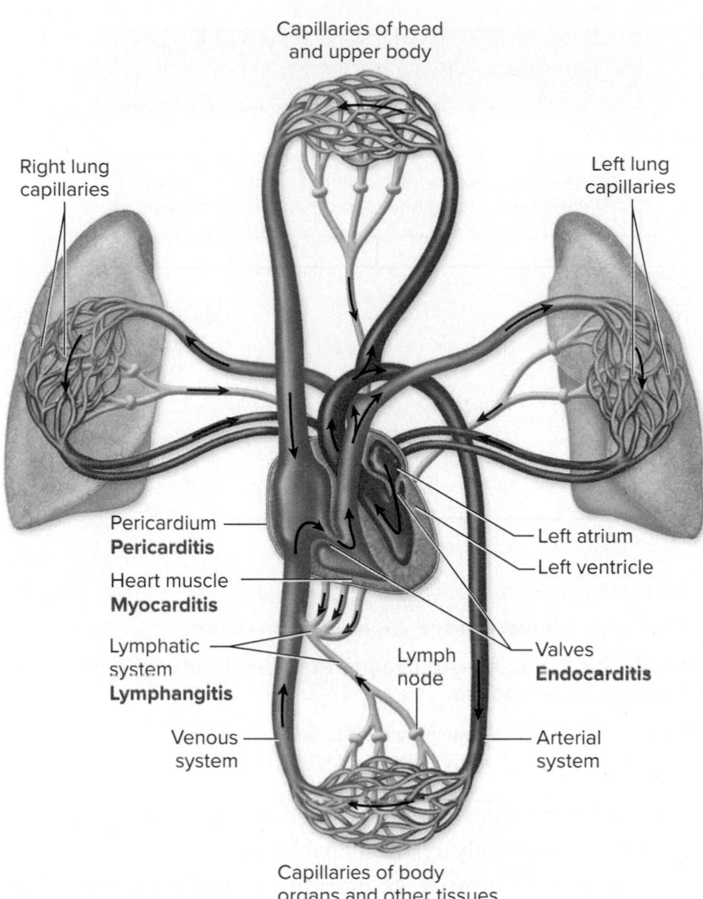

FIGURE 25.1 The Cardiovascular and Lymphatic Systems The red color of the blood vessels indicates oxygenated blood; the blue color indicates deoxygenated blood. Disease conditions are indicated in red type. For simplicity, the spleen is not shown.

? In this illustration, what color are the vessels of the arterial system?

(small lymphatic vessels) (see figure 15.6). Various substances are carried along with that fluid, including invading microbes and their products; leukocytes enter the lymphatic capillaries as well.

Lymphatic Vessels

As with blood flow in veins, lymph in the lymphatic vessels moves in one direction due to valves and the pumping action of skeletal muscle contractions. Unlike blood, however, lymph does not circulate within a single system. Instead, the various lymphatic capillaries converge to form lymphatic vessels, and these eventually converge near the heart to form ducts through which the lymph enters the bloodstream. Thus, the fluid constantly cycles from blood, to tissues, to lymph, and back to blood.

Lymph Nodes

As lymph moves through the lymphatic vessels, it passes through lymph nodes, allowing dendritic cells, lymphocytes, and macrophages gathered there to inspect and respond to the contents (see figure 15.7). When a hand or a foot is infected, a visible red streak may spread up the limb from the infection site toward the nearest lymph node; this is due to inflammation of one or more lymphatic vessels, a condition called **lymphangitis** (**figure 25.2**).

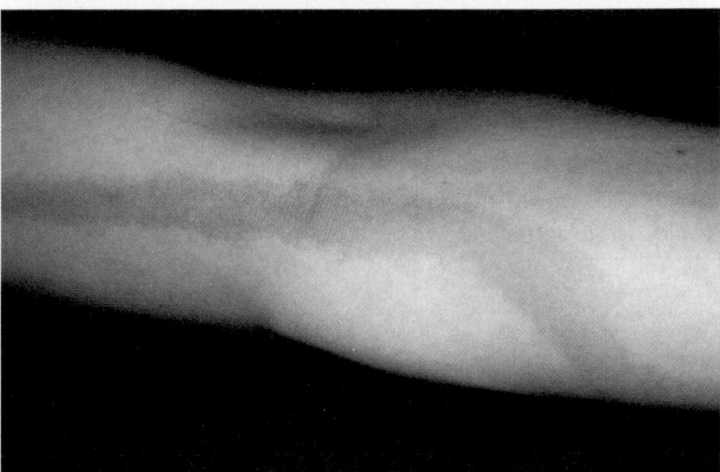

FIGURE 25.2 Lymphangitis Notice the red streak extending up the arm. SPL/Science Source

? What causes lymphangitis?

Spleen

The spleen is a fist-sized organ found on the left side of the abdominal cavity, behind the stomach. It contains two kinds of tissue: white pulp (lymphoid tissue) and red pulp (multiple blood-filled passageways). The white pulp functions as a lymph node for blood: Dendritic cells, lymphocytes, and macrophages gather there, ready to respond to microbial invaders. The red pulp cleans the blood of debris; resident macrophages there remove various materials, including aging or damaged RBCs, bacteria, and other foreign materials. In addition, the red pulp stores a reserve of monocytes, and it can produce new blood cells in rare situations where the bone marrow cannot make enough.

MicroAssessment 25.1

The heart pumps blood through arteries and veins, potentially carrying infectious agents and their toxins throughout the body. The lymphatic system carries lymph through lymph nodes, which contain immune cells that remove foreign materials, before returning it to the bloodstream. Blood is exposed to immune cells as it passes through the spleen.

1. What is the function of valves in the heart?
2. What are the functions of the spleen?
3. Why do the ankles of a person sitting still for an extended length of time sometimes swell?

25.2 ■ Bacterial Diseases of the Cardiovascular and Lymphatic Systems

Learning Outcomes

2. Explain the role of the inflammatory response in bacterial sepsis.
3. Outline the events that lead to infective endocarditis.
4. Compare and contrast plague, Lyme disease, *Vibrio vulnificus* infection, tularemia, and brucellosis.

Bacteria that cause diseases of the cardiovascular and lymphatic systems usually originate from an infected area in the tissues. From there, they enter the lymph and can be carried to the bloodstream. In addition to the bacterial diseases discussed in this section, additional examples that affect the cardiovascular and lymphatic systems include Rocky Mountain spotted fever and enteric fever, which were covered in earlier chapters.

Bacterial Sepsis

As mentioned in the chapter introduction, **sepsis** is life-threatening tissue damage and organ dysfunction resulting from an overwhelming host response to an infection. This definition has changed over the years as scientists learned more about the complex features that characterize the condition. Although any microbial infection can lead to sepsis, bacterial sepsis is most common and is typically healthcare-associated.

Signs and Symptoms

The signs and symptoms of sepsis include violent shaking, chills, and fever, often with rapid breathing and feelings of anxiety and confusion. As the disease progresses, urine output drops, respiration and pulse speed up, and the arms and legs become cool and dusky-colored. Bruising and bleeding can also occur.

Causative Agents

Systemic infection by any bacterium can lead to sepsis, as various bacterial components—including lipopolysaccharide (endotoxin) and peptidoglycan—can potentially trigger an exaggerated host response. The most common causes of bacterial sepsis depend on various factors such as patient characteristics (underlying health and age), source of infection (community-acquired versus hospital-acquired), and geographical area.

Pathogenesis

Bacterial sepsis usually starts from an infection somewhere in the body other than the bloodstream. When normal host defenses are compromised, particularly by medical treatments such as surgery, intravenous (IV) catheterization, or some medications, even microorganisms that normally have little invasive ability can overwhelm the local defenses or spread in the bloodstream. A combination of factors involving the pathogen and the patient's immune system then results in an out-of-control and ultimately damaging host response.

Sepsis involves the steps of a normal inflammatory response (see figure 14.15), but rather than being coordinated and highly controlled, the process is exaggerated and dysfunctional. When an infection is systemic, macrophages and other immune cells throughout the body detect the invading microbes and respond by releasing pro-inflammatory cytokines (**figure 25.3**). A substantial cytokine release—a cytokine storm—may overwhelm the normal mechanisms that control the inflammatory response; note that

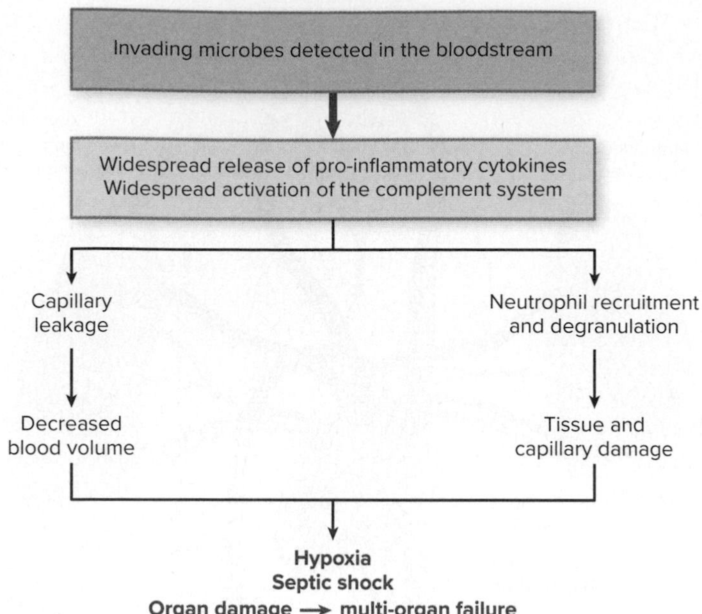

FIGURE 25.3 Sepsis—An Exaggerated and Dysfunctional Inflammatory Response

? How does neutrophil degranulation lead to tissue damage?

a cytokine storm does not always lead to sepsis but is an early indicator of a potentially dysfunctional immune response. Systemic microbes not only provoke an excessive inflammatory response, but they also activate the complement system and can overwhelm its normal control systems as well. The activated complement proteins then amplify the response, which becomes widespread and self-stimulating. The overburdened immune system may eventually fail, resulting in immunodeficiency. As an analogy for what happens during sepsis, imagine the chaos that would occur if every emergency alarm in a city were activated all at once: The uncoordinated responses themselves could cause significant damage. In the case of sepsis, the damage is associated with:

- **Capillary leakage, leading to decreased blood volume.** As described in section 14.8, the normal inflammatory response causes the capillaries to become more permeable. When this happens throughout the body, the extensive leakiness leads to decreased blood volume.
- **Neutrophil recruitment and degranulation, leading to tissue and capillary damage.** Neutrophils recruited to the tissues may degranulate there or in the capillaries, releasing their damaging enzymes and other products. Programmed cell death pathways, including apoptosis and pyroptosis, may also be triggered.
- **Hypoxia.** Hypoxia, meaning low levels of O_2 in body tissues, can occur as a result of lung damage. The lungs are particularly susceptible to inflammatory damage, both from toxic products released by phagocytes and from fluid accumulation; the overall result is compromised O_2 exchange, resulting in hypoxemia (low O_2 level in the blood), which, in turn, leads to hypoxia.

- **Septic shock. Shock** is a condition associated with a drop in blood pressure that results in inadequate blood flow to the organs. Shock associated with sepsis is called septic shock and is particularly severe because of the other damaging effects of sepsis.
- **Organ damage, leading to multi-organ failure.** The widespread cellular damage resulting from inflammation, coupled with hypoxia, results in the organ dysfunction that characterizes sepsis. In turn, this can lead to multi-organ failure.

As a complication of sepsis, widespread activation of the clotting system can lead to **disseminated intravascular coagulation (DIC).** Clots form in small blood vessels throughout the body, potentially resulting in organ failure (due to hypoxia from the blockage of blood circulation) and hemorrhage (due to depletion of clotting proteins and platelets).

Epidemiology

Bacterial sepsis is mainly healthcare-associated, reflecting the high incidence of bloodstream infections in hospitalized patients with impaired host defenses; it is most common in older populations (over 65 years of age) and those with diabetes and obesity. Bacterial sepsis rates have been increasing, likely due to longer life spans, antibiotic suppression of normal microbiota, the use of immunosuppressive medications, and the use of invasive medical devices (such as ventilators, intravenous (IV) catheters, and urinary catheters) on which biofilms readily develop.

Treatment and Prevention

Prompt treatment of bacterial sepsis is essential. One or more broad-spectrum antimicrobial medications are used to control the underlying infection, and immediate supportive care is given (including supplemental O_2 to reverse the hypoxia, IV fluids to restore blood volume, and organ dysfunction support). Care must be taken with respect to antimicrobial choice because bactericidal antibiotics often lyse the cells, thereby increasing the release of inflammation-inducing microbial components.

Sepsis can be prevented by quickly identifying and effectively treating localized infections, particularly in people with impaired immune defenses. Also, predisposing conditions such as bedsores and pyelonephritis (inflammation of the kidneys)—which commonly lead to sepsis in patients with cancer and diabetes—can usually be prevented. **Table 25.1** gives the main features of bacterial sepsis.

TABLE 25.1 Bacterial Sepsis

Signs and Symptoms	Shaking, chills, fever, rapid breathing and pulse, confusion, and anxiety; if shock develops, drop in urine output, bruising, bleeding, and organ failure
Incubation Period	Variable
Causative Agents	Any bacteria
Pathogenesis	Macrophages release pro-inflammatory cytokines in response to widespread bacterial components and tissue damage; complement system is activated; uncontrolled inflammatory response results, leading to organ dysfunction.
Epidemiology	Mainly involves a healthcare-associated infection
Treatment and Prevention	Treatment: One or more antimicrobial medications, as well as supportive care. Prevention: prompt identification and treatment of localized infections.

Infective Endocarditis

Infective endocarditis (IE) is an infection of an inner surface of the heart, usually involving heart valves. Some types of IE develop slowly but others have a more rapid progression.

Signs and Symptoms

People with IE usually have a fever and noticeable fatigue. They often become ill gradually and slowly lose energy over a period of weeks or months, and may experience various other symptoms including aching joints or muscles, headache, shortness of breath, and nausea. Most IE patients have a detectable heart murmur (an abnormal sound between heartbeats). Later in the disease, small purplish hemorrhagic lesions may develop; these are called **petechiae** if in the skin and splinter hemorrhages if under the nails. Strokes and heart failure can be life-threatening complications.

Causative Agent

A number of different pathogens cause IE, including *Staphylococcus aureus, Staphylococcus epidermidis,* enterococci, and various *Streptococcus* species, such as viridans group streptococci normally found in the mouth; the latter are less common causes but significant because of their increasing antibiotic resistance. Organisms causing IE are usually shed from the heart valves into the bloodstream and can be identified by cultivating samples taken from an arm vein.

Pathogenesis

The bacteria that cause IE may enter the bloodstream from another infected body site or during dental procedures, toothbrushing, or trauma. These microbes adhere to heart tissue, particularly if they become trapped in the thin blood clots that often form around deformed heart valves or areas with disturbed blood flow. The bacteria multiply, creating a biofilm that protects them from phagocytosis and antimicrobial medications. Although the bacteria usually have little invasive ability, when they grow to large numbers they can sometimes penetrate the heart tissue—resulting in abscesses or damaged valve tissue.

A complication of IE involves the clots that trap the bacteria. The clot grows larger around the multiplying organisms, gradually building up a fragile mass. Bacteria continually wash off the mass into the circulation, and pieces of infected clot (septic emboli) can break off. Emboli can block blood vessels, leading to tissue death in the regions supplied by the vessel, and they can also cause a vessel to weaken and balloon out, forming an aneurysm.

People with IE often have high levels of antibodies against the causative agent. These can be harmful because they may bind to soluble antigens to form immune complexes that lodge in the skin, the eyes, or other body structures, triggering an inflammatory response. In turn, this may result in the hemorrhagic lesions in the skin or conjunctiva or, if the complexes lodge in the kidneys, glomerulonephritis.

Epidemiology

Predisposing factors for IE include indwelling intravenous (IV) catheters, underlying heart defects such as deformed or damaged valves (possibly caused by rheumatic fever, a post-streptococcal illness), or injected drug abuse. In recent years, the number of IE cases due to prosthetic (artificial) heart valves or IV catheters has risen as the devices have become more common. Increased abuse of injected drugs also results in higher rates of the disease.

Treatment and Prevention

IE is treated with antibacterial medications, chosen according to the susceptibility of the pathogen, usually for a duration of more than a month. Devices covered with resistant biofilms must be replaced. Surgery is sometimes needed to remove an infected clot.

The only routine measure to help prevent IE is to maintain good oral health along with regular dental care. The highest-risk patients, however, such as those with an artificial heart valve or a history of IE, may be given an antibacterial medication shortly before dental or other bacteremia-causing procedures. Healthcare-associated IE can be prevented by strict use of aseptic technique when inserting IV catheters, maintaining the indwelling catheters properly, and removing them as soon as possible. Such precautions help prevent bacterial colonization of the catheters and consequent bacteremia that could lead to heart valve infection. The main characteristics of IE are presented in **table 25.2.**

TABLE 25.2 Infective Endocarditis

Signs and Symptoms	Fever, loss of energy over a period of weeks or months; sometimes, stroke or heart failure
Incubation Period	Poorly defined, usually weeks
Causative Agents	*Staphylococcus aureus* or *S. epidermidis;* enterococci; streptococci including oral viridans group streptococci
Pathogenesis	Normal microbiota or organisms from an infected body site enter bloodstream; disturbed blood flow causes a thin clot to form that traps circulating organisms; a biofilm forms, protecting the organisms from phagocytes; pieces of clot break off, blocking blood vessels, leading to tissue death.
Epidemiology	Risk factors include indwelling IV catheters, heart abnormalities (including artificial valves or damage from rheumatic fever), and injected drug abuse.
Treatment and Prevention	Treatment: antibiotics. Prevention: maintain good oral health; for highest-risk patients, an antimicrobial medication immediately before anticipated bacteremia, such as before dental work; proper procedures for inserting and maintaining IV catheters.

Plague ("Black Death")

Medically, the term *plague* refers to the disease once known as the "black death," which killed approximately a quarter of the population of Europe between 1346 and 1350. Crowded conditions in the cities and a large rat population played important roles in the spread of this zoonotic disease. Recent evidence suggests that human body lice may have played a crucial role as well. Plague is a potential bioterrorism threat, listed as a Category A agent (see Focus Your Perspective 19.1).

Signs and Symptoms

Three general types of plague occur, with different signs and symptoms:

- **Bubonic plague.** This form is by far the most common type today and typically occurs as a result of a bite by an infected flea. Signs and symptoms develop 2 to 8 days after the bite—the person characteristically develops high fever, chills, malaise, headache, and significantly swollen and tender lymph nodes called **buboes** in the region that receives lymph drainage from the area of the flea bite. Between 50% and 80% of people with bubonic plague die if not treated, often because the bacterium spreads to the bloodstream.
- **Septicemic plague.** This can develop with or without a preceding bubo when the organism spreads via the bloodstream, typically after a flea bite. The systemic response to the bloodstream infection results in high fever, shock, and DIC. The latter causes bleeding into the skin and the organs, leading to red or black patches that probably inspired the common name "black death." Clots may also interrupt blood flow in small blood vessels, especially those of the fingers and toes, which may lead to tissue death and gangrene in those areas. The bacterium may spread to the lungs, resulting in pneumonic plague. About 90% of people with septicemic plague die if not treated.

- **Pneumonic plague.** This results when respiratory droplets from an infected person or animal are inhaled, or when organisms in the bloodstream spread to the lungs. It typically develops 1 to 3 days after the bacterium enters the lungs, and is characterized by high fever, headache, a productive cough, and pneumonia accompanied by shortness of breath, chest pain, and possibly bloody sputum. Untreated pneumonic plague is almost always fatal within a few days.

Causative Agent

Plague is caused by *Yersinia pestis,* a facultatively anaerobic, Gram-negative rod that is a member of the Enterobacteriaceae (**figure 25.4**). It is non-motile and grows best at 28°C. The organism resembles a safety pin when stained with certain dyes because the ends of the bacterium stain more intensely than the middle.

Pathogenesis

Y. pestis forms biofilms in the digestive tract of infected fleas, often blocking the tract (**figure 25.5**). This both starves the flea—increasing the likelihood that it will try to feed again—and causes bacteria to be regurgitated (vomited) into the bite wound as the flea attempts to feed. Infected fleas may also release the bacterial cells in their feces, and these can then be introduced into human tissue when a person scratches the flea bite.

Y. pestis has multiple virulence factors, making it extremely well equipped for avoiding mammalian host defenses. Once the organism enters via a flea bite, its protein-degrading protease (Pla) inactivates certain complement system components and promotes clearing of clots from the lymphatics and capillaries, allowing it to spread. The lymphatic vessels carry the bacterial cells to the regional lymph nodes, where they are taken up by macrophages. Inside a macrophage, *Y. pestis* senses the intracellular conditions and responds by turning on various genes required for resisting the killing effects of the macrophage and for multiplying within it. Genes for other products that allow the organism to survive in the host are also activated. After several days, an acute inflammatory response develops in the lymph nodes, producing the enlargement and marked tenderness that characterize the bubos of bubonic plague.

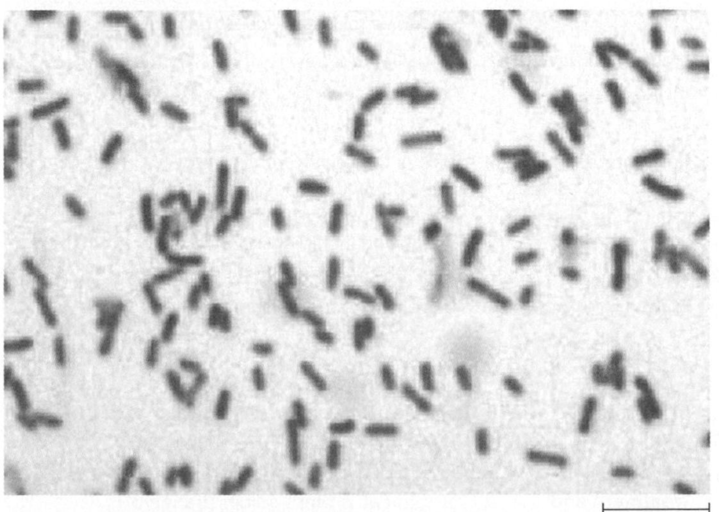

FIGURE 25.4 ***Yersinia pestis*** Gram stain Larry Stauffer, Oregon State Public Health Laboratory/CDC

? **What is unusual about the appearance of *Yersinia pestis* after staining with certain dyes?**

(a)

(b)

FIGURE 25.5 Fleas Following a Blood Meal (a) Healthy flea with fresh blood in gut. **(b)** Flea with obstruction due to *Yersinia pestis* infection. a: Science Source; b: Daniel Cooper/E+/Getty Images

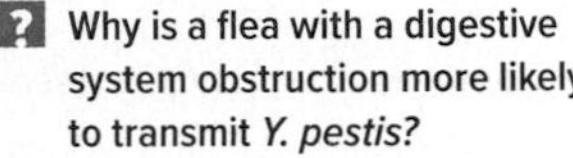

? **Why is a flea with a digestive system obstruction more likely to transmit *Y. pestis?***

The infected macrophages eventually die, releasing bacteria that are now "armed" to withstand host defenses. For example, a capsule allows *Y. pestis* to avoid phagocytosis, and an iron acquisition system allows the organism to trap and take up iron from the host. The bacteria also make a type III secretion system to inject proteins called YOPS (*Yersinia* outer proteins) into host cells; YOPS have multiple effects, including disrupting the host cell cytoskeleton, inhibiting phagocytosis, and blocking the production of pro-inflammatory cytokines, thereby reducing the inflammatory response. The swollen nodes may become necrotic, allowing large numbers of the bacterial cells to spill into the bloodstream. In turn, this provokes a systemic inflammatory response that leads to the signs and symptoms of septicemic plague.

In 10% to 20% of plague cases, the infection spreads to the lungs, resulting in pneumonic plague. A person with pneumonic plague can transmit that disease to others in respiratory droplets; organisms acquired this way are already expressing the genes required for full virulence and, therefore, are especially dangerous. The virulence factors of *Y. pestis* are summarized in **table 25.3.**

Epidemiology

Plague is a zoonotic disease, endemic in rodent populations in many regions of the world; occasional epidemics in those animals (epizootics) can spill over into humans. Globally,

TABLE 25.3 Virulence Factors of *Yersinia pestis*

Factor	Coded by	Action
Pla (protease)	pPCP1 plasmid	Destroys certain complement components and promotes clearance of clots
YOPs (proteins)	pCD1 plasmid	Interfere with phagocytosis and the immune response
V antigen	pCD1 plasmid	Controls type III secretion system that delivers YOPs (proteins)
F1	pMT1 plasmid	Forms antiphagocytic capsule at 37°C
PsaA (adhesin)	Chromosome	Allows attachment to host cells
Complement resistance	Chromosome	Protects against lysis by activated complement
Iron acquisition	Chromosome	Traps iron-containing substances; stores iron compounds intracellularly

most cases are in Africa, Asia, and South America. In the United States, the disease mostly occurs in the western states. Transmission of plague is usually via bites of infected fleas (resulting in bubonic plague); the main reservoirs are wild rodents such as rock squirrels and prairie dogs, as well as their fleas, but rats, rabbits, dogs, and cats can also be hosts. Other modes of transmission include contact with tissues or fluids of infected animals (resulting in septicemic plague) and inhalation of respiratory droplets produced by coughing (resulting in pneumonic plague). Recent evidence suggests that body lice played a role in historical plague epidemics.

Treatment and Prevention

Plague can be effectively treated with prompt administration of an appropriate antimicrobial medication (such as a fluoroquinolone), but delays can significantly affect the outcome because the disease progresses so rapidly. Supportive care (such as rehydration and mechanical ventilation) may also be required.

Plague epidemics can be prevented by using rodent control measures such as disposing of garbage properly, rodent-proofing buildings, and exterminating rats; extermination programs must be combined with insecticide use to kill any fleas that had been feeding on the rodents because fleas can remain infectious for a year or more. People at risk of flea bites should use an insecticide containing DEET, and anyone handling potentially infected animals should wear gloves. An appropriate antimicrobial medication can be given as a post-exposure prophylaxis (PEP) to someone exposed to plague. No vaccine is currently licensed to prevent plague, but several candidates are in clinical trials. The main features of plague are presented in **table 25.4.**

Lyme Disease

Lyme disease is a zoonotic disease transmitted by infected ticks. Approximately 30,000 cases are reported to the CDC each year, making it the most common vector-borne disease in the United States. Surprisingly, it was not recognized until the 1970s, when a cluster of cases occurred in Lyme, Connecticut; we now know that the disease was widespread long before then.

TABLE 25.4 Plague ("Black Death")

1. In most cases, *Yersinia pestis* is acquired via the bite of an infected flea.
2. The bacteria are carried to regional lymph nodes.
3. Phagocytes ingest the bacteria, but the intracellular conditions activate genes responsible for virulence.
4. Fully virulent bacteria break out of the phagocytes and infect the lymph nodes, resulting in the buboes that characterize bubonic plague.
5. The bacteria may enter the bloodstream, causing septicemic plague.
6. The infection may spread to the lungs, resulting in the highly contagious and lethal pneumonic plague.
7. Bacteria exit with coughing in those with pneumonic plague.

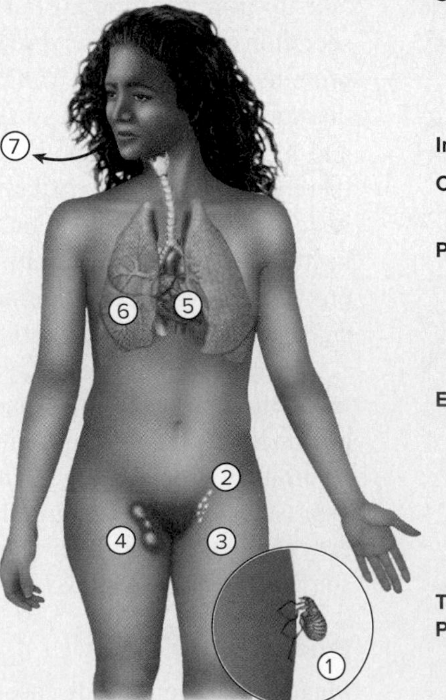

Signs and Symptoms	Bubonic plague: high fever, swollen lymph nodes called buboes. Septicemic plague: high fever, shock, DIC leading to red or black patches. Pneumonic plague: high fever, cough, shortness of breath, chest pain, bloody sputum.
Incubation Period	Usually 1 to 8 days
Causative Agent	*Yersinia pestis*, a Gram-negative rod
Pathogenesis	Enters the body via the bite of an infected flea or by inhalation; bacteria taken up by macrophages, but survive and begin producing multiple virulence factors that interfere with host immune defenses.
Epidemiology	Endemic in rodents and their fleas; in the United States, mainly in the western states. Bubonic plague is transmitted by fleas; septicemic plague can be transmitted by contact with contaminated tissues and fluids; pneumonic plague can be transmitted person to person in respiratory droplets.
Treatment and Prevention	Treatment: prompt administration of antimicrobial medication. Prevention: avoid contact with potentially infected rodents and their fleas; use insecticides and rat control; post-exposure prophylaxis; no vaccine currently available.

Signs and Symptoms

Untreated Lyme disease can be divided into three general stages, although individual patients may be asymptomatic in one or more of these:

- **Early localized.** This begins a few days to several weeks after a bite by an infected tick, and is often characterized by swollen lymph nodes near the bite and a circular skin rash called erythema migrans (EM). The rash begins as a red spot or bump at the bite site and slowly expands to reach a diameter of up to 15 cm (6 inches). Its advancing edge is bright red, leaving behind an area of fading redness as the lesion gets bigger, creating an appearance like the center of a target (**figure 25.6**). This characteristic bull's-eye rash—the hallmark of Lyme disease—is present in most but not all cases (about 75% of people develop it). Moreover, a circular rash may appear after the bite of a type of tick that does not transmit Lyme disease. Most of the other signs and symptoms that occur during this stage are flu-like: malaise, chills, fever, headache, fatigue, joint and muscle pains, and backache.
- **Early disseminated.** This begins days to months after the initial infection and is characterized by nervous system involvement as well as rashes on various parts of the body. Patients may experience one or more of the following: facial paralysis, severe headache, stiff neck, pain when moving the eyes, fatigue, and numbness or pain in the legs or arms. In rare cases, the bacterial cells enter heart tissues, interfering with normal electrical conduction between the chambers (Lyme carditis), which leads to dizzy spells or fainting, heart palpitations, and chest pain. A temporary pacemaker may be required to maintain a normal heartbeat.
- **Late.** This begins months to years after the skin rash and is characterized by joint pain and swelling, usually of large joints such as the knee. These symptoms slowly disappear over subsequent years. Chronic nervous system problems such as localized pain and paralysis can occur.

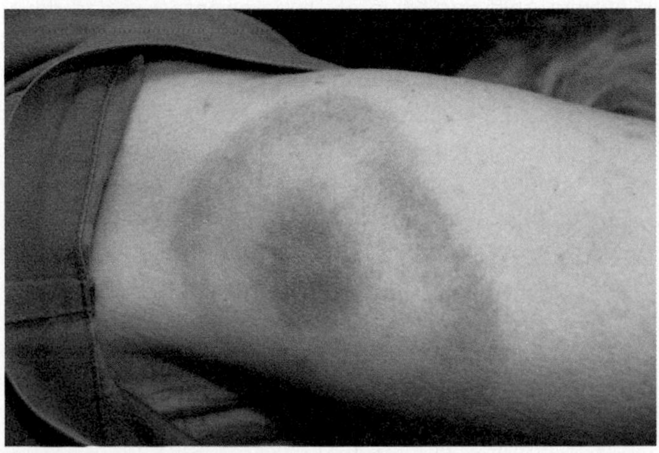

FIGURE 25.6 Erythema Migrans (EM) The characteristic rash of Lyme disease. James Gathany/CDC

? **What causes the bull's-eye pattern of this rash?**

Causative Agent

In the United States, Lyme disease is nearly always caused by *Borrelia burgdorferi,* a Gram-negative, microaerophilic spirochete (**figure 25.7**); a newly recognized species, *B. mayonii,* has been identified as an occasional cause in the Upper Midwest. In Europe and other parts of the world, species closely related to *B. burgdorferi* are common causes, and the disease is often called Lyme borreliosis.

Pathogenesis

The bacterial cells are introduced into the skin when an infected tick feeds, and they multiply there, where tick saliva components temporarily suppress the inflammatory response. The bacteria then begin moving away from the bite site, at which point the host response to lipopolysaccharide (endotoxin) results in an inflammatory reaction, causing the expanding rash. The bacterial cells then enter the bloodstream and begin spreading to other other parts of the body, leading to the flu-like symptoms of the first stage. An intense inflammatory response against the bacterial antigens develops, resulting in the rashes as well as nervous system and occasional cardiac effects of the second stage. The third stage of Lyme disease appears to involve an excessive pro-inflammatory response; some joints have high concentrations of reactive immune cells and immune complexes.

Epidemiology

Lyme disease is a widespread zoonosis in the United States, with humans an accidental host (**figure 25.8**); its incidence depends on complex ecological factors, including temperature, rainfall, and plant density, among others. Most of the infections are in the eastern and north-central states and transmitted by *Ixodes scapularis* (commonly called the blacklegged tick or

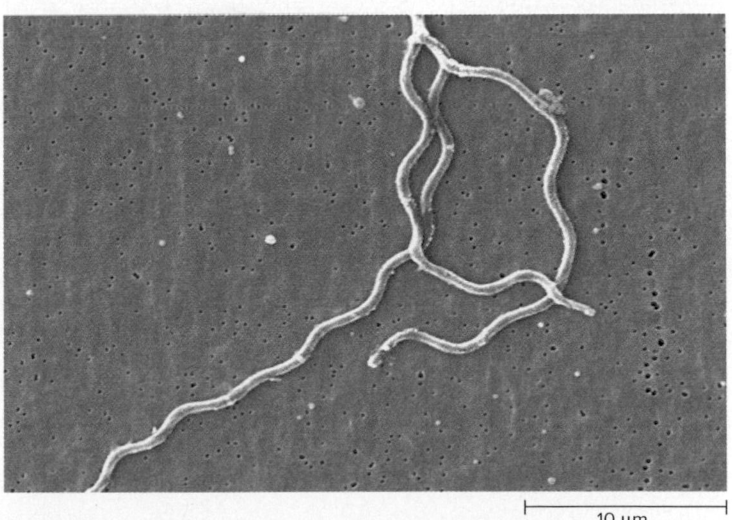

FIGURE 25.7 ***Borrelia burgdorferi,* the Cause of Lyme Disease** Color-enhanced SEM. Janice Haney Carr/CDC

? **How would you describe the shape of this organism?**

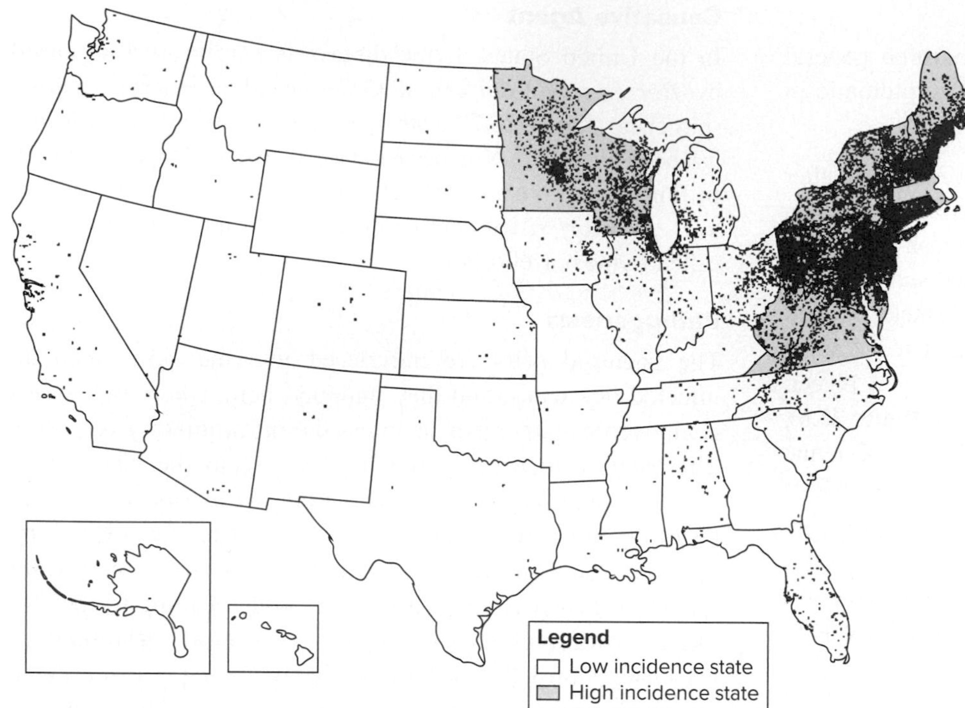

FIGURE 25.8 Reported Cases of Lyme Disease Each dot represents the county of residence of one confirmed case, 2019
Source: Data Compiled from CDC pub.data.

? **Why might the incidence of Lyme disease be influenced by ecological factors?**

deer tick) (**figure 25.9**); in states along the Pacific coast, particularly in California, *I. pacificus* (the western blacklegged tick) transmits the disease.

Ixodes ticks mature during a multi-year life cycle that involves an egg and three active stages: larva (plural: larvae), nymph, and adult. Each of these requires a blood meal from a new host, a process that can transmit *Borrelia* species. Starting with an adult female tick laying eggs, the steps of the *I. scapularis* life cycle and how it relates to *B. burgdorferi* transmission can be summarized as follows (**figure 25.10**):

FIGURE 25.9 The Black-Legged (Deer) Tick, *Ixodes scapularis* This tick is the most important vector of Lyme disease in the eastern and north-central United States. Scott Camazine/Science Source

? **Why are people often unaware that they have been bitten by this tick?**

1. **Adult female lays eggs.** The eggs are generally uninfected, even if they originated from an infected female.
2. **Eggs hatch, releasing larvae.** Larvae emerge from eggs in the summer.
3. **Larvae feed on first host and later develop into nymphs.** Larvae must each take a blood meal to develop into the next stage. To do this, a larva attaches to any of a variety of animals, with the most common reservoir of *B. burgdorferi* being the white-footed mouse. The larva inserts a feeding tube through the host's skin to ingest blood over a period of several days, and if that host is infected with *B. burgdorferi*, the larva may acquire the infection during the feeding process. Once a larva has obtained enough blood, it drops off the animal and then remains inactive during the winter. The following spring it molts (sheds its outer covering) to become a nymph; an infected larva becomes an infected nymph.
4. **Nymphs feed on second host and then develop into adults.** The nymphs become active when the weather warms in the spring and, like the larvae, they must each take a blood meal to develop into the next stage. When a *B. burgdorferi*-infected nymph feeds, it may transmit the infection to its host. If the infected nymph feeds on a mouse, that mouse typically becomes infected and develops a sustained bacteremia, thus becoming a source of infection for other ticks. If a nymph feeds on a human, that person might become infected and develop Lyme disease. For this to occur, however, the nymph must remain attached for least 2 days—a situation that often occurs because the nymph is very small (similar to a poppy seed) and is thus often unnoticed and therefore not removed; about two-thirds of Lyme disease patients are unable to recall a tick bite. The nymphs are most active from May to September, corresponding to the peak incidence of Lyme disease. After obtaining enough blood, the nymph falls off the animal and then molts to become an adult; infected nymphs develop into infected adults.
5. **Female adults feed on third host, mate, and then release eggs.** The adults become active in the fall and may remain so throughout the winter. During that time, each adult female tick must take a blood meal from an animal, often a white-tailed deer. Although deer do not

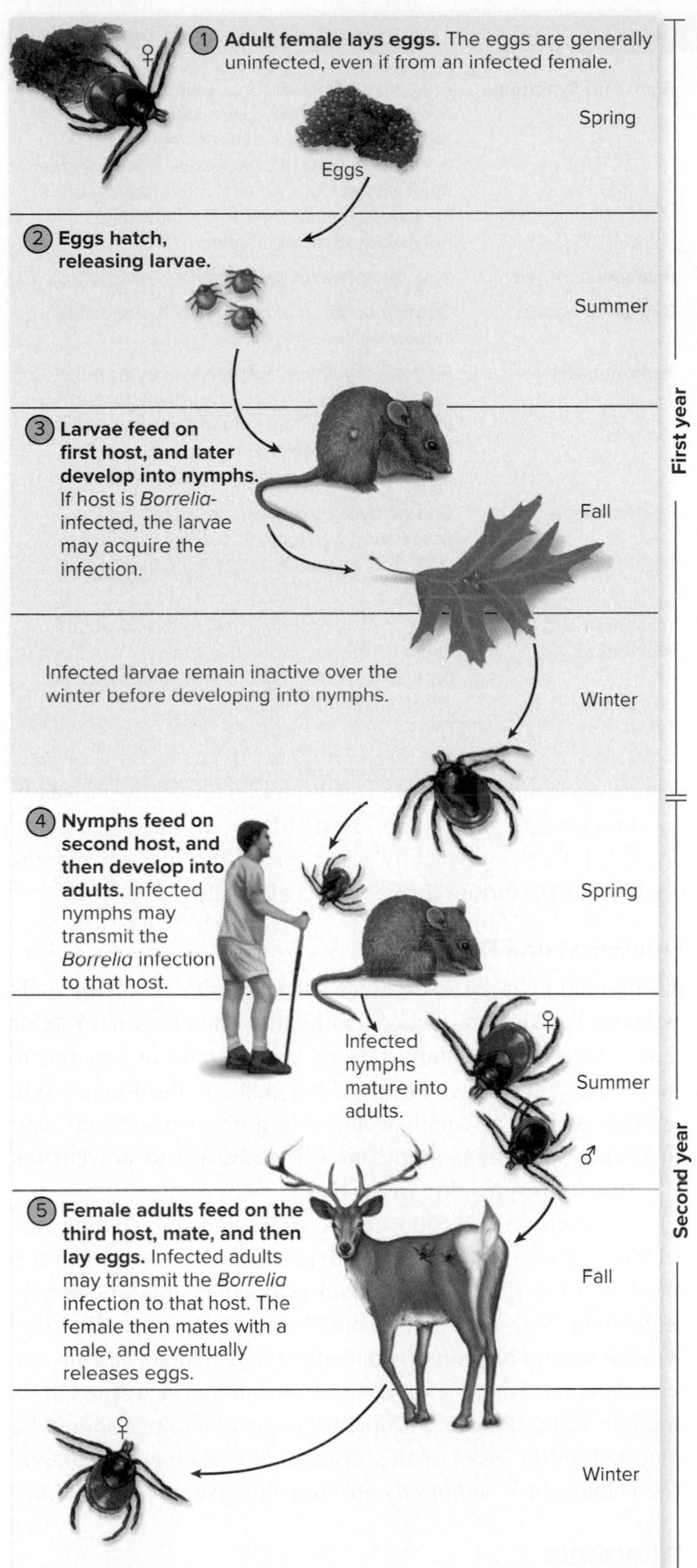

FIGURE 25.10 Life Cycle of the Black-Legged (Deer) Tick, *Ixodes scapularis* This illustration emphasizes the steps that relate to the transmission of *B. burgdorferi*. The life cycle of *I. pacificus* is similar but spans 3 years.

? Which stage of the Ixodes life cycle is usually responsible for transmitting Lyme disease and why?

appear to become infected, they are important because their movement spreads the attached ticks. Adult ticks that feed on humans can transmit Lyme disease, but they are more noticeable and therefore are often removed before transmission can occur. After the female feeds, male and female ticks mate, and the life cycle starts once again as the female tick then releases eggs.

Treatment and Prevention

Early antibiotic treatment prevents Lyme disease progression. For early localized disease, an oral antibiotic such as doxycycline is given. Treatment of later stages depends on the symptoms, but an extended course of doxycycline is often used.

General measures to prevent Lyme disease include avoiding exposure to ticks, wearing protective clothing, and using tick repellents. Post-exposure prophylaxis (PEP) with an oral dose of doxycycline taken within 72 hours of a high-risk tick bite reduces the risk of developing the disease. No vaccine is currently available. **Table 25.5** summarizes some features of Lyme disease.

Vibrio vulnificus Infection

Vibrio vulnificus was first isolated in 1976, from blood samples submitted to the CDC. Since then, the organism has become increasingly common in coastal environments of the United States.

Signs and Symptoms

Signs and symptoms of *V. vulnificus* infection vary, depending on the portal of entry and host factors. The two general routes of initial entry typically lead to different outcomes:

- **Oral.** When *V. vulnificus* is ingested, an otherwise healthy person may develop only a self-limiting case of gastroenteritis, with symptoms including fever, abdominal pain, vomiting, and diarrhea. In a person who has predisposing conditions such as high iron levels, liver disease, or immunodeficiency, the situation is quite different: The ingested bacterium may invade the bloodstream, leading to fevers, chills, severe skin blistering, and sepsis, which is often fatal.
- **Broken skin.** When *V. vulnificus* enters through broken skin, large blood-filled blisters develop; these rapidly progress to cellulitis and sometimes necrotizing fasciitis. The organism may enter the bloodstream, particularly in people who have predisposing factors, leading to the often fatal progression described above.

Causative Agent

V. vulnificus is a Gram-negative, motile, curved, rod-shaped bacterium. It is a halophile, found in warm coastal waters. Virulent strains have a polysaccharide capsule.

TABLE 25.5 Lyme Disease

① The bite of an infected tick introduces the bacteria into the skin.

② The bacteria multiply and move outward from the bite, resulting in an expanding red rash (erythema migrans).

③ The bacteria spread in the bloodstream; intense inflammatory response affects function of the nervous system and occasionally the heart.

④ Chronic symptoms may develop.

Signs and Symptoms	*Early localized infection:* expanding rash that resembles a bull's-eye develops at the site of the bite; swollen lymph nodes near the bite, flu-like symptoms. *Early disseminated infection:* heart and nervous system involvement. *Late persistent infection*: chronic arthritis and nervous system impairment.
Incubation Period	Several days to several weeks
Causative Agent	*Borrelia burgdorferi* (spirochetes) and certain closely related species
Pathogenesis	Bacteria introduced into the skin by an infected tick multiply and move outward; bacteria enter the bloodstream and spread throughout the body; immune reaction to bacterial antigens leads to tissue damage.
Epidemiology	Spread by infected ticks; usually found in association with animals such as white-footed mice and white-tailed deer living in wooded areas
Treatment and Prevention	Treatment: appropriate antibiotic. Prevention: avoid ticks; wear protective clothing; use tick repellents; post-exposure prophylaxis; no vaccine currently available

Pathogenesis

V. vulnificus enters the body orally or through broken skin. It produces a wide range of virulence factors that allow it to evade the host immune responses and multiply. The capsule appears to be essential for pathogenicity, protecting the bacterial cells from the effects of complement proteins. Iron is also very important, which is why anyone with an underlying disease that elevates serum iron levels is much more susceptible to severe infection. Normal iron levels in the body help limit the growth of *V. vulnificus*, even though the organism produces siderophores (that scavenge iron) as well as a hemolysin (that lyses blood cells to release iron-containing hemoglobin).

V. vulnificus directly damages tissues by producing a wide range of degradative enzymes. One protease in particular—metalloproteinase—damages the basement membrane of blood vessels and destroys fibrin, leading to the blistering. The organism also produces a type of multi-functional toxin that forms pores in host cell membranes, through which the toxin's lethal effector components then enter cells. The damage, as well as the inflammatory response to the infection, leads to systemic symptoms.

Epidemiology

V. vulnificus is a naturally occurring bacterium in coastal regions worldwide, found in water, sediments, and shellfish. Most infections in the United States are associated with eating contaminated oysters or other seafood, but the bacterium can also infect through cuts, abrasions, or other open wounds. Anyone can be at risk of *V. vulnificus* infection, but those with underlying conditions affecting the immune system—particularly people with high iron levels, liver disease, or immunodeficiency—are most likely to develop severe disease. Infections typically occur during the warmer months because the bacterium thrives in warmer water.

Treatment and Prevention

Prompt and aggressive treatment of *V. vulnificus* infections is essential because any delay significantly increases the risk of death. Patients are usually given a combination of synergistic antibiotics (such as doxycycline and a third-generation cephalosporin). Infected wounds require debridement and cleansing; sometimes amputation is necessary to prevent the infection from spreading in the tissues.

V. vulnificus infection can be prevented by avoiding raw seafood—especially oysters and particularly during warmer months—or by thoroughly cooking seafood. People who handle seafood in high-risk areas should wear protective clothing. Wounds should be promptly and effectively treated, and anyone with an open wound should avoid swimming in warm salt or brackish water. People with predisposing conditions should be educated on the risks of this disease. No vaccine is available. The features of *V. vulnificus* infection are given in **table 25.6.**

Tularemia

Tularemia, a zoonotic disease, can be transmitted from various wild animals to humans in multiple ways, resulting in very different disease manifestations. Although named after Tulare County, California, it occurs in many parts of the world. It is also a potential bioterrorism threat, listed as a Category A agent (see Focus Your Perspective 19.1).

TABLE 25.6	*Vibrio vulnificus* Infection
Signs and Symptoms	When the bacterium is ingested, it causes fever, vomiting, diarrhea, and abdominal pain; when it enters through broken skin, blood-filled blisters and cellulitis develop. Regardless of the entry route, people with predisposing conditions may develop a bloodstream infection that leads to skin blistering and sepsis.
Causative Agent	*Vibrio vulnificus,* a Gram-negative, halophilic, curved rod
Incubation Period	1 to 4 days
Pathogenesis	Encapsulated bacterium avoids phagocytosis and multiplies using iron from the host; it produces a variety of degradative enzymes as well as toxins that cause host cell damage and death.
Epidemiology	Found worldwide in warmer coastal habitats; typically acquired by eating raw or undercooked seafood, especially oysters; can also enter through broken skin; people with predisposing conditions are at greater risk of developing severe disease.
Treatment and Prevention	Treatment: combination of antibiotics; for wound infections, also debridement or amputation. Prevention: avoid raw or undercooked seafood and wear protective clothing when handling seafood in high-risk areas; no vaccine available.

Signs and Symptoms

Signs and symptoms of tularemia typically appear 2 to 5 days after exposure; they usually clear in 1 to 4 weeks but sometimes last for months. Several forms of the disease occur, depending on how the causative organism enters the body:

- **Ulceroglandular.** This, the most common type, develops after the bacterium is transmitted via direct contact or an arthropod bite, resulting in a characteristic ulcer at the entry site (**figure 25.11**). Regional lymph nodes enlarge, and fever, chills, headaches, and muscle aches occur.

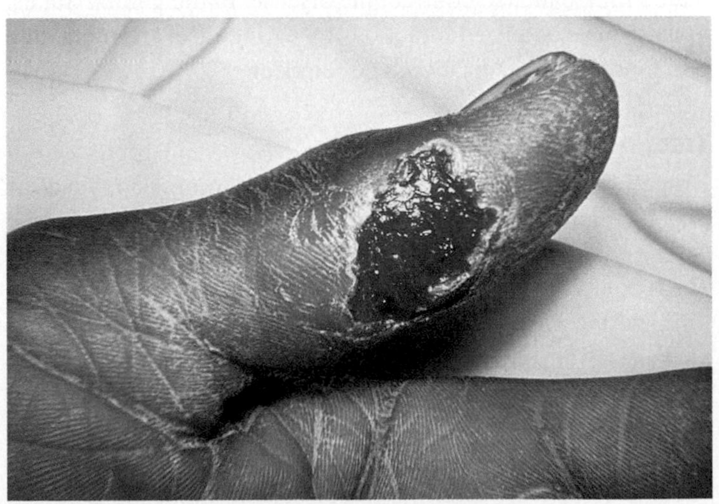

FIGURE 25.11 Tularemic Ulcer of the Thumb CDC/Science Source

? How can tularemia be transmitted?

- **Glandular.** This is similar to the ulceroglandular form, but without ulcers.
- **Oculoglandular.** This develops when the bacterium is transmitted to the eye, resulting in eye inflammation and swelling of regional lymph nodes.
- **Oropharyngeal.** This develops when the bacterium is ingested, resulting in sore throat, mouth ulcers, and swollen lymph nodes in the neck.
- **Pneumonic.** This develops when the bacterium is inhaled or spreads to the lungs from the bloodstream, resulting in pneumonia, with cough, chest pain, and breathing difficulty.

Causative Agent

Tularemia is caused by *Francisella tularensis,* a fastidious, non-motile, aerobic, Gram-negative rod. The organism requires a special medium enriched with the amino acid cysteine in order to grow.

Pathogenesis

F. tularensis enters tissues through a break in the skin or through a mucous membrane. The inflammatory response to the invasion is initially weak, partly due to altered surface components of the bacterial cells that limit detection by certain pattern recognition receptors of the innate immune system. When ingested by a macrophage, the bacteria escape from the phagosomes and grow within the phagocyte's cytosol, protected from antibodies. Infected macrophages then carry the bacteria to the regional lymph nodes, which become large, tender, and often filled with pus. From there, the bacteria spread to other parts of the body via the lymphatics and the blood vessels. In the lungs, infected alveolar macrophages release large quantities of pro-inflammatory cytokines, and this cytokine storm can result in capillary leakage, tissue injury, and lethal organ failure.

Epidemiology

Tularemia is a zoonosis associated with wild animals—including rabbits, hares, and rodents. Person-to-person transmission does not occur. The disease occurs in many areas of the Northern Hemisphere, including all the states of the United States except Hawaii. *F. tularensis* is highly infectious, with an infectious dose as low as 10 to 50 cells. Infections typically result from the bites of infected ticks, although in the western United States, they also result from the bites of infected deer flies. The bacterium can also be acquired from handling or eating contaminated animals and meat, or inhaling contaminated aerosol droplets or dust. While most tularemia cases occur in summer, some appear in the winter months in people skinning rabbits. Pneumonic tularemia, although rare, is serious and has a case-fatality rate as high as 30% if untreated; outbreaks have been associated with activities such as mowing lawns and entering rodent-infested buildings.

Treatment and Prevention

Tularemia is treated with antibiotics, but the medication choice usually depends on disease severity. Mild or moderate cases are often treated with an oral fluoroquinolone, but severe cases are usually treated with an aminoglycoside.

Tularemia can be prevented by using insect repellents and protective clothing to protect against insect and tick vectors, and by using rubber gloves and goggles or face shields when skinning or handling wild animals. The main features of tularemia are summarized in **table 25.7.**

Brucellosis

Brucellosis, like tularemia, is a zoonotic disease and potential bioterrorism threat. It is listed as a Category B agent (see Focus Your Perspective 19.1).

Signs and Symptoms

Brucellosis usually starts gradually 2 to 4 weeks after exposure, with vague signs and symptoms including mild fever, sweating, weakness, aches and pains, swollen lymph nodes, fatigue, and weight loss. The fevers sometimes recur over weeks or months, a characteristic that gave rise to the alternative name of the disease, "undulant fever." Most people recover within a few months even without treatment, but some people develop chronic illness.

Causative Agent

Brucella species are small, aerobic, non-motile, Gram-negative rods with complex nutritional requirements. At least four varieties cause brucellosis in humans. DNA studies indicate that many varieties fall into a single species, but different species names are assigned, some of which indicate the preferred host: *B. abortus* infects cattle; *B. canis,* dogs; *B. melitensis,* goats and sheep; and *B. suis,* pigs. The distinctions among the different varieties are mainly useful epidemiologically.

TABLE 25.7 Tularemia

Signs and Symptoms	Ulcer at site of entry, swollen regional lymph nodes, fever, chills, aches; pneumonia if causative agent is inhaled
Incubation Period	1 to 10 days; usually 2 to 5 days
Causative Agent	*Francisella tularensis,* a Gram-negative rod
Pathogenesis	Organisms are ingested by macrophages, grow within these cells, and then spread throughout body.
Epidemiology	Present among wildlife in Northern Hemisphere; most commonly acquired through the bite of an infected tick or insect but also by handling wildlife.
Treatment and Prevention	Treatment: appropriate antibacterial medication. Prevention: avoiding bites of ticks and insects; wearing rubber gloves and goggles or face shields when skinning animals.

Pathogenesis

Brucella cells enter tissues through mucous membranes (following ingestion or inhalation) or through wounds. The initial inflammatory response to the invasion is relatively weak, probably because some of the bacterial cells' surface features have structural modifications that limit detection by pattern recognition receptors of the innate immune system. When the bacterial cells are taken up by macrophages, they prevent phagosome-lysosome fusion and multiply within the phagocytic cell, protected from antibodies. Infected macrophages carry the bacteria to other parts of the body, including lymph nodes, spleen, liver, bone marrow, heart, and kidneys, where the bacteria then infect other mononuclear phagocytes (see figure 14.6). The case-fatality rate of brucellosis is low; most deaths are due to endocarditis or meningitis. Osteomyelitis (bone infection) is a serious but rare complication.

Epidemiology

Brucellosis is typically a chronic infection of domestic animals. It contaminates their milk and can cause abortions in the affected animals. Worldwide, brucellosis is a major problem in animals used for food, causing millions of dollars of losses each year.

Most human brucellosis cases involve occupational exposure to the causative agents; veterinarians, slaughterhouse workers, and butchers, for example, are at higher risk because of their contact with animals or their meat. People drinking unpasteurized milk or eating soft cheeses made from contaminated milk can also contract the disease. In the United States, hunters have acquired infections by eating meat from infected elk, moose, bison, caribou, and reindeer.

Treatment and Prevention

Brucellosis can usually be treated using an oral doxycycline plus rifampin for at least 6 weeks. In complicated or chronic cases, extended treatment with different antibiotics may be needed.

The most important control measures for brucellosis are pasteurization of dairy products and inspection of domestic animals for evidence of the disease. Veterinarians, butchers, and slaughterhouse workers should use protective gear such as goggles or face shields and rubber gloves. An attenuated vaccine effectively controls the disease in domestic animals. **Table 25.8** gives the main features of brucellosis.

TABLE 25.8	Brucellosis
Signs and Symptoms	Fever, body aches, weight loss, swollen lymph nodes; symptoms usually resolve within a few months even without treatment but may become chronic.
Incubation Period	5 days to 6 months (usually 2 to 4 weeks)
Causative Agent	*Brucella* species; small Gram-negative rods
Pathogenesis	Organisms enter through mucous membranes and are carried to heart, kidneys, and other parts of the body via the blood and lymphatic system; they resist killing by macrophages and grow within these cells.
Epidemiology	Main sources of human infections are domestic animals; veterinarians, slaughterhouse workers, butchers, and those who drink unpasteurized milk are at risk. Disease also occurs in wild animals.
Treatment and Prevention	Treatment: appropriate antibiotics. Prevention: vaccinate domestic animals; pasteurize milk and milk products; use gloves and face shields.

MicroAssessment 25.2

A systemic bacterial infection may lead to bacterial sepsis, life-threatening tissue damage, and organ dysfunction resulting from an overwhelming host response to to an infection. People with indwelling IV catheters or heart abnormalities, or who abuse intravenous drugs, are at increased risk of developing infective endocarditis. Plague is endemic in rodents and is transmitted to humans by flea bites. Lyme disease is the most common vector-borne disease in the United States. *Vibrio vulnificus* infection may cause severe disease, particularly in people with high iron levels or other predisposing conditions. Tularemia and brucellosis are zoonotic diseases.

4. How is bacterial sepsis different from bacteremia?
5. What is the most common way to contract *Vibrio vulnificus* infection?
6. Why might clots on the heart valves make microorganisms colonizing there inaccessible to phagocytic killing?

25.3 ■ Viral Diseases of the Cardiovascular and Lymphatic Systems

Learning Outcomes

5. Describe the characteristics of infectious mononucleosis.
6. Compare and contrast Ebola virus disease and Marburg virus disease.
7. Compare and contrast dengue and severe dengue, chikungunya, Zika virus disease, and yellow fever.

This section will focus on viruses that primarily infect the cardiovascular and/or lymphatic system, causing widespread effects. In addition to these viruses, others (such as SARS-CoV-2) typically infect different body systems but from there can sometimes spread systemically. Regardless of their origins, these systemic infections may elicit an excessive production of cytokines (a cytokine storm), which can potentially lead to viral sepsis. Relatively little is known about the mechanisms or the frequency of viral sepsis, but the observation that the signs and symptoms of severe COVID-19 have similarities to bacterial sepsis has led to research that will hopefully provide a better understanding.

Of the diseases described in this section, infectious mononucleosis, Ebola virus disease, and Marburg virus disease are transmitted from person to person in body fluids. In contrast, dengue and severe dengue, chikungunya, Zika virus disease, and yellow fever are transmitted by mosquitoes; the viruses that cause these mosquito-borne diseases are part of a group referred to as **arboviruses,** meaning **ar**thropod **bo**rne.

Infectious Mononucleosis ("Mono" or "Kissing Disease")

Infectious mononucleosis ("mono") is a disease familiar to many students because of its high incidence among young people. The term *mononucleosis* refers to the fact that proliferating infected B cells accumulate in large numbers. Their abundance and appearance sometimes lead to the incorrect diagnosis of leukemia—this is disproved when the person spontaneously recovers.

Signs and Symptoms

Signs and symptoms of infectious mononucleosis typically appear after a long incubation period, usually 30 to 60 days. They include fatigue, fever, a sore throat, swollen lymph nodes, white patches of pus near the tonsils, and sometimes an enlarged spleen. In extremely rare situations, the spleen ruptures, a potentially fatal complication. People generally fully recover within 4 weeks, although some suffer severe exhaustion and difficulty concentrating for months.

Causative Agent

Infectious mononucleosis is caused by the Epstein-Barr virus (EBV), an enveloped double-stranded DNA virus of the family *Herpesviridae*. Like other members of this family, EBV can remain latent within host cells, thereby causing a lifelong infection.

MicroByte

EBV was discovered in the 1960s by Michael Anthony Epstein and Yvonne Barr, who isolated it from Burkitt's lymphoma, a malignant tumor derived from B lymphocytes.

Pathogenesis

EBV initially infects cells in the mouth and throat, causing a sore throat. After replicating there, the virus is carried to the lymph nodes, where it infects B lymphocytes (**figure 25.12**). The B-cell infection can be:

- **Productive.** The virus replicates, resulting in death of the B cell.
- **Latent.** This is the most common outcome; the virus does not replicate, but instead is maintained in the infected B cell as either a plasmid or a provirus (integrated into the host cell chromosome). EBV can activate the latently infected cells, causing them to proliferate and produce immunoglobulins, including what are called *heterophile antibodies*; these have no pathological significance but provide the foundation for a diagnostic test called the monospot test, which relies on an agglutination reaction that occurs between the antibodies and horse red blood cells.

Effector cytotoxic T cells respond to the infection by destroying productively infected B cells. In blood smears, those effector T cells look different from normal resting lymphocytes—they are larger and have an unusually shaped nucleus—and thus are referred to as atypical lymphocytes (**figure 25.13**). Proliferation of the lymphocytes causes lymph node swelling and spleen enlargement.

EBV infection is linked to the development of certain malignancies (cancers) and autoimmune diseases. Most people are latently infected with EBV, however, so although the viral genome is detectable in almost all cases of Burkitt's lymphoma, multiple sclerosis, and nasopharyngeal carcinoma, other factors must also be involved.

Epidemiology

EBV is distributed worldwide, and humans are the only reservoir. People with infectious mononucleosis typically shed the virus in saliva for up to 18 months after recovery and may shed it intermittently for life as the latent virus can reactivate; people with AIDS or other immunodeficiencies may shed the virus continuously. Thus, anyone who has ever been infected with EBV, even without symptoms, can be a source of the virus. Mouth-to-mouth kissing is an important mode of transmission in young adults, leading to the name "kissing disease."

Children living in low socioeconomic environments often become infected with EBV at an early age, usually with no obvious symptoms. In more affluent populations, the primary infections occur later in life (such as when students go to college), often resulting in infectious mononucleosis; even then, however, only about half of newly infected young adults are symptomatic. By middle age, most people have antibodies to the virus.

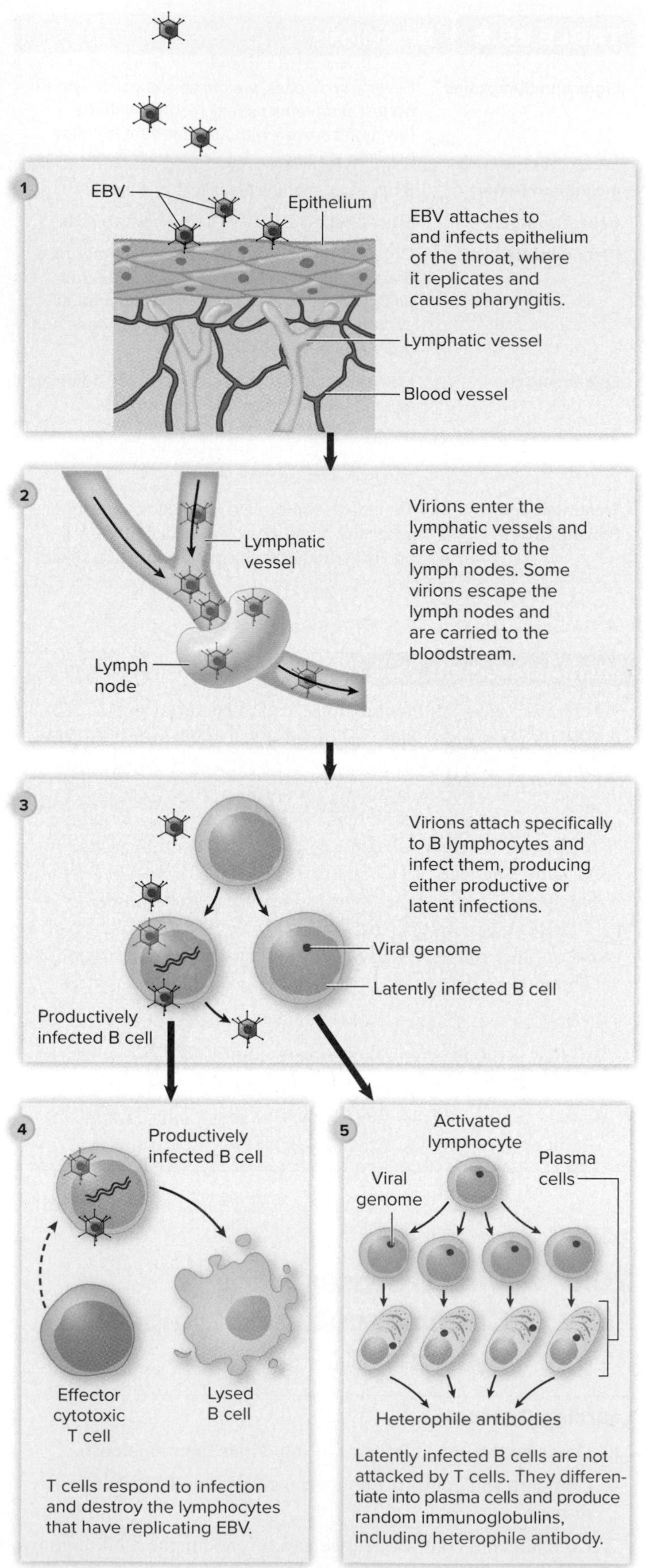

FIGURE 25.12 Pathogenesis of Infectious Mononucleosis

? What is the importance of heterophile antibodies in infectious mononucleosis?

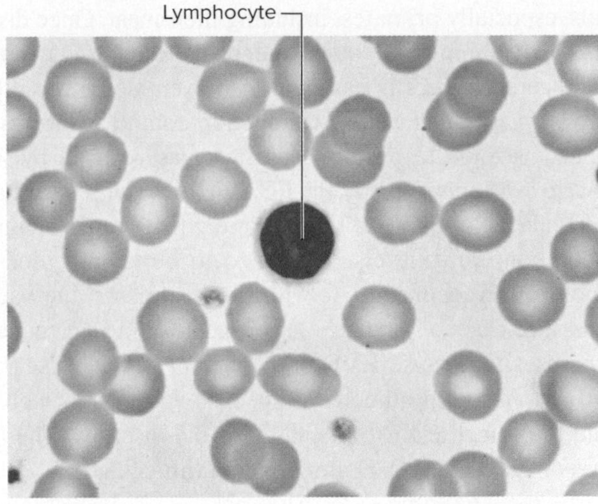

(a)

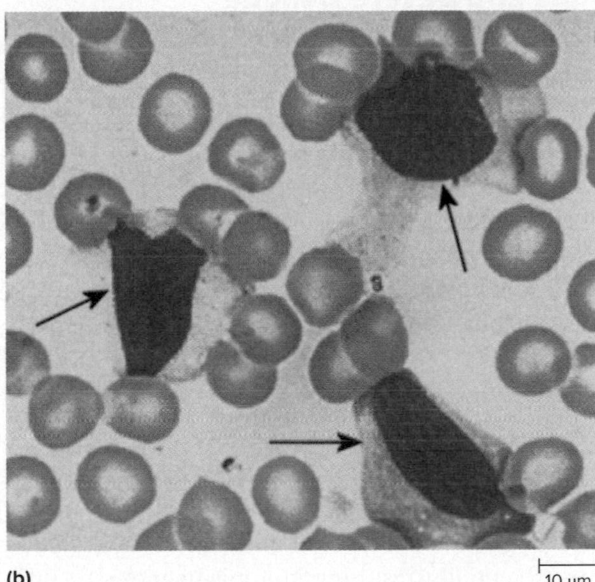

(b)

FIGURE 25.13 Normal and Atypical Lymphocytes (a) A normal lymphocyte. **(b)** Atypical lymphocytes characteristic of infectious mononucleosis. a: jarun011/Getty Images; b: Chamberlain, Medical Microbiology: The Big Picture

? Why is infectious mononucleosis sometimes misdiagnosed as leukemia?

Treatment and Prevention

No specific treatment is available for infectious mononucleosis, but supportive care such as pain medication can relieve symptoms. The disease can be prevented by avoiding the saliva of another person; this includes not sharing toothbrushes, drinking glasses, or any other object that may be contaminated with saliva. There is no vaccine for this disease. **Table 25.9** gives the main features of infectious mononucleosis.

TABLE 25.9	Infectious Mononucleosis ("Mono" or "Kissing Disease")
Signs and Symptoms	Fatigue, fever, sore throat, swollen lymph nodes, and sometimes enlarged spleen
Incubation Period	Usually 1 to 2 months
Causative Agent	Epstein-Barr virus (EBV), an enveloped double-stranded DNA virus
Pathogenesis	Productive infection of cells in the throat and mouth; productive and latent infections of B lymphocytes
Epidemiology	Spread by saliva; once infected, lifelong recurrent shedding of the virus into saliva
Treatment and Prevention	Treatment: no specific treatment available. Prevention: avoiding saliva of other people. No vaccine currently available.

Ebola Virus Disease and Marburg Virus Disease

Ebola virus disease (EBV) and Marburg virus disease (MVD) are both emerging diseases. Typically found in Africa, their person-to-person transmission and high case-fatality rates make them global concerns; they are both considered Category A bioterrorism threats (see Focus Your Perspective 19.1).

The names of EVD and MVD have changed over time, which creates confusion but provides an excellent example of the difficulties in communicating about newly recognized diseases. EVD, for example, was once called Ebola hemorrhagic fever, but because many cases do not involve hemorrhaging, the name was changed to Ebola virus disease (reflecting the common name of the causative agent—Ebola virus). Further complicating matters was the recognition that a group of related viruses cause clinically indistinguishable diseases. Some scientists now use the term *Ebola virus disease* to refer to disease caused by any of the related viruses (a practice we will use here), but others instead reserve that term for illness caused specifically by Ebola virus; these scientists use *Ebola disease* as a general term to refer to any of the clinically indistinguishable diseases. Although those distinctions may seem trivial, available vaccines and medications are virus-specific, so awareness of the terminology differences can be important.

Signs and Symptoms

Signs and symptoms of EVD and MVD both appear about 5 to 10 days after infection (range 2 to 21 days). Patients initially develop flu-like symptoms, including fever, headache, muscle pain, and sometimes a sore throat, and then abdominal pain, diarrhea, and a diffuse macular rash (small flat red spots). Hemorrhaging in the internal organs can also occur, in addition to bleeding from mucous membranes, puncture sites, and body orifices, particularly the eyes. The case-fatality rates of the diseases vary considerably, ranging from 25% to 90%, depending on the infecting strain and the supportive treatment available.

Causative Agents

The viruses that cause EVD and MVD belong to the family *Filoviridae*, a group of filamentous, enveloped, single-stranded

RNA viruses. Because of this, both EVD and MVD are within the category of Filovirus disease (FVD).

EVD is caused by ebolaviruses, meaning members of the genus *Ebolavirus* (named after the Ebola River Valley in the Democratic Republic of the Congo, formerly Zaire). At least three ebolaviruses have caused outbreaks in humans: *Zaire ebolavirus* (also called Ebola virus), *Sudan ebolavirus* (also called Sudan virus), and *Bundibugyo ebolavirus* (also called Bundibugyo virus); of these, most outbreaks have been caused by *Zaire ebolavirus*. MVD is caused by marburgviruses (named for the German town of Marburg), usually Marburg virus.

Pathogenesis

The viruses that cause EVD and MVD primarily infect mononuclear phagocytes, including dendritic cells and macrophages, but other cell types can be infected as well, and they spread via the lymphatic system and the bloodstream. Infection initially interferes with the innate and adaptive immune responses, allowing the viruses to replicate to high levels. The infection then triggers an excessive release of pro-inflammatory cytokines—a cytokine storm—resulting in leaky capillaries and disrupted blood clotting, which can lead to hemorrhaging, organ dysfunction, and shock.

Epidemiology

EVD and MVD are both endemic in Africa. Egyptian fruit bats (*Rousettus aegyptiacus*) were recently confirmed to be a reservoir host for Marburg virus, and bats are also thought to be the reservoir of the EVD-associated viruses. The diseases have been found in non-human primates, but the high mortality rate in these infected animals makes it unlikely that they are natural reservoirs. People can become infected by handling an infected animal, but most cases are acquired from an infected person. Human-to-human transmission occurs through direct contact with body fluids of a symptomatic or deceased patient, a situation that puts healthcare professionals and family members of infected individuals at greatest risk. Social and economic conditions in endemic areas often make outbreaks difficult to contain. The viruses can also be spread to hosts in other countries by infected travelers as well as in imported monkeys.

Treatment and Prevention

Patients with EVD or MVD are given supportive care, including fluids and supplemental O_2. To treat EVD caused by *Zaire ebolavirus* (Ebola virus), two monoclonal antibody preparations were approved by the FDA in 2020: one is a cocktail of three monoclonal antibodies and the other is derived from an antibody produced by a lymphocyte obtained from a survivor of the disease. No medications are currently approved for treating EVD caused by the other viruses or for treating MVD.

To reduce the risk of animal-to-human transmission of EVD and MVD, people living in or visiting areas where either disease has been reported should avoid contact with living or dead wild animals, especially primates, including their meat. Once disease occurs in a community, quick diagnosis and rapid response by healthcare workers is essential to prevent spread to others, and the workers must use strict infection-control measures. An attenuated vaccine to protect against disease caused by *Zaire ebolavirus* was approved by the FDA in 2020; ring vaccination is used once a case is identified, meaning that family members, neighbors, and anyone else identified to be at increased risk of exposure is vaccinated. As with the treatments, the vaccine is virus-specific; no vaccine is licensed to protect against the other viruses that cause EVD or to prevent MVD. Research is underway to develop other virus-specific vaccines as well as a vaccine to protect against all causes of the diseases. The main features of EVD and MVD are shown in **table 25.10.**

Dengue and Severe Dengue

Dengue (pronounced DENG-ee), a mosquito-borne disease, is also known as "break-bone fever" because it causes severe joint and muscle pain. It is the most common vector-borne viral disease in the world; the World Health Organization lists it as a neglected tropical disease (NTD) and estimates that about half the global population lives in dengue endemic areas. Most infections by the causative agent are asymptomatic or subclinical, but sometimes life-threatening severe dengue develops, primarily in people infected previously.

Signs and Symptoms

Dengue symptoms usually begin 4 to 7 days after infection, with sudden onset of a fever that lasts up to a week; the patient may also experience headaches with eye pain, joint and muscle pain, and a rash. In mild cases, the rash does not develop, potentially resulting in a misdiagnosis of influenza

TABLE 25.10	Ebola Virus Disease (EVD) and Marburg Virus Disease (MVD)
Signs and Symptoms	Fever, headache, muscle pain, abdominal pain, diarrhea, a rash, and sometimes bleeding from internal organs and mucous membranes
Incubation Period	Usually 5 to 10 days
Causative Agents	Ebolaviruses cause EBV and marburgviruses cause MVD; filamentous, enveloped, single-stranded RNA viruses
Pathogenesis	Primarily infect mononuclear cells; interfere with immune responses and blood clotting
Epidemiology	Endemic to Africa; fruit bats are the reservoir of Marburg virus and possibly the other disease-causing filoviruses; direct human-to-human transmission
Treatment and Prevention	Treatment: supportive care; for EVD caused by *Zaire ebolavirus*, monoclonal antibody treatments. Prevention: strict infection control measures; vaccine to prevent EVD caused by *Zaire ebolavirus*.

or another viral disease. Although dengue is generally self-limiting and rarely fatal, a critical phase begins as the patient's fever subsides. At this point, the patient may recover uneventfully, but if certain warning signs develop—including abdominal pain, persistent vomiting, mucosal bleeding, and restlessness—life-threatening severe dengue may follow.

Severe dengue begins with signs of plasma leaking from blood vessels. The patient initially develops a weak, rapid pulse, and the disease then progresses quickly. Petechiae (tiny purplish spots due hemorrhage), bruising, and sometimes massive gastrointestinal and vaginal bleeding occur. Blood pressure rapidly declines, potentially resulting in irreversible shock and organ damage. Widespread blood clotting and disseminated intravascular coagulation (DIC) may develop.

Causative Agent

Dengue and severe dengue are caused by dengue fever virus (DENV), an enveloped single-stranded RNA arbovirus of the *Flaviviridae.* Four widely spread serotypes of the virus are known (DENV1, DENV2, DENV3, and DENV4), all of which are transmitted by the bite of infected mosquitoes of the *Aedes* family. The mosquitoes are biological vectors, meaning that the infectious agents they transmit replicate within them.

Pathogenesis

The details of dengue pathogenesis are not well understood because of lack of suitable animal models in which to study the disease. Research indicates, however, that once DENV is injected by a mosquito, it infects local dendritic cells and macrophages at the bite site. Infected dendritic cells then migrate to the lymph nodes, where monocytes and macrophages are recruited and also become infected. DENV multiplies in those cells, and the resulting viral particles may spread in the bloodstream leading to general symptoms of a systemic viral infection.

Recovery from an infection by a given DENV serotype usually provides long-lasting immunity to that serotype. It provides only transient immunity to the other serotypes, however, and actually increases the chance of severe dengue if infection by a different serotype occurs later. The susceptibility is thought to be the result of **antibody-dependent enhancement (ADE).** In this model, pre-existing anti-dengue antibodies from the first dengue infection recognize and bind to antigens on the DENV of the second infection but fail to neutralize the virus. Instead, the antibodies facilitate viral entry into cells that express Fc receptors, especially monocytes and macrophages, where the virus then freely replicates. ADE thus leads to increased numbers of infected cells and a high viral load.

In severe dengue, infected monocytes and macrophages produce high levels of pro-inflammatory cytokines, resulting in a cytokine storm. This, along with activated complement proteins and cytokines released by T cells, increases vascular permeability, leading to plasma leakage that may result in decreased blood pressure and respiratory distress; the crucial feature of severe dengue is the plasma leakage. Capillaries become fragile, resulting in the observed hemorrhages (petechiae, easy bruising, and gastrointestinal and vaginal bleeding). Infected monocytes and macrophages die and release toxic products that cause blood clotting and DIC.

Epidemiology

Dengue is the fastest-spreading mosquito-borne viral disease in the world, and is now endemic in tropical and subtropical regions globally. It is most common in urban and semi-urban areas that have high rainfall and accumulations of fresh water in which the mosquito vectors can breed.

Dengue's epidemiology is largely explained by the distribution and feeding habits of its mosquito vectors. Its main vector, *Aedes aegypti,* thrives in tropical and subtropical urban areas and preferentially feeds on humans (**figure 25.14**). Another vector, *Aedes albopictus*, is significant because it tolerates a wide range of climates and is rapidly expanding its range, resulting in a corresponding spread of dengue; the species is native to Southeast Asia but has now spread to parts of Europe, the Americas, Africa, and the Pacific. *A. albopictus* also bites more frequently than *A. aegypti* and feeds on a range of animals. Geographic spread of mosquitoes—and thus mosquito-borne diseases—is linked to international travel, breakdown of vector control measures, climate change, and an increase in mosquito-breeding habitats (such as used tires and containers in which water has accumulated). Infected humans are a primary reservoir of dengue, but non-human primates and possibly other animals appear to be reservoir hosts as well.

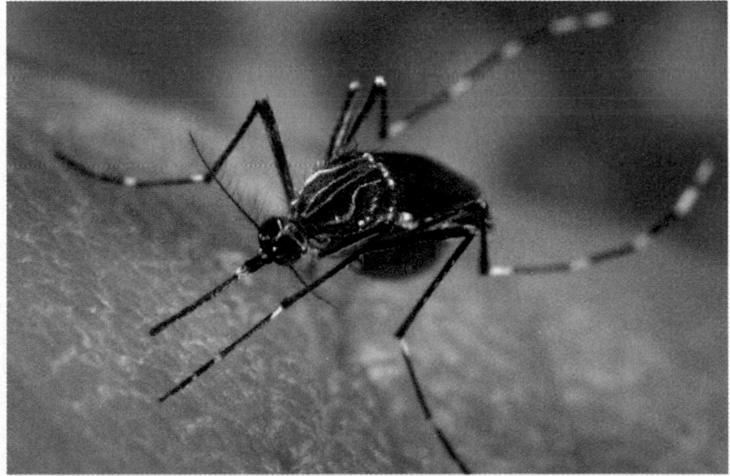

FIGURE 25.14 ***Aedes aegypti* Mosquito Taking a Blood Meal** Female mosquitoes require blood for their eggs to develop. During a bite, the female pumps saliva through her hollow mouthpart (proboscis), which increases blood flow and prevents clotting. Infectious agents picked up from the host's capillaries replicate within her body and can be transferred to a new host in a later bite. James Gathany/CDC

? **Why is dengue spreading geographically?**

Treatment and Prevention

There is no specific medication to treat dengue. Supportive care, including fluid replacement, helps relieve symptoms and is required for severe dengue. Aspirin and nonsteroidal anti-inflammatory drugs (NSAIDs) should be avoided, however, because they can worsen bleeding. Efforts are underway to develop antiviral medications.

Prevention and control of dengue and severe dengue are particularly important given the diseases' expanding geographic distribution and increasing incidence. Personal protection—including wearing long sleeves and pants, using an effective insect repellent, and sleeping under an insecticide-treated bed net (ITN)—helps prevent mosquito bites. ITNs referred to as long-lasting insecticide-treated nets (LLINs) maintain their effectiveness for at least 3 years or 20 washes. Additional control measures rely on limiting or eradicating mosquito populations by destroying indoor and outdoor breeding sites (standing water) and using effective insecticides, although insecticide resistance is increasing. Some countries have successfully used biological control methods (for example, using fish that eat the mosquito larvae). An exciting new approach is the release of lab-grown mosquitoes within which dengue virus replicates poorly; these mosquitoes are infected with *Wolbachia pipientis*, a bacterium that interferes with not only dengue virus replication, but also the replication of the other arboviruses described in this section. Although the bacterium naturally infects a wide range of insects, it must be introduced artificially into *Aedes aegypti;* once infected, however, the mosquitoes can breed with wild mosquitoes, and *W. pipientis* is transmitted to the resulting offspring.

A recombinant attenuated dengue vaccine that includes components from all four dengue strains is available in a number dengue-endemic countries for people ages 9 to 45 years old who have laboratory-confirmed evidence of a previous dengue infection. Vaccination is limited to those with prior infection because they are at risk of severe dengue; it is not used for people who have not been infected because it might increase their risk of developing the severe disease. The FDA has approved the vaccine for children 9 to 16 years old with laboratory-confirmed previous infection and living in certain U.S. territories where the disease is common. A different recombinant attenuated vaccine has been approved in several countries for use even in people who have not had a previous infection. The characteristics of dengue fever are compared to other mosquito-borne viral diseases at the end of this section (see table 25.11).

Chikungunya

Chikungunya (pronounced chik-en-gun-ye), another mosquito-borne neglected tropical disease, got its name from an African

FOCUS ON A CASE 25.1

The patients were two American boys, 9 and 11 years old, living in Thailand, who developed irritated eyes and slightly runny noses, progressing to fever, headache, and severe muscle pain. It hurt them to move their eyes, and they refused to walk because of pain in their legs. The symptoms subsided after a few days, then recurred, although less intensely. No treatment was given, and they recovered quickly. Their illness was diagnosed as dengue.

1. Why was it significant that the boys were living in Thailand rather than the United States?
2. What would be the risk if the boys became infected with a different strain of dengue virus?
3. What can the boys do to prevent reinfection with a dengue virus?
4. How does the spread of dengue illustrate the importance of the One Health approach to disease control described in section 19.3?

Discussion

1. Dengue, which is spread by mosquitoes, is endemic in large areas of the tropics and subtropics around the world. It is not common in the United States—most reported cases in this country have occurred in travelers infected elsewhere. The boys, being in Thailand, were living in a country in which dengue is endemic and were thus at risk of contracting the disease.
2. If either boy were to become infected with a different strain of the virus, he would be at risk of developing severe dengue, which is much more serious than regular dengue. People who have severe dengue suffer from capillary leakage, bleeding, and bruising. Blood pressure drops and disseminated intravascular coagulation (DIC) may occur—these can lead to shock and death.
3. The boys can protect themselves from getting re-infected by taking precautions to avoid mosquito bites; these precautions include sleeping under an insecticide-treated bed net, wearing long sleeves and pants, and avoiding areas where mosquitoes breed. Because of their ages and the fact that they have had a confirmed dengue infection, both boys can receive the dengue vaccine.
4. Because of the complex nature of dengue's spread, effective control programs must rely on the coordinated expertise of people working in diverse fields. Among the many examples of contributing professions are: climate analysts (to predict changes in mosquito range and behavior), molecular biologists (to develop improved diagnostics), information technologists (to manage data), vector control specialists (to develop methods for mosquito control), environmental engineers (to eliminate mosquito breeding areas), virologists (to study viral evolution), and policymakers/governments (to design and implement regulations).

word meaning "that which bends up" because people with the disease show bent or contorted posture due to severe joint pain. Many infections involving the causative agent, however, are asymptomatic or subclinical. Although chikungunya is not a new disease, it is considered an emerging disease because of several recent outbreaks.

Signs and Symptoms

Chikungunya signs and symptoms usually begin 3 to 7 days after infection and include sudden onset of fever, along with severe, sometimes debilitating, joint pain; the pain usually involves many joints, frequently affects the extremities (hands, feet, wrists, and ankles), and can persist for weeks or months, or even longer. Patients often develop a rash. Nonspecific symptoms include headache, conjunctivitis ("pink eye") and photophobia, back pain, nausea, and general malaise. Symptoms usually resolve within a week, and the disease is seldom fatal.

Causative Agent

Chikungunya is caused by chikungunya virus (CHIKV), an enveloped, single-stranded RNA arbovirus of the *Togaviridae* family. The virus is mainly transmitted to humans by the bite of infected *Aedes* mosquitoes, the biological vector.

Pathogenesis

The disease mechanism of CHIKV is largely unknown, but its damaging effects appear to be related to the immune response. Infection stimulates significant release of cytokines and an eventual decrease of helper T cells. Severe joint pain may be due to inflammatory damage to bone and joint tissue as a result of viral infection.

Epidemiology

The epidemiology of chikungunya is similar to that of dengue: It is found in tropical and subtropical regions that have suitable mosquito vectors. The disease is transmitted by *Aedes* mosquito species, mostly *A. aegypti* and *A. albopictus,* and is spreading as a result of the expanding range of *A. albopictus.*

Treatment and Prevention

There is no specific medication to treat chikungunya, but supportive treatment, including fluid replacement, relieves symptoms. Aspirin and non-steroidal pain relievers (NSAIDs) should be avoided until the possibility of dengue has been eliminated because they could worsen the hemorrhaging associated with that disease.

As with most other mosquito-borne diseases, chikungunya can be prevented by avoiding mosquito bites and using the vector control methods described in the section on dengue. Although no chikungunya vaccine is currently available, one candidate is being reviewed by the FDA, and several others are in clinical trials. The characteristics of chikungunya are compared to other mosquito-borne viral diseases at the end of this section (see table 25.11).

Zika Virus Disease

Zika virus disease, a mosquito-borne disease, dominated media reports in May 2015 when it was associated with a severe birth defect called microcephaly (smaller than normal head, usually due to incomplete brain development). The viral cause was first isolated in 1947 in monkeys of the Zika Forest of Uganda, and then found in humans five years later. Documented cases of the disease were rare before 2007, when an outbreak in Micronesia signaled its emerging status. Once introduced into the Western Hemisphere in 2015, the disease rapidly appeared across the Americas; most infections by the causative agent, however, are asymptomatic or subclinical. Fortunately, the incidence has since decreased, but outbreaks still occur.

Signs and Symptoms

Zika virus disease signs and symptoms usually begin 2 to 14 days after infection, characterized by fever, rash, joint pain, and conjunctivitis; these are usually mild and fade in 2 to 7 days. In rare cases, patients develop an autoimmune complication called Guillain-Barré syndrome, in which the immune system damages certain nerves.

A baby born to a woman infected with Zika virus during pregnancy may have congenital Zika syndrome, a characteristic pattern of neurological abnormalities.

Causative Agent

Zika virus (ZIKV) is an enveloped, single-stranded RNA arbovirus in the family *Flaviviridae*, transmitted by the bite of infected *Aedes* mosquitoes, the biological vector.

Pathogenesis

Pathogenesis of ZIKV is poorly understood. Studies indicate that when the virus enters the host, it infects local dendritic cells and macrophages, triggering a host response that may lead to general symptoms of a viral infection.

Significantly, ZIKV can infect a developing fetus, resulting in congenital Zika syndrome. ZIKV preferentially infects neural cells in the fetus, and in particular, neural stem cells from which the brain develops; these cells are present throughout fetal development, so infection during any trimester of pregnancy can damage the brain. Even newborns with normal head size but infected in utero can show developmental delays and neurological abnormalities.

Epidemiology

Zika virus disease has been found in parts of the Americas, Africa, Asia, and the Pacific, transmitted by the bite of infected *Aedes* mosquitoes (the biological vector). Most cases involve *A. aegypti*, but the observation that it might also be transmitted by *A. albopictus* is a concern because of the expanding range of that species. ZIKV can pass from a pregnant woman to the developing fetus and can also be sexually transmitted.

Treatment and Prevention

No specific medication exists to treat Zika virus disease, but supportive treatment, including fluid replacement, relieves symptoms. Aspirin and non-steroidal pain relievers (NSAIDS) should be avoided until the possibility of dengue has been eliminated because it could worsen the hemorrhaging associated with that disease. The U.S. Zika Pregnancy and Infant Registry, a program of the CDC, tracks Zika infections in pregnant women in the United States to learn more about the outcomes, with the goal of obtaining information that will help scientists make recommendations to people who might be affected.

No approved vaccine for Zika virus disease is currently available, but several are in clinical trials. Completing those trials is now challenging, however, because the number of ZIKV infections has dropped dramatically since 2017, thereby making it difficult to determine a vaccine's effectiveness. The best preventive measures are avoiding mosquito bites and controlling the mosquito vector, as described in the section on dengue. People with ZIKV infection and those who have had possible exposure (such as anyone living in or visiting areas with active Zika transmission) should take precautions to avoid sexual transmission. The characteristics of ZIKV are compared to other mosquito-borne viral diseases at the end of this section (see table 25.11).

Yellow Fever

Yellow fever is a vaccine-preventable mosquito-borne disease. Most infections by the causative agent are asymptomatic or subclinical, but severe forms of the disease have a high case-fatality rate. After hundreds of people died in a significant yellow fever outbreak in Africa in 2016, the WHO launched an initiative called Eliminate Yellow Fever Epidemics (EYE) with the goal of ending epidemics of the disease by 2026.

Signs and Symptoms

Signs and symptoms of yellow fever usually begin 3 to 6 days after infection and are similar to many other viral infections: fever, muscle pain, headache, and malaise. In most cases, these disappear after a day or two. Some patients, however, then progress to severe disease, experiencing high fever, nausea, bleeding, "black vomit" (from gastrointestinal bleeding), and jaundice (hence, the name *yellow fever*). The case-fatality rate of severe yellow fever can reach 50% or more, with death due to organ failure and shock.

Causative Agent

Yellow fever is caused by yellow fever virus (YFV), an enveloped, single-stranded RNA arbovirus of the family *Flaviviridae*. The virus is mainly transmitted to humans by the bite of infected *Aedes* mosquitoes, the biological vector.

Pathogenesis

Once injected by a mosquito, YFV multiplies in local host cells, enters the bloodstream, and is carried to the liver and other parts of the body. The host response to the infection may lead to general symptoms of a systemic viral infection.

In the case of severe yellow fever, liver damage results in jaundice and decreased production of clotting proteins. Injury to small blood vessels produces petechiae throughout the body. The immune response to infection results in a cytokine storm, which leads to tissue and organ-damaging effects. Kidney failure is a common consequence of blood loss and low blood pressure. Disseminated intravascular coagulation (DIC) may also occur.

Epidemiology

Yellow fever is mainly found in sub-Saharan Africa and regions of Central and South America, where infected primates living in the jungles are a significant reservoir. Although the mosquito species that transmit YFV in these wild animals do not normally bite humans, people are occasionally bitten and develop disease. In turn, these infected people may spread the disease from the jungle reservoir to urban areas in the tropics or subtropics, where *Aedes aegypti* then transmits it among humans, sometimes resulting in significant epidemics.

Treatment and Prevention

There is no specific medication to treat yellow fever, but severe cases require supportive care including hydration. As with dengue and severe dengue, aspirin and nonsteroidal anti-inflammatory drugs (NSAIDs) should be avoided because they can worsen bleeding.

Yellow fever can be prevented in urban areas by avoiding mosquito bites and controlling the mosquito vector, as described in the section on dengue. In the jungle, controlling yellow fever is almost impossible because of the monkey reservoir. An attenuated vaccine is available to immunize people who might become exposed, including travelers to endemic areas. **Table 25.11** compares the characteristics of the arboviral diseases discussed in this section.

MicroAssessment 25.3

Infectious mononucleosis occurs worldwide and is transmitted from person to person by saliva. Ebola virus disease (EVD) and Marburg virus disease (MVD) are both severe emerging diseases of global concern because they can result in massive bleeding and multi-organ failure. Dengue is the most common vector-borne viral disease in the world; severe dengue can be fatal. Chikungunya is a newly emerging disease that, like dengue, causes severe joint pain. Zika virus disease is usually mild but can cause severe birth defects. Yellow fever can result in life-threatening damage to the liver and other organs.

7. What characteristic changes occur in the blood of patients with infectious mononucleosis?
8. Compare and contrast the global spread of dengue, chikungunya, and Zika virus disease.
9. Why must a dengue vaccine be effective against all strains of the virus?

TABLE 25.11 Dengue and Severe Dengue, Chikungunya, Zika Virus Disease, and Yellow Fever Compared

	Dengue and Severe Dengue	Chikungunya	Zika Virus Disease	Yellow Fever
Signs and Symptoms	Fever, headache, joint pain, and sometimes a rash; in severe dengue, weak rapid pulse and bleeding; shock and organ damage can occur.	Similar to dengue, but the joint pain is sometimes debilitating.	Usually mild disease with fever, rash, joint pain, conjunctivitis; complicated by Guillain-Barré syndrome on rare occasions; congenital Zika syndrome.	Often mild; severe cases characterized by high fever, nausea, bleeding, black vomit, and jaundice; shock and organ damage can occur.
Incubation Period	Usually 4 to 7 days	Usually 3 to 7 days	Usually 2 to 14 days	Usually 3 to 6 days
Causative Agents	Dengue virus; four serotypes; enveloped single-stranded RNA arbovirus	Chikungunya virus; enveloped single-stranded RNA arbovirus	Zika virus; enveloped single-stranded RNA arbovirus	Yellow fever virus; enveloped, single-stranded RNA arbovirus
Pathogenesis	Host response to initial infection leads to general symptoms. Severe dengue: usually involves ADE resulting from a previous infection; cytokine storm leads to tissue and organ-damaging effects.	Host response leads to general symptoms; infection damages bone and joint tissue.	Host response leads to general symptoms; virus can infect fetus, affecting neural stem cells.	Host response leads to general symptoms; virus targets liver cells, leading to death of those cells; in severe cases, cytokine storm leads to tissue and organ-damaging effects.
Epidemiology	Mosquito-borne; transmitted by *A. aegypti* and *A. albopictus.*	Similar to dengue.	Mosquito-borne; transmitted by *A. aegypti* and perhaps *A. albopictus;* transmitted from pregnant female to the developing fetus; also sexually transmitted.	Mosquito-borne; infected non-human primates are an important reservoir; transmitted among humans by *A. aegypti*
Treatment and Prevention	Treatment: no specific treatment. Prevention: vector control; vaccine to prevent severe dengue in certain groups.	Treatment: no specific treatment. Prevention: vector control.	Treatment: no specific treatment. Prevention: vector control; precautions to avoid sexual transmission.	Treatment: no specific treatment. Prevention: attenuated vaccine; vector control.

25.4 ■ Protozoan Diseases of the Cardiovascular and Lymphatic Systems

Learning Outcomes

8. Outline the *Plasmodium vivax* life cycle and explain why malaria has been so difficult to control.
9. Describe leishmaniasis.

Protozoa infect the circulatory systems of millions of people globally, particularly in developing countries. Malaria and leishmaniasis are examples of debilitating protozoan diseases.

Malaria

Malaria is a leading cause of morbidity and mortality in many parts of the world. The name means "bad air," but the disease is not related to air quality—rather, it is spread by mosquitoes, which are commonly found in unpleasant-smelling swampy areas.

Global malaria control has been a struggle for decades. In the hopes of wiping out the disease, the World Health Organization (WHO) began the Global Malaria Eradication Program in 1955. Although that helped eliminate malaria from some parts of the world, the goal of eradication was never achieved, and the program was abandoned in 1969. Nearly 30 years later, a new disease-fighting initiative called Roll Back Malaria was started, with the goal of reducing malaria deaths by improving access to medical care, strengthening local healthcare facilities, supplying medications, and distributing long-lasting insecticide-treated mosquito nets. The support of the Global Fund (a private/public partnership that facilitates funding for AIDS, tuberculosis, and malaria prevention) and the Bill and Melinda Gates Foundation has since provided an additional focus on eradication. Although the combined control efforts resulted in a 60% decline in global malaria death rates between 2000 and 2015, that reduction then stalled, a problem recently exacerbated by the COVID-19 pandemic. According to the World Malaria Report, 2020 had the highest number of malarial deaths in a decade: 627,000.

MicroByte

Most of the people who die from malaria each year are children under the age of 5 who live in sub-Saharan Africa.

Signs and Symptoms

The first symptoms of malaria are flu-like, with fever, headache, and pain in the joints and muscles. These generally begin about 1 to 2 weeks after being bitten by an infected mosquito but in some cases start many weeks later. These vague symptoms continue for 2 or 3 weeks and then begin to occur in three phases:

- **Cold stage.** The patient feels cold and develops shaking chills that last for up to an hour.
- **Hot stage.** The patient's temperature rises sharply for 3 to 8 hours, often reaching 40°C (104°F) or more.
- **Sweat stage.** The patient's temperature falls; drenching sweating occurs for 2 to 4 hours, leaving the patient exhausted.

This cycle of intense symptoms—chills, fevers, and sweats—is called a **paroxysm.** Except for extreme fatigue, the patient feels well until 24 or 48 hours later, when the next paroxysm occurs. Occasionally, infection with a malarial parasite damages the brain or other organs, resulting in what is referred to as severe malaria—a life-threatening form of the disease. Severe malaria is considered a medical emergency because it can progress rapidly.

Causative Agent

The five main *Plasmodium* species that cause malaria—*P. vivax, P. falciparum, P. malariae, P. ovale,* and *P. knowlesi*—differ in microscopic appearance, certain life cycle features, and disease outcomes; *P. falciparum* causes most cases of severe malaria. All of these species are transmitted to humans by the bite of an infected female *Anopheles* mosquito, the biological vector.

Plasmodium species are apicomplexan parasites. As described in chapter 12, apicomplexans have complex life cycles involving sexual and asexual reproduction in different hosts.

Pathogenesis

The parasite's life cycle involves two stages within the human host: (1) a liver stage and (2) a red blood cell stage—referred to as the erythrocytic stage. In stained preparations of blood, the parasites can be seen microscopically within infected RBCs, so the red blood cell stage is important in diagnosis as well as in the disease process. Furthermore, distinct physical features of the different *Plasmodium* species become apparent as the parasites mature in the RBCs, and these can be used to help identify the infecting species.

Malarial parasites multiply asexually in the human host using the process of **schizogony** (multiple fission). To do this, the parasite enlarges and then develops into a multinucleated form called a **schizont,** which undergoes schizogony to simultaneously produce many progeny called **merozoites.** The complex life cycle includes the following steps (**figure 25.15**):

1. **Transmission to a human.** A person becomes infected when an infected mosquito injects **sporozoites**—the infectious form of the protozoan—while taking a blood meal. The motile sporozoites are then carried by the bloodstream to the liver.
2. **Liver stage.** Sporozoites in the bloodstream quickly infect hepatocytes (liver cells). Within a liver cell, a parasite enlarges, forms a schizont, and then reproduces by schizogony, producing tens of thousands of merozoites. In the case of some *Plasmodium* species, not all the sporozoites reproduce immediately; some will instead form a dormant stage called a hypnozoite, which can survive for years before becoming active and producing merozoites. Regardless, once merozoites have been made, packets of them bud from the liver cells into the bloodstream to begin the erythrocytic stage.
3. **Erythrocytic stage.** Merozoites enter RBCs and use the hemoglobin as a source of nutrients for growth. Upon entering an RBC, the parasite has the appearance of a ring and is thus referred to as a **ring form.** It quickly develops into an active feeding form called a **trophozoite** and then, after enlarging, develops into a schizont that gives rise to about 6 to 24 merozoites (the number depends on the *Plasmodium* species); the RBC eventually bursts, releasing the merozoites. These merozoites then infect other RBCs, repeating the cycle. Each cycle takes 24 to 72 hours, depending on the *Plasmodium* species. Not all parasites with the RBCs give rise to more merozoites; some develop into male or female **gametocytes,** the forms required for sexual reproduction in a mosquito host.
4. **Transmision to a mosquito.** When a female *Anopheles* mosquito feeds on an infected person's blood, it ingests RBCs, some of which contain gametocytes.
5. **Mosquito stage.** Once in the mosquito midgut, the gametocytes are released from the RBCs and develop into gametes. Fusion of a male gamete and a female gamete forms a zygote, which burrows into the midgut wall and develops into an oocyst. The oocyst then enlarges and its nucleus divides repeatedly, giving rise to thousands of sporozoites. These invade the mosquito's salivary glands, from which they may then be injected into a new human host.

The characteristic feature of malaria—recurrent paroxysms—results from the repeating cycles of RBC infections that release merozoites into the bloodstream. The infections in the millions of RBCs become nearly synchronous so that all infected RBCs rupture at about the same time. This rapidly releases the parasite's antigens into the bloodstream, triggering the initial symptoms of the paroxysm.

Lysis of the infected RBCs has damaging effects beyond antigen release. The RBC destruction results in anemia, a situation compounded by the parasite's conversion of iron in hemoglobin to a form not readily recycled in the body. Meanwhile, the overworked spleen enlarges as it removes the high levels of foreign material and abnormal RBCs from circulation and, as a result, may rupture. The overburdened immune system may also fail, resulting in immunodeficiency.

P. falciparum infections are typically more severe than other types of malaria. One reason is that this species can

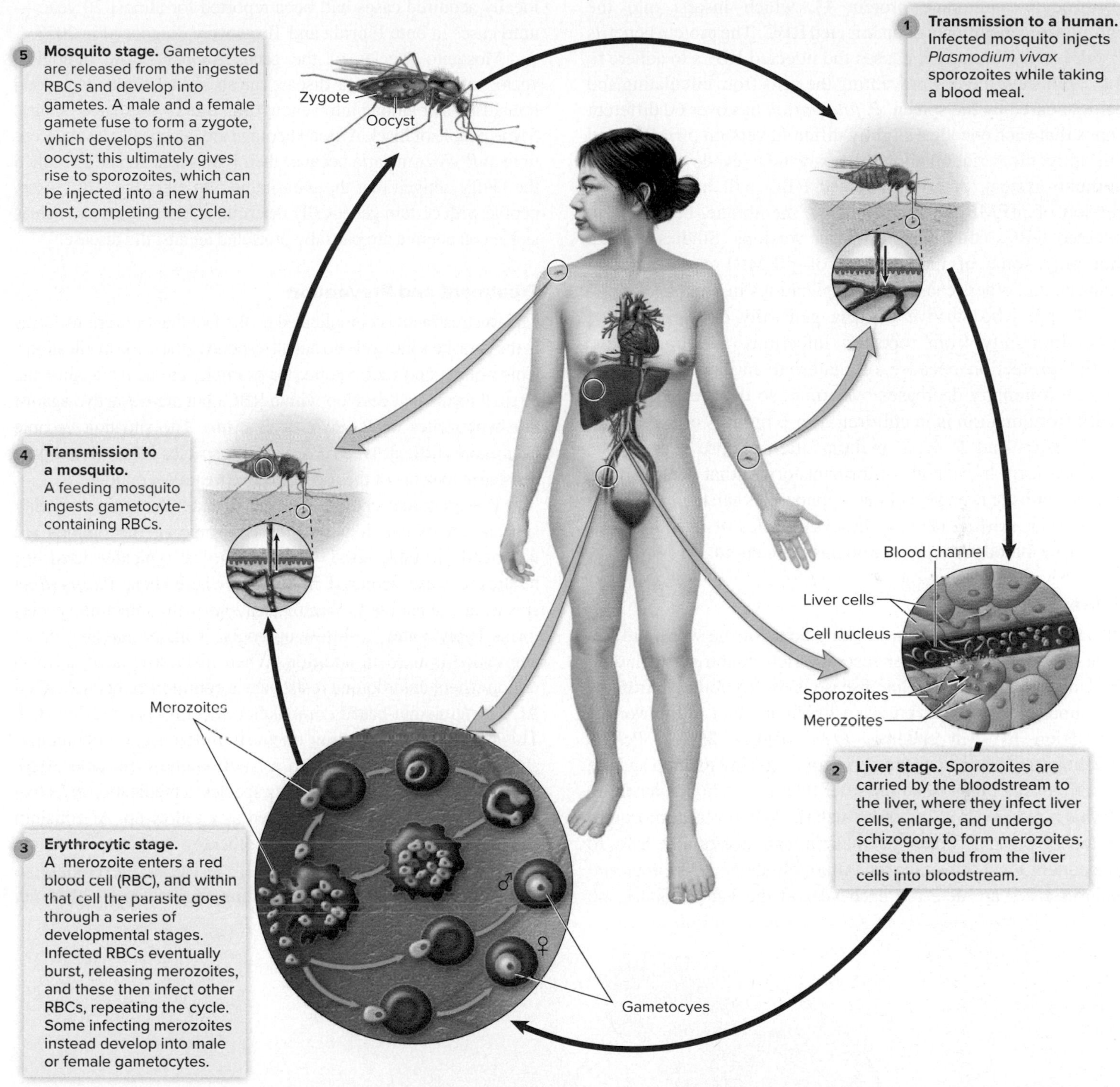

1 Transmission to a human. Infected mosquito injects *Plasmodium vivax* sporozoites while taking a blood meal.
Blood channel
Liver cells
Cell nucleus
Sporozoites
Merozoites
2 Liver stage. Sporozoites are carried by the bloodstream to the liver, where they infect liver cells, enlarge, and undergo schizogony to form merozoites; these then bud from the liver cells into bloodstream.
Merozoites
Gametocyes
♂
♀
3 Erythrocytic stage. A merozoite enters a red blood cell (RBC), and within that cell the parasite goes through a series of developmental stages. Infected RBCs eventually burst, releasing merozoites, and these then infect other RBCs, repeating the cycle. Some infecting merozoites instead develop into male or female gametocytes.
4 Transmission to a mosquito. A feeding mosquito ingests gametocyte-containing RBCs.
5 Mosquito stage. Gametocytes are released from the ingested RBCs and develop into gametes. A male and a female gamete fuse to form a zygote, which develops into an oocyst; this ultimately gives rise to sporozoites, which can be injected into a new human host, completing the cycle.
Zygote
Oocyst

FIGURE 25.15 Life Cycle of *Plasmodium vivax*

With respect to the number of merozoites produced, why is a single infected liver cell more significant than a single infected red blood cell?

infect any RBC, leading to very high levels of parasitemia. In contrast, most others can infect only one stage of an RBC's life cycle—either immature RBCs (reticulocytes) or mature RBCs, depending on the infecting *Plasmodium* species—thus limiting the levels of parasitemia. Perhaps more significantly, *P. falciparum* causes the infected RBCs to stick to the walls of capillaries, an action that blocks blood flow and thereby deprives the local tissues of O_2. This leads to the organ damage that characterizes severe malaria; if it occurs in the brain, life-threatening cerebral malaria develops. The RBCs can also clog the capillaries in the placenta of a pregnant woman, leading to devastating outcomes for the developing fetus.

The stickiness of *P. falciparum*–infected RBCs is due to a parasite-encoded protein referred to as pfEMP1 (*P. falciparum* erythrocyte membrane protein 1), which inserts into the cytoplasmic membrane of an infected RBC. The protein benefits *P. falciparum* because it causes the infected RBCs to adhere to the capillaries, thereby preventing the cells from circulating and being cleared by the spleen. *P. falciparum* has over 60 different genes that each encode a slightly different version pfEMP1, and this antigenic variation allows the parasite to evade the adaptive immune system. A given infected RBC will have only one version of pfEMP1 inserted into the membrane, but different infected RBCs could have different versions. Studies suggest that only some of the versions of pfEMP1 cause cerebral malaria, and others cause the complications in pregnancy.

People who survive malaria generally develop at least some immunity from repeated infections. Newborns have partial protection because of maternal antibodies, but that passive immunity decreases over time, so the greatest risk of death from malaria is in children over 6 months of age.

P. vivax and *P. ovale* malaria often relapse because the parasites form hypnozoites (dormant forms) that persist in the liver. Months or even years later, hypnozoites can begin growing in the liver, starting new erythrocytic cycles of infection after the earlier bloodstream infection has been cured.

Epidemiology

Malaria was once common in many areas of the world but has been eliminated from most resource-rich countries, including the United States. Today, the disease is mostly found in tropical and subtropical regions, but even so, almost half of the world's population lives in endemic areas (**figure 25.16**). People traveling to those regions are at risk of acquiring malaria and are sometimes not diagnosed until after their return home, resulting in what is referred to as imported malaria. Wherever the mosquito vector is present, imported malaria can potentially lead to subsequent local transmission. Although about 2,000 imported malaria cases are detected each year in the United States, no locally acquired cases had been reported for almost 20 years—until cases in both Florida and Texas were diagnosed in 2023.

Mosquito species of the genus *Anopheles* are biological vectors of malaria, but the disease can also be transmitted by blood transfusions or among intravenous drug users who share syringes. Some people of black African heritage are genetically less susceptible to *P. vivax* malaria because their RBCs lack an antigen called the Duffy antigen that the parasite uses to enter the RBC. Also, people with certain genetically determined blood diseases such as sickle cell anemia are partially protected against the disease.

Treatment and Prevention

Malaria treatment is complicated by the fact that the various forms in the parasite's life cycle do not all respond to the same medications. Chloroquine and mefloquine, for example, are active against the asexual forms that develop within RBCs but are not active against the hypnozoites of *P. vivax* or *P. ovale.* The situation became even more difficult when *Plasmodium* species began developing resistance to some of the most effective treatment options.

The treatment choice for malaria depends on the infecting species and where it was acquired; some of the options are described in table 20.5. Chloroquine is typically used for treating disease acquired in regions where most *Plasmodium* species are sensitive to that medication; if the infecting species forms hypnozoites, a follow-up course with primaquine or an alternative is used in addition. When malaria is acquired in a region where chloroquine resistance is common, a therapy called ACT (artemisinin-based combination therapy) is typically used. This combines a fast-acting artemisinin derivative with another anti-malarial medication such as mefloquine (see table 20.5). Again, depending on the infecting species, a medication effective against hypnozoites may be needed as a follow-up. Artemisinin should never be used as a monotherapy (meaning a single medication) because of the risk of resistance developing to this important medication. In certain regions, particularly in

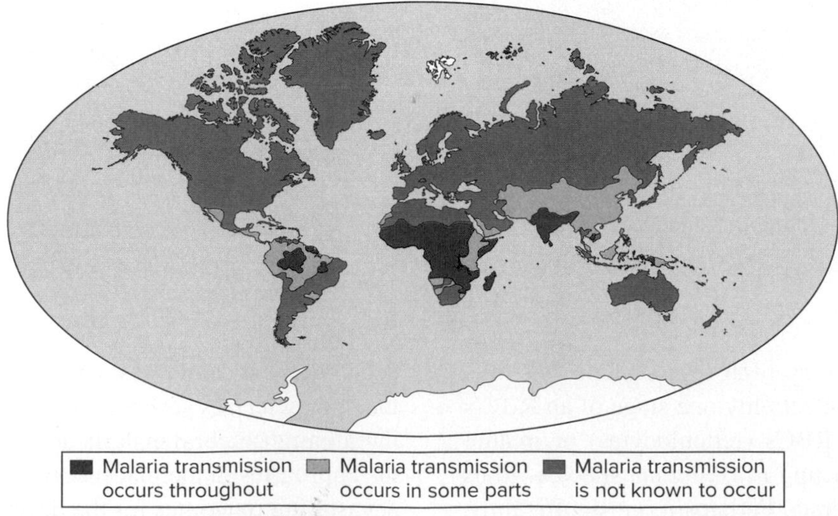

FIGURE 25.16 Distribution of Malaria Source: www.cdc.gov/malaria/malaria_worldwide/impact.html

? **How might someone diagnosed with malaria in the United States have acquired the disease?**

Southeast Asia, artemisinin resistance has already developed, in which case other combination therapies are used. Although artemisinin resistance is currently confined to certain regions, scientists are concerned it will spread; studies have already detected parasites with reduced susceptibility in Africa.

Preventing malaria in endemic areas depends largely on controlling the mosquito population, particularly by eliminating mosquito breeding areas. Indoor spraying with insecticides also helps eliminate the vector. As described in the section on dengue, personal protection—including wearing long sleeves and pants, using an effective repellent, and sleeping under an insecticide-treated net (ITN)—helps prevent mosquito bites. For travelers, some of the same medications used for malaria treatment can also be used for prevention (chemoprophylaxis). The best way to prevent malaria, however, will be through vaccination. In 2021, WHO began recommending the recombinant vaccine RTS,S for children from 5 months of age in regions with moderate to high *P. falciparum* transmission. That vaccine targets the sporozoite stage of the parasite, and although its use results in only a 30% decrease in severe malaria among recipients, it should save lives and also reduce disease transmission. In 2023, Ghana and Nigeria approved the vaccine R21 for use in children ages 5 to 36 months—a decision reached even before the final clinical testing results were published. That vaccine, which also targets the sporozoite stage, seems to be more effective than RTS,S, so the approval was in anticipation of WHO recommending R21 once the clinical results are scrutinized. **Table 25.12** describes the main features of malaria.

TABLE 25.12	**Malaria**
Signs and Symptoms	Flu-like symptoms initially. In untreated infections, recurrent paroxysms of intense chills, fever, and sweats, alternating with feeling healthy; severe malaria is associated with life-threatening organ damage.
Incubation Period	Varies with species; 6 days to a month or longer
Causative Agents	Five species of *Plasmodium;* severe malaria is mostly due to *P. falciparum.*
Pathogenesis	*Plasmodium* life cycle consists of a liver stage and a red blood cell stage; paroxysms resulting from repeating cycles of RBC infections that release merozoites into the bloodstream, triggering fever; spleen enlarges as it removes high levels of foreign material and abnormal blood cells from the circulation; with *P. falciparum* infection, red blood cells stick to walls of capillaries, blocking vessels and depriving tissue of O_2, leading to organ damage.
Epidemiology	Mosquito-borne; transmitted by *Anopheles* mosquitoes; mostly in tropical and subtropical regions.
Treatment and Prevention	Treatment: usually ACTs; other medicines if sensitivity known. Prevention: vector control; for travelers, chemoprophylaxis; newly approved vaccine for children living in *P. falciparum*-endemic regions,

Leishmaniasis

Leishmaniasis, a disease transmitted by the bite of a sandfly, is associated with malnutrition, poverty, and immunosuppression. The WHO lists it as a neglected tropical disease.

Signs and Symptoms

The signs and symptoms of leishmaniasis vary considerably, depending on both the virulence of the parasite and the immune response of the host; most infections do not result in disease. Clinical forms include:

- **Cutaneous leishmaniasis (CL).** This is the most common form and has an incubation period of weeks to months, and sometimes years. Sores develop at exposed sites on the skin where sandflies have bitten and usually remain localized, starting out as papules (small raised bumps) that gradually progress to nodules (relatively large, elevated, solid lesions). These eventually ulcerate to form raised edges surrounding a central crater (like a volcano) and are generally painless unless secondary bacterial infection occurs; if untreated they can last for months to years. Satellite lesions (additional lesions surrounding the primary ones) may also appear, along with swelling of regional lymph nodes. The lesions usually remain localized due to an effective cellular immune response, but if that response is weak, then the skin lesions spread widely, resulting in more severe forms of CL.
- **Mucocutaneous leishmaniasis (MCL).** Also referred to as mucosal leishmaniasis or espundia, this is a form of CL that results from an excessive inflammatory response that gradually destroys the mucous membranes of the mouth, nose, and throat.
- **Visceral leishmaniasis (VL).** Sometimes referred to as kala-azar (meaning black fever in Hindi, indicating the darkening of the skin that can occur), this usually has an incubation period of several months but can develop years after infection. The signs and symptoms reflect organ damage, and include fever, weight loss, enlarged spleen and liver, and panocytopenia (low levels of all types of blood cells and their derivatives). Without treatment, VL is typically fatal.

Causative Agent

Leishmaniasis is caused by more than 20 species of protozoa in the genus *Leishmania*. These hemoflagellates can be transmitted to humans and certain other mammals by the bite of infected sandflies in the genera *Phlebotomus* and *Lutzomyia,* the biological vectors.

Leishmania species are obligate intracellular parasites with complex life cycles, involving alternating stages in humans (or other reservoir hosts) and sandflies. Several different *Leishmania* species cause CL, but those found in the Eastern Hemisphere are often different from those in the Americas and are different still from the species that cause VL. The various species are identical microscopically but can be distinguished using monoclonal antibodies and molecular methods.

Pathogenesis

When an infected sandfly takes a blood meal, it injects a form of the leishmania parasite called a promastigote, and these are then taken up by macrophages in the skin. Once inside the phagocyte's phagolysosome, the parasite differentiates into a form called an amastigote and multiplies. Eventually the host cell ruptures, releasing the amastigotes to infect other macrophages. A sandfly can ingest the parasite in an infected macrophage while taking a blood meal, thus completing the parasite's life cycle.

The outcome of infection depends not only on the species of parasite but also on characteristics of the patient's helper T cell response, which directs the various components of the immune system (see figure 15.1). In localized CL, the cell-mediated response predominates, limiting the spread of the intracellular parasite; in the more severe forms of CL, the cell-mediated response is weak and fails to stop the parasite within infected host cells. MCL involves a cell-mediated response that controls the infection but also results in mucosa-damaging excessive inflammation. Like patients with severe forms of CL, those who develop VL fail to develop an adequate cell-mediated response against the infecting parasite. The multiplying parasites then spread, accumulating in the spleen, liver, and other organs that have high concentrations of macrophages (see figure 14.5).

Epidemiology

Most leishmaniasis cases occur in tropical and subtropical regions, but the disease is also endemic in southern Europe. In the United States, locally acquired CL cases are occasionally reported in Texas and Oklahoma, but because the disease is not nationally notifiable, its true incidence is not known. Travelers and military personnel who return from endemic areas can also develop leishmaniasis.

Various types of mammals are the primary reservoirs for many *Leishmania* species, but humans are the most important reservoir for some species. Transmission requires a sandfly vector, but the bite often goes unnoticed because it is sometimes painless and the insects are very small.

Treatment and Prevention

The sores that characterize localized CL often heal on their own, but may be treated to decrease scarring, prevent the more severe forms of disease, and lessen the number of infected reservoir hosts. Other forms of leishmaniasis are routinely treated. The medication chosen depends on the infecting species and the host characteristics; options include IV liposomal amphotericin for VL and oral miltefosine for all types of leishmaniasis.

No vaccines or medications are available to prevent leishmaniasis, so people in endemic areas should avoid being bitten by sandflies. Outdoor activities should be avoided during times when sandflies typically feed, which is usually from dusk to dawn. The sandflies cannot bite through clothing, so long-sleeved shirts and long pants tucked into socks should be worn. Effective insect repellents and—unless the home is well screened—insecticide-treated nets (ITNs) should be used as well. Preventive public health measures include treating or removing reservoir hosts, including domestic dogs. **Table 25.13** describes the main features of leishmaniasis.

TABLE 25.13 Leishmaniasis

Signs and Symptoms	Considerable variation; the most common form, cutaneous leishmaniasis (CL), usually results in localized skin lesions but can result in widespread lesions; mucocutaneous leishmaniasis (ML) results in destruction of mucous membranes of the mouth, nose, and throat; visceral leishmaniasis (VL) results in symptoms associated with organ damage.
Incubation Period	Variable—months to years, or even decades
Causative Agents	Various species of *Leishmania*
Pathogenesis	Infects macrophages; disease outcome depends on the infecting species but also on characteristics of the immune response: an effective cell-mediated immune response is needed to control the parasite.
Epidemiology	Sandfly-borne; transmitted by *Phlebotomus* and *Lutzomyia* sandflies; mostly in tropical and subtropical regions; disease is also endemic in southern Europe.
Treatment and Prevention	Treatment: medication depends on the infecting species and the host characteristics. Prevention: prevent sandfly bites by avoiding outdoor activities from dusk until dawn, wearing protective clothing, and using effective insecticides and insecticide-treated nets; treat or remove animal reservoirs.

MicroAssessment 25.4

Malaria, a leading cause of morbidity and mortality worldwide, is transmitted by mosquitoes; *Plasmodium falciparum* causes most cases of severe malaria. Leishmaniasis is transmitted by sandflies and occurs in three clinical forms—cutaneous, mucocutaneous, and visceral.

10. What causes the recurrent paroxysms that characterize malaria?

11. How do localized cutaneous leishmaniasis, mucocutaneous leishmaniasis, and visceral leishmaniasis differ with respect to their symptoms and underlying characteristics of the host immune response?

12. Vaccinating against which stage of a malarial parasite would prevent transmission from a human to a mosquito?

The key features of the diseases covered in this chapter are highlighted in the **Diseases in Review 25.1** table that follows.

Diseases in Review 25.1

Cardiovascular and Lymphatic System Infections

Disease	Causative Agent	Comment	Summary Table
BACTERIAL INFECTIONS			
Sepsis	Various microbes	Life-threatening tissue damage and organ dysfunction resulting from an overwhelming host response to an infection.	Table 25.1
Infective endocarditis	Pathogens from another infected site, or normal microbiota of the skin or the mucous membranes	People with indwelling intravenous catheters, artificial heart valves, or heart abnormalities, or who abuse intravenous drugs are most at risk.	Table 25.2
Plague ("black death")	*Yersinia pestis*	Spread by flea bites (bubonic plague) and aerosols (pneumonic plague); septicemic plague occurs when the bacteria enter the bloodstream; endemic in rodent populations; Category A bioterrorism agent.	Table 25.4
Lyme disease	*Borrelia burgdorferi* and certain other related species	Spread by ticks; often leads to "bull's-eye rash"; later symptoms include injury to heart and nervous system, and arthritis.	Table 25.5
***Vibrio vulnificus* infection**	*Vibrio vulnificus*	Bloodstream infections can lead to sepsis, particularly in people with predisposing conditions; acquired by eating contaminated seafood or through breaks in the skin.	Table 25.6
Tularemia	*Francisella tularensis*	Zoonotic disease acquired multiple ways, often by direct contact with an infected animal; Category A bioterrorism agent.	Table 25.7
Brucellosis	*Brucella species*	Zoonotic disease acquired via contact with an infected animal or consumption of infected milk or meat; Category B bioterrorism agent.	Table 25.8
VIRAL INFECTIONS			
Infectious mononucleosis ("mono")	Epstein-Barr virus (EBV)	Long incubation period, followed by symptoms that can last for months; transmitted in saliva; causes both productive and latent infections in B cells.	Table 25.9
Ebola virus disease and Marburg virus disease	Ebolaviruses and marburgviruses	Severe, often fatal diseases characterized by fever and often internal and external bleeding; spread person to person by body fluids.	Table 25.10
Dengue and severe dengue	Dengue virus	Transmitted by mosquitoes; fever and joint pain; life-threatening severe dengue occurs more frequently with second infection. Vaccine used to prevent severe disease.	Table 25.11
Chikungunya	Chikungunya virus	Transmitted by mosquitoes; characterized by fever and severe joint pain, which may become chronic.	Table 25.11
Zika virus disease	Zika virus	Transmitted by mosquitoes or through sexual contact; infection during pregnancy can lead to congenital Zika syndrome.	Table 25.11
Yellow fever	Yellow fever virus	Transmitted by mosquitoes; often mild, but severe form is characterized by fever, bleeding, black vomit, and jaundice, and often fatal. Preventive vaccine available.	Table 25.11
PROTOZOAN INFECTIONS			
Malaria	*Plasmodium* species	Transmitted by mosquitoes; flu-like symptoms that gradually develop into paroxysms of chills, fevers, and sweats; *P. falciparum* infections are life-threatening, even in otherwise healthy people. Vaccine lessens chance of severe disease.	Table 25.12
Leishmaniasis	*Leishmania* species	Transmitted by sandflies. Clinical forms include cutaneous, mucocutaneous, and visceral.	Table 25.13

Summary

25.1 ■ Anatomy, Physiology, and Ecology of the Cardiovascular and Lymphatic Systems (figure 25.1)

The Cardiovascular System

The heart pumps blood throughout the body, allowing O_2 to be delivered to tissues. Deoxygenated blood enters the right side of the heart, where it is then pumped to the lungs to become oxygenated before it is returned to the left side of the heart. Arteries carry blood away from the heart. In the capillaries, gas and fluid exchange occurs between the blood and the tissues. Veins carry blood toward the heart.

The Lymphatic System (figure 25.2)

Lymphatic vessels carry lymph, which contains phagocytes, lymphocytes, and antigens. As lymph moves through the lymphatic vessels, it passes through lymph nodes, allowing dendritic cells, lymphocytes, and macrophages that gather there to inspect and respond to the contents. The white pulp of the spleen functions as a lymph node for blood; the red pulp cleans the blood of debris, stores a reserve of monocytes, and can make RBCs if needed.

25.2 ■ Bacterial Diseases of the Cardiovascular and Lymphatic Systems

Bacterial Sepsis (table 25.1; figure 25.3)

Sepsis is life-threatening tissue damage and organ dysfunction resulting from an overwhelming host response to an infection. It is commonly a healthcare-associated illness, and most often occurs in older populations and those with diabetes and obesity.

Infective Endocarditis (table 25.2)

Infective endocarditis is often caused by members of the normal microbiota in people with indwelling intravenous catheters or heart abnormalities, or who abuse intravenous drugs. It also may be caused by pathogens that have spread from another infected site.

Plague ("Black Death") (table 25.4)

Plague is caused by *Yersinia pestis,* which has many virulence factors that interfere with phagocytosis and immunity (table 25.3, figure 25.4). **Bubonic plague** is transmitted to humans by fleas (figure 25.5). If the bacteria enter the bloodstream, **septicemic plague** may develop and can give rise to **pneumonic plague,** which is transmitted from person to person by aerosols.

Lyme Disease (table 25.5; figure 25.8)

Lyme disease is caused by the spirochete *Borrelia burgdorferi* and related species (figure 25.7). A bull's-eye rash often develops during the early stage of the disease (figure 25.6); systemic symptoms develop as the bacteria disseminate. The causative agents are transmitted by certain ticks (figures 25.9, 25.10).

Vibrio vulnificus Infection (table 25.6)

Vibrio vulnificus is typically ingested in raw or undercooked contaminated seafood or enters through wounds. If it enters the bloodstream, skin blistering and sepsis may result, particularly in those with predisposing conditions.

Tularemia (table 25.7, figure 25.11)

Tularemia, caused by *Francisella tularensis,* is usually transmitted from wild animals to humans by exposure to the animal's blood or via insects and ticks.

Brucellosis (table 25.8)

Brucellosis, caused by *Brucella* species, is usually acquired from cattle or other domestic animals, sometimes from wild animals. The organisms can infect via mucous membranes and minor skin injuries.

25.3 ■ Viral Diseases of the Cardiovascular and Lymphatic Systems

Infectious Mononucleosis ("Mono" or "Kissing Disease") (table 25.9)

Infectious mononucleosis is caused by the Epstein-Barr virus (EBV), which establishes a lifelong latent infection of B lymphocytes. The disease has a high incidence in young people and can cause exhaustion that lasts for months (figures 25.12, 25.13).

Ebola Virus Disease and Marburg Virus Disease (table 25.10)

Ebola virus disease (EVD) and Marburg virus disease (EVD) are both severe emerging diseases of global concern. They start with flu-like symptoms but can result in massive bleeding and multi-organ failure.

Dengue and Severe Dengue (table 25.11)

Dengue is an arboviral disease that can result in joint pain, giving the common name "breakbone fever." Its most common vector is *Aedes aegypti* (figure 25.14), but *A. albopictus* also transmits it, which is a concern because of the expanding range of that species. Severe dengue is usually thought to be a result of **antibody-dependent enhancement (ADE),** a situation that can occur with a second dengue infection. For certain people who have had a previous dengue infection, a vaccine is available.

Chikungunya (table 25.11)

Chikungunya is an arboviral disease that can cause debilitating joint pain, which sometimes becomes chronic. Like dengue, its most common vector is *A. aegypti,* but *A. albopictus* also transmits it.

Zika Virus Disease (table 25.11)

Zika virus disease is an arboviral disease that is typically mild, but can result in congenital Zika syndrome in a baby born to to female infected during pregnancy. Its most common vector is *A. aegypti.*

Yellow Fever (table 25.11)

Yellow fever is an arboviral disease characterized by fever, jaundice, and hemorrhaging. Infected primates living in jungles are a primary reservoir, but it can become epidemic in humans where a suitable *Aedes* mosquito vector is present. A highly effective attenuated vaccine is available for preventing the disease.

25.4 ■ Protozoan Diseases of the Cardiovascular and Lymphatic Systems

Malaria (table 25.12)

Malaria symptoms are initially flu-like, but then begin to occur as paroxysms. The disease is caused by five species of *Plasmodium* and is transmitted by certain species of *Anopheles* mosquitoes, its biological vector. The *Plasmodium* life cycle involves a liver stage and a red blood cell stage (figure 25.15). *P. falciparum* malaria is the most life-threatening. Malaria treatment is complicated by the fact that the various stages in the parasite's life cycle do not all respond to the same medications.

Leishmaniasis (table 25.13)

Leishmaniasis is caused by various *Leishmania* species and transmitted by sandflies. Cutaneous leishmaniasis is the most common form; other clinical forms include mucocutaneous leishmaniasis and visceral leishmaniasis.

Review Questions

Short Answer

1. Describe disseminated intravascular coagulation (DIC).
2. What is the significance of immune complex formation in infective endocarditis?
3. Why might the *Yersinia pestis* from a patient with pneumonic plague be more dangerous than the same organism from fleas?
4. Why might rodent burrows be a source of plague months after they are abandoned?
5. What activities of humans are likely to expose them to tularemia?
6. Why is brucellosis a threat to big-game hunters?
7. What type of leukocyte does EBV infect?
8. Why is a second infection with dengue virus more serious than the first?
9. Travelers to and from which areas of the world should be vaccinated against yellow fever?
10. Which *Plasmodium* species causes the most dangerous form of malaria?

Multiple Choice

1. Which of the following infection fighters are found in lymph?
   a) Leukocytes
   b) Antibodies
   c) Complement
   d) Interferon
   e) All of the above
2. Which of the following statements about the spleen is *false*?
   a) It routinely produces new blood cells.
   b) It cleanses the blood of foreign material and damaged cells.
   c) It contains B cells.
   d) It consists of red pulp and white pulp.
   e) It enlarges in a number of infectious diseases.
3. Choose the one *true* statement about sepsis.
   a) It is a rare healthcare-associated disease.
   b) The output of urine increases if shock develops.
   c) It can be caused only by anaerobic bacteria.
   d) Antibiotic treatment reliably cures the patient.
   e) Lung damage is an important cause of death.
4. Which statement about *Yersinia pestis* is *false*?
   a) Conditions inside human phagocytes activate virulence genes.
   b) The bacterium can form biofilms in the flea digestive system.
   c) YOPs (proteins) increase phagocytosis by macrophages.
   d) The organism resembles a safety pin in certain stained preparations.
   e) It was responsible for the "black death" in Europe during the 1300s.
5. Which of these statements about tularemia is *false*?
   a) It can be contracted from muskrats and bobcats.
   b) Biting insects and ticks can transmit the disease.
   c) The causative organism has growth requirements similar to those of *E. coli.*
   d) Formation of an ulcer at the bacterial entry site occurs in some forms of the disease.
   e) It can be treated effectively with antibiotics.
6. Which of the following statements about brucellosis is *false*?
   a) Fevers that come and go over a long period of time are reflected in the name "undulant fever."
   b) The causative agent can infect via mucous membranes.
   c) The causative agent is readily killed by phagocytes.
   d) The disease in cattle is characterized by chronic infection.
   e) Butchers are advised to wear goggles or a face shield to help protect against the disease.
7. Which of the following statements about yellow fever is *false*?
   a) Humans are the only reservoir.
   b) The name "yellow" reflects that many victims have jaundice.
   c) No proven therapy exists.
   d) An attenuated vaccine is widely used to prevent the disease.
8. Which of these infections is *not* transmitted by mosquitoes?
   a) Malaria
   b) Yellow fever
   c) Ebola virus disease
   d) Chikungunya
   e) Dengue
9. The malarial form infectious for mosquitoes is called a
   a) gametocyte.
   b) trophozoite.
   c) sporozoite.
   d) schizont.
   e) merozoite.

10. Which of the following statements about malaria is *true*?
   a) Transmission cannot occur in temperate climates.
   b) Transmission usually occurs with the bite of a male *Anopheles* mosquito.
   c) The disease is currently well controlled in tropical Africa.
   d) *P. falciparum* infects only old RBCs and therefore causes milder disease than other *Plasmodium* species.
   e) The characteristic recurrent fevers are associated with the release of merozoites from RBCs.

Applications

1. Patients with an artificial heart valve or who have had infective endocarditis are often given an antibacterial medication shortly before dental procedures or surgery. What is the rationale for this approach?
2. A healthcare worker in country where yellow fever is endemic was tasked with implementing a strategy to prevent epidemics of the disease. A key focus of the plan was vaccinating people who live on the outskirts of jungle areas. Why would vaccinating this group be particularly important?

Critical Thinking

1. At least four emerging mosquito-borne diseases have recently spread to the Western Hemisphere and now have global distribution. If you were assigned to control the spread of these diseases, would you focus on the mosquito population or on the causative agent that the mosquitoes carry? Explain your rationale and how you would implement your plan.
2. Transfusion of blood from a malaria-infected donor can result in transfusion-transmitted malaria (TTM). Why would a person who develops *Plasmodium vivax* TTM require treatment effective against the blood stage forms of the parasite but not against the hypnozoites?

26 | Nervous System Infections

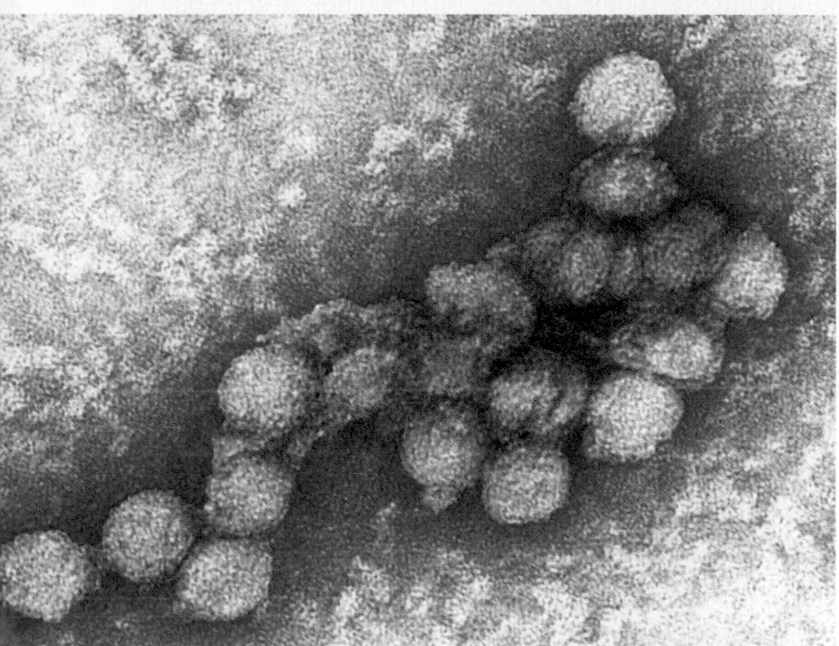
Transmission electron micrograph of West Nile virus particles. (*Cynthia Goldsmith/CDC*)

KEY TERMS

Blood-Brain Barrier Cells that function together to create a protective semipermeable border that separates the CNS from the bloodstream.

Central Nervous System (CNS) Brain and spinal cord.

Cerebrospinal Fluid (CSF) Fluid produced in the brain that flows within and around the CNS.

Encephalitis Inflammation of the brain.

Meninges Membranes covering the brain and spinal cord.

Meningitis Inflammation of the meninges.

Peripheral Nervous System (PNS) Division of the nervous system that carries information to and from the CNS.

Transmissible Spongiform Encephalopathy (TSE) Chronic degenerative brain disease caused by prions; characterized by spongy appearance of brain tissue.

A Glimpse of History

Today it is hard to appreciate the fear and loathing once attached to leprosy (lepros, meaning "scaly"). Many historical and religious texts refer to several disfiguring skin diseases, including leprosy, and portray those suffering from the diseases as unclean and sinful. Lepers were regularly segregated from mainstream society.

Gerhard Henrik Armauer Hansen (1841–1912) was a Norwegian physician with many interests, ranging from science to religion to polar exploration. After graduating from medical school, he went to work with Dr. Daniel C. Danielson, a leading authority on leprosy. Danielson believed that leprosy was a hereditary disease, but Hansen disproved that hypothesis in careful studies conducted over a number of years. He found a unique bacterium associated with the disease in every leprosy patient he studied. His 1873 report of the findings marked the first time that a specific bacterium was linked to a disease—almost a decade before Robert Koch detemined the cause of tuberculosis.

In the United States, even during the first half of the twentieth century, people diagnosed with leprosy risked having their houses burned to destroy the source of infection. Their names were changed to avoid embarrassing their families, and they were sent to a leprosarium such as the one at Carville, Louisiana, which was surrounded by a 12-foot fence topped with barbed wire. Those afflicted were separated from spouses and children and were denied the right to marry or vote. Those who tried to escape were captured and brought back in handcuffs. The Carville leprosarium was finally closed and converted to a military-style academy in 1999.

Because the word leprosy carries centuries of grim overtones, many people prefer to use the term Hansen's disease, a name that honors the discoverer of the causative bacterium. Today, the disease can be treated.

Nervous system infections are frightening. They threaten a person's ability to move, feel, or even think. Consider polio, which can result in a paralyzed limb or the inability to breathe without mechanical assistance. Hansen's disease (leprosy) can result in loss of fingers or toes or deformity of the face. Infections of the brain or the membranes covering it can render a child deaf or intellectually disabled. Before the discovery of antibiotics, bacterial infections of the nervous system were often fatal. Fortunately, these infections are uncommon.

26.1 ■ Anatomy, Physiology, and Ecology of the Nervous System

Learning Outcomes

1. Describe how information flows through and between neurons.
2. Differentiate between the central nervous system and the peripheral nervous system.
3. Explain how bone, cerebrospinal fluid, meninges, and the blood-brain barrier protect the central nervous system.

Nerve cells work together, transmitting electrical impulses throughout the body like a highly sophisticated circuit board. Each nerve cell, or **neuron,** has three functionally distinct regions: (1) branching projections called dendrites, (2) the cell body, which contains the nucleus, and (3) a single long, thin extension called an axon (**figure 26.1a**). Dendrites receive information from other neurons and then convey it to the cell body; from there the information is transmitted by the axon to another cell. In many neurons, the axons are surrounded by a fatty sheath called myelin that aids in signal transmission. The tiny gap at the junction between two cells where information transmission takes place is called a synapse. Most neurons communicate using **neurotransmitters,** chemicals stored in vesicles at the end of the axon. When a neurotransmitter is released, the molecules diffuse across the synapse to a neighboring cell; that cell has receptors for the neurotransmitter, allowing it to receive and respond to the signal.

For purposes of discussion, the nervous system is divided into the **central nervous system (CNS)** and the **peripheral nervous system (PNS).** The CNS includes the brain and spinal cord, both of which are enclosed by bone—the brain by the skull and the spinal cord by the vertebral column (**figure 26.1b**). Different parts of the brain have distinct functions; in fact, physicians can sometimes determine the location of a brain abscess or tumor by evaluating any associated loss of function. The spinal cord extends from the brain down the vertebral column. It conveys sensory information from receptors to the brain, and motor commands from the brain to the muscles and internal organs; it also contains neural circuits for simple reflexes. Inflammation of the brain is called **encephalitis;** inflammation of the spinal cord is called myelitis. The PNS is made up of nerves—each composed of bundles of axons—that carry information to and from the CNS. Motor nerves carry messages from the CNS to different parts of the body and cause them to respond; sensory nerves detect external stimuli such as heat, light, and pressure and transmit them to the CNS.

Cerebrospinal fluid (CSF) produced deep inside the brain provides cushion and support for the brain and also transports nutrients and other materials throughout the CNS. It is continuously secreted into four fluid-filled cavities in the brain called ventricles and then circulates over the surface

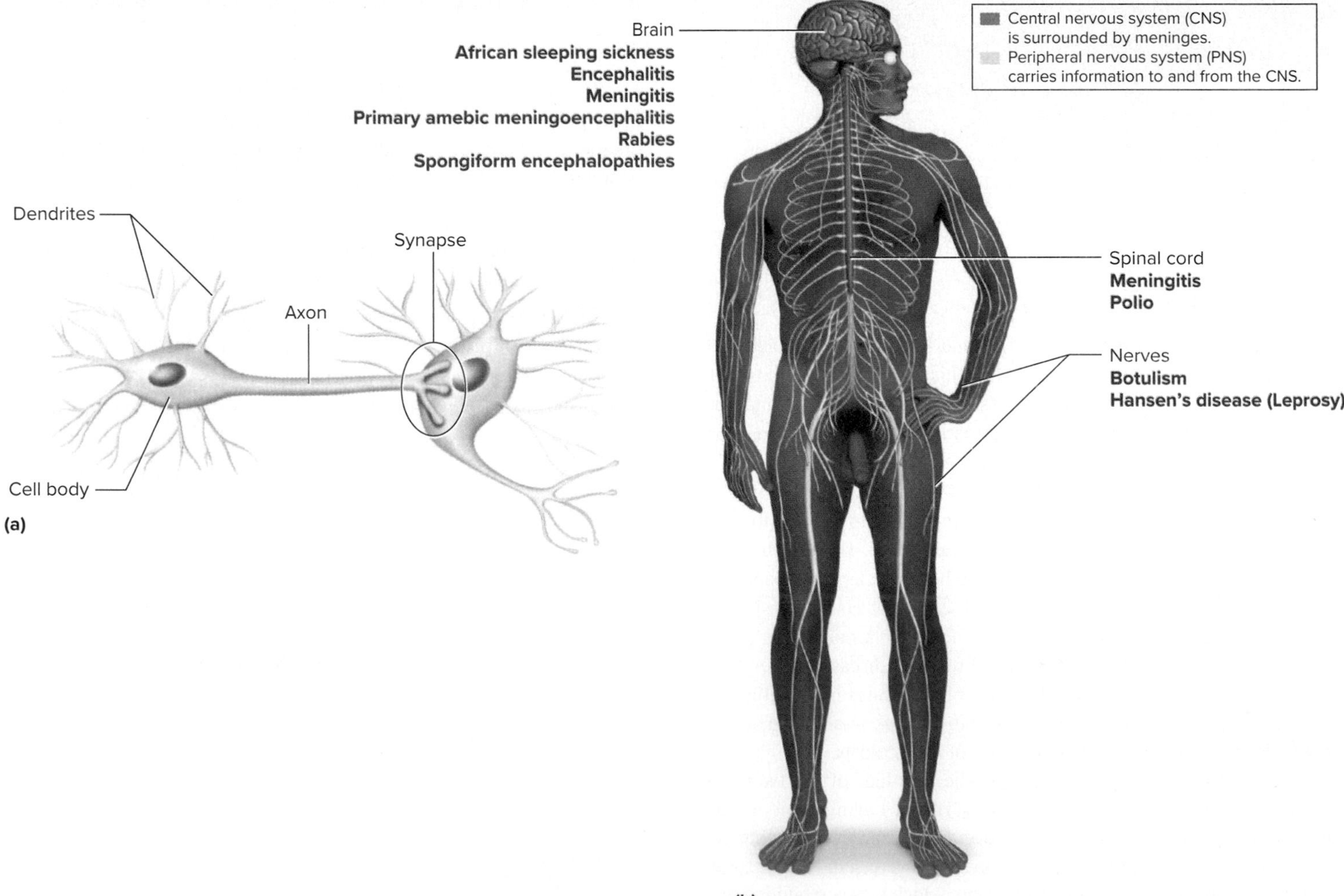

FIGURE 26.1 The Nervous System **(a)** A neuron receives information via dendrites and then transmits it through the axon. **(b)** The central nervous system is composed of the brain and spinal cord. Peripheral nerves carry information to and from the CNS. Nervous system diseases are shown in red.

? What are the respective roles of the dendrites and the axon in a neuron?

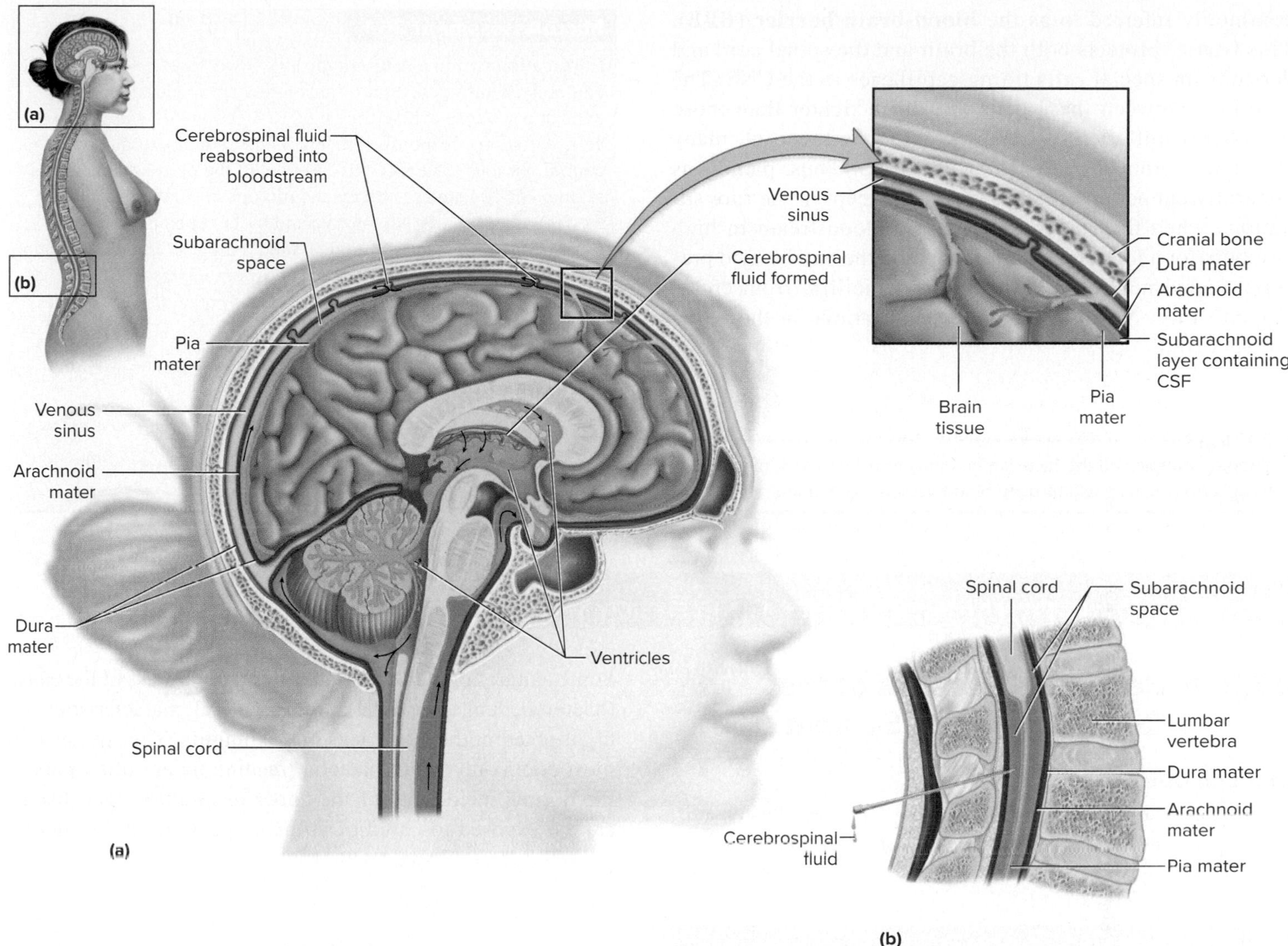

FIGURE 26.2 Meninges and Cerebrospinal Fluid **(a)** Three meninges—the dura mater, arachnoid mater, and pia mater—surround the CNS. CSF is formed in the ventricles, flows within the subarachnoid space, and is reabsorbed into the bloodstream. **(b)** A sample of cerebrospinal fluid can be withdrawn and examined.

? What is the purpose of a lumbar puncture?

of the brain and spinal cord. Some of that fluid is absorbed into the bloodstream at specialized sites within a venous sinus (channel for blood flow from the brain). CSF is normally sterile, so the presence of microbes indicates an infection.

Three layers of membranes called **meninges** cover the surface of the brain and spinal cord (**figure 26.2a**). The outer dura mater is tough and provides a barrier to the spread of infection from bones surrounding the central nervous system. It adheres closely to the skull and the vertebrae, and in some parts of the brain encloses a blood-filled venous sinus. The two inner membranes, the arachnoid mater and the pia mater, are separated by the subarachnoid space where CSF flows. The pia mater is the innermost membrane and adheres to the brain and spinal cord. Inflammation or infection of these membranes is called **meningitis.** When both the meninges and the brain are infected, the condition is meningoencephalitis. Meningitis is diagnosed using a sample of CSF obtained through a procedure called a lumbar puncture (or spinal tap). A needle is inserted into the subarachnoid space between lumbar vertebrae where the spinal cord has tapered to a thread-like structure, and a sample of CSF is withdrawn (**figure 26.2b**).

The nervous system lies entirely within body tissues and has no normal microbiota, although scientists believe that metabolites released by intestinal microbiota (such as short chain fatty acids, SCFAs) could be affecting brain function. The primary source of CNS infections is the bloodstream; infectious agents are rarely able to spread from the blood to the brain or spinal cord, however, because of what is commonly referred to as the **blood-brain barrier (BBB).** This barrier protects both the brain and the spinal cord and depends on special cells lining capillaries in the CNS. The junctions between these cells are much tighter than those in other capillaries, restricting the movement of many substances from the blood into the CNS. Thus, pathogens generally cannot enter nervous tissue except in the rare situations when they are present in the bloodstream in high numbers for a long time. Unfortunately, the barrier also prevents many medications, including penicillin, from crossing into the CNS unless their concentrations in the blood are very high.

MicroByte

Popping pimples on the face can be dangerous because it may lead to infections that spread through blood vessels to the brain.

MicroAssessment 26.1

Information within a neuron typically flows from dendrites to the cell body to the axon. Neurotransmitters are released from the axon at a synapse, where they bind to receptors on a neighboring cell, initiating a response. The brain and spinal cord make up the central nervous system (CNS). Nerves in the peripheral nervous system (PNS) convey sensory and motor information to and from the CNS. The CNS is protected by bone, by circulating cerebrospinal fluid (CSF), and by the meninges. The blood-brain barrier (BBB) prevents passage of most pathogens from the bloodstream into the CNS, but it also can prevent antimicrobial medications from accessing the CNS.

1. What molecules are released at the synapse, and what is their function?
2. Distinguish between the terms *encephalitis* and *meningitis.*
3. Why can encephalitis usually be detected in CSF obtained from the lower back?

CENTRAL NERVOUS SYSTEM INFECTIONS

26.2 ■ Bacterial Diseases of the Central Nervous System

Learning Outcomes

4. Compare and contrast the various types of bacterial meningitis.
5. Explain how newborns most often develop meningitis.

Some features of meningitis are similar regardless of the cause (bacterial, fungal, or viral), so the general characteristics are highlighted in the box Focus on Meningitis. The species that most commonly cause bacterial meningitis are often part of the normal microbiota of the upper respiratory tract; babies can be exposed to certain causative agents from the mother

FOCUS ON MENINGITIS

Meningitis—inflammation of the meninges—most often results from a microbial infection.

Signs and Symptoms

Signs and symptoms of meningitis typically begin a few days to a week after infection and can include the abrupt onset of a characteristic severe, throbbing headache and a stiff neck, often accompanied by fever, light sensitivity (photophobia), nausea, and vomiting. Seizures can occur, particularly in younger children. Deafness, confusion, loss of consciousness, and coma may develop. The case-fatality rate of untreated bacterial meningitis can be almost 100%, and death may occur within hours of the first symptoms. People who recover sometimes suffer permanent disabilities, including deafness, blindness, paralysis, or mental impairment. Viral meningitis is more common than bacterial meningitis but is typically less severe and causes little permanent damage in those with normal immunity.

To diagnose meningitis, a sample of cerebrospinal fluid (CSF) obtained by lumbar puncture is examined (see figure 26.2b). Bacterial meningitis is a medical emergency, so specimens are examined immediately. A person with bacterial meningitis has elevated CSF pressure (measured during lumbar puncture), and the CSF is cloudy and opaque, rather than a clear and pale yellow. It typically contains neutrophils, high protein levels, and low glucose levels compared to the blood. As part of the process of identifying the pathogen, a concentrated sample of the specimen is used to make a stained smear for microscopic examination and to inoculate appropriate laboratory media. Viruses are the most common cause of **aseptic meningitis,** meaning meningitis without detectable bacteria in the CSF using routine culture methods. CSF in people with viral meningitis is clear but typically contains many lymphocytes.

Pathogenesis

Most bacteria that cause meningitis are normal microbiota of the upper respiratory tract. These organisms can occasionally enter the bloodstream and, once there, may cross the blood-brain barrier (BBB) into the CSF. Certain types of trauma, such as skull fractures that allow contact between the sinuses or middle ear and the CSF, can predispose a patient to meningitis.

Much of the damage associated with bacterial meningitis is due to the inflammatory response to the invading pathogen. The vasodilation that characterizes inflammation results in fluid accumulation, causing brain swelling and nerve damage. The

(Continued)

FOCUS ON MENINGITIS *(continued)*

vasodilation also disrupts the BBB, allowing proteins and neutrophils to enter the CSF. Although neutrophils may help clear an invader, they can also cause life-threatening tissue damage in sensitive areas such as the brain and the spinal cord. In addition, the clots that form in the capillaries can block the blood supply, leading to cell death. Inflammation can also obstruct the normal outflow of CSF, causing the brain to be squeezed against the skull by the buildup of internal pressure.

Viral meningitis also induces inflammation of the meninges but is much less severe and typically causes little neurological damage. The disease usually resolves in a few days.

Epidemiology

Meningitis is a relatively rare disease, probably because pathogens generally cannot cross the BBB that protects the CNS. As with many diseases, people with lowered immunity such as newborns and the elderly are at greatest risk. Different bacterial causes of meningitis tend to affect specific age groups—either newborns, young children, young adults or older adults. Some of the bacteria are transmitted through respiratory droplets, so young adults living in crowded, confined environments such as military barracks or college dormitories are at increased risk. Some viral causes of meningitis are transmitted by the fecal-oral route rather than by respiratory droplets; not only are these viruses (enteroviruses) shed in the feces of infected individuals for weeks, but they are relatively stable in the environment.

Treatment and Prevention

Because of the life-threatening nature and rapid progression of bacterial meningitis, **empiric treatment** is begun immediately; an antimicrobial medication targeting the most likely causative agent based on the patient's age and other host factors is typically used. Once the bacterium causing the infection has been identified, then the treatment may be modified to target that organism. In most cases, no specific treatments exist for viral meningitis.

Vaccines are now available to prevent the most significant types of bacterial meningitis, resulting in a dramatic drop in the incidence of the disease in recent years. Unvaccinated people who have been in close contact with an infected person may be given an antimicrobial medication to help prevent the disease (post-exposure prophylaxis). There are no vaccines for viral meningitis, so routine prevention relies on handwashing and avoiding crowded areas.

while in the uterus or during passage through the birth canal. A table at the end of this section compares the main causes of bacterial meningitis (see table 26.3).

Pneumococcal Meningitis

Pneumococcal meningitis is caused by *Streptococcus pneumoniae,* part of the normal microbiota in the throat of many healthy people. Although best known as a cause of pneumonia, *S. pneumoniae* is also the leading cause of meningitis in adults.

Signs and Symptoms

Like other types of bacterial meningitis, symptoms of pneumococcal meningitis usually includes a severe, throbbing headache, followed by stiffness in the neck and back. Fever, nausea, and vomiting may also occur. Deafness, confusion, loss of consciousness, and coma may develop, and the disease can be rapidly fatal.

Causative Agent

S. pneumoniae, often called pneumococcus, is a Gram-positive diplococcus—the lancet-shaped cells are typically in pairs (see figure 21.11). Many strains are protected from phagocytosis by a polysaccharide capsule; antigenic differences in the capsules result in over 100 recognized serotypes of the organism.

Pathogenesis

S. pneumoniae is a common cause of otitis media (ear infection), sinusitis, and pneumonia—any of these can precede pneumococcal meningitis. Encapsulated strains that resist phagocytosis may enter the bloodstream of an infected person and then cross the blood-brain barrier into the cerebrospinal fluid, causing meningitis; the associated damage is largely due to the inflammatory response. Fluid accumulation (cerebral edema) and clots that form in the capillaries can lead to tissue death. Obstruction of the normal outflow of CSF can cause a buildup of internal pressure that squeezes the brain against the skull. Mortality and neurological damage are more common with pneumococcal meningitis, especially among the elderly.

Epidemiology

S. pneumoniae causes over half of the cases of bacterial meningitis in the United States, most of them in adults. It is also the leading cause of meningitis in children under 5. The case-fatality rate is less than 10% in children, but more than twice that in adults. The pathogen is spread by respiratory droplets, but very few who are infected will develop meningitis.

Treatment and Prevention

Penicillin remains the medication of choice for susceptible pneumococcal strains. If the strain is penicillin-resistant, a third-generation cephalosporin such as ceftriaxone is administered, with the possible addition of vancomycin.

The vaccines described in the discussion of pneumococcal pneumonia offer protection against the *S. pneumoniae* serotypes that cause the most serious pneumococcal diseases. The vaccines include:

- **Pneumococcal conjugate vaccine (PCV).** The three types of PCV currently available, PCV13, PCV15, and PCV20, use polysaccharides from 13, 15, and 20 pneumococcal serotypes, respectively. Recall that conjugate vaccines consist of capsular polysaccharides attached to bacterial proteins, thereby converting what would otherwise be T-independent antigens to T-dependent antigens; this is important because children younger than age 2 do not produce an effective immune response against T-independent antigens. PCV13 or PCV15 is recommended for children under 5 years old and for high-risk children through age 18. PCV15 or PCV20 is recommended for adults 65 years or older and for some high-risk adults 19 years or older.

- **Pneumococcal polysaccharide vaccine (PPSV).** The one version available, PPSV23, contains polysaccharides from 23 pneumococcal serotypes. It is recommended for high-risk children 2 through 18 years old and anyone age 19 or older who gets PCV15.

Meningococcal Disease

Neisseria meningitidis, often called meningococcus, is commonly part of the normal respiratory microbiota of healthy individuals but can also cause meningitis. Most cases of meningococcal meningitis are accompanied by meningococcemia (meningococcus in the bloodstream), so the term meningococcal disease includes both.

Signs and Symptoms

Symptoms of meningococcal disease appear after 1 to 7 days and are similar to those of pneumococcal meningitis but often include purplish spots on the skin called **petechiae** (**figure 26.3**); these indicate capillary damage, in this case due to meningococcemia.

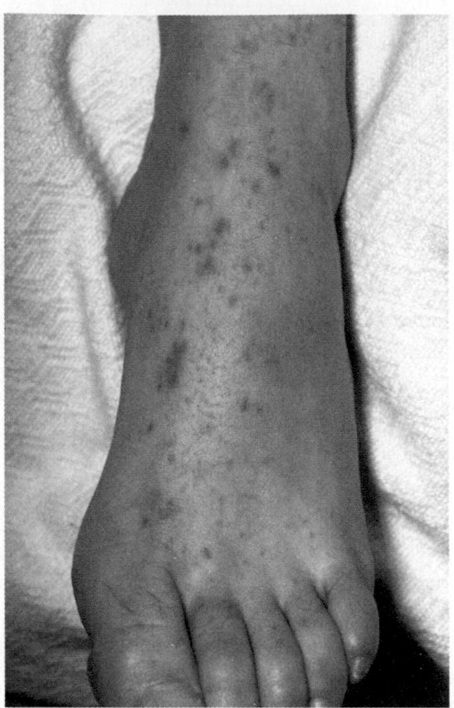

FIGURE 26.3 Petechiae of Meningococcal Disease These spots are a symptom of meningococcal disease. Medical-on-Line/Alamy Stock Photo

? What causes the petechiae associated with meningococcal disease?

Causative Agent

Meningococcal disease is caused by *N. meningitidis,* a Gram-negative encapsulated diplococcus. Its outer membrane differs from that of most Gram-negative bacteria in that it contains lipooligosaccharide (LOS) instead of lipopolysaccharide (LPS); like LPS, LOS contains lipid A and is therefore referred to as endotoxin. Meningococci can be grouped into at least 13 distinct serotypes, and most serious infections are due to serotypes A, B, C, W, X, and Y. Groups B, C, and Y are most common in the United States.

Pathogenesis

When airborne droplets containing meningococci are inhaled, the bacteria attach via pili to mucous membranes and then multiply. Specific proteins in the bacterial outer membrane allow the cells to pass though the respiratory mucosa and into the bloodstream. As is the case for many pathogens, the *N. meningitidis* capsule helps the cells avoid phagocytosis and protects them from complement system proteins.

As the *N. meningitidis* cells grow, they release outer membrane vesicles (called blebs), which contain LOS. The lipid A component of LOS is recognized by the host cell's pattern recognition receptors (PRRs), triggering the release of pro-inflammatory cytokines. In turn, this leads to systemic vasodilation and capillary leakage, sometimes resulting in petechiae and a drop in blood pressure. Septic shock results when blood pressure becomes so low that circulation cannot adequately supply O_2 to vital body tissues.

When meningococci invade the CSF, large numbers of neutrophils respond, but the bacteria multiply faster than they can be destroyed. As in other types of bacterial meningitis, tissue damage and cerebral edema (fluid accumulation) often result from inflammation.

Epidemiology

N. meningitidis can be transmitted by a person with meningococcal disease or by an asymptomatic carrier. The bacterium causes meningitis in all age groups, but infants, adolescents, and young adults are at particular risk. In the United States, the disease is relatively rare, but outbreaks may arise if the bacteria spread via respiratory droplets through crowded populations such as at universities; between 2013 and 2018, 10 university-based outbreaks of meningococcal serotype B disease were reported in 7 states, resulting in a total of 39 cases and 2 deaths.

The highest incidence of meningococcal meningitis worldwide occurs in the "meningitis belt" in sub-Saharan Africa (**figure 26.4**). During the dry season (between December and June), incidence there has reached 1,000 cases per 100,000 people (compared with 0.3 to 3 cases per 100,000 in the rest of the world). Epidemics in that region were primarily due to serogroup

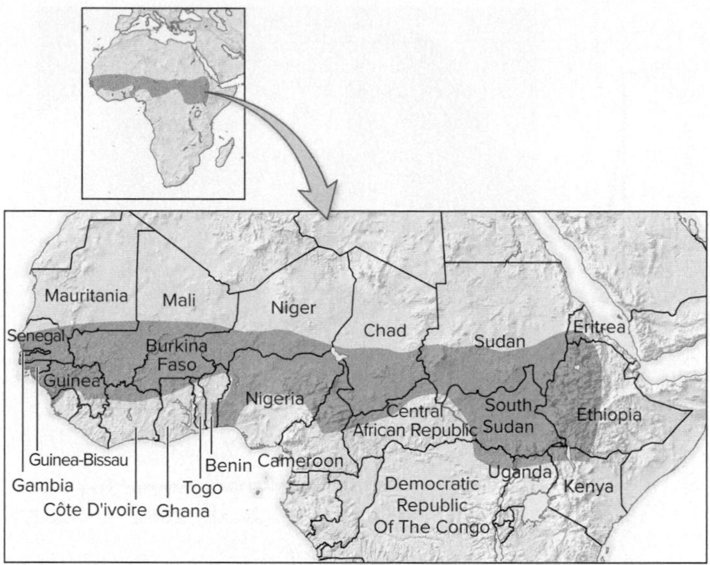

FIGURE 26.4 The Meningitis Belt CDC

? What is the "meningitis belt"?

A, but use of a meningococcal serogroup A conjugate vaccine starting in 2010 has dramatically decreased the number of cases; outbreaks now are primarily due to serogroups C and W.

Treatment and Prevention

Meningococcal meningitis can be effectively treated with prompt administration of antibiotics such as ceftriaxone, a third generation cephalosporin, and high doses of penicillin. With timely treatment, the case-fatality rate is less than 10%, and most patients recover without permanent nervous system damage.

Two types of vaccines are generally used to prevent meningococcal disease:

- **MenACWY.** This is a conjugate vaccine (bacterial polysaccharides attached to an immunogenic protein) that protects against serotypes A, C, W, and Y. It is recommended for all 11- to 12-year-olds with a booster at age 16, and for children and adults at increased risk of meningococcal disease.
- **MenB.** This is a recombinant subunit vaccine that protects against serotype B; it is recommended for adolescents and young adults between 16 and 23 who are at an increased risk for serotype B meningococcal disease, particularly in the event of an outbreak or if they have certain medical conditions. Ideally, the vaccine should be given between ages 16 and 18 when the risk of acquiring the disease is highest.

To prevent the disease from developing in close contacts of a meningococcal disease patient, an appropriate antibiotic can be used for post-exposure prophylaxis. **Table 26.1** describes the main features of meningococcal disease.

Haemophilus influenzae Meningitis

Haemophilus influenzae was once the leading cause of meningitis in young children, with about 1 out of 200 children

TABLE 26.1 Meningococcal Disease

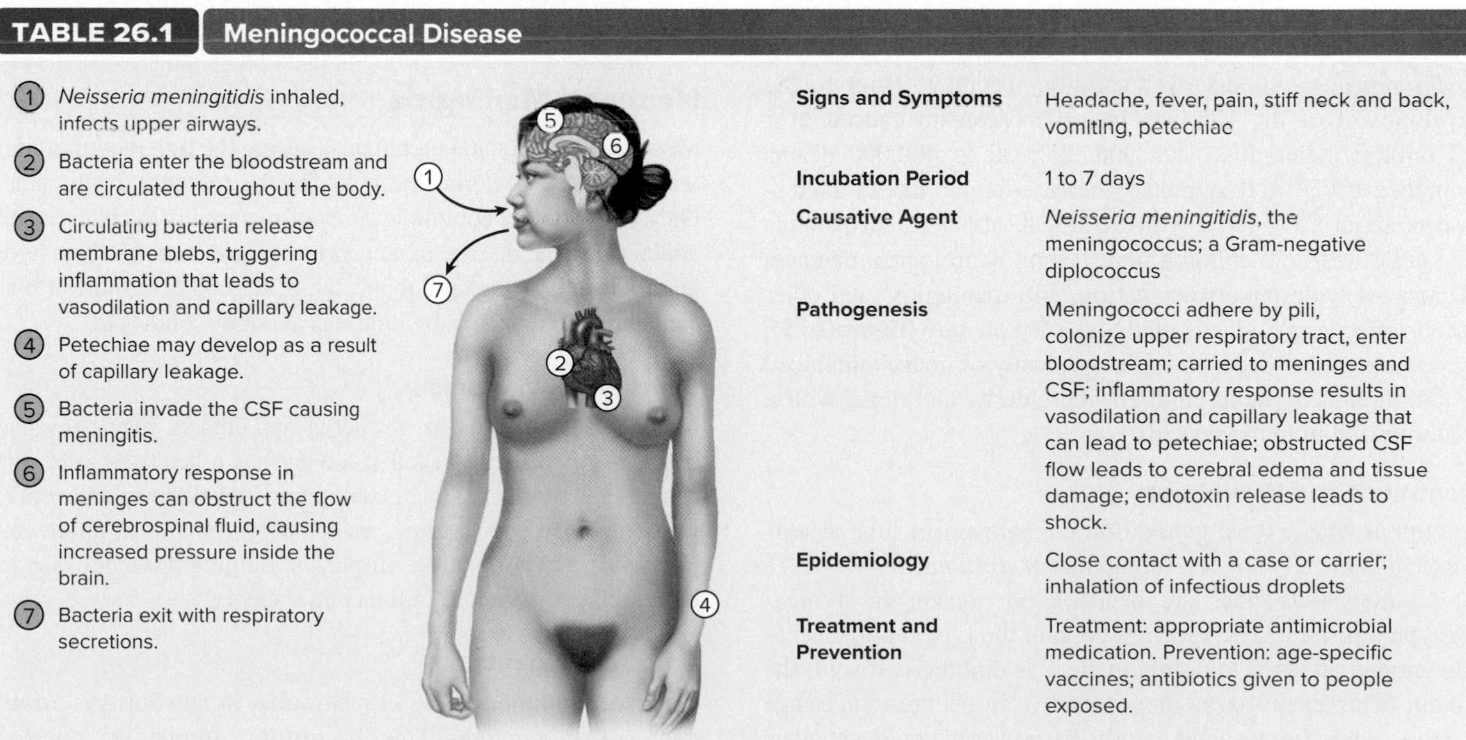

1. *Neisseria meningitidis* inhaled, infects upper airways.
2. Bacteria enter the bloodstream and are circulated throughout the body.
3. Circulating bacteria release membrane blebs, triggering inflammation that leads to vasodilation and capillary leakage.
4. Petechiae may develop as a result of capillary leakage.
5. Bacteria invade the CSF causing meningitis.
6. Inflammatory response in meninges can obstruct the flow of cerebrospinal fluid, causing increased pressure inside the brain.
7. Bacteria exit with respiratory secretions.

Signs and Symptoms	Headache, fever, pain, stiff neck and back, vomiting, petechiae
Incubation Period	1 to 7 days
Causative Agent	*Neisseria meningitidis*, the meningococcus; a Gram-negative diplococcus
Pathogenesis	Meningococci adhere by pili, colonize upper respiratory tract, enter bloodstream; carried to meninges and CSF; inflammatory response results in vasodilation and capillary leakage that can lead to petechiae; obstructed CSF flow leads to cerebral edema and tissue damage; endotoxin release leads to shock.
Epidemiology	Close contact with a case or carrier; inhalation of infectious droplets
Treatment and Prevention	Treatment: appropriate antimicrobial medication. Prevention: age-specific vaccines; antibiotics given to people exposed.

under 5 years old developing the disease; vaccines against the most serious antigenic type have decreased the incidence in the United States by over 99%.

Signs and Symptoms

As with other types of meningitis, signs and symptoms of *Haemophilus influenzae* meningitis include severe headache, fever, and vomiting. Older children may report a stiff neck. Infants may show a bulging fontanel (the "soft spot" on the top of an infant's skull). The disease may progress rapidly to coma and death.

Causative Agent

H. influenzae, a Gram-negative, non-motile coccobacillus (very short rod), was so named because in the 1890s it was mistakenly thought to be the cause of influenza. Encapsulated strains cause most disease in children and are distinguishable by serotype (a to f); *H. influenzae* type b (abbreviated as Hib) is usually the most virulent.

MicroByte

In 1991, *H. influenzae* earned the distinction of being the first free-living organism to have its genome completely sequenced.

Pathogenesis

Inhaled encapsulated *H. influenzae* use pili and other adhesins to bind to epithelial cells in the upper respiratory tract, and then penetrate the epithelial layer to enter the bloodstream. Bacteremia can result in meningitis or other infections such as epiglottitis or cellulitis. Unencapsulated strains typically cause local infections such as sinusitis and otitis media, although they sometimes cause invasive disease.

Epidemiology

H. influenzae is spread by respiratory droplets. Prior to the development of the Hib vaccine, the organism caused over 2.2 million meningitis cases and 300,000 to 400,000 deaths globally every year. If untreated, the case-fatality rate of the disease is about 90%. Even with treatment, about 5% of patients die, and 10–30% of children have lasting neurological damage. Because of widespread vaccination, Hib meningitis (and other serious infections) in the United States is now rare (**figure 26.5**); the few cases that occur today are typically in underimmunized or unimmunized young children, the elderly, and those with a compromised immune system.

Treatment and Prevention

Treatment with a third-generation cephalosporin such as ceftriaxone is usually successful against *H. influenzae.*

Conjugate vaccines are available to prevent *H. influenzae* type b disease. The vaccines contain the type b polysaccharide antigen attached to a protein such as diphtheria toxoid; the protein component makes them effective in children under age 2, who otherwise respond poorly to polysaccharide antigens.

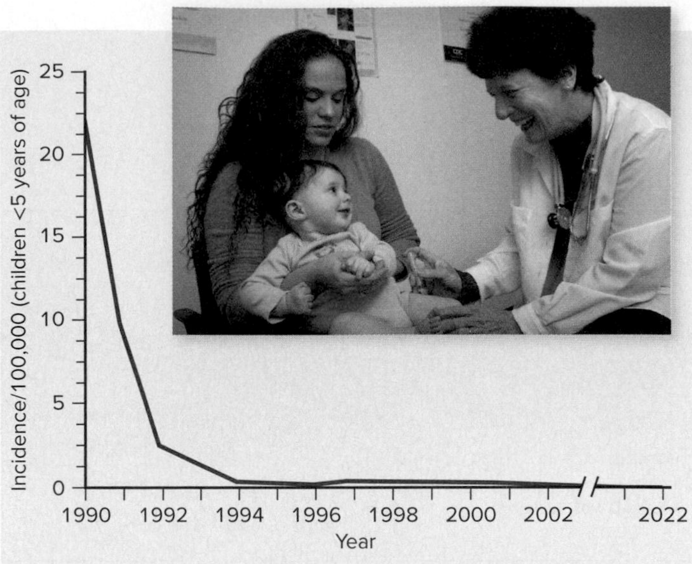

FIGURE 26.5 Incidence of Serious *Haemophilus influenzae* Disease per 100,000 Children Under Age 5, United States, 1990–2022 Before the availability of conjugate vaccines in late 1987, *H. influenzae* type b was the most common cause of bacterial meningitis in preschool children. James Gathany/CDC

? **What are the components of the Hib conjugate vaccine?**

Infants in the United States are routinely vaccinated against Hib beginning at 2 months of age (see figure 26.5); older children and adults with certain medical conditions are also vaccinated. The combination vaccine DTaP-IPV/Hib protects infants from Hib meningitis as well as tetanus, diphtheria, whooping cough, and polio; the DTaP-IPV/Hib/HepB vaccine additionally protects against hepatitis B. To prevent the disease from developing in close contacts of a patient with *H. influenzae* type b disease, an appropriate antibiotic can be used for post-exposure prophylaxis.

Neonatal Meningitis

Most cases of neonatal meningitis (during the first month of life) are caused by bacteria that colonize the mother's birth canal. Only rarely do the common causes of meningitis in children and adults cause the disease in newborns because most mothers have antibodies (IgG) against them; these antibodies cross the placenta and protect the baby until it is about 6 months old.

Signs and Symptoms

Meningitis symptoms in newborns and infants may be vague because it is difficult to tell if a baby has a headache or a stiff neck. Signs of meningitis in babies may include fever and vomiting, irritability, poor feeding, and lethargy. As with *H. influenzae* infection, newborns may display a bulging fontanel due to increased pressure within the cranial cavity.

Causative Agents

The most common cause of meningitis in newborns is *Streptococcus agalactiae.* This bacterium, frequently referred

FOCUS ON A CASE 26.1

The patient was a 31-month-old girl admitted to the hospital because of fever, headache, drowsiness, and vomiting. She had been well until 12 hours before admission, when she developed a runny nose, fever, and loss of appetite.

Her birth and development were normal, but her routine immunizations had been neglected. On examination, her temperature was 40°C (104°F), her neck was stiff, and she did not respond to verbal commands. There was no history of head trauma.

The child's white blood cell (WBC) count was elevated and showed a marked increase in the percentage of neutrophils. Her blood sugar was in the normal range. A spinal tap revealed cloudy CSF containing 18,000 WBCs per microliter (normally, there are few or none), a markedly elevated protein level, and a markedly low glucose level. Gram stain of the fluid showed tiny, Gram-negative coccobacilli.

1. What was the diagnosis, and what was the causative agent?
2. What was the prognosis in this case?
3. What age group is most susceptible to this illness?

Discussion

1. The patient had bacterial meningitis caused by *Haemophilus influenzae* serotype b.
2. With treatment, the fatality rate is approximately 5%. Unfortunately, even with treatment, up to one-third of those who survive are left with permanent damage to the nervous system, such as deafness or paralysis of facial nerves. Prompt diagnosis and correct choice of antibacterial treatment minimize the chance of permanent damage.
3. The peak incidence of this disease is in the age range of 6 to 18 months, corresponding to the time when maternal antibodies have diminished and adaptive immunity has not yet fully developed. Since 1987, vaccines consisting of type b capsular antigen conjugated with a protein have been used to immunize infants and thus eliminate the immunity gap. As a result, meningitis caused by *H. influenzae* type b is now rare.

to as group B streptococcus (GBS), colonizes the vagina of many women. Certain Gram-negative rods, such as encapsulated strains of *E. coli,* originate from the mother's intestinal tract and can also cause neonatal meningitis. Other cases are caused by *Listeria monocytogenes* from the bloodstream of an infected mother (discussed next).

Pathogenesis

Infection of the meninges is usually preceded by bacteremia in the newborn. Inflammation increases intracranial pressure that can block CSF flow, causing hydrocephalus (abnormal buildup of CSF that may compress and damage brain tissue). Disruption of blood flow damages nervous tissue. Infection may also lead to the development of a brain abscess.

Epidemiology

Neonates usually acquire the infecting bacteria from the mother's genital tract shortly before or during birth; premature or low birth weight babies are at most risk. Neonatal meningitis causes death in 5–20% of affected newborns. Those who survive often have long-lasting consequences, such as hearing loss or mental disability.

Treatment and Prevention

Treatment of neonatal meningitis includes intravenous administration of ampicillin and gentamicin, or another combination of antibacterial medications effective against both group B streptococci and *E. coli.* Other medications may be used after the causative agent is identified.

The Centers for Disease Control and Prevention (CDC) recommends that the vagina and rectum of pregnant women be tested for group B streptococci late in pregnancy. Women with positive cultures can then be treated with an appropriate antibacterial medication during labor. Screening and subsequent treatment has significantly decreased the incidence of serious group B streptococcal disease.

Listeriosis

Listeriosis, a systemic foodborne illness caused by *Listeria monocytogenes*, can lead to meningoencephalitis or meningitis. Most *L. monocytogenes* infections are asymptomatic or mild, but in certain populations the consequences can be severe.

Signs and Symptoms

Listeriosis is usually characterized by fever and muscle aches, and sometimes nausea or diarrhea. Most people requiring medical attention have meningitis or meningoencephalitis with fever, headache, stiff neck, and vomiting. Pregnant women who become infected often miscarry or deliver terminally ill infants. Babies infected at birth usually develop meningitis after an incubation period of 1 to 4 weeks; when a baby is infected in utero, the incubation period is shorter.

Causative Agent

L. monocytogenes is a motile, non-spore-forming, facultatively anaerobic, Gram-positive rod. It can grow at 4°C, even in foods that are vacuum-packaged.

Pathogenesis

L. monocytogenes is primarily a foodborne pathogen. The organism can induce its own phagocytosis and then survive and reproduce within many different host cell types, where it is protected from the immune system. It can also cause host cell actin to polymerize, forming a "tail" that propels the organism with enough force to push it into an adjoining cell (see Focus Your Perspective 3.1). When the organism enters the bloodstream, the resulting bacteremia can progress to meningitis. In pregnant women, *L. monocytogenes* crosses the placenta and produces widespread abscesses in fetal tissues.

Epidemiology

L. monocytogenes is widespread in natural waters and vegetation and can be carried in the intestines of asymptomatic humans and other animals. Pregnant women, the elderly, and those with underlying illnesses such as immunodeficiency, diabetes, cancer, or liver disease are especially susceptible to listeriosis. Outbreaks have resulted from *L. monocytogenes*-contaminated foods, including unpasteurized milk, soft cheeses, processed meats, and fresh produce. Because the organisms can grow in commercially prepared food stored at refrigeration temperatures, thousands of infections can come from a single food-processing plant.

Treatment and Prevention

Most *L. monocytogenes* strains remain susceptible to antibacterial medications such as penicillin. Even though the disease is often mild in pregnant women, prompt diagnosis and treatment are important to protect the fetus.

L. monocytogenes can be killed by thoroughly cooking foods. To reduce the risk of cross-contamination, uncooked meats should not be kept with other foods; countertops and utensils should be cleaned after food preparation; raw fruits and vegetables should be thoroughly washed before eating. Pregnant women and others at high risk are advised to avoid soft cheeses, refrigerated meat spreads, and raw or smoked seafood. They should also heat cold cuts and hot dogs before eating them and avoid the fluids that may be in the packaging. The U.S. Food and Drug Administration (FDA) has approved a food additive consisting of a mixture of bacteriophage strains that lyse *L. monocytogenes* (**figure 26.6**). The additive can be sprayed on a variety of meats during the production process, and its presence is recorded on the food label. The bacteriophages destroy contaminating *L. monocytogenes* on the treated products, thereby making the products safer to eat. Some of the main features of listeriosis are presented in **table 26.2.** The main causes of bacterial meningitis are compared in **table 26.3.**

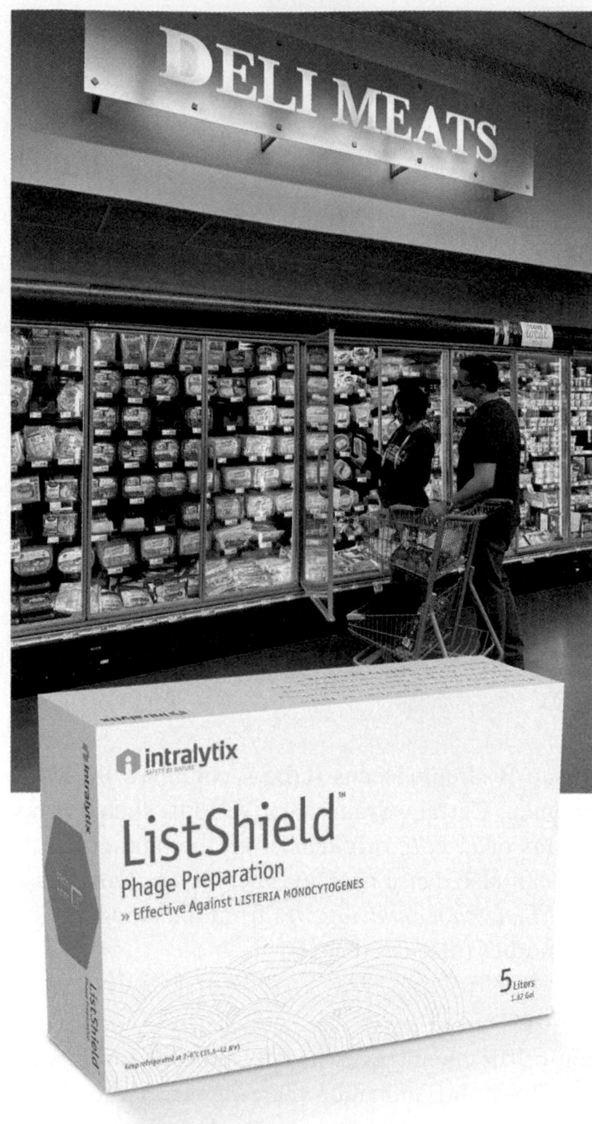

FIGURE 26.6 Bacteriophage in Food Safety When sprayed onto ready-to-eat foods, bacteriophage preparations significantly reduce the risk of listeriosis. (top): Mira Beins; (bottom): Intralytix, Inc.

How does the phage preparation reduce the risk of listeriosis?

MicroAssessment 26.2

Bacterial infections of the nervous system are uncommon but have a relatively high case-fatality rate. Most cases of bacterial meningitis are caused by species commonly carried in the upper respiratory tract (*Streptococcus pneumoniae, Neisseria meningitidis,* and *Haemophilus influenzae*). Organisms from the mother's birth canal are frequent causes in neonates. Meningoencephalitis and meningitis are complications of listeriosis, caused by the foodborne pathogen *Listeria monocytogenes.*

4. Why are most neonates unlikely to contract pneumococcal, meningococcal, or *Haemophilus influenzae* meningitis?
5. How would a Gram stain distinguish *Streptococcus pneumoniae* from *Neisseria meningitidis*?
6. Why would you expect to find relatively high levels of protein and low levels of glucose in the CSF of someone with bacterial meningitis?

TABLE 26.2 Listeriosis

1. *Listeria monocytogenes* is ingested with food such as soft cheeses, unpasteurized milk, raw fruits and vegetables, hot dogs, or smoked fish.
2. The bacteria rapidly penetrate the intestinal epithelium and establish bacteremia, especially in pregnant women, the elderly, and the immunodeficient.
3. In pregnant women, circulating *L. monocytogenes* cells cross the placenta and infect the fetus; the bacteria can also be transmitted to the baby at birth. The mother usually does not have a serious illness.
4. In older people and those with underlying diseases, *L. monocytogenes* infects the brain and meninges.

Signs and Symptoms	Fever and muscle aches, with or without gastrointestinal symptoms; headache and stiff neck mark the onset of meningitis or meningoencephalitis.
Incubation Period	A few days to 2 to 3 months; in newborn babies, 1 to 4 weeks
Causative Agent	*Listeria monocytogenes*, a non-spore-forming Gram-positive rod able to grow at 4°C
Pathogenesis	Ingested *L. monocytogenes* cells penetrate the intestinal epithelium and enter the bloodstream; the resulting bacteremia spreads to the CNS, causing meningoencephalitis or meningitis.
Epidemiology	Outbreaks from contaminated foods. Pregnant women develop bacteremia that may result in fetal infection or miscarriage. Babies contract infection in the uterus or at birth. Elderly and those with immunodeficiency, cancer, and diabetes also at high risk.
Treatment and Prevention	Treatment: appropriate antimicrobial medication. Prevention: care in handling, cooking of raw meats; thorough washing of vegetables; reheating of cold cuts, hot dogs, and refrigerated leftovers.

TABLE 26.3 Main Causes of Acute Bacterial Meningitis Compared

Agent	Characteristics	Source	Vaccine?	Age Group Usually Affected
Streptococcus pneumoniae	Gram-positive diplococcus	Respiratory system	Yes, against multiple serotypes	Mostly adults, also infants and young children
Neisseria meningitidis	Gram-negative diplococcus	Respiratory system	Yes, against multiple serotypes	Mostly infants, adolescents, and young adults; can cause epidemics
Haemophilus influenzae	Gram-negative coccobacillus	Respiratory system	Yes, against type b	Infants and young children
Streptococcus agalactiae	Gram-positive coccus; forms chains	Colon, vagina	No	Mostly neonates
Escherichia coli	Gram-negative rod	Colon	No	Mostly neonates
Listeria monocytogenes	Gram-positive rod; multiplies at refrigerator temperatures	Environment; contaminated cheeses, cold cuts, other foods	No	Pregnant women, neonates, elderly

26.3 ■ Viral Diseases of the Central Nervous System

Learning Outcomes

6. Compare and contrast viral meningitis with viral encephalitis.
7. Explain the history of polio and why it has been targeted for global elimination.
8. Outline the steps in the pathogenesis of rabies in wild animals, and explain how the disease can be prevented in humans.

A wide variety of viruses are neuroinvasive, meaning they can infect the nervous system to cause meningitis or encephalitis. Polio, now targeted for global elimination, and rabies, a disease normally found in animals, result from viral infection of neurons. Many viruses that typically affect other body systems can occasionally infect the central nervous system (CNS). These include the mumps, rubeola, varicella zoster, and herpes simplex viruses, Epstein-Barr virus, and, most commonly, human enteroviruses.

Viral Meningitis

Viral meningitis, sometimes referred to as aseptic meningitis, is more common than bacterial meningitis. It usually does not require specific treatment, and affected individuals generally recover in 7 to 10 days. Certain characteristics were described in Focus on Meningitis.

Signs and Symptoms

Signs and symptoms of viral meningitis typically begin abruptly within 1 to 4 weeks after infection, with fever, severe headache, sensitivity to light, a stiff neck, and often vomiting. Depending on the causative agent, a sore throat, chest pain, swollen parotid salivary glands, or a skin rash may also be present.

Causative Agents

Enteroviruses, non-enveloped RNA viruses that are members of the *Picornaviridae* family, are responsible for most cases of viral meningitis. Of these, the most common offenders are coxsackieviruses and echoviruses.

Pathogenesis

Enteroviruses characteristically infect the throat, intestinal epithelium, and lymphoid tissue, and then spread through the bloodstream to other sites, including the meninges. The inflammatory response differs from bacterial meningitis in that fewer white blood cells usually enter the CSF. Viral meningitis is typically less severe than bacterial meningitis and rarely causes lasting neurological damage.

Epidemiology

Enteroviral meningitis is transmitted by the fecal-oral route, and the feces of infected persons often contain viral particles for weeks. The viruses are relatively stable in the environment, sometimes even persisting in chlorinated swimming pools.

Treatment and Prevention

Most people recover from viral meningitis without treatment in 7 to 10 days. Handwashing and avoiding crowded swimming pools are reasonable preventive measures during outbreaks. There are no vaccines against most causes of viral meningitis. The main features of the disease are presented in **table 26.4.**

TABLE 26.4 Viral Meningitis

Signs and Symptoms	Abrupt onset, fever, severe headache, stiff neck, often vomiting; sometimes sore throat, rash, or chest pain
Incubation Period	Usually 1 to 4 weeks
Causative Agents	Most cases: small non-enveloped RNA enteroviruses of *Picornaviridae*, usually coxsackieviruses or echoviruses
Pathogenesis	Viruses spread through the bloodstream to the meninges. Fewer leukocytes enter CSF than with bacterial infections.
Epidemiology	Enteroviruses are transmitted by the fecal-oral route.
Treatment and Prevention	Treatment: no specific treatment in most cases. Prevention: handwashing, avoiding crowded swimming pools during enterovirus outbreaks.

West Nile and Other Types of Viral Encephalitis

Viral encephalitis (inflammation of the brain) is rare but can be deadly; it is often accompanied by viral meningitis, as the causative agents spread from the meninges to brain tissue. In cases where the causative agent is identified, the most common cause is reactivation of latent herpesvirus infections. **Arboviruses** (**ar**thropod-**bo**rne viruses) are also relatively common causes, depending on the geographic region; distribution of these cases is influenced by climate patterns that affect the range and activity of the arthropod vectors (typically mosquitoes). Herpesviruses are covered in section 24.3, so this section will mainly focus on arboviral encephalitis, particularly disease caused by West Nile virus.

Signs and Symptoms

Most arboviral infections are asymptomatic. If arboviral encephalitis occurs, early signs and symptoms typically develop within 1 week of infection and include fever, headache, and vomiting. Disorientation, dizziness, tremors, and seizures are also possible. Most people recover, but some are left with permanent damage, such as epilepsy, paralysis, memory loss, or mental impairment. Coma and death can occur in rare cases.

Causative Agents

Arboviral encephalitis is caused by a number of different arboviruses, a group of enveloped, single-stranded RNA viruses. The leading cause in the United States is West Nile virus (WNV), but other causes include La Crosse encephalitis virus (LACV), St. Louis encephalitis virus (SLEV), and eastern equine encephalitis virus (EEEV) (**table 26.5**). These viruses are all transmitted by mosquitoes; the vector of several arboviruses, including WNV, is the *Culex* mosquito that feeds at dusk and dawn.

TABLE 26.5 Arboviruses That Cause Encephalitis in the United States

Virus	Family
Eastern equine encephalitis virus (EEEV)	*Togaviridae*
La Crosse encephalitis virus (LACV)	*Bunyaviridae*
St. Louis encephalitis virus (SLEV)	*Flaviviridae*
West Nile virus (WNV)	*Flaviviridae*

Pathogenesis

West Nile and other arboviruses replicate initially in the dendritic cells at the site of a mosquito bite and are then carried in these cells to regional lymph nodes; from there, viremia occurs, allowing spread to other tissues and organs. The route of viral entry into the CNS is not clear, but appears to involve multiple mechanisms. Once in the CNS, they can cause meningitis, encephalitis, or meningoencephalitis. Brain tissue destruction is caused by both viral replication in neurons and by the damaging effects of an uncontrolled inflammatory response.

Epidemiology

West Nile and other types of arboviral encephalitis are zoonoses; the natural reservoir of the arboviruses is in birds or other small animals. Humans are an accidental or dead-end host—they can acquire the virus from a mosquito bite but do not develop sufficient viremia to transmit it back to the arthropod vector. WNV and SLEV have wild bird reservoirs. When WNV was introduced into New York from the Middle East in the summer of 1999, migrating birds quickly spread the virus across the country. As a result, neuroinvasive disease due to WNV infections is now widespread across the United States (**figure 26.7**). Birds are also the natural host for EEEV, but that virus can cause disease in equines (such as horses and donkeys) as well as humans, thus the name. Squirrels and chipmunks are the natural host of LACV.

The likelihood of disability due to arboviral encephalitis depends largely on the kind of virus and the age of the patient; as with many diseases, the very young, the elderly, and the immunocompromised are most affected. Case-fatality rates range from about <1% with La Crosse encephalitis to approximately 30% with eastern equine encephalitis. In 2022, the CDC reported 79 deaths in the United States resulting from West Nile neuroinvasive disease (which includes encephalitis), with a case-fatality rate of <11%.

Treatment and Prevention

There are no proven antiviral therapies for arboviral encephalitis.

Currently, the best preventive measure for arboviral encephalitis is to avoid mosquito bites. Sentinel chickens housed in mosquito-prevalent areas are used to track virus circulation in that environment; the blood of these caged birds is tested weekly for antibodies to certain arboviruses—if the results are positive, authorities know that the pathogen is present in that area, triggering a community alert. These alerts warn people to avoid outdoor activities during the peak hours of biting for mosquito vectors, to wear long sleeves and pants when outdoors, to make sure windows and porches are properly screened, and to use insect repellents and insecticides. Sentinel chickens are not made ill by these viruses, but because they are often bitten by the mosquito vectors, they provide a very sensitive detection and tracking system. Although there are no vaccines that prevent the types of viral encephalitis found in the United States, one is available that protects against a type called Japanese encephalitis (JE), which occurs in certain regions of Asia. The vaccine is recommended for people living in a JE-endemic country, as well as travelers to those areas. The main features of arboviral encephalitis are presented in **table 26.6.**

Polio (Poliomyelitis)

The characteristic feature of polio (also called poliomyelitis or paralytic poliomyelitis) is motor neuron destruction, resulting in paralysis of muscle groups, such as those of an arm or a leg. Jonas Salk and Albert Sabin each developed a highly effective vaccine; together these vaccines are key to controlling this terrifying disease. After the introduction of the vaccines in the mid 1950s and early 1960s, the United States and other resource-rich countries saw a dramatic drop in the number of polio cases. In 1988, the WHO launched the Global Polio Eradication Initiative, with the aim of eliminating polio worldwide by enhancing vaccination efforts in all countries. As a result, the disease is now endemic in only two

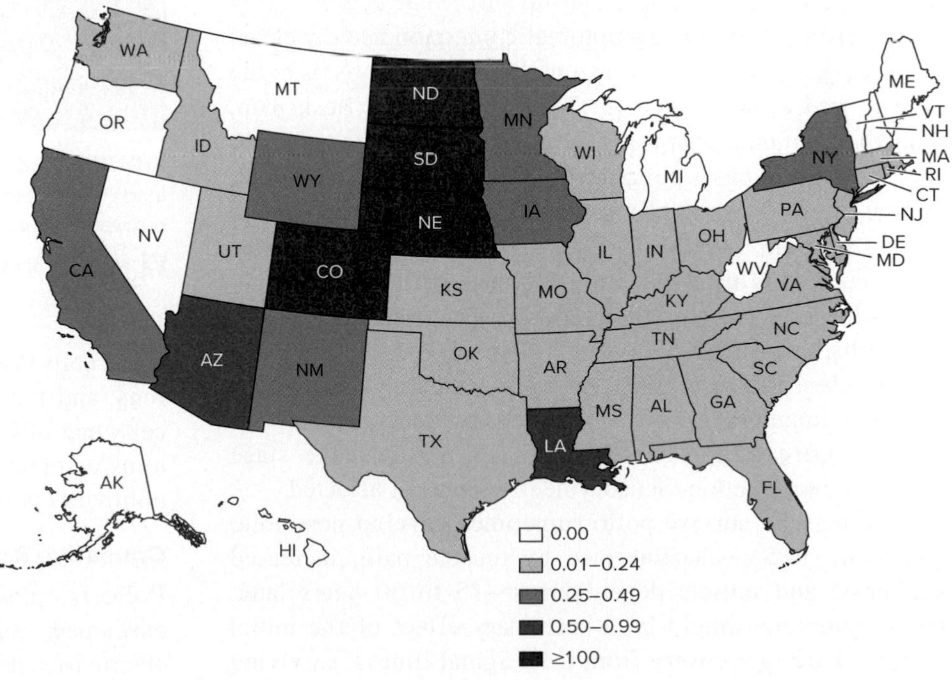

FIGURE 26.7 Incidence of West Nile Virus Neuroinvasive disease per 100,000, United States, 2022 (preliminary). ArboNET, Arboviral Diseases Branch, Centers for Disease Control and Prevention. https://www.cdc.gov/westnile/statsmaps/preliminarymapsdata2022/incidencestate-2022.html

? **What is the best way to prevent West Nile virus infection?**

TABLE 26.6 Arboviral Encephalitis

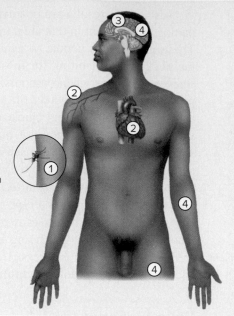

1. Infected mosquito introduces encephalitis virus.
2. Virus replicates locally, establishes brief low-level viremia.
3. Virus crosses blood-brain barrier and enters the CNS.
4. Destruction of brain tissue causes death or permanent disabilities such as paralysis, or mental impairment.

There is no exit for the virus; thus, humans are dead-end hosts.

Signs and Symptoms	Abrupt onset, fever, headache, vomiting, disorientation, paralysis, seizures, and hearing loss
Incubation Period	About a week
Causative Agent	Several arboviruses including West Nile virus, La Crosse encephalitis virus, St. Louis encephalitis virus or eastern equine encephalitis virus
Pathogenesis	Virus replicates in dendritic cells at the site of the mosquito bite, spread to lymph nodes, and then invades brain tissue. Neurons in the brain are destroyed.
Epidemiology	Mosquitoes transmit the viruses to humans from animal reservoirs.
Treatment and Prevention	Treatment: no proven treatment. Prevention: sentinel chickens to warn of arbovirus presence; insecticides and other anti-mosquito measures; vaccine for Japanese encephalitis.

countries—Afghanistan and Pakistan. Eradication is proving to be a very difficult challenge, however, because of several factors, including political instability within the endemic countries as well as interruption of vaccination programs during the height of the COVID-19 pandemic.

Signs and Symptoms

The outcome of infection by the virus that causes polio is highly variable, with only a fraction of people developing the disease. Up to 90% of infected individuals are asymptomatic, but some develop a flu-like illness (fever, sore throat, headache, stomach pain, nausea, and malaise) that begins 3 to 6 days after infection. The symptomatic infection usually clears within about a week, but occasionally the virus spreads to the CNS to cause meningitis, resulting in back and neck pain, headache, and muscle spasms. This usually resolves within 10 days, but in rare cases polio develops, characterized by weak, floppy, poorly controlled muscles, and sometimes complete loss of muscle function—called **flaccid paralysis;** when this occurs, it is usually within 3 weeks of infection. Muscles in the affected area then shrink over a period of weeks to months, and children's bones do not develop normally; in severe cases, the respiratory muscles are paralyzed, and air must be pumped in and out of the lungs by a mechanical ventilator (**figure 26.8**). Those who survive this acute stage recover some function. Sensory neurons are not affected.

People who survive polio sometimes develop post-polio syndrome (PPS)—characterized by muscle pain, increased weakness, and muscle degeneration—15 to 50 years later. These symptoms might be a secondary effect of the initial damage: During recovery from the original illness, surviving nerve cells branch out to take over the functions of killed neurons, and the "neural fatigue" (deterioration) of these nerve cells that did double duty for years likely leads to the symptoms associated with PPS. Another hypothesis is that some poliovirus persists in the CNS and is later reactivated.

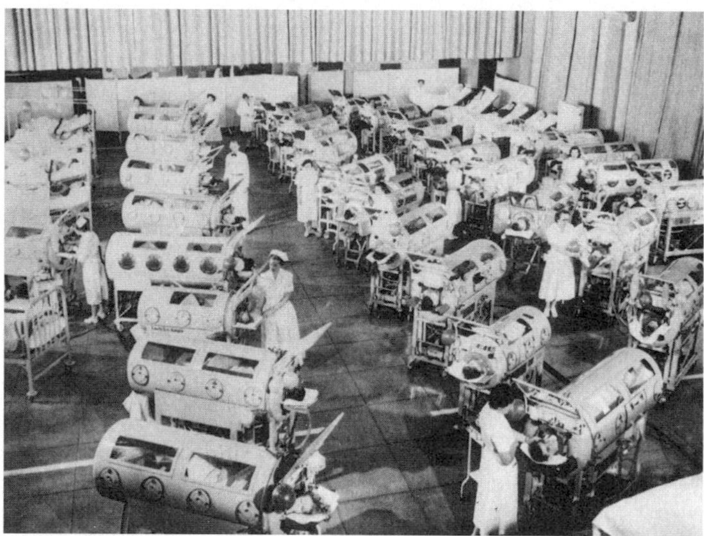

FIGURE 26.8 The Horror of Polio These early respirators ("iron lungs") were used during polio epidemics to keep people with paralyzed respiratory muscles alive. Everett Collection Historical/Alamy Stock Photo

? What type of paralysis occurs in some people with polio?

Causative Agent

Polio is caused by three serotypes of poliovirus, a non-enveloped, single-stranded RNA virus belonging to the enterovirus subgroup of the family *Picornaviridae.* Of those serotypes (designated 1, 2, and 3), only type 1 is still circulating; type 2 was declared eradicated in 2015 (after not being detected anywhere in the world since 1999), and type 3 was declared eradicated in 2019 (after not being detected since 2012).

MicroByte

Acute flaccid myelitis (AFM) in children resembles polio. Several viruses can cause AFM, including other non-polio enteroviruses such as Enterovirus D68 (EV-D68).

Pathogenesis

Poliovirus enters the body orally and infects cells that line the throat and intestinal tract. The virus is then carried by immune cells to regional lymph nodes, and from there it enters the bloodstream, resulting in viremia. In most cases, the virus is eliminated by the host immune response, leading to complete recovery. Occasionally, however, the circulating virus enters the CNS, either by crossing the blood-brain barrier or via peripheral nerves. Once in the CNS, the virus binds to specific receptors on motor neurons, replicates in the cell bodies, and destroys the cells when mature virus is released. Muscles innervated by the affected motor neurons no longer function properly, resulting in polio.

Epidemiology

Poliovirus is transmitted by the fecal-oral route and is quite stable in the environment, so it spreads easily if sanitation is poor. In areas where polio is endemic, babies become infected while still protected by the maternal antibodies they acquired as a fetus (see figure 15.20). In this situation, maternal antipolio antibodies can neutralize the virus in a baby's bloodstream before it can infect the motor neurons. In contrast, when a person is first infected with poliovirus later in life, maternal antibodies are no longer present, so paralytic disease is more likely to result. Because most people infected with the virus do not develop nervous system involvement, a single case of polio means that the virus is already widespread.

Treatment and Prevention

There are currently no antiviral treatments for polio, but two types of vaccines are used to prevent polio, and each has an important role in disease control:

- **Inactivated polio vaccine (IPV).** This contains inactivated forms of all three poliovirus types and is administered by injection in four doses (at ages 2 months, 4 months, 6 to 18 months, and 5 to 6 years); it is the vaccine currently administered in the United States as well as most other countries where polio has been eliminated. Although IPV effectively protects against polio, it is unreliable at eliciting mucosal immunity (secretory IgA response); recall that sIgA provides mucosal immunity because it can neutralize viral particles that would otherwise infect mucous membranes. Thus, even in IPV-vaccinated individuals, the virus can still potentially replicate in cells that line the digestive tract, be shed in the feces, and then be transmitted to someone else.
- **Oral polio vaccine (OPV).** This contains attenuated poliovirus administered as drops on the tongue (see chapter 17 opener photo) in four doses (at birth and ages 6, 10, and 14 weeks); a single dose of IPV is also given to children at least 14 weeks old. OPV is generally only used in countries where polio is still circulating or where it may be reintroduced; the vaccine is not licensed in the United States. In the past, OPV was trivalent, meaning that it contained all three poliovirus types. After type 2 was eradicated, however, the trivalent version was replaced by bivalent OPV (bOPV), which contains only types 1 and 3; monovalent options against those types are also available. In addition to protecting from polio, OPV generates mucosal immunity, thereby preventing the spread of poliovirus in a population. Because of this, polio eradication requires the use of OPV in regions where the virus is endemic or is likely to be reintroduced, as described in chapter 17.

The importance of OPV was demonstrated in 2013 when poliovirus was found in sewage in Israel. That country had been free of poliovirus and had stopped using OPV in 2004, so the only polio vaccine the children had received was IPV. The vaccine protected the recipients from polio, but it did not prevent reintroduced poliovirus from spreading within the population. After the virus was detected, Israel started an intensive immunization program using OPV, successfully eliminating the virus once again.

The drawback of OPV is that a vaccine strain can mutate as it replicates in the recipient, potentially developing virulence and giving rise to vaccine-associated polio. In fact, after poliovirus was eliminated in the United States, the rare cases of polio that occurred were not due to wild poliovirus (WPV) but to vaccine-derived poliovirus (now called VDPV) (**figure 26.9**); those cases generally occurred in people who had recently received their first OPV dose or in their unvaccinated contacts. Once WPV was eliminated from the Western Hemisphere, the small risk of developing polio due to VDPV led the United States to change to the routine use of IPV only.

As the geographic distribution of WPV is limited and detection methods are now more powerful, the source of infecting strains can be traced more easily. In turn, this makes it easier for public health officials to target effective vaccination campaigns. Testing of sewage and other environments also helps officials monitor the spread of the virus. The task of eradication is difficult, however, because even a seemingly low number of polio cases indicates perhaps 100 times as many infections—thus 50 polio cases suggests that over 5,000 people were infected!

Another factor complicating eradication efforts is VDPV strains referred to as circulating VDPV (cVDPV), which

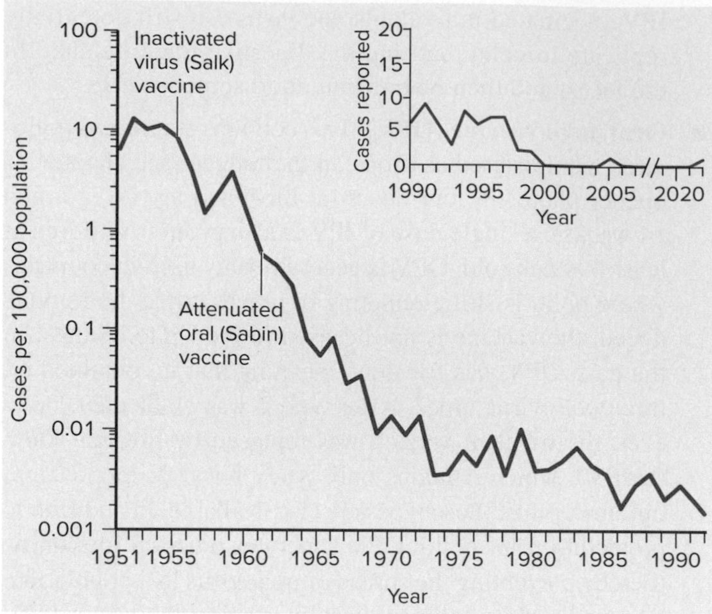

FIGURE 26.9 Incidence of Polio in the United States, 1951–2022 Wild poliovirus (WPV) was eliminated from the United States by 1980 and from the entire Western Hemisphere by 1991. Polio cases acquired in the United States after 1980 were caused by vaccine-derived poliovirus (VDPV).

? What was the cause of the polio case reported in 2005?

spread in populations that have inadequate mucosal immunity. Just as public health officials must eradicate WPV, they must also prevent the spread of cVDPV. Most cVDPV stains are derived from type 2 poliovirus, which IPV protects against, but because that vaccine does not elicit strong mucosal immunity, it might not prevent cVDPV from spreading. The problem was illustrated by the detection of that virus in wastewater samples in several New York counties in 2022, after a case of polio caused by cVDPV was reported in an unvaccinated person in one of those counties. A new, more genetically stable attenuated form of poliovirus type 2 has been developed to use as a vaccine against cVDPV. This vaccine, novel oral polio vaccine type 2 (nOPV2), is authorized for emergency use by the WHO when outbreaks of type 2 cVDPV occur.

In addition to controlling poliovirus through vaccination, prevention relies on breaking the transmission cycle by providing access to safe, clean water supplies, having adequate sanitation (such as treating sewage), and promoting handwashing. The main features of polio are summarized in **table 26.7.**

Rabies

Rabies is a classic zoonotic disease—one normally found in animals but that can be transmitted to humans. In the United States, immunization of dogs and cats against rabies has practically eliminated them as a source of human disease. Still, rabies has multiple wild animal hosts, and these remain a constant threat to non-immunized domestic animals and humans.

Signs and Symptoms

Signs and symptoms of rabies typically begin about 2 to 3 months after infection, but can vary considerably beyond that range (from a few days to several years). A characteristic paresthesia (pins and needles sensation) often occurs at the site of viral entry, usually an animal bite. Non-specific symptoms then occur, including fever, head and muscle aches, sore throat, fatigue, and nausea. In the later stage of the

TABLE 26.7 Polio (Poliomyelitis)

1. Poliovirus enters the body orally.
2. The virus replicates in cells lining the throat and the intestinal tract.
3. In some cases, the virus spreads via the bloodstream, causing mild symptoms, including fever, headache, and nausea.
4. In a small percentage of cases, the virus enters the CNS and may then cause the paralysis that characterizes polio.
5. Poliovirus exits the body with feces.

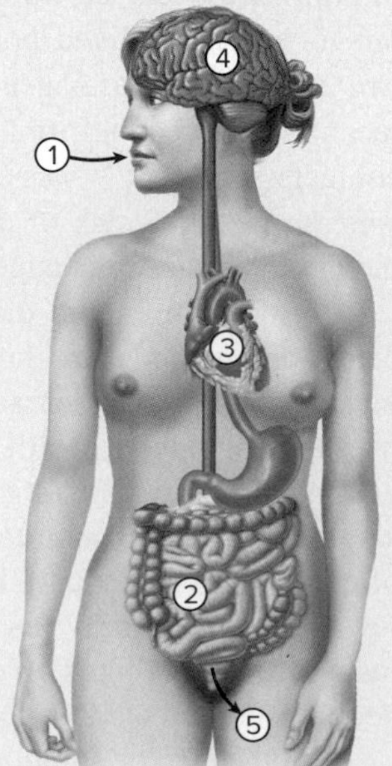

Signs and Symptoms	Headache, fever, stiff neck, nausea, pain, muscle spasm, followed by paralysis
Incubation Period	7 to 14 days
Causative Agent	Poliovirus serotypes 1, 2, 3; member of *Picornaviridae*; only type 1 is still circulating
Pathogenesis	Virus infects the throat and the intestinal tract, circulates via the bloodstream, and enters some motor nerve cells of the CNS; infected nerve cells are destroyed upon mature virus release.
Epidemiology	Spreads by the fecal-oral route; asymptomatic and non-paralytic cases common
Treatment and Prevention	Treatment: artificial ventilation for respiratory paralysis; physical therapy and rehabilitation. Prevention: vaccination; adequate sanitation and safe, clean water supplies.

disease, the virus spreads to the CNS, resulting in agitation, confusion, hallucinations, seizures, and increased sensitivity. That later stage has two recognized forms:

- **Encephalitic (furious) rabies.** This is characterized by increased salivation and difficulty in swallowing, causing the "frothing at the mouth" classically associated with rabid animals. Swallowing, or even the sight of fluids, often leads to severe spasms of the throat and respiratory muscles; the historic name for rabies is "hydrophobia" (fear of water). In some cases, the spasms lead to "aerophobia" (fear of air). Soon the individual goes into a coma and dies of respiratory failure or cardiac arrest.
- **Paralytic rabies.** This is characterized by flaccid paralysis, and is less dramatic than encephalitic rabies and persists longer before death occurs. It is also less common, accounting for about 20% of rabies cases.

Causative Agent

The cause of rabies is the rabies virus (**figure 26.10a**), a member of the family *Rhabdoviridae*. This bullet-shaped virus is enveloped and has a single-stranded RNA genome. Viral replication in brain tissue results in the formation of characteristic inclusions—the sites of viral replication—called **Negri bodies** (**figure 26.10b**).

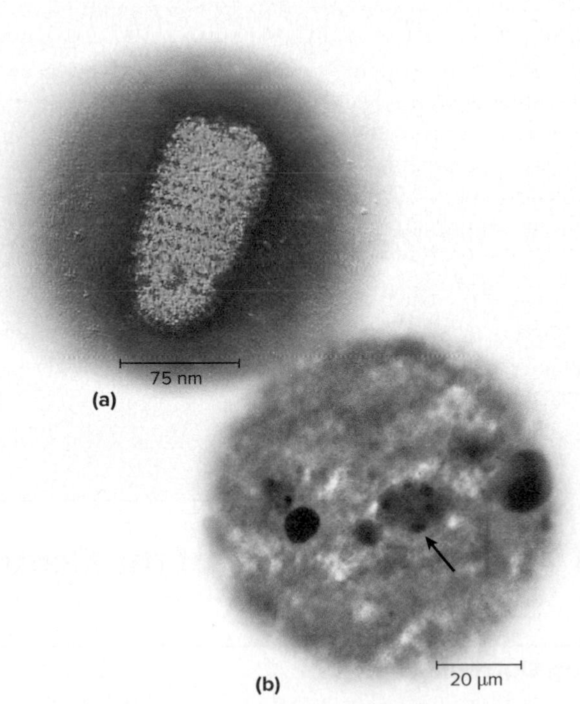

FIGURE 26.10 Rabies Virus (a) Color-enhanced transmission electron micrograph of rabies virus. Notice the bullet shape. **(b)** Stained smear of brain tissue from a rabid dog, showing a Negri body (arrow).
a: Tektoff-BM/CNRI/SPL/Science Source; b: Evans Roberts

? What inclusion body confirms the presence of rabies virus in brain tissue?

Pathogenesis

Rabies virus typically replicates in cells at the site of introduction for some time before entering a sensory neuron. It then travels by a mechanism called retrograde transport up the axon to the neuron cell body, and from there to the spinal cord and eventually the brain. The length of time before signs and symptoms develop depends on the location of the bite, the amount of virus introduced, and the condition of the host. Individuals with head wounds, for example, tend to show symptoms sooner than those with leg wounds.

Once in brain tissue, the virus replicates extensively, causing the symptoms of encephalitis. From the brain, the virus spreads outward via the nerves to various body tissues, notably the salivary glands, eyes, and fatty tissue under the skin, as well as to the heart and other vital organs. Rabies can be diagnosed before death by identifying the virus in stained smears collected from the surface of the eyes.

Epidemiology

Rabies is most commonly transmitted to humans via the bite of a rabid animal. In the United States, dogs are routinely vaccinated for rabies, so the main reservoir is wild animals such as raccoons, bats, skunks, and foxes (**figure 26.11**); over 5,000 wild animal cases are reported each year. Raccoons lead the list of wildlife cases, but the few human cases reported per year (one to three) are usually due to contact with infected bats.

Globally, about 50,000 to 70,000 people die of rabies every year, mostly in Asia and Africa due to dog bites in areas where dogs are not routinely vaccinated. A person bitten by an infected dog has about a 30% risk of developing rabies. Most rabid dogs excrete the virus in their saliva, sometimes even a few days before they get sick. Therefore, if an unvaccinated dog bites a person, the animal should be confined for 10 days to see if it develops symptoms of rabies.

Treatment and Prevention

There is no effective treatment for rabies, and very few people are known to have recovered from the disease once symptoms appear.

Although rabies is a rare disease, prevention is particularly important because it is nearly always fatal. The prevention involves multiple approaches:

- **Post exposure.** If a person has contact with a possibly rabid animal, post-exposure prophylaxis (PEP) should be administered. As part of rabies PEP, passive immunity is provided by injecting human rabies immune globulin (HRIG) at the wound site to neutralize free virus. In addition, multiple doses of the rabies vaccine are given to elicit active immunity; this post-exposure vaccination is effective because of the long incubation period before the rabies virus enters neurons. The first vaccine dose is given as soon as possible after exposure, and the others

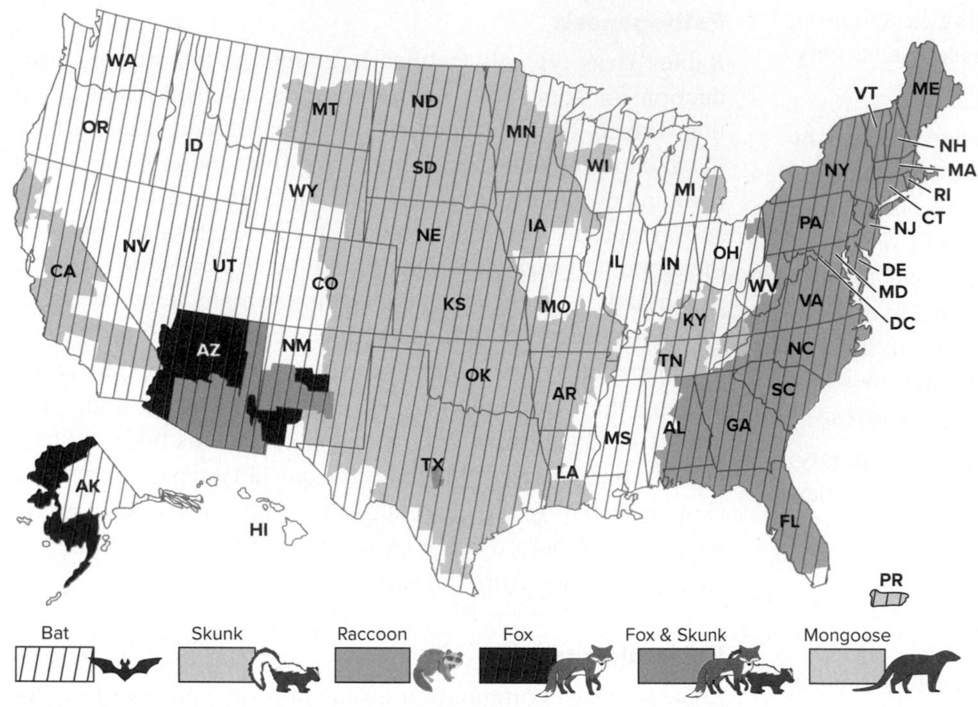

FIGURE 26.11 Wildlife Reservoirs of Rabies in the United States https://www.cdc.gov/rabies/location/usa/surveillance/wild_animals.html

What animal is responsible for most human cases of rabies in the United States?

at 3, 7, and 14 days after the first. If a person is immunocompromised, an additional dose is given at 28 days.

- **High risk pre-exposure.** People at high risk of exposure (such as those who work with the rabies virus or with animals that could have rabies) should receive the rabies vaccine as pre-exposure prophylaxis (PrEP). If someone who has received PrEP is then exposed to rabies, a modified version of PEP is administered.

- **Reservoir control.** A rabies vaccine is available for animals, so pets in most countries should be vaccinated (an exception is certain rabies-free countries); imported animals are often quarantined. Although vaccinating wild animals can be a difficult goal, programs are underway in the United States to administer an oral rabies vaccine to wild animal populations by placing the vaccine into a bait food, such as peanut butter or fish meal; many eastern states have participated in air-drop programs to vaccinate raccoons, and Texas has a similar program targeting foxes and coyotes. Animals suspected of having rabies should be avoided.

Some features of rabies are summarized in **table 26.8.**

TABLE 26.8	**Rabies**
Signs and Symptoms	Fever, head and muscle aches, sore throat, fatigue, and nausea at onset. Encephalitic rabies: spasms of the muscles of mouth and throat, and coma. Paralytic rabies: flaccid paralysis and coma.
Incubation Period	Varies, but usually 2 to 3 months
Causative Agent	Rabies virus, single-stranded RNA; bullet shape
Pathogenesis	During the incubation period, virus replicates in cells at bite site, then travels via nerves to the central nervous system; replicates and spreads outward via nerves to infect heart and other organs.
Epidemiology	Bite of rabid animal
Treatment and Prevention	Treatment: no effective treatment once symptoms begin. Prevention: immediate post-exposure prophylaxis with rabies vaccine and human rabies immunoglobulin; vaccine for people at high risk of exposure; immunize pets; avoid suspect animals

MicroAssessment 26.3

Although many different kinds of viruses can attack the nervous system, they generally do so in only a small percentage of infected people. Most viral meningitis cases are caused by the enterovirus subgroup of picornaviruses, which are generally spread by the fecal-oral route. Viral meningitis is usually mild, but viral encephalitis often causes permanent disability. Polio, characterized by paralysis of one or more muscle groups, is caused by poliovirus. Rabies is a widespread zoonosis, usually transmitted by animal bites, and almost always fatal for humans.

7. Explain why some types of viral encephalitis are considered zoonoses.
8. What are the advantages and disadvantages of OPV relative to IPV?
9. Why is rabies now rare in humans when it is still so common in wildlife?

26.4 ■ Fungal Diseases of the Central Nervous System

Learning Outcome

9. Compare and contrast the development of cryptococcal meningoencephalitis in AIDS patients and in healthy individuals.

Fungal cells in the environment are often inhaled but are generally destroyed by phagocytic cells and thus seldom cause disease. Occasionally, however, fungi survive phagocytosis and are then carried within the phagocytic cells via the bloodstream to the brain, leading to inflammation of the brain and the meninges.

Cryptococcal Meningoencephalitis

Cryptococcal meningoencephalitis was uncommon before the onset of the AIDS epidemic but is now among the most important HIV-related opportunistic infections. Because the causative agent is difficult to clear without T-cell involvement, AIDS patients with this infection are sometimes maintained indefinitely on antifungal medications. Between 1999 and 2007, the incidence of cryptococcal meningoencephalitis increased among healthy individuals in the western United States and Canada due to an emerging pathogen, *Cryptococcus gattii*.

Signs and Symptoms

In apparently healthy people, symptoms of cryptococcal meningoencephalitis develop gradually and generally consist of confusion, dizziness, intermittent headache, and sometimes a slight fever. After weeks or months, symptoms may progress to vomiting, weight loss, seizures, paralysis, and coma. In immunocompromised people, the disease generally progresses much faster; without treatment, the outcome can be fatal in as little as 2 weeks.

Causative Agent

Cryptococcal meningoencephalitis can be caused by either *Cryptococcus neoformans* or *C. gattii,* both of which are encapsulated yeasts (**figure 26.12**). *C. neoformans* is an opportunistic pathogen that usually causes disease only in immunocompromised persons, whereas *C. gattii* causes disease even in healthy individuals.

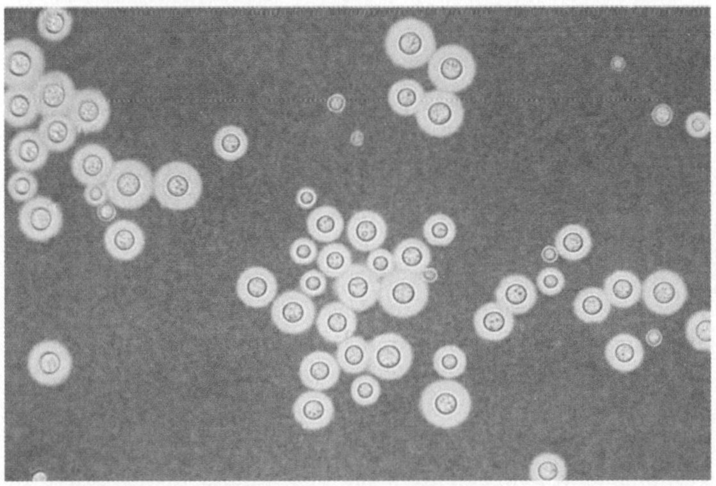

FIGURE 26.12 ***Cryptococcus neoformans*** This India ink stain reveals the capsule surrounding the yeast cells. Dr. Leanor Haley/CDC

How does a capsule allow an organism to establish an infection?

Pathogenesis

The *Cryptococcus* capsule is a key virulence factor that plays two seemingly opposite roles—preventing phagocytosis yet also allowing the cells to survive within macrophages. The contrasting roles are at least partly due to the complex and dynamic nature of the capsule, a structure that can vary significantly in size, even among cells of a pure culture.

When *C. neoformans* and *C. gattii* are inhaled into the lungs, their capsules are smaller, likely because they are desiccated (dried out); as a result, the cells are more easily engulfed by resident alveolar macrophages. Once inside the phagocytes, some organisms not only survive, but also multiply and eventually escape. Factors that aid intracellular survival include protection by the capsule and production of both melanin (a pigment that protects cells from oxidizing environments) and urease (an enzyme that generates acid-neutralizing urea). Within a day of infection, many of the yeast cells have developed the characteristic large antiphagocytic capsules.

If the lung infection is not controlled by an effective cell-mediated immune response, *C. neoformans* and *C. gattii* can spread to the bloodstream and into the meninges and brain. How the yeast cells cross the blood-brain barrier is not entirely understood, but multiple mechanisms may be involved, such as (1) inducing host cells that line the barrier to move them across it and (2) being carried across by phagocytes. Capsular material can be detected in spinal fluid and in serum, aiding diagnosis.

Epidemiology

C. neoformans is distributed worldwide in soil and vegetation contaminated with bird droppings. Humans and other animals may inhale the organism, but subsequent onset of disease is rare. Symptomatic infection is sometimes the first indication of AIDS. Person-to-person transmission of the disease does not occur.

C. gattii was once associated with eucalyptus trees in tropical or subtropical regions of the world, but it emerged in 1999 in British Columbia, Canada. Pathogenic strains have since spread in the U.S. Pacific Northwest and Canada. The appearance of more virulent strains has increased the case-fatality rate from 5% to about 25%. These emerging strains also cause more cases of fungal pneumonia.

Treatment and Prevention

Treatment of cryptococcal meningoencephalitis with the antifungal medication amphotericin B is often effective, particularly if given with flucytosine followed by the oral medicine fluconazole. Amphotericin B must be given intravenously and the dose carefully regulated to minimize its toxic effects. Because amphotericin B does not reliably cross the blood-brain barrier, the medication may be administered through a plastic tube inserted through the skull into a ventricle of the brain. Treatment is successful in about 70% of cases except in

AIDS patients, who respond poorly, most likely because they lack T-cell-dependent killing that normally assists the action of the antifungal medications. Unless their T-cell function can be restored by treatment, AIDS patients are rarely cured of their infection.

There is no vaccine or other measures to prevent cryptococcal meningoencephalitis. The main features of the disease are summarized in **table 26.9.**

MicroAssessment 26.4

Cryptococcal meningoencephalitis is caused by *Cryptococcus neoformans* and *Cryptococcus gattii,* encapsulated yeasts. *C. neoformans* can be found worldwide in soil and vegetation contaminated with bird droppings; it causes disease mainly in immunocompromised people such as AIDS patients. In contrast, the disease caused by *C. gattii* affects healthy individuals; it emerged in the late 1990s in the Pacific Northwest.

10. Why is it difficult to cure AIDS patients of a fungal infection?
11. Why are AIDS patients so vulnerable to cryptococcal meningoencephalitis?
12. What might cause the spread of new strains of *C. gattii* in the northwest United States?

TABLE 26.9 Cryptococcal Meningoencephalitis

Signs and Symptoms	Headache, vomiting, confusion, and weight loss; slight or no fever; symptoms may progress to seizures, paralysis, coma
Incubation Period	Widely variable, weeks to months
Causative Agent	*Cryptococcus neoformans* and *Cryptococcus gattii*—encapsulated yeasts
Pathogenesis	Organisms are inhaled and phagocytosed by alveolar macrophages; encapsulated organisms survive and multiply within phagocytes, eventually escaping to enter the bloodstream, from where they are carried to the meninges and brain.
Epidemiology	Inhalation of material contaminated with the fungus
Treatment and Prevention	Treatment: antifungal medications. Prevention: none.

26.5 ■ Protozoan Diseases of the Central Nervous System

Learning Outcome

10. Compare and contrast African trypanosomiasis, toxoplasmosis, and primary amebic meningoencephalitis.

Protozoan diseases of the nervous system are difficult to treat because the causative agents are eukaryotes, and available medications may also affect fragile neurons.

African Trypanosomiasis ("African Sleeping Sickness")

African trypanosomiasis—also known as African sleeping sickness—is transmitted by the tsetse fly, the causative agent's biological vector. The disease can persist for years, producing the symptoms for which it is named, and is always fatal if not treated. It can be contracted by residents and visitors in a wide area across the middle of the African continent.

Signs and Symptoms

African trypanosomiasis can be chronic or acute, depending on the subspecies of the causative agent. Signs and symptoms of both types progress in two stages:

- **Early stage.** This begins a week or so after infection. In the chronic form, disease progression is slow, sometimes taking 3 years to advance to the late stage; signs and symptoms are often relatively mild and intermittent, but patients may experience fatigue, headaches, muscle and joint aches, fever, itchiness, and swollen lymph nodes. By contrast, the acute form progresses rapidly and has pronounced signs and symptoms: A painful chancre (sore) develops at the pathogen entry site, followed by most of the same symptoms as the chronic form, but often more severe (extreme fatigue, intense headaches, muscle and joint pains, fever, and swollen lymph nodes). The acute disease often reaches the late stage within months.
- **Late stage.** This occurs after the organism enters the central nervous system (CNS). The patient begins exhibiting confusion, changes in behavior, poor coordination, decreased activity, and indifference to food. Sleep cycle disturbances typically develop—the patient falls asleep even while eating or standing, has drooping eyelids, and slurred speech. Eventually the patient goes into a coma and dies.

Causative Agent

African trypanosomiasis is caused by the flagellated protozoan *Trypanosoma brucei.* This organism (referred to as a trypanosome) is slender and has a wavy, undulating membrane and a characteristic flagellum (**figure 26.13**). Two subspecies are identical in appearance: *T. b. gambiense* causes chronic disease (West African trypanosomiasis), and *T. b. rhodesiense* causes acute disease (East African trypanosomiasis). Both subspecies are transmitted by tsetse flies, large biting insects of the genus *Glossina.*

Pathogenesis

Trypanosoma brucei enters the bite wound in the saliva of an infected tsetse fly. The parasite multiplies at the site and eventually enters the circulation, causing the first systemic signs and symptoms. The immune response initially controls the parasite, but within about a week and at roughly weekly

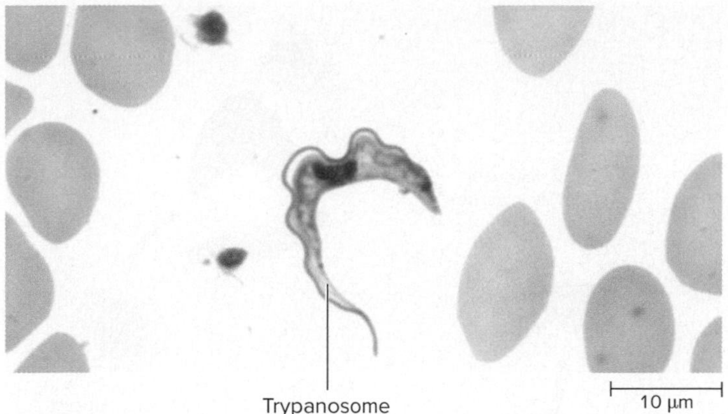

FIGURE 26.13 ***Trypanosoma brucei* in a Blood Smear** Notice the slender protozoan among the red blood cells of a person with African trypanosomiasis. CDC/ Blaine Mathison

? **What is the common name for the biological vector of *Trypanosoma brucei*?**

intervals thereafter, the number of parasites in the blood increases. Each burst coincides with the appearance of a new surface glycoprotein on the trypanosome cells. The parasite can make more than a thousand different versions of that glycoprotein—each encoded by a different gene—but only one is expressed at a time. Whenever another version is made, a new immune response is needed to produce appropriate antibodies. Because of this antigenic variation, recurrent cycles of parasitemia and antibody production continue until the patient is treated or dies.

With *T. b. gambiense* infections, progression of the disease is slow. Years may pass before death occurs, often from secondary infection. Damage to the host is partly due to the formation of immune complexes that then react with complement system proteins. In *T. b. rhodesiense* infections, disease can progress rapidly. The parasite enters the heart and the brain within 6 weeks of infection. Irritability, personality changes, and mental dullness result from brain involvement, but the patient usually dies from heart failure within 6 months.

Epidemiology

The distribution of African trypanosomiasis depends on the presence of the tsetse fly vector as well as reservoir hosts. The estimated population at risk is 65 million people, most of whom live in remote, rural areas in certain sub-Saharan countries. Almost 40,000 cases were reported in 1998; however, successful WHO-coordinated efforts since then have led to a substantial decrease—only 663 new cases were reported in 2020, a record low.

The chronic form of African trypanosomiasis (caused by *T. b. gambiense*) is found in the forested areas of central Africa and parts of West Africa; it is by far the most common form, accounting for the vast majority of cases. Humans are the main reservoir. The acute form (caused by *T. b. rhodesiense*) is found in the savannahs of eastern and southeastern Africa. It is a zoonosis; the main reservoirs are cattle and wild animals, making this form more difficult to control.

Treatment and Prevention

Early diagnosis and treatment of African trypanosomiasis is important to prevent damage to the nervous system; this is particularly true for the acute form because the disease progresses so rapidly. The medication choice depends on the type and stage of the disease. Suramin is used to treat early stages of the acute form of the illness; pentamidine is used to treat early stages of the chronic form. Medications that can cross the blood-brain barrier must be used to treat the late stage of either form of the disease; these medications are relatively toxic, however, and can be difficult to administer. Eflornithine, sometimes in combination with nifurtimox, is used to treat the late stage of the chronic form; melarsoprol is used to treat the late stage of the acute form. A new oral medication, fexinidazole, is effective against both stages of the chronic form of the disease.

Preventing African trypanosomiasis relies on controlling the tsetse fly vectors by using insect repellents and wearing protective clothing to prevent bites, clearing brush to reduce fly breeding habitats, and using baited traps containing insecticides to kill the flies. The main features of the disease are presented in **table 26.10.**

Toxoplasmosis

The protozoan parasite that causes toxoplasmosis can be a serious problem for those who are immunocompromised, as well as for fetuses. Even in healthy people, the immune

TABLE 26.10	African Trypanosomiasis ("African Sleeping Sickness")
Signs and Symptoms	Early: fatigue, headache, fever, swollen lymph nodes, joint and muscle aches Late: involvement of the central nervous system, confusion, withdrawal, uncontrollable sleepiness, coma, death
Incubation Period	Days or weeks (acute) to months or years (chronic)
Causative Agent	*Trypanosoma brucei*, a flagellated protozoan. *T. b. gambiense* causes chronic disease; *T. b. rhodesiense* causes acute disease
Pathogenesis	The protozoa multiply at the site of a tsetse fly bite, then enter the circulation; the parasites can change a surface glycoprotein, making previously formed antibodies ineffective; entry of the parasite into the CNS is fatal.
Epidemiology	Bites of infected tsetse flies transmit the trypanosomes through saliva; humans are the primary reservoir for *T. b. gambiense*; cattle and wild animals are the reservoir for *T. b. rhodesiense*.
Treatment and Prevention	Treatment: antiprotozoal medications. Prevention: protective clothing, insecticides, clearing brush where flies breed.

system may not clear the organism, resulting in a latent infection that can reactivate when immunity declines.

Signs and Symptoms

In otherwise healthy people, infection by the parasite that causes toxoplasmosis is generally asymptomatic. Some people, however, develop signs and symptoms—including sore throat, fever, and swollen lymph nodes and enlarged spleen—similar to those of other mild systemic infections. These subside over weeks or months and do not require treatment.

Toxoplasmosis in immunocompromised people commonly occurs due to latent infection reactivation. In these patients, the systemic infection can progress to life-threatening encephalitis, manifested by confusion, weakness, impaired coordination, seizures, stiff neck, paralysis, and coma. The patient may also develop brain masses much like tumors, which cause headaches and other neurological signs and symptoms such as poor coordination, seizures, sensory loss, and weakness. Infection of the retina (the light-sensitive part of the eye) is also common.

Fetal toxoplasmosis can be acquired across the placenta if a woman is infected with the protozoan for the first time either during or just before her pregnancy; if she was infected previously, her IgG protects the fetus. Infection of the fetus during the first trimester is most severe, often resulting in miscarriage or stillbirth; babies born live may have serious birth defects. Infections during the last trimester usually have less severe effects; most of these infants appear normal at birth, but retinitis (inflammation of the retina in the eye) may occur later in life, resulting in recurrent episodes of pain, sensitivity to light, and blurred vision. Intellectual disability and epilepsy may develop.

Causative Agent

Toxoplasmosis is caused by the apicomplexan *Toxoplasma gondii,* an obligate intracellular protozoan that infects a wide variety of animals. Domestic cats and other felines are the definitive hosts in which the organism reproduces sexually. When a newly infected cat defecates, millions of *T. gondii* oocysts are released daily into soil or cat litter for a few weeks. After release, an oocyst matures over a period of 1 to 5 days, becoming infectious to cats, humans, or other animals that inadvertently eat it. The mature oocysts can remain viable for up to a year, contaminating soil and water and, secondarily, hands and food.

Each mature *T. gondii* oocyst contains two sporocysts, and within each of these are several sporozoites (**figure 26.14a**). Once an oocyst is ingested, the sporozoites are released; these invade the cells of the small intestine, developing into a rapidly multiplying form called a tachyzoite (*tachy* means "fast") (**figure 26.14b**). The intestinal infection spreads by the lymphatics and blood vessels throughout the host tissues, infecting heart, brain, and muscle cells. As host immunity

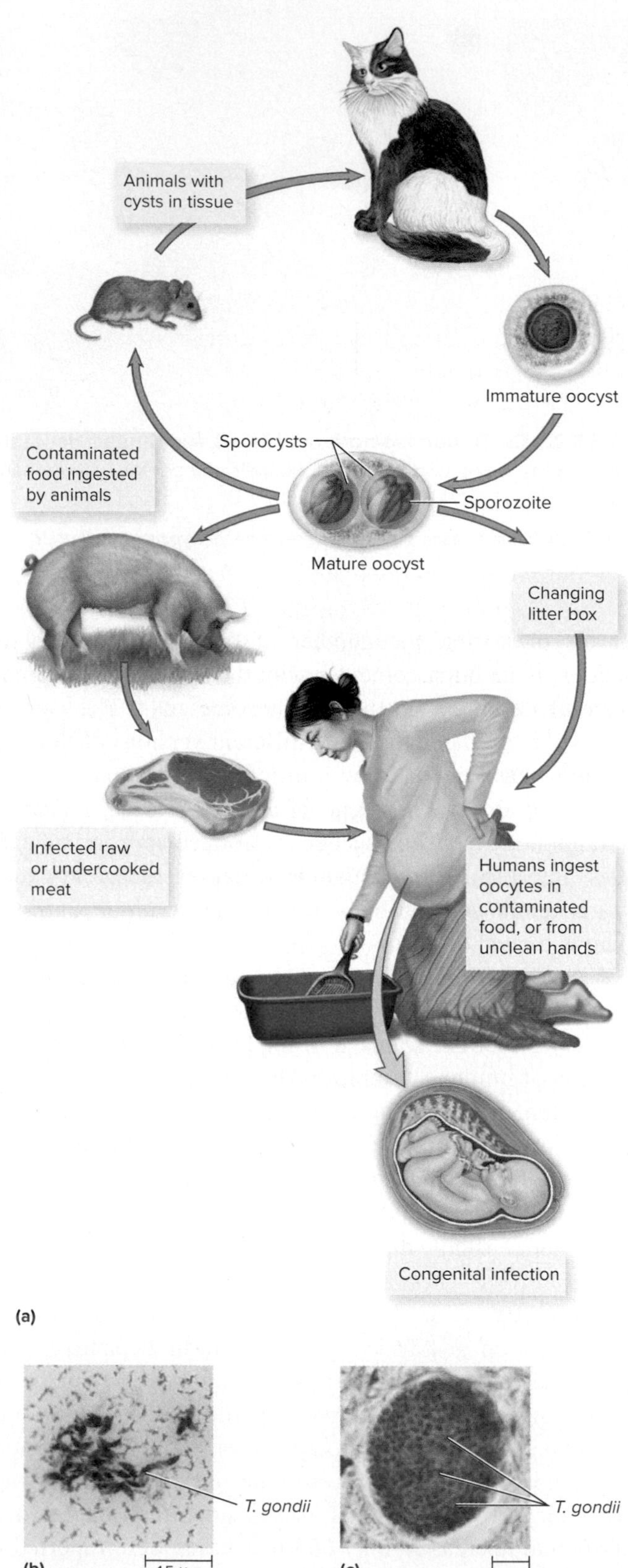

FIGURE 26.14 ***Toxoplasma gondii*** **(a)** Life cycle. Oocysts from cat feces and cysts from raw or inadequately cooked meat can infect humans and many other animals. **(b)** Tachyzoites. **(c)** Tissue cyst containing bradyzoites. b: CDC/ Dr. L.L. Moore, Jr.; c: Dr. Thomas Fritsche

? **Which form of *T. gondii* is found during a latent infection?**

develops, multiplication of the parasite slows. A tough, fibrous capsule forms around infected host cells, forming a tissue cyst (**figure 26.14c**). Each cyst contains large numbers of a smaller form of *T. gondii* called a bradyzoite (*brady* means "slow"), which survive for months or years, resulting in a latent infection. Tissue cysts, like oocysts, are infectious to humans, cats, and other animals when eaten.

Pathogenesis

A person acquires *T. gondii* by accidentally ingesting mature oocysts after handling contaminated cat litter or by consuming inadequately cooked meat containing tissue cysts. The organism can infect any kind of nucleated cell. Within a host cell, a tachyzoite develops and then proliferates, eventually destroying the infected cell. This process is normally brought under control by the immune response, but the tissue cysts persist. In immunocompromised individuals, activation of latent infections as well as newly acquired infections can result in widespread and uncontrolled multiplication of the parasite, producing many areas of tissue necrosis. Severe toxoplasmosis can damage the brain, the eyes, or other organs.

Epidemiology

T. gondii is distributed worldwide, particularly in areas where eating raw or undercooked meat is common; many people have latent infections. In the United States, an estimated 6% of the population has been exposed to this pathogen. Globally, as many as 60% of some populations are infected with *T. gondii;* the incidence is highest in warm, humid, low-altitude countries because the oocysts survive better in these conditions. Animal-to-person transmission occurs, as seen with cats; gardening in areas where cats hunt birds or rodents and then defecate in the soil can result in *T. gondii* contamination of hands or produce. Horizontal person-to-person transmission (via organ transplants or blood transfusions) rarely occurs.

Treatment and Prevention

Toxoplasmosis is not treated in otherwise healthy people. In immunocompromised patients, the disease is typically treated with medications such as pyrimethamine (interferes with folate metabolism) in combination with sulfadiazine (a sulfa drug). An additional medication called folinic acid is also given to prevent bone marrow toxicity associated with pyrimethamine. Treatment is not curative, however, because it does not destroy the tissue cysts; for this reason, maintenance therapy using lower doses of the medications are given to prevent relapses. If PCR tests of amniotic fluid indicate that a fetus is infected, various medications may be used, depending on the stage of pregnancy.

Measures to prevent *T. gondii* infection are especially important for immunocompromised people and pregnant women. Immunocompromised patients who test positive for antibody to *T. gondii* are presumed to have a latent infection and are given prophylactic antimicrobial therapy. The organisms can be avoided by washing hands after touching raw meat, soil, or cat litter. Meat, especially lamb, pork, and venison (deer meat), should be cooked thoroughly. Fruits and vegetables should be washed before eating. Litter boxes should be cleaned regularly, before oocysts have a chance to mature. Cats should not be allowed to hunt birds and rodents, nor should they be fed undercooked or raw meat. The main features of toxoplasmosis are presented in **table 26.11.**

Primary Amebic Meningoencephalitis (PAM)

Naegleria fowleri, the cause of primary amebic meningoencephalitis (PAM), is commonly found in warm fresh water and soils. Nevertheless, fewer than 200 cases have been reported in the United States (typically less than five per year), making this a rare disease, but one that is usually fatal.

Signs and Symptoms

Signs and symptoms of PAM appear within a week of infection and are similar to those of bacterial meningitis. Early indications include headache, fever, stiff neck, and vomiting. Once neurological symptoms appear—such as confusion or seizures—death quickly follows in the vast majority of cases.

TABLE 26.11	Toxoplasmosis
Signs and Symptoms	Sore throat, fever, and enlarged lymph nodes. Encephalitis may develop in immunodeficient individuals, resulting in confusion, poor coordination, weakness, paralysis, seizures, and coma. Fetal infections can result in stillbirth, birth defects, seizures, intellectual disability, and retinitis.
Incubation Period	Usually unknown
Causative Agent	*Toxoplasma gondii*, an apicomplexan protozoan
Pathogenesis	*T. gondii* is acquired by ingesting mature oocysts or by eating inadequately cooked meat containing tissue cysts. Tachyzoite develops within a host cell, killing that cell. Immunity brings the infection under control but tissue cysts persist, resulting in a latent infection. Organisms are released from tissue cysts if immunity is impaired.
Epidemiology	Worldwide distribution, particularly in warm, humid countries. Infected cats discharge oocysts with their feces.
Treatment and Prevention	Treatment: not usually treated in healthy adults; antimicrobial medications if the patient is pregnant or immunocompromised. Prevention: avoid inadequately cooked meat and contact with material that may contain oocysts from cat feces. Immunocompromised individuals with latent *T. gondii* infections are given prophylactic antimicrobial medications.

Causative Agent

N. fowleri is one of only a few free-living protozoa pathogenic for humans. It is found in soil and in warm freshwater lakes and rivers (**figure 26.15**). The organism has three forms: an amoeboid trophozoite ("amoeboid" means the shape changes as the organism moves), a temporary flagellated cell that can swim to new locations, and a cyst that can survive environmental stress. Structures on the trophozoite called food cups allow the organism to engulf bacteria, yeast, and other food sources.

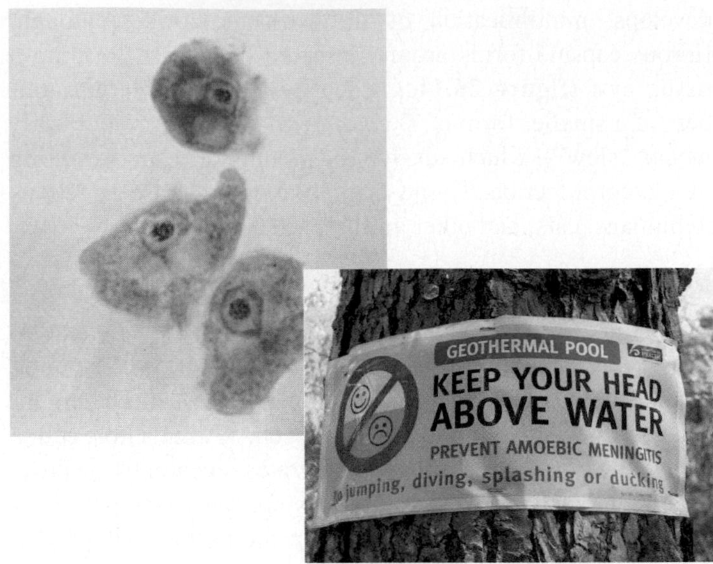

FIGURE 26.15 ***Naegleria fowleri*** Dr. George Healy/CDC; (bottom):Dr. Jay Stahl-Herz

? Which form of *Naegleria fowleri* can survive environmental stress?

Pathogenesis

N. fowleri enters the body when contaminated water is forced or snorted into the nasal passages and the organism penetrates the nasal mucosa. The trophozoite then moves along the nerves serving the nasal mucosa. From there, it travels to the olfactory bulbs (structures containing nerve cells involved with the sense of smell) in the brain, where the parasite uses its food cups to literally "eat the brain" by "nibbling off" pieces of nerve cells, a process called trogocytosis. An intense innate immune response occurs as a result, resulting in further nervous tissue destruction. Hemorrhage, coma, and death occur within a week.

Epidemiology

N. fowleri rarely causes disease; for every case of PAM, many millions of people are exposed to the organism without harm. The organism is usually acquired when people swim or engage in other water sports in warm, natural fresh water, including hot springs, lakes, and rivers; it has even been acquired from poorly chlorinated swimming pools. While commonly found in the southern United States, the organism's geographic range is expanding; it has now been reported in some northern states. A few cases of PAM have been reported from the use of untreated water for sinus and nasal rinsing, but *N. fowleri* is not acquired by drinking contaminated water, nor is it transmitted from person to person.

MicroByte

A person is over 10,000 times more likely to die from drowning in natural waters than to die from PAM.

Treatment and Prevention

A key factor in treating PAM is rapid diagnosis, which is a challenge because the signs and symptoms resemble those of bacterial meningitis. Several medications were tried with little success in the past, but an early, aggressive, and relatively new approach shows promise. In fact, three children have survived the disease—the first U.S. survivors of PAM in 35 years—and two of them made a full neurologic recovery. The treatment involves administering a drug called miltefosine in combination with various other medications, and also minimizing any brain swelling, partly by inducing hypothermia (cooling the body to below normal temperature).

There is no vaccine for PAM. Because the disease is so rare, prevention recommendations are based on common sense rather than data. With respect to swimming and other water sports, the only reliable prevention method is to refrain from any activity in warm fresh water, particularly in southern states. People who do enter such waters should keep their heads above the water, use nose clips, or hold their nose shut to prevent water from entering. In situations where a saline solution is needed to flush the nasal passages, sterile or distilled water should be used in the preparation; tap water should not be used.

The main characteristics of primary amebic meningoencephalitis are presented in **table 26.12.**

TABLE 26.12	Primary Amebic Meningoencephalitis (PAM)
Signs and Symptoms	Headache, fever, stiff neck, vomiting, confusion, seizures
Incubation Period	Usually less than a week
Causative Agent	*Naegleria fowleri*
Pathogenesis	Destroys brain tissue
Epidemiology	Rare incidence after swimming or diving in warm, natural waters
Treatment and Prevention	Treatment: miltefosine combined with other medications and treatments. Prevention: no vaccine; avoiding getting water up the nose when swimming.

MicroAssessment 26.5

Residents and visitors to a wide region of Africa are at risk of contracting African trypanosomiasis, caused by the protozoan parasite *Trypanosoma brucei* and transmitted by the tsetse fly. The parasites may penetrate the CNS, causing sleepiness, coma, and death. Toxoplasmosis is caused by the apicomplexan parasite, *Toxoplasma gondii*. Infection in healthy individuals is typically asymptomatic, but it can lead to encephalitis in immunocompromised persons. Fetal infections may result in stillbirth or in a variety of nervous system disorders. *Naegleria fowleri* is a free-living protozoan that causes meningoencephalitis; this is rare but almost always fatal.

13. Explain two ways in which a person might become infected with *Toxoplasma gondii.*
14. How likely is it that a person who swims in warm fresh water will contract primary amebic meningoencephalitis?
15. How can one explain repeated, abrupt increases in *T. brucei* in the blood of African trypanosomiasis patients?

26.6 ■ Diseases Caused by Prions

Learning Outcomes

11. Explain how prions differ from other infectious agents.
12. Describe how prions can cause disease.

As described in section 13.9, prions cause a rare and mysterious group of chronic, degenerative brain diseases in wild animals (mink, elk, and deer), domestic animals (sheep, goats, and cattle), and humans. Scrapie, a prion disease of sheep and goats, was named because affected animals had difficulty standing and therefore "scraped" along fences for support. Affected brain tissue has a spongy appearance, which is why these diseases are called spongiform encephalopathies. Cattle, presumably when fed meat and bone meal from infected sheep, developed bovine spongiform encephalopathy (BSE), or "mad cow disease." BSE gained attention in the 1990s when an outbreak of a spongiform encephalopathy in humans in the United Kingdom was related to an earlier outbreak of mad cow disease.

Transmissible Spongiform Encephalopathies in Humans

Transmissible spongiform encephalopathies (TSEs) are rare in humans, with an incidence of less than one case per million people; the most common of these are Creutzfeldt-Jakob disease (CJD), which mainly affects individuals over 45 years old and sometimes occurs in families. Because the disease acquired by eating affected animals is similar to CJD yet distinctively different, it is called variant CJD (vCJD). Cases of CJD and vCJD are always fatal.

MicroByte

Kuru is a TSE associated with cannibalism in New Guinea, where local people ate deceased relatives as a sign of respect.

Signs and Symptoms

Early symptoms of TSEs include vague behavioral changes, anxiety, insomnia, and fatigue. These progress to characteristic muscle jerks, lack of coordination, memory loss, and dementia. Although the incubation period may last for years, patients generally die within a year of developing symptoms.

Causative Agent

TSEs are caused by proteinaceous infectious particles, or **prions.** Much smaller than a virus, a prion (PrP^{Sc}) appears to be a misfolded form of a normal cellular protein (PrP^{c}). That cellular protein is encoded by a normal human gene, but post-translational misfolding gives rise to PrP^{Sc}; the misfolding results in a protease-resistant protein, whereas the normal protein is protease-sensitive.

Pathogenesis

Once in the body, prions increase in quantity as the misfolded protein (PrP^{Sc}) acts as a template that promotes misfolding of the normal cellular protein (PrP^{c}) on the surface of neurons (see figure 13.26). The accumulating prions then aggregate in insoluble masses in the brain, causing tissue damage. They may be taken up by neurons or phagocytic cells, but they cannot be degraded by cellular proteases. Death of neurons produces the spongy appearance of affected brain tissue. Not all prions, however, act the same. They differ in host range, incubation period, and their nervous system targets. TSEs can be differentiated from encephalitis because they typically do not induce an immune response.

Epidemiology

CJD generally occurs in people older than 45 years. Transmission from human to human has occurred through corneal transplants, contaminated surgical instruments, and injections of contaminated human growth hormone. It can be transmitted experimentally to chimpanzees.

Current evidence indicates that consumption of BSE-contaminated beef can result in variant Creutzfeldt-Jakob disease (vCJD), distinguished from CJD by differences in symptoms, brain pathology, and age of onset. The median age of individuals with vCJD is only 28 years. If an animal is suspected of having BSE, radical measures may be used to prevent (1) spread of the disease and (2) the potential transmission of the causative agent to humans. In 2003, for example, the United States banned the importation of cattle from Canada when the disease was identified in several animals there. To date, BSE is the only example of a TSE agent being transmitted to humans from wild or domestic animals.

Treatment and Prevention

There is no treatment for TSEs, and they are always fatal. It is important to avoid eating the meat of any animals that show neurological symptoms. Prions are highly resistant to disinfectants (including formaldehyde), heat, and radiation. They appear to be inactivated by extended autoclaving in 1M (molar) sodium hydroxide. The main features of TSEs are presented in **table 26.13.**

MicroAssessment 26.6

TSEs are degenerative nervous system diseases that are rare in humans but can sometimes be transmitted from animals via ingested tissues. These diseases appear to be caused by prions—infectious agents consisting only of protein and highly resistant to inactivation by heat, radiation, and disinfectants. TSEs are always fatal.

16. What is a prion?

17. What is the best way to prevent TSEs?

18. If you were an eye surgeon, which age would you prefer for a cornea transplant donor: under 35 or over 45?

TABLE 26.13	Transmissible Spongiform Encephalopathies
Signs and Symptoms	Behavioral changes, anxiety, insomnia, fatigue, progressing to muscle jerks, lack of coordination, dementia
Incubation Period	Usually many years
Causative Agents	Prions, which are protease-resistant misfolded versions of normal proteins
Pathogenesis	Prions increase in quantity by converting normal protein to more prions; aggregate into masses in the brain; tissue damage and death of neurons.
Epidemiology	Human-to-human transmission by corneal transplantation and by contaminated surgical instruments; probable transmission of cattle prions to humans through consumption of contaminated beef.
Treatment and Prevention	Treatment: none, invariably fatal. Prevention: autoclave contaminated materials in concentrated sodium hydroxide.

PERIPHERAL NERVOUS SYSTEM INFECTIONS

26.7 ■ Bacterial Diseases of the Peripheral Nervous System

Learning Outcomes

13. Explain how Hansen's disease can cause peripheral nerve damage.
14. Explain how botulism affects the nervous system.

Hansen's Disease (Leprosy)

Hansen's disease, or leprosy, is an ancient disease that has been widely feared throughout history (see A Glimpse of History). In reality, even long-term exposure to Hansen's disease patients rarely results in disease, and it can now be easily treated. The prevalence of Hansen's disease has dropped by 99% in the last three decades, mostly due to effective treatment programs established by the World Health Organization (WHO). Still, over 6,000 people in the United States continue to be treated for the long-lasting effects of this disease, and 150 to 250 new cases are reported annually by the National Hansen's Disease Program.

Signs and Symptoms

Hansen's disease begins gradually, usually about 3 years after exposure (range: 3 months to 20 years). Early signs and symptoms include numbness and tingling in the hands and feet, along with skin lesions characterized by pigmentation changes and decreased sensation (**figure 26.16a**.) The disease progression varies significantly, however, depending largely on characteristics of the patient's immune response; treatment options depend on that progression, so various classification schemes have been developed to describe the clinical appearance of a given case. At one end of the disease spectrum is **tuberculoid leprosy** (also called paucibacillary Hansen's disease), a form well controlled by cell-mediated immunity and characterized by the formation of relatively few and well-defined skin lesions. At the other end is **lepromatous leprosy** (also called multibacillary Hansen's disease), which occurs when the immune response fails to control the disease, resulting in extensive skin involvement and peripheral nerve damage. The affected person may have multiple unnoticed injuries because of loss of nerve activity; secondary infections can lead to ulceration, tissue loss, and malformation of fingers and toes as cartilage in the affected digits is absorbed back into the body (**figure 26.16b**). In more severe cases, changes are most obvious in the face, with thickening of the nose and ears and deep wrinkling of the facial skin. Collapse of the supporting cartilage in the nose (called saddle nose) leads to congestion and bleeding. Most cases of the disease fall somewhere between these two extremes.

Causative Agent

Two species of *Mycobacterium* cause leprosy—*M. leprae* and *M. lepromatosis*—and these are referred to as the *Mycobacterium leprae* complex. Both organisms are aerobic,

FIGURE 26.16 Effects of Leprosy (a) An early symptom of tuberculoid leprosy is a change in skin pigmentation. **(b)** Notice the absence of fingers as a result of lepromatous disease.
a: Biophoto Associates/Science Source; b: Palo Koch/Science Source

? What causes the loss of fingers, toes, and the nose bridge in those affected by leprosy?

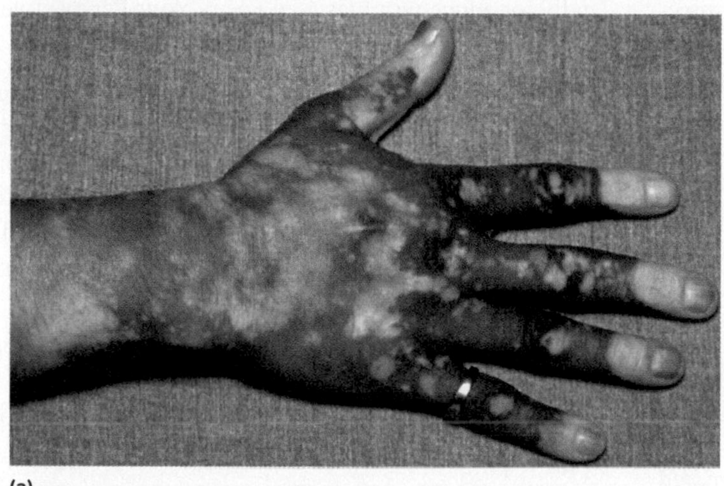

(a)

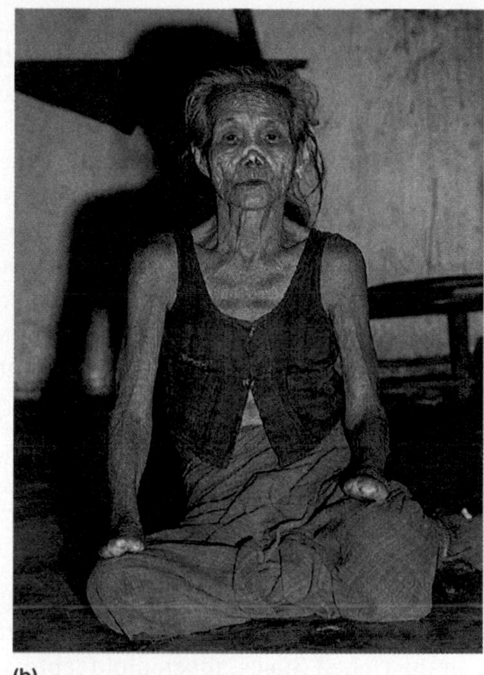

(b)

rod-shaped, acid-fast bacteria (**figure 26.17**). They grow very slowly (generation time 12–13 days) and prefer the slightly cooler temperatures of the body's extremities. Hansen's disease is usually diagnosed based on clinical findings but confirmed by skin biopsy. Research on these organisms has been challenging because they cannot grow in standard laboratory media; instead they are grown in animal models, including the footpads of mice and in armadillos.

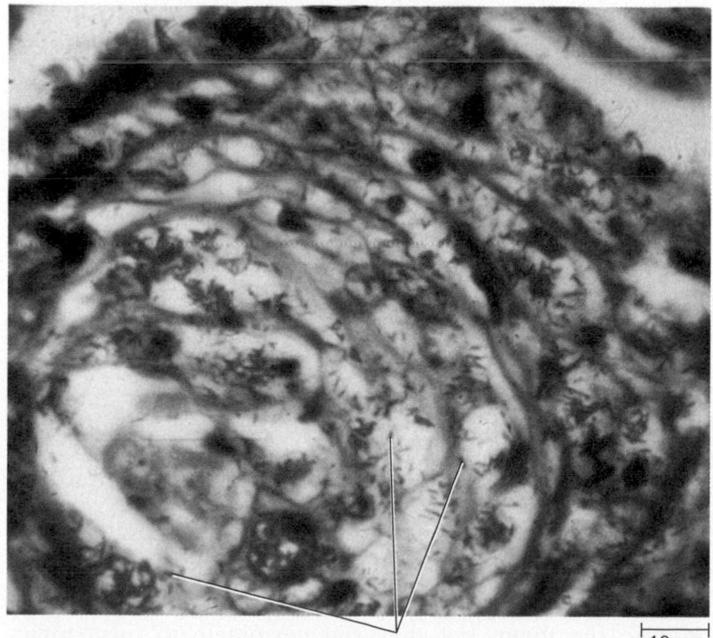

FIGURE 26.17 ***Mycobacterium leprae*** **in a Biopsy Specimen** Acid-fast stain reveals dense red masses of the bacterial cells, a typical finding in lepromatous leprosy. CDC/Arthur E. Kaye

? Why does *M. leprae* preferentially infect a person's extremities?

Pathogenesis

M. leprae and *M. lepromatosis* are the only known human pathogens that preferentially infect cells of the peripheral nervous system. From there, the course of the infection depends on the immune response of the host. In tuberculoid leprosy, cell-mediated immunity develops against the invading bacteria, and activated macrophages limit their spread. The chronically infected cells, however, are progressively damaged by attacking immune cells. The disease often spontaneously stops progressing, and the nerve damage, although permanent, does not worsen. People with tuberculoid leprosy rarely, if ever, transmit the disease to others.

When cell-mediated immunity to the pathogen fails to develop or is suppressed, unrestricted growth occurs, leading to lepromatous leprosy. The bacteria first multiply in the cooler tissues of the body, notably in skin macrophages and cells of the peripheral nervous system, and later spread throughout the body. The tissues and mucous membranes contain billions of *M. leprae* or *M. lepromatosis,* but almost no inflammatory response results. Mucus of the nose and throat contains high numbers of the pathogen, which can easily be transmitted to others. Tissue damage leads to disabling deformities, resorption of cartilage and bone, and skin ulcerations that characterize the disease.

Epidemiology

M leprae and *M. lepromatosis* are transmitted person-to-person, perhaps by direct contact with nasal secretions of lepromatous cases; the disease develops in only a very small percentage of those exposed, however, being controlled by immune defenses in the rest. Natural infections with *M. leprae* also occur in wild armadillos. Interaction with these animals may explain cases of leprosy seen in people who have neither traveled to a part of the world where the disease is considered endemic nor had contact with those who have.

Although Hansen's disease is difficult to eradicate due to the causative agent's long generation time, elimination of the disease as a public health problem (defined as less than one case per 10,000 individuals) has been achieved in most countries. In countries where the disease is still endemic, the WHO is encouraging the development and implementation of country-specific "zero-leprosy roadmaps" as part of it's Global Leprosy Strategy 2021-2030. Goals of the strategy include interrupting transmission via active detection and prompt treatment of new cases and their close contacts as well as expanding the management of leprosy to include the mental well-being of the patient.

MicroByte

During much of history, Hansen's disease was so feared that "lepers" were forced to carry a bell or horn to warn others of their presence.

Treatment and Prevention

Early treatment using combination therapy keeps Hansen's disease from progressing and prevents drug-resistant strains from developing. In the United States, tuberculoid leprosy is treated using a combination of dapsone and rifampin for 12 months. For lepromatous leprosy, a third drug, clofazimine, is added, and treatment is extended to 2 years. Although no vaccine is currently available to control the disease, several are in clinical trials. Prophylactic treatment with an anti-mycobacterial agent is recommended for close contacts of patients. Some features of Hansen's disease are shown in **table 26.14.**

Botulism

Botulism is not a nervous system infection but an intoxication with botulinum toxin (BTX), also called botulinum neurotoxin (BoNT), which is produced by *Clostridium botulinum* and strains of a few similar species; the disease is considered in this chapter because its key symptom is paralysis. BTX is also significant because its paralytic effects have medical and cosmetic applications: Commercial versions such as Botox are used to prevent headaches and provide relief from a variety of conditions involving muscle contractions, and they are used cosmetically to reduce wrinkles (see Focus Your Perspective 26.1).

TABLE 26.14 Hansen's Disease (Leprosy)

Signs and Symptoms	Skin lesions that lack sensation; limited disease progression in tuberculoid leprosy, but lepromatous leprosy can result in deformed facial features and loss of fingers or toes.
Incubation Period	Usually about 3 years
Causative Agent	*Mycobacterium* leprae complex (*M. leprae* and *M. lepromatosis*); acid-fast rods
Pathogenesis	Infection of macrophages and cells of the peripheral nervous system; in tuberculoid leprosy, activated macrophages limit the growth of the bacterium, but the damage to infected nerve cells leads to disabling deformities; in lepromatous leprosy, cell-mediated immunity fails, allowing unrestrained growth of organisms.
Epidemiology	Likely direct contact with *M. leprae* or *M. lepromatosis* from mucous membrane secretions; possibly also via contact with armadillos.
Treatment and Prevention	Treatment: antimicrobial medications. Prevention: no vaccine; prophylactic treatment of patient's close contacts recommended.

Botulism has several different forms:

- **Foodborne botulism.** This results from consuming BTX-contaminated food, and was the first form of botulism recognized. In fact, the name "botulism" comes from *botulus*, meaning "sausages," chosen because some of the earliest cases occurred in people who ate contaminated sausages. Additional characteristics of foodborne botulism, and the events leading to it, are described in chapter 30.
- **Infant botulism.** This results from ingesting endospores of a BTX-producing *Clostridium* species that germinate into vegetative cells, which then temporarily colonize the intestine of infants and produce toxin there.
- **Wound botulism.** This results from a BTX-producing *Clostridium* species colonizing a wound and producing toxin there.
- **Adult intestinal toxemia botulism.** This is similar to infant botulism but occurs in adults and is much rarer.
- **Iatrogenic botulism.** Iatrogenic (*iatro* means "healer, physician" and *genic* means "generated") results when too much BTX is injected for either cosmetic or medical reasons.

Signs and Symptoms

The incubation period of botulism depends on the form of the disease. In the case of foodborne botulism, symptoms usually begin 18 to 36 hours after a BTX-contaminated food is ingested; the incubation time for other forms is variable or unknown. Most cases begin with dizziness, dry mouth, and blurred or double vision, indicating eye muscle weakness. Abdominal symptoms—including pain, nausea, vomiting, and diarrhea or constipation—can also occur. Infants with botulism may drool, have trouble sucking, and have drooping eyelids. Over time, muscles that allow a person to move become progressively paralyzed; respiratory paralysis is the most common cause of death.

Causative Agent

BTX is produced mainly by *C. botulinum*, so we will focus our discussion on that species; however, a few strains of *Clostridium butyricum* and *Clostridium baratii* also produce the toxin. *C. botulinum* is an anaerobic, endospore-forming Gram-positive rod, found in soils and aquatic sediments worldwide (**figure 26.18**). The endospores, which are dormant, do not produce the toxin, but they can germinate to form toxin-producing vegetative cells if the environment is favorable (nutrient-rich, anaerobic, a pH above 4.5, and a temperature above 4°C). Seven main types of BTX are known (A to G), but types A, B, and E are responsible for most human disease.

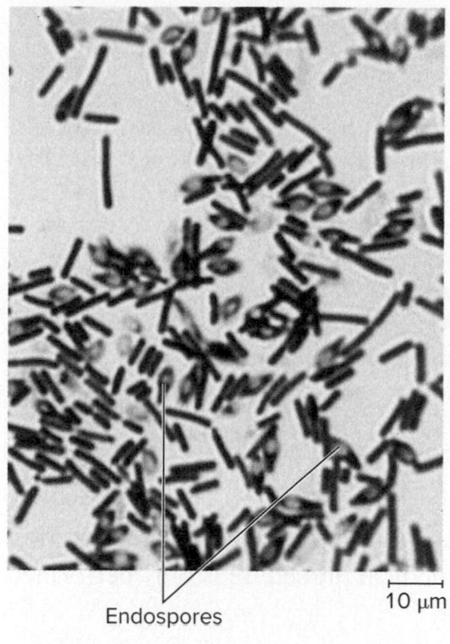

FIGURE 26.18 ***Clostridium botulinum*** Notice the lighter-colored endospores in this Gram stain. Richard Gross/McGraw Hill

? **Do *C. botulinum* endospores cause disease? Explain.**

Pathogenesis

The signs and symptoms of botulism are due entirely to the effects of BTX. It is one of the most powerful toxins known—a few milligrams could kill the entire population of a large city! Foodborne botulism occurs when a person eats food contaminated with the toxin, which passes through the stomach and into the small intestine, where it is absorbed into the bloodstream; it has even resulted from eating a single contaminated string bean! In contrast, infant botulism occurs when endospores are ingested; these germinate in the child's intestinal tract to become vegetative cells that multiply and produce the toxin, which is then absorbed. In wound botulism, the bacterial cells multiply and produce the toxin in necrotic (dead) tissue; the toxin then diffuses into the bloodstream. The pathogenesis of adult intestinal toxemia botulism is similar to infant botulism; it rarely occurs in healthy adults, presumably because the organism cannot compete with the normal microbiota in the adult intestinal tract. Iatrogenic botulism results from direct injection of purified BTX preparations for either cosmetic or medical use, in most cases at concentrations higher than recommended.

BTX is an A-B toxin; the B portion binds to specific receptors on motor nerve endings, and the A portion enters the nerve cell, where it inactivates proteins that regulate the release of neurotransmitters. Unlike tetanus toxin, which results in spastic muscle contraction and paralysis (figure 23.9), BTX stops muscle contraction, resulting in flaccid paralysis.

Epidemiology

The epidemiology of botulism depends on the form:

- **Foodborne botulism.** This is mostly associated with eating improperly home-canned foods; outbreaks involving commercially canned foods are now rare because of strict processing controls. Cases have also resulted from eating improperly stored foods such as stew that had been left at room temperature (see Focus on a Case 30.1). Fortunately, foodborne botulism is relatively rare, usually with less than 30 cases per year in the United States (**figure 26.19**).
- **Infant botulism.** This generally occurs in infants 6 weeks to 6 months old, likely because members of their normal intestinal microbiota cannot inhibit the growth of many bacteria, including BTX-producing *Clostridium* species. Honey—a source of *Clostridium* endospores—has been implicated in several infant botulism cases, leading to the recommendation that honey not be given to children under 1 year old. Although infant botulism is the most common form of botulism in the United States, fewer than 170 cases generally occur per year.
- **Wound botulism.** This is rare in the United States, and most cases are due to dirty wounds associated with abuse of injected drugs.
- **Adult intestinal toxemia botulism.** This is rare but can occur in immunocompromised persons whose normal microbiota may have been suppressed by antibiotic treatment.
- **Iatrogenic botulism.** Relatively few cases have occurred, and the most severe ones involved the use of unlicensed BTX preparations.

Treatment and Prevention

Botulism is treated by intravenous administration of antitoxin as soon as possible, but the formulation depends on the age of the patient. A formulation called BabyBIG is used for infants and protects against the two toxin types most commonly associated with infant botulism; HBAT (heptavalent botulinum antitoxin) is used for anyone older than 1 year of age and protects against seven toxin types. Antitoxin, however, can only neutralize toxin

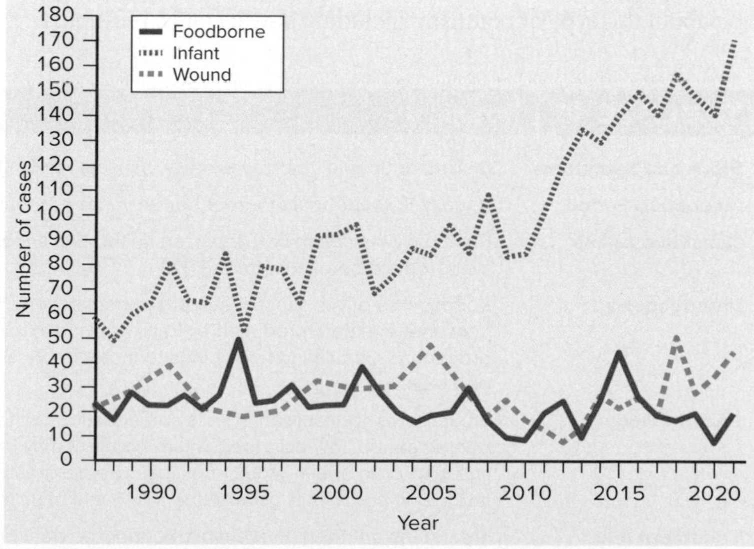

FIGURE 26.19 Botulism in the United States, 1992–2021

? **What is the most common form of botulism in the United States?**

FOCUS YOUR PERSPECTIVE 26.1

Botulinum Toxin for Beauty and Pain Relief

Botulinum toxin (BTX) can be very dangerous, but it can also be used as a medication to treat various conditions associated with muscle contractions. An FDA-approved dilute preparation of BTX (Botox, for example) can be injected directly into affected muscles, where the medication remains effective for 3 to 4 months, after which time the treatment can be repeated. One popular use of BTX injection is a cosmetic treatment to remove frown lines and other facial wrinkles. By paralyzing the muscles involved, the lines are effectively erased. The treatment is also used to relieve a number of very painful and disabling conditions. It gives relief from cervical dystonia (severe muscle cramping in the neck and shoulders), a disorder characterized by jerky movements, muscle pain, and tremors (shaking). It also lessens some symptoms of Parkinson's disease, a condition in which certain nerve cells are lost, resulting in tremors, slow movement, stiffness, and in some cases, severe cramping. So, in spite of BTX's powerful and dangerous properties, it can also be a useful therapeutic agent.

circulating in the bloodstream. The nerves already affected recover slowly, over weeks or months. If a patient shows signs of respiratory failure, mechanical ventilation may be required. For wound botulism, surgical removal of dirt and dead tissue helps to eliminate any organisms and unabsorbed toxin.

Prevention of botulism depends on the form:

- **Foodborne.** This can be prevented by properly sterilizing and sealing the food as it is canned. Contaminated food does not always have a spoiled smell, taste, or appearance, but fortunately, the toxin is heat-labile (destroyed by heat); thus, as a safety measure, home-canned, low-acid foods (pH over 4.5) should be heated to 100°C for 10 minutes just before serving.
- **Infant botulism.** This is difficult to prevent, but because honey can contain *C. botulinum* endospores, it should not be fed to infants younger than 1 year of age.
- **Wound botulism.** This is prevented through prompt and appropriate wound care and by discouraging injection-drug abuse.
- **Adult intestinal toxemia botulism.** Very little is known about this type of botulism, including how it can be prevented.
- **Iatrogenic botulism.** This can be prevented by ensuring that any cosmetic or medical procedure is only performed by a licensed provider.

Immunity does not develop in response to the low levels of BTX, so a person may get the disease more than once. The main features of botulism are presented in **table 26.15**.

The key features of the diseases covered in this chapter are highlighted in the Diseases in Review 26.1 table that follows.

MicroAssessment 26.7

Hansen's disease (leprosy) is caused by *Mycobacterium leprae* and *M. lepromatosis,* which preferentially invade peripheral neurons. Botulism, a toxin-mediated disease characterized by flaccid paralysis, is caused mainly by *Clostridium botulinum*. Forms of botulism include foodborne, infant, wound, adult intestinal toxemia, and iatrogenic.

19. Why is blurred vision an early sign of botulism?
20. Differentiate between the two types of Hansen's disease.
21. Why is it difficult to determine an incubation period for *Clostridium botulinum* in cases of wound botulism seen in chronic injection-drug abusers?

TABLE 26.15 Botulism

Signs and Symptoms	Blurred or double vision, weakness, nausea, vomiting, diarrhea or constipation; generalized paralysis and respiratory insufficiency
Incubation Period	Usually 18 to 36 hours for foodborne; variable or unknown for other types
Causative Agent	Primarily *Clostridium botulinum*, an anaerobic, endospore-forming, Gram-positive rod; also strains of other *Clostridium* species that produce botulinum toxin (BTX)
Pathogenesis	Endospores of the causative agent germinate, and the resulting vegetative cells multiply, releasing BTX. The toxin may be consumed in food, produced by cells growing in the intestinal tract or an infected wound, or injected in a medical or cosmetic procedure. Circulating toxin binds to motor nerve endings and stops the transmission of signals to the muscles, producing flaccid paralysis.
Epidemiology	Endospores widespread in soils and aquatic sediments. Foodborne botulism is often due to consuming improperly processed home-canned, low-acid food. Infant botulism may involve eating endospore-containing honey. Wound botulism often involves injection-drug abuse. Adult intestinal toxemia botulism may affect immunocompromised persons after antibiotic treatment. Iatrogenic botulism is usually due to the use of unlicensed preparations of BTX.
Treatment and Prevention	Treatment: antitoxin. Prevention: Foodborne botulism, proper home-canning; infant botulism, do not feed honey to infants under 1 year of age; wound botulism, prompt wound care; adult intestinal toxemia botulism, unknown; iatrogenic botulism, use licensed providers for procedures involving the toxin.

Diseases in Review 26.1

Nervous System Diseases

Disease	Causative Agent	Comment	Summary Table
BACTERIAL DISEASES OF THE CENTRAL NERVOUS SYSTEM			
Pneumococcal meningitis	*Streptococcus pneumoniae* (pneumococcus)	Leading cause of meningitis in adults; vaccines that protect against multiple serotypes are available	Table 26.3
Meningococcal disease	*Neisseria meningitidis* (meningococcus)	Associated with meningitis epidemics; can lead to septic shock; formation of petechiae due to capillary leakage; vaccines available	Table 26.1
***Haemophilus influenzae* meningitis**	*Haemophilus influenzae*	Once the most common cause of infant meningitis, now largely controlled by a conjugate vaccine	Table 26.3
Neonatal meningitis	*Streptococcus agalactiae, Escherichia coli*	Newborns acquire causative agent from the mother's birth canal; pregnant women in the United States are screened for *S. agalactiae* before delivery and, if positive, treated with antibiotics.	Table 26.3
Listeriosis	*Listeria monocytogenes*	Contaminated foods cause outbreaks; multiplies at refrigeration temperatures; infected pregnant women often miscarry or deliver terminally ill infants.	Table 26.2
VIRAL DISEASES OF THE CENTRAL NERVOUS SYSTEM			
Viral meningitis	Usually enteroviruses	Aseptic meningitis; more common and much milder than bacterial meningitis; often transmitted by fecal-oral route	Table 26.4
Viral encephalitis	Herpesviruses; arboviruses	More likely than viral meningitis to cause death or disability; arboviruses transmitted by mosquitoes; fatality rate varies, depending upon type of virus	Table 26.6
Polio	Poliovirus	Destruction of motor neurons leads to paralysis; fecal-oral transmission, but infection rarely leads to disease; vaccination has eliminated the disease in most parts of the world, and the disease is now targeted for eradication.	Table 26.7
Rabies	Rabies virus	Zoonotic disease; virus travels from bite wound through sensory neurons to CNS, causing a generally fatal infection; post-exposure prophylaxis prevents disease.	Table 26.8
FUNGAL DISEASES OF THE CENTRAL NERVOUS SYSTEM			
Cryptococcal meningoencephalitis	*Cryptococcus neoformans* and *C. gattii*	Inhaled fungal cells access the brain; can be an early sign of AIDS, but also seen in healthy individuals where *C. gattii* is found.	Table 26.9
PROTOZOAN DISEASES OF THE CENTRAL NERVOUS SYSTEM			
African trypanosomiasis	*Trypanosoma brucei*	Chronic (West African) or acute (East African); two subspecies of protozoa transmitted by tsetse fly; loss of interest, sleepiness, coma, death	Table 26.10
Toxoplasmosis	*Toxoplasma gondii*	Acquired by ingestion of oocysts (in cat feces) or tissue cysts (in raw meat); infection usually asymptomatic in healthy individuals, but fetal infections may cause stillbirth, and encephalitis may develop in the immunocompromised.	Table 26.11
Primary amebic meningoencephalitis (PAM)	*Naegleria fowleri*	Swimming or diving in water containing the organism transmits this rare but deadly disease; travels to brain via olfactory neurons, where it damages brain tissue.	Table 26.12
PRION DISEASES			
Transmissible spongiform encephalopathies	Prions	Caused by an accumulation of destruction-resistant misfolded proteins, resulting in death of neurons; may come from contaminated animal tissues; no treatment; always fatal	Table 26.13
BACTERIAL DISEASES OF THE PERIPHERAL NERVOUS SYSTEM			
Hansen's disease (leprosy)	*Mycobacterium leprae* complex	Infects peripheral nerves; course of disease determined by cell-mediated immune response; long incubation period; loss of limbs, blindness	Table 26.14
Botulism	Botulinum toxin-producing *Clostridium* species: *C. botulinum* and certain strains of *C. butyricum* and *C. baratii*	Botulinum toxin causes flaccid paralysis; foodborne botulism is the most common form worldwide; infant botulism is the most common form in the United States; wound botulism can also occur; adult intestinal toxemia botulism and iatrogenic botulism are both rare.	Table 26.15

Summary

26.1 ■ Anatomy, Physiology, and Ecology of the Nervous System (figure 26.1)

The brain and spinal cord make up the **central nervous system (CNS)**; the **peripheral nervous system (PNS)** includes nerves that carry information to and from the CNS. **Cerebrospinal fluid (CSF)** is produced in ventricles in the brain and flows out over the brain and spinal cord (figure 26.2). **Meninges** are the membranes that cover the brain and spinal cord.

CENTRAL NERVOUS SYSTEM INFECTIONS

Focus on Meningitis

Infection of the membranes surrounding the brain and spinal cord causes **meningitis.** Bacteria that cause meningitis are often part of the normal microbiota, transmitted through respiratory droplets. Enteroviruses that cause meningitis are transmitted by the fecal-oral route. Signs and symptoms include the abrupt onset of fever, severe headache, pain and stiffness of the neck and back, nausea, and vomiting. Immunization has greatly decreased the incidence of bacterial meningitis, particularly among children. Viral meningitis is more common but typically less severe than bacterial meningitis, which can have a fatality rate of nearly 100% if left untreated.

26.2 ■ Bacterial Diseases of the Central Nervous System

Pneumococcal Meningitis (table 26.3)

Streptococcus pneumoniae, or pneumococcus, is the most common cause of meningitis in adults.

Meningococcal Disease (figure 26.4; tables 26.1, 26.3)

Meningococcal disease caused by *Neisseria meningitidis* is associated with meningitis epidemics. Small hemorrhages in the skin (figure 26.3), deafness, and coma can occur. Septic shock can result from the release of endotoxin into the bloodstream.

Haemophilus influenzae Meningitis (table 26.3)

Haemophilus influenzae, once the leading cause of childhood bacterial meningitis, is largely sporadic and mostly controlled by the Hib vaccine (figure 26.5).

Neonatal Meningitis (table 26.3)

Newborns most often acquire meningitis-causing bacteria from the mother's genital tract shortly before or during birth. Infants who survive often face long-lasting consequences of their infection.

Listeriosis (tables 26.2, 26.3)

Listeriosis is caused by *Listeria monocytogenes,* a non-spore-forming, Gram-positive rod usually associated with foodborne illness. The bacterium sometimes contaminates refrigerated foods such as nonpasteurized milk, cold cuts, and soft cheeses (figure 26.6). The bacterial cells can enter the bloodstream, and infect the meninges.

26.3 ■ Viral Diseases of the Central Nervous System

Viral Meningitis (table 26.4)

Viral meningitis is much more common than bacterial meningitis. It is generally a mild disease for which there is no specific treatment.

West Nile and Other Types of Viral Encephalitis (table 26.6)

Viral encephalitis can have a high case-fatality rate and may leave survivors with permanent disabilities. Herpes simplex virus reactivation is the most common cause; encephalitis caused by **arboviruses** are also relatively common (figure 26.7; table 26.5).

Polio (Poliomyelitis) (figures 26.8, 26.9; table 26.7)

Infection of motor neurons in the CNS leads to nerve degeneration and results in paralysis and muscle atrophy. Post-polio syndrome occurs years after polio, and it is probably caused by the death of nerve cells that had taken over for those killed by poliovirus.

Rabies (figure 26.10; table 26.8)

Rabies is a widespread zoonosis transmitted to humans mainly through the bite of an infected animal (figure 26.11). Once signs and symptoms appear, the disease is almost always fatal. If a person has had contact with a possibly rabid animal, post-exposure prophylaxis using both human rabies immune globulin (HRIG) and rabies vaccine should be used.

26.4 ■ Fungal Diseases of the Central Nervous System

Cryptococcal Meningoencephalitis (table 26.9)

Infection begins in the lung after a person inhales spores of *Cryptococcus neoformans* or *Cryptococcus gattii,* encapsulated yeasts that resist phagocytosis (figure 26.12). *C. neoformans* usually causes disease only in immunocompromised individuals, but *C. gattii* can cause disease in healthy people.

26.5 ■ Protozoan Diseases of the Central Nervous System

African Trypanosomiasis ("African Sleeping Sickness") (table 26.10)

African trypanosomiasis is a major health problem in a wide area across equatorial Africa. In its late stages, it is marked by indifference, sleepiness, coma, and death. The disease is caused by *Trypanosoma brucei* (figure 26.13) transmitted by its biological vector, the tsetse fly.

Toxoplasmosis (figure 26.14; table 26.11)

Toxoplasmosis, caused by *Toxoplasma gondii,* is contracted by ingesting oocytes discharged in the feces of infected cats or tissue cysts in inadequately cooked meat from an infected animal. Symptoms are usually mild in otherwise healthy people, but infections can lead to encephalitis, brain masses, and other nervous system problems among immunocompromised people. Infection of the fetus, especially during the first trimester, results in miscarriage, stillbirth, or birth defects.

Primary Amebic Meningoencephalitis (PAM) (table 26.12)

PAM is rare, but usually fatal. Infection with *Naegleria fowleri* (figure 26.15) can occur after swimming in warm fresh water. The parasite gains entry into the CNS via the nasal passages.

26.6 ■ Diseases Caused by Prions

Prions—abnormal proteins that are resistant to heat, radiation, and disinfectants—cause the spongiform encephalopathies, rare but fatal diseases characterized by a sponge-like appearance of brain tissue resulting from loss of nerve cells.

Transmissible Spongiform Encephalopathies in Humans (table 26.13)

Examples of **transmissible spongiform encephalopathies (TSEs)** include "mad cow disease," Creutzfeldt-Jakob disease (CJD), and variant Creutzfeldt-Jakob disease (vCJD), which can be transmitted to humans from infected meat. There is no treatment for these diseases, and they are invariably fatal.

PERIPHERAL NERVOUS SYSTEM INFECTIONS

26.7 ■ Bacterial Diseases of the Peripheral Nervous System

Hansen's Disease (Leprosy) (figure 26.16; table 26.14)

Hansen's disease is characterized by infection of cells in the peripheral nervous system by acid-fast rods in the *Mycobacterium leprae* complex (figure 26.17). The disease occurs in two main forms—tuberculoid and lepromatous—depending on the immune status of the individual.

Botulism (table 26.15)

Botulism is caused by a toxin produced by *Clostridium botulinum* (and a few other *Clostridium* species), obligate anaerobes that form heat-resistant endospores (figure 26.18). **Foodborne botulism** results from ingesting botulinum toxin (BTX), typically in inadequately processed canned foods. **Infant botulism** occurs when endospores are ingested and then germinate to form vegetative cells that colonize the infant's intestine (figure 26.19). **Wound botulism** results when endospores contaminate a wound and then germinate to form vegetative cells that multiply. Adult intestinal toxemia botulism, which is similar to infant botulism but can occur in immunocompromised individuals, and iatrogenic botulism, which can occur due to improper administration of purified BTX preparations in procedures, are both rare.

Review Questions

Short Answer

1. What sign would differentiate meningococcal disease from pneumococcal meningitis?
2. Name and describe the organism that is the leading cause of bacterial meningitis in adults.
3. What measures can be undertaken to prevent neonatal meningitis?
4. Why is listeriosis so important to prevent in pregnant women even though it usually causes them few symptoms?
5. Give two ways in which viral meningitis usually differs from bacterial meningitis.
6. Why are humans considered dead-end hosts when they have a West Nile virus infection?
7. Which type of polio vaccine is important in preventing the spread of poliovirus in a population?
8. Why is it possible to prevent rabies by vaccinating after exposure?
9. If you contract African trypanosomiasis on a visit to West Africa, are you more likely to have an acute form or a chronic form?
10. Can botulism be spread from person to person?

Multiple Choice

1. Which is the best way to prevent meningococcal meningitis in individuals immediately after they have been intimately exposed to the disease?
   a) Vaccinate them against *Neisseria meningitidis.*
   b) Give them an antimicrobial agent such as rifampin.
   c) Culture their throat and hospitalize them for observation.
   d) Withdraw a sample of spinal fluid and begin antibacterial treatment if the cell count is high and the glucose level is low.
   e) Have them return to their usual activities, but seek medical evaluation if symptoms of meningitis occur.
2. Which of these statements concerning the causative agent of listeriosis is *false*?
   a) It can cause meningitis during the first month of life.
   b) It is a Gram-positive rod that can grow in refrigerated food.
   c) It is usually transmitted by the respiratory route.
   d) Infection can result in bacteremia.
   e) It is widespread in natural waters and vegetation.
3. Which of these statements concerning Hansen's disease is *false*?
   a) It is not common in the United States.
   b) An early symptom is loss of sensation, sweating, and hair in a localized patch of skin.
   c) The incubation period is usually less than 1 month.
   d) Treatment should include more than one antimicrobial medication given at the same time.
   e) The form the disease takes depends on the individual's immune status.
4. Which of the following statements about viral meningitis is *true*?
   a) Vaccines are generally available to protect against the disease.
   b) The main symptom is muscle paralysis.
   c) Transmission is often by the fecal-oral route.
   d) The causative agents do not survive well in the environment.
   e) Recovery is rarely complete.
5. Which of these statements concerning arboviral encephalitis is *false*?
   a) It is likely to occur in epidemics.
   b) Mosquitoes can be an important vector.
   c) Epilepsy, paralysis, and thinking difficulties are among the possible sequels to the disease.
   d) Use of sentinel chickens helps warn about the disease.
   e) In the United States, the disease is primarily a zoonosis involving cattle.
6. Which of these statements concerning polio is *false*?
   a) The sensory nerves are usually involved.
   b) It is now endemic in only two countries in the world.
   c) Only a small fraction of those infected will develop the disease.
   d) The disease is transmitted via the fecal-oral route.
   e) Post-polio syndrome can develop years after recovery from the original illness.
7. Which of these statements concerning cryptococcal meningoencephalitis is *true*?
   a) It is caused by a yeast that can produce a large capsule.
   b) It is a disease of trees transmissible to humans.
   c) It typically attacks the meninges but spares the brain.
   d) Person-to-person transmission commonly occurs.
   e) It is seen only in persons who are immunocompromised.

8. Which of these statements concerning African trypanosomiasis is *true*?
   a) It is transmitted by a species of biting mosquito.
   b) It is a threat to visitors to Africa.
   c) The onset of sleepiness is usually within 2 weeks of contracting the disease.
   d) It is caused by free-living protozoa.
   e) Distribution of the disease is determined mainly by the presence of standing water.
9. Which of these statements concerning Creutzfeldt-Jakob disease (CJD) and vCJD is *true*?
   a) CJD occurs in children; vCJD occurs in adults over 45.
   b) CJD and vCJD are fatal in under 50% of cases.
   c) CJD is caused by prions; vCJD is a viral infection.
   d) Only humans suffer from diseases like CJD and vCJD.
   e) Both CJD and vCJD produce a spongy appearance in affected brain tissue.
10. Which of these statements concerning foodborne botulism is *false*?
    a) It is not a central nervous system infection.
    b) Blurred vision is an early symptom.
    c) Food can taste normal but still cause botulism.
    d) Treatment is based on choosing the correct antibiotic.
    e) Control of the disease depends largely on proper food-canning techniques.

Applications

1. An outbreak of viral meningitis in a small eastern city was linked epidemiologically to a group who swam in a non-chlorinated pool in an abandoned quarry outside of town. What might public health officials surmise about the probable cause of the outbreak?
2. Two microbiologists are writing a textbook, but they cannot agree on where to place the discussion of botulism. One favored the chapter on nervous system infections, whereas the other insisted on the chapter covering digestive system infections. Where do you think the discussion should be placed, and why?

Critical Thinking

1. A pathologist stated that it was much easier to determine the causative agent of meningitis than that of an infection of the skin or of the intestine. Is her statement valid? Why or why not? Why is it important to learn about rabies when only a few cases occur in the entire United States each year?

www.mcgrawhillconnect.com

Enhance your study of this chapter with study tools and practice tests. Also ask your instructor about the resources available through Connect, including the media-rich eBook, interactive learning tools, and animations.

27 Genitourinary Tract Infections

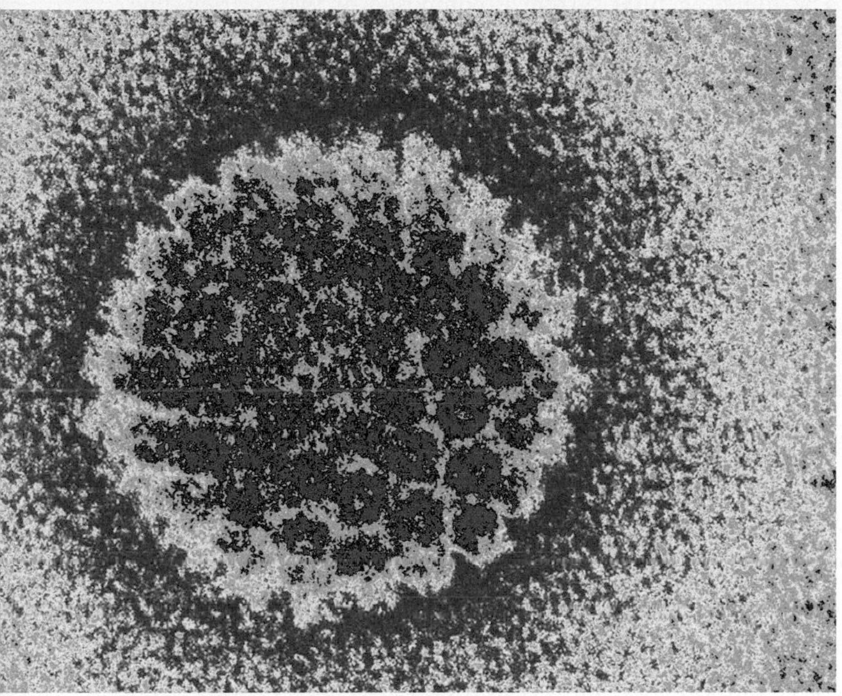

Human papillomavirus particle (color-enhanced TEM). *Science Photo Library—PASIEKA/Getty Images*

KEY TERMS

Catheter A flexible tube inserted into the bladder or other body location to drain or deliver fluids.

Chancre Ulcerating sore that develops during the initial stage of syphilis.

Cystitis Inflammation of the bladder, usually due to an infection.

Papilloma Abnormal tissue growth that projects outward; a wart is an example.

Pelvic Inflammatory Disease (PID) Infection of the fallopian tubes, uterus, or ovaries.

Pyelonephritis Inflammation of one or both kidneys, usually due to an infection.

Toxic Shock Syndrome Potentially life-threatening drop in blood pressure due to a circulating bacterial toxin.

A Glimpse of History

Syphilis, an ancient disease that was once very common, was first named "French pox" or the "Neapolitan disease" because people thought it came from France or Italy. In 1530, the Italian physician Girolamo Fracastoro wrote a poem featuring a shepherd named Syphilis, who had ulcerating sores all over his body. The description matched the signs and symptoms of syphilis, and from then on, the disease was known by the shepherd's name.

Syphilis is sexually transmitted. In 1905, the German biologist Fritz Schaudinn examined fluid from a syphilitic sore and observed faintly visible motile organisms that were very pale and slender. Using dark-field microscopy, he could see the organisms more clearly and noted that they looked like a corkscrew without a handle—bacteria that are now known as spirochetes. The spirochetes were found in specimens from other cases of syphilis, and Schaudinn felt certain he had discovered the cause of the disease, even though the organism could not be cultured using laboratory media. Schaudinn named the organisms *Spirochaeta pallida*, "the pale spirochete." It is now called *Treponema pallidum*, from the Greek words meaning "turning thread."

Infections of the urinary and reproductive tracts are very common and sometimes cause great discomfort; some are self-limiting or easily treated, while others can have serious consequences and may even be life-threatening. Urinary tract infections (UTIs) are the most frequent healthcare-associated infections (HAIs) and an important origin of fatal bloodstream invasions. Healthcare-associated uterine infections such as puerperal fever, once a common cause of maternal deaths from childbirth, still require medical awareness to be prevented (see A Glimpse of History, chapter 19). Sexually transmitted infections (STIs) are widespread; the World Health Organization (WHO) estimates that more than 1 million people worldwide acquire an STI each day. Although most STIs are transmitted by sexual contact, some can also be transmitted by contaminated needles, shared objects, childbirth, or breast feeding.

27.1 ■ Anatomy, Physiology, and Ecology of the Genitourinary System

Learning Outcome

1. Describe the anatomy, physiology, and ecology of the urinary and genital systems.

The genital and urinary systems are referred to as the *genitourinary system* because they are positioned close to each other. Both systems are often affected by the same pathogens.

The Urinary System

The urinary system (or tract) consists of the kidneys, ureters, bladder, and urethra (**figure 27.1**). The kidneys act as a multistage blood cleansing system, removing waste materials and excess fluid from the blood to produce urine. Each kidney is connected to the bladder by a ureter. The bladder acts as a holding tank for urine; once full, it empties through the urethra, using sphincter muscles at both ends of the urethra to control the flow.

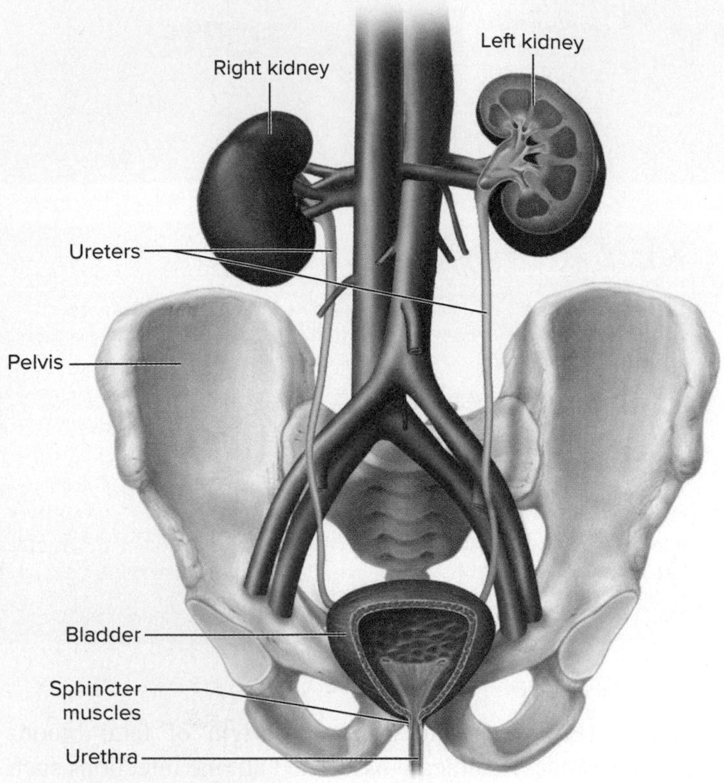

FIGURE 27.1 Anatomy of the Urinary System Urine flows from the kidneys, down the ureters, and into the bladder, which empties through the urethra.

? What is the function of the kidneys?

The lower urethra has a population of resident microbes, including species of *Lactobacillus, Staphylococcus, Corynebacterium, Haemophilus, Streptococcus,* and *Bacteroides.* Standard culture results indicate that urine is generally sterile, but studies using nucleic acid amplification tests to detect bacterial DNA suggest the presence of a urinary microbiome. Some bacteria detected by DNA studies have now been cultivated using modifications of the standard culture methods, and their role in urinary tract health and disease is being investigated.

The urinary system is protected from infection by a number of mechanisms. The sphincter muscles keep the system closed most of the time and help prevent infections by stopping bacteria from ascending to the bladder. The downward flow of urine also helps clean the system by washing away many microorganisms before they have a chance to multiply and cause infection. Recent evidence indicates that the most abundant protein in urine—a kidney-produced filamentous glycoprotein—binds bacterial adhesins, thus entangling bacterial cells before they can attach to the the bladder wall. In addition, normal urine contains antimicrobial substances such as host defense peptides (antimicrobial peptides) and antibodies. The length and position of the urethra also play a role in preventing infection. Males have a relatively long urethra (20 cm) emptying far from the anus, and because of this, they develop significantly fewer urinary tract infections (UTIs) than females, who have a shorter urethra (4 cm) emptying close to the anus.

MicroByte

Alkaline urine suggests that a bacterium causing a UTI produces urease, an enzyme that converts urea to ammonia.

The Genital System

The female genital system is composed of two ovaries, two fallopian tubes, a uterus, the vagina, and the external genitalia (vulva), including the labia and the clitoris (**figure 27.2a**). In women of childbearing age, an egg (ovum) is released from one of the two ovaries each month during ovulation and swept into the adjacent fallopian tube. When fertilization occurs, it is normally in the fallopian tube. Ciliated epithelium of the tube then moves the fertilized ovum to the uterus, where it implants in the epithelial lining, developing into an embryo and then a fetus. If fertilization does not occur, the epithelial lining of the uterus sloughs off, producing a menstrual period. The lower end of the uterus is the cervix, which is filled with antimicrobial mucus, except during menstruation. The cervix opens to the vagina, which leads to the external genitalia.

The vagina is a portal of entry for a number of pathogens, particularly those that are sexually transmitted. Infection of the cervix frequently results in **cervicitis,** meaning inflammation of the cervix. If the infection spreads to the fallopian tubes, the inflammation there may lead to scarring and destruction of the ciliated epithelium; that damage may then interfere with efficient movement of ova to the uterus, leading to infertility. The fallopian tubes are open at both ends, so microorganisms that infect the tubes may also move into the abdominal cavity, where they can infect the liver and other organs.

Estrogen hormones acting on vaginal epithelial cells affect the normal microbiota of the female genital tract. When estrogens are present, the cells produce glycogen, which lactobacilli can then convert to lactic acid; the resulting acidic pH inhibits the growth of many potential pathogens. Lactobacilli may also release hydrogen peroxide, an inhibitor of some anaerobic bacteria. Thus, the normal microbiota and resistance to infection of the female genital tract vary considerably with the person's hormonal status.

In males, the reproductive organs include a pair of testes (testicles) within the scrotum, a variety of tubes, ducts, and glands, and the penis (**figure 27.2b**). Sperm cells from each testis collect in a tightly coiled tubule called the epididymis and are carried to the urethra by a long tube, the vas deferens. At the junction of these tubes are glands called the seminal vesicles. Surrounding the urethra just below the bladder is another gland, the prostate. During ejaculation, sperm is released from the body via the urethra as semen—a whitish fluid composed of millions of sperm along with secretions from the prostate gland and seminal vesicles. Sexually transmitted pathogens commonly infect the urethra, causing **urethritis** (inflammation of the urethra). Although prostate secretions have antimicrobial properties, the prostate

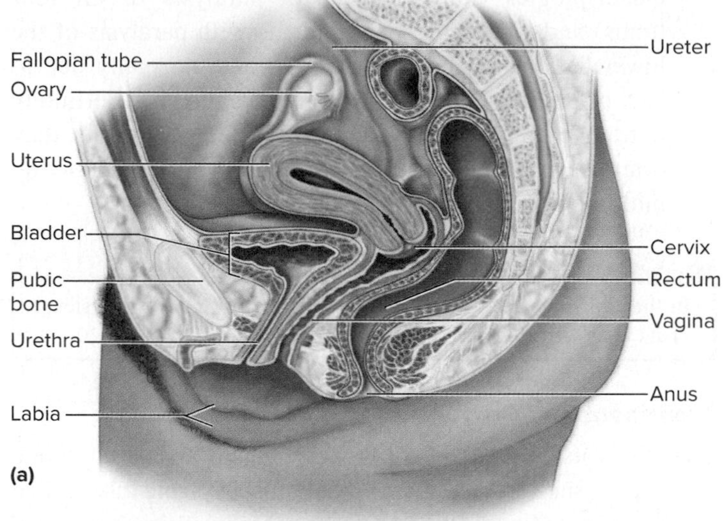

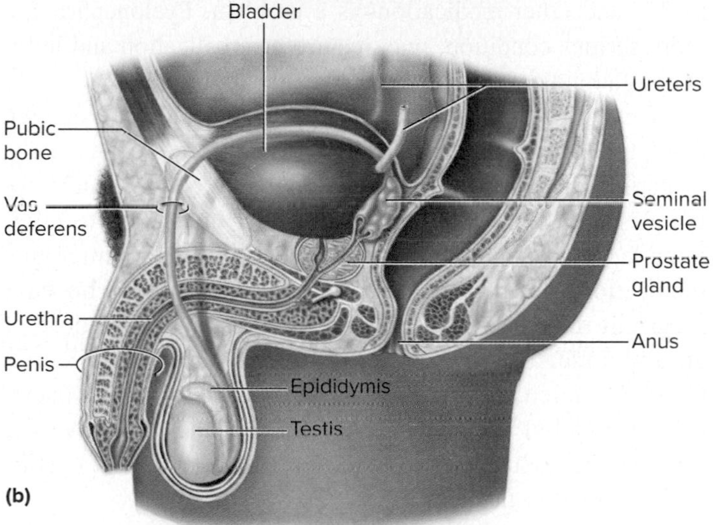

FIGURE 27.2 Anatomy of the Genital System **(a)** Female. **(b)** Male.

What is the role of the ciliated epithelium in the fallopian tubes?

gland can be infected by urinary or sexually transmitted pathogens. In older males, the prostate often enlarges, compressing the urethra and slowing the flow of urine, making it more likely that a UTI will develop.

MicroAssessment 27.1

The genitourinary system is one of the portals of entry for pathogens. The urinary system includes the kidneys, ureters, bladder, and urethra. The female reproductive system includes the vagina, cervix, uterus, fallopian tubes, and ovaries. The male reproductive system includes the testes, various glands, and the penis.

1. Describe how the bladder is protected from pathogens.
2. Why do females get urinary tract infections more often than males?
3. What changes might occur in the vagina if lactobacilli were eliminated?

27.2 ■ Urinary Tract Infections

Learning Outcomes

2. Describe the features of bacterial cystitis.
3. Outline the characteristics of leptospirosis.

Urinary tract infections (UTIs), which can involve the urethra, bladder, and/or kidneys, account for several million visits to the doctor's office each year in the United States. UTIs normally involve intestinal microbes that gain access via the urethral opening, but some pathogens enter the body through another site, infect the blood, and then accumulate in the urine.

Bacterial Cystitis ("Bladder Infection")

Cystitis, meaning inflammation of the bladder, is typically considered synonymous with bladder infection because the inflammation is usually due to infection. It is the most common type of UTI, particularly in otherwise healthy women. Bladder infections are also frequent healthcare-associated infections, connected to the use of urinary **catheters**—tubes inserted into the bladder to drain urine into a bag.

Signs and Symptoms

Bacterial cystitis is sometimes asymptomatic, especially among children and the elderly, and also in catheter-associated cases. When symptoms occur, they typically start suddenly and include a burning pain during urination, an urgent need to urinate, and frequent release of small amounts of urine. The urine often has a bad odor, may be cloudy due to leukocyte accumulation, and sometimes has a pale red color due to the presence of blood. In addition, the area above the pubic bone may be painful because of the underlying inflamed bladder.

If not treated, a bladder infection can potentially spread to the kidneys, leading to **pyelonephritis** (meaning kidney inflammation, typically due to infection); symptoms then include fever, chills, vomiting, and back pain overlying the kidneys. From there, the pathogen can spread to the blood, sometimes leading to sepsis, a life-threatening condition. Furthermore, pyelonephritis can result in permanent kidney damage if not treated promptly, and recurrent episodes increase the risk of kidney failure.

Causative Agents

Bladder infections are usually caused by members of the normal intestinal microbiota. Most community-acquired cases are due to specific uropathogenic strains of *Escherichia coli,* but other Gram-negative Enterobacteriaceae such as *Klebsiella* species are also causes, as is Gram-positive *Staphylococcus saprophyticus*. A wider range of organisms causes healthcare-associated UTIs, including *Pseudomonas aeruginosa* and *Enterococcus faecalis*. Standard urine culture methods are designed to detect all these organisms.

Pathogenesis

The bacteria that cause cystitis generally reach the bladder by moving up the urethra. In the case of uropathogenic *E. coli* (UPEC) strains, some of the bacteria multiply in the urine, whereas others attach via pili to receptors on bladder epithelial cells. Once attached, the bacteria may be taken in by the epithelial cells, where they then multiply rapidly to create densely populated biofilm-like intracellular bacterial communities (IBCs); bacterial cells can later detach from these and move to surrounding epithelium, creating new IBCs. A few of the bacteria within the IBCs develop into filamentous forms, and although these may simply be a stress response to conditions in the bladder, they might help the bacteria avoid phagocytosis or attach to other epithelial cells. Eventually, some bacterial cells may establish a dormant intracellular reservoir that resists antibiotics, potentially leading to chronic or recurrent UTIs. Bacteria able to move up the ureters to the kidneys can cause pyelonephritis.

Epidemiology

Cystitis can develop in anyone if urine is allowed to accumulate in the bladder, such as when someone is too busy to urinate, but is much more common in females because of the following predisposing factors:

- **Short urethra.** The length and position of the urethra put it at risk for fecal contamination and colonization with intestinal bacteria. From the urethra, the distance to the bladder is only a few centimeters.
- **Sexual intercourse.** UTIs in sexually active females are commonly associated with intercourse, and many women develop them after having sex for the first time. The massaging effect of sexual intercourse on the urethra moves bacteria from the urethra into the bladder.
- **Changes in vaginal microbiota.** Disruption of the naturally occurring microbes in the vagina can predispose women to UTIs. Spermicides used alone, with condoms, or with diaphragms can lead to such changes. Additionally, changes can be attributed to menopause, when monthly menstruation no longer occurs.

Other factors involved in the development of UTIs include:

- **Enlarged prostate.** Males over age 50 are at risk of UTIs because an enlarged prostate can compress the urethra, slowing urine flow and making it difficult for the bladder to empty completely.
- **Indwelling urinary catheters.** Medical conditions may require a urinary catheter to drain the bladder, and unfortunately the catheter (1) gives bacteria direct access to the bladder and (2) provides surfaces for biofilm formation; recall that bacteria growing within a biofilm are difficult or impossible to kill with antibacterial medications.

  Quadriplegics (individuals with paralysis of all four limbs) and paraplegics (individuals with paralysis of the lower half of the body) cannot urinate normally due to lack of bladder control and require a catheter indefinitely to transfer their urine to a container. Because of this, they often have chronic or recurrent UTIs, often involving multiple species.

MicroByte

In the United States, UTIs result in more than 7 million physician visits yearly, costing over $1 billion.

Treatment and Prevention

Cystitis is usually successfully treated with an antimicrobial medication such as trimethoprim-sulfamethoxazole (also called co-trimoxazole or TMP-SMX), although emerging resistance to this and other medications is a problem. Pyelonephritis, a more serious condition, may require hospitalization and intravenous administration of an appropriate antibiotic.

General ways to prevent UTIs include drinking enough fluid to ensure urinating at least four or five times daily, urinating immediately after sexual intercourse, and, for females, wiping from front to back after defecation to minimize fecal contamination of the urethra. Many antimicrobial medications accumulate in the urine, so women who have recurrent infections are sometimes prescribed a low dose of an antibiotic to be taken daily or after sexual intercourse. Cranberry juice and green tea are often used to prevent bacterial cystitis, but the scientific evidence of their effectiveness remains inconclusive. The main features of bacterial cystitis are presented in **table 27.1.**

Leptospirosis

Leptospirosis is a zoonotic disease that occurs worldwide but is more common in warm climates. The causative bacterium enters the body through a mucous membrane or a wound and is then carried by the bloodstream to the urinary system. The outcome of infection is highly variable—most infections are asymptomatic or result in mild symptoms, but some lead to life-threatening organ damage.

Signs and Symptoms

Signs and symptoms of leptospirosis begin about 5 to 14 days (range 2 to 30 days) after exposure. Most cases are mild, with abrupt onset of flu-like symptoms that can include headache, fever, chills, eye redness, muscle pain, dry cough, and vomiting. Symptoms usually resolve within a week, but a second phase sometimes occurs either immediately or within a few days after apparent recovery. This phase, referred to as the immune phase, is marked by renewed fever, muscle pain, and intense headaches. Although most patients recover, some develop Weil's disease, a life-threatening form of leptospirosis that involves

TABLE 27.1	Bacterial Cystitis
Signs and Symptoms	Sudden onset, burning pain on urination, frequent need to urinate, cloudy, red-colored urine that smells bad. Fever, chills, back pain, and vomiting suggest pyelonephritis.
Incubation Period	Usually 1 to 3 days
Causative Agents	Most commonly *E. coli,* but also other Enterobacteriaceae as well as *Staphylococcus saprophyticus*; in the case of healthcare-associated infections, also *Pseudomonas* and *Enterococcus* species
Pathogenesis	Bacteria ascend the urethra, enter the bladder, and attach by pili to receptors on epithelium; attachment causes uptake by epithelial cells within which intracellular bacterial communities develop; bacterial spread to the kidneys can occur via the ureters, causing pyelonephritis.
Epidemiology	Common in females, promoted by a relatively short urethra, sexual intercourse, and changes in the microbiota. Older males are at risk of infection because an enlarged prostate gland partially obstructs their urethra; catheterization increases the risk of infection.
Treatment and Prevention	Treatment: antimicrobial medication; longer treatment for pyelonephritis. Prevention: drinking enough liquid to urinate at least four to five times daily; in females, urinating after sexual intercourse, wiping from front to back; a low dose of antibiotic daily or after sexual intercourse may help prevent recurrent UTIs.

FOCUS ON A CASE 27.1

A 32-year-old married woman complained of 1 week of burning pain on urination and frequent release of small amounts of bloody urine. About 8 days earlier, she had completed 3 days of treatment with co-trimoxazole (TMP-SMX) for similar symptoms; urine culture results before that treatment showed *E. coli,* resistant only to amoxicillin. When her symptoms returned, she began drinking 12 ounces of cranberry juice three times daily but felt only slightly better. She did not have other symptoms such as chills, fever, back pain, nausea, or vomiting, and the rest of her physical examination results were normal.

Her medical history showed that she had suffered two or three similar episodes of urinary symptoms every year for several years. Sometimes the symptoms would go away when she forced herself to drink more, but when they persisted she would obtain medical evaluation and treatment with an antibacterial drug. On one occasion several years before the present illness, she had chills, fever, back pain, nausea, and vomiting with her other urinary symptoms, and she was hospitalized for a "kidney infection." She had no history suggesting any underlying disease such as diabetes, cancer, or immunodeficiency.

The results of her laboratory tests included normal leukocyte count and kidney function tests. Microscopic examination of her urine showed many red and white blood cells, including neutrophils. A stained smear of the urine showed numerous Gram-negative rods, later identified as *E. coli.* A culture of the urine revealed more than 100,000 colony-forming units (CFUs) per milliliter. The bacterium was resistant to amoxicillin but sensitive to other antibacterials used to treat UTIs.

1. What is the diagnosis?
2. What is the treatment?
3. What is the prognosis?
4. What future preventive measures would be advisable?

Discussion

1. This woman's signs and symptoms led to a diagnosis of bacterial cystitis, but clues in the presentation suggest that it was not as simple as first thought. First, her signs and symptoms recurred only 1 day after completing her medication; most patients with uncomplicated bacterial cystitis are cured by 3 days of appropriate an antibacterial drug. Second, she gave a history of being hospitalized for pyelonephritis. In addition, she had signs and symptoms for a full week before seeing her physician, potentially increasing the possibility of developing serious complications. These clues make it highly likely that she has subclinical pyelonephritis, meaning her bladder infection has spread to her kidneys but has not yet resulted in symptoms. As many as 30% of cystitis patients have subclinical pyelonephritis, depending on risk factors such as the clues mentioned in this case. The diagnosis of a kidney infection can be confirmed using imaging studies, which allow the physician to view the kidney and see areas of inflammation or structural abnormalities.
2. The patient can be treated with co-trimoxazole, the same medication given before. Because kidney infections take much longer to cure than bladder infections, however, the treatment must be continued for 2 weeks or longer. A urine culture must be done 1 week after treatment is completed, to be sure the infection is truly cleared.
3. The outlook is good for a full recovery without any permanent kidney damage. The patient should be advised, however, that additional kidney infections could lead to scarring and potential kidney failure if not treated quickly.
4. In the future, this patient should use all the usual methods for preventing UTIs, including drinking enough to ensure urinating at least four or five times daily, avoiding delays in emptying the bladder, and urinating promptly after intercourse. Because she has recurrent infections, a preventive antimicrobial drug may be appropriate.

damage to the kidneys, liver, and sometimes other organs. The patient may have high fever, abnormal heartbeat, difficulty breathing, jaundice (yellowing of skin and eyes due to liver failure), and/or hemorrhaging, depending on which organs are affected. Pulmonary hemorrhage is most often associated with fatal outcomes.

Causative Agent

Leptospirosis is caused by certain *Leptospira* species—most commonly *L. interrogans*—thin, motile, aerobic Gram-negative spirochetes with hooked ends (**figure 27.3**). As with all spirochetes, their endoflagella (flagella contained within the periplasm) propel the cells in a corkscrew-like motion, allowing them to burrow through mucous membranes and into tissue.

Pathogenesis

Pathogenic *Leptospira* species enter the body through mucous membranes, eyes, and breaks in the skin. The cells then multiply and spread throughout the body by way of the bloodstream, infecting many organs. Although an effective antibody response controls the infection in most tissues, the bacteria may continue to multiply in the kidneys. The mechanisms that lead to Weil's disease are not well understood, but some studies suggest that certain events in the infection process may trigger a poorly regulated inflammatory response, with excessive production of pro-inflammatory cytokines (a cytokine storm); recall that this response has effects on the circulatory system that can lead to life-threatening tissue damage and organ failure.

Epidemiology

Leptospirosis occurs worldwide but is more common in tropical and subtropical climates. Pathogenic *Leptospira* species infect many types of wild and domestic animals, usually causing little or no apparent illness; rodents are the most important reservoir hosts. The bacteria accumulate in the animal's kidneys and are subsequently shed in the urine (the main mode of transmission to other hosts); urine spots on the ground remain infectious for weeks to months. Humans contract leptospirosis from water, soil, or food contaminated with infected animal urine. Because the bacteria may survive in mud or water for months, outbreaks typically occur after heavy rainfall; cases have been connected to wading through flooded areas or swimming in contaminated lakes. Infected humans may excrete the organisms in their urine for up to several months, but person-to-person transmission is rare.

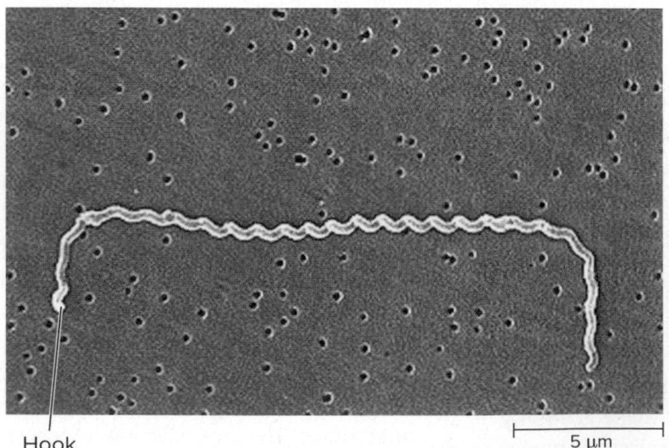

FIGURE 27.3 ***Leptospira interrogans*** This spirochete is the most common cause of leptospirosis, a zoonosis (SEM). Janice Haney Carr/NCID/HIP/CDC

? What is the importance of the endoflagella of spirochetes?

Treatment and Prevention

Treatment of leptospirosis depends on the severity of the disease. For mild cases, doxycycline is typically recommended; if the disease is severe enough to require hospitalization, intravenous antimicrobial medication is given, usually penicillin.

The most important measure for preventing leptospirosis is to avoid animal urine. Vaccines can prevent infections in domestic animals, which, in turn, reduces disease occurrence in humans. The main features of leptospirosis are shown in **table 27.2.**

MicroAssessment 27.2

Bladder infections are common, especially in women, and are usually caused by normal intestinal bacteria ascending from the urethra. Situations that affect normal urine flow predispose a person to UTIs. Pyelonephritis is a serious complication that can lead to kidney failure. In leptospirosis, a widespread zoonosis spread by urine, the causative organism enters the urinary system from the bloodstream; leptospirosis is usually mild, but a form called Weil's disease can be life-threatening.

4. What organism causes the majority of bladder infections in otherwise healthy females? From where does this organism come?
5. How do people become infected with *Leptospira interrogans?*
6. Treatment of leptospirosis with penicillin or other bactericidal antibiotics sometimes results in a Jarisch-Herxheimer reaction, characterized by fever, chills, headache, malaise, tachycardia, and rarely, death. What might be the cause of this reaction?

27.3 ■ Genital System Diseases

Learning Outcome

4. Compare and contrast bacterial vaginosis (BV), vulvovaginal candidiasis (VVC), and staphylococcal toxic shock syndrome.

The genital tract is the portal of entry for many infectious diseases, both sexually and non-sexually transmitted. This section discusses some genital system diseases that are not classified as sexually transmitted.

TABLE 27.2 Leptospirosis

1. Urine from an animal infected with certain *Leptospira* species contacts mucous membranes or damaged skin, usually via contaminated soil or food.
2. The bacteria enter the bloodstream and are carried throughout all body tissue, sometimes causing fever and intense pain.
3. Signs and symptoms resolve and bacteria disappear from blood and tissues, except kidneys.
4. Weil's disease is associated with severe damage to liver and kidneys.
5. Complete recovery occurs if organ failure is avoided or effectively treated.
6. Bacteria continue to be excreted in urine.

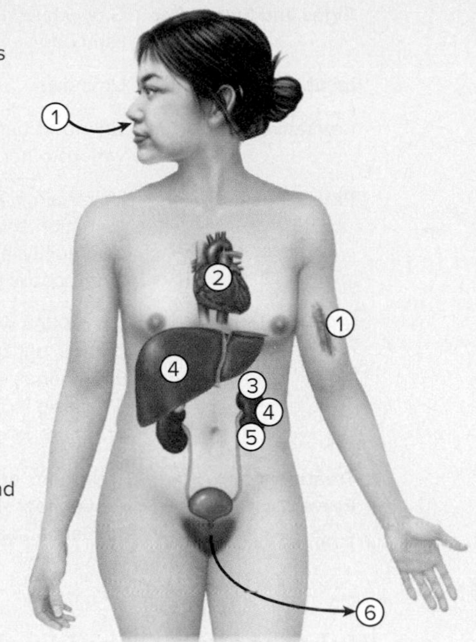

Signs and Symptoms	Most infections are asymptomatic. If symptoms occur they are usually mild, with fever, headache, muscle pain, and bloodshot eyes; a second phase may follow with renewed fever and headaches; severe form (Weil's disease) may lead to organ failure.
Incubation Period	Usually 5 to 14 days
Causative Agent	Certain *Leptospira* species, most commonly *L. interrogans*
Pathogenesis	The bacterial cells penetrate mucous membranes or breaks in the skin and are carried in the bloodstream to all parts of the body. Weil's disease may be due to a cytokine storm.
Epidemiology	Worldwide distribution; wide range of animal hosts chronically excrete the bacteria in their urine, contaminating natural waters and soils.
Treatment and Prevention	Treatment: antibacterial medication. Prevention: avoiding contact with animal urine; vaccines prevent infection in domestic animals.

Bacterial Vaginosis (BV)

Bacterial vaginosis (BV) is the most common vaginal disease of females, particularly in their childbearing years. It is termed vaginosis rather than vaginitis because there are no inflammatory changes. BV is a particular concern for pregnant women because it is linked to premature delivery and low-birth-weight babies.

Signs and Symptoms

About half of BV cases are asymptomatic. Symptomatic BV is characterized by a thin, grayish-white, slightly bubbly vaginal discharge that has a characteristic strong fish-like smell. The bacteria associated with BV may spread to the uterus or the fallopian tubes, causing **pelvic inflammatory disease (PID),** which can lead to infertility.

Causative Agent

The cause or causes of BV are unknown, but the disease is characterized by an imbalance in the normal vaginal microbiota; recall that an imbalance in a microbial population is called dysbiosis. The commonly dominant *Lactobacillus* population declines, accompanied by an increase in the numbers of various facultative and anaerobic bacteria, including *Gardnerella vaginalis* and members of the genera *Mobiluncus, Prevotella,* and *Peptostreptococcus*. PCR-based methods suggest the additional involvement of three novel species termed BV-associated bacteria (BVAB1, BVAB2, and BVAB3). Although numerous studies implicate *G. vaginalis* in the disease, the role of other bacteria is not clear (they could simply be bystanders). In addition, other factors are likely involved because women voluntarily inoculated with cultures of the implicated organisms do not always develop the disease, and low numbers of each species can occur in vaginal secretions of healthy women.

MicroByte

Although BVA-associated bacteria have yet to be cultured, researchers pieced together BVAB1's entire genome sequence using data obtained by analyzing secretions from females diagnosed with BV.

Pathogenesis

BV is thought to start with the development of a *G. vaginalis* biofilm, which then facilitates the development of the associated polymicrobial population. No inflammation occurs unless another concurrent vaginal infection is present. BV is associated with characteristic changes in the vagina, including a loss of acidity of the vaginal secretions (normally pH 3.8–4.2), disruption of the normal microbiota, and substantial increase in the numbers of what are called clue cells—vaginal epithelial cells that have a roughened appearance because they are covered with bacteria (**figure 27.4**). The strong fishy odor that characterizes BV is caused by metabolic products of the anaerobic bacteria; this can be detected using the whiff test in which a drop of potassium hydroxide is added to a slide containing a vaginal discharge sample—in females with BV, the fish-like smell is produced.

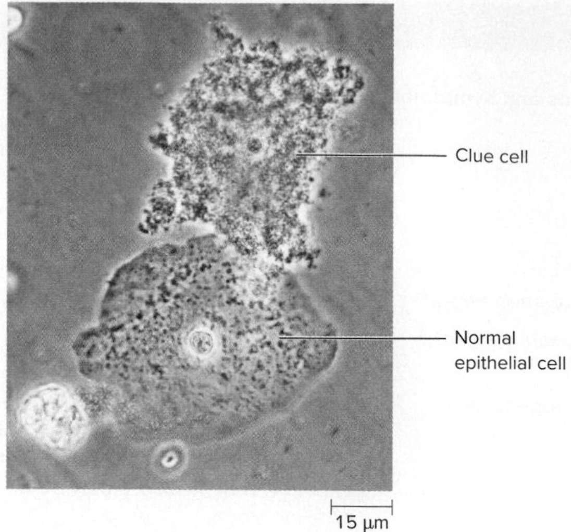

FIGURE 27.4 Clue Cell from a Female with Bacterial Vaginosis The clue cell is covered with bacteria, giving it a roughened appearance. M. Rein/CDC

? **Why does the clue cell appear different in texture from normal epithelial cells?**

Epidemiology

The epidemiology of BV is not well understood. Although there is no proof that BV is sexually transmitted, people who have never had sex seldom get the disease. It is most common among sexually active women and sometimes occurs in children who have been sexually abused. Females who douche, have multiple sex partners, have sex with other women, have a new sex partner, or use an intrauterine device (IUD) also have an increased risk. Perhaps because of the vaginal dysbiosis, women with BV are at higher risk of acquiring certain STIs, including chlamydia, gonorrhea, HIV, and trichomoniasis. Pregnant women are at increased risk of BV, and the disease is linked to premature delivery and low-birth-weight babies.

Treatment and Prevention

Most cases of BV respond quickly to treatment with an antimicrobial medication such as metronidazole or clindamycin, administered intravaginally (meaning within the vagina) or orally. Many females with BV have a recurrence several months after initial treatment, requiring an additional course of medication; continuous therapy may be required to prevent multiple recurrences. Treating pregnant women who have BV is important because the disease increases the risk of premature delivery. Studies on using yogurt or probiotics orally or intravaginally to restore lactobacilli gave conflicting results.

BV can be prevented by abstinence, limiting the number of sex partners, and avoiding vaginal douching. Treatment of the partners of patients with BV does not prevent recurrences. The main features of BV are summarized in **table 27.3.**

TABLE 27.3 Bacterial Vaginosis

Signs and Symptoms	Gray-white vaginal discharge with unpleasant fishy odor
Incubation Period	Unknown
Causative Agent	Unknown but involves *Gardnerella vaginalis* and various other bacteria
Pathogenesis	Uncertain, but includes *G. vaginalis* biofilm formation and development of a polymicrobial community; appearance of clue cells; odor due to metabolic products of anaerobic bacteria
Epidemiology	Associated with many sexual partners or a new partner, but can occur in the absence of sexual intercourse; increases risk of acquiring other STIs; also increases risk of premature delivery in pregnant women.
Treatment and Prevention	Treatment: appropriate antimicrobial medication. Prevention: abstinence, limiting number of sexual partners, avoiding douching.

Vulvovaginal Candidiasis (VVC)

Vulvovaginal candidiasis (VVC), a fungal infection, is the second most common cause of vaginal symptoms after BV. Like BV, it seems to involve dysbiosis. As the name indicates, VVC often involves not only the vagina but the vulva as well.

Signs and Symptoms

The most common signs and symptoms of VVC are intense itching and burning sensation in the vagina or vulva, which appear red and swollen. Large amounts of thick, clumpy, whitish vaginal discharge resembling cottage cheese are typically produced.

Causative Agent

VVC is caused by *Candida albicans,* a yeast that is part of the normal vaginal microbiota in about a third of all females (**figure 27.5**). Because *C. albicans* is a fungus, it has a eukaryotic cell structure. It typically grows as yeast cells but can also form hyphae or pseudohyphae.

Pathogenesis

C. albicans is often found in the vagina without causing symptoms, its growth limited by the immune system and competition from the normal vaginal lactobacilli. When the normal microbiota balance is disturbed—as occurs during menstruation, pregnancy, when using oral contraceptives, or taking antibiotics—*C. albicans* can multiply freely, triggering an inflammatory response; the signs and symptoms occur within about 10 days.

Epidemiology

Most women with VVC have no known predisposing circumstances, but factors that increase the likelihood of *Candida*

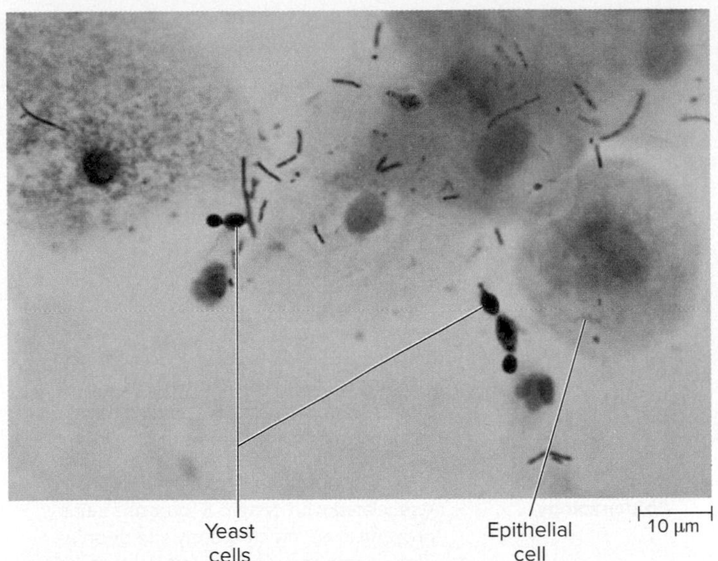

FIGURE 27.5 ***Candida albicans*** Stained smear of vaginal discharge from a female with vulvovaginal candidiasis, showing *C. albicans.* Dr Stuart Brown/CDC

? What is the most common structural form of *C. albicans* in VVC?

infection are the late stage of pregnancy, poorly controlled diabetes, and the use of antibiotics or oral contraceptives. Hormone replacement therapy may also be a predisposing factor. The disease does not spread from person to person.

Treatment and Prevention

Vulvovaginal candidiasis can be treated with one of various azoles taken orally or applied intravaginally. A single dose of oral fluconazole or one of several formulations of intravaginal creams, ointments, and suppositories is often prescribed. Over-the-counter intravaginal medications are also available, but self-diagnosis and treatment can lead to the development of drug-resistant organisms.

Preventing VVC depends on minimizing the use of antibacterial medications and on the effective treatment of underlying conditions such as diabetes. As with BV, insufficient evidence supports the use of oral or intravaginal yogurt for treatment. The main features of VVC are presented in **table 27.4.**

TABLE 27.4	**Vulvovaginal Candidiasis**
Signs and Symptoms	Itching, burning sensation; thick, white vaginal discharge; redness; and swelling
Incubation Period	Usually unknown, but generally within about 10 days of microbiota balance disturbance
Causative Agent	*Candida albicans*, a yeast
Pathogenesis	Inflammatory response to *C. albicans* overgrowth
Epidemiology	Not contagious; associated with antibacterial therapy, use of oral contraceptives, pregnancy, and uncontrolled diabetes, but most cases have no identifiable predisposing factor.
Treatment and Prevention	Treatment: oral or intravaginal antifungal medication. Prevention: minimizing antibiotic use and treating known predisposing conditions.

Staphylococcal Toxic Shock Syndrome

Toxic shock syndrome was first described in the late 1970s in several children with staphylococcal infections. In 1980, it became epidemic in young, healthy, menstruating women using a brand of high-absorbency tampon that has since been removed from the market. The term *toxic shock syndrome* was used to describe the illness. Now that we know its cause, it is called *staphylococcal toxic shock syndrome.* This is not a new disease but one that emerged in a new form and became much more common as a result of changes in technology and human behavior.

Signs and Symptoms

Staphylococcal toxic shock syndrome is characterized by the sudden development of high temperature, headache, muscle aches, bloodshot eyes, vomiting, diarrhea, a sunburn-like rash, and confusion. Typically, the skin peels about a week after symptoms begin. Without treatment, the blood pressure may drop, leading to multi-organ failure, coma, and sometimes death.

Causative Agent

Staphylococcal toxic shock syndrome is caused by strains of *Staphylococcus aureus* that produce toxic shock syndrome toxin-1 (TSST-1) or related exotoxins.

Pathogenesis

Tampon-associated toxic shock syndrome results from *S. aureus* growing in a tampon used during menstruation. The bacteria rarely spread throughout the body, but as they multiply they produce TSST-1 or related exotoxins that can be absorbed into the bloodstream. The toxins are superantigens that cause the activation of large numbers of helper T cells, leading to a massive release of cytokines (cytokine storm). This in turn causes a drop in blood pressure, sometimes resulting in life-threatening multi-organ failure.

Epidemiology

Staphylococcal toxic shock syndrome can occur after infection of any number of body sites with a *S. aureus* strain that produces one of the responsible exotoxins. For females, improper handling of tampons increases the likelihood of bacterial contamination, which may result in staphylococcal toxic shock. Higher-absorbency tampons that allow *S. aureus* to grow to high numbers and persist over time can pose a greater threat. The use of intravaginal contraceptive sponges also increases risk. People who have recovered from the disease are not always immune to it. Since 1990, there has been a slow, steady decline in the number of cases in the United States, now usually less than 50 per year (**figure 27.6**).

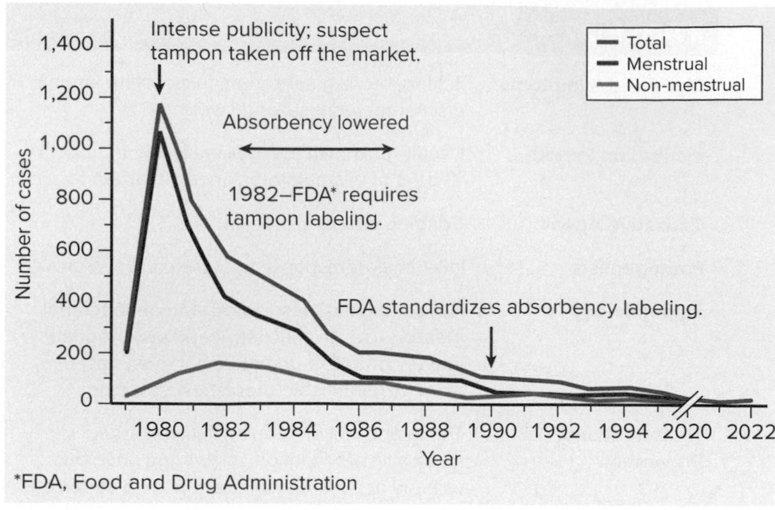

FIGURE 27.6 Staphylococcal Toxic Shock Syndrome, United States A sharp drop in cases occurred when a brand of high-absorbency tampon was taken off the market.

Does staphylococcal toxic shock syndrome only occur in females? Explain.

Treatment and Prevention

Staphylococcal toxic shock syndrome is a severe disease and requires hospitalization. It can be effectively treated with antibacterial medication active against the infecting *S. aureus* strain, along with supportive measures to prevent shock and organ failure. The infection source, such as tampons or wound dressings, should be eliminated if possible. Although most people recover fully in 2 to 3 weeks, the disease can be fatal within a few hours.

Toxic shock syndrome associated with the use of tampons can be prevented by the appropriate use of tampons, including washing hands thoroughly before and after inserting a tampon, using tampons with the lowest practical absorbency, changing tampons at least every 6 hours, and using a pad instead of a tampon while sleeping. It is also important to avoid trauma to the vagina when inserting tampons, to recognize the symptoms of the disease, and to remove a tampon immediately if symptoms occur. Women who have had the disease previously should not use tampons. **Table 27.5** describes the main features of staphylococcal toxic shock syndrome.

TABLE 27.5 Staphylococcal Toxic Shock Syndrome

Signs and Symptoms	Fever, vomiting, diarrhea, muscle aches, a rash that peels, and low blood pressure that may lead to multi-organ failure and sometimes death
Incubation Period	3 to 7 days
Causative Agent	Certain toxin-producing strains of *Staphylococcus aureus*
Pathogenesis	Toxin (TSST-1 and others) produced by the infecting bacterium is absorbed into the bloodstream; the toxins are superantigens, triggering a cytokine storm and drop in blood pressure.
Epidemiology	Associated with certain *S. aureus* strains growing in an infected body site or on a high-absorbency tampon left in place for an extended time period.
Treatment and Prevention	Treatment: antimicrobial medication effective against the causative *S. aureus* strain; supportive therapies. Prevention: being aware of symptoms; promptly treating *S. aureus* infections; care when using tampons.

MicroAssessment 27.3

Bacterial vaginosis (BV), the most common vaginal disease for females in their childbearing years, is characterized by a significant change in the composition of the normal vaginal microbiota. Vulvovaginal candidiasis (VVC) may arise for unknown reasons but can occur as a result of conditions that suppress the normal vaginal microbiota. Staphylococcal toxic shock syndrome is caused by strains of *Staphylococcus aureus* that produce certain exotoxins, which are absorbed into the bloodstream, causing the massive release of cytokines responsible for shock.

7. What is the causative agent of bacterial vaginosis (BV)?
8. Why was the incidence of staphylococcal toxic shock syndrome higher in menstruating women than in other people during the early 1980s?
9. Why would using antibiotics predispose a woman to vulvovaginal candidiasis (VVC)?

SEXUALLY TRANSMITTED INFECTIONS

27.4 ■ Bacterial STIs

Learning Outcomes

5. Compare and contrast chlamydia, gonorrhea, and *Mycoplasma genitalium* infection.
6. Compare and contrast syphilis and chancroid.

In general, all STIs have similar epidemiologic characteristics and require the same prevention measures; these topics are covered in the box Focus on Sexually Transmitted Infections. Three of the most common bacterial STIs—chlamydia, gonorrhea, and *Mycoplasma genitalium* infection—are caused by distinctly different pathogens yet all have similar signs and symptoms as well as epidemiology, so to avoid too

FOCUS ON SEXUALLY TRANSMITTED INFECTIONS

The terms sexually transmitted infection (STI) and sexually transmitted disease (STD) are often used interchangeably but their definitions have a subtle difference: STI refers to colonization with a sexually transmitted pathogen, whereas STD implies a disease state as a result of the infection. Confusion sometimes arises, however, because people assume that "disease state" equates to obvious symptoms, yet many sexually transmitted pathogens can damage the reproductive tract without causing noticeable symptoms; in addition, these silent infections are transmissible. Thus, as a means to avoid giving the impression that STDs are usually accompanied by obvious symptoms, many public health experts prefer the term STI.

Signs and Symptoms

When signs and symptoms of an STI occur, they vary according to the pathogen and infection site but may include vaginal or urethral discharge, pain with urination, genital sores, rashes, abdominal pain, or abnormal vaginal bleeding. Some STIs become systemic, potentially causing flu-like symptoms, and others may result in cancer.

The most common sexually transmitted microorganisms can ascend the female reproductive tract to cause **pelvic inflammatory disease (PID),** an infection of the fallopian tubes, uterus, or ovaries that can result in chronic pelvic pain. PID sometimes results in infertility or an increased risk of ectopic pregnancy—a pregnancy in which the embryo develops in the fallopian tube or elsewhere outside of the uterus, potentially leading to life-threatening internal hemorrhaging.

Pathogenesis

Microbes that cause STIs typically do not survive long in the environment, so intimate contact is required for their transmission. They generally enter through mucous membranes; recall that pathogens infect mucous membranes much more easily than they penetrate skin. Most commonly, the microbes enter through the vagina or urethra, but they can also enter through the rectum or the oral cavity. In females, genital infections sometimes spread to the rectum because of the proximity of the two body sites. The ease with which a pathogen can infect a membrane varies, but even a small amount of body fluid sometimes contains enough infectious cells or viral particles to put a person at high risk of acquiring that microbe.

Because sexually transmitted pathogens do not persist well in the outside environment, they must be able to survive for relatively long periods in their host by avoiding immune defenses. The mechanisms vary according to the pathogen but often include avoiding destruction by phagocytes and complement system proteins.

The signs and symptoms associated with bacterial STIs are usually a result of damage due to inflammation, while those of viral STIs are often due to direct damage to infected cells. In some cases, the resulting scarring may cause permanent problems. For example, infertility and risk of ectopic pregnancy associated with PID are due to scarring in the fallopian tubes that can interfere with the progression of a fertilized egg to the uterus.

Epidemiology

STIs are typically spread through intimate contact with body fluids such as urethral discharges, vaginal secretions, and sometimes saliva. Because of this, they are usually transmitted through unprotected sexual contact—including vaginal, oral, and anal sex—but they can also be acquired through other intimate activities that result in close contact between mucous membranes or fluids. Notably, in the United States, almost half of new STIs are in the age group 15–24 years old.

The risk of acquiring an STI through unprotected sex increases when multiple partners are involved (**box figure 27.1**). Even infected persons who are asymptomatic and therefore unaware that they have an STI can potentially transmit the pathogen—a

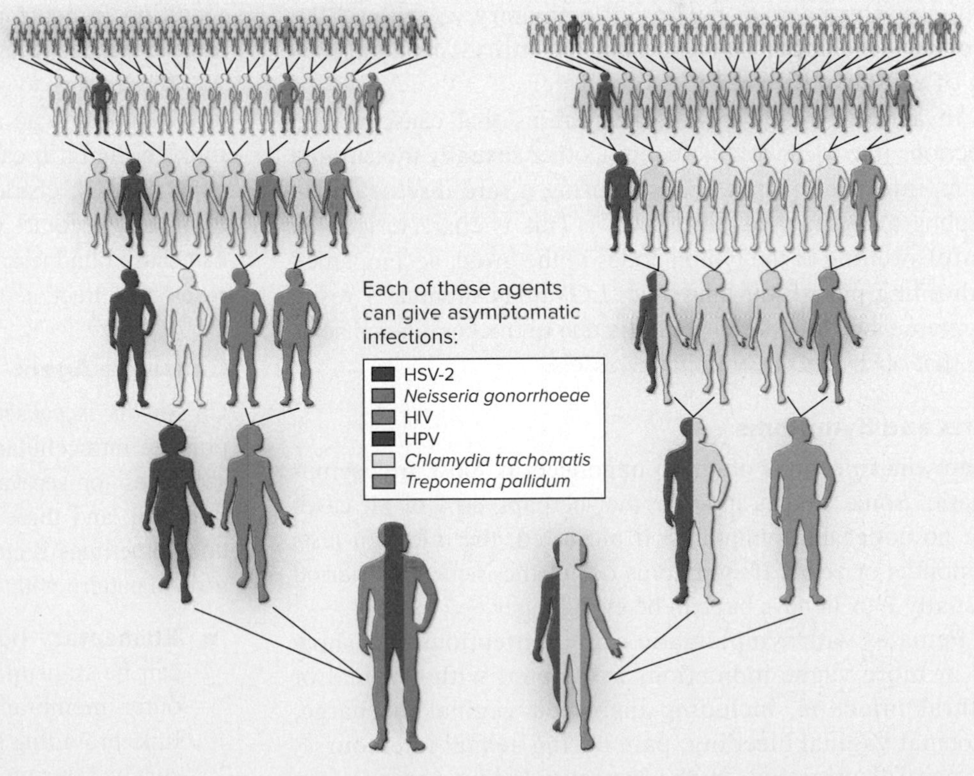

BOX FIGURE 27.1 The Possible Risk of Acquiring STIs in People Having Unprotected Sex Each partner had two previous sexual partners, and each of these partners had two previous partners, and so on. This risk of contracting any number of STIs rises with the number of sexual partners. (Examples provided: HSV-2—herpes simplex virus type 2, HIV—human immunodeficiency virus, HPV—human papillomavirus)

continued

major factor in the spread of STIs. Another concern is that any STI that results in open sores (even those too small to be noticeable) or inflammation increases the risk of contracting HIV; the sores allow HIV access to tissues, and the inflammation recruits cell types that HIV infects (helper T cells and macrophages).

Fetuses and newborns are at risk of acquiring certain genital pathogens through vertical transmission—via the placenta or during the birth process. In infants, the infections may have serious consequences such as blindness or even death.

Treatment and Prevention

Common bacterial STIs are currently curable with antimicrobial medications, although some infectious agents are now resistant to most available options. Viral STIs cannot be cured with medications, but some antivirals reduce the severity and duration of symptoms. To prevent further transmission of STIs, sex partners of anyone diagnosed with one should also be notified so that they can be evaluated and, if necessary, treated.

The most reliable ways to avoid contracting an STI include not having sex or having a monogamous relationship with an uninfected person; not sharing sexual devices also limits transmission. Consistent, correct condom use significantly reduces the chances of acquiring an STI but does not guarantee full protection. Some human papillomavirus (HPV) infections can be prevented through vaccination, but there are no vaccines against other STI agents. Male circumcision significantly reduces the risk of acquiring some sexually transmitted viruses such as HIV. Ideally, all sexually active people should be screened regularly for asymptomatic infections, but due to cost, logistics, and the risk of anxiety from false positives, testing may not be feasible. Instead, screening is recommended for people at higher risk for acquiring an STI, particularly if long-term consequences may develop.

much repetition in this section, we will cover chlamydia in the greatest detail and then refer back to that information where appropriate as we describe the others.

Chlamydia

Chlamydia is the most common bacterial STI in the United States. Like many other STIs, it is often a silent disease because many people do not show symptoms. This is particularly significant for females due to the potential long-term consequences if pelvic inflammatory disease (PID) develops—including chronic pain, infertility, and increased risk of ectopic pregnancies.

In addition to the chlamydial strains that cause genital infections (the focus of this section), other sexually transmitted strains infect the lymph nodes, causing a rare disease called lymphogranuloma venereum (LGV). This is characterized by painful swelling of the lymph nodes in the groin, accompanied by draining pus; if left untreated, LGV may eventually result in extreme swelling of the genitalia due to thickening and scarring that obstructs the lymphatic vessels.

Signs and Symptoms

Chlamydia infections often go unnoticed as most are asymptomatic. Some studies indicate that perhaps 80% of all cases have no noticeable symptoms; if untreated, the infection lasts for months or years. If symptoms occur, the incubation period is usually 7 to 14 days but can be even longer.

Females with symptomatic genital infections may show one or more vague indications associated with cervical or urethral infection, including increased vaginal discharge, abnormal vaginal bleeding, pain during sexual intercourse, and painful urination. If the causative agent spreads into the upper reproductive tract, PID may develop—potentially leading to the associated complications. Also, the bacteria may spread from the fallopian tubes into the abdominal cavity, where they can affect the liver or other abdominal organs.

In males, the infection usually involves the urethra; if symptomatic, this results in a thin, gray-white discharge from the urethra and painful urination. The inflammatory response to infection can result in scar tissue accumulation that partially obstructs the urethra, slowing urination and creating a predisposition to urinary tract infections (UTIs). In addition, the bacteria sometimes spread to the epididymis, resulting in acute pain and swelling of the scrotum. Some studies suggest that the infection can decrease fertility.

Babies born to women who have symptomatic or asymptomatic chlamydia can acquire the causative agent during passage through the birth canal. Infection of the eyes leads to neonatal conjunctivitis, characterized by a discharge from the eye that eventually becomes thick with pus. If left untreated the infection can cause blindness. If the organism enters the newborn's lower respiratory tract, neonatal chlamydial pneumonia may result.

Causative Agent

Chlamydia is caused by *Chlamydia trachomatis,* a spherical, obligate intracellular Gram-negative bacterium. Eight different serotypes (or serovars) are associated with genital chlamydia infection, and three others cause lymphogranuloma venereum and sometimes rectal infections. The bacterium has a unique growth pattern with two developmental forms (see figure 11.26):

- **Elementary body (EB).** This is the infectious form that can be transmitted to nearby cells or another host. The outer membrane has cysteine-rich proteins that cross-link, providing structural stability that enhances the cell's survival during transmission.
- **Reticulate body (RB).** This is the active form that replicates within infected host cells.

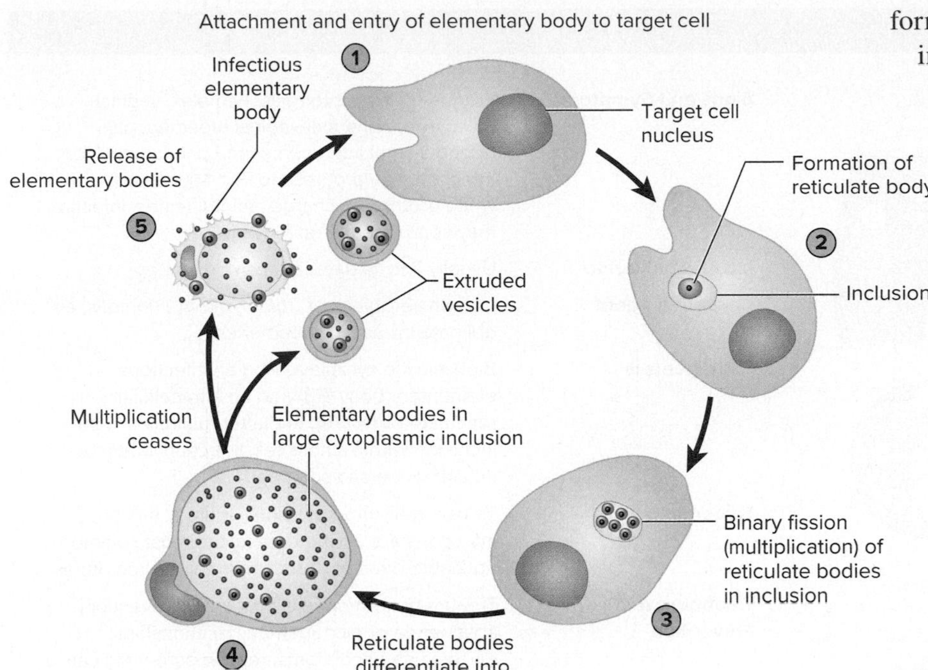

FIGURE 27.7 ***Chlamydia trachomatis*** **Life Cycle**

Which is the infectious form of *C. trachomatis*?

Pathogenesis

Infection begins when an EB of *C. trachomatis* attaches to a mucosal epithelial cell. From there, it begins its unique biphasic life cycle (**figure 27.7**):

1. Once attached to a target cell, the EB uses a type III secretion system to deliver effector proteins into the host cell that induce uptake of the EB by endocytosis.
2. Inside the host cell, two important processes occur: The bacterium directs modifications of the endosome to form a protective compartment called an inclusion; the EB then undergoes changes within the inclusion to become a larger, active RB.
3. Within the inclusion, the RB divides rapidly by binary fission, obtaining many nutrients from the host cell. The inclusion can take up much of the cytoplasmic space of the host cell.
4. While still in the inclusion, the RBs stop multiplying and then differentiate into infectious EBs.
5. The EBs are released from the infected cells, either when the host cell bursts or by extrusion in vesicles through the host cell membrane. The EBs can then infect additional nearby host cells, or be transmitted in semen or vaginal fluid to another host.

Chlamydia can persist for long periods of time in a host, indicating that the bacterium is well equipped to avoid the host defenses. One way it does this while multiplying within host cells is by preventing the infected cells from presenting antigens to T cells; the bacterium synthesizes a protein that interferes with MHC synthesis and MHC-antigen complex formation. *C. trachomatis* can also inhibit apoptosis in infected cells, keeping the host cells alive long enough for the RBs to replicate and develop into EBs. In addition, the microbe interferes with the formation of neutrophil extracellular traps.

The damage associated with chlamydia is largely due to the inflammatory response that results when infected cells produce pro-inflammatory cytokines. Neutrophils recruited to the infection site release various substances that kill bacteria but also harm tissues in the process; further tissue damage results from the cell-mediated immune response. In the case of PID, scarring in the fallopian tubes as a result of tissue repair processes increases the risk of ectopic pregnancy or infertility.

Epidemiology

Chlamydia is the most common of all notifiable bacterial infectious diseases in the United States, with over 1.4 million reported cases in 2022; many infections are asymptomatic, however, so the actual number is thought to be at least twice that. The number of reported cases is influenced not only by the incidence of infections, but also the effectiveness of screening programs as well as sensitivity of diagnostic tests used. As with other STIs, the infection can be transmitted by vaginal, oral, or anal sex as well as other intimate activities that result in close contact with mucous membranes or body fluids. *C. trachomatis* can also be passed from mother to baby during vaginal childbirth.

Treatment and Prevention

Chlamydia is typically treated with doxycycline, but other options are available. Neonatal conjunctivitis and pneumonia are usually treated with oral erythromycin.

Like all STIs, chlamydia can be prevented by abstinence, monogamous relationships, not sharing sexual devices, and barrier methods such as condoms used consistently and correctly. Sexual partners should be screened and, if infected, treated. Neonatal conjunctivitis is prevented by screening pregnant females for chlamydia and treating those who are infected. The main features of chlamydia are summarized in **table 27.6.**

Gonorrhea

Gonorrhea is the second most common bacterial STI in the United States, and it shares significant characteristics with chlamydia: it is often asymptomatic, and even silent infections can lead to long-term consequences. The name of the disease originates from the Greek words *gonos* meaning "seed" and *rhoia* meaning "to flow," probably because the pus produced during infection looks like semen.

TABLE 27.6 Chlamydia

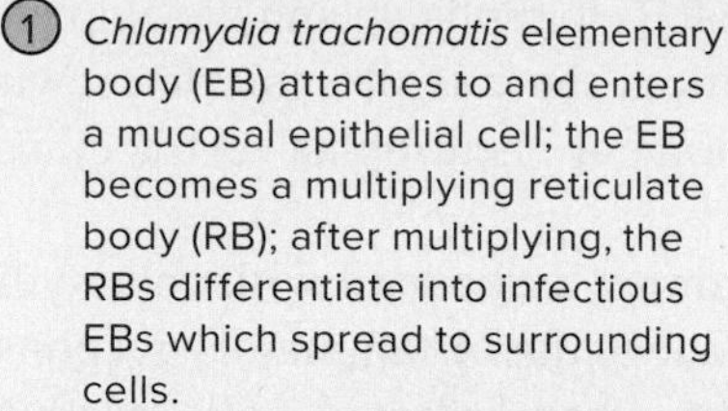

1. *Chlamydia trachomatis* elementary body (EB) attaches to and enters a mucosal epithelial cell; the EB becomes a multiplying reticulate body (RB); after multiplying, the RBs differentiate into infectious EBs which spread to surrounding cells.

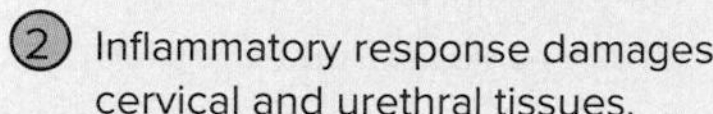

2. Inflammatory response damages cervical and urethral tissues.
3. Infection of the fallopian tubes may result in scarring, which can lead to infertility or ectopic pregnancy.
4. Urethral infection in males may result in urethral discharge and painful urination; scarring may slow urination, thus predisposing to UTIs.
5. Bacteria may spread to the epididymis, resulting in acute pain and swelling of the scrotum.

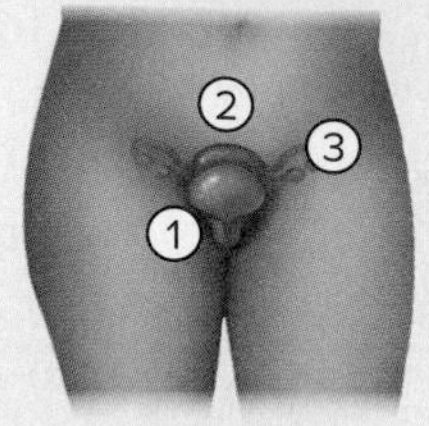

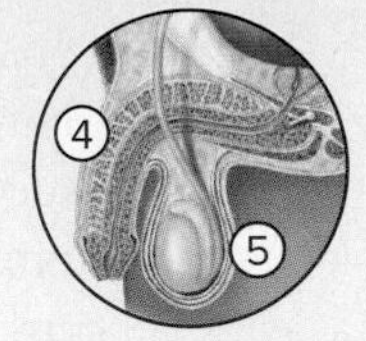

Signs and Symptoms	Frequently asymptomatic. Females: vaginal discharge, abnormal vaginal bleeding, pain during sexual intercourse, and painful urination. Infection may progress to PID. Males: thin, gray-white urethral discharge, painful testes. Infection may spread to the epididymis.
Incubation Period	Usually 7 to 14 days, but can be longer
Causative Agent	Certain serotypes of *Chlamydia trachomatis*, an obligate intracellular bacterium
Pathogenesis	Biphasic life cycle involving an infectious elementary body (EB) and an intracellular reticulate body (RB); the latter multiplies in an inclusion within a host cell. Infection leads to inflammation and scarring.
Epidemiology	Transmitted through sexual contact; most infections are asymptomatic; the most common notifiable bacterial infection in the United States.
Treatment and Prevention	Treatment: appropriate antibiotic. Prevention: abstinence or monogamous relationships; correct use of condoms; regular screening can identify infections, thus limiting spread to sexual partners.

Signs and Symptoms

Gonorrhea is often asymptomatic in both females and males, but if signs and symptoms occur, they typically appear within 14 days of infection and are similar to those of chlamydia. Symptomatic females may notice indications associated with cervical or urethral infection, including vaginal itchiness, increased vaginal discharge, abnormal vaginal bleeding, pain with sexual intercourse, and painful urination. The infection can progress upward through the reproductive tract, causing PID; occasionally, it spreads into the abdominal cavity. In males, the infection usually involves the urethra and, if symptomatic, typically results in painful urination and a thick, pus-containing urethral discharge (**figure 27.8**); scar tissue can potentially accumulate, partially obstructing the urethra and thus increasing the likelihood of UTIs. The bacteria sometimes spread to the epididymis, resulting in pain and swelling of the scrotum.

Certain strains of the causative agent can cause **disseminated gonococcal infection (DGI),** a systemic infection characterized by fever, rash, and arthritis caused by growth of the pathogen within the joint spaces. DGI can also lead to infective endocarditis and meningitis. The disseminated infection is not usually preceded by urogenital symptoms.

Babies born to mothers who have gonorrhea (symptomatic or asymptomatic) can develop neonatal conjunctivitis, as described for chlamydia.

Causative Agent

Gonorrhea is caused by *Neisseria gonorrhoeae,* commonly called gonococcus (GC), a fastidious Gram-negative diplococcus (**figure 27.9**). Due to the bacterium's fastidious nature, it requires a rich medium such as chocolate agar with various antibiotics added for cultivation. The outer membrane of GC has the general structure of a typical Gram-negative outer membrane

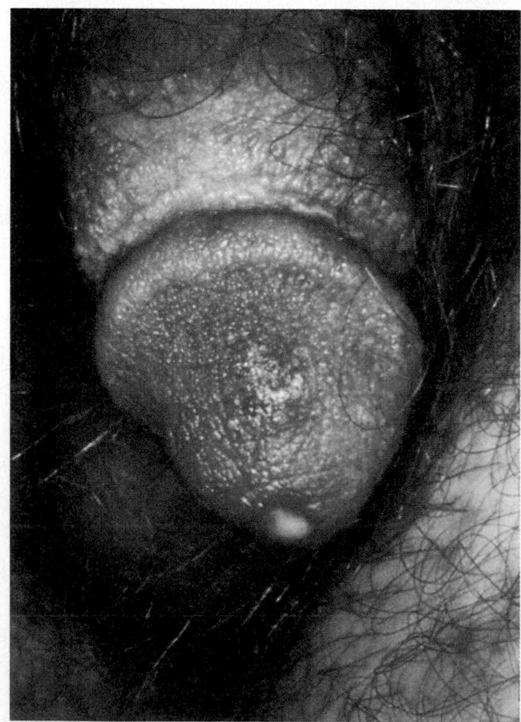

FIGURE 27.8 Urethral Discharge in a Male with Gonorrhea
While symptoms of gonorrhea in males are noticeable when present, the disease is often asymptomatic. CDC/ Dr. Gavin Hart

? Compare the signs and symptoms of gonorrhea in males and females.

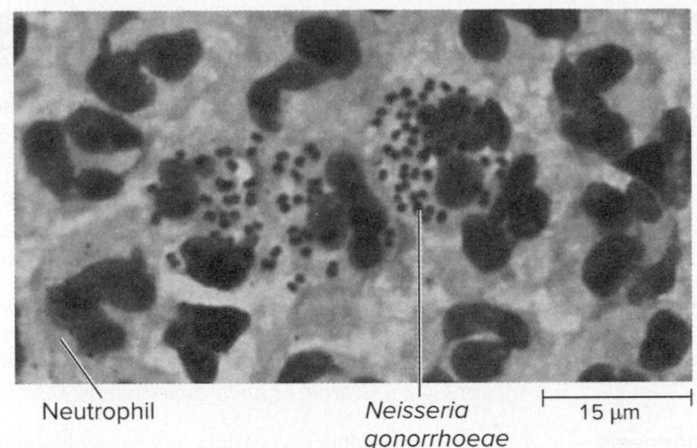

FIGURE 27.9 ***Neisseria gonorrhoeae*** Stained smear of pus from the urethra showing *N. gonorrhoeae*. Bill Schwartz/CDC

? **What term describes the morphology of *N. gonorrhoeae*?**

but contains lipooligosaccharide (LOS) instead of lipopolysaccharide (LPS); like LPS, LOS contains lipid A and is therefore referred to as endotoxin. The organism produces pili used for (1) attachment to host cells and (2) twitching motility (**figure 27.10**).

Pathogenesis

N. gonorrhoeae is well adapted to grow within the human host. Like other STI agents, it is a human-specific pathogen that survives poorly in the environment and therefore must be able to avoid the host defenses in order to survive.

Infection begins when GC attaches by means of its pili to receptors on mucosal epithelial cells; some of the bacterial cells grow on the surface, but others direct the host cells to engulf them. The intracellular bacteria can then multiply within the protected environment of the epithelial cell or are released to the other side of the mucosal barrier (the submucosa).

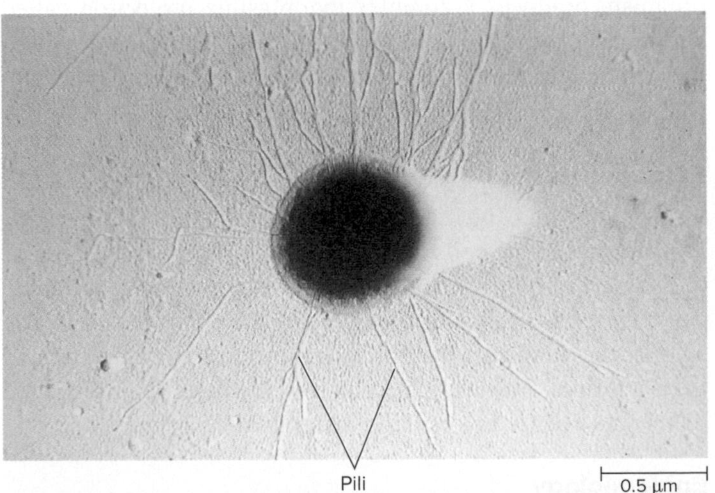

FIGURE 27.10 Pili on Single Coccus of *Neisseria gonorrhoeae* (SEM) Stanley Falkow

? **What are the functions of the pili?**

The symptoms of gonorrhea are primarily due to the intense inflammatory response that results from infection. As the bacterial cells grow, they release outer membrane vesicles (called blebs), which contain LOS. The lipid A component of LOS is recognized by the host cell's pattern recognition receptors (PRRs), leading to the release of pro-inflammatory cytokines. The complement system is also activated. Neutrophils are recruited to the site of infection as a result, and the microbe-destroying substances they release contribute to tissue damage. In addition, peptidoglycan fragments released from gonococcal cells are toxic to nearby ciliated mucosal epithelial cells.

GC persists and multiplies in the host, using several mechanisms to avoid the immune response. Antigenic variation and phase variation of three important antigens—pili, a membrane protein called Opa (for opacity), and the carbohydrate portion of LOS—allow the bacterial cells to avoid being tagged by antibodies. By the time antibodies are produced against one version of any of these antigens, some gonococcal cells will already have switched to making alternative versions. This genetic plasticity not only helps GC evade the immune response, it also allows the bacterial cell population to survive different host environments; for example, certain LOS versions make it easier for GC to attach to host cells. In addition to varying their antigens, the bacterial cells prevent efficient phagocytosis by binding to the host's complement regulatory proteins; recall that these proteins normally protect host cells by inactivating C3b, which could otherwise opsonize those cells (see figure 14.13).

Gonococci have a variety of other features important for survival. Although the bacteria have several ways to avoid phagocytosis, they also appear to stimulate their own uptake by neutrophils. Once engulfed, many of the bacterial cells are killed, but some survive and even seem to multiply within neutrophils, an impressive characteristic considering that neutrophils are the most powerful of the body's phagocytic cells. Several outer membrane proteins allow gonococci to attach to helper T cells, preventing the activation and proliferation of those important cells. The bacteria also produce IgA protease, which destroys secretory IgA and seems to interfere with phagosome maturation. To obtain the iron required for growth, gonococci produce membrane proteins that "steal" the element directly from the host's storage forms (transferrin and lactoferrin). Also, gonococcal cells secrete DNA and are naturally competent, which means they continually exchange genetic information through DNA-mediated transformation. Some strains carry conjugative plasmids—another mechanism of horizontal gene transfer.

Epidemiology

Gonorrhea is one of the most common notifiable bacterial diseases in the United States, second only to chlamydia. The number of reported cases each year has increased over the last decade, with about 700,000 cases in 2021, but the actual numbers are thought to be at least double that. Little or no long-lasting immunity to the disease develops, so a person can contract gonorrhea repeatedly.

N. gonorrhoeae infects only humans, living mainly on the mucous membranes of the genital tract. Most strains are susceptible to UV light, desiccation, and cold temperatures and thus do not survive well outside the host. For this reason, gonorrhea is transmitted almost exclusively by vaginal, oral, and anal sex, as well as other intimate activities that result in close contact with mucous membranes or body fluids. *N. gonorrhoeae* can also be passed from mother to baby during vaginal childbirth.

Treatment and Prevention

N. gonorrhoeae strains are now resistant to many antibacterial medications, including penicillins, tetracyclines, sulfonamides, and fluoroquinolones. Because of this, the CDC lists drug-resistant *N. gonorrhoeae* as an urgent threat (see table 20.2). A single dose of intramuscular ceftriaxone is used for treatment.

Preventing gonorrhea depends on abstinence, monogamous relationships, not sharing sexual devices, and consistent, correct use of condoms. Rapid identification and treatment of sexual contacts help minimize the spread of the disease by decreasing the overall number of infected individuals. No vaccine for preventing gonorrhea has been developed due to antigenic variation in GC and the lack of a suitable animal model for studying the immune response to the disease.

Neonatal conjunctivitis is prevented by screening pregnant women for *N. gonorrhoeae* infection, treating those who are infected, and administering erythromycin-containing ointment into the eyes of newborn infants within 1 hour of birth. As a result of these important preventive measures, this disease of newborns is now unusual in the United States. **Table 27.7** describes the main features of gonorrhea.

TABLE 27.7 Gonorrhea

Signs and Symptoms	Frequently asymptomatic. Females: abnormal vaginal discharge, painful urination; complications can include pelvic pain, infertility, ectopic pregnancy, and arthritis. Males: pain on urination, urethral discharge; complications may include impaired urinary flow and swelling of scrotum.
Incubation Period	Typically 14 days or less
Causative Agent	*Neisseria gonorrhoeae* (also called gonococcus or GC), a Gram-negative diplococcus
Pathogenesis	Organisms attach to certain mucosal epithelial cells by pili; phase and antigenic variation in surface proteins and pili allows attachment to different host cells and escape from immune mechanisms; infection leads to inflammation and scarring.
Epidemiology	Transmitted through sexual contact; many infections are asymptomatic; little or no immunity following recovery.
Treatment and Prevention	Treatment: appropriate antibiotic; widespread resistance limits medication options. Prevention: abstinence or monogamous relationships; correct use of condoms; regular screening can identify infections, thus limiting spread to sexual partners.

Mycoplasma genitalium Infection

Mycoplasma genitalium infection is an emerging STI in men and women worldwide. The causative agent was discovered and identified only in the early 1980s, partly because it cannot be detected with the Gram stain, but also because it is very difficult to culture. Little was known about its disease association until nucleic acid amplification tests (NAATs) were developed for its detection.

Signs and Symptoms

Like many STIs, *M. genitalium* infection is often asymptomatic. The incubation period is not well established, but when signs and symptoms do occur, they are similar to those of chlamydia and gonorrhea and in women include increased vaginal discharge. The infection can progress upward through the female reproductive tract, causing PID. In men, infections typically involve the urethra and, if symptoms occur, typically include urethral discharge and pain during urination.

Causative Agent

M. genitalium, like all mycoplasmas, does not have a cell wall. The organism has a very small genome; as a result, it is fastidious and obtains many of its required nutrients from the host. It grows very slowly, with a generation time of at least 16 hours in the lab, and can take several months to culture in vitro. Even with that small genome, however, the organism produces a complex cytoplasmic projection called a *terminal organelle* at one end of the cell. The structure gives the cells a characteristic bottle shape and plays an important role in gliding motility and colonization. (**figure 27.11**).

Pathogenesis

Using adhesins on its terminal organelle, *M. genitalium* attaches to the cells lining the genitourinary tract. Similar to *Neisseria gonorrhoeae*, the bacterium uses both antigenic variation and phase variation to avoid an immune response against the adhesins. Pattern recognition receptors recognize the bacterium, however, and much of the infection-associated damage is due to the resulting inflammatory response.

Epidemiology

M. genitalium infection may be more common than gonorrhea and chlamydia but is often undiagnosed, partly due to less awareness of the problem; that may start changing now that an

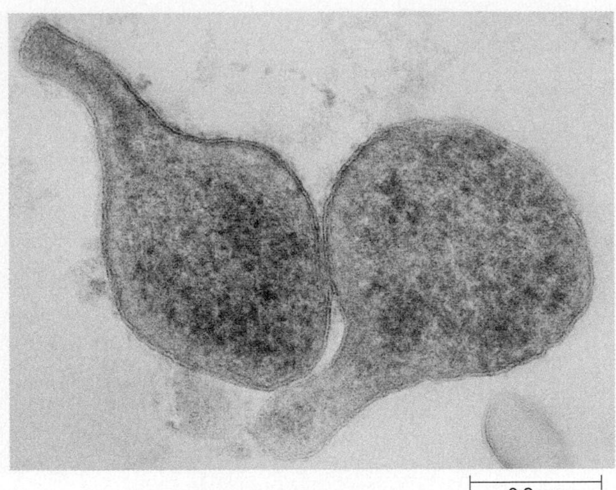

FIGURE 27.11 ***Mycoplasma genitalium*** Notice the distinctive shape caused by the terminal organelle. Thomas Deerinck, NCMIR/Science Source

? Why is it difficult to grow this organism in culture?

FDA-approved commercial diagnostic test is available. As with other sexually transmitted infections, *M. genitalium* is acquired through direct mucosal contact, typically during unprotected sex.

Treatment and Prevention

Suspected *M. genitalium* infections are initially treated with doxycycline, followed by azithromycin; an alternative must be used if the bacterium is resistant to azithromycin. The organism lacks a cell wall, so it has intrinsic resistance to peptidoglycan synthesis inhibitors such as penicillin. As with other STIs, prevention is by abstinence, monogamy, and proper use of condoms. **Table 27.8** summarizes the disease's features.

Syphilis

Syphilis is an example of both the successes and setbacks in disease control. During the first half of the twentieth century, the disease was a major cause of mental illness and blindness, and a significant cause of heart disease and stroke (see A Glimpse of History, chapter 20). Once antibiotics became widely available, however, the number of cases plummeted; in the United States by the mid-1950s, the disease was almost eliminated through a successful program aimed at locating and treating all people with syphilis and their sexual contacts. Unfortunately, the number of cases then increased in the late 1980s due to factors that included prostitution and drug use. Renewed efforts in education, case finding, and treatment reversed that trend so that by the end of 1998, syphilis incidence was at the lowest level since it became a notifiable disease in the early 1940s. Unfortunately, the rates have risen steadily since then, often associated with exchanging sex for drugs, and are now at the highest they have been in three decades; the disease is also emerging as a global health problem.

Signs and Symptoms

Syphilis causes so many different signs and symptoms that it is easily confused with other diseases and is often called "the great imitator." Symptoms occur in defined clinical stages:

- **Primary syphilis.** The stage is characterized by a firm, round, painless red ulcer called a **chancre** (pronounced "shanker") that appears at the site of infection about 3 weeks after exposure (**figure 27.12**). Chancres usually develop on the genitalia but may occur anywhere on the body. Because they are painless, they often go unnoticed, particularly if small or hidden from view in the vagina or the rectum. Lymph nodes near the chancre may swell. The chancre usually heals within 3 to 6 weeks even without treatment.
- **Secondary syphilis.** The symptoms of secondary syphilis usually appear 2 to 10 weeks after the primary stage. The

TABLE 27.8	*Mycoplasma genitalium* Infections
Signs and Symptoms	Frequently asymptomatic. Males: urethral discharge, painful urination. Females: vaginal discharge.
Incubation Period	Not well established
Causative Agent	*Mycoplasma genitalium*, a fastidious bacterium that lacks a cell wall and grows very slowly in culture.
Pathogenesis	Bacterium attaches to host cells by adhesins on terminal organelle. Infection leads to inflammation.
Epidemiology	Transmitted through sexual contact; most infections are asymptomatic.
Treatment and Prevention	Treatment: appropriate antibiotic. Prevention: abstinence or monogamous relationships; correct use of condoms.

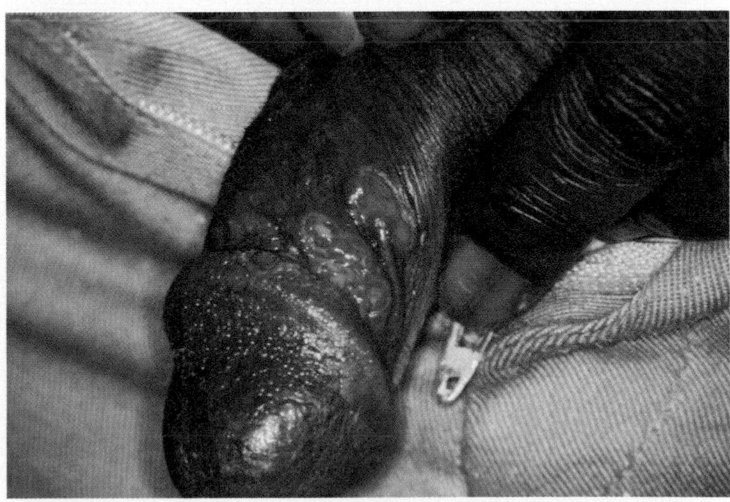

FIGURE 27.12 Syphilitic Chancre A chancre develops at the site where *Treponema pallidum* entered the body, as on the shaft of this penis. M. Rein, VD/CDC

? What is the risk associated with an open ulcer, as shown here?

infection has spread systemically by this time, resulting in an immune response that gives rise to diverse signs and symptoms. The most common is a body-wide rash that involves the palms and soles (**figure 27.13**); white areas referred to as *mucous patches* develop on the mucous membranes. The patient may also experience aches and pains, sore throat, fever, malaise, and weight loss. Occasionally, hepatitis (liver disease) and renal (kidney) disease develop.

- **Latent syphilis.** As the term implies, no signs or symptoms are apparent during this stage; infected people, however, have antibodies to the causative agent. Most people either recover or remain latently infected, but some progress to the next stage of the disease.
- **Tertiary syphilis.** After the latent period that can last for many years, this stage sometimes occurs, resulting in variable signs and symptoms that depend on the area or organ affected. In gummatous syphilis, localized regions of tissue damage develop as a result of prolonged inflammatory responses (**figure 27.14**). These chronic lesions, called **gummas,** are granulomas that can occur anywhere in the body. Cardiovascular syphilis is characterized by aneurysms (bulging of the blood vessel wall) that form in the ascending aorta (a large blood vessel that comes out of the heart).

At any stage of syphilis, involvement of the central nervous system (CNS) can lead to **neurosyphilis.** Early symptoms include headache, nausea, and vomiting, with later development of personality changes, emotional instability, forgetfulness, and sometimes strokes. As CNS damage progresses, damage to the brain results in delusions and hallucinations, speech defects, and abnormalities of the pupils of the eye; spinal cord damage results in difficulty walking and extreme pain in the legs.

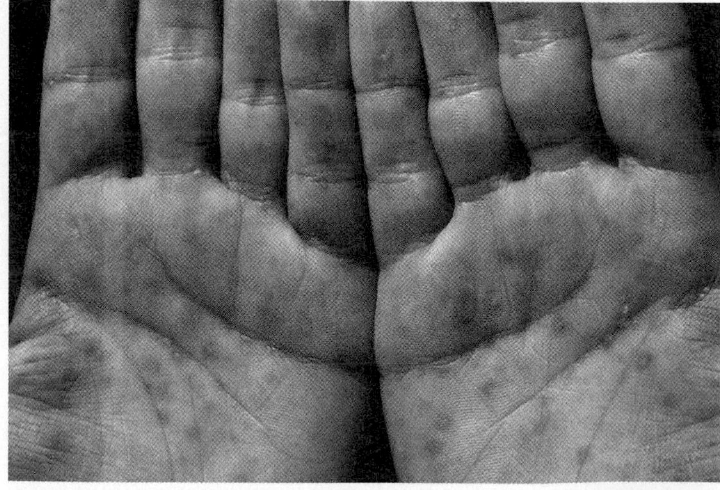

FIGURE 27.13 Secondary Syphilis The most common symptom is a body-wide rash that involves the palms of the hands. Martin M. Rotker/Science Source

What causes the manifestations of secondary syphilis?

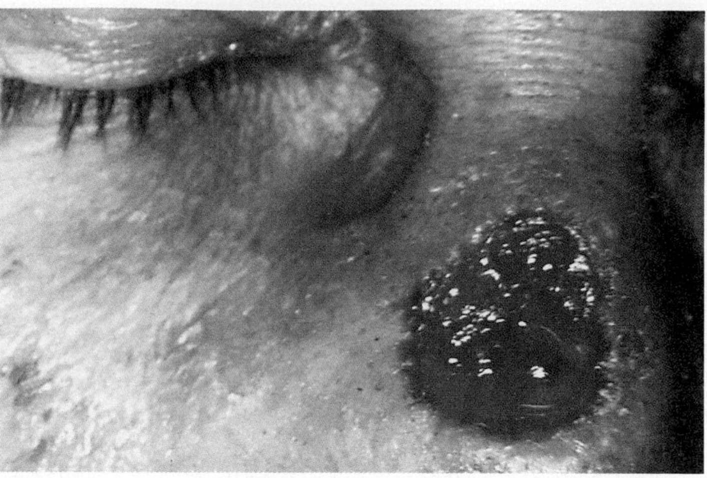

FIGURE 27.14 A Gumma of Tertiary Syphilis Gummas are formed by an inflammatory mass that can perforate tissue, as in this example. J. Pledger/CDC

Where in the body do gummas occur?

In pregnant females, the bacterium that causes syphilis easily crosses the placenta and infects the fetus, causing devastating outcomes including spontaneous abortion, stillbirth, and neonatal death. Newborns with **congenital syphilis** (meaning present at birth) are often asymptomatic, but those that survive eventually develop distinctive signs and symptoms. Some infected infants develop early manifestations within weeks, including liver enlargement, rashes, and severe nasal discharge (snuffles) (**figure 27.15**), but others remain asymptomatic until much later. Without treatment, the child's bones, cartilage, and teeth can become deformed over time, leading to outcomes including bowing of the large bones of the lower legs, sunken nasal bridge (saddle nose), and small, widely spaced, notched permanent teeth (Hutchinson teeth). Eye inflammation and deafness often occur.

Causative Agent

Syphilis is caused by *Treponema pallidum,* a very slender, highly motile spirochete. The organism cannot easily be seen using the Gram stain. Instead, immunofluorescence or various specialized stains that contain silver nitrate are used to make the bacterial cells visible (**figure 27.16**); the organism can also be detected in wet mounts by dark-field microscopy. Like all spirochetes, *T. pallidum* cells have endoflagella that propel them in a corkscrew-like motion, and this allows them to penetrate a wide variety of tissues and organs. In nature, *T. pallidum* infects only humans.

Until recently, *T. pallidum* could not be grown in vitro. Researchers discovered, however, that the bacterium would grow continuously within tissue culture rabbit epithelial cells bathed in a nutrient-rich medium. This method makes studying the organism easier, which should provide better insight into its pathogenic nature.

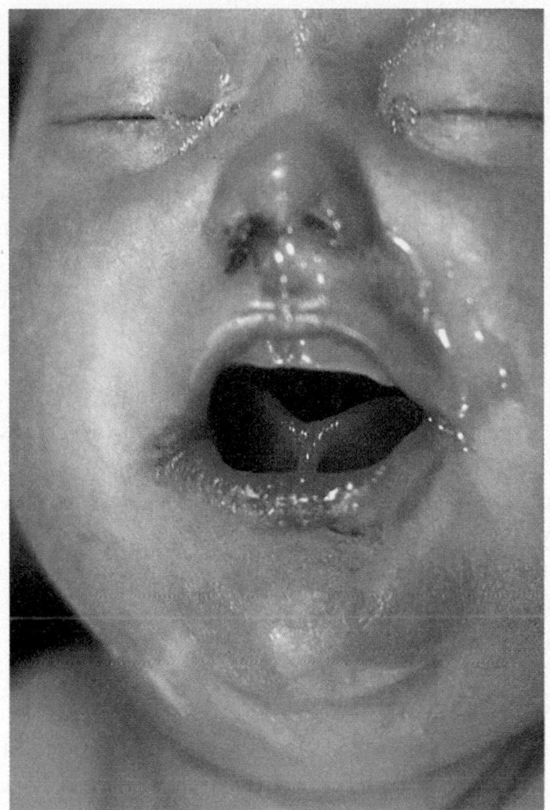

FIGURE 27.15 Congenital Syphilis Severe nasal discharge (snuffles) is one of the early symptoms in infected infants. Centers for Disease Control and Prevention

? What are some other manifestations of congenital syphilis?

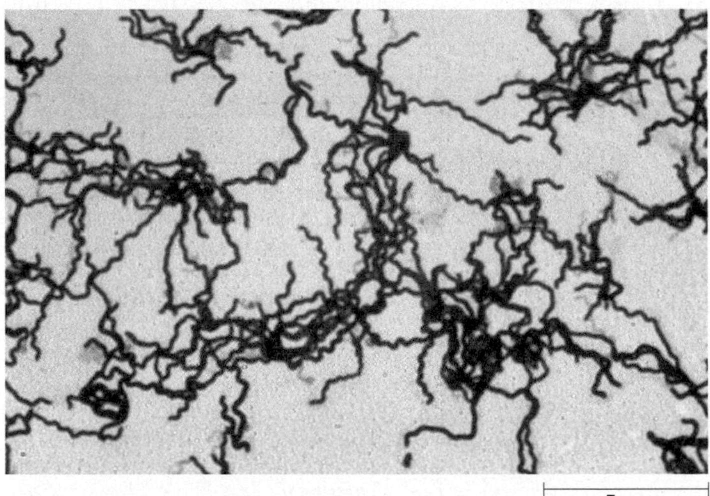

FIGURE 27.16 *Treponema pallidum* Silver stain Melba Photo Agency/Alamy Stock Photo

? In wet mounts, why is dark-field microscopy needed to observe *T. pallidum*?

Pathogenesis

Treponema pallidum easily penetrates mucous membranes and damaged skin. The infectious dose is very low—less than 100 cells. In primary syphilis, *T. pallidum* multiplies in a localized area at the site of infection, usually the genitalia, spreading from there to the lymph nodes and the bloodstream. The chancre that forms is caused by an intense inflammatory response to high numbers of bacterial cells in the lesion. Although the chancre disappears, the organism avoids destruction by the body's defenses, and disease progression can continue for years. The *T. pallidum* outer membrane lacks LPS, the molecule that normally triggers an immune response, and also has few membrane-bound proteins to which opsonizing antibodies could otherwise bind.

Many of the signs and symptoms of secondary syphilis are due to immune complexes that form as specific antibodies bind to circulating *T. pallidum.* By this time, the spirochetes have become systemic, and infectious lesions occur on the skin and mucous membranes in various locations, especially in the mouth.

Tertiary syphilis results from a hypersensitivity (delayed-type hypersensitivity) reaction to small numbers of *T. pallidum* that grow and persist in the tissues. In this stage, the patient is no longer infectious. The organisms may be present in almost any part of the body, and the outcome depends on where the hypersensitivity reactions occur. If they occur in the skin, the bones, or other non-vital areas, the disease is not life-threatening. If, however, they occur within the walls of a major blood vessel such as the aorta, the vessel may become weakened and even rupture, resulting in death. The main characteristics of the stages of syphilis are summarized in **table 27.9.**

Epidemiology

In 1990, the number of reported cases of primary and secondary syphilis in the United States was just over 50,000. Effective control efforts over the next 10 years then demonstrated the power of prevention, with just less than 6,000 reported cases in 2000—a nearly 90% decrease! The disease has now rebounded, however, with over 52,000 reported cases of primary and secondary syphilis in 2021; as with other notifiable diseases, the actual number is thought to be at least twice that. Alarmingly, the number of congenital syphilis cases in the United States

TABLE 27.9 Stages of Syphilis

Stage of Disease	Main Characteristics	Infectious?
Primary	Firm, painless ulcer (chancre) at site of infection; lymph node enlargement	Yes
Secondary	Aches and pains; rash and mucous membrane lesions	Yes
Latent	No signs or symptoms	Early—yes; Late—no
Tertiary	Gummas occur anywhere in the body; damage to large blood vessels	No

has also been on the rise, with over 2,600 reported cases in 2021.

Syphilis is most commonly transmitted by sexual intercourse, but infection can occur from any contact with a *T. pallidum*-containing lesion; thus, kissing or touching a chancre, skin rash, or mucous patch of an infected person can result in transmission.

Syphilis-associated fetal risk is affected by several factors, many of which are poorly understood. Fetal infections are most likely to occur when high numbers of *T. pallidum* circulate in the maternal bloodstream, typically during the primary, secondary, or early latent stages of the mother's disease. Fetal infections acquired at any stage of the pregnancy can lead to spontaneous abortion or stillbirth, and those acquired during later stages can result in the damage associated with signs and symptoms of congenital syphilis; that damage is thought to be due to the fetal immune response.

Treatment and Prevention

All stages and types of syphilis can be treated with one or more injections of penicillin. Primary, secondary, and latent syphilis are easily treated, but tertiary syphilis treatment takes longer and only limits disease progression—it has little effect on the existing tissue damage. If congenital syphilis is treated within the first 3 months of birth, some (but not all) of the later manifestations are prevented; the infants and children are also monitored to assess their development.

Abstinence, monogamous relationships, and correct use of condoms decrease the risk of contracting the disease. Congenital syphilis can be prevented by diagnosing and treating the mother's syphilis. The CDC recommends screening all pregnant women at the first prenatal visit, with repeated screenings for high-risk women during pregnancy and just before delivery.

Quick identification and treatment of sexual contacts are important in limiting the spread of the disease. **Table 27.10** summarizes the main features of syphilis.

Chancroid

Chancroid is another bacterial STI that, like syphilis, causes genital sores. The disease is most common in resource-limited countries, usually associated with sex workers. Certain strains of the bacterium that causes chancroid were recently found to be a frequent cause of ulcerative skin sores in children and young adults in some tropical countries; the relationship of these strains to ones that cause chancroid is still being studied.

Signs and Symptoms

Chancroid is characterized by one or more painful soft genital ulcers (**figure 27.17**). Typically, these begin as a small pimple at the site of bacterial entry through the skin, but they then ulcerate and enlarge. The lymph nodes in the groin also become large and tender. Sometimes the nodes are pus-filled and may rupture, releasing the pus. In females, ulcers typically occur on the thighs or labia, causing pain during urination and sexual intercourse.

TABLE 27.10 Syphilis

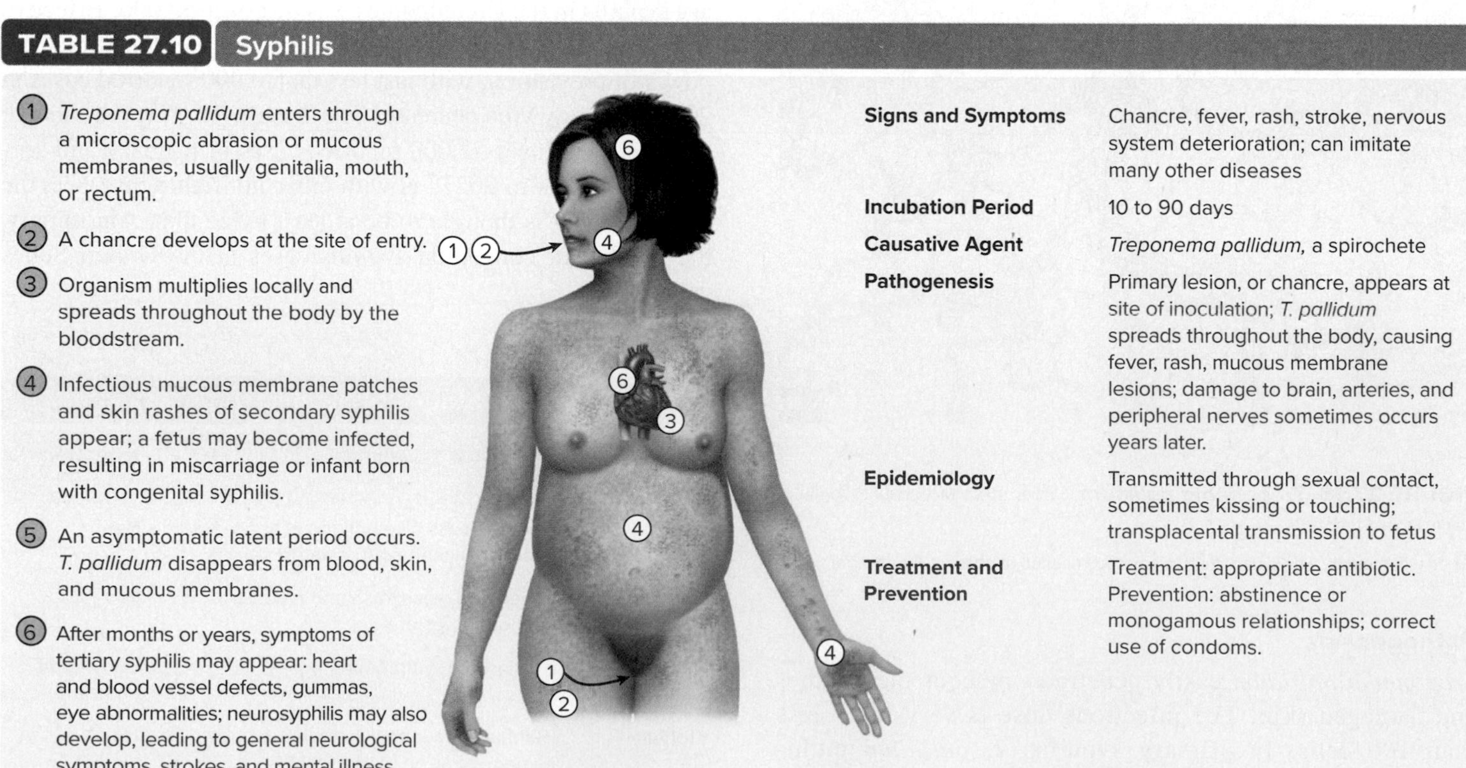

1 Treponema pallidum enters through a microscopic abrasion or mucous membranes, usually genitalia, mouth, or rectum.
2 A chancre develops at the site of entry.
3 Organism multiplies locally and spreads throughout the body by the bloodstream.
4 Infectious mucous membrane patches and skin rashes of secondary syphilis appear; a fetus may become infected, resulting in miscarriage or infant born with congenital syphilis.
5 An asymptomatic latent period occurs. T. pallidum disappears from blood, skin, and mucous membranes.
6 After months or years, symptoms of tertiary syphilis may appear: heart and blood vessel defects, gummas, eye abnormalities; neurosyphilis may also develop, leading to general neurological symptoms, strokes, and mental illness.
6
1 2
4
6
3
4
4
1
2

Signs and Symptoms	Chancre, fever, rash, stroke, nervous system deterioration; can imitate many other diseases
Incubation Period	10 to 90 days
Causative Agent	*Treponema pallidum*, a spirochete
Pathogenesis	Primary lesion, or chancre, appears at site of inoculation; *T. pallidum* spreads throughout the body, causing fever, rash, mucous membrane lesions; damage to brain, arteries, and peripheral nerves sometimes occurs years later.
Epidemiology	Transmitted through sexual contact, sometimes kissing or touching; transplacental transmission to fetus
Treatment and Prevention	Treatment: appropriate antibiotic. Prevention: abstinence or monogamous relationships; correct use of condoms.

FOCUS YOUR PERSPECTIVE 27.1

The Painful Past of Syphilis Research

For 40 years (1932 to 1972), the U.S. Public Health Service conducted a study (now known as the Tuskegee Syphilis Experiment) of 399 Black men in Alabama with advanced syphilis. The study was done to assess the natural progression of the disease in Black men, but the men were neither told of their diagnosis nor were they educated about the disease and its transmission. For the first half of the study period, there were no effective medications against syphilis, and so the men remained untreated. Even after effective therapy became available, however, the study was continued without treating the men, in a blatant mix of bad science and racism. Twenty-eight of the men died of the disease, and another hundred died from its complications. Forty of their wives became infected, and 19 children were born with congenital syphilis. Many people remain bitter and distrustful of the U.S. Public Health Service as a result. In 1997, President Clinton, speaking for the government, formally apologized to the surviving eight subjects.

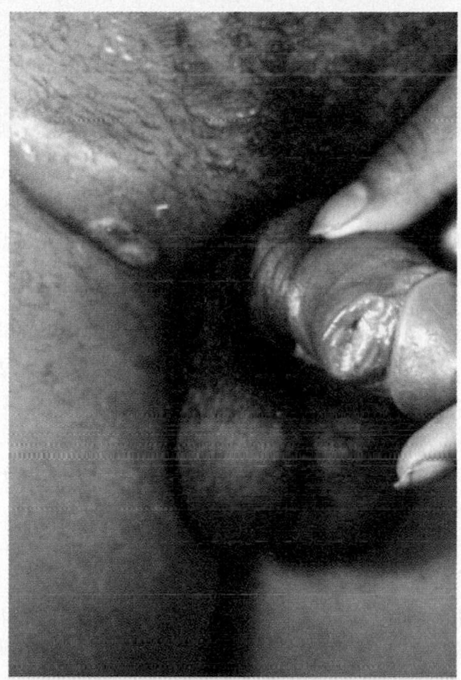

FIGURE 27.17 Chancroid A genital ulcer and a ruptured lymph node can be seen in this patient. Dr. Pirozzi/CDC

? How are chancroid ulcers different from syphilis chancres?

Causative Agent

Chancroid is caused by *Haemophilus ducreyi,* a fastidious, pleomorphic, Gram-negative coccobacillus that can be cultivated only on a rich medium such as chocolate agar.

Pathogenesis

H. ducreyi infection causes an intense inflammatory response, with accumulation of neutrophils and macrophages in the area. The bacterial cells prevent the phagocytes from engulfing them, however, and also interfere with the function of activated complement proteins. In addition, the bacteria produce a cytotoxin that kills local epithelial cells, an action that, together with the inflammatory response, results in the characteristic ulcers.

Epidemiology

Although only 10 or fewer cases of chancroid are reported each year in the United States, outbreaks associated with prostitution have occurred. The disease may occur more often than recorded because the bacterium is difficult to detect, potentially leading to misdiagnosis.

As with other sexually transmitted infections, *H. ducreyi* is acquired through direct contact, typically during unprotected sex. Although endemic in many regions of the world, the global incidence is unknown because of the lack of simple culturing and diagnostic techniques.

Treatment and Prevention

Chancroid is typically treated with a single oral dose of azithromycin or a single injection of ceftriaxone. As with other STIs, chancroid can be prevented by abstinence from sexual intercourse, monogamous relationships, and correct use of condoms. **Table 27.11** gives the main features of the disease.

TABLE 27.11 Chancroid

Signs and Symptoms	One or more painful, gradually enlarging, soft genital ulcers; large, tender regional lymph nodes
Incubation Period	3 to 10 days
Causative Agent	*Haemophilus ducreyi*, a pleomorphic, fastidious Gram-negative rod requiring special growth media
Pathogenesis	A small pimple appears first, which ulcerates and gradually enlarges; multiple lesions may join together; lymph nodes enlarge and may rupture.
Epidemiology	Transmitted through sexual contact
Treatment and Prevention	Treatment: appropriate antimicrobial medication. Prevention: abstinence or monogamous relationships, correct use of condoms.

MicroAssessment 27.4

Transmission of STIs usually requires direct person-to-person contact. Even asymptomatic infections can cause genital tract damage (including PID) and can be spread to other people. STIs in pregnant women can be a serious risk to fetuses and newborns. Chlamydia, gonorrhea, and *Mycoplasma genitalium* infections are often asymptomatic but can cause similar signs and symptoms, including genital discharge and painful urination. Syphilis can imitate other diseases; untreated, it may lead to stroke and nervous system deterioration. Chancroid is associated mainly with sex workers but is rare in the United States.

10. Can a baby have congenital syphilis without its mother ever having had signs and symptoms of syphilis? Explain.
11. Why is it difficult to diagnose *Mycoplasma genitalium* infections?
12. Why does scarring of a fallopian tube increase the risk of an ectopic pregnancy?

27.5 ■ Viral STIs

Learning Outcomes

7. Compare the infections that cause genital herpes, genital warts, and certain cancers.
8. Outline the relationship between HIV infection and AIDS.

Viral STIs are more common than bacterial STIs and cannot be cured with medications. As mentioned in the last section, however, all STIs have similar epidemiology and prevention, topics covered in the box Focus on Sexually Transmitted Infections.

Human Papillomavirus STIs: Genital Warts and Cancers

Sexually transmitted human papillomaviruses (HPVs) are the most common STI agents. Some HPV types cause **papillomas**—warty growths of the external and internal genitalia; other types can cause cervical, penile, anal, and oral cancer.

Signs and Symptoms

Genital HPV infections are usually asymptomatic and self-limiting, but if symptoms do occur, warts are the most easily recognized. The warts can appear months after infection—developing at the vaginal opening, around the anus, or on the shaft or head of the penis (**figure 27.18**). Depending on the viral type and the location of the infection, the lesions can be flat or raised, cauliflower-like, or hidden within the epithelium; occasionally, they become inflamed and bleed. Even if warts are removed, HPV persists in surrounding normal-looking epithelium and can cause additional lesions. Warts sometimes partly block the urethra or, if they are very large, the birth canal.

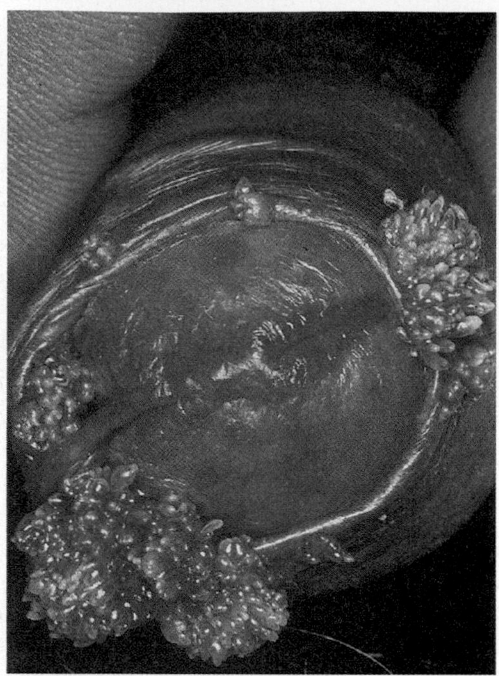

FIGURE 27.18 Genital Warts Warts on the penis are the result of human papillomavirus infection. Biophoto Associates/Science Source

Why do individuals sometimes get additional warts after initial warts are removed?

Certain HPV types can cause cervical cancer. Precancerous lesions (areas of abnormal cell growth associated with the development of cancer) are generally asymptomatic and can be detected only by examining the cervical tissues. The HPV types that cause cervical cancer can also cause cancers of the mouth, penis, vagina, and anus.

Causative Agents

Human papillomaviruses are non-enveloped, double-stranded DNA viruses of the *Papillomaviridae* family. The more than 100 HPV types are tissue-specific; for example, HPV types that cause the common dermal warts of the hands and feet generally do not infect the genitalia. At least 40 HPV types are transmitted by sexual contact. Those that cause genital warts rarely cause cancer and thus are called low-risk HPVs; they include types 6 and 11. The types associated with cancer of the cervix, rectum, vagina, penis, and throat are known as oncogenic (cancer-causing) high-risk HPVs (hrHPV) and include types 16, 18, 31, and 45.

Pathogenesis

HPVs enter and infect the deeper layers of epithelium through microscopic abrasions, where they can infect a self-renewing cell type called a basal cell. As these cells divide, they can either become replacement cells (self-renewing basal cells) or differentiate to become part of the superficial epithelial layer. The viruses replicate in the differentiating cell type—sometimes giving rise to warts—and are shed along with the outermost layer. The viruses can also exist as a persistent form in the self-renewing basal cells, usually as extrachromosomal, closed circular DNA. Thus, the viruses can give rise to more viral particles (in the differentiating cells) as well as persist in the host (in the self-renewing basal cells). Most HPV infections are cleared by the immune response, but sometimes the genome of hrHPVs integrates into the host cell chromosome to become a permanent part of that cell. The viruses can be oncogenic because they code for a protein that allows excessive cell growth. Although hrHPVs are present in more than 90% of cervical cancers, only a small percentage of HPV infections result in cancer, indicating that other factors must be involved.

Epidemiology

HPV infection is not a notifiable disease in the United States, but studies indicate that in 2018, about 13 million people in the age group 15 to 59 became infected with a sexually transmitted HPV type associated with genital warts or cancer. An estimated 42.5 million people in that age group had such an infection—new or existing—at any given point during that year.

HPVs can be spread by vaginal, anal, or oral sex, and even with close skin-to-skin sexual contact. Significantly, transmission can occur even when symptoms are not present. The infection is the most common reason for an abnormal Pap smear (discussed shortly) in teenage females. Approximately 80% of all cancers identified in the oropharynx or anogenital regions are linked to a current HPV infection. The risk of transmission of HPV to a neonate during birth appears to be low.

Treatment and Prevention

Warts may spontaneously clear over time, but several treatments are available to remove them or hasten their disappearance. Unfortunately, none of these methods cures the infection, so warts may still recur. Dermal and external genital warts can be removed by laser treatment, freezing with liquid nitrogen, surgical excision, or acid treatment using trichloroacetic acid. External genital warts can also be treated with topical medications that either stimulate the host cell antiviral response (imiquimod) or interfere with cell replication (podofilox and sinecatechin).

Treatments for HPV-associated cancers include surgery, radiation, and chemotherapy, but a more effective strategy is to identify and remove precancerous lesions. The CDC recommends that females with an average risk of acquiring HPV should have regular cervical cancer screenings starting at age 21. In the past, the only available test was the Pap smear, a cytology-based test (cytology is the study of cells) in which cells are removed from the cervix, stained, and examined for any abnormalities that indicate precancerous lesions (**figure 27.19**). More recently, however, a nucleic acid–based test to detect hrHPVs became available. Depending on the situation, the two types of tests may be used alone, at the same time (co-testing), or sequentially. Patients with abnormal results should be monitored, and the lesion may be removed to reduce the cancer risk.

HPV infections can be prevented by abstinence or monogamous relationships. Condoms do not provide complete protection against HPVs because the viruses can be transmitted by exposure to areas not covered by the condom.

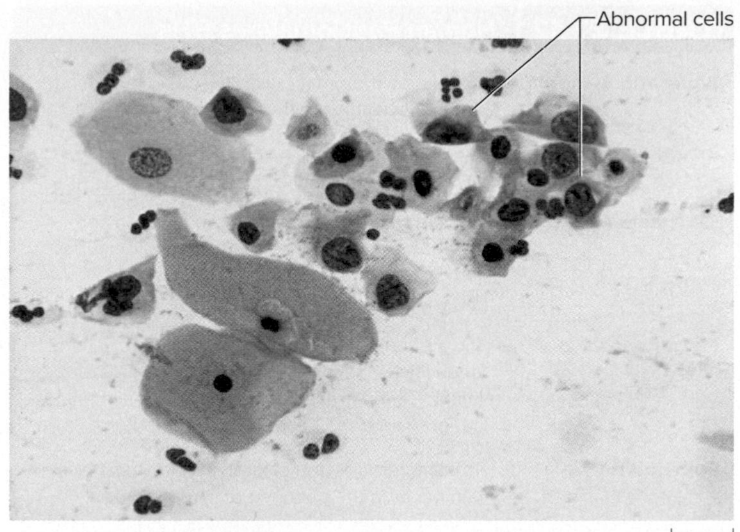

FIGURE 27.19 Abnormal Pap Smear The pink and blue cells are squamous epithelial cells; abnormalities include double nuclei that have a clear area around them. Frederick C. Skvara, M.D.

What is the most frequent cause of an abnormal Pap smear in young women?

The vaccine Gardasil 9 (9vHPV) helps protect against nine different HPV types, including the four types identified in most HPV-associated cancers (HPV types 16, 18, 31, and 45) and the two types that cause more than 90% of genital warts (HPV types 6 and 11). The vaccine is part of the routine immunization schedule for adolescents at age 11 or 12, although it can be given to children as young as 9 years old. It is also recommended for teens and young adults up to age 26 who have either not started or not completed the vaccine series. The number of doses given is age-dependent; a two-dose series is used for people under age 15, and a three-dose series is used for people 15 and older. **Table 27.12** summarizes some important features of human papillomavirus STIs.

MicroByte

Within 10 years after the first HPV vaccine was introduced in 2006, the number of infections caused by the HPV types in that vaccine decreased by about 86% in females ages 14 to 19 years old.

Genital Herpes

Genital herpes is one of the more frequently identified STIs. Unlike most HPV infections, however, genital herpes cannot be cleared by the immune system and therefore remains for life.

Signs and Symptoms

Infections by the viruses that cause genital herpes are often asymptomatic or so mild that they go unnoticed. When symptoms do occur during the initial (primary) infection, they usually begin about a week after exposure (range 2 to 12 days)

TABLE 27.12 Human Papillomavirus STIs

Signs and Symptoms	Usually asymptomatic; warts of the external and internal genitalia are the most noticeable symptom.
Incubation Period	Weeks or months for warts; perhaps years for HPV-associated cancers
Causative Agent	Human papillomaviruses (many types): non-enveloped DNA viruses of the papillomavirus family; the various types are tissue- and lesion-specific.
Pathogenesis	Virus enters epithelium through abrasions, infects deep layer of epithelium to replicate and establish latency; cancer-associated viral types sometimes integrate into the host cell chromosome.
Epidemiology	Transmitted through sexual contact; transmission can occur in the absence of symptoms.
Treatment and Prevention	Treatment: remove warts or hasten their disappearance; remove precancerous lesions. Prevention: abstinence or monogamous relationships; correct use of condoms to minimize transmission; vaccine to protect against the most common cancer- and wart-causing strains.

and include itching and a burning sensation at the site of infection and sometimes pain. Groups of small red bumps may then appear on or around the genitalia or anus, depending on the site of infection. These bumps become vesicles (blisters) surrounded by redness (**figure 27.20**). The vesicles break in a few days, leaving an ulcerated area that slowly dries and becomes crusted before eventually healing. Some cases are accompanied by urethral or vaginal discharge, headache, fever, muscle pain, and in the case of females, painful urination. The symptoms usually disappear within 3 weeks, but the virus becomes latent and can reactivate to again cause lesions; some people never experience these recurrences, but others have them throughout life. The repeated episodes are usually not as severe as the primary infection, are of shorter duration, and may occur less frequently over time.

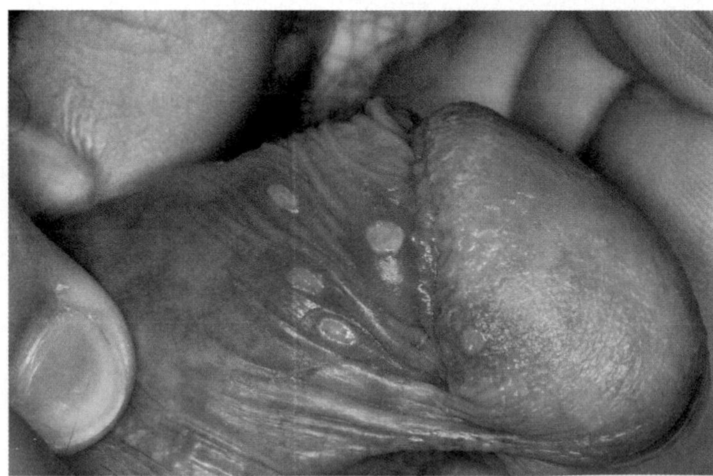

FIGURE 27.20 Genital Herpes Lesions on the Shaft of the Penis Dr. P. Marazzi/Science Source

? Why can the signs and symptoms of genital herpes recur?

One of the most serious complications of genital herpes is neonatal herpes. If a pregnant female near the time of delivery has genital herpes, particularly a primary infection, the infectious agent can potentially be transmitted as the baby passes through the birth canal. The outcome of neonatal herpes is variable, but about one-quarter of those infections become systemic. Disseminated neonatal herpes is often fatal, and children who survive may be left with permanent neurological abnormalities, including developmental delay and blindness.

Causative Agents

Genital herpes is often caused by herpes simplex virus type 2 (HSV-2), an enveloped, double-stranded DNA virus of the *Herpesviridae* family. That type has long been regarded as the most common cause, but herpes simplex virus type 1 (HSV-1)—the cause of cold sores—is increasingly associated with genital herpes as well; recent studies have implicated it in over half of new cases. The two virus types look the same, and their genomes are about 50% homologous.

Pathogenesis

Genital herpes lesions begin when groups of infected epithelial cells lyse following viral replication, creating small, fluid-filled vesicles containing numerous infectious virions. The vesicles burst, producing painful ulcers. The virus is shed from the genital region during the primary infection and continues to be shed intermittently thereafter, even in the absence of symptoms. The frequency of shedding decreases over time, and is affected by the viral type (HSV-1 appears to be shed less frequently than HSV-2).

HSV remains latent in a cell by preventing transcription of most viral genes; the only gene expressed encodes a small RNA segment—a microRNA (miRNA)—that allows the virus to remain silent. In this state, the viral DNA exists within nerve cells in a circular, non-infectious form, causing no symptoms. At times, however, the entire viral chromosome can be transcribed and complete infectious virions produced. This reactivation can lead to viral shedding with or without symptom recurrence.

Epidemiology

HSV infection is not a notifiable disease in the United States, but estimates indicate that about 572,000 people in the age group 15 to 49 years acquired an HSV-2 infection in 2018. At any given point during that year, an estimated 18.6 million people in that age group had an HSV-2 infection—new or existing. The number of new and existing cases of genital herpes caused by HSV-1 is not known.

Sexual transmission of HSV is most likely to occur during the first few days of symptomatic disease but can also happen in the absence of symptoms; once infected, a person is forever at risk of transmitting the virus to someone else.

HSV remains infectious for short periods on fomites or in water, but non-sexual transmission is rare.

The risk of transmitting HSV to a neonate during birth is greatest during a primary infection. If the mother has recurrent symptoms, the risk is very low, presumably because maternal anti-HSV antibodies protect the baby.

Treatment and Prevention

Herpes has no cure, but anti-HSV medications such as acyclovir and valacyclovir can decrease the severity of the primary infection and the incidence of recurrences.

Genital herpes can be prevented by abstaining from sex or maintaining a monogamous relationship with a partner who is known to be uninfected. Consistently using condoms along with HSV-inactivating spermicide, while also avoiding sexual activities during active signs or symptoms, reduces exposure. However, using a condom does not always give full protection because herpes lesions can occur on parts of the genitalia that a condom does not cover.

Neonatal herpes can be prevented by eliminating contact with the infected birth canal, so babies of women who have a primary infection are delivered by cesarean section. The main features of genital herpes are presented in **table 27.13.**

HIV Infection and AIDS

AIDS was first recognized in the United States in 1981 when a number of young, previously healthy homosexual and bisexual males developed *Pneumocystis* pneumonia, a disease caused by a fungus of such low virulence that it seldom affects healthy people. As more people developed unusual opportunistic infections for no apparent reason (**table 27.14**), the CDC began using the acronym AIDS (acquired immunodeficiency syndrome) to describe the unexplained increased susceptibility that indicated underlying immunosuppression. Fairly soon after AIDS was recognized, scientists identified the causative agent: human immunodeficiency virus (HIV).

TABLE 27.13 Genital Herpes

Signs and Symptoms	Itching, burning pain at the site of infection, painful urination, tiny vesicles; vesicles break, leaving painful superficial ulcers, which heal without scarring; recurrences may occur. In neonates, herpes infections can become systemic.
Incubation Period	About 1 week (range, 2 to 12 days)
Causative Agents	Herpes simplex virus type 2 and increasingly, herpes simplex type 1 (the cold sore virus); herpesviruses are enveloped, double-stranded DNA viruses.
Pathogenesis	Lysis of infected epithelial cells results in fluid-filled vesicles containing infectious virions; vesicles burst, causing a painful ulcer; the acute infection is controlled by immune defenses; viral genome persists within nerve cells in a non-infectious form; reactivated virus may be shed with or without symptoms.
Epidemiology	Transmission through sexual contact; transmission can occur in the absence of symptoms.
Treatment and Prevention	Treatment: no cure; anti-HSV medications help prevent recurrences, shorten duration of symptoms. Prevention: abstinence or monogamous relationships; correct use of condoms to minimize transmission.

TABLE 27.14 Opportunistic Infections Characteristic of AIDS Patients

Infection	Causative Agent
Bacterial Infections	
Disseminated *Mycobacterium avium* complex (MAC) infection	*Mycobacterium avium* complex
Tuberculosis	*Mycobacterium tuberculosis*
Fungal Infections	
Candidiasis (thrush)	*Candida albicans*
Coccidioidomycosis	*Coccidioides immitis*
Cryptococcosis	*Cryptococcus* species
Histoplasmosis	*Histoplasma capsulatum*
Pneumocystis pneumonia	*Pneumocystis jirovecii*
Protozoan Infections	
Cryptosporidiosis	*Cryptosporidium parvum*
Toxoplasmosis	*Toxoplasma gondii*
Viral Infections	
Cytomegalovirus disease	Cytomegalovirus

Signs and Symptoms

The terms HIV and AIDS are not synonymous, and the distinctions are very important. AIDS is the last of several stages of HIV infection (**figure 27.21**):

- **Acute infection.** This initial stage includes the first 6 weeks of infection, a time during which HIV replicates to high levels. The virus infects helper T cells (CD4 lymphocytes) and certain other cells that have the CD4 protein on their surface, leading to the destruction of some of these cells. During this stage, some patients experience **acute retroviral syndrome (ARS)**—characterized by temporary flu-like symptoms, including fever, head and muscle aches, sore throat, enlarged lymph nodes, and a generalized rash. The immune system eventually begins controlling the infection, leading to decreased viral levels; at the same time, the number of CD4 lymphocytes in the bloodstream increases as the cells are replaced.
- **Clinical latency.** During this stage, the virus persists without causing symptoms—a period of chronic but

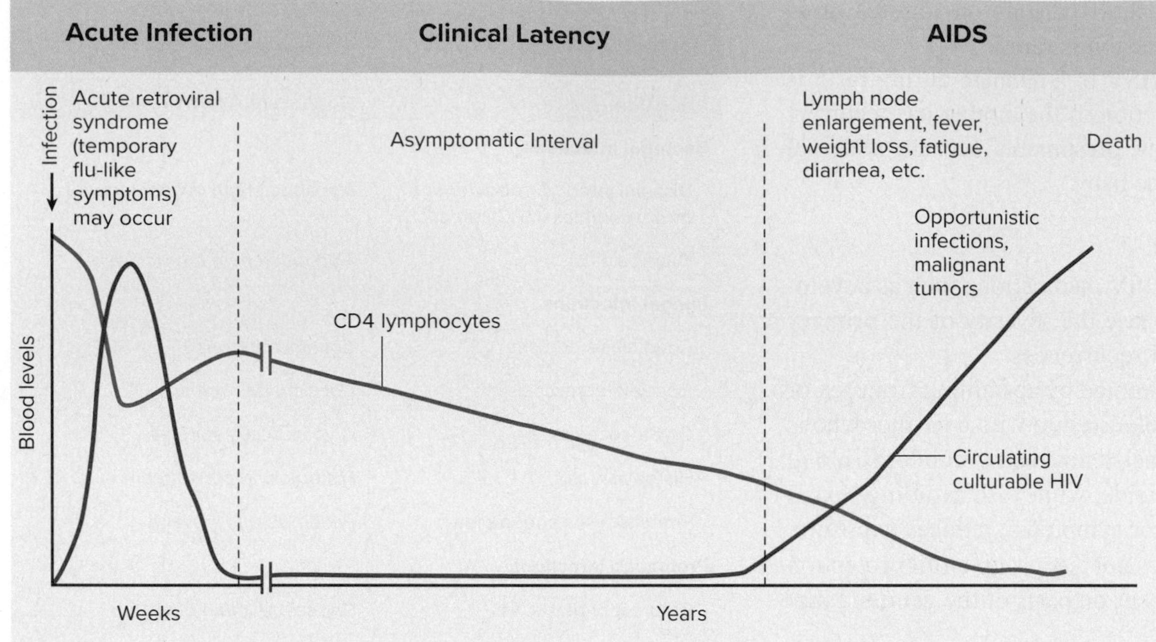

FIGURE 27.21 Stages of HIV Infection During acute infection, viral levels are very high. In the clinical latency period, the disease continues to progress, as shown by the falling CD4 lymphocyte (helper T cell) count. The onset of AIDS is marked by a very low CD4 cell count.

Why would a high viral set point lead to faster progression of the disease?

clinically silent HIV infection. The viral load (amount of virus in the bloodstream) remains at a relatively stable level during this stage; that level, called the **HIV (viral) set point,** serves as a predictor of disease progression—the higher the viral set point, the more quickly AIDS usually develops. Throughout clinical latency, the virus continues infecting cells, resulting in the gradual loss of CD4 lymphocytes, which is significant because of the central role these cells play in the body's adaptive immune response (see figure 15.1).

- **AIDS.** When the number of CD4 lymphocytes drops to a very low level (below 200 cells per microliter of blood), the severe immunodeficiency that characterizes AIDS results. As the immune system stops functioning properly, the viral load increases dramatically, and the patient begins experiencing a variety of cancers and opportunistic infections (see table 27.14). Most people with AIDS also suffer from weight loss, fever, fatigue, and diarrhea. AIDS is fatal without treatment.

Causative Agent

The human immunodeficiency virus (HIV) is a retrovirus, the common name for members of the *Retroviridae* family. Like all retroviruses, HIV is enveloped and has duplicate copies of a single-stranded RNA genome. Two distinct types of HIV occur (HIV-1 and HIV-2), and these have only about 50% genetic similarity. Most AIDS cases are caused by HIV-1, so we will focus on that type (HIV-2 is found mainly in West Africa and causes a less severe form of disease). Based on HIV-1 genome sequence comparisons, many different strains exist, and these can be divided into at least four general groups; by far the most common is group M (for "major"), and within this group are subtypes (also called clades).

Various proteins make up the HIV structure, and these are important for understanding the HIV replication cycle, particularly the role of the proteins that anti-HIV medications target. Some of the proteins also provide the basis for antigen and antibody tests used to diagnose infection (see figure 17.11). The HIV proteins are often referred to by abbreviations that indicate the functional role or location of the protein or a term that indicates the size of the protein (or glycoprotein) in kilodaltons (**figure 27.22**). The most important HIV protein structure from the standpoint of initial infection is Env—the spike that projects from the viral surface. It is made up of two types of glycoprotein subunits: gp41, which anchors the spike to the envelope, and gp120, which rests on gp41 like a cap and attaches to host cells. The capsid protein (CA or p24), which makes up the viral capsid (shell), is the most abundant protein in the virus and the target of a recently approved anti-HIV medication; in addition, its serum levels can be measured to detect early HIV infection. Sandwiched between the capsid shell and the envelope is a layer of matrix protein (MA or p17), which functions in stabilizing the virion. HIV viral particles carry three important virally encoded enzymes that are targets of anti-HIV medications:

- **Reverse transcriptase (RT).** This is an RNA-dependent DNA polymerase that uses the viral RNA genome as a template as part of the process to make a double-stranded DNA version of that viral genome (see figure 13.13).
- **Integrase (IN).** This inserts the double-stranded DNA copy of the viral genome into the host cell genome.
- **HIV protease (PR).** This cleaves virally encoded polyproteins into individual functional components that are required to produce more viral particles.

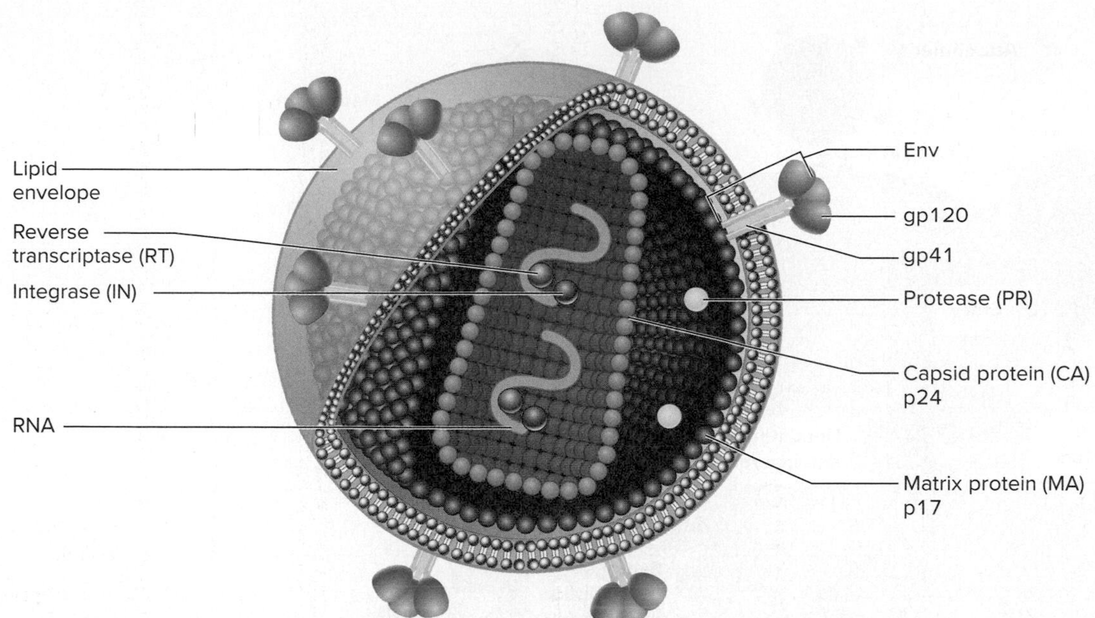

FIGURE 27.22 Human Immunodeficiency Virus Type 1 (HIV-1) Diagrammatic representation of a viral particle showing important components.

? **What is the function of reverse transcriptase?**

The functions of the HIV components are summarized in **table 27.15.**

Pathogenesis

HIV can infect several human cell types, but the most significant are helper T cells. These cells, also referred to as CD4 T cells, have the CD4 surface protein that HIV uses as a receptor for attachment. Monocytes, including macrophages and dendritic cells, also have CD4 and can therefore be targeted by HIV as well. **Figure 27.23** illustrates the key steps of the HIV replication cycle:

1. **Attachment.** HIV uses Env (the spike) to bind to the host cell CD4 protein. That interaction causes the shape of Env to change slightly so that it can then bind a co-receptor—either CCR5 or CXCR4 (cytokine receptors on the host cell surface). CCR5 is generally most important in the initial infection, and people who have lower than normal amounts on their cells are much less susceptible to HIV infection.

TABLE 27.15 HIV Components

Protein	Function
Capsid protein (CA or p24)	Multiple copies together form the capsid shell that protects viral nucleic acid
Env; composed of gp120 and gp41	Contacts the host cell and facilitates viral entry into the cell
Matrix protein (MA or p17)	Stabilizes the virion
Reverse transcriptase (RT)	Enzyme that uses the viral RNA genome as a template to make DNA
Integrase (IN)	Enzyme that inserts a DNA copy of the viral genome into the host genome
Protease (PR)	Enzyme involved in viral protein processing

2. **Entry.** Binding of Env to a co-receptor allows a component of Env to insert into the host cell membrane. In turn, this allows the viral envelope to fuse with the host cell membrane, thus releasing the nucleocapsid (capsid together with the nucleic acid) into the host cell. The nucleocapsid, or at least most of it, then begins its transport to the host cell nucleus.
3. **Reverse transcription.** As the nucleocapsid moves to the host cell nucleus, the viral enzyme reverse transcriptase uses the HIV single-stranded RNA genome as a template to make a double-stranded DNA version. This process—reverse transcription—occurs within the HIV capsid, thus shielding the viral RNA genome from detection by the host cell's cytoplasmic defenses. Because reverse transcriptase makes frequent mistakes during the process, the nucleotide sequence of the DNA copy is slightly different from that of the parent molecule.
4. **Uncoating and insertion.** Uncoating (disassembly of the capsid shell) releases the DNA copy, but details of the process are still being investigated; early studies suggested that uncoating occurs in the host cell cytoplasm, but recent evidence indicates that the final steps occur in the host cell nucleus. The viral enzyme integrase inserts the DNA copy of the HIV genome randomly into the host cell's DNA; the provirus (integrated DNA copy of the viral genome) is a permanent part of the cell's genome.
5. **Synthesis. (a)** In active host cells, transcription of the provirus produces viral mRNA and viral genome RNA. **(b)** Translation of the viral mRNA then generates viral proteins and polyproteins; the polyproteins must be cleaved by viral protease to release the functional proteins.

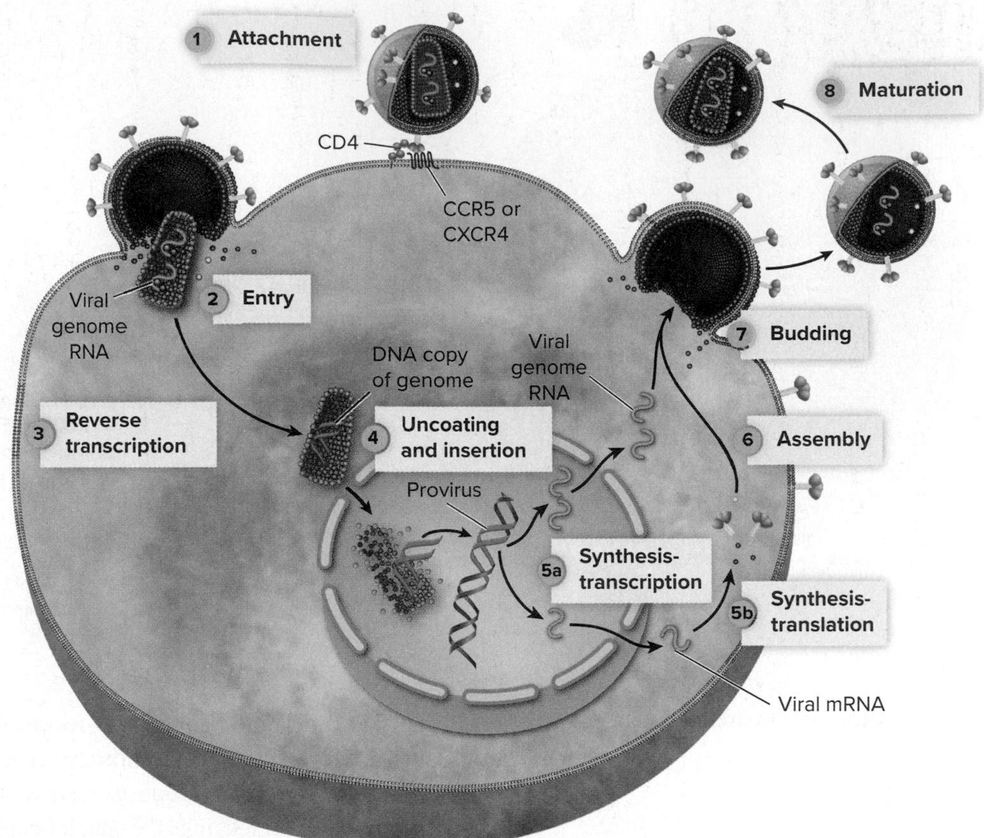

FIGURE 27.23 HIV Replication (1) HIV attaches to CD4 and then either CCR5 or CXCR4. **(2)** The viral envelope fuses with the host membrane, thus releasing the HIV nucleocapsid into the host cell. **(3)** Within the nucleocapsid, reverse transcriptase makes a DNA copy of the HIV RNA genome. **(4)** The final steps of uncoating release the DNA copy of the HIV genome into the host cell nucleus; integrase inserts that copy into the host cell genome, creating a provirus. **(5a)** Transcription of the provirus creates HIV genome RNA and HIV mRNA; **(5b)** translation of the HIV mRNA makes HIV proteins. **(6)** HIV proteins come together to make immature HIV particles, and HIV spikes embed in the host cell membrane. **(7)** Immature HIV particles gain their spike-embedded envelope as they exit the cell. **(8)** Viral protease cleaves a viral polyprotein, releasing capsid proteins that form the viral capsid.

? **Why are there so many different strains of HIV?**

⑥ **Assembly.** Viral envelope proteins (spikes) embed in the host cell membrane. At the same time, immature viral particles are formed when the other viral components—including the genome, structural proteins, polyproteins, and enzymes—come together.

⑦ **Budding.** The immature virions bud from the host cell, gaining their spike-embedded envelope as they do so.

⑧ **Maturation.** Outside of the host cell, the viral particle matures. Viral protease cleaves copies of a crucial polyprotein to release individual proteins, including the capsid protein (CA); the CA copies self-assemble, forming a cone-shaped structure (the capsid shell) around the viral genome and key enzymes. Because of the reverse transcriptase–associated mutations, the newly produced mature viral particles will be slightly different from the parent virus that entered the cell.

The patient's initial immune response to the infection is strong but unable to clear the virus because key HIV antigens change rapidly due to mutations introduced by the error-prone reverse transcriptase; antibodies produced against one antigenic variant may not protect against another. Helper T cells typically die as a result of HIV infection, and although the body can replace hundreds of billions of them, their number slowly declines over many months or years. Infected macrophages and dendritic cells usually survive, but they release new virions over long periods of time and thus become reservoirs for replicating HIV. The virus also remains latent in memory T cells, providing yet another reservoir. Meanwhile, the infected macrophages and dendritic cells do not function normally, further weakening the immune response. Eventually, the immune system is no longer strong enough respond to infections or cancers.

About half of untreated HIV patients progress to AIDS within 10 years. Rapid progressors (patients who have high viral set points) develop AIDS within a few years of initial infection. Near the other end of the spectrum are long-term non-progressors (patients who have very low viral set points). These people show no decrease in CD4 cell numbers and maintain high levels of anti-HIV antibodies and HIV-specific $CD8^+$ cytotoxic T cells over many years. At the extreme end is a group called elite controllers, who maintain very low viral set-point levels (below the detection limit of conventional clinical assays) without the use of any antiviral medications.

Epidemiology

More than 40 million people worldwide have died of AIDS over the last three decades, and in 2021, an estimated 38.4 million are living with HIV (**figure 27.24**). The most affected region is sub-Saharan Africa—home to about two-thirds of all people living with HIV. In the United States, the rate of new infections has decreased by about two-thirds since the mid-1980s, with an estimated 38,400 new infections occurring in 2019. Despite the significant advances made in understanding and controlling the disease, however, it still cannot be cured. Nonetheless, the rate of AIDS-related deaths worldwide is decreasing as effective treatments become more accessible.

HIV is present in blood, semen, and vaginal secretions of infected individuals; common mechanisms of person-to-person transmission include:

- **Sexual contact.** HIV can be transmitted through unprotected anal or vaginal sex as well as shared sexual devices. The risk of spread by anal sex is greater than that of vaginal sex, probably because the rectal mucosa is thin and easily damaged, allowing viral entry into the blood. The likelihood of transmission is also increased when a sexual partner has an additional STI.
- **Blood and blood products.** HIV can be spread among IV (intravenous) drug abusers who share hypodermic needles. It can potentially be transmitted to healthcare workers by needle sticks, but the risk of contracting the infection this way is extremely low (less than 0.3% per exposure). Whole blood and blood products such as plasma and clotting factors are now screened for antibodies against HIV; the presence of antibodies suggests that the donor was infected, so the product is not used.
- **Vertical transmission.** HIV can be transmitted vertically (from mother to fetus or newborn) during pregnancy, during passage through the birth canal, and through breastfeeding.

The virus is not spread by insect bites or by casual contact, and no evidence indicates that it is transmitted through sweat, urine, tears, or saliva, even though these fluids may contain viral particles.

Treatment and Prevention

As scientists learned more about HIV replication and transmission, treatment and prevention options improved dramatically.

Treatment Anti-HIV medications, called **antiretrovirals (ARVs),** are designed to block replication of the virus. However, they do not affect viral nucleic acid already integrated

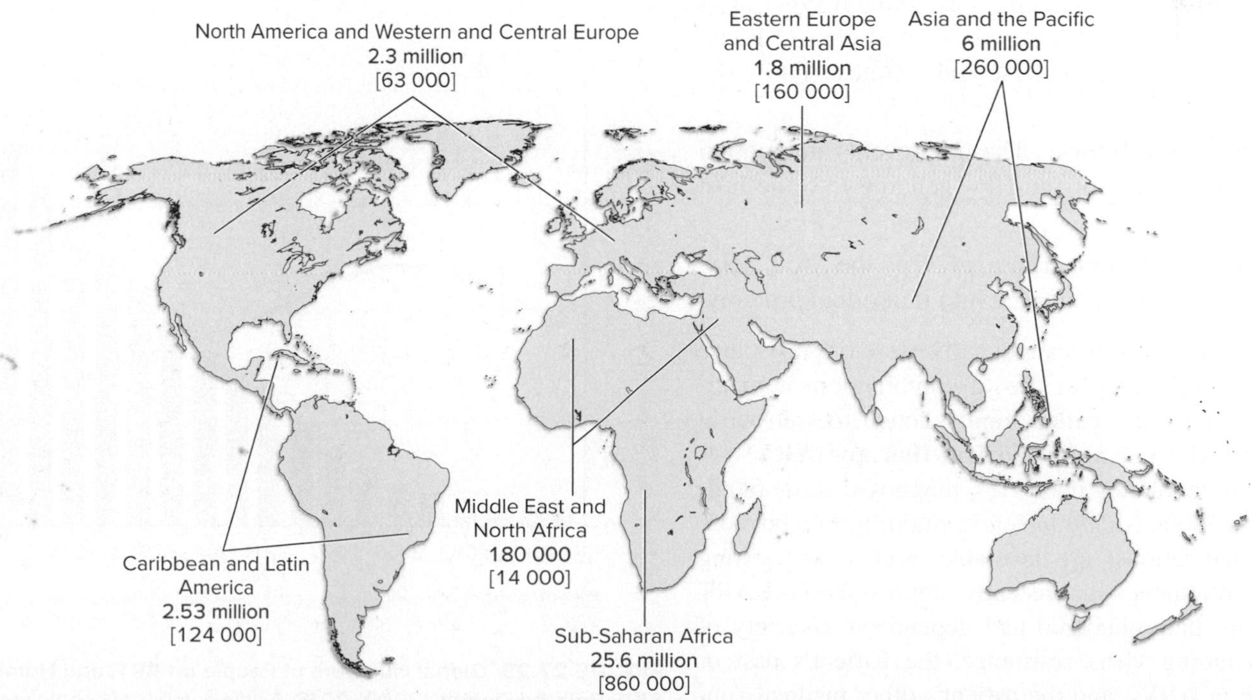

FIGURE 27.24 The Global HIV/AIDS Epidemic at the End of 2021 Altogether, over 38 million people are living with HIV/AIDS (red numbers). An estimated 1.5 million people were newly infected during the year (numbers in brackets).

? Considering that the rate of new infections has decreased considerably since the 1980s, why is the number of people living with HIV/AIDS worldwide still so high?

TABLE 27.16 Antiretroviral Medications

Category	Action	Examples
Entry inhibitors	Interfere with viral entry into a host cell by blocking viral attachment or fusion with the host cell membrane	Fostemsavir (FTR), enfuviritide (ENF), maraviroc (MVC), ibalizumab
Capsid inhibitors	Interfere with disassembly and assembly of the capsid shell	Lenacapavir (LEN)
NRTIs (nucleoside reverse transcriptase inhibitors)	Terminate synthesis of the DNA copy of the HIV RNA genome	Abacavir (ABC), emtricitabine (FTC), lamivudine (3TC), tenofovir alafenamide (TAF), zidovudine (AZT or ZDV).
NNRTIs (non-nucleoside reverse transcriptase inhibitors)	Inhibit the activity of reverse transcriptase	Doravirine (DOR), efavirenz (EFV), etravirine (ETR), nevirapine (NVP), rilpivirine (RPV)
Integrase inhibitors	Prevent HIV genome from integrating into host genome	Dolutegravir (DTG), elvitegravir (EVG), raltegravir (RAL)
Protease inhibitors	Prevent virions from maturing	Atazanavir (ATV), darunavir (DRV), ritonavir (RTV)

into the genome of host cells. The various categories of these medications were described in chapter 20, but their characteristics can be summarized as follows (**table 27.16**):

- **Entry inhibitors.** These interfere with viral entry by blocking attachment of the virus to the host cell or by preventing fusion of the viral envelope with the host membrane.
- **Capsid inhibitors.** The first example of this new drug class binds to capsid proteins that make up the capsid shell; in doing so, it interferes with the shell's disassembly (during uncoating) as well as assembly (during maturation to create new virions).
- **Reverse transcriptase inhibitors.** These prevent the viral enzyme from using the HIV genome as a template to make a DNA copy. Two general types are used: nucleoside reverse transcriptase inhibitors (NRTIs) and non-nucleoside reverse transcriptase inhibitors (NNRTIs).
- **Integrase inhibitors.** These prevent integration of the DNA copy of the HIV genome into the host genome.
- **Protease inhibitors.** These prevent the viral polyproteins from being cleaved into individual proteins.

Unfortunately, HIV can quickly develop resistance to ARVs. To help prevent this, a combination of medications—a "cocktail" called highly active antiretroviral therapy (HAART) or **antiretroviral therapy (ART)**—is prescribed immediately upon HIV diagnosis; some fixed-dose combinations (single tablets containing two or more different medications) are available, and a long-lasting injectable treatment was recently approved. The ART regimens are individualized and depend on a variety of factors, including virus resistance, the patient's past or current use of ARVs, and the patient's other medical conditions. The regimens typically include three medications: two NRTIs and either a protease inhibitor or an NNRTI. In many cases, ART lowers the viral set point to below detectable levels, stops progression of the disease, and allows partial recovery of immune function. Even with ART, however, HIV strains can still develop resistance to the medications. To treat multi-drug resistant HIV infections, three recently developed novel drugs are available for use in combination with other ARVs: fostemsavir, an entry inhibitor; lenacapavir, a capsid inhibitor; and ibalizumab, a humanized monoclonal antibody (mAb) that binds CD4 molecules on host cells in a way that prevents the virus from then binding to a co-receptor (see figure 27.23).

HIV treatments, as well as better management of opportunistic infections, have significantly slowed the progression of HIV infection and decreased the number of AIDS-related deaths (**figure 27.25**). By lowering the viral set point, ART

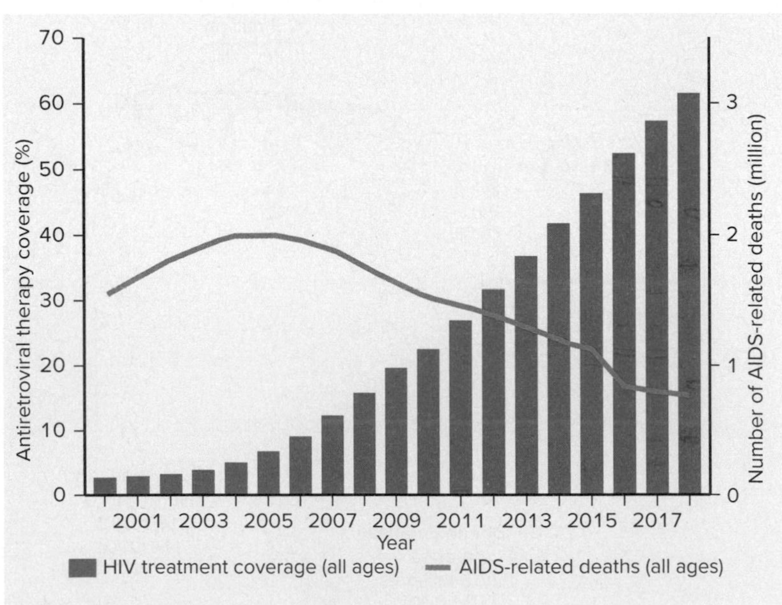

FIGURE 27.25 Global Numbers of People on ART, and Number of AIDS-Related Deaths, 2000–2018 As the number of people receiving antiretroviral therapy has increased worldwide (green bars), the number of people dying of AIDS-related conditions has decreased (blue line).

What caused the drop in AIDS-related deaths after 2005?

improves the quality of the patient's life, reduces the risk of other complicating diseases, and minimizes transmission. It also decreases the chance of resistance developing because fewer replicating viral particles means less opportunity for mutation to drug resistance. Unfortunately, disruptions due to the COVID-19 pandemic interfered with access to some therapies, raising concerns that death rates may rise again.

ART does not cure HIV infection, and several factors limit its use. For one thing, many patients experience side effects that range from uncomfortable (such as nausea, diarrhea, and fatigue) to severe (liver damage or painful peripheral nerve injury). The medications are also expensive, and thus unaffordable in many parts of the world. Fortunately, efforts such as negotiated price decreases, the U.S. FDA approval of generic medications, and efforts by the Global Fund (a private/public partnership that facilitates funding for AIDS, tuberculosis, and malaria prevention) have made HIV treatment more available worldwide. Still, many people lack access to therapies. Also, it is important to remember that ART does not eliminate HIV provirus integrated into host cell genomes; if medications are stopped, the viral set point rebounds to pretreatment levels.

Preventing Transmission via Sexual Contact Educating people about HIV transmission is a powerful weapon against the AIDS epidemic. The virus is not highly contagious, and the risk of contracting and spreading it can be eliminated or greatly reduced by avoiding activities that might transmit it. Latex condom use during sex significantly reduces HIV transmission; both male and female condoms are available, but male condoms are less expensive and more readily available. People who are unsure of their HIV status, and especially those at increased risk of contracting the virus, should get tested; those who are HIV-positive should receive ART because this lowers the viral set point and therefore reduces transmissibility.

The World Health Organization reports that circumcision (surgical removal of the penis foreskin) can reduce HIV acquisition by up to 60% and should also be included in prevention strategies. Circumcision changes the anatomy of the penis, reducing the number of certain cell types that HIV targets. It also changes the penis microbiota composition, reducing the number of certain anaerobic species that enhance virus stability. Several countries that have high HIV infection rates have developed programs promoting voluntary medical male circumcision (VMMC) as a preventive measure.

Individuals at high risk of HIV exposure can prevent infection by taking certain antiretroviral medications either before or immediately after exposure. Pre-exposure prophylaxis (PrEP) involves taking a pill daily that contains a fixed dose of two reverse transcriptase inhibitors or an injection every two months of cabotegravir (CAB-LA), a medication approved in 2021. PrEP is not 100% effective, however, so it should be used in combination with other prevention methods, including proper use of condoms. In contrast to PrEP, post-exposure prophylaxis (PEP) is an emergency measure used within 3 days after HIV exposure as a means to prevent the virus from establishing infection.

Preventing Transmission via Blood and Blood Products All blood and blood products should be handled as though they contain HIV. Transmission via blood transfusion decreased dramatically once potential blood donors were screened for risk factors and donated blood was tested for antibodies to HIV. Screening tests have also significantly reduced the risk of HIV transmission from artificial insemination and organ transplantation. HIV in or on blood-contaminated material is easily inactivated by high-level disinfectants and heat at 56°C or more for 30 minutes.

Transmission of HIV and other blood-borne diseases among injected-drug abusers can be prevented through needle and syringe exchange programs that give people sterile syringes and needles in exchange for used ones. Programs should also provide drug rehabilitation efforts and education about condom use and other safer sex practices.

Preventing Vertical Transmission Methods to prevent vertical transmission of HIV depend partly on the level of resources available. In resource-rich regions such as the United States, the transmission is significantly reduced by treating the mother with ART, delivering the baby by cesarean section, treating the baby with a short course of ART, and advising the mother to feed the baby with formula rather than breastfeed. In resource-poor countries, prevention is more difficult. In those settings, ART is recommended for the mother and a short course for the baby. However, delivering the baby by cesarean section is often not practical. Also, breastfeeding is important in resource-poor countries because breast milk contains antibodies that protect against many other diseases common in the regions; in addition, clean water may not be available for making infant formula safely. Regardless of the obstacles, methods to prevent vertical HIV transmission have been so successful that the Global Alliance to End AIDS in Children—an international effort involving the World Health Organization and various other government and non-government groups—was recently launched to end AIDS in children by 2030.

Vaccine Prospects A vaccine against HIV would prevent new infections, but creating an effective option has proven extremely difficult, largely because of the virus's rapid evolution that results from the error-prone reverse transcriptase—the key HIV antigens change rapidly due to mutations. Despite the challenges, development of an effective vaccine offers the best hope for preventing HIV infection. The successes have been limited, but the fact that a small subset of infected people have an immune response that appears to either prevent or control the infection encourages scientists to continue trying; various options are being explored and include the nucleic acid–based vaccines described in section 17.2. **Table 27.17** summarizes the main features of HIV infection and AIDS.

TABLE 27.17 HIV Infection and AIDS

Signs and Symptoms	During acute infection, patients may develop temporary flu-like symptoms, including fever, sore throat, head and muscle aches, rash, and enlarged lymph nodes. After a period of clinical latency, AIDS develops, leading to various unusual cancers and opportunistic infections.
Incubation Period	Within weeks for acute infection; without treatment, about half of patients progress to AIDS within 10 years.
Causative Agent	Mainly human immunodeficiency virus type 1 (HIV-1), many subtypes and strains; HIV-2 mostly in West Africa
Pathogenesis	HIV primarily infects CD4 T lymphocytes, macrophages, and dendritic cells. The T cells are killed—their numbers slowly decline until the immune system can no longer resist infections or tumor development.
Epidemiology	Three main routes of transmission: sexual contact, transfer of blood or blood products, and from mother to child around the time of childbirth.
Treatment and Prevention	Treatment: ART slows disease progression. Prevention: abstinence or monogamous relationships; correct use of condoms; male circumcision reduces risk of acquisition; pre-exposure or post-exposure prophylaxis; vertical transmission prevented with anti-HIV medications and, in resource-rich countries, cesarean section and not breastfeeding; other effective preventive measures include educating people about transmission and needle exchange programs for drug users; no vaccine yet available.

MicroAssessment 27.5

Viral STIs are as common as bacterial STIs, but cannot be cured with medications. Human papillomaviruses cause genital warts; some play a significant role in cancer of the cervix as well as cancers of the vagina, penis, anus, and throat. Genital herpes is common and, like most other STIs, can be transmitted in the absence of signs or symptoms. HIV infection generally ends in AIDS, but treatment can significantly improve longevity and quality of life.

13. What are the possible consequences of sexually transmitted human papillomavirus infections for males, females, and babies?
14. Why are circumcised males less likely to contract HIV?
15. Why should a person be concerned about genital herpes—is it not just a cold sore on the genitals? Explain.

27.6 ■ Protozoan STIs

Learning Outcome

9. List the distinctive characteristics of the organism that causes trichomoniasis.

Many intestinal protozoa can be spread through oral–anal sexual contact. Trichomoniasis, however, is a sexually transmitted protozoan disease that involves the genital system; its epidemiology and prevention are similar to other STI agents, as covered in the box Focus on Sexually Transmitted Infections.

Trichomoniasis ("Trich")

Trichomoniasis is not a reportable disease in the United States, but estimates suggest that almost 7 million people per year develop the disease.

Signs and Symptoms

Trichomoniasis is often asymptomatic. The incubation period is unknown, but if symptoms occur they seem to develop 4 to 20 days after infection. Symptomatic infections in females are characterized by a burning sensation and itching in the genitals and a frothy, sometimes smelly, yellowish-green vaginal discharge. In some cases, burning pain occurs with urination or during sex. The vulva and vaginal wall are red and slightly swollen. Pinpoint hemorrhages may develop on the cervix, a characteristic referred to as strawberry cervix. In pregnant women, trichomoniasis sometimes leads to preterm birth. Symptomatic men may have urethral discharge, burning pain with urination, painful testes, and a tender prostate gland.

Causative Agent

Trichomoniasis is caused by *Trichomonas vaginalis,* a motile protozoan that has four anterior flagella and a posterior flagellum attached to an undulating membrane (**figure 27.26**). It also has a slender, posteriorly protruding rigid structure called an axostyle that is thought to be used for attachment. Unlike most other pathogenic protozoa, *T. vaginalis* lacks a cyst form, so it does not survive long outside the host. Although the organism is a eukaryote, it does not have mitochondria; instead, it has cytoplasmic organelles called hydrogenosomes that generate hydrogen gas during anaerobic metabolism.

Pathogenesis

The pathogenesis of *T. vaginalis* is not fully understood. The organism produces a variety of adherence factors that allow it to attach to the urogenital epithelium. The contact damages the epithelium, perhaps a result of mechanical trauma by the axostyle of the moving protozoan

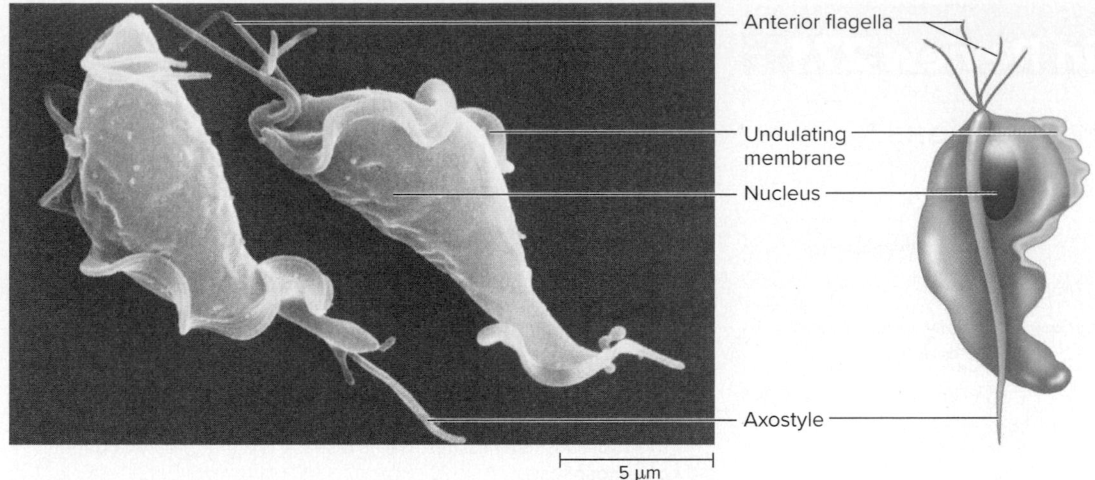

FIGURE 27.26 ***Trichomonas vaginalis*** Scanning electron micrograph and diagram of *T. vaginalis,* a common cause of vaginitis. David M. Phillips/Science Source

What is the presumed role of the axostyle in the pathogenesis of *T. vaginalis*?

cells. Cytotoxic molecules and other factors may also be involved, some of which are delivered to host cells in membrane-bound vesicles. The inflammatory response to infection causes many of the symptoms, and the hydrogen gas produced by the organism contributes to the frothiness of the vaginal discharge. *T. vaginalis* also appears to contribute to vaginal dysbiosis, which in turn, might favor the additional development of bacterial vaginosis (BV).

Epidemiology

T. vaginalis is a widespread human parasite that has no other reservoirs and is spread through sexual intercourse or other intimate activities. Many women have trichomoniasis and BV concurrently, so one of these diseases might create a predisposition for the other.

Treatment and Prevention

Most strains of *T. vaginalis* respond quickly to treatment with metronidazole or tinidazole; if resistance occurs, cases are often treated with higher doses of metronidazole due to a lack of alternatives. As with other STIs, transmission is prevented by abstinence, monogamy, and the use of condoms. **Table 27.18** gives the main features of trichomoniasis.

The key features of the diseases covered in this chapter are highlighted in the **Diseases in Review 27.1** table that follows.

TABLE 27.18 Trichomoniasis

Signs and Symptoms	Frequently asymptomatic. Females: burning sensation and itching in the genitals, vaginal redness, and frothy yellow-green discharge; preterm births may be a complication in pregnant females. Males: urethral discharge, burning pain on urination, painful testes, and tender prostate.
Incubation Period	Appears to be about 4 to 20 days
Causative Agent	*Trichomonas vaginalis*, a motile protozoan that lacks a cyst form.
Pathogenesis	Not fully understood; involves adherence factors, host cell damage, and the inflammatory response. Appears to contribute to vaginal dysbiosis.
Epidemiology	Transmitted through sexual contact.
Treatment and Prevention	Treatment: appropriate antimicrobial agent. Prevention: abstinence, monogamy, and consistent use of condoms.

MicroAssessment 27.6

Trichomoniasis is caused by a flagellated protozoan that does not form cysts and lacks mitochondria.

16. What significance does the lack of a cyst form of *Trichomonas vaginalis* have on the epidemiology of trichomoniasis?

17. What is the relationship between being “easily killed by drying” and “transmission usually by sexual contact”?

Diseases in Review 27.1

Genitourinary Infections

Disease	Causative Agent	Comment	Summary Table
UROGENITAL INFECTIONS			
Cystitis (bladder infection)	Usually uropathogenic *Escherichia coli*	Most common in females because of their relatively short urethra; characterized by painful urination; can progress to kidney infection.	Table 27.1
Leptospirosis	Certain *Leptospira* species	Spread by water, soil, or food contaminated with the urine of infected animals; enters via mucous membranes or breaks in skin then spreads to all tissues. Weil's disease is a life-threatening form of leptospirosis.	Table 27.2
Bacterial vaginosis (BV)	Unknown, involves vaginal dysbiosis	Characterized by thin, gray-white discharge and fishy odor; most common in sexually active women and pregnant women.	Table 27.3
Vulvovaginal candidiasis (VVC)	*Candida albicans* (a fungal disease)	Characterized by intense itching and thick, white discharge; associated with conditions that disrupt the normal vaginal microbiota, allowing the overgrowth of *C. albicans*.	Table 27.4
Staphylococcal toxic shock	Toxin-producing strains of *Staphylococcus aureus*	Toxin is a superantigen and causes cytokine release, leading to a drop in blood pressure.	Table 27.5
BACTERIAL STIs			
Chlamydia	*Chlamydia trachomatis*	Most common STI in the United States; infections are often asymptomatic; signs and symptoms include genital discharge and painful urination; can progress to PID.	Table 27.6
Gonorrhea	*Neisseria gonorrhoeae*	Infections are often asymptomatic, particularly in females; signs and symptoms include genital discharge and painful urination; can progress to PID.	Table 27.7
***Mycoplasma genitalium* infections**	*Mycoplasma genitalium*	Infections often asymptomatic; signs and symptoms resemble those of chlamydia and gonorrhea; can progress to PID.	Table 27.8
Syphilis	*Treponema pallidum*	Primary syphilis is characterized by a painless chancre, secondary by a widespread rash, and tertiary by cardiovascular damage; neurosyphilis and congenital infections also occur.	Table 27.9 Table 27.10
Chancroid	*Haemophilus ducreyi*	Characterized by painful genital sores and enlarged groin lymph nodes.	Table 27.11
VIRAL STIs			
Human papillomavirus infections	Human papillomaviruses (HPVs)	Some strains cause genital warts while others cause cancers. Vaccines protect against the most significant strains.	Table 27.12
Genital herpes	Herpes simplex virus, usually type 2 (HSV-2)	Virus can cause painful blisters; virus becomes latent, so infections are lifelong.	Table 27.13
HIV infection and AIDS	Human immunodeficiency virus, usually type 1 (HIV-1)	Without treatment, HIV infection progresses to AIDS; virus infects helper T cells, resulting in immunodeficiency; ART slows disease progression but is not curative.	Table 27.17
PROTOZOAN STI			
Trichomoniasis ("trich")	*Trichomonas vaginalis*	Characterized by itching, discharge, and urinary discomfort.	Table 27.18

Summary

27.1 ■ Anatomy, Physiology, and Ecology of the Genitourinary System

The Urinary System (figure 27.1)

The urinary system is composed of the kidneys, ureters, bladder, and urethra. Infections occur more frequently in women than in men because of the shortness of the female urethra and its proximity to the anus.

The Genital System (figure 27.2)

The vagina is a portal of entry for a number of infections. The normal vaginal microbiota, including lactobacilli, protect the vagina from colonization by pathogens. The fallopian tubes provide a passageway for infections to enter the abdominal cavity. In males, prostate secretions have antimicrobial properties.

27.2 ■ Urinary Tract Infections

The urinary system usually becomes infected by organisms ascending from the urethra, but it can also be infected from the bloodstream.

Bacterial Cystitis ("Bladder Infection") (table 27.1)

Most UTIs in healthy people are caused by *Escherichia coli* or other Enterobacteriaceae members from the person's own normal intestinal microbiota. Healthcare-associated UTIs are common and are caused by *Pseudomonas aeruginosa* and *Enterococcus faecalis,* which are often resistant to many antibiotics. Kidney infection (**pyelonephritis**) may complicate a bladder infection when pathogens ascend through the ureters to the kidneys.

Leptospirosis (table 27.2)

In leptospirosis, the urinary system is infected by *Leptospira interrogans* from the bloodstream (figure 27.3). The outcome of infection is highly variable—most infections are asymptomatic or have mild symptoms, but some result in life-threatening organ damage. Weil's disease is a sometimes fatal form of leptospirosis that affects the liver and kidneys.

27.3 ■ Genital System Diseases

Bacterial Vaginosis (BV) (table 27.3, figure 27.4)

Bacterial vaginosis is the most common vaginal disease. Signs and symptoms include a gray-white vaginal discharge and a strong fishy odor. BV is characterized by an imbalance in the normal microbiota and appears to involve development of a *Gardnerella vaginalis* biofilm.

Vulvovaginal Candidiasis (VVC) (table 27.4)

Vulvovaginal candidiasis signs and symptoms include itching, burning, vulvar redness and swelling, and a thick, white discharge. The causative agent is a yeast, *Candida albicans,* that is commonly part of the normal vaginal microbiota (figure 27.5). Antibacterial treatment, uncontrolled diabetes, and oral contraceptives are predisposing factors, but in most cases no such factor can be identified.

Staphylococcal Toxic Shock Syndrome (table 27.5)

Staphylococcal toxic shock syndrome became widely known with a 1980 epidemic in menstruating females who used a certain kind of tampon that has since been removed from the market (figure 27.6). Signs and symptoms include sudden fever, headache, muscle aches, bloodshot eyes, vomiting, diarrhea, a sunburn-like rash that later peels, and confusion. Blood pressure drops, and without treatment organ failure and death may occur.

SEXUALLY TRANSMITTED INFECTIONS

Focus on Sexually Transmitted Infections

STIs are typically spread through intimate contact with body fluids. Many are asymptomatic; signs and symptoms often include genital discharge and pain during urination. Inflammatory reaction to the infection may cause scarring, which can partially obstruct the fallopian tubes in females, leading to potential infertility or ectopic pregnancy. Simple measures for controlling STIs include abstinence from sexual activities, a monogamous relationship with an uninfected person, and consistent and correct use of latex or polyurethane condoms.

27.4 ■ Bacterial STIs

Chlamydia (table 27.6, figure 27.7)

Chlamydia, caused by *Chlamydia trachomatis,* is reported more often than any other STI. Asymptomatic infections are common and easily transmitted. Signs and symptoms include genital discharge and pain.

Gonorrhea (table 27.7; figures 27.9, 27.10)

Gonorrhea, caused by *Neisseria gonorrhoeae,* is often asymptomatic. Symptomatic females may notice discharge and pain associated with cervical or urethral infection. Symptomatic males typically experience painful urination and thick pus draining from the urethra (figure 27.8).

***Mycoplasma genitalium* Infection** (table 27.8)

Mycoplasma genitalium (figure 27.11) infections are often asymptomatic and are becoming increasingly common. Signs and symptoms, when present, are similar to those of chlamydia and gonorrhea. Diagnosis without a commercially available test is difficult because the fastidious causative agent grows very slowly in vitro.

Syphilis (tables 27.9, 27.10)

Syphilis is caused by the spirochete *Treponema pallidum* (figure 27.16). Primary syphilis is characterized by a painless, firm ulceration called a hard **chancre** (figure 27.12). In secondary syphilis, the immune response gives rise to diverse signs and symptoms, but the most common is a body-wide rash that involves the palms and soles (figure 27.13); a latent period of months or years occurs between the secondary and tertiary phases of the disease. Tertiary syphilis is the result of a prolonged inflammatory response, causing chronic lesions called **gummas** that can involve any part of the body (figure 27.14). Syphilis in pregnant women can spread across the placenta to involve the fetus, resulting in congenital syphilis (figure 27.15).

Chancroid (table 27.11)

Chancroid is caused by *Haemophilus ducreyi,* and is characterized by single or multiple soft, tender genital ulcers, and enlarged, painful groin lymph nodes (figure 27.17).

27.5 ■ Viral STIs

Viral STIs are more common than bacterial STIs, and they are not yet curable.

Human Papillomavirus STIs: Genital Warts and Cancers (table 27.12)

Human papillomavirus STIs are caused by human papillomaviruses (HPVs). Low-risk HPV types cause warts on or near the genitalia (figure 27.18); high-risk types are associated with various cancers, including those of the cervix, vagina, penis, anus, and throat.

Asymptomatic precancerous lesions can be detected with a Pap smear (figure 27.19) and a nucleic acid–based test to identify high-risk HPV. Gardasil 9 (9vHPV) is a vaccine that protects against the most common cancer- and wart-causing HPV types.

Genital Herpes (table 27.13)

Genital herpes is caused by herpes simplex viruses (usually HSV type 2, but also HSV type 1). Signs and symptoms include vesicles (blisters) with itching, burning, or pain; these break, leaving ulcers (figure 27.20). HSV establishes a latent infection in sensory nerves and cannot be cured; the virus can reactivate, causing recurrent symptoms. Genital herpes can be transmitted in the absence of signs or symptoms, but the risk is greatest when lesions are present.

HIV Infection and AIDS (tables 27.15, 27.16, 27.17)

HIV infection has three stages: acute infection, clinical latency, and AIDS (figure 27.21). HIV is a retrovirus; three virally encoded enzymes—reverse transcriptase, integrase, and protease—are targets of anti-HIV medications (figures 27.22). The HIV replication is complex but helps explain the mechanisms of anti-HIV medications (Figure 27.23). Worldwide, millions of new HIV infections and deaths from AIDS occur each year (figure 27.24). No HIV vaccine or medical cure is yet available, but antiretroviral therapy (ART) reduces viral load and maintains health in infected people (figure 27.25). Pre- and post-exposure prophylaxis is available. Methods to prevent HIV infections are grouped by transmission route (sexual contact, blood and blood products, and vertical).

27.6 ■ Protozoan STIs

Trichomoniasis ("Trich") (table 27.18)

Trichomoniasis is caused by *Trichomonas vaginalis* (figure 27.26). Most infections are asymptomatic. In females, signs and symptoms may include itching, burning, swelling, and redness of the vagina; frothy, sometimes smelly, yellow-green discharge, and burning on urination. Males may have discharge from the penis and burning on urination, sometimes accompanied by painful testes and a tender prostate gland.

Review Questions

Short Answer

1. How do lactobacilli help protect the vagina from potential pathogens?
2. List four factors that increase the likelihood of developing a urinary tract infection.
3. What are the typical signs and symptoms of cystitis?
4. What danger might possibly be found in a spot on the ground where an animal urinated 1 week earlier?
5. What is a clue cell?
6. List three diseases caused by different antigenic types of *Chlamydia trachomatis*.
7. What is neonatal conjunctivitis?
8. Why is silver-staining used to visualize *Treponema pallidum*?
9. Give two ways in which the chancre of chancroid differs from the chancre of syphilis.
10. What is the relationship between AIDS and HIV infection?

Multiple Choice

1. Which of the following about bacterial cystitis is *false*?
   a) About one-third of all women will have it at some time during their lives.
   b) Catheterization of the bladder markedly increases the risk of contracting the disease.
   c) Bladder infections may lead to pyelonephritis.
   d) Bladder infections occur as often in males as they do in females.
   e) Bladder infections can be asymptomatic.
2. Choose the one *true* statement about leptospirosis.
   a) Humans are the only reservoir.
   b) Most infections produce severe symptoms.
   c) Transmission is by the fecal-oral route.
   d) Some cases can have two phases of signs and symptoms.
   e) An effective vaccine is generally available for preventing human disease.
3. Which one of the following statements about bacterial vaginosis is *false*?
   a) It is the most common vaginal disease in women of childbearing age.
   b) In pregnant women, it is associated with premature delivery.
   c) Inflammation of the vagina is a constant feature of the disease.
   d) The vaginal microbiota shows a decrease in lactobacilli and an increase in anaerobic bacteria.
   e) The microbial cause of bacterial vaginosis is unknown.
4. Pick the one *false* statement about vulvovaginal candidiasis.
   a) It often involves the external genitalia.
   b) It is readily transmitted by sexual intercourse.
   c) It is caused by a yeast present among the normal vaginal microbiota.
   d) It is associated with prolonged antibiotic use.
   e) Its predisposing factors include the late stage of pregnancy.
5. All of the following statements about staphylococcal toxic shock are true *except*
   a) it can lead to kidney failure.
   b) the causative organism usually does not enter the bloodstream.
   c) it occurs only in tampon users.
   d) the causative organism produces superantigens.
   e) person-to-person spread does not occur.
6. Which one of these statements about chlamydia is *false*?
   a) The incubation period is usually shorter than in gonorrhea.
   b) Infected cells develop inclusions.
   c) Infection can progress to pelvic inflammatory disease and affect the liver.
   d) Tissue damage largely results from an inflammatory response.
   e) Fallopian tube damage may occur, potentially leading to infertility.

7. Which of the following statements about gonorrhea is *false*?
   a) The incubation period is less than two weeks.
   b) Disseminated gonococcal infection (DGI) is almost invariably preceded by prominent urogenital symptoms.
   c) Disseminated gonococcal infection can result in arthritis of joints.
   d) Phase and antigenic variation help the causative organism evade the immune response.
   e) Pelvic inflammatory disease (PID) is common in untreated women.
8. Which symptom is least likely to occur as a result late stage syphilis (tertiary or neurosyphilis)?
   a) Gummas
   b) White patches on mucous membranes
   c) Aneurysms
   d) Stroke
   e) Speech defects
9. Which of the following statements about HIV infection and AIDS is *false*?
   a) HIV is transmitted by body fluids such as blood and semen.
   b) Antiretroviral medications can cure AIDS.
   c) Unusual cancers and infections indicate the start of AIDS.
   d) HIV can be transmitted transplacentally to a fetus.
   e) HIV infects white blood cells, damaging the immune system.
10. All of the following are true of "trich" (trichomoniasis) *except*
    a) it can cause burning pain on urination and painful testes in men.
    b) it occurs worldwide.
    c) asymptomatic carriers are rare.
    d) transmission can be prevented by the proper use of condoms.
    e) individuals with multiple sex partners are at high risk of contracting the disease.

Applications

1. Religious restrictions of a small North African community are preventing a World Health Organization project from reducing the incidence of gonorrhea. The community will not allow the testing of females for the disease. They can be treated, however, if they show outward evidence of the disease. Only males are allowed to participate fully in the project, with testing for the disease and treatment. The village elders argue that eradicating the disease from males would eventually remove it from the population. What would be the impact of these restrictions on the success of the project?
2. Former President Ronald Reagan once commented at a press conference that the best way to combat the spread of AIDS in the United States was to prohibit everyone from having sexual contact for 5 years. What would be the success of such a program if it were possible to carry it out?

Critical Thinking

1. The middle curve of figure 27.6 shows the occurrence of staphylococcal toxic shock syndrome in menstruating women from 1979 to 2013. What aspect of these data argues that high-absorbency tampons were not the only cause of staphylococcal toxic shock syndrome associated with menstruation?
2. In early attempts to identify and isolate the cause of syphilis, various bacteria in the discharge from syphilitic lesions in experimental animals were isolated in pure culture. None of them, however, caused the disease when used in attempts to infect healthy animals. Why was it considered a critical step to have the cultivated bacteria reproduce the disease in healthy animals?

28 | Microbial Ecology

Farming relies on the activities of microorganisms. *Tim McCabe/Natural Resource Conservation Service/USDA*

KEY TERMS

Consumers Organisms that eat primary producers or other consumers.

Decomposers Organisms that digest the remains of primary producers and consumers.

Ecology Study of interactions of organisms with one another and with their environment.

Eutrophic A nutrient-rich environment that supports the excessive growth of algae and other organisms.

Hydrothermal Vents Undersea geysers that spew out mineral-laden hot water.

Hypoxic An environment very low in dissolved O_2.

Microbial Mat A type of microbial community characterized by distinct layers of different groups of microbes that together make up a thick, dense, highly organized structure.

Nitrogen Fixation Conversion of nitrogen gas to ammonia.

Oligotrophic A nutrient-poor environment.

Primary Producers Organisms that convert CO_2 into organic compounds, sustaining other forms of life.

Rhizosphere Zone around plant roots containing organic materials secreted by the roots.

A Glimpse of History

Although many microbiologists in the late 1800s were particularly interested in studying disease-causing bacteria, the Russian microbiologist Sergei Winogradsky (1856–1953) was more intrigued by microorganisms in soil and water. Considered by some to be the founder of microbial ecology, he was the first to recognize what he called chemosynthesis: the biological conversion of inorganic to organic chemicals.

Winogradsky realized that although the pure culture techniques are useful for studying medically important bacteria, they are not adequate for studying environmental microorganisms. So instead he tried to mimic in the laboratory the natural conditions in which organisms grow. A device called a Winogradsky column illustrates his method. This in vitro ecosystem can be built using a cylindrical glass container filled with mud, carbon and sulfur sources, and water from a pond or lake. The column is covered to prevent evaporation and is placed in sunlight. Eventually, layers of different colors corresponding to microbial populations with different physiological capabilities will appear in the column. The top layer consists of algae and cyanobacteria that produce O_2. Below that are the green and purple sulfur bacteria that oxidize H_2S, producing sulfur. Near the bottom of the column are the sulfate-reducers, producing H_2S and precipitating iron.

Winogradsky's work was instrumental in describing the global nutrient cycles described in this chapter. In recognition of his contributions, the bacterium *Nitrobacter winogradskyi* was named in his honor.

Microorganisms have existed on Earth for about 4 billion years. By all evidence they occupy every possible environment, from ocean depths to hot springs and from rich agricultural soil to the human gut. In fact, without microbial activities, life on Earth could not exist—we depend on them to capture energy, cycle nutrients, maintain fertile soil, and decompose dead organisms and pollutants. Without microorganisms, the Earth would quickly become buried by tons of wastes, and nutrients would be depleted, halting growth and reproduction.

In view of the crucial functions microorganisms perform, it seems we should know a great deal about the diverse microbial species in the environment. Quite the opposite is true, however, as less than 1% have been grown in culture. In chapter 4, we discussed how organisms are cultivated in the laboratory under controlled conditions that promote optimal growth. In nature, however, organisms generally grow in close associations with many different species, often in biofilms. Nutrients may be in short supply, quite unlike the rich conditions in growth media.

When researchers work with environmental samples in the laboratory, the microorganisms that grow in culture are not necessarily the most abundant or the most important in their natural setting; they are simply the ones able to grow best under the conditions we provide. Even if all microorganisms could be cultivated, their behavior in culture might not accurately reflect their environmental role. Molecular biology techniques provide much more extensive and reliable information through the direct analysis of DNA from environmental samples, offering vital clues to the diversity and roles of microbes in the natural world. As part of the Earth BioGenome Project, efforts are currently underway to sequence microbial genomes from around the globe.

Chapter 11 presented the great diversity of physiological mechanisms that prokaryotes use to survive in terrestrial and aquatic environments. This chapter will expand on some of those concepts and describe microbial activities that are essential to life.

28.1 ■ Principles of Microbial Ecology

Learning Outcomes

1. Compare and contrast the roles of primary producers, consumers, and decomposers.
2. Explain how some microbes can grow in low-nutrient environments.
3. Describe the significance of microbial competition.
4. Describe how environmental changes can result in alterations in a microbial community.
5. Describe the structural organization of a microbial mat.

Ecology is the study of interactions of organisms with one another and with their physical environment. Organisms of the same type in a given location make up a **population.** All of the different organisms in the location form a **community.** An **ecosystem** consists of a community of organisms and the non-living environment with which they interact. Major ecosystems include the oceans, fresh waters (rivers and lakes), deserts, marshes, grasslands, forests, and tundra. Each ecosystem possesses characteristic organisms and physical conditions. All of the ecosystems on Earth make up the **biosphere.**

Within the biosphere, ecosystems vary in both **biodiversity** (number and variety of species present and their evenness of distribution) and **biomass** (the weight of all organisms present). Microorganisms play a major role in most ecosystems, and many ecosystems host microbes unique to themselves. The role an organism plays in a particular ecosystem is called its **ecological niche.**

The environment immediately surrounding an individual microbe—the **microenvironment**—is most relevant to that cell, but because microorganisms are so small, the microenvironment is difficult to identify and measure. The more easily measured gross environment—the **macroenvironment**—may be very different from the microenvironment. Consider a bacterial cell living within a biofilm (see figure 4.3); growth of aerobic organisms in the biofilm can deplete O_2, creating tiny zones where obligate anaerobes can grow. Fermenters can produce organic acids that may then be metabolized by other organisms in the biofilm. In addition, various growth factors and even genes can be transferred between organisms. Thus, microorganisms that might be unexpected in a given macroenvironment may actually thrive there within specialized microenvironments.

As described in chapter 16, living organisms often interact with one another in long-lasting and intimate relationships called symbioses. In a mutualistic relationship involving two organisms, both benefit from the relationship. Bacteria in the human gut benefit from a steady flow of available nutrients, but they also provide the body with the many benefits described in section 16.2. In a commensalistic relationship, one organism benefits and the other is unaffected. We have this type of relationship with many of the microorganisms found on our skin. In a parasitic relationship, one organism benefits, but the other is harmed. Well-adapted parasites do not usually kill their host, unlike predators.

Nutrient Acquisition

Organisms are categorized according to their trophic level (source of food), which is intimately related to nutrient cycling. Three general trophic levels occur in a food chain (**figure 28.1**):

- **Primary producers.** These are autotrophs; they convert CO_2 into organic materials. Producers include both photoautotrophs (use sunlight for energy) and chemolithoautotrophs (oxidize inorganic chemicals for energy). Primary producers serve as a food source for consumers and decomposers.
- **Consumers.** These are heterotrophs that eat primary producers or other consumers. Herbivores, which eat plants or algae, are primary consumers. Carnivores that eat herbivores are secondary consumers; carnivores that eat other carnivores are tertiary consumers. A chain of consumption is a food chain; interacting food chains form a food web.
- **Decomposers.** These are heterotrophs that digest the waste products and remains of primary producers and consumers. Decomposers specialize in degrading complex materials such as cellulose, converting them into small molecules that can more easily be used by other organisms. The complete breakdown of organic molecules into inorganic molecules such as ammonia, sulfates, phosphates, and carbon dioxide is called mineralization. Microorganisms, particularly bacteria and fungi, play a major role in decomposition because of their unique metabolic capabilities.

Microbes in Low-Nutrient Environments

Low-nutrient environments such as lakes, rivers, and streams are common in nature, so microorganisms that can grow in dilute aqueous solutions are widespread. Most microbial growth in these settings is in biofilms, and the cells are shed from the biofilm into the aqueous solution.

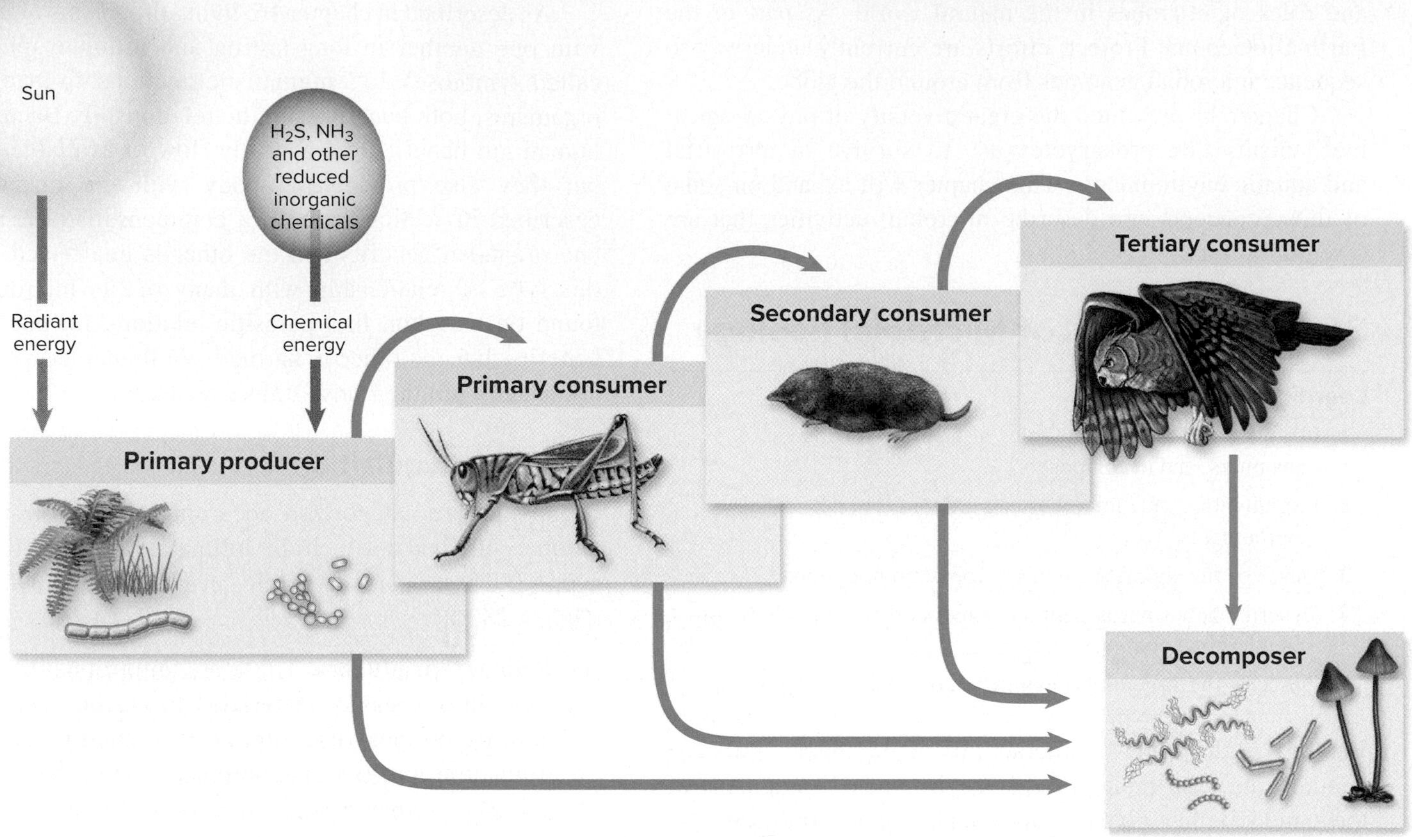

FIGURE 28.1 Trophic Levels in a Food Chain

? Which organisms in the diagram are autotrophs?

Even distilled water has trace amounts of nutrients absorbed from the air, so microorganisms can grow in distilled-water reservoirs used in laboratories. Although the organisms multiply slowly, they can reach concentrations as high as 10^7 per milliliter. This cell concentration is not high enough to result in an obviously cloudy solution, so the growth usually goes unnoticed, but it can have serious consequences for the success of laboratory experiments that depend on water purity.

Oligotrophs (organisms that can grow in dilute environments) contain highly efficient transport systems for moving nutrients into the cell. Other mechanisms that bacteria use to thrive in dilute aquatic environments are described in chapter 11.

Microbial Competition

Perhaps nowhere in the living world is competition more quickly evident than among microorganisms. The ability of a microorganism to compete successfully for a habitat is generally related to the rate at which it multiplies, as well as to its ability to withstand adverse environmental conditions. Because bacteria multiply logarithmically, even small differences in their generation times can be significant; the species that multiplies fastest under the given conditions yields the largest population (**figure 28.2**).

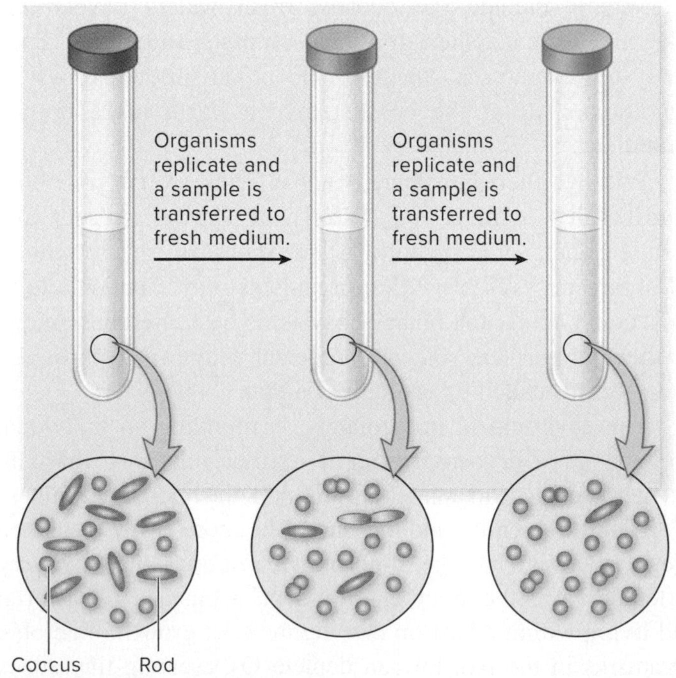

FIGURE 28.2 Competition The species that multiplies faster yields the larger population.

? Why does transferring only some of the culture make it easier to see the effects of competition?

The outcome of microbial competition is not always determined by the replication rate; some microorganisms actively compete using antagonistic strategies. In the soil, for example, certain microbes resort to a type of chemical warfare, producing antimicrobial compounds such as **bacteriocins** (bacterial proteins that kill closely related strains). Antibiotics produced by *Streptomyces* species also inhibit competing organisms, but their exact roles in nature are still poorly understood.

Microorganisms and Environmental Changes

Environmental changes often result in changes in a community. Those organisms that have adapted to live several inches beneath the surface of an untilled field will probably not be well suited to growth in that field if it is plowed, fertilized, and irrigated. In addition to external sources of environmental change, the growth and metabolism of organisms themselves can alter the environment dramatically. Nutrients may become depleted, and toxic waste products accumulate.

In some environments, the changing conditions bring about a highly ordered and predictable succession of bacterial species. An example occurs in unpasteurized milk, which usually contains various species of microbes associated with cows. Initially, the dominant bacterium is *Lactococcus lactis,* which breaks down the milk sugar lactose, forming lactic acid as a fermentation end product (**figure 28.3**). This sours the milk and denatures the proteins, causing the milk to curdle. The acid inhibits most other organisms in the milk, and eventually enough acid is produced to inhibit *L. lactis*. *Lactobacillus* species can multiply in this highly acidic environment, however, and these bacteria metabolize any remaining sugar, forming more acid until their growth is also inhibited. Yeasts and molds, which tolerate even more acidic conditions, then become the dominant group. They oxidize the lactic acid, which raises the pH. Most of the sugar has already been used at this point, but milk protein (casein) is still available, and it can be used by protease-producing members of the endospore-forming genus *Bacillus*. This breakdown of protein, known as putrefaction, yields a completely clear and foul-smelling product. The milk thus goes through a succession of changes with time, first souring and finally putrefying.

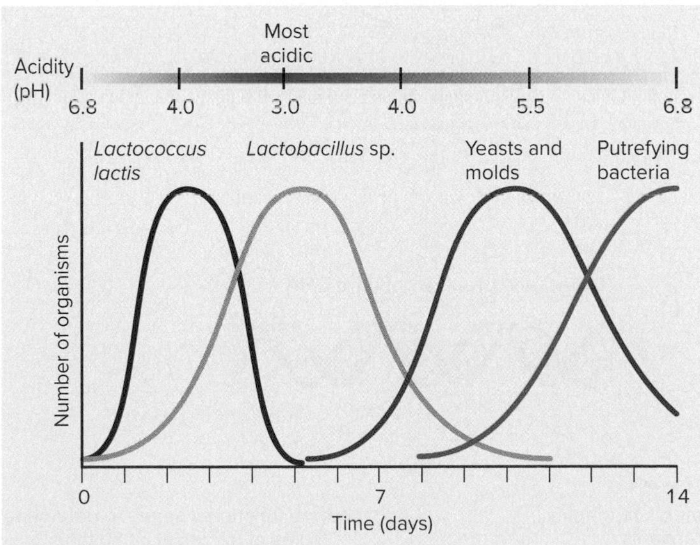

FIGURE 28.3 Growth of Microbial Populations in Unpasteurized Raw Milk at Room Temperature Production of acid causes souring and encourages the growth of yeasts and molds. Eventually, bacteria digest the proteins, causing putrefaction.

? **Why does the pH of the milk first decrease and then increase?**

Microbial Communities

Microorganisms most often grow as biofilms attached to solid surfaces or at air–water interfaces. General aspects of biofilms were described in detail in chapter 4. In this section, we focus on a specific type of biofilm: a microbial mat.

A **microbial mat** is a thick, dense, highly organized structure composed of distinct layers. Often the layers are green, reddish-pink, and black, which indicate the growth of different microbial groups (**figure 28.4**). The top green layer is typically composed of various species of cyanobacteria, and the color is due to their photosynthetic pigments. Directly below the green layer is a reddish-pink layer consisting of purple sulfur bacteria. The pigments of these anoxygenic phototrophs collect wavelengths of light not used by the cyanobacteria. At the bottom is a black layer, resulting from iron molecules reacting with hydrogen sulfide produced by a group of bacteria called sulfate-reducers. These obligate anaerobes oxidize the organic compounds produced by the photosynthetic microbes growing in the mat's upper layers, using sulfate as a terminal electron acceptor.

Although microbial mats can be found in many areas, those near hot springs in Yellowstone National Park are some of the most intensively studied. The mats in these extreme

FIGURE 28.4 A Microbial Mat This dense structure is composed of distinct layers of different microbial communities. Reut S. Abramovich

? **Why are the layers of the mat different colors?**

areas are undisturbed by grazing eukaryotic organisms and, consequently, provide an important model for studying microbial interactions.

MicroAssessment 28.1

Microorganisms play a major role in most ecosystems. Organisms are categorized as primary producers, consumers, or decomposers. Many microorganisms live in low-nutrient environments, where they form biofilms. Competition among microorganisms in a habitat can be intense. Microbial mats have distinct layers.

1. What are the roles of primary producers, consumers, and decomposers?
2. Describe the microbial succession that occurs as unpasteurized milk spoils.
3. What aspects of an ecosystem would increase its biodiversity?

28.2 ■ Studying Microbial Ecology

Learning Outcome

6. Explain the role of cultivation-independent methods to study microorganisms in the environment.

Estimates vary widely, but one gram of soil may contain between 10^4 and 10^7 different microbial species. Studying these is a daunting task by any method, and impossible using culture techniques (as mentioned earlier, less than 1% of microorganisms can be expected to grow under laboratory conditions). Still, however, scientists strive to understand the diversity of microorganisms on the planet, including their roles in nature and how they affect the environment. Researchers in microbial ecology often rely on **metagenomics,** the cultivation-independent study of communities or their members by analyzing genetic material taken directly from an environment (**figure 28.5**). Fragments of DNA are extracted from soil or another natural sample, and then one of a variety of methods is used to determine the nucleotide sequences of that DNA. The sequences are then compared with others available in public databases. One goal of current research is to expand and improve the existing databases because the power of metagenomics depends on the quality of the databases.

Metagenomic data provide two separate but related types of information about a microbial community: taxonomic diversity and genetic capabilities. Taxonomic diversity is studied by examining either the nucleotide sequence of 16S rRNA genes (18S rRNA genes for eukaryotes) or the deduced amino acid sequences of ribosomal proteins; by doing this, scientists can classify and identify the community's microbial inhabitants. The genetic capabilities of the community as a whole are determined by analyzing the entire spectrum of sequenced genes in a sample. A significant challenge is to link the two types of information, allowing researchers to determine which uncultivated microbes have a given genetic capability.

Culture-independent methods are rapidly advancing the study of microbial ecology because sequence information learned from one species can be applied to others. For example, researchers found that variations of a gene coding for bacterial rhodopsin, a light-sensitive pigment that provides a mechanism for harvesting the energy of sunlight, are widespread in marine bacteria. This gene provides bacteria with a mechanism for phototrophy that does not require chlorophyll and might be an important mechanism for energy accumulation in ocean environments.

Additional methods to study microbial ecology are available as well. These include DNA microarrays designed to study gene expression (see figure 9.18) and fluorescence in situ hybridization (FISH), which uses probe technology in combination with fluorescence microscopy to determine if a specific species or group is present in a sample (see figure 9.17).

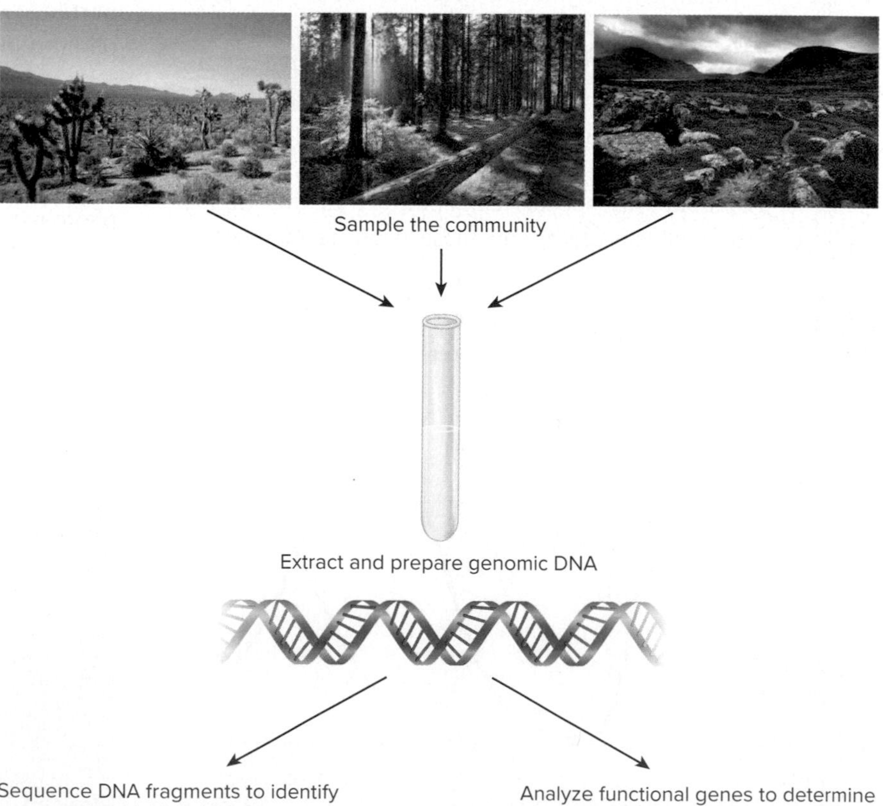
Sample the community
Extract and prepare genomic DNA
Sequence DNA fragments to identify microbial inhabitants
Analyze functional genes to determine roles of microbial inhabitants

FIGURE 28.5 Metagenomics Used to Characterize Microbial Communities Gene sequences in DNA from environmental samples are compared to known gene sequences in current databases. (top left): John & Karen Hollingsworth/U.S. Fish & Wildlife Service; (top middle): konradlew/Vetta/Getty Images; (top right): ajliikala/Getty Images;

What gene sequences are used to identify the organisms present in an environmental sample?

MicroAssessment 28.2

Metagenomics is the study of genetic material taken from an environment. Such cultivation-independent techniques allow researchers to better understand natural microbial communities.

4. What can researchers learn about a microbial community using metagenomics?
5. Why are molecular techniques important in studying microbial ecology?
6. How could FISH (fluorescence in situ hybridization) be used to determine the relative proportions of archaea and bacteria in a community?

28.3 ■ Aquatic Habitats

Learning Outcome

7. Compare and contrast the habitats provided by marine, freshwater, and specialized aquatic environments.

Marine environments such as open oceans cover more than 70% of Earth's surface. They are the most abundant aquatic habitat, representing about 95% of the global water. The freshwater environments—lakes and rivers—represent only a small fraction of the total water.

Deep lakes and oceans have characteristic zones that influence the distribution of microbial populations. The uppermost layer—where sufficient light penetrates—supports the growth of photosynthetic microorganisms, including algae and cyanobacteria. These primary producers synthesize organic material that is then metabolized by heterotrophs as it gradually sinks.

The number of microorganisms in waters is influenced by the nutrient content. In **oligotrophic** waters, meaning nutrient poor, the lack of phosphate, nitrate, and iron limits the growth of photosynthetic organisms and other autotrophs. When waters are **eutrophic** (nutrient-rich), the availability of inorganic nutrients allows photosynthetic organisms to flourish, often forming a visible layer on the surface (**figure 28.6**). In turn, photosynthesis produces organic compounds that foster the growth of heterotrophs in lower layers. The heterotrophs consume dissolved O_2 as they metabolize the organic material. Because O_2 consumption can outpace the slow rate of diffusion of atmospheric O_2 into the waters, the environment can become **hypoxic** (very low in dissolved O_2). Insufficient O_2 leads to the death of resident fish and other aquatic animals.

Marine Environments

Marine environments range from the deep sea, where nutrients are scarce, to the shallower coastal regions, where nutrients may be abundant due to runoff from the land. Seawater contains about 3.5% salt, compared with about 0.05% for fresh water. Consequently, it supports the growth of halophilic organisms, which prefer or require high salt concentrations, and halotolerant ones. Temperatures often vary widely at the surface, but they decrease with depth until reaching about 2°C in the deeper waters; an exception is the areas around hydrothermal vents.

Ocean waters are typically oligotrophic, limiting the growth of microorganisms. The small amounts of organic compounds produced by photosynthetic organisms are quickly consumed as they sink, so few nutrients reach the sediments below. Even in the deep sea, marine water is O_2-saturated due to mixing associated with tides, currents, and wind action.

The ecology of inshore areas is not as stable as the deep sea and can be dramatically affected by nutrient-rich runoff. An unfortunate example is a **dead zone** (a region in a large body of water that lacks fish and other marine life) that forms in the Gulf of Mexico (**figure 28.7**). The Mississippi River—carrying nutrients accumulated as it runs through agricultural, industrial, and urbanized regions—feeds into the Gulf. As a consequence of the excess nitrate and phosphate, algae and cyanobacteria grow to high numbers (referred to as a *bloom*) in the spring and summer when sunlight is also plentiful. Heterotrophic microbes then metabolize the organic compounds synthesized by these primary producers, consuming dissolved O_2 in the

FIGURE 28.6 Eutrophication in a Polluted Stream Photosynthetic organisms flourish in the nutrient-rich water. The organic compounds they produce foster the growth of heterotrophs, which deplete O_2 as they metabolize organic compounds. Geography Photos/Getty Images

? What is the likely fate of a fish living in a eutrophic stream?

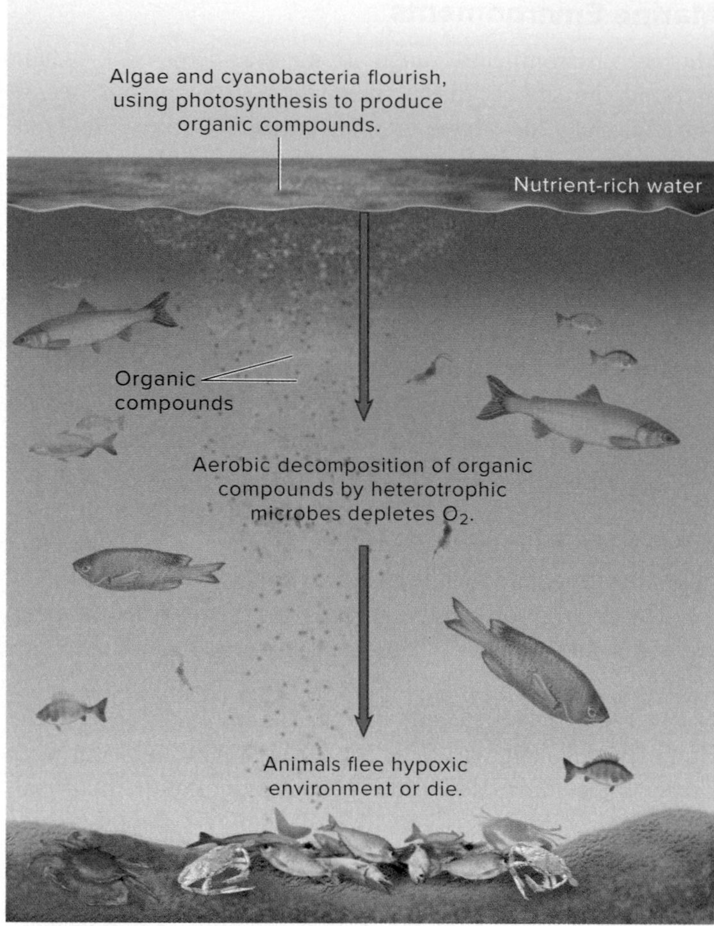
Algae and cyanobacteria flourish, using photosynthesis to produce organic compounds.
Nutrient-rich water
Organic compounds
Aerobic decomposition of organic compounds by heterotrophic microbes depletes O_2.
Animals flee hypoxic environment or die.

FIGURE 28.7 Dead Zone Formation

Why does decomposition by microbes deplete O_2 in water?

process. This causes a large region in the Gulf—sometimes in excess of 7,000 square miles—to become hypoxic. Animals in the water either move away or die. Dead zones are a problem in other waters as well, including fresh waters such as Lake Erie (see Focus on a Case 11.1). Nutrient enrichment of coastal waters also contributes to the excessive growth of toxin-producing photosynthetic microorganisms. When an algal bloom has a harmful effect—such as creating a dead zone or toxic conditions—it is referred to as a harmful algal bloom (HAB).

Freshwater Environments

As with marine environments, the types and relative numbers of microbes inhabiting fresh waters depend on multiple factors, including light, concentration of dissolved O_2 and nutrients, and temperature.

Oligotrophic lakes in temperate climates may have anaerobic layers due to thermal stratification resulting from seasonal temperature changes. During the summer months, the surface water warms. This decreases the density of the water, causing it to form a distinct layer that does not mix with the cooler, denser water below. The upper layer, called the epilimnion, is generally O_2 rich due to the activities of photosynthetic organisms. In contrast, the lower layer, the hypolimnion, may be anaerobic due to the consumption of O_2 by heterotrophs. Separating these two layers is the thermocline, a zone of rapid temperature change. As the weather cools, the waters mix, delivering O_2 to the deep water.

Rapidly moving waters, such as rivers and streams, are very different from lakes. They are usually shallow and turbulent, facilitating O_2 circulation, so they are generally aerobic. Light may penetrate to their bottoms, making photosynthesis possible. Sheathed bacteria such as *Sphaerotilus* and *Leptothrix* species commonly adhere to rocks and other solid structures, where they then use nutrients that flow by.

Specialized Aquatic Environments

Specialized aquatic environments include salt lakes, such as the Great Salt Lake in Utah, which have no outlets. As water in these lakes evaporates, the salt concentrations become much higher than that in seawater. Extreme halophiles thrive in this environment.

Other specialized habitats include iron springs that contain large quantities of ferrous ions; these springs are habitats for species of *Gallionella* and *Sphaerotilus*. Sulfur springs support the growth of both photosynthetic and non-photosynthetic sulfur bacteria. Other aquatic environments include groundwater, stagnant ponds, swimming pools, and drainage ditches, each offering its own opportunity for microbial growth.

MicroAssessment 28.3

Aquatic habitats include marine, freshwater, and specialized environments. Nutrient-rich waters can become hypoxic. Lakes often exhibit thermal stratification during summer months.

7. Compare the salt content of seawater with that of fresh water and salt lakes.
8. Explain how nutrient-rich runoff can cause waters to become hypoxic.
9. Why would the nutrient content of a body of water be more homogeneous than that of a terrestrial environment?

28.4 ■ Terrestrial Habitats

Learning Outcome

8. Describe soil as a microbial habitat.

Although microorganisms can adhere to and grow on a variety of objects on land, the focus in this section is soil—a crucial component of terrestrial ecosystems. Extreme terrestrial habitats, such as volcanic vents and fissures, and some of the extremophiles that inhabit them are described in chapter 11.

Soil is composed of finely ground rock, decaying organic material, air, and water. It is full of life, including bacteria, fungi, algae, protozoa, worms, insects, and plant roots. The top 6 inches of fertile soil may contain more than 2 tons of bacteria and fungi per acre! Scientists have known for years that the microorganisms are essential to the health and productivity of plants, but DNA sequencing projects such as the Earth BioGenome Project now provide an exciting chance to learn much more. Just as it is difficult to unravel the relationships in the complex human microbiome, however, so it will be with soil communities.

Microorganisms in soil are important for reasons beyond their ecological niche. The various species of *Streptomyces* produce over 500 different antibiotic substances, at least 50 of which have useful applications in medicine, agriculture, and industry. The pharmaceutical industry has tested many thousands of soil microorganisms in search of those that produce useful antibiotics. In addition, soil microbes are being investigated for their ability to degrade toxic chemicals, an application of microbiology called bioremediation. Probably no other habitat represents such a wide range of biosynthetic and biodegradative capabilities as that of soil.

MicroByte

The number of microorganisms in a cup of fertile soil is greater than the number of people who have ever lived on Earth.

Characteristics of Soil

Soil forms as rock weathers. Water, temperature changes, windblown particles, and other physical forces gradually cause the rock to crack and break. Photosynthetic organisms growing on the rock surfaces synthesize organic compounds, which various microorganisms then use as carbon and energy sources; as a result, acids and other chemicals are produced that gradually decompose the rocks. As soil slowly forms, some plants begin to grow. When these die and decay, the residual organic material functions as a sponge, retaining water and thus allowing more plants to grow. Over time, more organic compounds accumulate, forming a slowly degrading complex polymeric substance called humus.

Soil represents an environment that can change abruptly and dramatically. Heavy rains, for example, can cause a soil to rapidly become waterlogged. Trees dropping their leaves can suddenly enrich the soil with organic nutrients. Farmers and gardeners rapidly change the nutrient mix by applying fertilizers.

The texture of the soil influences how much air and water can flow through it. Finely textured soils, such as clay soils, are more likely to become waterlogged and anaerobic. In contrast, sandy soils that dry quickly allow water to pass through and are generally aerobic.

Microorganisms in Soil

The density and composition of the soil microbiome are dramatically affected by environmental conditions. Wet soils, for example, are unfavorable for aerobic microbes because the spaces fill up with water, lowering the amount of air in the soil. When soils become very dry, as during a drought or in a desert environment, the metabolic activity and number of soil microorganisms decrease. Many microbes that live in soil produce desiccation-resistant survival forms such as endospores and cysts. Other environmental influences that affect soil microbes include acidity, temperature, and nutrient supply. For example, acidity suppresses bacterial growth, allowing fungi to thrive with less competition for nutrients. This is why mushrooms often appear in a lawn fertilized with an acid-producing fertilizer such as ammonium chloride.

An especially important part of the soil microbiome is found in the **rhizosphere** (the zone of soil that adheres to plant roots). The concentration of microbes there, particularly Gram-negative bacteria, is relatively high because the root cells secrete organic molecules that serve as a nutrient source. Specific types of bacteria appear to preferentially interact with particular plants. For example, the rhizosphere of certain grasses can have high concentrations of *Azospirillum* species, which fix nitrogen (meaning they convert nitrogen gas into ammonia, a form that can be incorporated into cellular material).

Prokaryotes are the most numerous soil inhabitants. Their physiological diversity allows them to colonize all types of soil. In general, Gram-positive bacteria are more abundant in soils than Gram-negative bacteria (except around the rhizosphere). Among the most common Gram-positive bacteria are members of the genus *Bacillus*; these form endospores, allowing them to survive long periods of adverse conditions such as drought or extreme heat. *Streptomyces* species produce conidia, which are desiccation-resistant structures, and they also produce metabolites called geosmins, which give soil its characteristic musty odor. As already mentioned, *Streptomyces* species produce many medically useful antibiotics. Other bacteria adapted to thrive in terrestrial environments—including myxobacteria and species of *Clostridium, Azotobacter, Agrobacterium,* and *Rhizobium*—were discussed in chapter 11.

Although prokaryotes are the most numerous soil microbes, the biomass of fungi is much greater. Most fungi are aerobes, so they usually grow in the top 10 cm of soil. The soil fungi degrade complex macromolecules such as lignin (the major component of cell walls of woody plants) and cellulose. Some soil fungi are free-living, and others live in symbiotic relationships. The latter include **mycorrhizas** ("fungus root"), which are fungi growing in a symbiotic relationship with the roots of certain types of plants. The fungal partners in the relationships do more than assist individual plants; they create hyphal networks that link various plants, thereby providing the plants with a mechanism to share signals. These important examples of symbiosis are discussed later in the chapter.

In addition to bacteria and fungi, various algae and protozoa are found in most soils. Algae depend on sunlight for energy, so they mostly live on or near the soil surface. Most protozoa require O_2, so they too are found near the surface, typically where microbes on which they feed are plentiful.

MicroAssessment 28.4

The density and composition of soil are dramatically affected by environmental conditions. The concentration of microbes in the rhizosphere is generally much higher than that of the surrounding soil.

10. What is the significance of the rhizosphere?
11. Why are wet soils unfavorable for aerobic organisms?
12. How can the biomass of fungi in soil be greater than the biomass of bacteria, considering that bacteria are far more numerous?

28.5 ■ Biogeochemical Cycling and Energy Flow

Learning Outcomes

9. Diagram the carbon, nitrogen, sulfur, and phosphorus cycles, and describe some of the important microbial contributors.
10. Compare and contrast energy cycling in environments with sunlight versus those far removed from sunlight.

Biogeochemical cycles are the cyclical paths that elements take as they flow through living (biotic) and non-living (abiotic) components of ecosystems. These cycles are important because a fixed and limited amount of the elements that make up living cells exists on Earth and in the atmosphere. Thus, in order for an ecosystem to sustain its characteristic life-forms, elements must continually be recycled. For example, the organic carbon that animals use as an energy source is exhaled as carbon dioxide (CO_2); if this inorganic carbon were not eventually converted back to an organic form, we would run out of the organic carbon needed for growth. The carbon and nitrogen cycles are particularly important because they involve stable gaseous forms (carbon dioxide and nitrogen gas), which enter the atmosphere and thus have global impacts.

Although elements continually cycle in an ecosystem, energy does not. Instead, energy must be continually added to an ecosystem, fueling the activities required for life.

Understanding the cycling of nutrients and the flow of energy is particularly important because human activities can affect the environment. For example, industrial processes that convert nitrogen gas (N_2) into ammonia-containing fertilizers have boosted food production substantially, but they also alter the nitrogen cycle by increasing the amount of nitrogen available in organic compounds. When these nitrogen sources pollute lakes and coastal areas, eutrophication can result. As a consequence, the water's dissolved O_2 is depleted, leading to the death of aquatic animals. Burning of coal, oil, and other carbon-rich fossil fuels provides energy for our daily activities, but it also releases CO_2 and other carbon-containing gases into the atmosphere. Fossil fuels, the ancient remains of partially decomposed plants and animals, are nutrient reservoirs that would be unavailable without human intervention and therefore would not normally participate in biogeochemical cycles. Carbon-containing gases in the atmosphere absorb infrared radiation and reflect it back to Earth, raising global temperatures—a phenomenon called the greenhouse effect.

When studying biogeochemical cycles, it is helpful to consider the role of a given element in a particular organism's metabolism. Elements are used for three general purposes:

- **Biosynthesis (biomass production).** All organisms require elements for biosynthesis. As an example, nitrogen is required to produce amino acids. Plants and many prokaryotes assimilate nitrogen by incorporating ammonia (NH_3) to synthesize the amino acid glutamate (see figure 6.30). Some prepare for this step by converting nitrate (NO_3^-) to ammonia. Once glutamate has been synthesized, the amino group can then be transferred to other carbon compounds to produce the other amino acids. Animals cannot incorporate ammonia, so instead they must consume amino acids in their diet. Some prokaryotes can reduce atmospheric nitrogen to form ammonia—the process of nitrogen fixation. The ammonia can then be incorporated into cellular material.
- **Energy source.** Reduced carbon compounds such as sugars, lipids, and amino acids are used as energy sources by chemoorganotrophs. Chemolithotrophs can use reduced inorganic molecules such as hydrogen sulfide (H_2S), ammonia (NH_3), and hydrogen gas (H_2) (see table 4.5).
- **Terminal electron acceptor (TEA).** In aerobic conditions, O_2 is used as a TEA in respiration. In anaerobic conditions, some prokaryotes can use nitrate (NO_3^-), nitrite (NO_2^-), sulfate (SO_4^-), or carbon dioxide (CO_2) as a TEA.

Carbon Cycle

All organisms are composed of organic molecules such as proteins, lipids, and carbohydrates. The carbon in these molecules travels through the food chain when primary producers are eaten by primary consumers that are then eaten by secondary consumers. Decomposers then use organic molecules in the remains of primary producers and consumers.

Carbon Fixation

A fundamental part of the carbon cycle is **carbon fixation,** the defining characteristic of primary producers (**figure 28.8**). Fixation incorporates chemical elements into molecules that can be directly used by living organisms. Primary producers convert CO_2 into an organic form, using mechanisms described in chapter 6. Without primary producers, no other organisms, including humans, could exist. We depend on them to generate the organic carbon compounds we use as an energy source and for biosynthesis.

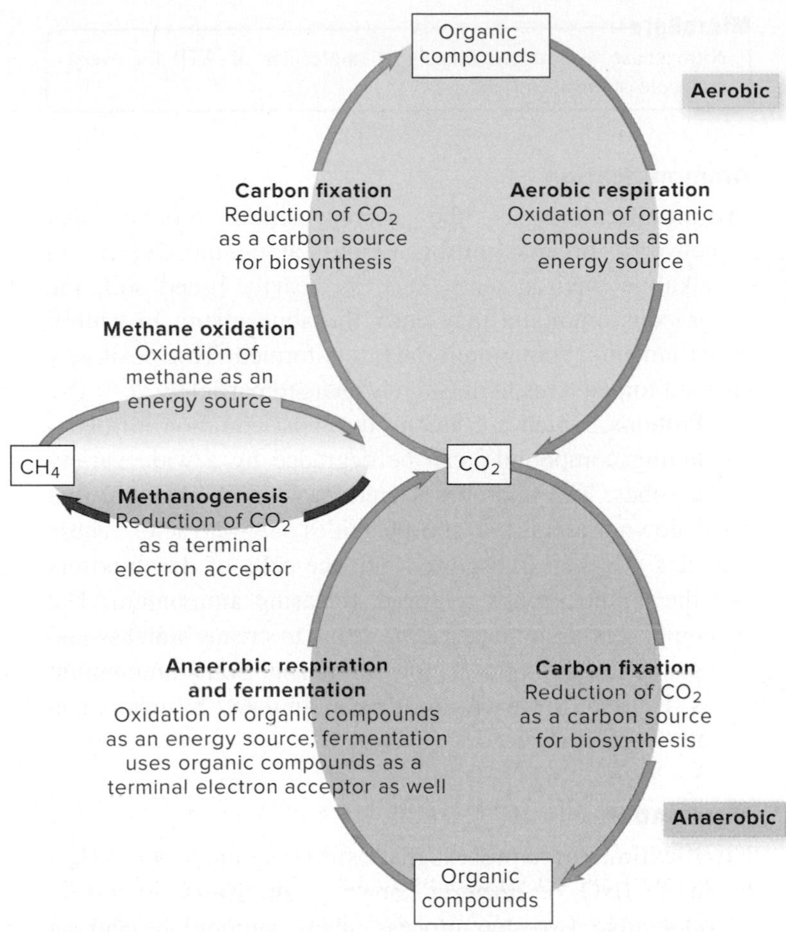

FIGURE 28.8 Carbon Cycle Blue arrows represent steps where carbon compounds are used as energy sources, red arrows represent steps where carbon compounds are used as terminal electron acceptors, and green arrows represent steps where inorganic carbon is used in biosynthesis.

? **How does the burning of fossil fuels affect the carbon cycle?**

Respiration and Fermentation

When heterotrophs consume organic material, they break it down using respiration and/or fermentation to release energy, which is captured to make ATP. These processes usually also make CO_2.

The type of organic material helps dictate which species degrade it. A wide variety of organisms use sugars, amino acids, and proteins as energy sources, but rapidly multiplying bacteria often play the dominant role in decomposing these substances. In contrast, relatively few organisms (primarily fungi) can break down lignin, a major component of wood (**figure 28.9**). Aerobic conditions are required for this degradation, so wood at the bottom of marshes resists decay.

The O_2 supply has a strong influence on the carbon cycle. Not only does O_2 allow degradation of certain compounds such as lignin, it also helps determine the types of carbon-containing gases produced. When organic matter is degraded aerobically, a great deal of CO_2 is produced. When the O_2 level is low, however, as is the case in marshes, swamps, and manure piles, the degradation is incomplete, generating some CO_2 and also a variety of other products.

Methanogenesis and Methane Oxidation

In anaerobic environments, CO_2 is used by methanogens. These archaea obtain energy by oxidizing hydrogen gas, using CO_2 as a terminal electron acceptor, and generating methane (CH_4). Methane that enters the atmosphere is oxidized by ultraviolet light and chemical ions, forming carbon monoxide (CO) and CO_2. Microorganisms called methylotrophs can use methane as an energy source, oxidizing it to produce CO_2.

Nitrogen Cycle

Nitrogen is a component of proteins and nucleic acids. As consumers ingest plants and animals for carbon and energy, they also obtain their required nitrogen. Prokaryotes, as a group, are far more diverse in their use of nitrogen-containing compounds. Some use oxidized nitrogen compounds such as nitrate (NO_3^-) and nitrite (NO_2^-) as terminal electron acceptors; others use reduced nitrogen compounds such as ammonium (NH_4^+) as energy sources. These metabolic activities represent essential steps in the nitrogen cycle (**figure 28.10**).

Nitrogen Fixation

Nitrogen fixation is the process in which nitrogen gas (N_2) is reduced to form ammonia (NH_3), which can then be incorporated into cellular material. The process, catalyzed by the enzyme complex nitrogenase, requires a tremendous amount of energy because N_2 has a very stable triple covalent bond. Organisms that can fix nitrogen are referred to as **diazotrophs** (*di* means "two," *azo* means "nitrogen," and *troph* means "nourishment"). Although our atmosphere consists of approximately 80% N_2, relatively few organisms are diazotrophs, and all of these are prokaryotes. Thus, just as humans and other animals depend on primary producers to fix carbon, they rely on prokaryotes to convert atmospheric nitrogen to a form they can assimilate to create biomass.

FIGURE 28.9 Wood-Degrading Fungus These fungi digest lignin, the major cell wall component of woody plants. Nature Picture Library/ Alamy Stock Photo

? **Dead trees submerged in marshes resist decay. Why?**

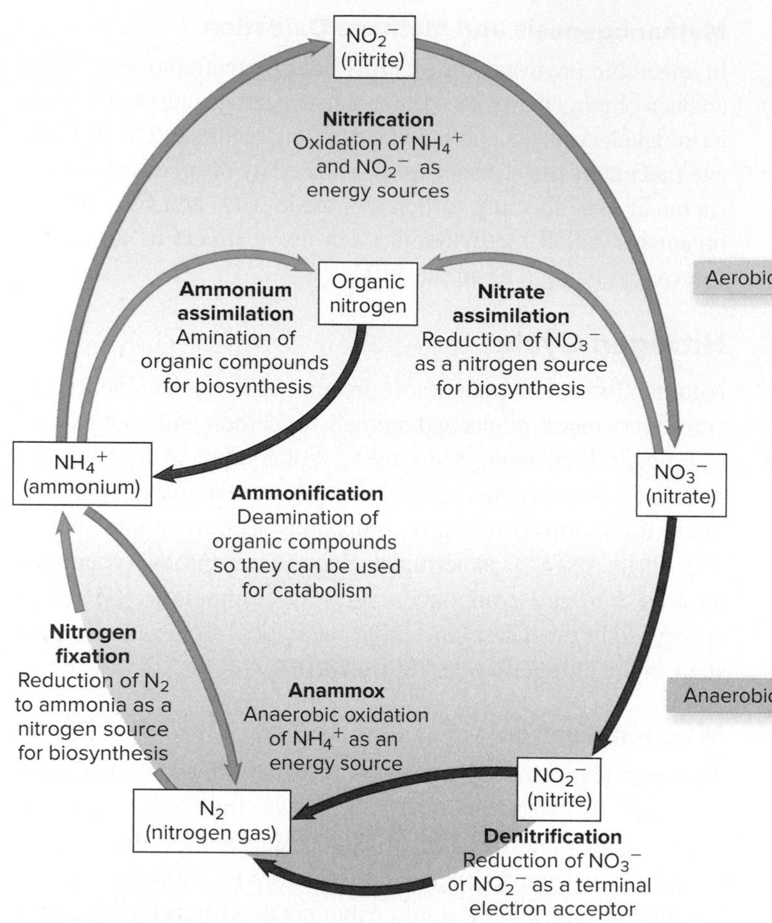

FIGURE 28.10 Nitrogen Cycle Blue arrows represent steps where nitrogen compounds are used as energy sources, red arrows represent steps where nitrogen compounds are used as terminal electron acceptors, and green arrows represent steps where inorganic nitrogen is used in biosynthesis. The black arrow represents decomposition reactions. Not all parts of the cycle are shown.

? **The only organisms that can fix nitrogen are prokaryotes. Why does this make them crucial in the nitrogen cycle?**

Some diazotrophs are free-living, whereas others form symbiotic associations with higher organisms, particularly certain plants. Among the free-living examples are members of the genus *Azotobacter.* These heterotrophic, aerobic, Gram-negative rods may be the chief suppliers of fixed nitrogen in ecosystems such as grasslands that lack plants with nitrogen-fixing symbionts. The dominant free-living, anaerobic, soil diazotrophs are certain members of the genus *Clostridium.* Some cyanobacteria are diazotrophs; these photosynthetic bacteria use both nitrogen and carbon from the atmosphere. Symbiotic diazotrophs will be discussed later in the chapter.

Energy-expensive chemical processes can fix nitrogen to make fertilizers, and these are playing an increasingly larger role in the nitrogen cycle. In fact, fixed nitrogen sources associated with human intervention, including fertilizer production and planting crops that foster the growth of symbiotic nitrogen-fixers, now surpass natural biological nitrogen fixation.

MicroByte

Nitrogenase uses approximately 16 molecules of ATP for every molecule of nitrogen fixed.

Ammonification

Ammonification is the decomposition process that converts organic nitrogen into ammonia (NH_3). In alkaline environments, such as heavily limed soil, the gaseous ammonia may enter the atmosphere. In neutral environments, ammonium (NH_4^+) is formed. This positively charged ion adheres to negatively charged particles.

Proteins, which are among the most common nitrogen-containing compounds, can be degraded by a wide variety of microbes. The microbes secrete proteolytic enzymes that break down proteins into short peptides or amino acids. These products are then transported into the cells of decomposers and the amino groups removed, releasing ammonium. The decomposers assimilate much of this to create biomass and use the remaining parts for catabolism. Some ammonium is released, however, and then assimilated by plants and other organisms.

Nitrification

Nitrification is the process that oxidizes ammonium (NH_4^+) to nitrate (NO_3^-). Bacteria known as **nitrifiers** do this in a cooperative two-step process, using ammonium and an intermediate, nitrite (NO_2^-), as energy sources. Nitrifiers are obligate aerobes, using O_2 as a terminal electron acceptor. Consequently, nitrification does not occur in waterlogged soils or in anaerobic regions of aquatic environments.

Nitrification has some important consequences with respect to agricultural practices and pollution. Farmers often apply ammonium-containing compounds to soils as a source of nitrogen for plants. The ammonium is retained by soils because its positive charge causes it to adhere to negatively charged soil particles. Nitrification converts the ammonium to nitrate, a nitrogen form more readily used by plants but rapidly leached from soil by rainwater. To slow nitrification, certain chemicals can be added to ammonium-fertilized soils.

Leaching of nitrate and nitrite from soil is a health concern if the compounds contaminate drinking water. Nitrite is toxic because it can combine with hemoglobin of the blood, reducing blood's O_2-carrying capacity. Even nitrate, which in itself is not very toxic, can be dangerous if high levels are ingested because some intestinal bacteria can use it as a terminal electron acceptor, converting it to nitrite.

Denitrification

Denitrification is the process that reduces nitrate (NO_3^-), converting it to gaseous forms such as nitrous oxide (N_2O) and molecular nitrogen (N_2). This happens when prokaryotes anaerobically respire using nitrate as a terminal electron acceptor.

Denitrification can have negative environmental and economic consequences. Under anaerobic conditions in wet soils, for example, denitrifying bacteria will reduce the oxidized nitrogen compounds of fertilizers, releasing gaseous nitrogen to the atmosphere. In some areas, this process may represent 80% of nitrogen lost from fertilized soil, a considerable economic loss to the farmer. In addition, nitrous oxide contributes to global warming.

Denitrification is not always undesirable. The process can be actively fostered in certain steps of wastewater treatment as a means to remove nitrate. This compound could otherwise act as a fertilizer in the waters to which the wastewater is discharged, promoting algal growth.

Anammox

Certain bacteria oxidize ammonium under anaerobic conditions, using nitrite as a terminal electron acceptor. This reaction, called **anammox** (for **an**oxic **amm**onia **ox**idation), forms N_2 and might provide an economical means of removing nitrogen compounds during wastewater treatment.

Sulfur Cycle

Sulfur is found in all living matter, primarily as a component of the amino acids methionine and cysteine. Like the nitrogen cycle, key steps of the sulfur cycle depend on the activities of prokaryotes (**figure 28.11**).

Sulfur Assimilation and Decomposition

Most plants and microorganisms assimilate sulfur as sulfate (SO_4^{2-}), reducing it and then incorporating the reduced form into biomass. Like nitrogen, organic sulfur is present chiefly as a part of the amino acids that make up proteins. Decomposition of the sulfur-containing amino acids releases hydrogen sulfide (H_2S), a gas.

Sulfur Oxidation

Hydrogen sulfide (H_2S) and elemental sulfur (S^0) can both serve as an energy source for certain chemolithotrophs. Sulfur-oxidizing prokaryotes, including species of *Beggiatoa, Thiothrix,* and *Thiobacillus,* oxidize these molecules to sulfate (SO_4^{2-}). Certain prokaryotes in anaerobic marine environments can oxidize elemental sulfur, using nitrate as a terminal electron acceptor. As discussed in chapter 11, these organisms, including *Thioploca* species and *Thiomargarita namibiensis,* have unusual mechanisms to cope with the fact that their energy source and terminal electron acceptor are found in two different environments.

Hydrogen sulfide and elemental sulfur are oxidized anaerobically by photosynthetic purple sulfur and green bacteria. These bacteria harvest energy from sunlight but require reduced molecules as a source of electrons to generate reducing power. Like the chemolithotrophs that use hydrogen sulfide and elemental sulfur, the photosynthetic sulfur-oxidizers produce sulfate.

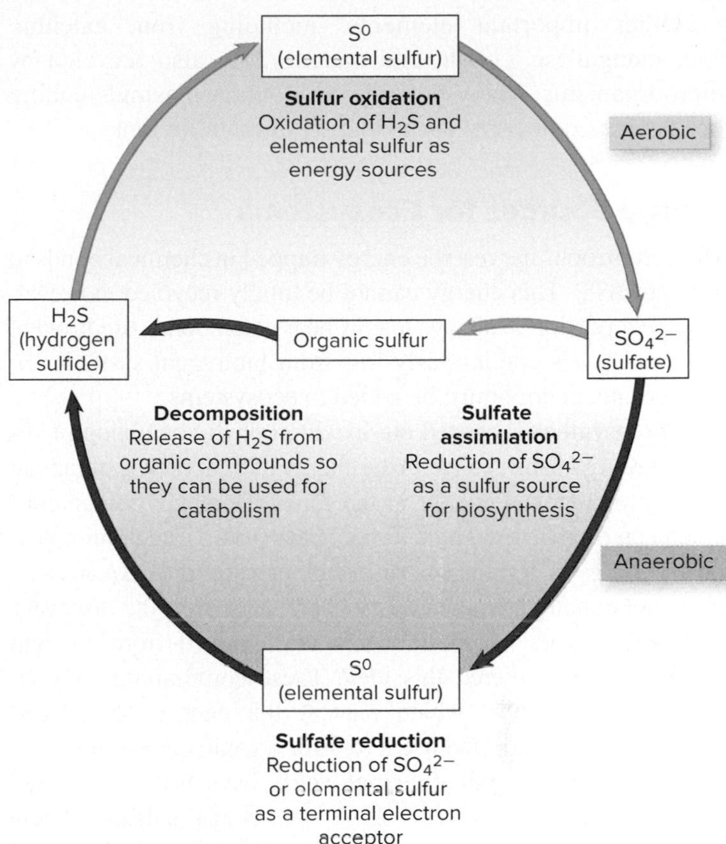

FIGURE 28.11 Sulfur Cycle Blue arrows represent steps where sulfur compounds are used as energy sources, red arrows represent steps where sulfur compounds are used as terminal electron acceptors, and green arrows represent steps where inorganic sulfur is used in biosynthesis. The black arrow represents decomposition reactions.

? Organisms that oxidize sulfur use it for what purpose in their metabolism?

Sulfur Reduction

Under anaerobic conditions, sulfate generated by the sulfur-oxidizers can then be used as a terminal electron acceptor by certain organisms. The sulfur- and sulfate-reducing bacteria and archaea use sulfate in the process of anaerobic respiration, reducing it to hydrogen sulfide (H_2S). In addition to its unpleasant odor, the H_2S is a problem because it reacts with metals, resulting in corrosion.

Phosphorus Cycle and Other Cycles

Phosphorus is a component of several critical biological compounds, including nucleic acids, phospholipids, and ATP. Most plants and microorganisms can take up phosphorus as orthophosphate (PO_4^{3-}), incorporating it into biomass. From there, the phosphorus is passed along the food web. When plants and animals die, decomposers convert organic phosphate back to the inorganic form.

In many aquatic habitats, the growth of algae and cyanobacteria (the primary producers) is limited by low concentrations of phosphorus. When phosphates are added from sources such as agricultural runoff, phosphate-containing detergents, and wastewater, eutrophication can result.

Other important elements, including iron, calcium, zinc, manganese, cobalt, and mercury, are also recycled by microorganisms. Many prokaryotes contain plasmids coding for enzymes that carry out oxidation of metallic ions.

Energy Sources for Ecosystems

All chemotrophs harvest the energy trapped in chemical bonds to generate ATP. This energy cannot be totally recycled, however, because a portion is always lost as heat when bonds are broken. Thus, energy is continuously lost from biological systems. To compensate, energy must be added to ecosystems.

Photosynthesis, carried out by chlorophyll-containing plants and microorganisms, converts the energy of light (a type of radiant energy) to chemical energy in the form of organic compounds, which can then be used by chemoorganotrophs. The requirement for the energy of light has traditionally been used to explain why life is not equally abundant everywhere. However, the discovery of different types of communities far removed from sunlight has dramatically altered this idea. These communities rely on chemolithoautotrophs, which harvest the energy of reduced inorganic compounds and use it to form organic compounds.

A number of **hydrothermal vents** have been discovered some thousands of meters below the ocean surface. These vents form when water seeps into cracks in the ocean floor and becomes heated by molten rock, finally spewing out in the form of mineral-laden undersea geysers. Hydrogen sulfide discharged from the vents supports thriving deep-sea communities—oases in the otherwise desolate ocean floor (**figure 28.12**). Large numbers of sulfur-oxidizing chemolithoautotrophs are found in and around the vents. Many of these bacteria and archaea are free-living, but some live in symbiotic association with the large tube worms and clams that inhabit the areas. The chemolithoautotrophs obtain energy by oxidizing hydrogen sulfide, and they fix CO_2, providing the animals with a source of both carbon and energy.

Microbial populations have been found almost 3 km under the ground and in iron-rich volcanic rocks from nearly a thousand meters below the surface of the Columbia River. These organisms gain energy from hydrogen (H_2) produced in the subsurface. It has been estimated that if (and it is a big "if" at this point) most similar rocks contain microbes, there could be as much as 2×10^{14} tons of underground microorganisms—equivalent to a layer 1.5 meters thick over the entire land surface of the earth!

MicroAssessment 28.5

Elements are recycled as organisms incorporate them to produce biomass, oxidize the reduced forms as energy sources, and reduce the oxidized forms as terminal electron acceptors. Carbon fixation incorporates atmospheric CO_2 into organic material; the CO_2 is regenerated during respiration and some fermentations. Prokaryotes are essential for several steps of the nitrogen cycle, including nitrogen fixation, nitrification, denitrification, and anammox. Prokaryotes are also essential for several steps of the sulfur cycle, including sulfur reduction and sulfate oxidation.

13. What are the three general roles of carbon-containing compounds in metabolism?
14. Why do farmers try to prevent nitrification?
15. Although chemoautotrophs serve as the primary producers near hydrothermal vents, animals there still ultimately depend on the photosynthetic activities of plants and cyanobacteria. Why?

(a)

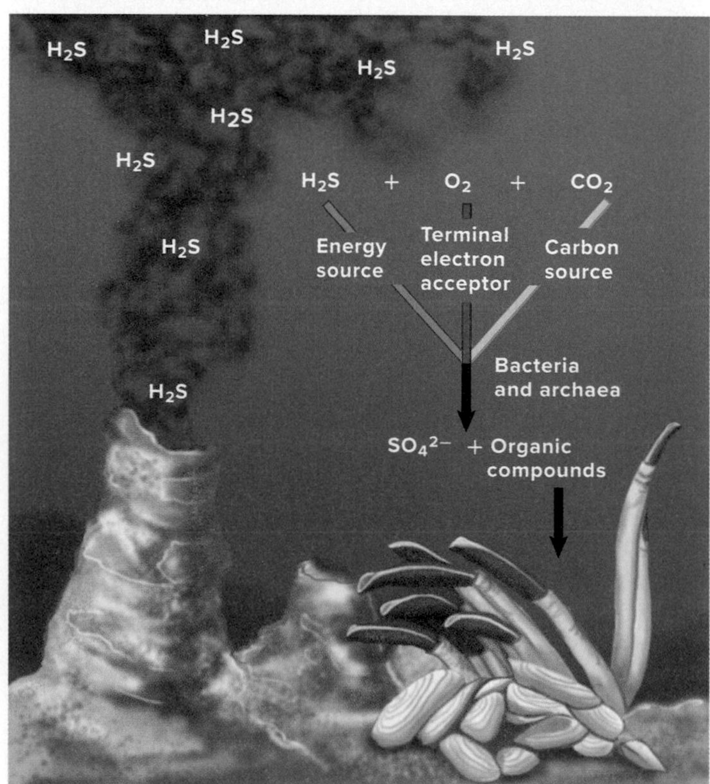

(b)

FIGURE 28.12 Hydrothermal Vent Community (a) This diverse community is supported by the metabolic activities of chemolithoautotrophs. **(b)** Water escaping from the vent is rich in reduced compounds, including hydrogen sulfide, that can serve as energy sources. a: Kim Juniper/AFP/Newscom

? Photosynthetic organisms are normally considered the most important primary producers; which organisms are primary producers in hydrothermal vent communities?

28.6 ■ Mutualistic Relationships Between Microorganisms and Eukaryotes

Learning Outcome

11. Describe the mutualistic relationships between fungi and plant roots, symbiotic nitrogen-fixers and plants, and microorganisms and herbivores.

As described in chapter 16, mutualism is a symbiotic association in which both partners benefit. A variety of other ecologically significant symbiotic relationships exist, but mutualistic relationships highlight the vital importance of microorganisms to life on this planet.

Mycorrhizas

Mycorrhizas (*myco* means "fungus"; *rhizo* means "root") are fungi growing in symbiotic relationships with plant roots. They enhance the competitiveness of plants by helping them take up phosphorus, water, and other substances from the soil. In turn, the fungi gain sugars and other nutrients for their own growth from root secretions. Some mycorrhizal fungi even form networks that link various plants, perhaps allowing plants to exchange nutrients or transmit stress signals. Over 85% of vascular plants (plants with specialized water- and food-conducting tissues) are thought to have mycorrhizas.

The two general types of mycorrhizal relationships are (**figure 28.13**):

- **Endomycorrhizas.** The fungi penetrate the root cell walls to create invaginations in the cytoplasmic membrane; this is by far the most common mycorrhizal relationship and is formed with most herbaceous plants. Relatively few fungal species form this type of relationship, perhaps only 100 or so, and most appear to be obligate symbionts. The relationship for some plants is also obligate; for example, most orchid seeds will not germinate without the activities of a fungal partner. Of the several categories of endomycorrhizas, the fungal partner in the most common type grows as bush-like masses called arbuscules; the relationship is called an arbuscular mycorrhiza.
- **Ectomycorrhizas.** The fungi do not penetrate the root cell wall and instead grow around the plant cells, forming a sheath around the root tip and a network of hypha referred to as a "Hartig net" surrounding some of the root cells. Ecotomycorrhizal fungi associate mainly with certain trees, including conifers, beeches, and oaks; many of the over 5,000 species of fungi involved form a partnership with only with a single type of plant. Chanterelles and truffles are examples of commercially valuable ectomycorrhizal fungi.

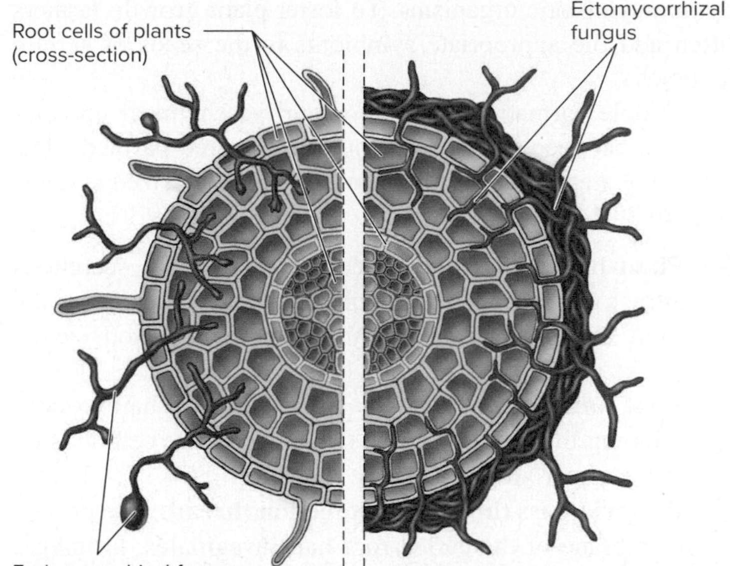

(a) Diagram of Mycorrhizas

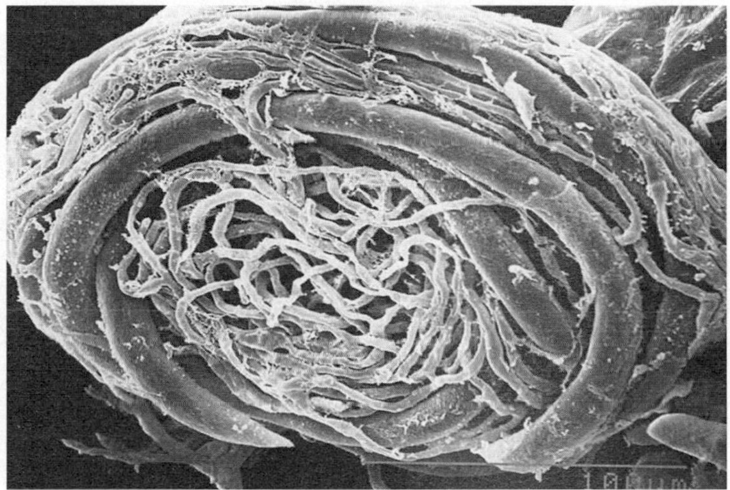

(b) Endomycorrhiza

FIGURE 28.13 Mycorrhizas (a) Diagram illustrating the two types of mycorrhizas. **(b)** In an endomycorrhizal relationship, the fungi penetrate root cells and grow within them. b: Stephan Imhof

? **Which type of mycorrhizal relationship is common in herbaceous plants?**

Symbiotic Nitrogen-Fixers and Plants

Although some free-living bacteria can add fixed nitrogen to the soil, symbiotic nitrogen-fixing organisms are far more significant in benefiting plant growth and crop production. They are important in both terrestrial and aquatic habitats.

Rhizobia

Members of a diverse group of genera collectively referred to as **rhizobia** are the most agriculturally important symbiotic nitrogen-fixing bacteria. These grow as endosymbionts within specialized organs called **nodules** on the roots of legumes (plants that bear seeds in pods), including alfalfa, clover, peas, beans, and peanuts. The input of soil nitrogen from rhizobia may be about 10 times the annual rate of nitrogen fixation by non-symbiotic organisms. To foster plant growth, farmers often add the appropriate symbionts to the seeds of certain legumes.

Nodule formation involves extensive chemical communication between the rhizobia and the legume partners. The following steps occur in the most well-characterized systems (**figure 28.14**):

① **Plant-bacterial communication.** Plant root secretions attract the appropriate rhizobia, which then colonize the roots and produce chemical signals called Nod factors (NFs).

② **Root hairs curl.** The NFs cause the root hair to curl, trapping the bacterial cells; they also induce cell division in the underlying plant cells.

③ **Bacteria pass through an infection thread.** The plasma membrane of the curled root hair invaginates, forming a tube-like structure called an infection thread. The rhizobia cells travel through this structure to underlying plant tissues and are then engulfed by the dividing plant cells, forming membrane-bound structures called symbiosomes.

④ **Nitrogen-fixing nodule forms.** Within a symbiosome, a rhizobial cell changes in shape and function to become a specialized nitrogen-fixing cell called a bacteroid.

The plant cells make an O_2-binding protein called leghemoglobin. This protein, which is somewhat similar to hemoglobin, maintains low O_2 concentration in the nodule, thereby protecting the O_2-sensitive nitrogenase. The plant also provides various nutrients to the endosymbionts. Meanwhile, the endosymbionts provide fixed nitrogen to the plant.

Although the relationship between the plant and the bacterium is not obligate, it offers a competitive advantage to both partners. The rhizobia do not fix nitrogen in soils lacking legumes, and they compete poorly with other microbes, slowly disappearing from soils in which legumes are not grown. Likewise, legumes compete poorly against other plants in heavily fertilized soils.

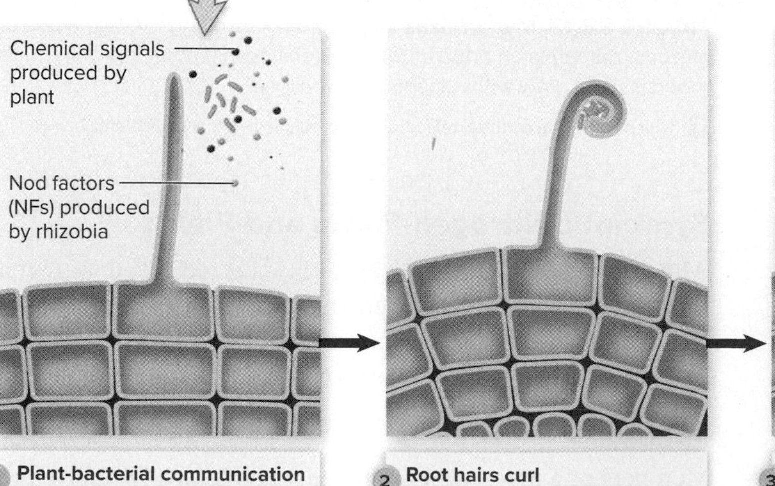

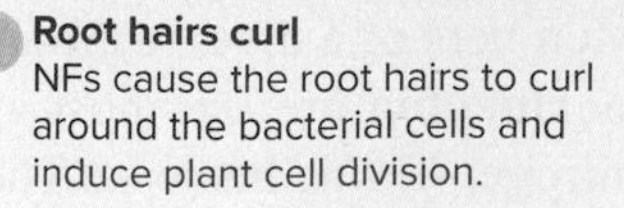

2 Root hairs curl
NFs cause the root hairs to curl around the bacterial cells and induce plant cell division.

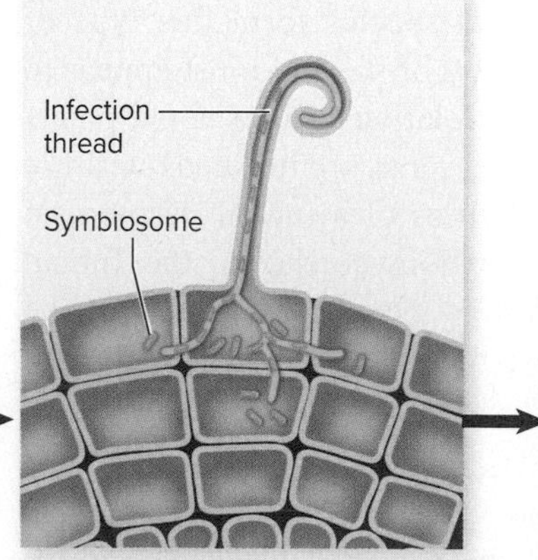

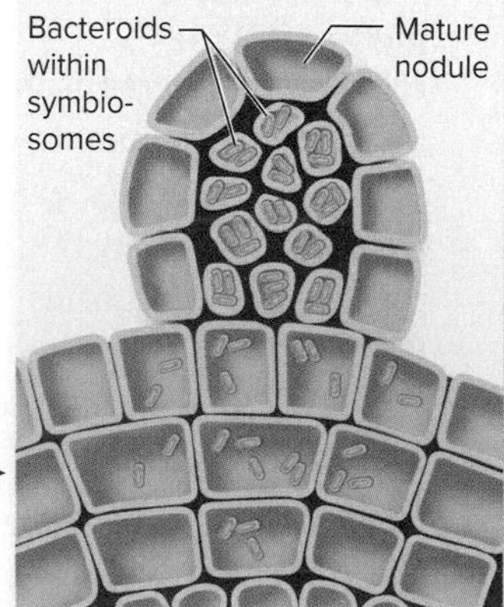

FIGURE 28.14 Root Nodules The major steps leading to nodule formation by *Rhizobium* species. a: Lisa Burgess/McGraw Hill

? The root shown in the photograph could be from any of which group of plants?

Other Nitrogen-Fixing Symbionts

Several genera of non-leguminous trees, including alder, have nitrogen-fixing root nodules at some stages of their life cycle. The bacteria involved in the symbiosis are members of the genus *Frankia.*

In aquatic environments, the most significant nitrogen-fixers are cyanobacteria. They are especially important in flooded soils such as rice paddies. In fact, rice has been cultivated successfully for centuries without the addition of nitrogen-containing fertilizer because of the symbiotic relationship between the cyanobacterium *Anabaena azollae* and the aquatic fern *Azolla.* The bacterium grows in specialized sacs in the leaves of the fern, providing nitrogen to the fern. Before planting rice, the farmer allows the flooded rice paddy to overgrow with *Azolla* ferns. Then, as the rice grows, it eventually crowds out the ferns. As the ferns die and decompose, their nitrogen is released into the water.

MicroByte

Scientists recently discovered that an ancient variety of corn obtains nitrogen by producing syrup-coated aerial roots that nourish nitrogen-fixing bacteria.

Microorganisms and Herbivores

Another mutualistic relationship occurs between microbes and certain herbivores. In order to live on grass and other plant material, herbivores such as cattle and horses rely on microbial communities that inhabit a specialized digestive compartment within the animal. The microbes digest cellulose and hemicellulose, two of the major components of plant material, releasing compounds that can then serve as a nutrient source for the animal. The specialized digestive compartment in ruminants such as cattle, sheep, and deer is called a rumen, and it precedes the true stomach. Non-ruminant herbivores such as horses and rabbits have a compartment called a cecum, which serves a similar purpose but lies between the small intestine and the large intestine.

The rumen functions as an anaerobic fermentation vessel to which nutrients in the form of plant materials are intermittently added. Each milliliter of rumen content contains approximately 10^{10} bacteria, 10^6 protozoa, and 10^3 fungi; of the over 200 species identified, no single one accounts for more than 3% of the total number of cells. The microorganisms in the rumen degrade the ingested material, releasing sugars that are then fermented to produce various organic acids. Those acids are absorbed by the animal, providing an energy and nutrient source. Large quantities of gas—produced as a result of fermentation—are discharged when the animal belches.

After being digested in the rumen, the food mass enters another compartment (the omasum), eventually reaching the acidic true stomach (abomasum). There, organic acids

FOCUS ON A CASE 28.1

A novice farmer noticed his two steers behaving strangely; one was staggering and the other was bellowing and urinating. Both looked as if someone had blown them up like balloons. The farmer had fed the animals a diet of hay for several weeks at the end of winter, but then moved the animals into a pasture with newly sprouted grass and lush alfalfa. The animals greedily grazed on the tender plants, but then an hour later the problems began. The farmer ran to call the veterinarian and desperately tried to remember what he had been told about something called "frothy bloat"—a type of gas buildup in the rumen.

1. Why are ruminants subject to bloat?
2. What was the likely cause of bloat in the farmer's animals?
3. Why is bloat dangerous for ruminants?
4. What can be done to prevent bloat?
5. What can be done to treat animals with bloat?

Discussion

1. When ruminants swallow freshly eaten vegetation, it enters the rumen. There, microorganisms ferment cellulose and other structural carbohydrates, producing 200 to 400 liters of gas per day (largely CO_2 and methane). This gas is expelled regularly by a type of belching called eructation that healthy cattle do about once per minute. If not released, the gas remains in the rumen, causing bloat in as little as 15 minutes.
2. Microorganisms in the rumen can digest high-protein alfalfa much faster than high-fiber grass or hay, producing greater volumes of gases, small particles, and slime associated with biofilms. Foam or froth forms, trapping the gas in the mass of small bubbles that cannot rise freely to be released by belching.
3. Trapped gas inflates the rumen, causing it to press against vital organs such as the heart and lungs. This can kill the animal, often by suffocation. In addition, rapid breakdown of the easily digested alfalfa causes a buildup of organic acids, the end products of fermentation. These can diffuse into the bloodstream and cause systemic acidosis.
4. Bloat can be prevented by allowing only gradual changes in the animal's diet so that microbial communities in the rumen can adjust to new foods.
5. Walking an animal may cause it to belch and release gas. Oils that act as antifoaming agents can also be administered to break the bubbles that otherwise prevent gas from escaping. Inserting a hose down the animal's esophagus into the rumen may physically allow gas to leave under pressure through the hose. In extreme cases, a tube may be inserted directly into the rumen through the skin. This offers immediate relief for the animal but poses a risk of infection.

are absorbed. In addition, lysozyme is secreted, allowing the animal to lyse and then digest members of the microbial population—a source of even more nutrients. A crucial feature of ruminants is that the microbial population gets the first opportunity to use the ingested nutrients. The animal then uses the metabolic end products as well as the microbial cells themselves.

The cecum of non-ruminant herbivores serves a function similar to a rumen. Microbes in the cecum, however, cannot be used as a food source because of the cecum's location relative to the stomach. The benefit of the location is that the animal can digest and absorb readily available nutrients without competition from microbes.

MicroAssessment 28.6

Mycorrhizal fungi gain nutrients from plant root secretions while helping plants take up substances from soil. Symbiotic nitrogen-fixers provide plants with a source of usable nitrogen while being provided with an exclusive habitat. Microorganisms in the rumen and cecum of herbivores digest cellulose and hemicellulose, allowing the animal to subsist on plant material.

16. Describe the differences between an endomycorrhiza and an ectomycorrhiza.
17. Describe the differences between a rumen and a cecum.
18. Gardeners sometimes plant clover between productive growing seasons. Why would this practice be beneficial?

Summary

28.1 ■ Principles of Microbial Ecology

Ecosystems vary in their biodiversity and biomass. The microenvironment is most relevant to a microorganism's survival and growth.

Nutrient Acquisition

Primary producers convert CO_2 into organic material that can be used by **consumers**; **decomposers** digest the remains of primary producers and consumers (figure 28.1).

Microbes in Low-Nutrient Environments

Microorganisms able to grow in dilute aqueous solutions are common in nature; often they grow in biofilms.

Microbial Competition

Microorganisms in the environment compete for the same limited pool of nutrients (figure 28.2). A species can competitively exclude others or produce compounds that inhibit others.

Microorganisms and Environmental Changes

Environmental changes are common and often result in a different population becoming dominant (figure 28.3).

Microbial Communities

A **microbial mat** is a thick, dense, highly organized biofilm composed of distinct colored layers of different groups of microbes (figure 28.4).

28.2 ■ Studying Microbial Ecology

Microbial ecology has been difficult to study because so few environmental prokaryotes can be successfully grown in the laboratory. **Metagenomics** allows researchers to characterize microbial communities (figure 28.5).

28.3 ■ Aquatic Habitats

Oligotrophic waters are nutrient poor; **eutrophic** waters are nutrient rich (figure 28.6). Excessive growth of aerobic heterotrophs may cause an aquatic environment to become **hypoxic,** resulting in the death of fish and other aquatic animals.

Marine Environments

Ocean waters are generally oligotrophic and aerobic, but inshore areas can be dramatically affected by nutrient-rich runoff (figure 28.7).

Freshwater Environments

Oligotrophic lakes may have anaerobic layers due to thermal stratification. Shallow, turbulent streams are generally aerobic.

Specialized Aquatic Environments

Salt lakes and mineral-rich springs support the growth of microbes specifically adapted to thrive in these specialized environments.

28.4 ■ Terrestrial Habitats

Characteristics of Soil

Soil represents an environment that can change abruptly and dramatically.

Microorganisms in Soil

Environmental conditions affect the density and composition of the soil microbiome. The concentration of microbes in the **rhizosphere** is generally much higher than that in the surrounding soil.

28.5 ■ Biogeochemical Cycling and Energy Flow

Organisms use elements for biosynthesis, as sources of energy, and as terminal electron acceptors.

Carbon Cycle (figure 28.8)

One of the fundamental aspects of the carbon cycle is **carbon fixation.** As consumers and decomposers degrade organic materials, CO_2 is released during respiration and some fermentations (figure 28.9).

Nitrogen Cycle (figure 28.10)

The steps of the nitrogen cycle include **nitrogen fixation, ammonification, nitrification, denitrification,** and **anammox.**

Sulfur Cycle (figure 28.11)

Sulfur reduction and sulfur oxidation are steps of the sulfur cycle that depend on the activities of prokaryotes.

Phosphorus Cycle and Other Cycles

Most plants and microorganisms take up orthophosphate, incorporating it into biomass. Iron, calcium, zinc, manganese, cobalt, and mercury are recycled by microorganisms.

Energy Sources for Ecosystems

Photosynthetic organisms convert the energy of light to chemical bond energy in the form of organic compounds. Chemolithoautotrophs harvest energy from reduced inorganic chemicals (figure 28.12).

28.6 ■ Mutualistic Relationships Between Microorganisms and Eukaryotes

Mycorrhizas

Mycorrhizas help plants take up phosphorus and other substances from soil; in turn the fungal partners gain nutrients for their own growth (figure 28.13). Endomycorrhizal fungi penetrate the root cell wall; ectomycorrhizal fungi grow around root cells.

Symbiotic Nitrogen-Fixers and Plants

Rhizobia reside as nitrogen-fixing endosymbionts in nodules on legume roots (figure 28.14). *Frankia* species fix nitrogen in nodules of alder. A species of cyanobacteria fixes nitrogen in specialized sacs in the leaves of the *Azolla* fern.

Microorganisms and Herbivores

In order to live on a diet of grass and other plant material, herbivores rely on a community of microbes that inhabit a specialized digestive compartment, either a rumen or a cecum.

Review Questions

Short Answer

1. Describe why a microbial mat has green, reddish-pink, and black layers.
2. How do bacteriocins benefit bacteria growing in their natural habitat?
3. Why is there a high concentration of microbes in the rhizosphere?
4. What is the role of publicly available genetic databases in metagenomic studies of environmental samples?
5. Why does wood resting at the bottom of a bog resist decay?
6. What is the importance of nitrogen fixation?
7. Describe the relationship between ammonia oxidizers and nitrite oxidizers.
8. How do hydrothermal vents support thriving communities of microbes, clams, and tube worms?
9. Give examples of free-living and symbiotic nitrogen-fixing microorganisms. Are these prokaryotic or eukaryotic?
10. Describe the steps that lead to the formation of the symbiotic relationship between rhizobia and legumes.

Multiple Choice

1. Cyanobacteria are
   a) primary producers.
   b) consumers.
   c) herbivores.
   d) decomposers.
   e) more than one of the above.
2. Which of the following is *false*?
   a) Culture techniques are an accurate way of determining which members in a microbial community are most common.
   b) Fluorescence in situ hybridization (FISH) can be used to distinguish subsets of prokaryotes that contain a specific nucleotide sequence.
   c) Taxonomic diversity of prokaryotes can be studied by examining 16S rRNA sequences.
   d) Studying the genome of one organism can give insights into the characteristics of another.
3. Which of the following pairs that relate to aquatic environments does not match?
   a) Oligotrophic—nutrient poor
   b) Hypoxic—O_2 poor
   c) Epilimnion—O_2 poor
   d) Hypolimnion—lower layer
   e) Eutrophic—nutrient rich
4. Adding high levels of nutrients to coastal waters would have all of the following effects in that environment *except*
   a) death of clams and crabs.
   b) increased growth of heterotrophic microbes.
   c) increased growth of photosynthetic organisms.
   d) increased levels of dissolved O_2.
5. Which of the following is *not* a matching pair?
   a) Soil—minimal biodiversity
   b) *Bacillus*—endospores
   c) *Streptomyces*—geosmin production
   d) Fungi—lignin degradation
   e) Rhizosphere—soil that adheres to plant root
6. Atmospheric nitrogen can be used
   a) directly by all living organisms.
   b) only by aerobic bacteria.
   c) only by anaerobic bacteria.
   d) in symbiotic relationships between rhizobia and plants.
   e) in photosynthesis.
7. Which process converts ammonium (NH_4^+) into nitrate (NO_3^-)?
   a) Nitrogen fixation
   b) Ammonification
   c) Nitrification
   d) Denitrification
   e) Anammox
8. Energy for ecosystems can come from
   a) sunlight via photosynthesis.
   b) oxidation of reduced inorganic chemicals by chemoautotrophs.
   c) both a and b.
9. Mycorrhizas represent associations between plant roots and microorganisms that
   a) are antagonistic.
   b) help plants take up phosphorus and other nutrients from soil.
   c) involve algae in the association with plant roots.
   d) form nodules on the plant's leaves.
   e) lead to the production of antibiotics.
10. In symbiotic nitrogen fixation by rhizobia and legumes,
   a) the amount of nitrogen fixed is much greater than by non-symbiotic organisms.
   b) neither the bacteria nor the legume can exist independently.
   c) the bacteria enter the leaves of the legume.
   d) the bacteria operate independently of the legume.

Applications

1. A farmer who was growing soybeans, a type of legume, saw an Internet site advertising an agricultural product for safely killing soil bacteria. The ad claimed that soil bacteria were responsible for most crop losses. The farmer called the agricultural extension office at a local university for advice. Explain what the extension office adviser most likely told the farmer about the usefulness of the product.
2. Recent reports suggest that human activities, such as the generous use of nitrogen fertilizers, have doubled the rate at which elemental nitrogen is fixed, raising concerns about environmental overload of nitrogen. What problems could arise from too much fixed nitrogen, and what could be done about this situation?

Critical Thinking

1. Each colony growing on an agar plate arises from a single cell (see photo). Colonies growing close together are much smaller than those that are well separated. Why would this be so?
2. An entrepreneur found an economically feasible way of collecting large amounts of sulfur from underwater hot vents in the Pacific Ocean. The sulfur will be harvested from the microorganisms found in the vent areas. A group of ecologists argued that the project would destroy the fragile ecosystem by depleting it of usable sulfur. The entrepreneur argued that the environment would not be harmed because the vents produce more than enough sulfur for the clams and tube worms in the area. Explain who is correct.

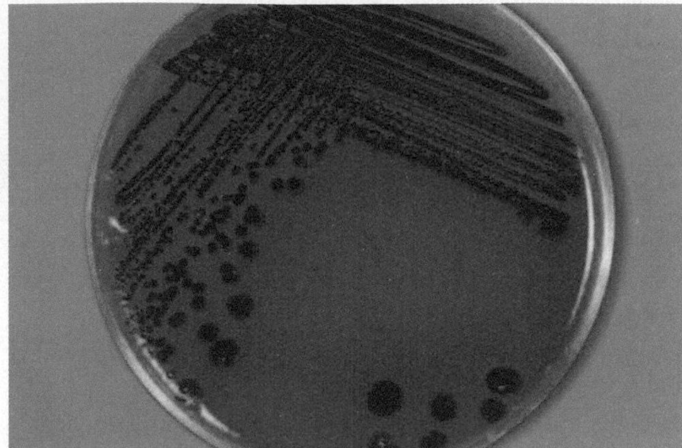

Dr. V.R. Dowell, Jr/CDC

www.mcgrawhillconnect.com

Enhance your study of this chapter with study tools and practice tests. Also ask your instructor about the resources available through Connect, including the media-rich eBook, interactive learning tools, and animations.

29 Environmental Microbiology: Treatment of Water, Wastes, and Polluted Habitats

A pristine mountain lake. *Robert Glusic/Getty Images*

KEY TERMS

Advanced Treatment In wastewater treatment, any physical, chemical, or biological purification process beyond secondary treatment.

Biochemical Oxygen Demand (BOD) The amount of O_2 required for the microbial decomposition of organic matter in a sample.

Bioremediation Process that uses microorganisms to degrade harmful chemicals.

Compost Material that results from the controlled decomposition of solid organic material.

Indicator Organisms Microbes whose presence in an environment suggests fecal contamination.

Primary Treatment A wastewater treatment process used to remove material that settles out.

Sanitary Landfill A site used for disposal of non-hazardous solid wastes in a manner that minimizes damage to human health and the environment.

Secondary Treatment A wastewater treatment process that uses microorganisms to convert suspended solids to inorganic compounds and removable cell mass.

Septic System An individual wastewater treatment system in which the solid portion settles out and is degraded by microorganisms while the liquid portion slowly passes through a gravel-containing drain field.

Sludge The solid portion of wastewater that settles to the bottom of sedimentation tanks during primary and secondary treatment.

Wastewater Any water that has been used by people; includes water that has gone down drains or toilets in homes or businesses or was used in an industrial process.

A Glimpse of History

Delivering fresh water to urban areas and removing human wastes have been practiced at least since Roman times—the ruins of ancient aqueducts can still be seen today in many parts of Europe. In the 1800s, the desire for clean, clear water led to the use of sand filtration systems in London and elsewhere. Later, Robert Koch showed that this kind of filtration not only yields clear water but also removes most bacteria.

In the 1840s, Edwin Chadwick, an English activist, proposed a system to flush solid wastes away from a city. His plan was to use narrow, smooth ceramic pipes and water under pressure to move the wastes to a distant site, where they then could be used to produce fertilizer to sell to farmers. Concern about cholera in 1848 prompted the newly formed Board of Health in England to begin installing a sewage system somewhat similar to that envisioned by Chadwick. New York City did the same in 1866, again in response to a threatened cholera epidemic. By the end of the nineteenth century, most large European and U.S. cities had sewer systems to remove and treat waste materials as well as water systems to deliver safe drinking water. Cholera in the industrialized nations of Europe and North America virtually disappeared.

Most people living in developed countries take for granted that their tap water is safe to drink, that their wastes will reliably disappear into sewers or landfills for proper disposal, and that pollutants (substances that are harmful or injurious) will not accumulate in the environment. They seldom consider the role of microbes in these essential aspects of modern life.

Microorganisms are important in the treatment processes described in this chapter for two very distinct reasons: (1) they are the ultimate recyclers, playing an essential role in the decomposition of our wastes and (2) some are pathogens and must be removed from sewage before it is discharged and from drinking water before it is safe for human consumption. Recreational waters such as swimming pools, water parks, lakes, rivers, and shorelines are also monitored to ensure they do not have harmful levels of certain pathogens.

Treatment of water, waste, and polluted habitats is a significant challenge, particularly in densely populated areas. Consider that every day the average American uses about 100 gallons of water and generates 80 gallons of wastewater and 5 pounds of garbage. This means that a city with only 1 million inhabitants is faced with the disposal of approximately 30 billion gallons of wastewater and a million tons of garbage each year!

29.1 ■ Microbiology of Wastewater Treatment

Learning Outcomes

1. Describe the concept of biochemical oxygen demand (BOD).
2. Compare and contrast primary treatment, secondary treatment, advanced treatment, and anaerobic digestion.
3. Describe how septic systems function.

Wastewater is any water that has has been used by people; it includes water that has gone down drains or toilets in homes or businesses or was used in an industrial process. The term **sewage** refers specifically to wastewater contaminated with human feces or urine. Municipal wastewater flows through a sewer system to a treatment plant. In many cities, stormwater runoff that flows into street drains enters wastewater treatment systems as well.

The most obvious reason sewage must be treated before discharge into the environment is that pathogenic microbes—including those that cause diarrheal diseases and hepatitis—can be transmitted in feces. If untreated sewage is released into a river or lake that is then used as a source of drinking water, disease can easily spread. If marine waters become contaminated in a similar manner, then eating the local shellfish can result in disease. Shellfish are filter feeders, and they concentrate microbes from the waters in which they live.

The high nutrient content of wastewater can also be damaging to the receiving water. When any nutrient-rich substance is added to an aqueous environment, microorganisms quickly use the compounds as energy sources. As a result, microbes that aerobically respire consume available O_2 in the water, using it as a terminal electron acceptor (see figure 6.21). The amount of dissolved O_2 in lakes and rivers is limited and can easily be depleted during the microbial breakdown of nutrients. Fish and other aquatic animals in the environment die because they require O_2 for respiration (see figure 28.7). Thus, effective wastewater treatment must decrease the level of organic compounds substantially, in addition to eliminating pathogens and pollutants.

Biochemical Oxygen Demand (BOD)

An important goal of wastewater treatment is to decrease the environmental impact by reducing the **biochemical oxygen demand (BOD)**—the amount of O_2 required for the microbial decomposition of organic matter in a given sample. High BOD values indicate that large amounts of degradable materials are present, resulting in the consumption of correspondingly large amounts of O_2 during biological degradation of the material. The BOD of raw sewage is approximately 300 to 400 mg/L, whereas the dissolved O_2 content of natural waters is generally 5 to 10 mg/L. Thus, adding raw sewage to a lake could easily deplete the dissolved O_2 in that water.

To determine the BOD of a sample, the O_2 level is first measured. The sample is then incubated in a sealed container in the dark under standard conditions of time and temperature, usually 5 days at 20°C. During that time, microbes in the sample will degrade the organic material. The O_2 level is then determined again; the difference between the amount of dissolved O_2 at the beginning of the test and the amount at the end reflects the BOD. Samples that have a very high BOD must be diluted as part of the procedure; otherwise, the O_2 in the water could be depleted before the sample has been completely decomposed.

Municipal Wastewater Treatment Methods

Large-scale wastewater treatment plants in the United States use a series of two processes—primary and secondary treatment—as required by the Clean Water Act. Once treated, the **effluent** (the liquid portion) can be discharged into the receiving water. Additional steps are used to treat **sludge** (the solid portion). Industrial wastes, including organic solvents and heavy metals, can interfere with standard wastewater treatment processes, so industries must have facilities to pretreat their wastewater; these are specific to the types of wastes generated and may remove organic solvents, heavy metals, and other harmful chemicals. **Figure 29.1** provides an overview of the municipal wastewater treatment process.

Primary Treatment

Primary treatment is a physical process designed to remove large objects and material that will settle out. Raw sewage is first passed through a series of screens to remove items such as sticks, rags, and trash (see figure 29.1). Skimmers then remove scum and other floating materials, including grease and oil. After that, the sewage sits in a sedimentation tank, allowing more solids to settle out. Primary treatment removes approximately 50% of the solids and 25% of the BOD from raw wastewater. Once the settling period is complete, the material at the bottom of the tank is scraped out and forms primary sludge. The sludge is pumped to an anaerobic sludge digester (described later in this section), and the remaining liquid waste is sent for secondary treatment.

Secondary Treatment

Secondary treatment is mainly a biological process that converts most of the suspended solids to inorganic compounds and cell mass that can then be removed, eliminating as much as 95% of the remaining BOD (see figure 29.1). Microbial growth is actively encouraged during secondary treatment, allowing aerobic organisms to oxidize the biologically degradable organic material to CO_2 and H_2O. Because secondary treatment relies on the metabolic activities of microorganisms, the processes could be devastated if too much toxic industrial waste or too many hazardous household materials were dumped into wastewater systems, killing the microbial population.

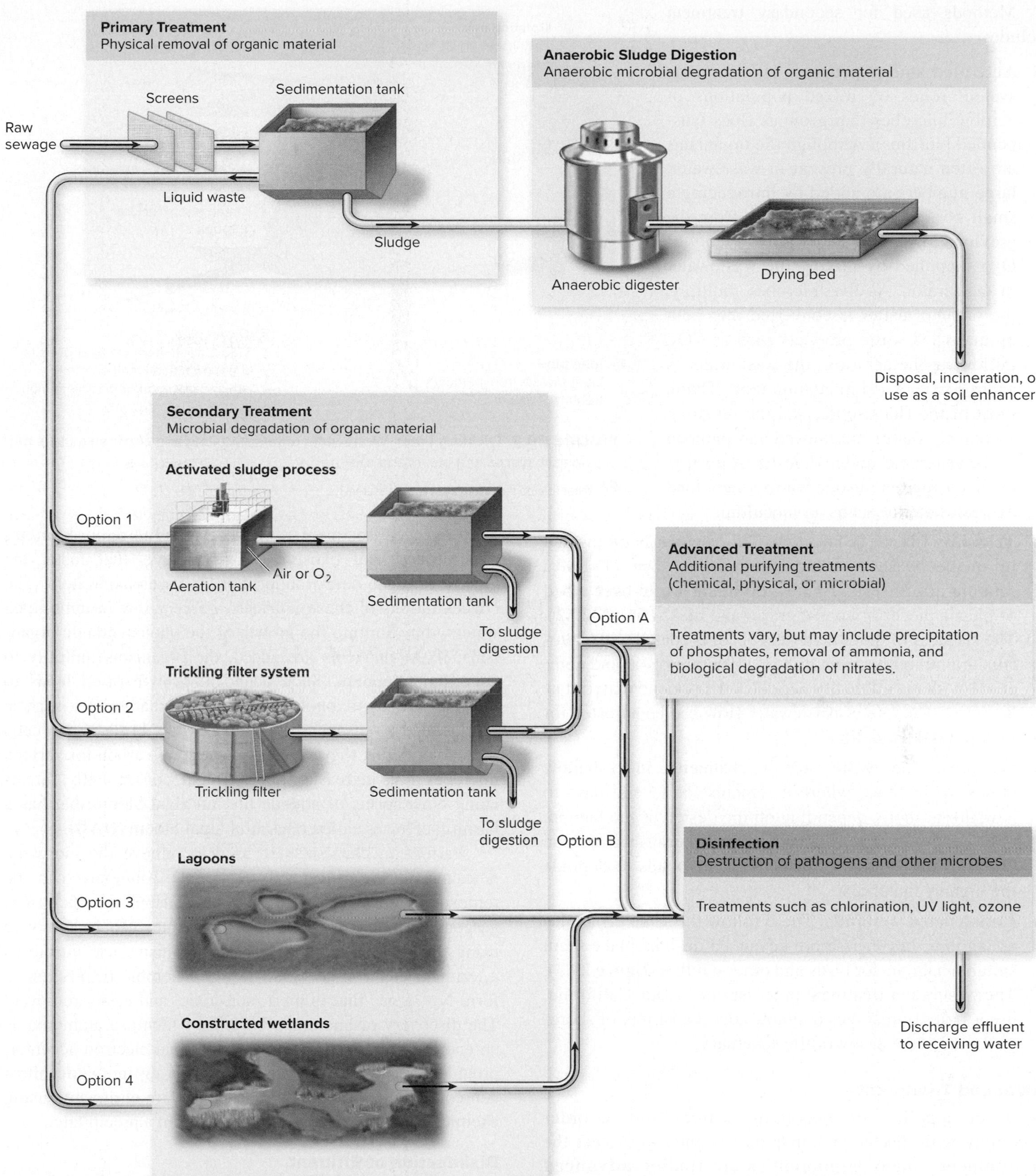

FIGURE 29.1 Municipal Wastewater Treatment The processes consist of primary treatment, secondary treatment, advanced treatment (optional), disinfection, and anaerobic sludge digestion.

? What is sludge?

Methods used for secondary treatment include:

- **Activated sludge process.** This common system relies on mixed populations of aerobic microbes that grow as flocs (suspended biofilms). Although the organisms are often naturally present in wastewater, large numbers are added by introducing a small portion of leftover sludge from the previous load of treated wastes. Plenty of O_2 is supplied by mixing the wastewater in an aerator. As the microbes multiply, the organic matter is converted into both biomass and waste products such as CO_2. Following the aeration, the wastewater is again sent to a sedimentation tank. There, most of the flocs settle, and the resulting secondary sludge is removed and pumped to the anaerobic sludge digester. A portion of this sludge is introduced to a new load of wastewater to act as an inoculum.

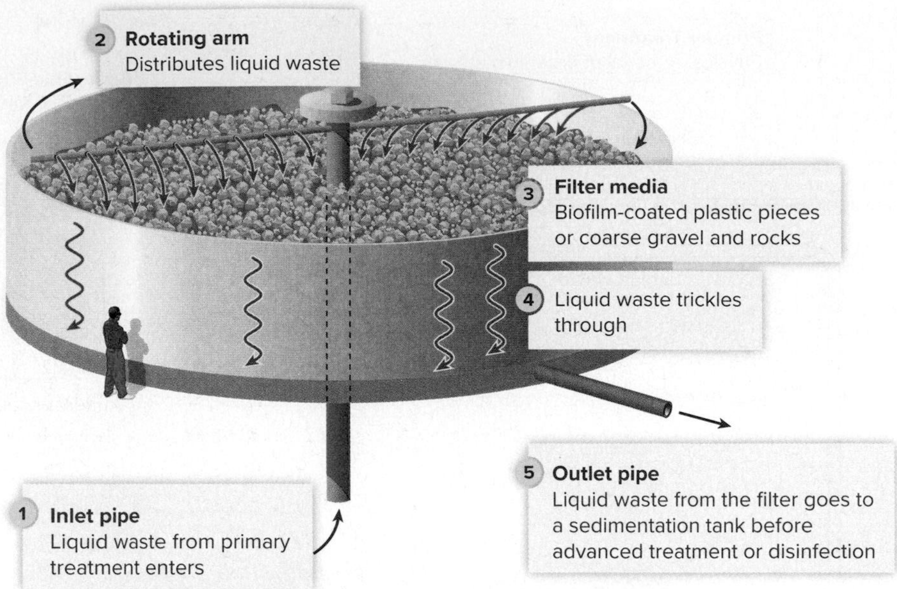
2 Rotating arm
Distributes liquid waste
3 Filter media
Biofilm-coated plastic pieces
or coarse gravel and rocks
4 Liquid waste trickles
through
5 Outlet pipe
Liquid waste from the filter goes to
a sedimentation tank before
advanced treatment or disinfection
1 Inlet pipe
Liquid waste from primary
treatment enters

FIGURE 29.2 Trickling Filter Wastewater enters a rotating arm and then trickles out onto plastic pieces or a gravel and rock bed.

? What role do biofilms play in trickling filters?

- **Trickling filter (TF) system.** This method is often used in smaller wastewater treatment plants. The TF has a rotating arm that distributes the liquid waste over a bed of plastic pieces or coarse gravel and rocks (**figure 29.2**). The surfaces of these materials become coated with a biofilm—a mix of bacteria, fungi, algae, protozoa, and nematodes—that aerobically degrades the organic material as it passes. The rate of wastewater flow can be adjusted for maximum degradation.
- **Lagoons.** The wastewater is channeled into shallow ponds, or lagoons, where it remains for several days to a month or more, depending on the design of the lagoon. Algae and cyanobacteria that grow at the surface provide O_2, allowing aerobic organisms in the ponds to degrade the organic materials.
- **Constructed wetlands.** These follow the same principles as lagoons, but their more advanced designs make them suitable habitats for birds and other wildlife (**figure 29.3**). The wastewater treatment processes in Arcata, California, use a series of marshes that now attract a variety of shorebirds and serve as a wildlife sanctuary.

Advanced Treatment

As water supplies are becoming scarce—and in order to comply with discharge standards designed to protect the environment—many communities are finding **advanced treatment** necessary. This includes any purification process beyond secondary treatment, and it may involve physical, chemical, or biological processes, or any combination of these. Advanced treatment is expensive, however, and has not been common in the past.

Advanced treatment is often designed to remove ammonia, nitrates, and phosphates—compounds that foster the growth of algae and cyanobacteria in receiving waters. The concentrations of those nutrients are very low in unpolluted waters, thus limiting the growth of the photosynthetic organisms; if the nutrients are added, the organisms multiply to very high numbers. Such abundant growth often leads to buoyant masses of photosynthetic cells that form a surface scum called a nuisance bloom (see figure 11.6). When cells in the bloom die, they serve as a source of carbon for various microbes, thus increasing the BOD and, consequently, threatening other forms of aquatic life; an algal bloom that has a harmful effect is called a harmful algal bloom (HAB).

Various mechanisms are used to remove the inorganic nutrients during advanced treatment. Ammonia can be removed by a process called ammonia stripping, which frees gaseous ammonia from the water. Nitrates can be removed using denitrifying bacteria. These organisms use nitrate as a terminal electron acceptor during anaerobic respiration to form N_2—a gas that is inert, non-toxic, and easily removed. The discovery of anammox bacteria, which use ammonia as an energy source and nitrite as a terminal electron acceptor, provides an alternative means of removing inorganic nitrogen-containing compounds. Phosphates are eliminated using chemicals that combine with them to form a precipitate.

Disinfection of Effluent

Before discharge into the receiving water, the effluent is disinfected with chlorine, ozone, or UV light to decrease the numbers of microorganisms and viruses. If chlorine is used, the disinfected water can then be dechlorinated to avoid releasing excessive amounts of the toxic chemical into the environment.

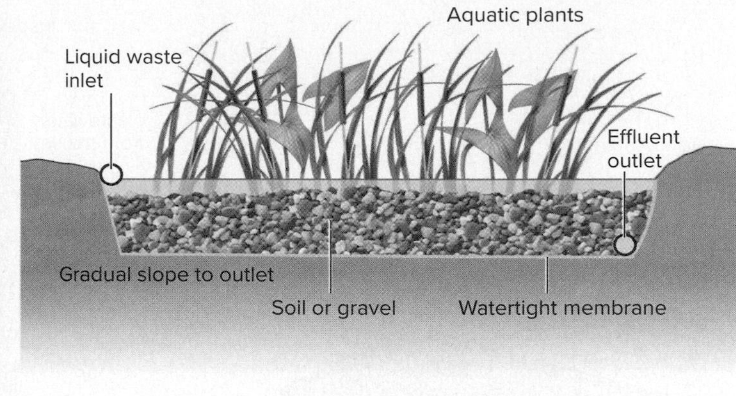

(a) (b)

FIGURE 29.3 Constructed Wetland (a) Wastewater may be turned into a natural resource. **(b)** General components of a constructed wetland.

a: Carrie Garcia/Alamy Stock Photo

? What advantage does a constructed wetland have over a lagoon?

Anaerobic Digestion of Sludge

Sludge obtained during the sedimentation steps of primary and secondary treatment is transferred to a tank for **anaerobic digestion.** There, various anaerobic microbes convert much of the organic material to methane using the following sequence of reactions:

- Organic compounds → organic acids, CO_2, H_2
- Organic acids → acetate, CO_2, H_2
- Acetate, CO_2, H_2 → methane (CH_4)

Many wastewater treatment plants are equipped to use the methane generated, thereby avoiding the cost of other sources of energy to run their equipment.

After anaerobic digestion, water is removed from the remaining sludge, generating a nutrient-rich product called stabilized sludge. This can be incinerated or disposed of in landfills, but biosolids—a term for stabilized sludge that meets certain standards—may be used to improve soils and promote plant growth. An increasing number of wastewater treatment facilities are finding ways to recycle their treated sludge. Concerns exist, however, about heavy metals and other pollutants that can sometimes be concentrated in the product.

MicroByte

Sludge generated by the city of Milwaukee, Wisconsin, is used to produce fertilizer for lawns, gardens, golf courses, and playfields.

Individual Wastewater Treatment Systems

Rural homes usually rely on **septic systems** for wastewater treatment (**figure 29.4**). These small-scale on-site systems move household wastewater to a large tank, where much of the solid material settles and is degraded by anaerobic microorganisms. The effluent—which has a high BOD—flows to a series of perforated pipes, where it percolates (slowly passes) through a gravel-containing drain field. The drain field is designed to allow aerobic microorganisms to oxidize the organic material. Sludge that remains in the septic tank must be pumped out from time to time.

Septic systems must be properly designed and monitored to ensure that they function properly. Circumstances that prevent efficient drainage—such as a clay soil under a drain field—allow anaerobic conditions to develop, preventing the organic material from being oxidized adequately. In addition, certain materials, including antibacterial soaps and bleach, can inhibit microbial activity in the system. Drainage from a septic tank may contain pathogens, so the tank must never be allowed to drain where it can contaminate water supplies.

MicroAssessment 29.1

Primary treatment of wastewater removes material that settles out. Secondary treatment converts suspended material into inorganic compounds and microbial biomass. Advanced treatment is often designed to remove ammonia, phosphates, and nitrates. Anaerobic digestion converts much of the organic matter to methane. Septic systems are small-scale on-site wastewater treatment systems.

1. Why is it important that the BOD of wastewater is decreased before discharge?
2. What is the advantage of removing phosphates and nitrates from wastewater?
3. Why is denitrification unlikely to occur in an aeration tank used during secondary treatment?

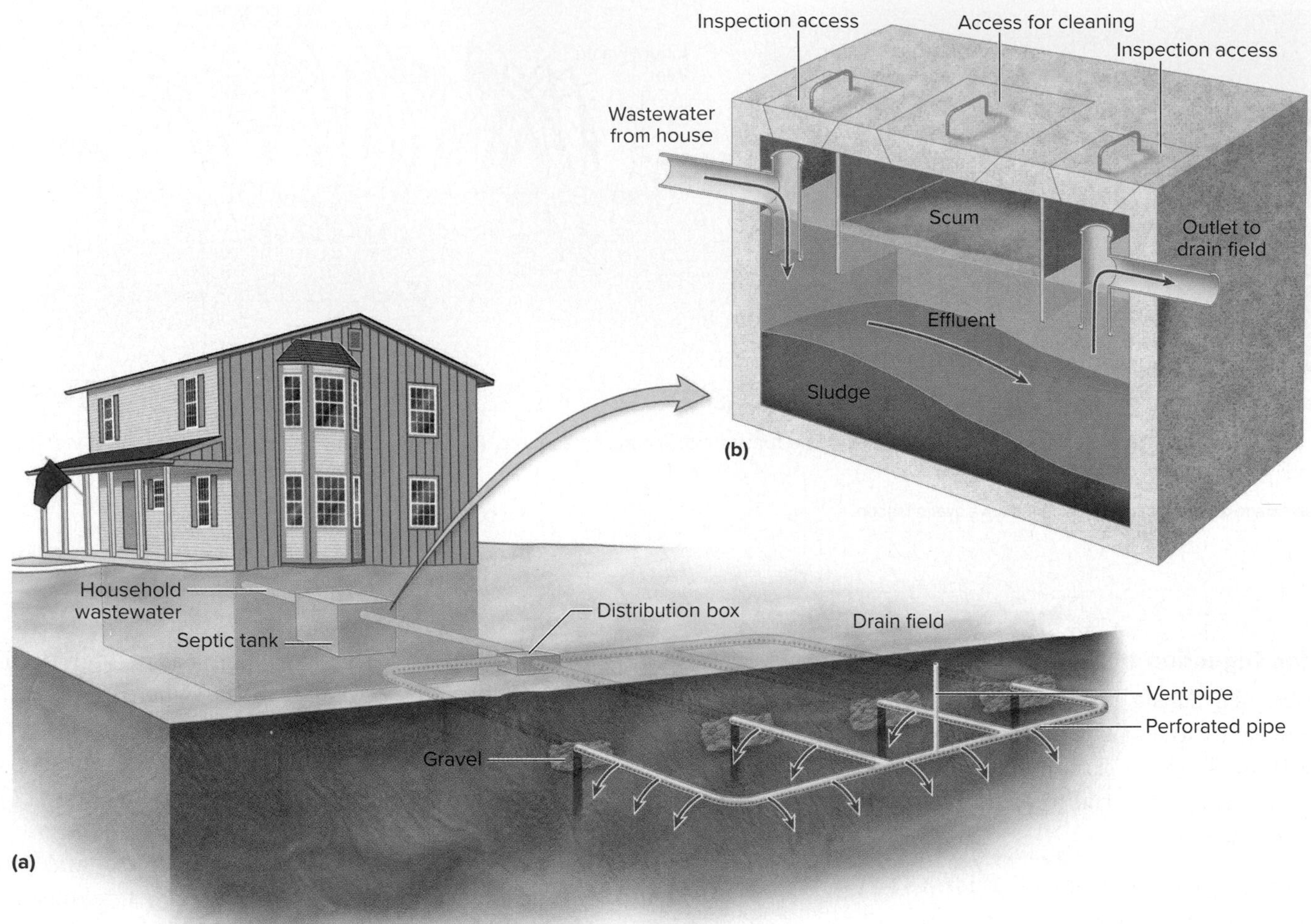

FIGURE 29.4 Septic System **(a)** The components of a septic system. **(b)** Septic tank, where anaerobic degradation takes place.

Why might a farm family avoid using bleach in their laundry?

29.2 ■ Drinking Water Treatment and Testing

Learning Outcomes

4. Describe how drinking water is typically treated.
5. Describe why and how drinking water is tested for total coliforms.

Large cities generally obtain their drinking water from surface waters such as lakes or rivers. Because surface water may also have served as the receiving water for another city's wastewater effluent, drinking water treatment is intimately connected to wastewater treatment. The quality of the surface water is also affected by the characteristics of the watershed (the land over which water flows into the river or lake). Even pristine rivers are likely contaminated with feces of animals that inhabit the watershed.

Smaller communities often use groundwater—pumped from a well—as a source of drinking water. Groundwater is in aquifers (water-containing layers of rock, sand, and gravel) and is replenished as water from rain and other sources seeps through the soil. Because aquifers are not directly exposed to animals and the atmosphere, they are somewhat protected from contamination. However, poorly located or maintained septic systems and sewer lines, as well as sludge or other fertilizers, can lead to groundwater contamination (**figure 29.5**).

Public water systems in the United States are regulated under the Safe Drinking Water Act. This gives the Environmental Protection Agency (EPA) the authority to set standards in order to control the level of contaminants in drinking water. Standards are modified in response to new concerns; for example, regulations now govern the maximum levels of *Cryptosporidium* oocysts, *Giardia* cysts, and enteric (intestinal) viruses in drinking water.

Water Treatment Processes

Community drinking water treatment is designed to provide safe or **potable** water that lacks pathogenic microbes as well

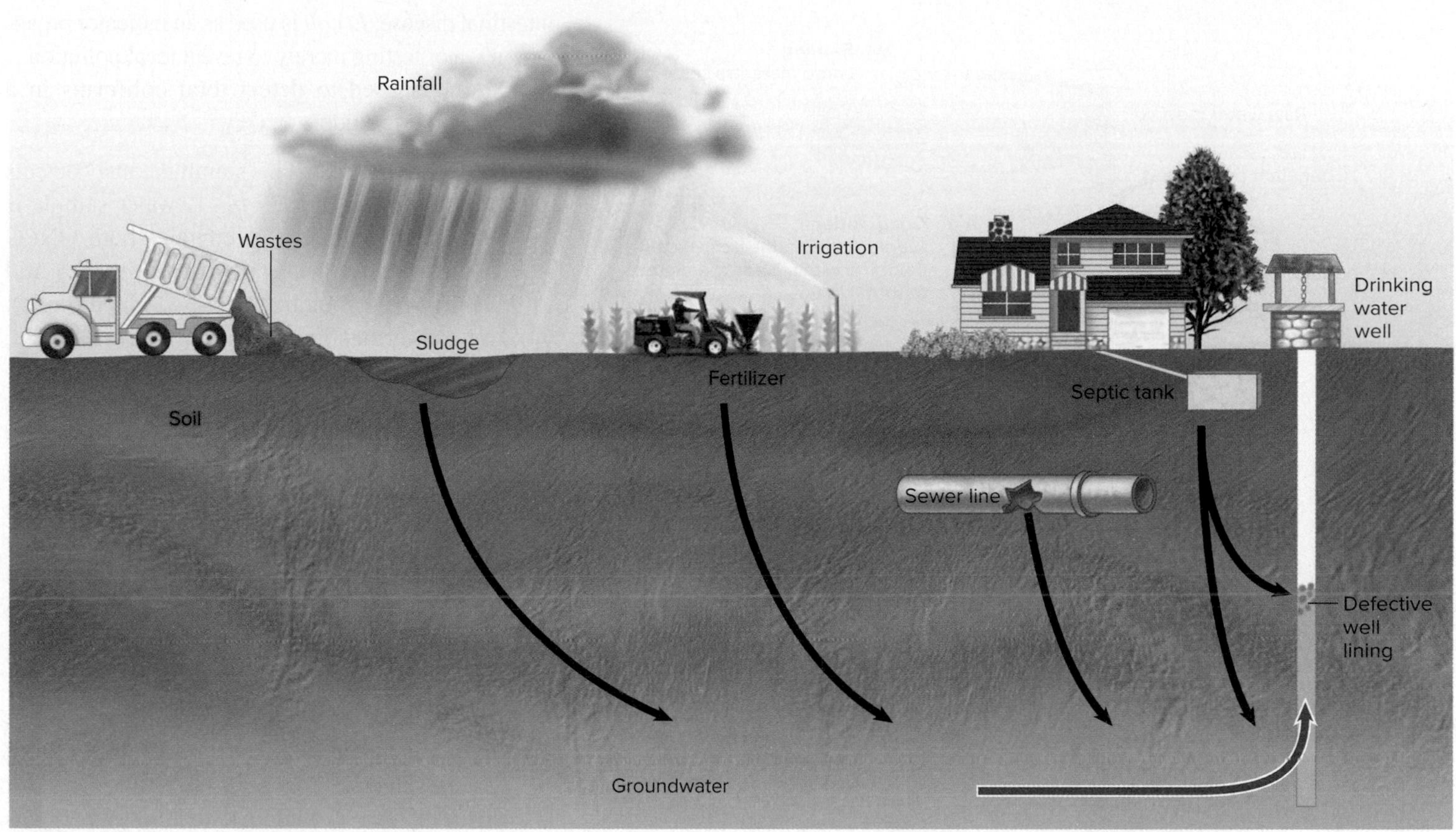

FIGURE 29.5 Groundwater Contamination Poorly located or maintained septic tanks and sewer lines, as well as sludge or other fertilizers, can lead to groundwater contamination.

? How is water in the aquifer replenished?

as harmful chemicals. Municipal drinking water treatment involves the following steps (**figure 29.6**):

1. **Settling.** Water flows into a reservoir where it is allowed to stand long enough for material to settle out. The water is then transferred to a flocculation tank.
2. **Coagulation.** The water is mixed with a chemical that causes suspended materials to coagulate (flocculate; form small clumps); an example of a coagulant is alum (aluminum sulfate). The mixture then flows to a sedimentation tank.
3. **Sedimentation.** The coagulated materials are allowed to slowly sink to the bottom of the tank. As they settle, some microbes and other substances become trapped and are thereby removed as well. The water then flows to a filtration unit.
4. **Filtration.** The water is filtered—often through a thick bed of sand and gravel—to remove various microbes, including bacteria and protozoan cysts and oocysts. Organic chemicals that may be harmful or give undesirable tastes and odors can be removed by additional filtration using an activated charcoal filter, which adsorbs dissolved chemicals. An added benefit of filtration is that microbes growing on the filter materials use carbon from the water as it passes, thus lowering the water's organic carbon content, which, in turn, results in less microbial growth in pipes delivering the water.
5. **Disinfection and storage.** The water is treated with chlorine or other disinfectants to kill or inactivate harmful microorganisms and viruses that might remain. A concern with using chlorine, however, is that some of the disinfection by-products (DBPs) might be carcinogenic. In response to this concern, ultraviolet radiation and ozone are increasingly being used as alternatives, but a small amount of chlorine must still be added to prevent problems associated with post-treatment contamination. Note that disinfection of waters with a high organic content requires more chlorine because organic compounds consume free chlorine.

Water Testing

A primary concern regarding the safety of drinking water is the possibility that it might be contaminated with any of a wide variety of intestinal pathogens, such as those discussed in chapter 24. It is not practical to test for all of the pathogens, however, so **indicator organisms** are used—these microbes are routinely found in feces, survive longer than intestinal

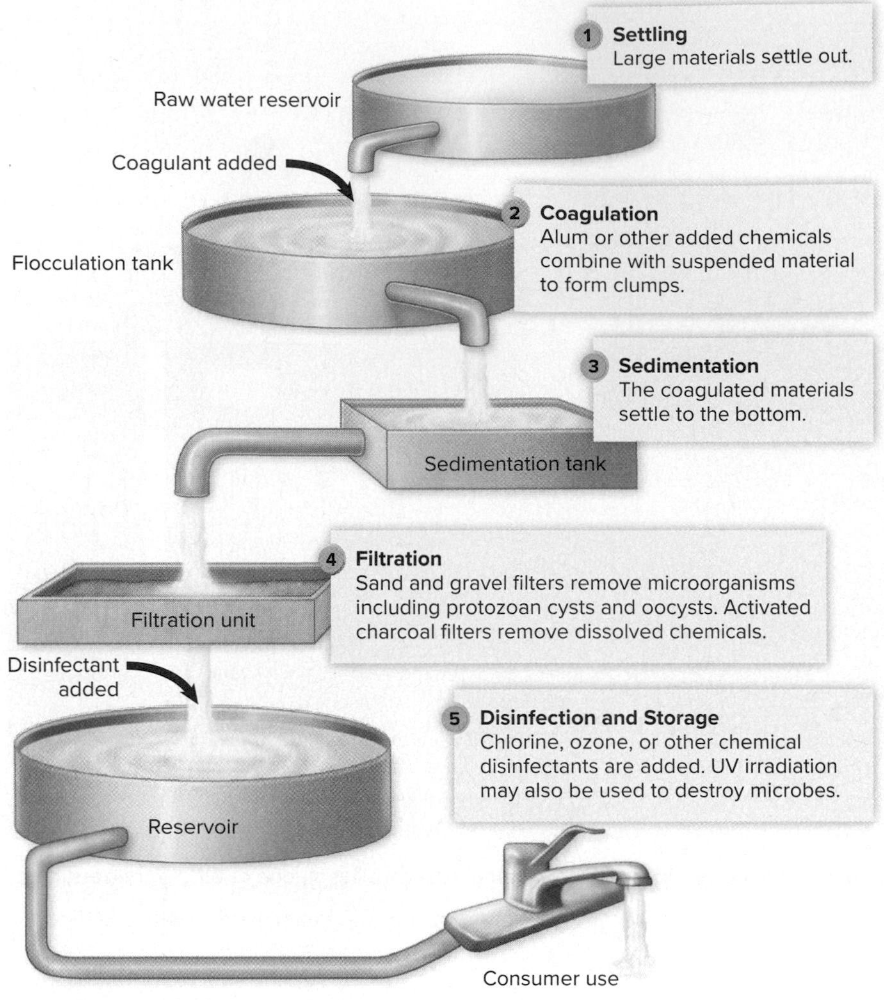
1 Settling
Large materials settle out.
Raw water reservoir
Coagulant added
2 Coagulation
Alum or other added chemicals combine with suspended material to form clumps.
Flocculation tank
3 Sedimentation
The coagulated materials settle to the bottom.
Sedimentation tank
4 Filtration
Sand and gravel filters remove microorganisms including protozoan cysts and oocysts. Activated charcoal filters remove dissolved chemicals.
Filtration unit
Disinfectant added
5 Disinfection and Storage
Chlorine, ozone, or other chemical disinfectants are added. UV irradiation may also be used to destroy microbes.
Reservoir
Consumer use

FIGURE 29.6 Municipal Drinking Water Treatment

? Considering that *Cryptosporidium* oocysts are resistant to chlorine, which step in water treatment protects us from them?

pathogens, and are relatively easy to detect. Finding them suggests fecal contamination and therefore a greater chance that intestinal pathogens are also present.

The most common bacterial group used as indicator organisms in the United States is **total coliforms**—lactose-fermenting members of the family Enterobacteriaceae, including *Escherichia coli.* The group is functionally defined as facultatively anaerobic, Gram-negative, rod-shaped, non-spore-forming bacteria that ferment lactose, forming acid and gas within 48 hours at 35°C.

Although total coliforms are routinely present in the intestinal contents of warm-blooded animals, certain species can also thrive in soils and on plant material. Because of this, their presence does not necessarily imply fecal pollution. To compensate for this shortcoming, **fecal coliforms,** a subset of total coliforms more likely to be of intestinal origin, are also used as indicator organisms. The most common fecal coliform is *E. coli.* Note that although some *E. coli* strains can cause intestinal disease, *E. coli* is used as an indicator organism in water testing merely to reveal fecal pollution.

Methods used to detect total coliforms in a water sample include:

- **ONPG/MUG test.** This simultaneously detects total coliforms and *E. coli*. A water sample is added to a medium that contains both ONPG (*o*-nitrophenyl-β-D-galactopyranoside) and MUG (4-methyl-umbelliferyl-β-D-glucuronide). Lactose-fermenting bacteria hydrolyze ONPG, generating a yellow compound; thus, if a coliform is in the sample, the medium will turn yellow during incubation (**figure 29.7**). *E. coli* is usually the only coliform that produces an enzyme that hydrolyzes MUG, releasing a compound that fluoresces blue under long wave ultraviolet (UV) light (365 nm; see figure 5.7); thus, if *E. coli* is in the sample, the incubated medium will fluoresce blue in addition to being yellow.
- **Presence/absence test.** A 100 mL water sample is added to a lactose-containing broth that is selective for Gram-negative rods. An inverted tube is included to trap gas. If gas is present after incubation, the broth is then tested to confirm that coliforms are present.
- **Most probable number (MPN) method.** This statistical assay of cell numbers uses successive dilutions to determine the most probable number of bacteria in a sample (see figure 4.21). To test drinking water, the

FIGURE 29.7 ONPG/MUG Test *E. coli* hydrolyzes MUG, generating a compound that fluoresces blue under long wave UV light (bottle on the left). No color change, indicating that coliforms are not present (bottle in the center). Coliforms hydrolyze ONPG, yielding a yellow-colored compound (bottle on the right). IDEXX Laboratories, Inc.

? Which of these tested bottles would you prefer to contain a sample of your drinking water?

FOCUS ON A CASE 29.1

A large lake that serves as a popular recreation site was found to have *Escherichia coli* levels well above the U.S. Environmental Protection Agency's suggested maximum limit. In response, authorities closed a number of public beaches surrounding the lake due to health concerns. Local business owners and homeowners were angry about the closure, worrying about the effect on tourism and property values. Prospective homebuyers were hesitant to invest in the region because of the controversy.

Discussions about the source of the problem caused considerable confusion, as no one wanted to take responsibility. Some people felt that the results were meaningless because they followed periods of heavy rain and the resulting runoff. Others blamed the region's small wastewater treatment facilities that served local business and housing developments or the hundreds of aging septic systems that served individual homes. Yet others felt that because the lake water was clear, there was no cause for concern.

In an attempt to find the source of the high *E. coli* levels, authorities began inspecting the local wastewater treatment facilities. Over 30% of the facilities were found to be in violation of the conditions agreed to in their permits. The state then instituted a strict policy of testing and enforcement to improve the water quality. In addition, officials began exploring longer-term solutions, including developing a regional sewage treatment system to take the place of smaller treatment facilities and septic systems.

1. Would you expect the level of *E. coli* in the lake to fall to zero before swimmers are allowed on the beach?
2. What are the possible sources of increased *E. coli* found in the lake?
3. List three possible causes of failure of wastewater facilities.
4. How can individuals improve the quality of the lake water?

Discussion

1. Natural waters are expected to contain a low level of *E. coli* because wild mammals and birds can defecate in the lake and the surrounding area; runoff from precipitation can carry feces and bacteria into the water from surrounding soil. The Environmental Protection Agency recommends that *E. coli* levels stay below 235 CFUs (colony-forming units) per 100 mL of water at public beaches. The water in this case contained over 2,000 CFUs per 100 mL.
2. *E. coli* contamination may result from a specific source such as a faulty septic system serving a private home or a wastewater treatment facility serving a larger population, such as in a campground or condominium development. It may also come from non-specific sources such as surface runoff of rainfall flowing through agricultural land or storm water from city streets.
3. Possible causes for wastewater treatment failure could include: (1) allowing filters to become clogged with sludge or vegetation; (2) leaks in wastewater tanks that allow wastewater to seep into the soil; (3) uncontrolled overflow of wastewater tanks or sewers due to storms or flooding; (4) not upgrading the capacity of the facility to keep up with population growth; or simply (5) allowing sewage to bypass the system and flow directly into the lake.
4. People with septic systems can regularly maintain and test their systems to make sure they function properly and comply with local regulations. People with lakefront property can plant vegetation at the shoreline to minimize erosion and runoff and reduce their use of pesticides, fertilizers, and other chemical pollutants. They can also avoid dumping yard waste into the water and keep pets from defecating in or near the lake. Residents and tourists can pressure any businesses that have been repeatedly cited for water quality violations to comply with regulations.

broth used is similar to that in the presence/absence test, and each tube includes a small inverted tube to trap gas. MPN tubes that show gas production are further tested to confirm that they contain coliforms.

- **Membrane filtration.** A water sample is passed through a filter that retains bacteria, thereby collecting the organisms from a known volume of water (see figure 4.20). The filter is then placed on a lactose-containing selective and differential agar medium.

As part of the Safe Drinking Water Act, the Revised Total Coliform Rule (RTCR) outlines the sample collection requirements for water testing. It also establishes monitoring and corrective actions if samples are positive for total coliforms or *E. coli*.

Because total coliform and fecal coliform assays cannot always predict contamination with protozoan cysts and oocysts, alternatives are being explored. Other microbes that can be used as indicators of fecal pollution include enterococci, some *Clostridium* species, and certain types of bacteriophages.

MicroAssessment 29.2

Drinking water may be obtained from surface water or groundwater. Treatment of drinking water is designed to eliminate pathogens and harmful chemicals. In the United States, total and fecal coliforms are the most commonly used indicator organisms.

4. What is the purpose of coagulation in drinking water treatment?
5. Describe two methods of water testing.
6. Which would be more likely to cause illness and why—a water sample that tested positive for fecal coliforms or one that tested positive for *E. coli* O157:H7? Explain your answer.

29.3 ■ Microbiology of Solid Waste Treatment

Learning Outcome

6. Compare and contrast sanitary landfills and composting programs.

In addition to ridding our environment of wastes in water, we must dispose of the solid wastes (garbage) generated each day. Eliminating these has become an increasingly complex problem.

Sanitary Landfills

Sanitary landfills are widely used to dispose of non-hazardous solid wastes in a manner that minimizes damage to human health and the environment. Federal standards dictate that they must be located away from wetlands, earthquake-prone faults, floodplains, or other sensitive areas. Before sanitary landfills were developed, solid wastes were often piled up on the ground in open-burning dumps, attracting insects and rodents and causing aesthetic and public health problems.

Sanitary landfills are constructed by excavating (digging out) a site and then lining it with plastic sheets or a special membrane on top of a thick layer of clay to prevent contaminants from seeping out into the surrounding environment. A layer of sand with drainage pipes is placed on top of this. When wastes are added to the site, they are compacted and covered with a layer of soil. Once a landfill is full, a layer of material that minimizes the amount of stormwater that will flow through is added; that is then covered with soil and plants so that the area can be used for recreation and eventually as a site for construction.

Sanitary landfills have several disadvantages with respect to waste management. For one thing, only a limited number of sites are available near urban and suburban areas. In addition, the methane gas must be removed as the organic waste material anaerobically decomposes, a process that may continue for years after the landfill is full. The methane is burned or recovered for use; if it is not removed, disastrous explosions can occur. If pollutants such as heavy metals and pesticides leak from a landfill, they may reach underground aquifers, and purifying these once they have become contaminated is difficult.

Sanitary landfills have traditionally been a low-cost method of handling large quantities of solid waste, but because of increased costs and decreased availability of land, cities are looking for alternatives. Programs to recycle paper, plastics, glass, and metal have been implemented in many regions with great success. In some cities, garbage collection fees are based on the size of the container collected—the smaller the can, the lower the cost—offering an incentive to recycle while hopefully raising awareness of the amount of solid waste generated.

Municipal and Backyard Composting—Alternative to Landfills

Composting is the process of making **compost,** the material that results from the controlled decomposition of solid organic material. Not only does composting reduce the amount of organic waste added to landfills, but the product can be used to improve garden soils.

Backyard composting usually starts with a supply of organic material such as leaves, grass clippings, and food scraps (**figure 29.8**); some soil and water are often added to facilitate the process. If composting is done correctly, the inside of the pile heats up due to the metabolic reactions of the multiplying microorganisms. Achieving high temperatures (55°C to 66°C) is important to kill pathogens that might be present in the solid wastes. Thermophilic organisms thrive

FIGURE 29.8 Backyard Composting (a) A home compost bin. **(b)** Garden debris and many kitchen organic wastes can be composted. a: Wave Royalty Free/Design Pics/Alamy Stock Photo; b: Sharon Dominick/iStock Exclusive/Getty Images

? **Why should a compost pile be turned frequently?**

at the high temperature, so they continue decomposing the material. Frequent aeration of the pile—accomplished by physically stirring and turning it as well as keeping it moist—speeds the process so that composting can be completed in as little as 6 weeks.

Composting on a large scale offers cities a way to reduce the amount of solid wastes sent to their landfills while also generating a commercially valuable product. In many areas, yard wastes are collected separately from the main garbage; these are then processed using machinery such as grinders that help make the composting more efficient by increasing the surface area for microbial action (**figure 29.9**).

MicroAssessment 29.3

Sanitary landfills are used to dispose of non-hazardous solid wastes. Composting dramatically reduces the need for large landfills and generates a useful product.

7. What are the advantages and disadvantages of landfills?
8. List the steps in successful composting.
9. Why would adding soil and water facilitate composting?

29.4 ■ Microbiology of Bioremediation

Learning Outcomes

7. Describe why pollutants are a problem, and why some persist in the environment.
8. Compare and contrast the bioremediation strategies of biostimulation and bioaugmentation.

Bioremediation is the use of microorganisms to degrade or detoxify pollutants in a given environment. It generally takes advantage of organisms already present, but in some cases specific organisms are added to the polluted environment.

Pollutants

Most naturally occurring organic compounds are biodegradable, meaning they can be degraded by one or more microbial species. As oil spills dramatically demonstrate, however, some natural materials can cause devastating effects before degradation occurs.

Synthetic compounds are more likely to be biodegradable if their chemical composition is similar to that of naturally occurring substances. In contrast, xenobiotics (synthetic compounds quite different from any in nature) usually persist in the environment because microorganisms are unlikely to have suitable enzymes to break them down. Such enzymes would not give a microbe a competitive advantage in a natural situation.

Relatively slight molecular changes significantly alter the biodegradability of a compound. Perhaps the best-studied example involves the herbicides 2,4-dichlorophenoxyacetic acid (2,4-D) and 2,4,5-trichlorophenoxyacetic acid (2,4,5-T). The only difference between these two compounds is the additional chlorine atom on the latter. When 2,4-D is applied to the soil, it disappears within several weeks, as a result of its degradation by soil microbes. When 2,4,5-T is applied, however, it is often still present more than a year later (**figure 29.10**). The additional chlorine atom of 2,4,5-T blocks the enzyme that makes the initial change in 2,4-D.

Most herbicides and insecticides not only are toxic to their target, but also have damaging effects on fish, birds, and other animals. For example, the pesticide DDT accumulates in the fat of predatory birds through biological magnification (**table 29.1**). Small amounts of the pollutant in water can be concentrated in the tiny plankton, which are eaten by minnows, so that even more accumulates in the fish. When large birds eat the fish, the amount of chemical in tissues is

FIGURE 29.9 Municipal Composting Compost is turned for aeration at a municipal composting site. FLPA/Alamy Stock Photo

What is the purpose of grinding yard waste before putting it in piles?

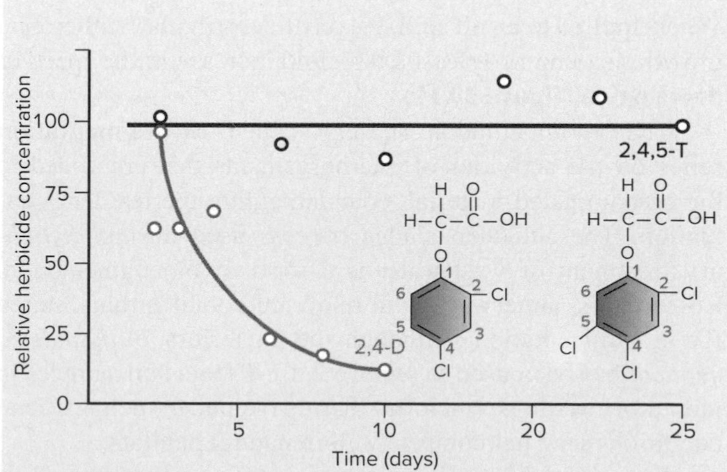

FIGURE 29.10 Comparison of the Rates of Disappearance of Two Structurally Related Herbicides, 2,4-D and 2,4,5-T

Several countries have banned the use of 2,4,5-T. Why would this be the case?

TABLE 29.1 Biological Magnification of DDT

Parts per Million DDT	Source
0.00005	Water
0.04	Plankton
0.23	Minnow
3.57	Heron
22.8	Merganser (fish-eating duck)

amplified. The continuing ingestion of DDT, which accumulates in fat, results in an ever-greater concentration of the pollutant as it passes upward through the food chain. DDT interferes with the reproductive process of birds, leading to the production of fragile eggs, which break before the young can hatch. Although banned in the United States, DDT is still used in other countries, particularly to control mosquitoes that transmit malaria.

Strategies of Bioremediation

Many factors influence the degradation rate of pollutants; as a general rule, however, any practice that favors microbial growth increases the rate. Thus, providing adequate nutrients, maintaining a nearly neutral pH, raising the temperature, and providing an optimal amount of moisture are all likely to promote the degradation.

The bioremediation strategy called **biostimulation** enhances the growth of resident microbes in a contaminated site by providing additional nutrients. Petroleum-degrading bacteria are naturally present in seawater, but they degrade oil at a very slow rate because the low levels of certain nutrients, including nitrogen and phosphorus, limit their growth. To enhance bioremediation of oil spills, a fertilizer containing these nutrients—and which adheres to oil—was developed. When applied to an oil spill, the fertilizer stimulates microbial growth, leading to at least a threefold increase in the speed of degradation (**figure 29.11**).

The bioremediation strategy called **bioaugmentation** relies on the activities of microorganisms that are added to the contaminated material, complementing the resident population. The activated sludge process used during secondary treatment of wastewater is a form of bioaugmentation. Research is underway to develop microbial strains suited for bioaugmentation. Although the bacterium *Burkholderia cepacia* has been used to remove 2,4,5-T from soil samples in laboratory settings, microbes that thrive under such artificial conditions may not compete well in natural habitats.

Successful bioremediation may also involve controlling metabolic processes by manipulating the growth conditions. Aerobic conditions are important when degrading trichloroethylene (TCE), a solvent used to clean metal parts, because

FIGURE 29.11 Oil Spill Bioremediation Nutrients being added to stimulate the growth of resident oil-degrading microorganisms. Accent Alaska.com/Alamy Stock Photo

What characteristic described in the legend puts this bioremediation in the category of biostimulation?

anaerobic degradation results in the accumulation of vinyl chloride, a compound more toxic than TCE. Some pollutants are degraded only when specific substrates are made available to the microbes. This phenomenon, called co-metabolism, occurs because the enzyme produced by the microbe to degrade the additional substrate degrades the pollutant as well. Co-metabolism explains why the presence of methane enhances the degradation of TCE: Enzymes produced by some microbes to degrade methane also degrade TCE.

Bioremediation may be done either in situ ("in place") or off-site. In situ bioremediation generally relies on biostimulation and is less disruptive. Technicians can add oxygen (O_2) to contaminated groundwater and soil either by injecting hydrogen peroxide, which rapidly decomposes to release O_2 and water, or by pumping air into soil. Off-site processes may be performed using a bioreactor, a large tank designed to accelerate microbial processes. Nutrients and O_2 may be added to facilitate microbial growth and metabolism, while the slurry is mixed to ensure that the microbes remain in contact with the contaminants. A slower process involves mounding the contaminated soil over a layer that traps seeping chemicals. To provide O_2, the soil can be turned occasionally or air can be forced through it.

MicroAssessment 29.4

Bioremediation uses microorganisms to degrade pollutants. Biostimulation and bioaugmentation are two methods of bioremediation.

10. Why do xenobiotics often persist in the environment?

11. How is biostimulation different from bioaugmentation?

12. In treating an oil spill, why might biostimulation be preferred over bioaugmentation?

Summary

29.1 ■ Microbiology of Wastewater Treatment

Biochemical Oxygen Demand (BOD)

An important goal of **wastewater** treatment is the reduction of the **biochemical oxygen demand (BOD).**

Municipal Wastewater Treatment Methods (figures 29.1, 29.3)

Primary treatment is a physical process designed to remove materials that settle out. **Secondary treatment** is chiefly a biological process designed to convert most of the suspended solids to inorganic compounds and microbial biomass, removing most of the BOD. **Advanced treatment** is often designed to remove ammonia, nitrates, and phosphates. Biosolids that result from **anaerobic digestion** of **sludge** can be used to improve soils and promote plant growth.

Individual Wastewater Treatment Systems

Rural homes often use **septic systems** for wastewater treatment (figure 29.4).

29.2 ■ Drinking Water Treatment and Testing

Water Treatment Processes

Community drinking water is treated to remove particulate and suspended matter, various microorganisms, and organic chemicals (figure 29.6). Chlorine or other disinfectants are then used to destroy harmful microbes.

Water Testing (figure 29.7)

Total coliforms and **fecal coliforms** are used as **indicator organisms**; their presence suggests the possible presence of pathogens.

29.3 ■ Microbiology of Solid Waste Treatment

Sanitary Landfills

Landfills are used to dispose of solid wastes near towns and cities. Increased recycling of solid wastes decreases the demand on landfills.

Municipal and Backyard Composting—Alternative to Landfills

Composting reduces the amount of garbage sent to landfills (figures 29.8, 29.9).

29.4 ■ Microbiology of Bioremediation

Pollutants

Synthetic compounds are more likely to be biodegradable if they have a chemical composition similar to that of naturally occurring compounds (figure 29.10).

Strategies of Bioremediation

Biostimulation can be used to increase the effectiveness of oil degradation by naturally occurring microorganisms (figure 29.11). **Bioaugmentation** involves the addition of other microbes to the contaminated area.

Review Questions

Short Answer

1. Describe how the BOD of a water sample is determined.
2. Which step of wastewater treatment removes most of the BOD?
3. Compare and contrast the activated sludge process and the trickling filter system used in secondary treatment of wastewater.
4. What is effluent?
5. How does a septic system work?
6. What is an aquifer?
7. Why do water-testing procedures look for coliforms rather than pathogens?
8. How does the ONPG/MUG test allow a sample to be assayed simultaneously for the presence of both total coliforms and *E. coli*?
9. Describe two advantages of composting.
10. Describe the use of bioremediation in the cleanup of oil spills.

Multiple Choice

1. A significant decrease in BOD during secondary treatment indicates
   a) lack of O_2 during treatment.
   b) effective aerobic decomposition during treatment.
   c) effective anaerobic decomposition during treatment.
   d) removal of all pathogenic bacteria.
   e) removal of all toxic chemicals.
2. Advanced treatment is often designed to remove
   a) BOD.
   b) nitrates and phosphates.
   c) bacteria.
   d) protozoa.
   e) methane.
3. Which of the following is *not* a matching pair?
   a) Potable water—presence of pathogens
   b) High BOD—high organic content
   c) Stabilized sludge—fertilizer
   d) Primary treatment—removal of material that settles
4. Which of the following is *false*?
   a) Trickling filter systems are used for primary treatment.
   b) Artificial wetlands provide a habitat for wildlife.
   c) Removal of nitrates by microorganisms requires anaerobic conditions.
   d) Methane is a by-product of anaerobic digestion.
   e) Ozone can be used to disinfect water.
5. Which of the following is *not* a matching pair?
   a) Surface water—watershed
   b) Groundwater—aquifer
   c) Sand and gravel filters—remove organic chemicals
   d) Alum—causes suspended material to coagulate
   e) Disinfection—chlorine

6. Septic tanks should be placed
   a) as close to the well as possible.
   b) near a lagoon.
   c) under the house.
   d) in deep clay soil.
   e) where the outflow cannot contaminate any water supply.
7. Which of the following about coliform testing methods is *true*?
   a) They all determine the number of *E. coli* present in a sample.
   b) The MPN procedure precisely indicates the concentration of coliforms.
   c) The culture media used in these tests check for the ability to ferment lactose.
   d) A positive test indicates that pathogens are definitely present in the sample.
   e) All coliforms hydrolyze ONPG and MUG.
8. Landfills are often used to dispose of
   a) household wastewater.
   b) commercial wastewater.
   c) solid wastes.
   d) petroleum wastes.
   e) wastewater effluent.
9. Backyard composting is an excellent way to dispose of
   a) cooking fats.
   b) garden debris.
   c) spoiled meats.
   d) insecticides.
   e) cleaning supplies.
10. Synthetic compounds are most likely to be biodegradable if they
    a) are totally different from anything found in nature.
    b) have three chlorine atoms per molecule.
    c) are plastics.
    d) are present in very large amounts.
    e) are chemically similar to naturally occurring substances.

Applications

1. A developer is interested in building vacation homes on 150 acres of oceanfront property. A priority is to retain as much natural beauty of the area as possible. Safe and effective wastewater treatment must be part of the plan. What advantages and disadvantages of each of the following options must the developer consider before selecting one?
   a) Individual septic systems for each home
   b) Trickling filter system
   c) Constructed wetlands
2. A public health official is investigating waterborne diseases in Illinois. She notes that over half of the cases of waterborne diseases originating from drinking water were caused by *Giardia lamblia.* Other data showed that most cases of gastroenteritis attributed to exposure to recreational waters were caused by *Cryptosporidium hominis.* What does this suggest about controlling waterborne diseases?

Critical Thinking

1. Why is oil not degraded when in a natural habitat underground, yet it is susceptible to bioremediation in an oil spill?
2. The accompanying figure shows the effects of different treatments of drinking water on the incidence of typhoid fever in Philadelphia, 1890–1935. If filtration of drinking water caused such a dramatic decrease in the disease incidence, was it necessary to introduce chlorination a few years later? Why or why not?

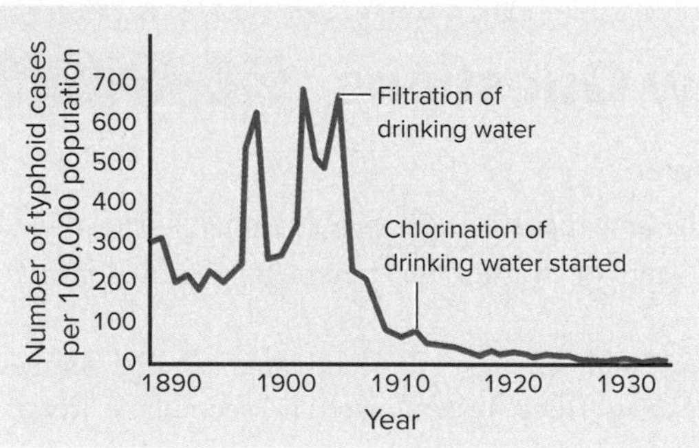

30 Food Microbiology

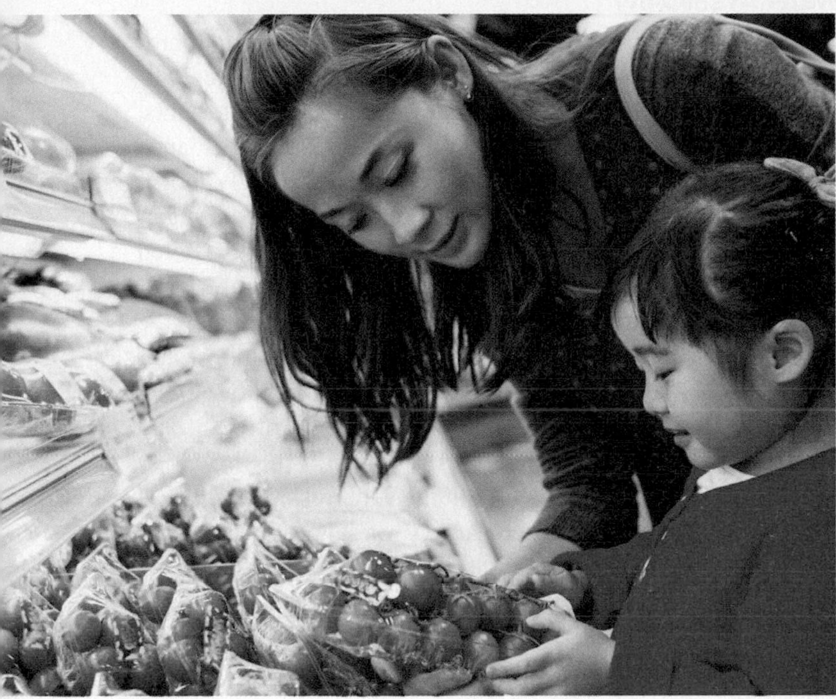

Refrigeration is a method of food preservation. *Images By Tang Ming Tung/Getty Images*

KEY TERMS

Fermented Foods Foods intentionally altered during production by encouraging the activity of bacteria, yeasts, or molds.

Food Preservation The process of increasing the shelf life of foods by preventing the growth and activities of undesirable microorganisms.

Food Spoilage Undesirable biochemical changes in foods.

Foodborne Infection An illness that results from consuming a food product contaminated with microbes that colonize the host and cause disease.

Foodborne Intoxication An illness that results from consuming an exotoxin produced by a microorganism growing in a food product.

Water Activity (a_w) The relative amount of water available for microbial growth; pure water has an a_w of 1.0.

A Glimpse of History

Alice Catherine Evans (1881–1975), the first female president of the Society of American Bacteriology (now the American Society for Microbiology), helped show that unpasteurized milk could be a source of disease in humans. A graduate of both Cornell University and the University of Wisconsin, Evans worked for the U.S. Department of Agriculture, seeking out the sources of microbial contamination of dairy products. In 1917, she reported that cases of human brucellosis were related to finding *Brucella abortus* in cows' milk.

Evans's conclusion that *B. abortus* could be transmitted from cows to humans through milk conflicted with the common view of a number of prominent scientists, including Robert Koch. In 1900, Koch had declared that bovine tuberculosis and brucellosis could not be transmitted to humans. As a result, many scientists and dairy workers would not accept the increasing evidence that milk could be a source of the diseases. In the late 1930s, after a number of children of dairy workers had died of brucellosis, the problem was finally acknowledged. Today, milk is routinely pasteurized, and only very small amounts of unpasteurized milk are sold in the United States.

Everyone has experienced finding moldy bread, sour milk, or perhaps a container of long-forgotten green goo in the back of the refrigerator. Although disappointed with the fate of our food, we are seldom surprised. Just as our foods provide nutrition for our bodies, they also provide a rich nutrient source for ever-present populations of microbes.

Microorganisms on or in foods are not necessarily undesirable. Sometimes, their growth results in pleasant flavors or textures. Foods intentionally altered during production by carefully controlling the activity of bacteria, yeasts, or molds are called **fermented foods** (**figure 30.1**). The processes that cause food to spoil are often the same ones involved in the production of fermented foods. The souring of milk, for example, involves the same microbial processes that produce the agreeable acidic flavor of sour cream, and the mold growing on bread may be related to the one that causes the blue veining in Gorgonzola cheese. In fact, a food considered by one cultural population to be fermented may be thought of as spoiled by another. Strictly speaking, the term *fermentation* refers to only those metabolic activities that use pyruvate or another organic compound as an electron acceptor, with the result that alcohols and acids are produced (see figure 6.24). Food scientists, however, often use the term more generally, to encompass any desirable change that a microorganism contributes to food.

Foods like fresh vegetables and meats that easily support the growth of microorganisms are considered perishable. Microbes growing on these foods can lead to **food spoilage,** meaning a deterioration in the food's quality (**figure 30.2**). Spoilage can be slowed by adjusting the food's properties so that microbes are less able to grow or by changing the conditions under which the food is stored or prepared. Slowing microbial growth maintains food quality for a longer period, and it also helps prevent foodborne illnesses (diseases that result from eating contaminated food).

FIGURE 30.1 Fermented Foods John Thoeming/McGraw Hill

How is a fermented food different from a spoiled food?

30.1 ■ Factors Influencing the Growth of Microorganisms in Foods

Learning Outcome

1. Describe five intrinsic factors and two extrinsic factors that influence the growth of microorganisms in foods.

Understanding the factors that influence microbial growth is essential to maintaining food quality. As a general rule, bacteria will predominate in fresh meats and other moist, pH-neutral, nutrient-rich foods. Yeasts and molds can also grow in these foods, but the more rapid increase of bacteria overwhelms the competitors. When conditions such as low moisture or high acidity restrict the growth of bacteria, fungi predominate despite their relatively slow growth.

Intrinsic Factors

Characteristics of the food itself—such as water availability, acidity, and nutrient level—are called **intrinsic factors.** They affect which microbes predominate on the product.

Water Availability

Foods vary in terms of how much water is accessible to microorganisms. Fresh meats and milk, for example, have plenty of water to support the growth of many microbes. Bread, nuts, and dried foods, on the other hand, are relatively dry. Jams, jellies, and some other sugar-rich foods are seemingly moist, but most of that water is chemically interacting with the sugar, making it unavailable for use by microbes. Highly salted foods, for similar reasons, have little available moisture.

The term **water activity (a_w)** is used to indicate the amount of water available in foods. By definition, pure water has an a_w of 1.0. Most fresh foods have an a_w above 0.98, whereas ham has an a_w of 0.91, jam has an a_w of 0.85, and some cakes have an a_w of 0.70. Most bacteria require an a_w above 0.90 for growth, which explains why fresh, moist foods spoil more quickly than dried, sugary, or salted foods. Fungi can grow at an a_w as low as 0.80, so forgotten bread, cheese, jam, and dried foods often become moldy.

Staphylococcus species, which are adapted to grow on the dry, salty surfaces of skin, can grow at an a_w of 0.86, lower than the minimum required by most common spoilage bacteria. Staphylococci normally do not compete well with other bacteria, but on salty products such as ham and other cured meats, they can multiply with little competition. Ham is a common vehicle for *S. aureus* food poisoning.

pH

Many bacterial species, including most pathogens, are inhibited by acidic conditions and cannot grow at a pH below 4.5. An exception is the lactic acid bacteria, which can grow at a pH as low as 3.5.

FIGURE 30.2 Examples of Spoiled Food
(a) Rotten apples. **(b)** Moldy bread. **(c)** An ear of corn and a lemon after several weeks in the refrigerator.
a: Melbourne Etc/Alamy Stock Photo; b: Westend61/Alamy Stock Photo; c: Martha Neste

Why would refrigeration slow spoilage but not stop it?

(a)

(b)

(c)

These bacteria produce lactic acid as a result of fermentative metabolism and are used to make yogurt, sauerkraut, and some other fermented foods. Acidic end products preserve fermented foods by preventing the growth of other microorganisms. Those same end products are undesirable in many foods, however, so lactic acid bacteria are also common spoilage bacteria.

Fungi can grow at a lower pH than most spoilage bacteria, so acidic foods may eventually become moldy. The low pH of lemons (approximately 2.2) inhibits the growth of bacteria, including the lactic acid group, but some fungi can grow at that pH.

The pH of a food product can also determine whether toxins can be produced. *Clostridium botulinum* (the bacterium that causes botulism) does not grow or produce toxin below pH 4.5, so it is not considered a danger in highly acidic foods. This is why the canning process for acidic fruits and pickles does not call for as high a temperature as that required for canning less acidic foods. Some newer varieties of tomatoes have a higher pH than older types, so acid must be added if they are to be safely canned using the lower temperatures.

Nutrients

Microorganisms requiring a particular vitamin cannot grow in a food lacking that vitamin, but other organisms can grow if conditions are favorable. Members of the genus *Pseudomonas* often spoil foods because they can synthesize essential nutrients and can multiply in various environments, including refrigeration.

Biological Barriers

Rinds, shells, and other coverings help protect foods from invasion by microorganisms. Eggs, for example, retain their quality much longer with intact shells. Whole lemons keep longer than slices. Even so, microorganisms will eventually break down these coverings and cause spoilage.

Antimicrobial Chemicals

Some foods contain natural antimicrobial chemicals that help prevent spoilage. Egg white, for instance, is rich in the enzyme lysozyme. If lysozyme-susceptible bacteria get through the protective shell of an egg, the enzyme destroys them before they can cause spoilage. Other examples of naturally occurring antimicrobial chemicals are benzoic acid in cranberries and allicin in garlic.

Extrinsic Factors

Environmental conditions under which a food is held—such as temperature and atmosphere—are called **extrinsic factors.** These can greatly affect the extent of microbial growth on food.

Storage Temperature

The storage temperature affects the growth rate of microorganisms. At low temperatures above freezing, many enzymatic reactions are either very slow or nonexistent, with the result that microorganisms multiply slowly, if at all. Microorganisms that grow on refrigerated foods are most likely psychrophiles or psychrotrophs, such as some members of the genus *Pseudomonas.* Most pathogens prefer the warmer temperatures of the human body and do not grow well under refrigeration.

Some foods can be stored for longer periods of time at freezing temperatures. Temperatures below about −23°C (about −10°F) stop the growth of most organisms because water is solid and therefore unavailable for microbial growth. Even frozen foods, however, should not be stored indefinitely.

Atmosphere

The presence or absence of O_2 affects the type of microbial population able to grow in food. Obligate aerobes cannot grow in foods stored under conditions that exclude their required O_2. Keeping O_2 out of food, however, may permit other bacteria to grow, including the obligate anaerobe *Clostridium botulinum.* People have developed botulism after eating homemade stew left at room temperature overnight; the cooking process does not destroy *C. botulinum* endospores, and it drives off the O_2, thereby creating anaerobic conditions in which the endospores can germinate and the resulting vegetative cells can multiply and produce toxin.

MicroAssessment 30.1

Intrinsic factors (such as available moisture, pH, and the presence of antimicrobial chemicals) and extrinsic factors (including storage temperature and atmosphere) influence the types of microorganisms that grow and predominate in a food product.

1. Why is *Staphylococcus aureus* more likely to be found in high numbers on ham than on fresh meat?
2. Which is important to refrigerate: fresh stew or bread? Why?
3. A camper carries dried meat, shelled pecans, and cranberries on a week-long hike. What characteristics of the foods will prevent spoilage during that time?

30.2 ■ Microorganisms in Food and Beverage Production

Learning Outcomes

2. Explain the role of the lactic acid bacteria in the production of fermented milk products, pickled vegetables, and fermented meats.
3. Compare and contrast the production of wine, beer, distilled spirits, and vinegar.
4. Describe the production of soy sauce.

Fermented foods and beverages, such as yogurt, cheese, pickled vegetables, wine, and beer, are perceived as having a pleasant taste. In addition, various compounds produced by the fermenting microbes inhibit the growth of many spoilage organisms as well as of foodborne pathogens. Thus, fermentation historically has been, and continues to be today, an important method of food preservation.

To maintain the quality of fermentations, **starter cultures** containing one or more strains of the desired microorganisms are often used. These strains are carefully selected to produce the most pleasant flavors and textures. Precious starter cultures must be carefully maintained and protected against contamination, particularly by viruses that can damage or destroy them. The species used in starter cultures are genetically diverse, and so they are often taxonomically divided into subspecies; for example, *Streptococcus lactis* subsp. *lactis* and *S. lactis* subsp. *cremoris* are both used in cheese production.

Coagulation. Lactic acid production and rennin activity cause the milk proteins to coagulate. The coagulated mixture is then cut so that the liquid whey will start separating from the solid curd.

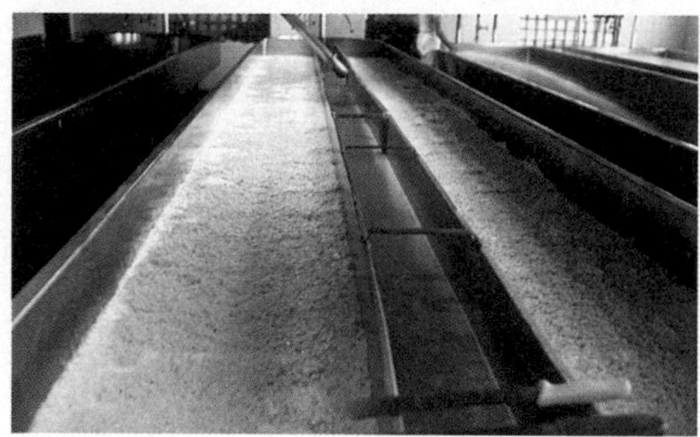

Separation of curds from whey. The curd is heated and cut into small pieces. The liquid whey is removed by draining.

Ripening. Curds are salted and pressed into blocks or wheels for aging.

FIGURE 30.3 Commercial Production of Cheese Courtesy of the Wisconsin Milk Marketing Board

? **Why does a longer ripening process give rise to a sharper cheese?**

Lactic Acid Fermentations by the Lactic Acid Bacteria

The tart taste of yogurt, pickles, sharp cheeses, and some sausages is due to the production of lactic acid by one or more members of the lactic acid bacteria. These bacteria—including members of the genera *Lactobacillus, Lactococcus, Streptococcus, Leuconostoc,* and *Pediococcus*—are obligate fermenters that characteristically use the lactic acid fermentation pathway to ferment monosaccharides (simple sugars), producing lactic acid as the primary end product (see figure 6.23a). Some also produce flavorful and aromatic compounds that contribute to the overall quality of fermented foods.

Cheese, Yogurt, and Other Fermented Milk Products

Lactic acid bacteria ferment the main sugar in milk, lactose, producing lactic acid. As a result, the growth of many other microbes is inhibited because: (1) lactose is no longer available as a nutrient source and: (2) the accumulated lactic acid lowers the pH. Aesthetic features of the milk change as well because the low pH causes the milk proteins to coagulate or curdle, and it sours the flavor.

Cheese Cottage cheese is one of the simplest cheeses to make. Pasteurized milk is inoculated with a starter culture, usually containing *Streptococcus lactis* subsp. *lactis* and *S. lactis* subsp. *cremoris,* and then incubated until fermentation products cause the proteins in milk to coagulate. The coagulated proteins, or **curd,** are heated and cut into small pieces to make it easier to drain the liquid portion (called whey). Unlike most cheeses that undergo further microbial processes called ripening or curing, cottage cheese is unripened.

The initial steps of ripened cheese production are the same as those of cottage cheese, except the enzyme rennin is added to the fermenting milk to speed protein coagulation (**figure 30.3**). After the proteins coagulate, the whey is removed. The curds are then salted, pressed, and shaped into the traditional forms, usually bricks or wheels. The cheese is then ripened (aged), resulting in characteristic textural and flavor changes due to the metabolic activities of naturally occurring or starter lactic acid bacteria. Depending on the type of cheese, ripening can take from several weeks to years. Longer ripening creates more acidic, sharper cheeses.

Some cheeses are inoculated with additional bacteria or fungi that give characteristics particular to the kind of cheese. The bacterium *Propionibacterium freudenreichii* subsp. *shermanii* helps ripen Swiss cheese and gives it the characteristic holes and a nutty flavor. This bacterium ferments organic compounds to produce propionic acid and CO_2; the CO_2 gas causes the holes in the cheese, while the propionic acid gives the typical flavor. Propionic acid also inhibits spoilage organisms. Roquefort, Gorgonzola, and Stilton cheeses obtain their distinctive-bluish green veins from the growth of the fungus *Penicillium roquefortii* along cracks in the cheese. The white coating on Brie and Camembert is due to surface growth of a fungus such as *Penicillium camemberti*, which produces enzymes that alter the texture and flavor of the cheeses.

Yogurt To produce yogurt, pasteurized milk is concentrated slightly by evaporation and then inoculated with a starter culture containing *Streptococcus thermophilus* and *Lactobacillus delbrueckii* subsp. *bulgaricus.* The mixture is incubated at 40°C to 45°C for several hours, during which time these thermophilic bacteria grow rapidly. They produce lactic acid and other end products that contribute to the flavor. Carefully controlled incubation conditions favoring the balanced growth of the two species ensure the proper levels of acid and flavor compounds.

MicroByte

The enzyme rennin is found in calves' stomachs, where it helps digest the mother's milk. Today, genetically engineered microbes make it.

Pickled Vegetables

The fermentation process known as pickling originated as a way to preserve vegetables such as cucumbers and cabbage. Today, pickled products such as sauerkraut (cabbage), pickles (cucumbers), and olives are valued for their flavor. The pickling typically relies on lactic acid bacteria naturally on the vegetables rather than starter cultures.

One of the most well-studied natural fermentations is the process that makes sauerkraut. To make sauerkraut, shredded cabbage is mixed with salt and then firmly packed into a container to provide an anaerobic environment. The salt draws water and nutrients from the cabbage, creating a brine that inhibits most microbes other than the lactic acid bacteria. Under the correct conditions, natural successions of lactic acid bacteria grow. These bacteria—*Leuconostoc mesenteroides, Lactobacillus brevis,* and *Lactobacillus plantarum*—produce lactic acid, which lowers the pH, further inhibiting undesired microbes. The lactic acid and other fermentation end products that develop over a 2- to 4-week period give sauerkraut its characteristic tangy taste. When the desired flavor has developed, the sauerkraut is often canned. Similar processes are used to make other vegetable products, including olives and some pickles.

Fermented Meat Products

Fermented meat products—such as salami, pepperoni, and summer sausage—were traditionally produced by allowing the small numbers of lactic acid bacteria naturally present on or in the meats to multiply to the point of dominance. This approach is inherently risky, however, because the incubation conditions can potentially support the growth and toxin production of pathogens such as *Staphylococcus aureus* and *Clostridium botulinum.* Starter cultures are now used because they ensure that lactic acid is rapidly produced, thereby inhibiting the growth of pathogens and also improving flavor development. Starter cultures used by U.S. sausage-makers typically contain *Lactobacillus* and/or *Pediococcus* species.

To make fermented sausages, meat is ground and combined with a starter culture and other ingredients, including sugar, salt, and nitrite. The sugar serves as a substrate for fermentation because meat does not naturally contain enough fermentable carbohydrate to produce adequate lactic acid. Salt and nitrite contribute to the flavor and also inhibit the growth of spoilage microorganisms. More important, they inhibit *C. botulinum.* After thorough blending, the mixture is stuffed into a casing and incubated from one to several days. The product can then be smoked or otherwise heated to kill bacteria. Finally, it is dried.

Table 30.1 summarizes the characteristics of fermented milk products, vegetables, and meats that are produced using lactic acid bacteria.

TABLE 30.1 Foods Produced Using Lactic Acid Bacteria

Food	Characteristic
Milk Products	
Cheese (unripened)	Uses a starter culture, usually containing *Lactococcus lactis* subsp. *cremoris* and *L. lactis* subsp. *lactis*
Cheese (ripened)	Uses rennin and a starter culture containing *Lactococcus lactis* subsp. *cremoris* and *L. lactis* subsp. *lactis;* ripened for weeks to years; other bacteria and/or fungi may be added to enhance flavor development.
Yogurt	Uses a starter culture containing *Streptococcus thermophilus* and *Lactobacillus delbrueckii* subsp. *bulgaricus*
Vegetables	
Sauerkraut (fermented cabbage)	Relies on a succession of naturally occurring bacteria including *Leuconostoc mesenteroides, Lactobacillus brevis,* and *Lactobacillus plantarum*
Pickles (fermented cucumbers)	Relies on naturally occurring bacteria
Poi (fermented taro root)	Relies on naturally occurring bacteria
Olives (fermented green olives)	Relies on naturally occurring bacteria
Kimchee (fermented cabbage and other vegetables)	Relies on naturally occurring bacteria
Meats	
Dry and semidry sausages	Use a starter culture containing species of *Lactobacillus* and *Pediococcus;* meat is stuffed into casings, incubated, heated, and then dried.

Ethanol Fermentations by Yeast

Some yeasts, such as members of the genus *Saccharomyces,* use the ethanol fermentation pathway to ferment simple sugars, producing ethanol and CO_2 (see figure 6.23b). They are used to make alcoholic beverages as well as vinegar and bread.

Wine

Wine production relies on fermentation of naturally occurring sugars in the juices of grapes or other fruits. The steps of commercial wine production include (**figure 30.4**):

1. **Crushing.** Grapes are crushed in a tank. The resulting solids and juices, called must, are collected.
2. **Fermentation.** If white wine is to be made, only the clear juices are fermented. For red wine, the entire must of red grapes is put into the fermentation vat, and the solids removed once the desired amount of color and flavor compounds have been extracted; the color and complex flavors of these wines are derived from components of the grape skin and seeds. Rose wines get their light pink color from a short fermentation of the entire crushed red grape (about 1 day), after which the juice is fermented alone. For the fermentation, a specially selected strain of *Saccharomyces cerevisiae* is inoculated. Sulfur dioxide (SO_2) is generally added to inhibit the growth of the natural microbial population of the grape, especially acetic acid bacteria; these can spoil wine because they convert alcohol to acetic acid (vinegar). The *S. cerevisiae* strains used to make wine are more resistant to the antimicrobial action of SO_2 and produce a higher alcohol content than naturally occurring yeasts. Fermentation is carried out at a carefully controlled temperature for a period ranging from a few days to several weeks. During fermentation, most of the sugar is converted to ethanol and CO_2, generally resulting in a final alcohol content of less than 14%. Dry wines result from the complete fermentation of the sugar, whereas sweet wines contain residual sugar.
3. **Settling.** The wine is transferred to a settling tank, where fermented solids can settle out.
4. **Aging.** Most red wines and some white wines are then aged in oak barrels, contributing to the complexity of the flavor.
5. **Bottling.** Wine is clarified by filtration and then bottled. The CO_2 produced during fermentation is usually released before the wine is bottled, resulting in a "still" (non-carbonated) wine. Other processes are used to prepare carbonated wines such as champagne.

The Japanese wine sake depends on several microbial fermentation reactions. First, cooked rice is inoculated with the fungus *Aspergillus oryzae,* which produces amylase—an enzyme that degrades the rice starch to sugar. Then a strain of *S. cerevisiae* is added to convert the sugar to alcohol and CO_2. Lactic acid bacteria contribute to the flavor by producing lactic acid and other fermentation end products.

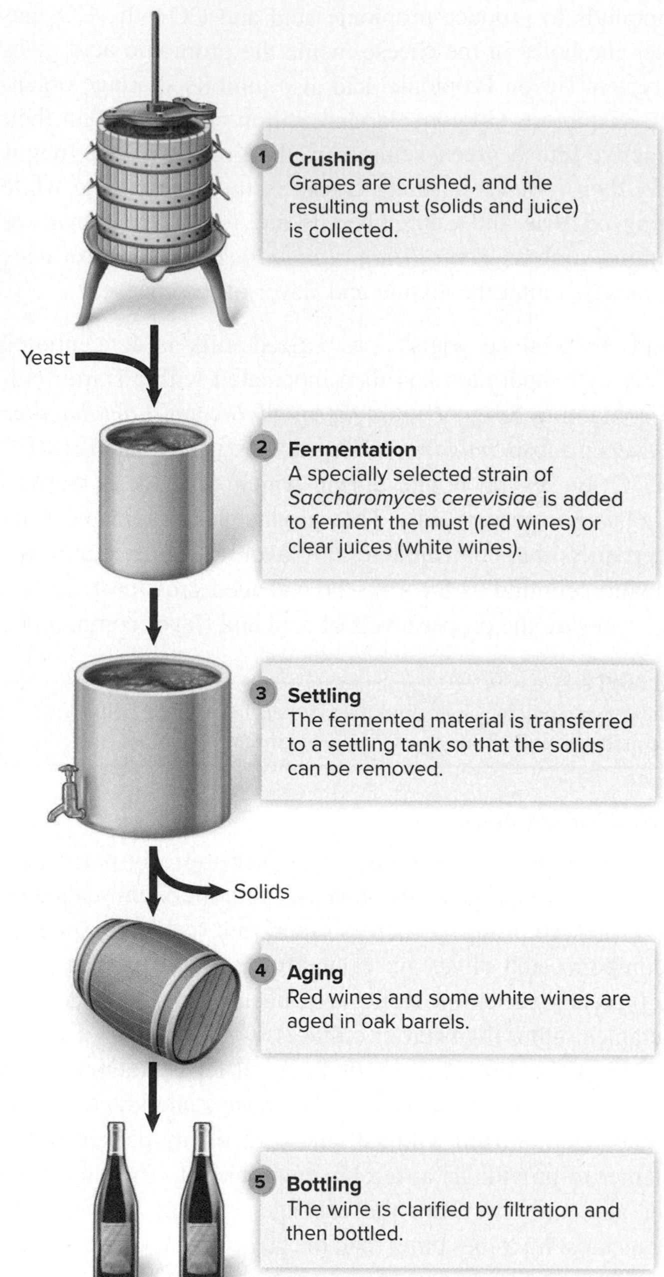

FIGURE 30.4 Commercial Production of Wine

? How are the initial steps of making white wines different from those of making red wines?

Beer

Beer production is a multistep process designed to break down the starches of grains such as barley to produce simple sugars, which are then fermented by yeast (**figure 30.5**). Yeasts alone cannot convert grain to alcohol because they lack the enzymes that degrade starch, the primary carbohydrate of grain. Sprouted or germinated barley, however, known as malted barley or malt,

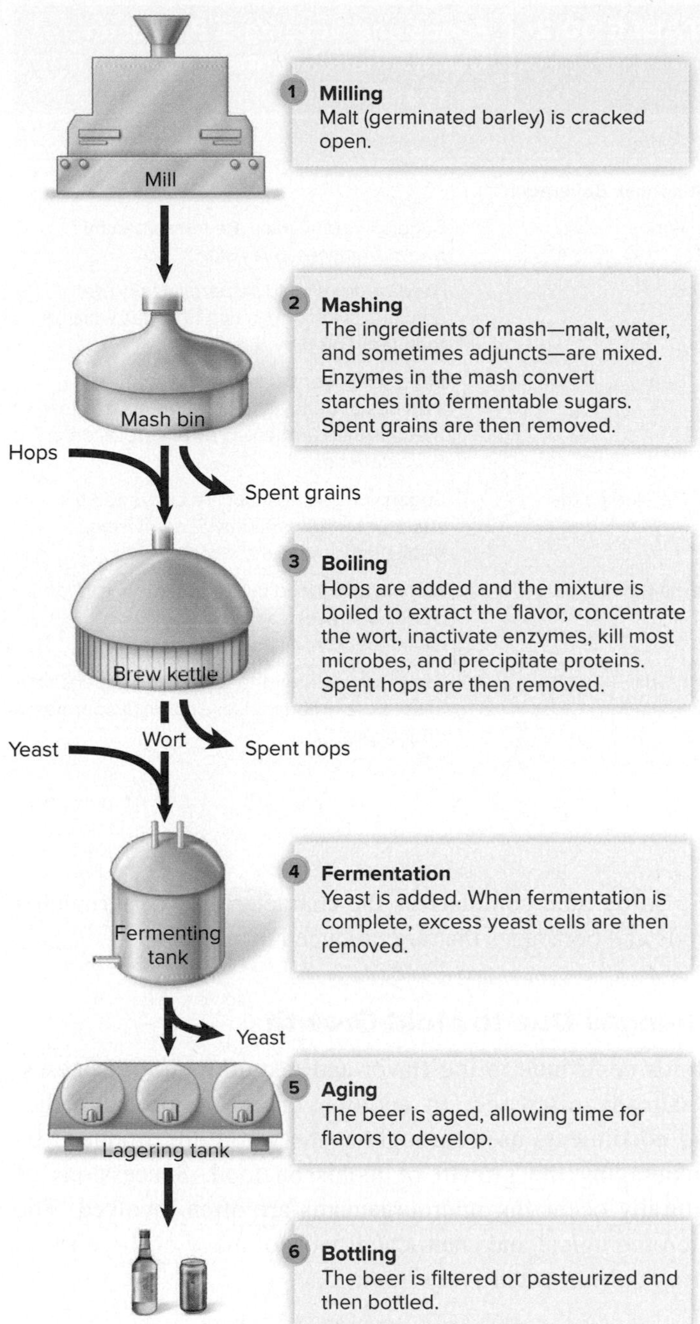

Mill
1 Milling
Malt (germinated barley) is cracked open.
Mash bin
2 Mashing
The ingredients of mash—malt, water, and sometimes adjuncts—are mixed. Enzymes in the mash convert starches into fermentable sugars. Spent grains are then removed.
Hops
Spent grains
Brew kettle
3 Boiling
Hops are added and the mixture is boiled to extract the flavor, concentrate the wort, inactivate enzymes, kill most microbes, and precipitate proteins. Spent hops are then removed.
Yeast
Wort
Spent hops
Fermenting tank
4 Fermentation
Yeast is added. When fermentation is complete, excess yeast cells are then removed.
Yeast
Lagering tank
5 Aging
The beer is aged, allowing time for flavors to develop.
6 Bottling
The beer is filtered or pasteurized and then bottled.

FIGURE 30.5 Commercial Production of Beer

? **How are the yeasts used to make ales different from those used to make lagers?**

naturally contains these and other important enzymes. Steps of commercial beer production include:

① **Milling.** The malt is dried and milled (ground) in preparation for the brewing process. The ground malt is transferred to the mash bin.

② **Mashing.** The malt is mixed with adjuncts (starches, sugars, or whole grains such as rice, corn, or sorghum) and then soaked in warm water, allowing enzymes of the malt to act on the starches, converting them to fermentable sugars. The final characteristics of the beer, such as color, flavor, and foam, are derived entirely from compounds in the malt. The adjuncts simply serve as less expensive sources of carbohydrates for alcohol production. The spent grains (residual solids) are removed to yield the sugary liquid called wort that is then transferred to the brew kettle for boiling.

③ **Boiling.** Hops, the flowers of the vine-like hop plant, are added to the wort to contribute a desirable bitter flavor and provide antibacterial substances. The mixture is boiled to extract the flavor components of hops, concentrate the wort, inactivate enzymes, kill most microbes, and precipitate proteins so that they can be removed more easily. The wort is then centrifuged to remove the solids and cooled before being transferred to the fermentation tank.

④ **Fermentation.** For the fermentation, special strains of **brewer's yeasts** (as opposed to baker's yeasts) are added. When making lager-style beers, *Saccharomyces pastorianus* strains are used. These are "bottom yeasts," meaning that they tend to form clumps that sink to the bottom of the fermentation tank. They ferment best at relatively low temperatures (between 6°C and 12°C), usually completing the fermentation in about a week or two. In contrast, when making ales, porters, or stouts, *S. cerevisiae* strains are used. These are called "top-fermenting yeasts" because they are carried to the top of the tank by the rising CO_2. They ferment at higher temperatures (14°C to 23°C) and over a shorter period (5 to 7 days). The fermentation process generates beer with an alcohol content ranging from 3.4% to 6%. Most of the yeast cells settle out following fermentation and are removed; they can be sold as flavor and dietary supplements. The beer is then transferred to a lagering tank for aging.

⑤ **Aging.** During aging, unwanted flavor compounds are metabolized by remaining yeast cells or settle out. Cask-conditioned beer undergoes a second fermentation, which generates CO_2 in the cask. Other beers must be carbonated to replace the CO_2 that escapes during fermentation.

⑥ **Bottling.** Microorganisms are removed or killed using membrane filtration or pasteurization, and the product is packaged.

Distilled Spirits

The manufacturing of distilled spirits such as scotch, whiskey, and gin is initially similar to that of beer, except the wort is not boiled. Consequently, enzymes in the wort continue breaking down the starches during fermentation. When fermentation is complete, the ethanol is collected by distillation.

Different types of spirits are made in different ways. Rum production uses sugar cane or molasses as the fermentation substrate. Malt scotch whiskey production uses barley as the fermentation substrate, and the distilled product is then aged for several years in oak sherry casks; the wood and the residual sherry contribute both flavor and color to the whiskey as it ages. To make sour-mash whiskey, lactic acid bacteria are used to produce lactic acid in grain mash; a yeast strain then ferments the mash to form alcohol. Tequila production involves fermentation of the juices of cooked agave plant leaves by the bacterium *Zymomonas mobilis,* which uses a fermentation pathway similar to that of yeast to produce ethanol and CO_2.

Vinegar

Vinegar, an aqueous solution of at least 4% acetic acid, is the product of ethanol oxidation by the acetic acid bacteria—*Acetobacter* and *Gluconobacter* species. Acetic acid bacteria are acid-tolerant, strictly aerobic, Gram-negative rods, characterized by their ability to carry out a number of oxidations.

Alcohol is commercially converted to vinegar using processes that provide O_2 to speed the oxidation reaction. One method uses a vinegar generator, which sprays alcohol onto loosely packed wood shavings that have a biofilm of acetic acid bacteria. As the alcohol trickles through the bacteria-coated shavings, it is oxidized to acetic acid. In principle, the vinegar generator operates much like the trickling filter used in wastewater treatment by providing a large surface area for aerobic metabolism. Another method uses a submerged culture reactor, an enclosed system that continuously pumps small air bubbles into alcohol that has been inoculated with acetic acid bacteria.

TABLE 30.2 Foods and Beverages Produced Using Ethanol Fermentation by Yeast

Product	Characteristic
Alcoholic Beverages	
Wine	Sugars in grape juice are fermented by *Saccharomyces cerevisiae.*
Sake	Amylase from mold *(Aspergillus oryzae)* converts the starch in rice to sugar, which is then fermented by *S. cerevisiae.*
Beer	Enzymes in germinated barley convert starches of barley and other grains to sugar, which is then fermented by *S. cerevisiae* or *S. pastorianus.*
Distilled spirits	Sugars, or starches that are converted to sugars, are fermented by *S. cerevisiae;* distillation purifies the alcohol.
Vinegar	Alcohol (produced by fermentation) is oxidized to acetic acid by species of *Gluconobacter* or *Acetobacter.*
Breads	*S. cerevisiae* ferments sugar; CO_2 expansion causes the bread to rise; alcohol evaporates during baking.

Bread

Yeast bread rises through the action of **baker's yeast**—strains of *Saccharomyces cerevisiae* carefully selected for the commercial baking industry. The CO_2 produced during fermentation causes the bread to rise, producing the spongy texture characteristic of yeast breads (**figure 30.6**). The alcohol evaporates during baking.

Yeast bread is made from a mixture of flour, sugar, salt, milk or water, yeast, and sometimes butter or oil. Packaged baker's yeast that can be reconstituted in warm water is readily available as pressed cakes or dried granules. An excess of yeast is added to allow adequate production of CO_2 in a time period too short to permit multiplication of spoilage bacteria.

Sourdough bread is made with a combination of yeast and lactic acid bacteria. Lactic acid is produced, as well as alcohol and CO_2, giving the bread its sour flavor.

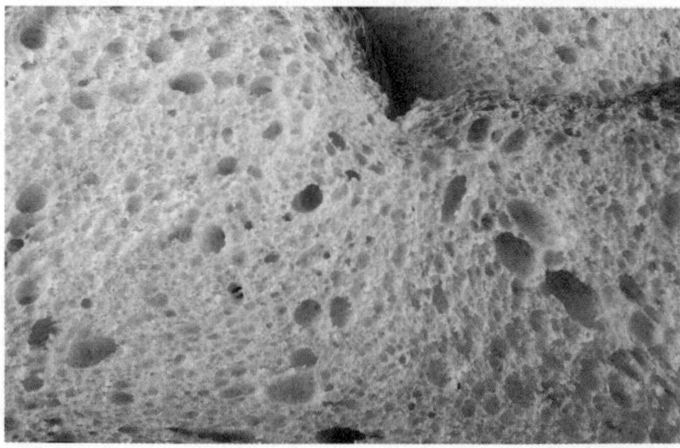

FIGURE 30.6 Bread Pixtal/age fotostock

? What creates the air pockets in bread?

Table 30.2 summarizes the characteristics of fermented foods and beverages that are produced using yeast.

Changes Due to Mold Growth

Molds contribute to the flavor and texture of some cheeses, as already discussed. In addition, many traditional dishes and condiments used throughout the world are produced by encouraging the growth of molds on food. Successions of naturally occurring microorganisms are often involved. The microbiological and chemical aspects of many of these foods have not been extensively studied.

Soy Sauce

Soy sauce is made by first inoculating equal parts of cooked soybeans and roasted cracked wheat with a culture of either *Aspergillus oryzae* or *Aspergillus sojae*. This is allowed to ferment for several days, during which time carbohydrates and proteins are broken down to produce a yellow-green liquid called koji that contains fermentable sugars, peptides, and amino acids. The koji is then put into a large container with an 18% NaCl solution (brine), where salt-tolerant microorganisms including certain lactobacilli, pediococci, and yeasts grow and produce flavor changes. After this brine mixture is allowed to ferment for 6 to 12 months, the liquid soy sauce is separated from the residual solids, and the latter are used as animal feed. **Table 30.3** summarizes the characteristics of soy sauce and other foods produced using molds.

TABLE 30.3 Foods Produced Using Molds

Food	Characteristic
Soy sauce	Cooked soybeans and cracked wheat are inoculated with *Aspergillus oryzae* or A. sojae and allowed to ferment; the product is then added to a brine and fermented for months.
Tempeh	Cooked soybeans are inoculated with a species of the mold *Rhizopus* and allowed to ferment; Indonesia.
Miso	Cooked rice, soybeans, or barley are inoculated with *Aspergillus oryzae* and allowed to ferment; Asia.
Cheeses	
Roquefort, Gorgonzola, and Stilton	Curd is inoculated with *Penicillium roqueforti* and allowed to ferment
Brie and Camembert	Wheels of cheese are inoculated with selected species of *Penicillium* and allowed to ferment.

MicroAssessment 30.2

Lactic acid bacteria are used to produce a variety of foods including cheese, yogurt, sauerkraut, and some sausages. Yeasts are used to produce alcoholic beverages and breads. Some cheeses and many traditional foods owe their characteristics to changes caused by molds.

4. Describe how the metabolism of lactic acid bacteria differs from that of most other microorganisms that can grow aerobically.
5. How does the use of starter cultures improve the safety of fermented meat products?
6. How could cottage cheese be produced without bacteria?

30.3 ■ Food Spoilage

Learning Outcome

5. Distinguish between fermented foods and spoiled foods.

Spoilage microorganisms produce metabolites that have undesirable tastes and odors, resulting in **food spoilage.** Although these microbes damage the quality of the food, they are generally not harmful. This is not surprising when microbial growth requirements are considered: Most human pathogens grow best near normal body temperatures of 37°C, whereas most foods are usually stored at temperatures well below that. Similarly, the nutrients available in fruits, vegetables, and other foods are generally not suitable for the optimum growth of human pathogens. As a result, the non-pathogens typically outgrow the pathogens when competing for the same nutrients. Spoiled foods are considered unsafe to eat, however, because high numbers of spoilage organisms indicate that foodborne pathogens may be present as well.

Common Spoilage Bacteria

Many types of bacteria are important in food spoilage. *Pseudomonas* species can degrade a wide variety of compounds and grow on and spoil numerous kinds of foods, including meats and vegetables. Psychrotrophic species are notorious for spoiling refrigerated foods. Members of the genus *Erwinia* produce enzymes that degrade pectin, and so they commonly cause soft rot of fruits and vegetables. *Acetobacter* species transform ethanol to acetic acid, the principal acid of vinegar. Although this is beneficial to commercial producers of vinegar, it presents a great problem to wine producers. Milk products are sometimes spoiled by *Alcaligenes* species that form a glycocalyx, causing strings of slime, or "ropiness," in raw milk. The lactic acid bacteria, including species of *Streptococcus, Leuconostoc,* and *Lactobacillus,* all produce lactic acid. Anyone who has unexpectedly consumed sour milk knows that this can be disagreeable. *Bacillus* and *Clostridium* species are particularly troublesome causes of food spoilage because their heat-resistant endospores survive cooking and, in some cases, canning. *Bacillus coagulans* and *Bacillus stearothermophilus* spoil some canned foods.

Common Spoilage Fungi

A wide variety of fungi, including species of *Rhizopus, Alternaria, Penicillium, Aspergillus,* and *Botrytis,* spoil foods. Because fungi grow in acidic as well as low-moisture environments, fruits and breads are more likely to be spoiled by fungi than by bacteria. *Aspergillus flavus* grows on peanuts and other grains, producing **aflatoxin,** a potent carcinogen monitored by the U.S. Food and Drug Administration (FDA).

MicroAssessment 30.3

The metabolites of microbes can spoil foods by generating undesirable flavors, odors, and textures.

7. What characteristics of *Pseudomonas* species allow them to spoil such a wide variety of foods?
8. Why is a grapefruit more likely to be spoiled by fungi than by bacteria?
9. How can disinfection of milking equipment by farmers postpone spoilage of milk in your refrigerator?

30.4 ■ Foodborne Illness

Learning Outcome

6. Distinguish between foodborne intoxication and foodborne infection, and give two examples of each.

Foodborne illness, commonly referred to as food poisoning, occurs when a pathogen, or a toxin it produced, is consumed in a food product. Food production is carefully regulated in the United States, with federal, state, and local agencies cooperating in inspections to help enforce laws that help prevent foodborne illness. In spite of strict controls, millions of food poisoning cases occur each year from foods prepared commercially, at home, or in institutions such as hospitals and schools. The vast majority could have been prevented with proper storage, sanitation, and preparation.

To monitor trends in foodborne illness in the United States, a government program called **FoodNet** (Foodborne Disease Active Surveillance Network) collects data on laboratory-confirmed cases of diarrheal illness in 10 states, covering approximately 15% of the population (www.cdc.gov/foodnet). The aim is to better understand the epidemiology of foodborne diseases so that these diseases can be prevented more easily.

Foodborne Intoxication

Foodborne intoxication is an illness that results from consuming an exotoxin produced by a microorganism growing in a food product. The toxin causes illness, not the living organisms. *Staphylococcus aureus* and *Clostridium botulinum* are two examples of bacteria that cause foodborne intoxication.

Staphylococcus aureus

Recall that *S. aureus* causes a wide variety of infectious diseases, many involving skin and wound infections (see table 22.3). A very different set of circumstances leads to what is referred to as "Staph food poisoning," an illness that results from ingesting a toxin made by some *S. aureus* strains; the toxin, an enterotoxin, causes nausea and vomiting. *S. aureus* does not compete well with most spoilage organisms, but it thrives in moist, nutrient-rich foods in which other organisms have been killed or their growth inhibited. It can grow with little competition on unrefrigerated creamy pastries and starchy salads made with cooked ingredients or salty products such as ham (a_w of 0.91).

The source of *S. aureus* is usually a human carrier who has not followed adequate hygiene procedures (such as handwashing) before preparing the food. If *S. aureus* is inoculated into a food that supports its growth, and the food is left at room temperature for several hours, the bacterial cells can multiply and produce the toxin. Unlike most exotoxins, *S. aureus* toxin is heat-stable, so cooking the food will not destroy it.

Clostridium botulinum

Foodborne botulism is a deadly paralytic disease caused by ingesting botulinum toxin (BTX), a neurotoxin most commonly produced by *C. botulinum*—a strictly anaerobic, endospore-forming, Gram-positive rod; some strains of *Clostridium butyricum* and *Clostridium baratii* also produce BTX but the disease principles are the same so we will focus our discussion on *C. botulinum*. The toxin, one of the most powerful poisons known, blocks transmission of the signals that trigger muscle contraction; the result is a potentially fatal flaccid paralysis. A few milligrams of this toxin are enough to kill the entire population of a large city. Characteristics of foodborne botulism were covered with nervous system diseases (chapter 26), so this section will instead focus on the events that lead to the disease.

C. botulinum is common in soils and marine sediments, so it is found naturally on many foods. The endospores survive usual cooking methods and inadequate canning procedures. If conditions in the food are favorable (anaerobic, nutrient-rich, a pH above 4.5, and a temperature above 4°C), the endospores can germinate, and the resulting vegetative cells may then multiply and produce the toxin (see Focus on a Case 30.1). Canning processes for low-acid foods are specifically designed to destroy the *C. botulinum* endospores. Manufacturing errors are rare in commercially canned foods, so most cases of foodborne botulism are due to improperly processed home-canned foods. As an added safety measure, low-acid, home-canned foods should be boiled for at least 10 minutes immediately before serving. The toxin is heat-labile (sensitive), so the heat treatment will destroy any toxin that may have been produced.

Unfortunately, the growth of *C. botulinum* and toxin production may not result in any noticeable changes in the taste or appearance of the food. Cans of foods that are damaged should be discarded, because they might have small holes through which the organism could enter. Bulging cans should also be discarded, because the swelling could be due to pressure from gases produced by microorganisms growing within the can.

Figure 30.7 summarizes the typical events that lead to foodborne intoxications by *S. aureus* and *C. botulinum.*

Foodborne Infection

Unlike foodborne intoxication, **foodborne infection** requires ingestion of live organisms. The signs and symptoms of the illness, which generally appear at least a day after eating the contaminated food, usually include diarrhea. Thorough cooking of food immediately before consuming it will kill the organisms, thereby preventing infection. *Salmonella* species, *Campylobacter* species, and Shiga toxin–producing *Escherichia coli* (STEC) are examples of organisms that cause foodborne infection. Characteristics of these bacteria and the diseases they cause were covered with digestive system infections (chapter 24), so this section will focus on the events that lead to foodborne infection.

Salmonella and *Campylobacter*

Salmonella and *Campylobacter* are two genera commonly associated with poultry products such as chicken, turkey, and eggs. Inadequate cooking of these products can result in foodborne infection. In addition, if a cutting board on which raw

Staphylococcus aureus

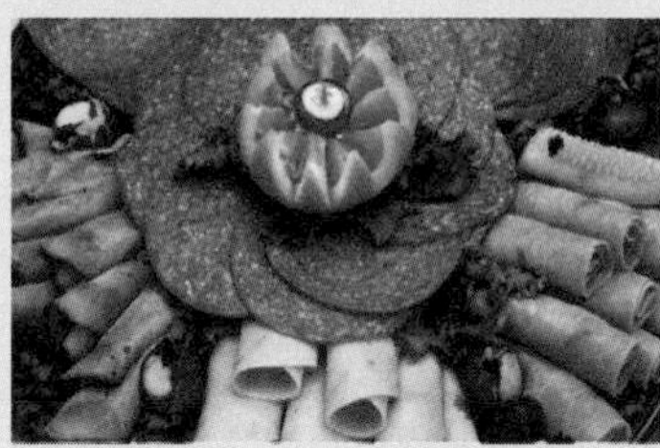

- Most bacteria that normally compete with *Staphylococcus aureus* are either killed by cooking or inhibited by high salt conditions.
- A food handler inadvertently transfers *S. aureus* onto food.
- *S. aureus* grows and produces toxin when food is allowed to slowly cool or is stored at room temperature.
- A person ingests the toxin-containing food. Symptoms of staph food poisoning—nausea, abdominal cramping, and vomiting—begin after 4 to 6 hours.

Clostridium botulinum

- *Clostridium botulinum* endospores, common in soil and marine sediments, contaminate many different foods.
- Endospores survive inadequate canning processes. Canned foods are anaerobic.
- Surviving *C. botulinum* endospores germinate; the resulting vegetative cells grow and produce toxin in low-acid canned foods.
- A person ingests the toxin-containing food. Symptoms of botulism, including weakness, double vision, and progressive inability to speak, swallow, and breathe, begin in 12 to 36 hours.

FIGURE 30.7 Typical Events Leading to Foodborne Intoxications by *Staphylococcus aureus* and *Clostridium botulinum* (left): Scott Bauer/ARS/U.S. Department of Agriculture; (right): Gaetano/Corbis

? Why is botulism primarily a problem in canned foods rather than fresh ones?

chicken was cut is then immediately used to slice vegetables for a salad, that salad can become contaminated as a result of cross-contamination.

Shiga Toxin–Producing *Escherichia coli* (STEC)

Escherichia coli O157:H7 and other Shiga toxin–producing *E. coli* (STEC) cause bloody diarrhea. Infection sometimes results in hemolytic uremic syndrome (HUS), a life-threatening condition.

STEC strains can colonize the intestinal tract of healthy cattle and other livestock, and are then shed in their feces. Because of this, meats can easily become contaminated. The initial contamination normally occurs on the meat surface, and the bacterial cells are easily destroyed by searing the exterior of meats such as steaks. Grinding the meat to create beef patties distributes the pathogen throughout the product. Prevention then involves more thorough cooking so that enough heat reaches the center to kill all the *E. coli* cells, which is why hamburgers are a particularly troublesome source of infection. Outbreaks have also been linked to unpasteurized milk and various kinds of produce that were contaminated with animal manure.

Figure 30.8 summarizes the typical events leading to foodborne infection by *Salmonella*, *Campylobacter*, and Shiga toxin–producing *E. coli*.

Some of the microbes that cause foodborne illness are listed in **table 30.4**.

Salmonella, Campylobacter

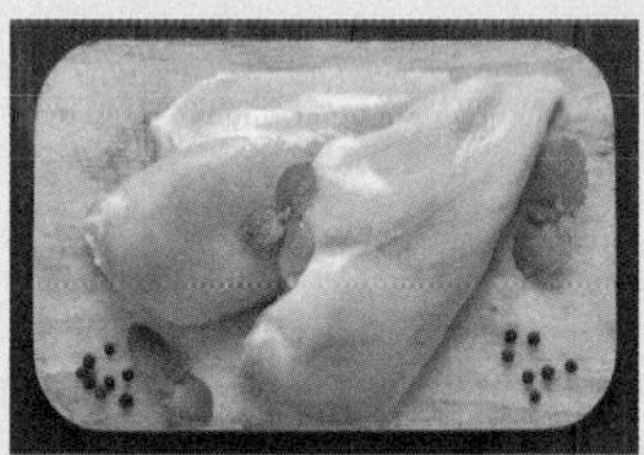

- Incomplete cooking fails to kill all pathogens. Surviving *Salmonella* and/or *Campylobacter* can multiply as food is cooled slowly or stored at room temperature.
- Live organisms are ingested. They multiply in the intestinal tract and cause disease. Symptoms include diarrhea, abdominal pain, and vomiting.

Shiga Toxin–Producing *E. coli*

- Incomplete cooking fails to kill all pathogens. Even low numbers of surviving Shiga toxin–producing *E. coli* can cause illness.
- Live organisms are ingested. They multiply in the intestinal tract and cause disease. Symptoms include severe abdominal pain and bloody diarrhea.

FIGURE 30.8 Typical Events Leading to Foodborne Infections by *Salmonella, Campylobacter,* and Shiga Toxin–Producing *E. coli* (STEC) (left): villagemon/iStock/Getty Images; (right): Olga Khomyakova/alisali/123RF

? Why are ground meats more common than steaks as a source of STEC infection?

TABLE 30.4 Common Foodborne Illnesses

Microbe	Signs and Symptoms	Foods Commonly Implicated
Intoxication		
Bacillus cereus	Nausea, vomiting, abdominal cramping	Rice
Clostridium botulinum	Blurred or double vision, progressive muscle weakness, nausea, vomiting, diarrhea, and inability to breathe	Low-acid canned foods such as vegetables and meats
Staphylococcus aureus	Nausea, vomiting, abdominal cramping	Cured meats, creamy salads, cream-filled pastries
Infection		
Bacillus cereus	Nausea, diarrhea, abdominal cramping	Meats, rice, fresh vegetables
Campylobacter species	Diarrhea, fever, abdominal pain, vomiting	Poultry, unpasteurized milk
Clostridium perfringens	Intense abdominal cramps, watery diarrhea	Meats, meat products
Cryptosporidium species	Watery diarrhea, nausea, abdominal cramping, fever, loss of appetite	Fresh produce, unpasteurized juices
Cyclospora species	Watery diarrhea, loss of appetite, bloating, vomiting, fatigue, weight loss	Fresh produce
Shiga toxin–producing *Escherichia coli* (STEC)	Severe abdominal pain, bloody diarrhea; sometimes hemolytic uremic syndrome	Ground beef, fresh vegetables, unpasteurized juices
Listeria monocytogenes	Fever, muscle aches; may progress to sepsis, meningitis	Unpasteurized milk, cheese, meats, fresh produce
Norovirus	Nausea, vomiting, diarrhea, abdominal cramps	Shellfish, salads, fresh produce
Salmonella species	Vomiting, abdominal cramps, diarrhea, fever, headache	Poultry, eggs, milk, meat
Shigella species	Abdominal cramps; diarrhea with blood, pus, or mucus; fever; vomiting, headache	Salads, fresh vegetables
Vibrio vulnificus	Fever, nausea, muscle pain, abdominal cramps	Oysters
Vibrio parahaemolyticus	Diarrhea, abdominal cramps, nausea, vomiting, headache, fever, chills	Fish and shellfish

FOCUS ON A CASE 30.1

A 37-year-old man was admitted to the hospital with dizziness, dry mouth, blurred vision, slurred speech, and nausea. He had difficulty breathing, so he was placed on a mechanical ventilator. Based on the signs and symptoms, the physician suspected botulism, and botulinum antitoxin was immediately administered intravenously.

When asked about the foods he had consumed the previous day, the patient had reported eating homemade stew and home-canned green beans. As the physician interviewed the patient, he noticed a partially healed wound on the man's elbow.

A stool sample and a specimen from the wound were obtained and cultured anaerobically for *Clostridium botulinum;* the stool was positive, but the wound was negative. In addition, leftover stew and green beans, as well as the stool sample, were tested for botulinum toxin, which was detected in the stew and in the stool sample, but not in the green beans. These results confirmed the diagnosis of foodborne botulism.

The patient was hospitalized for 6 weeks before being discharged, but he required many months at home for a complete recovery.

1. How did the antitoxin help the patient, and why was it administered before the diagnosis was confirmed?
2. Why is it surprising that the implicated food was homemade stew rather than home-canned green beans, and what could have led to the problem?
3. How could this case have been prevented?
4. Why was the patient's wound cultured along with the other samples?

Discussion

1. The antitoxin is a mixture of antibodies that bind to and thereby neutralize the different serotypes of botulinum toxin. Botulism is a life-threatening disease, and the toxin is one of the most potent known, so it is essential that antitoxin be administered as soon as possible when botulism is suspected.
2. Foodborne botulism typically involves home-canned low-acid foods that have not been adequately heat-processed to destroy the endospores. Although stews and certain other foods can also support the germination of *C. botulinum* endospores and the subsequent growth of the vegetative cells, refrigeration usually prevents this from occurring. In this case, however, the stew had been cooked in a large pot, covered with a heavy lid, and then left at room temperature for 3 days. Cooking does not destroy the endospores, and the extended time at room temperature—along with an environment that provided anaerobic conditions—resulted in the opportunity for those spores to germinate and the resulting vegetative cells to multiply to high numbers and produce botulinum toxin.

continued

3. If the stew had been transferred to smaller containers and refrigerated shortly after cooking, the endospores would not have germinated, nor would any vegetative cells have multiplied and produced toxin. Also, if the stew had been boiled for 10 minutes immediately before it was eaten, the heat would have destroyed any botulinum toxin that had been produced. Heating the stew to destroy the toxin would have been a risky option in this case, however, because if the food also contained *Staphylococcus aureus,* the room temperature storage would have allowed its growth as well. Recall that some *S. aureus* strains produce a toxin not destroyed by heat.
4. Not all botulism is foodborne. Wound botulism also occurs, so this was another possible cause of the man's signs and symptoms.

MicroAssessment 30.4

Foodborne intoxication results from consuming toxins produced by microbes growing in a food. Foodborne infection results from consuming living organisms.

10. How does boiling a home-canned food immediately prior to serving it prevent botulism?

11. Which foodborne pathogen can cause hemolytic uremic syndrome?

12. Why would a large number of competing microorganisms in a food sample result in lack of sensitivity of culture methods for detecting pathogens?

30.5 ■ Food Preservation

Learning Outcome

7. Describe the methods used to preserve foods.

Food preservation—the process of increasing the shelf life of foods by preventing the growth and metabolic activities of undesirable microorganisms—preserves the quality of food. The major methods are briefly summarized here and described in more detail in chapter 5.

- **Canning.** The canning process destroys all spoilage and pathogenic organisms capable of growth at normal storage temperatures. Low-acid foods are processed using steam under pressure (autoclaving) in order to reach temperatures high enough to destroy the endospores of *Clostridium botulinum.* Acidic foods do not require such high heat because *C. botulinum* cannot grow and produce toxin in those foods.
- **Pasteurization.** Heating foods under controlled conditions at high temperatures for short periods of time destroys non-spore-forming pathogens and reduces the numbers of spoilage organisms without significantly altering the flavor of food.
- **Cooking.** Cooking, like pasteurization, can destroy non-spore-forming organisms. Cooking, however, obviously alters the characteristics of food.
- **Refrigeration.** Refrigeration preserves food by slowing the growth rate of microbes. Many common organisms, including most pathogens, cannot multiply at low temperatures.
- **Freezing.** Freezing stops microbial growth because water in the form of ice is unavailable for biological reactions. Some of the microbial cells will be killed by damage caused by ice crystals, but those remaining can grow and spoil food once it is thawed.
- **Drying/reducing the *a*w.** Drying foods or adding high concentrations of sugars or salts inhibits microbial growth by decreasing the available moisture. Molds might grow, however, although slowly.
- **Lowering the pH.** Lowering the pH, either by adding acids or by encouraging fermentation by lactic acid bacteria, inhibits a wide range of spoilage organisms and pathogens.
- **Adding antimicrobial chemicals.** Organic acids such as propionic acid, benzoic acid, and sorbic acid are naturally occurring antimicrobial chemicals that are added to a variety of foods to inhibit fungal growth. Nitrates are added to cured meats to inhibit the growth of *Clostridium botulinum* and other organisms. Wine, fruit juices, and other products are often preserved by adding sulfur dioxide.
- **Irradiation.** Gamma radiation destroys microorganisms without significantly altering the flavor of foods such as spices and meats.

MicroAssessment 30.5

Food spoilage can be eliminated or delayed by destroying microorganisms or altering conditions to inhibit their growth.

13. Why are the process temperatures for canning low-acid foods higher than ones for acidic foods?

14. Why are nitrates added to cured meats?

15. Microorganisms are often grouped according to their optimum growth temperatures. Which groups are most likely to spoil refrigerated foods?

Summary

30.1 ■ Factors Influencing the Growth of Microorganisms in Foods

Intrinsic Factors

Bacteria require a high a_w. Fungi often grow when the a_w is too low to support bacterial growth. Many bacterial species, including most pathogens, are inhibited by acidic conditions. The nutritional content of a food determines the kinds of organisms that can grow in it. Rinds, shells, and other coverings aid in protecting some foods from invasion by microorganisms. Some foods contain natural antimicrobial chemicals that help prevent spoilage.

Extrinsic Factors

Low temperatures stop or inhibit the growth of most foodborne microorganisms. Psychrophiles and psychrotrophs, however, grow at refrigeration temperatures. The presence or absence of O_2 affects the type of microbial population able to grow in or on a food.

30.2 ■ Microorganisms in Food and Beverage Production

Not only are fermented foods perceived as pleasant tasting, the acids inhibit the growth of many spoilage organisms and foodborne pathogens. **Starter cultures** are sometimes used.

Lactic Acid Fermentations by the Lactic Acid Bacteria (table 30.1)

The tart taste of yogurt, pickles, sharp cheese, and some sausages is due to the metabolic products of the lactic acid bacteria. Other bacteria or fungi are sometimes added to cheese to give characteristic flavors or textures (figure 30.3). Pickling relies on naturally occurring lactic acid bacteria. Commercial sausage production uses starter cultures to rapidly decrease the pH and prevent the growth of pathogens.

Ethanol Fermentations by Yeast (table 30.2)

Wine is the product of the fermentation of sugars in fruit juices (figure 30.4). Beer production is a multistep process designed to break down the starches of grains such as barley to produce simple sugars, which can then serve as a substrate for ethanol fermentation by yeast (figure 30.5). Distilled spirits are produced using distillation to collect the alcohol generated during fermentation. Vinegar is the product of the oxidation of alcohol by the acetic acid bacteria. In bread-making, the CO_2 produced by yeast causes bread to rise; the alcohol is lost to evaporation (figure 30.6).

Changes Due to Mold Growth (table 30.3)

Some cheeses and other foods are produced by encouraging the growth of molds on foods. Soy sauce is made by allowing an *Aspergillus* species to degrade a mixture of soybeans and wheat, which is then fermented in brine.

30.3 ■ Food Spoilage

Food spoilage is most often due to the metabolic activities of microorganisms as they grow and use the nutrients in the food.

Common Spoilage Bacteria

Pseudomonas, Erwinia, Acetobacter, Alcaligenes, lactic acid bacteria, and endospore-forming bacteria are important causes of food spoilage.

Common Spoilage Fungi

Fungi grow readily in acidic as well as low-moisture environments; therefore, fruits and breads are more likely to be spoiled by fungi than by bacteria.

30.4 ■ Foodborne Illness (table 30.4)

Foodborne Intoxication

Foodborne intoxication results from consuming a toxin produced by a microorganism growing in a food product (figure 30.7). Many strains of *Staphylococcus aureus* produce a toxin that, when ingested, causes nausea and vomiting. Botulism is caused by ingestion of a neurotoxin produced by the anaerobic, spore-forming, Gram-positive rod *Clostridium botulinum* or certain closely related bacteria. As an added safety measure, low-acid home-canned foods should be boiled at least 10 minutes immediately before serving to destroy any botulinum toxin that could be present.

Foodborne Infection

Foodborne infection requires the consumption of living organisms (figure 30.8). Thorough cooking of food immediately before eating it will kill bacteria, thereby preventing infection. *Salmonella* and *Campylobacter* species are commonly associated with poultry products. Some outbreaks of Shiga toxin–producing *E. coli* have been traced to undercooked contaminated hamburger patties and produce contaminated with manure.

30.5 ■ Food Preservation

Food spoilage can be eliminated or delayed by destroying microorganisms or altering conditions to inhibit their growth. Methods used to preserve foods include canning, pasteurization, cooking, refrigeration, freezing, reducing the a_w, lowering the pH, adding antimicrobial chemicals, and irradiation.

Review Questions

Short Answer

1. What is the purpose of rennin in cheese-making?
2. What causes bluish-green veins to form in some types of cheese?
3. What causes the holes to form in Swiss cheese?
4. Why are starter cultures used when making semidry sausages?
5. What is the purpose of the mashing step in beer-making?
6. Explain how *Alcaligenes* species cause "ropiness" in raw milk.
7. Explain the significance of *Aspergillus flavus* in grain products.
8. Explain the typical sequence of events that lead to foodborne botulism.
9. Explain the typical sequence of events that lead to staphylococcal food poisoning.
10. How does canning differ from pasteurization?

Multiple Choice

1. The a_w of a food product reflects which of the following?
   a) Acidity of the food
   b) Presence of antimicrobial chemicals such as lysozyme
   c) Amount of water available
   d) Storage atmosphere
   e) Nutrient content
2. Most spoilage bacteria cannot grow below an a_w of
   a) 0.3.
   b) 0.5.
   c) 0.7.
   d) 0.9.
   e) 1.0.
3. What is a generally minimum pH for growth and toxin production by *Clostridium botulinum* and other foodborne pathogens?
   a) 8.5
   b) 7.0
   c) 6.5
   d) 4.5
   e) 2.0
4. Benzoic acid is an antimicrobial chemical naturally found in which of the following foods?
   a) Apples
   b) Cranberries
   c) Eggs
   d) Milk
   e) Yogurt
5. Which of the following is often added to wine to inhibit the growth of the natural microbial populations on grapes?
   a) Benzoic acid
   b) Lactic acid
   c) Carbon dioxide
   d) Sulfur dioxide
   e) Oxygen
6. In the brewing process, the sugar and nutrient extract obtained by soaking germinated grain in warm water is called
   a) baker's yeast.
   b) hops.
   c) malt.
   d) must.
   e) wort.
7. Members of which of the following genera are used in bread, wine, and beer production?
   a) *Lactobacillus*
   b) *Pseudomonas*
   c) *Saccharomyces*
   d) *Streptococcus*
   e) *Staphylococcus*
8. Which group of organisms most commonly spoils breads, fruits, and dried foods?
   a) *Acetobacter*
   b) Fungi
   c) Lactic acid bacteria
   d) *Pseudomonas*
   e) *Saccharomyces*
9. Which of the following types of bacteria causes foodborne intoxication?
   a) Shiga toxin–producing *E. coli*
   b) *Campylobacter* species
   c) *Lactobacillus* species
   d) *Salmonella* species
   e) *Staphylococcus aureus*
10. Canned pickles require less stringent heat processing than canned beans because pickles
    a) contain fewer nutrients.
    b) are more acidic.
    c) have a lower a_w.
    d) contain antimicrobial chemicals.
    e) are less likely to be contaminated with endospores.

Applications

1. A small cheese-manufacturing company in Wisconsin is looking for ways to reduce the costs of disposing of whey, a cheese by-product. As a food microbiologist, what would you suggest that the company do with the thousands of liters of whey being produced per month as a means to profit from it?
2. A microbiologist is troubleshooting a batch of home-brewed ale that did not ferment properly. She noticed that the alcohol content was only 2%, well below the desired level. Microscopic examination showed numerous yeast cells. Chemical analysis indicated low levels of sugar, high levels of CO_2, and large amounts of protein in the liquid. What did the microbiologist conclude was the probable cause of the problem when the beer did not come out properly?

Critical Thinking

1. It has been argued that the nature of the growth of fungi in Roquefort cheese, indicated by the appearance of bluish-green veins, is evidence that these fungi require O_2 for growth. How does this evidence lead to the conclusion?
2. In the production of sauerkraut, a natural succession of lactic acid bacteria is observed growing in the product. What causes the succession? What does this tell you about the optimal growth conditions of the different species of lactic acid bacteria?

Appendix I
Microbial Mathematics

Because prokaryotes are very tiny and can multiply to very large numbers of cells in short time periods, convenient and simple ways are used to indicate their numbers without resorting to many zeros before or after the number. This is one reason it is important to understand the metric system, which is used in scientific measurements.

The basic unit of measure is the meter, which is equal to about 39 inches. All other units are fractions of a meter:

1 decimeter is one tenth = 0.1 meter

1 centimeter is one hundredth = 0.01 meter

1 millimeter is one thousandth = 0.001 meter

Because prokaryotes are much smaller than a millimeter, even smaller units of measure are used. A millionth of a meter is a micrometer = 0.000001 meter, and is abbreviated μm. This is the most frequently used size measurement in microbiology because most bacteria are about 1 μm wide. For comparison, a human hair is about 75 μm wide.

Because it is inconvenient to write so many zeros in front of the 1, an easier way of indicating the same number is through the use of superscript, or exponential, numbers (exponents). One hundred dollars can be written 10^2 dollars. The 10 is called the base number and the 2 is the exponent. Conversely, one hundredth of a dollar is 10^{-2} dollars; thus the exponent is negative. The base most commonly used in biology is 10 (which is designated as $\log_{10}$). The preceding information can be summarized as follows:

1 millimeter = 1 mm = 0.001 meter = 10^{-3} meter

1 micrometer = 1 μm = 0.000001 meter = 10^{-6} meter

1 nanometer = 1 nm = 0.000000001 meter = 10^{-9} meter

The same prefix designations can be used for weights. The basic unit of weight is the gram, abbreviated g. Approximately 450 grams are in a pound.

1 milligram = 1 mg = 0.001 gram = 10^{-3} g

1 microgram = 1 μg = 0.000001 gram = 10^{-6} g

1 nanogram = 1 ng = 0.000000001 gram = 10^{-9} g

1 picogram = 1 pg = 0.000000000001 gram = 10^{-12} g

Note that the number of zeros before the 1 is one less than the exponent.

The value of the number is obtained by multiplying the base by itself the number of times indicated by the exponent.

Thus, $10^1 = 10 \times 1 = 10$

$10^2 = 10 \times 10 = 100$

$10^3 = 10 \times 10 \times 10 = 1{,}000$

When the exponent is negative, the base and exponent are divided into 1.

For example, $10^{-2} = 1/10 \times 1/10 = 1/100 = 0.01$

When multiplying numbers having exponents to the same base, the exponents are added.

For example, $10^3 \times 10^2 = 10^5$ (not 10^6)

When dividing numbers having exponents to the same base, the exponents are subtracted.

For example, $10^5 \div 10^2 = 10^3$

In both cases, only if the bases are the same can the exponents be added or subtracted.

Appendix II
Pronunciation Key for Bacterial, Fungal, Protozoan, and Viral Names

A

Acetobacter (a-see′-toe-back-ter)
Acinetobacter (a-sin-et′-oh-back-ter)
Actinomycetes (ak-tin-oh-my′-seats)
Adenovirus (ad′-eh-no-vi-rus)
Agrobacterium tumefaciens (ag-rho-bak-teer′-ee-um too-meh-faysh′-ee-enz)
Alcaligenes (al-ka-li′-jen-ease)
Aliivibrio fischeri (ali-vib′-ree-oh fish′-er-i)
Amoeba (ah-mee′-bah)
Arbovirus (are′-bow-vi-rus)
Aspergillus niger (ass-per-jill′-us nye′-jer)
Aspergillus oryzae (ass-per-jill′-us or-eye′-zee)
Azolla (aye-zol′-lah)
Azotobacter (ay-zoh′-toe-back-ter)

B

Bacillus anthracis (bah-sill′-us an-thra′-siss)
Bacillus cereus (bah-sill′-us seer′-ee-us)
Bacillus fastidiosus (bah-sill′-us fas-tid-ee-oh′-sus)
Bacillus subtilis (bah-sill′-us sut′-ill-us)
Bacillus thuringiensis (bah-sill′-us thur′-in-jee-en-sis)
Bacteroides (back′-ter-oid′-eez)
Bdellovibrio (del′-o-vib′-re-oh)
Beggiatoa (beg-gee-ah-toe′-ah)
Beijerinckia (by-yer-ink′-ee-ah)
Bordetella pertussis (bor-deh-tell′-ah per-tuss′-iss)
Borrelia burgdorferi (bor-real′-ee-ah berg-dor′-fir-ee)
Bradyrhizobium (bray-dee-rye-zoe′-bee-um)
Brucella abortus (bru-sell′-ah ah-bore′-tus)

C

Campylobacter jejuni (kam′-peh-low-back-ter je-june′-ee)
Candida albicans (kan′-did-ah al′-bi-kanz)
Caulobacter (caw′-loh-back-ter)
Chikungunya virus (chik-en-gun-ye vi-rus)
Chlamydia trachomatis (klah-mid′-ee-ah trah-ko-ma′-tiss)
Chlamydophila pneumoniae (klah-mid′-o-fil-ah new-moan′-ee-ee)
Claviceps purpurea (kla′-vi-seps purr-purr′-ee-ah)
Clostridioides difficile (kloss-trid′-ee-oid-eez dif′-fi-seal)
Clostridium botulinum (kloss-trid′-ee-um bot-you-line′-um)
Clostridium perfringens (kloss-trid′-ee-um per-frin′-gens)
Clostridium tetani (kloss-trid′-ee-um tet′-an-ee)
Coccidioides immitis (cock-sid-ee-oid′-eez im′-mi-tiss)
Coronavirus (kor-oh′-nah-vi-rus)
Corynebacterium diphtheriae (koh-ryne′-nee-bak-teer-ee-um dif-theer′-ee-ee)
Coxsackievirus (cock-sack-ee′-vi-rus)
Cryptococcus neoformans (krip-toe-cock′-us knee-oh-for′-manz)
Cryptosporidium parvum (krip-toe-spo-rid-ee-um par-vum)
Cutibacterium acnes (kyoo-tee-bak-teer-ee-um ak-neez)
Cyclospora (sigh-clo-spo-rah)

D

Dengue virus (deng-ee vi-rus)
Desulfovibrio (dee-sul-foh-vib′-ree-oh)

E

Entamoeba histolytica (en-ta-mee′-bah his-toh-lit′-ik-ah)
Enterobacter (en′-ter-oh-back-ter)
Enterococcus faecalis (en′-ter-oh-kock′-us fee-ka′-liss)
Enterovirus (en′-ter-oh-vi-rus)
Epidermophyton (eh-pee-der′-moh-fy-ton)
Epulopiscium (ep′-you-low-pis-se-um)
Escherichia coli (esh-er-ee′-she-ah koh′-lee)

F

Flavivirus (flay′-vih-vi-rus)
Francisella tularensis (fran-siss-sell′-ah tu-lah-ren′-siss)
Frankia (frank′-ee-ah)

G

Gallionella (gal-ee-oh-nell′-ah)
Gardnerella vaginalis (gard-nee-rel′-lah va-jin-al′-is)
Geobacillus stearothermophilus (gee′-oh-bah-sill′-us steer-oh-ther-maw′-fill-us)
Giardia lamblia (jee-are′-dee-ah lamb′-lee-ah)
Gluconobacter (glue-kon-oh-back′-ter)
Gonyaulax (gon-ee-ow′-lax)

H

Haemophilus influenzae (hee-moff′-ill-us in-flew-en′-zee)
Helicobacter pylori (he′-lih-koh-back-ter pie-lore′-ee)
Hepadnavirus (hep-ad′-nah-vi-rus)
Hepatitis virus (hep-ah-ti′-tis vi-rus)
Herpes simplex (her′-peas sim′-plex)
Herpes zoster (her′-peas zoh′-ster)
Histoplasma capsulatum (his-toh-plaz′-mah cap-su-lah′-tum)
Hyphomicrobium (high-foh-my-krow′-bee-um)

I

Influenza virus (in-flew-en′-za vi-rus)

K

Klebsiella pneumoniae (kleb-see-ell′-ah new-moan′-ee-ee)

L

Lactobacillus brevis (lack-toe-ba-sil′-lus bre′-vis)
Lactobacillus bulgaricus (lack-toe-ba-sil′-lus bull-gair′-i-kus)
Lactobacillus casei (lack-toe-ba-sil′-us kay′-see-ee)
Lactobacillus plantarum (lack-toe-ba-sil′-us plan-tar′-um)
Lactobacillus thermophilus (lack-toe-ba-sil′-us ther-mo′-fil-us)
Lactococcus lactis (lack-toe-kock′-us lak′-tiss)
Legionella pneumophila (lee-jon-ell′-ah new-moh′-fill-ah)
Leptospira interrogans (lep-toe-spire′-ah in-ter-roh′-ganz)
Leuconostoc mesenteroides (lew-kow-nos′-tok mes-en-te-roi-deze)
Listeria monocytogenes (lis-tear′-ee-ah mon′-oh-sigh-to- jen′-eze)

M

Malassezia (mal-as-seez′-e-ah)
Methanobacterium (me-than′-oh-bak-teer-ee-um)
Methanococcus (me-than-oh-ko′-kus)
Microsporum (my-kroh-spore′-um)
Mobiluncus (moh-bi-lun′-kus)
Moraxella catarrhalis (more-ax-ell′-ah kah-tah-rah′-liss)
Moraxella lacunata (more-ax-ell′-ah lak-u-nah′-tah)
Mucor (mu′-kor)
Mycobacterium leprae (my-koh-bak-teer′-ee-um lep-ree)
Mycobacterium tuberculosis (my-koh-bak-teer′-ee-um too-ber-kew-loh′-siss)
Mycoplasma pneumoniae (my-koh-plaz′-mah new-moan′-ee-ee)
Mycoplasma genitalium (my-koh-plaz′-mah gen-ih-tale′-ee-um)

N

Neisseria gonorrhoeae (nye-seer′-ee-ah gahn-oh-ree′-ee)
Neisseria meningitidis (nye-seer′-ee-ah men-in-jit′-id-iss)
Neurospora (new-rah′-spor-ah)

O

Orthomyxovirus (or-thoe-mix′-oh-vi-rus)
Oscillatoria (os-sil-la-tor′-ee-ah)

P

Papillomavirus (pap-il-oh′-ma-vi-rus)
Parainfluenza virus (par-ah-in-flew-en′-zah vi-rus)
Paramecium (pair′-ah-mee-see-um)
Paramyxovirus (par-ah-mix′-oh-vi-rus)
Parvovirus (par′-vo-vi-rus)
Pasteurella multocida (pass-ture-ell′-ah mul-toe-sid′-ah)
Pediococcus (ped-ih-oh-ko′-kus)
Penicillium camemberti (pen-eh-sill′-ee-um cam-em-bare′-tee)
Penicillium roqueforti (pen-eh-sill′-ee-um rok-e-for′-tee)
Peptostreptococcus (pep′-to-strep-to-ko-kus)
Phytophthora infestans (fy′-toe-fy-thor-ah in-fes′-tanz)
Picornavirus (pi-kor′-na-vi-rus)
Plasmodium falciparum (plaz-moh′-dee-um fall-sip′-air-um)
Plasmodium malariae (plaz-moh′-dee-um ma-lair′-ee-ee)
Plasmodium ovale (plaz-moh′-dee-um oh-vah′-lee)
Plasmodium vivax (plaz-moh′-dee-um vye′-vax)
Pneumocystis jirovecii (new-mo-sis′-tis yee′-row′-vet-ze)
Poliovirus (poe′-lee-oh-vi-rus)
Polyomavirus (po-lee-oh′-mah vi-rus)
Propionibacterium shermanii (proh-pee-ah-nee-bak-teer′-ee-um sher-man′-ee-ee)
Proteus mirabilis (proh′-tee-us mee-rab′-il-us)
Pseudomonas aeruginosa (sue-dough-moan′-ass aye-rue-gin-o′-sa)

R

Rabies virus (ray′-bees vi-rus)
Retrovirus (re′-trow-vi-rus)
Rhabdovirus (rab′-doh-vi-rus)
Rhinovirus (rye′-no-vi-rus)
Rhizobium (rye-zoh′-bee-um)
Rhizopus stolonifer (rise′-oh-pus stoh′-lon-ih-fer)
Rickettsia rickettsii (rik-kett′-see-ah rik-kett′-see-ee)
Rotavirus (row′-tah-vi-rus)
Rubella virus (rue-bell′-ah vi-rus)
Rubeola virus (rue-bee-oh′-la vi-rus)

S

Saccharomyces carlsbergensis (sack-ah-row-my′-sees karls-ber-gen′-siss)
Saccharomyces cerevisiae (sack-ah-row-my′-sees sara-vis′-ee-ee)
Salmonella enterica (sall-moh-nell′-ah en-ter′-ih-kah)
Serratia marcescens (ser-ray′-sha mar-sess′-sens)
Shigella dysenteriae (shig-ell′-ah diss-en-tair′-ee-ee)
Spirillum volutans (spy-rill′-um vol′-u-tanz)
Sporothrix schenckii (spore′-oh-thrix shenk′-ee-ee)
Staphylococcus aureus (staff-ill-oh-kok′-us aw′-ree-us)
Staphylococcus epidermidis (staff-ill-oh-kok′-us epi-der′-mid-iss)
Streptobacillus moniliformis (strep-tow-bah-sill′-us mon-ill-i-form′-is)
Streptococcus agalactiae (strep-toe-kock′-us a-ga-lac′-tee-ee)
Streptococcus cremoris (strep-toe-kock′-us kre-more′-iss)
Streptococcus mutans (strep-toe-kock′-us mew′-tanz)
Streptococcus pneumoniae (strep-toe-kock′-us new-moan′-ee-ee)
Streptococcus pyogenes (strep-toe-kock′-us pie-ah′-gen-ease)
Streptococcus salivarius (strep-toe-kock′-us sal-ih-vair′-ee-us)
Streptococcus thermophilis (strep-toe-kock′-us ther-moh′-fill-us)
Streptomyces griseus (strep-toe-my′-seez gree′-see-us)

T

Thiobacillus (thigh-oh-bah-sill′-us)
Treponema pallidum (tre-poh-nee′-mah pal′-ih-dum)
Trichomonas vaginalis (trick-oh-moan′-as vag-in-al′-iss)
Trichophyton (trick-oh-phye′-ton)
Trypanosoma brucei (tri-pan′-oh-soh-mah bru′-see-ee)

V

Varicella-zoster virus (var-ih-sell′-ah zoh′-ster vi-rus)
Vibrio cholerae (vib′-ree-oh kahl′-er-ee)
Vibrio vulnificus (vib′-ree-oh vul-ni′-fi-kus)

W

Wolbachia (wol-bach′-ee-ah)

Y

Yersinia enterocolitica (yer-sin′-ee-ah en-ter-oh-koh-lih′-tih-kah)
Yersinia pestis (yer-sin′-ee-ah pess′-tiss)

Appendix III

Metabolic Pathways

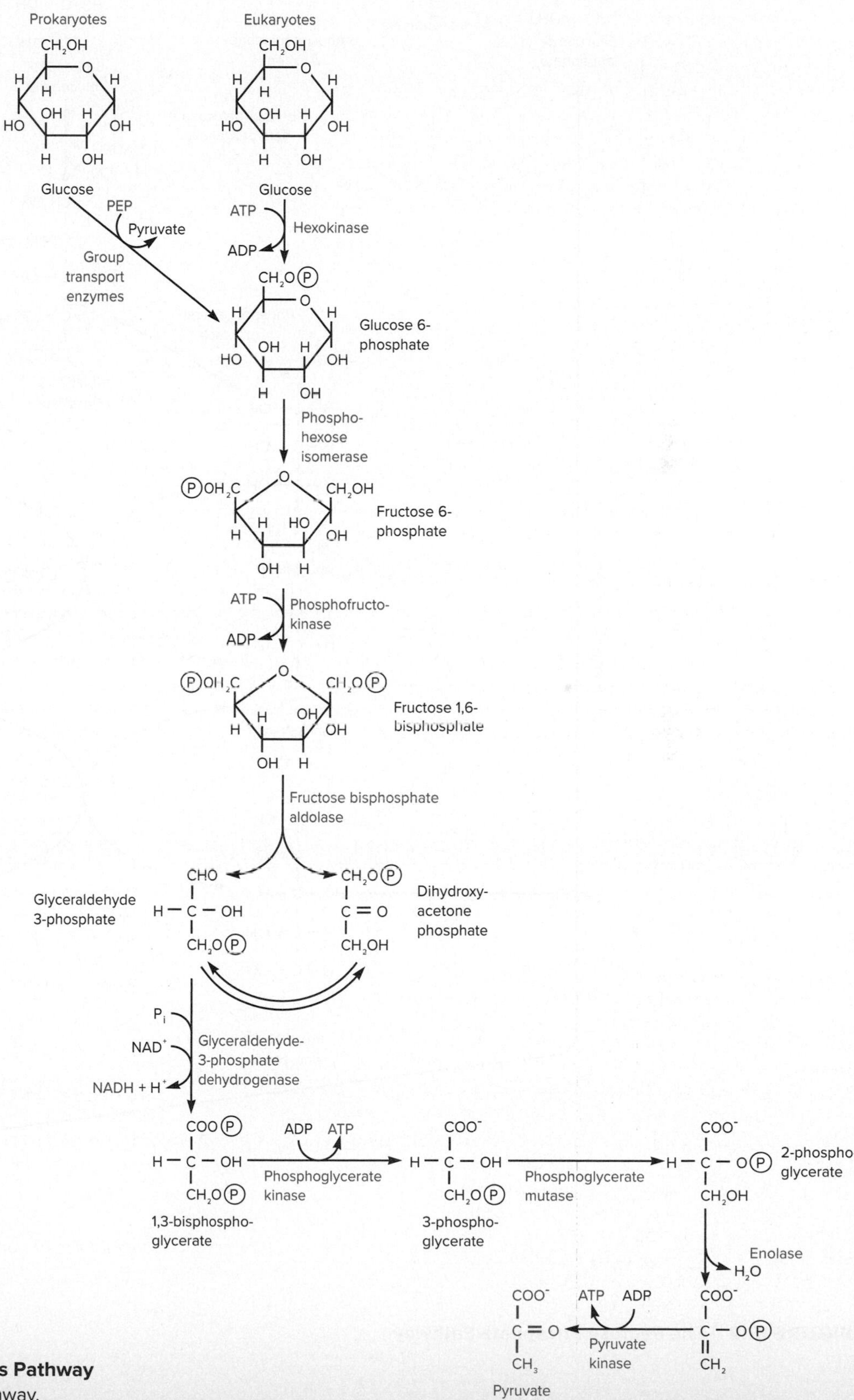

FIGURE III.1 The Embden-Meyerhof-Parnas Pathway
Commonly called glycolysis or the glycolytic pathway.

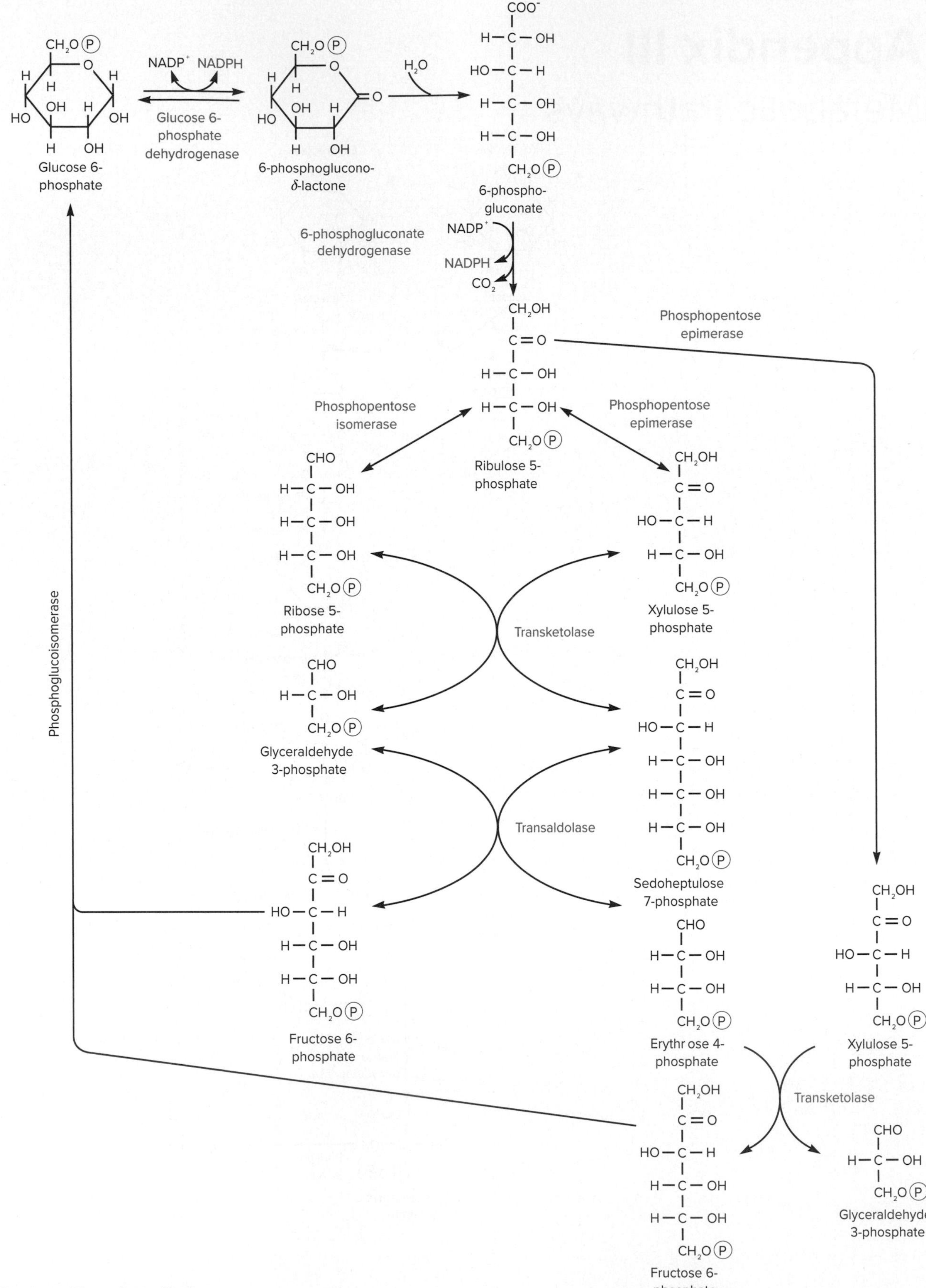

FIGURE III.2 The Pentose Phosphate Pathway

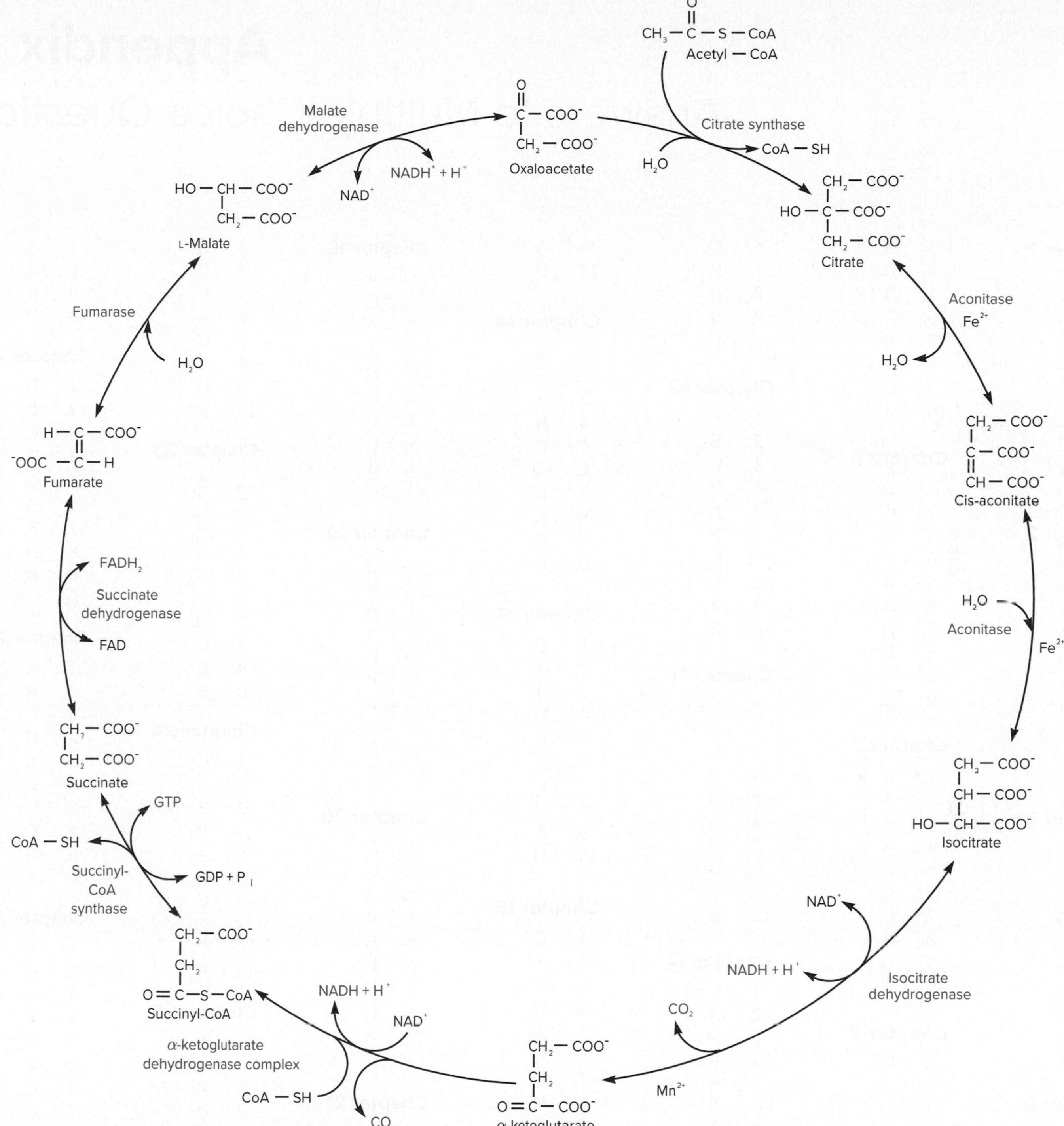

FIGURE III.3 The Tricarboxylic Acid Cycle (TCA Cycle)
Also referred to as the Krebs cycle or the citric acid cycle.

Appendix IV

Answers to Multiple Choice Questions

Chapter 1
1. C
2. C
3. C
4. A
5. C
6. C
7. C
8. C
9. A
10. D

Chapter 2
1. C
2. B
3. B
4. A
5. D
6. E
7. A
8. C
9. D
10. E

Chapter 3
1. C
2. B
3. E
4. D
5. E
6. B
7. D
8. A
9. A
10. A

Chapter 4
1. D
2. A
3. B
4. B
5. B
6. E
7. C
8. A
9. C
10. B

Chapter 5
1. C
2. D
3. A
4. B
5. B
6. A
7. E
8. D
9. C
10. A

Chapter 6
1. D
2. C
3. D
4. E
5. B
6. A
7. D
8. D
9. D
10. E

Chapter 7
1. B
2. A
3. A
4. A
5. C
6. C
7. C
8. B
9. D
10. D

Chapter 8
1. C
2. A
3. D
4. A
5. B
6. D
7. A
8. B
9. A
10. B

Chapter 9
1. C
2. C
3. E
4. B
5. B
6. C
7. B
8. B
9. B
10. D

Chapter 10
1. E
2. B
3. E
4. B
5. D
6. B
7. C
8. D
9. B
10. E

Chapter 11
1. E
2. B
3. E
4. B
5. D
6. A
7. D
8. B
9. C
10. D

Chapter 12
1. C
2. A
3. D
4. B
5. B
6. C
7. A
8. A
9. C
10. B

Chapter 13
1. C
2. A
3. C
4. D
5. E
6. C
7. D
8. A
9. A
10. D

Chapter 14
1. B
2. E
3. A
4. B
5. C
6. A
7. E
8. A
9. C
10. D

Chapter 15
1. D
2. A
3. B
4. D
5. E
6. E
7. D
8. D
9. B
10. D

Chapter 16
1. C
2. A
3. D
4. D
5. A
6. C
7. C
8. D
9. D
10. A

Chapter 17
1. C
2. D
3. D
4. E
5. D
6. A
7. C
8. D
9. A
10. C

Chapter 18
1. A
2. E
3. A
4. B
5. D
6. E
7. C
8. C
9. A
10. C

Chapter 19
1. A
2. B
3. A
4. B
5. E
6. B
7. B
8. E
9. C
10. E

Chapter 20
1. D
2. C
3. A
4. B
5. A
6. B
7. D
8. D
9. A
10. C

Chapter 21
1. E
2. C
3. E
4. A
5. B
6. A
7. A
8. C
9. D
10. D

Chapter 22
1. E
2. E
3. A
4. D
5. E
6. A
7. C
8. A
9. D
10. D

Chapter 23
1. B
2. E
3. C
4. A
5. A
6. C
7. D
8. C
9. E
10. B

Chapter 24
1. C
2. A
3. B
4. D
5. A
6. D
7. C
8. D
9. C
10. E

Chapter 25
1. E
2. A
3. E
4. C
5. C
6. C
7. A
8. C
9. A
10. E

Chapter 26
1. B
2. C
3. C
4. C
5. E
6. A
7. A
8. B
9. E
10. D

Chapter 27
1. D
2. D
3. C
4. B
5. C
6. A
7. B
8. B
9. B
10. C

Chapter 28
1. A
2. A
3. C
4. D
5. A
6. D
7. C
8. C
9. B
10. A

Chapter 29
1. B
2. B
3. A
4. A
5. C
6. E
7. C
8. C
9. B
10. E

Chapter 30
1. C
2. D
3. D
4. B
5. D
6. E
7. C
8. B
9. E
10. B

Appendix V
Microbial Terminology

Singular and Plural Forms

Singular	Plural	Singular	Plural
alg*a*	alg*ae*	fung*us*	fung*i*
amoeb*a*	amoeb*ae*	hydroly*sis*	hydroly*ses*
bacill*us*	bacill*i*	hyph*a*	hyph*ae*
bacteri*um*	bacteri*a*	inocul*um*	inocul*a*
cili*um*	cili*a*	medi*um*	medi*a*
clostridi*um*	clostridi*a*	mucos*a*	mucos*ae*
cocc*us*	cocc*i*	myceli*um*	myceli*a*
conidi*um*	conidi*a*	myco*sis*	myco*ses*
dat*um*	dat*a*	phyl*um*	phyl*a*
diagno*sis*	diagno*ses*	pil*us*	pil*i*
fimbri*a*	fimbri*ae*	nucle*us*	nucle*i*
flagell*um*	flagell*a*	sept*um*	sept*a*
foc*us*	foc*i*	synthe*sis*	synthe*ses*

Meanings of Prefixes and Suffixes

Prefix or Suffix	Meaning	Example
a-, an-	not, without	avirulent (lacking virulence), anaerobic (without air)
aer-	air	aerobic
anti-	against	antiseptic
-ase	enzyme	penicillinase
chlor-	green	chlorophyll
chrom	color	metachromatic (staining differently with the same dye)
-cide	causing death	germicide
co-, com-, con-	together	coenzyme
-cyan-	blue	pyocyanin (a blue bacterial pigment)
de-	down, from	dehydrate (remove water)
-dem-	people, district	epidemic
endo-	within	endospore (spore within a cell)
-enter-	intestine	enteritis (inflammation of the intestine)
epi-	upon	epidermis
erythro-	red	erythocyte (red blood cell)
eu-	well, normal	eukaryotic (true nucleus)
exo-	outside	exoenzyme (enzyme that acts outside the cell that produced it)
extra-	outside of	extracellular
flav-	yellow	flavoprotein (a protein containing a yellow compound)
-gen	produce, originate	antigen (a substance that induces a production of antibodies)
glyc-	sweet	glycemia (the presence of sugar in the blood)
hetero-	other	heterotroph (organism that obtains carbon from organic compounds)
homo-	common, same	homologous (similar in structure or origin)
hydr-	water	dehydrate
hyper-	excessive, above	hypersensitive
hypo-	under	hypotonic (having low osmotic pressure)

Prefix or Suffix	Meaning	Example
iso-	same, equal	isotonic (having the same osmotic pressure)
-itis	inflammation	appendicitis, meningitis
leuko-	white	leukocyte (white blood cell)
ly-, -lys, -lyt-	loosen; dissolve	bacteriolysis (dissolution of bacteria)
meso-	middle	mesophilic (preferring moderate temperatures)
meta-	after, beyond, changed	metachromatic (staining differently with the same dye)
micro-	small; one-millionth part	microscopic
milli-	one-thousandth part	millimeter (10^{-3} meter)
mito-	thread	mitochondrion (small, rod-shaped or granular organelle)
mono-	single	monotrichous (having a single flagellum)
multi-	many	multinuclear (having many nuclei)
myc-	fungus	mycotic (caused by fungus)
myx-	mucus	myxomycete (slime mold)
-oid	resembling	lymphoid (resembling lymphocytes)
-ose	a sugar	lactose (milk sugar)
-osis	disease of	coccidioidomycosis (disease caused by *Coccidioides*)
pan-	all	pandemic (widespread epidemic)
para-	beside	parasite (an organism that feeds in and at the expense of the host)
patho-	disease	pathogenic (producing disease)
peri-	around	peritrichous (having flagella on all sides)
-phag-	eat	phagocyte (a cell that ingests other cell substances)
-phil	like, having affinity for	eosinophilic (staining with the dye eosin)
-phot-	light	photosynthesis
-phyll	leaf	chlorophyll (green leaf pigment)
pleo-	more	pleomorphic (occurring in more than one form)
poly-	many	polymorphonuclear (having a many-shaped nucleus)
post-	after	postnatal (after birth)
pyo-	pus	pyogenic (producing pus)
-sta-	stop	bacteriostatic (inhibiting bacterial multiplication)
sym-, syn-	together	symbiosis (life together)
thermo-	heat	thermophilic (liking heat)
tox-	poison	toxin
trans-	through, across	transfusion
-trich-	hair	monotrichous (having a single flagellum)
-troph	nourishment	autotroph (organism that obtains carbon from CO_2)
zym-	ferment	enzyme

Glossary/Index

Page numbers followed by *t* and *f* indicate tables and figures respectively. Bolded terms are defined here.

A

B

D

E

F

griseofulvin, 67, 529, 530t
gRNA, 240
groundwater contamination, 798, 799f
group A streptococcal flesh-eating disease, 612–613, 613t
group A streptococcus (GAS), 541
group B streptococcus (GBS), 711
group translocation Type of transport process that chemically alters a molecule during its passage through the cytoplasmic membrane, 49, 49f, 49t
growth curve Growth pattern observed when cells are grown in a closed system; consists of five stages—lag phase, log phase (or exponential phase), stationary phase, death phase, and the phase of prolonged decline, 95–96, 95f
growth factors Compounds that a particular bacterium requires in a growth medium because it cannot synthesize them, 101
guanine, 38, 39f
Guillain-Barré syndrome, 653, 691
gum infection, 634f
gumma Localized area of chronic inflammation and necrosis in tertiary syphilis, 754, 754f
gummatous syphilis, 754, 754f
gunshot wounds, 609
gut microbiome, 416

H

HA. *See* hemagglutinin (HA)
Haemophilus, 294
Haemophilus ducreyi, 294, 757
Haemophilus influenzae, 294, 487, 565
  conjunctivitis, 539
  genomic sequence, 199
  scientific name, 12t
  throat, 536
  T-independent antigen, 406
Haemophilus influenzae meningitis, 709–710, 713t
  case study, 711
  causative agent, 710
  epidemiology, 710, 710f
  pathogenesis, 710
  signs and symptoms, 710
  treatment and prevention, 710
Haemophilus influenzae type b disease, 441, 443t
hair follicle infections, 584–588, 588t
  carbuncle, 585
  causative agents, 585, 585t
  epidemiology, 587
  folliculitis, 584–585
  furuncle, 585
  pathogenesis, 585–587, 586f
  signs and symptoms, 584–585
  summary/overview, 588t
  treatment and prevention, 587–588
HAIs. *See* healthcare-associated infections (HAIs)
half-life, 506
halitosis, 631
Halobacterium, 298
halofantrine, 531t
halogens, 130–131, 132t
halophile Organism that prefers or requires a high salt (NaCl) medium to grow, 100
Halorubrum, 298
halotolerant Describing an organism that can grow in relatively high salt concentrations, up to 10% NaCl, 100
HA-MRSA. *See* healthcare-associated MRSA (HA-MRSA)
hand-foot-and-mouth disease (HFMD), 602
hand hygiene, 498
hand sanitizers, alcohol-based, 129
hand scrubbing, 497
handwashing, 118, 120, 548
Hansen, Gerhard Henrik Armauer, 703
Hansen's disease (leprosy), 98, 398, 417, 703, 728–730
  causative agent, 30, 728–729, 729f
  effects of leprosy, 729f
  epidemiology, 729–730
  lepromatous leprosy, 728
  pathogenesis, 729
  signs and symptoms, 728, 729f
  summary/overview, 730t
  treatment and prevention, 730
  tuberculoid leprosy, 728
  WHO's Global Leprosy Strategy, 730
Hantaan virus, 8f
hantaviruses, 571, 572f
hantavirus pulmonary syndrome (HPS), 571–572
  causative agent, 571–572, 572f
  epidemiology, 572
  pathogenesis, 572
  signs and symptoms, 571
  summary/overview, 572t
  treatment and prevention, 572
harmful algal bloom (HAB), 780, 796
Hartig net, 787
Hatfull, Graham, 338
HAV. *See* hepatitis A virus (HAV)
hay fever Allergic rhinitis; sneezing, runny nose, teary eyes resulting from exposure of a sensitized person to inhaled antigen; an IgE-mediated allergic reaction, 464
HBV. *See* hepatitis B virus (HBV)
HCV. *See* hepatitis C virus (HCV)
HDN. *See* hemolytic disease of the newborn (HDN)
HDPs. *See* host defense peptides (HDPs)
healer macrophages, 376
healthcare-associated IE, 676
healthcare-associated infections (HAIs) Infections acquired while receiving treatment in a hospital or other healthcare facility, 118, 242, 495–498, 519, 520, 737. *See also* urinary tract infections
  bacterial causes, 496t
  common types, 497t
  prevention, 498
  reservoirs of infectious agents, 496
  transmission of infectious agents, 496–497
healthcare-associated MRSA (HA-MRSA), 520, 588, 612
healthcare-associated pneumonias (HCAPs), 550
Healthcare Infection Control Practices Advisory Committee (HICPAC), 498
healthcare worker, 496
heart, 672
heart muscle, 672
heart valves, 672
heavy chain The two higher-molecular-weight polypeptide chains that make up an antibody molecule; the type of heavy chain determines the class of antibody molecule, 402f, 411
HeLa cell line, 351
helical viruses, 329, 331f
Helicobacter, 294
Helicobacter pylori, 9, 57, 98, 99, 260, 294, 638
  urea breath test (UBT), 260
Helicobacter pylori gastritis, 638–639
  case study, 638
  causative agent, 638
  epidemiology, 639
  pathogenesis, 638–639, 639f
  signs and symptoms, 638
  summary/overview, 639t
  treatment and prevention, 639
Heliobacterium, 277
helminths Worms; parasitic helminths often have complex life cycles, 15, 16t, 320–324
  diseases caused by, 324t
  flatworms, 321
  flukes, 321, 323–324, 324t
  life cycles and transmission, 321
  nematodes, 321–322, 324t
  pathogenesis, 433
  tapeworms, 321, 322–323, 323f, 324t
helper T cell Type of lymphocyte programmed to activate B cells and macrophages, and assist other parts of the adaptive immune response, 385
helper virus, 334
hemagglutination, 352
hemagglutination assay In virology, a test that relies on clumping of red blood cells to determine the relative concentration of viral particles, 352, 352f
hemagglutinin (HA), 564, 598
hematopoiesis The formation and development of blood cells, 364
hematopoietic stem cells Bone marrow cells that give rise to all blood cells, 364
hemoflagellate A flagellated protozoan blood parasite, 314–315
hemolysin Substance that lyses red blood cell, 426, 682
hemolytic disease of the newborn (HDN) Disease of the fetus or newborn caused by transplacental passage of maternal antibodies against the baby's red blood cells, resulting in red cell destruction; also called erythroblastosis fetalis, 467–469, 468f
hemolytic transfusion reaction Occurs when a person receives a transfusion of erythrocytes, or red blood cells (RBCs) with different antigens from his or her own, and it results in destruction of the cells, 467, 467t
hemolytic uremic syndrome (HUS) Serious condition characterized by red blood cell breakdown and kidney failure, 647, 648, 817
Hepadnaviridae, 658
HEPA filters. *See* high-efficiency particulate air (HEPA) filters
hepatitis A, 417, 443t, 490, 657–658
  causative agent, 658
  epidemiology, 658
  pathogenesis, 658
  post-exposure prophylaxis (PEP), 658
  signs and symptoms, 657
  summary/overview, 661t
  treatment and prevention, 658
  vaccination, 658
hepatitis A virus (HAV), 658
hepatitis B, 443t, 658–660
  causative agent, 658 659, 658f
  epidemiology, 659–660
  pathogenesis, 659, 659f
  signs and symptoms, 658
  summary/overview, 661t
  treatment and prevention, 660
hepatitis B core antigen (HBcAg), 659
hepatitis B e antigen (HBeAg), 659
hepatitis B immune globulin (HBIG), 438, 660
hepatitis B surface antigen (HBsAg), 658–659
hepatitis B virus (HBV), 344, 347t, 349t, 658
hepatitis B virus vaccine, 441
hepatitis C, 660–661
  causative agent, 660
  country where disease first appeared, 8f
  epidemiology, 660
  pathogenesis, 660
  signs and symptoms, 660
  summary/overview, 661t
  treatment and prevention, 660–661
hepatitis C virus (HCV), 347t, 349t, 660
hepatitis D, 8f
hepatitis Inflammation of the liver, 657
heptavalent botulinum antitoxin (HBAT), 731
HER2. *See* human epidermal growth factor receptor 2 (HER2)
herd immunity Protection of an entire population based upon a critical concentration of immune hosts that prevents the spread of an infectious agent, 439, 486
hereditary angioedema, 475
hereditary transthyretin amyloid, 199
herpes labialis, 640
herpes simplex virus type 1 (HSV-1), 347–348, 347t, 640, 641, 760
  genetically engineered, 349
herpes simplex virus type 2 (HSV-2), 347t, 640, 760
Herpesviridae, 593, 685, 760
herpesviruses, 347t, 432, 525
herpesvirus infection, 526
herpetic whitlow, 641
Hesse, Fannie, 90
heterocysts, 279, 279f
heterophile antibodies, 686
heterotroph Organism that obtains carbon from an organic compound such as glucose, 100, 775, 779
hexachlorophene, 131
Hfr cells Cells that have the F plasmid integrated into their chromosome, allowing them to begin transferring the chromosome by conjugation; stands for "high frequency of recombination cells," 221, 222f
HGT. *See* horizontal gene transfer (HGT)
Hib vaccine, 710
high-efficiency particulate air (HEPA) filters Special filters that remove from air nearly all particles, including microorganisms, that have a diameter of 0.3 μm or larger, 125, 483
high-energy phosphate bond Bond that joins a phosphate group to a compound and is easily broken to release energy that can drive cellular reactions; indicated by the symbol~, 41
high-level disinfectant Chemical used to destroy all viruses and vegetative cells, but not endospores, 129
highly active antiretroviral therapy (HAART), 766
highly resistant microbes, 121
high-pressure processing (HPP), 126
high-temperature–short-time (HTST) method Most common pasteurization protocol; using this method, milk is pasteurized by holding it at 72°C for 15 seconds, 123
high-throughput sequencing Highly automated DNA sequencing methods that generate huge amounts of data relatively quickly , 243
histones, 59
Histoplasma capsulatum, 310, 574, 575f
histoplasmosis, 574–575
  causative agent, 574, 575f
  epidemiology, 574, 575f
  pathogenesis, 574
  signs and symptoms, 574
  summary/overview, 575t
  treatment and prevention, 574–575
historical overview, 1, 4f
  golden age of microbiology, 3, 4f
  major milestones and timeline, 4f
  Pasteur, 2, 3f
  scientific method, 3, 5
  spontaneous generation, 2
  Tyndall, 2–3
Hitchings, George, 505
HIV-1 (human immunodeficiency virus type 1), 8
HIV/AIDS, 346, 347t, 761–768. *See also* acquired immunodeficiency syndrome (AIDS)
  acute infection, 761
  AIDS onset, 762, 764
  AIDS-related deaths, 766, 766f
  antiretrovirals (ARVs), 765–766, 766t
  antiretroviral therapy (ART), 766–767, 766t
  blood and blood products, 765, 767
  causative agents, 762–763, 763f
  CD4 lymphocytes, 761
  clinical latency, 761–762
  elite controllers, 765
  epidemiology, 765
  functions of HIV components, 763t
  global HIV/AIDS epidemic, 765f
  HIV replication, 763–764, 764f
  immune response to infection, 764
  multi-drug resistant, 525, 766
  non-progressors, 765
  pathogenesis, 763–765
  person-to-person transmission, 765
  pre-exposure prophylaxis, 767
  rapid progressors, 764
  sexual contact, 765, 767
  signs and symptoms, 761–762
  stages of infection, 761–762, 762f
  summary/overview, 768t

I

M

N

O

P

T

U

V

W

X

Y

Z